TRIP GENERATION MANUAL

9th Edition • Volume 3: Data

Trip Generation Rates, Plots and Equations

- Institutional (Land Uses 500–599)
- Medical (Land Uses 600–699)
- Office (Land Uses 700–799)
- Retail (Land Uses 800–899)
- Services (Land Uses 900–999)

ite
Institute of Transportation Engineers

Trip Generation Manual, 9th Edition

Volume 3: Data

The Institute of Transportation Engineers is an international educational and scientific association of transportation professionals who are responsible for meeting mobility and safety needs. ITE facilitates the application of technology and scientific principles to research, planning, functional design, implementation, operation, policy development and management for any mode of ground transportation. Through its products and services, ITE promotes professional development of its members, supports and encourages education, stimulates research, develops public awareness programs and serves as a conduit for the exchange of professional information.

Founded in 1930, ITE is a community of transportation professionals including, but not limited to transportation engineers, transportation planners, consultants, educators and researchers. Through meetings, seminars, publications and a network of nearly 17,000 members, working in more than 90 countries, ITE is your source for expertise, knowledge and ideas.

ite

Institute of Transportation Engineers
1627 Eye Street, NW, Suite 600
Washington, DC 20006 USA
Telephone: +1 202-785-0060
Fax: +1 202-785-0609
www.ite.org

Copyright © 2012 Institute of Transportation Engineers. Use of the *Trip Generation Manual* is governed by the license agreement that is bound with the three volumes of the *Trip Generation Manual* and is also available at www.ite.org/Trip Generation Manual License Agreement.
Publication No. IR-016G
1500/BH/AGS/1212
Second Printing

ISBN-13: 978-1-933452-64-7
ISBN-10: 1-933452-64-1
Printed in the United States of America

Table of Contents (9th Edition)
Volume 3: Data

Trip Generation Rates, Plots and Equations

Preface vii

Institutional (Land Uses 500–599)

CODE	LAND USE	PAGE
501	Military Base	967
520	Elementary School	978
522	Middle School/Junior High School	991
530	High School	1004
534	Private School (K-8)	1029
536	Private School (K-12)	1036
540	Junior/Community College	1047
550	University/College	1075
560	Church	1088
561	Synagogue	1103
562	Mosque	1112
565	Day Care Center	1114
566	Cemetery	1143
571	Prison	1151
580	Museum	1160
590	Library	1162
591	Lodge/Fraternal Organization	1181

Medical (Land Uses 600–699)

CODE	LAND USE	PAGE
610	Hospital	1184
620	Nursing Home	1214
630	Clinic	1236
640	Animal Hospital/Veterinary Clinic	1247

Office (Land Uses 700–799)

CODE	LAND USE	PAGE
710	General Office Building	1250
714	Corporate Headquarters Building	1266
715	Single Tenant Office Building	1277

720	Medical-Dental Office Building	1284
730	Government Office Building	1303
731	State Motor Vehicles Department	1306
732	United States Post Office	1325
733	Government Office Complex	1344
750	Office Park	1352
760	Research and Development Center	1374
770	Business Park	1396

Retail (Land Uses 800–899)

810	Tractor Supply Store	1412
811	Construction Equipment Rental Store	1415
812	Building Materials and Lumber Store	1418
813	Free-Standing Discount Superstore	1437
814	Variety Store	1449
815	Free-Standing Discount Store	1455
816	Hardware/Paint Store	1475
817	Nursery (Garden Center)	1503
818	Nursery (Wholesale)	1531
820	Shopping Center	1557
823	Factory Outlet Center	1568
826	Specialty Retail Center	1578
841	Automobile Sales	1588
842	Recreational Vehicle Sales	1602
843	Automobile Parts Sales	1604
848	Tire Store	1611
849	Tire Superstore	1628
850	Supermarket	1643
851	Convenience Market (Open 24 Hours)	1654
852	Convenience Market (Open 15-16 Hours)	1664
853	Convenience Market with Gasoline Pumps	1669
854	Discount Supermarket	1688
857	Discount Club	1710
860	Wholesale Market	1730
861	Sporting Goods Superstore	1735
862	Home Improvement Superstore	1739

863	Electronics Superstore	1749
864	Toy/Children's Superstore	1755
865	Baby Superstore	1758
866	Pet Supply Superstore	1760
867	Office Supply Superstore	1764
868	Book Superstore	1766
869	Discount Home Furnishing Superstore	1770
872	Bed and Linen Superstore	1776
875	Department Store	1778
876	Apparel Store	1787
879	Arts and Crafts Store	1791
880	Pharmacy/Drugstore without Drive-Through Window	1794
881	Pharmacy/Drugstore with Drive-Through Window	1801
890	Furniture Store	1808
896	DVD/Video Rental Store	1827
897	Medical Equipment Store	1831

Services (Land Uses 900–999)

911	Walk-in Bank	1833
912	Drive-in Bank	1835
918	Hair Salon	1857
920	Copy, Print and Express Ship Store	1859
925	Drinking Place	1861
931	Quality Restaurant	1864
932	High-Turnover (Sit-Down) Restaurant	1883
933	Fast-Food Restaurant without Drive-Through Window	1903
934	Fast-Food Restaurant with Drive-Through Window	1910
935	Fast-Food Restaurant with Drive-Through Window and No Indoor Seating	1932
936	Coffee/Donut Shop without Drive-Through Window	1935
937	Coffee/Donut Shop with Drive-Through Window	1942
938	Coffee/Donut Shop with Drive-Through Window and No Indoor Seating	1956
939	Bread/Donut/Bagel Shop without Drive-Through Window	1963
940	Bread/Donut/Bagel Shop with Drive-Through Window	1966
941	Quick Lubrication Vehicle Shop	1969
942	Automobile Care Center	1973
943	Automobile Parts and Service Center	1981

944	Gasoline/Service Station	1983
945	Gasoline/Service Station with Convenience Market	1991
946	Gasoline/Service Station with Convenience Market and Car Wash	2003
947	Self-Service Car Wash	2010
948	Automated Car Wash	2014
950	Truck Stop	2016

Preface

Trip Generation Manual is a publication of the Institute of Transportation Engineers (ITE).

Volume 1 of the publication, the *User's Guide and Handbook*, contains definitions of the independent variables and terms used in this manual. The *User's Guide and Handbook* also provides general instructional material on statistical data and helps users understand the data plots contained in the second and third volumes. With this edition, *Trip Generation Handbook, 2nd Edition: An ITE Recommended Practice*, is being included as part of Volume 1. *Trip Generation Handbook*, Second Edition (Publication Number RP-028B), has two primary purposes: to provide instruction and guidance in the proper use of data presented in *Trip Generation* and to provide information on supplemental issues of importance in estimating trip generation for development sites. Some additional topics covered in the *Trip Generation Handbook*, Second Edition include primary/pass-by/diverted link trips, multi-use developments, truck trip generation and transportation demand management programs.

Volumes 2 and 3: Data, are prepared for informational purposes only and do not include ITE recommendations on the best course of action or the preferred application of the data. The information in these volumes is based on trip generation studies submitted voluntarily to ITE by public agencies, developers, consulting firms and associations. Users are encouraged to review and become familiar with the *User's Guide and Handbook* prior to using the data contained in Volumes 2 and 3.

Land Use: 501
Military Base

Description

A military base is a complex that serves one division of the armed forces of the United States. It typically contains offices and training, housing, dining and recreational facilities.

Additional Data

The independent variable, vehicles, used in this land use refers to the number of vehicles authorized to enter the facility.

Most of the sites were surveyed at air force bases between the 1970s and the 1990s at installations throughout the United States.

Source Numbers

18, 285, 406

Military Base
(501)

Average Vehicle Trip Ends vs: Employees
On a: Weekday

Number of Studies: 7
Avg. Number of Employees: 7,747
Directional Distribution: 50% entering, 50% exiting

Trip Generation per Employee

Average Rate	Range of Rates	Standard Deviation
1.78	1.00 - 4.18	1.64

Data Plot and Equation

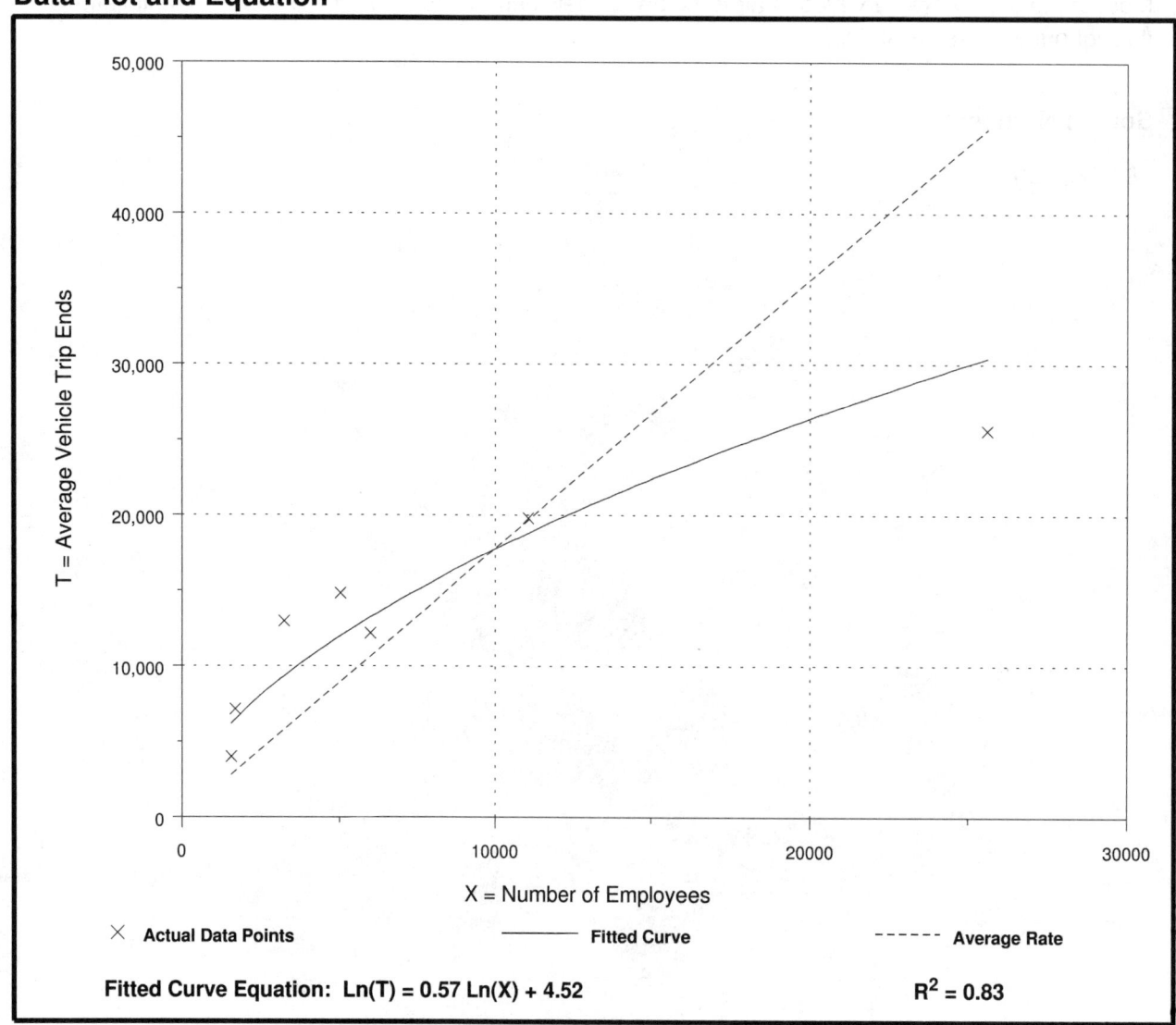

Fitted Curve Equation: $Ln(T) = 0.57\ Ln(X) + 4.52$ $R^2 = 0.83$

Military Base
(501)

Average Vehicle Trip Ends vs: Employees
On a: Weekday,
Peak Hour of Adjacent Street Traffic,
One Hour Between 7 and 9 a.m.

Number of Studies: 6
Avg. Number of Employees: 6,028
Directional Distribution: Not available

Trip Generation per Employee

Average Rate	Range of Rates	Standard Deviation
0.39	0.26 - 0.55	0.64

Data Plot and Equation

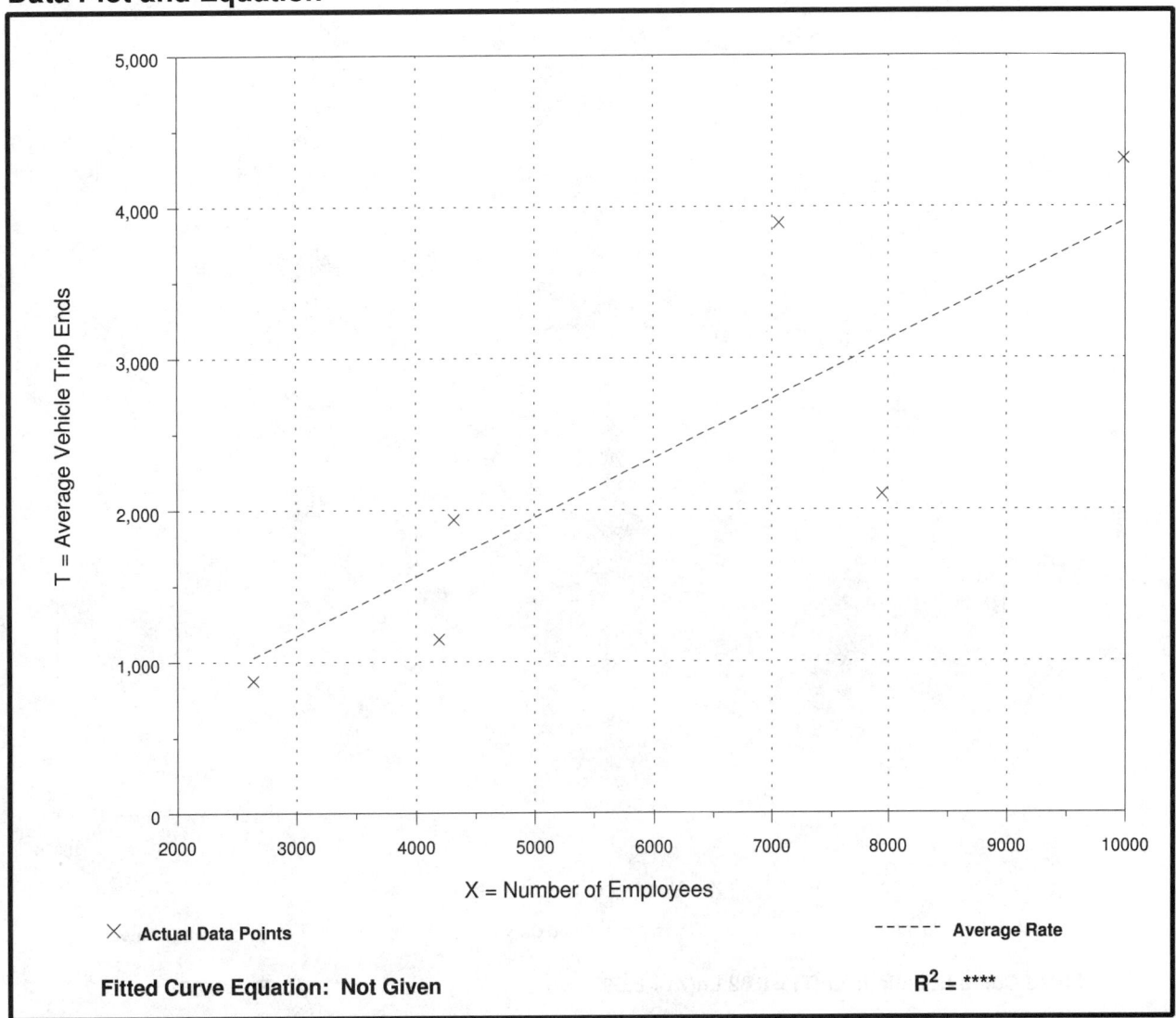

Fitted Curve Equation: Not Given $R^2 = ****$

Military Base
(501)

Average Vehicle Trip Ends vs: Employees
On a: Weekday,
Peak Hour of Adjacent Street Traffic,
One Hour Between 4 and 6 p.m.

Number of Studies: 6
Avg. Number of Employees: 6,028
Directional Distribution: Not available

Trip Generation per Employee

Average Rate	Range of Rates	Standard Deviation
0.39	0.31 - 0.49	0.63

Data Plot and Equation

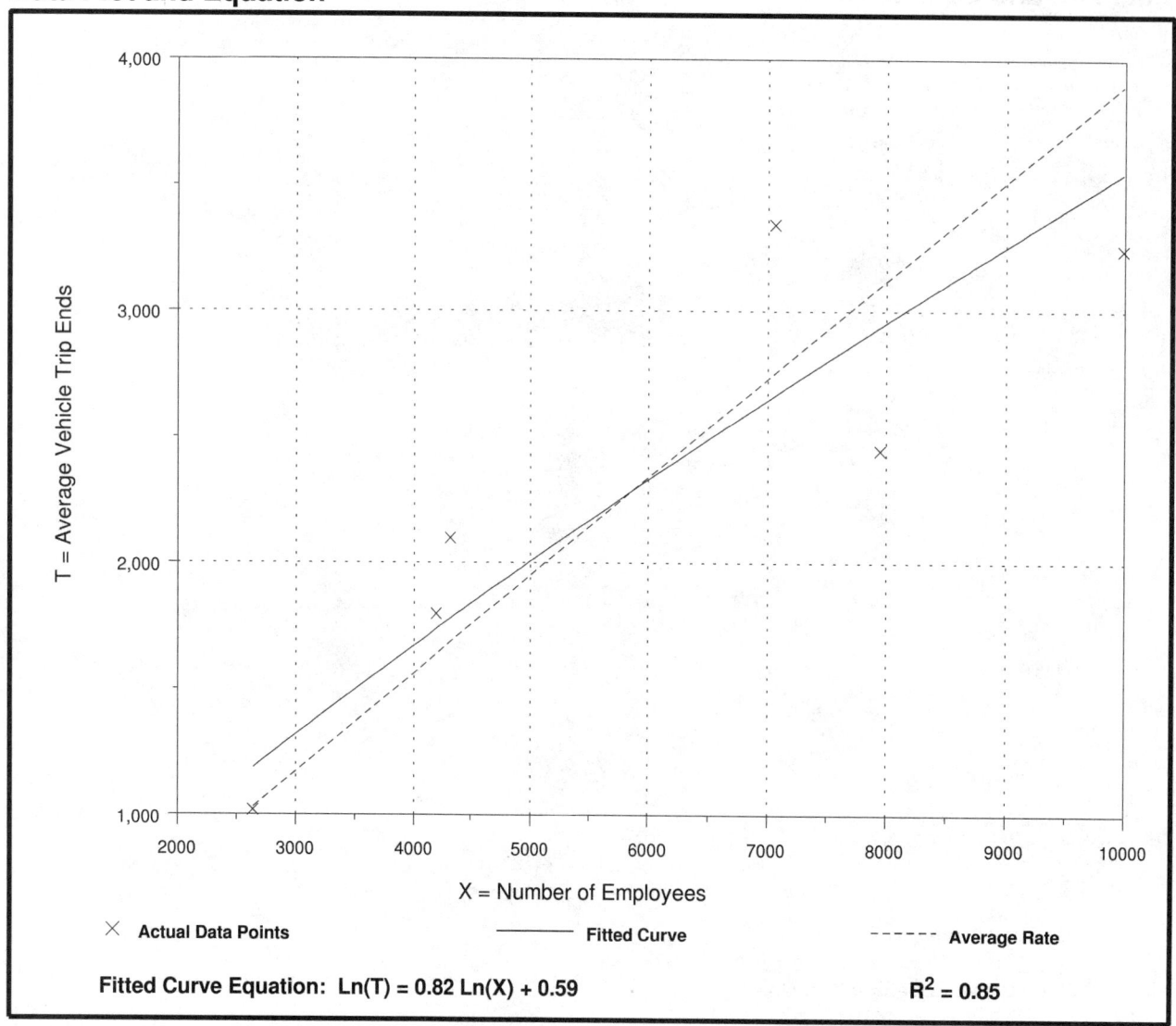

Fitted Curve Equation: $Ln(T) = 0.82 \, Ln(X) + 0.59$ $R^2 = 0.85$

Military Base
(501)

Average Vehicle Trip Ends vs: Employees
On a: Weekday,
A.M. Peak Hour of Generator

Number of Studies: 8
Avg. Number of Employees: 6,156
Directional Distribution: 88% entering, 12% exiting

Trip Generation per Employee

Average Rate	Range of Rates	Standard Deviation
0.37	0.25 - 0.55	0.62

Data Plot and Equation

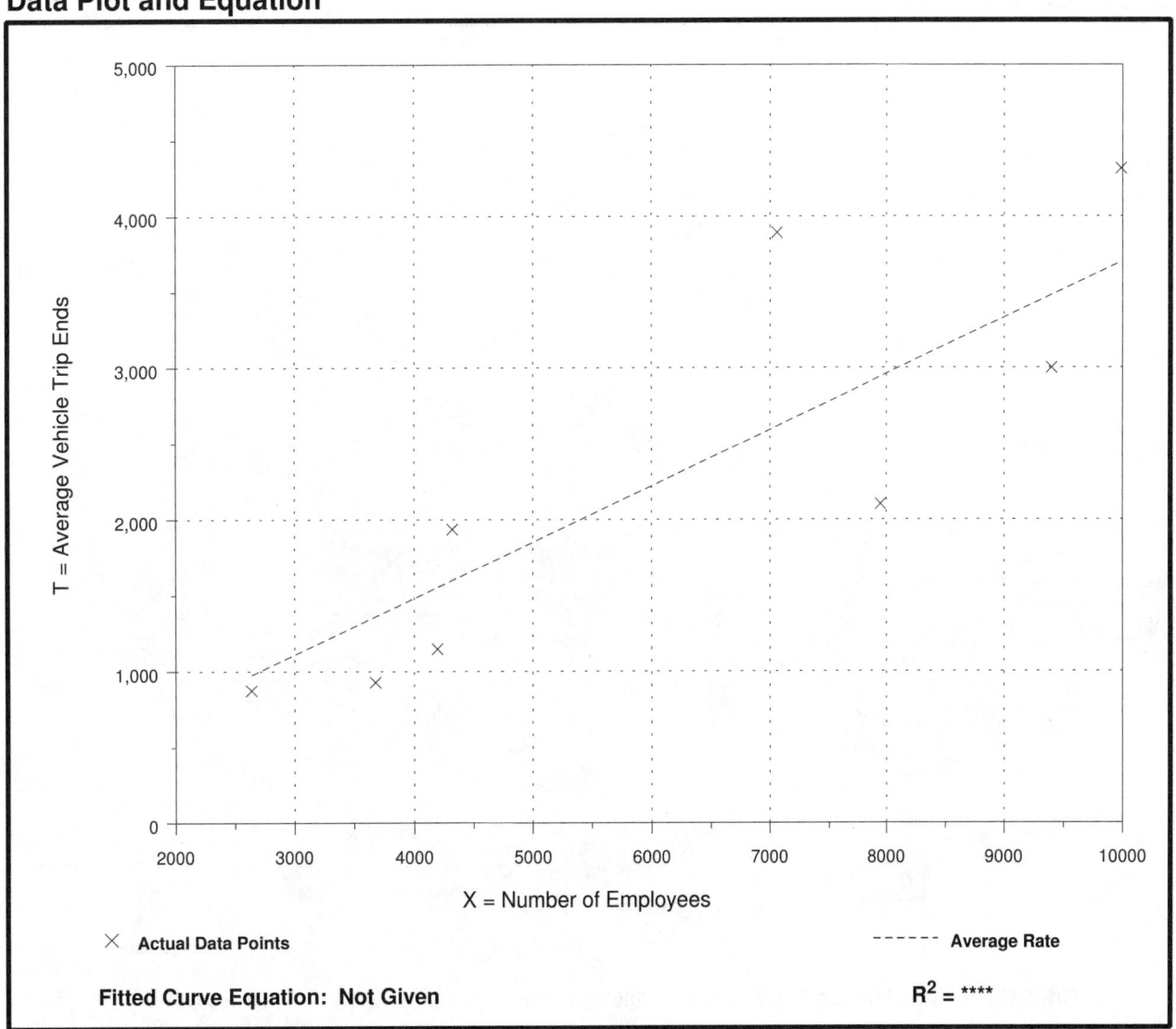

X = Number of Employees

× Actual Data Points   ----- Average Rate

Fitted Curve Equation: Not Given $R^2 = ****$

Military Base
(501)

Average Vehicle Trip Ends vs: Employees
On a: Weekday,
P.M. Peak Hour of Generator

Number of Studies: 8
Avg. Number of Employees: 6,156
Directional Distribution: 25% entering, 75% exiting

Trip Generation per Employee

Average Rate	Range of Rates	Standard Deviation
0.37	0.30 - 0.49	0.61

Data Plot and Equation

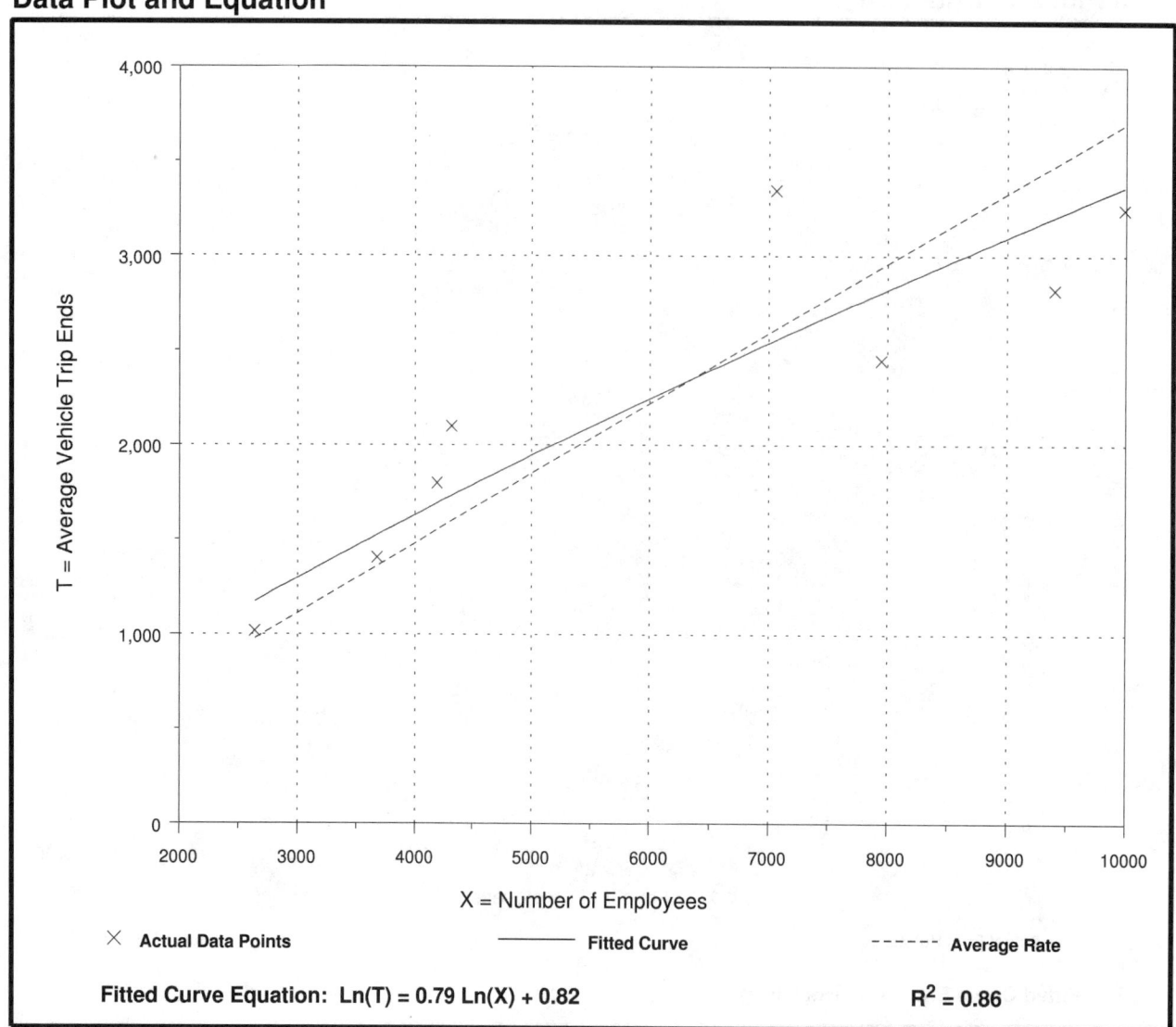

Fitted Curve Equation: $Ln(T) = 0.79 \, Ln(X) + 0.82$ $R^2 = 0.86$

Military Base
(501)

Average Vehicle Trip Ends vs: Employees
On a: Saturday

Number of Studies: 7
Avg. Number of Employees: 5,692
Directional Distribution: 50% entering, 50% exiting

Trip Generation per Employee

Average Rate	Range of Rates	Standard Deviation
2.64	1.08 - 4.88	2.16

Data Plot and Equation

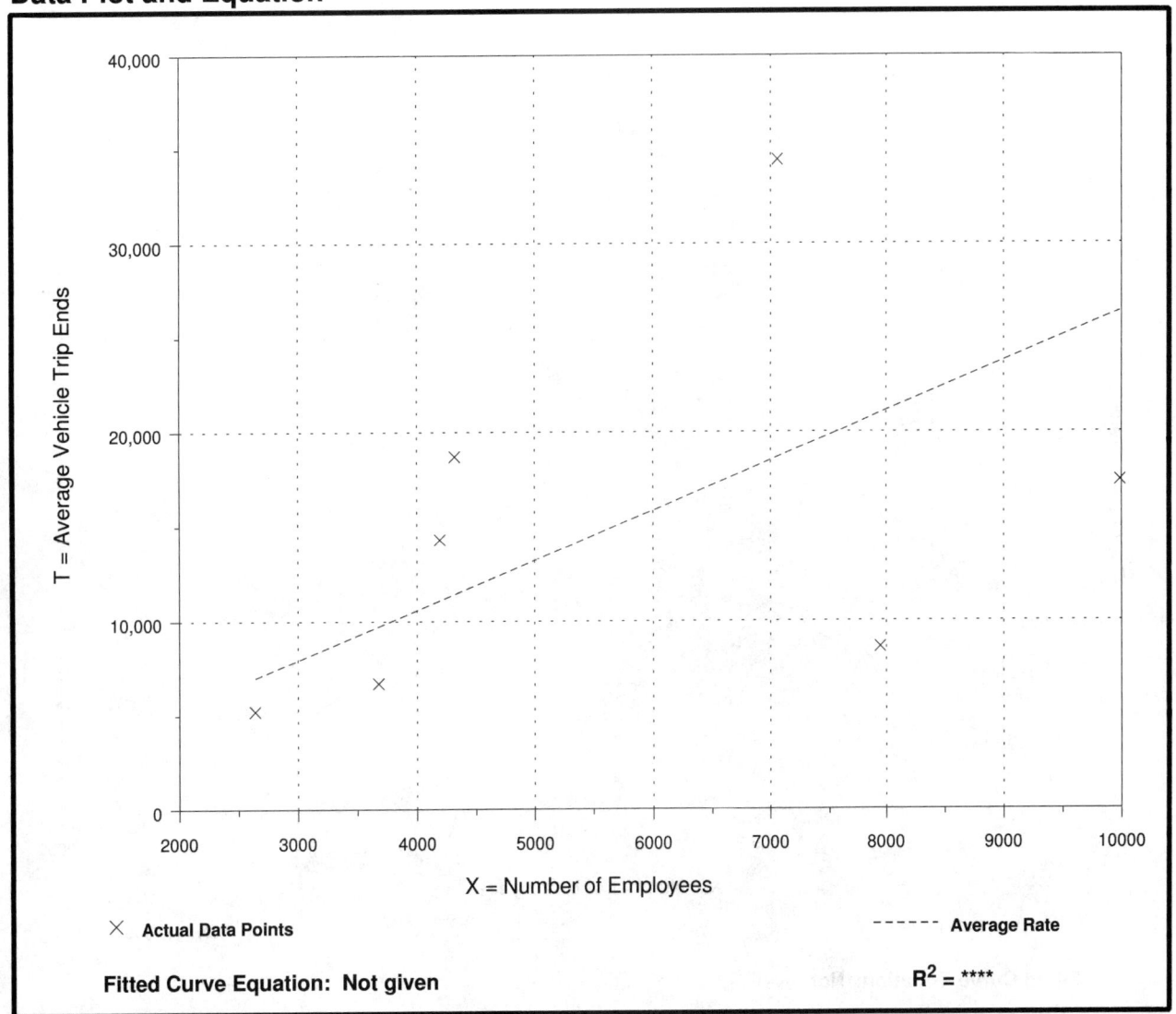

Fitted Curve Equation: Not given $R^2 = ****$

Military Base
(501)

Average Vehicle Trip Ends vs: Employees
On a: Saturday,
Peak Hour of Generator

Number of Studies: 7
Avg. Number of Employees: 5,692
Directional Distribution: Not available

Trip Generation per Employee

Average Rate	Range of Rates	Standard Deviation
0.26	0.11 - 0.50	0.53

Data Plot and Equation

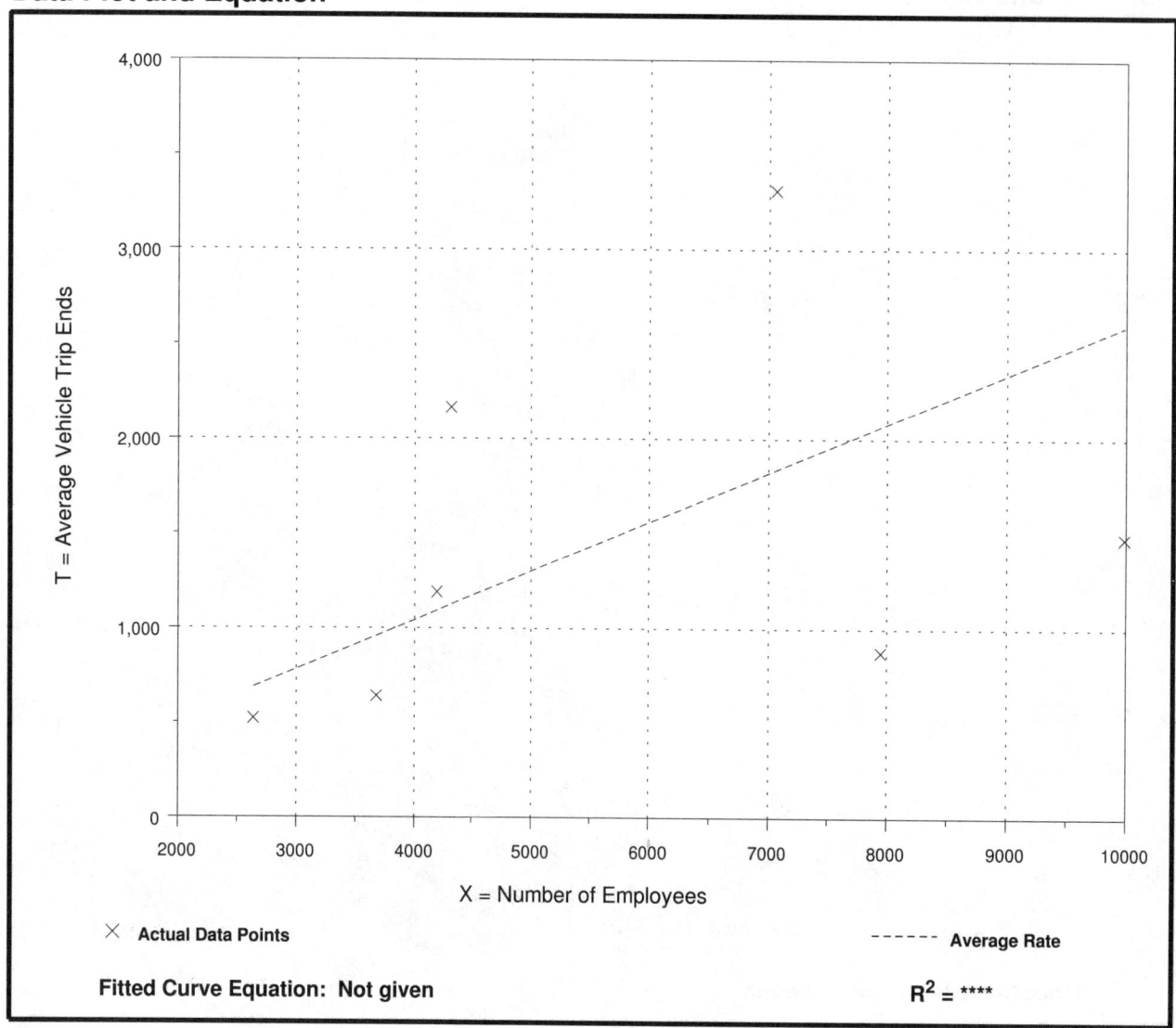

Fitted Curve Equation: Not given $R^2 = ****$

Military Base
(501)

Average Vehicle Trip Ends vs: Employees
On a: Sunday

Number of Studies: 7
Avg. Number of Employees: 5,692
Directional Distribution: 50% entering, 50% exiting

Trip Generation per Employee

Average Rate	Range of Rates	Standard Deviation
1.67	0.72 - 3.31	1.56

Data Plot and Equation

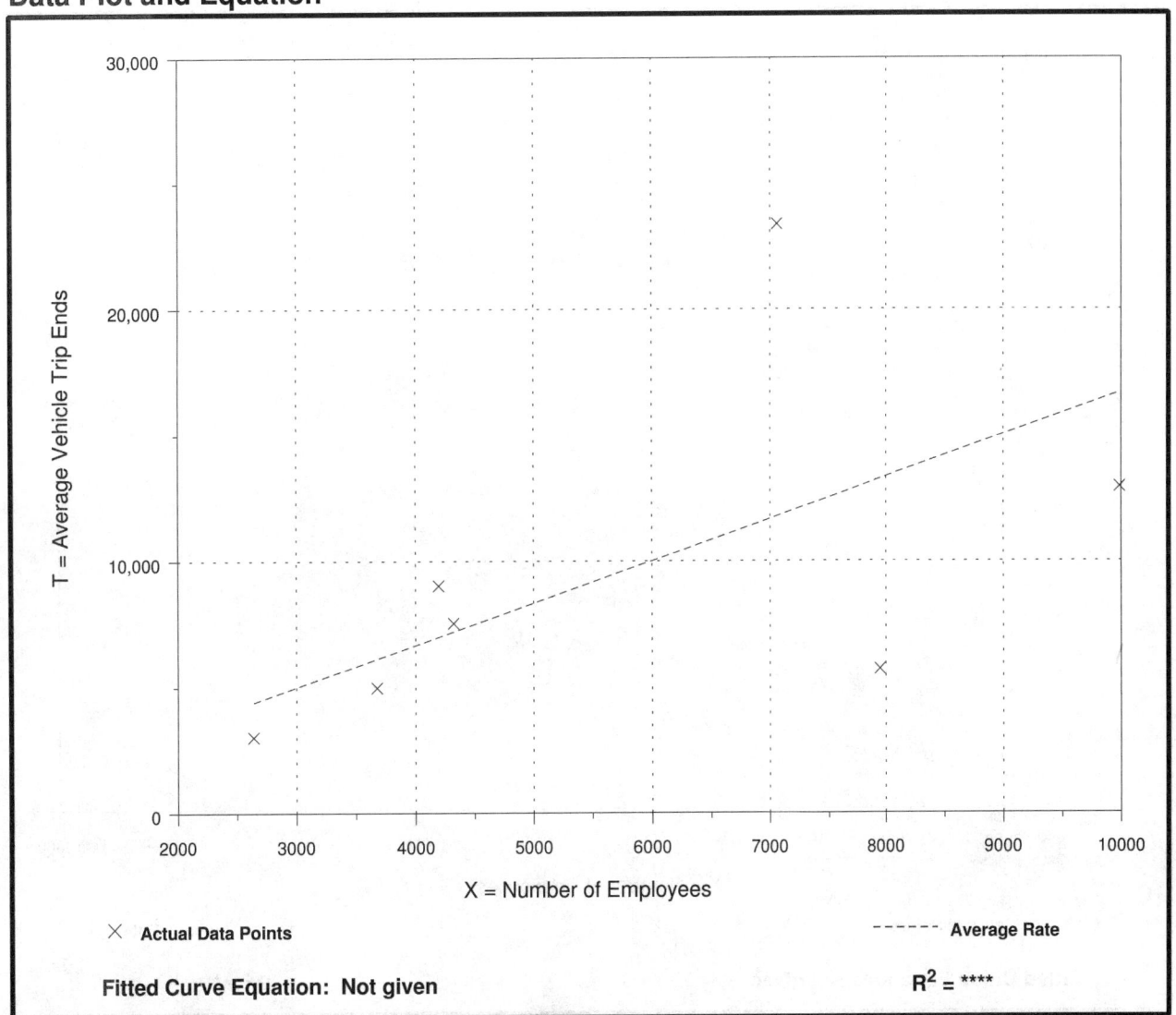

Fitted Curve Equation: Not given $R^2 = ****$

Military Base
(501)

Average Vehicle Trip Ends vs: Employees
On a: Sunday,
Peak Hour of Generator

Number of Studies: 7
Avg. Number of Employees: 5,692
Directional Distribution: Not available

Trip Generation per Employee

Average Rate	Range of Rates	Standard Deviation
0.18	0.06 - 0.45	0.45

Data Plot and Equation

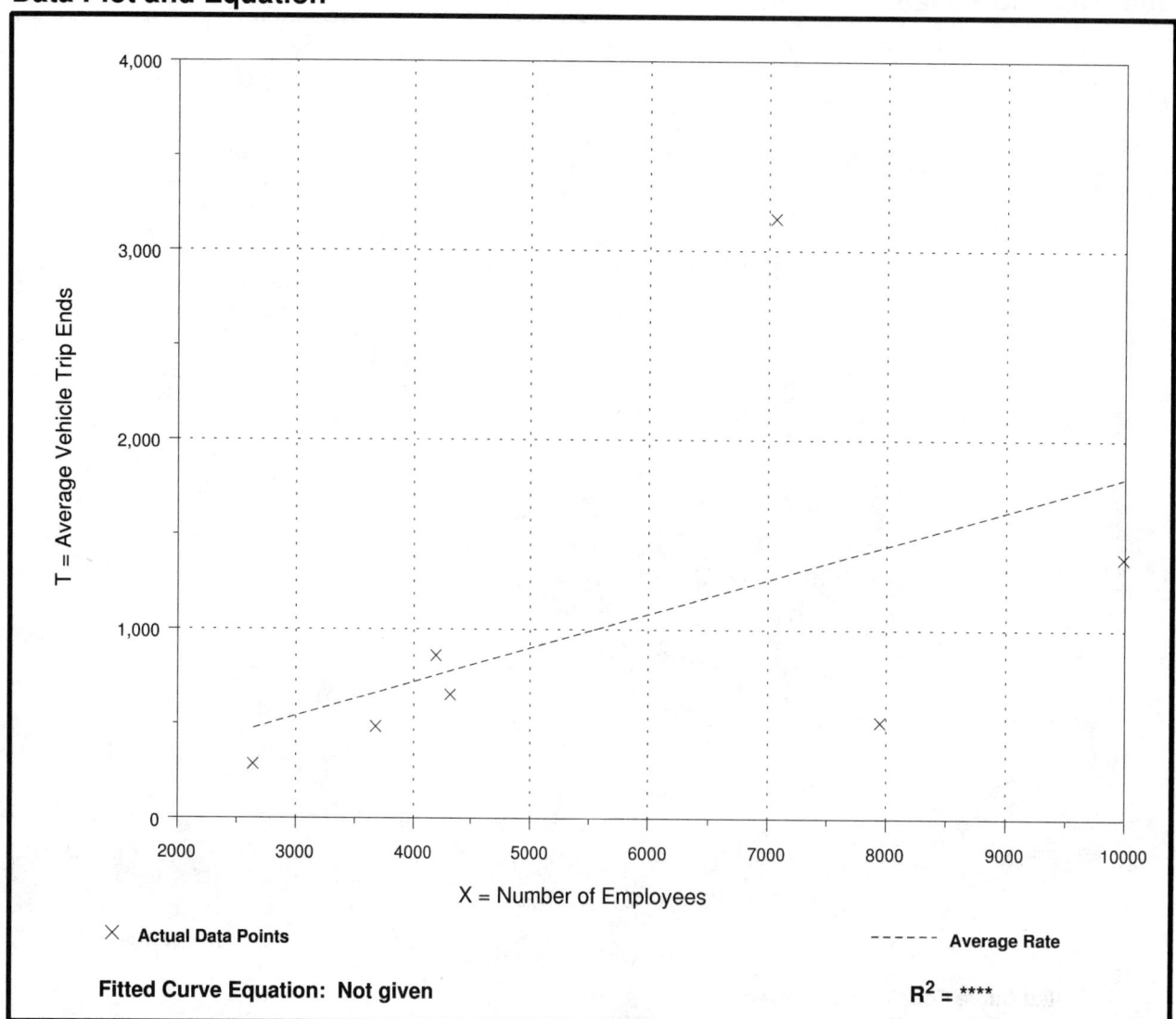

Fitted Curve Equation: Not given $R^2 = ****$

Military Base
(501)

Average Vehicle Trip Ends vs: Vehicles
On a: Weekday

Number of Studies: 7
Average Number of Vehicles: 16,002
Directional Distribution: 50% entering, 50% exiting

Trip Generation per Vehicle

Average Rate	Range of Rates	Standard Deviation
0.86	0.64 - 2.33	0.99

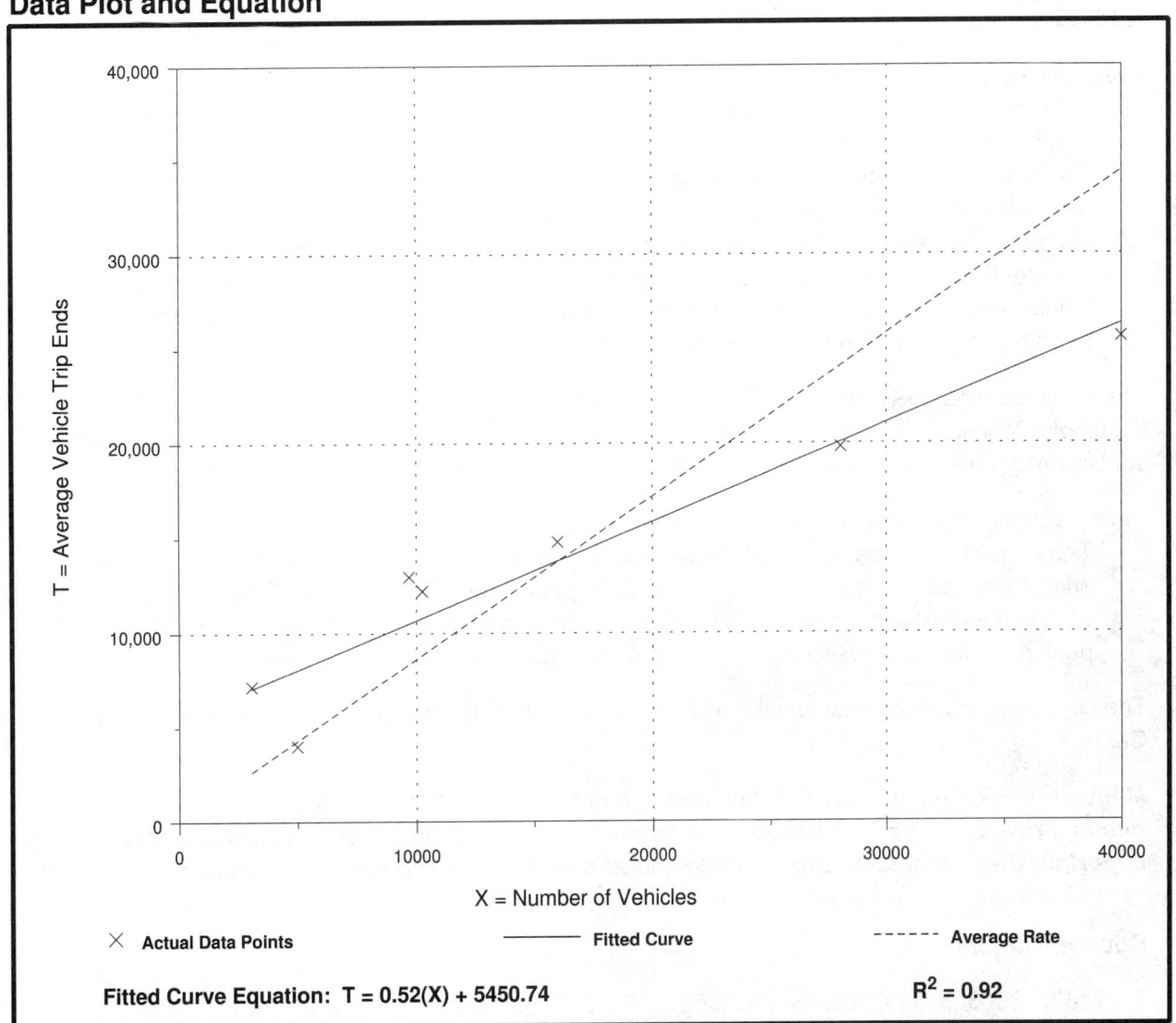

Data Plot and Equation

Fitted Curve Equation: $T = 0.52(X) + 5450.74$ $R^2 = 0.92$

Land Use: 520
Elementary School

Description

Elementary schools typically serve students attending kindergarten through the fifth or sixth grade. Elementary schools are usually centrally located in residential communities in order to facilitate student access and have no student drivers. This land use consists of schools where bus service is usually provided to students living beyond a specified distance from the school. Both public and private elementary schools are included in this land use. Middle school/junior high school (Land Use 522), high school (Land Use 530), private school (K-8) (Land Use 534) and private school (K-12) (Land Use 536) are related uses.

Additional Data

Average weekday transit trip ends—
Elementary school students generally used school buses more than regular transit and were dropped off and picked up more than high school students, who were apt to walk longer distances, ride bicycles, or, in some cases, drive to school. The percentage of students at the sites who were transported to school via bus varied considerably. Some sites experienced higher than average trip rates because many students did not utilize the available school bus service. Due to the varied transit and school bus usage at these sites, it is desirable that future studies report additional detail on the percentage of students who were bused to school and the percentage that were dropped off and picked up.

The elementary schools surveyed exhibited significant variations in terms of facilities provided. Because the ratio of floor space to student population varied widely among the schools surveyed, the number of students may be a more reliable independent variable on which to establish trip generation rates.

Peak hours of the generator—
The weekday A.M. peak hour of the generator typically coincided with the peak hour of the adjacent street traffic; therefore, only one A.M. peak hour, which represents both the peak hour of the generator and the peak hour of the adjacent street traffic, is displayed. The weekday P.M. peak hour varied between 2:00 p.m. and 4:00 p.m.

The sites were surveyed between the mid-1970s and the 2000s throughout the United States and Canada.

Many of the studies included in this land use did not indicate if the sites surveyed were public, private, or charter schools. To assist in the future analysis of this land use, it is important that this information be collected and included in trip generation data submissions.

Source Numbers

7, 32, 86, 186, 383, 390, 395, 444, 533, 536, 572, 579, 583, 609, 611, 612, 613, 632, 707

Elementary School
(520)

Average Vehicle Trip Ends vs: Students
On a: Weekday

Number of Studies: 33
Average Number of Students: 620
Directional Distribution: 50% entering, 50% exiting

Trip Generation per Student

Average Rate	Range of Rates	Standard Deviation
1.29	0.45 - 2.45	1.26

Data Plot and Equation

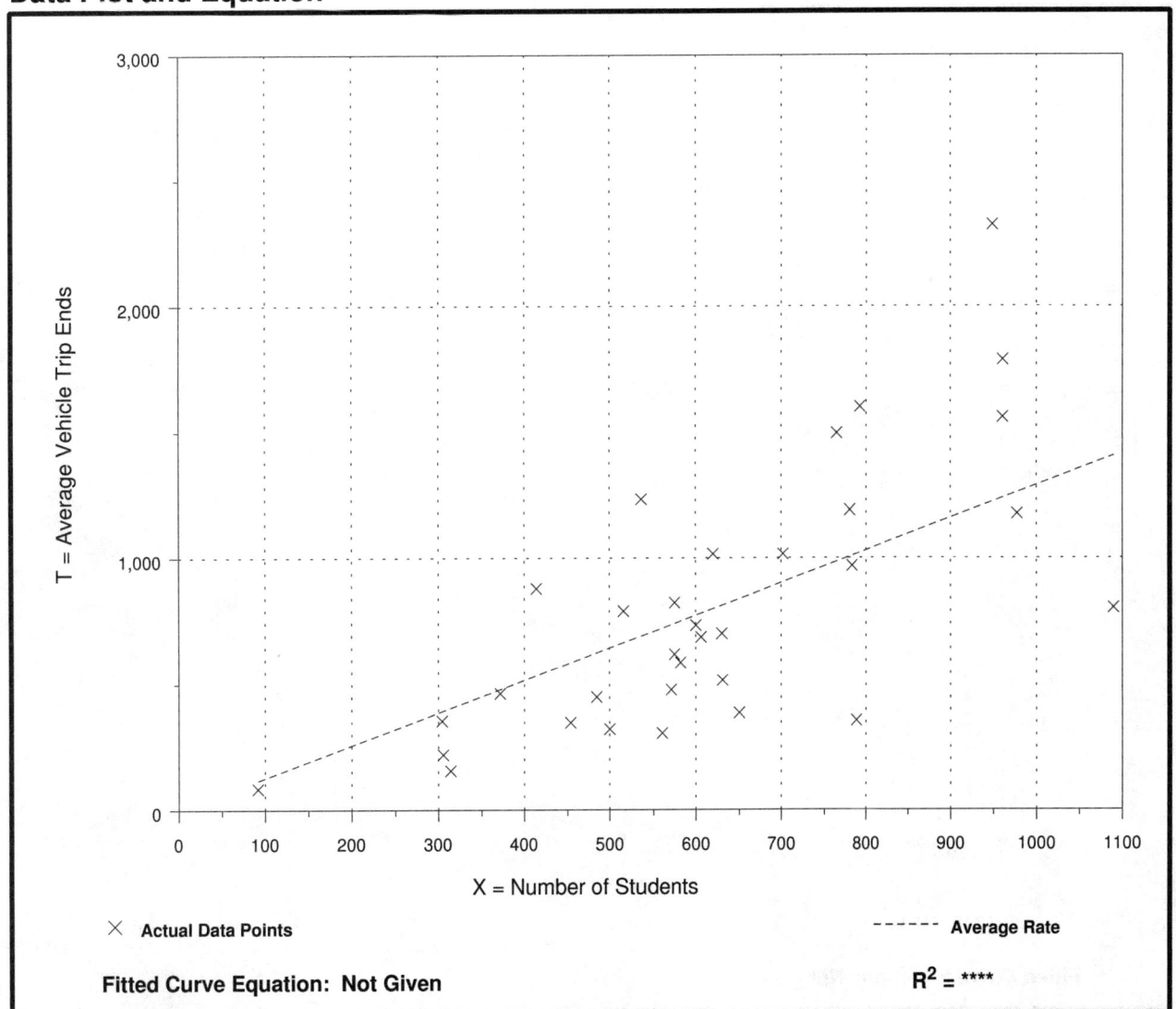

Fitted Curve Equation: Not Given　　　　　　　　$R^2 = ****$

Elementary School
(520)

Average Vehicle Trip Ends vs: Students
On a: Weekday,
A.M. Peak Hour

Number of Studies: 49
Average Number of Students: 630
Directional Distribution: 55% entering, 45% exiting

Trip Generation per Student

Average Rate	Range of Rates	Standard Deviation
0.45	0.11 - 0.92	0.70

Data Plot and Equation

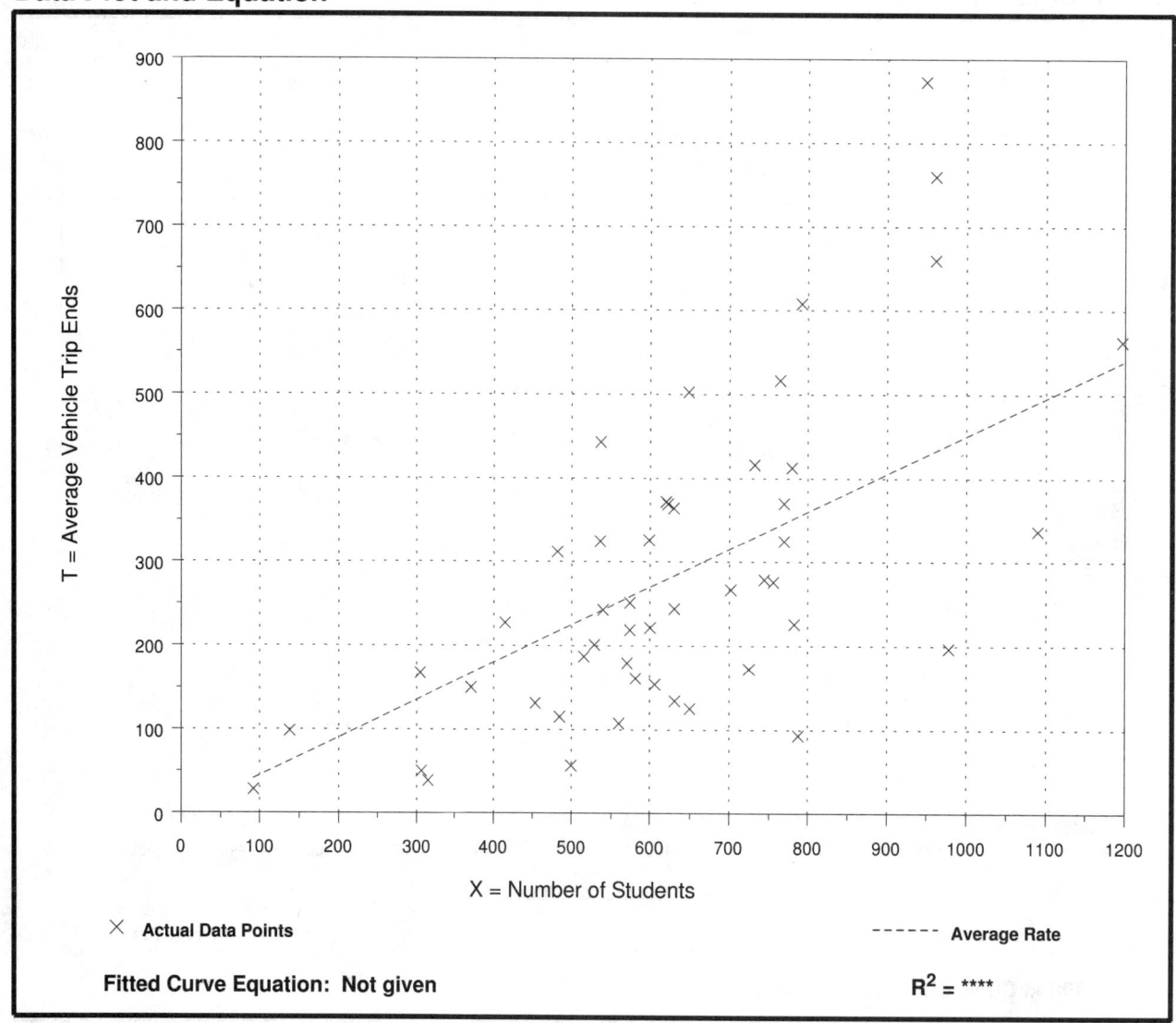

Fitted Curve Equation: Not given $R^2 = ****$

Elementary School
(520)

Average Vehicle Trip Ends vs: Students
On a: Weekday,
Peak Hour of Adjacent Street Traffic,
One Hour Between 4 and 6 p.m.

Number of Studies: 21
Average Number of Students: 684
Directional Distribution: 49% entering, 51% exiting

Trip Generation per Student

Average Rate	Range of Rates	Standard Deviation
0.15	0.05 - 0.37	0.40

Data Plot and Equation

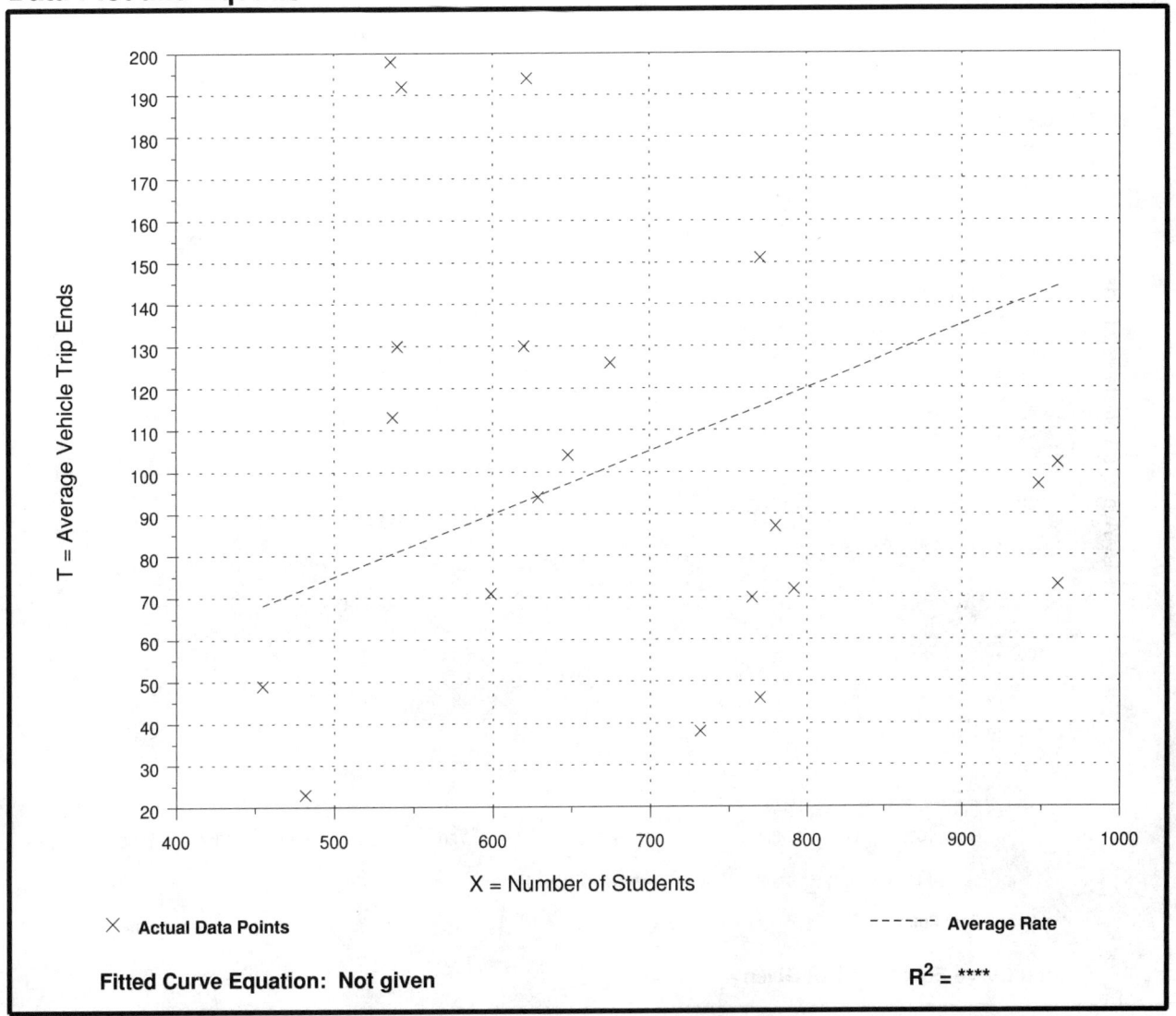

Fitted Curve Equation: Not given $R^2 = ****$

Elementary School
(520)

Average Vehicle Trip Ends vs: Students
On a: Weekday,
P.M. Peak Hour of Generator

Number of Studies: 45
Average Number of Students: 642
Directional Distribution: 45% entering, 55% exiting

Trip Generation per Student

Average Rate	Range of Rates	Standard Deviation
0.28	0.09 - 0.50	0.54

Data Plot and Equation

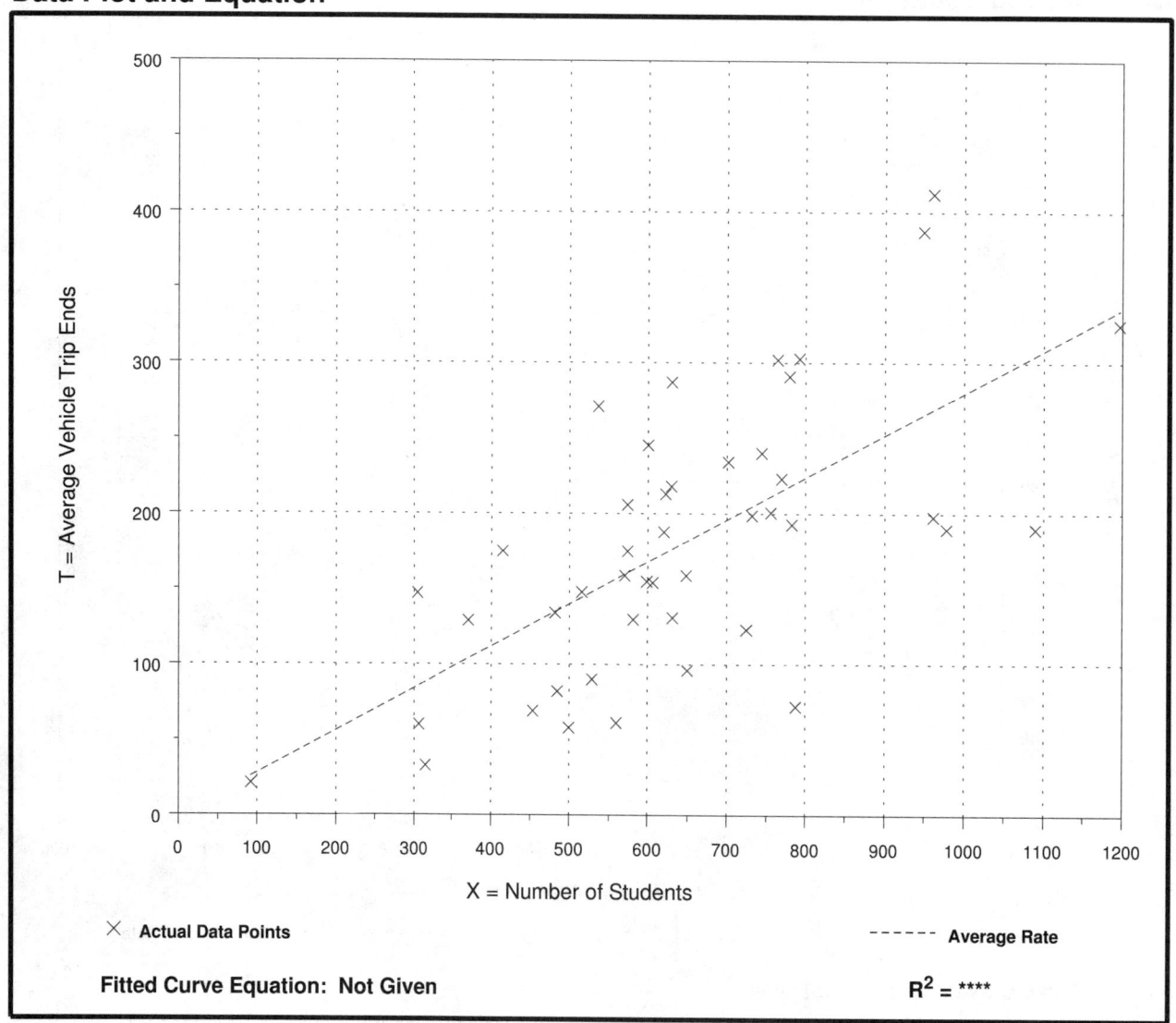

Fitted Curve Equation: Not Given $R^2 = ****$

Elementary School
(520)

Average Vehicle Trip Ends vs: Employees
On a: Weekday

Number of Studies: 31
Avg. Number of Employees: 50
Directional Distribution: 50% entering, 50% exiting

Trip Generation per Employee

Average Rate	Range of Rates	Standard Deviation
15.71	4.47 - 26.49	6.99

Data Plot and Equation

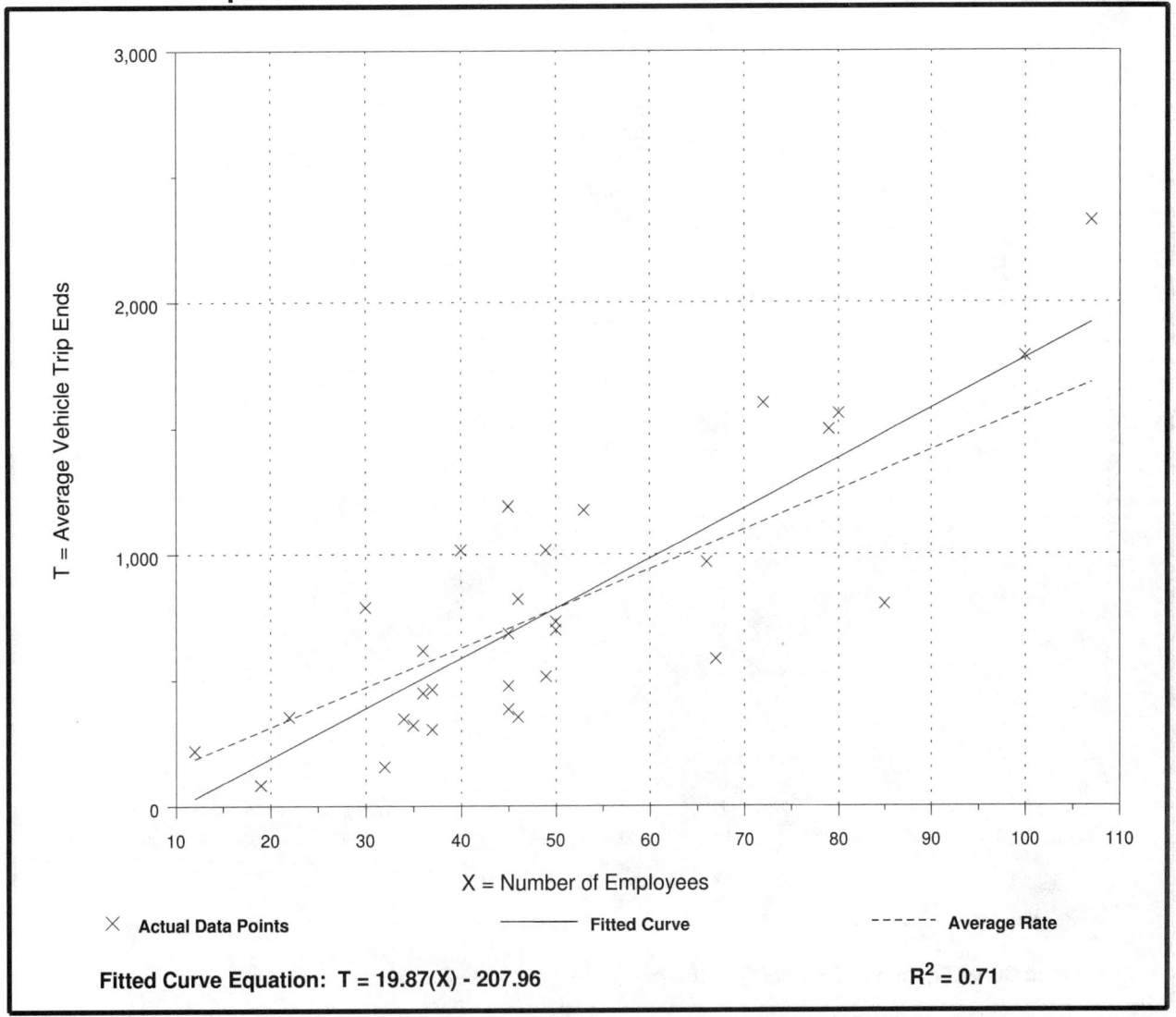

Fitted Curve Equation: T = 19.87(X) - 207.96 $R^2 = 0.71$

Trip Generation, 9th Edition • Institute of Transportation Engineers

Elementary School
(520)

Average Vehicle Trip Ends vs: Employees
On a: Weekday,
A.M. Peak Hour

Number of Studies: 36
Avg. Number of Employees: 51
Directional Distribution: 54% entering, 46% exiting

Trip Generation per Employee

Average Rate	Range of Rates	Standard Deviation
5.33	1.22 - 10.27	3.30

Data Plot and Equation

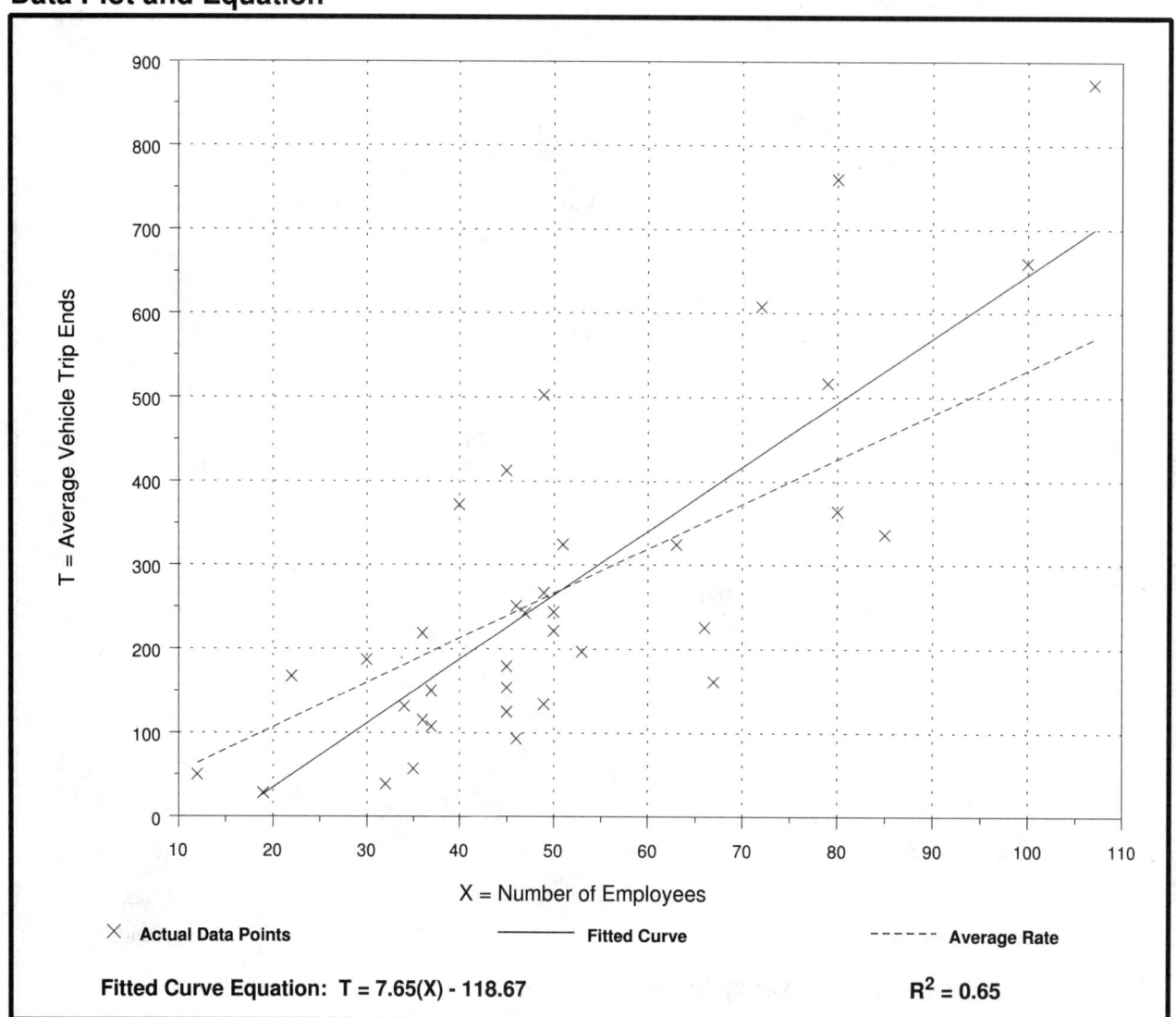

Fitted Curve Equation: $T = 7.65(X) - 118.67$ $R^2 = 0.65$

Elementary School
(520)

Average Vehicle Trip Ends vs: Employees
On a: Weekday,
Peak Hour of Adjacent Street Traffic,
One Hour Between 4 and 6 p.m.

Number of Studies: 14
Avg. Number of Employees: 66
Directional Distribution: 49% entering, 51% exiting

Trip Generation per Employee

Average Rate	Range of Rates	Standard Deviation
1.76	0.73 - 3.88	1.66

Data Plot and Equation

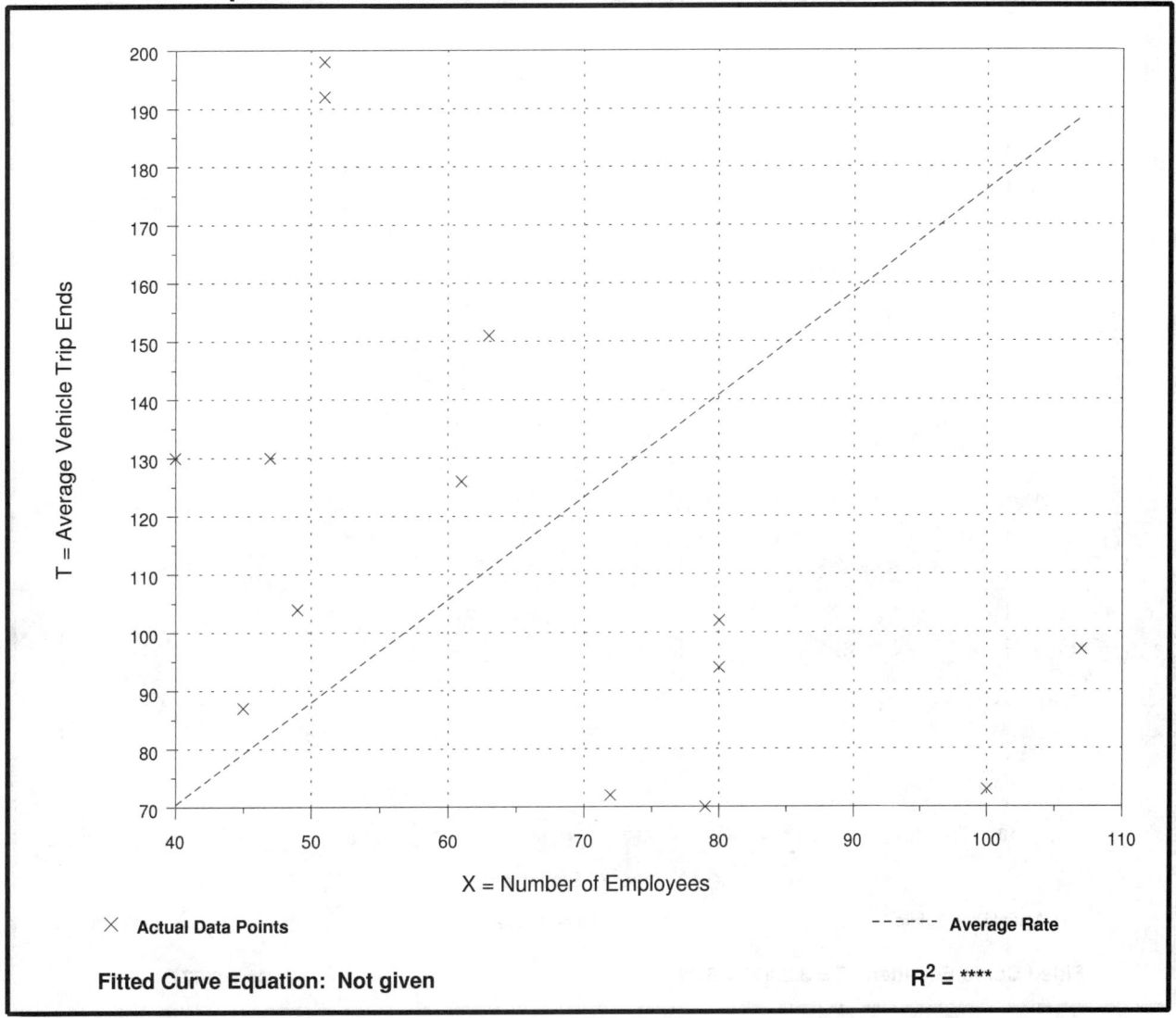

Fitted Curve Equation: Not given $R^2 = ****$

Elementary School
(520)

Average Vehicle Trip Ends vs: Employees
On a: Weekday,
P.M. Peak Hour of Generator

Number of Studies: 33
Avg. Number of Employees: 51
Directional Distribution: 44% entering, 56% exiting

Trip Generation per Employee

Average Rate	Range of Rates	Standard Deviation
3.41	1.03 - 6.68	2.24

Data Plot and Equation

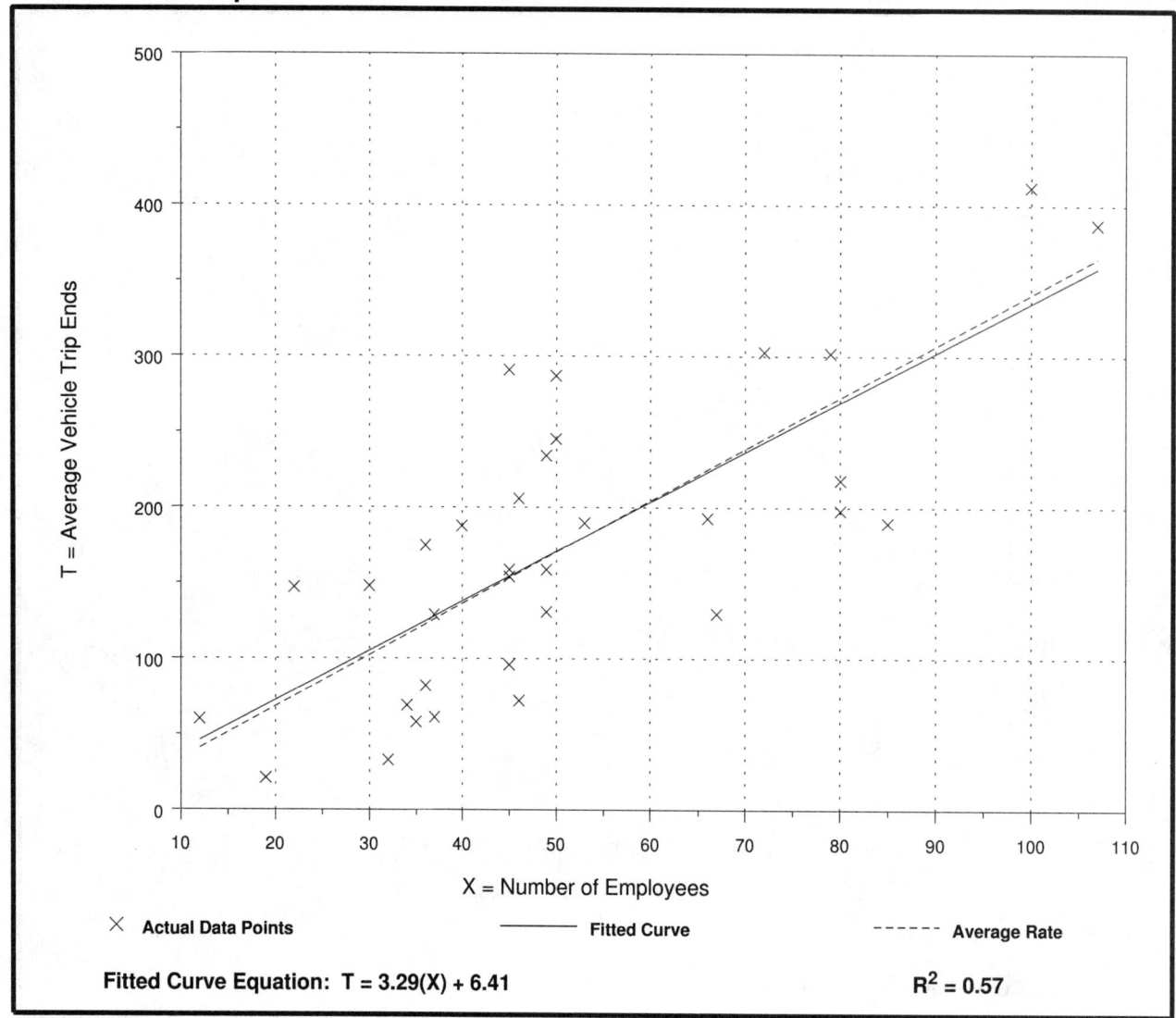

Fitted Curve Equation: T = 3.29(X) + 6.41 $R^2 = 0.57$

Elementary School
(520)

Average Vehicle Trip Ends vs: 1000 Sq. Feet Gross Floor Area
On a: Weekday

Number of Studies: 34
Average 1000 Sq. Feet GFA: 57
Directional Distribution: 50% entering, 50% exiting

Trip Generation per 1000 Sq. Feet Gross Floor Area

Average Rate	Range of Rates	Standard Deviation
15.43	4.69 - 30.15	7.81

Data Plot and Equation

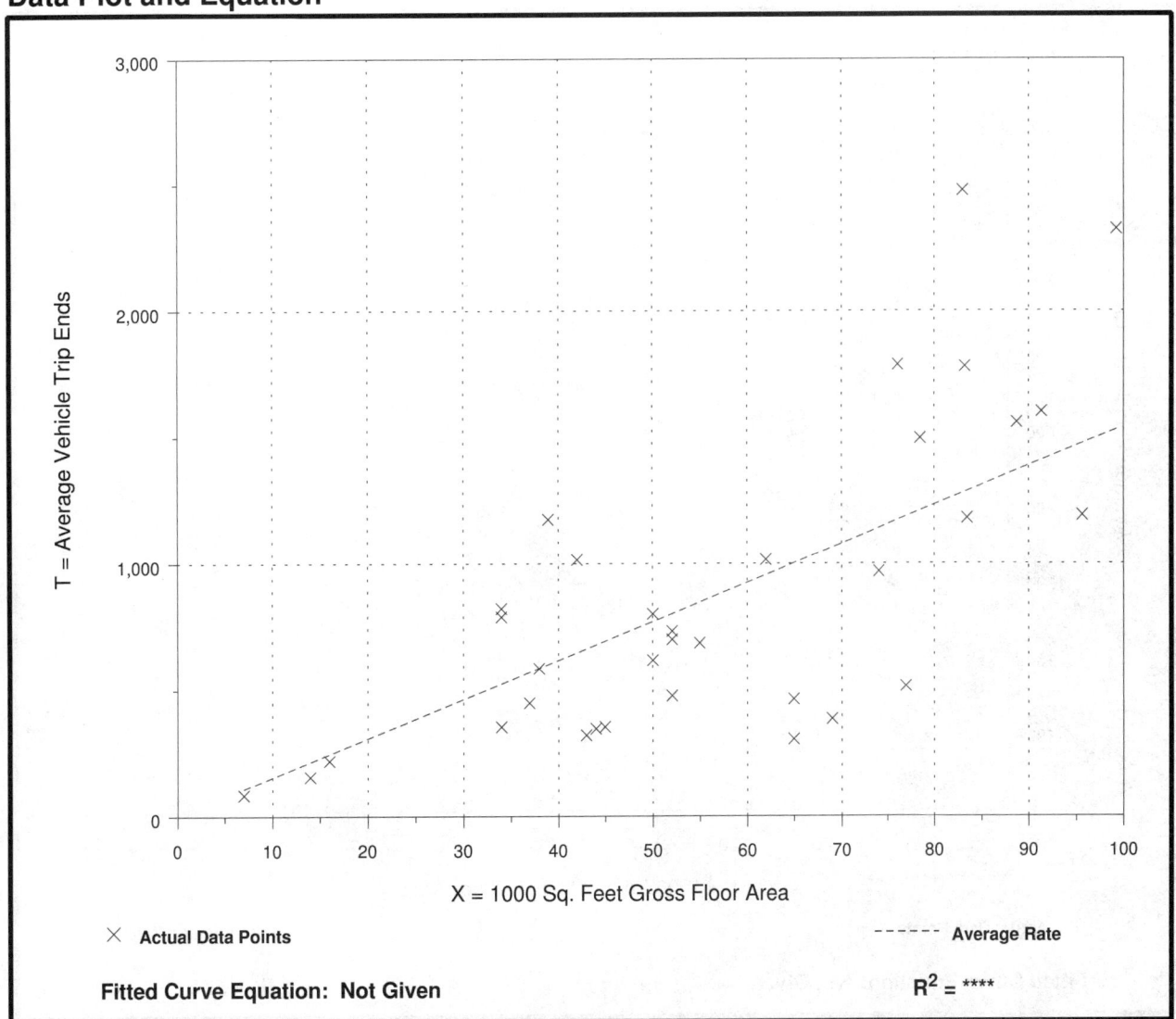

Fitted Curve Equation: Not Given

$R^2 =$ ****

Elementary School
(520)

Average Vehicle Trip Ends vs: 1000 Sq. Feet Gross Floor Area
On a: Weekday,
A.M. Peak Hour

Number of Studies: 35
Average 1000 Sq. Feet GFA: 58
Directional Distribution: 56% entering, 44% exiting

Trip Generation per 1000 Sq. Feet Gross Floor Area

Average Rate	Range of Rates	Standard Deviation
5.20	1.33 - 11.95	3.54

Data Plot and Equation

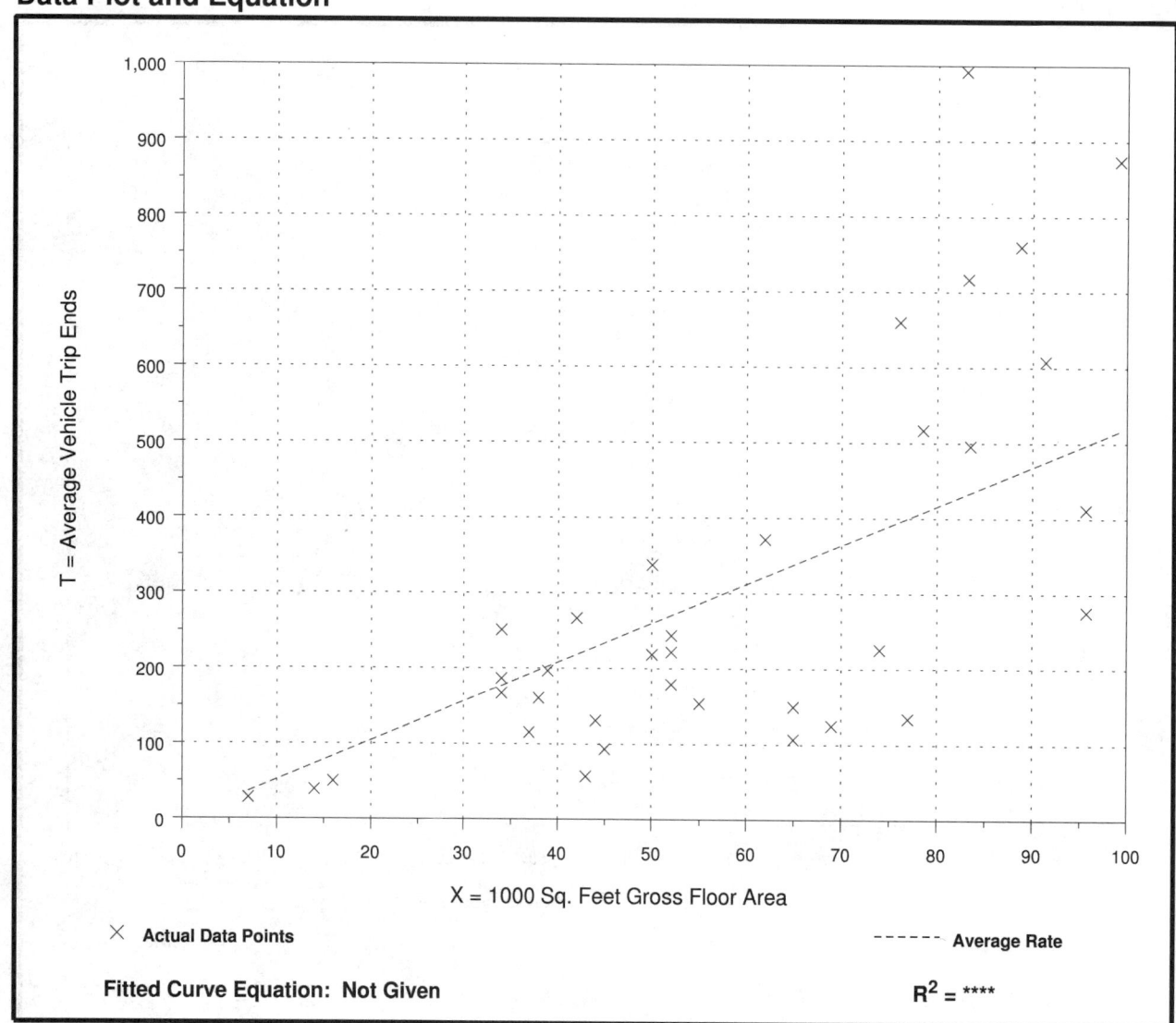

Fitted Curve Equation: Not Given $R^2 = ****$

Elementary School
(520)

Average Vehicle Trip Ends vs: 1000 Sq. Feet Gross Floor Area
On a: Weekday,
Peak Hour of Adjacent Street Traffic,
One Hour Between 4 and 6 p.m.

Number of Studies: 10
Average 1000 Sq. Feet GFA: 84
Directional Distribution: 45% entering, 55% exiting

Trip Generation per 1000 Sq. Feet Gross Floor Area

Average Rate	Range of Rates	Standard Deviation
1.21	0.79 - 2.25	1.20

Data Plot and Equation

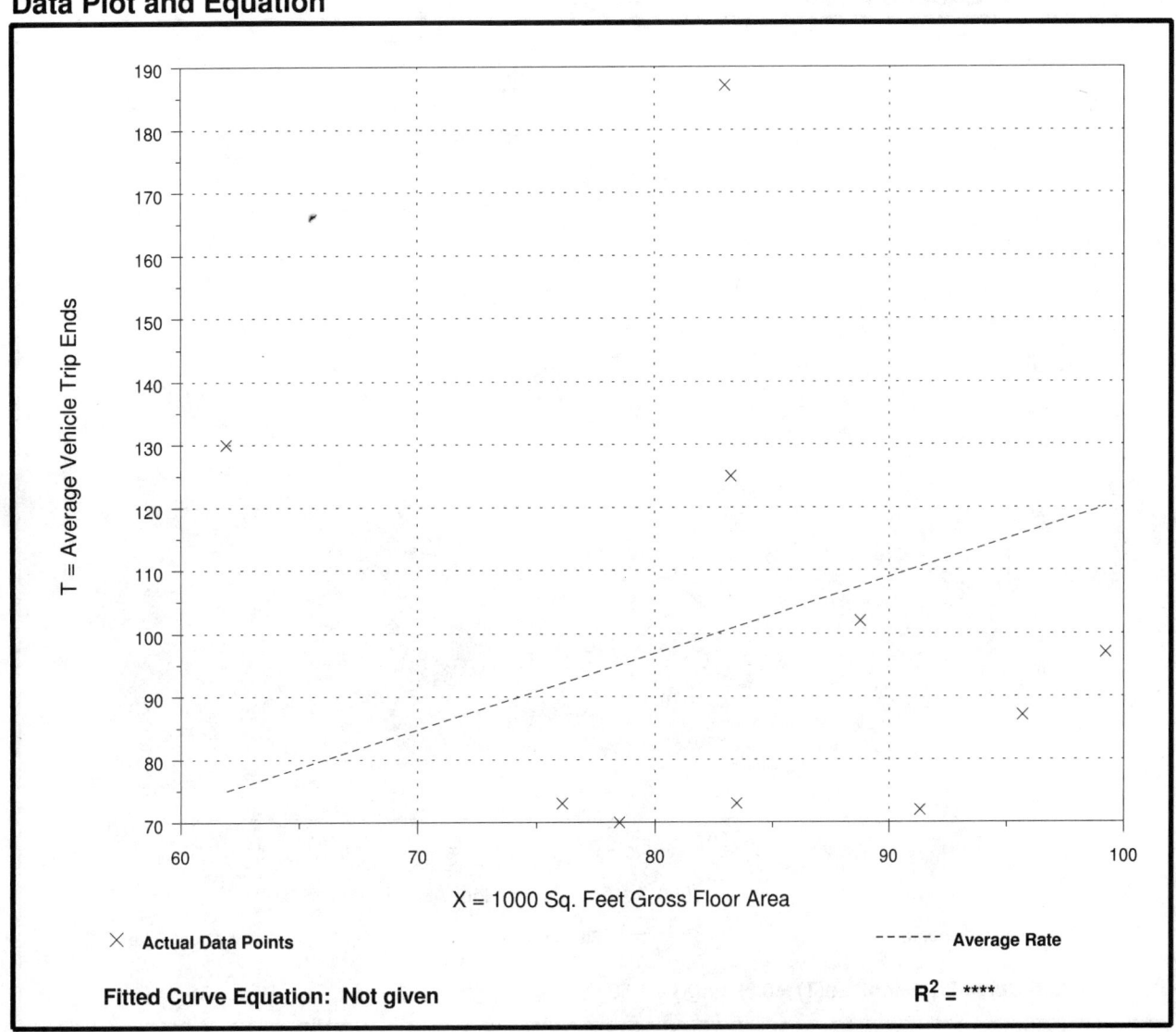

× Actual Data Points ----- Average Rate

Fitted Curve Equation: Not given $R^2 = ****$

Trip Generation, 9th Edition • Institute of Transportation Engineers

Elementary School
(520)

Average Vehicle Trip Ends vs: 1000 Sq. Feet Gross Floor Area
On a: Weekday,
P.M. Peak Hour of Generator

Number of Studies: 35
Average 1000 Sq. Feet GFA: 58
Directional Distribution: 44% entering, 56% exiting

Trip Generation per 1000 Sq. Feet Gross Floor Area

Average Rate	Range of Rates	Standard Deviation
3.11	0.94 - 6.06	2.17

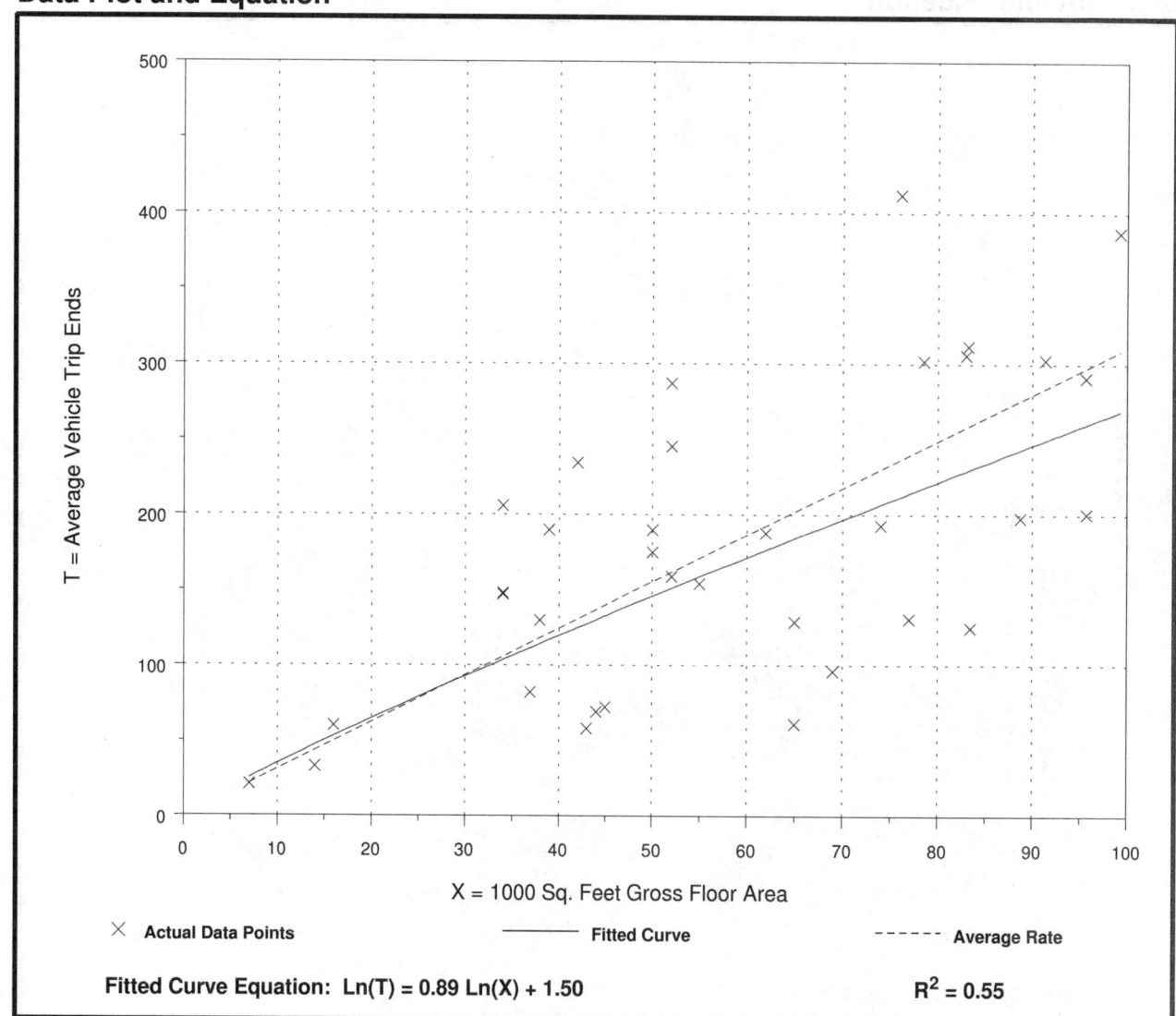

Data Plot and Equation

Fitted Curve Equation: $Ln(T) = 0.89 \, Ln(X) + 1.50$ $R^2 = 0.55$

Land Use: 522
Middle School/Junior High School

Description

Middle or junior high schools serve students who have completed elementary school and have not yet entered high school. Both public and private middle schools/junior high schools are included in this land use. Elementary school (Land Use 520), high school (Land Use 530), private school (K-8) (Land Use 534) and private school (K-12) (Land Use 536) are related uses.

Additional Data

Average weekday transit trip ends—
> The percentage of students at the sites who were transported to school via bus varied considerably. Due to the varied transit and school bus usage at these sites, it is desirable that future studies include additional detail on the percentage of students who were bused to school and the percentage that were dropped off and picked up.

Because the ratio of floor space to student population varies widely among the schools surveyed, the number of students may be a more reliable independent variable on which to establish trip generation rates.

Peak hours of the generator—
> The weekday A.M. peak hour of the generator typically coincided with the peak hour of the adjacent street traffic; therefore, only one A.M. peak hour, which represents both the peak hour of the generator and the peak hour of the adjacent street traffic, is displayed. The weekday P.M. peak hour varied between 2:00 p.m. and 4:00 p.m.

The sites were surveyed between the 1990s and the 2000s throughout the United States.

Many of the studies included in this land use did not indicate if the sites surveyed were public, private, or charter schools. To assist in the future analysis of this land use, it is important that this information be collected and included in trip generation data submissions.

Source Numbers

86, 422, 431, 444, 534, 536, 564, 579, 592, 611, 719

Middle School/Junior High School
(522)

Average Vehicle Trip Ends vs: Students
On a: Weekday

Number of Studies: 20
Average Number of Students: 904
Directional Distribution: 50% entering, 50% exiting

Trip Generation per Student

Average Rate	Range of Rates	Standard Deviation
1.62	0.72 - 2.81	1.45

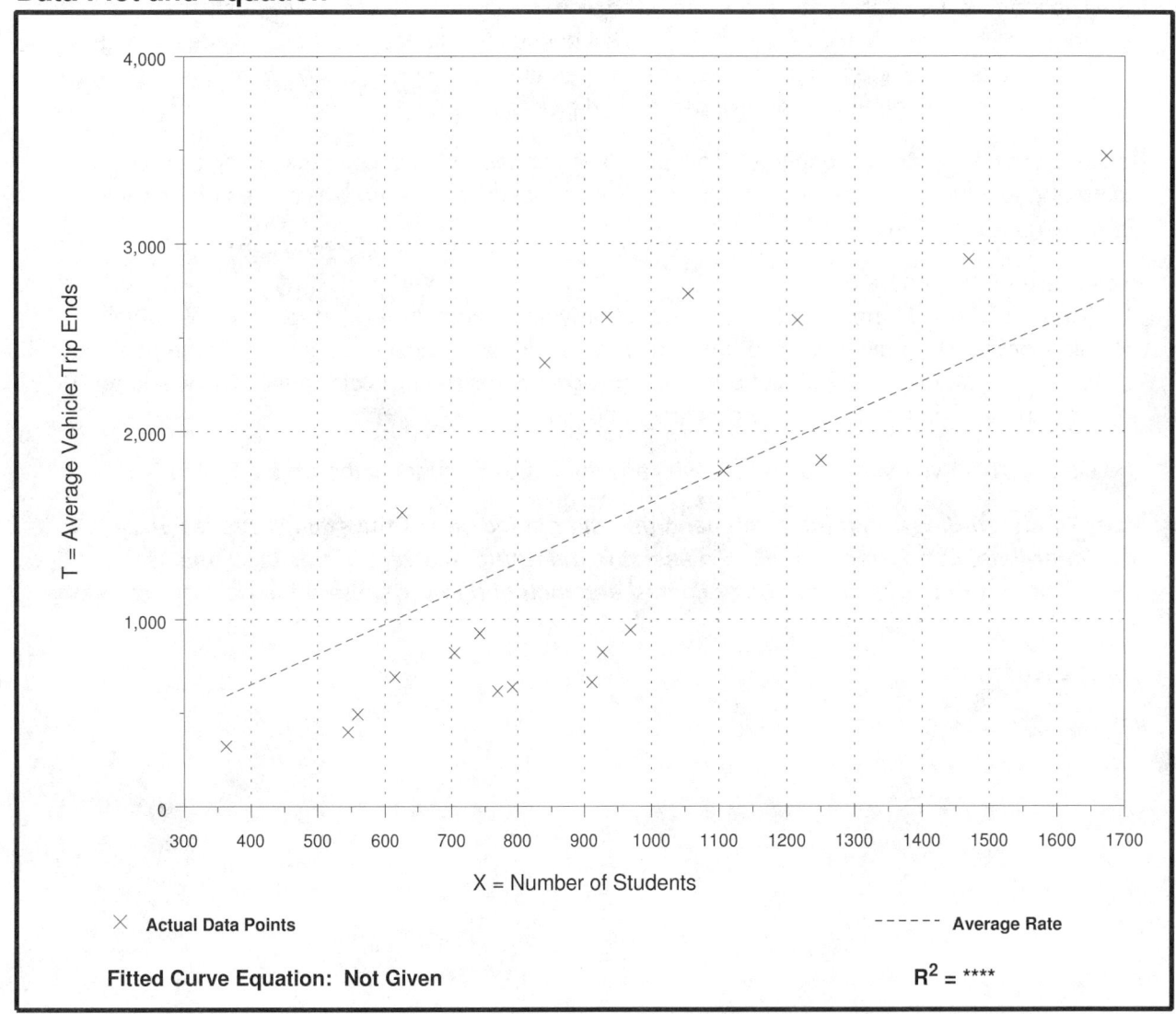

Data Plot and Equation

Fitted Curve Equation: Not Given $R^2 = ****$

Middle School/Junior High School
(522)

Average Vehicle Trip Ends vs: Students
On a: Weekday,
A.M. Peak Hour

Number of Studies: 25
Average Number of Students: 876
Directional Distribution: 55% entering, 45% exiting

Trip Generation per Student

Average Rate	Range of Rates	Standard Deviation
0.54	0.14 - 1.29	0.80

Data Plot and Equation

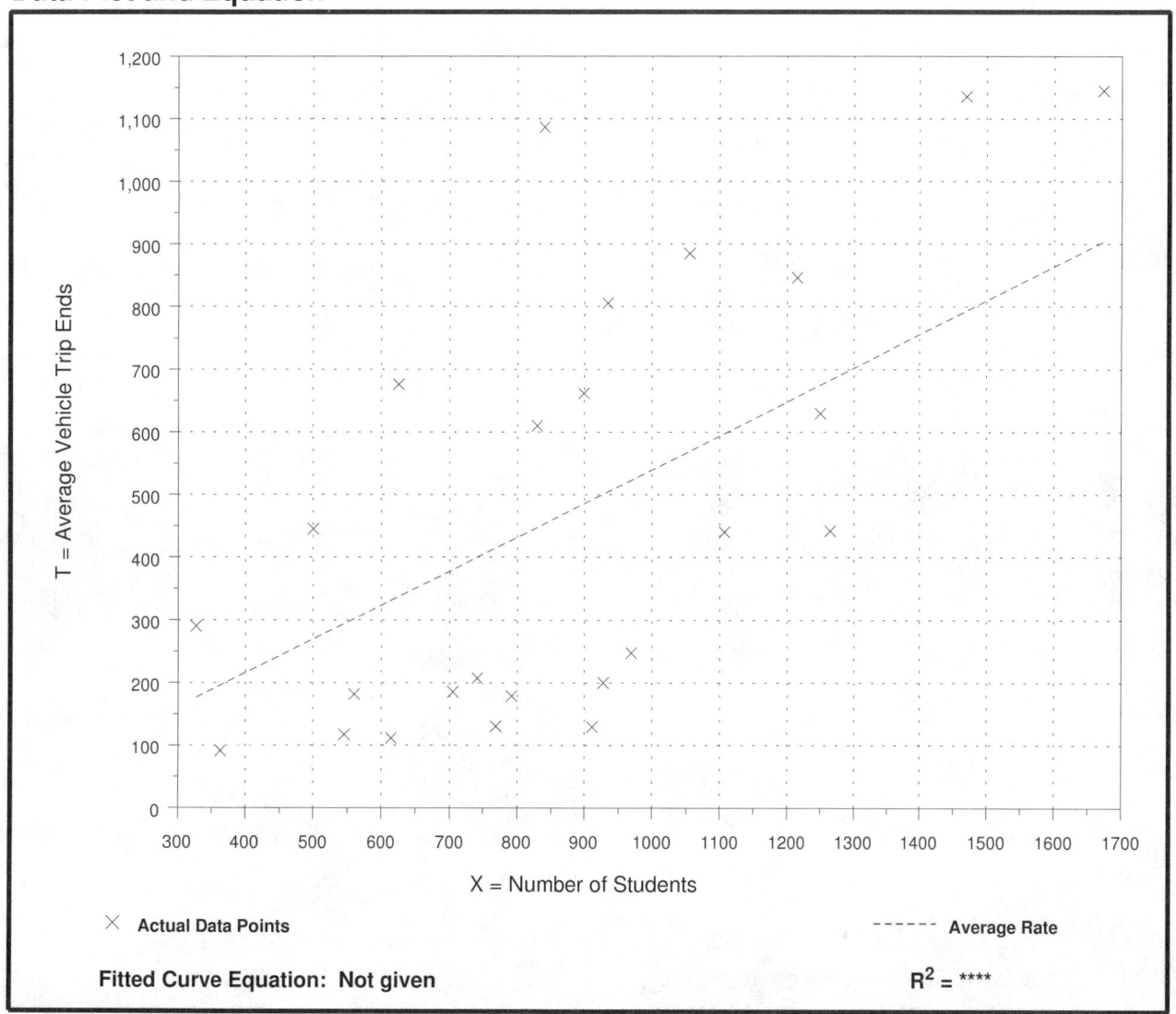

Fitted Curve Equation: Not given $R^2 = ****$

Middle School/Junior High School
(522)

Average Vehicle Trip Ends vs: Students
On a: Weekday,
Peak Hour of Adjacent Street Traffic,
One Hour Between 4 and 6 p.m.

Number of Studies: 16
Average Number of Students: 982
Directional Distribution: 49% entering, 51% exiting

Trip Generation per Student

Average Rate	Range of Rates	Standard Deviation
0.16	0.06 - 0.36	0.40

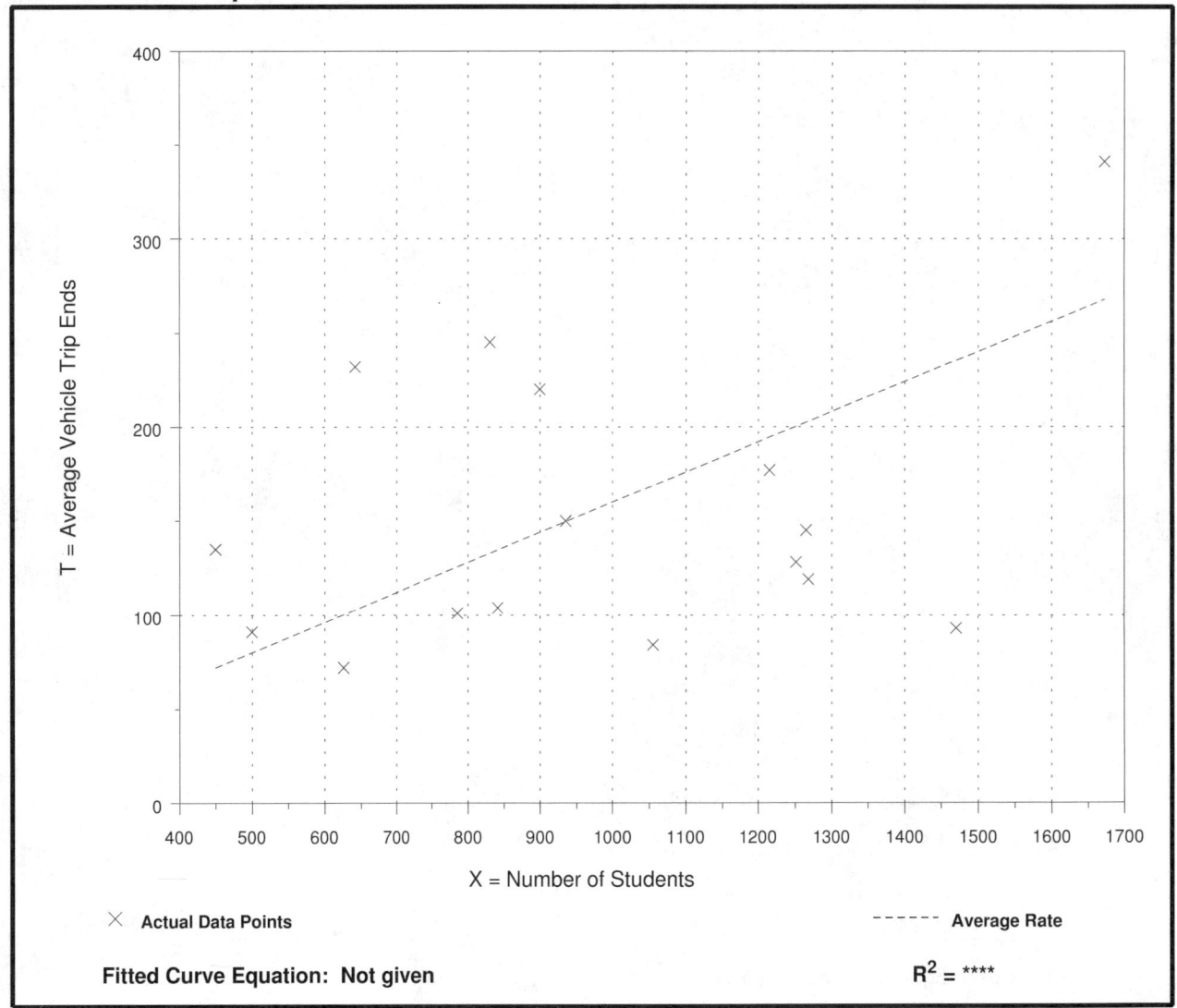

Data Plot and Equation

Fitted Curve Equation: Not given $R^2 =$ ****

Middle School/Junior High School
(522)

Average Vehicle Trip Ends vs: Students
On a: Weekday,
P.M. Peak Hour of Generator

Number of Studies: 25
Average Number of Students: 843
Directional Distribution: 45% entering, 55% exiting

Trip Generation per Student

Average Rate	Range of Rates	Standard Deviation
0.30	0.12 - 0.63	0.56

Data Plot and Equation

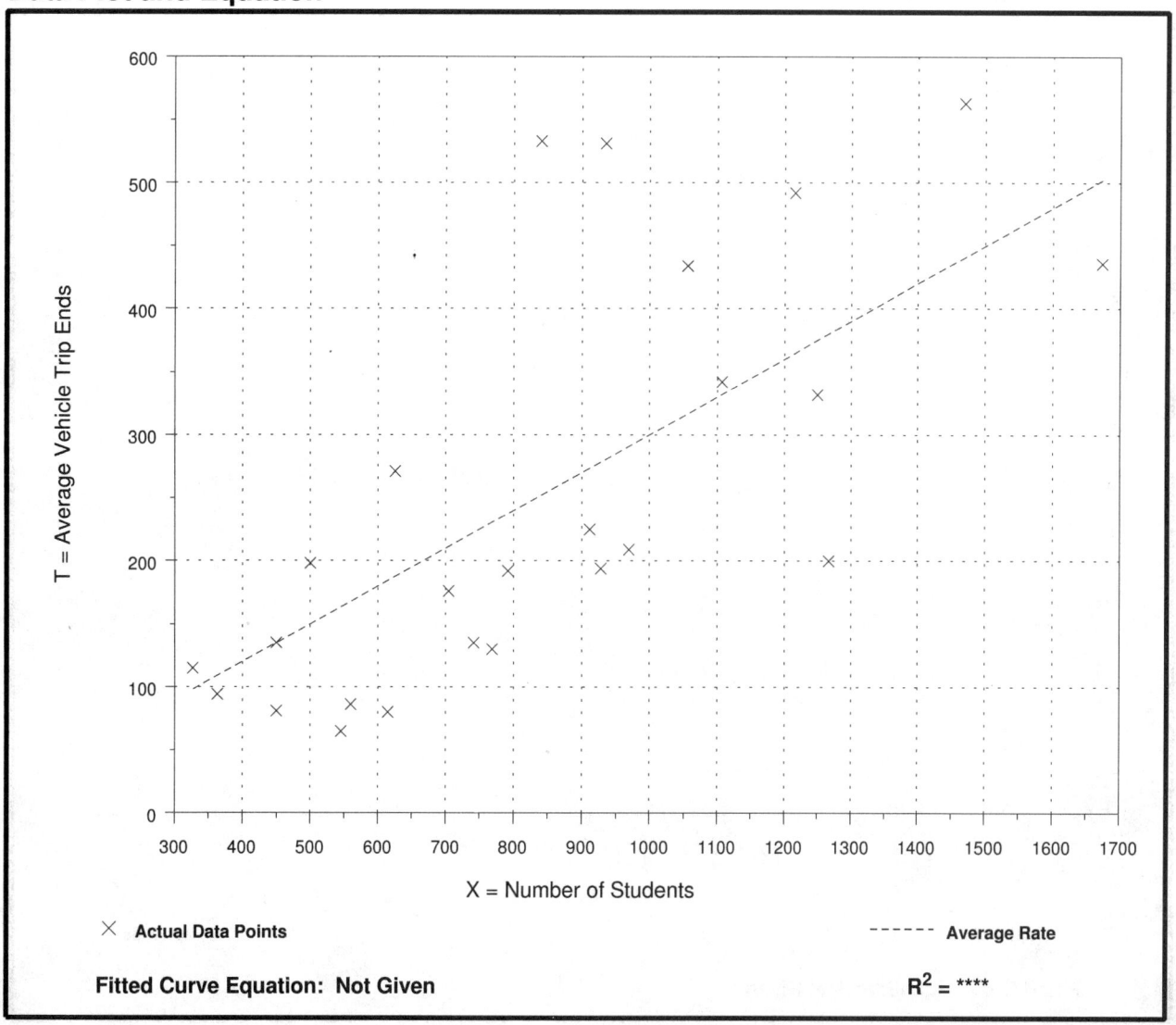

Fitted Curve Equation: Not Given $R^2 = ****$

Middle School/Junior High School
(522)

Average Vehicle Trip Ends vs: Employees
On a: Weekday

Number of Studies: 16
Avg. Number of Employees: 77
Directional Distribution: 50% entering, 50% exiting

Trip Generation per Employee

Average Rate	Range of Rates	Standard Deviation
16.39	6.96 - 28.47	8.48

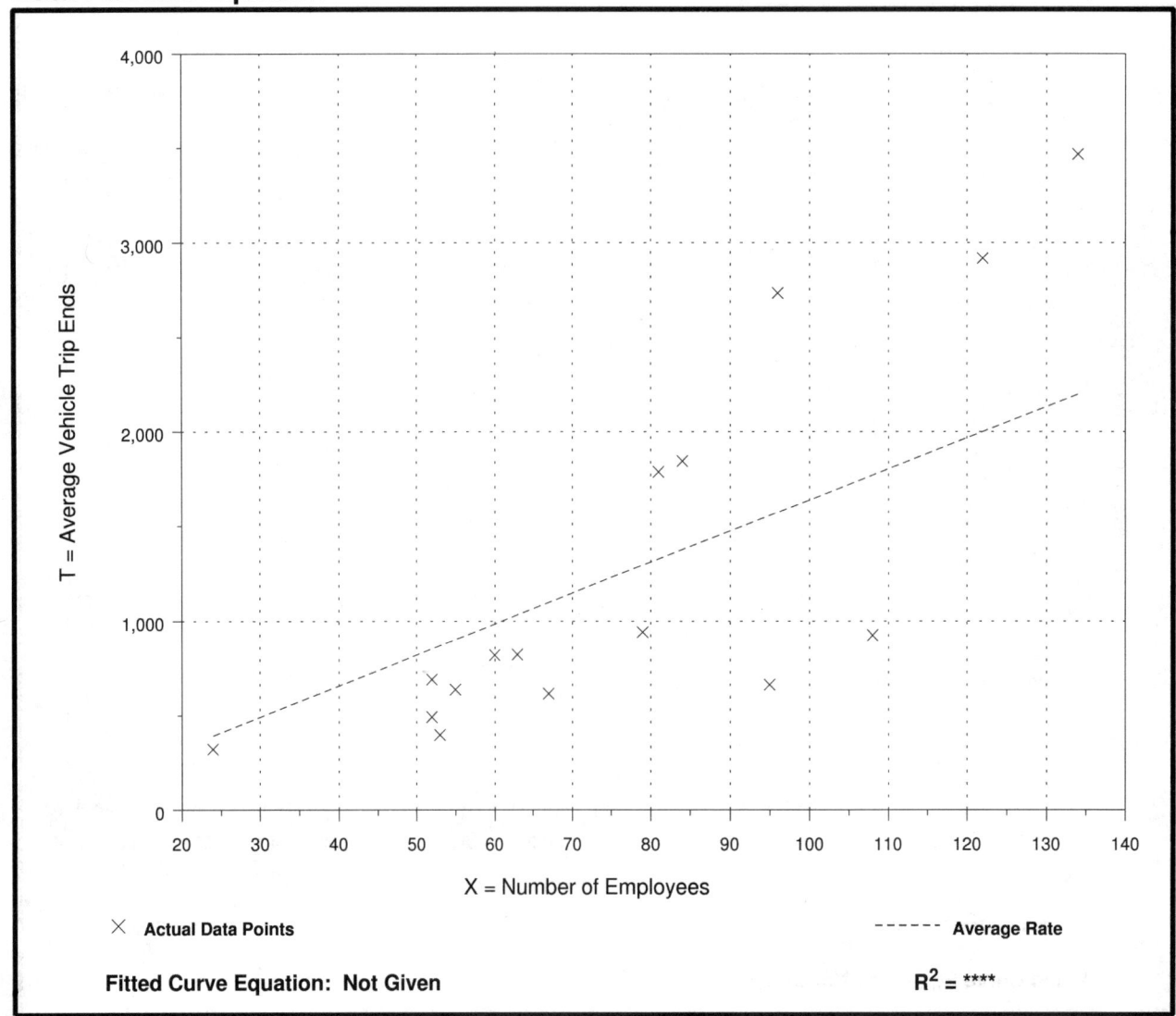

Data Plot and Equation

Fitted Curve Equation: Not Given $R^2 = ****$

Middle School/Junior High School
(522)

Average Vehicle Trip Ends vs: Employees
On a: Weekday,
A.M. Peak Hour

Number of Studies: 21
Avg. Number of Employees: 76
Directional Distribution: 54% entering, 46% exiting

Trip Generation per Employee

Average Rate	Range of Rates	Standard Deviation
5.30	1.37 - 9.30	3.64

Data Plot and Equation

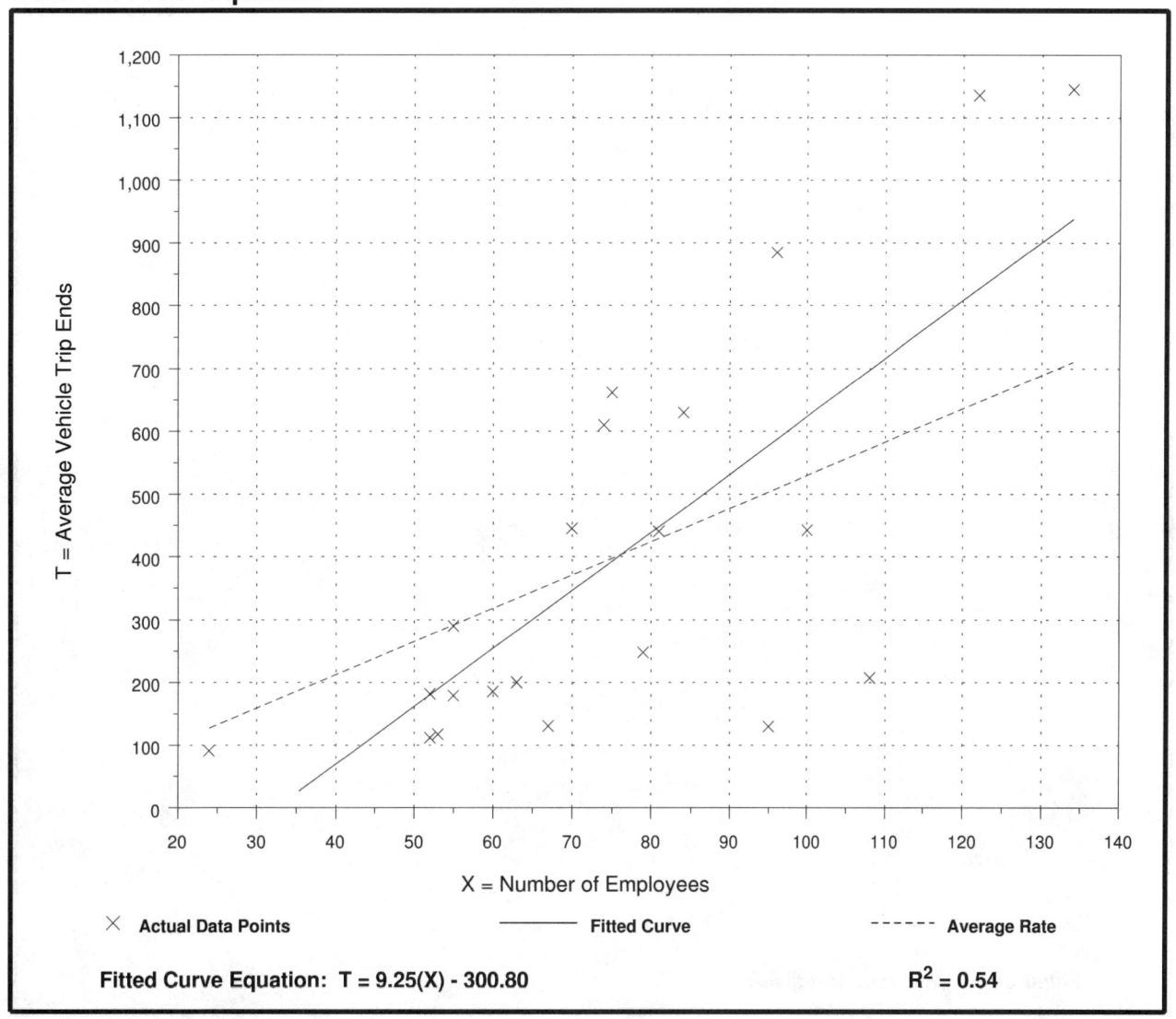

Fitted Curve Equation: T = 9.25(X) - 300.80 $R^2 = 0.54$

Middle School/Junior High School
(522)

Average Vehicle Trip Ends vs: Employees
On a: Weekday,
Peak Hour of Adjacent Street Traffic,
One Hour Between 4 and 6 p.m.

Number of Studies: 9
Avg. Number of Employees: 90
Directional Distribution: 50% entering, 50% exiting

Trip Generation per Employee

Average Rate	Range of Rates	Standard Deviation
1.94	0.76 - 3.93	1.72

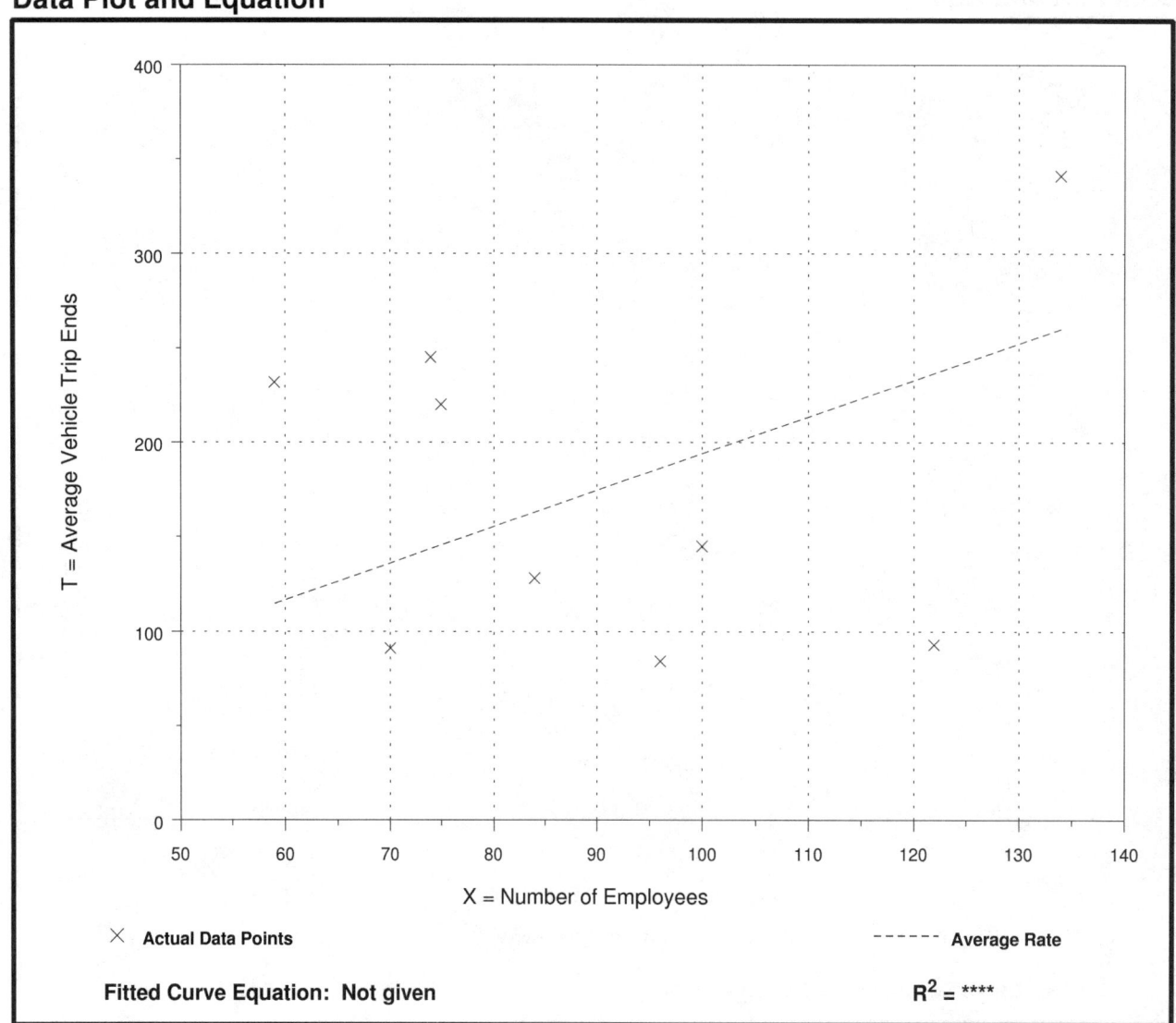

Data Plot and Equation

Fitted Curve Equation: Not given $R^2 = ****$

Middle School/Junior High School
(522)

Average Vehicle Trip Ends vs: Employees
On a: Weekday,
P.M. Peak Hour of Generator

Number of Studies: 18
Avg. Number of Employees: 75
Directional Distribution: 38% entering, 62% exiting

Trip Generation per Employee

Average Rate	Range of Rates	Standard Deviation
2.97	1.23 - 4.61	2.04

Data Plot and Equation

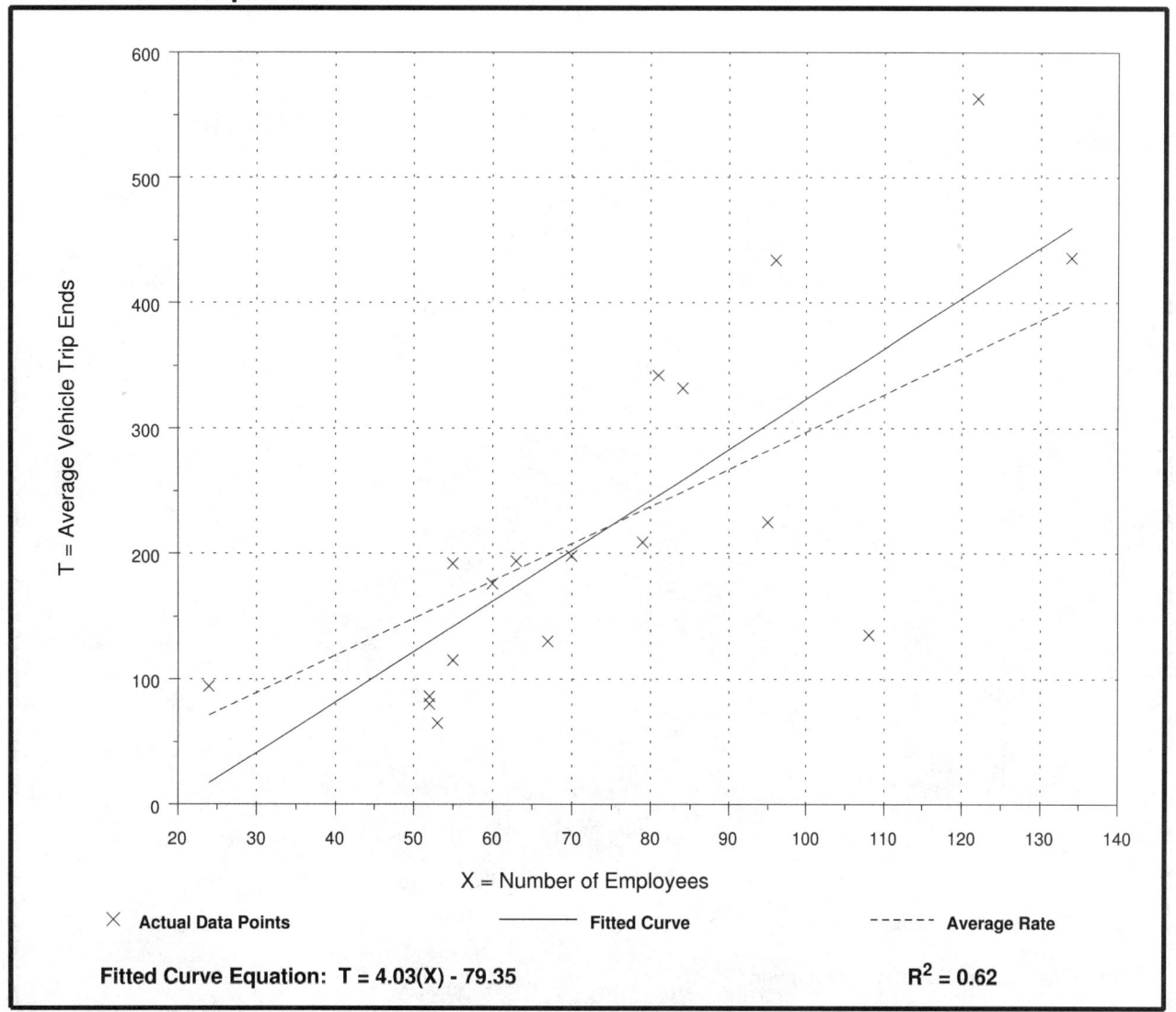

Fitted Curve Equation: $T = 4.03(X) - 79.35$ $R^2 = 0.62$

Middle School/Junior High School
(522)

Average Vehicle Trip Ends vs: 1000 Sq. Feet Gross Floor Area
On a: Weekday

Number of Studies: 20
Average 1000 Sq. Feet GFA: 106
Directional Distribution: 50% entering, 50% exiting

Trip Generation per 1000 Sq. Feet Gross Floor Area

Average Rate	Range of Rates	Standard Deviation
13.78	3.89 - 48.31	9.07

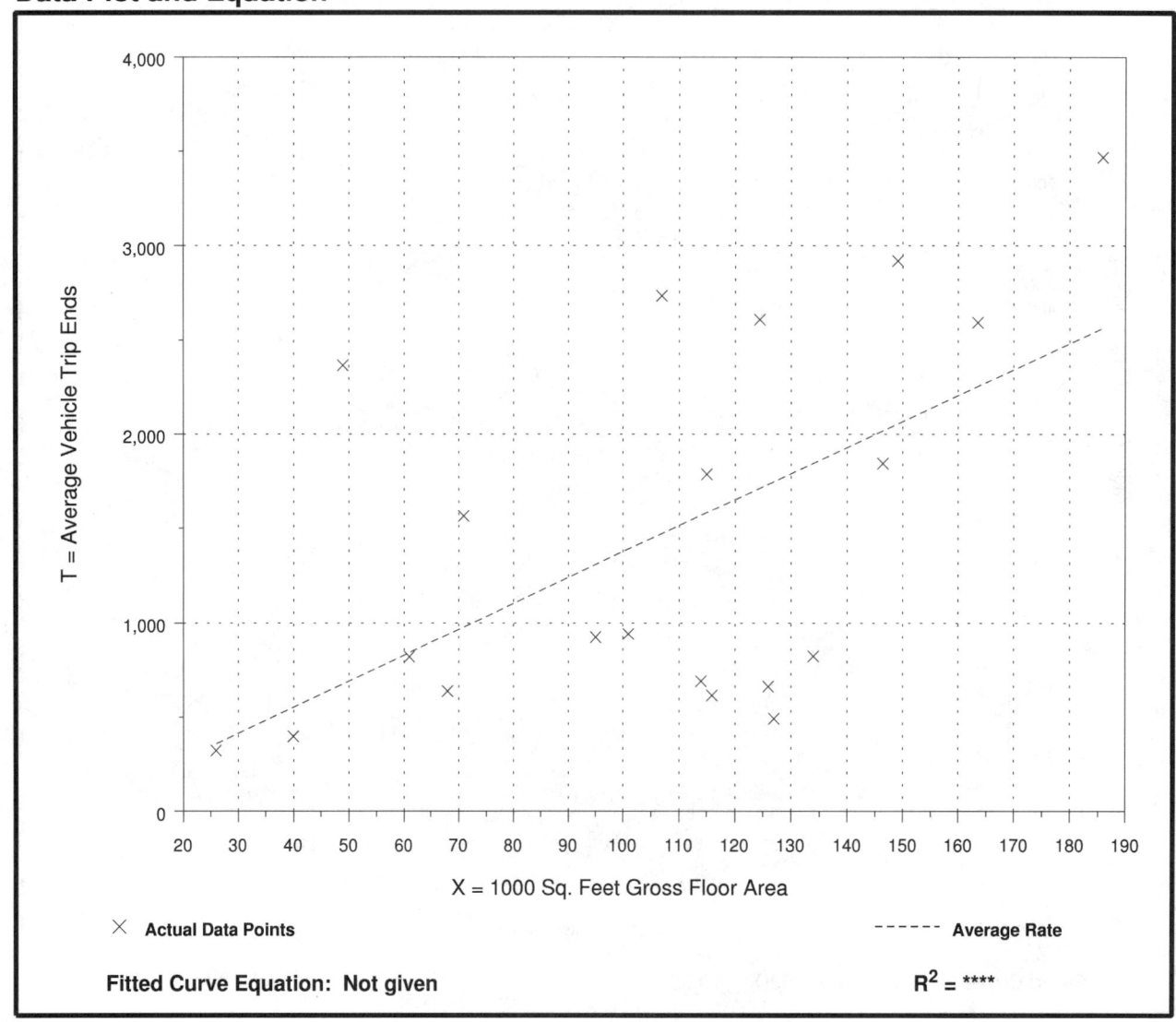

Data Plot and Equation

Fitted Curve Equation: Not given $R^2 = ****$

Middle School/Junior High School
(522)

Average Vehicle Trip Ends vs: 1000 Sq. Feet Gross Floor Area
On a: Weekday,
A.M. Peak Hour

Number of Studies: 21
Average 1000 Sq. Feet GFA: 107
Directional Distribution: 55% entering, 45% exiting

Trip Generation per 1000 Sq. Feet Gross Floor Area

Average Rate	Range of Rates	Standard Deviation
4.35	0.98 - 22.16	4.19

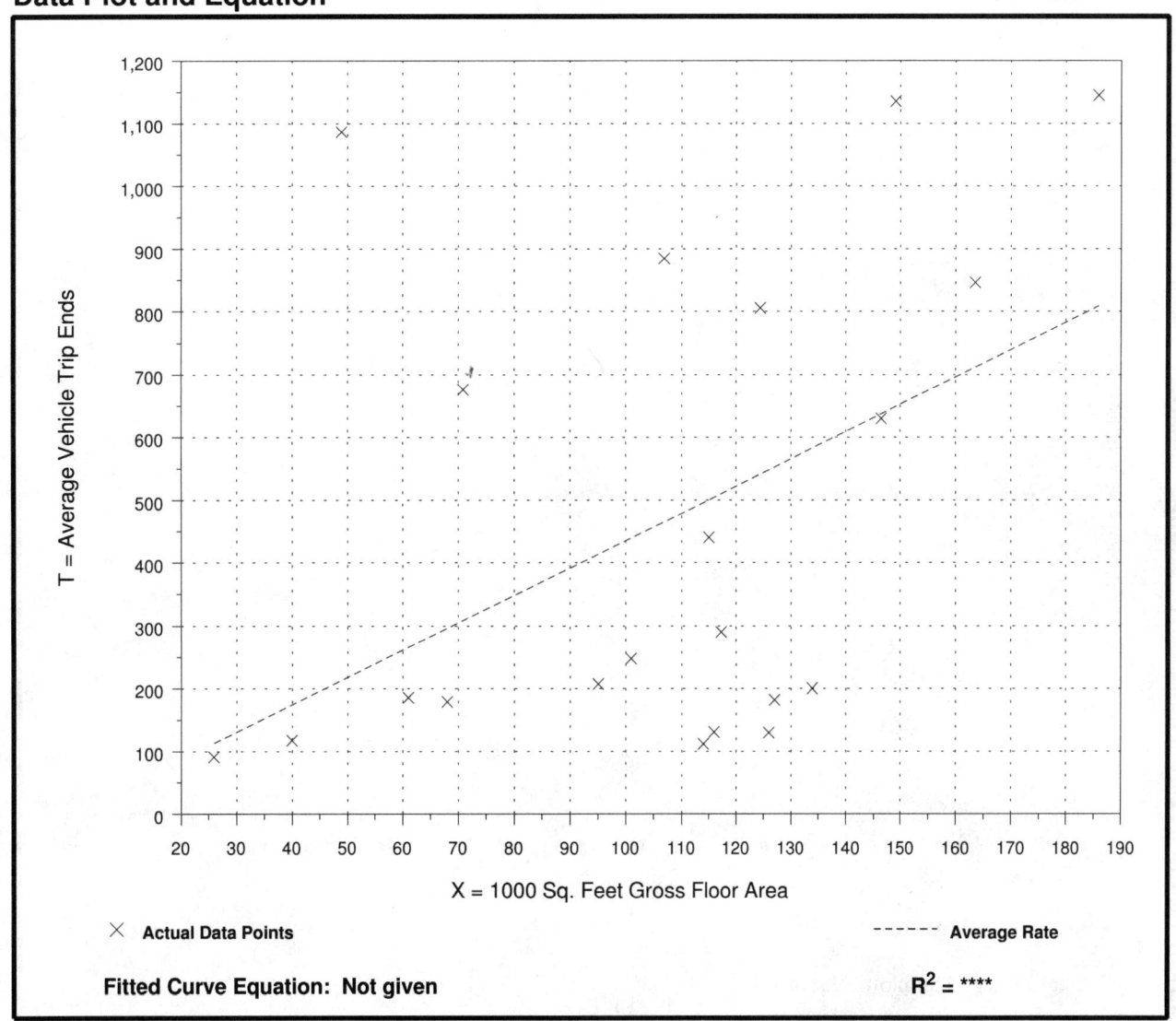

Data Plot and Equation

Fitted Curve Equation: Not given $R^2 = ****$

Middle School/Junior High School
(522)

Average Vehicle Trip Ends vs: 1000 Sq. Feet Gross Floor Area
On a: Weekday,
Peak Hour of Adjacent Street Traffic,
One Hour Between 4 and 6 p.m.

Number of Studies: 9
Average 1000 Sq. Feet GFA: 116
Directional Distribution: 52% entering, 48% exiting

Trip Generation per 1000 Sq. Feet Gross Floor Area

Average Rate	Range of Rates	Standard Deviation
1.19	0.62 - 2.12	1.19

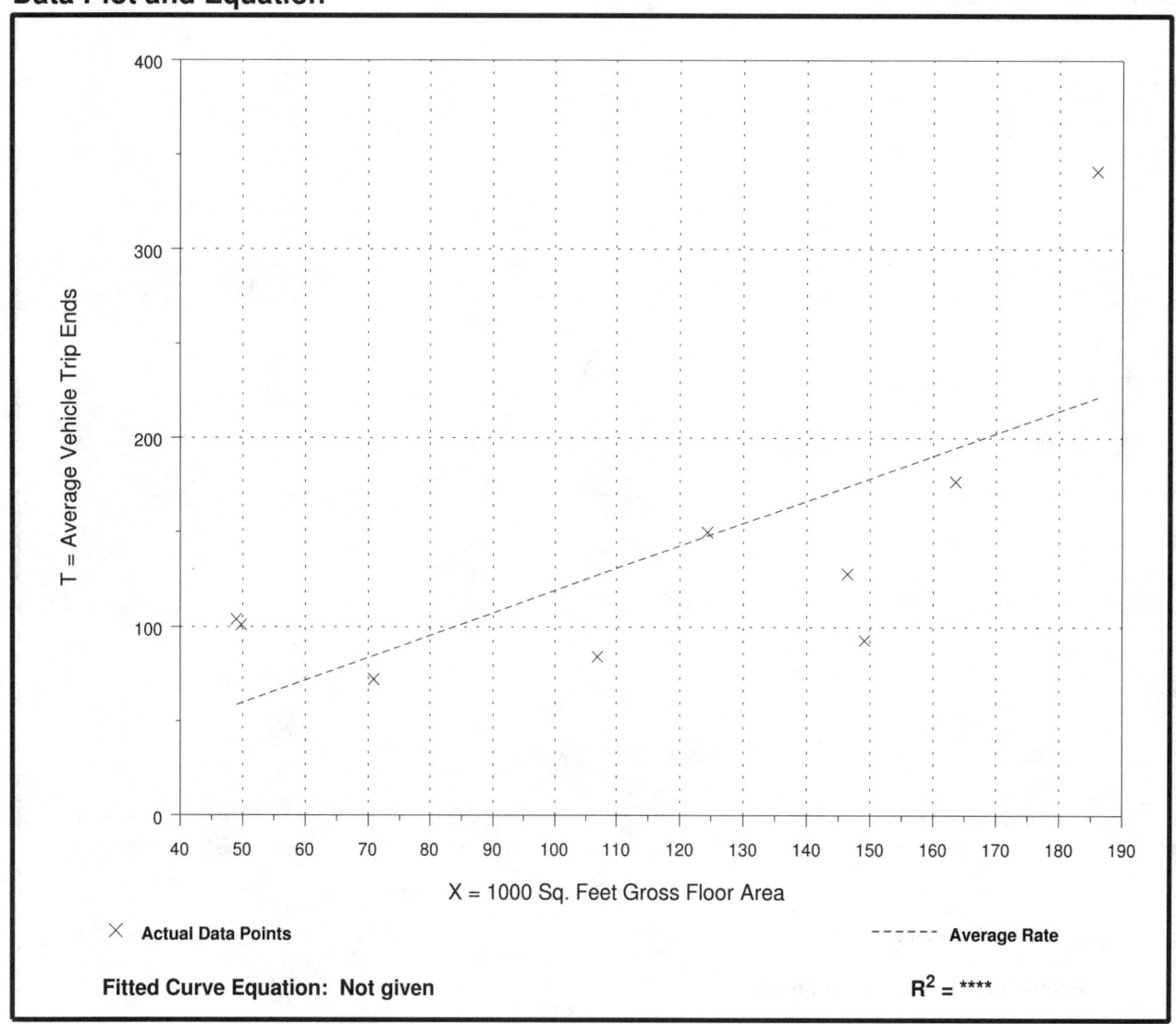

Data Plot and Equation

Fitted Curve Equation: Not given $R^2 =$ ****

Middle School/Junior High School
(522)

Average Vehicle Trip Ends vs: 1000 Sq. Feet Gross Floor Area
On a: Weekday,
P.M. Peak Hour of Generator

Number of Studies: 21
Average 1000 Sq. Feet GFA: 107
Directional Distribution: 45% entering, 55% exiting

Trip Generation per 1000 Sq. Feet Gross Floor Area

Average Rate	Range of Rates	Standard Deviation
2.52	0.68 - 10.88	2.30

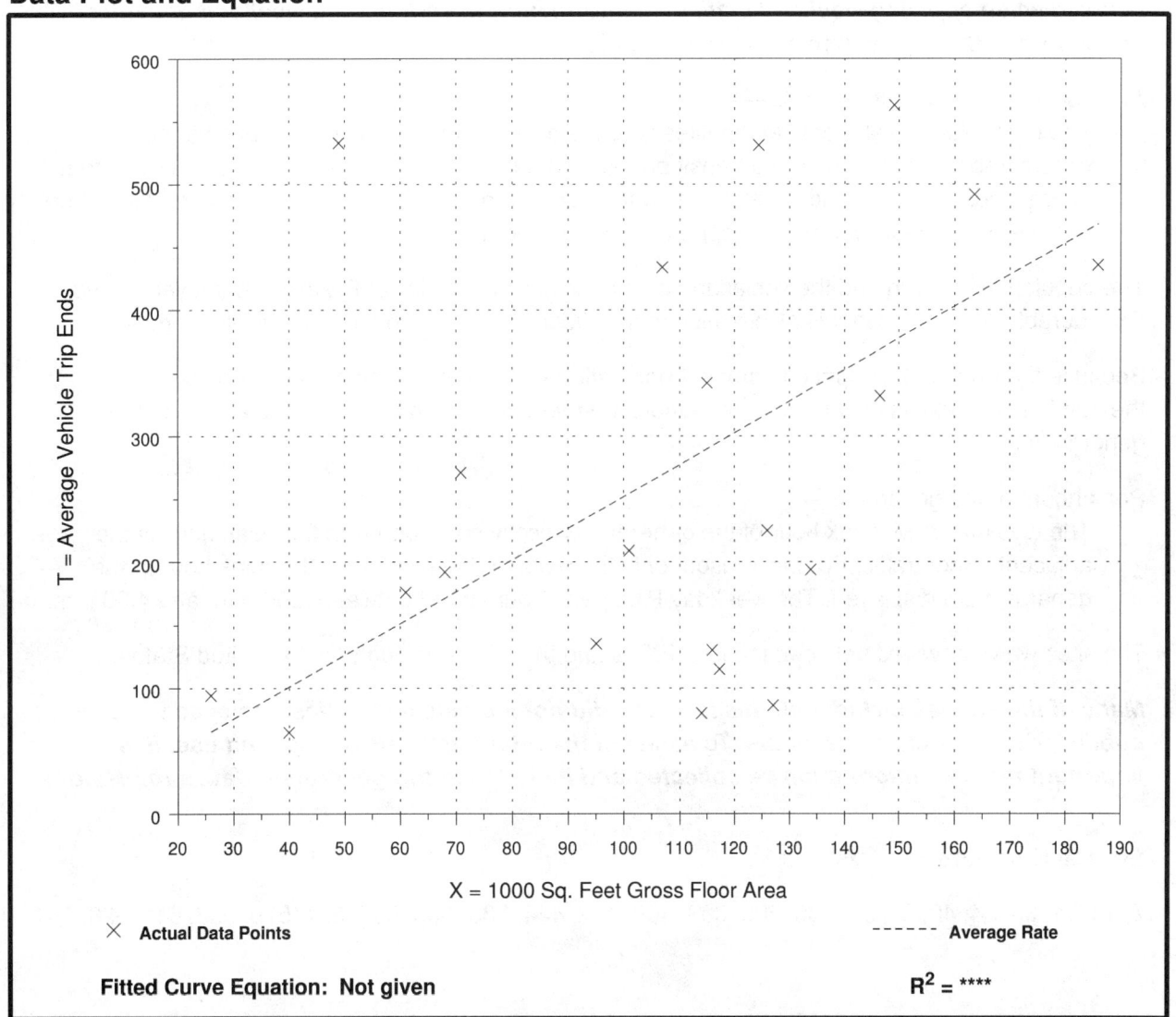

Data Plot and Equation

X   Actual Data Points   ----- Average Rate

Fitted Curve Equation: Not given $R^2 = ****$

Land Use: 530
High School

Description

High schools serve students who have completed middle or junior high school. Both public and private high schools are included in this land use. Elementary school (Land Use 520), middle school/junior high school (Land Use 522), private school (K-8) (Land Use 534) and private school (K-12) (Land Use 536) are related uses.

Additional Data

The trip generation for weekend time periods varied considerably; therefore, caution should be used when applying weekend statistics. Information describing the weekend activities conducted at the high schools was not available.

Average weekday transit trip ends—
   The percentage of students at the sites who were transported to school via bus varied considerably. Due to the varied transit and school bus usage at these sites, it is desirable that future studies include additional detail on the percentage of students who were bused to school and the percentage who were dropped off and picked up.

The populations served and the social and economic characteristics of the areas surveyed varied considerably. The high schools also exhibited significant variations in terms of facilities provided.

Because the ratio of floor space to student population varied widely among the schools surveyed, the number of students may be a more reliable independent variable on which to establish trip generation rates.

Peak hours of the generator—
   The weekday A.M. peak hour of the generator typically coincided with the peak hour of the adjacent street traffic. For this reason, only the plots for the weekday P.M. peak hour of the generator are displayed. The weekday P.M. peak hour varied between 2:00 p.m. and 4:00 p.m.

The sites were surveyed between the late 1960s and the 2000s throughout the United States.

Many of the studies included in this land use did not indicate if the sites surveyed were public, private, or charter schools. To assist in the future analysis of this land use, it is important that this information be collected and included in trip generation data submissions.

Source Numbers

7, 10, 31, 33, 34, 40, 86, 91, 186, 293, 383, 409, 422, 444, 533, 536, 550, 564, 579, 598, 611, 620, 751

High School
(530)

Average Vehicle Trip Ends vs: Students
On a: Weekday

Number of Studies: 51
Average Number of Students: 1,382
Directional Distribution: 50% entering, 50% exiting

Trip Generation per Student

Average Rate	Range of Rates	Standard Deviation
1.71	0.71 - 3.96	1.49

Data Plot and Equation

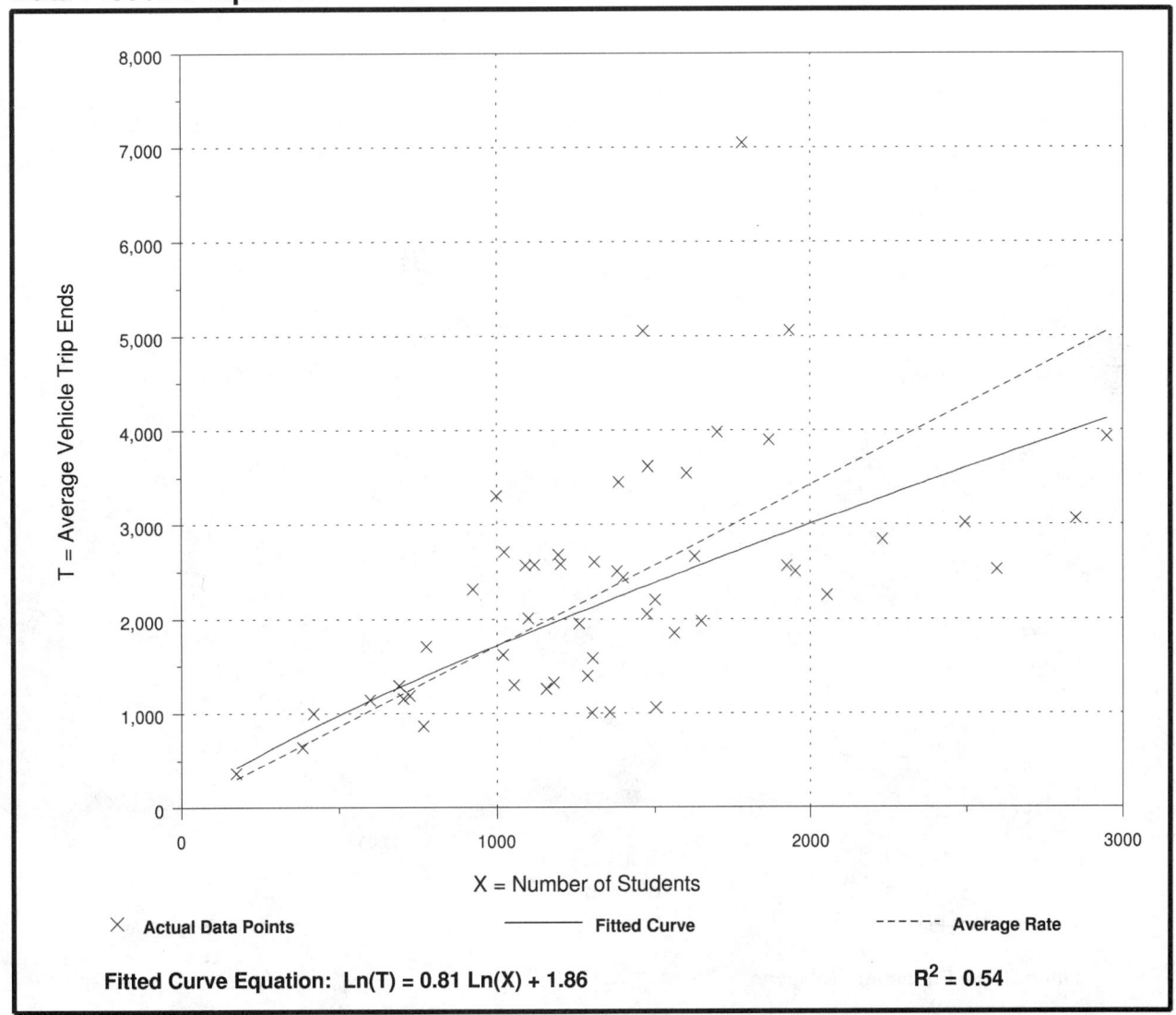

Fitted Curve Equation: $Ln(T) = 0.81 \, Ln(X) + 1.86$ $R^2 = 0.54$

High School
(530)

Average Vehicle Trip Ends vs: Students
On a: Weekday,
A.M. Peak Hour

Number of Studies: 75
Average Number of Students: 1,231
Directional Distribution: 68% entering, 32% exiting

Trip Generation per Student

Average Rate	Range of Rates	Standard Deviation
0.43	0.14 - 1.15	0.69

Data Plot and Equation

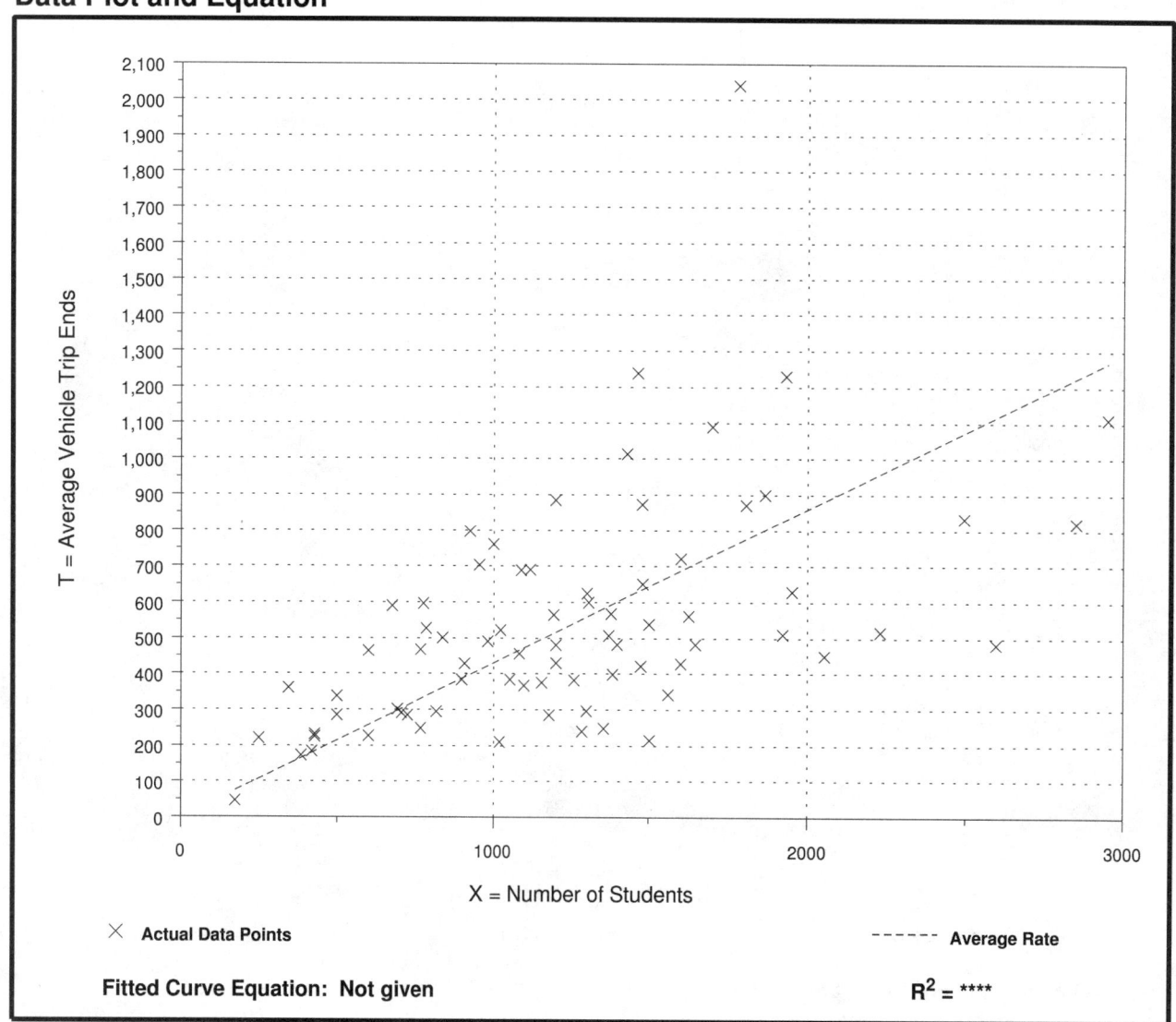

X = Actual Data Points
----- Average Rate

Fitted Curve Equation: Not given $R^2 = ****$

High School
(530)

Average Vehicle Trip Ends vs: Students
On a: Weekday,
Peak Hour of Adjacent Street Traffic,
One Hour Between 4 and 6 p.m.

Number of Studies: 40
Average Number of Students: 1,352
Directional Distribution: 47% entering, 53% exiting

Trip Generation per Student

Average Rate	Range of Rates	Standard Deviation
0.13	0.03 - 0.38	0.37

Data Plot and Equation

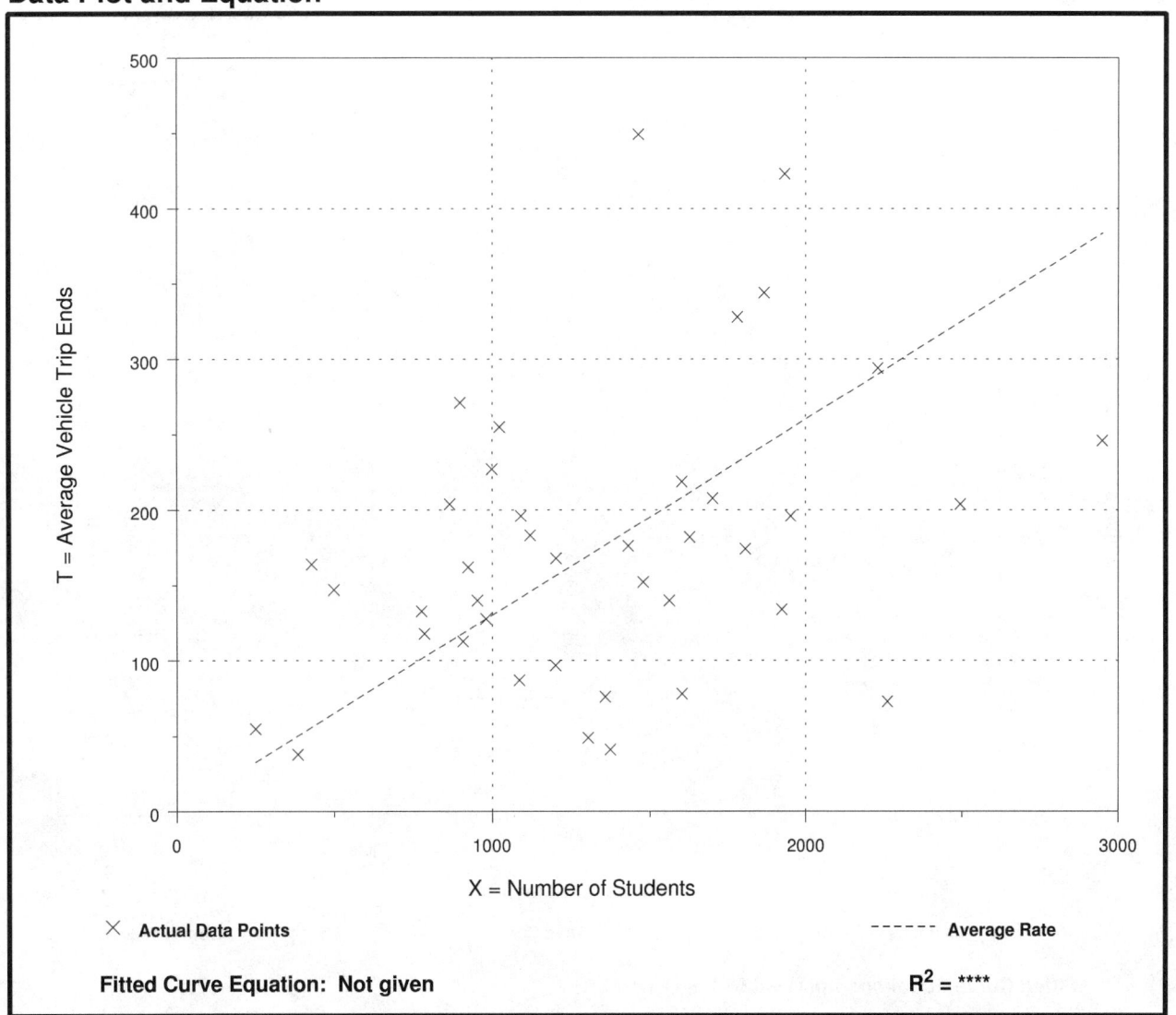

Fitted Curve Equation: Not given $R^2 = ****$

High School
(530)

Average Vehicle Trip Ends vs: Students
On a: Weekday,
P.M. Peak Hour of Generator

Number of Studies: 74
Average Number of Students: 1,235
Directional Distribution: 33% entering, 67% exiting

Trip Generation per Student

Average Rate	Range of Rates	Standard Deviation
0.29	0.10 - 0.74	0.55

Data Plot and Equation

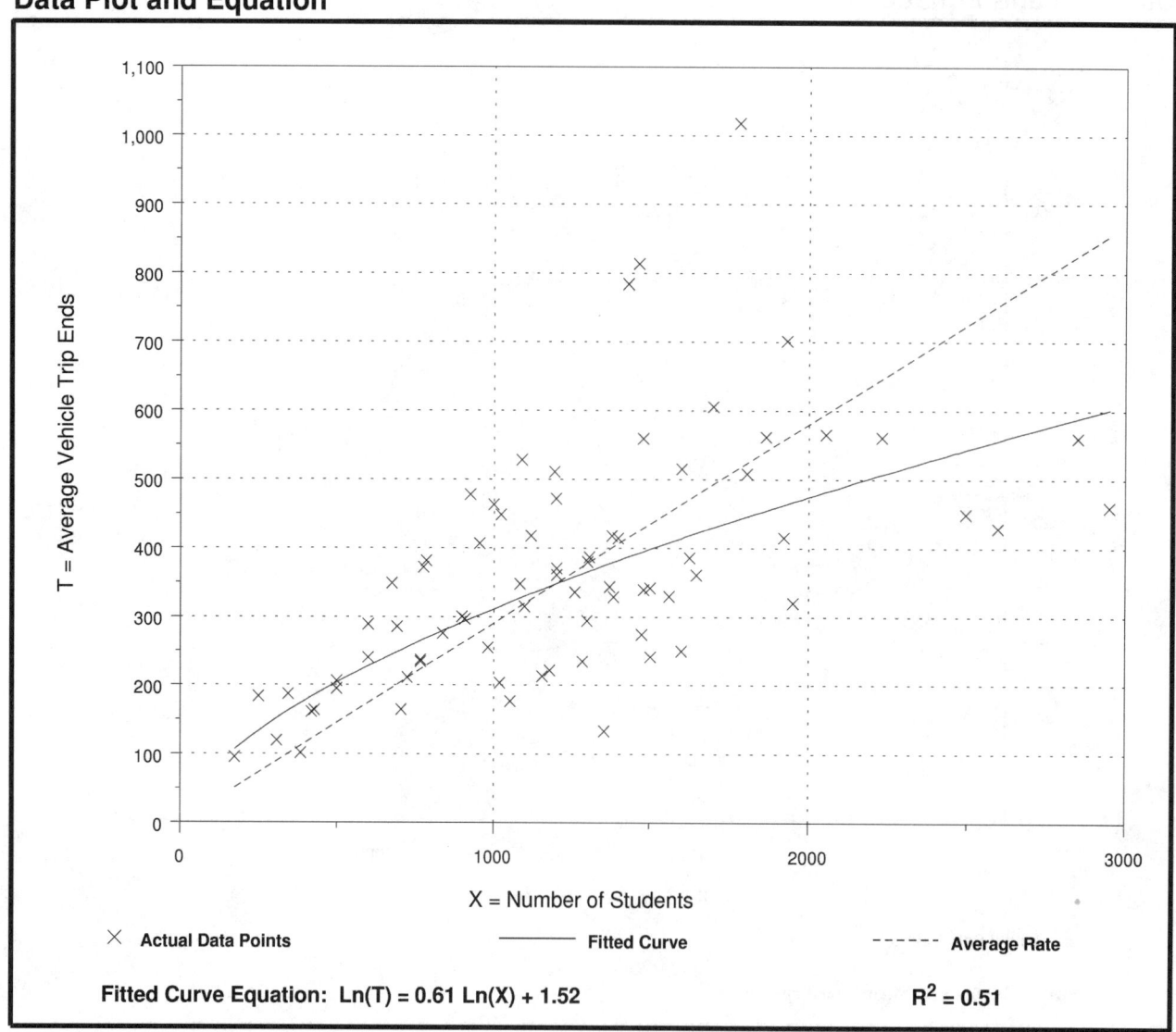

Fitted Curve Equation: $\text{Ln}(T) = 0.61 \, \text{Ln}(X) + 1.52$ $R^2 = 0.51$

High School
(530)

Average Vehicle Trip Ends vs: Students
On a: Saturday

Number of Studies: 20
Average Number of Students: 1,523
Directional Distribution: 50% entering, 50% exiting

Trip Generation per Student

Average Rate	Range of Rates	Standard Deviation
0.61	0.08 - 1.62	0.88

Data Plot and Equation

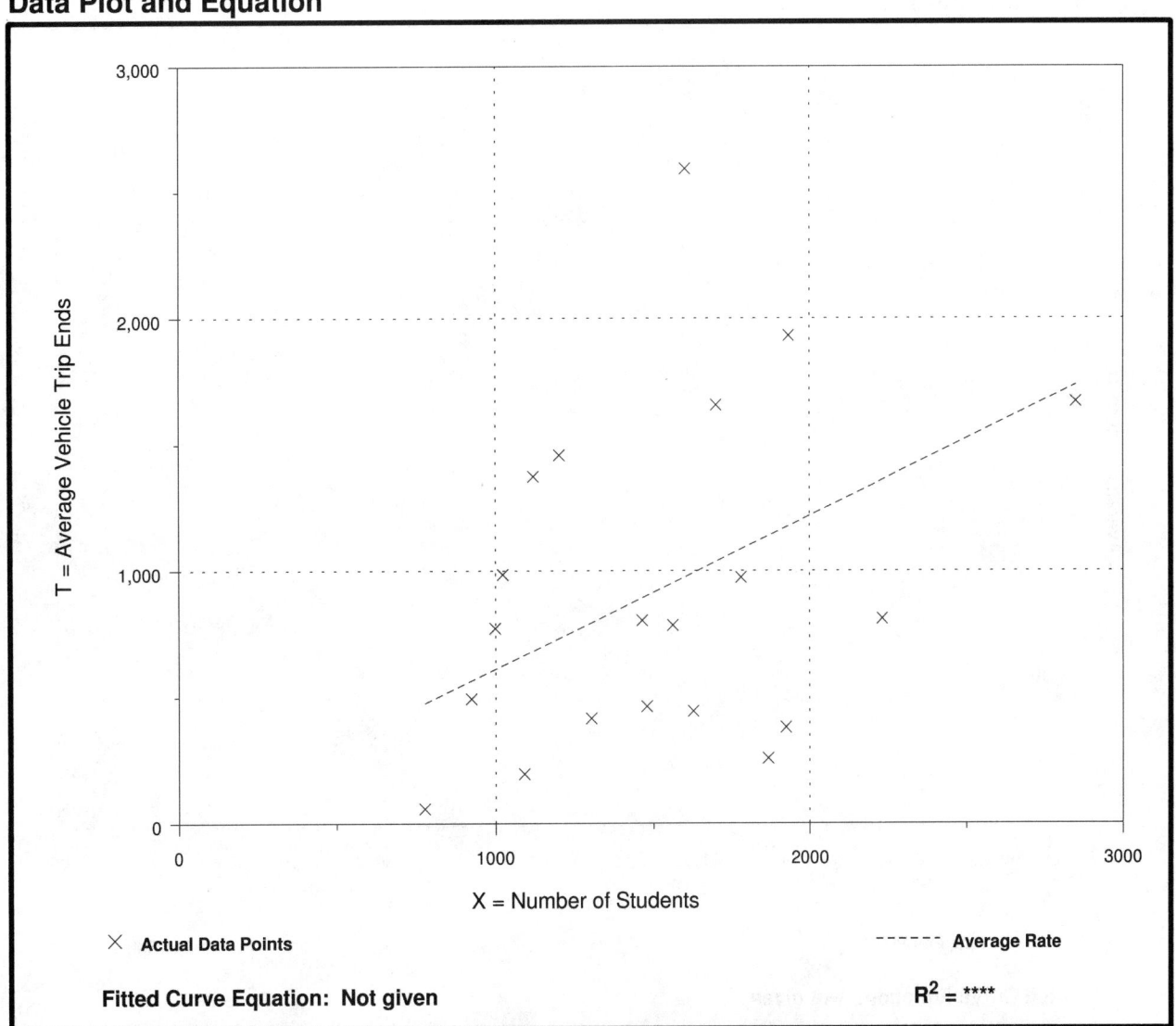

X = Number of Students
T = Average Vehicle Trip Ends

X Actual Data Points
----- Average Rate

Fitted Curve Equation: Not given $R^2 = ****$

High School
(530)

Average Vehicle Trip Ends vs: Students
On a: Saturday,
        Peak Hour of Generator

Number of Studies: 20
Average Number of Students: 1,523
Directional Distribution: 64% entering, 36% exiting

Trip Generation per Student

Average Rate	Range of Rates	Standard Deviation
0.11	0.02 - 0.24	0.34

Data Plot and Equation

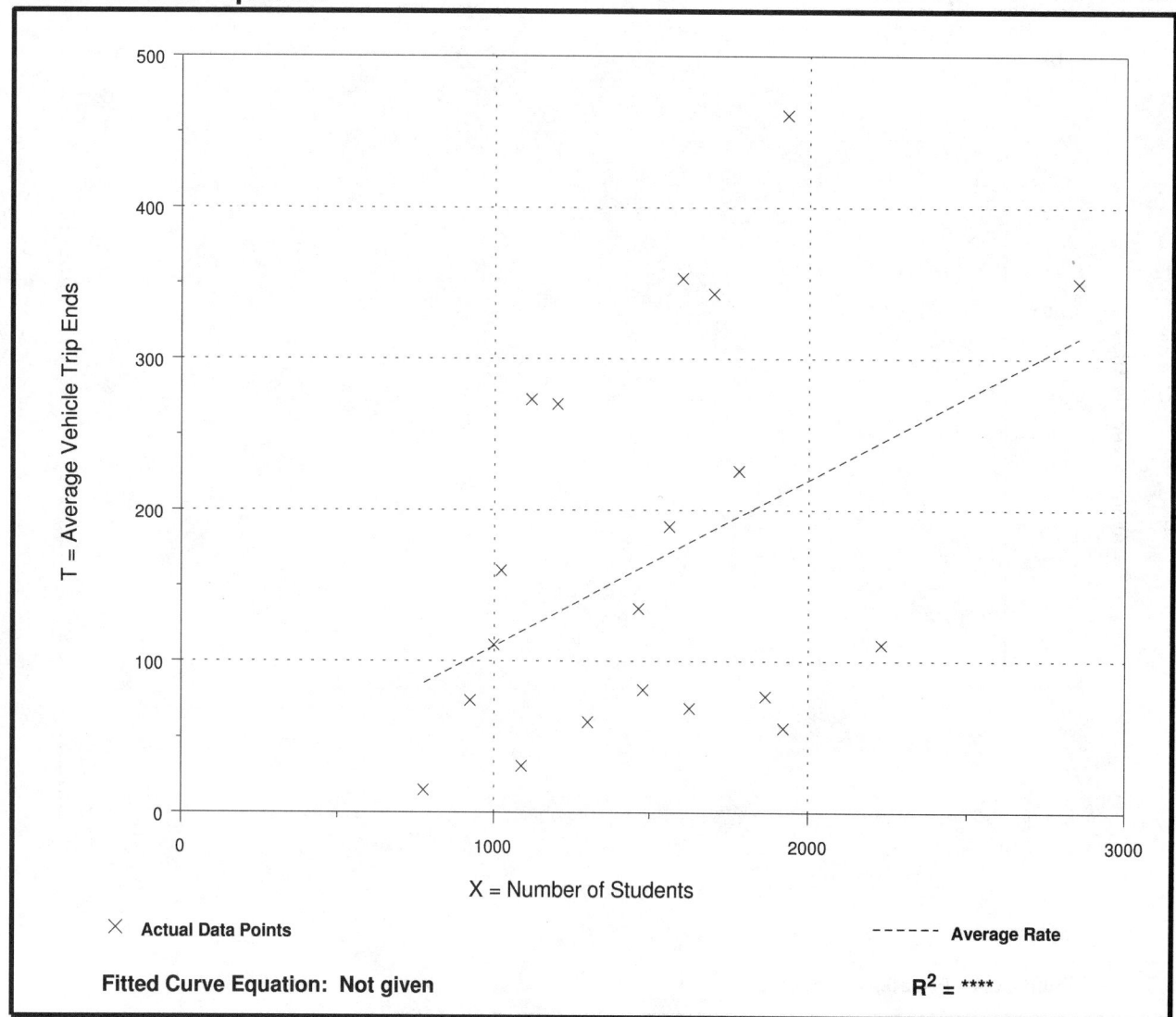

Fitted Curve Equation: Not given $R^2 =$ ****

High School
(530)

Average Vehicle Trip Ends vs: Students
On a: Sunday

Number of Studies: 20
Average Number of Students: 1,523
Directional Distribution: 50% entering, 50% exiting

Trip Generation per Student

Average Rate	Range of Rates	Standard Deviation
0.25	0.04 - 0.92	0.54

Data Plot and Equation

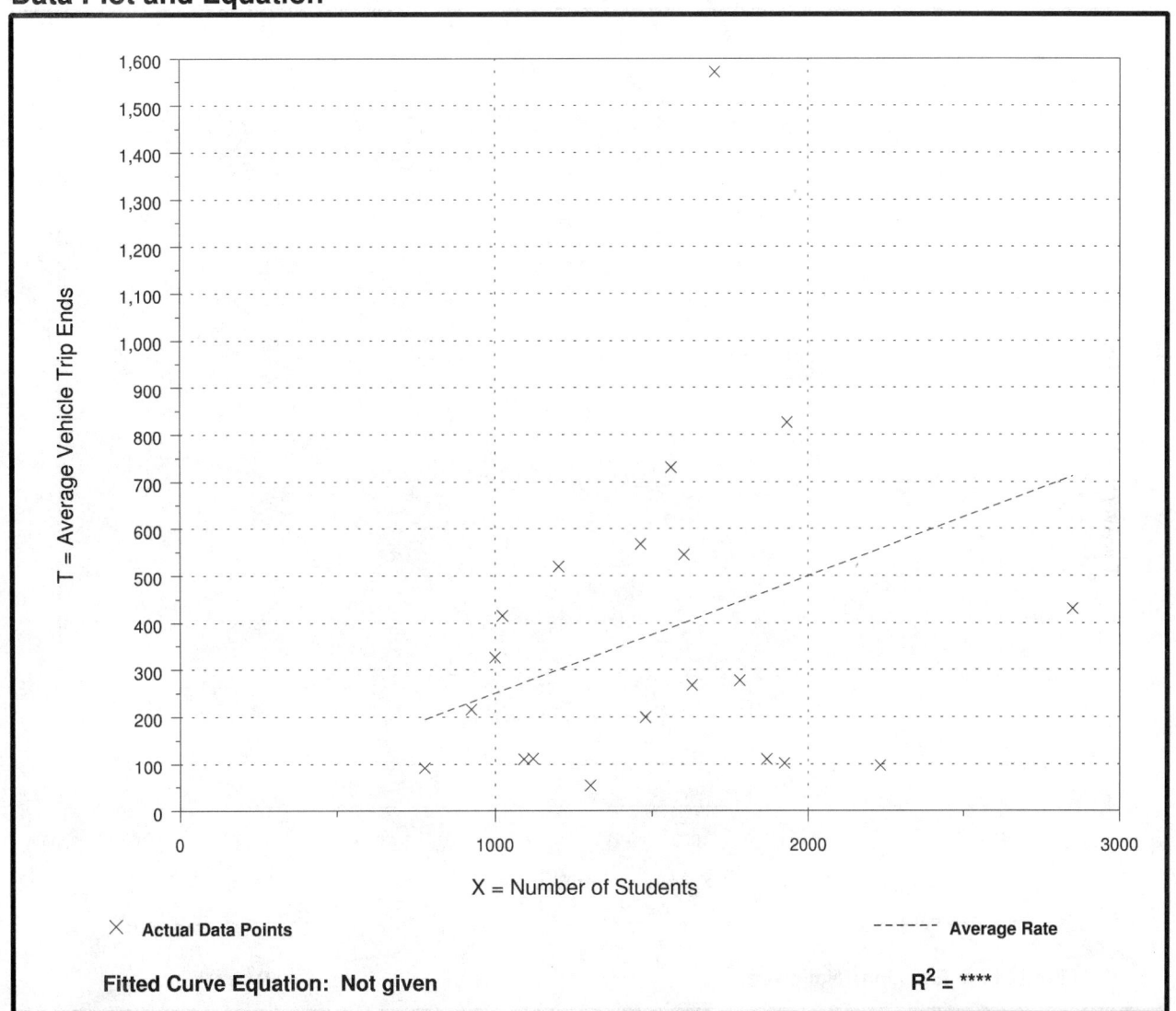

× Actual Data Points   - - - - - Average Rate

Fitted Curve Equation: Not given $R^2 = ****$

High School
(530)

Average Vehicle Trip Ends vs: Students
On a: Sunday,
Peak Hour of Generator

Number of Studies: 20
Average Number of Students: 1,523
Directional Distribution: 41% entering, 59% exiting

Trip Generation per Student

Average Rate	Range of Rates	Standard Deviation
0.04	0.01 - 0.20	0.22

Data Plot and Equation

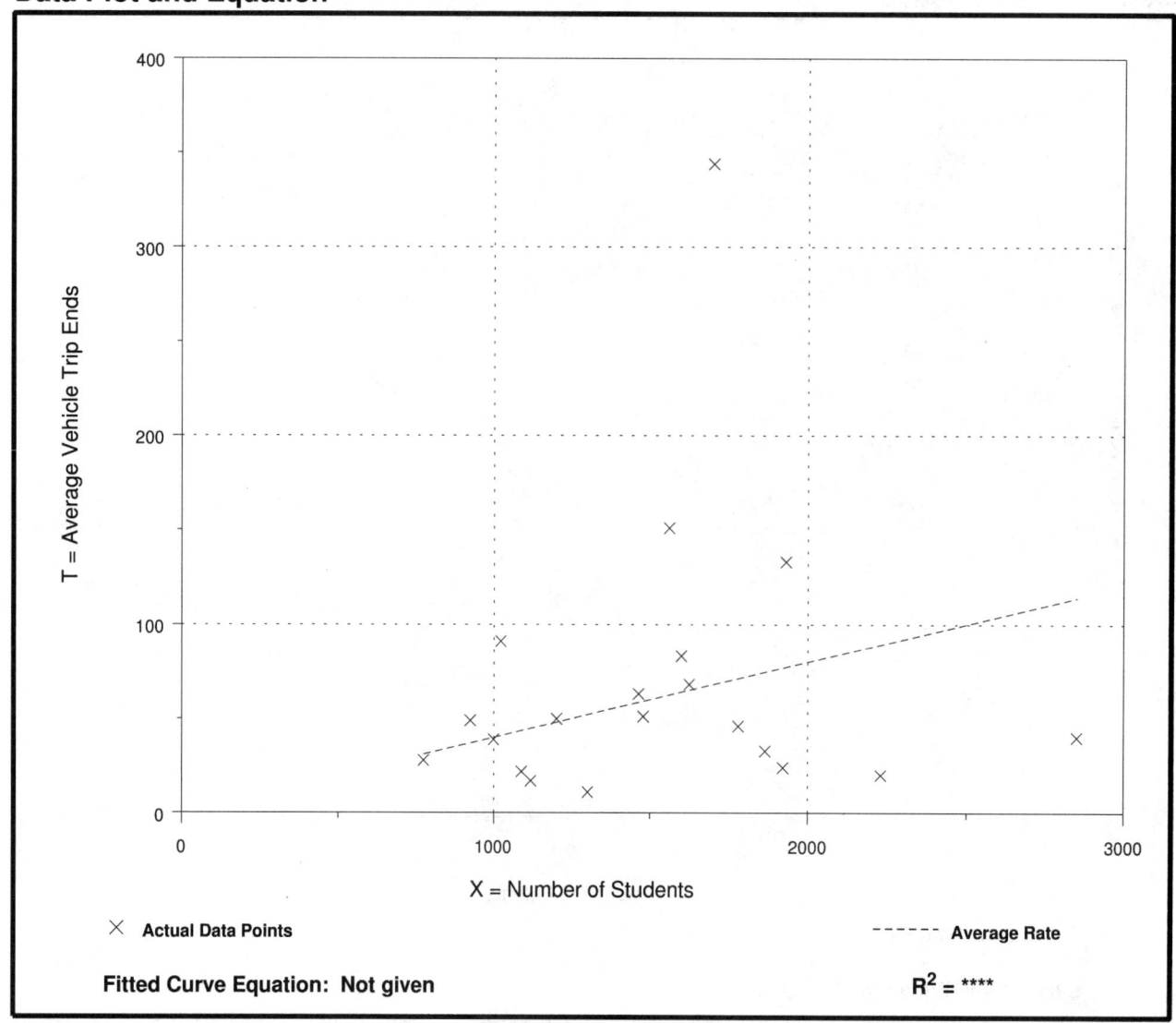

Fitted Curve Equation: Not given $R^2 = ****$

High School
(530)

Average Vehicle Trip Ends vs: Employees
On a: Weekday

Number of Studies: 51
Avg. Number of Employees: 119
Directional Distribution: 50% entering, 50% exiting

Trip Generation per Employee

Average Rate	Range of Rates	Standard Deviation
19.74	4.28 - 35.26	7.95

Data Plot and Equation

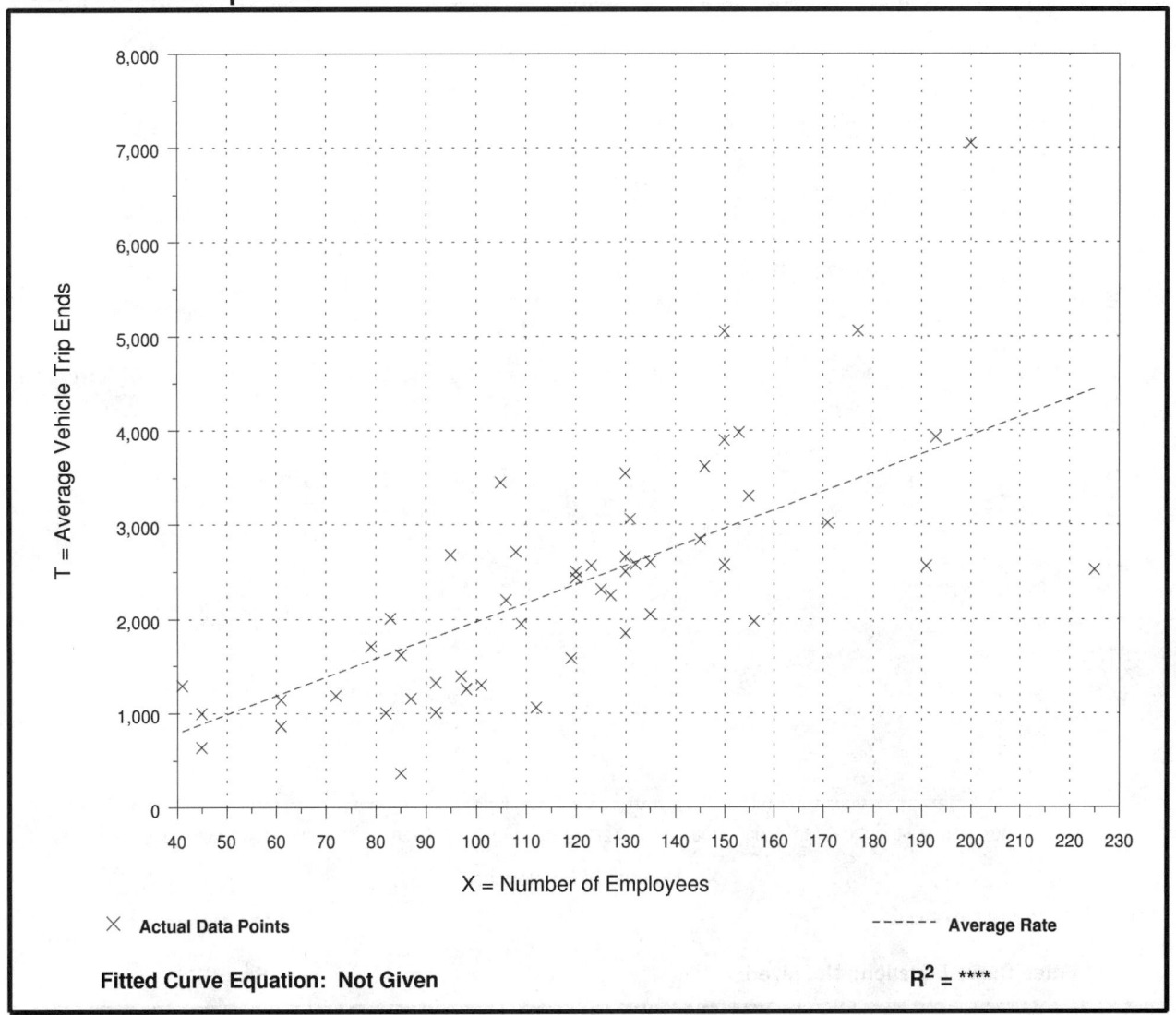

Fitted Curve Equation: Not Given $R^2 = ****$

High School
(530)

Average Vehicle Trip Ends vs: Employees
On a: Weekday,
A.M. Peak Hour

Number of Studies: 53
Avg. Number of Employees: 118
Directional Distribution: 70% entering, 30% exiting

Trip Generation per Employee

Average Rate	Range of Rates	Standard Deviation
4.68	0.54 - 10.20	2.88

Data Plot and Equation

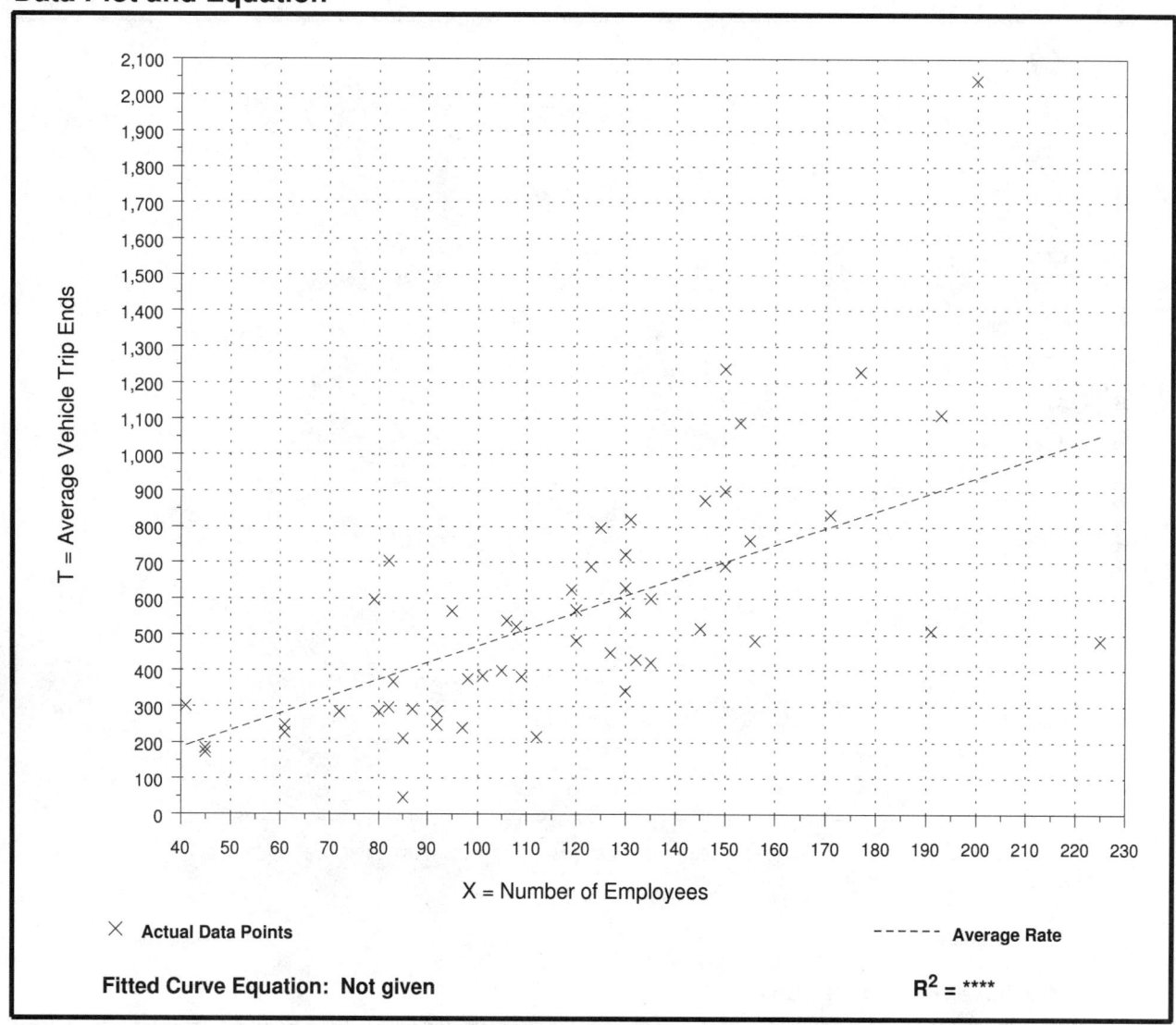

Fitted Curve Equation: Not given $R^2 = ****$

High School
(530)

Average Vehicle Trip Ends vs: Employees
On a: Weekday,
Peak Hour of Adjacent Street Traffic,
One Hour Between 4 and 6 p.m.

Number of Studies: 25
Avg. Number of Employees: 134
Directional Distribution: 54% entering, 46% exiting

Trip Generation per Employee

Average Rate	Range of Rates	Standard Deviation
1.55	0.41 - 3.00	1.39

Data Plot and Equation

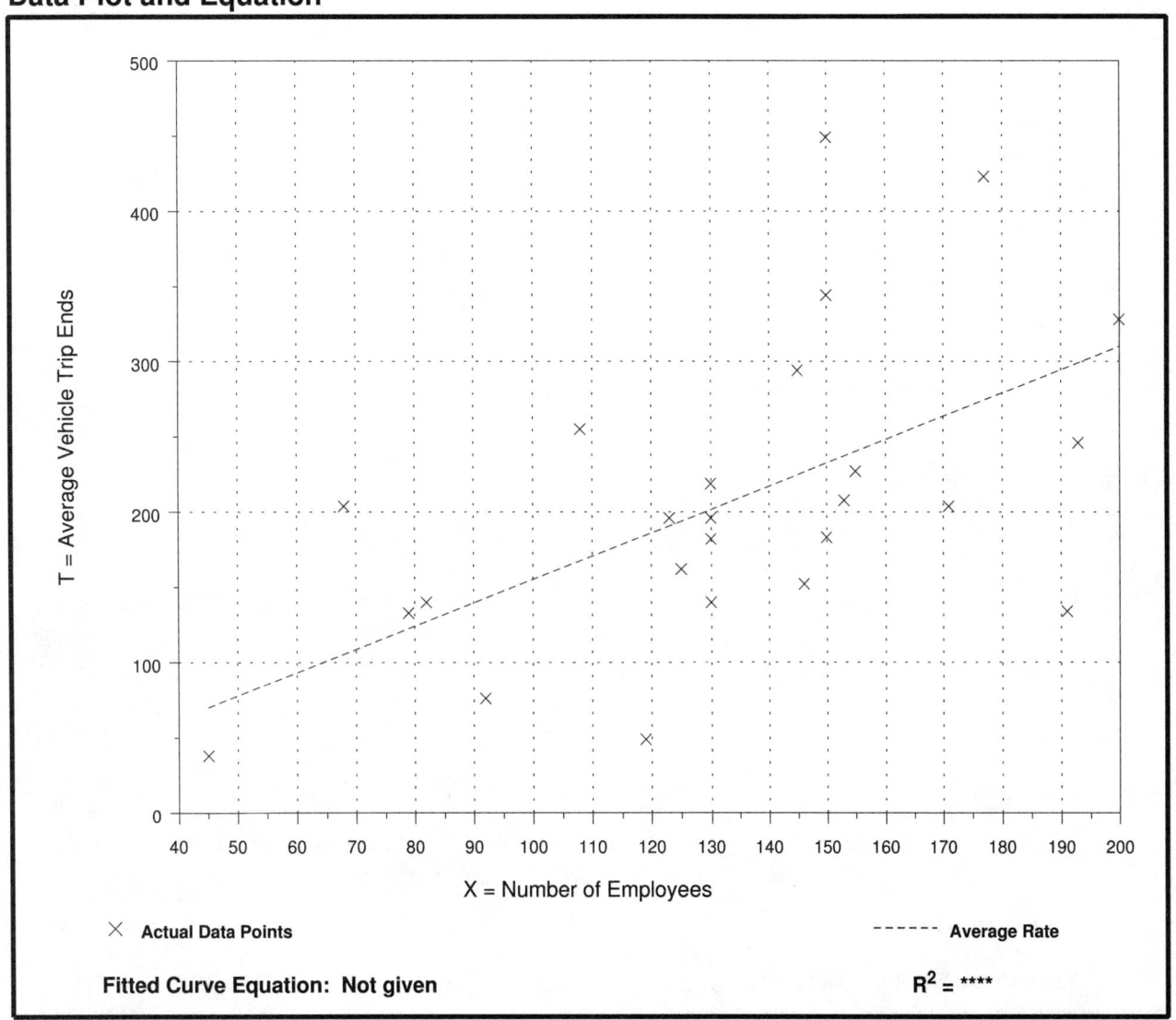

Fitted Curve Equation: Not given $R^2 = ****$

High School
(530)

Average Vehicle Trip Ends vs: Employees
On a: Weekday,
P.M. Peak Hour of Generator

Number of Studies: 53
Avg. Number of Employees: 118
Directional Distribution: 31% entering, 69% exiting

Trip Generation per Employee

Average Rate	Range of Rates	Standard Deviation
3.23	1.13 - 6.98	2.08

Data Plot and Equation

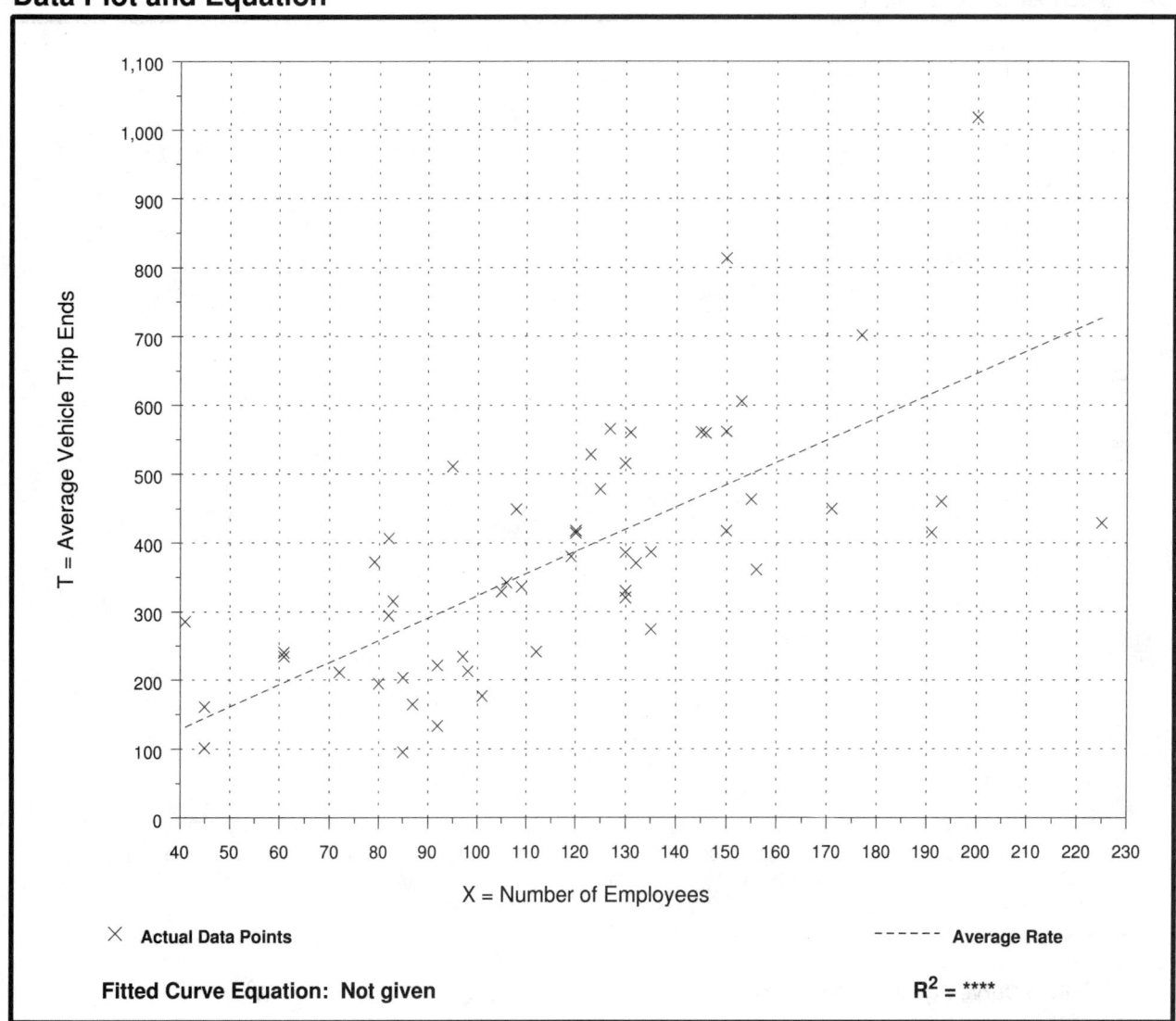

X   Actual Data Points   ----- Average Rate

Fitted Curve Equation: Not given $R^2 = ****$

High School
(530)

Average Vehicle Trip Ends vs: Employees
On a: Saturday

Number of Studies: 20
Avg. Number of Employees: 141
Directional Distribution: 50% entering, 50% exiting

Trip Generation per Employee

Average Rate	Range of Rates	Standard Deviation
6.57	0.75 - 19.95	5.21

Data Plot and Equation

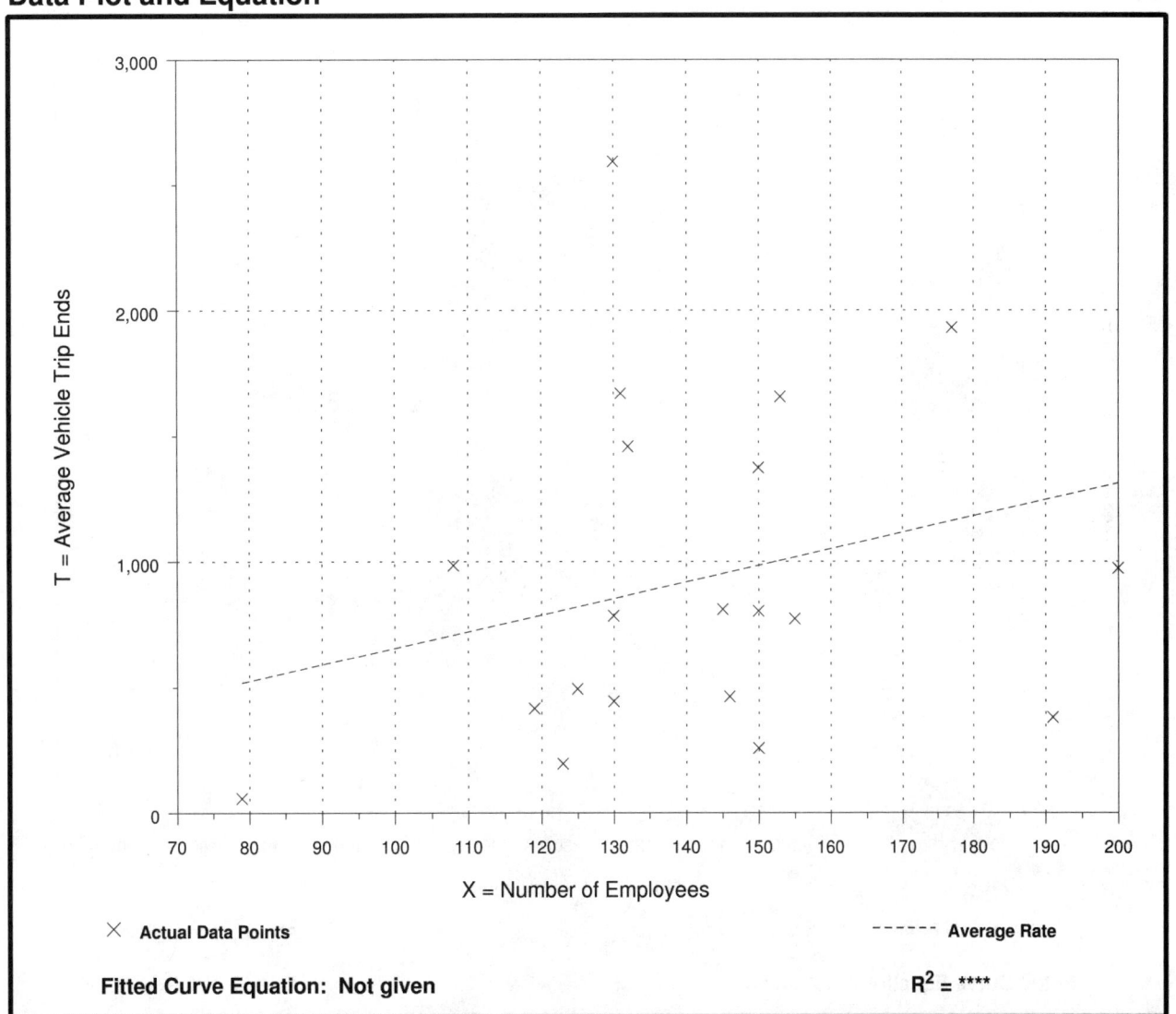

Fitted Curve Equation: Not given $R^2 = ****$

High School
(530)

Average Vehicle Trip Ends vs: Employees
On a: Saturday,
Peak Hour of Generator

Number of Studies: 20
Avg. Number of Employees: 141
Directional Distribution: 64% entering, 36% exiting

Trip Generation per Employee

Average Rate	Range of Rates	Standard Deviation
1.22	0.19 - 2.72	1.38

Data Plot and Equation

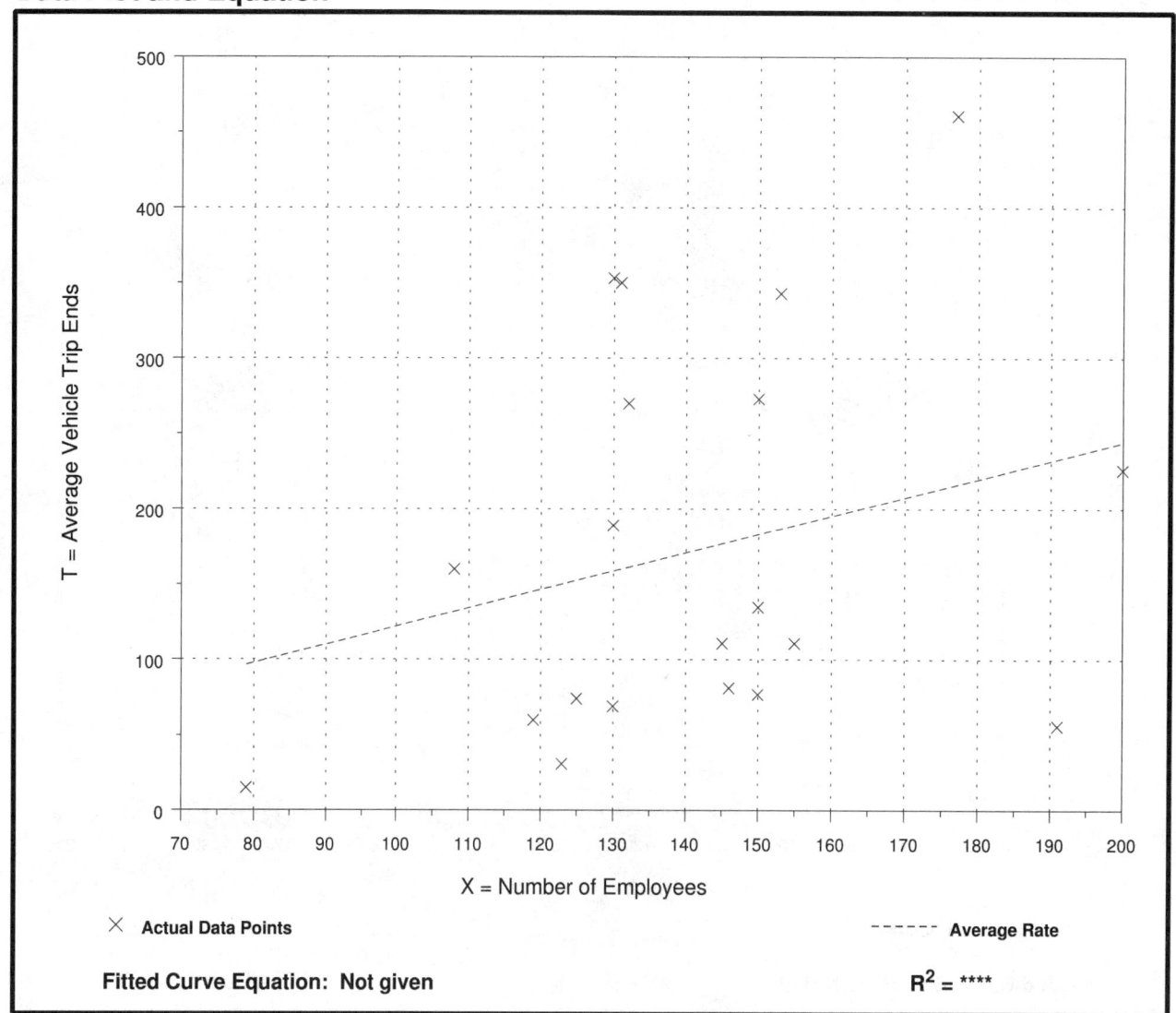

Fitted Curve Equation: Not given $R^2 = ****$

High School
(530)

Average Vehicle Trip Ends vs: Employees
On a: Sunday

Number of Studies: 20
Avg. Number of Employees: 141
Directional Distribution: 50% entering, 50% exiting

Trip Generation per Employee

Average Rate	Range of Rates	Standard Deviation
2.68	0.45 - 10.27	2.89

Data Plot and Equation

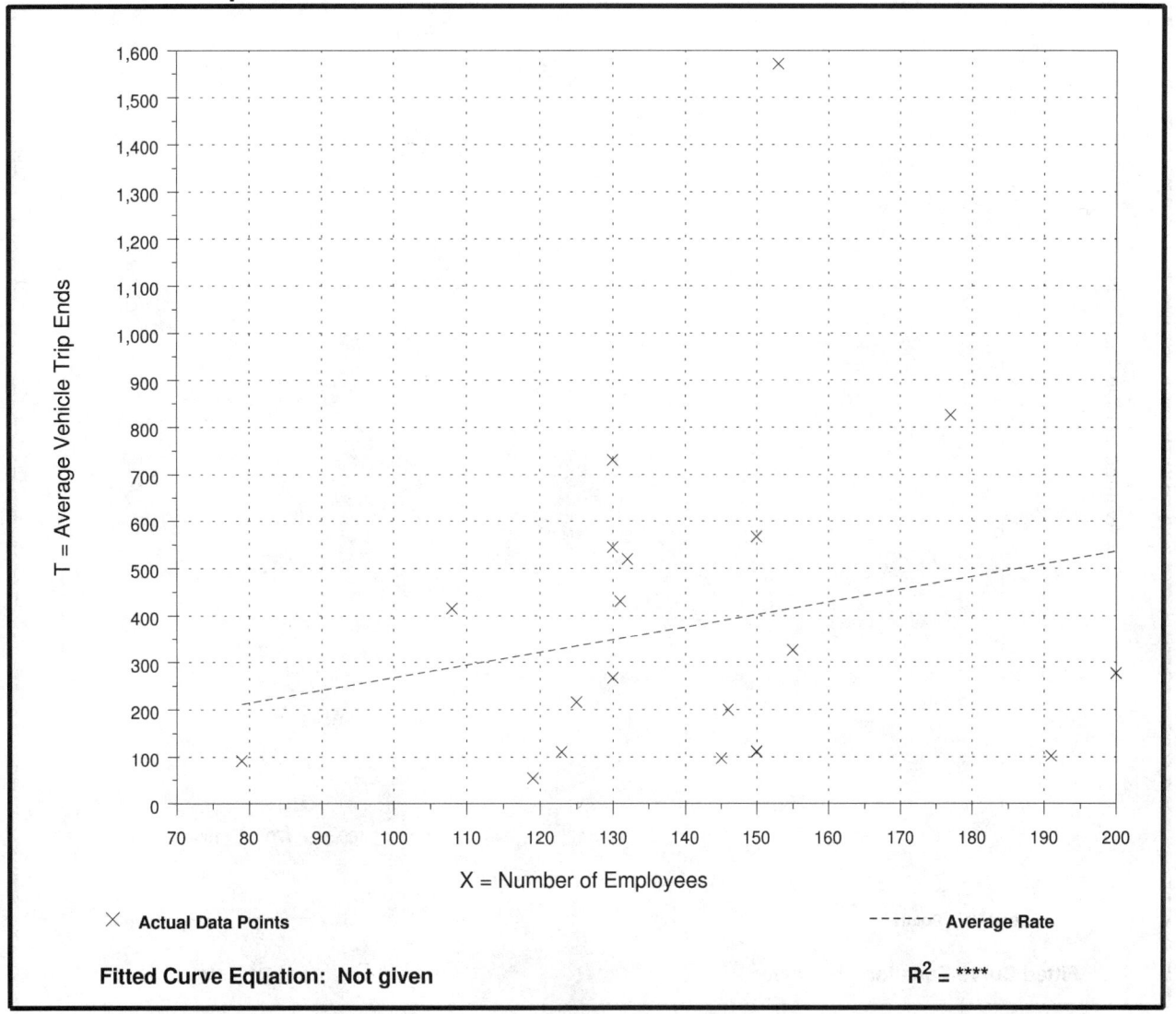

Fitted Curve Equation: Not given $R^2 = ****$

High School
(530)

Average Vehicle Trip Ends vs: Employees
On a: Sunday,
Peak Hour of Generator

Number of Studies: 20
Avg. Number of Employees: 141
Directional Distribution: 41% entering, 59% exiting

Trip Generation per Employee

Average Rate	Range of Rates	Standard Deviation
0.48	0.09 - 2.25	0.85

Data Plot and Equation

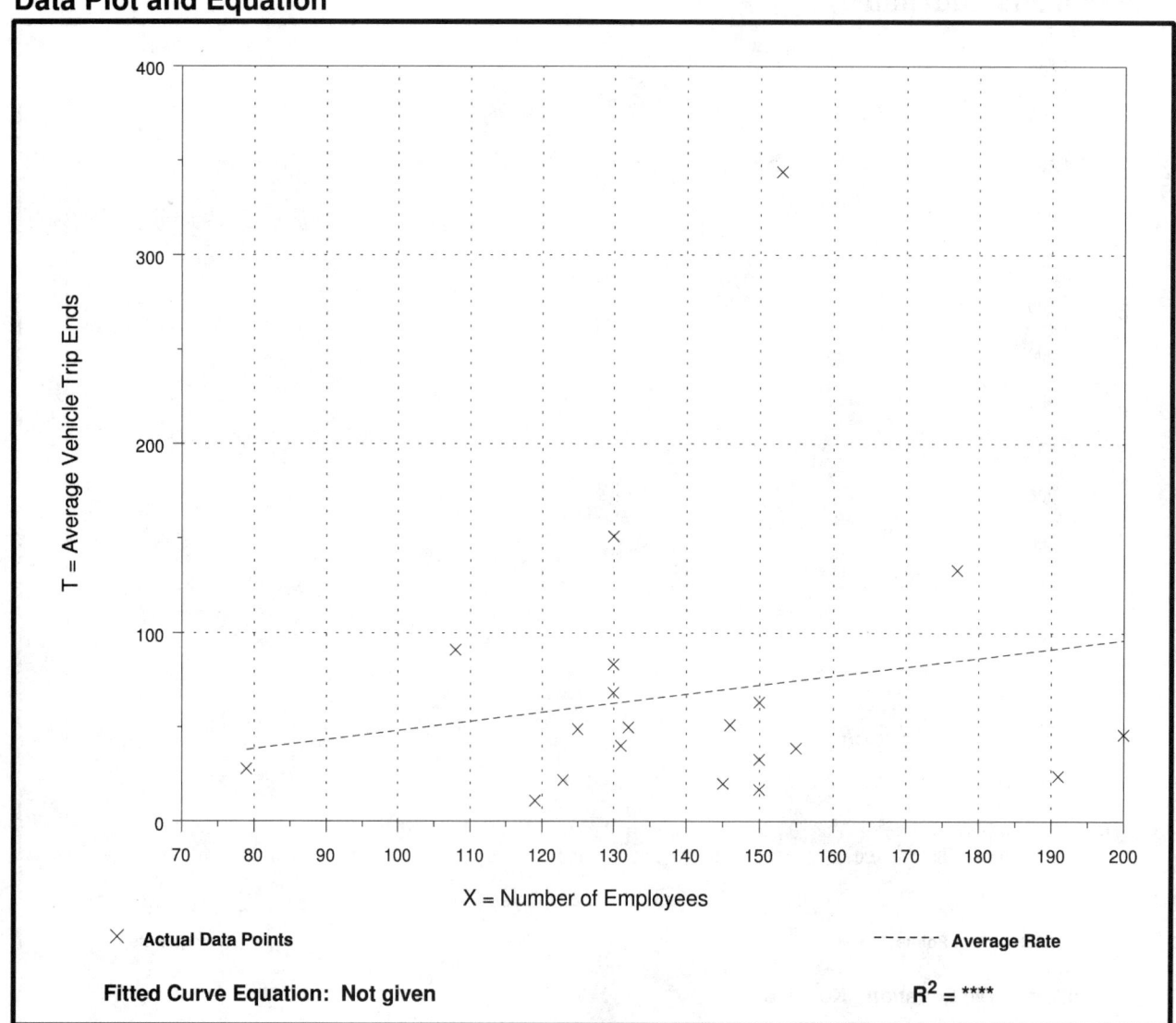

Fitted Curve Equation: Not given $R^2 =$ ****

High School
(530)

Average Vehicle Trip Ends vs: 1000 Sq. Feet Gross Floor Area
On a: Weekday

Number of Studies: 43
Average 1000 Sq. Feet GFA: 196
Directional Distribution: 50% entering, 50% exiting

Trip Generation per 1000 Sq. Feet Gross Floor Area

Average Rate	Range of Rates	Standard Deviation
12.89	4.00 - 34.06	7.17

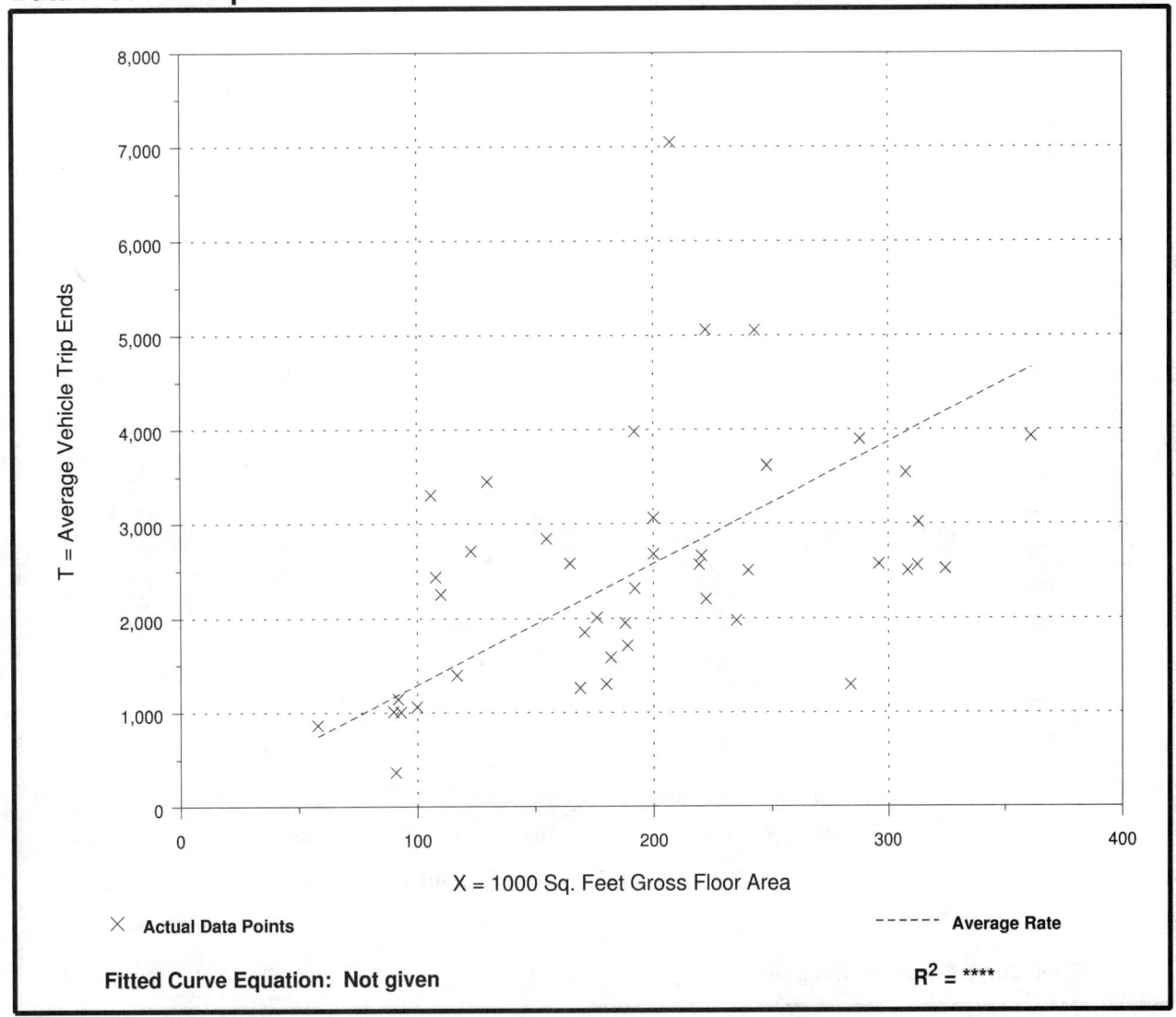

Data Plot and Equation

Fitted Curve Equation: Not given $R^2 = ****$

High School
(530)

Average Vehicle Trip Ends vs: 1000 Sq. Feet Gross Floor Area
On a: Weekday,
A.M. Peak Hour

Number of Studies: 44
Average 1000 Sq. Feet GFA: 194
Directional Distribution: 71% entering, 29% exiting

Trip Generation per 1000 Sq. Feet Gross Floor Area

Average Rate	Range of Rates	Standard Deviation
3.06	0.51 - 9.86	2.36

Data Plot and Equation

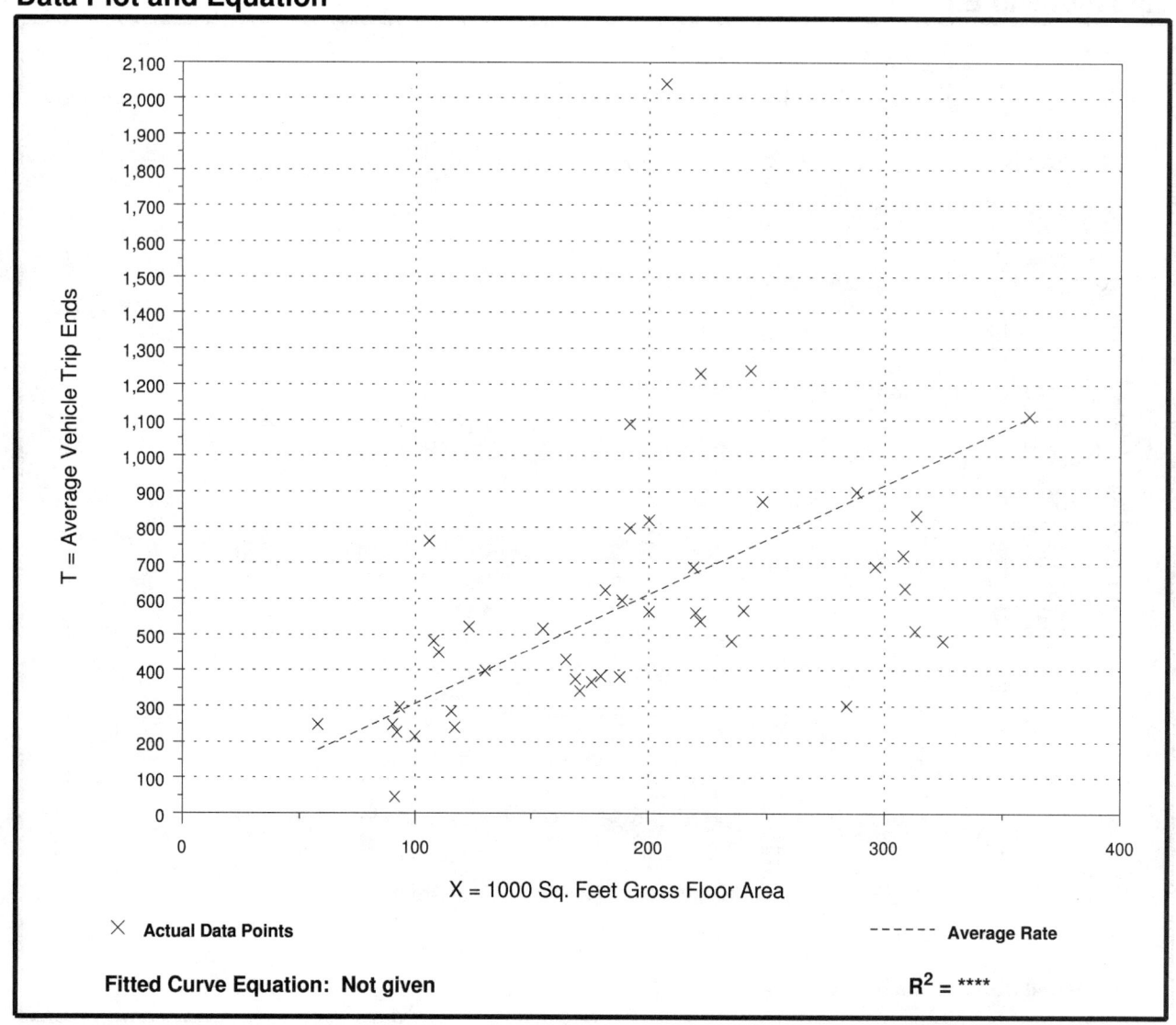

Fitted Curve Equation: Not given $R^2 = ****$

High School
(530)

Average Vehicle Trip Ends vs: 1000 Sq. Feet Gross Floor Area
On a: Weekday,
Peak Hour of Adjacent Street Traffic,
One Hour Between 4 and 6 p.m.

Number of Studies: 22
Average 1000 Sq. Feet GFA: 225
Directional Distribution: 54% entering, 46% exiting

Trip Generation per 1000 Sq. Feet Gross Floor Area

Average Rate	Range of Rates	Standard Deviation
0.97	0.27 - 2.14	1.11

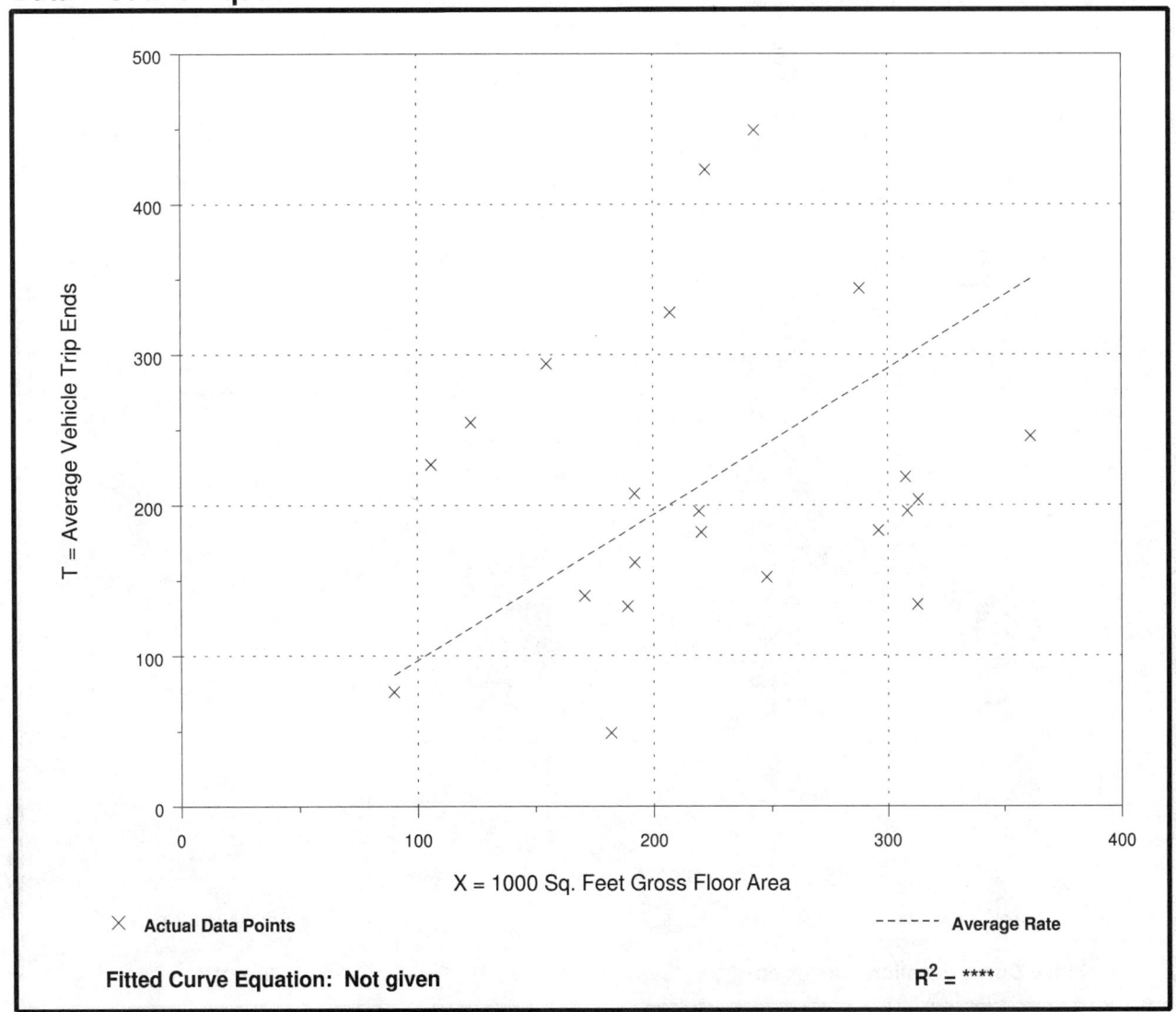

Data Plot and Equation

Fitted Curve Equation: Not given $R^2 = ****$

High School
(530)

Average Vehicle Trip Ends vs: 1000 Sq. Feet Gross Floor Area
On a: Weekday,
P.M. Peak Hour of Generator

Number of Studies: 44
Average 1000 Sq. Feet GFA: 194
Directional Distribution: 31% entering, 69% exiting

Trip Generation per 1000 Sq. Feet Gross Floor Area

Average Rate	Range of Rates	Standard Deviation
2.12	0.98 - 5.14	1.74

Data Plot and Equation

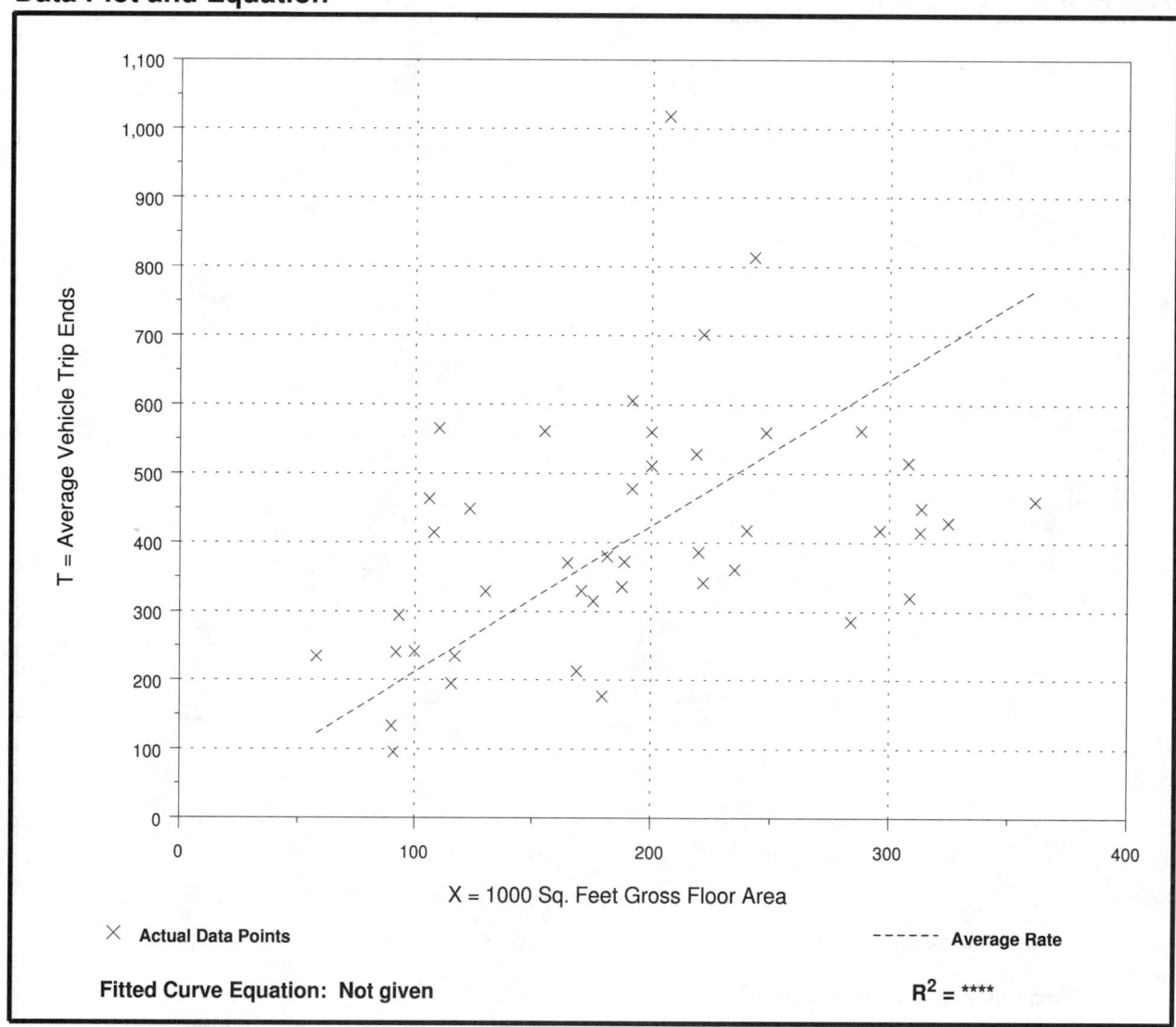

Fitted Curve Equation: Not given $R^2 = ****$

High School
(530)

Average Vehicle Trip Ends vs: 1000 Sq. Feet Gross Floor Area
On a: Saturday

Number of Studies: 20
Average 1000 Sq. Feet GFA: 212
Directional Distribution: 50% entering, 50% exiting

Trip Generation per 1000 Sq. Feet Gross Floor Area

Average Rate	Range of Rates	Standard Deviation
4.37	0.31 - 8.85	3.67

Data Plot and Equation

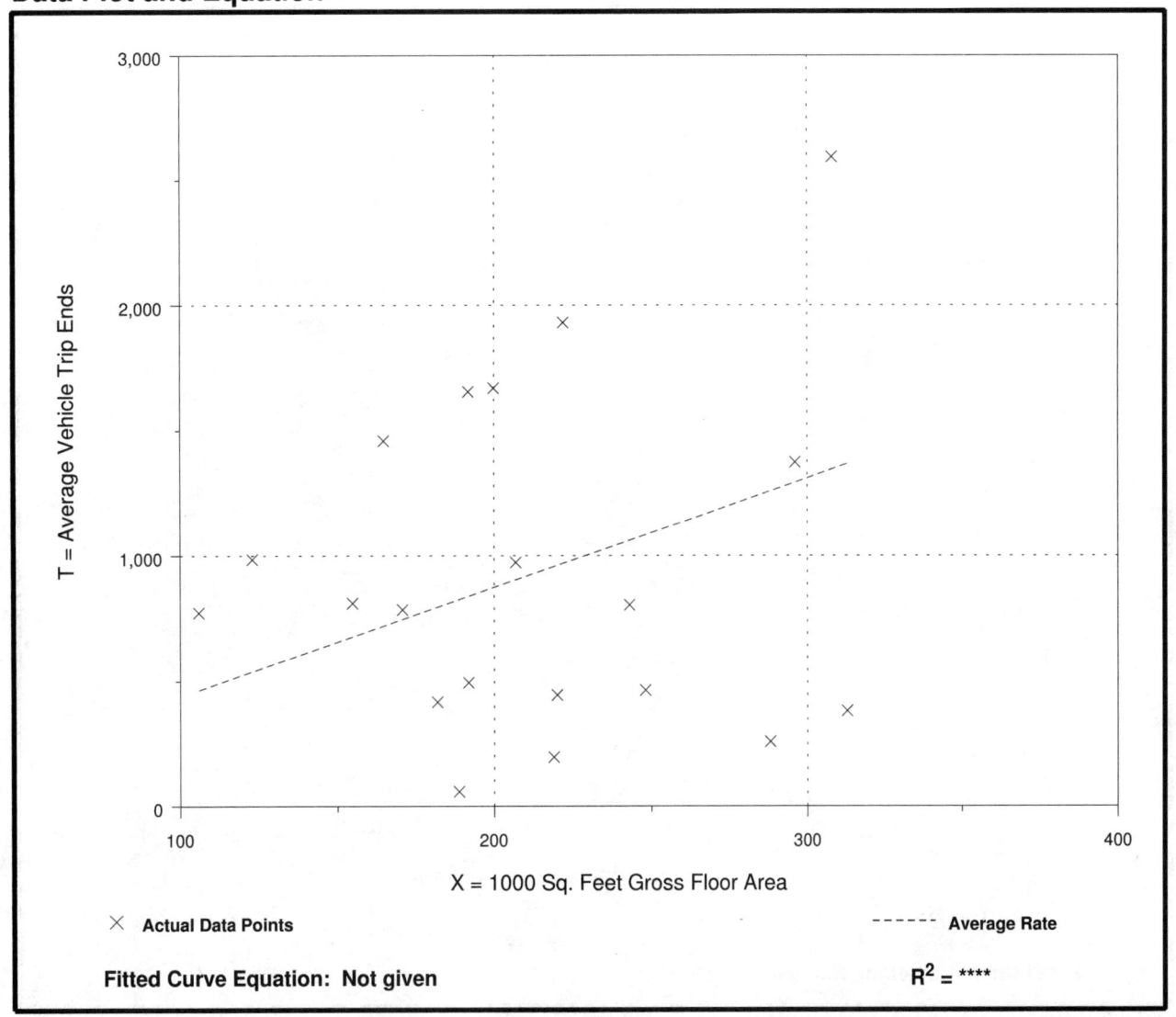

X Actual Data Points   ----- Average Rate

Fitted Curve Equation: Not given $R^2 = ****$

Trip Generation, 9th Edition • Institute of Transportation Engineers

High School
(530)

Average Vehicle Trip Ends vs: 1000 Sq. Feet Gross Floor Area
On a: Saturday,
Peak Hour of Generator

Number of Studies: 20
Average 1000 Sq. Feet GFA: 212
Directional Distribution: 64% entering, 36% exiting

Trip Generation per 1000 Sq. Feet Gross Floor Area

Average Rate	Range of Rates	Standard Deviation
0.81	0.08 - 2.08	1.08

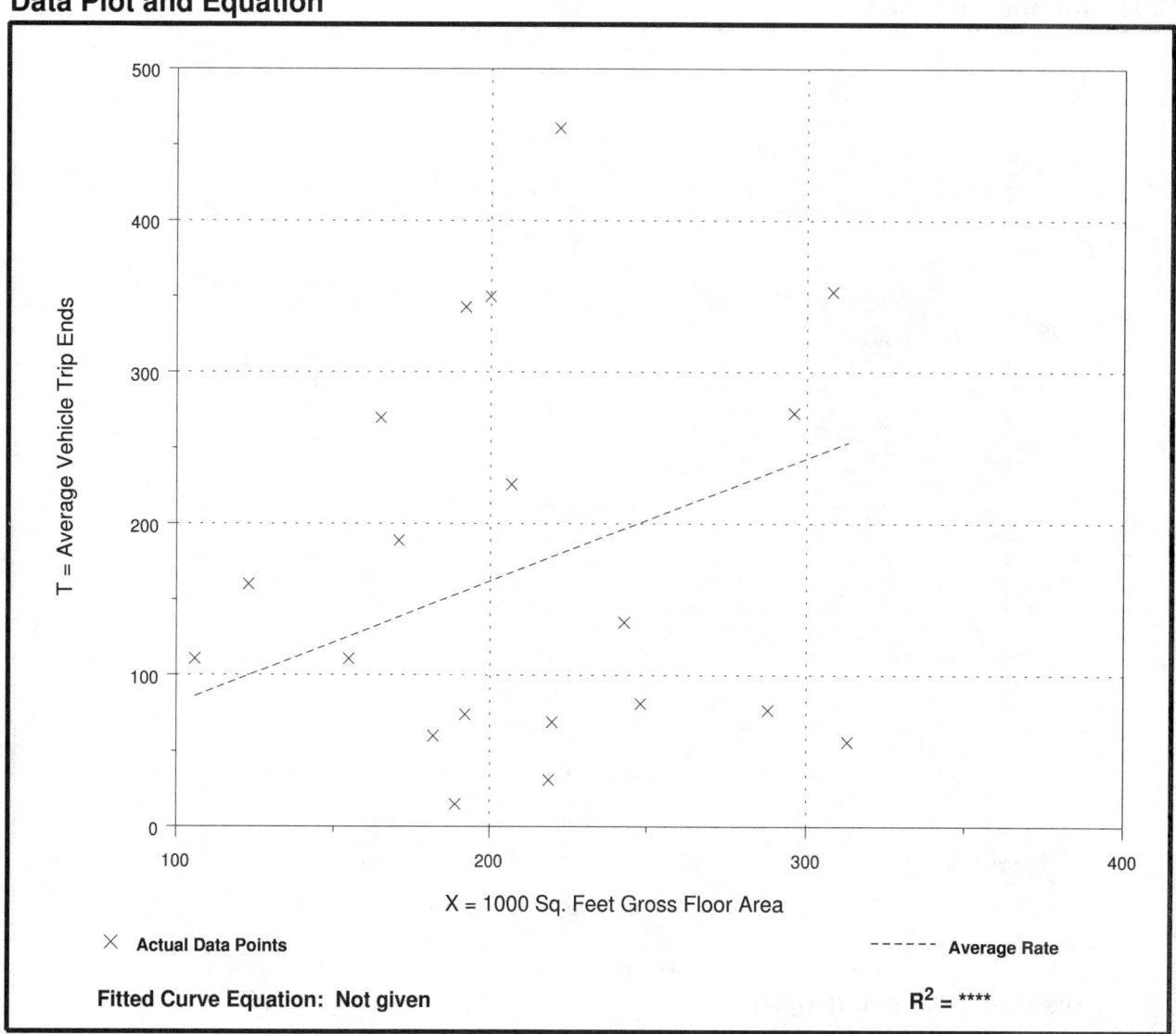

Data Plot and Equation

Fitted Curve Equation: Not given $R^2 = ****$

High School
(530)

Average Vehicle Trip Ends vs: 1000 Sq. Feet Gross Floor Area
On a: Sunday

Number of Studies: 20
Average 1000 Sq. Feet GFA: 212
Directional Distribution: 50% entering, 50% exiting

Trip Generation per 1000 Sq. Feet Gross Floor Area

Average Rate	Range of Rates	Standard Deviation
1.79	0.30 - 8.19	2.26

Data Plot and Equation

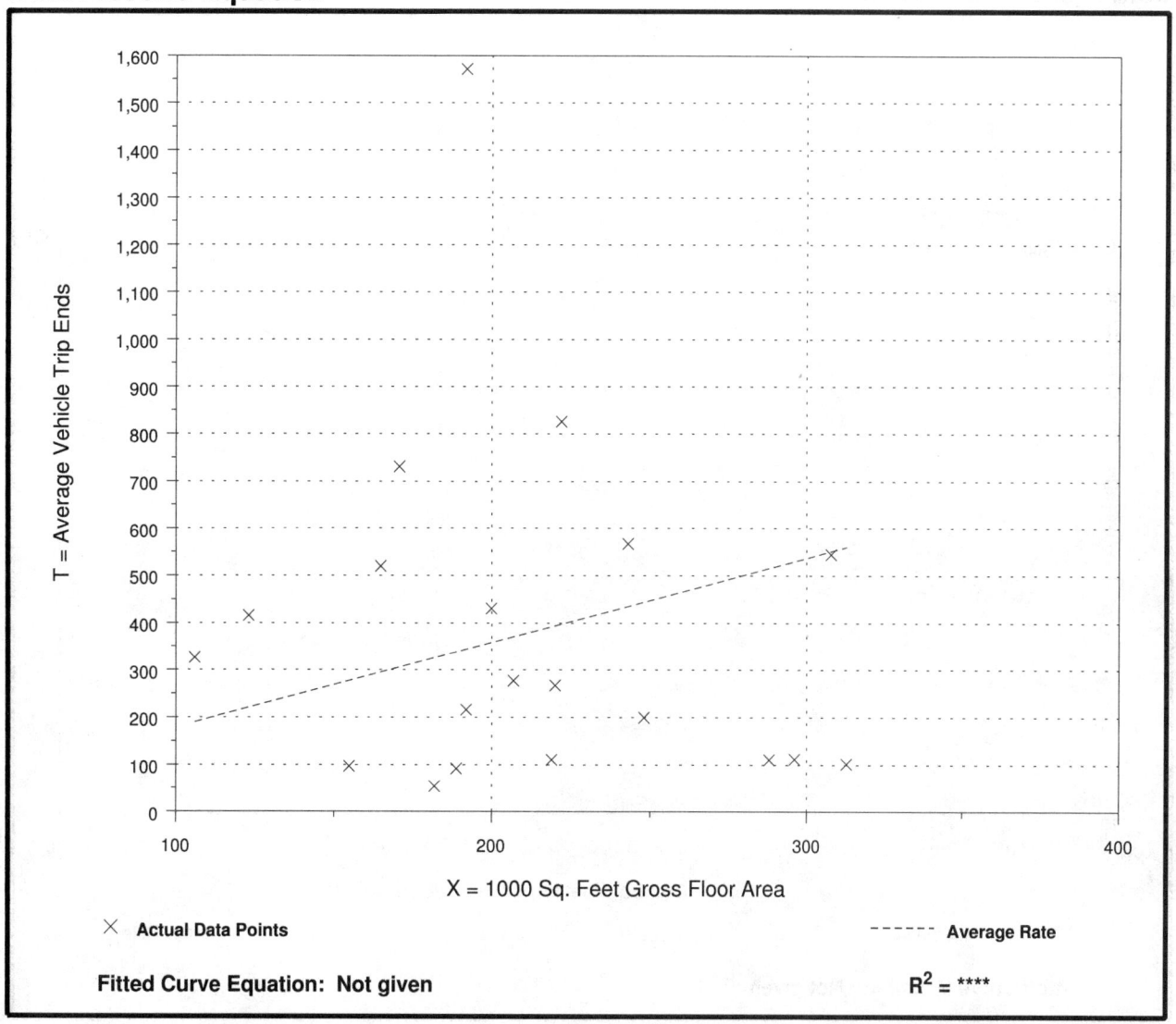

Fitted Curve Equation: Not given $R^2 = ****$

High School
(530)

Average Vehicle Trip Ends vs: 1000 Sq. Feet Gross Floor Area
On a: Sunday,
Peak Hour of Generator

Number of Studies: 20
Average 1000 Sq. Feet GFA: 212
Directional Distribution: 41% entering, 59% exiting

Trip Generation per 1000 Sq. Feet Gross Floor Area

Average Rate	Range of Rates	Standard Deviation
0.32	0.06 - 1.79	0.68

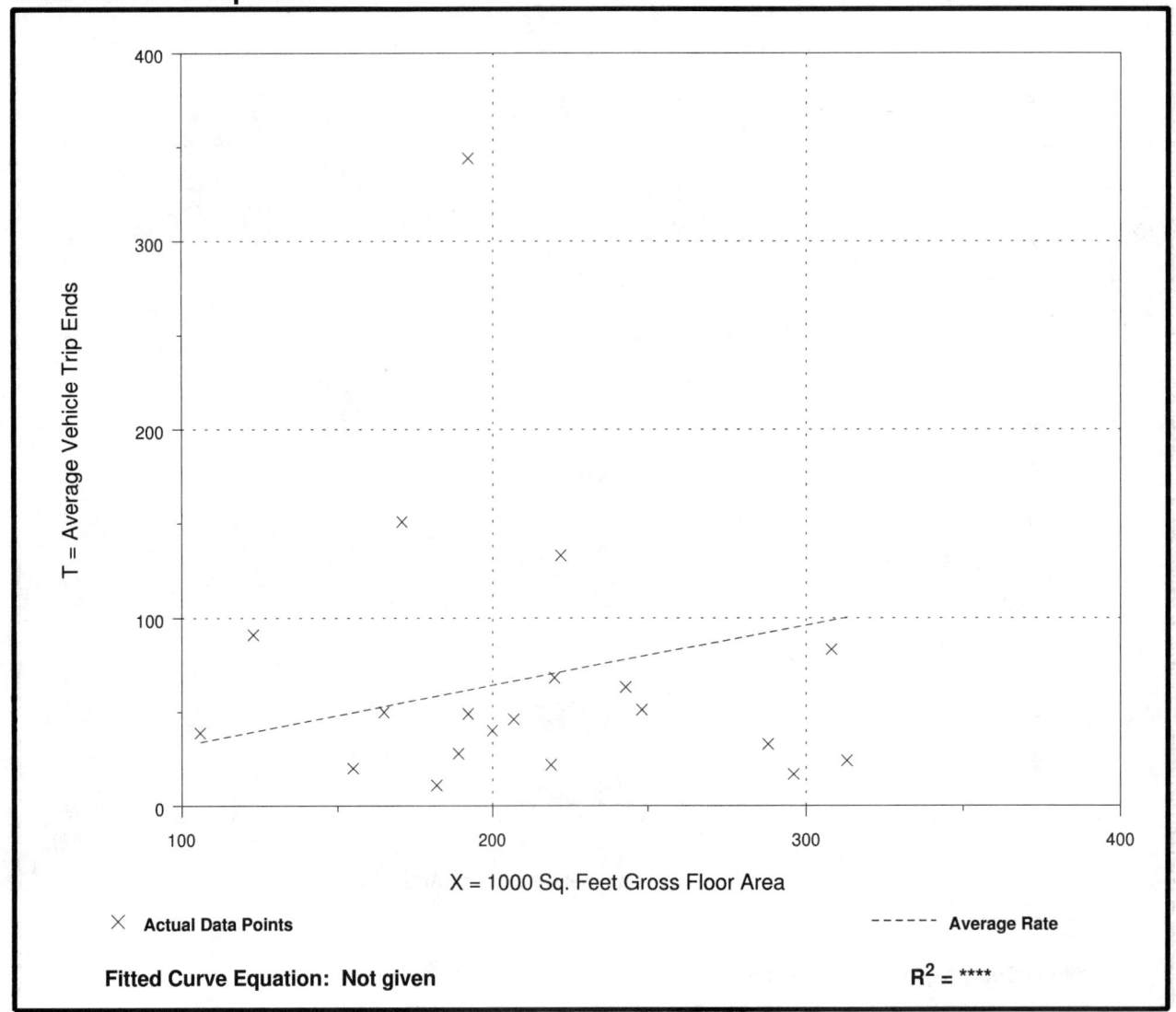

Data Plot and Equation

Fitted Curve Equation: Not given

$R^2 = ****$

Land Use: 534
Private School (K-8)

Description

Private schools in this land use category primarily serve students attending kindergarten through the eighth grade but may also include students beginning with pre–K classes. These schools may also offer extended care and day care. Students may travel a long distance to get to private schools. Elementary school (Land Use 520), middle school/junior high school (Land Use 522), high school (Land Use 530) and private school (K-12) (Land Use 536) are related uses.

Additional Data

Peak hours of the generator—
    The weekday A.M. peak hour of the generator typically coincided with the peak hour of the adjacent street traffic; therefore, only one A.M. peak hour, which represents both the peak hour of the generator and the peak hour of the adjacent street traffic, is displayed. The weekday P.M. peak hour varied between 2:00 p.m. and 4:00 p.m.

The sites were surveyed in the early 1990s and the 2000s in Florida, Maryland, Texas, Pennsylvania and Oregon.

Source Numbers

355, 444, 516, 536, 634

Private School (K-8)
(534)

Average Vehicle Trip Ends vs: Students
On a: Weekday,
A.M. Peak Hour

Number of Studies: 10
Average Number of Students: 322
Directional Distribution: 55% entering, 45% exiting

Trip Generation per Student

Average Rate	Range of Rates	Standard Deviation
0.90	0.76 - 1.58	0.96

Data Plot and Equation

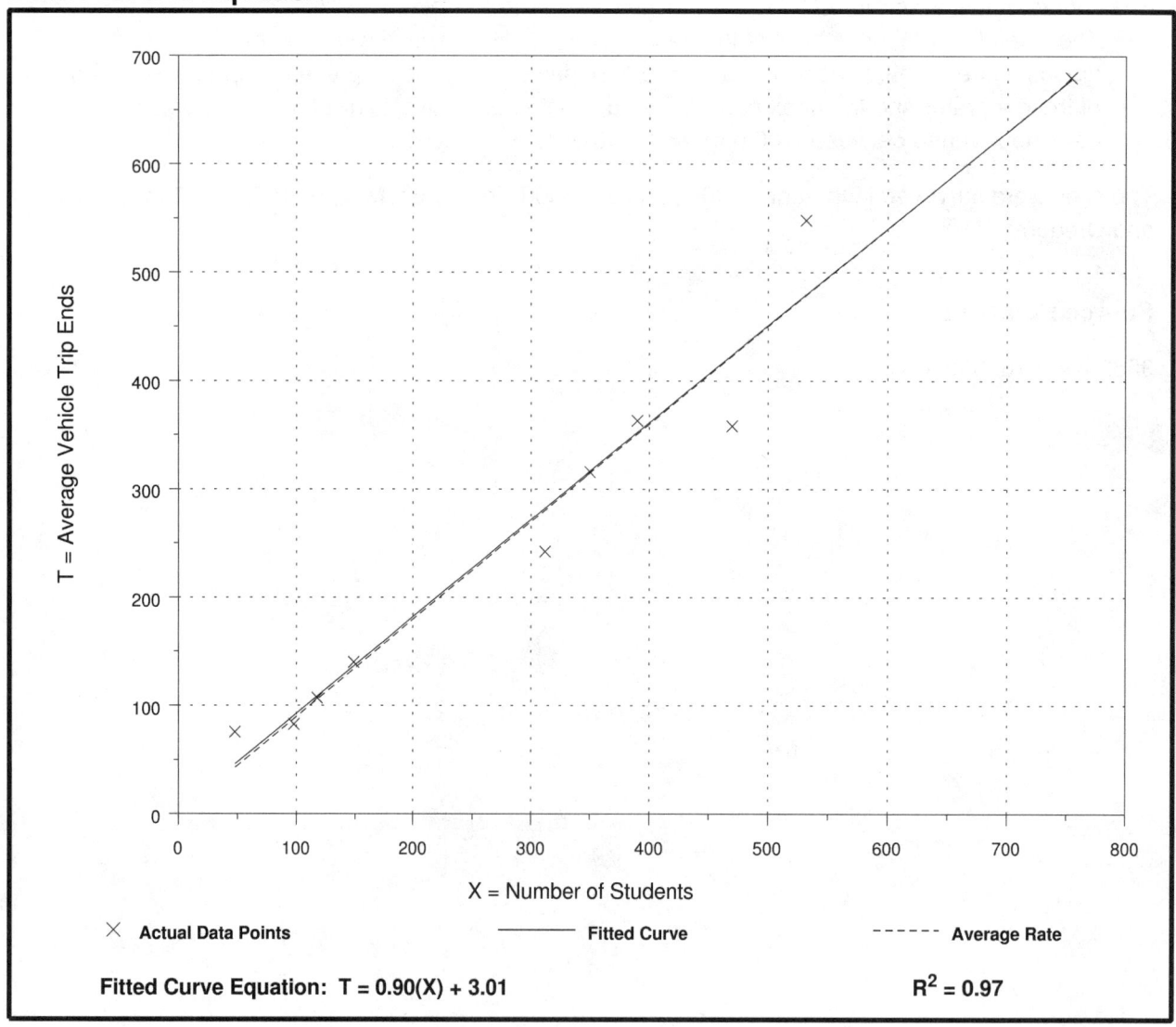

Fitted Curve Equation: $T = 0.90(X) + 3.01$ $R^2 = 0.97$

Private School (K-8)
(534)

Average Vehicle Trip Ends vs: Students
On a: Weekday,
P.M. Peak Hour of Generator

Number of Studies: 8
Average Number of Students: 340
Directional Distribution: 47% entering, 53% exiting

Trip Generation per Student

Average Rate	Range of Rates	Standard Deviation
0.60	0.42 - 0.75	0.78

Data Plot and Equation

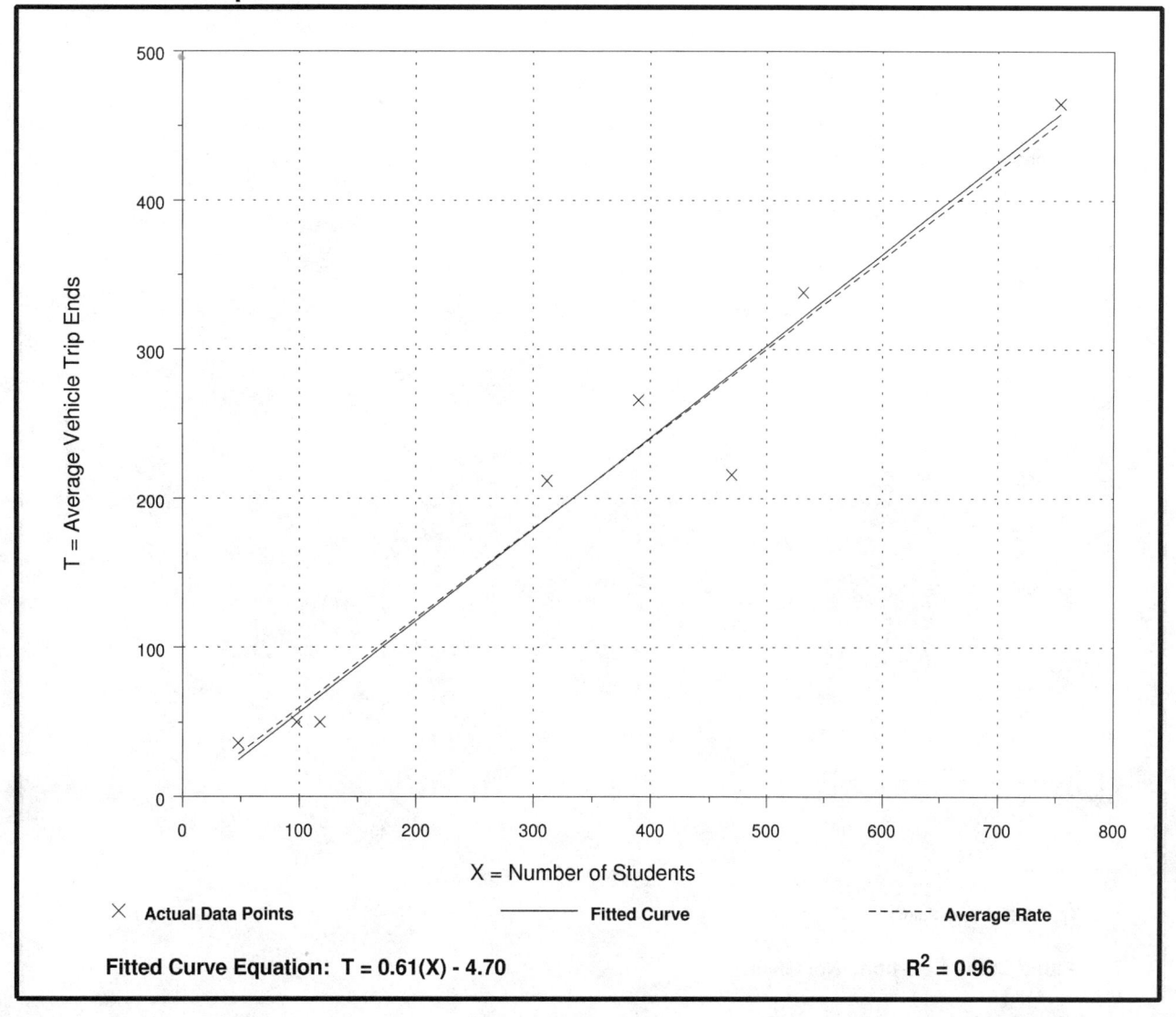

Fitted Curve Equation: $T = 0.61(X) - 4.70$ $R^2 = 0.96$

Private School (K-8)
(534)

Average Vehicle Trip Ends vs: Employees
On a: Weekday,
A.M. Peak Hour

Number of Studies: 8
Avg. Number of Employees: 33
Directional Distribution: 54% entering, 46% exiting

Trip Generation per Employee

Average Rate	Range of Rates	Standard Deviation
8.88	3.96 - 14.17	4.57

Data Plot and Equation

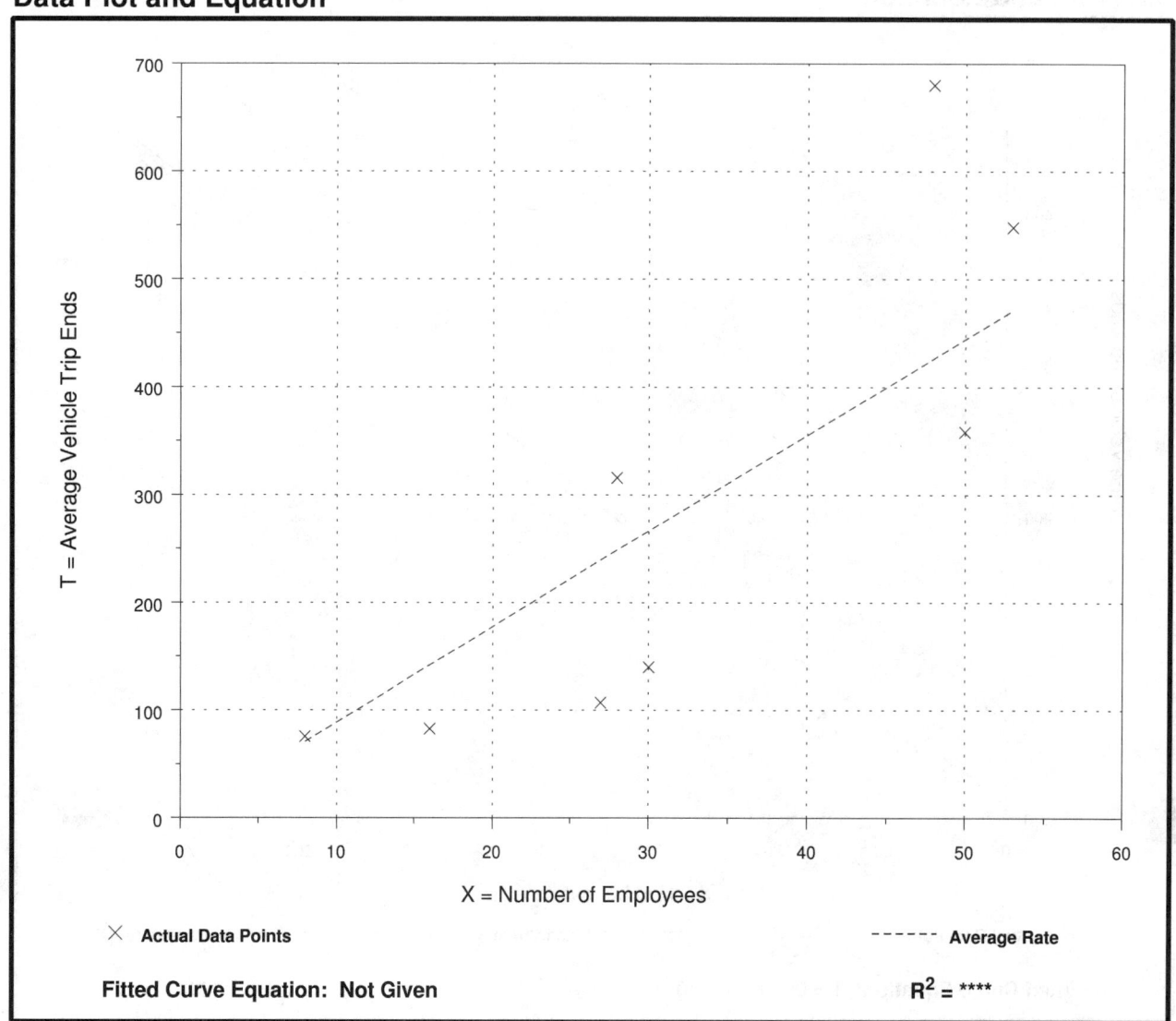

Fitted Curve Equation: Not Given $R^2 = ****$

Private School (K-8)
(534)

Average Vehicle Trip Ends vs: Employees
On a: Weekday,
P.M. Peak Hour of Generator

Number of Studies: 6
Avg. Number of Employees: 34
Directional Distribution: 47% entering, 53% exiting

Trip Generation per Employee

Average Rate	Range of Rates	Standard Deviation
5.72	1.85 - 9.69	3.54

Data Plot and Equation

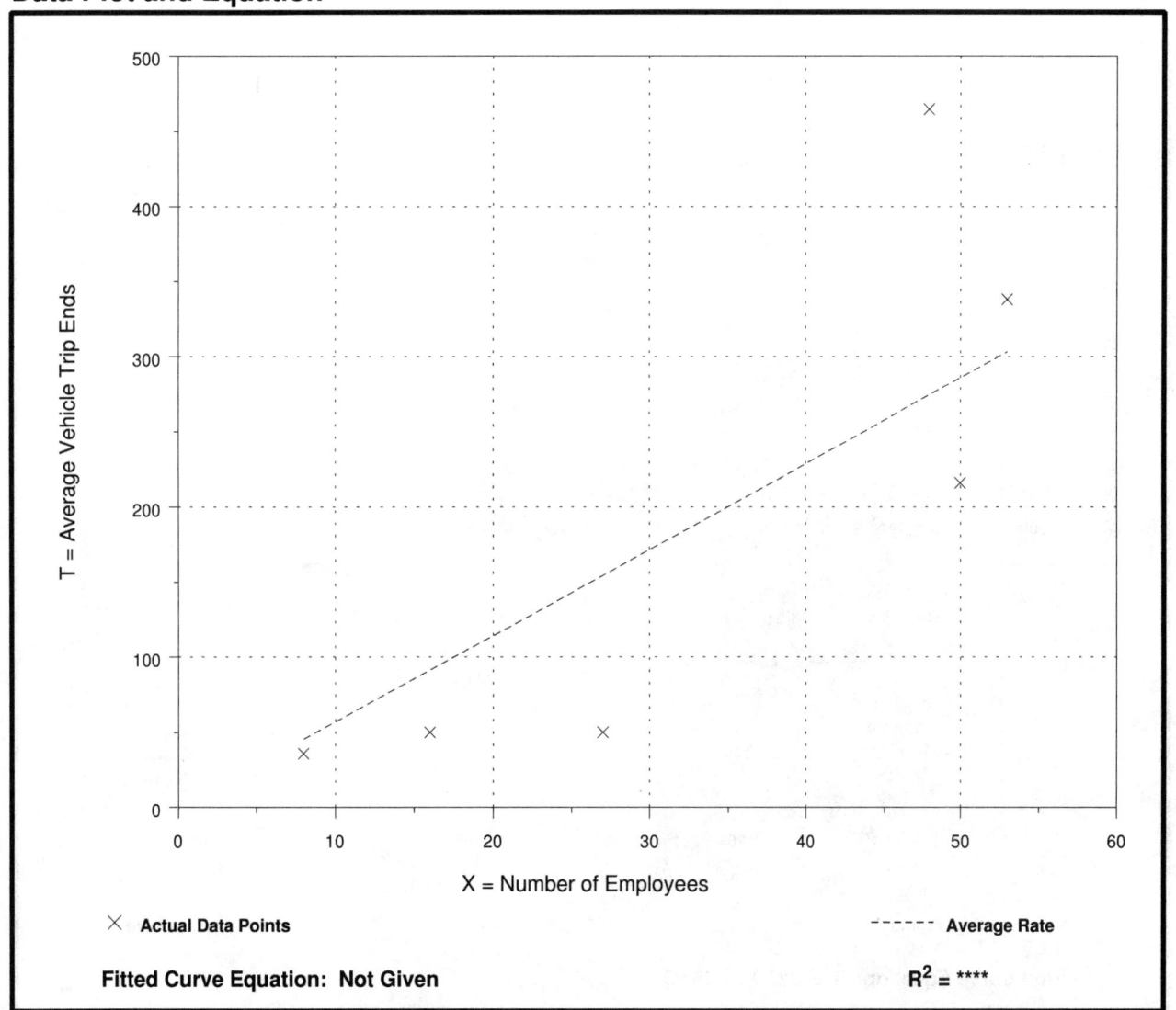

Fitted Curve Equation: Not Given $R^2 = ****$

Private School (K-8)
(534)

Average Vehicle Trip Ends vs: 1000 Sq. Feet Gross Floor Area
On a: Weekday,
A.M. Peak Hour

Number of Studies: 4
Average 1000 Sq. Feet GFA: 25
Directional Distribution: 55% entering, 45% exiting

Trip Generation per 1000 Sq. Feet Gross Floor Area

Average Rate	Range of Rates	Standard Deviation
11.59	6.92 - 20.65	5.53

Data Plot and Equation

Caution - Use Carefully - Small Sample Size

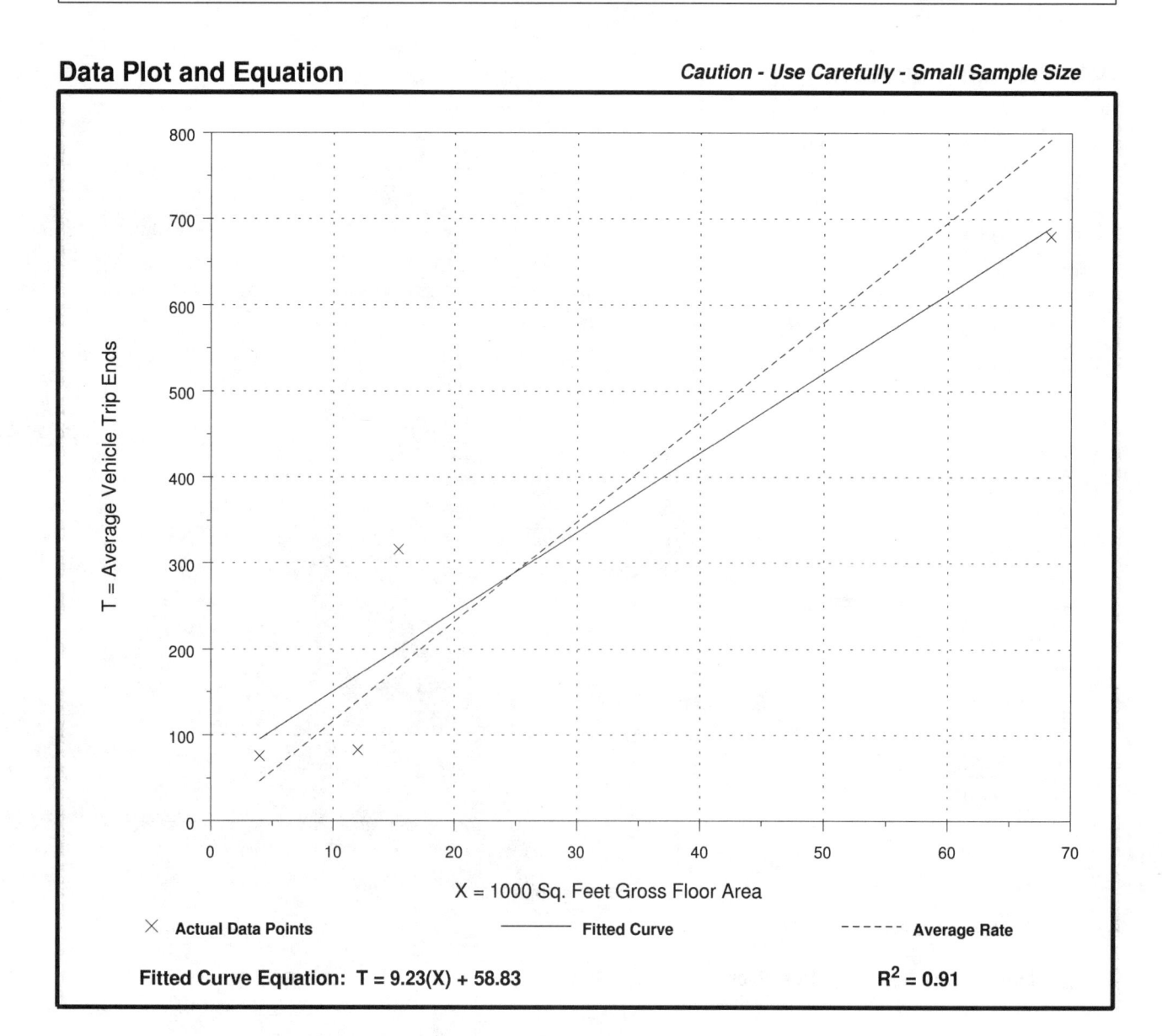

Fitted Curve Equation: T = 9.23(X) + 58.83 $R^2 = 0.91$

Private School (K-8)
(534)

Average Vehicle Trip Ends vs: 1000 Sq. Feet Gross Floor Area
On a: Weekday,
P.M. Peak Hour of Generator

Number of Studies: 3
Average 1000 Sq. Feet GFA: 28
Directional Distribution: 49% entering, 51% exiting

Trip Generation per 1000 Sq. Feet Gross Floor Area

Average Rate	Range of Rates	Standard Deviation
6.53	4.17 - 9.00	2.74

Data Plot and Equation

Caution - Use Carefully - Small Sample Size

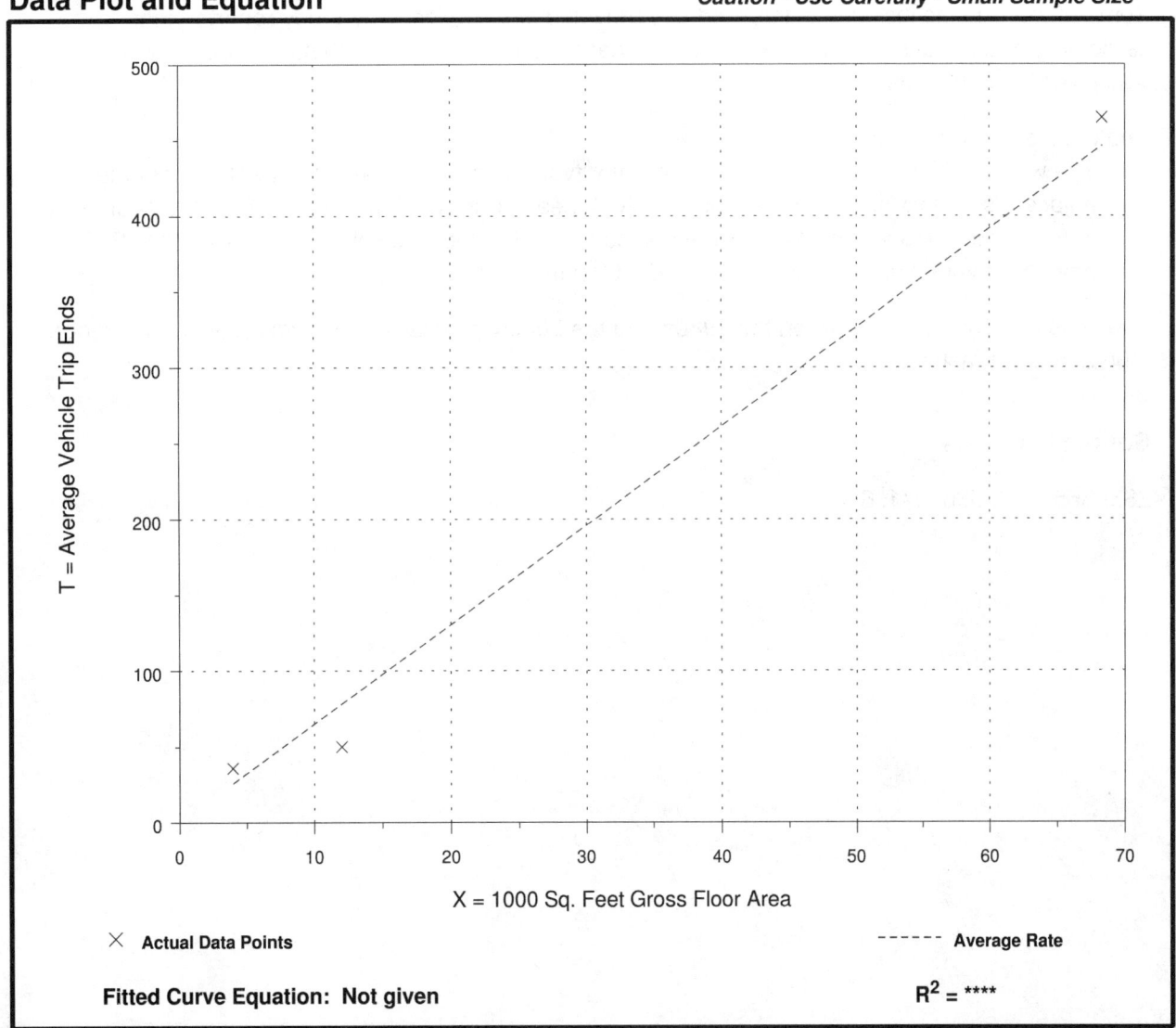

Fitted Curve Equation: Not given $R^2 = ****$

Land Use: 536
Private School (K-12)

Description

Private schools in this land use category primarily serve students attending kindergarten through the 12th grade but may also include those beginning with pre–K classes. These schools may also offer extended care and day care. Students may travel a long distance to get to private schools. Elementary school (Land Use 520), middle school/junior high school (Land Use 522), high school (Land Use 530) and private school (K-8) (Land Use 534) are related uses.

Additional Data

Some of the schools included in this land use provided bus service. One study reported that carpooling was used instead of bus service. At least one study indicated that no public bus service was provided to the school.

Peak hours of the generator—
    The weekday A.M. peak hour of the generator typically coincided with the peak hour of the adjacent street traffic; therefore, only one A.M. peak hour, which represents both the peak hour of the generator and peak hour of the adjacent street traffic, is displayed. The weekday P.M. peak hour varied between 2:30 p.m. and 4:00 p.m.

The sites were surveyed between the 1980s and the 2000s in Delaware, Florida, Maryland, Oregon, Montana and Washington.

Source Numbers

283, 355, 370, 536, 571, 613

Land Use: 536
Private School (K-12)

Independent Variables with One Observation

The following trip generation data are for independent variables with only one observation. This information is shown in this table only; there are no related plots for these data.

Users are cautioned to use data with care because of the small sample size.

Independent Variable	Trip Generation Rate	Size of Independent Variable	Number of Studies	Directional Distribution
1,000 Square Feet Gross Floor Area				
Weekday P.M. Peak Hour of Generator	5.50	45	1	28% entering, 72% exiting

Private School (K-12)
(536)

Average Vehicle Trip Ends vs: Students
On a: Weekday

Number of Studies: 2
Average Number of Students: 537
Directional Distribution: 50% entering, 50% exiting

Trip Generation per Student

Average Rate	Range of Rates	Standard Deviation
2.48	1.74 - 3.12	*

Data Plot and Equation

Caution - Use Carefully - Small Sample Size

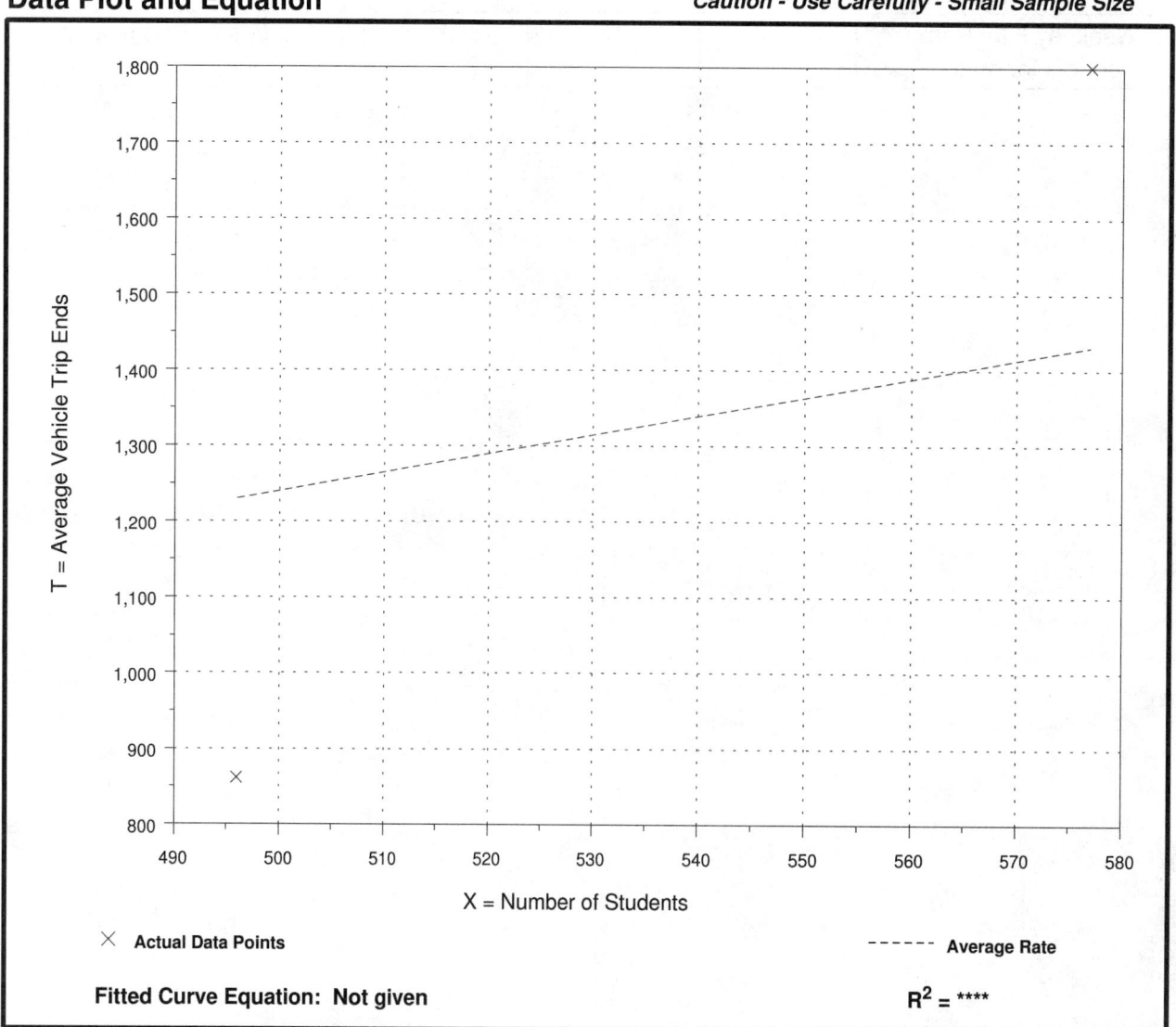

Fitted Curve Equation: Not given

$R^2 = ****$

Private School (K-12)
(536)

Average Vehicle Trip Ends vs: Students
On a: Weekday,
A.M. Peak Hour

Number of Studies: 5
Average Number of Students: 460
Directional Distribution: 61% entering, 39% exiting

Trip Generation per Student

Average Rate	Range of Rates	Standard Deviation
0.81	0.52 - 0.96	0.91

Data Plot and Equation

Caution - Use Carefully - Small Sample Size

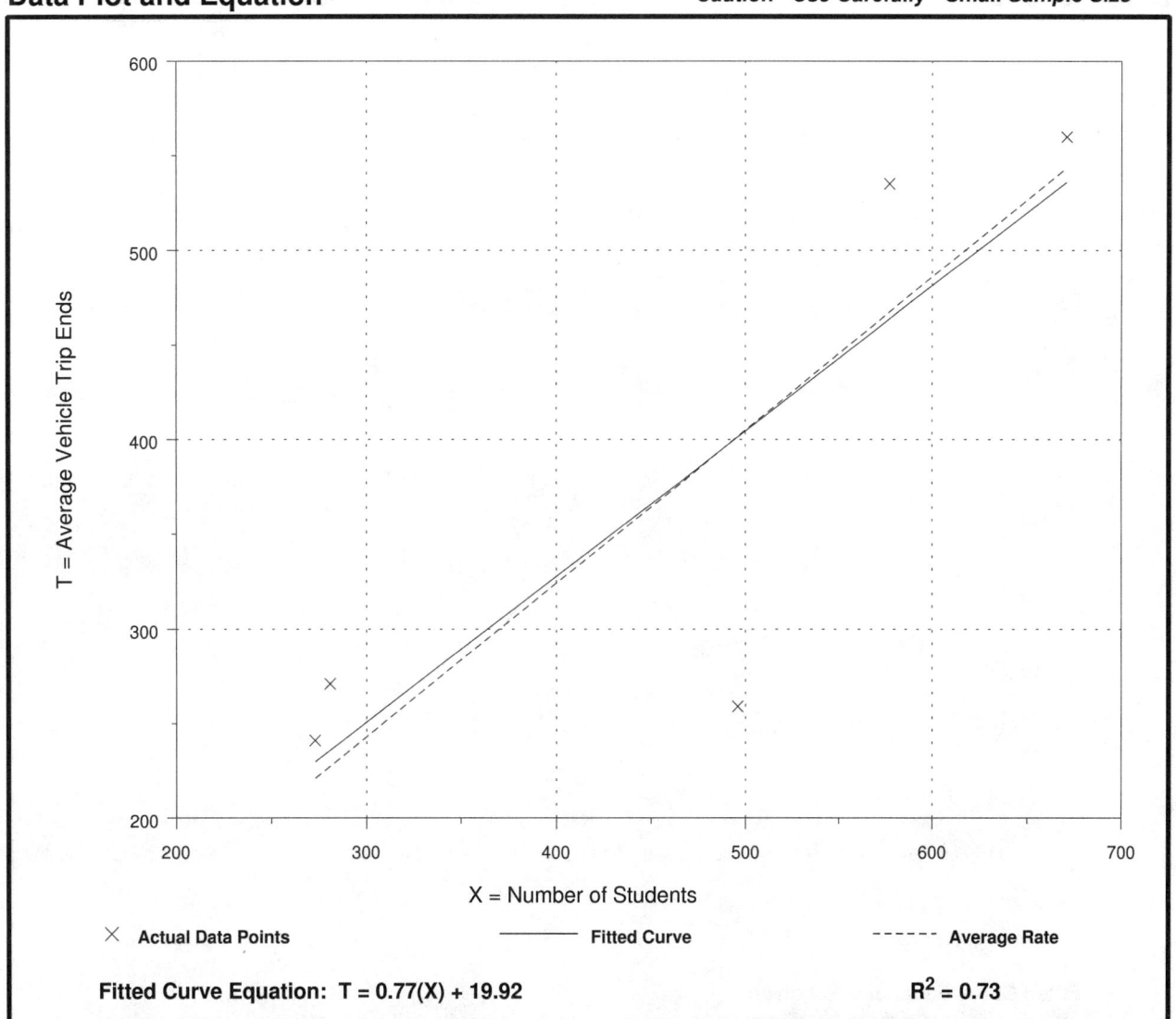

Fitted Curve Equation: $T = 0.77(X) + 19.92$ $R^2 = 0.73$

Private School (K-12)
(536)

Average Vehicle Trip Ends vs: Students
On a: Weekday,
　　　　Peak Hour of Adjacent Street Traffic,
　　　　One Hour Between 4 and 6 p.m.

Number of Studies: 3
Average Number of Students: 581
Directional Distribution: 43% entering, 57% exiting

Trip Generation per Student

Average Rate	Range of Rates	Standard Deviation
0.17	0.13 - 0.23	0.41

Data Plot and Equation

Caution - Use Carefully - Small Sample Size

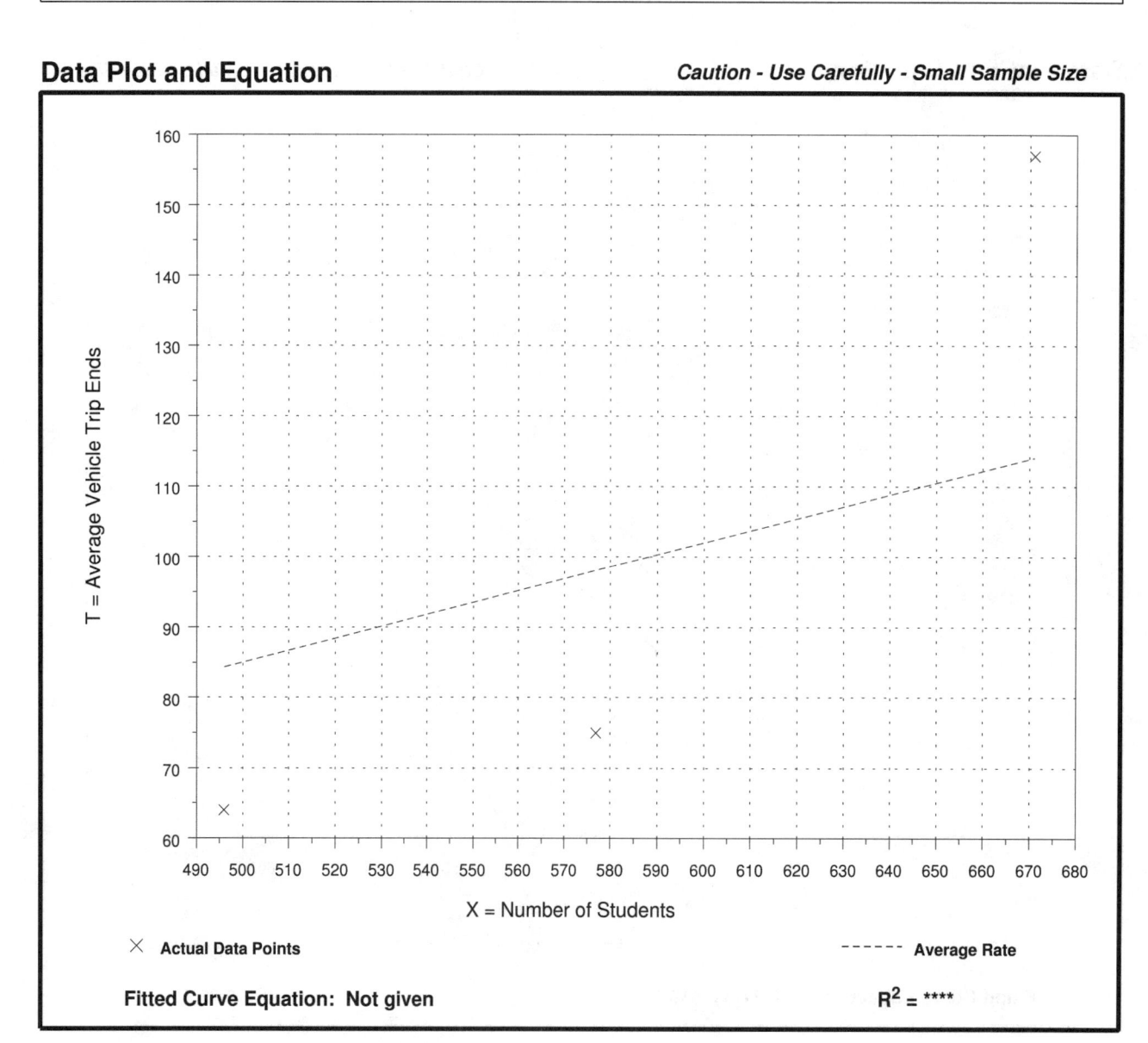

Fitted Curve Equation: Not given　　　　　$R^2 = ****$

Private School (K-12)
(536)

Average Vehicle Trip Ends vs: Students
On a: Weekday,
P.M. Peak Hour of Generator

Number of Studies: 4
Average Number of Students: 506
Directional Distribution: 42% entering, 58% exiting

Trip Generation per Student

Average Rate	Range of Rates	Standard Deviation
0.58	0.46 - 0.79	0.77

Data Plot and Equation

Caution - Use Carefully - Small Sample Size

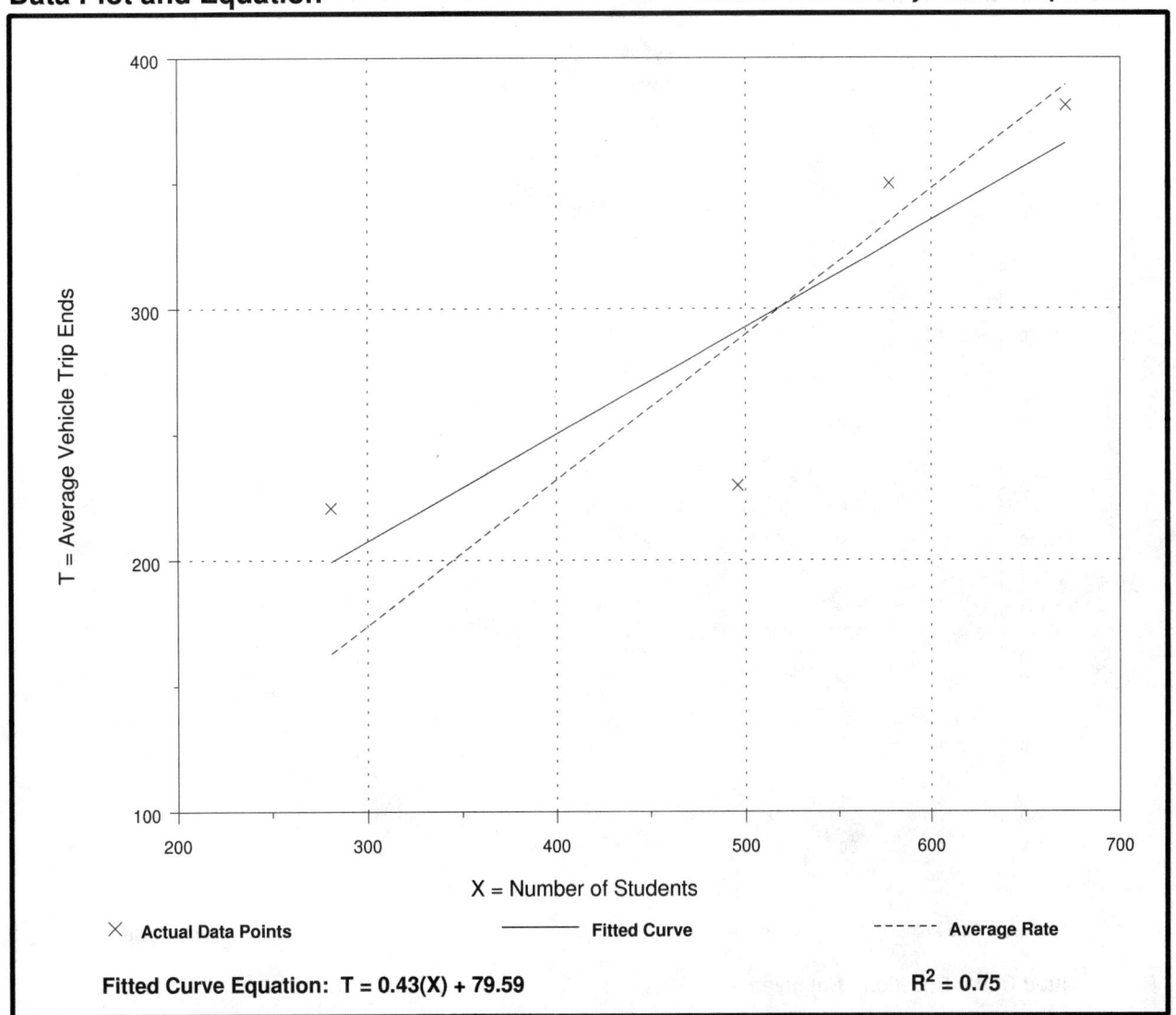

Fitted Curve Equation: T = 0.43(X) + 79.59 $R^2 = 0.75$

Private School (K-12)
(536)

Average Vehicle Trip Ends vs: Employees
On a: Weekday

Number of Studies: 2
Avg. Number of Employees: 81
Directional Distribution: 50% entering, 50% exiting

Trip Generation per Employee

Average Rate	Range of Rates	Standard Deviation
16.43	16.36 - 16.58	*

Data Plot and Equation

Caution - Use Carefully - Small Sample Size

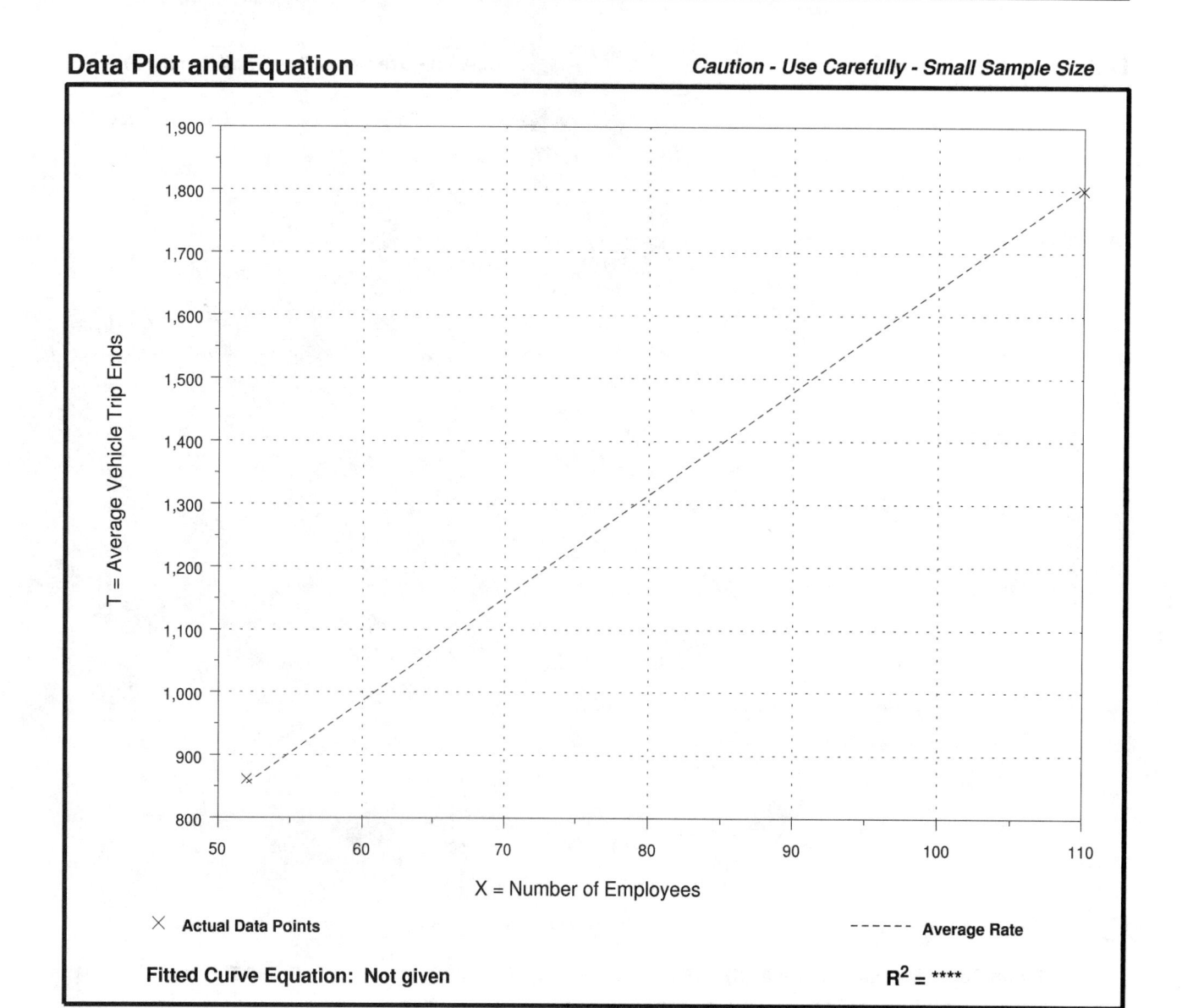

Fitted Curve Equation: Not given $R^2 = ****$

Private School (K-12)
(536)

Average Vehicle Trip Ends vs: Employees
On a: Weekday,
A.M. Peak Hour

Number of Studies: 4
Avg. Number of Employees: 62
Directional Distribution: 61% entering, 39% exiting

Trip Generation per Employee

Average Rate	Range of Rates	Standard Deviation
5.31	4.86 - 8.03	2.51

Data Plot and Equation

Caution - Use Carefully - Small Sample Size

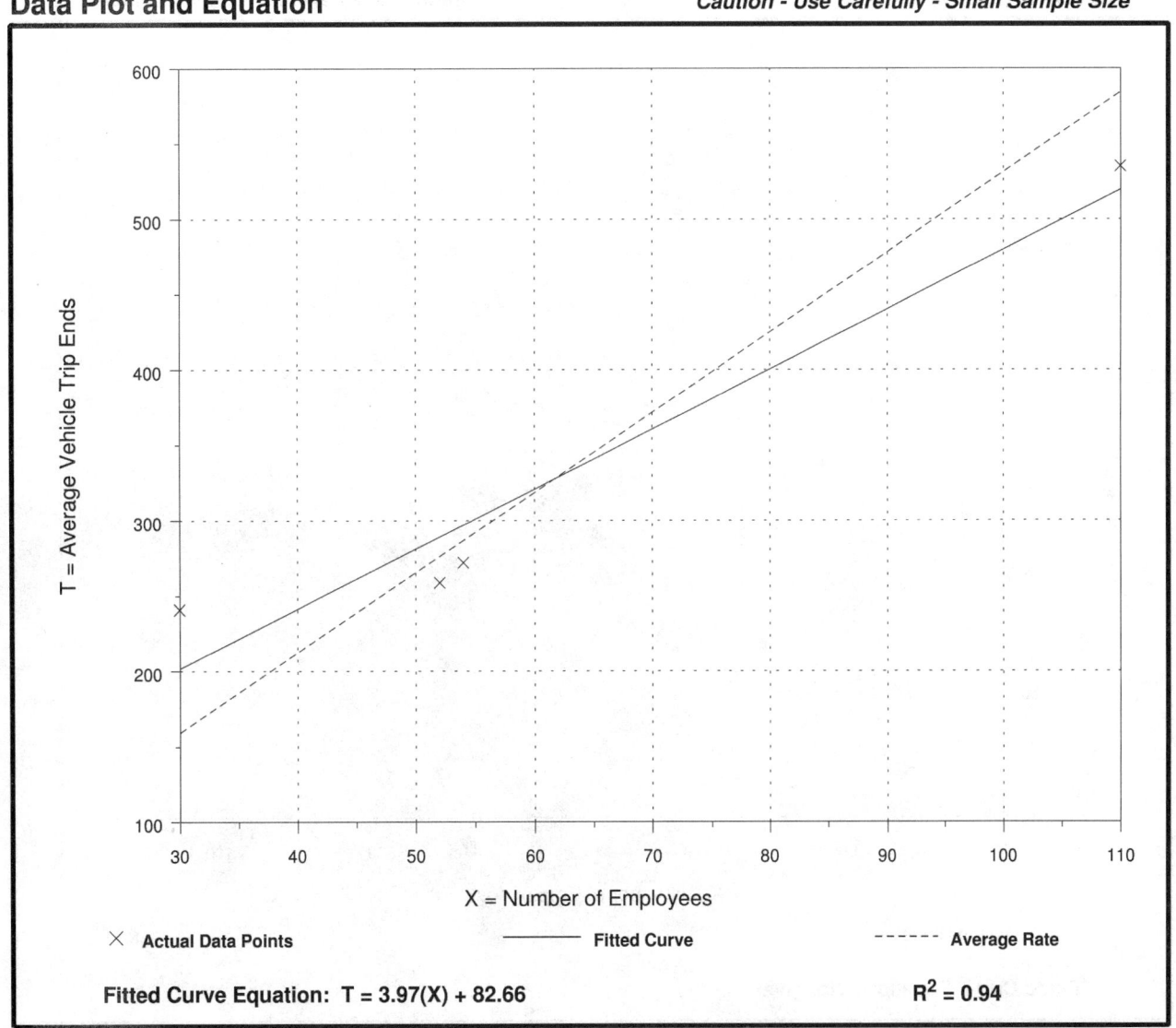

Fitted Curve Equation: T = 3.97(X) + 82.66 $R^2 = 0.94$

Trip Generation, 9th Edition • Institute of Transportation Engineers

Private School (K-12)
(536)

Average Vehicle Trip Ends vs: Employees
On a: Weekday,
Peak Hour of Adjacent Street Traffic,
One Hour Between 4 and 6 p.m.

Number of Studies: 2
Avg. Number of Employees: 81
Directional Distribution: 38% entering, 62% exiting

Trip Generation per Employee

Average Rate	Range of Rates	Standard Deviation
0.86	0.68 - 1.23	*

Data Plot and Equation

Caution - Use Carefully - Small Sample Size

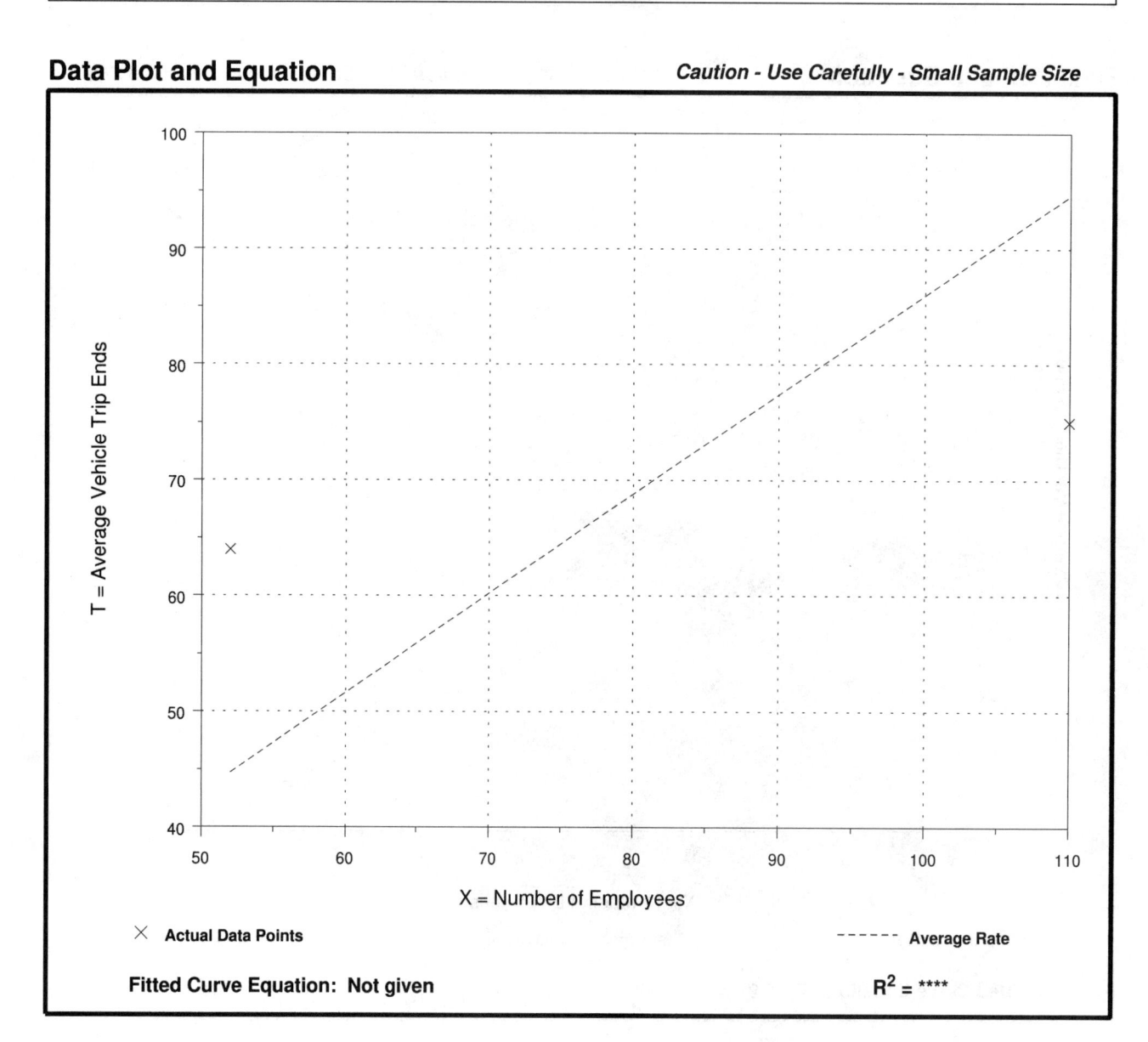

Fitted Curve Equation: Not given $R^2 = ****$

Private School (K-12)
(536)

Average Vehicle Trip Ends vs: Employees
On a: Weekday,
P.M. Peak Hour of Generator

Number of Studies: 3
Avg. Number of Employees: 72
Directional Distribution: 39% entering, 61% exiting

Trip Generation per Employee

Average Rate	Range of Rates	Standard Deviation
3.82	3.18 - 4.56	2.05

Data Plot and Equation

Caution - Use Carefully - Small Sample Size

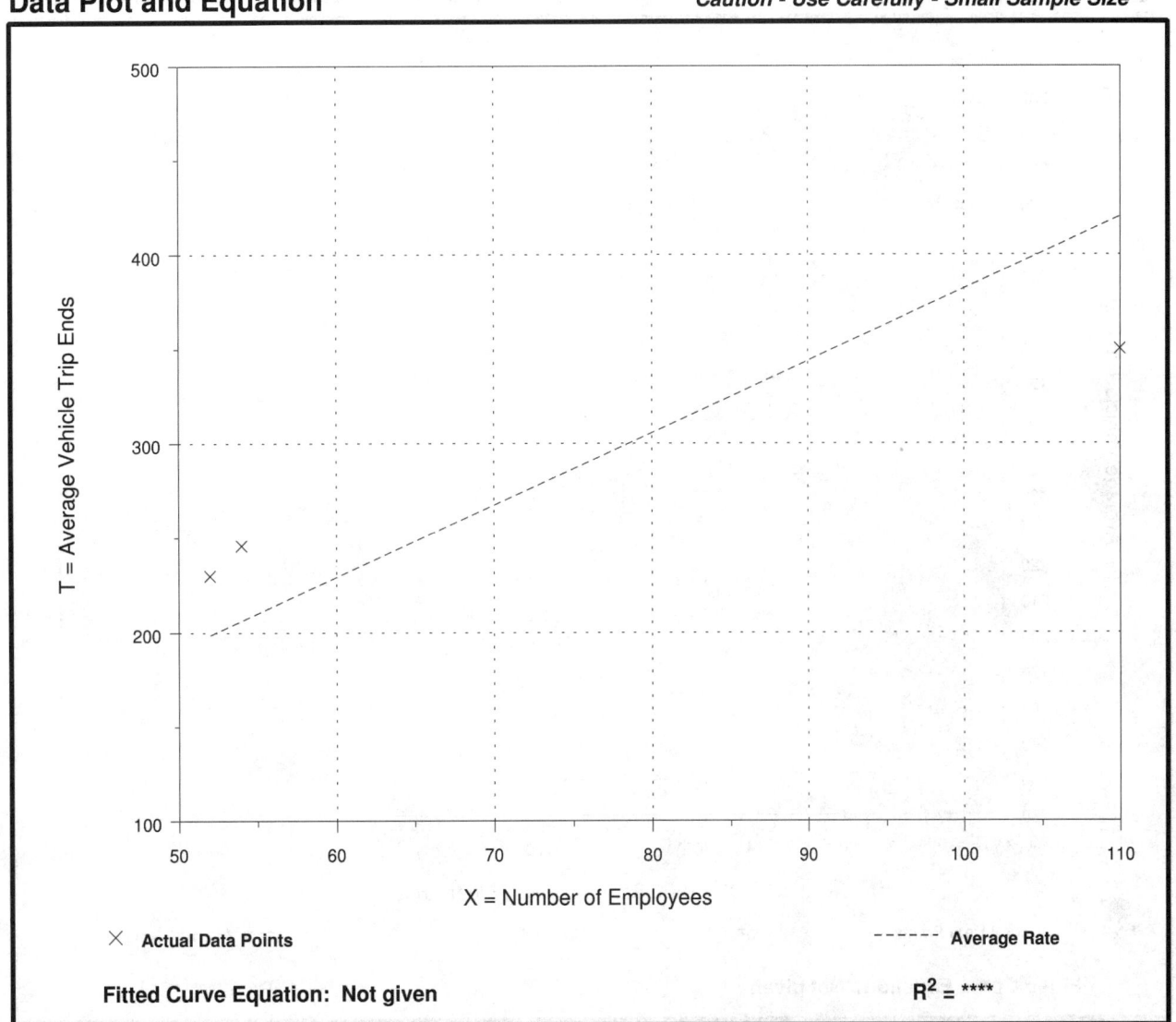

Fitted Curve Equation: Not given $R^2 = ****$

Private School (K-12)
(536)

Average Vehicle Trip Ends vs: 1000 Sq. Feet Gross Floor Area
On a: Weekday,
A.M. Peak Hour

Number of Studies: 2
Average 1000 Sq. Feet GFA: 72
Directional Distribution: 63% entering, 37% exiting

Trip Generation per 1000 Sq. Feet Gross Floor Area

Average Rate	Range of Rates	Standard Deviation
3.54	2.41 - 6.08	*

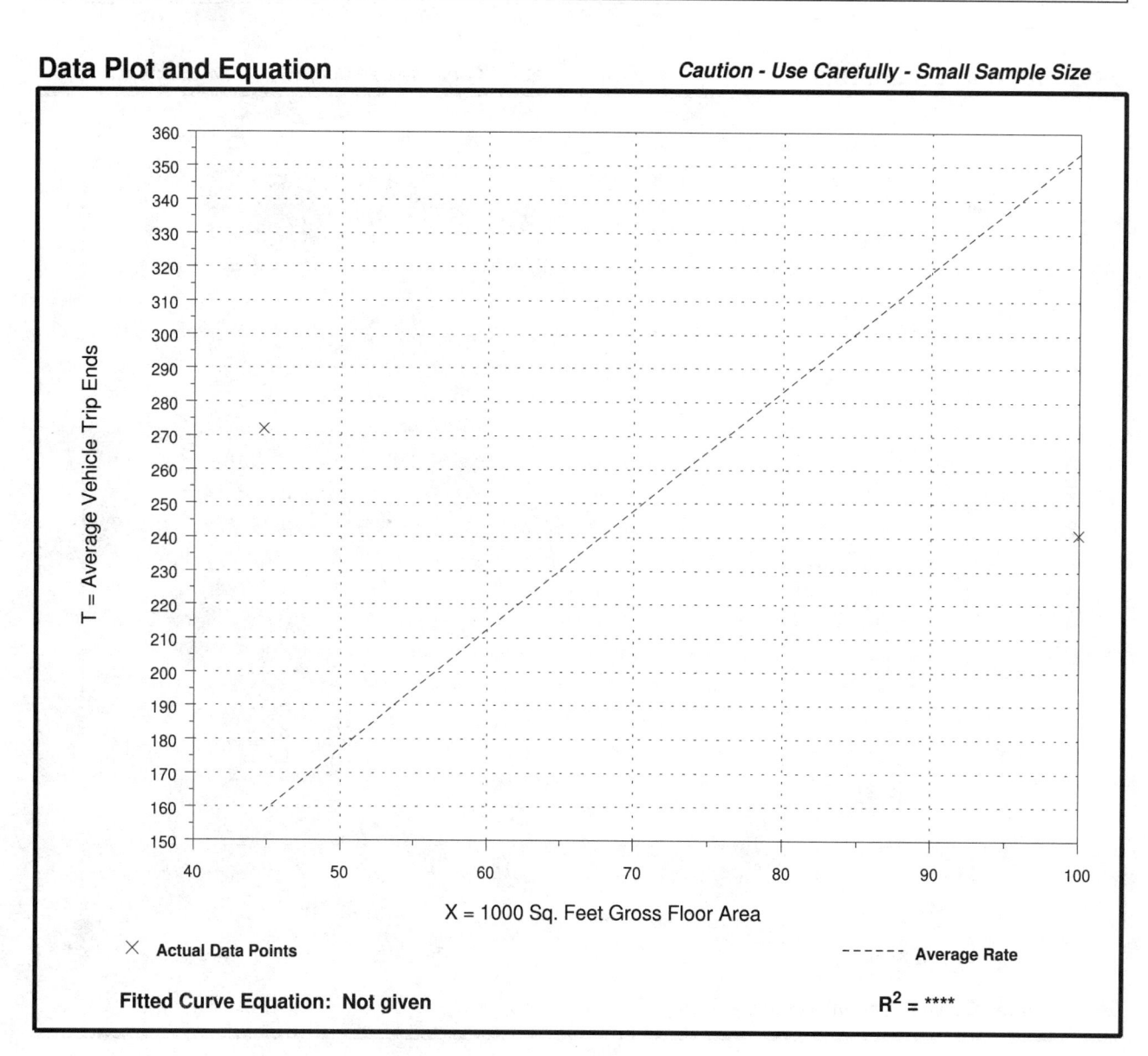

Data Plot and Equation

Caution - Use Carefully - Small Sample Size

Fitted Curve Equation: Not given $R^2 = ****$

Land Use: 540
Junior/Community College

Description

This land use includes two-year junior, community, or technical colleges. Four-year (or more) colleges or universities are described in university/college (Land Use 550). A number of two-year institutions have sizable evening programs.

Additional Data

The trip generation for weekend time periods varied considerably; therefore, caution should be used when applying weekend statistics. Information describing the weekend activities conducted at junior/community colleges was not available.

Two of the studies indicated that transit centers were located within close proximity of the colleges.

Acreage, floor space, staff and parking accommodations varied widely with the populations served and the social and economic characteristics of the area; thus, the number of students may be a more reliable independent variable on which to establish trip generation rates.

The sites were surveyed between the late 1970s and the 2000s throughout the United States.

Many of the studies included in this land use did not indicate if significant transit service was available within close proximity of the site and did not specify the type of transit service. To assist in the future analysis of this land use, it is important that this information be collected and included in trip generation data submissions.

Source Numbers

86, 111, 396, 413, 533, 628

Junior/Community College
(540)

Average Vehicle Trip Ends vs: Students
On a: Weekday

Number of Studies: 7
Average Number of Students: 11,864
Directional Distribution: 50% entering, 50% exiting

Trip Generation per Student

Average Rate	Range of Rates	Standard Deviation
1.23	0.93 - 2.16	1.16

Data Plot and Equation

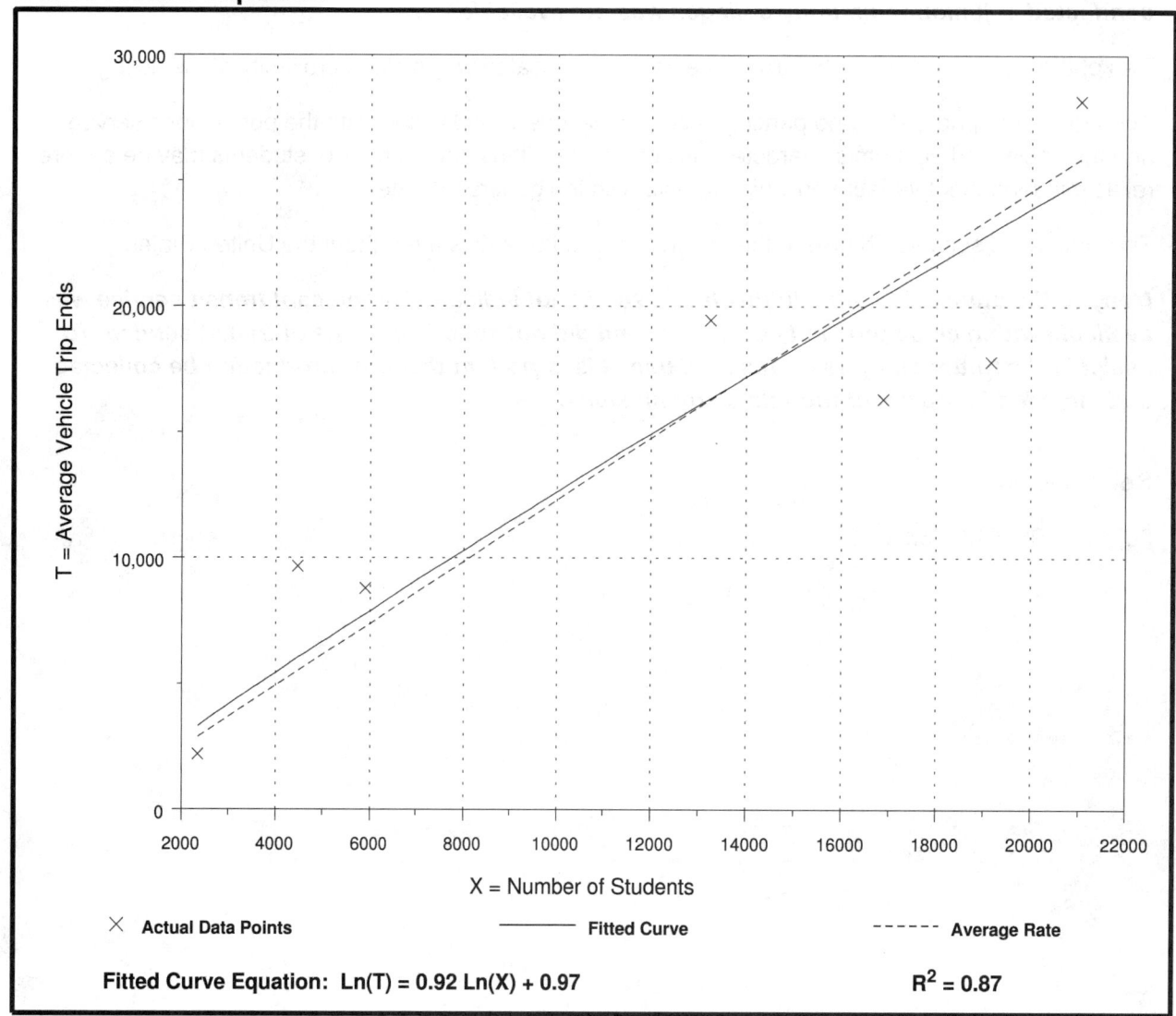

Fitted Curve Equation: $Ln(T) = 0.92\ Ln(X) + 0.97$ $R^2 = 0.87$

Junior/Community College
(540)

Average Vehicle Trip Ends vs: Students
On a: Weekday,
Peak Hour of Adjacent Street Traffic,
One Hour Between 7 and 9 a.m.

Number of Studies: 6
Average Number of Students: 13,450
Directional Distribution: 84% entering, 16% exiting

Trip Generation per Student

Average Rate	Range of Rates	Standard Deviation
0.12	0.10 - 0.17	0.35

Data Plot and Equation

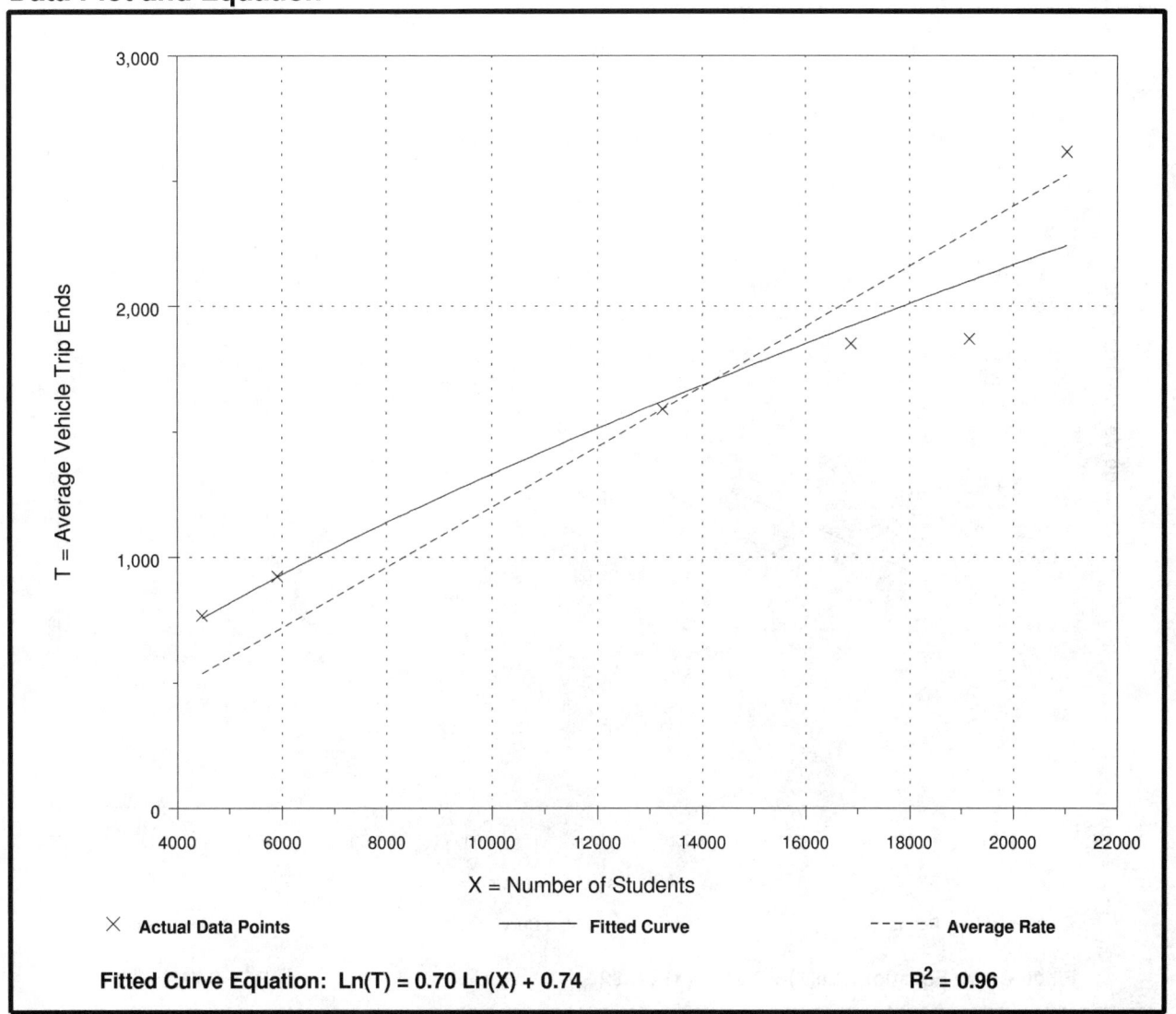

Fitted Curve Equation: $Ln(T) = 0.70 \, Ln(X) + 0.74$ $R^2 = 0.96$

Junior/Community College
(540)

Average Vehicle Trip Ends vs: Students
On a: Weekday,
Peak Hour of Adjacent Street Traffic,
One Hour Between 4 and 6 p.m.

Number of Studies: 6
Average Number of Students: 13,450
Directional Distribution: 63% entering, 37% exiting

Trip Generation per Student

Average Rate	Range of Rates	Standard Deviation
0.12	0.08 - 0.19	0.35

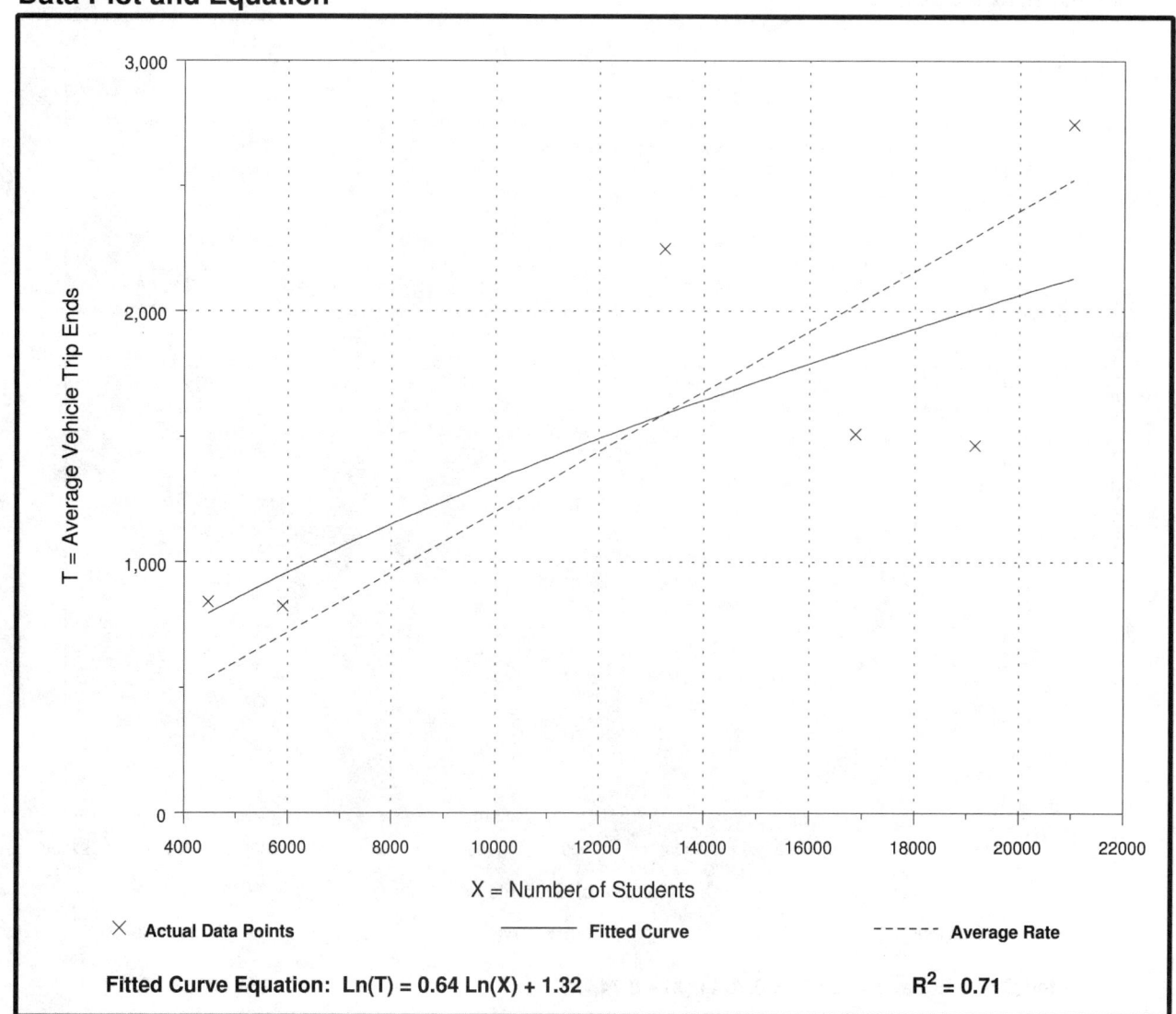

Data Plot and Equation

Fitted Curve Equation: $Ln(T) = 0.64\ Ln(X) + 1.32$ $R^2 = 0.71$

Junior/Community College
(540)

Average Vehicle Trip Ends vs: Students
On a: Weekday,
A.M. Peak Hour of Generator

Number of Studies: 6
Average Number of Students: 10,336
Directional Distribution: 65% entering, 35% exiting

Trip Generation per Student

Average Rate	Range of Rates	Standard Deviation
0.12	0.10 - 0.17	0.35

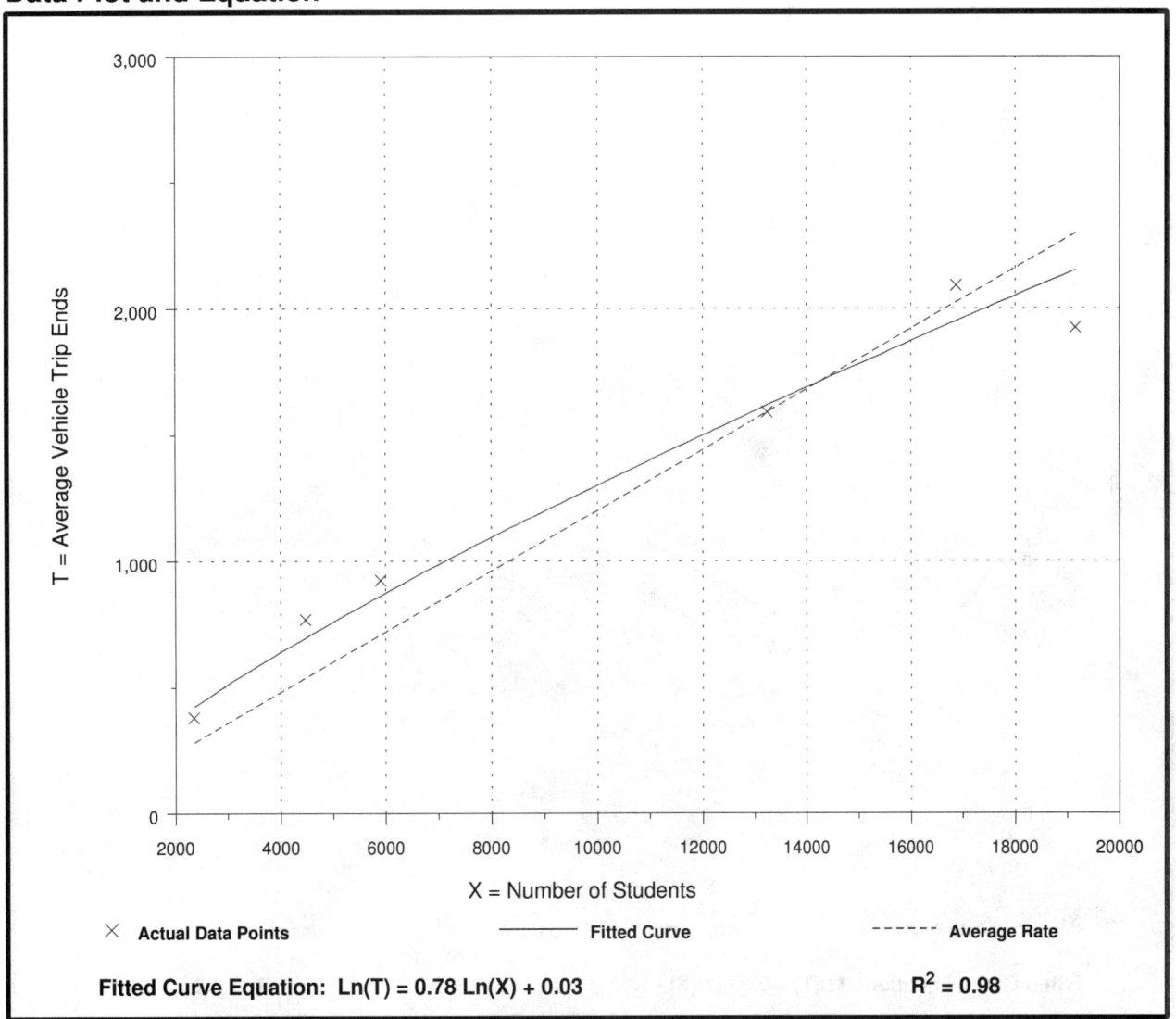

Data Plot and Equation

Fitted Curve Equation: $\text{Ln}(T) = 0.78\, \text{Ln}(X) + 0.03$ $R^2 = 0.98$

Junior/Community College
(540)

Average Vehicle Trip Ends vs: Students
On a: Weekday,
P.M. Peak Hour of Generator

Number of Studies: 6
Average Number of Students: 10,336
Directional Distribution: 55% entering, 45% exiting

Trip Generation per Student

Average Rate	Range of Rates	Standard Deviation
0.12	0.08 - 0.20	0.35

Data Plot and Equation

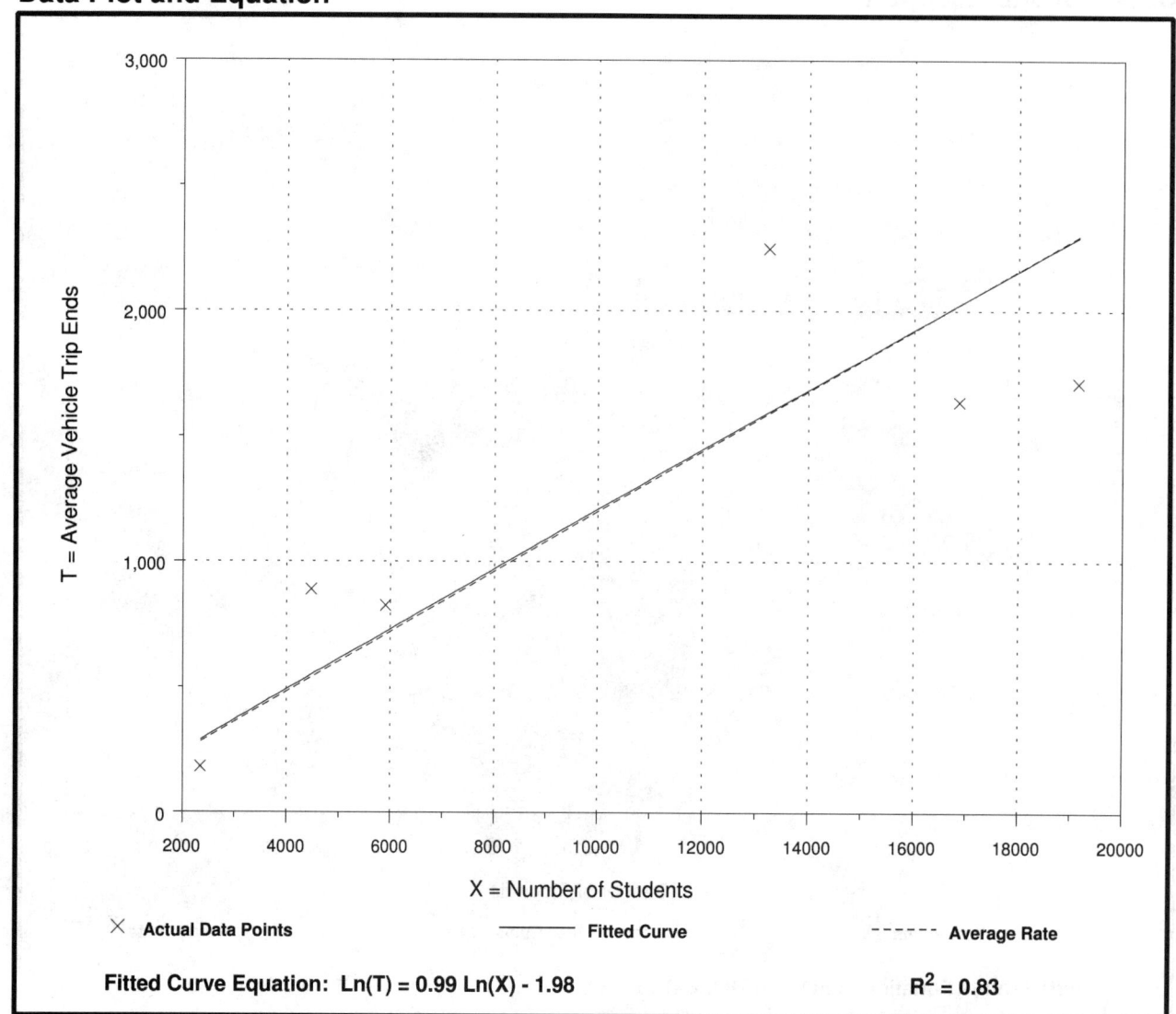

Fitted Curve Equation: $\text{Ln}(T) = 0.99\ \text{Ln}(X) - 1.98$ $R^2 = 0.83$

Junior/Community College
(540)

Average Vehicle Trip Ends vs: Students
On a: Saturday

Number of Studies: 3
Average Number of Students: 13,503
Directional Distribution: 50% entering, 50% exiting

Trip Generation per Student

Average Rate	Range of Rates	Standard Deviation
0.42	0.29 - 0.71	0.66

Data Plot and Equation

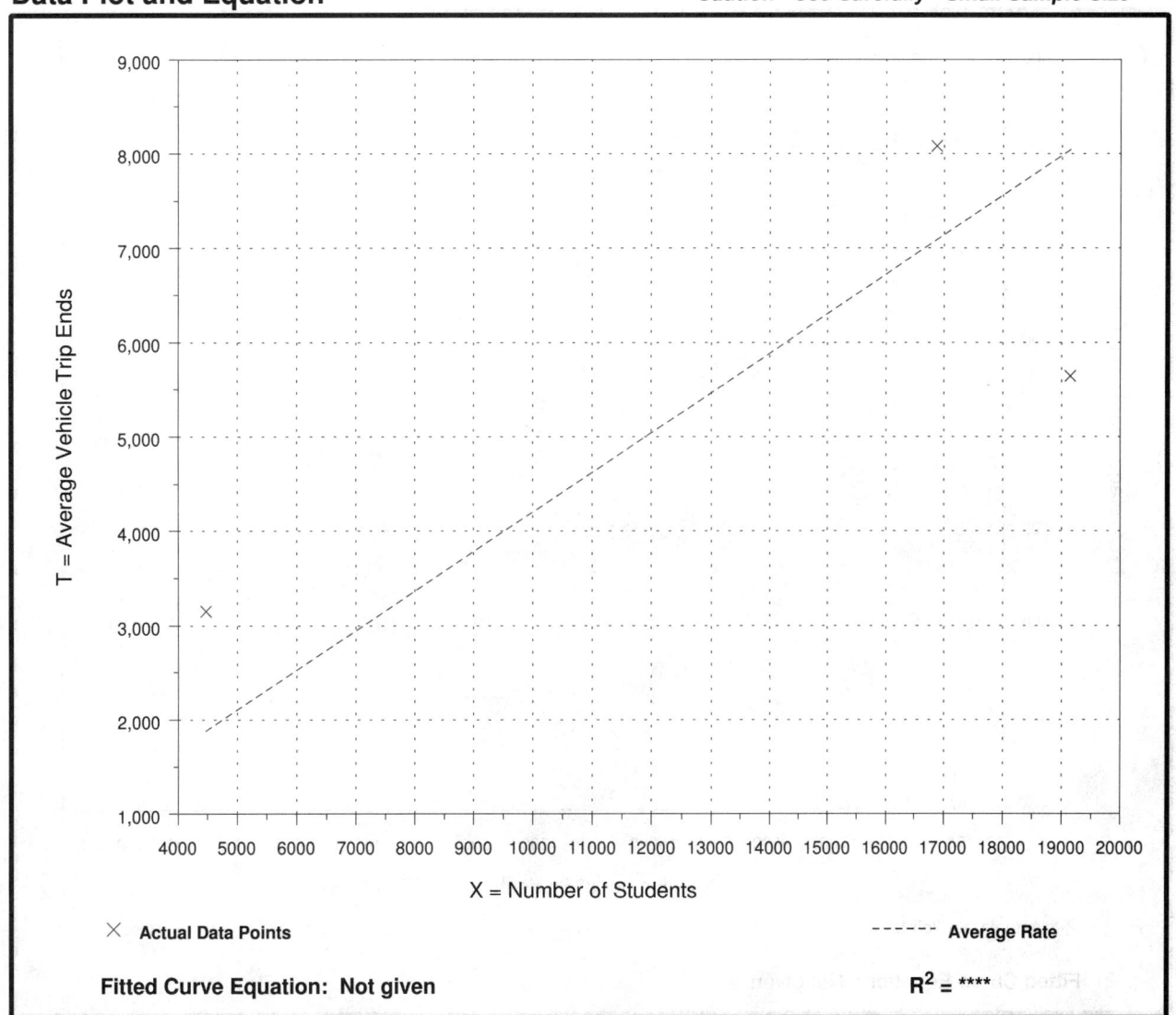

Caution - Use Carefully - Small Sample Size

Fitted Curve Equation: Not given $R^2 = ****$

Junior/Community College
(540)

Average Vehicle Trip Ends vs: Students
On a: Saturday,
Peak Hour of Generator

Number of Studies: 3
Average Number of Students: 13,503
Directional Distribution: 57% entering, 43% exiting

Trip Generation per Student

Average Rate	Range of Rates	Standard Deviation
0.05	0.04 - 0.08	0.23

Data Plot and Equation

Caution - Use Carefully - Small Sample Size

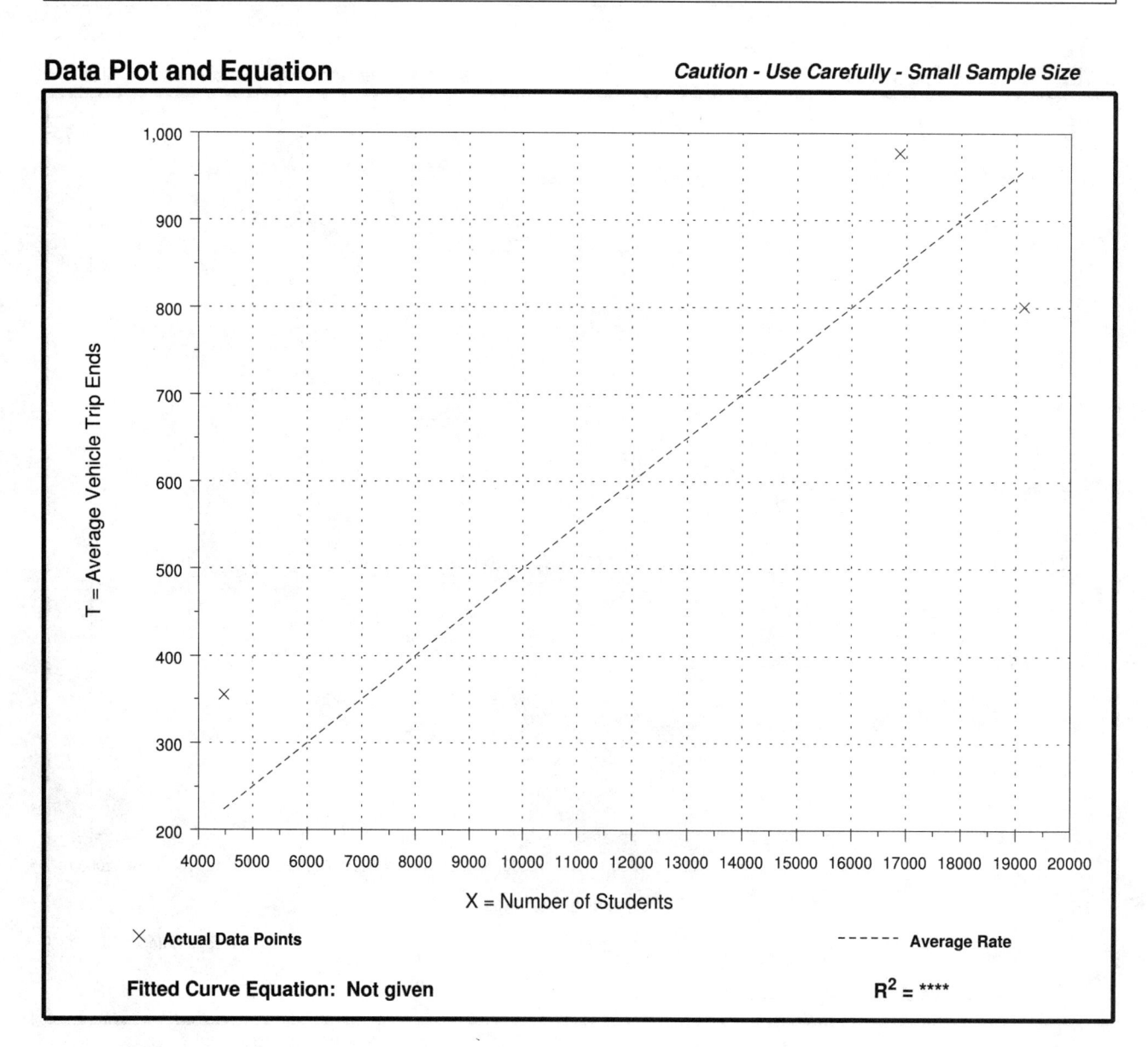

Fitted Curve Equation: Not given

$R^2 =$ ****

Junior/Community College
(540)

Average Vehicle Trip Ends vs: Students
On a: Sunday

Number of Studies: 3
Average Number of Students: 13,503
Directional Distribution: 50% entering, 50% exiting

Trip Generation per Student

Average Rate	Range of Rates	Standard Deviation
0.04	0.03 - 0.16	0.22

Data Plot and Equation

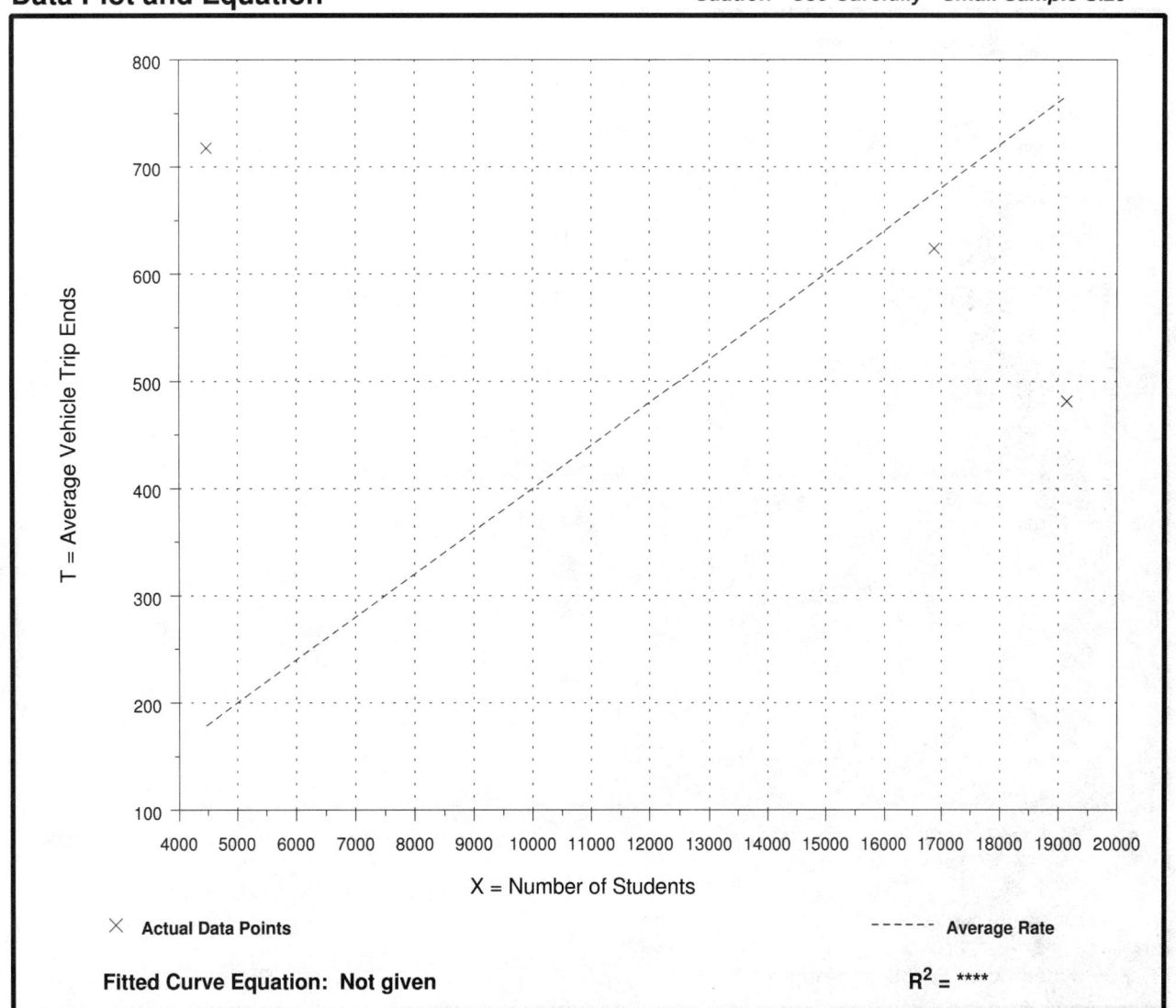

Caution - Use Carefully - Small Sample Size

Fitted Curve Equation: Not given $R^2 = ****$

Junior/Community College
(540)

Average Vehicle Trip Ends vs: Students
On a: Sunday,
Peak Hour of Generator

Number of Studies: 3
Average Number of Students: 13,503
Directional Distribution: 46% entering, 54% exiting

Trip Generation per Student

Average Rate	Range of Rates	Standard Deviation
0.01	0.00 - 0.03	0.09

Data Plot and Equation

Caution - Use Carefully - Small Sample Size

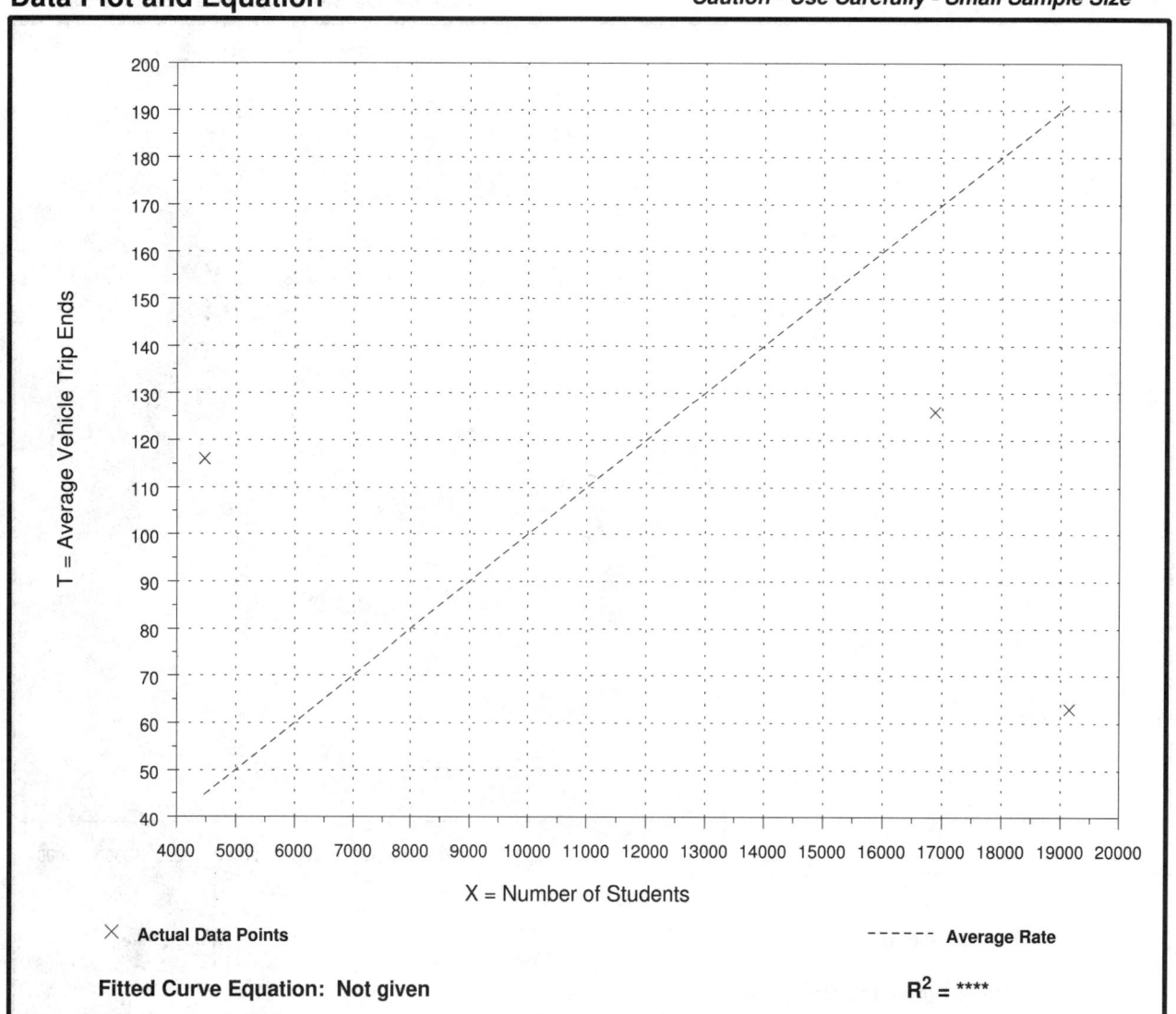

Fitted Curve Equation: Not given $R^2 = ****$

Junior/Community College
(540)

Average Vehicle Trip Ends vs: Employees
On a: Weekday

Number of Studies: 4
Avg. Number of Employees: 740
Directional Distribution: 50% entering, 50% exiting

Trip Generation per Employee

Average Rate	Range of Rates	Standard Deviation
15.55	10.06 - 35.79	7.62

Data Plot and Equation

Caution - Use Carefully - Small Sample Size

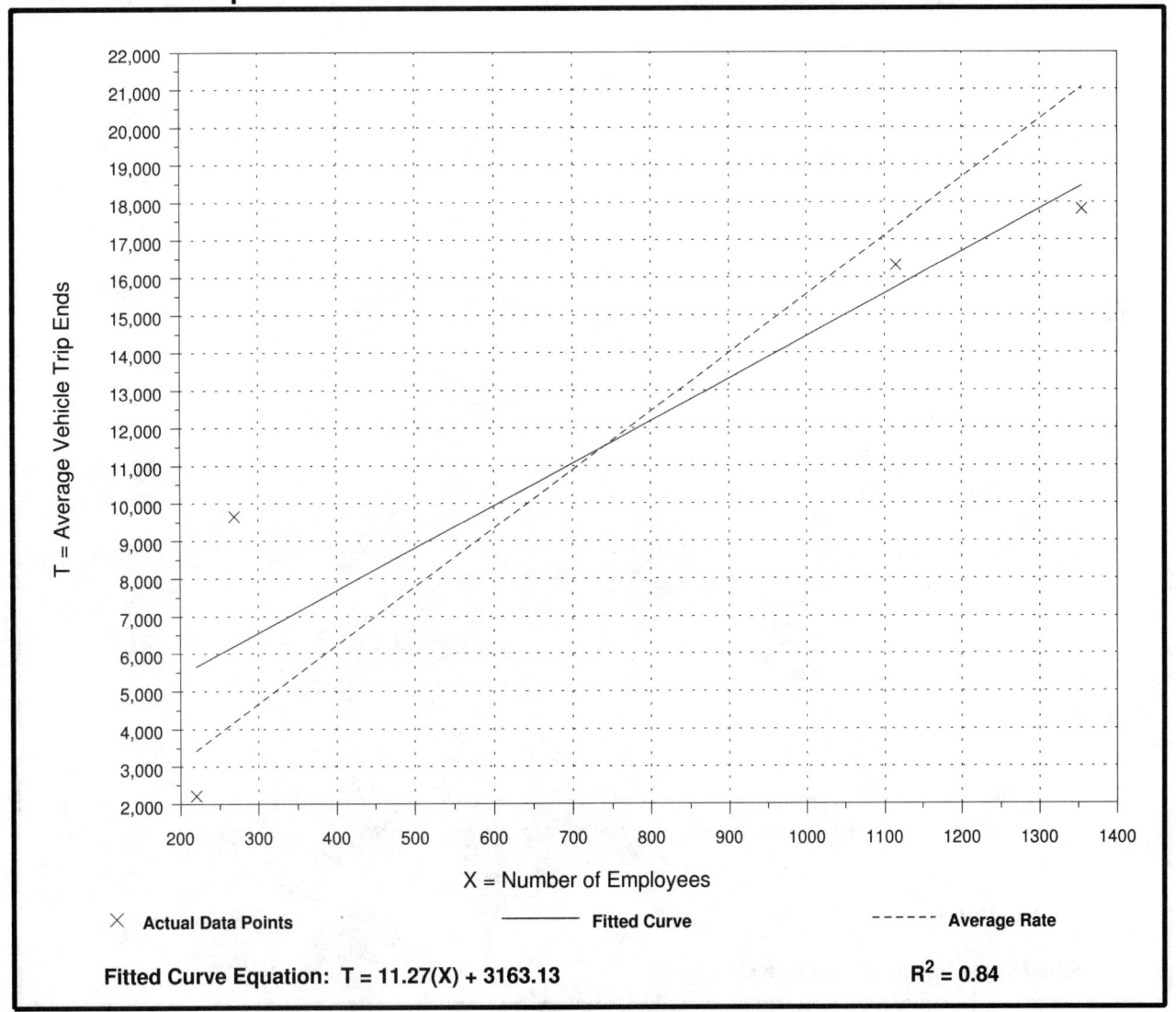

Fitted Curve Equation: $T = 11.27(X) + 3163.13$ $R^2 = 0.84$

Junior/Community College
(540)

Average Vehicle Trip Ends vs: Employees
On a: Weekday,
Peak Hour of Adjacent Street Traffic,
One Hour Between 7 and 9 a.m.

Number of Studies: 3
Avg. Number of Employees: 913
Directional Distribution: 74% entering, 26% exiting

Trip Generation per Employee

Average Rate	Range of Rates	Standard Deviation
1.64	1.38 - 2.85	1.35

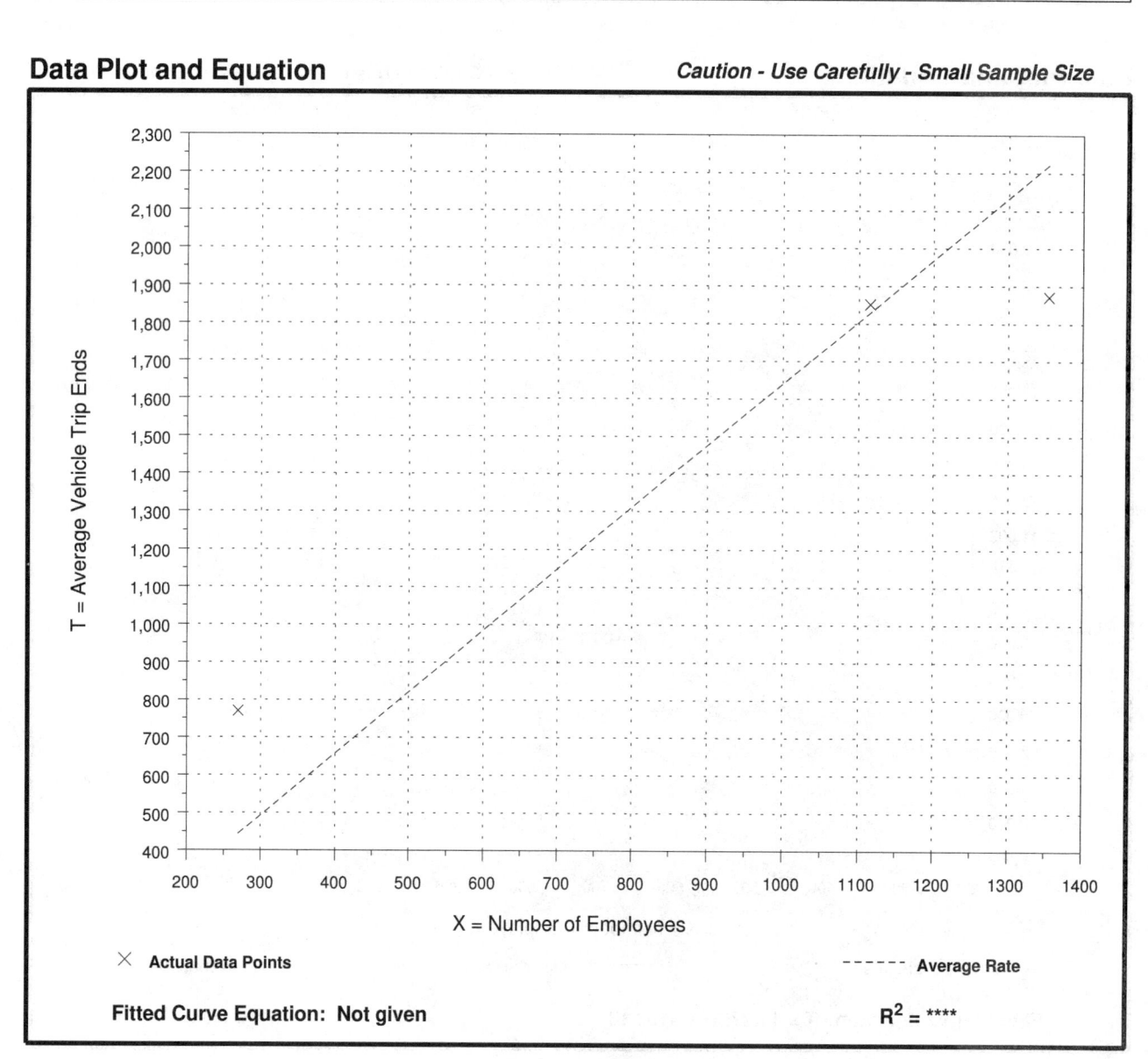

Data Plot and Equation

Caution - Use Carefully - Small Sample Size

Fitted Curve Equation: Not given $R^2 = ****$

Junior/Community College
(540)

Average Vehicle Trip Ends vs: Employees
On a: Weekday,
Peak Hour of Adjacent Street Traffic,
One Hour Between 4 and 6 p.m.

Number of Studies: 3
Avg. Number of Employees: 913
Directional Distribution: 58% entering, 42% exiting

Trip Generation per Employee

Average Rate	Range of Rates	Standard Deviation
1.39	1.08 - 3.12	1.32

Data Plot and Equation

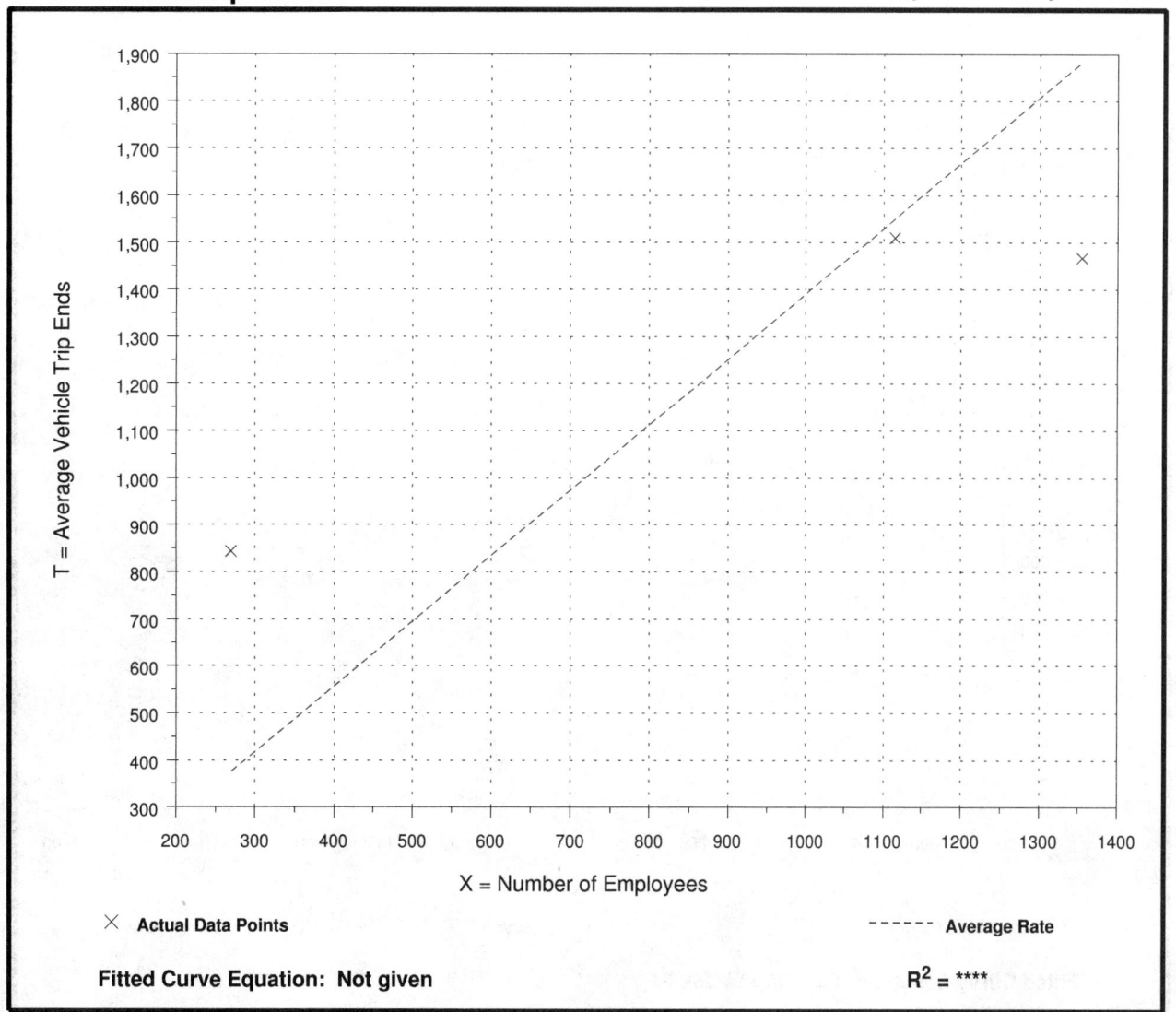

Caution - Use Carefully - Small Sample Size

Fitted Curve Equation: Not given $R^2 = ****$

Junior/Community College
(540)

Average Vehicle Trip Ends vs: Employees
On a: Weekday,
A.M. Peak Hour of Generator

Number of Studies: 4
Avg. Number of Employees: 740
Directional Distribution: 50% entering, 50% exiting

Trip Generation per Employee

Average Rate	Range of Rates	Standard Deviation
1.75	1.42 - 2.85	1.38

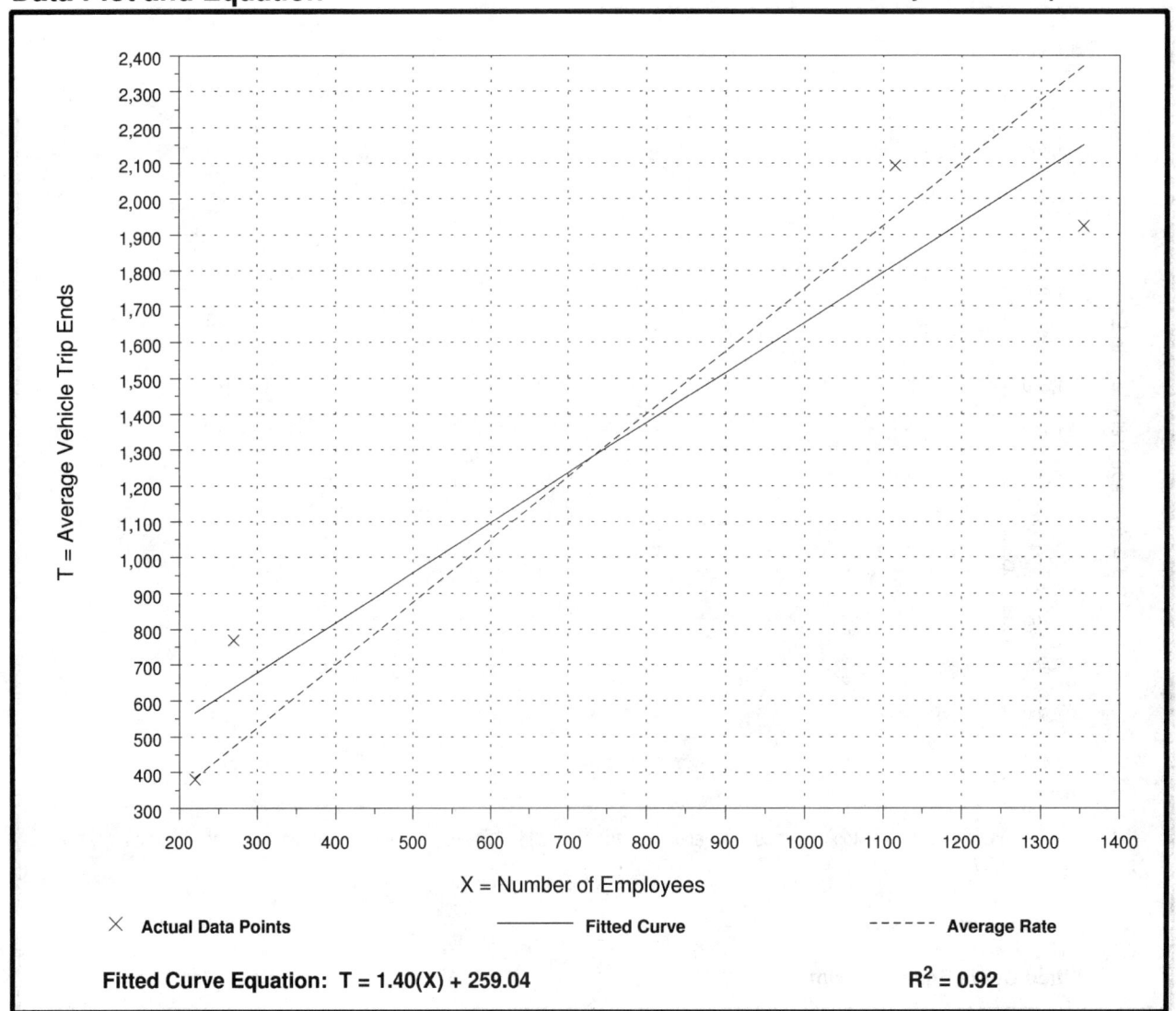

Data Plot and Equation

Caution - Use Carefully - Small Sample Size

Fitted Curve Equation: T = 1.40(X) + 259.04 $R^2 = 0.92$

Junior/Community College
(540)

Average Vehicle Trip Ends vs: Employees
On a: Weekday,
P.M. Peak Hour of Generator

Number of Studies: 4
Avg. Number of Employees: 740
Directional Distribution: 44% entering, 56% exiting

Trip Generation per Employee

Average Rate	Range of Rates	Standard Deviation
1.49	0.83 - 3.29	1.36

Data Plot and Equation

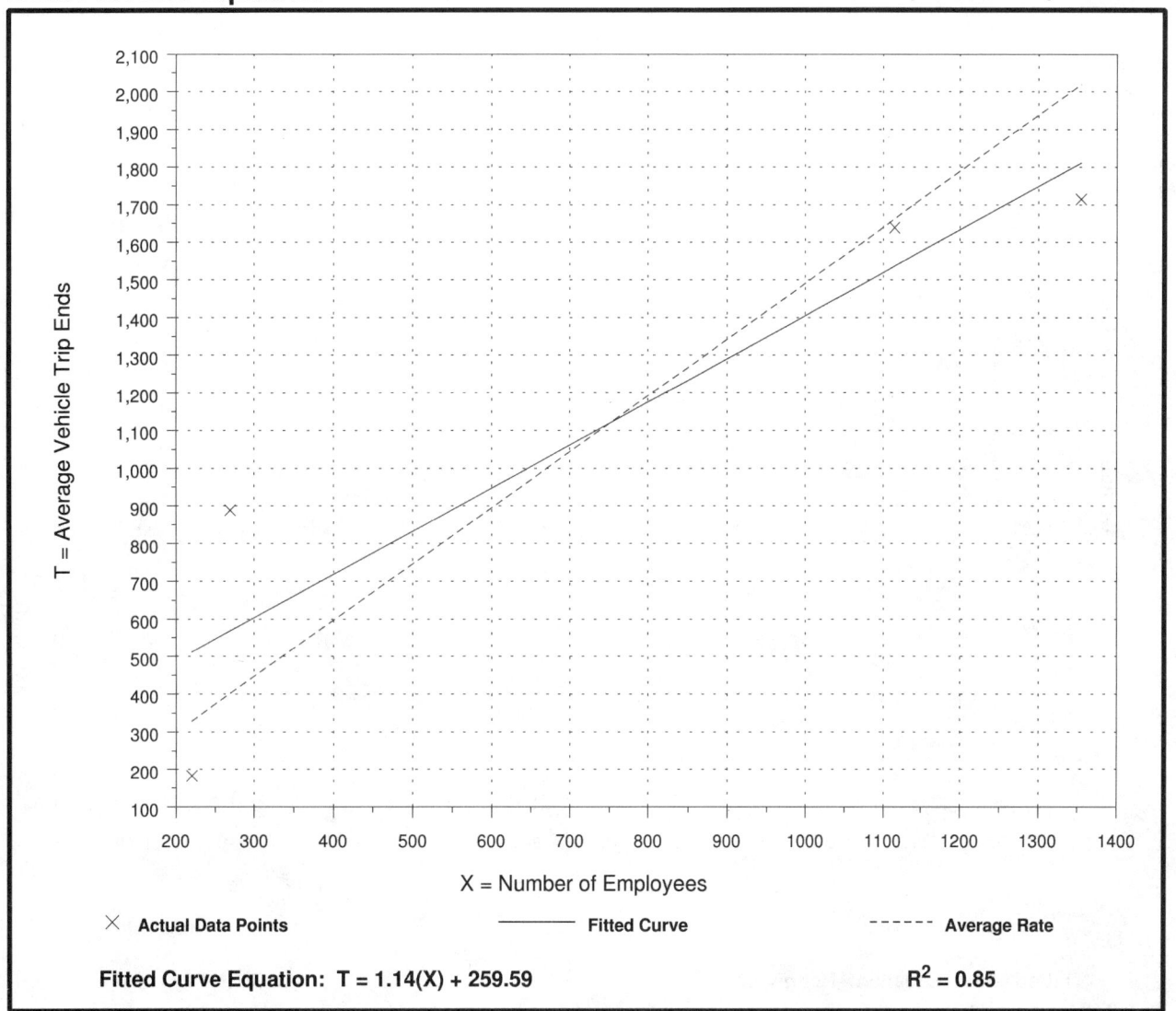

Caution - Use Carefully - Small Sample Size

Fitted Curve Equation: $T = 1.14(X) + 259.59$ $\qquad$ $R^2 = 0.85$

Junior/Community College
(540)

Average Vehicle Trip Ends vs: Employees
On a: Saturday

Number of Studies: 3
Avg. Number of Employees: 913
Directional Distribution: 50% entering, 50% exiting

Trip Generation per Employee

Average Rate	Range of Rates	Standard Deviation
6.16	4.17 - 11.68	3.41

Data Plot and Equation

Caution - Use Carefully - Small Sample Size

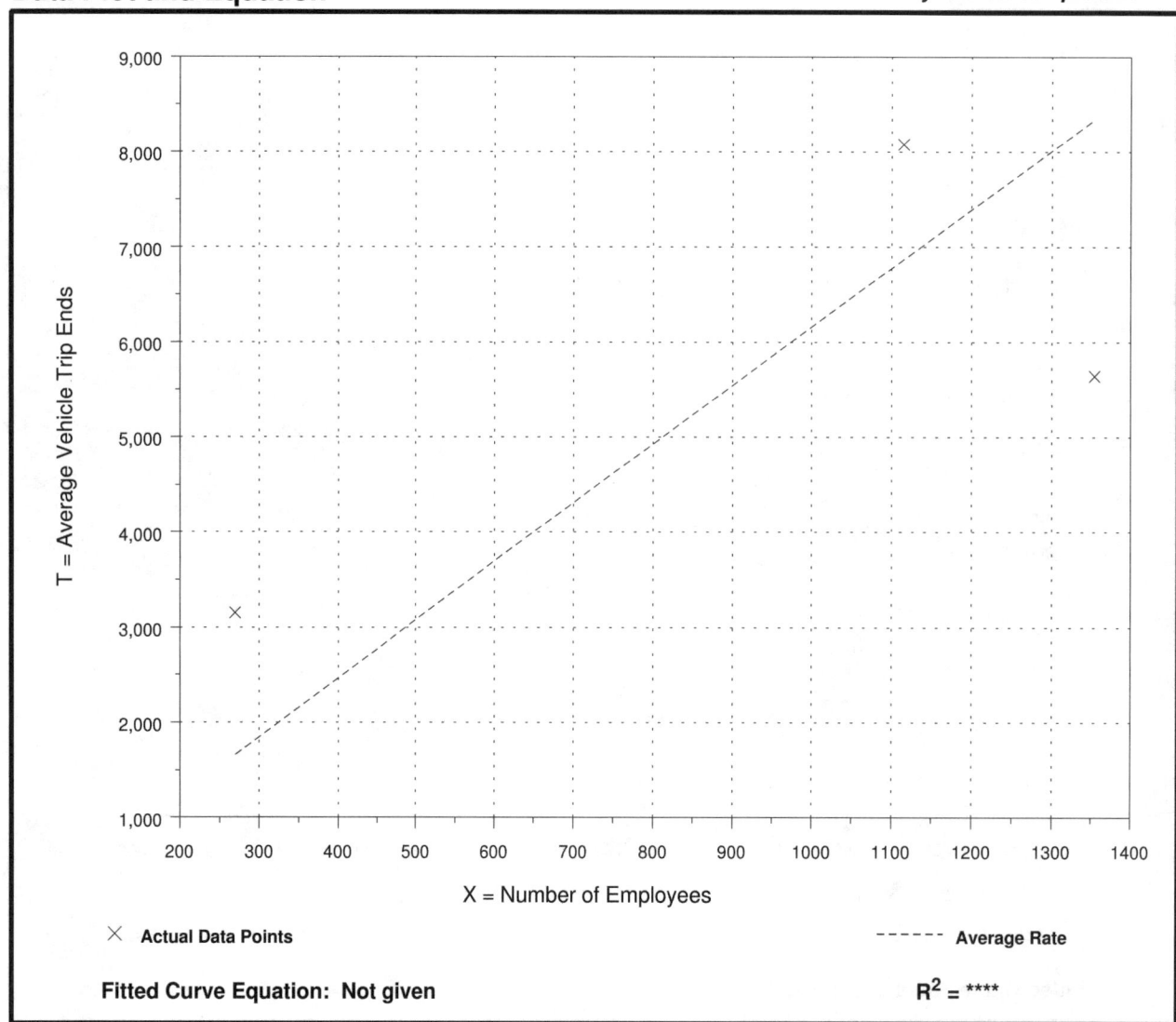

Fitted Curve Equation: Not given $R^2 = ****$

Junior/Community College
(540)

Average Vehicle Trip Ends vs: Employees
On a: Saturday,
Peak Hour of Generator

Number of Studies: 3
Avg. Number of Employees: 913
Directional Distribution: 57% entering, 43% exiting

Trip Generation per Employee

Average Rate	Range of Rates	Standard Deviation
0.78	0.59 - 1.31	0.91

Data Plot and Equation

Caution - Use Carefully - Small Sample Size

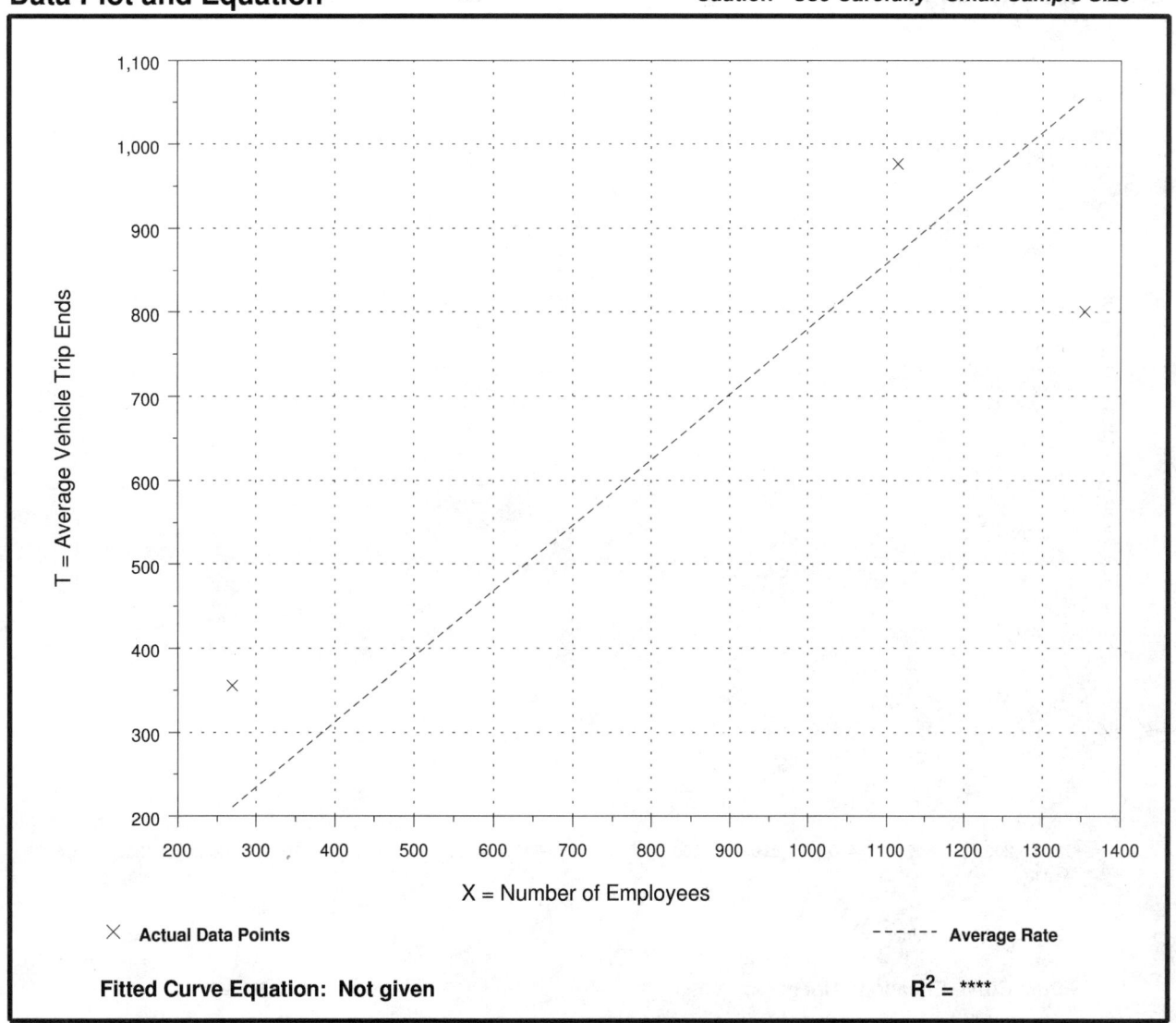

Fitted Curve Equation: Not given $R^2 = ****$

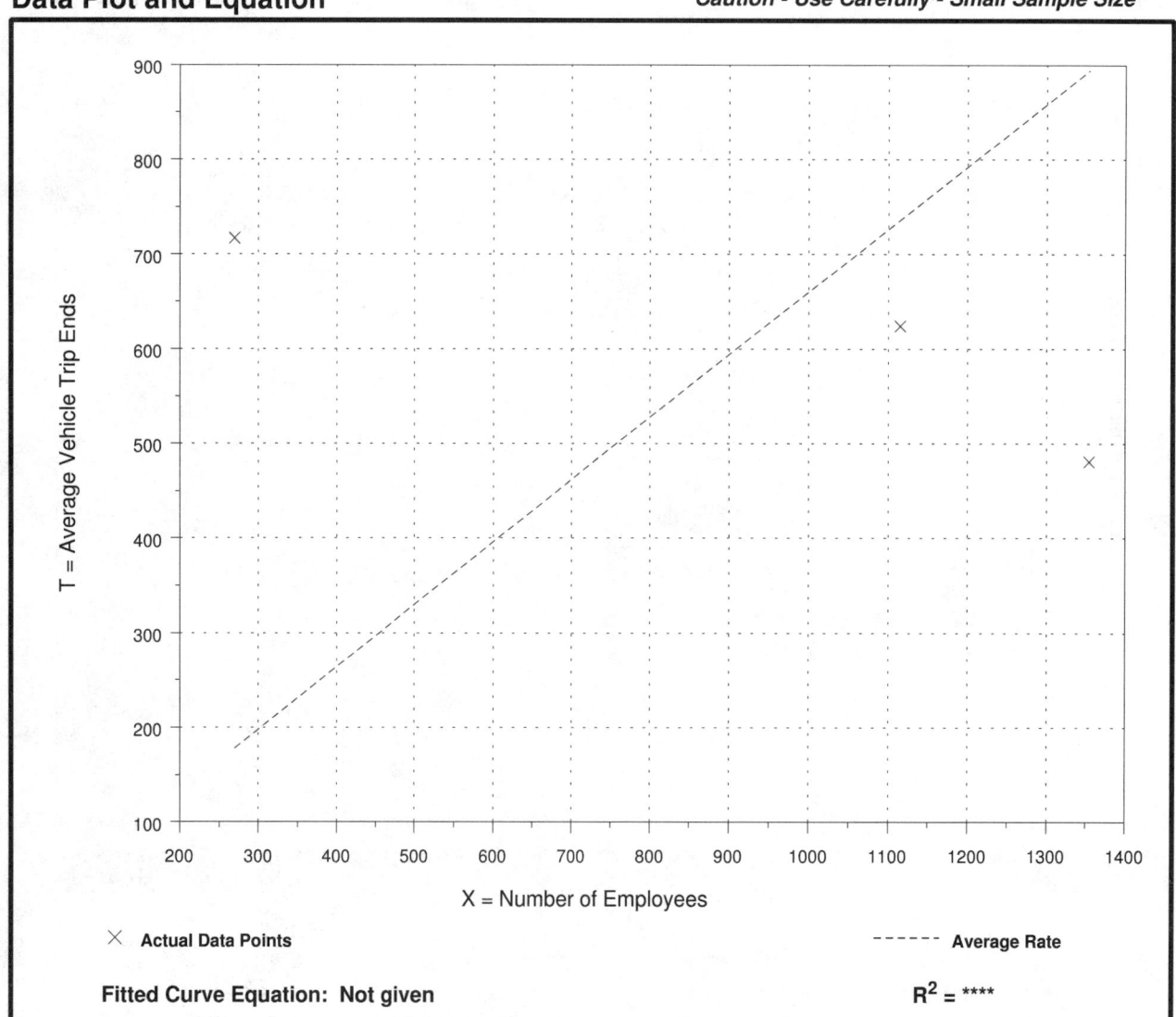

Junior/Community College
(540)

Average Vehicle Trip Ends vs: Employees
On a: Sunday

Number of Studies: 3
Avg. Number of Employees: 913
Directional Distribution: 50% entering, 50% exiting

Trip Generation per Employee

Average Rate	Range of Rates	Standard Deviation
0.66	0.35 - 2.66	1.05

Data Plot and Equation *Caution - Use Carefully - Small Sample Size*

Fitted Curve Equation: Not given $R^2 = ****$

Junior/Community College
(540)

Average Vehicle Trip Ends vs: Employees
On a: Sunday,
Peak Hour of Generator

Number of Studies: 3
Avg. Number of Employees: 913
Directional Distribution: 46% entering, 54% exiting

Trip Generation per Employee

Average Rate	Range of Rates	Standard Deviation
0.11	0.05 - 0.43	0.35

Data Plot and Equation

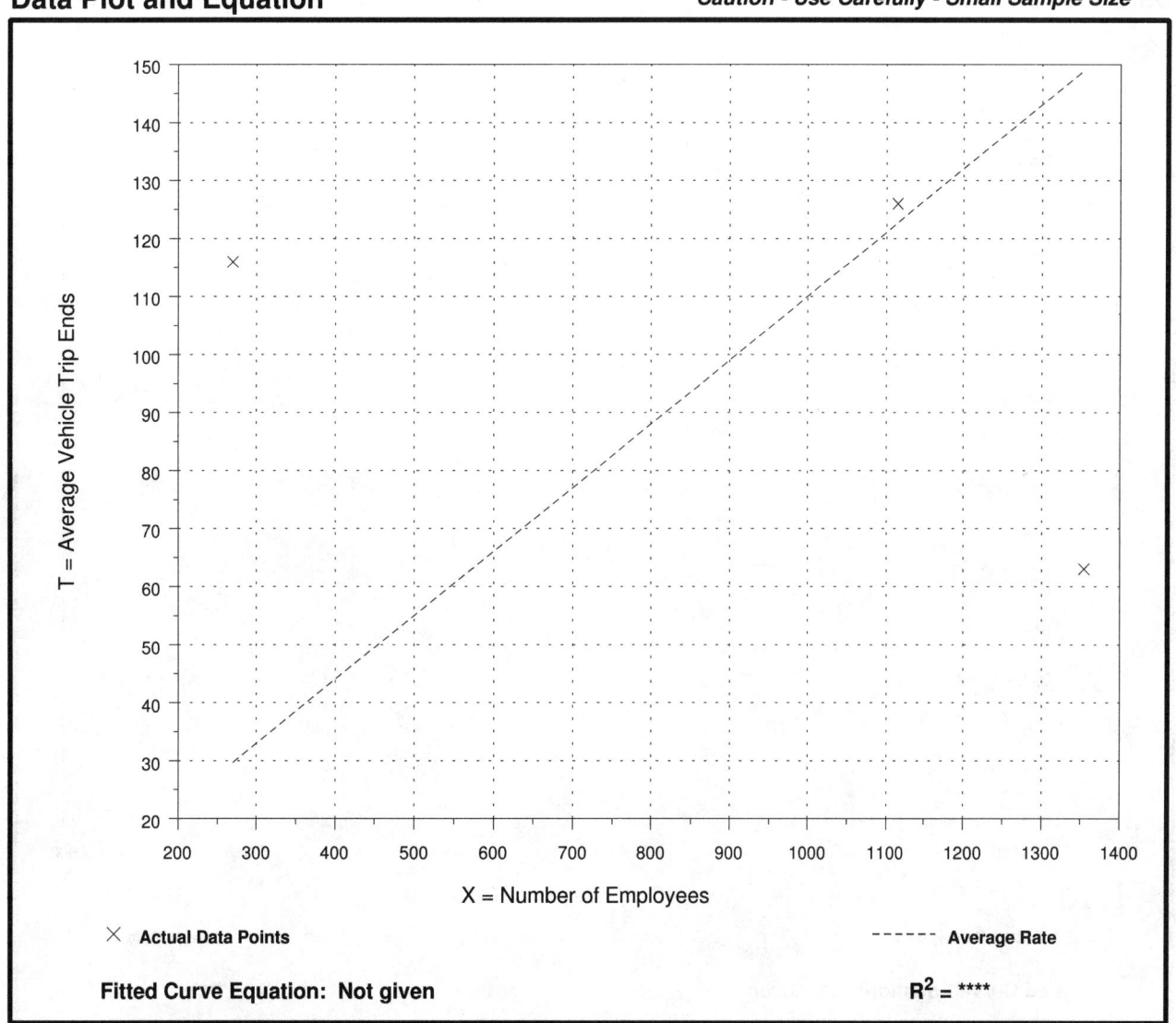

Caution - Use Carefully - Small Sample Size

Fitted Curve Equation: Not given $R^2 = ****$

Junior/Community College
(540)

Average Vehicle Trip Ends vs: 1000 Sq. Feet Gross Floor Area
On a: Weekday

Number of Studies: 4
Average 1000 Sq. Feet GFA: 419
Directional Distribution: 50% entering, 50% exiting

Trip Generation per 1000 Sq. Feet Gross Floor Area

Average Rate	Range of Rates	Standard Deviation
27.49	12.87 - 35.98	9.41

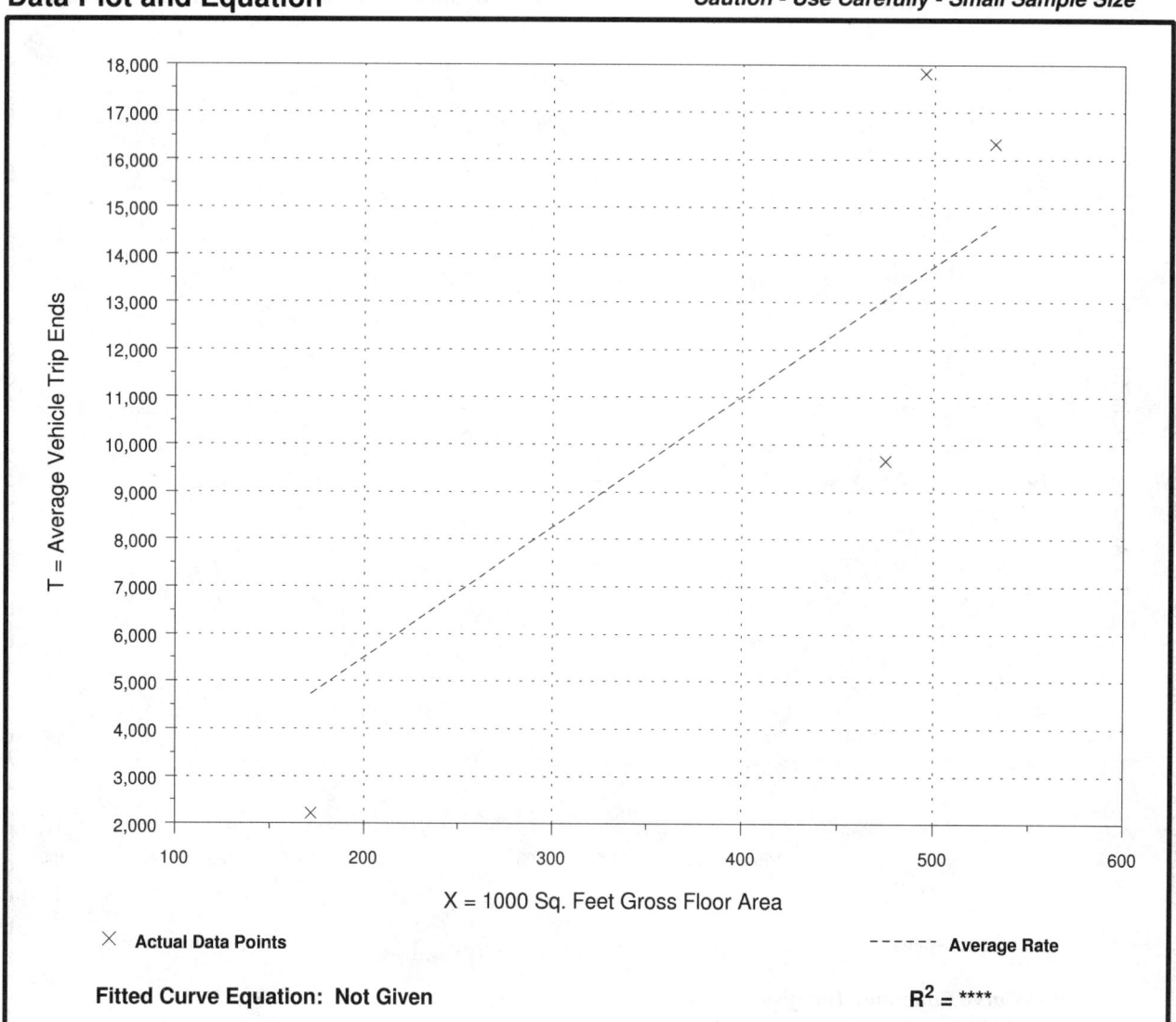

Data Plot and Equation

Caution - Use Carefully - Small Sample Size

× Actual Data Points   ----- Average Rate

Fitted Curve Equation: Not Given $R^2 = ****$

Junior/Community College
(540)

Average Vehicle Trip Ends vs: 1000 Sq. Feet Gross Floor Area
On a: Weekday,
Peak Hour of Adjacent Street Traffic,
One Hour Between 7 and 9 a.m.

Number of Studies: 3
Average 1000 Sq. Feet GFA: 501
Directional Distribution: 74% entering, 26% exiting

Trip Generation per 1000 Sq. Feet Gross Floor Area

Average Rate	Range of Rates	Standard Deviation
2.99	1.62 - 3.78	1.97

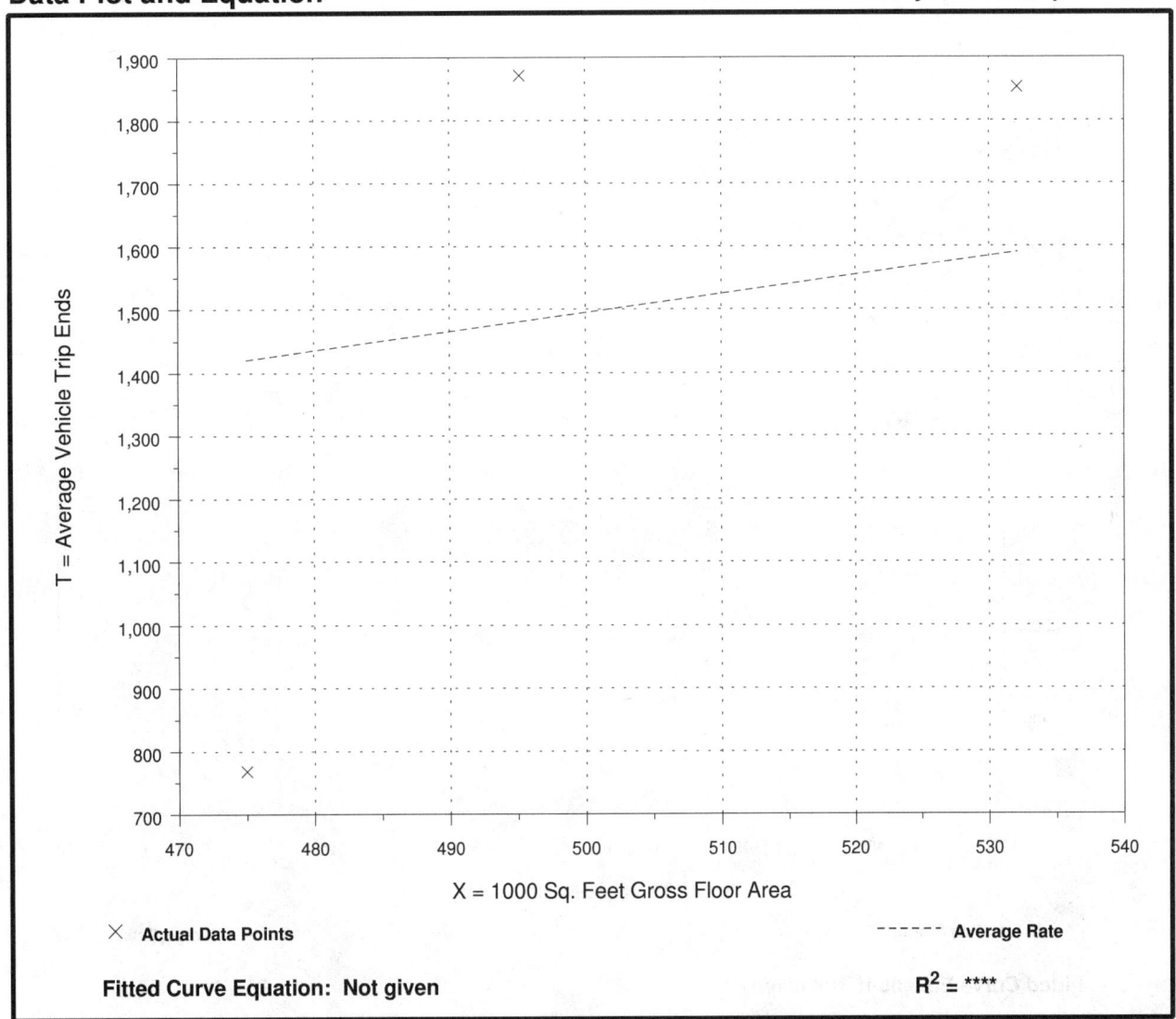

Data Plot and Equation

Caution - Use Carefully - Small Sample Size

× Actual Data Points
-------- Average Rate

Fitted Curve Equation: Not given $R^2 = ****$

Junior/Community College
(540)

Average Vehicle Trip Ends vs: 1000 Sq. Feet Gross Floor Area
On a: Weekday,
Peak Hour of Adjacent Street Traffic,
One Hour Between 4 and 6 p.m.

Number of Studies: 3
Average 1000 Sq. Feet GFA: 501
Directional Distribution: 58% entering, 42% exiting

Trip Generation per 1000 Sq. Feet Gross Floor Area

Average Rate	Range of Rates	Standard Deviation
2.54	1.77 - 2.96	1.68

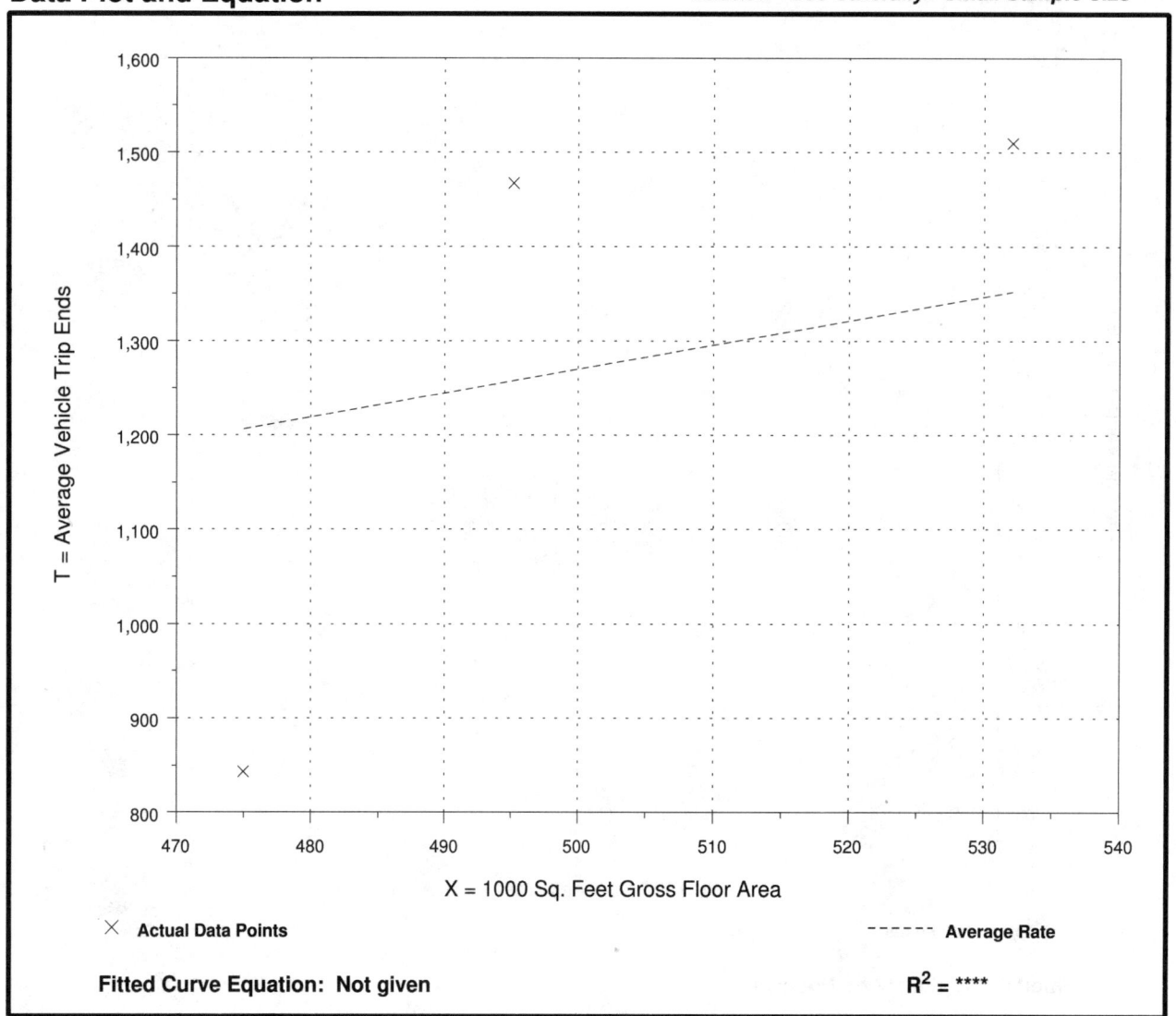

Data Plot and Equation

Caution - Use Carefully - Small Sample Size

Fitted Curve Equation: Not given $R^2 = ****$

Junior/Community College
(540)

Average Vehicle Trip Ends vs: 1000 Sq. Feet Gross Floor Area
On a: Weekday,
A.M. Peak Hour of Generator

Number of Studies: 4
Average 1000 Sq. Feet GFA: 419
Directional Distribution: 50% entering, 50% exiting

Trip Generation per 1000 Sq. Feet Gross Floor Area

Average Rate	Range of Rates	Standard Deviation
3.09	1.62 - 3.93	2.05

Data Plot and Equation

Caution - Use Carefully - Small Sample Size

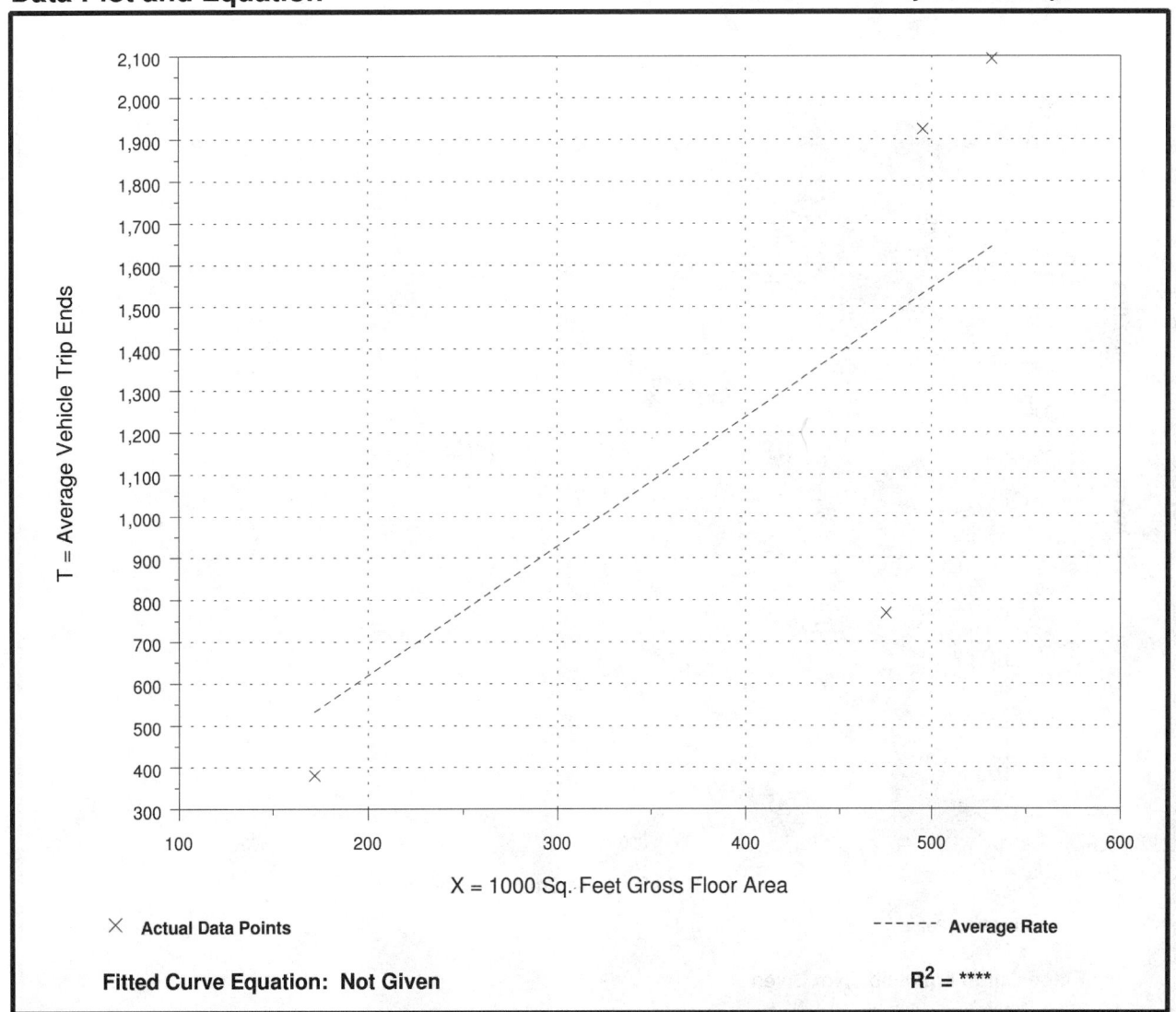

Fitted Curve Equation: Not Given $R^2 = ****$

Junior/Community College
(540)

Average Vehicle Trip Ends vs: 1000 Sq. Feet Gross Floor Area
On a: Weekday,
P.M. Peak Hour of Generator

Number of Studies: 4
Average 1000 Sq. Feet GFA: 419
Directional Distribution: 44% entering, 56% exiting

Trip Generation per 1000 Sq. Feet Gross Floor Area

Average Rate	Range of Rates	Standard Deviation
2.64	1.06 - 3.46	1.82

Data Plot and Equation

Caution - Use Carefully - Small Sample Size

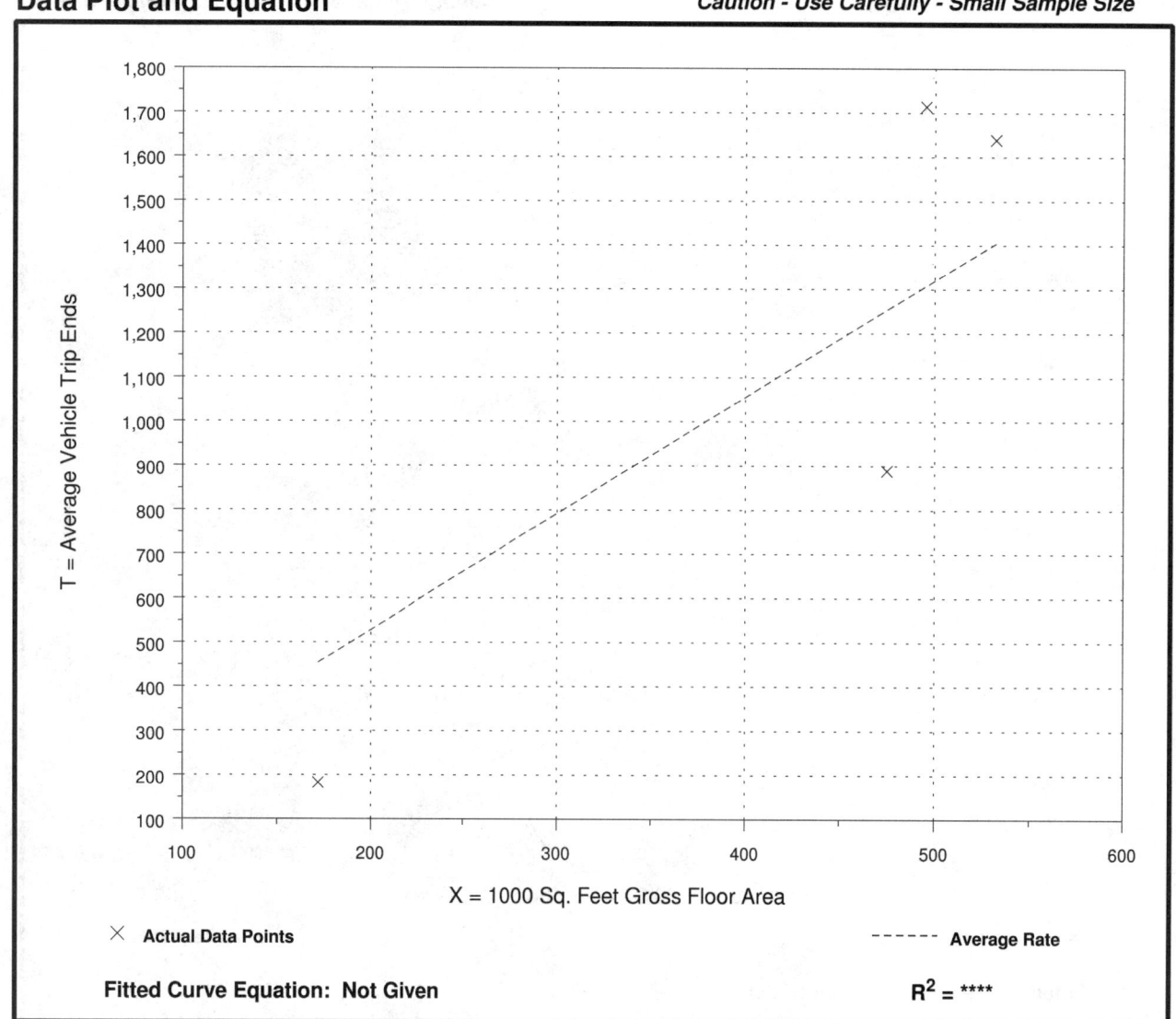

Fitted Curve Equation: Not Given $R^2 = ****$

Junior/Community College
(540)

Average Vehicle Trip Ends vs: 1000 Sq. Feet Gross Floor Area
On a: Saturday

Number of Studies: 3
Average 1000 Sq. Feet GFA: 501
Directional Distribution: 50% entering, 50% exiting

Trip Generation per 1000 Sq. Feet Gross Floor Area

Average Rate	Range of Rates	Standard Deviation
11.23	6.64 - 15.18	4.84

Data Plot and Equation

Caution - Use Carefully - Small Sample Size

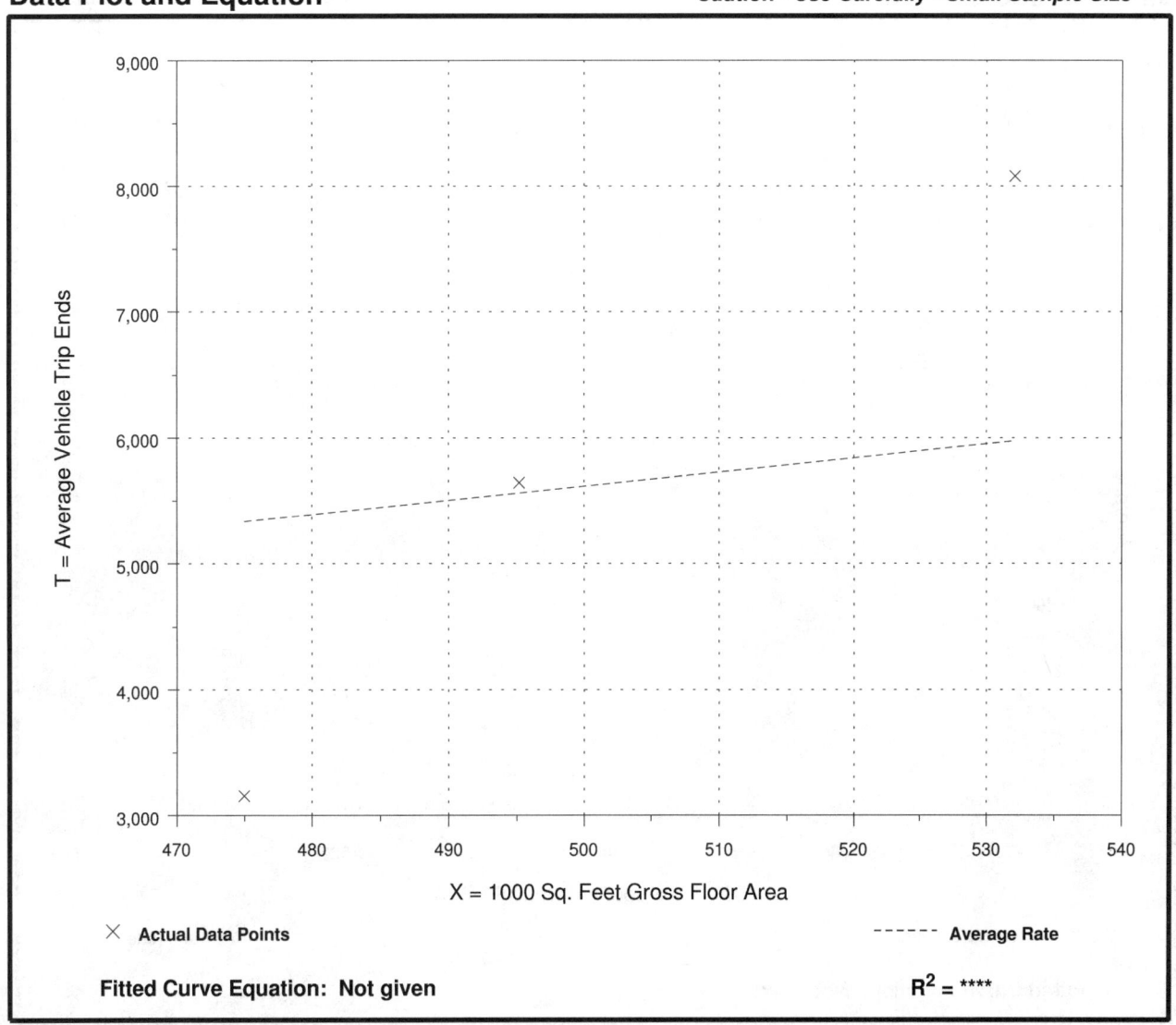

× Actual Data Points   ----- Average Rate

Fitted Curve Equation: Not given $R^2 =$ ****

Junior/Community College
(540)

Average Vehicle Trip Ends vs: 1000 Sq. Feet Gross Floor Area
On a: Saturday,
Peak Hour of Generator

Number of Studies: 3
Average 1000 Sq. Feet GFA: 501
Directional Distribution: 57% entering, 43% exiting

Trip Generation per 1000 Sq. Feet Gross Floor Area

Average Rate	Range of Rates	Standard Deviation
1.42	0.75 - 1.84	1.28

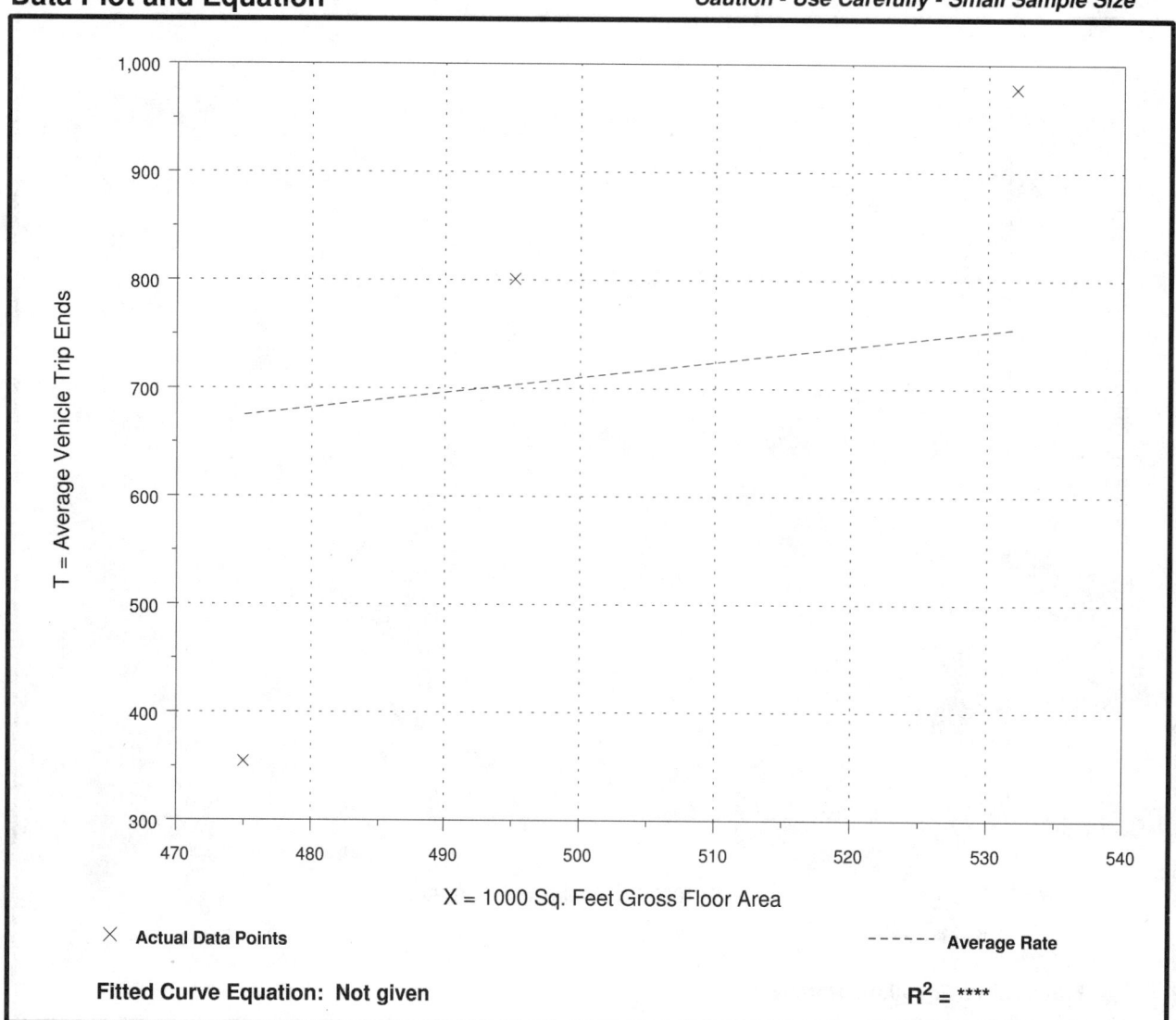

Data Plot and Equation

Caution - Use Carefully - Small Sample Size

Fitted Curve Equation: Not given $R^2 = ****$

Junior/Community College
(540)

Average Vehicle Trip Ends vs: 1000 Sq. Feet Gross Floor Area
On a: Sunday

Number of Studies: 3
Average 1000 Sq. Feet GFA: 501
Directional Distribution: 50% entering, 50% exiting

Trip Generation per 1000 Sq. Feet Gross Floor Area

Average Rate	Range of Rates	Standard Deviation
1.21	0.97 - 1.51	1.12

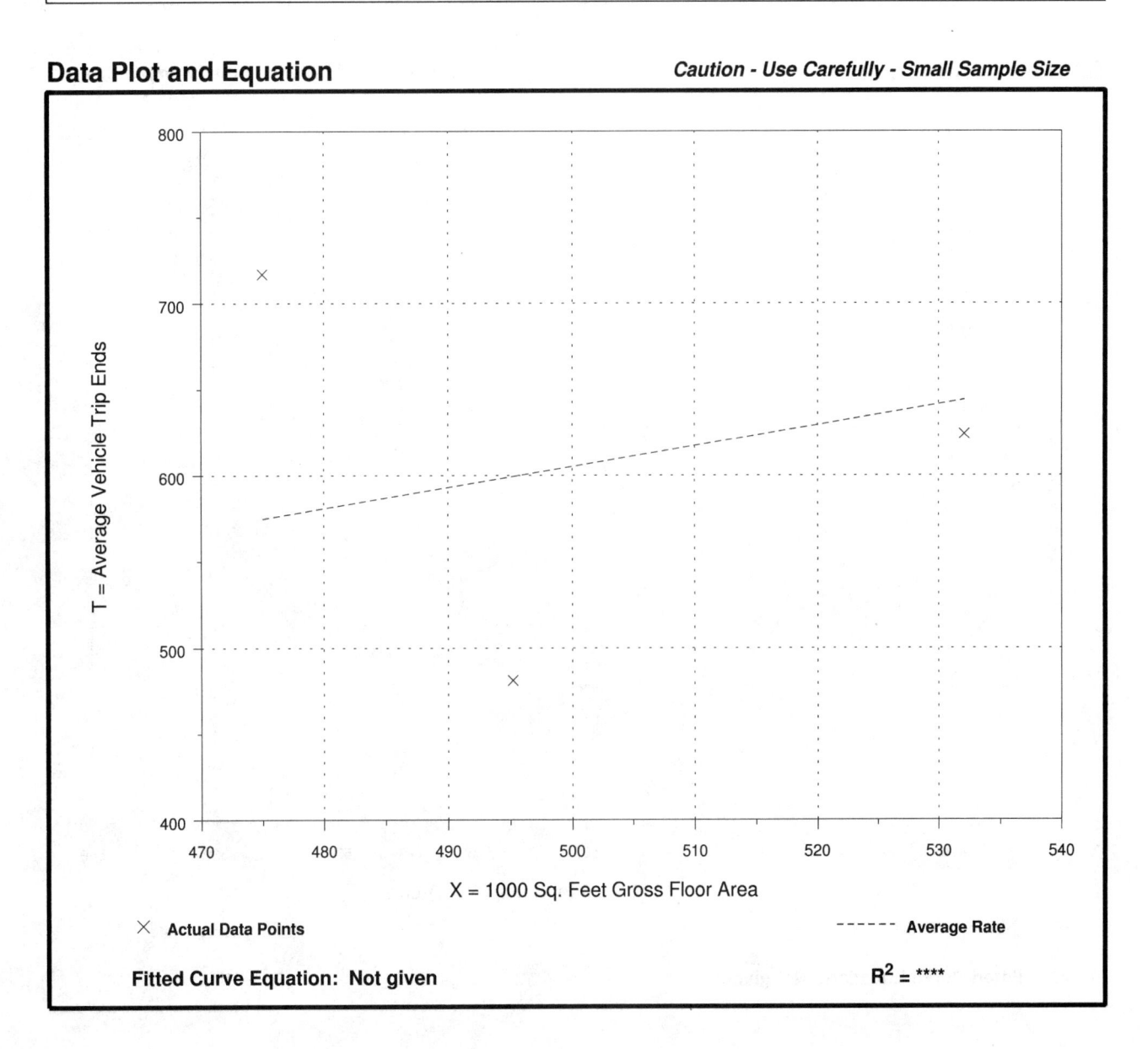

Data Plot and Equation

Caution - Use Carefully - Small Sample Size

Fitted Curve Equation: Not given $R^2 = ****$

Junior/Community College
(540)

Average Vehicle Trip Ends vs: 1000 Sq. Feet Gross Floor Area
On a: Sunday,
Peak Hour of Generator

Number of Studies: 3
Average 1000 Sq. Feet GFA: 501
Directional Distribution: 46% entering, 54% exiting

Trip Generation per 1000 Sq. Feet Gross Floor Area

Average Rate	Range of Rates	Standard Deviation
0.20	0.13 - 0.24	0.45

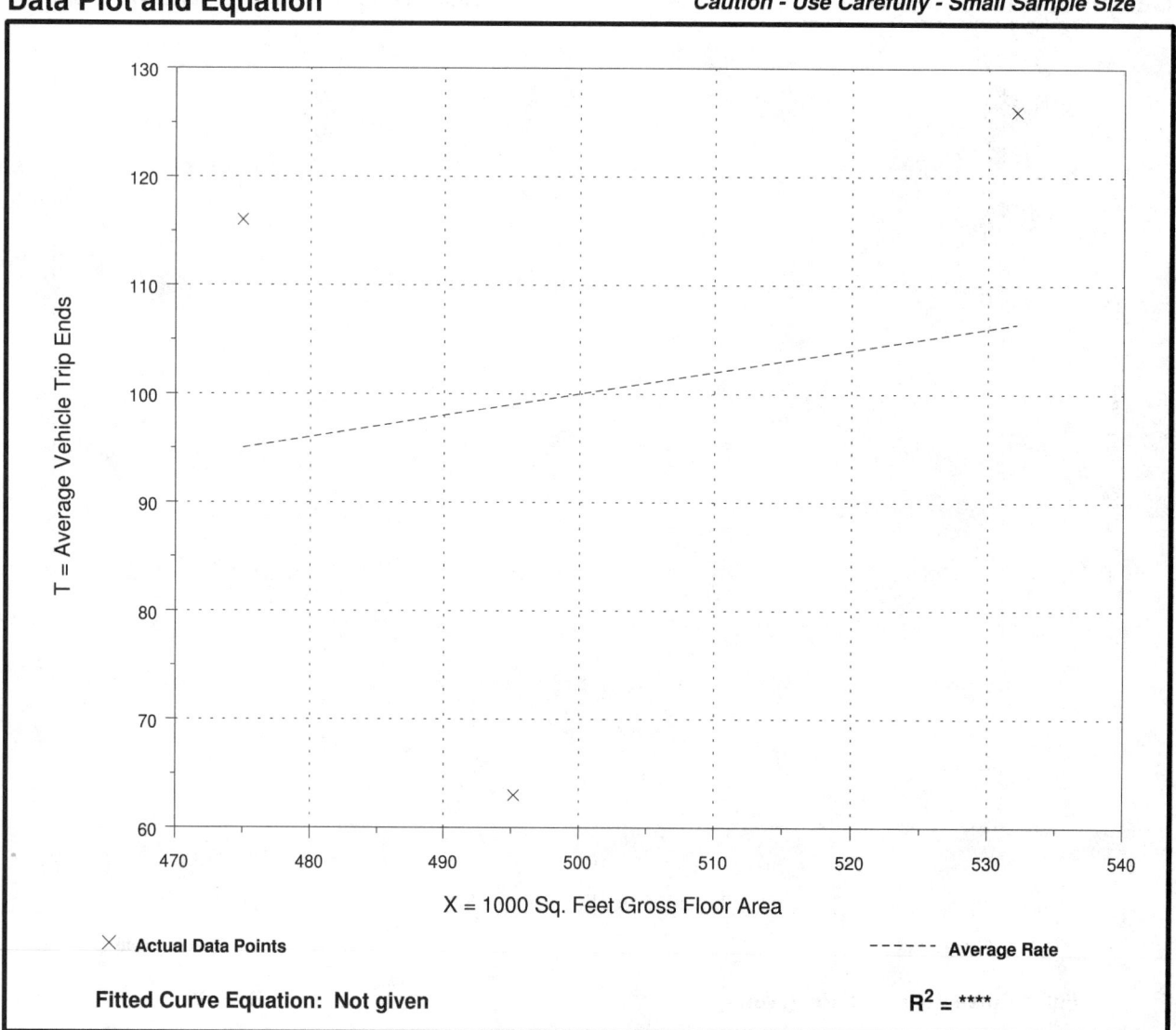

Data Plot and Equation

Caution - Use Carefully - Small Sample Size

Fitted Curve Equation: Not given $R^2 = ****$

Land Use: 550
University/College

Description

This land use includes four-year universities or colleges that may or may not offer graduate programs. Two-year junior, community, or technical colleges are described in junior/community college (Land Use 540).

Additional Data

The trip generation for weekend time periods varied considerably; therefore, caution should be used when applying weekend statistics. Information describing the weekend activities conducted at universities/colleges was not available.

Two studies indicated that transit service was available within close proximity of the college. Trip generation rates were generally lower at these sites compared to the average trip generation rates of this land use. **Due to the varied transit usage at the sites included in this land use,** *caution should be used when applying trip generation rates.*

Acreage, floor space, staff and parking accommodations varied widely with the populations served and the social and economic characteristics of the area; thus, the number of students may be a more reliable independent variable on which to establish trip generation rates.

The sites were surveyed between the late 1970s and the 2000s throughout the United States.

Many of the studies included in this land use did not indicate if significant transit service was available within close proximity of the site and did not specify the type of transit service. To assist in the future analysis of this land use, it is important that this information be collected and included in trip generation data submissions.

Source Numbers

86, 365, 423, 440, 612, 702

University/College
(550)

Average Vehicle Trip Ends vs: Students
On a: Weekday

Number of Studies: 9
Average Number of Students: 7,858
Directional Distribution: 50% entering, 50% exiting

Trip Generation per Student

Average Rate	Range of Rates	Standard Deviation
1.71	1.25 - 3.31	1.39

Data Plot and Equation

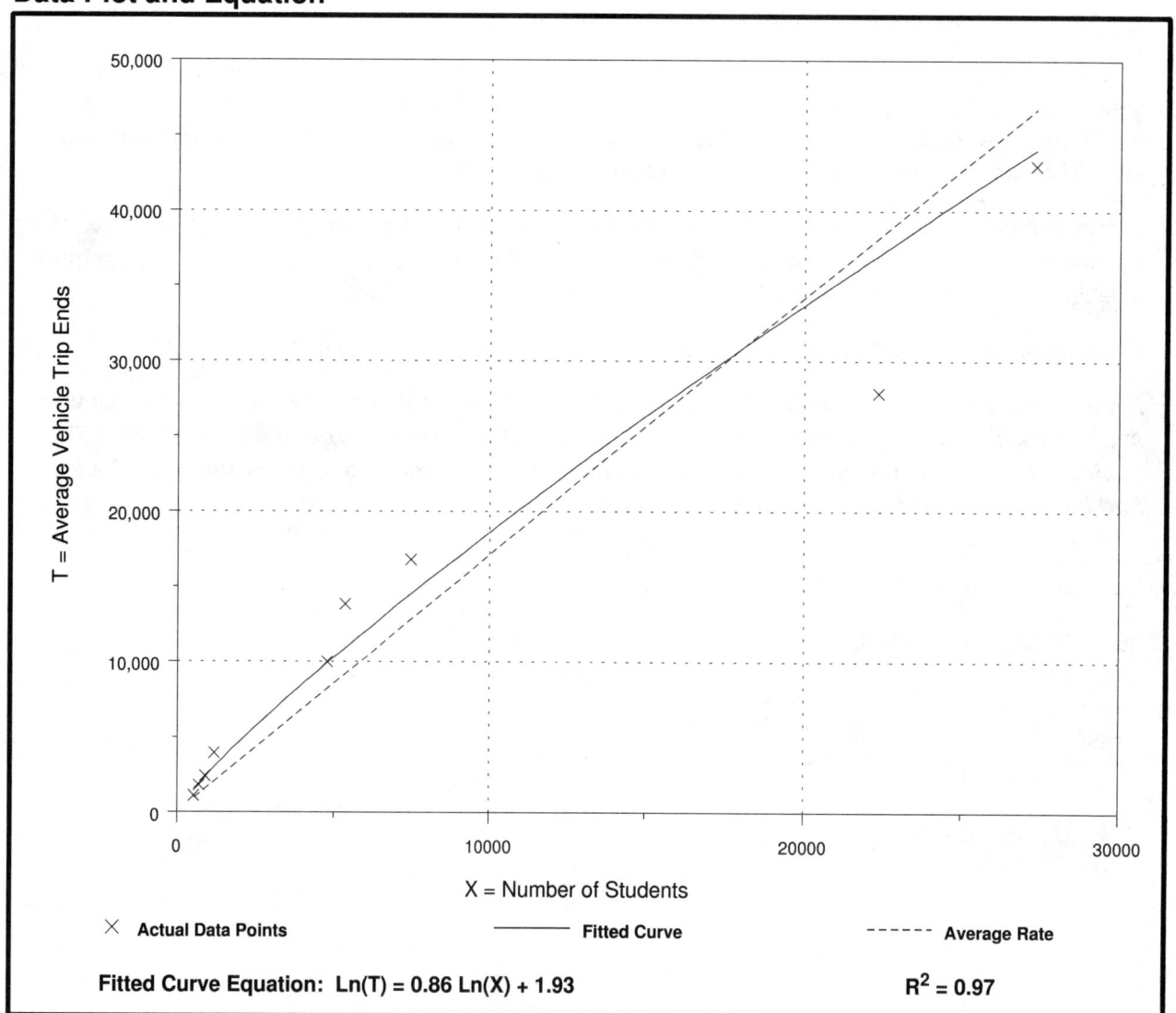

Fitted Curve Equation: Ln(T) = 0.86 Ln(X) + 1.93 $R^2 = 0.97$

University/College
(550)

Average Vehicle Trip Ends vs: Students
On a: Weekday,
Peak Hour of Adjacent Street Traffic,
One Hour Between 7 and 9 a.m.

Number of Studies: 8
Average Number of Students: 13,372
Directional Distribution: 78% entering, 22% exiting

Trip Generation per Student

Average Rate	Range of Rates	Standard Deviation
0.17	0.09 - 0.26	0.41

Data Plot and Equation

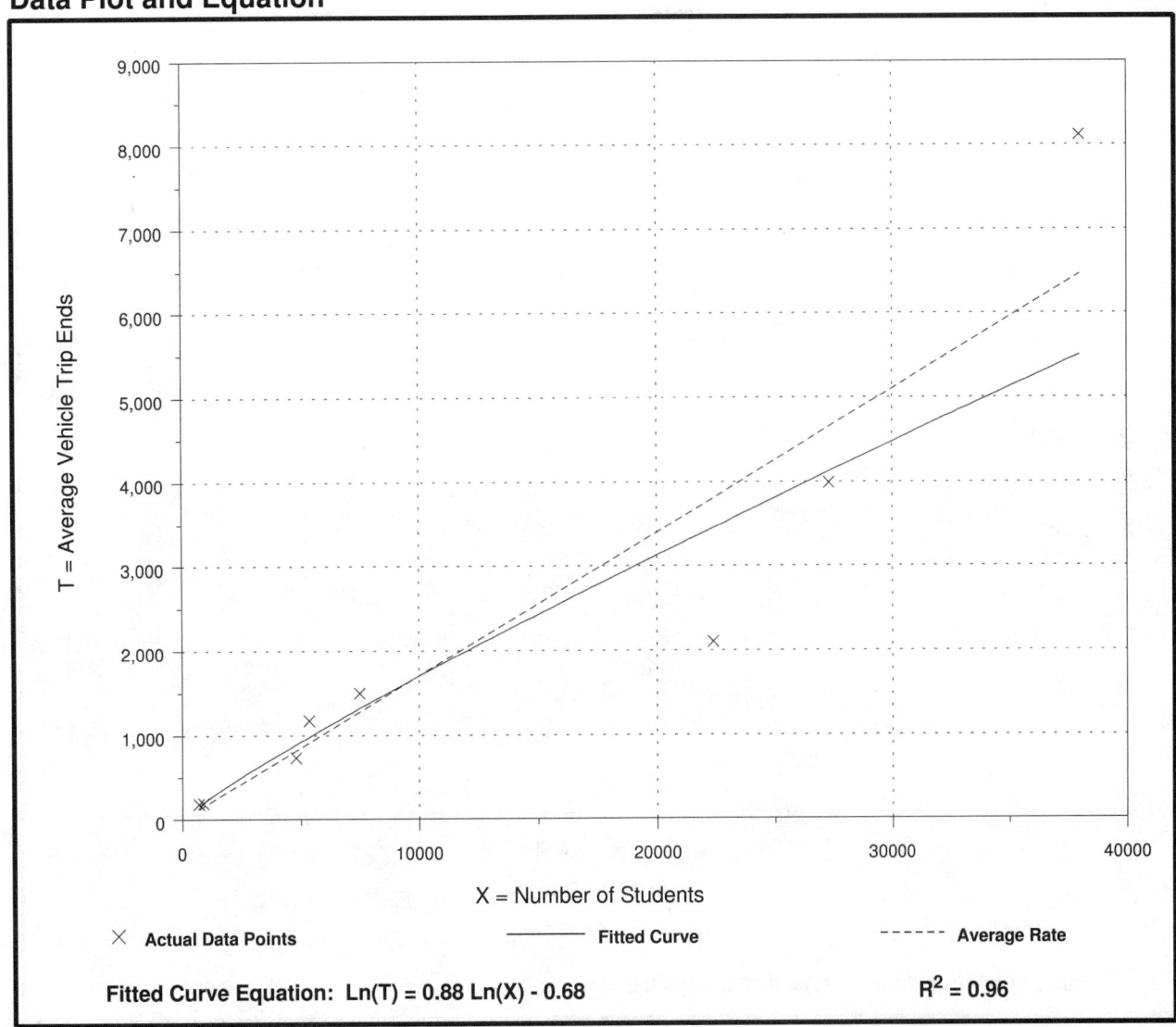

Fitted Curve Equation: $Ln(T) = 0.88 \, Ln(X) - 0.68$ $R^2 = 0.96$

University/College
(550)

Average Vehicle Trip Ends vs: Students
On a: Weekday,
Peak Hour of Adjacent Street Traffic,
One Hour Between 4 and 6 p.m.

Number of Studies: 10
Average Number of Students: 10,832
Directional Distribution: 32% entering, 68% exiting

Trip Generation per Student

Average Rate	Range of Rates	Standard Deviation
0.17	0.10 - 0.77	0.42

Data Plot and Equation

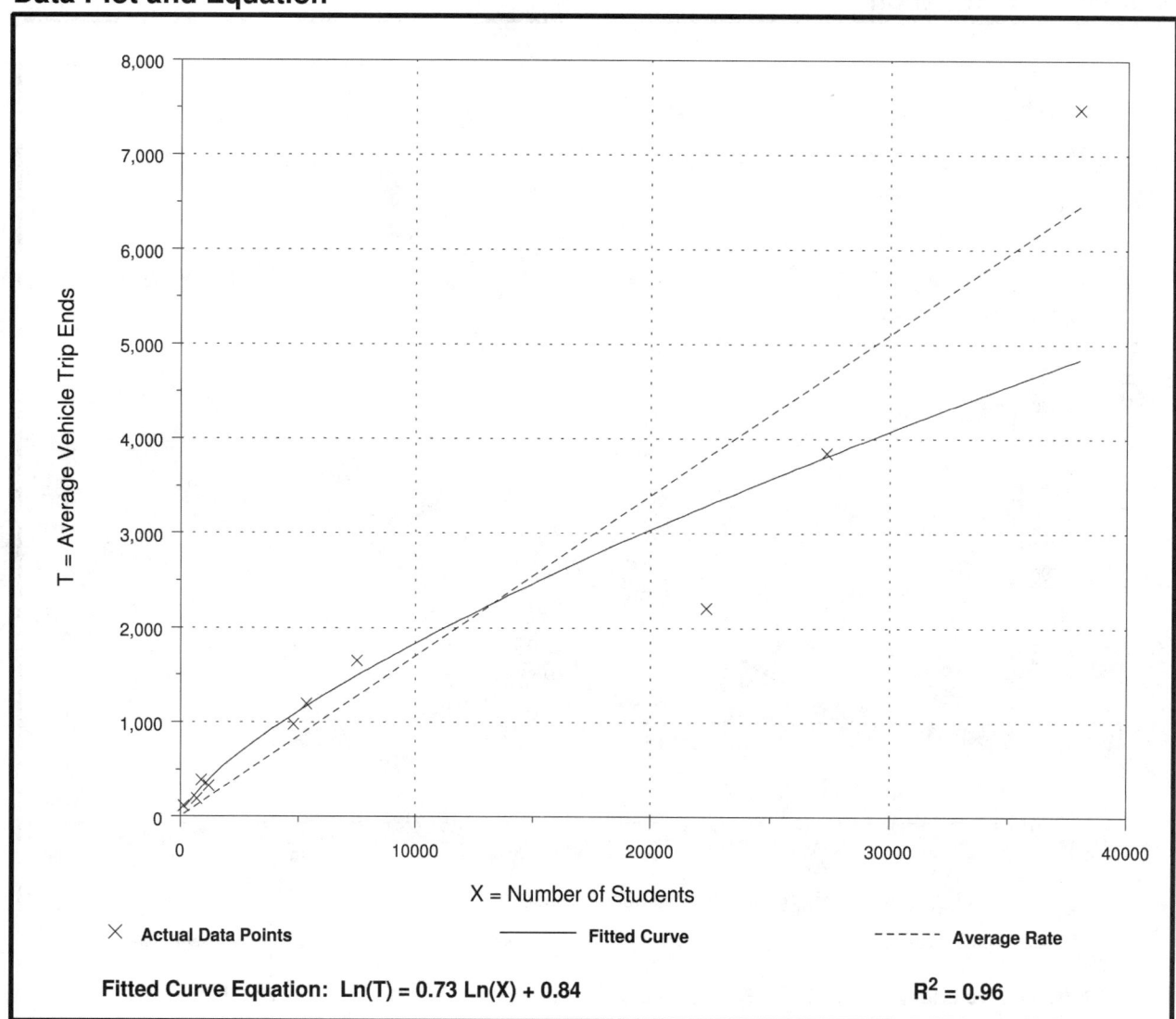

Fitted Curve Equation: Ln(T) = 0.73 Ln(X) + 0.84 $R^2 = 0.96$

University/College
(550)

Average Vehicle Trip Ends vs: Students
On a: Weekday,
A.M. Peak Hour of Generator

Number of Studies: 7
Average Number of Students: 8,860
Directional Distribution: 74% entering, 26% exiting

Trip Generation per Student

Average Rate	Range of Rates	Standard Deviation
0.14	0.10 - 0.30	0.38

Data Plot and Equation

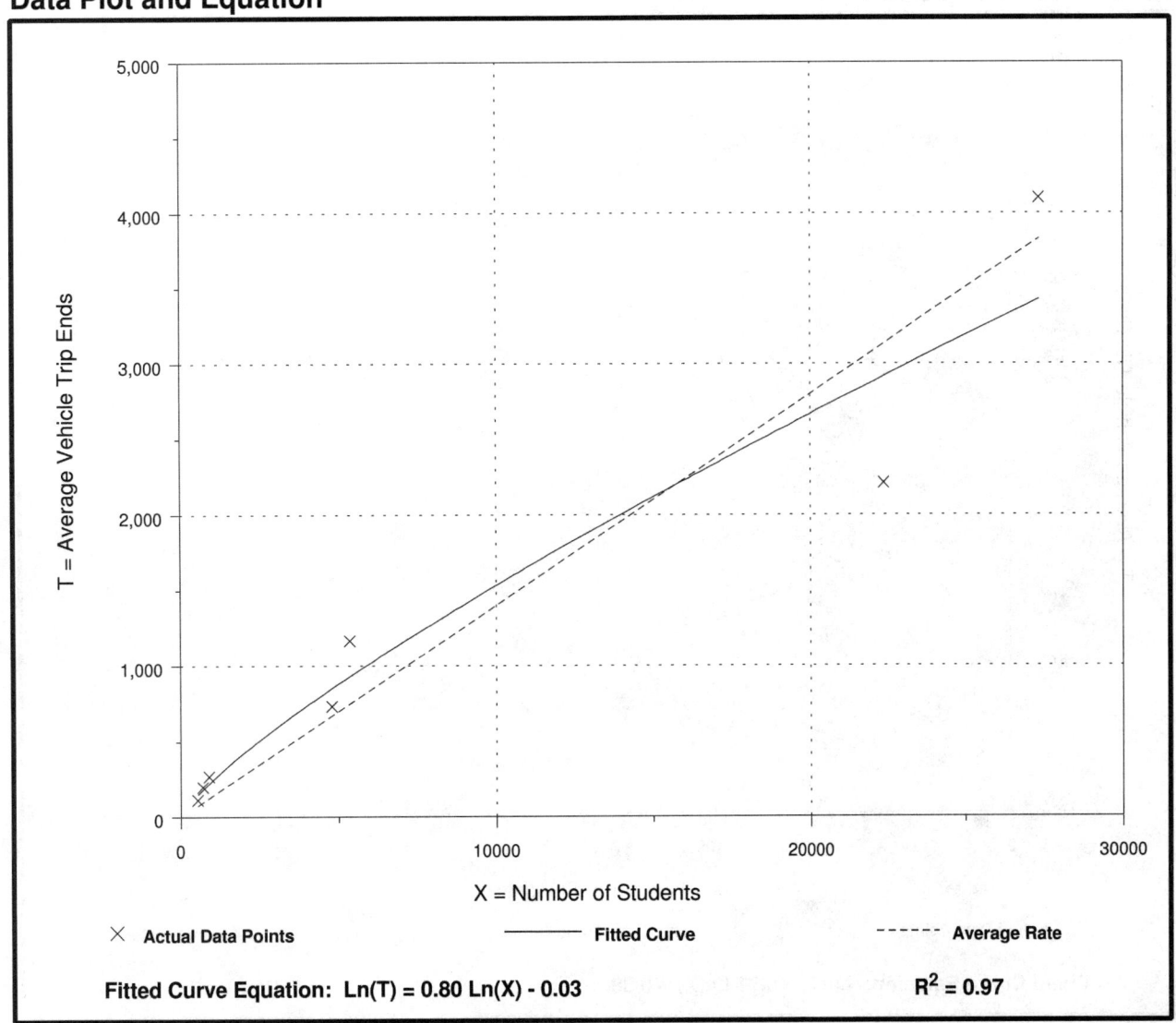

Fitted Curve Equation: $Ln(T) = 0.80\ Ln(X) - 0.03$ $R^2 = 0.97$

University/College
(550)

Average Vehicle Trip Ends vs: Students
On a: Weekday,
P.M. Peak Hour of Generator

Number of Studies: 7
Average Number of Students: 8,860
Directional Distribution: 32% entering, 68% exiting

Trip Generation per Student

Average Rate	Range of Rates	Standard Deviation
0.15	0.11 - 0.44	0.39

Data Plot and Equation

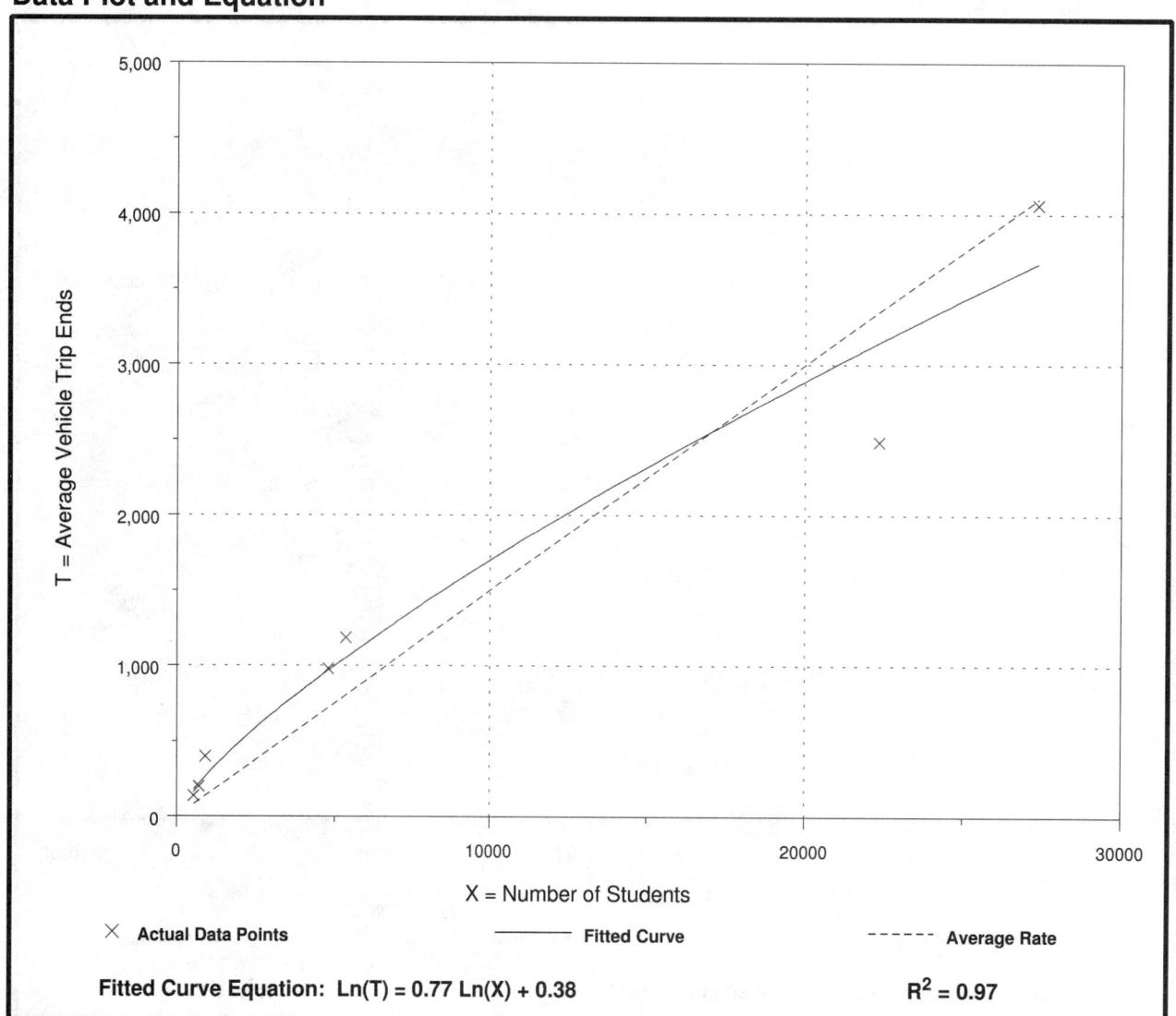

Fitted Curve Equation: $Ln(T) = 0.77 \, Ln(X) + 0.38$ $R^2 = 0.97$

University/College
(550)

Average Vehicle Trip Ends vs: Students
On a: Saturday

Number of Studies: 2
Average Number of Students: 2,749
Directional Distribution: 50% entering, 50% exiting

Trip Generation per Student

Average Rate	Range of Rates	Standard Deviation
1.30	1.08 - 2.83	*

Data Plot and Equation

Caution - Use Carefully - Small Sample Size

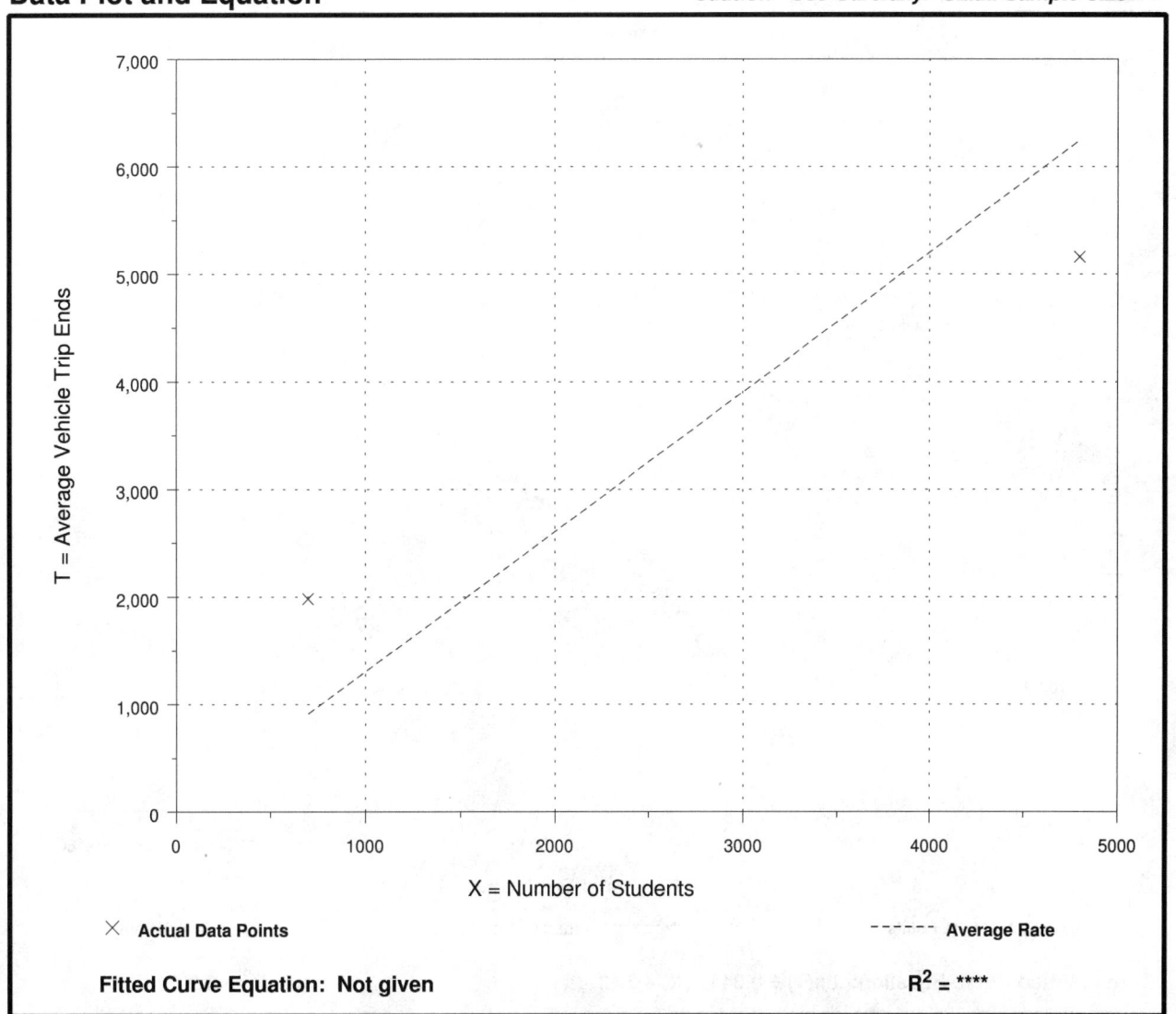

Fitted Curve Equation: Not given $R^2 =$ ****

University/College
(550)

Average Vehicle Trip Ends vs: Employees
On a: Weekday

Number of Studies: 7
Avg. Number of Employees: 1,596
Directional Distribution: 50% entering, 50% exiting

Trip Generation per Employee

Average Rate	Range of Rates	Standard Deviation
8.96	4.96 - 19.70	4.98

Data Plot and Equation

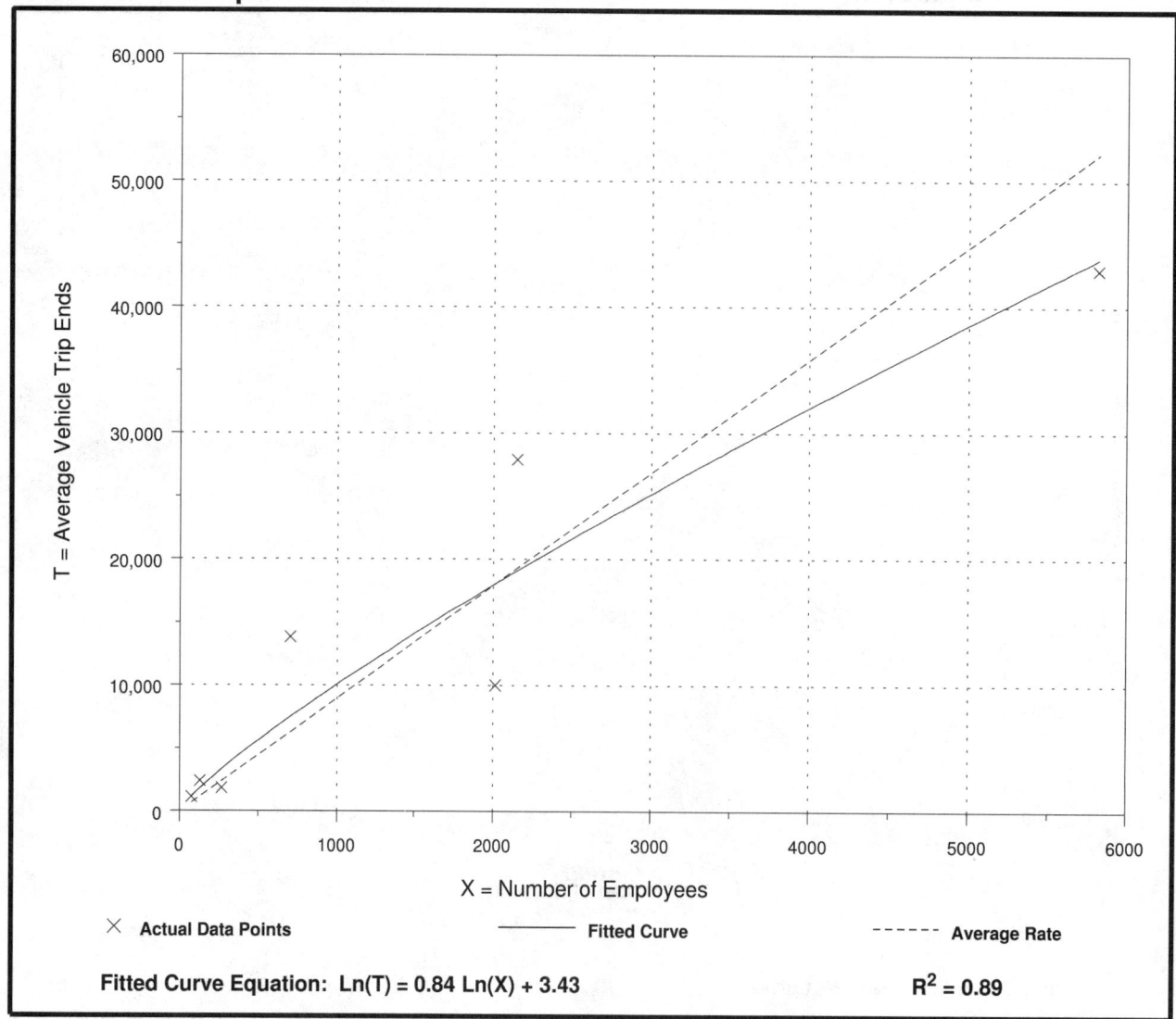

Fitted Curve Equation: $Ln(T) = 0.84\ Ln(X) + 3.43$ $R^2 = 0.89$

Trip Generation, 9th Edition • Institute of Transportation Engineers

University/College
(550)

Average Vehicle Trip Ends vs: Employees
On a: Weekday,
Peak Hour of Adjacent Street Traffic,
One Hour Between 7 and 9 a.m.

Number of Studies: 6
Avg. Number of Employees: 1,849
Directional Distribution: 76% entering, 24% exiting

Trip Generation per Employee

Average Rate	Range of Rates	Standard Deviation
0.75	0.36 - 1.67	0.92

Data Plot and Equation

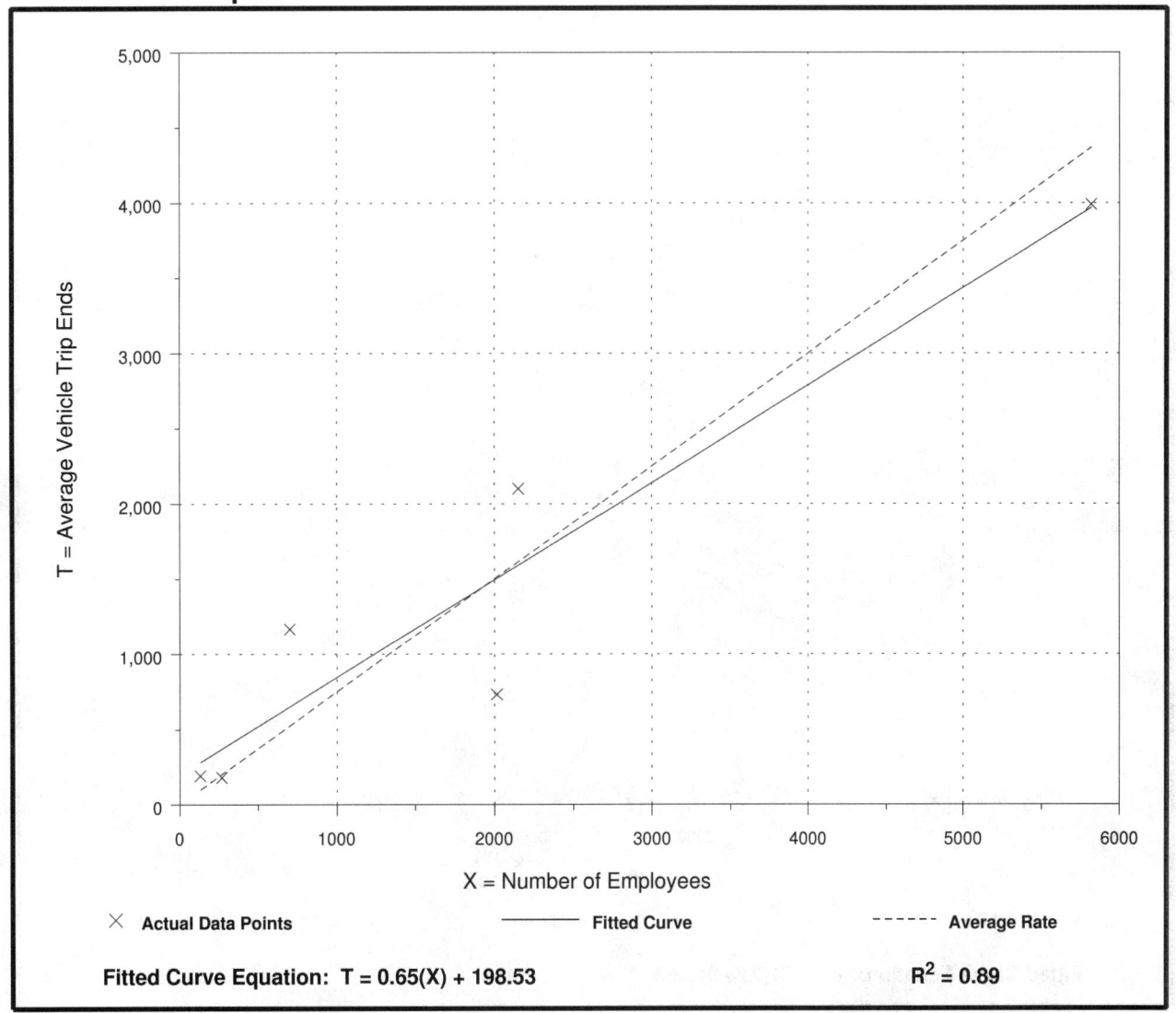

Fitted Curve Equation: $T = 0.65(X) + 198.53$ $R^2 = 0.89$

University/College
(550)

Average Vehicle Trip Ends vs: Employees
On a: Weekday,
Peak Hour of Adjacent Street Traffic,
One Hour Between 4 and 6 p.m.

Number of Studies: 6
Avg. Number of Employees: 1,849
Directional Distribution: 34% entering, 66% exiting

Trip Generation per Employee

Average Rate	Range of Rates	Standard Deviation
0.79	0.49 - 3.00	0.97

Data Plot and Equation

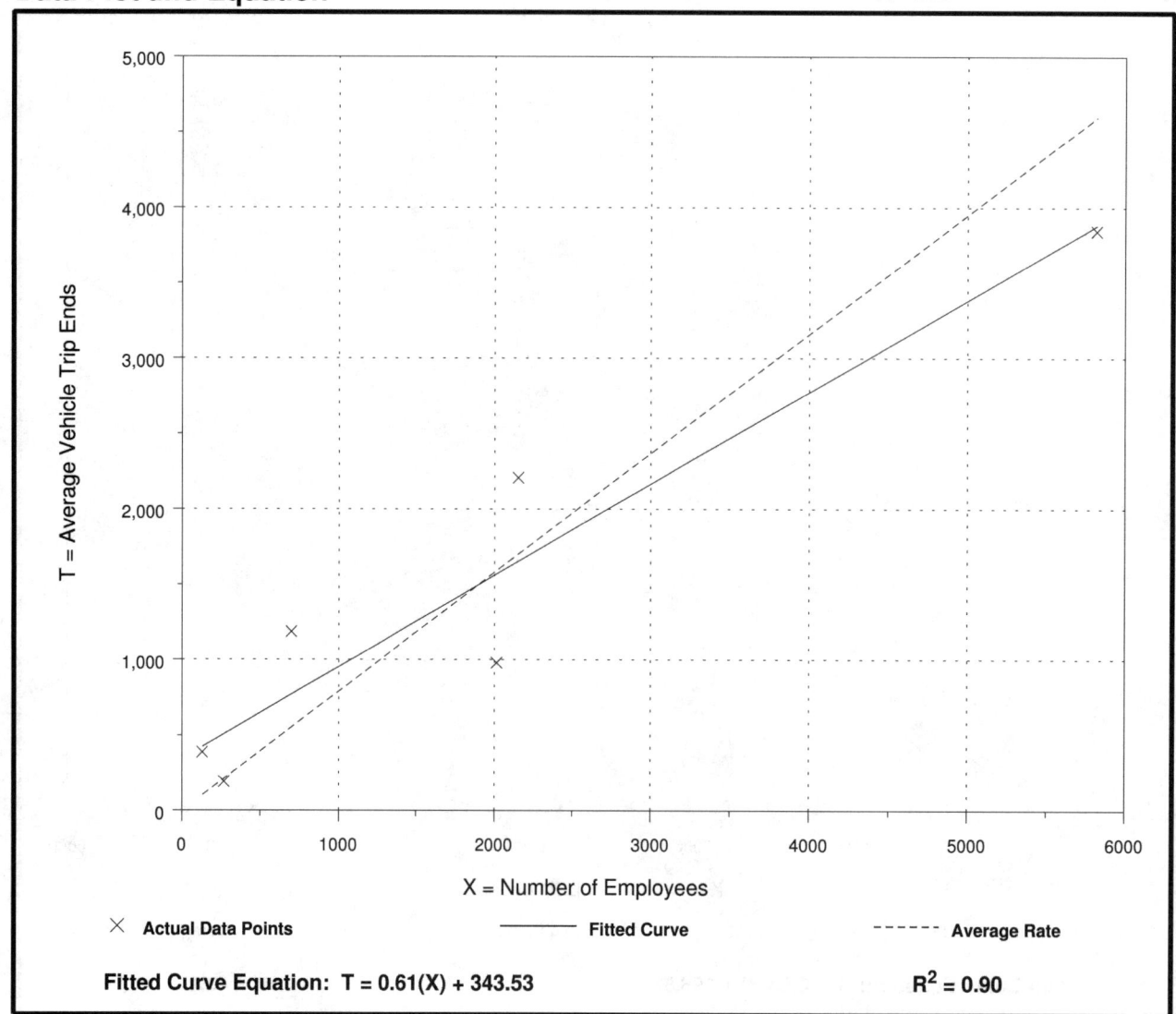

Fitted Curve Equation: T = 0.61(X) + 343.53 $R^2 = 0.90$

University/College
(550)

Average Vehicle Trip Ends vs: Employees
On a: Weekday,
A.M. Peak Hour of Generator

Number of Studies: 7
Avg. Number of Employees: 1,596
Directional Distribution: 74% entering, 26% exiting

Trip Generation per Employee

Average Rate	Range of Rates	Standard Deviation
0.79	0.36 - 2.08	0.95

Data Plot and Equation

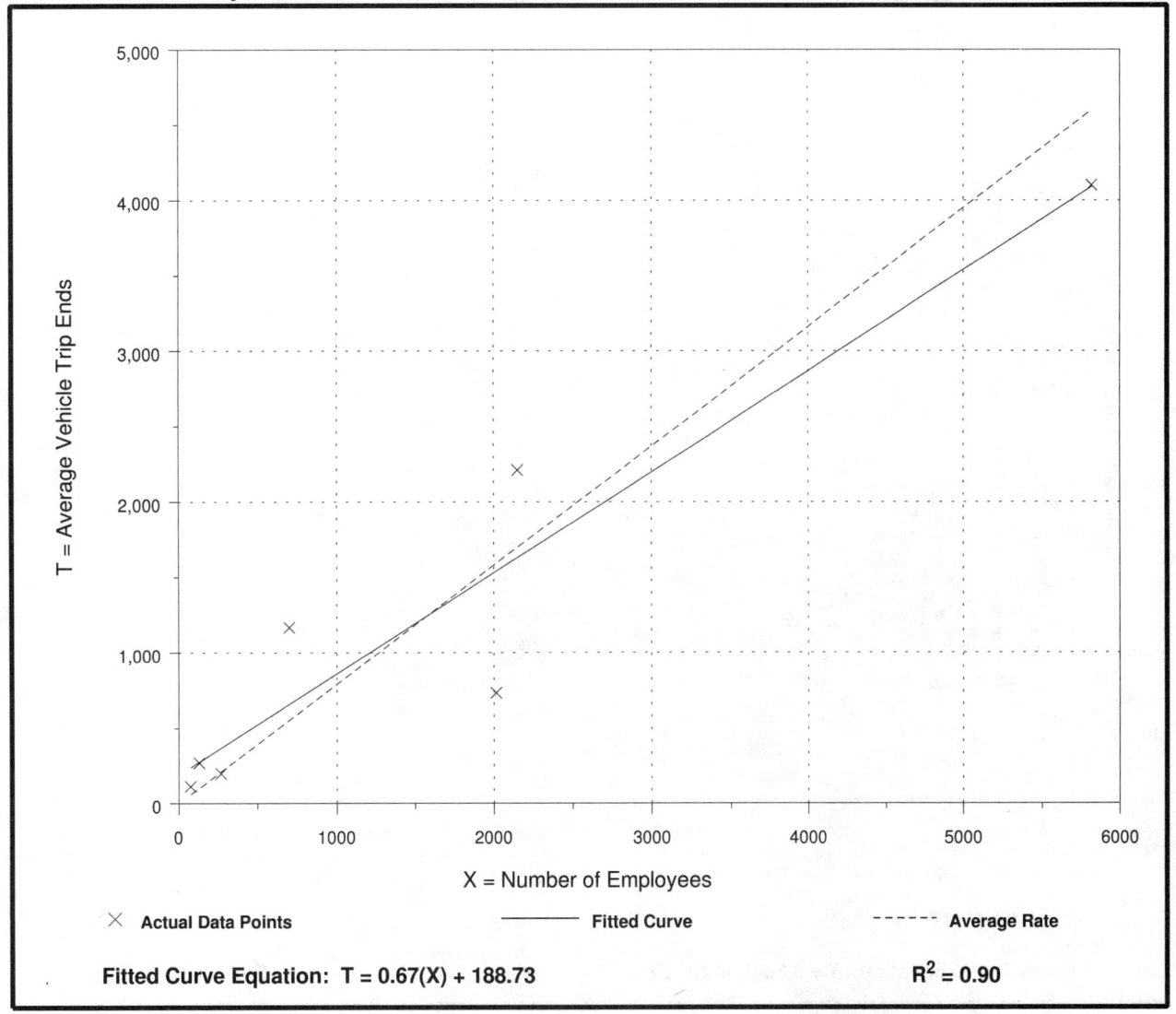

Fitted Curve Equation: T = 0.67(X) + 188.73 $R^2 = 0.90$

University/College
(550)

Average Vehicle Trip Ends vs: Employees
On a: Weekday,
P.M. Peak Hour of Generator

Number of Studies: 7
Avg. Number of Employees: 1,596
Directional Distribution: 32% entering, 68% exiting

Trip Generation per Employee

Average Rate	Range of Rates	Standard Deviation
0.85	0.49 - 3.08	1.00

Data Plot and Equation

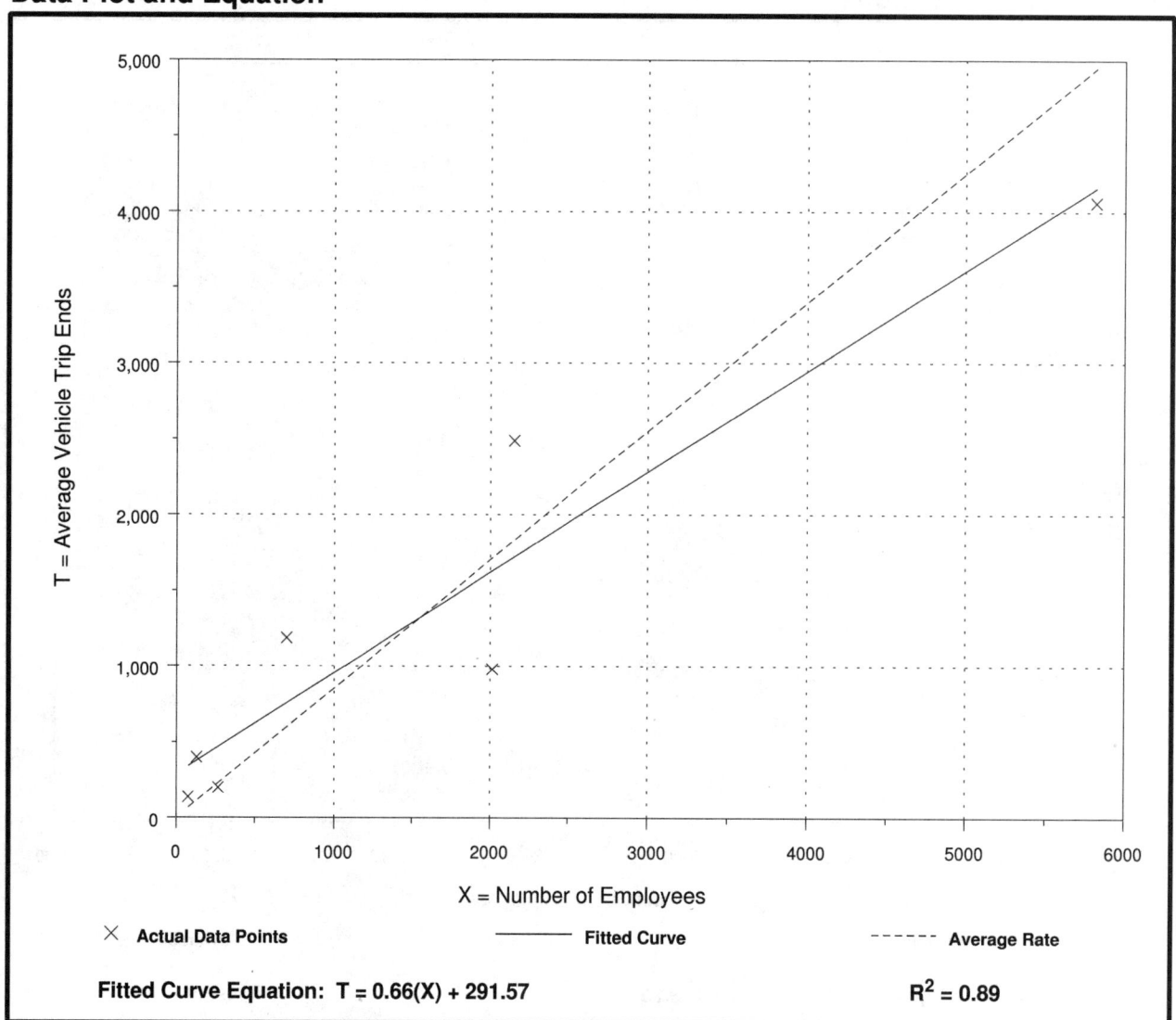

Fitted Curve Equation: T = 0.66(X) + 291.57 $R^2 = 0.89$

University/College
(550)

Average Vehicle Trip Ends vs: Employees
On a: Saturday

Number of Studies: 2
Avg. Number of Employees: 1,143
Directional Distribution: 50% entering, 50% exiting

Trip Generation per Employee

Average Rate	Range of Rates	Standard Deviation
3.12	2.56 - 7.39	*

Data Plot and Equation

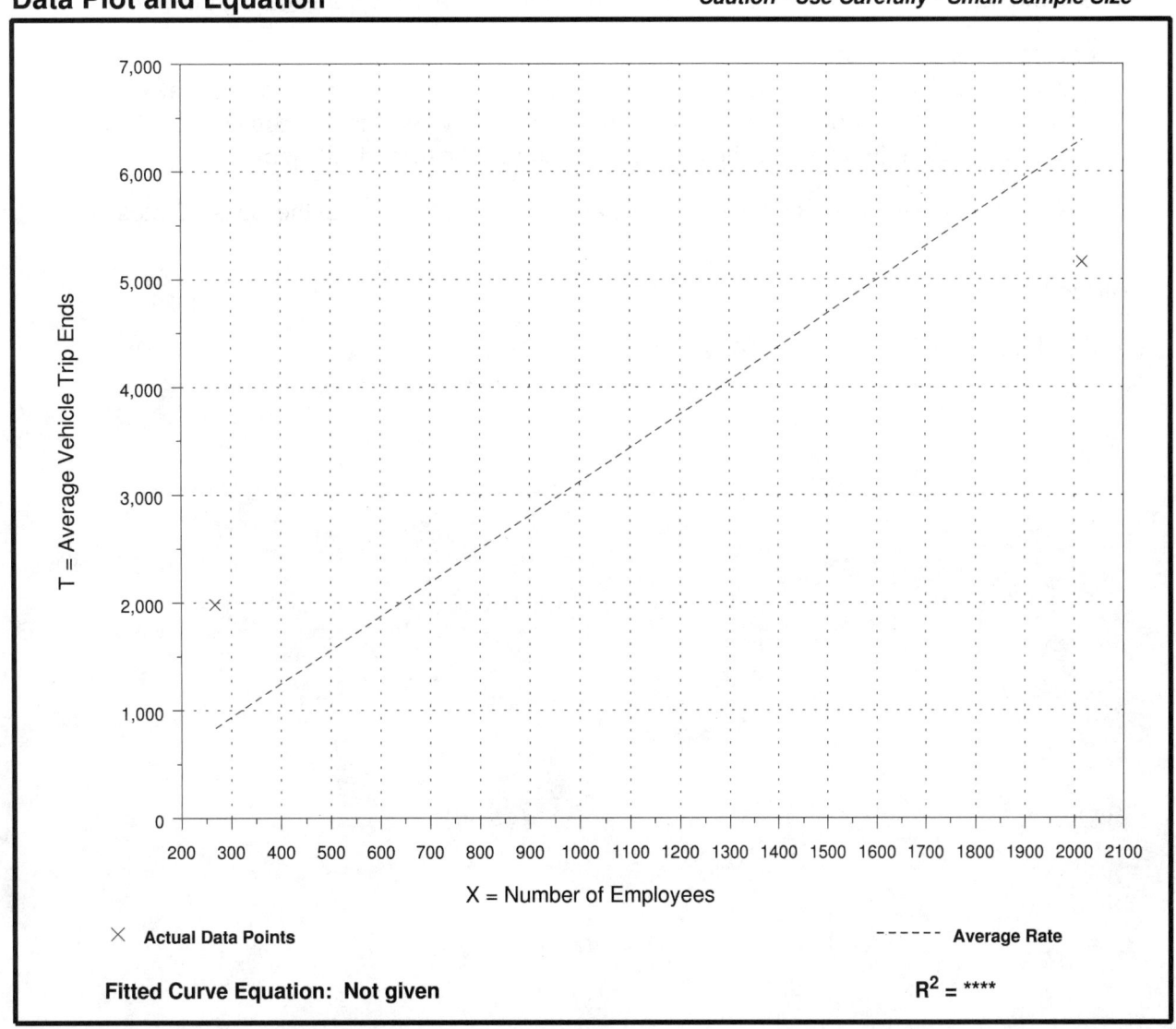

Caution - Use Carefully - Small Sample Size

Fitted Curve Equation: Not given $R^2 = ****$

Land Use: 560
Church

Description

A church is a building in which public worship services are held. A church houses an assembly hall or sanctuary; it may also house meeting rooms, classrooms and, occasionally, dining, catering, or party facilities. Synagogue (Land Use 561) and mosque (Land Use 562) are related uses.

Additional Data

Worship services are typically held on Sundays.

Some of the surveyed churches offered day care or extended care programs during the week.

Peak hours of the generator—
The weekday A.M. peak hour varied between 10:00 a.m. and 12:00 p.m. The weekday P.M. peak hour varied between 7:00 p.m. and 11:00 p.m. The Saturday peak hour varied between 5:00 p.m. and 8:00 p.m. The Sunday peak hour varied between 9:00 a.m. and 1:00 p.m.

The sites were surveyed between the late 1970s and the 2000s throughout the United States.

Source Numbers

90, 120, 169, 170, 423, 428, 436, 554, 571, 583, 629, 631, 704

Church
(560)

Average Vehicle Trip Ends vs: 1000 Sq. Feet Gross Floor Area
On a: Weekday

Number of Studies: 8
Average 1000 Sq. Feet GFA: 19
Directional Distribution: 50% entering, 50% exiting

Trip Generation per 1000 Sq. Feet Gross Floor Area

Average Rate	Range of Rates	Standard Deviation
9.11	4.35 - 30.20	7.20

Data Plot and Equation

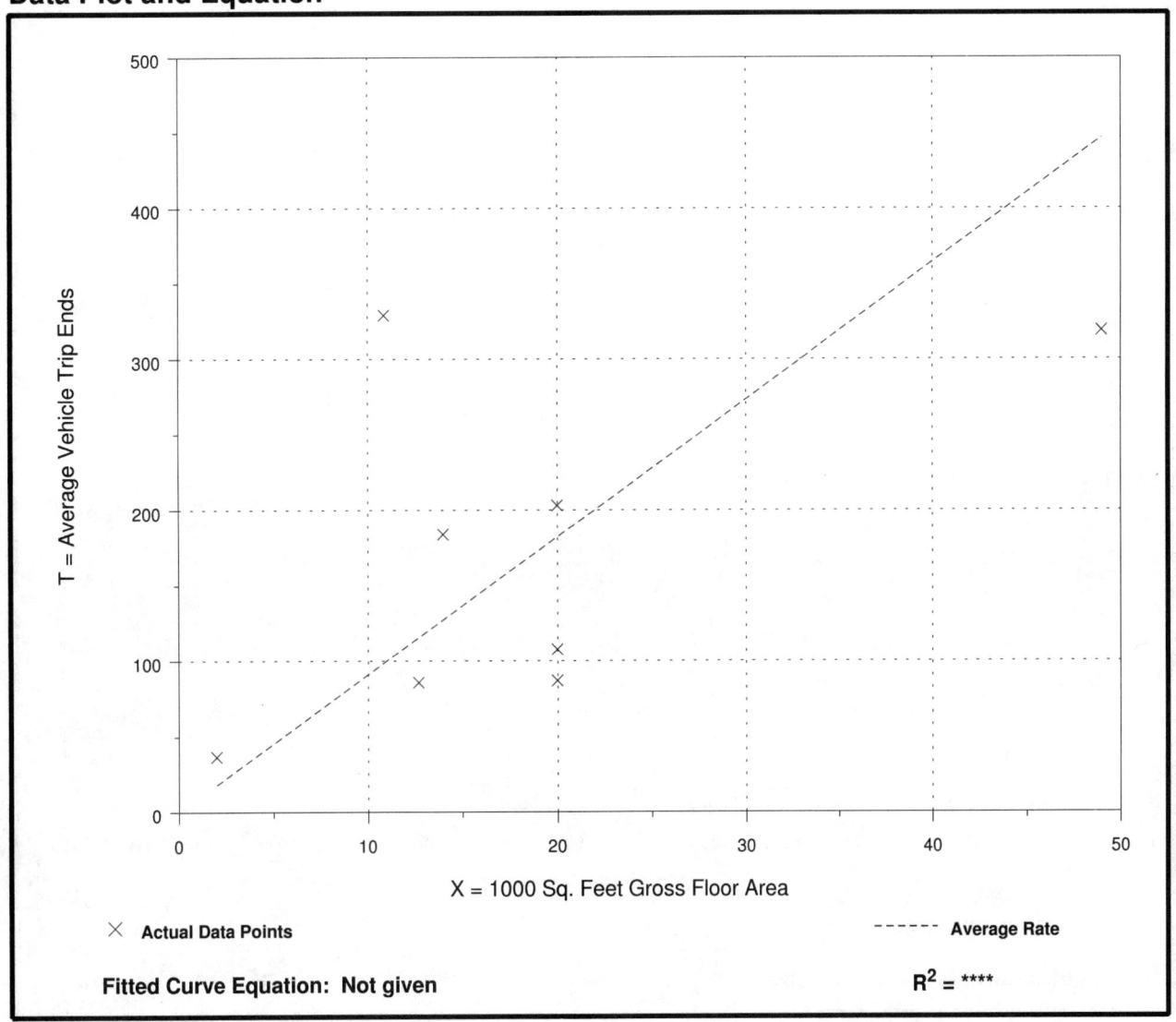

T = Average Vehicle Trip Ends
X = 1000 Sq. Feet Gross Floor Area

× Actual Data Points
------ Average Rate

Fitted Curve Equation: Not given　　　　$R^2 = ****$

Church
(560)

Average Vehicle Trip Ends vs: 1000 Sq. Feet Gross Floor Area
On a: Weekday,
Peak Hour of Adjacent Street Traffic,
One Hour Between 7 and 9 a.m.

Number of Studies: 9
Average 1000 Sq. Feet GFA: 30
Directional Distribution: 62% entering, 38% exiting

Trip Generation per 1000 Sq. Feet Gross Floor Area

Average Rate	Range of Rates	Standard Deviation
0.56	0.08 - 6.61	1.45

Data Plot and Equation

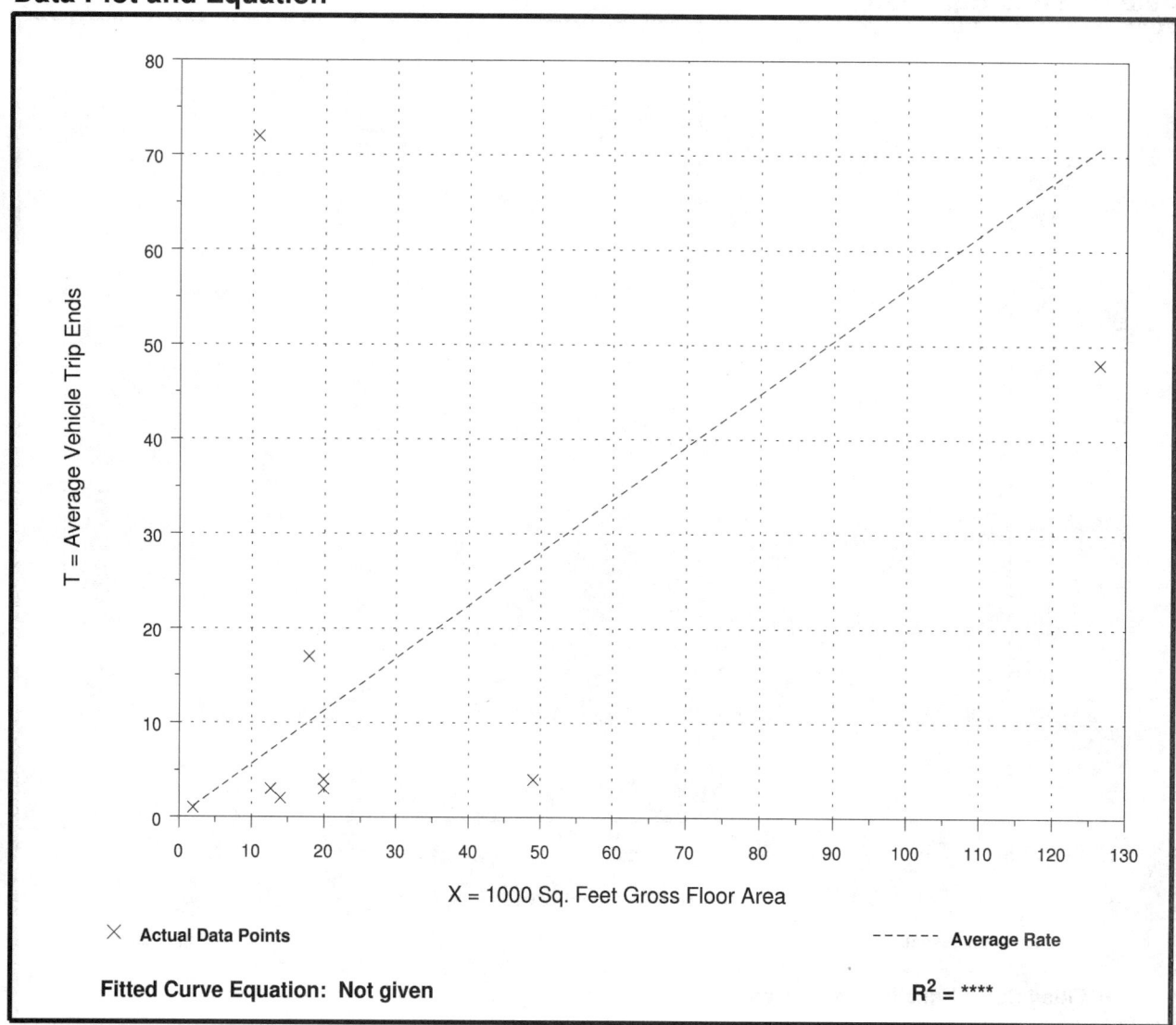

Fitted Curve Equation: Not given $R^2 = ****$

Church
(560)

Average Vehicle Trip Ends vs: 1000 Sq. Feet Gross Floor Area
On a: Weekday,
Peak Hour of Adjacent Street Traffic,
One Hour Between 4 and 6 p.m.

Number of Studies: 12
Average 1000 Sq. Feet GFA: 26
Directional Distribution: 48% entering, 52% exiting

Trip Generation per 1000 Sq. Feet Gross Floor Area

Average Rate	Range of Rates	Standard Deviation
0.55	0.21 - 2.10	0.87

Data Plot and Equation

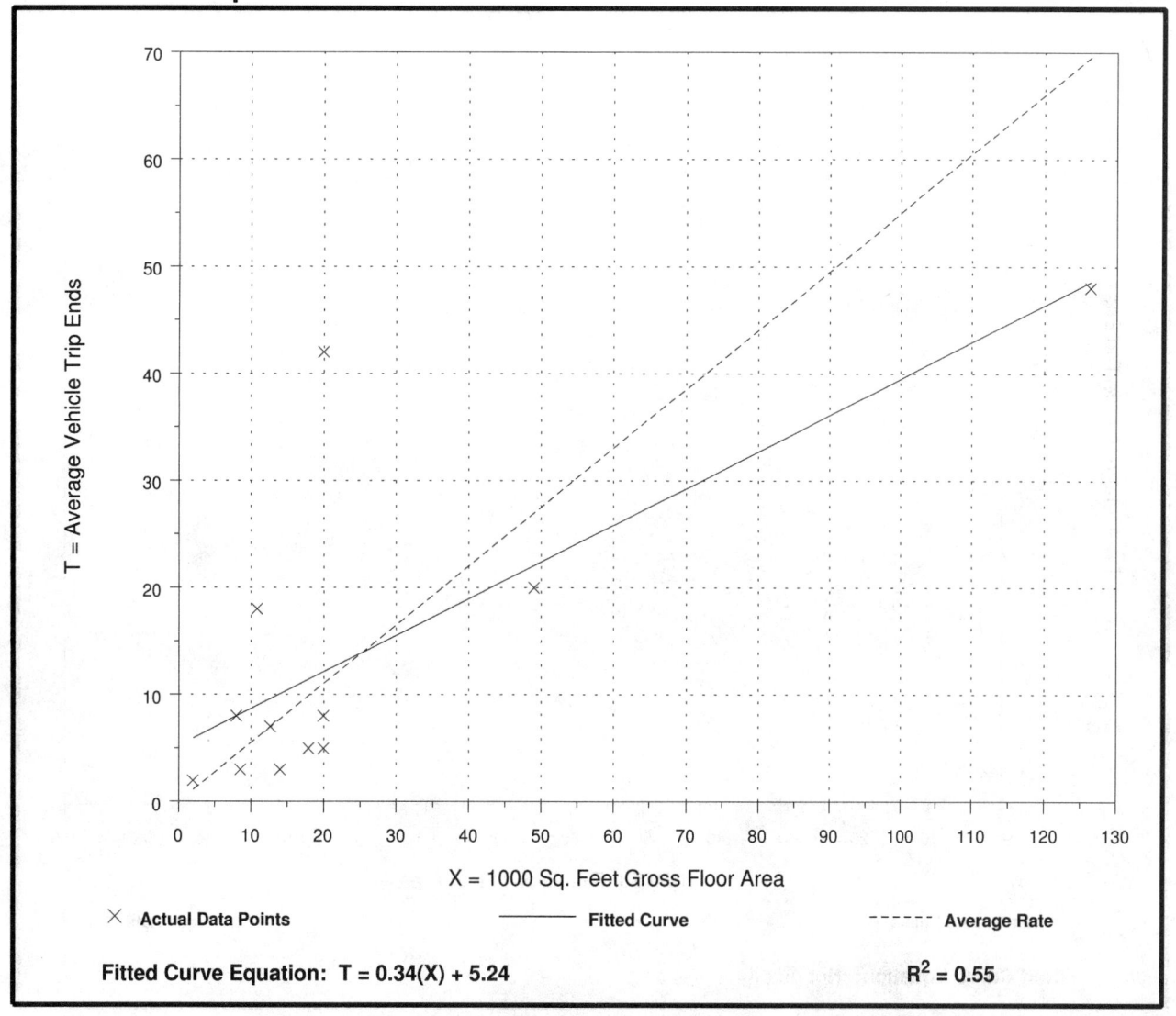

Fitted Curve Equation: $T = 0.34(X) + 5.24$ $R^2 = 0.55$

Church
(560)

Average Vehicle Trip Ends vs: 1000 Sq. Feet Gross Floor Area
On a: Weekday,
A.M. Peak Hour of Generator

Number of Studies: 9
Average 1000 Sq. Feet GFA: 31
Directional Distribution: 55% entering, 45% exiting

Trip Generation per 1000 Sq. Feet Gross Floor Area

Average Rate	Range of Rates	Standard Deviation
0.87	0.30 - 6.61	1.57

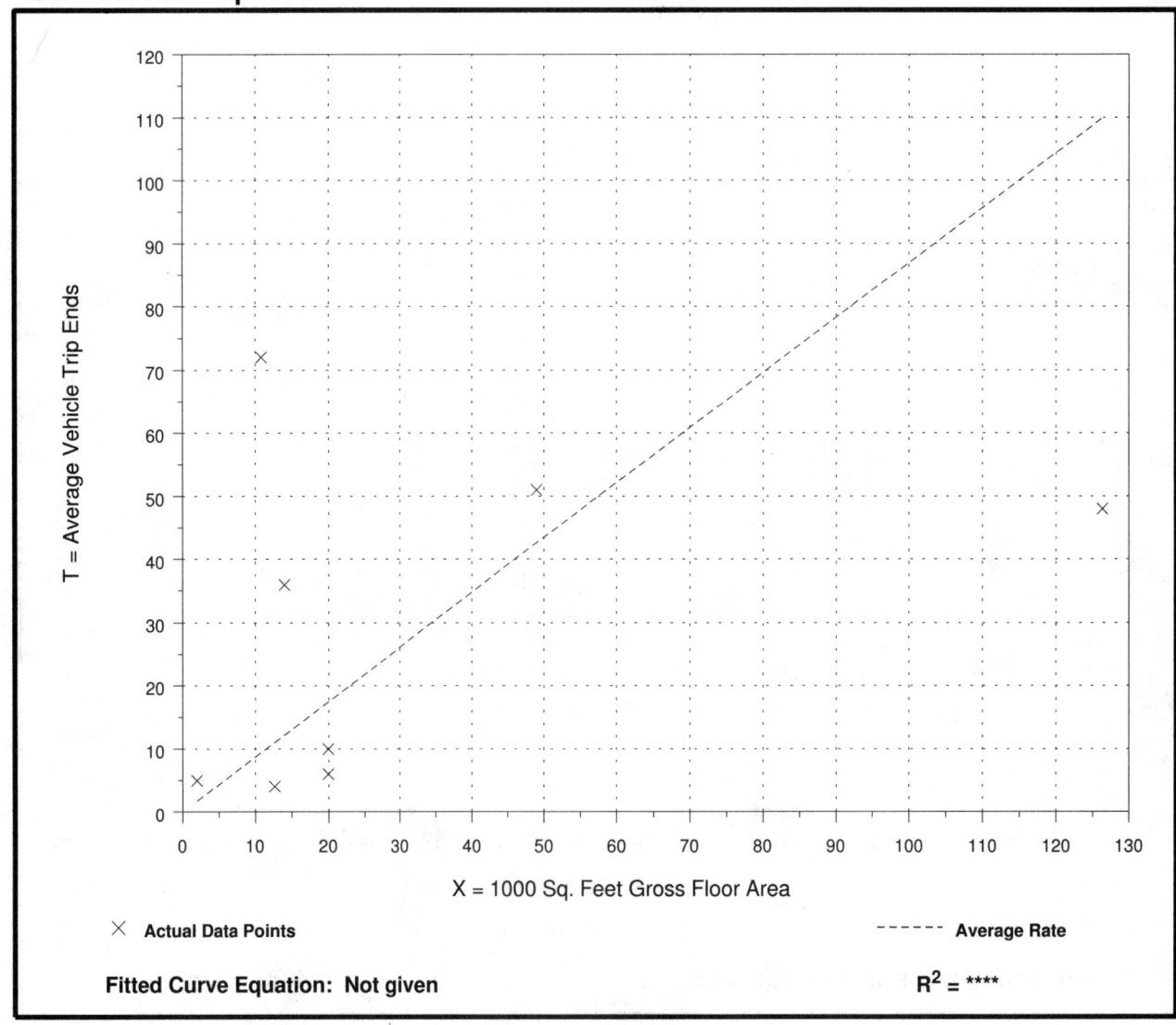

Data Plot and Equation

Fitted Curve Equation: Not given $R^2 = ****$

Church
(560)

Average Vehicle Trip Ends vs: 1000 Sq. Feet Gross Floor Area
On a: Weekday,
P.M. Peak Hour of Generator

Number of Studies: 9
Average 1000 Sq. Feet GFA: 31
Directional Distribution: 54% entering, 46% exiting

Trip Generation per 1000 Sq. Feet Gross Floor Area

Average Rate	Range of Rates	Standard Deviation
0.94	0.38 - 4.04	1.26

Data Plot and Equation

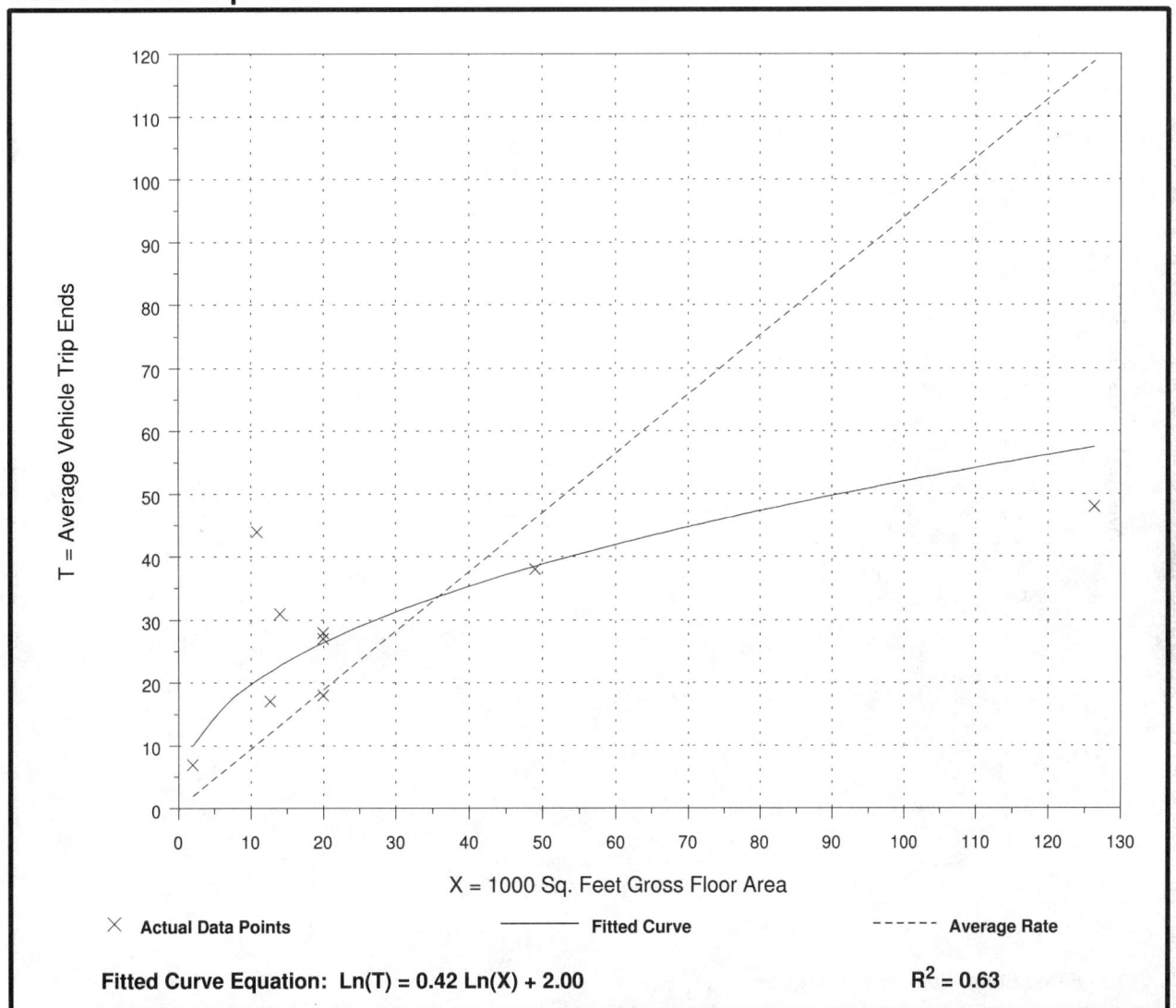

Fitted Curve Equation: $Ln(T) = 0.42\, Ln(X) + 2.00$ $R^2 = 0.63$

Church
(560)

Average Vehicle Trip Ends vs: 1000 Sq. Feet Gross Floor Area
On a: Saturday

Number of Studies: 6
Average 1000 Sq. Feet GFA: 19
Directional Distribution: 50% entering, 50% exiting

Trip Generation per 1000 Sq. Feet Gross Floor Area

Average Rate	Range of Rates	Standard Deviation
10.37	2.35 - 56.83	16.74

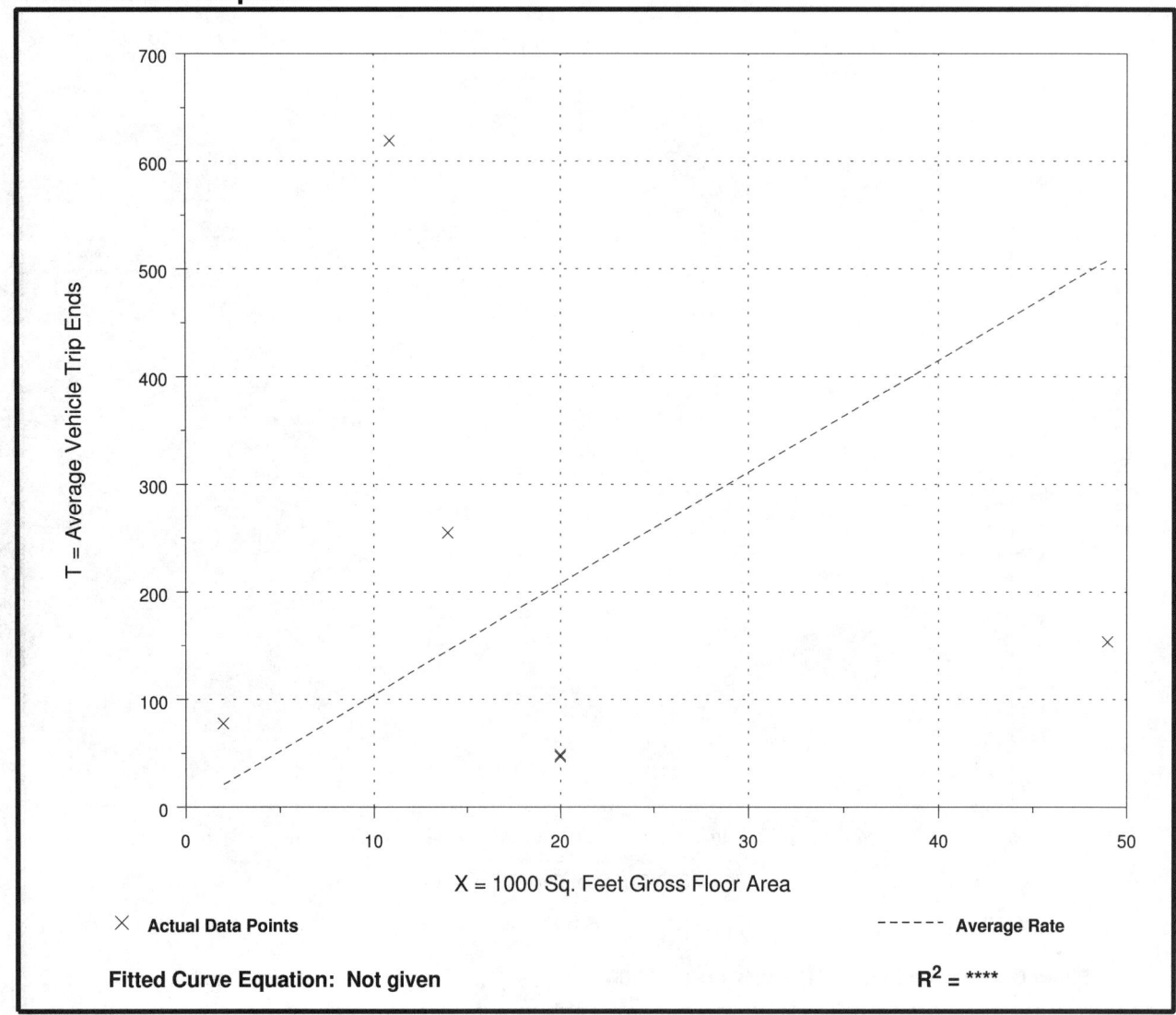

Data Plot and Equation

Fitted Curve Equation: Not given $R^2 = ****$

Church
(560)

Average Vehicle Trip Ends vs: 1000 Sq. Feet Gross Floor Area
On a: Saturday,
Peak Hour of Generator

Number of Studies: 6
Average 1000 Sq. Feet GFA: 19
Directional Distribution: 71% entering, 29% exiting

Trip Generation per 1000 Sq. Feet Gross Floor Area

Average Rate	Range of Rates	Standard Deviation
3.54	0.40 - 23.32	6.87

Data Plot and Equation

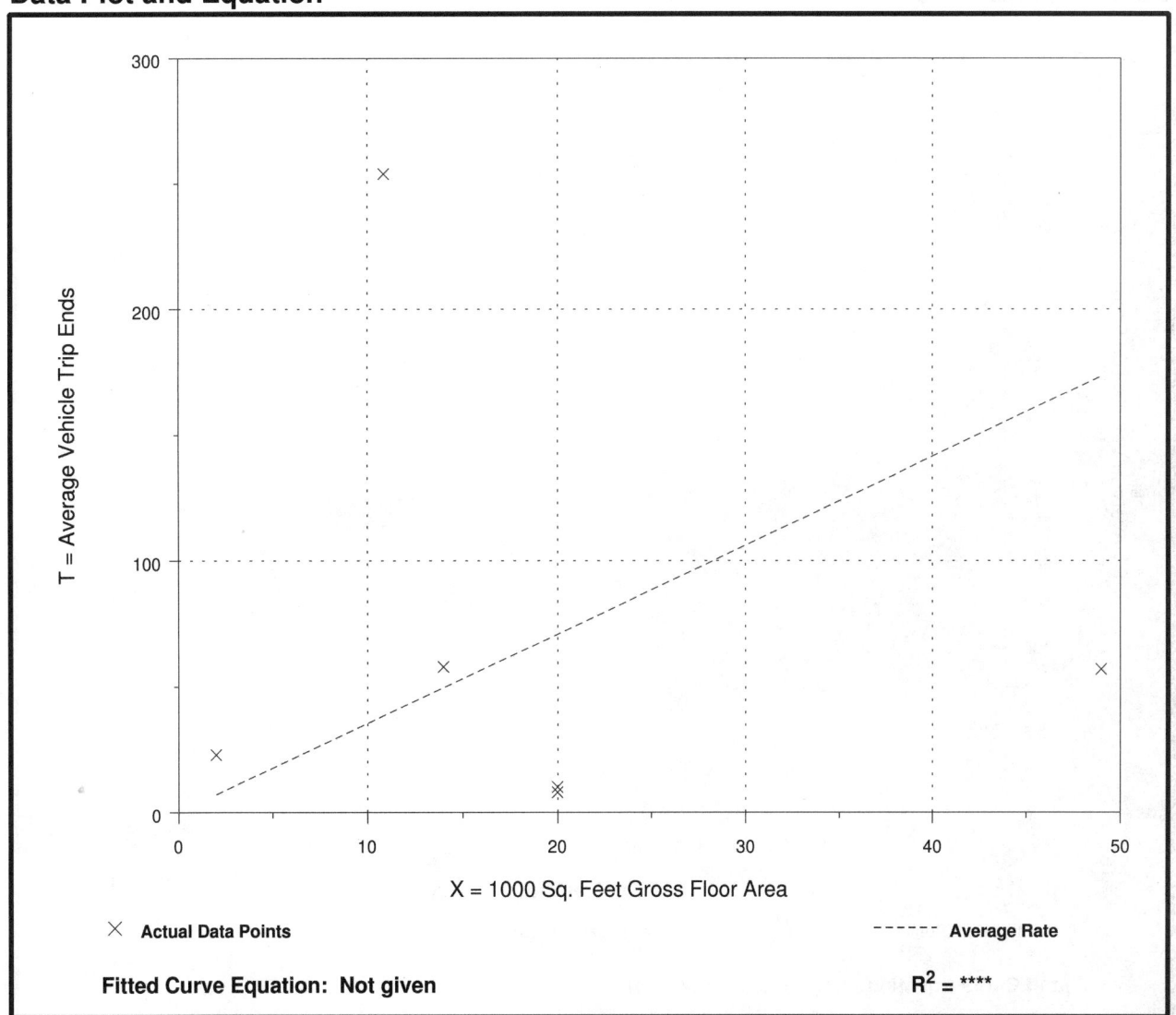

Fitted Curve Equation: Not given $R^2 = ****$

Church
(560)

Average Vehicle Trip Ends vs: 1000 Sq. Feet Gross Floor Area
On a: Sunday

Number of Studies: 7
Average 1000 Sq. Feet GFA: 19
Directional Distribution: 50% entering, 50% exiting

Trip Generation per 1000 Sq. Feet Gross Floor Area

Average Rate	Range of Rates	Standard Deviation
36.63	19.15 - 113.38	29.13

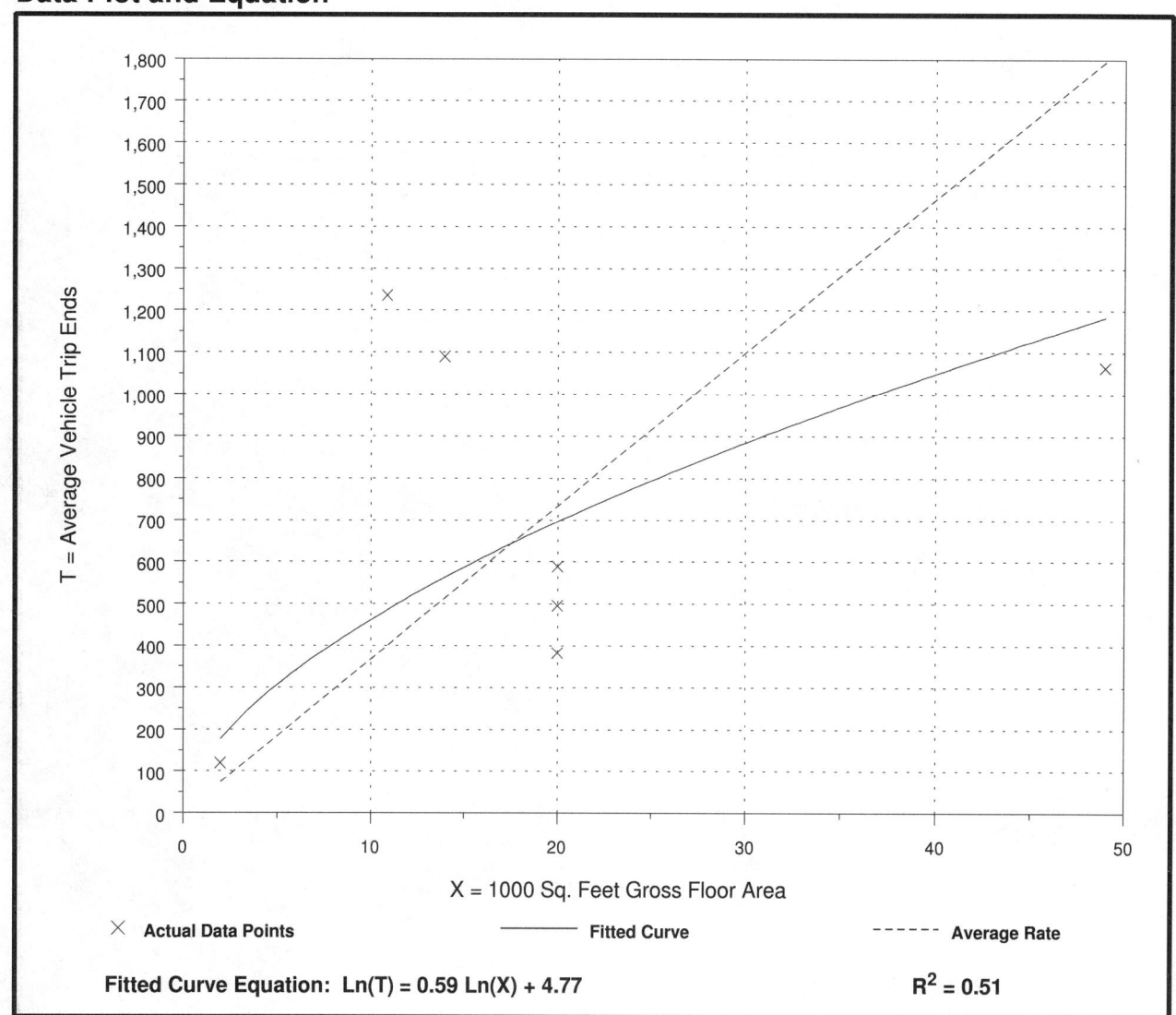

Data Plot and Equation

Fitted Curve Equation: $Ln(T) = 0.59 \, Ln(X) + 4.77$ $R^2 = 0.51$

Church
(560)

Average Vehicle Trip Ends vs: 1000 Sq. Feet Gross Floor Area
On a: Sunday,
Peak Hour of Generator

Number of Studies: 15
Average 1000 Sq. Feet GFA: 32
Directional Distribution: 49% entering, 51% exiting

Trip Generation per 1000 Sq. Feet Gross Floor Area

Average Rate	Range of Rates	Standard Deviation
12.04	5.17 - 51.27	8.72

Data Plot and Equation

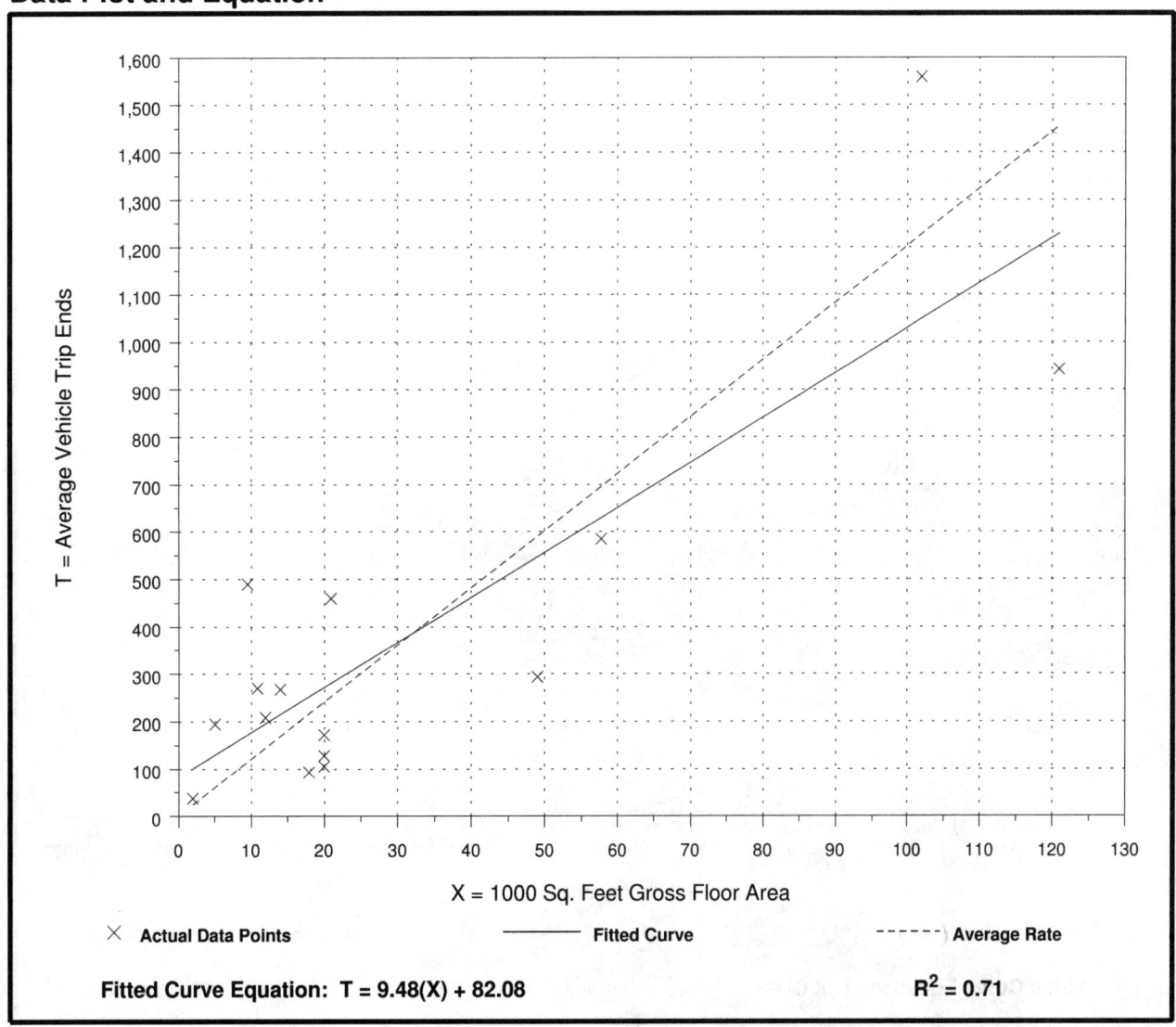

Fitted Curve Equation: $T = 9.48(X) + 82.08$ $R^2 = 0.71$

Church
(560)

Average Vehicle Trip Ends vs: Seats
On a: Weekday

Number of Studies: 4
Average Number of Seats: 534
Directional Distribution: 50% entering, 50% exiting

Trip Generation per Seat

Average Rate	Range of Rates	Standard Deviation
0.61	0.21 - 0.84	0.82

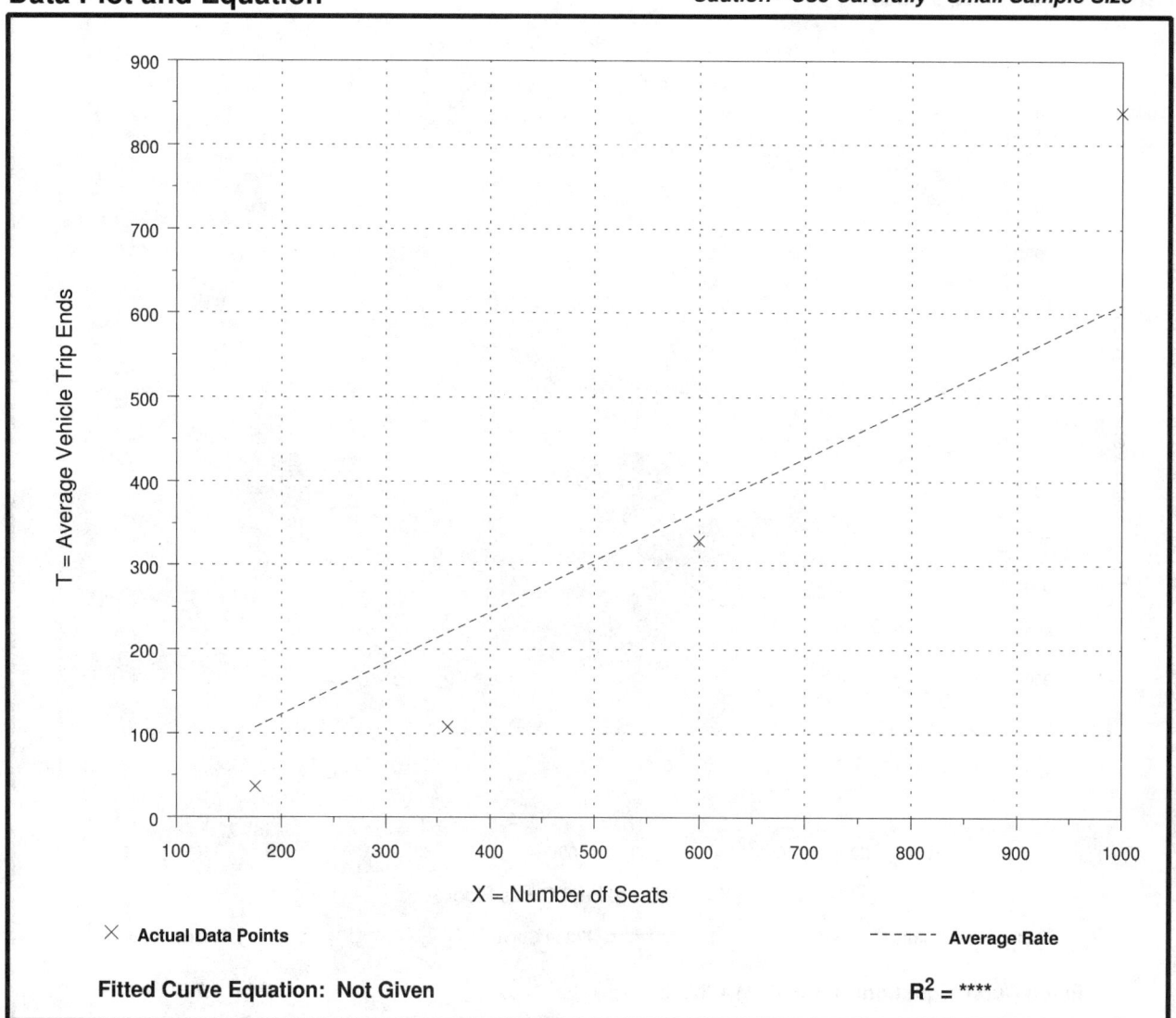

Data Plot and Equation

Caution - Use Carefully - Small Sample Size

Fitted Curve Equation: Not Given $R^2 = ****$

Church
(560)

Average Vehicle Trip Ends vs: Seats
On a: Saturday

Number of Studies: 2
Average Number of Seats: 388
Directional Distribution: 50% entering, 50% exiting

Trip Generation per Seat

Average Rate	Range of Rates	Standard Deviation
0.90	0.45 - 1.03	*

Data Plot and Equation

Caution - Use Carefully - Small Sample Size

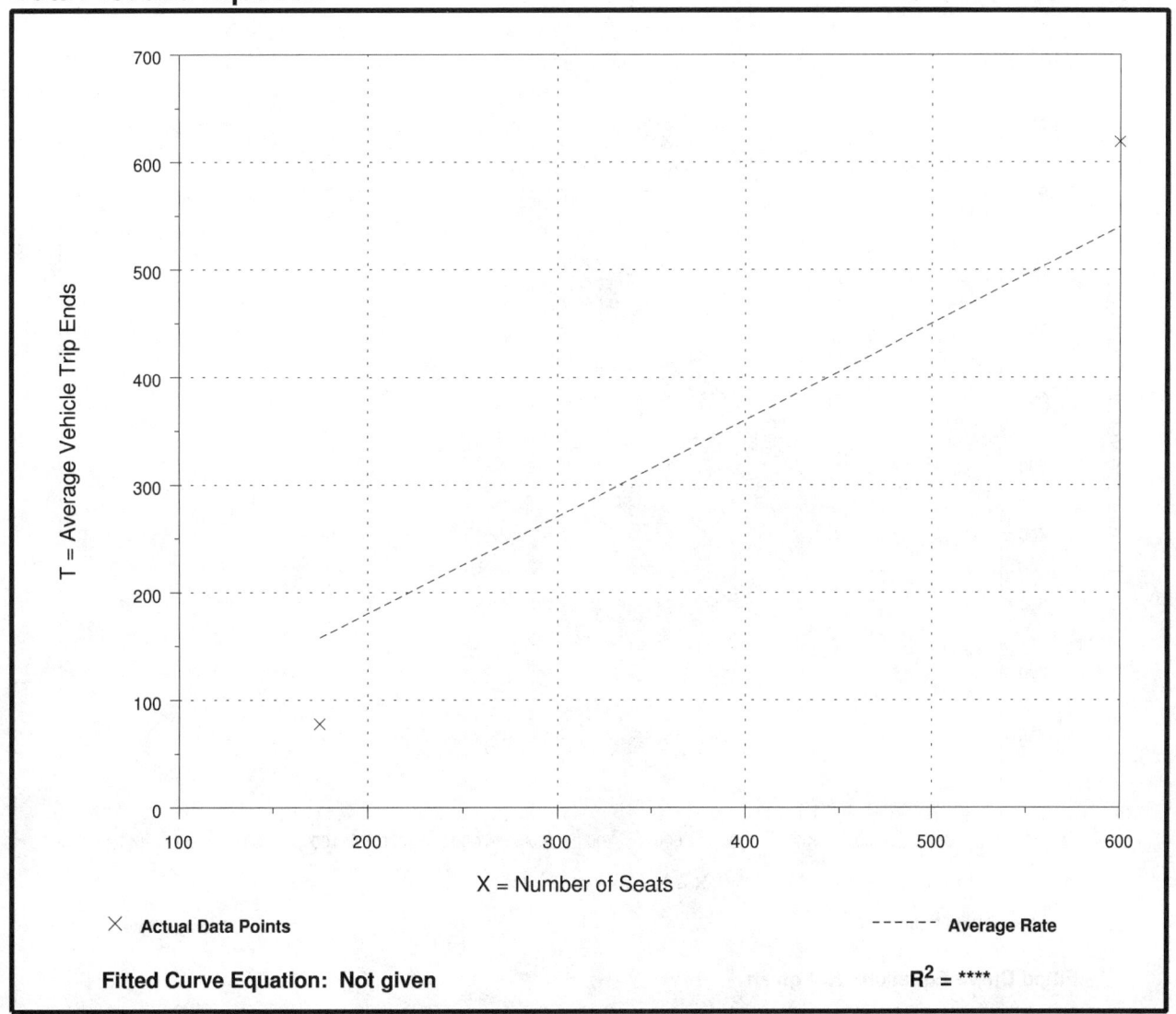

Fitted Curve Equation: Not given $R^2 = ****$

Church
(560)

Average Vehicle Trip Ends vs: Seats
On a: Saturday,
Peak Hour of Generator

Number of Studies: 3
Average Number of Seats: 758
Directional Distribution: 43% entering, 57% exiting

Trip Generation per Seat

Average Rate	Range of Rates	Standard Deviation
0.60	0.13 - 0.72	0.79

Data Plot and Equation

Caution - Use Carefully - Small Sample Size

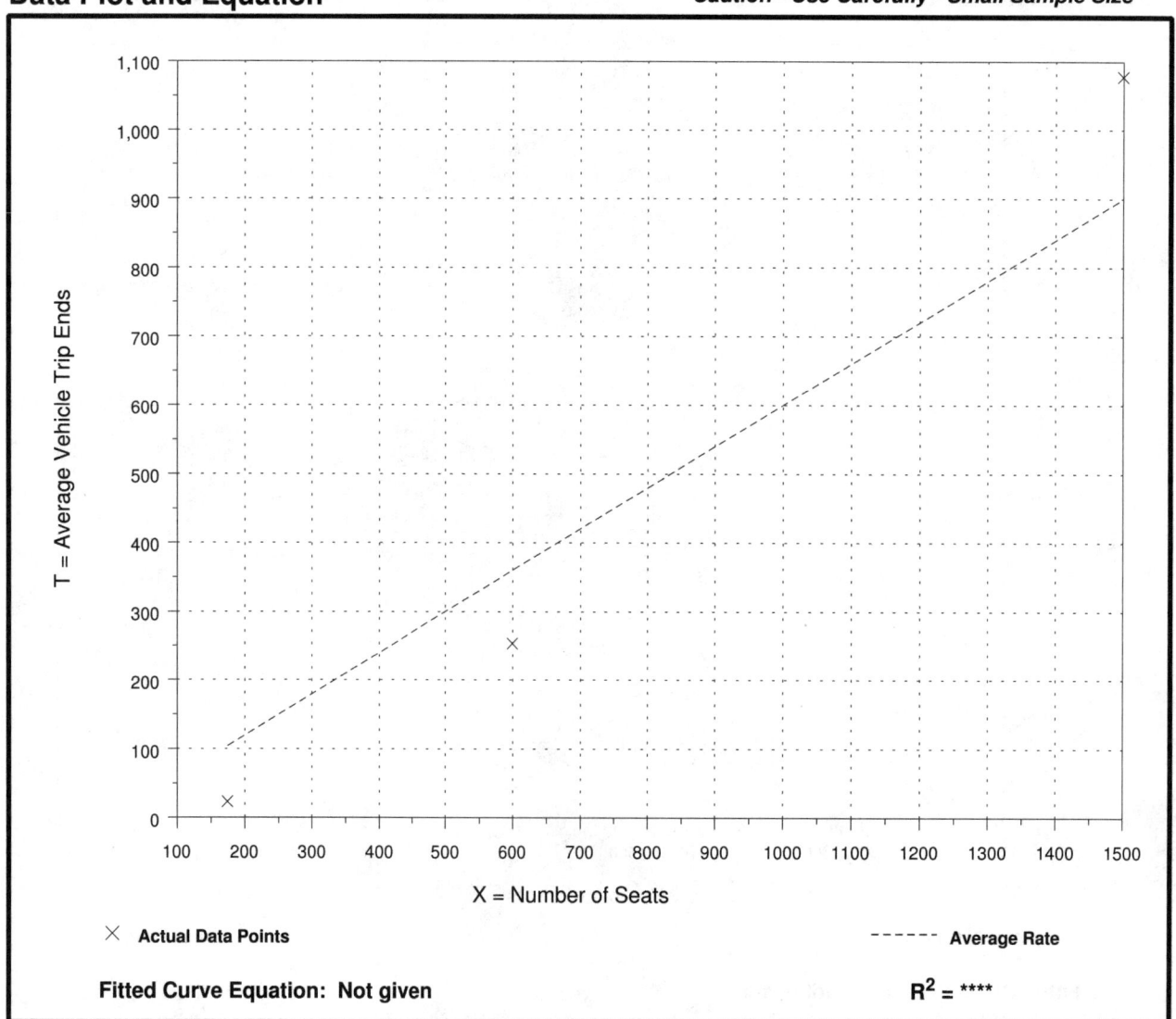

Fitted Curve Equation: Not given $R^2 = ****$

Church
(560)

Average Vehicle Trip Ends vs: Seats
On a: Sunday

Number of Studies: 4
Average Number of Seats: 534
Directional Distribution: 50% entering, 50% exiting

Trip Generation per Seat

Average Rate	Range of Rates	Standard Deviation
1.85	0.69 - 2.21	1.46

Data Plot and Equation

Caution - Use Carefully - Small Sample Size

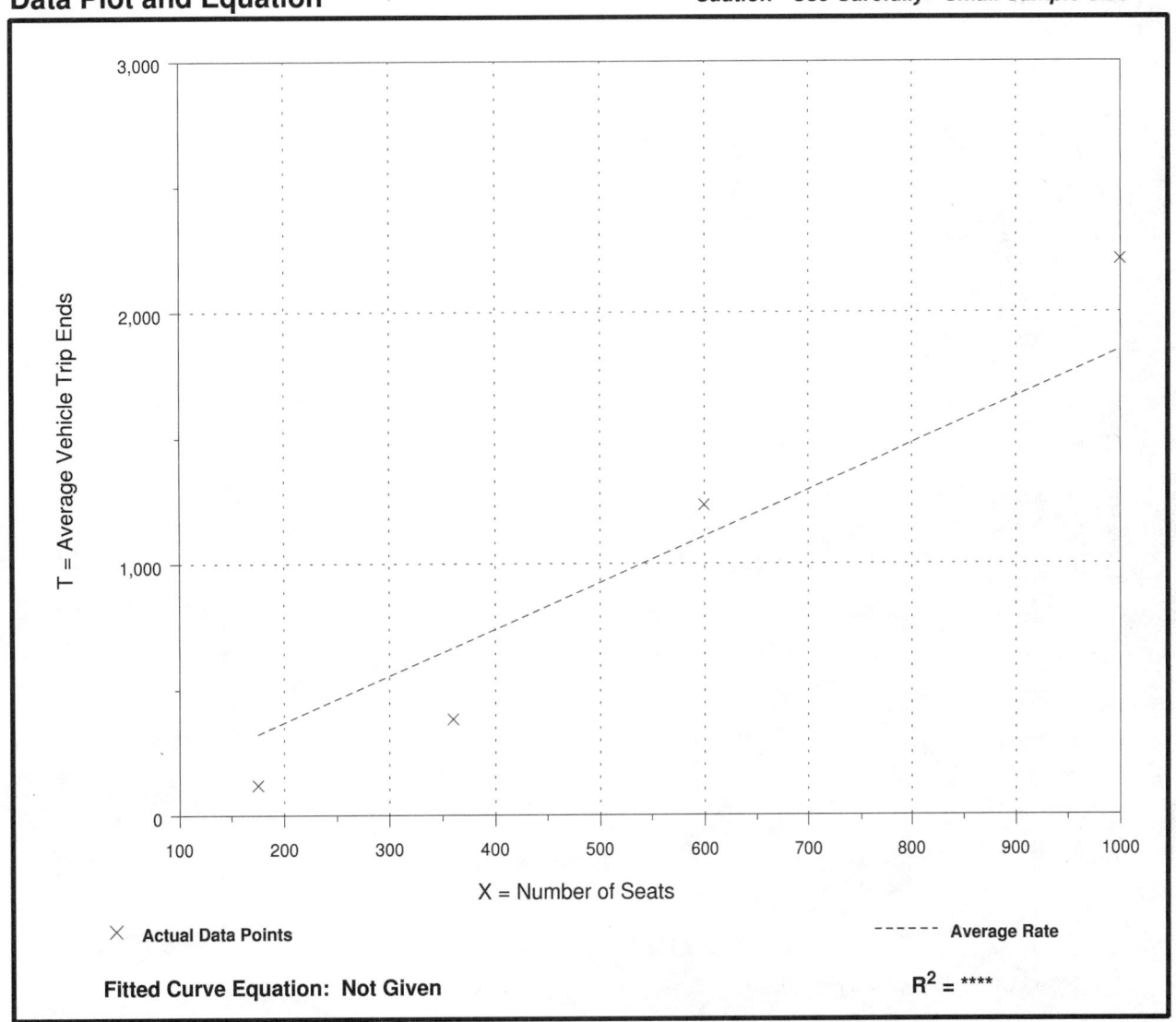

Fitted Curve Equation: Not Given $R^2 = ****$

Trip Generation, 9th Edition • Institute of Transportation Engineers

Church
(560)

Average Vehicle Trip Ends vs: Seats
On a: Sunday,
Peak Hour of Generator

Number of Studies: 13
Average Number of Seats: 903
Directional Distribution: 50% entering, 50% exiting

Trip Generation per Seat

Average Rate	Range of Rates	Standard Deviation
0.61	0.21 - 1.14	0.81

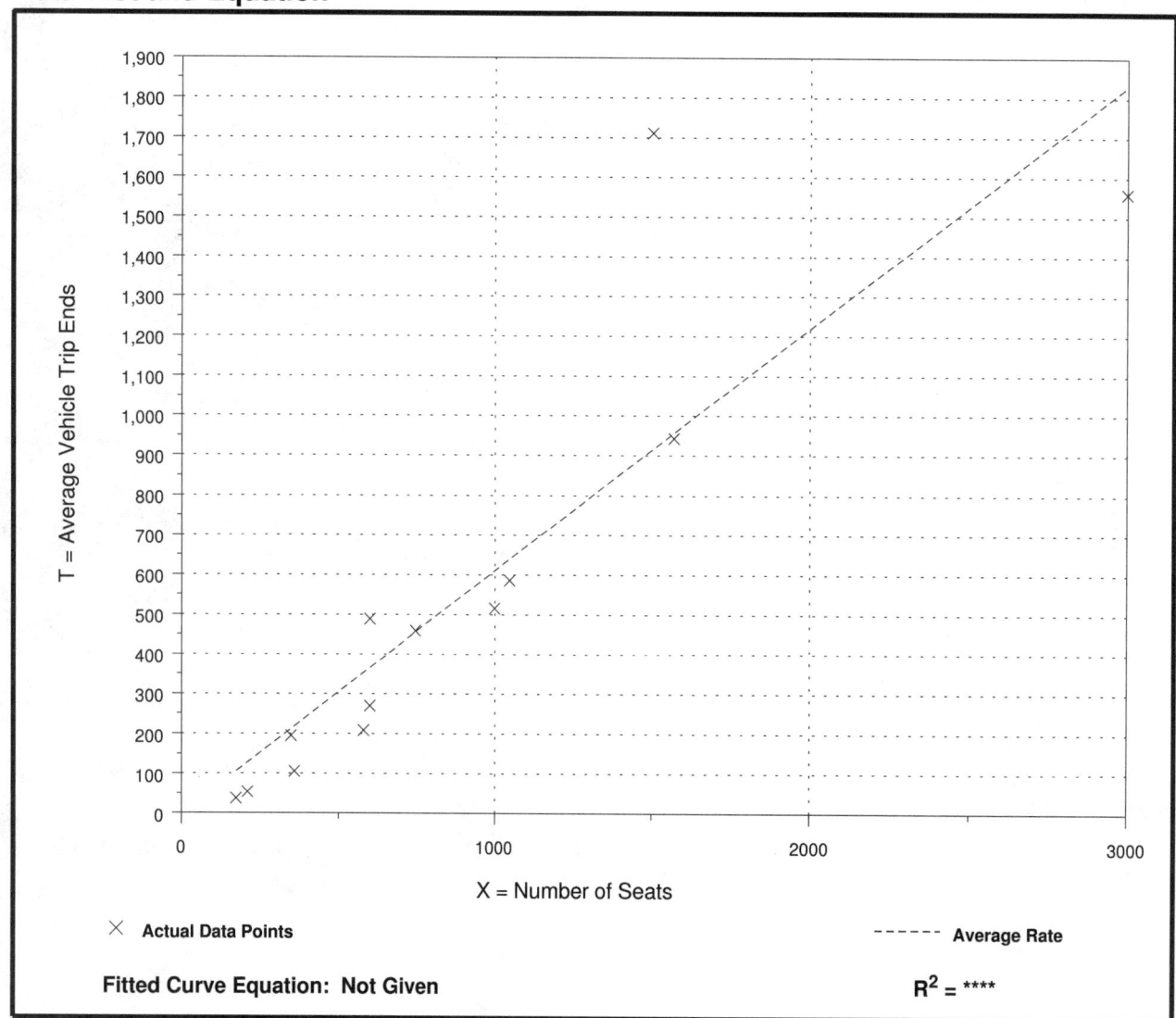

Data Plot and Equation

Fitted Curve Equation: Not Given $R^2 = ****$

Land Use: 561
Synagogue

Description

A synagogue is a building in which public worship services are held. A synagogue may also house a sanctuary, meeting rooms, classrooms and, occasionally, dining, catering, or party facilities. Church (Land Use 560) and mosque (Land Use 562) are related uses.

Additional Data

One study reported trip counts taken during the Yom Kippur holiday. The 30,000-square foot site had 20 employees and 1,040 family members. The information collected during this period is presented in the table below and was excluded from the data plots.

Independent Variable	Trip Generation Rates			
	Weekday Peak Hour of Adjacent Street Traffic		Weekday Peak Hour of Generator	
	A.M.	P.M.	A.M.	P.M.
1,000 Square Feet GFA	14.87	10.27	29.03	25.73
Employees	22.30	15.40	43.55	38.60
Family Members	0.43	0.30	0.84	0.74

Peak hours of the generator—
   The weekday A.M. peak hour was between 10:00 a.m. and 11:00 a.m. The weekday P.M. peak hour was between 4:00 p.m. and 5:00 p.m. The Saturday peak hour was between 4:00 p.m. and 5:00 p.m. The Sunday peak hour of the generator was between 11:00 a.m. and 12:00 p.m.

The sites were surveyed in 1976 and 2001 in California and Pennsylvania, respectively.

Source Numbers

90, 571

Land Use: 561
Synagogue
Independent Variables with One Observation

The following trip generation data are for independent variables with only one observation. This information is shown in this table only; there are no related plots for these data.

Users are cautioned to use data with care because of the small sample size.

Independent Variable	Trip Generation Rate	Size of Independent Variable	Number of Studies	Directional Distribution
Employees				
Weekday	20.64	11	1	50% entering, 50% exiting
Weekday A.M. Peak Hour of Adjacent Street Traffic	0.27	11	1	100% entering, 0% exiting
Weekday P.M. Peak Hour of Adjacent Street Traffic	3.27	11	1	47% entering, 53% exiting
Weekday A.M. Peak Hour of Generator	0.82	11	1	56% entering, 44% exiting
Weekday P.M. Peak Hour of Generator	3.27	11	1	47% entering, 53% exiting
Saturday	11.45	11	1	50% entering, 50% exiting
Sunday	43.64	11	1	50% entering, 50% exiting
Family Members				
Weekday	0.47	487	1	50% entering, 50% exiting
Weekday A.M. Peak Hour of Adjacent Street Traffic	0.01	487	1	Not available
Weekday P.M. Peak Hour of Adjacent Street Traffic	0.07	487	1	47% entering, 53% exiting
Weekday A.M. Peak Hour of Generator	0.02	487	1	56% entering, 44% exiting
Weekday P.M. Peak Hour of Generator	0.07	487	1	47% entering, 53% exiting
Saturday	0.26	487	1	50% entering, 50% exiting
Sunday	0.99	487	1	50% entering, 50% exiting

Land Use: 561
Synagogue
Independent Variables with One Observation

1,000 Square Feet Gross Floor Area

Weekday	10.64	21	1	50% entering, 50% exiting
Weekday A.M. Peak Hour of Adjacent Street Traffic	0.14	21	1	Not available
Weekday P.M. Peak Hour of Adjacent Street Traffic	1.69	21	1	47% entering, 53% exiting
Weekday A.M. Peak Hour of Generator	0.42	21	1	56% entering, 44% exiting
Weekday P.M. Peak Hour of Generator	1.69	21	1	47% entering, 53% exiting
Saturday	5.91	21	1	50% entering, 50% exiting
Sunday	22.50	21	1	50% entering, 50% exiting

Synagogue
(561)

Average Vehicle Trip Ends vs: Employees
On a: Saturday,
Peak Hour of Generator

Number of Studies: 2
Avg. Number of Employees: 16
Directional Distribution: 42% entering, 58% exiting

Trip Generation per Employee

Average Rate	Range of Rates	Standard Deviation
4.52	2.18 - 5.80	*

Data Plot and Equation

Caution - Use Carefully - Small Sample Size

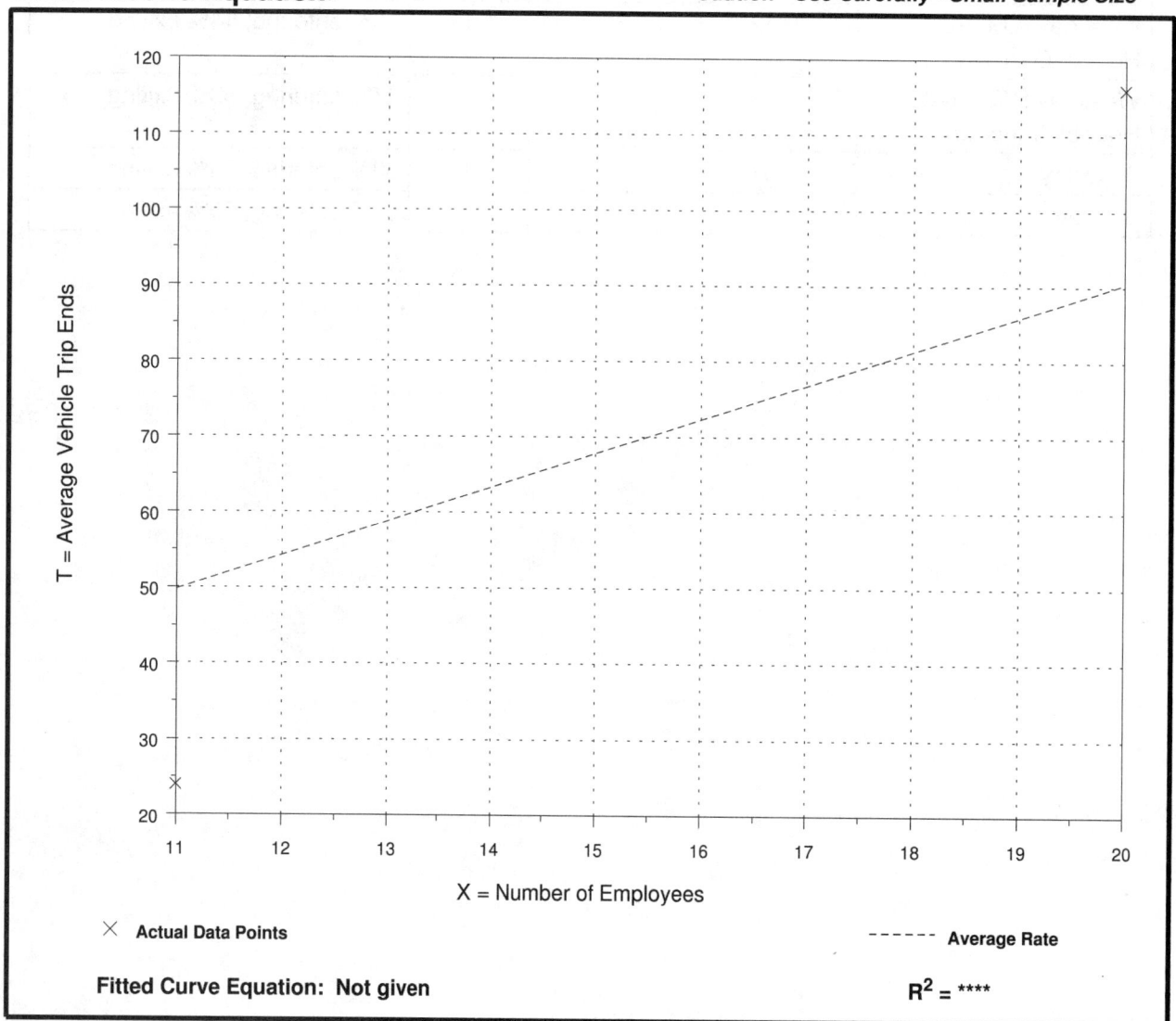

Fitted Curve Equation: Not given $R^2 = ****$

Synagogue
(561)

Average Vehicle Trip Ends vs: Employees
On a: Sunday,
Peak Hour of Generator

Number of Studies: 2
Avg. Number of Employees: 16
Directional Distribution: 49% entering, 51% exiting

Trip Generation per Employee

Average Rate	Range of Rates	Standard Deviation
12.55	11.75 - 14.00	*

Data Plot and Equation

Caution - Use Carefully - Small Sample Size

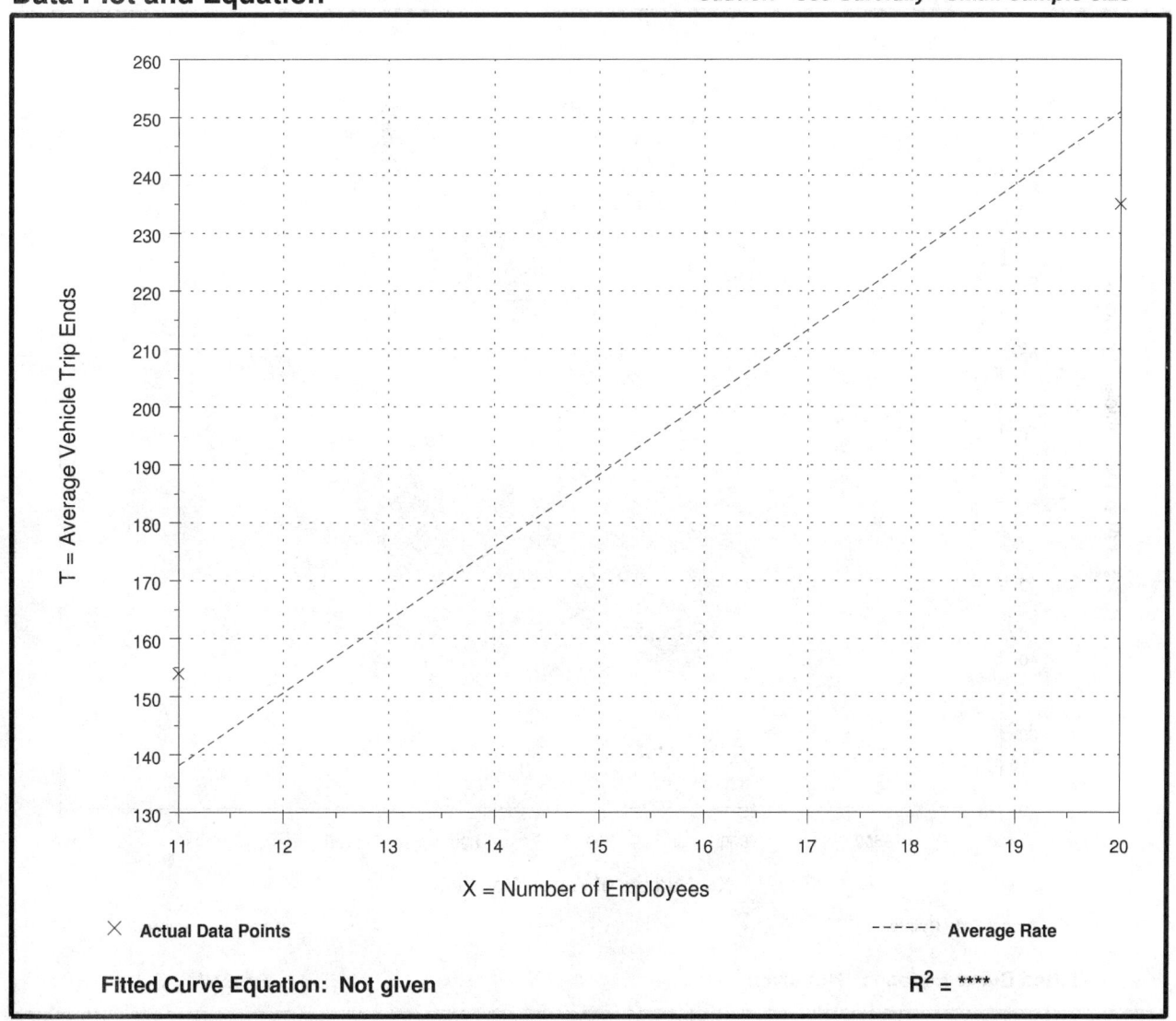

Fitted Curve Equation: Not given

R^2 = ****

Synagogue
(561)

Average Vehicle Trip Ends vs: Family Members
On a: Saturday,
Peak Hour of Generator

Number of Studies: 2
Average Number of Family Members: 764
Directional Distribution: 42% entering, 58% exiting

Trip Generation per Family Member

Average Rate	Range of Rates	Standard Deviation
0.09	0.05 - 0.11	*

Data Plot and Equation

Caution - Use Carefully - Small Sample Size

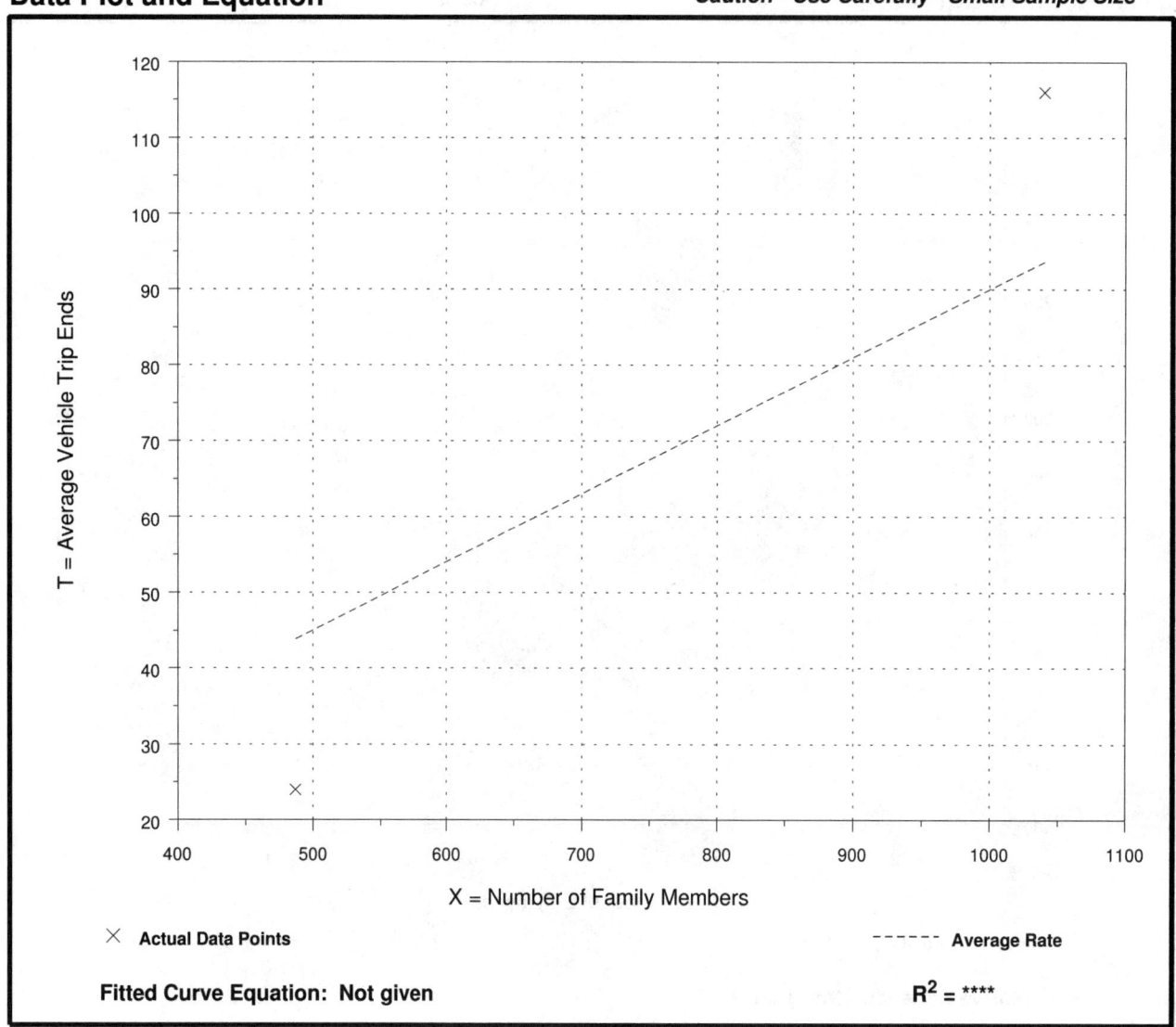

Fitted Curve Equation: Not given $R^2 = ****$

Synagogue
(561)

Average Vehicle Trip Ends vs: Family Members
On a: Sunday,
Peak Hour of Generator

Number of Studies: 2
Average Number of Family Members: 764
Directional Distribution: 49% entering, 51% exiting

Trip Generation per Family Member

Average Rate	Range of Rates	Standard Deviation
0.25	0.23 - 0.32	*

Data Plot and Equation

Caution - Use Carefully - Small Sample Size

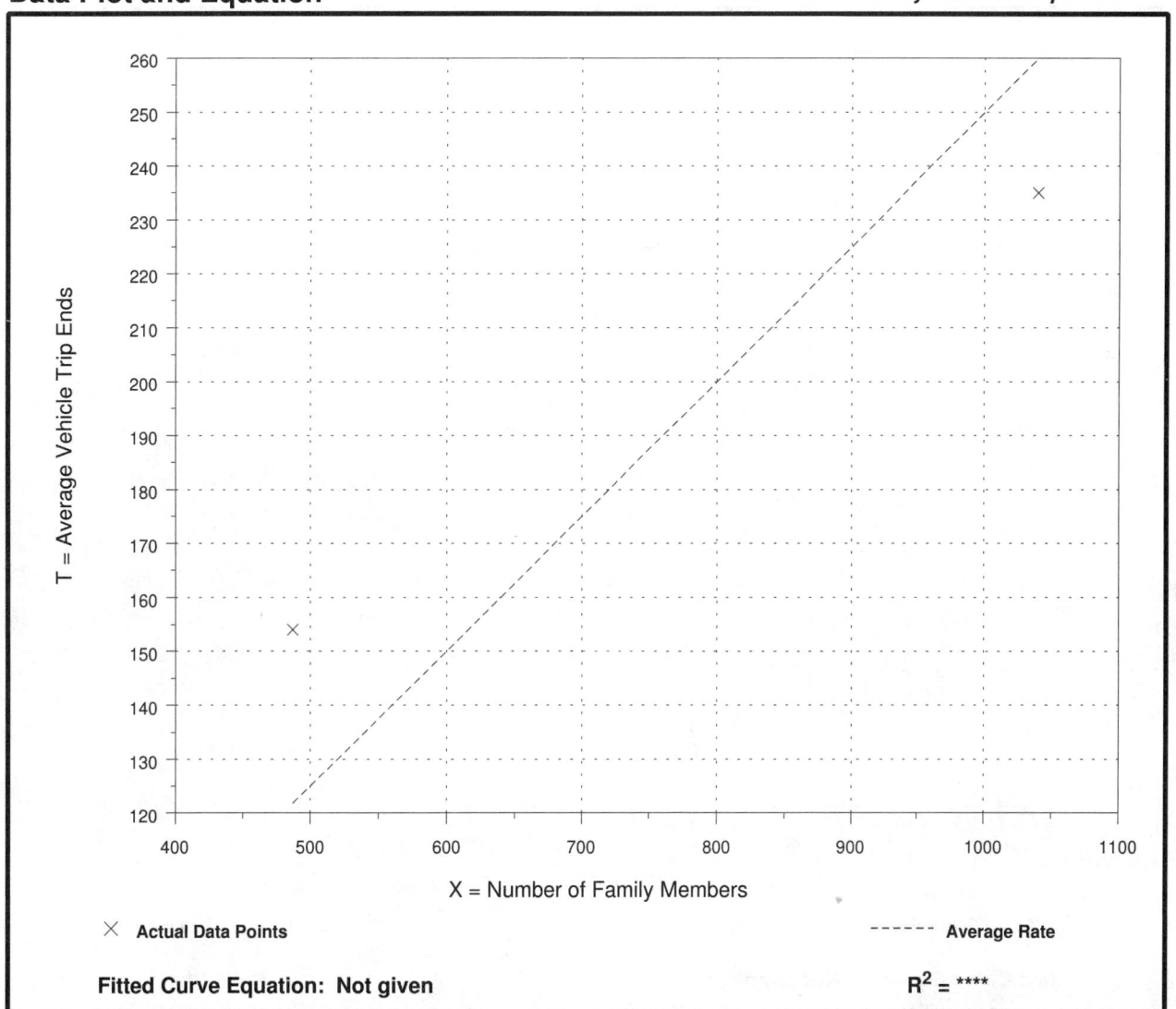

Fitted Curve Equation: Not given $R^2 = ****$

Synagogue
(561)

Average Vehicle Trip Ends vs: 1000 Sq. Feet Gross Floor Area
On a: Saturday,
Peak Hour of Generator

Number of Studies: 2
Average 1000 Sq. Feet GFA: 26
Directional Distribution: 42% entering, 58% exiting

Trip Generation per 1000 Sq. Feet Gross Floor Area

Average Rate	Range of Rates	Standard Deviation
2.73	1.12 - 3.87	*

Data Plot and Equation

Caution - Use Carefully - Small Sample Size

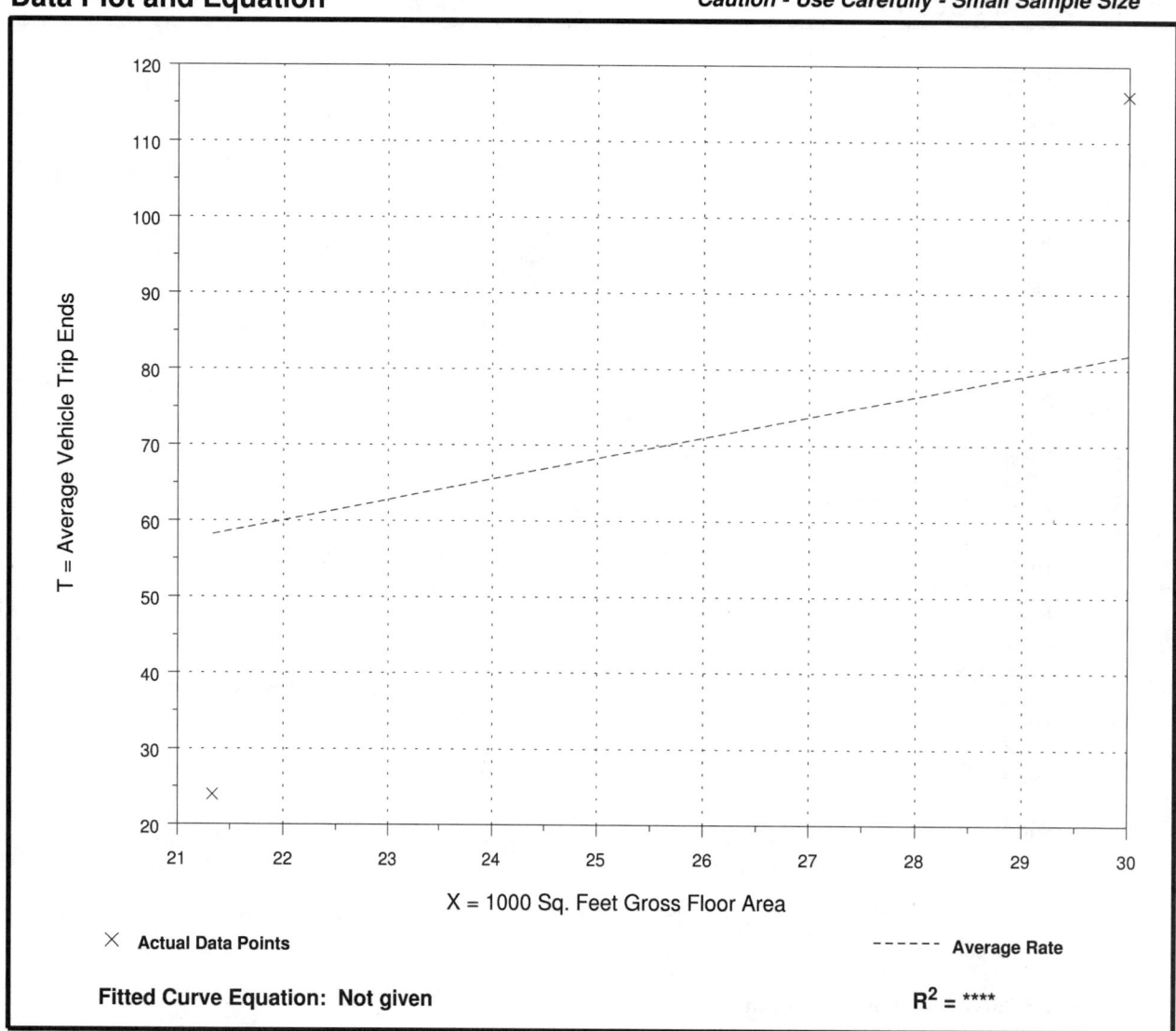

Fitted Curve Equation: Not given $R^2 = ****$

Synagogue
(561)

Average Vehicle Trip Ends vs: 1000 Sq. Feet Gross Floor Area
On a: Sunday,
Peak Hour of Generator

Number of Studies: 2
Average 1000 Sq. Feet GFA: 26
Directional Distribution: 49% entering, 51% exiting

Trip Generation per 1000 Sq. Feet Gross Floor Area

Average Rate	Range of Rates	Standard Deviation
7.58	7.22 - 7.83	*

Data Plot and Equation

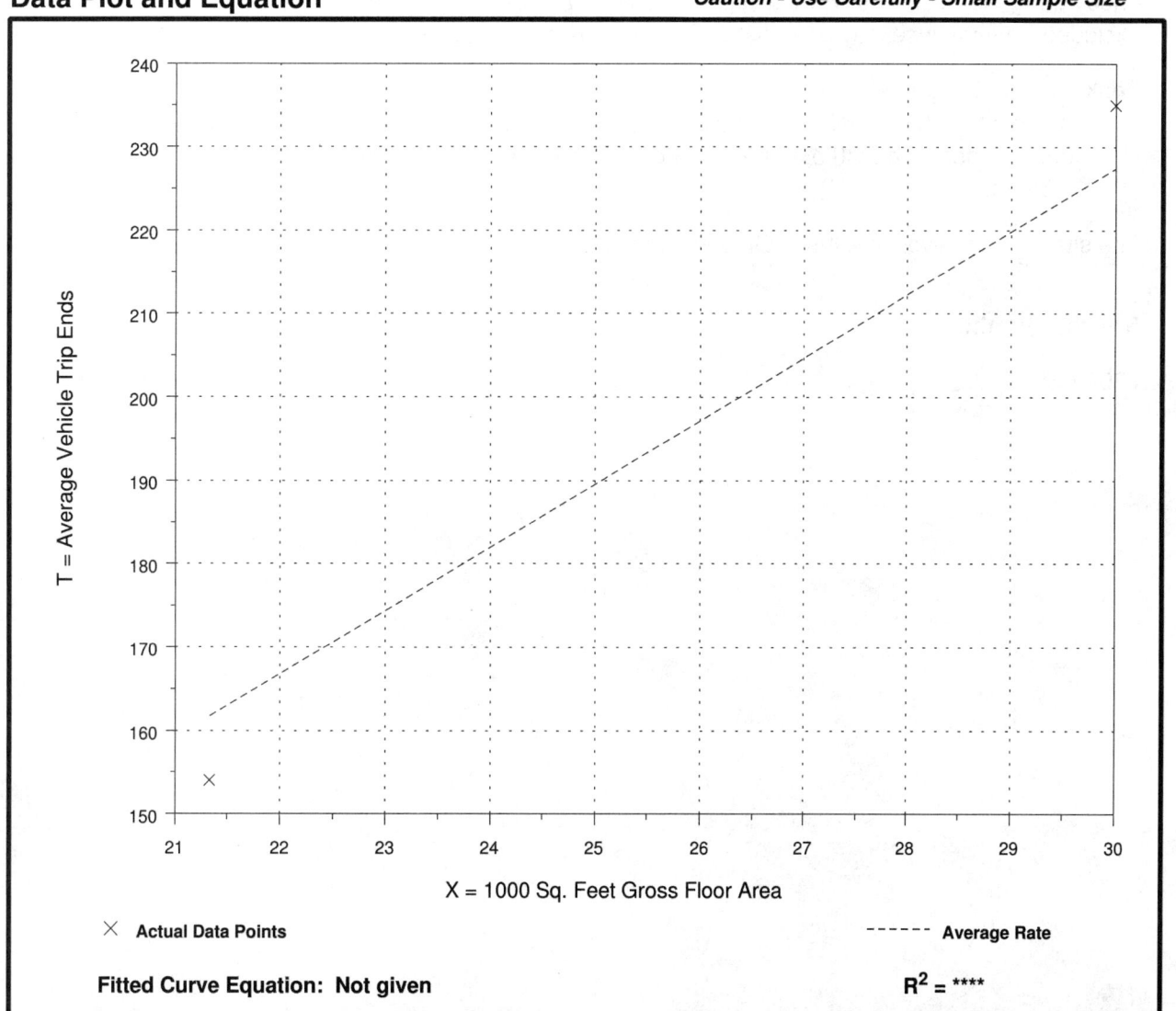

Caution - Use Carefully - Small Sample Size

Fitted Curve Equation: Not given $R^2 = ****$

Land Use: 562
Mosque

Description

A mosque is a building in which public worship services are held. A mosque houses an assembly hall and sanctuary; it may also house meeting rooms, classrooms and dining facilities. Church (Land Use 560) and synagogue (Land Use 561) are related uses.

Additional Data

Worship services are typically held on Fridays.

Friday trip generation was typically higher than other weekdays. Therefore, Friday's data have been excluded from the weekday analysis and have been independently studied.

Peak hours of the generator—
    The weekday A.M. peak hour was between 6:00 a.m. and 7:00 a.m. The weekday P.M. peak hour was between 7:30 p.m. and 8:30 p.m. The Friday peak hour was between 12:15 p.m. and 1:15 p.m.

The site was surveyed in 2009 in Ontario, Canada.

Source Number

730

Land Use: 562
Mosque

Independent Variables with One Observation

The following trip generation data are for independent variables with only one observation. This information is shown in this table only; there are no related plots for these data.

Users are cautioned to use data with care because of the small sample size.

Independent Variable	Trip Generation Rate	Size of Independent Variable	Number of Studies	Directional Distribution
1,000 Square Feet Gross Floor Area				
Weekday A.M. Peak Hour of Generator	1.63	7	1	Not available
Weekday P.M. Peak Hour of Generator	11.02	7	1	67% entering, 33% exiting
Friday Peak Hour of Generator	18.37	7	1	96% entering, 4% exiting

Land Use: 565
Day Care Center

Description

A day care center is a facility where care for pre-school age children is provided, normally during the daytime hours. Day care facilities generally include classrooms, offices, eating areas and playgrounds. Some centers also provide after-school care for school-age children.

Additional Data

Peak hours of the generator—
   The weekday A.M. and P.M. peak hours of the generator typically coincided with the peak hours of the adjacent street traffic.

Information on approximate hourly variation in day care center traffic is shown in the following table. It should be noted, however, that the information contained in this table is based on a limited sample size. Therefore, caution should be exercised when applying the data. Also, some information provided in the table may conflict with the results obtained by applying the average rate or regression equations. When this occurs, it is suggested that the results from the average rate or regression equations be used, as they are based on a larger number of studies.

| Hourly Variation in Day Care Center Traffic ||||||
| Time | Average Weekday[a] || Average Saturday[b] || Average Sunday[c] ||
	Percent of 24-Hour Entering Traffic	Percent of 24-Hour Exiting Traffic	Percent of 24-Hour Entering Traffic	Percent of 24-Hour Exiting Traffic	Percent of 24-Hour Entering Traffic	Percent of 24-Hour Exiting Traffic
6 a.m.–7 a.m.	1.4	0.8	0.0	0.0	0.0	0.0
7 a.m.–8 a.m.	15.6	11.8	0.0	0.0	0.0	0.0
8 a.m.–9 a.m.	19.4	15.8	6.0	6.5	4.5	5.0
9 a.m.–10 a.m.	6.9	7.9	2.0	2.2	9.1	0.0
10 a.m.–11 a.m.	3.0	3.0	4.0	4.3	13.6	10.0
11 a.m.–12 p.m.	5.2	4.7	14.0	13.0	4.5	0.0
12 p.m.–1 p.m.	4.0	3.3	2.0	2.2	9.1	10.0
1 p.m.–2 p.m.	2.4	2.6	16.0	8.7	0.0	10.0
2 p.m.–3 p.m.	2.4	2.7	18.0	17.4	0.0	10.0
3 p.m.–4 p.m.	5.9	5.5	2.0	2.2	4.5	10.0
4 p.m.–5 p.m.	8.4	9.1	4.0	2.2	9.1	0.0
5 p.m.–6 p.m.	15.0	17.0	12.0	2.2	18.2	15.0
6 p.m.–7 p.m.	7.8	12.6	4.0	4.3	13.6	5.0
7 p.m.–8 p.m.	1.0	1.5	0.0	2.2	4.5	5.0
8 p.m.–9 p.m.	0.4	0.5	0.0	4.3	0.0	5.0
9 p.m.–10 p.m.	0.6	0.8	4.0	0.0	9.1	5.0
10 p.m.–6 a.m.	0.6	0.5	12.0	28.3	0.0	10.0

Sites ranged in size from 3,800 to 28,000 Square Feet Gross Floor Area and from 52 to 210 students
[a] Source numbers – 208, Southern New Hampshire Planning Commission; based on three studies
[b] Source numbers – Southern New Hampshire Planning Commission; based on two studies
[c] Source numbers – Southern New Hampshire Planning Commission; based on two studies

The sites were surveyed between the mid-1980s and the 2000s throughout the United States.

Source Numbers

169, 208, 216, 253, 335, 336, 337, 355, 418, 423, 536, 550, 562, 583, 633, 734

Day Care Center
(565)

Average Vehicle Trip Ends vs: Employees
On a: Weekday

Number of Studies: 11
Avg. Number of Employees: 13
Directional Distribution: 50% entering, 50% exiting

Trip Generation per Employee

Average Rate	Range of Rates	Standard Deviation
26.73	11.08 - 50.43	11.43

Data Plot and Equation

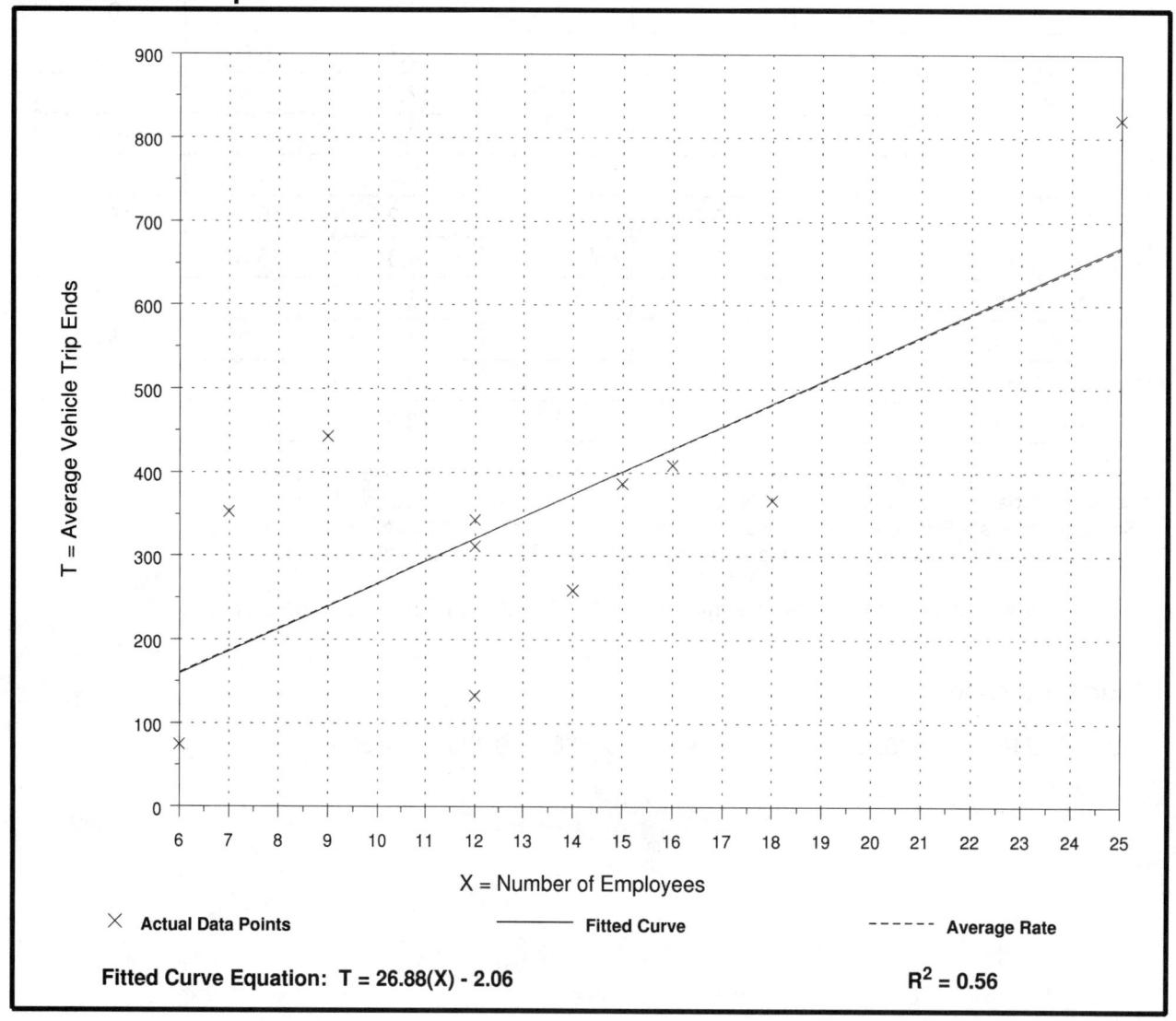

Fitted Curve Equation: T = 26.88(X) - 2.06 $R^2 = 0.56$

Day Care Center
(565)

Average Vehicle Trip Ends vs: Employees
On a: Weekday,
Peak Hour of Adjacent Street Traffic,
One Hour Between 7 and 9 a.m.

Number of Studies: 60
Avg. Number of Employees: 10
Directional Distribution: 53% entering, 47% exiting

Trip Generation per Employee

Average Rate	Range of Rates	Standard Deviation
4.85	1.83 - 16.33	3.04

Data Plot and Equation

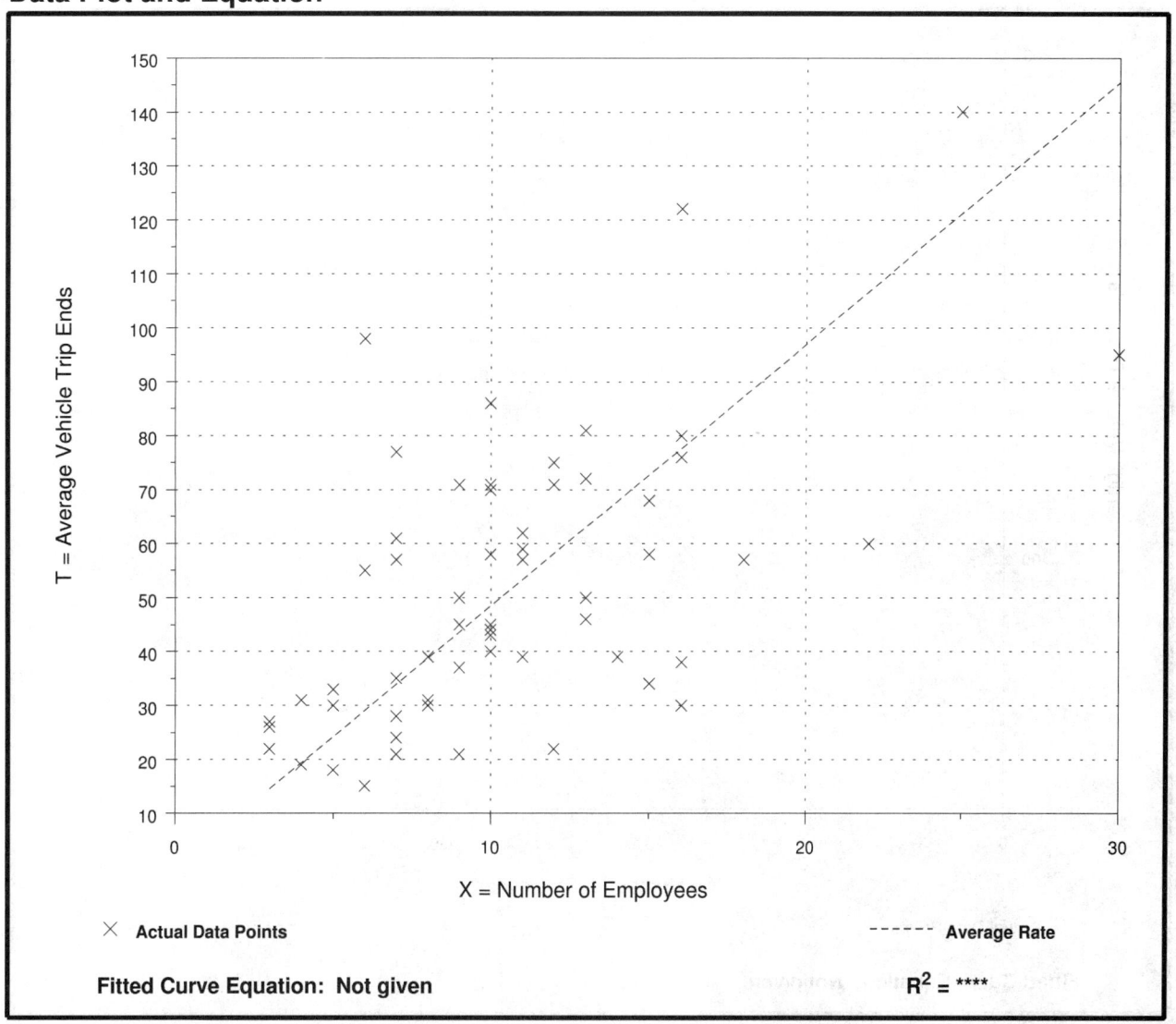

Fitted Curve Equation: Not given $R^2 = ****$

Day Care Center
(565)

Average Vehicle Trip Ends vs: Employees
On a: Weekday,
Peak Hour of Adjacent Street Traffic,
One Hour Between 4 and 6 p.m.

Number of Studies: 61
Avg. Number of Employees: 11
Directional Distribution: 47% entering, 53% exiting

Trip Generation per Employee

Average Rate	Range of Rates	Standard Deviation
4.73	0.69 - 14.00	3.20

Data Plot and Equation

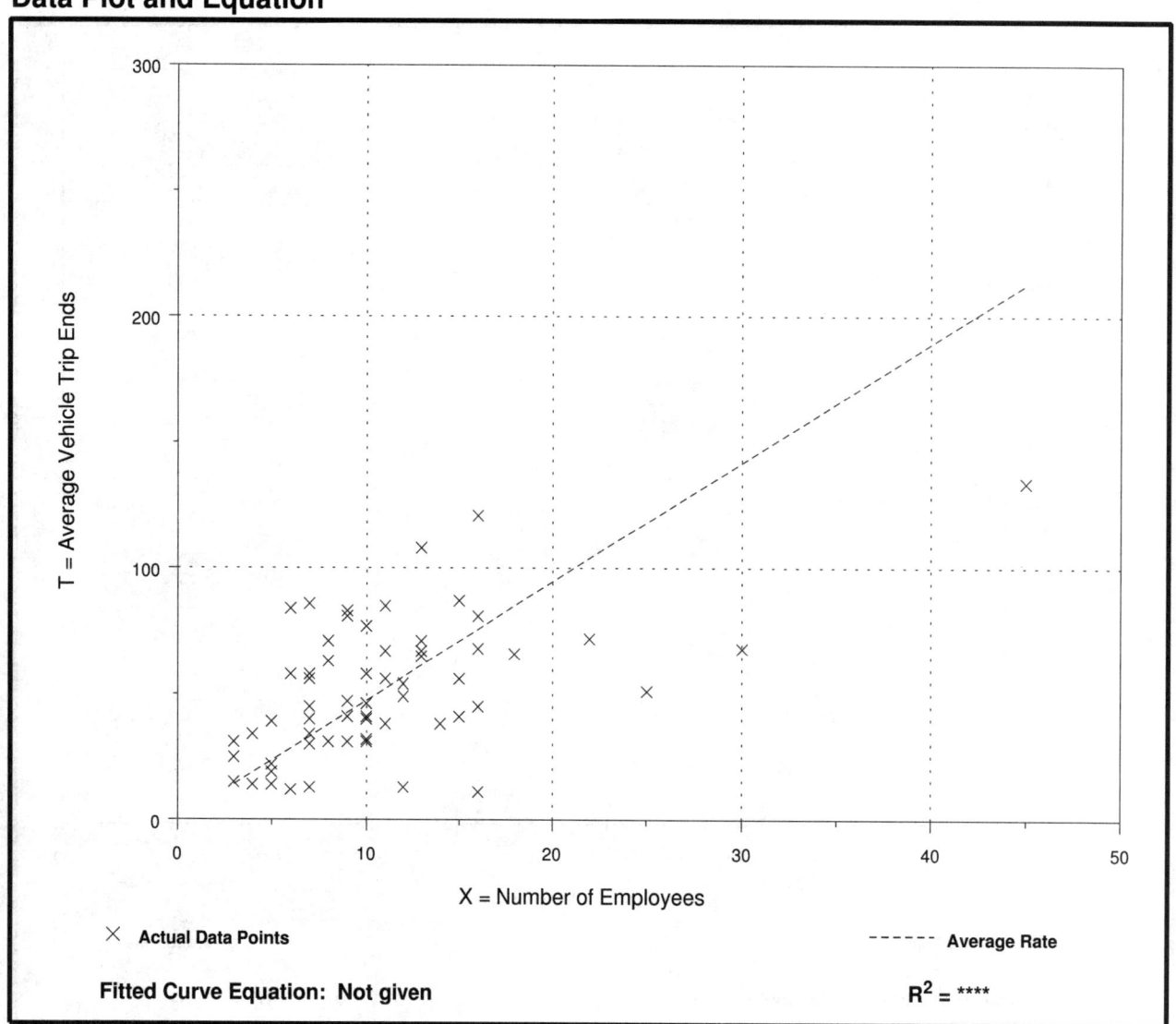

X Actual Data Points   ----- Average Rate

Fitted Curve Equation: Not given $R^2 = ****$

Day Care Center
(565)

Average Vehicle Trip Ends vs: Employees
On a: Weekday,
A.M. Peak Hour of Generator

Number of Studies: 60
Avg. Number of Employees: 10
Directional Distribution: 53% entering, 47% exiting

Trip Generation per Employee

Average Rate	Range of Rates	Standard Deviation
5.08	1.83 - 16.33	3.14

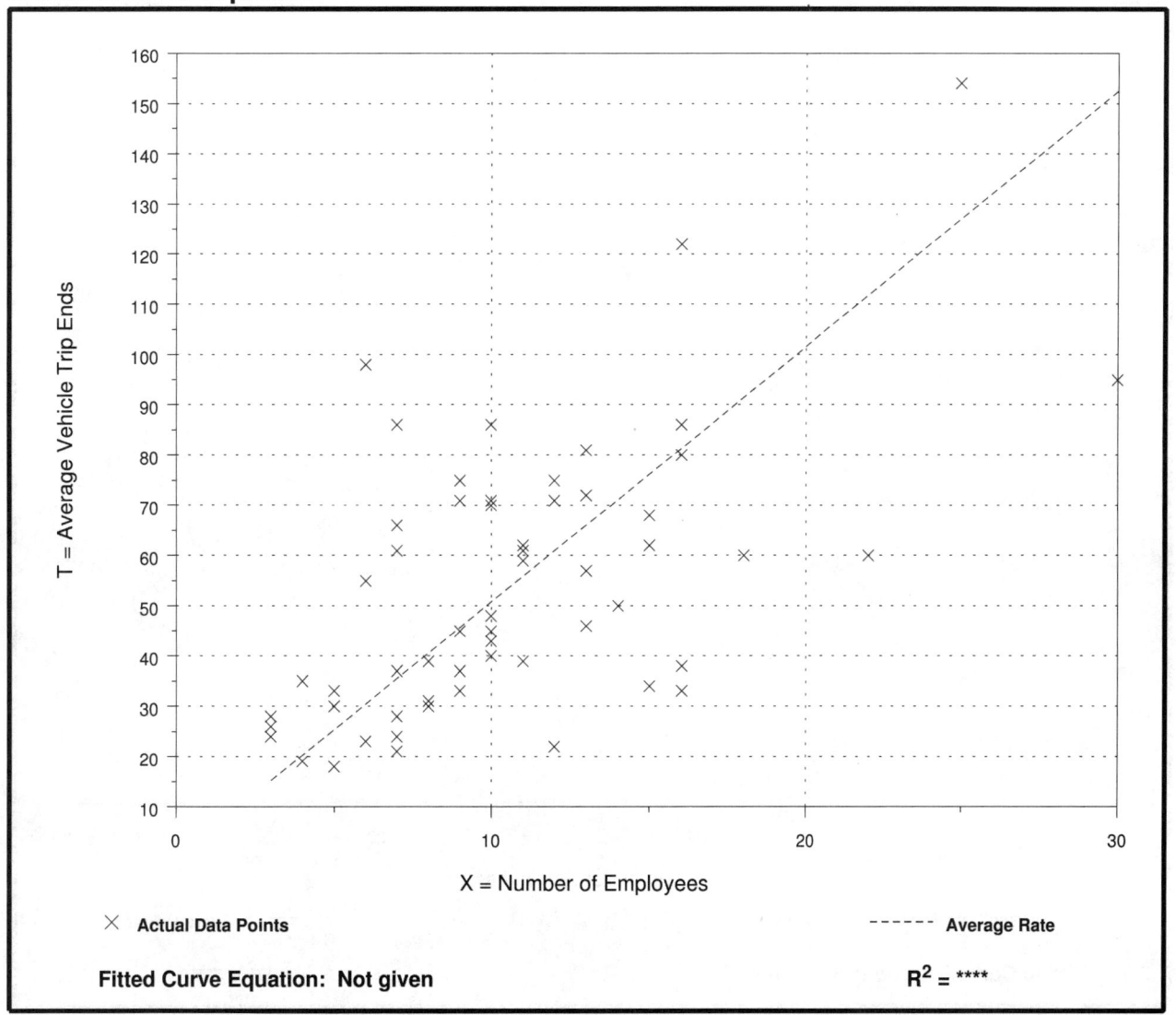

Data Plot and Equation

× Actual Data Points   ----- Average Rate

Fitted Curve Equation: Not given $R^2 = ****$

Day Care Center
(565)

Average Vehicle Trip Ends vs: Employees
On a: Weekday,
P.M. Peak Hour of Generator

Number of Studies: 60
Avg. Number of Employees: 10
Directional Distribution: 47% entering, 53% exiting

Trip Generation per Employee

Average Rate	Range of Rates	Standard Deviation
5.12	1.13 - 14.00	3.24

Data Plot and Equation

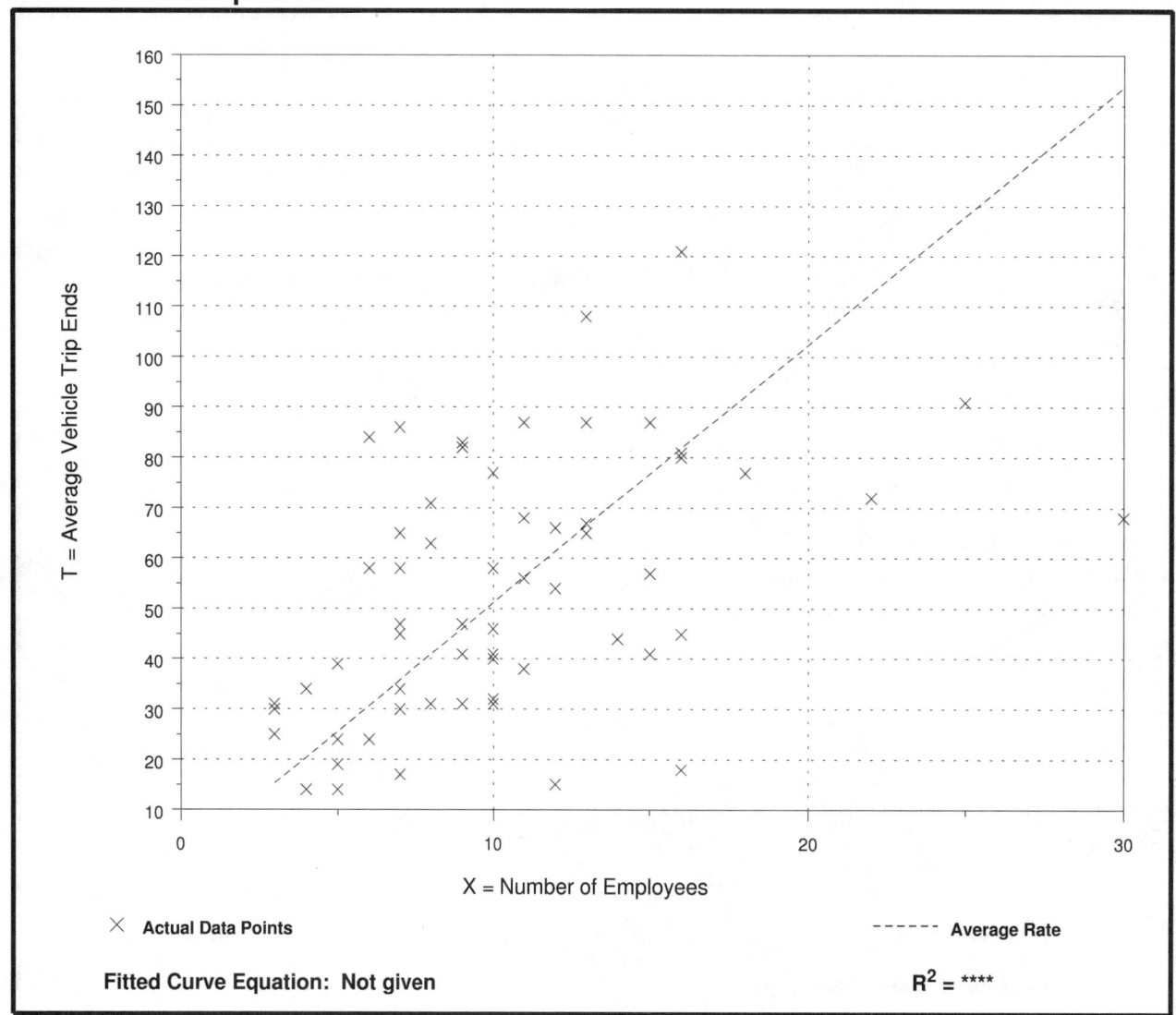

Fitted Curve Equation: Not given $R^2 = ****$

Day Care Center
(565)

Average Vehicle Trip Ends vs: Employees
On a: Saturday

Number of Studies: 5
Avg. Number of Employees: 11
Directional Distribution: 50% entering, 50% exiting

Trip Generation per Employee

Average Rate	Range of Rates	Standard Deviation
2.61	0.67 - 8.00	3.07

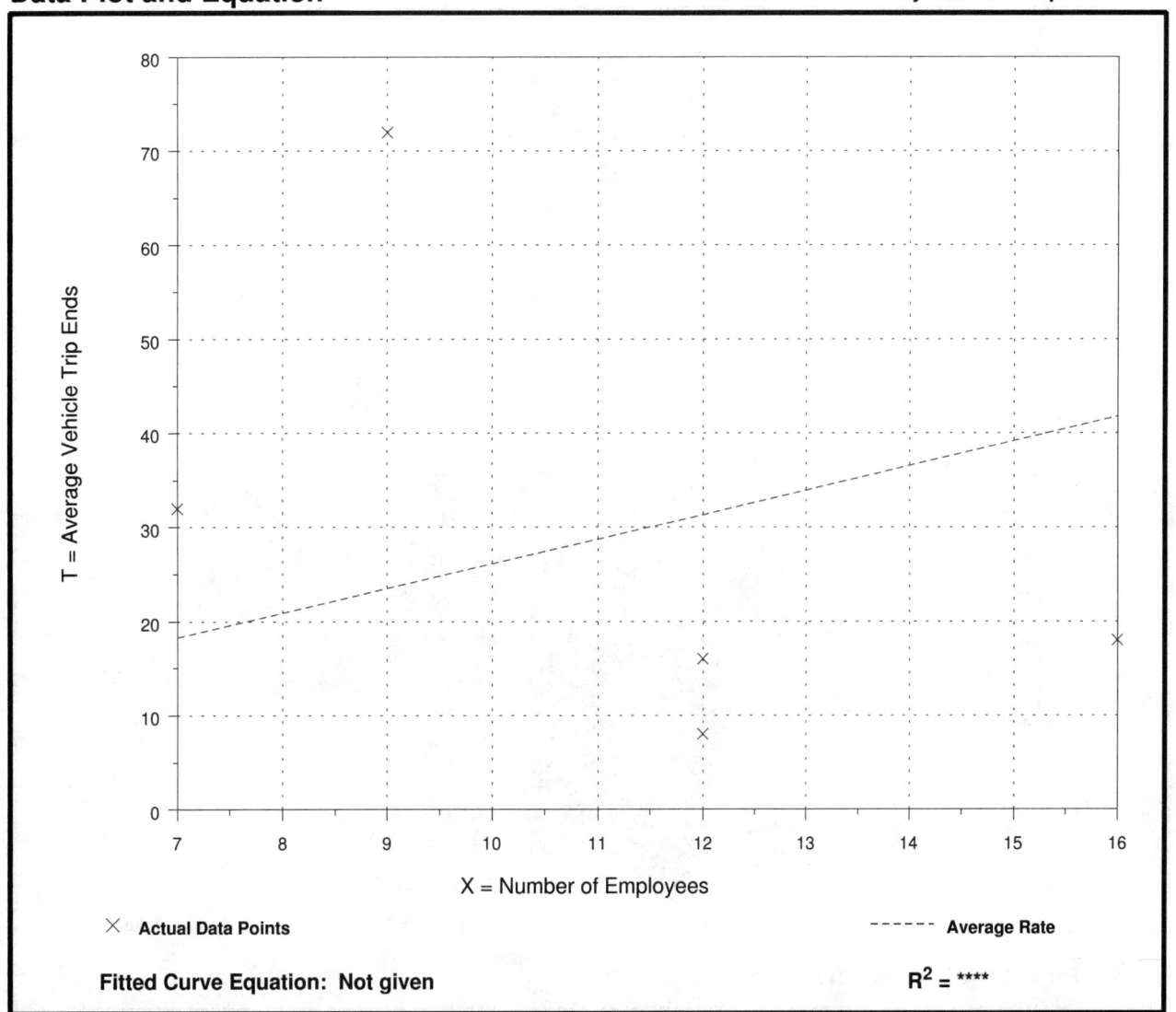

Data Plot and Equation

Caution - Use Carefully - Small Sample Size

Fitted Curve Equation: Not given $R^2 = ****$

Day Care Center
(565)

Average Vehicle Trip Ends vs: Employees
On a: Saturday,
Peak Hour of Generator

Number of Studies: 5
Avg. Number of Employees: 11
Directional Distribution: 63% entering, 37% exiting

Trip Generation per Employee

Average Rate	Range of Rates	Standard Deviation
0.71	0.17 - 2.22	1.09

Data Plot and Equation

Caution - Use Carefully - Small Sample Size

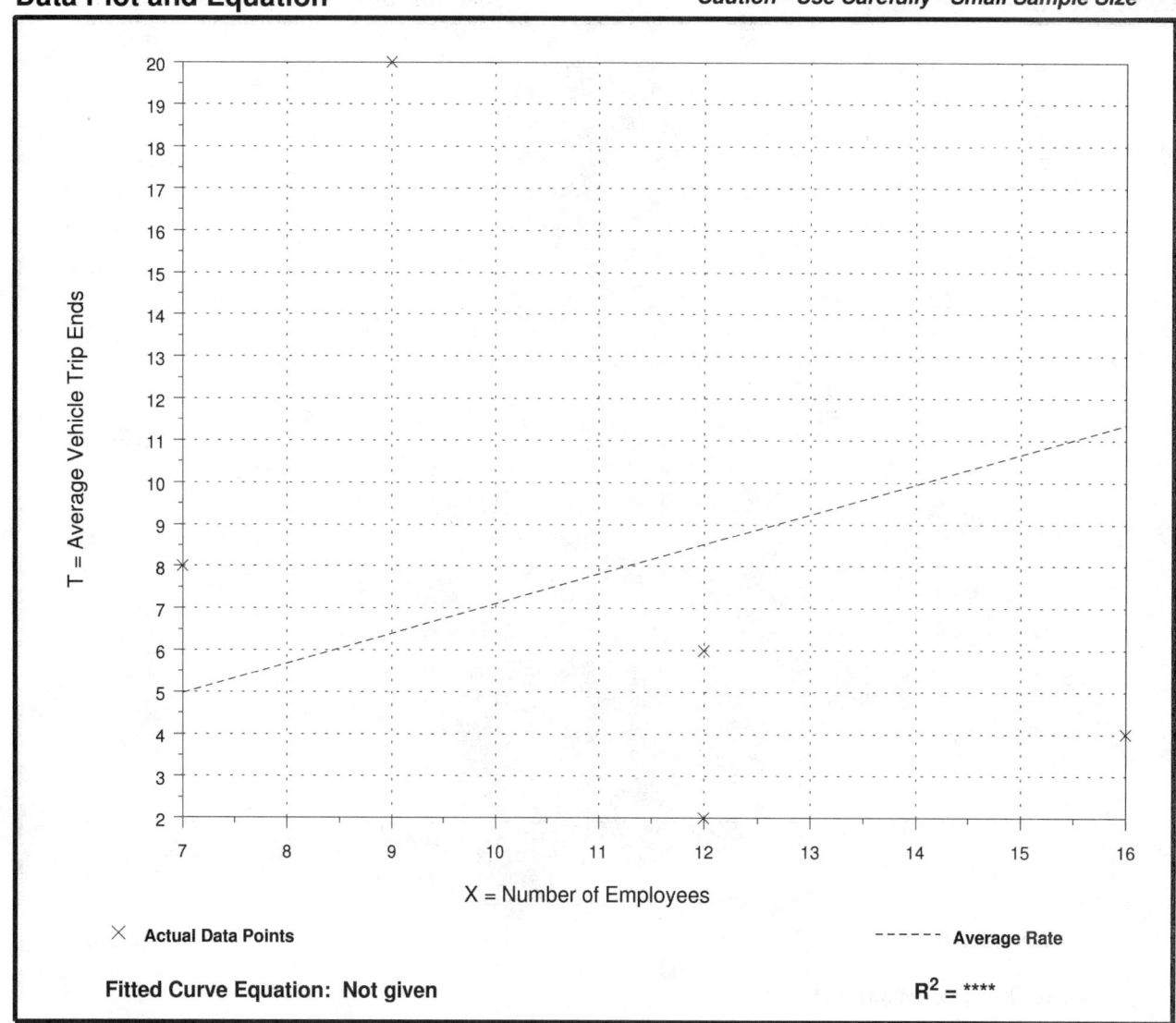

Fitted Curve Equation: Not given $R^2 = ****$

Day Care Center
(565)

Average Vehicle Trip Ends vs: Employees
On a: Sunday

Number of Studies: 5
Avg. Number of Employees: 11
Directional Distribution: 50% entering, 50% exiting

Trip Generation per Employee

Average Rate	Range of Rates	Standard Deviation
2.45	1.13 - 7.00	2.56

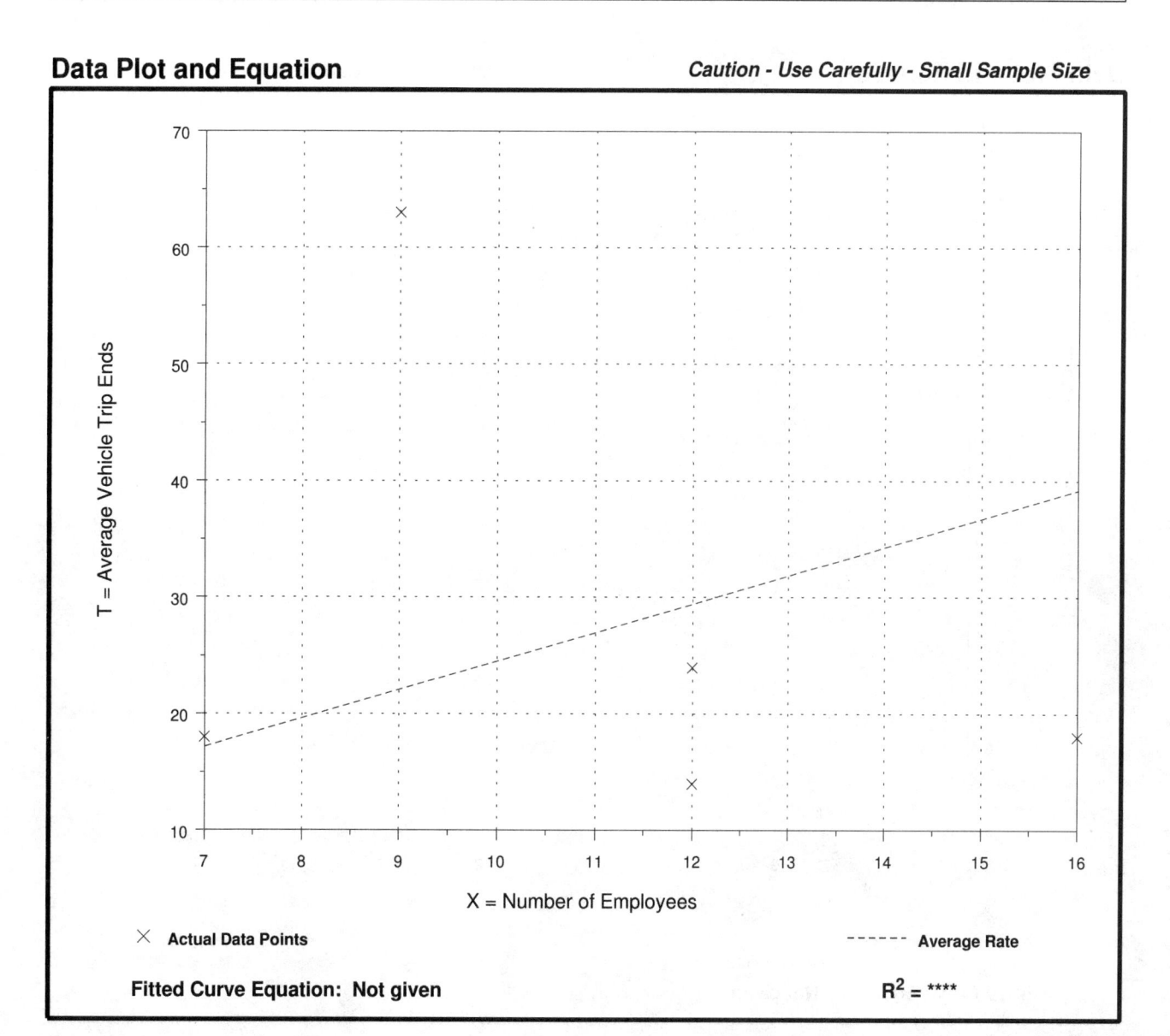

Data Plot and Equation

Caution - Use Carefully - Small Sample Size

Fitted Curve Equation: Not given $R^2 = ****$

Day Care Center
(565)

Average Vehicle Trip Ends vs: Employees
On a: Sunday,
Peak Hour of Generator

Number of Studies: 5
Avg. Number of Employees: 11
Directional Distribution: 54% entering, 46% exiting

Trip Generation per Employee

Average Rate	Range of Rates	Standard Deviation
0.73	0.38 - 1.44	0.90

Data Plot and Equation

Caution - Use Carefully - Small Sample Size

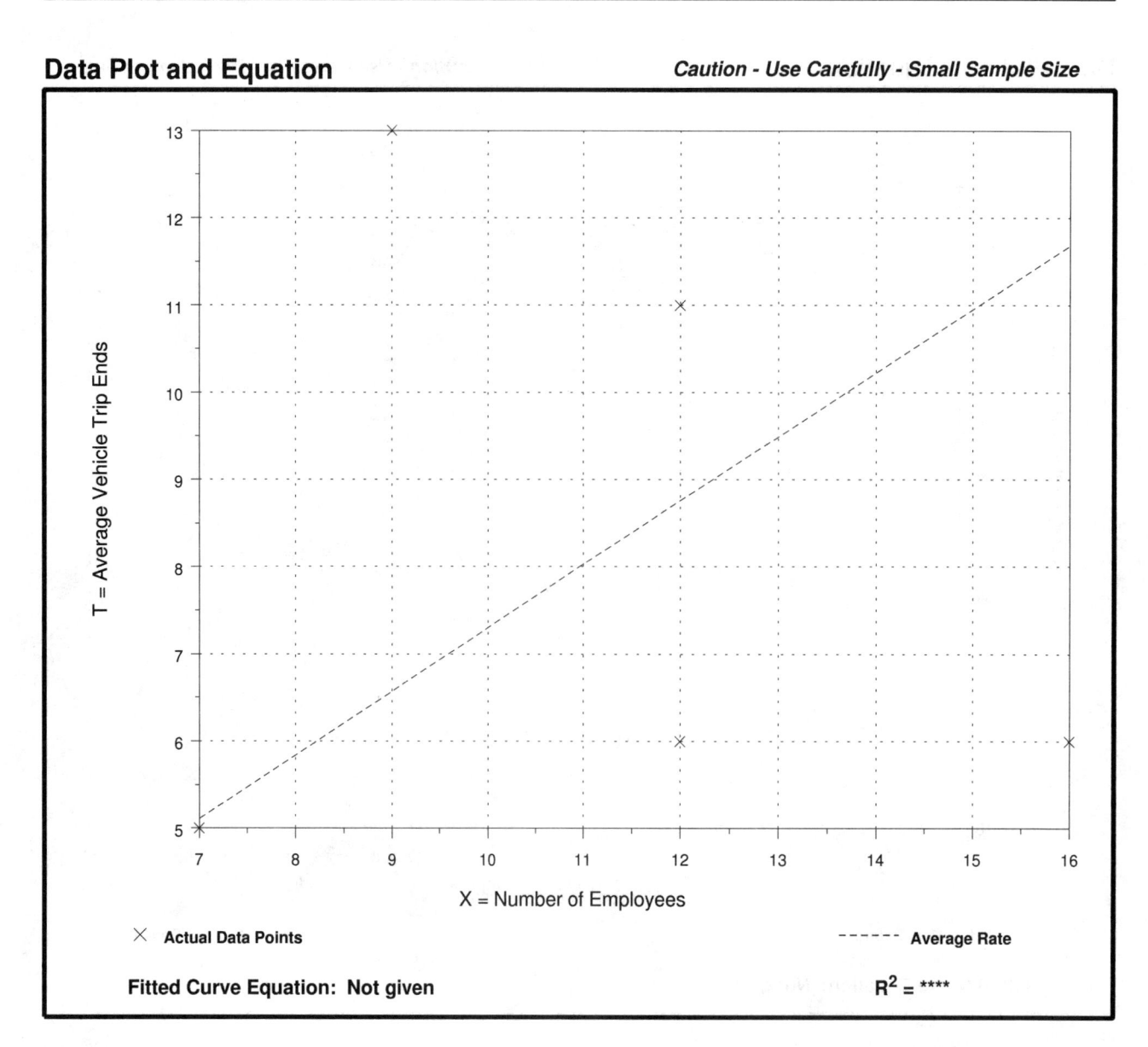

Fitted Curve Equation: Not given $R^2 = ****$

Day Care Center
(565)

Average Vehicle Trip Ends vs: 1000 Sq. Feet Gross Floor Area
On a: Weekday

Number of Studies: 7
Average 1000 Sq. Feet GFA: 5
Directional Distribution: 50% entering, 50% exiting

Trip Generation per 1000 Sq. Feet Gross Floor Area

Average Rate	Range of Rates	Standard Deviation
74.06	35.00 - 126.07	24.53

Data Plot and Equation

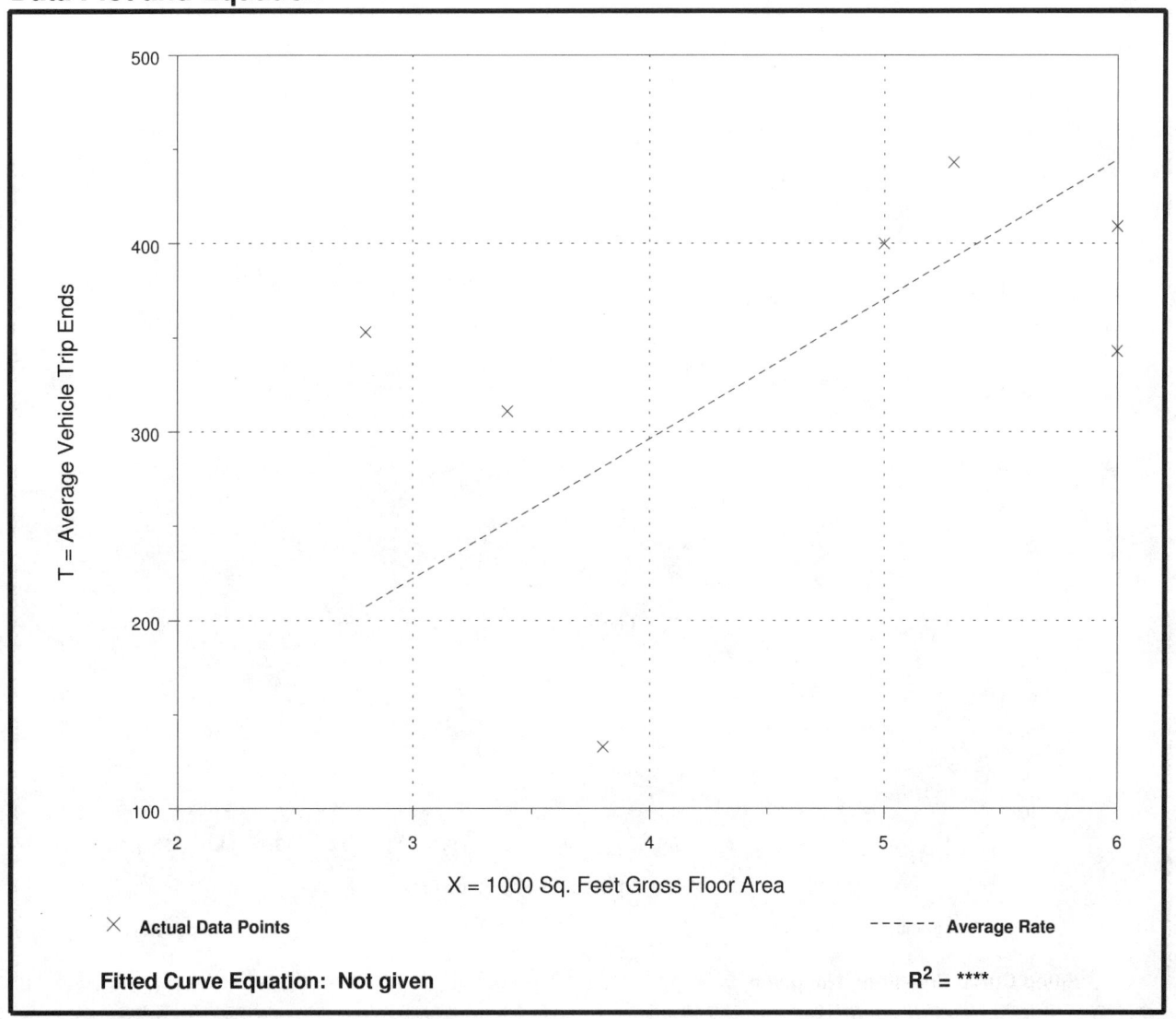

X = 1000 Sq. Feet Gross Floor Area

× Actual Data Points   ----- Average Rate

Fitted Curve Equation: Not given R^2 = ****

Day Care Center
(565)

Average Vehicle Trip Ends vs: 1000 Sq. Feet Gross Floor Area
On a: Weekday,
Peak Hour of Adjacent Street Traffic,
One Hour Between 7 and 9 a.m.

Number of Studies: 67
Average 1000 Sq. Feet GFA: 4
Directional Distribution: 53% entering, 47% exiting

Trip Generation per 1000 Sq. Feet Gross Floor Area

Average Rate	Range of Rates	Standard Deviation
12.18	4.43 - 34.92	6.40

Data Plot and Equation

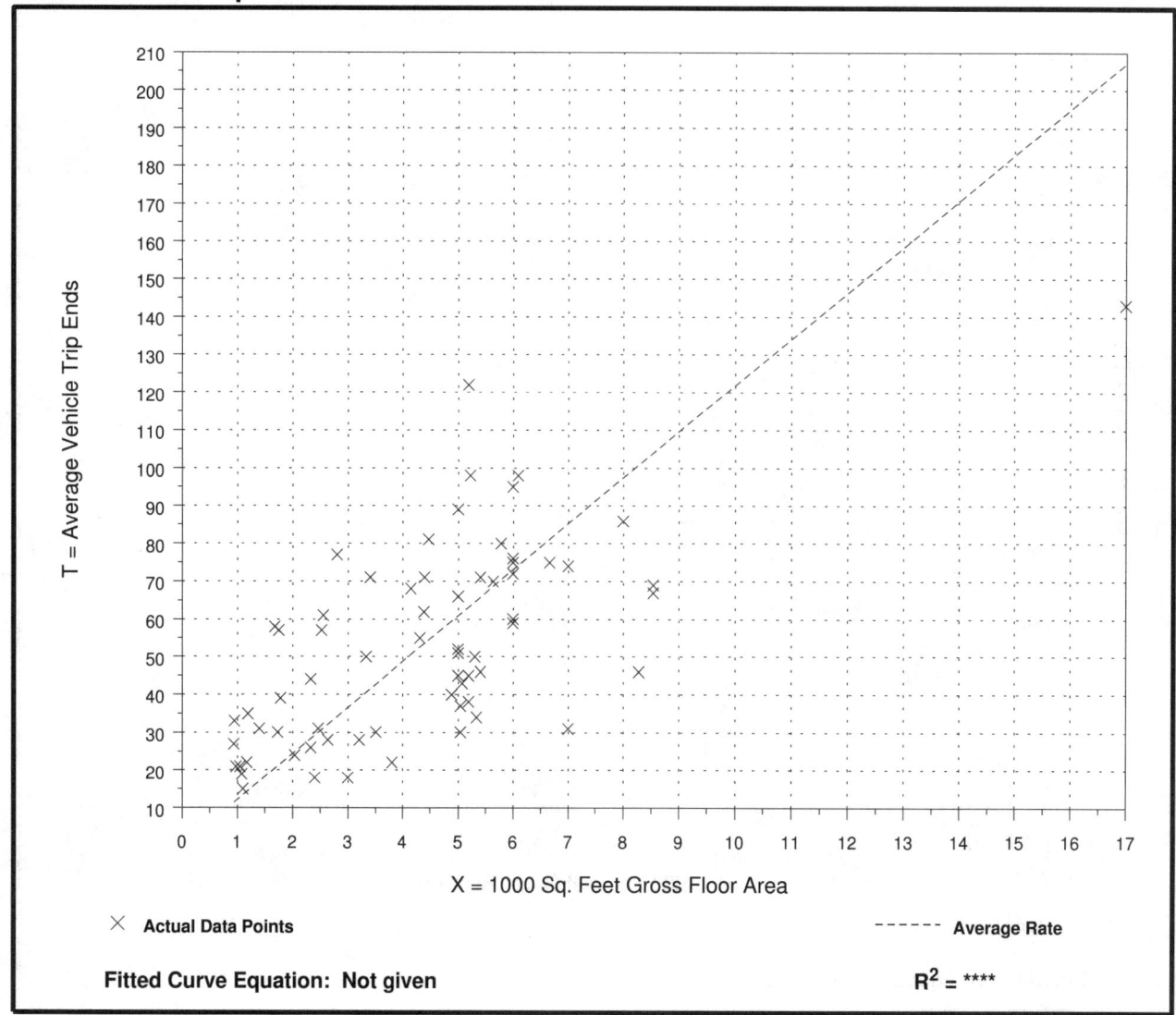

Fitted Curve Equation: Not given $R^2 = ****$

Day Care Center
(565)

Average Vehicle Trip Ends vs: 1000 Sq. Feet Gross Floor Area
On a: Weekday,
Peak Hour of Adjacent Street Traffic,
One Hour Between 4 and 6 p.m.

Number of Studies: 68
Average 1000 Sq. Feet GFA: 4
Directional Distribution: 47% entering, 53% exiting

Trip Generation per 1000 Sq. Feet Gross Floor Area

Average Rate	Range of Rates	Standard Deviation
12.34	2.66 - 33.66	6.93

Data Plot and Equation

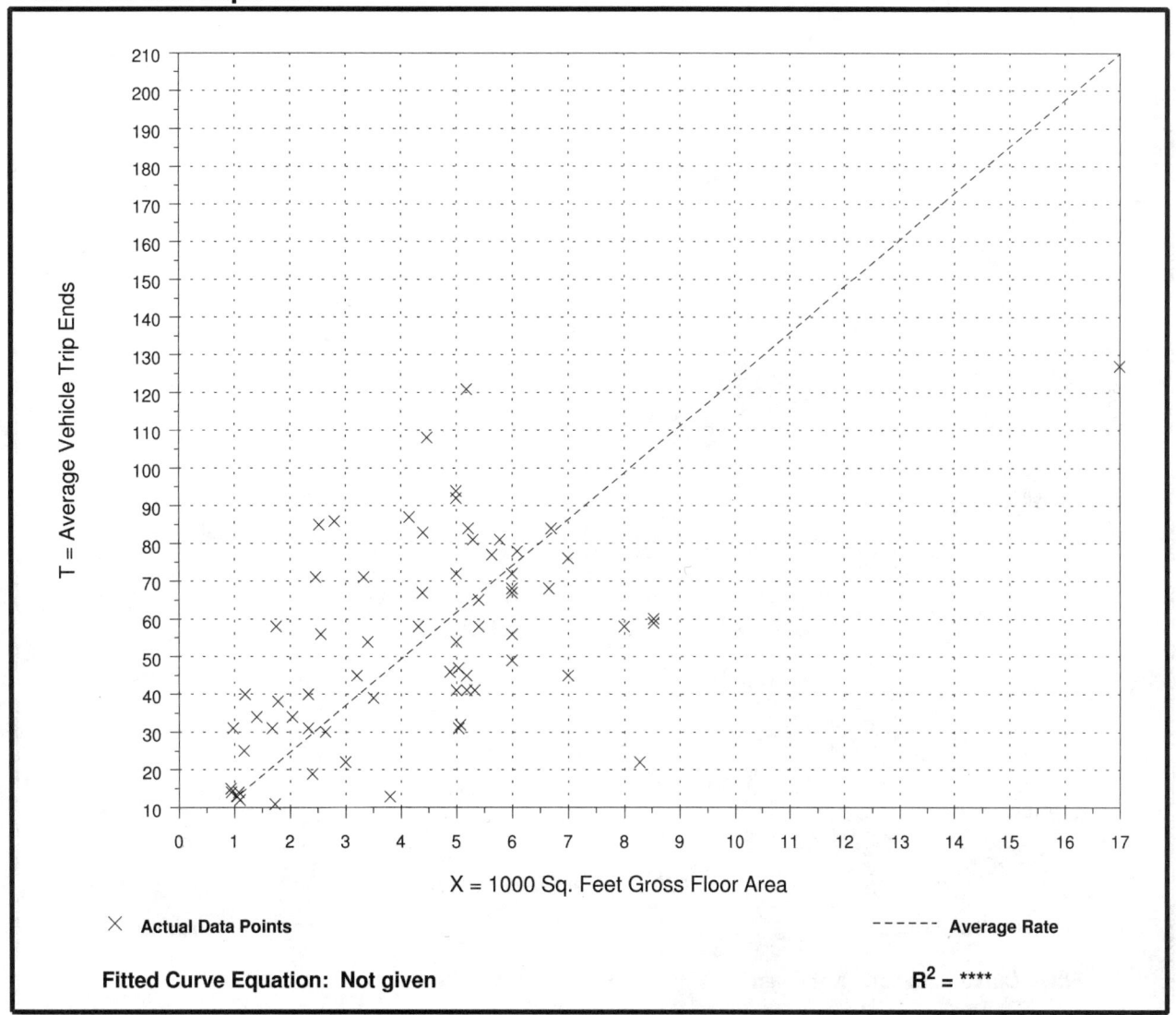

Fitted Curve Equation: Not given $R^2 = ****$

Day Care Center
(565)

Average Vehicle Trip Ends vs: 1000 Sq. Feet Gross Floor Area
On a: Weekday,
A.M. Peak Hour of Generator

Number of Studies: 62
Average 1000 Sq. Feet GFA: 4
Directional Distribution: 53% entering, 47% exiting

Trip Generation per 1000 Sq. Feet Gross Floor Area

Average Rate	Range of Rates	Standard Deviation
13.44	4.43 - 41.57	7.17

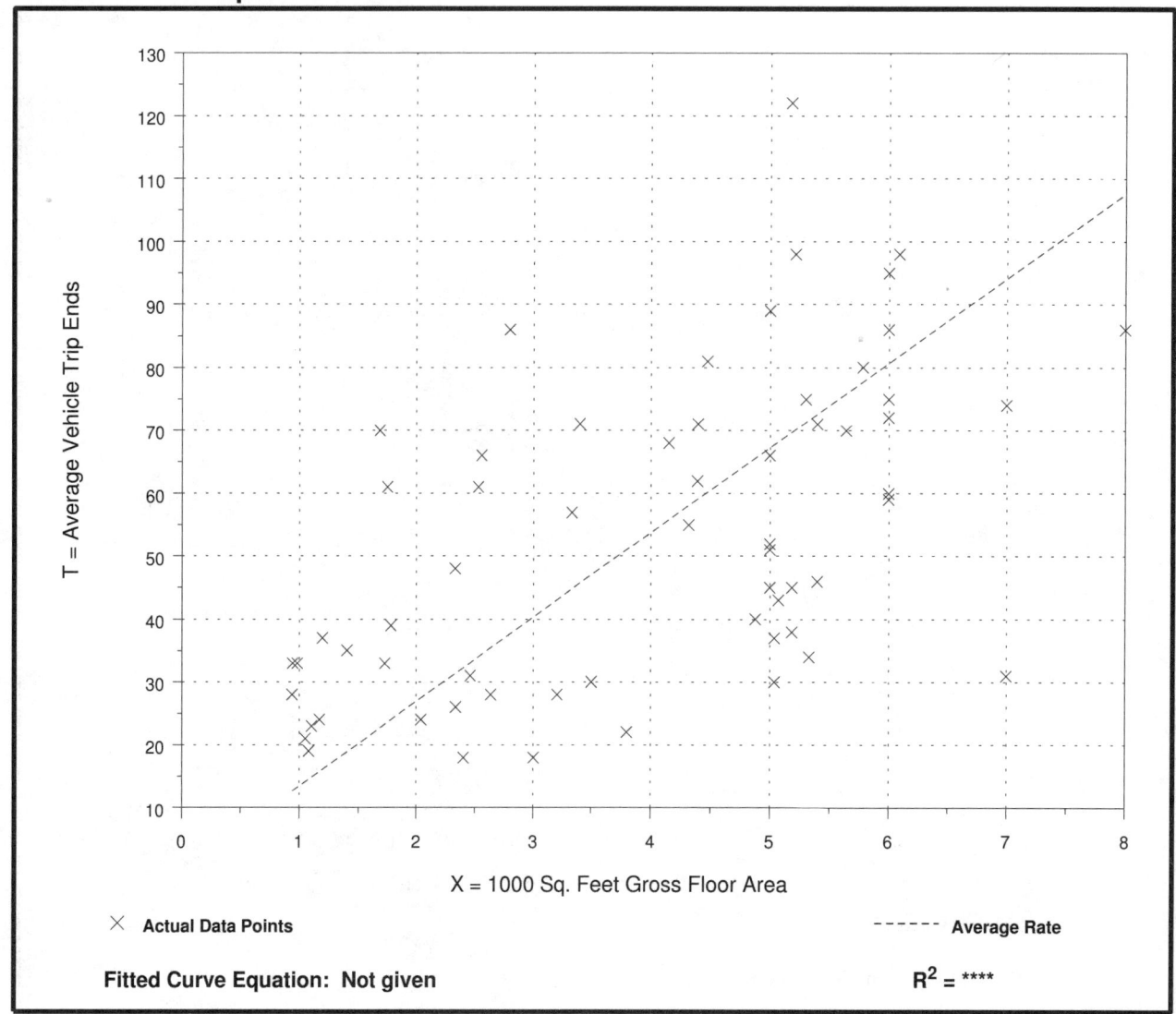

Data Plot and Equation

Fitted Curve Equation: Not given $R^2 = ****$

Day Care Center
(565)

Average Vehicle Trip Ends vs: 1000 Sq. Feet Gross Floor Area
On a: Weekday,
P.M. Peak Hour of Generator

Number of Studies: 62
Average 1000 Sq. Feet GFA: 4
Directional Distribution: 47% entering, 53% exiting

Trip Generation per 1000 Sq. Feet Gross Floor Area

Average Rate	Range of Rates	Standard Deviation
13.75	3.95 - 39.17	7.45

Data Plot and Equation

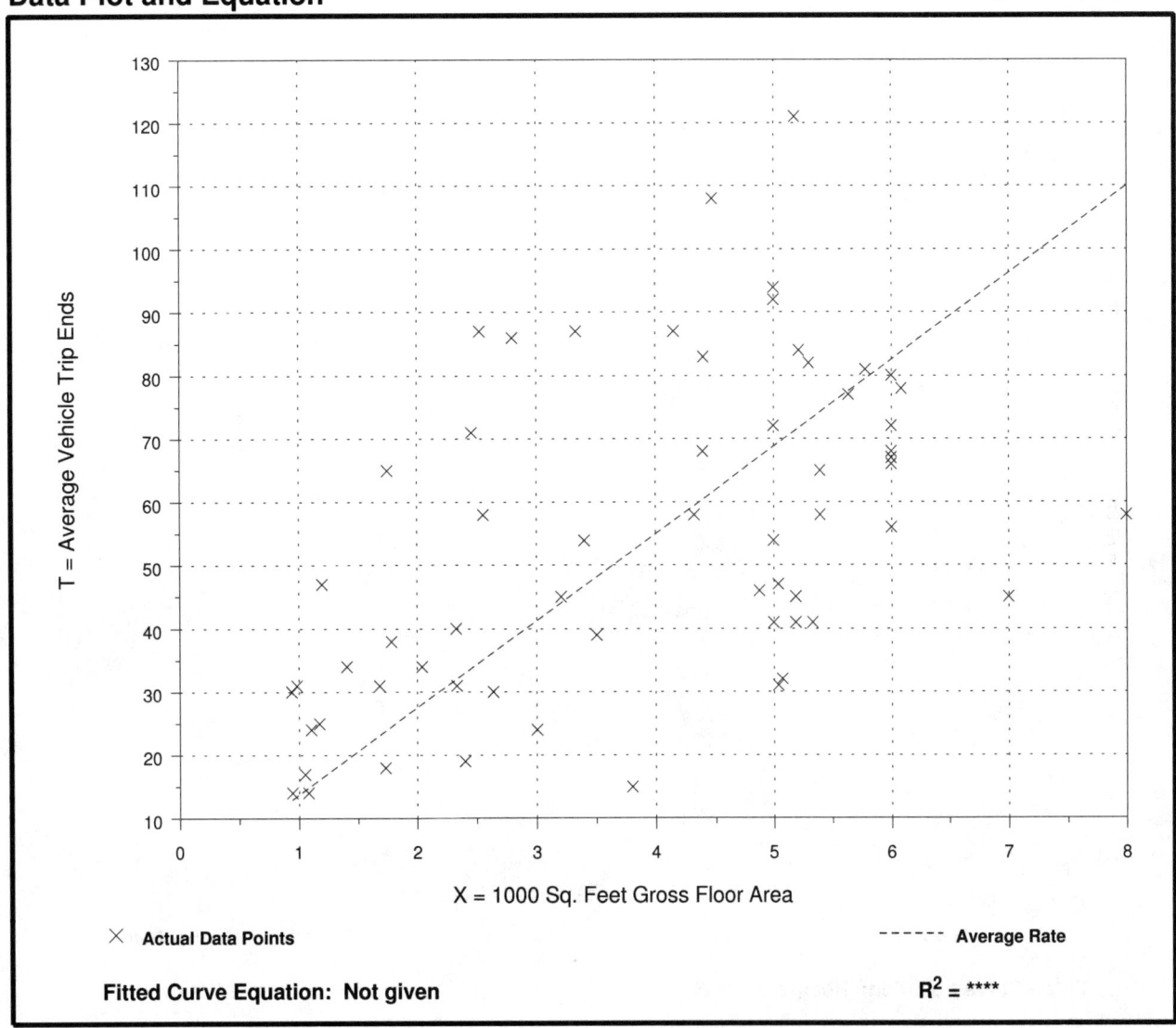

Fitted Curve Equation: Not given $R^2 = ****$

Day Care Center
(565)

Average Vehicle Trip Ends vs: 1000 Sq. Feet Gross Floor Area
On a: Saturday

Number of Studies: 5
Average 1000 Sq. Feet GFA: 5
Directional Distribution: 50% entering, 50% exiting

Trip Generation per 1000 Sq. Feet Gross Floor Area

Average Rate	Range of Rates	Standard Deviation
6.21	2.35 - 13.58	5.44

Data Plot and Equation

Caution - Use Carefully - Small Sample Size

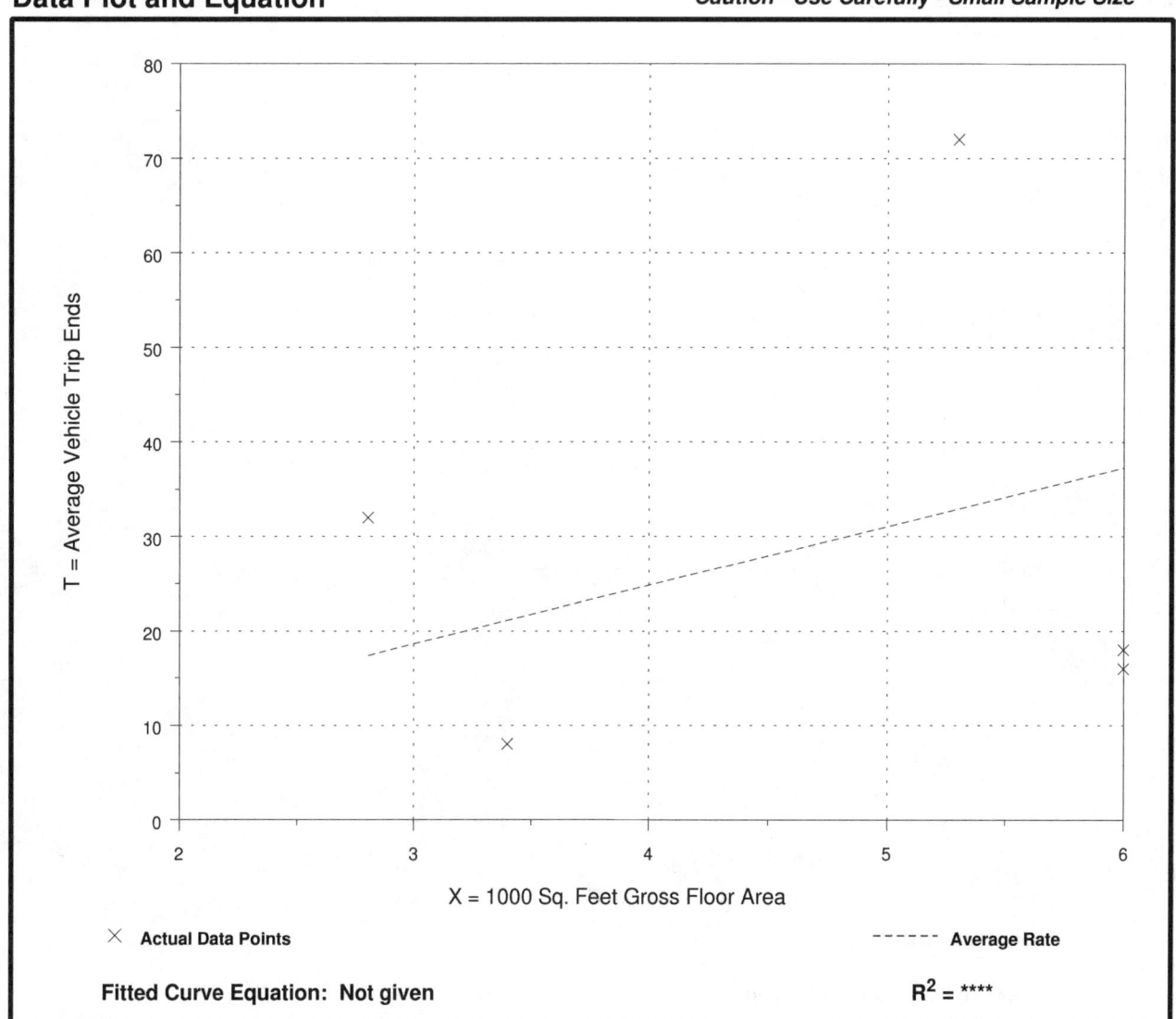

Fitted Curve Equation: Not given $R^2 = ****$

Day Care Center
(565)

Average Vehicle Trip Ends vs: 1000 Sq. Feet Gross Floor Area
On a: Saturday,
Peak Hour of Generator

Number of Studies: 5
Average 1000 Sq. Feet GFA: 5
Directional Distribution: 63% entering, 37% exiting

Trip Generation per 1000 Sq. Feet Gross Floor Area

Average Rate	Range of Rates	Standard Deviation
1.70	0.59 - 3.77	1.78

Data Plot and Equation

Caution - Use Carefully - Small Sample Size

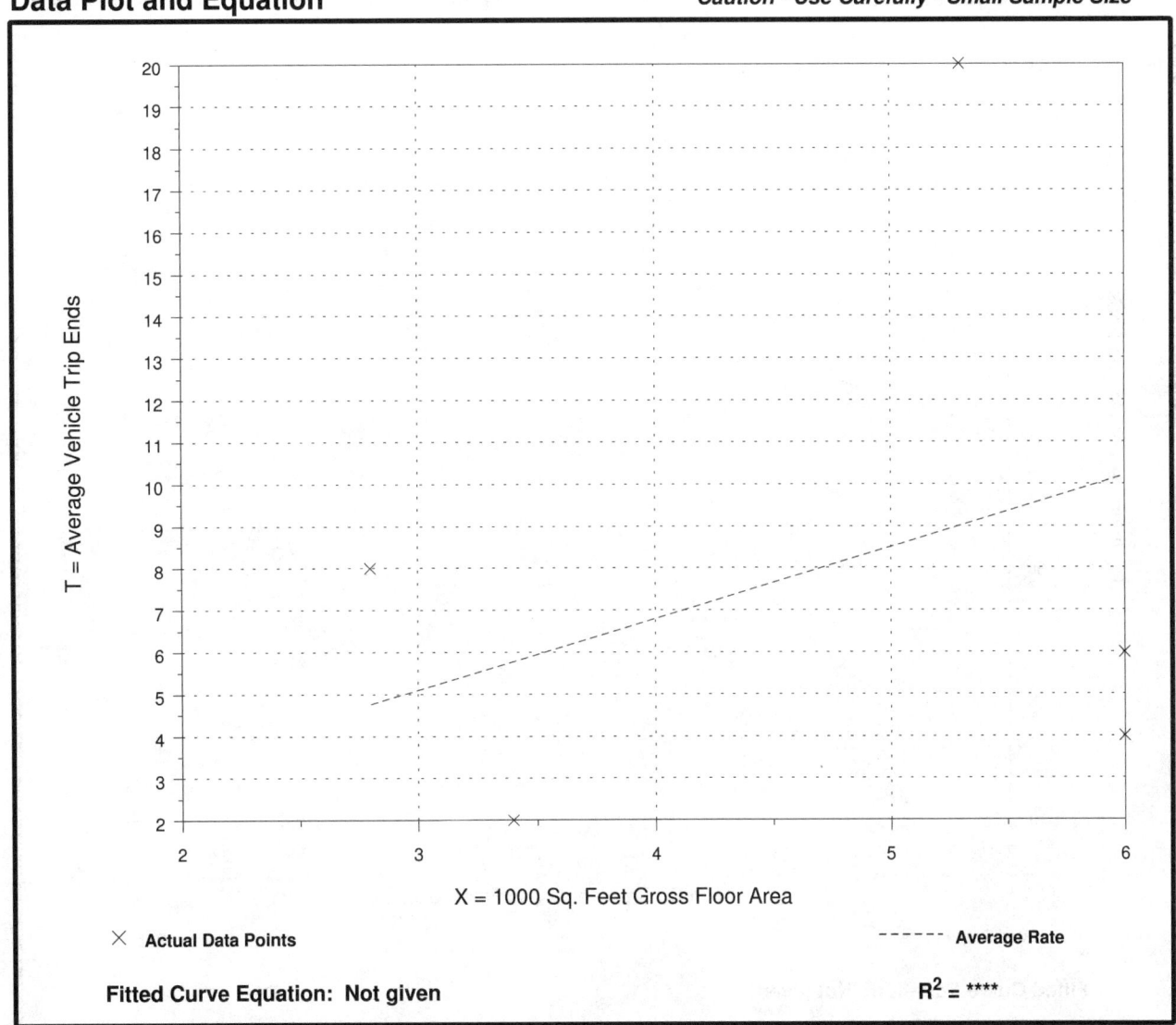

× Actual Data Points   ----- Average Rate

Fitted Curve Equation: Not given $R^2 =$ ****

Day Care Center
(565)

Average Vehicle Trip Ends vs: 1000 Sq. Feet Gross Floor Area
On a: Sunday

Number of Studies: 5
Average 1000 Sq. Feet GFA: 5
Directional Distribution: 50% entering, 50% exiting

Trip Generation per 1000 Sq. Feet Gross Floor Area

Average Rate	Range of Rates	Standard Deviation
5.83	3.00 - 11.89	4.12

Data Plot and Equation

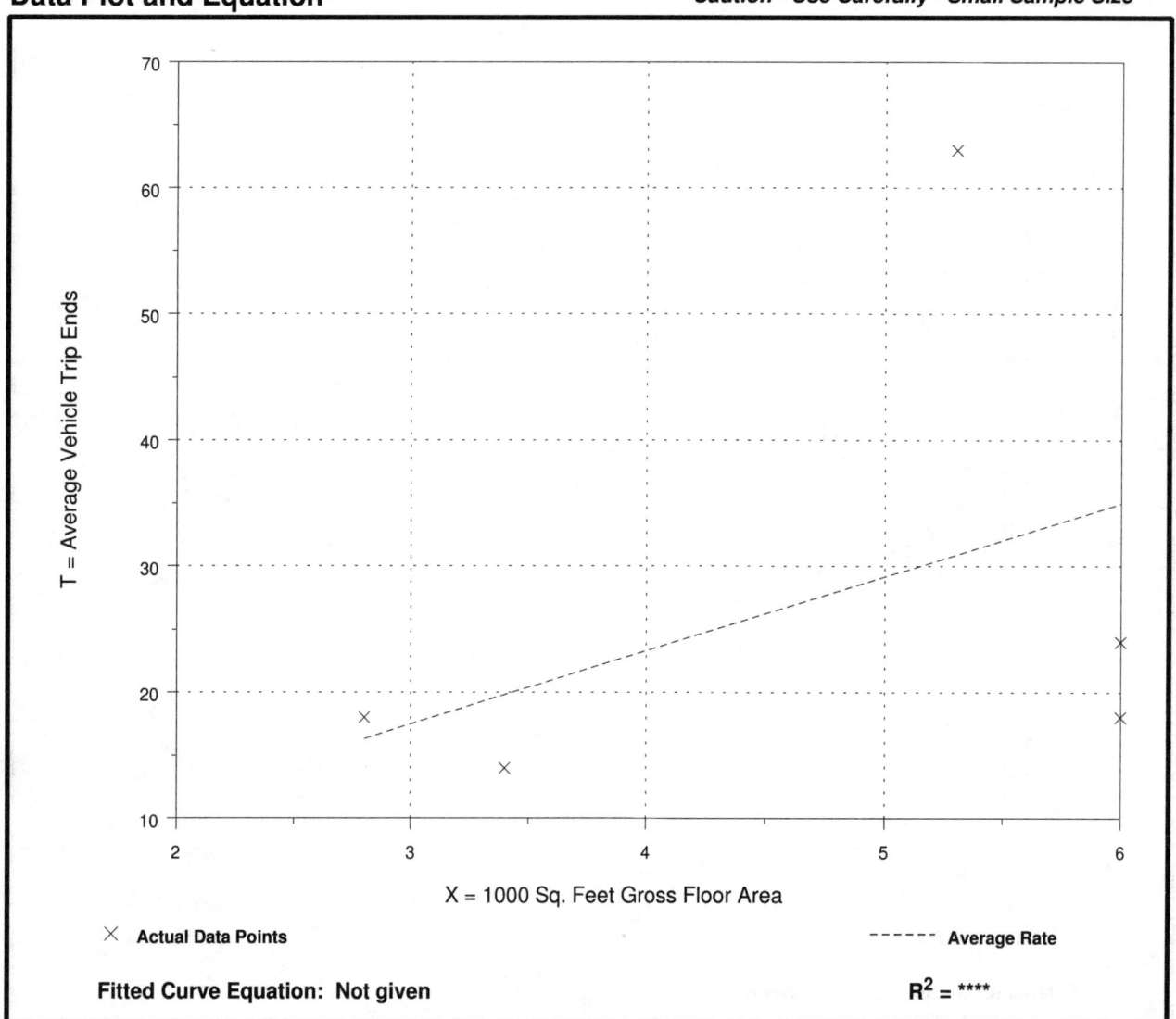

Caution - Use Carefully - Small Sample Size

Fitted Curve Equation: Not given $R^2 = ****$

1132   *Trip Generation*, 9th Edition • Institute of Transportation Engineers

Day Care Center
(565)

Average Vehicle Trip Ends vs: 1000 Sq. Feet Gross Floor Area
On a: Sunday,
Peak Hour of Generator

Number of Studies: 5
Average 1000 Sq. Feet GFA: 5
Directional Distribution: 54% entering, 46% exiting

Trip Generation per 1000 Sq. Feet Gross Floor Area

Average Rate	Range of Rates	Standard Deviation
1.74	1.00 - 2.45	1.30

Data Plot and Equation

Caution - Use Carefully - Small Sample Size

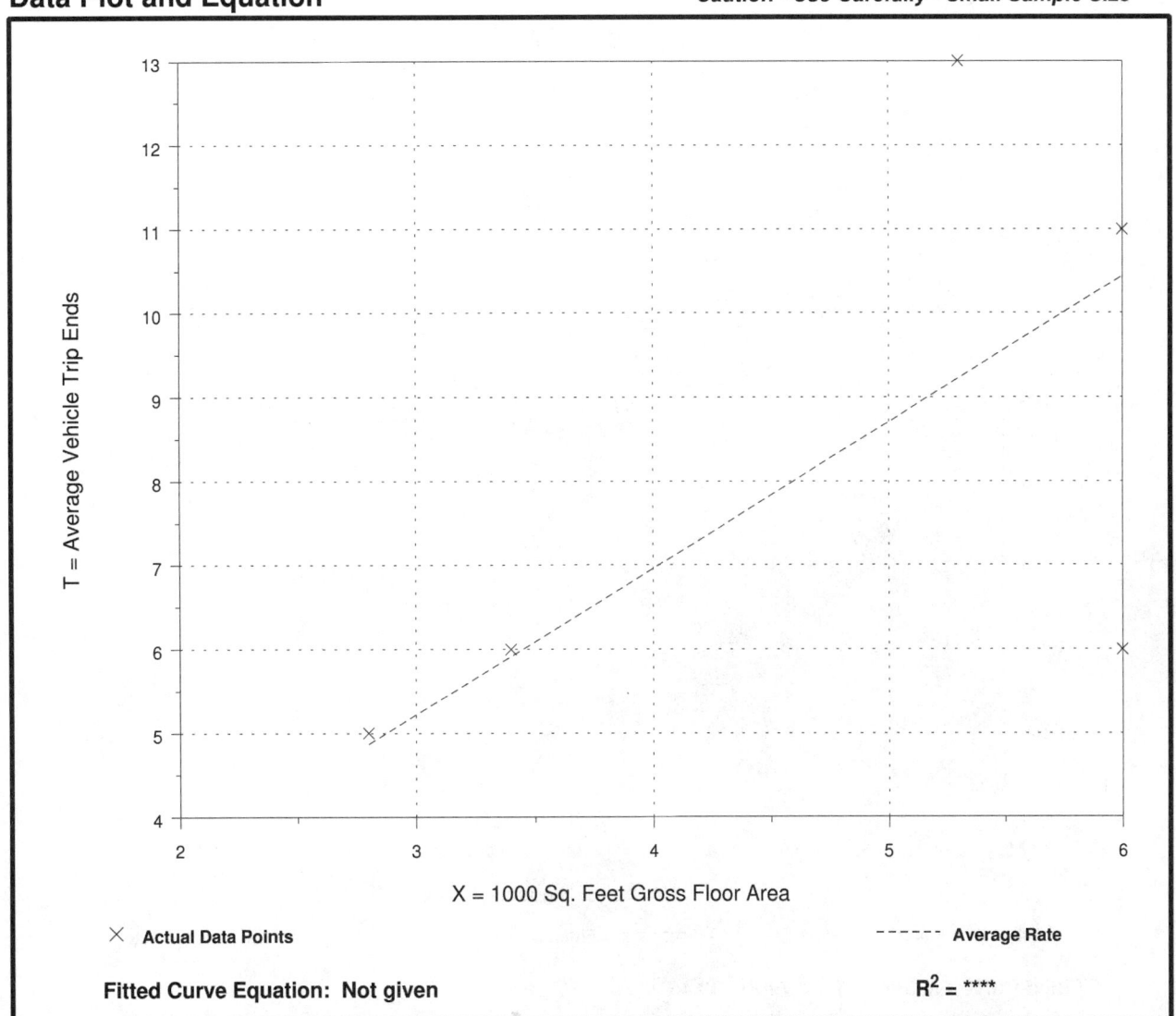

Fitted Curve Equation: Not given $R^2 = ****$

Day Care Center
(565)

Average Vehicle Trip Ends vs: Students
On a: Weekday

Number of Studies: 12
Average Number of Students: 82
Directional Distribution: 50% entering, 50% exiting

Trip Generation per Student

Average Rate	Range of Rates	Standard Deviation
4.38	2.50 - 7.06	2.37

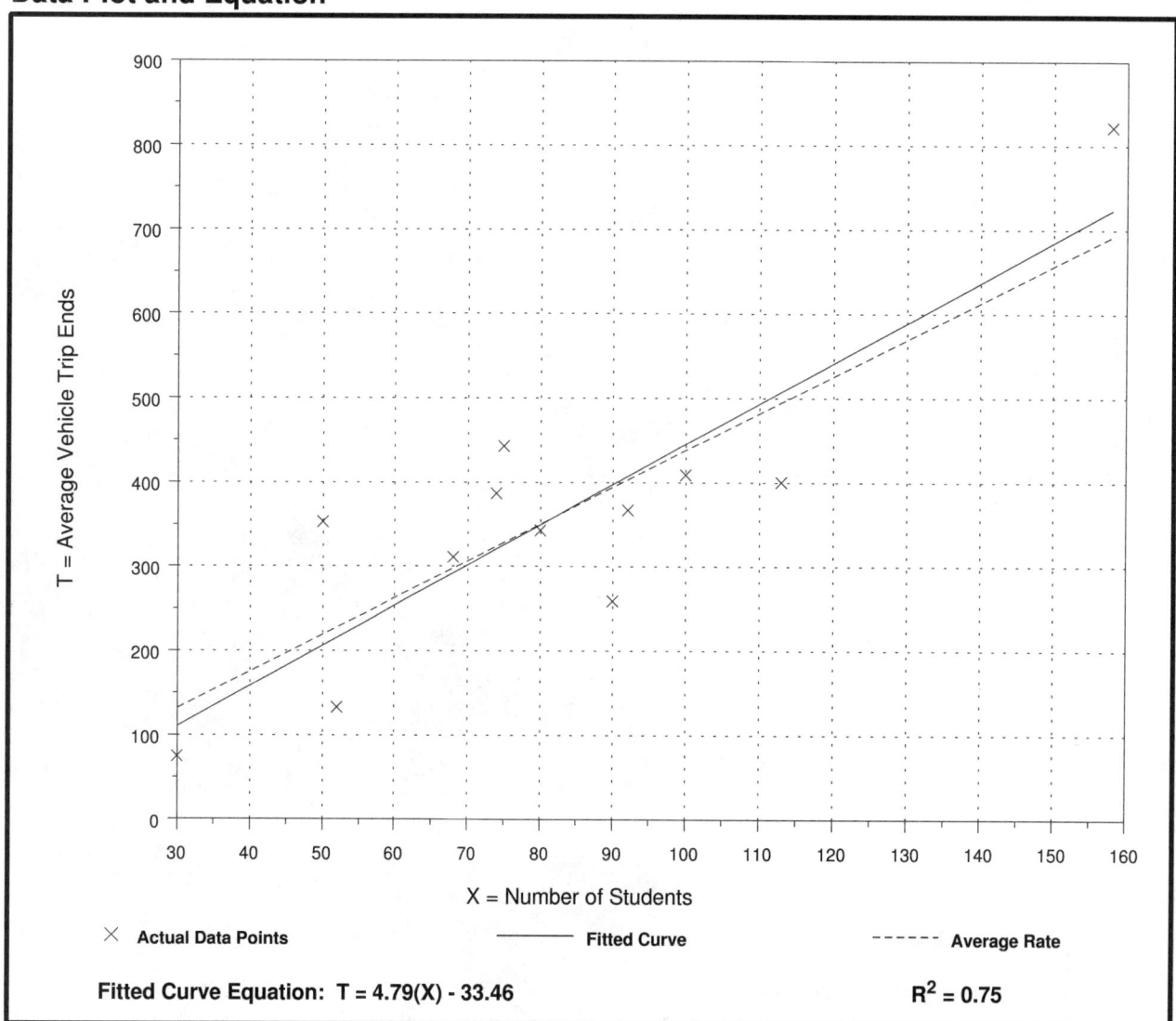

Data Plot and Equation

Fitted Curve Equation: $T = 4.79(X) - 33.46$ $R^2 = 0.75$

Day Care Center
(565)

Average Vehicle Trip Ends vs: Students
On a: Weekday,
Peak Hour of Adjacent Street Traffic,
One Hour Between 7 and 9 a.m.

Number of Studies: 71
Average Number of Students: 67
Directional Distribution: 53% entering, 47% exiting

Trip Generation per Student

Average Rate	Range of Rates	Standard Deviation
0.80	0.39 - 1.78	0.92

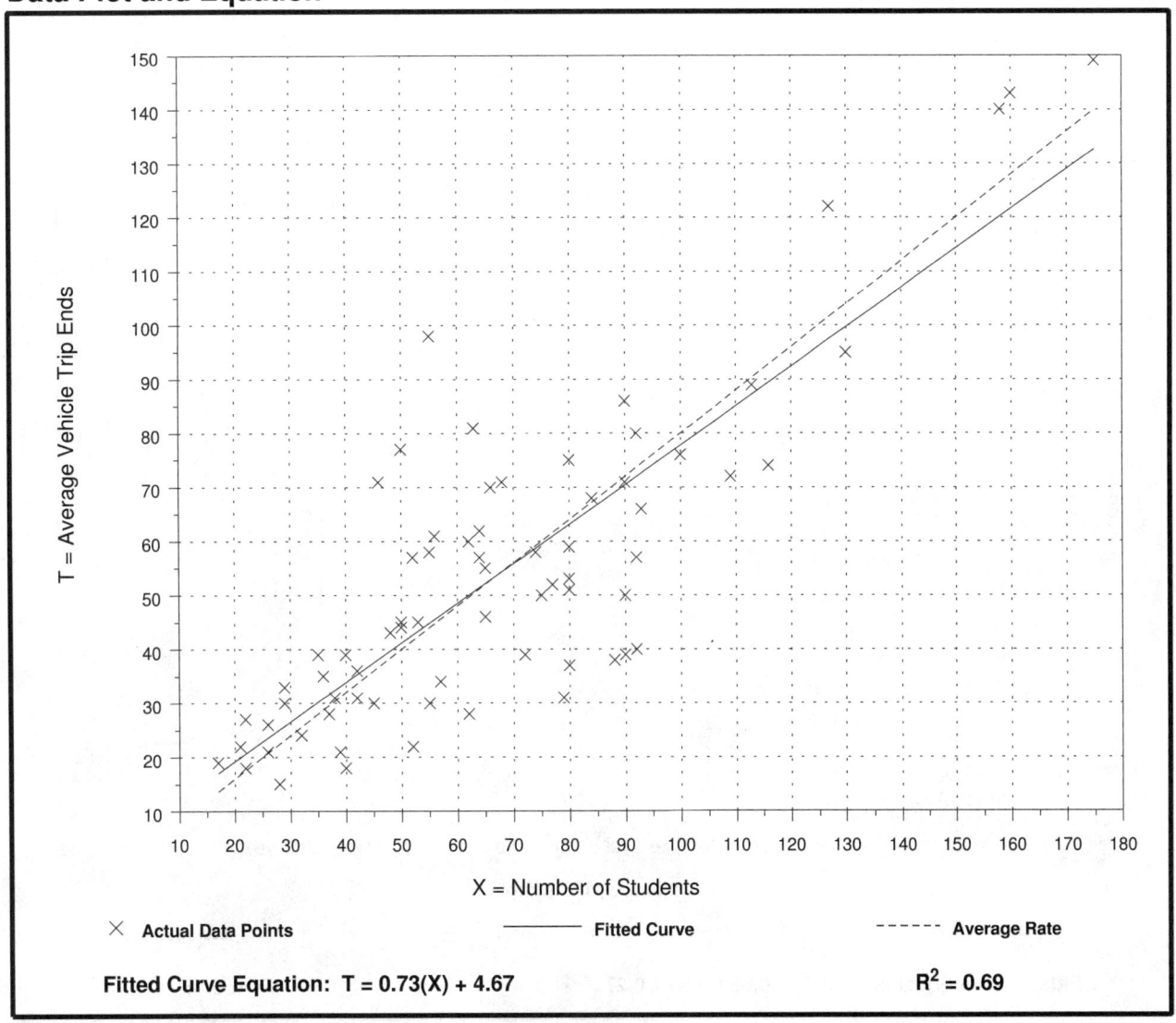

Data Plot and Equation

Fitted Curve Equation: $T = 0.73(X) + 4.67$ $R^2 = 0.69$

Day Care Center
(565)

Average Vehicle Trip Ends vs: Students
On a: Weekday,
Peak Hour of Adjacent Street Traffic,
One Hour Between 4 and 6 p.m.

Number of Studies: 72
Average Number of Students: 69
Directional Distribution: 47% entering, 53% exiting

Trip Generation per Student

Average Rate	Range of Rates	Standard Deviation
0.81	0.24 - 1.72	0.94

Data Plot and Equation

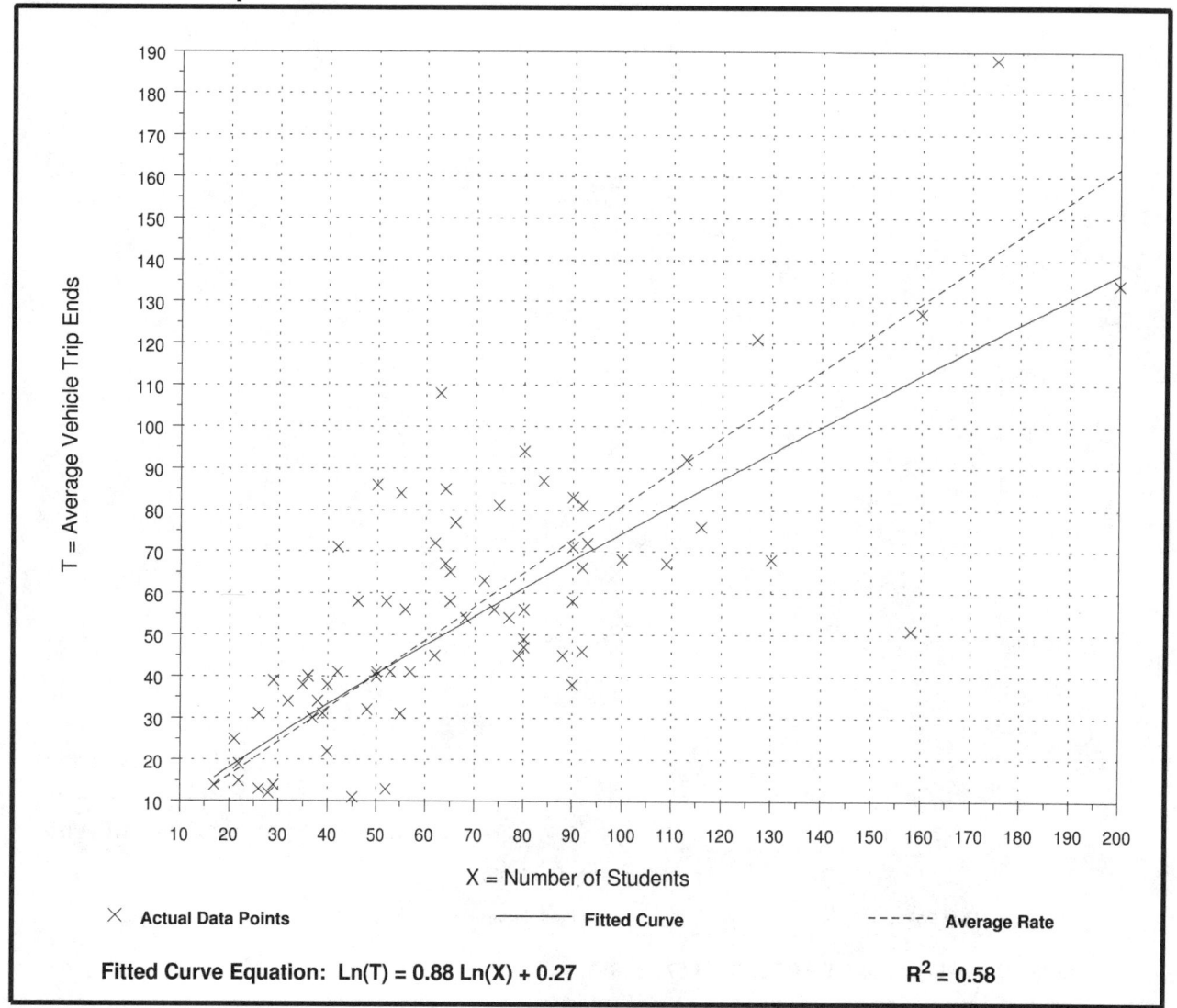

Fitted Curve Equation: $Ln(T) = 0.88 \, Ln(X) + 0.27$ $R^2 = 0.58$

Day Care Center
(565)

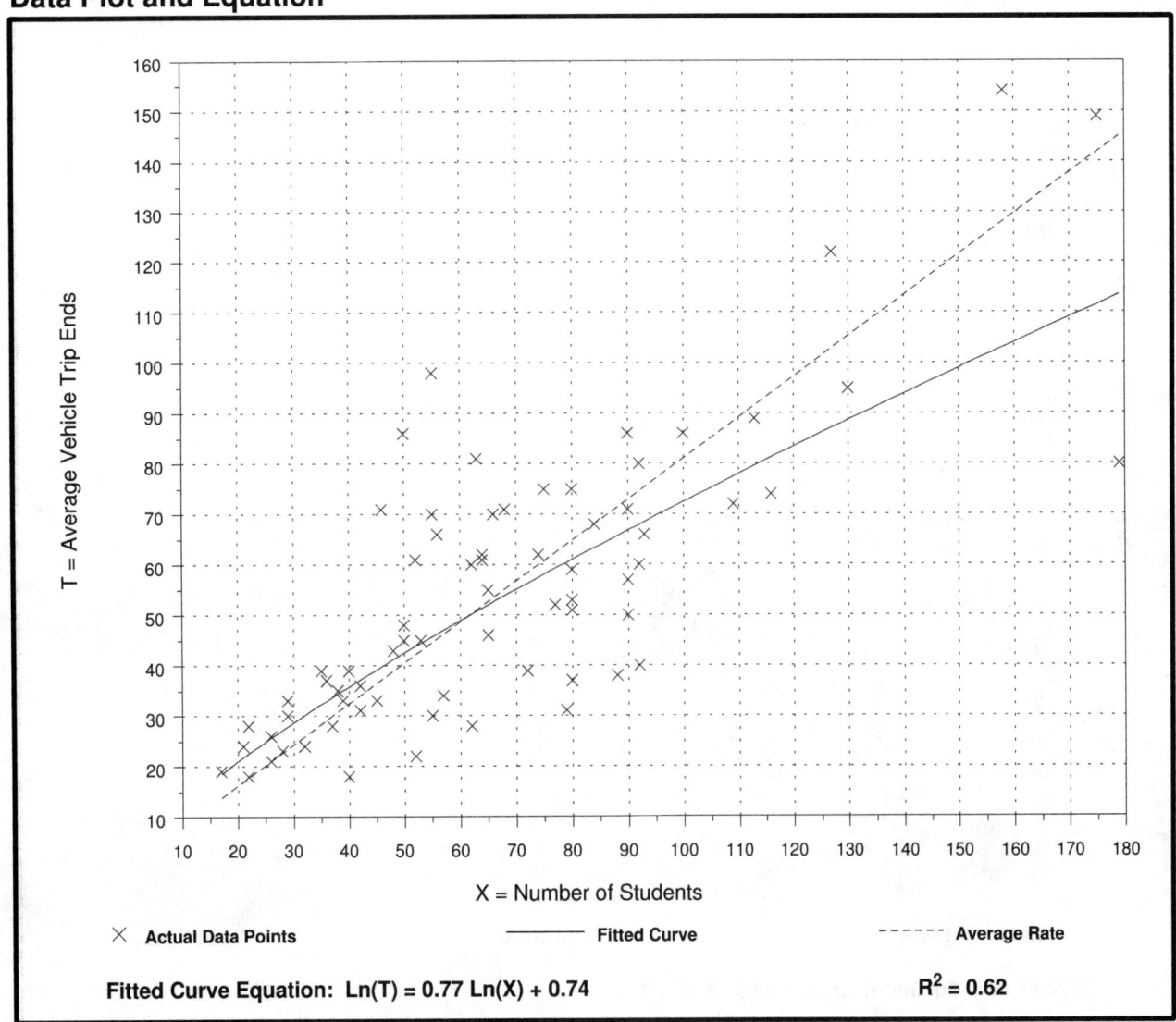

Average Vehicle Trip Ends vs: Students
On a: Weekday,
A.M. Peak Hour of Generator

Number of Studies: 71
Average Number of Students: 67
Directional Distribution: 53% entering, 47% exiting

Trip Generation per Student

Average Rate	Range of Rates	Standard Deviation
0.81	0.39 - 1.78	0.93

Data Plot and Equation

Fitted Curve Equation: $Ln(T) = 0.77\, Ln(X) + 0.74$ $R^2 = 0.62$

Day Care Center
(565)

Average Vehicle Trip Ends vs: Students
On a: Weekday,
P.M. Peak Hour of Generator

Number of Studies: 71
Average Number of Students: 67
Directional Distribution: 47% entering, 53% exiting

Trip Generation per Student

Average Rate	Range of Rates	Standard Deviation
0.84	0.29 - 1.72	0.96

Data Plot and Equation

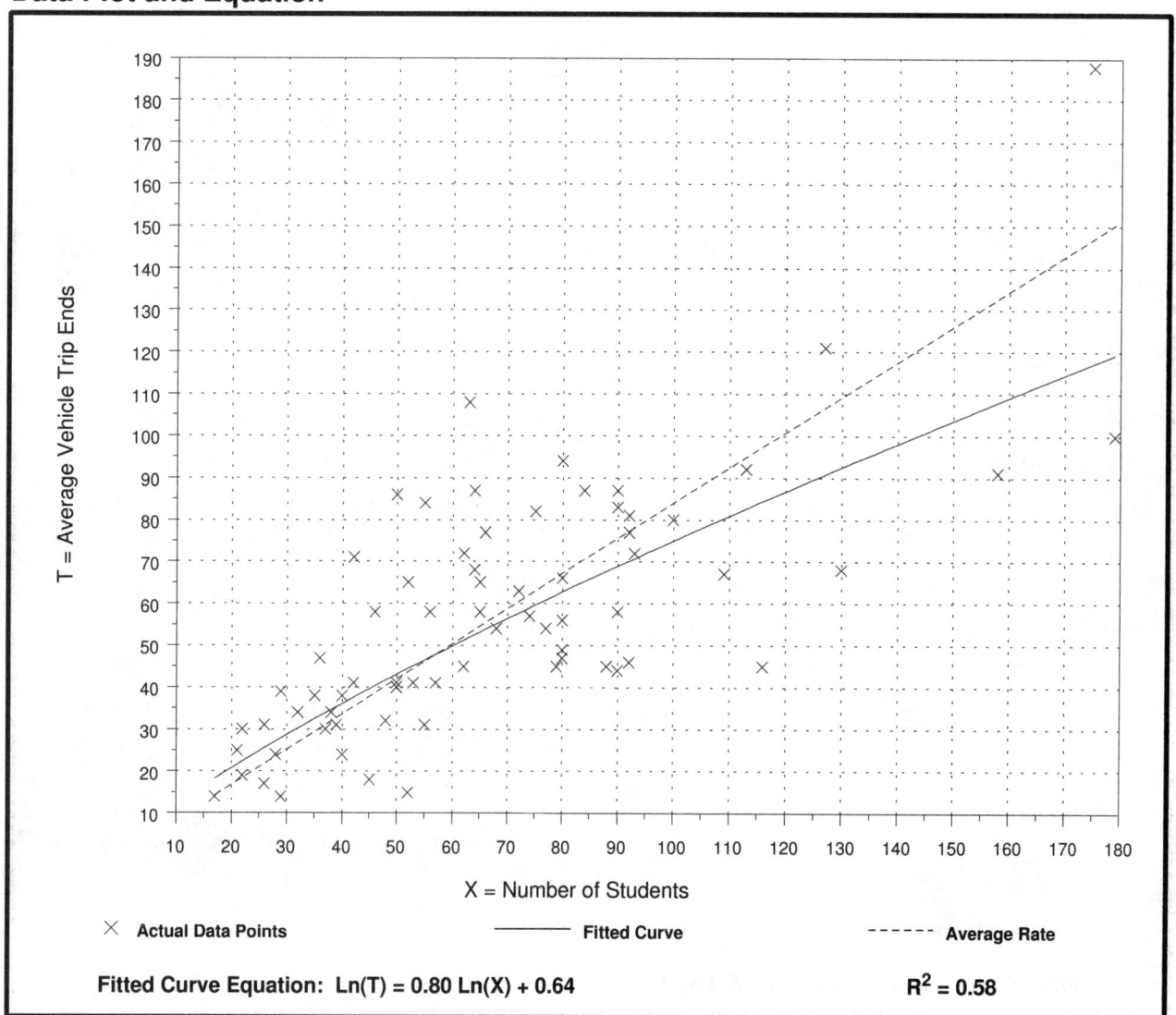

Fitted Curve Equation: $\text{Ln}(T) = 0.80 \, \text{Ln}(X) + 0.64$ $R^2 = 0.58$

Day Care Center
(565)

Average Vehicle Trip Ends vs: Students
On a: Saturday

Number of Studies: 5
Average Number of Students: 75
Directional Distribution: 50% entering, 50% exiting

Trip Generation per Student

Average Rate	Range of Rates	Standard Deviation
0.39	0.12 - 0.96	0.70

Data Plot and Equation

Caution - Use Carefully - Small Sample Size

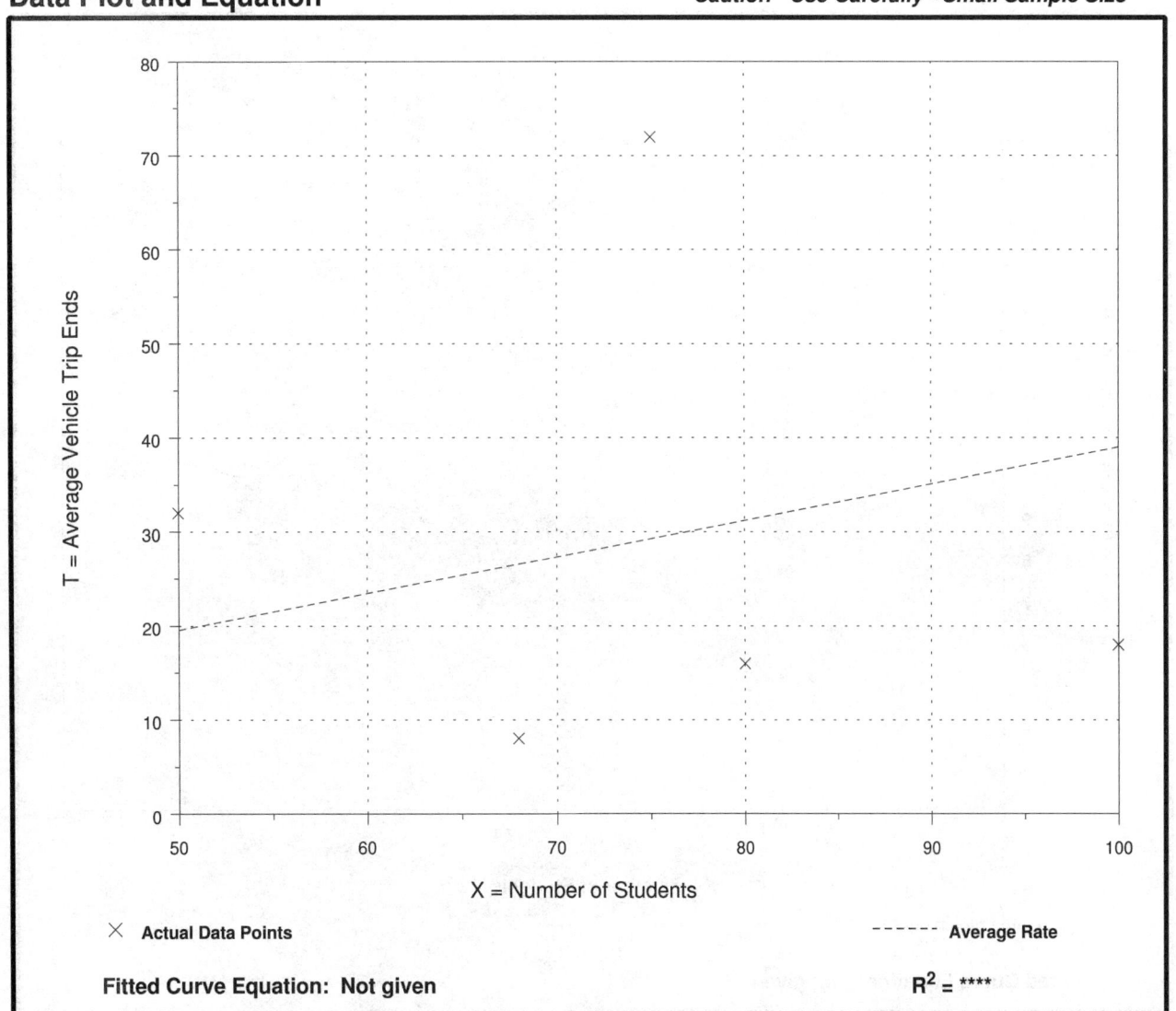

Fitted Curve Equation: Not given $R^2 = ****$

Day Care Center
(565)

Average Vehicle Trip Ends vs: Students
On a: Saturday,
Peak Hour of Generator

Number of Studies: 5
Average Number of Students: 75
Directional Distribution: 63% entering, 37% exiting

Trip Generation per Student

Average Rate	Range of Rates	Standard Deviation
0.11	0.03 - 0.27	0.34

Data Plot and Equation

Caution - Use Carefully - Small Sample Size

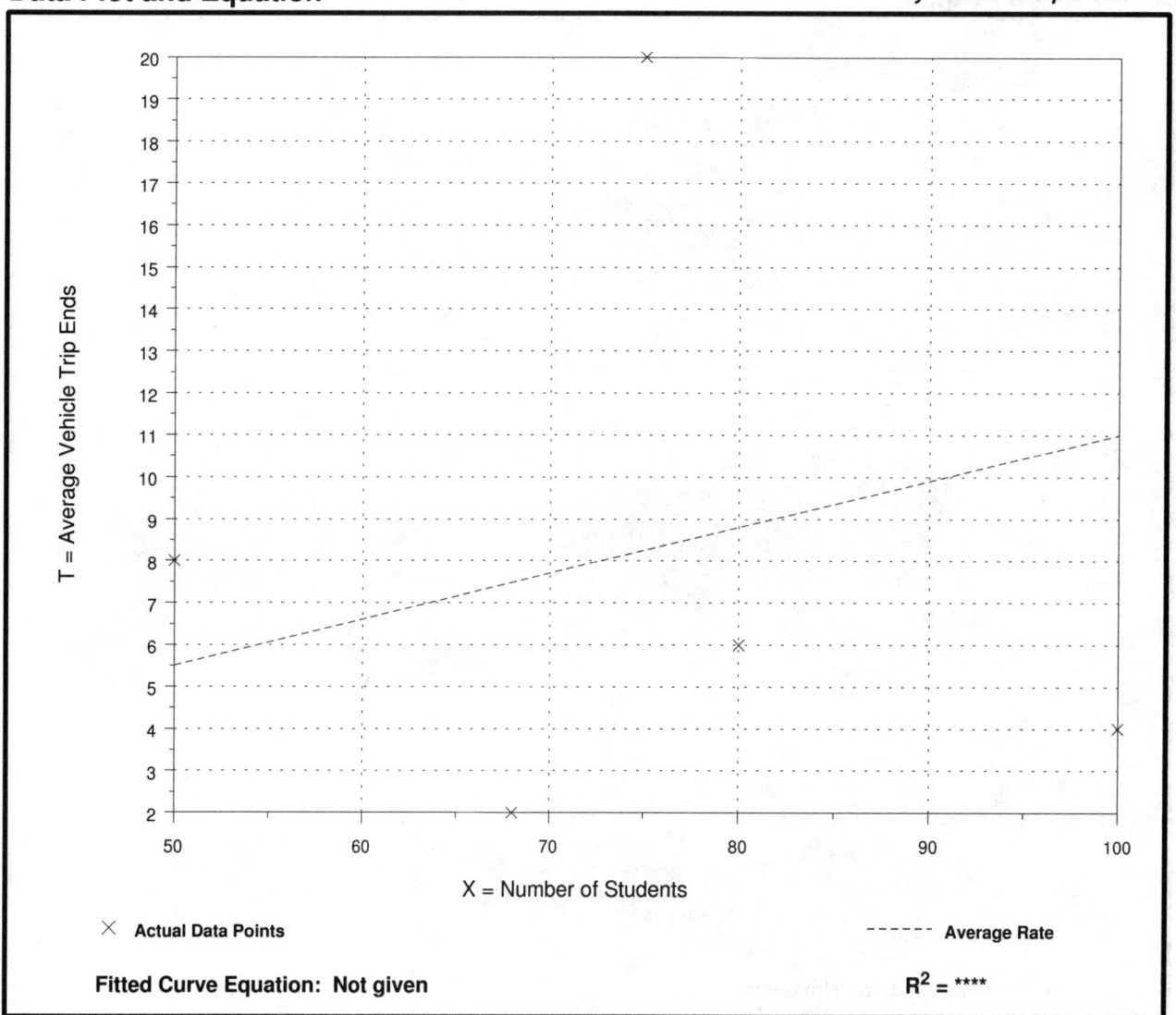

Fitted Curve Equation: Not given $R^2 = ****$

Day Care Center
(565)

Average Vehicle Trip Ends vs: Students
On a: Sunday

Number of Studies: 5
Average Number of Students: 75
Directional Distribution: 50% entering, 50% exiting

Trip Generation per Student

Average Rate	Range of Rates	Standard Deviation
0.37	0.18 - 0.84	0.65

Data Plot and Equation

Caution - Use Carefully - Small Sample Size

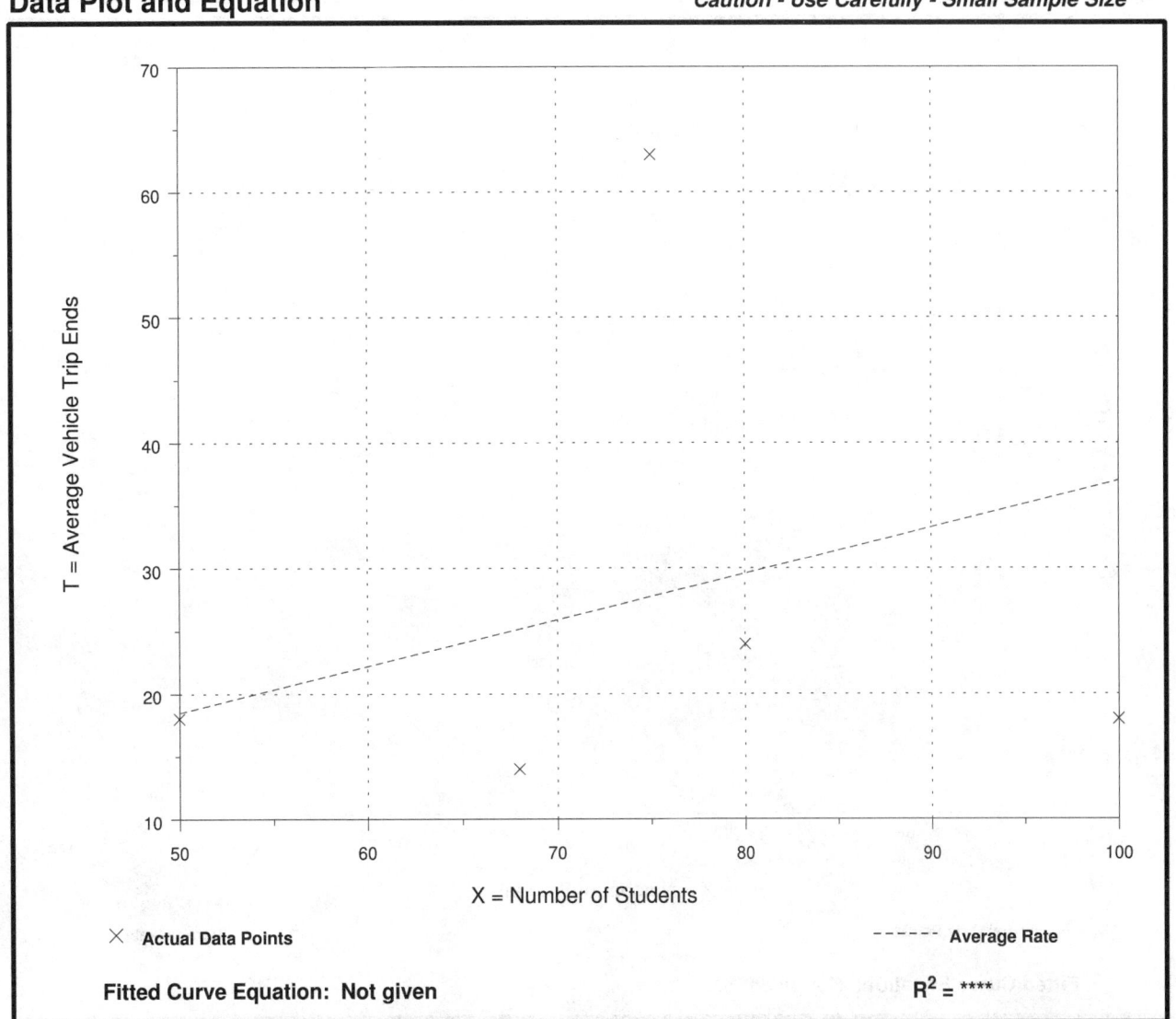

Fitted Curve Equation: Not given $R^2 = ****$

Day Care Center
(565)

Average Vehicle Trip Ends vs: Students
On a: Sunday,
Peak Hour of Generator

Number of Studies: 5
Average Number of Students: 75
Directional Distribution: 54% entering, 46% exiting

Trip Generation per Student

Average Rate	Range of Rates	Standard Deviation
0.11	0.06 - 0.17	0.33

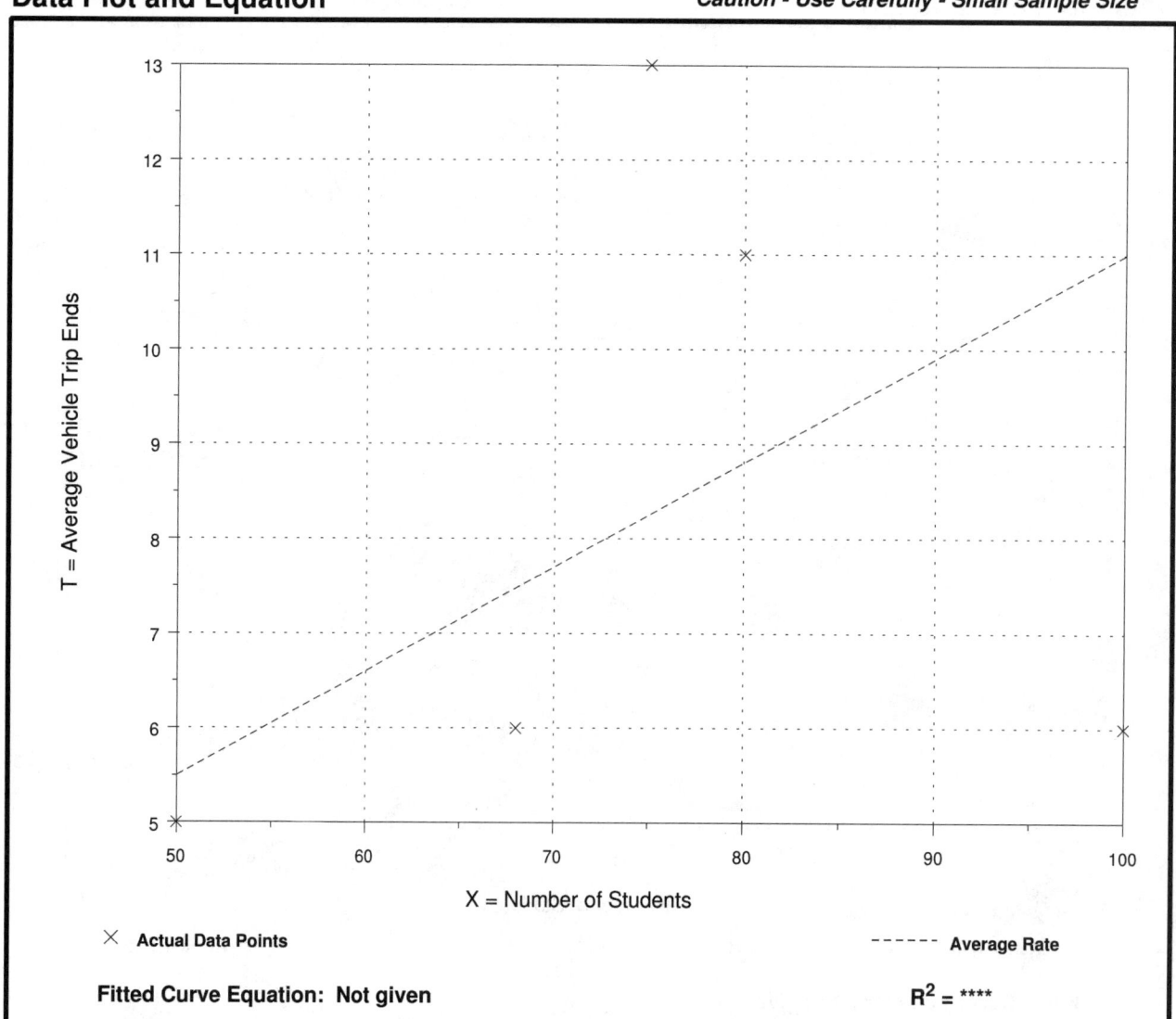

Data Plot and Equation

Caution - Use Carefully - Small Sample Size

Fitted Curve Equation: Not given $R^2 = ****$

Land Use: 566
Cemetery

Description

A cemetery is a place for burying the deceased, possibly including buildings used for funeral services, a mausoleum and a crematorium.

Additional Data

The sites were surveyed between the 1970s and the mid-1990s in California.

Source Numbers

214, 392, 430

Land Use: 566
Cemetery
Independent Variables with One Observation

The following trip generation data are for independent variables with only one observation. This information is shown in this table only; there are no related plots for these data.

Users are cautioned to use data with care because of the small sample size.

Independent Variable	Trip Generation Rate	Size of Independent Variable	Number of Studies	Directional Distribution
Acres				
Weekday A.M. Peak Hour of Adjacent Street Traffic	0.17	58	1	70% entering, 30% exiting
Weekday P.M. Peak Hour of Adjacent Street Traffic	0.84	58	1	33% entering, 67% exiting
Weekday A.M. Peak Hour of Generator	0.76	58	1	48% entering, 52% exiting
Weekday P.M. Peak Hour of Generator	1.64	58	1	75% entering, 25% exiting
Saturday Peak Hour of Generator	3.09	58	1	51% entering, 49% exiting
Sunday Peak Hour of Generator	6.21	58	1	48% entering, 52% exiting
Employees				
Weekday A.M. Peak Hour of Adjacent Street Traffic	1.43	7	1	70% entering, 30% exiting
Weekday P.M. Peak Hour of Adjacent Street Traffic	7.00	7	1	33% entering, 67% exiting
Weekday A.M. Peak Hour of Generator	6.29	7	1	48% entering, 52% exiting
Weekday P.M. Peak Hour of Generator	13.57	7	1	75% entering, 25% exiting
Saturday Peak Hour of Generator	25.57	7	1	51% entering, 49% exiting
Sunday Peak Hour of Generator	51.43	7	1	48% entering, 52% exiting

Cemetery
(566)

Average Vehicle Trip Ends vs: Acres
On a: Weekday

Number of Studies: 5
Average Number of Acres: 108
Directional Distribution: 50% entering, 50% exiting

Trip Generation per Acre

Average Rate	Range of Rates	Standard Deviation
4.73	1.67 - 9.40	3.92

Data Plot and Equation

Caution - Use Carefully - Small Sample Size

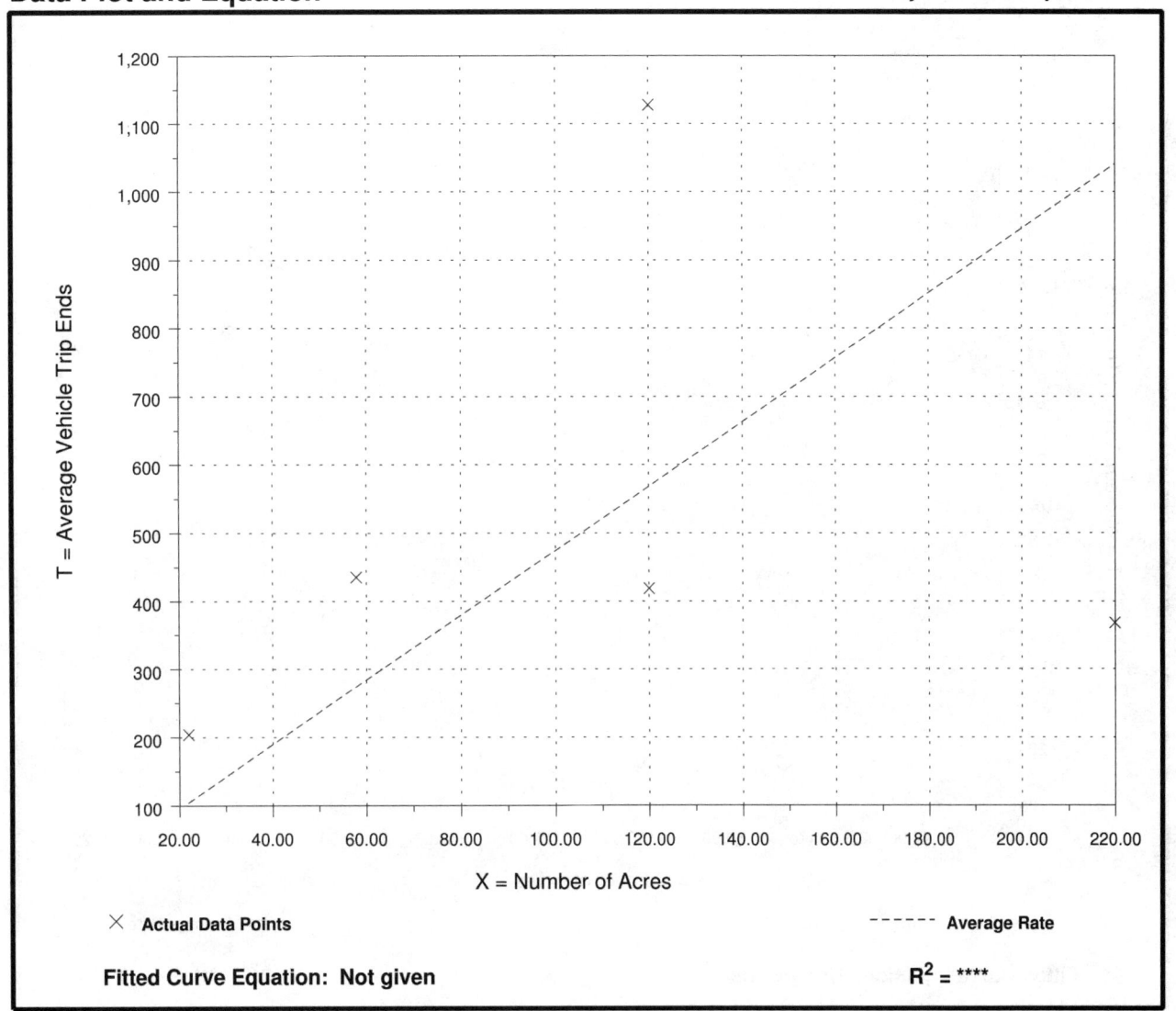

Fitted Curve Equation: Not given $R^2 =$ ****

Cemetery
(566)

Average Vehicle Trip Ends vs: Acres
On a: Saturday

Number of Studies: 5
Average Number of Acres: 108
Directional Distribution: 50% entering, 50% exiting

Trip Generation per Acre

Average Rate	Range of Rates	Standard Deviation
5.94	1.27 - 18.52	5.70

Data Plot and Equation

Caution - Use Carefully - Small Sample Size

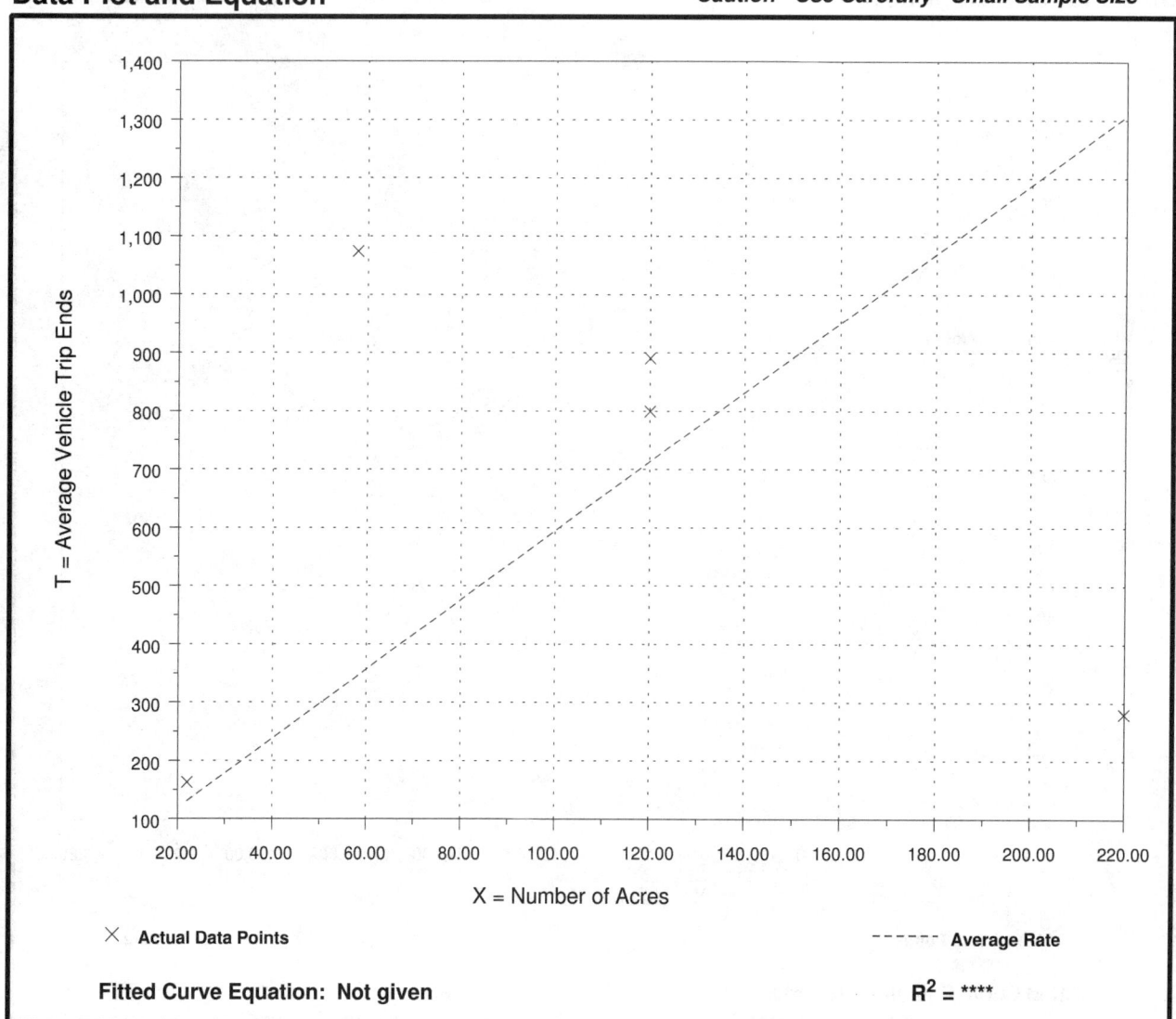

X = Number of Acres

× Actual Data Points   ----- Average Rate

Fitted Curve Equation: Not given $R^2 =$ ****

Cemetery
(566)

Average Vehicle Trip Ends vs: Acres
On a: Sunday

Number of Studies: 5
Average Number of Acres: 108
Directional Distribution: 50% entering, 50% exiting

Trip Generation per Acre

Average Rate	Range of Rates	Standard Deviation
7.62	2.00 - 35.71	10.35

Data Plot and Equation

Caution - Use Carefully - Small Sample Size

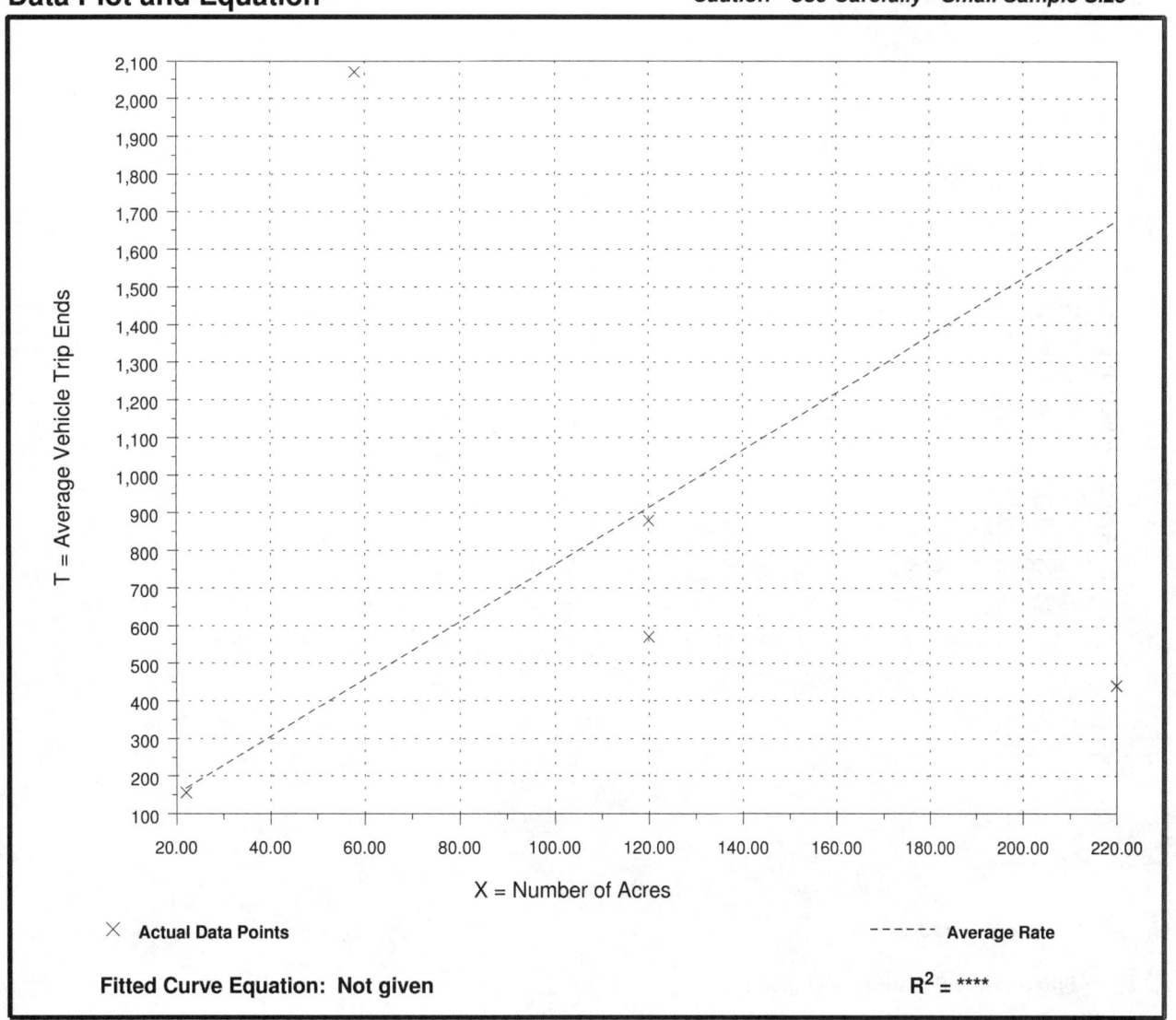

× Actual Data Points   ------ Average Rate

Fitted Curve Equation: Not given $R^2 =$ ****

Cemetery
(566)

Average Vehicle Trip Ends vs: Employees
On a: Weekday

Number of Studies: 2
Avg. Number of Employees: 6
Directional Distribution: 50% entering, 50% exiting

Trip Generation per Employee

Average Rate	Range of Rates	Standard Deviation
58.09	51.00 - 62.14	*

Data Plot and Equation

Caution - Use Carefully - Small Sample Size

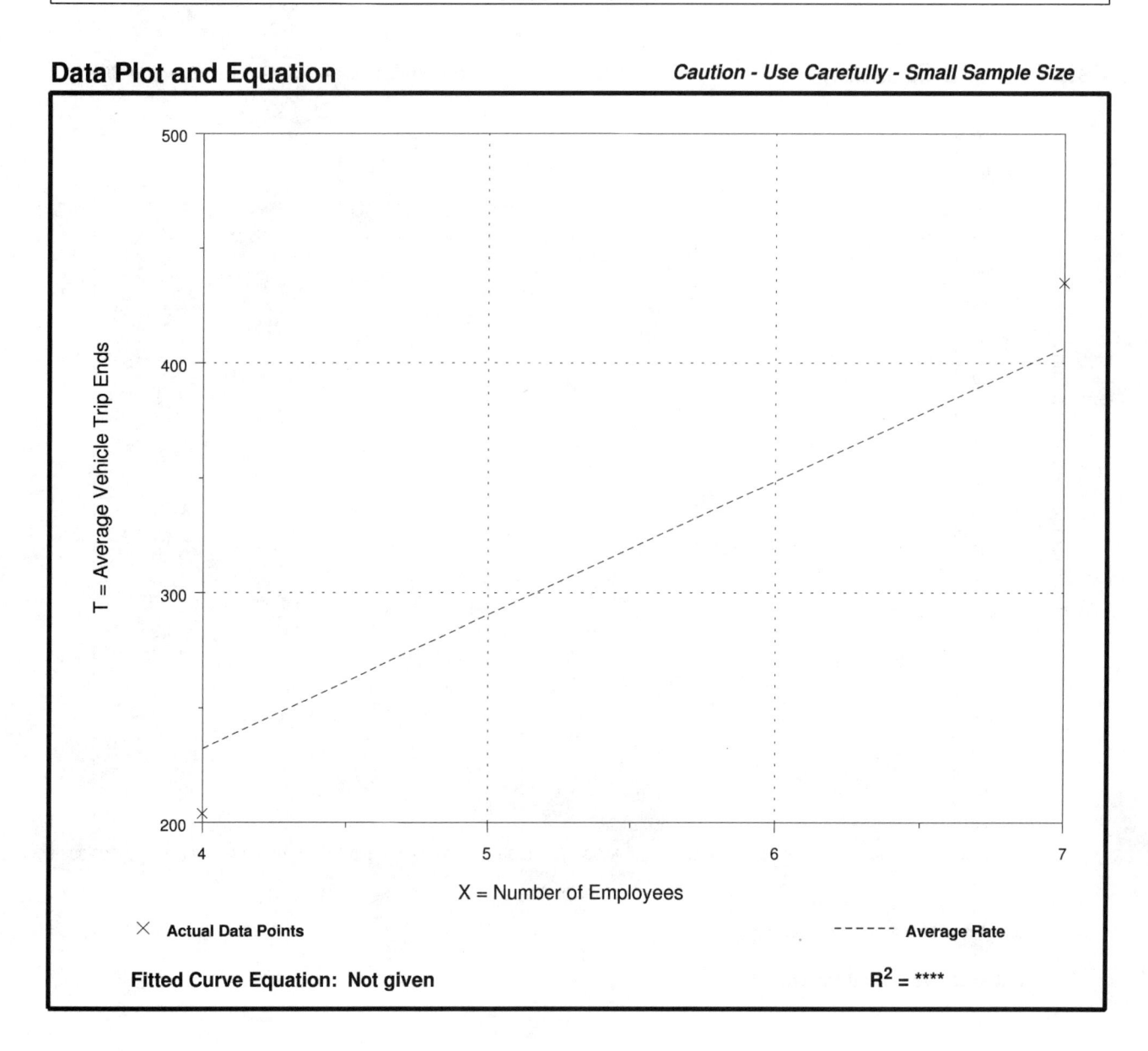

Fitted Curve Equation: Not given $R^2 = ****$

Cemetery
(566)

Average Vehicle Trip Ends vs: Employees
On a: Saturday

Number of Studies: 2
Avg. Number of Employees: 6
Directional Distribution: 50% entering, 50% exiting

Trip Generation per Employee

Average Rate	Range of Rates	Standard Deviation
112.45	40.75 - 153.43	*

Data Plot and Equation

Caution - Use Carefully - Small Sample Size

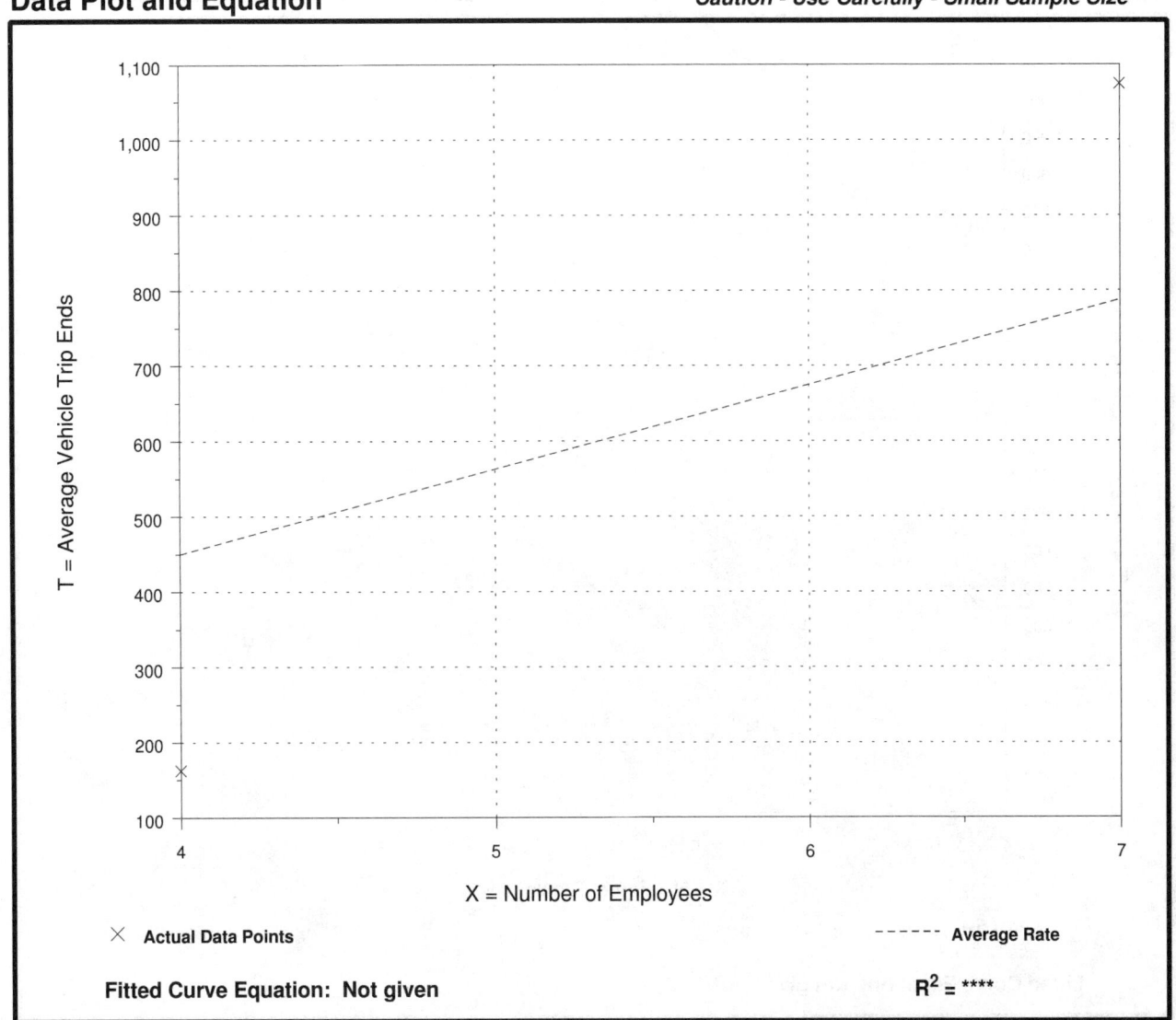

Fitted Curve Equation: Not given

$R^2 = ****$

Cemetery
(566)

Average Vehicle Trip Ends vs: Employees
On a: Sunday

Number of Studies: 2
Avg. Number of Employees: 6
Directional Distribution: 50% entering, 50% exiting

Trip Generation per Employee

Average Rate	Range of Rates	Standard Deviation
202.45	39.00 - 295.86	*

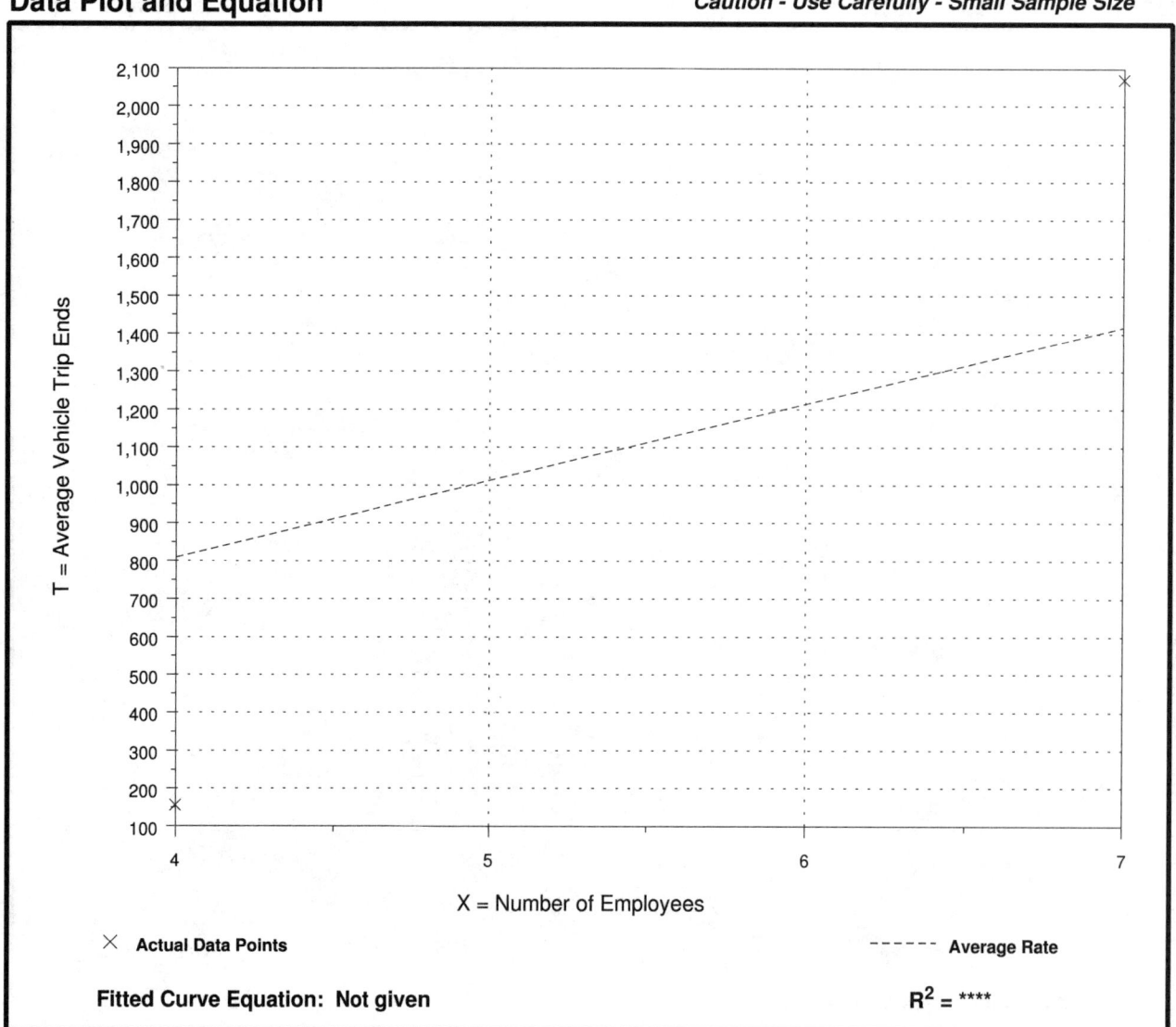

Data Plot and Equation

Caution - Use Carefully - Small Sample Size

Fitted Curve Equation: Not given R^2 = ****

Land Use: 571
Prison

Description

A prison is a building where persons who have been convicted of a crime or are awaiting trial are confined. A prison usually consists of cells, dining and food preparation facilities, limited recreational facilities, work areas and offices.

Additional Data

Peak hours of the generator—
    The weekend peak hours of the generator varied between 9:00 a.m. and 11:00 a.m. and 2:00 p.m. and 3:00 p.m.

These sites were surveyed in 1990 and 1996 in Florida, Connecticut and Oregon.

Source Numbers

247, 326, 583

Land Use: 571
Prison

Independent Variables with One Observation

The following trip generation data are for independent variables with only one observation. This information is shown in this table only; there are no related plots for these data.

Users are cautioned to use data with care because of the small sample size.

Independent Variable	Trip Generation Rate	Size of Independent Variable	Number of Studies	Directional Distribution

Occupied Beds

Independent Variable	Trip Generation Rate	Size of Independent Variable	Number of Studies	Directional Distribution
Weekday A.M. Peak Hour of Adjacent Street Traffic	0.78	77	1	Not available
Weekday P.M. Peak Hour of Adjacent Street Traffic	0.31	77	1	Not available
Weekday A.M. Peak Hour of Generator	1.30	77	1	Not available
Weekday P.M. Peak Hour of Generator	1.22	77	1	Not available
Saturday	0.70	77	1	50% entering, 50% exiting

1,000 Square Feet Gross Floor Area

Independent Variable	Trip Generation Rate	Size of Independent Variable	Number of Studies	Directional Distribution
Weekday A.M. Peak Hour of Adjacent Street Traffic	7.27	8	1	Not available
Weekday P.M. Peak Hour of Adjacent Street Traffic	2.91	8	1	Not available
Weekday A.M. Peak Hour of Generator	12.11	8	1	Not available
Weekday P.M. Peak Hour of Generator	11.39	8	1	Not available
Saturday	6.54	8	1	50% entering, 50% exiting

Prison
(571)

Average Vehicle Trip Ends vs: Employees
On a: Weekday,
Peak Hour of Adjacent Street Traffic,
One Hour Between 7 and 9 a.m.

Number of Studies: 2
Avg. Number of Employees: 185
Directional Distribution: 66% entering, 34% exiting

Trip Generation per Employee

Average Rate	Range of Rates	Standard Deviation
0.42	0.29 - 1.20	*

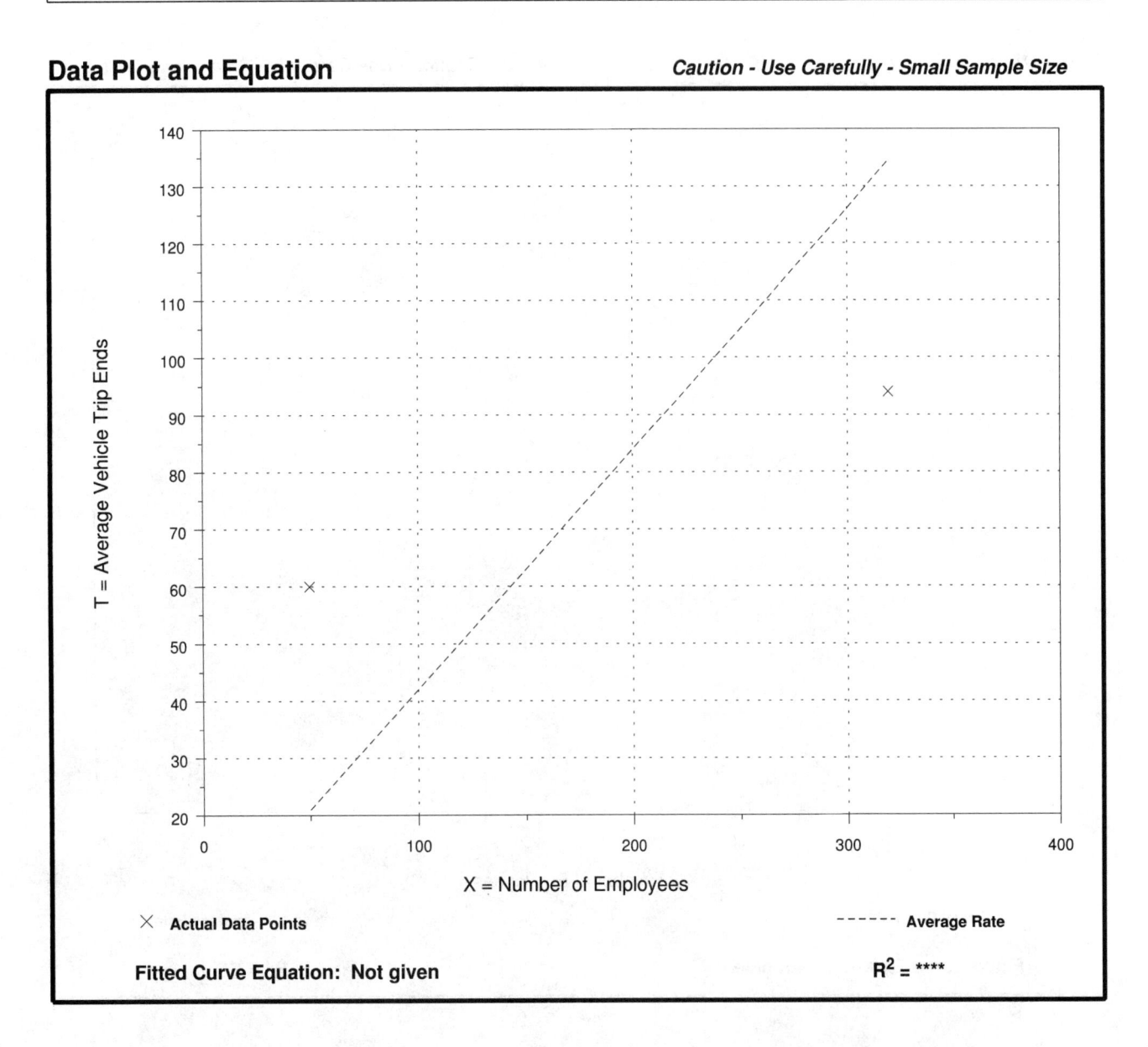

Data Plot and Equation

Caution - Use Carefully - Small Sample Size

Fitted Curve Equation: Not given $R^2 = ****$

Prison
(571)

Average Vehicle Trip Ends vs: Employees
On a: Weekday,
  Peak Hour of Adjacent Street Traffic,
  One Hour Between 4 and 6 p.m.

Number of Studies: 2
Avg. Number of Employees: 185
Directional Distribution: 28% entering, 72% exiting

Trip Generation per Employee

Average Rate	Range of Rates	Standard Deviation
0.23	0.19 - 0.48	*

Data Plot and Equation

Caution - Use Carefully - Small Sample Size

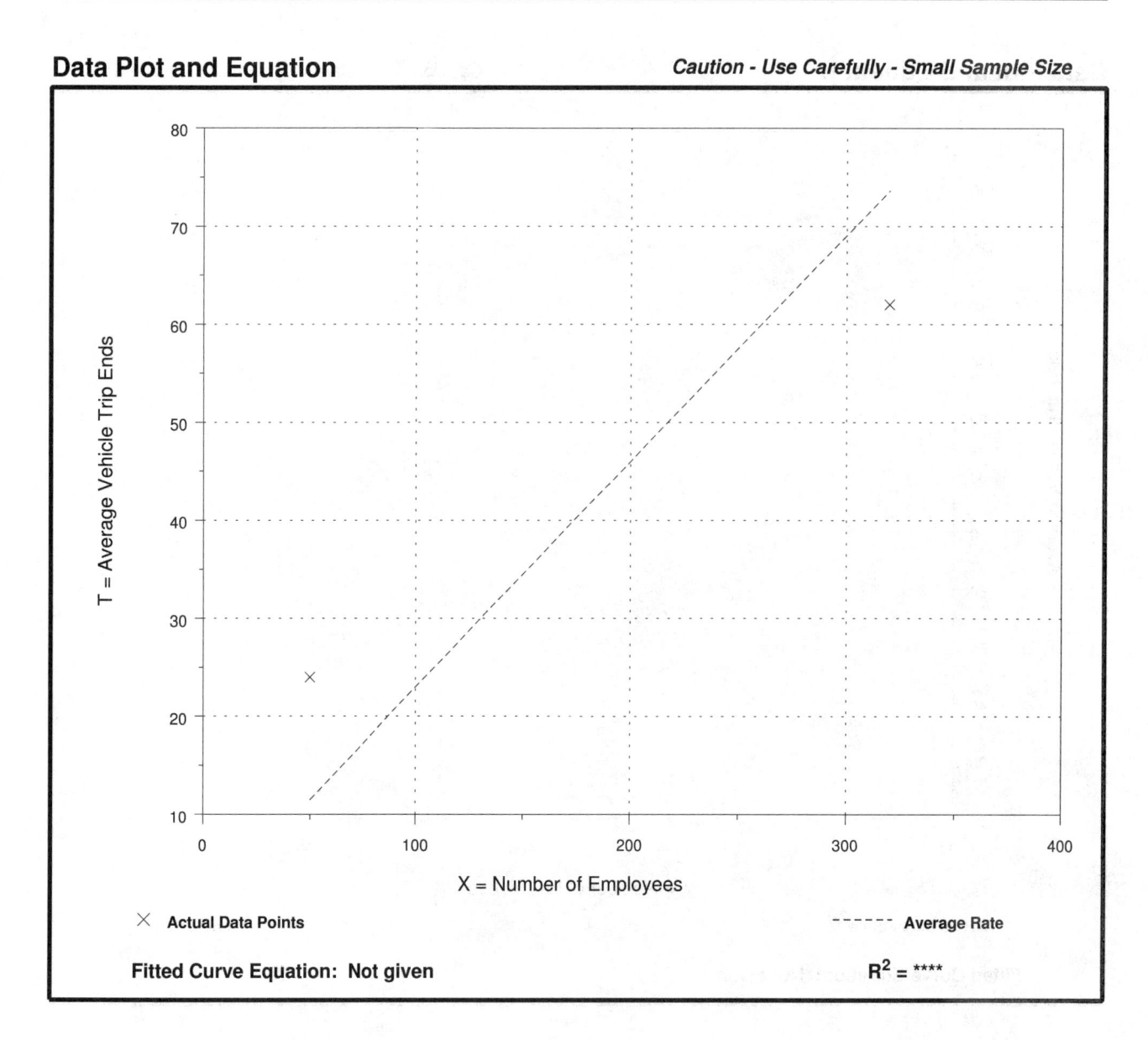

Fitted Curve Equation: Not given $R^2 =$ ****

Prison
(571)

Average Vehicle Trip Ends vs: Employees
On a: Weekday,
A.M. Peak Hour of Generator

Number of Studies: 2
Avg. Number of Employees: 185
Directional Distribution: 62% entering, 38% exiting

Trip Generation per Employee

Average Rate	Range of Rates	Standard Deviation
0.52	0.29 - 2.00	*

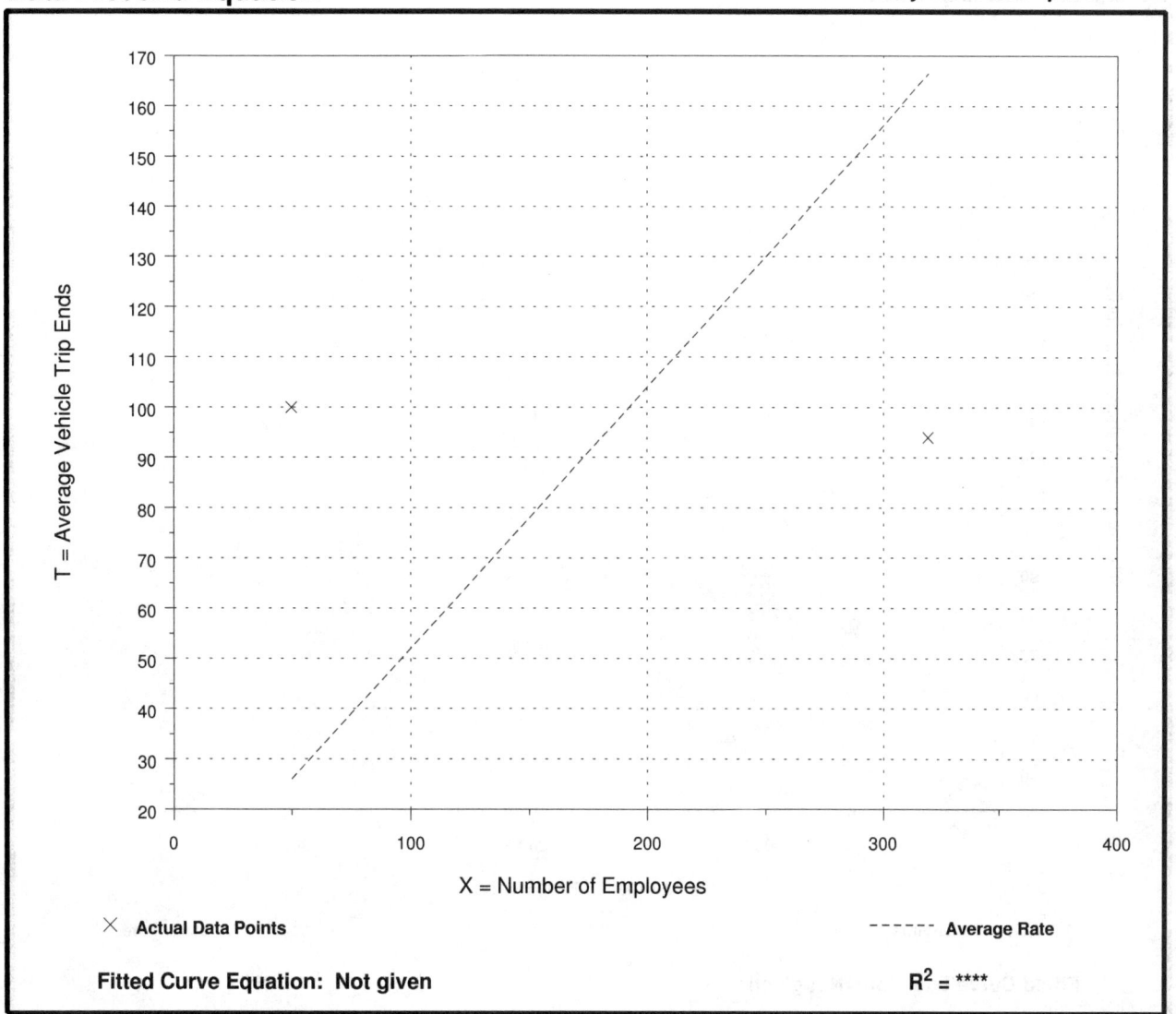

Data Plot and Equation

Caution - Use Carefully - Small Sample Size

Fitted Curve Equation: Not given $R^2 = ****$

Prison
(571)

Average Vehicle Trip Ends vs: Employees
On a: Weekday,
P.M. Peak Hour of Generator

Number of Studies: 2
Avg. Number of Employees: 185
Directional Distribution: 27% entering, 73% exiting

Trip Generation per Employee

Average Rate	Range of Rates	Standard Deviation
0.68	0.50 - 1.88	*

Data Plot and Equation

Caution - Use Carefully - Small Sample Size

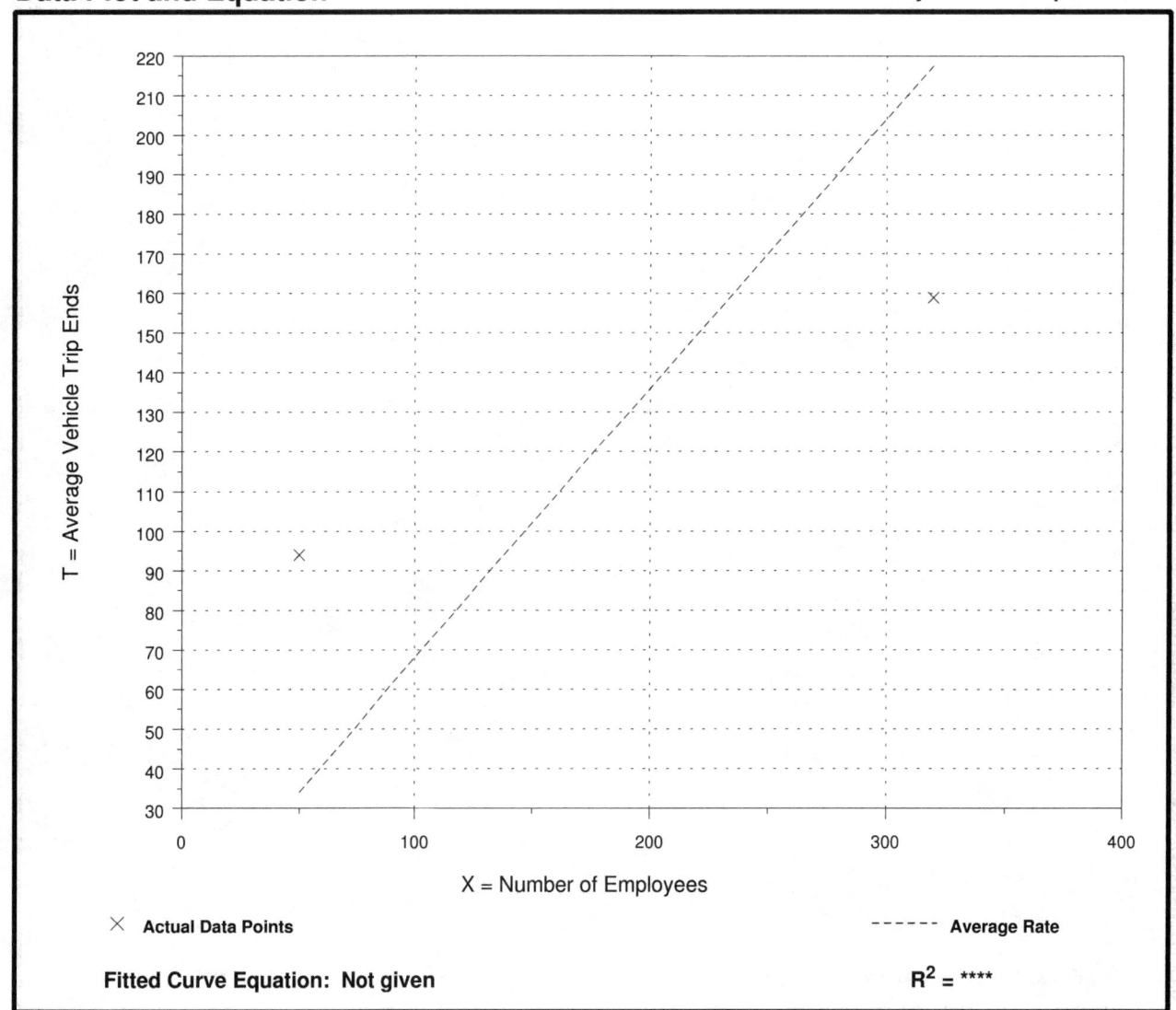

Fitted Curve Equation: Not given $R^2 = ****$

Prison
(571)

Average Vehicle Trip Ends vs: Employees
On a: Saturday

Number of Studies: 2
Avg. Number of Employees: 185
Directional Distribution: 50% entering, 50% exiting

Trip Generation per Employee

Average Rate	Range of Rates	Standard Deviation
1.80	1.08 - 1.91	*

Data Plot and Equation

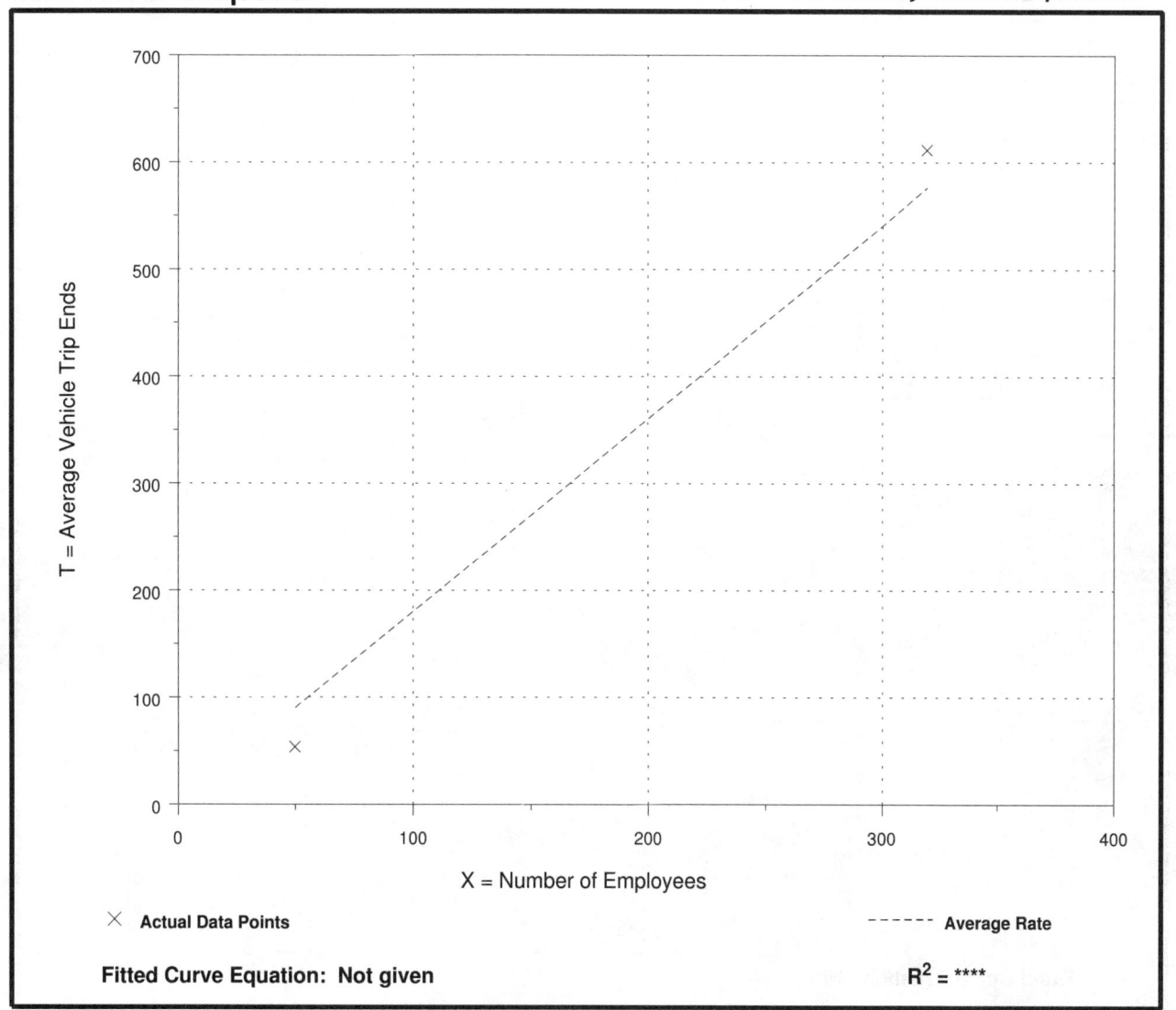

Caution - Use Carefully - Small Sample Size

Fitted Curve Equation: Not given $R^2 = ****$

Prison
(571)

Average Vehicle Trip Ends vs: Beds
On a: Weekday,
Peak Hour of Adjacent Street Traffic,
One Hour Between 7 and 9 a.m.

Number of Studies: 2
Average Number of Beds: 347
Directional Distribution: 54% entering, 46% exiting

Trip Generation per Bed

Average Rate	Range of Rates	Standard Deviation
0.10	0.03 - 0.30	*

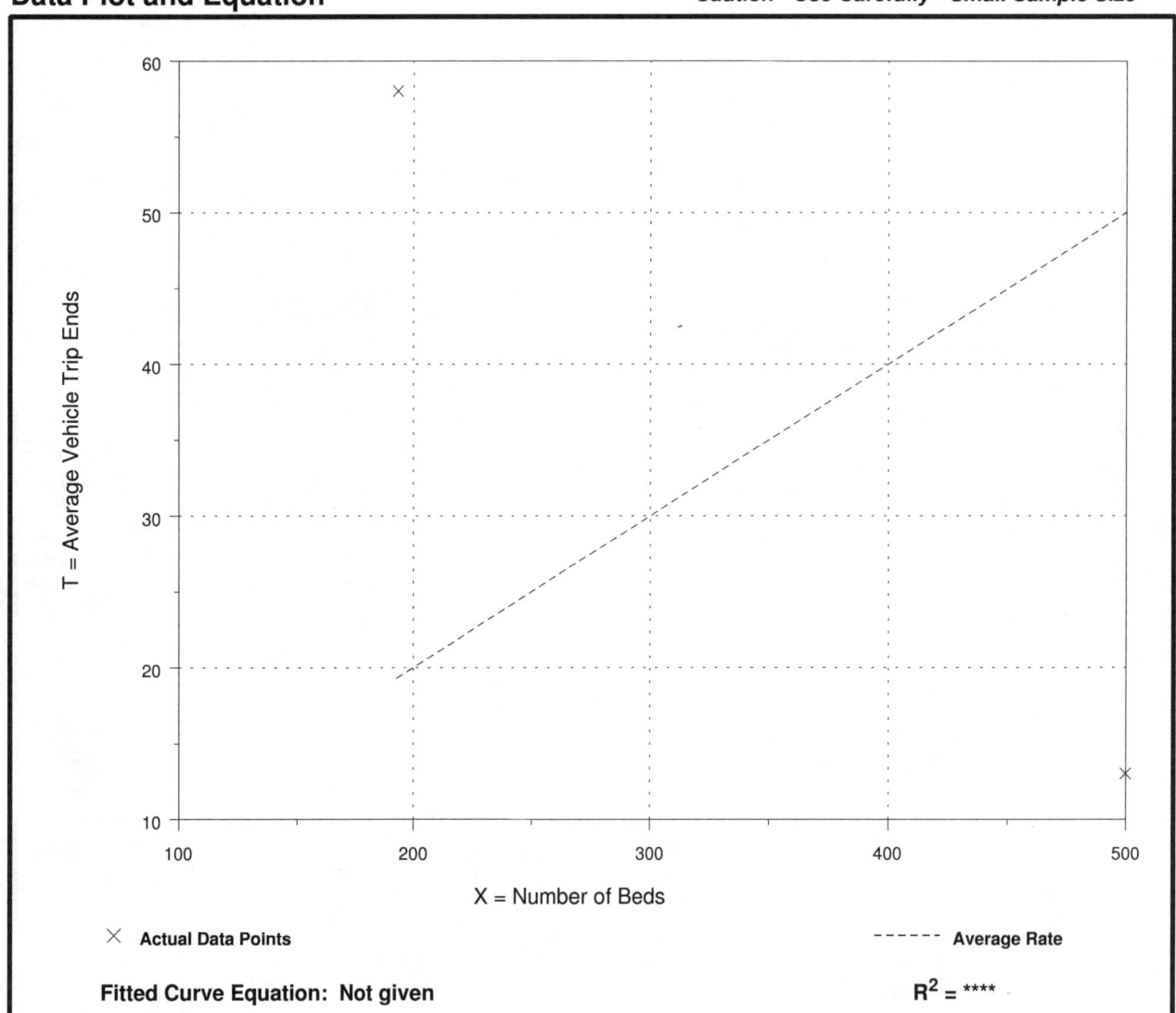

Data Plot and Equation

Caution - Use Carefully - Small Sample Size

Fitted Curve Equation: Not given $R^2 = ****$

Prison
(571)

Average Vehicle Trip Ends vs: Beds
On a: Weekday,
Peak Hour of Adjacent Street Traffic,
One Hour Between 4 and 6 p.m.

Number of Studies: 2
Average Number of Beds: 347
Directional Distribution: 10% entering, 90% exiting

Trip Generation per Bed

Average Rate	Range of Rates	Standard Deviation
0.05	0.02 - 0.13	*

Data Plot and Equation

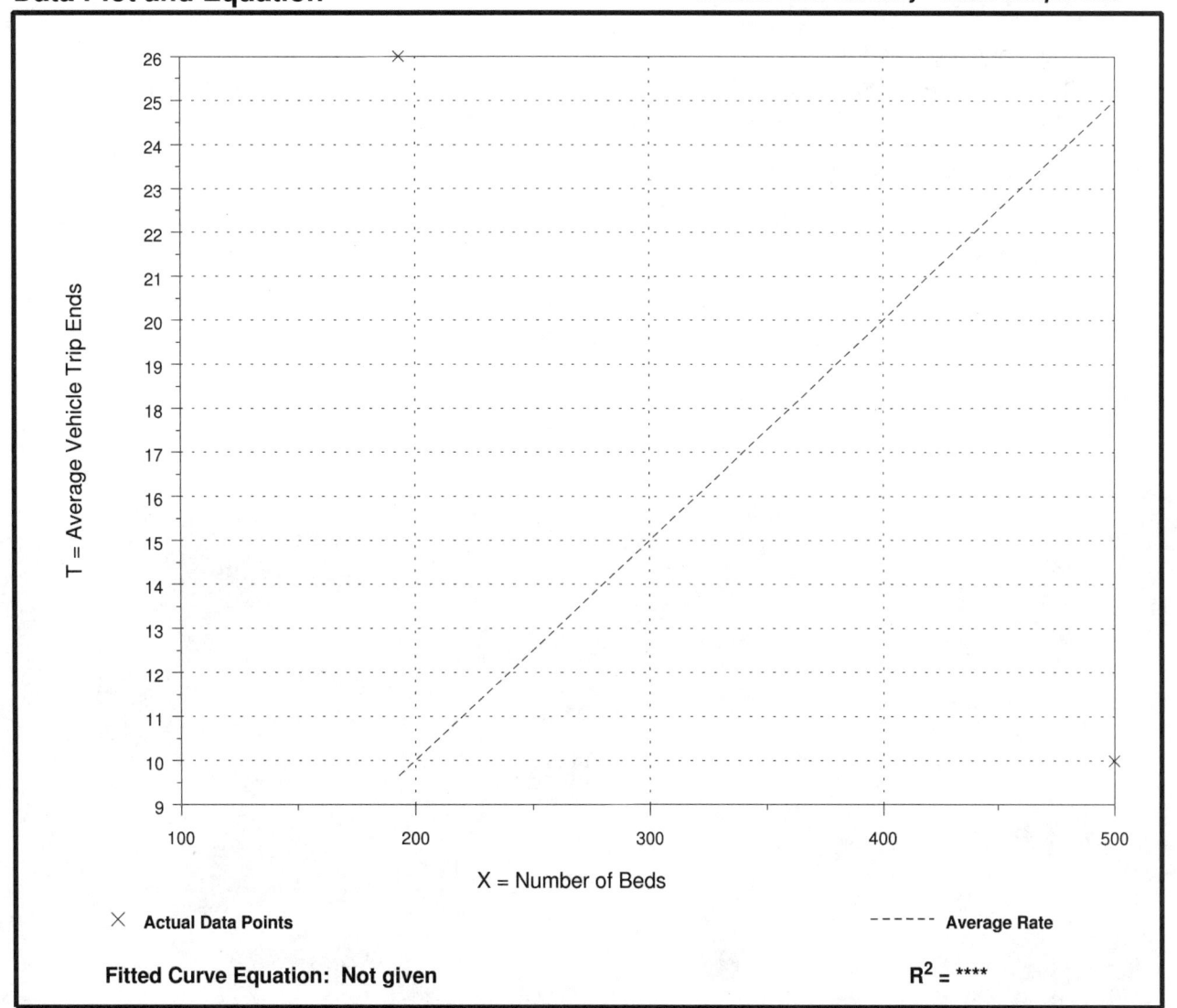

Caution - Use Carefully - Small Sample Size

× Actual Data Points ----- Average Rate

Fitted Curve Equation: Not given $R^2 = ****$

Land Use: 580
Museum

Description

Museums are facilities that include displays, shows, exhibits and/or demonstration of historical, science, nature, art, entertainment, or other cultural significance.

Additional Data

Due to variation in type of museums, caution should be exercised when using the trip generation rates for this land use because they may not be appropriate for all museum types.

The site surveyed had 45,000 square feet of exhibition space.

Peak hours of the generator—
   The weekday A.M. peak hour of the generator was between 11:00 a.m. and 12:00 p.m. The Saturday peak hour of the generator was between 1:00 p.m. and 2:00 p.m.

The site was surveyed in 2010 in Tennessee.

Source Number

725

Land Use: 580
Museum

Independent Variables with One Observation

The following trip generation data are for independent variables with only one observation. This information is shown in this table only; there are no related plots for these data.

Users are cautioned to use data with care because of the small sample size.

Independent Variable	Trip Generation Rate	Size of Independent Variable	Number of Studies	Directional Distribution
1,000 Square Feet Gross Floor Area				
Weekday A.M. Peak Hour of Adjacent Street Traffic	0.28	176	1	86% entering, 14% exiting
Weekday P.M. Peak Hour of Adjacent Street Traffic	0.18	176	1	16% entering, 84% exiting
Weekday A.M. Peak Hour of Generator	0.35	176	1	40% entering, 60% exiting
Saturday Peak Hour of Generator	0.66	176	1	71% entering, 29% exiting
Employees				
Weekday A.M. Peak Hour of Adjacent Street Traffic	0.89	55	1	86% entering, 14% exiting
Weekday P.M. Peak Hour of Adjacent Street Traffic	0.58	55	1	16% entering, 84% exiting
Weekday A.M. Peak Hour of Generator	1.13	55	1	40% entering, 60% exiting
Saturday Peak Hour of Generator	2.11	55	1	71% entering, 29% exiting

Land Use: 590
Library

Description

A library can be either a public or private facility that consists of shelved books, reading rooms or areas and, sometimes, meeting rooms.

Additional Data

The sites were surveyed between the late 1960s and the 2000s in California, Florida, New Jersey, Oregon and Georgia.

Source Numbers

10, 12, 88, 113, 275, 407, 415, 444, 590

Library
(590)

Average Vehicle Trip Ends vs: Employees
On a: Weekday

Number of Studies: 10
Avg. Number of Employees: 25
Directional Distribution: 50% entering, 50% exiting

Trip Generation per Employee

Average Rate	Range of Rates	Standard Deviation
52.52	19.80 - 109.00	22.70

Data Plot and Equation

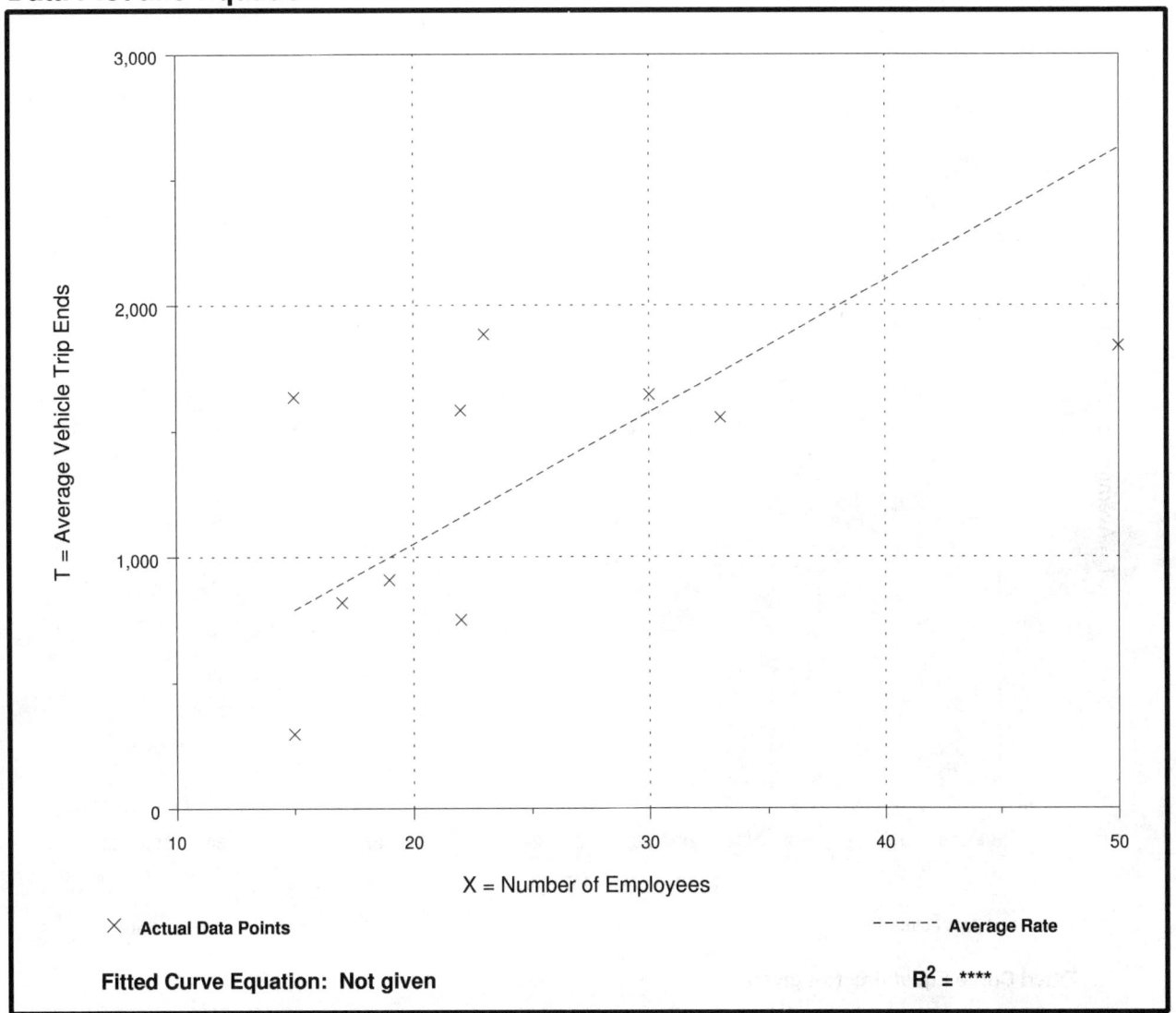

Fitted Curve Equation: Not given $R^2 = ****$

Library
(590)

Average Vehicle Trip Ends vs: Employees
On a: Weekday,
Peak Hour of Adjacent Street Traffic,
One Hour Between 7 and 9 a.m.

Number of Studies: 5
Avg. Number of Employees: 24
Directional Distribution: 69% entering, 31% exiting

Trip Generation per Employee

Average Rate	Range of Rates	Standard Deviation
1.03	0.53 - 1.36	1.05

Data Plot and Equation

Caution - Use Carefully - Small Sample Size

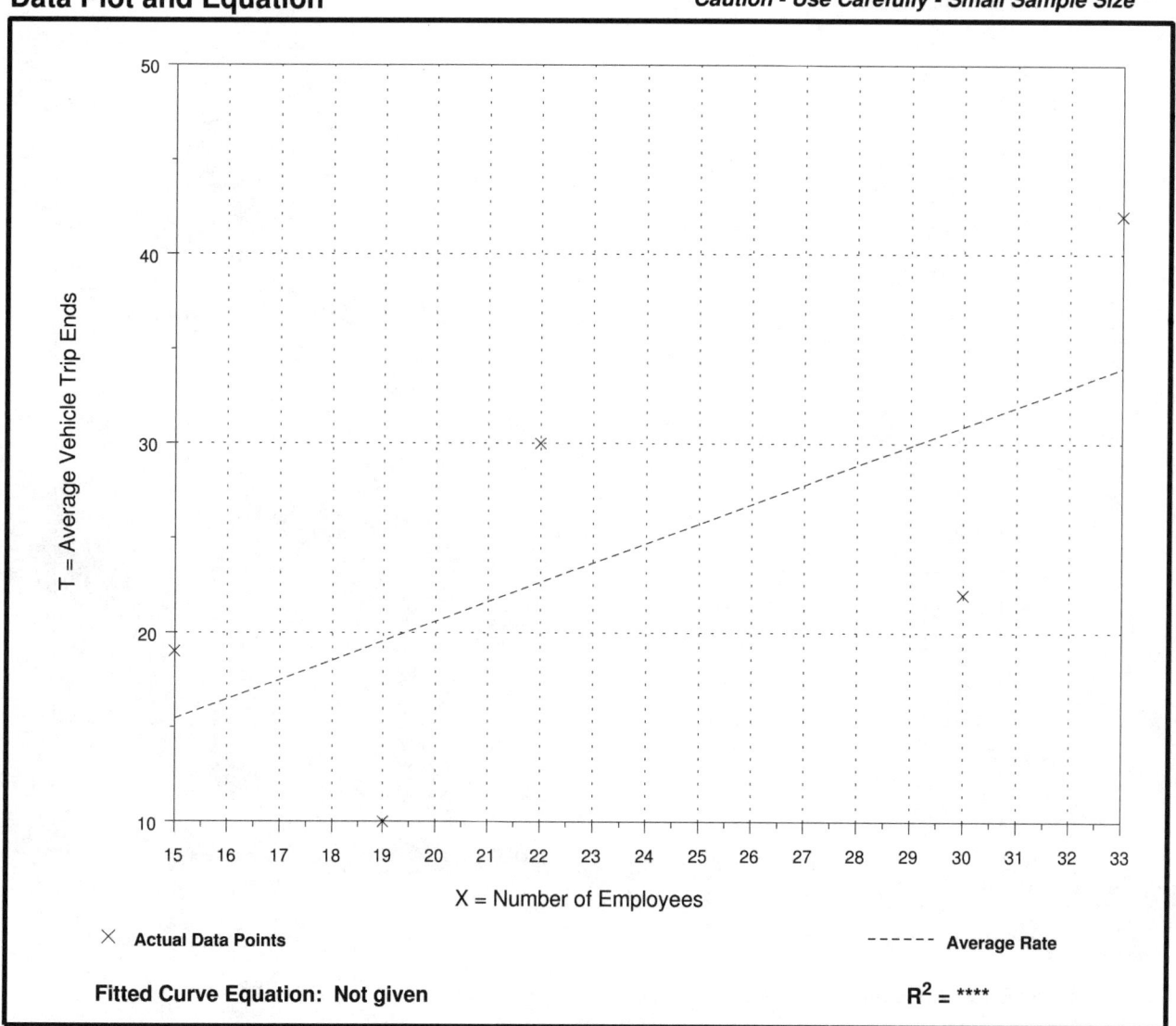

Fitted Curve Equation: Not given $R^2 = ****$

Library
(590)

Average Vehicle Trip Ends vs: Employees
On a: Weekday,
Peak Hour of Adjacent Street Traffic,
One Hour Between 4 and 6 p.m.

Number of Studies: 8
Avg. Number of Employees: 22
Directional Distribution: 47% entering, 53% exiting

Trip Generation per Employee

Average Rate	Range of Rates	Standard Deviation
5.40	1.60 - 12.33	3.50

Data Plot and Equation

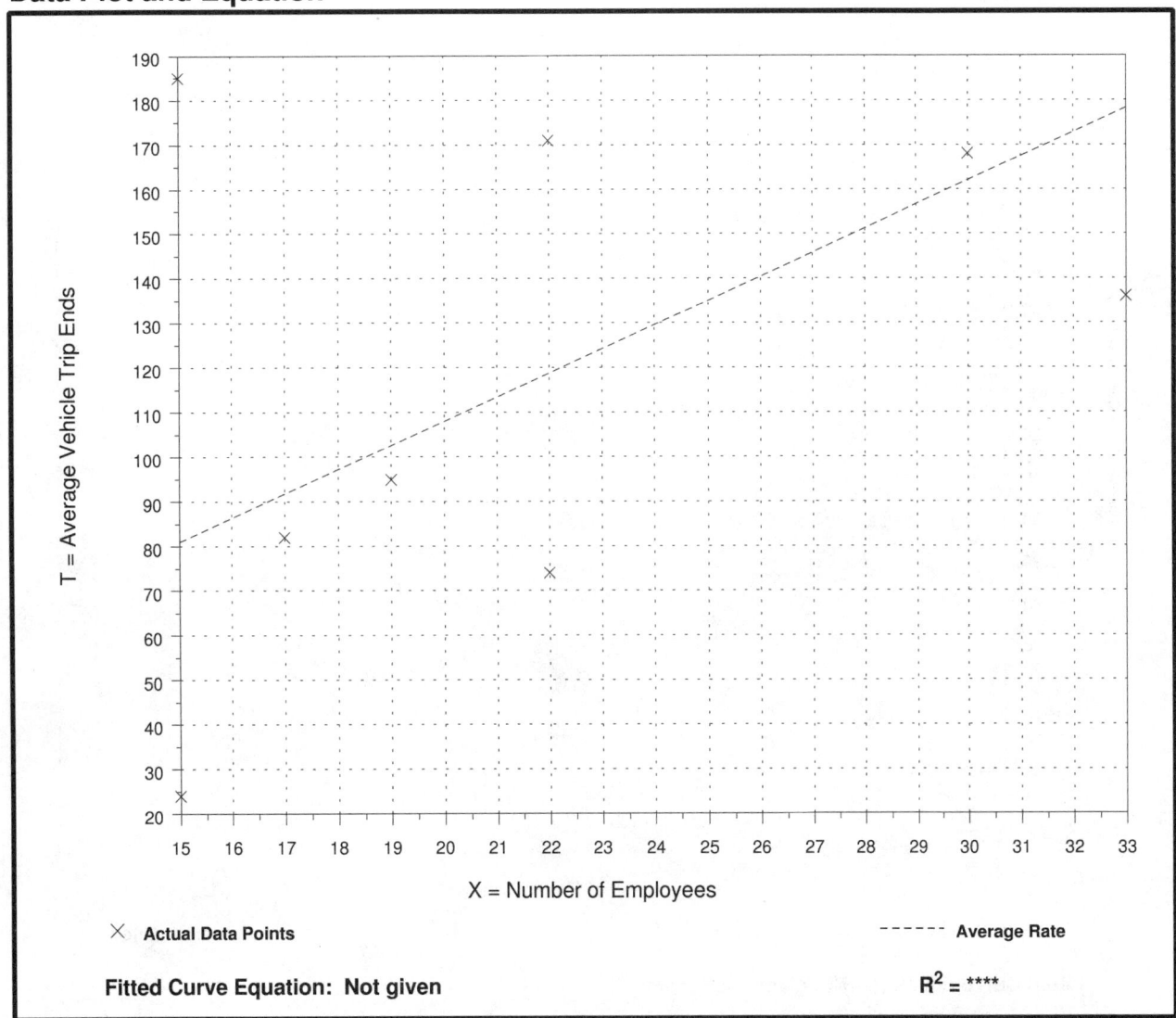

X Actual Data Points   ----- Average Rate

Fitted Curve Equation: Not given $R^2 = ****$

Library
(590)

Average Vehicle Trip Ends vs: Employees
On a: Weekday,
A.M. Peak Hour of Generator

Number of Studies: 10
Avg. Number of Employees: 25
Directional Distribution: 50% entering, 50% exiting

Trip Generation per Employee

Average Rate	Range of Rates	Standard Deviation
4.17	2.13 - 8.47	2.67

Data Plot and Equation

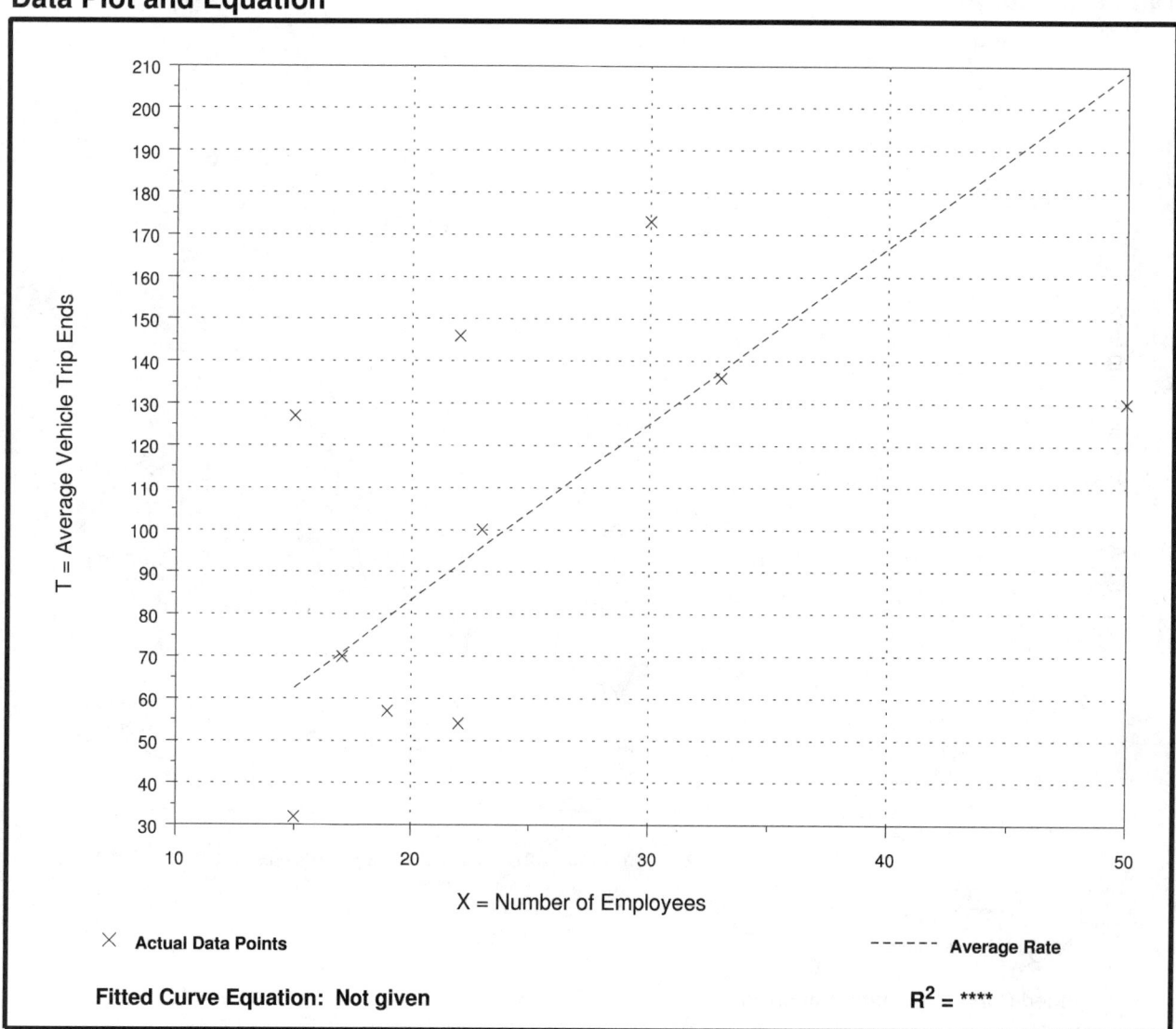

Fitted Curve Equation: Not given $R^2 = ****$

Library
(590)

Average Vehicle Trip Ends vs: Employees
On a: Weekday,
P.M. Peak Hour of Generator

Number of Studies: 10
Avg. Number of Employees: 25
Directional Distribution: 51% entering, 49% exiting

Trip Generation per Employee

Average Rate	Range of Rates	Standard Deviation
6.78	3.13 - 12.73	3.82

Data Plot and Equation

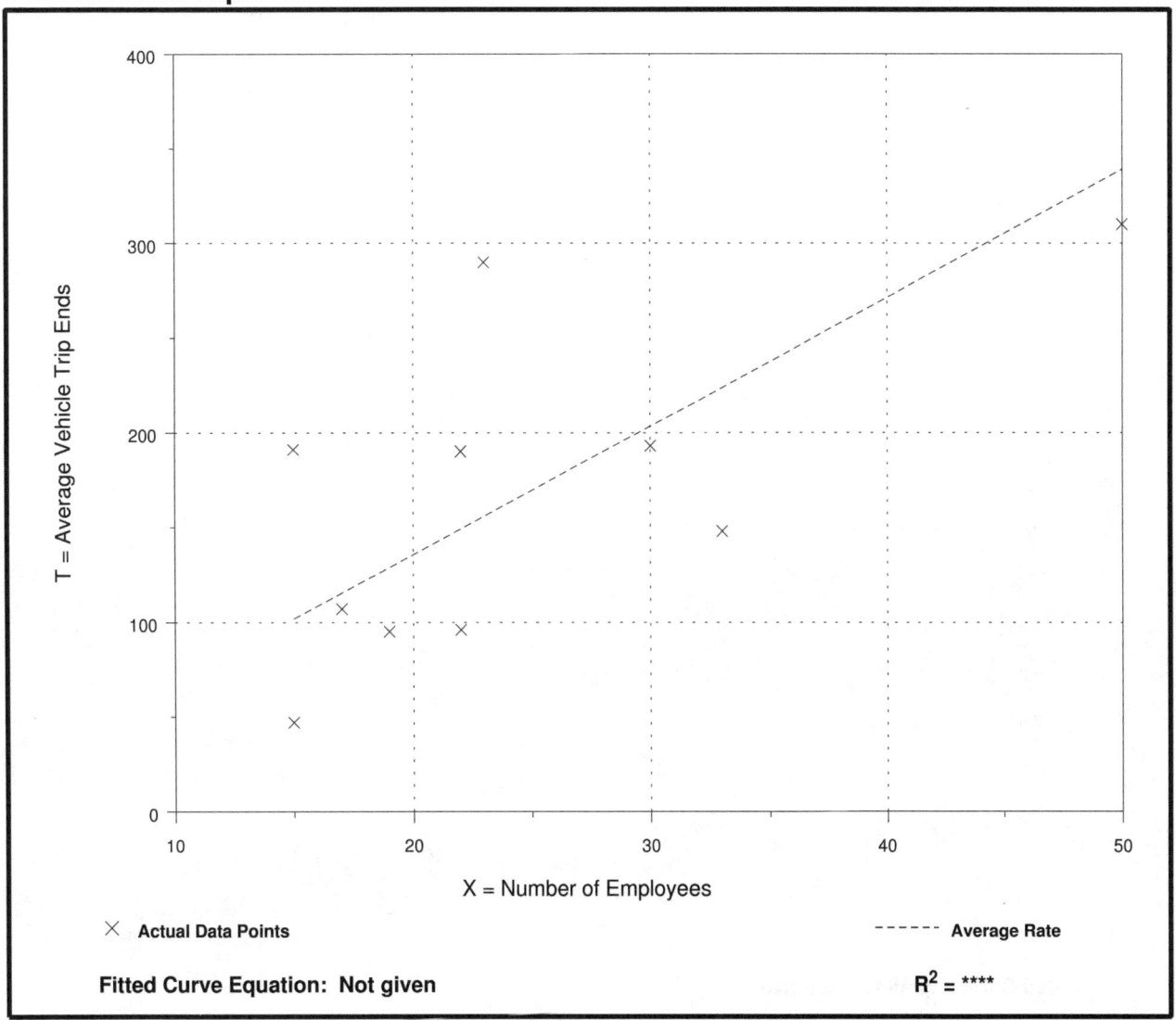

Fitted Curve Equation: Not given $R^2 = ****$

Library
(590)

Average Vehicle Trip Ends vs: Employees
On a: Saturday

Number of Studies: 6
Avg. Number of Employees: 30
Directional Distribution: 50% entering, 50% exiting

Trip Generation per Employee

Average Rate	Range of Rates	Standard Deviation
47.68	30.91 - 80.45	17.80

Data Plot and Equation

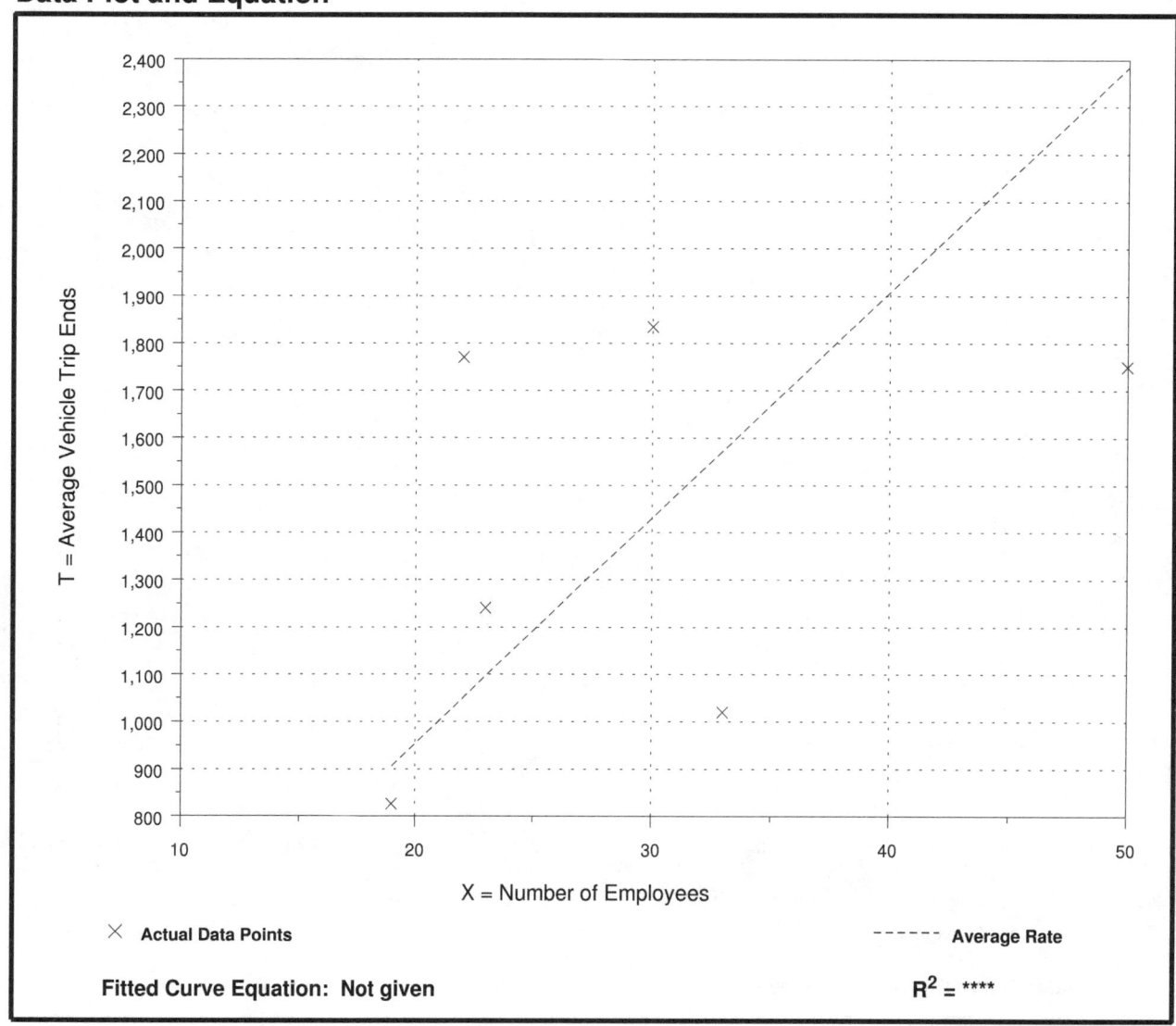

Fitted Curve Equation: Not given $R^2 =$ ****

Library
(590)

Average Vehicle Trip Ends vs: Employees
On a: Saturday,
Peak Hour of Generator

Number of Studies: 6
Avg. Number of Employees: 30
Directional Distribution: 53% entering, 47% exiting

Trip Generation per Employee

Average Rate	Range of Rates	Standard Deviation
6.91	3.80 - 13.14	4.00

Data Plot and Equation

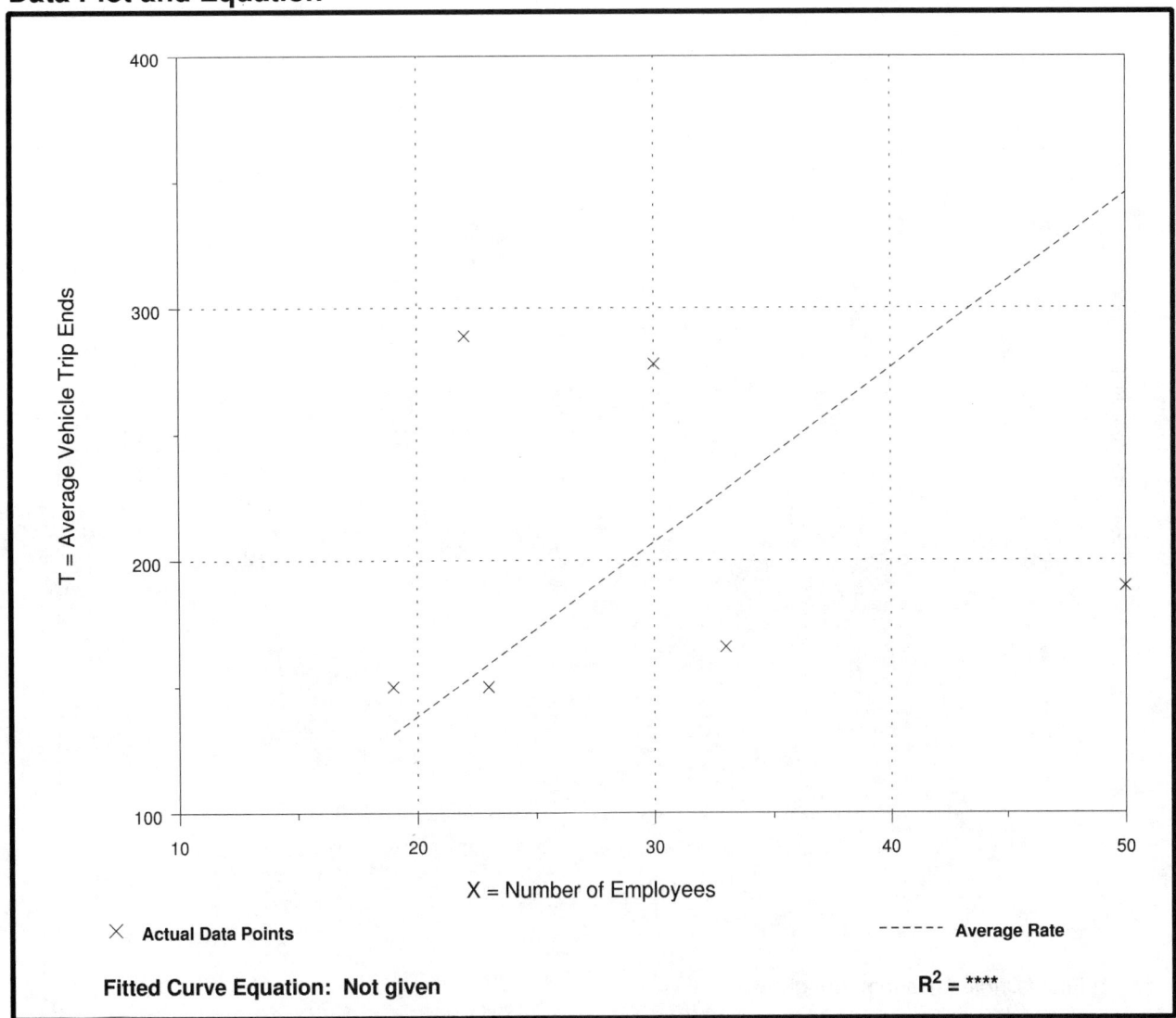

Fitted Curve Equation: Not given $R^2 = ****$

Library
(590)

Average Vehicle Trip Ends vs: Employees
On a: Sunday

Number of Studies: 5
Avg. Number of Employees: 25
Directional Distribution: 50% entering, 50% exiting

Trip Generation per Employee

Average Rate	Range of Rates	Standard Deviation
23.54	4.70 - 41.73	16.54

Data Plot and Equation

Caution - Use Carefully - Small Sample Size

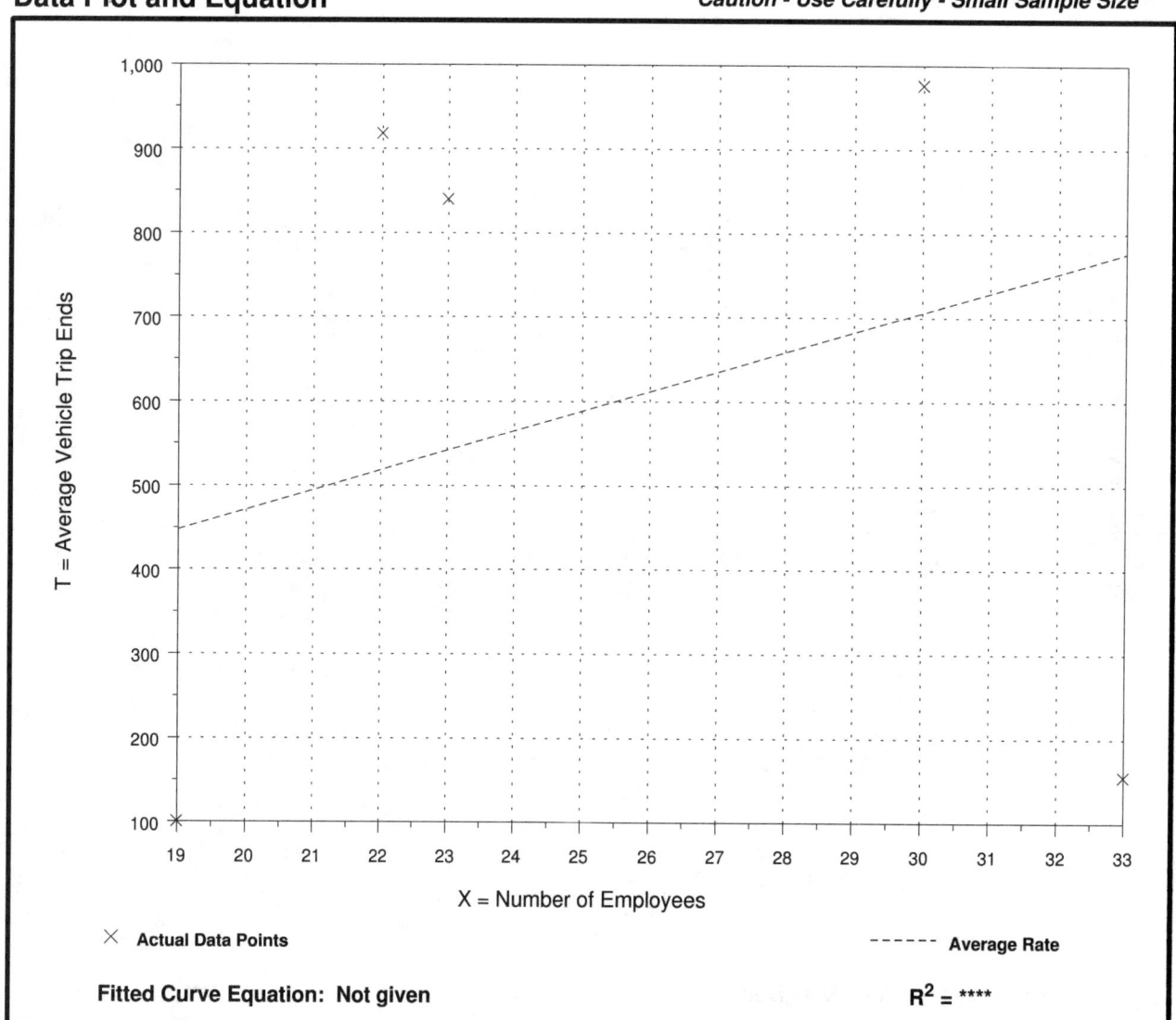

[Plot: T = Average Vehicle Trip Ends vs. X = Number of Employees; × Actual Data Points; ----- Average Rate]

Fitted Curve Equation: Not given $R^2 =$ ****

Library
(590)

Average Vehicle Trip Ends vs: Employees
On a: Sunday,
Peak Hour of Generator

Number of Studies: 5
Avg. Number of Employees: 25
Directional Distribution: 53% entering, 47% exiting

Trip Generation per Employee

Average Rate	Range of Rates	Standard Deviation
4.75	1.06 - 8.91	3.85

Data Plot and Equation

Caution - Use Carefully - Small Sample Size

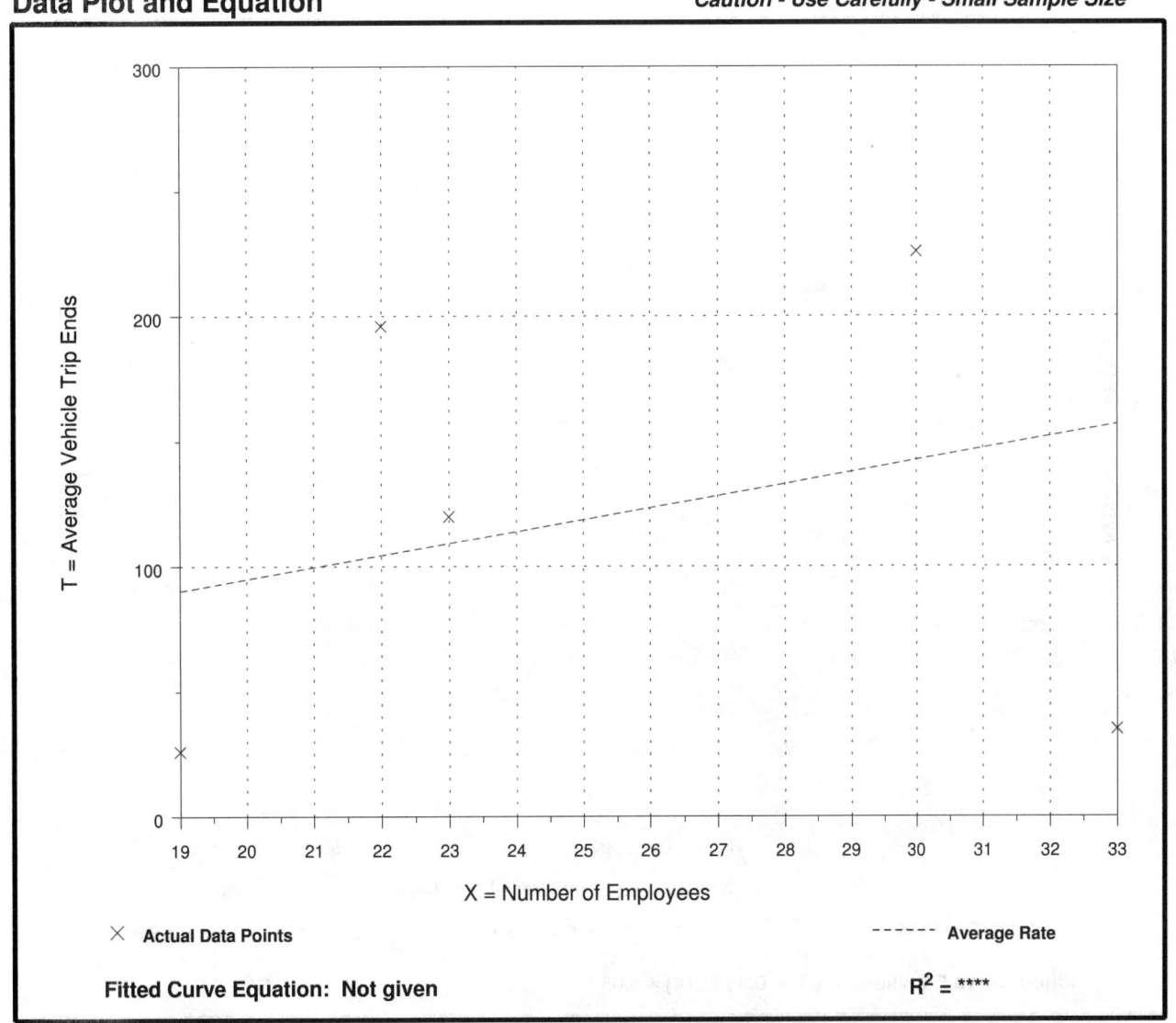

Fitted Curve Equation: Not given $R^2 = ****$

Library
(590)

Average Vehicle Trip Ends vs: 1000 Sq. Feet Gross Floor Area
On a: Weekday

Number of Studies: 10
Average 1000 Sq. Feet GFA: 23
Directional Distribution: 50% entering, 50% exiting

Trip Generation per 1000 Sq. Feet Gross Floor Area

Average Rate	Range of Rates	Standard Deviation
56.24	28.75 - 88.25	22.45

Data Plot and Equation

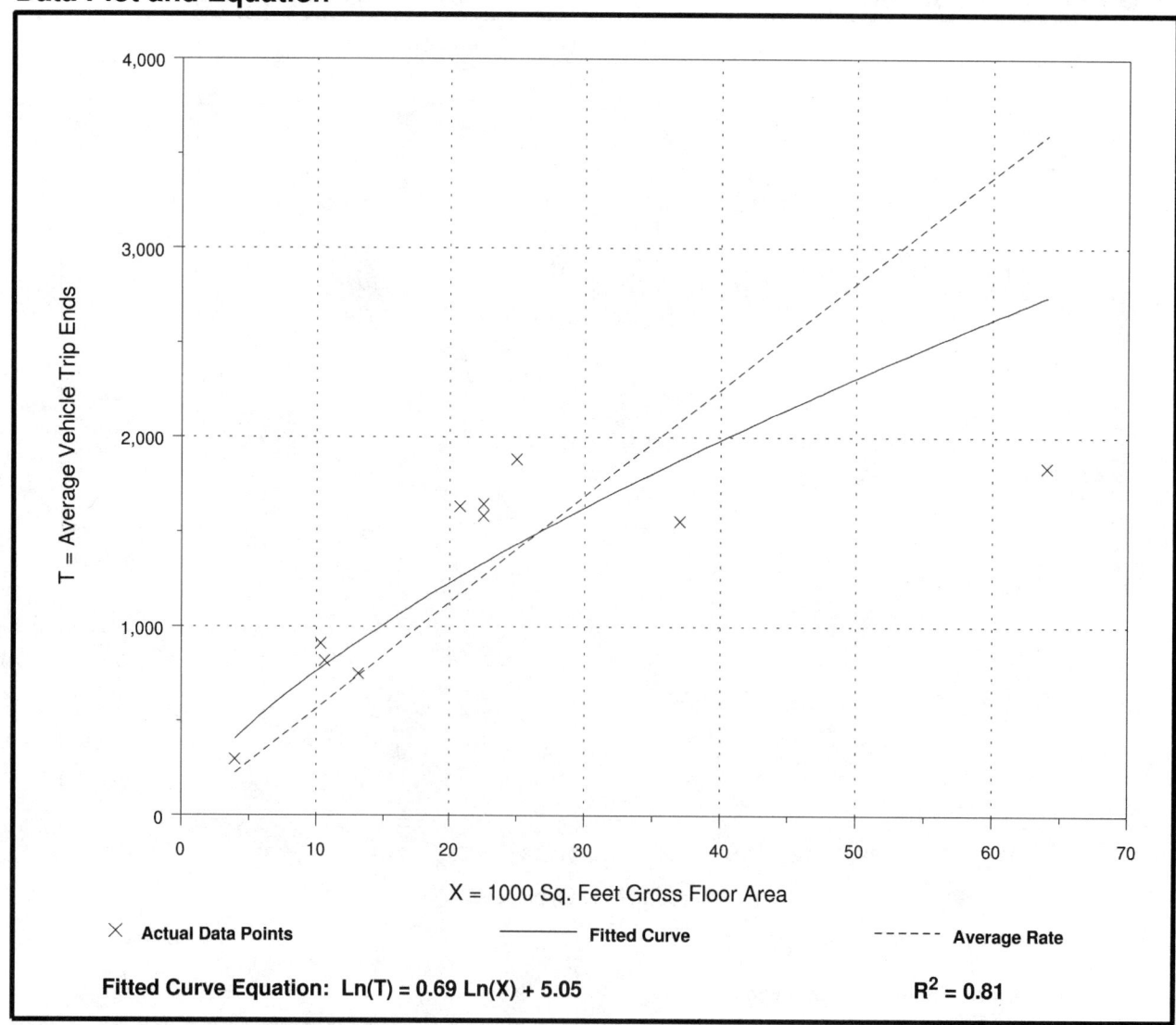

Fitted Curve Equation: $Ln(T) = 0.69\, Ln(X) + 5.05$ $R^2 = 0.81$

Library
(590)

Average Vehicle Trip Ends vs: 1000 Sq. Feet Gross Floor Area
On a: Weekday,
Peak Hour of Adjacent Street Traffic,
One Hour Between 7 and 9 a.m.

Number of Studies: 6
Average 1000 Sq. Feet GFA: 21
Directional Distribution: 71% entering, 29% exiting

Trip Generation per 1000 Sq. Feet Gross Floor Area

Average Rate	Range of Rates	Standard Deviation
1.04	0.52 - 1.33	1.02

Data Plot and Equation

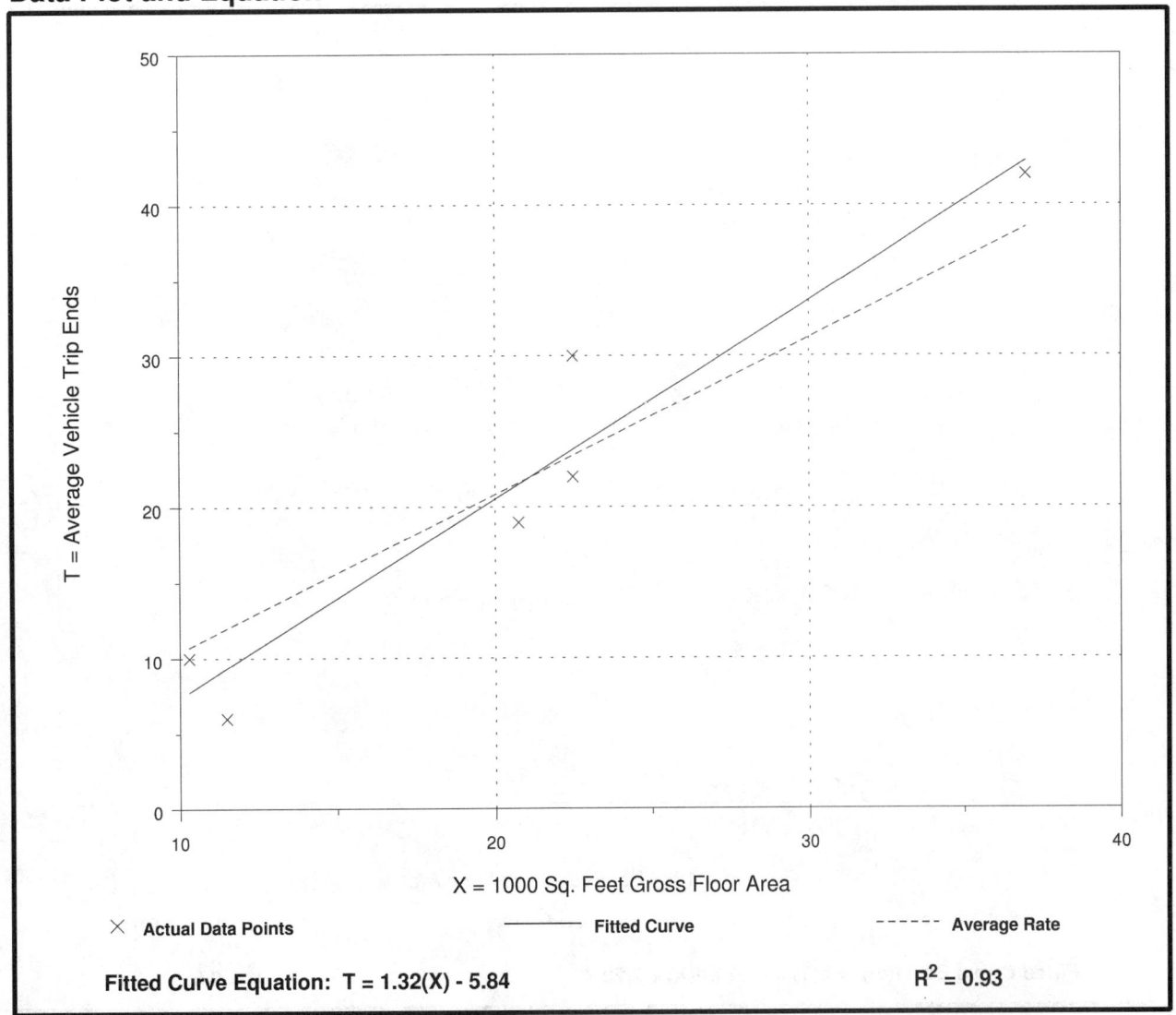

Fitted Curve Equation: $T = 1.32(X) - 5.84$ $R^2 = 0.93$

Library
(590)

Average Vehicle Trip Ends vs: 1000 Sq. Feet Gross Floor Area
On a: Weekday,
Peak Hour of Adjacent Street Traffic,
One Hour Between 4 and 6 p.m.

Number of Studies: 11
Average 1000 Sq. Feet GFA: 16
Directional Distribution: 48% entering, 52% exiting

Trip Generation per 1000 Sq. Feet Gross Floor Area

Average Rate	Range of Rates	Standard Deviation
7.30	3.68 - 12.25	3.81

Data Plot and Equation

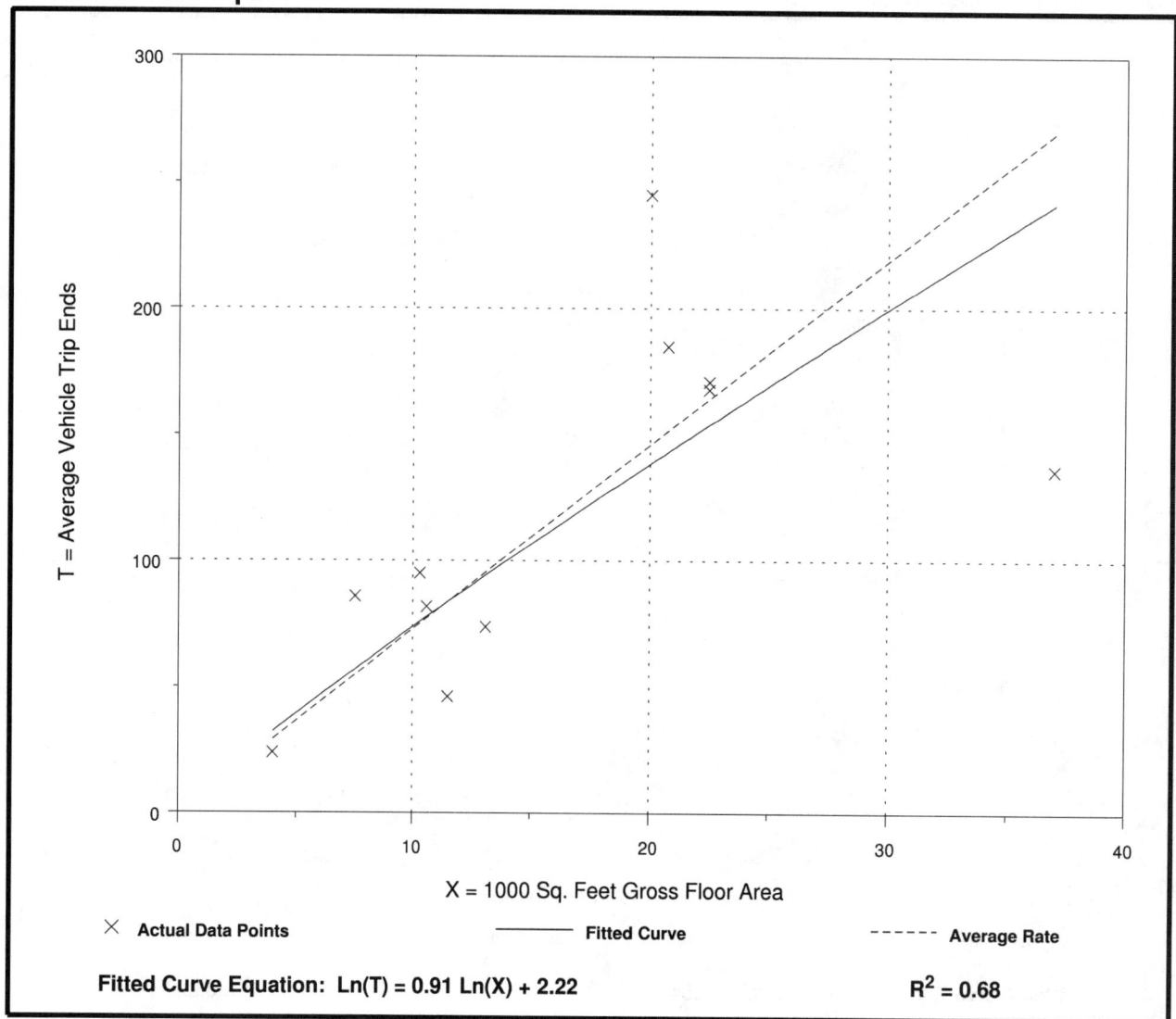

Fitted Curve Equation: $Ln(T) = 0.91\, Ln(X) + 2.22$ $R^2 = 0.68$

Library
(590)

Average Vehicle Trip Ends vs: 1000 Sq. Feet Gross Floor Area
On a: Weekday,
A.M. Peak Hour of Generator

Number of Studies: 11
Average 1000 Sq. Feet GFA: 22
Directional Distribution: 49% entering, 51% exiting

Trip Generation per 1000 Sq. Feet Gross Floor Area

Average Rate	Range of Rates	Standard Deviation
4.47	2.03 - 8.00	2.83

Data Plot and Equation

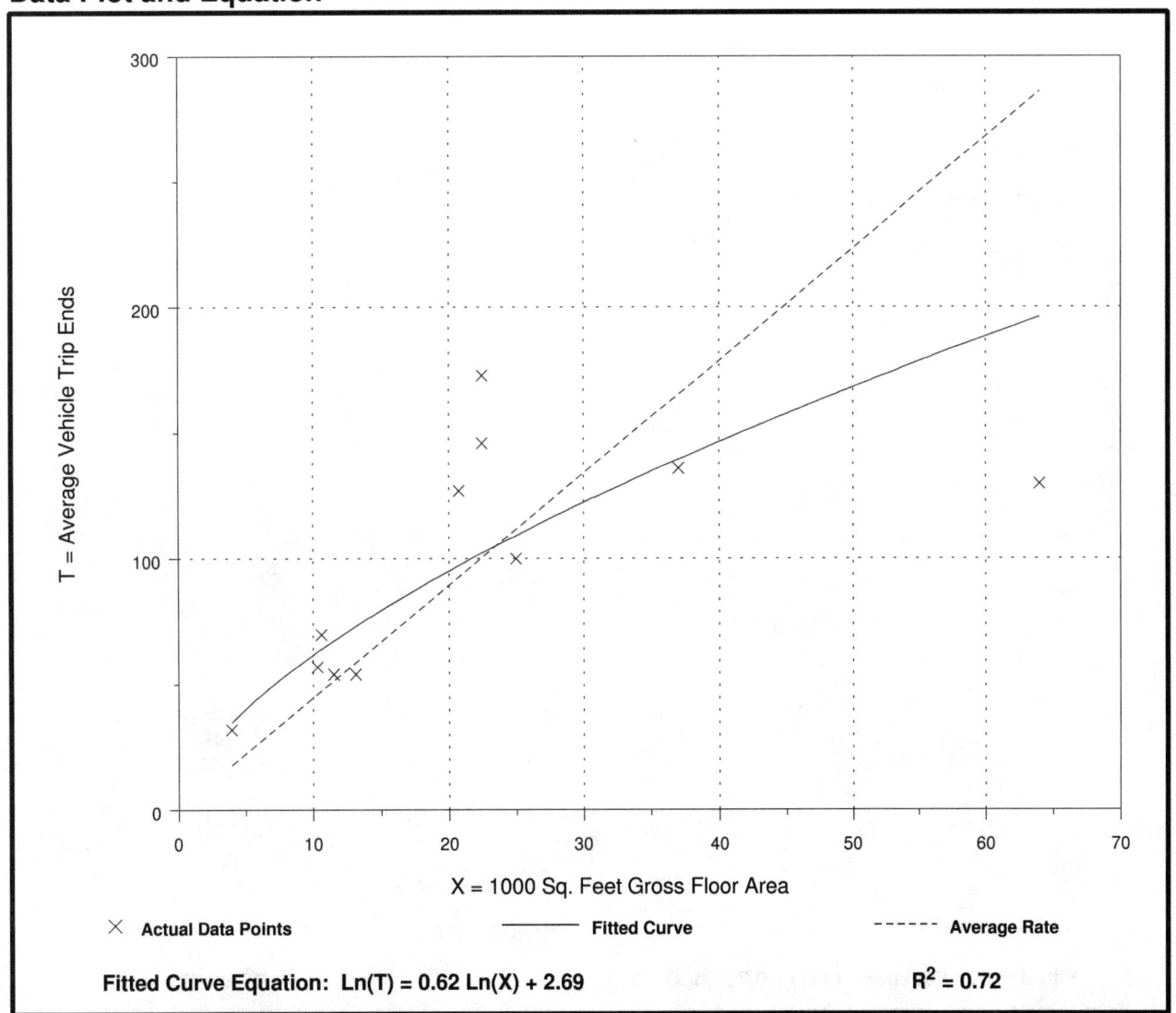

Fitted Curve Equation: $Ln(T) = 0.62 \, Ln(X) + 2.69$ $R^2 = 0.72$

Library
(590)

Average Vehicle Trip Ends vs: 1000 Sq. Feet Gross Floor Area
On a: Weekday,
P.M. Peak Hour of Generator

Number of Studies: 11
Average 1000 Sq. Feet GFA: 22
Directional Distribution: 52% entering, 48% exiting

Trip Generation per 1000 Sq. Feet Gross Floor Area

Average Rate	Range of Rates	Standard Deviation
7.20	4.00 - 11.75	3.68

Data Plot and Equation

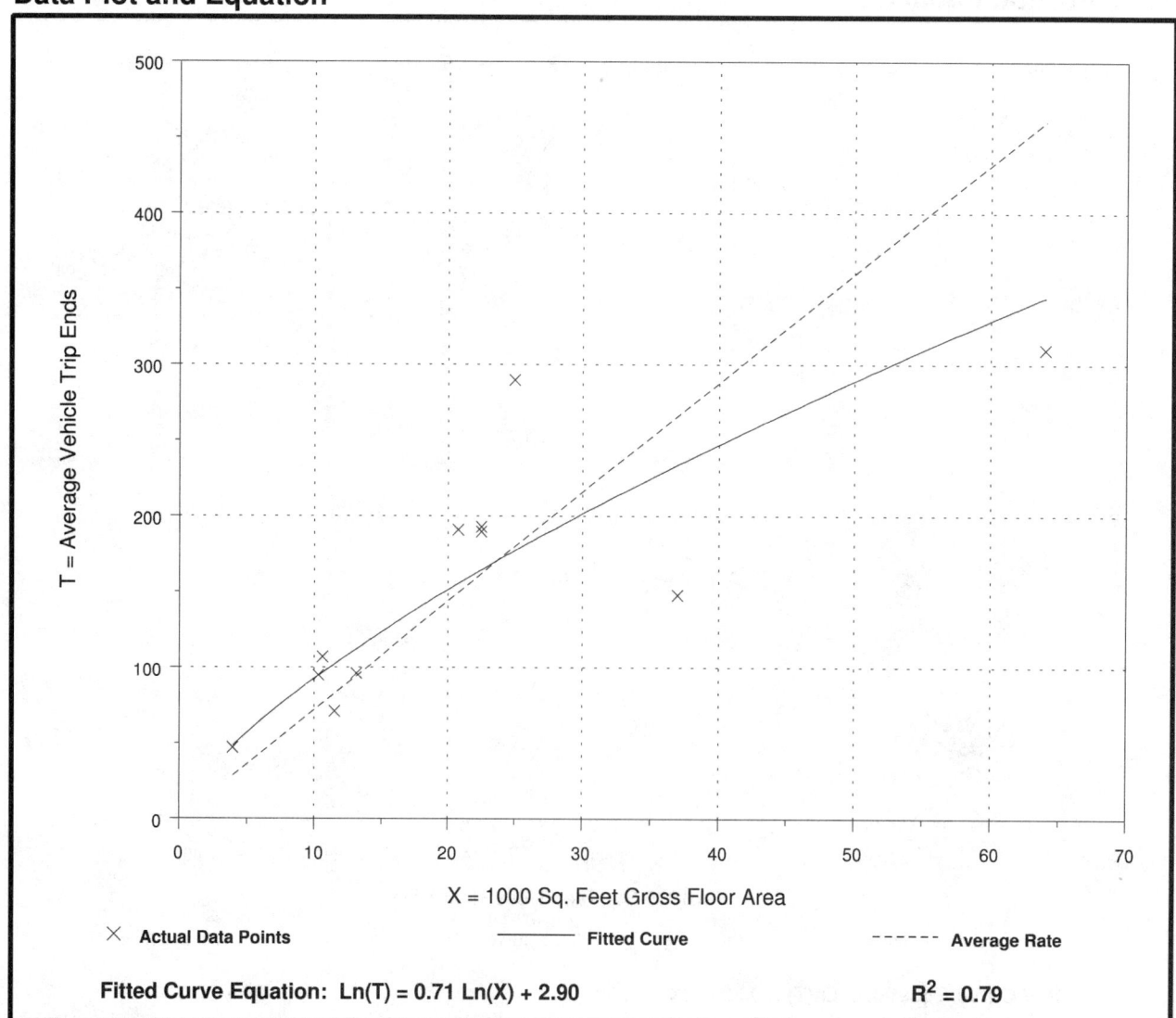

Fitted Curve Equation: Ln(T) = 0.71 Ln(X) + 2.90 $R^2 = 0.79$

Library
(590)

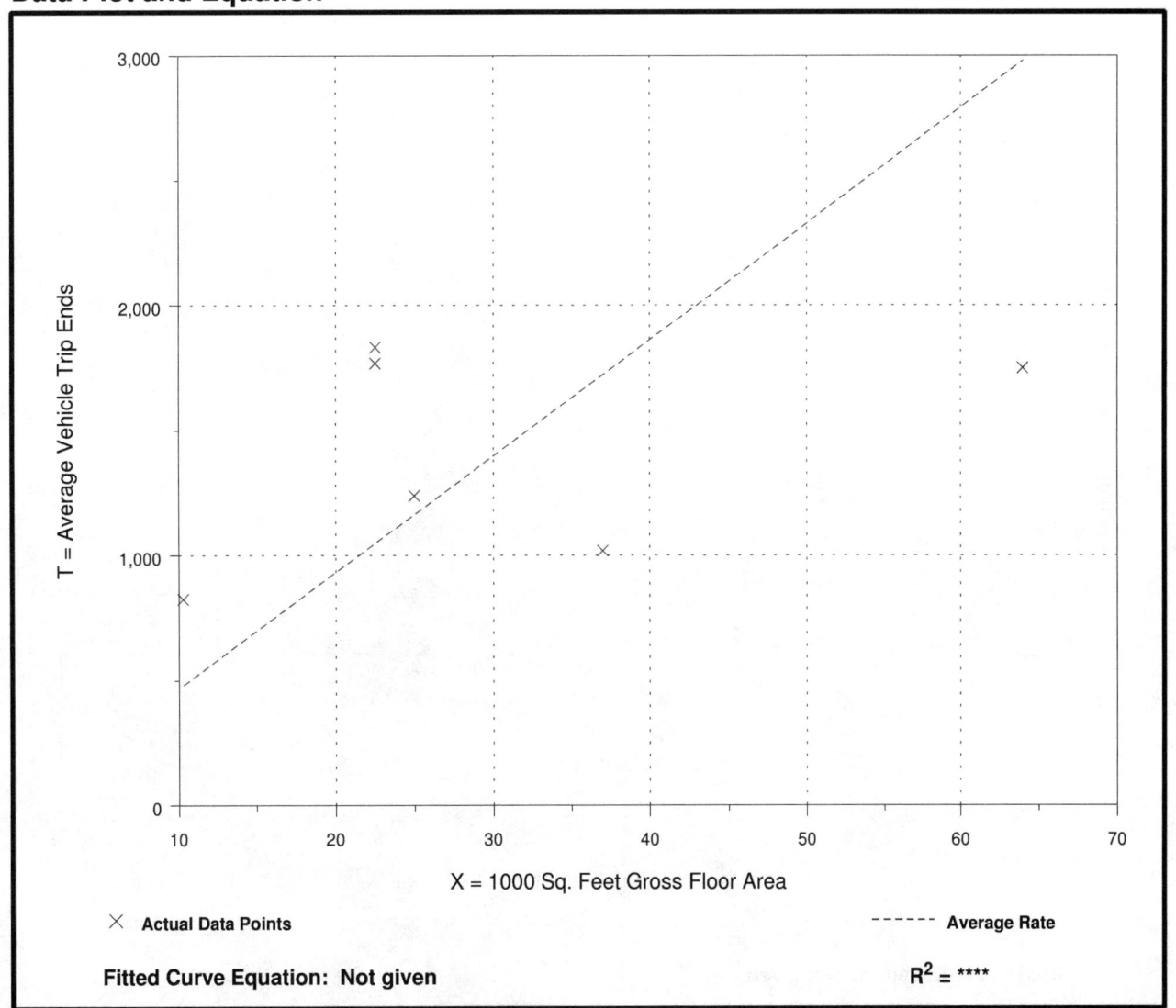

Average Vehicle Trip Ends vs: 1000 Sq. Feet Gross Floor Area
On a: Saturday

Number of Studies: 6
Average 1000 Sq. Feet GFA: 30
Directional Distribution: 50% entering, 50% exiting

Trip Generation per 1000 Sq. Feet Gross Floor Area

Average Rate	Range of Rates	Standard Deviation
46.55	27.34 - 81.51	24.43

Data Plot and Equation

Fitted Curve Equation: Not given $R^2 = ****$

Library
(590)

Average Vehicle Trip Ends vs: 1000 Sq. Feet Gross Floor Area
On a: Saturday,
Peak Hour of Generator

Number of Studies: 6
Average 1000 Sq. Feet GFA: 30
Directional Distribution: 53% entering, 47% exiting

Trip Generation per 1000 Sq. Feet Gross Floor Area

Average Rate	Range of Rates	Standard Deviation
6.75	2.97 - 14.56	4.97

Data Plot and Equation

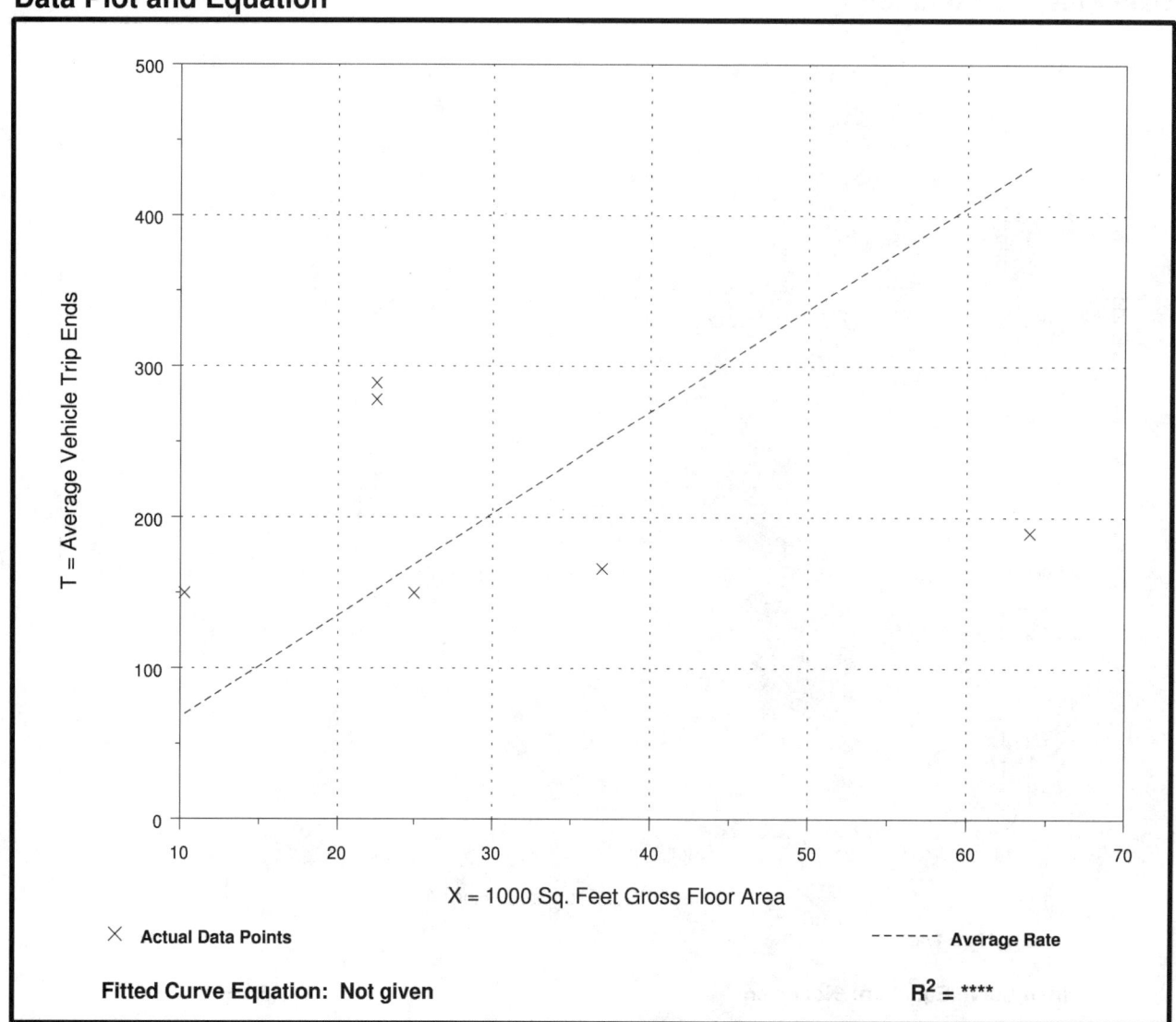

Fitted Curve Equation: Not given $R^2 = ****$

Library
(590)

Average Vehicle Trip Ends vs: 1000 Sq. Feet Gross Floor Area
On a: Sunday

Number of Studies: 5
Average 1000 Sq. Feet GFA: 23
Directional Distribution: 50% entering, 50% exiting

Trip Generation per 1000 Sq. Feet Gross Floor Area

Average Rate	Range of Rates	Standard Deviation
25.49	4.19 - 43.38	17.67

Data Plot and Equation

Caution - Use Carefully - Small Sample Size

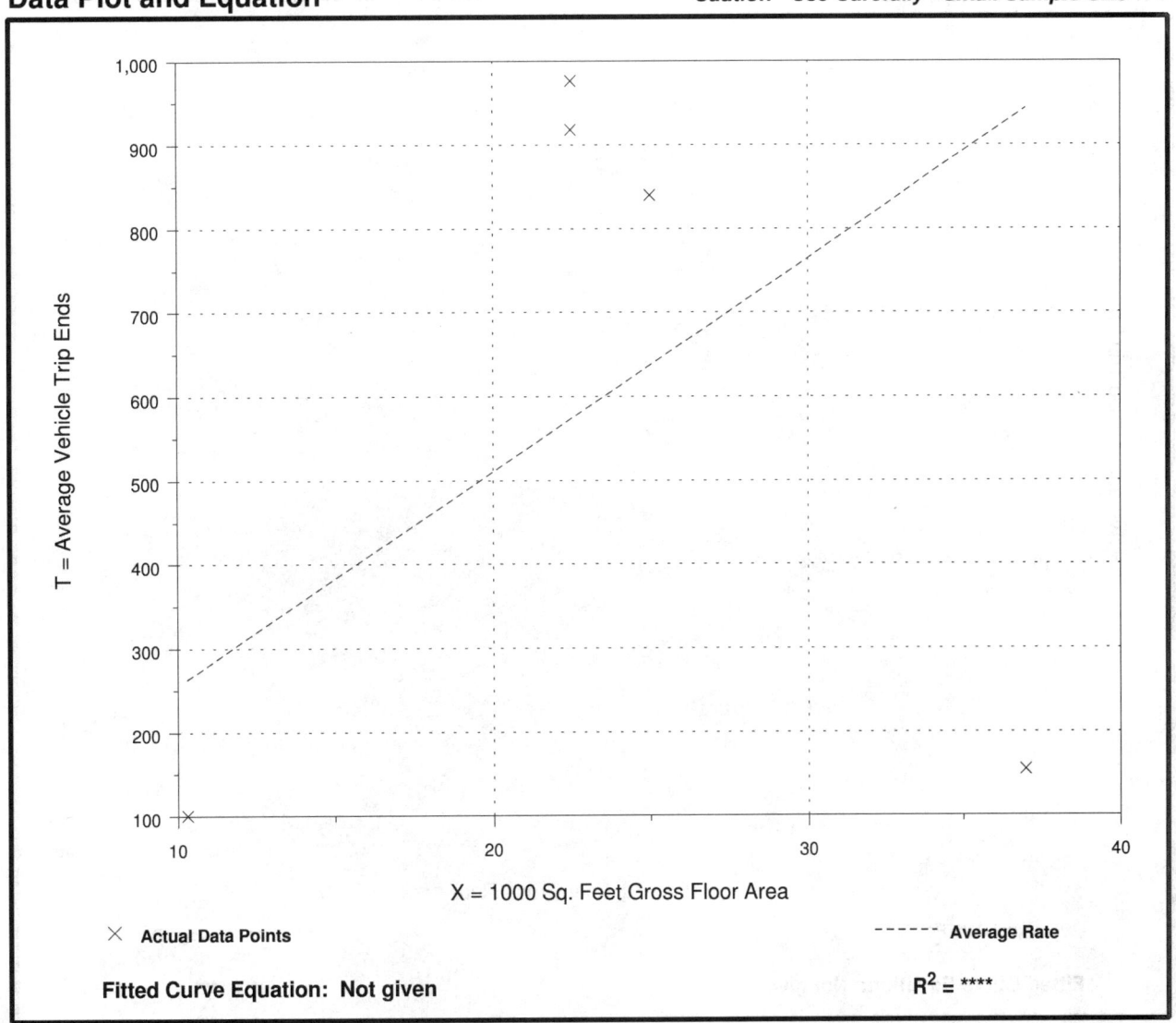

[Plot: T = Average Vehicle Trip Ends vs X = 1000 Sq. Feet Gross Floor Area]

× Actual Data Points
----- Average Rate

Fitted Curve Equation: Not given $R^2 =$ ****

Library
(590)

Average Vehicle Trip Ends vs: 1000 Sq. Feet Gross Floor Area
On a: Sunday,
Peak Hour of Generator

Number of Studies: 5
Average 1000 Sq. Feet GFA: 23
Directional Distribution: 53% entering, 47% exiting

Trip Generation per 1000 Sq. Feet Gross Floor Area

Average Rate	Range of Rates	Standard Deviation
5.14	0.95 - 10.04	4.28

Data Plot and Equation

Caution - Use Carefully - Small Sample Size

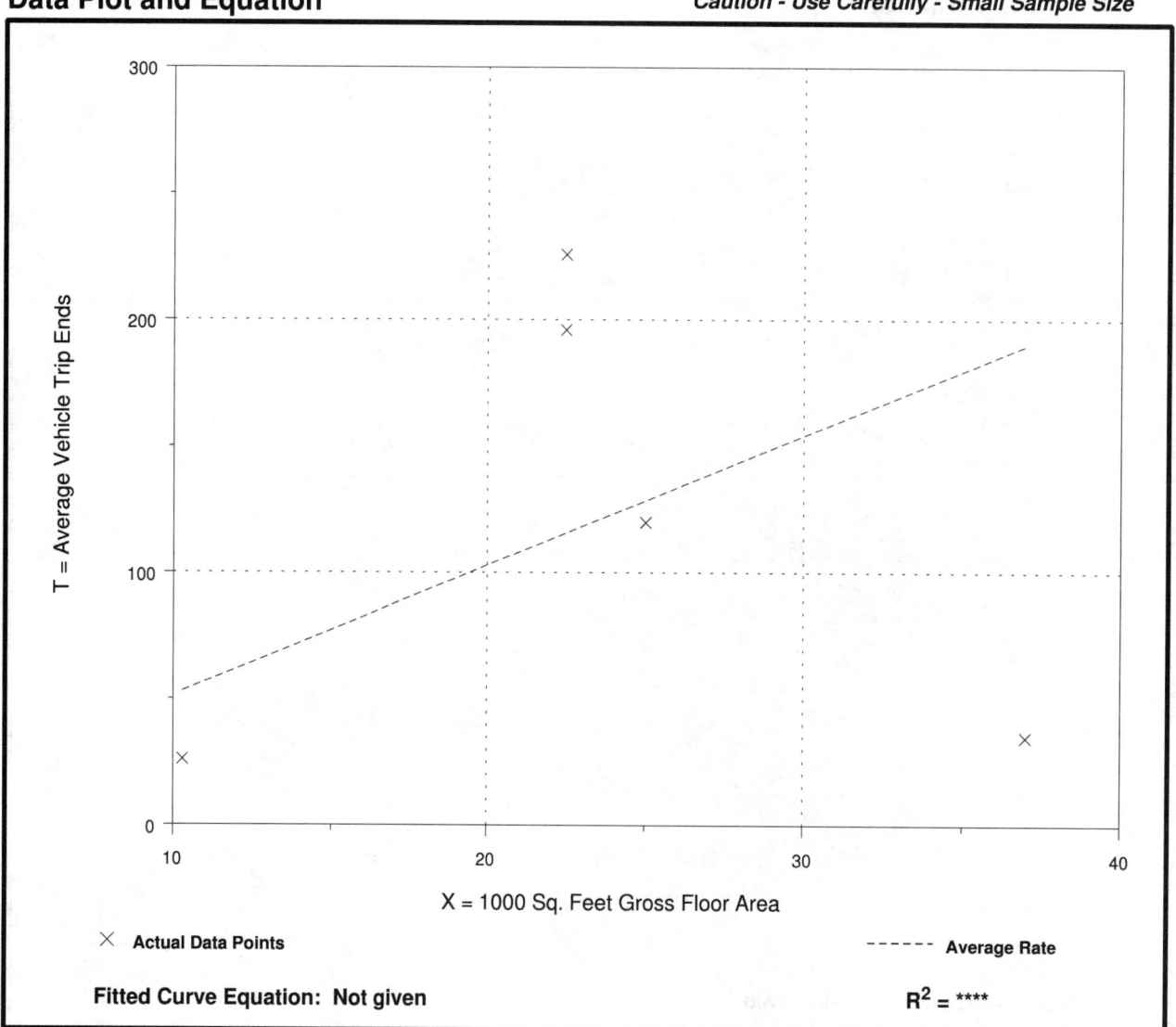

Fitted Curve Equation: Not given $R^2 = ****$

Land Use: 591
Lodge/Fraternal Organization

Description

A lodge or fraternal organization typically includes a clubhouse with dining and drinking facilities, recreational and entertainment areas and meeting rooms.

Additional Data

Peak hours of the generator—
The weekday A.M. peak hour was between 11:00 a.m. and 12:00 p.m. The weekday P.M. peak hour was between 3:00 p.m. and 4:00 p.m. The Saturday peak hour was between 12:00 p.m. and 1:00 p.m. The Sunday peak hour was between 5:00 p.m. and 6:00 p.m.

This site was surveyed in 1977 at a lodge in California with 3,200 members, 20 employees and 246 parking spaces. On-site facilities included a clubhouse with three dining rooms and two bars, tennis and handball courts, bowling lanes, a billiard room, swimming pool, exercise room and steam room.

Source Number

113

Land Use: 591
Lodge/Fraternal Organization
Independent Variables with One Observation

The following trip generation data are for independent variables with only one observation. This information is shown in this table only; there are no related plots for these data.

Users are cautioned to use data with care because of the small sample size.

Independent Variable	Trip Generation Rate	Size of Independent Variable	Number of Studies	Directional Distribution
Employees				
Weekday	46.90	20	1	50% entering, 50% exiting
Weekday A.M. Peak Hour of Adjacent Street Traffic	2.10	20	1	Not available
Weekday P.M. Peak Hour of Adjacent Street Traffic	4.05	20	1	Not available
Weekday A.M. Peak Hour of Generator	4.30	20	1	Not available
Weekday P.M. Peak Hour of Generator	4.05	20	1	Not available
Saturday	29.55	20	1	50% entering, 50% exiting
Saturday Peak Hour of Generator	3.10	20	1	Not available
Sunday	29.10	20	1	50% entering, 50% exiting
Sunday Peak Hour of Generator	3.75	20	1	Not available

Independent Variable	Trip Generation Rate	Size of Independent Variable	Number of Studies	Directional Distribution
Members				
Weekday	0.29	3,200	1	50% entering, 50% exiting
Weekday A.M. Peak Hour of Adjacent Street Traffic	0.01	3,200	1	Not available
Weekday P.M. Peak Hour of Adjacent Street Traffic	0.03	3,200	1	Not available
Weekday A.M. Peak Hour of Generator	0.03	3,200	1	Not available
Weekday P.M. Peak Hour of Generator	0.03	3,200	1	Not available

Saturday	0.18	3,200	1	50% entering, 50% exiting
Saturday Peak Hour of Generator	0.02	3,200	1	Not available
Sunday	0.18	3,200	1	50% entering, 50% exiting
Sunday Peak Hour of Generator	0.02	3,200	1	Not available

Land Use: 610
Hospital

Description

A hospital is any institution where medical or surgical care and overnight accommodations are provided to non-ambulatory and ambulatory patients. However, the term "hospital" does not refer to medical clinics (facilities that provide diagnoses and outpatient care only) or nursing homes (facilities devoted to the care of persons unable to care for themselves), which are covered elsewhere in this report. Clinic (Land Use 630) is a related use.

Additional Data

Peak hours of the generator—
    The weekday A.M. peak hour varied between 6:45 a.m. and 11:00 a.m. The weekday P.M. peak hour varied between 1:00 p.m. and 5:00 p.m.

The sites were surveyed between the 1960s and the 2000s throughout the United States.

Specialized Land Use Data

A 2009 study provided data on a research and training-type medical center in Alberta, Canada. The size and trip generation characteristics of this site differed from sites included in this land use; therefore, trip generation information for this site is presented in the following tables and was excluded from the data plots. The A.M. peak hour of the generator was from 7:15 a.m. to 8:15 a.m. The P.M. peak hour of generator was from 3:45 p.m. to 4:45 p.m.

Independent Variable	Average Trip Generation Rate	Size of Independent Variable	Number of Studies	Directional Distribution
Employees				
Weekday	2.76	9,682	1	Not available
Weekday A.M. Peak Hour of Adjacent Street Traffic	0.25	9,682	1	77% entering, 23% exiting
Weekday P.M. Peak Hour of Adjacent Street Traffic	0.23	9,682	1	25% entering, 75% exiting
Weekday A.M. Peak Hour of Generator	0.26	9,682	1	74% entering, 26% exiting
Weekday P.M. Peak Hour of Generator	0.24	9,682	1	26% entering, 74% exiting

Source: 702

Independent Variable	Average Trip Generation Rate	Size of Independent Variable	Number of Studies	Directional Distribution

1,000 Square Feet Gross Floor Area

Weekday	6.95	3,843	1	Not available
Weekday A.M. Peak Hour of Adjacent Street Traffic	0.63	3,843	1	77% entering, 23% exiting
Weekday P.M. Peak Hour of Adjacent Street Traffic	0.60	3,843	1	25% entering, 75% exiting
Weekday A.M. Peak Hour of Generator	0.65	3,843	1	74% entering, 26% exiting
Weekday P.M. Peak Hour of Generator	0.61	3,843	1	26% entering, 74% exiting

Source: 702

Independent Variable	Average Trip Generation Rate	Size of Independent Variable	Number of Studies	Directional Distribution

Beds

Weekday	28.16	949	1	Not available
Weekday A.M. Peak Hour of Adjacent Street Traffic	2.54	949		77% entering, 23% exiting
Weekday P.M. Peak Hour of Adjacent Street Traffic	2.39	949	1	25% entering, 75% exiting
Weekday A.M. Peak Hour of Generator	2.62	949	1	74% entering, 26% exiting
Weekday P.M. Peak Hour of Generator	2.46	949	1	26% entering, 74% exiting

Source: 702

Another study, which was conducted in 2008, provided data on a research hospital in Baltimore, Maryland. The trip generation characteristics of this site differed from sites included in this land use; therefore, trip generation information for this site is presented in the following tables and was excluded from the data plots.

Independent Variable	Average Trip Generation Rate	Size of Independent Variable	Number of Studies	Directional Distribution

Employees

Independent Variable	Average Trip Generation Rate	Size of Independent Variable	Number of Studies	Directional Distribution
Weekday A.M. Peak Hour of Adjacent Street Traffic	0.21	5,500	1	81% entering, 19% exiting
Weekday P.M. Peak Hour of Adjacent Street Traffic	0.20	5,500	1	31% entering, 69% exiting

Source: 749

Independent Variable	Average Trip Generation Rate	Size of Independent Variable	Number of Studies	Directional Distribution

1,000 Square Feet Gross Floor Area

Independent Variable	Average Trip Generation Rate	Size of Independent Variable	Number of Studies	Directional Distribution
Weekday A.M. Peak Hour of Adjacent Street Traffic	0.42	2,800	1	81% entering, 19% exiting
Weekday P.M. Peak Hour of Adjacent Street Traffic	0.39	2,800	1	31% entering, 69% exiting

Source: 749

Source Numbers

2, 6, 14, 28, 88, 98, 110, 112, 186, 241, 253, 262, 423, 429, 533, 573, 591, 601, 630, 702, 719, 749

Hospital
(610)

Average Vehicle Trip Ends vs: Employees
On a: Weekday

Number of Studies: 21
Avg. Number of Employees: 1,137
Directional Distribution: 50% entering, 50% exiting

Trip Generation per Employee

Average Rate	Range of Rates	Standard Deviation
4.50	2.17 - 10.48	2.83

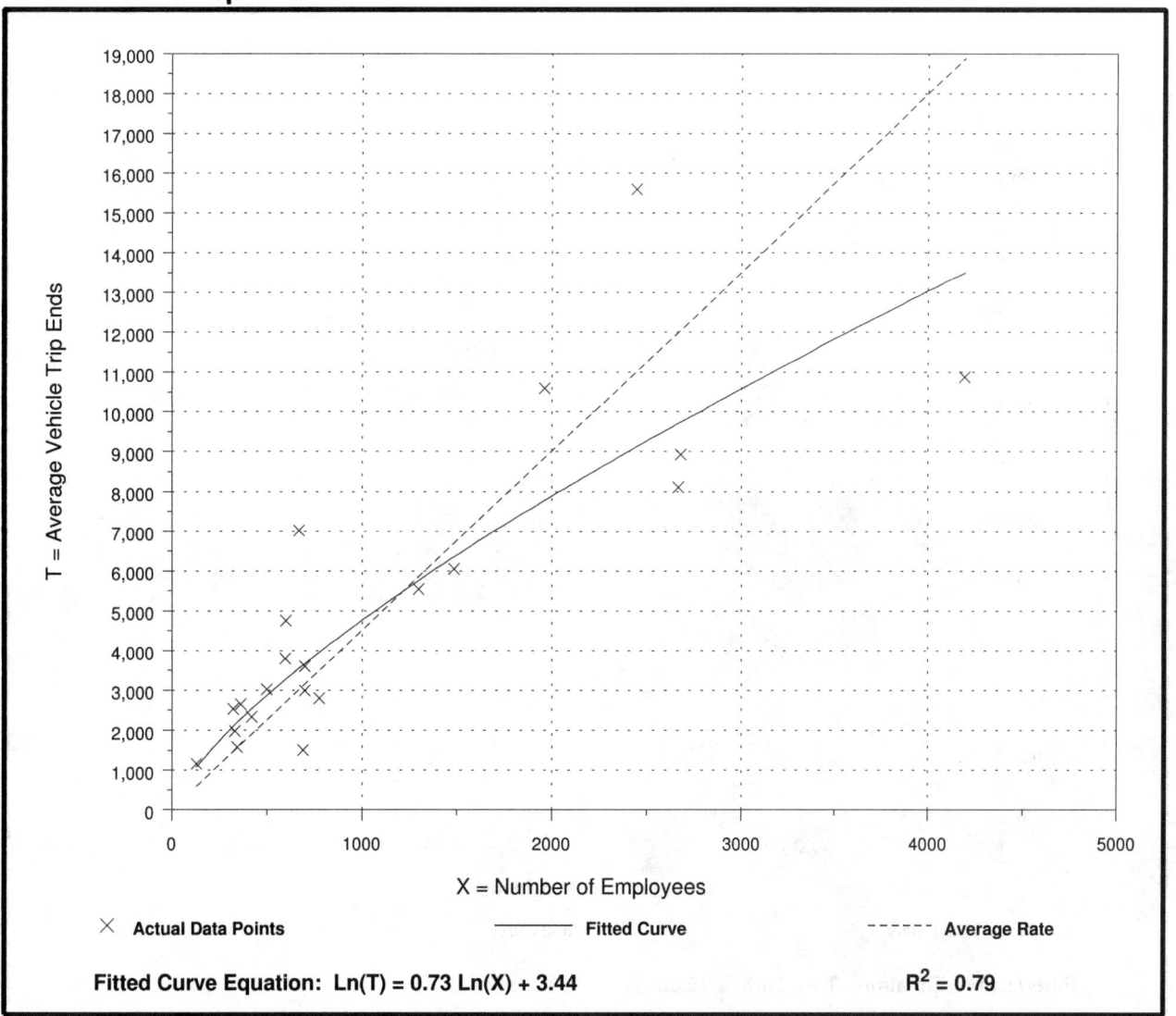

Data Plot and Equation

Fitted Curve Equation: $Ln(T) = 0.73\ Ln(X) + 3.44$ $R^2 = 0.79$

Hospital
(610)

Average Vehicle Trip Ends vs: Employees
On a: Weekday,
Peak Hour of Adjacent Street Traffic,
One Hour Between 7 and 9 a.m.

Number of Studies: 12
Avg. Number of Employees: 1,625
Directional Distribution: 72% entering, 28% exiting

Trip Generation per Employee

Average Rate	Range of Rates	Standard Deviation
0.31	0.12 - 0.85	0.57

Data Plot and Equation

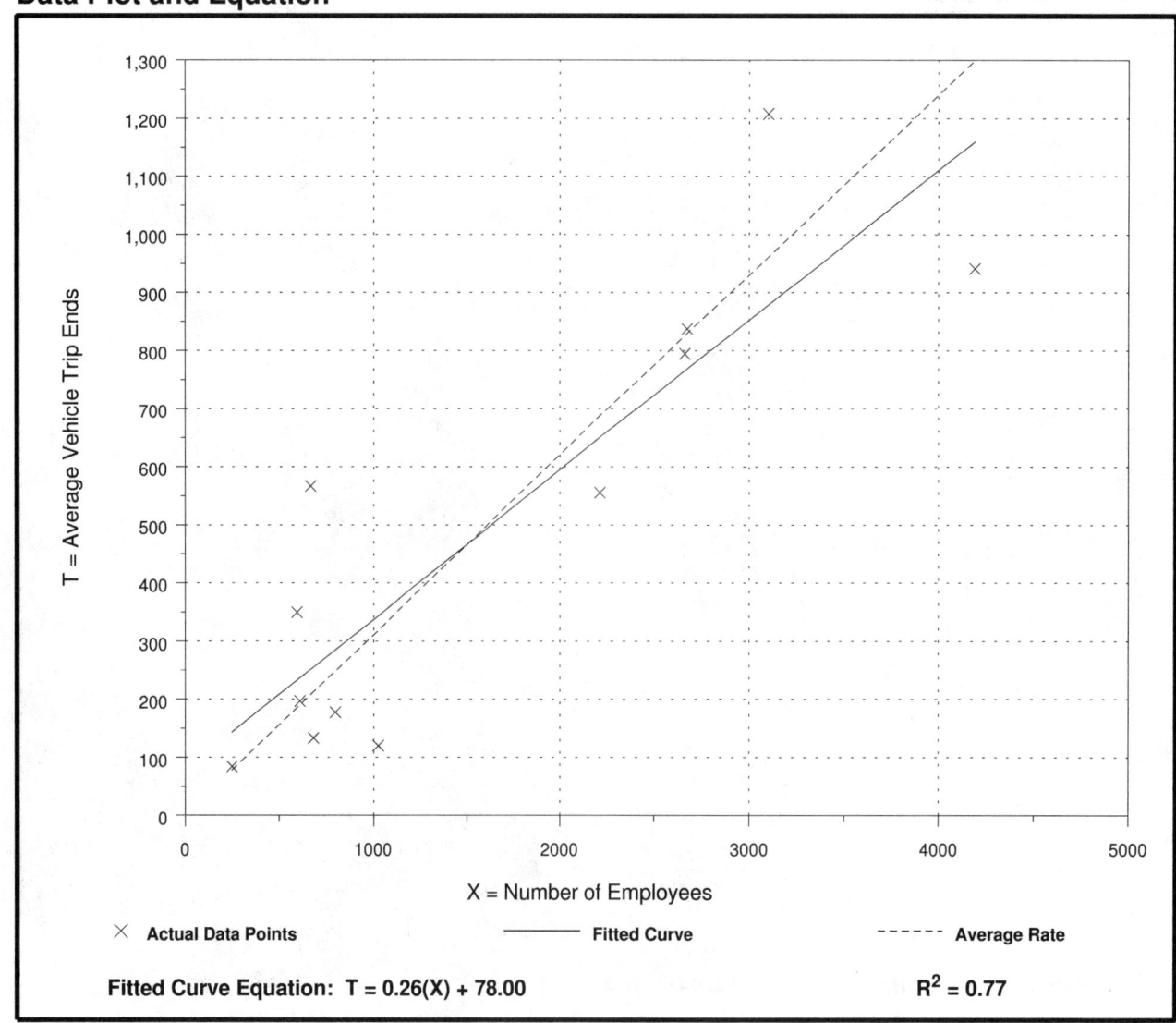

Fitted Curve Equation: T = 0.26(X) + 78.00 $R^2 = 0.77$

Hospital
(610)

Average Vehicle Trip Ends vs: Employees
On a: Weekday,
Peak Hour of Adjacent Street Traffic,
One Hour Between 4 and 6 p.m.

Number of Studies: 11
Avg. Number of Employees: 1,718
Directional Distribution: 29% entering, 71% exiting

Trip Generation per Employee

Average Rate	Range of Rates	Standard Deviation
0.29	0.15 - 1.08	0.56

Data Plot and Equation

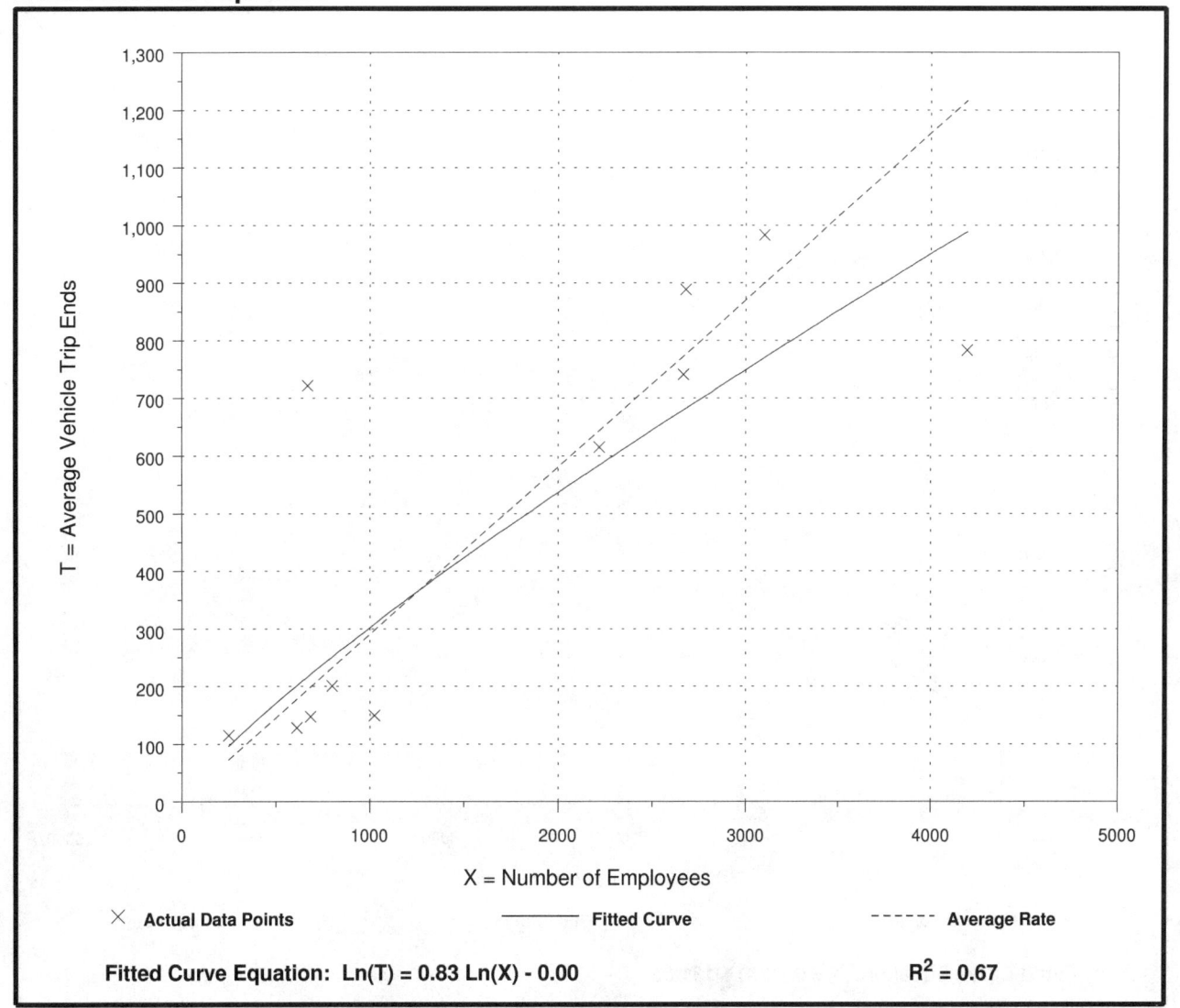

Fitted Curve Equation: $Ln(T) = 0.83\, Ln(X) - 0.00$ $R^2 = 0.67$

Hospital
(610)

Average Vehicle Trip Ends vs: Employees
On a: Weekday,
A.M. Peak Hour of Generator

Number of Studies: 10
Avg. Number of Employees: 1,659
Directional Distribution: 69% entering, 31% exiting

Trip Generation per Employee

Average Rate	Range of Rates	Standard Deviation
0.34	0.23 - 0.89	0.59

Data Plot and Equation

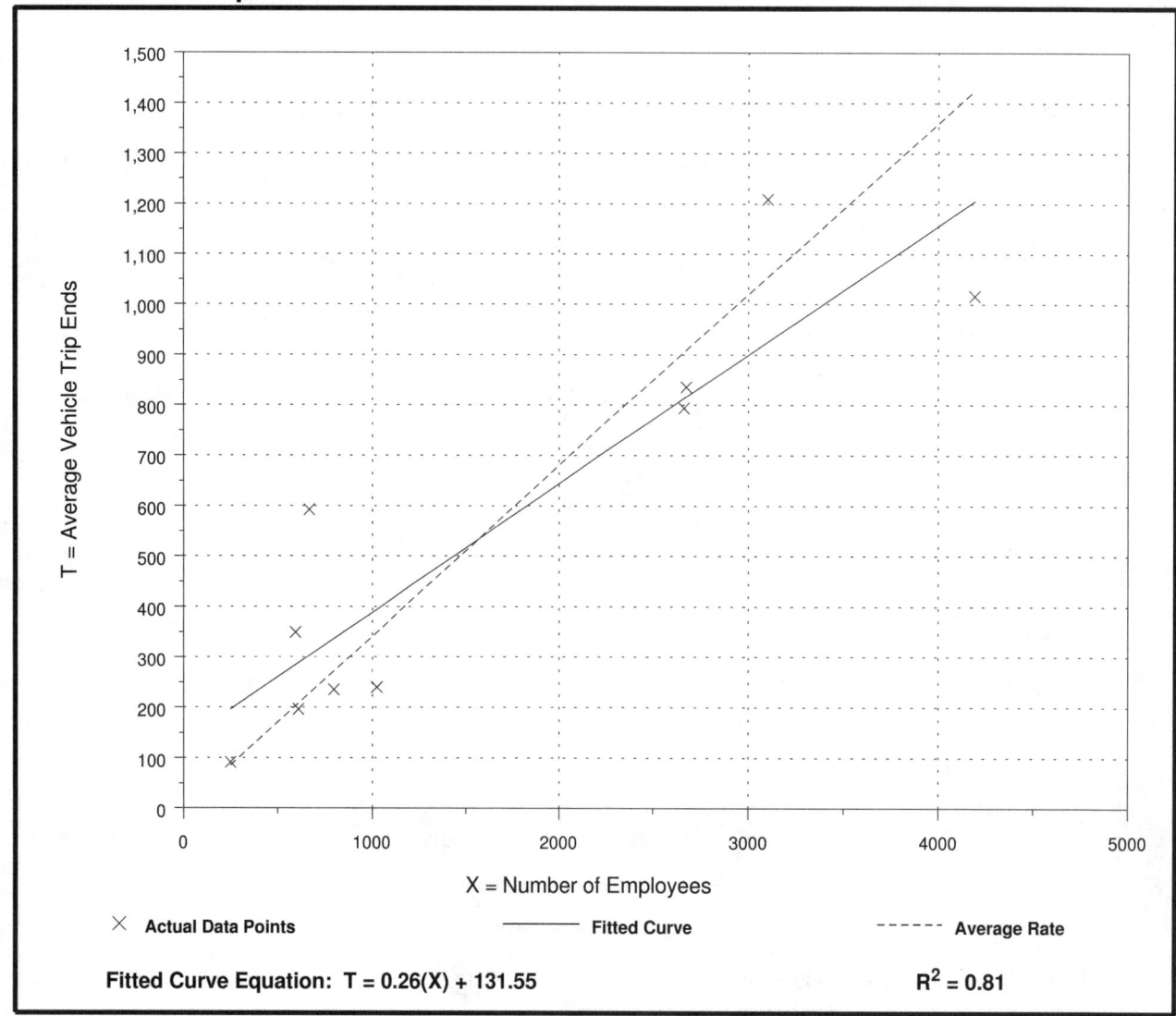

Fitted Curve Equation: $T = 0.26(X) + 131.55$ $R^2 = 0.81$

Hospital
(610)

Average Vehicle Trip Ends vs: Employees
On a: Weekday,
P.M. Peak Hour of Generator

Number of Studies: 18
Avg. Number of Employees: 1,157
Directional Distribution: 33% entering, 67% exiting

Trip Generation per Employee

Average Rate	Range of Rates	Standard Deviation
0.41	0.21 - 1.19	0.67

Data Plot and Equation

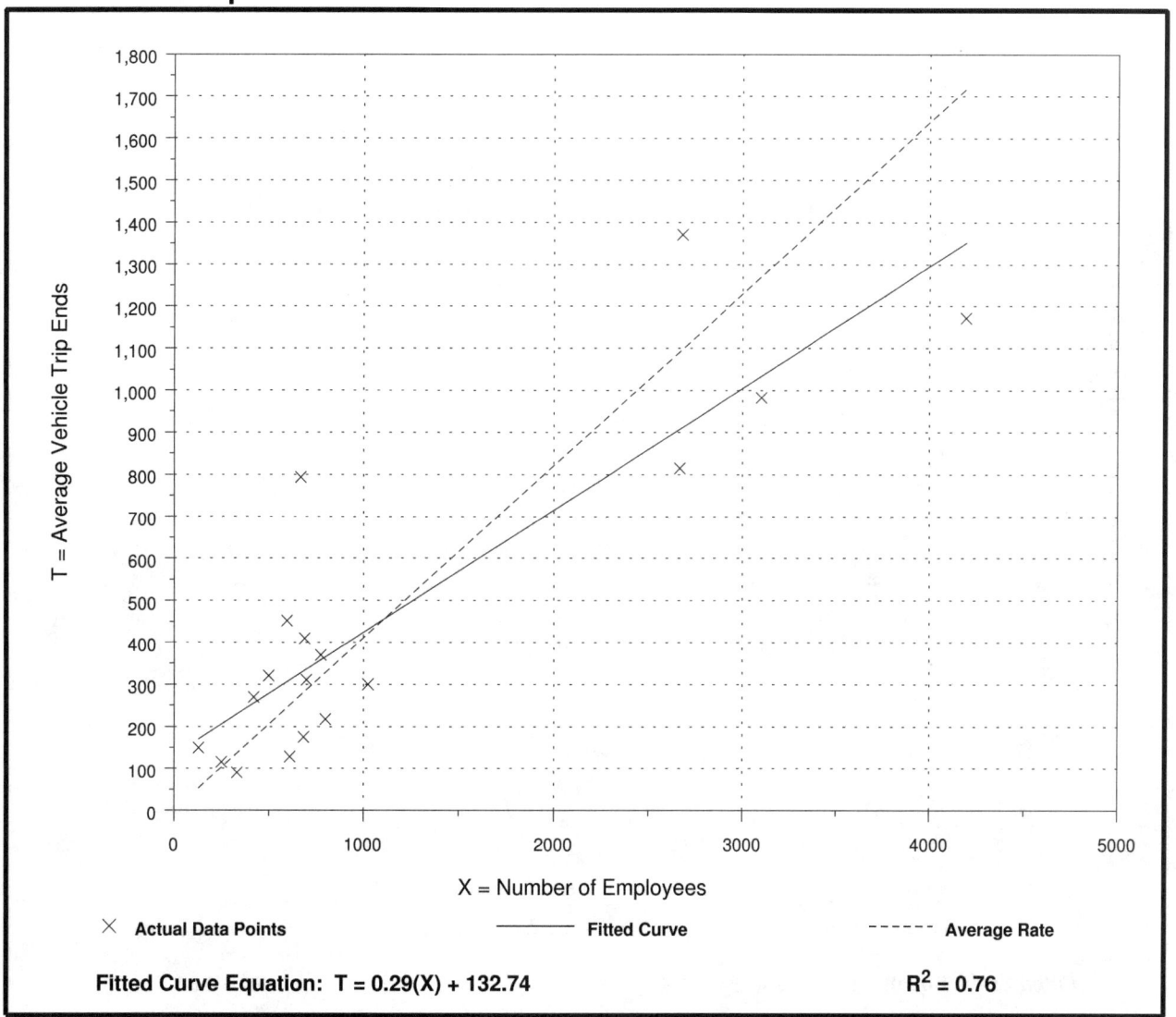

× Actual Data Points —— Fitted Curve ----- Average Rate

Fitted Curve Equation: $T = 0.29(X) + 132.74$ $R^2 = 0.76$

Trip Generation, 9th Edition • Institute of Transportation Engineers

Hospital
(610)

Average Vehicle Trip Ends vs: Employees
On a: Saturday

Number of Studies: 15
Avg. Number of Employees: 835
Directional Distribution: 50% entering, 50% exiting

Trip Generation per Employee

Average Rate	Range of Rates	Standard Deviation
3.78	1.60 - 7.98	2.27

Data Plot and Equation

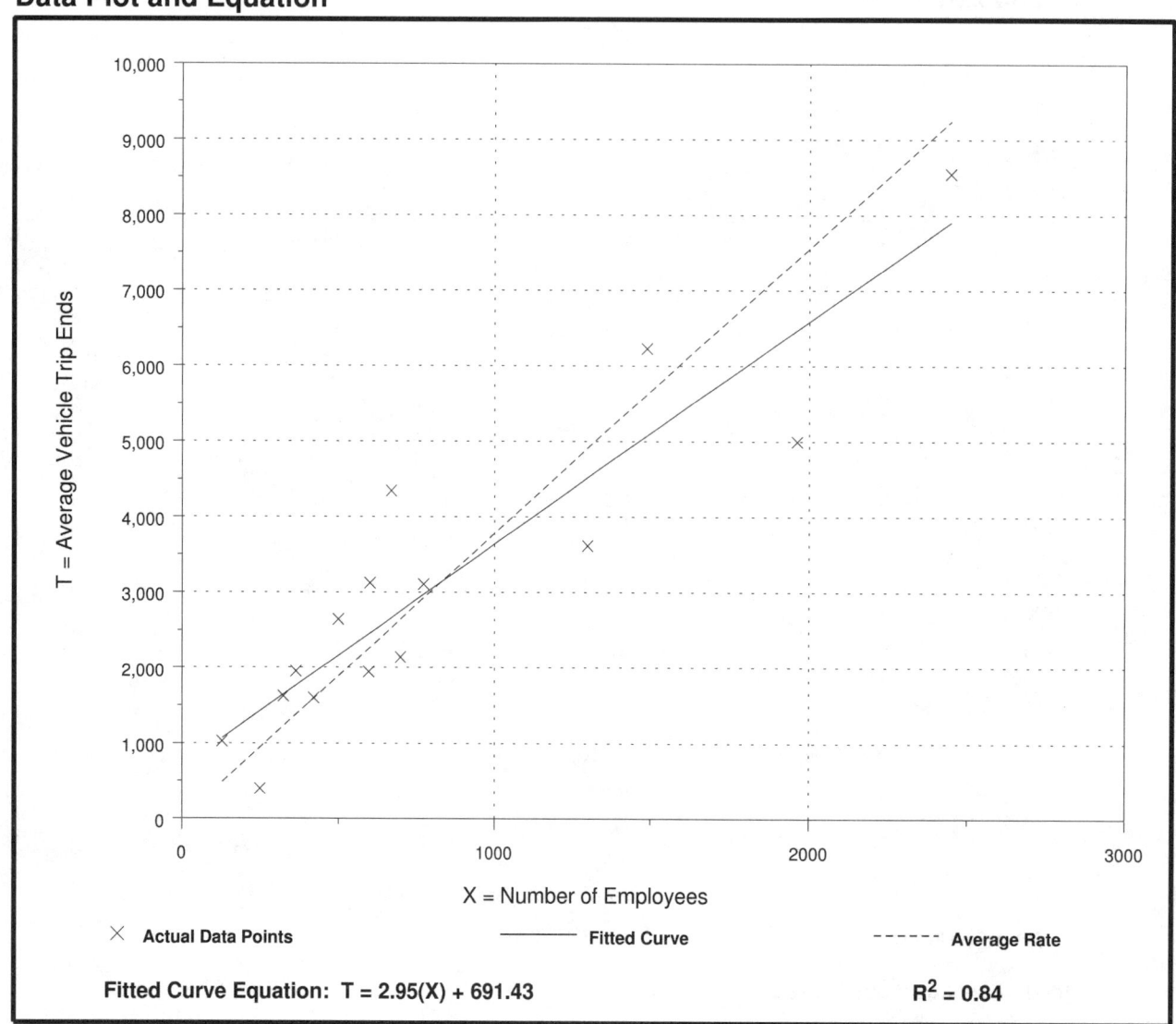

Fitted Curve Equation: $T = 2.95(X) + 691.43$ $\qquad R^2 = 0.84$

Hospital
(610)

Average Vehicle Trip Ends vs: Employees
On a: Saturday,
Peak Hour of Generator

Number of Studies: 4
Avg. Number of Employees: 502
Directional Distribution: 49% entering, 51% exiting

Trip Generation per Employee

Average Rate	Range of Rates	Standard Deviation
0.53	0.18 - 0.93	0.80

Data Plot and Equation

Caution - Use Carefully - Small Sample Size

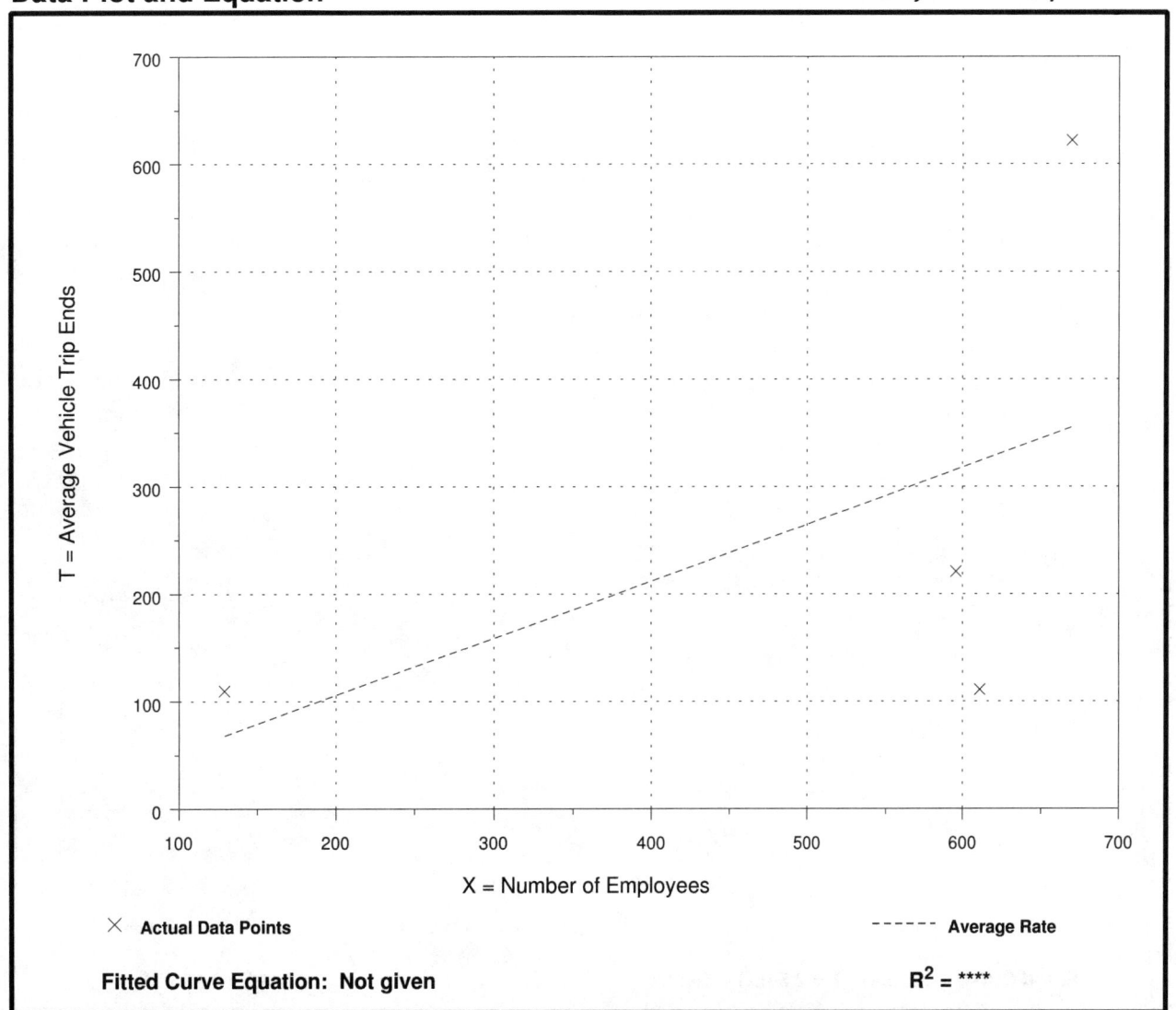

Fitted Curve Equation: Not given $R^2 = ****$

Hospital
(610)

Average Vehicle Trip Ends vs: Employees
On a: Sunday

Number of Studies: 14
Avg. Number of Employees: 852
Directional Distribution: 50% entering, 50% exiting

Trip Generation per Employee

Average Rate	Range of Rates	Standard Deviation
3.34	1.59 - 6.28	2.11

Data Plot and Equation

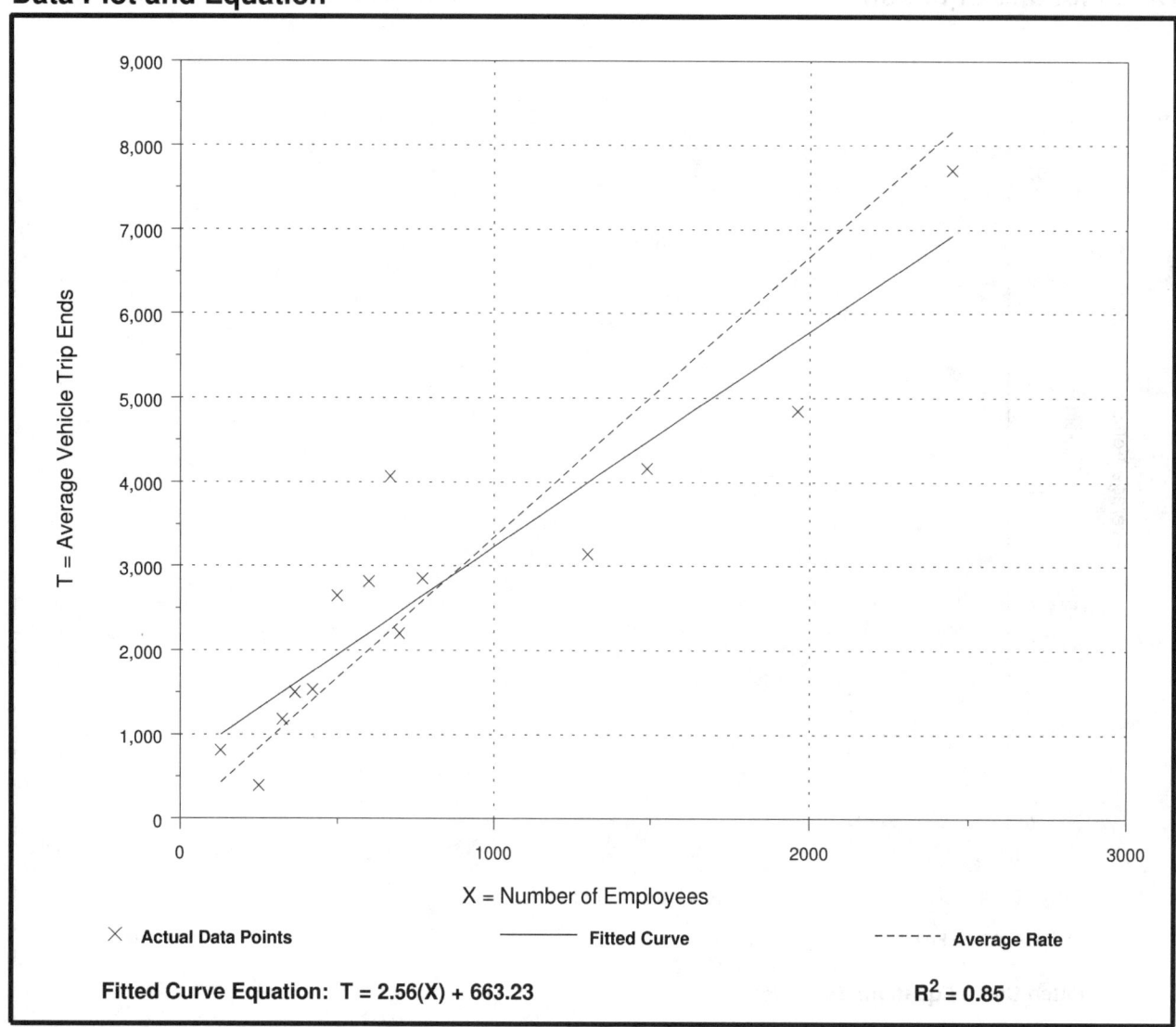

Fitted Curve Equation: T = 2.56(X) + 663.23 $R^2 = 0.85$

Hospital
(610)

Average Vehicle Trip Ends vs: Employees
On a: Sunday,
Peak Hour of Generator

Number of Studies: 6
Avg. Number of Employees: 515
Directional Distribution: 44% entering, 56% exiting

Trip Generation per Employee

Average Rate	Range of Rates	Standard Deviation
0.55	0.34 - 0.85	0.76

Data Plot and Equation

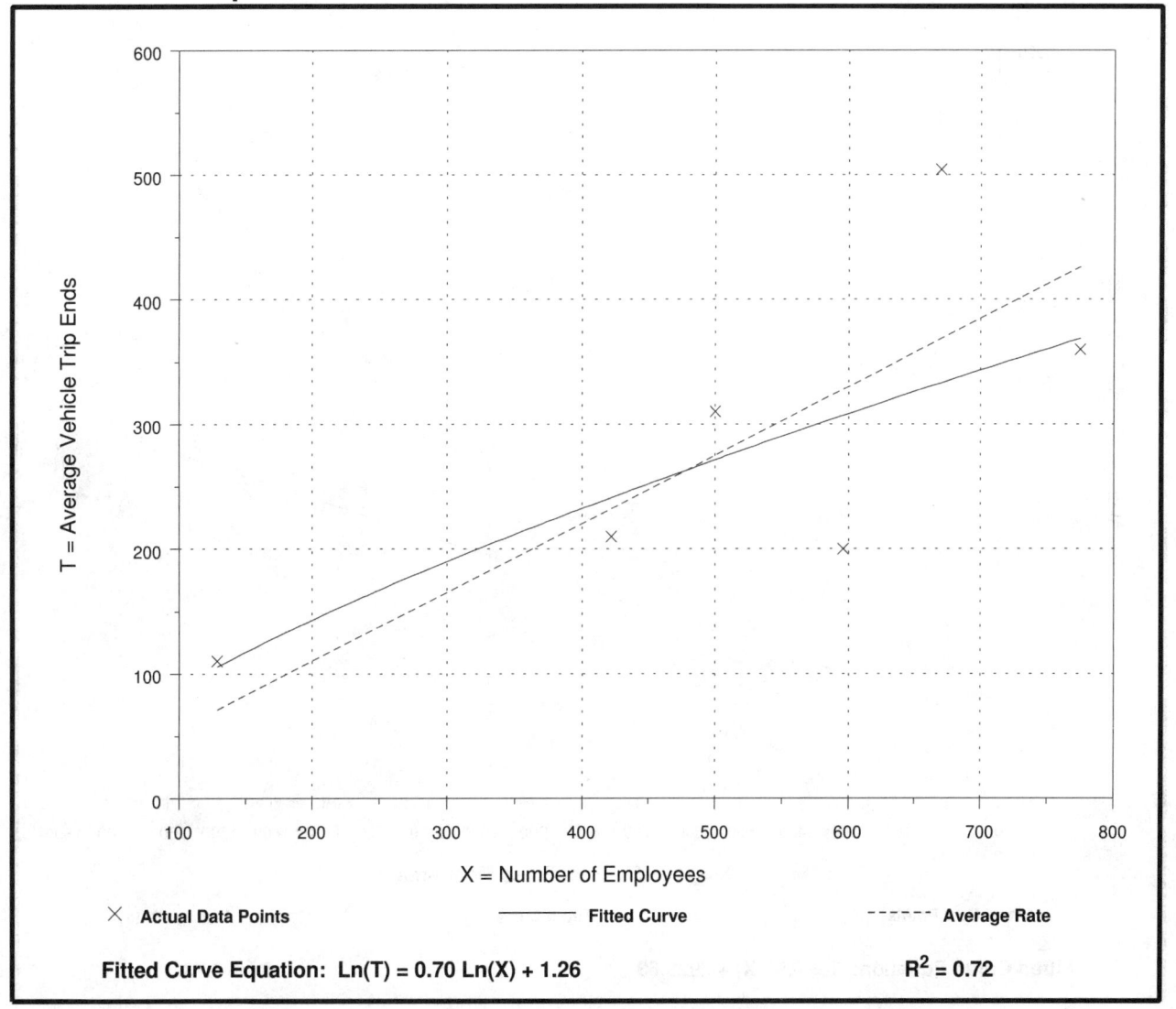

Fitted Curve Equation: $Ln(T) = 0.70\ Ln(X) + 1.26$ $R^2 = 0.72$

Hospital
(610)

Average Vehicle Trip Ends vs: 1000 Sq. Feet Gross Floor Area
On a: Weekday

Number of Studies: 17
Average 1000 Sq. Feet GFA: 463
Directional Distribution: 50% entering, 50% exiting

Trip Generation per 1000 Sq. Feet Gross Floor Area

Average Rate	Range of Rates	Standard Deviation
13.22	6.12 - 67.52	10.09

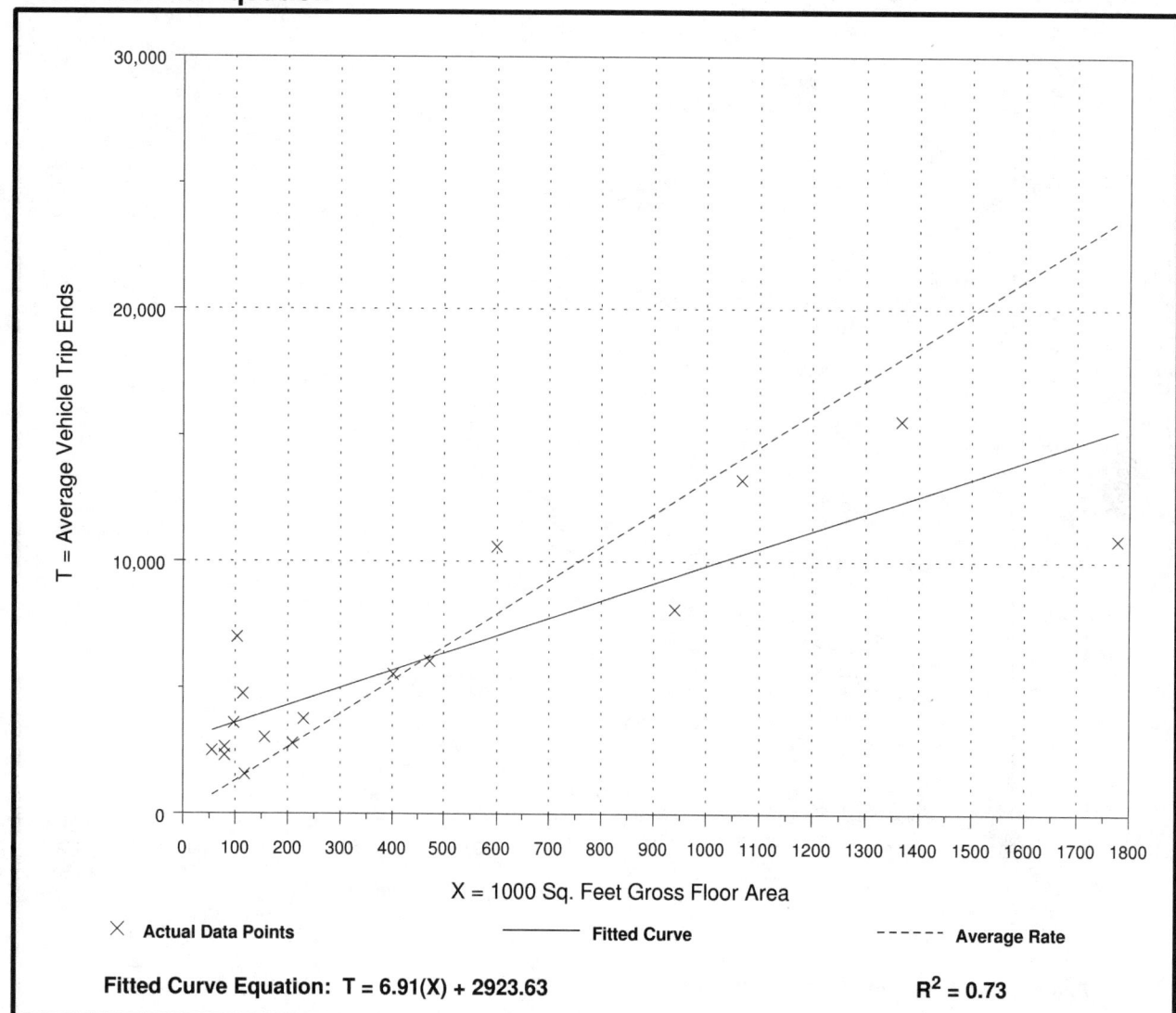

Data Plot and Equation

Fitted Curve Equation: T = 6.91(X) + 2923.63 $R^2 = 0.73$

Hospital
(610)

Average Vehicle Trip Ends vs: 1000 Sq. Feet Gross Floor Area
On a: Weekday,
Peak Hour of Adjacent Street Traffic,
One Hour Between 7 and 9 a.m.

Number of Studies: 13
Average 1000 Sq. Feet GFA: 599
Directional Distribution: 63% entering, 37% exiting

Trip Generation per 1000 Sq. Feet Gross Floor Area

Average Rate	Range of Rates	Standard Deviation
0.95	0.53 - 5.45	1.15

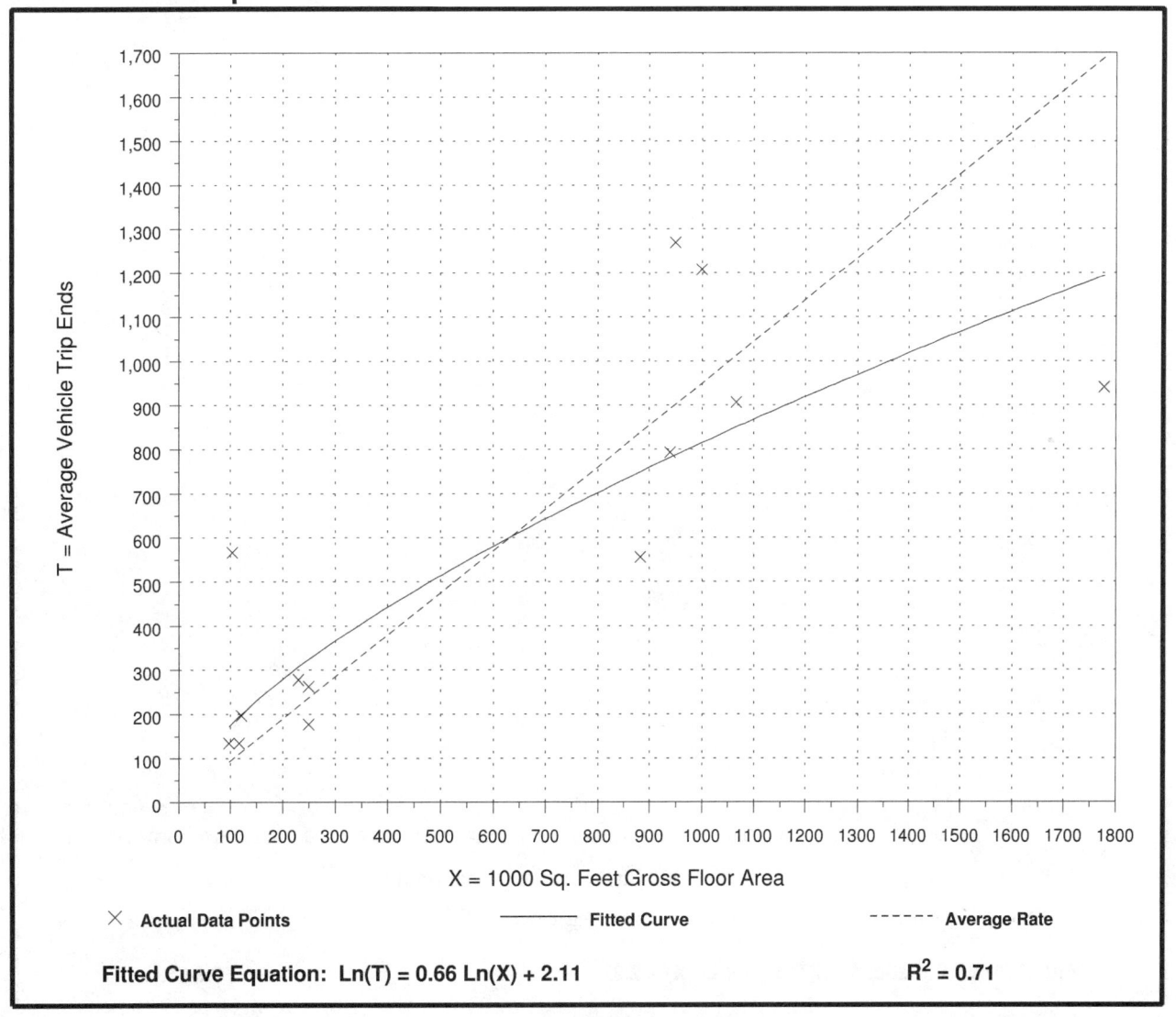

Data Plot and Equation

Fitted Curve Equation: $Ln(T) = 0.66\ Ln(X) + 2.11$ $R^2 = 0.71$

Hospital
(610)

Average Vehicle Trip Ends vs: 1000 Sq. Feet Gross Floor Area
On a: Weekday,
Peak Hour of Adjacent Street Traffic,
One Hour Between 4 and 6 p.m.

Number of Studies: 13
Average 1000 Sq. Feet GFA: 599
Directional Distribution: 38% entering, 62% exiting

Trip Generation per 1000 Sq. Feet Gross Floor Area

Average Rate	Range of Rates	Standard Deviation
0.93	0.44 - 6.94	1.23

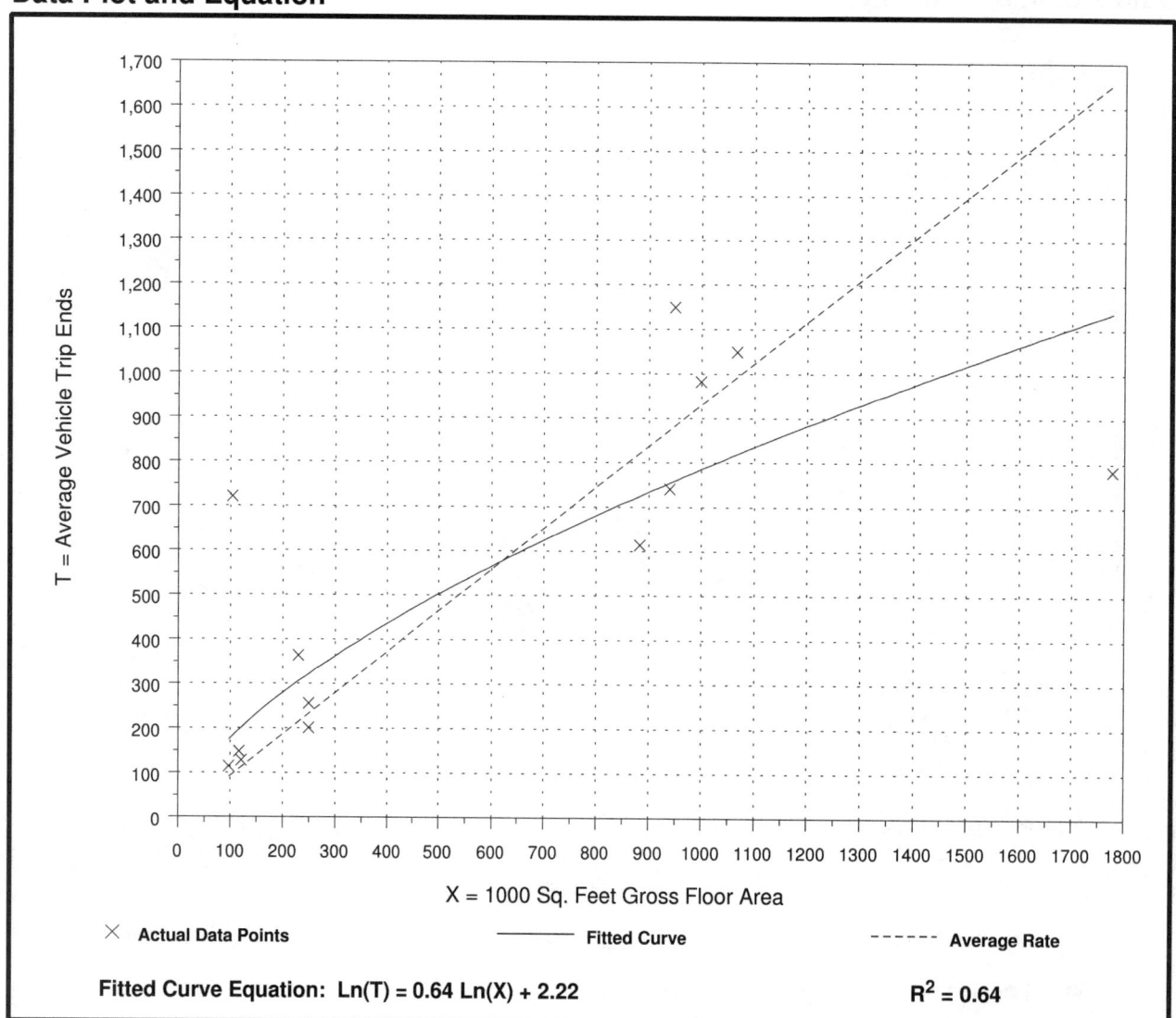

Data Plot and Equation

Fitted Curve Equation: $Ln(T) = 0.64\ Ln(X) + 2.22$ $R^2 = 0.64$

Hospital
(610)

Average Vehicle Trip Ends vs: 1000 Sq. Feet Gross Floor Area
On a: Weekday,
A.M. Peak Hour of Generator

Number of Studies: 8
Average 1000 Sq. Feet GFA: 686
Directional Distribution: 59% entering, 41% exiting

Trip Generation per 1000 Sq. Feet Gross Floor Area

Average Rate	Range of Rates	Standard Deviation
0.96	0.57 - 5.70	1.21

Data Plot and Equation

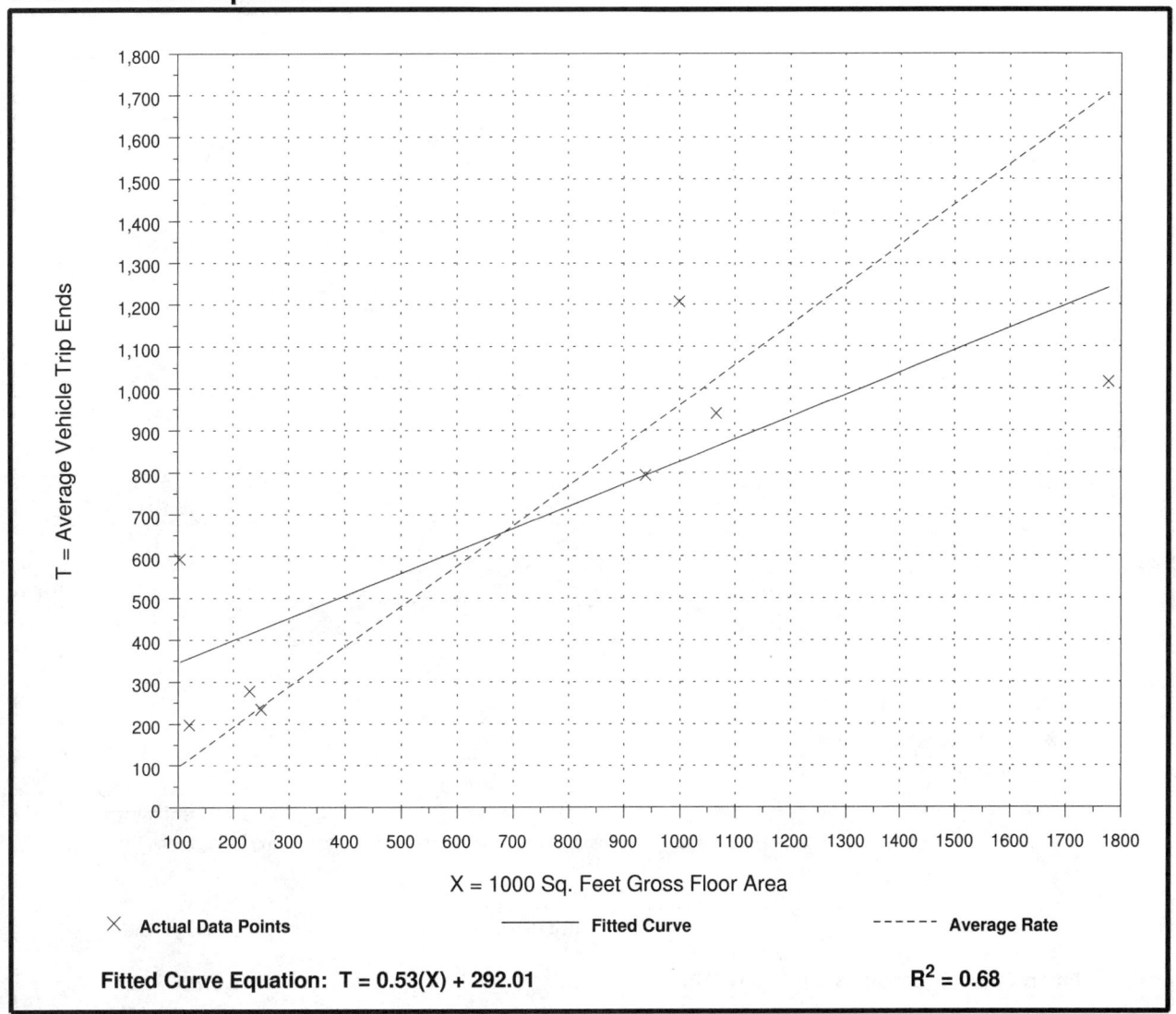

Fitted Curve Equation: $T = 0.53(X) + 292.01$ $R^2 = 0.68$

Hospital
(610)

Average Vehicle Trip Ends vs: 1000 Sq. Feet Gross Floor Area
On a: Weekday,
P.M. Peak Hour of Generator

Number of Studies: 13
Average 1000 Sq. Feet GFA: 505
Directional Distribution: 40% entering, 60% exiting

Trip Generation per 1000 Sq. Feet Gross Floor Area

Average Rate	Range of Rates	Standard Deviation
1.16	0.66 - 7.63	1.42

Data Plot and Equation

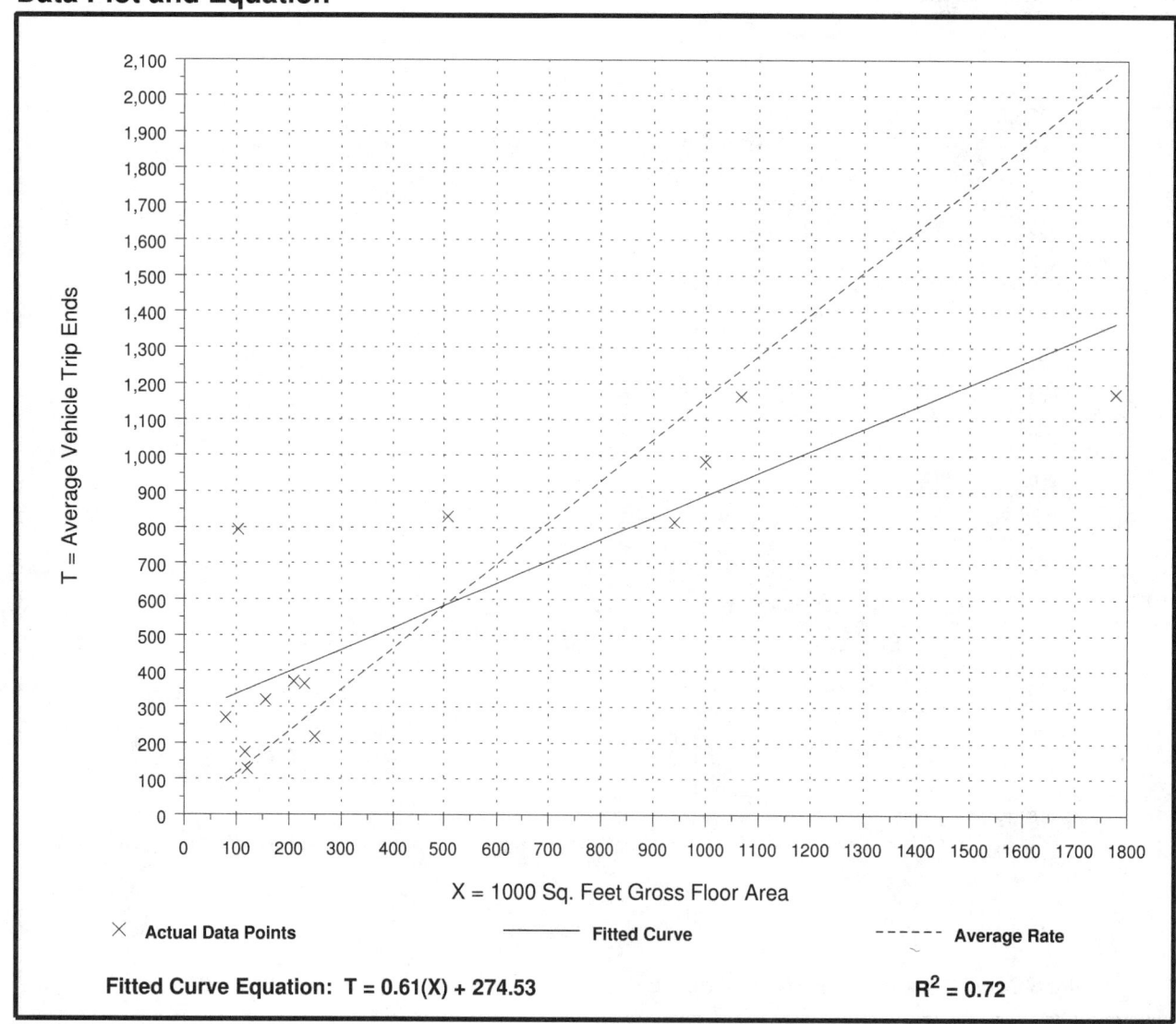

Fitted Curve Equation: $T = 0.61(X) + 274.53$ $R^2 = 0.72$

Hospital
(610)

Average Vehicle Trip Ends vs: 1000 Sq. Feet Gross Floor Area
On a: Saturday

Number of Studies: 14
Average 1000 Sq. Feet GFA: 360
Directional Distribution: 50% entering, 50% exiting

Trip Generation per 1000 Sq. Feet Gross Floor Area

Average Rate	Range of Rates	Standard Deviation
10.18	4.40 - 41.80	7.98

Data Plot and Equation

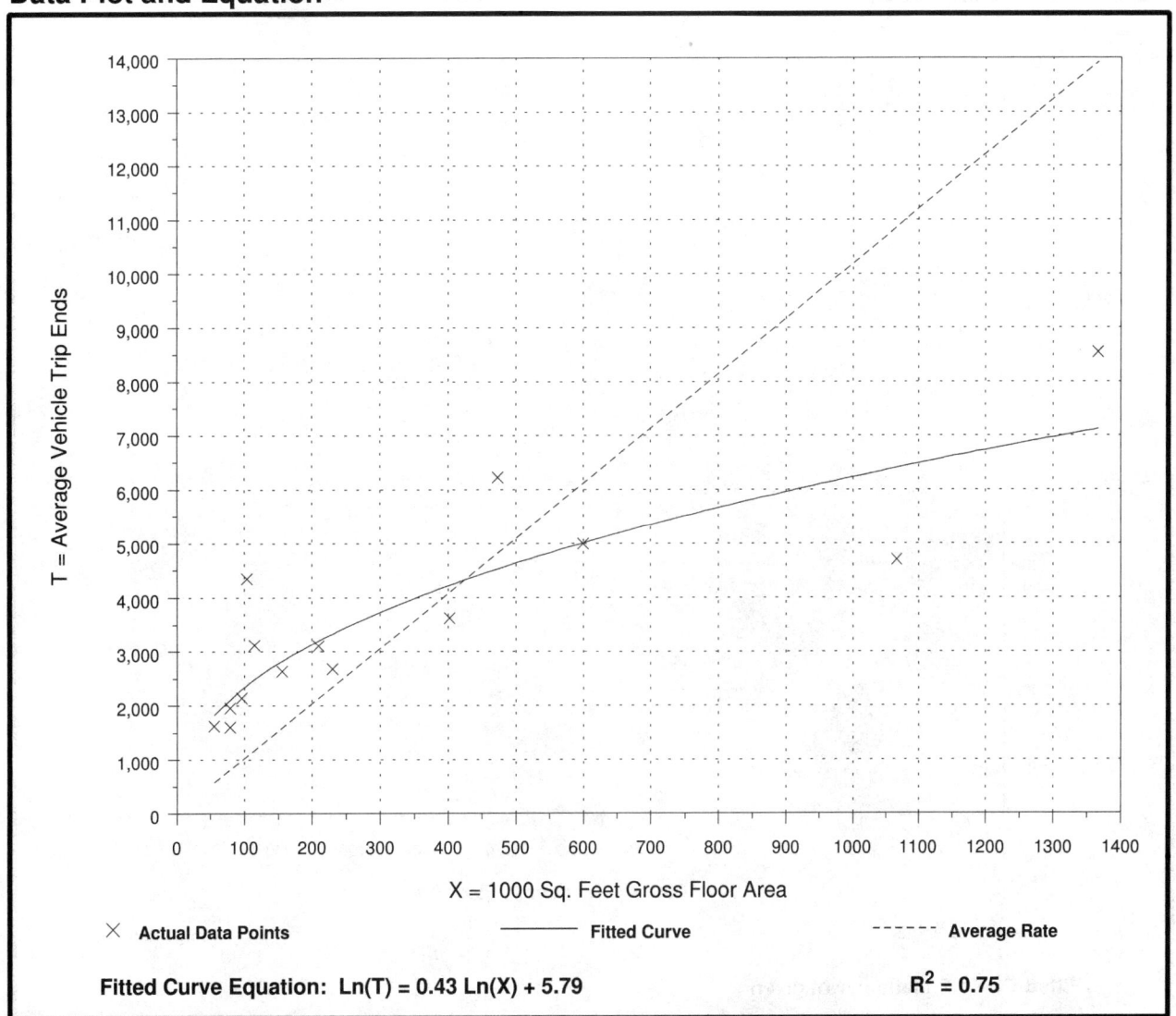

Fitted Curve Equation: $Ln(T) = 0.43\,Ln(X) + 5.79$ $R^2 = 0.75$

Hospital
(610)

Average Vehicle Trip Ends vs: 1000 Sq. Feet Gross Floor Area
On a: Saturday,
Peak Hour of Generator

Number of Studies: 3
Average 1000 Sq. Feet GFA: 152
Directional Distribution: 50% entering, 50% exiting

Trip Generation per 1000 Sq. Feet Gross Floor Area

Average Rate	Range of Rates	Standard Deviation
2.26	0.92 - 5.98	2.53

Data Plot and Equation

Caution - Use Carefully - Small Sample Size

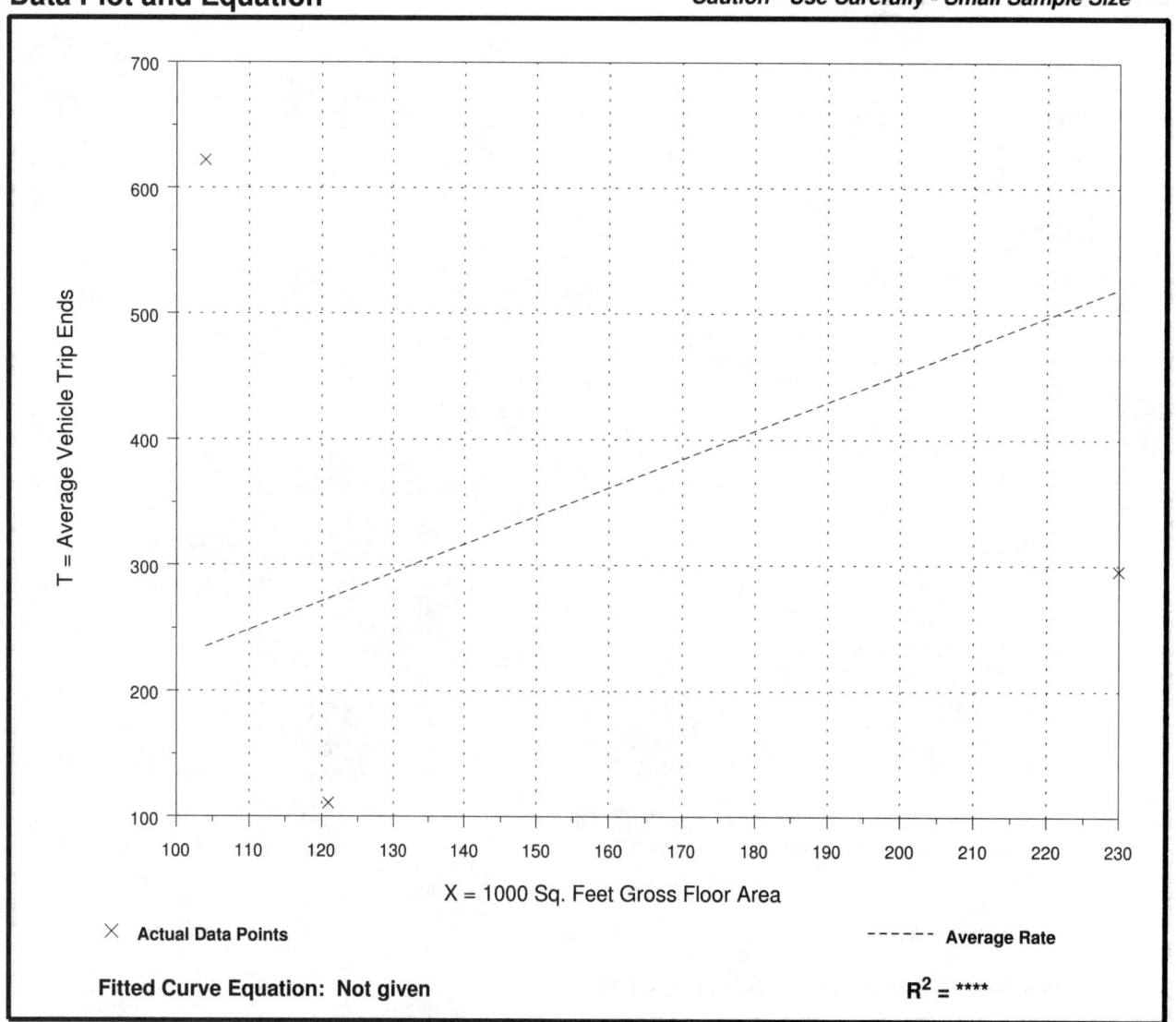

Fitted Curve Equation: Not given $R^2 = ****$

Hospital
(610)

Average Vehicle Trip Ends vs: 1000 Sq. Feet Gross Floor Area
On a: Sunday

Number of Studies: 14
Average 1000 Sq. Feet GFA: 360
Directional Distribution: 50% entering, 50% exiting

Trip Generation per 1000 Sq. Feet Gross Floor Area

Average Rate	Range of Rates	Standard Deviation
8.91	3.62 - 39.13	7.34

Data Plot and Equation

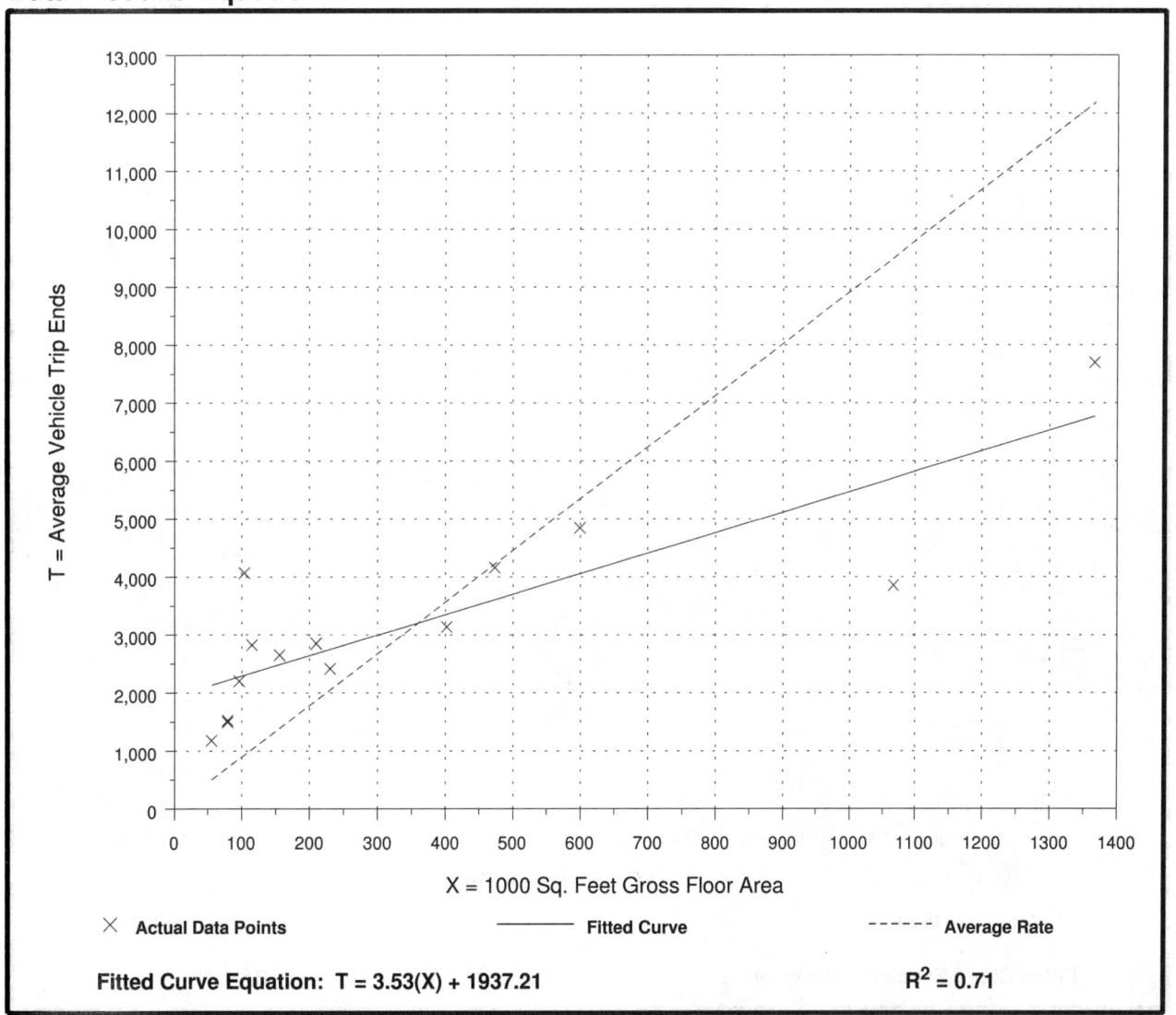

Fitted Curve Equation: $T = 3.53(X) + 1937.21$ $R^2 = 0.71$

Hospital
(610)

Average Vehicle Trip Ends vs: 1000 Sq. Feet Gross Floor Area
On a: Sunday,
Peak Hour of Generator

Number of Studies: 5
Average 1000 Sq. Feet GFA: 156
Directional Distribution: 45% entering, 55% exiting

Trip Generation per 1000 Sq. Feet Gross Floor Area

Average Rate	Range of Rates	Standard Deviation
2.13	1.20 - 4.85	1.85

Data Plot and Equation

Caution - Use Carefully - Small Sample Size

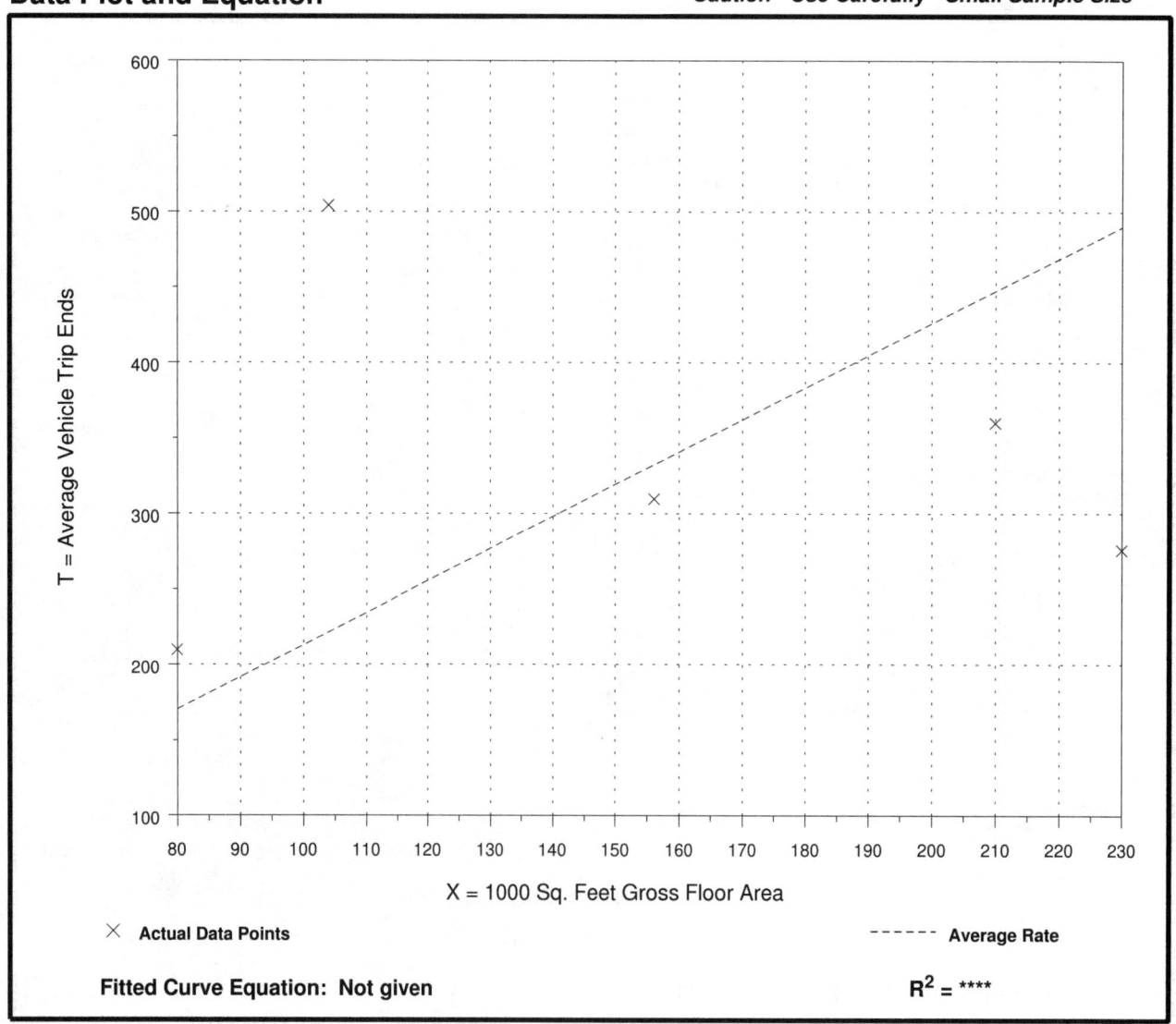

Fitted Curve Equation: Not given $R^2 = ****$

Hospital
(610)

Average Vehicle Trip Ends vs: Beds
On a: Weekday

Number of Studies: 22
Average Number of Beds: 395
Directional Distribution: 50% entering, 50% exiting

Trip Generation per Bed

Average Rate	Range of Rates	Standard Deviation
12.94	3.00 - 57.13	9.07

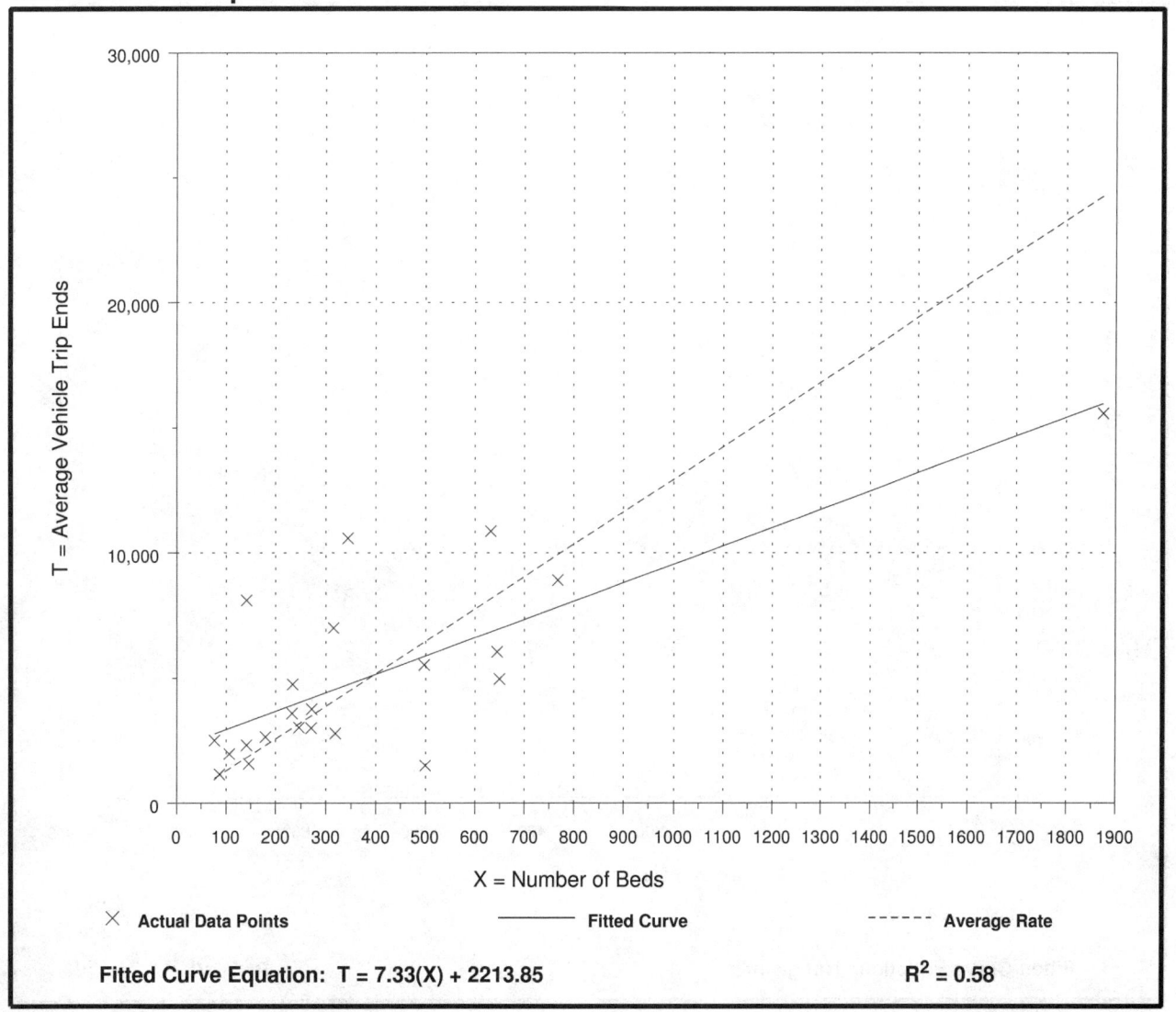

Data Plot and Equation

Fitted Curve Equation: $T = 7.33(X) + 2213.85$ $R^2 = 0.58$

Hospital
(610)

Average Vehicle Trip Ends vs: Beds
On a: Weekday,
Peak Hour of Adjacent Street Traffic,
One Hour Between 7 and 9 a.m.

Number of Studies: 11
Average Number of Beds: 430
Directional Distribution: 72% entering, 28% exiting

Trip Generation per Bed

Average Rate	Range of Rates	Standard Deviation
1.32	0.32 - 5.59	1.43

Data Plot and Equation

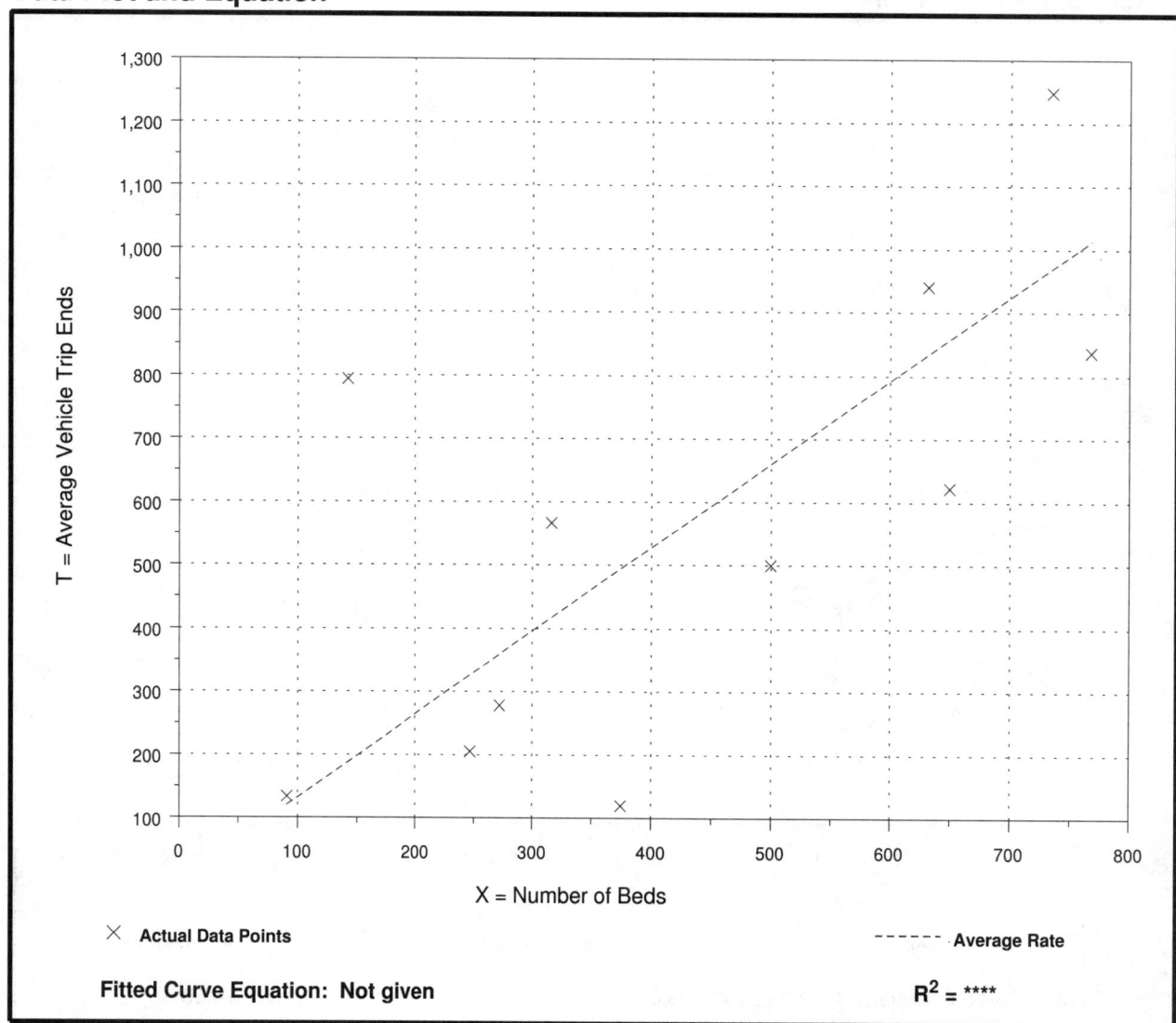

Fitted Curve Equation: Not given $R^2 = ****$

Hospital
(610)

Average Vehicle Trip Ends vs: Beds
On a: Weekday,
Peak Hour of Adjacent Street Traffic,
One Hour Between 4 and 6 p.m.

Number of Studies: 11
Average Number of Beds: 430
Directional Distribution: 33% entering, 67% exiting

Trip Generation per Bed

Average Rate	Range of Rates	Standard Deviation
1.42	0.40 - 5.22	1.44

Data Plot and Equation

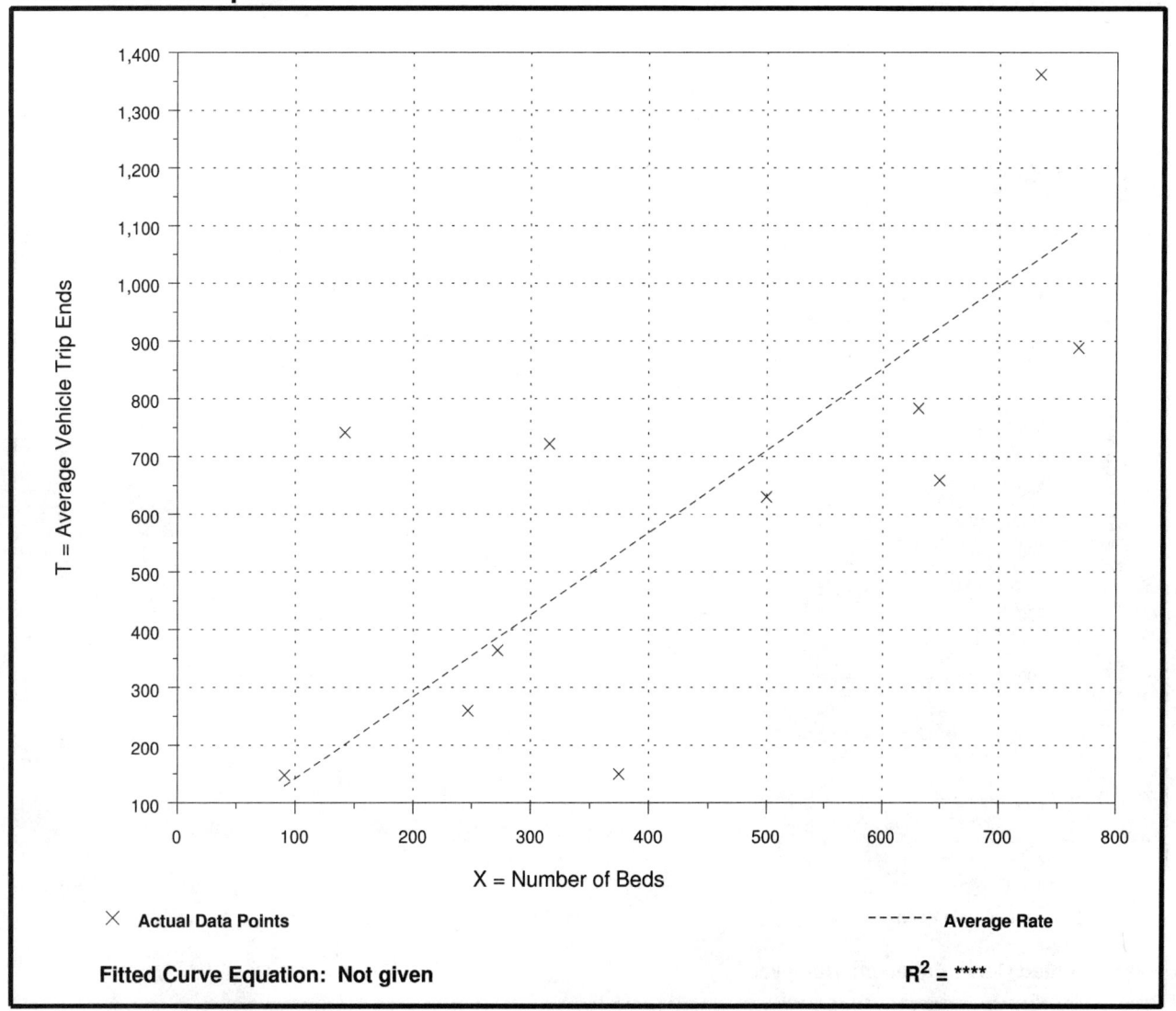

$\times$ Actual Data Points ----- Average Rate

Fitted Curve Equation: Not given $R^2 = ****$

Hospital
(610)

Average Vehicle Trip Ends vs: Beds
On a: Weekday,
A.M. Peak Hour of Generator

Number of Studies: 9
Average Number of Beds: 460
Directional Distribution: 69% entering, 31% exiting

Trip Generation per Bed

Average Rate	Range of Rates	Standard Deviation
1.45	0.64 - 5.59	1.49

Data Plot and Equation

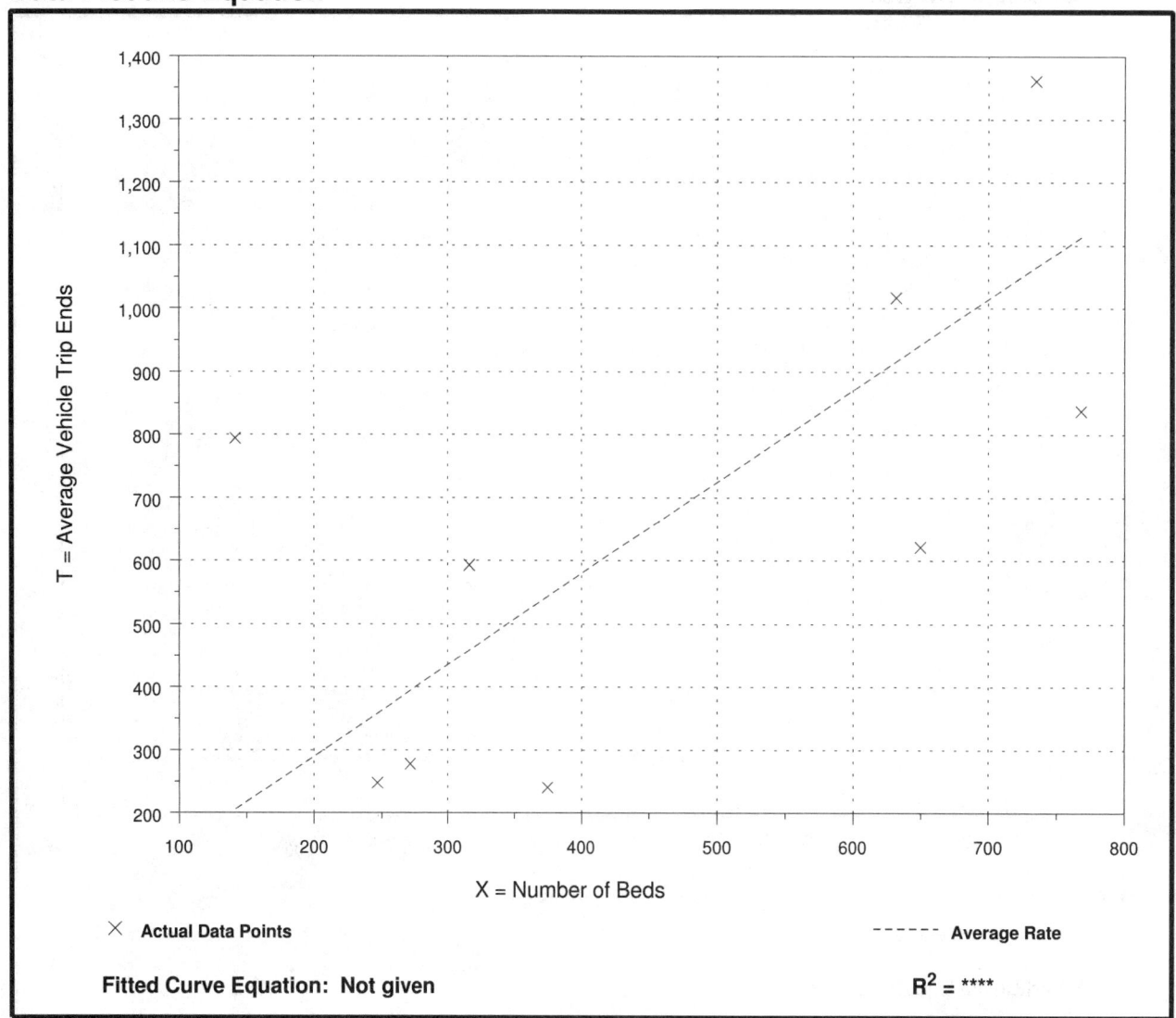

Fitted Curve Equation: Not given $R^2 = ****$

Hospital
(610)

Average Vehicle Trip Ends vs: Beds
On a: Weekday,
P.M. Peak Hour of Generator

Number of Studies: 17
Average Number of Beds: 329
Directional Distribution: 34% entering, 66% exiting

Trip Generation per Bed

Average Rate	Range of Rates	Standard Deviation
1.60	0.80 - 5.74	1.52

Data Plot and Equation

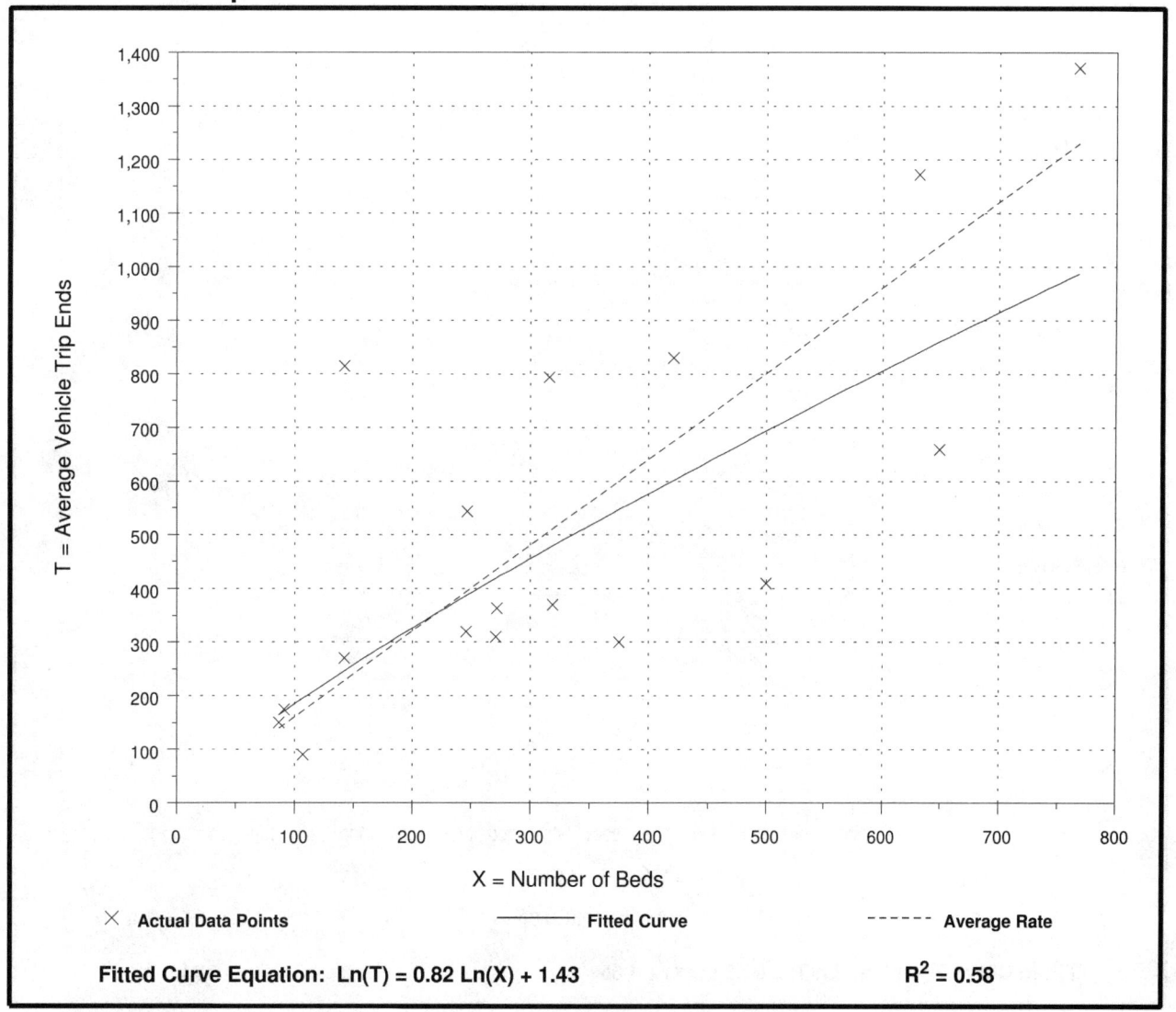

Fitted Curve Equation: Ln(T) = 0.82 Ln(X) + 1.43 $R^2 = 0.58$

Hospital
(610)

Average Vehicle Trip Ends vs: Beds
On a: Saturday

Number of Studies: 15
Average Number of Beds: 408
Directional Distribution: 50% entering, 50% exiting

Trip Generation per Bed

Average Rate	Range of Rates	Standard Deviation
8.14	3.35 - 21.04	4.80

Data Plot and Equation

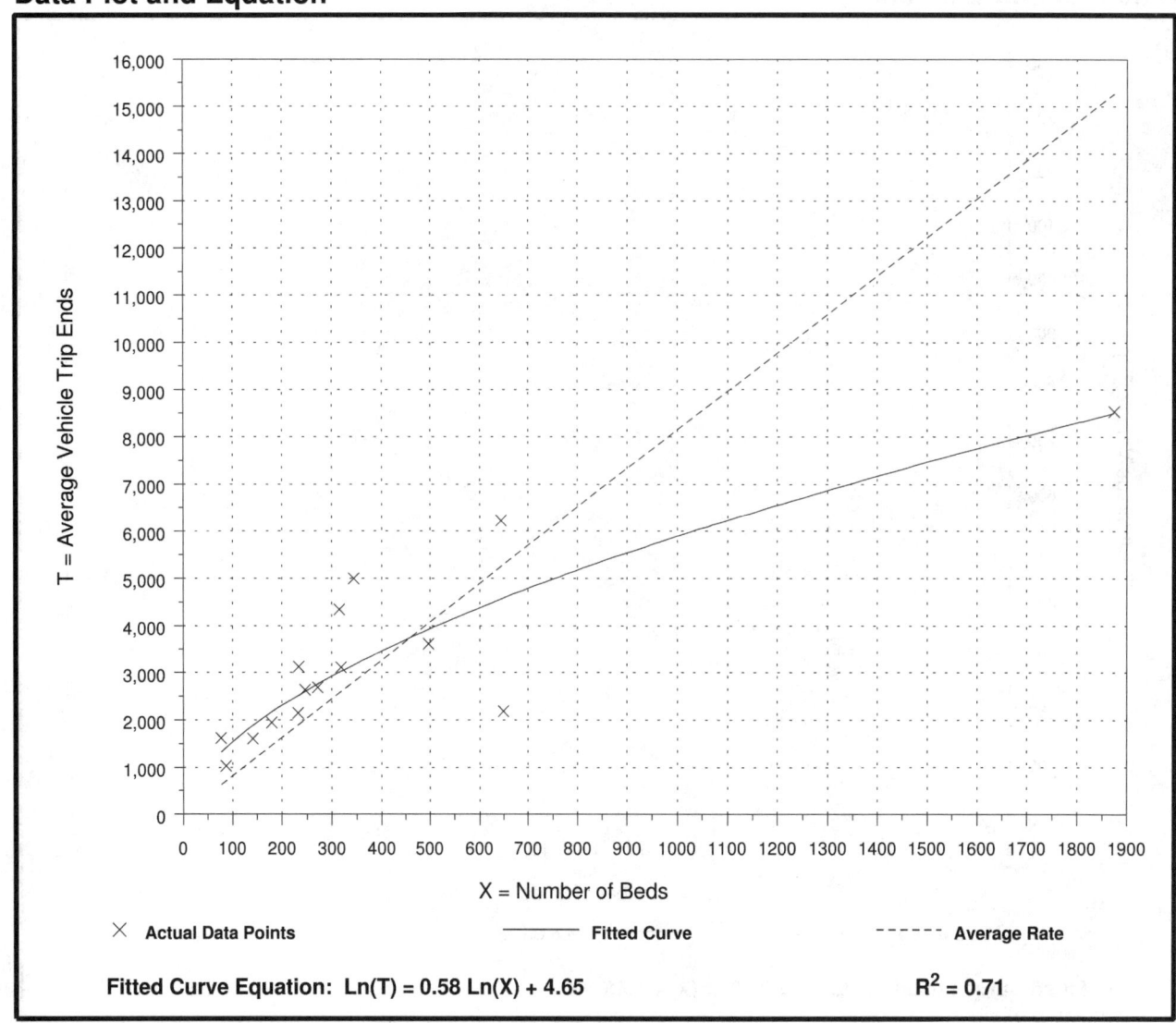

Fitted Curve Equation: $Ln(T) = 0.58\ Ln(X) + 4.65$ $R^2 = 0.71$

Hospital
(610)

Average Vehicle Trip Ends vs: Beds
On a: Saturday,
Peak Hour of Generator

Number of Studies: 4
Average Number of Beds: 331
Directional Distribution: 47% entering, 53% exiting

Trip Generation per Bed

Average Rate	Range of Rates	Standard Deviation
1.00	0.45 - 1.97	1.17

Data Plot and Equation

Caution - Use Carefully - Small Sample Size

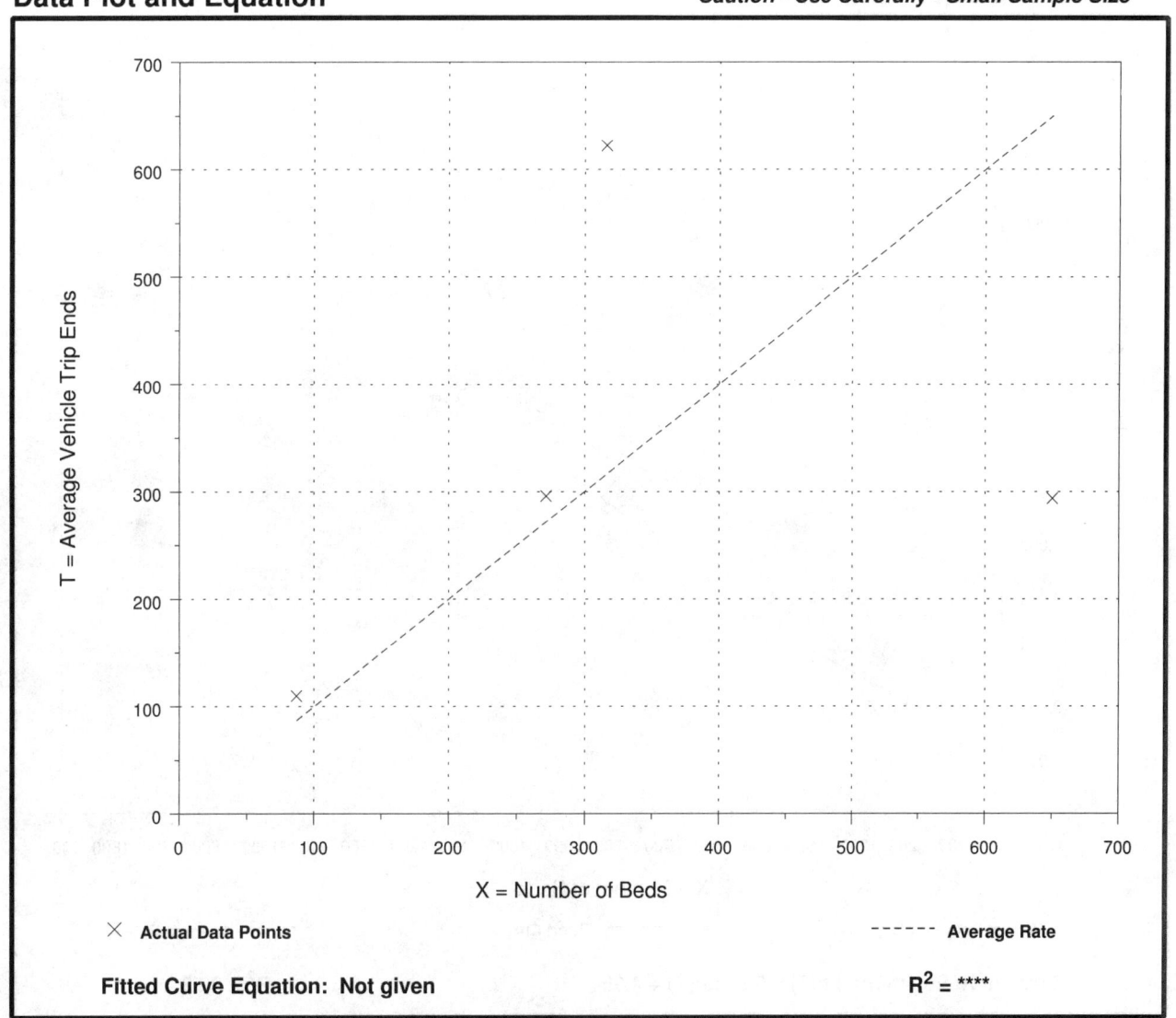

Fitted Curve Equation: Not given $R^2 = ****$

Hospital
(610)

Average Vehicle Trip Ends vs: Beds
On a: Sunday

Number of Studies: 15
Average Number of Beds: 408
Directional Distribution: 50% entering, 50% exiting

Trip Generation per Bed

Average Rate	Range of Rates	Standard Deviation
7.19	3.22 - 15.32	4.40

Data Plot and Equation

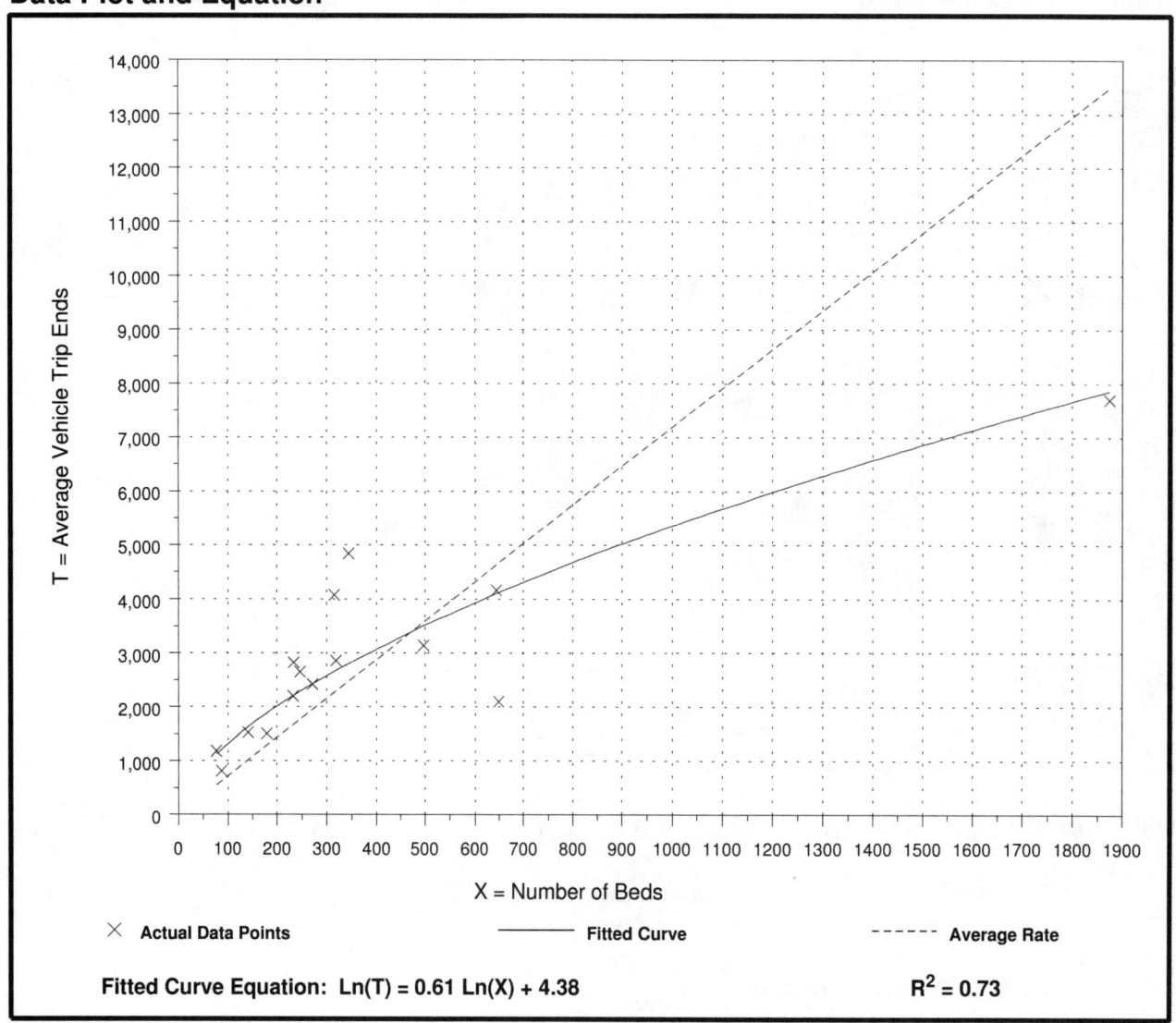

Fitted Curve Equation: $Ln(T) = 0.61\ Ln(X) + 4.38$ $R^2 = 0.73$

Hospital
(610)

Average Vehicle Trip Ends vs: Beds
On a: Sunday,
Peak Hour of Generator

Number of Studies: 7
Average Number of Beds: 290
Directional Distribution: 45% entering, 55% exiting

Trip Generation per Bed

Average Rate	Range of Rates	Standard Deviation
1.03	0.50 - 1.59	1.09

Data Plot and Equation

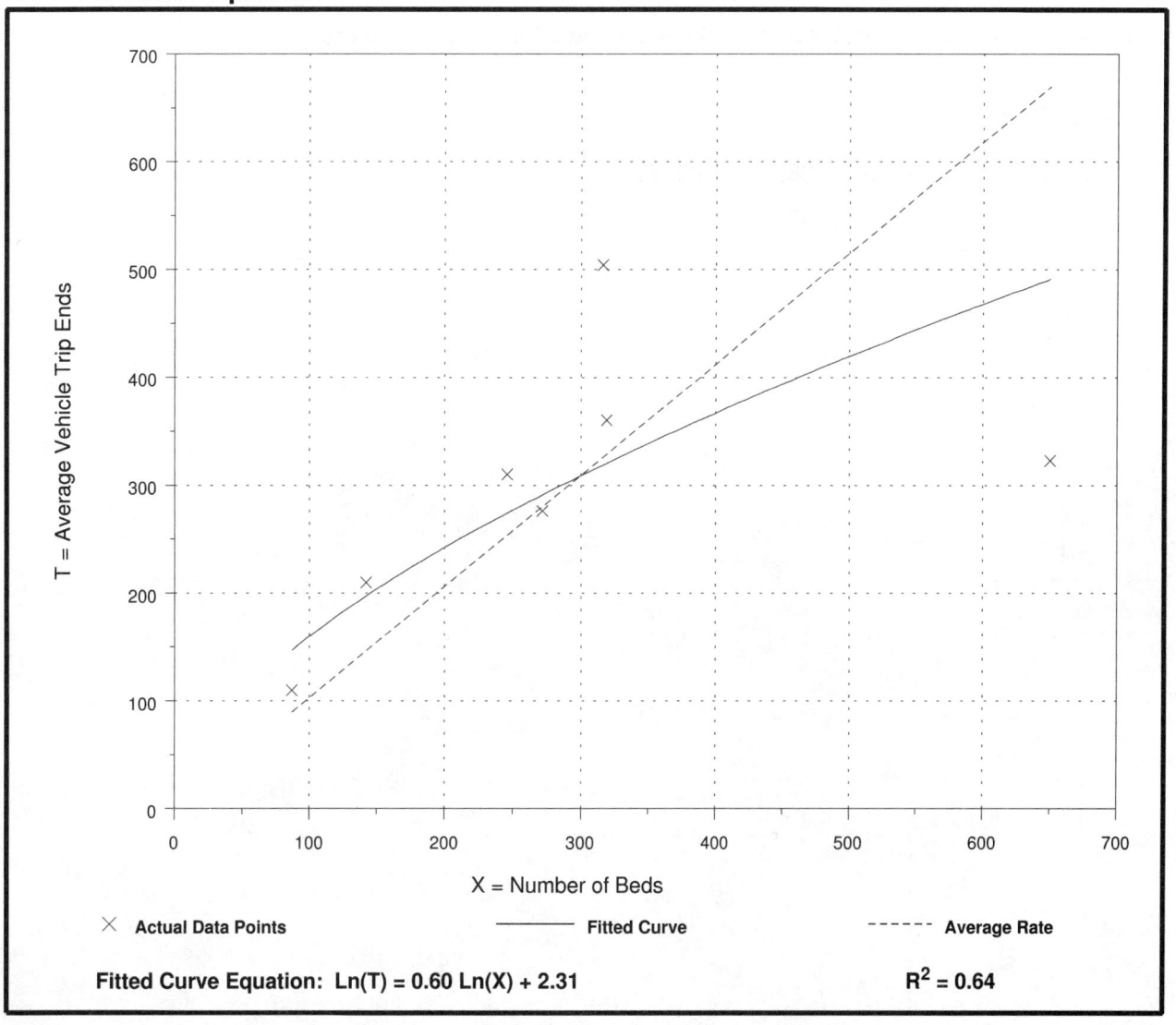

Fitted Curve Equation: $Ln(T) = 0.60 \, Ln(X) + 2.31$ $R^2 = 0.64$

Land Use: 620
Nursing Home

Description

A nursing home is any facility whose primary function is to provide care for persons who are unable to care for themselves. Examples of such facilities include rest homes and chronic care and convalescent homes. Skilled nurses and nursing aides are present 24 hours a day at these sites. Nursing homes are occupied by residents who do little or no driving; traffic is primarily generated by employees, visitors and deliveries. Assisted living (Land Use 254) and continuing care retirement community (Land Use 255) are related uses.

Additional Data

The sites were surveyed between the 1960s and the 2000s throughout the United States.

Source Numbers

91, 221, 237, 245, 253, 397, 436, 502, 598, 734

Nursing Home
(620)

Average Vehicle Trip Ends vs: Employees
On a: Weekday

Number of Studies: 4
Avg. Number of Employees: 113
Directional Distribution: 50% entering, 50% exiting

Trip Generation per Employee

Average Rate	Range of Rates	Standard Deviation
3.26	2.60 - 8.72	2.40

Data Plot and Equation

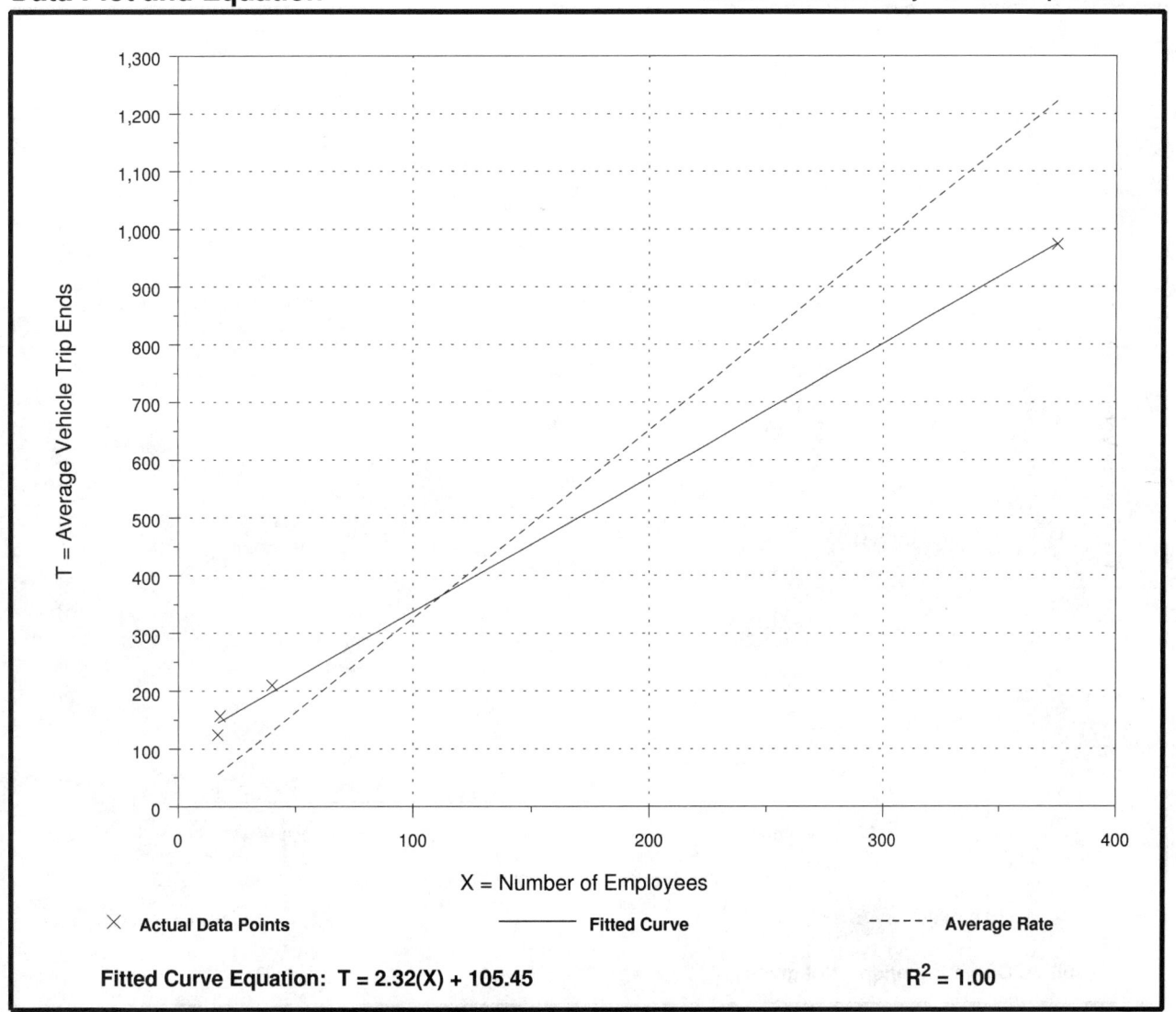

Caution - Use Carefully - Small Sample Size

Fitted Curve Equation: $T = 2.32(X) + 105.45$ $R^2 = 1.00$

Nursing Home
(620)

Average Vehicle Trip Ends vs: Employees
On a: Weekday,
A.M. Peak Hour of Generator

Number of Studies: 2
Avg. Number of Employees: 222
Directional Distribution: 69% entering, 31% exiting

Trip Generation per Employee

Average Rate	Range of Rates	Standard Deviation
0.23	0.19 - 0.24	*

Data Plot and Equation

Caution - Use Carefully - Small Sample Size

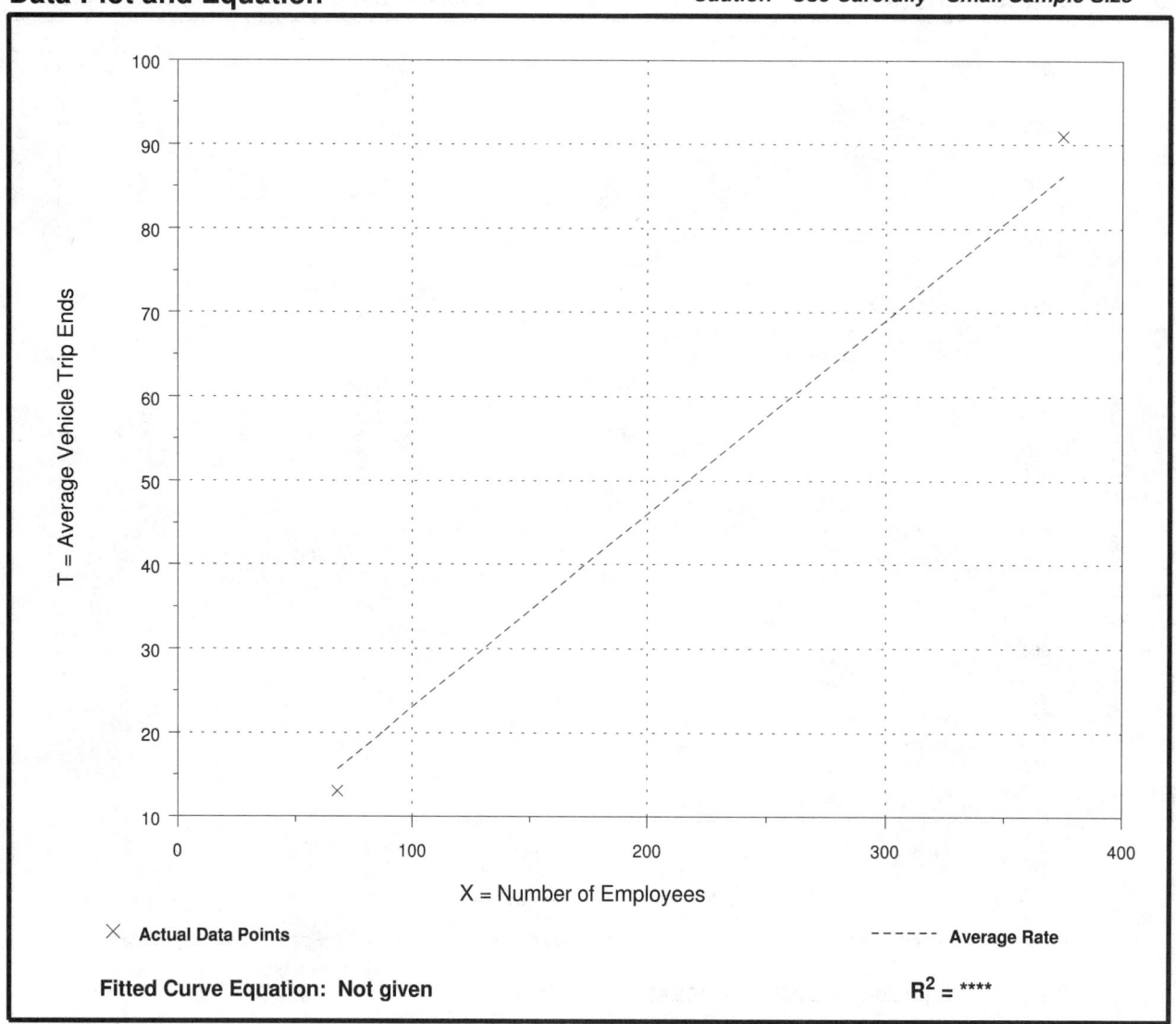

Fitted Curve Equation: Not given R^2 = ****

Nursing Home
(620)

Average Vehicle Trip Ends vs: Employees
On a: Weekday,
P.M. Peak Hour of Generator

Number of Studies: 4
Avg. Number of Employees: 113
Directional Distribution: Not available

Trip Generation per Employee

Average Rate	Range of Rates	Standard Deviation
0.47	0.41 - 0.94	0.70

Data Plot and Equation

Caution - Use Carefully - Small Sample Size

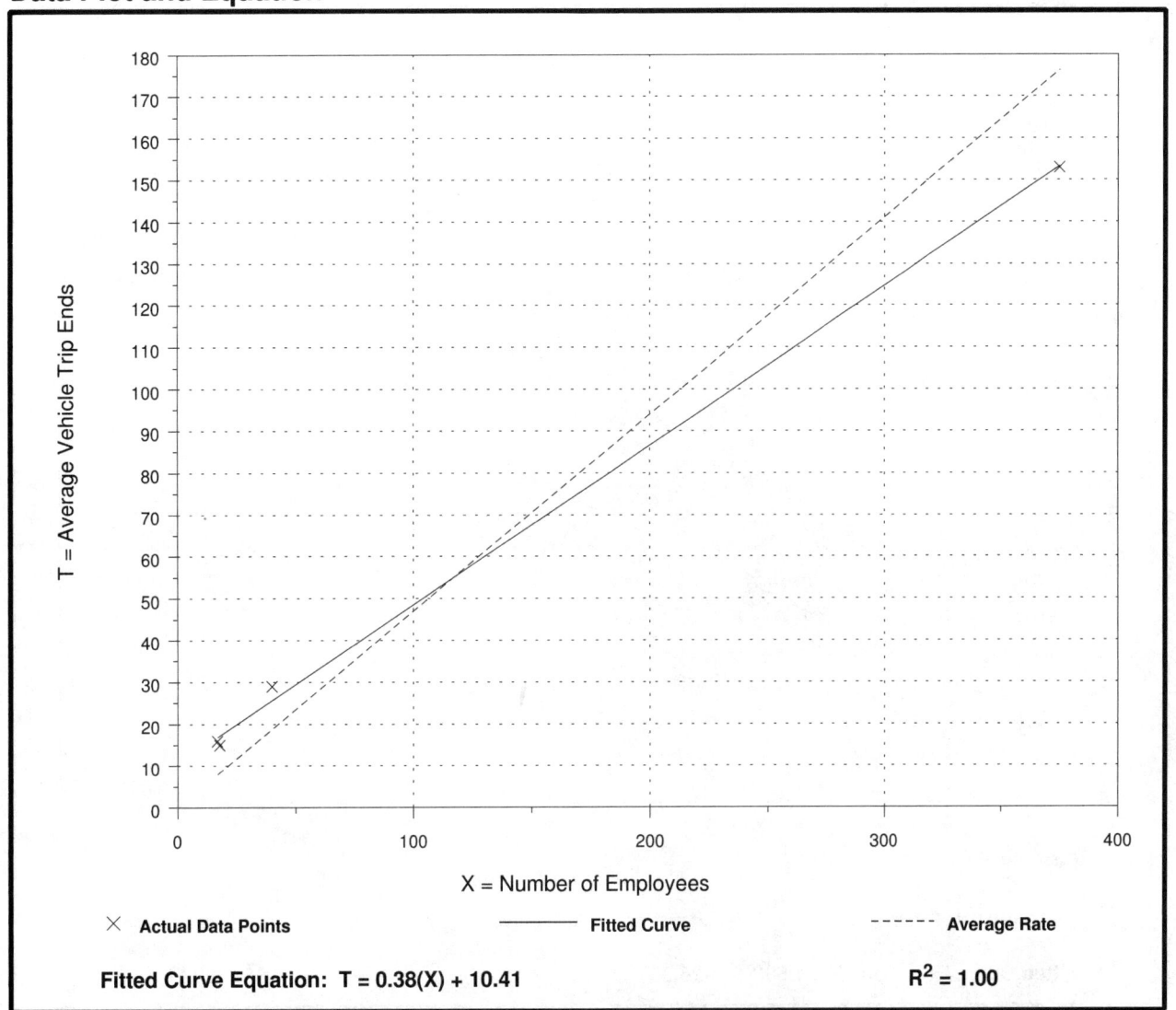

Fitted Curve Equation: $T = 0.38(X) + 10.41$ $R^2 = 1.00$

Nursing Home
(620)

Average Vehicle Trip Ends vs: Employees
On a: Saturday

Number of Studies: 4
Avg. Number of Employees: 113
Directional Distribution: 50% entering, 50% exiting

Trip Generation per Employee

Average Rate	Range of Rates	Standard Deviation
2.65	1.86 - 8.83	2.46

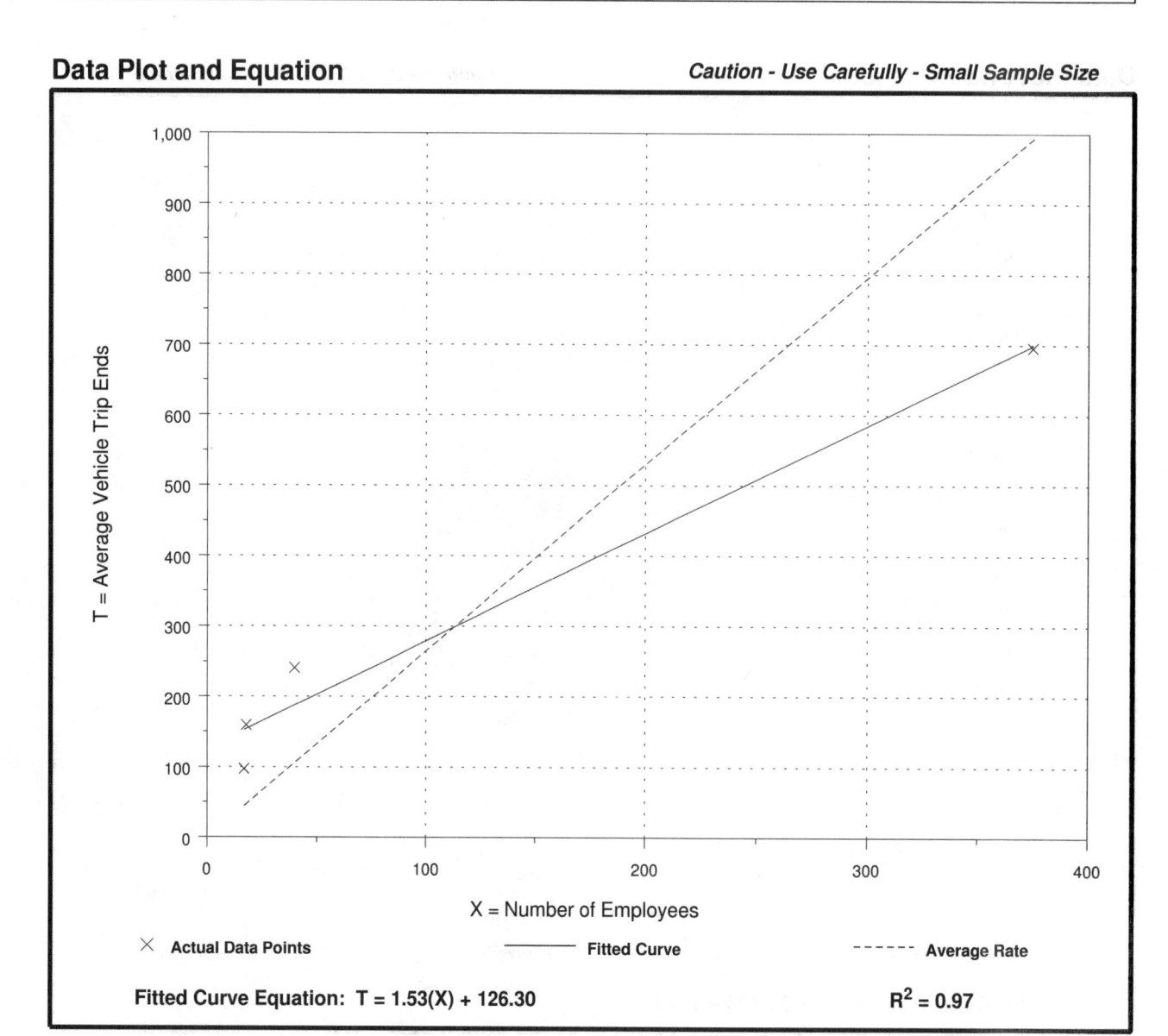

Data Plot and Equation

Caution - Use Carefully - Small Sample Size

Fitted Curve Equation: T = 1.53(X) + 126.30 $R^2 = 0.97$

Nursing Home
(620)

Average Vehicle Trip Ends vs: Employees
On a: Saturday,
Peak Hour of Generator

Number of Studies: 4
Avg. Number of Employees: 113
Directional Distribution: Not available

Trip Generation per Employee

Average Rate	Range of Rates	Standard Deviation
0.45	0.29 - 2.06	0.78

Data Plot and Equation

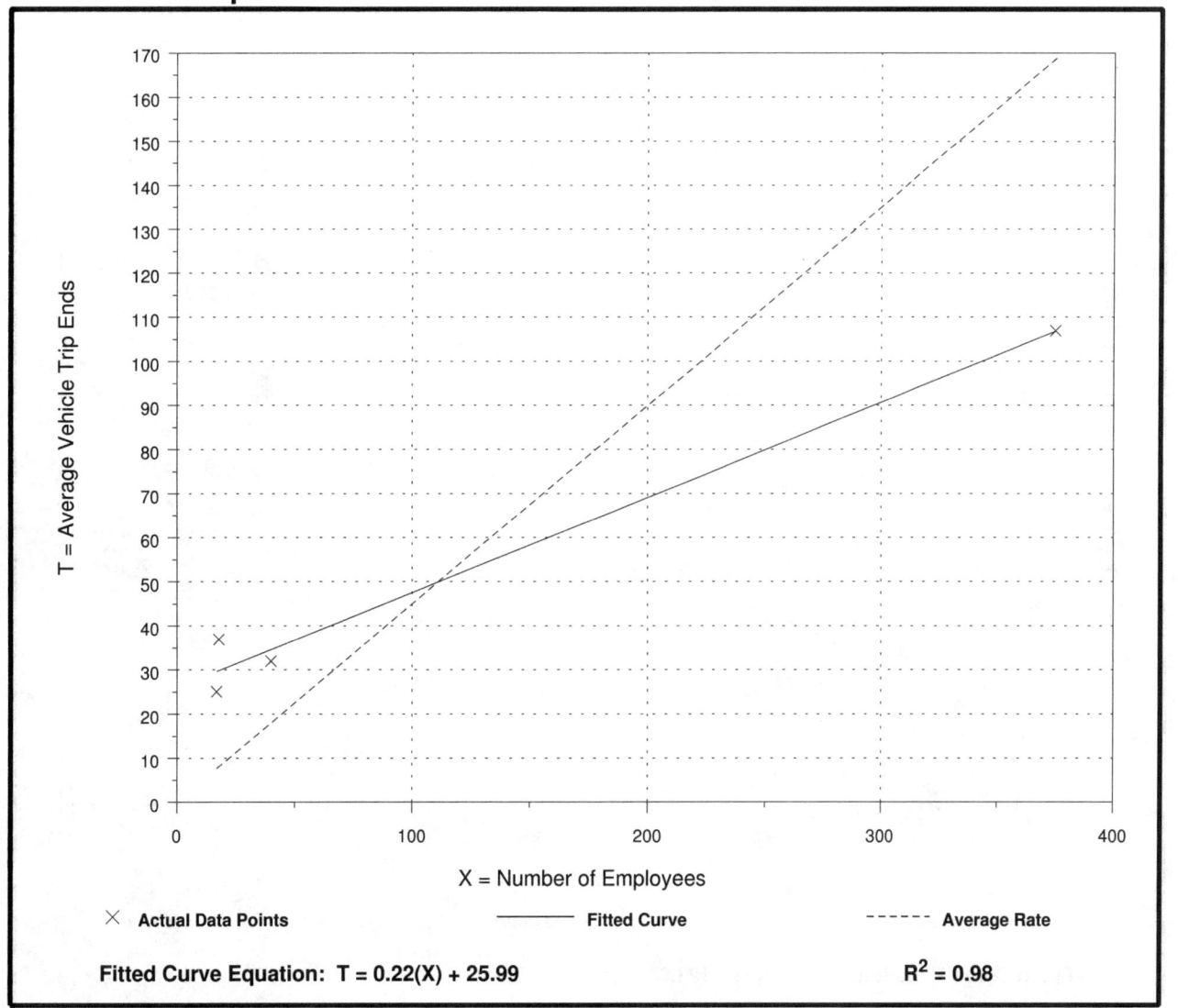

Caution - Use Carefully - Small Sample Size

Fitted Curve Equation: T = 0.22(X) + 25.99 $R^2 = 0.98$

Nursing Home
(620)

Average Vehicle Trip Ends vs: Employees
On a: Sunday

Number of Studies: 4
Avg. Number of Employees: 113
Directional Distribution: 50% entering, 50% exiting

Trip Generation per Employee

Average Rate	Range of Rates	Standard Deviation
2.70	1.93 - 10.28	2.53

Data Plot and Equation

Caution - Use Carefully - Small Sample Size

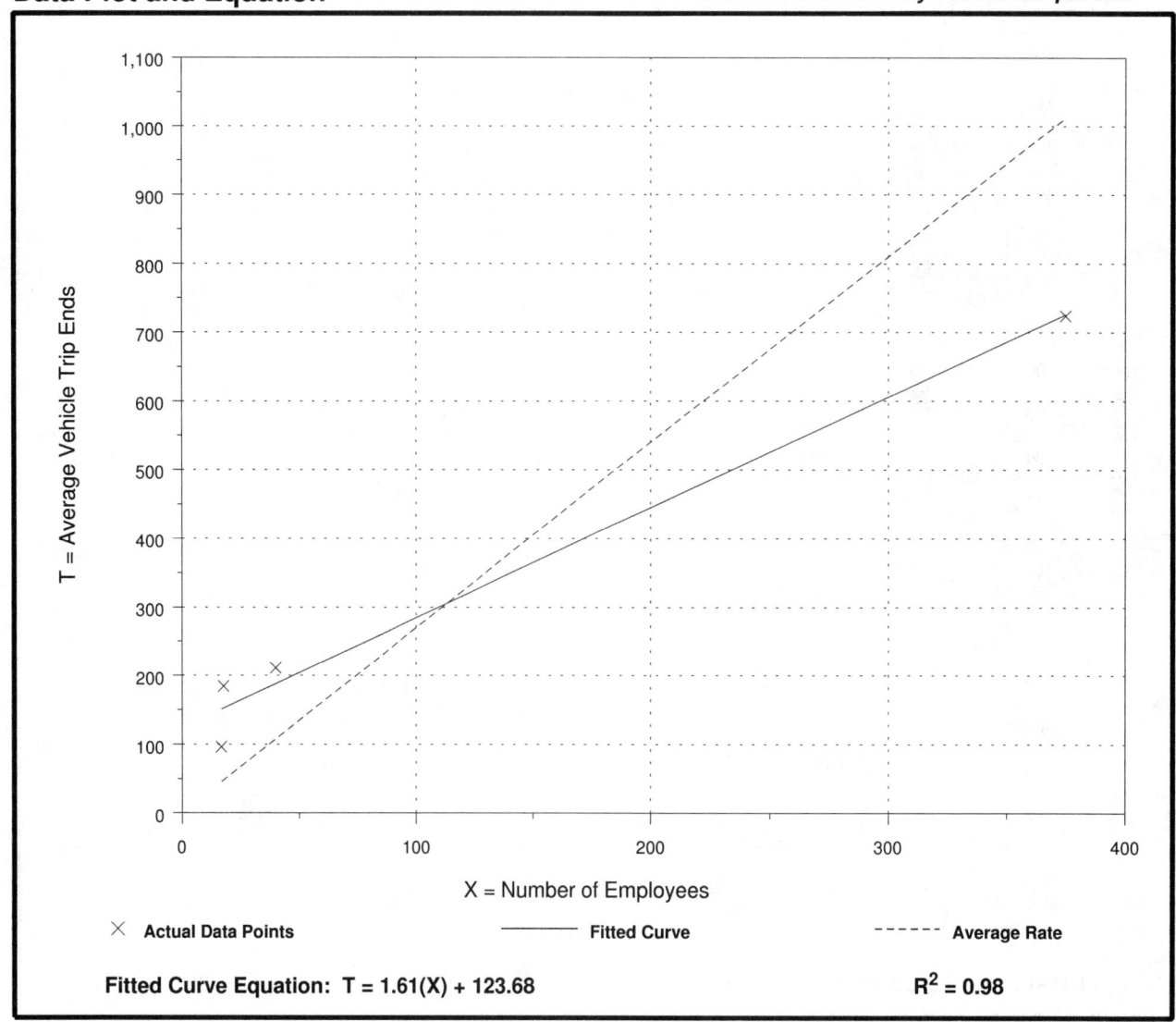

Fitted Curve Equation: T = 1.61(X) + 123.68 $R^2 = 0.98$

Nursing Home
(620)

Average Vehicle Trip Ends vs: Employees
On a: Sunday,
Peak Hour of Generator

Number of Studies: 5
Avg. Number of Employees: 104
Directional Distribution: 66% entering, 34% exiting

Trip Generation per Employee

Average Rate	Range of Rates	Standard Deviation
0.53	0.32 - 2.28	0.83

Data Plot and Equation

Caution - Use Carefully - Small Sample Size

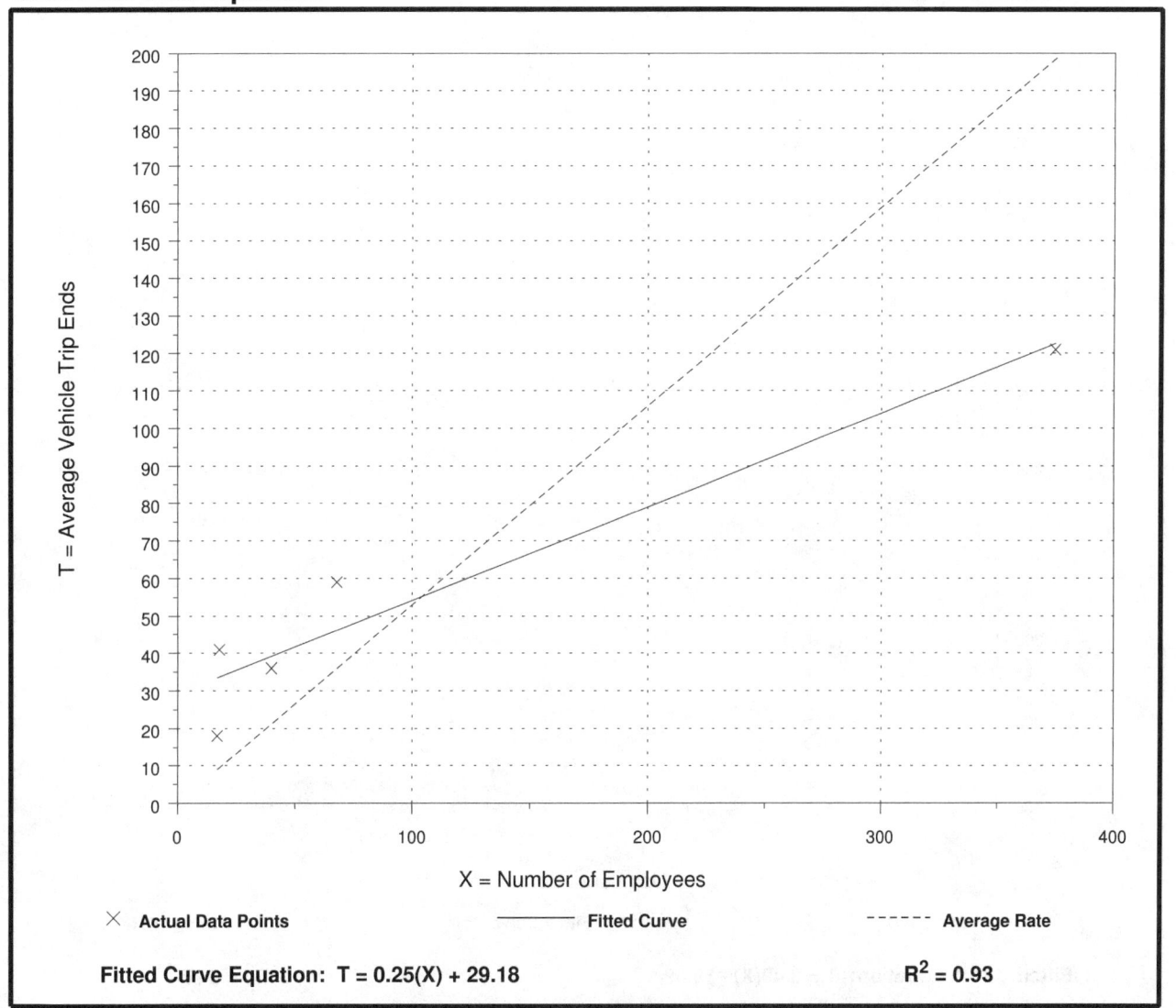

Fitted Curve Equation: $T = 0.25(X) + 29.18$ $R^2 = 0.93$

Nursing Home
(620)

Average Vehicle Trip Ends vs: Beds
On a: Weekday

Number of Studies: 6
Average Number of Beds: 119
Directional Distribution: 50% entering, 50% exiting

Trip Generation per Bed

Average Rate	Range of Rates	Standard Deviation
2.74	2.00 - 3.25	1.73

Data Plot and Equation

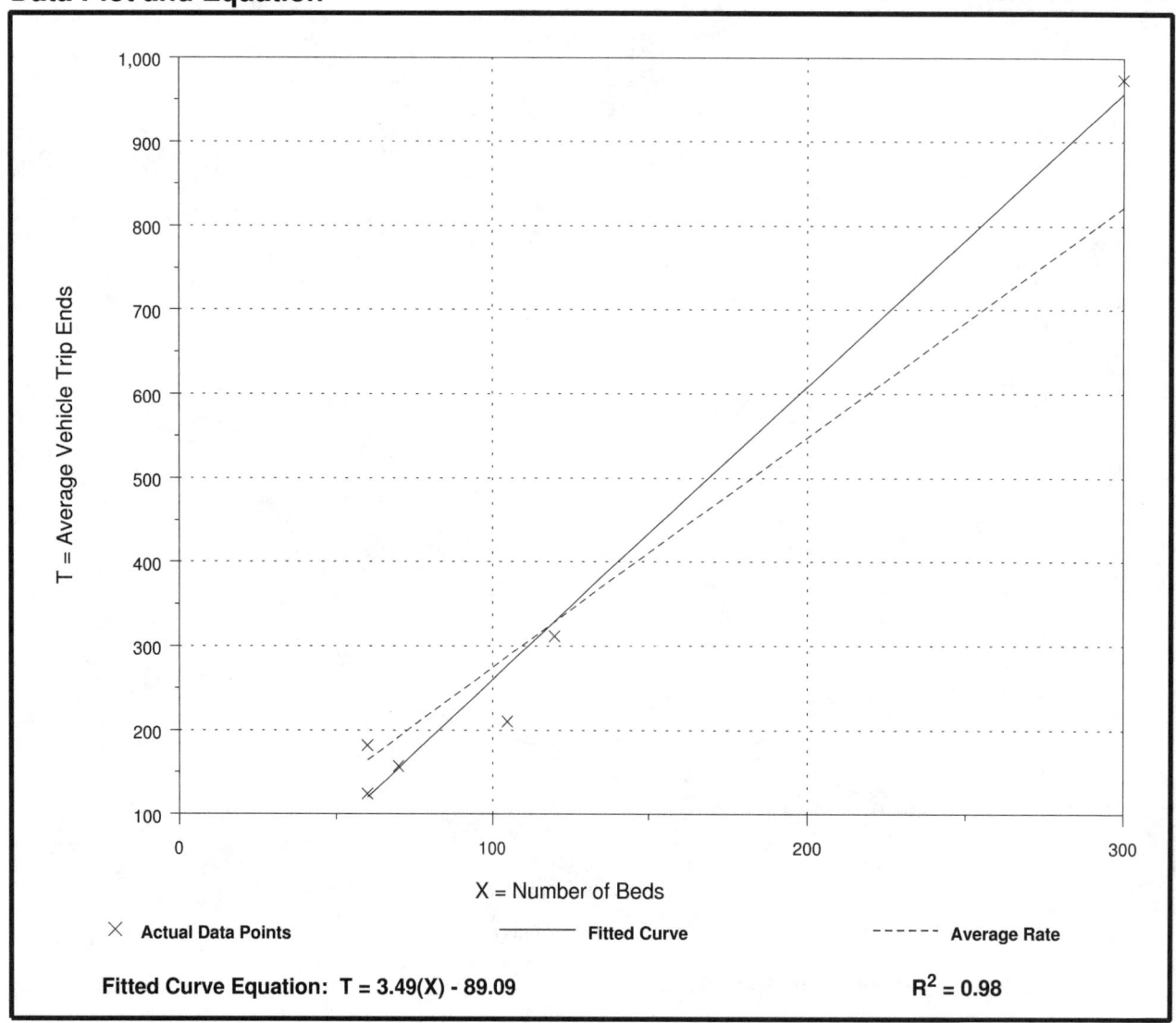

Fitted Curve Equation: T = 3.49(X) - 89.09 $R^2 = 0.98$

Nursing Home
(620)

Average Vehicle Trip Ends vs: Beds
On a: Weekday,
Peak Hour of Adjacent Street Traffic,
One Hour Between 7 and 9 a.m.

Number of Studies: 2
Average Number of Beds: 90
Directional Distribution: Not available

Trip Generation per Bed

Average Rate	Range of Rates	Standard Deviation
0.17	0.16 - 0.20	*

Data Plot and Equation

Caution - Use Carefully - Small Sample Size

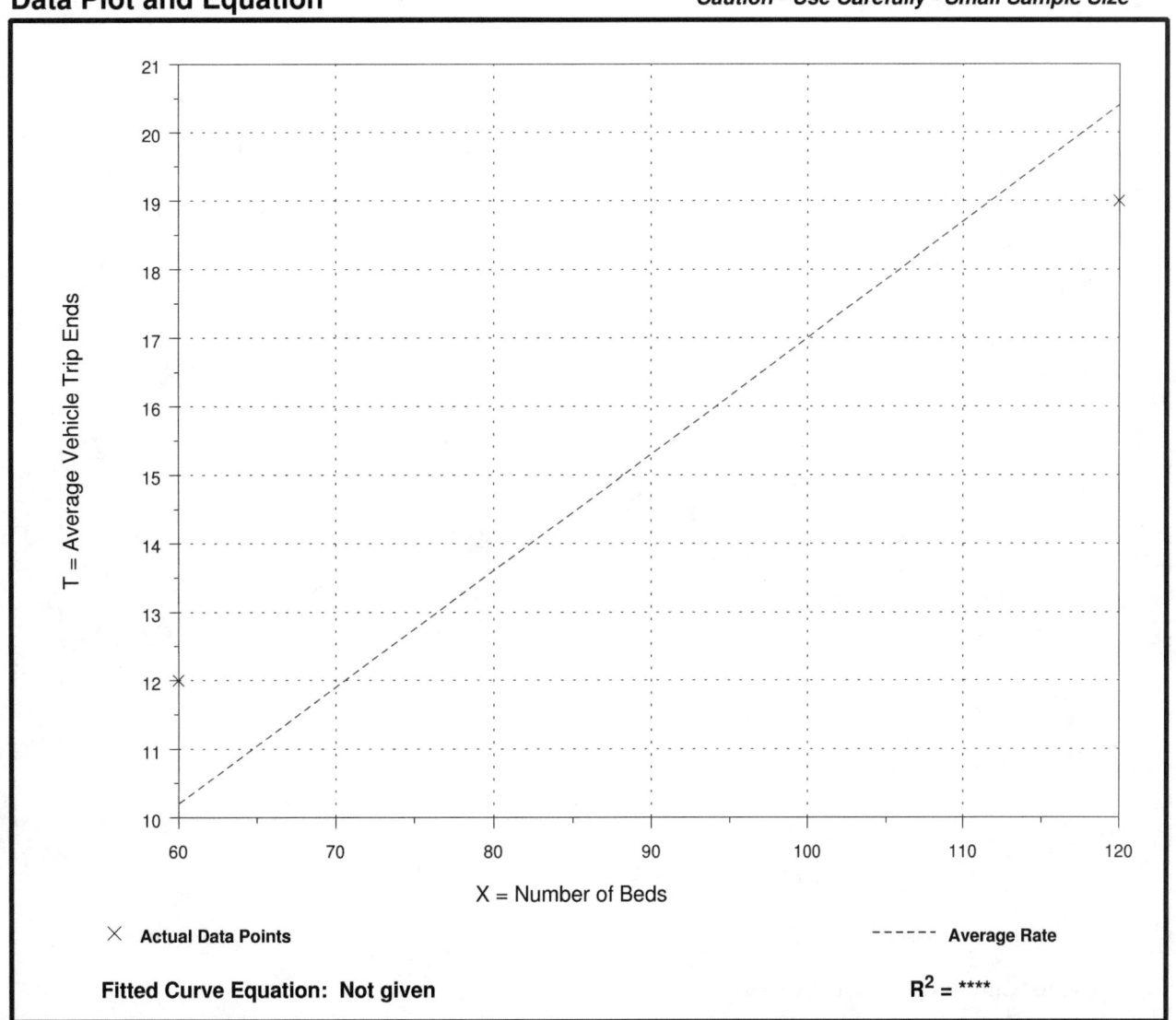

Fitted Curve Equation: Not given $R^2 = ****$

Nursing Home
(620)

Average Vehicle Trip Ends vs: Beds
On a: Weekday,
Peak Hour of Adjacent Street Traffic,
One Hour Between 4 and 6 p.m.

Number of Studies: 4
Average Number of Beds: 99
Directional Distribution: 33% entering, 67% exiting

Trip Generation per Bed

Average Rate	Range of Rates	Standard Deviation
0.22	0.12 - 0.27	0.47

Data Plot and Equation

Caution - Use Carefully - Small Sample Size

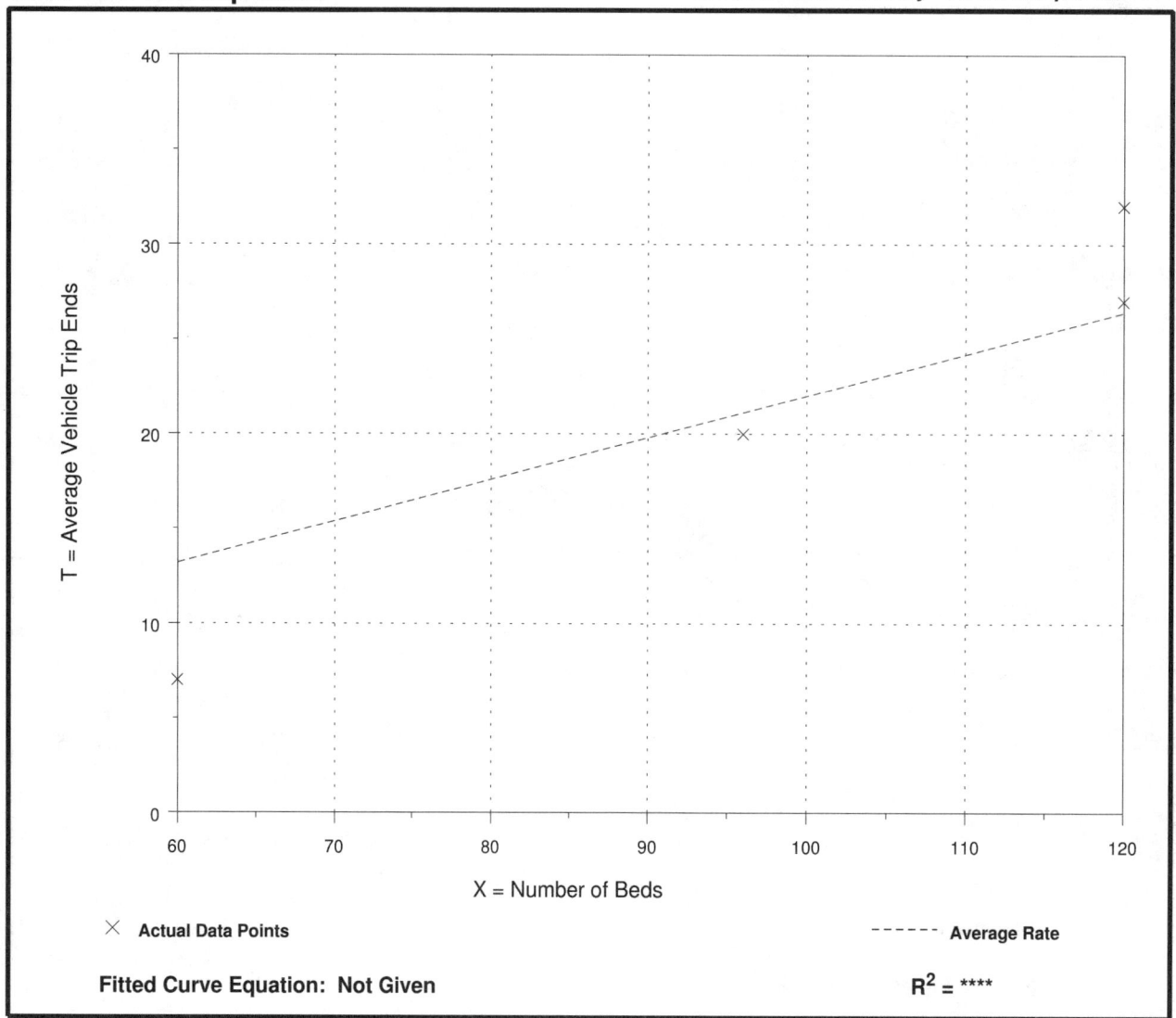

Fitted Curve Equation: Not Given $R^2 = ****$

Nursing Home
(620)

Average Vehicle Trip Ends vs: Beds
On a: Weekday,
A.M. Peak Hour of Generator

Number of Studies: 4
Average Number of Beds: 171
Directional Distribution: 69% entering, 31% exiting

Trip Generation per Bed

Average Rate	Range of Rates	Standard Deviation
0.20	0.06 - 0.30	0.46

Data Plot and Equation

Caution - Use Carefully - Small Sample Size

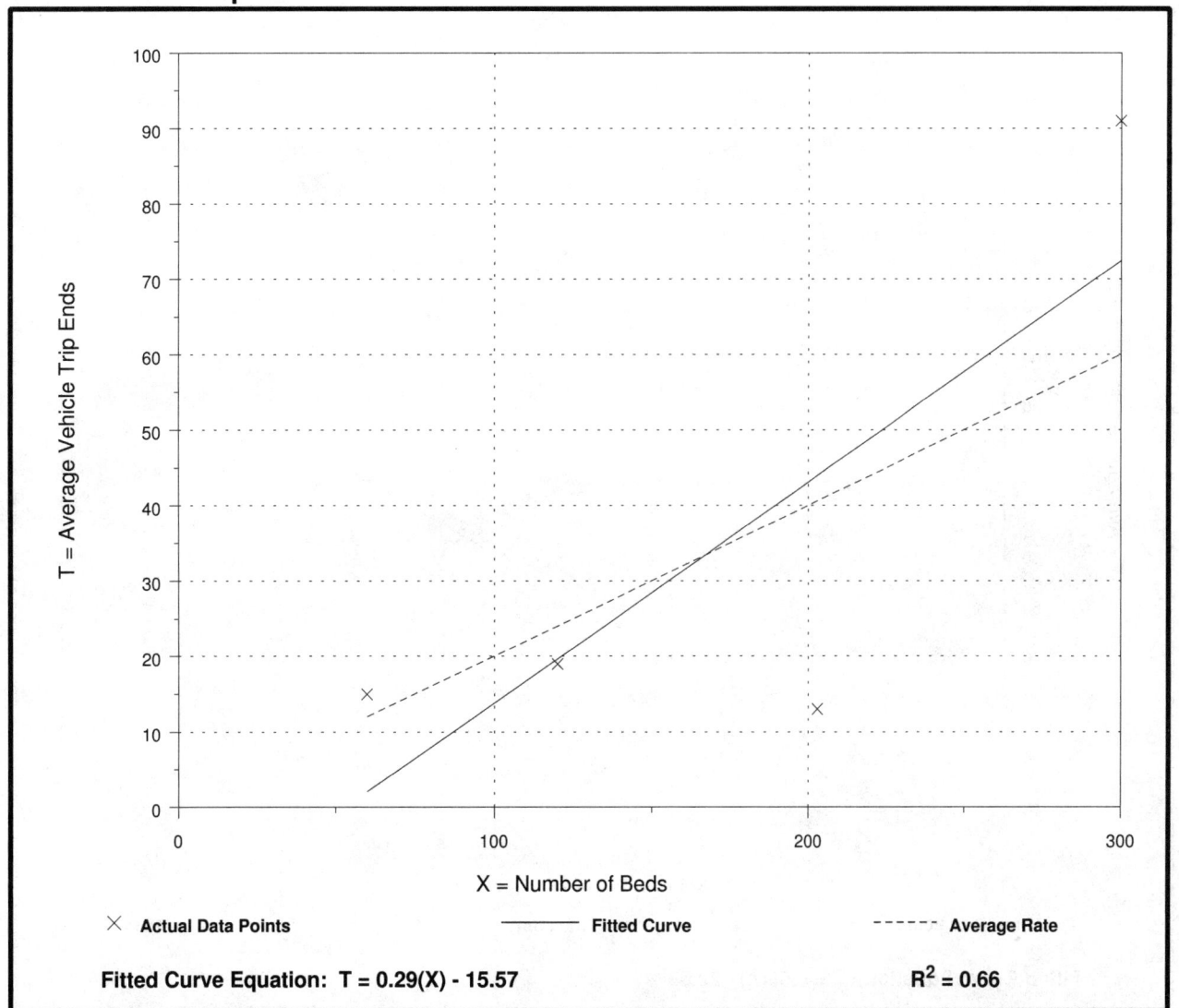

Fitted Curve Equation: T = 0.29(X) - 15.57 $R^2 = 0.66$

Nursing Home
(620)

Average Vehicle Trip Ends vs: Beds
On a: Weekday,
P.M. Peak Hour of Generator

Number of Studies: 8
Average Number of Beds: 116
Directional Distribution: 40% entering, 60% exiting

Trip Generation per Bed

Average Rate	Range of Rates	Standard Deviation
0.37	0.21 - 0.51	0.62

Data Plot and Equation

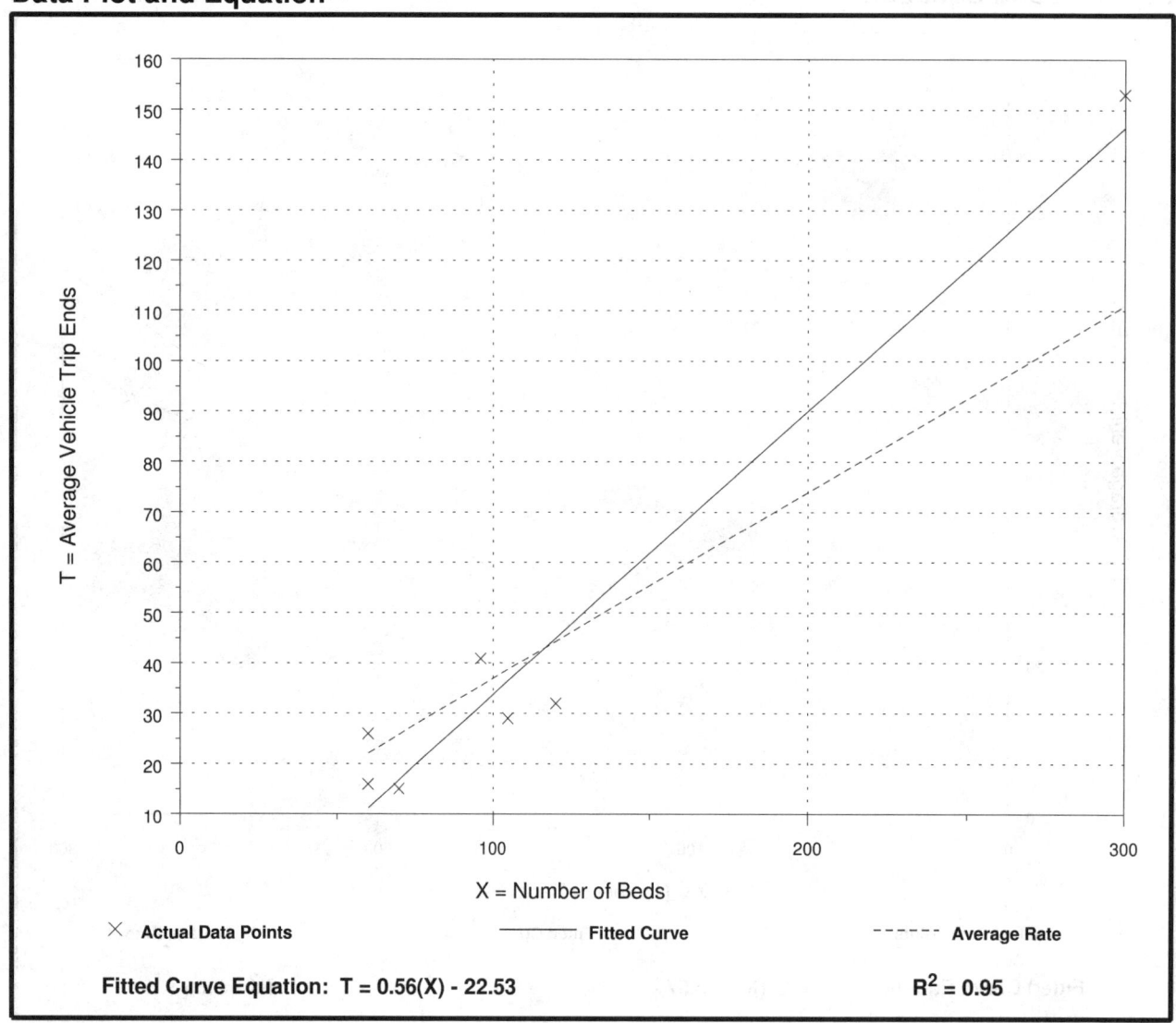

Fitted Curve Equation: $T = 0.56(X) - 22.53$ $R^2 = 0.95$

Nursing Home
(620)

Average Vehicle Trip Ends vs: Beds
On a: Saturday

Number of Studies: 4
Average Number of Beds: 134
Directional Distribution: 50% entering, 50% exiting

Trip Generation per Bed

Average Rate	Range of Rates	Standard Deviation
2.23	1.62 - 2.32	1.51

Data Plot and Equation

Caution - Use Carefully - Small Sample Size

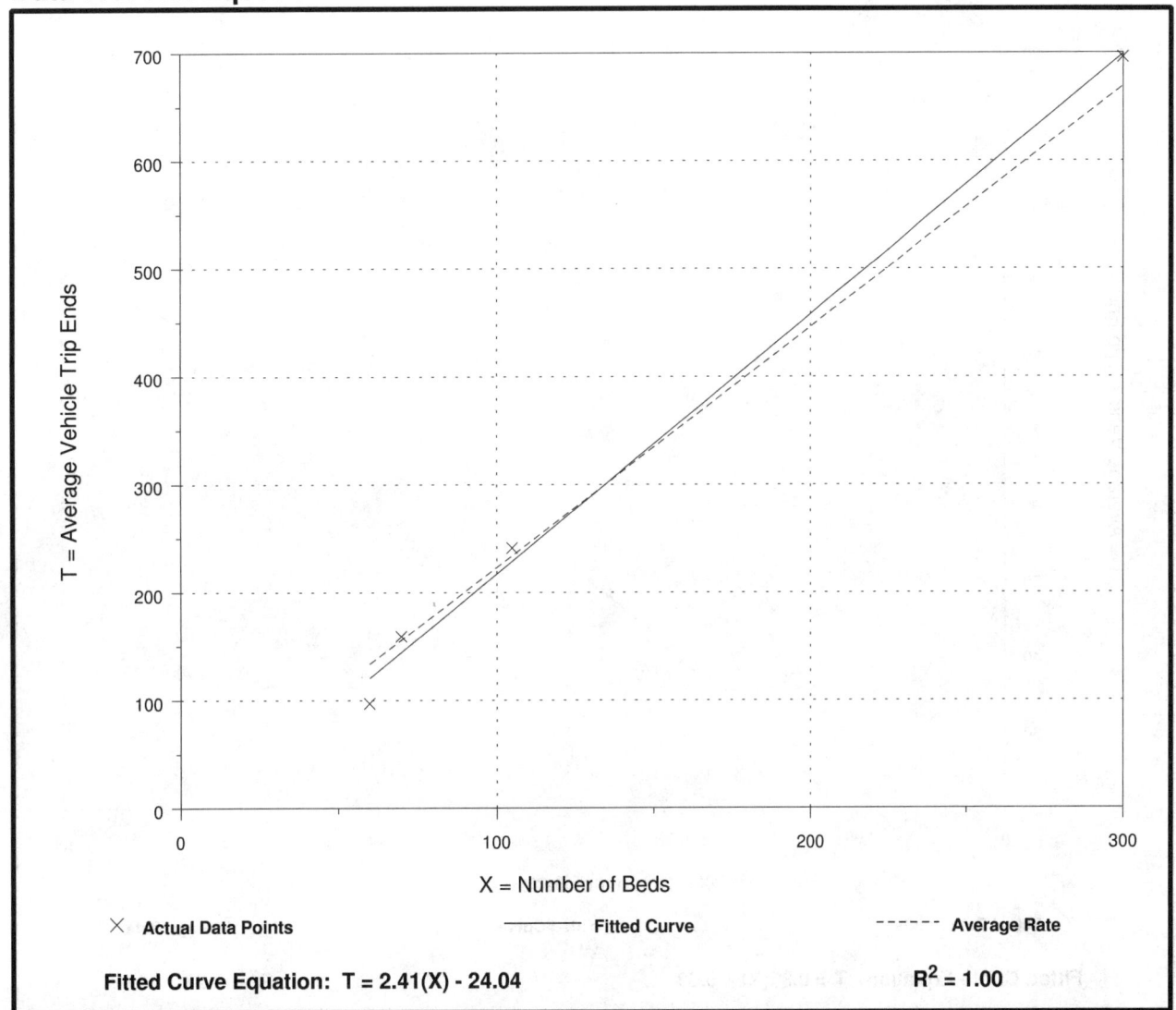

Fitted Curve Equation: $T = 2.41(X) - 24.04$ $R^2 = 1.00$

Nursing Home
(620)

Average Vehicle Trip Ends vs: Beds
On a: Saturday,
Peak Hour of Generator

Number of Studies: 4
Average Number of Beds: 134
Directional Distribution: Not available

Trip Generation per Bed

Average Rate	Range of Rates	Standard Deviation
0.38	0.30 - 0.53	0.61

Data Plot and Equation

Caution - Use Carefully - Small Sample Size

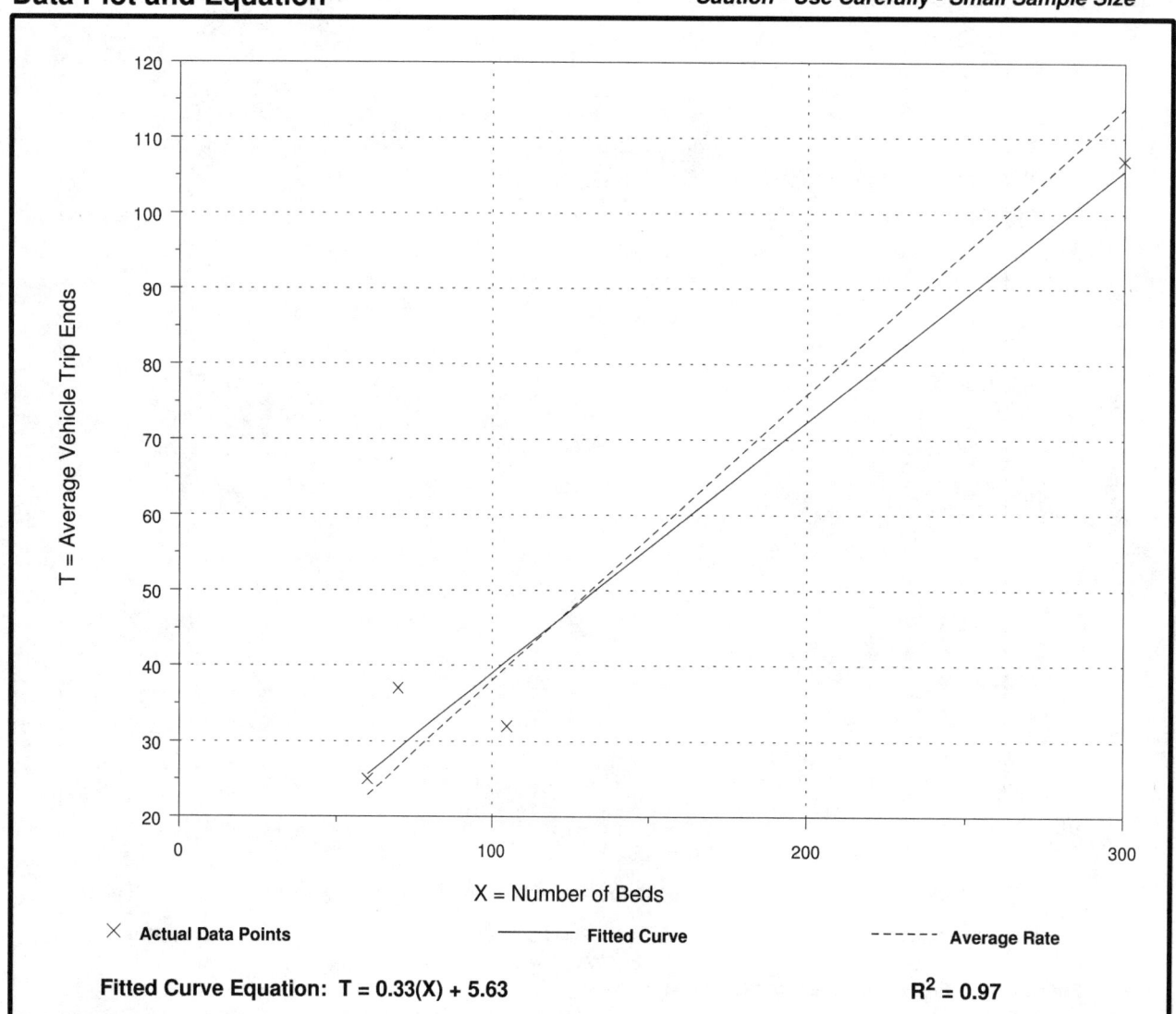

Fitted Curve Equation: $T = 0.33(X) + 5.63$ $R^2 = 0.97$

Nursing Home
(620)

Average Vehicle Trip Ends vs: Beds
On a: Sunday

Number of Studies: 4
Average Number of Beds: 134
Directional Distribution: 50% entering, 50% exiting

Trip Generation per Bed

Average Rate	Range of Rates	Standard Deviation
2.27	1.62 - 2.64	1.53

Data Plot and Equation

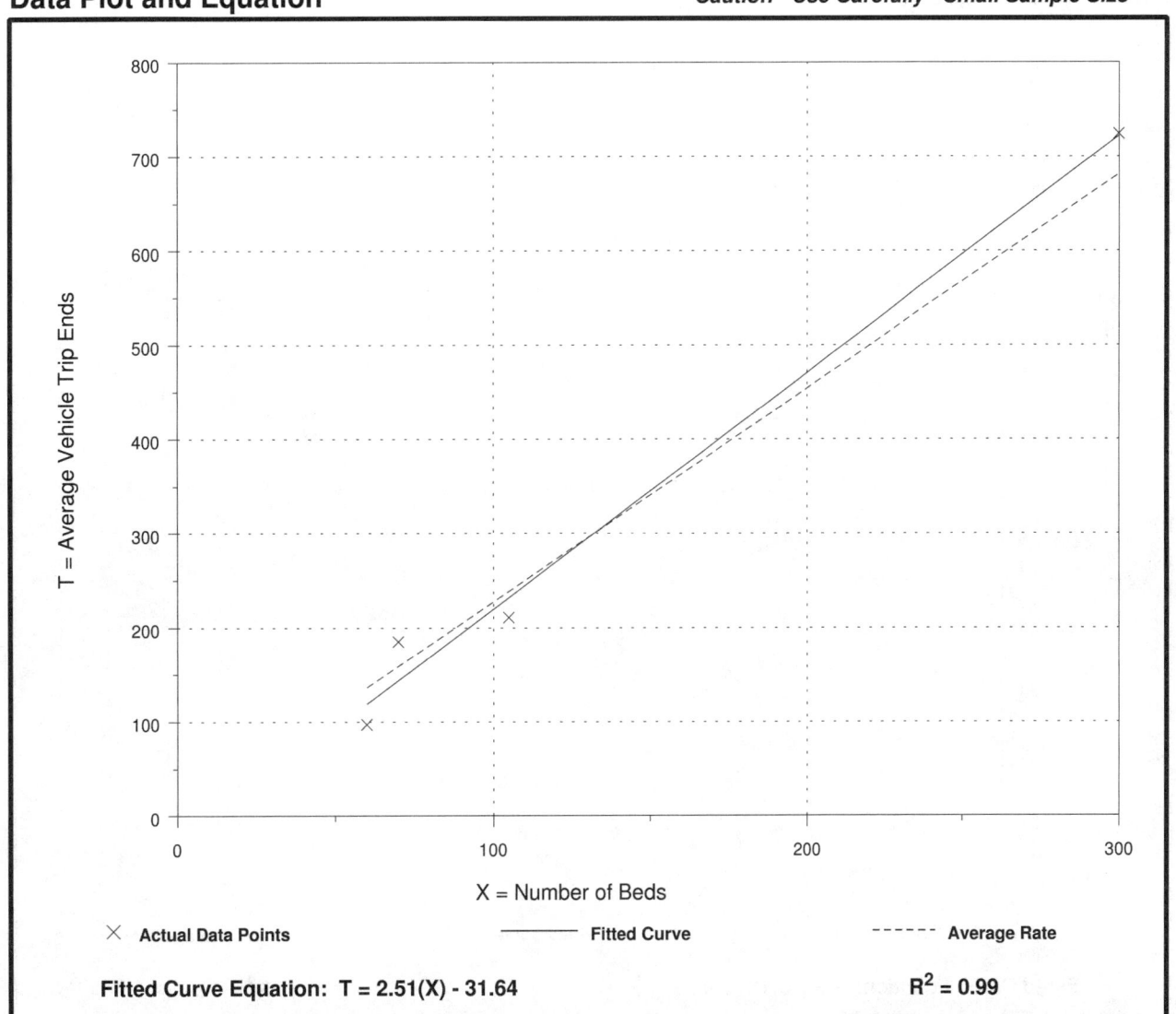

Caution - Use Carefully - Small Sample Size

Fitted Curve Equation: $T = 2.51(X) - 31.64$ $R^2 = 0.99$

Nursing Home
(620)

Average Vehicle Trip Ends vs: Beds
On a: Sunday,
Peak Hour of Generator

Number of Studies: 6
Average Number of Beds: 139
Directional Distribution: 57% entering, 43% exiting

Trip Generation per Bed

Average Rate	Range of Rates	Standard Deviation
0.37	0.29 - 0.59	0.61

Data Plot and Equation

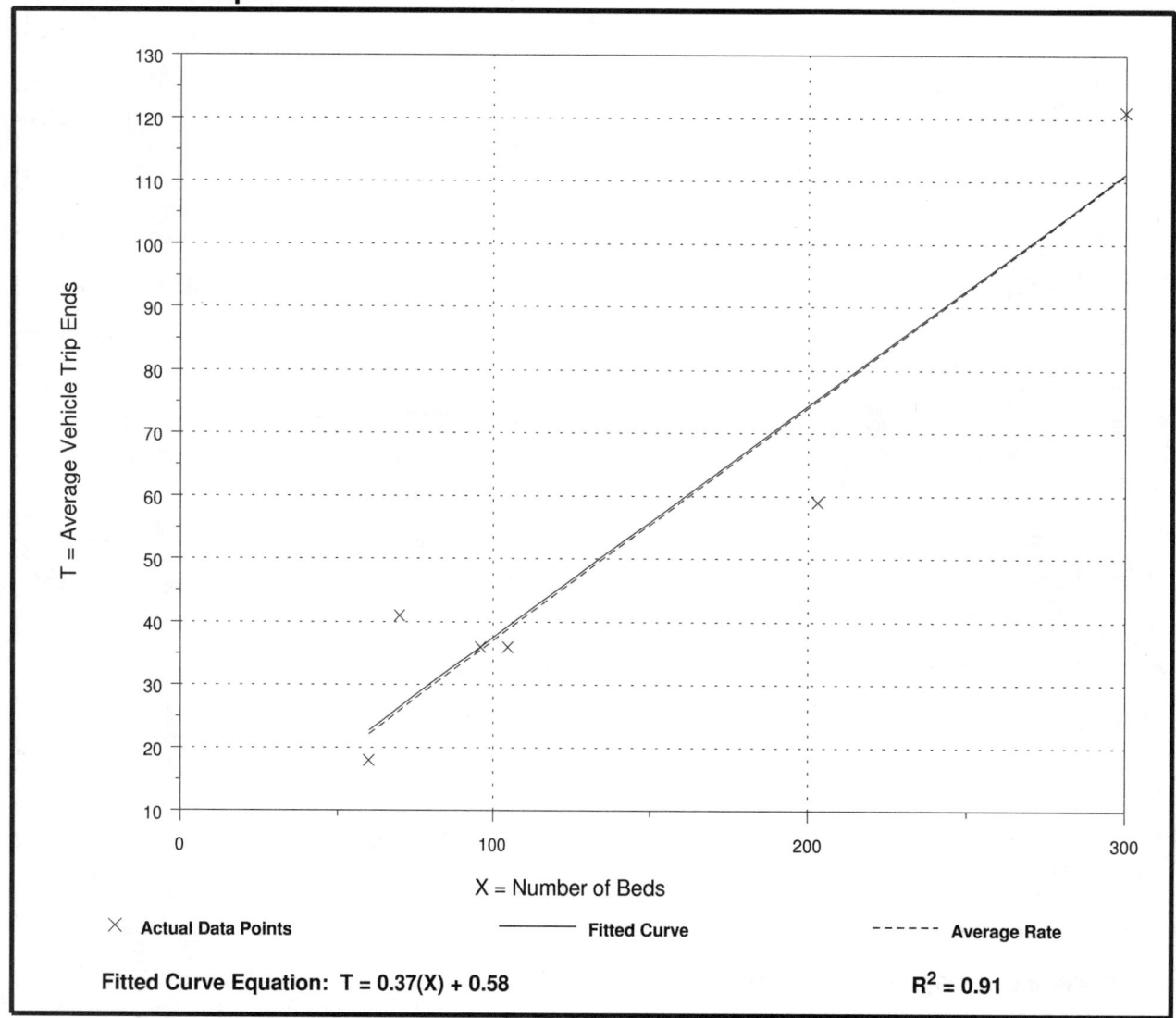

Fitted Curve Equation: T = 0.37(X) + 0.58 $R^2 = 0.91$

Nursing Home
(620)

Average Vehicle Trip Ends vs: 1000 Sq. Feet Gross Floor Area
On a: Weekday

Number of Studies: 4
Average 1000 Sq. Feet GFA: 63
Directional Distribution: 50% entering, 50% exiting

Trip Generation per 1000 Sq. Feet Gross Floor Area

Average Rate	Range of Rates	Standard Deviation
7.60	5.67 - 10.31	3.10

Data Plot and Equation

Caution - Use Carefully - Small Sample Size

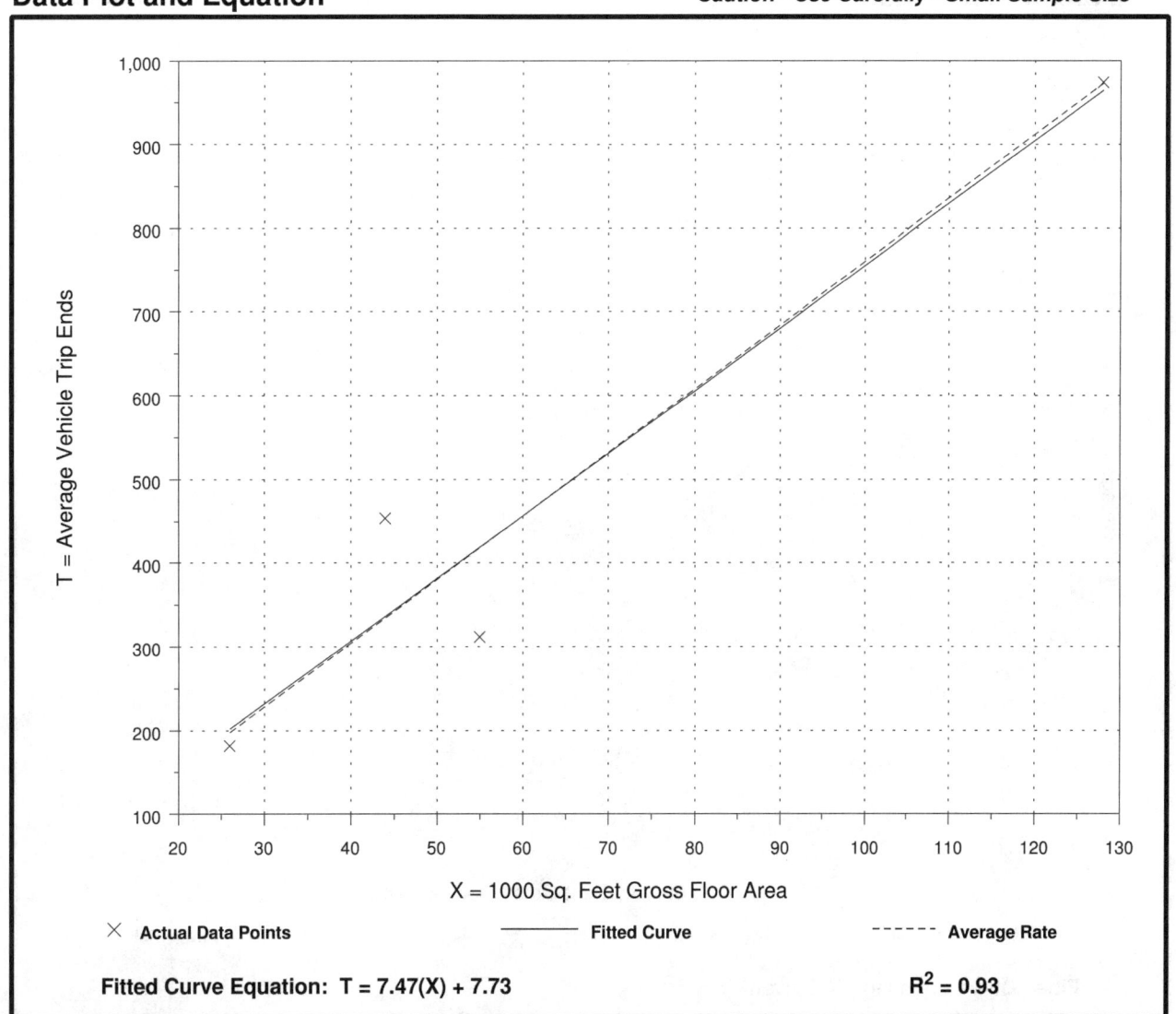

Fitted Curve Equation: $T = 7.47(X) + 7.73$ $R^2 = 0.93$

Nursing Home
(620)

Average Vehicle Trip Ends vs: 1000 Sq. Feet Gross Floor Area
On a: Weekday,
Peak Hour of Adjacent Street Traffic,
One Hour Between 7 and 9 a.m.

Number of Studies: 3
Average 1000 Sq. Feet GFA: 42
Directional Distribution: 71% entering, 29% exiting

Trip Generation per 1000 Sq. Feet Gross Floor Area

Average Rate	Range of Rates	Standard Deviation
0.55	0.35 - 0.86	0.77

Data Plot and Equation

Caution - Use Carefully - Small Sample Size

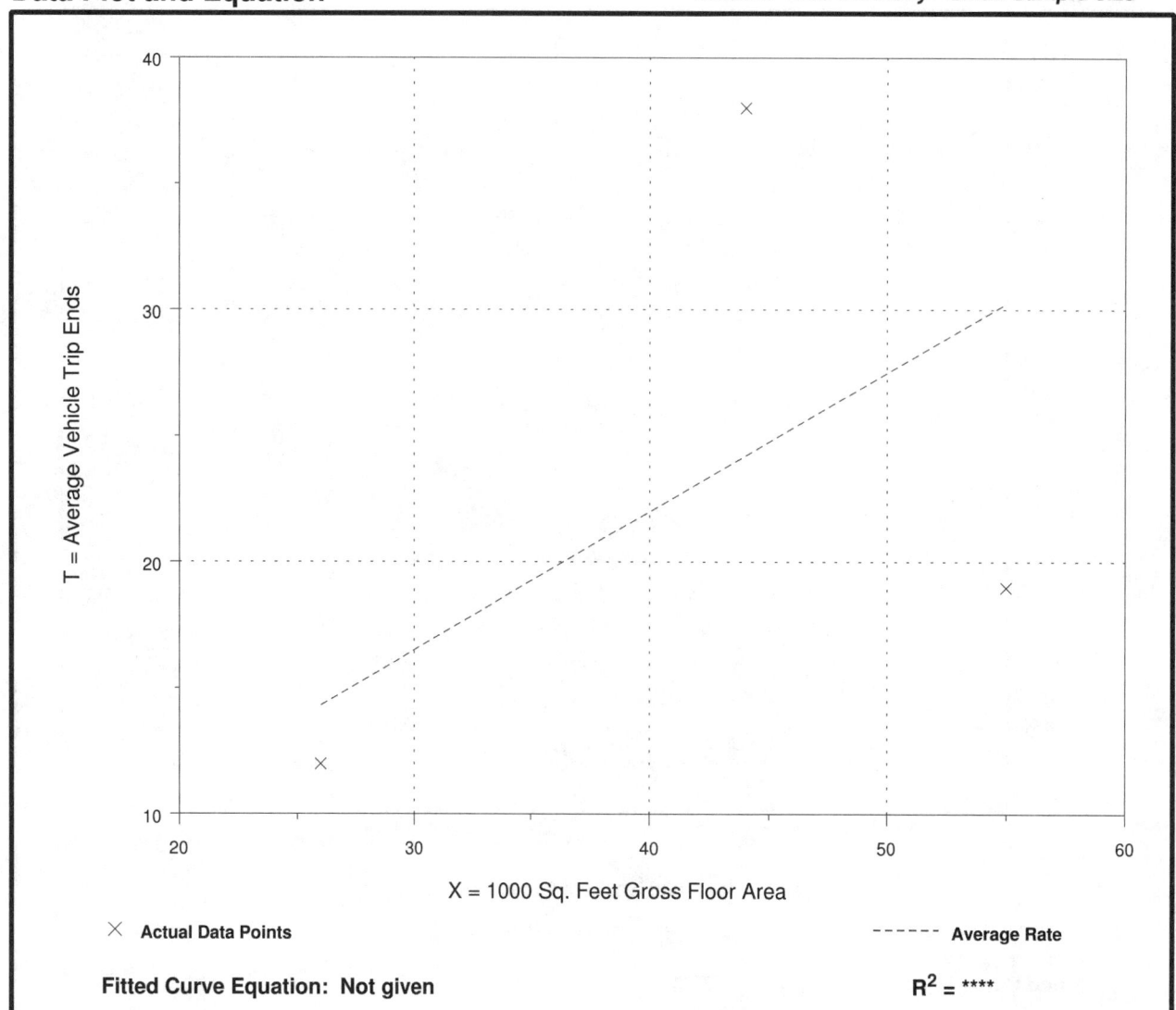

Fitted Curve Equation: Not given $R^2 = ****$

Nursing Home
(620)

Average Vehicle Trip Ends vs: 1000 Sq. Feet Gross Floor Area
On a: Weekday,
Peak Hour of Adjacent Street Traffic,
One Hour Between 4 and 6 p.m.

Number of Studies: 3
Average 1000 Sq. Feet GFA: 42
Directional Distribution: 52% entering, 48% exiting

Trip Generation per 1000 Sq. Feet Gross Floor Area

Average Rate	Range of Rates	Standard Deviation
0.74	0.27 - 1.32	0.96

Data Plot and Equation

Caution - Use Carefully - Small Sample Size

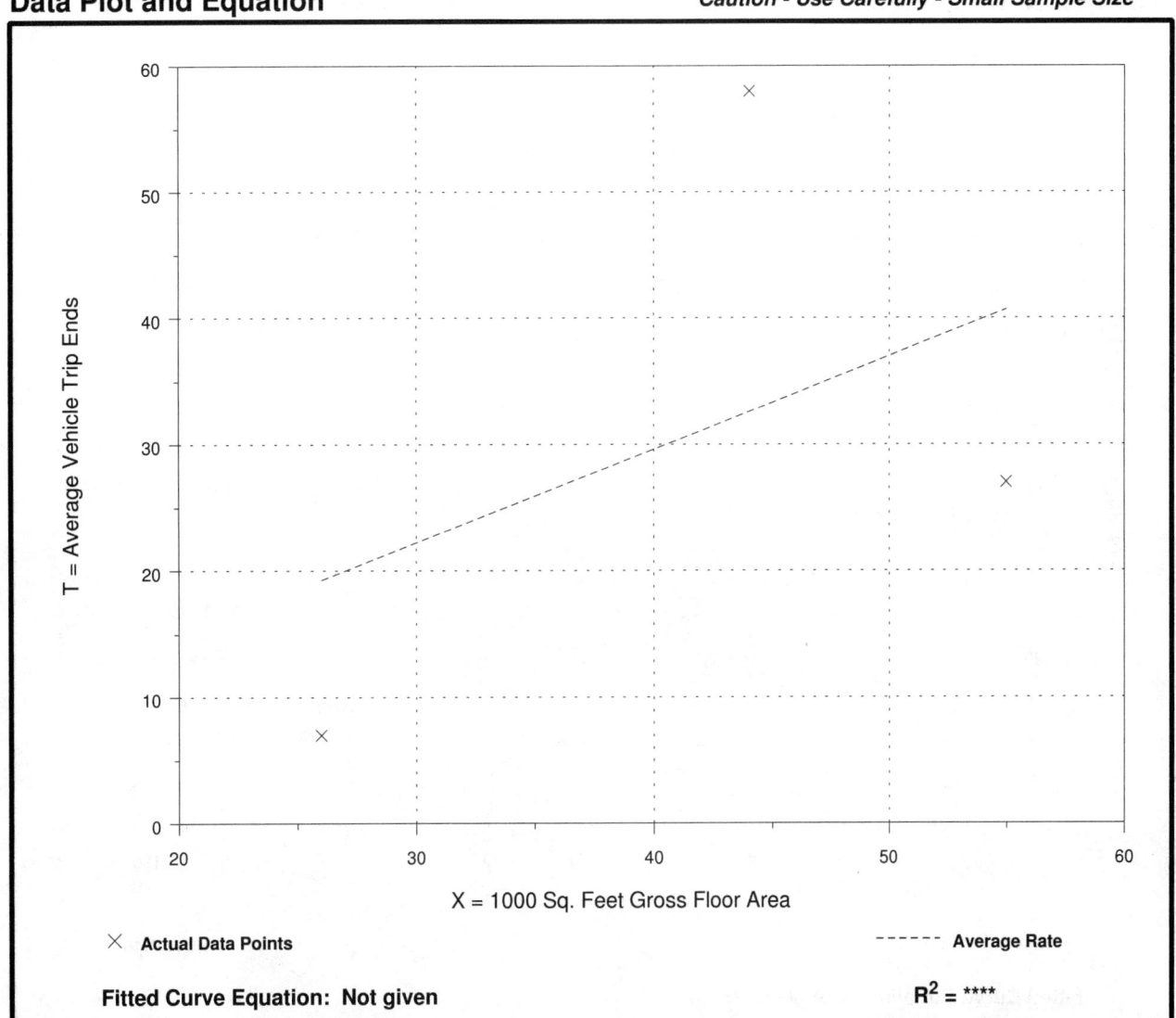

Fitted Curve Equation: Not given $R^2 = ****$

Nursing Home
(620)

Average Vehicle Trip Ends vs: 1000 Sq. Feet Gross Floor Area
On a: Weekday,
A.M. Peak Hour of Generator

Number of Studies: 3
Average 1000 Sq. Feet GFA: 70
Directional Distribution: Not available

Trip Generation per 1000 Sq. Feet Gross Floor Area

Average Rate	Range of Rates	Standard Deviation
0.60	0.35 - 0.71	0.79

Data Plot and Equation

Caution - Use Carefully - Small Sample Size

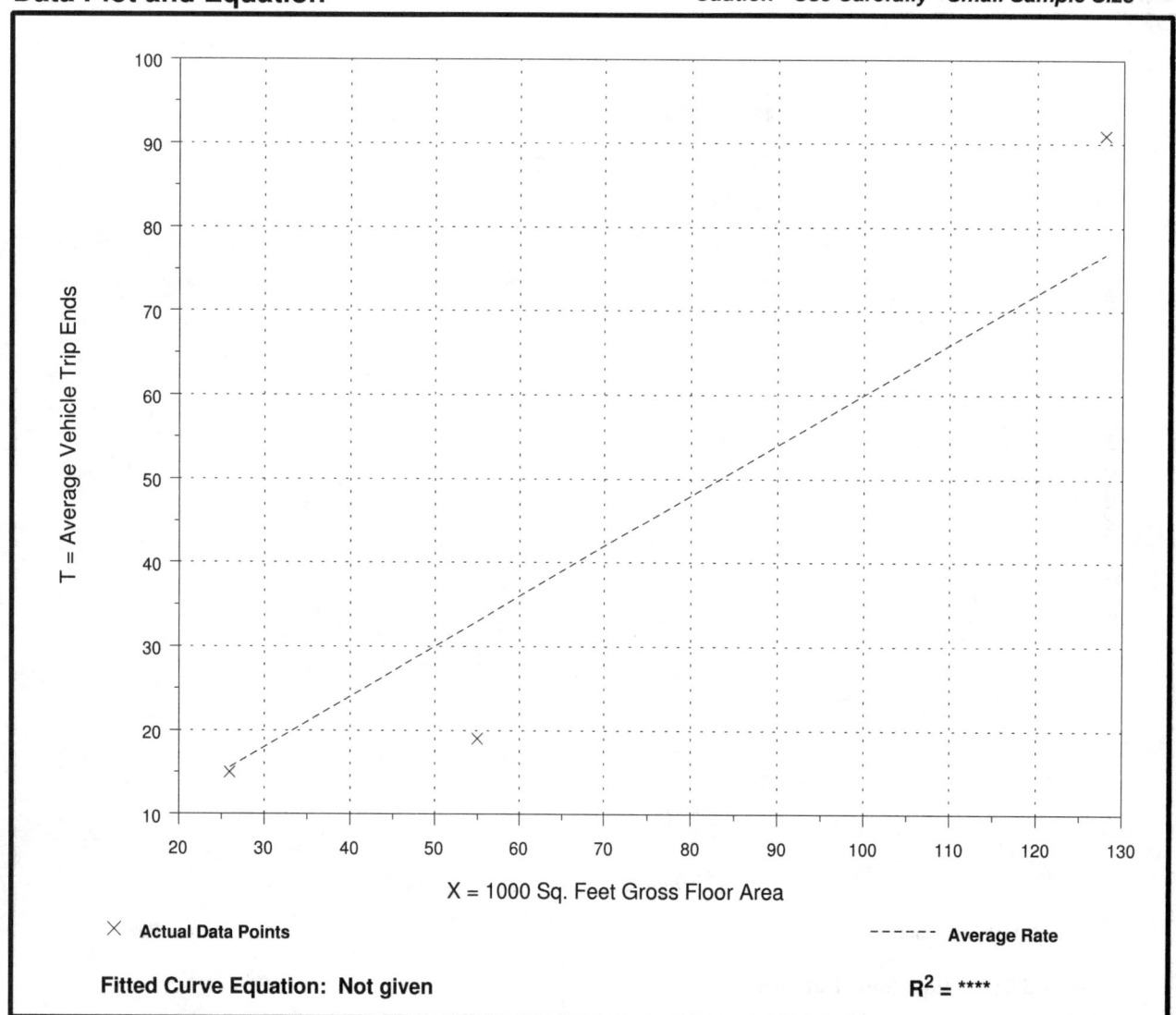

Fitted Curve Equation: Not given $R^2 = ****$

Nursing Home
(620)

Average Vehicle Trip Ends vs: 1000 Sq. Feet Gross Floor Area
On a: Weekday,
P.M. Peak Hour of Generator

Number of Studies: 3
Average 1000 Sq. Feet GFA: 70
Directional Distribution: 45% entering, 55% exiting

Trip Generation per 1000 Sq. Feet Gross Floor Area

Average Rate	Range of Rates	Standard Deviation
1.01	0.58 - 1.20	1.03

Data Plot and Equation

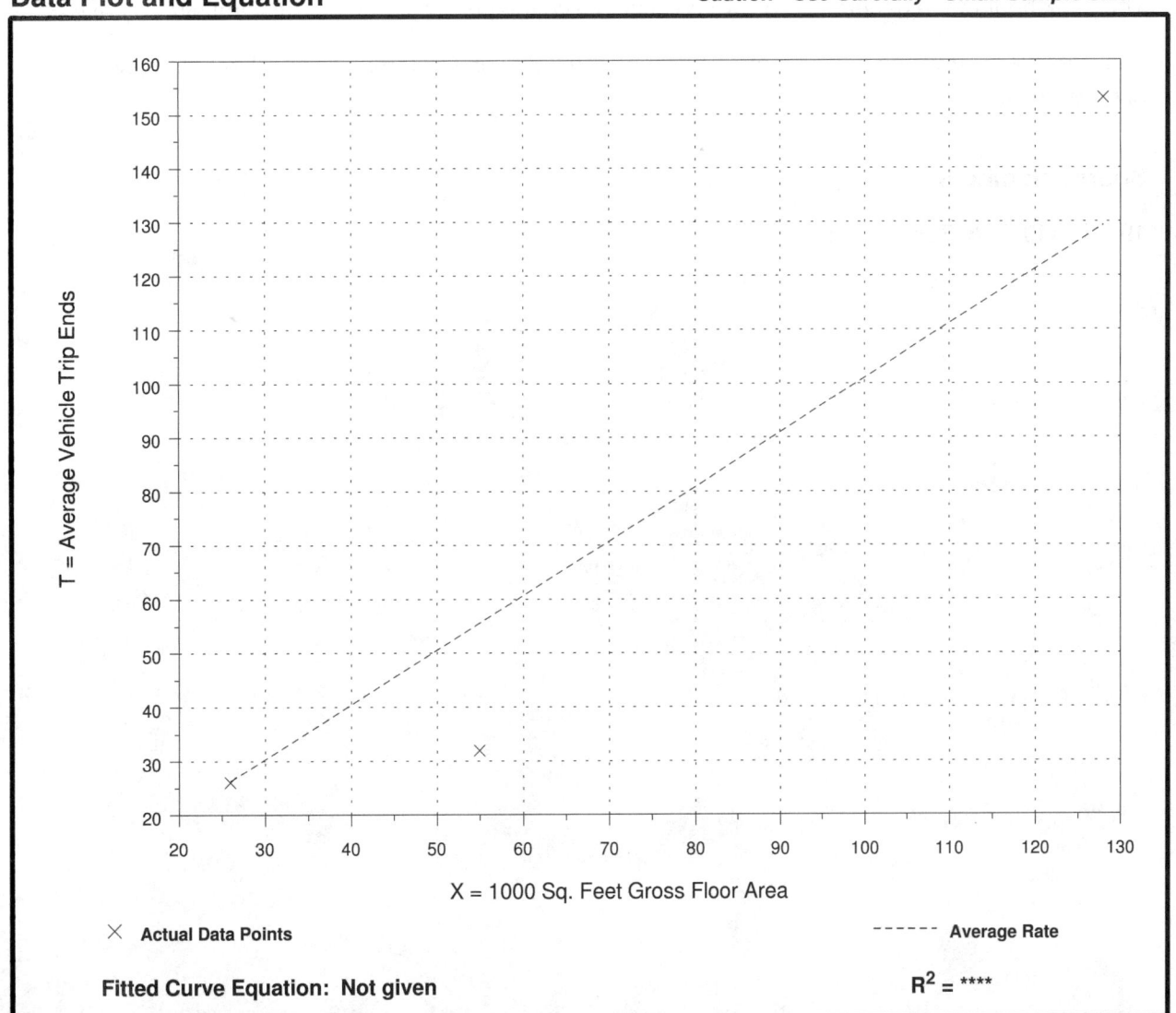

Caution - Use Carefully - Small Sample Size

Fitted Curve Equation: Not given $R^2 = ****$

Land Use: 630
Clinic

Description

A clinic is any facility that provides limited diagnostic and outpatient care but is unable to provide prolonged in-house medical and surgical care. Clinics commonly have lab facilities, supporting pharmacies and a wide range of services (compared to the medical office, which may only have specialized or individual physicians). Hospital (Land Use 610) and medical-dental office building (Land Use 720) are related uses.

Additional Data

The peak hour of the generator typically coincided with the peak hour of the adjacent street traffic.

The sites were surveyed between the 1960s and the 2000s in California, Illinois, New Hampshire and Vermont.

Source Numbers

19, 98, 440, 728, 734

Land Use: 630
Clinic

Independent Variables with One Observation

The following trip generation data are for independent variables with only one observation. This information is shown in this table only; there are no related plots for these data.

Users are cautioned to use data with care because of the small sample size.

Independent Variable	Trip Generation Rate	Size of Independent Variable	Number of Studies	Directional Distribution
Full-Time Doctors				
Weekday A.M. Peak Hour of Generator	3.60	5	1	50% entering, 50% exiting
1,000 Square Feet Gross Floor Area				
Weekday P.M. Peak Hour of Adjacent Street Traffic	5.18	64	1	Not available
Saturday	13.54	161	1	50% entering, 50% exiting
Sunday	24.10	161	1	50% entering, 50% exiting

Clinic
(630)

Average Vehicle Trip Ends vs: Employees
On a: Weekday

Number of Studies: 3
Avg. Number of Employees: 454
Directional Distribution: 50% entering, 50% exiting

Trip Generation per Employee

Average Rate	Range of Rates	Standard Deviation
8.01	5.89 - 12.33	3.72

Data Plot and Equation

Caution - Use Carefully - Small Sample Size

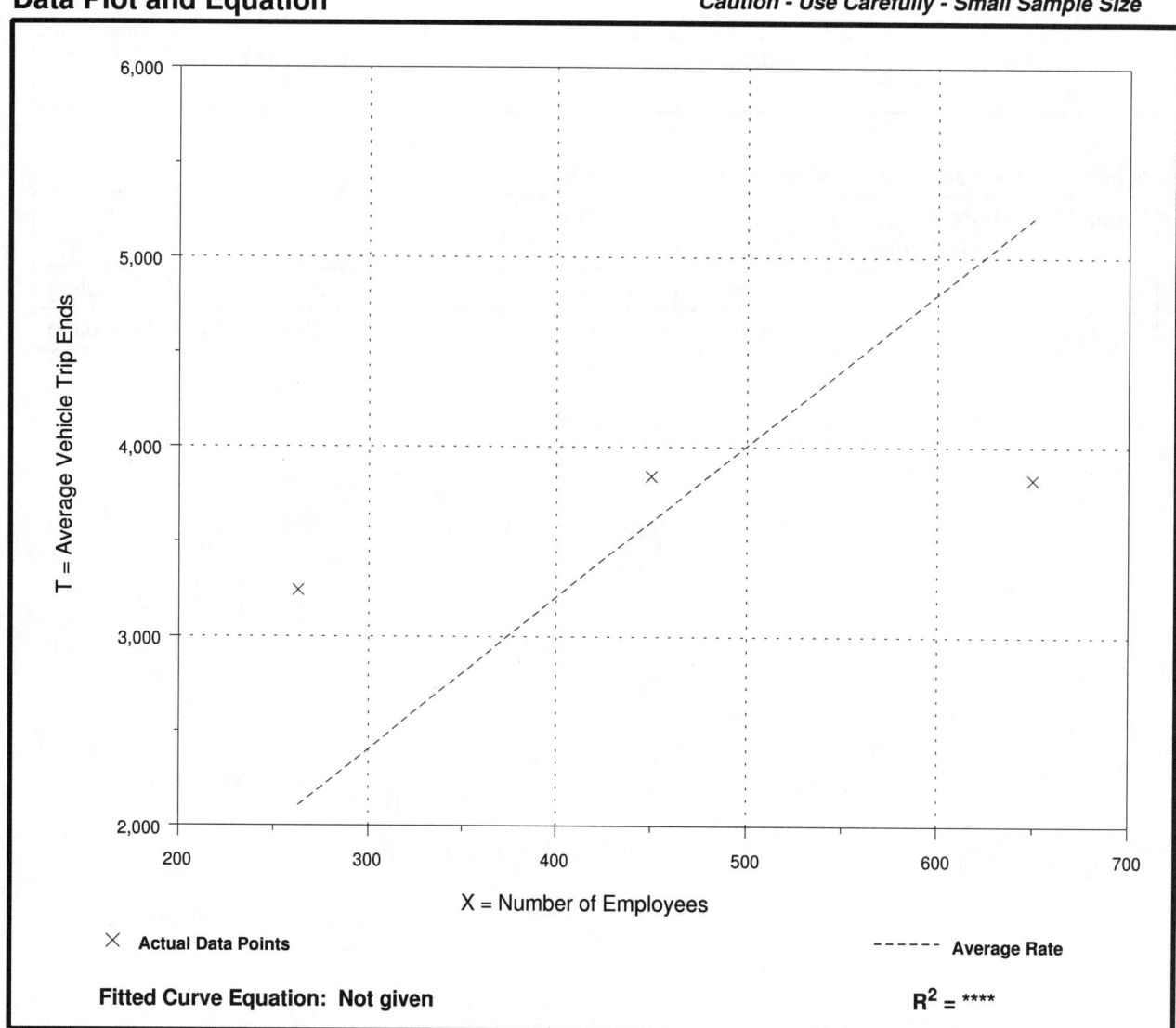

Fitted Curve Equation: Not given $R^2 = ****$

Clinic
(630)

Average Vehicle Trip Ends vs: Employees
On a: Weekday,
Peak Hour of Adjacent Street Traffic,
One Hour Between 4 and 6 p.m.

Number of Studies: 4
Avg. Number of Employees: 198
Directional Distribution: 41% entering, 59% exiting

Trip Generation per Employee

Average Rate	Range of Rates	Standard Deviation
0.96	0.76 - 1.26	1.01

Data Plot and Equation

Caution - Use Carefully - Small Sample Size

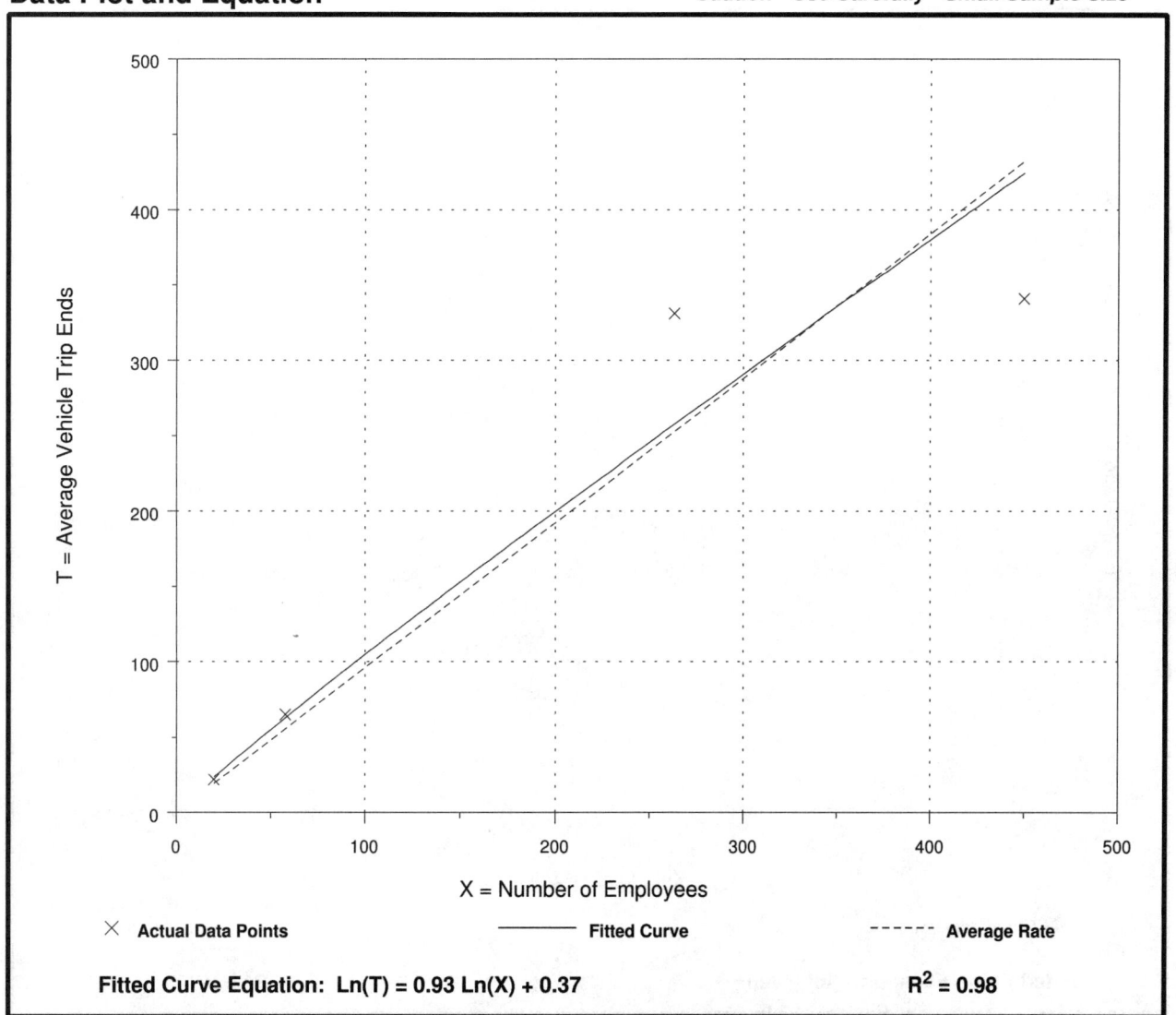

Fitted Curve Equation: $Ln(T) = 0.93 Ln(X) + 0.37$ $R^2 = 0.98$

Clinic
(630)

Average Vehicle Trip Ends vs: Employees
On a: Weekday,
A.M. Peak Hour of Generator

Number of Studies: 2
Avg. Number of Employees: 235
Directional Distribution: 50% entering, 50% exiting

Trip Generation per Employee

Average Rate	Range of Rates	Standard Deviation
0.81	0.80 - 0.90	*

Data Plot and Equation

Caution - Use Carefully - Small Sample Size

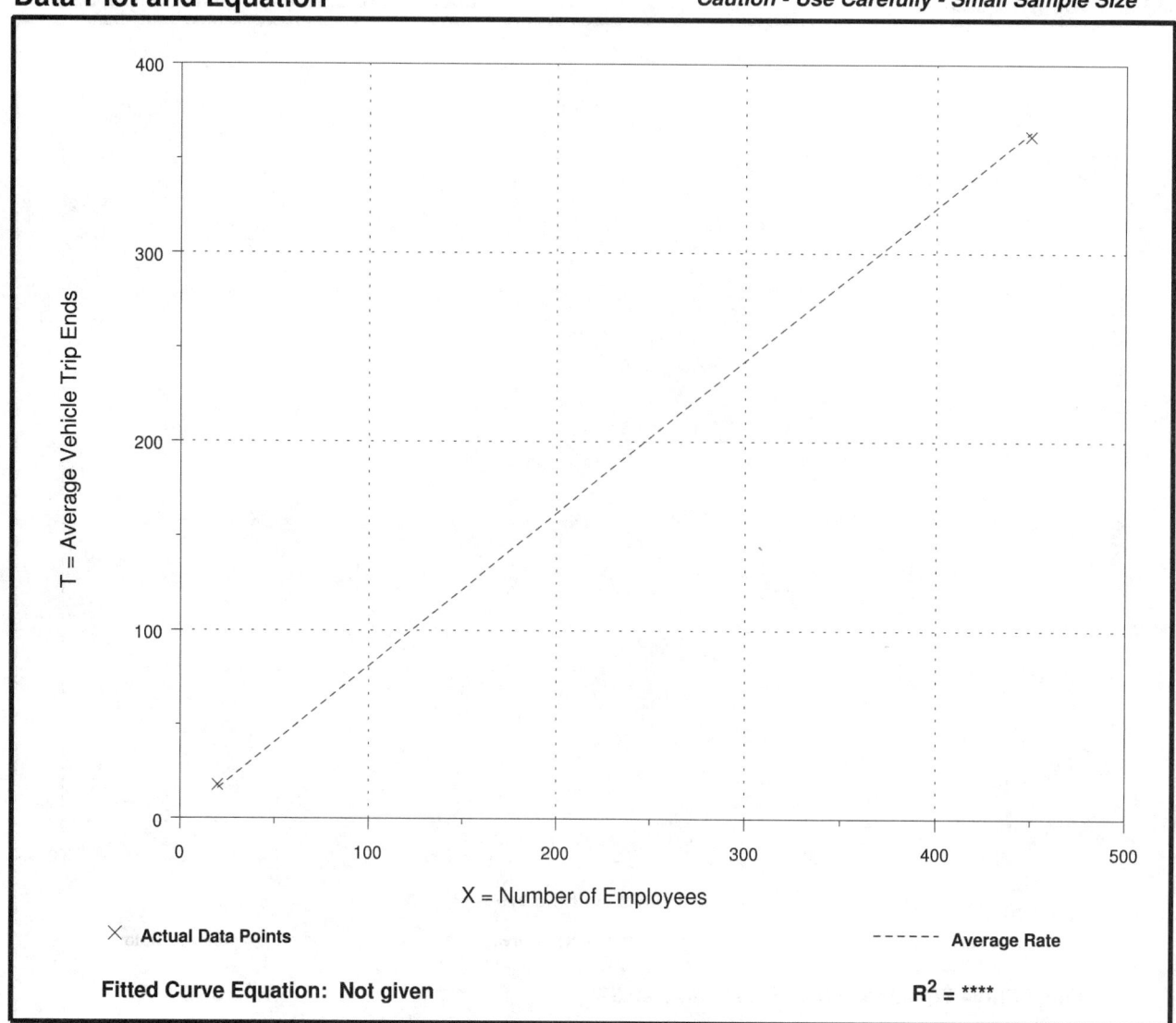

Fitted Curve Equation: Not given

R^2 = ****

Clinic
(630)

Average Vehicle Trip Ends vs: Employees
On a: Weekday,
P.M. Peak Hour of Generator

Number of Studies: 3
Avg. Number of Employees: 176
Directional Distribution: 50% entering, 50% exiting

Trip Generation per Employee

Average Rate	Range of Rates	Standard Deviation
0.86	0.78 - 1.38	0.95

Data Plot and Equation

Caution - Use Carefully - Small Sample Size

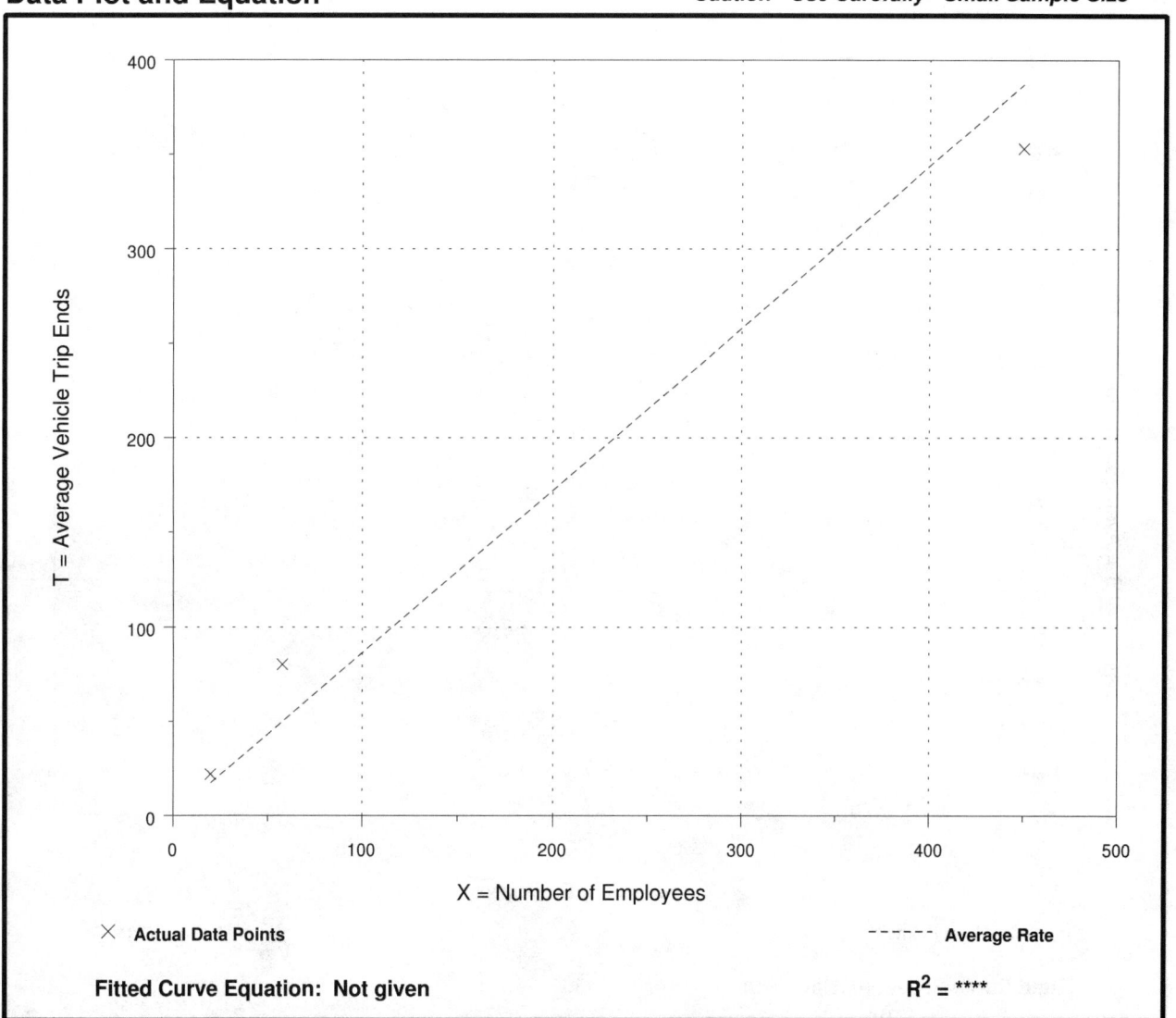

Fitted Curve Equation: Not given $R^2 = ****$

Clinic
(630)

Average Vehicle Trip Ends vs: Employees
On a: Saturday

Number of Studies: 2
Avg. Number of Employees: 550
Directional Distribution: 50% entering, 50% exiting

Trip Generation per Employee

Average Rate	Range of Rates	Standard Deviation
2.53	1.33 - 3.35	*

Data Plot and Equation

Caution - Use Carefully - Small Sample Size

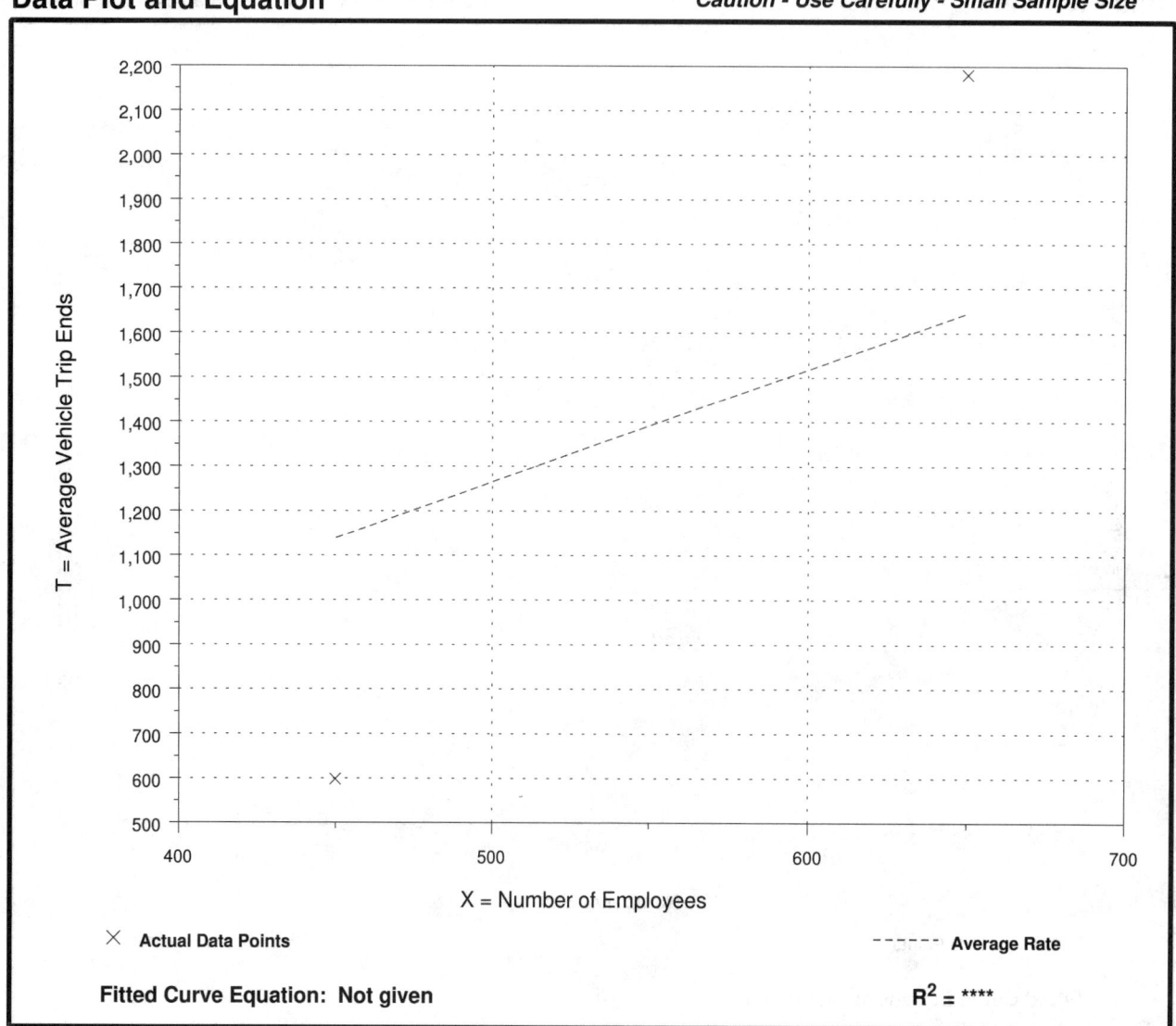

Fitted Curve Equation: Not given

$R^2 = ****$

Clinic
(630)

Average Vehicle Trip Ends vs: Employees
On a: Sunday

Number of Studies: 2
Avg. Number of Employees: 550
Directional Distribution: 50% entering, 50% exiting

Trip Generation per Employee

Average Rate	Range of Rates	Standard Deviation
3.89	0.89 - 5.97	*

Data Plot and Equation

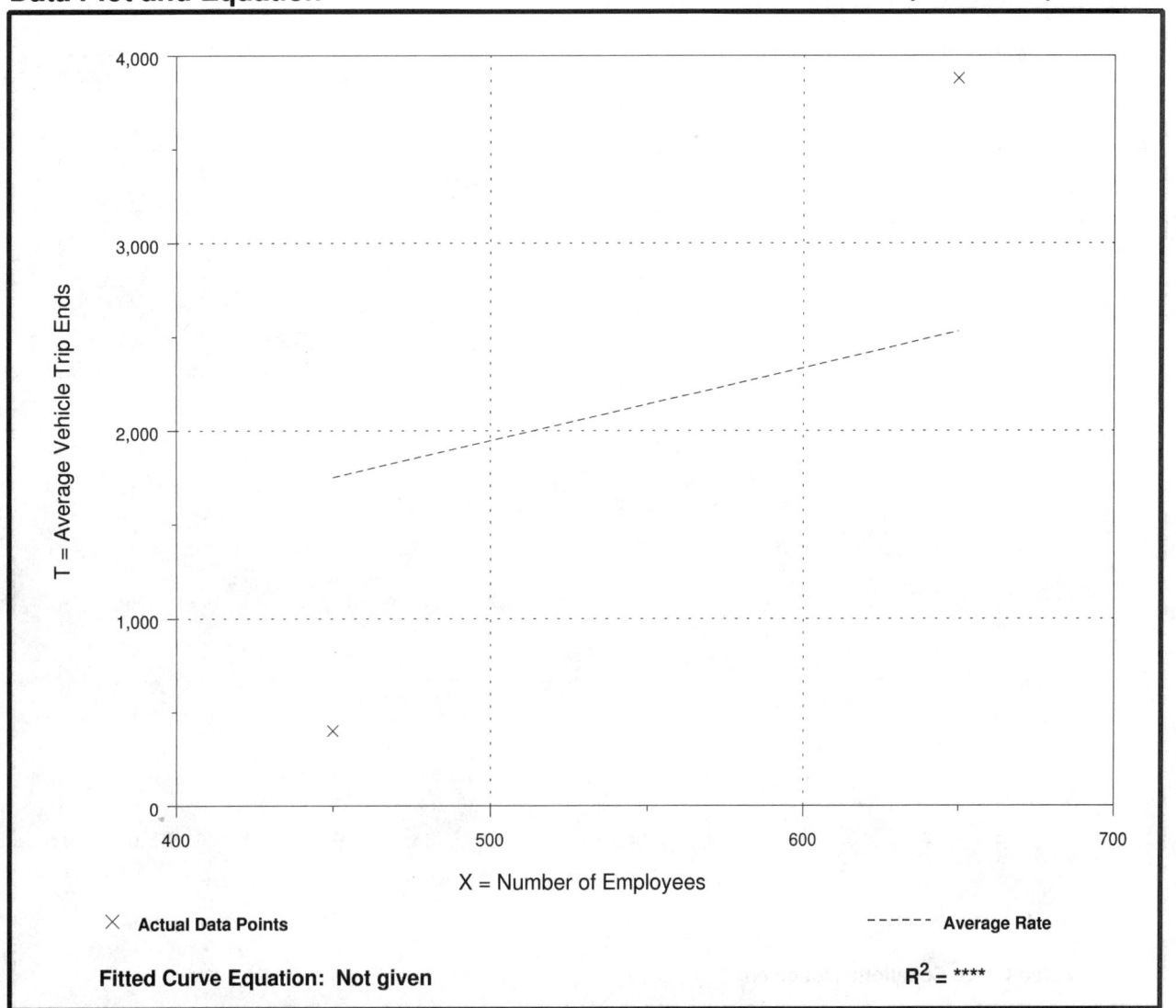

Caution - Use Carefully - Small Sample Size

[Scatter plot: T = Average Vehicle Trip Ends vs X = Number of Employees, with two data points and dashed average rate line]

Fitted Curve Equation: Not given $R^2 = ****$

Clinic
(630)

Average Vehicle Trip Ends vs: Full-time Doctors
On a: Weekday,
Peak Hour of Adjacent Street Traffic,
One Hour Between 4 and 6 p.m.

Number of Studies: 2
Avg. Number of Full-time Doctors: 12
Directional Distribution: 41% entering, 59% exiting

Trip Generation per Full-time Doctor

Average Rate	Range of Rates	Standard Deviation
3.78	3.61 - 4.40	*

Data Plot and Equation

Caution - Use Carefully - Small Sample Size

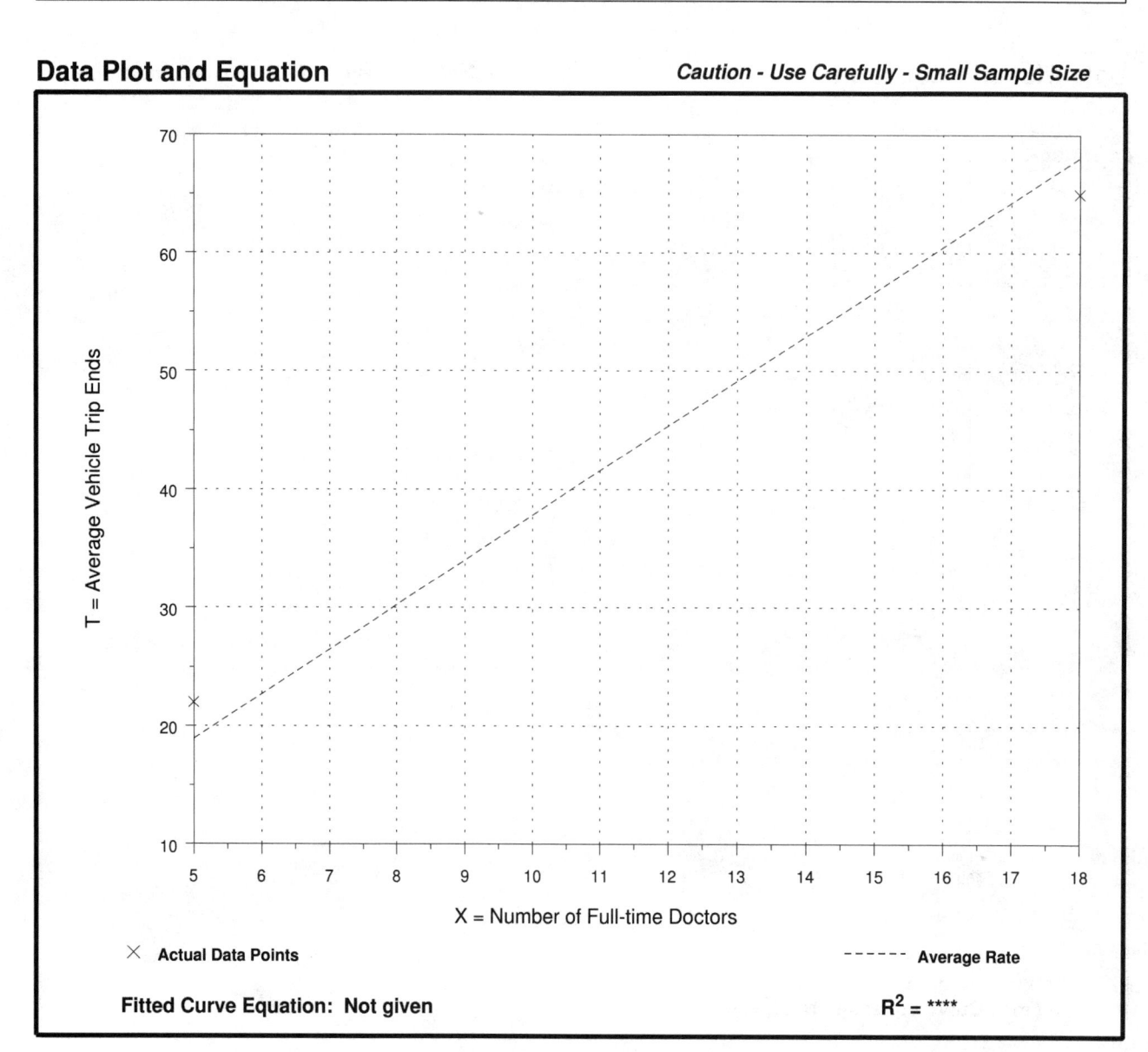

Fitted Curve Equation: Not given $R^2 = ****$

Clinic
(630)

Average Vehicle Trip Ends vs: Full-time Doctors
On a: Weekday,
P.M. Peak Hour of Generator

Number of Studies: 2
Avg. Number of Full-time Doctors: 12
Directional Distribution: 50% entering, 50% exiting

Trip Generation per Full-time Doctor

Average Rate	Range of Rates	Standard Deviation
4.43	4.40 - 4.44	*

Data Plot and Equation

Caution - Use Carefully - Small Sample Size

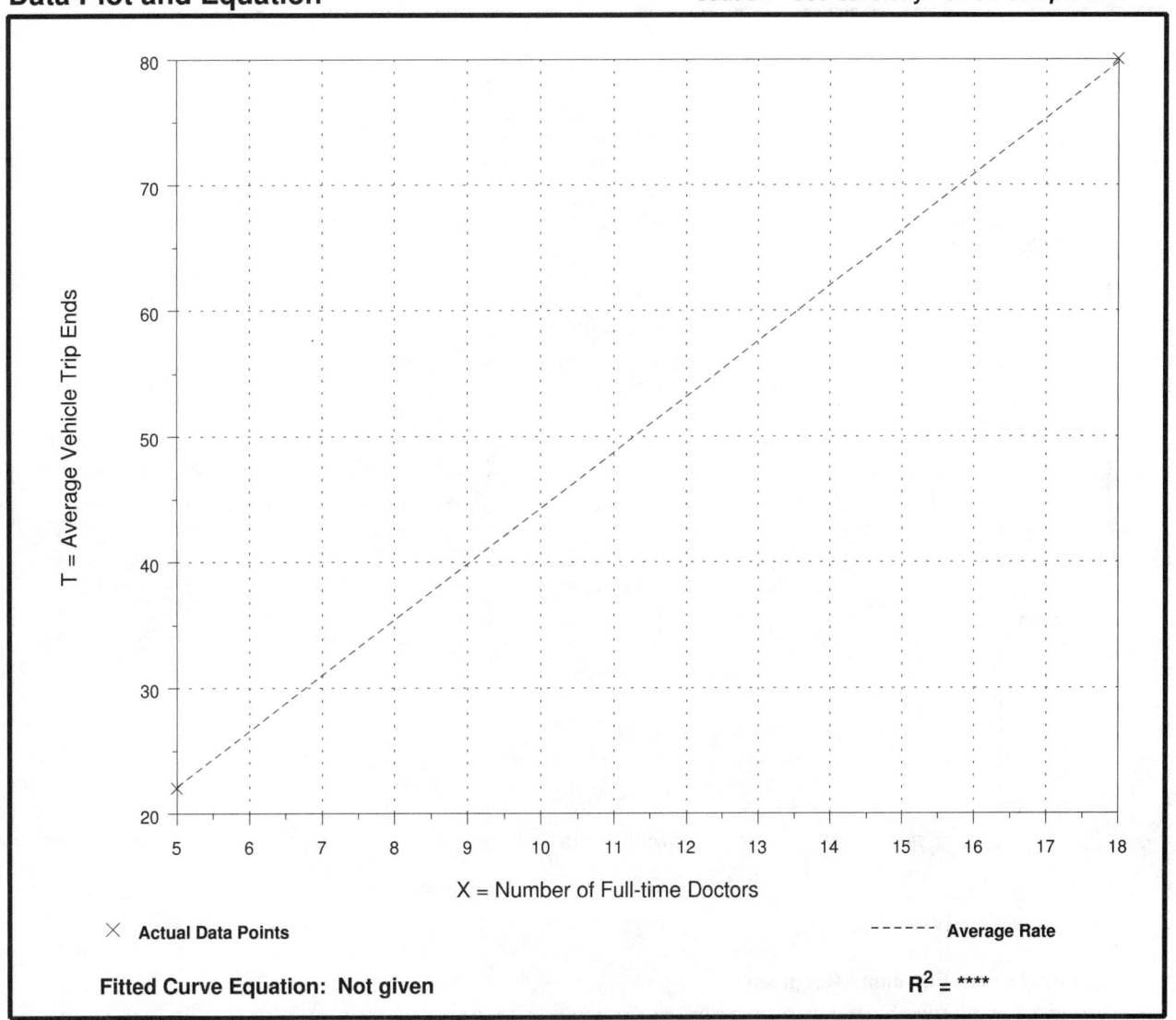

× Actual Data Points
------ Average Rate

Fitted Curve Equation: Not given R^2 = ****

Clinic
(630)

Average Vehicle Trip Ends vs: 1000 Sq. Feet Gross Floor Area
On a: Weekday

Number of Studies: 2
Average 1000 Sq. Feet GFA: 112
Directional Distribution: 50% entering, 50% exiting

Trip Generation per 1000 Sq. Feet Gross Floor Area

Average Rate	Range of Rates	Standard Deviation
31.45	23.79 - 50.74	*

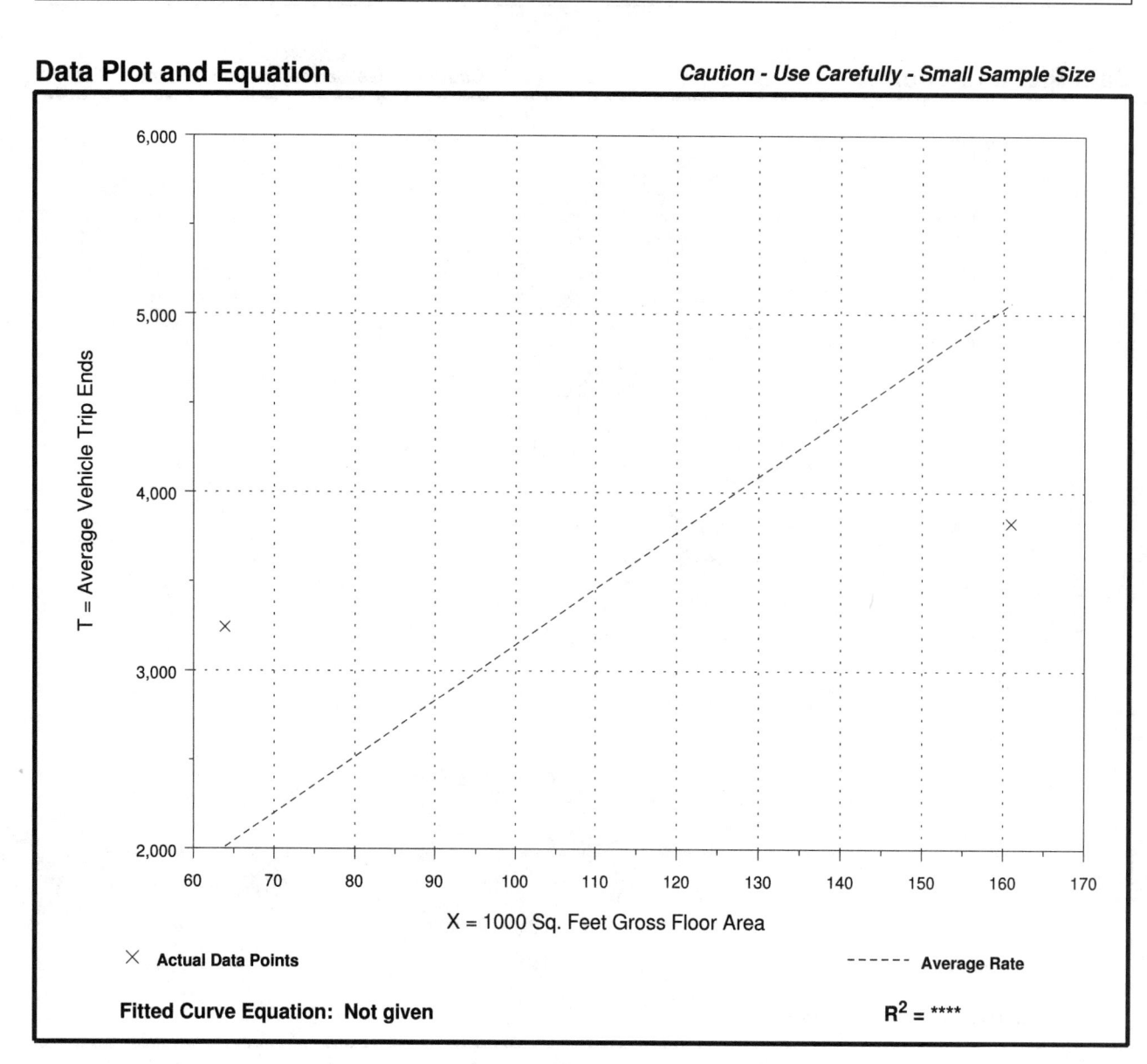

Data Plot and Equation

Caution - Use Carefully - Small Sample Size

Fitted Curve Equation: Not given $R^2 = ****$

Land Use: 640
Animal Hospital/Veterinary Clinic

Description

An animal hospital or veterinary clinic is a facility that specializes in the medical care and treatment of animals.

Additional Data

One of the sites surveyed was located in a rural area.

The sites were surveyed in the 2000s in California.

Source Numbers

597, 662

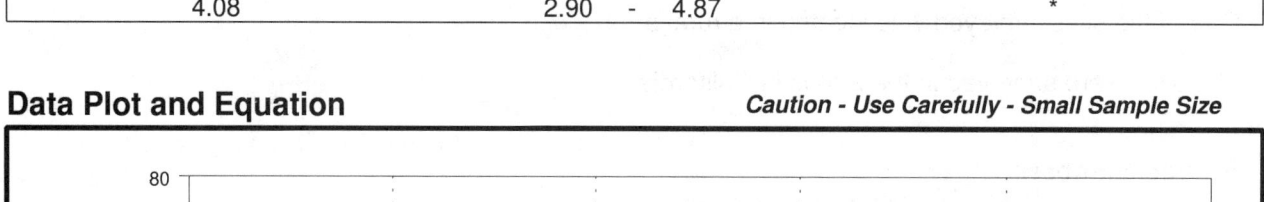

Animal Hospital/Veterinary Clinic
(640)

Average Vehicle Trip Ends vs: 1000 Sq. Feet Gross Floor Area
On a: Weekday,
Peak Hour of Adjacent Street Traffic,
One Hour Between 7 and 9 a.m.

Number of Studies: 2
Average 1000 Sq. Feet GFA: 13
Directional Distribution: 72% entering, 28% exiting

Trip Generation per 1000 Sq. Feet Gross Floor Area

Average Rate	Range of Rates	Standard Deviation
4.08	2.90 - 4.87	*

Data Plot and Equation

Caution - Use Carefully - Small Sample Size

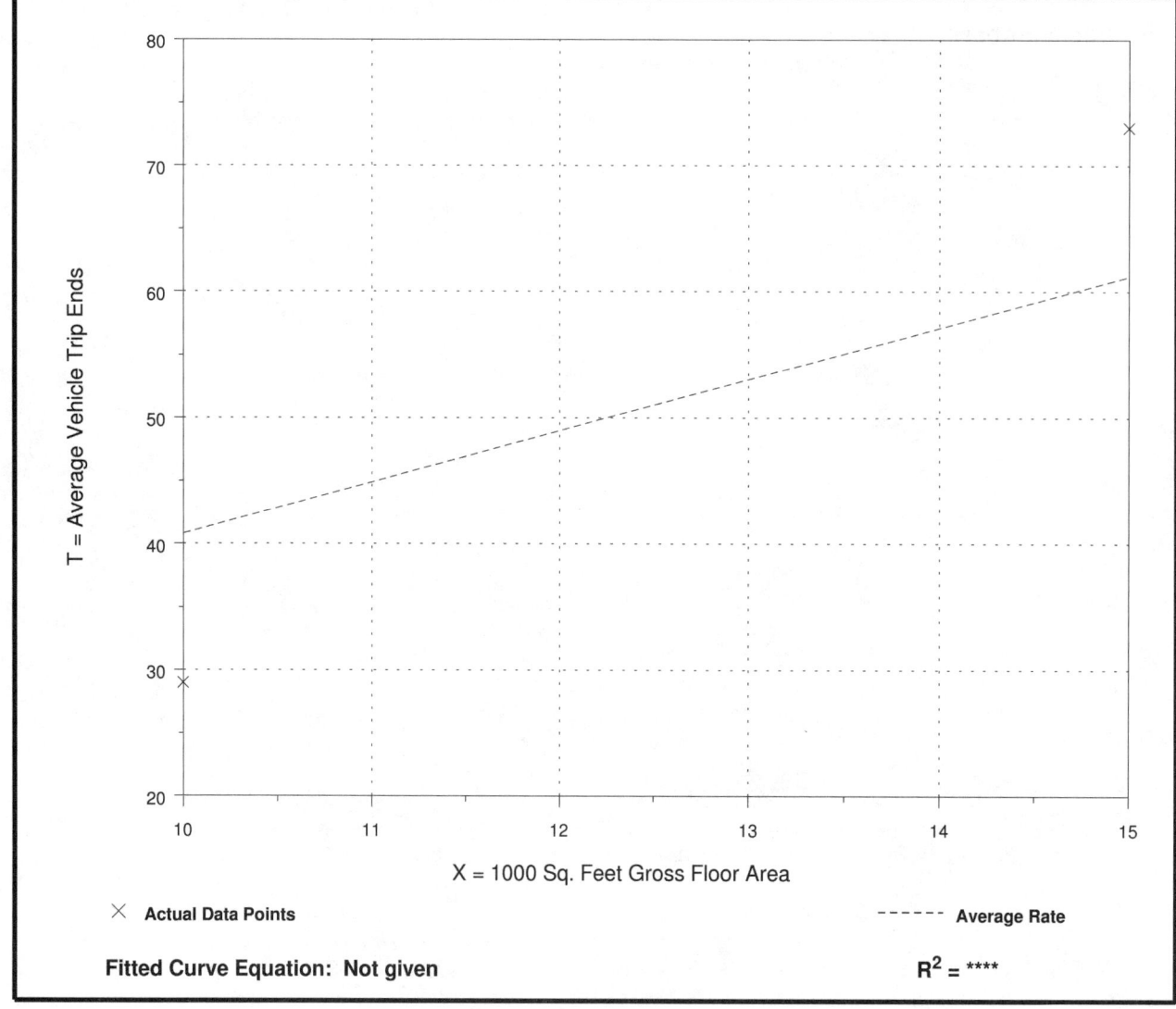

Fitted Curve Equation: Not given $R^2 = ****$

Animal Hospital/Veterinary Clinic
(640)

Average Vehicle Trip Ends vs: 1000 Sq. Feet Gross Floor Area
On a: Weekday,
Peak Hour of Adjacent Street Traffic,
One Hour Between 4 and 6 p.m.

Number of Studies: 2
Average 1000 Sq. Feet GFA: 13
Directional Distribution: 39% entering, 61% exiting

Trip Generation per 1000 Sq. Feet Gross Floor Area

Average Rate	Range of Rates	Standard Deviation
4.72	4.60 - 4.90	*

Data Plot and Equation

Caution - Use Carefully - Small Sample Size

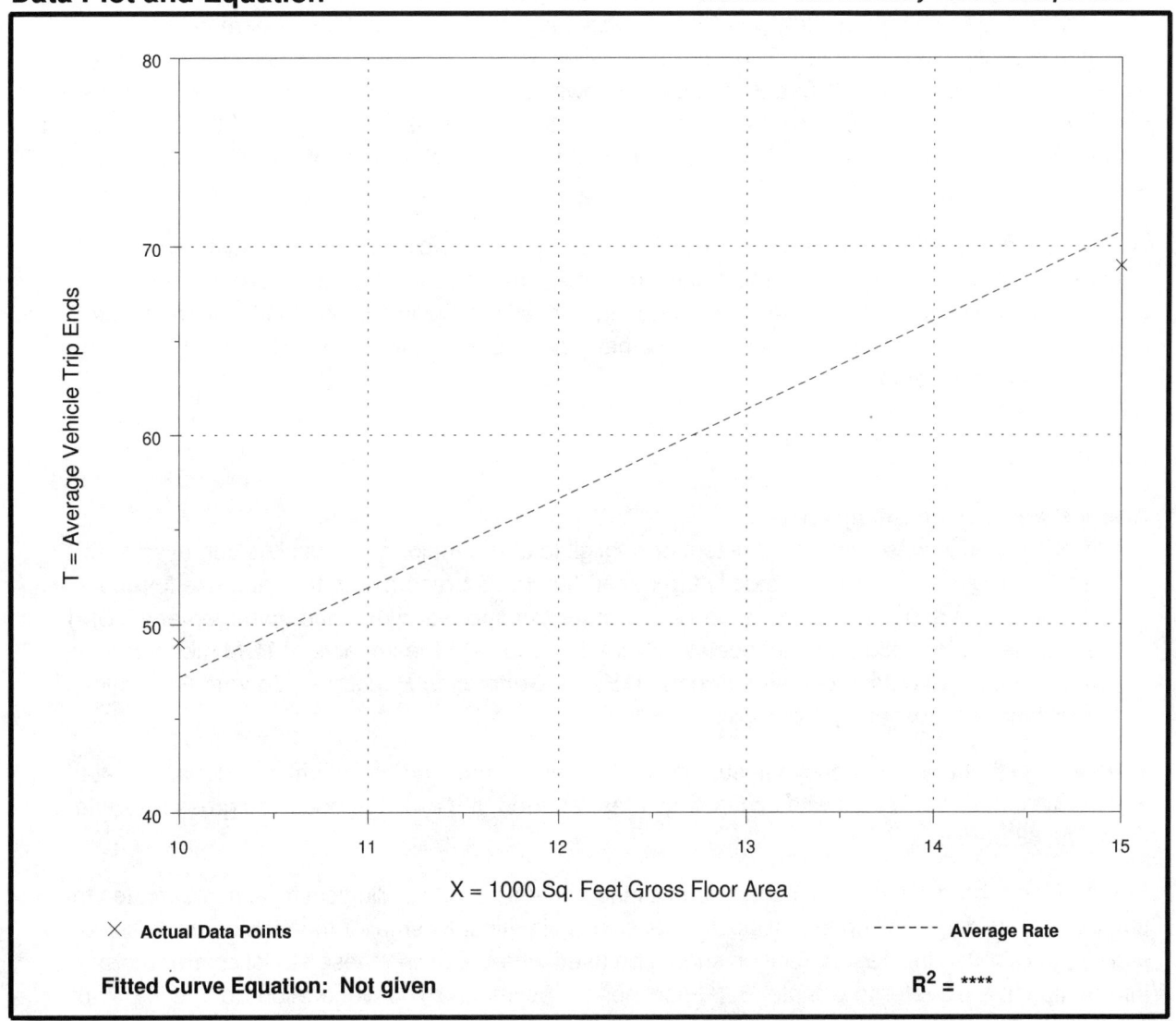

Fitted Curve Equation: Not given $R^2 = ****$

Land Use: 710
General Office Building

Description

A general office building houses multiple tenants; it is a location where affairs of businesses, commercial or industrial organizations, or professional persons or firms are conducted. An office building or buildings may contain a mixture of tenants including professional services, insurance companies, investment brokers and tenant services, such as a bank or savings and loan institution, a restaurant or cafeteria and service retail facilities. Corporate headquarters building (Land Use 714), single tenant office building (Land Use 715), office park (Land Use 750), research and development center (Land Use 760) and business park (Land Use 770) are related uses.

If information is known about individual buildings, it is suggested that the general office building category be used rather than office parks when estimating trip generation for one or more office buildings in a single development. The office park category is more general and should be used when a breakdown of individual or different uses is not known. If the general office building category is used and if additional buildings, such as banks, restaurants, or retail stores, are included in the development, the development should be treated as a multiuse project. On the other hand, if the office park category is used, internal trips are already reflected in the data and do not need to be considered.

When the buildings are interrelated (defined by shared parking facilities or the ability to easily walk between buildings) or house one tenant, it is suggested that the total area or employment of all the buildings be used for calculating the trip generation. When the individual buildings are isolated and not related to one another, it is suggested that trip generation be calculated for each building separately and then summed.

Additional Data

Average weekday transit trip ends—
Transit service was either nonexistent or negligible at the majority of the sites surveyed in this land use. Users may wish to modify trip generation rates presented in this land use to reflect the presence of public transit, carpools and other transportation demand management (TDM) strategies. Information has not been analyzed to document the impacts of TDM measures on the total trip generation of a site. See the ITE *Trip Generation Handbook*, Second Edition for additional information on this topic.

The average building occupancy varied considerably within the studies for which occupancy data were provided. For buildings with occupancy rates reported, the average occupied gross leasable area was 88 percent.

Some of the regression curves plotted for this land use may produce illogical trip-end estimates for small office buildings. When the proposed site size is significantly smaller than the average-sized facility published in this report, caution should be used when applying these statistics. For more information, please refer to Chapter 3, "Guidelines for Estimating Trip Generation," of the ITE *Trip Generation Handbook*, Second Edition.

In some regions, peaking may occur earlier or later and may last somewhat longer than the traditional 7:00 a.m. to 9:00 a.m. and 4:00 p.m. to 6:00 p.m. peak period time frames.

The sites were surveyed between the 1960s and the 2000s throughout the United States.

Trip Characteristics

The trip generation for the A.M. and P.M. peak hours of the generator typically coincided with the peak hours of the adjacent street traffic; therefore, only one A.M. peak hour and one P.M. peak hour, which represent both the peak hour of the generator and the peak hour of the adjacent street traffic, are shown for general office buildings.

Source Numbers

2, 5, 20, 21, 51, 53, 54, 72, 88, 89, 92, 95, 98, 100, 159, 161, 172, 175, 178, 183, 184, 185, 189, 193, 207, 212, 217, 247, 253, 257, 260, 262, 279, 295, 297, 298, 300, 301, 302, 303, 304, 321, 322, 323, 324, 327, 404, 407, 408, 418, 419, 423, 562, 734

General Office Building
(710)

Average Vehicle Trip Ends vs: Employees
On a: Weekday

Number of Studies: 62
Avg. Number of Employees: 610
Directional Distribution: 50% entering, 50% exiting

Trip Generation per Employee

Average Rate	Range of Rates	Standard Deviation
3.32	1.59 - 7.28	2.16

Data Plot and Equation

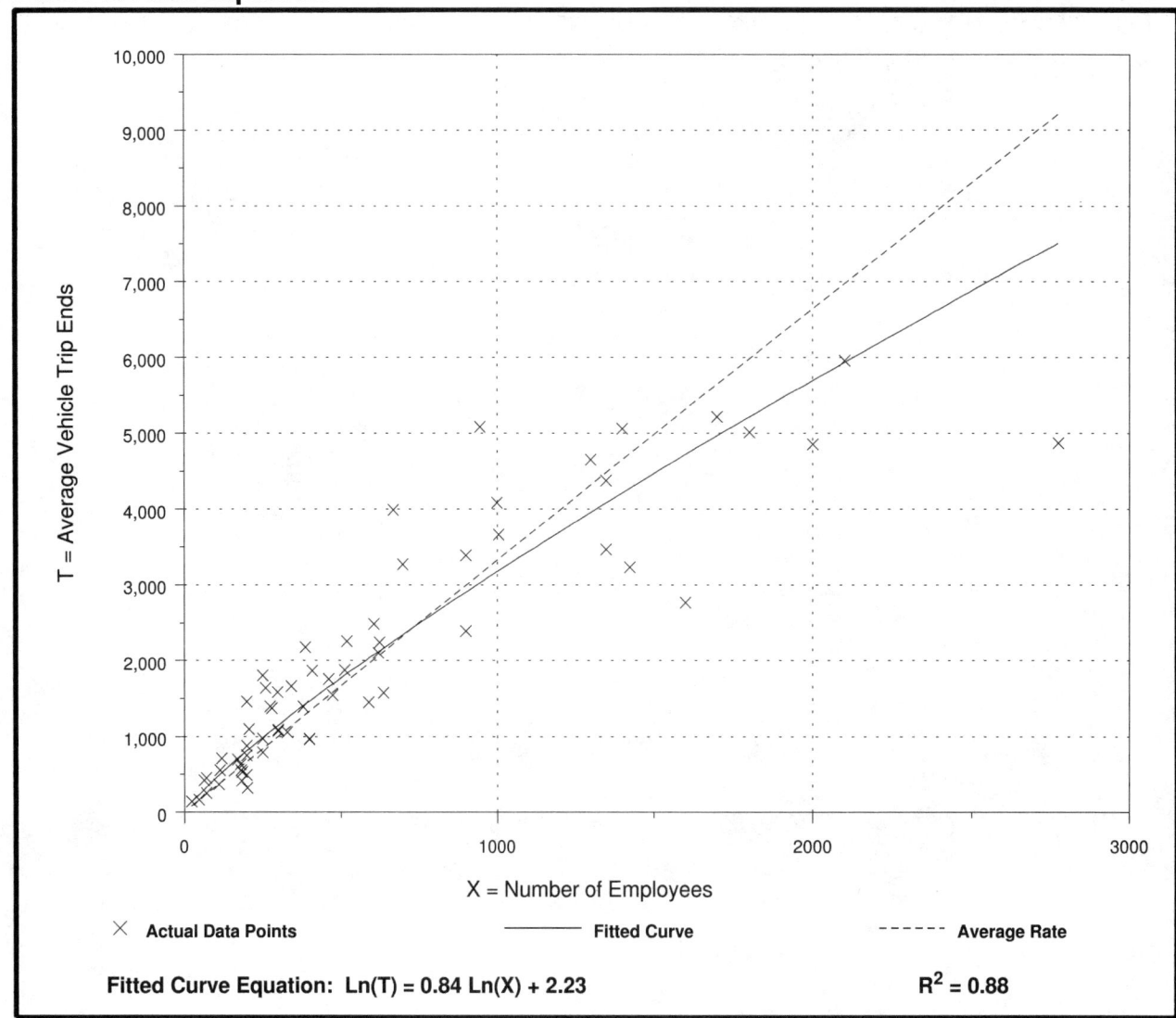

Fitted Curve Equation: $Ln(T) = 0.84\ Ln(X) + 2.23$ $R^2 = 0.88$

General Office Building
(710)

Average Vehicle Trip Ends vs: Employees
On a: Weekday,
A.M. Peak Hour

Number of Studies: 163
Avg. Number of Employees: 695
Directional Distribution: 88% entering, 12% exiting

Trip Generation per Employee

Average Rate	Range of Rates	Standard Deviation
0.48	0.20 - 1.62	0.71

Data Plot and Equation

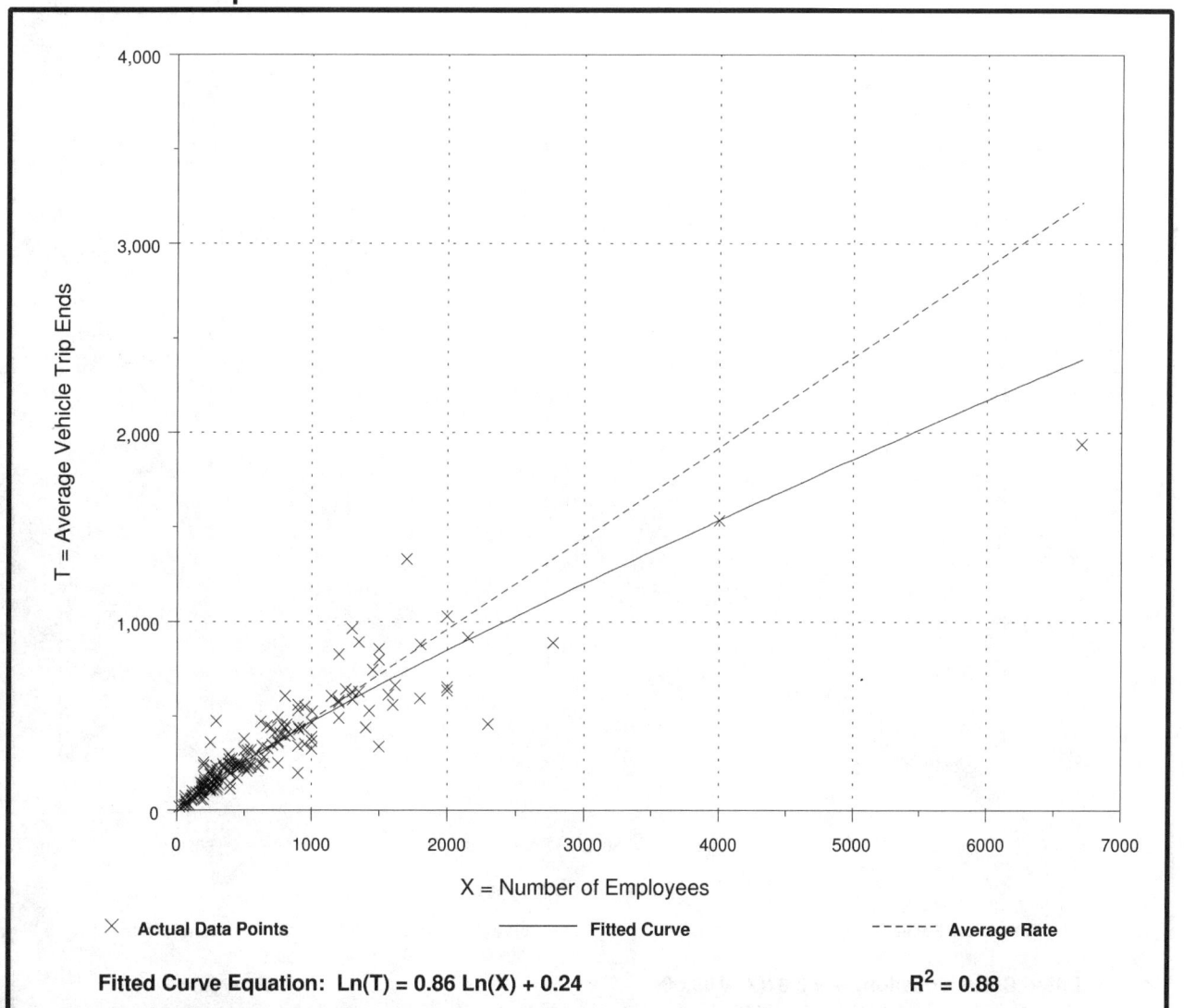

Fitted Curve Equation: $Ln(T) = 0.86\ Ln(X) + 0.24$ $R^2 = 0.88$

Trip Generation, 9th Edition • Institute of Transportation Engineers

General Office Building
(710)

Average Vehicle Trip Ends vs: Employees
On a: Weekday,
P.M. Peak Hour

Number of Studies: 173
Avg. Number of Employees: 688
Directional Distribution: 17% entering, 83% exiting

Trip Generation per Employee

Average Rate	Range of Rates	Standard Deviation
0.46	0.16 - 3.12	0.70

Data Plot and Equation

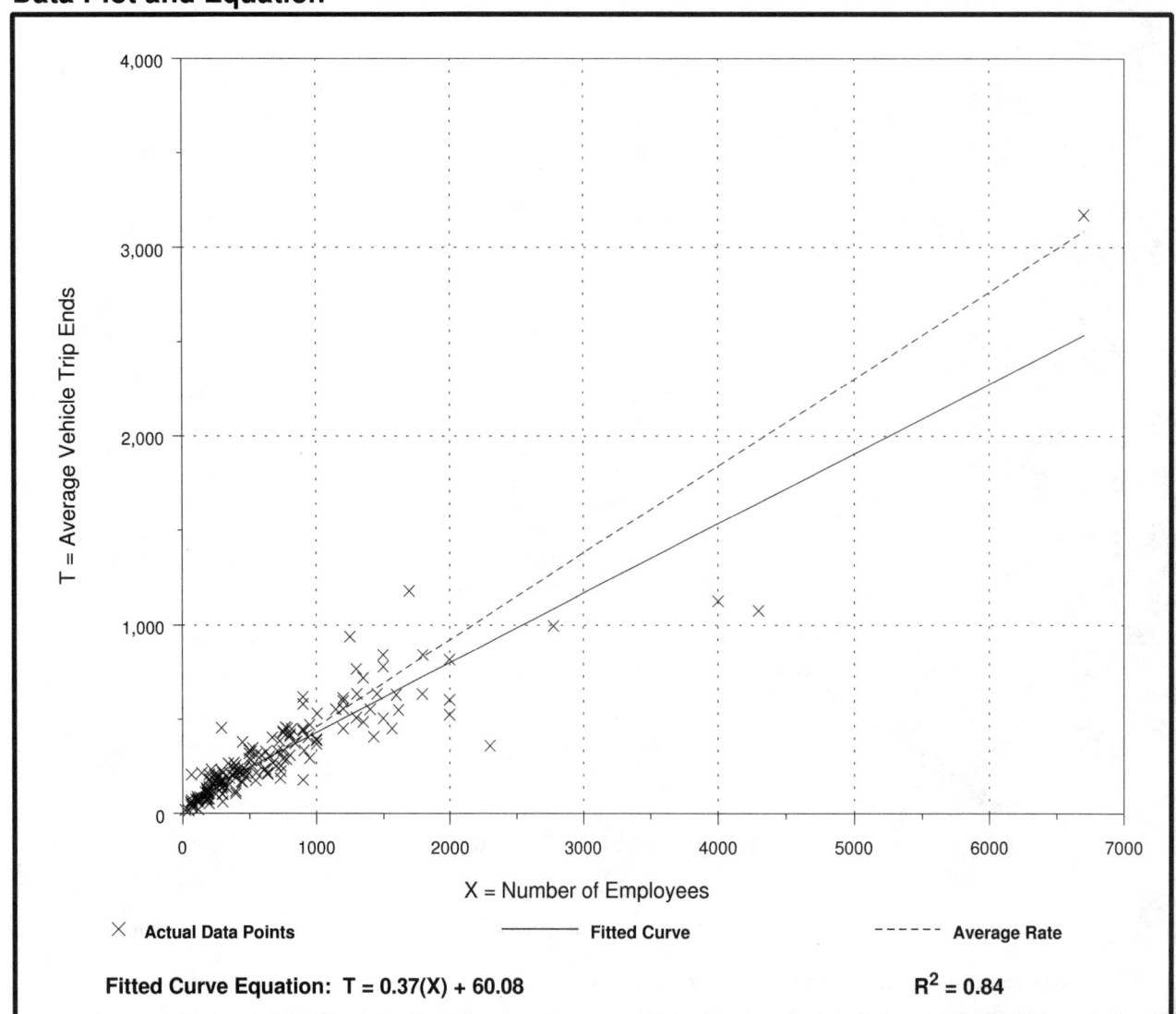

Fitted Curve Equation: $T = 0.37(X) + 60.08$ $R^2 = 0.84$

General Office Building
(710)

Average Vehicle Trip Ends vs: Employees
On a: Saturday

Number of Studies: 17
Avg. Number of Employees: 346
Directional Distribution: 50% entering, 50% exiting

Trip Generation per Employee

Average Rate	Range of Rates	Standard Deviation
0.54	0.12 - 3.83	0.80

Data Plot and Equation

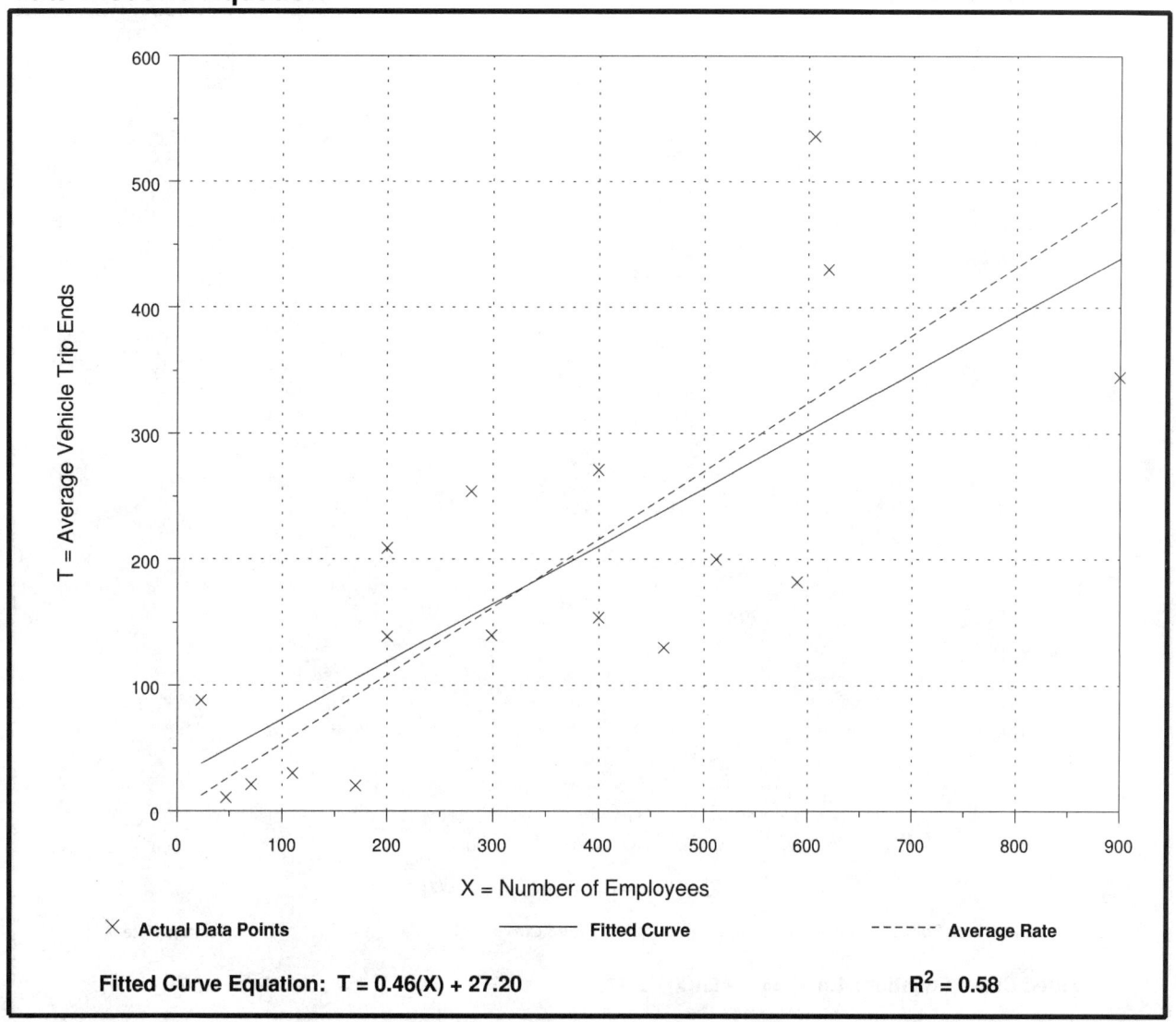

Fitted Curve Equation: T = 0.46(X) + 27.20 $R^2 = 0.58$

General Office Building
(710)

Average Vehicle Trip Ends vs: Employees
On a: Saturday,
Peak Hour of Generator

Number of Studies: 10
Avg. Number of Employees: 427
Directional Distribution: 54% entering, 46% exiting

Trip Generation per Employee

Average Rate	Range of Rates	Standard Deviation
0.09	0.04 - 0.22	0.31

Data Plot and Equation

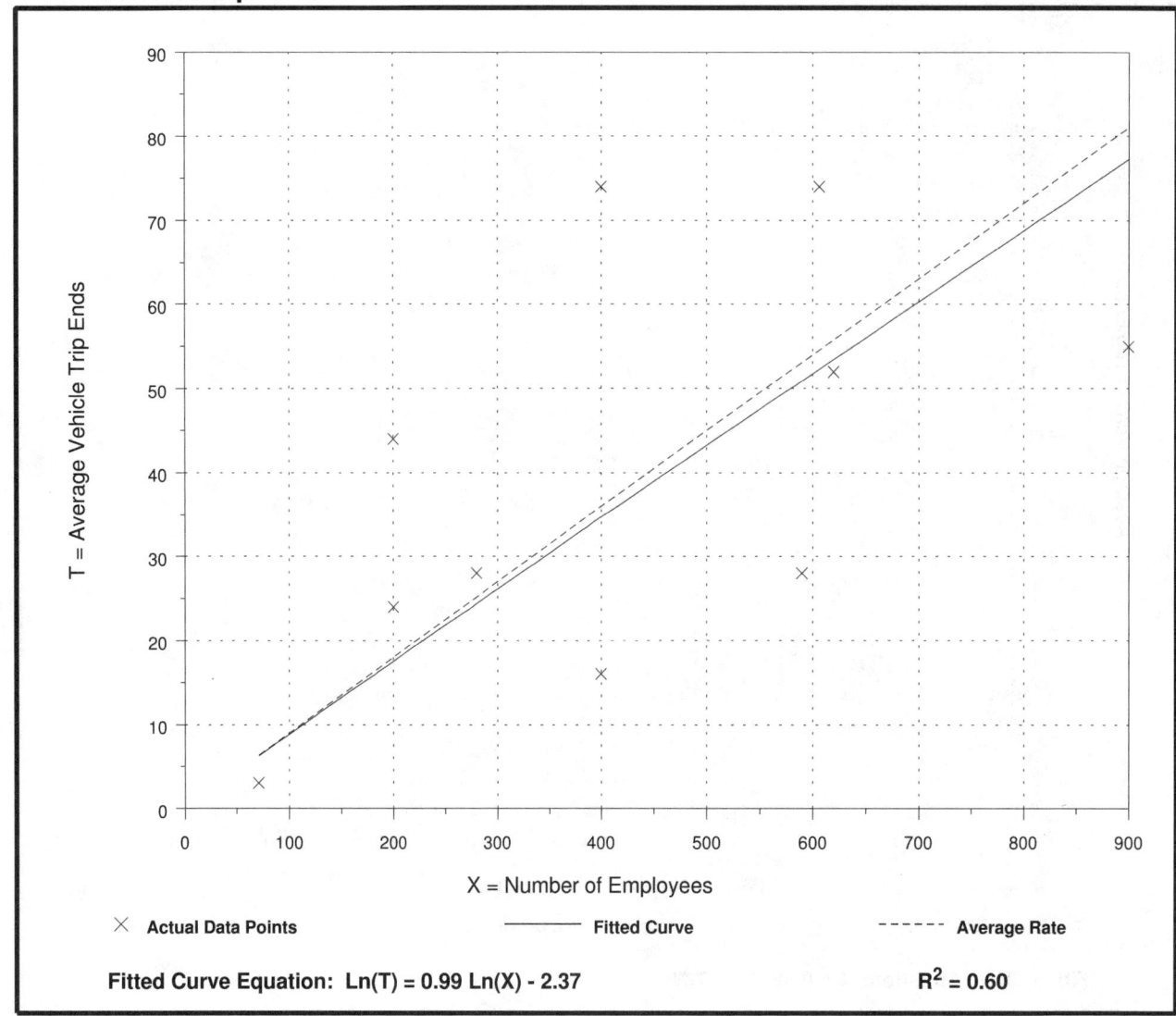

Fitted Curve Equation: $Ln(T) = 0.99\ Ln(X) - 2.37$ $R^2 = 0.60$

General Office Building
(710)

Average Vehicle Trip Ends vs: Employees
On a: Sunday

Number of Studies: 17
Avg. Number of Employees: 346
Directional Distribution: 50% entering, 50% exiting

Trip Generation per Employee

Average Rate	Range of Rates	Standard Deviation
0.22	0.04 - 1.91	0.51

Data Plot and Equation

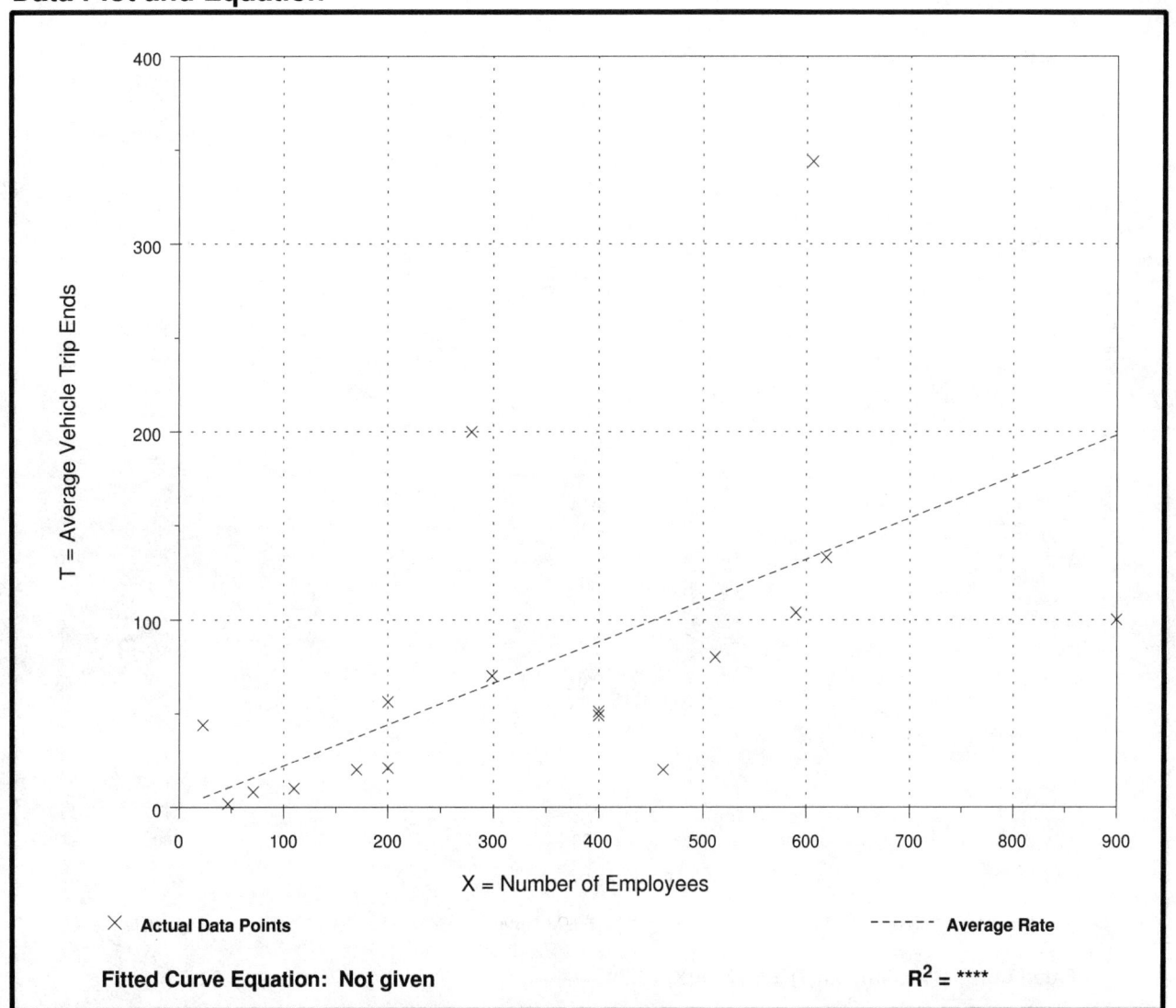

Fitted Curve Equation: Not given $R^2 = ****$

General Office Building
(710)

Average Vehicle Trip Ends vs: Employees
On a: Sunday,
Peak Hour of Generator

Number of Studies: 10
Avg. Number of Employees: 427
Directional Distribution: 58% entering, 42% exiting

Trip Generation per Employee

Average Rate	Range of Rates	Standard Deviation
0.03	0.02 - 0.08	0.18

Data Plot and Equation

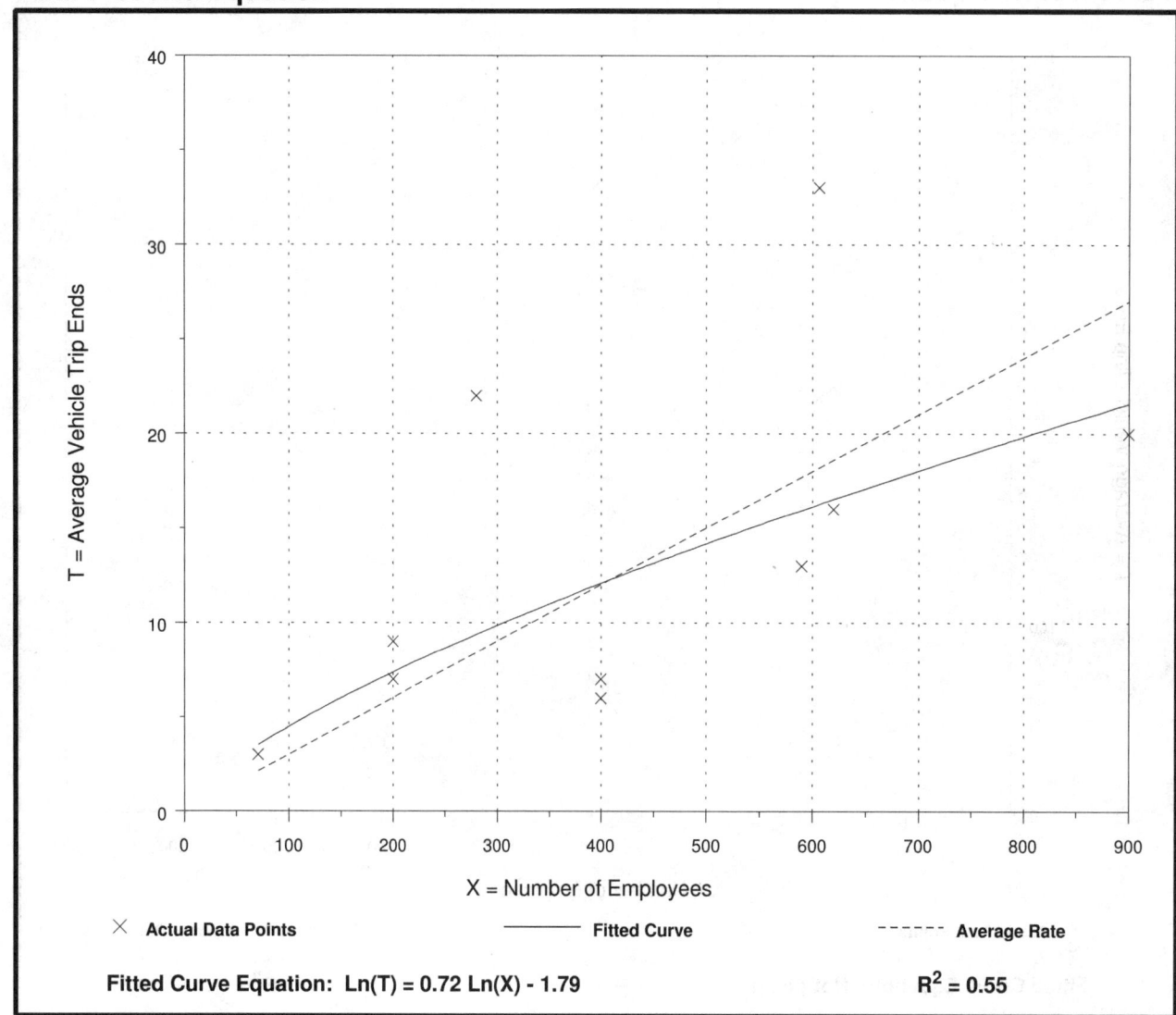

Fitted Curve Equation: $Ln(T) = 0.72 \, Ln(X) - 1.79$ $R^2 = 0.55$

General Office Building
(710)

Average Vehicle Trip Ends vs: 1000 Sq. Feet Gross Floor Area
On a: Weekday

Number of Studies: 79
Average 1000 Sq. Feet GFA: 197
Directional Distribution: 50% entering, 50% exiting

Trip Generation per 1000 Sq. Feet Gross Floor Area

Average Rate	Range of Rates	Standard Deviation
11.03	3.58 - 28.80	6.15

Data Plot and Equation

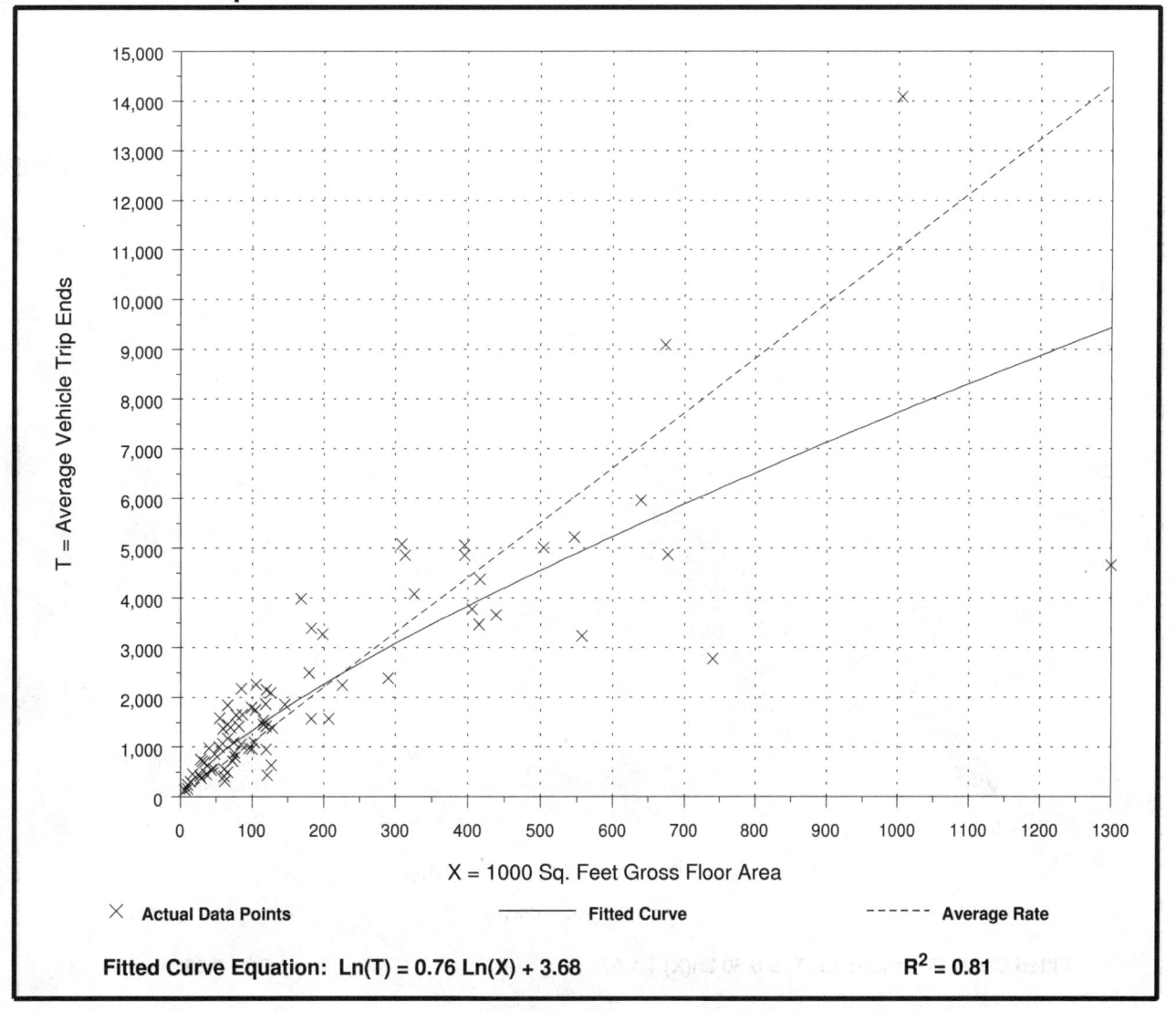

Fitted Curve Equation: $Ln(T) = 0.76\, Ln(X) + 3.68$ $R^2 = 0.81$

Trip Generation, 9th Edition • Institute of Transportation Engineers

General Office Building
(710)

Average Vehicle Trip Ends vs: 1000 Sq. Feet Gross Floor Area
On a: Weekday,
A.M. Peak Hour

Number of Studies: 218
Average 1000 Sq. Feet GFA: 222
Directional Distribution: 88% entering, 12% exiting

Trip Generation per 1000 Sq. Feet Gross Floor Area

Average Rate	Range of Rates	Standard Deviation
1.56	0.60 - 5.98	1.40

Data Plot and Equation

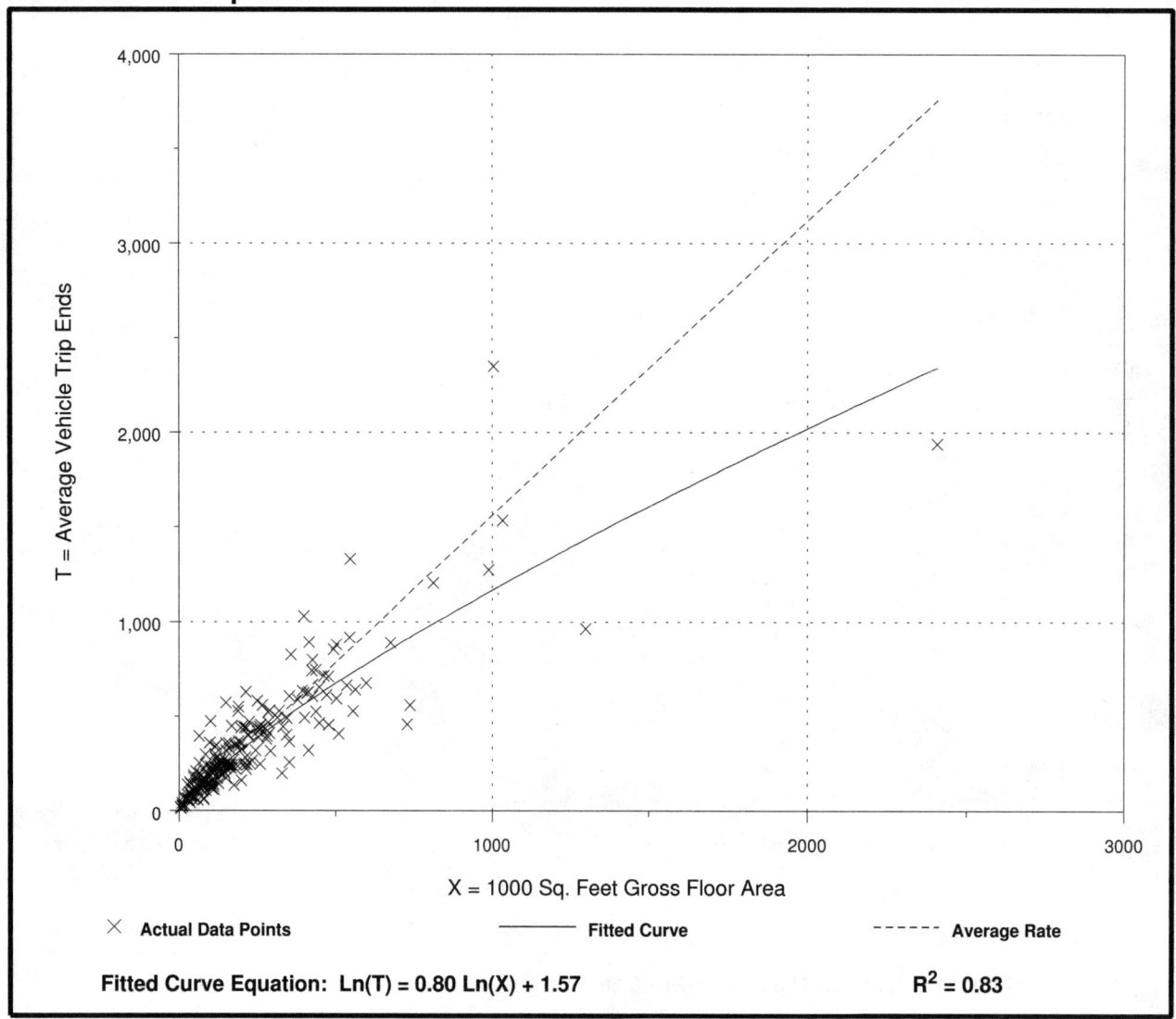

Fitted Curve Equation: $Ln(T) = 0.80 \, Ln(X) + 1.57$ $R^2 = 0.83$

General Office Building
(710)

Average Vehicle Trip Ends vs: 1000 Sq. Feet Gross Floor Area
On a: Weekday,
P.M. Peak Hour

Number of Studies: 236
Average 1000 Sq. Feet GFA: 215
Directional Distribution: 17% entering, 83% exiting

Trip Generation per 1000 Sq. Feet Gross Floor Area

Average Rate	Range of Rates	Standard Deviation
1.49	0.49 - 6.39	1.37

Data Plot and Equation

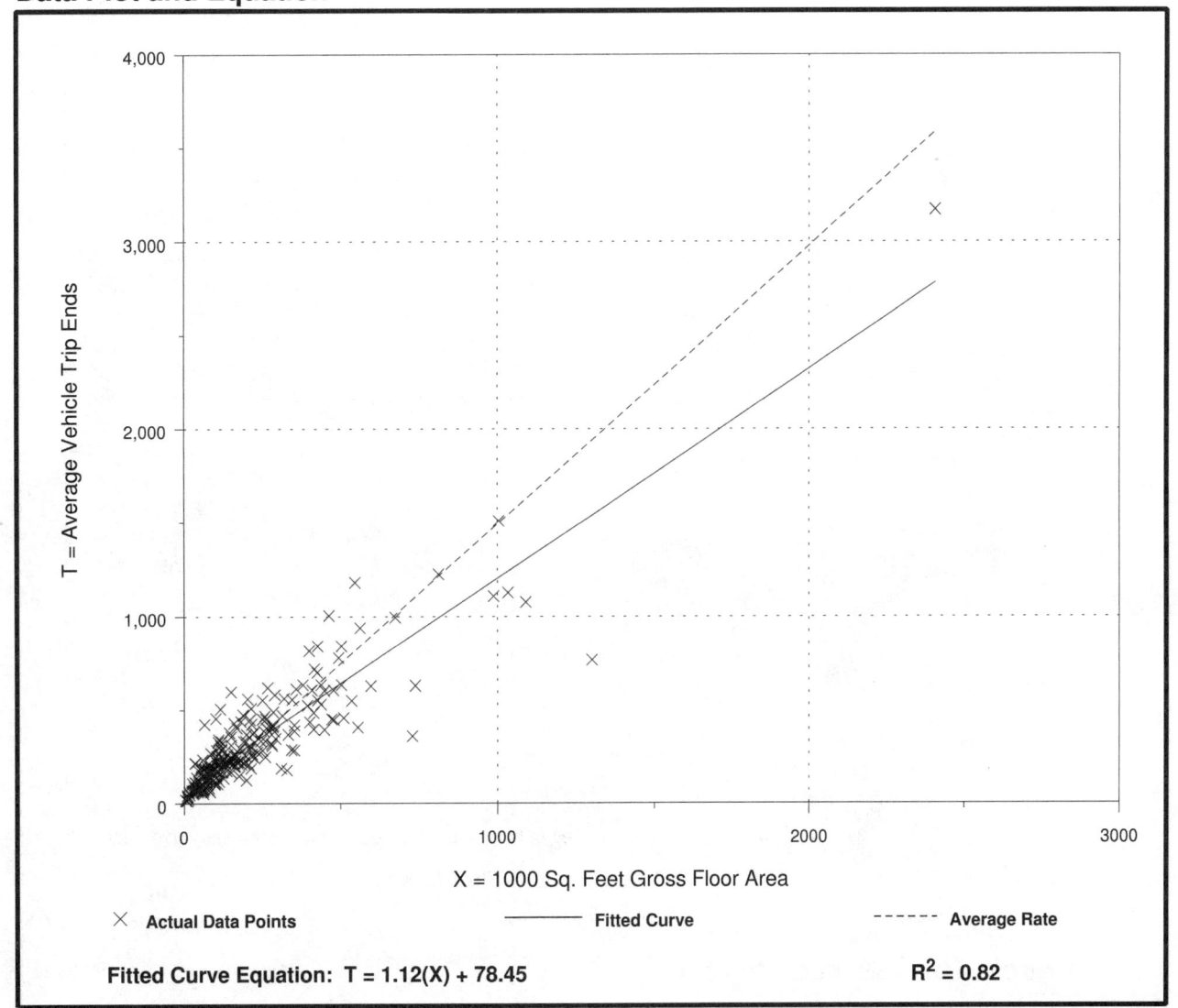

Fitted Curve Equation: $T = 1.12(X) + 78.45$ $\qquad$ $R^2 = 0.82$

General Office Building
(710)

Average Vehicle Trip Ends vs: 1000 Sq. Feet Gross Floor Area
On a: Saturday

Number of Studies: 18
Average 1000 Sq. Feet GFA: 75
Directional Distribution: 50% entering, 50% exiting

Trip Generation per 1000 Sq. Feet Gross Floor Area

Average Rate	Range of Rates	Standard Deviation
2.46	0.59 - 14.67	2.21

Data Plot and Equation

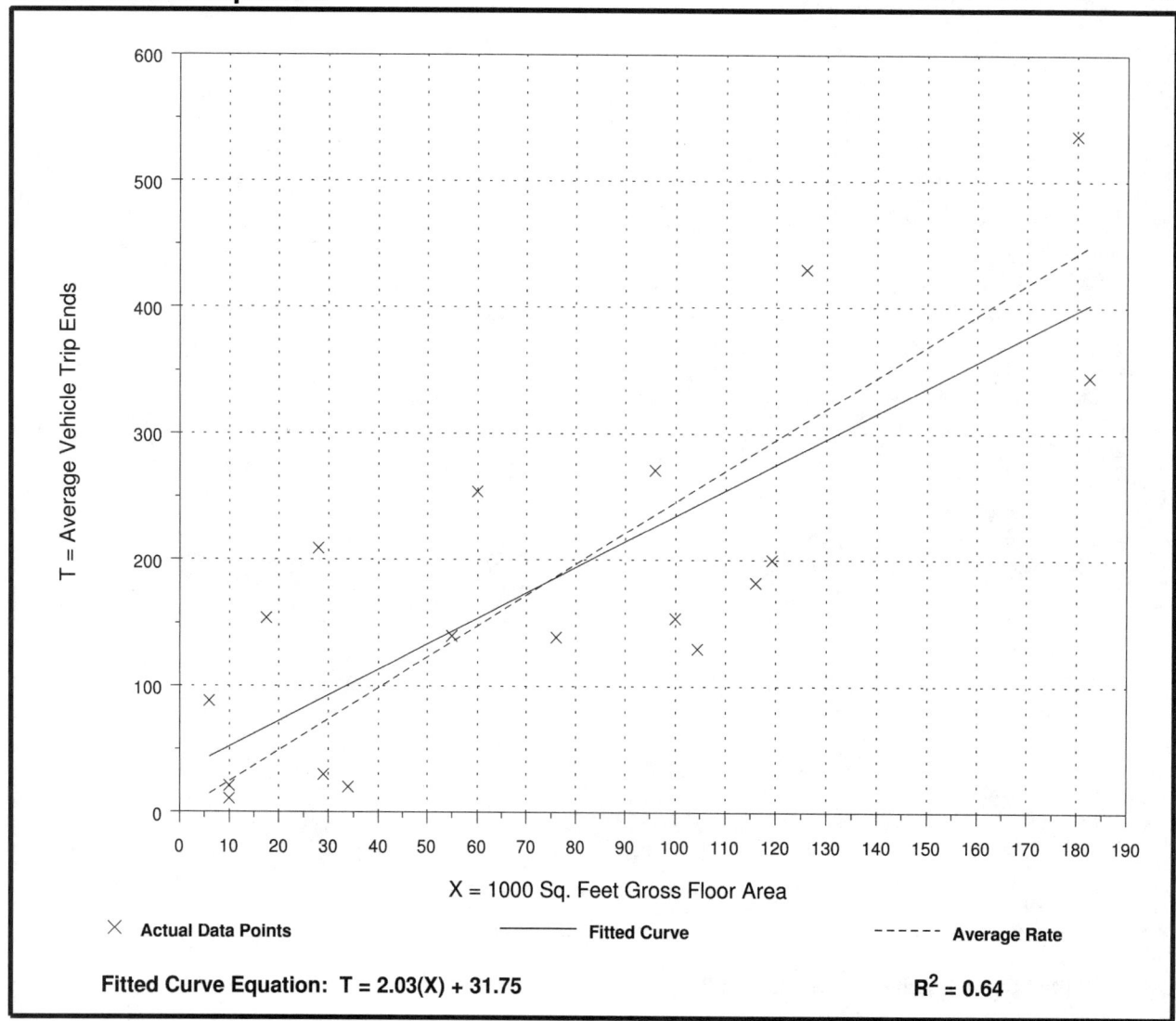

Fitted Curve Equation: T = 2.03(X) + 31.75 $R^2 = 0.64$

General Office Building
(710)

Average Vehicle Trip Ends vs: 1000 Sq. Feet Gross Floor Area
On a: Saturday,
Peak Hour of Generator

Number of Studies: 11
Average 1000 Sq. Feet GFA: 90
Directional Distribution: 54% entering, 46% exiting

Trip Generation per 1000 Sq. Feet Gross Floor Area

Average Rate	Range of Rates	Standard Deviation
0.43	0.16 - 1.77	0.72

Data Plot and Equation

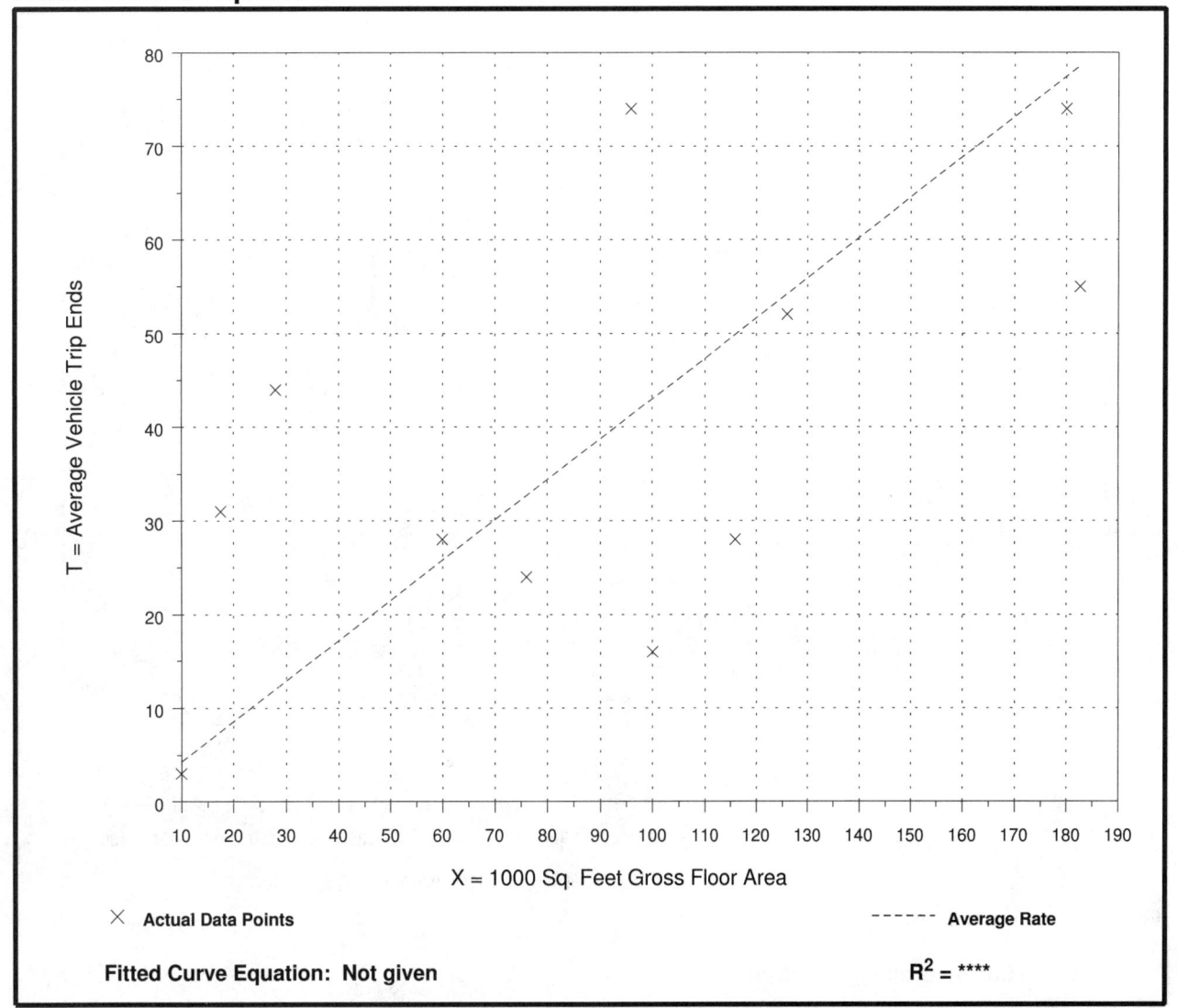

X Actual Data Points   ----- Average Rate

Fitted Curve Equation: Not given $R^2 = ****$

Trip Generation, 9th Edition • Institute of Transportation Engineers

General Office Building
(710)

Average Vehicle Trip Ends vs: 1000 Sq. Feet Gross Floor Area
On a: Sunday

Number of Studies: 18
Average 1000 Sq. Feet GFA: 75
Directional Distribution: 50% entering, 50% exiting

Trip Generation per 1000 Sq. Feet Gross Floor Area

Average Rate	Range of Rates	Standard Deviation
1.05	0.19 - 7.33	1.43

Data Plot and Equation

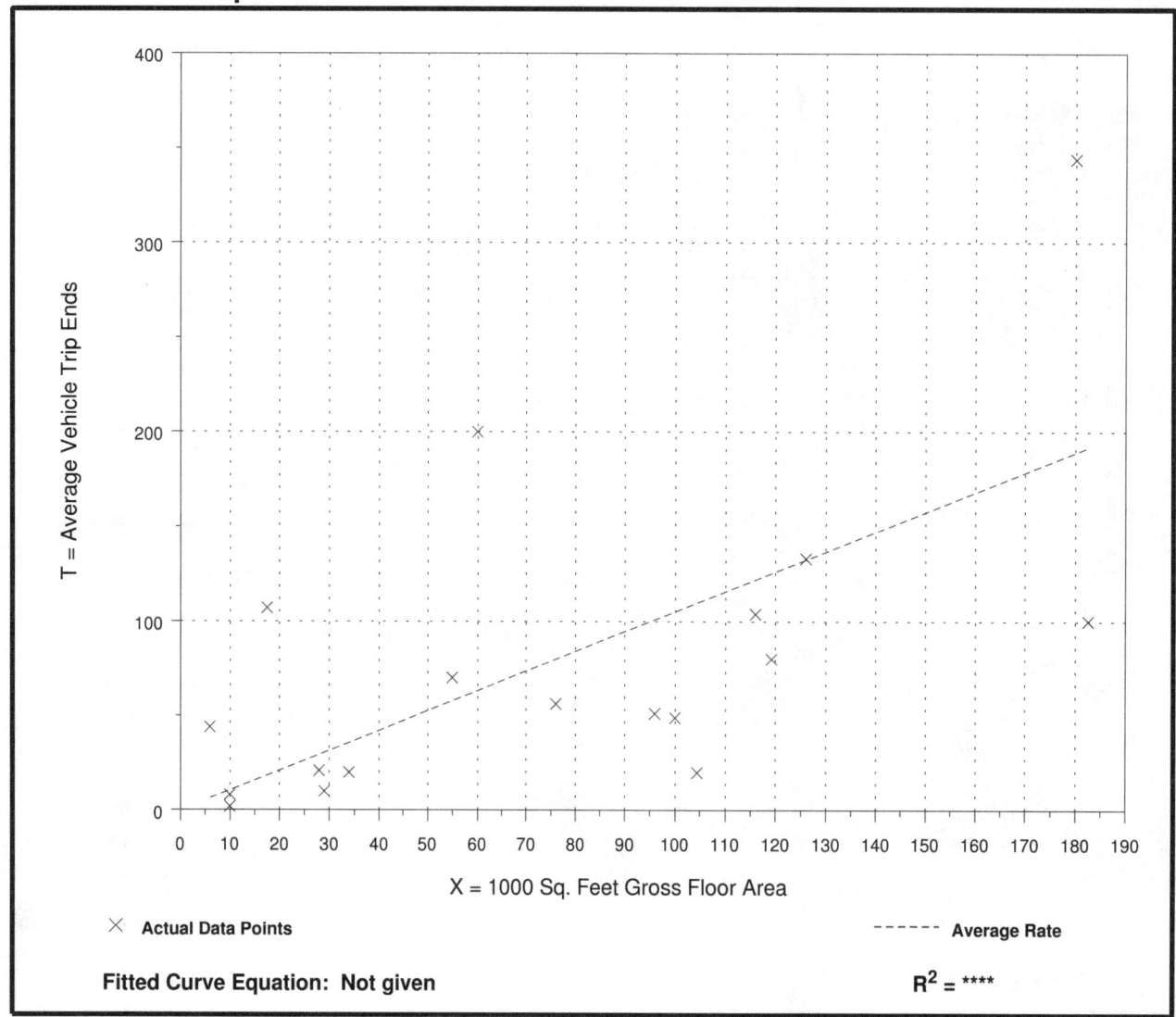

Fitted Curve Equation: Not given $R^2 = ****$

General Office Building
(710)

Average Vehicle Trip Ends vs: 1000 Sq. Feet Gross Floor Area
On a: Sunday,
Peak Hour of Generator

Number of Studies: 11
Average 1000 Sq. Feet GFA: 90
Directional Distribution: 58% entering, 42% exiting

Trip Generation per 1000 Sq. Feet Gross Floor Area

Average Rate	Range of Rates	Standard Deviation
0.16	0.06 - 1.37	0.44

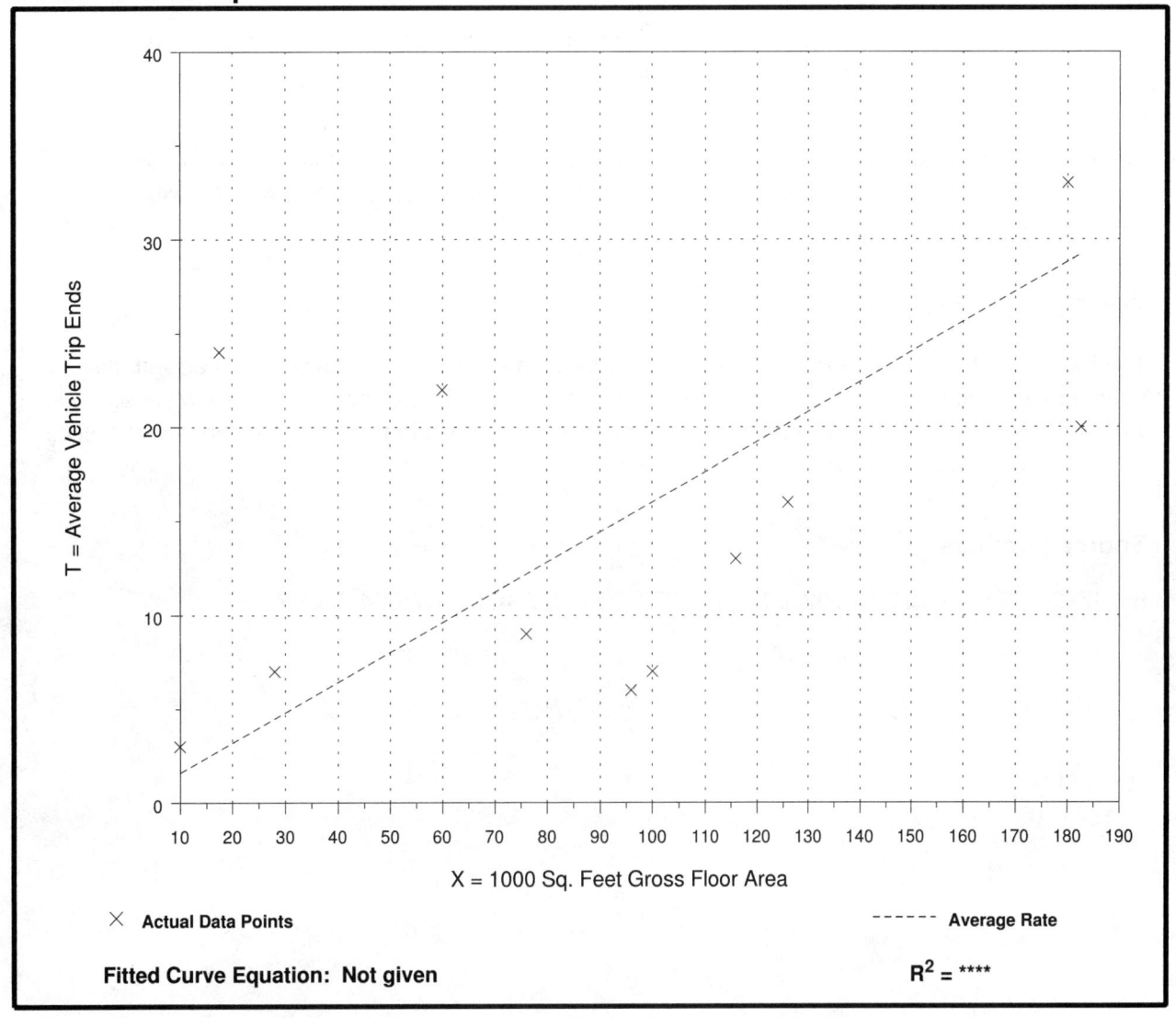

Data Plot and Equation

Fitted Curve Equation: Not given $R^2 = ****$

Land Use: 714
Corporate Headquarters Building

Description

A corporate headquarters building is a single tenant office building that houses the corporate headquarters of a company or organization, which generally consists of offices, meeting rooms, space for file storage and data processing, a restaurant or cafeteria and other service functions. General office building (Land Use 710), single tenant office building (Land Use 715), office park (Land Use 750), research and development center (Land Use 760) and business park (Land Use 770) are related uses.

Additional Data

The average vehicle occupancy for the 10 studies for which information was submitted was approximately 1.2 persons per automobile. The vehicle occupancy rates ranged from 1.03 to 1.74 persons per automobile.

The majority of the sites were surveyed between the late 1980s and the late 1990s throughout the United States, with nearly half conducted in the Washington, DC and Atlanta, Georgia metropolitan areas.

Trip Characteristics

The trip generation for the A.M. and P.M. peak hours of the generator typically coincided with the peak hours of the adjacent street traffic; therefore, only one A.M. peak hour and one P.M. peak hour, which represent both the peak hour of the generator and the peak hour of the adjacent street traffic, are shown for a corporate headquarters building.

Source Numbers

9, 247, 251, 262, 273, 298, 302, 303, 304, 323, 324, 327, 406, 444, 524, 552, 717

Corporate Headquarters Building
(714)

Average Vehicle Trip Ends vs: Employees
On a: Weekday

Number of Studies: 7
Avg. Number of Employees: 602
Directional Distribution: 50% entering, 50% exiting

Trip Generation per Employee

Average Rate	Range of Rates	Standard Deviation
2.33	2.01 - 3.53	1.61

Data Plot and Equation

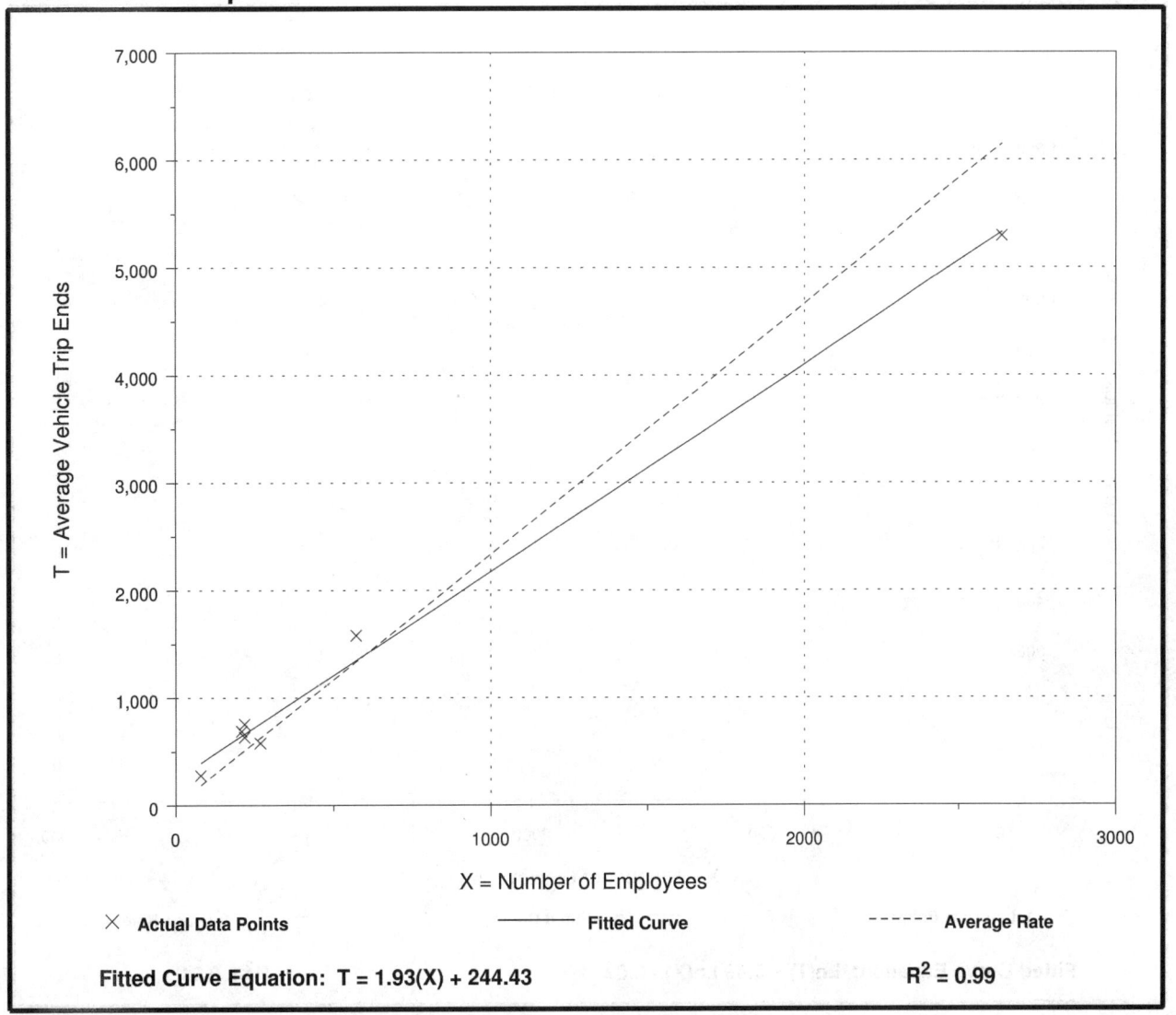

Fitted Curve Equation: T = 1.93(X) + 244.43 $R^2 = 0.99$

Corporate Headquarters Building
(714)

Average Vehicle Trip Ends vs: Employees
On a: Weekday,
A.M. Peak Hour

Number of Studies: 18
Avg. Number of Employees: 939
Directional Distribution: 93% entering, 7 % exiting

Trip Generation per Employee

Average Rate	Range of Rates	Standard Deviation
0.45	0.18 - 0.66	0.68

Data Plot and Equation

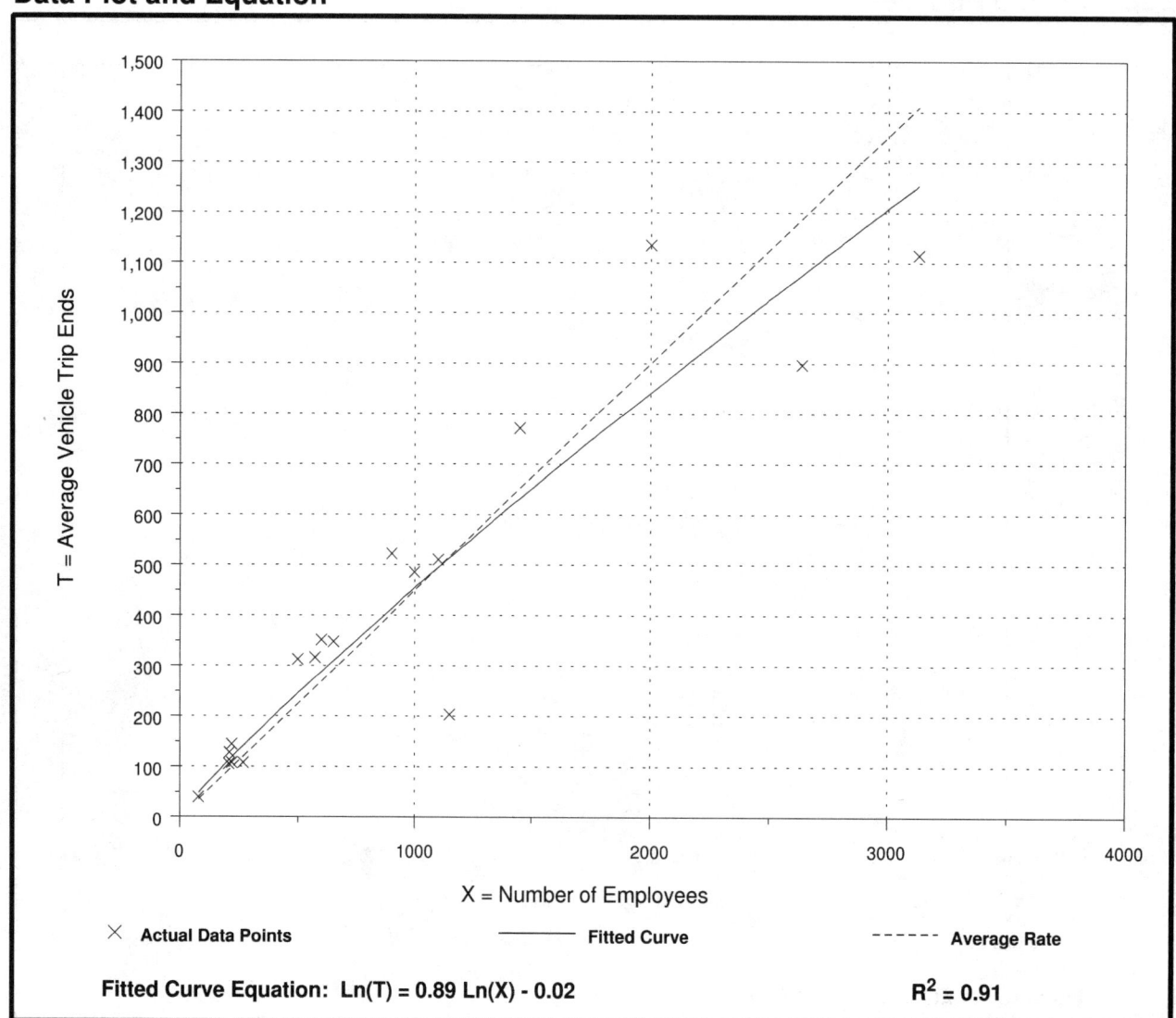

Fitted Curve Equation: $Ln(T) = 0.89 \, Ln(X) - 0.02$ $R^2 = 0.91$

Corporate Headquarters Building
(714)

Average Vehicle Trip Ends vs: Employees
On a: Weekday,
P.M. Peak Hour

Number of Studies: 20
Avg. Number of Employees: 1,045
Directional Distribution: 11% entering, 89% exiting

Trip Generation per Employee

Average Rate	Range of Rates	Standard Deviation
0.38	0.20 - 1.00	0.63

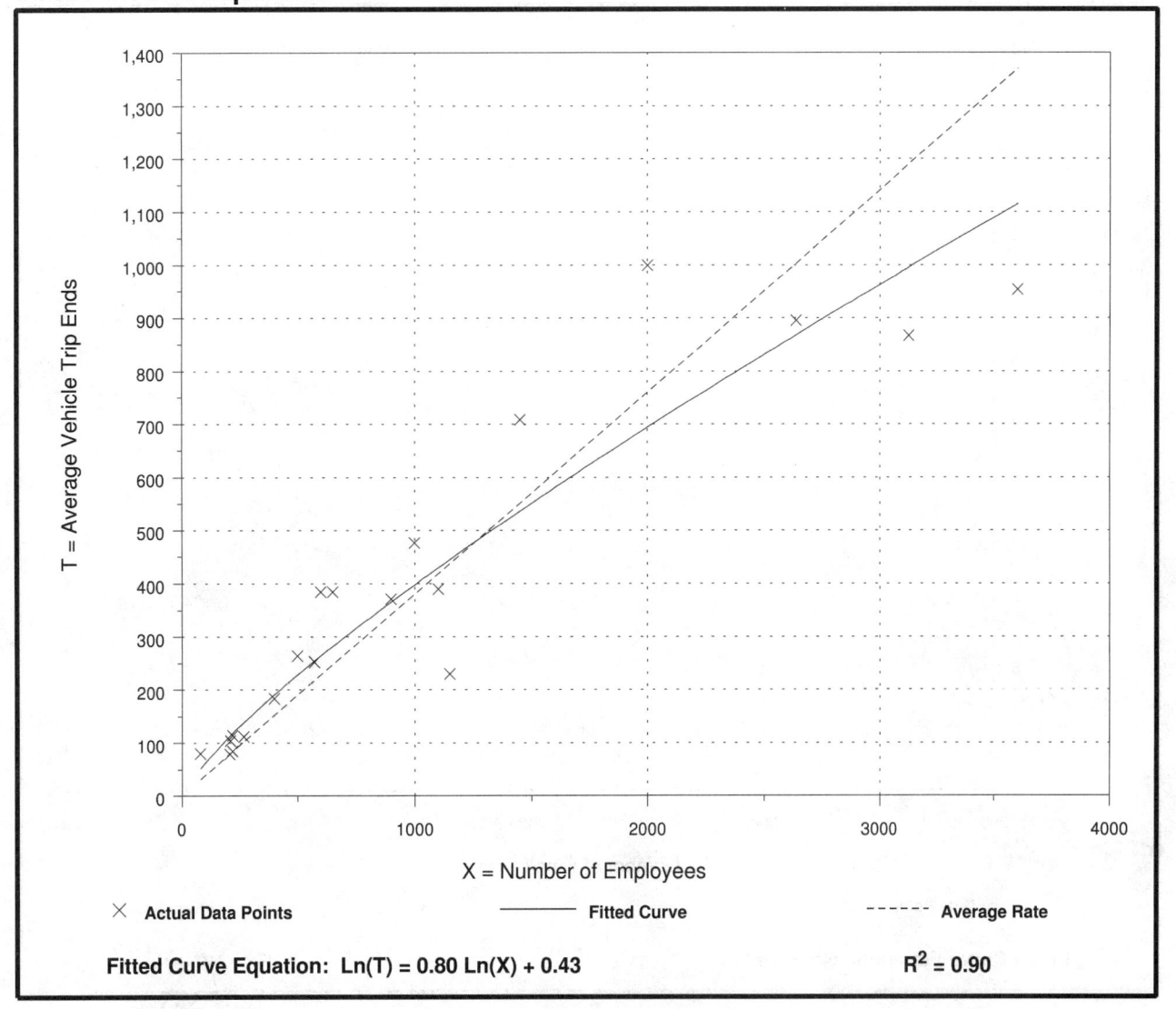

Data Plot and Equation

Fitted Curve Equation: $Ln(T) = 0.80\ Ln(X) + 0.43$ $\qquad R^2 = 0.90$

Corporate Headquarters Building
(714)

Average Vehicle Trip Ends vs: Employees
On a: Saturday,
Peak Hour of Generator

Number of Studies: 2
Avg. Number of Employees: 397
Directional Distribution: 44% entering, 56% exiting

Trip Generation per Employee

Average Rate	Range of Rates	Standard Deviation
0.03	0.03 - 0.03	*

Data Plot and Equation

Caution - Use Carefully - Small Sample Size

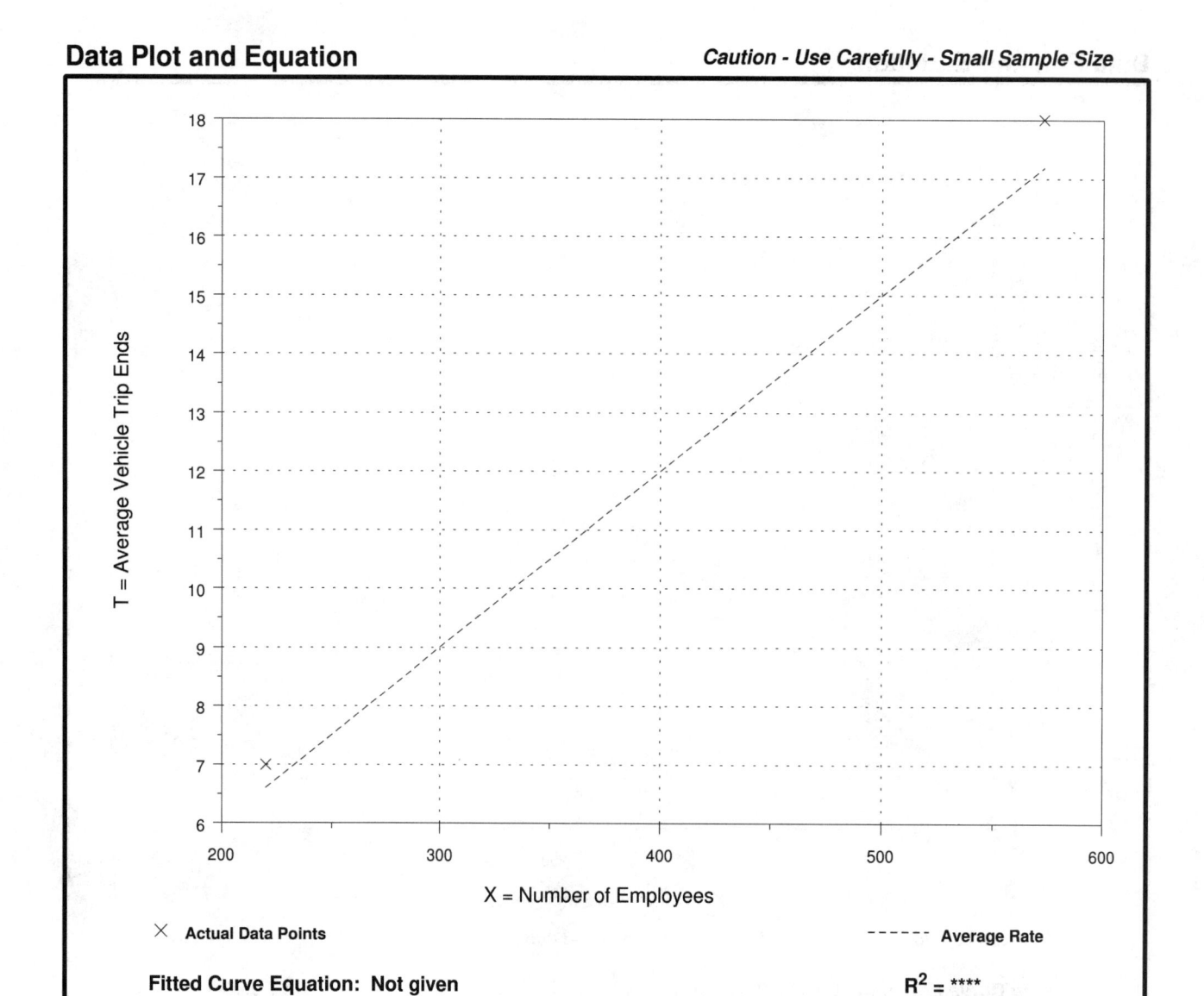

Fitted Curve Equation: Not given $R^2 = ****$

Corporate Headquarters Building
(714)

Average Vehicle Trip Ends vs: Employees
On a: Sunday,
Peak Hour of Generator

Number of Studies: 2
Avg. Number of Employees: 397
Directional Distribution: 63% entering, 37% exiting

Trip Generation per Employee

Average Rate	Range of Rates	Standard Deviation
0.03	0.02 - 0.03	*

Data Plot and Equation   *Caution - Use Carefully - Small Sample Size*

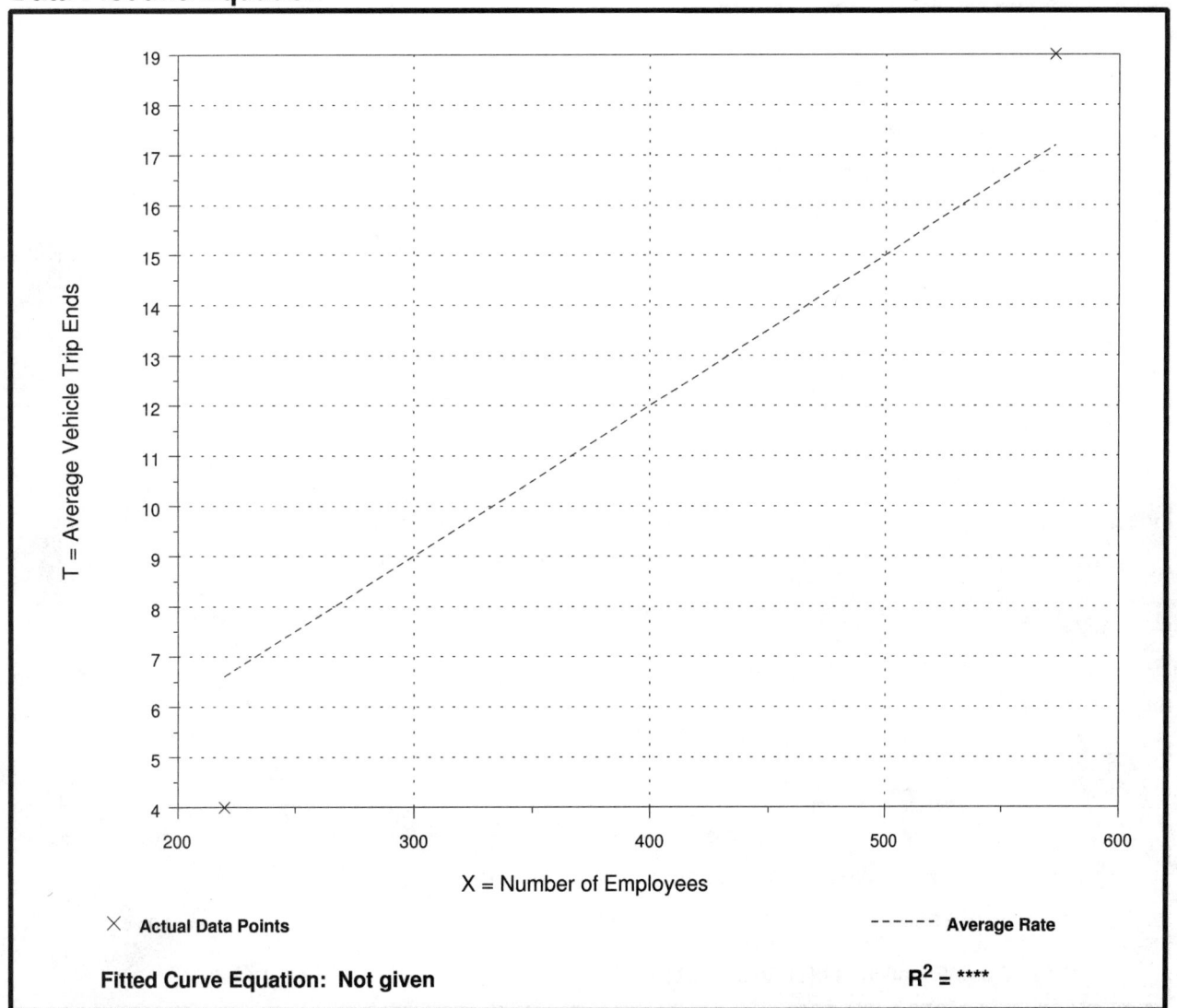

Fitted Curve Equation: Not given $R^2 = ****$

Corporate Headquarters Building
(714)

Average Vehicle Trip Ends vs: 1000 Sq. Feet Gross Floor Area
On a: Weekday

Number of Studies: 8
Average 1000 Sq. Feet GFA: 229
Directional Distribution: 50% entering, 50% exiting

Trip Generation per 1000 Sq. Feet Gross Floor Area

Average Rate	Range of Rates	Standard Deviation
7.98	5.87 - 12.39	3.89

Data Plot and Equation

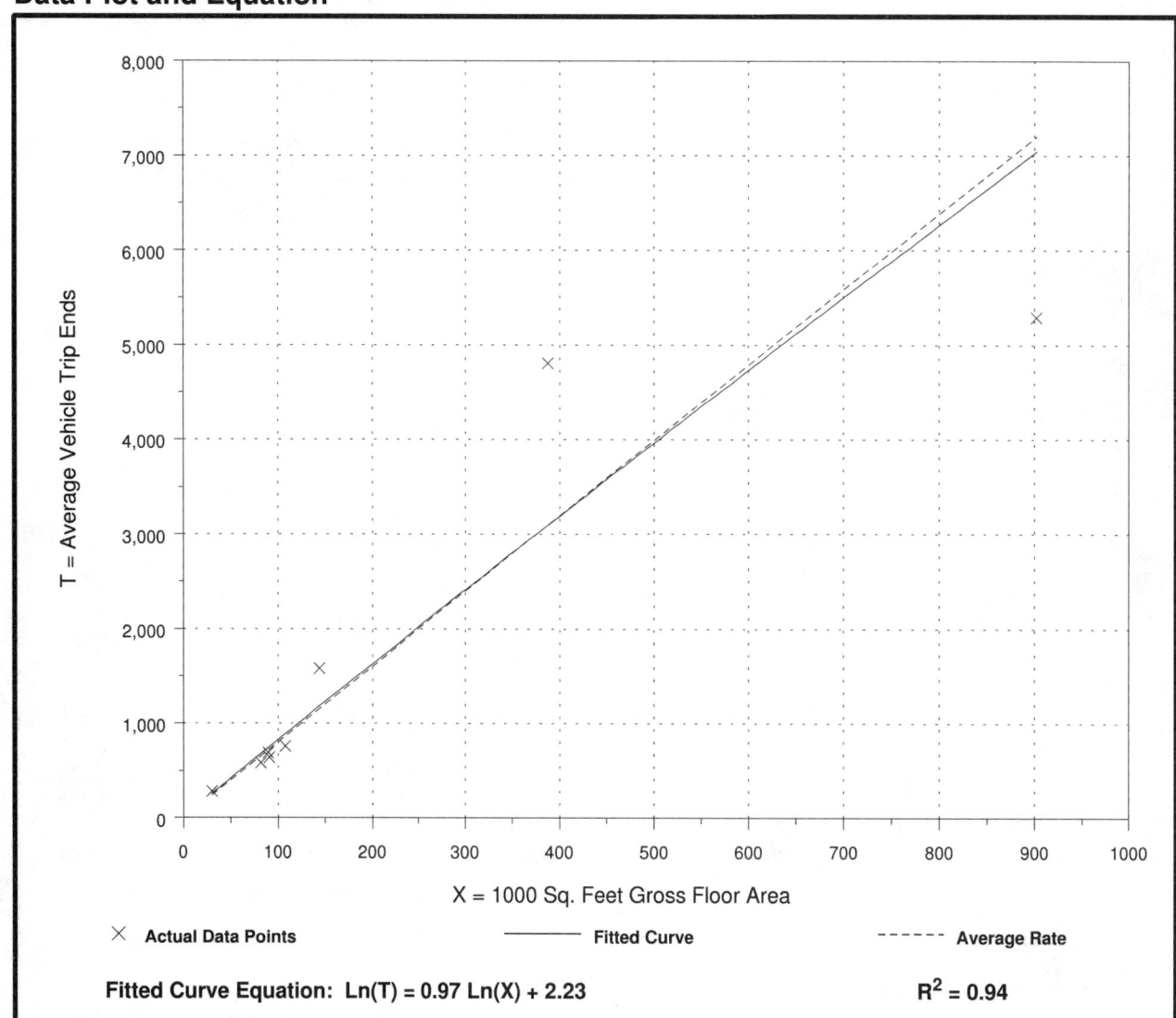

Fitted Curve Equation: $Ln(T) = 0.97 Ln(X) + 2.23$ $R^2 = 0.94$

Corporate Headquarters Building
(714)

Average Vehicle Trip Ends vs: 1000 Sq. Feet Gross Floor Area
On a: Weekday,
A.M. Peak Hour

Number of Studies: 20
Average 1000 Sq. Feet GFA: 277
Directional Distribution: 93% entering, 7 % exiting

Trip Generation per 1000 Sq. Feet Gross Floor Area

Average Rate	Range of Rates	Standard Deviation
1.52	0.46 - 3.01	1.37

Data Plot and Equation

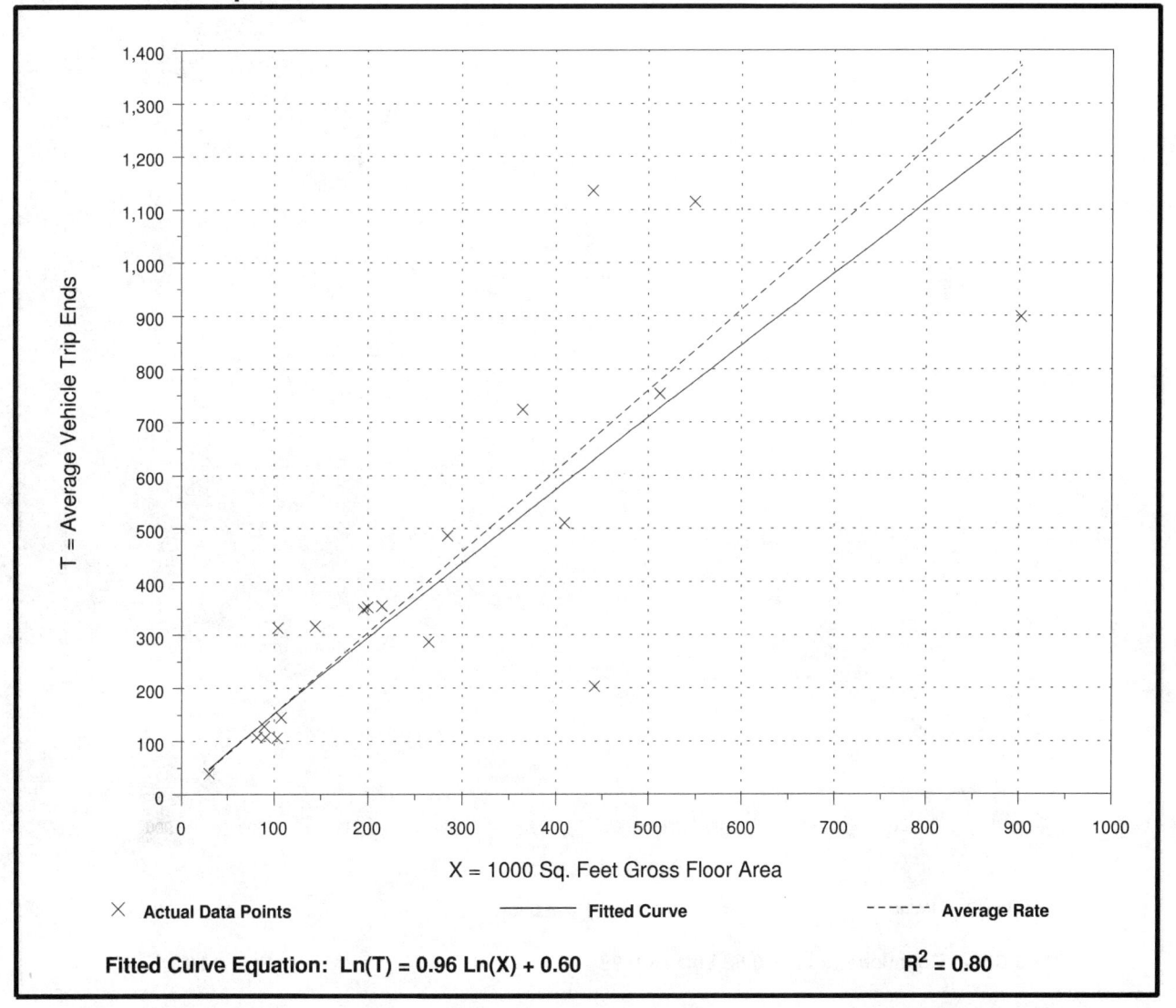

Fitted Curve Equation: $\text{Ln}(T) = 0.96 \, \text{Ln}(X) + 0.60$ $R^2 = 0.80$

Trip Generation, 9th Edition • Institute of Transportation Engineers

Corporate Headquarters Building
(714)

Average Vehicle Trip Ends vs: 1000 Sq. Feet Gross Floor Area
On a: Weekday,
P.M. Peak Hour

Number of Studies: 22
Average 1000 Sq. Feet GFA: 283
Directional Distribution: 10% entering, 90% exiting

Trip Generation per 1000 Sq. Feet Gross Floor Area

Average Rate	Range of Rates	Standard Deviation
1.41	0.52 - 2.67	1.28

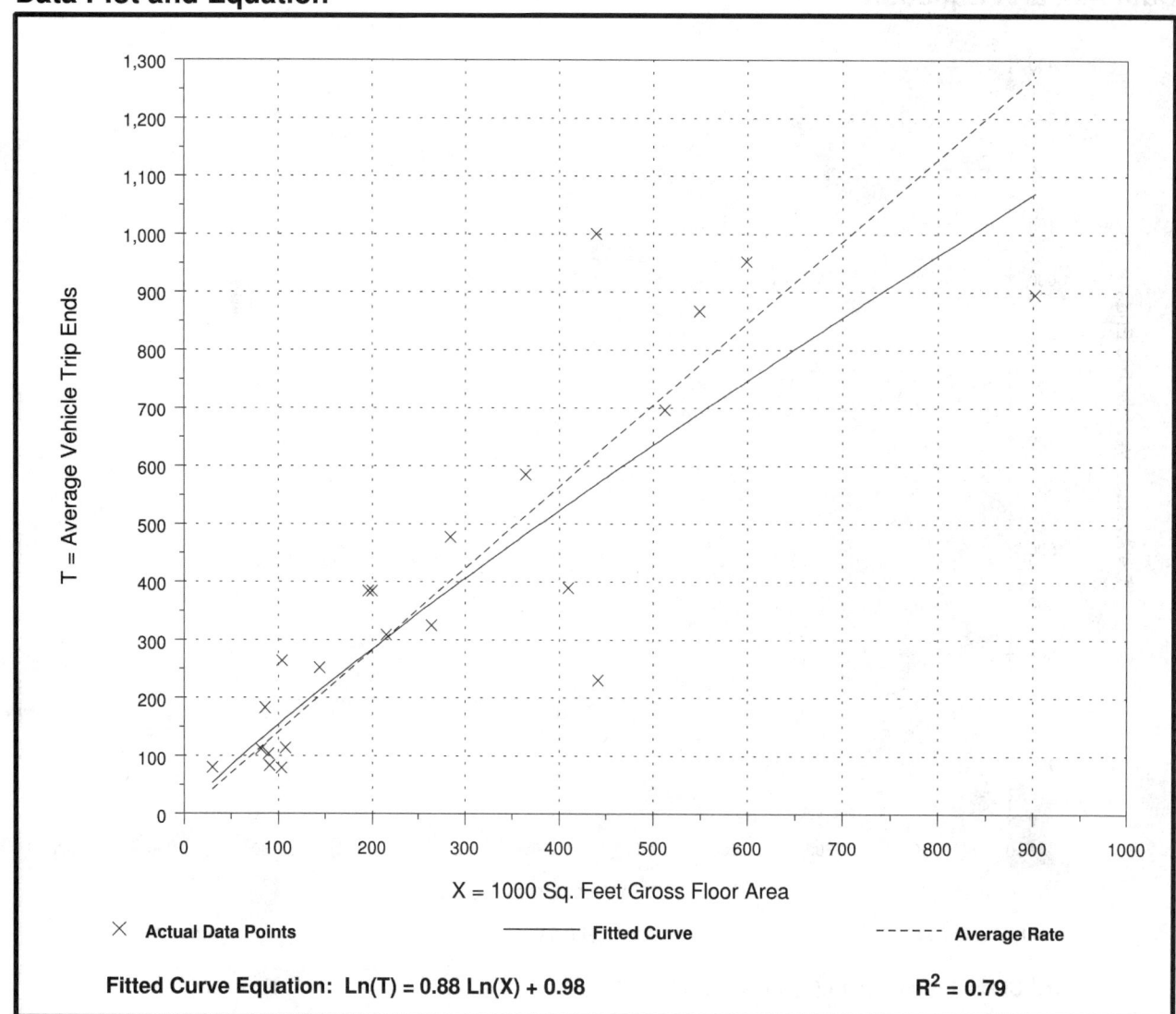

Data Plot and Equation

Fitted Curve Equation: $Ln(T) = 0.88 \, Ln(X) + 0.98$ **$R^2 = 0.79$**

Corporate Headquarters Building
(714)

Average Vehicle Trip Ends vs: 1000 Sq. Feet Gross Floor Area
On a: Saturday,
Peak Hour of Generator

Number of Studies: 2
Average 1000 Sq. Feet GFA: 126
Directional Distribution: 44% entering, 56% exiting

Trip Generation per 1000 Sq. Feet Gross Floor Area

Average Rate	Range of Rates	Standard Deviation
0.10	0.06 - 0.13	*

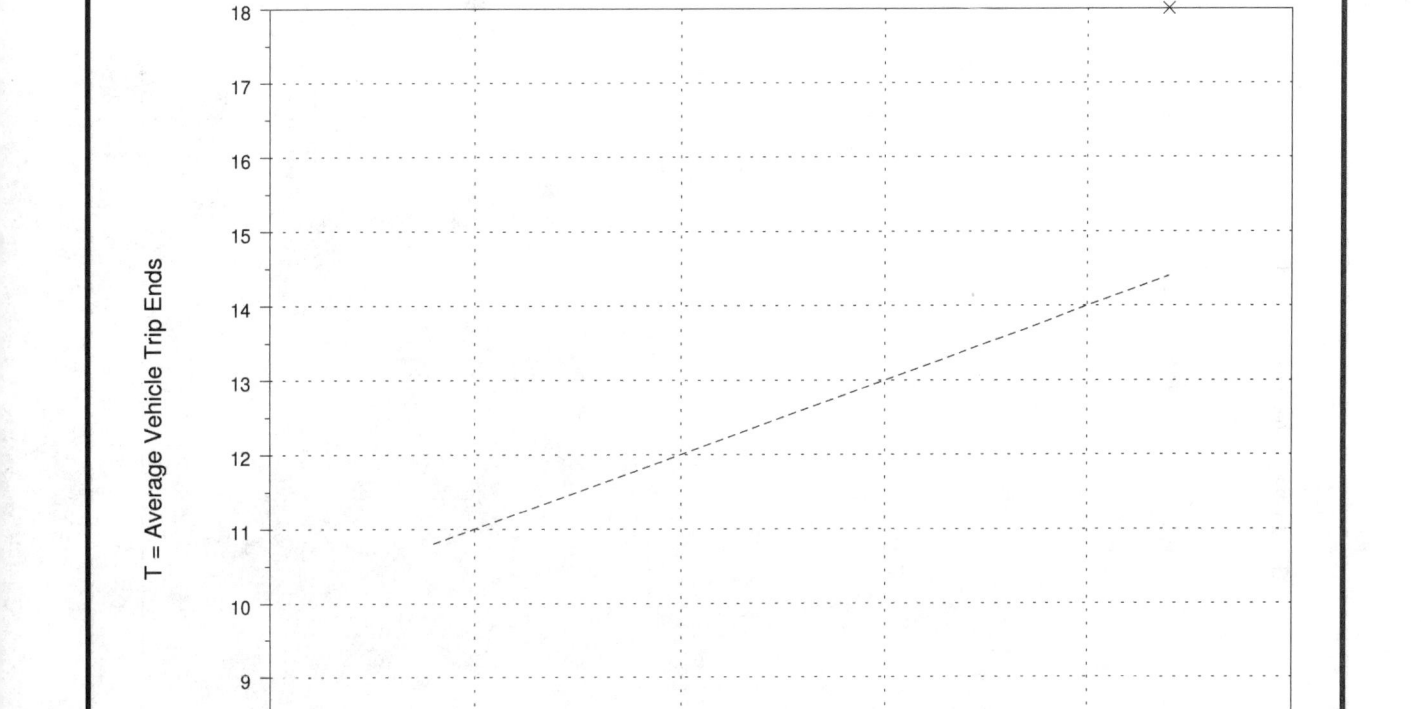

Data Plot and Equation

Caution - Use Carefully - Small Sample Size

Fitted Curve Equation: Not given $R^2 = ****$

Corporate Headquarters Building
(714)

Average Vehicle Trip Ends vs: 1000 Sq. Feet Gross Floor Area
On a: Sunday,
Peak Hour of Generator

Number of Studies: 2
Average 1000 Sq. Feet GFA: 126
Directional Distribution: 63% entering, 37% exiting

Trip Generation per 1000 Sq. Feet Gross Floor Area

Average Rate	Range of Rates	Standard Deviation
0.09	0.04 - 0.13	*

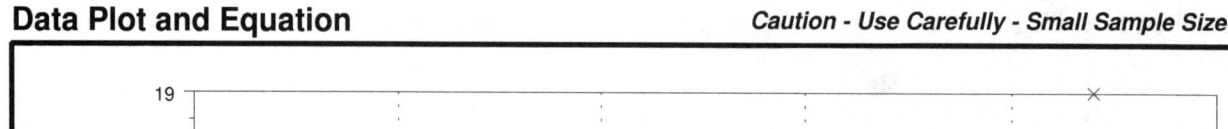

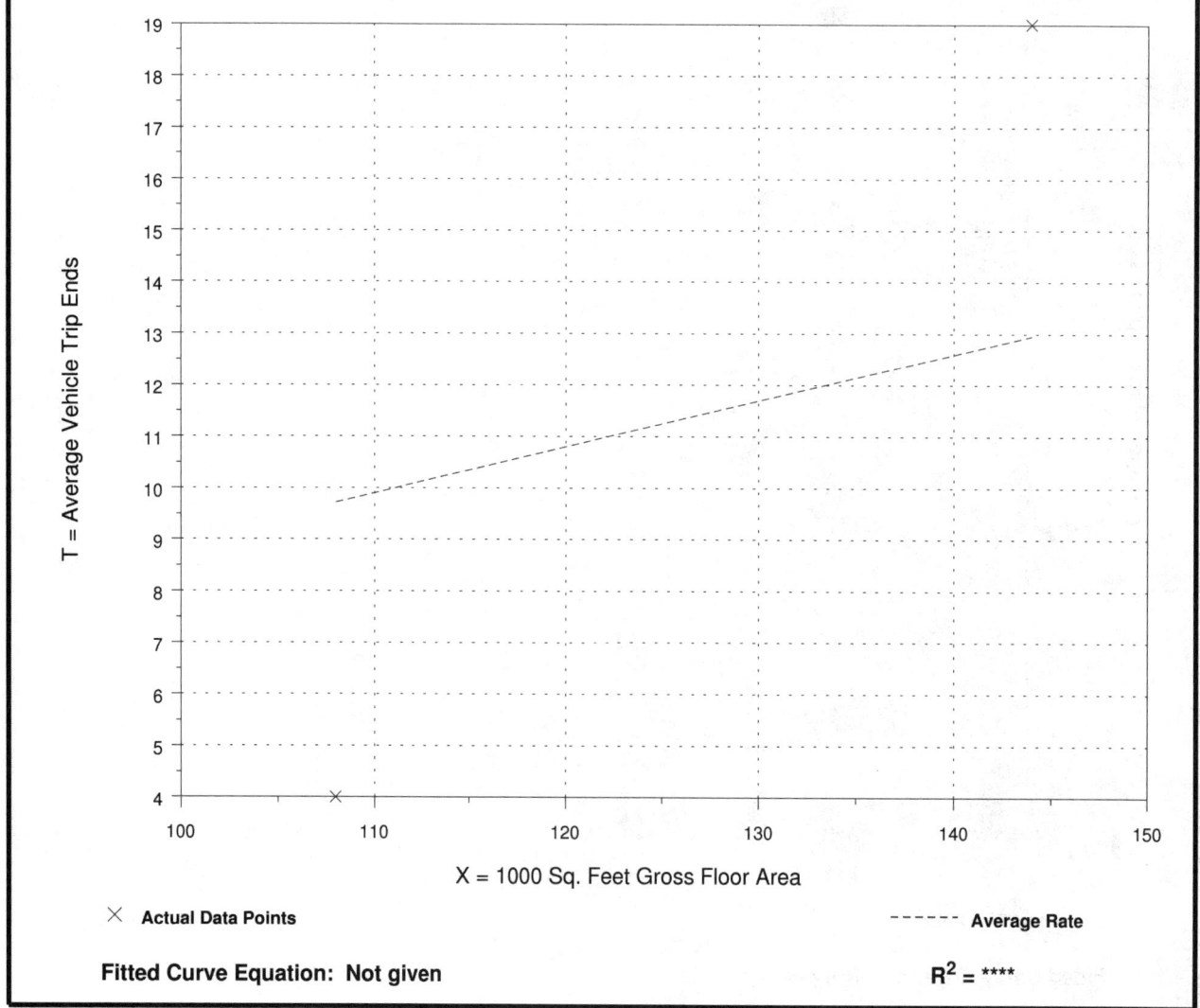

Data Plot and Equation

Caution - Use Carefully - Small Sample Size

Fitted Curve Equation: Not given $R^2 = ****$

Land Use: 715
Single Tenant Office Building

Description

A single tenant office building generally contains offices, meeting rooms and space for file storage and data processing of a single business or company and possibly other service functions including a restaurant or cafeteria. General office building (Land Use 710), corporate headquarters building (Land Use 714), office park (Land Use 750), research and development center (Land Use 760) and business park (Land Use 770) are related uses.

Additional Data

The average vehicle occupancy for the 10 studies for which information was submitted was approximately 1.1 persons per automobile. The vehicle occupancy rates ranged from 1.03 to 1.14 persons per automobile.

The sites were surveyed between the 1970s and the late 2000s throughout the United States.

Trip Characteristics

The trip generation for the A.M. and P.M. peak hours of the generator typically coincided with the peak hours of the adjacent street traffic; therefore, only one A.M. peak hour and one P.M. peak hour, which represent both the peak hour of the generator and the peak hour of the adjacent street traffic, are shown for single tenant office buildings.

Source Numbers

89, 92, 212, 262, 273, 279, 303, 304, 322, 323, 324, 327, 407, 510, 701

Single Tenant Office Building
(715)

Average Vehicle Trip Ends vs: Employees
On a: Weekday

Number of Studies: 15
Avg. Number of Employees: 330
Directional Distribution: 50% entering, 50% exiting

Trip Generation per Employee

Average Rate	Range of Rates	Standard Deviation
3.70	2.09 - 7.75	2.48

Data Plot and Equation

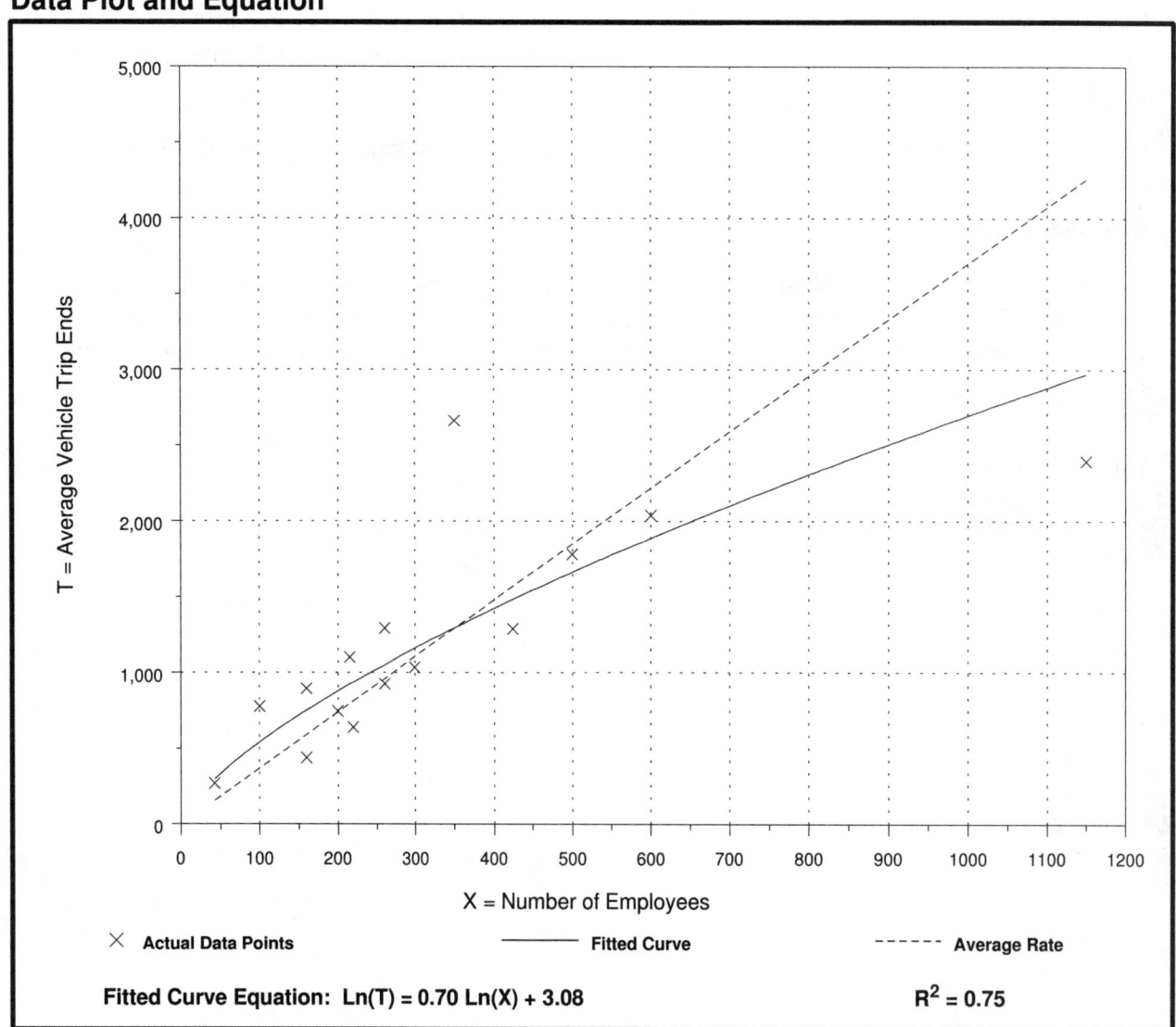

Fitted Curve Equation: $Ln(T) = 0.70 Ln(X) + 3.08$ $R^2 = 0.75$

Single Tenant Office Building
(715)

Average Vehicle Trip Ends vs: Employees
On a: Weekday,
A.M. Peak Hour

Number of Studies: 39
Avg. Number of Employees: 545
Directional Distribution: 89% entering, 11% exiting

Trip Generation per Employee

Average Rate	Range of Rates	Standard Deviation
0.53	0.27 - 1.01	0.75

Data Plot and Equation

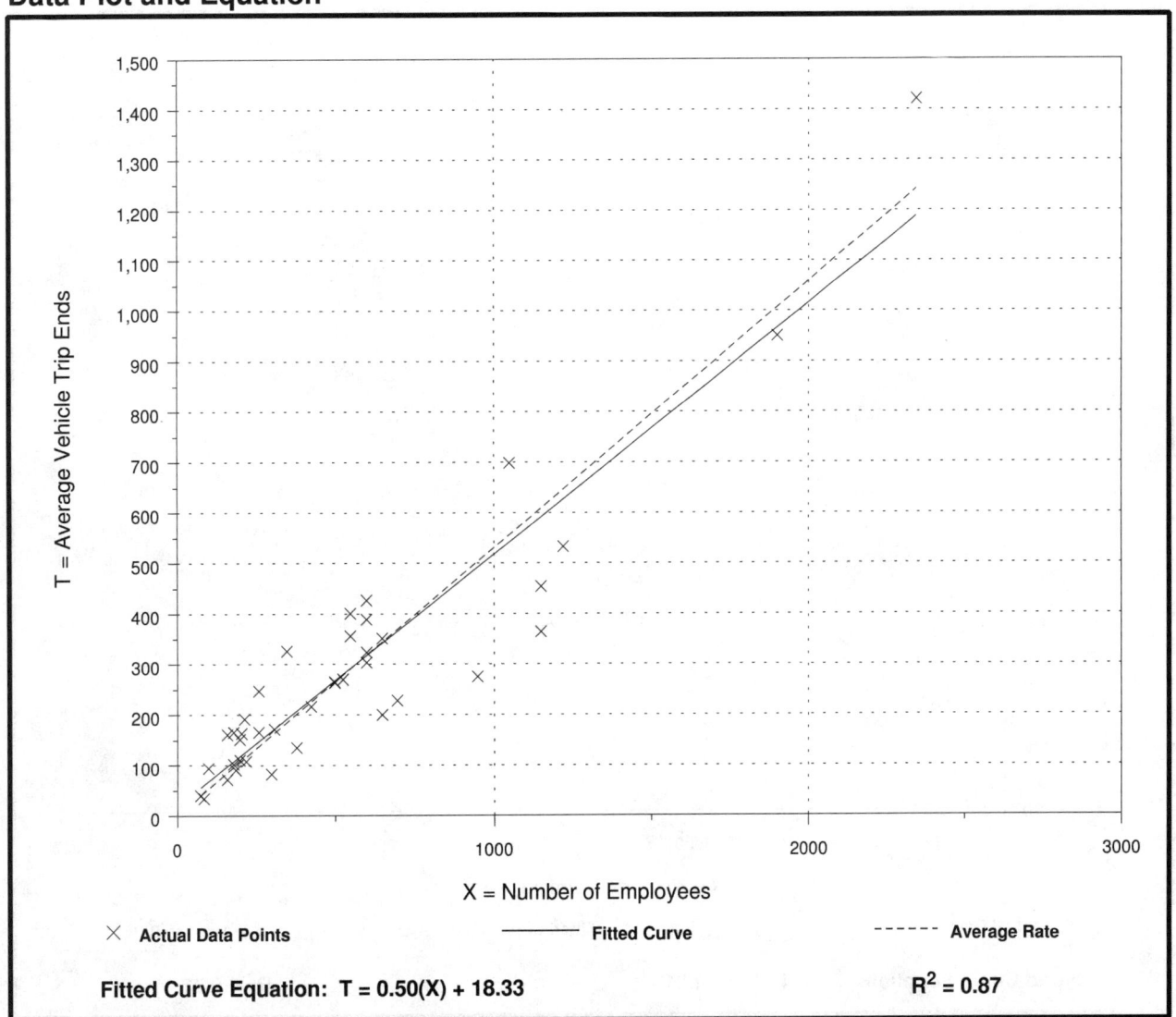

Fitted Curve Equation: $T = 0.50(X) + 18.33$ $R^2 = 0.87$

Single Tenant Office Building
(715)

Average Vehicle Trip Ends vs: Employees
On a: Weekday,
P.M. Peak Hour

Number of Studies: 39
Avg. Number of Employees: 545
Directional Distribution: 15% entering, 85% exiting

Trip Generation per Employee

Average Rate	Range of Rates	Standard Deviation
0.51	0.29 - 1.14	0.73

Data Plot and Equation

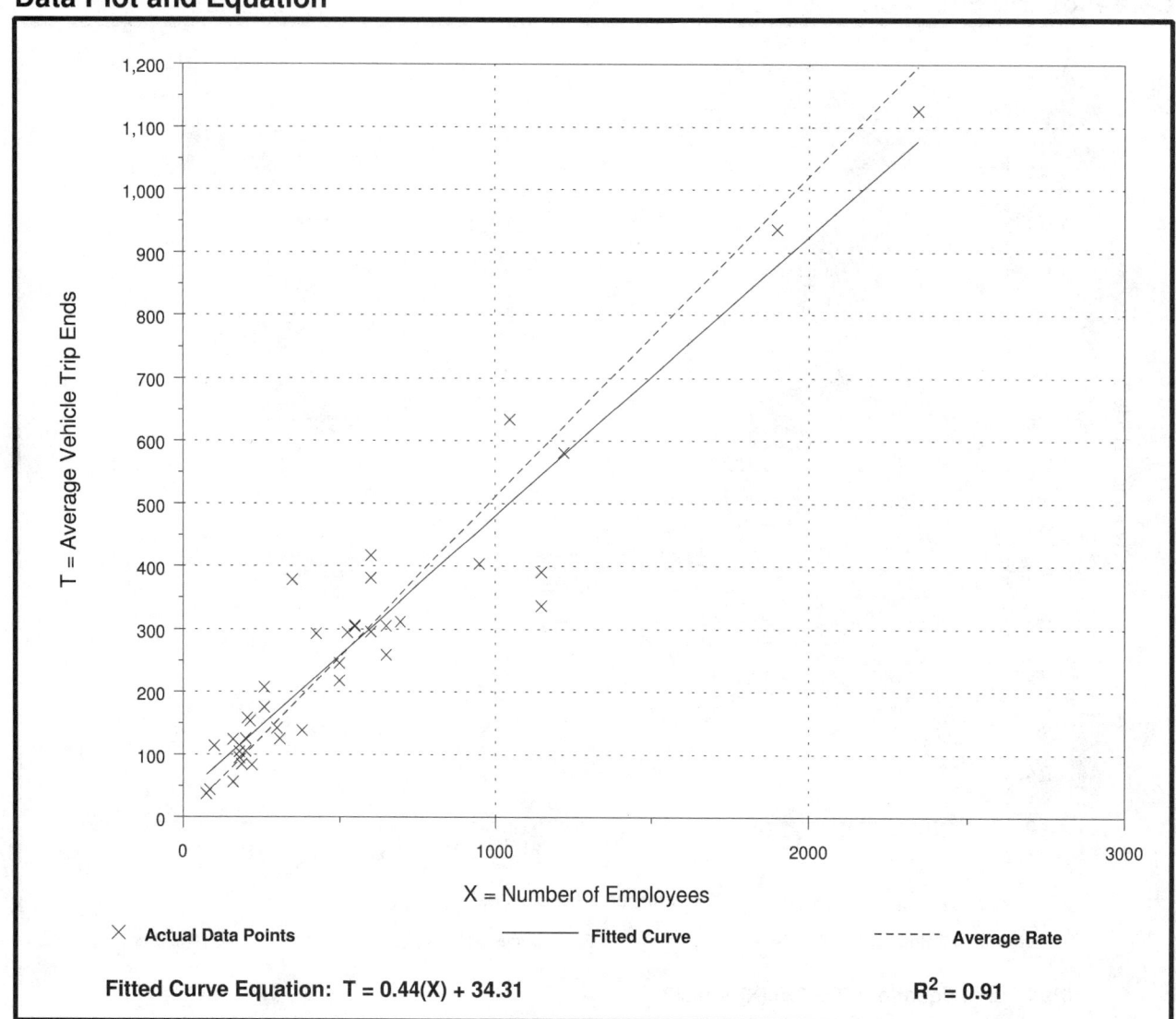

Fitted Curve Equation: $T = 0.44(X) + 34.31$ $R^2 = 0.91$

Single Tenant Office Building
(715)

Average Vehicle Trip Ends vs: 1000 Sq. Feet Gross Floor Area
On a: Weekday

Number of Studies: 15
Average 1000 Sq. Feet GFA: 105
Directional Distribution: 50% entering, 50% exiting

Trip Generation per 1000 Sq. Feet Gross Floor Area

Average Rate	Range of Rates	Standard Deviation
11.65	5.33 - 35.68	8.19

Data Plot and Equation

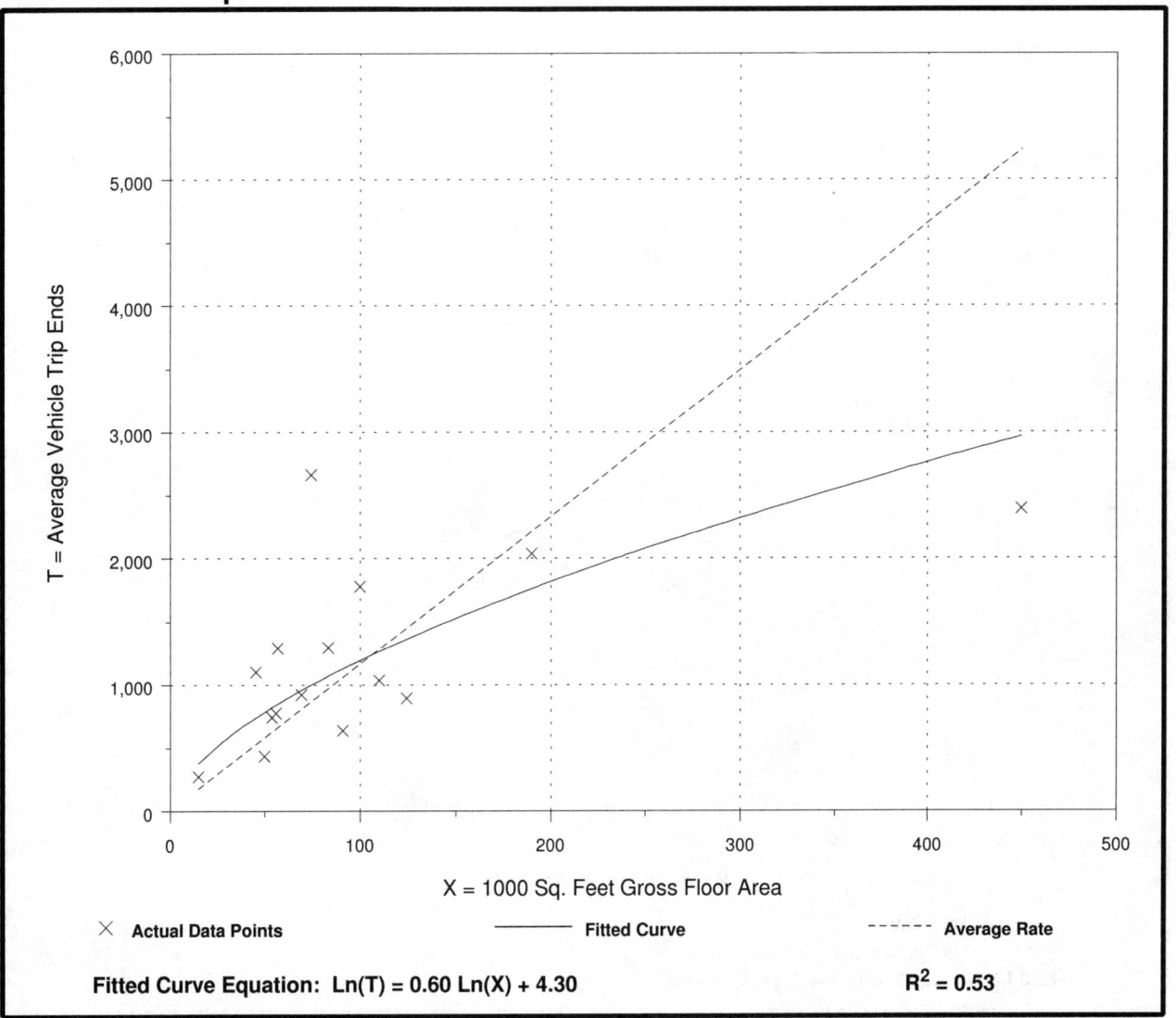

Fitted Curve Equation: $Ln(T) = 0.60\, Ln(X) + 4.30$ $R^2 = 0.53$

Trip Generation, 9th Edition • Institute of Transportation Engineers

Single Tenant Office Building
(715)

Average Vehicle Trip Ends vs: 1000 Sq. Feet Gross Floor Area
On a: Weekday,
A.M. Peak Hour

Number of Studies: 43
Average 1000 Sq. Feet GFA: 162
Directional Distribution: 89% entering, 11% exiting

Trip Generation per 1000 Sq. Feet Gross Floor Area

Average Rate	Range of Rates	Standard Deviation
1.80	0.75 - 4.57	1.51

Data Plot and Equation

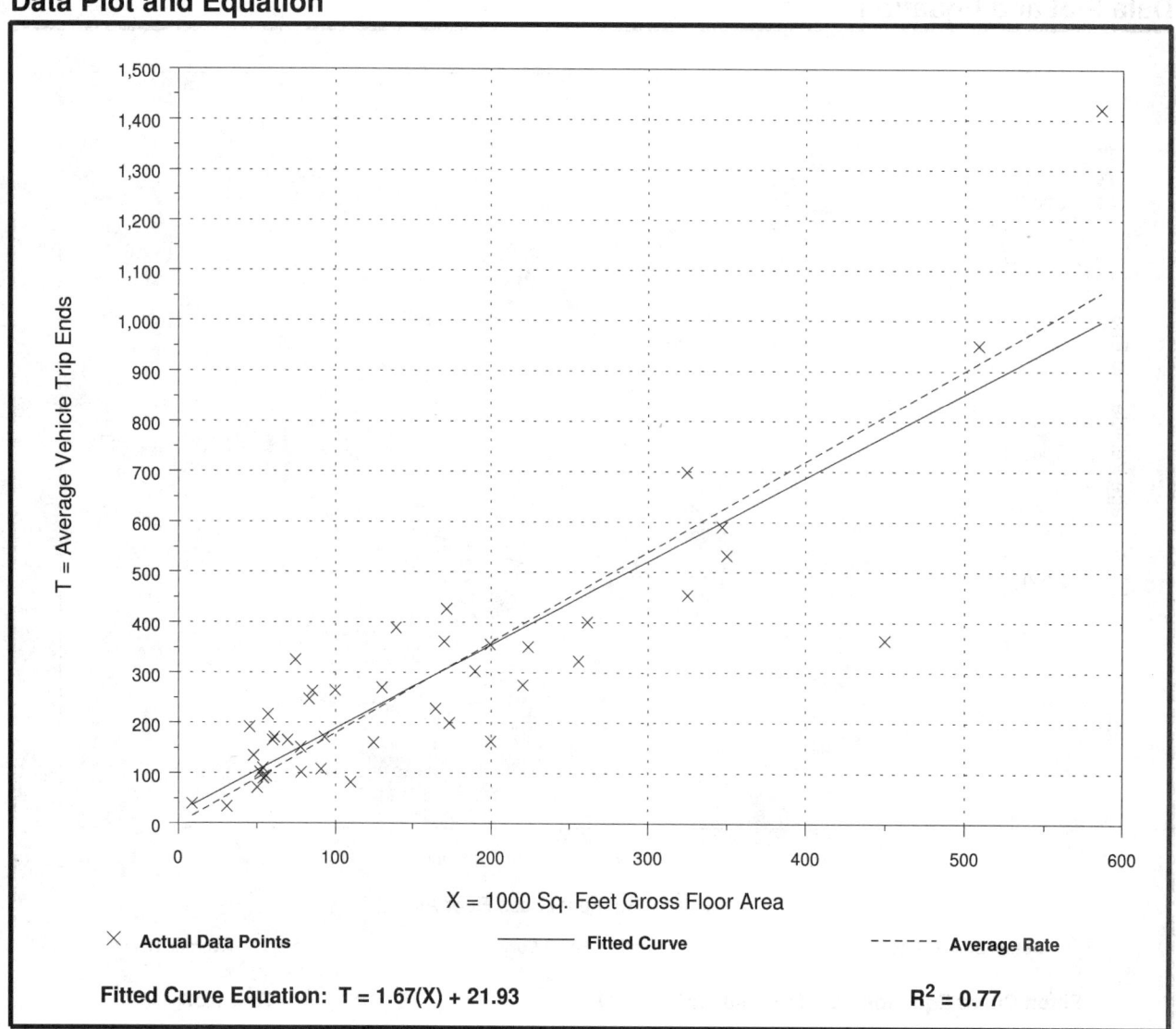

Fitted Curve Equation: T = 1.67(X) + 21.93 $R^2 = 0.77$

Single Tenant Office Building
(715)

Average Vehicle Trip Ends vs: 1000 Sq. Feet Gross Floor Area
On a: Weekday,
P.M. Peak Hour

Number of Studies: 43
Average 1000 Sq. Feet GFA: 162
Directional Distribution: 15% entering, 85% exiting

Trip Generation per 1000 Sq. Feet Gross Floor Area

Average Rate	Range of Rates	Standard Deviation
1.74	0.79 - 5.14	1.49

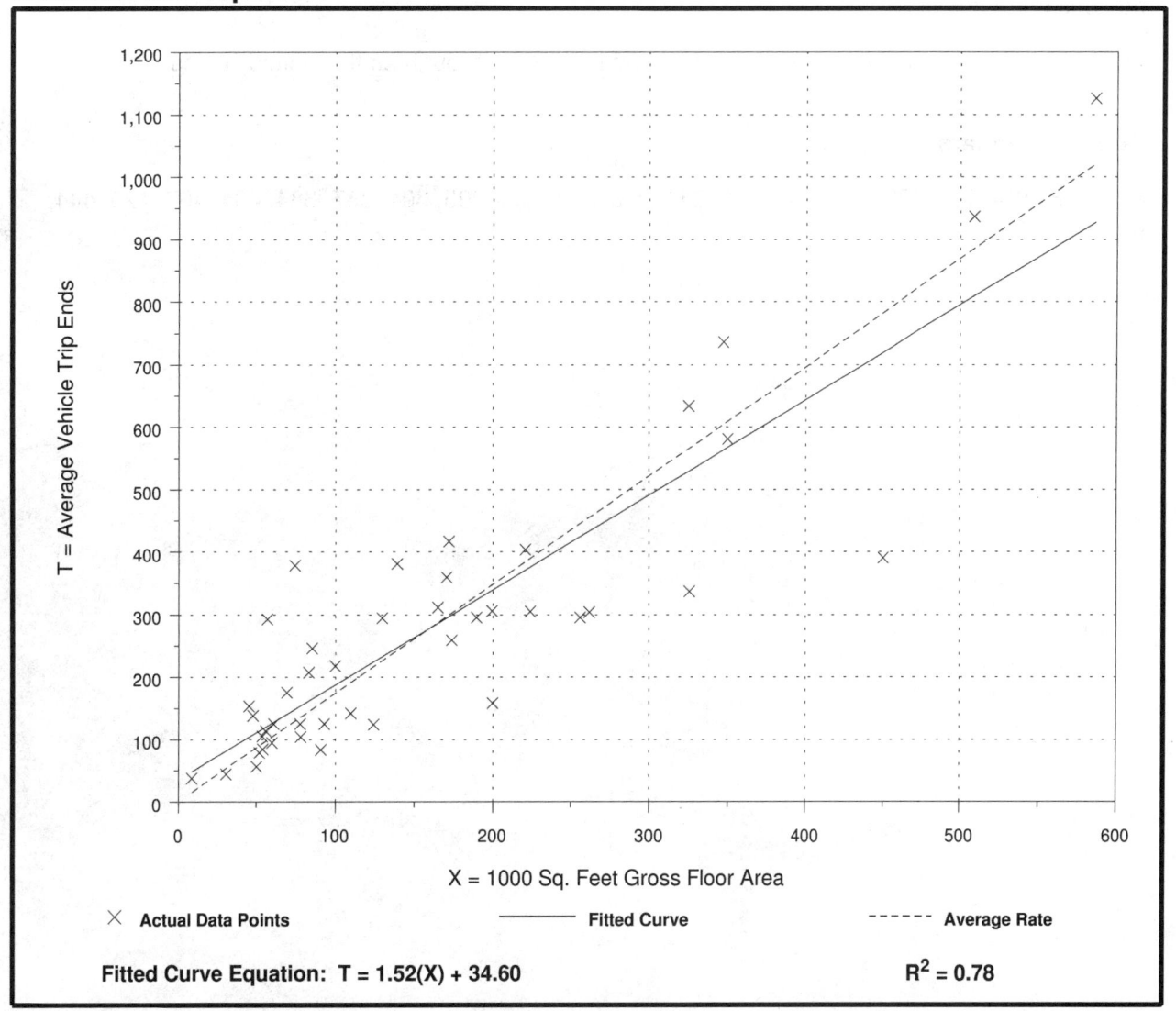

Data Plot and Equation

Fitted Curve Equation: $T = 1.52(X) + 34.60$ $R^2 = 0.78$

Land Use: 720
Medical-Dental Office Building

Description

A medical-dental office building is a facility that provides diagnoses and outpatient care on a routine basis but is unable to provide prolonged in-house medical and surgical care. One or more private physicians or dentists generally operate this type of facility. Clinic (Land Use 630) is a related use.

Additional Data

The average vehicle occupancy for the six studies for which information was submitted was approximately 1.37 persons per automobile. The vehicle occupancy rates ranged from 1.32 to 1.44 persons per automobile.

The sites were surveyed between the 1980s and the 2000s throughout the United States.

Source Numbers

8, 19, 98, 104, 109, 120, 157, 184, 209, 211, 253, 287, 294, 295, 304, 357, 384, 404, 407, 423, 444, 509, 601, 715

Medical-Dental Office Building
(720)

Average Vehicle Trip Ends vs: Employees
On a: Weekday

Number of Studies: 5
Avg. Number of Employees: 127
Directional Distribution: 50% entering, 50% exiting

Trip Generation per Employee

Average Rate	Range of Rates	Standard Deviation
8.91	5.69 - 13.03	3.95

Data Plot and Equation

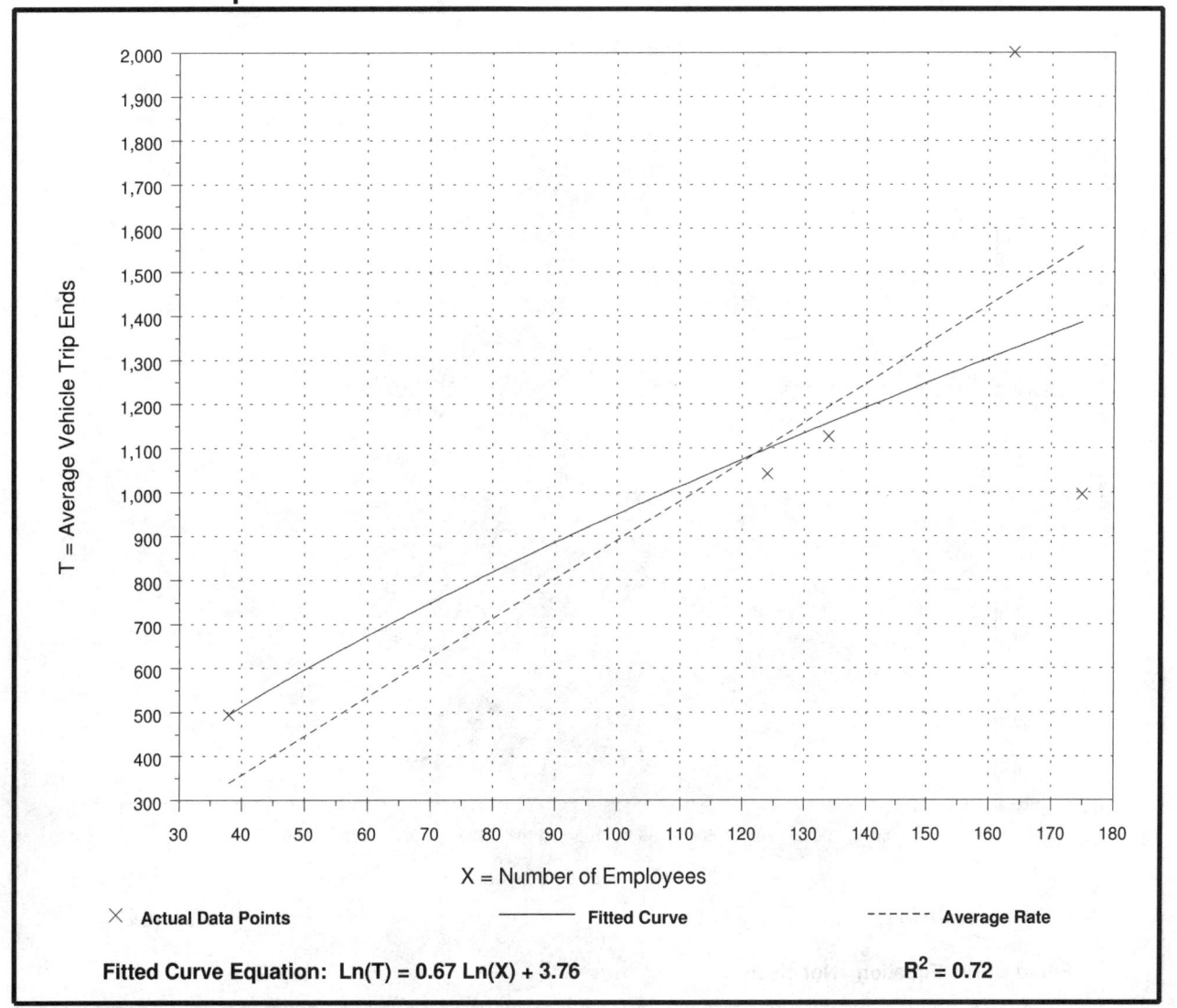

Caution - Use Carefully - Small Sample Size

Fitted Curve Equation: $Ln(T) = 0.67\ Ln(X) + 3.76$ $R^2 = 0.72$

Medical-Dental Office Building
(720)

Average Vehicle Trip Ends vs: Employees
On a: Weekday,
Peak Hour of Adjacent Street Traffic,
One Hour Between 7 and 9 a.m.

Number of Studies: 10
Avg. Number of Employees: 120
Directional Distribution: 79% entering, 21% exiting

Trip Generation per Employee

Average Rate	Range of Rates	Standard Deviation
0.53	0.24 - 1.14	0.76

Data Plot and Equation

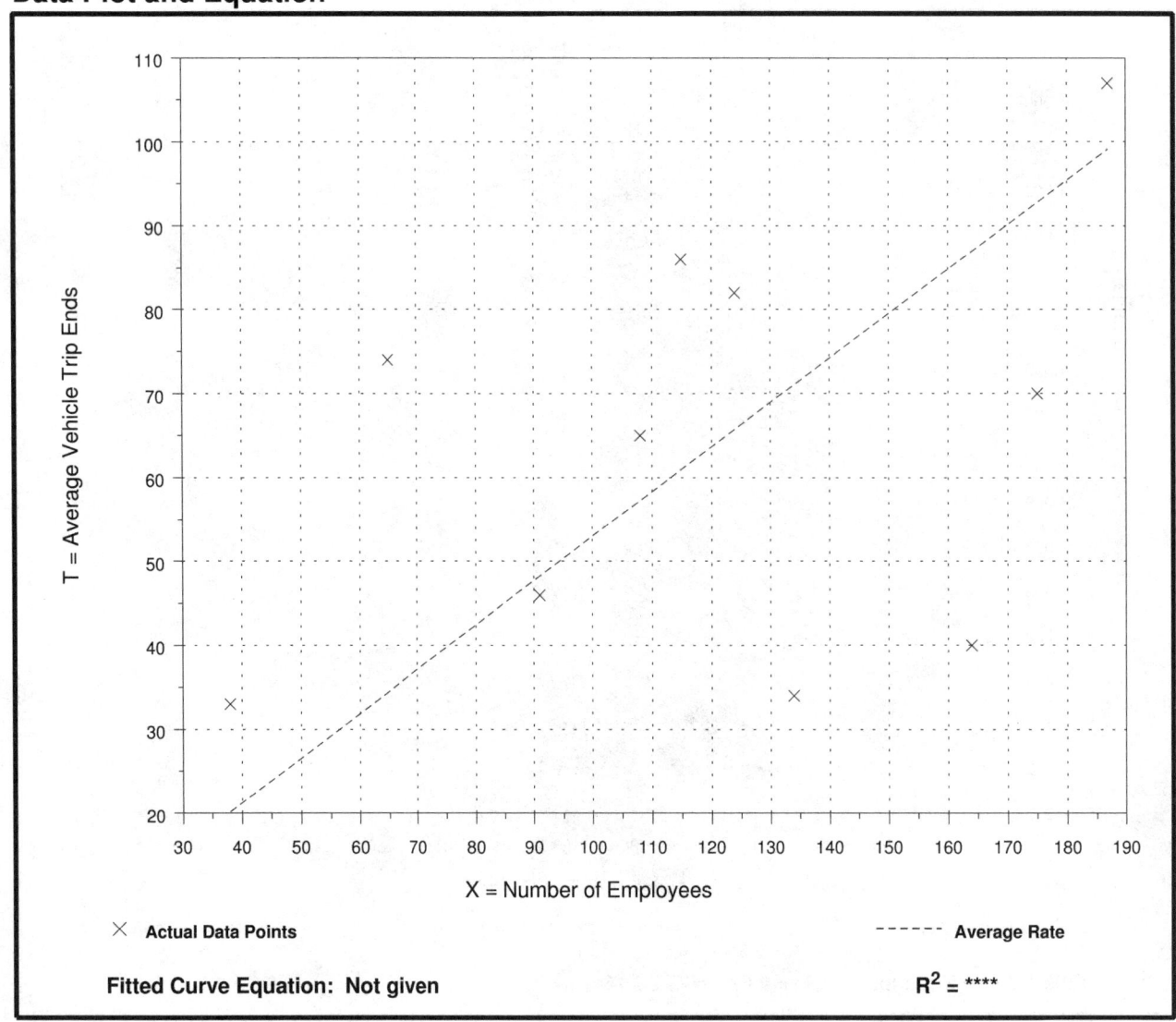

× Actual Data Points
------ Average Rate

Fitted Curve Equation: Not given $R^2 =$ ****

Medical-Dental Office Building
(720)

Average Vehicle Trip Ends vs: Employees
On a: Weekday,
Peak Hour of Adjacent Street Traffic,
One Hour Between 4 and 6 p.m.

Number of Studies: 15
Avg. Number of Employees: 88
Directional Distribution: 34% entering, 66% exiting

Trip Generation per Employee

Average Rate	Range of Rates	Standard Deviation
1.06	0.58 - 1.75	1.08

Data Plot and Equation

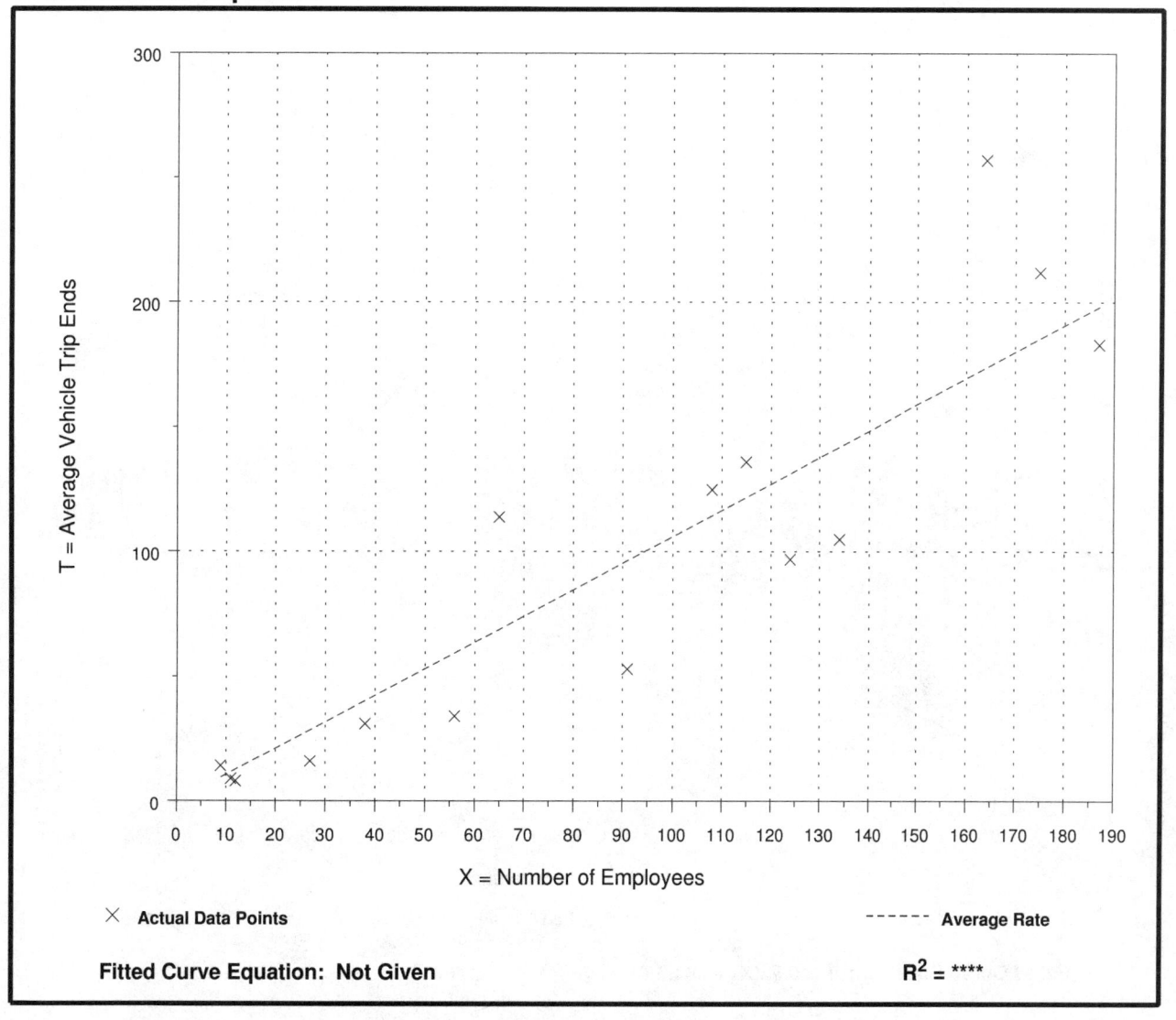

Fitted Curve Equation: Not Given $R^2 = ****$

Medical-Dental Office Building
(720)

Average Vehicle Trip Ends vs: Employees
On a: Weekday,
A.M. Peak Hour of Generator

Number of Studies: 11
Avg. Number of Employees: 209
Directional Distribution: 65% entering, 35% exiting

Trip Generation per Employee

Average Rate	Range of Rates	Standard Deviation
0.80	0.40 - 2.03	0.98

Data Plot and Equation

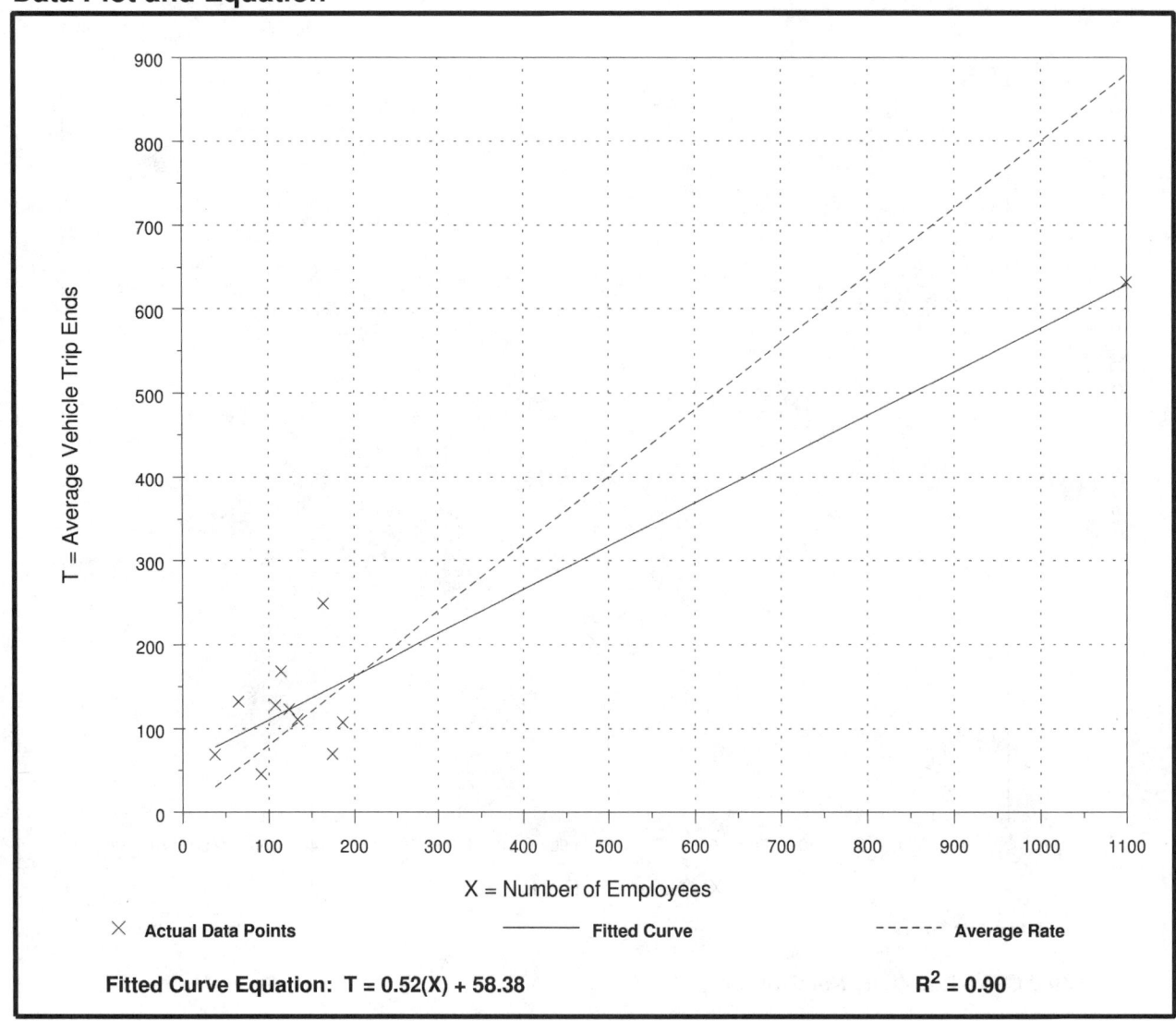

Fitted Curve Equation: $T = 0.52(X) + 58.38$ $R^2 = 0.90$

Medical-Dental Office Building
(720)

Average Vehicle Trip Ends vs: Employees
On a: Weekday,
P.M. Peak Hour of Generator

Number of Studies: 16
Avg. Number of Employees: 151
Directional Distribution: 39% entering, 61% exiting

Trip Generation per Employee

Average Rate	Range of Rates	Standard Deviation
0.97	0.58 - 2.06	1.06

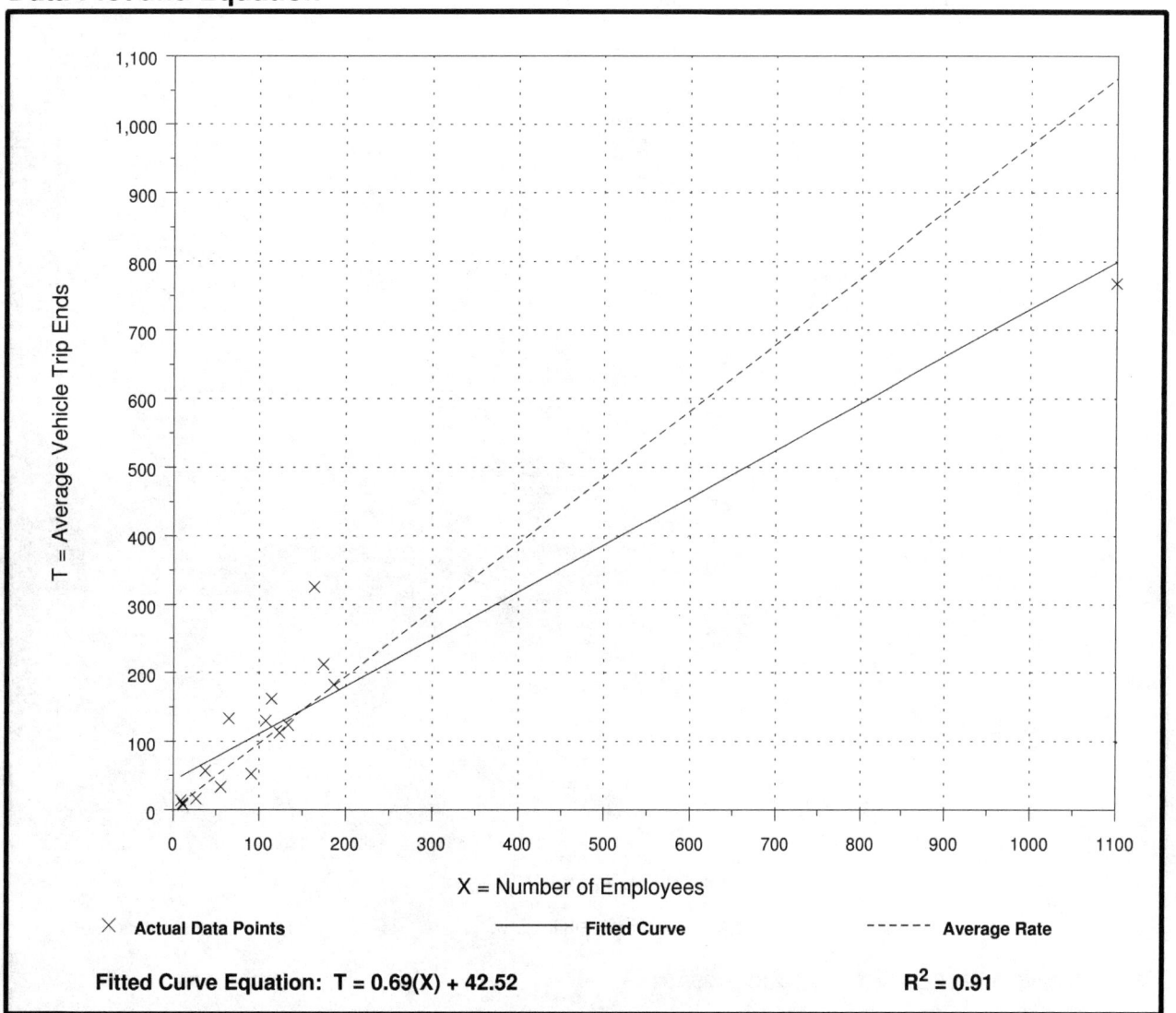

Data Plot and Equation

Fitted Curve Equation: $T = 0.69(X) + 42.52$ $R^2 = 0.91$

Medical-Dental Office Building
(720)

Average Vehicle Trip Ends vs: Employees
On a: Saturday

Number of Studies: 4
Avg. Number of Employees: 116
Directional Distribution: 50% entering, 50% exiting

Trip Generation per Employee

Average Rate	Range of Rates	Standard Deviation
4.02	1.95 - 5.39	2.41

Data Plot and Equation

Caution - Use Carefully - Small Sample Size

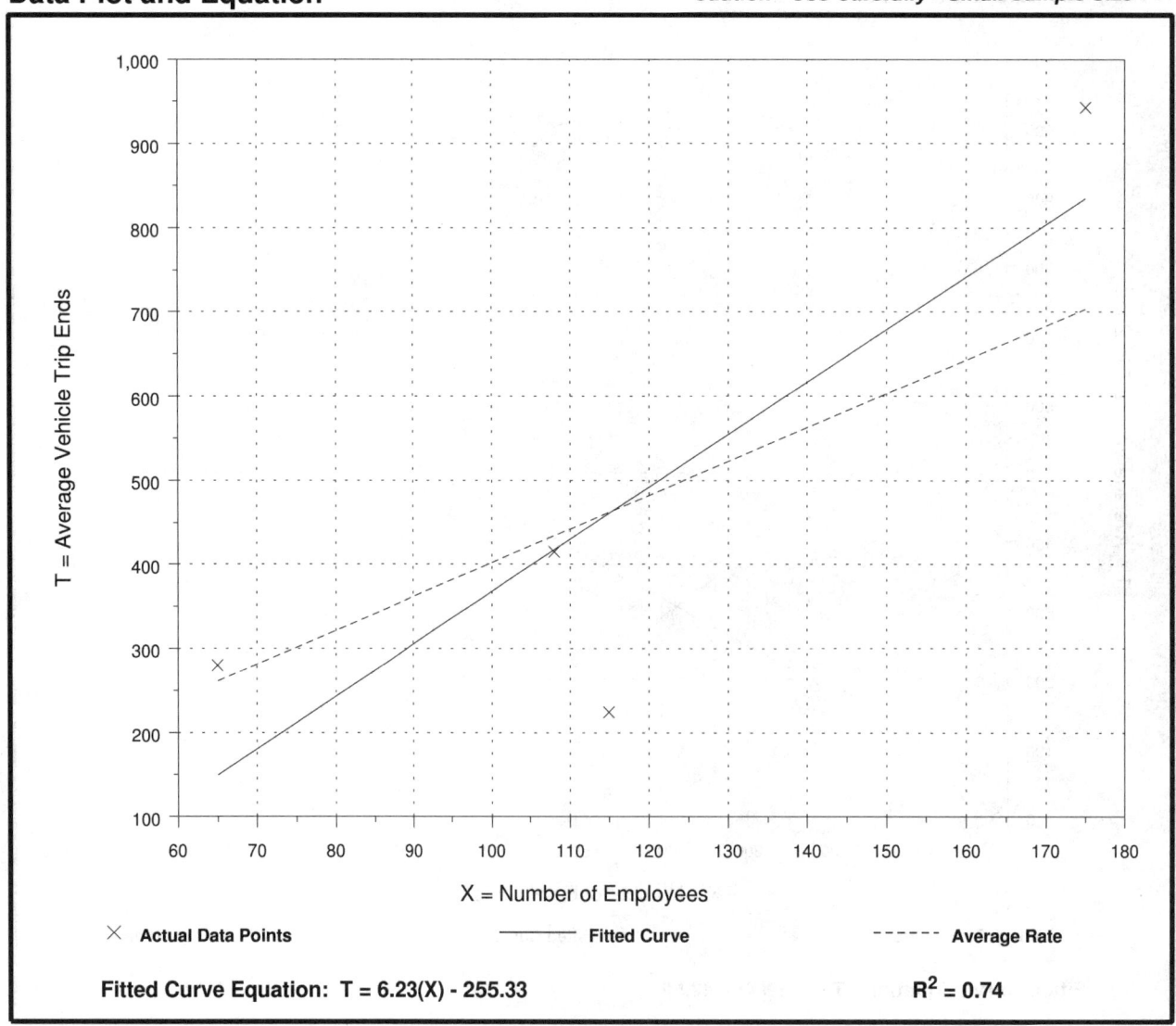

Fitted Curve Equation: T = 6.23(X) - 255.33 $R^2 = 0.74$

Medical-Dental Office Building
(720)

Average Vehicle Trip Ends vs: Employees
On a: Saturday,
Peak Hour of Generator

Number of Studies: 3
Avg. Number of Employees: 116
Directional Distribution: 57% entering, 43% exiting

Trip Generation per Employee

Average Rate	Range of Rates	Standard Deviation
0.88	0.69 - 0.99	0.95

Data Plot and Equation

Caution - Use Carefully - Small Sample Size

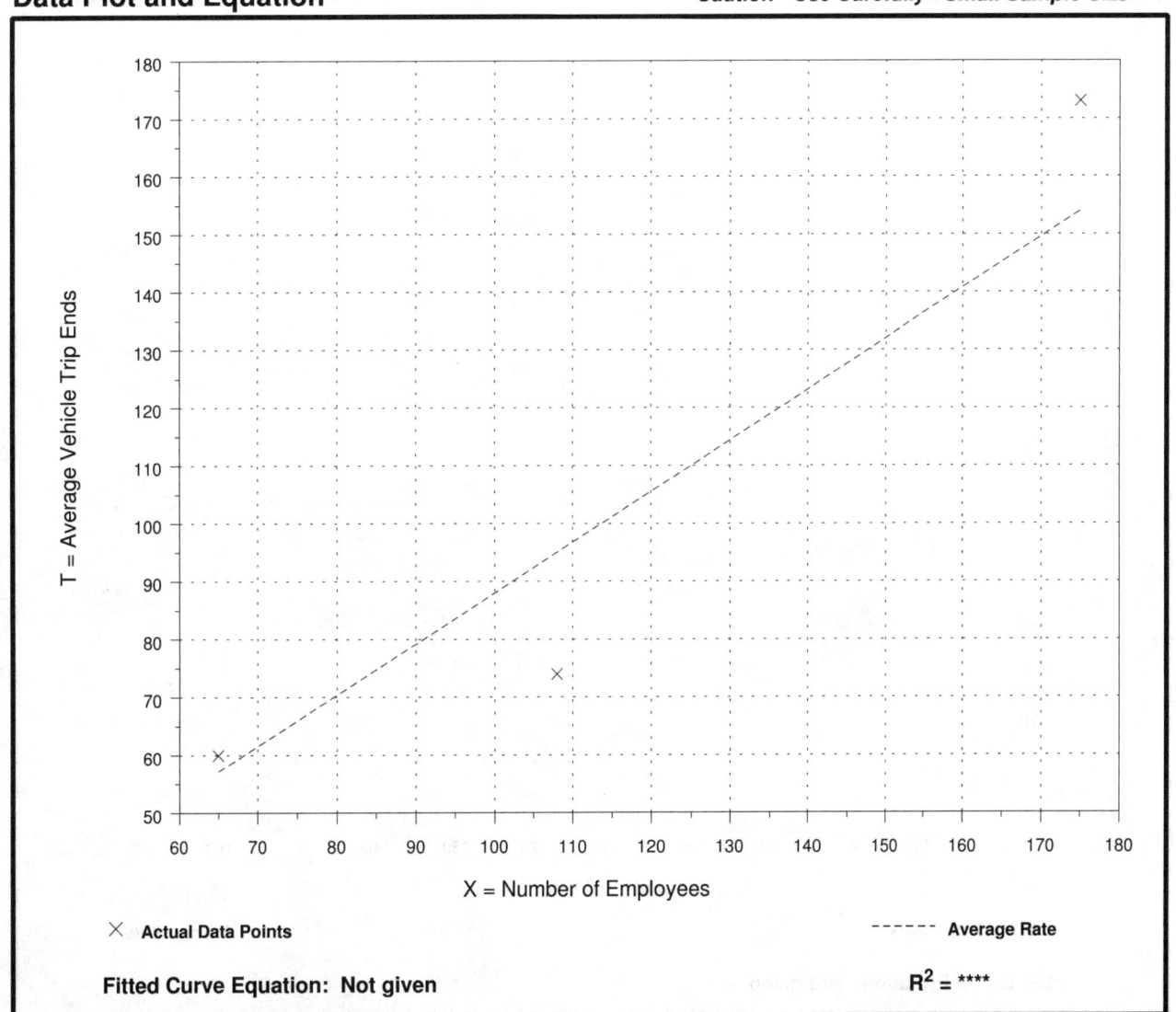

Fitted Curve Equation: Not given $R^2 = ****$

Medical-Dental Office Building
(720)

Average Vehicle Trip Ends vs: Employees
On a: Sunday

Number of Studies: 3
Avg. Number of Employees: 116
Directional Distribution: 50% entering, 50% exiting

Trip Generation per Employee

Average Rate	Range of Rates	Standard Deviation
0.64	0.39 - 1.38	0.88

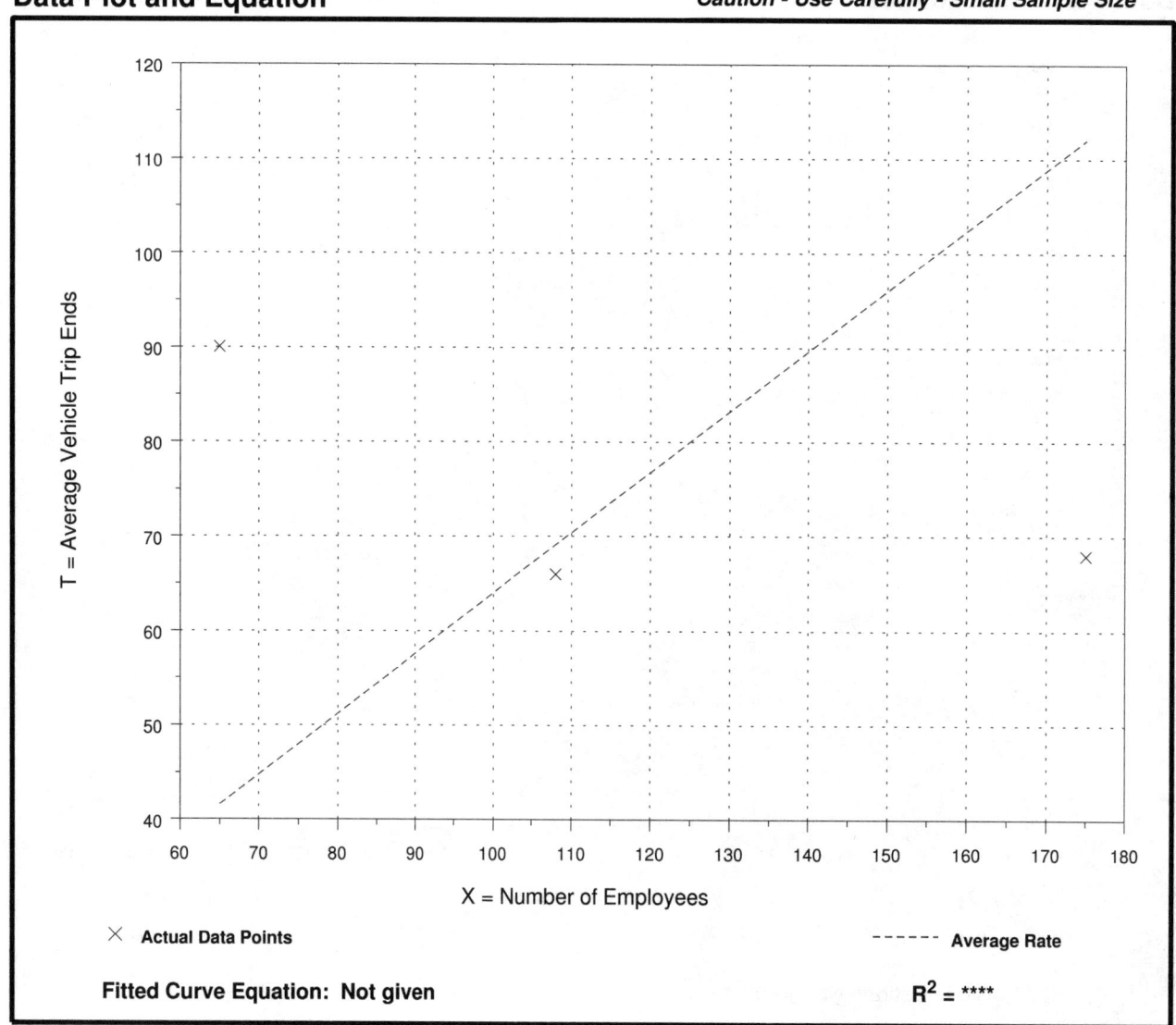

Data Plot and Equation

Caution - Use Carefully - Small Sample Size

Fitted Curve Equation: Not given $R^2 = ****$

Medical-Dental Office Building
(720)

Average Vehicle Trip Ends vs: Employees
On a: Sunday,
Peak Hour of Generator

Number of Studies: 2
Avg. Number of Employees: 142
Directional Distribution: 52% entering, 48% exiting

Trip Generation per Employee

Average Rate	Range of Rates	Standard Deviation
0.10	0.07 - 0.14	*

Data Plot and Equation

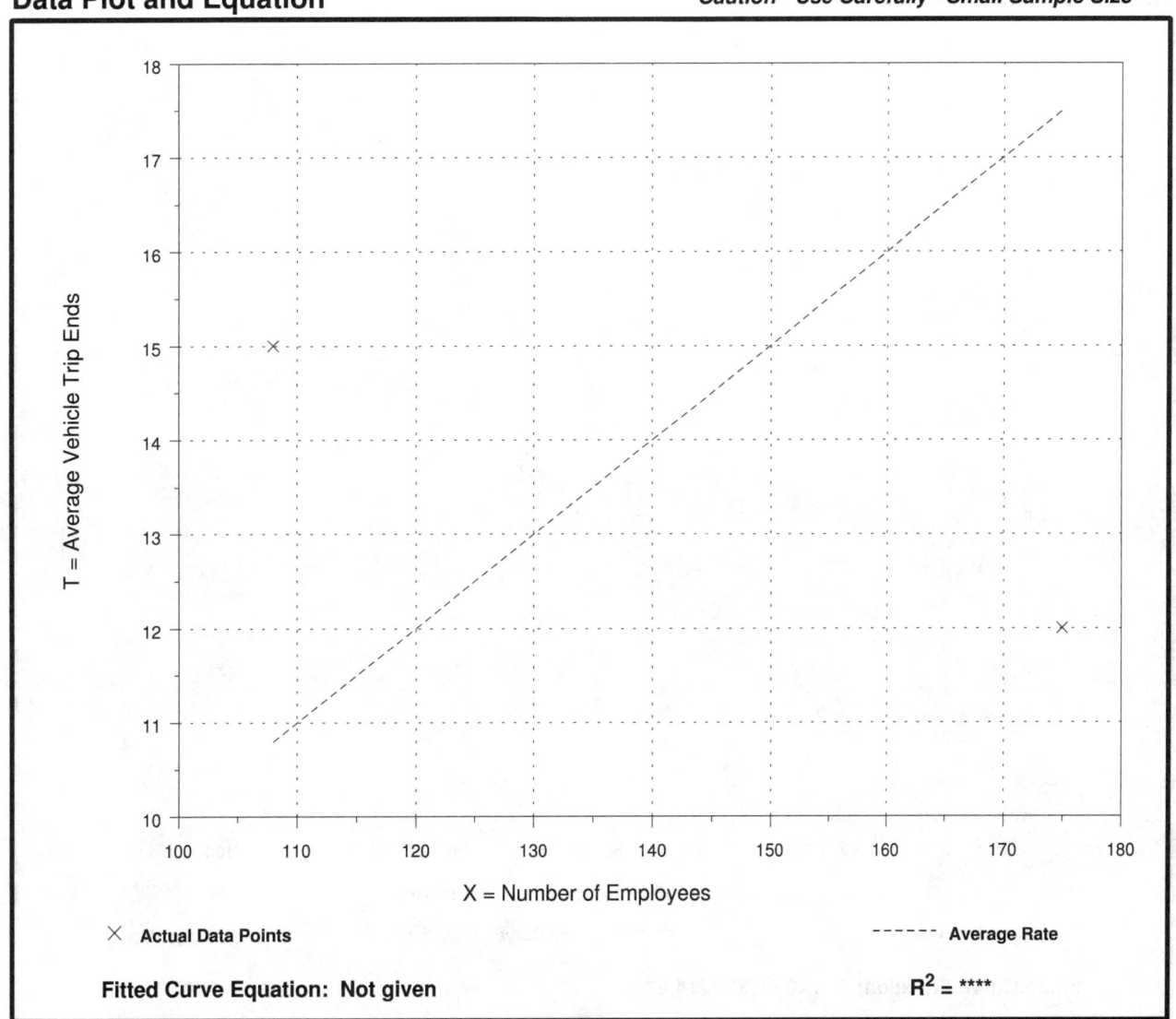

Caution - Use Carefully - Small Sample Size

Fitted Curve Equation: Not given $R^2 = ****$

Medical-Dental Office Building
(720)

Average Vehicle Trip Ends vs: 1000 Sq. Feet Gross Floor Area
On a: Weekday

Number of Studies: 10
Average 1000 Sq. Feet GFA: 45
Directional Distribution: 50% entering, 50% exiting

Trip Generation per 1000 Sq. Feet Gross Floor Area

Average Rate	Range of Rates	Standard Deviation
36.13	23.16 - 50.51	10.18

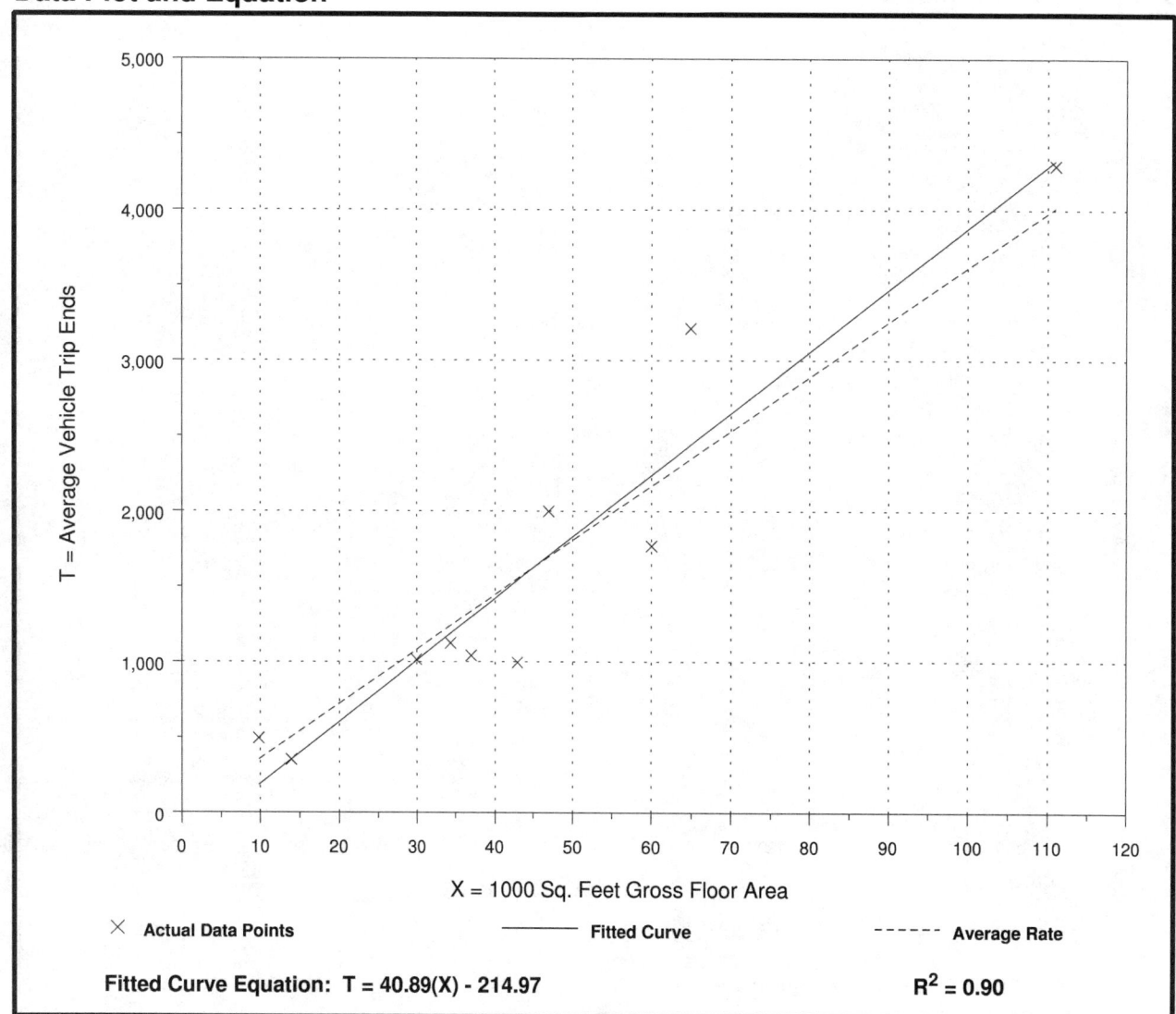

Data Plot and Equation

Fitted Curve Equation: T = 40.89(X) - 214.97 $R^2 = 0.90$

Medical-Dental Office Building
(720)

Average Vehicle Trip Ends vs: 1000 Sq. Feet Gross Floor Area
On a: Weekday,
Peak Hour of Adjacent Street Traffic,
One Hour Between 7 and 9 a.m.

Number of Studies: 23
Average 1000 Sq. Feet GFA: 41
Directional Distribution: 79% entering, 21% exiting

Trip Generation per 1000 Sq. Feet Gross Floor Area

Average Rate	Range of Rates	Standard Deviation
2.39	0.85 - 4.79	1.89

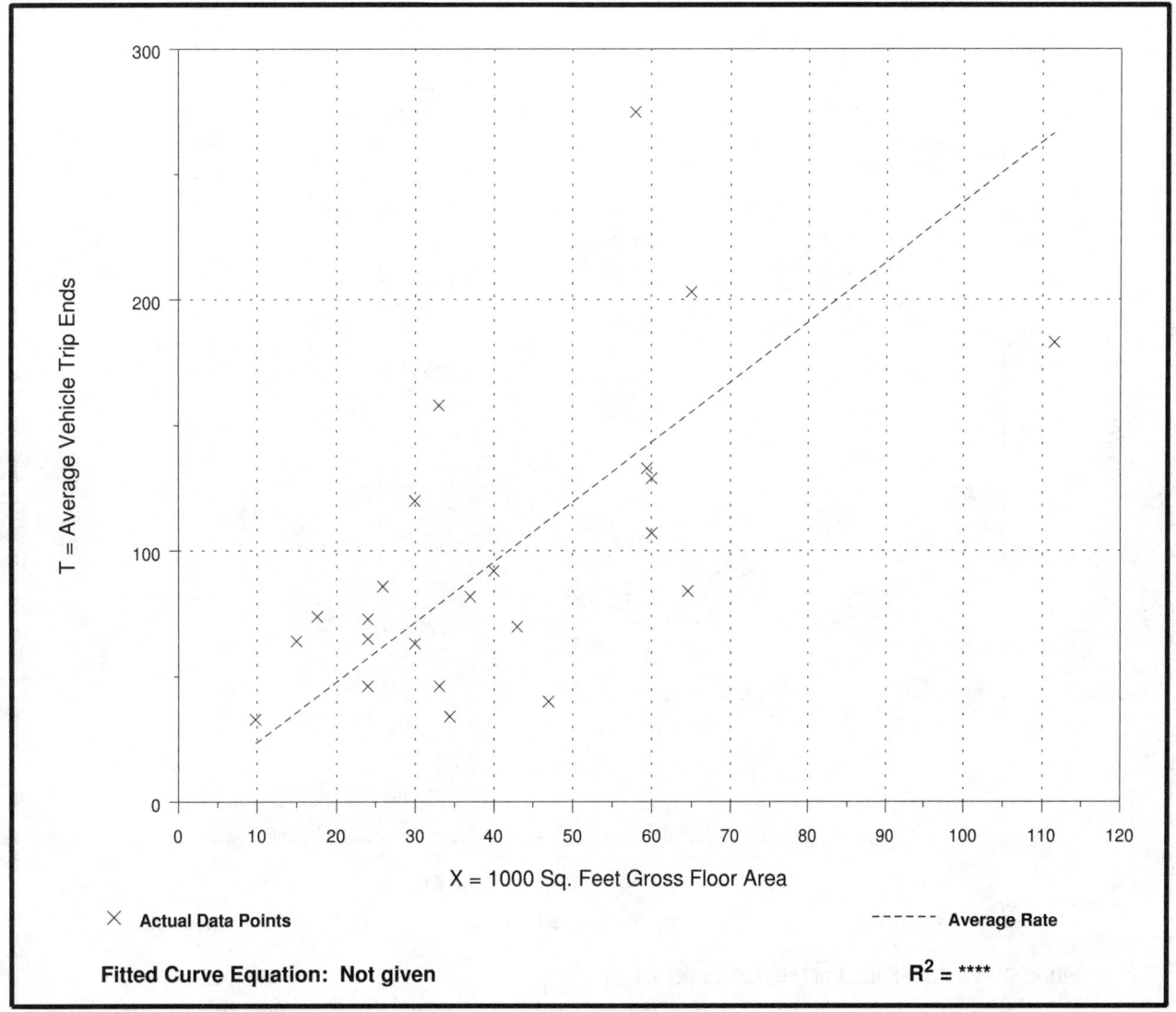

Data Plot and Equation

Fitted Curve Equation: Not given $R^2 = ****$

Medical-Dental Office Building
(720)

Average Vehicle Trip Ends vs: 1000 Sq. Feet Gross Floor Area
On a: Weekday,
Peak Hour of Adjacent Street Traffic,
One Hour Between 4 and 6 p.m.

Number of Studies: 43
Average 1000 Sq. Feet GFA: 31
Directional Distribution: 28% entering, 72% exiting

Trip Generation per 1000 Sq. Feet Gross Floor Area

Average Rate	Range of Rates	Standard Deviation
3.57	0.97 - 8.86	2.47

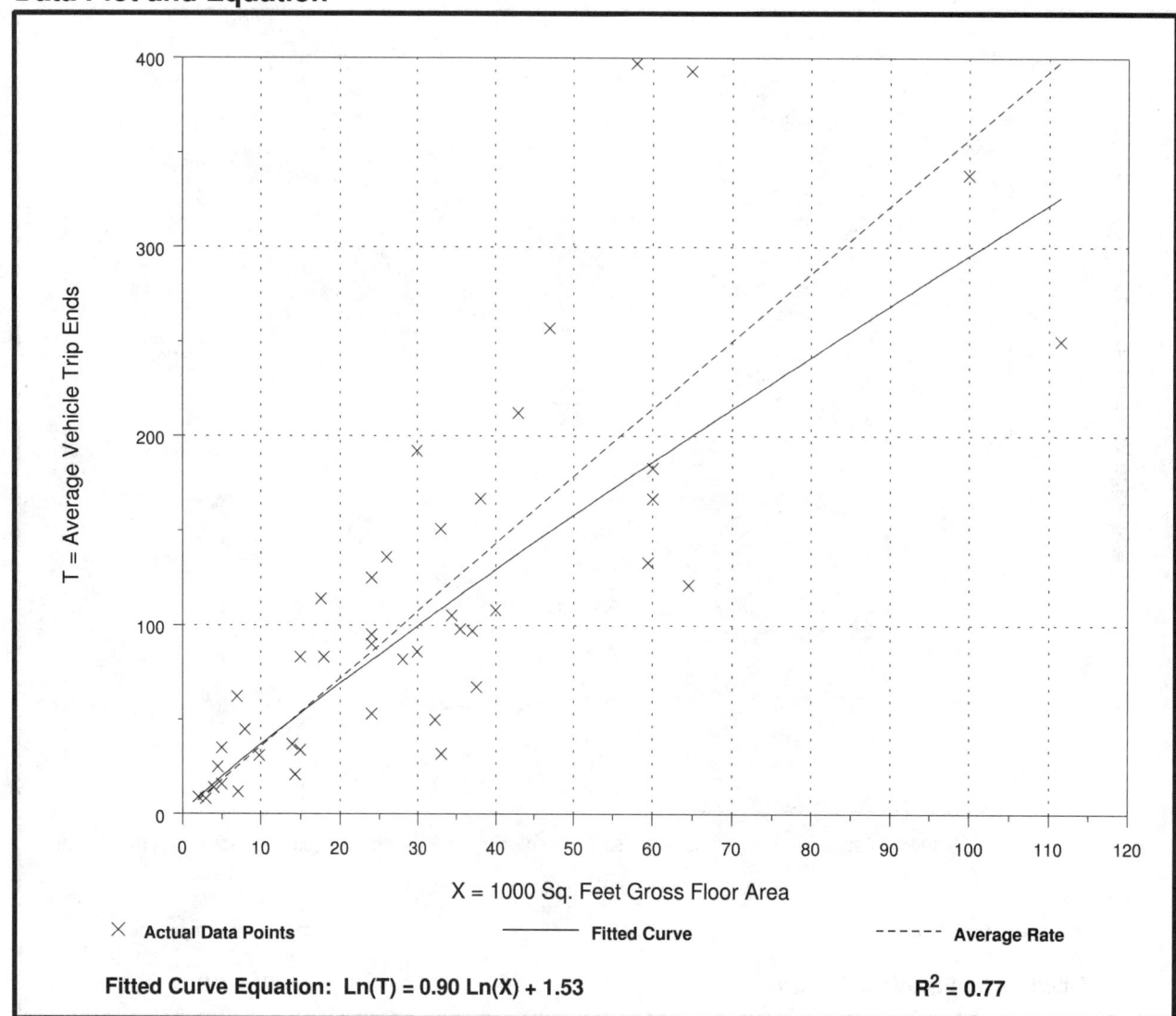

Data Plot and Equation

Fitted Curve Equation: $Ln(T) = 0.90\, Ln(X) + 1.53$ $R^2 = 0.77$

Medical-Dental Office Building
(720)

Average Vehicle Trip Ends vs: 1000 Sq. Feet Gross Floor Area
On a: Weekday,
A.M. Peak Hour of Generator

Number of Studies: 17
Average 1000 Sq. Feet GFA: 41
Directional Distribution: 67% entering, 33% exiting

Trip Generation per 1000 Sq. Feet Gross Floor Area

Average Rate	Range of Rates	Standard Deviation
3.50	1.21 - 7.49	2.35

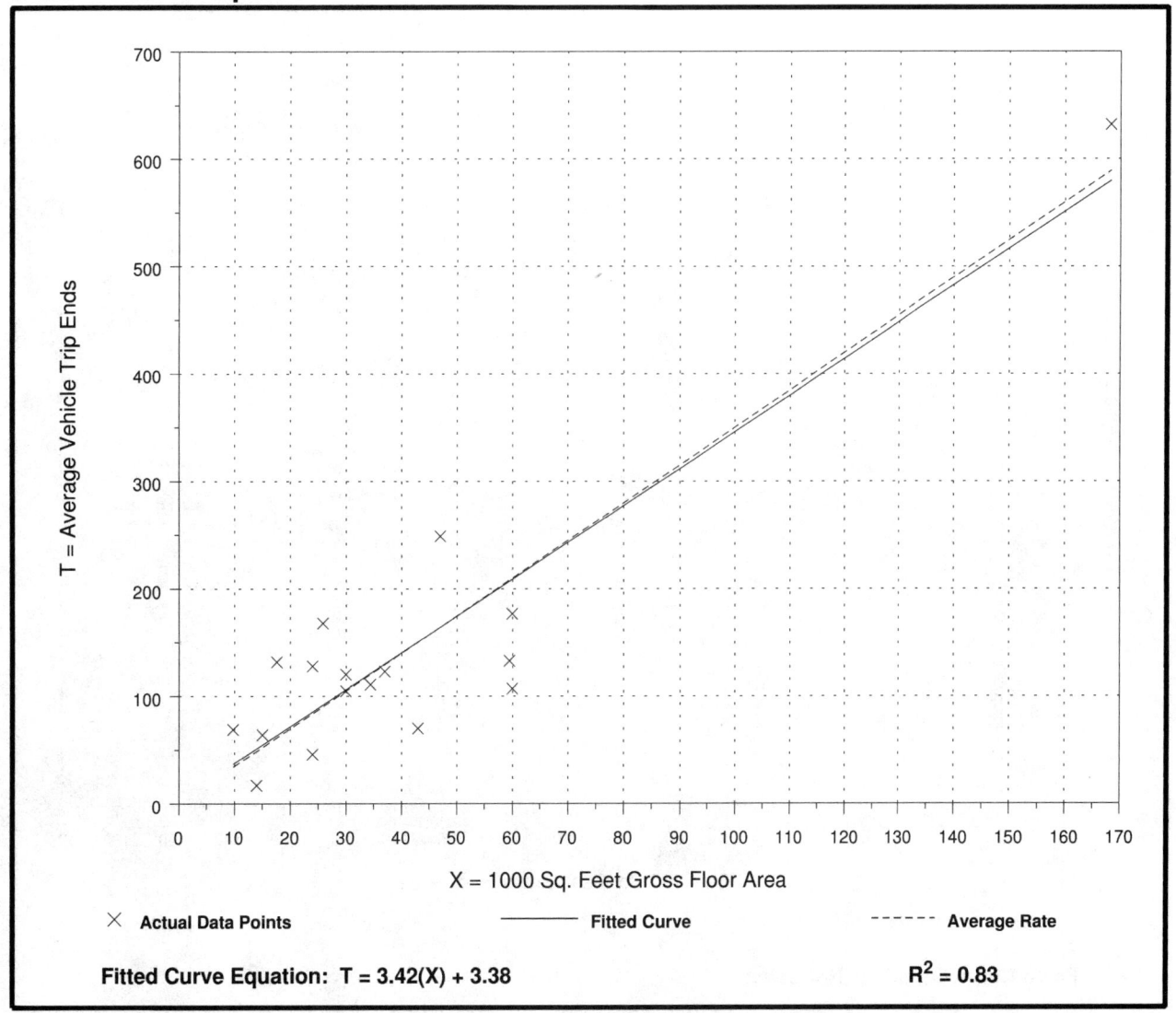

Data Plot and Equation

Fitted Curve Equation: T = 3.42(X) + 3.38 $R^2 = 0.83$

Medical-Dental Office Building
(720)

Average Vehicle Trip Ends vs: 1000 Sq. Feet Gross Floor Area
On a: Weekday,
P.M. Peak Hour of Generator

Number of Studies: 22
Average 1000 Sq. Feet GFA: 33
Directional Distribution: 39% entering, 61% exiting

Trip Generation per 1000 Sq. Feet Gross Floor Area

Average Rate	Range of Rates	Standard Deviation
4.27	2.21 - 7.60	2.50

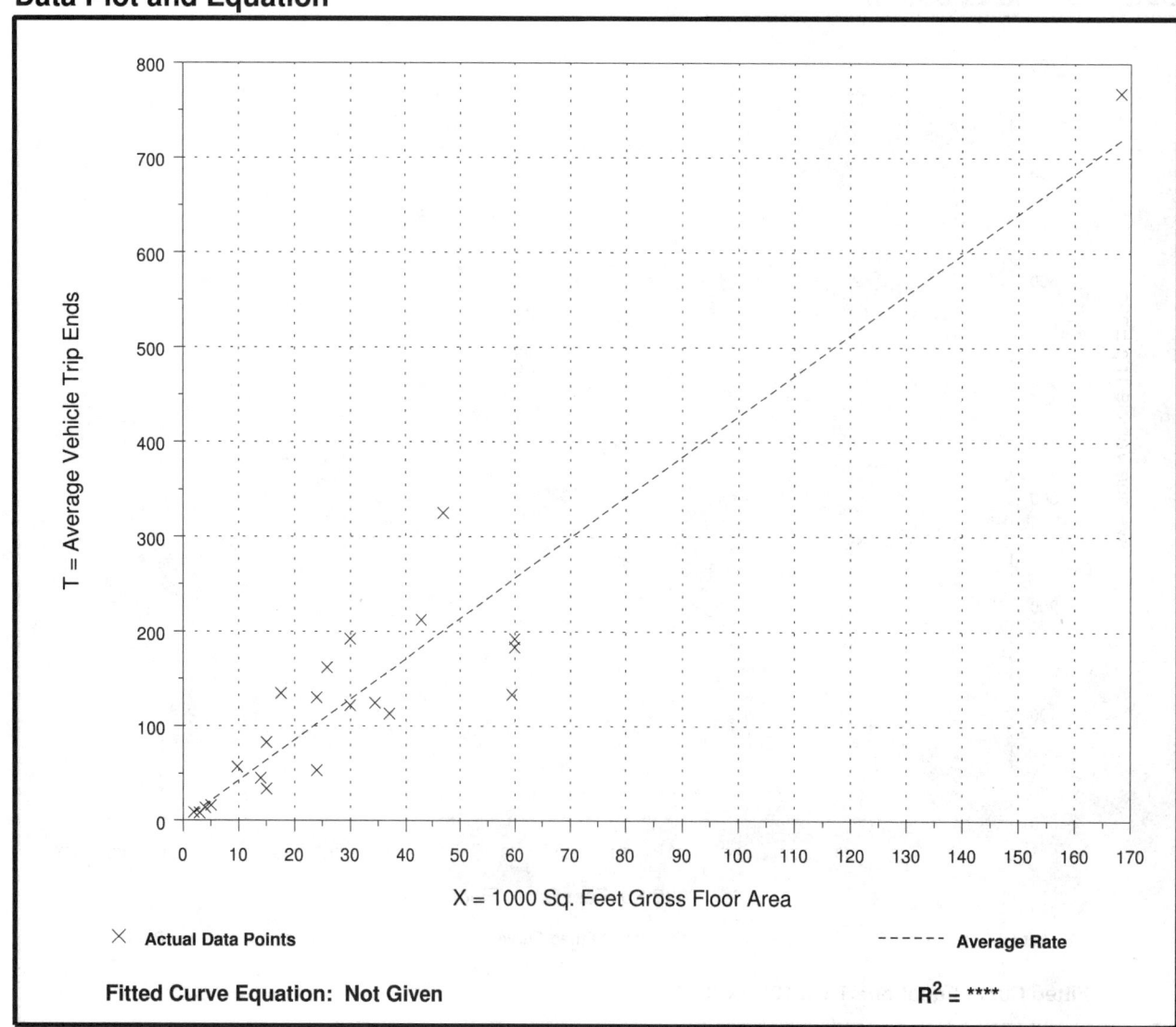

Data Plot and Equation

Fitted Curve Equation: Not Given $R^2 = ****$

Medical-Dental Office Building
(720)

Average Vehicle Trip Ends vs: 1000 Sq. Feet Gross Floor Area
On a: Saturday

Number of Studies: 5
Average 1000 Sq. Feet GFA: 44
Directional Distribution: 50% entering, 50% exiting

Trip Generation per 1000 Sq. Feet Gross Floor Area

Average Rate	Range of Rates	Standard Deviation
8.96	1.10 - 21.93	9.17

Data Plot and Equation

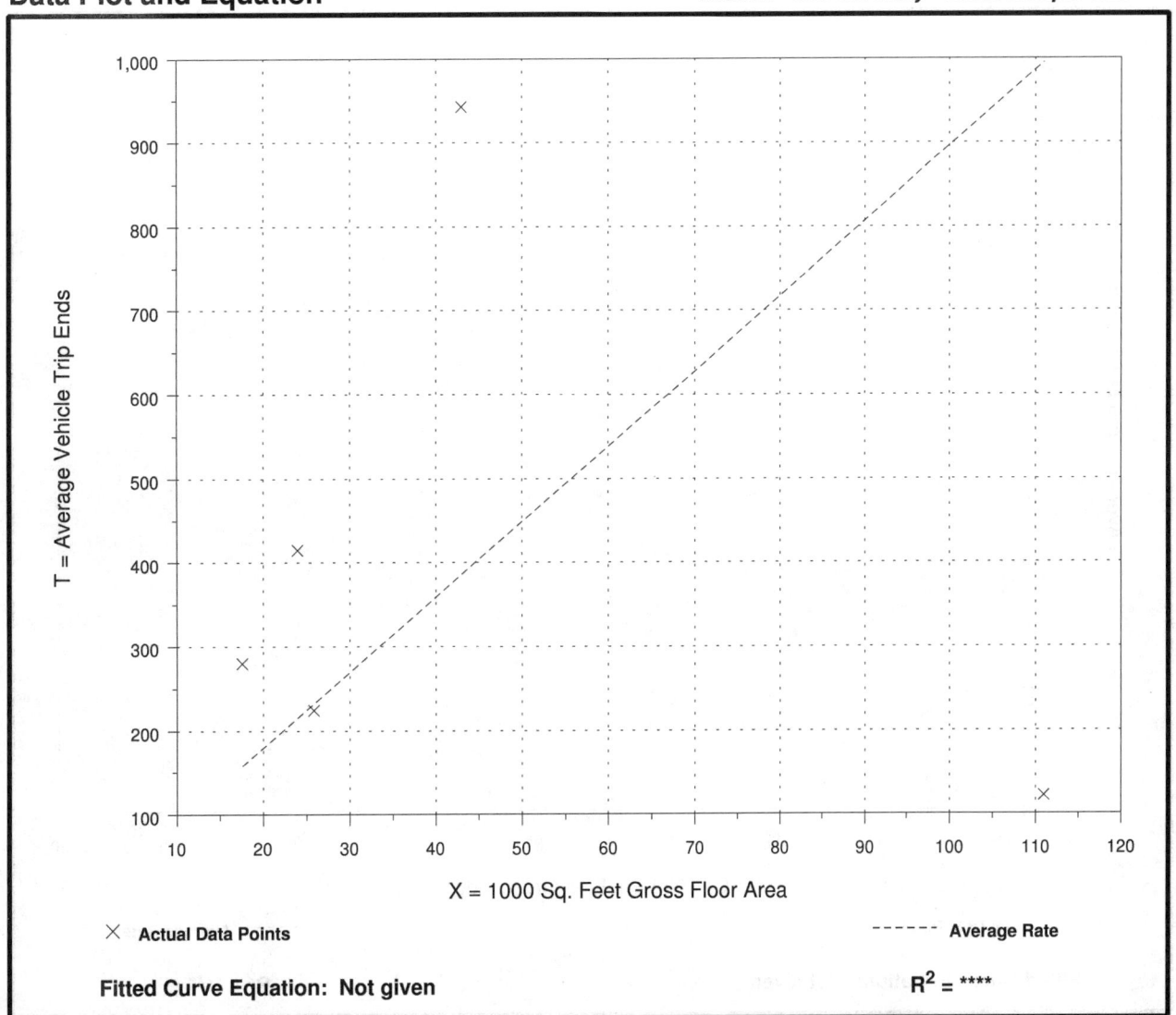

Caution - Use Carefully - Small Sample Size

X Actual Data Points   ----- Average Rate

Fitted Curve Equation: Not given R^2 = ****

Trip Generation, 9th Edition • Institute of Transportation Engineers

Medical-Dental Office Building
(720)

Average Vehicle Trip Ends vs: 1000 Sq. Feet Gross Floor Area
On a: Saturday,
Peak Hour of Generator

Number of Studies: 3
Average 1000 Sq. Feet GFA: 28
Directional Distribution: 57% entering, 43% exiting

Trip Generation per 1000 Sq. Feet Gross Floor Area

Average Rate	Range of Rates	Standard Deviation
3.63	3.08 - 4.02	1.93

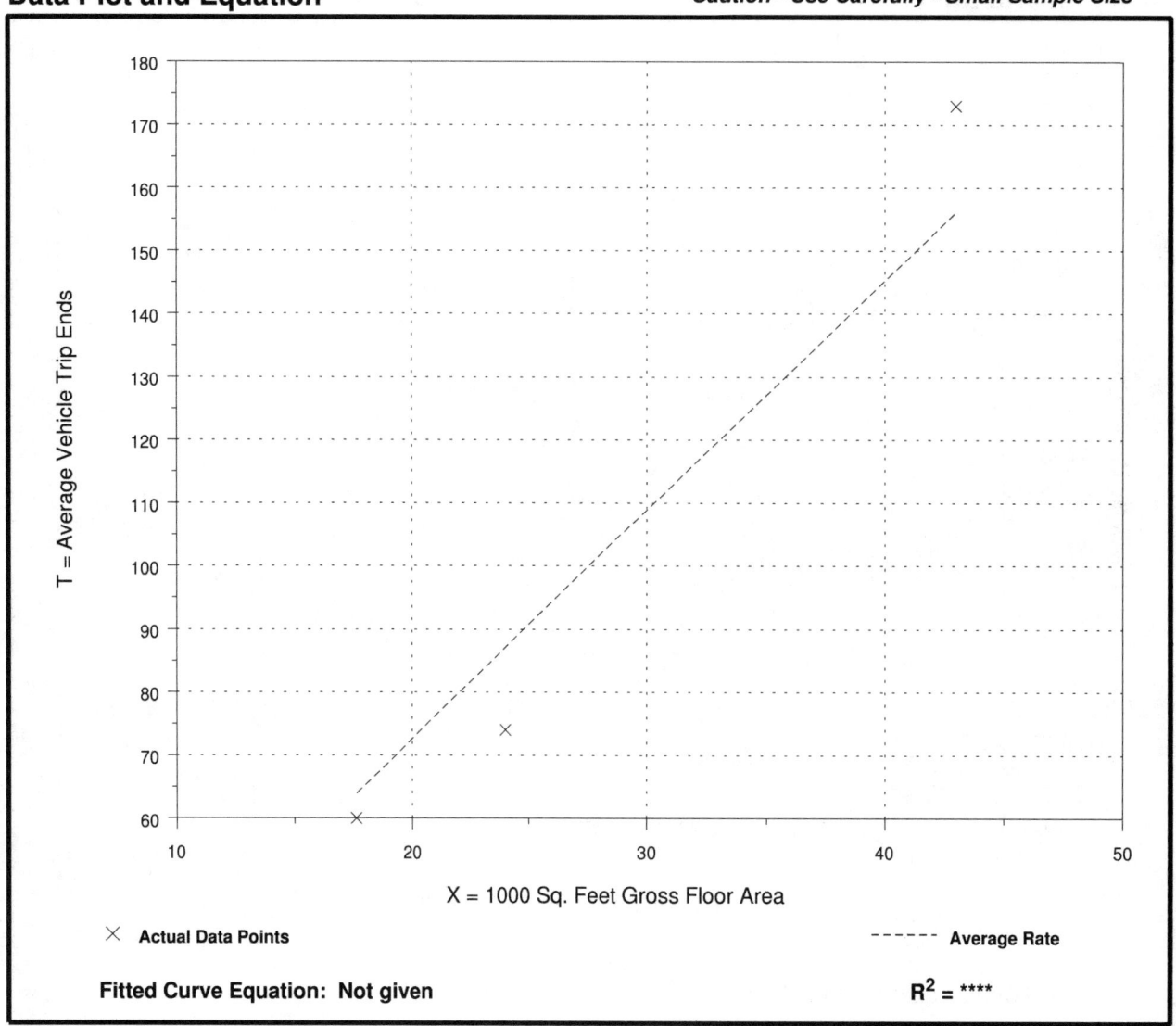

Data Plot and Equation

Caution - Use Carefully - Small Sample Size

Fitted Curve Equation: Not given $R^2 = ****$

Medical-Dental Office Building
(720)

Average Vehicle Trip Ends vs: 1000 Sq. Feet Gross Floor Area
On a: Sunday

Number of Studies: 4
Average 1000 Sq. Feet GFA: 49
Directional Distribution: 50% entering, 50% exiting

Trip Generation per 1000 Sq. Feet Gross Floor Area

Average Rate	Range of Rates	Standard Deviation
1.55	0.71 - 5.11	1.80

Data Plot and Equation

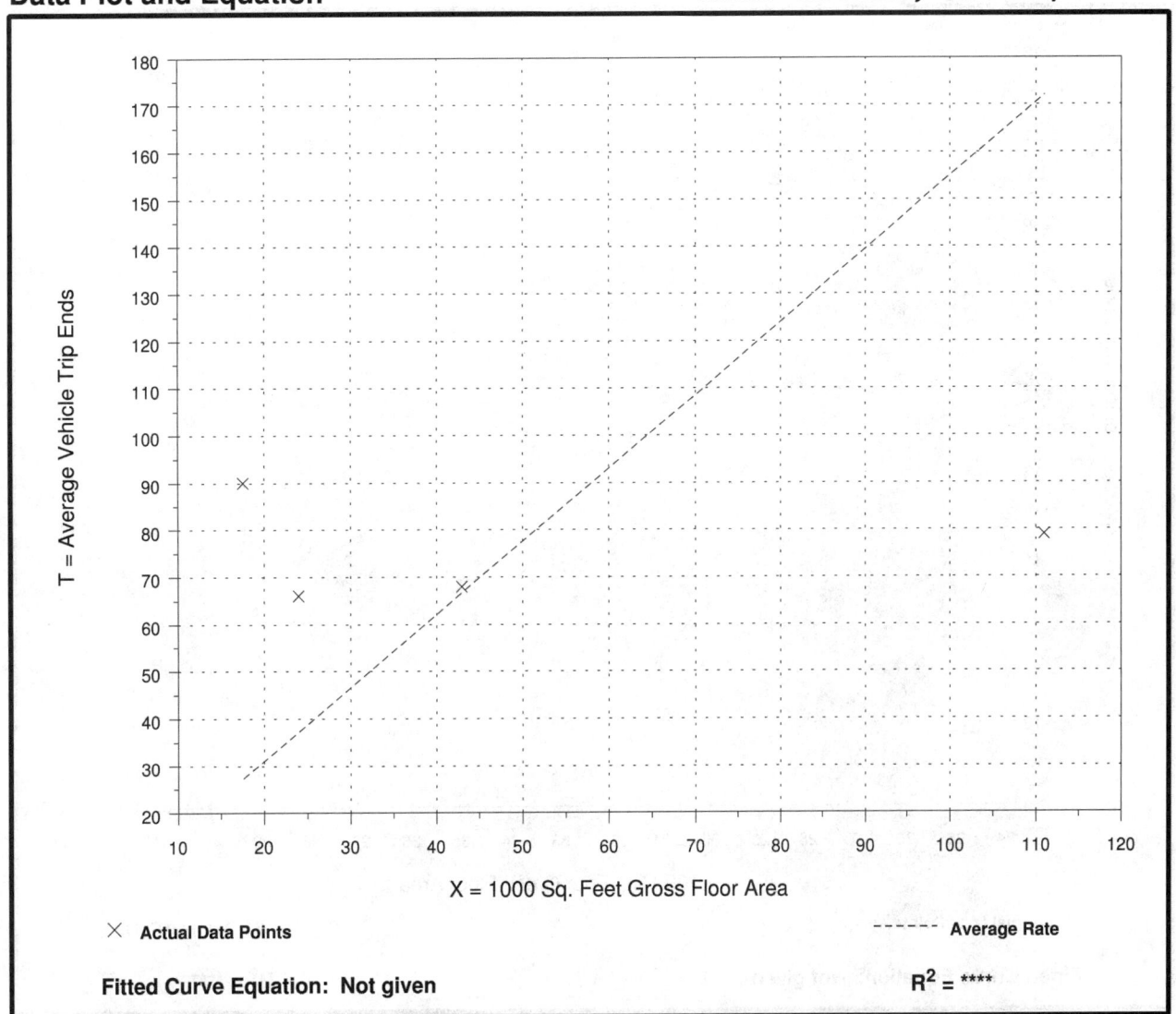

Caution - Use Carefully - Small Sample Size

[Plot: T = Average Vehicle Trip Ends vs. X = 1000 Sq. Feet Gross Floor Area]

× Actual Data Points
----- Average Rate

Fitted Curve Equation: Not given $R^2 = $ ****

Medical-Dental Office Building
(720)

Average Vehicle Trip Ends vs: 1000 Sq. Feet Gross Floor Area
On a: Sunday,
Peak Hour of Generator

Number of Studies: 2
Average 1000 Sq. Feet GFA: 34
Directional Distribution: 52% entering, 48% exiting

Trip Generation per 1000 Sq. Feet Gross Floor Area

Average Rate	Range of Rates	Standard Deviation
0.40	0.28 - 0.63	*

Data Plot and Equation

Caution - Use Carefully - Small Sample Size

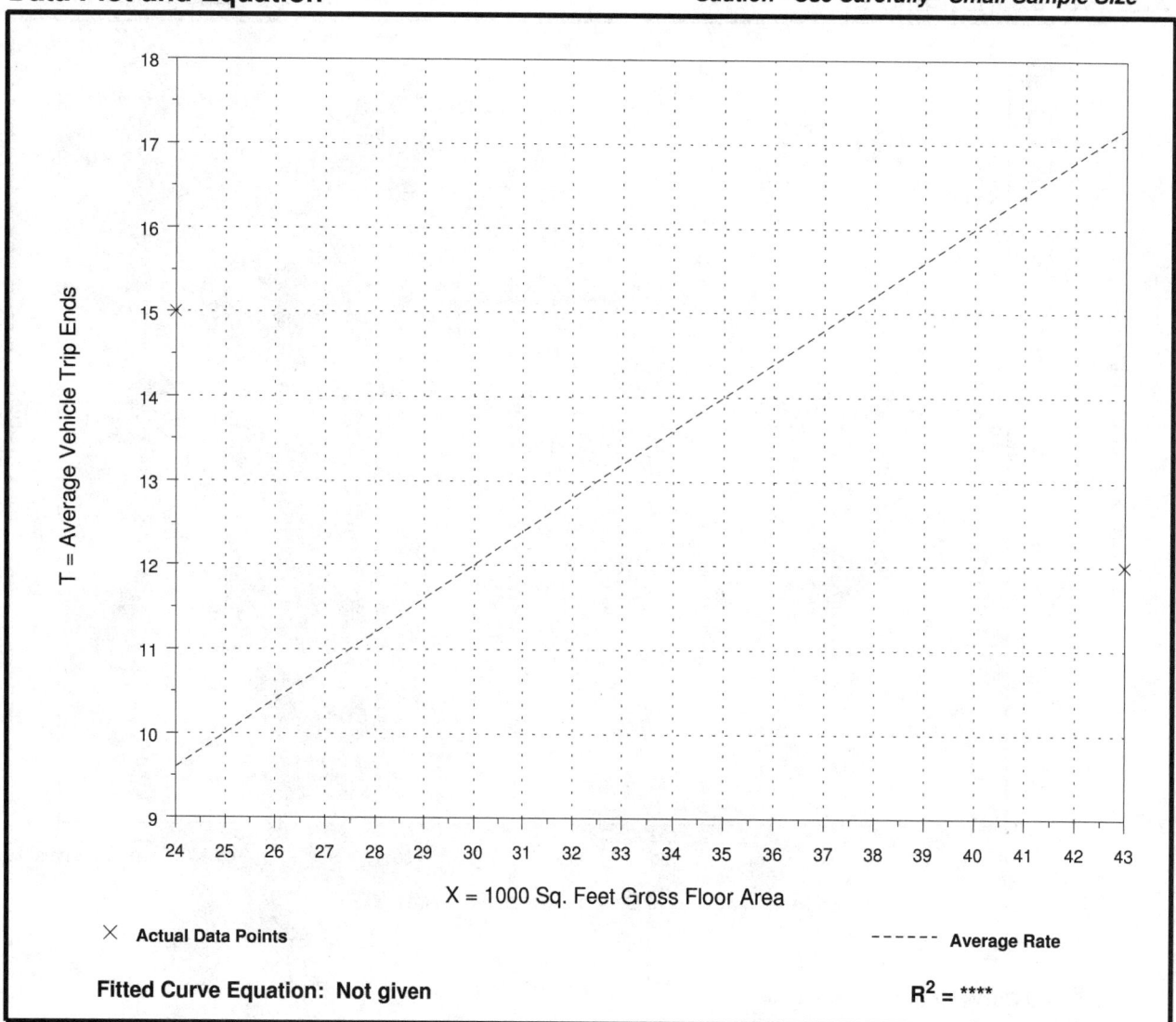

Fitted Curve Equation: Not given $R^2 = ****$

Land Use: 730
Government Office Building

Description

A government office building is an individual building containing either the entire function or simply one agency of a city, county, state, federal, or other governmental unit. This type of building differs from a government office complex (Land Use 733) in that it is not a group of buildings that are interconnected by pedestrian walkways.

Additional Data

Peak hours of the generator—
    The weekday A.M. peak hour typically coincided with the peak hour of the adjacent street traffic.
    The weekday P.M. peak hour was between 1:00 p.m. and 2:00 p.m.

The sites were surveyed in 1970 and 2002 in California and Oregon, respectively. Two of the sites were city halls.

Source Numbers

11, 579

Land Use: 730
Government Office Building
Independent Variables with One Observation

The following trip generation data are for independent variables with only one observation. This information is shown in this table only; there are no related plots for these data.

Users are cautioned to use data with care because of the small sample size.

Independent Variable	**Average Trip Generation Rate**	**Size of Independent Variable**	**Number of Studies**	**Directional Distribution**
Employees				
Weekday	11.95	102	1	50% entering, 50% exiting
Weekday A.M. Peak Hour of Adjacent Street Traffic	1.02	102	1	84% entering, 16% exiting
Weekday A.M. Peak Hour of Generator	1.02	102	1	84% entering, 16% exiting
Weekday P.M. Peak Hour of Generator	1.91	102	1	74% entering, 26% exiting
1,000 Square Feet Gross Floor Area				
Weekday	68.93	18	1	50% entering, 50% exiting
Weekday A.M Peak Hour of Adjacent Street Traffic	5.88	18	1	84% entering, 16% exiting
Weekday A.M. Peak Hour of Generator	5.88	18	1	84% entering, 16% exiting
Weekday P.M. Peak Hour of Generator	11.03	18	1	74% entering, 26% exiting

Government Office Building
(730)

Average Vehicle Trip Ends vs: 1000 Sq. Feet Gross Floor Area
On a: Weekday,
Peak Hour of Adjacent Street Traffic,
One Hour Between 4 and 6 p.m.

Number of Studies: 2
Average 1000 Sq. Feet GFA: 52
Directional Distribution: 31% entering, 69% exiting

Trip Generation per 1000 Sq. Feet Gross Floor Area

Average Rate	Range of Rates	Standard Deviation
1.21	1.17 - 1.22	*

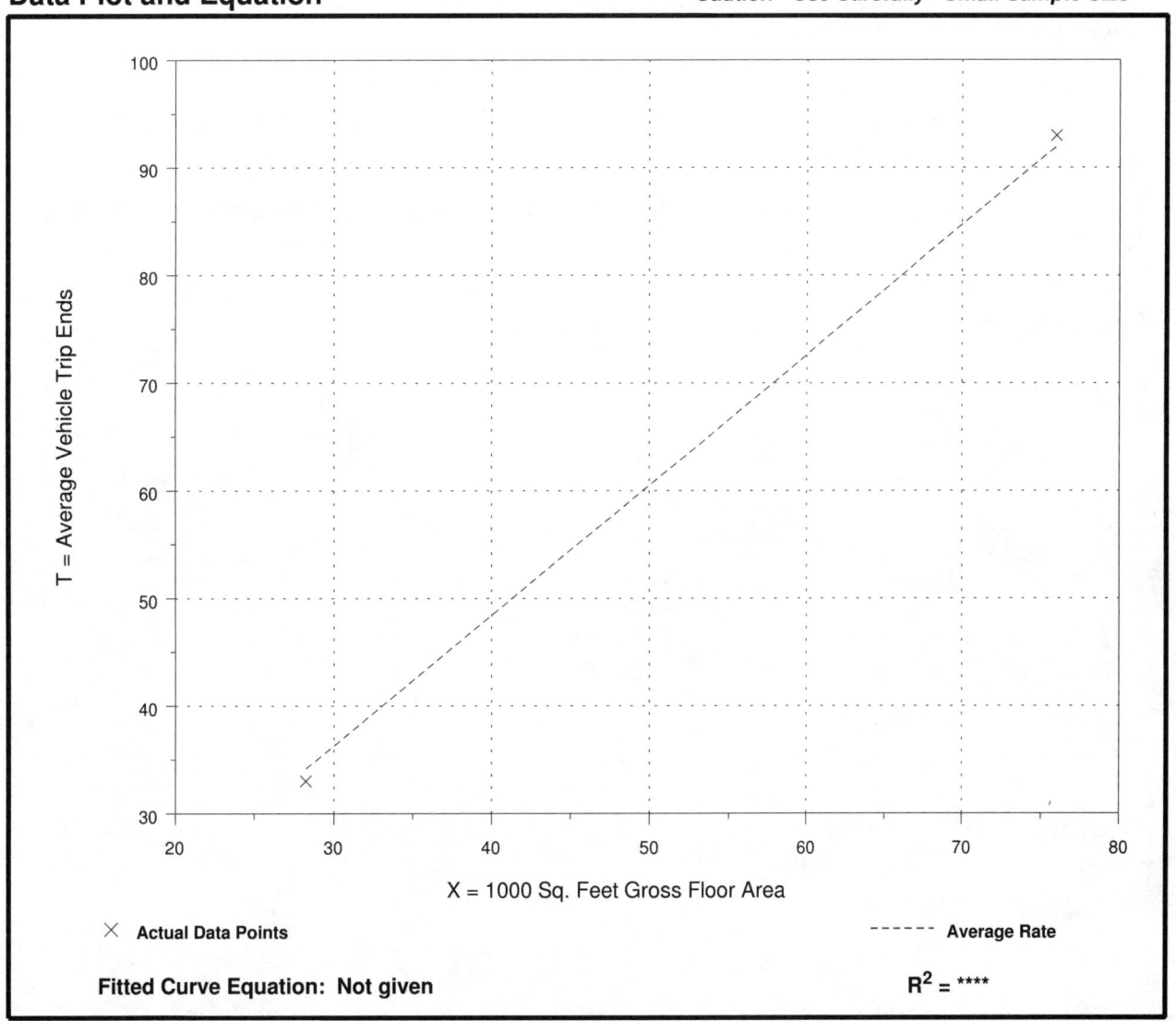

Data Plot and Equation

Caution - Use Carefully - Small Sample Size

× Actual Data Points
----- Average Rate

Fitted Curve Equation: Not given $R^2 =$ ****

Trip Generation, 9th Edition • Institute of Transportation Engineers

Land Use: 731
State Motor Vehicles Department

Description

A state motor vehicles department is an office-type building where driver license testing, vehicle registration and other related functions are administered.

Additional Data

Truck trips accounted for 0.44 percent of the weekday traffic at the motor vehicles departments surveyed (range of 0.12 percent to 0.85 percent).

The average vehicle occupancy was 1.38 persons per automobile. The vehicle occupancy rates ranged from 1.30 to 1.48 persons per automobile.

Peak hours of the generator—
　　The weekday peak hours varied between 10:00 a.m. and 4:00 p.m. The Saturday peak hour varied between 10:00 a.m. and 2:00 p.m.

The sites were surveyed between the mid- and late 1970s in California. The sites had an average of 121 parking spaces.

Source Numbers

88, 113

State Motor Vehicles Department
(731)

Average Vehicle Trip Ends vs: Employees
On a: Weekday

Number of Studies: 8
Avg. Number of Employees: 38
Directional Distribution: 50% entering, 50% exiting

Trip Generation per Employee

Average Rate	Range of Rates	Standard Deviation
44.54	28.16 - 60.83	11.02

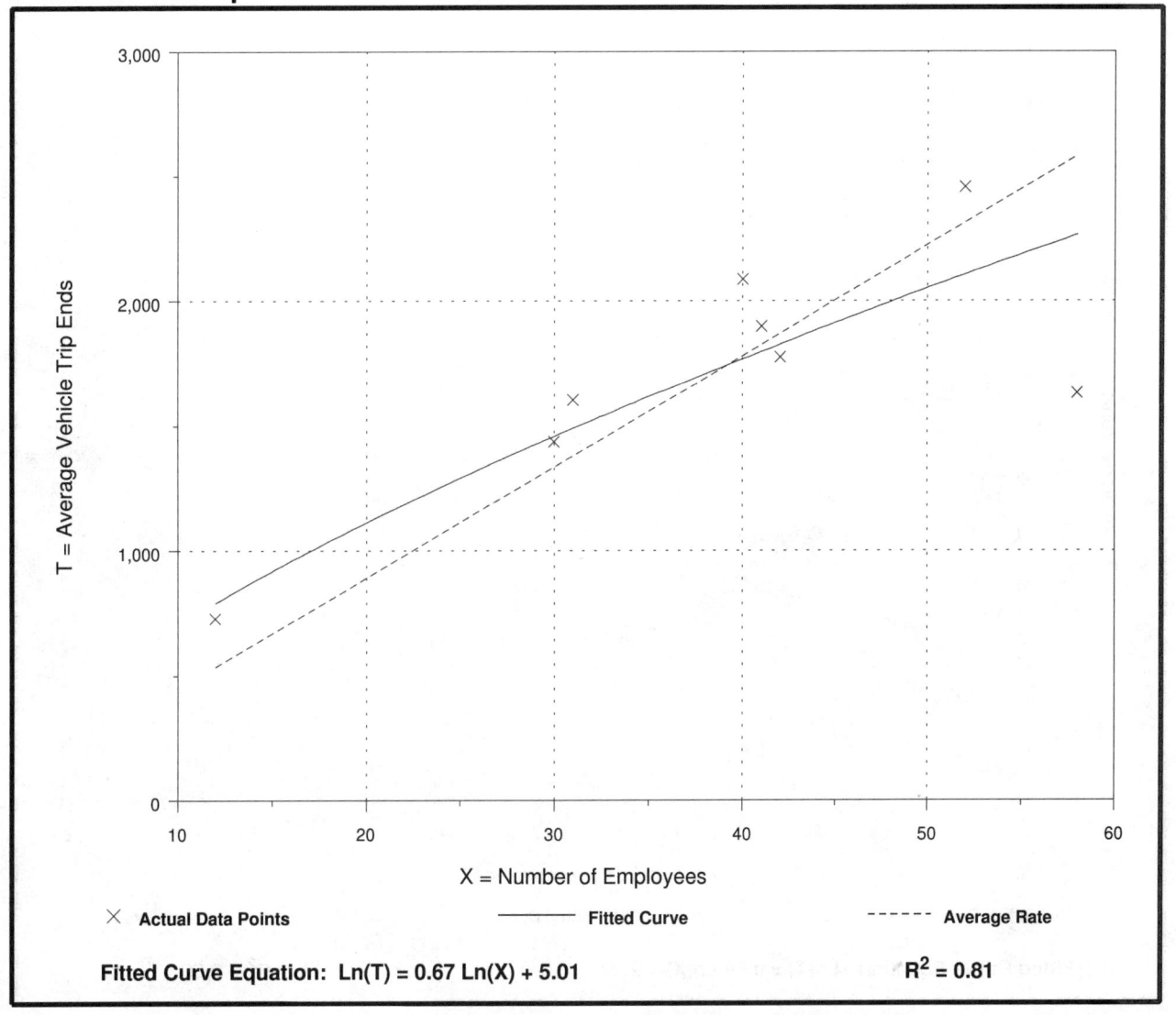

Data Plot and Equation

Fitted Curve Equation: $\text{Ln}(T) = 0.67 \, \text{Ln}(X) + 5.01$ $R^2 = 0.81$

Trip Generation, 9th Edition • Institute of Transportation Engineers

State Motor Vehicles Department
(731)

Average Vehicle Trip Ends vs: Employees
On a: Weekday,
Peak Hour of Adjacent Street Traffic,
One Hour Between 7 and 9 a.m.

Number of Studies: 8
Avg. Number of Employees: 38
Directional Distribution: Not available

Trip Generation per Employee

Average Rate	Range of Rates	Standard Deviation
2.64	1.62 - 4.05	1.81

Data Plot and Equation

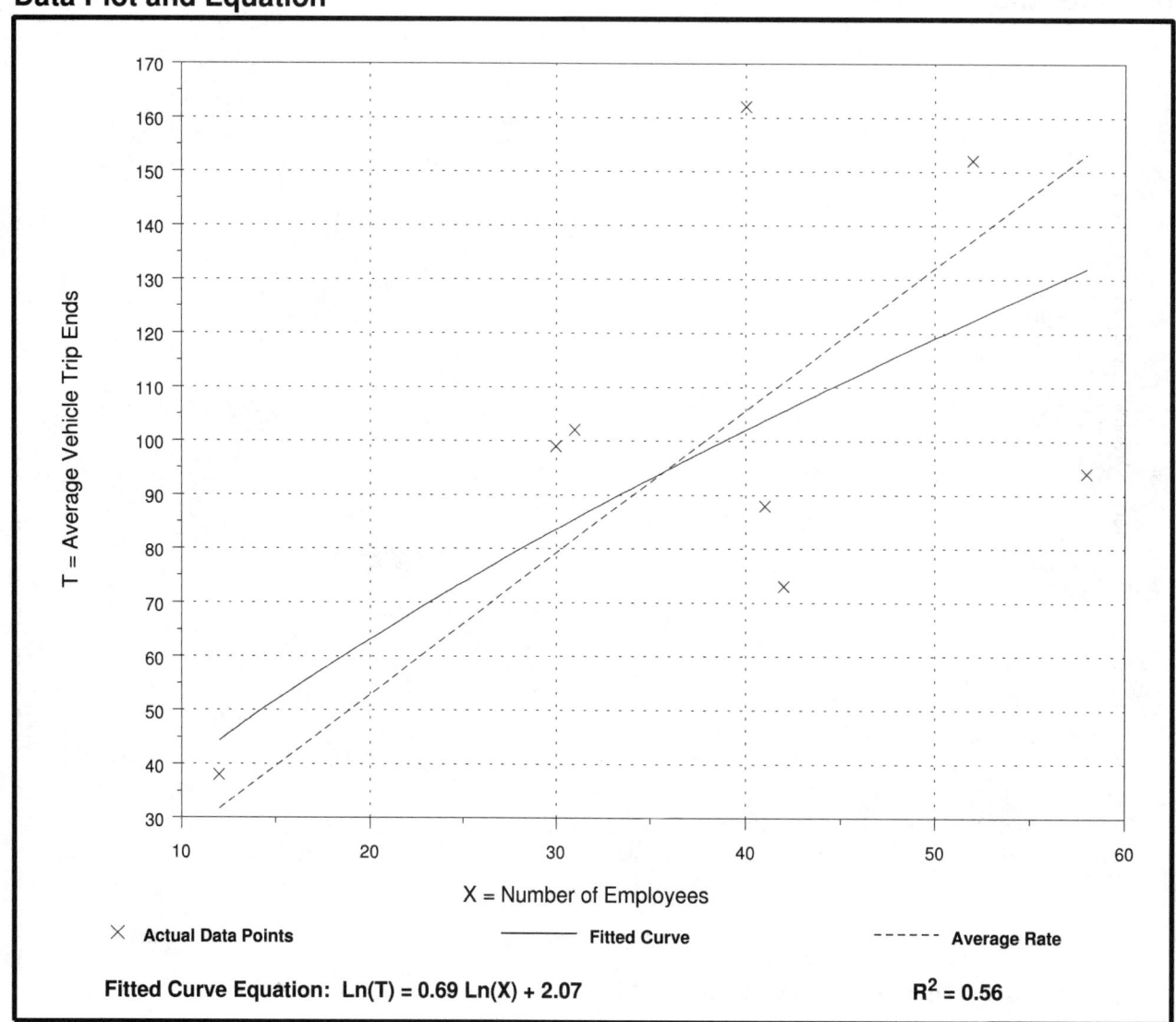

Fitted Curve Equation: $\text{Ln}(T) = 0.69 \, \text{Ln}(X) + 2.07$ $R^2 = 0.56$

State Motor Vehicles Department
(731)

Average Vehicle Trip Ends vs: Employees
On a: Weekday,
Peak Hour of Adjacent Street Traffic,
One Hour Between 4 and 6 p.m.

Number of Studies: 8
Avg. Number of Employees: 38
Directional Distribution: Not available

Trip Generation per Employee

Average Rate	Range of Rates	Standard Deviation
4.58	3.14 - 7.58	2.43

Data Plot and Equation

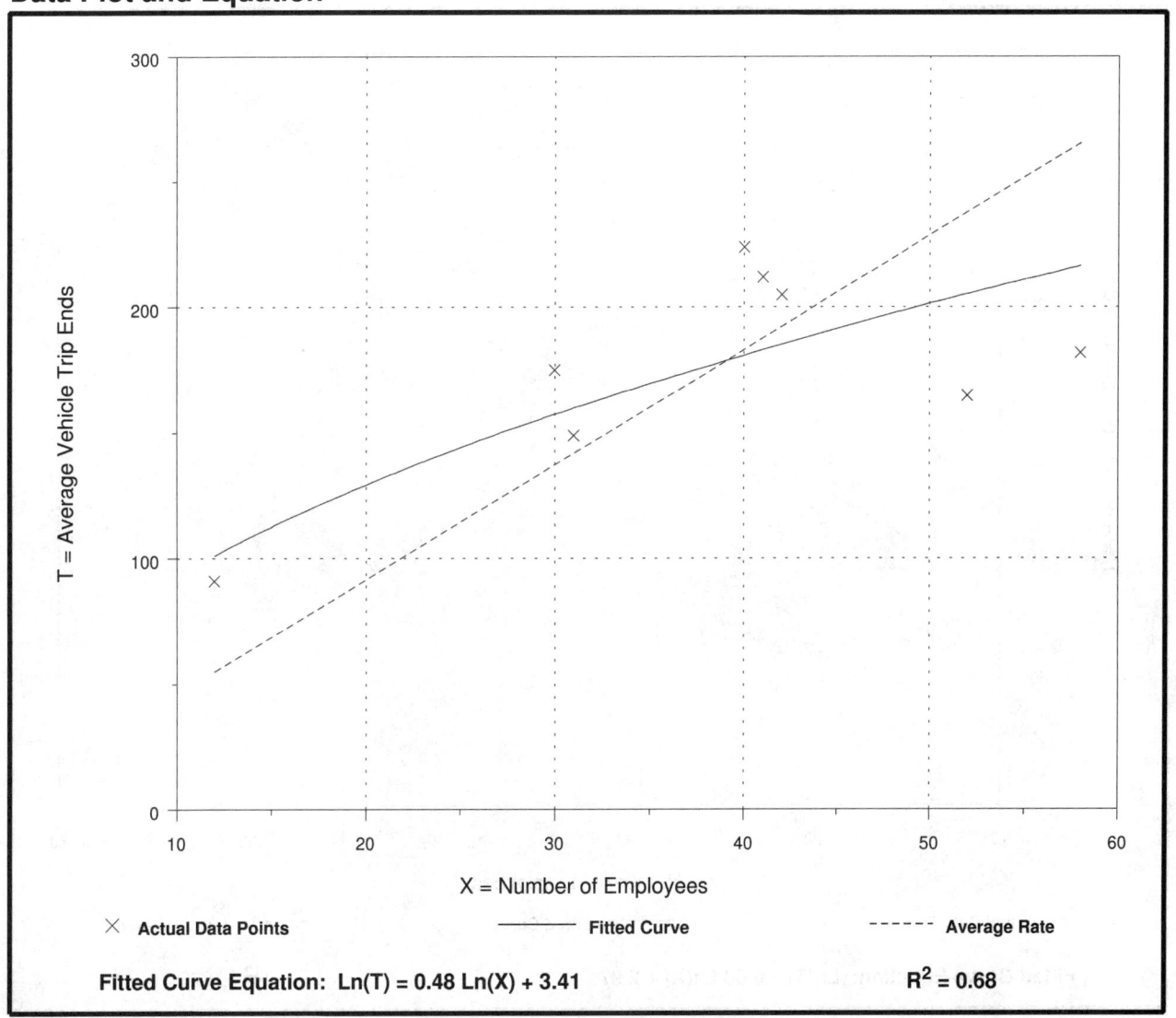

Fitted Curve Equation: $Ln(T) = 0.48\, Ln(X) + 3.41$ $R^2 = 0.68$

State Motor Vehicles Department
(731)

Average Vehicle Trip Ends vs: Employees
On a: Weekday,
A.M. Peak Hour of Generator

Number of Studies: 8
Avg. Number of Employees: 38
Directional Distribution: Not available

Trip Generation per Employee

Average Rate	Range of Rates	Standard Deviation
4.97	3.12 - 7.17	2.46

Data Plot and Equation

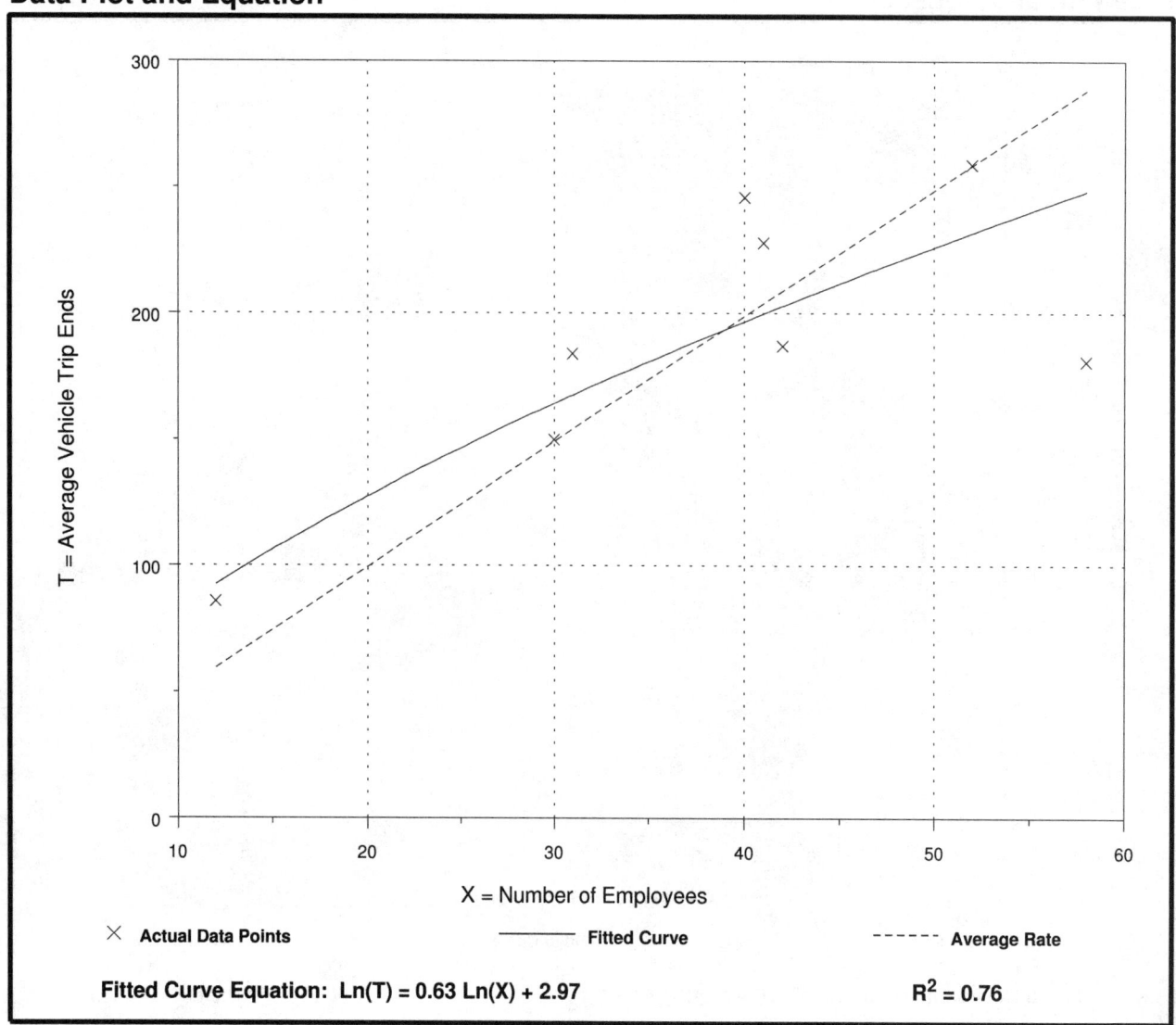

Fitted Curve Equation: $Ln(T) = 0.63\ Ln(X) + 2.97$ $R^2 = 0.76$

State Motor Vehicles Department
(731)

Average Vehicle Trip Ends vs: Employees
On a: Weekday,
P.M. Peak Hour of Generator

Number of Studies: 8
Avg. Number of Employees: 38
Directional Distribution: Not available

Trip Generation per Employee

Average Rate	Range of Rates	Standard Deviation
5.35	3.24 - 7.58	2.55

Data Plot and Equation

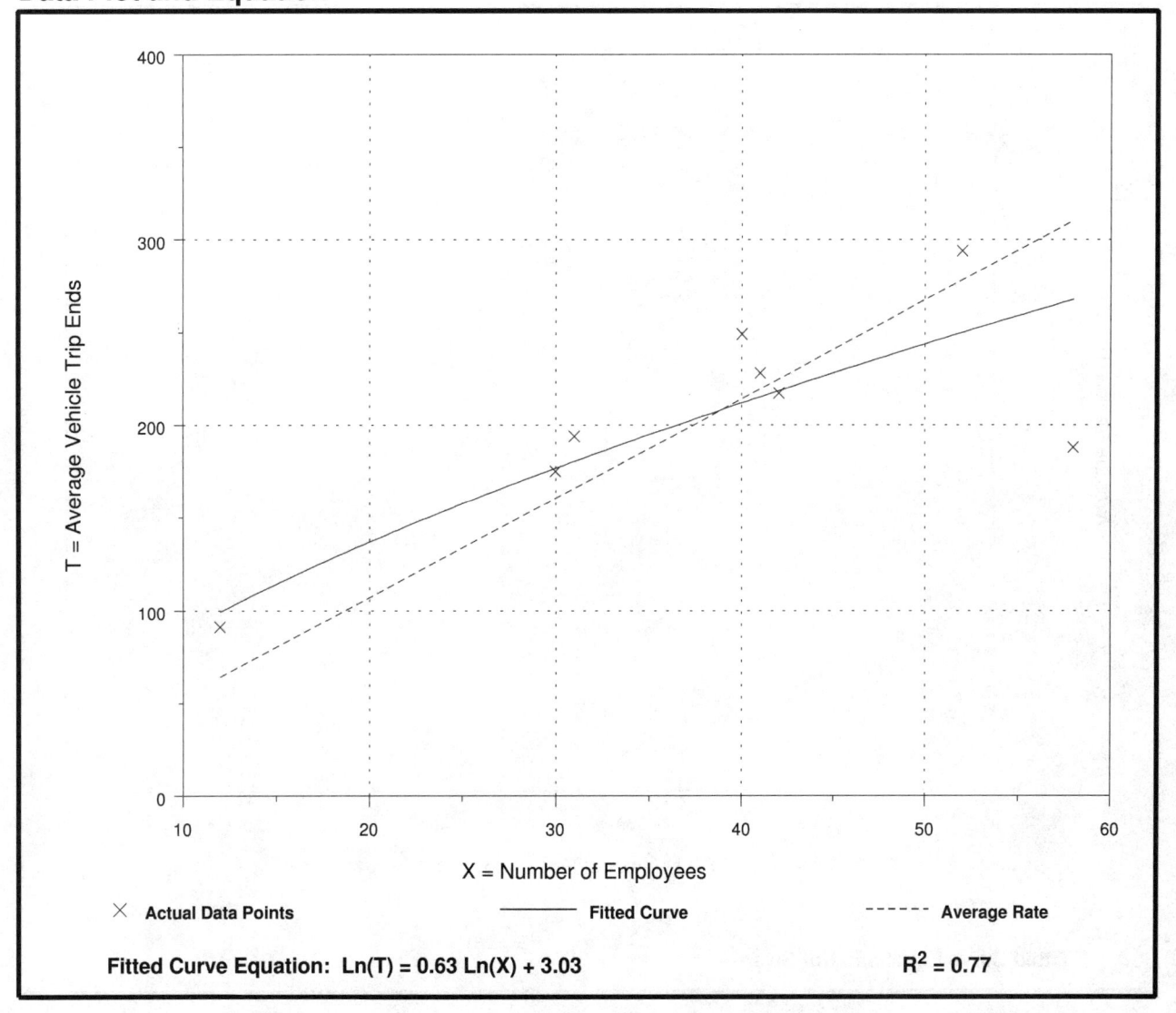

[Plot: T = Average Vehicle Trip Ends vs X = Number of Employees]

× Actual Data Points — Fitted Curve ----- Average Rate

Fitted Curve Equation: $Ln(T) = 0.63 Ln(X) + 3.03$ $R^2 = 0.77$

State Motor Vehicles Department
(731)

Average Vehicle Trip Ends vs: Employees
On a: Saturday

Number of Studies: 7
Avg. Number of Employees: 38
Directional Distribution: 50% entering, 50% exiting

Trip Generation per Employee

Average Rate	Range of Rates	Standard Deviation
2.39	0.83 - 4.27	1.82

Data Plot and Equation

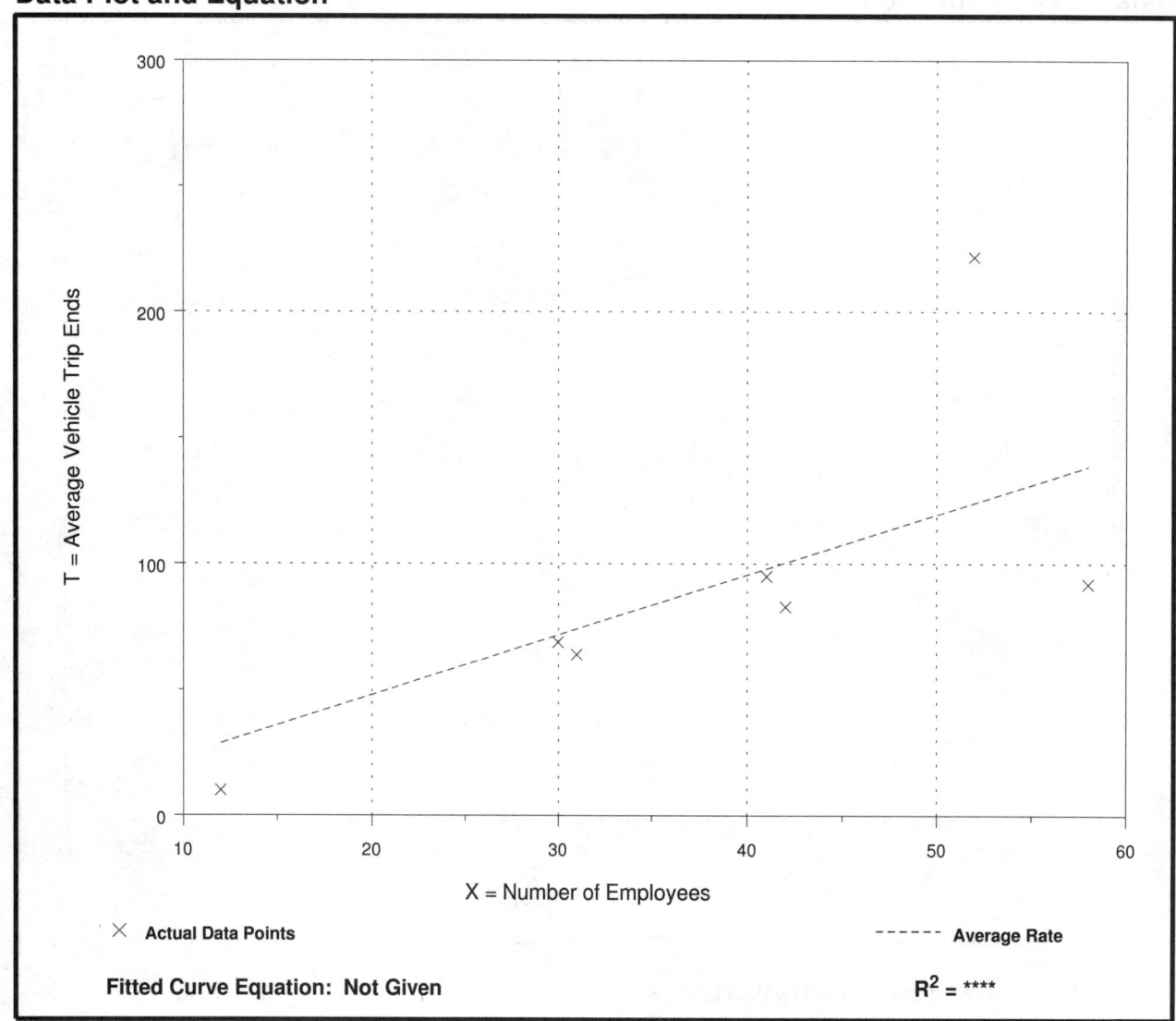

[Plot: T = Average Vehicle Trip Ends vs X = Number of Employees]

× Actual Data Points   ----- Average Rate

Fitted Curve Equation: Not Given R^2 = ****

State Motor Vehicles Department
(731)

Average Vehicle Trip Ends vs: **Employees**
On a: **Saturday,**
Peak Hour of Generator

Number of Studies: 7
Avg. Number of Employees: 38
Directional Distribution: Not available

Trip Generation per Employee

Average Rate	Range of Rates	Standard Deviation
0.35	0.17 - 0.50	0.60

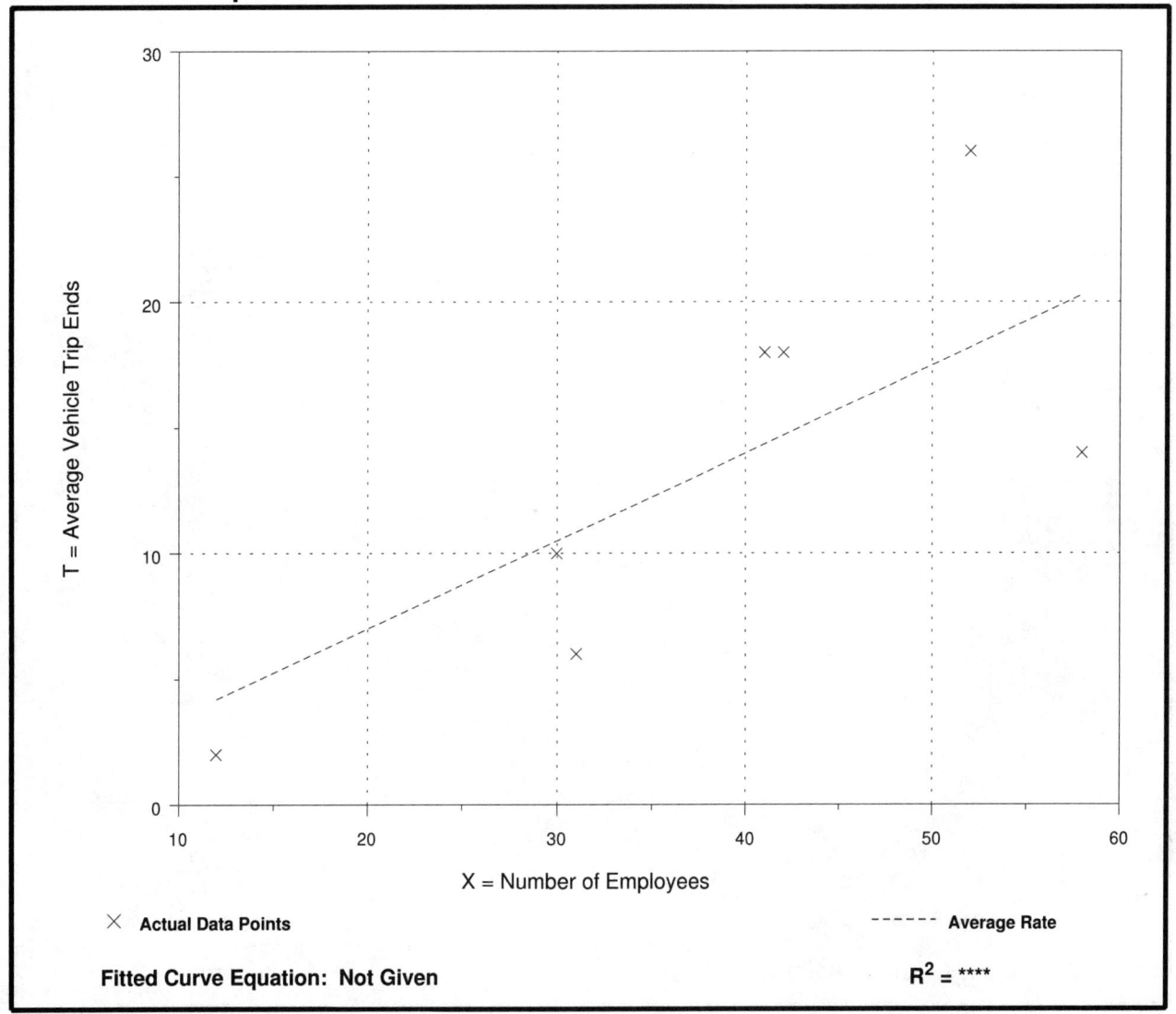

Data Plot and Equation

$\times$ Actual Data Points
$----$ Average Rate

Fitted Curve Equation: Not Given $R^2 = ****$

State Motor Vehicles Department
(731)

Average Vehicle Trip Ends vs: Employees
On a: Sunday

Number of Studies: 7
Avg. Number of Employees: 38
Directional Distribution: 50% entering, 50% exiting

Trip Generation per Employee

Average Rate	Range of Rates	Standard Deviation
1.70	0.42 - 3.15	1.53

Data Plot and Equation

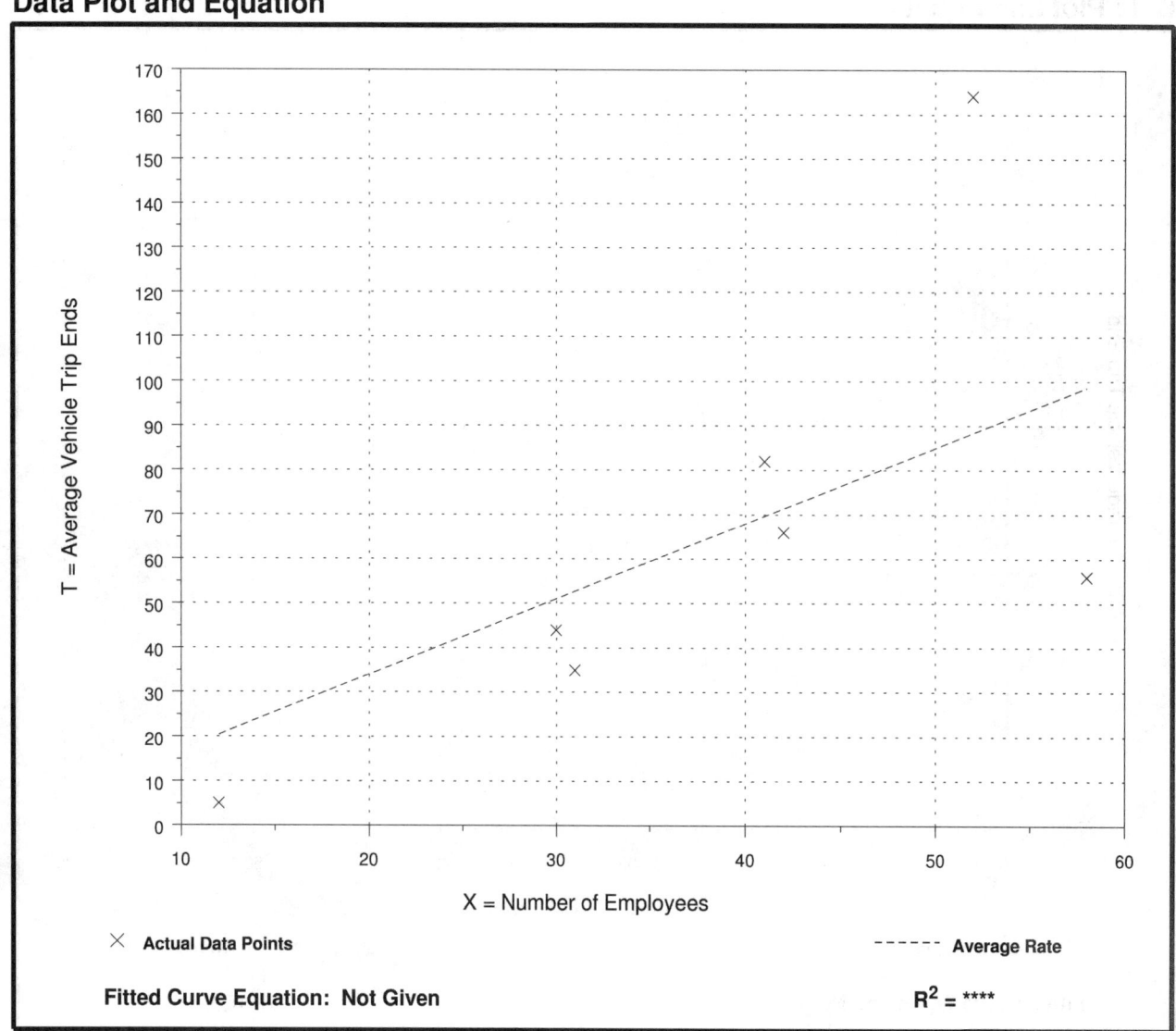

Fitted Curve Equation: Not Given $R^2 = ****$

State Motor Vehicles Department
(731)

Average Vehicle Trip Ends vs: Employees
On a: Sunday,
Peak Hour of Generator

Number of Studies: 7
Avg. Number of Employees: 38
Directional Distribution: Not available

Trip Generation per Employee

Average Rate	Range of Rates	Standard Deviation
0.31	0.16 - 0.51	0.57

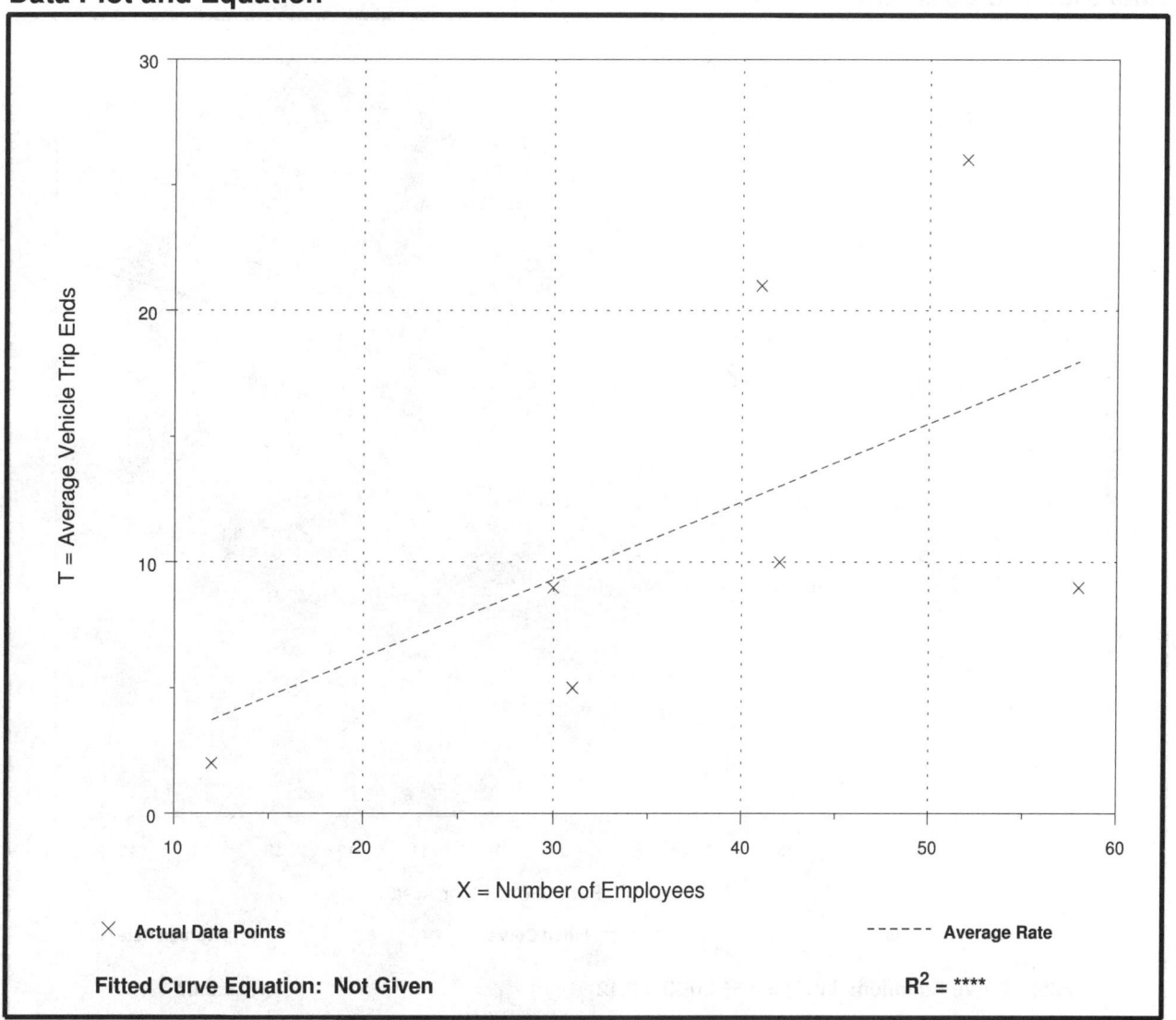

Data Plot and Equation

Fitted Curve Equation: Not Given $R^2 = ****$

State Motor Vehicles Department
(731)

Average Vehicle Trip Ends vs: 1000 Sq. Feet Gross Floor Area
On a: Weekday

Number of Studies: 8
Average 1000 Sq. Feet GFA: 10
Directional Distribution: 50% entering, 50% exiting

Trip Generation per 1000 Sq. Feet Gross Floor Area

Average Rate	Range of Rates	Standard Deviation
166.02	113.23 - 272.17	49.39

Data Plot and Equation

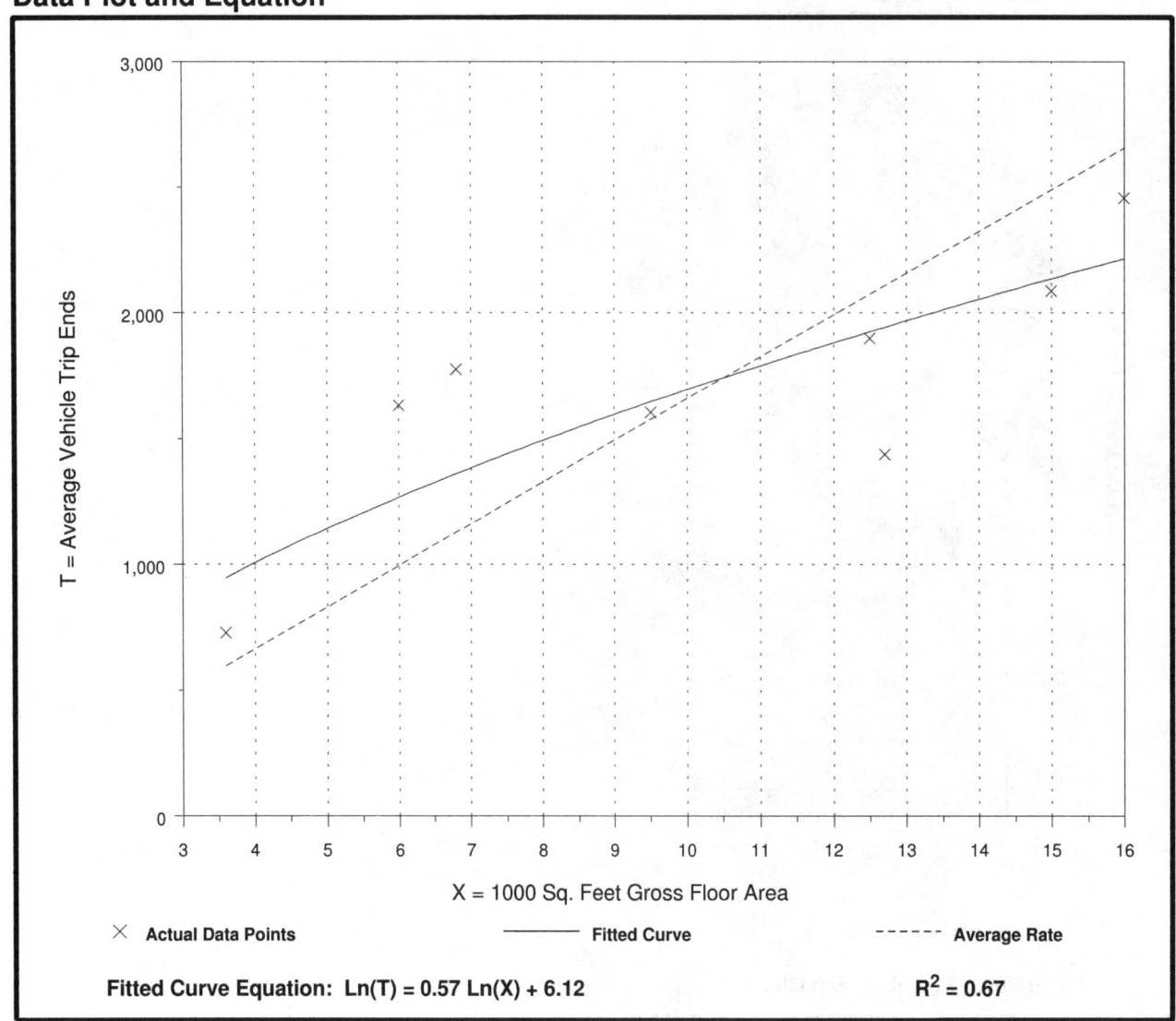

Fitted Curve Equation: $Ln(T) = 0.57\ Ln(X) + 6.12$ $R^2 = 0.67$

State Motor Vehicles Department
(731)

Average Vehicle Trip Ends vs: 1000 Sq. Feet Gross Floor Area
On a: Weekday,
Peak Hour of Adjacent Street Traffic,
One Hour Between 7 and 9 a.m.

Number of Studies: 8
Average 1000 Sq. Feet GFA: 10
Directional Distribution: Not available

Trip Generation per 1000 Sq. Feet Gross Floor Area

Average Rate	Range of Rates	Standard Deviation
9.84	7.04 - 15.67	3.70

Data Plot and Equation

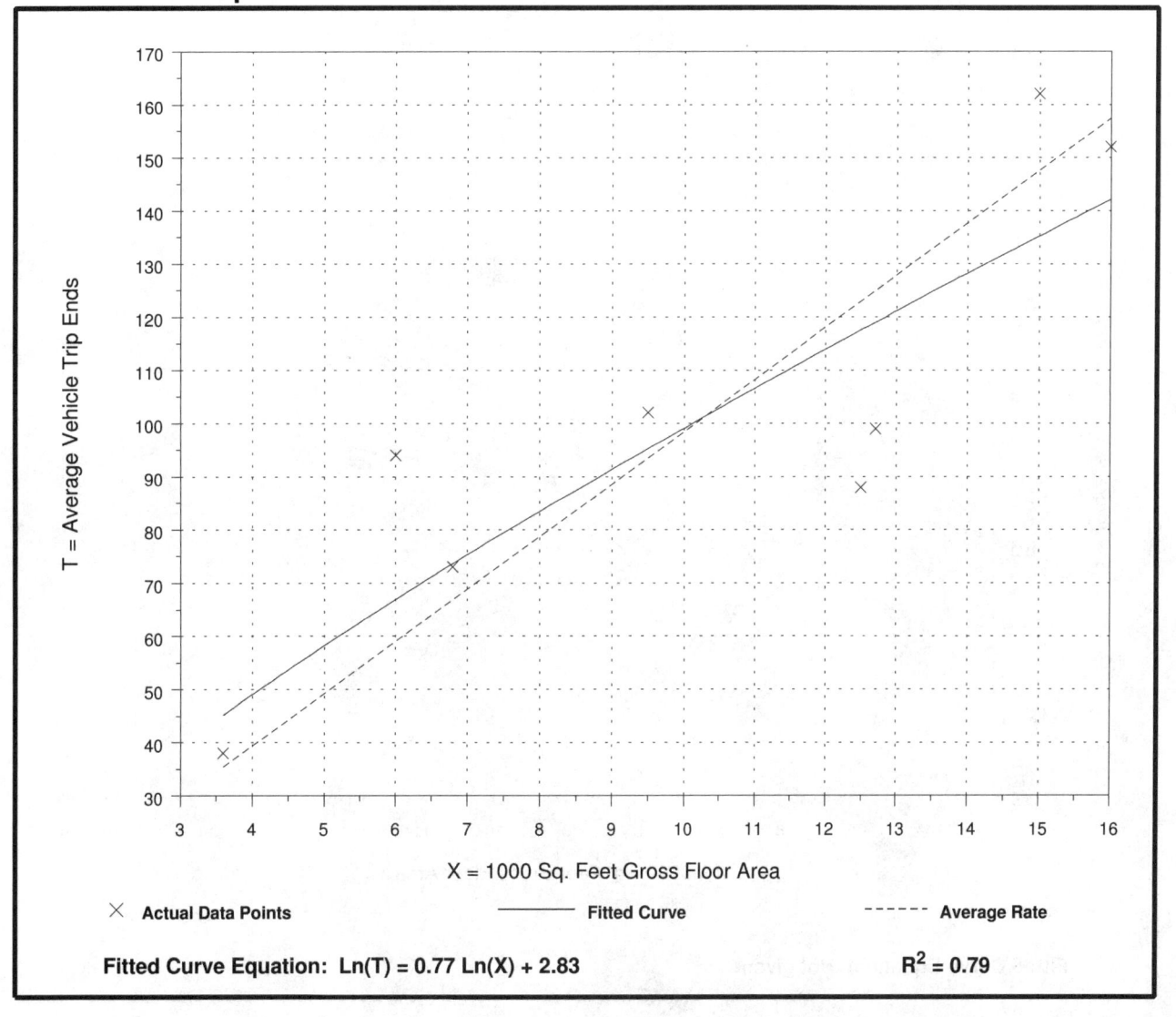

Fitted Curve Equation: $\text{Ln}(T) = 0.77 \, \text{Ln}(X) + 2.83$ $R^2 = 0.79$

Trip Generation, 9th Edition • Institute of Transportation Engineers

State Motor Vehicles Department
(731)

Average Vehicle Trip Ends vs: 1000 Sq. Feet Gross Floor Area
On a: Weekday,
Peak Hour of Adjacent Street Traffic,
One Hour Between 4 and 6 p.m.

Number of Studies: 8
Average 1000 Sq. Feet GFA: 10
Directional Distribution: Not available

Trip Generation per 1000 Sq. Feet Gross Floor Area

Average Rate	Range of Rates	Standard Deviation
17.09	10.31 - 30.33	7.58

Data Plot and Equation

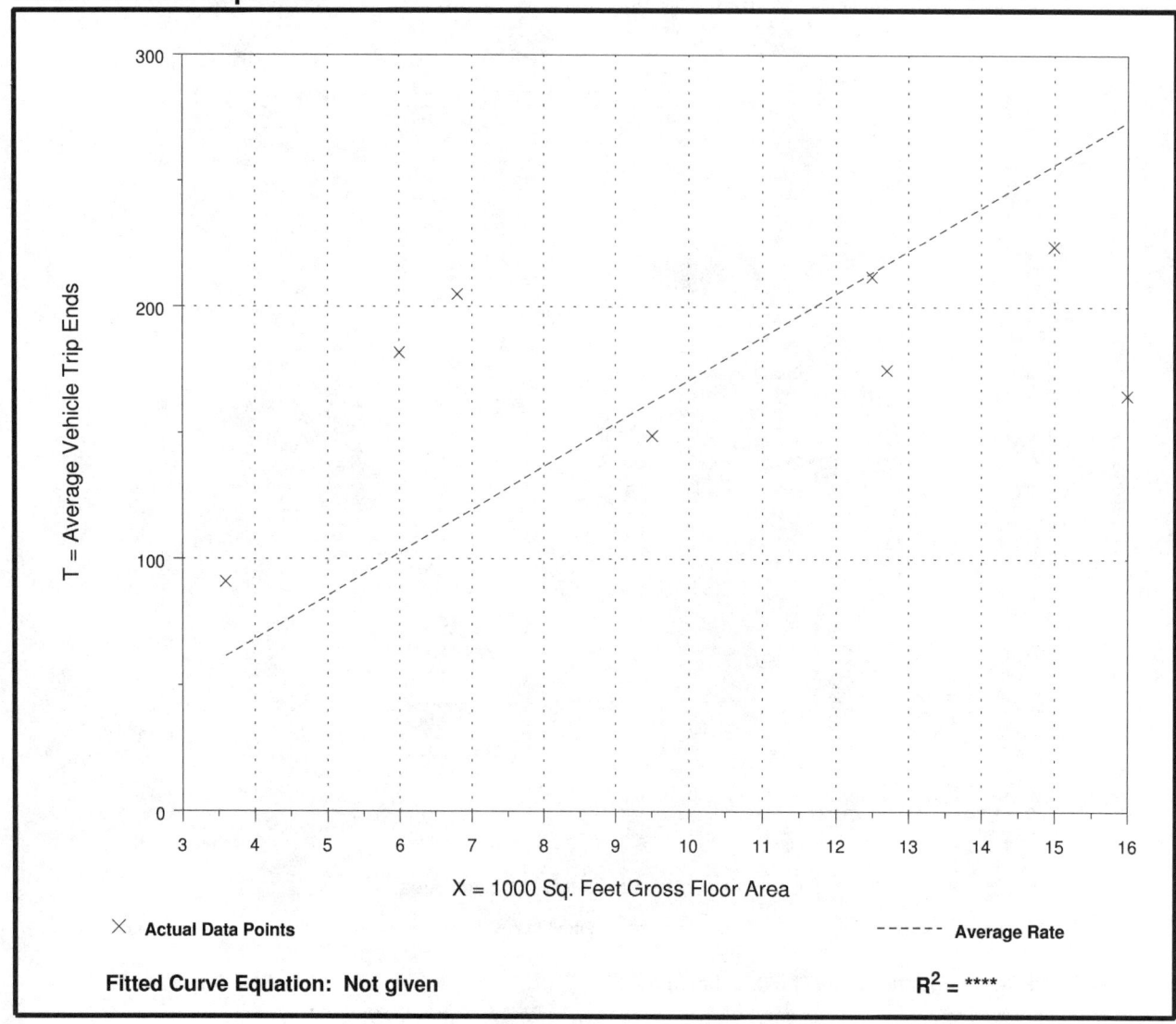

Fitted Curve Equation: Not given $R^2 = ****$

State Motor Vehicles Department
(731)

Average Vehicle Trip Ends vs: 1000 Sq. Feet Gross Floor Area
On a: Weekday,
A.M. Peak Hour of Generator

Number of Studies: 8
Average 1000 Sq. Feet GFA: 10
Directional Distribution: Not available

Trip Generation per 1000 Sq. Feet Gross Floor Area

Average Rate	Range of Rates	Standard Deviation
18.53	11.81 - 30.17	6.62

Data Plot and Equation

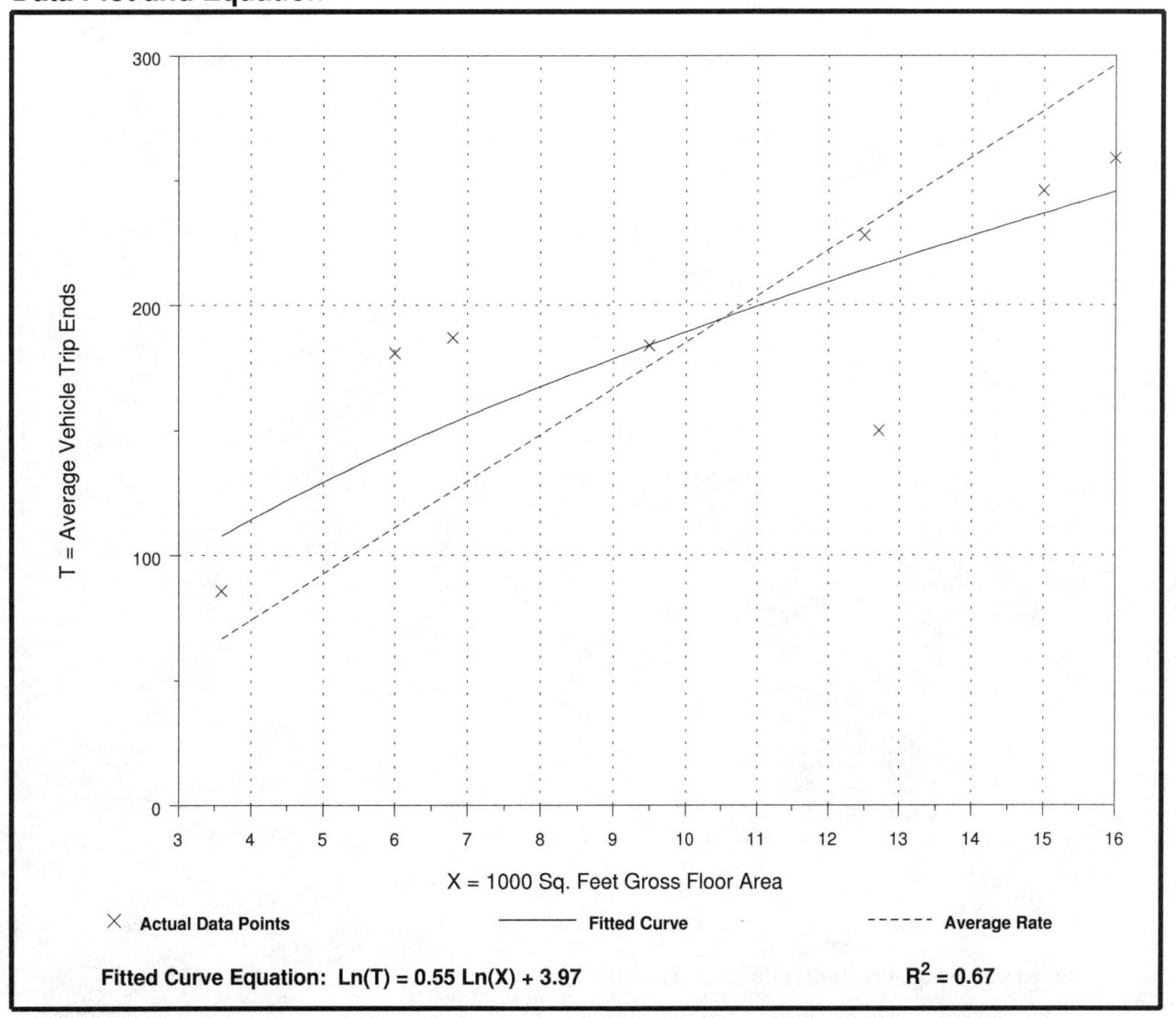

Fitted Curve Equation: $Ln(T) = 0.55\ Ln(X) + 3.97$ $R^2 = 0.67$

State Motor Vehicles Department
(731)

Average Vehicle Trip Ends vs: 1000 Sq. Feet Gross Floor Area
On a: Weekday,
P.M. Peak Hour of Generator

Number of Studies: 8
Average 1000 Sq. Feet GFA: 10
Directional Distribution: Not available

Trip Generation per 1000 Sq. Feet Gross Floor Area

Average Rate	Range of Rates	Standard Deviation
19.93	13.78 - 31.91	7.06

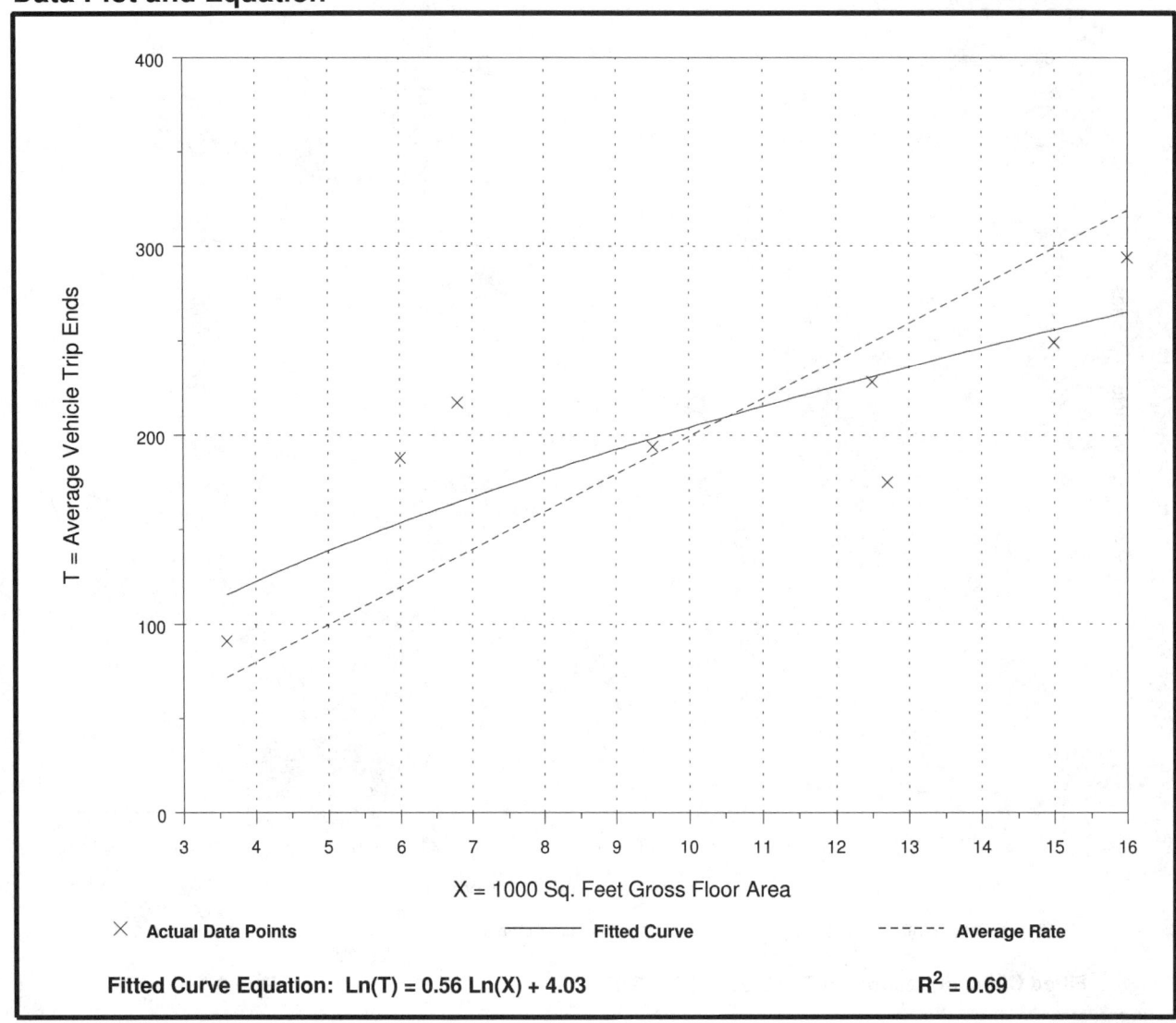

Data Plot and Equation

Fitted Curve Equation: $Ln(T) = 0.56\, Ln(X) + 4.03$ $R^2 = 0.69$

State Motor Vehicles Department
(731)

Average Vehicle Trip Ends vs: 1000 Sq. Feet Gross Floor Area
On a: Saturday

Number of Studies: 7
Average 1000 Sq. Feet GFA: 10
Directional Distribution: 50% entering, 50% exiting

Trip Generation per 1000 Sq. Feet Gross Floor Area

Average Rate	Range of Rates	Standard Deviation
9.46	2.78 - 15.33	4.95

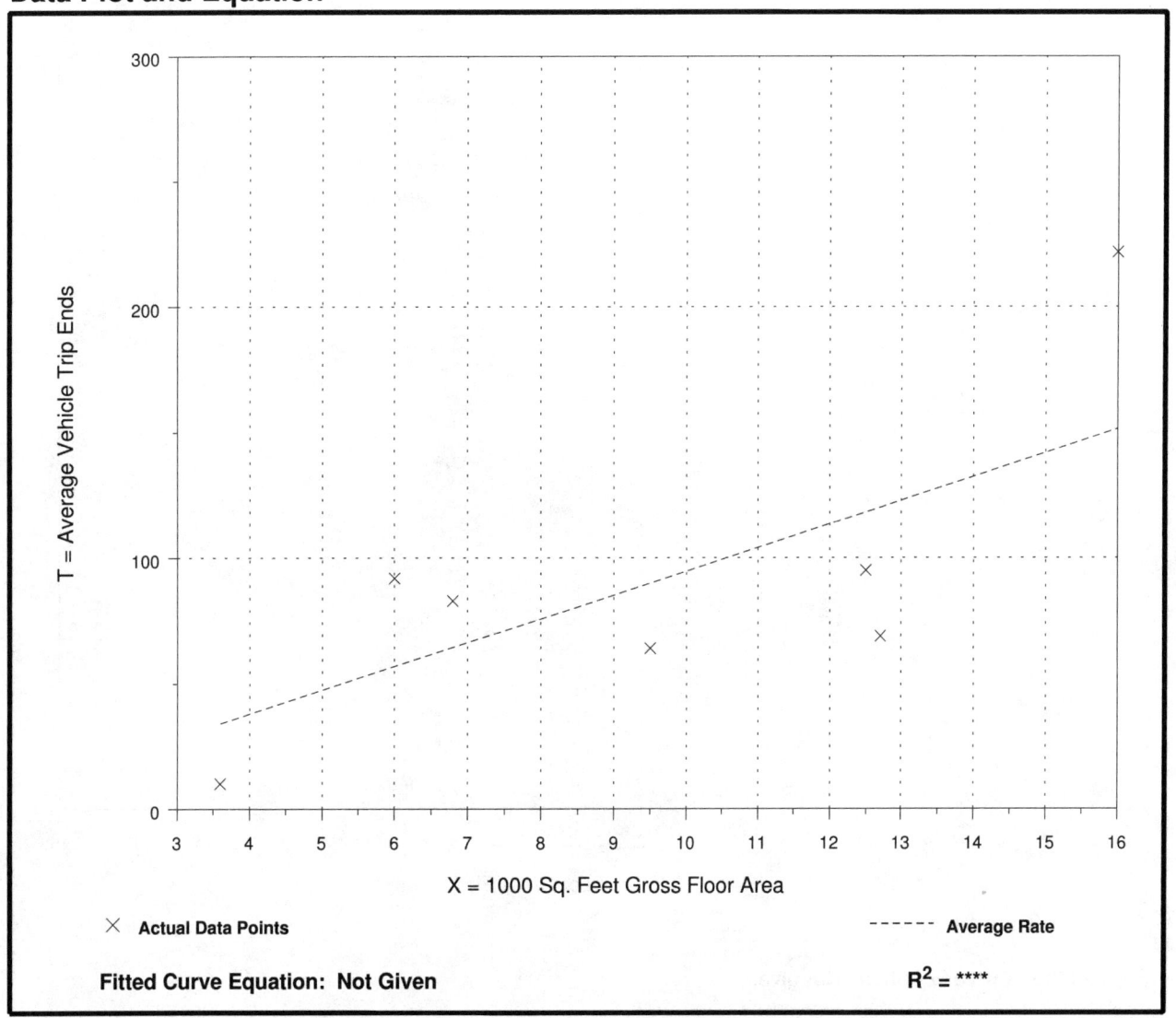

Data Plot and Equation

X   Actual Data Points   - - - - - Average Rate

Fitted Curve Equation: Not Given $R^2 = ****$

Trip Generation, 9th Edition • Institute of Transportation Engineers

State Motor Vehicles Department
(731)

Average Vehicle Trip Ends vs: 1000 Sq. Feet Gross Floor Area
On a: Saturday,
Peak Hour of Generator

Number of Studies: 7
Average 1000 Sq. Feet GFA: 10
Directional Distribution: Not available

Trip Generation per 1000 Sq. Feet Gross Floor Area

Average Rate	Range of Rates	Standard Deviation
1.40	0.56 - 2.65	1.31

Data Plot and Equation

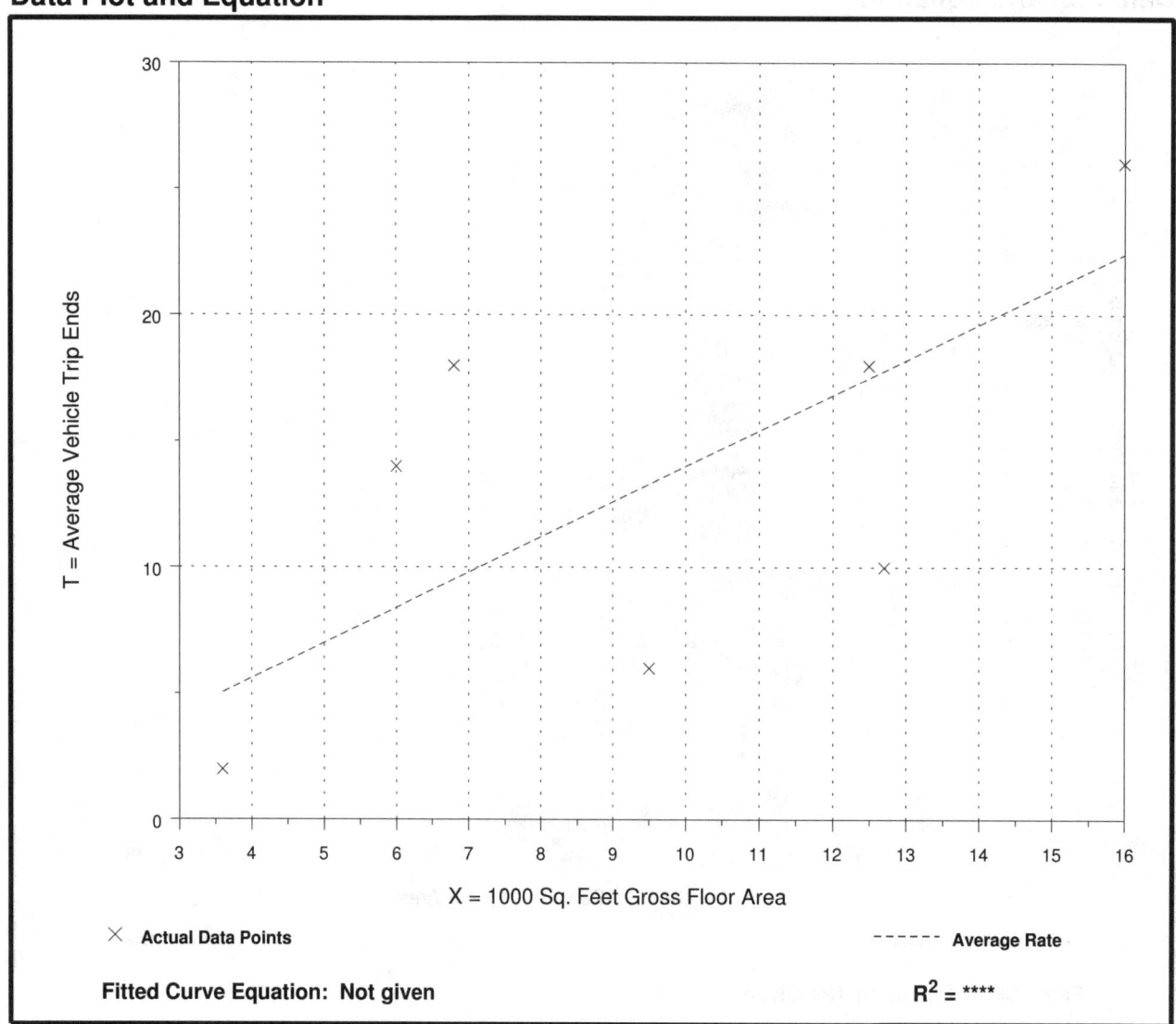

Fitted Curve Equation: Not given $R^2 = ****$

Trip Generation, 9th Edition • Institute of Transportation Engineers

State Motor Vehicles Department
(731)

Average Vehicle Trip Ends vs: 1000 Sq. Feet Gross Floor Area
On a: Sunday

Number of Studies: 7
Average 1000 Sq. Feet GFA: 10
Directional Distribution: 50% entering, 50% exiting

Trip Generation per 1000 Sq. Feet Gross Floor Area

Average Rate	Range of Rates	Standard Deviation
6.74	1.39 - 10.25	3.95

Data Plot and Equation

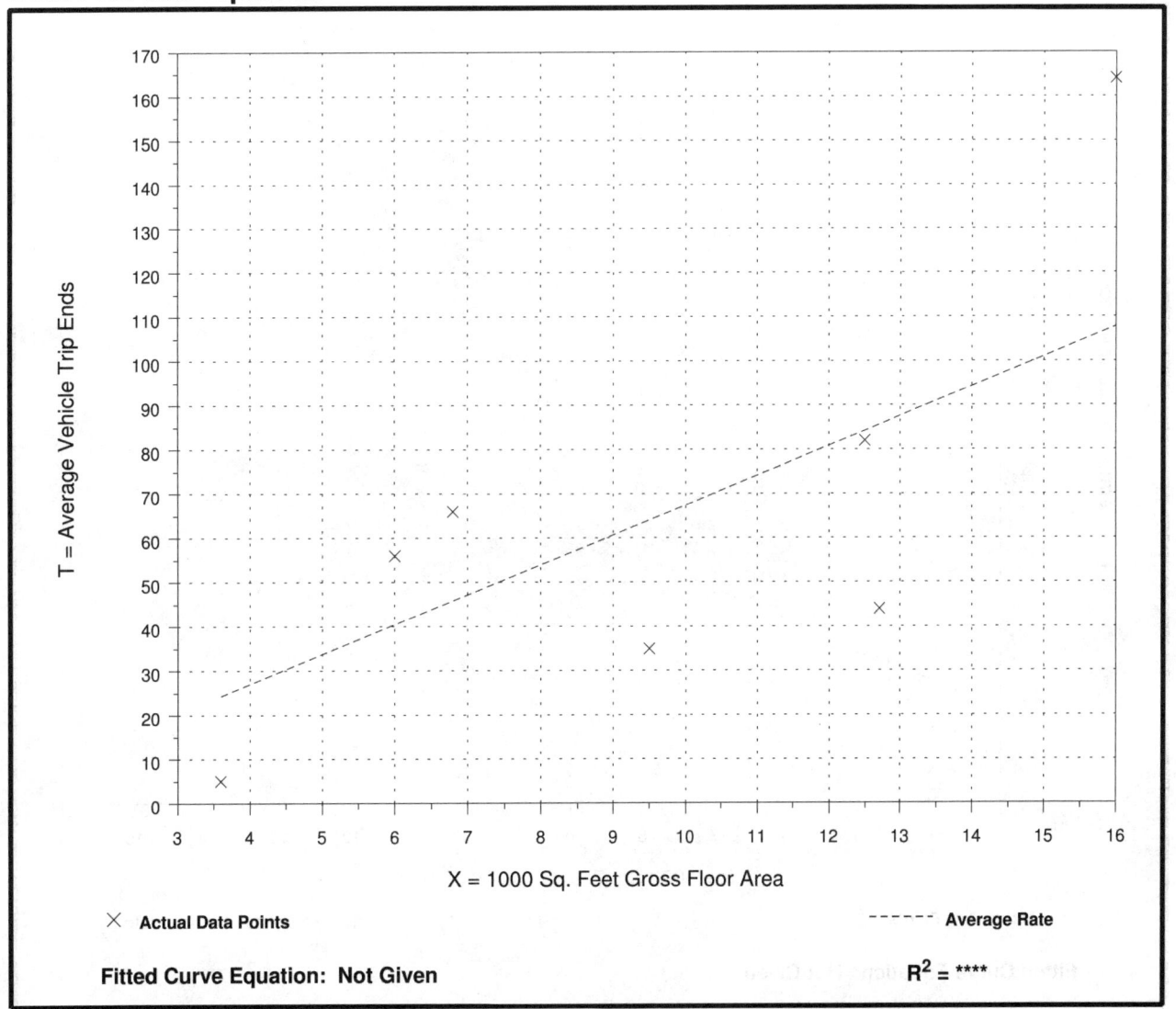

Fitted Curve Equation: Not Given $R^2 = ****$

State Motor Vehicles Department
(731)

Average Vehicle Trip Ends vs: 1000 Sq. Feet Gross Floor Area
On a: Sunday,
Peak Hour of Generator

Number of Studies: 7
Average 1000 Sq. Feet GFA: 10
Directional Distribution: Not available

Trip Generation per 1000 Sq. Feet Gross Floor Area

Average Rate	Range of Rates	Standard Deviation
1.22	0.53 - 1.68	1.16

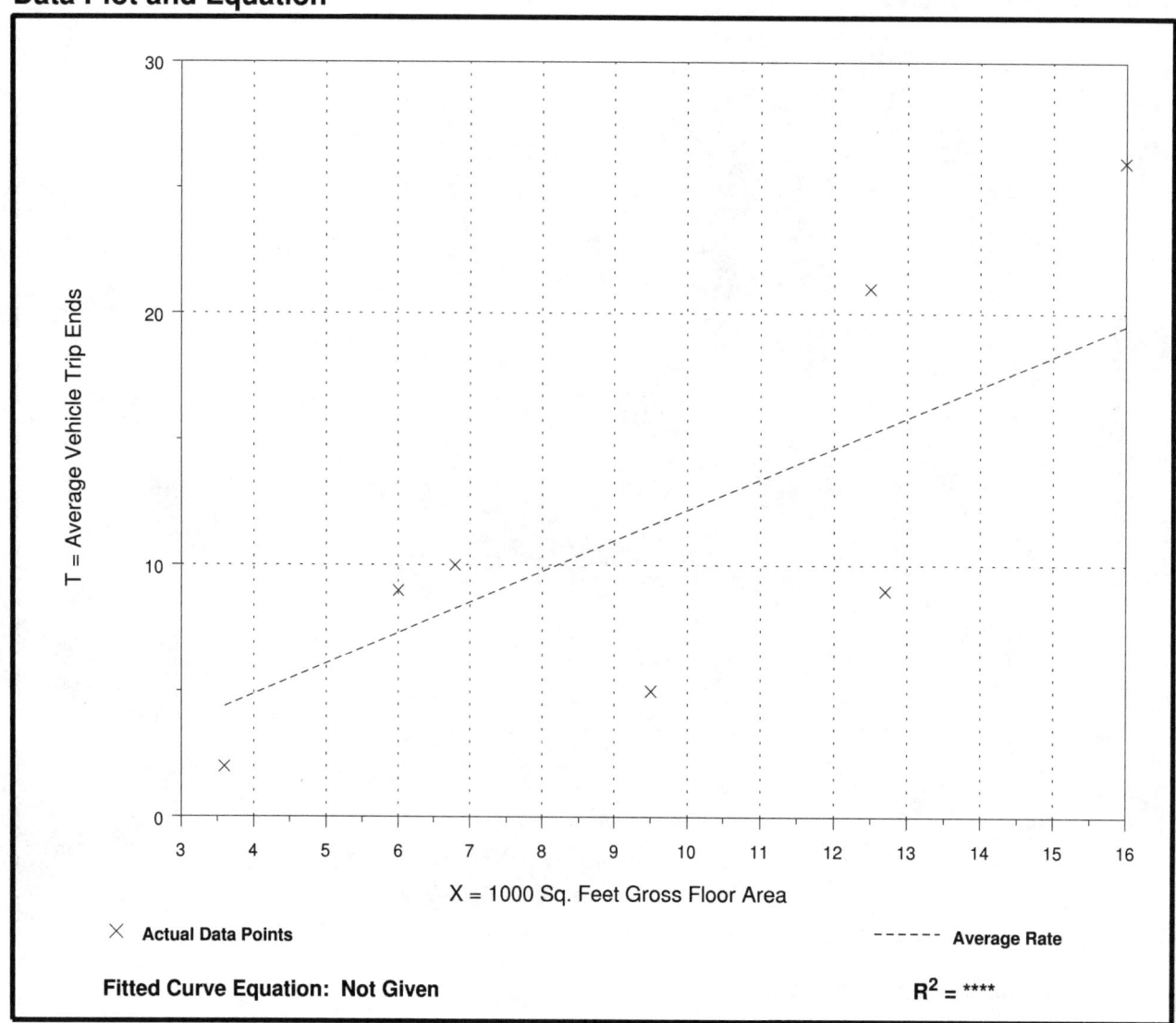

Data Plot and Equation

Fitted Curve Equation: Not Given $R^2 = ****$

Land Use: 732
United States Post Office

Description

A U.S. post office is a federal building that contains service windows for mailing packages and letters, post office boxes, offices, sorting and distributing facilities for mail and vehicle storage areas.

Additional Data

Truck trips accounted for 1.2 percent of the weekday traffic at the post offices surveyed.

The average vehicle occupancy for the four studies for which information was submitted was approximately 1.14 persons per automobile.

Peak hours of the generator—
    The weekday A.M. peak hour was between 9:00 a.m. and 10:00 a.m. The weekday P.M. peak hour was between 3:00 p.m. and 4:00 p.m. The Saturday peak hour was between 11:00 a.m. and 12:00 p.m.

The sites were surveyed between the 1980s and the 2000s throughout the United States.

Source Numbers

88, 170, 248, 250, 262, 269, 275, 357, 435, 444, 579, 609, 732, 734

United States Post Office
(732)

Average Vehicle Trip Ends vs: Employees
On a: Weekday

Number of Studies: 9
Avg. Number of Employees: 156
Directional Distribution: 50% entering, 50% exiting

Trip Generation per Employee

Average Rate	Range of Rates	Standard Deviation
25.37	13.33 - 92.79	19.61

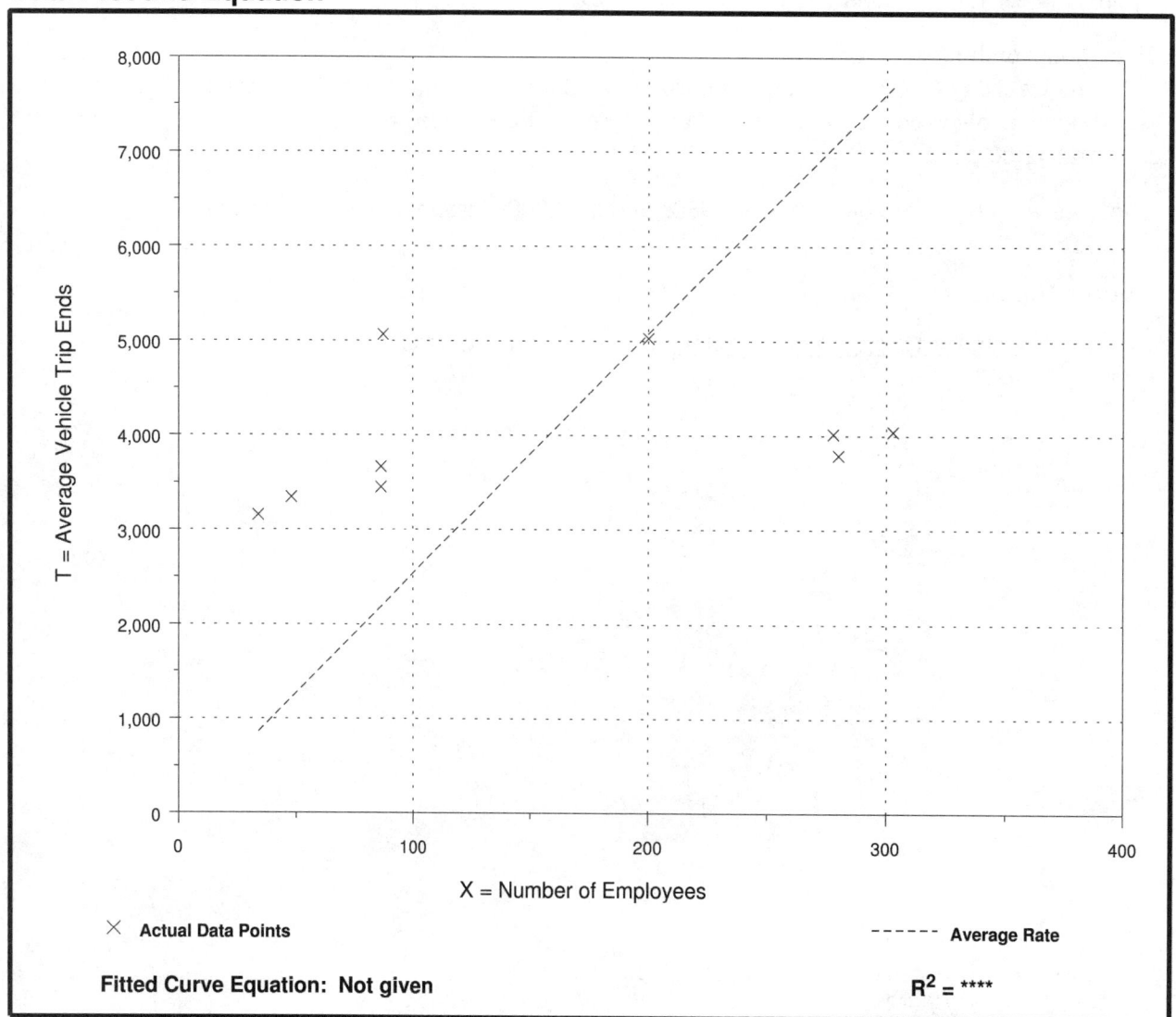

Data Plot and Equation

Fitted Curve Equation: Not given $R^2 = ****$

United States Post Office
(732)

Average Vehicle Trip Ends vs: Employees
On a: Weekday,
Peak Hour of Adjacent Street Traffic,
One Hour Between 7 and 9 a.m.

Number of Studies: 11
Avg. Number of Employees: 129
Directional Distribution: 52% entering, 48% exiting

Trip Generation per Employee

Average Rate	Range of Rates	Standard Deviation
1.78	0.85 - 22.90	2.76

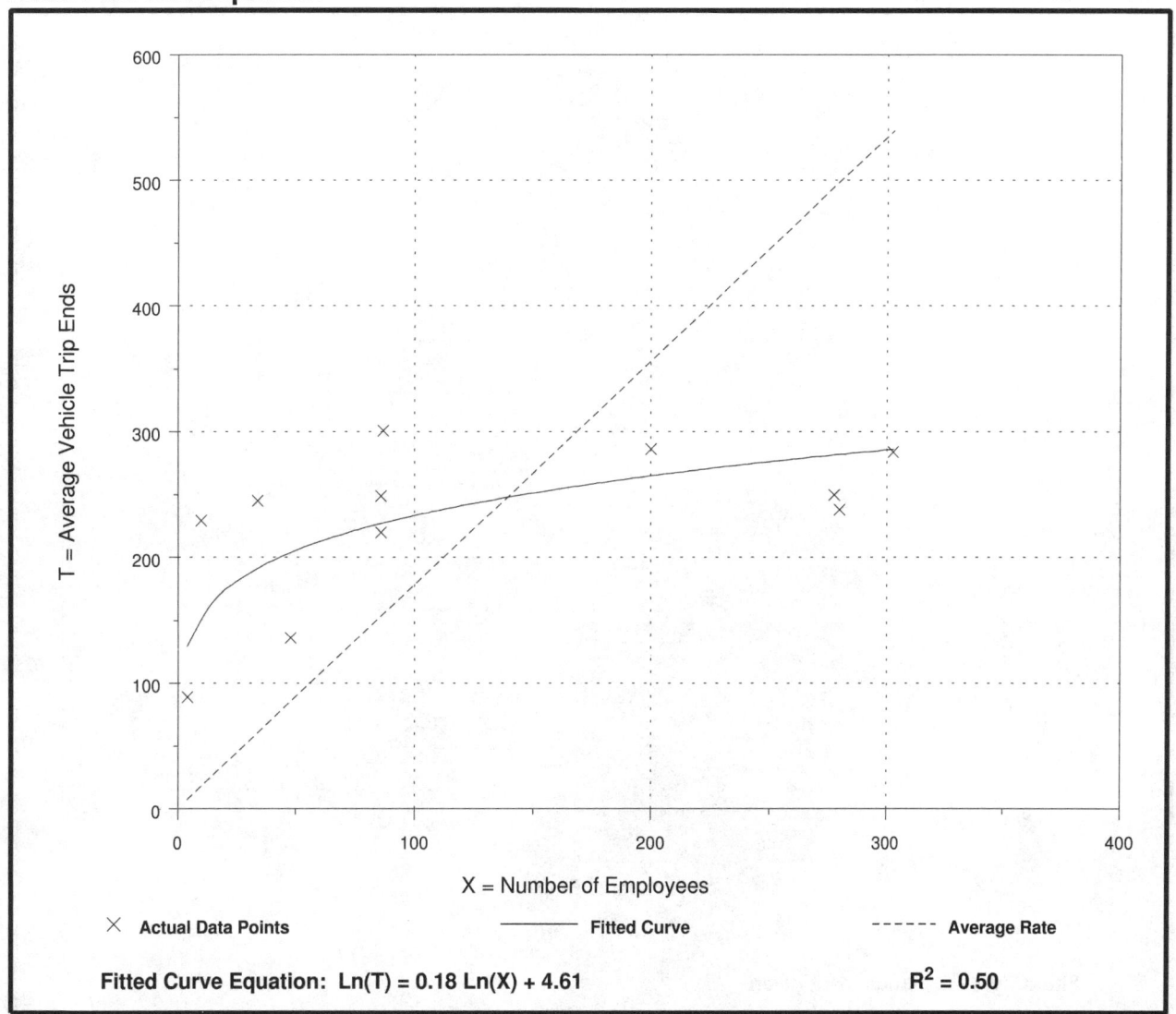

Data Plot and Equation

Fitted Curve Equation: $Ln(T) = 0.18\ Ln(X) + 4.61$ $R^2 = 0.50$

United States Post Office
(732)

Average Vehicle Trip Ends vs: Employees
On a: Weekday,
Peak Hour of Adjacent Street Traffic,
One Hour Between 4 and 6 p.m.

Number of Studies: 11
Avg. Number of Employees: 129
Directional Distribution: 51% entering, 49% exiting

Trip Generation per Employee

Average Rate	Range of Rates	Standard Deviation
2.50	0.90 - 40.40	4.33

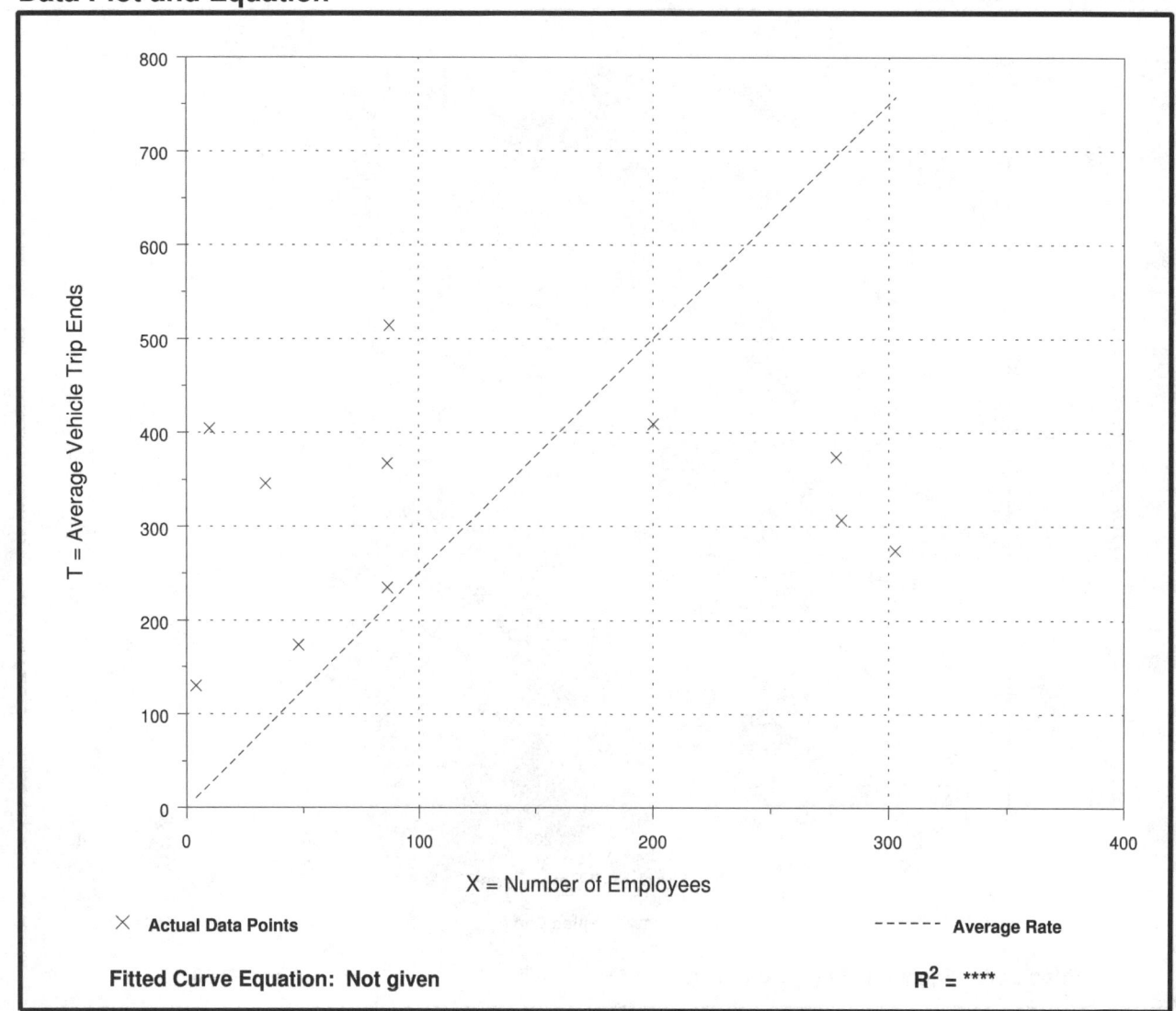

Data Plot and Equation

Fitted Curve Equation: Not given $R^2 =$ ****

United States Post Office
(732)

Average Vehicle Trip Ends vs: Employees
On a: Weekday,
A.M. Peak Hour of Generator

Number of Studies: 11
Avg. Number of Employees: 128
Directional Distribution: 48% entering, 52% exiting

Trip Generation per Employee

Average Rate	Range of Rates	Standard Deviation
2.71	0.94 - 27.80	3.71

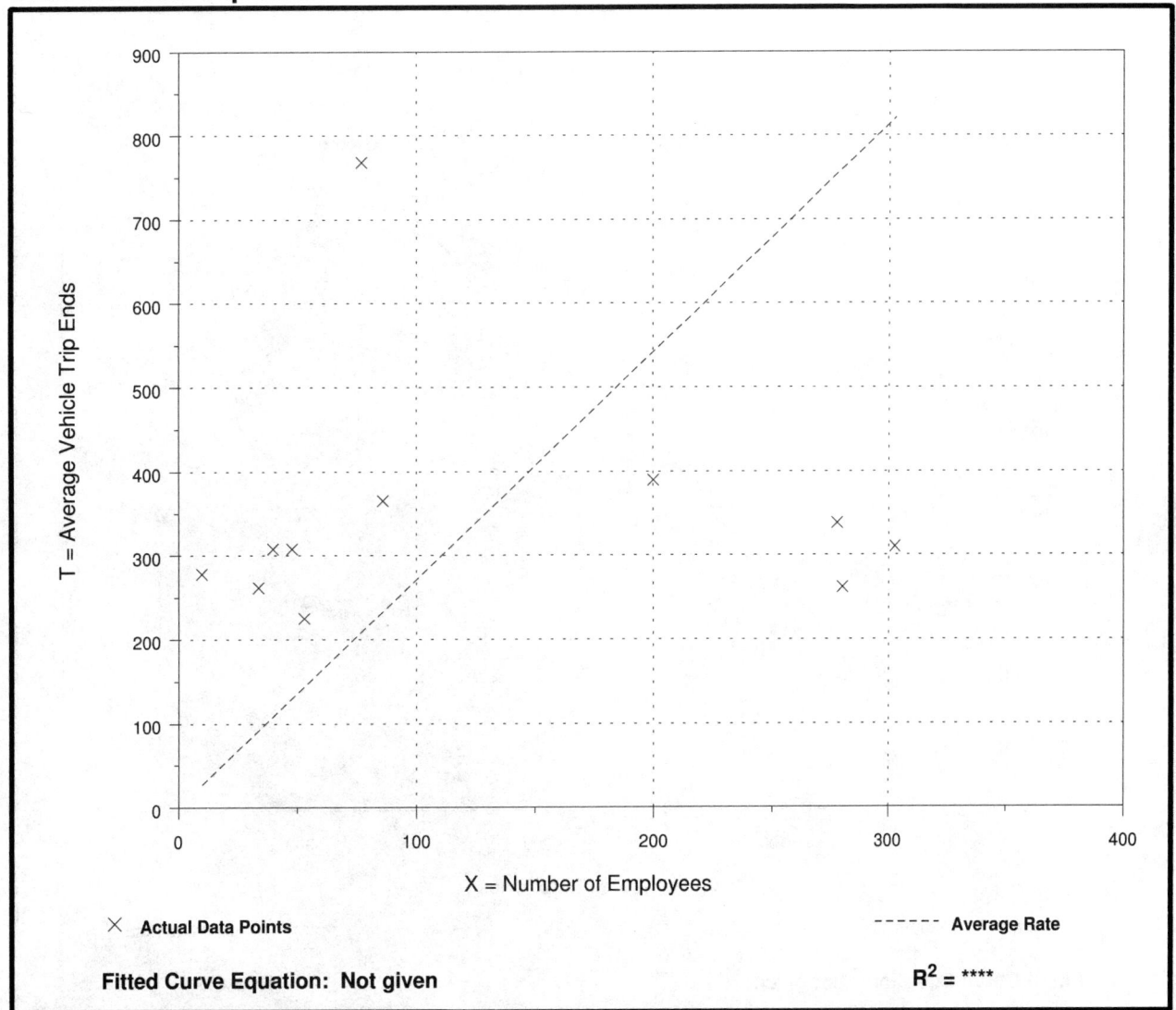

Data Plot and Equation

Fitted Curve Equation: Not given $R^2 = ****$

Trip Generation, 9th Edition • Institute of Transportation Engineers

United States Post Office
(732)

Average Vehicle Trip Ends vs: Employees
On a: Weekday,
P.M. Peak Hour of Generator

Number of Studies: 11
Avg. Number of Employees: 128
Directional Distribution: 51% entering, 49% exiting

Trip Generation per Employee

Average Rate	Range of Rates	Standard Deviation
3.11	0.97 - 40.40	4.70

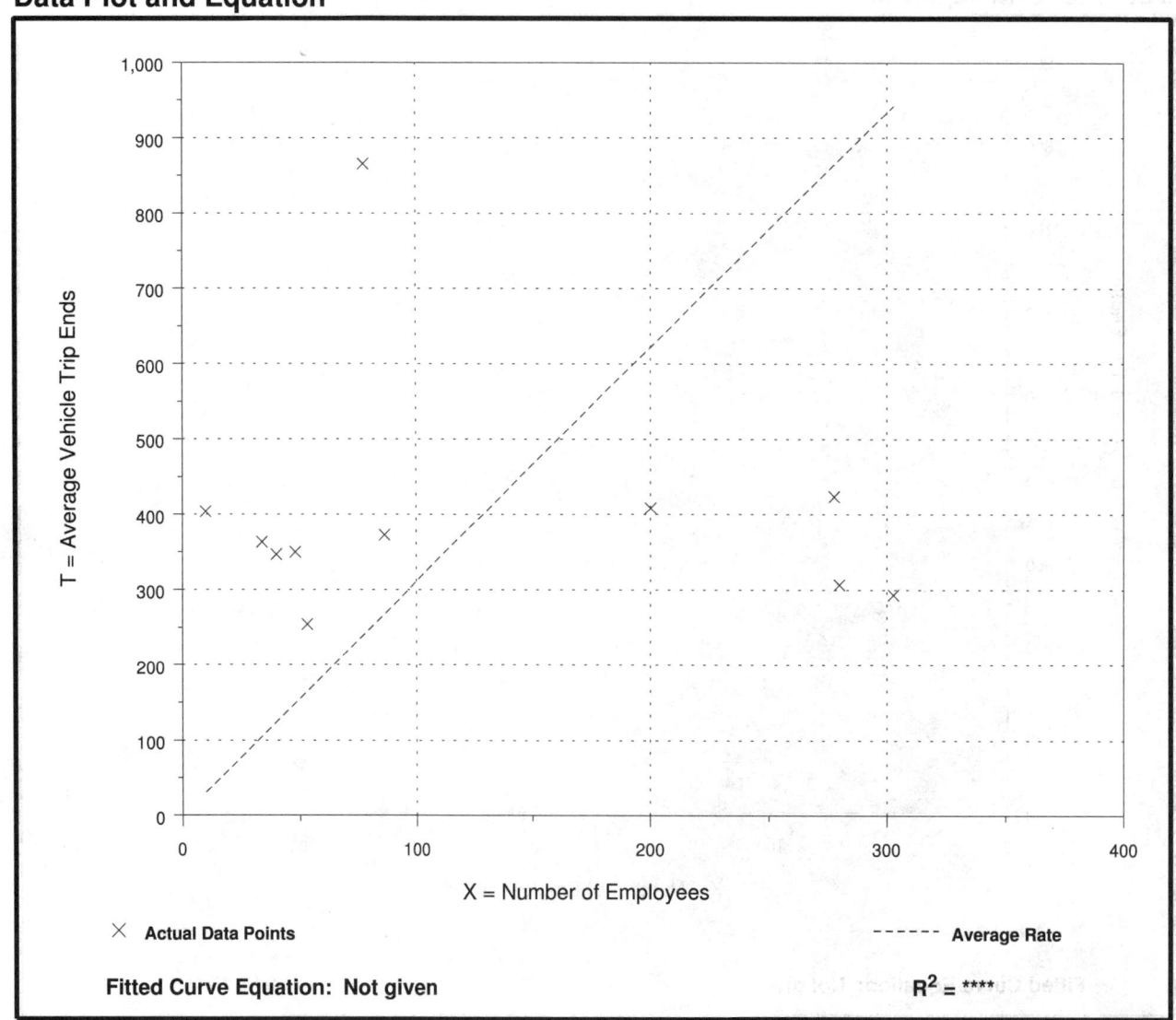

Data Plot and Equation

Fitted Curve Equation: Not given $R^2 = ****$

United States Post Office
(732)

Average Vehicle Trip Ends vs: Employees
On a: Saturday

Number of Studies: 7
Avg. Number of Employees: 176
Directional Distribution: 50% entering, 50% exiting

Trip Generation per Employee

Average Rate	Range of Rates	Standard Deviation
13.02	7.46 - 46.09	8.93

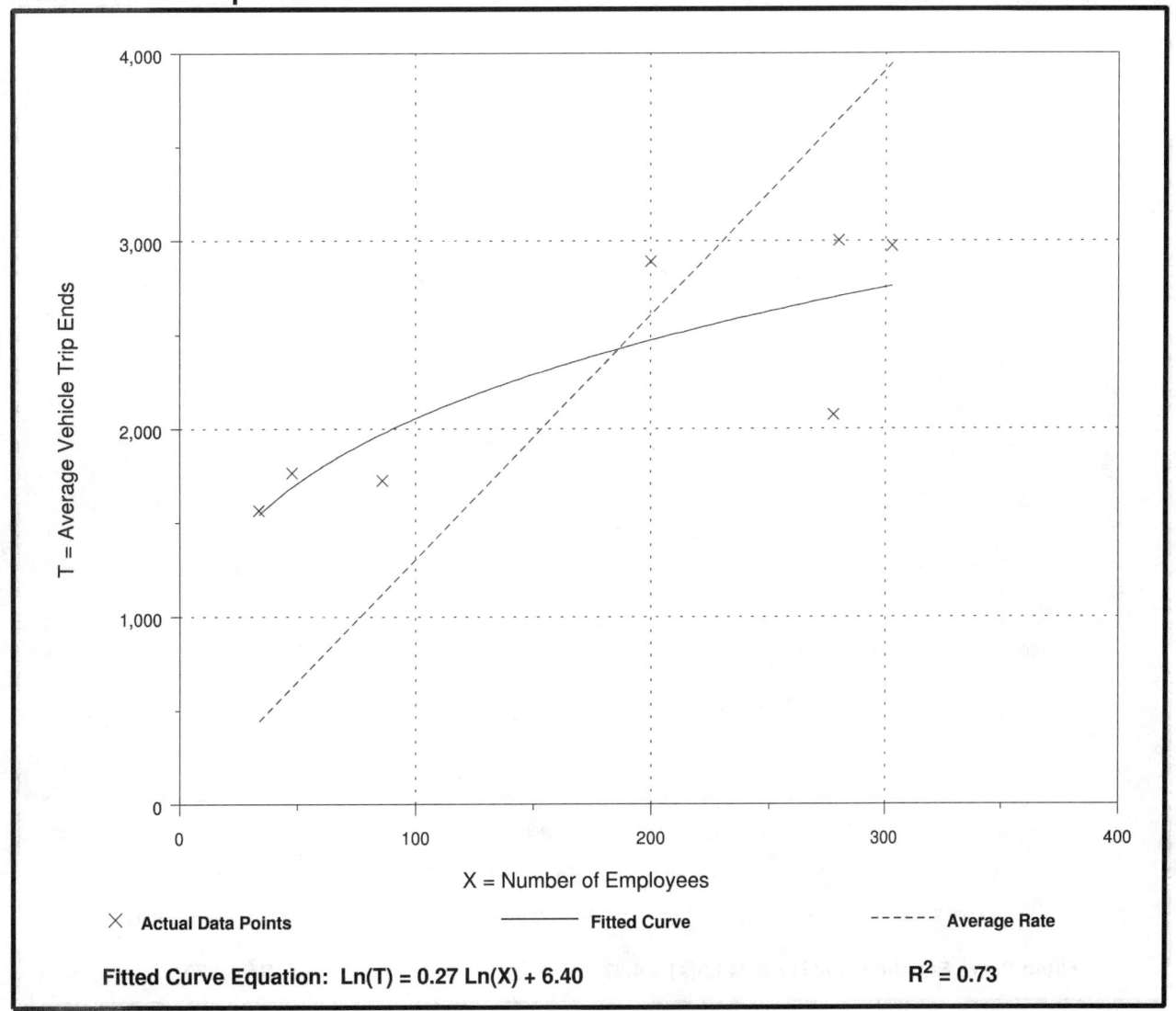

Data Plot and Equation

Fitted Curve Equation: $Ln(T) = 0.27\ Ln(X) + 6.40$ $R^2 = 0.73$

United States Post Office
(732)

Average Vehicle Trip Ends vs: Employees
On a: Saturday,
Peak Hour of Generator

Number of Studies: 7
Avg. Number of Employees: 176
Directional Distribution: 55% entering, 45% exiting

Trip Generation per Employee

Average Rate	Range of Rates	Standard Deviation
1.61	0.97 - 6.00	1.65

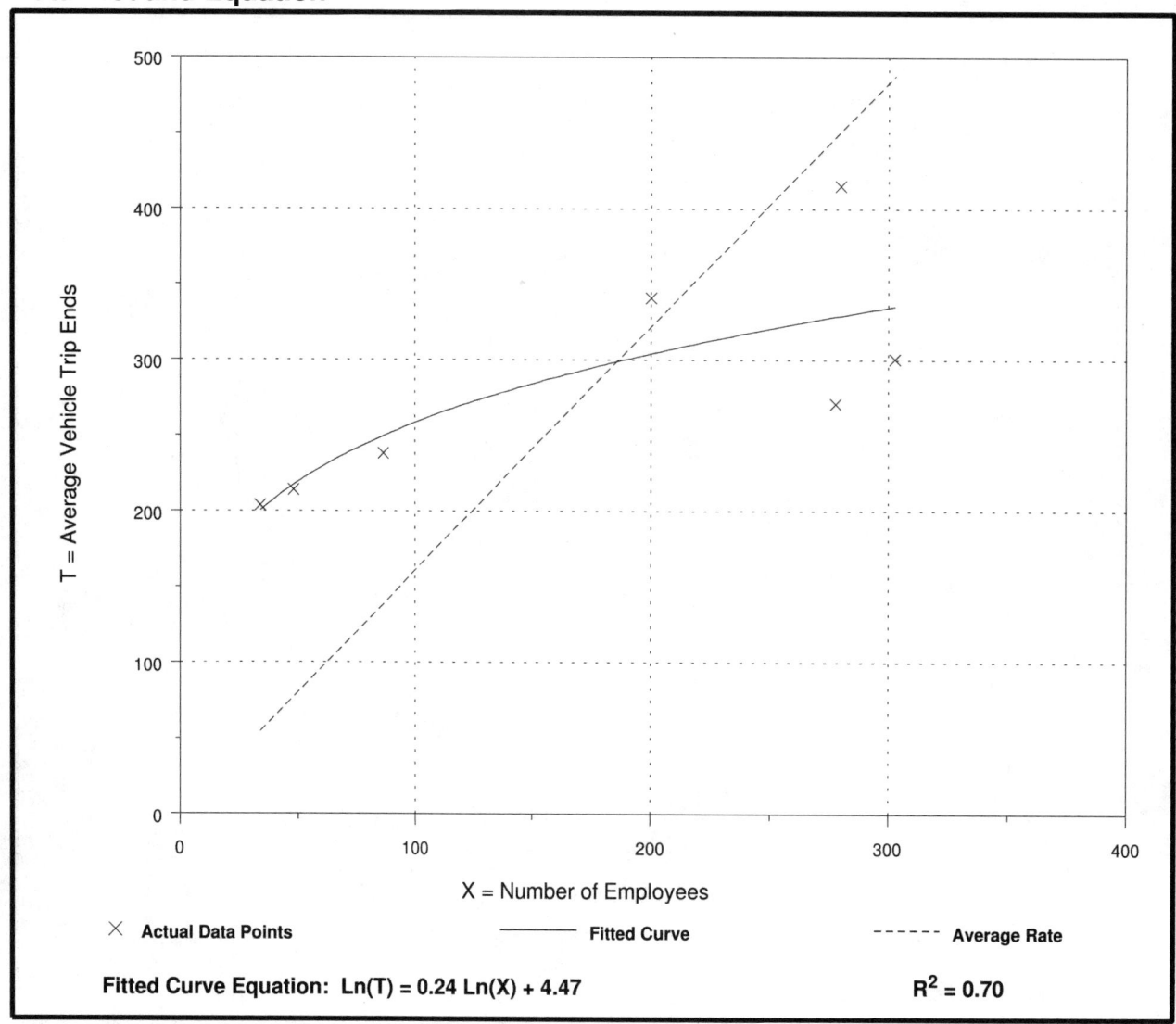

Data Plot and Equation

Fitted Curve Equation: $Ln(T) = 0.24\ Ln(X) + 4.47$ $R^2 = 0.70$

United States Post Office
(732)

Average Vehicle Trip Ends vs: Employees
On a: Sunday

Number of Studies: 7
Avg. Number of Employees: 176
Directional Distribution: 50% entering, 50% exiting

Trip Generation per Employee

Average Rate	Range of Rates	Standard Deviation
7.76	5.36 - 26.44	5.17

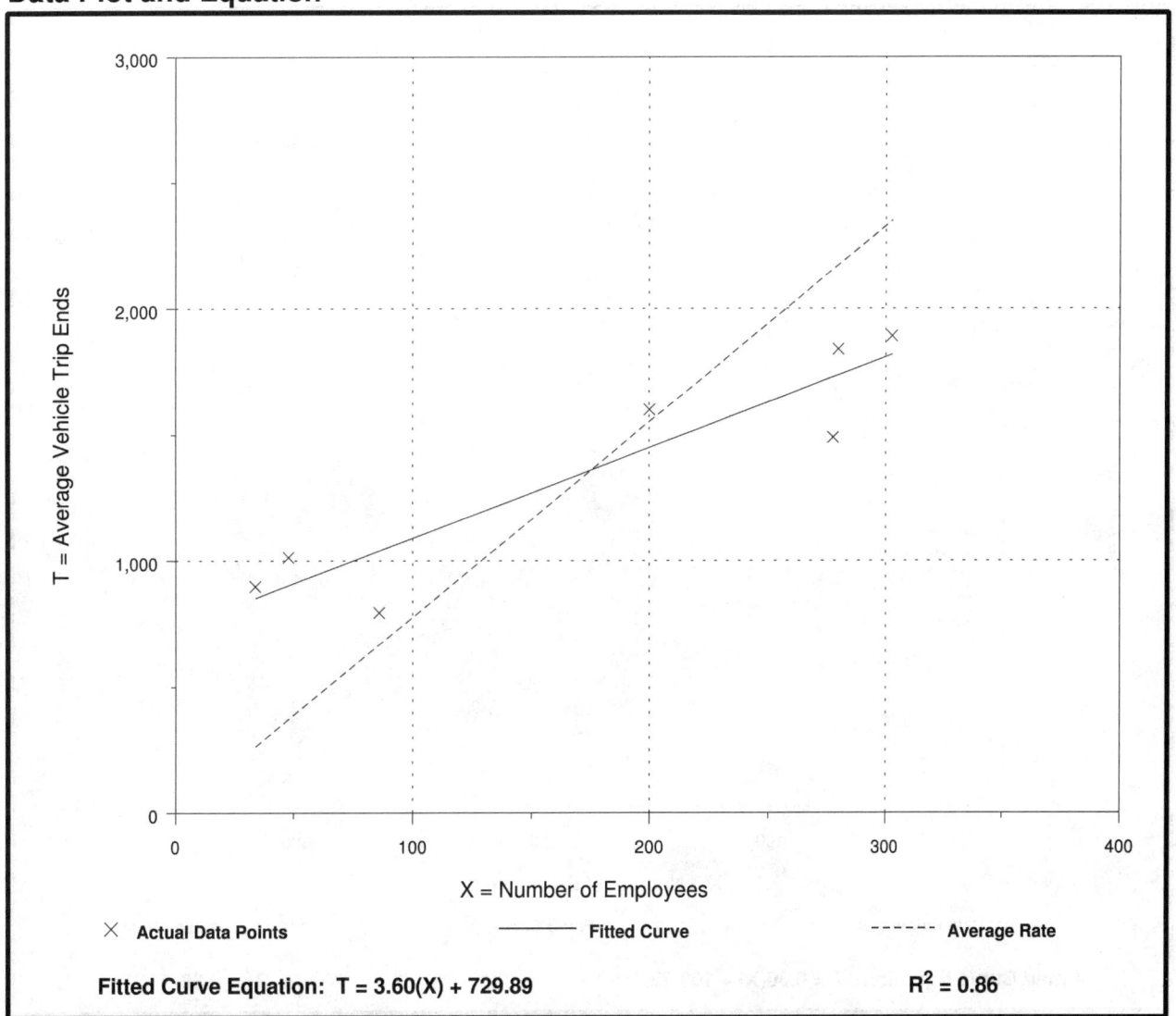

Data Plot and Equation

Fitted Curve Equation: T = 3.60(X) + 729.89 $R^2 = 0.86$

United States Post Office
(732)

Average Vehicle Trip Ends vs: Employees
On a: Sunday,
Peak Hour of Generator

Number of Studies: 7
Avg. Number of Employees: 176
Directional Distribution: 55% entering, 45% exiting

Trip Generation per Employee

Average Rate	Range of Rates	Standard Deviation
0.89	0.61 - 3.50	1.12

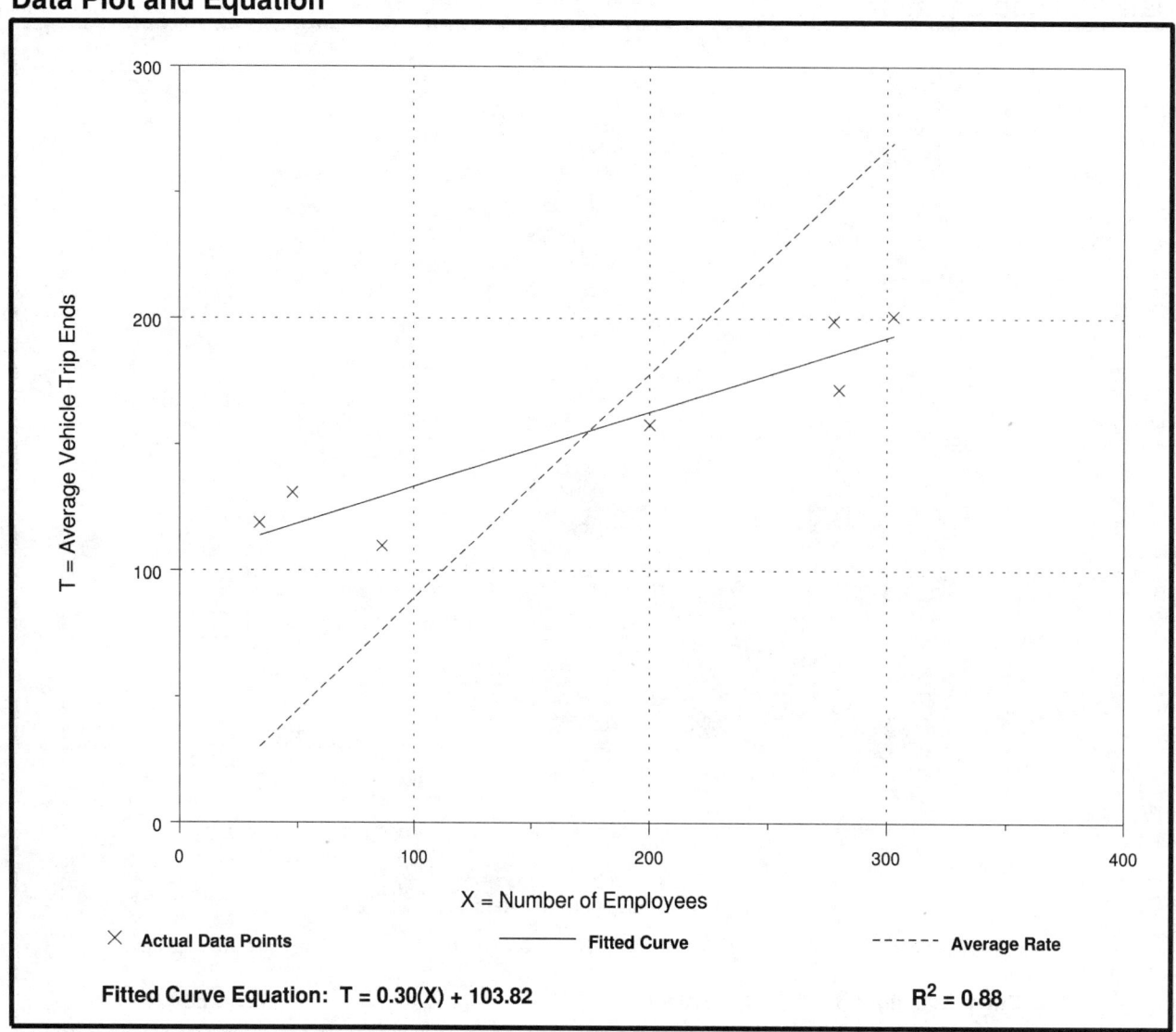

Data Plot and Equation

Fitted Curve Equation: T = 0.30(X) + 103.82 $R^2 = 0.88$

United States Post Office
(732)

Average Vehicle Trip Ends vs: 1000 Sq. Feet Gross Floor Area
On a: Weekday

Number of Studies: 8
Average 1000 Sq. Feet GFA: 37
Directional Distribution: 50% entering, 50% exiting

Trip Generation per 1000 Sq. Feet Gross Floor Area

Average Rate	Range of Rates	Standard Deviation
108.19	35.57 - 352.42	110.21

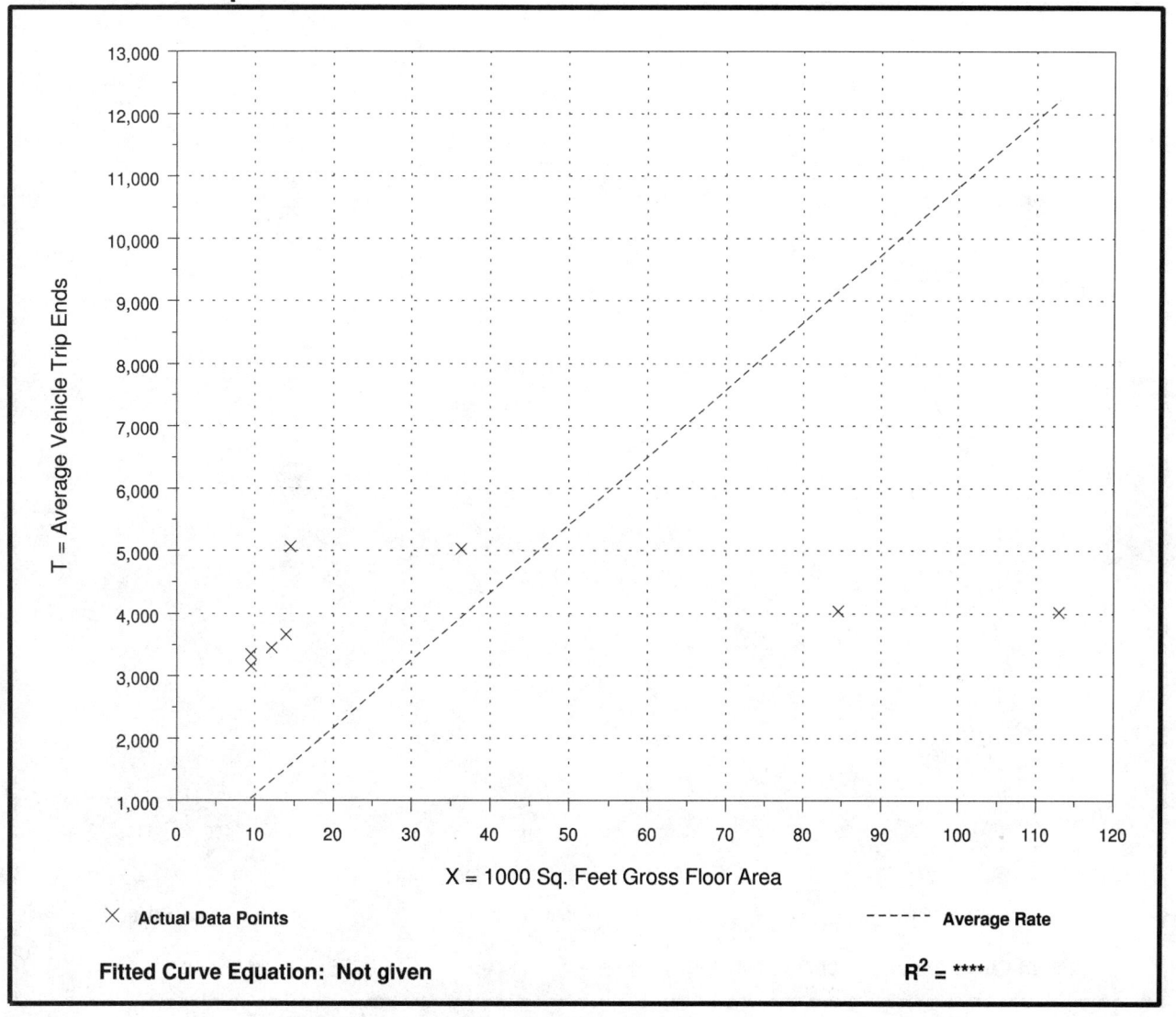

Data Plot and Equation

Fitted Curve Equation: Not given $R^2 = ****$

United States Post Office
(732)

Average Vehicle Trip Ends vs: 1000 Sq. Feet Gross Floor Area
On a: Weekday,
Peak Hour of Adjacent Street Traffic,
One Hour Between 7 and 9 a.m.

Number of Studies: 13
Average 1000 Sq. Feet GFA: 25
Directional Distribution: 52% entering, 48% exiting

Trip Generation per 1000 Sq. Feet Gross Floor Area

Average Rate	Range of Rates	Standard Deviation
8.23	2.21 - 38.17	9.14

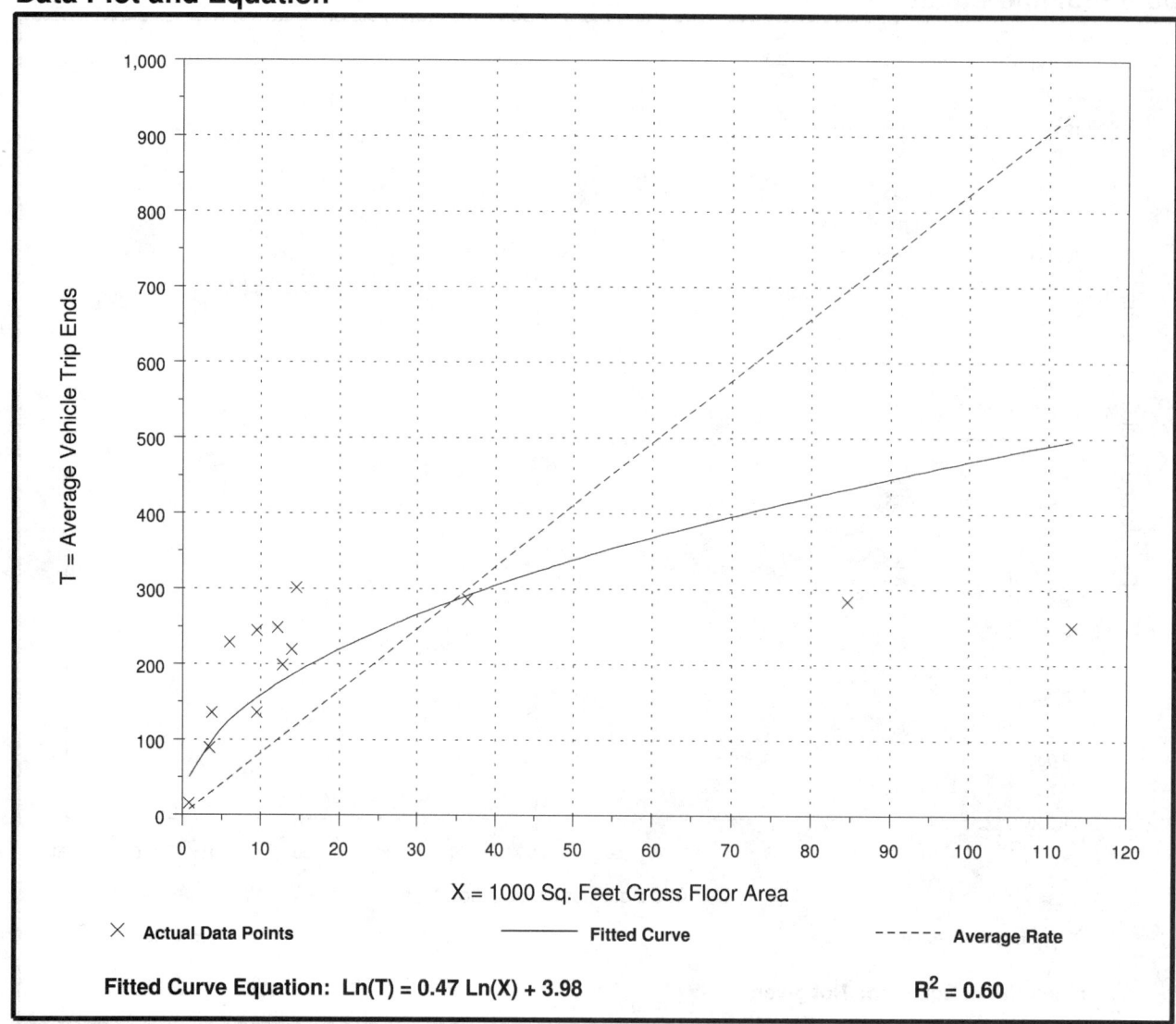

Data Plot and Equation

Fitted Curve Equation: $Ln(T) = 0.47\ Ln(X) + 3.98$ $R^2 = 0.60$

United States Post Office
(732)

Average Vehicle Trip Ends vs: 1000 Sq. Feet Gross Floor Area
On a: Weekday,
Peak Hour of Adjacent Street Traffic,
One Hour Between 4 and 6 p.m.

Number of Studies: 15
Average 1000 Sq. Feet GFA: 27
Directional Distribution: 51% entering, 49% exiting

Trip Generation per 1000 Sq. Feet Gross Floor Area

Average Rate	Range of Rates	Standard Deviation
11.22	3.24 - 80.00	14.25

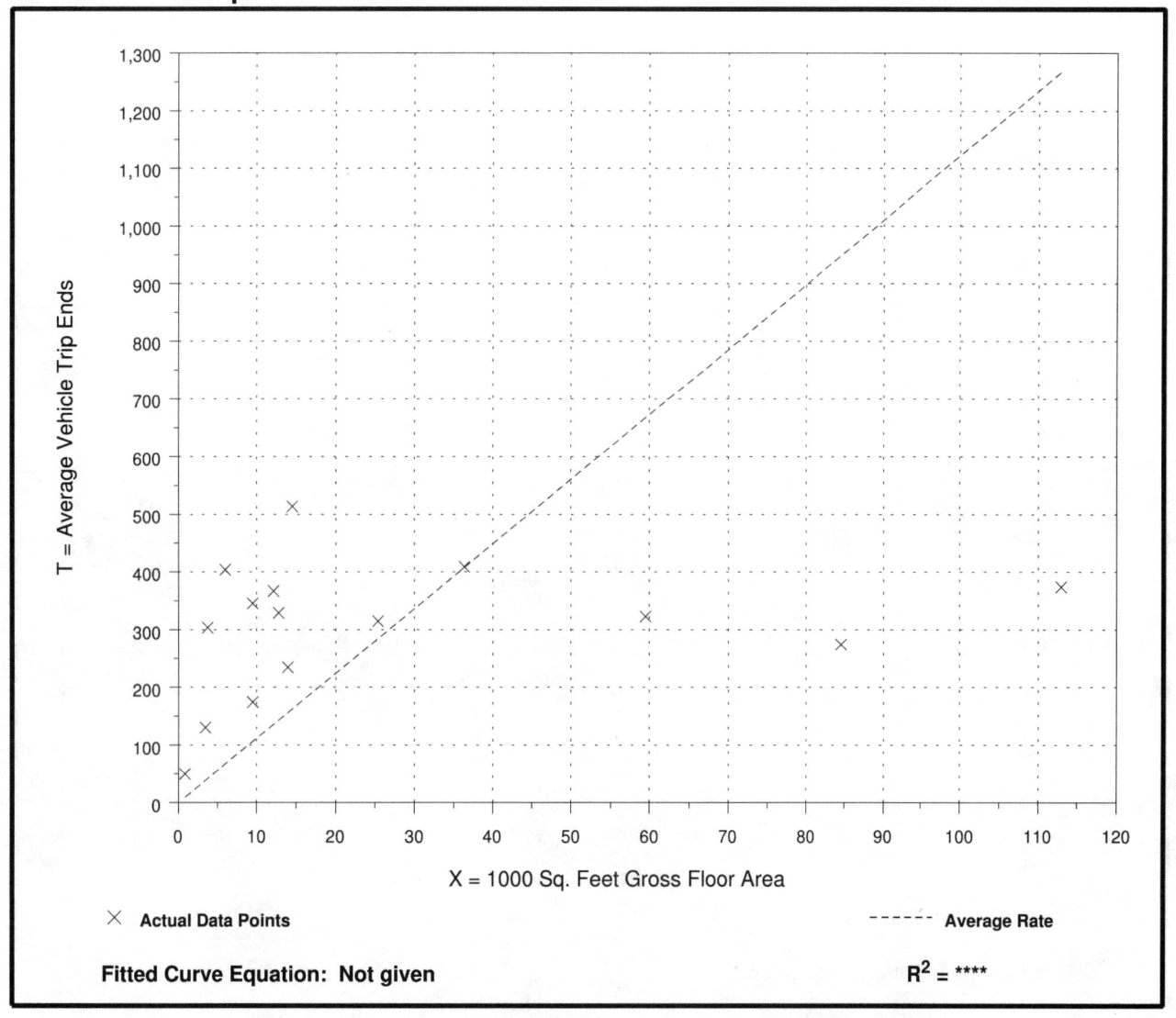

Data Plot and Equation

Fitted Curve Equation: Not given $R^2 =$ ****

United States Post Office
(732)

Average Vehicle Trip Ends vs: 1000 Sq. Feet Gross Floor Area
On a: Weekday,
A.M. Peak Hour of Generator

Number of Studies: 11
Average 1000 Sq. Feet GFA: 29
Directional Distribution: 49% entering, 51% exiting

Trip Generation per 1000 Sq. Feet Gross Floor Area

Average Rate	Range of Rates	Standard Deviation
12.19	2.99 - 51.20	14.41

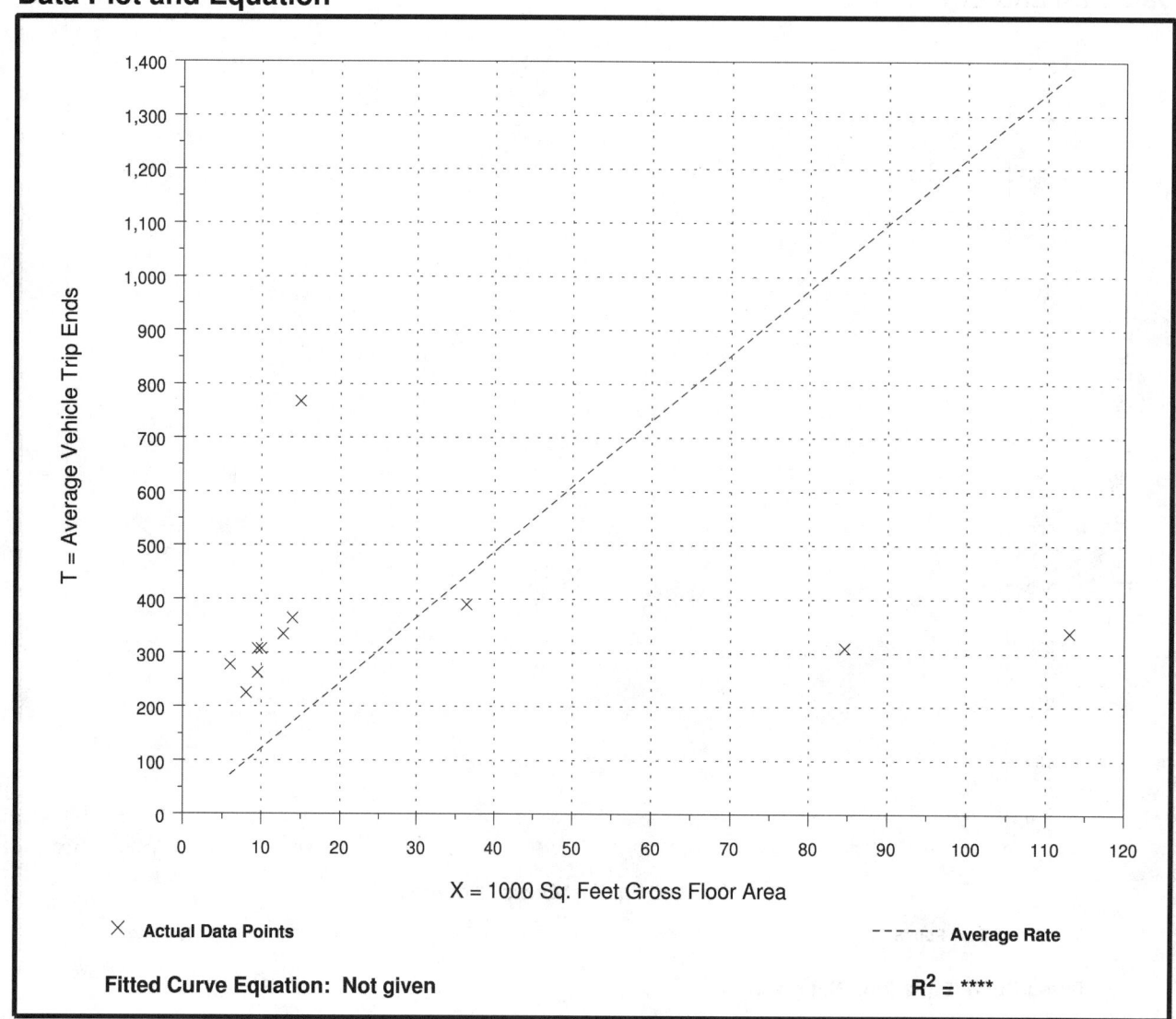

Data Plot and Equation

X   Actual Data Points   ----- Average Rate

Fitted Curve Equation: Not given $R^2 = ****$

United States Post Office
(732)

Average Vehicle Trip Ends vs: 1000 Sq. Feet Gross Floor Area
On a: Weekday,
P.M. Peak Hour of Generator

Number of Studies: 12
Average 1000 Sq. Feet GFA: 27
Directional Distribution: 51% entering, 49% exiting

Trip Generation per 1000 Sq. Feet Gross Floor Area

Average Rate	Range of Rates	Standard Deviation
14.67	3.46 - 82.89	18.56

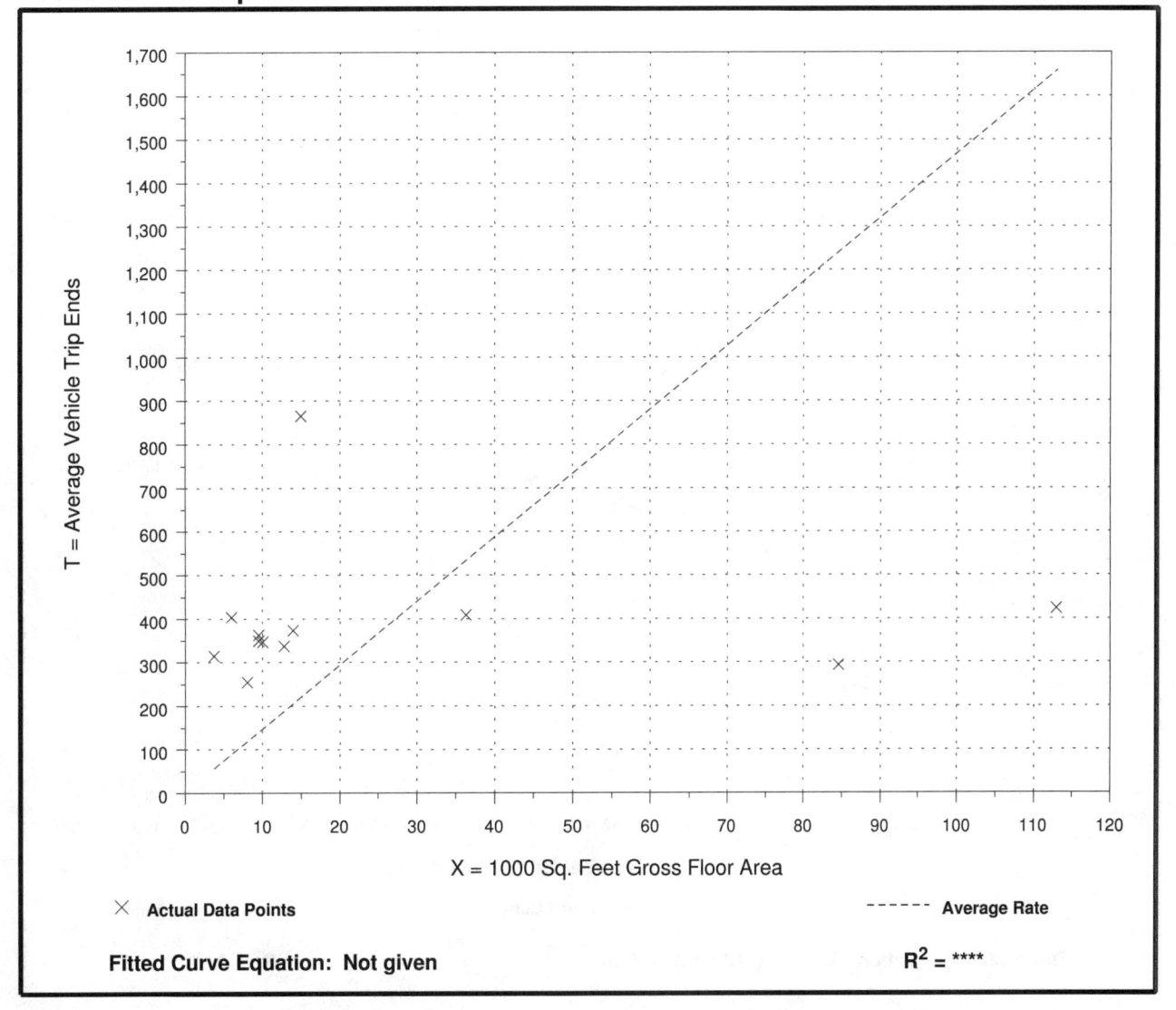

Data Plot and Equation

Fitted Curve Equation: Not given $R^2 = ****$

United States Post Office
(732)

Average Vehicle Trip Ends vs: 1000 Sq. Feet Gross Floor Area
On a: Saturday

Number of Studies: 6
Average 1000 Sq. Feet GFA: 44
Directional Distribution: 50% entering, 50% exiting

Trip Generation per 1000 Sq. Feet Gross Floor Area

Average Rate	Range of Rates	Standard Deviation
48.69	18.35 - 185.89	45.54

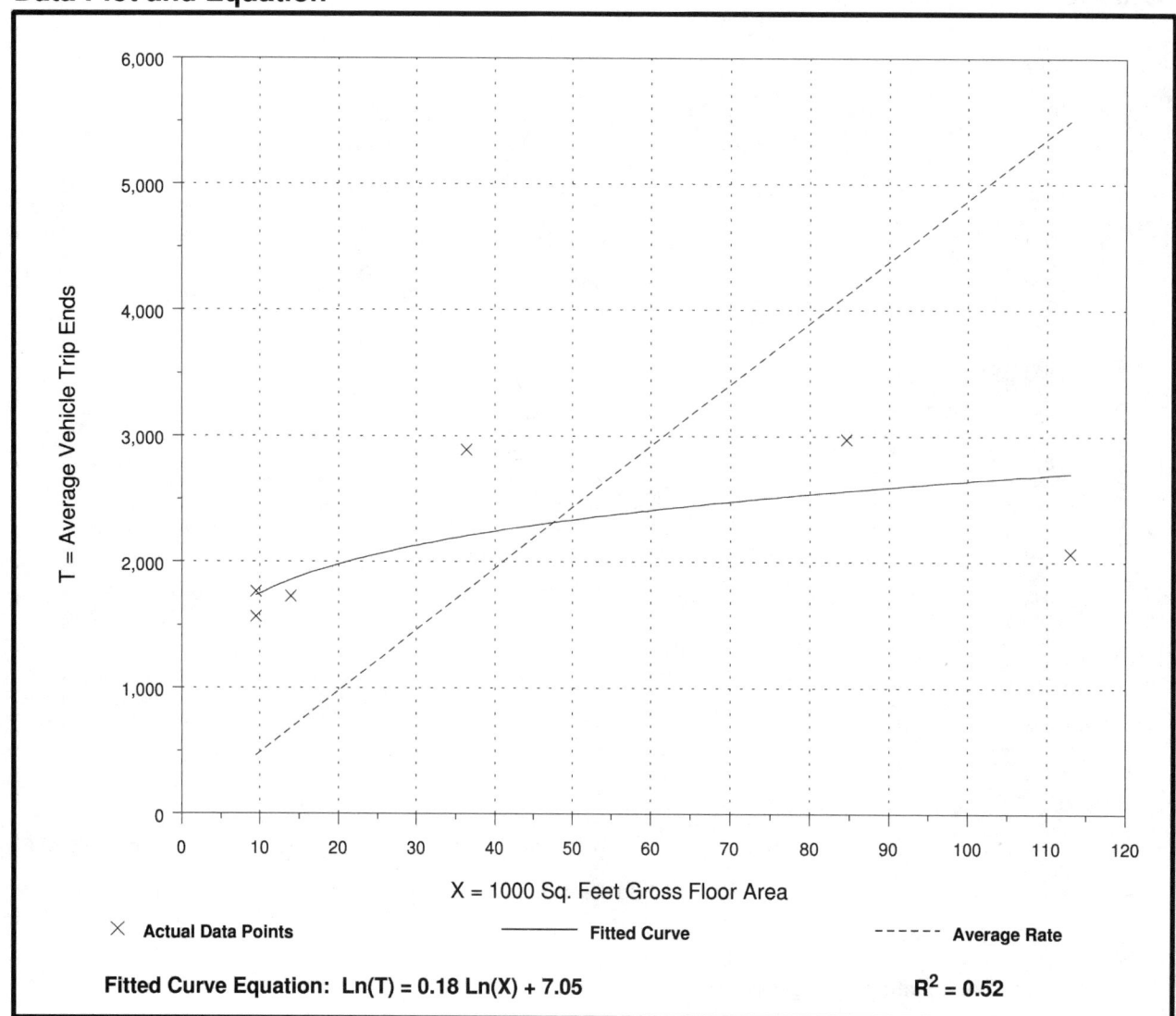

Data Plot and Equation

Fitted Curve Equation: $Ln(T) = 0.18\, Ln(X) + 7.05$ $R^2 = 0.52$

United States Post Office
(732)

Average Vehicle Trip Ends vs: 1000 Sq. Feet Gross Floor Area
On a: Saturday,
Peak Hour of Generator

Number of Studies: 6
Average 1000 Sq. Feet GFA: 44
Directional Distribution: 55% entering, 45% exiting

Trip Generation per 1000 Sq. Feet Gross Floor Area

Average Rate	Range of Rates	Standard Deviation
5.88	2.40 - 22.53	6.27

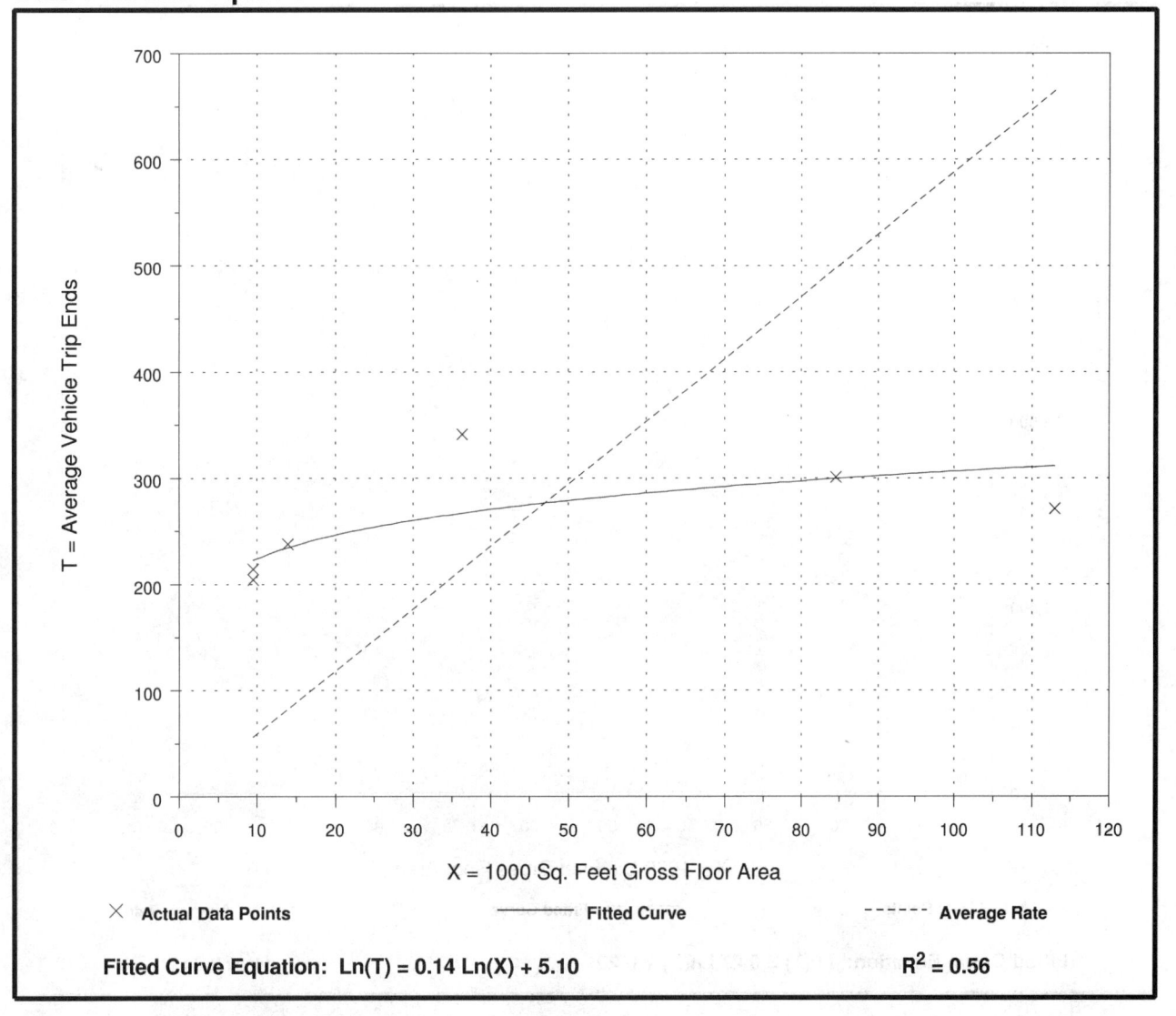

Data Plot and Equation

Fitted Curve Equation: $Ln(T) = 0.14\ Ln(X) + 5.10$ $R^2 = 0.56$

United States Post Office
(732)

Average Vehicle Trip Ends vs: 1000 Sq. Feet Gross Floor Area
On a: Sunday

Number of Studies: 6
Average 1000 Sq. Feet GFA: 44
Directional Distribution: 50% entering, 50% exiting

Trip Generation per 1000 Sq. Feet Gross Floor Area

Average Rate	Range of Rates	Standard Deviation
28.81	13.18 - 106.74	24.28

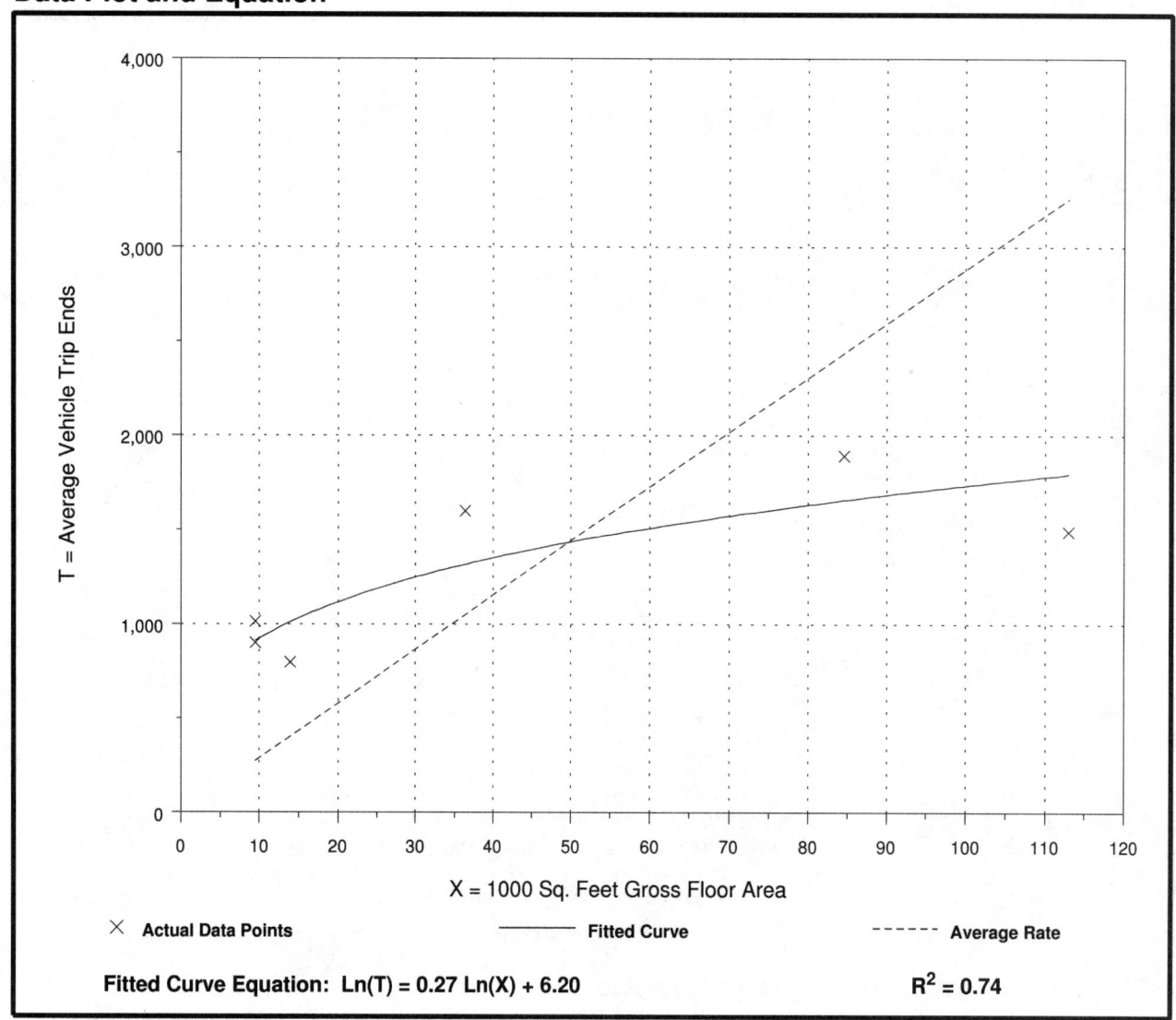

Data Plot and Equation

Fitted Curve Equation: $Ln(T) = 0.27\, Ln(X) + 6.20$ $R^2 = 0.74$

United States Post Office
(732)

Average Vehicle Trip Ends vs: 1000 Sq. Feet Gross Floor Area
On a: Sunday,
Peak Hour of Generator

Number of Studies: 6
Average 1000 Sq. Feet GFA: 44
Directional Distribution: 55% entering, 45% exiting

Trip Generation per 1000 Sq. Feet Gross Floor Area

Average Rate	Range of Rates	Standard Deviation
3.44	1.76 - 13.79	3.57

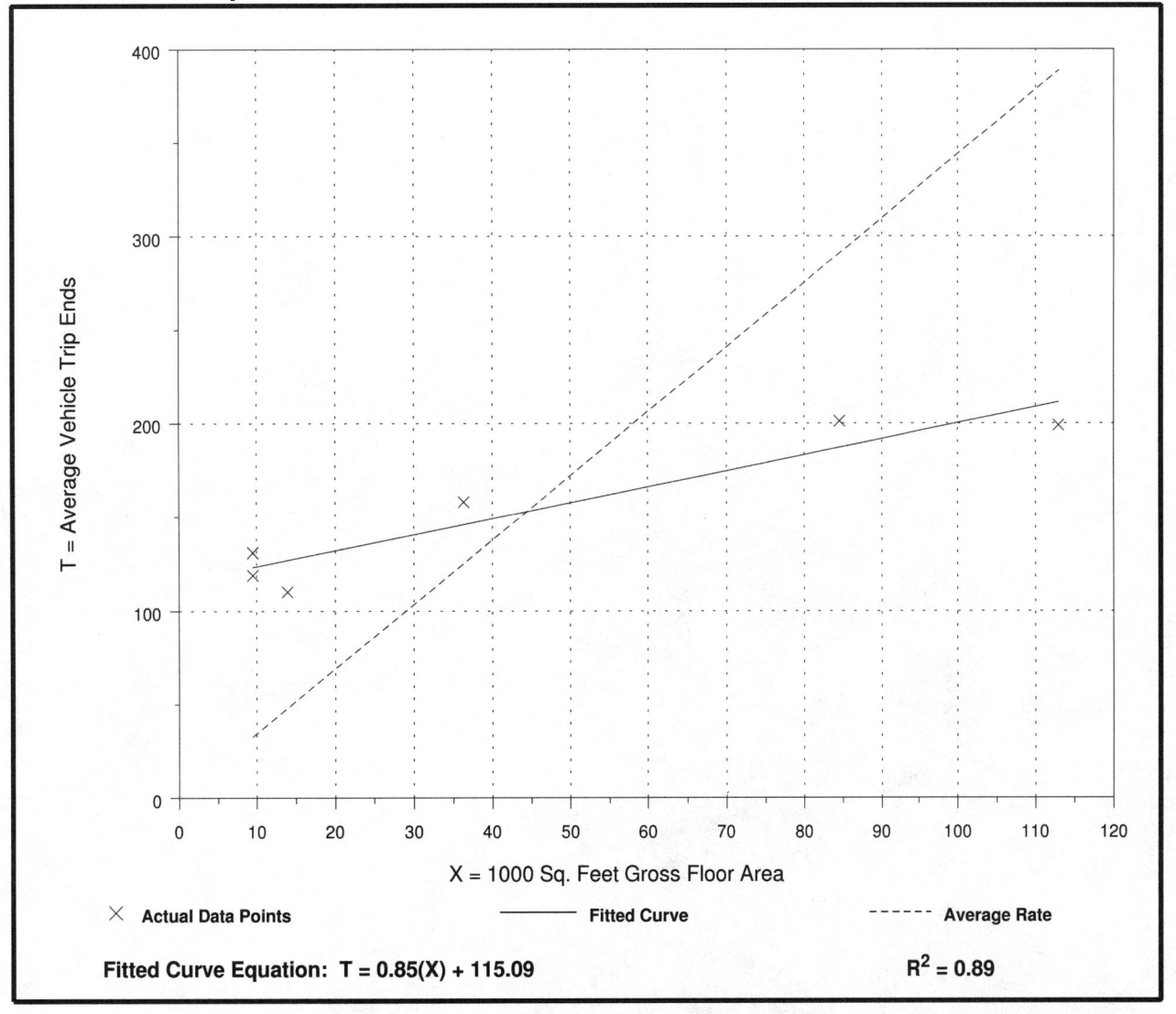

Data Plot and Equation

Fitted Curve Equation: $T = 0.85(X) + 115.09$ $R^2 = 0.89$

Land Use: 733
Government Office Complex

Description

A government office complex is a related group of buildings where a variety of functions of a city, county, state, federal, other governmental unit, or multiple governmental units are carried out. This complex differs from a government office building (Land Use 730) in that it is a group of buildings that are interconnected by pedestrian walkways.

Additional Data

The sites were surveyed in 1967 and 1999 in California.

Source Numbers

8, 508

Land Use: 733
Government Office Complex
Independent Variables with One Observation

The following trip generation data are for independent variables with only one observation. This information is shown in this table only; there are no related plots for these data.

Users are cautioned to use data with care because of the small sample size.

Independent Variable	Average Trip Generation Rate	Size of Independent Variable	Number of Studies	Directional Distribution
Employees				
Weekday A.M. Peak Hour of Generator	1.13	173	1	Not available
1,000 Square Feet Gross Floor Area				
Weekday P.M. Peak Hour of Generator	3.59	68	1	Not available

Government Office Complex
(733)

Average Vehicle Trip Ends vs: Employees
On a: Weekday

Number of Studies: 2
Avg. Number of Employees: 374
Directional Distribution: 50% entering, 50% exiting

Trip Generation per Employee

Average Rate	Range of Rates	Standard Deviation
7.75	6.09 - 13.29	*

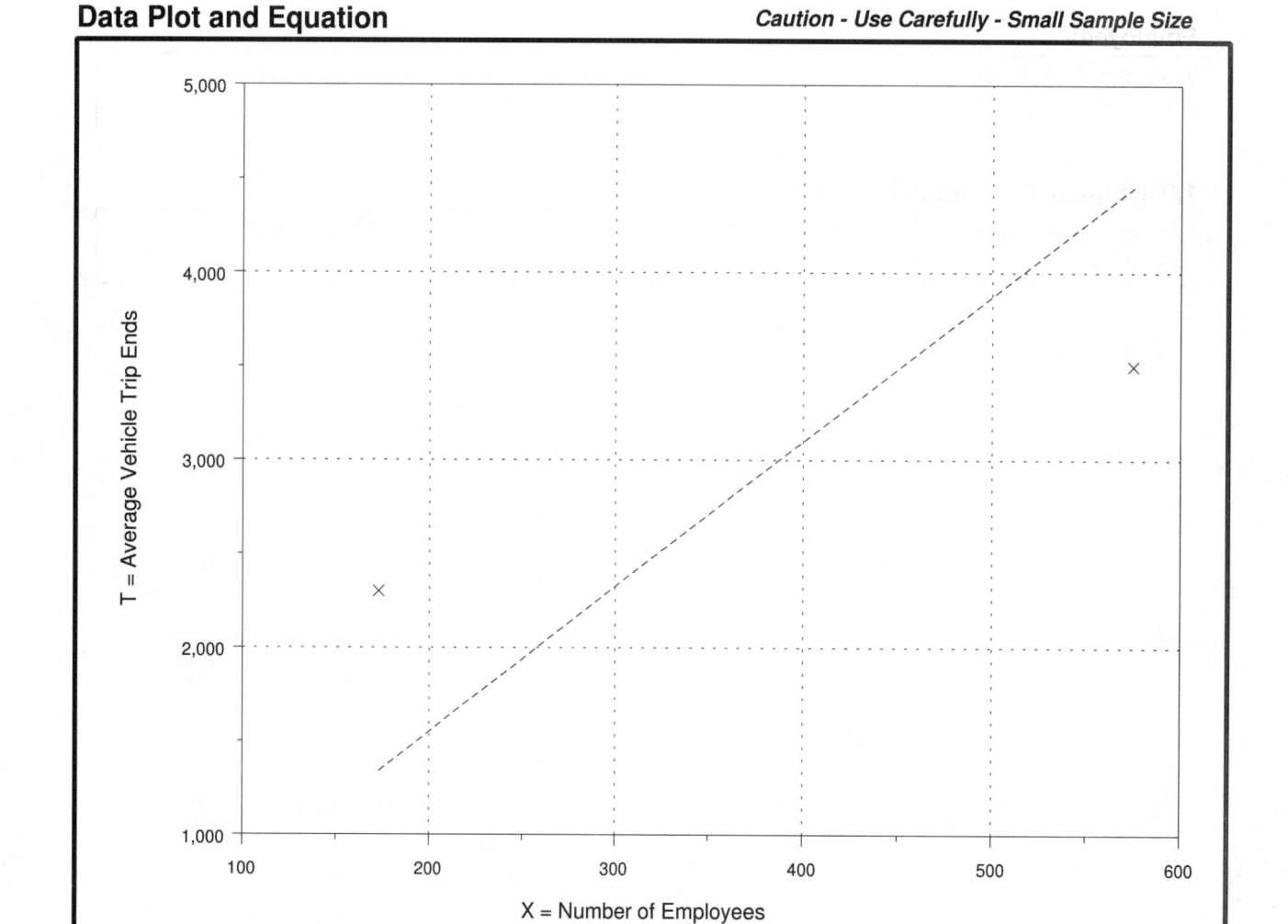

Data Plot and Equation

Caution - Use Carefully - Small Sample Size

Fitted Curve Equation: Not given $R^2 = ****$

Government Office Complex
(733)

Average Vehicle Trip Ends vs: Employees
On a: Weekday,
Peak Hour of Adjacent Street Traffic,
One Hour Between 7 and 9 a.m.

Number of Studies: 2
Avg. Number of Employees: 374
Directional Distribution: 89% entering, 11% exiting

Trip Generation per Employee

Average Rate	Range of Rates	Standard Deviation
0.61	0.55 - 0.83	*

Data Plot and Equation

Caution - Use Carefully - Small Sample Size

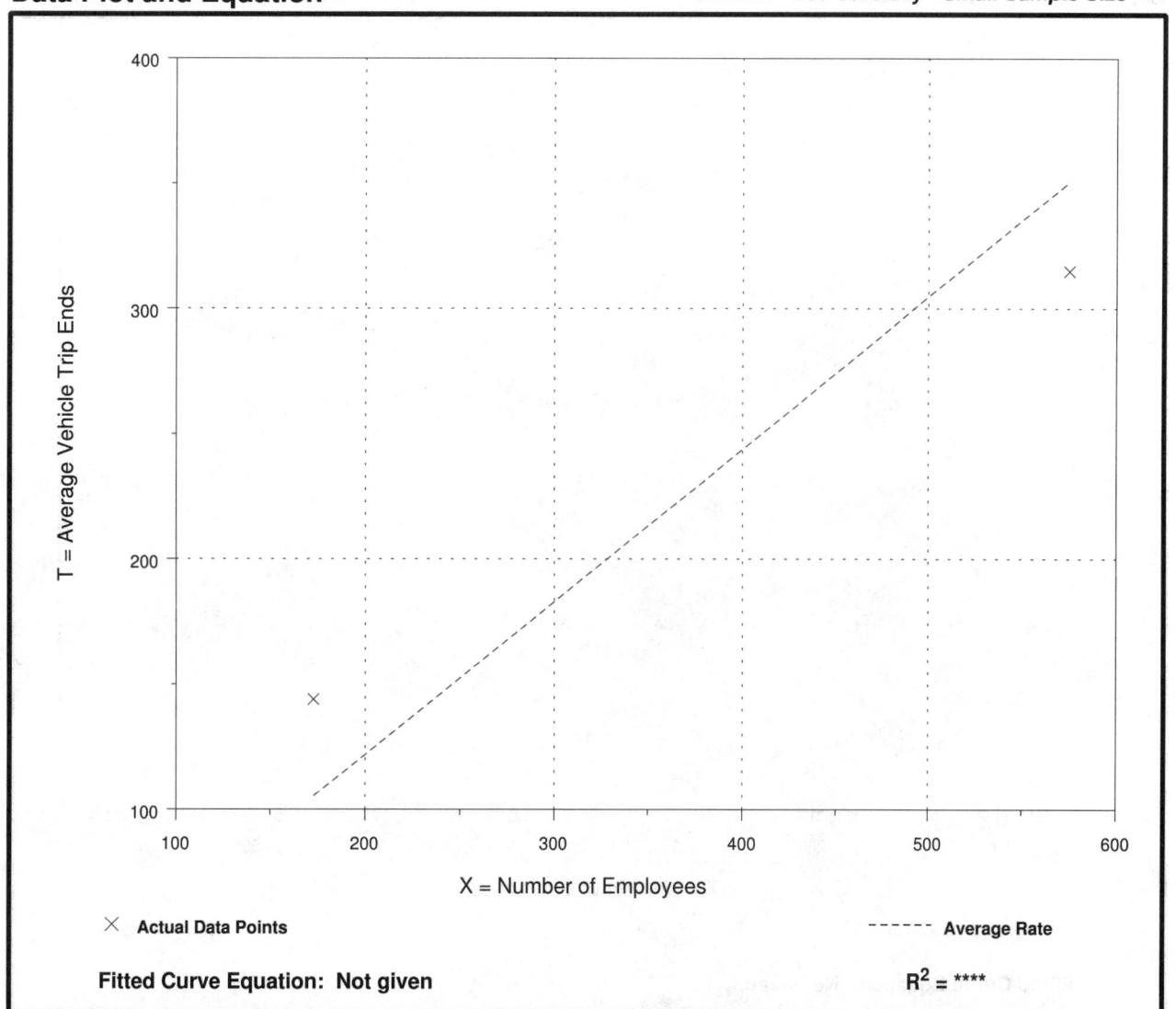

Fitted Curve Equation: Not given R^2 = ****

Government Office Complex
(733)

Average Vehicle Trip Ends vs: Employees
On a: Weekday,
Peak Hour of Adjacent Street Traffic,
One Hour Between 4 and 6 p.m.

Number of Studies: 2
Avg. Number of Employees: 374
Directional Distribution: 31% entering, 69% exiting

Trip Generation per Employee

Average Rate	Range of Rates	Standard Deviation
0.79	0.70 - 1.10	*

Data Plot and Equation

Caution - Use Carefully - Small Sample Size

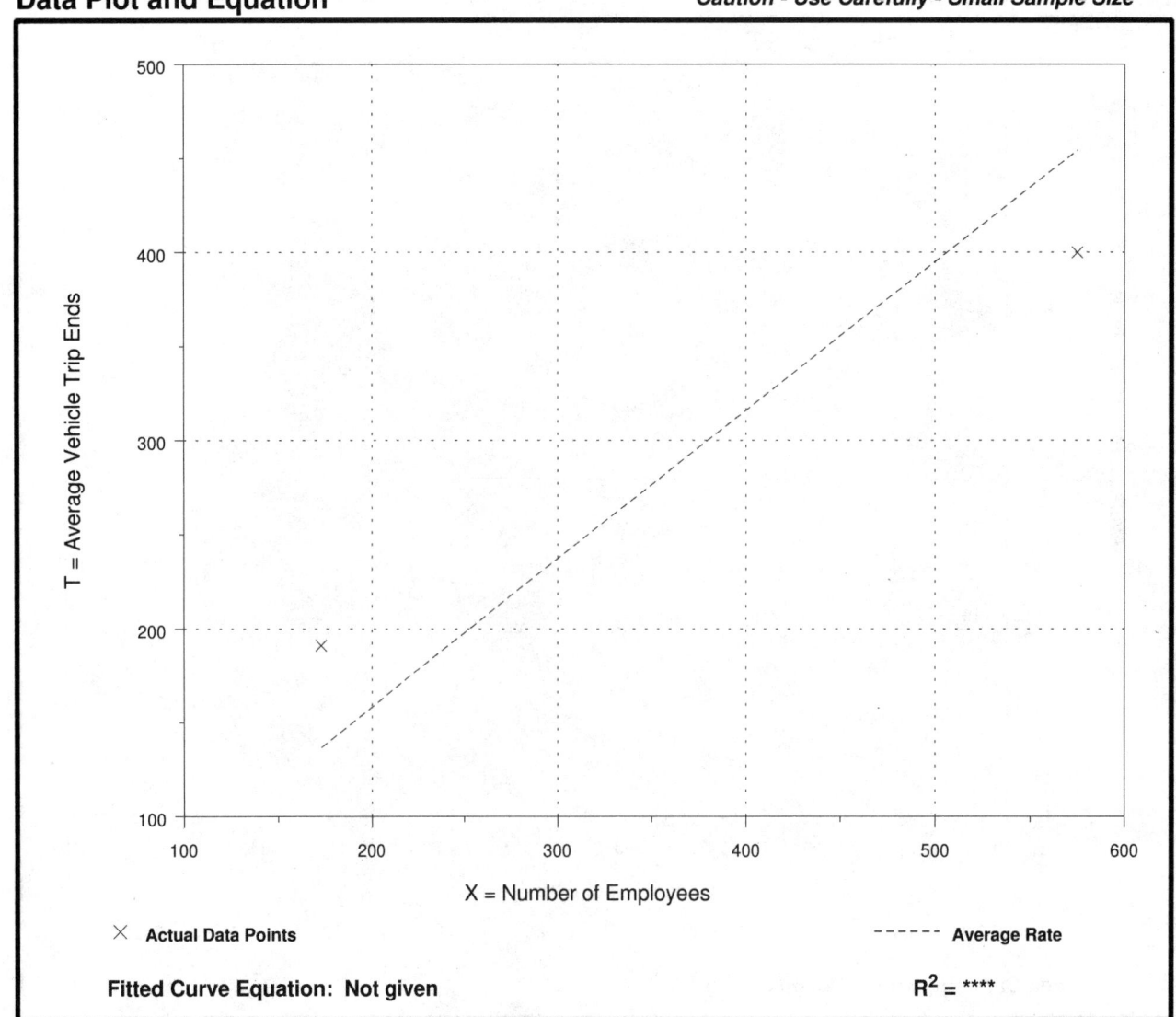

Fitted Curve Equation: Not given $R^2 = ****$

Government Office Complex
(733)

Average Vehicle Trip Ends vs: 1000 Sq. Feet Gross Floor Area
On a: Weekday

Number of Studies: 2
Average 1000 Sq. Feet GFA: 104
Directional Distribution: 50% entering, 50% exiting

Trip Generation per 1000 Sq. Feet Gross Floor Area

Average Rate	Range of Rates	Standard Deviation
27.92	25.00 - 33.97	*

Data Plot and Equation

Caution - Use Carefully - Small Sample Size

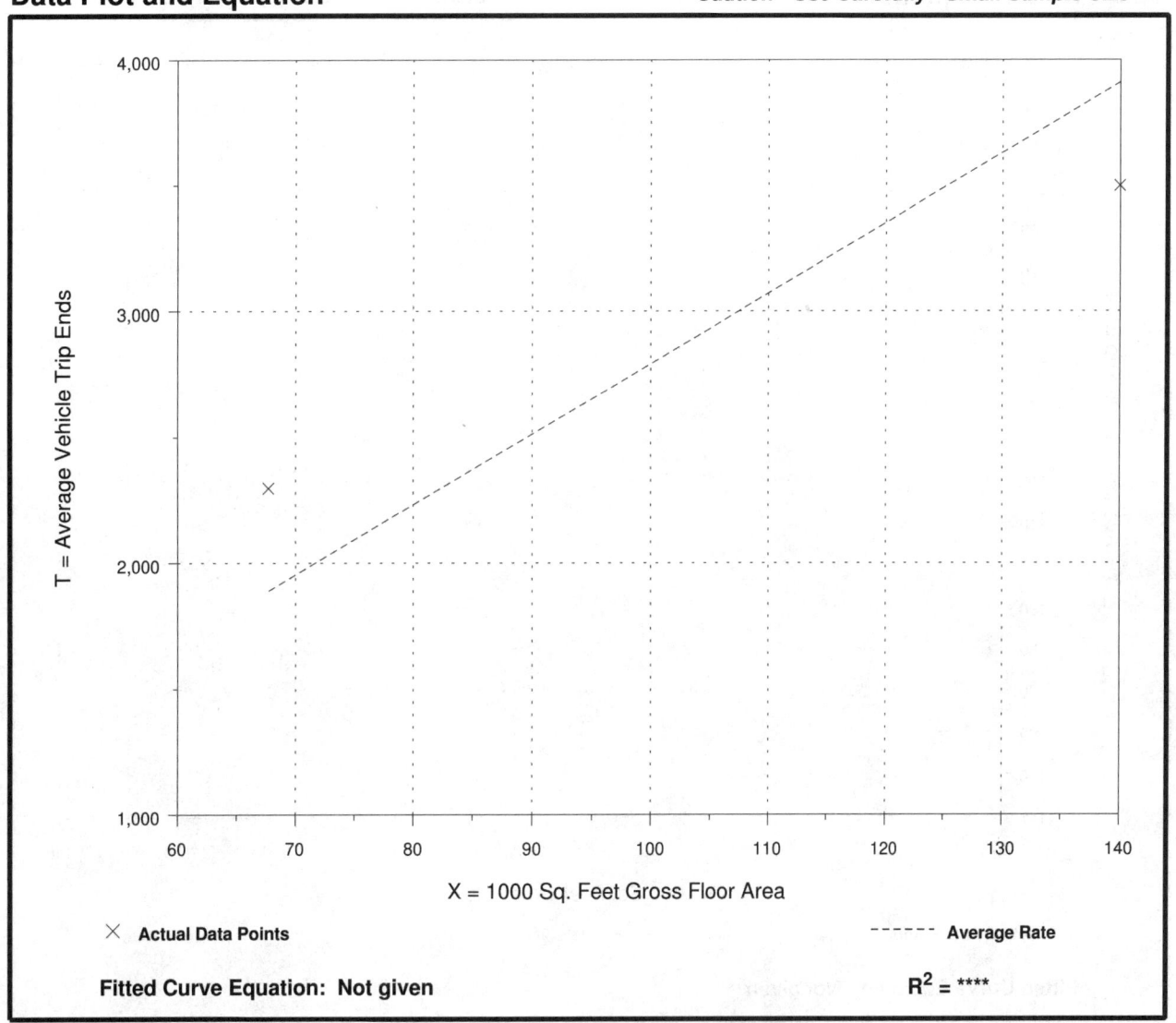

Fitted Curve Equation: Not given $R^2 = ****$

Trip Generation, 9th Edition • Institute of Transportation Engineers

Government Office Complex
(733)

Average Vehicle Trip Ends vs: 1000 Sq. Feet Gross Floor Area
On a: Weekday,
Peak Hour of Adjacent Street Traffic,
One Hour Between 7 and 9 a.m.

Number of Studies: 2
Average 1000 Sq. Feet GFA: 104
Directional Distribution: 89% entering, 11% exiting

Trip Generation per 1000 Sq. Feet Gross Floor Area

Average Rate	Range of Rates	Standard Deviation
2.21	2.13 - 2.25	*

Data Plot and Equation

Caution - Use Carefully - Small Sample Size

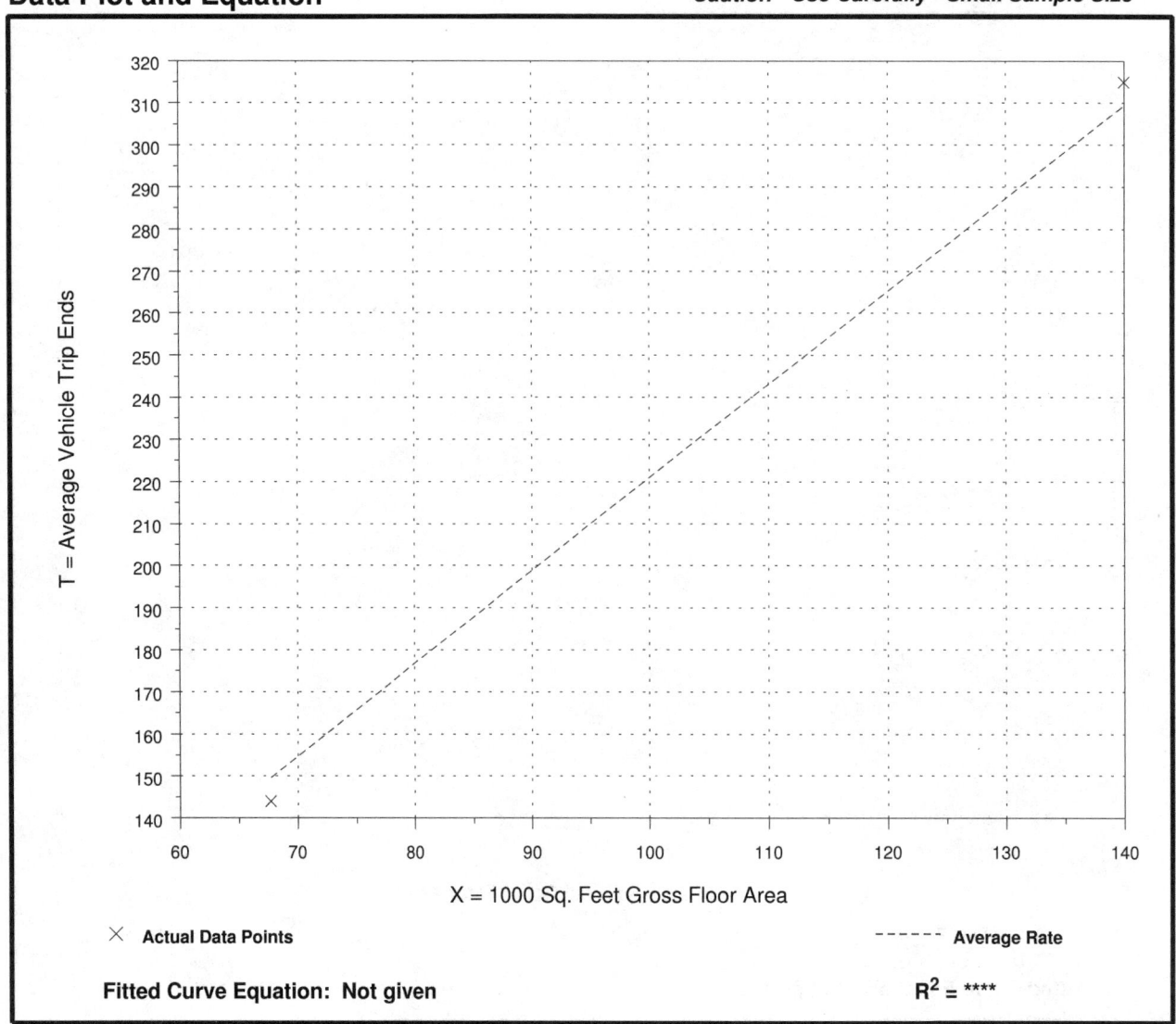

× Actual Data Points

- - - - - Average Rate

Fitted Curve Equation: Not given $R^2 =$ ****

Government Office Complex
(733)

Average Vehicle Trip Ends vs: 1000 Sq. Feet Gross Floor Area
On a: Weekday,
Peak Hour of Adjacent Street Traffic,
One Hour Between 4 and 6 p.m.

Number of Studies: 2
Average 1000 Sq. Feet GFA: 104
Directional Distribution: 31% entering, 69% exiting

Trip Generation per 1000 Sq. Feet Gross Floor Area

Average Rate	Range of Rates	Standard Deviation
2.85	2.82 - 2.86	*

Data Plot and Equation

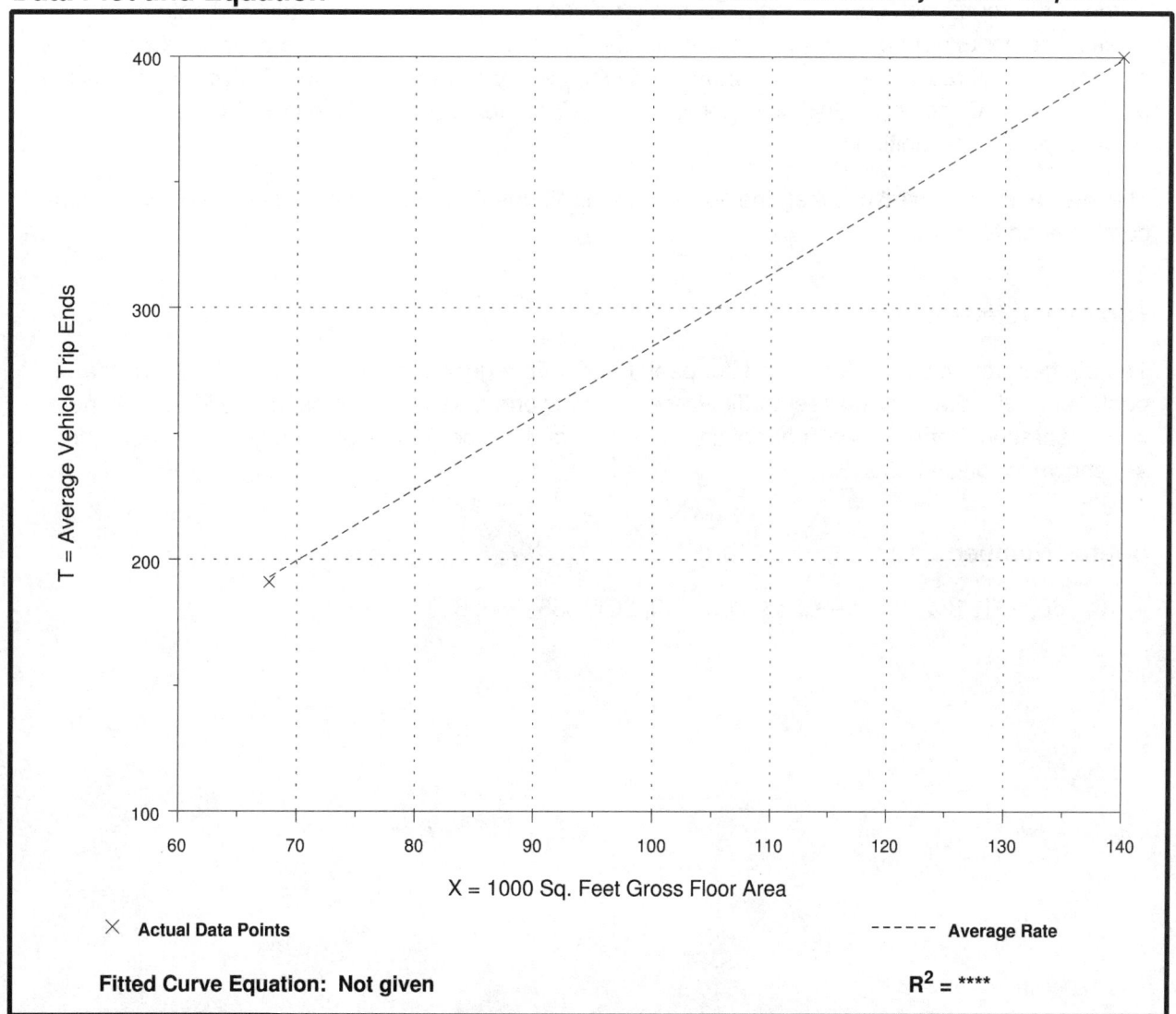

Caution - Use Carefully - Small Sample Size

Fitted Curve Equation: Not given $R^2 = ****$

Land Use: 750
Office Park

Description

Office parks are usually suburban subdivisions or planned unit developments containing general office buildings and support services, such as banks, restaurants and service stations, arranged in a park- or campus-like atmosphere. General office building (Land Use 710), corporate headquarters building (Land Use 714), single tenant office building (Land Use 715), research and development center (Land Use 760) and business park (Land Use 770) are related uses.

Additional Data

Some of the regression curves plotted for this land use may produce illogical trip-end estimates for small office parks. When the proposed site size is significantly smaller than the average-sized facility published in this report, caution should be used when applying these statistics. For more information, please refer to Chapter 3, "Guidelines for Estimating Trip Generation," of the ITE *Trip Generation Handbook*, Second Edition.

The sites were surveyed between the 1970s and the 2000s throughout the United States, with many conducted in New York.

Trip Characteristics

The trip generation for the A.M. and P.M. peak hours of the generator typically coincided with the peak hours of the adjacent street traffic; therefore, only one A.M. peak hour and one P.M. peak hour, which represent both the peak hour of the generator and the peak hour of the adjacent street traffic, are shown for office parks.

Source Numbers

4, 15, 160, 161, 184, 185, 193, 253, 268, 300, 301, 356, 550, 618

Office Park
(750)

Average Vehicle Trip Ends vs: Employees
On a: Weekday

Number of Studies: 4
Avg. Number of Employees: 1,461
Directional Distribution: 50% entering, 50% exiting

Trip Generation per Employee

Average Rate	Range of Rates	Standard Deviation
3.50	2.92 - 3.85	1.88

Data Plot and Equation

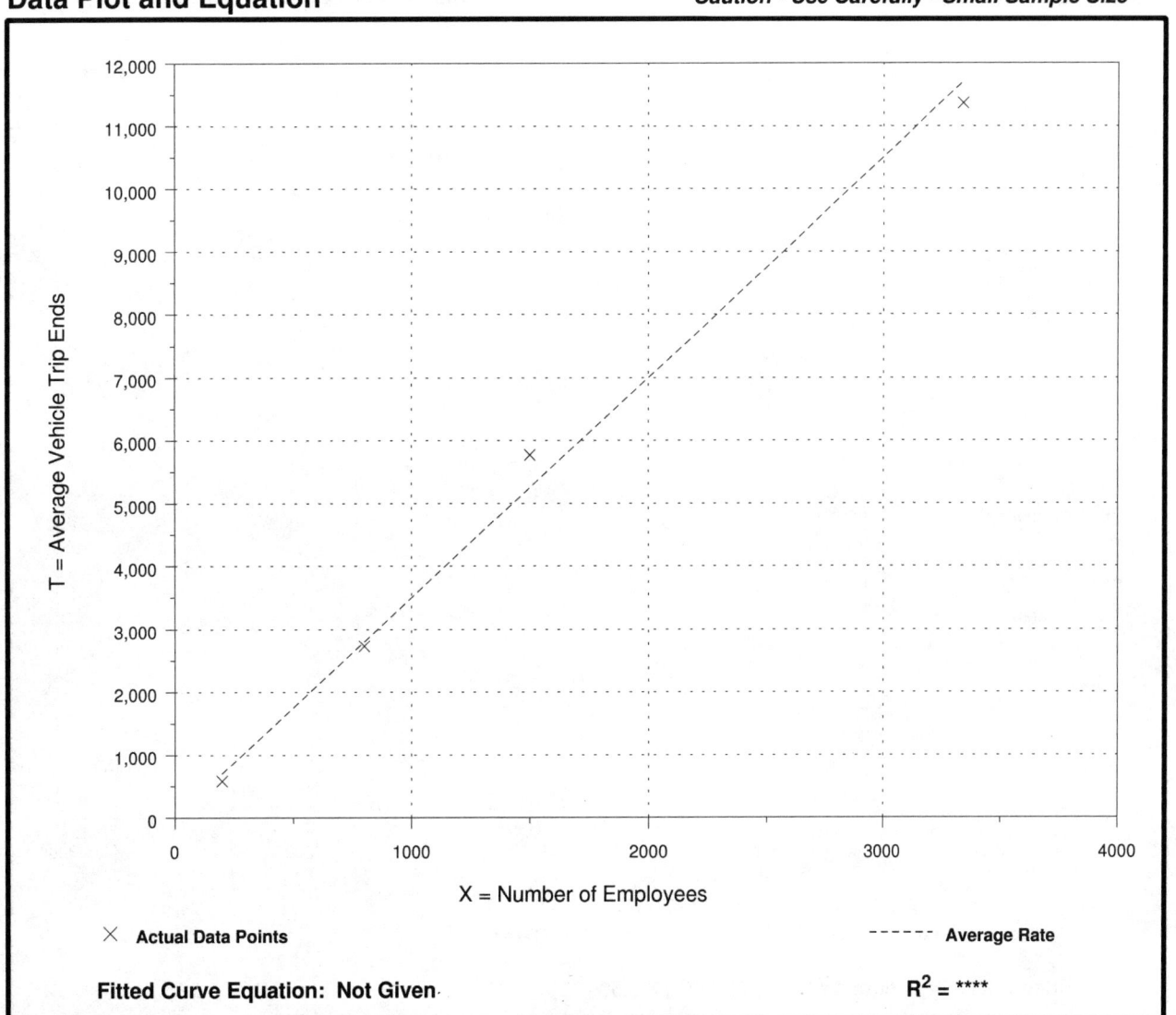

Caution - Use Carefully - Small Sample Size

Fitted Curve Equation: Not Given $R^2 = ****$

Trip Generation, 9th Edition • Institute of Transportation Engineers

Office Park
(750)

Average Vehicle Trip Ends vs: Employees
On a: Weekday,
A.M. Peak Hour

Number of Studies: 5
Avg. Number of Employees: 1,469
Directional Distribution: 92% entering, 8 % exiting

Trip Generation per Employee

Average Rate	Range of Rates	Standard Deviation
0.43	0.31 - 0.66	0.67

Data Plot and Equation

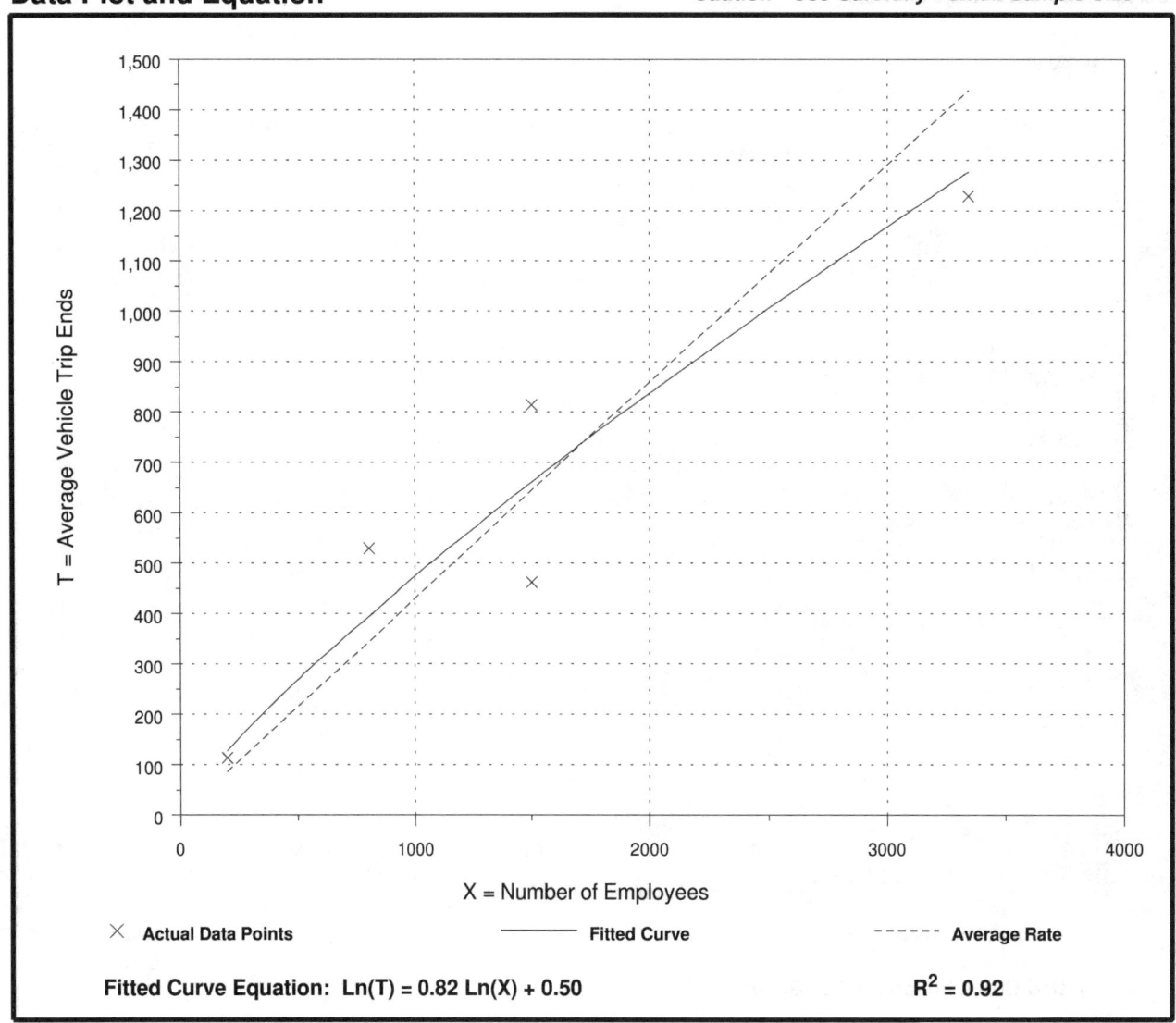

Caution - Use Carefully - Small Sample Size

Fitted Curve Equation: $Ln(T) = 0.82 \, Ln(X) + 0.50$ $R^2 = 0.92$

Office Park
(750)

Average Vehicle Trip Ends vs: Employees
On a: Weekday,
P.M. Peak Hour

Number of Studies: 5
Avg. Number of Employees: 1,469
Directional Distribution: 15% entering, 85% exiting

Trip Generation per Employee

Average Rate	Range of Rates	Standard Deviation
0.39	0.31 - 0.51	0.63

Data Plot and Equation

Caution - Use Carefully - Small Sample Size

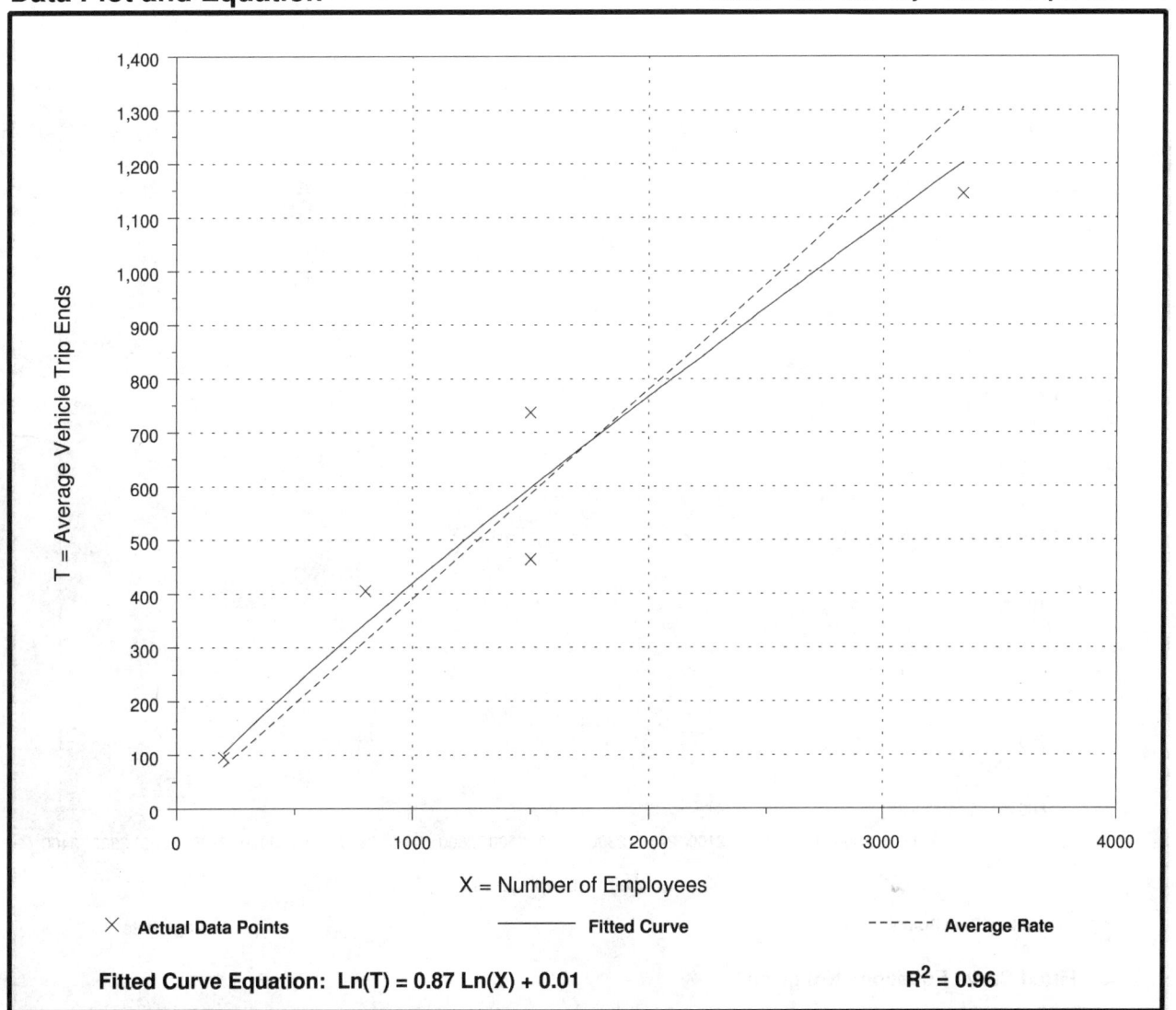

Fitted Curve Equation: Ln(T) = 0.87 Ln(X) + 0.01 $R^2 = 0.96$

Office Park
(750)

Average Vehicle Trip Ends vs: Employees
On a: Saturday

Number of Studies: 2
Avg. Number of Employees: 2,423
Directional Distribution: 50% entering, 50% exiting

Trip Generation per Employee

Average Rate	Range of Rates	Standard Deviation
0.56	0.45 - 0.82	*

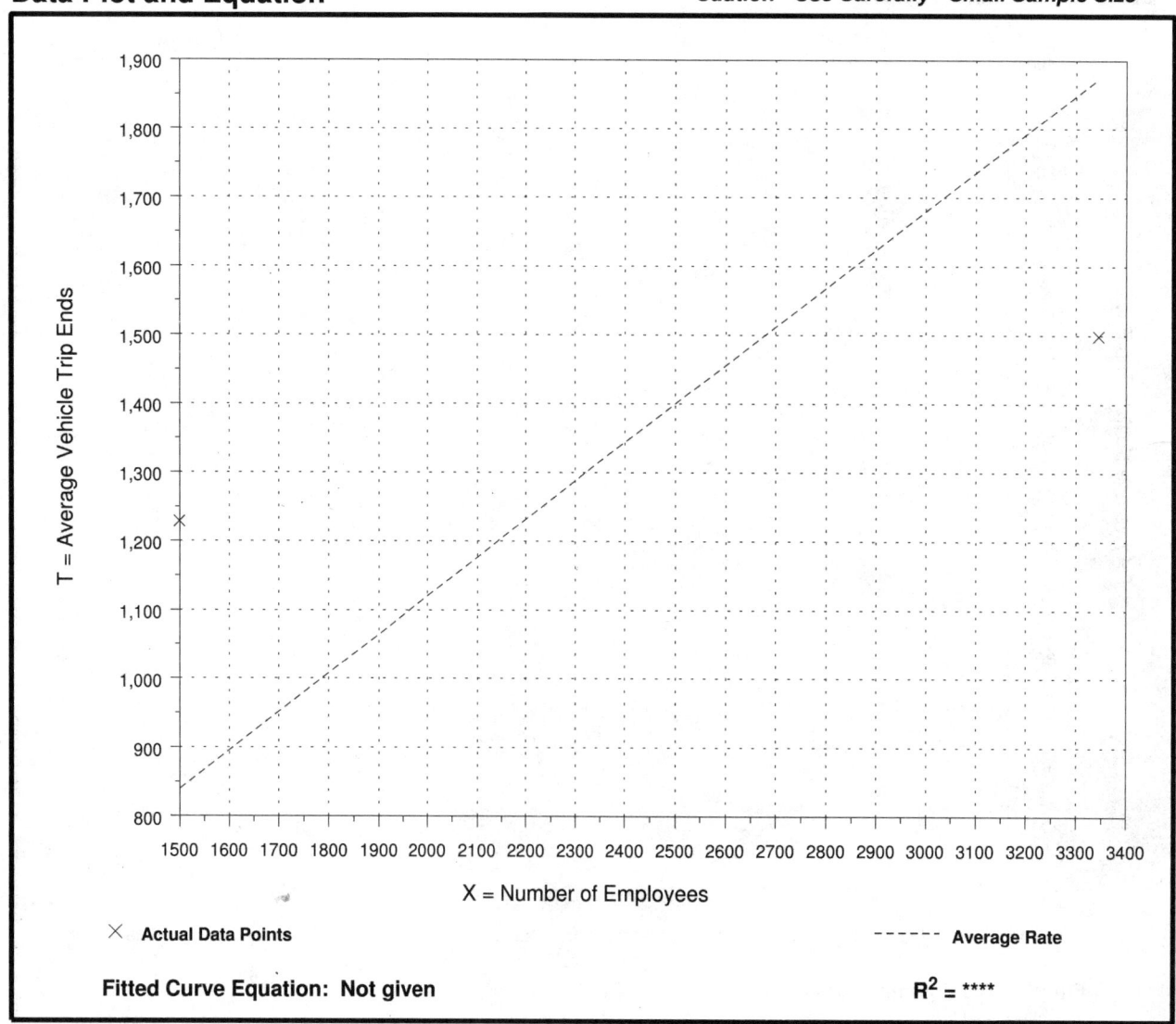

Data Plot and Equation

Caution - Use Carefully - Small Sample Size

[Plot: T = Average Vehicle Trip Ends vs X = Number of Employees, ranging from 1500 to 3400 employees. Two actual data points shown, with dashed average rate line.]

Fitted Curve Equation: Not given

$R^2 = ****$

Office Park
(750)

Average Vehicle Trip Ends vs: Employees
On a: Saturday,
Peak Hour of Generator

Number of Studies: 2
Avg. Number of Employees: 2,423
Directional Distribution: 78% entering, 22% exiting

Trip Generation per Employee

Average Rate	Range of Rates	Standard Deviation
0.05	0.04 - 0.05	*

Data Plot and Equation

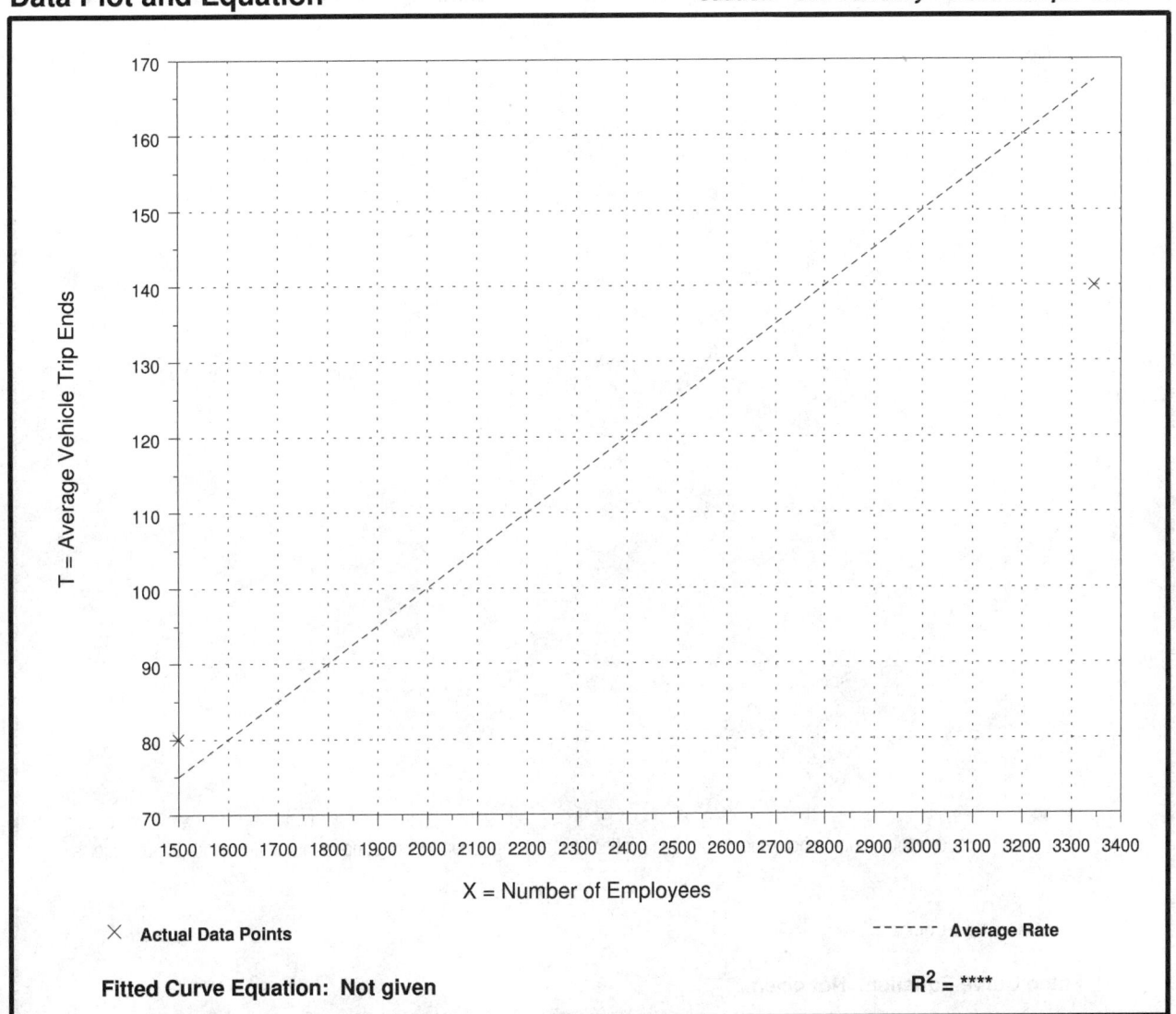

Caution - Use Carefully - Small Sample Size

[Plot: T = Average Vehicle Trip Ends vs. X = Number of Employees, with Actual Data Points marked by X and dashed line for Average Rate]

Fitted Curve Equation: Not given $R^2 = ****$

Trip Generation, 9th Edition • Institute of Transportation Engineers   1357

Office Park
(750)

Average Vehicle Trip Ends vs: Employees
On a: Sunday

Number of Studies: 2
Avg. Number of Employees: 2,423
Directional Distribution: 50% entering, 50% exiting

Trip Generation per Employee

Average Rate	Range of Rates	Standard Deviation
0.26	0.17 - 0.47	*

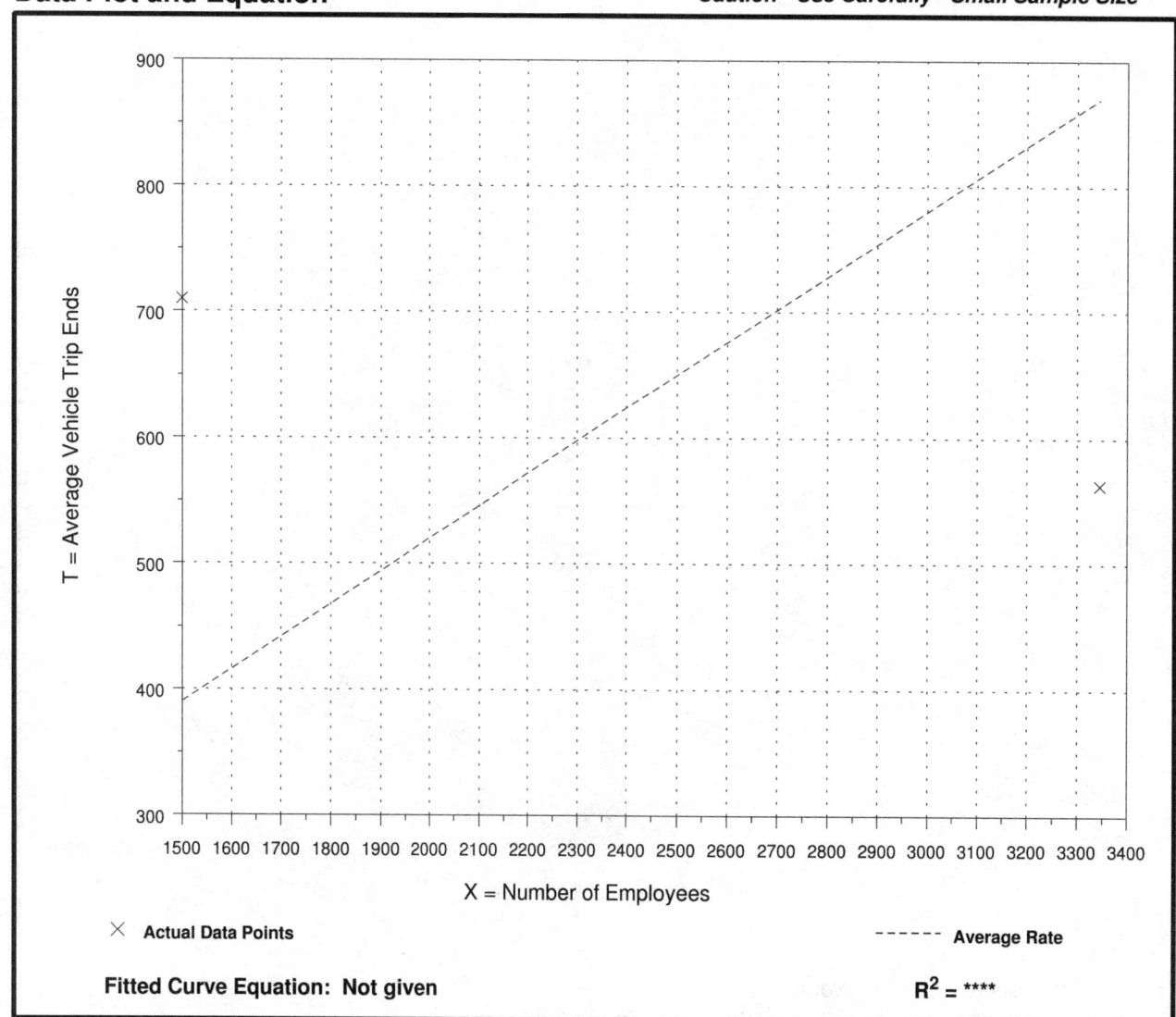

Data Plot and Equation

Caution - Use Carefully - Small Sample Size

Fitted Curve Equation: Not given $R^2 = ****$

Office Park
(750)

Average Vehicle Trip Ends vs: Employees
On a: Sunday,
Peak Hour of Generator

Number of Studies: 2
Avg. Number of Employees: 2,423
Directional Distribution: 41% entering, 59% exiting

Trip Generation per Employee

Average Rate	Range of Rates	Standard Deviation
0.04	0.03 - 0.07	*

Data Plot and Equation

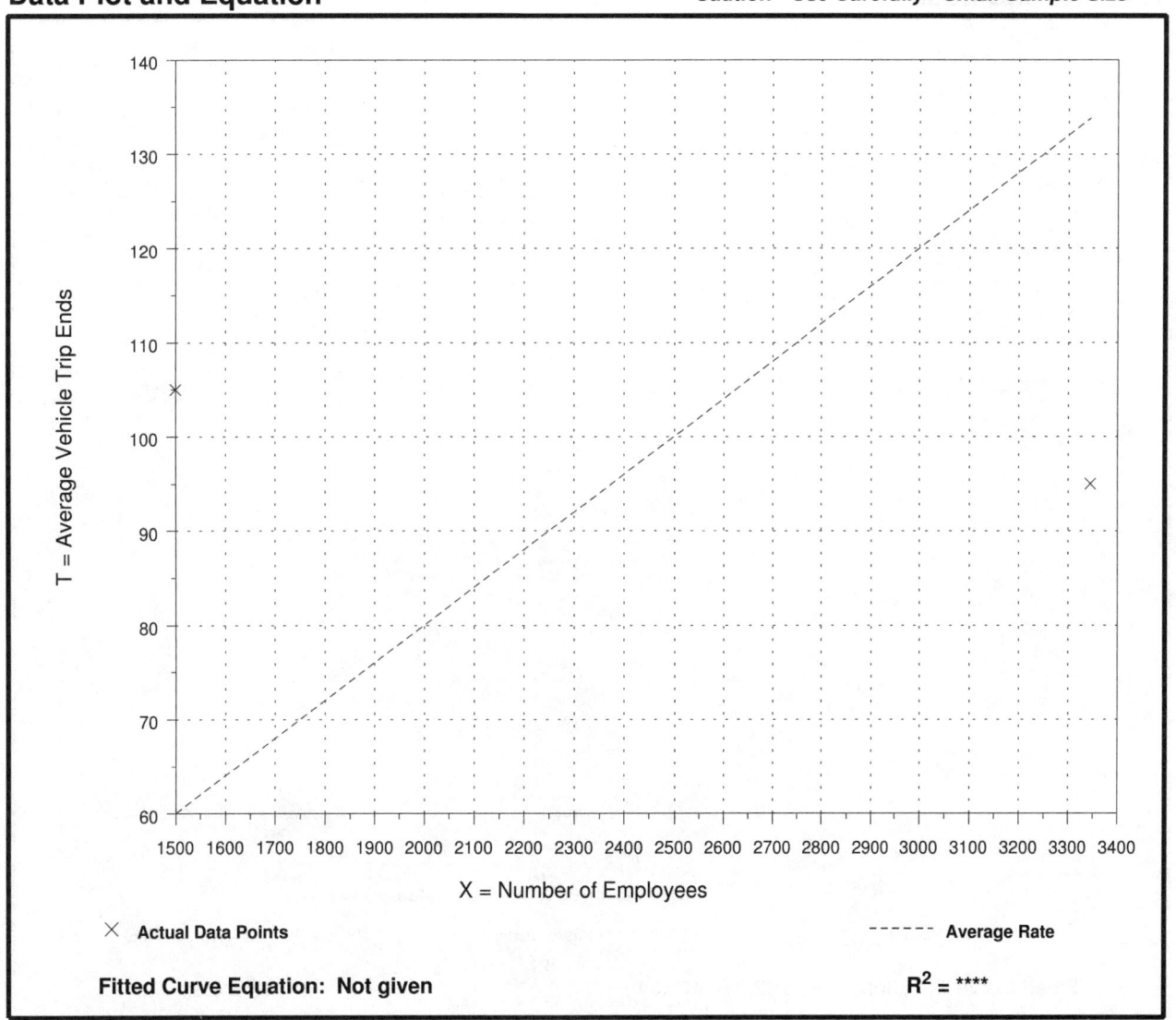

Caution - Use Carefully - Small Sample Size

Fitted Curve Equation: Not given

R^2 = ****

Office Park
(750)

Average Vehicle Trip Ends vs: 1000 Sq. Feet Gross Floor Area
On a: Weekday

Number of Studies: 12
Average 1000 Sq. Feet GFA: 412
Directional Distribution: 50% entering, 50% exiting

Trip Generation per 1000 Sq. Feet Gross Floor Area

Average Rate	Range of Rates	Standard Deviation
11.42	7.56 - 30.30	4.69

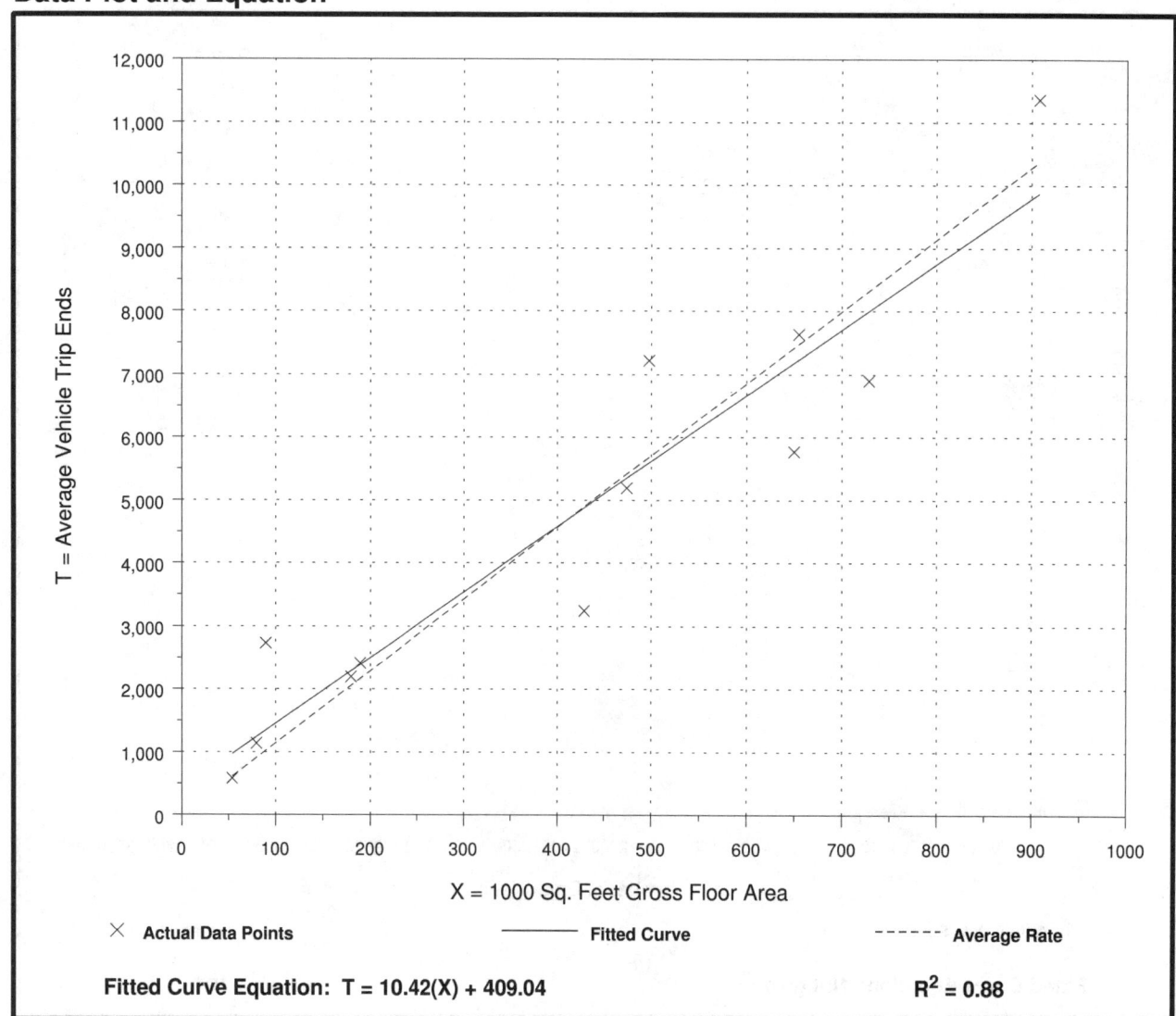

Data Plot and Equation

Fitted Curve Equation: T = 10.42(X) + 409.04 $R^2 = 0.88$

Office Park
(750)

Average Vehicle Trip Ends vs: 1000 Sq. Feet Gross Floor Area
On a: Weekday,
A.M. Peak Hour

Number of Studies: 31
Average 1000 Sq. Feet GFA: 362
Directional Distribution: 89% entering, 11% exiting

Trip Generation per 1000 Sq. Feet Gross Floor Area

Average Rate	Range of Rates	Standard Deviation
1.71	0.60 - 5.89	1.46

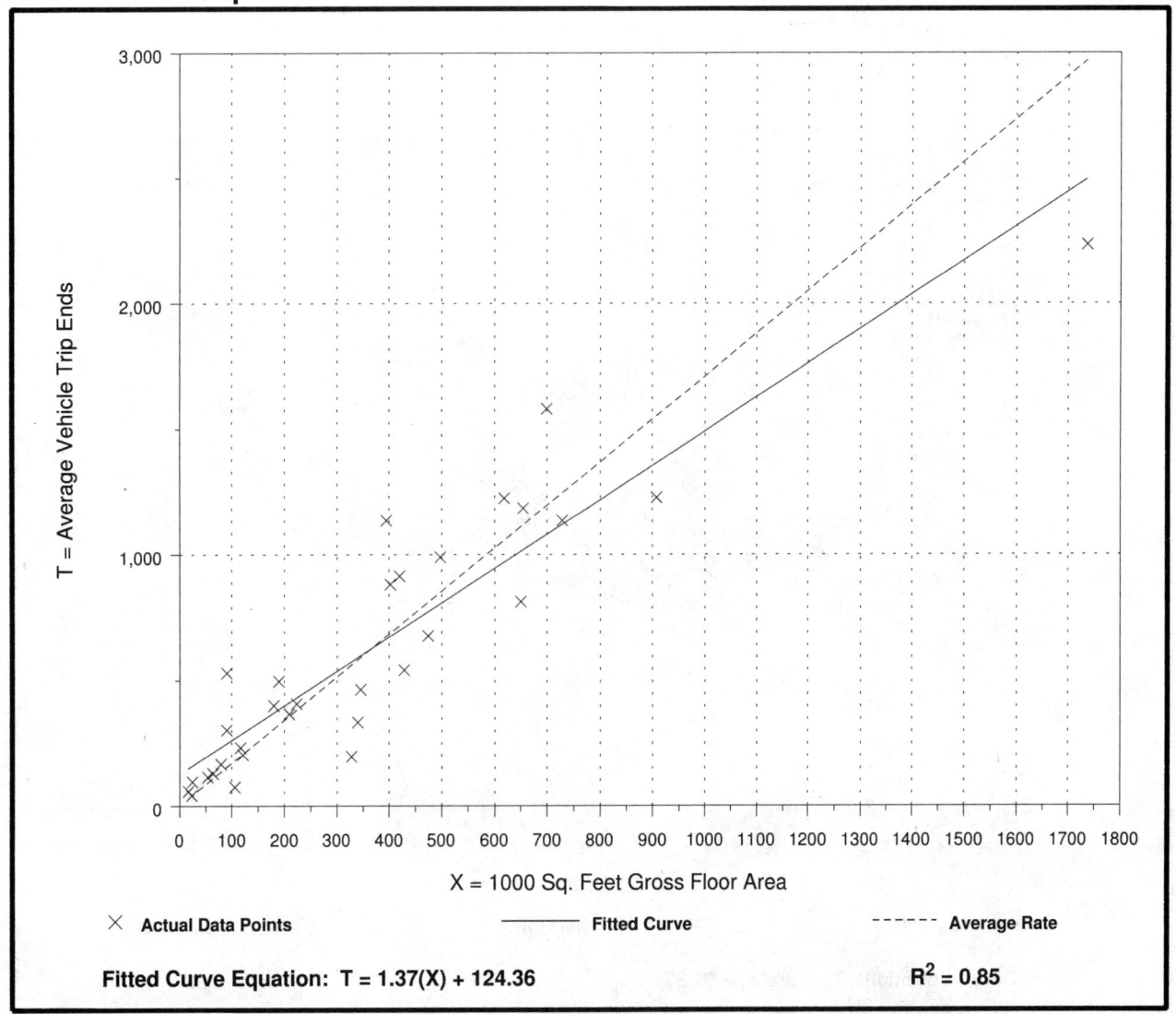

Data Plot and Equation

Fitted Curve Equation: $T = 1.37(X) + 124.36$ $R^2 = 0.85$

Office Park
(750)

Average Vehicle Trip Ends vs: 1000 Sq. Feet Gross Floor Area
On a: Weekday,
P.M. Peak Hour

Number of Studies: 32
Average 1000 Sq. Feet GFA: 369
Directional Distribution: 14% entering, 86% exiting

Trip Generation per 1000 Sq. Feet Gross Floor Area

Average Rate	Range of Rates	Standard Deviation
1.48	0.64 - 4.50	1.31

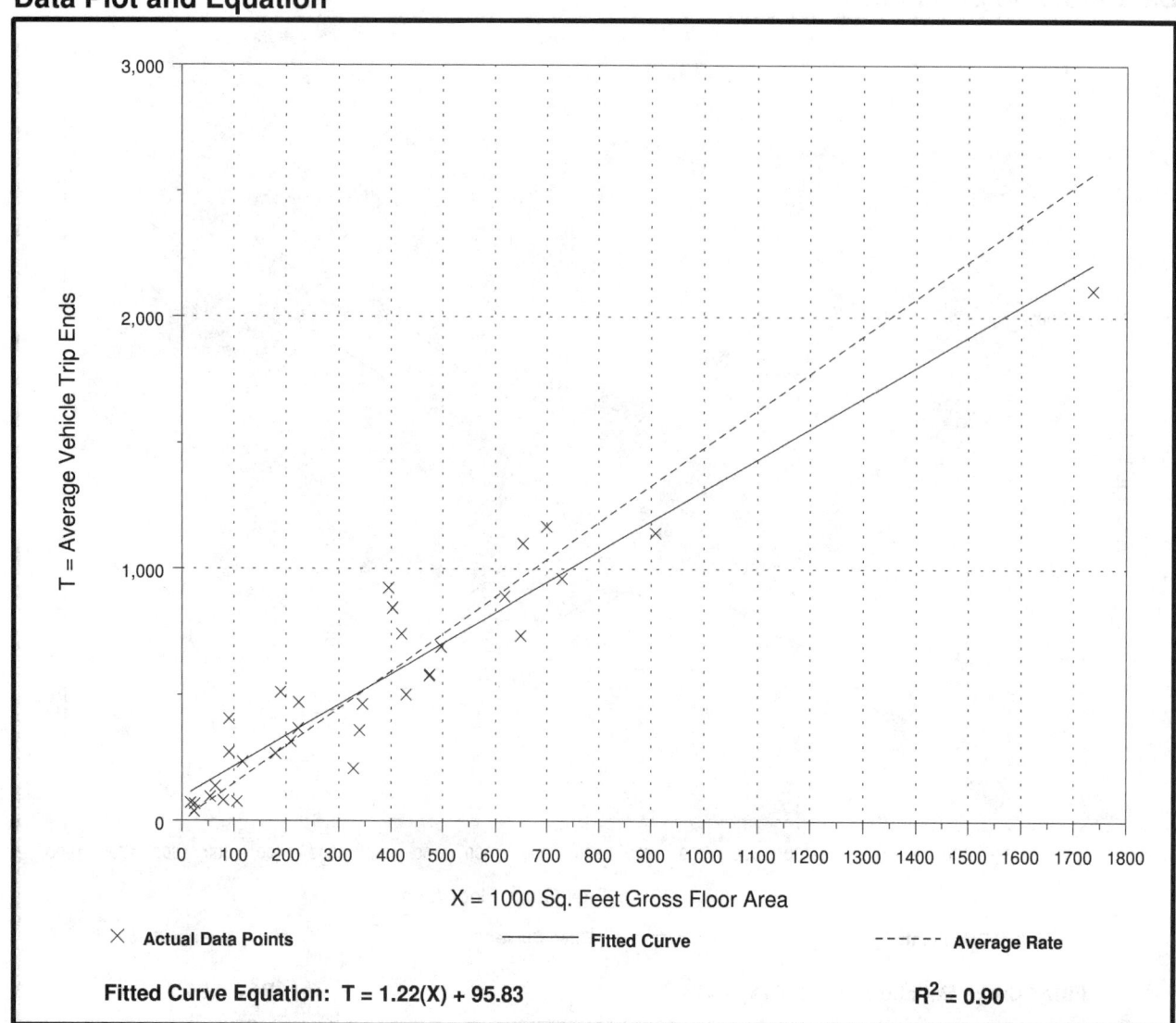

Data Plot and Equation

Fitted Curve Equation: T = 1.22(X) + 95.83 $R^2 = 0.90$

Office Park
(750)

Average Vehicle Trip Ends vs: 1000 Sq. Feet Gross Floor Area
On a: Saturday

Number of Studies: 3
Average 1000 Sq. Feet GFA: 583
Directional Distribution: 50% entering, 50% exiting

Trip Generation per 1000 Sq. Feet Gross Floor Area

Average Rate	Range of Rates	Standard Deviation
1.64	0.73 - 1.89	1.32

Data Plot and Equation

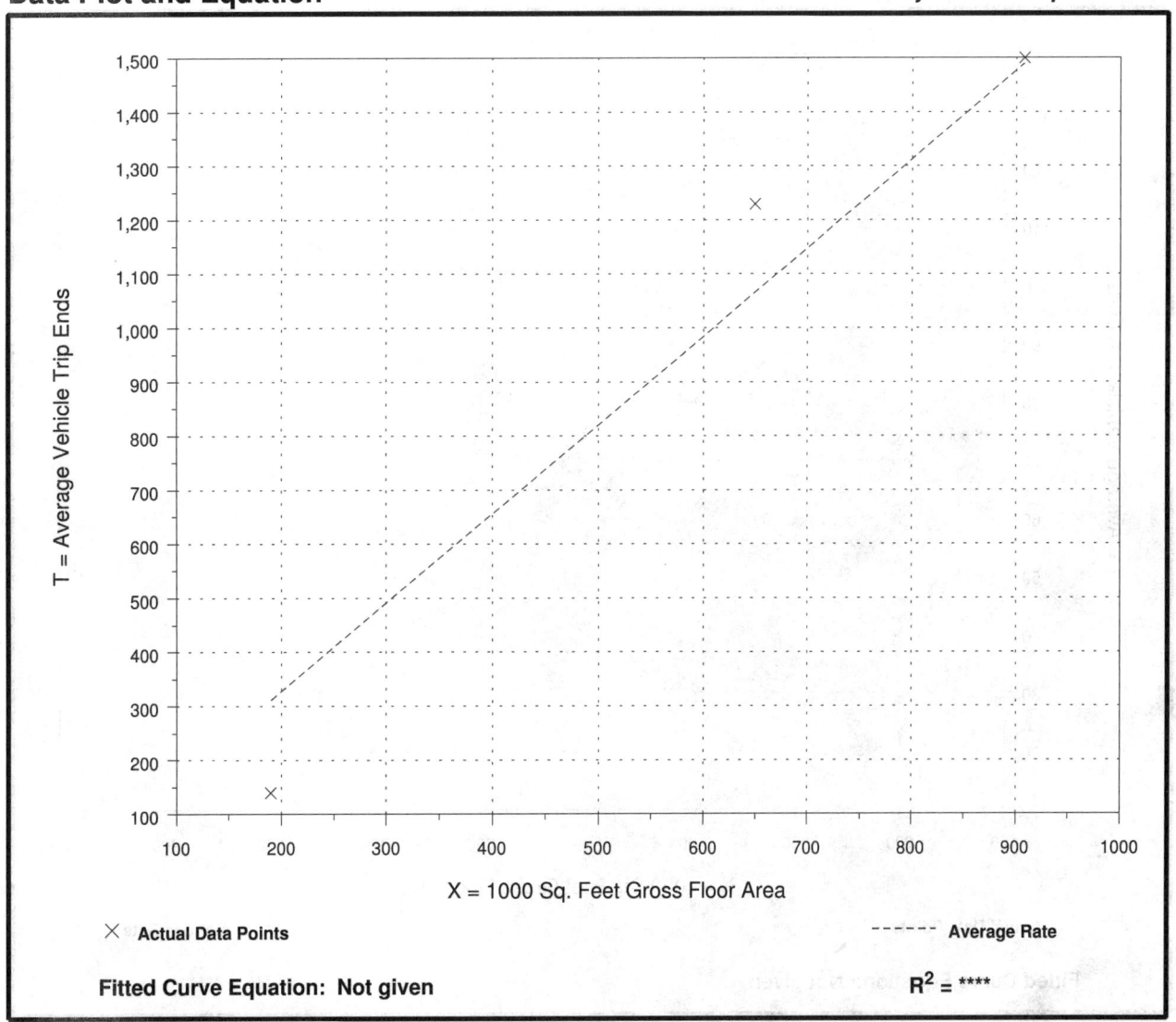

Caution - Use Carefully - Small Sample Size

Fitted Curve Equation: Not given $R^2 = ****$

Office Park
(750)

Average Vehicle Trip Ends vs: 1000 Sq. Feet Gross Floor Area
On a: Saturday,
Peak Hour of Generator

Number of Studies: 3
Average 1000 Sq. Feet GFA: 583
Directional Distribution: 74% entering, 26% exiting

Trip Generation per 1000 Sq. Feet Gross Floor Area

Average Rate	Range of Rates	Standard Deviation
0.14	0.09 - 0.15	0.37

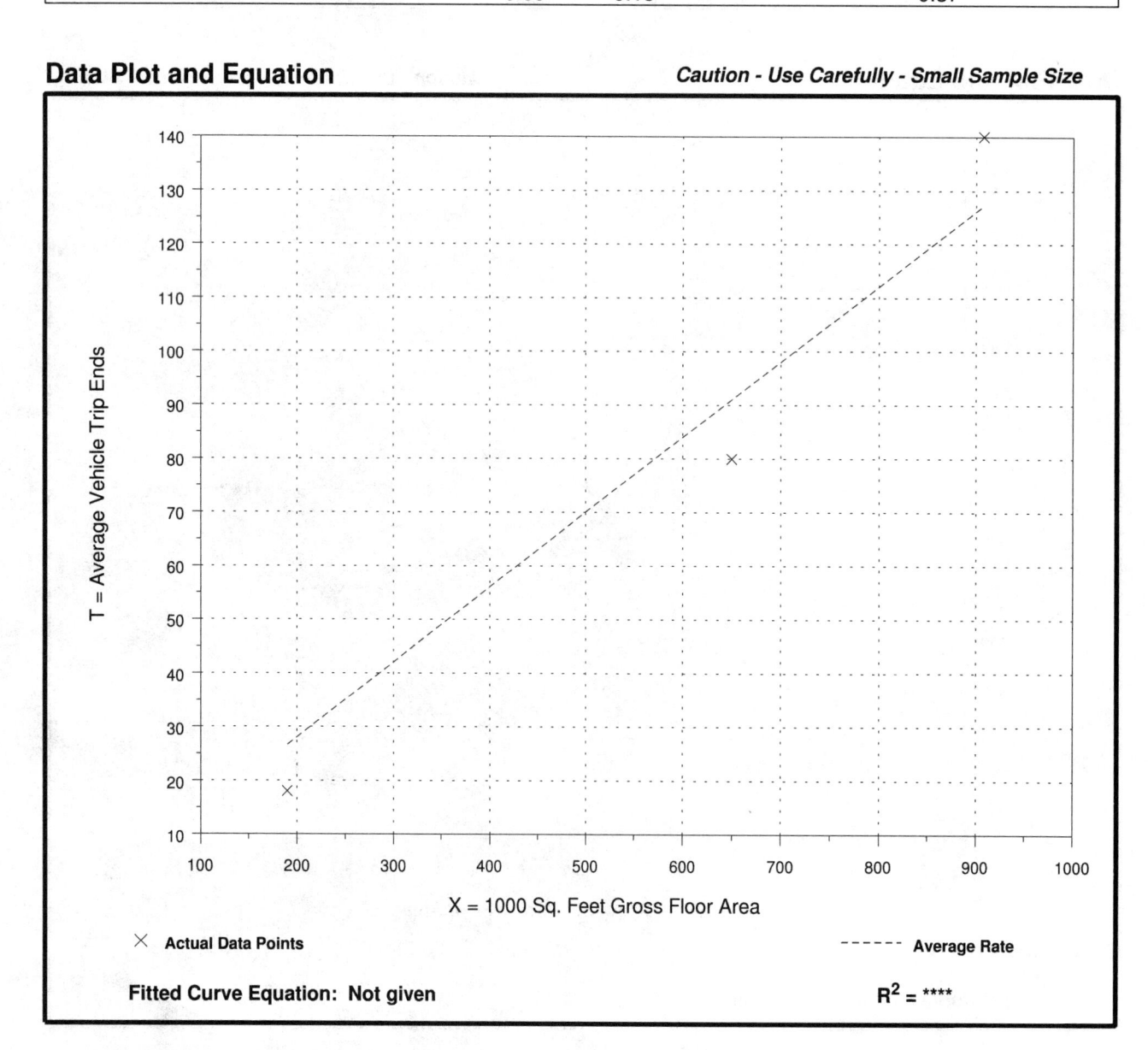

Data Plot and Equation

Caution - Use Carefully - Small Sample Size

Fitted Curve Equation: Not given $R^2 = ****$

Office Park
(750)

Average Vehicle Trip Ends vs: 1000 Sq. Feet Gross Floor Area
On a: Sunday

Number of Studies: 3
Average 1000 Sq. Feet GFA: 583
Directional Distribution: 50% entering, 50% exiting

Trip Generation per 1000 Sq. Feet Gross Floor Area

Average Rate	Range of Rates	Standard Deviation
0.76	0.25 - 1.09	0.91

Data Plot and Equation

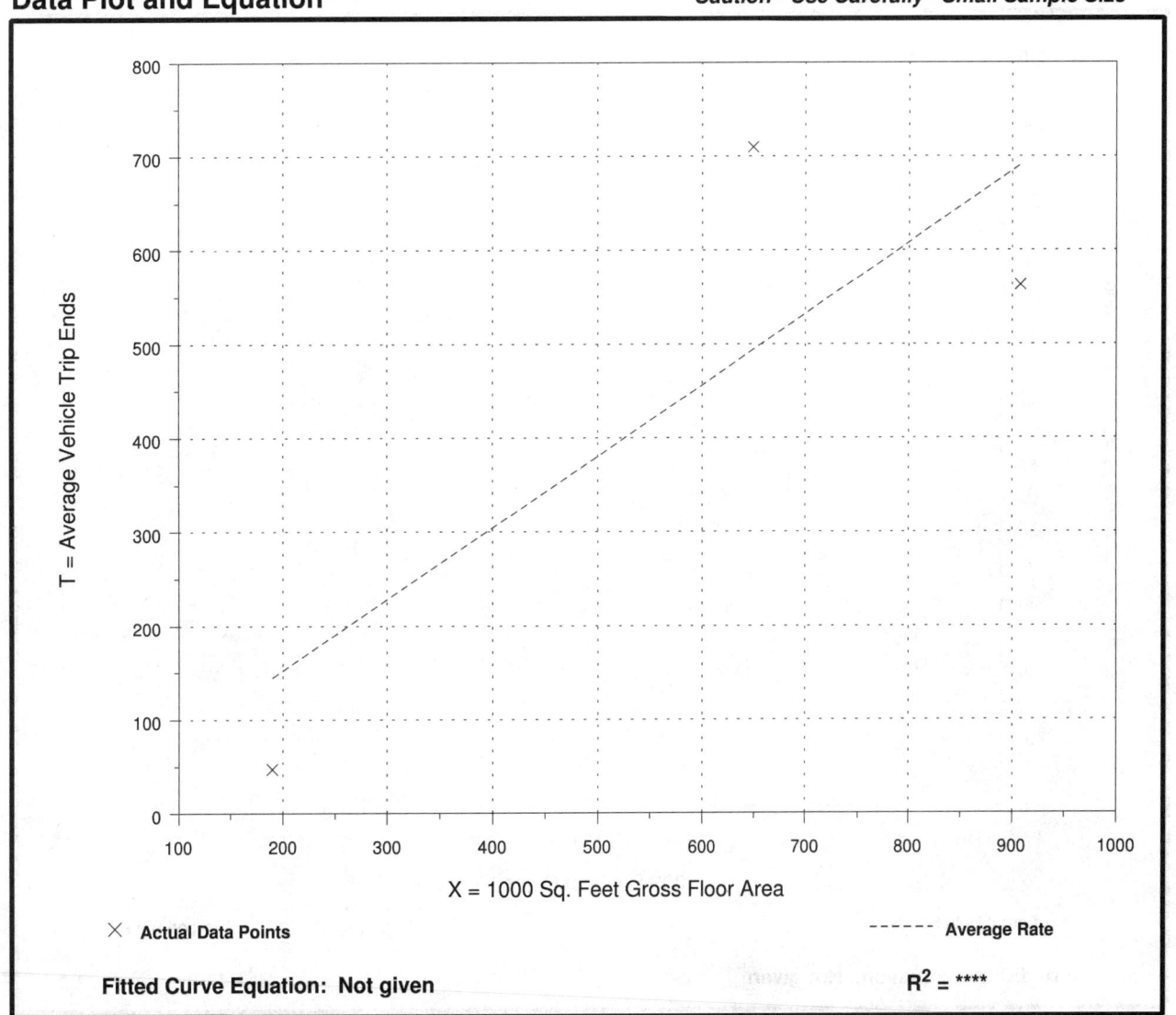

Caution - Use Carefully - Small Sample Size

Fitted Curve Equation: Not given $R^2 = ****$

Office Park
(750)

Average Vehicle Trip Ends vs: 1000 Sq. Feet Gross Floor Area
On a: Sunday,
Peak Hour of Generator

Number of Studies: 3
Average 1000 Sq. Feet GFA: 583
Directional Distribution: 42% entering, 58% exiting

Trip Generation per 1000 Sq. Feet Gross Floor Area

Average Rate	Range of Rates	Standard Deviation
0.12	0.04 - 0.16	0.35

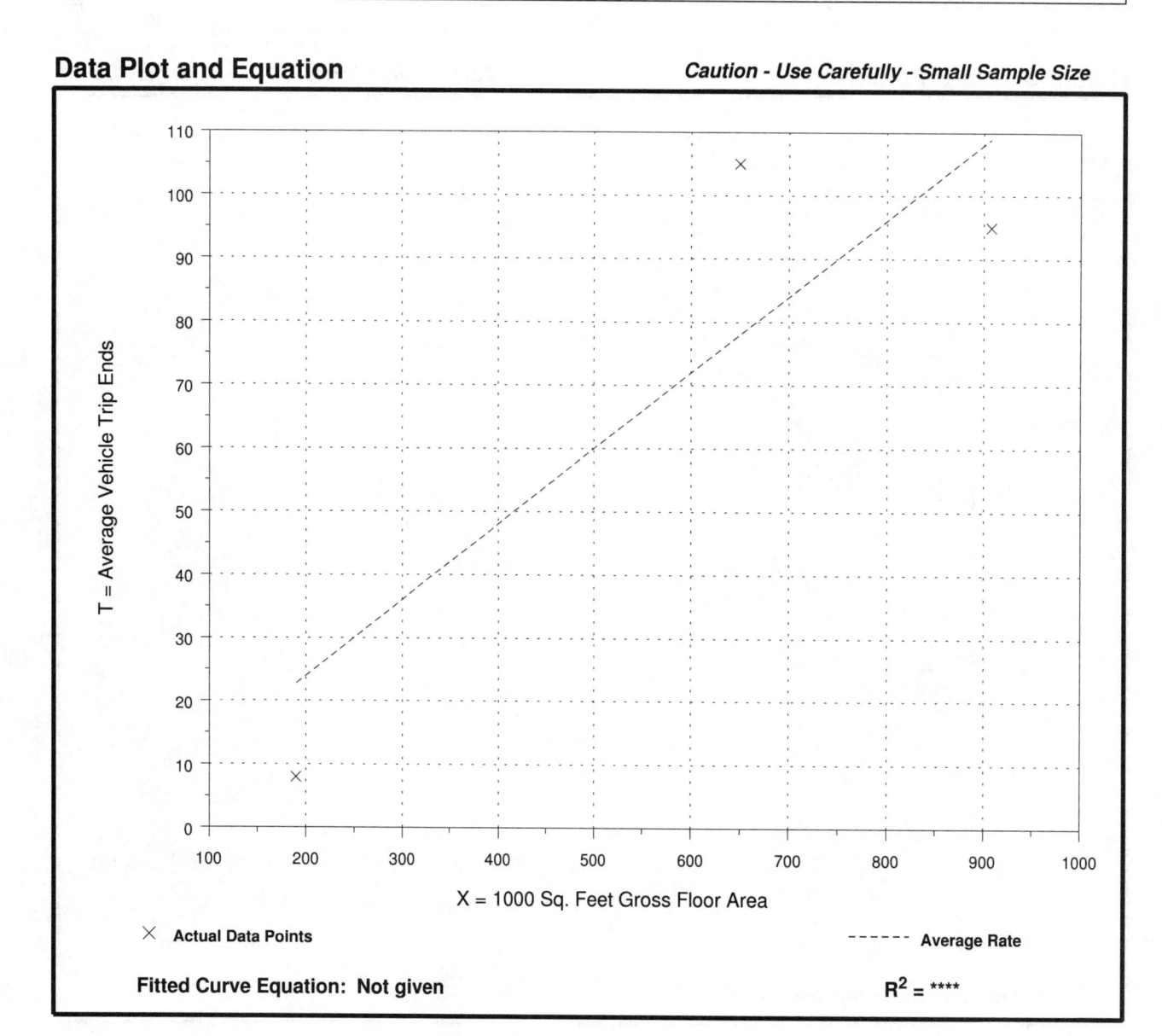

Data Plot and Equation

Caution - Use Carefully - Small Sample Size

Fitted Curve Equation: Not given $R^2 = ****$

Office Park
(750)

Average Vehicle Trip Ends vs: Acres
On a: Weekday

Number of Studies: 4
Average Number of Acres: 26
Directional Distribution: 50% entering, 50% exiting

Trip Generation per Acre

Average Rate	Range of Rates	Standard Deviation
195.11	151.51 - 340.88	77.46

Data Plot and Equation

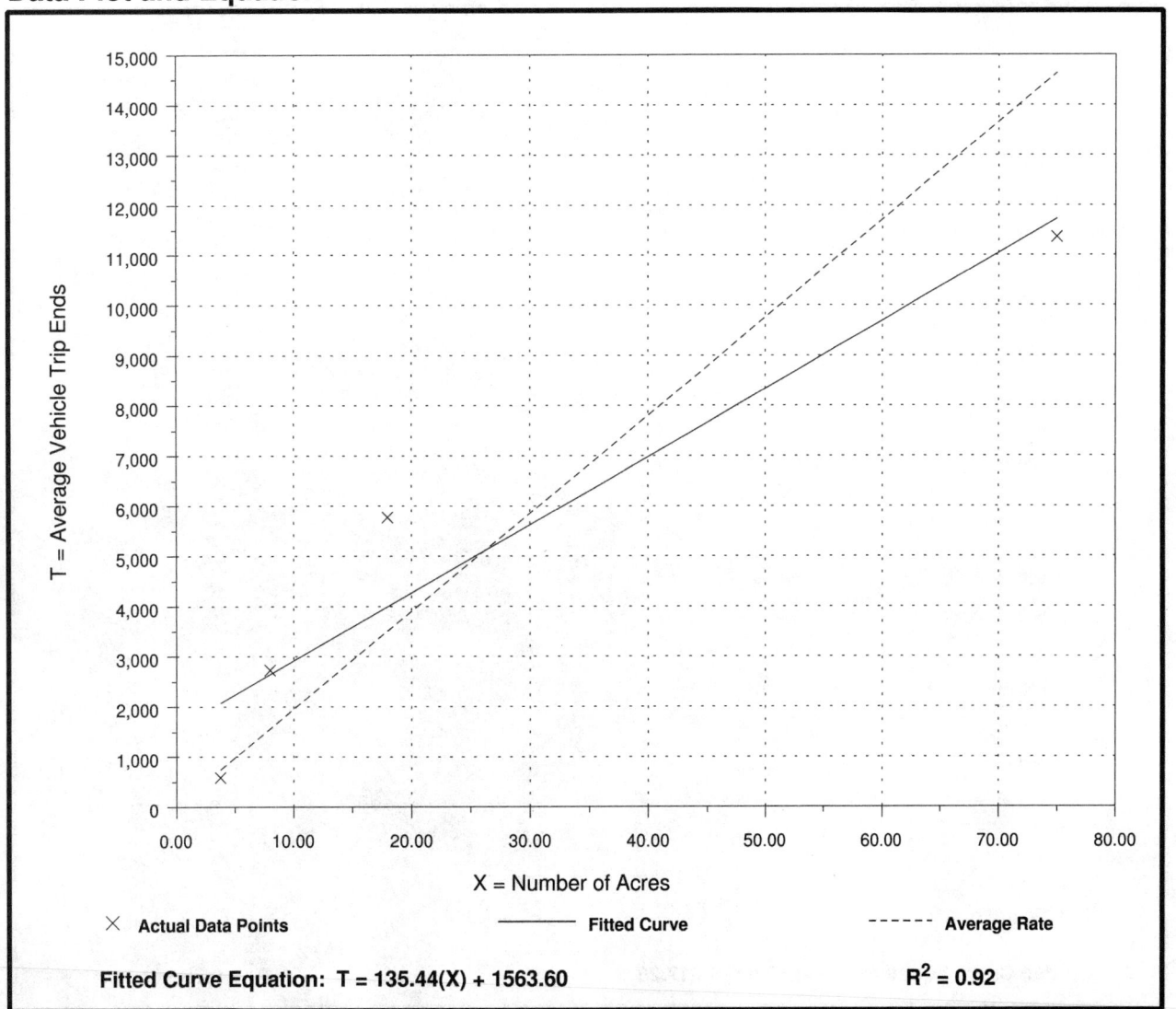

Caution - Use Carefully - Small Sample Size

Fitted Curve Equation: T = 135.44(X) + 1563.60 $R^2 = 0.92$

Office Park
(750)

Average Vehicle Trip Ends vs: Acres
On a: Weekday,
A.M. Peak Hour

Number of Studies: 5
Average Number of Acres: 25
Directional Distribution: 92% entering, 8 % exiting

Trip Generation per Acre

Average Rate	Range of Rates	Standard Deviation
25.65	16.39 - 66.25	15.58

Data Plot and Equation

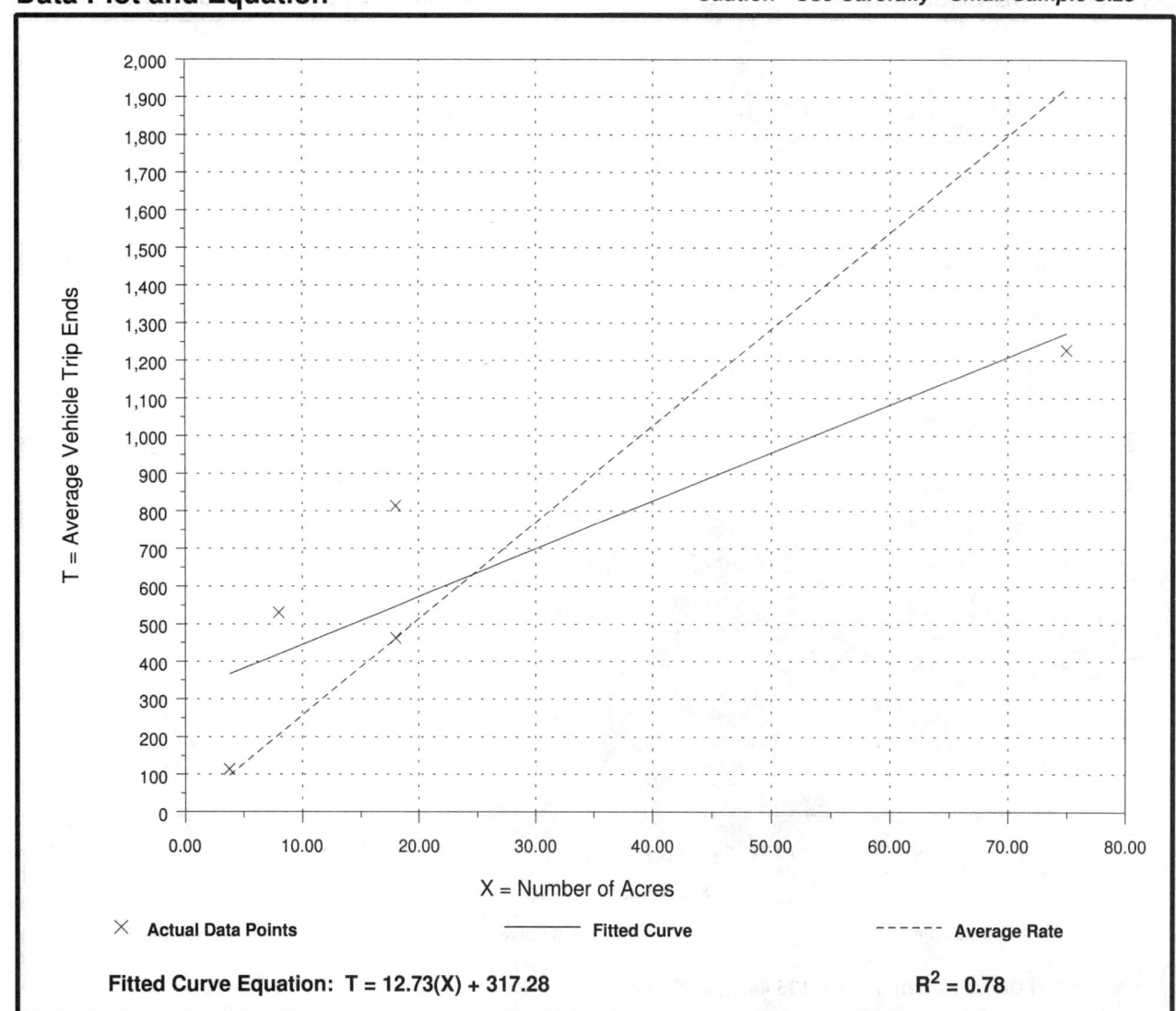

Caution - Use Carefully - Small Sample Size

Fitted Curve Equation: $T = 12.73(X) + 317.28$ $R^2 = 0.78$

Office Park
(750)

Average Vehicle Trip Ends vs: Acres
On a: Weekday,
P.M. Peak Hour

Number of Studies: 7
Average Number of Acres: 21
Directional Distribution: 15% entering, 85% exiting

Trip Generation per Acre

Average Rate	Range of Rates	Standard Deviation
28.28	15.25 - 88.40	20.57

Data Plot and Equation

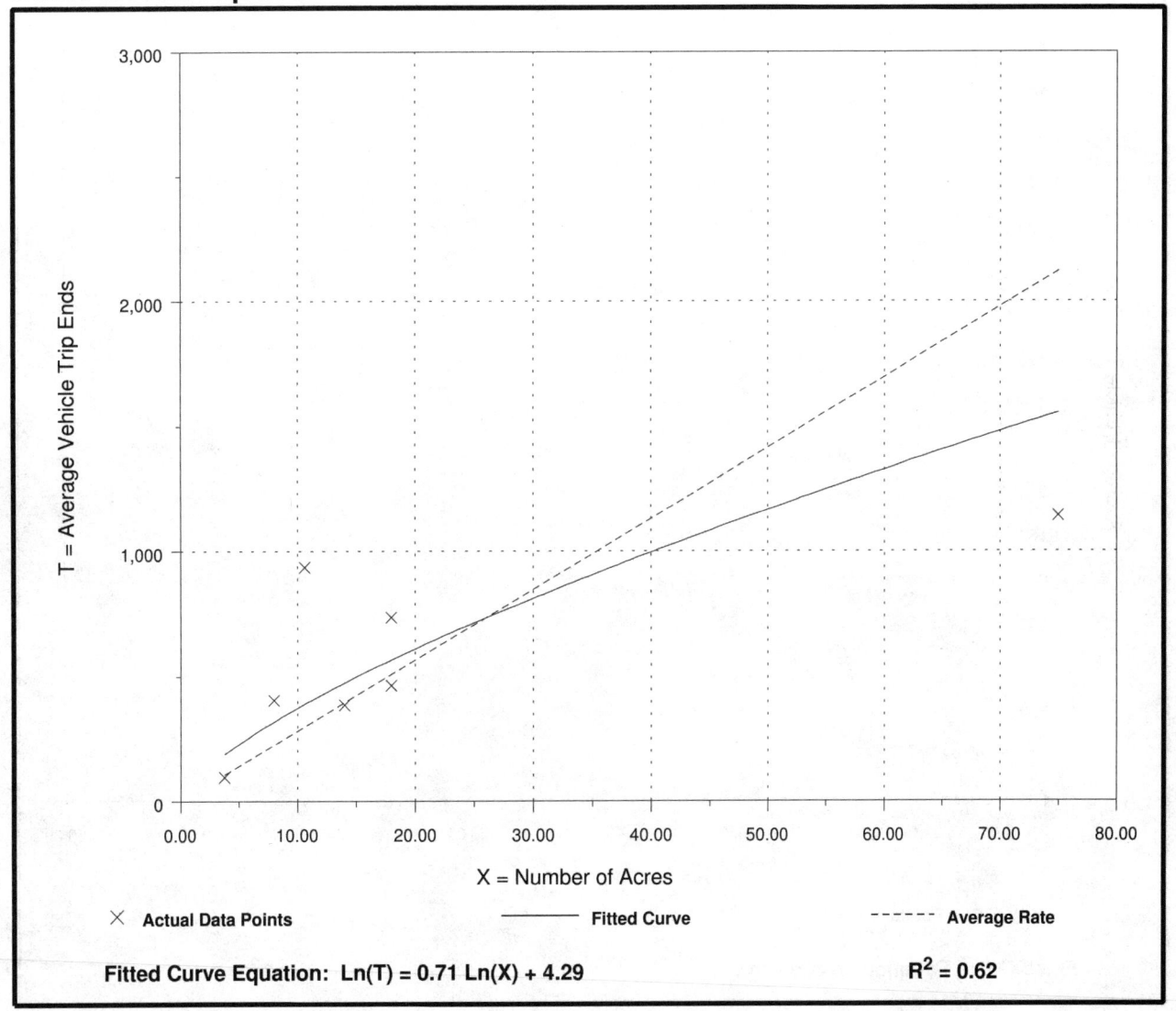

Fitted Curve Equation: $Ln(T) = 0.71\ Ln(X) + 4.29$ $R^2 = 0.62$

Office Park
(750)

Average Vehicle Trip Ends vs: Acres
On a: Saturday

Number of Studies: 2
Average Number of Acres: 47
Directional Distribution: 50% entering, 50% exiting

Trip Generation per Acre

Average Rate	Range of Rates	Standard Deviation
29.33	19.99 - 68.28	*

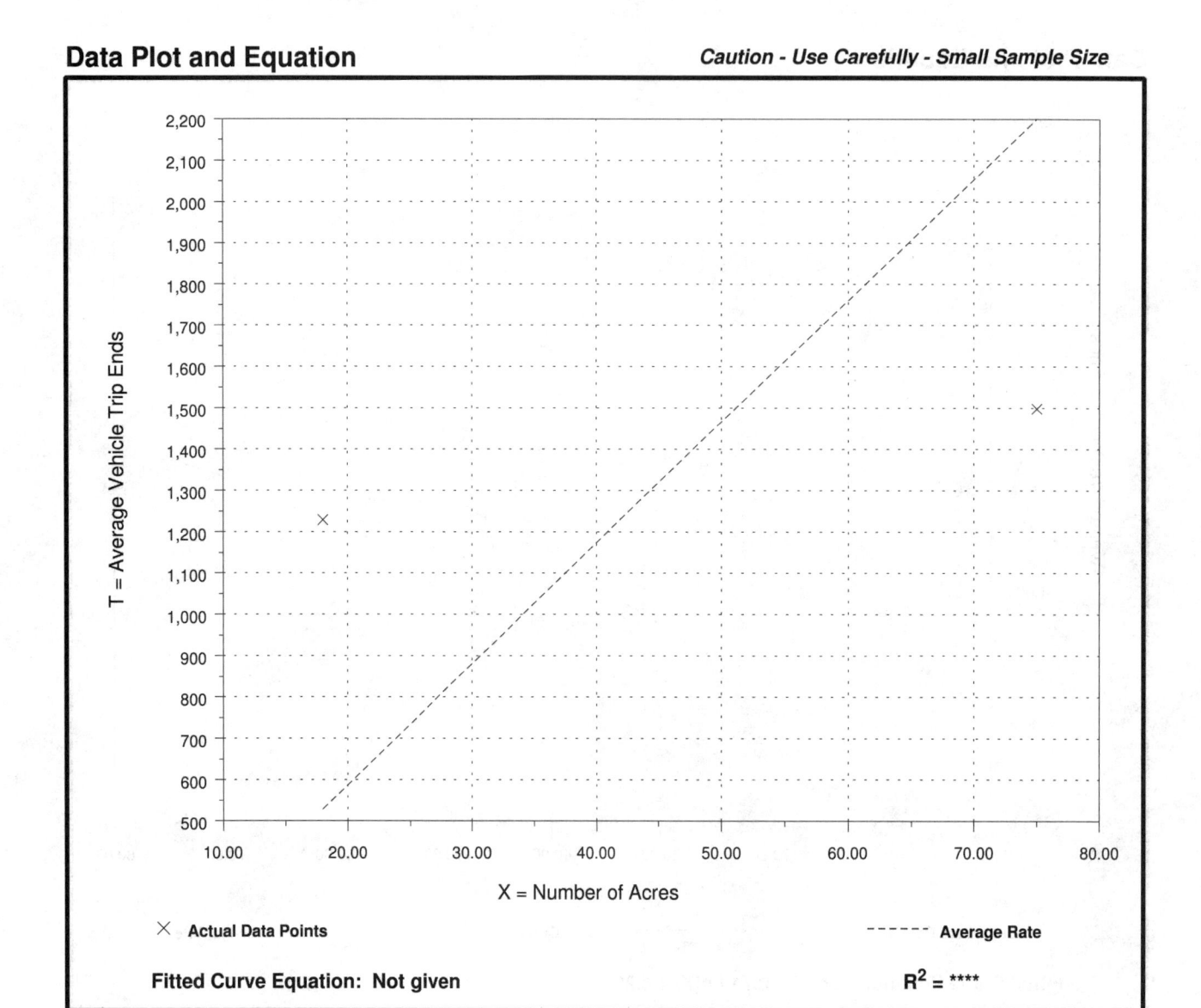

Data Plot and Equation

Caution - Use Carefully - Small Sample Size

Fitted Curve Equation: Not given $R^2 = ****$

Office Park
(750)

Average Vehicle Trip Ends vs: Acres
On a: Saturday,
Peak Hour of Generator

Number of Studies: 2
Average Number of Acres: 47
Directional Distribution: 78% entering, 22% exiting

Trip Generation per Acre

Average Rate	Range of Rates	Standard Deviation
2.37	1.87 - 4.44	*

Data Plot and Equation

Caution - Use Carefully - Small Sample Size

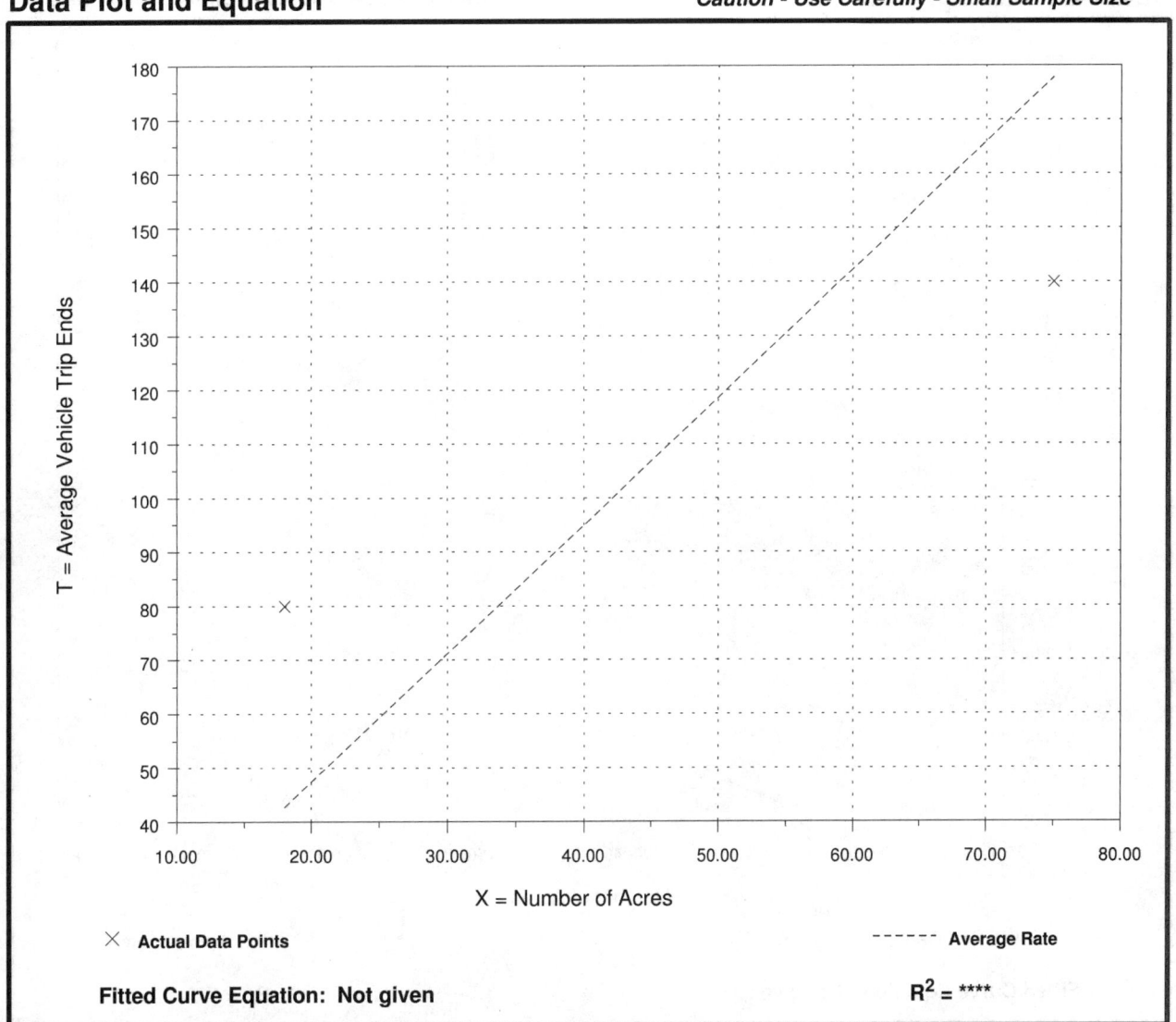

Fitted Curve Equation: Not given $R^2 = ****$

Office Park
(750)

Average Vehicle Trip Ends vs: Acres
On a: Sunday

Number of Studies: 2
Average Number of Acres: 47
Directional Distribution: 50% entering, 50% exiting

Trip Generation per Acre

Average Rate	Range of Rates	Standard Deviation
13.69	7.51 - 39.44	*

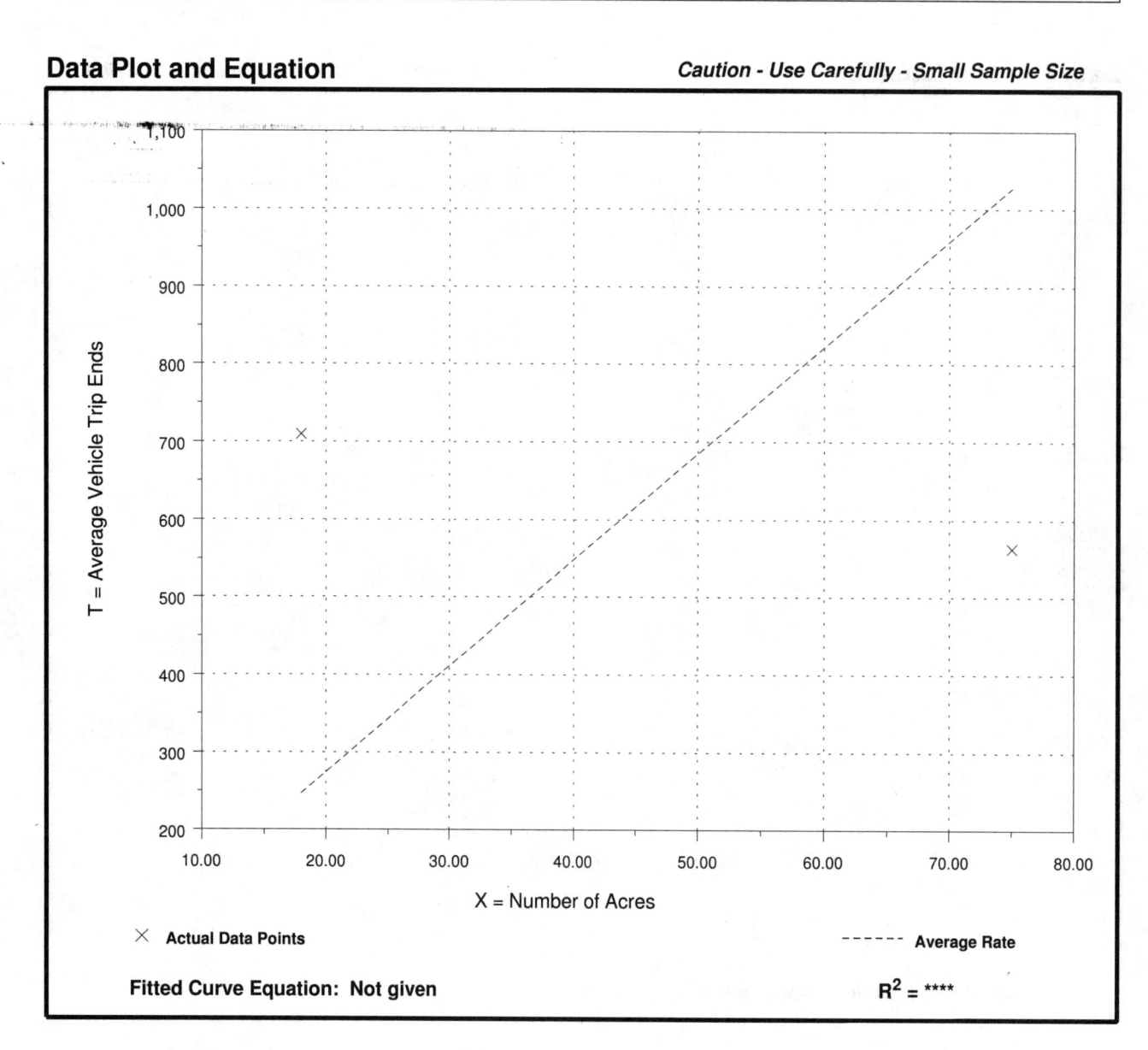

Data Plot and Equation

Caution - Use Carefully - Small Sample Size

Fitted Curve Equation: Not given $R^2 = ****$

Office Park
(750)

Average Vehicle Trip Ends vs: Acres
On a: Sunday,
Peak Hour of Generator

Number of Studies: 2
Average Number of Acres: 47
Directional Distribution: 41% entering, 59% exiting

Trip Generation per Acre

Average Rate	Range of Rates	Standard Deviation
2.15	1.27 - 5.83	*

Data Plot and Equation

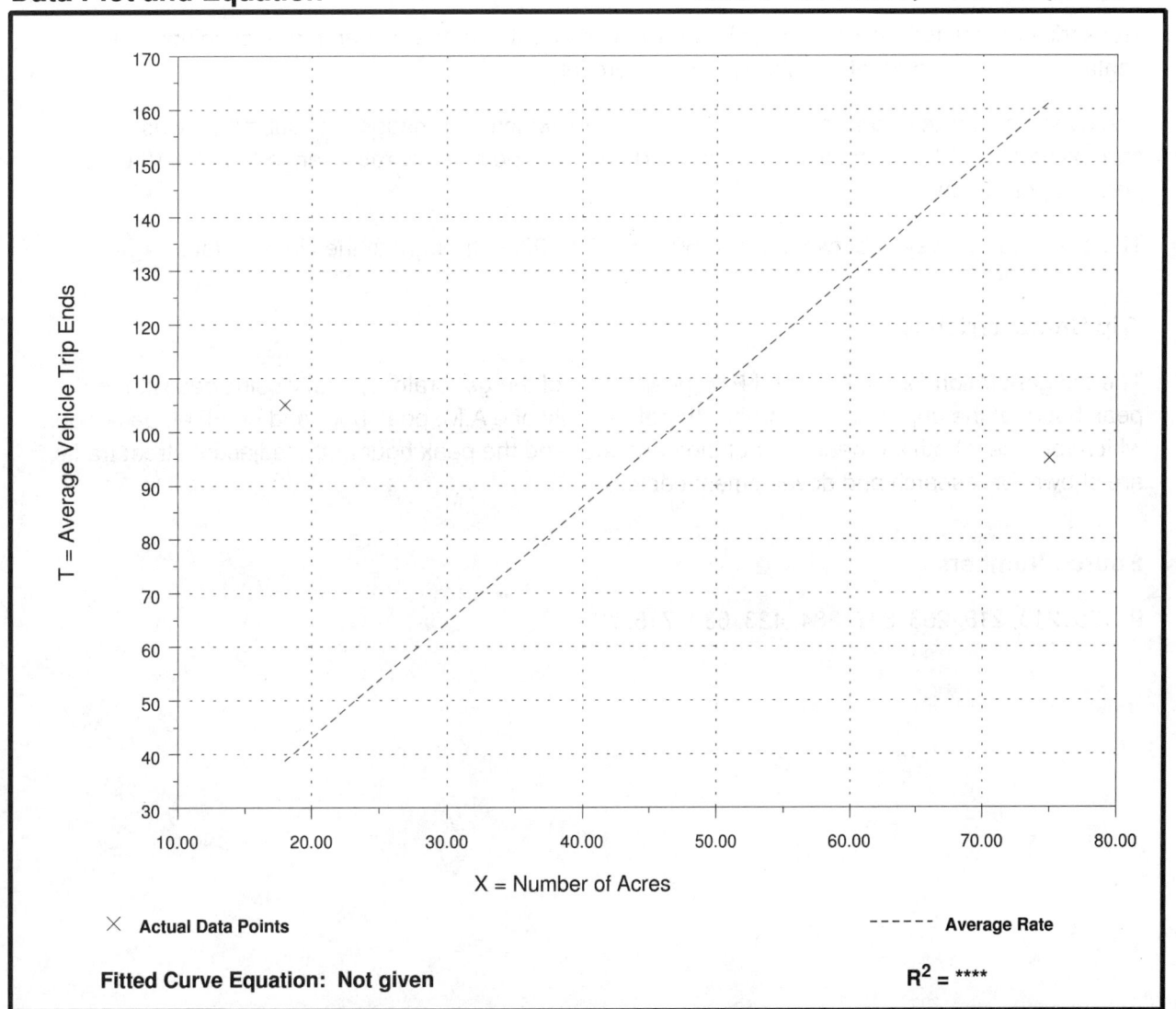

Caution - Use Carefully - Small Sample Size

Fitted Curve Equation: Not given $R^2 = ****$

Land Use: 760
Research and Development Center

Description

Research and development centers are facilities or groups of facilities devoted almost exclusively to research and development activities. The range of specific types of businesses contained in this land use category varies significantly. Research and development centers may contain offices and light fabrication areas. General office building (Land Use 710), corporate headquarters building (Land Use 714), single tenant office building (Land Use 715), office park (Land Use 750) and business park (Land Use 770) are related uses.

Additional Data

Truck trips accounted for 1.84 percent of the weekday traffic at the research and development centers surveyed (range of 0.4 percent to 4.0 percent).

The average vehicle occupancy for the 13 studies for which information was submitted was approximately 1.19 persons per automobile. The vehicle occupancy rates ranged from 1.10 to 1.33 persons per automobile.

The sites were surveyed between the 1960s and the 2000s throughout the United States.

Trip Characteristics

The trip generation for the A.M. and P.M. peak hours of the generator typically coincided with the peak hours of the adjacent street traffic; therefore, only one A.M. peak hour and one P.M. peak hour, which represent both the peak hour of the generator and the peak hour of the adjacent street traffic, are shown for research and development centers.

Source Numbers

9, 105, 213, 218, 253, 332, 384, 423, 630, 715, 723

Research and Development Center
(760)

Average Vehicle Trip Ends vs: Employees
On a: Weekday

Number of Studies: 27
Avg. Number of Employees: 1,022
Directional Distribution: 50% entering, 50% exiting

Trip Generation per Employee

Average Rate	Range of Rates	Standard Deviation
2.77	0.96 - 10.63	2.09

Data Plot and Equation

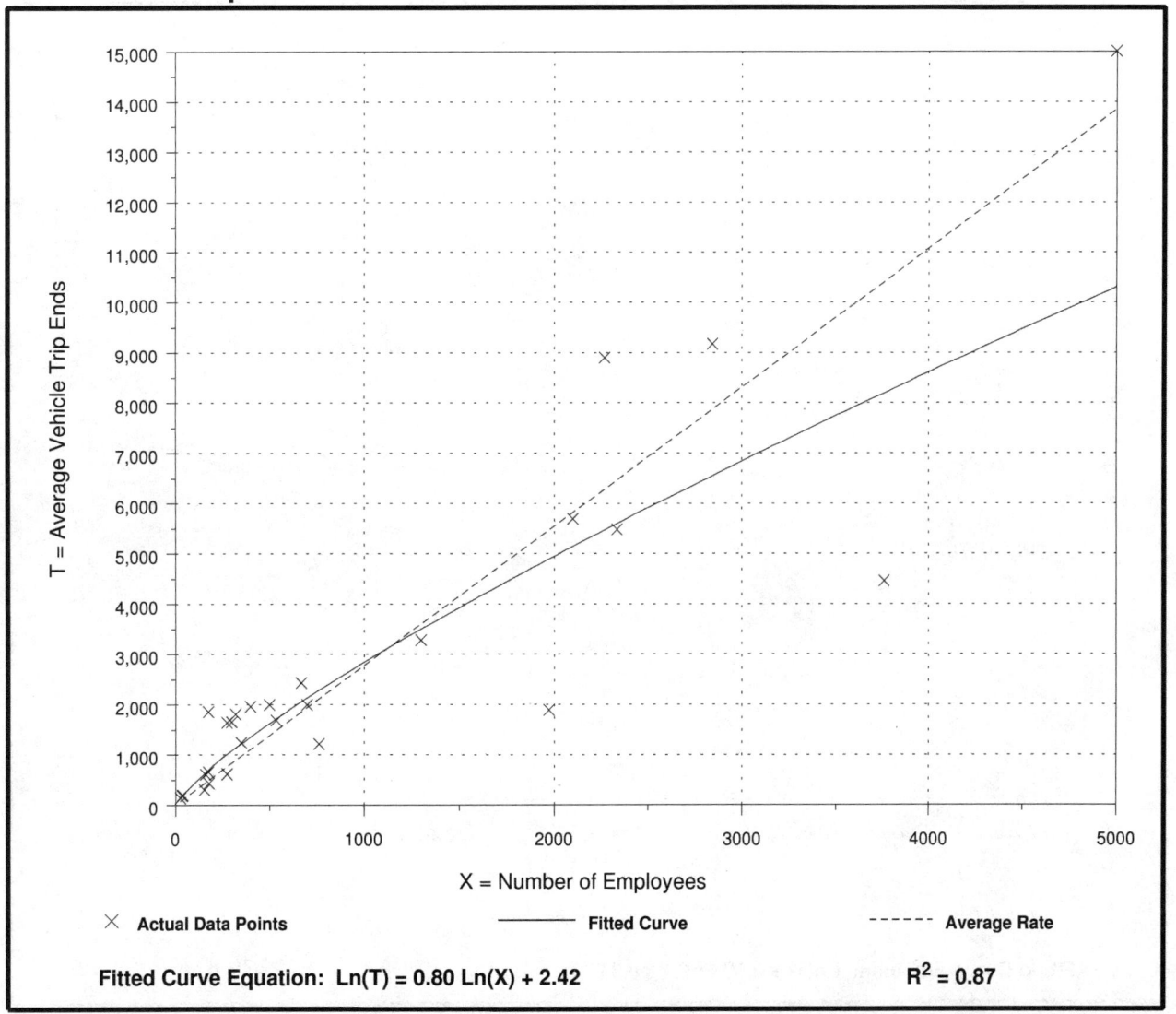

Fitted Curve Equation: $Ln(T) = 0.80\ Ln(X) + 2.42$ $R^2 = 0.87$

Research and Development Center
(760)

Average Vehicle Trip Ends vs: Employees
On a: Weekday,
A.M. Peak Hour

Number of Studies: 28
Avg. Number of Employees: 1,038
Directional Distribution: 86% entering, 14% exiting

Trip Generation per Employee

Average Rate	Range of Rates	Standard Deviation
0.43	0.20 - 1.39	0.67

Data Plot and Equation

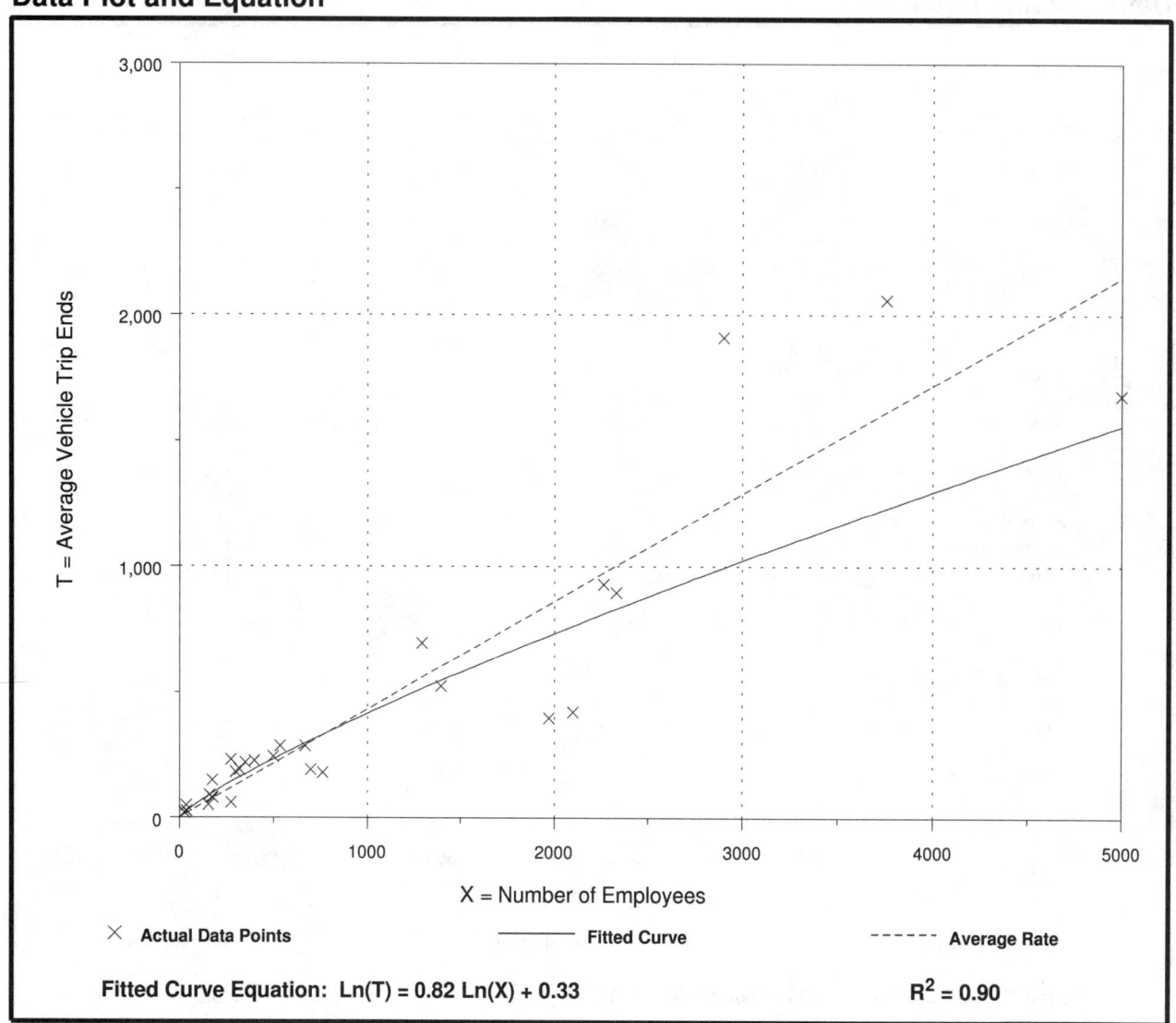

Fitted Curve Equation: $Ln(T) = 0.82\ Ln(X) + 0.33$ $R^2 = 0.90$

Research and Development Center
(760)

Average Vehicle Trip Ends vs: Employees
On a: Weekday,
P.M. Peak Hour

Number of Studies: 29
Avg. Number of Employees: 1,049
Directional Distribution: 10% entering, 90% exiting

Trip Generation per Employee

Average Rate	Range of Rates	Standard Deviation
0.41	0.18 - 1.39	0.66

Data Plot and Equation

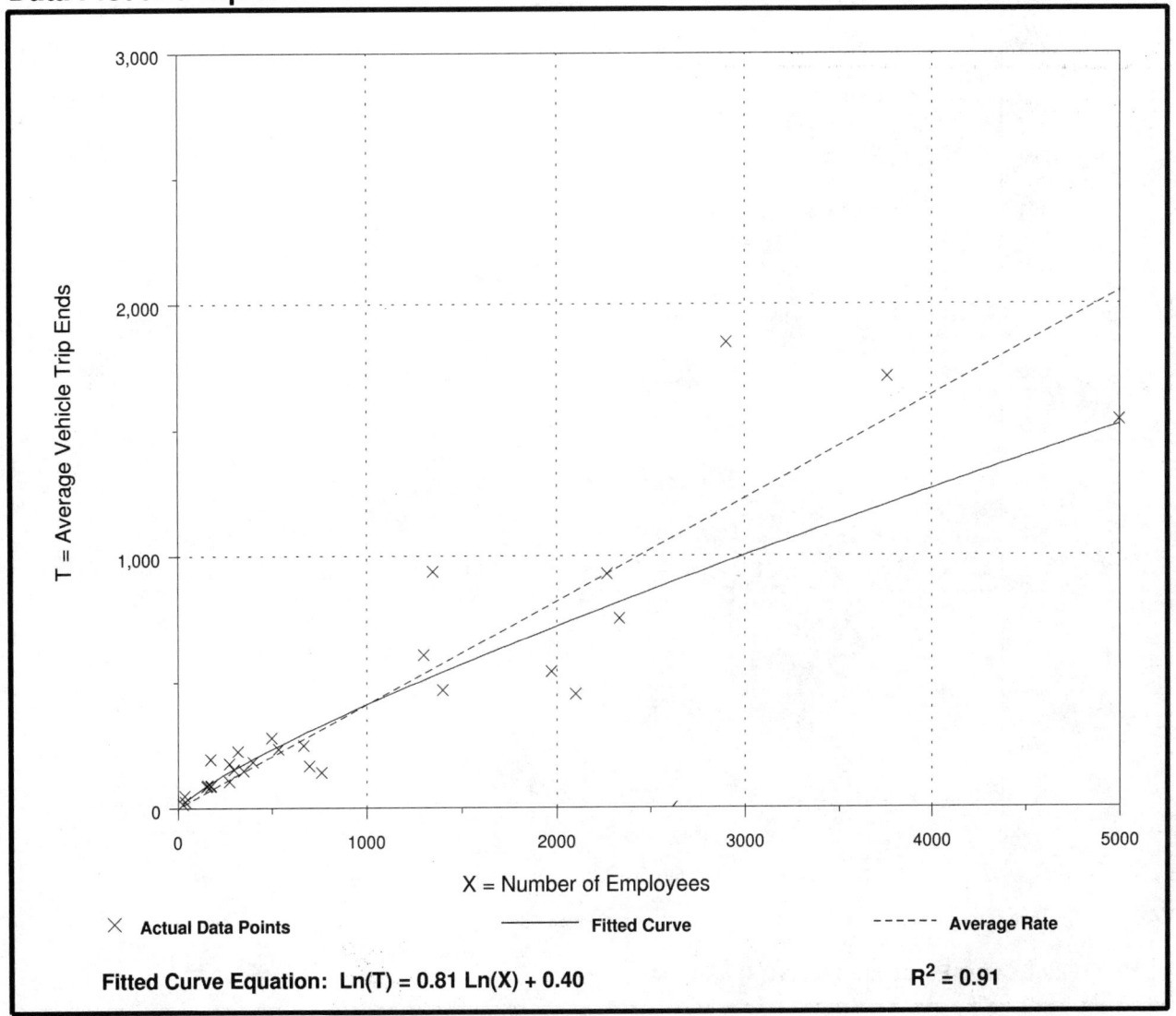

Fitted Curve Equation: $Ln(T) = 0.81\, Ln(X) + 0.40$ $R^2 = 0.91$

Research and Development Center
(760)

Average Vehicle Trip Ends vs: Employees
On a: Saturday

Number of Studies: 20
Avg. Number of Employees: 572
Directional Distribution: 50% entering, 50% exiting

Trip Generation per Employee

Average Rate	Range of Rates	Standard Deviation
0.57	0.03 - 2.97	0.88

Data Plot and Equation

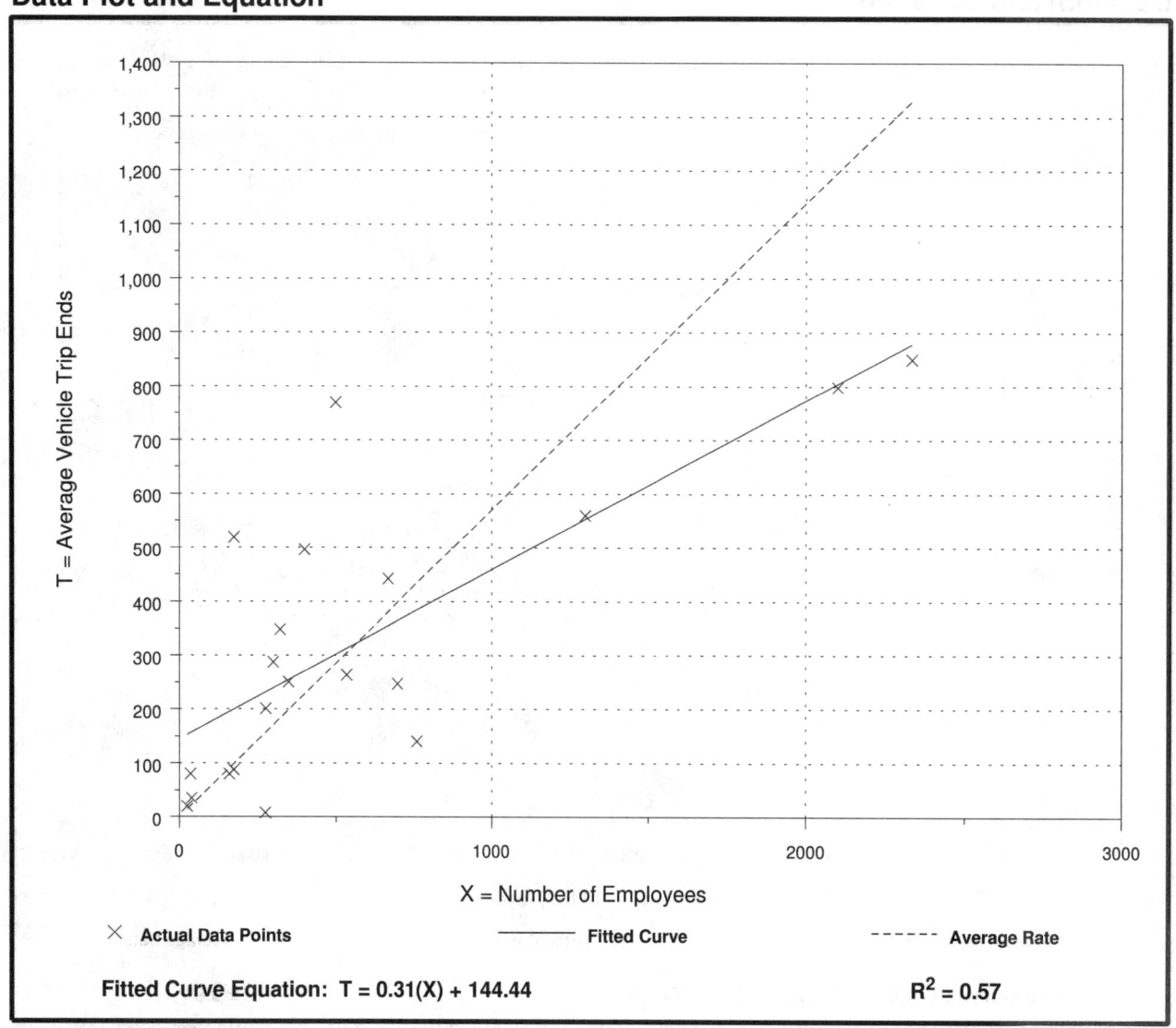

Fitted Curve Equation: T = 0.31(X) + 144.44 $R^2 = 0.57$

Research and Development Center
(760)

Average Vehicle Trip Ends vs: Employees
On a: Saturday,
Peak Hour of Generator

Number of Studies: 13
Avg. Number of Employees: 518
Directional Distribution: Not available

Trip Generation per Employee

Average Rate	Range of Rates	Standard Deviation
0.07	0.01 - 0.56	0.26

Data Plot and Equation

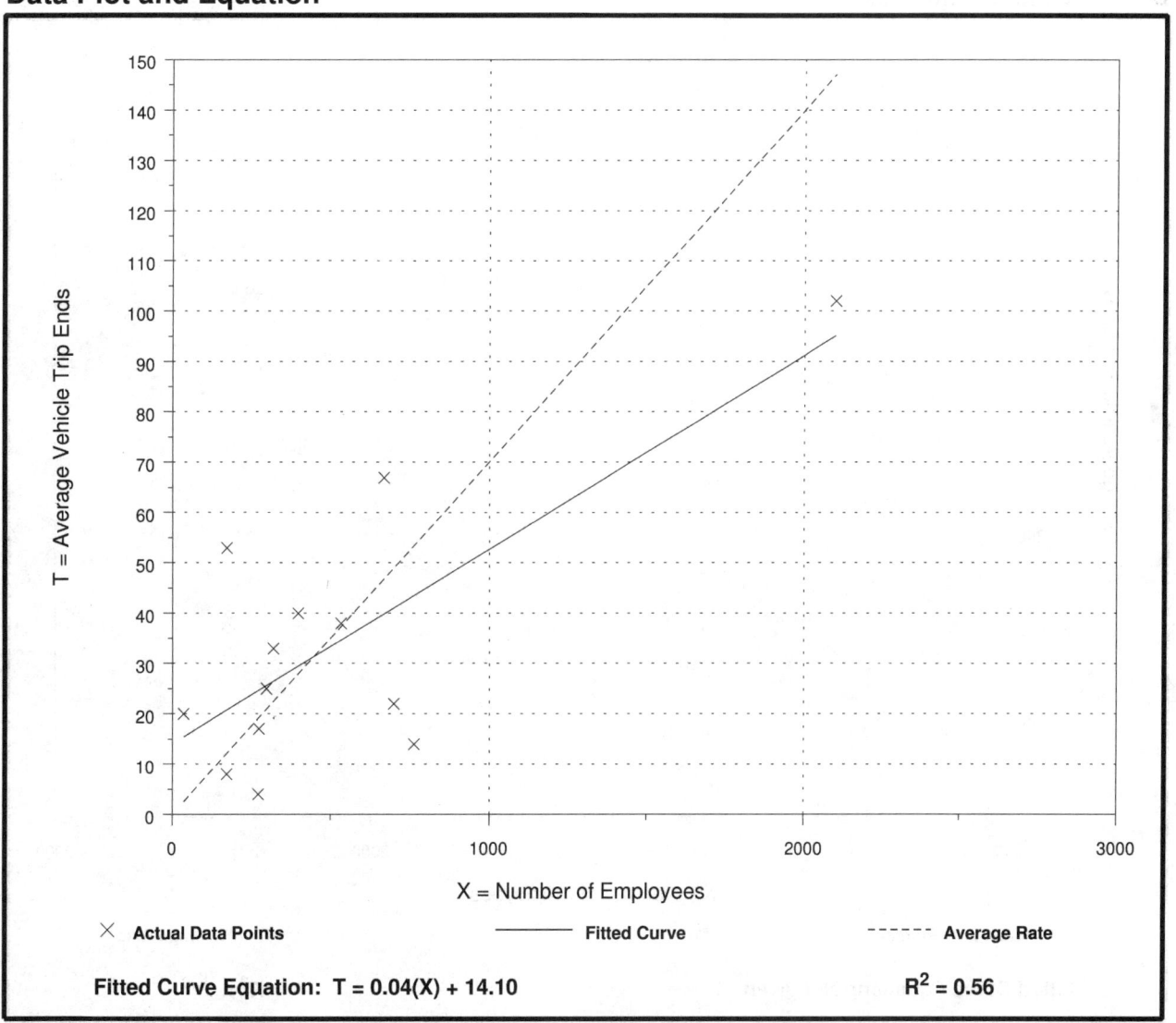

Fitted Curve Equation: $T = 0.04(X) + 14.10$ $R^2 = 0.56$

Research and Development Center
(760)

Average Vehicle Trip Ends vs: Employees
On a: Sunday

Number of Studies: 19
Avg. Number of Employees: 600
Directional Distribution: 50% entering, 50% exiting

Trip Generation per Employee

Average Rate	Range of Rates	Standard Deviation
0.33	0.02 - 1.78	0.67

Data Plot and Equation

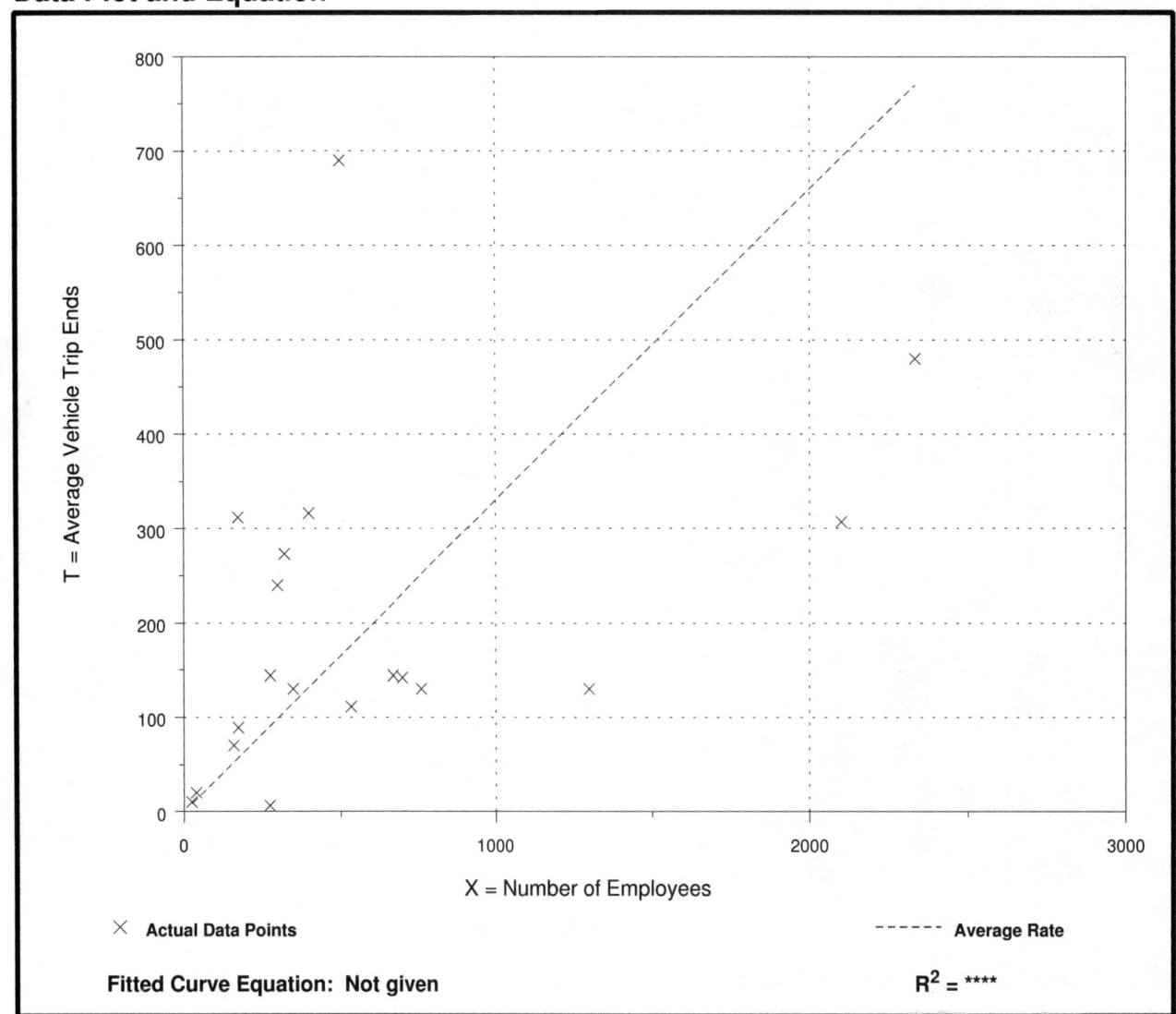

Fitted Curve Equation: Not given $R^2 = ****$

Research and Development Center
(760)

Average Vehicle Trip Ends vs: Employees
On a: Sunday,
Peak Hour of Generator

Number of Studies: 12
Avg. Number of Employees: 558
Directional Distribution: Not available

Trip Generation per Employee

Average Rate	Range of Rates	Standard Deviation
0.04	0.01 - 0.23	0.21

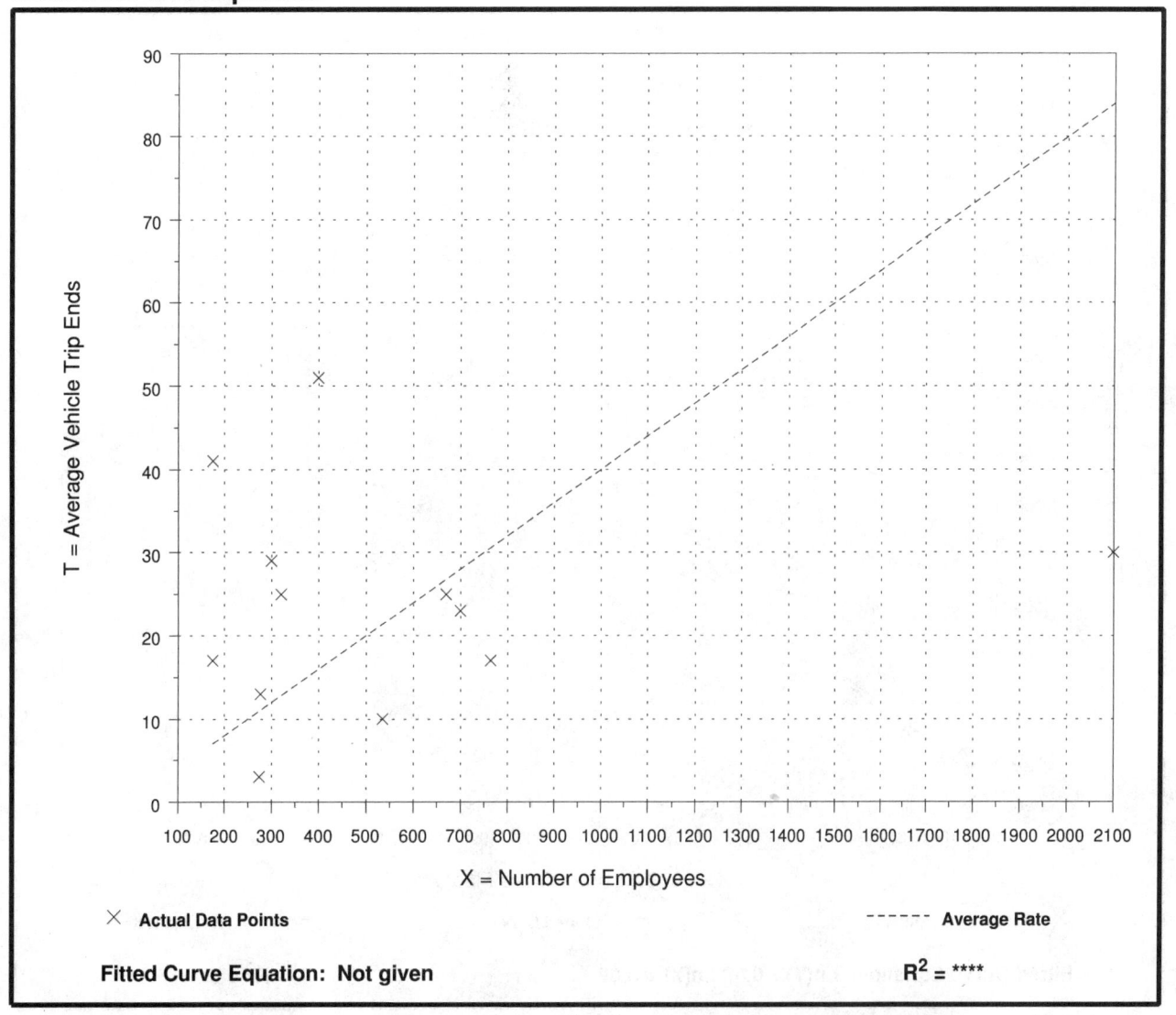

Data Plot and Equation

Fitted Curve Equation: Not given $R^2 = ****$

Research and Development Center
(760)

Average Vehicle Trip Ends vs: 1000 Sq. Feet Gross Floor Area
On a: Weekday

Number of Studies: 29
Average 1000 Sq. Feet GFA: 298
Directional Distribution: 50% entering, 50% exiting

Trip Generation per 1000 Sq. Feet Gross Floor Area

Average Rate	Range of Rates	Standard Deviation
8.11	1.78 - 24.95	5.84

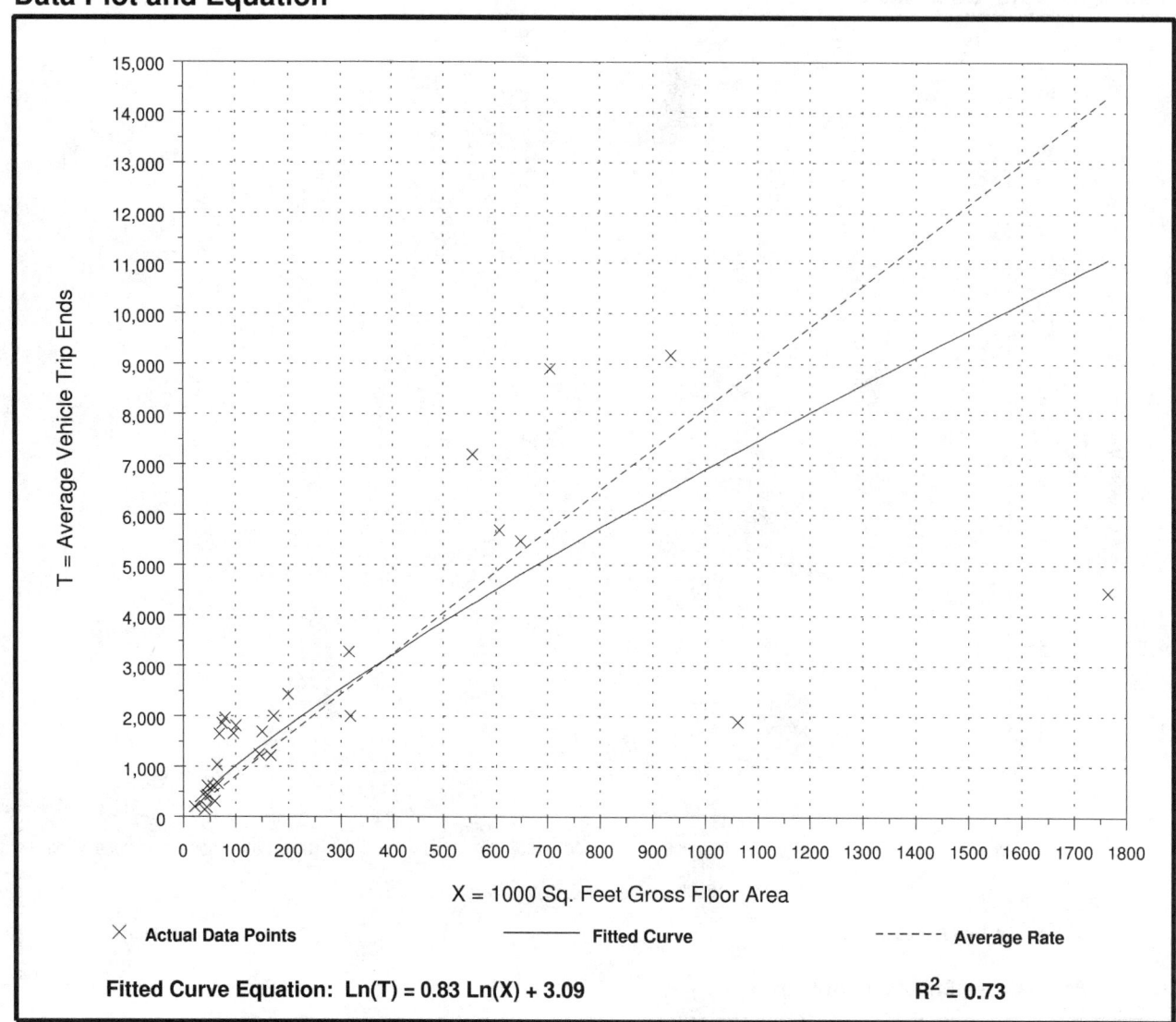

Data Plot and Equation

Fitted Curve Equation: $Ln(T) = 0.83\ Ln(X) + 3.09$ $R^2 = 0.73$

Research and Development Center
(760)

Average Vehicle Trip Ends vs: 1000 Sq. Feet Gross Floor Area
On a: Weekday,
A.M. Peak Hour

Number of Studies: 34
Average 1000 Sq. Feet GFA: 274
Directional Distribution: 83% entering, 17% exiting

Trip Generation per 1000 Sq. Feet Gross Floor Area

Average Rate	Range of Rates	Standard Deviation
1.22	0.37 - 3.73	1.31

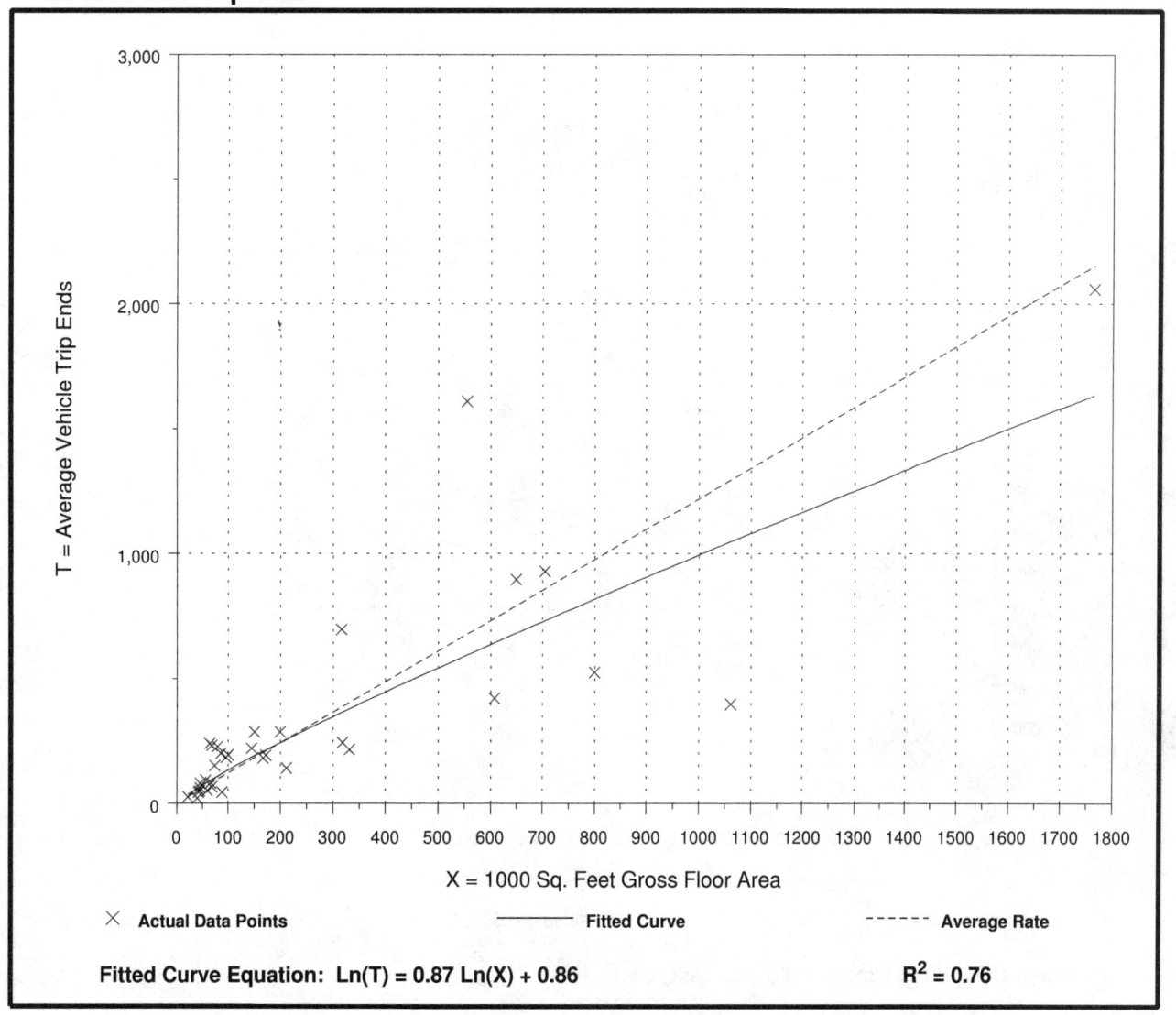

Data Plot and Equation

Fitted Curve Equation: $Ln(T) = 0.87\, Ln(X) + 0.86$ $R^2 = 0.76$

Research and Development Center
(760)

Average Vehicle Trip Ends vs: 1000 Sq. Feet Gross Floor Area
On a: Weekday,
P.M. Peak Hour

Number of Studies: 36
Average 1000 Sq. Feet GFA: 299
Directional Distribution: 15% entering, 85% exiting

Trip Generation per 1000 Sq. Feet Gross Floor Area

Average Rate	Range of Rates	Standard Deviation
1.07	0.40 - 4.13	1.18

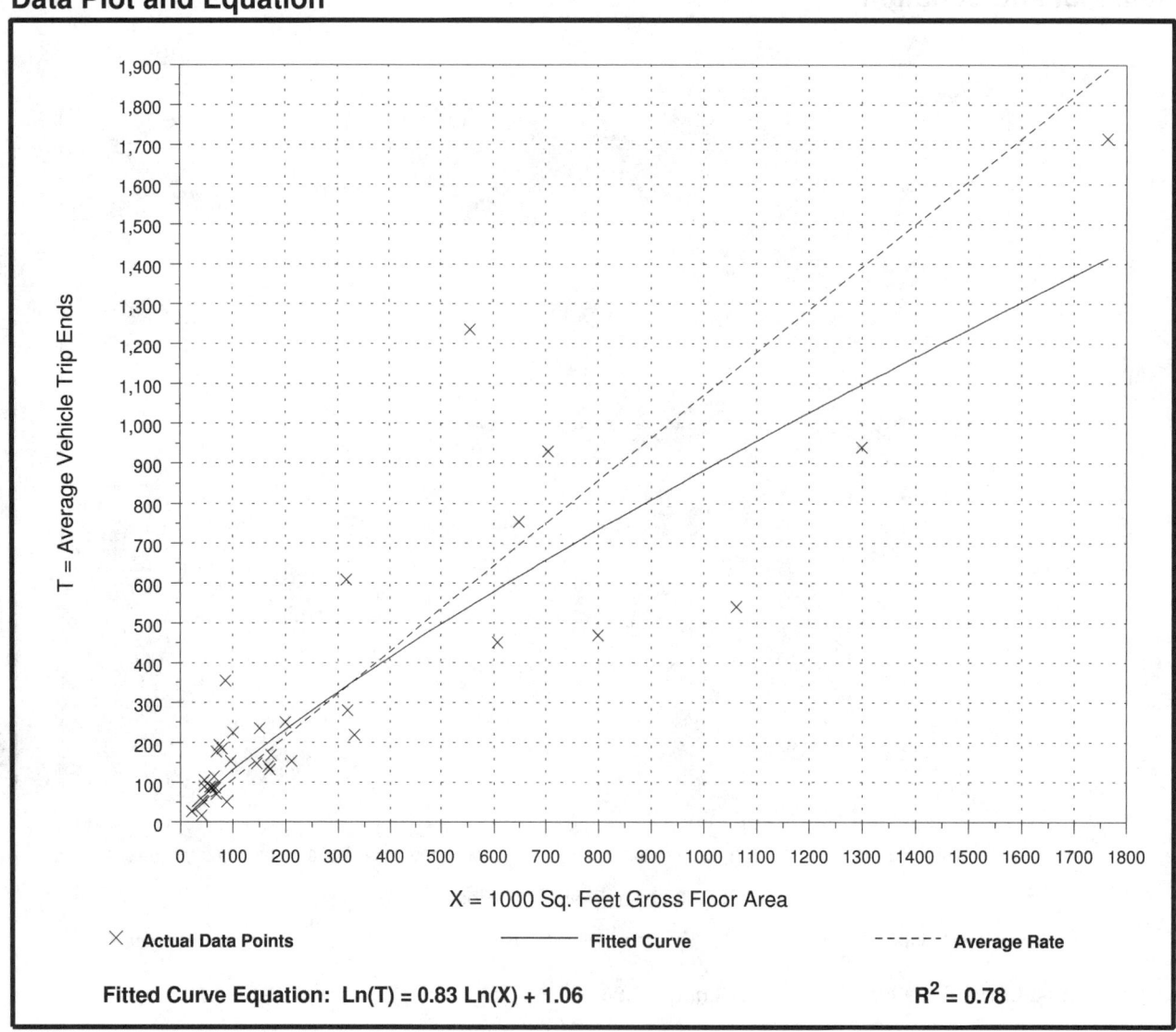

Data Plot and Equation

Fitted Curve Equation: $Ln(T) = 0.83 \, Ln(X) + 1.06$ $R^2 = 0.78$

Research and Development Center
(760)

Average Vehicle Trip Ends vs: 1000 Sq. Feet Gross Floor Area
On a: Saturday

Number of Studies: 21
Average 1000 Sq. Feet GFA: 166
Directional Distribution: 50% entering, 50% exiting

Trip Generation per 1000 Sq. Feet Gross Floor Area

Average Rate	Range of Rates	Standard Deviation
1.90	0.18 - 6.96	1.81

Data Plot and Equation

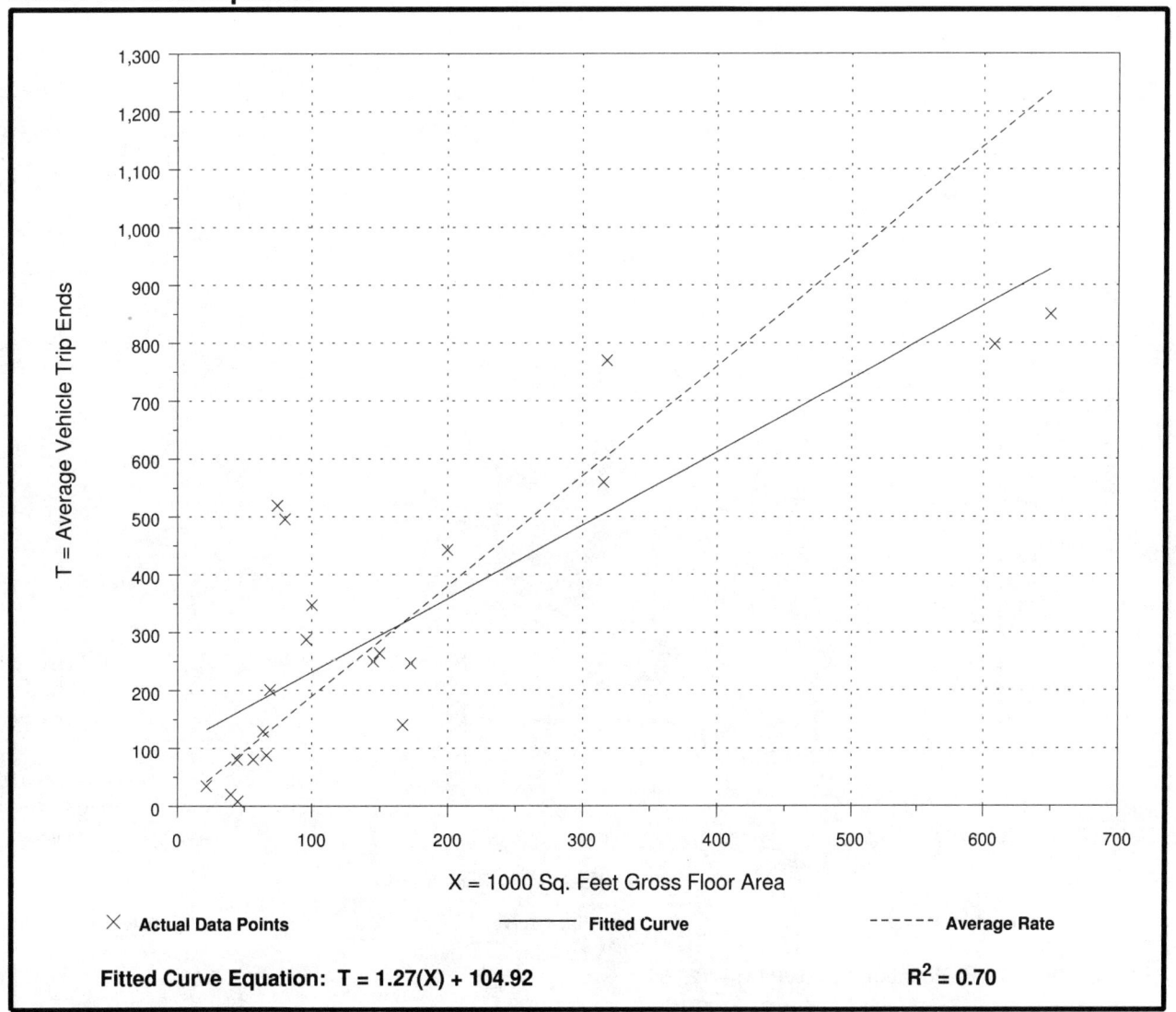

Fitted Curve Equation: $T = 1.27(X) + 104.92$ $R^2 = 0.70$

Research and Development Center
(760)

Average Vehicle Trip Ends vs: 1000 Sq. Feet Gross Floor Area
On a: Saturday,
Peak Hour of Generator

Number of Studies: 14
Average 1000 Sq. Feet GFA: 138
Directional Distribution: Not available

Trip Generation per 1000 Sq. Feet Gross Floor Area

Average Rate	Range of Rates	Standard Deviation
0.24	0.08 - 0.71	0.51

Data Plot and Equation

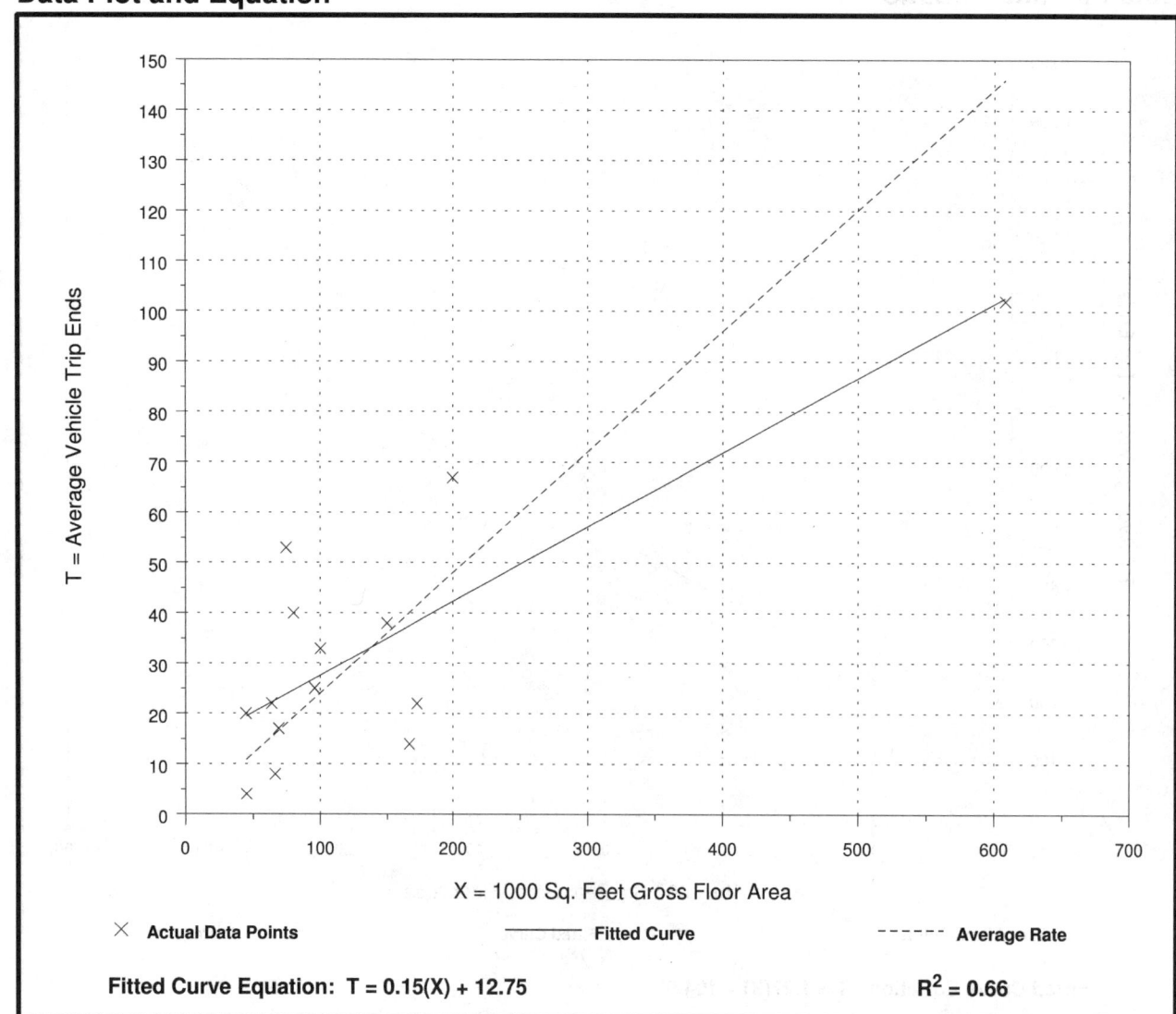

Fitted Curve Equation: T = 0.15(X) + 12.75 $R^2 = 0.66$

Research and Development Center
(760)

Average Vehicle Trip Ends vs: 1000 Sq. Feet Gross Floor Area
On a: Sunday

Number of Studies: 20
Average 1000 Sq. Feet GFA: 172
Directional Distribution: 50% entering, 50% exiting

Trip Generation per 1000 Sq. Feet Gross Floor Area

Average Rate	Range of Rates	Standard Deviation
1.11	0.13 - 4.18	1.39

Data Plot and Equation

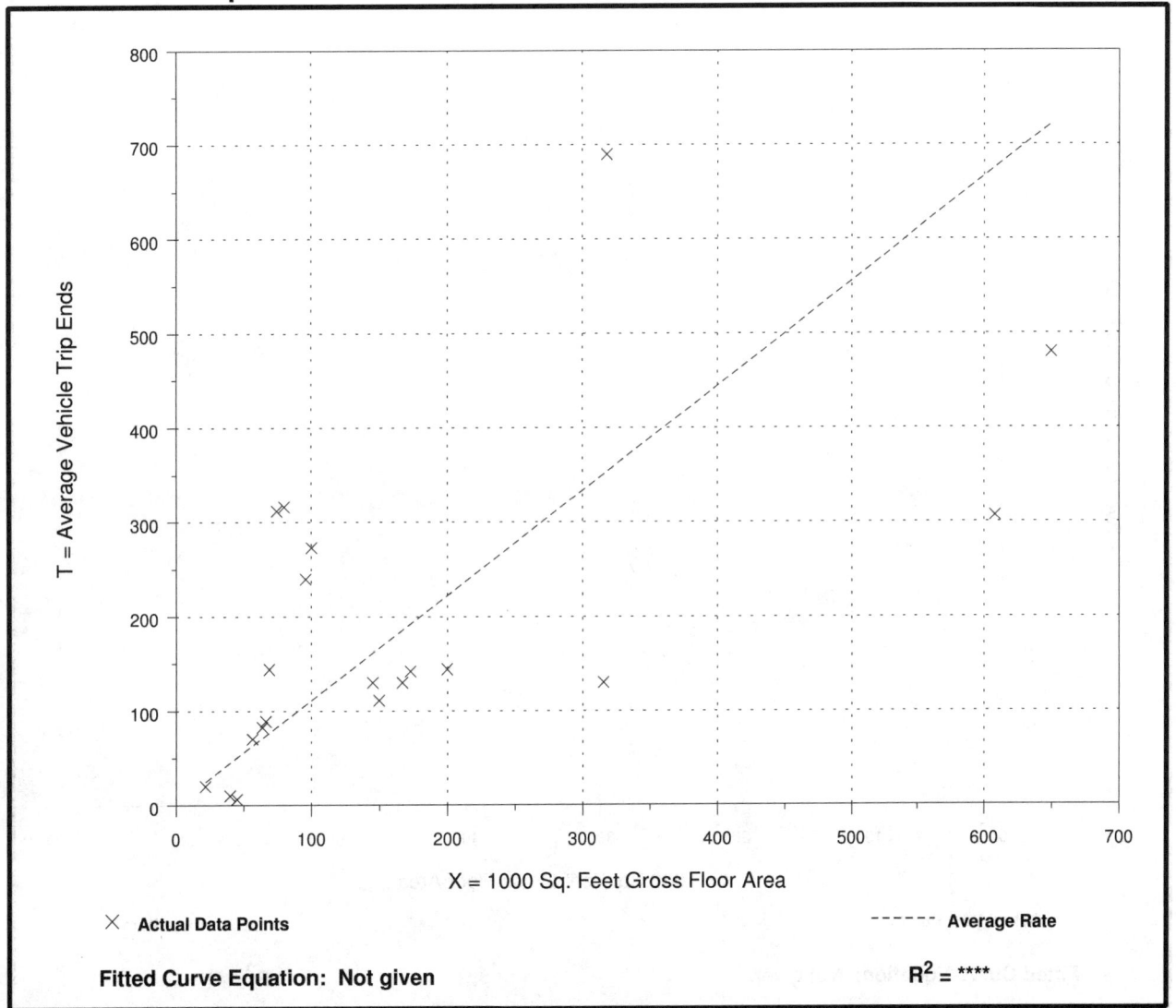

Fitted Curve Equation: Not given

R^2 = ****

Research and Development Center
(760)

Average Vehicle Trip Ends vs: 1000 Sq. Feet Gross Floor Area
On a: Sunday,
Peak Hour of Generator

Number of Studies: 13
Average 1000 Sq. Feet GFA: 146
Directional Distribution: Not available

Trip Generation per 1000 Sq. Feet Gross Floor Area

Average Rate	Range of Rates	Standard Deviation
0.16	0.05 - 0.64	0.42

Data Plot and Equation

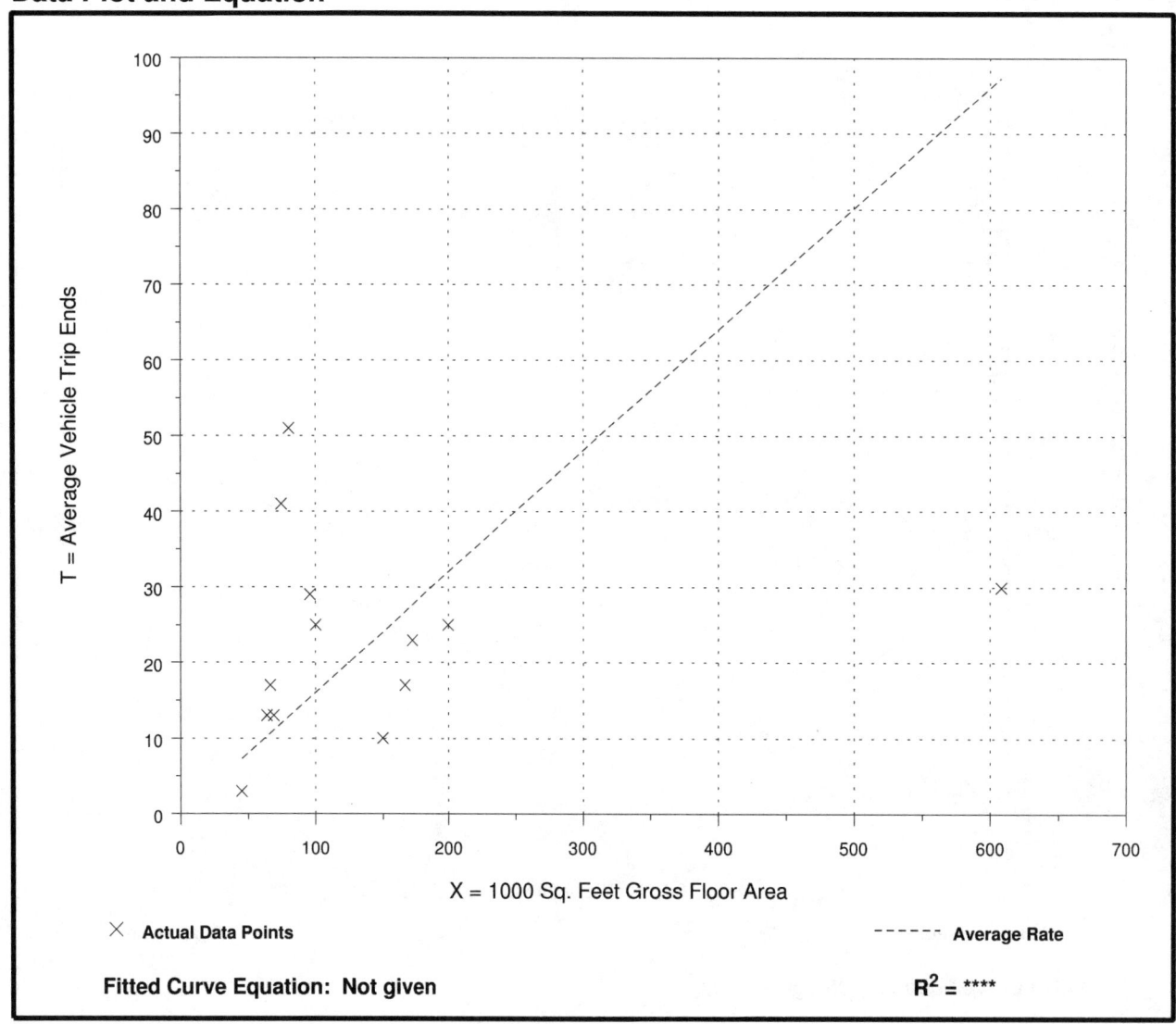

Fitted Curve Equation: Not given $R^2 =$ ****

1388   *Trip Generation*, 9th Edition • Institute of Transportation Engineers

Research and Development Center
(760)

Average Vehicle Trip Ends vs: Acres
On a: Weekday

Number of Studies: 25
Average Number of Acres: 22
Directional Distribution: 50% entering, 50% exiting

Trip Generation per Acre

Average Rate	Range of Rates	Standard Deviation
79.61	15.61 - 876.00	88.46

Data Plot and Equation

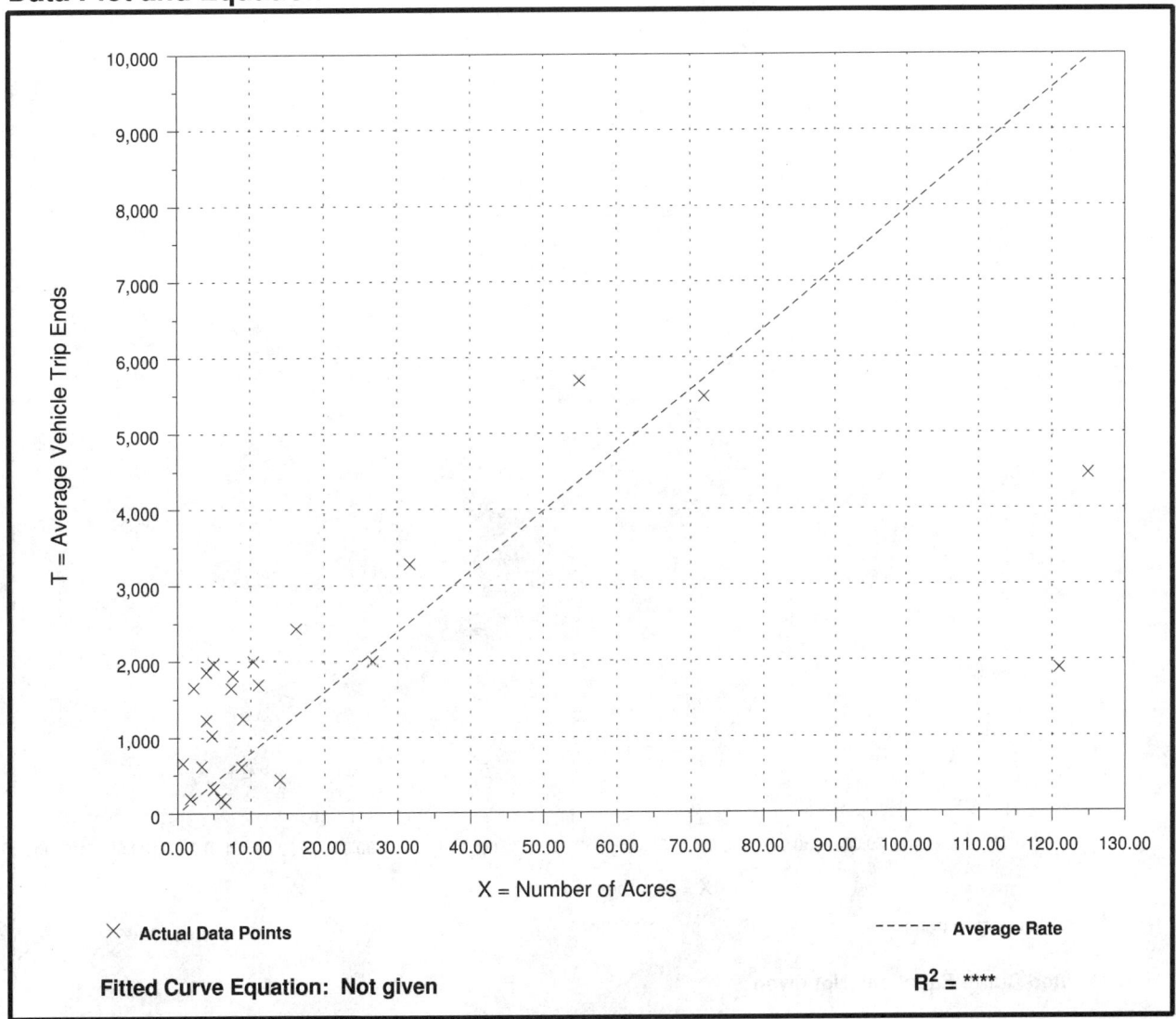

[Plot: T = Average Vehicle Trip Ends vs. X = Number of Acres; X marks = Actual Data Points; dashed line = Average Rate]

Fitted Curve Equation: Not given $R^2 = ****$

Research and Development Center
(760)

Average Vehicle Trip Ends vs: Acres
On a: Weekday,
A.M. Peak Hour

Number of Studies: 26
Average Number of Acres: 22
Directional Distribution: 84% entering, 16% exiting

Trip Generation per Acre

Average Rate	Range of Rates	Standard Deviation
16.77	3.03 - 293.85	31.93

Data Plot and Equation

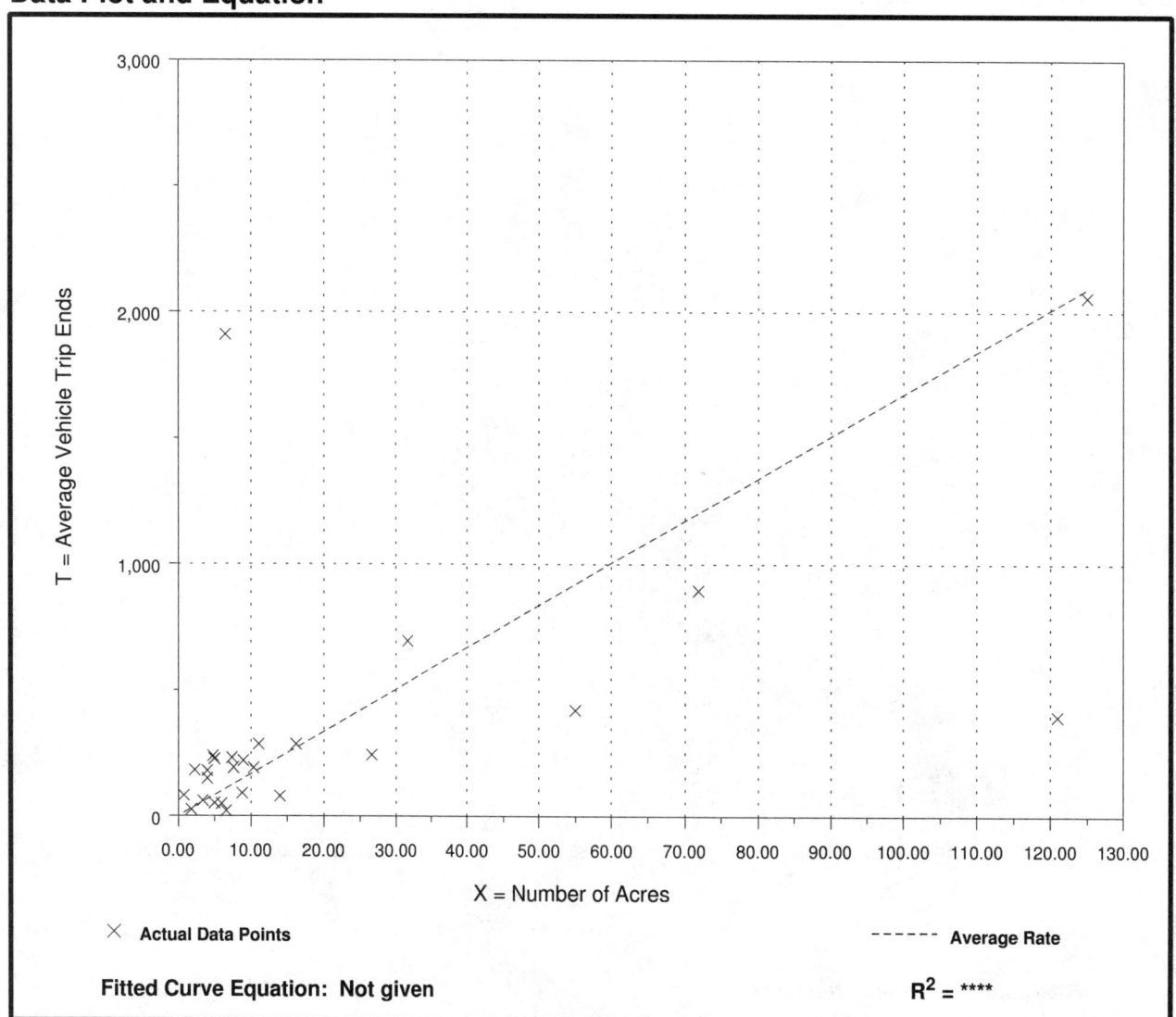

Fitted Curve Equation: Not given $R^2 = ****$

Research and Development Center
(760)

Average Vehicle Trip Ends vs: Acres
On a: Weekday,
P.M. Peak Hour

Number of Studies: 26
Average Number of Acres: 22
Directional Distribution: 12% entering, 88% exiting

Trip Generation per Acre

Average Rate	Range of Rates	Standard Deviation
15.44	2.42 - 284.62	30.56

Data Plot and Equation

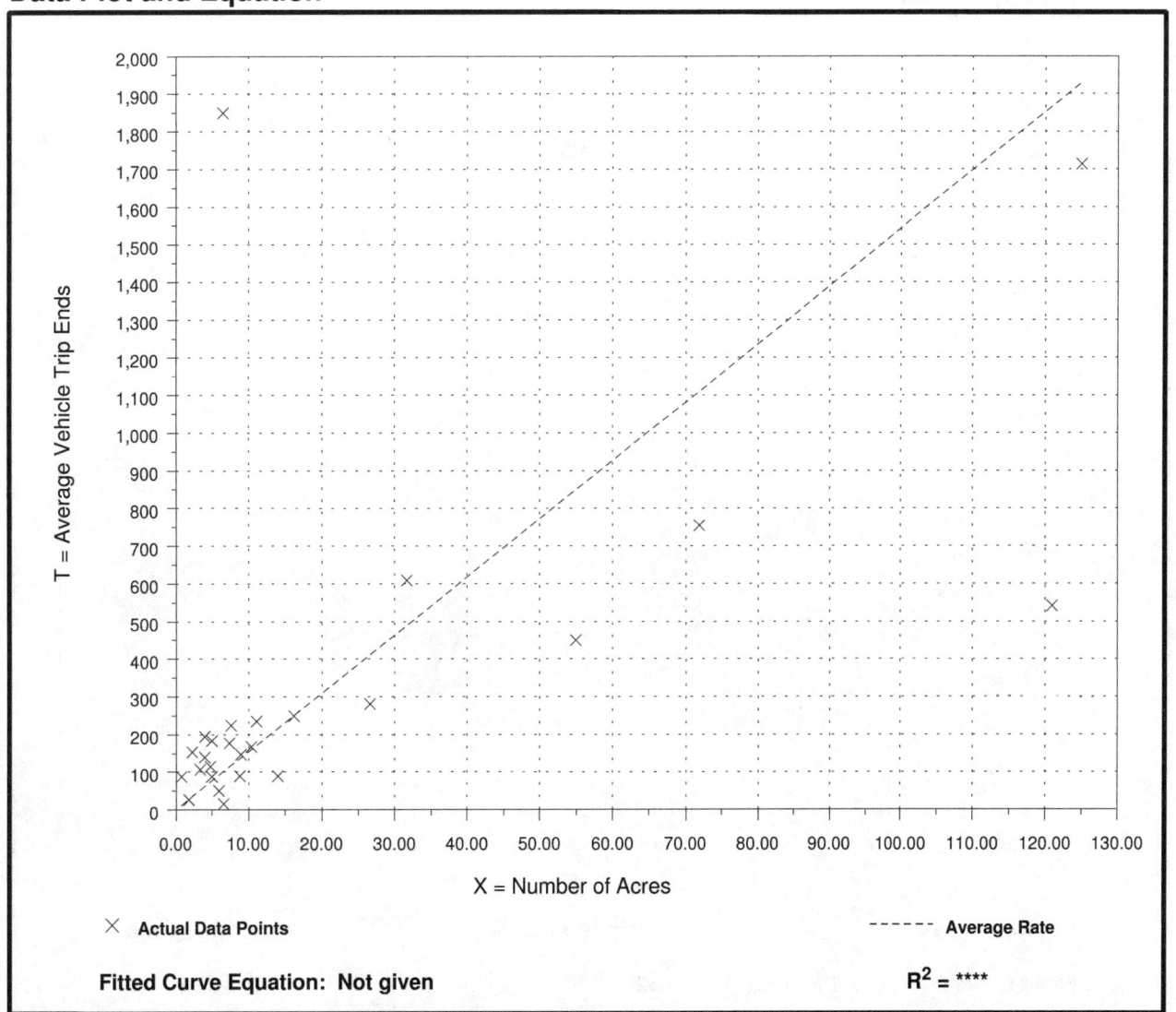

Fitted Curve Equation: Not given $R^2 = ****$

Research and Development Center
(760)

Average Vehicle Trip Ends vs: Acres
On a: Saturday

Number of Studies: 21
Average Number of Acres: 14
Directional Distribution: 50% entering, 50% exiting

Trip Generation per Acre

Average Rate	Range of Rates	Standard Deviation
22.47	2.35 - 128.78	21.60

Data Plot and Equation

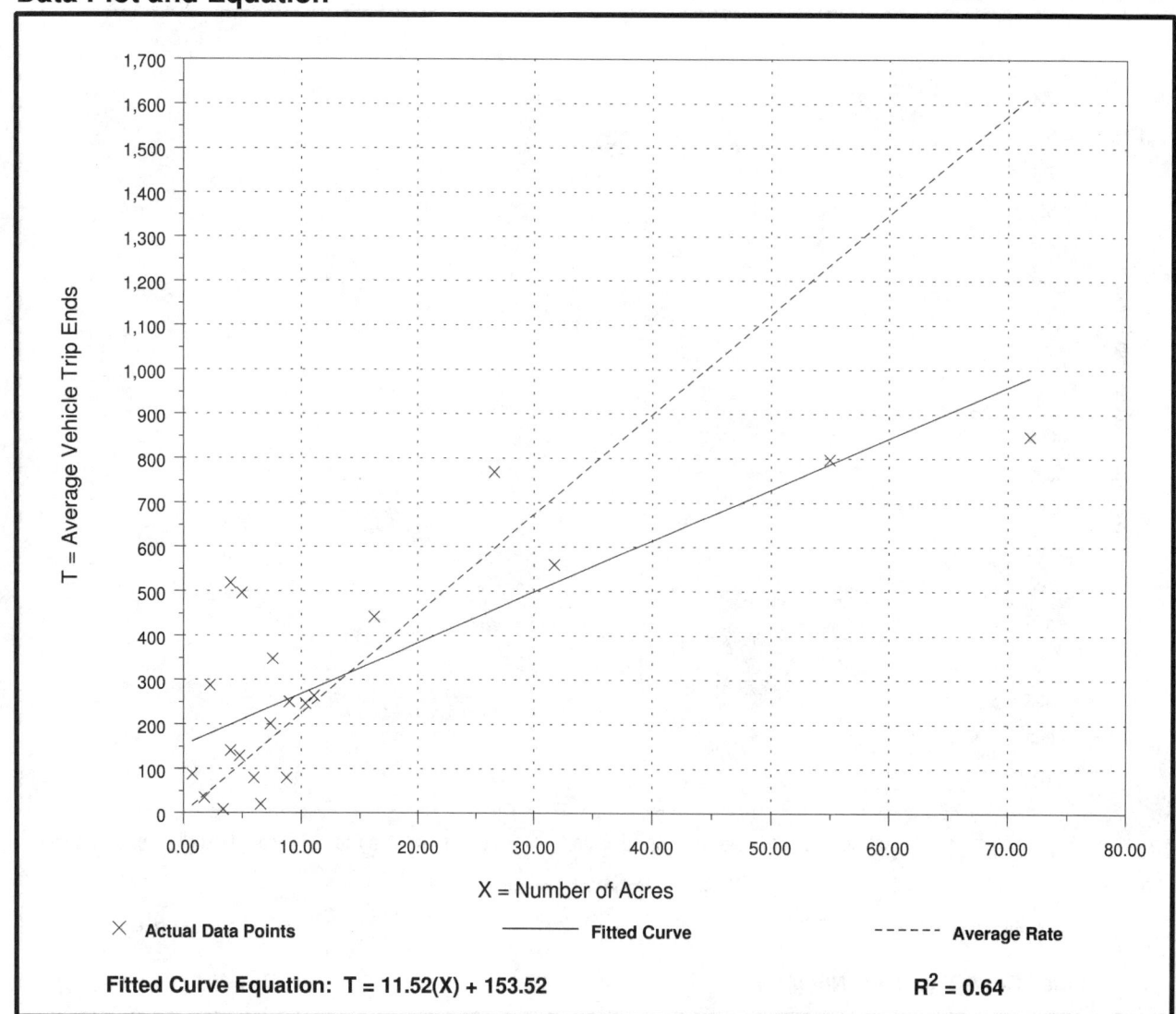

Fitted Curve Equation: $T = 11.52(X) + 153.52$ $R^2 = 0.64$

Research and Development Center
(760)

Average Vehicle Trip Ends vs: Acres
On a: Saturday,
Peak Hour of Generator

Number of Studies: 14
Average Number of Acres: 10
Directional Distribution: Not available

Trip Generation per Acre

Average Rate	Range of Rates	Standard Deviation
3.37	1.18 - 13.15	3.01

Data Plot and Equation

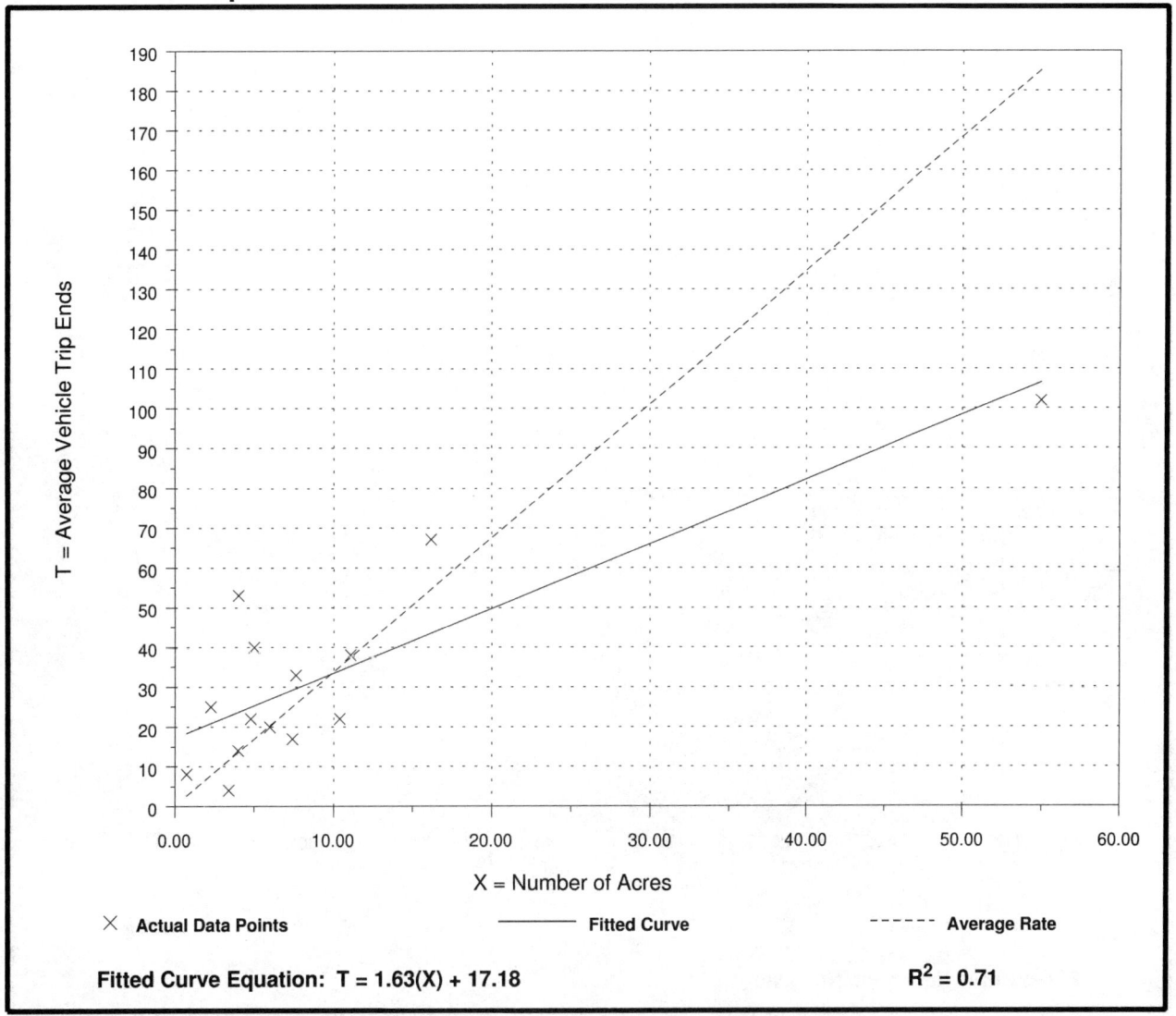

Fitted Curve Equation: T = 1.63(X) + 17.18 $R^2 = 0.71$

Trip Generation, 9th Edition • Institute of Transportation Engineers

Research and Development Center
(760)

Average Vehicle Trip Ends vs: Acres
On a: Sunday

Number of Studies: 20
Average Number of Acres: 14
Directional Distribution: 50% entering, 50% exiting

Trip Generation per Acre

Average Rate	Range of Rates	Standard Deviation
13.27	1.52 - 118.67	16.76

Data Plot and Equation

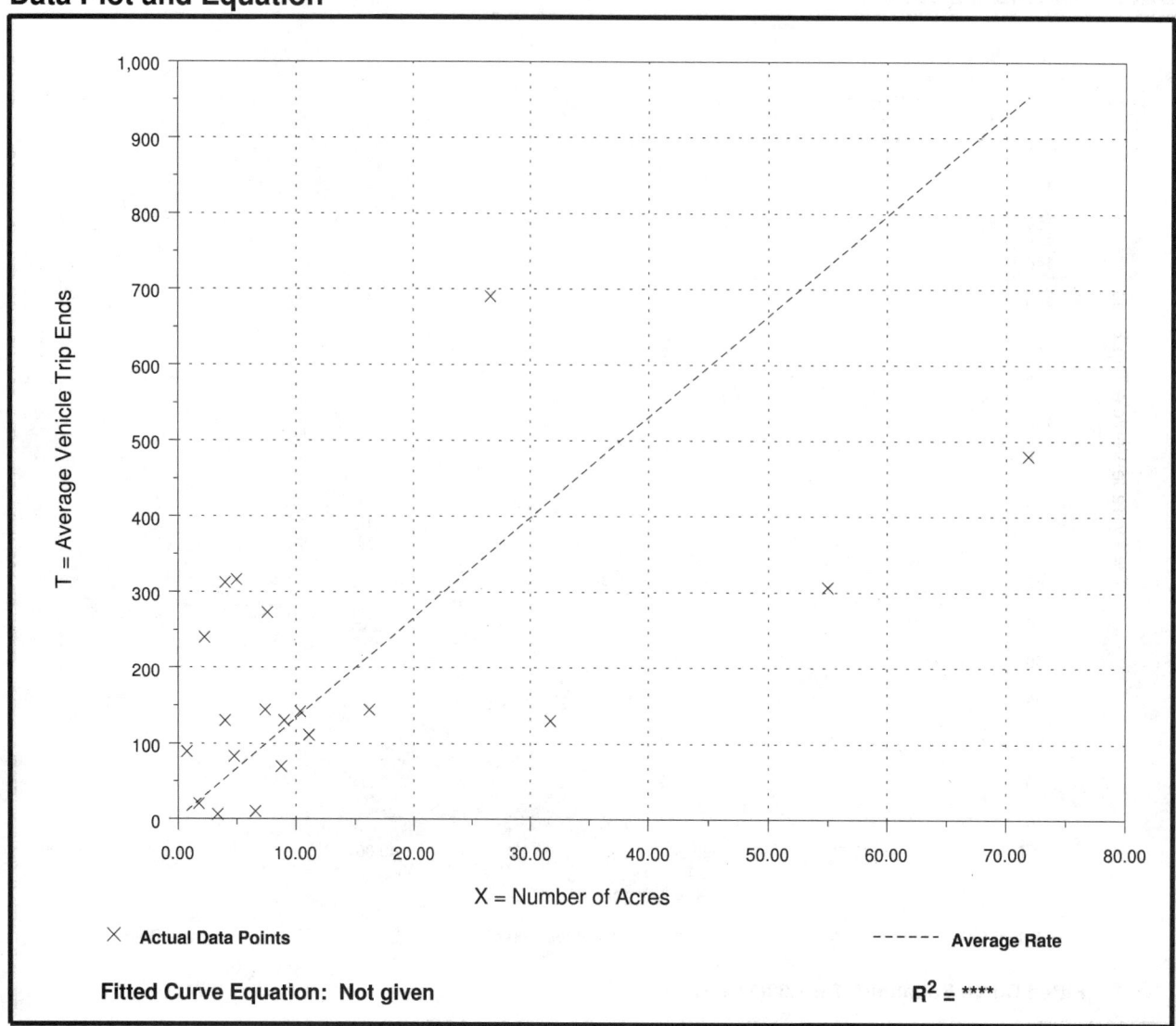

Fitted Curve Equation: Not given $R^2 = ****$

Research and Development Center
(760)

Average Vehicle Trip Ends vs: Acres
On a: Sunday,
Peak Hour of Generator

Number of Studies: 13
Average Number of Acres: 10
Directional Distribution: Not available

Trip Generation per Acre

Average Rate	Range of Rates	Standard Deviation
2.25	0.55 - 22.67	3.46

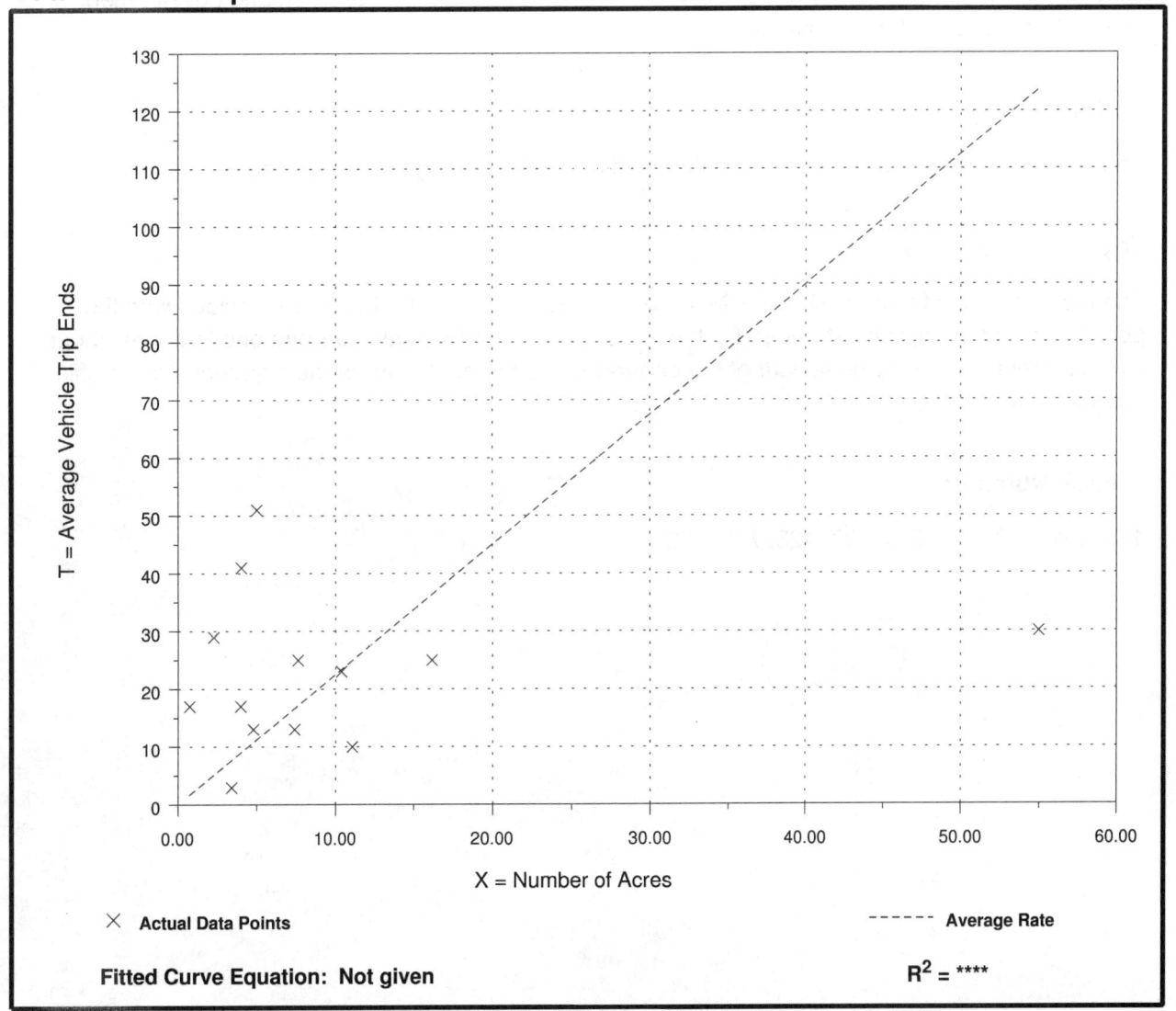

Data Plot and Equation

Fitted Curve Equation: Not given R^2 = ****

Trip Generation, 9th Edition • Institute of Transportation Engineers

Land Use: 770
Business Park

Description

Business parks consist of a group of flex-type or incubator one- or two-story buildings served by a common roadway system. The tenant space is flexible and lends itself to a variety of uses; the rear side of the building is usually served by a garage door. Tenants may be start-up companies or small mature companies that require a variety of space. The space may include offices, retail and wholesale stores, restaurants, recreational areas and warehousing, manufacturing, light industrial, or scientific research functions. The average mix is 20 to 30 percent office/commercial and 70 to 80 percent industrial/warehousing. Industrial park (Land Use 130), warehousing (Land Use 150), general office building (Land Use 710), corporate headquarters building (Land Use 714), single tenant office building (Land Use 715), office park (Land Use 750) and research and development center (Land Use 760) are related uses.

Additional Data

The sites were surveyed between the 1980s and the 2000s throughout the United States.

Trip Characteristics

The trip generation for the A.M. and P.M. peak hours of the generator typically coincided with the peak hours of the adjacent street traffic; therefore, only one A.M. peak hour and one P.M. peak hour, which represent both the peak hour of the generator and the peak hour of the adjacent street traffic, are shown for business parks.

Source Numbers

155, 211, 212, 213, 216, 407, 423, 715, 728

Business Park
(770)

Average Vehicle Trip Ends vs: Employees
On a: Weekday

Number of Studies: 12
Avg. Number of Employees: 1,097
Directional Distribution: 50% entering, 50% exiting

Trip Generation per Employee

Average Rate	Range of Rates	Standard Deviation
4.04	3.25 - 8.19	2.20

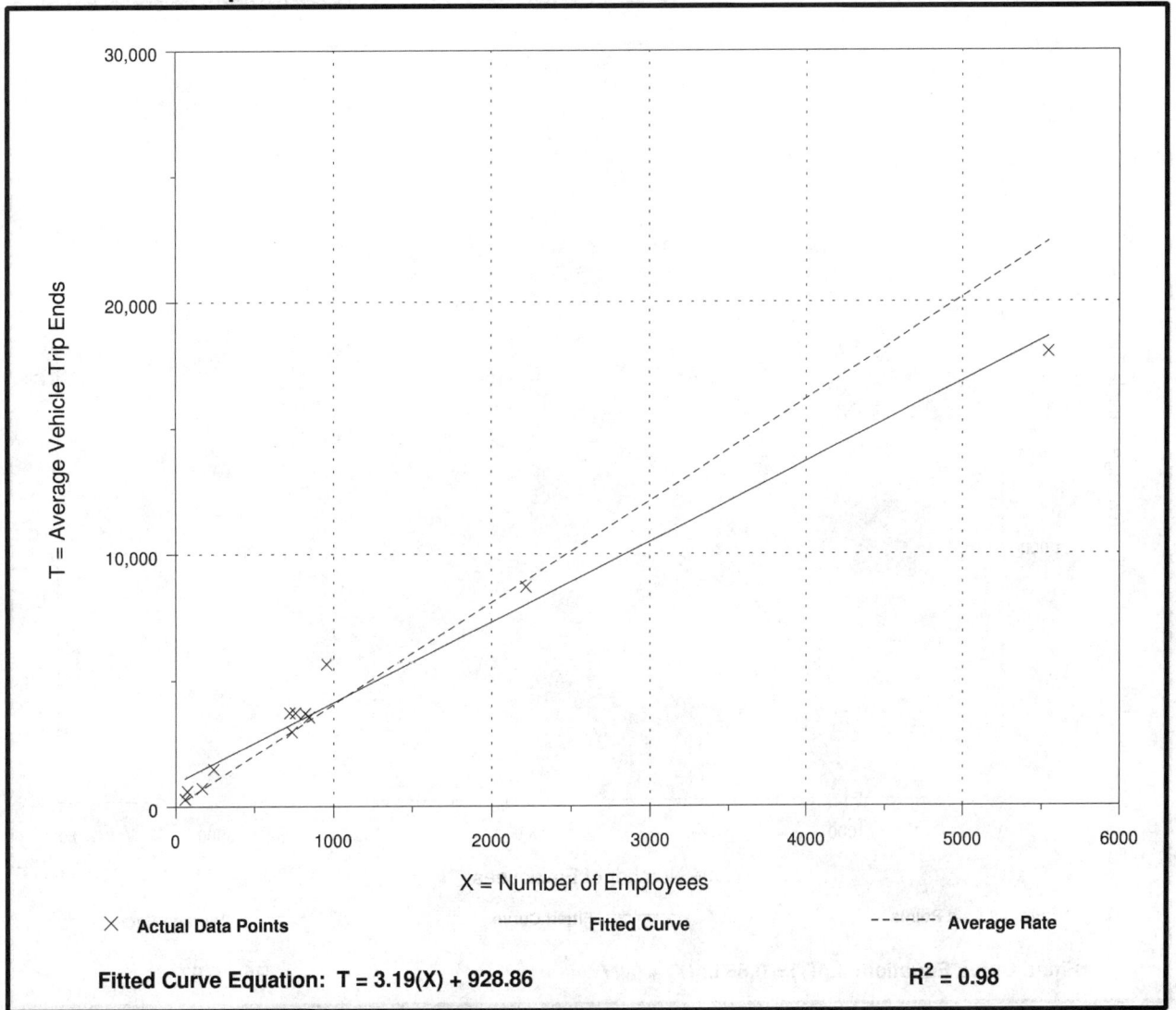

Data Plot and Equation

Fitted Curve Equation: $T = 3.19(X) + 928.86$ $R^2 = 0.98$

Business Park
(770)

Average Vehicle Trip Ends vs: Employees
On a: Weekday,
A.M. Peak Hour

Number of Studies: 12
Avg. Number of Employees: 1,097
Directional Distribution: 85% entering, 15% exiting

Trip Generation per Employee

Average Rate	Range of Rates	Standard Deviation
0.45	0.30 - 0.95	0.69

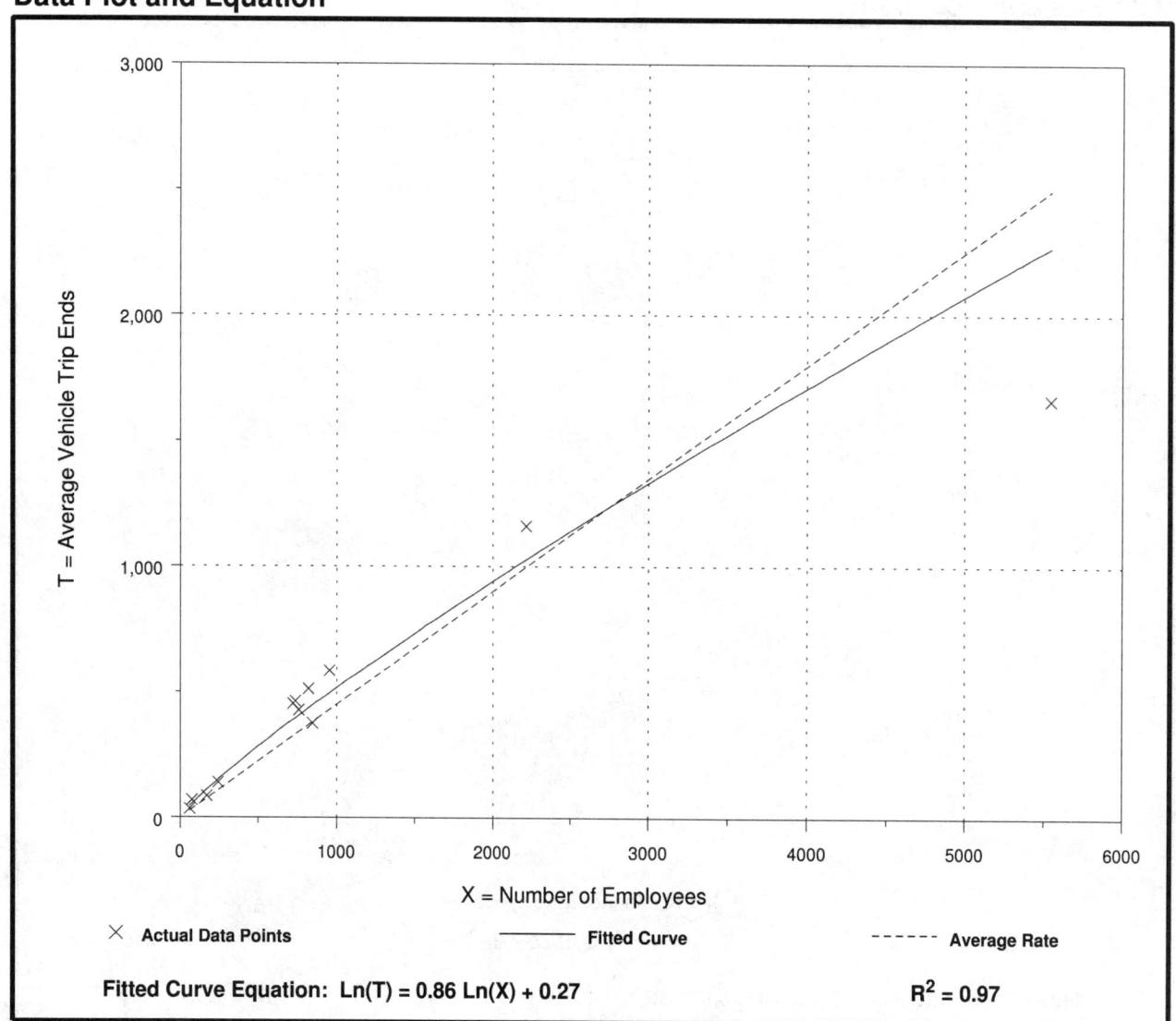

Data Plot and Equation

Fitted Curve Equation: $Ln(T) = 0.86 \, Ln(X) + 0.27$ $R^2 = 0.97$

Business Park
(770)

Average Vehicle Trip Ends vs: Employees
On a: Weekday, P.M. Peak Hour

Number of Studies: 13
Avg. Number of Employees: 1,163
Directional Distribution: 22% entering, 78% exiting

Trip Generation per Employee

Average Rate	Range of Rates	Standard Deviation
0.39	0.24 - 1.01	0.64

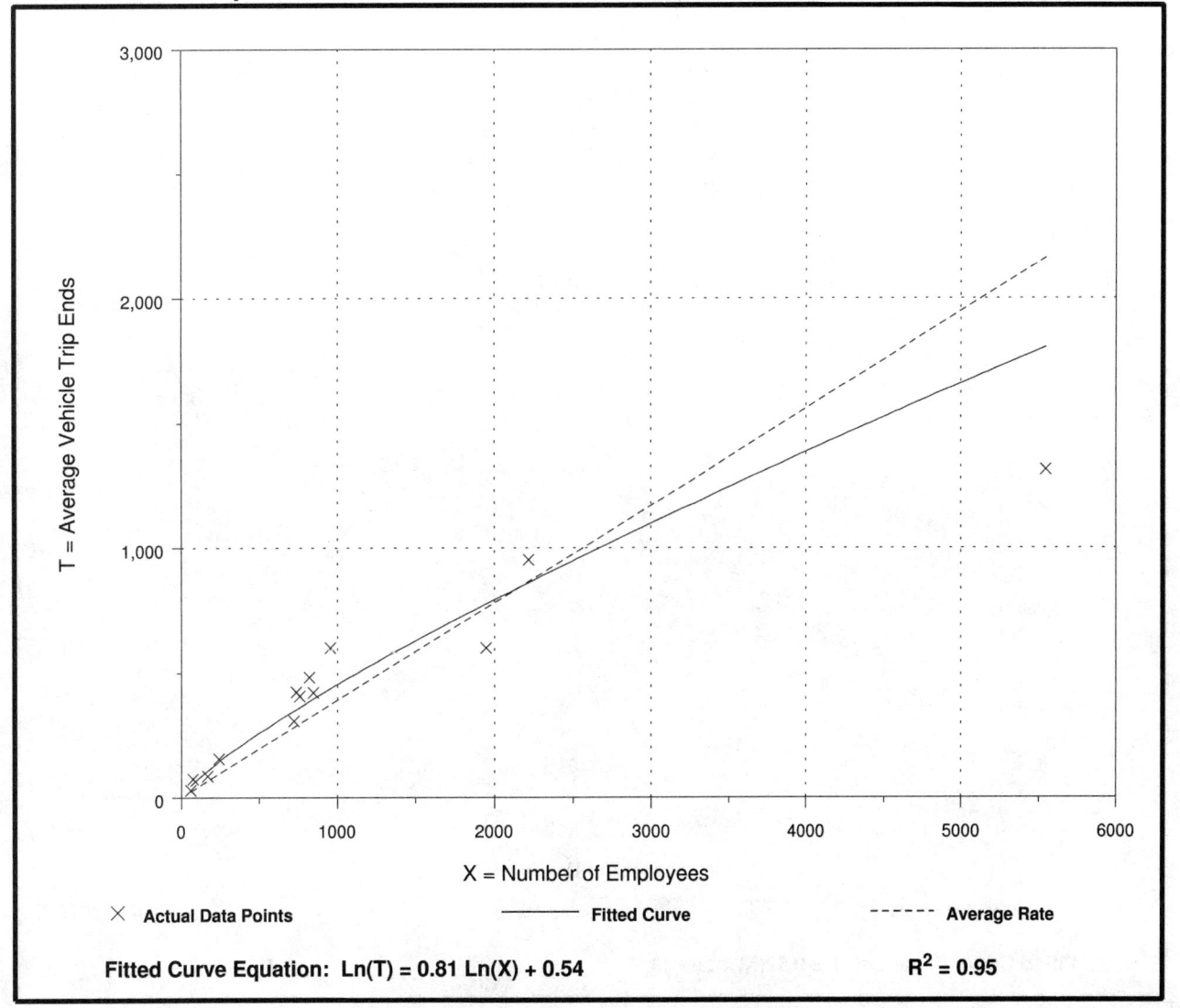

Data Plot and Equation

Fitted Curve Equation: $Ln(T) = 0.81\ Ln(X) + 0.54$ $R^2 = 0.95$

Business Park
(770)

Average Vehicle Trip Ends vs: Employees
On a: Saturday

Number of Studies: 9
Avg. Number of Employees: 1,429
Directional Distribution: 50% entering, 50% exiting

Trip Generation per Employee

Average Rate	Range of Rates	Standard Deviation
0.71	0.48 - 1.37	0.88

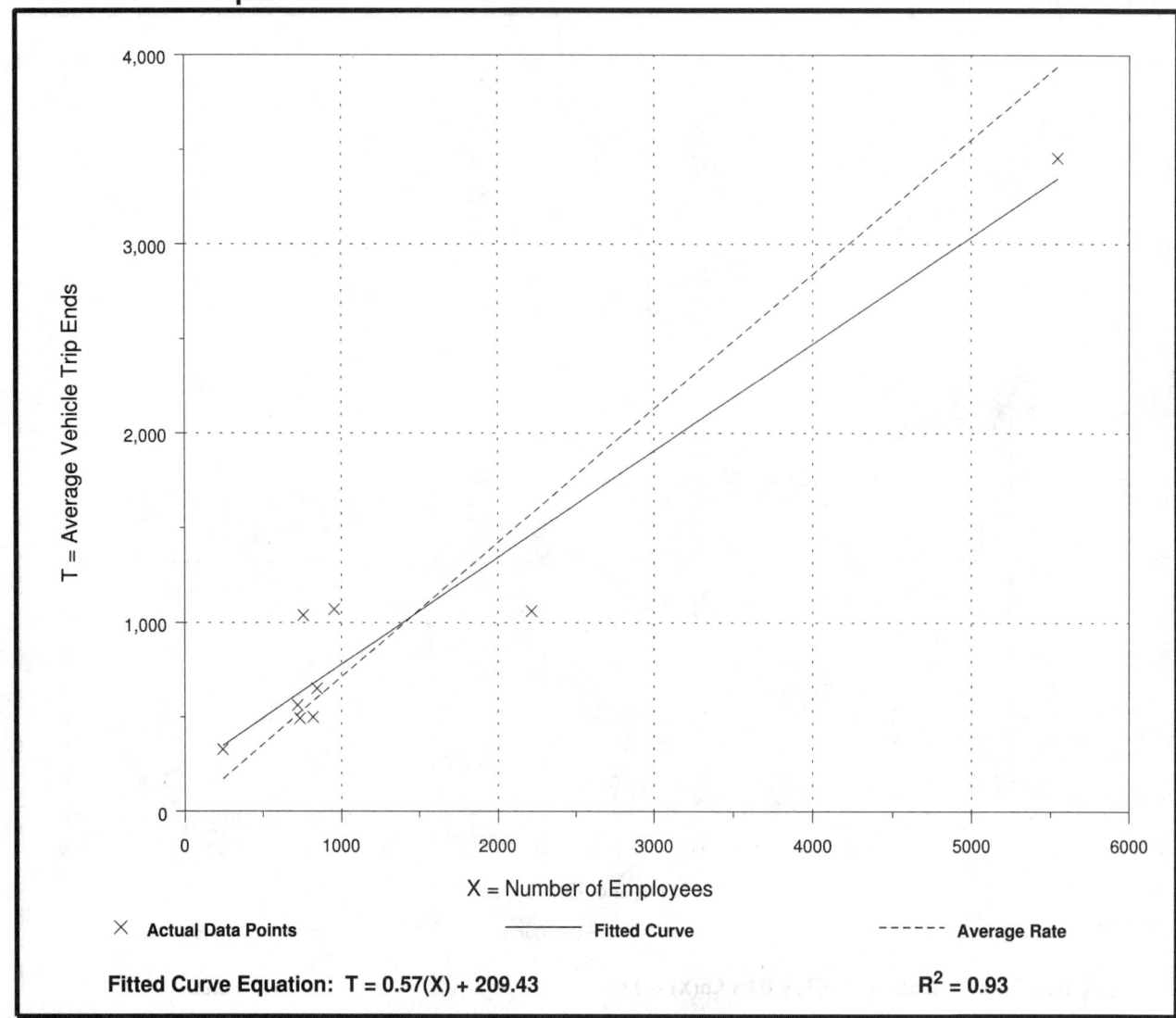

Data Plot and Equation

Fitted Curve Equation: T = 0.57(X) + 209.43 $R^2 = 0.93$

Business Park
(770)

Average Vehicle Trip Ends vs: Employees
On a: Sunday

Number of Studies: 9
Avg. Number of Employees: 1,429
Directional Distribution: 50% entering, 50% exiting

Trip Generation per Employee

Average Rate	Range of Rates	Standard Deviation
0.36	0.17 - 1.02	0.63

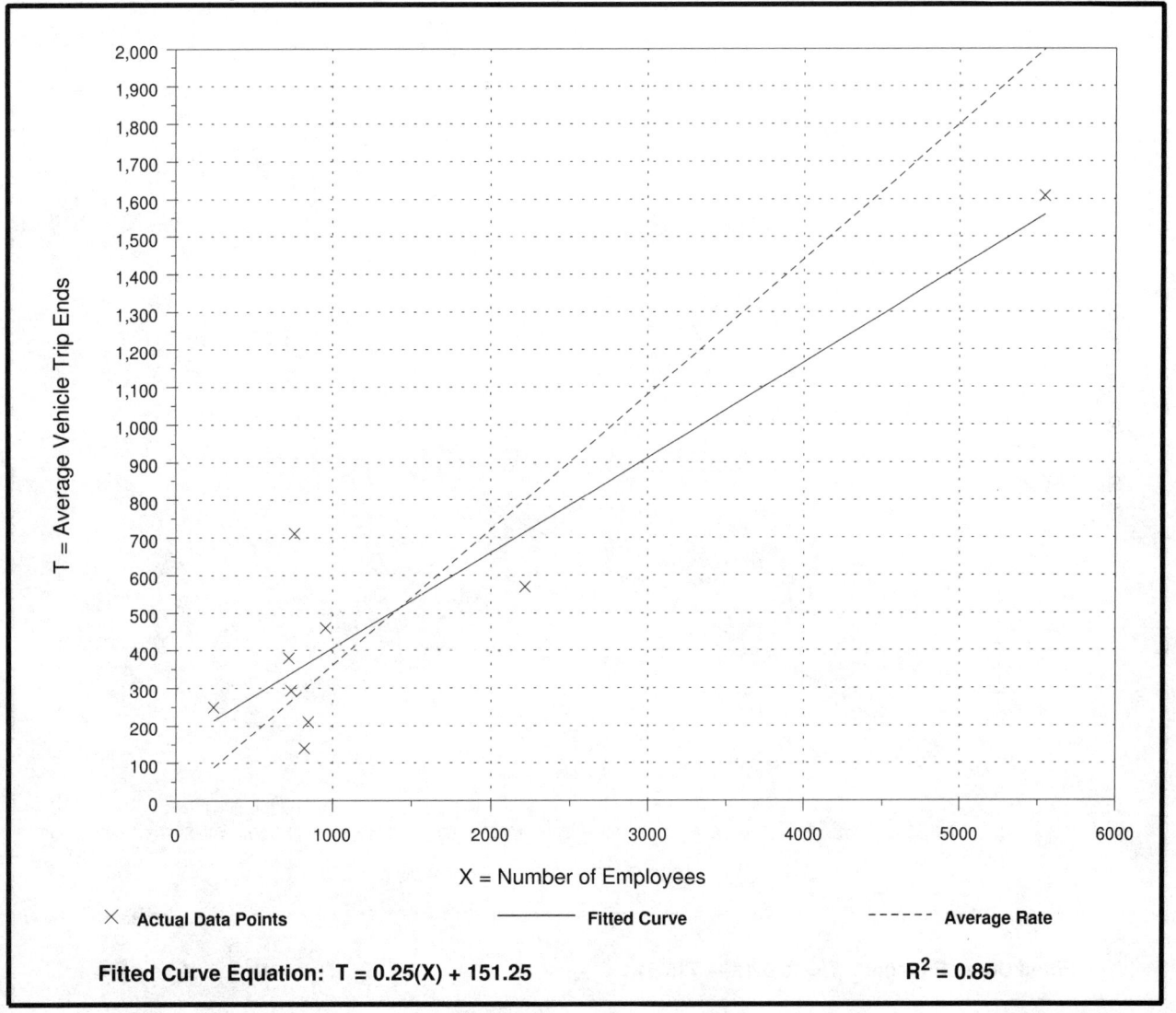

Data Plot and Equation

Fitted Curve Equation: T = 0.25(X) + 151.25 $R^2 = 0.85$

Trip Generation, 9th Edition ● Institute of Transportation Engineers

Business Park
(770)

Average Vehicle Trip Ends vs: 1000 Sq. Feet Gross Floor Area
On a: Weekday

Number of Studies: 16
Average 1000 Sq. Feet GFA: 393
Directional Distribution: 50% entering, 50% exiting

Trip Generation per 1000 Sq. Feet Gross Floor Area

Average Rate	Range of Rates	Standard Deviation
12.44	5.56 - 27.96	5.61

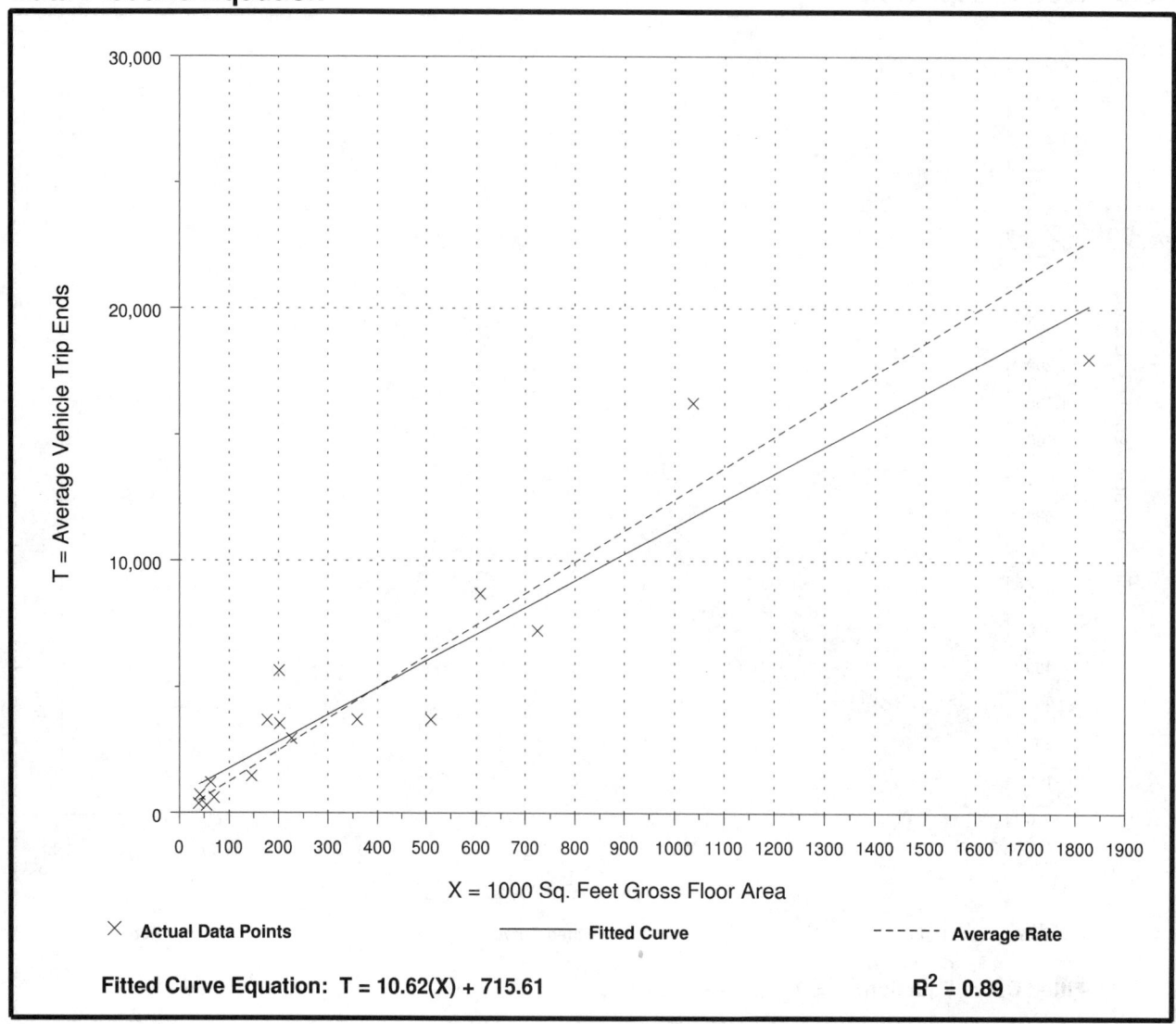

Data Plot and Equation

Fitted Curve Equation: T = 10.62(X) + 715.61 $R^2 = 0.89$

Business Park
(770)

Average Vehicle Trip Ends vs: 1000 Sq. Feet Gross Floor Area
On a: Weekday,
A.M. Peak Hour

Number of Studies: 20
Average 1000 Sq. Feet GFA: 384
Directional Distribution: 85% entering, 15% exiting

Trip Generation per 1000 Sq. Feet Gross Floor Area

Average Rate	Range of Rates	Standard Deviation
1.40	0.65 - 2.90	1.32

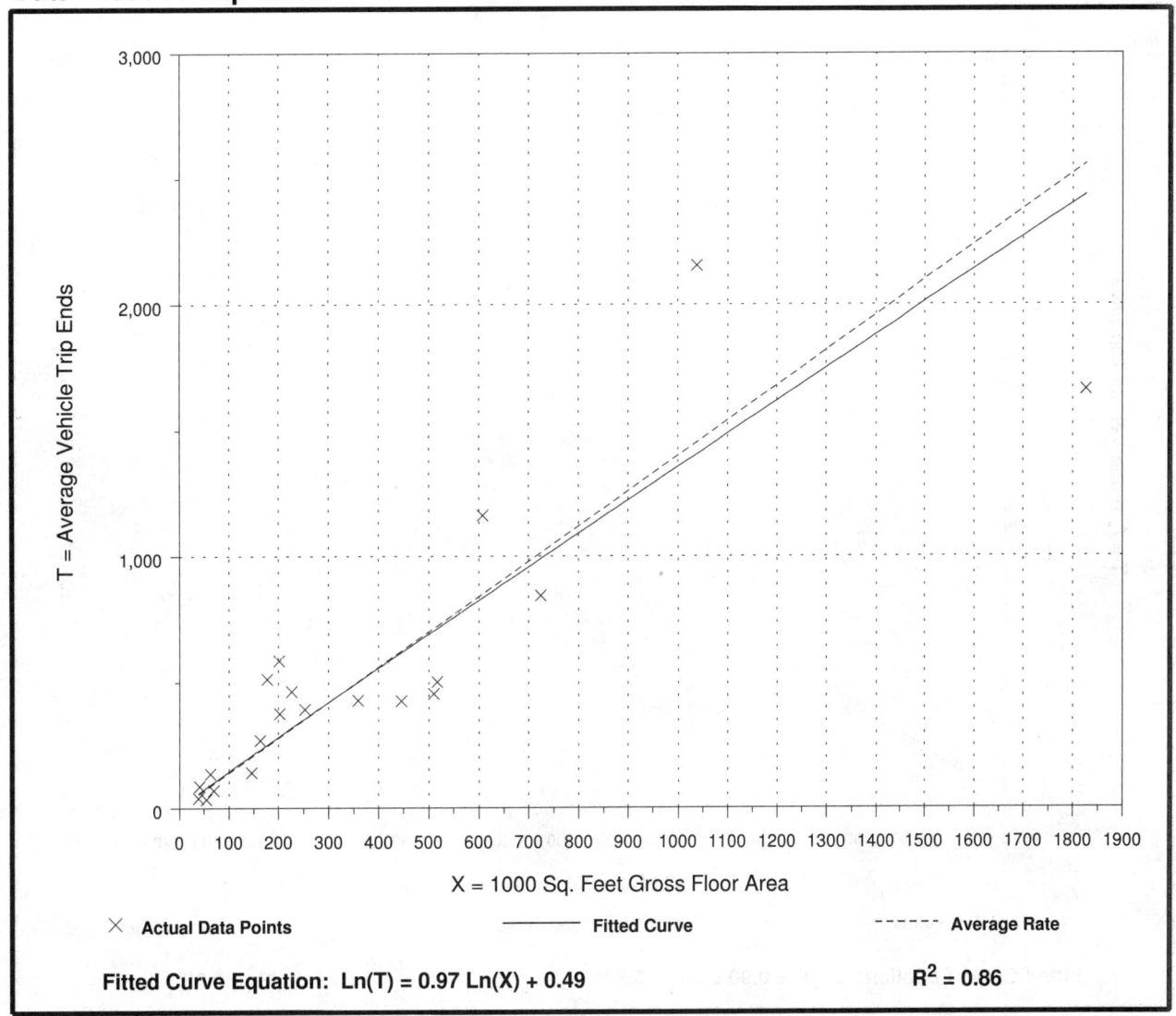

Data Plot and Equation

Fitted Curve Equation: $Ln(T) = 0.97 \, Ln(X) + 0.49$ $R^2 = 0.86$

Business Park
(770)

Average Vehicle Trip Ends vs: 1000 Sq. Feet Gross Floor Area
On a: Weekday,
P.M. Peak Hour

Number of Studies: 21
Average 1000 Sq. Feet GFA: 396
Directional Distribution: 26% entering, 74% exiting

Trip Generation per 1000 Sq. Feet Gross Floor Area

Average Rate	Range of Rates	Standard Deviation
1.26	0.55 - 2.97	1.26

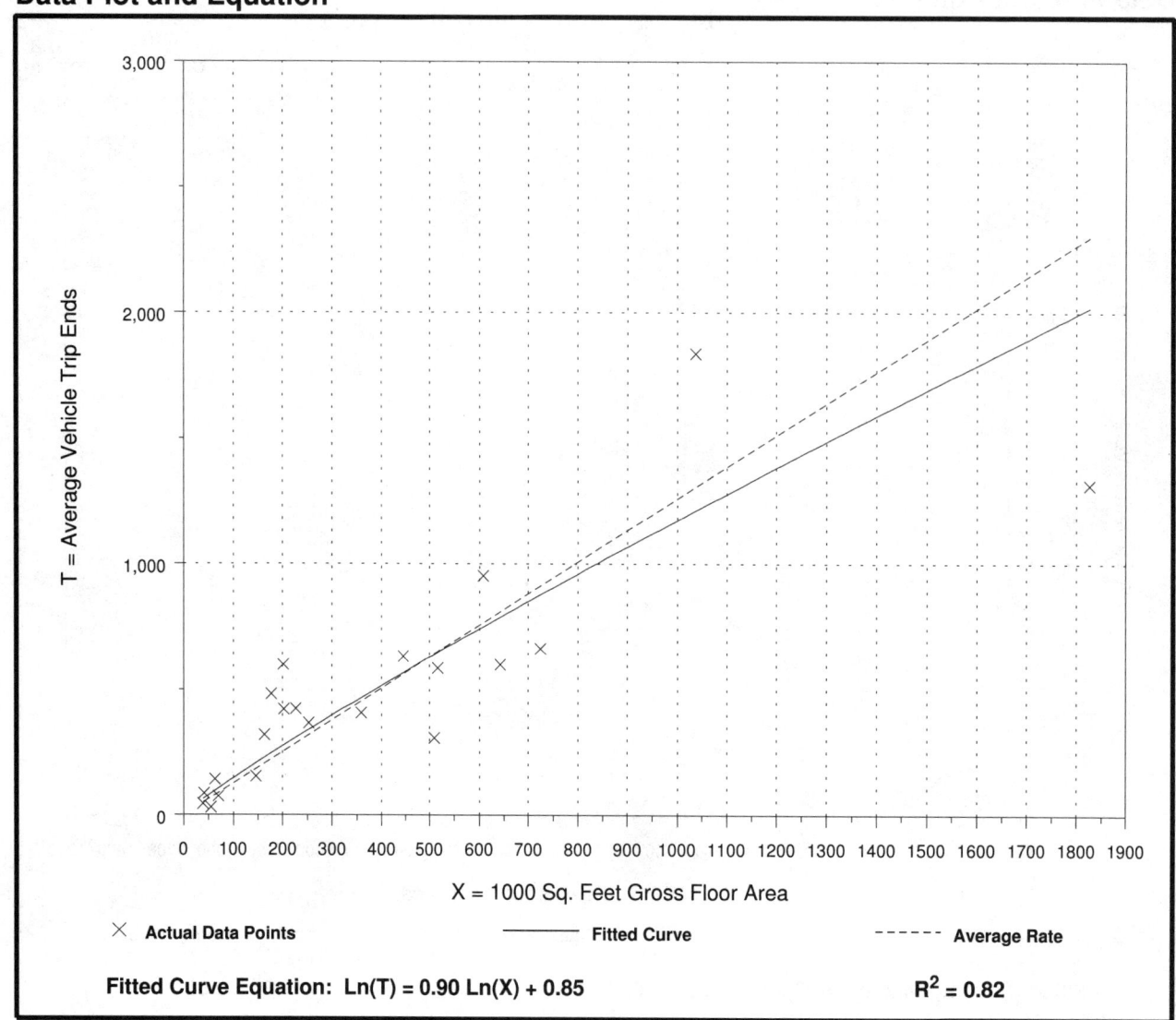

Data Plot and Equation

Fitted Curve Equation: $Ln(T) = 0.90 \, Ln(X) + 0.85$ $R^2 = 0.82$

Business Park
(770)

Average Vehicle Trip Ends vs: 1000 Sq. Feet Gross Floor Area
On a: Saturday

Number of Studies: 11
Average 1000 Sq. Feet GFA: 485
Directional Distribution: 50% entering, 50% exiting

Trip Generation per 1000 Sq. Feet Gross Floor Area

Average Rate	Range of Rates	Standard Deviation
2.56	1.10 - 5.30	1.96

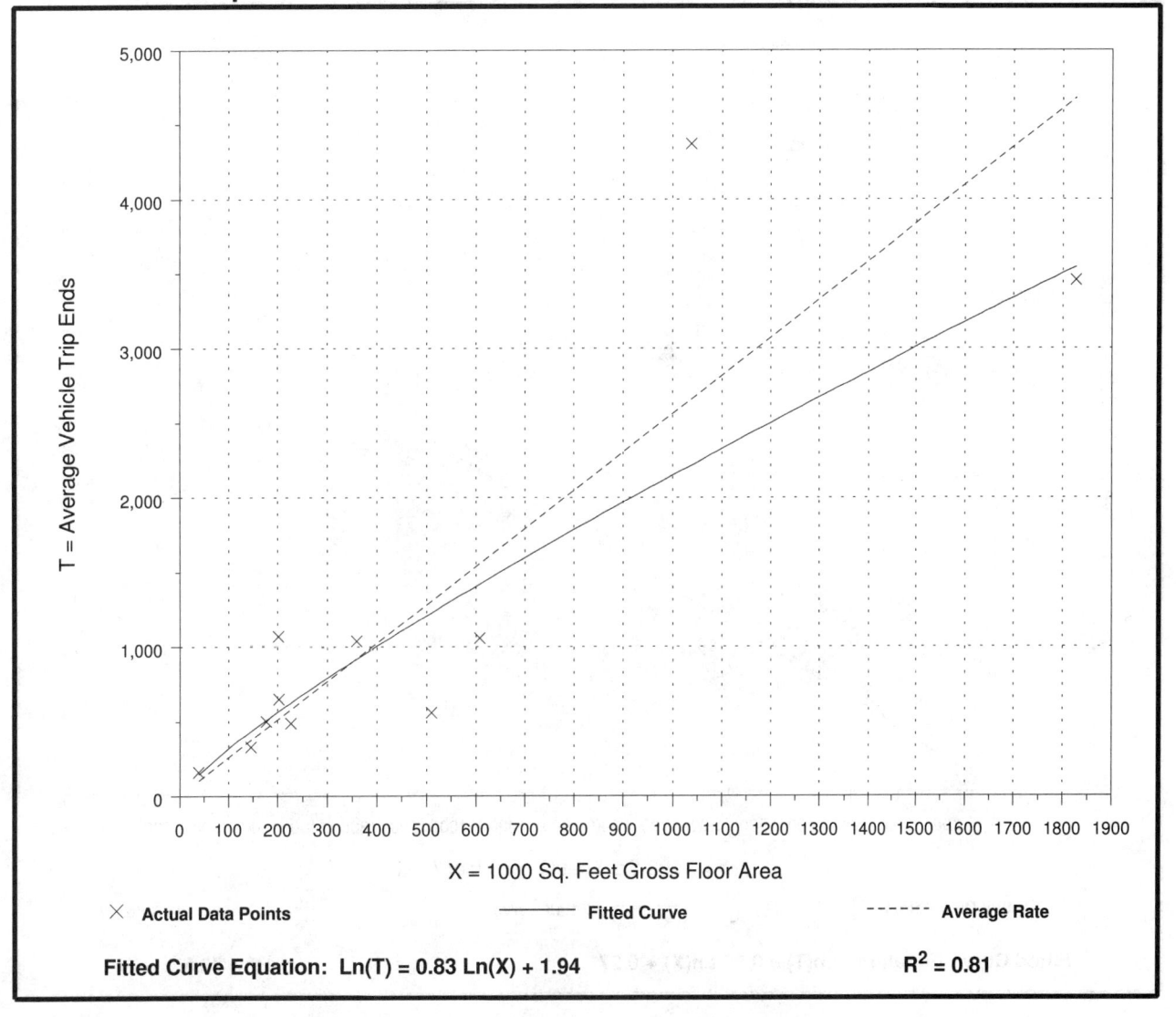

Data Plot and Equation

Fitted Curve Equation: $\operatorname{Ln}(T) = 0.83 \operatorname{Ln}(X) + 1.94$

$R^2 = 0.81$

Business Park
(770)

Average Vehicle Trip Ends vs: 1000 Sq. Feet Gross Floor Area
On a: Sunday

Number of Studies: 11
Average 1000 Sq. Feet GFA: 485
Directional Distribution: 50% entering, 50% exiting

Trip Generation per 1000 Sq. Feet Gross Floor Area

Average Rate	Range of Rates	Standard Deviation
1.29	0.74 - 2.28	1.27

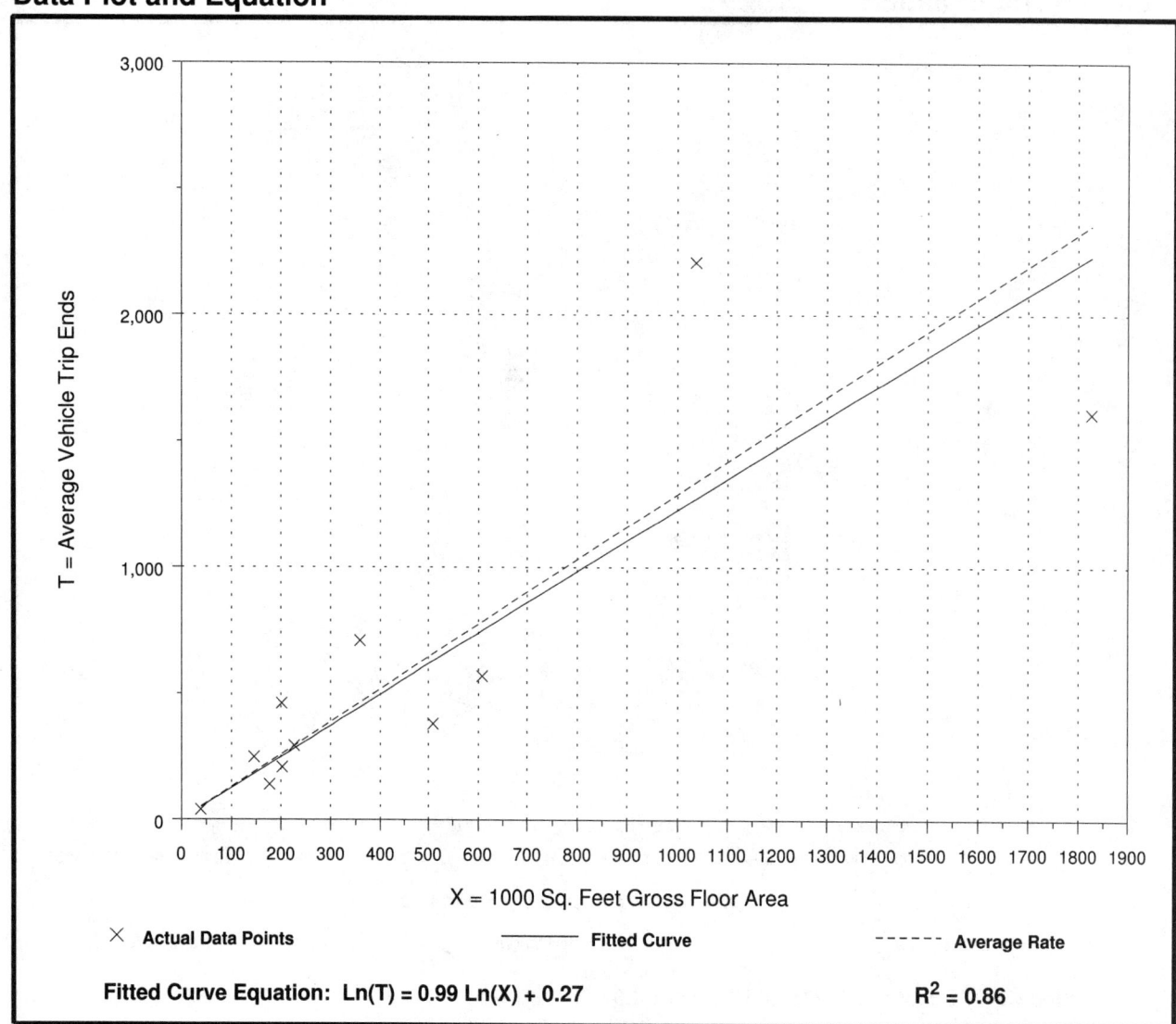

Data Plot and Equation

Fitted Curve Equation: Ln(T) = 0.99 Ln(X) + 0.27 $R^2 = 0.86$

Business Park
(770)

Average Vehicle Trip Ends vs: Acres
On a: Weekday

Number of Studies: 12
Average Number of Acres: 28
Directional Distribution: 50% entering, 50% exiting

Trip Generation per Acre

Average Rate	Range of Rates	Standard Deviation
149.79	23.54 - 276.76	72.38

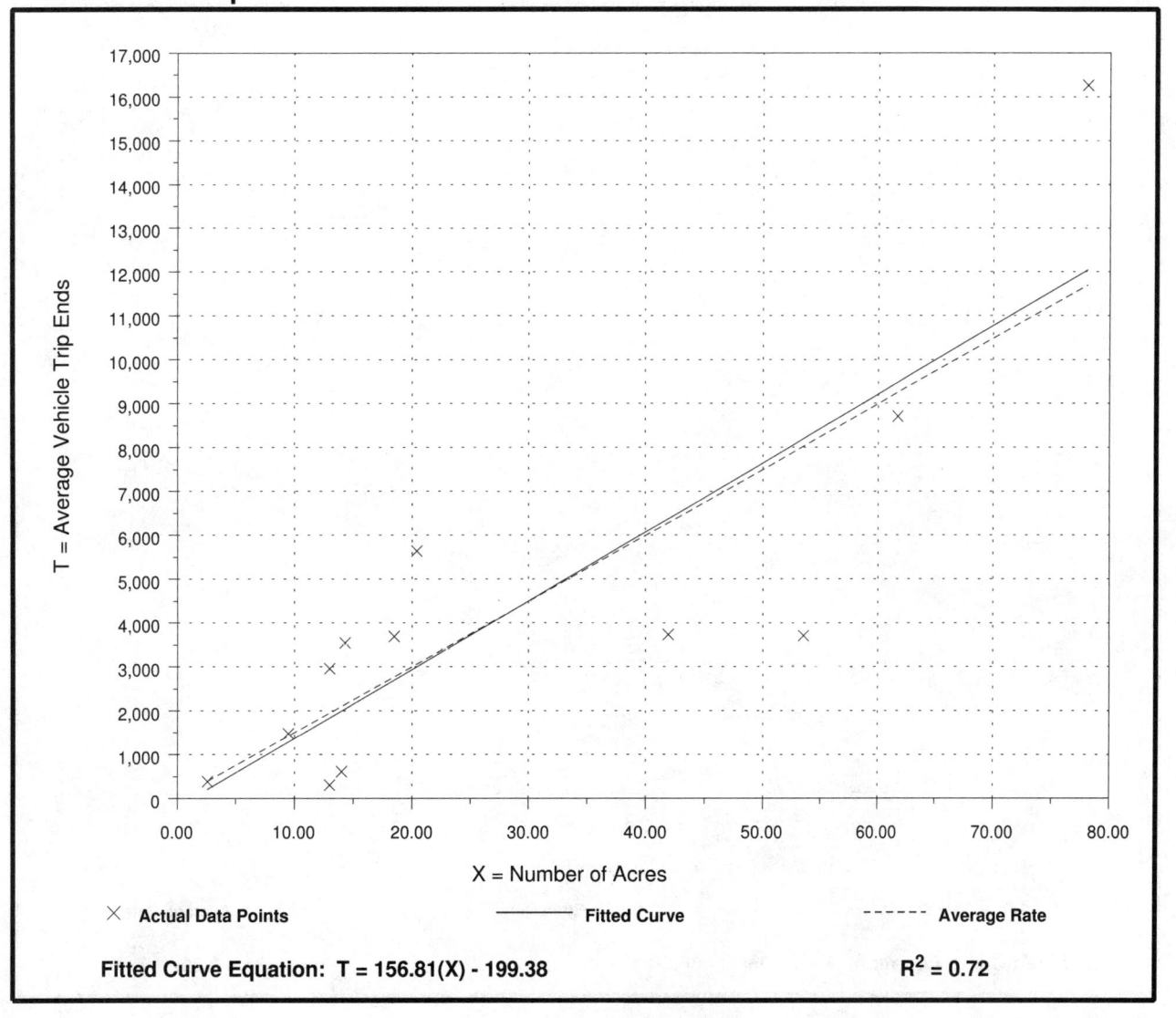

Data Plot and Equation

Fitted Curve Equation: T = 156.81(X) - 199.38 $R^2 = 0.72$

Business Park
(770)

Average Vehicle Trip Ends vs: Acres
On a: Weekday,
A.M. Peak Hour

Number of Studies: 12
Average Number of Acres: 28
Directional Distribution: 85% entering, 15% exiting

Trip Generation per Acre

Average Rate	Range of Rates	Standard Deviation
18.86	2.77 - 35.62	10.17

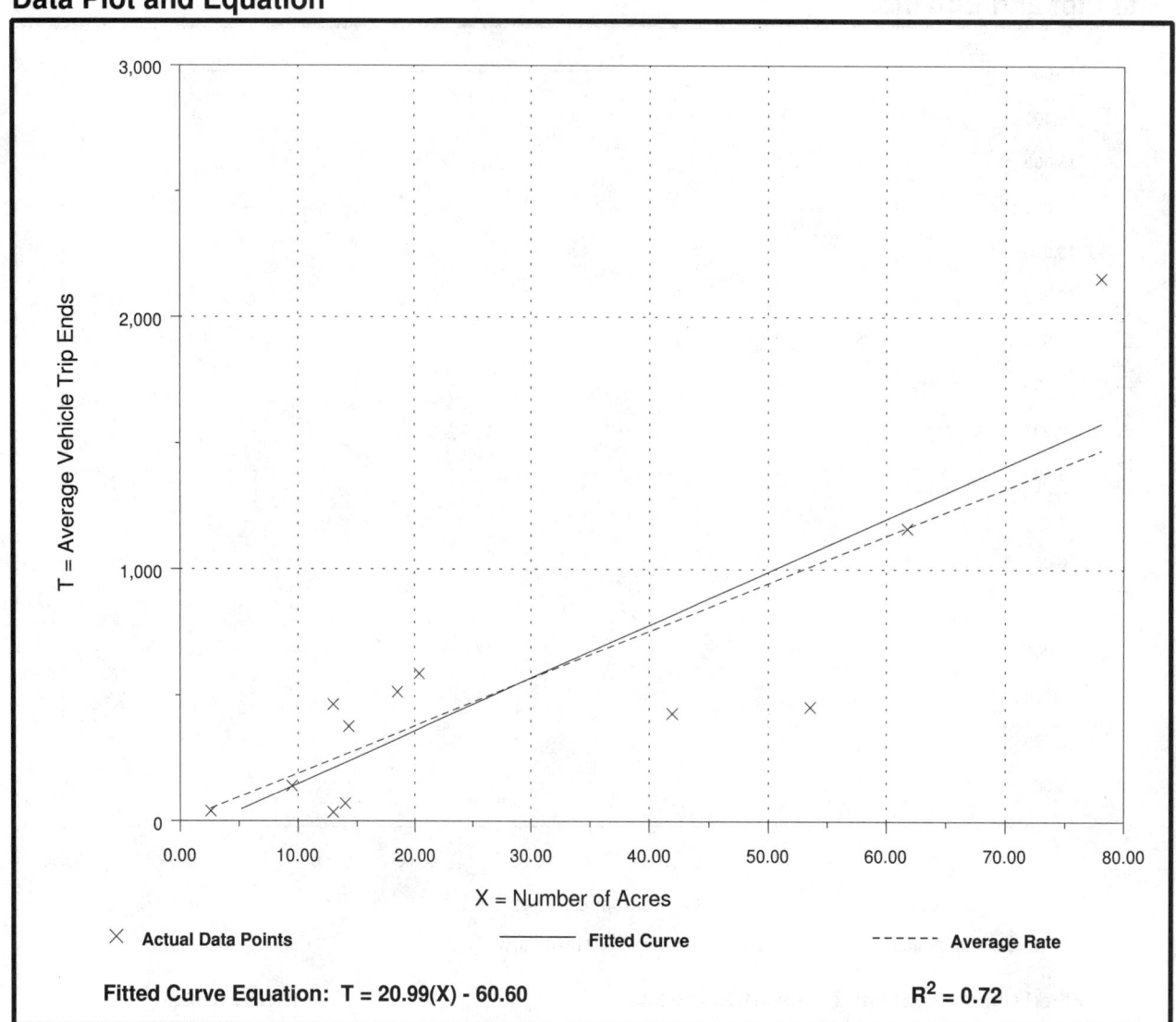

Data Plot and Equation

Fitted Curve Equation: $T = 20.99(X) - 60.60$ $R^2 = 0.72$

Business Park
(770)

Average Vehicle Trip Ends vs: Acres
On a: Weekday,
P.M. Peak Hour

Number of Studies: 12
Average Number of Acres: 28
Directional Distribution: 20% entering, 80% exiting

Trip Generation per Acre

Average Rate	Range of Rates	Standard Deviation
16.84	2.31 - 32.54	9.82

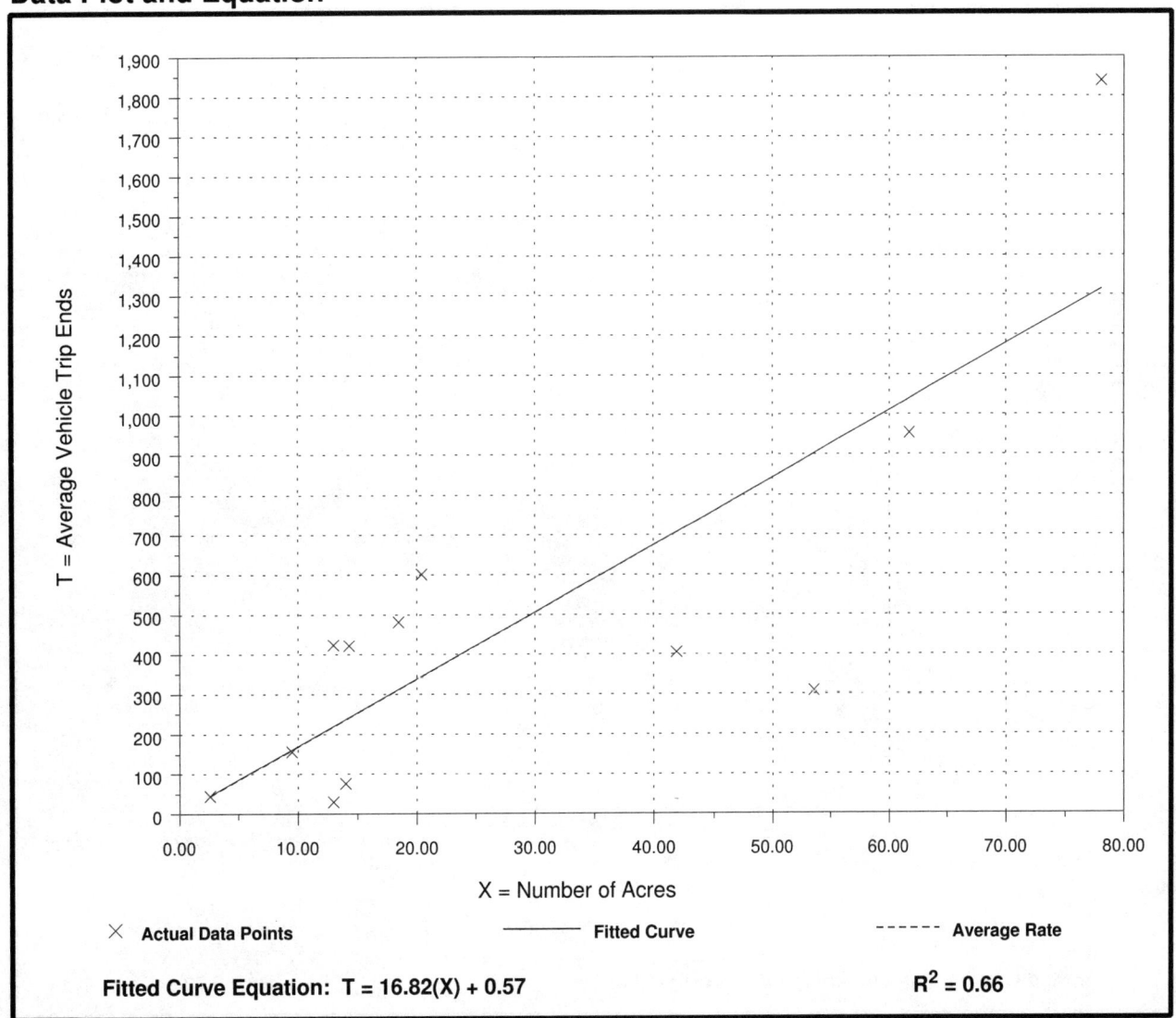

Data Plot and Equation

Fitted Curve Equation: $T = 16.82(X) + 0.57$ $R^2 = 0.66$

Business Park
(770)

Average Vehicle Trip Ends vs: Acres
On a: Saturday

Number of Studies: 10
Average Number of Acres: 31
Directional Distribution: 50% entering, 50% exiting

Trip Generation per Acre

Average Rate	Range of Rates	Standard Deviation
32.61	10.45 - 61.54	18.72

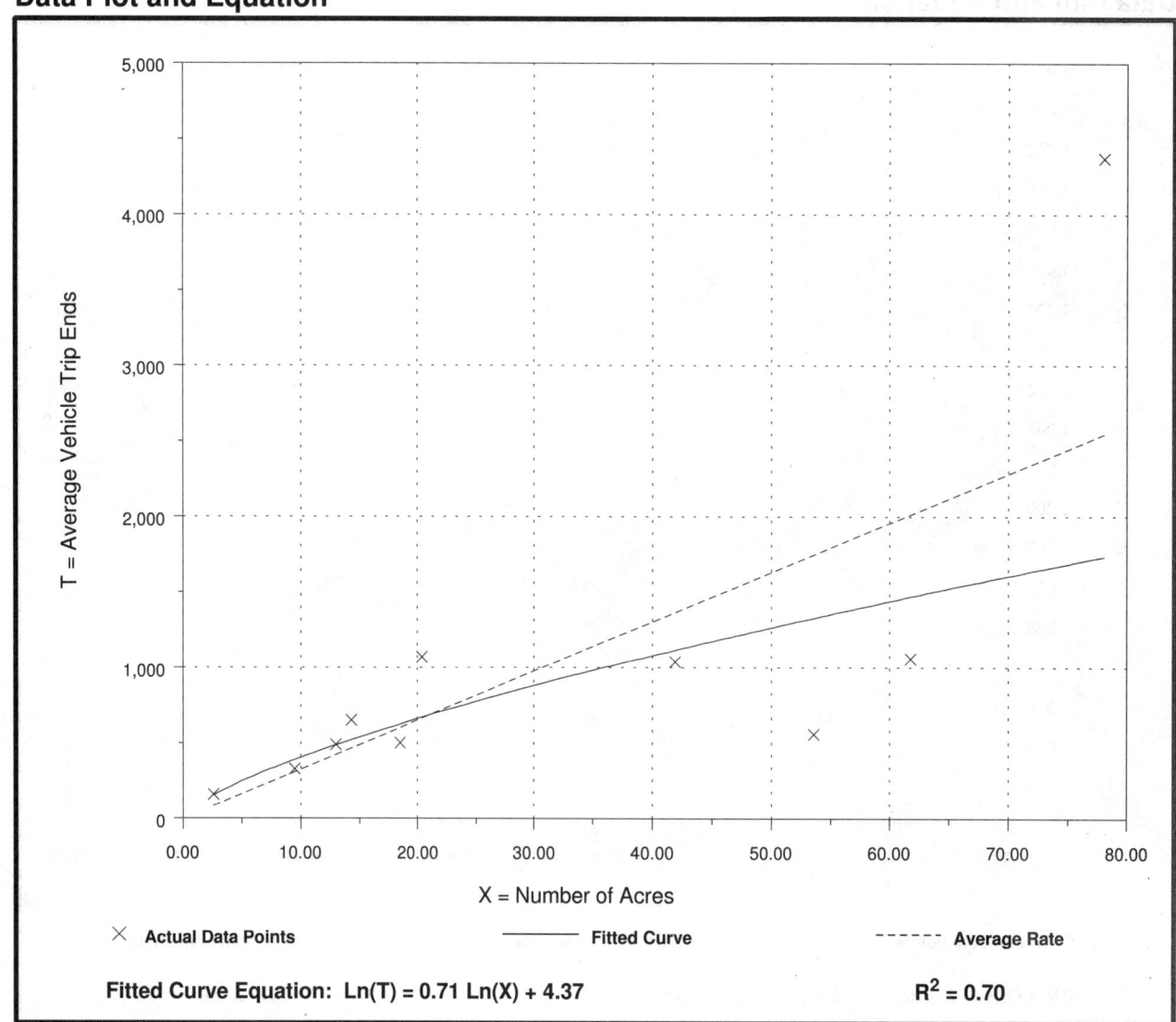

Data Plot and Equation

Fitted Curve Equation: $Ln(T) = 0.71\, Ln(X) + 4.37$ $R^2 = 0.70$

Business Park
(770)

Average Vehicle Trip Ends vs: Acres
On a: Sunday

Number of Studies: 10
Average Number of Acres: 31
Directional Distribution: 50% entering, 50% exiting

Trip Generation per Acre

Average Rate	Range of Rates	Standard Deviation
16.78	7.09 - 28.30	9.40

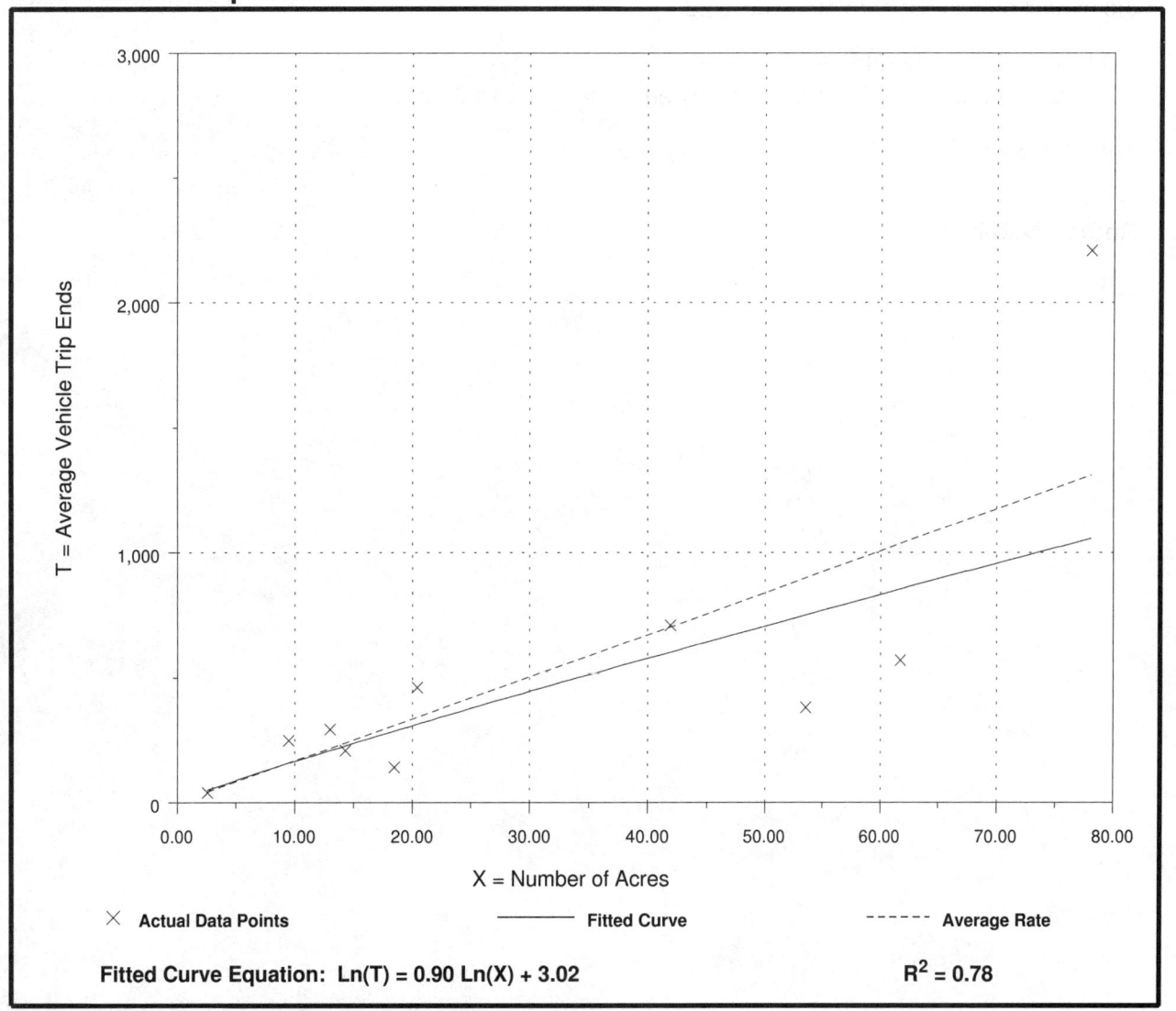

Data Plot and Equation

Fitted Curve Equation: Ln(T) = 0.90 Ln(X) + 3.02 $R^2 = 0.78$

Land Use: 810
Tractor Supply Store

Description

A tractor supply store is a free-standing facility that specializes in the sale of agricultural and garden equipment, power tools, vehicle maintenance parts and heavy-duty outdoor machinery. They may also offer ancillary items such as clothing, footwear and other accessories.

Additional Data

Outside storage areas are not included in the overall gross floor area measurements. However, if storage areas are located within the principal outside faces of the exterior walls, they are included in the overall gross floor area of the building.

Peak hour of the generator—
    The Saturday peak hour varied between 11:00 a.m. to 1:00 p.m.

The sites were surveyed in 2008 in New Jersey.

Source Number

737

Tractor Supply Store
(810)

Average Vehicle Trip Ends vs: 1000 Sq. Feet Gross Floor Area
On a: Weekday,
Peak Hour of Adjacent Street Traffic,
One Hour Between 4 and 6 p.m.

Number of Studies: 7
Average 1000 Sq. Feet GFA: 44
Directional Distribution: 47% entering, 53% exiting

Trip Generation per 1000 Sq. Feet Gross Floor Area

Average Rate	Range of Rates	Standard Deviation
1.40	0.75 - 1.83	1.24

Data Plot and Equation

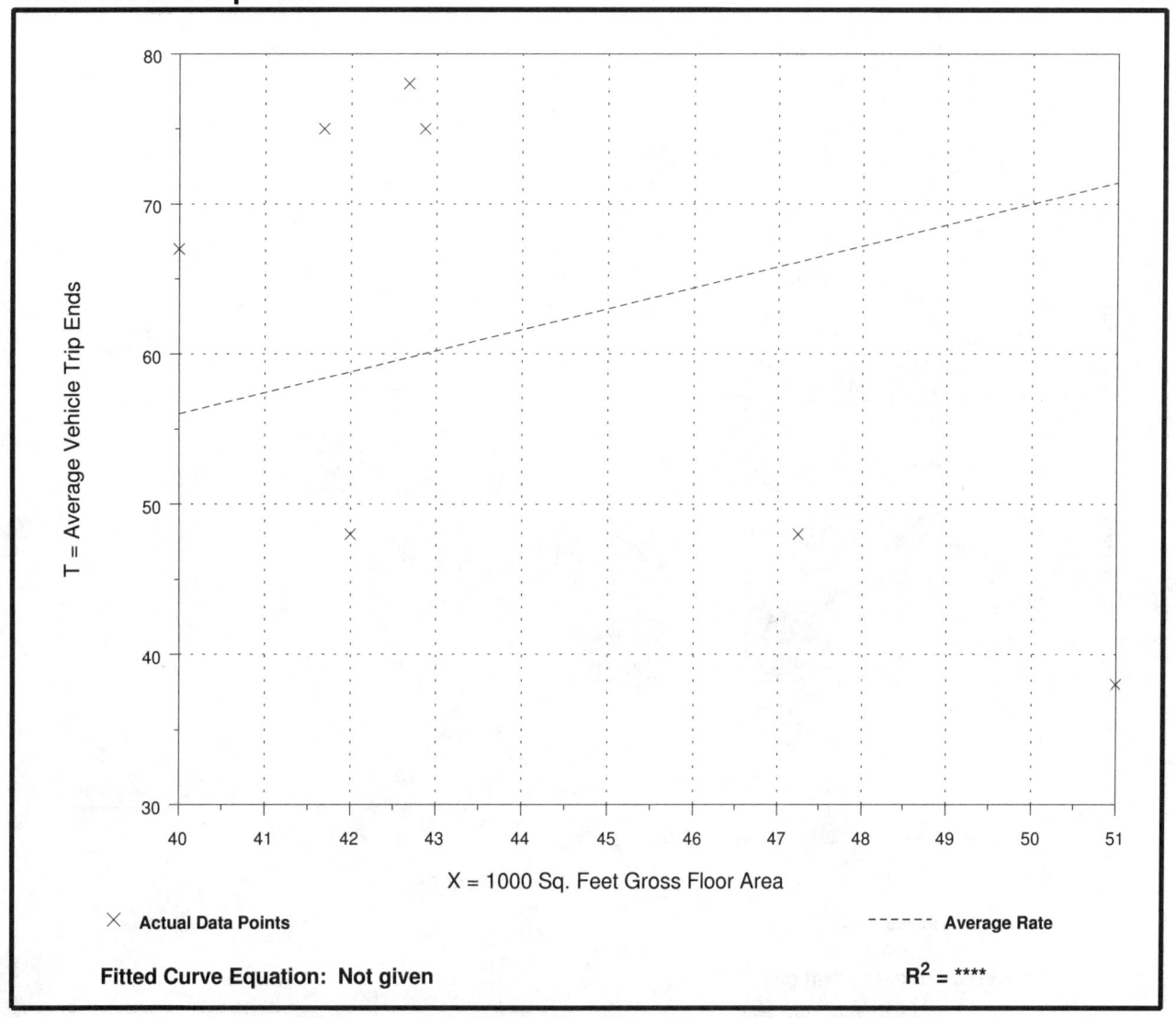

Fitted Curve Equation: Not given $R^2 = ****$

Tractor Supply Store
(810)

Average Vehicle Trip Ends vs: 1000 Sq. Feet Gross Floor Area
On a: Saturday,
Peak Hour of Generator

Number of Studies: 8
Average 1000 Sq. Feet GFA: 43
Directional Distribution: 49% entering, 51% exiting

Trip Generation per 1000 Sq. Feet Gross Floor Area

Average Rate	Range of Rates	Standard Deviation
3.17	2.12 - 4.90	1.93

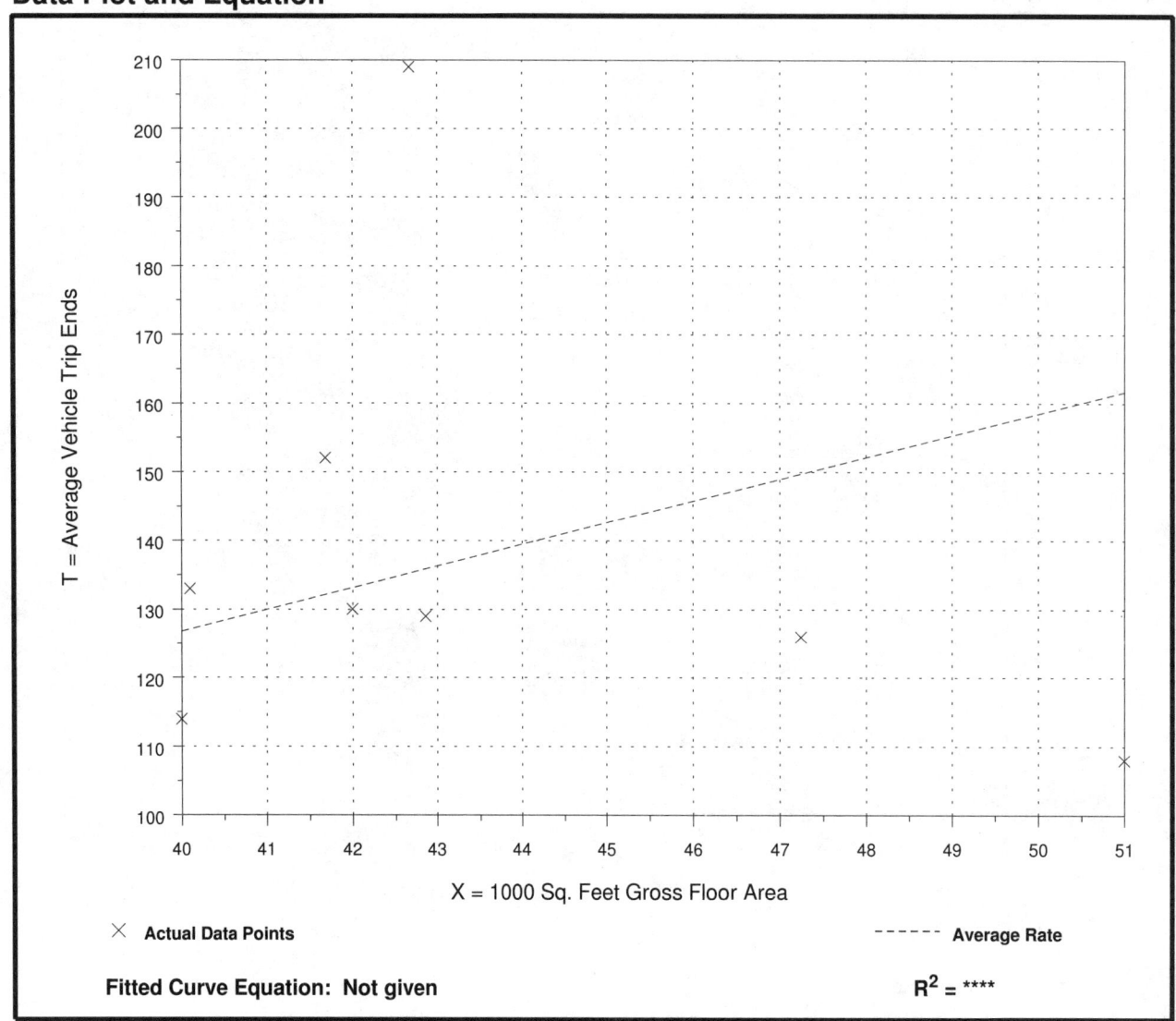

Data Plot and Equation

Fitted Curve Equation: Not given $R^2 = ****$

Land Use: 811
Construction Equipment Rental Store

Description

A construction equipment rental store is a business that specializes in the rental of construction equipment tools and supplies including, but not limited to, electrical and industrial tools, pumps, lawn and garden equipment, paving and earthmoving equipment, and safety equipment.

Additional Data

Outside storage areas are not included in the overall gross floor area measurements. However, if storage areas are located within the principal outside faces of the exterior walls, they are included in the overall gross floor area of the building.

The sites were surveyed in 2007 in Florida.

Source Number

721

Construction Equipment Rental Store
(811)

Average Vehicle Trip Ends vs: 1000 Sq. Feet Gross Floor Area
On a: Weekday,
Peak Hour of Adjacent Street Traffic,
One Hour Between 4 and 6 p.m.

Number of Studies: 3
Average 1000 Sq. Feet GFA: 20
Directional Distribution: 28% entering, 72% exiting

Trip Generation per 1000 Sq. Feet Gross Floor Area

Average Rate	Range of Rates	Standard Deviation
0.99	0.81 - 1.40	1.01

Data Plot and Equation

Caution - Use Carefully - Small Sample Size

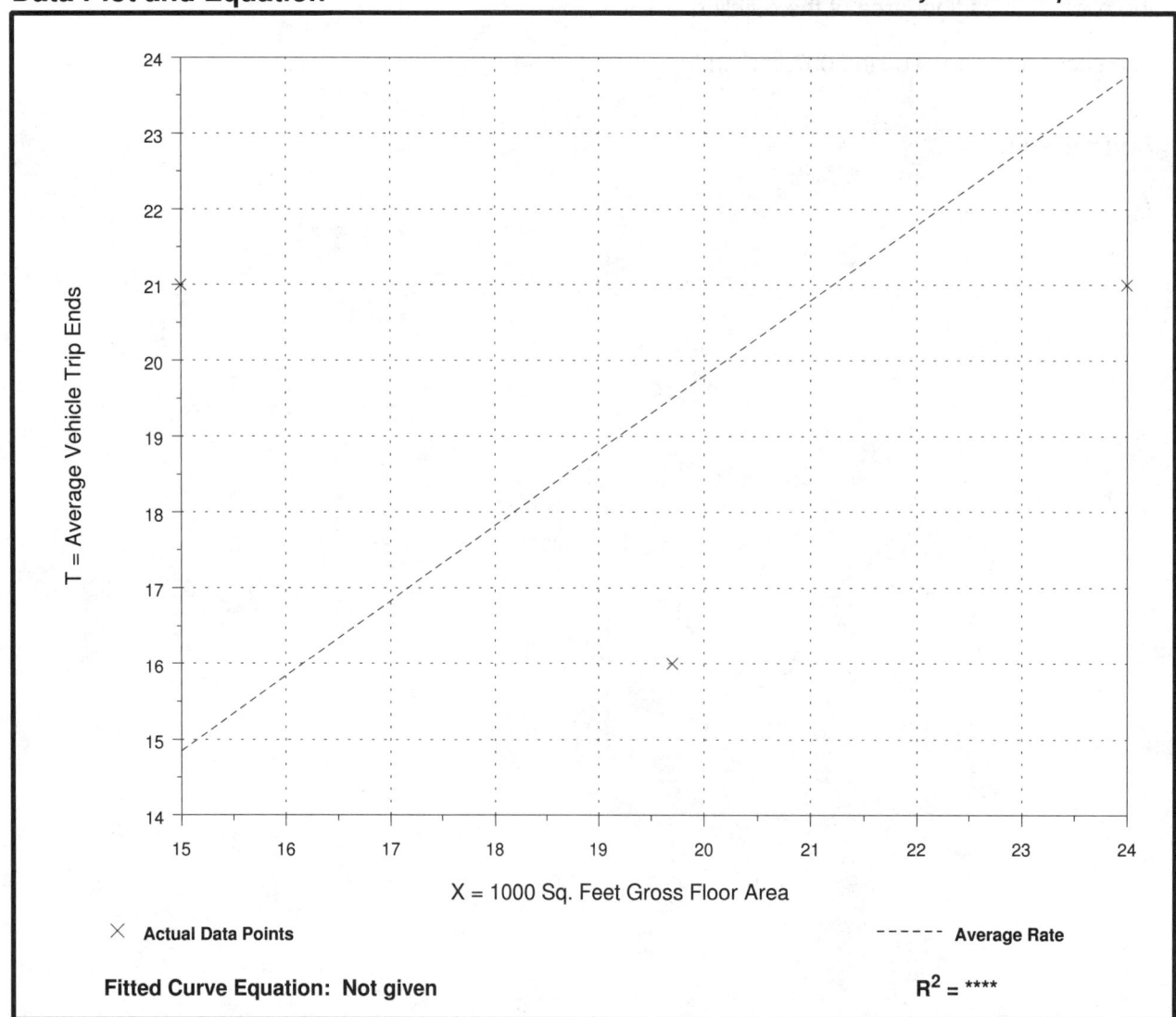

X Actual Data Points   ------ Average Rate

Fitted Curve Equation: Not given $R^2 = ****$

Construction Equipment Rental Store
(811)

Average Vehicle Trip Ends vs: Acres
On a: Weekday,
Peak Hour of Adjacent Street Traffic,
One Hour Between 4 and 6 p.m.

Number of Studies: 3
Average Number of Acres: 3
Directional Distribution: 28% entering, 72% exiting

Trip Generation per Acre

Average Rate	Range of Rates	Standard Deviation
6.44	4.67 - 12.31	3.42

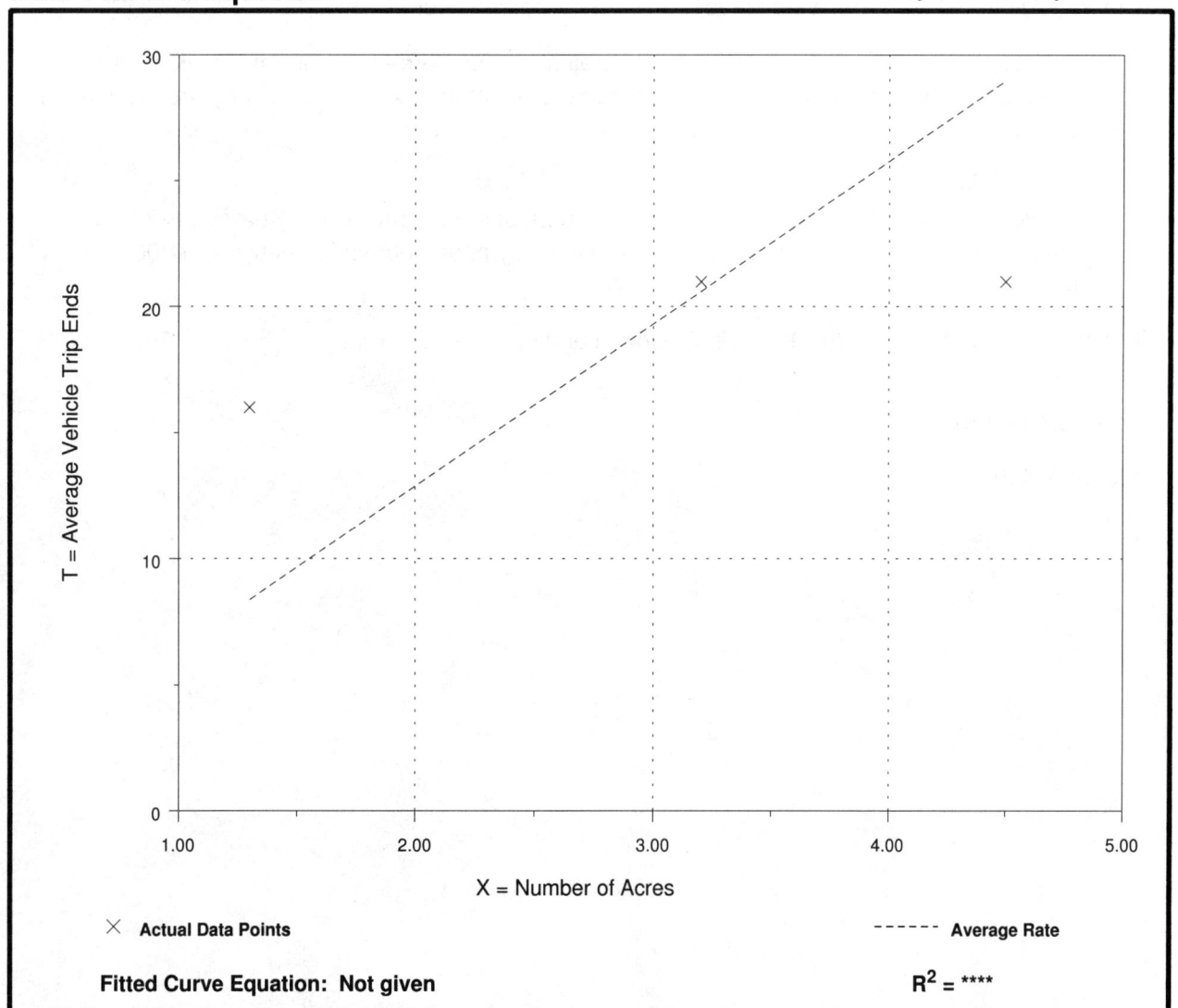

Data Plot and Equation

Caution - Use Carefully - Small Sample Size

Fitted Curve Equation: Not given $R^2 = ****$

Land Use: 812
Building Materials and Lumber Store

Description

A building materials and lumber store is a free-standing building that sells hardware, building materials and lumber. The lumber may be stored in the main building, yard, or storage shed. The buildings contained in this land use are less than 30,000 square feet gross floor area. Hardware/paint store (Land Use 816) and home improvement superstore (Land Use 862) are related uses.

Additional Data

Vehicle occupancy ranged from 1.10 to 1.21 persons per automobile on an average weekday. The average for all sites that were surveyed was 1.17 persons per automobile.

Outside storage areas are not included in the overall gross floor area measurements. However, if storage areas are located within the principal outside faces of the exterior walls, they are included in the overall gross floor area of the building.

Peak hours of the generator—
   The weekday peak hour varied between 10:00 a.m. and 3:00 p.m. The Saturday peak hour varied between 10:00 a.m. and 1:00 p.m. The Sunday peak hour varied between 11:00 a.m. and 1:00 p.m.

The sites were surveyed in the 1980s in California and upstate New York.

Source Numbers

126, 280, 449

Building Materials and Lumber Store
(812)

Average Vehicle Trip Ends vs: Employees
On a: Weekday

Number of Studies: 4
Avg. Number of Employees: 13
Directional Distribution: 50% entering, 50% exiting

Trip Generation per Employee

Average Rate	Range of Rates	Standard Deviation
32.12	25.72 - 52.75	11.32

Data Plot and Equation

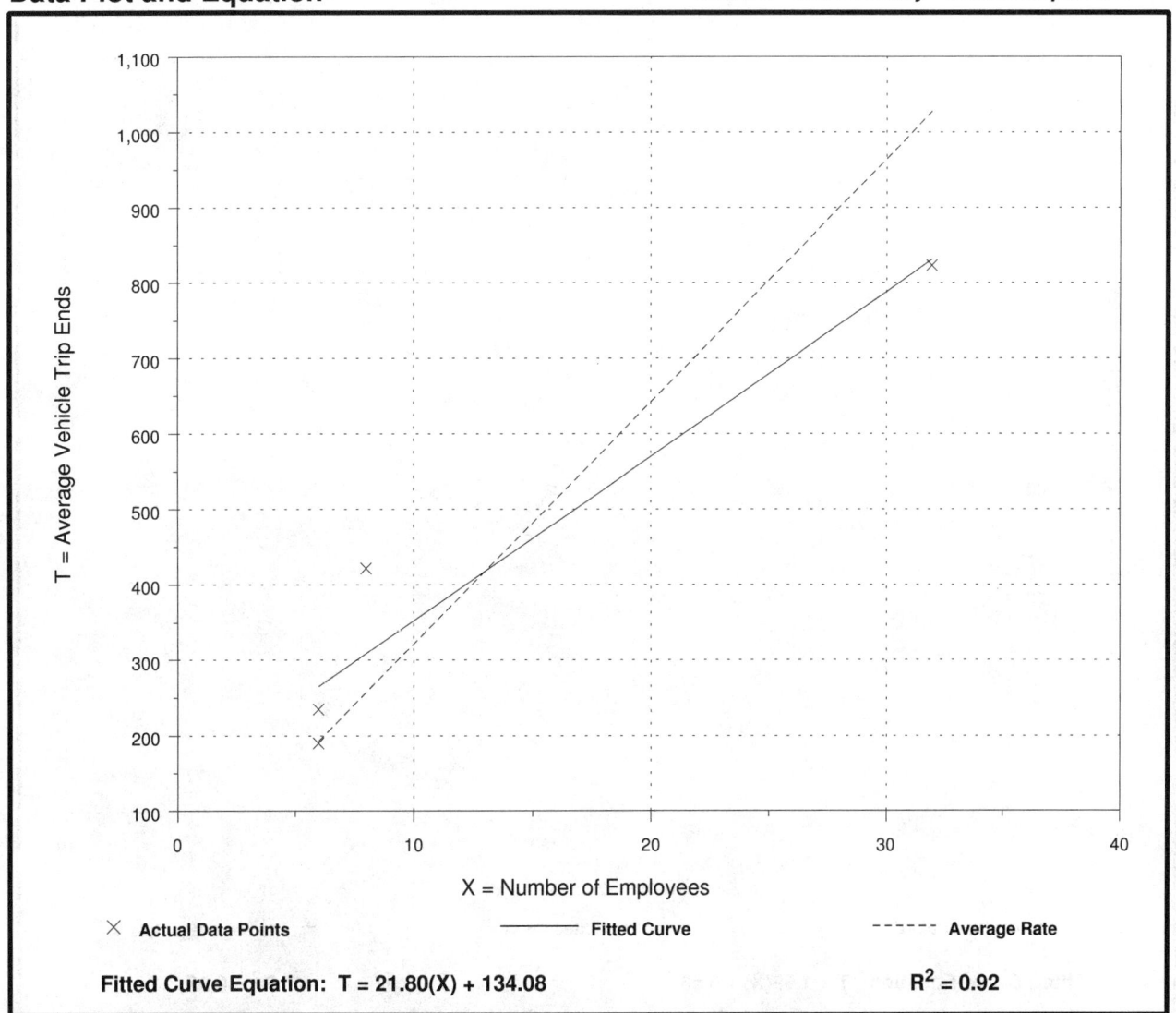

Caution - Use Carefully - Small Sample Size

Fitted Curve Equation: T = 21.80(X) + 134.08 $R^2 = 0.92$

Building Materials and Lumber Store
(812)

Average Vehicle Trip Ends vs: Employees
On a: Weekday,
Peak Hour of Adjacent Street Traffic,
One Hour Between 7 and 9 a.m.

Number of Studies: 4
Avg. Number of Employees: 13
Directional Distribution: 62% entering, 38% exiting

Trip Generation per Employee

Average Rate	Range of Rates	Standard Deviation
2.42	1.97 - 3.88	1.66

Data Plot and Equation

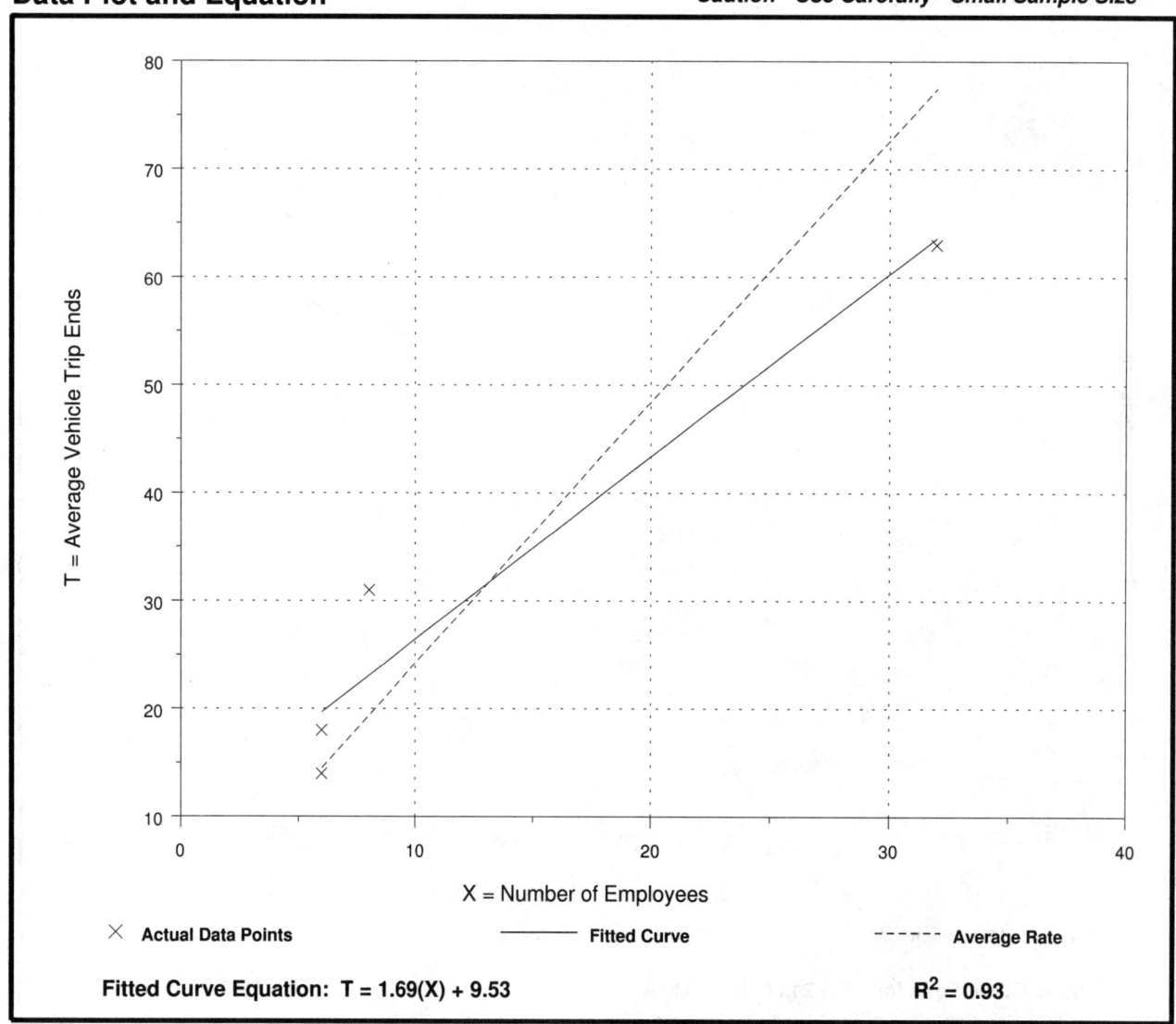

Caution - Use Carefully - Small Sample Size

Fitted Curve Equation: T = 1.69(X) + 9.53 $R^2 = 0.93$

Building Materials and Lumber Store
(812)

Average Vehicle Trip Ends vs: Employees
On a: Weekday,
Peak Hour of Adjacent Street Traffic,
One Hour Between 4 and 6 p.m.

Number of Studies: 4
Avg. Number of Employees: 13
Directional Distribution: 51% entering, 49% exiting

Trip Generation per Employee

Average Rate	Range of Rates	Standard Deviation
2.77	2.00 - 4.13	1.77

Data Plot and Equation

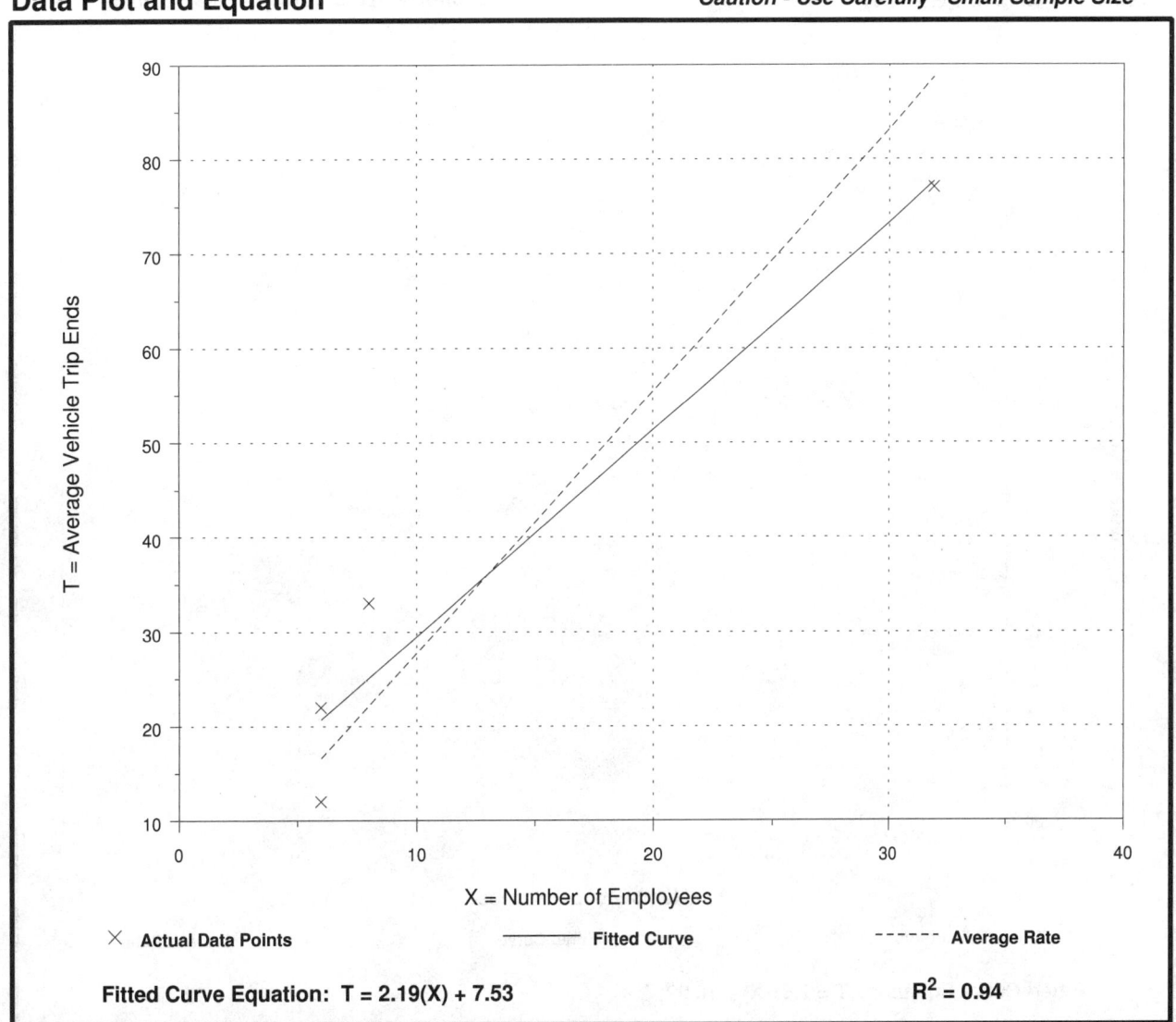

Caution - Use Carefully - Small Sample Size

Fitted Curve Equation: T = 2.19(X) + 7.53 $R^2 = 0.94$

Building Materials and Lumber Store
(812)

Average Vehicle Trip Ends vs: Employees
On a: Weekday,
A.M. Peak Hour of Generator

Number of Studies: 4
Avg. Number of Employees: 13
Directional Distribution: 54% entering, 46% exiting

Trip Generation per Employee

Average Rate	Range of Rates	Standard Deviation
3.94	3.06 - 6.38	2.30

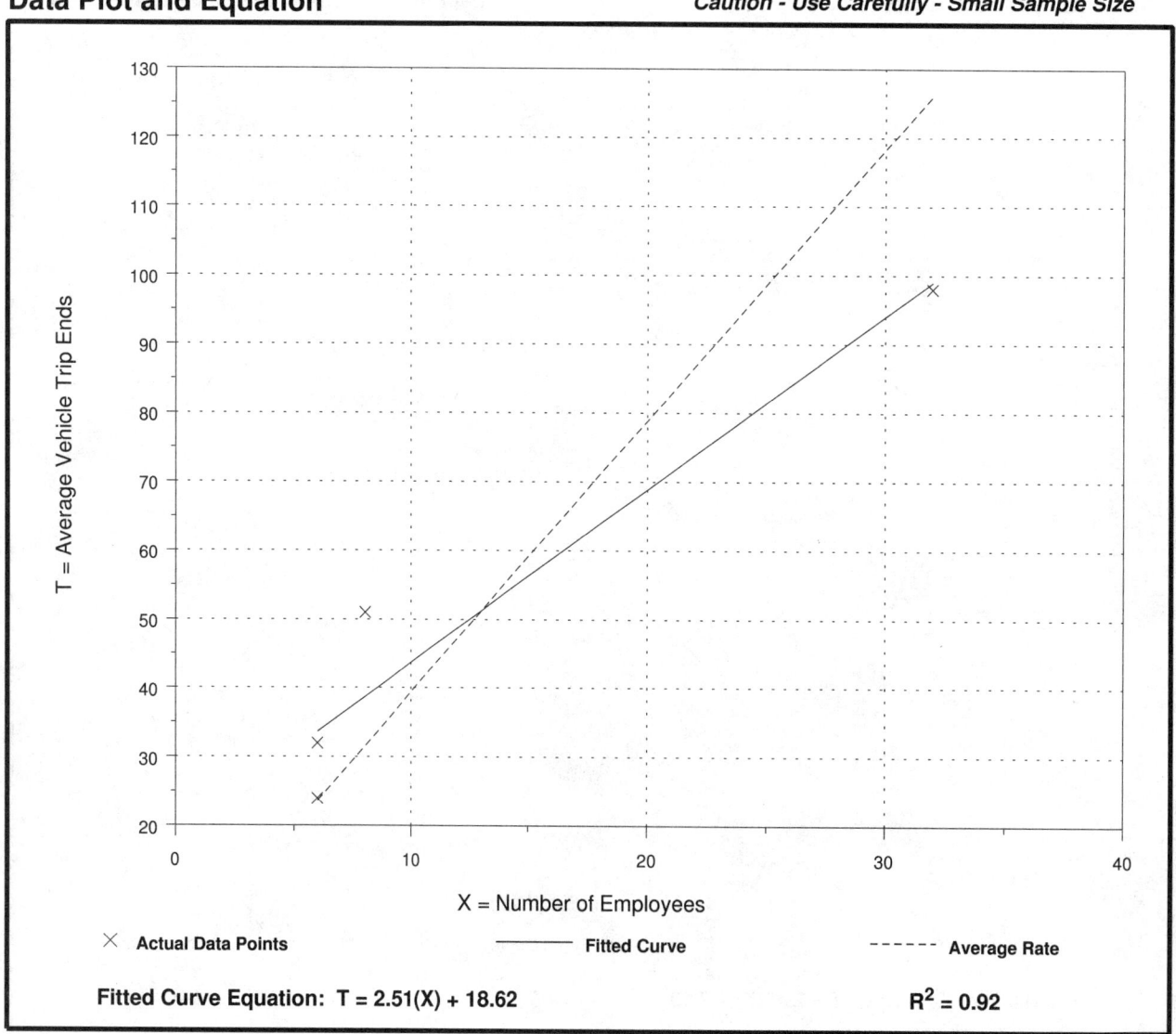

Data Plot and Equation

Caution - Use Carefully - Small Sample Size

Fitted Curve Equation: T = 2.51(X) + 18.62 $R^2 = 0.92$

Building Materials and Lumber Store
(812)

Average Vehicle Trip Ends vs: Employees
On a: Weekday,
P.M. Peak Hour of Generator

Number of Studies: 4
Avg. Number of Employees: 13
Directional Distribution: 50% entering, 50% exiting

Trip Generation per Employee

Average Rate	Range of Rates	Standard Deviation
3.83	3.19 - 5.75	2.11

Data Plot and Equation

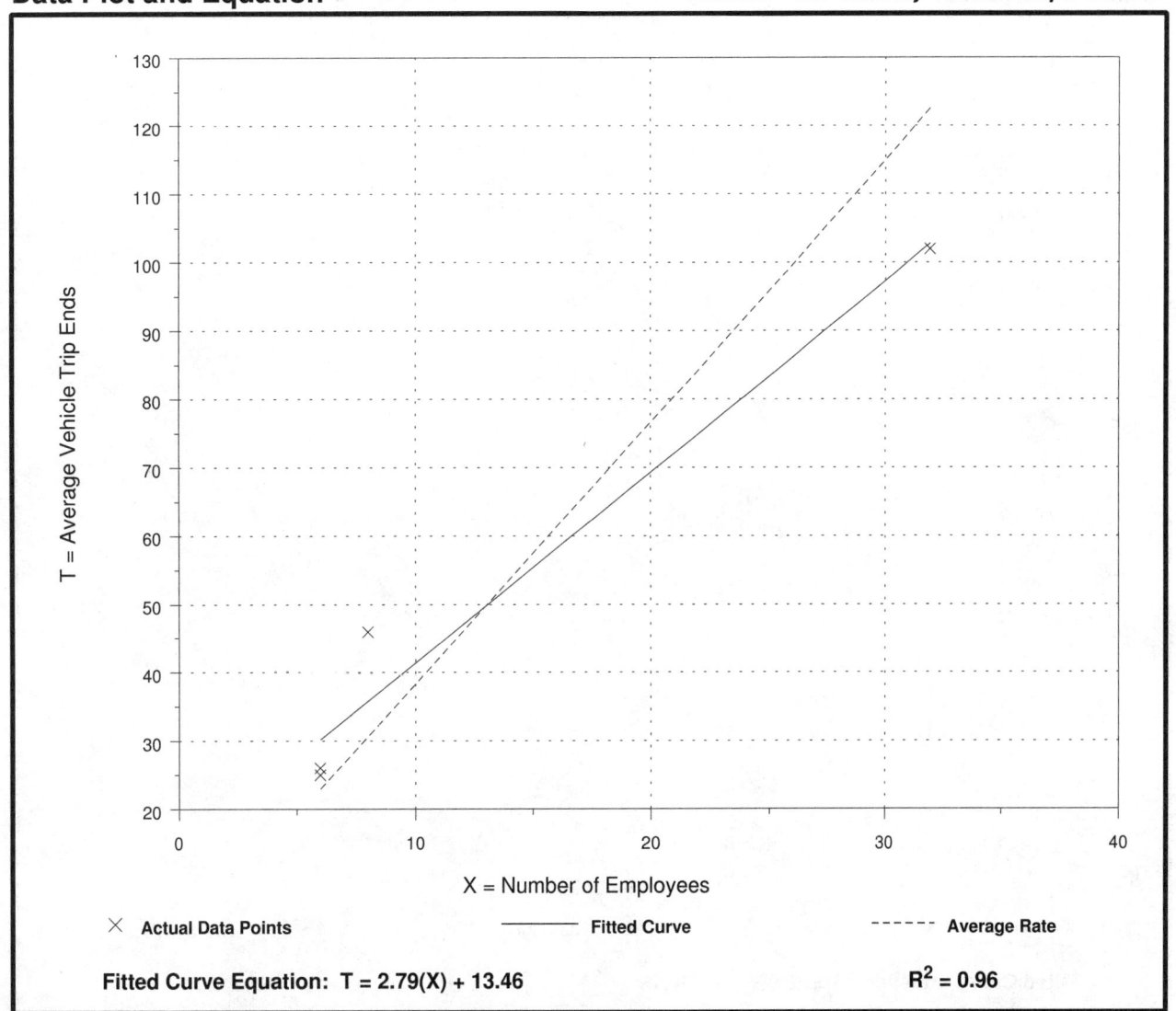

Caution - Use Carefully - Small Sample Size

Fitted Curve Equation: T = 2.79(X) + 13.46 $R^2 = 0.96$

Trip Generation, 9th Edition • Institute of Transportation Engineers

Building Materials and Lumber Store
(812)

Average Vehicle Trip Ends vs: Employees
On a: Saturday

Number of Studies: 4
Avg. Number of Employees: 13
Directional Distribution: 50% entering, 50% exiting

Trip Generation per Employee

Average Rate	Range of Rates	Standard Deviation
36.69	27.31 - 62.50	14.10

Data Plot and Equation

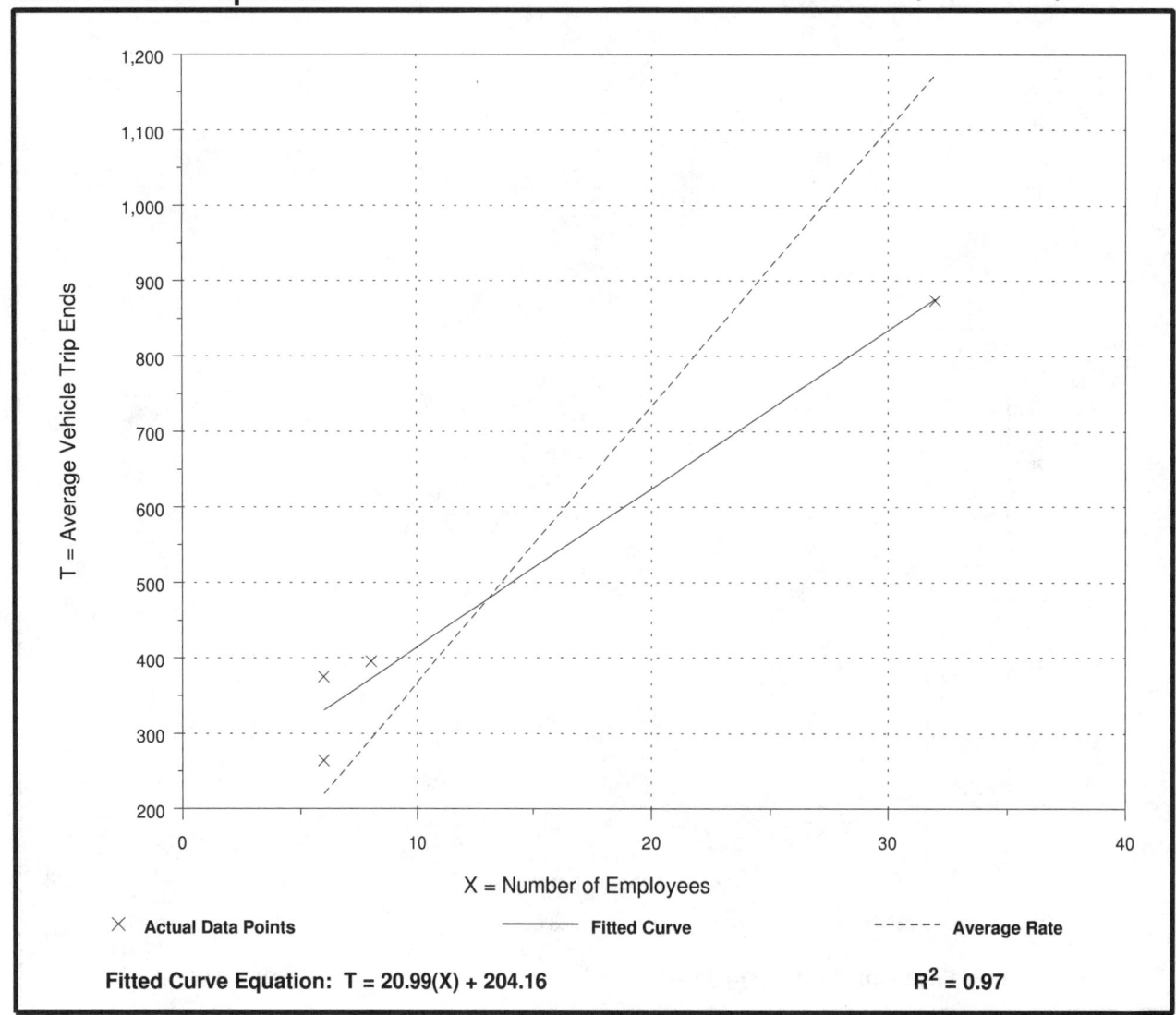

Caution - Use Carefully - Small Sample Size

Fitted Curve Equation: T = 20.99(X) + 204.16 $R^2 = 0.97$

Building Materials and Lumber Store
(812)

Average Vehicle Trip Ends vs: Employees
On a: Saturday,
Peak Hour of Generator

Number of Studies: 4
Avg. Number of Employees: 13
Directional Distribution: 51% entering, 49% exiting

Trip Generation per Employee

Average Rate	Range of Rates	Standard Deviation
5.23	4.13 - 8.17	2.66

Data Plot and Equation

Caution - Use Carefully - Small Sample Size

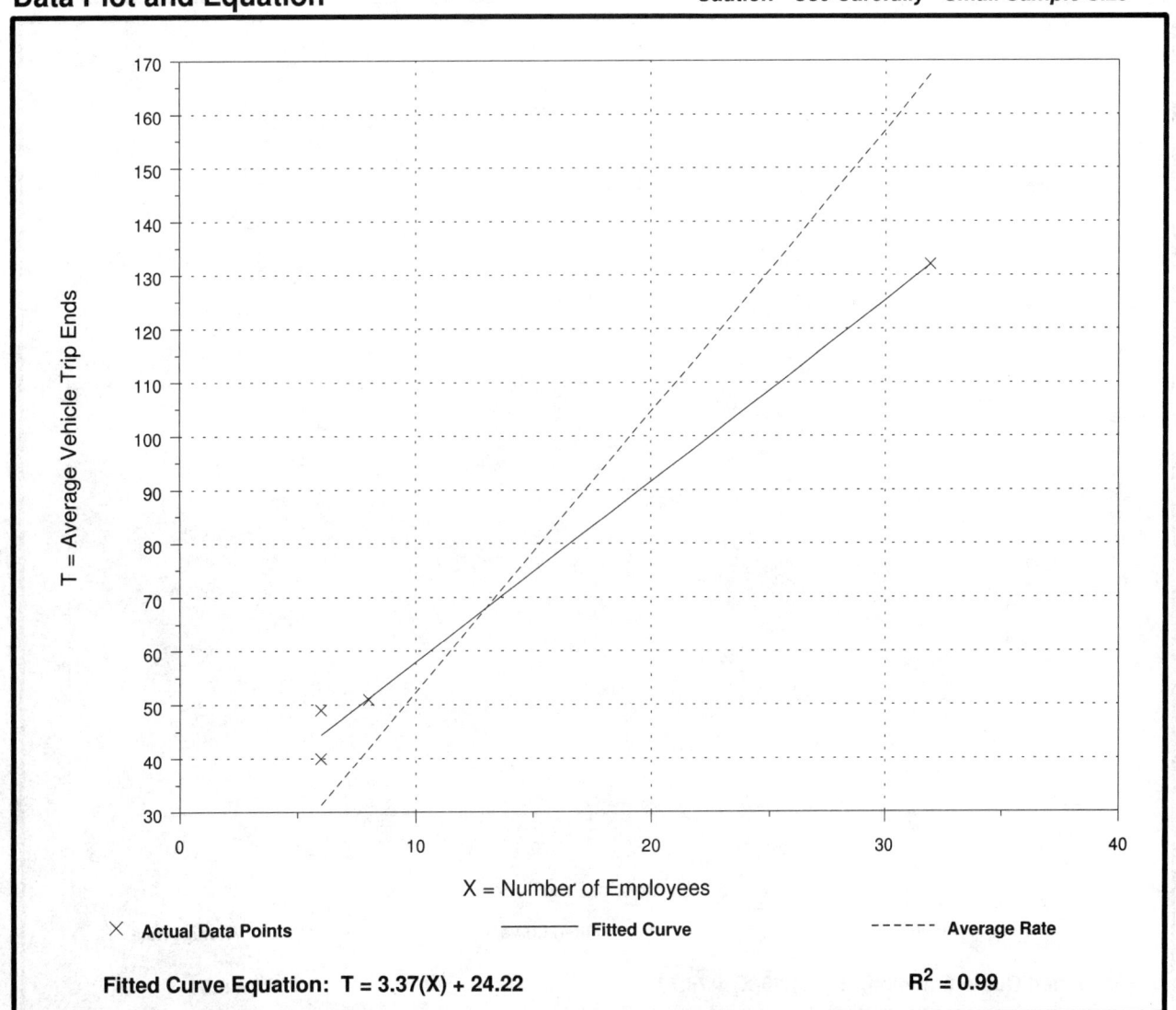

Fitted Curve Equation: T = 3.37(X) + 24.22 $R^2 = 0.99$

Building Materials and Lumber Store
(812)

Average Vehicle Trip Ends vs: Employees
On a: Sunday

Number of Studies: 4
Avg. Number of Employees: 13
Directional Distribution: 50% entering, 50% exiting

Trip Generation per Employee

Average Rate	Range of Rates	Standard Deviation
17.42	4.00 - 48.67	12.95

Data Plot and Equation

Caution - Use Carefully - Small Sample Size

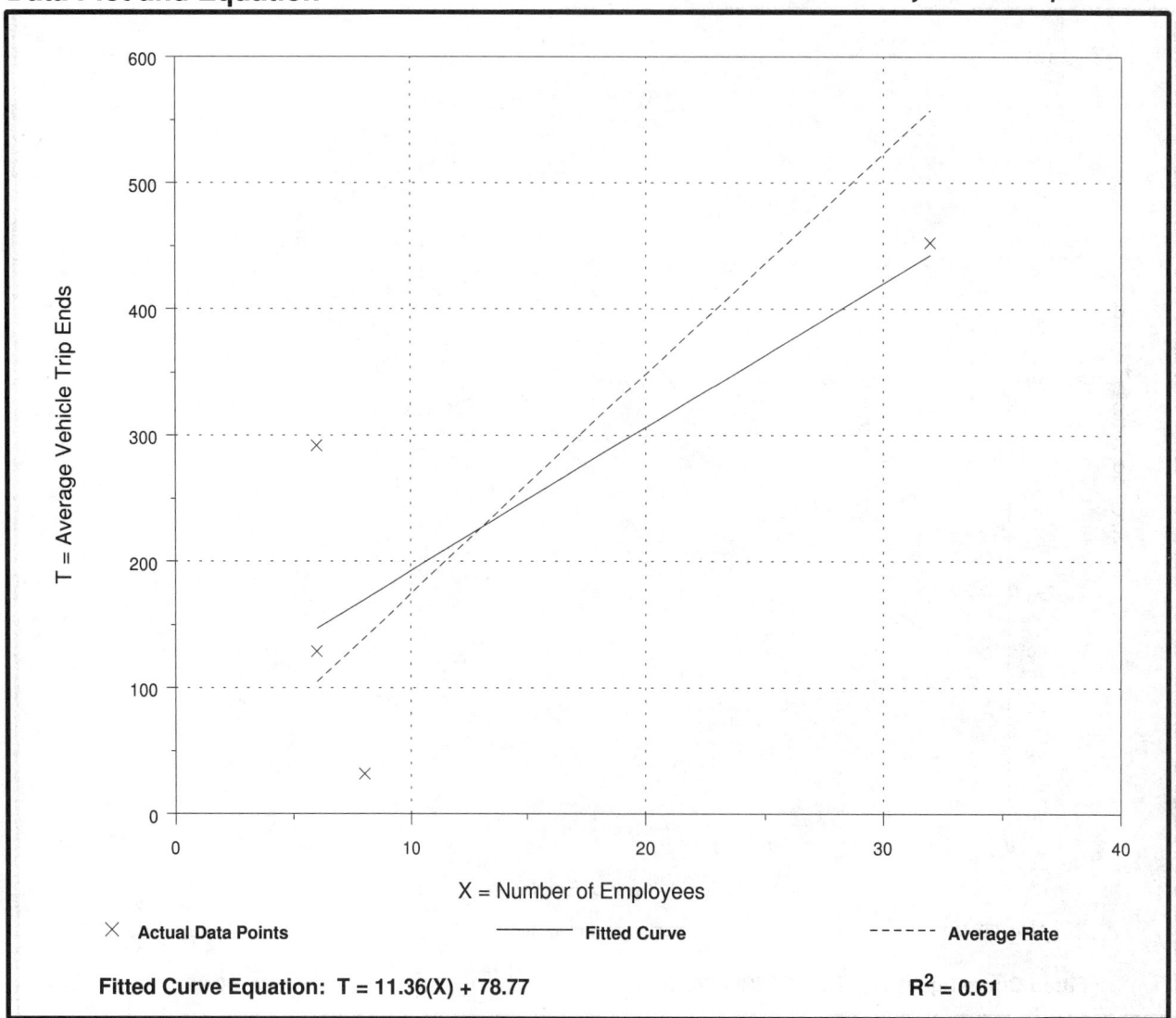

Fitted Curve Equation: T = 11.36(X) + 78.77 $R^2 = 0.61$

Building Materials and Lumber Store
(812)

Average Vehicle Trip Ends vs: Employees
On a: Sunday,
Peak Hour of Generator

Number of Studies: 4
Avg. Number of Employees: 13
Directional Distribution: 45% entering, 55% exiting

Trip Generation per Employee

Average Rate	Range of Rates	Standard Deviation
3.25	0.63 - 7.67	2.61

Data Plot and Equation

Caution - Use Carefully - Small Sample Size

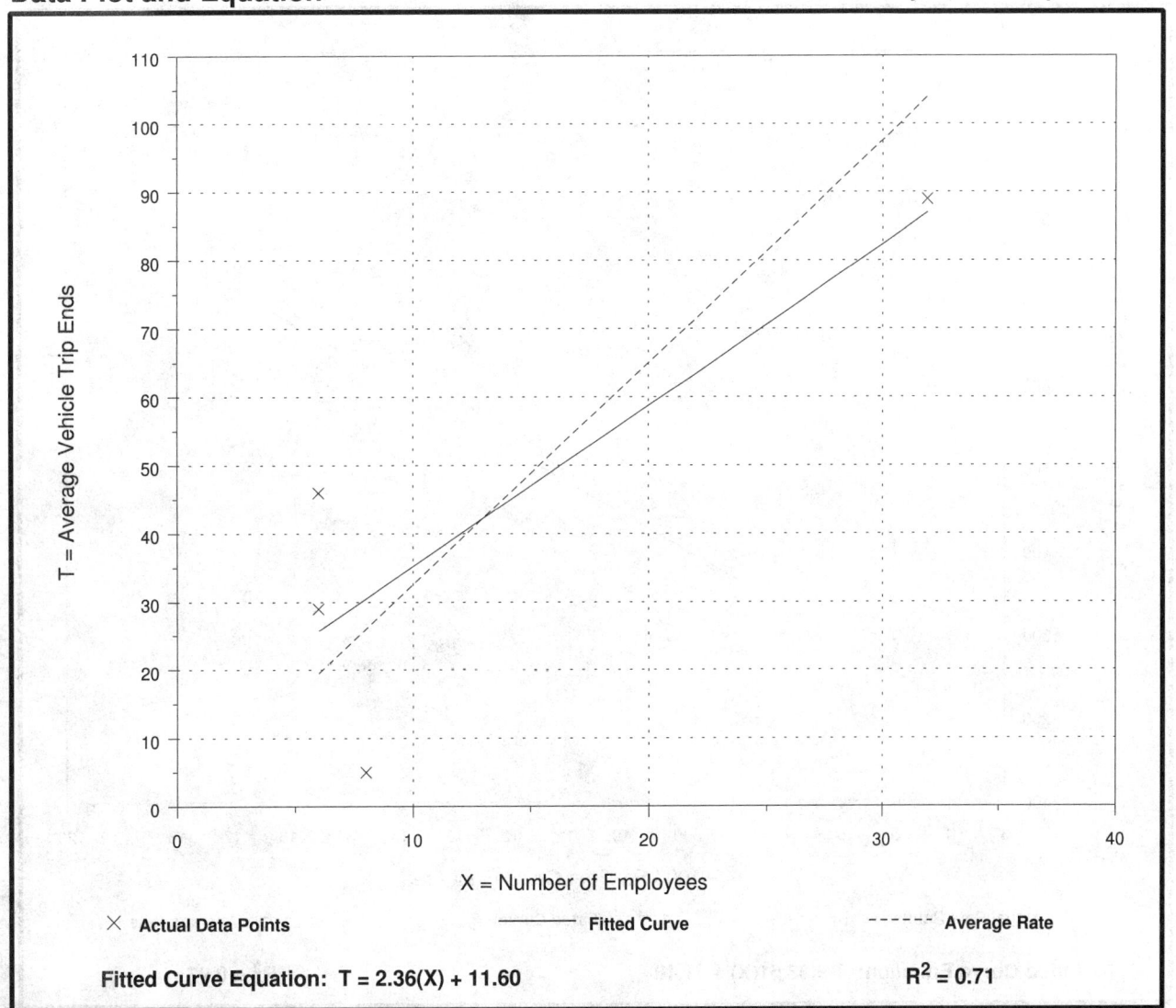

Fitted Curve Equation: $T = 2.36(X) + 11.60$ $R^2 = 0.71$

Trip Generation, 9th Edition • Institute of Transportation Engineers

Building Materials and Lumber Store
(812)

Average Vehicle Trip Ends vs: 1000 Sq. Feet Gross Floor Area
On a: Weekday

Number of Studies: 4
Average 1000 Sq. Feet GFA: 9
Directional Distribution: 50% entering, 50% exiting

Trip Generation per 1000 Sq. Feet Gross Floor Area

Average Rate	Range of Rates	Standard Deviation
45.16	39.17 - 56.27	9.51

Data Plot and Equation

Caution - Use Carefully - Small Sample Size

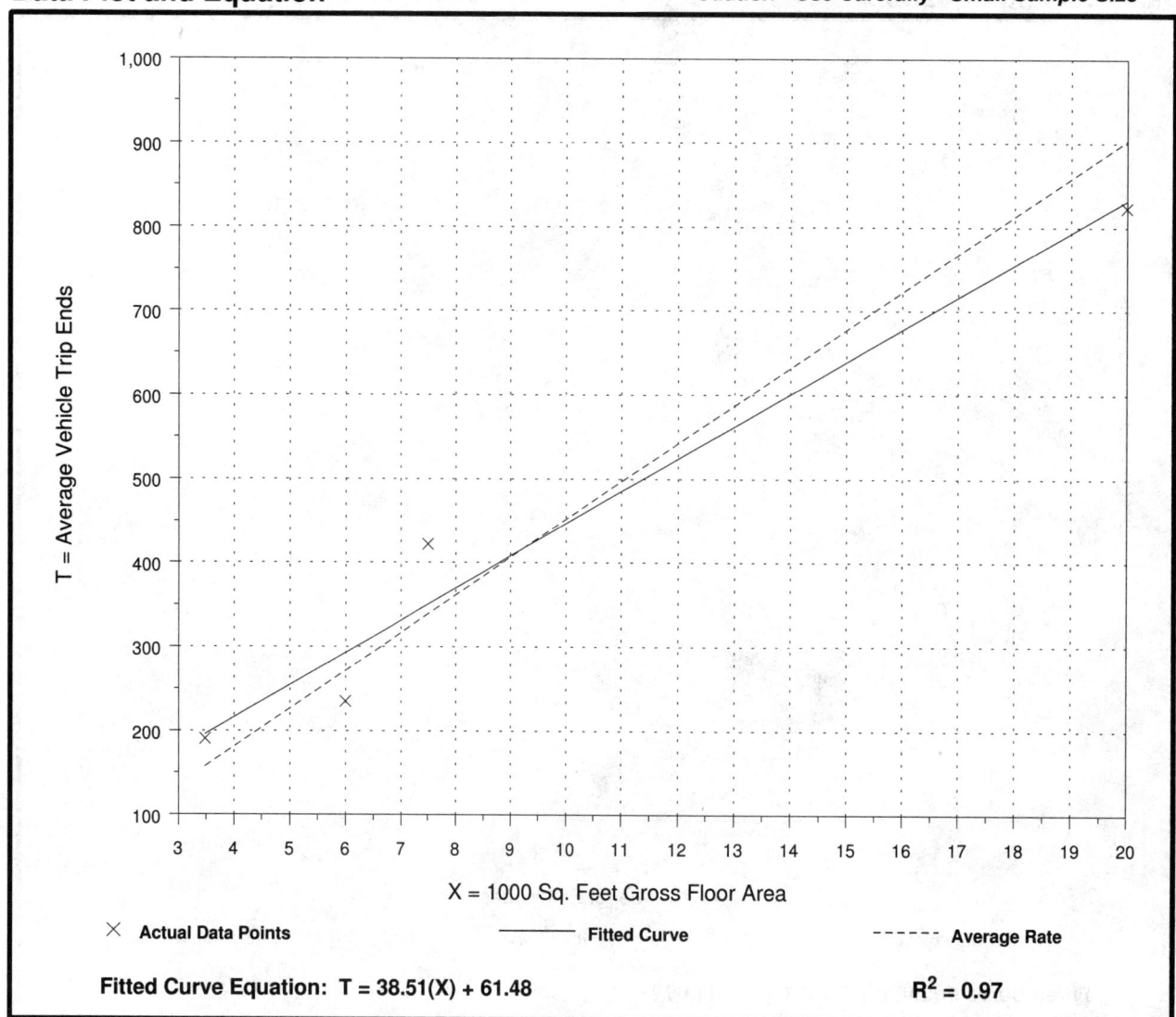

× Actual Data Points — Fitted Curve ----- Average Rate

Fitted Curve Equation: $T = 38.51(X) + 61.48$ $R^2 = 0.97$

Building Materials and Lumber Store
(812)

Average Vehicle Trip Ends vs: 1000 Sq. Feet Gross Floor Area
On a: Weekday,
Peak Hour of Adjacent Street Traffic,
One Hour Between 7 and 9 a.m.

Number of Studies: 6
Average 1000 Sq. Feet GFA: 11
Directional Distribution: 67% entering, 33% exiting

Trip Generation per 1000 Sq. Feet Gross Floor Area

Average Rate	Range of Rates	Standard Deviation
2.60	1.13 - 5.17	1.92

Data Plot and Equation

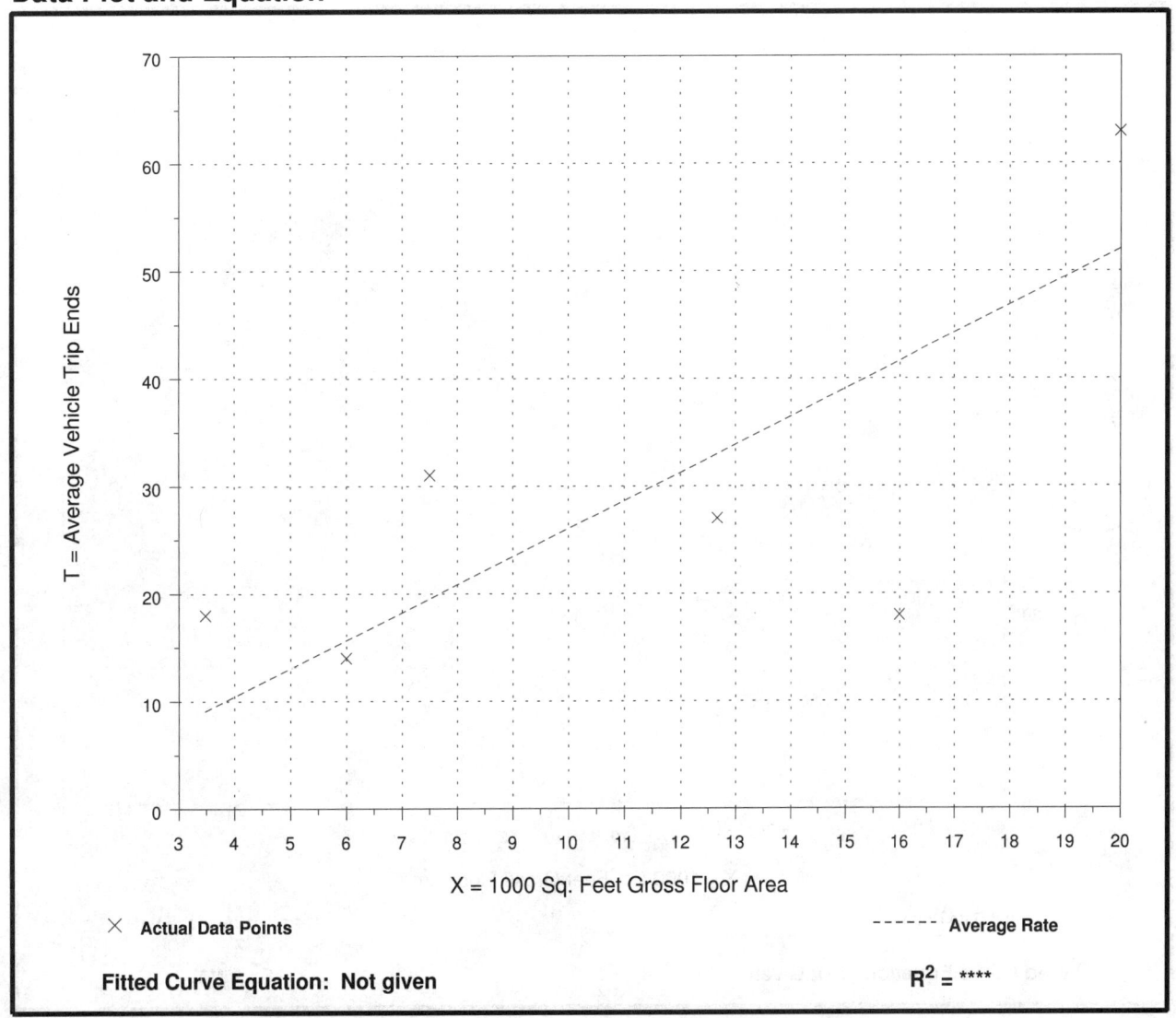

× Actual Data Points   ------ Average Rate

Fitted Curve Equation: Not given $R^2 = ****$

Trip Generation, 9th Edition • Institute of Transportation Engineers

Building Materials and Lumber Store
(812)

Average Vehicle Trip Ends vs: 1000 Sq. Feet Gross Floor Area
On a: Weekday,
Peak Hour of Adjacent Street Traffic,
One Hour Between 4 and 6 p.m.

Number of Studies: 6
Average 1000 Sq. Feet GFA: 11
Directional Distribution: 47% entering, 53% exiting

Trip Generation per 1000 Sq. Feet Gross Floor Area

Average Rate	Range of Rates	Standard Deviation
4.49	3.45 - 6.08	2.22

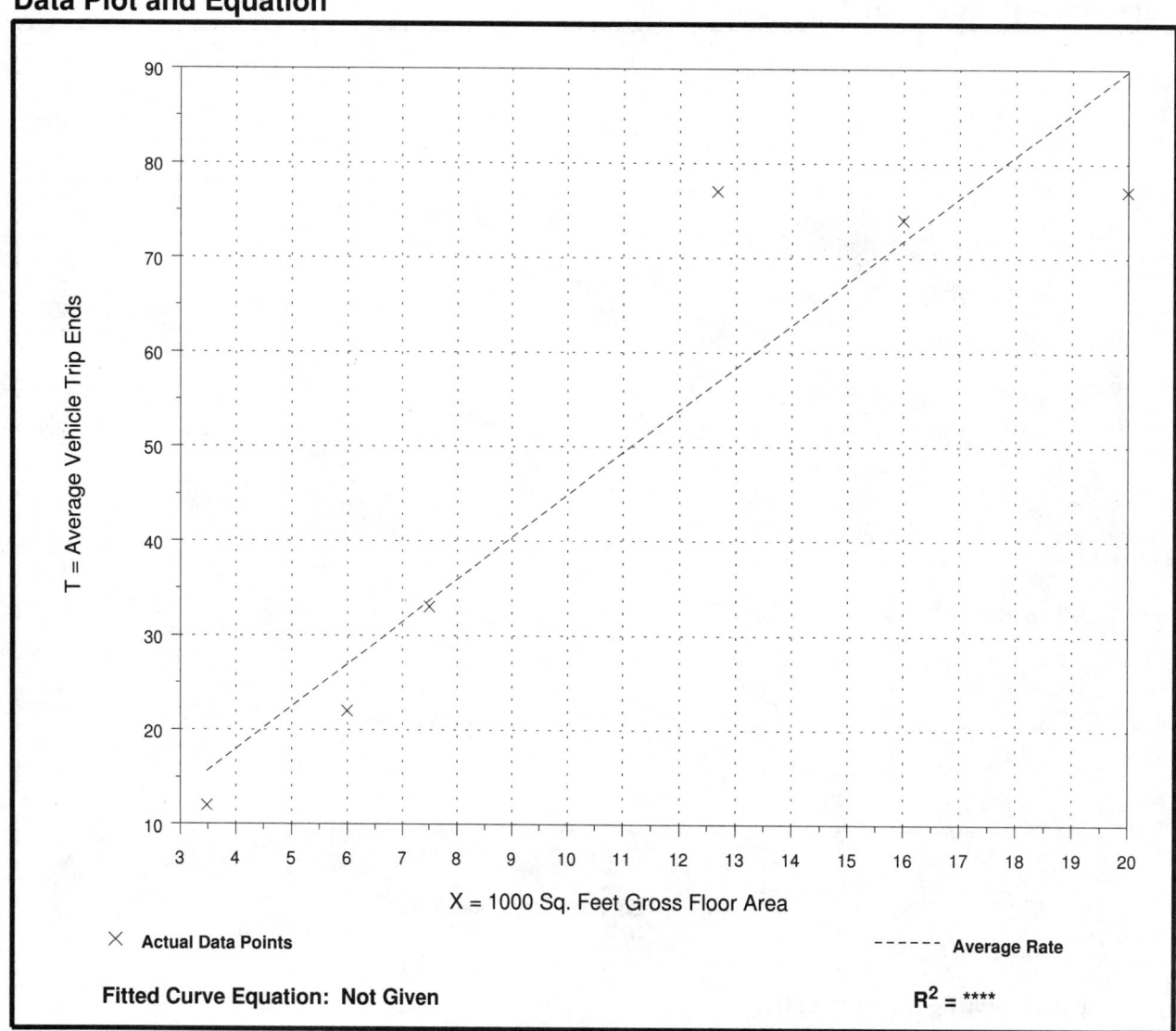

Data Plot and Equation

Fitted Curve Equation: Not Given $R^2 = ****$

Building Materials and Lumber Store
(812)

Average Vehicle Trip Ends vs: 1000 Sq. Feet Gross Floor Area
On a: Weekday,
A.M. Peak Hour of Generator

Number of Studies: 6
Average 1000 Sq. Feet GFA: 11
Directional Distribution: 55% entering, 45% exiting

Trip Generation per 1000 Sq. Feet Gross Floor Area

Average Rate	Range of Rates	Standard Deviation
4.16	1.75 - 9.20	2.78

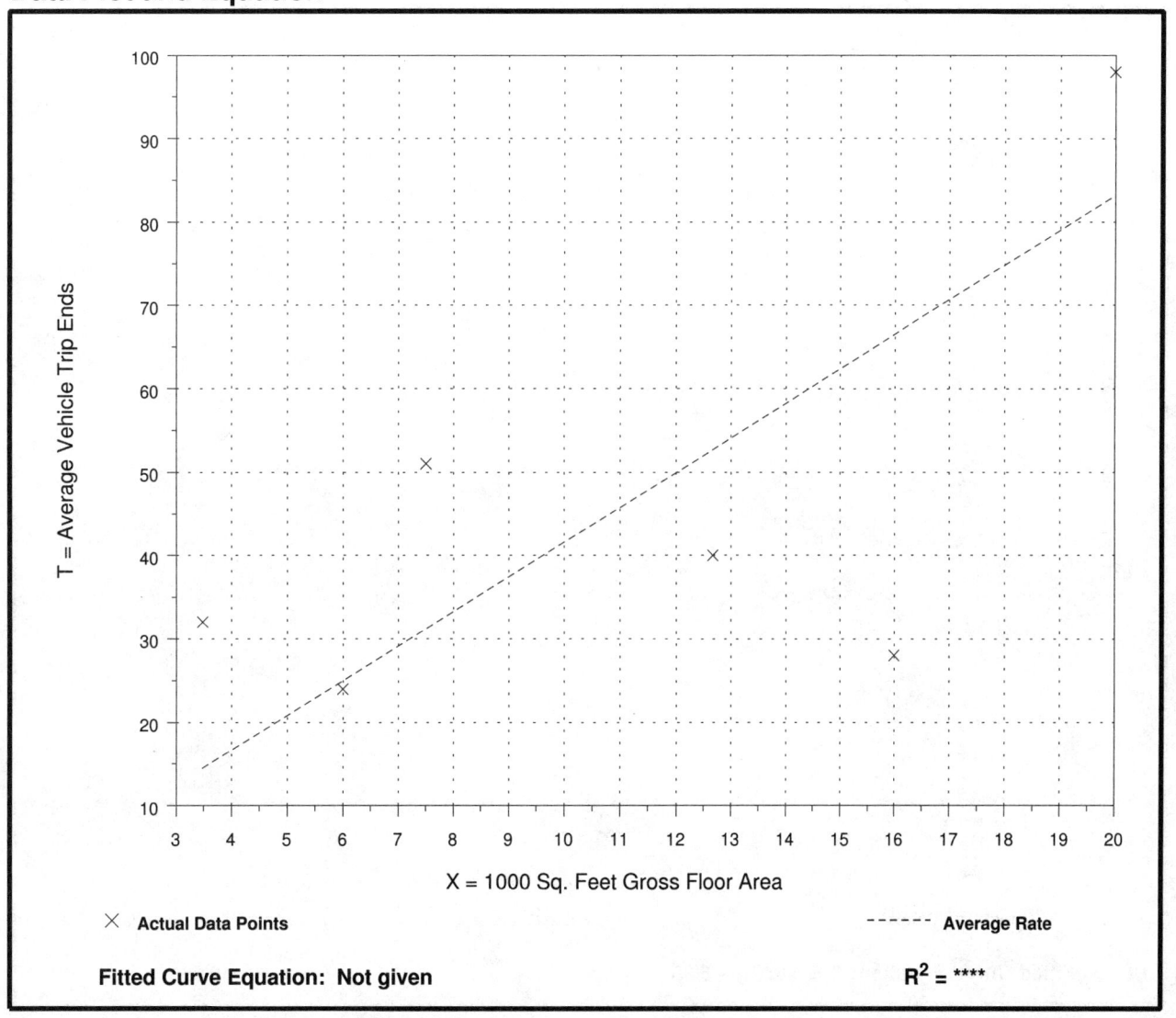

Data Plot and Equation

Fitted Curve Equation: Not given $R^2 = ****$

Building Materials and Lumber Store
(812)

Average Vehicle Trip Ends vs: 1000 Sq. Feet Gross Floor Area
On a: Weekday,
P.M. Peak Hour of Generator

Number of Studies: 6
Average 1000 Sq. Feet GFA: 11
Directional Distribution: 49% entering, 51% exiting

Trip Generation per 1000 Sq. Feet Gross Floor Area

Average Rate	Range of Rates	Standard Deviation
5.56	4.33 - 7.18	2.36

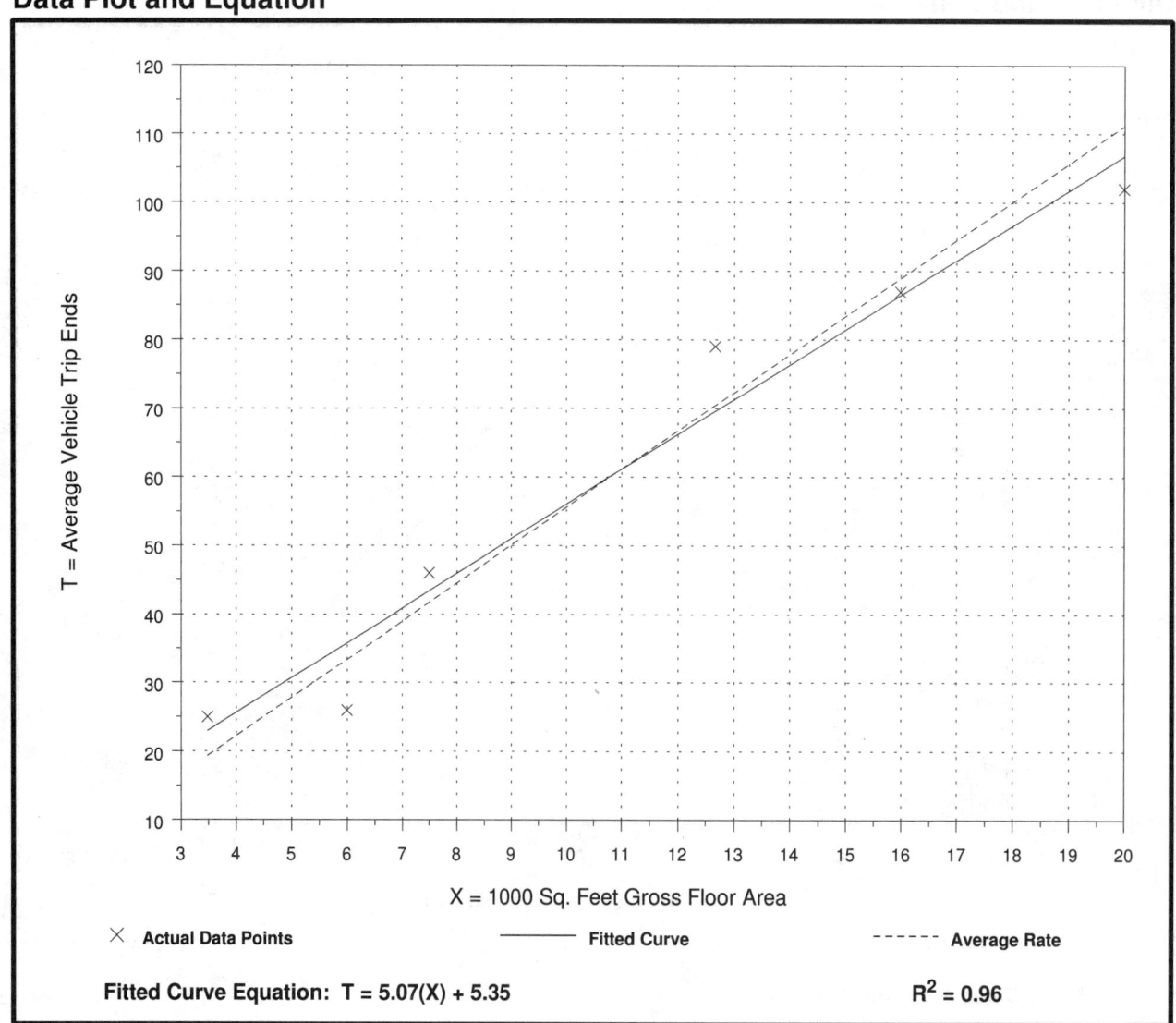

Data Plot and Equation

Fitted Curve Equation: $T = 5.07(X) + 5.35$ $R^2 = 0.96$

Building Materials and Lumber Store
(812)

Average Vehicle Trip Ends vs: 1000 Sq. Feet Gross Floor Area
On a: Saturday

Number of Studies: 4
Average 1000 Sq. Feet GFA: 9
Directional Distribution: 50% entering, 50% exiting

Trip Generation per 1000 Sq. Feet Gross Floor Area

Average Rate	Range of Rates	Standard Deviation
51.60	43.70 - 75.86	12.58

Data Plot and Equation

Caution - Use Carefully - Small Sample Size

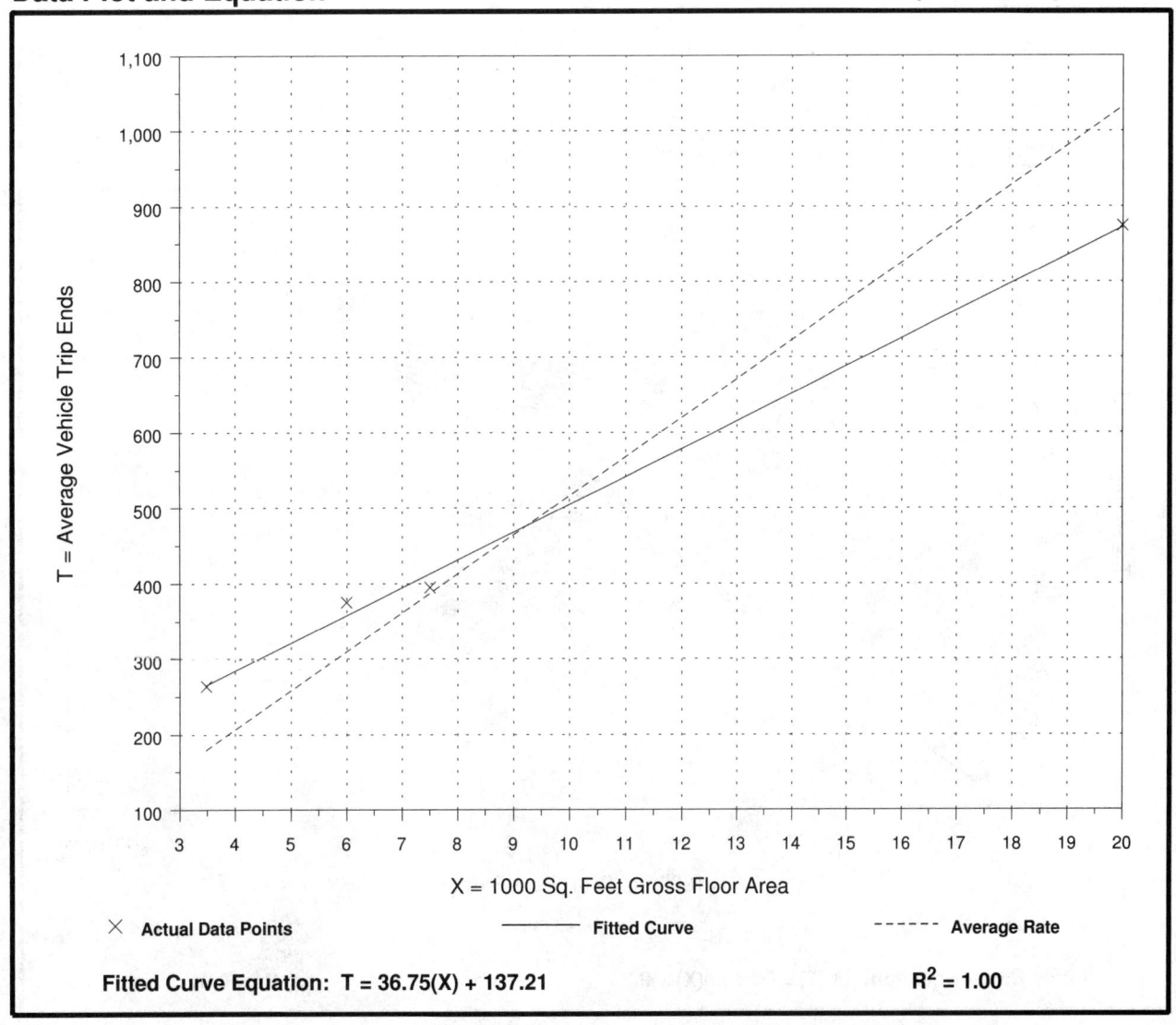

Fitted Curve Equation: T = 36.75(X) + 137.21 $R^2 = 1.00$

Building Materials and Lumber Store
(812)

Average Vehicle Trip Ends vs: 1000 Sq. Feet Gross Floor Area
On a: Saturday,
Peak Hour of Generator

Number of Studies: 6
Average 1000 Sq. Feet GFA: 11
Directional Distribution: 51% entering, 49% exiting

Trip Generation per 1000 Sq. Feet Gross Floor Area

Average Rate	Range of Rates	Standard Deviation
9.58	6.60 - 13.26	4.09

Data Plot and Equation

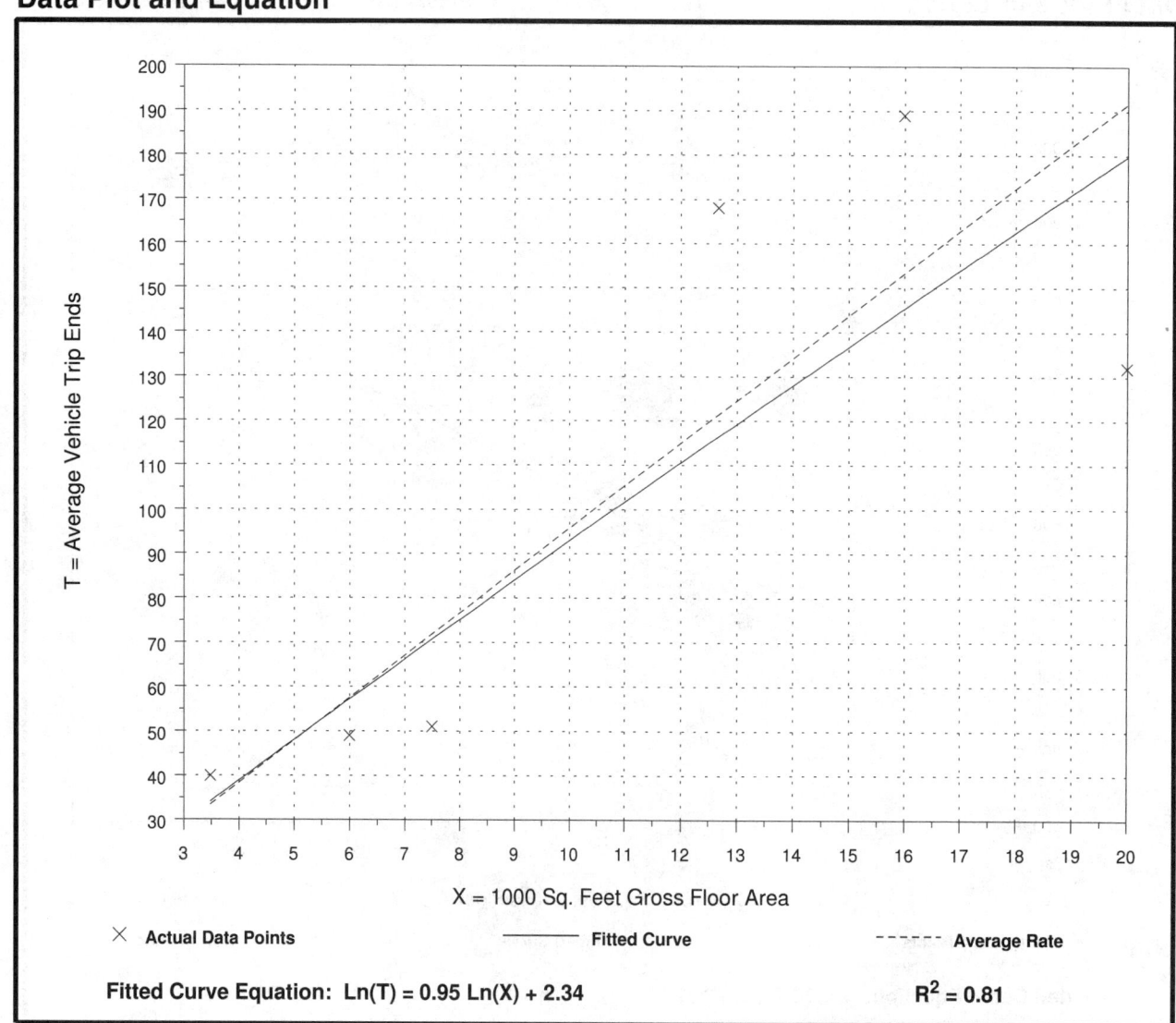

Fitted Curve Equation: $Ln(T) = 0.95 Ln(X) + 2.34$ $R^2 = 0.81$

Building Materials and Lumber Store
(812)

Average Vehicle Trip Ends vs: 1000 Sq. Feet Gross Floor Area
On a: Sunday

Number of Studies: 4
Average 1000 Sq. Feet GFA: 9
Directional Distribution: 50% entering, 50% exiting

Trip Generation per 1000 Sq. Feet Gross Floor Area

Average Rate	Range of Rates	Standard Deviation
24.50	4.27 - 48.67	14.90

Data Plot and Equation

Caution - Use Carefully - Small Sample Size

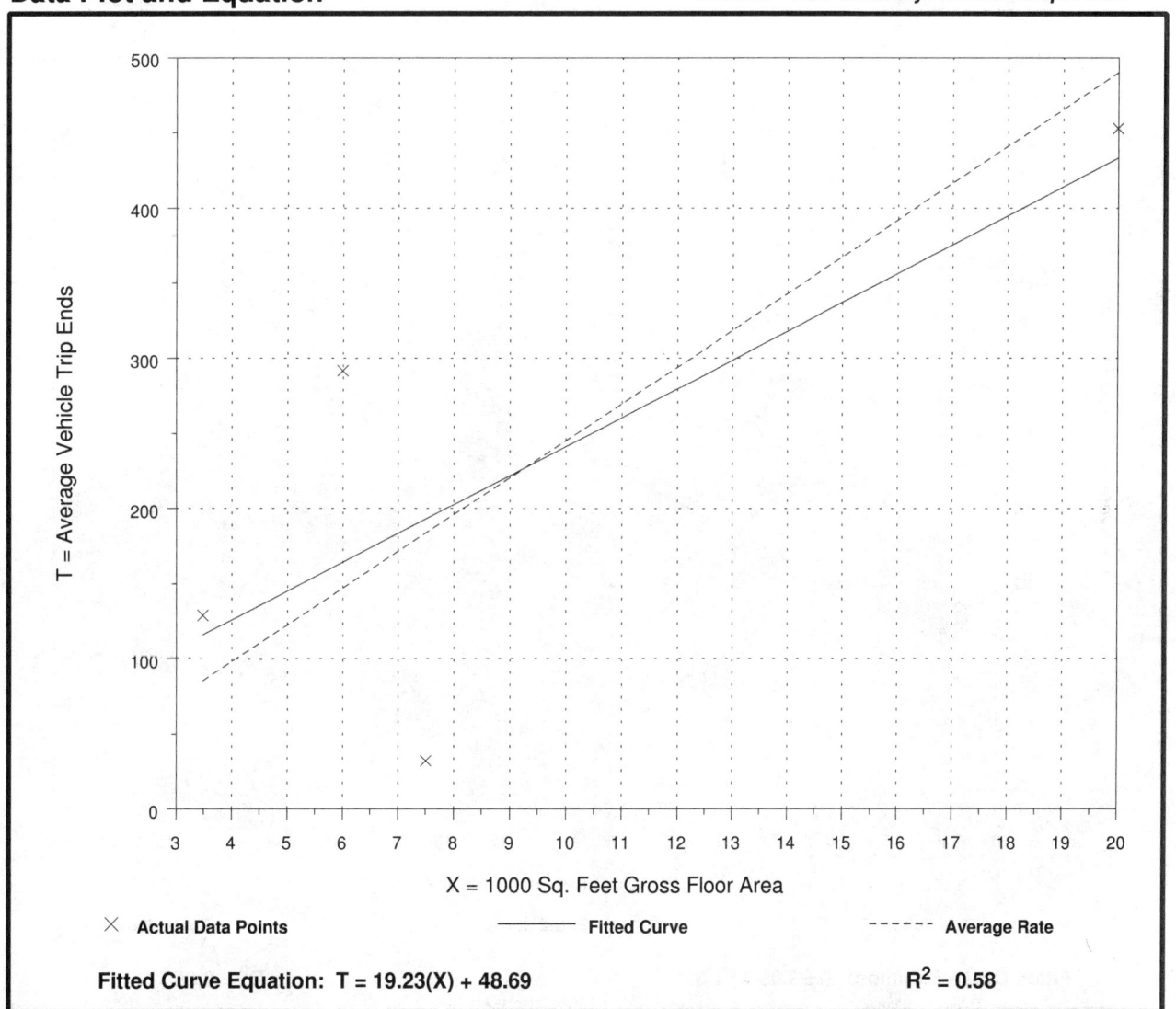

Fitted Curve Equation: T = 19.23(X) + 48.69 $R^2 = 0.58$

Building Materials and Lumber Store
(812)

Average Vehicle Trip Ends vs: 1000 Sq. Feet Gross Floor Area
On a: Sunday,
Peak Hour of Generator

Number of Studies: 4
Average 1000 Sq. Feet GFA: 9
Directional Distribution: 45% entering, 55% exiting

Trip Generation per 1000 Sq. Feet Gross Floor Area

Average Rate	Range of Rates	Standard Deviation
4.57	0.67 - 8.33	3.20

Data Plot and Equation

Caution - Use Carefully - Small Sample Size

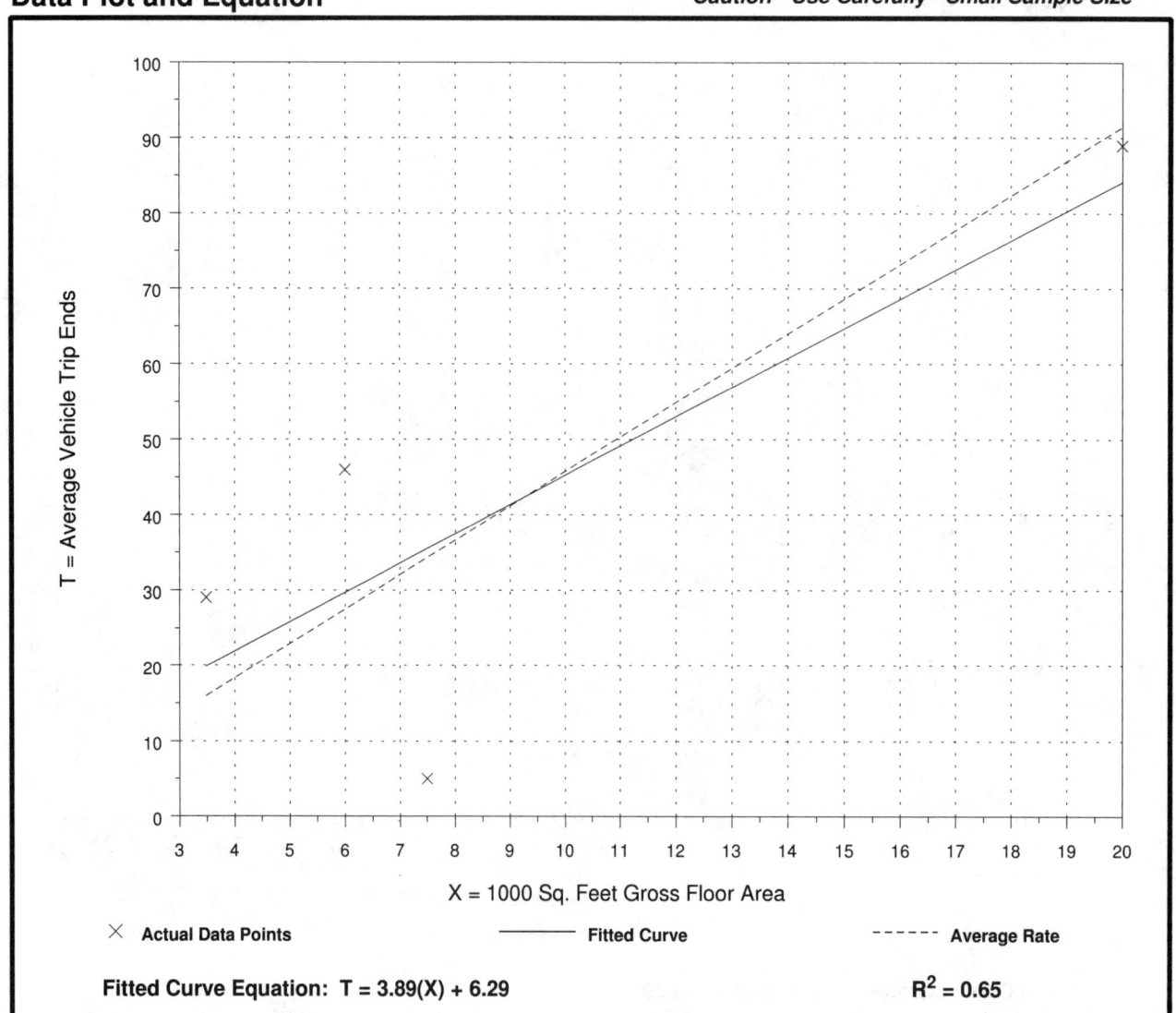

Fitted Curve Equation: $T = 3.89(X) + 6.29$ $R^2 = 0.65$

Land Use: 813
Free-Standing Discount Superstore

Description

The discount superstores in this category are similar to the free-standing discount stores described in Land Use 815 with the exception that they also contain a full service grocery department under the same roof that shares entrances and exits with the discount store area. The stores usually offer a variety of customer services, centralized cashiering and a wide range of products. They typically maintain long store hours 7 days a week. The stores included in this land use are often the only ones on the site, but they can also be found in mutual operation with a related or unrelated garden center and/or service station, or as a part of a shopping center, with or without their own dedicated parking area. Free-standing discount store (Land Use 815) is a related use.

Additional Data

Peak hours of the generator—
The weekday A.M. peak hour was generally between 10:00 a.m. and 11:00 a.m. The weekday P.M. peak hour varied between 12:00 p.m. and 5:00 p.m. The Saturday and Sunday peak hours varied between 12:00 p.m. and 5:00 p.m.

The weighted average truck trip generation rates from approximately 30 sites surveyed for this land use are summarized in the table below. The average gross floor area of these facilities is 206,000 square feet.

Day/Time Period	Weighted Average Truck Trip Generation Rate (trip ends per 1,000 square feet)
Weekday	0.87
Weekday A.M. Peak Hour of Adjacent Street Traffic	0.05
Weekday P.M. Peak Hour of Adjacent Street Traffic	0.03
Weekday A.M. Peak Hour of Generator	0.06
Weekday P.M. Peak Hour of Generator	0.04
Saturday	0.59
Saturday Peak Hour of Generator	0.04
Sunday	0.43
Sunday Peak Hour of Generator	0.02

One source provided information on trip generation rates for what the study defined as "typical" and "peak" seasons. These data indicated that weekday trip generation rates were similar in both seasons. However, trip generation rates on Saturdays during peak season were 13 to 20 percent higher than a typical season; Sunday rates were found to be 6 to 10 percent higher. For the purposes of this analysis, "peak" season was defined as the period between the week after

Thanksgiving and the week prior to Christmas; "typical" season was defined as September through mid-November when transactions are close to average. The seasonal trip generation information provided was based on a sample of five sites.

Information on approximate hourly variation in free-standing discount superstore traffic is shown in the table below. It should be noted, however, that the information contained in this table is based on a limited sample size. Therefore, caution should be exercised when applying the data. Also, some information provided in the table may conflict with the results obtained by applying the average rate or regression equations. When this occurs, it is suggested that the results from the average rate or regression equations be used, as they are based on a larger number of studies.

Hourly Variation in Free-Standing Discount Superstore Traffic

Time	Average Weekday[a] Percent of 24-Hour Entering Traffic	Average Weekday[a] Percent of 24-Hour Exiting Traffic	Average Saturday[b] Percent of 24-Hour Entering Traffic	Average Saturday[b] Percent of 24-Hour Exiting Traffic	Average Sunday[c] Percent of 24-Hour Entering Traffic	Average Sunday[c] Percent of 24-Hour Exiting Traffic
6 a.m.–7 a.m.	1.5	1.2	1.0	1.1	0.9	1.3
7 a.m.–8 a.m.	2.6	2.4	2.2	2.1	2.0	2.3
8 a.m.–9 a.m.	4.1	3.3	3.8	3.2	3.4	3.4
9 a.m.–10 a.m.	6.0	4.6	5.7	4.6	5.4	5.1
10 a.m.–11 a.m.	7.3	6.0	7.0	6.2	7.2	5.8
11 a.m.–12 p.m.	7.5	7.3	8.4	7.4	8.6	7.5
12 p.m.–1 p.m.	8.3	7.7	9.0	8.0	9.4	8.0
1 p.m.–2 p.m.	7.8	7.7	8.9	8.6	9.5	9.2
2 p.m.–3 p.m.	8.0	7.7	8.4	7.9	8.3	8.6
3 p.m.–4 p.m.	7.7	7.7	7.6	7.9	8.4	8.7
4 p.m.–5 p.m.	7.8	8.0	7.4	7.7	7.9	7.8
5 p.m.–6 p.m.	7.1	7.3	7.0	7.5	6.9	7.2
6 p.m.–7 p.m.	6.7	6.7	6.3	6.8	6.4	6.7
7 p.m.–8 p.m.	5.7	6.1	5.4	5.9	5.0	5.1
8 p.m.–9 p.m.	4.4	5.2	4.4	5.0	4.0	3.6
9 p.m.–10 p.m.	3.0	4.0	3.5	3.7	2.9	2.9
10 p.m.–6 a.m.	4.5	7.2	3.9	6.4	3.8	6.8

Sites ranged in size from 123,000 to 224,000 square feet gross floor area
[a] Source numbers – 354, 595 and 618; based on 11 studies
[b] Source numbers – 354 and 618; based on nine studies
[c] Source number – 354; based on eight studies

Garden centers contained within the principal outside faces of the exterior building walls were included in the gross square floor areas reported. Outdoor or fenced-in areas outside the principal faces of the exterior walls were excluded. Please refer to Volume 1, User's Guide, for a more detailed definition of gross floor area.

Several sites included in this land use indicated the presence of fenced/covered space.

The sites were surveyed between the 1990s and the 2000s throughout the United States.

To assist in the future analysis of this land use, it is important to collect and include information on the presence and size of garden centers, outdoor fenced-in space and service stations in trip generation data submissions.

Source Numbers

354, 522, 577, 595, 607, 609, 612, 618, 625, 630, 636, 651, 652, 661, 700, 731, 735

Free-Standing Discount Superstore
(813)

Average Vehicle Trip Ends vs: 1000 Sq. Feet Gross Floor Area
On a: Weekday

Number of Studies: 65
Average 1000 Sq. Feet GFA: 196
Directional Distribution: 50% entering, 50% exiting

Trip Generation per 1000 Sq. Feet Gross Floor Area

Average Rate	Range of Rates	Standard Deviation
50.75	21.39 - 85.01	14.73

Data Plot and Equation

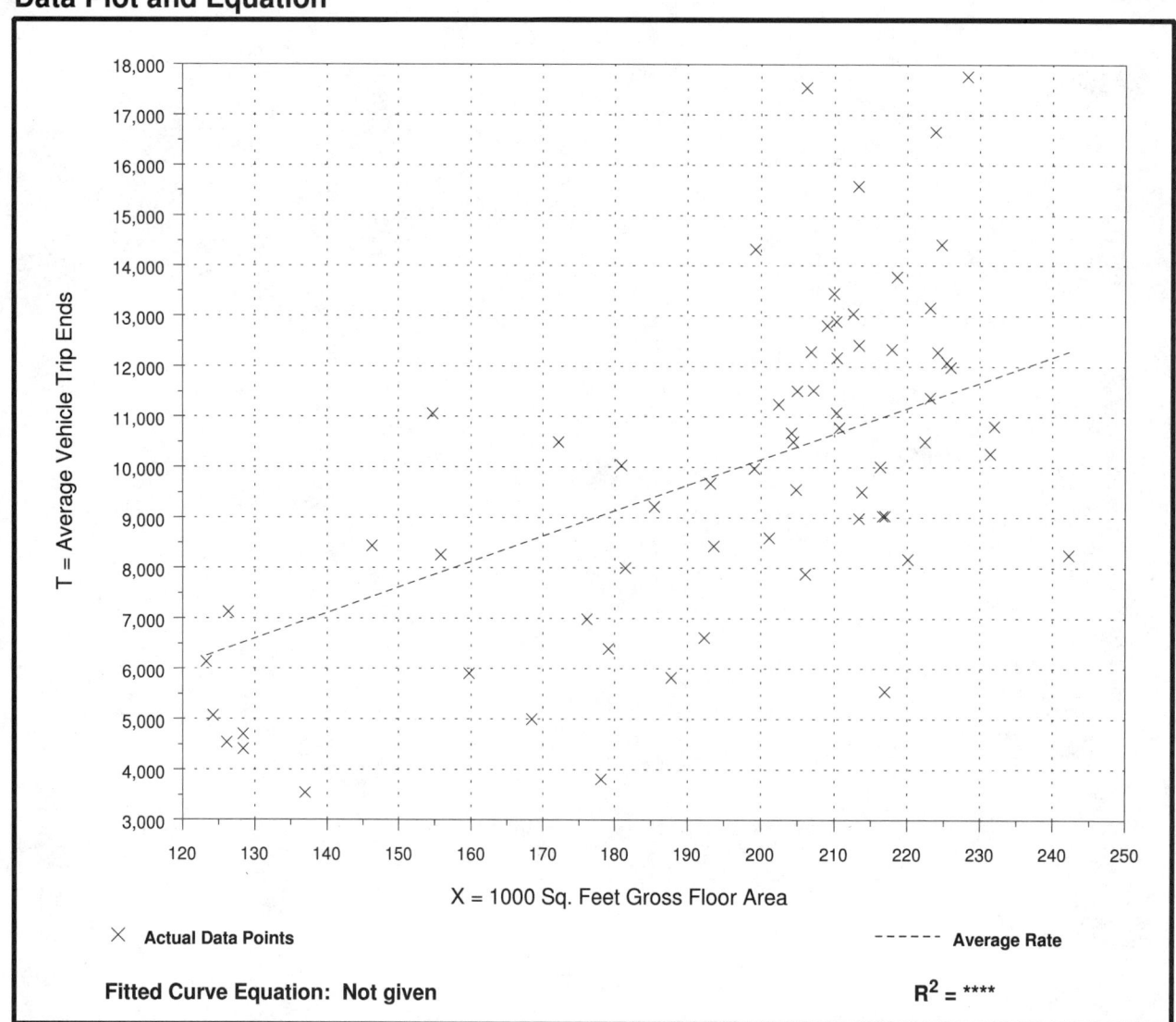

Fitted Curve Equation: Not given $R^2 = ****$

Free-Standing Discount Superstore
(813)

Average Vehicle Trip Ends vs: 1000 Sq. Feet Gross Floor Area
On a: Weekday,
Peak Hour of Adjacent Street Traffic,
One Hour Between 7 and 9 a.m.

Number of Studies: 67
Average 1000 Sq. Feet GFA: 196
Directional Distribution: 56% entering, 44% exiting

Trip Generation per 1000 Sq. Feet Gross Floor Area

Average Rate	Range of Rates	Standard Deviation
1.85	0.81 - 3.86	1.55

Data Plot and Equation

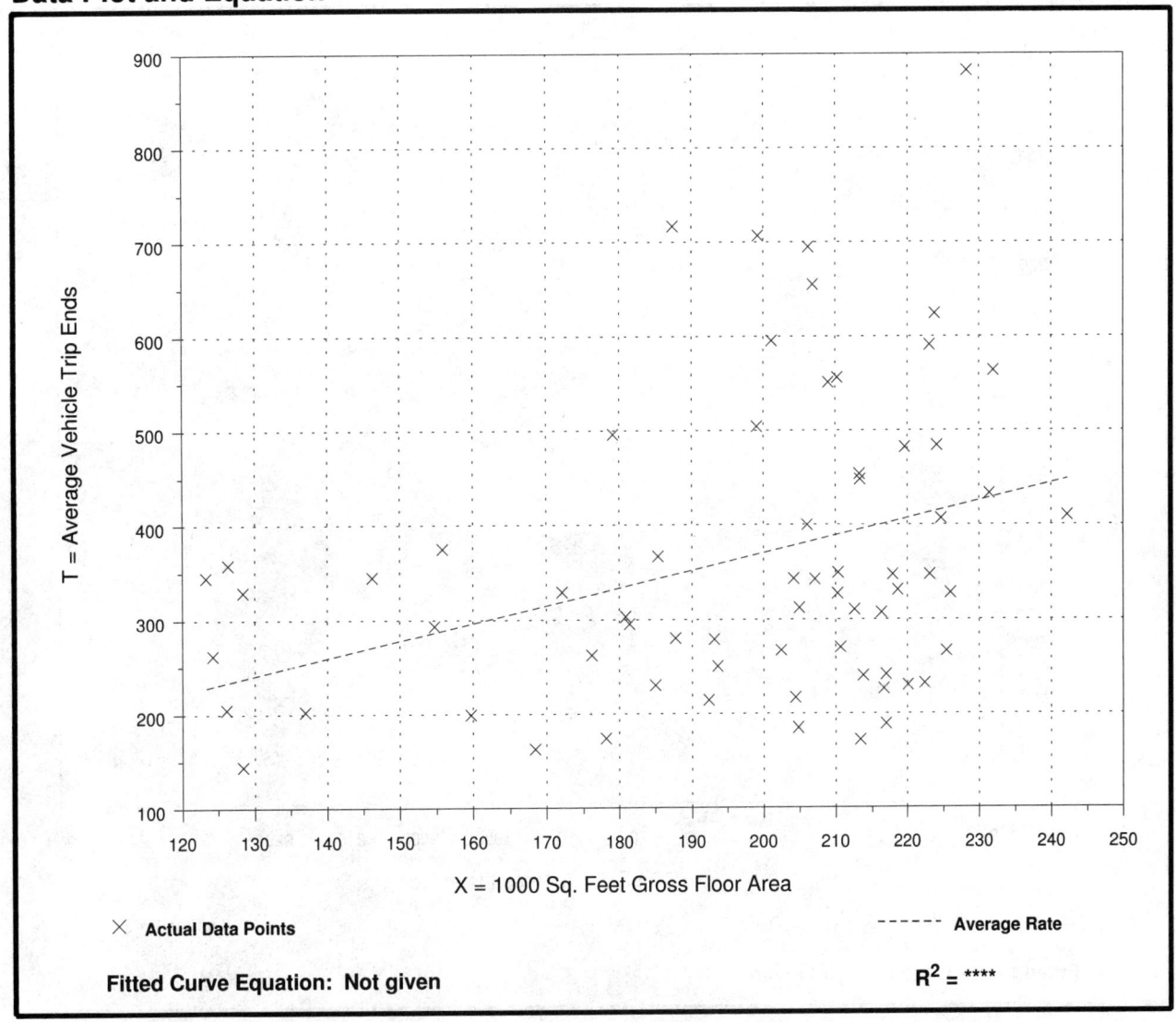

Fitted Curve Equation: Not given $R^2 = ****$

Trip Generation, 9th Edition • Institute of Transportation Engineers

Free-Standing Discount Superstore
(813)

Average Vehicle Trip Ends vs: 1000 Sq. Feet Gross Floor Area
On a: Weekday,
Peak Hour of Adjacent Street Traffic,
One Hour Between 4 and 6 p.m.

Number of Studies: 86
Average 1000 Sq. Feet GFA: 200
Directional Distribution: 49% entering, 51% exiting

Trip Generation per 1000 Sq. Feet Gross Floor Area

Average Rate	Range of Rates	Standard Deviation
4.35	1.83 - 7.40	2.36

Data Plot and Equation

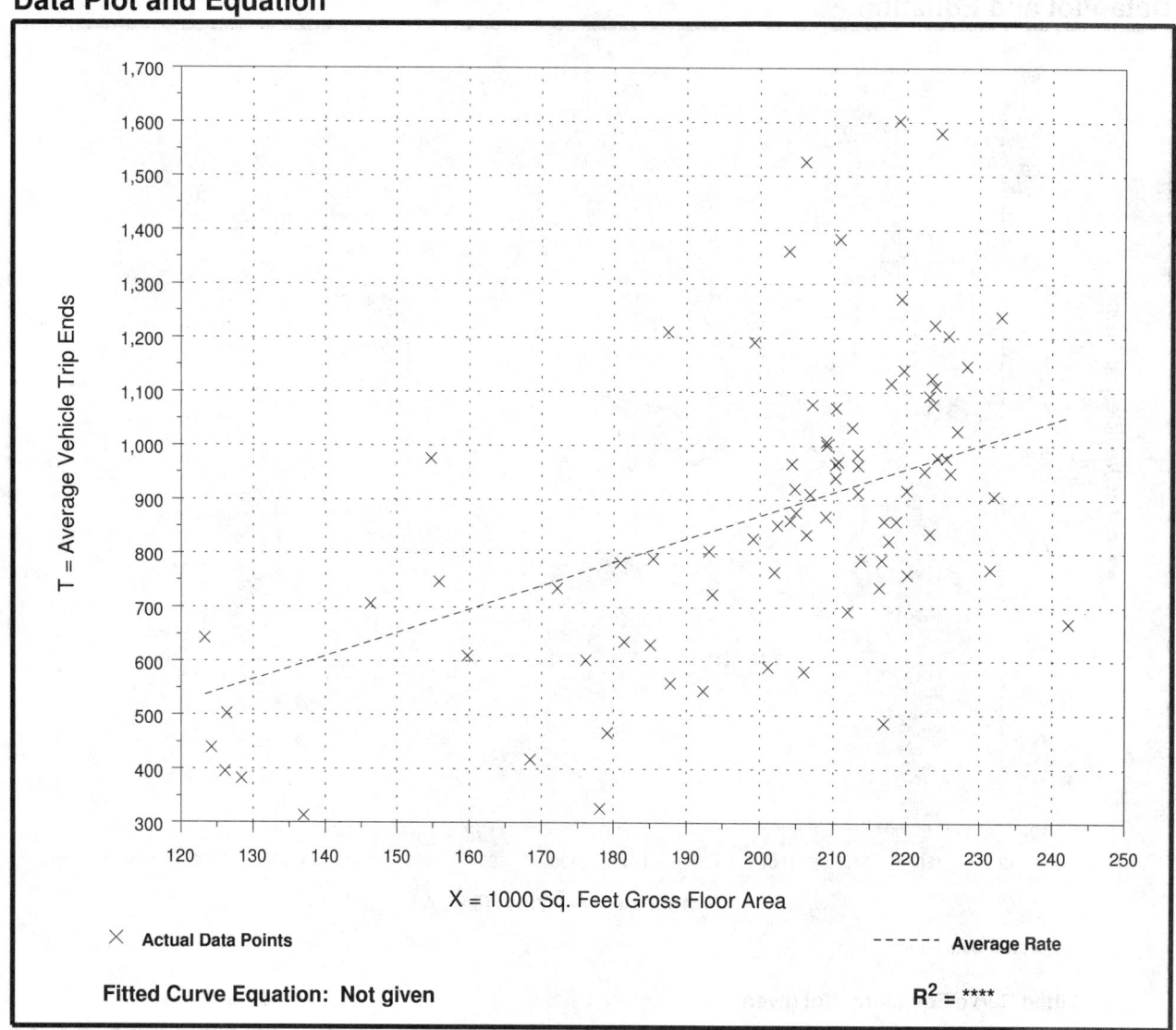

Fitted Curve Equation: Not given $R^2 = ****$

Free-Standing Discount Superstore
(813)

Average Vehicle Trip Ends vs: 1000 Sq. Feet Gross Floor Area
On a: Weekday,
A.M. Peak Hour of Generator

Number of Studies: 65
Average 1000 Sq. Feet GFA: 195
Directional Distribution: 52% entering, 48% exiting

Trip Generation per 1000 Sq. Feet Gross Floor Area

Average Rate	Range of Rates	Standard Deviation
3.08	0.98 - 5.67	2.03

Data Plot and Equation

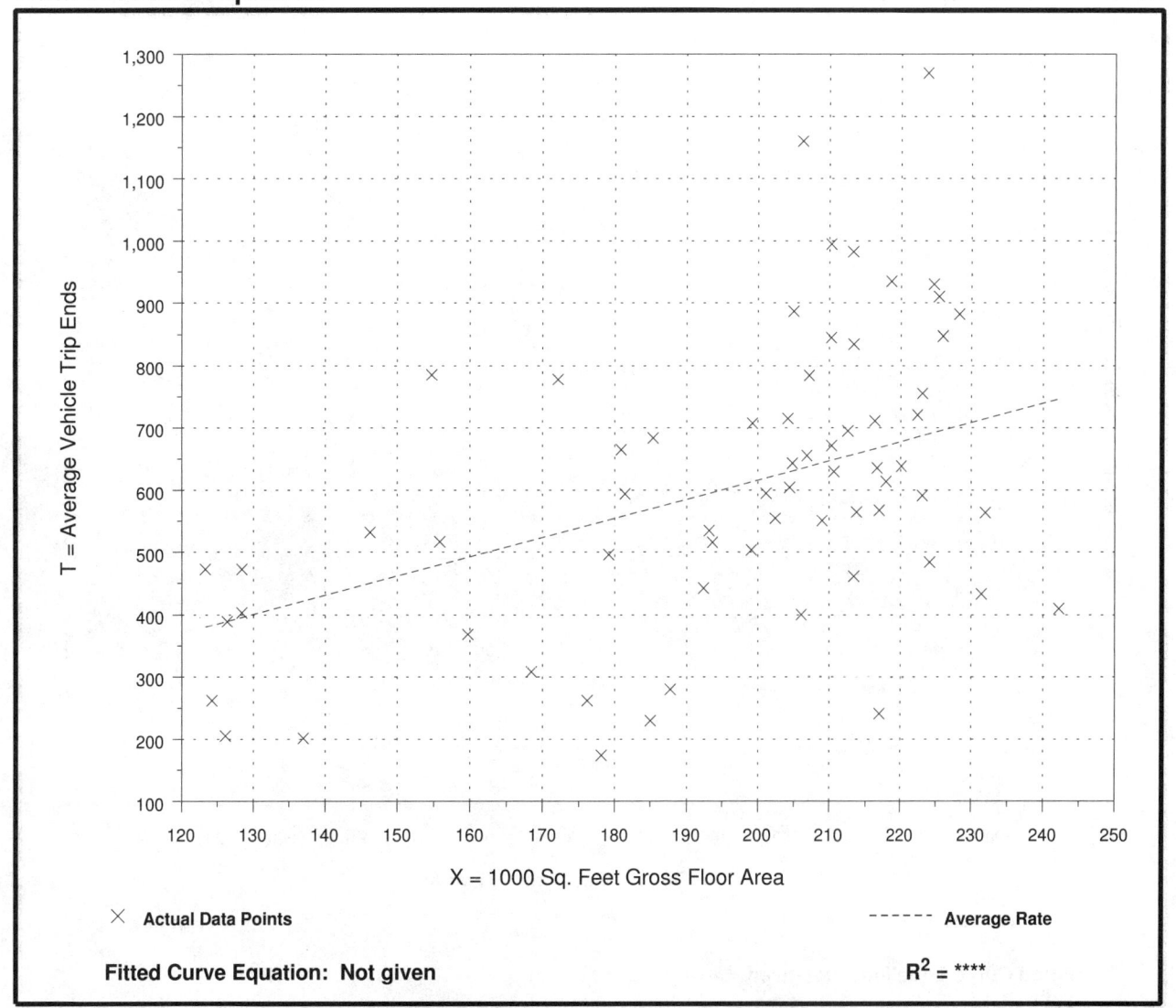

Fitted Curve Equation: Not given $R^2 = ****$

Free-Standing Discount Superstore
(813)

Average Vehicle Trip Ends vs: 1000 Sq. Feet Gross Floor Area
On a: Weekday,
P.M. Peak Hour of Generator

Number of Studies: 67
Average 1000 Sq. Feet GFA: 196
Directional Distribution: 50% entering, 50% exiting

Trip Generation per 1000 Sq. Feet Gross Floor Area

Average Rate	Range of Rates	Standard Deviation
4.40	2.05 - 7.40	2.37

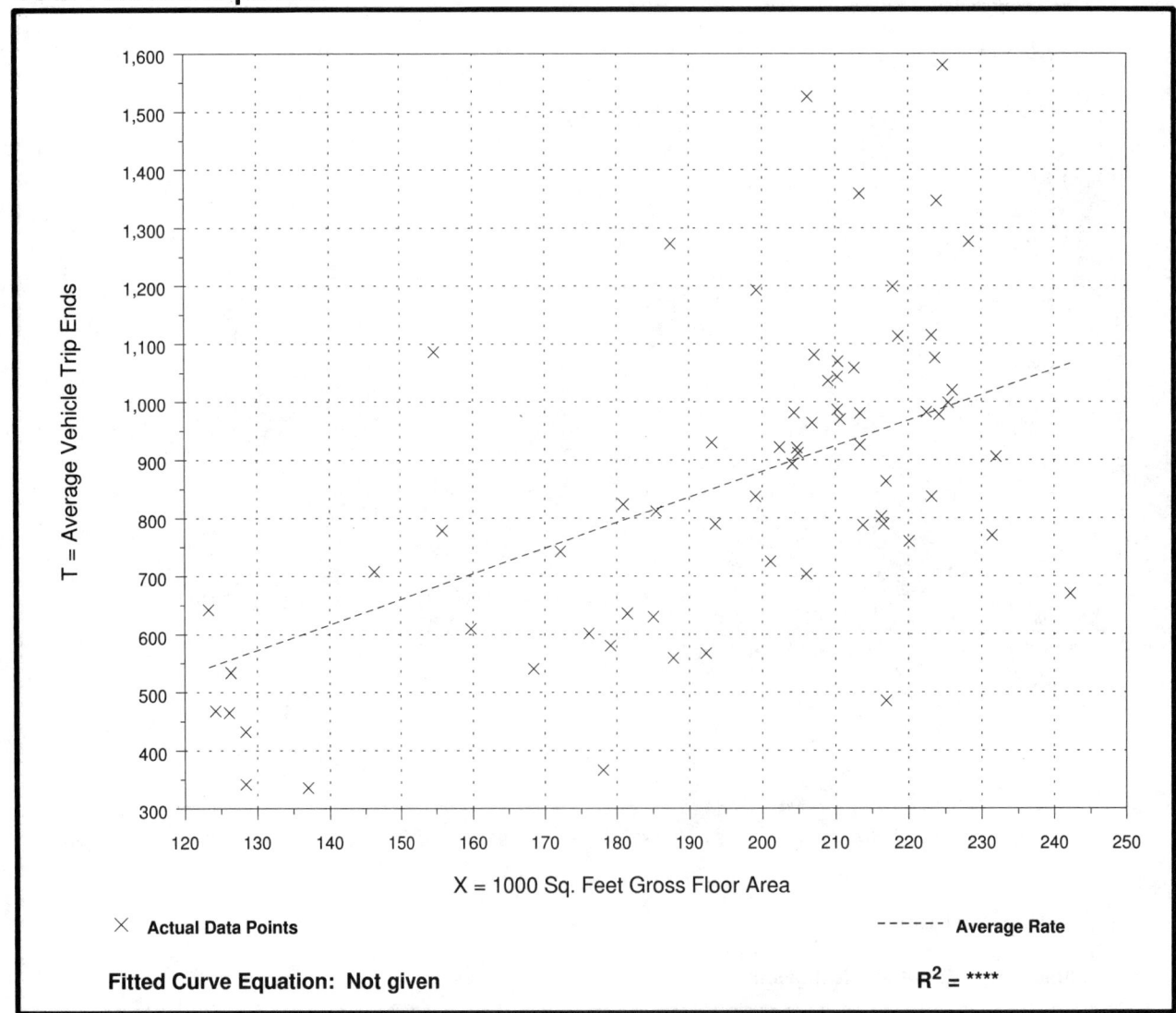

Data Plot and Equation

[Scatter plot: X = 1000 Sq. Feet Gross Floor Area (120 to 250); T = Average Vehicle Trip Ends (300 to 1,600); dashed line shows Average Rate]

× Actual Data Points   ------ Average Rate

Fitted Curve Equation: Not given $R^2 = ****$

Free-Standing Discount Superstore
(813)

Average Vehicle Trip Ends vs: 1000 Sq. Feet Gross Floor Area
On a: Saturday

Number of Studies: 43
Average 1000 Sq. Feet GFA: 195
Directional Distribution: 50% entering, 50% exiting

Trip Generation per 1000 Sq. Feet Gross Floor Area

Average Rate	Range of Rates	Standard Deviation
64.07	35.32 - 105.94	16.71

Data Plot and Equation

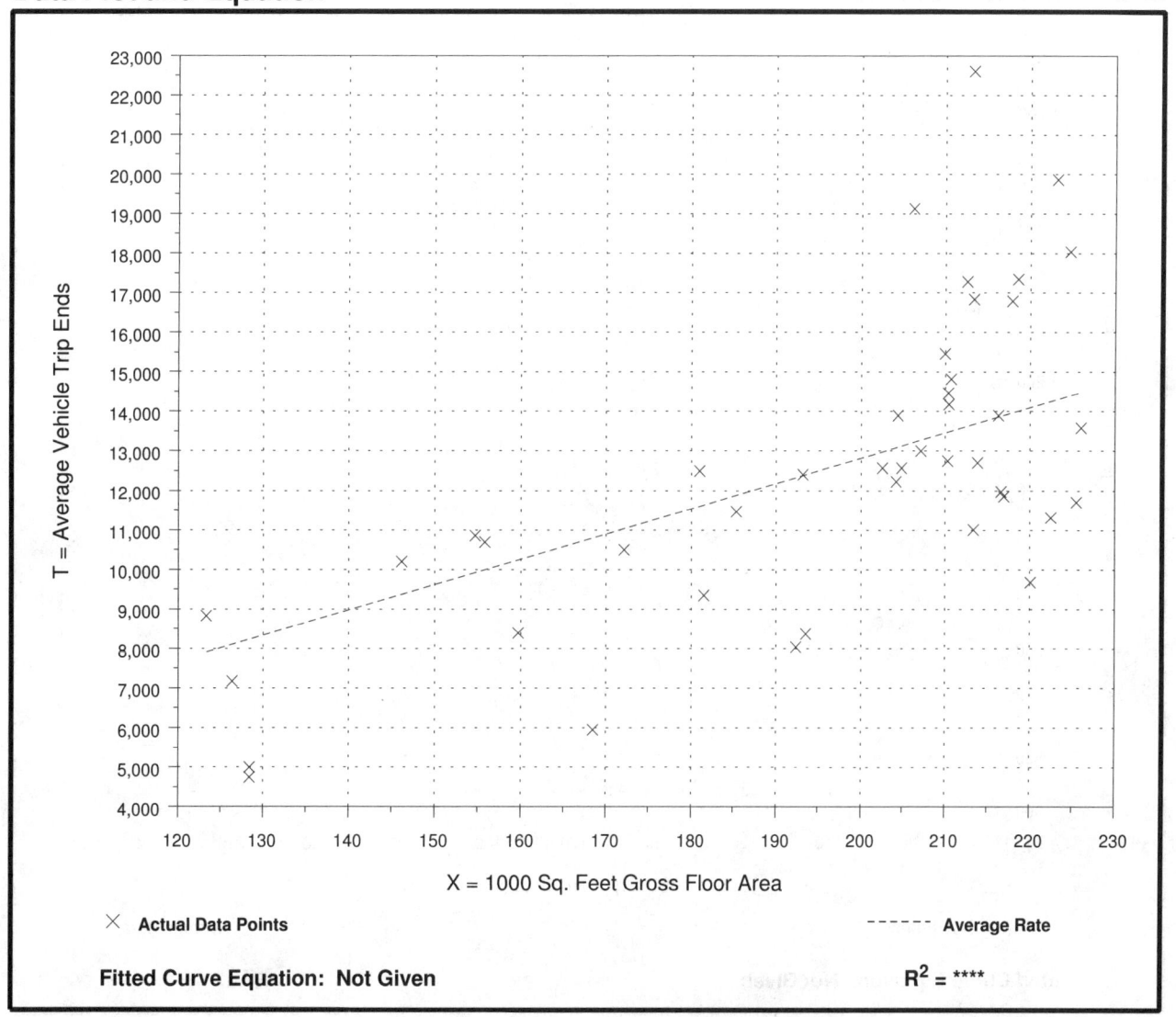

Fitted Curve Equation: Not Given $R^2 = ****$

Free-Standing Discount Superstore
(813)

Average Vehicle Trip Ends vs: 1000 Sq. Feet Gross Floor Area
On a: Saturday,
Peak Hour of Generator

Number of Studies: 52
Average 1000 Sq. Feet GFA: 198
Directional Distribution: 50% entering, 50% exiting

Trip Generation per 1000 Sq. Feet Gross Floor Area

Average Rate	Range of Rates	Standard Deviation
5.64	2.99 - 7.96	2.61

Data Plot and Equation

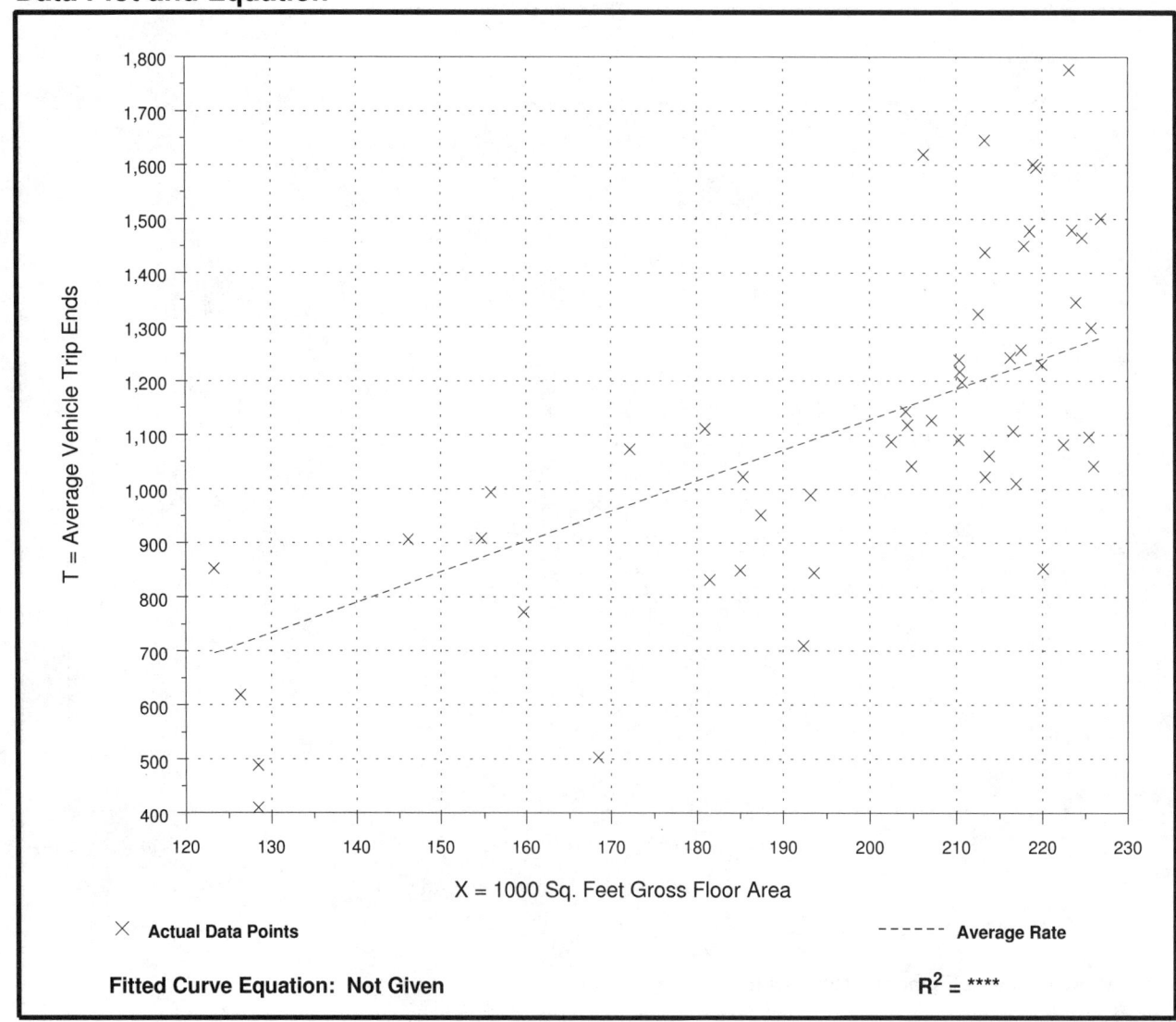

X = 1000 Sq. Feet Gross Floor Area

X   Actual Data Points   ------ Average Rate

Fitted Curve Equation: Not Given R^2 = ****

Free-Standing Discount Superstore
(813)

Average Vehicle Trip Ends vs: 1000 Sq. Feet Gross Floor Area
On a: Sunday

Number of Studies: 42
Average 1000 Sq. Feet GFA: 195
Directional Distribution: 50% entering, 50% exiting

Trip Generation per 1000 Sq. Feet Gross Floor Area

Average Rate	Range of Rates	Standard Deviation
56.12	27.61 - 99.78	16.49

Data Plot and Equation

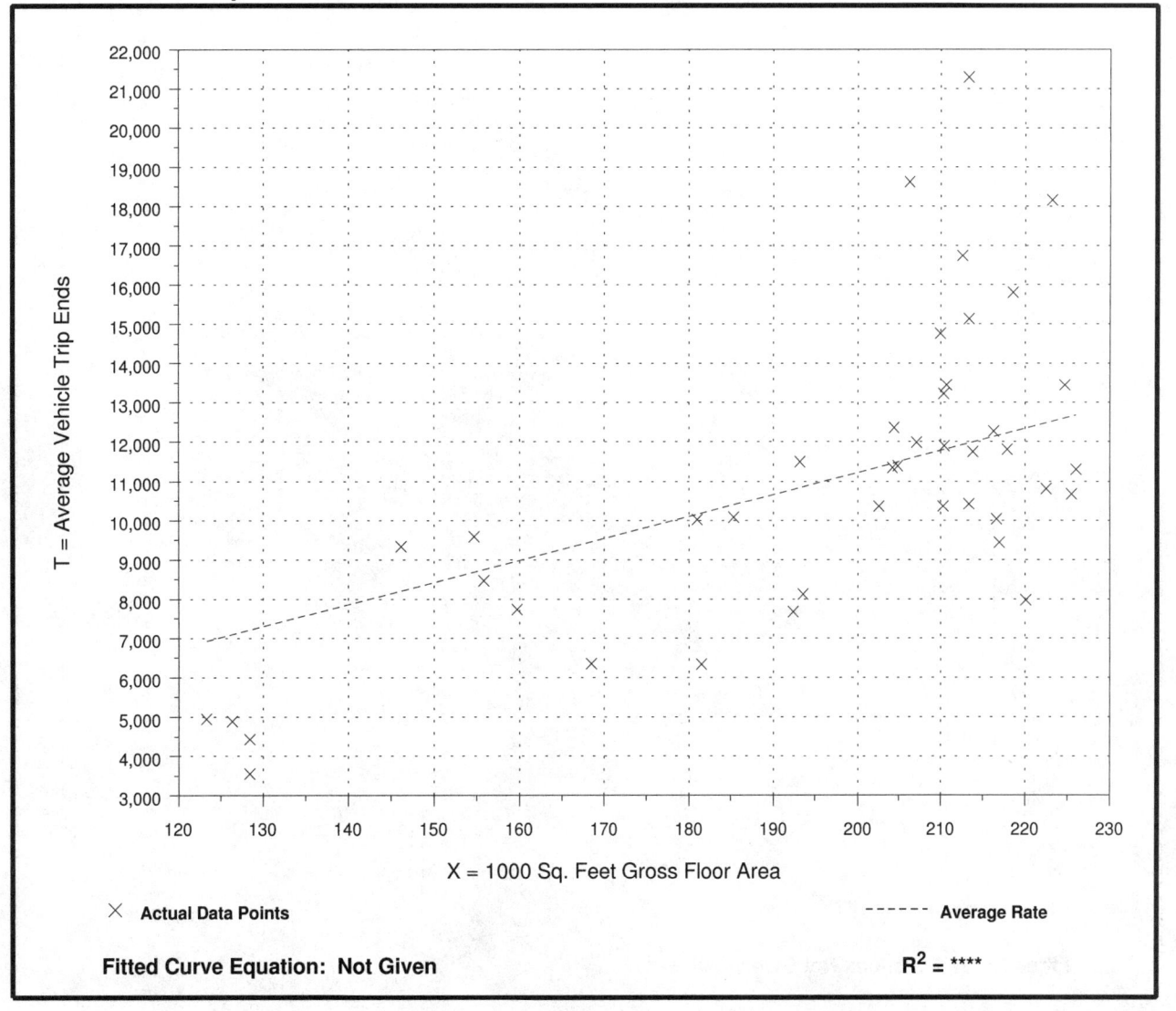

X Actual Data Points   ----- Average Rate

Fitted Curve Equation: Not Given $R^2 =$ ****

Trip Generation, 9th Edition • Institute of Transportation Engineers

Free-Standing Discount Superstore
(813)

Average Vehicle Trip Ends vs: 1000 Sq. Feet Gross Floor Area
On a: Sunday,
Peak Hour of Generator

Number of Studies: 42
Average 1000 Sq. Feet GFA: 195
Directional Distribution: 51% entering, 49% exiting

Trip Generation per 1000 Sq. Feet Gross Floor Area

Average Rate	Range of Rates	Standard Deviation
5.18	3.28 - 7.90	2.53

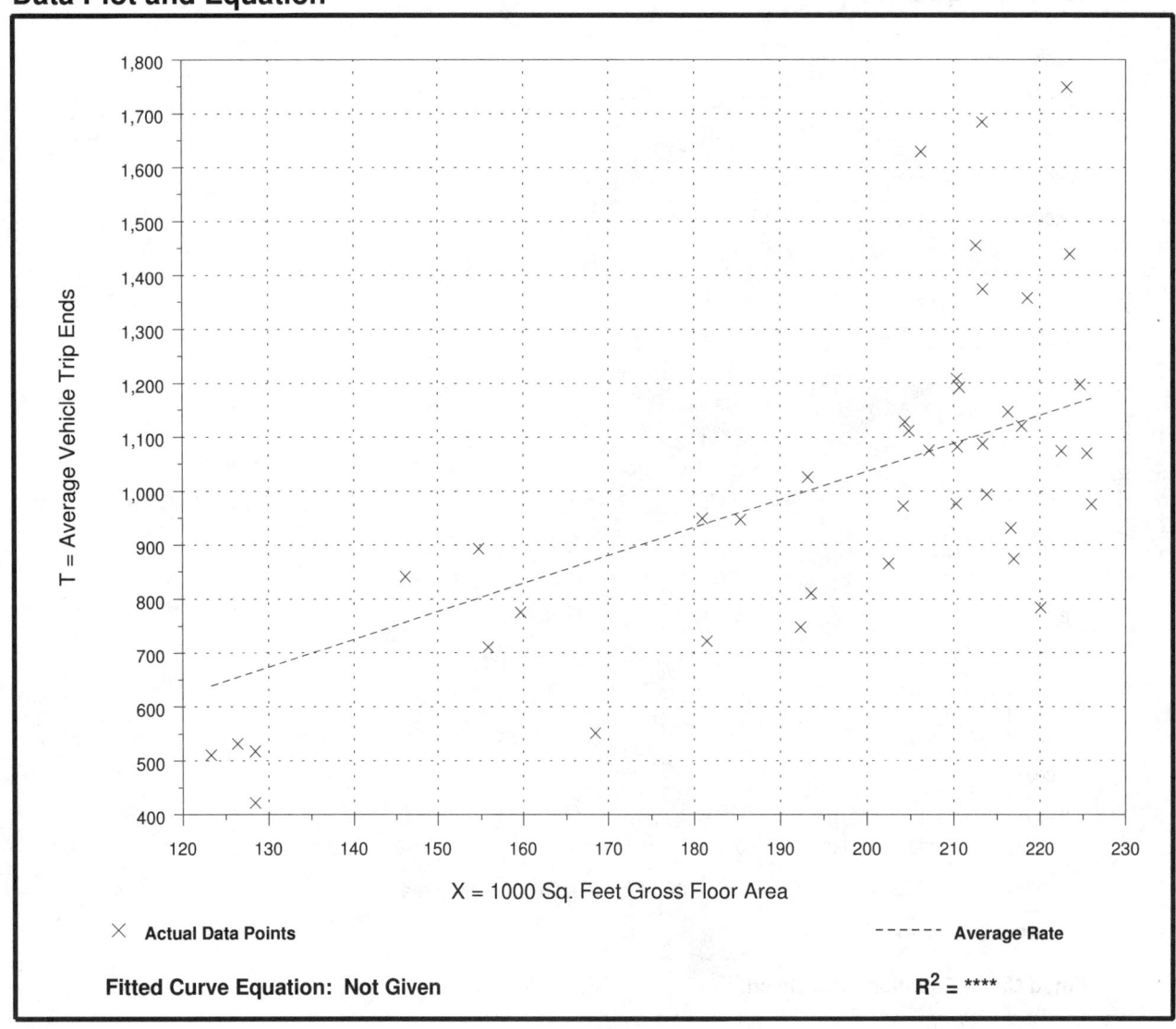

Data Plot and Equation

Fitted Curve Equation: Not Given $R^2 = ****$

Land Use: 814
Variety Store

Description

A variety store is a retail store that sells a broad range of inexpensive items often at a single price. These stores are typically referred to as "dollar stores." Items sold at these stores typically include kitchen supplies, cleaning products, home office supplies, food products, household goods, decorations and toys. These stores are sometimes stand-alone sites, but they may also be located in small strip shopping centers. Free-standing discount store (Land Use 815) is a related use.

Additional Data

The sites were surveyed in 2010 in Florida.

Source Number

731

Variety Store
(814)

Average Vehicle Trip Ends vs: 1000 Sq. Feet Gross Floor Area
On a: Weekday

Number of Studies: 15
Average 1000 Sq. Feet GFA: 10
Directional Distribution: 50% entering, 50% exiting

Trip Generation per 1000 Sq. Feet Gross Floor Area

Average Rate	Range of Rates	Standard Deviation
64.03	33.73 - 133.60	25.69

Data Plot and Equation

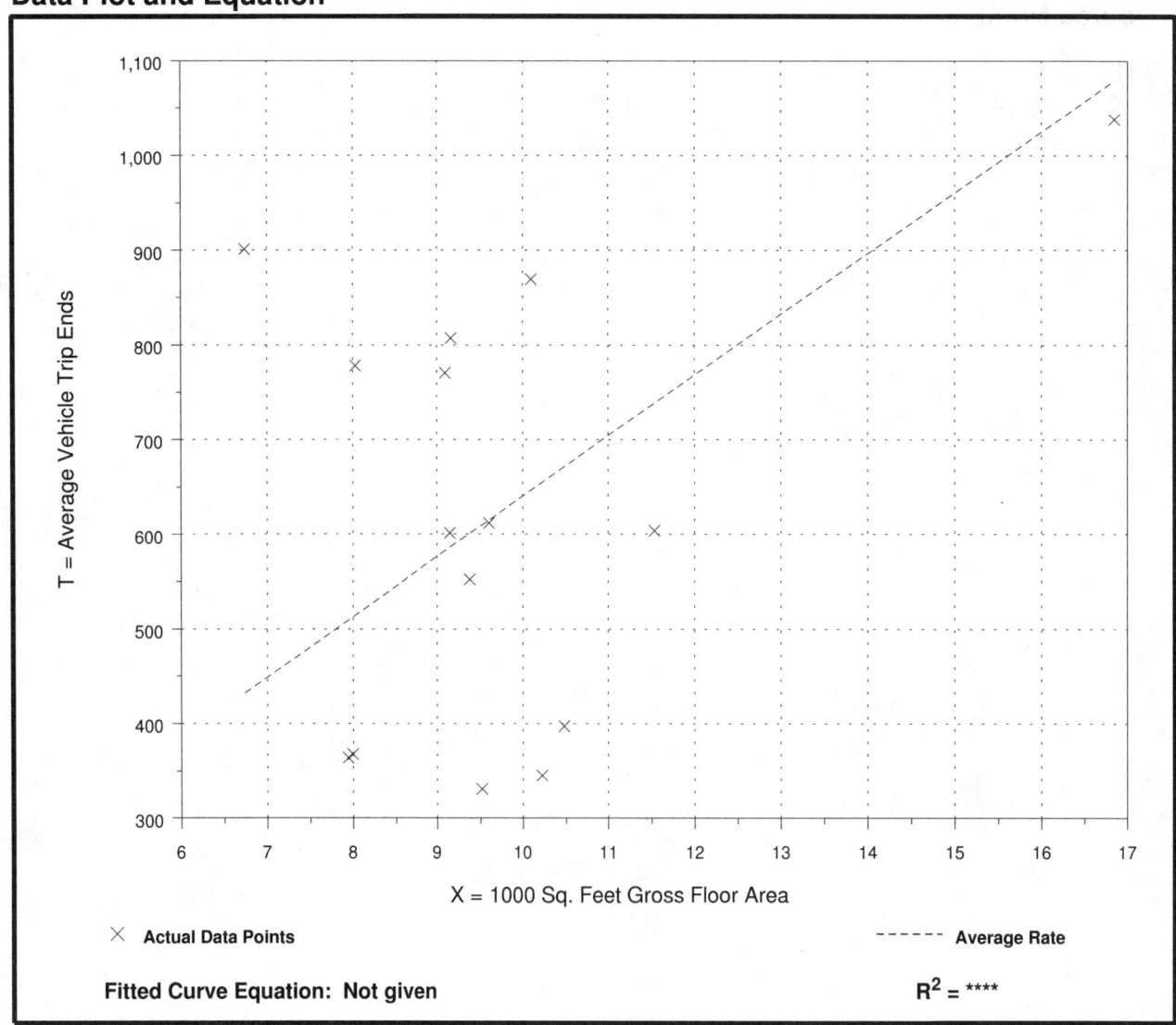

Fitted Curve Equation: Not given $R^2 = ****$

Trip Generation, 9th Edition • Institute of Transportation Engineers

Variety Store
(814)

Average Vehicle Trip Ends vs: 1000 Sq. Feet Gross Floor Area
On a: Weekday,
Peak Hour of Adjacent Street Traffic,
One Hour Between 7 and 9 a.m.

Number of Studies: 15
Average 1000 Sq. Feet GFA: 10
Directional Distribution: Not available

Trip Generation per 1000 Sq. Feet Gross Floor Area

Average Rate	Range of Rates	Standard Deviation
3.81	1.68 - 11.86	2.74

Data Plot and Equation

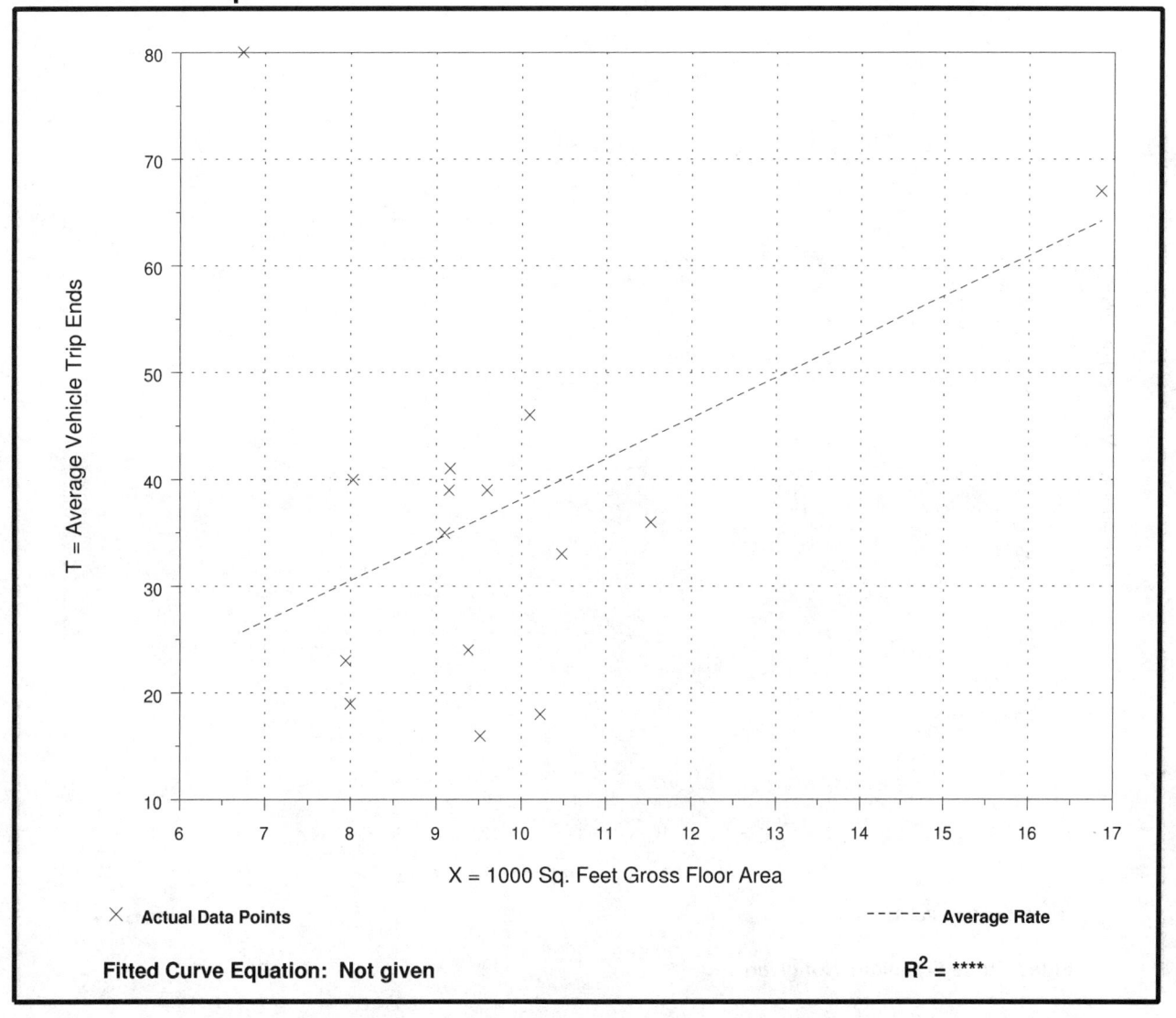

Fitted Curve Equation: Not given $R^2 = ****$

Variety Store
(814)

Average Vehicle Trip Ends vs: 1000 Sq. Feet Gross Floor Area
On a: Weekday,
Peak Hour of Adjacent Street Traffic,
One Hour Between 4 and 6 p.m.

Number of Studies: 15
Average 1000 Sq. Feet GFA: 10
Directional Distribution: Not available

Trip Generation per 1000 Sq. Feet Gross Floor Area

Average Rate	Range of Rates	Standard Deviation
6.82	3.15 - 13.94	3.80

Data Plot and Equation

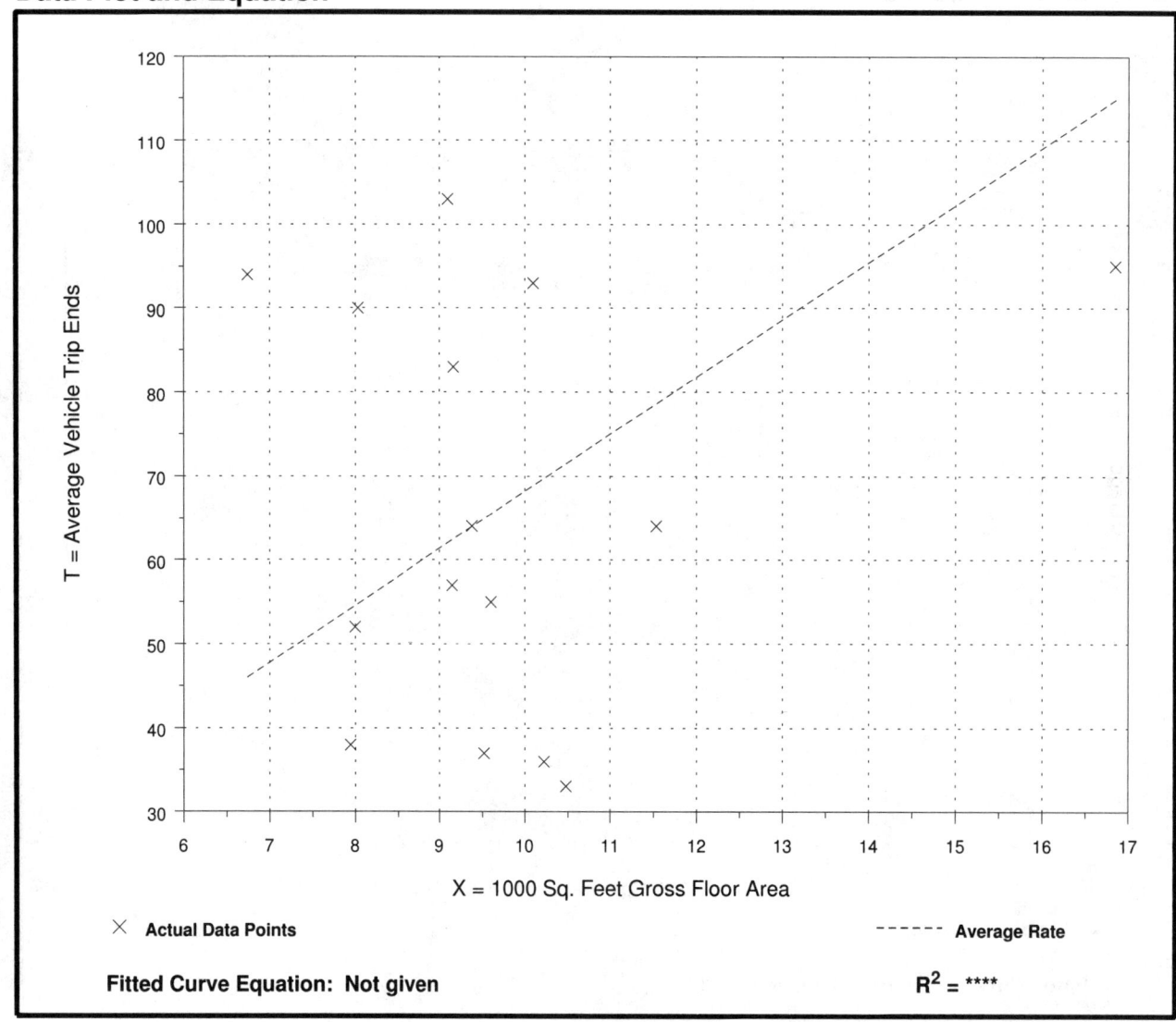

Fitted Curve Equation: Not given $R^2 = ****$

Variety Store
(814)

Average Vehicle Trip Ends vs: 1000 Sq. Feet Gross Floor Area
On a: Weekday,
A.M. Peak Hour of Generator

Number of Studies: 15
Average 1000 Sq. Feet GFA: 10
Directional Distribution: Not available

Trip Generation per 1000 Sq. Feet Gross Floor Area

Average Rate	Range of Rates	Standard Deviation
3.81	1.68 - 11.86	2.74

Data Plot and Equation

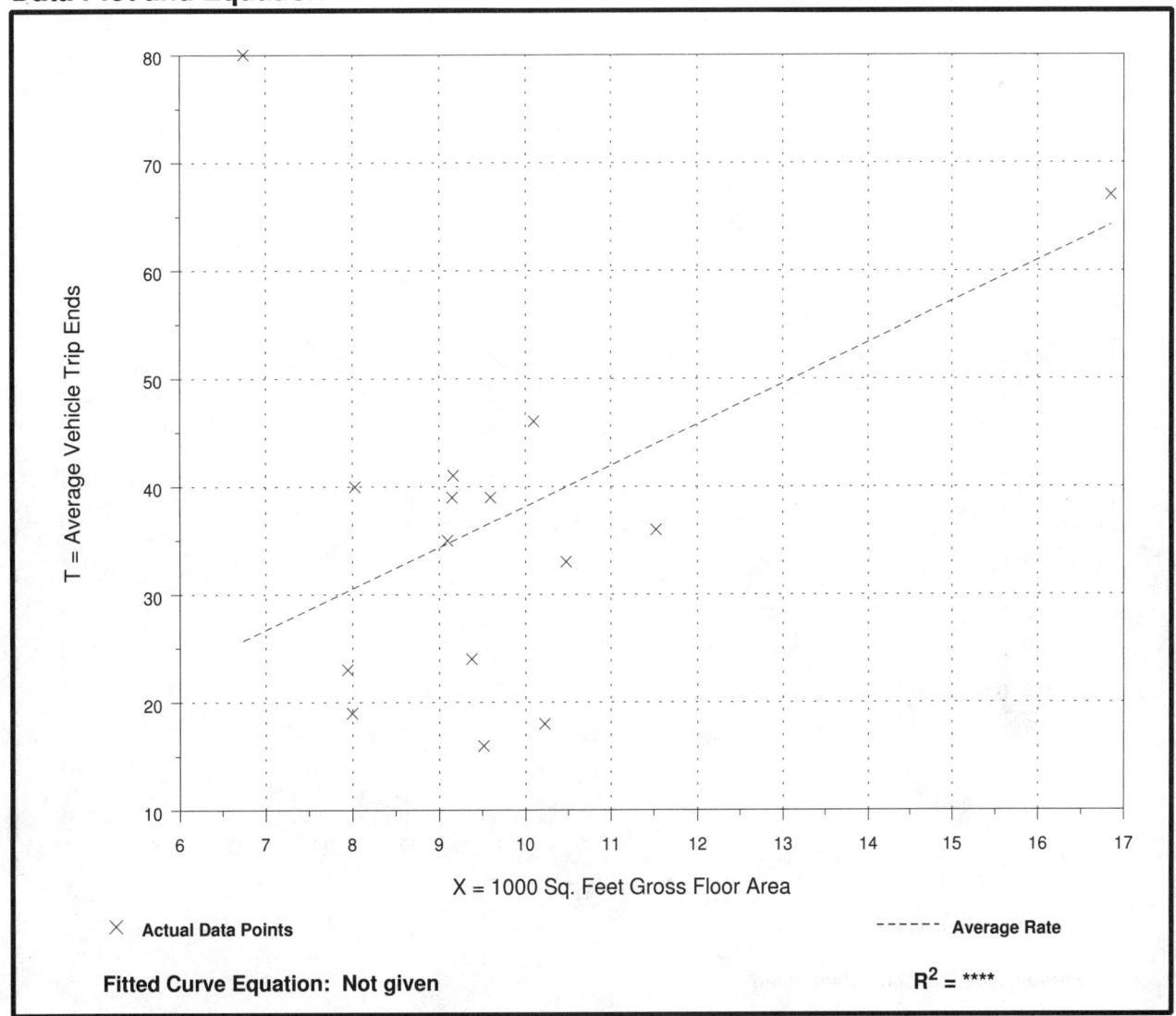

Fitted Curve Equation: Not given $R^2 = ****$

Variety Store
(814)

Average Vehicle Trip Ends vs: 1000 Sq. Feet Gross Floor Area
On a: Weekday,
P.M. Peak Hour of Generator

Number of Studies: 15
Average 1000 Sq. Feet GFA: 10
Directional Distribution: Not available

Trip Generation per 1000 Sq. Feet Gross Floor Area

Average Rate	Range of Rates	Standard Deviation
6.99	3.52 - 13.94	3.76

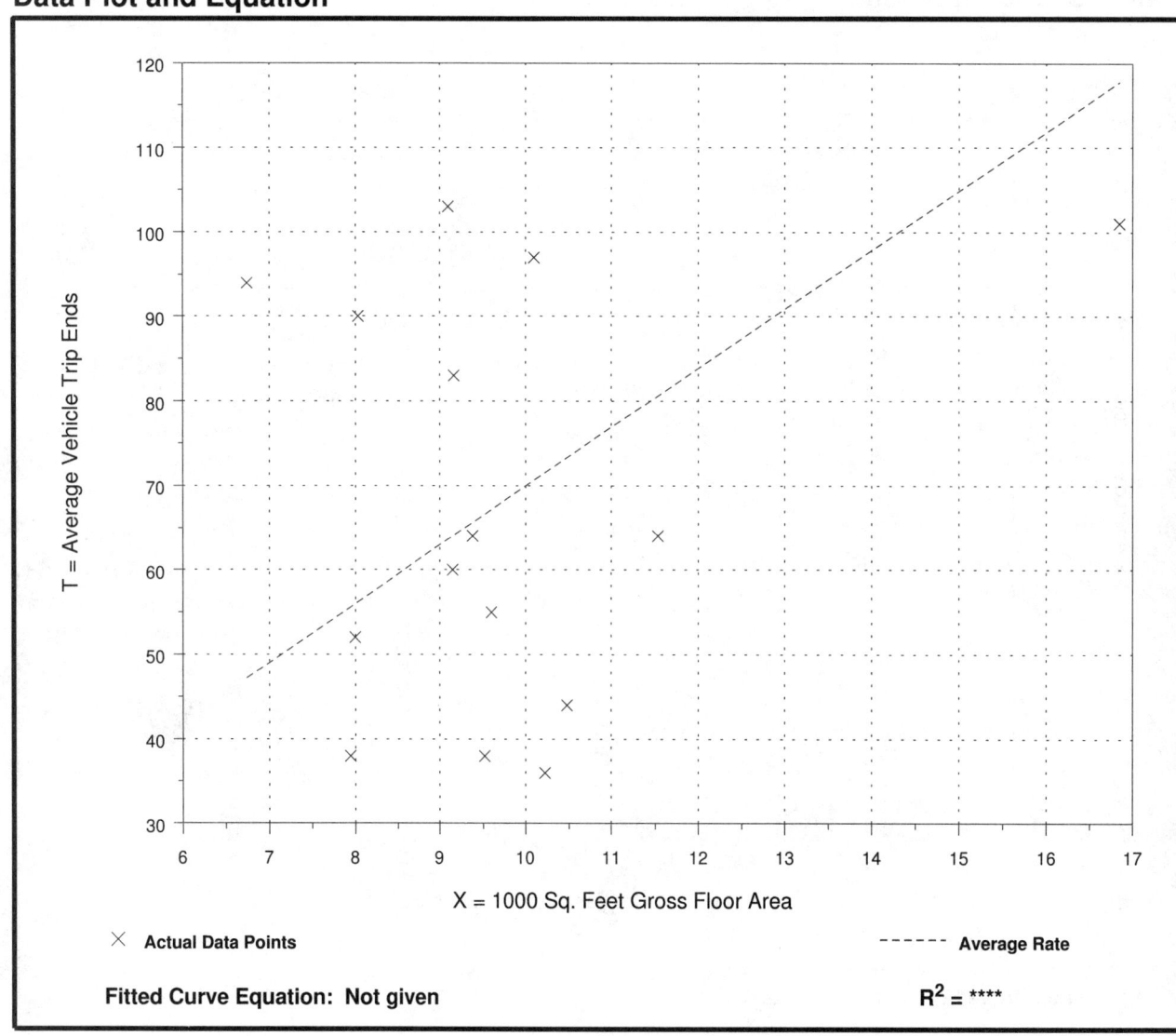

Data Plot and Equation

Fitted Curve Equation: Not given $R^2 = ****$

Land Use: 815
Free-Standing Discount Store

Description

The discount stores in this category are similar to the free-standing discount superstores described in Land Use 813 with the exception that they do not contain a full service grocery department. They are also similar to the department stores described in Land Use 875 with the exception that they generally offer centralized cashiering and sell products that are advertised at discount prices. These stores offer a variety of customer services and typically maintain long store hours 7 days a week. The stores included in this land use are often the only ones on the site, but they can also be found in mutual operation with a related or unrelated garden center and/or service station. Free-standing discount stores are also sometimes found as separate parcels within a retail complex, with or without their own dedicated parking. Free-standing discount superstore (Land Use 813), variety store (Land Use 814) and department store (Land Use 875) are related uses.

Additional Data

Peak hours of the generator—
    The weekday peak hour varied between 10:00 a.m. and 6:00 p.m. The weekend peak hour varied between 11:00 a.m. and 3:00 p.m.

Truck trips accounted for approximately 2 percent of the weekday traffic at one site.

Vehicle occupancy was 1.46 persons per automobile at one of the sites.

Information on approximate hourly variation in free-standing discount store traffic is shown in the following table. It should be noted, however, that the information contained in this table is based on a limited sample size. Therefore, caution should be exercised when applying the data. Also, some information provided in the table may conflict with the results obtained by applying the average rate or regression equations. When this occurs, it is suggested that the results from the average rate or regression equations be used, as they are based on a larger number of studies.

Hourly Variation in Free-Standing Discount Store Traffic			
Time	Average Weekday[a]		Average Saturday[b]	
	Percent of 24-Hour Entering Traffic	Percent of 24-Hour Exiting Traffic	Percent of 24-Hour Entering Traffic	Percent of 24-Hour Exiting Traffic
6 a.m.–7 a.m.	1.2	0.6	0.2	0.3
7 a.m.–8 a.m.	1.7	1.0	1.3	0.4
8 a.m.–9 a.m.	2.5	1.2	1.9	0.9
9 a.m.–10 a.m.	5.5	3.6	6.4	3.9
10 a.m.–11 a.m.	8.7	7.0	9.6	7.3
11 a.m.–12 p.m.	8.2	8.4	11.3	9.7
12 p.m.–1 p.m.	8.9	8.6	10.2	10.6
1 p.m.–2 p.m.	8.7	8.1	9.8	9.0
2 p.m.–3 p.m.	8.7	8.8	9.0	9.2
3 p.m.–4 p.m.	7.7	8.7	8.2	8.9
4 p.m.–5 p.m.	7.8	8.4	7.4	8.1
5 p.m.–6 p.m.	7.2	8.1	6.9	7.9
6 p.m.–7 p.m.	7.4	7.1	6.4	6.8
7 p.m.–8 p.m.	7.0	7.2	5.5	6.3
8 p.m.–9 p.m.	4.7	6.6	4.2	6.4
9 p.m.–10 p.m.	2.1	3.7	0.9	3.2
10 p.m.–6 a.m.	2.1	3.0	0.8	1.2

Sites ranged in size from 87,000 to 117,000 square feet gross floor area
[a] Source numbers – 376, 528 and 595; based on five studies
[b] Source numbers – 376 and 528; based on four studies

Garden centers contained within the principal outside faces of the exterior building walls were included in the gross square floor areas reported. Outdoor or fenced-in areas outside the principal outside faces of the exterior building walls were excluded. Please refer to Volume 1, User's Guide, for a more detailed definition of gross floor area.

The sites were surveyed between the 1970s and the 2000s throughout the United States.

To assist in the future analysis of this land use, it is important to collect and include information on the presence and size of garden centers, outdoor fenced-in space and service stations in trip generation data submissions.

Source Numbers

87, 113, 124, 245, 305, 340, 353, 358, 376, 386, 417, 504, 528, 579, 588, 595, 612, 630, 735

Free-Standing Discount Store
(815)

Average Vehicle Trip Ends vs: 1000 Sq. Feet Gross Floor Area
On a: Weekday

Number of Studies: 25
Average 1000 Sq. Feet GFA: 105
Directional Distribution: 50% entering, 50% exiting

Trip Generation per 1000 Sq. Feet Gross Floor Area

Average Rate	Range of Rates	Standard Deviation
57.24	25.53 - 106.88	19.54

Data Plot and Equation

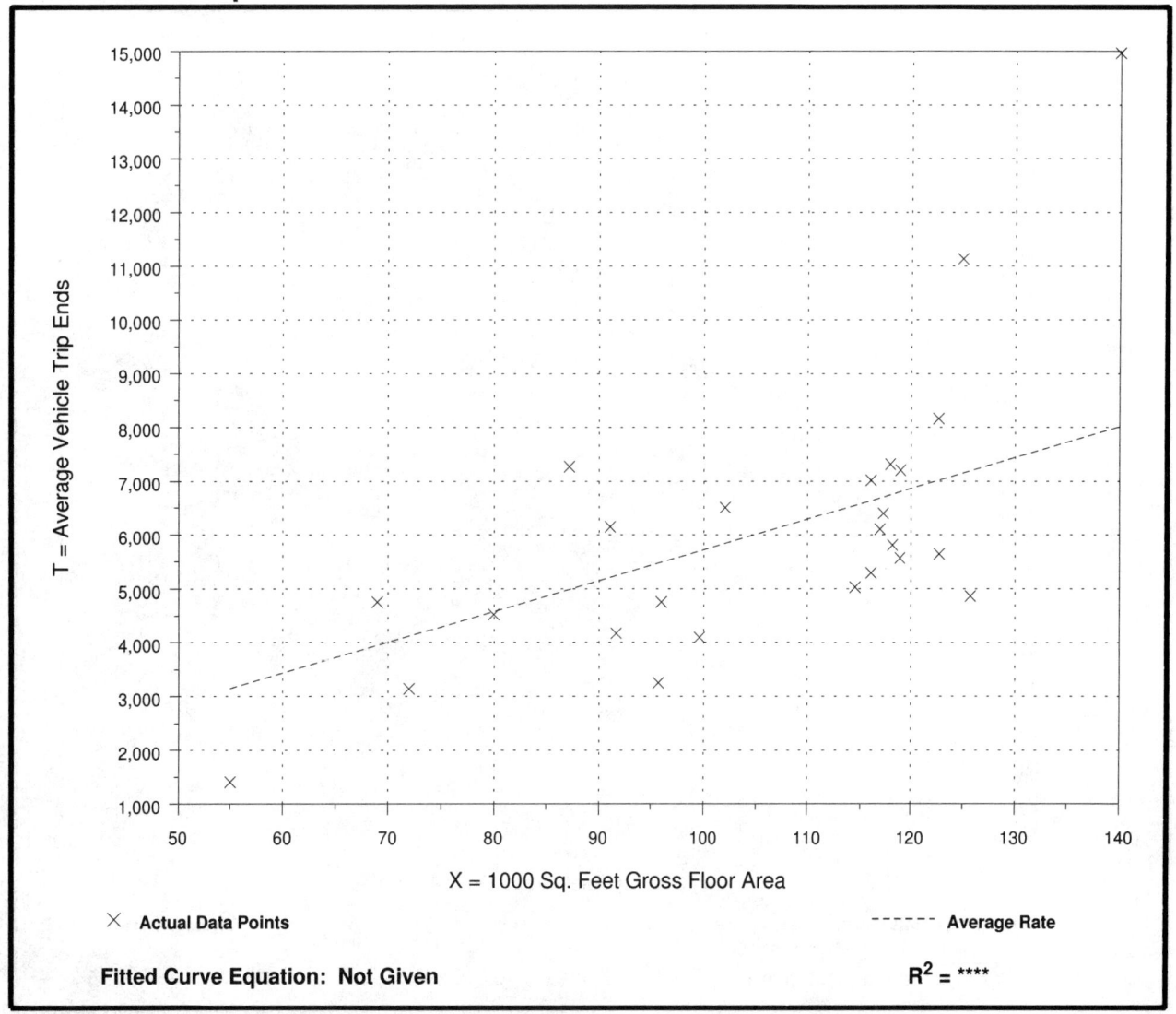

X   Actual Data Points   ----- Average Rate

Fitted Curve Equation: Not Given R^2 = ****

Trip Generation, 9th Edition • Institute of Transportation Engineers

Free-Standing Discount Store
(815)

Average Vehicle Trip Ends vs: 1000 Sq. Feet Gross Floor Area
On a: Weekday,
Peak Hour of Adjacent Street Traffic,
One Hour Between 7 and 9 a.m.

Number of Studies: 7
Average 1000 Sq. Feet GFA: 102
Directional Distribution: 68% entering, 32% exiting

Trip Generation per 1000 Sq. Feet Gross Floor Area

Average Rate	Range of Rates	Standard Deviation
1.06	0.51 - 2.64	1.22

Data Plot and Equation

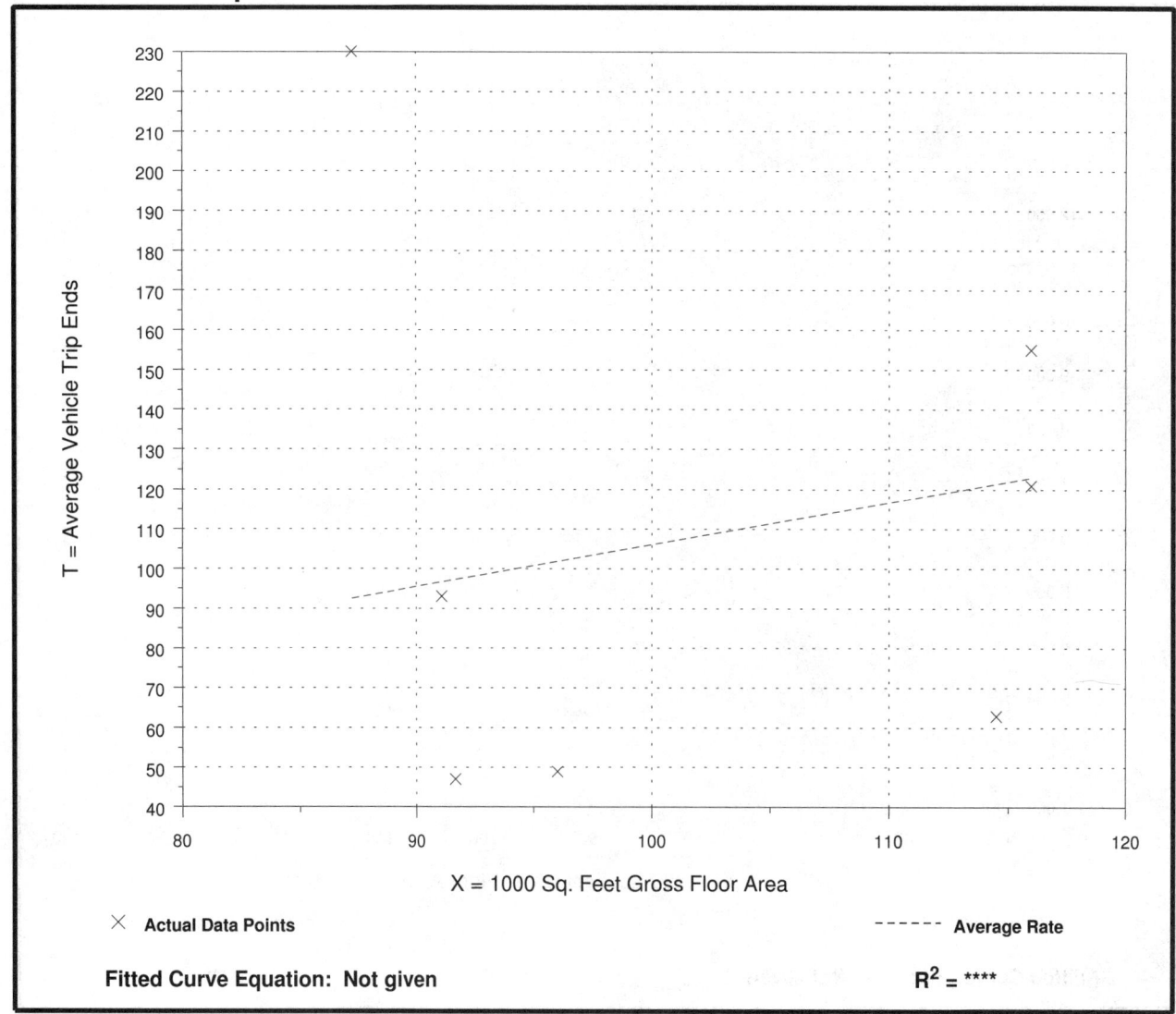

Fitted Curve Equation: Not given $R^2 = ****$

Free-Standing Discount Store
(815)

Average Vehicle Trip Ends vs: 1000 Sq. Feet Gross Floor Area
On a: Weekday,
Peak Hour of Adjacent Street Traffic,
One Hour Between 4 and 6 p.m.

Number of Studies: 53
Average 1000 Sq. Feet GFA: 116
Directional Distribution: 50% entering, 50% exiting

Trip Generation per 1000 Sq. Feet Gross Floor Area

Average Rate	Range of Rates	Standard Deviation
4.98	2.48 - 9.23	2.59

Data Plot and Equation

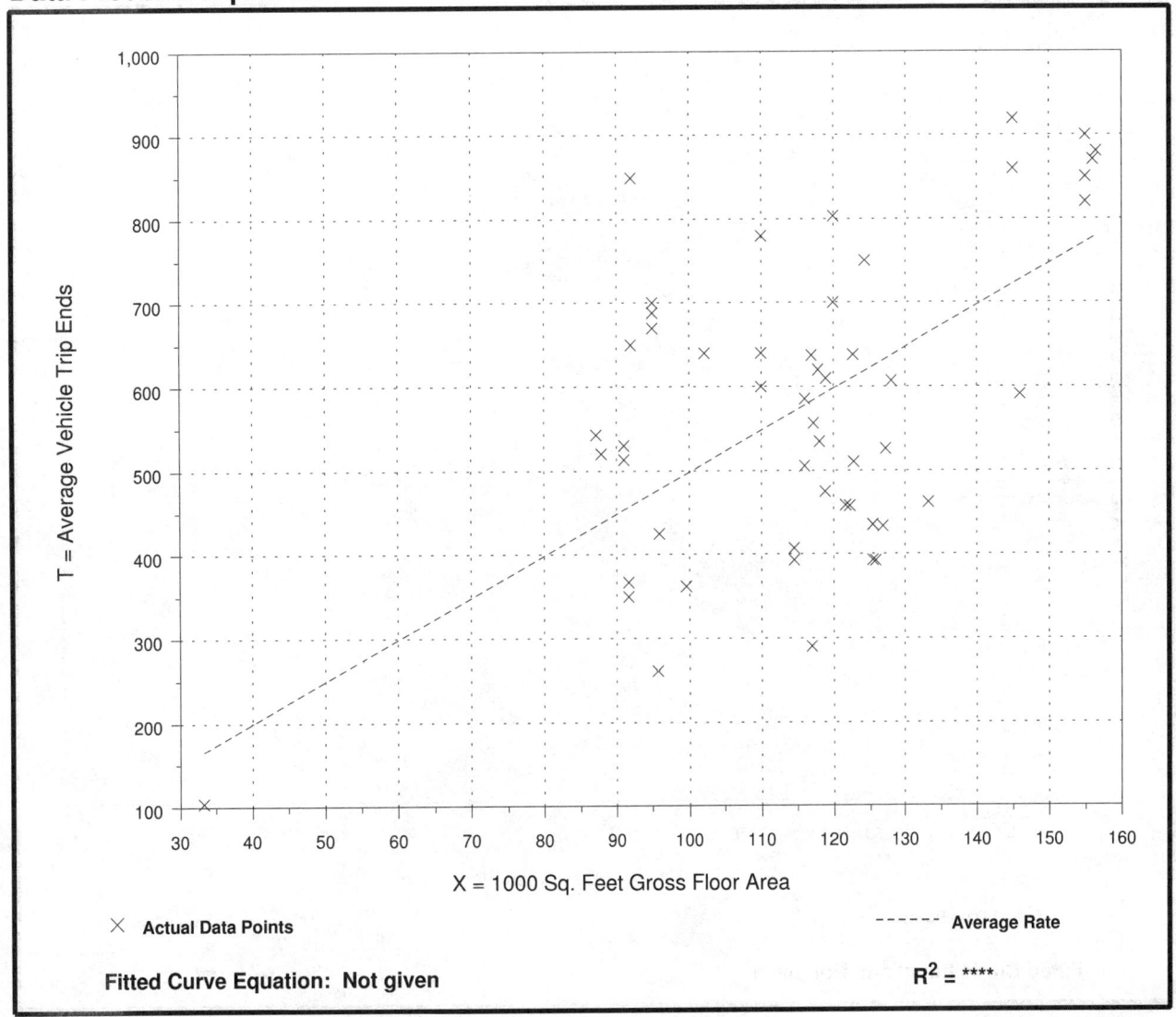

X - Actual Data Points
----- Average Rate

Fitted Curve Equation: Not given $R^2 = ****$

Free-Standing Discount Store
(815)

Average Vehicle Trip Ends vs: 1000 Sq. Feet Gross Floor Area
On a: Weekday,
A.M. Peak Hour of Generator

Number of Studies: 12
Average 1000 Sq. Feet GFA: 92
Directional Distribution: 51% entering, 49% exiting

Trip Generation per 1000 Sq. Feet Gross Floor Area

Average Rate	Range of Rates	Standard Deviation
5.48	2.90 - 9.05	3.03

Data Plot and Equation

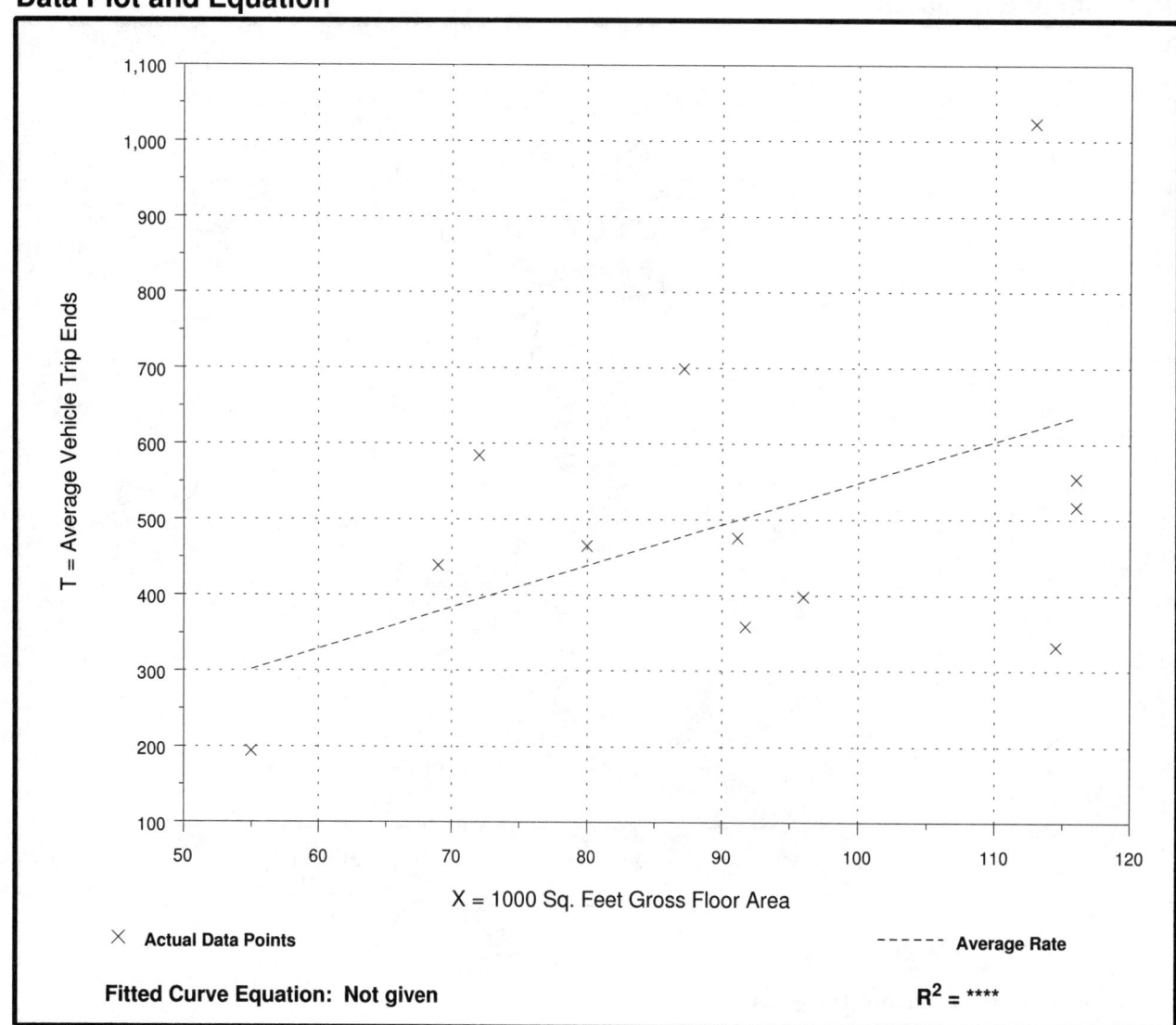

Fitted Curve Equation: Not given $R^2 = ****$

Free-Standing Discount Store
(815)

Average Vehicle Trip Ends vs: 1000 Sq. Feet Gross Floor Area
On a: Weekday,
P.M. Peak Hour of Generator

Number of Studies: 24
Average 1000 Sq. Feet GFA: 103
Directional Distribution: 50% entering, 50% exiting

Trip Generation per 1000 Sq. Feet Gross Floor Area

Average Rate	Range of Rates	Standard Deviation
5.57	3.17 - 9.44	2.82

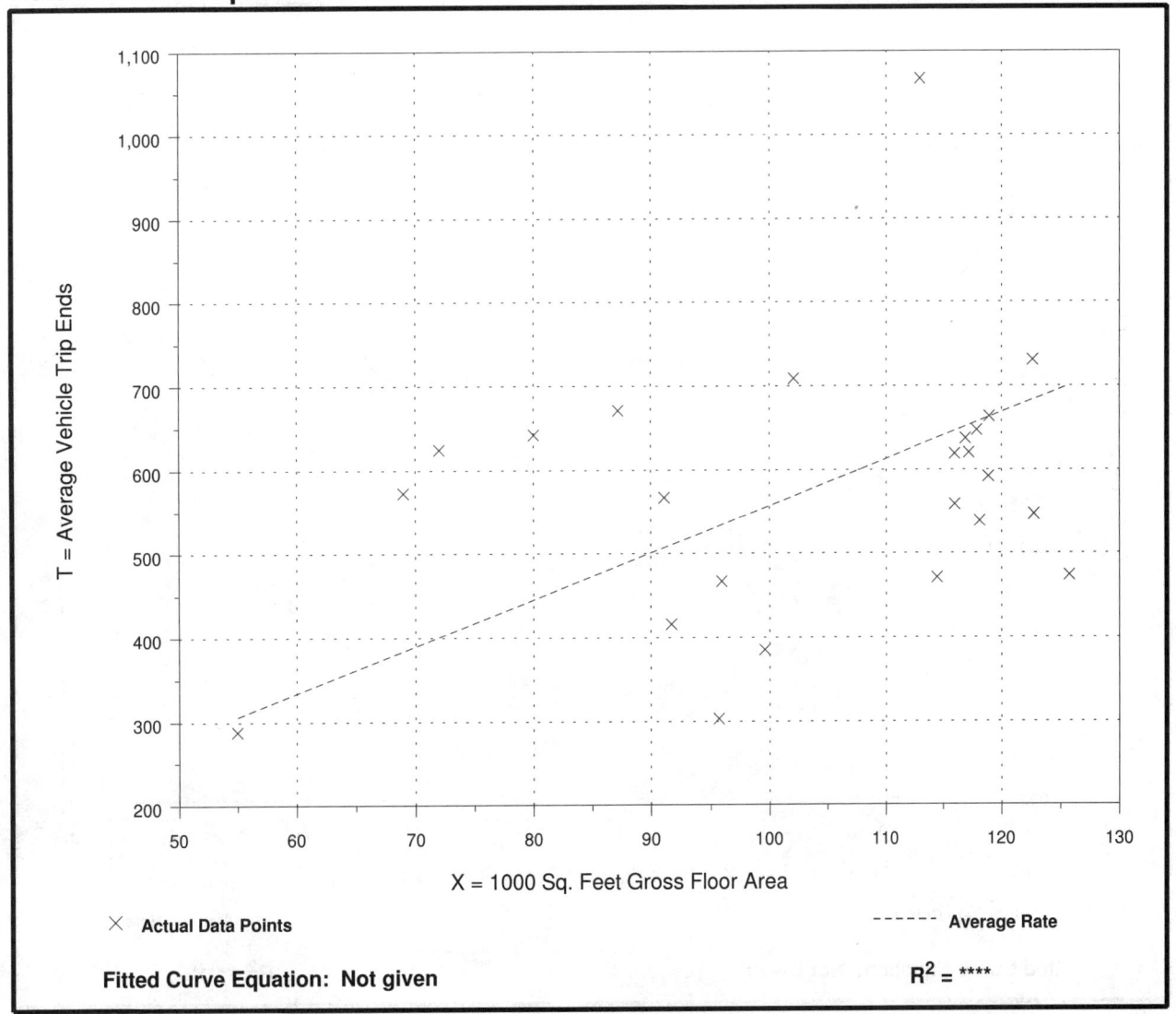

Data Plot and Equation

Fitted Curve Equation: Not given $R^2 = ****$

Free-Standing Discount Store
(815)

Average Vehicle Trip Ends vs: 1000 Sq. Feet Gross Floor Area
On a: Saturday

Number of Studies: 22
Average 1000 Sq. Feet GFA: 104
Directional Distribution: 50% entering, 50% exiting

Trip Generation per 1000 Sq. Feet Gross Floor Area

Average Rate	Range of Rates	Standard Deviation
71.07	45.42 - 92.59	15.44

Data Plot and Equation

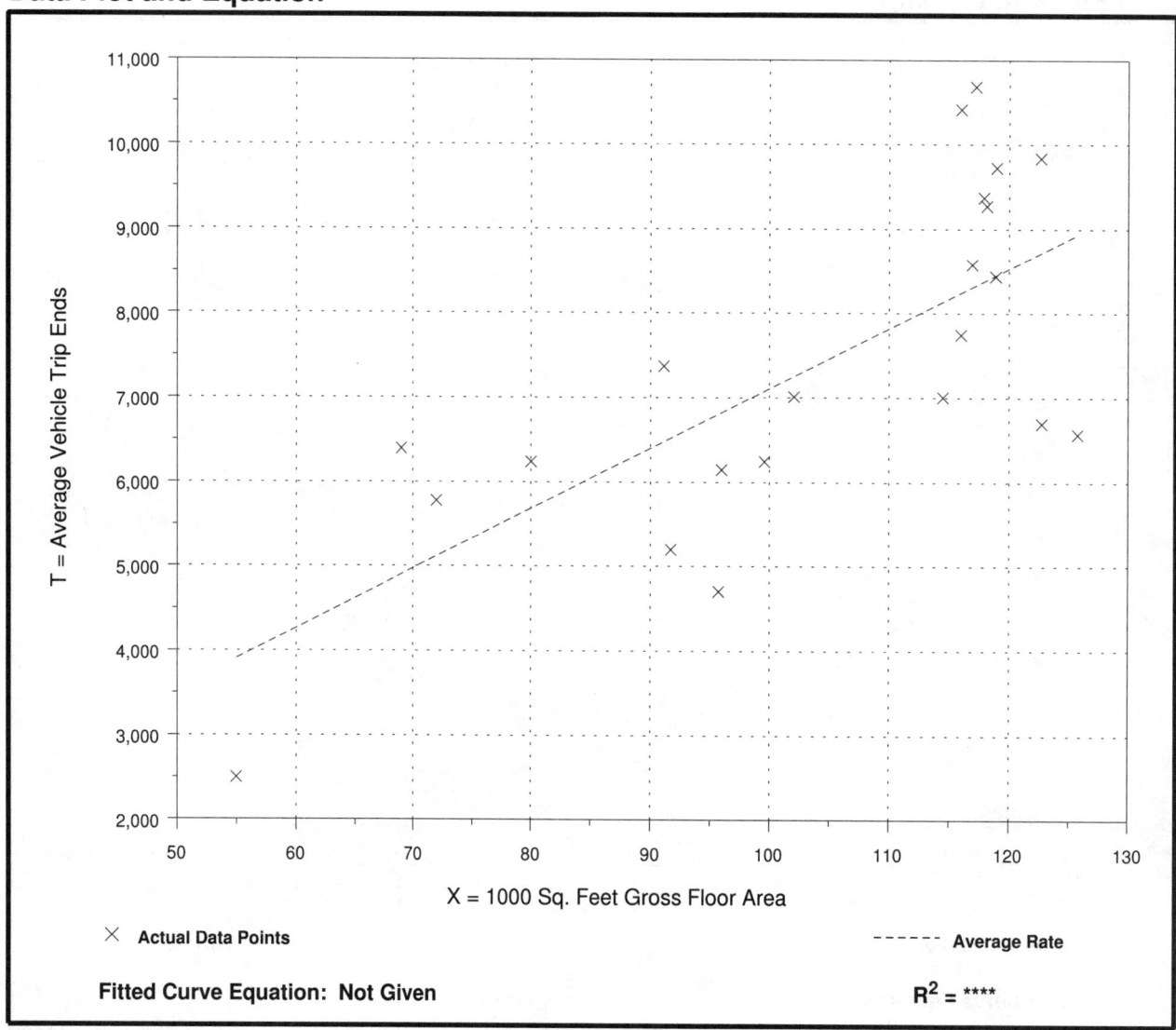

Fitted Curve Equation: Not Given $R^2 = ****$

Free-Standing Discount Store
(815)

Average Vehicle Trip Ends vs: 1000 Sq. Feet Gross Floor Area
On a: Saturday,
Peak Hour of Generator

Number of Studies: 29
Average 1000 Sq. Feet GFA: 114
Directional Distribution: 51% entering, 49% exiting

Trip Generation per 1000 Sq. Feet Gross Floor Area

Average Rate	Range of Rates	Standard Deviation
7.39	4.66 - 10.71	3.10

Data Plot and Equation

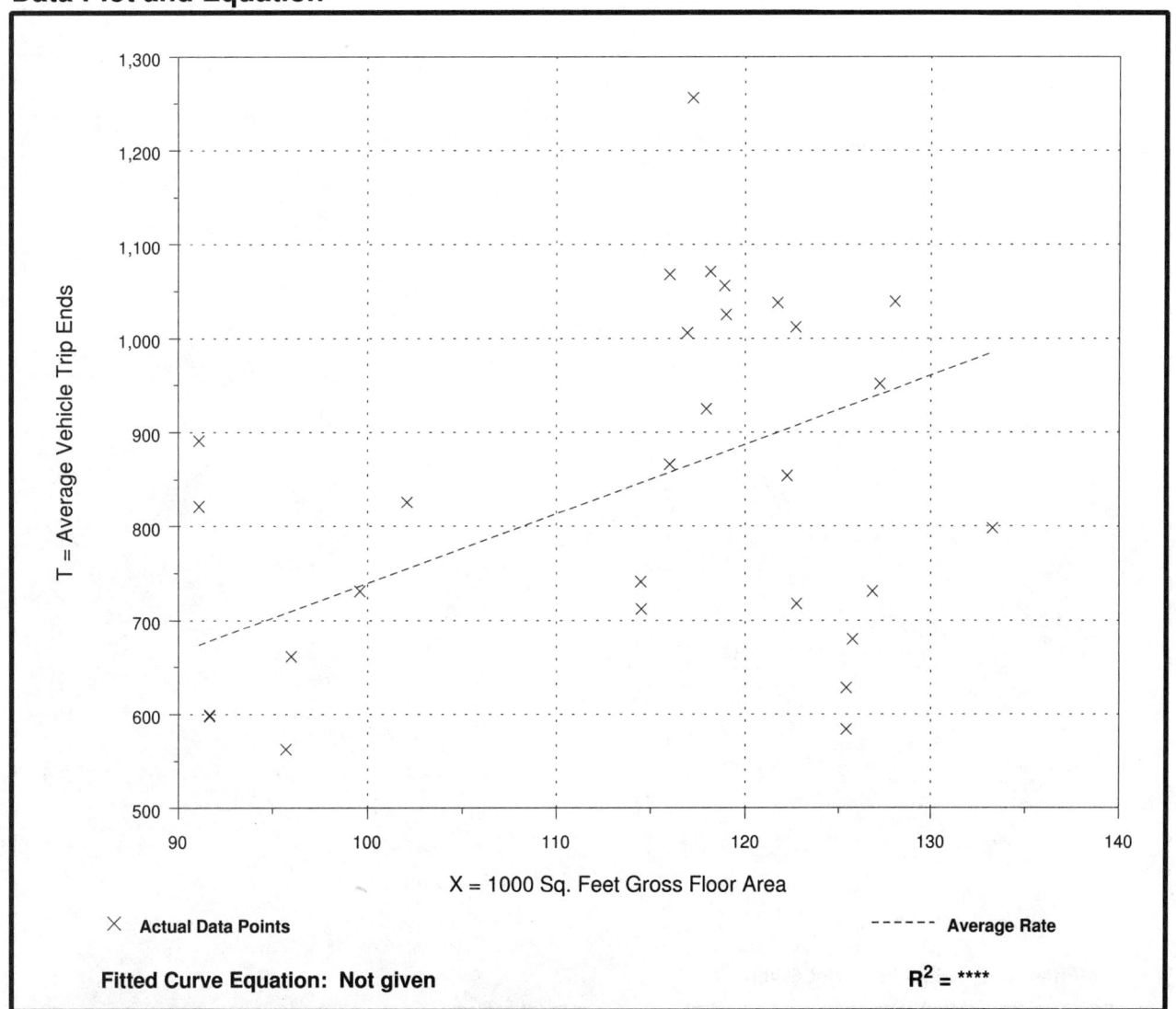

Fitted Curve Equation: Not given $R^2 = ****$

Free-Standing Discount Store
(815)

Average Vehicle Trip Ends vs: 1000 Sq. Feet Gross Floor Area
On a: Sunday

Number of Studies: 4
Average 1000 Sq. Feet GFA: 108
Directional Distribution: 50% entering, 50% exiting

Trip Generation per 1000 Sq. Feet Gross Floor Area

Average Rate	Range of Rates	Standard Deviation
56.36	42.95 - 64.27	10.89

Data Plot and Equation

Caution - Use Carefully - Small Sample Size

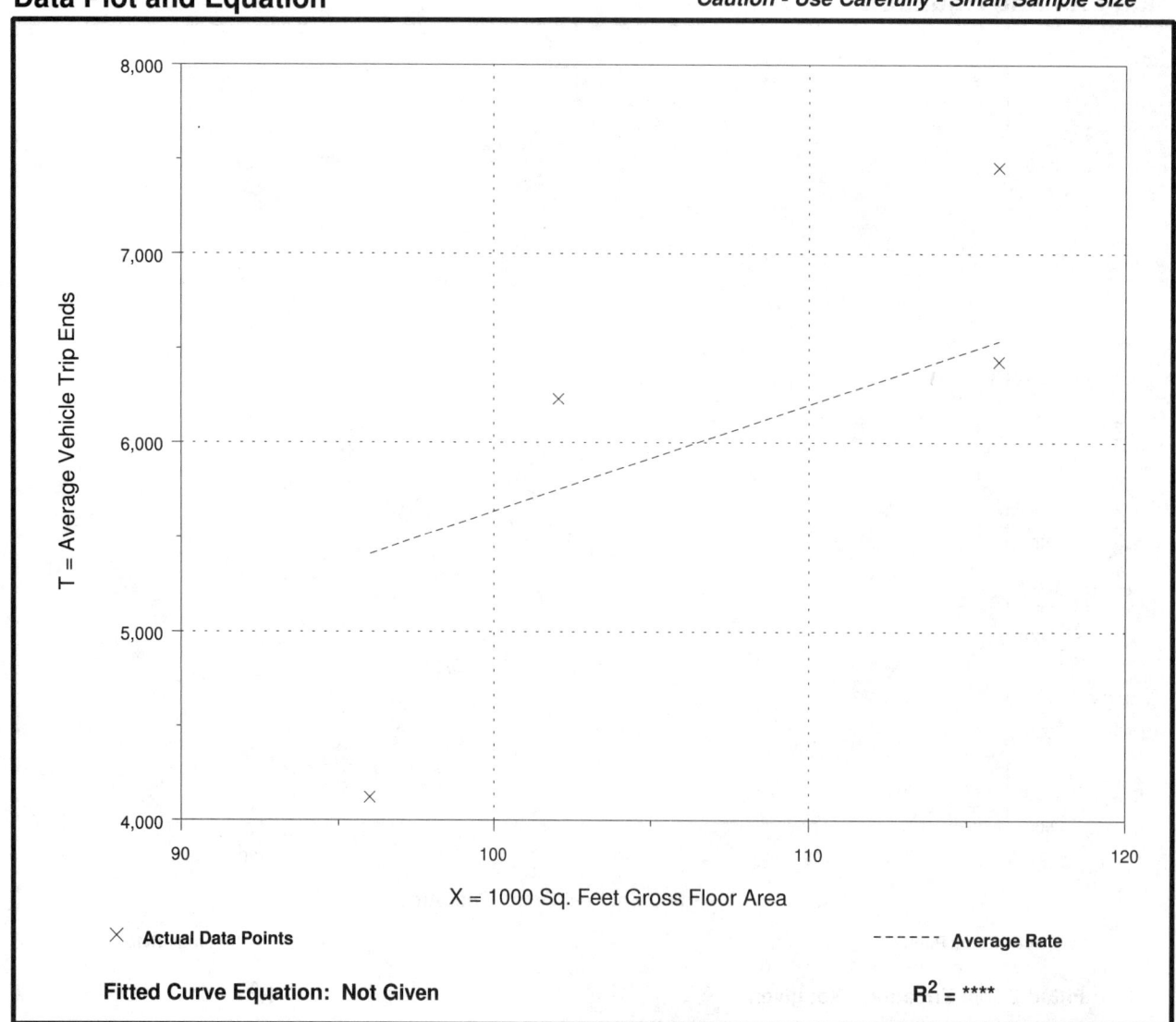

Fitted Curve Equation: Not Given $R^2 = ****$

Free-Standing Discount Store
(815)

Average Vehicle Trip Ends vs: 1000 Sq. Feet Gross Floor Area
On a: Sunday,
Peak Hour of Generator

Number of Studies: 3
Average 1000 Sq. Feet GFA: 109
Directional Distribution: 52% entering, 48% exiting

Trip Generation per 1000 Sq. Feet Gross Floor Area

Average Rate	Range of Rates	Standard Deviation
7.32	6.01 - 8.61	2.90

Data Plot and Equation

Caution - Use Carefully - Small Sample Size

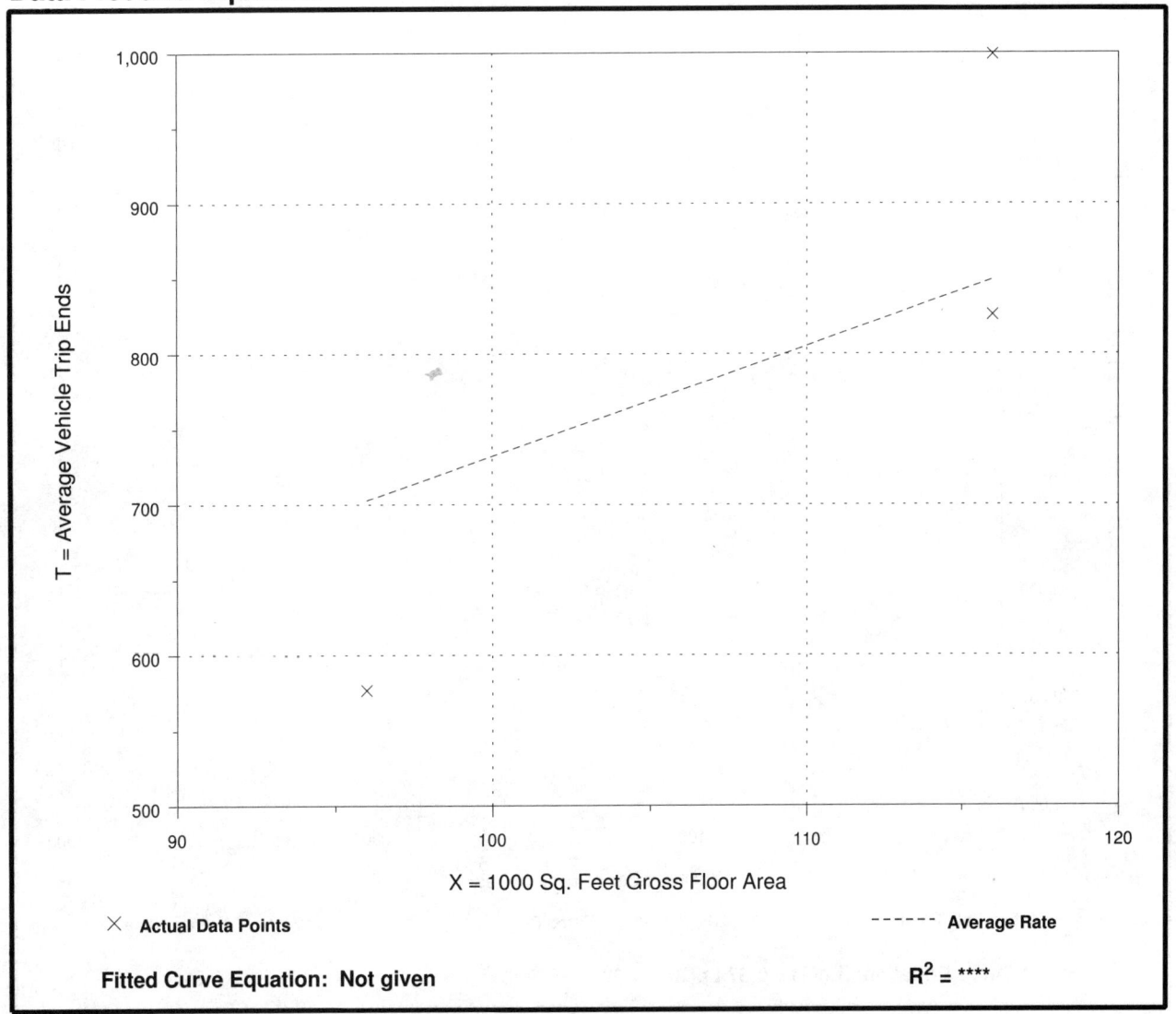

Fitted Curve Equation: Not given $R^2 = ****$

Free-Standing Discount Store
(815)

Average Vehicle Trip Ends vs: Employees
On a: Weekday

Number of Studies: 7
Avg. Number of Employees: 153
Directional Distribution: 50% entering, 50% exiting

Trip Generation per Employee

Average Rate	Range of Rates	Standard Deviation
28.84	21.18 - 35.46	7.28

Data Plot and Equation

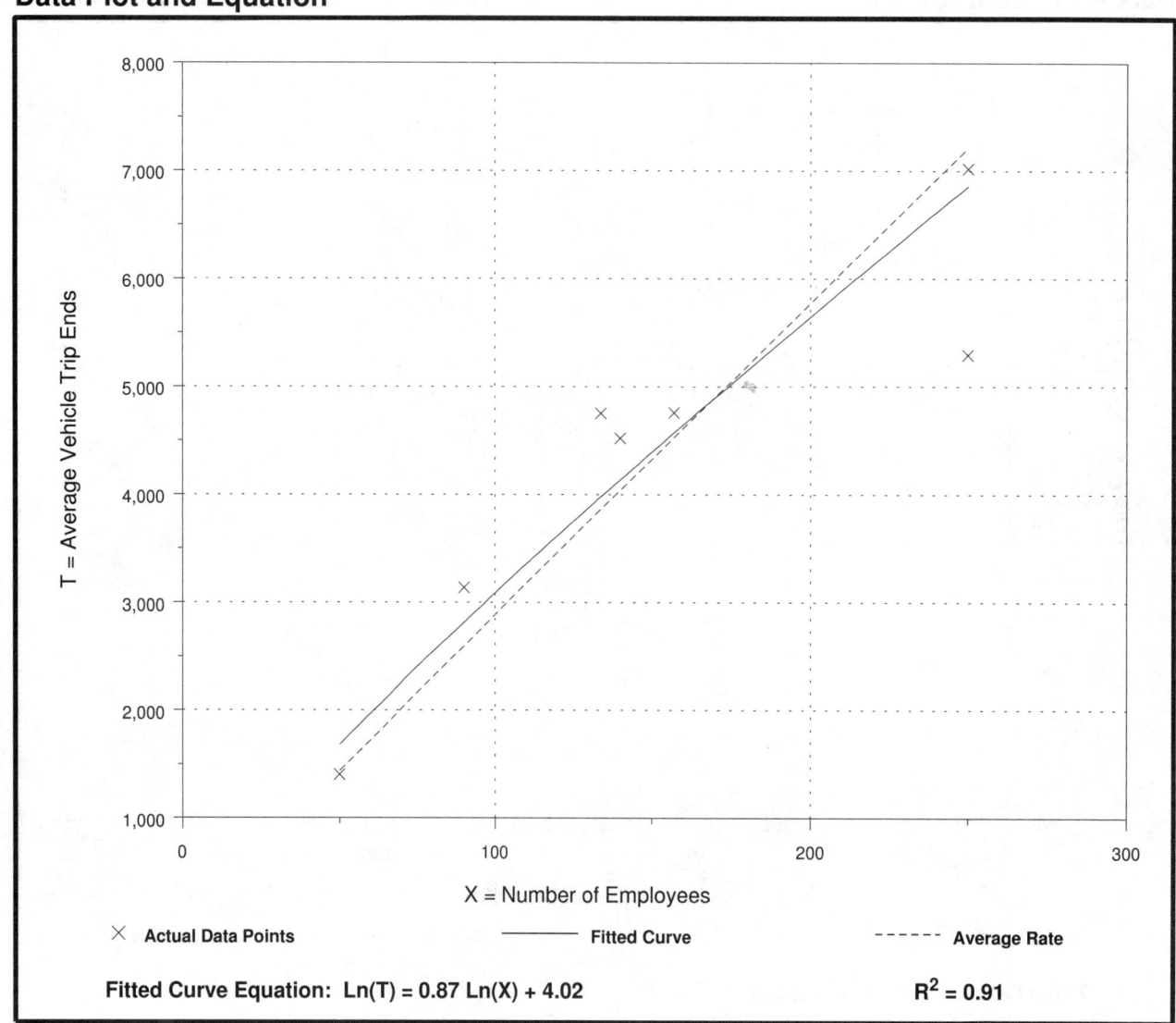

Fitted Curve Equation: $Ln(T) = 0.87\ Ln(X) + 4.02$ $R^2 = 0.91$

Free-Standing Discount Store
(815)

Average Vehicle Trip Ends vs: Employees
On a: Weekday,
Peak Hour of Adjacent Street Traffic,
One Hour Between 7 and 9 a.m.

Number of Studies: 3
Avg. Number of Employees: 211
Directional Distribution: 66% entering, 34% exiting

Trip Generation per Employee

Average Rate	Range of Rates	Standard Deviation
0.51	0.37 - 0.62	0.72

Data Plot and Equation

Caution - Use Carefully - Small Sample Size

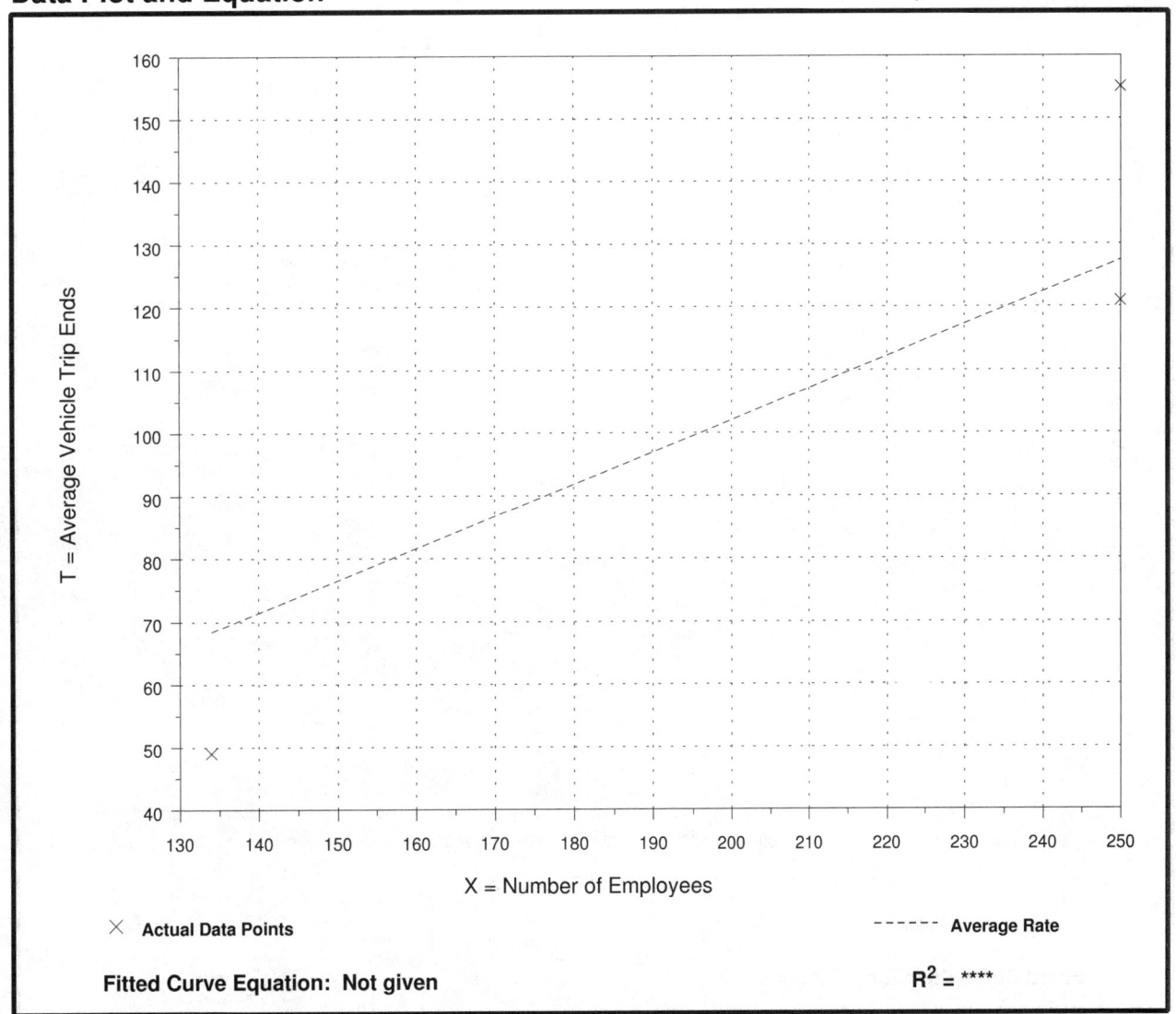

Fitted Curve Equation: Not given $R^2 = ****$

Free-Standing Discount Store
(815)

Average Vehicle Trip Ends vs: Employees
On a: Weekday,
Peak Hour of Adjacent Street Traffic,
One Hour Between 4 and 6 p.m.

Number of Studies: 21
Avg. Number of Employees: 212
Directional Distribution: 50% entering, 50% exiting

Trip Generation per Employee

Average Rate	Range of Rates	Standard Deviation
3.48	2.02 - 4.77	1.96

Data Plot and Equation

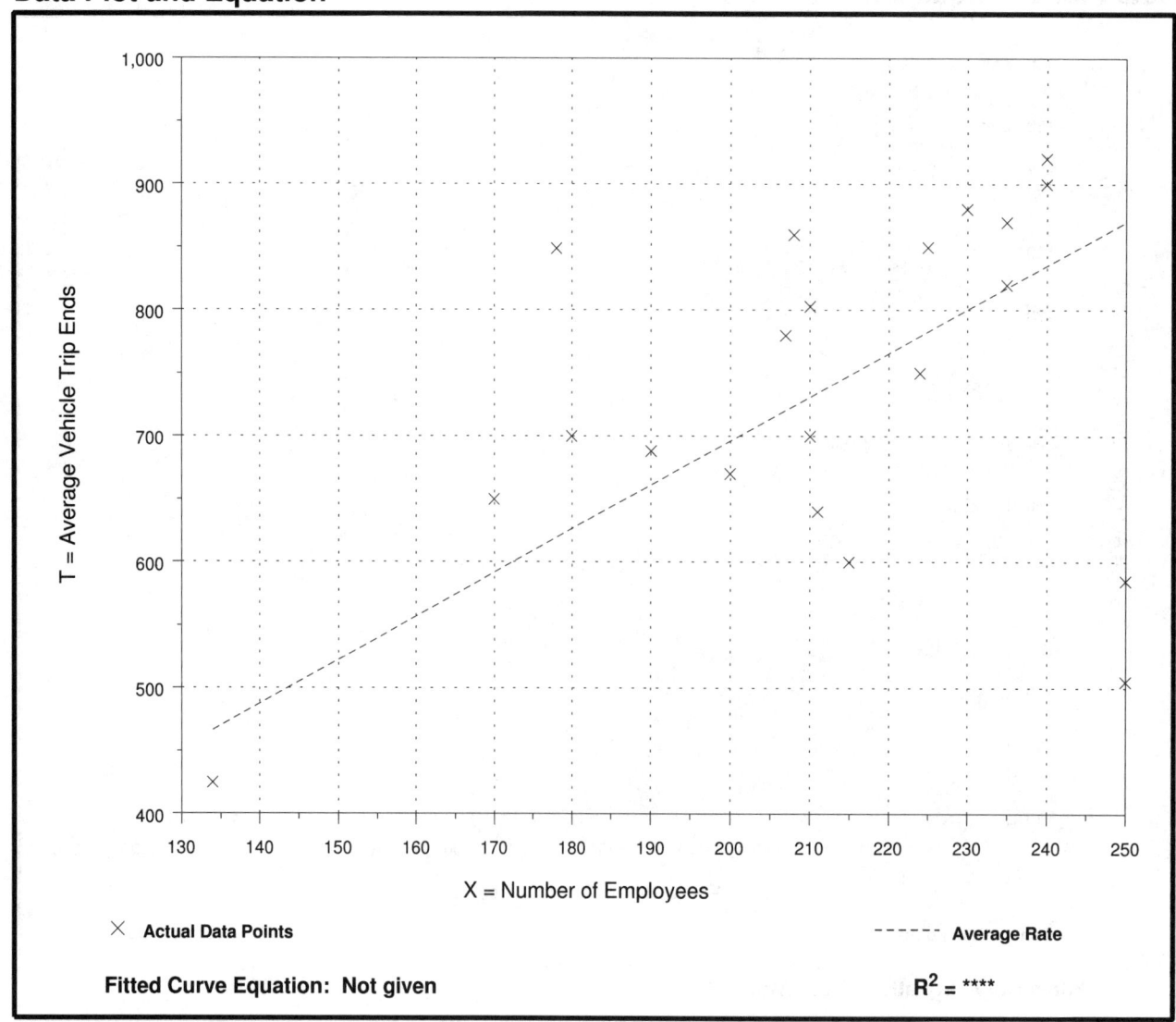

Fitted Curve Equation: Not given $R^2 = ****$

Free-Standing Discount Store
(815)

Average Vehicle Trip Ends vs: Employees
On a: Weekday,
A.M. Peak Hour of Generator

Number of Studies: 7
Avg. Number of Employees: 153
Directional Distribution: 45% entering, 55% exiting

Trip Generation per Employee

Average Rate	Range of Rates	Standard Deviation
2.94	2.07 - 6.49	2.08

Data Plot and Equation

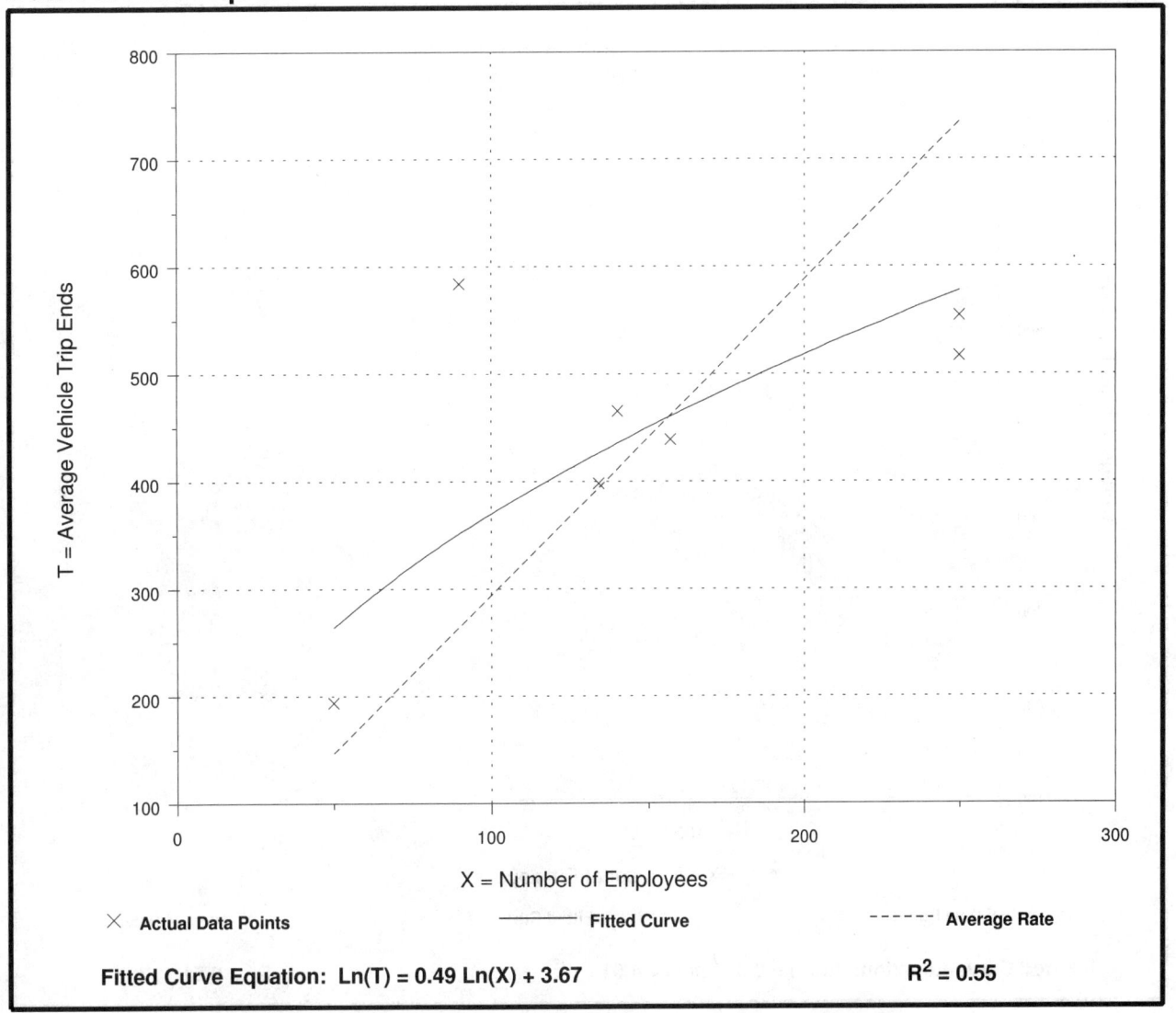

Fitted Curve Equation: $Ln(T) = 0.49 \, Ln(X) + 3.67$ $R^2 = 0.55$

Free-Standing Discount Store
(815)

Average Vehicle Trip Ends vs: Employees
On a: Weekday,
P.M. Peak Hour of Generator

Number of Studies: 7
Avg. Number of Employees: 153
Directional Distribution: 52% entering, 48% exiting

Trip Generation per Employee

Average Rate	Range of Rates	Standard Deviation
3.52	2.24 - 6.93	2.35

Data Plot and Equation

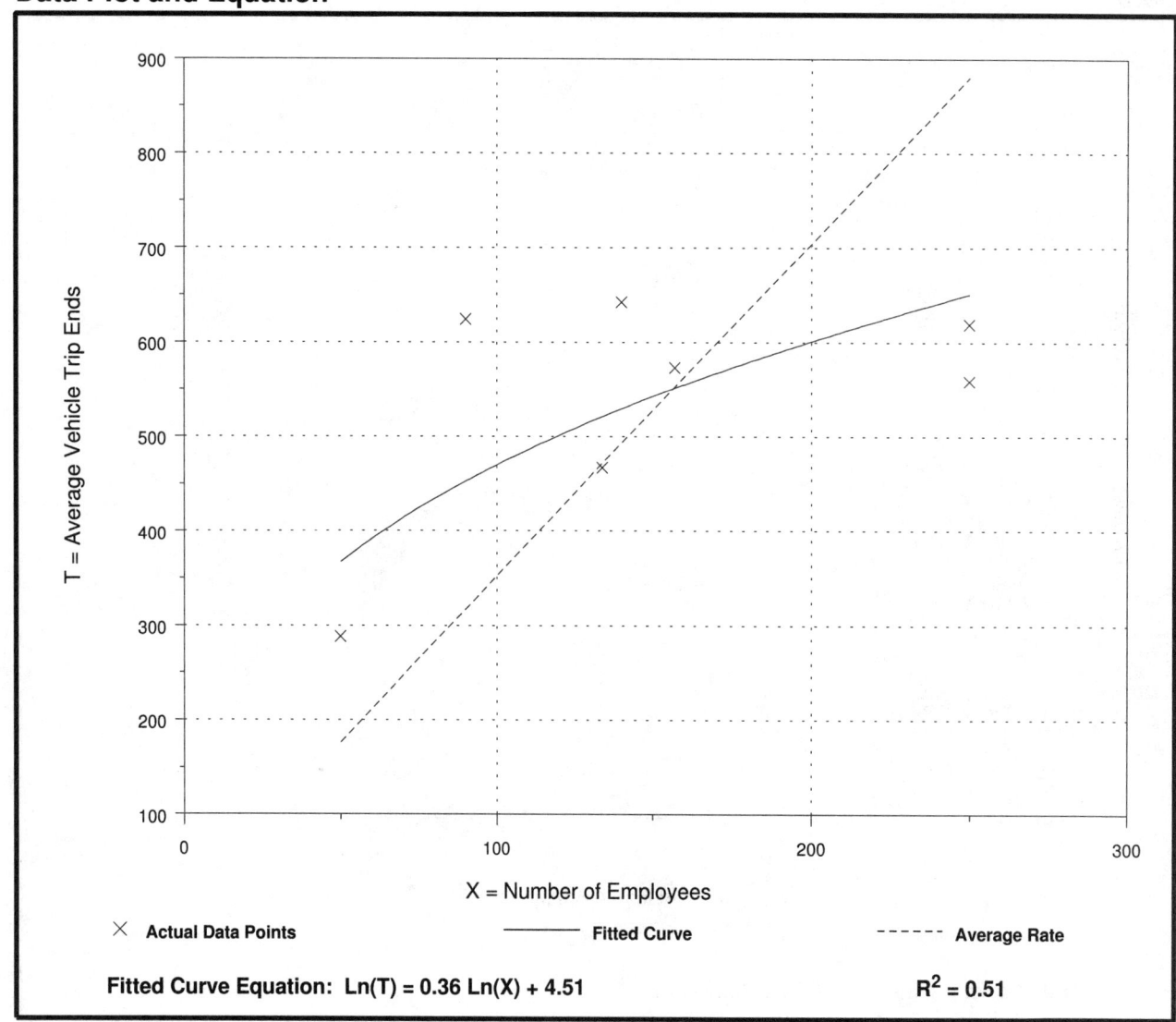

Fitted Curve Equation: $Ln(T) = 0.36\, Ln(X) + 4.51$ $R^2 = 0.51$

Free-Standing Discount Store
(815)

Average Vehicle Trip Ends vs: Employees
On a: Saturday

Number of Studies: 7
Avg. Number of Employees: 153
Directional Distribution: 50% entering, 50% exiting

Trip Generation per Employee

Average Rate	Range of Rates	Standard Deviation
42.20	30.98 - 64.19	10.85

Data Plot and Equation

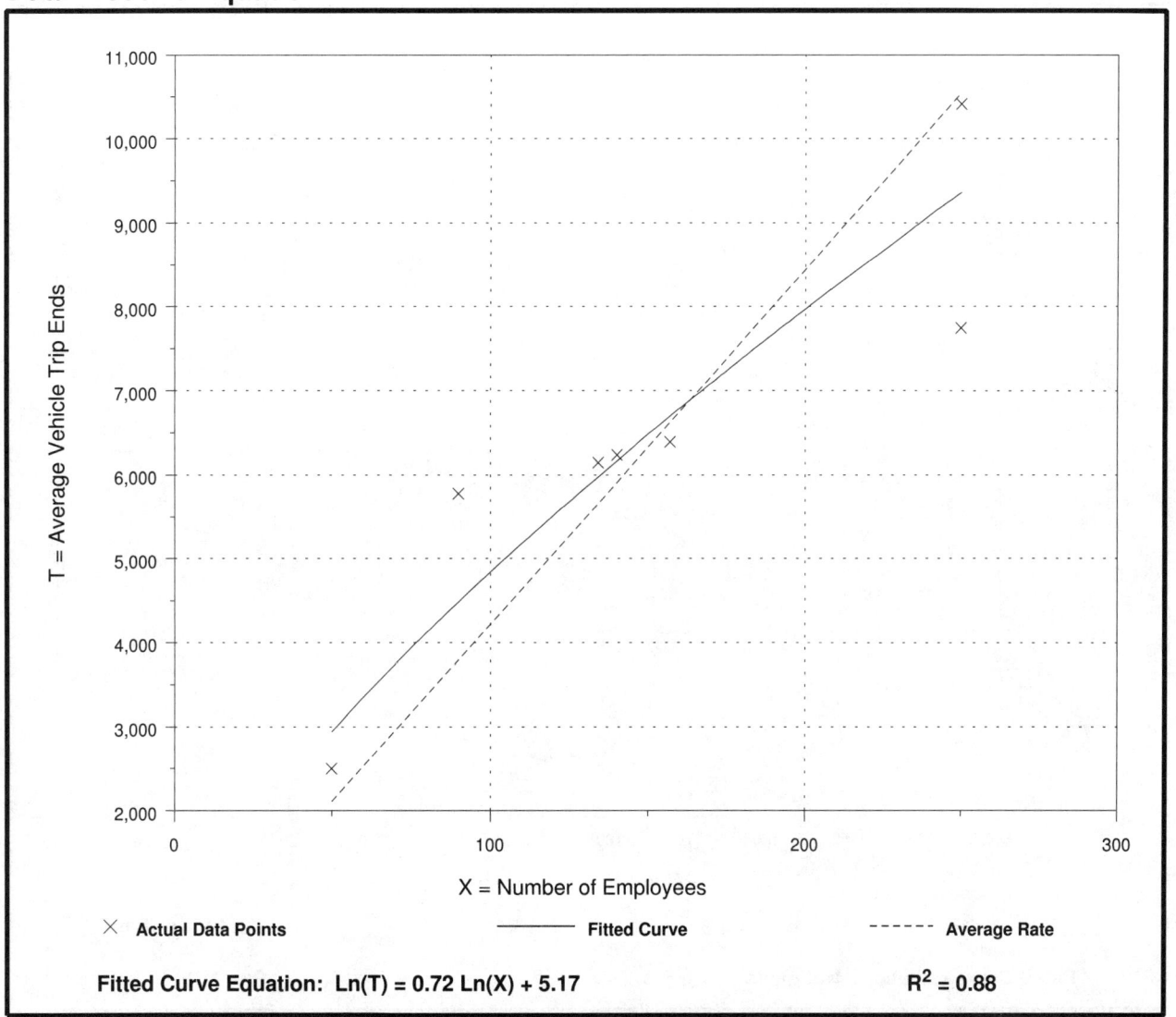

Fitted Curve Equation: $Ln(T) = 0.72\, Ln(X) + 5.17$ $\qquad R^2 = 0.88$

Free-Standing Discount Store
(815)

Average Vehicle Trip Ends vs: Employees
On a: Saturday,
Peak Hour of Generator

Number of Studies: 3
Avg. Number of Employees: 211
Directional Distribution: 54% entering, 46% exiting

Trip Generation per Employee

Average Rate	Range of Rates	Standard Deviation
4.09	3.46 - 4.93	2.10

Data Plot and Equation

Caution - Use Carefully - Small Sample Size

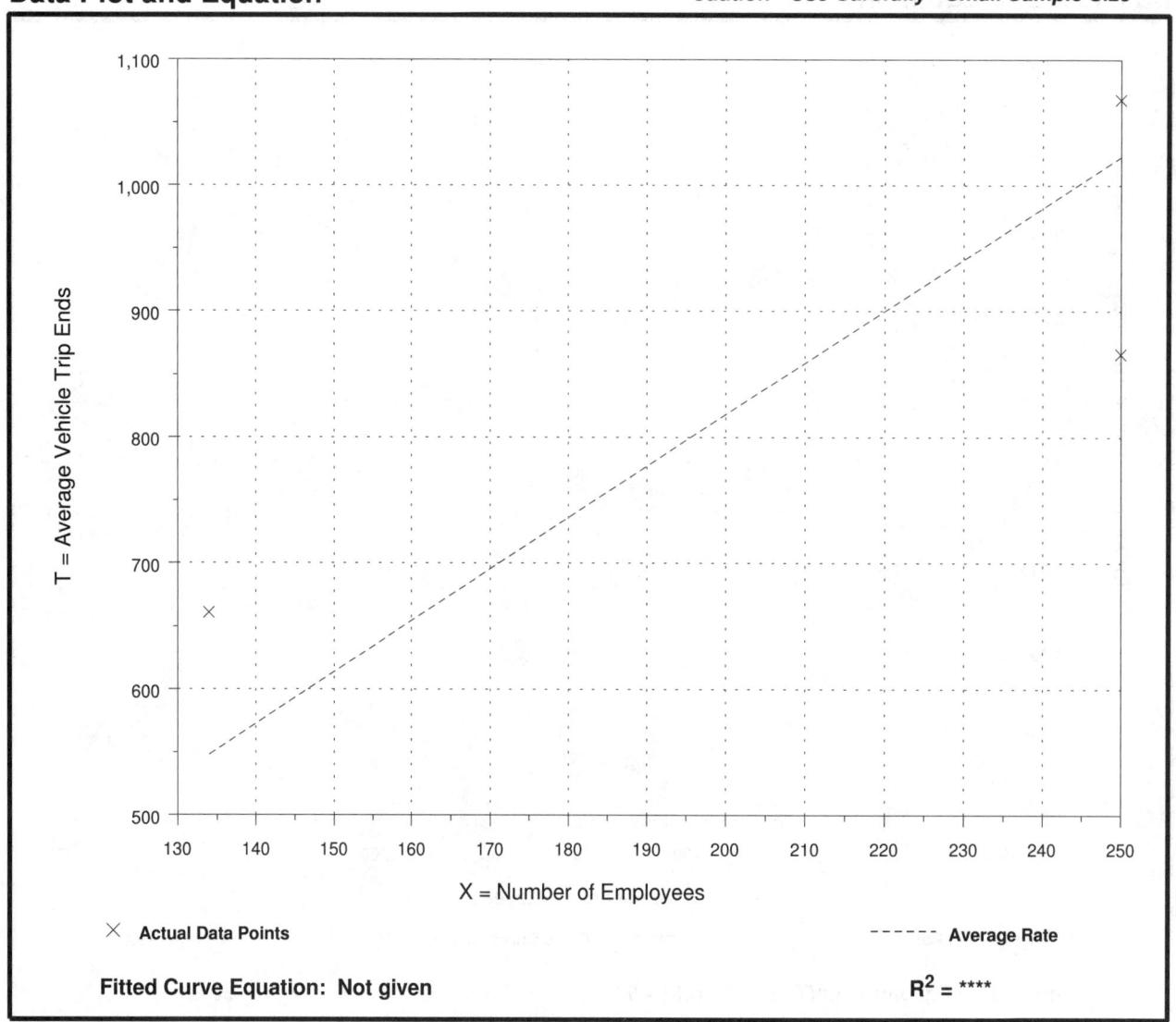

Fitted Curve Equation: Not given $R^2 = ****$

Free-Standing Discount Store
(815)

Average Vehicle Trip Ends vs: Employees
On a: Sunday

Number of Studies: 3
Avg. Number of Employees: 211
Directional Distribution: 50% entering, 50% exiting

Trip Generation per Employee

Average Rate	Range of Rates	Standard Deviation
28.40	25.72 - 30.77	5.76

Data Plot and Equation

Caution - Use Carefully - Small Sample Size

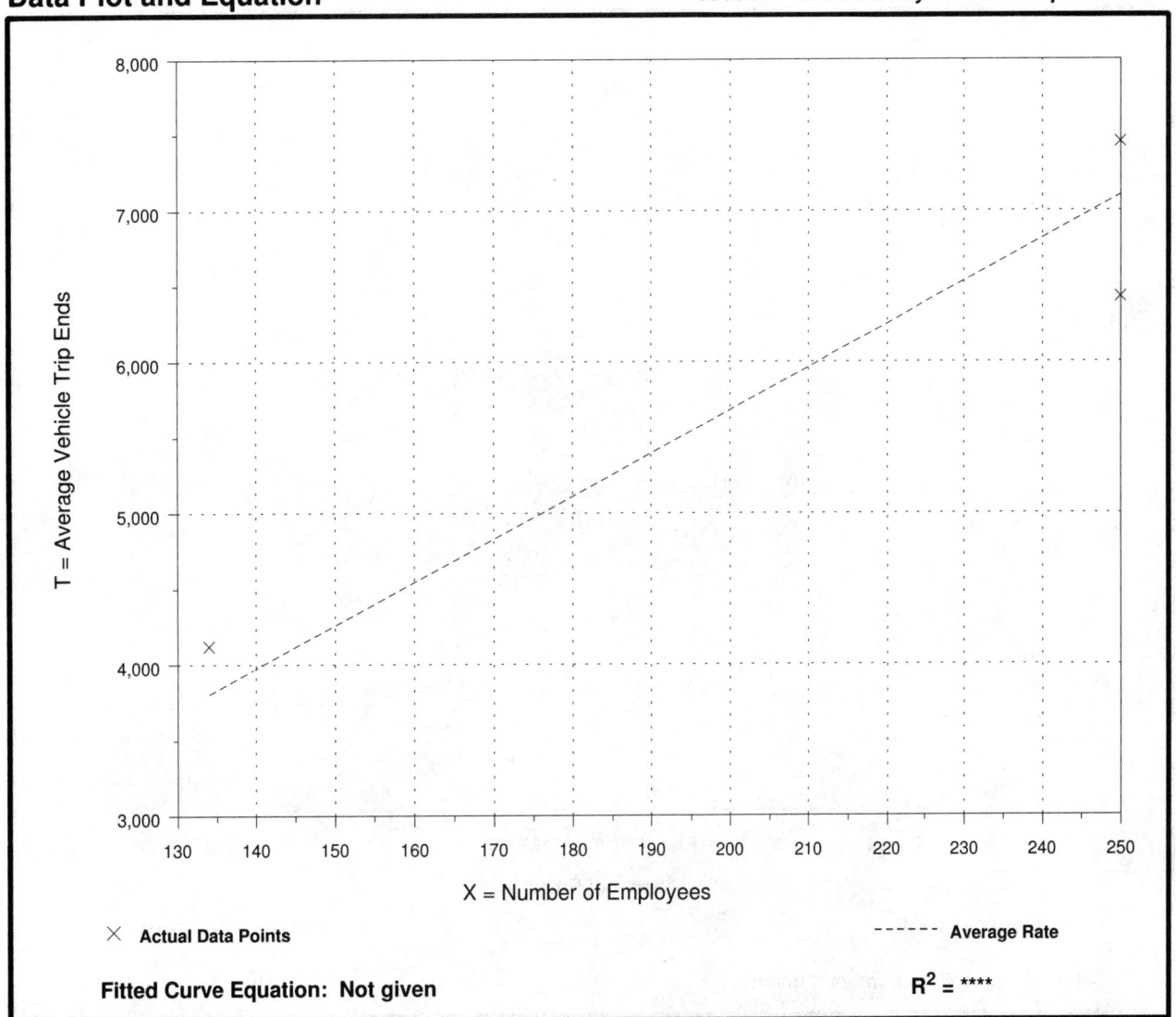

Fitted Curve Equation: Not given $R^2 = ****$

Free-Standing Discount Store
(815)

Average Vehicle Trip Ends vs: Employees
On a: Sunday,
Peak Hour of Generator

Number of Studies: 3
Avg. Number of Employees: 211
Directional Distribution: 52% entering, 48% exiting

Trip Generation per Employee

Average Rate	Range of Rates	Standard Deviation
3.79	3.30 - 4.31	1.99

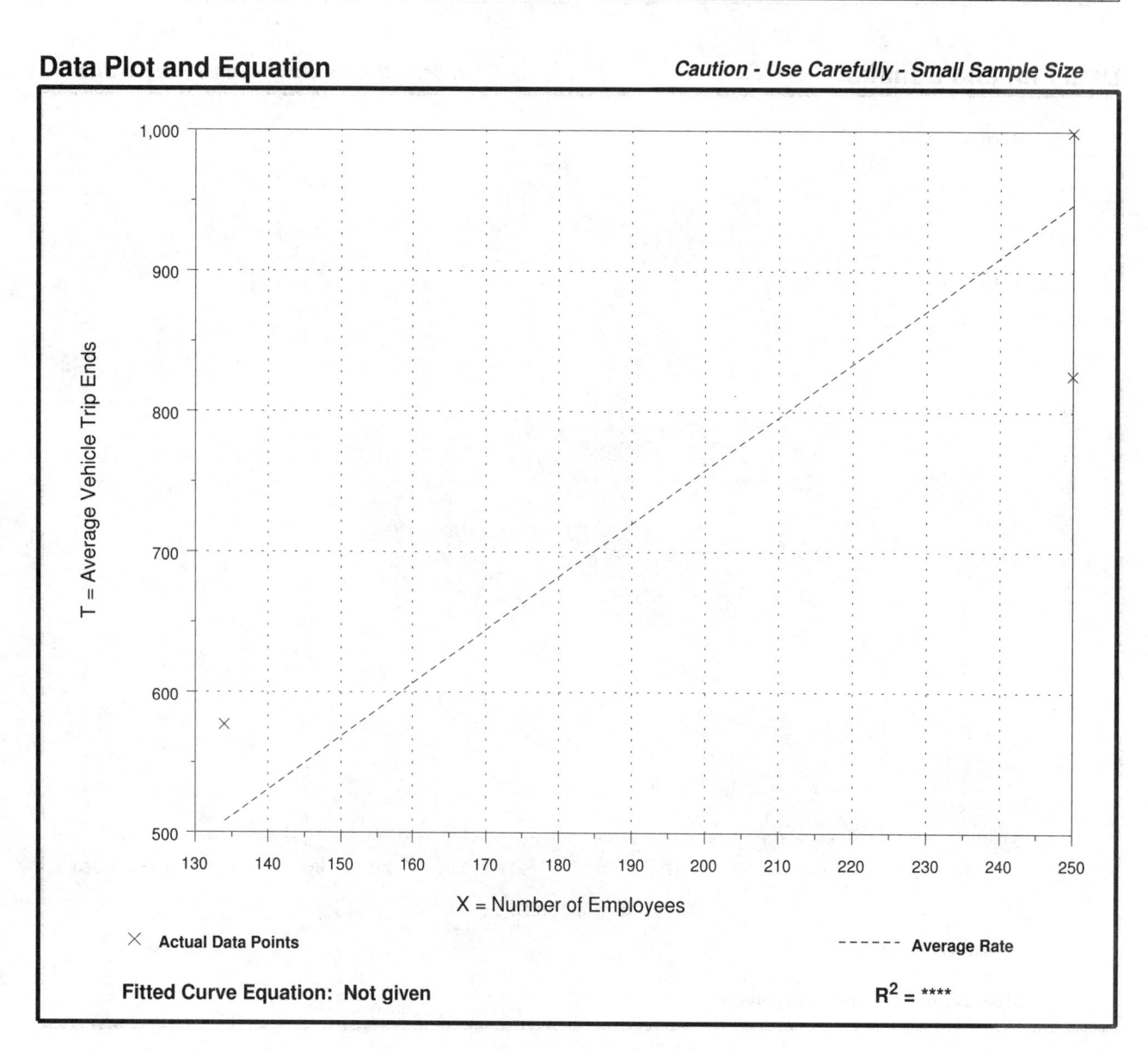

Data Plot and Equation

Caution - Use Carefully - Small Sample Size

Fitted Curve Equation: Not given $R^2 = ****$

Land Use: 816
Hardware/Paint Store

Description

Hardware/paint stores are generally free-standing buildings. Building materials and lumber store (Land Use 812) and home improvement superstore (Land Use 862) are related uses.

Additional Data

Truck trips accounted for approximately 1 to 3 percent of the weekday traffic at the sites surveyed. The average for all sites surveyed was approximately 2 percent.

Vehicle occupancy ranged from 1.15 to 1.39 persons per automobile on an average weekday. The average for all sites surveyed was 1.31 persons per automobile.

Peak hours of the generator—
    The weekday peak hour varied between 10:00 a.m. and 4:00 p.m. The weekend peak hour varied between 11:00 a.m. and 2:00 p.m.

To assist in the future analysis of this land use, it is important to specify if the store is a hardware store or paint store.

The sites were surveyed between the late 1970s and the 1990s in California, Oregon, South Dakota and Washington.

Source Numbers

90, 304, 358, 531

Hardware/Paint Store
(816)

Average Vehicle Trip Ends vs: Employees
On a: Weekday

Number of Studies: 3
Avg. Number of Employees: 27
Directional Distribution: 50% entering, 50% exiting

Trip Generation per Employee

Average Rate	Range of Rates	Standard Deviation
53.21	45.94 - 58.21	8.33

Data Plot and Equation

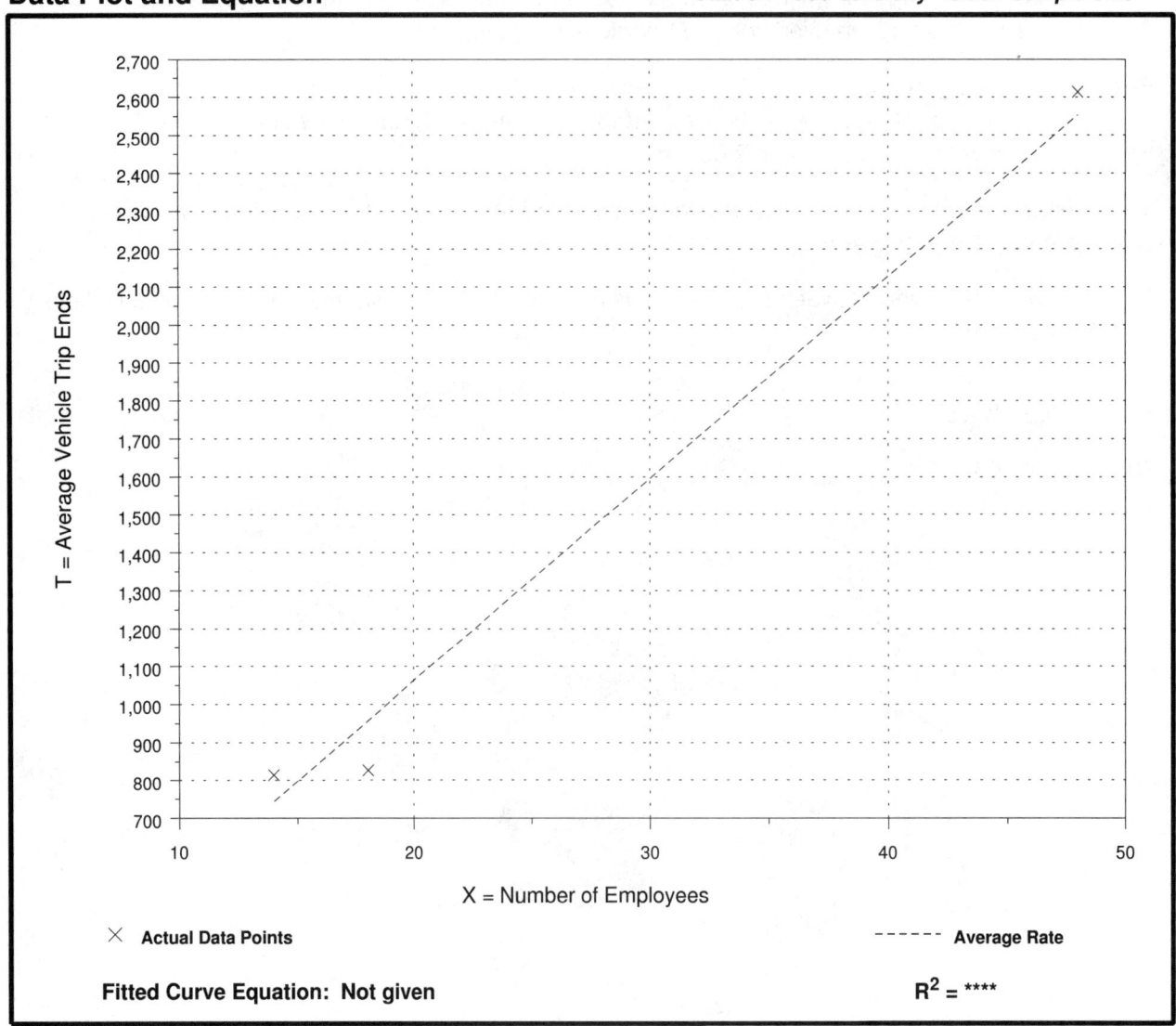

Caution - Use Carefully - Small Sample Size

Fitted Curve Equation: Not given R^2 = ****

Hardware/Paint Store
(816)

Average Vehicle Trip Ends vs: Employees
On a: Weekday,
Peak Hour of Adjacent Street Traffic,
One Hour Between 7 and 9 a.m.

Number of Studies: 3
Avg. Number of Employees: 27
Directional Distribution: Not available

Trip Generation per Employee

Average Rate	Range of Rates	Standard Deviation
1.13	0.52 - 2.33	1.30

Data Plot and Equation

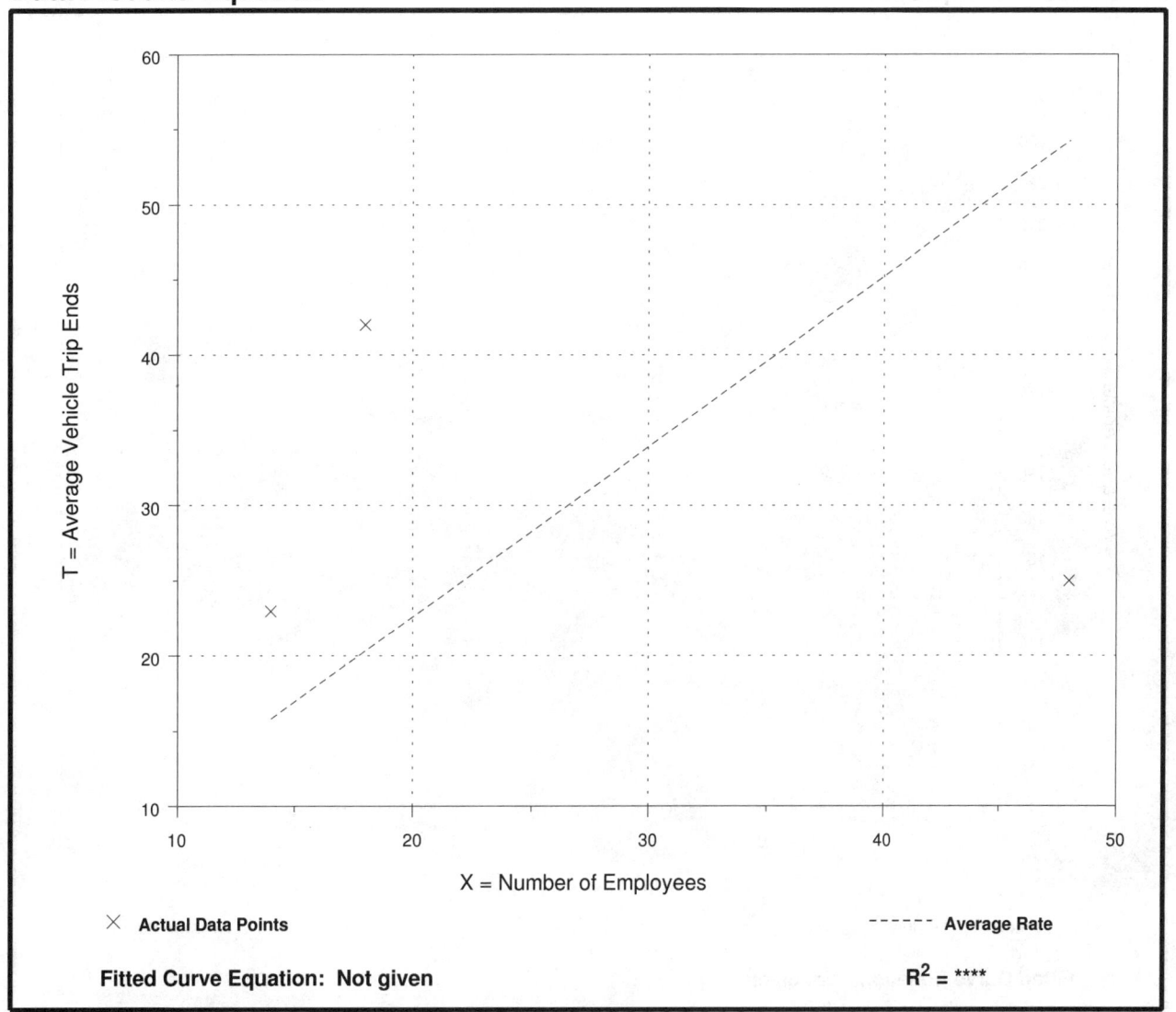

Caution - Use Carefully - Small Sample Size

Fitted Curve Equation: Not given $R^2 = ****$

Hardware/Paint Store
(816)

Average Vehicle Trip Ends vs: Employees
On a: Weekday,
Peak Hour of Adjacent Street Traffic,
One Hour Between 4 and 6 p.m.

Number of Studies: 3
Avg. Number of Employees: 27
Directional Distribution: Not available

Trip Generation per Employee

Average Rate	Range of Rates	Standard Deviation
5.05	4.83 - 5.79	2.24

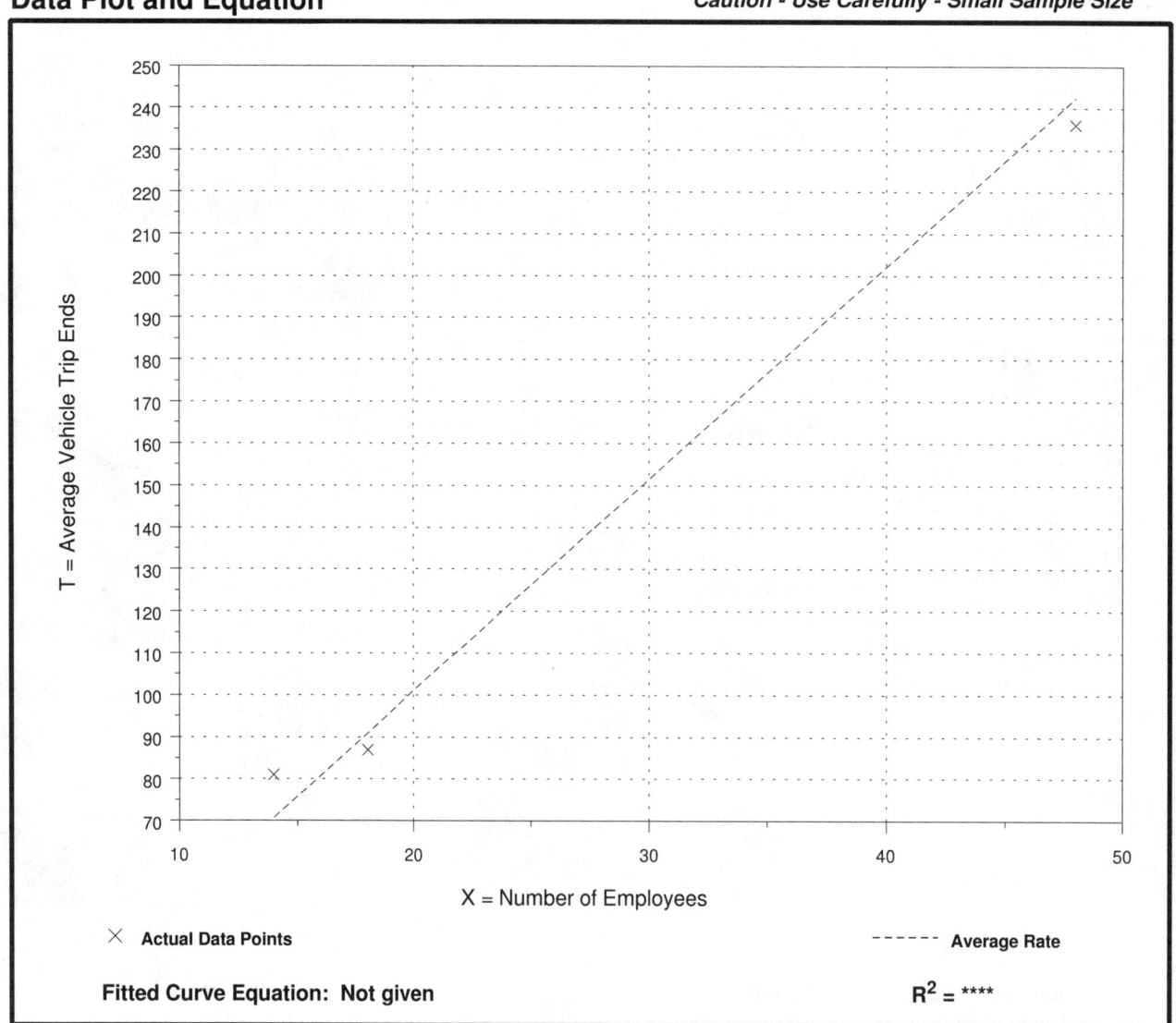

Data Plot and Equation

Caution - Use Carefully - Small Sample Size

× Actual Data Points   ------ Average Rate

Fitted Curve Equation: Not given R^2 = ****

Hardware/Paint Store
(816)

Average Vehicle Trip Ends vs: Employees
On a: Weekday,
A.M. Peak Hour of Generator

Number of Studies: 3
Avg. Number of Employees: 27
Directional Distribution: Not available

Trip Generation per Employee

Average Rate	Range of Rates	Standard Deviation
5.33	4.78 - 5.56	2.30

Data Plot and Equation

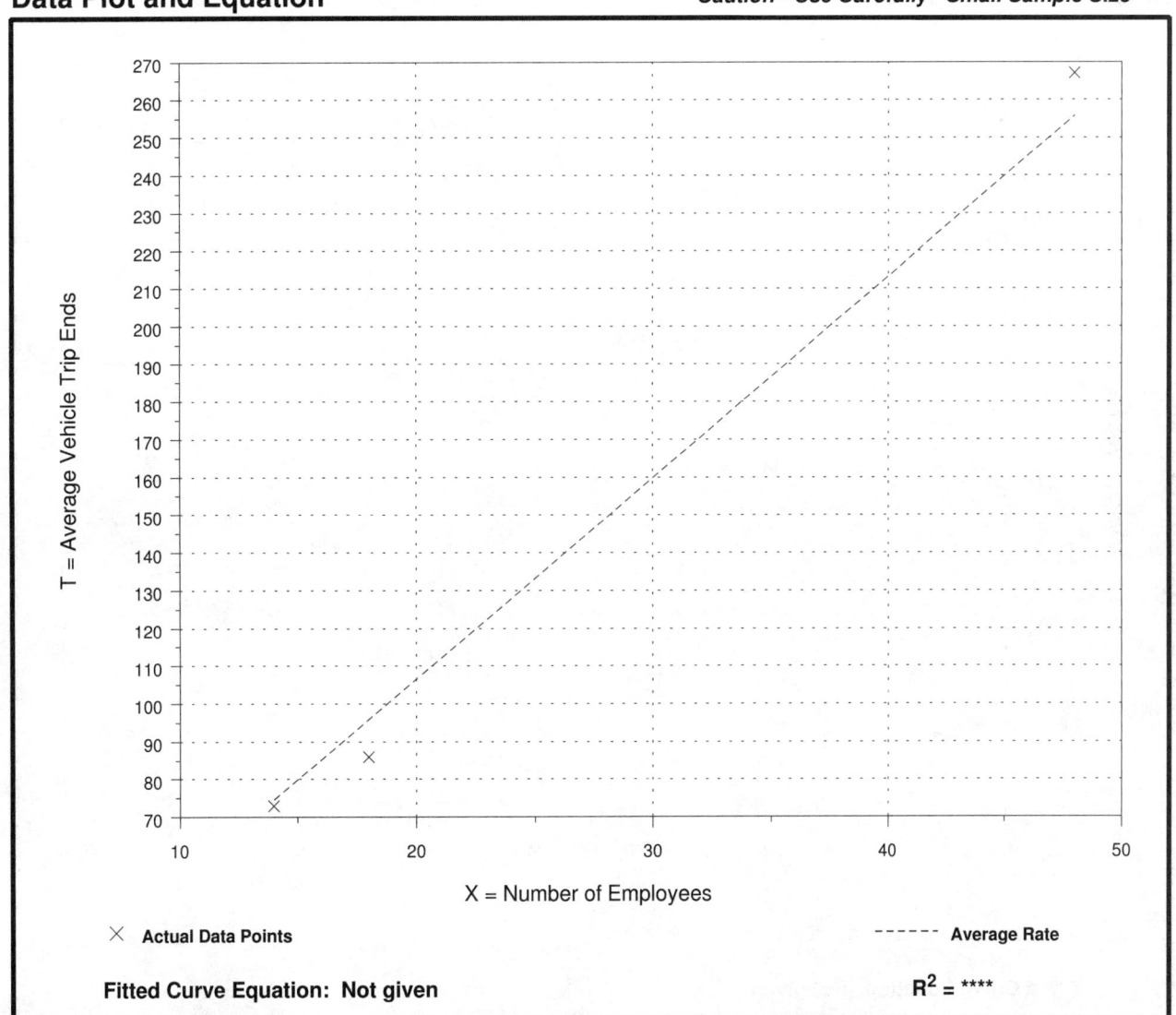

Caution - Use Carefully - Small Sample Size

Fitted Curve Equation: Not given $R^2 = ****$

Hardware/Paint Store
(816)

Average Vehicle Trip Ends vs: Employees
On a: Weekday,
P.M. Peak Hour of Generator

Number of Studies: 3
Avg. Number of Employees: 27
Directional Distribution: Not available

Trip Generation per Employee

Average Rate	Range of Rates	Standard Deviation
5.43	4.83 - 6.50	2.36

Data Plot and Equation

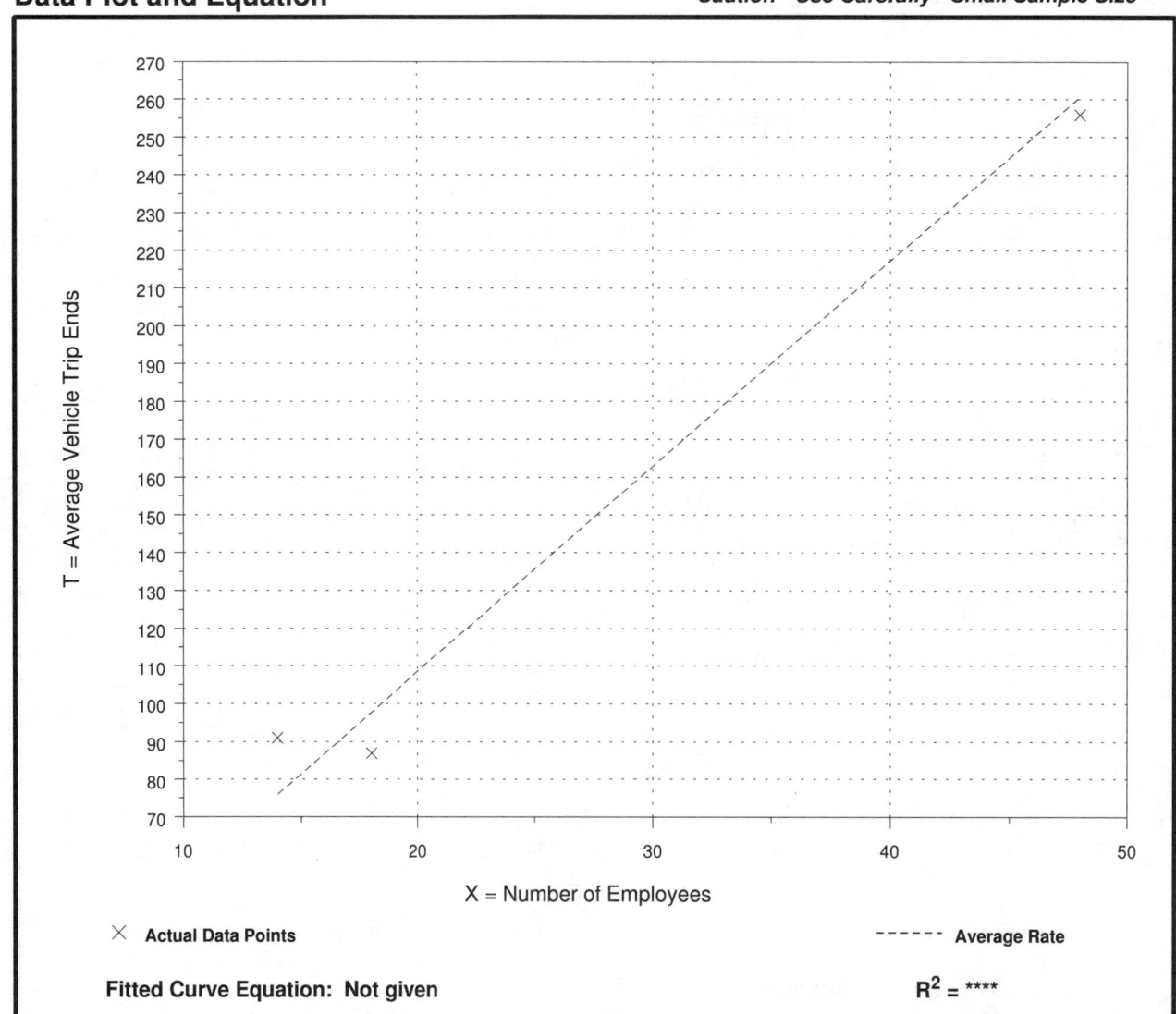

Caution - Use Carefully - Small Sample Size

Fitted Curve Equation: Not given $R^2 = ****$

Hardware/Paint Store
(816)

Average Vehicle Trip Ends vs: Employees
On a: Saturday

Number of Studies: 3
Avg. Number of Employees: 27
Directional Distribution: 50% entering, 50% exiting

Trip Generation per Employee

Average Rate	Range of Rates	Standard Deviation
85.61	62.83 - 94.13	15.68

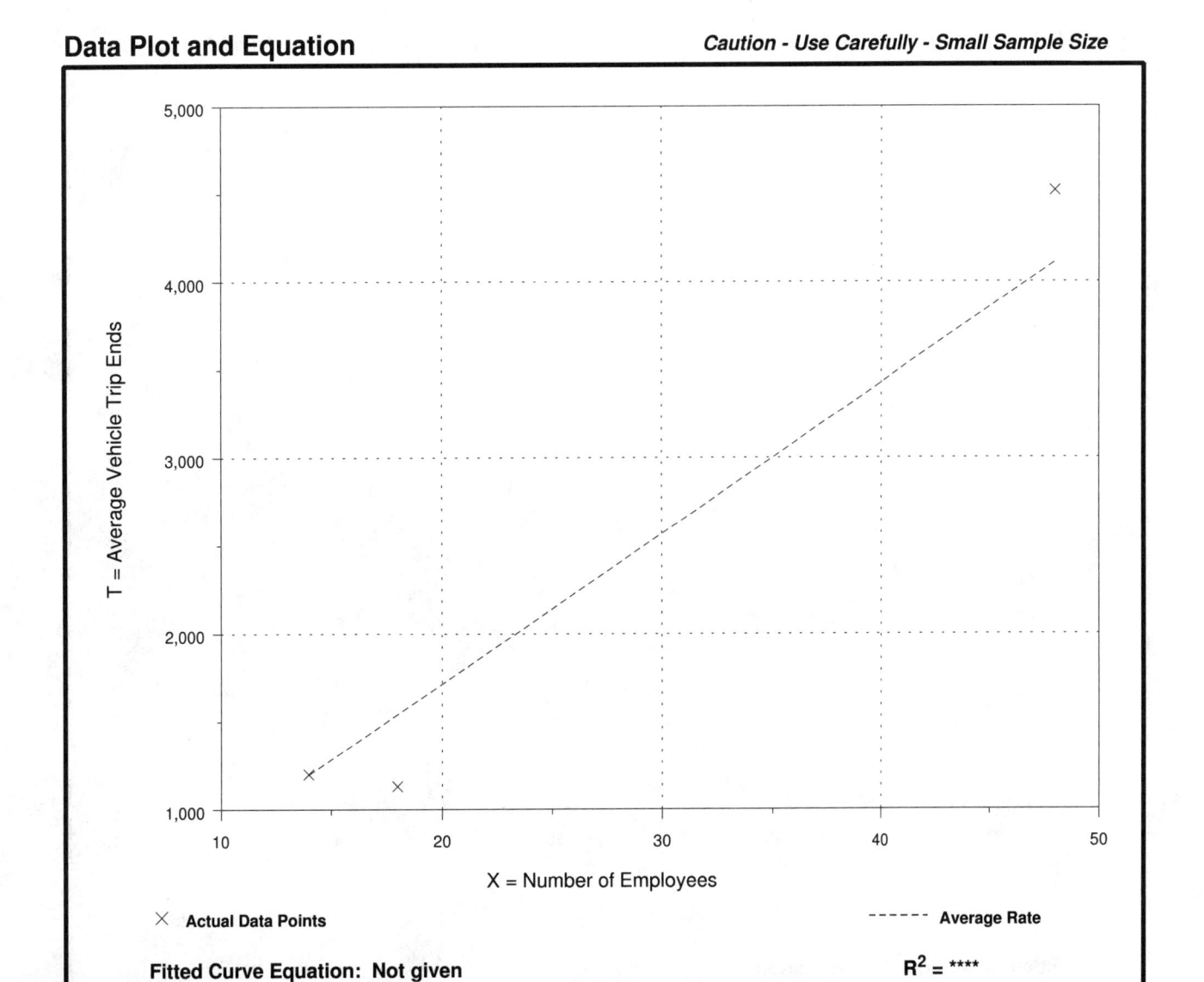

Data Plot and Equation

Caution - Use Carefully - Small Sample Size

Fitted Curve Equation: Not given

$R^2 = ****$

Hardware/Paint Store
(816)

Average Vehicle Trip Ends vs: Employees
On a: Saturday,
Peak Hour of Generator

Number of Studies: 3
Avg. Number of Employees: 27
Directional Distribution: Not available

Trip Generation per Employee

Average Rate	Range of Rates	Standard Deviation
11.60	8.28 - 12.92	3.86

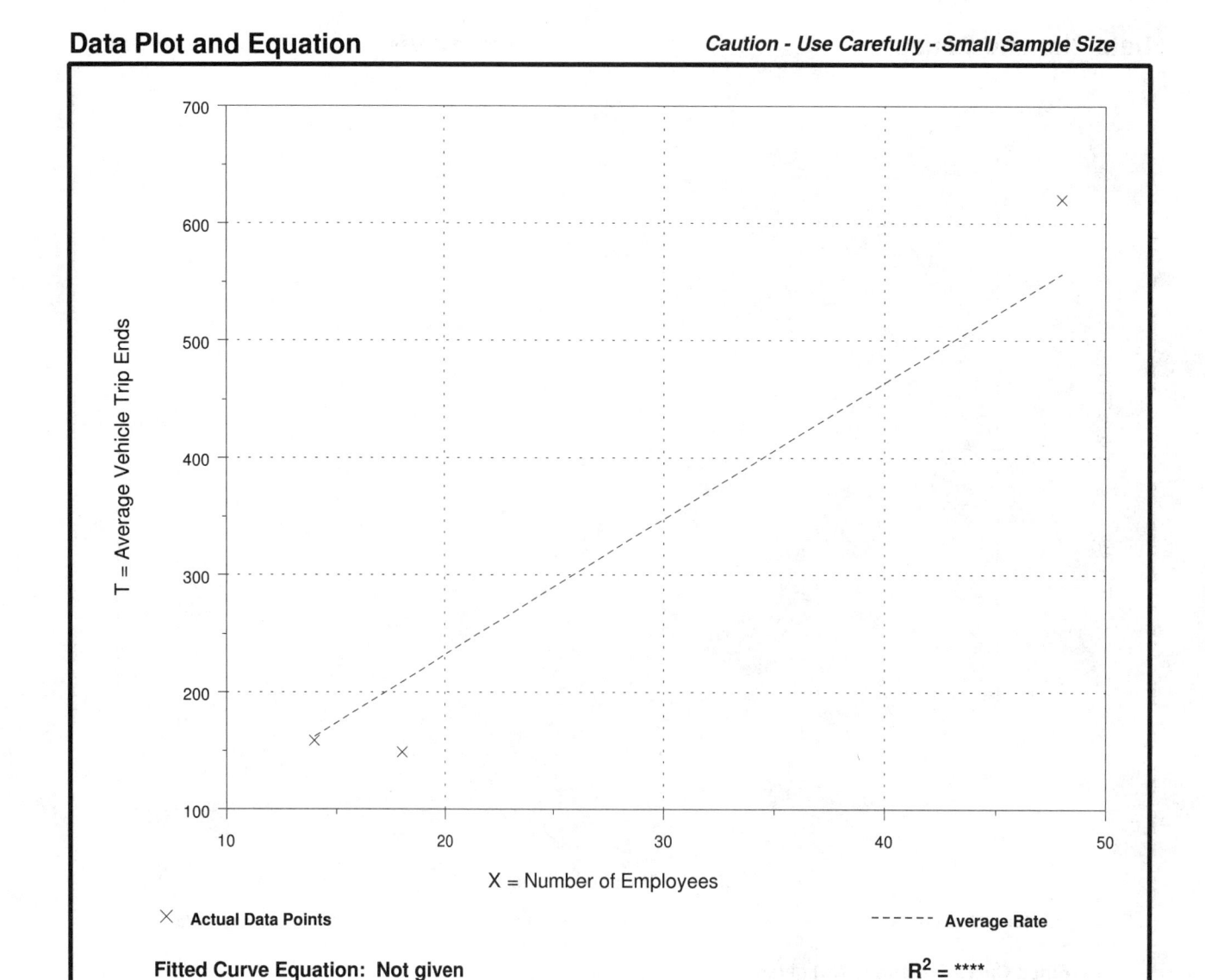

Data Plot and Equation

Caution - Use Carefully - Small Sample Size

Fitted Curve Equation: Not given $R^2 = ****$

Hardware/Paint Store
(816)

Average Vehicle Trip Ends vs: Employees
On a: Sunday

Number of Studies: 3
Avg. Number of Employees: 27
Directional Distribution: 50% entering, 50% exiting

Trip Generation per Employee

Average Rate	Range of Rates	Standard Deviation
71.23	49.78 - 80.00	14.91

Data Plot and Equation

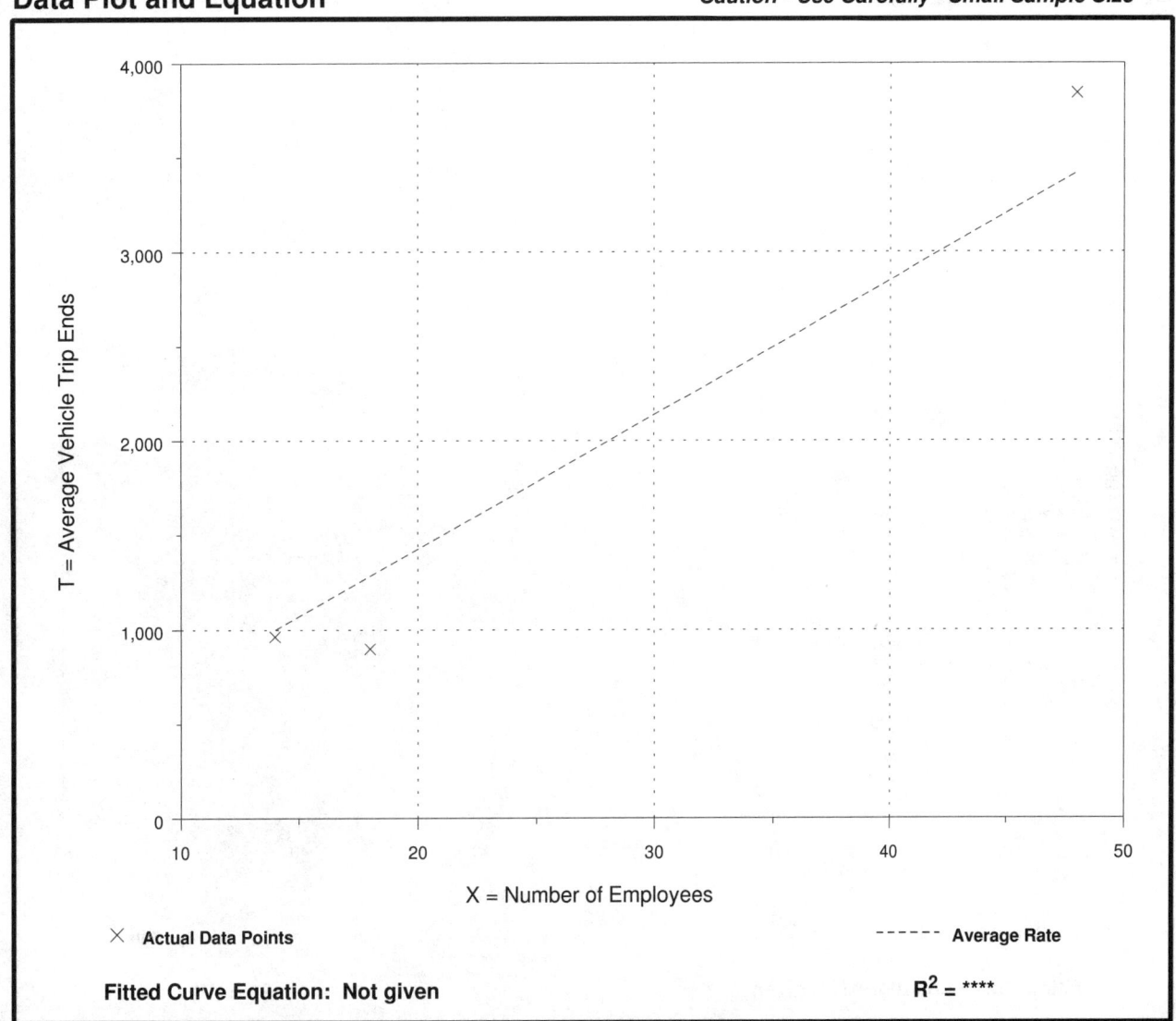

Caution - Use Carefully - Small Sample Size

X   Actual Data Points   ------ Average Rate

Fitted Curve Equation: Not given $R^2 = ****$

Hardware/Paint Store
(816)

Average Vehicle Trip Ends vs: Employees
On a: Sunday,
Peak Hour of Generator

Number of Studies: 3
Avg. Number of Employees: 27
Directional Distribution: Not available

Trip Generation per Employee

Average Rate	Range of Rates	Standard Deviation
10.18	7.89 - 10.96	3.39

Data Plot and Equation

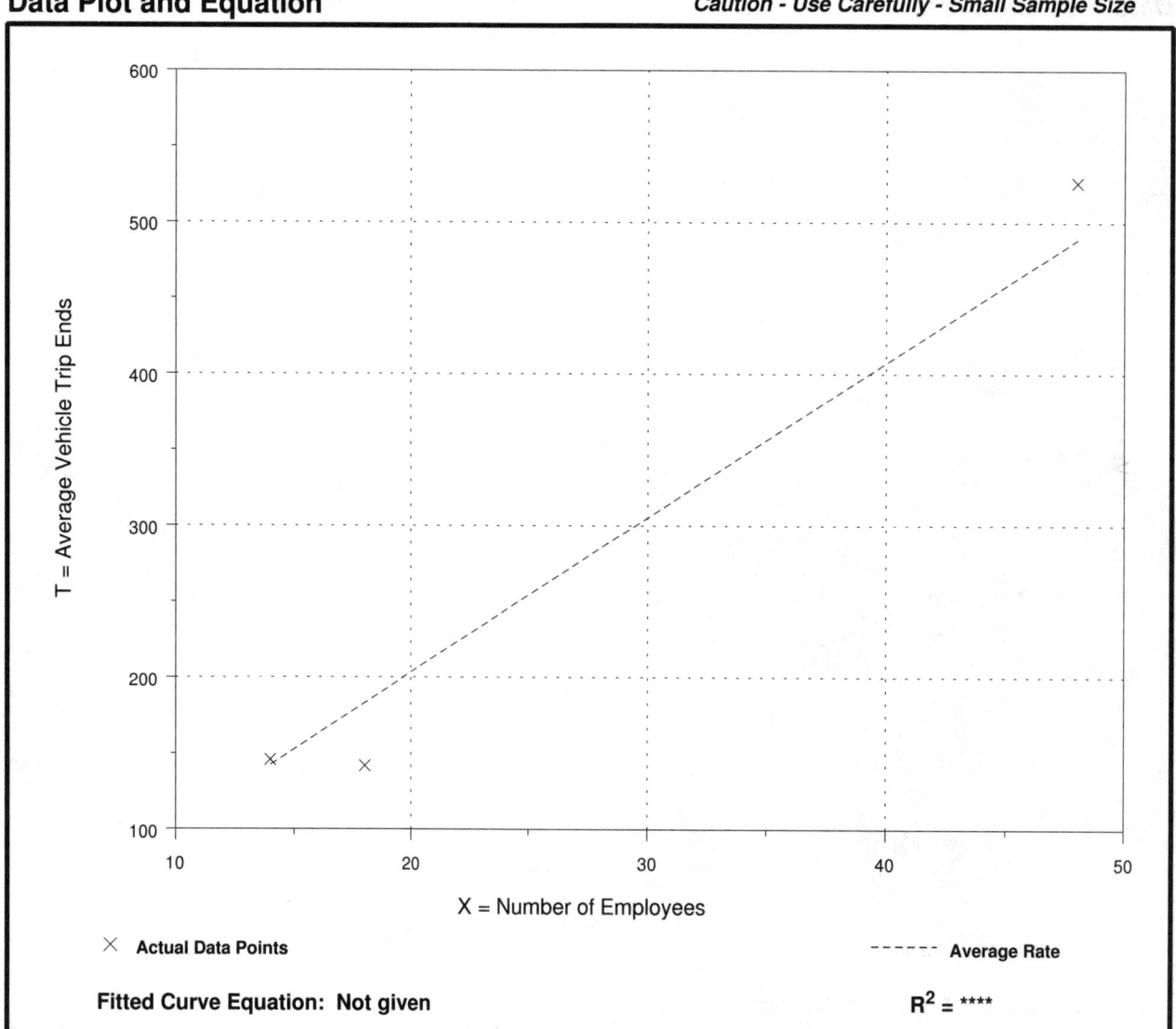

Caution - Use Carefully - Small Sample Size

Fitted Curve Equation: Not given $R^2 = ****$

Hardware/Paint Store
(816)

Average Vehicle Trip Ends vs: 1000 Sq. Feet Gross Floor Area
On a: Weekday

Number of Studies: 3
Average 1000 Sq. Feet GFA: 28
Directional Distribution: 50% entering, 50% exiting

Trip Generation per 1000 Sq. Feet Gross Floor Area

Average Rate	Range of Rates	Standard Deviation
51.29	43.58 - 74.09	14.43

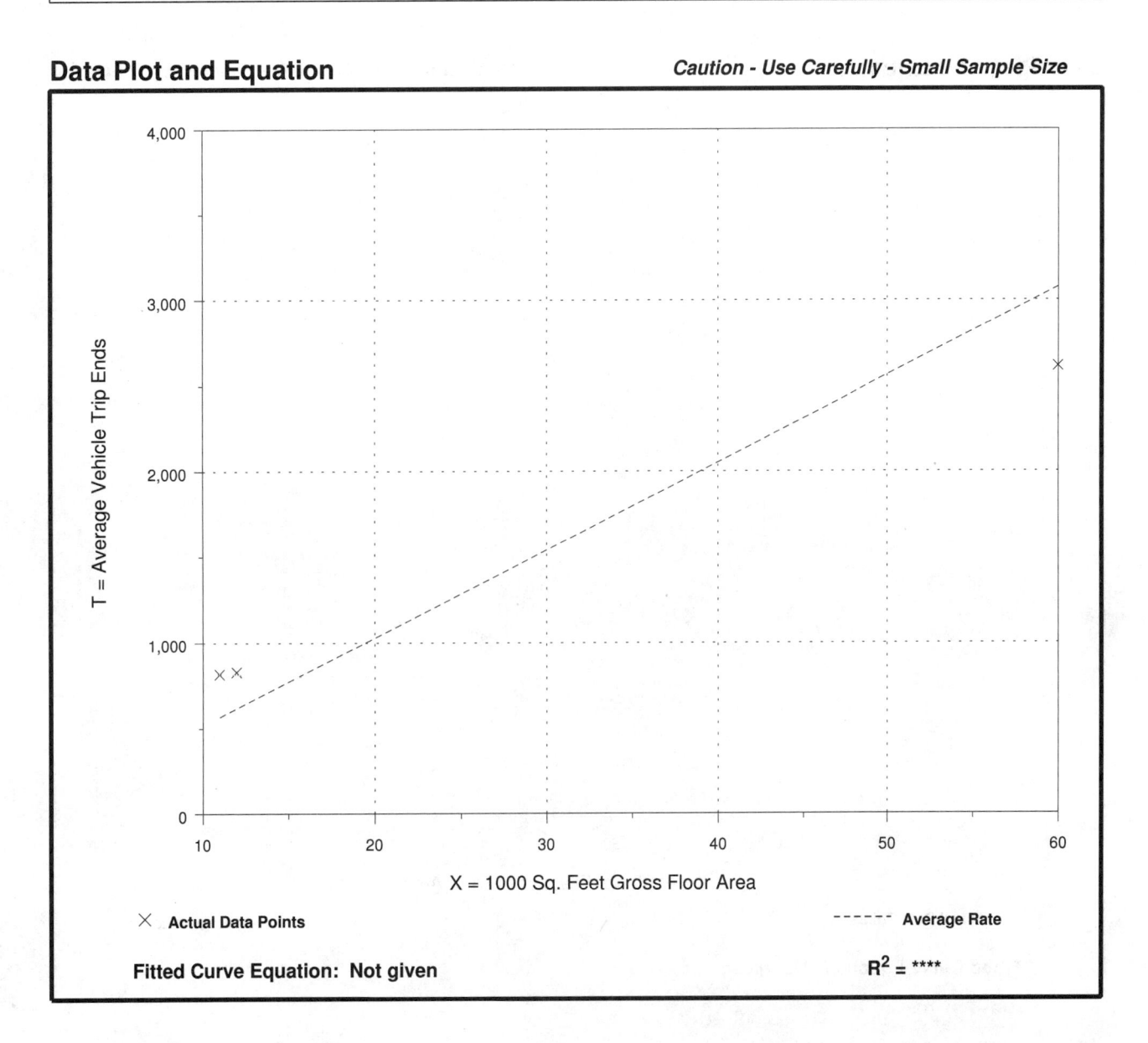

Data Plot and Equation

Caution - Use Carefully - Small Sample Size

Fitted Curve Equation: Not given $R^2 = ****$

Hardware/Paint Store
(816)

Average Vehicle Trip Ends vs: 1000 Sq. Feet Gross Floor Area
On a: Weekday,
Peak Hour of Adjacent Street Traffic,
One Hour Between 7 and 9 a.m.

Number of Studies: 3
Average 1000 Sq. Feet GFA: 28
Directional Distribution: Not available

Trip Generation per 1000 Sq. Feet Gross Floor Area

Average Rate	Range of Rates	Standard Deviation
1.08	0.42 - 3.50	1.53

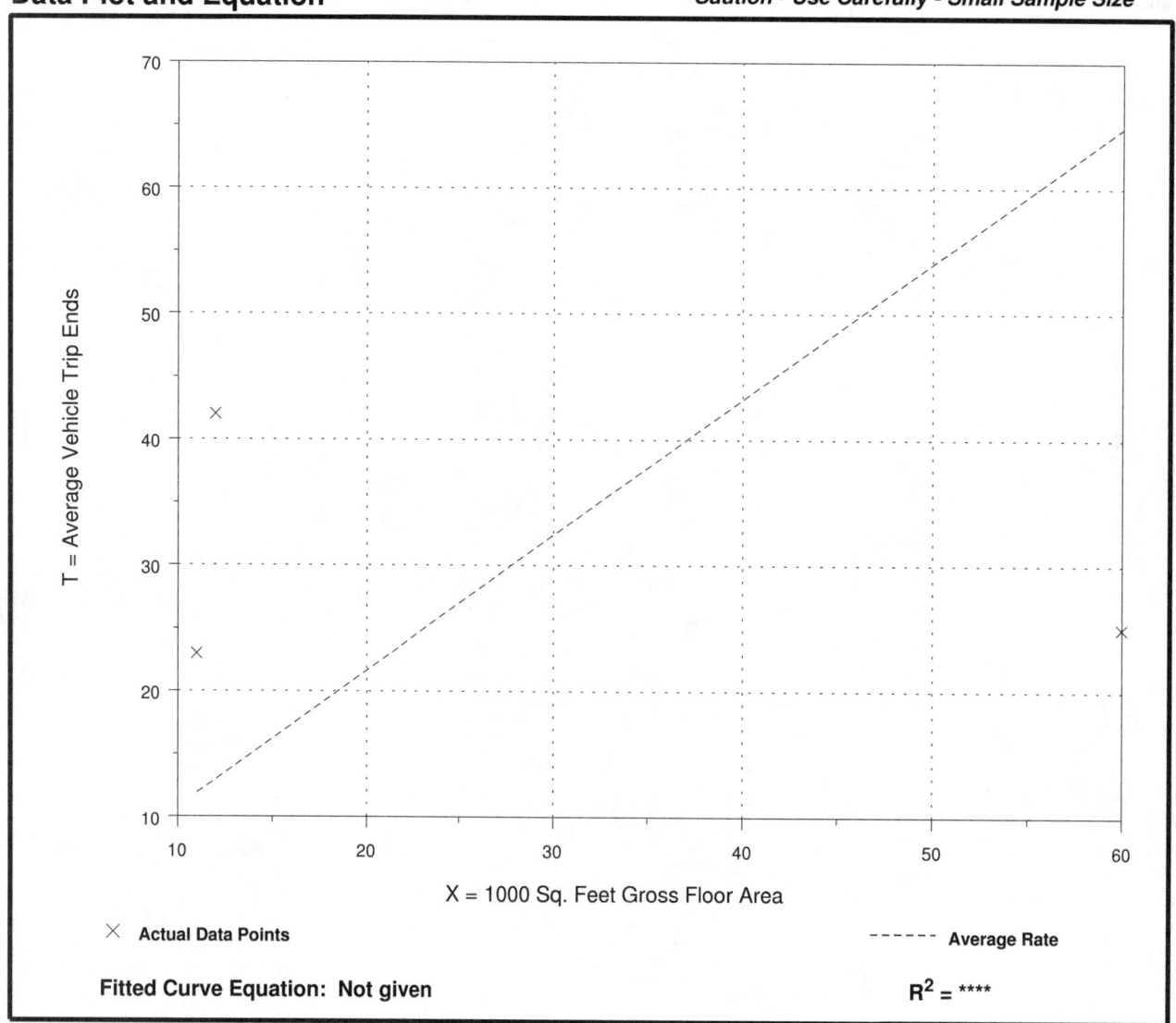

Data Plot and Equation

Caution - Use Carefully - Small Sample Size

Fitted Curve Equation: Not given $R^2 = ****$

Hardware/Paint Store
(816)

Average Vehicle Trip Ends vs: 1000 Sq. Feet Gross Floor Area
On a: Weekday,
Peak Hour of Adjacent Street Traffic,
One Hour Between 4 and 6 p.m.

Number of Studies: 8
Average 1000 Sq. Feet GFA: 18
Directional Distribution: 47% entering, 53% exiting

Trip Generation per 1000 Sq. Feet Gross Floor Area

Average Rate	Range of Rates	Standard Deviation
4.84	1.52 - 8.45	2.92

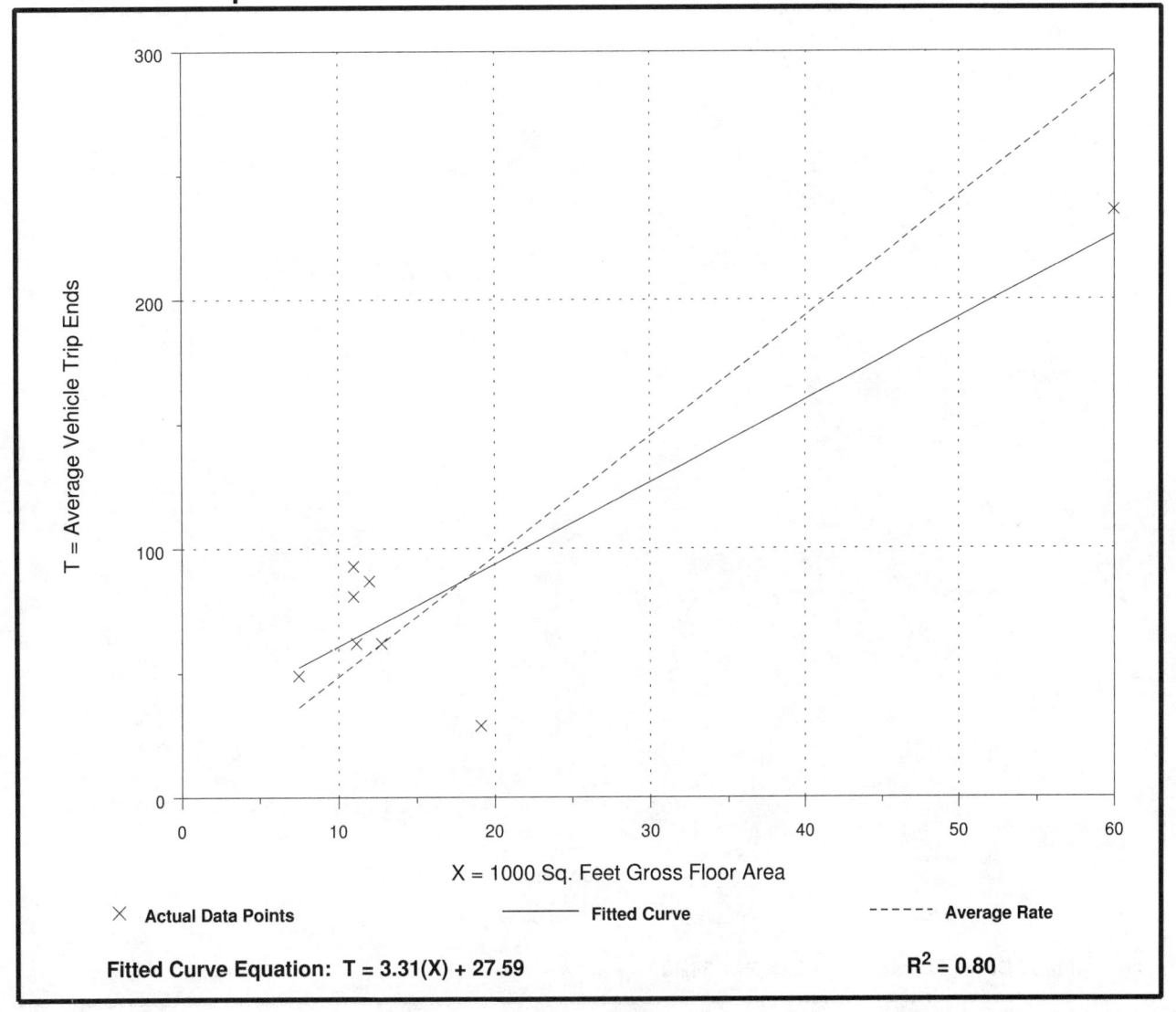

Data Plot and Equation

Fitted Curve Equation: $T = 3.31(X) + 27.59$ $\qquad R^2 = 0.80$

Hardware/Paint Store
(816)

Average Vehicle Trip Ends vs: 1000 Sq. Feet Gross Floor Area
On a: Weekday,
A.M. Peak Hour of Generator

Number of Studies: 4
Average 1000 Sq. Feet GFA: 34
Directional Distribution: 52% entering, 48% exiting

Trip Generation per 1000 Sq. Feet Gross Floor Area

Average Rate	Range of Rates	Standard Deviation
4.91	4.45 - 7.17	2.37

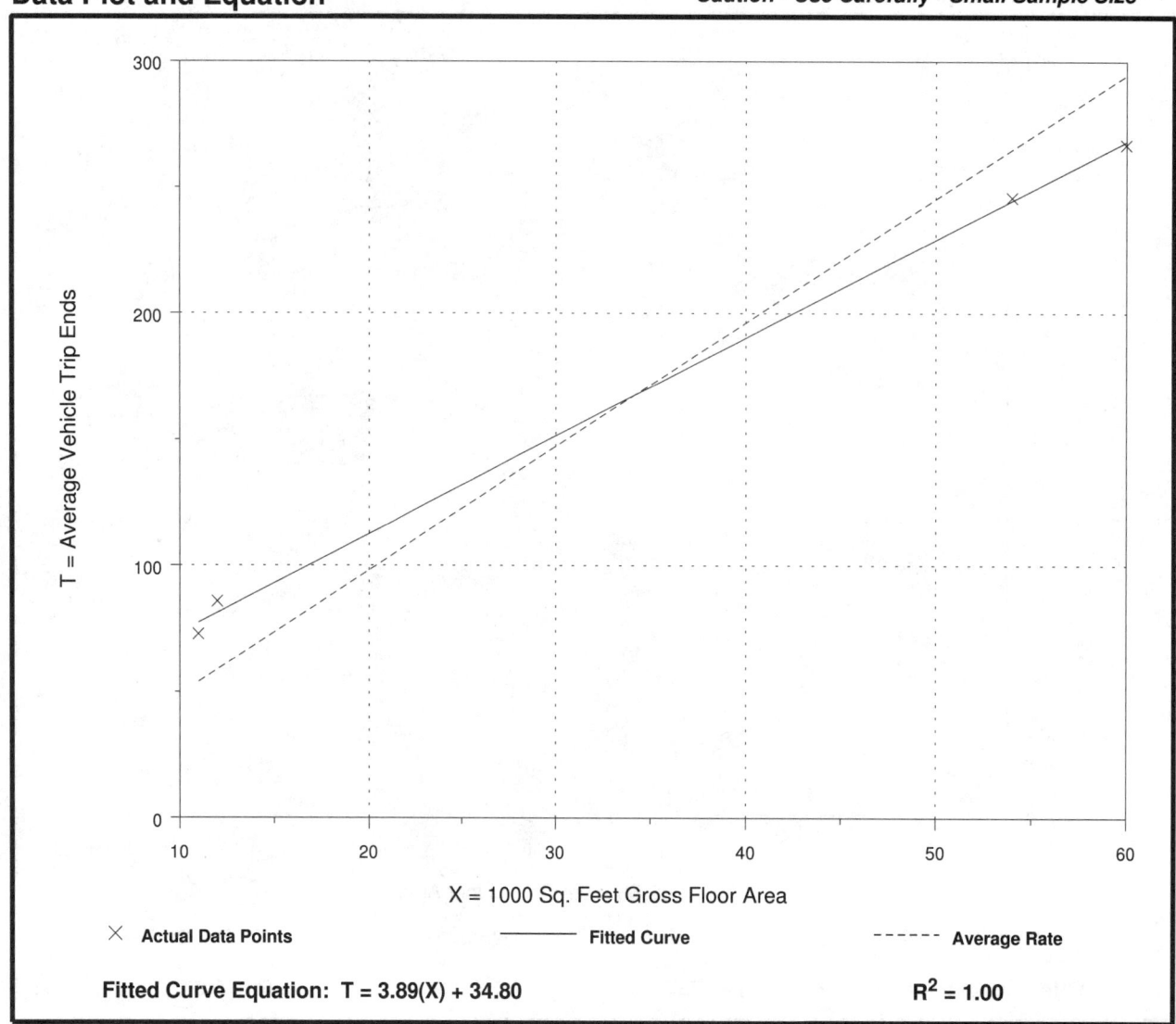

Data Plot and Equation

Caution - Use Carefully - Small Sample Size

Fitted Curve Equation: T = 3.89(X) + 34.80 $R^2 = 1.00$

Hardware/Paint Store
(816)

Average Vehicle Trip Ends vs: 1000 Sq. Feet Gross Floor Area
On a: Weekday,
P.M. Peak Hour of Generator

Number of Studies: 4
Average 1000 Sq. Feet GFA: 34
Directional Distribution: 54% entering, 46% exiting

Trip Generation per 1000 Sq. Feet Gross Floor Area

Average Rate	Range of Rates	Standard Deviation
4.74	3.98 - 8.27	2.55

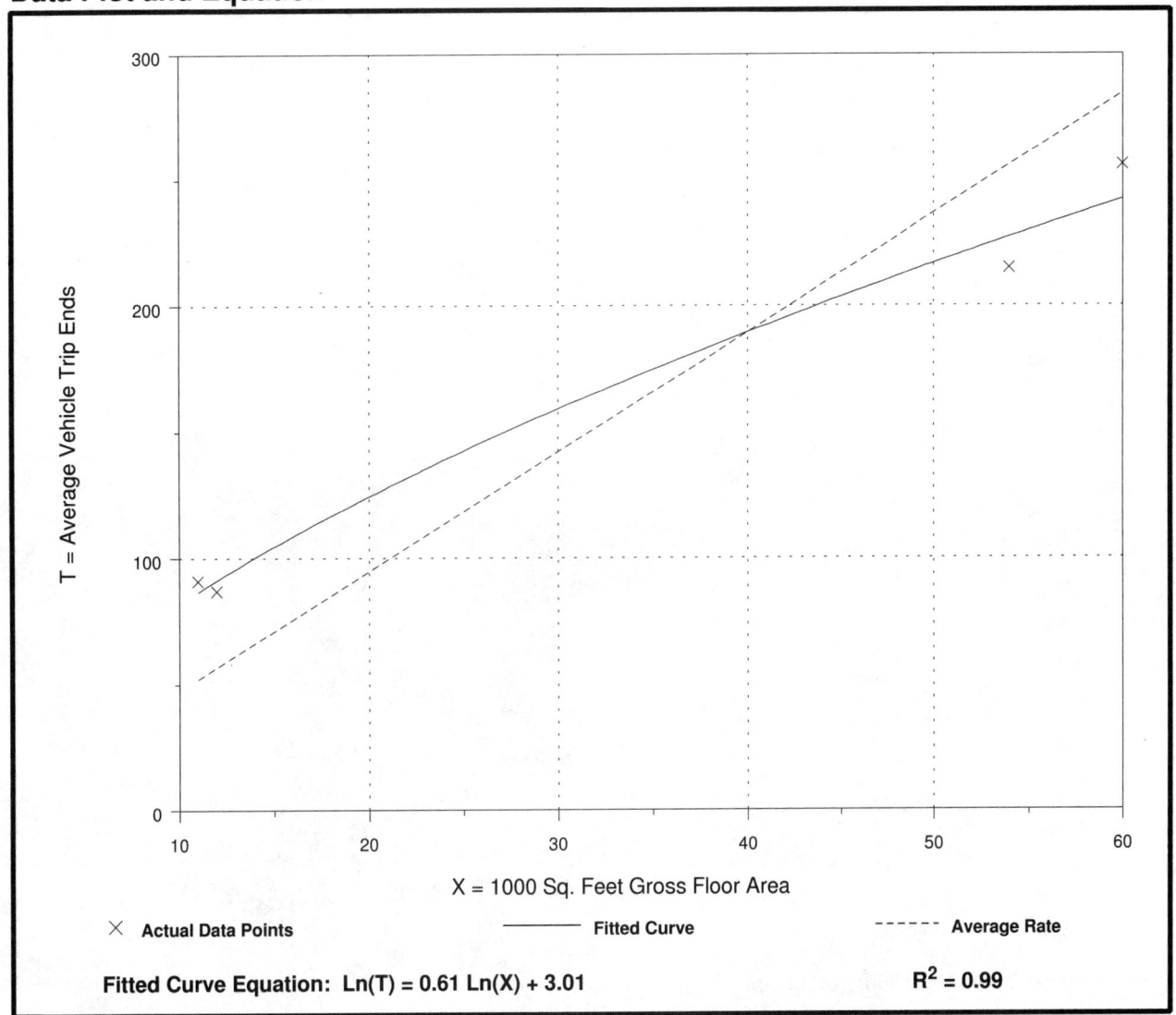

Data Plot and Equation

Caution - Use Carefully - Small Sample Size

Fitted Curve Equation: $Ln(T) = 0.61\ Ln(X) + 3.01$ $R^2 = 0.99$

Hardware/Paint Store
(816)

Average Vehicle Trip Ends vs: 1000 Sq. Feet Gross Floor Area
On a: Saturday

Number of Studies: 3
Average 1000 Sq. Feet GFA: 28
Directional Distribution: 50% entering, 50% exiting

Trip Generation per 1000 Sq. Feet Gross Floor Area

Average Rate	Range of Rates	Standard Deviation
82.52	75.30 - 109.09	15.27

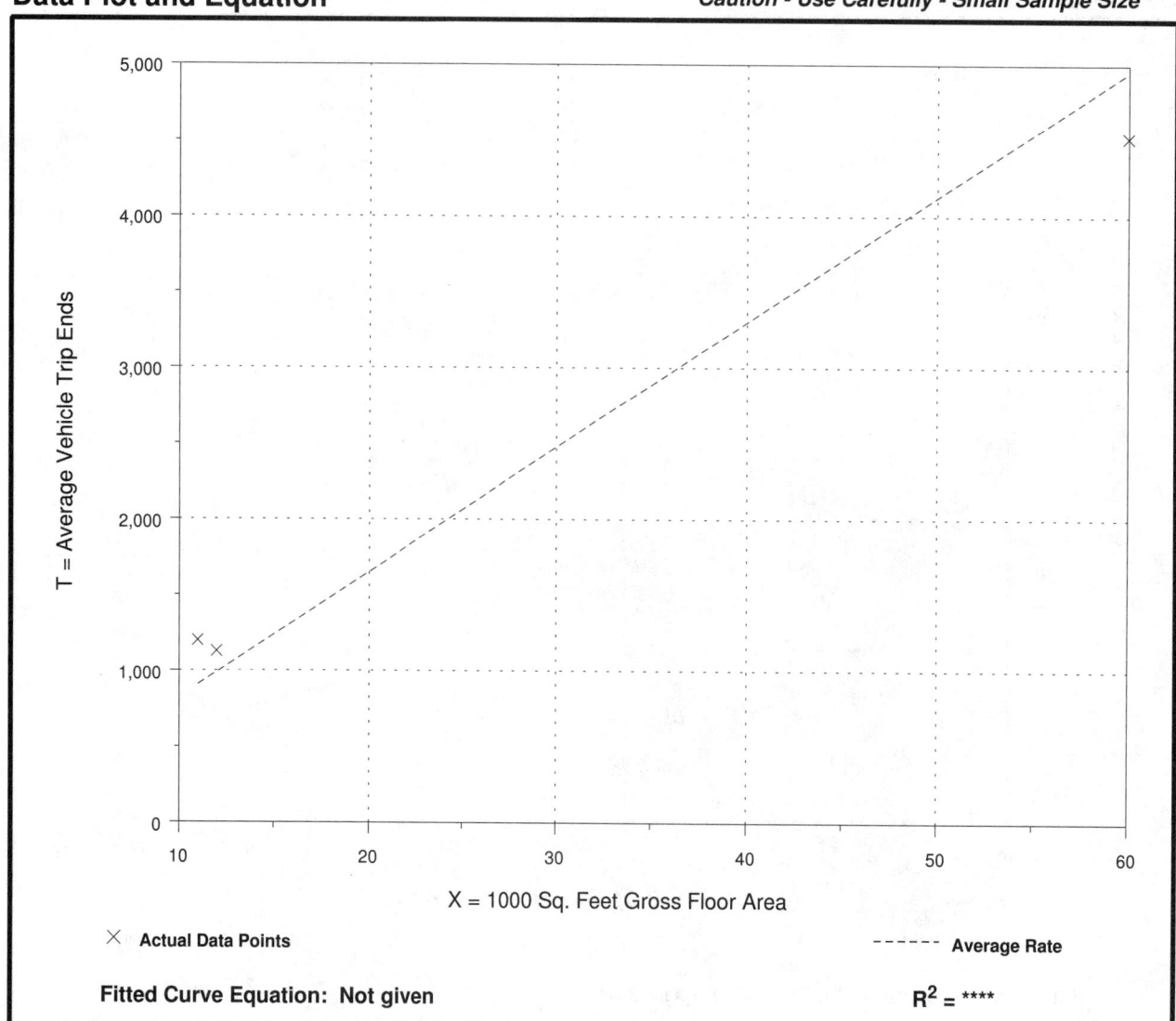

Data Plot and Equation

Caution - Use Carefully - Small Sample Size

Fitted Curve Equation: Not given $R^2 = ****$

Hardware/Paint Store
(816)

Average Vehicle Trip Ends vs: 1000 Sq. Feet Gross Floor Area
On a: Saturday,
Peak Hour of Generator

Number of Studies: 3
Average 1000 Sq. Feet GFA: 28
Directional Distribution: Not available

Trip Generation per 1000 Sq. Feet Gross Floor Area

Average Rate	Range of Rates	Standard Deviation
11.18	10.33 - 14.45	3.61

Data Plot and Equation

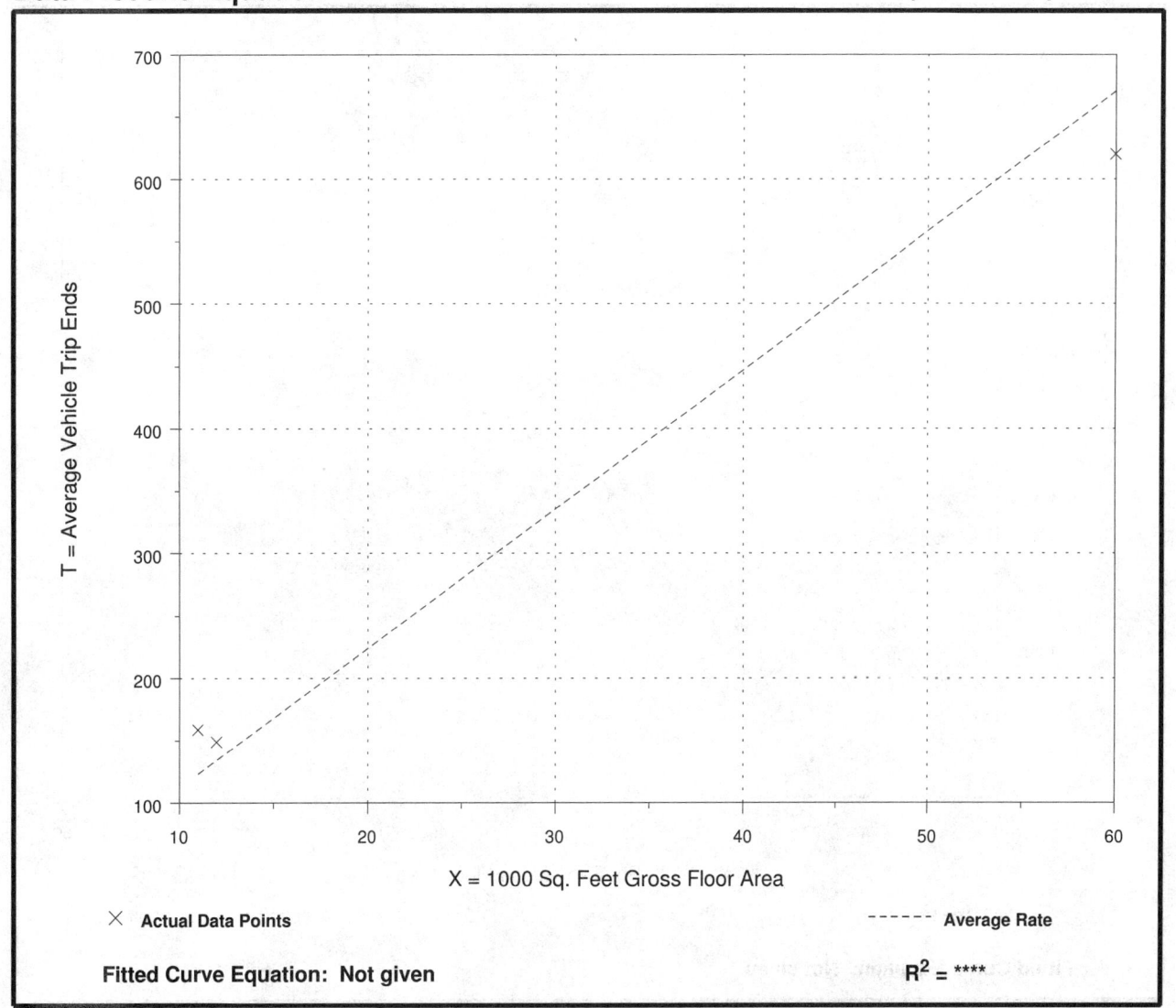

Caution - Use Carefully - Small Sample Size

Fitted Curve Equation: Not given $R^2 =$ ****

Hardware/Paint Store
(816)

Average Vehicle Trip Ends vs: 1000 Sq. Feet Gross Floor Area
On a: Sunday

Number of Studies: 3
Average 1000 Sq. Feet GFA: 28
Directional Distribution: 50% entering, 50% exiting

Trip Generation per 1000 Sq. Feet Gross Floor Area

Average Rate	Range of Rates	Standard Deviation
68.65	64.00 - 87.45	11.63

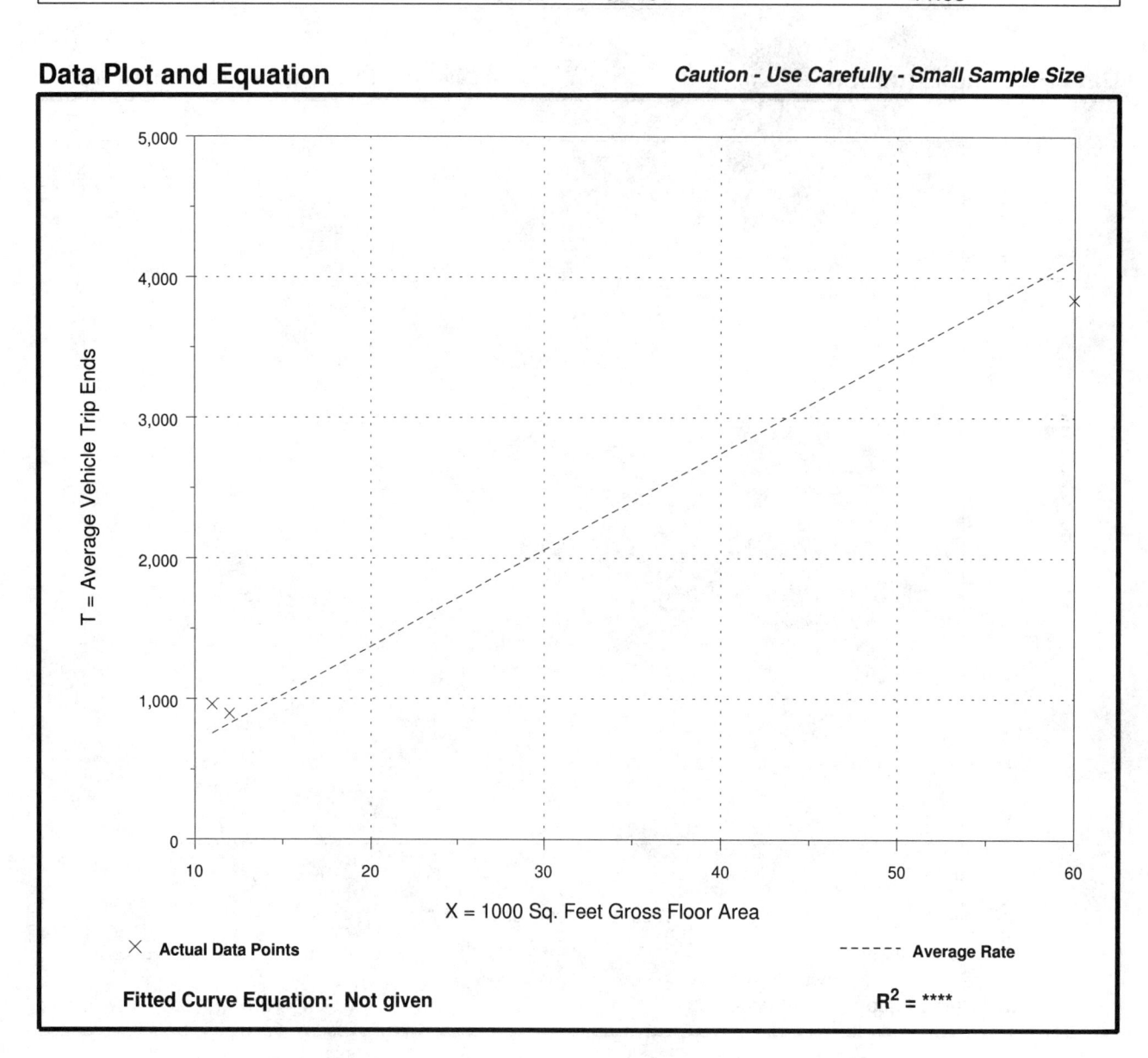

Data Plot and Equation

Caution - Use Carefully - Small Sample Size

Fitted Curve Equation: Not given $R^2 = ****$

Hardware/Paint Store
(816)

Average Vehicle Trip Ends vs: 1000 Sq. Feet Gross Floor Area
On a: Sunday,
Peak Hour of Generator

Number of Studies: 3
Average 1000 Sq. Feet GFA: 28
Directional Distribution: Not available

Trip Generation per 1000 Sq. Feet Gross Floor Area

Average Rate	Range of Rates	Standard Deviation
9.81	8.77 - 13.27	3.54

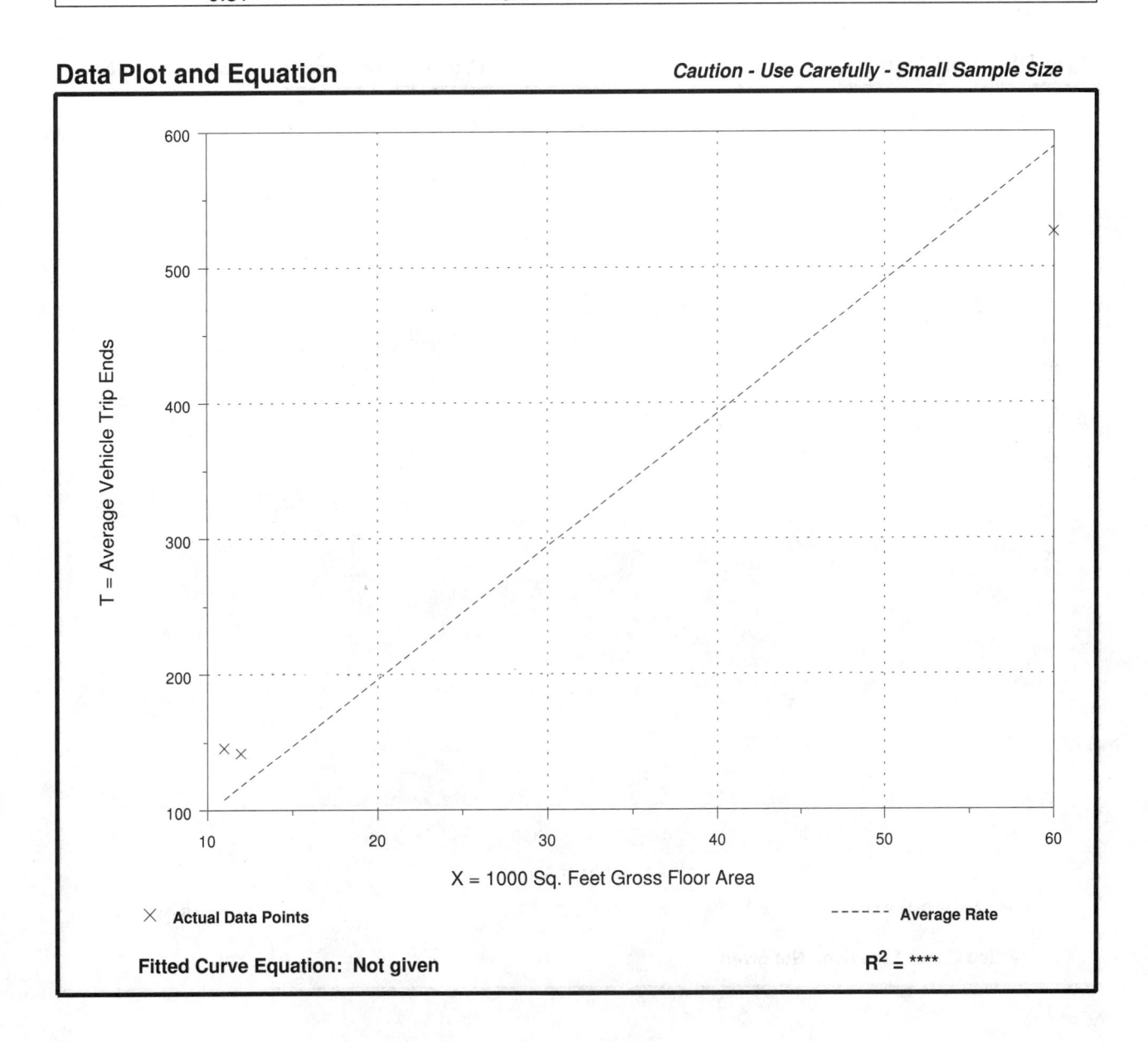

Data Plot and Equation *Caution - Use Carefully - Small Sample Size*

Fitted Curve Equation: Not given $R^2 = ****$

Hardware/Paint Store
(816)

Average Vehicle Trip Ends vs: Acres
On a: Weekday

Number of Studies: 3
Average Number of Acres: 3
Directional Distribution: 50% entering, 50% exiting

Trip Generation per Acre

Average Rate	Range of Rates	Standard Deviation
545.77	466.96 - 905.56	155.37

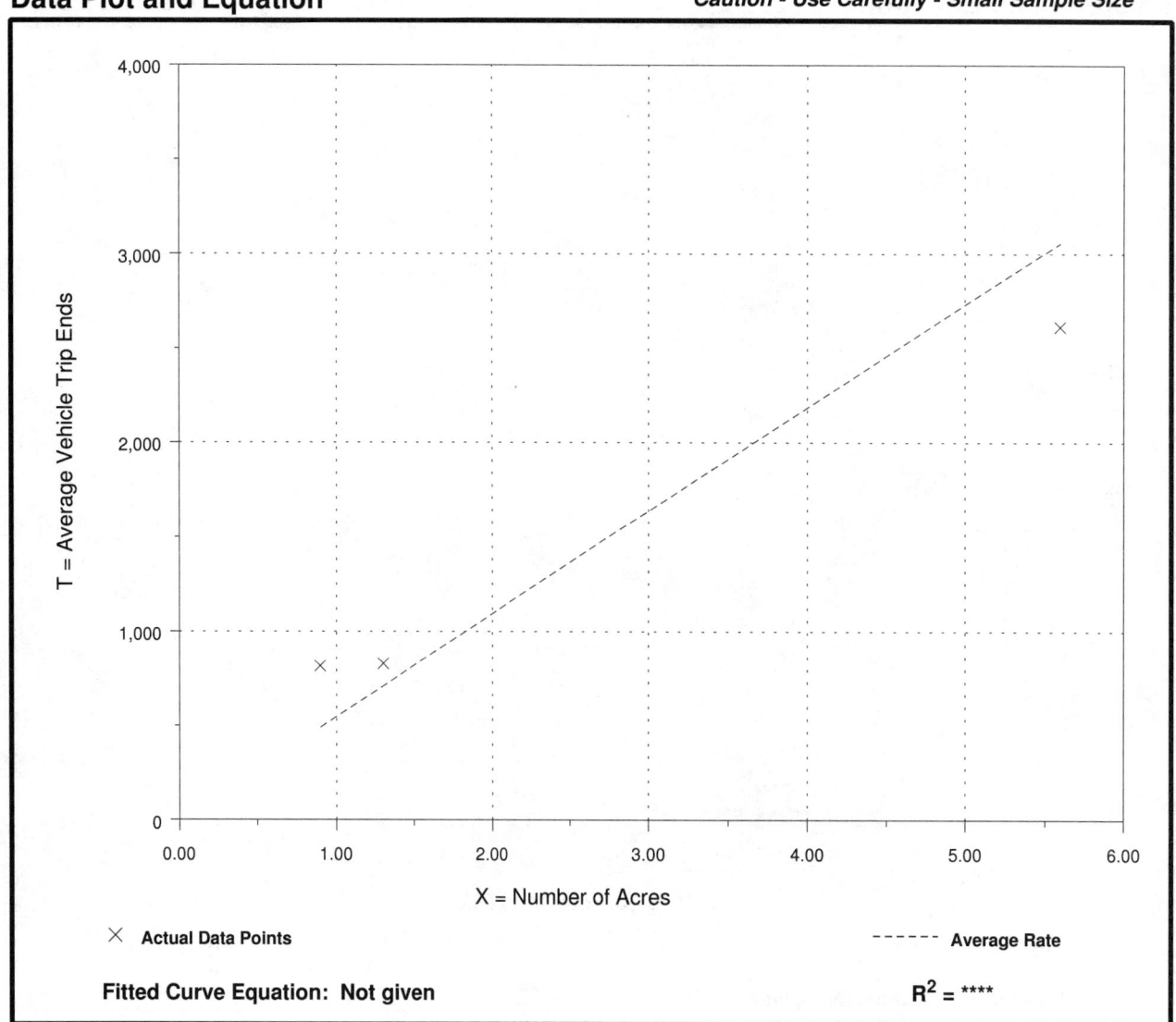

Data Plot and Equation

Caution - Use Carefully - Small Sample Size

Fitted Curve Equation: Not given $R^2 = ****$

Hardware/Paint Store
(816)

Average Vehicle Trip Ends vs: Acres
On a: Weekday,
Peak Hour of Adjacent Street Traffic,
One Hour Between 7 and 9 a.m.

Number of Studies: 3
Average Number of Acres: 3
Directional Distribution: Not available

Trip Generation per Acre

Average Rate	Range of Rates	Standard Deviation
11.54	4.46 - 32.31	12.40

Data Plot and Equation

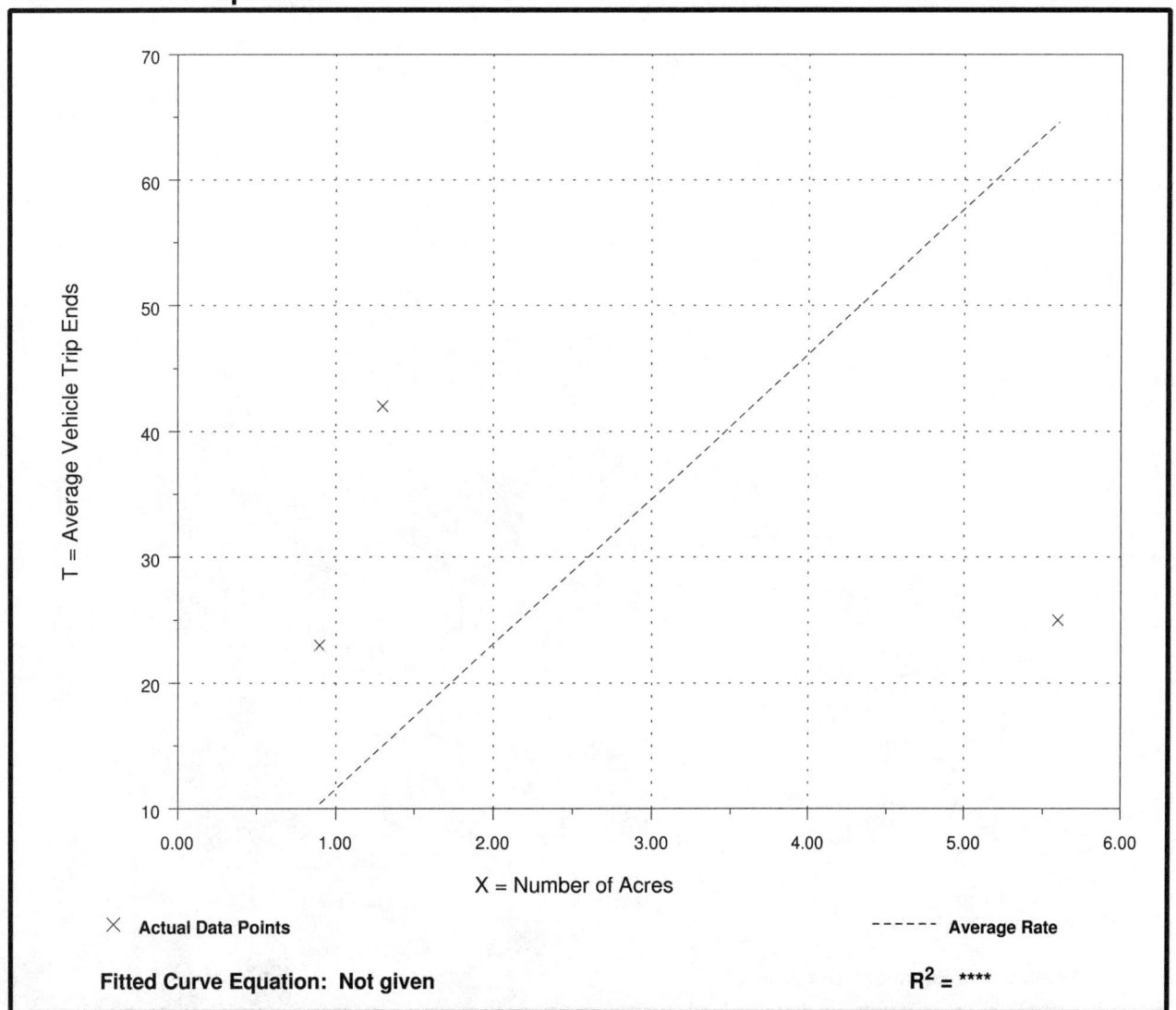

Caution - Use Carefully - Small Sample Size

Fitted Curve Equation: Not given $R^2 =$ ****

Hardware/Paint Store
(816)

Average Vehicle Trip Ends vs: Acres
On a: Weekday,
Peak Hour of Adjacent Street Traffic,
One Hour Between 4 and 6 p.m.

Number of Studies: 3
Average Number of Acres: 3
Directional Distribution: Not available

Trip Generation per Acre

Average Rate	Range of Rates	Standard Deviation
51.79	42.14 - 90.00	18.54

Data Plot and Equation

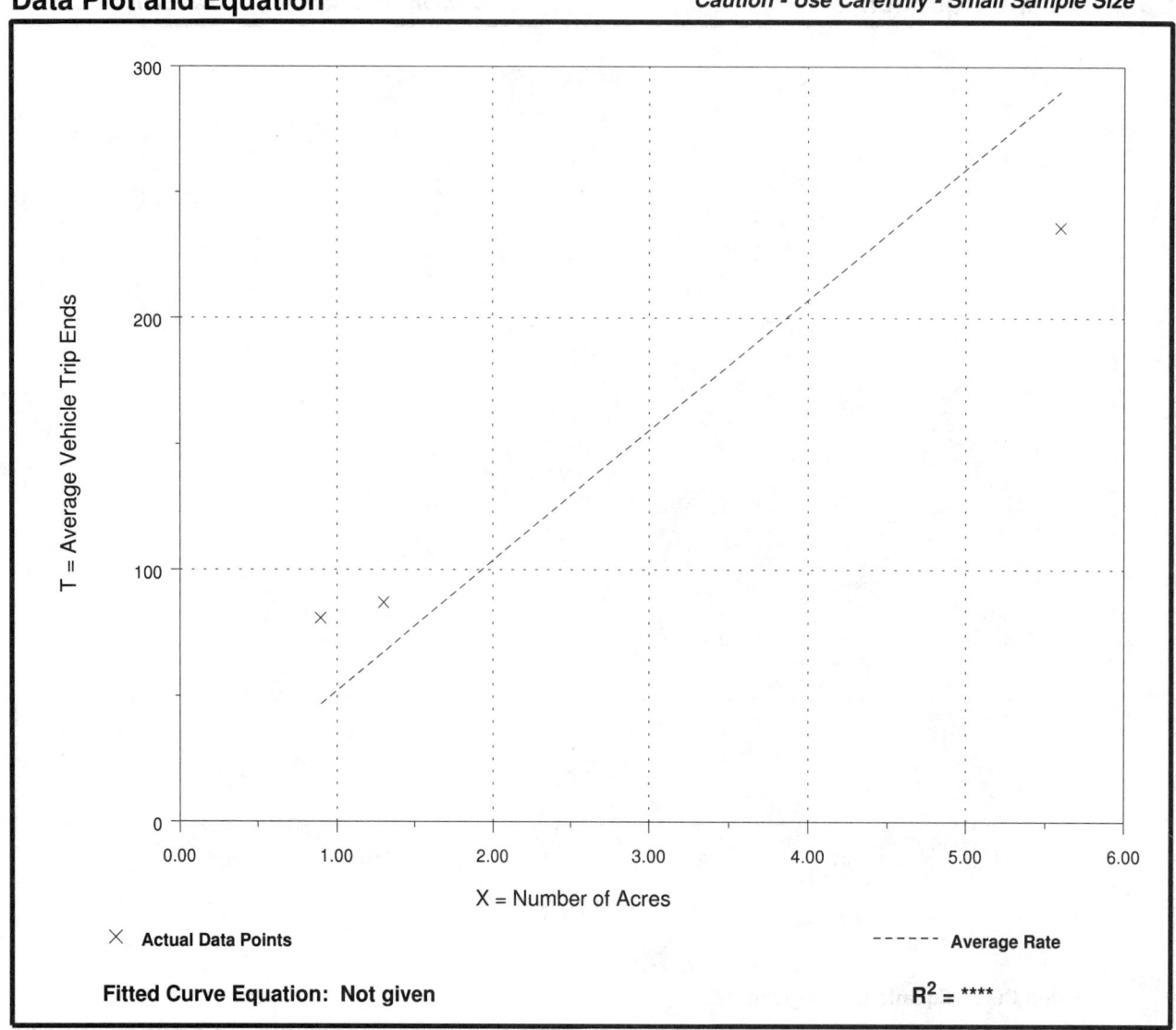

Caution - Use Carefully - Small Sample Size

Fitted Curve Equation: Not given $R^2 = ****$

Hardware/Paint Store
(816)

Average Vehicle Trip Ends vs: Acres
On a: Weekday,
A.M. Peak Hour of Generator

Number of Studies: 3
Average Number of Acres: 3
Directional Distribution: Not available

Trip Generation per Acre

Average Rate	Range of Rates	Standard Deviation
54.62	47.68 - 81.11	13.86

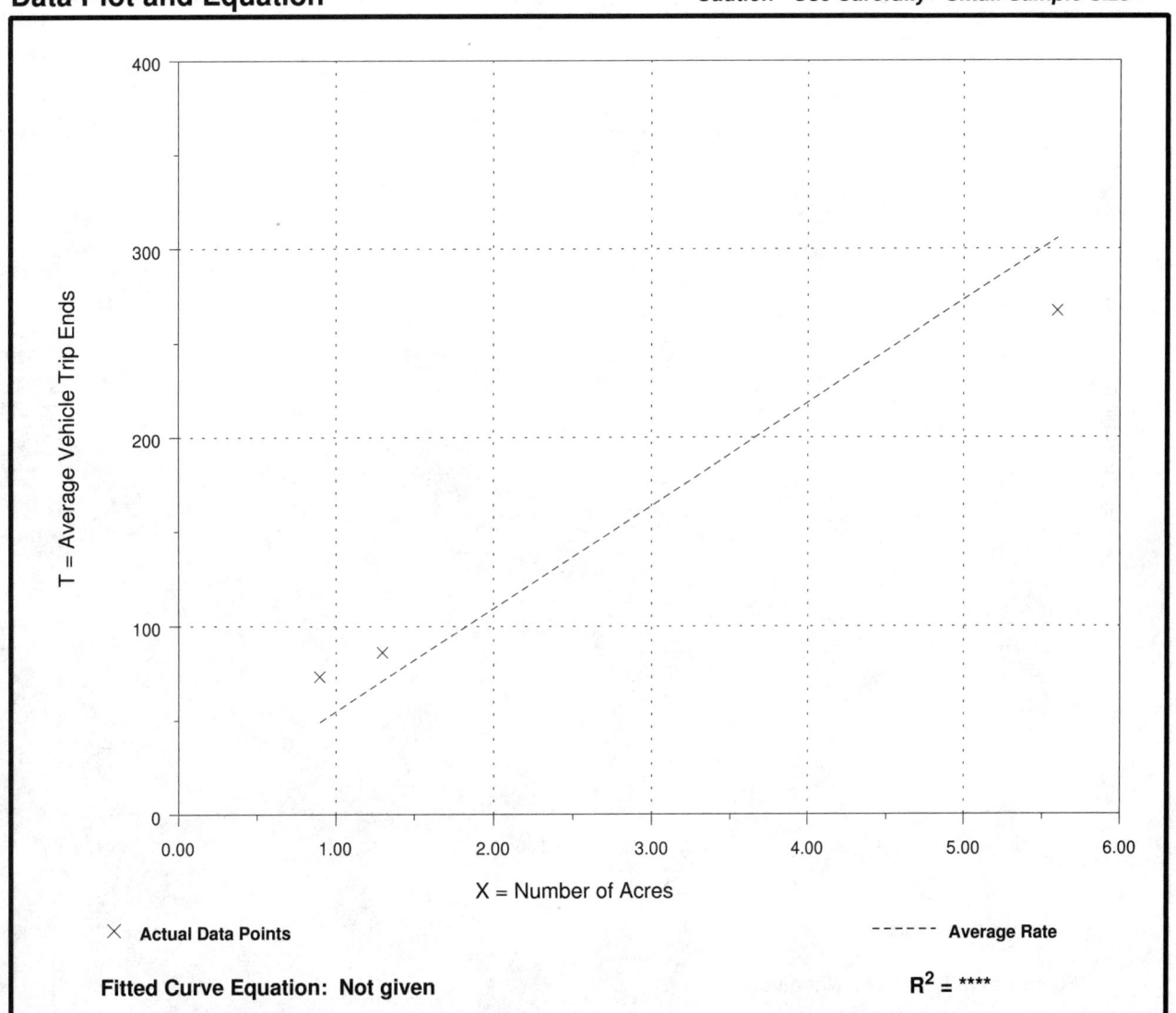

Data Plot and Equation

Caution - Use Carefully - Small Sample Size

Fitted Curve Equation: Not given $R^2 = ****$

Hardware/Paint Store
(816)

Average Vehicle Trip Ends vs: Acres
On a: Weekday,
P.M. Peak Hour of Generator

Number of Studies: 3
Average Number of Acres: 3
Directional Distribution: Not available

Trip Generation per Acre

Average Rate	Range of Rates	Standard Deviation
55.64	45.71 - 101.11	20.29

Data Plot and Equation

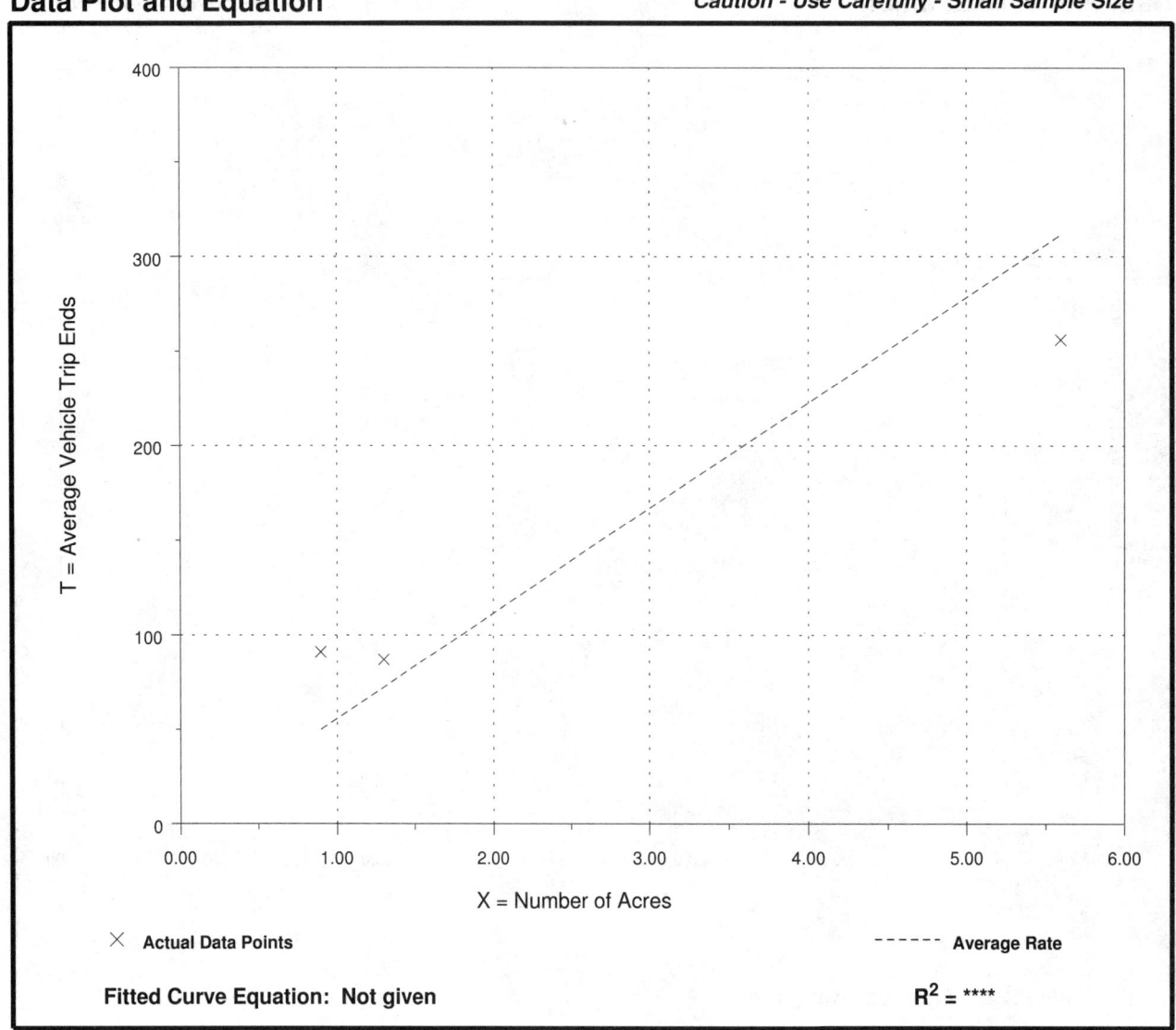

Caution - Use Carefully - Small Sample Size

Fitted Curve Equation: Not given $R^2 = ****$

Hardware/Paint Store
(816)

Average Vehicle Trip Ends vs: Acres
On a: Saturday

Number of Studies: 3
Average Number of Acres: 3
Directional Distribution: 50% entering, 50% exiting

Trip Generation per Acre

Average Rate	Range of Rates	Standard Deviation
878.08	806.79 - 1333.33	179.43

Data Plot and Equation

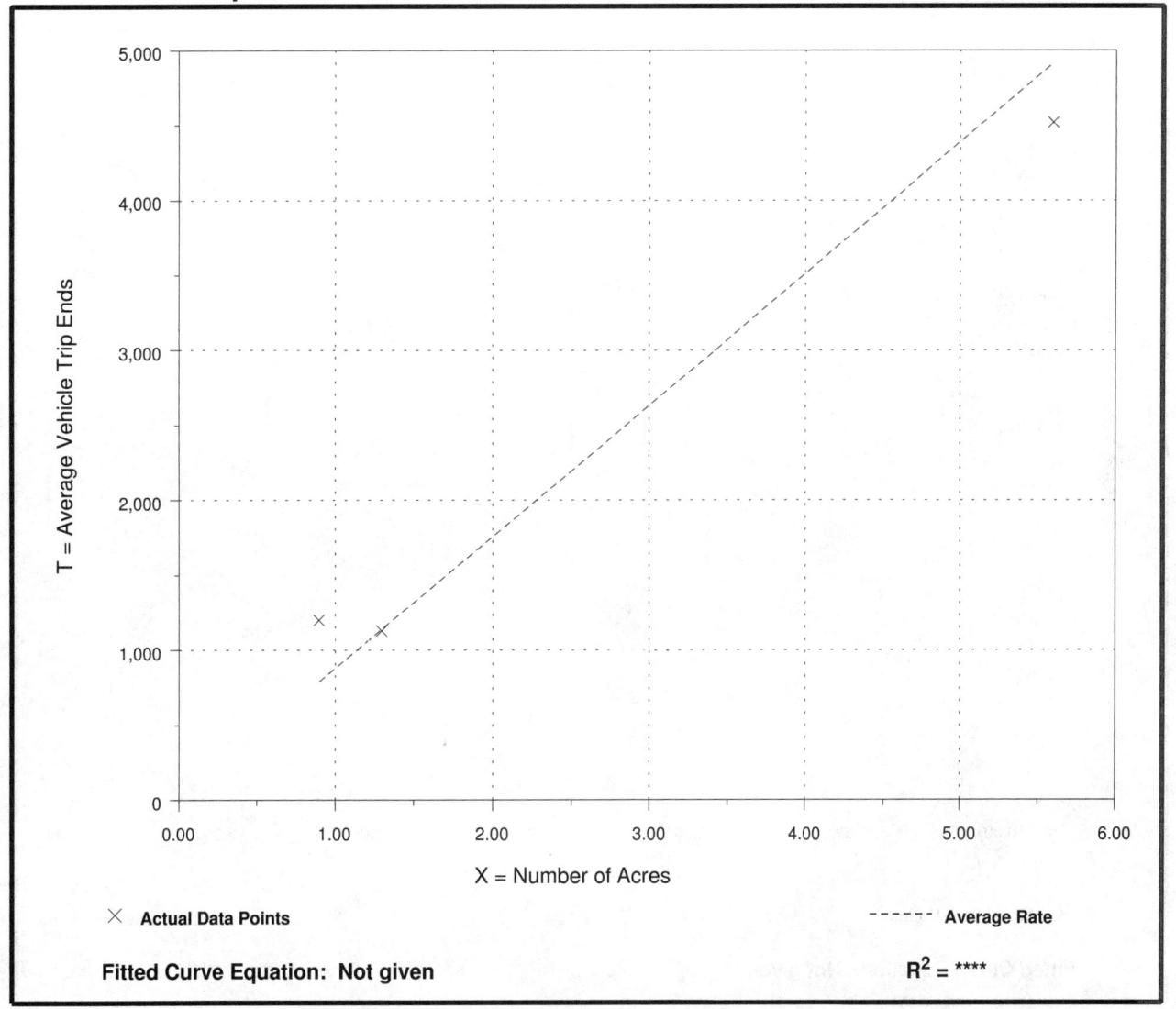

Caution - Use Carefully - Small Sample Size

Fitted Curve Equation: Not given $R^2 =$ ****

Hardware/Paint Store
(816)

Average Vehicle Trip Ends vs: Acres
On a: Saturday,
Peak Hour of Generator

Number of Studies: 3
Average Number of Acres: 3
Directional Distribution: Not available

Trip Generation per Acre

Average Rate	Range of Rates	Standard Deviation
118.97	110.71 - 176.67	24.04

Data Plot and Equation

Caution - Use Carefully - Small Sample Size

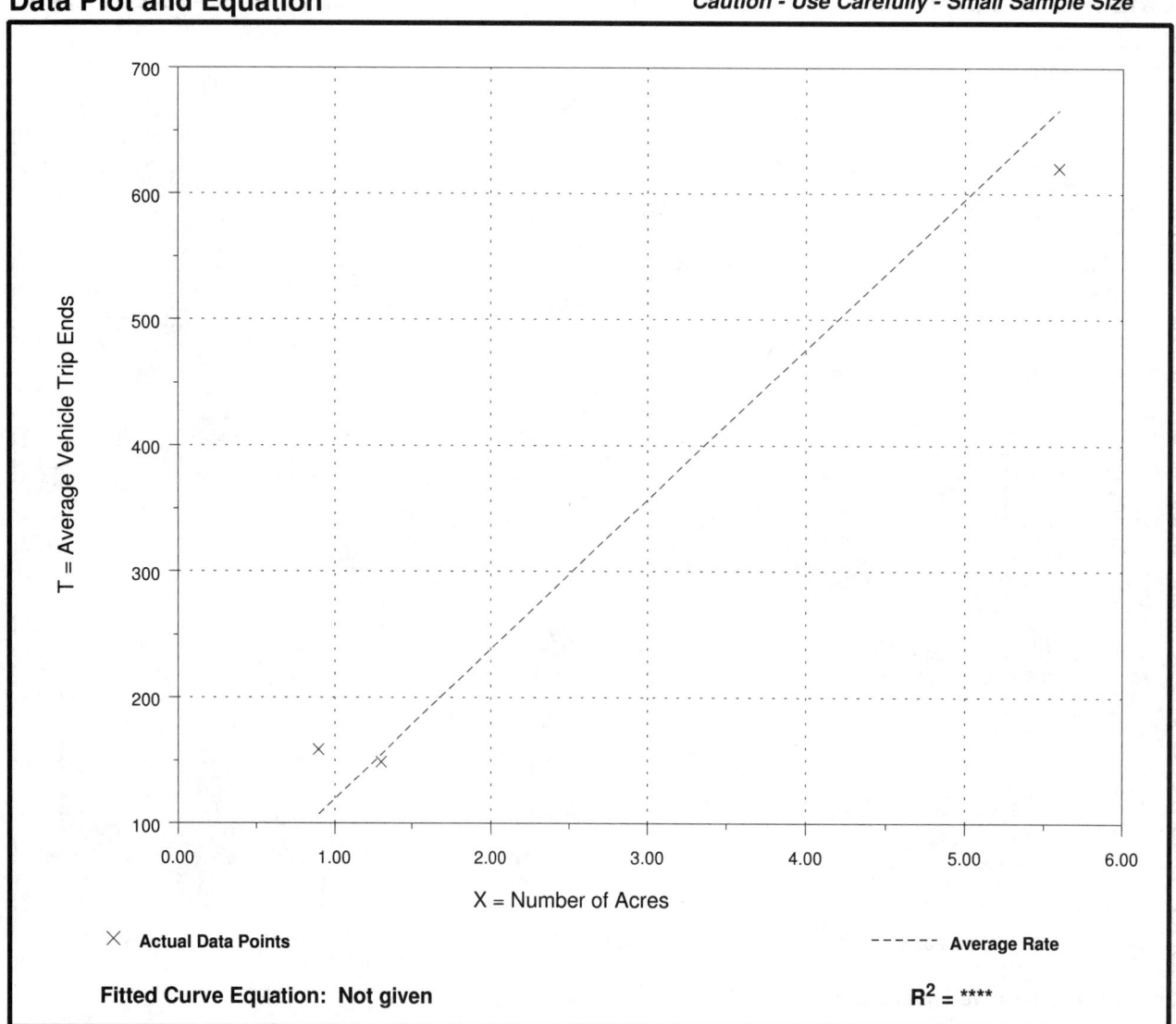

Fitted Curve Equation: Not given $R^2 = ****$

Hardware/Paint Store
(816)

Average Vehicle Trip Ends vs: Acres
On a: Sunday

Number of Studies: 3
Average Number of Acres: 3
Directional Distribution: 50% entering, 50% exiting

Trip Generation per Acre

Average Rate	Range of Rates	Standard Deviation
730.51	685.71 - 1068.89	132.71

Data Plot and Equation

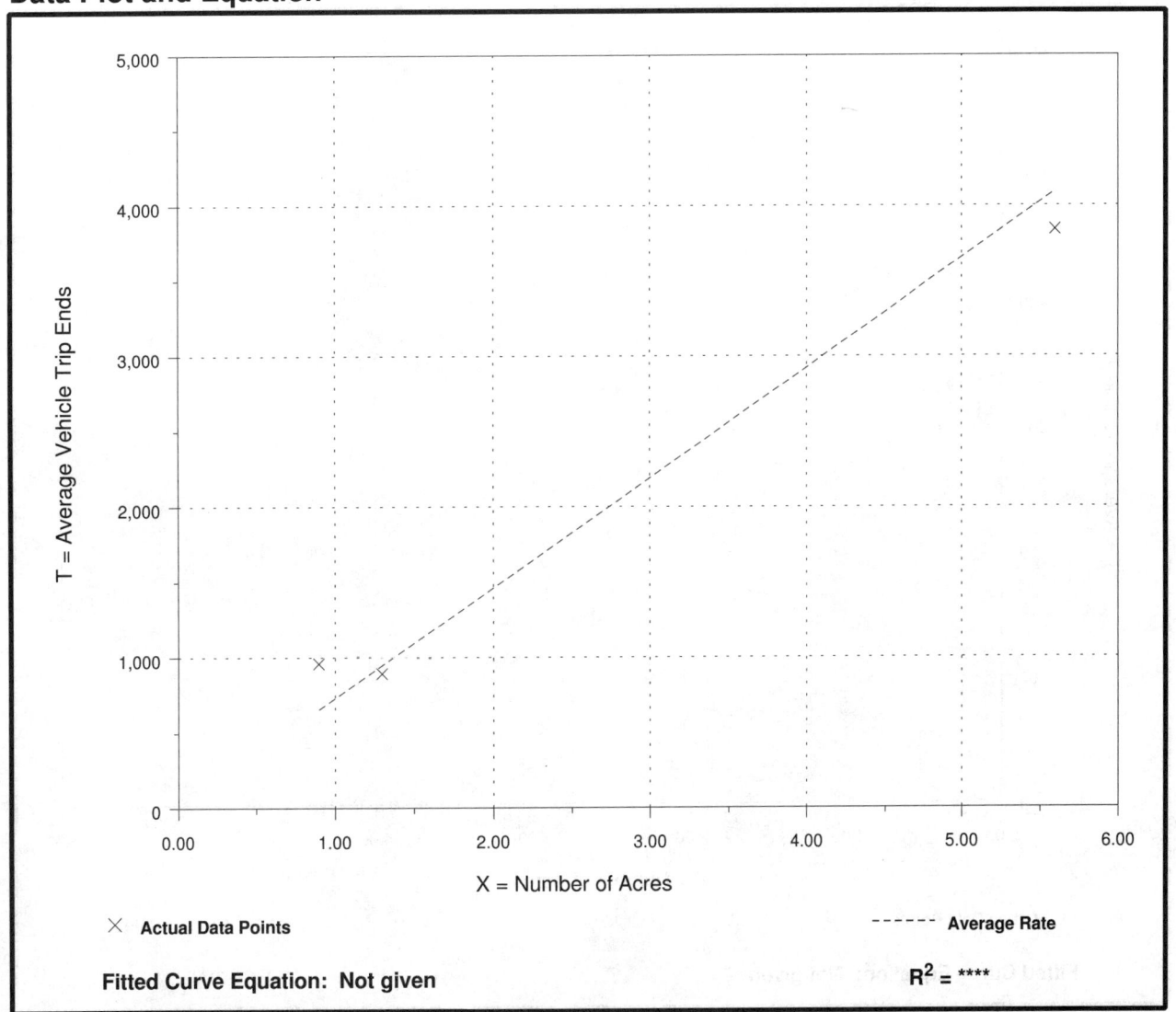

Caution - Use Carefully - Small Sample Size

Fitted Curve Equation: Not given $R^2 = ****$

Hardware/Paint Store
(816)

Average Vehicle Trip Ends vs: Acres
On a: Sunday,
Peak Hour of Generator

Number of Studies: 3
Average Number of Acres: 3
Directional Distribution: Not available

Trip Generation per Acre

Average Rate	Range of Rates	Standard Deviation
104.36	93.93 - 162.22	24.56

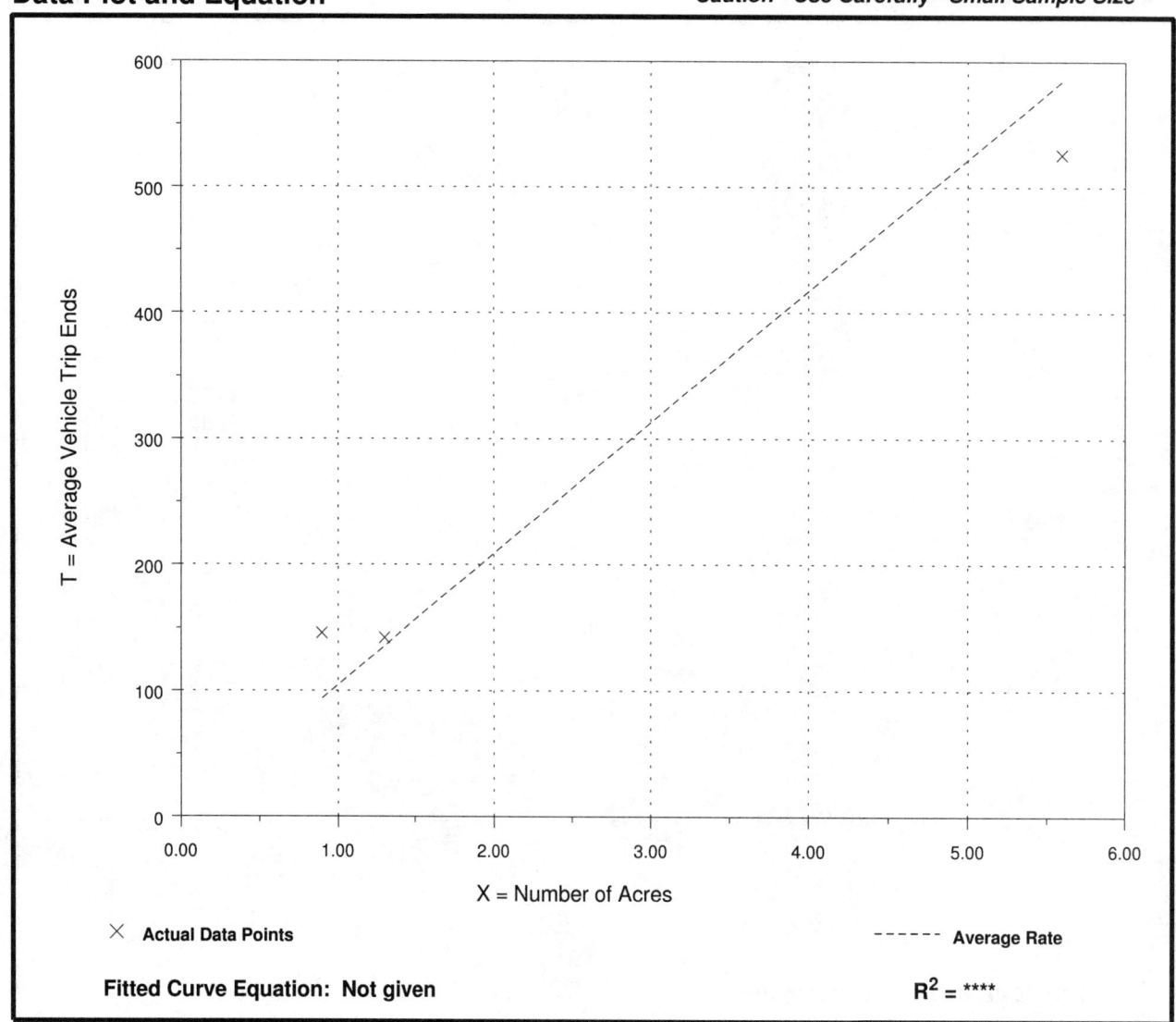

Data Plot and Equation

Caution - Use Carefully - Small Sample Size

Fitted Curve Equation: Not given $R^2 = ****$

Land Use: 817
Nursery (Garden Center)

Description

A nursery or garden center is a free-standing building with an outside storage area for planting or landscape stock. The nurseries surveyed primarily serve the general public. Some have large greenhouses and offer landscaping services. Most have office, storage and shipping facilities. Nurseries are characterized by seasonal variations in trip characteristics. Nursery (Wholesale) (Land Use 818) is a related use.

Additional Data

Outside storage areas are not included in the overall gross floor area measurements. However, if storage areas are located within the principal outside faces of the exterior walls, they are included in the overall gross floor area of the building.

The sites were surveyed in the 1980s in California.

Source Numbers

205, 240

Nursery (Garden Center)
(817)

Average Vehicle Trip Ends vs: Employees
On a: Weekday

Number of Studies: 10
Avg. Number of Employees: 16
Directional Distribution: 50% entering, 50% exiting

Trip Generation per Employee

Average Rate	Range of Rates	Standard Deviation
21.83	10.71 - 53.86	10.64

Data Plot and Equation

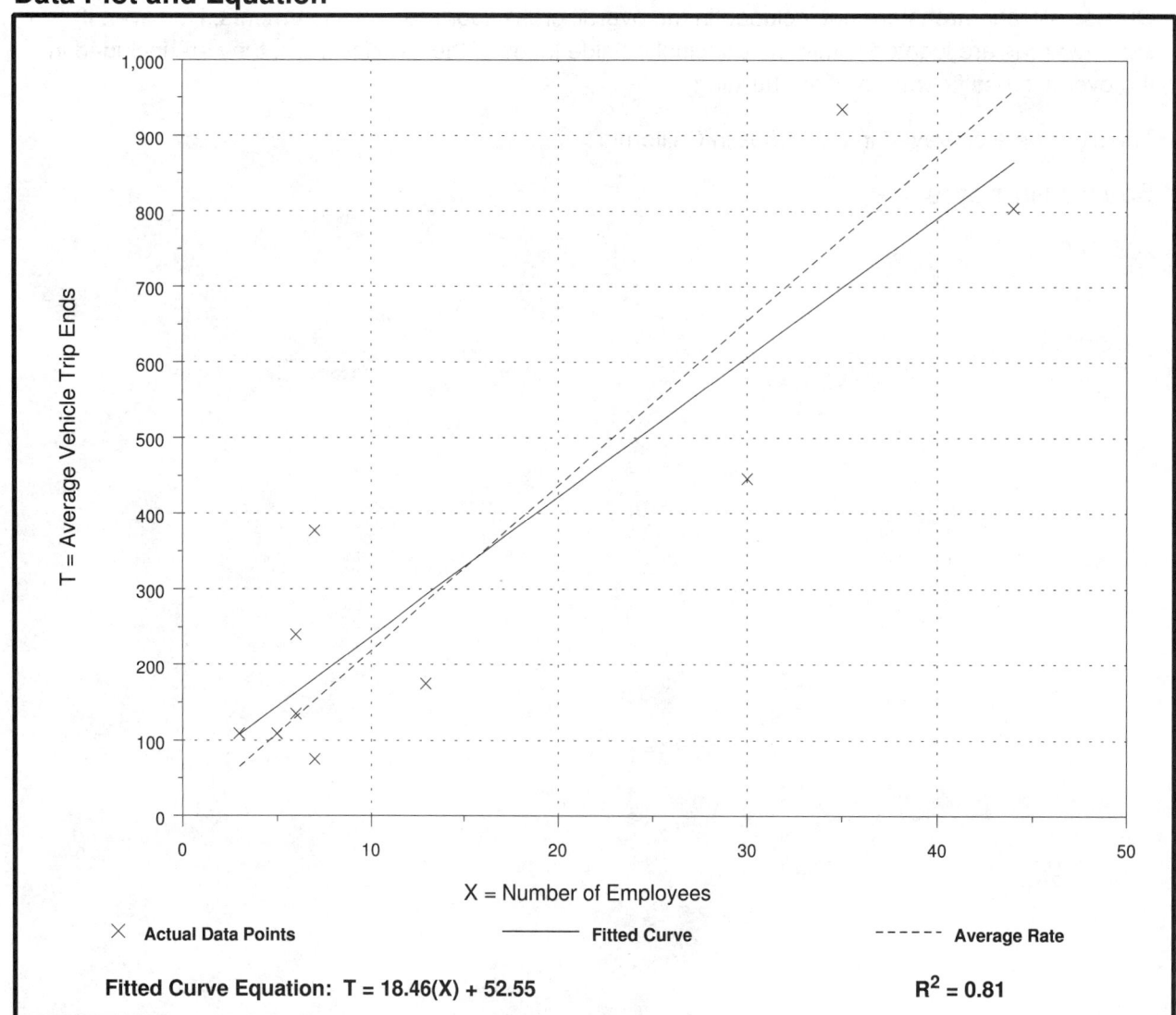

Fitted Curve Equation: $T = 18.46(X) + 52.55$ $R^2 = 0.81$

Nursery (Garden Center)
(817)

Average Vehicle Trip Ends vs: Employees
On a: Weekday,
Peak Hour of Adjacent Street Traffic,
One Hour Between 7 and 9 a.m.

Number of Studies: 11
Avg. Number of Employees: 17
Directional Distribution: Not available

Trip Generation per Employee

Average Rate	Range of Rates	Standard Deviation
0.69	0.08 - 1.67	0.87

Data Plot and Equation

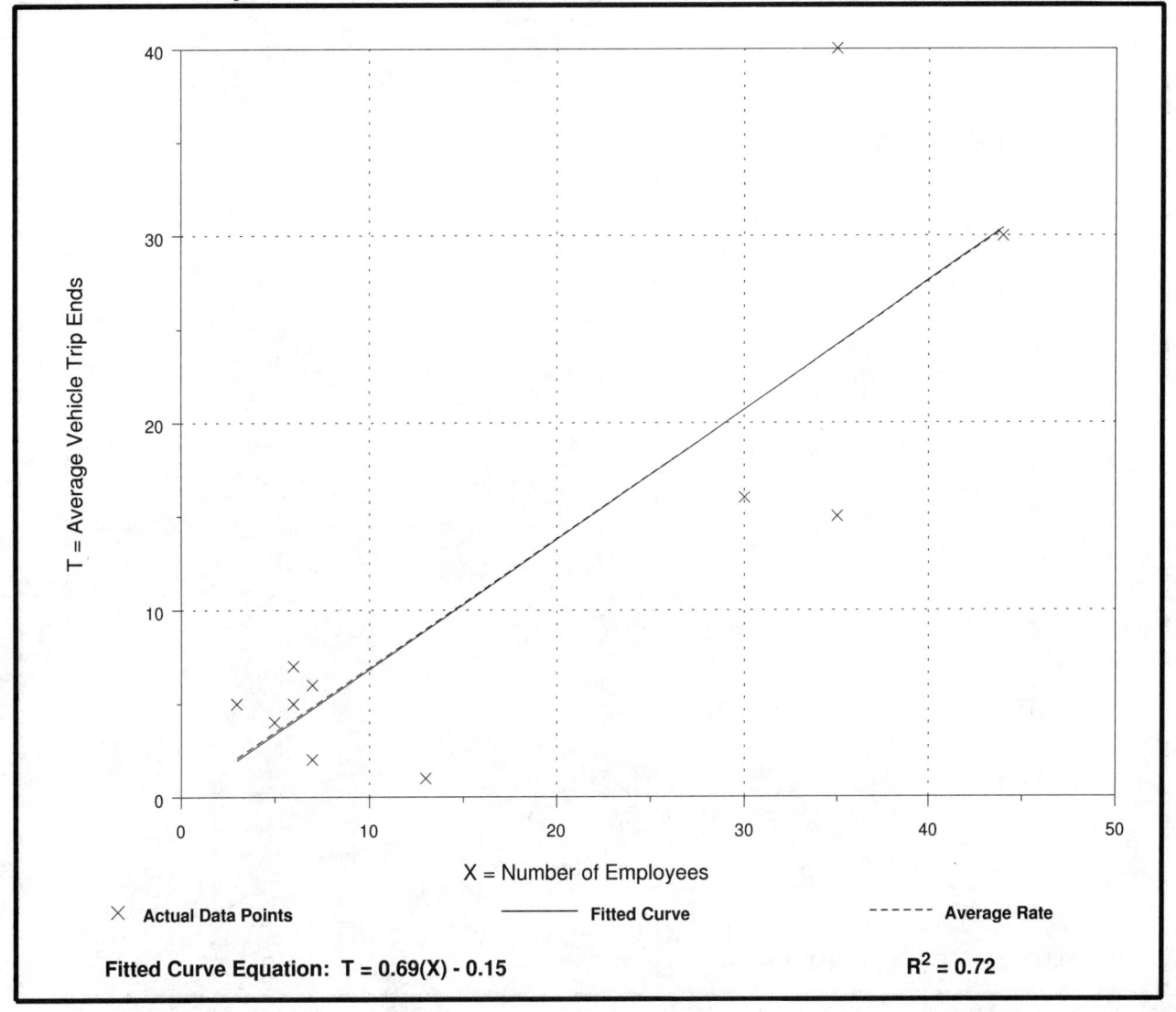

Fitted Curve Equation: $T = 0.69(X) - 0.15$ $R^2 = 0.72$

Nursery (Garden Center)
(817)

Average Vehicle Trip Ends vs: Employees
On a: Weekday,
Peak Hour of Adjacent Street Traffic,
One Hour Between 4 and 6 p.m.

Number of Studies: 11
Avg. Number of Employees: 17
Directional Distribution: Not available

Trip Generation per Employee

Average Rate	Range of Rates	Standard Deviation
1.96	0.06 - 7.43	2.07

Data Plot and Equation

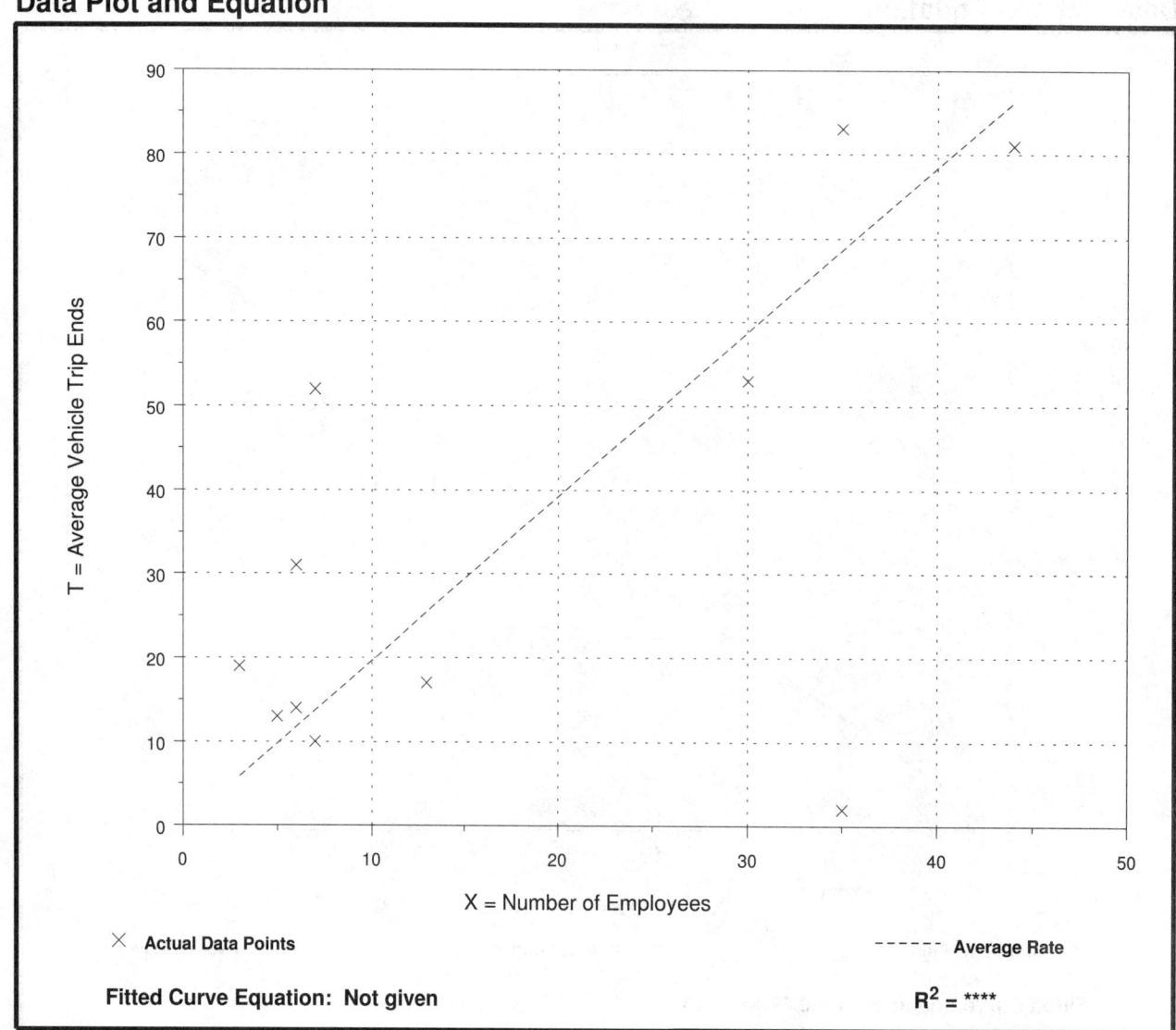

Fitted Curve Equation: Not given $R^2 = ****$

Nursery (Garden Center)
(817)

Average Vehicle Trip Ends vs: Employees
On a: Weekday,
A.M. Peak Hour of Generator

Number of Studies: 11
Avg. Number of Employees: 17
Directional Distribution: 52% entering, 48% exiting

Trip Generation per Employee

Average Rate	Range of Rates	Standard Deviation
2.26	1.14 - 7.00	1.92

Data Plot and Equation

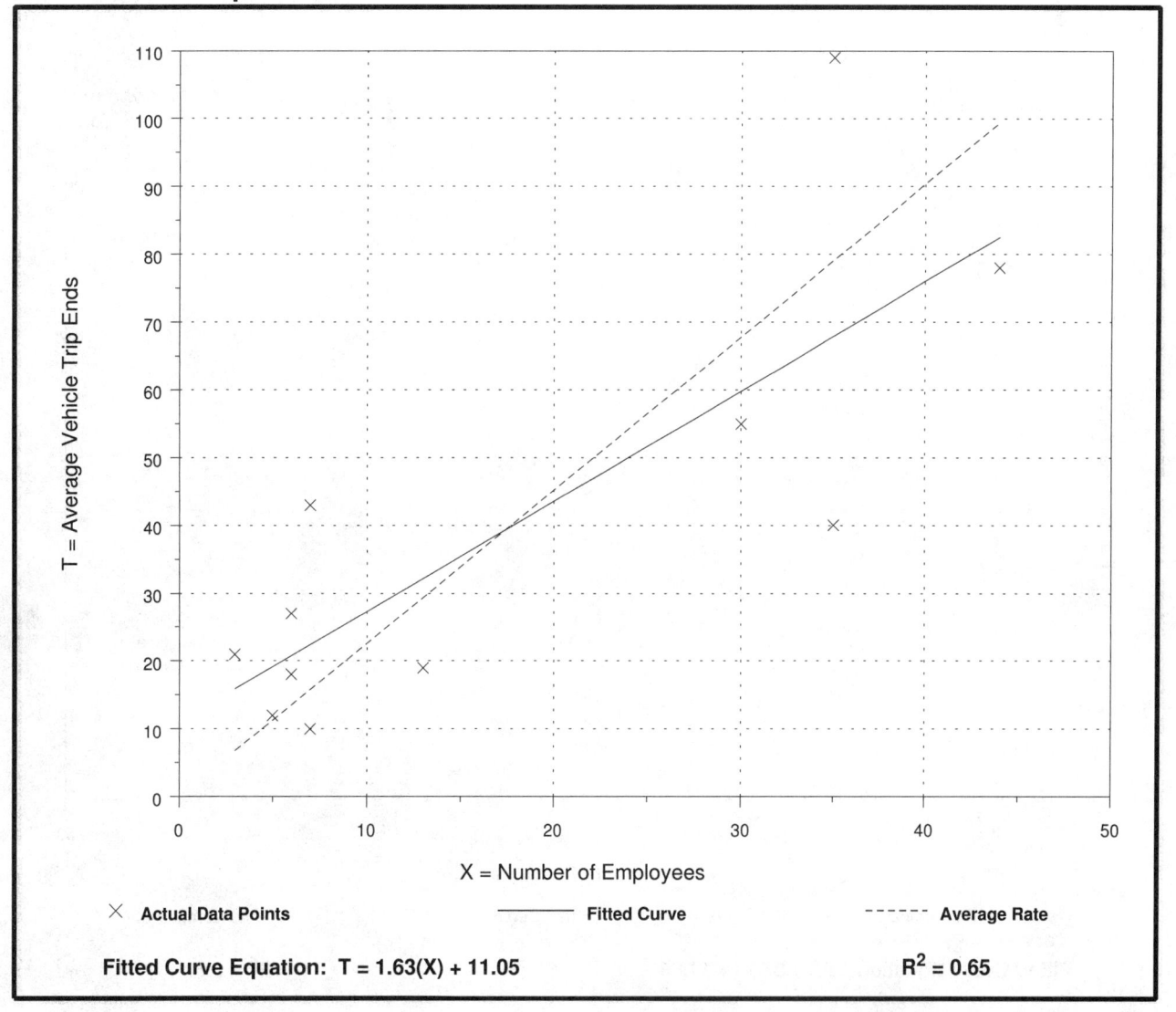

Fitted Curve Equation: $T = 1.63(X) + 11.05$ $R^2 = 0.65$

Nursery (Garden Center)
(817)

Average Vehicle Trip Ends vs: Employees
On a: Weekday,
P.M. Peak Hour of Generator

Number of Studies: 11
Avg. Number of Employees: 17
Directional Distribution: 49% entering, 51% exiting

Trip Generation per Employee

Average Rate	Range of Rates	Standard Deviation
2.55	1.03 - 7.43	2.10

Data Plot and Equation

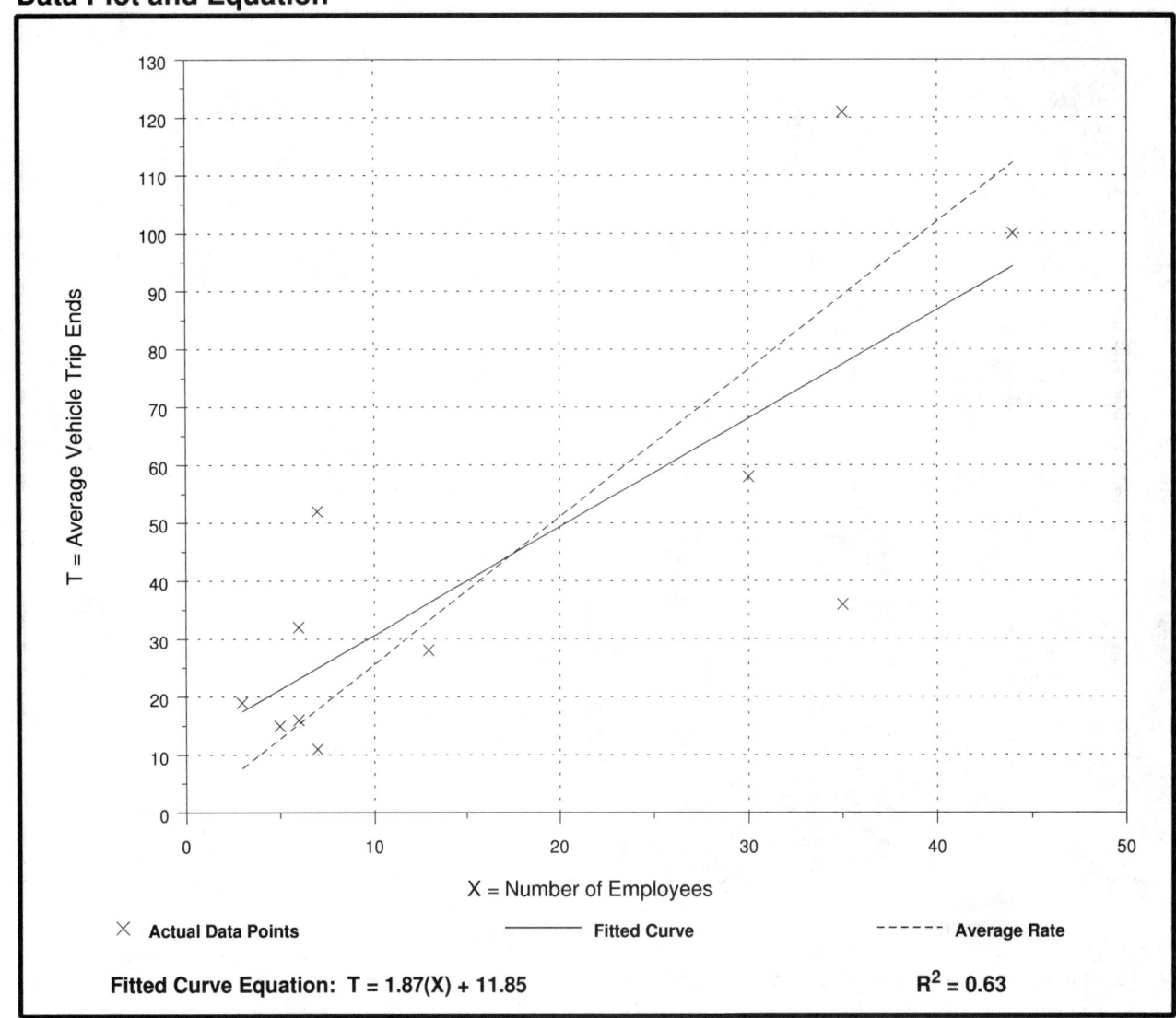

Fitted Curve Equation: T = 1.87(X) + 11.85 $R^2 = 0.63$

Nursery (Garden Center)
(817)

Average Vehicle Trip Ends vs: Employees
On a: Saturday

Number of Studies: 11
Avg. Number of Employees: 17
Directional Distribution: 50% entering, 50% exiting

Trip Generation per Employee

Average Rate	Range of Rates	Standard Deviation
37.69	3.89 - 160.14	32.57

Data Plot and Equation

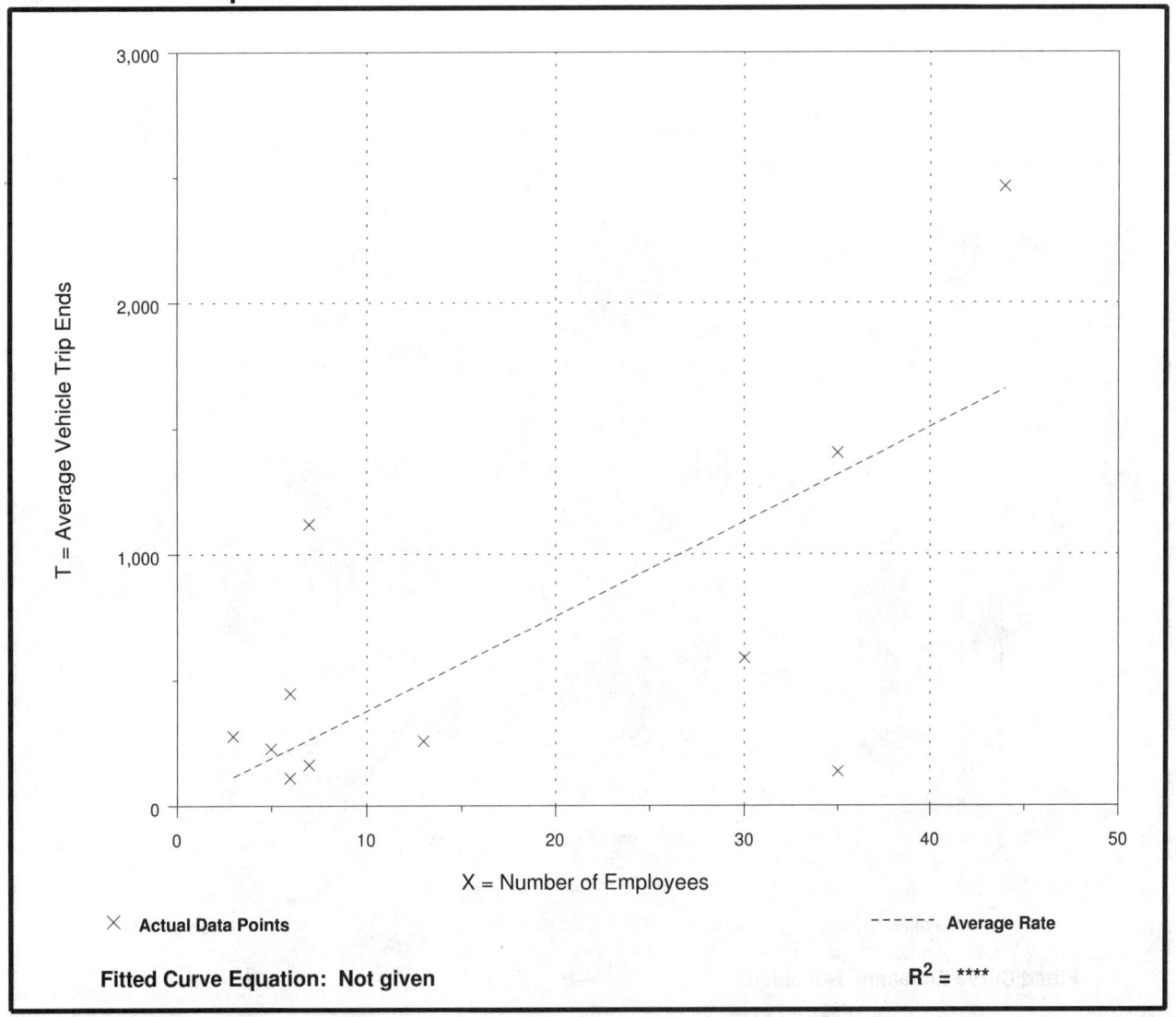

Fitted Curve Equation: Not given $R^2 = ****$

Nursery (Garden Center)
(817)

Average Vehicle Trip Ends vs: Employees
On a: Saturday,
Peak Hour of Generator

Number of Studies: 11
Avg. Number of Employees: 17
Directional Distribution: Not available

Trip Generation per Employee

Average Rate	Range of Rates	Standard Deviation
5.67	0.71 - 24.86	5.75

Data Plot and Equation

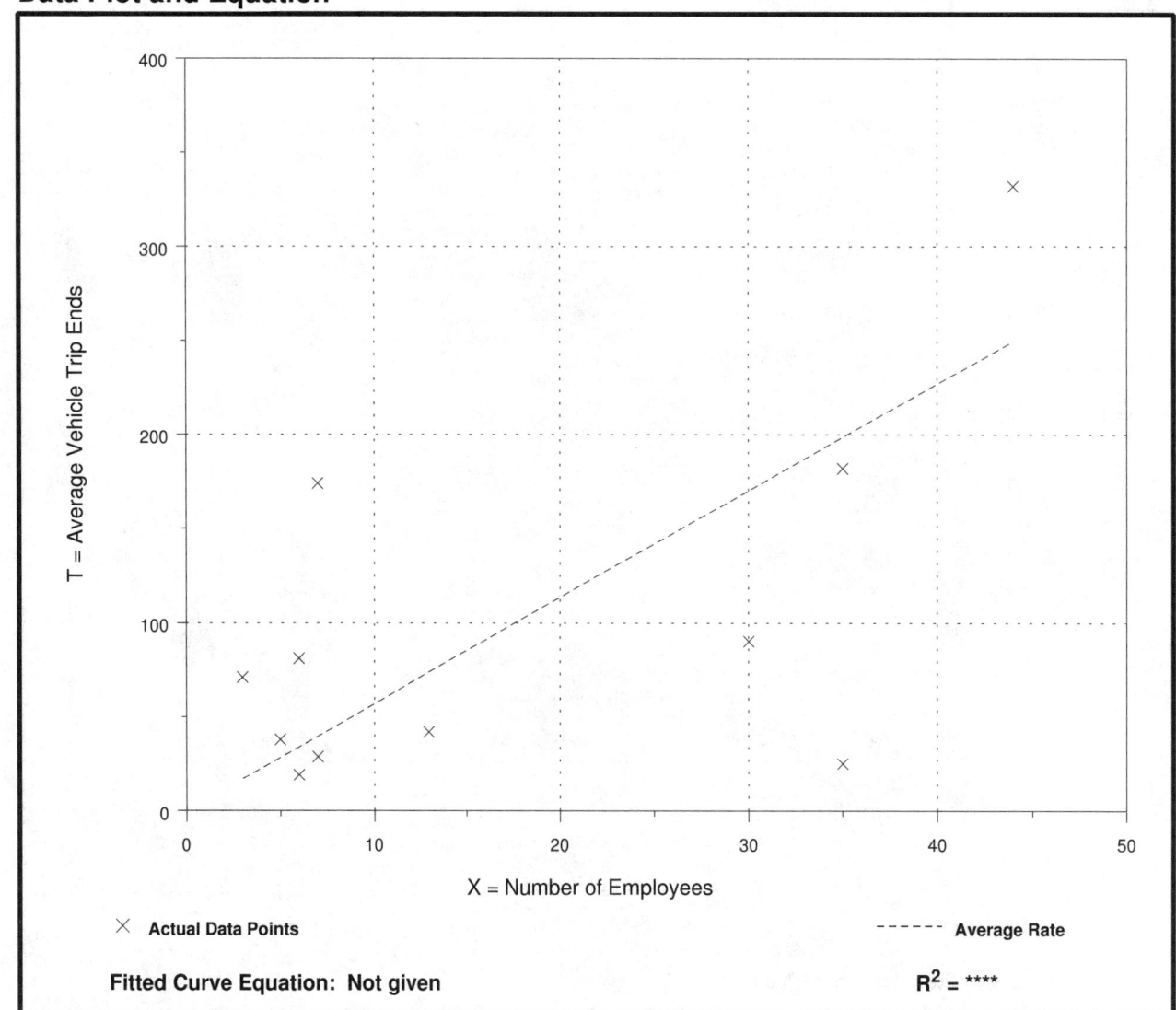

Fitted Curve Equation: Not given $R^2 = ****$

Nursery (Garden Center)
(817)

Average Vehicle Trip Ends vs: Employees
On a: Sunday

Number of Studies: 11
Avg. Number of Employees: 17
Directional Distribution: 50% entering, 50% exiting

Trip Generation per Employee

Average Rate	Range of Rates	Standard Deviation
30.03	0.31 - 160.29	31.54

Data Plot and Equation

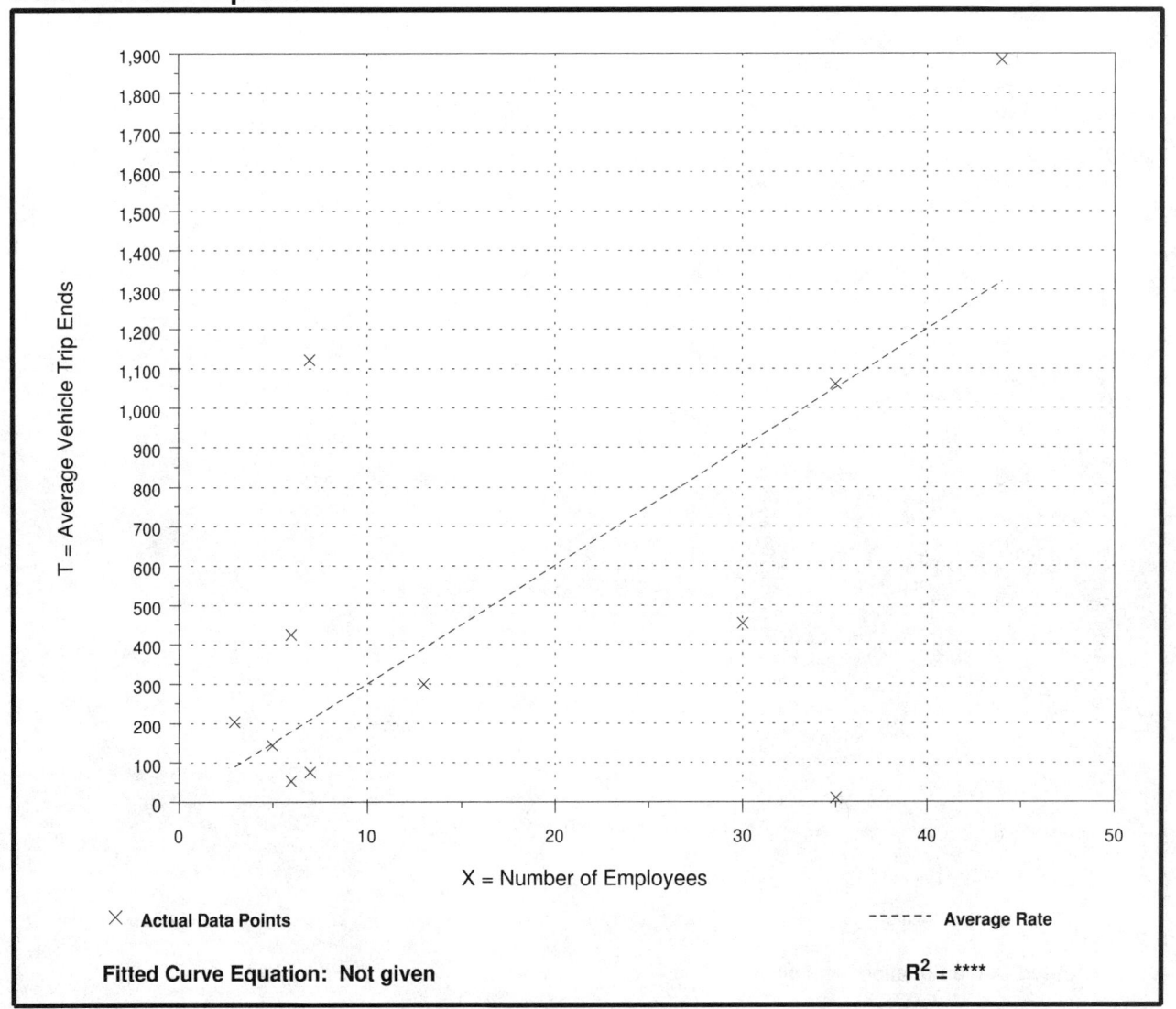

X = Number of Employees
T = Average Vehicle Trip Ends

× Actual Data Points   ----- Average Rate

Fitted Curve Equation: Not given $R^2 = ****$

Nursery (Garden Center)
(817)

Average Vehicle Trip Ends vs: Employees
On a: Sunday,
Peak Hour of Generator

Number of Studies: 11
Avg. Number of Employees: 17
Directional Distribution: Not available

Trip Generation per Employee

Average Rate	Range of Rates	Standard Deviation
5.30	0.26 - 30.14	6.21

Data Plot and Equation

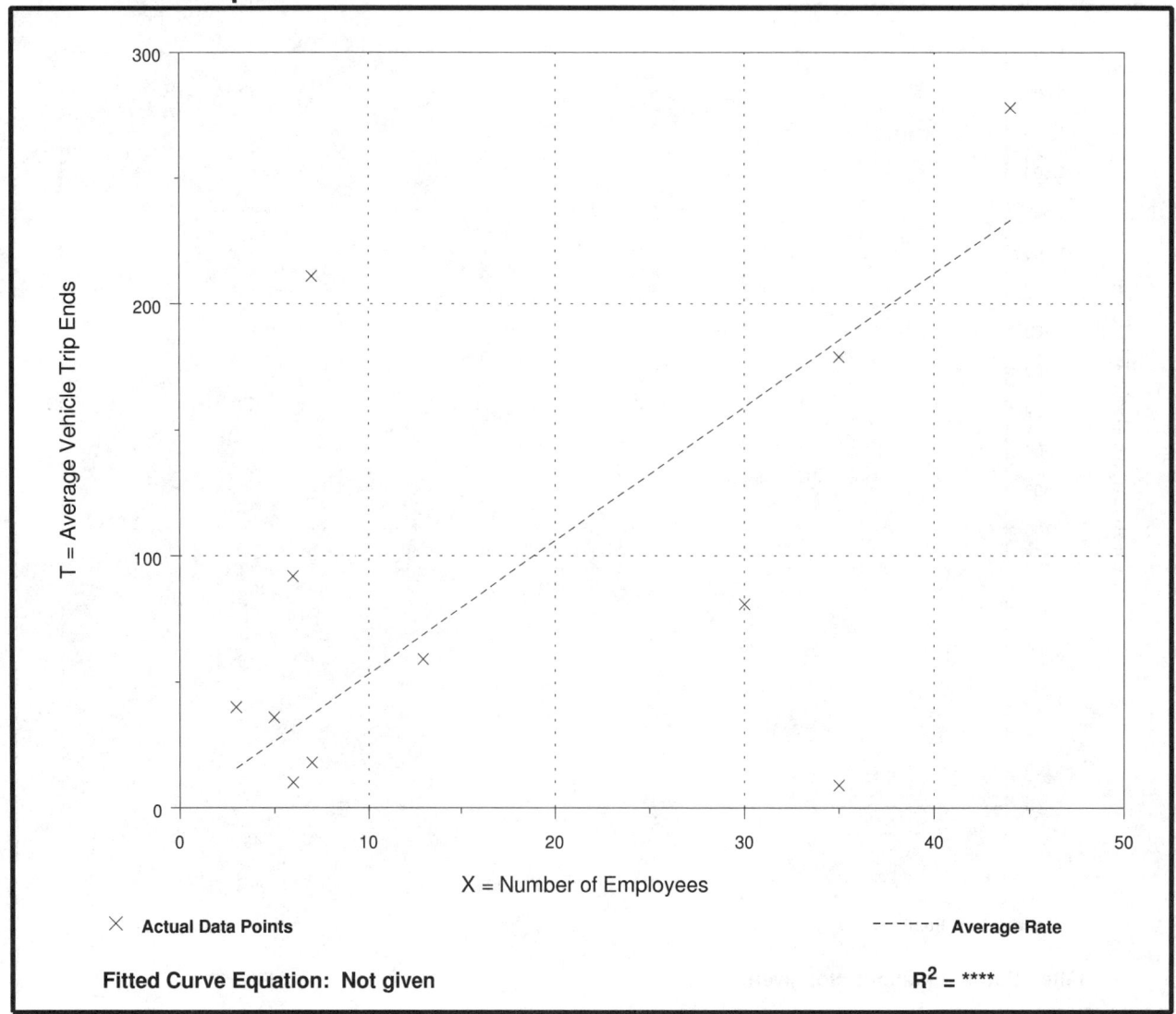

Fitted Curve Equation: Not given $R^2 = ****$

Nursery (Garden Center)
(817)

Average Vehicle Trip Ends vs: 1000 Sq. Feet Gross Floor Area
On a: Weekday

Number of Studies: 10
Average 1000 Sq. Feet GFA: 5
Directional Distribution: 50% entering, 50% exiting

Trip Generation per 1000 Sq. Feet Gross Floor Area

Average Rate	Range of Rates	Standard Deviation
68.10	18.46 - 233.75	58.80

Data Plot and Equation

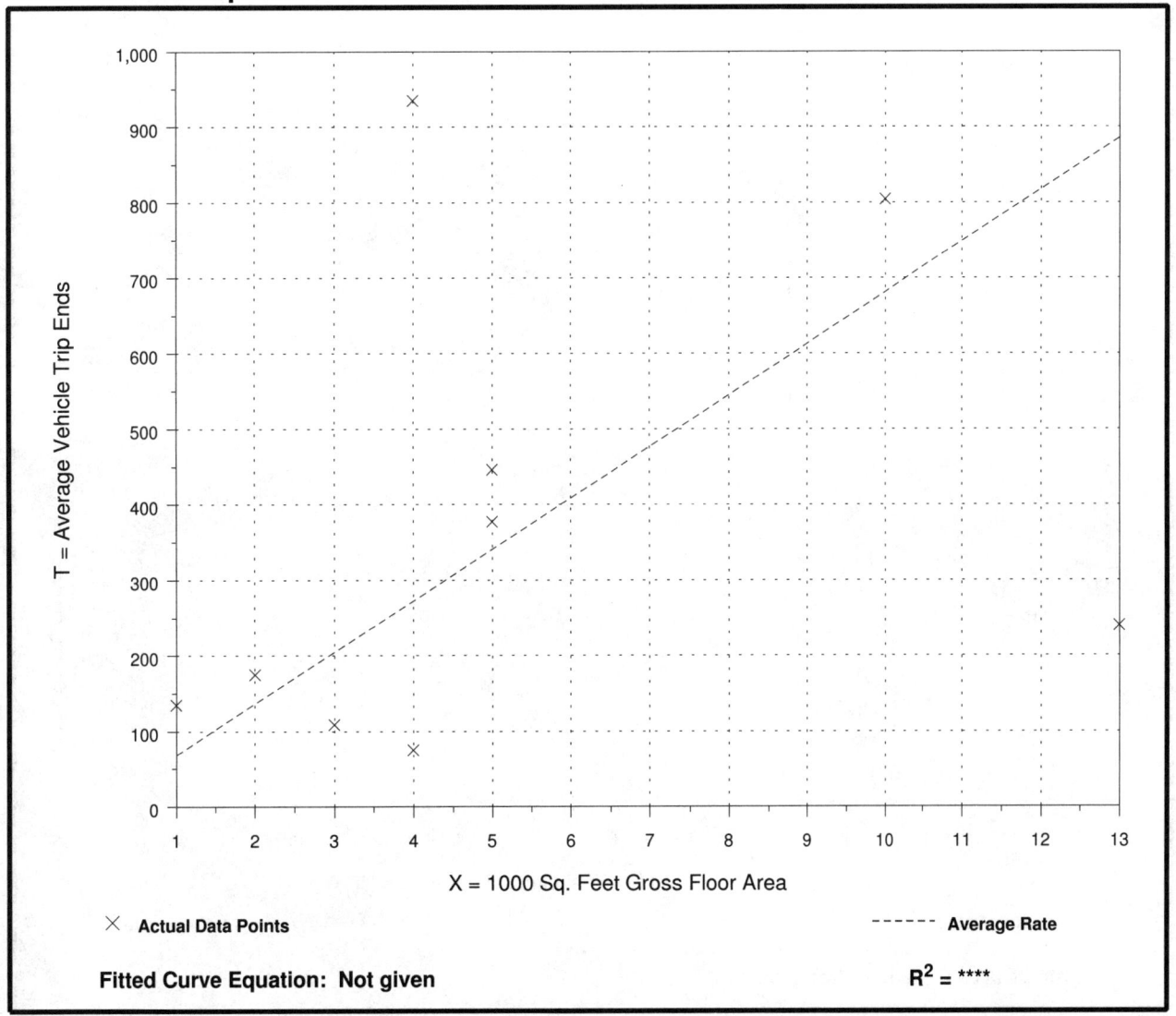

Fitted Curve Equation: Not given $R^2 = ****$

Nursery (Garden Center)
(817)

Average Vehicle Trip Ends vs: 1000 Sq. Feet Gross Floor Area
On a: Weekday,
Peak Hour of Adjacent Street Traffic,
One Hour Between 7 and 9 a.m.

Number of Studies: 11
Average 1000 Sq. Feet GFA: 5
Directional Distribution: Not available

Trip Generation per 1000 Sq. Feet Gross Floor Area

Average Rate	Range of Rates	Standard Deviation
2.43	0.38 - 10.00	2.92

Data Plot and Equation

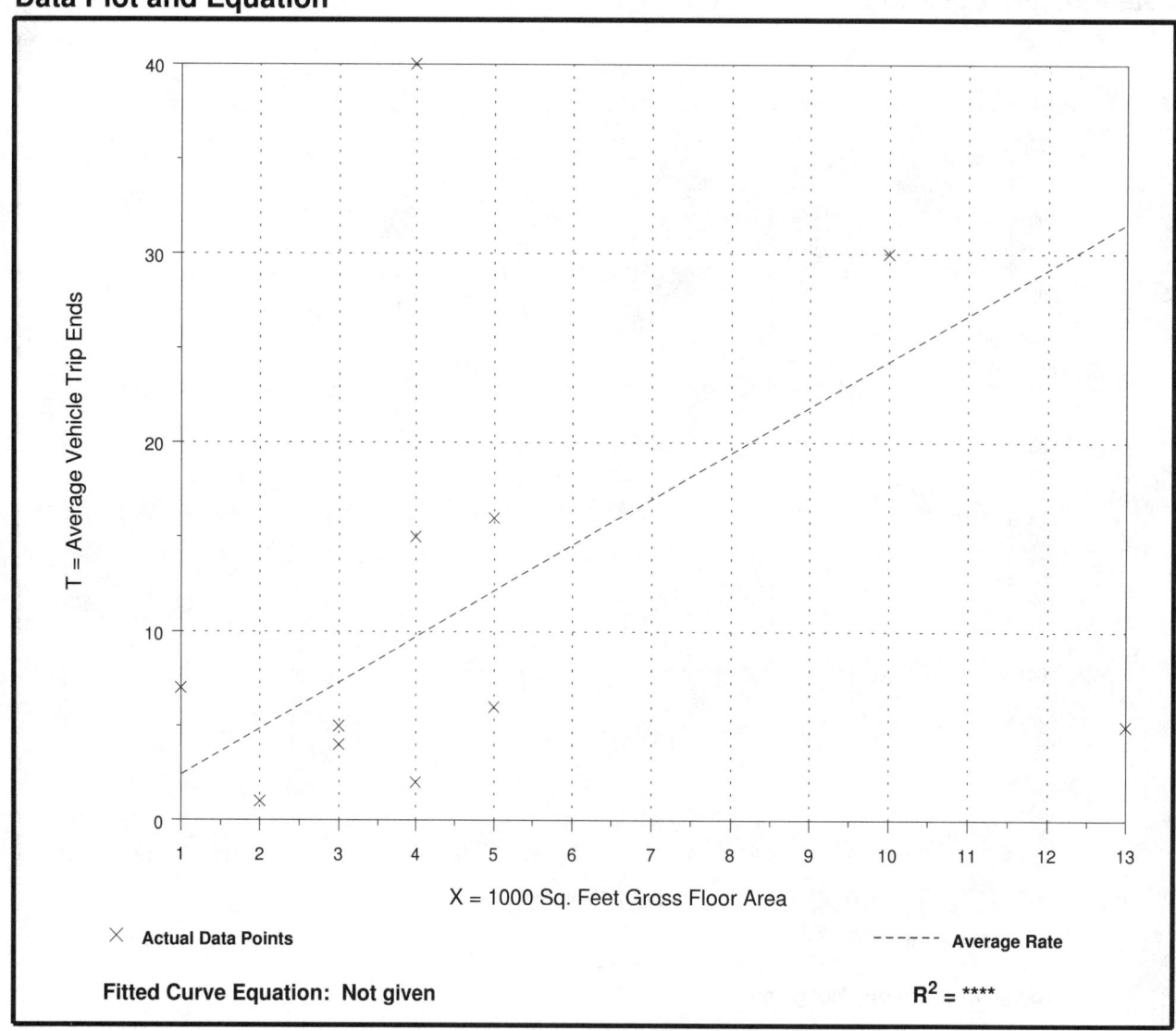

Fitted Curve Equation: Not given $R^2 = ****$

Nursery (Garden Center)
(817)

Average Vehicle Trip Ends vs: 1000 Sq. Feet Gross Floor Area
On a: Weekday,
Peak Hour of Adjacent Street Traffic,
One Hour Between 4 and 6 p.m.

Number of Studies: 11
Average 1000 Sq. Feet GFA: 5
Directional Distribution: Not available

Trip Generation per 1000 Sq. Feet Gross Floor Area

Average Rate	Range of Rates	Standard Deviation
6.94	0.50 - 20.75	5.80

Data Plot and Equation

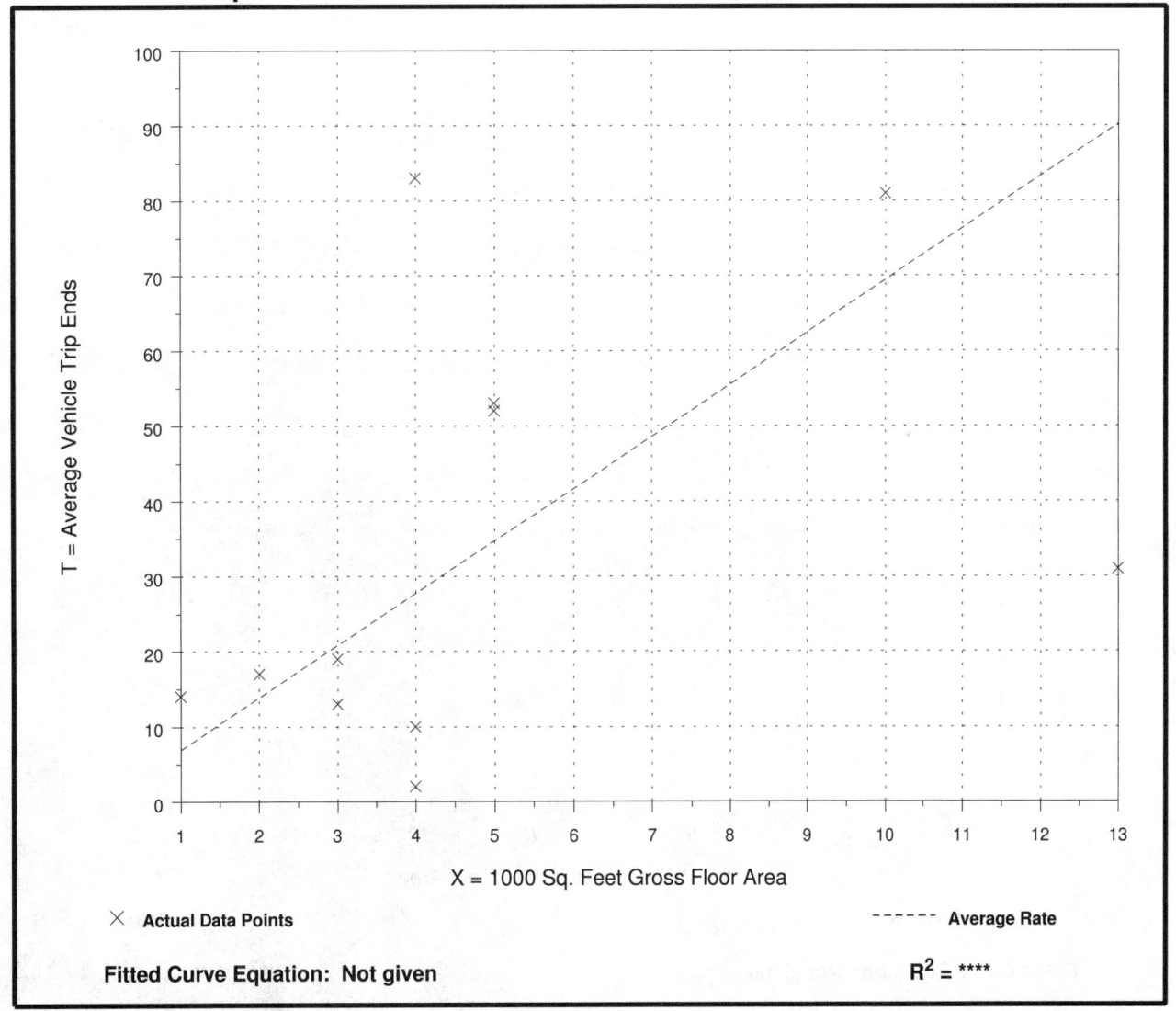

Fitted Curve Equation: Not given $R^2 = ****$

Nursery (Garden Center)
(817)

Average Vehicle Trip Ends vs: 1000 Sq. Feet Gross Floor Area
On a: Weekday,
A.M. Peak Hour of Generator

Number of Studies: 11
Average 1000 Sq. Feet GFA: 5
Directional Distribution: 52% entering, 48% exiting

Trip Generation per 1000 Sq. Feet Gross Floor Area

Average Rate	Range of Rates	Standard Deviation
8.00	2.08 - 27.25	7.01

Data Plot and Equation

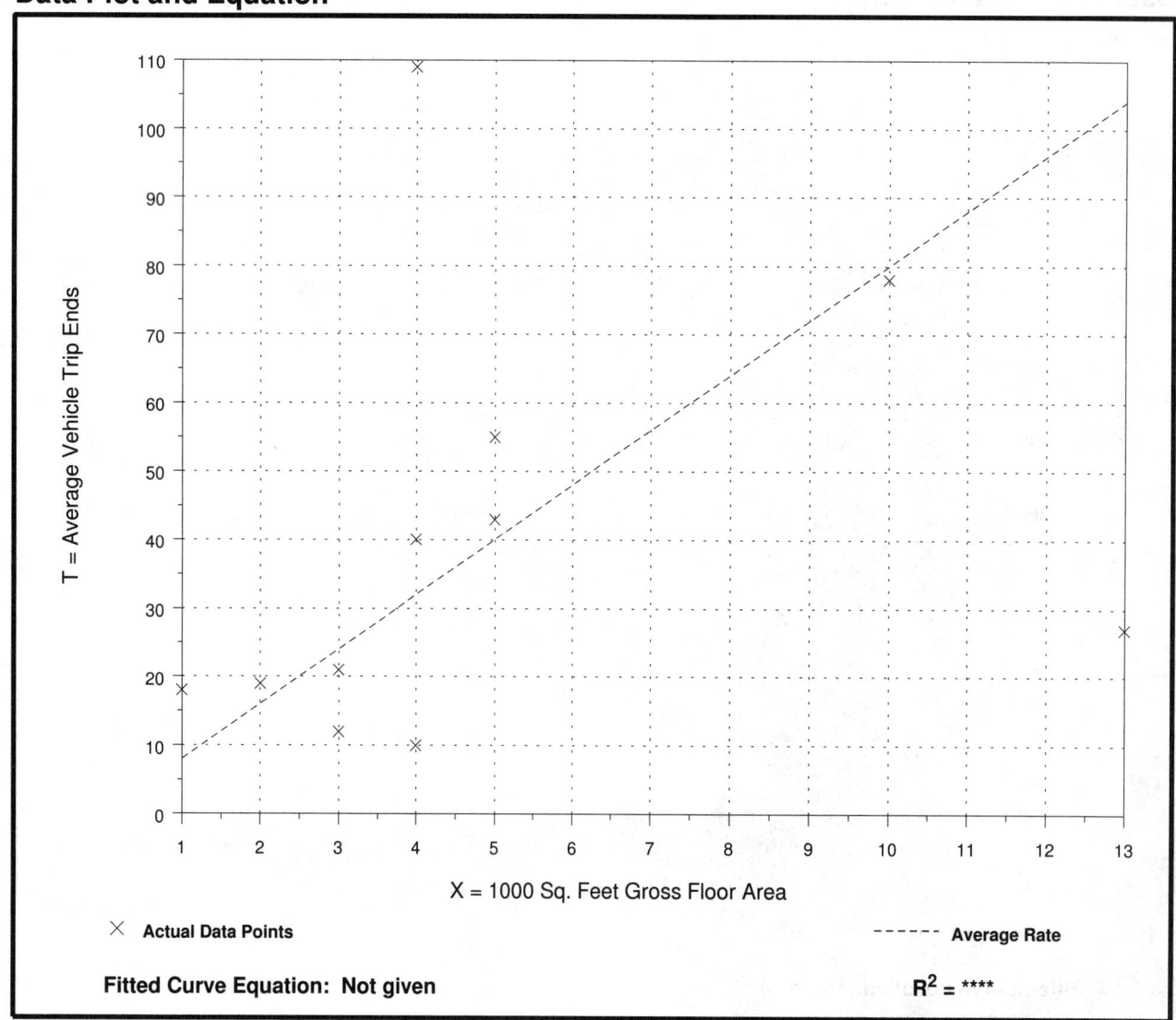

Fitted Curve Equation: Not given $R^2 = ****$

Nursery (Garden Center)
(817)

Average Vehicle Trip Ends vs: 1000 Sq. Feet Gross Floor Area
On a: Weekday,
P.M. Peak Hour of Generator

Number of Studies: 11
Average 1000 Sq. Feet GFA: 5
Directional Distribution: 49% entering, 51% exiting

Trip Generation per 1000 Sq. Feet Gross Floor Area

Average Rate	Range of Rates	Standard Deviation
9.04	2.46 - 30.25	7.66

Data Plot and Equation

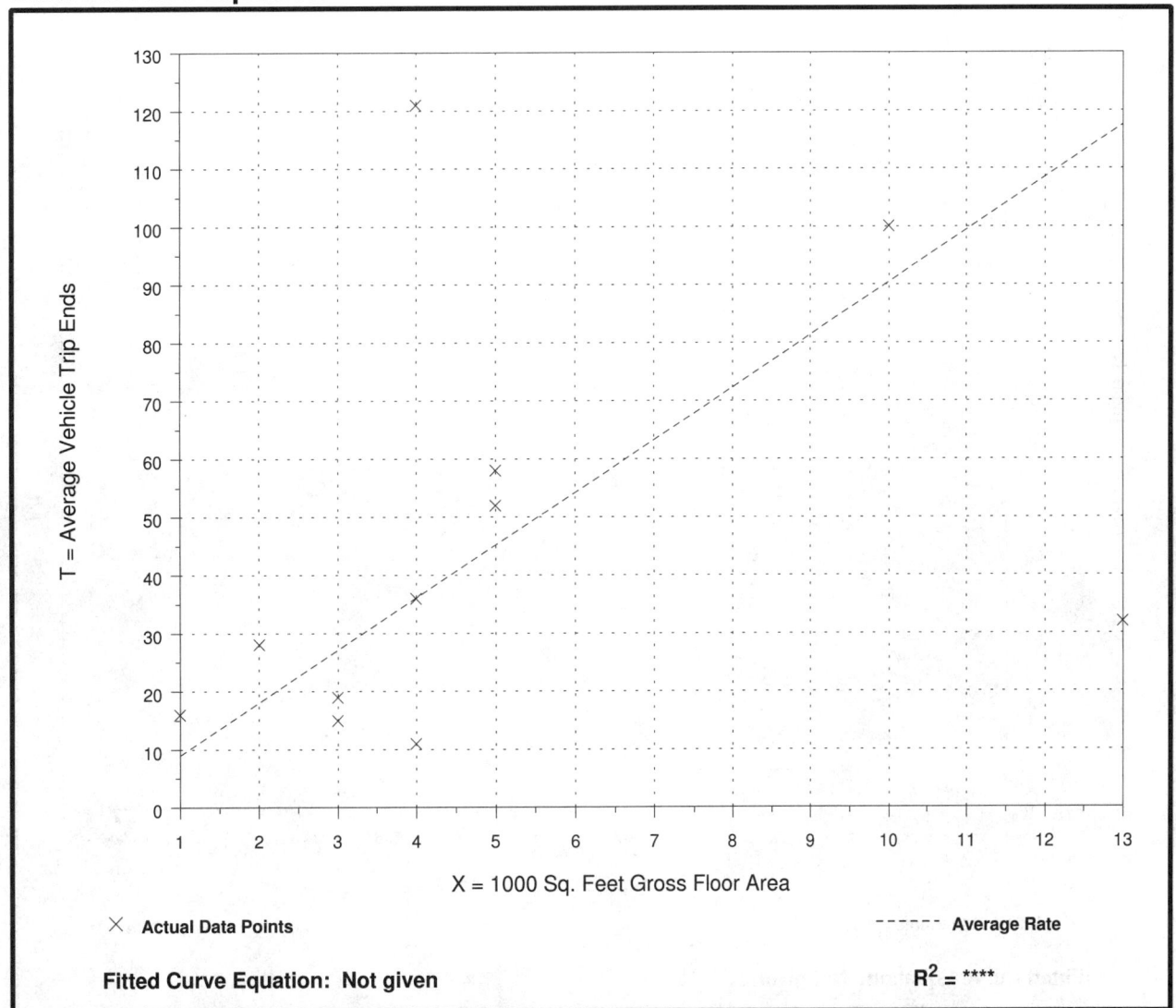

Fitted Curve Equation: Not given $R^2 = ****$

Nursery (Garden Center)
(817)

Average Vehicle Trip Ends vs: 1000 Sq. Feet Gross Floor Area
On a: Saturday

Number of Studies: 11
Average 1000 Sq. Feet GFA: 5
Directional Distribution: 50% entering, 50% exiting

Trip Generation per 1000 Sq. Feet Gross Floor Area

Average Rate	Range of Rates	Standard Deviation
133.31	34.00 - 351.25	104.85

Data Plot and Equation

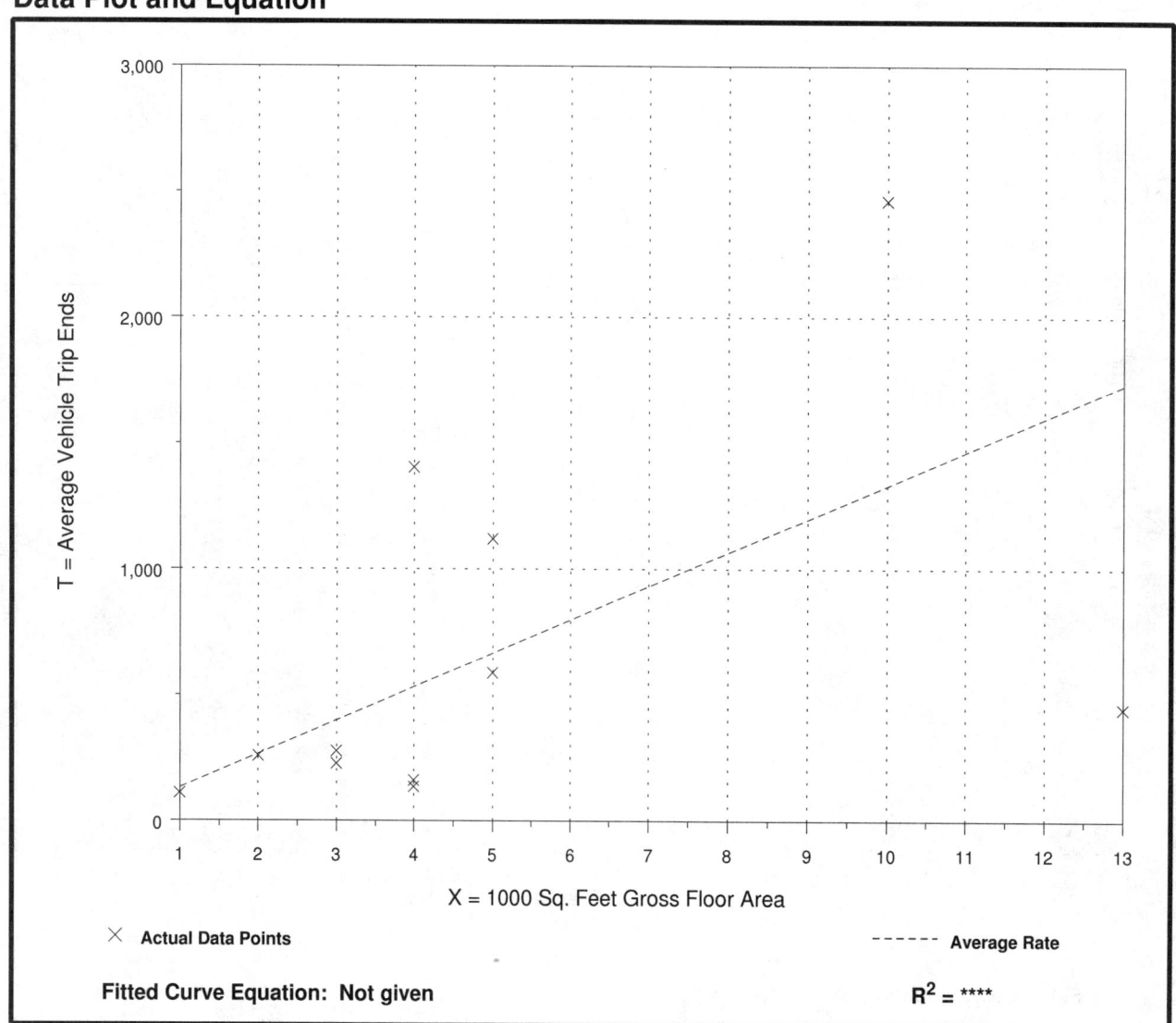

Fitted Curve Equation: Not given $R^2 = ****$

Nursery (Garden Center)
(817)

Average Vehicle Trip Ends vs: 1000 Sq. Feet Gross Floor Area
On a: Saturday,
Peak Hour of Generator

Number of Studies: 11
Average 1000 Sq. Feet GFA: 5
Directional Distribution: Not available

Trip Generation per 1000 Sq. Feet Gross Floor Area

Average Rate	Range of Rates	Standard Deviation
20.06	6.23 - 45.50	14.00

Data Plot and Equation

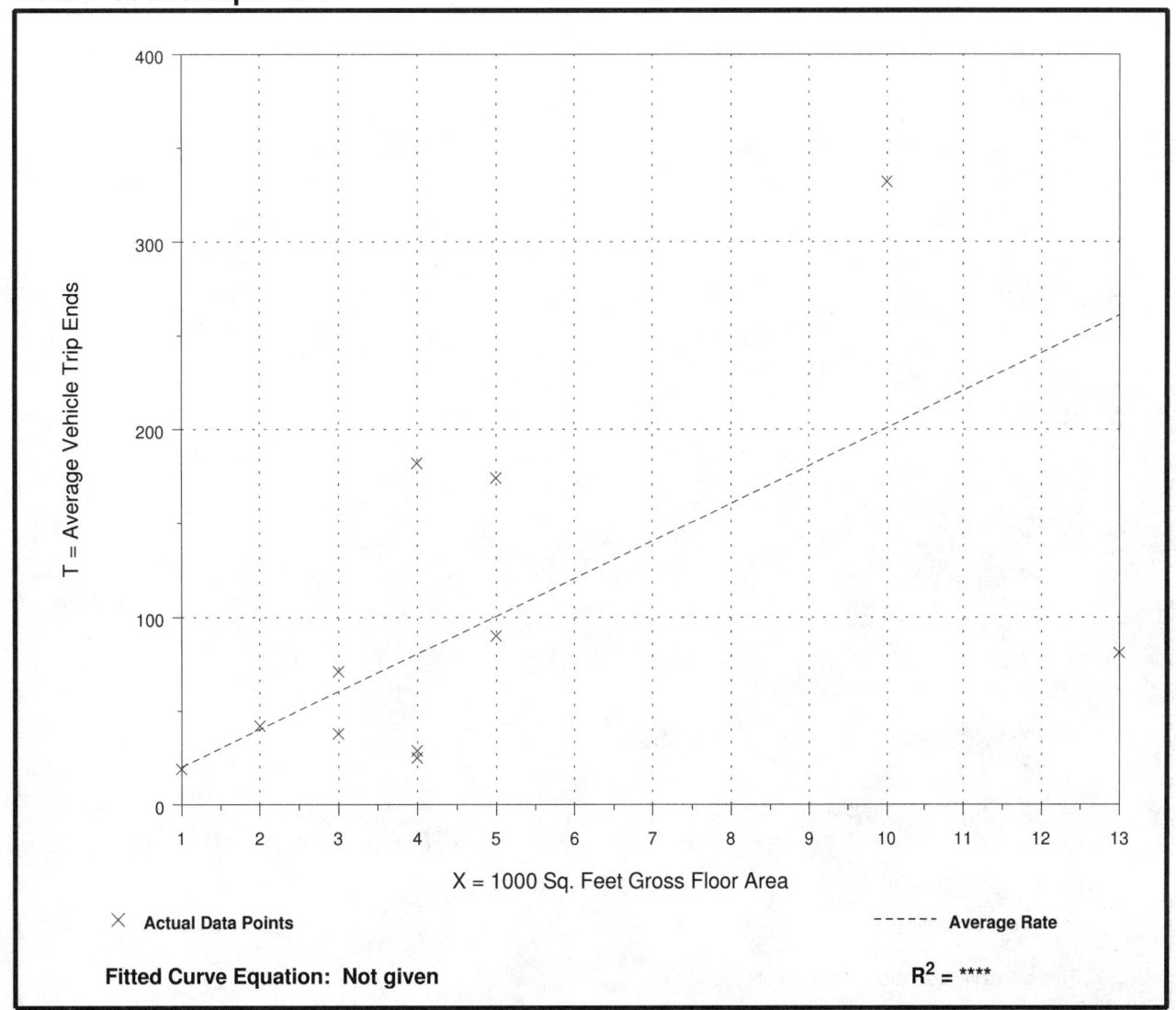

Fitted Curve Equation: Not given $R^2 = ****$

Nursery (Garden Center)
(817)

Average Vehicle Trip Ends vs: 1000 Sq. Feet Gross Floor Area
On a: Sunday

Number of Studies: 11
Average 1000 Sq. Feet GFA: 5
Directional Distribution: 50% entering, 50% exiting

Trip Generation per 1000 Sq. Feet Gross Floor Area

Average Rate	Range of Rates	Standard Deviation
106.20	2.75 - 265.25	87.83

Data Plot and Equation

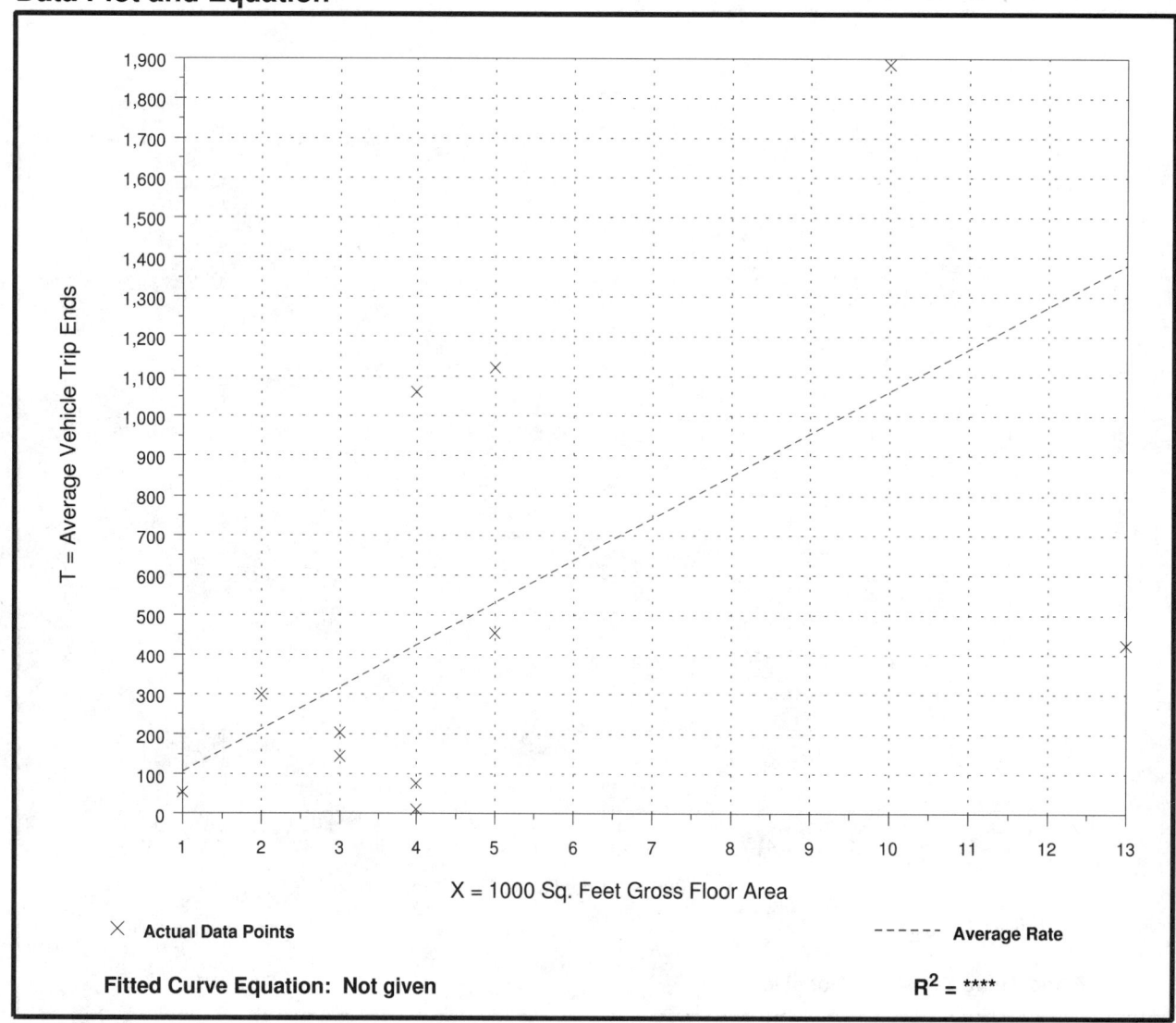

Fitted Curve Equation: Not given $R^2 = ****$

Nursery (Garden Center)
(817)

Average Vehicle Trip Ends vs: 1000 Sq. Feet Gross Floor Area
On a: Sunday,
Peak Hour of Generator

Number of Studies: 11
Average 1000 Sq. Feet GFA: 5
Directional Distribution: Not available

Trip Generation per 1000 Sq. Feet Gross Floor Area

Average Rate	Range of Rates	Standard Deviation
18.76	2.25 - 44.75	14.61

Data Plot and Equation

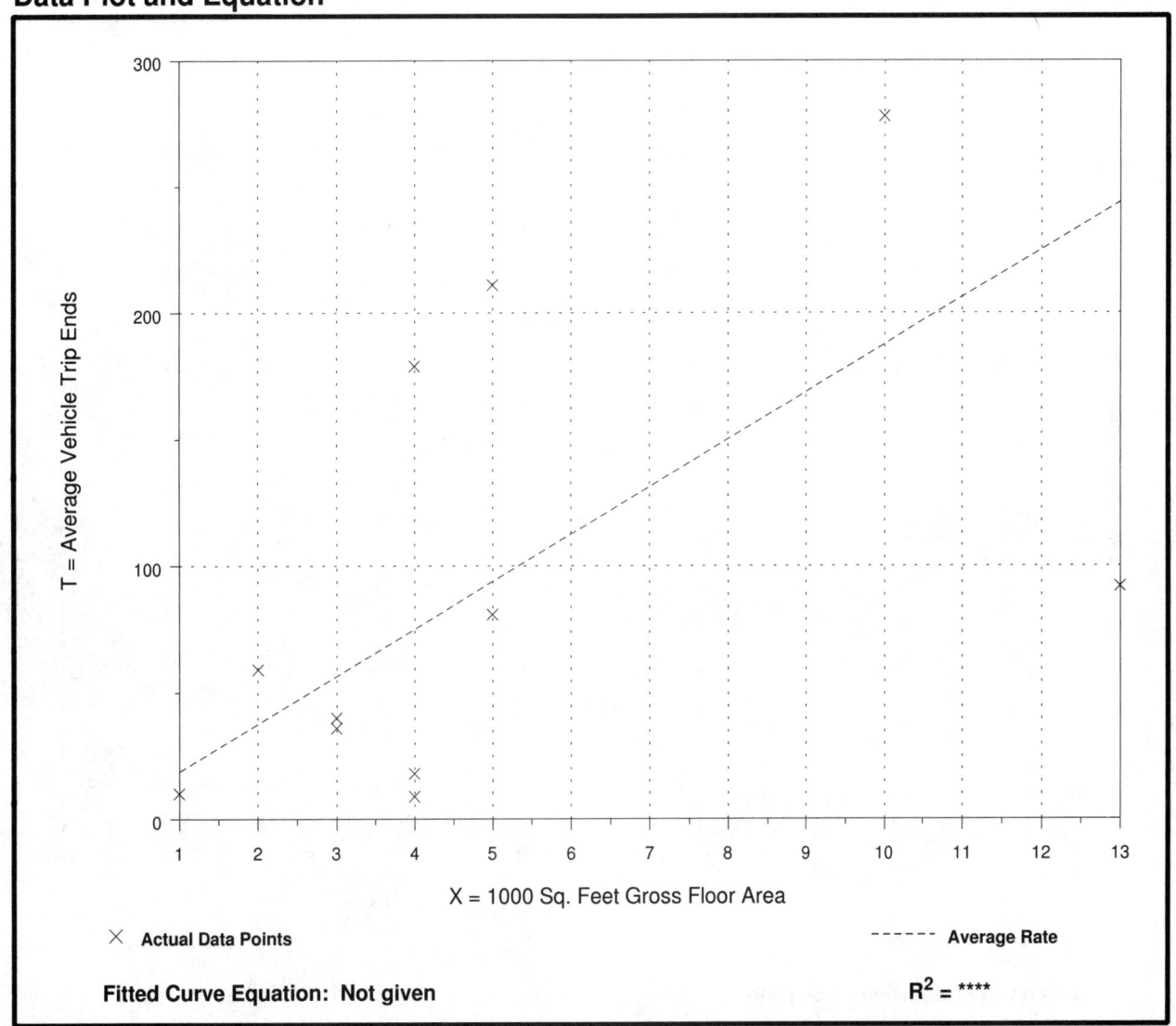

Fitted Curve Equation: Not given $R^2 = ****$

Nursery (Garden Center)
(817)

Average Vehicle Trip Ends vs: Acres
On a: Weekday

Number of Studies: 10
Average Number of Acres: 3
Directional Distribution: 50% entering, 50% exiting

Trip Generation per Acre

Average Rate	Range of Rates	Standard Deviation
108.10	25.00 - 335.00	91.85

Data Plot and Equation

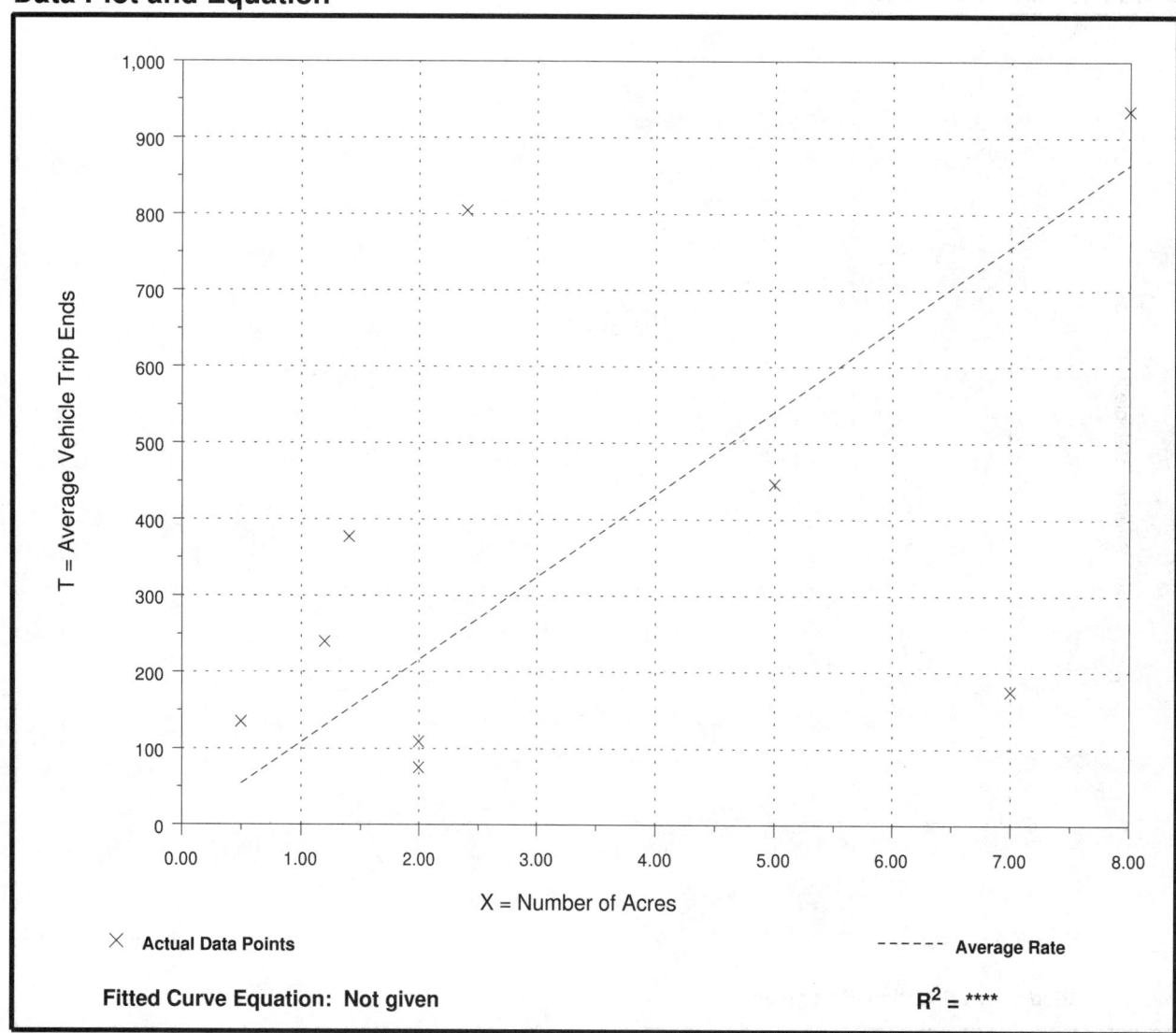

Fitted Curve Equation: Not given $R^2 = ****$

Nursery (Garden Center)
(817)

Average Vehicle Trip Ends vs: Acres
On a: Weekday,
Peak Hour of Adjacent Street Traffic,
One Hour Between 7 and 9 a.m.

Number of Studies: 11
Average Number of Acres: 4
Directional Distribution: Not available

Trip Generation per Acre

Average Rate	Range of Rates	Standard Deviation
2.82	0.14 - 14.00	3.38

Data Plot and Equation

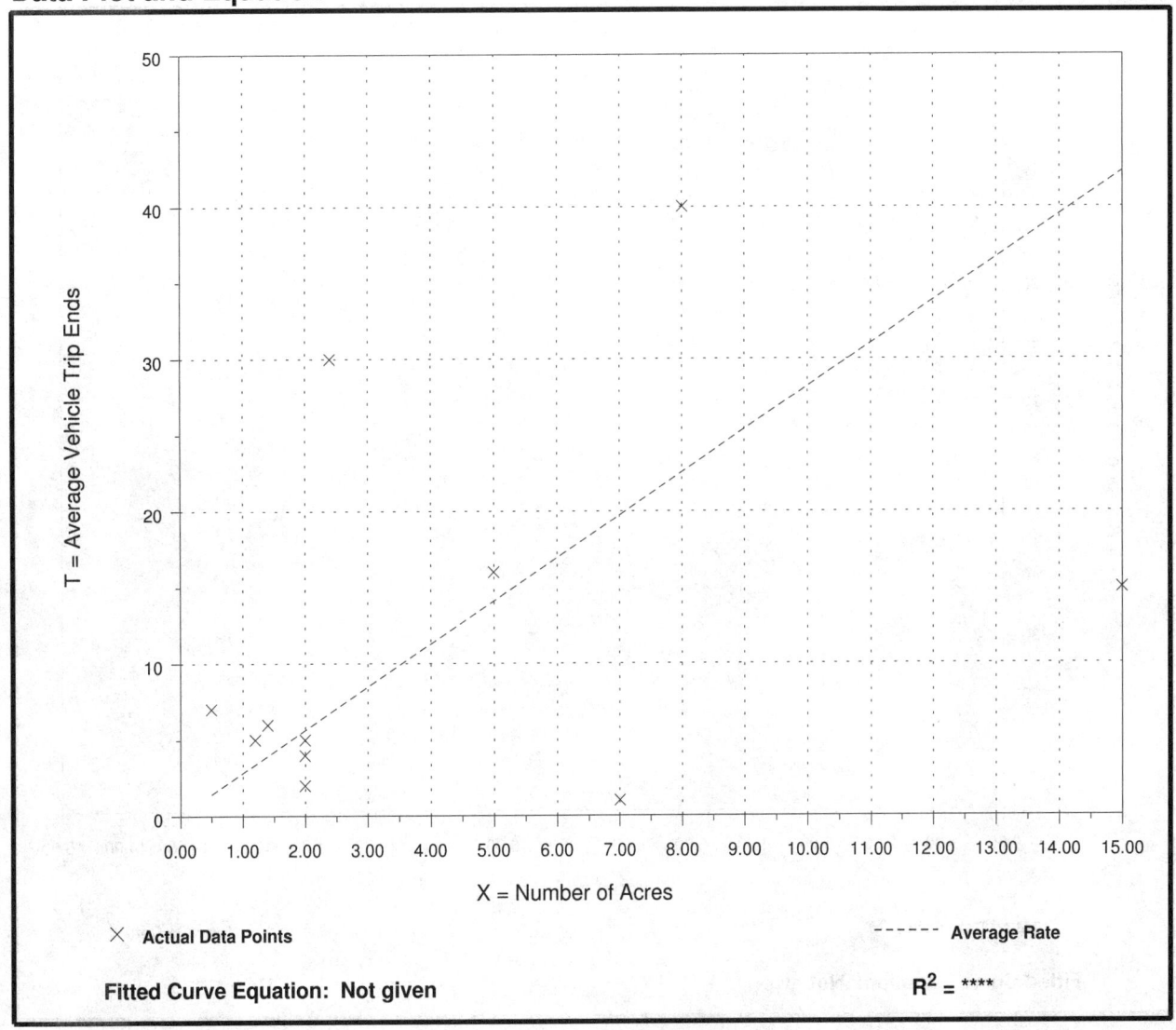

Fitted Curve Equation: Not given $R^2 = ****$

Nursery (Garden Center)
(817)

Average Vehicle Trip Ends vs: Acres
On a: Weekday,
Peak Hour of Adjacent Street Traffic,
One Hour Between 4 and 6 p.m.

Number of Studies: 11
Average Number of Acres: 4
Directional Distribution: Not available

Trip Generation per Acre

Average Rate	Range of Rates	Standard Deviation
8.06	0.13 - 37.14	10.29

Data Plot and Equation

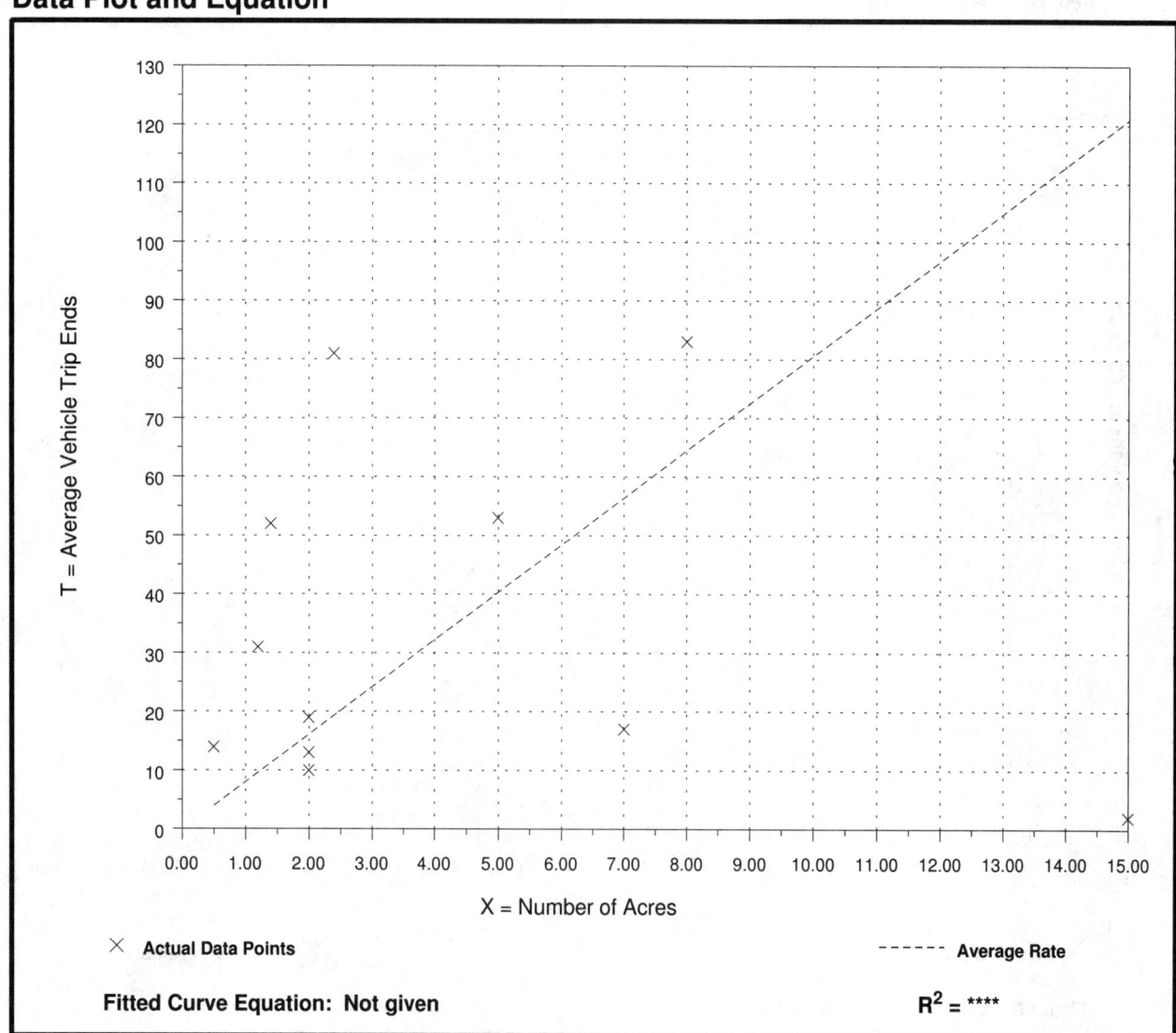

Fitted Curve Equation: Not given $R^2 = ****$

Nursery (Garden Center)
(817)

Average Vehicle Trip Ends vs: Acres
On a: Weekday,
A.M. Peak Hour of Generator

Number of Studies: 11
Average Number of Acres: 4
Directional Distribution: 52% entering, 48% exiting

Trip Generation per Acre

Average Rate	Range of Rates	Standard Deviation
9.29	2.67 - 36.00	9.31

Data Plot and Equation

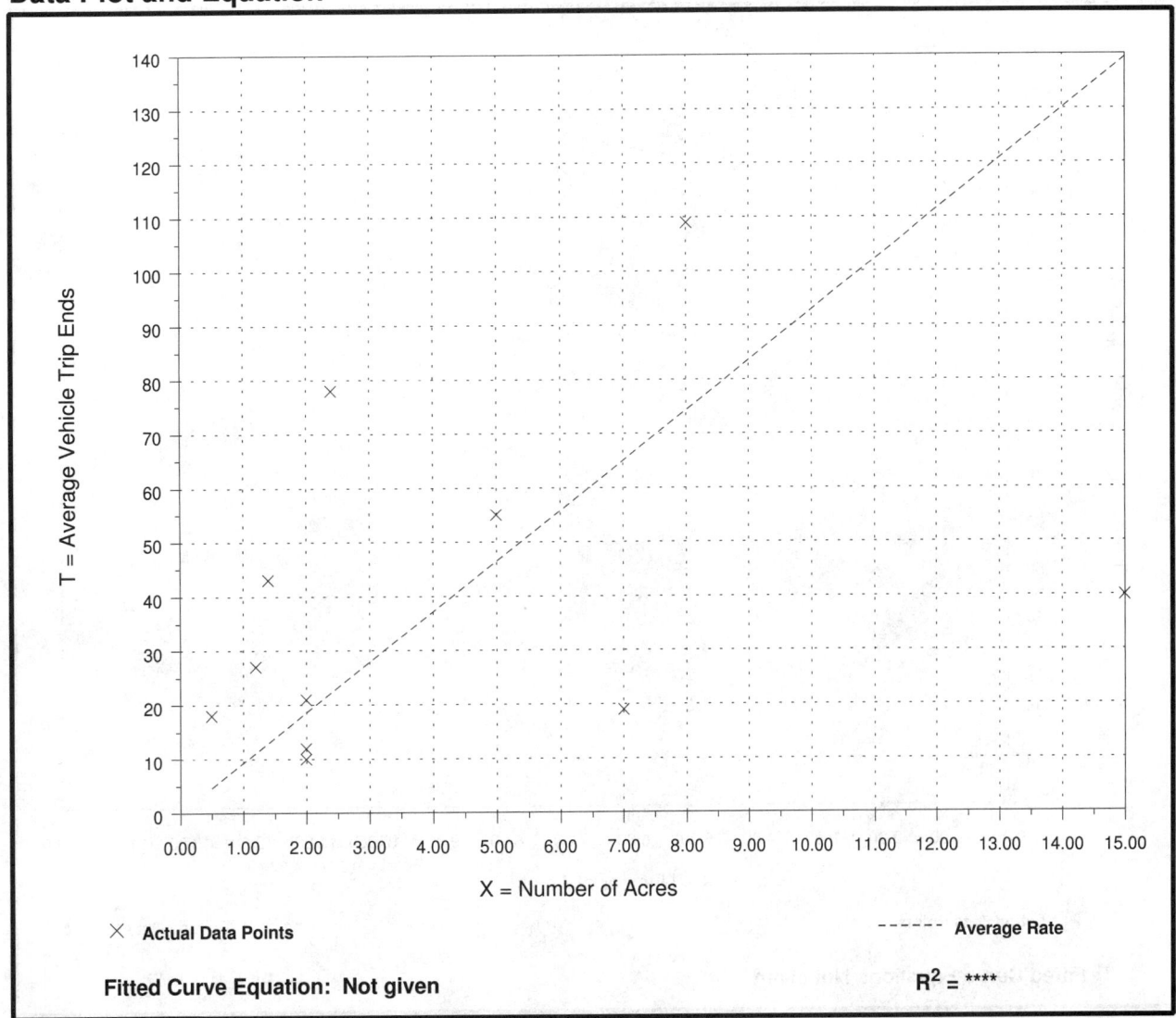

Fitted Curve Equation: Not given $R^2 = ****$

Nursery (Garden Center)
(817)

Average Vehicle Trip Ends vs: Acres
On a: Weekday,
P.M. Peak Hour of Generator

Number of Studies: 11
Average Number of Acres: 4
Directional Distribution: 49% entering, 51% exiting

Trip Generation per Acre

Average Rate	Range of Rates	Standard Deviation
10.49	2.40 - 41.67	11.18

Data Plot and Equation

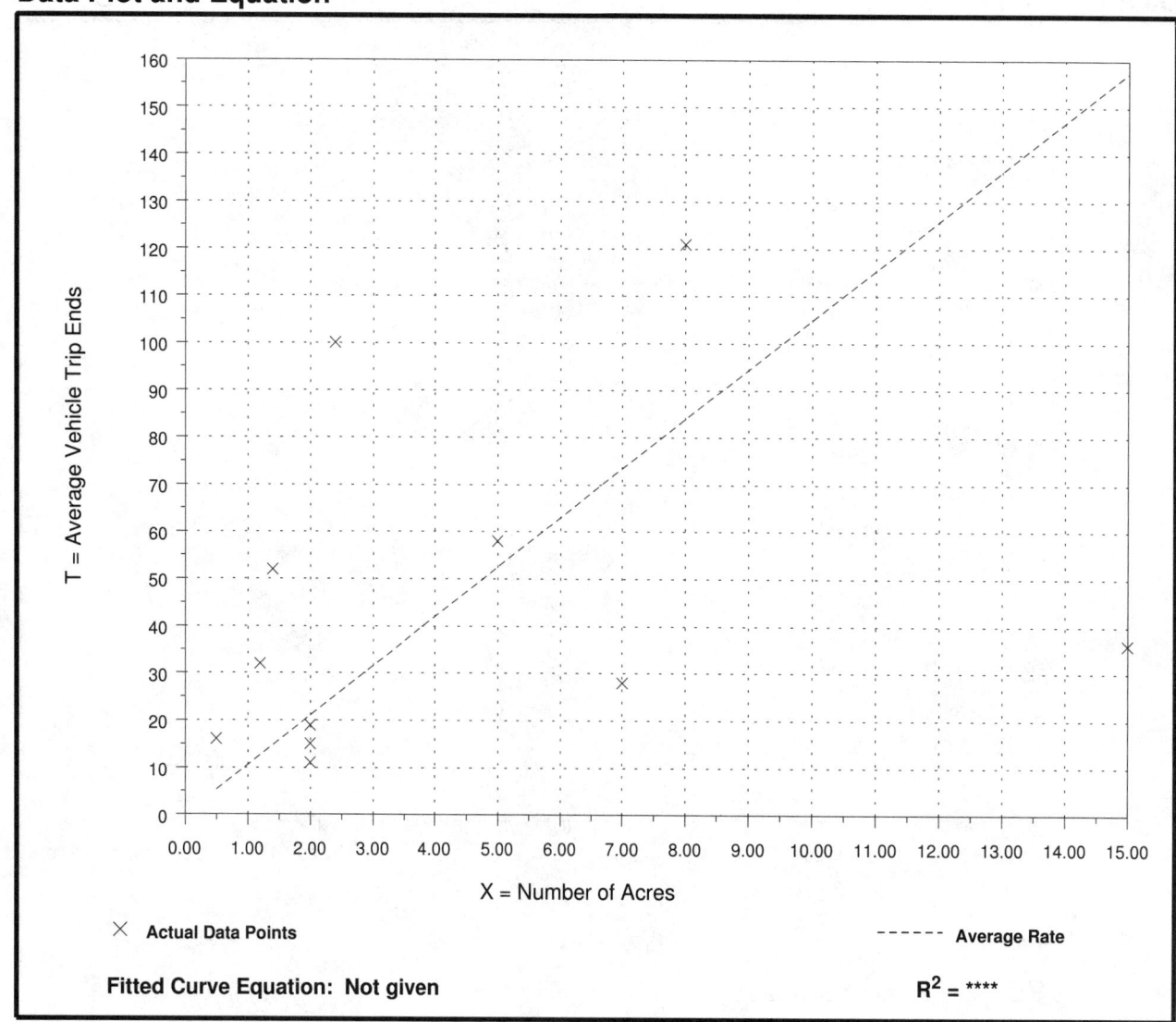

Fitted Curve Equation: Not given $R^2 = ****$

Nursery (Garden Center)
(817)

Average Vehicle Trip Ends vs: Acres
On a: Saturday

Number of Studies: 11
Average Number of Acres: 4
Directional Distribution: 50% entering, 50% exiting

Trip Generation per Acre

Average Rate	Range of Rates	Standard Deviation
154.82	9.07 - 1026.67	253.00

Data Plot and Equation

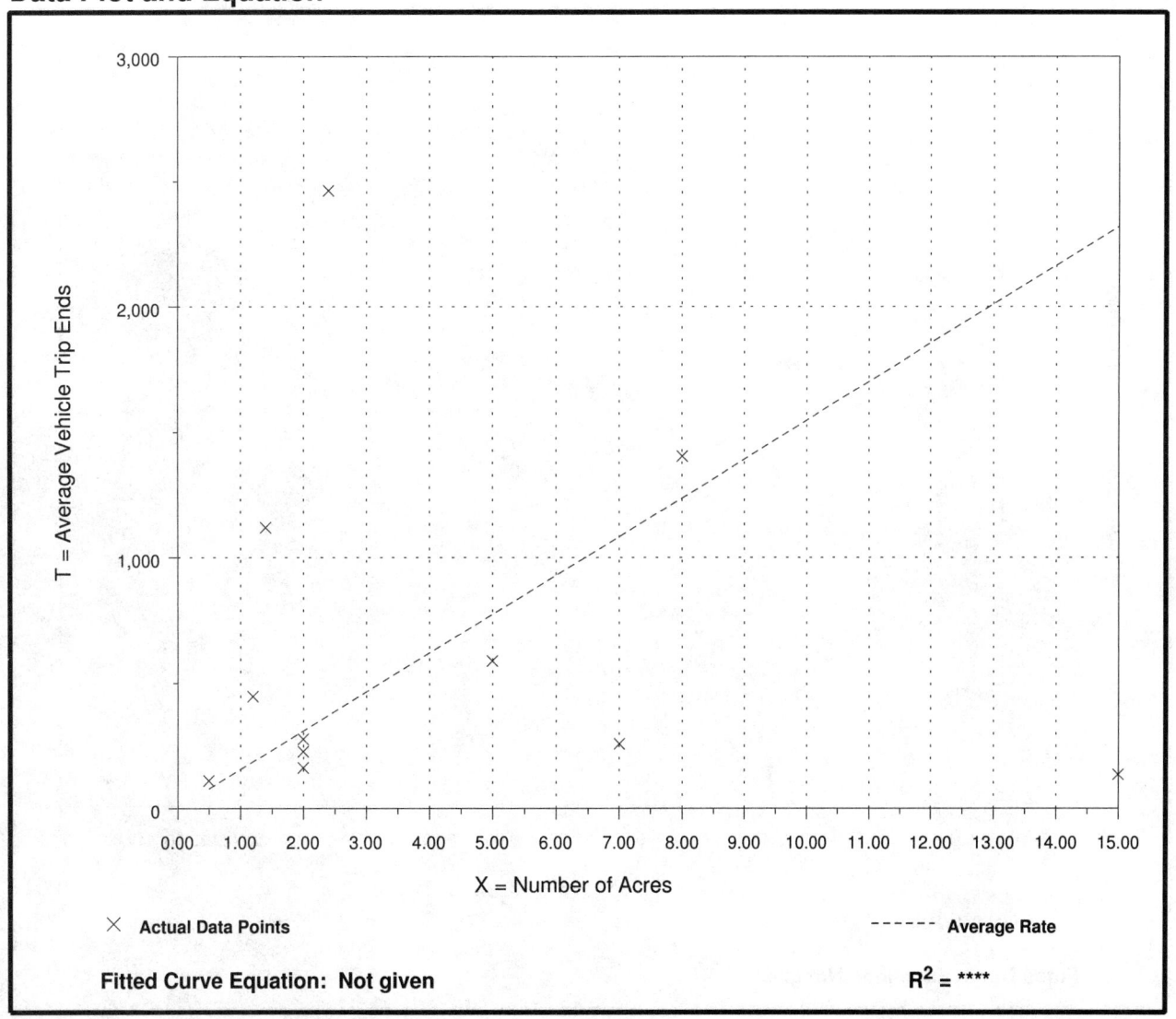

Fitted Curve Equation: Not given $R^2 =$ ****

Nursery (Garden Center)
(817)

Average Vehicle Trip Ends vs: Acres
On a: Saturday,
Peak Hour of Generator

Number of Studies: 11
Average Number of Acres: 4
Directional Distribution: Not available

Trip Generation per Acre

Average Rate	Range of Rates	Standard Deviation
23.29	1.67 - 138.33	35.96

Data Plot and Equation

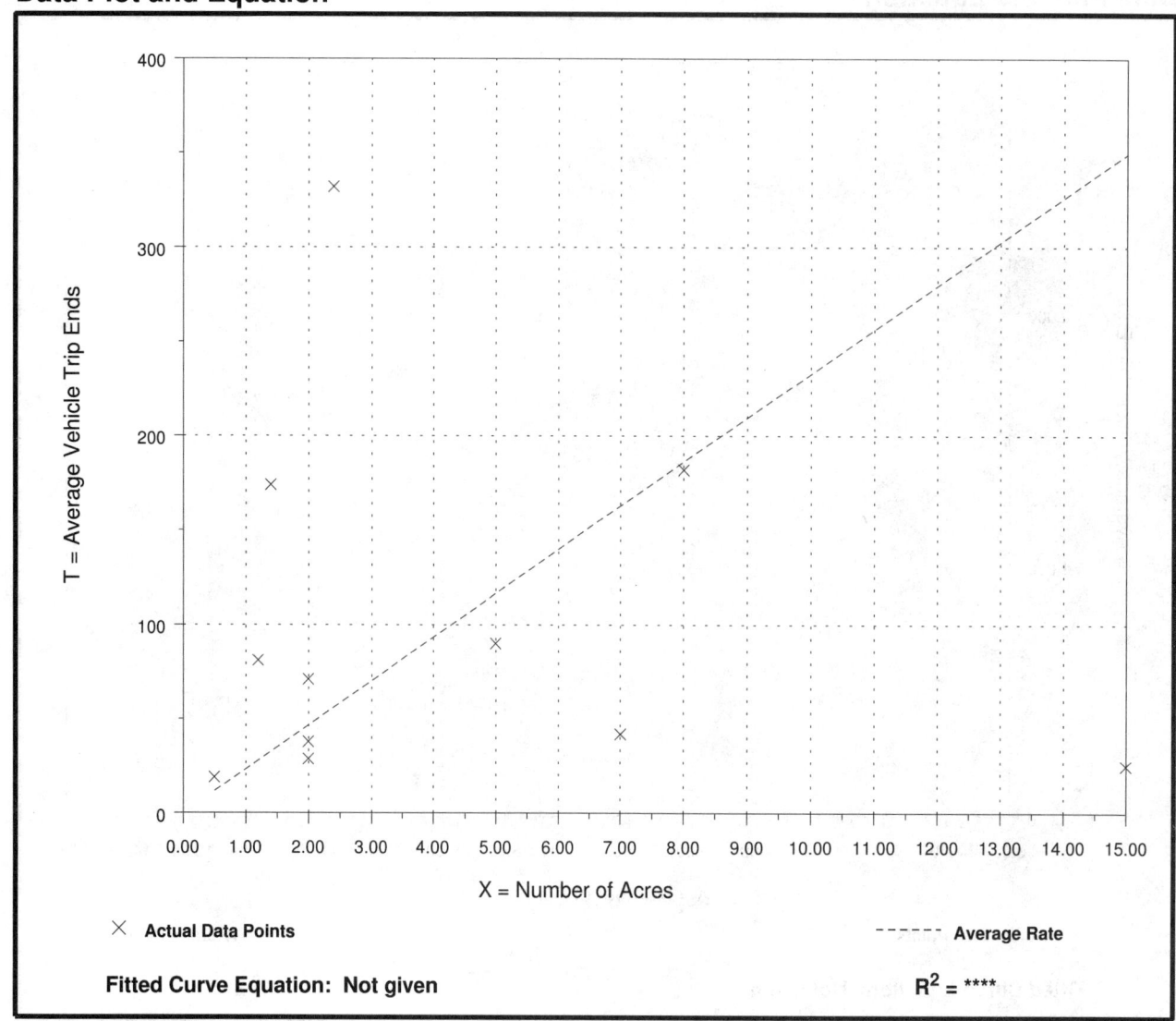

Fitted Curve Equation: Not given $R^2 = ****$

Nursery (Garden Center)
(817)

Average Vehicle Trip Ends vs: Acres
On a: Sunday

Number of Studies: 11
Average Number of Acres: 4
Directional Distribution: 50% entering, 50% exiting

Trip Generation per Acre

Average Rate	Range of Rates	Standard Deviation
123.33	0.73 - 801.43	212.75

Data Plot and Equation

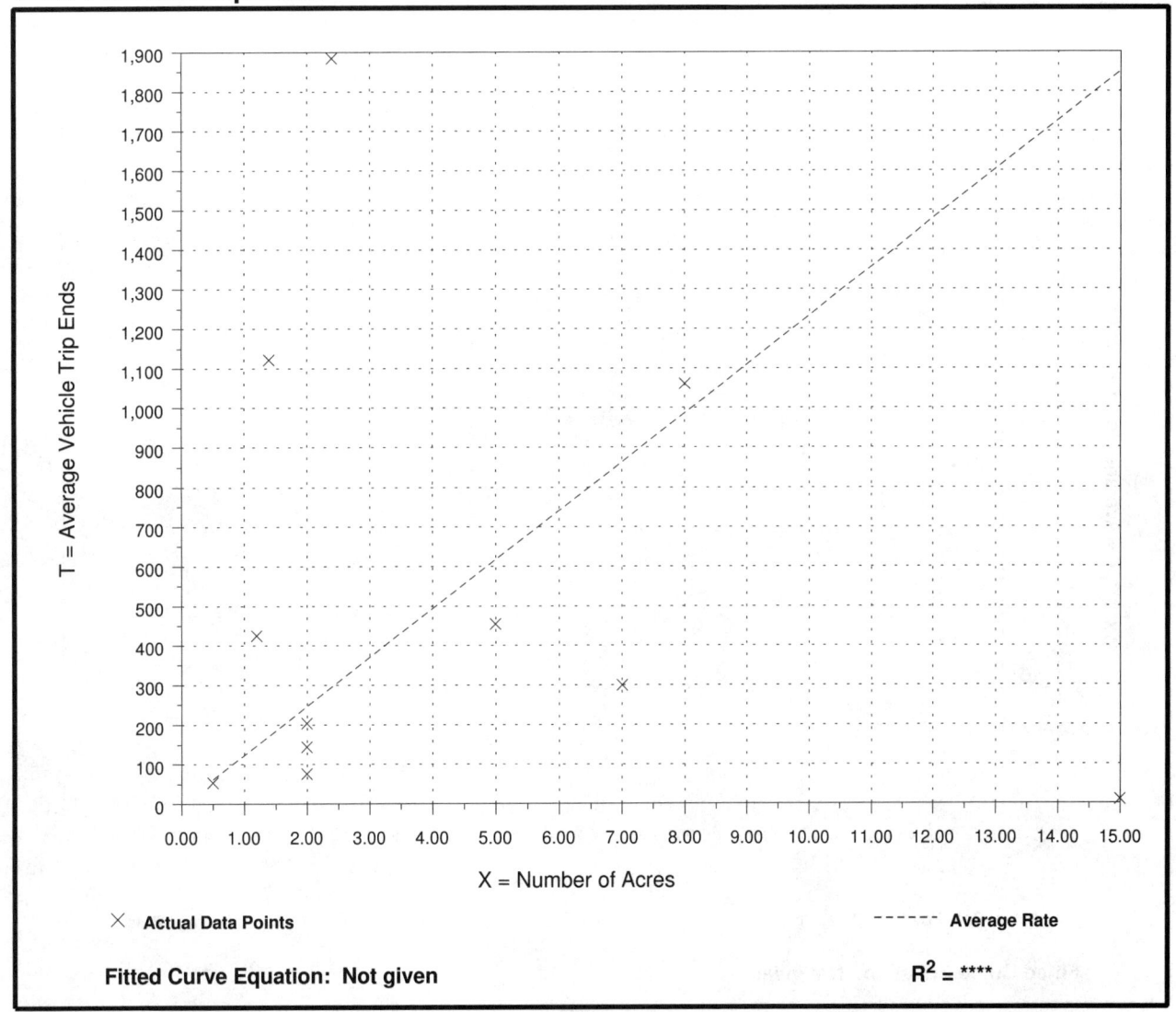

Fitted Curve Equation: Not given $R^2 = ****$

Nursery (Garden Center)
(817)

Average Vehicle Trip Ends vs: Acres
On a: Sunday,
Peak Hour of Generator

Number of Studies: 11
Average Number of Acres: 4
Directional Distribution: Not available

Trip Generation per Acre

Average Rate	Range of Rates	Standard Deviation
21.78	0.60 - 150.71	35.45

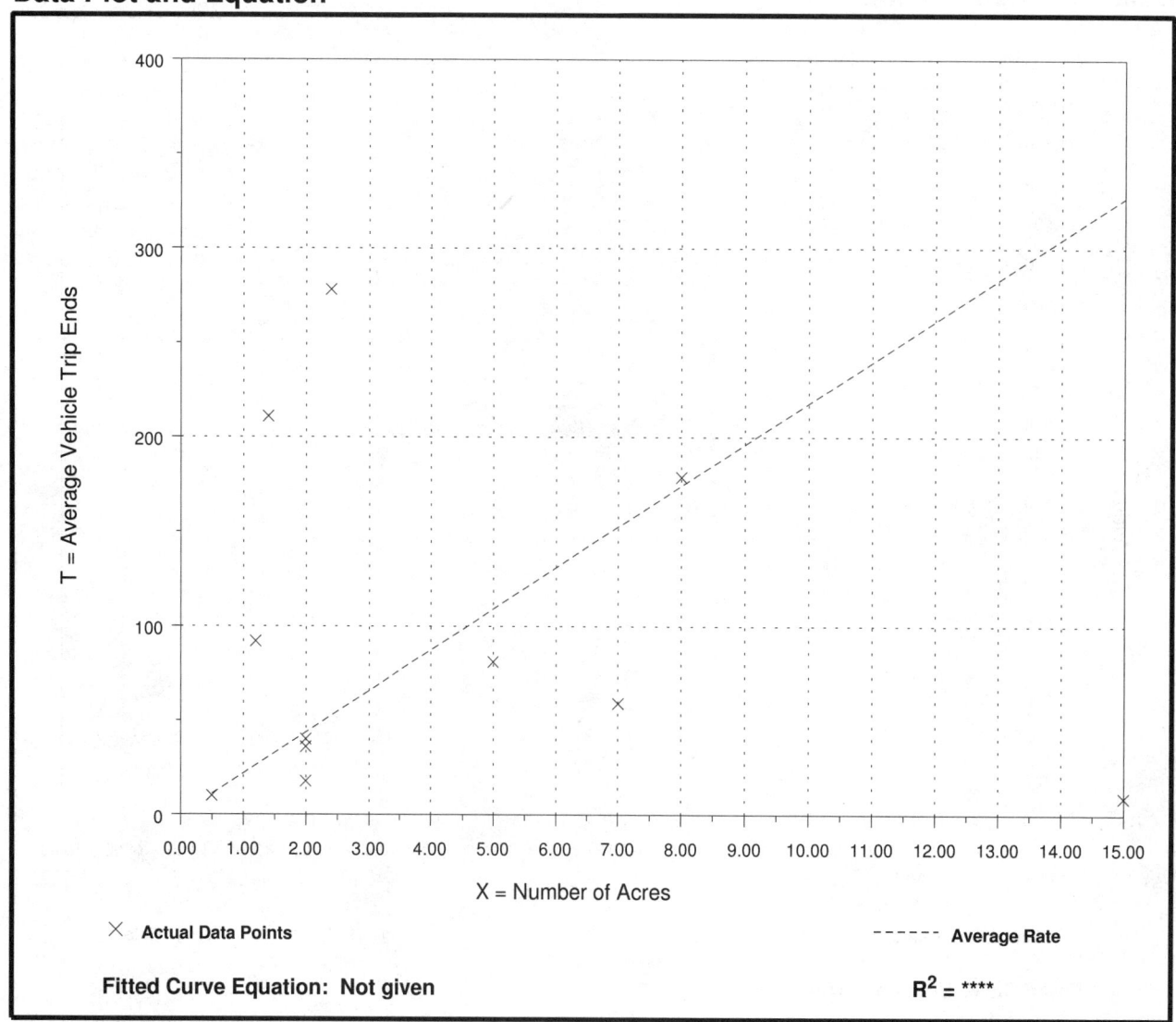

Data Plot and Equation

Fitted Curve Equation: Not given $R^2 = ****$

Land Use: 818
Nursery (Wholesale)

Description

A wholesale nursery is a free-standing building with an outside storage area for planting or landscape stock. The nurseries surveyed primarily serve contractors and suppliers. Some have large greenhouses and offer landscaping services. Most have office, storage and shipping facilities. Nurseries are characterized by seasonal variations in trip characteristics. Nursery (Garden Center) (Land Use 817) is a related use.

Additional Data

Outside storage areas are not included in the overall gross floor area measurements. However, if storage areas are located within the principal outside faces of the exterior walls, they are included in the overall gross floor area of the building.

The sites were surveyed in the 1980s in California.

Source Numbers

205, 240

Land Use: 818
Nursery (Wholesale)
Independent Variables with One Observation

The following trip generation data are for independent variables with only one observation. This information is shown in this table only; there are no related plots for these data.

Users are cautioned to use data with care because of the small sample size.

Independent Variable	Trip Generation Rate	Size of Independent Variable	Number of Studies	Directional Distribution
Employees				
Weekday	23.40	5	1	50% entering, 50% exiting
1,000 Square Feet Gross Floor Area				
Weekday	39.00	3	1	50% entering, 50% exiting
Acres				
Weekday	19.50	6	1	50% entering, 50% exiting

Nursery (Wholesale)
(818)

Average Vehicle Trip Ends vs: Employees
On a: Weekday,
Peak Hour of Adjacent Street Traffic,
One Hour Between 7 and 9 a.m.

Number of Studies: 8
Avg. Number of Employees: 19
Directional Distribution: Not available

Trip Generation per Employee

Average Rate	Range of Rates	Standard Deviation
0.34	0.15 - 1.40	0.64

Data Plot and Equation

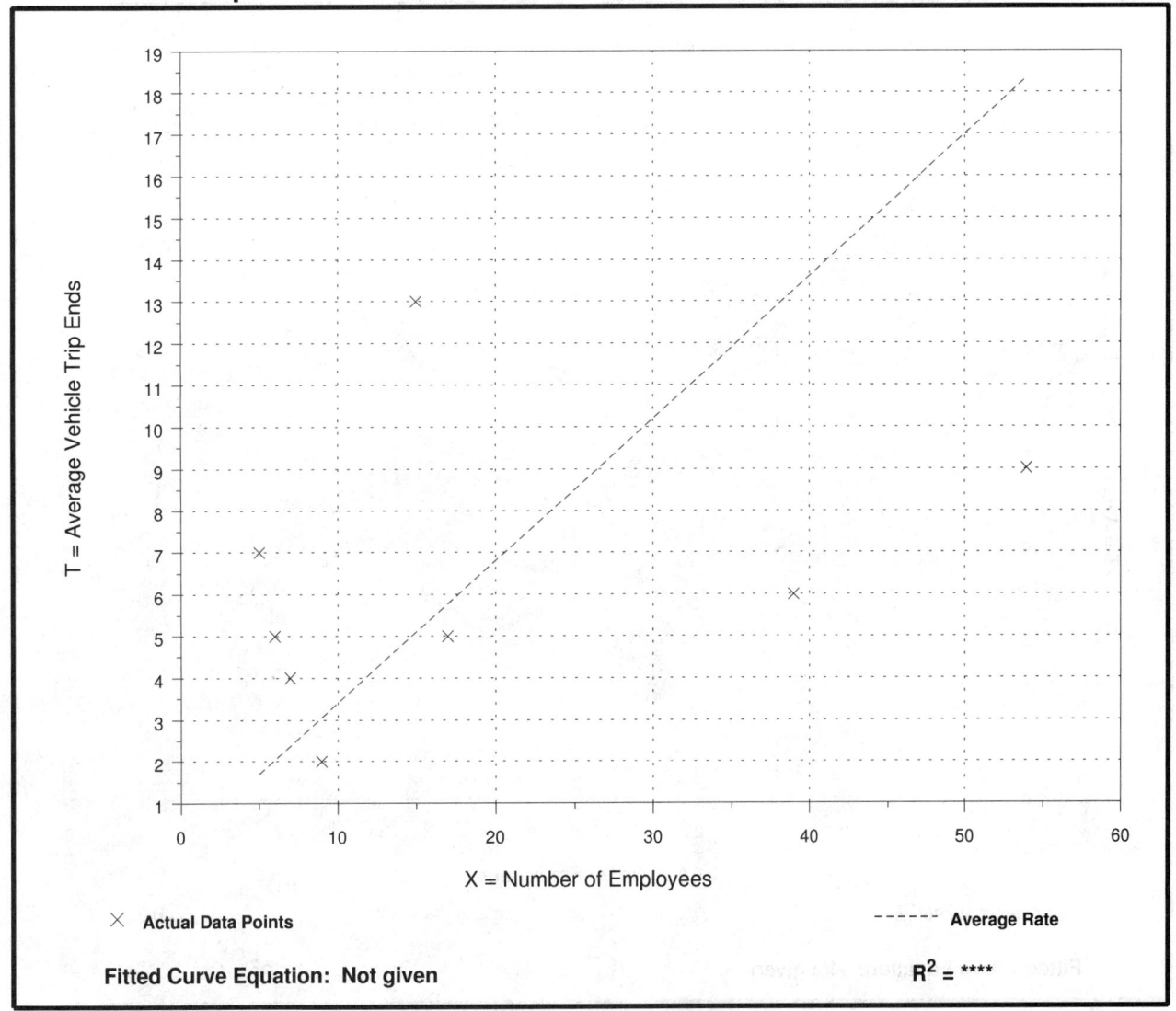

Fitted Curve Equation: Not given $R^2 = ****$

Nursery (Wholesale)
(818)

Average Vehicle Trip Ends vs: Employees
On a: Weekday,
Peak Hour of Adjacent Street Traffic,
One Hour Between 4 and 6 p.m.

Number of Studies: 6
Avg. Number of Employees: 22
Directional Distribution: Not available

Trip Generation per Employee

Average Rate	Range of Rates	Standard Deviation
0.47	0.15 - 3.00	0.85

Data Plot and Equation

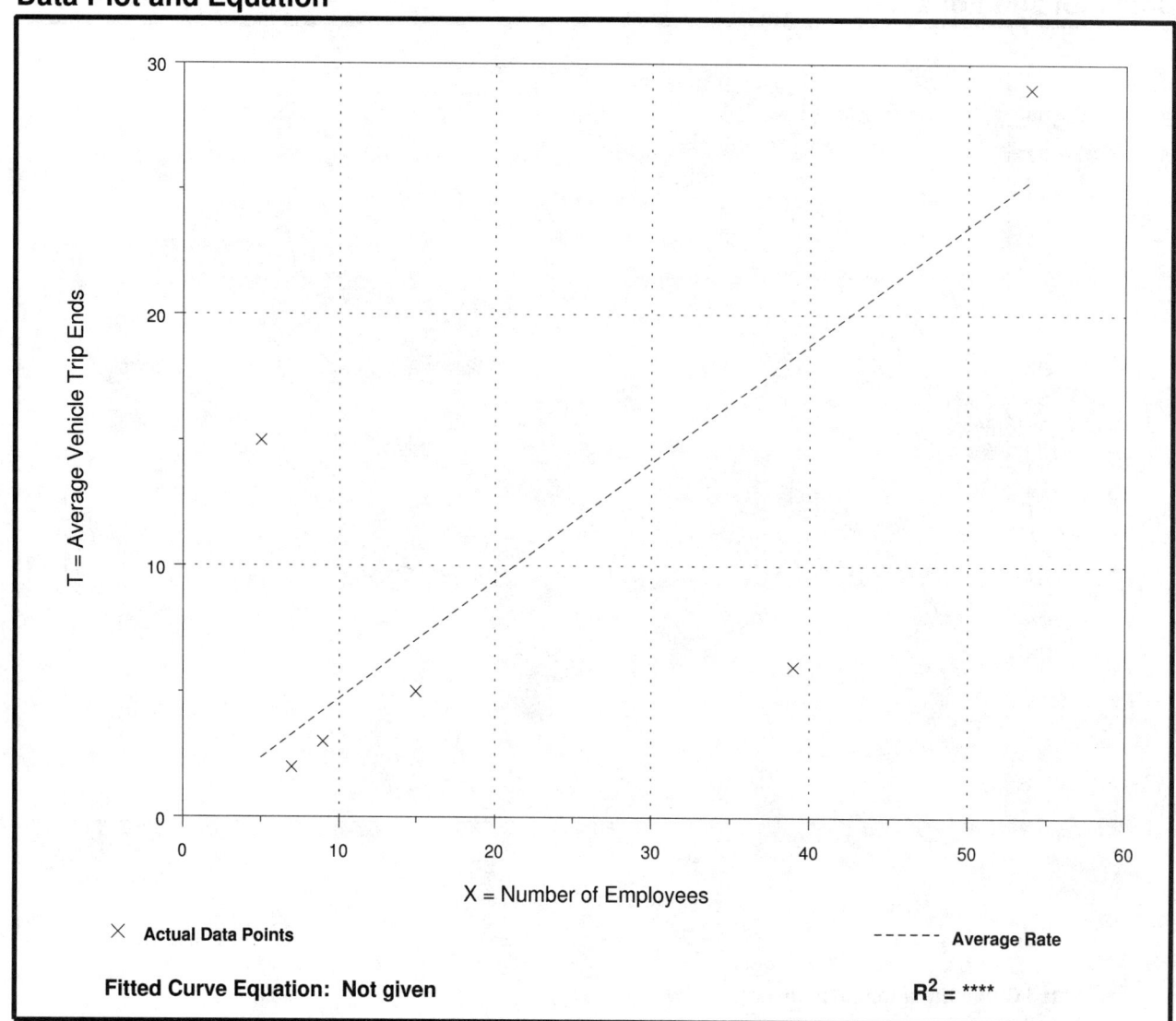

Fitted Curve Equation: Not given $R^2 = ****$

Nursery (Wholesale)
(818)

Average Vehicle Trip Ends vs: Employees
On a: Weekday,
A.M. Peak Hour of Generator

Number of Studies: 8
Avg. Number of Employees: 19
Directional Distribution: 43% entering, 57% exiting

Trip Generation per Employee

Average Rate	Range of Rates	Standard Deviation
0.43	0.15 - 2.40	0.78

Data Plot and Equation

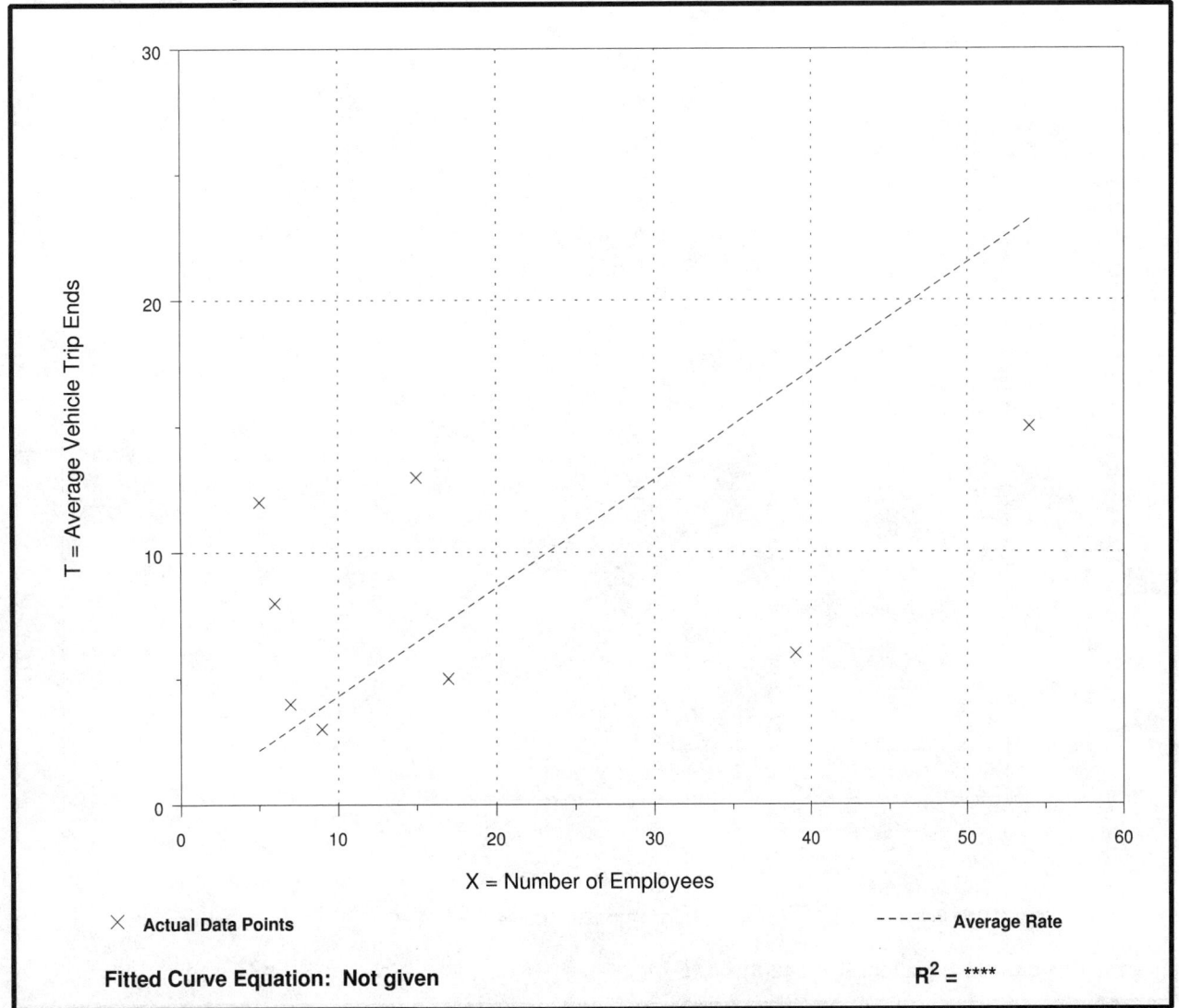

Fitted Curve Equation: Not given $R^2 = ****$

Nursery (Wholesale)
(818)

Average Vehicle Trip Ends vs: Employees
On a: Weekday,
P.M. Peak Hour of Generator

Number of Studies: 8
Avg. Number of Employees: 19
Directional Distribution: Not available

Trip Generation per Employee

Average Rate	Range of Rates	Standard Deviation
0.67	0.47 - 3.00	0.91

Data Plot and Equation

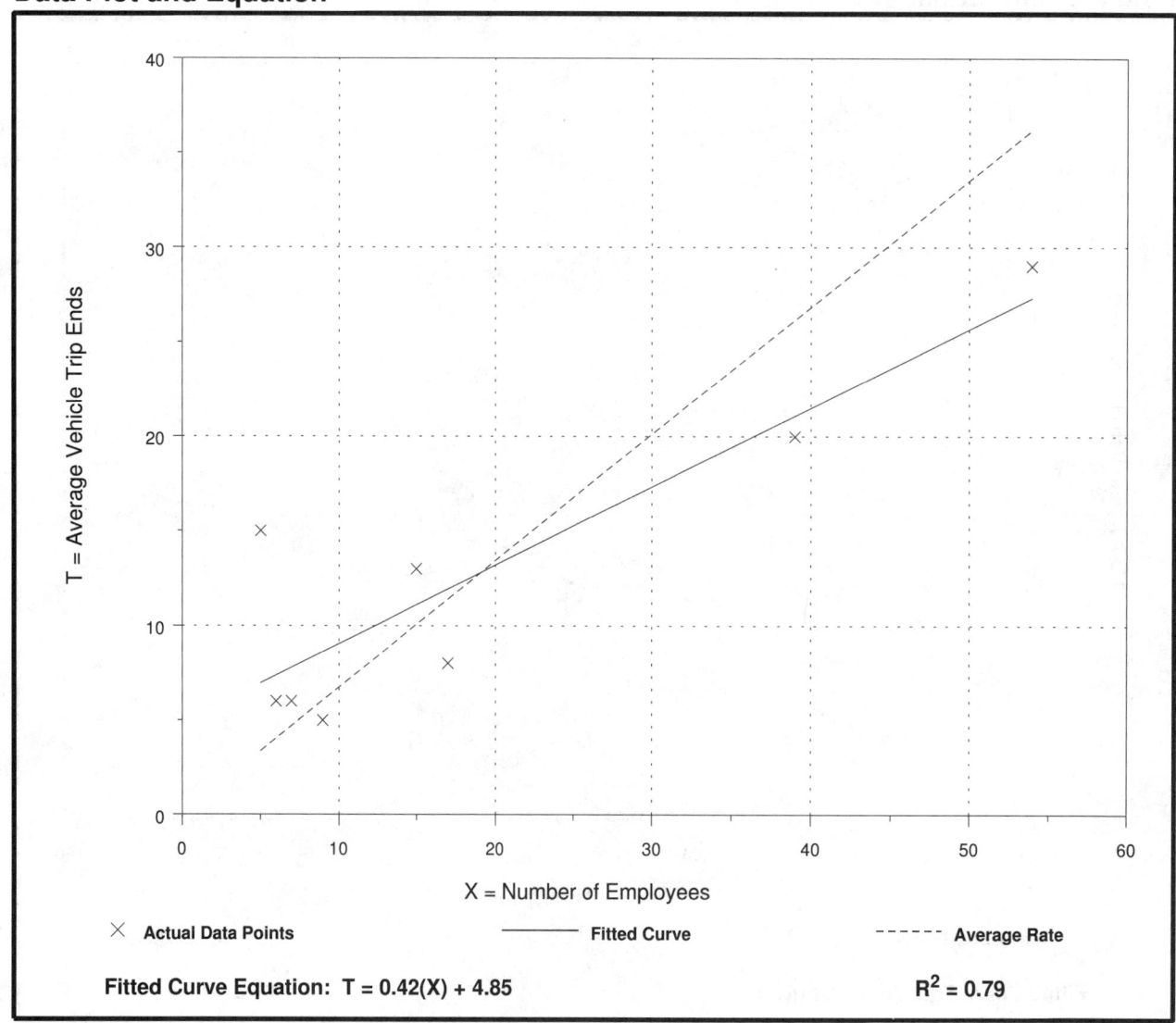

Fitted Curve Equation: T = 0.42(X) + 4.85 $R^2 = 0.79$

Nursery (Wholesale)
(818)

Average Vehicle Trip Ends vs: Employees
On a: Saturday

Number of Studies: 8
Avg. Number of Employees: 19
Directional Distribution: 50% entering, 50% exiting

Trip Generation per Employee

Average Rate	Range of Rates	Standard Deviation
3.97	1.78 - 27.40	5.05

Data Plot and Equation

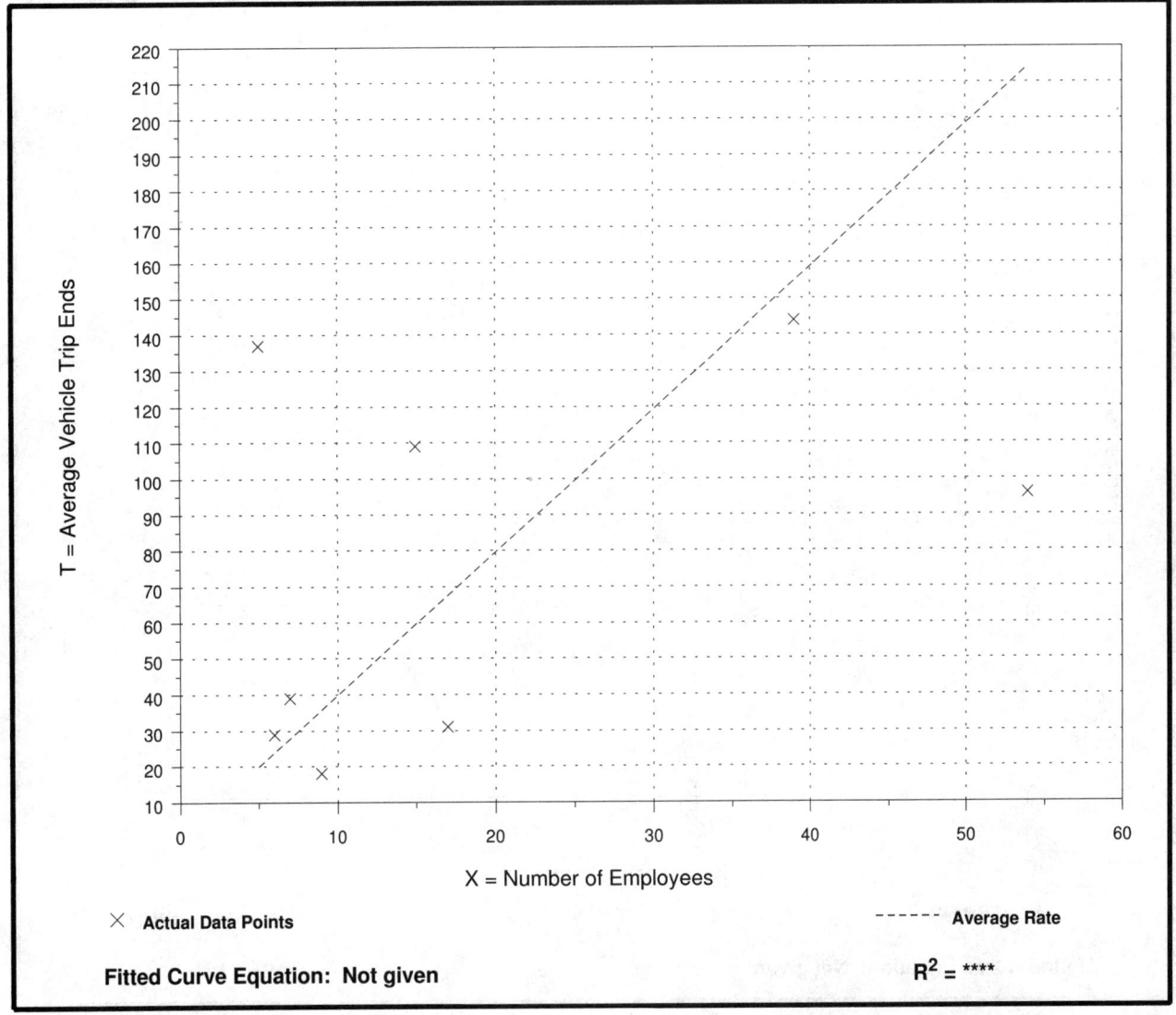

Fitted Curve Equation: Not given $R^2 = ****$

Nursery (Wholesale)
(818)

Average Vehicle Trip Ends vs: Employees
On a: Saturday,
Peak Hour of Generator

Number of Studies: 8
Avg. Number of Employees: 19
Directional Distribution: Not available

Trip Generation per Employee

Average Rate	Range of Rates	Standard Deviation
0.74	0.26 - 4.40	1.13

Data Plot and Equation

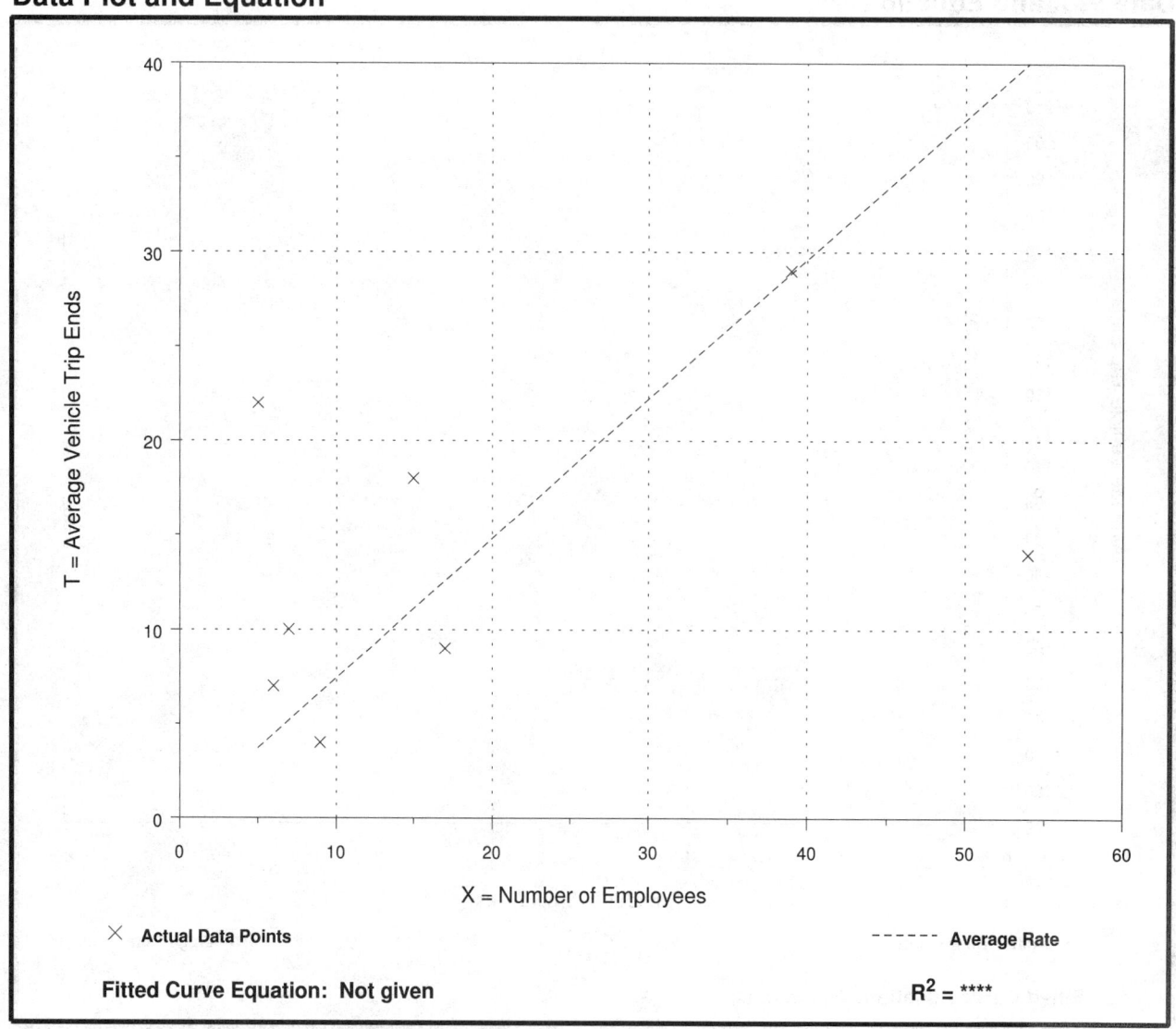

Fitted Curve Equation: Not given $R^2 = ****$

Nursery (Wholesale)
(818)

Average Vehicle Trip Ends vs: Employees
On a: Sunday

Number of Studies: 7
Avg. Number of Employees: 19
Directional Distribution: 50% entering, 50% exiting

Trip Generation per Employee

Average Rate	Range of Rates	Standard Deviation
2.35	0.49 - 27.60	5.24

Data Plot and Equation

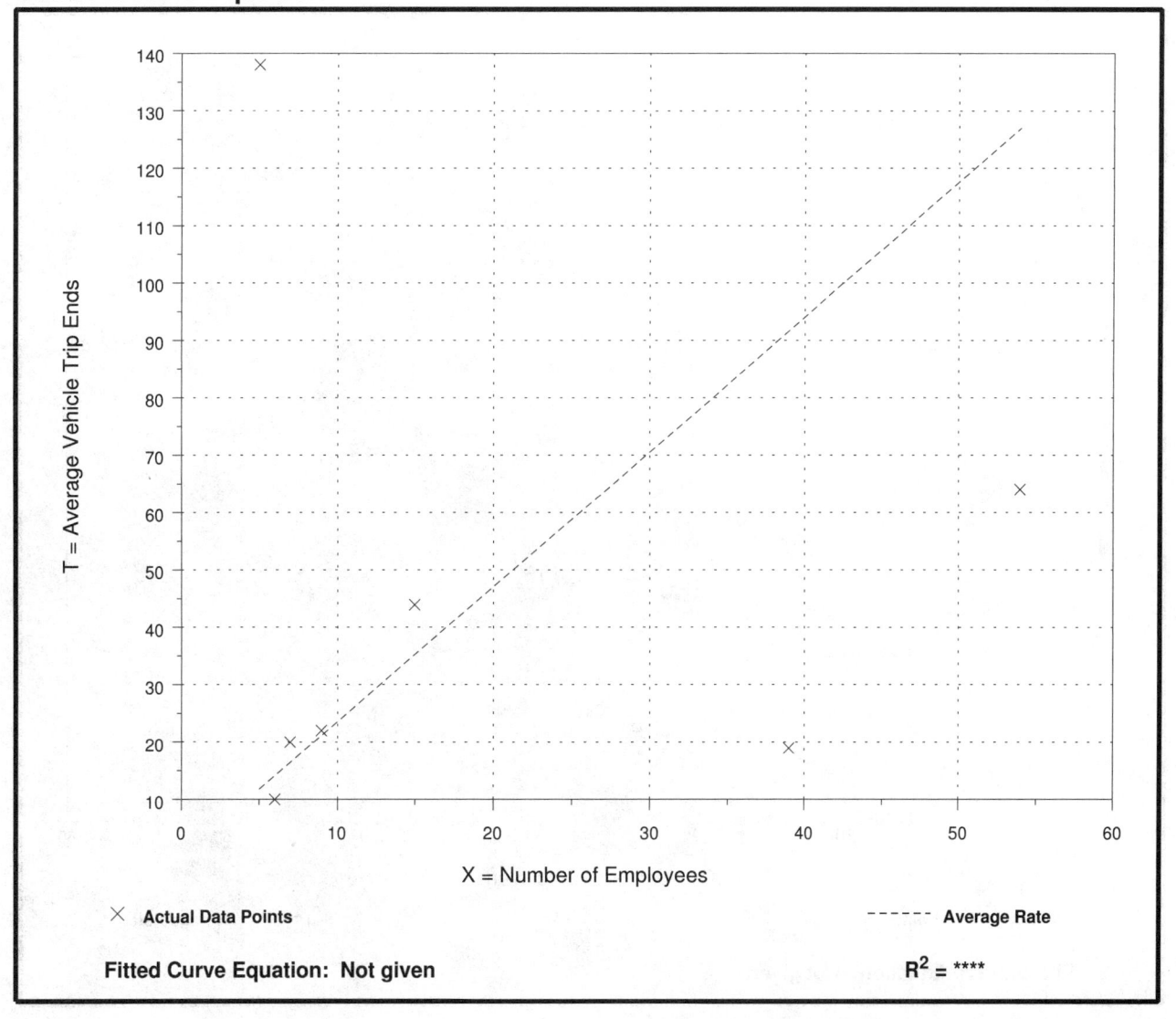

Fitted Curve Equation: Not given $R^2 = ****$

Nursery (Wholesale)
(818)

Average Vehicle Trip Ends vs: Employees
On a: Sunday,
Peak Hour of Generator

Number of Studies: 7
Avg. Number of Employees: 19
Directional Distribution: Not available

Trip Generation per Employee

Average Rate	Range of Rates	Standard Deviation
0.53	0.08 - 5.40	1.22

Data Plot and Equation

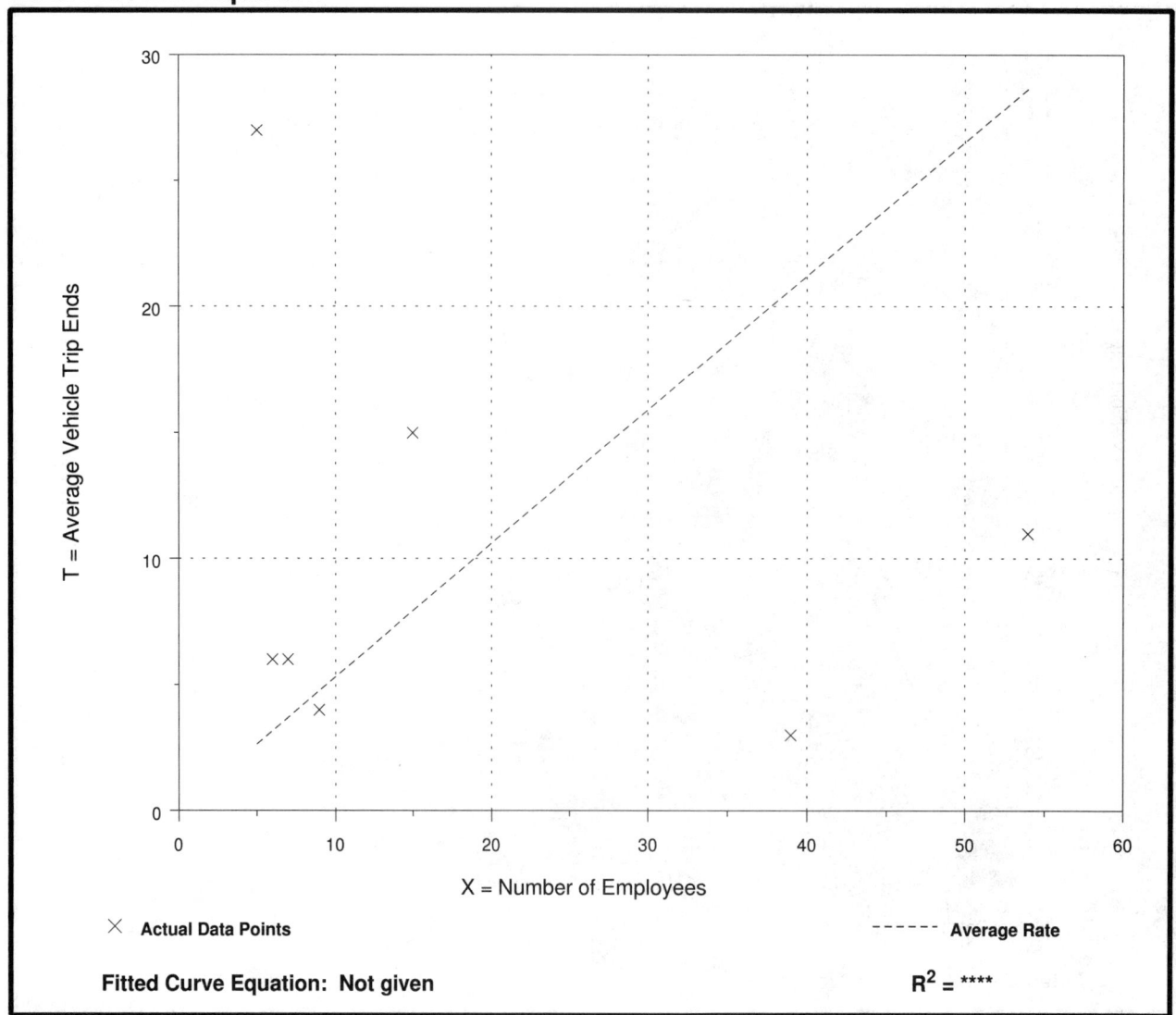

Fitted Curve Equation: Not given $R^2 = ****$

Nursery (Wholesale)
(818)

Average Vehicle Trip Ends vs: 1000 Sq. Feet Gross Floor Area
On a: Weekday,
Peak Hour of Adjacent Street Traffic,
One Hour Between 7 and 9 a.m.

Number of Studies: 7
Average 1000 Sq. Feet GFA: 3
Directional Distribution: Not available

Trip Generation per 1000 Sq. Feet Gross Floor Area

Average Rate	Range of Rates	Standard Deviation
2.40	0.66 - 9.00	2.31

Data Plot and Equation

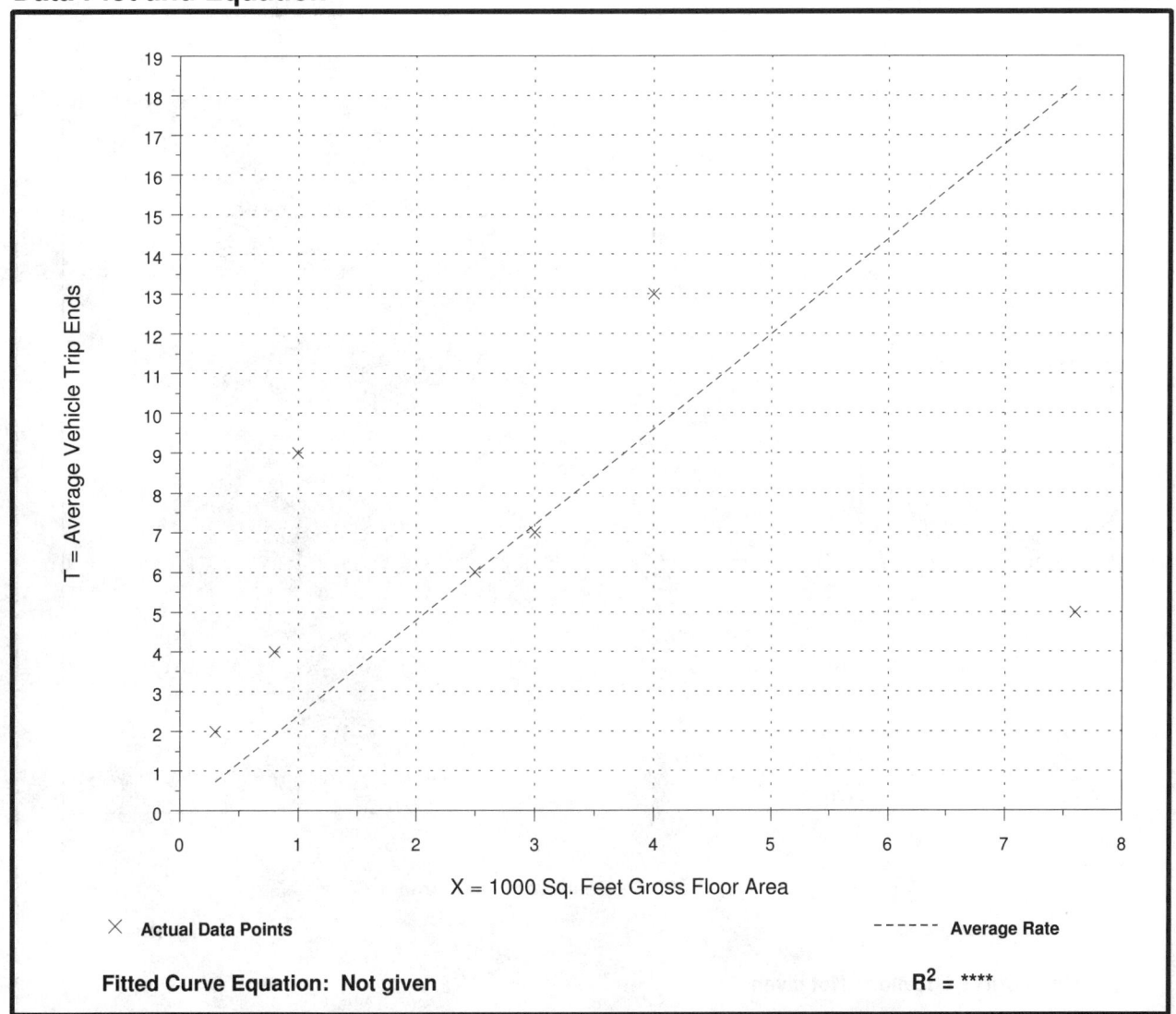

Fitted Curve Equation: Not given $R^2 = ****$

Nursery (Wholesale)
(818)

Average Vehicle Trip Ends vs: 1000 Sq. Feet Gross Floor Area
On a: Weekday,
Peak Hour of Adjacent Street Traffic,
One Hour Between 4 and 6 p.m.

Number of Studies: 6
Average 1000 Sq. Feet GFA: 2
Directional Distribution: Not available

Trip Generation per 1000 Sq. Feet Gross Floor Area

Average Rate	Range of Rates	Standard Deviation
5.17	1.25 - 29.00	7.96

Data Plot and Equation

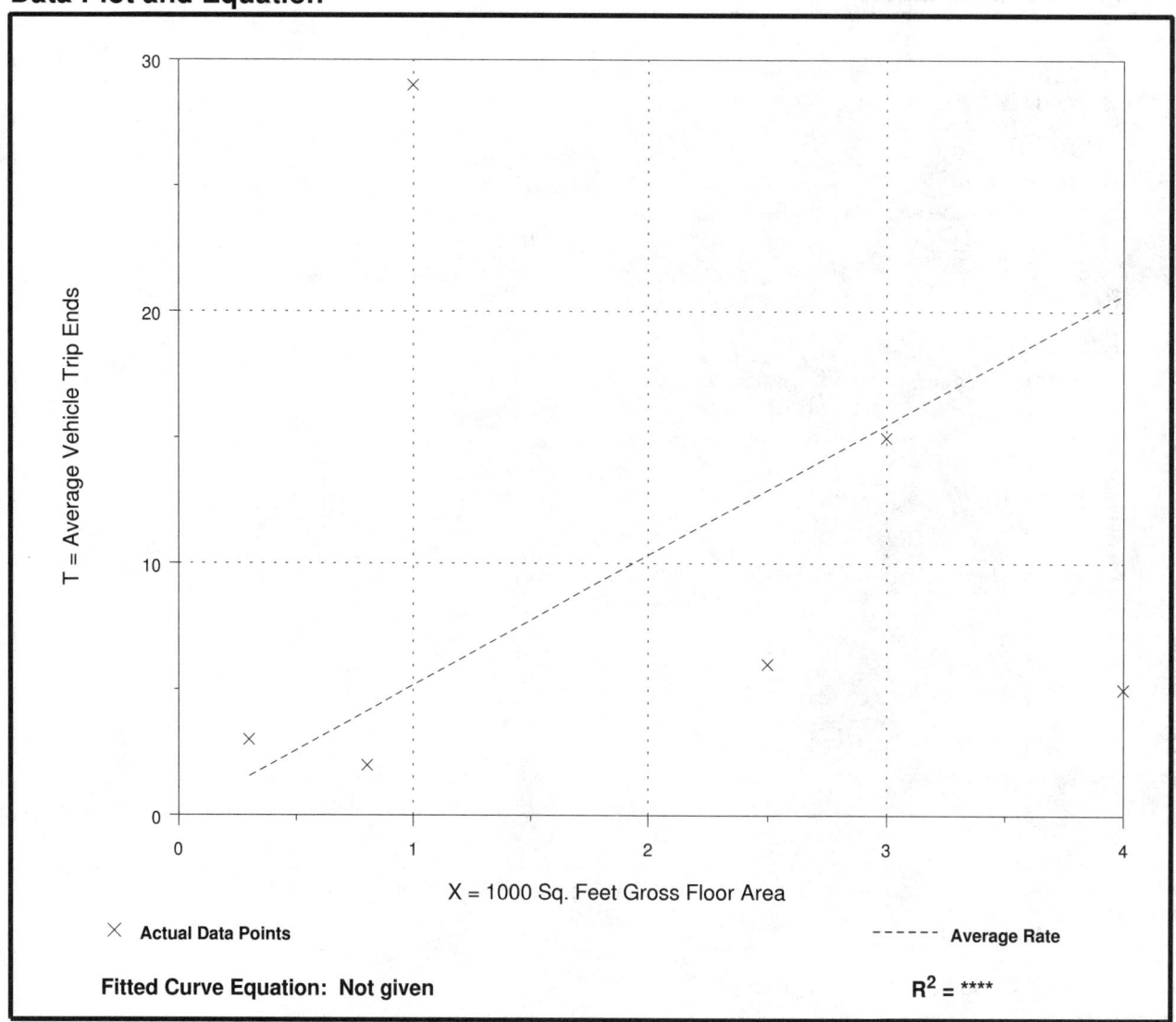

Fitted Curve Equation: Not given $R^2 = ****$

Nursery (Wholesale)
(818)

Average Vehicle Trip Ends vs: 1000 Sq. Feet Gross Floor Area
On a: Weekday,
A.M. Peak Hour of Generator

Number of Studies: 7
Average 1000 Sq. Feet GFA: 3
Directional Distribution: 43% entering, 57% exiting

Trip Generation per 1000 Sq. Feet Gross Floor Area

Average Rate	Range of Rates	Standard Deviation
3.02	0.66 - 15.00	3.52

Data Plot and Equation

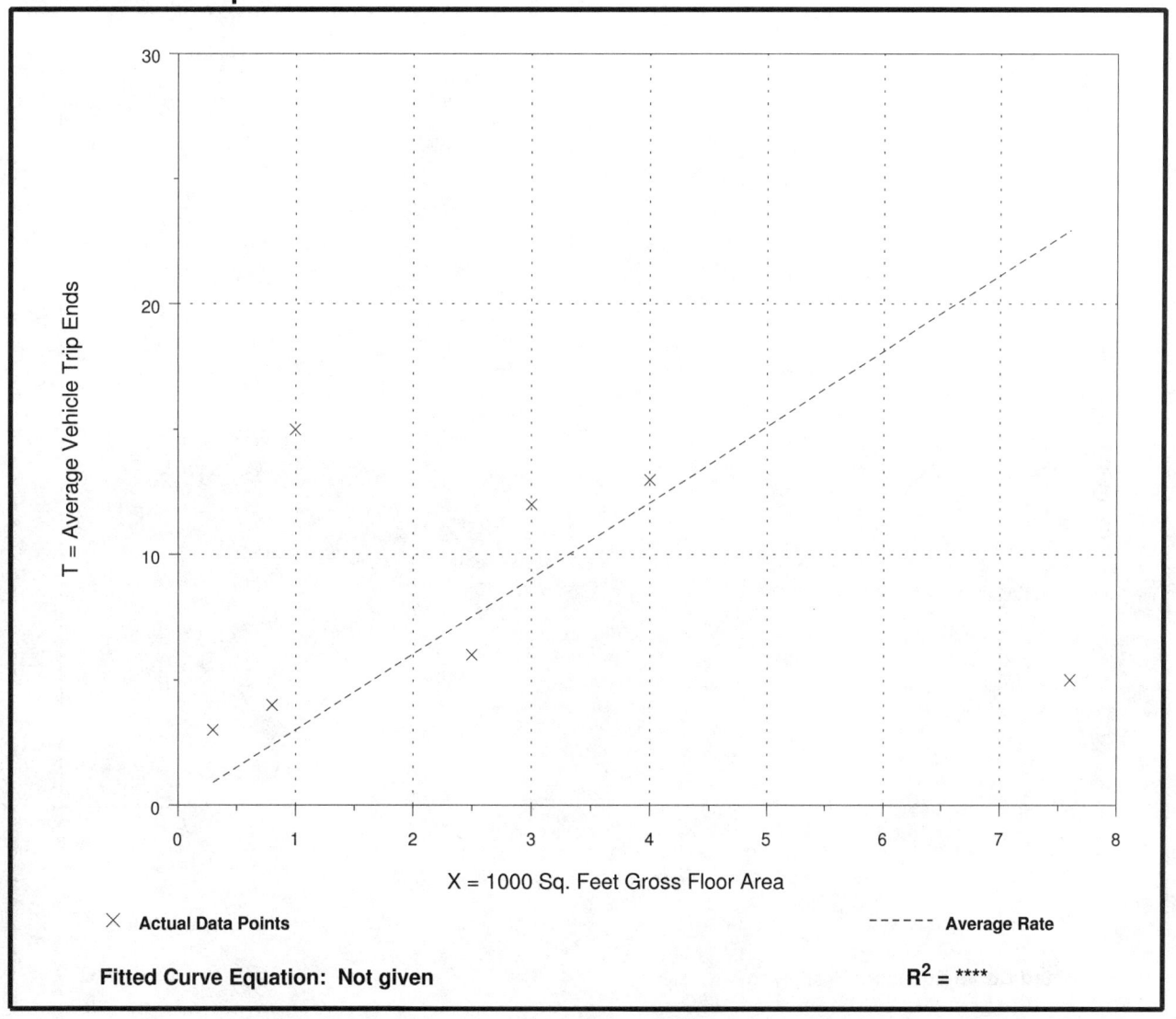

Fitted Curve Equation: Not given $R^2 = ****$

Nursery (Wholesale)
(818)

Average Vehicle Trip Ends vs: 1000 Sq. Feet Gross Floor Area
On a: Weekday,
P.M. Peak Hour of Generator

Number of Studies: 7
Average 1000 Sq. Feet GFA: 3
Directional Distribution: Not available

Trip Generation per 1000 Sq. Feet Gross Floor Area

Average Rate	Range of Rates	Standard Deviation
5.00	1.05 - 29.00	6.63

Data Plot and Equation

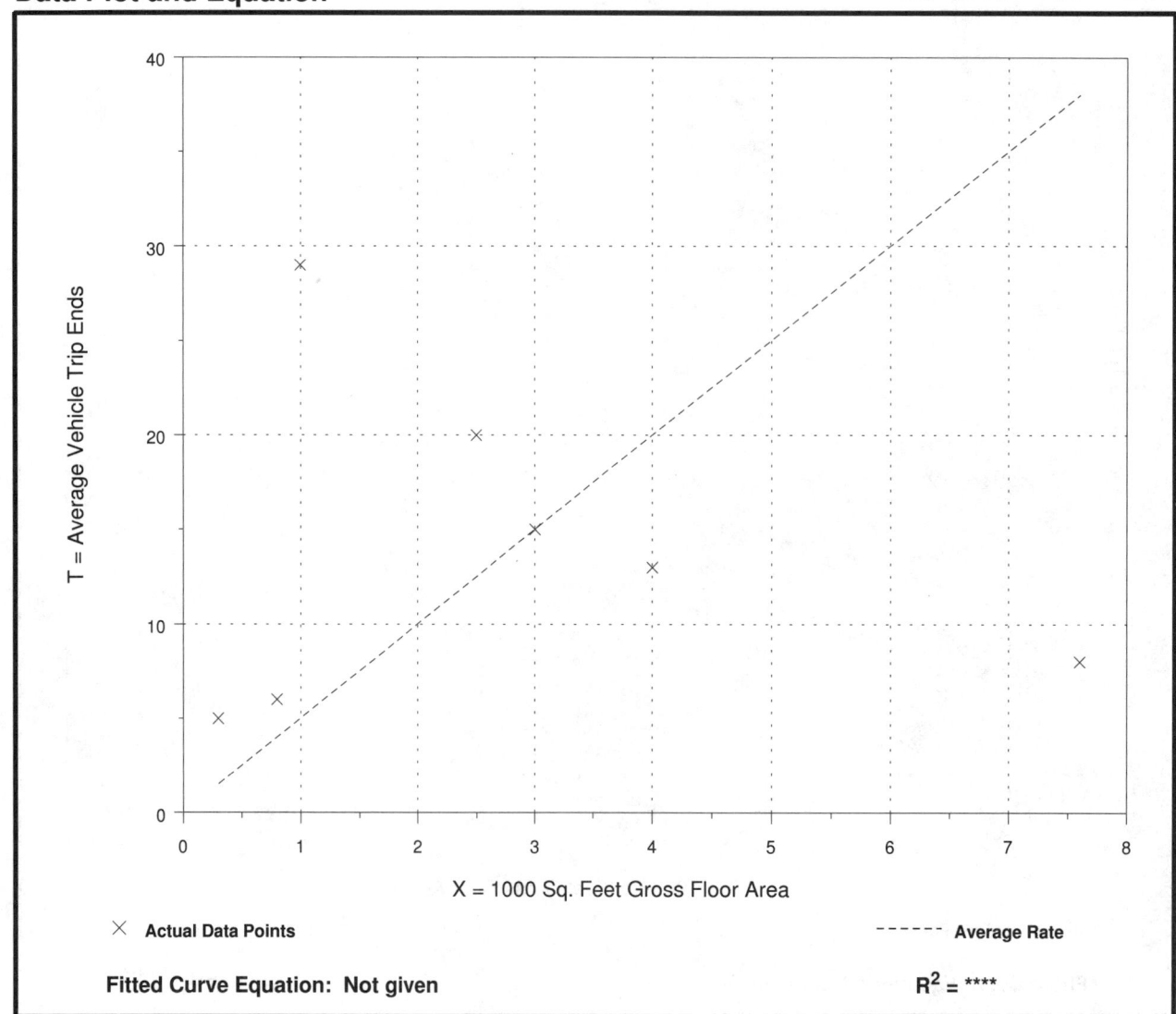

Fitted Curve Equation: Not given $R^2 = ****$

1544   *Trip Generation*, 9th Edition • Institute of Transportation Engineers

Nursery (Wholesale)
(818)

Average Vehicle Trip Ends vs: 1000 Sq. Feet Gross Floor Area
On a: Saturday

Number of Studies: 7
Average 1000 Sq. Feet GFA: 3
Directional Distribution: 50% entering, 50% exiting

Trip Generation per 1000 Sq. Feet Gross Floor Area

Average Rate	Range of Rates	Standard Deviation
29.90	4.08 - 96.00	26.64

Data Plot and Equation

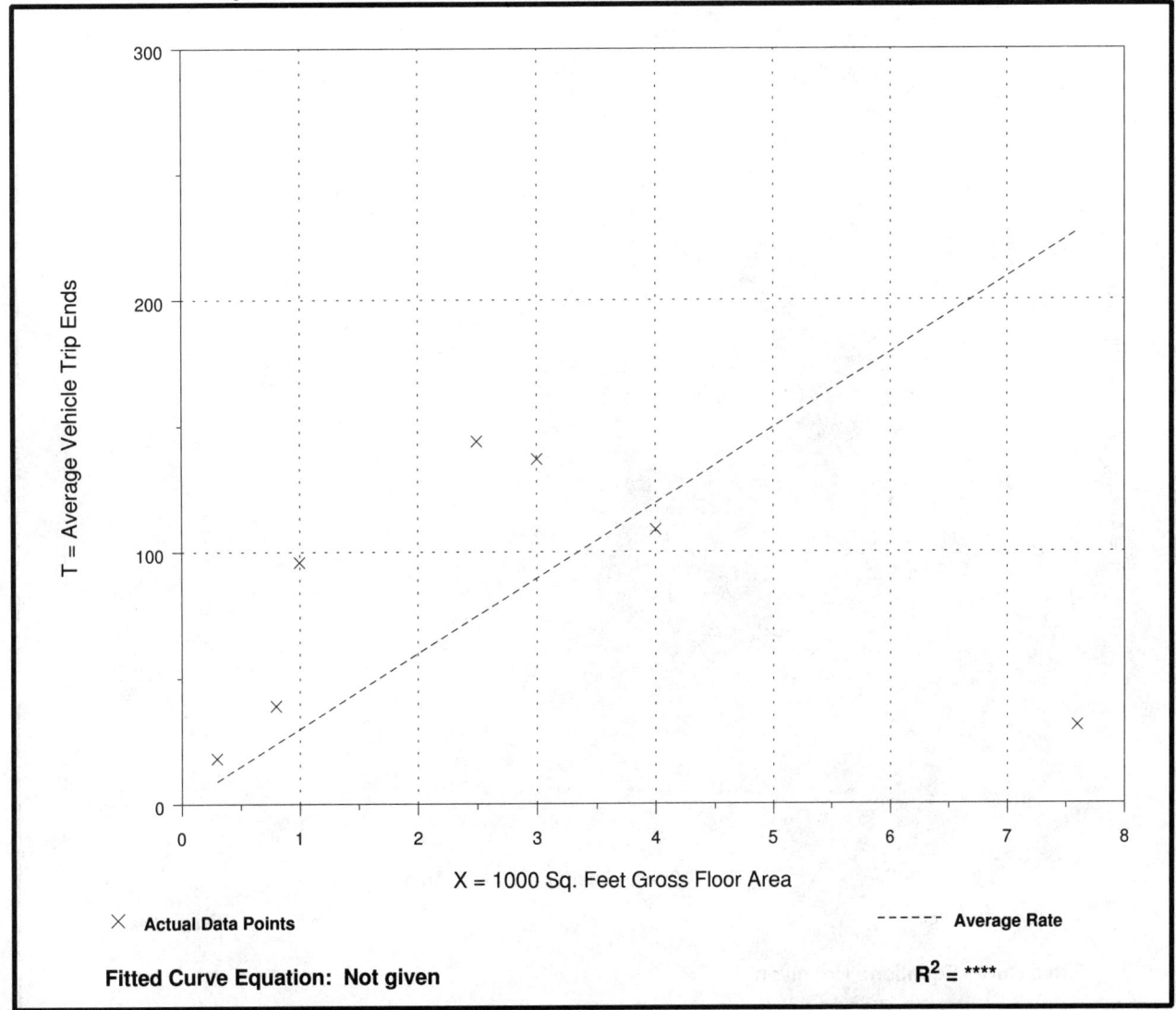

Fitted Curve Equation: Not given

$R^2 = ****$

Nursery (Wholesale)
(818)

Average Vehicle Trip Ends vs: 1000 Sq. Feet Gross Floor Area
On a: Saturday,
Peak Hour of Generator

Number of Studies: 7
Average 1000 Sq. Feet GFA: 3
Directional Distribution: Not available

Trip Generation per 1000 Sq. Feet Gross Floor Area

Average Rate	Range of Rates	Standard Deviation
5.52	1.18 - 14.00	4.80

Data Plot and Equation

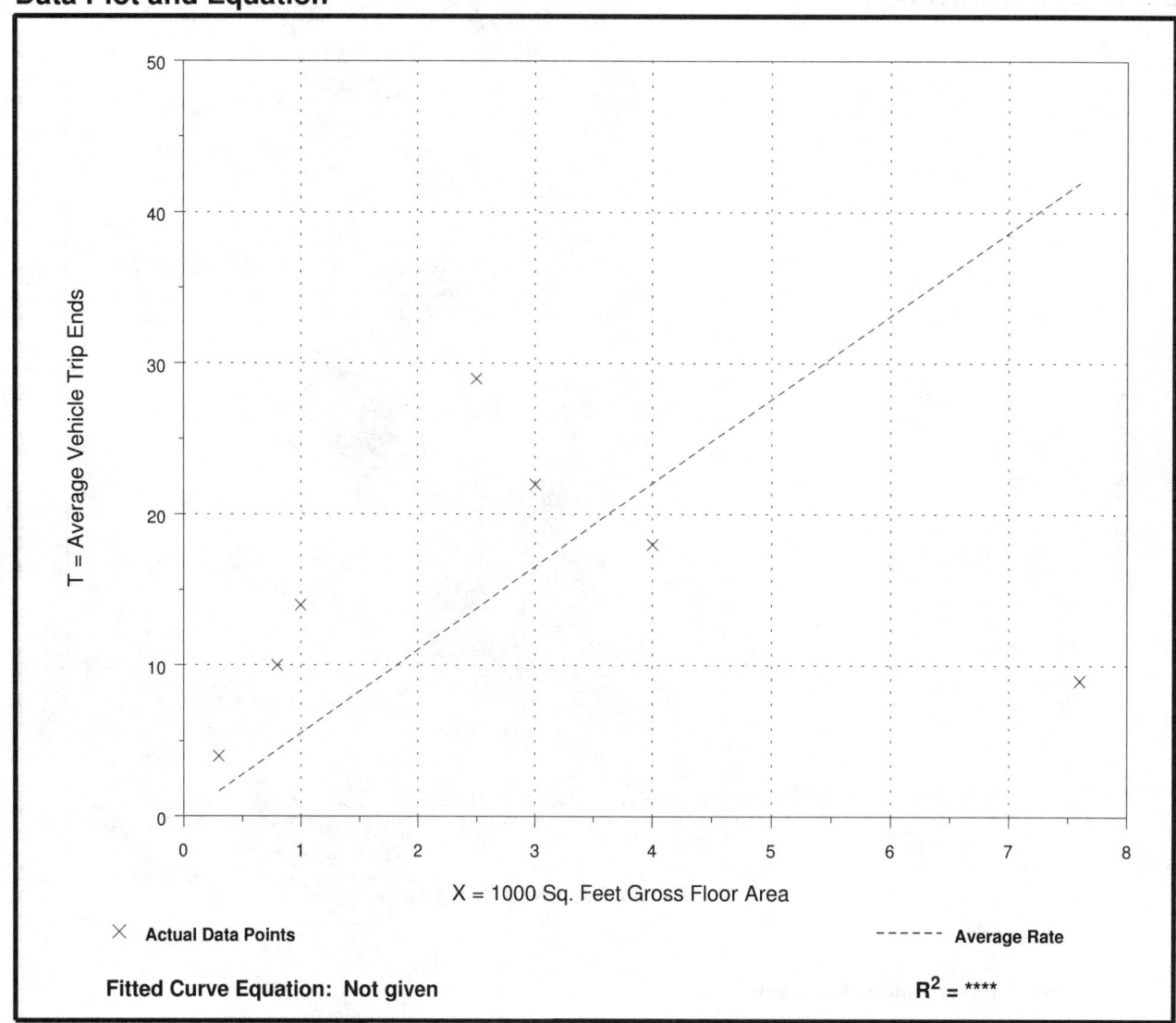

Fitted Curve Equation: Not given $R^2 = ****$

1546   *Trip Generation*, 9th Edition • Institute of Transportation Engineers

Nursery (Wholesale)
(818)

Average Vehicle Trip Ends vs: 1000 Sq. Feet Gross Floor Area
On a: Sunday

Number of Studies: 6
Average 1000 Sq. Feet GFA: 2
Directional Distribution: 50% entering, 50% exiting

Trip Generation per 1000 Sq. Feet Gross Floor Area

Average Rate	Range of Rates	Standard Deviation
26.47	7.60 - 73.33	22.02

Data Plot and Equation

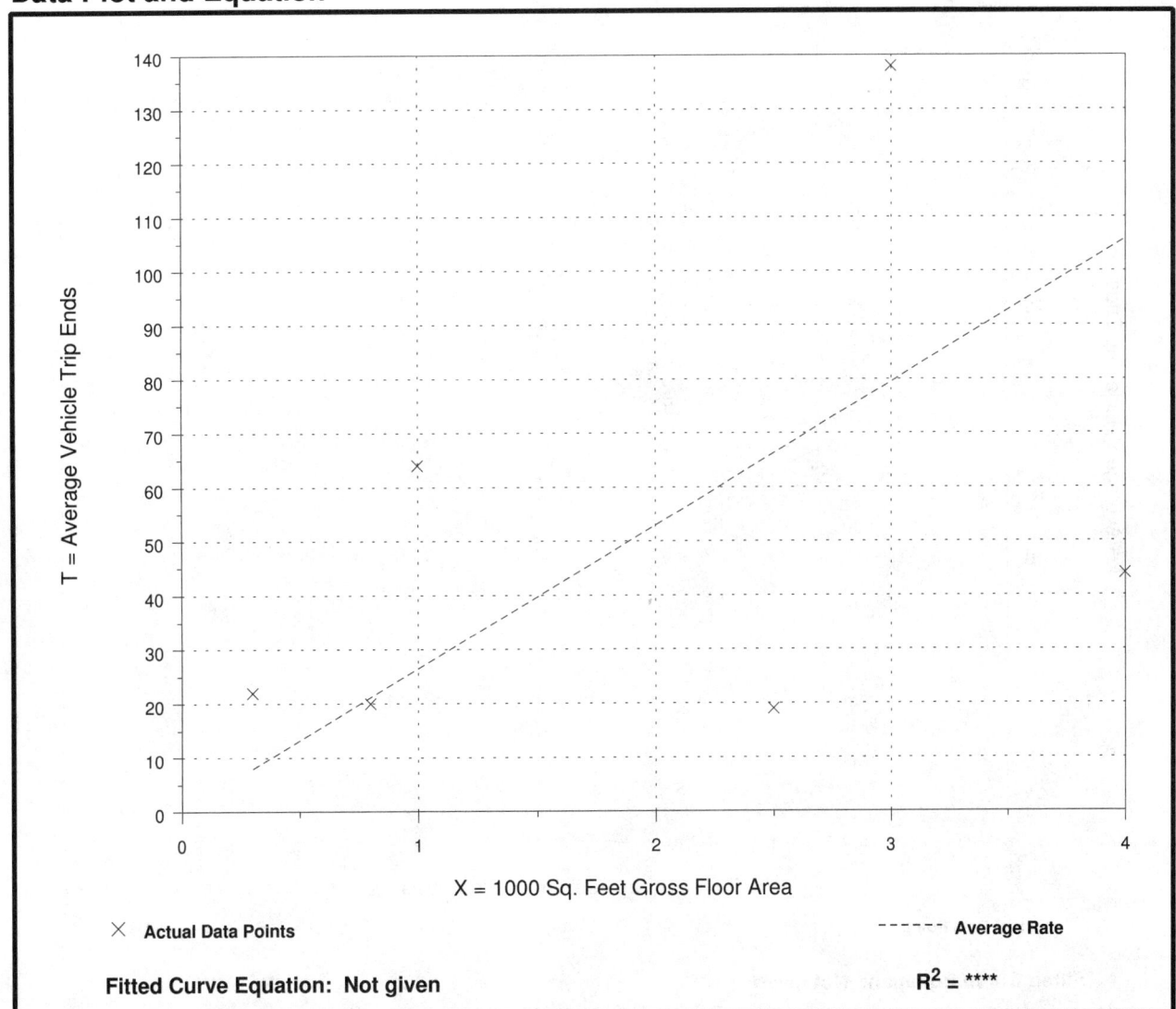

Fitted Curve Equation: Not given $R^2 = ****$

Nursery (Wholesale)
(818)

Average Vehicle Trip Ends vs: 1000 Sq. Feet Gross Floor Area
On a: Sunday,
Peak Hour of Generator

Number of Studies: 6
Average 1000 Sq. Feet GFA: 2
Directional Distribution: Not available

Trip Generation per 1000 Sq. Feet Gross Floor Area

Average Rate	Range of Rates	Standard Deviation
5.69	1.20 - 13.33	3.97

Data Plot and Equation

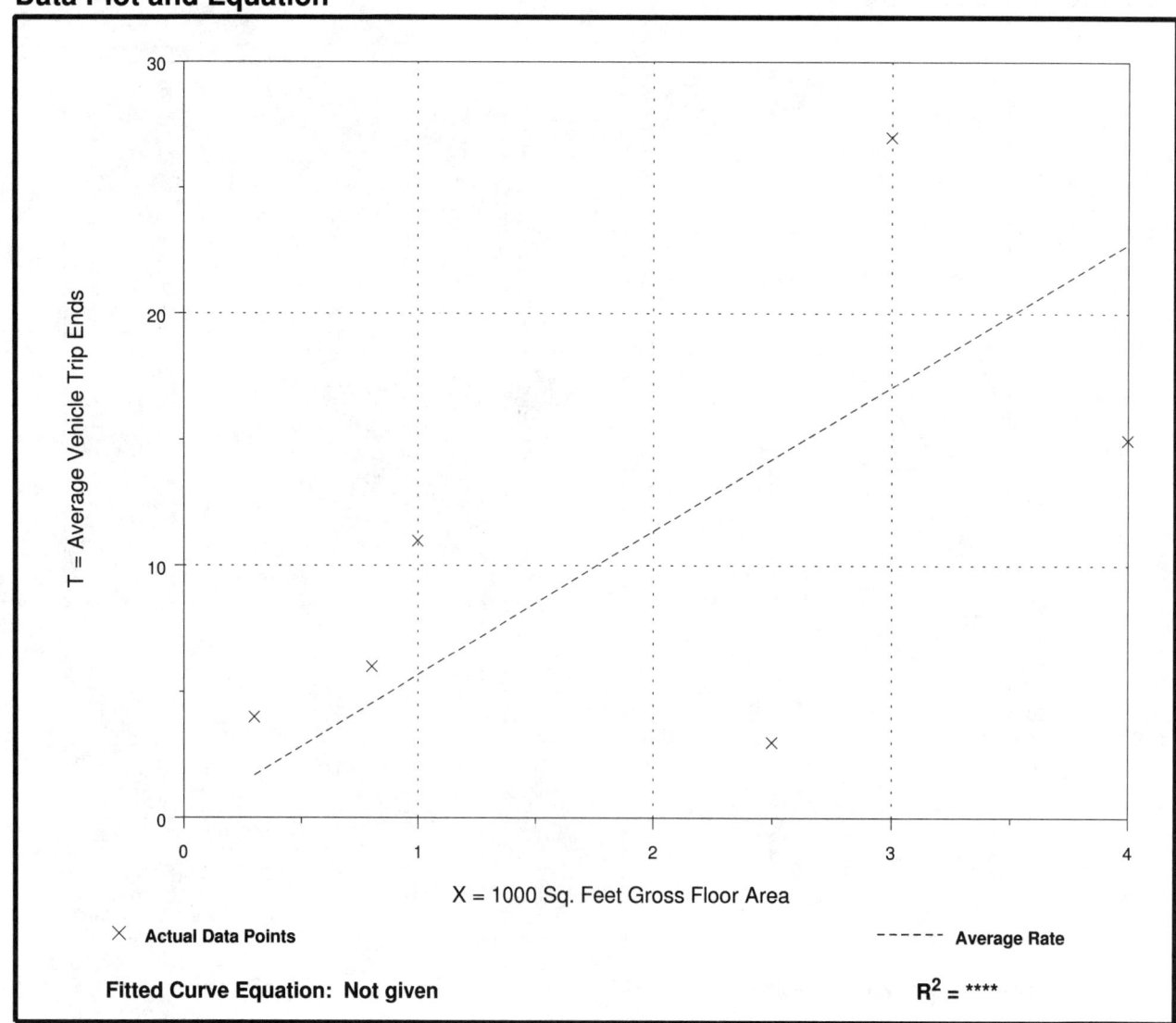

Fitted Curve Equation: Not given $R^2 = ****$

Nursery (Wholesale)
(818)

Average Vehicle Trip Ends vs: Acres
On a: Weekday,
Peak Hour of Adjacent Street Traffic,
One Hour Between 7 and 9 a.m.

Number of Studies: 8
Average Number of Acres: 24
Directional Distribution: Not available

Trip Generation per Acre

Average Rate	Range of Rates	Standard Deviation
0.26	0.10 - 1.17	0.56

Data Plot and Equation

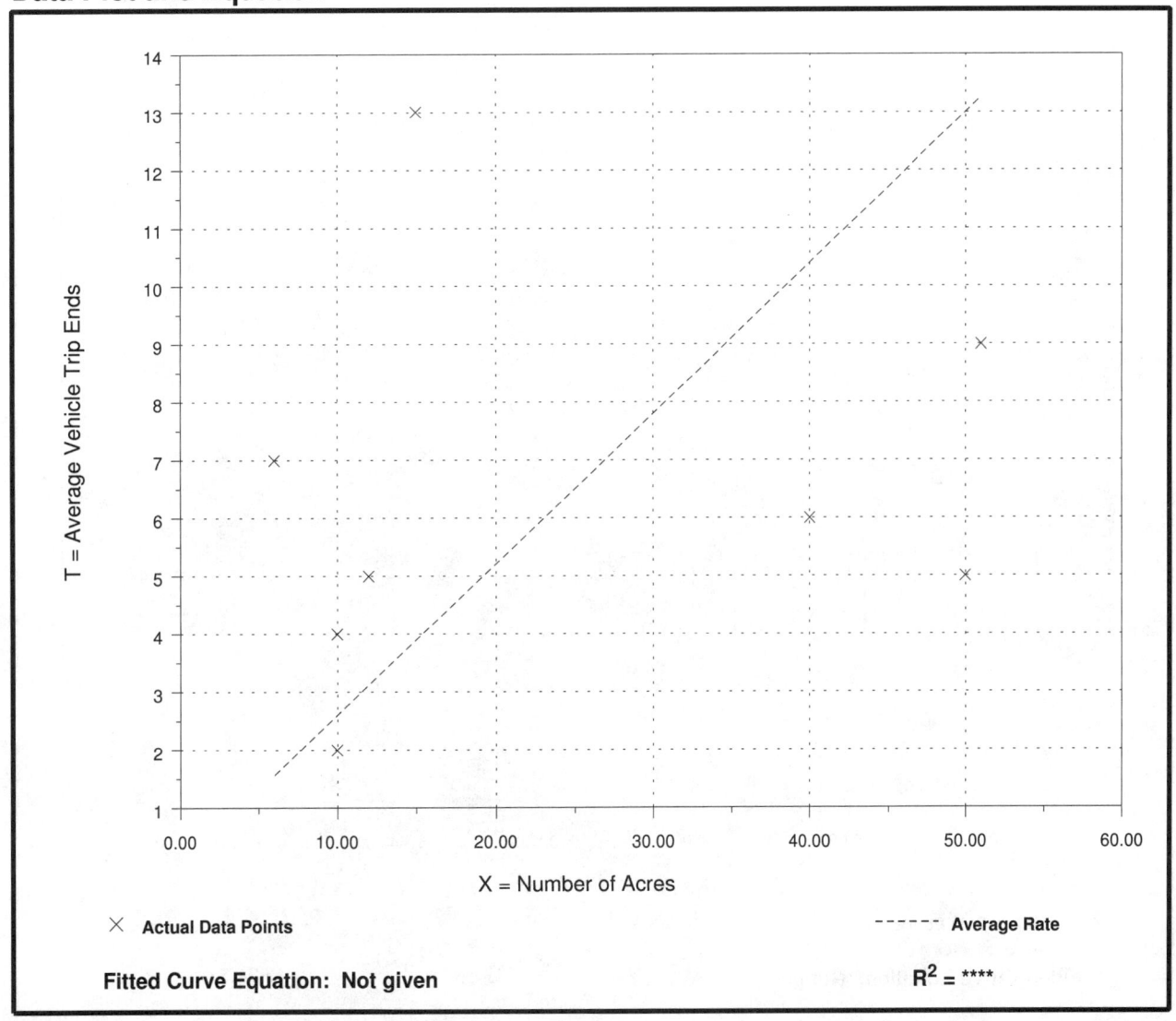

Fitted Curve Equation: Not given $R^2 = ****$

Nursery (Wholesale)
(818)

Average Vehicle Trip Ends vs: Acres
On a: Weekday,
Peak Hour of Adjacent Street Traffic,
One Hour Between 4 and 6 p.m.

Number of Studies: 6
Average Number of Acres: 22
Directional Distribution: Not available

Trip Generation per Acre

Average Rate	Range of Rates	Standard Deviation
0.45	0.15 - 2.50	0.81

Data Plot and Equation

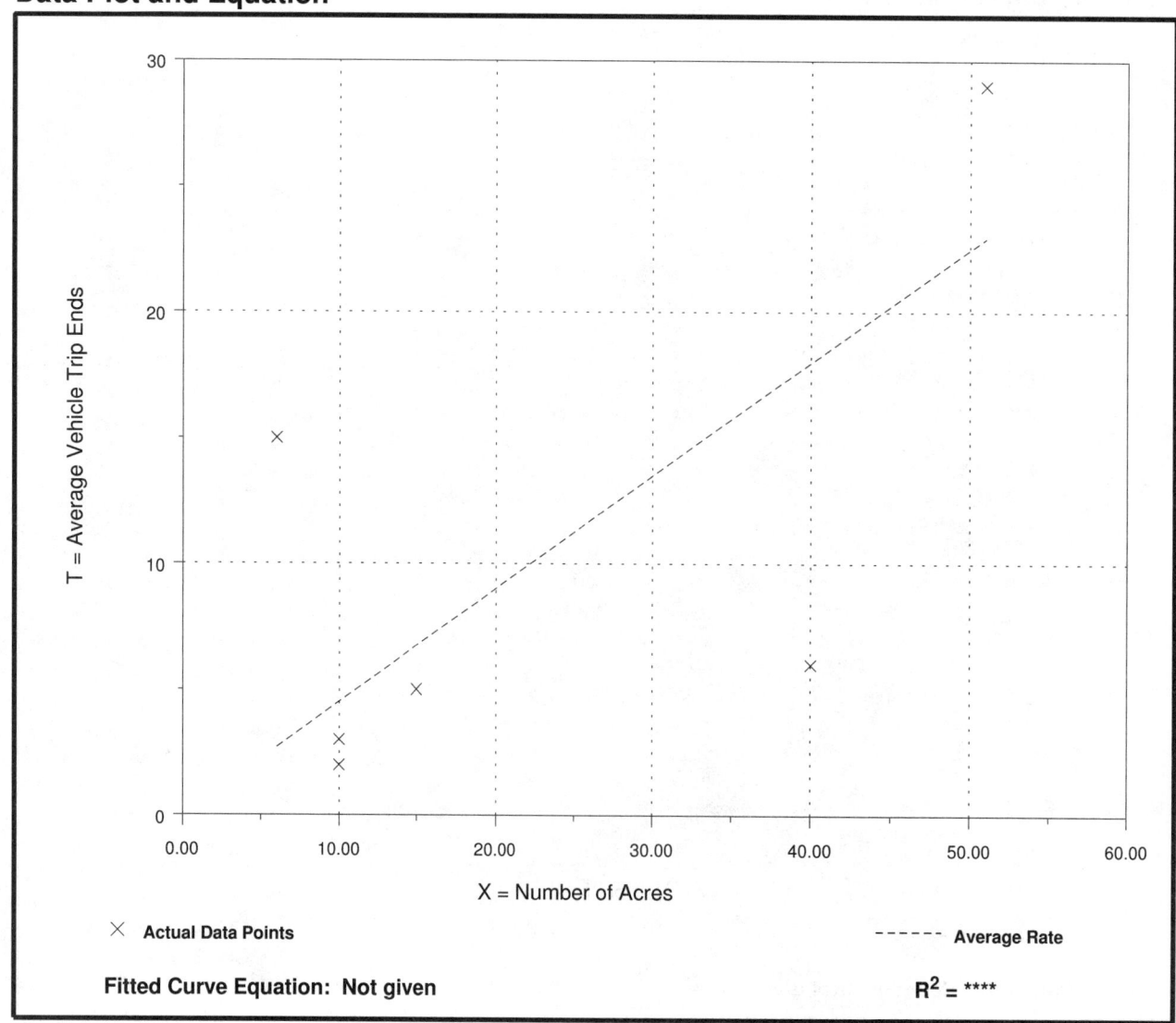

Fitted Curve Equation: Not given $R^2 = ****$

Nursery (Wholesale)
(818)

Average Vehicle Trip Ends vs: Acres
On a: Weekday,
A.M. Peak Hour of Generator

Number of Studies: 8
Average Number of Acres: 24
Directional Distribution: 43% entering, 57% exiting

Trip Generation per Acre

Average Rate	Range of Rates	Standard Deviation
0.34	0.10 - 2.00	0.67

Data Plot and Equation

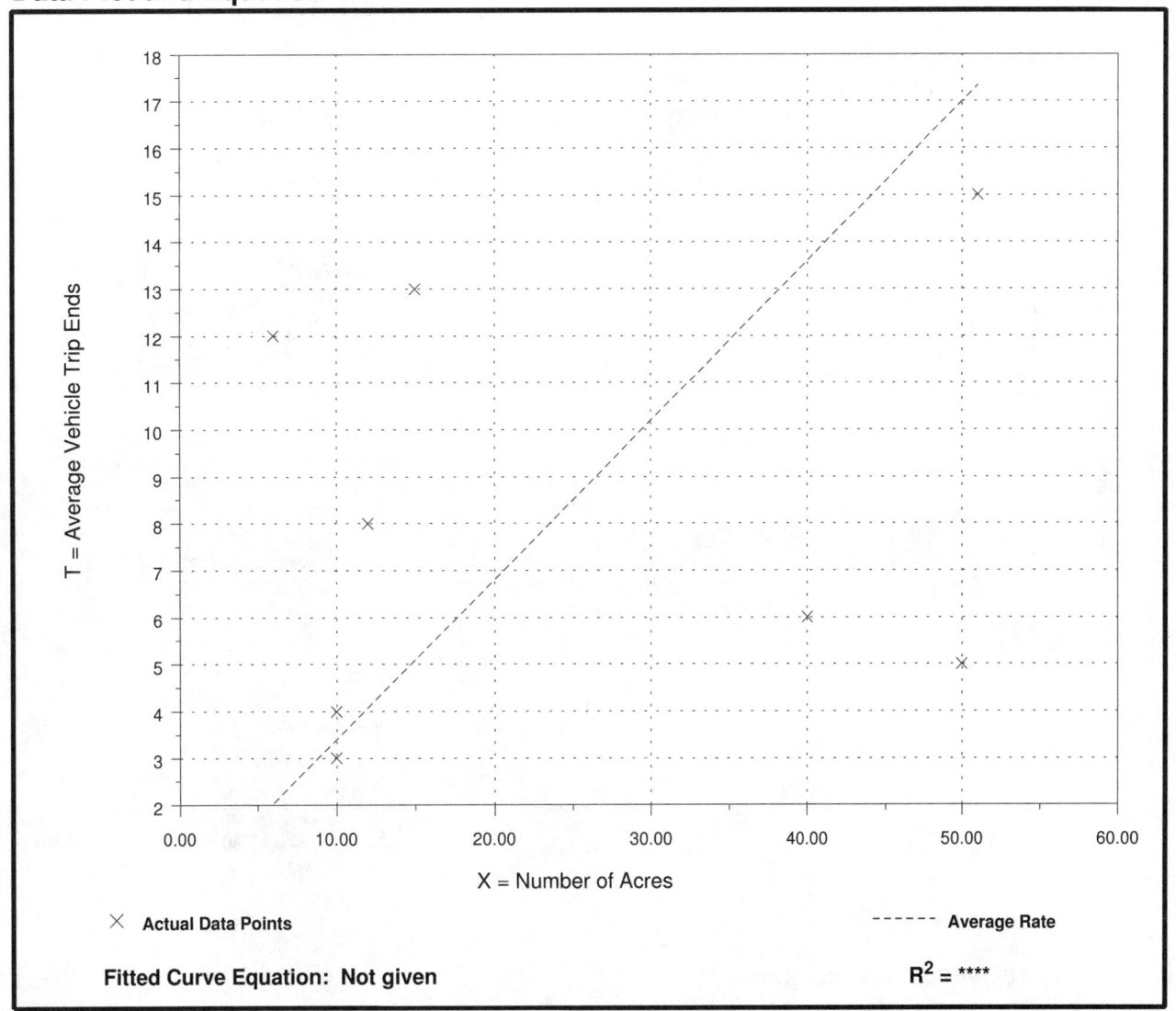

× Actual Data Points   ----- Average Rate

Fitted Curve Equation: Not given $R^2 = ****$

Nursery (Wholesale)
(818)

Average Vehicle Trip Ends vs: Acres
On a: Weekday,
P.M. Peak Hour of Generator

Number of Studies: 8
Average Number of Acres: 24
Directional Distribution: Not available

Trip Generation per Acre

Average Rate	Range of Rates	Standard Deviation
0.53	0.16 - 2.50	0.81

Data Plot and Equation

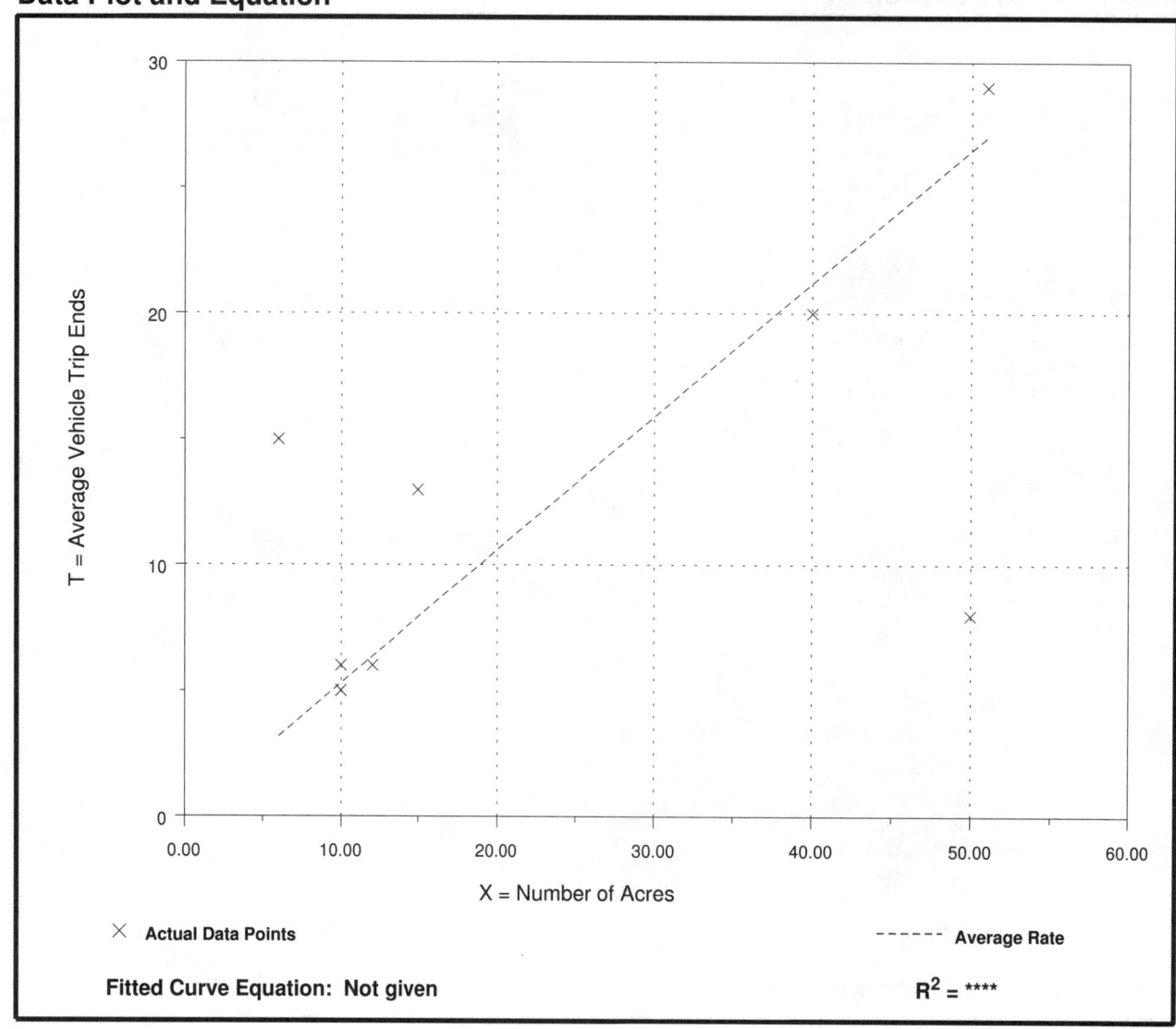

Fitted Curve Equation: Not given $R^2 = ****$

Nursery (Wholesale)
(818)

Average Vehicle Trip Ends vs: Acres
On a: Saturday

Number of Studies: 8
Average Number of Acres: 24
Directional Distribution: 50% entering, 50% exiting

Trip Generation per Acre

Average Rate	Range of Rates	Standard Deviation
3.11	0.62 - 22.83	4.31

Data Plot and Equation

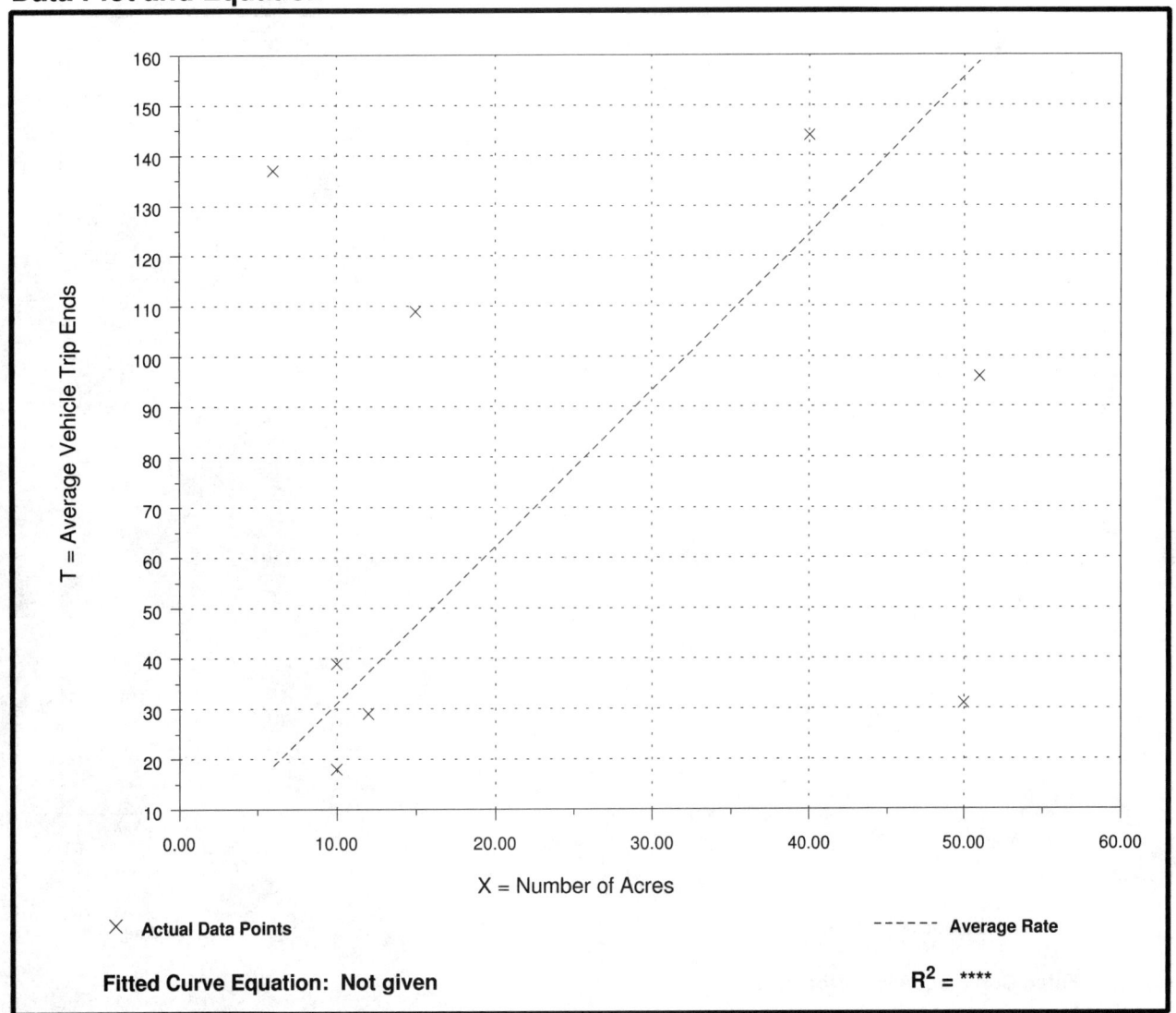

Fitted Curve Equation: Not given $R^2 = ****$

Nursery (Wholesale)
(818)

Average Vehicle Trip Ends vs: Acres
On a: Saturday,
Peak Hour of Generator

Number of Studies: 8
Average Number of Acres: 24
Directional Distribution: Not available

Trip Generation per Acre

Average Rate	Range of Rates	Standard Deviation
0.58	0.18 - 3.67	0.98

Data Plot and Equation

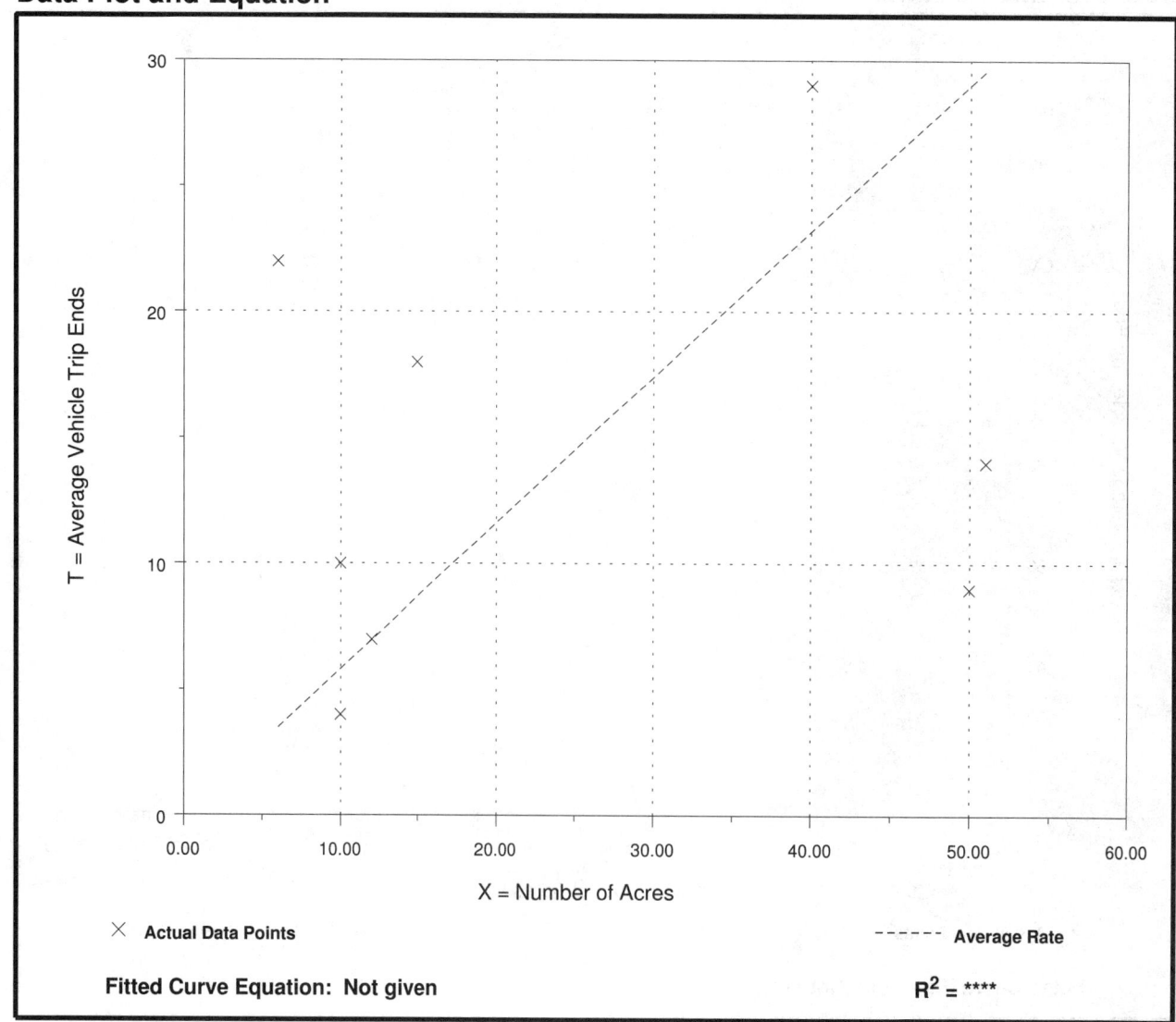

Fitted Curve Equation: Not given $R^2 = ****$

Nursery (Wholesale)
(818)

Average Vehicle Trip Ends vs: Acres
On a: Sunday

Number of Studies: 7
Average Number of Acres: 21
Directional Distribution: 50% entering, 50% exiting

Trip Generation per Acre

Average Rate	Range of Rates	Standard Deviation
2.20	0.48 - 23.00	4.64

Data Plot and Equation

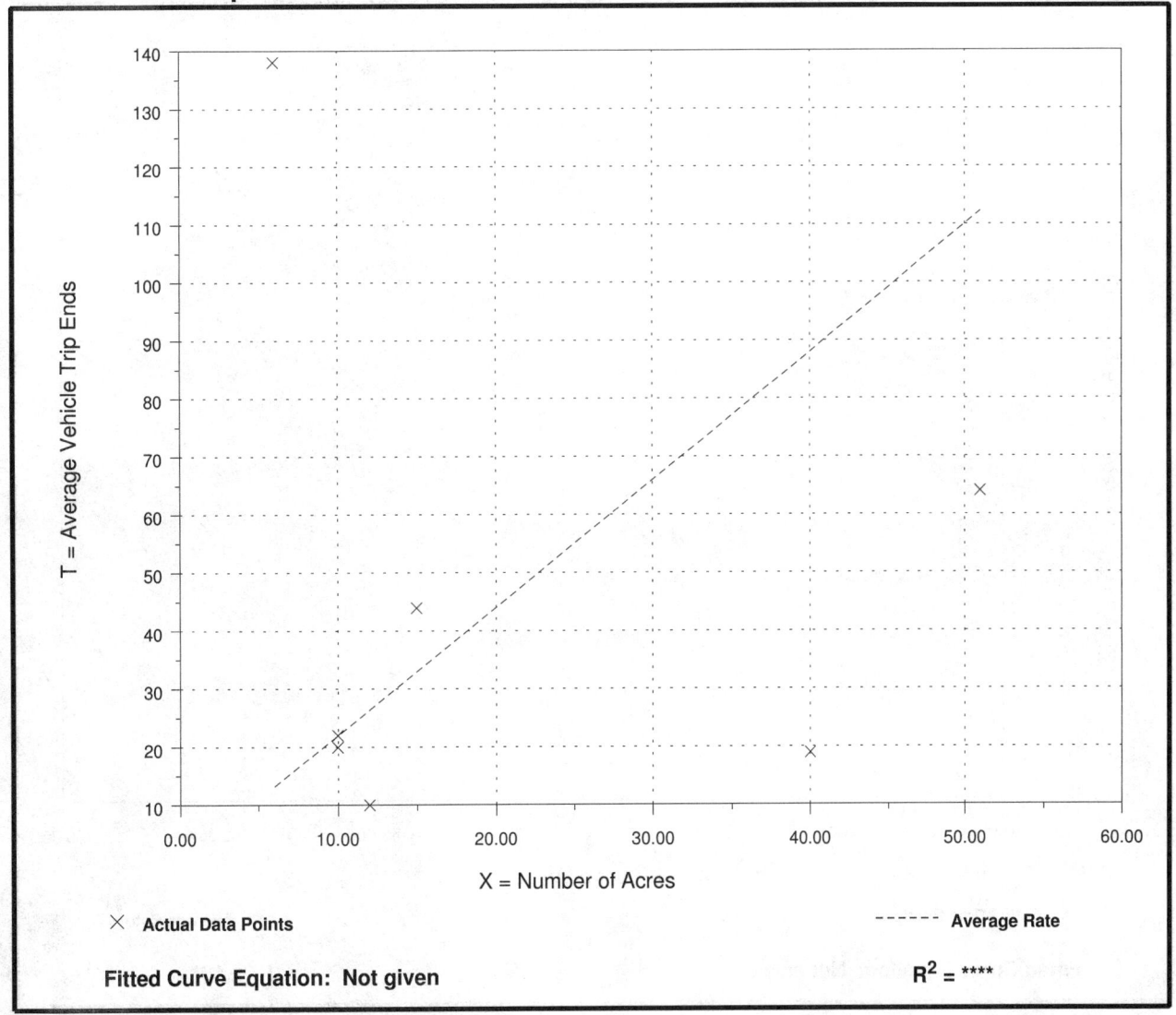

Fitted Curve Equation: Not given $R^2 = ****$

Nursery (Wholesale)
(818)

Average Vehicle Trip Ends vs: Acres
On a: Sunday,
Peak Hour of Generator

Number of Studies: 7
Average Number of Acres: 21
Directional Distribution: Not available

Trip Generation per Acre

Average Rate	Range of Rates	Standard Deviation
0.50	0.08 - 4.50	1.11

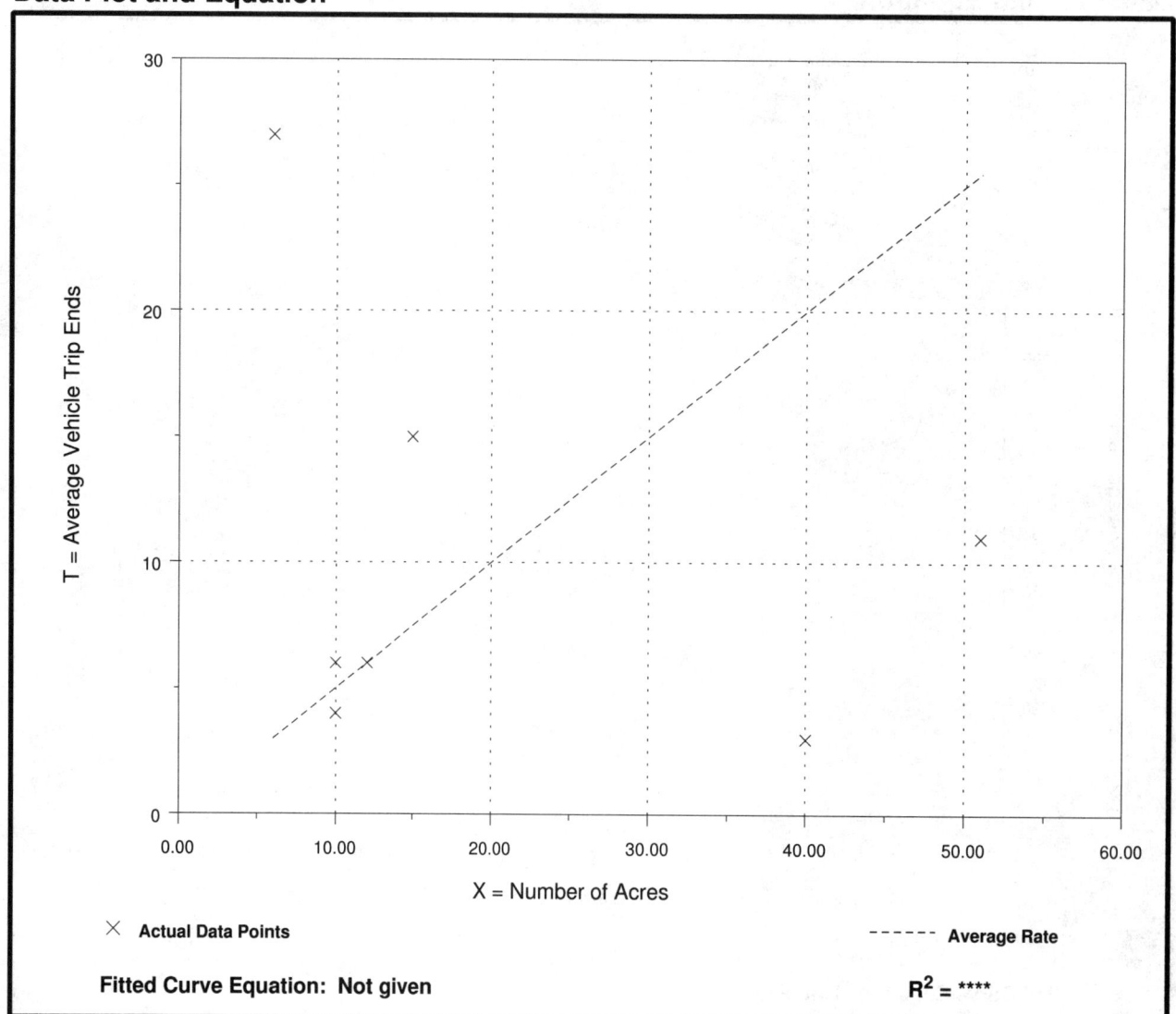

Data Plot and Equation

Fitted Curve Equation: Not given $R^2 = ****$

Land Use: 820
Shopping Center

Description

A shopping center is an integrated group of commercial establishments that is planned, developed, owned and managed as a unit. A shopping center's composition is related to its market area in terms of size, location and type of store. A shopping center also provides on-site parking facilities sufficient to serve its own parking demands. Specialty retail center (Land Use 826) and factory outlet center (Land Use 823) are related uses.

Additional Data

Shopping centers, including neighborhood centers, community centers, regional centers and super regional centers, were surveyed for this land use. Some of these centers contained non-merchandising facilities, such as office buildings, movie theaters, restaurants, post offices, banks, health clubs and recreational facilities (for example, ice skating rinks or indoor miniature golf courses). The centers ranged in size from 1,700 to 2.2 million square feet gross leasable area (GLA). The centers studied were located in suburban areas throughout the United States and, therefore, represent typical U.S. suburban conditions.

Many shopping centers, in addition to the integrated unit of shops in one building or enclosed around a mall, include outparcels (peripheral buildings or pads located on the perimeter of the center adjacent to the streets and major access points). These buildings are typically drive-in banks, retail stores, restaurants, or small offices. Although the data herein do not indicate which of the centers studied included peripheral buildings, it can be assumed that some of the data show their effect.

The vehicle trips generated at a shopping center are based upon the total GLA of the center. In cases of smaller centers without an enclosed mall or peripheral buildings, the GLA could be the same as the gross floor area of the building.

Separate equations have been developed for shopping centers during the Christmas shopping season. Plots were included for the weekday peak hour of adjacent street traffic and the Saturday peak hour of the generator.

Information on approximate hourly, monthly and daily variation in shopping center traffic is shown in Tables 1–3. It should be noted, however, that the information contained in these tables is based on a limited sample size. Therefore, caution should be exercised when applying the data. Also, some information provided in the tables may conflict with the results obtained by applying the average rate or regression equations. When this occurs, it is suggested that the results from the average rate or regression equations be used, as they are based on a larger number of studies.

Table 1 Hourly Variation in Shopping Center Traffic						
Time	Average Weekday[a]		Average Saturday[b]		Average Sunday[c]	
	Percent of 24-Hour Entering Traffic	Percent of 24-Hour Exiting Traffic	Percent of 24-Hour Entering Traffic	Percent of 24-Hour Exiting Traffic	Percent of 24-Hour Entering Traffic	Percent of 24-Hour Exiting Traffic
6 a.m.–7 a.m.	0.8	0.3	0.2	0.2	0.2	0.1
7 a.m.–8 a.m.	2.0	0.9	0.9	0.4	0.4	0.3
8 a.m.–9 a.m.	3.1	1.2	2.7	1.0	0.9	0.5
9 a.m.–10 a.m.	5.5	2.0	5.5	2.2	1.7	1.1
10 a.m.–11 a.m.	7.0	4.3	8.6	4.8	3.8	2.5
11 a.m.–12 p.m.	8.4	6.2	10.8	7.5	10.0	4.6
12 p.m.–1 p.m.	9.4	8.3	11.8	9.3	15.1	7.9
1 p.m.–2 p.m.	8.2	8.6	12.1	10.3	16.7	12.0
2 p.m.–3 p.m.	7.7	8.9	11.8	11.8	15.8	14.7
3 p.m.–4 p.m.	7.8	8.8	10.7	12.5	13.0	15.6
4 p.m.–5 p.m.	8.0	8.9	8.8	12.5	9.4	15.8
5 p.m.–6 p.m.	8.4	9.2	5.3	11.3	5.1	13.0
6 p.m.–7 p.m.	8.0	7.5	3.3	6.7	2.3	4.6
7 p.m.–8 p.m.	7.9	7.2	2.7	2.9	1.7	1.9
8 p.m.–9 p.m.	4.3	7.7	1.8	2.2	1.1	1.3
9 p.m.–10 p.m.	1.8	7.2	1.0	1.6	0.7	1.1
10 p.m.–6 a.m.	1.7	2.8	2.0	2.8	2.1	3.0

Sites ranged in size from 11,000 to 1,750,000 square feet gross leasable area
[a] Source numbers – 13, 73, 88, 190, 217, 220, 225 and 376; based on ten studies
[b] Source numbers – 13, 73, 88, 190, 220, 225 and 376; based on nine studies
[c] Source numbers – 13, 73, 88, 190, 220 and 225; based on eight studies

Table 2
Daily Variation in Shopping Center Traffic
Percentage of Average Weekday Volume (Monday through Friday)

Day	Less than 100,000 Square Feet GLA	100,000 to 300,000 Square Feet GLA	More than 300,000 Square Feet GLA	Discount Center
Sunday	45.2	65.4	77.4	82.1
Monday	97.3	96.8	96.8	95.1
Tuesday	92.9	103.1	97.1	91.4
Wednesday	92.7	99.1	93.6	94.8
Thursday	98.2	85.3	97.1	99.5
Friday	118.9	108.7	115.4	119.2
Saturday	128.5	113.4	128.0	151.0
Sample Size	6	8	17	2

Source numbers: 88, 124

Table 3
Monthly Variation in Shopping Center Traffic
Percentage of Average Month

Month	Percentage	Month	Percentage
January	85.3	July	100.8
February	78.1	August	102.1
March	92.0	September	94.8
April	93.2	October	98.9
May	105.4	November	101.5
June	106.0	December	141.8

Sample size: 2
Average gross leasable area: 938,000 square feet

The sites were surveyed between the 1960s and the 2000s throughout the United States and Canada.

Specialized Land Use Data

Two studies provided data on outdoor shopping centers in Illinois and Alberta, Canada. The trip generation characteristics of these sites varied from the other stores in this land use; therefore, the information collected for these facilities is presented in the following tables and was excluded from the data plots.

Independent Variable	Average Trip Generation Rate	Size of Independent Variable	Number of Studies	Directional Distribution
1,000 Square Feet Gross Leasable Area				
Weekday	66.64	797	2	Not available
Weekday A.M. Peak Hour of Adjacent Street Traffic	3.27	797	2	Not available
Weekday P.M. Peak Hour of Adjacent Street Traffic	5.46	797	2	Not available

Sources: 446, 702

Source Numbers

1, 2, 3, 4, 5, 6, 13, 14, 18, 19, 22, 26, 40, 42, 48, 49, 54, 59, 60, 61, 64, 65, 72, 73, 75, 76, 77, 78, 79, 87, 89, 90, 98, 99, 100, 105, 110, 124, 156, 159, 172, 186, 193, 194, 195, 196, 197, 198, 199, 202, 204, 211, 213, 260, 263, 269, 295, 299, 300, 301, 304, 305, 307, 308, 309, 310, 311, 312, 313, 314, 315, 316, 317, 318, 319, 358, 365, 376, 385, 390, 400, 404, 414, 420, 423, 428, 437, 440, 442, 444, 446, 507, 562, 563, 580, 598, 629, 658, 702, 715, 728

Shopping Center
(820)

Average Vehicle Trip Ends vs: 1000 Sq. Feet Gross Leasable Area
On a: Weekday

Number of Studies: 302
Average 1000 Sq. Feet GLA: 331
Directional Distribution: 50% entering, 50% exiting

Trip Generation per 1000 Sq. Feet Gross Leasable Area

Average Rate	Range of Rates	Standard Deviation
42.70	12.50 - 270.89	21.25

Data Plot and Equation

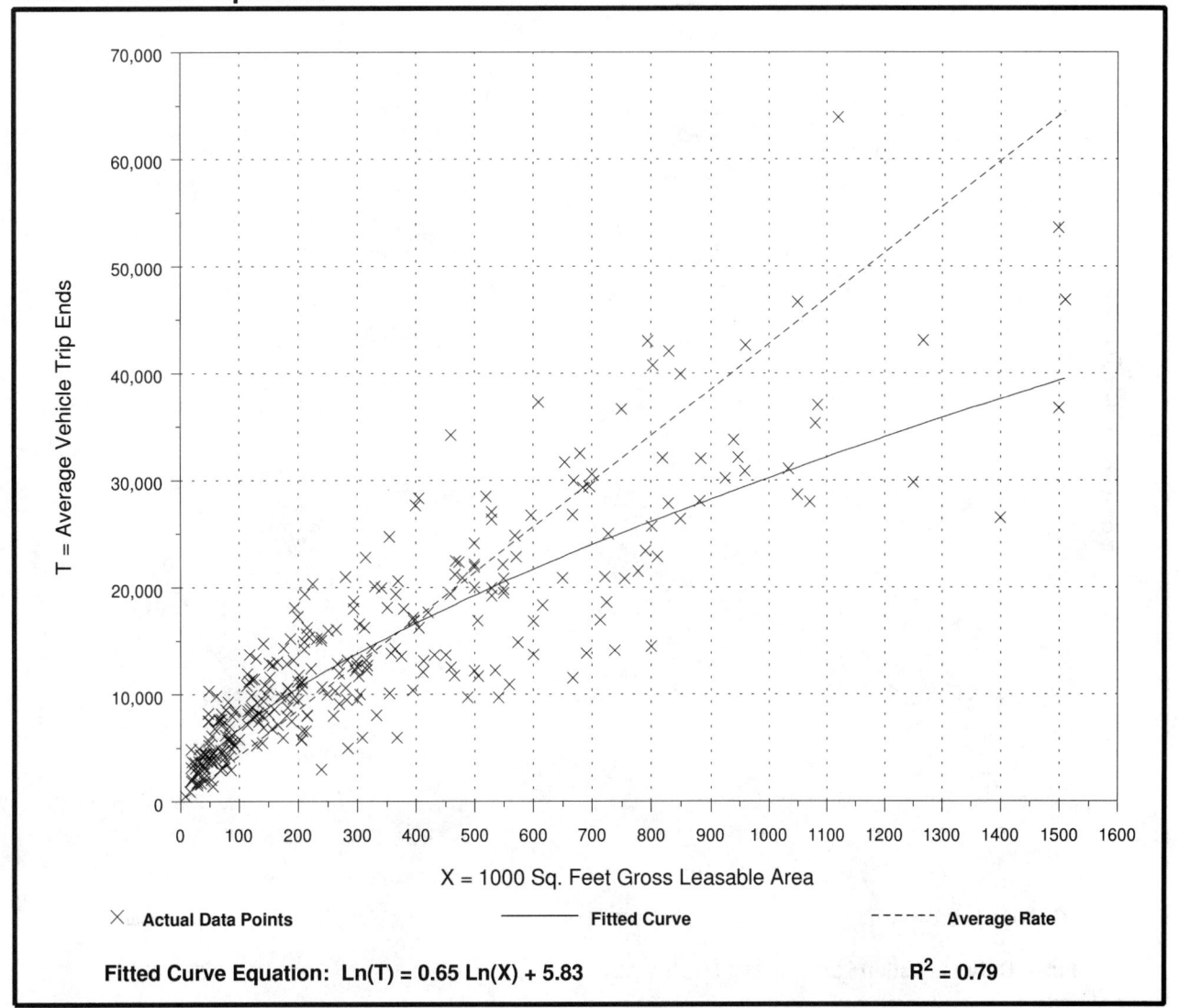

Fitted Curve Equation: $Ln(T) = 0.65\, Ln(X) + 5.83$ $R^2 = 0.79$

Trip Generation, 9th Edition • Institute of Transportation Engineers

Shopping Center
(820)

Average Vehicle Trip Ends vs: 1000 Sq. Feet Gross Leasable Area
On a: Weekday,
Peak Hour of Adjacent Street Traffic,
One Hour Between 7 and 9 a.m.

Number of Studies: 104
Average 1000 Sq. Feet GLA: 310
Directional Distribution: 62% entering, 38% exiting

Trip Generation per 1000 Sq. Feet Gross Leasable Area

Average Rate	Range of Rates	Standard Deviation
0.96	0.10 - 9.05	1.31

Data Plot and Equation

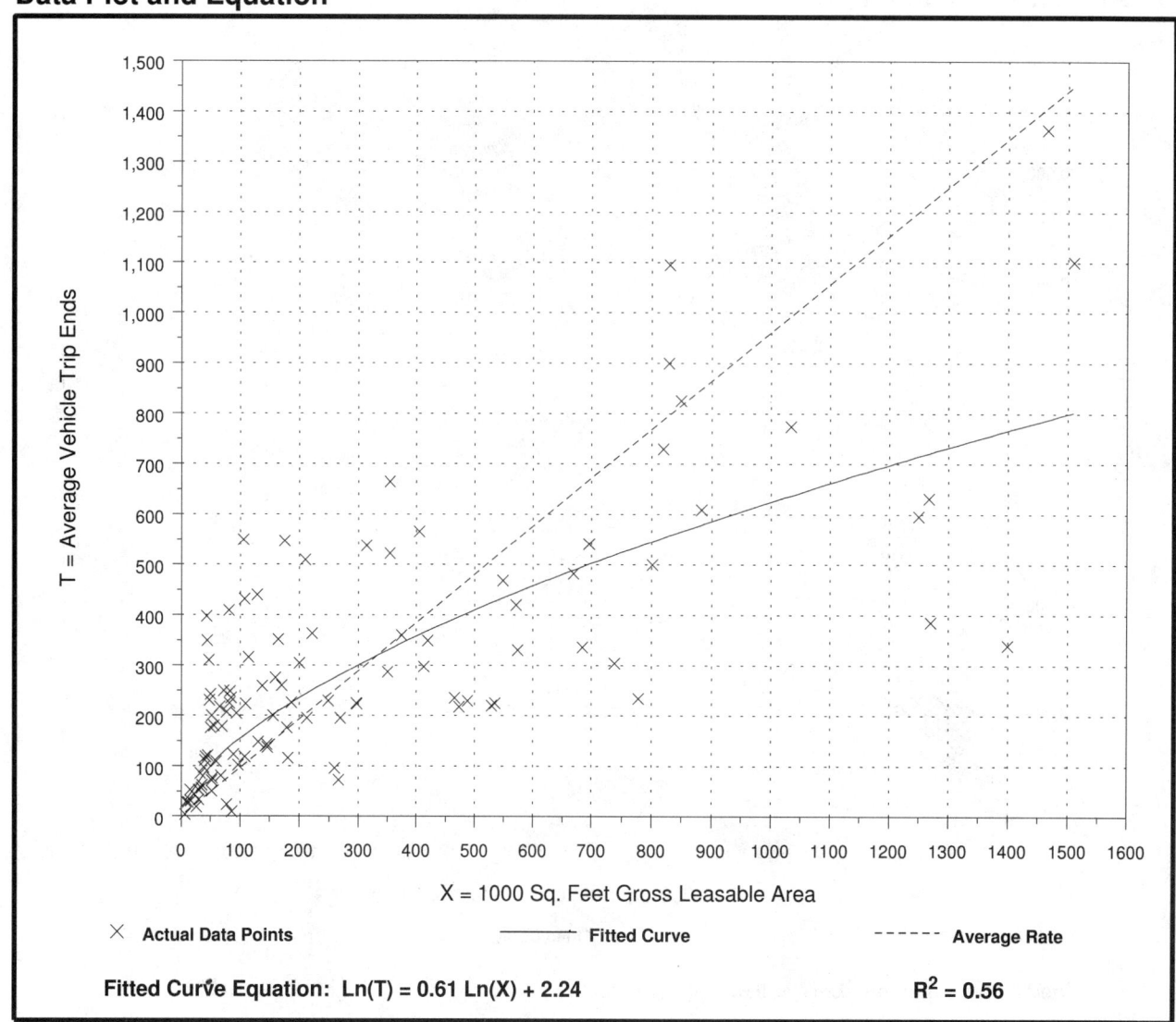

Fitted Curve Equation: $Ln(T) = 0.61\, Ln(X) + 2.24$ $R^2 = 0.56$

Shopping Center
(820)

Average Vehicle Trip Ends vs: 1000 Sq. Feet Gross Leasable Area
On a: Weekday,
Peak Hour of Adjacent Street Traffic,
One Hour Between 4 and 6 p.m.

Number of Studies: 426
Average 1000 Sq. Feet GLA: 376
Directional Distribution: 48% entering, 52% exiting

Trip Generation per 1000 Sq. Feet Gross Leasable Area

Average Rate	Range of Rates	Standard Deviation
3.71	0.68 - 29.27	2.74

Data Plot and Equation

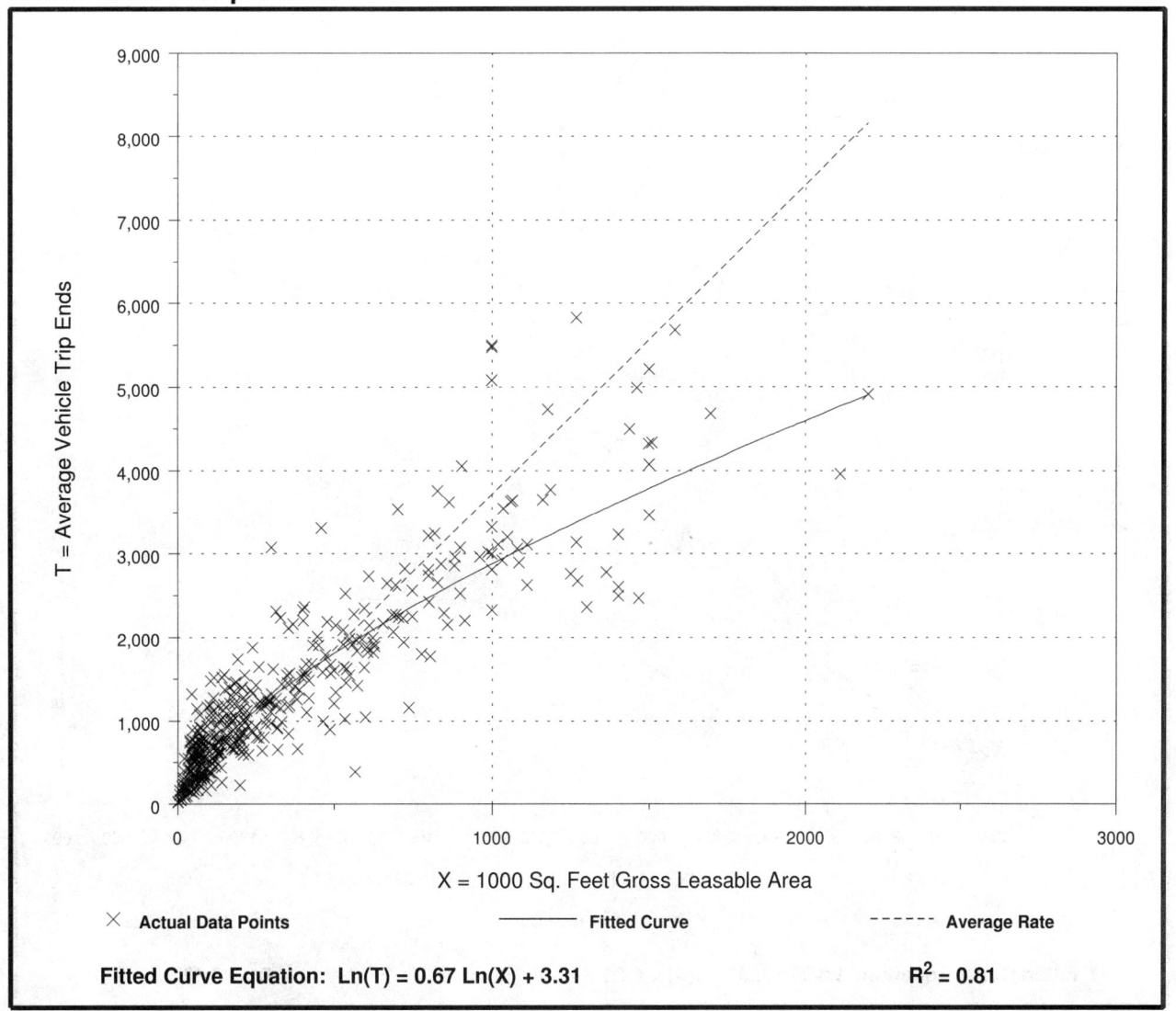

Fitted Curve Equation: $Ln(T) = 0.67\ Ln(X) + 3.31$ $R^2 = 0.81$

Shopping Center
(820)

Average Vehicle Trip Ends vs: 1000 Sq. Feet Gross Leasable Area
On a: Saturday

Number of Studies: 123
Average 1000 Sq. Feet GLA: 450
Directional Distribution: 50% entering, 50% exiting

Trip Generation per 1000 Sq. Feet Gross Leasable Area

Average Rate	Range of Rates	Standard Deviation
49.97	16.70 - 227.50	22.62

Data Plot and Equation

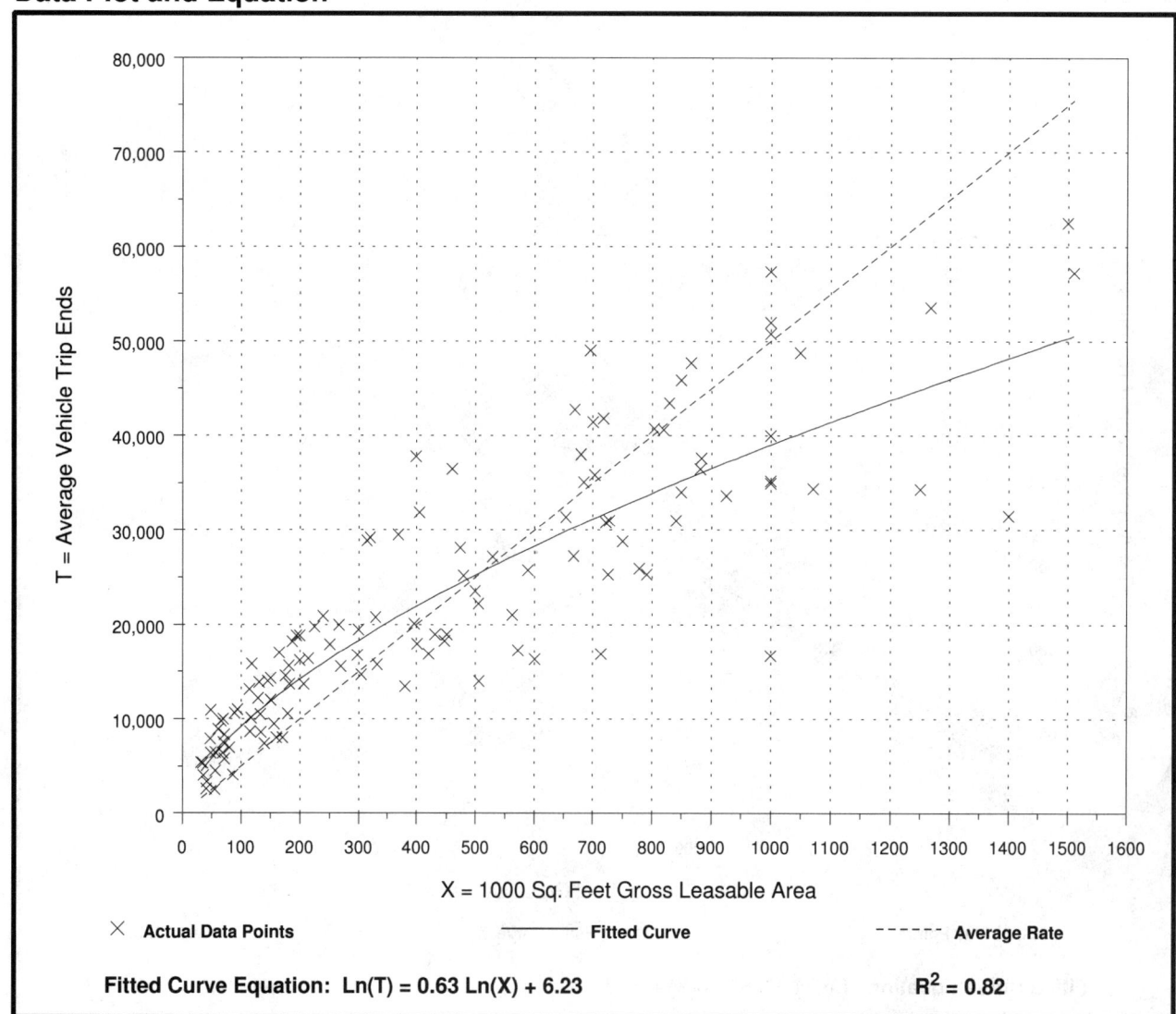

Fitted Curve Equation: $Ln(T) = 0.63\, Ln(X) + 6.23$ $\qquad R^2 = 0.82$

Shopping Center
(820)

Average Vehicle Trip Ends vs: 1000 Sq. Feet Gross Leasable Area
On a: Saturday,
Peak Hour of Generator

Number of Studies: 128
Average 1000 Sq. Feet GLA: 458
Directional Distribution: 52% entering, 48% exiting

Trip Generation per 1000 Sq. Feet Gross Leasable Area

Average Rate	Range of Rates	Standard Deviation
4.82	1.46 - 18.32	3.10

Data Plot and Equation

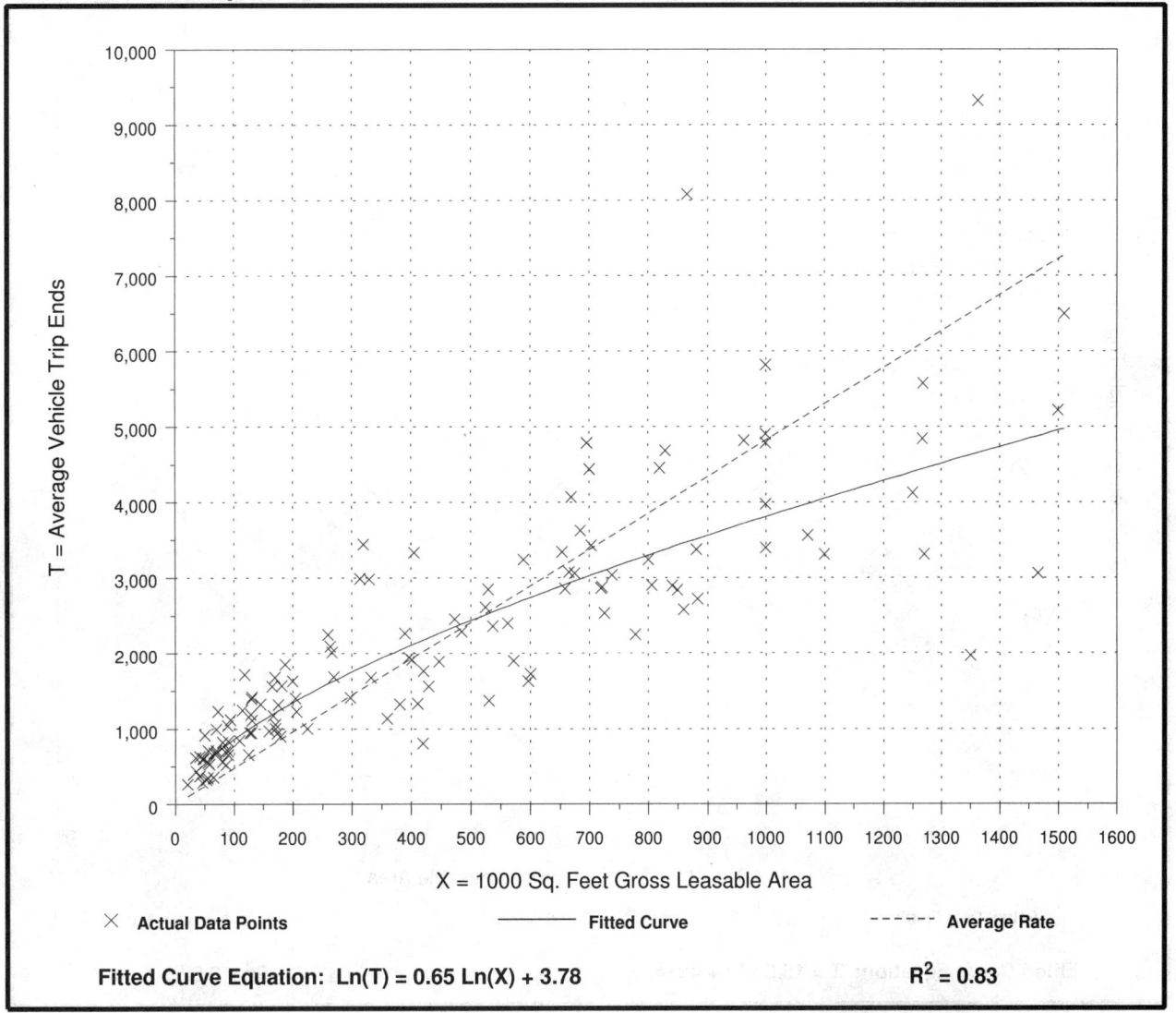

Fitted Curve Equation: $Ln(T) = 0.65 \, Ln(X) + 3.78$ $R^2 = 0.83$

Shopping Center
(820)

Average Vehicle Trip Ends vs: 1000 Sq. Feet Gross Leasable Area
On a: Sunday

Number of Studies: 77
Average 1000 Sq. Feet GLA: 439
Directional Distribution: 50% entering, 50% exiting

Trip Generation per 1000 Sq. Feet Gross Leasable Area

Average Rate	Range of Rates	Standard Deviation
25.24	4.15 - 148.15	17.23

Data Plot and Equation

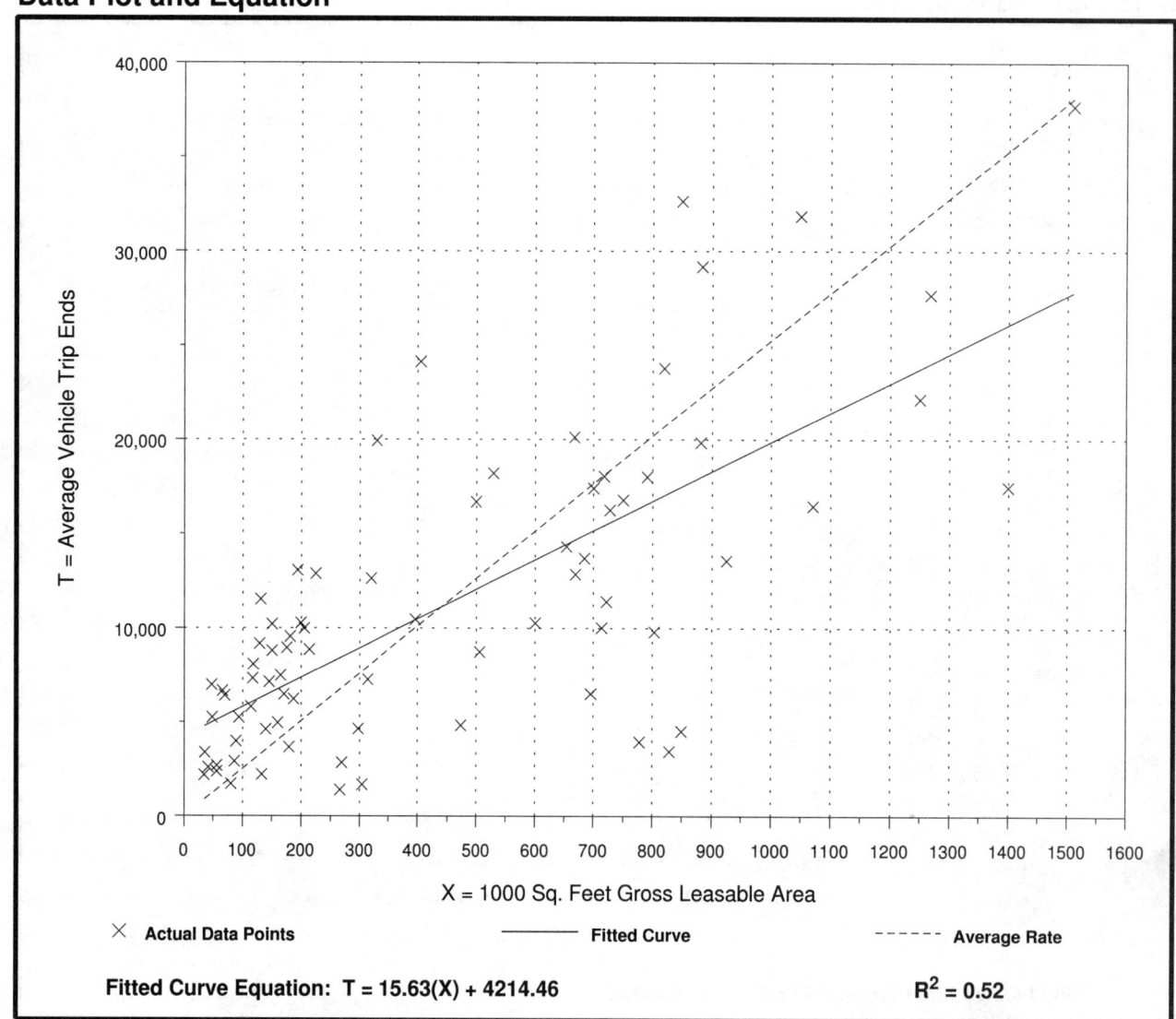

Fitted Curve Equation: T = 15.63(X) + 4214.46 $R^2 = 0.52$

Shopping Center
(820)

Average Vehicle Trip Ends vs: 1000 Sq. Feet Gross Leasable Area
On a: Sunday,
Peak Hour of Generator

Number of Studies: 39
Average 1000 Sq. Feet GLA: 369
Directional Distribution: 49% entering, 51% exiting

Trip Generation per 1000 Sq. Feet Gross Leasable Area

Average Rate	Range of Rates	Standard Deviation
3.12	0.39 - 12.40	2.78

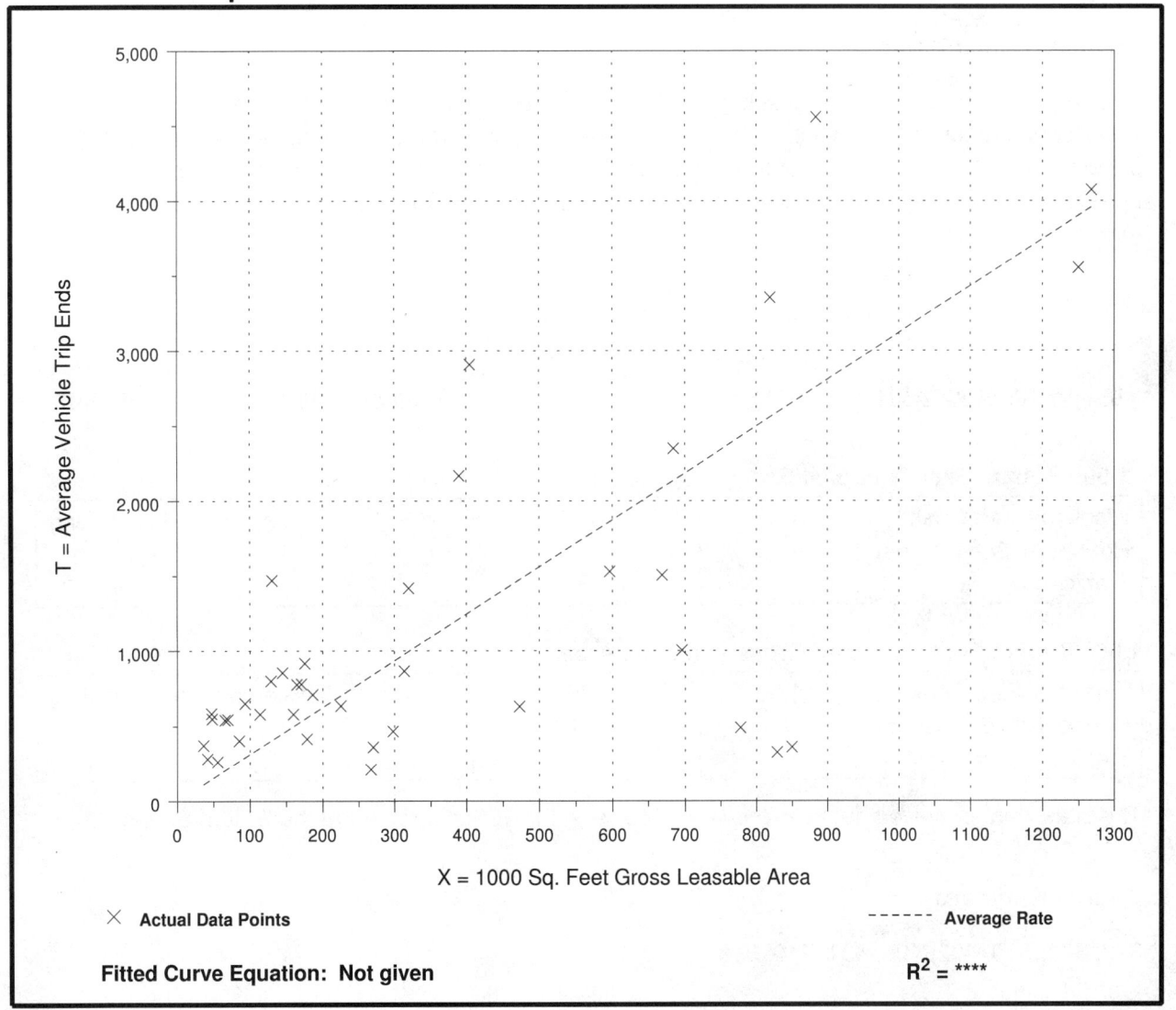

Data Plot and Equation

Fitted Curve Equation: Not given $R^2 = ****$

Land Use: 823
Factory Outlet Center

Description

A factory outlet center is a shopping center that primarily houses factory outlet stores, attracting customers from a wide geographic area, very often from a larger area than a regional shopping center. Shopping center (Land Use 820) is a related use.

Additional Data

Organized bus trips serve many factory outlet centers. Information on the number of trips served by tour bus or transit is not available.

Specialized Land Use Data

A survey conducted in 1998 was submitted for a large outlet shopping center containing more than 200 stores, numerous entertainment venues, full-service restaurants and a home improvement superstore. The size and scale of this site differs considerably from those contained in this land use. Therefore, the information collected for this facility is presented in the following tables and was excluded from the data plots.

Independent Variable	Trip Generation Rate	Size of Independent Variable	Number of Studies	Directional Distribution

1,000 Square Feet Occupied Gross Floor Area

Weekday P.M. Peak Hour of Adjacent Street Traffic	1.53	1,209	1	51% entering, 49% exiting

1,000 Square Feet Occupied Gross Leasable Area

Weekday P.M. Peak Hour of Adjacent Street Traffic	1.90	975	1	51% entering, 49% exiting

The sites were surveyed between the late 1980s and the 1990s throughout the United States.

Source Numbers

328, 379, 380, 402, 405, 431, 440, 518

Factory Outlet Center
(823)

Average Vehicle Trip Ends vs: 1000 Sq. Feet Gross Floor Area
On a: Weekday

Number of Studies: 11
Average 1000 Sq. Feet GFA: 137
Directional Distribution: 50% entering, 50% exiting

Trip Generation per 1000 Sq. Feet Gross Floor Area

Average Rate	Range of Rates	Standard Deviation
26.59	13.78 - 50.97	12.02

Data Plot and Equation

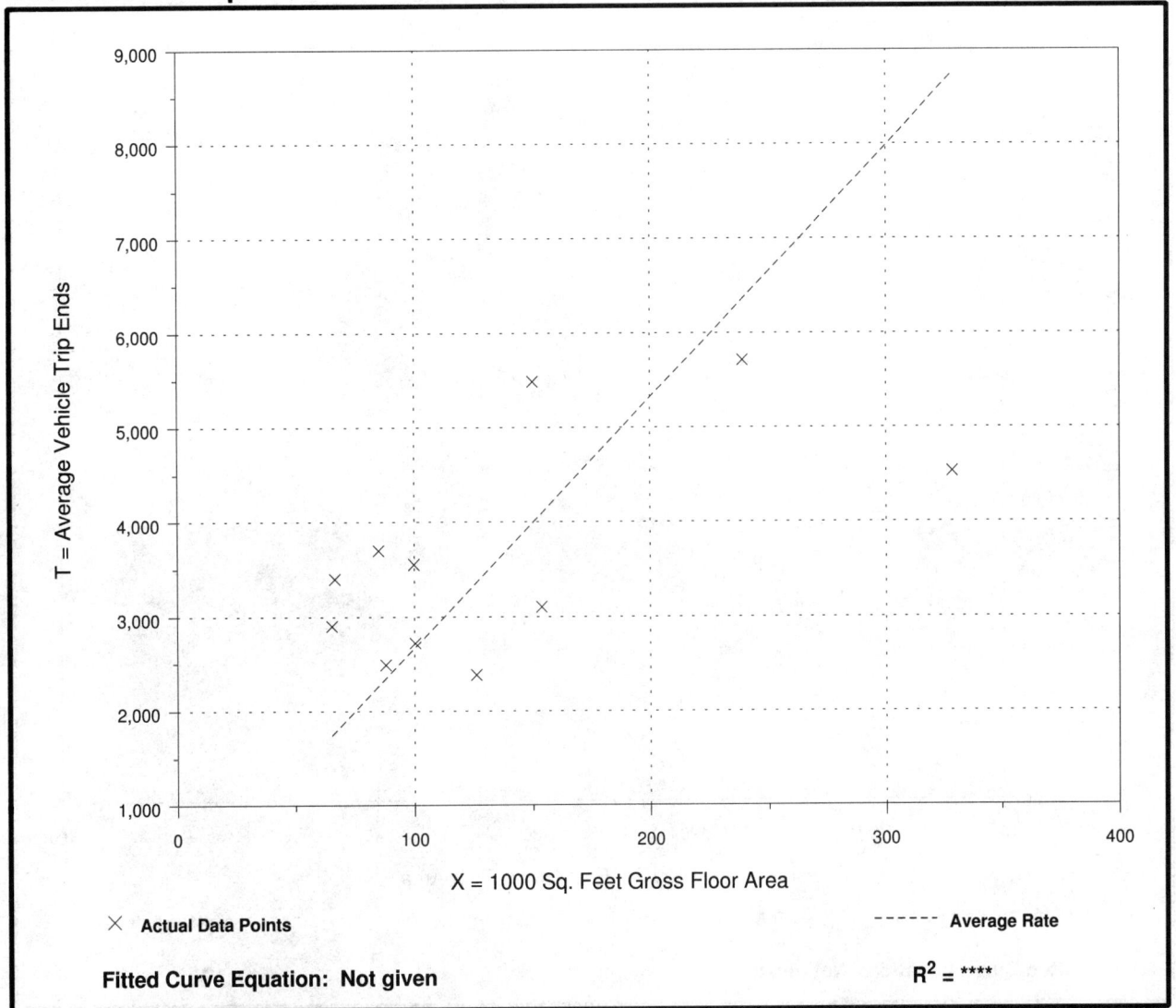

Fitted Curve Equation: Not given $R^2 =$ ****

Factory Outlet Center
(823)

Average Vehicle Trip Ends vs: 1000 Sq. Feet Gross Floor Area
On a: Weekday,
Peak Hour of Adjacent Street Traffic,
One Hour Between 7 and 9 a.m.

Number of Studies: 2
Average 1000 Sq. Feet GFA: 127
Directional Distribution: 73% entering, 27% exiting

Trip Generation per 1000 Sq. Feet Gross Floor Area

Average Rate	Range of Rates	Standard Deviation
0.67	0.47 - 0.97	*

Data Plot and Equation

Caution - Use Carefully - Small Sample Size

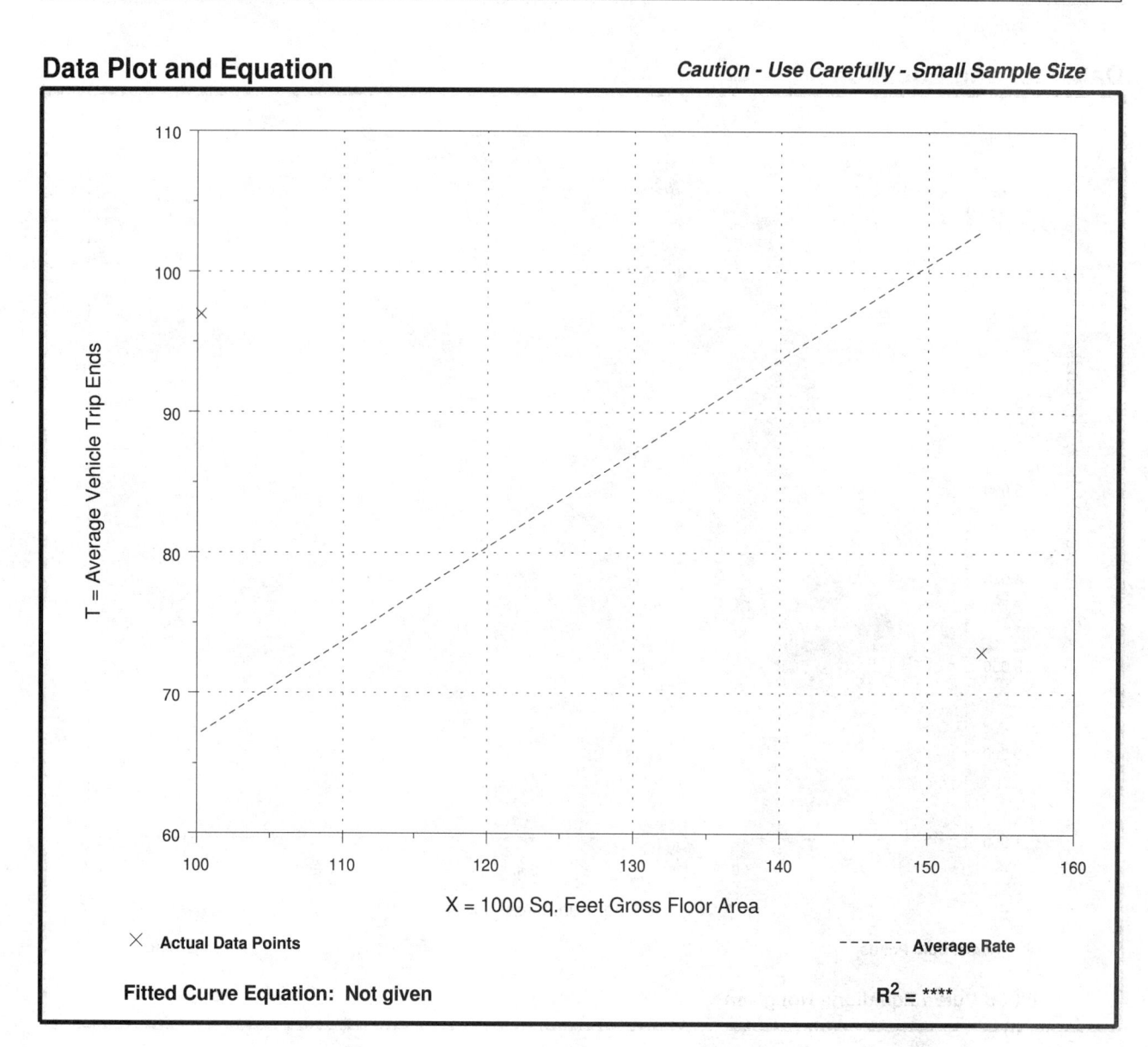

Fitted Curve Equation: Not given $R^2 = ****$

Factory Outlet Center
(823)

Average Vehicle Trip Ends vs: 1000 Sq. Feet Gross Floor Area
On a: Weekday,
Peak Hour of Adjacent Street Traffic,
One Hour Between 4 and 6 p.m.

Number of Studies: 14
Average 1000 Sq. Feet GFA: 146
Directional Distribution: 47% entering, 53% exiting

Trip Generation per 1000 Sq. Feet Gross Floor Area

Average Rate	Range of Rates	Standard Deviation
2.29	1.22 - 3.96	1.69

Data Plot and Equation

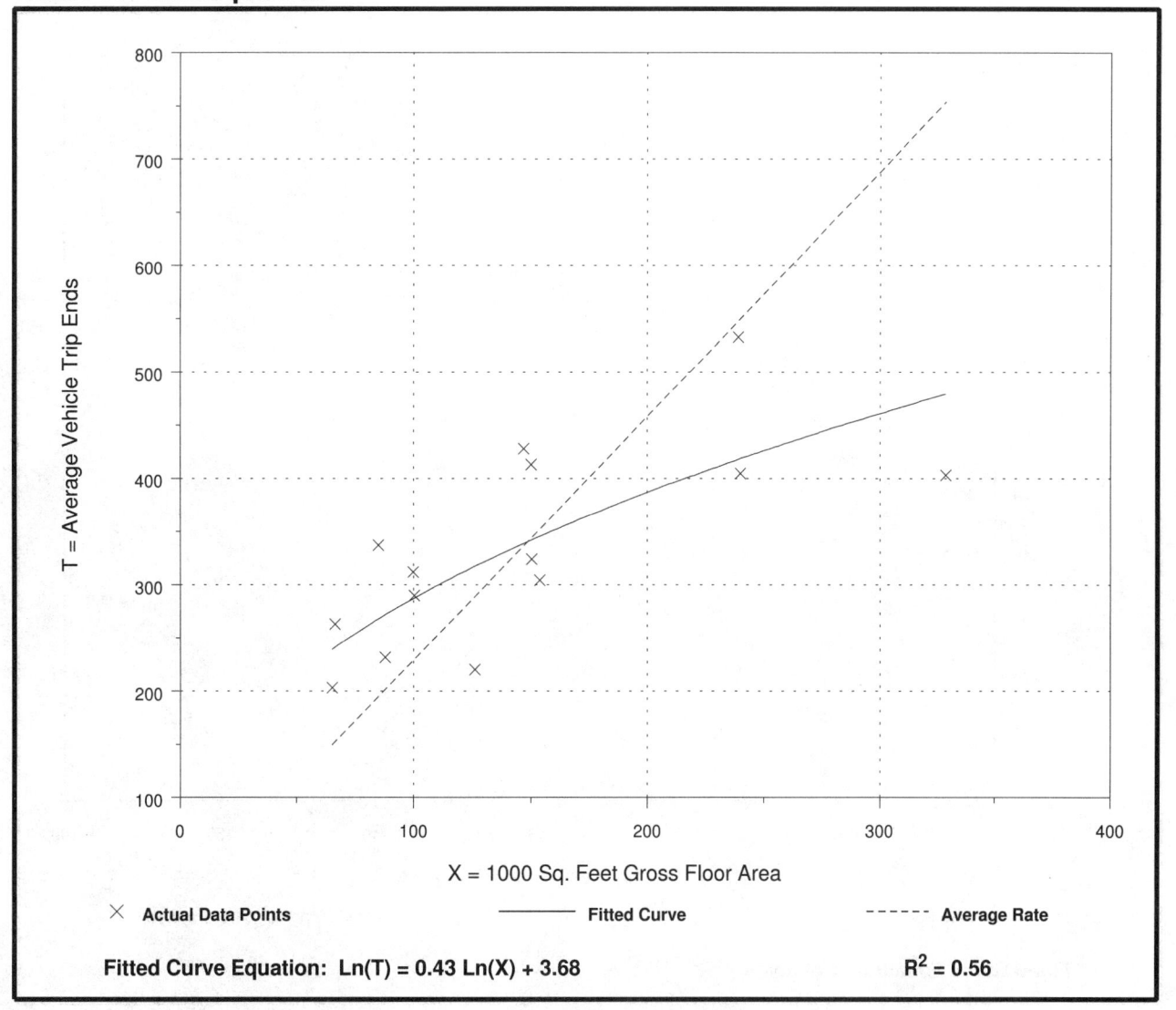

Fitted Curve Equation: $\operatorname{Ln}(T) = 0.43 \operatorname{Ln}(X) + 3.68$ $R^2 = 0.56$

Factory Outlet Center
(823)

Average Vehicle Trip Ends vs: 1000 Sq. Feet Gross Floor Area
On a: Weekday,
A.M. Peak Hour of Generator

Number of Studies: 2
Average 1000 Sq. Feet GFA: 127
Directional Distribution: 54% entering, 46% exiting

Trip Generation per 1000 Sq. Feet Gross Floor Area

Average Rate	Range of Rates	Standard Deviation
2.06	1.74 - 2.56	*

Data Plot and Equation

Caution - Use Carefully - Small Sample Size

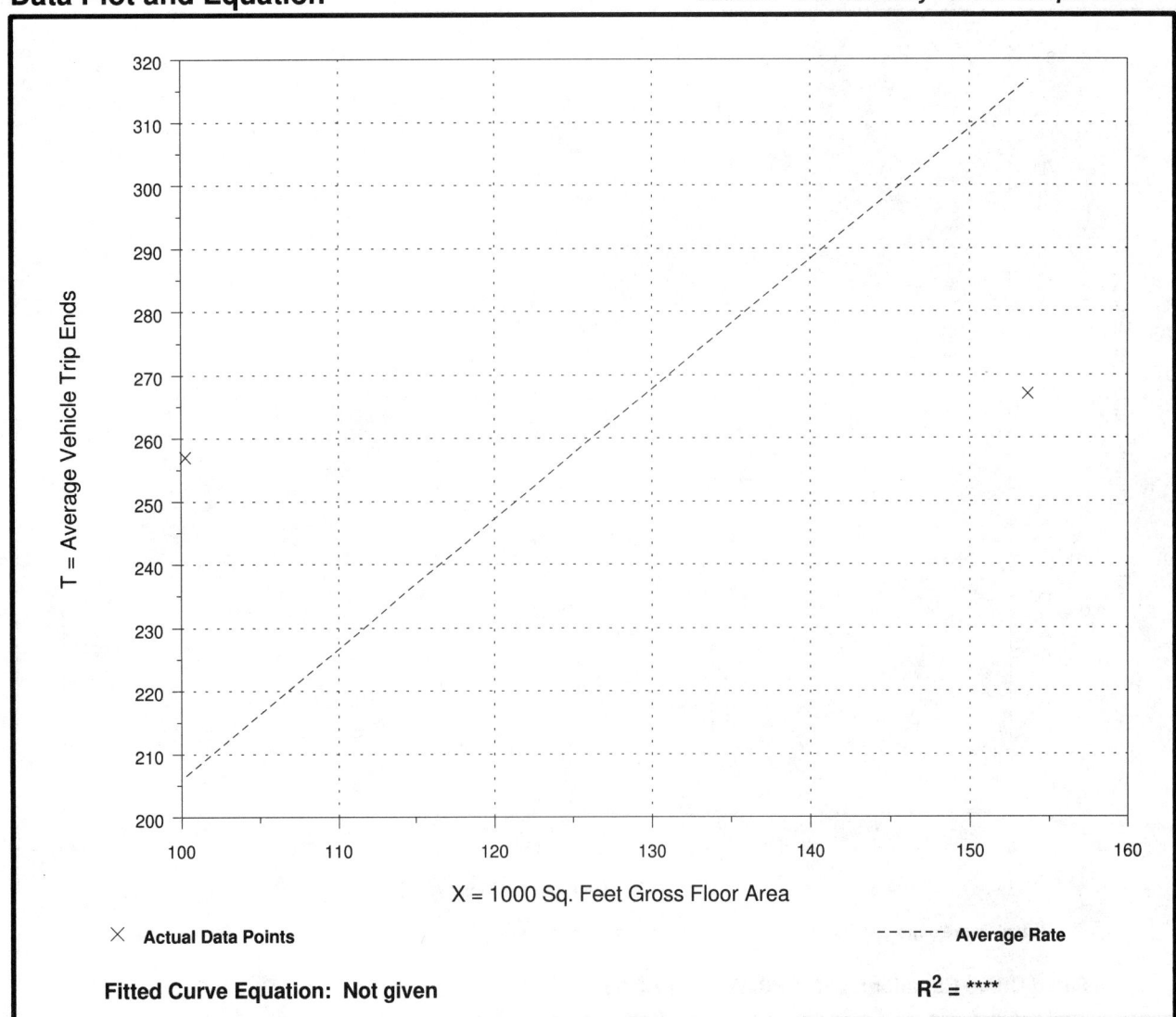

Fitted Curve Equation: Not given $R^2 = ****$

Factory Outlet Center
(823)

Average Vehicle Trip Ends vs: 1000 Sq. Feet Gross Floor Area
On a: Weekday,
P.M. Peak Hour of Generator

Number of Studies: 5
Average 1000 Sq. Feet GFA: 190
Directional Distribution: 51% entering, 49% exiting

Trip Generation per 1000 Sq. Feet Gross Floor Area

Average Rate	Range of Rates	Standard Deviation
1.94	1.57 - 3.20	1.47

Data Plot and Equation

Caution - Use Carefully - Small Sample Size

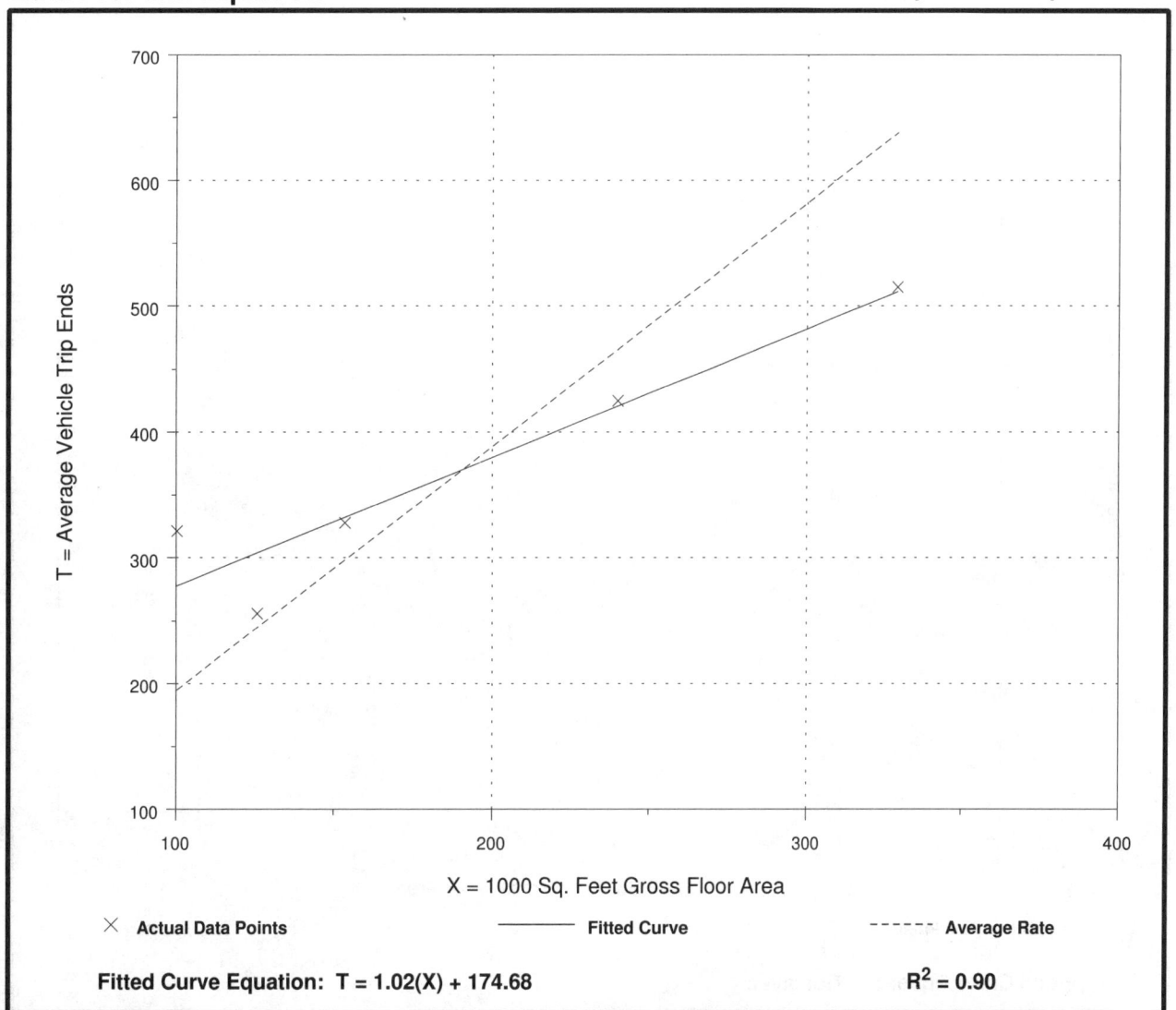

Fitted Curve Equation: T = 1.02(X) + 174.68 $R^2 = 0.90$

Factory Outlet Center
(823)

Average Vehicle Trip Ends vs: 1000 Sq. Feet Gross Floor Area
On a: Saturday

Number of Studies: 2
Average 1000 Sq. Feet GFA: 127
Directional Distribution: 50% entering, 50% exiting

Trip Generation per 1000 Sq. Feet Gross Floor Area

Average Rate	Range of Rates	Standard Deviation
40.97	29.38 - 58.73	*

Data Plot and Equation

Caution - Use Carefully - Small Sample Size

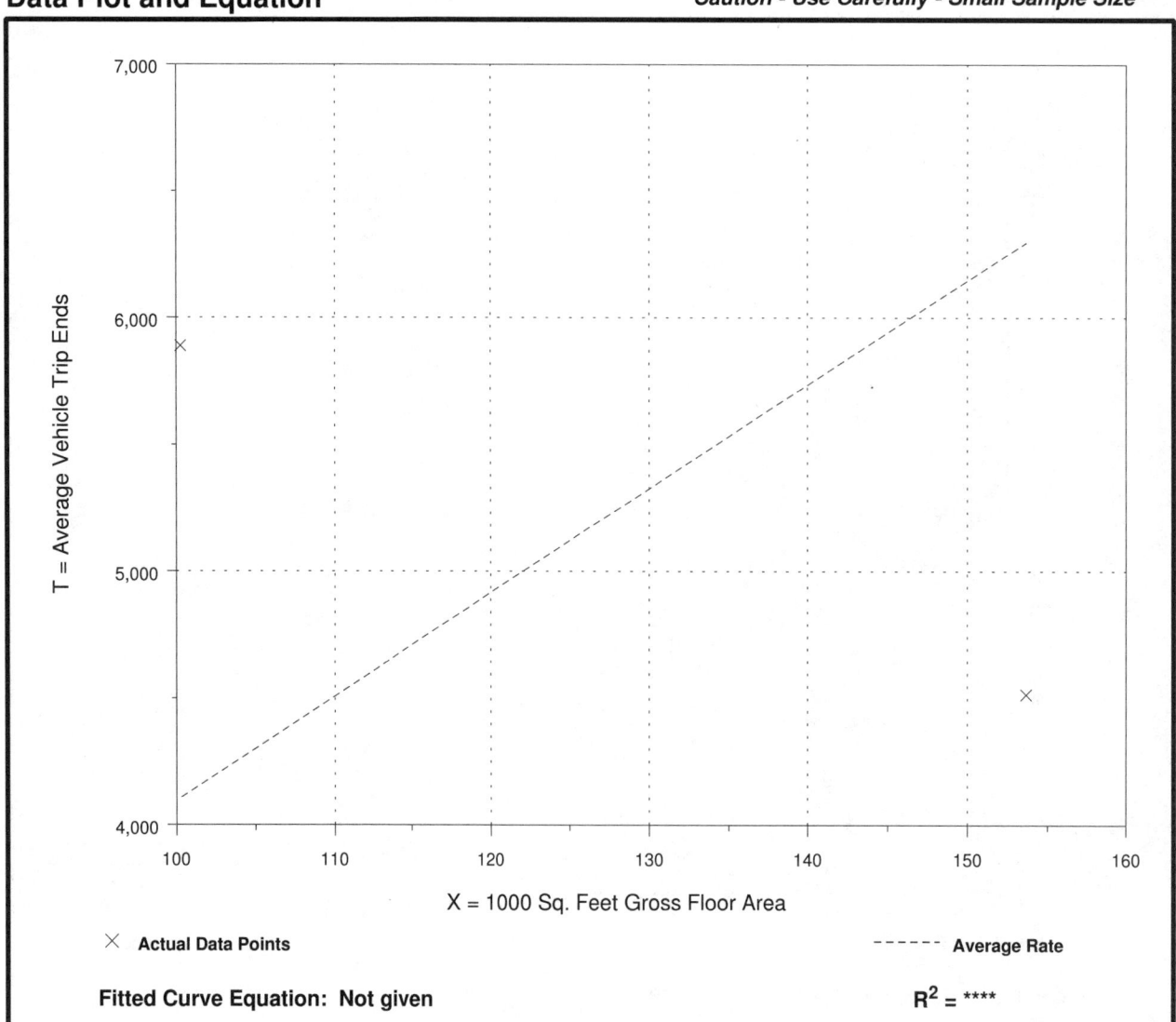

Fitted Curve Equation: Not given R^2 = ****

Factory Outlet Center
(823)

Average Vehicle Trip Ends vs: 1000 Sq. Feet Gross Floor Area
On a: Saturday,
Peak Hour of Generator

Number of Studies: 3
Average 1000 Sq. Feet GFA: 165
Directional Distribution: 51% entering, 49% exiting

Trip Generation per 1000 Sq. Feet Gross Floor Area

Average Rate	Range of Rates	Standard Deviation
3.79	2.95 - 6.49	2.38

Data Plot and Equation

Caution - Use Carefully - Small Sample Size

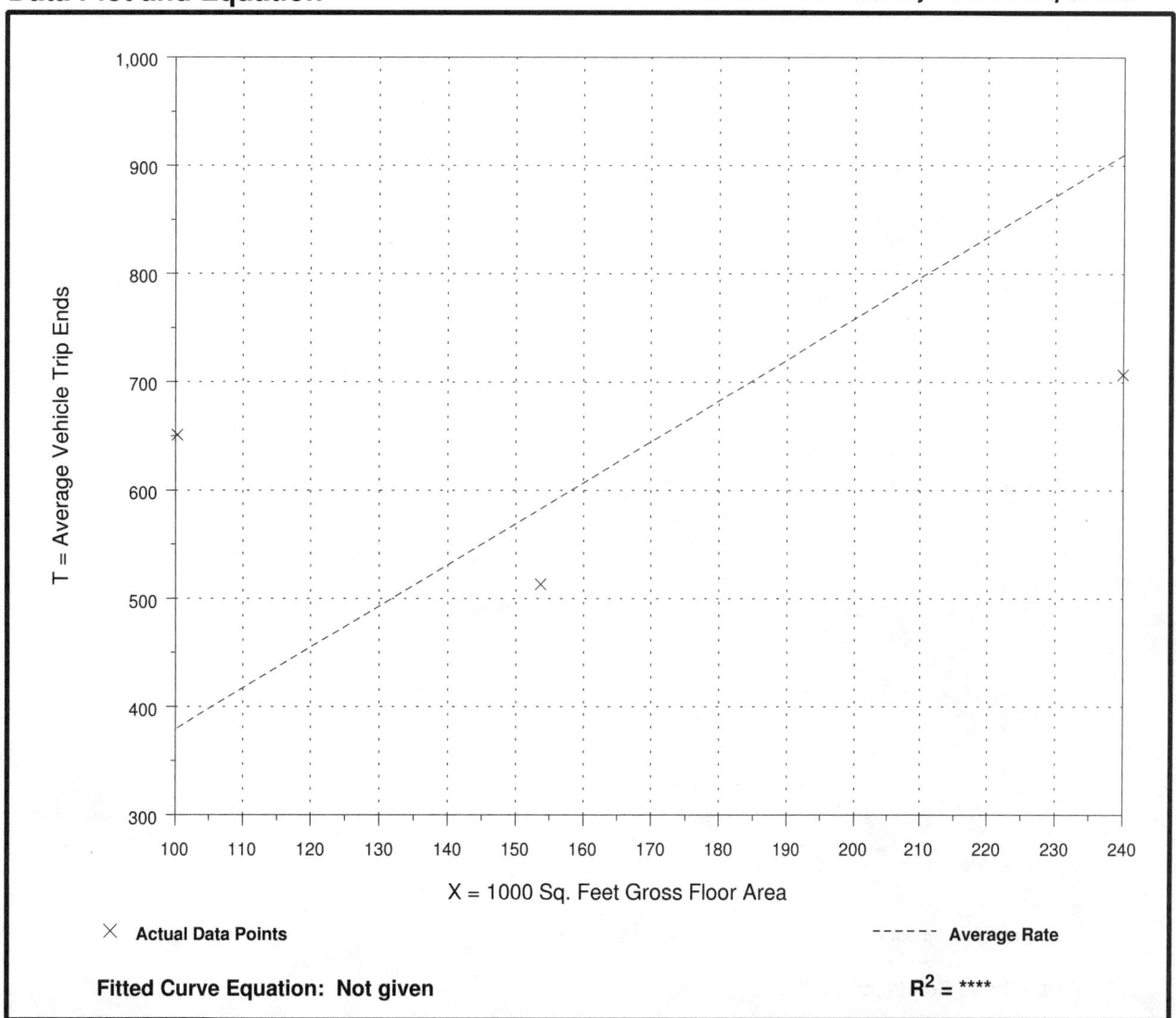

Fitted Curve Equation: Not given $R^2 = ****$

Factory Outlet Center
(823)

Average Vehicle Trip Ends vs: 1000 Sq. Feet Gross Floor Area
On a: Sunday

Number of Studies: 4
Average 1000 Sq. Feet GFA: 177
Directional Distribution: 50% entering, 50% exiting

Trip Generation per 1000 Sq. Feet Gross Floor Area

Average Rate	Range of Rates	Standard Deviation
26.16	19.82 - 45.95	10.14

Data Plot and Equation

Caution - Use Carefully - Small Sample Size

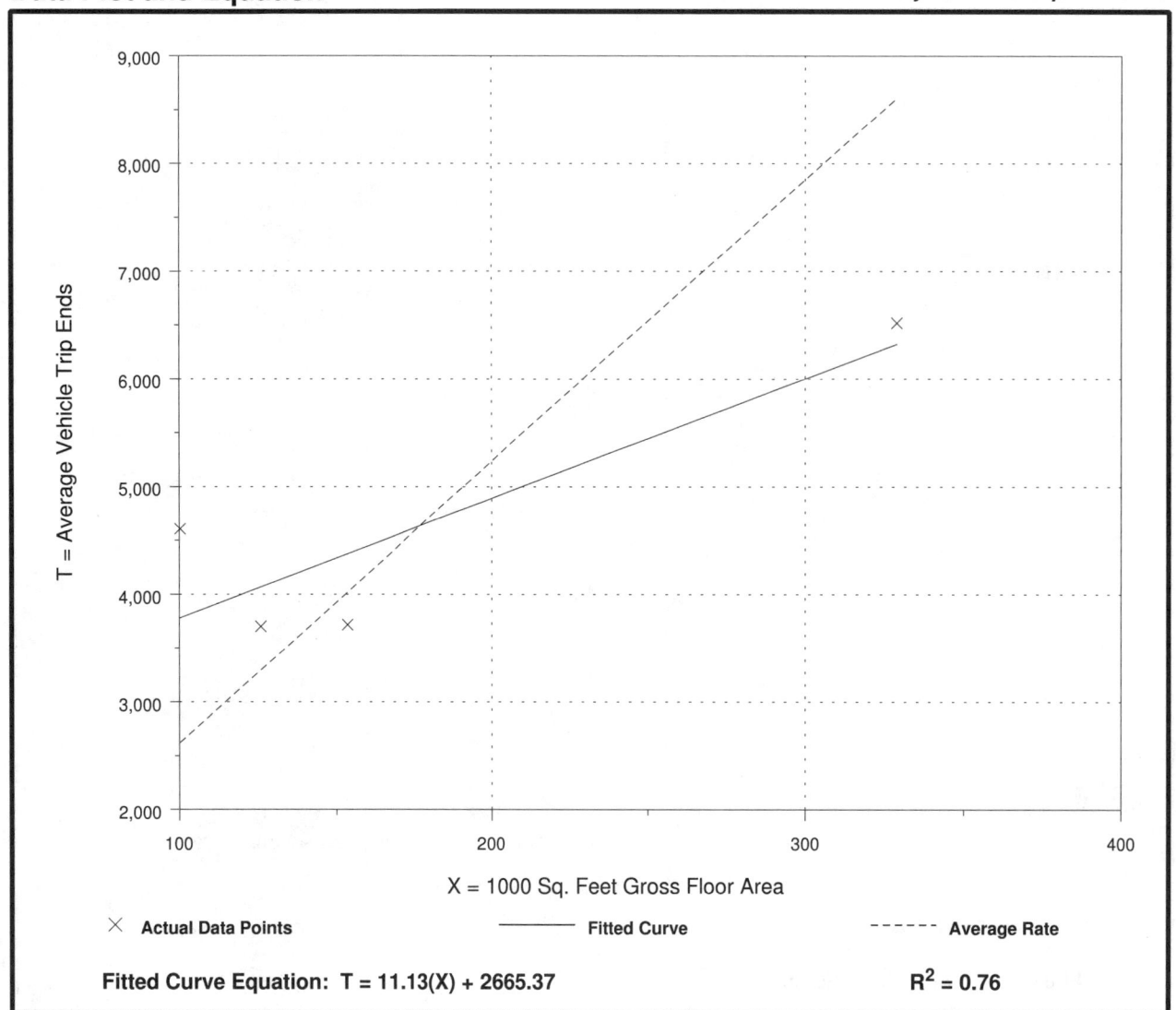

Fitted Curve Equation: T = 11.13(X) + 2665.37 $R^2 = 0.76$

Factory Outlet Center
(823)

Average Vehicle Trip Ends vs: 1000 Sq. Feet Gross Floor Area
On a: Sunday,
Peak Hour of Generator

Number of Studies: 4
Average 1000 Sq. Feet GFA: 177
Directional Distribution: 51% entering, 49% exiting

Trip Generation per 1000 Sq. Feet Gross Floor Area

Average Rate	Range of Rates	Standard Deviation
3.27	2.25 - 6.30	2.25

Data Plot and Equation

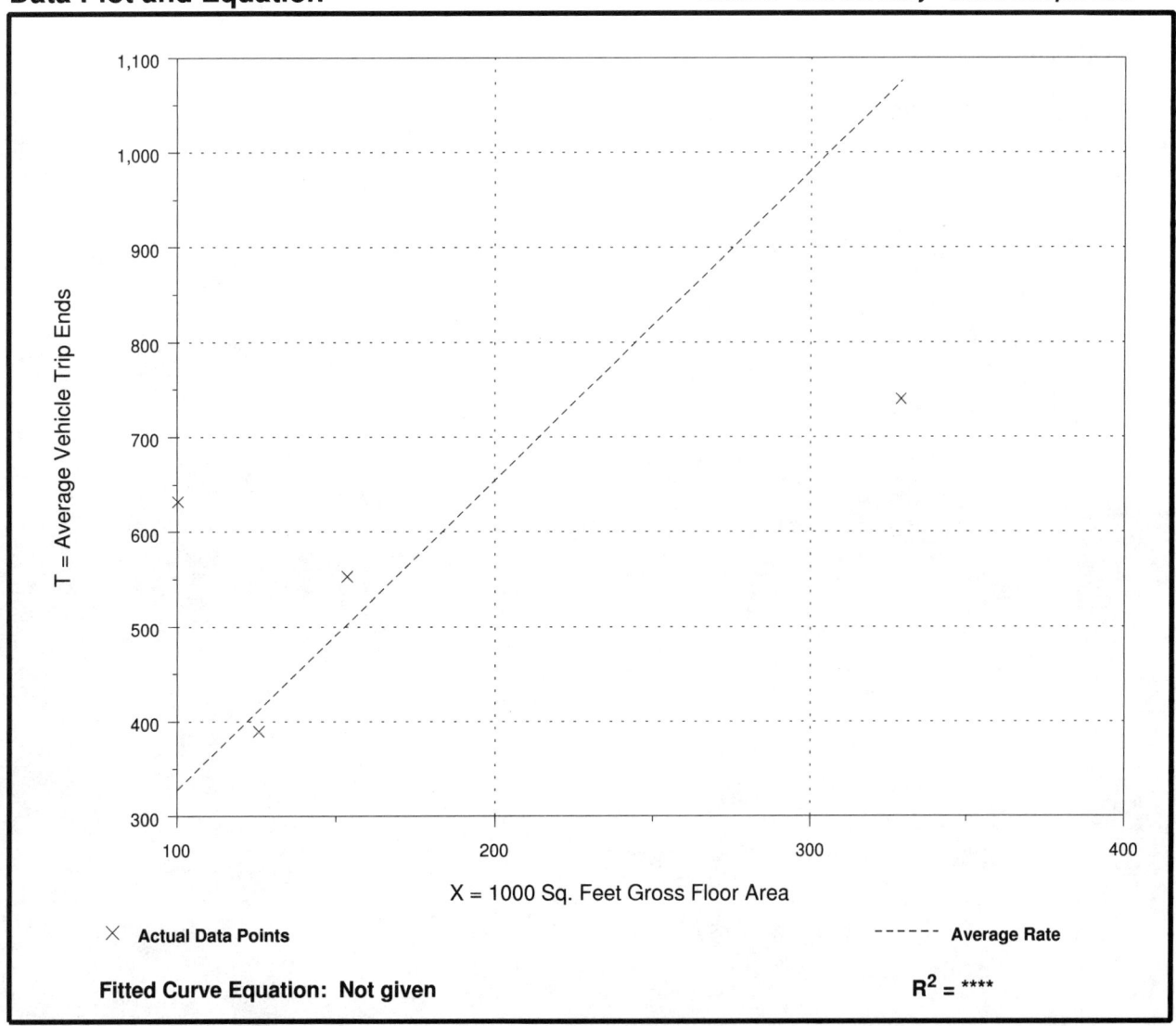

Caution - Use Carefully - Small Sample Size

Fitted Curve Equation: Not given $R^2 = ****$

Land Use: 826
Specialty Retail Center

Description

Specialty retail centers are generally small strip shopping centers that contain a variety of retail shops and specialize in quality apparel, hard goods and services, such as real estate offices, dance studios, florists and small restaurants. Shopping center (Land Use 820) is a related use.

Additional Data

The sites were surveyed between the late 1970s and the 2000s in California, Florida, Georgia, New York and Pennsylvania.

Source Numbers

100, 304, 305, 367, 423, 507, 577

Specialty Retail Center
(826)

Average Vehicle Trip Ends vs: 1000 Sq. Feet Gross Leasable Area
On a: Weekday

Number of Studies: 4
Average 1000 Sq. Feet GLA: 25
Directional Distribution: 50% entering, 50% exiting

Trip Generation per 1000 Sq. Feet Gross Leasable Area

Average Rate	Range of Rates	Standard Deviation
44.32	21.30 - 64.21	15.52

Data Plot and Equation

Caution - Use Carefully - Small Sample Size

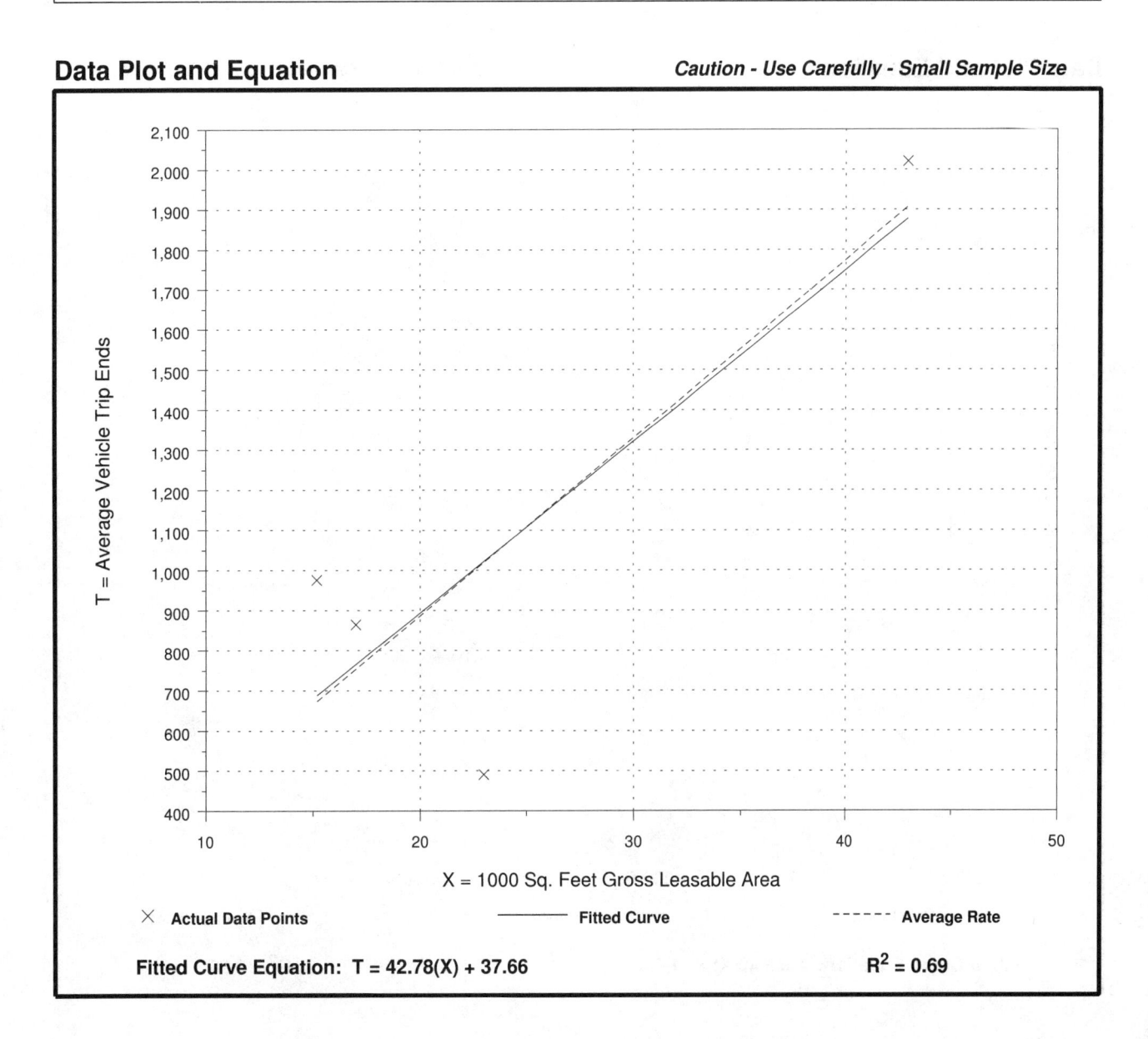

Fitted Curve Equation: T = 42.78(X) + 37.66 $R^2 = 0.69$

Specialty Retail Center
(826)

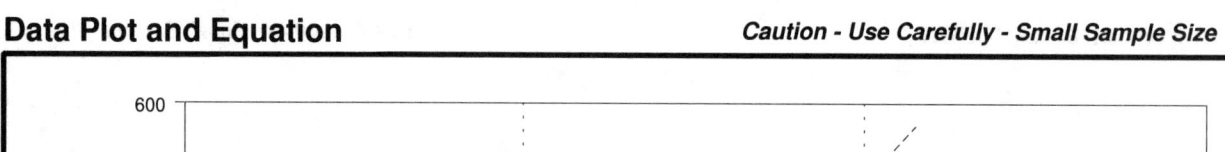

Average Vehicle Trip Ends vs: 1000 Sq. Feet Gross Leasable Area
On a: Weekday,
Peak Hour of Adjacent Street Traffic,
One Hour Between 4 and 6 p.m.

Number of Studies: 5
Average 1000 Sq. Feet GLA: 69
Directional Distribution: 44% entering, 56% exiting

Trip Generation per 1000 Sq. Feet Gross Leasable Area

Average Rate	Range of Rates	Standard Deviation
2.71	2.03 - 5.16	1.83

Data Plot and Equation

Caution - Use Carefully - Small Sample Size

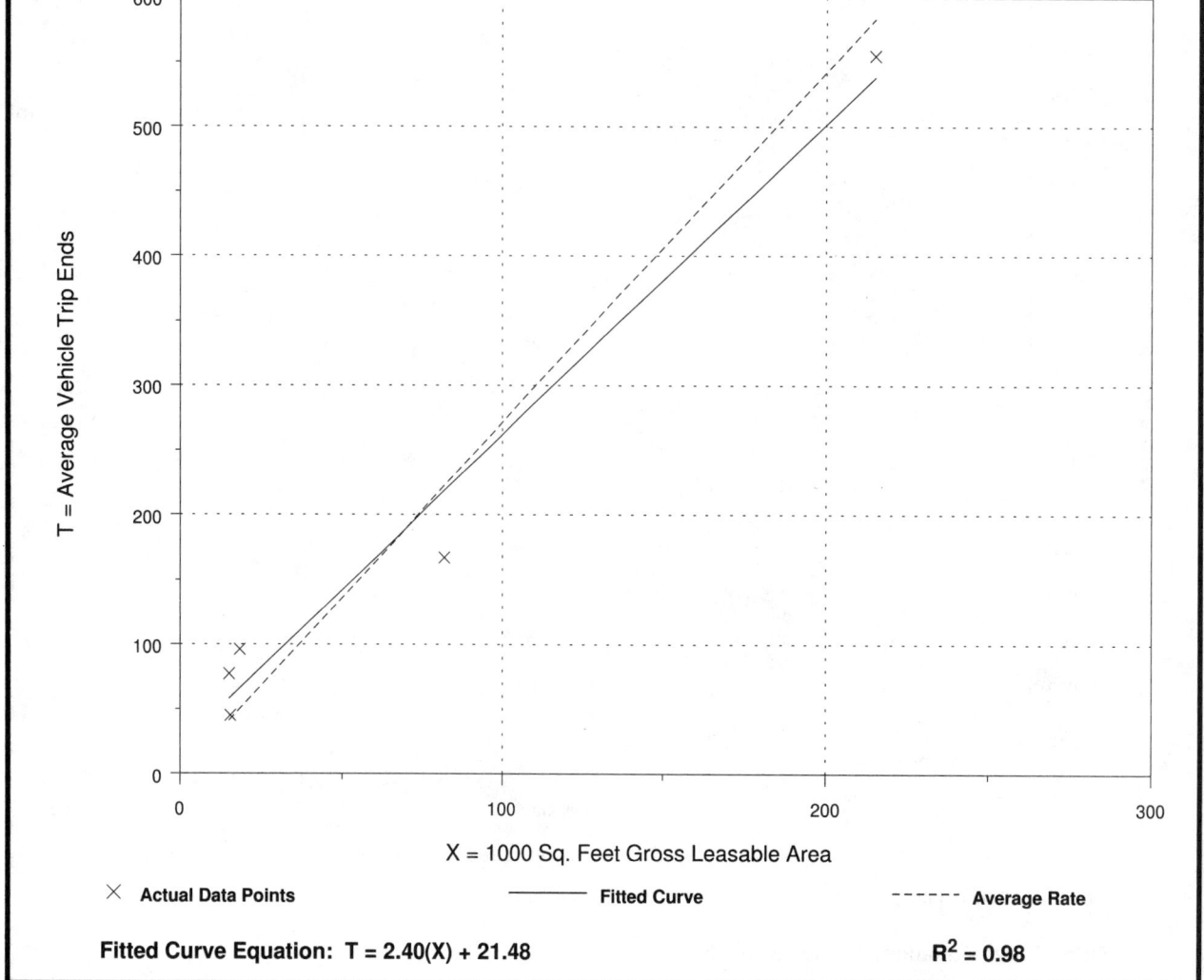

Fitted Curve Equation: $T = 2.40(X) + 21.48$ $R^2 = 0.98$

Specialty Retail Center
(826)

Average Vehicle Trip Ends vs: 1000 Sq. Feet Gross Leasable Area
On a: Weekday,
A.M. Peak Hour of Generator

Number of Studies: 4
Average 1000 Sq. Feet GLA: 60
Directional Distribution: 48% entering, 52% exiting

Trip Generation per 1000 Sq. Feet Gross Leasable Area

Average Rate	Range of Rates	Standard Deviation
6.84	5.33 - 14.08	3.55

Data Plot and Equation

Caution - Use Carefully - Small Sample Size

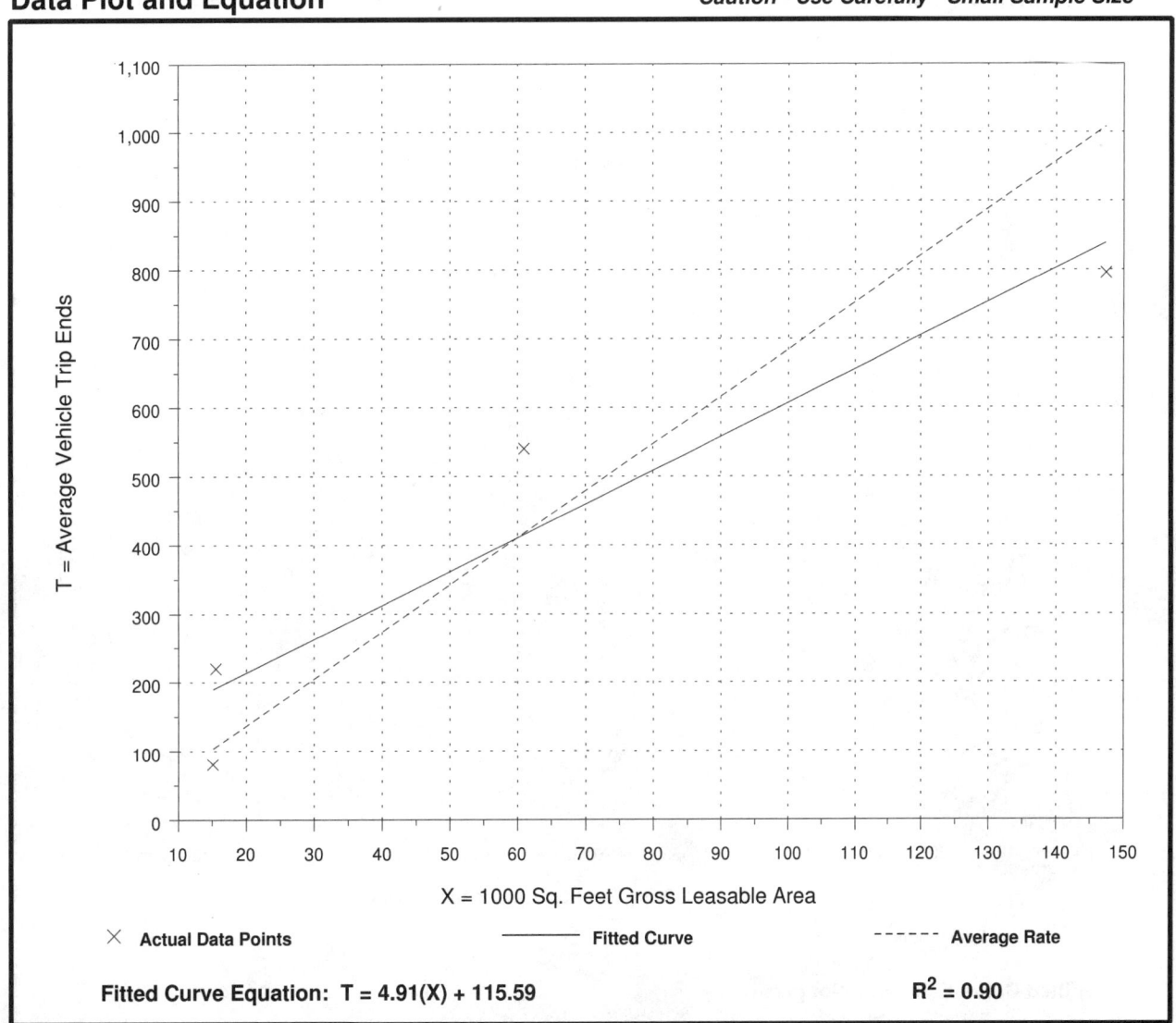

Fitted Curve Equation: T = 4.91(X) + 115.59 $R^2 = 0.90$

Specialty Retail Center
(826)

Average Vehicle Trip Ends vs: 1000 Sq. Feet Gross Leasable Area
On a: Weekday,
P.M. Peak Hour of Generator

Number of Studies: 3
Average 1000 Sq. Feet GLA: 75
Directional Distribution: 56% entering, 44% exiting

Trip Generation per 1000 Sq. Feet Gross Leasable Area

Average Rate	Range of Rates	Standard Deviation
5.02	4.59 - 6.18	2.31

Data Plot and Equation

Caution - Use Carefully - Small Sample Size

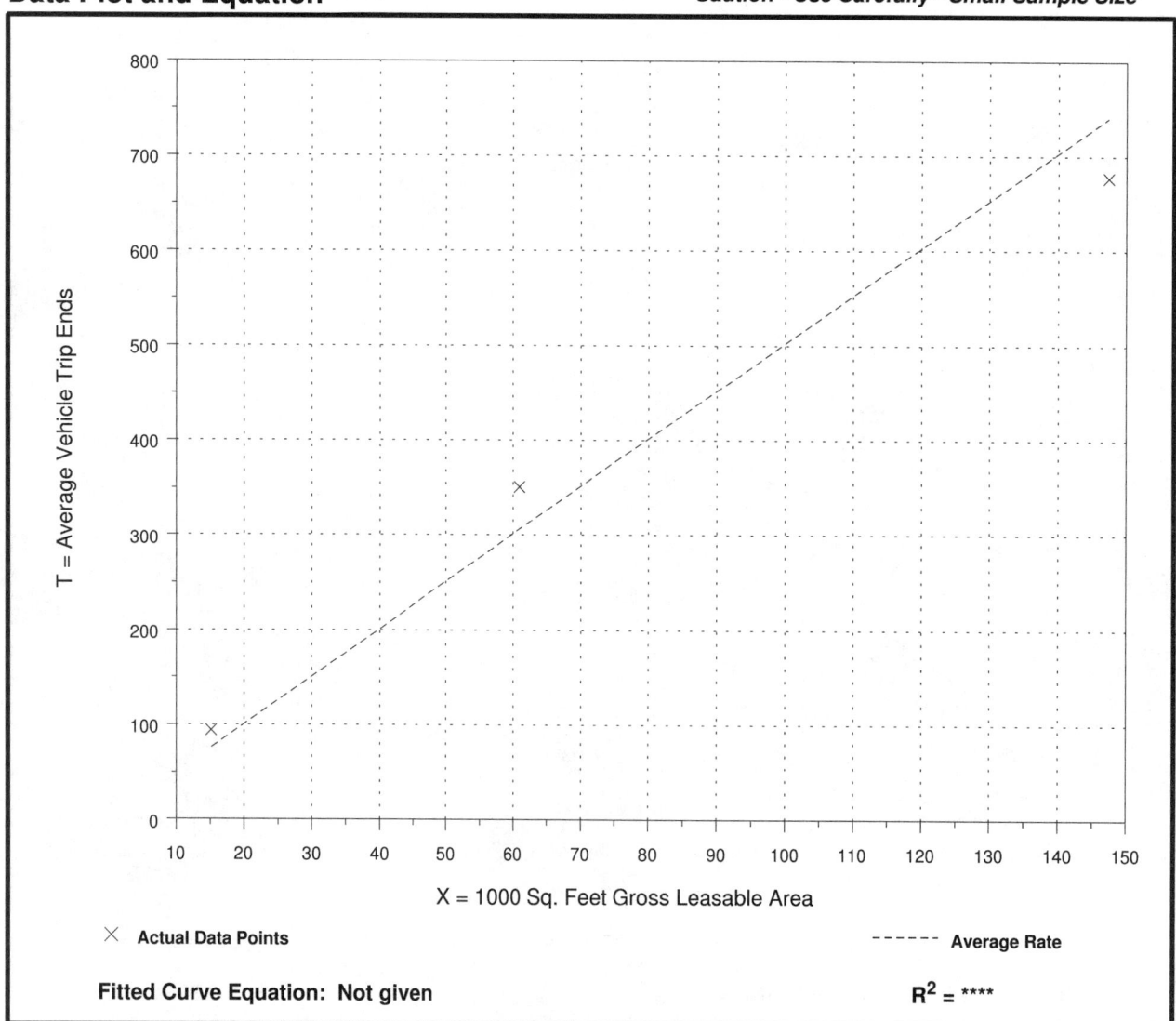

Fitted Curve Equation: Not given $R^2 = ****$

Specialty Retail Center
(826)

Average Vehicle Trip Ends vs: 1000 Sq. Feet Gross Leasable Area
On a: Saturday

Number of Studies: 3
Average 1000 Sq. Feet GLA: 28
Directional Distribution: 50% entering, 50% exiting

Trip Generation per 1000 Sq. Feet Gross Leasable Area

Average Rate	Range of Rates	Standard Deviation
42.04	22.57 - 54.47	13.97

Data Plot and Equation

Caution - Use Carefully - Small Sample Size

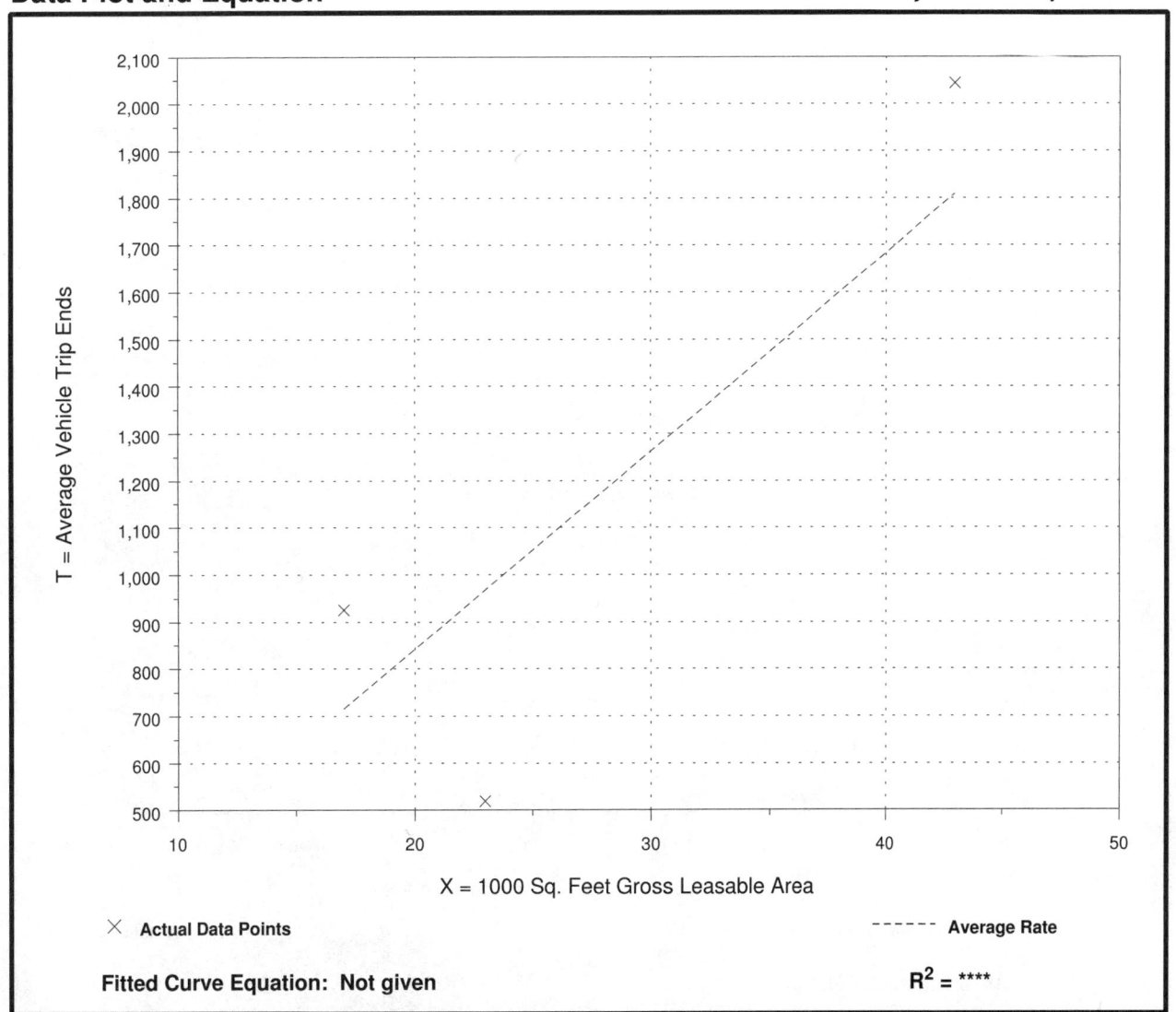

× Actual Data Points
----- Average Rate

Fitted Curve Equation: Not given R^2 = ****

Specialty Retail Center
(826)

Average Vehicle Trip Ends vs: 1000 Sq. Feet Gross Leasable Area
On a: Sunday

Number of Studies: 3
Average 1000 Sq. Feet GLA: 28
Directional Distribution: 50% entering, 50% exiting

Trip Generation per 1000 Sq. Feet Gross Leasable Area

Average Rate	Range of Rates	Standard Deviation
20.43	6.96 - 32.82	10.27

Data Plot and Equation

Caution - Use Carefully - Small Sample Size

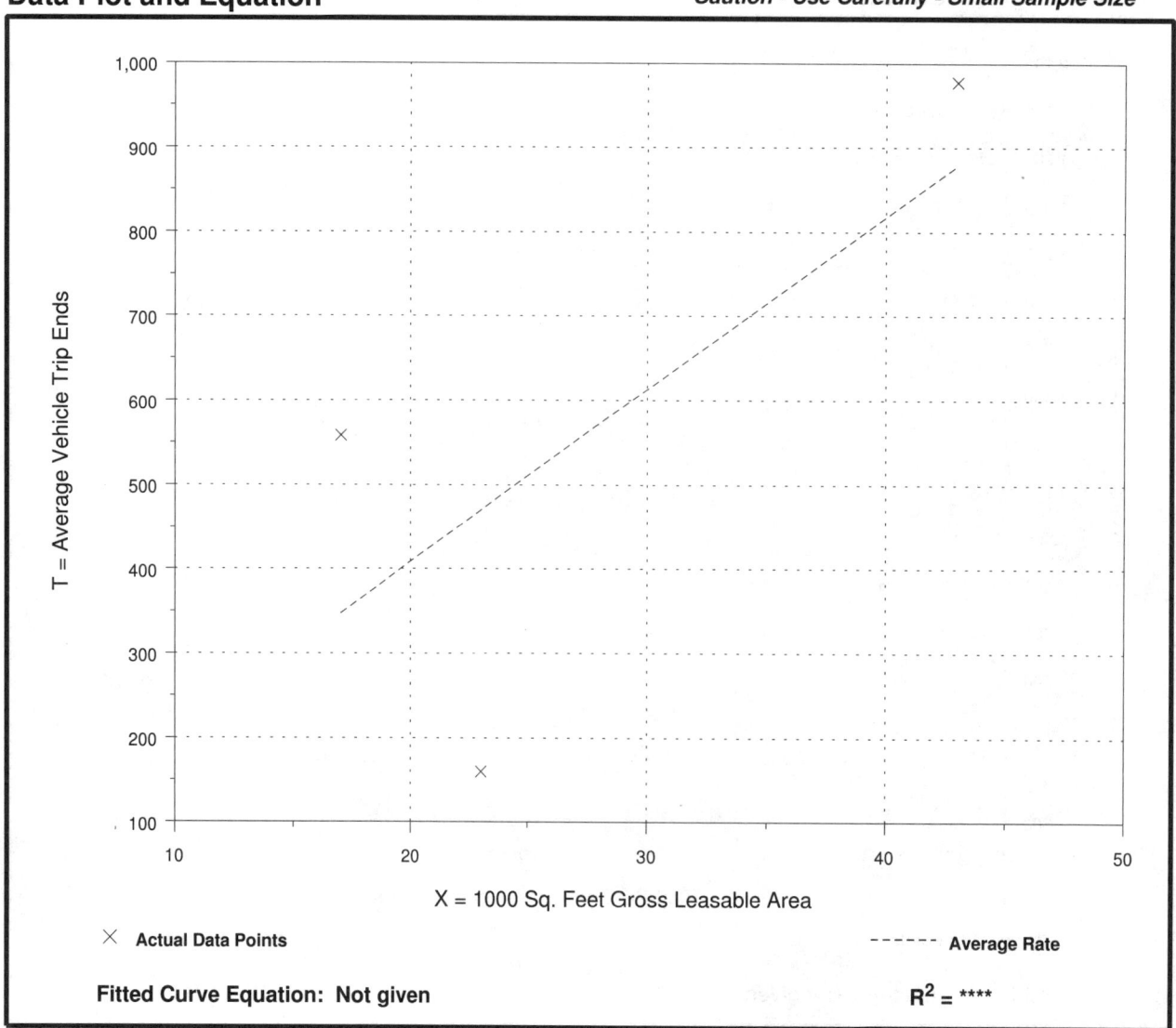

Fitted Curve Equation: Not given $R^2 = ****$

Specialty Retail Center
(826)

Average Vehicle Trip Ends vs: Employees
On a: Weekday

Number of Studies: 3
Avg. Number of Employees: 50
Directional Distribution: 50% entering, 50% exiting

Trip Generation per Employee

Average Rate	Range of Rates	Standard Deviation
22.36	21.96 - 24.50	4.77

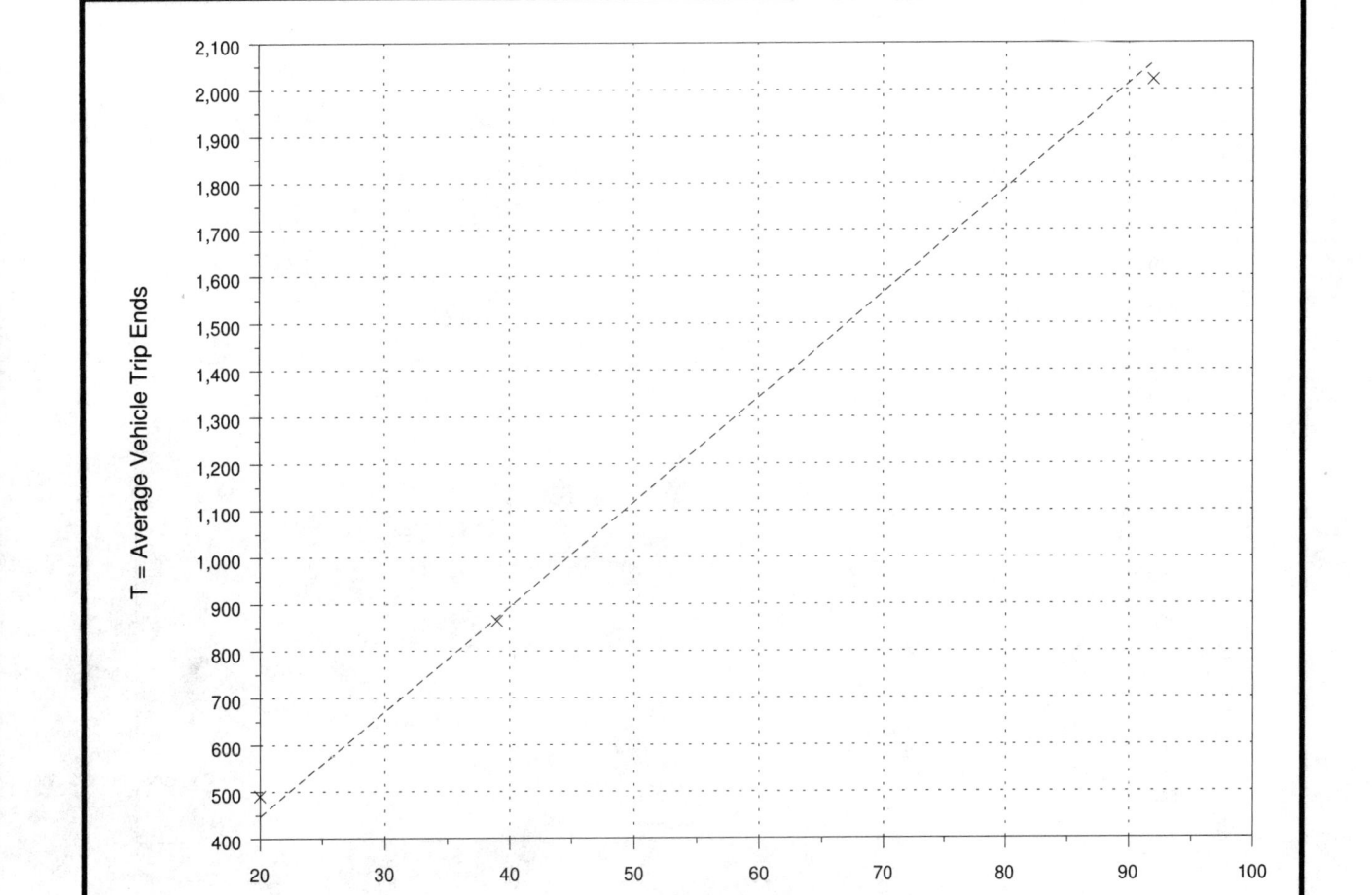

Data Plot and Equation

Caution - Use Carefully - Small Sample Size

Fitted Curve Equation: Not given $R^2 = ****$

Trip Generation, 9th Edition • Institute of Transportation Engineers

Specialty Retail Center
(826)

Average Vehicle Trip Ends vs: Employees
On a: Saturday

Number of Studies: 3
Avg. Number of Employees: 50
Directional Distribution: 50% entering, 50% exiting

Trip Generation per Employee

Average Rate	Range of Rates	Standard Deviation
23.11	22.22 - 25.95	4.94

Data Plot and Equation

Caution - Use Carefully - Small Sample Size

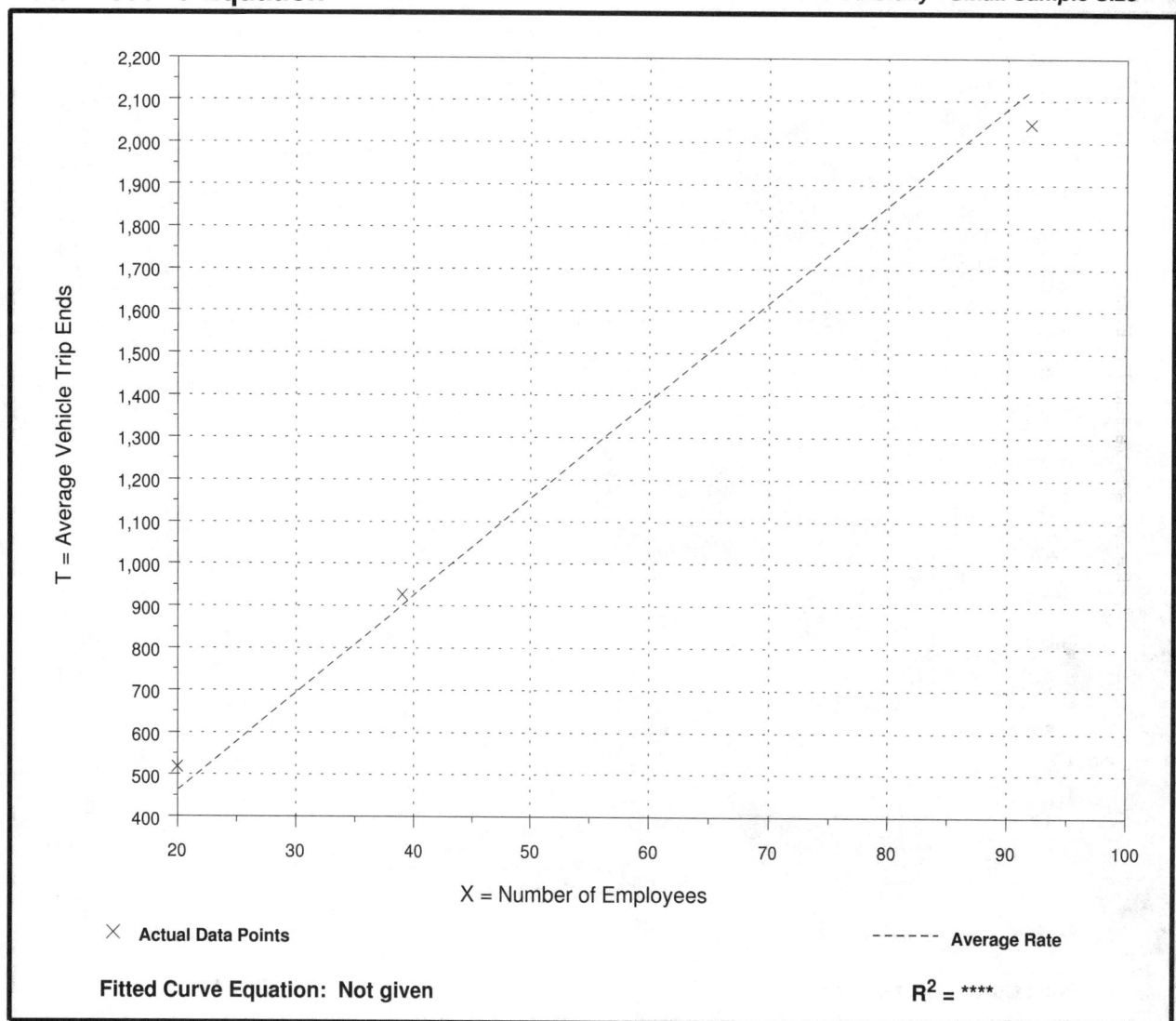

Fitted Curve Equation: Not given $R^2 = ****$

Specialty Retail Center
(826)

Average Vehicle Trip Ends vs: Employees
On a: Sunday

Number of Studies: 3
Avg. Number of Employees: 50
Directional Distribution: 50% entering, 50% exiting

Trip Generation per Employee

Average Rate	Range of Rates	Standard Deviation
11.23	8.00 - 14.31	3.89

Data Plot and Equation

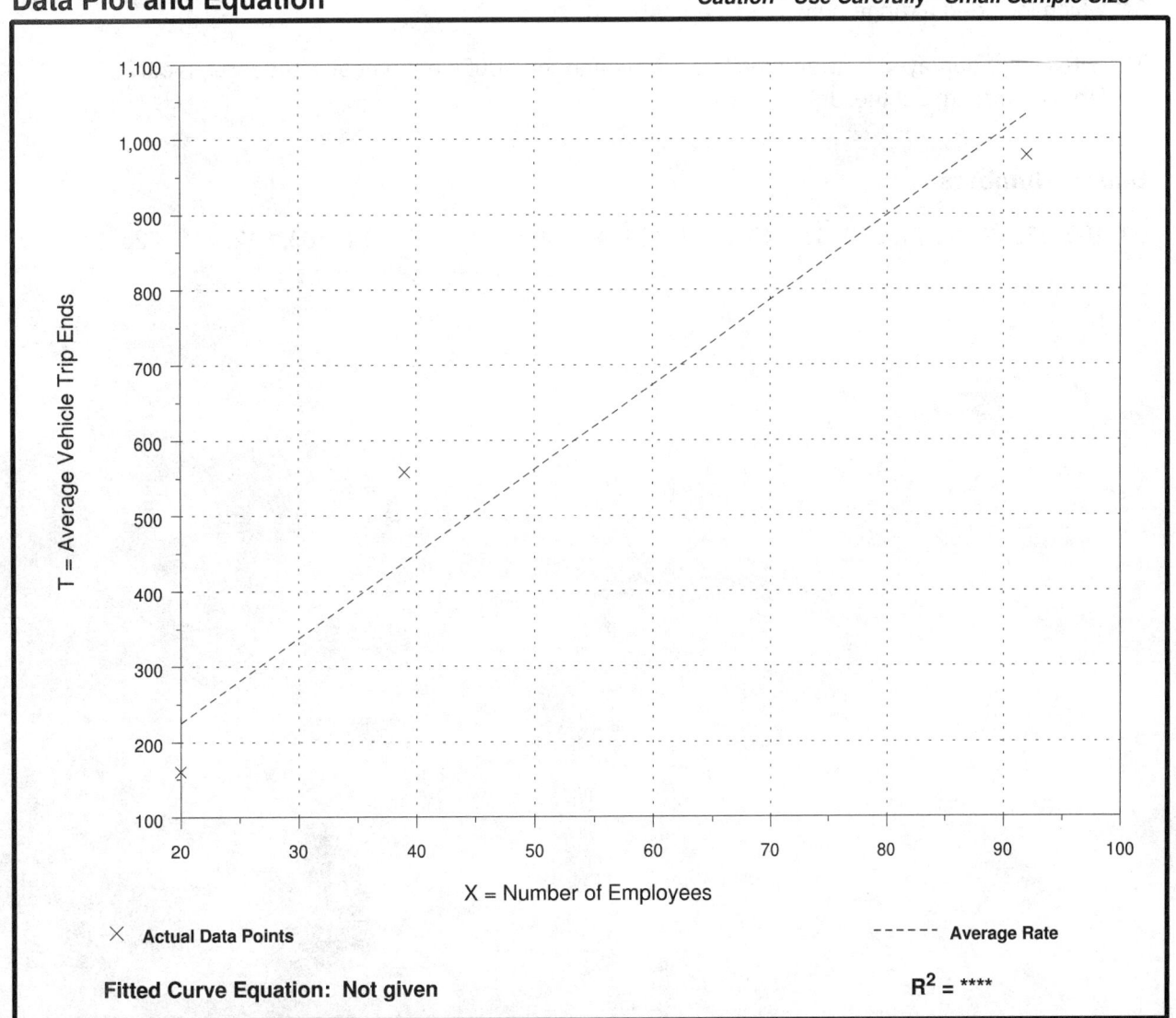

Caution - Use Carefully - Small Sample Size

Fitted Curve Equation: Not given R^2 = ****

Land Use: 841
Automobile Sales

Description

Automobile sales dealerships are typically located along major arterial streets characterized by abundant commercial development. Automobile services, parts sales and substantial used car sales may also be available. Some dealerships also include leasing options, truck sales and servicing. Recreational vehicle sales (Land Use 842) is a related use.

Additional Data

Three sites provided information on the number of service stalls. The average number of stalls was 34, which included a range between 12 and 49.

The sites were surveyed between the late 1960s and the 2000s at domestic and foreign car dealerships throughout the United States.

Source Numbers

98, 100, 172, 260, 271, 280, 328, 406, 414, 424, 427, 438, 440, 507, 571, 583, 612, 715, 728

Automobile Sales
(841)

Average Vehicle Trip Ends vs: Employees
On a: Weekday

Number of Studies: 4
Avg. Number of Employees: 62
Directional Distribution: 50% entering, 50% exiting

Trip Generation per Employee

Average Rate	Range of Rates	Standard Deviation
21.14	10.82 - 38.55	10.91

Data Plot and Equation

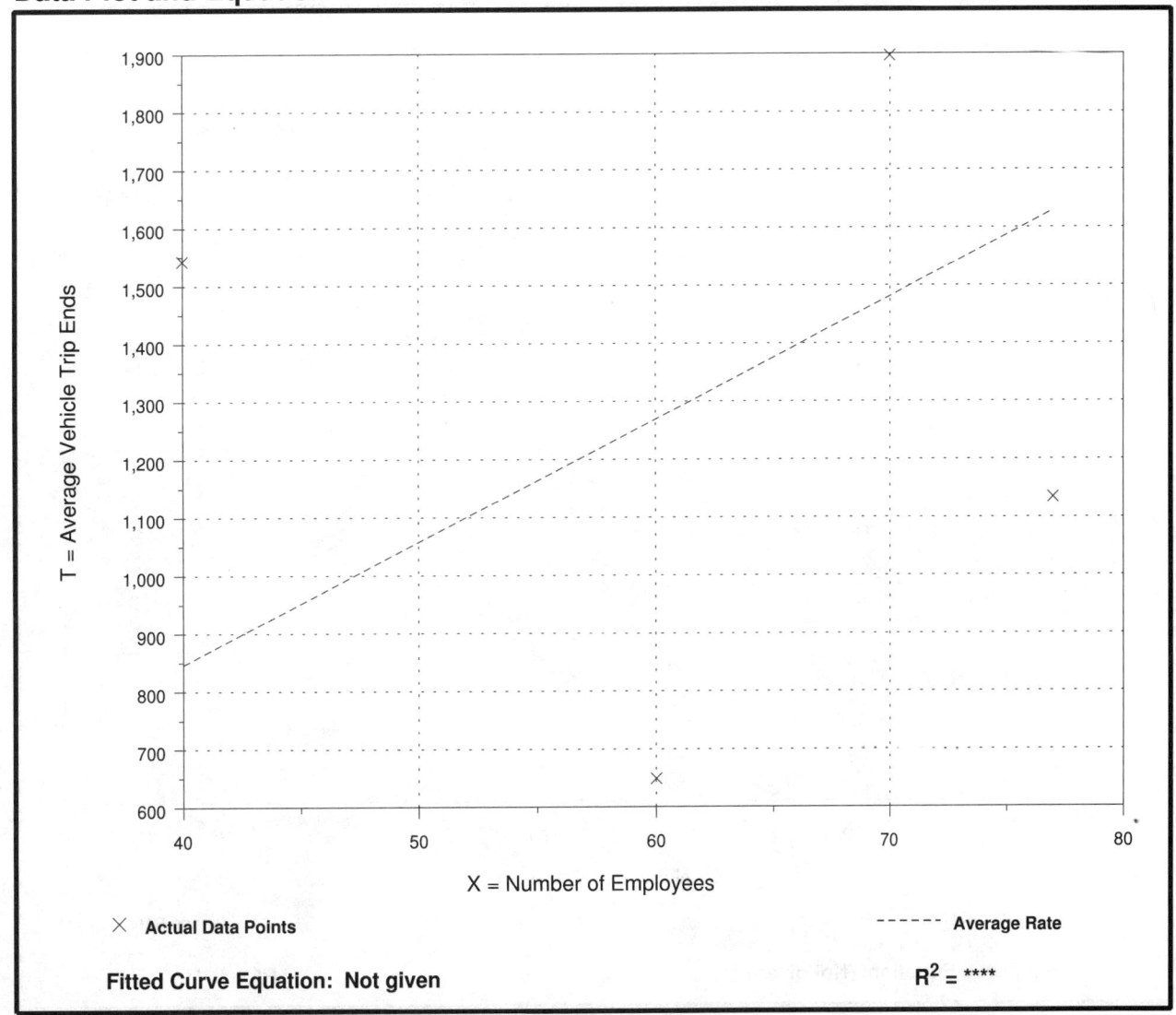

Caution - Use Carefully - Small Sample Size

Fitted Curve Equation: Not given $R^2 =$ ****

Automobile Sales
(841)

Average Vehicle Trip Ends vs: Employees
On a: Weekday,
A.M. Peak Hour of Generator

Number of Studies: 7
Avg. Number of Employees: 42
Directional Distribution: 44% entering, 56% exiting

Trip Generation per Employee

Average Rate	Range of Rates	Standard Deviation
0.67	0.35 - 1.13	0.85

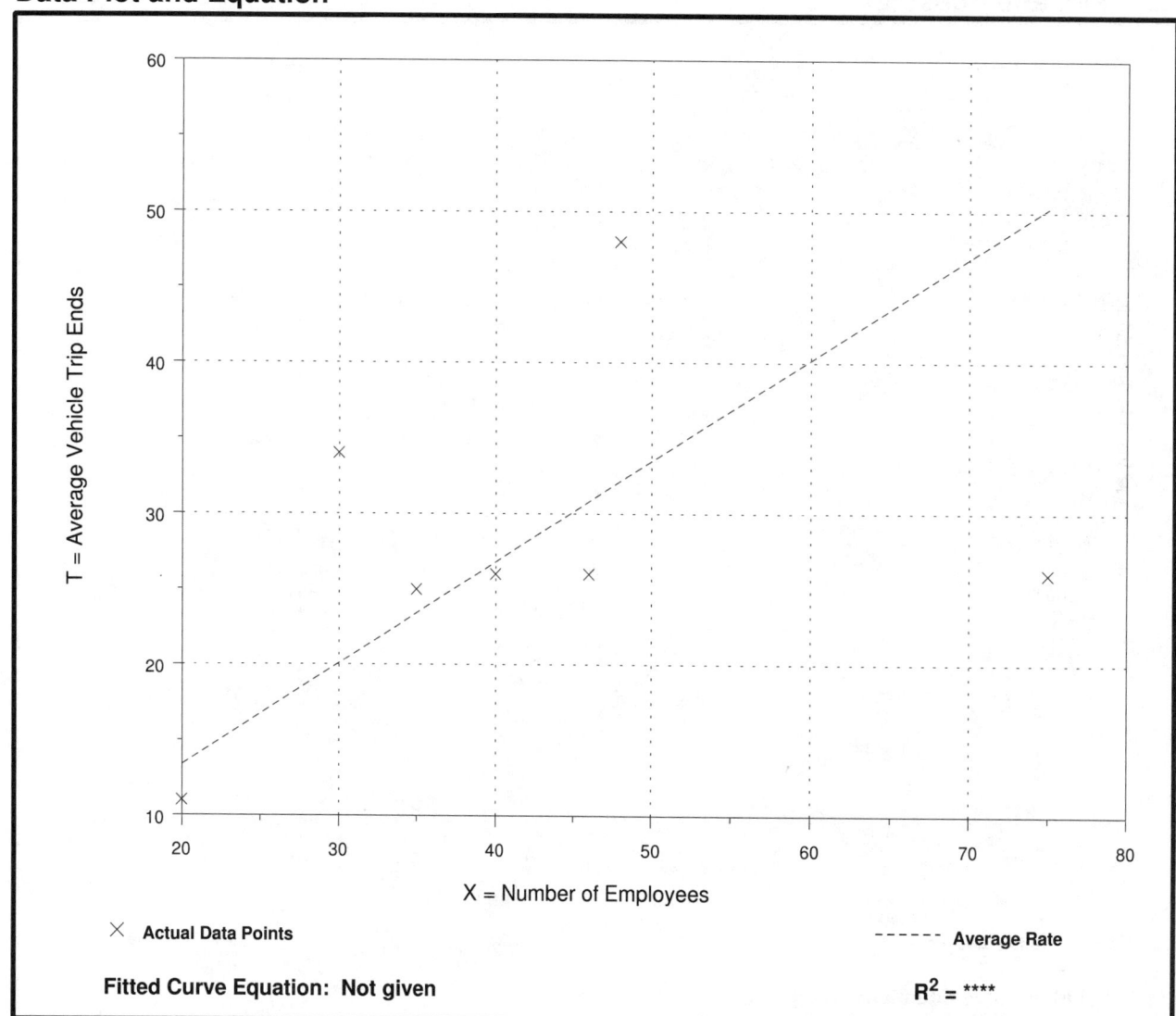

Data Plot and Equation

Fitted Curve Equation: Not given $R^2 = $ ****

Automobile Sales
(841)

Average Vehicle Trip Ends vs: Employees
On a: Weekday,
P.M. Peak Hour of Generator

Number of Studies: 7
Avg. Number of Employees: 42
Directional Distribution: 48% entering, 52% exiting

Trip Generation per Employee

Average Rate	Range of Rates	Standard Deviation
0.96	0.48 - 1.93	1.06

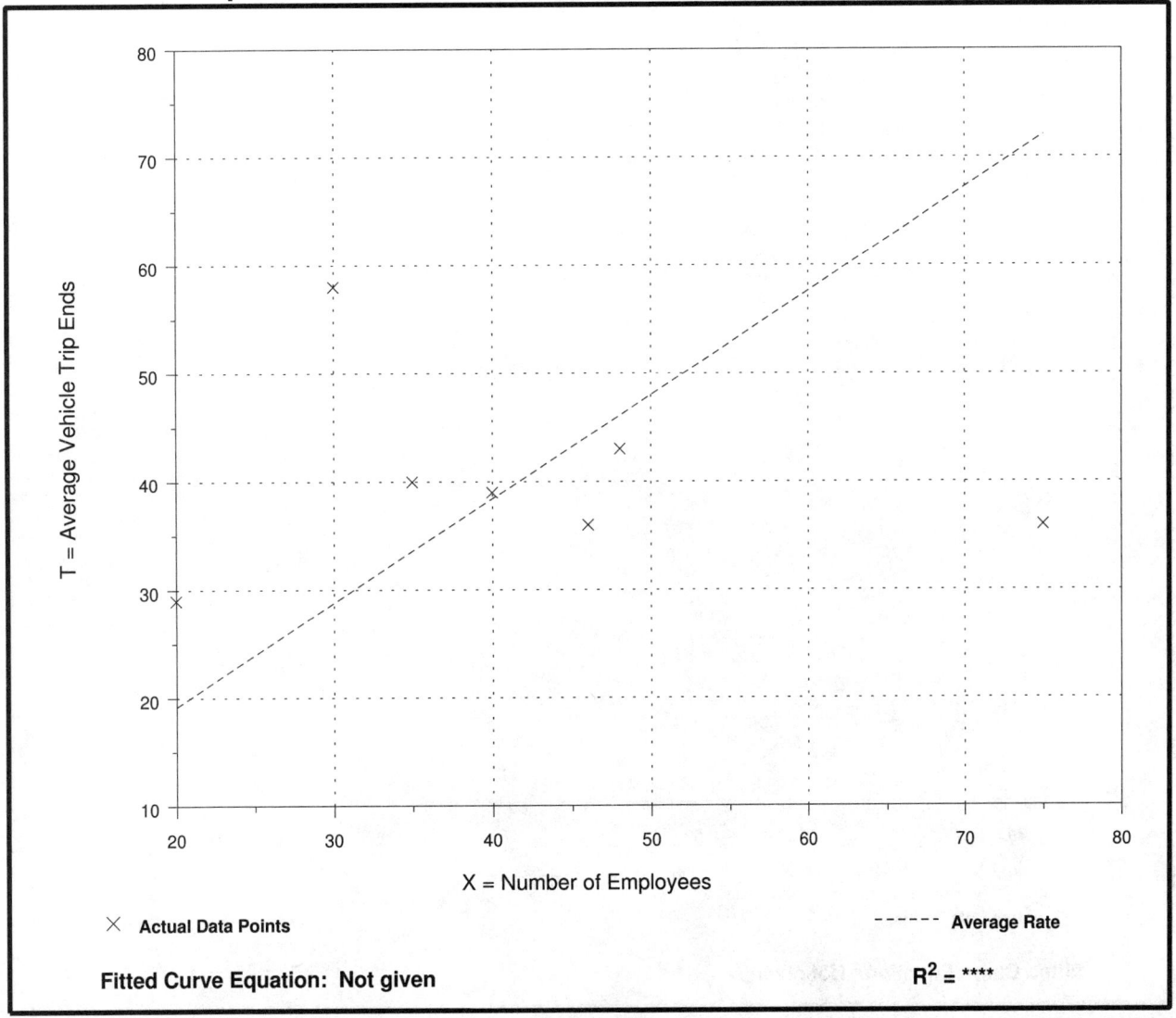

Data Plot and Equation

Fitted Curve Equation: Not given $R^2 = ****$

Automobile Sales
(841)

Average Vehicle Trip Ends vs: Employees
On a: Saturday

Number of Studies: 3
Avg. Number of Employees: 57
Directional Distribution: 50% entering, 50% exiting

Trip Generation per Employee

Average Rate	Range of Rates	Standard Deviation
10.55	8.50 - 11.60	3.45

Data Plot and Equation

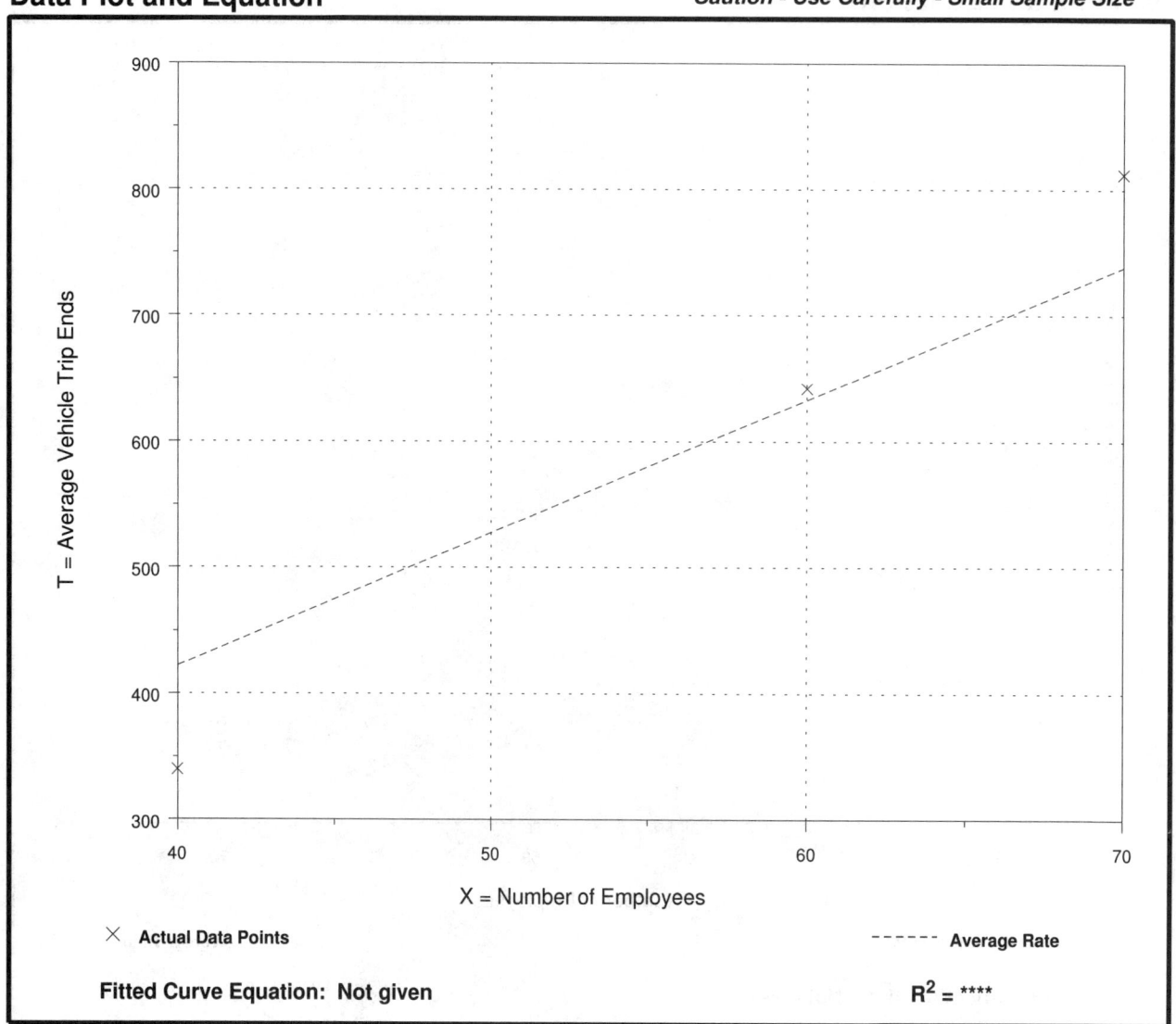

Caution - Use Carefully - Small Sample Size

Fitted Curve Equation: Not given $R^2 = ****$

Automobile Sales
(841)

Average Vehicle Trip Ends vs: Employees
On a: Sunday

Number of Studies: 3
Avg. Number of Employees: 57
Directional Distribution: 50% entering, 50% exiting

Trip Generation per Employee

Average Rate	Range of Rates	Standard Deviation
5.26	2.66 - 8.95	3.37

Data Plot and Equation

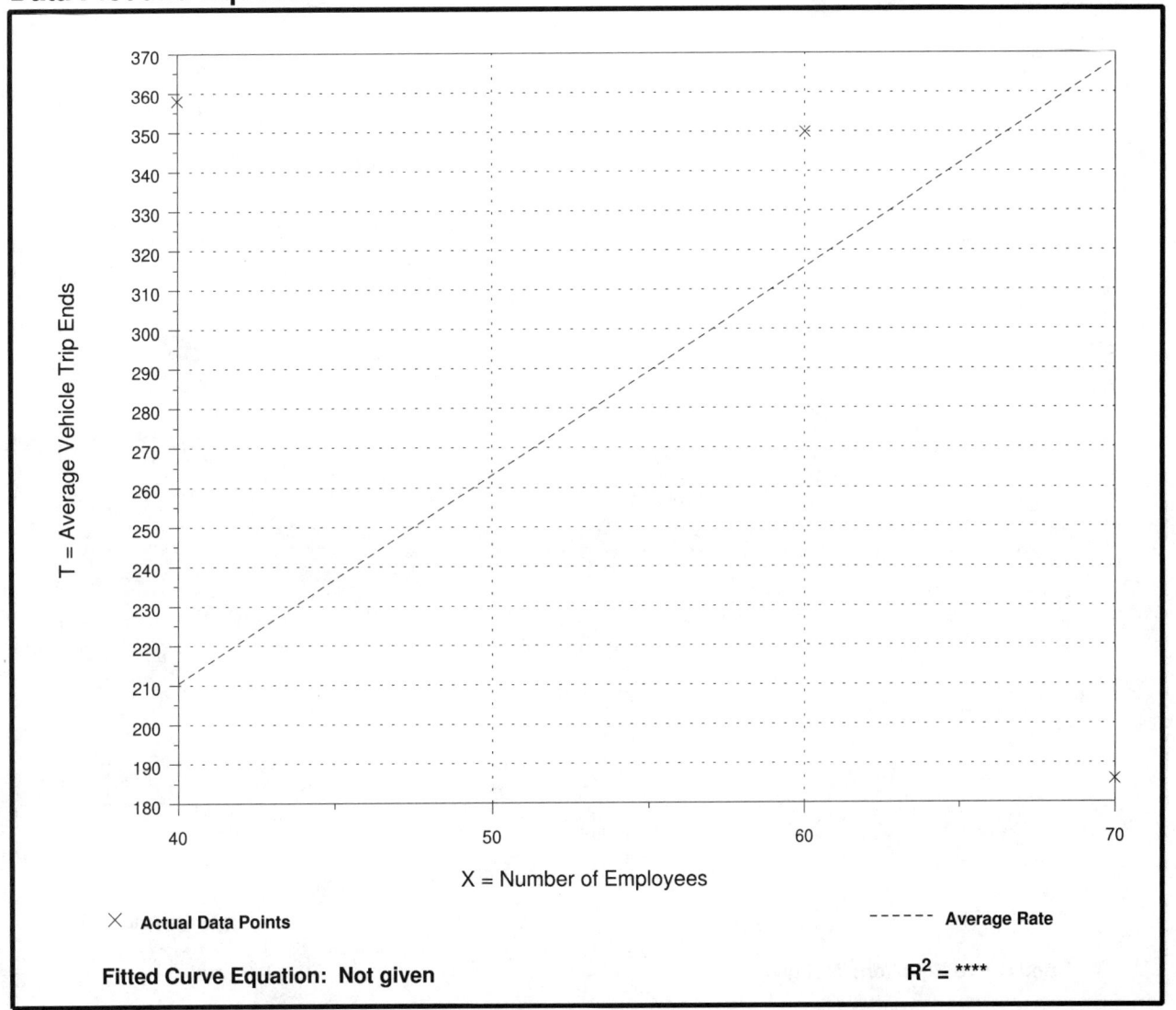

Caution - Use Carefully - Small Sample Size

X = Number of Employees
T = Average Vehicle Trip Ends

× Actual Data Points
----- Average Rate

Fitted Curve Equation: Not given $R^2 =$ ****

Automobile Sales
(841)

Average Vehicle Trip Ends vs: 1000 Sq. Feet Gross Floor Area
On a: Weekday

Number of Studies: 15
Average 1000 Sq. Feet GFA: 38
Directional Distribution: 50% entering, 50% exiting

Trip Generation per 1000 Sq. Feet Gross Floor Area

Average Rate	Range of Rates	Standard Deviation
32.30	15.64 - 79.66	15.70

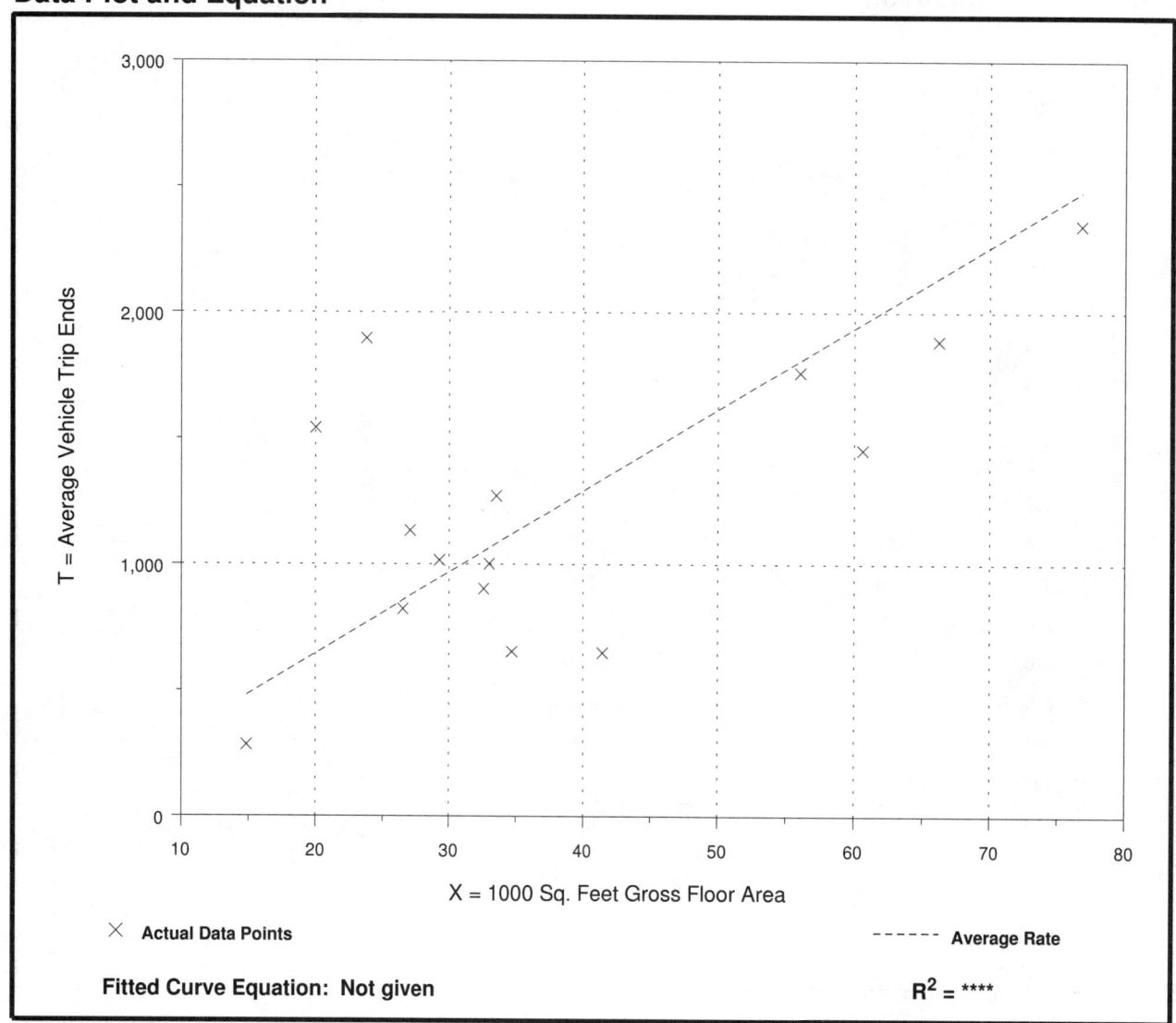

Data Plot and Equation

Fitted Curve Equation: Not given $R^2 = $ ****

Automobile Sales
(841)

Average Vehicle Trip Ends vs: 1000 Sq. Feet Gross Floor Area
On a: Weekday,
Peak Hour of Adjacent Street Traffic,
One Hour Between 7 and 9 a.m.

Number of Studies: 26
Average 1000 Sq. Feet GFA: 30
Directional Distribution: 75% entering, 25% exiting

Trip Generation per 1000 Sq. Feet Gross Floor Area

Average Rate	Range of Rates	Standard Deviation
1.92	0.59 - 6.17	1.72

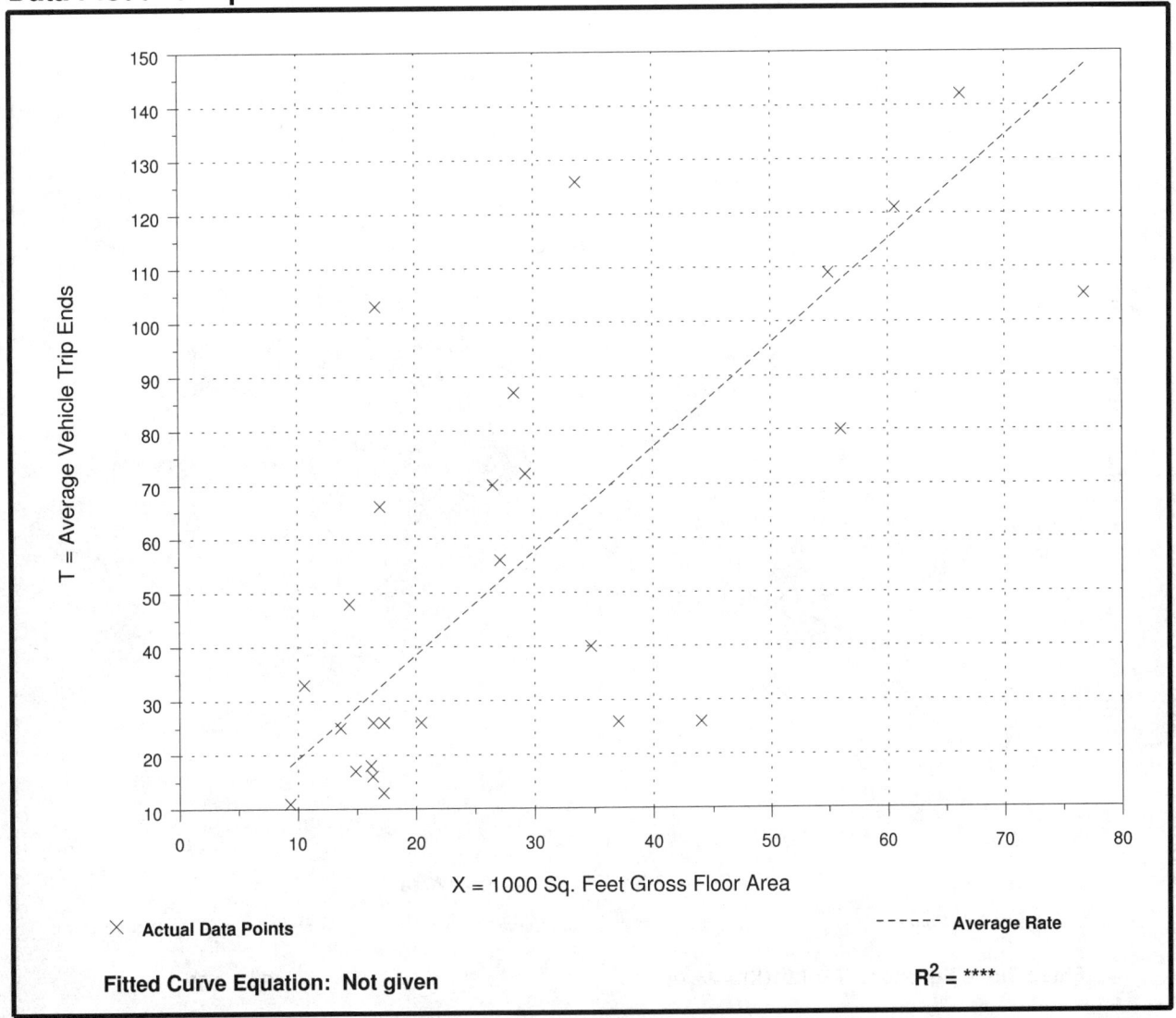

Data Plot and Equation

Fitted Curve Equation: Not given $R^2 = ****$

Automobile Sales
(841)

Average Vehicle Trip Ends vs: 1000 Sq. Feet Gross Floor Area
On a: Weekday,
Peak Hour of Adjacent Street Traffic,
One Hour Between 4 and 6 p.m.

Number of Studies: 41
Average 1000 Sq. Feet GFA: 33
Directional Distribution: 40% entering, 60% exiting

Trip Generation per 1000 Sq. Feet Gross Floor Area

Average Rate	Range of Rates	Standard Deviation
2.62	0.94 - 5.81	1.90

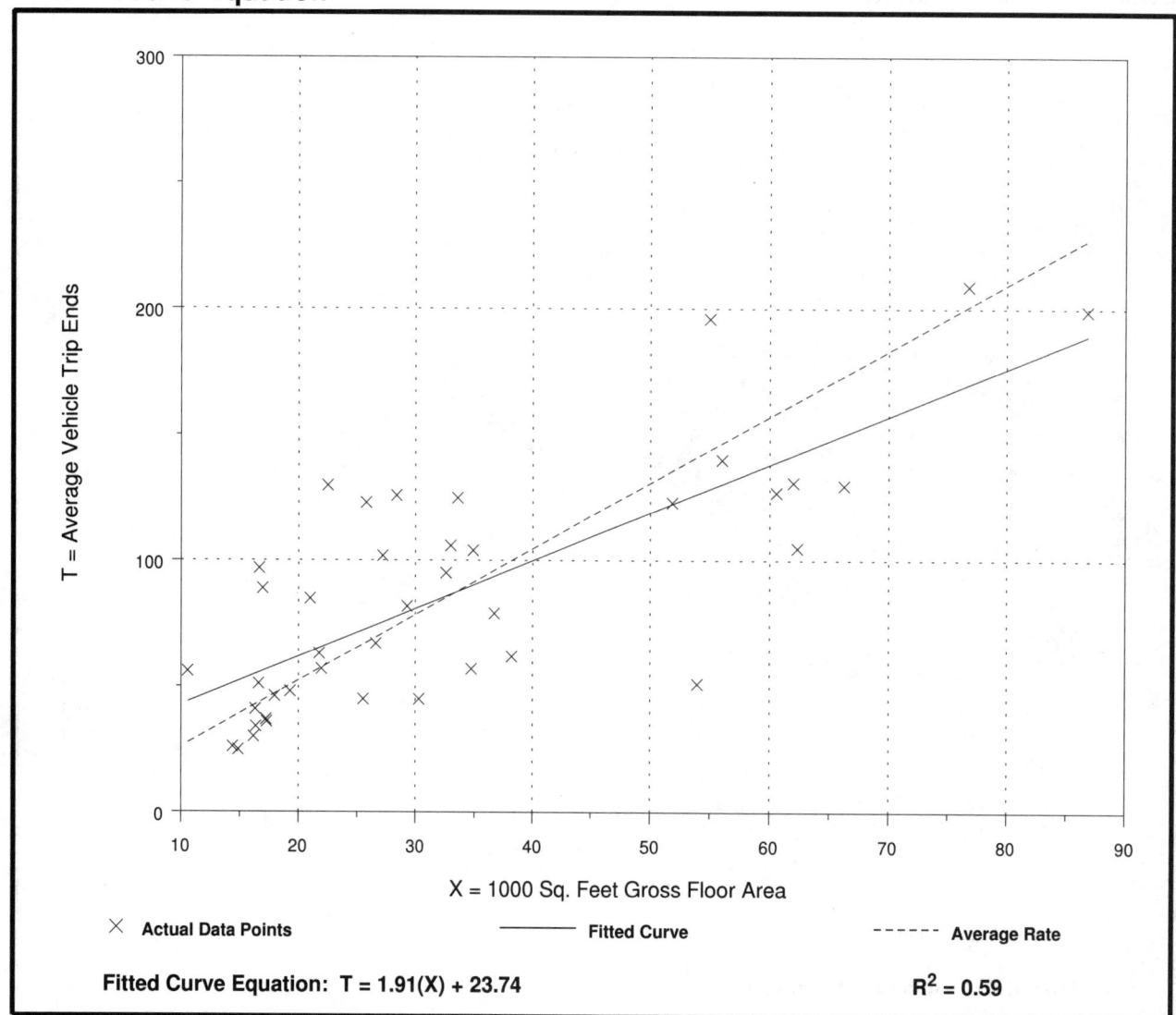

Data Plot and Equation

Fitted Curve Equation: $T = 1.91(X) + 23.74$ $R^2 = 0.59$

Automobile Sales
(841)

Average Vehicle Trip Ends vs: 1000 Sq. Feet Gross Floor Area
On a: Weekday,
A.M. Peak Hour of Generator

Number of Studies: 27
Average 1000 Sq. Feet GFA: 31
Directional Distribution: 55% entering, 45% exiting

Trip Generation per 1000 Sq. Feet Gross Floor Area

Average Rate	Range of Rates	Standard Deviation
2.22	0.59 - 6.00	1.76

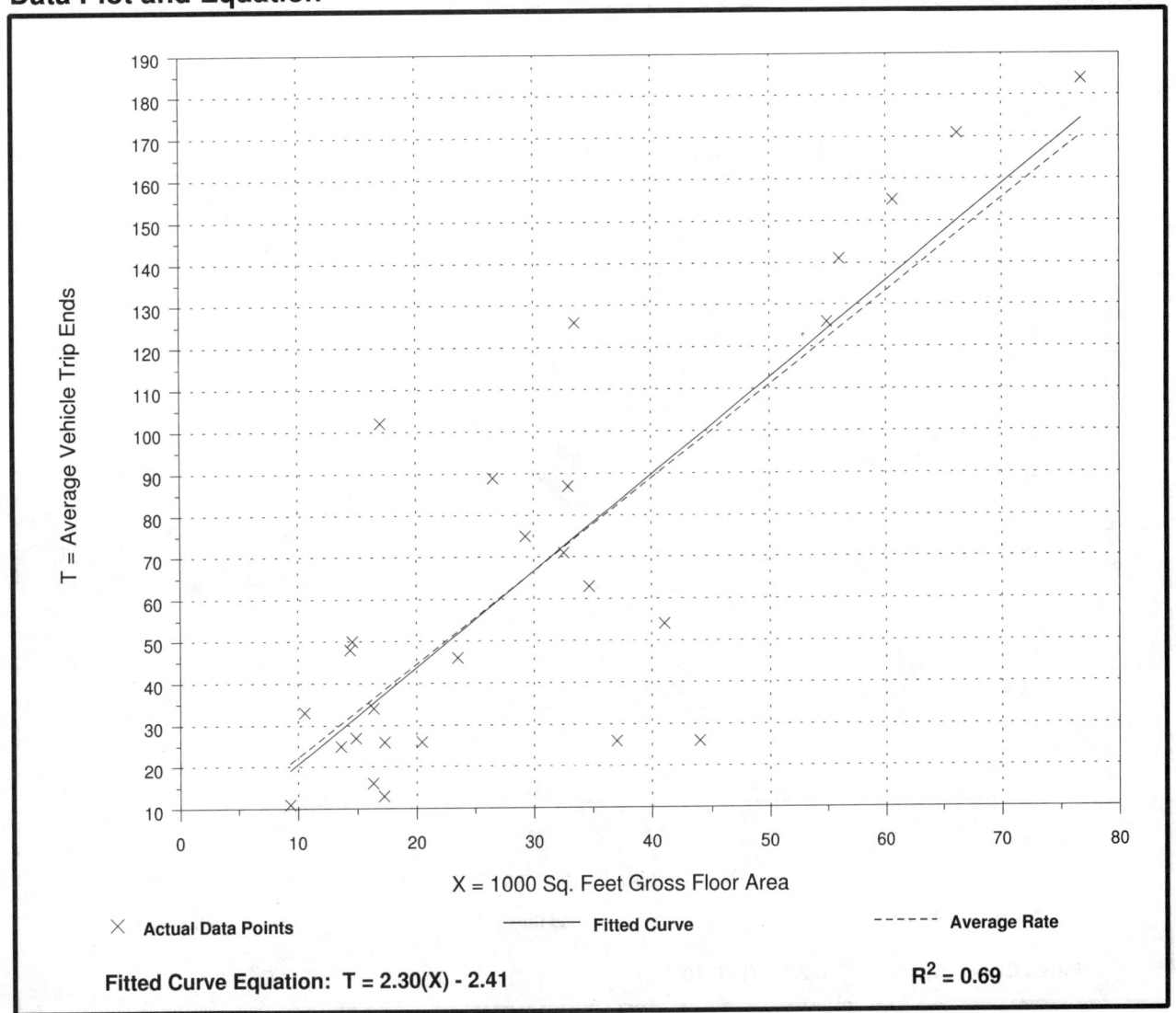

Data Plot and Equation

Fitted Curve Equation: T = 2.30(X) - 2.41 $R^2 = 0.69$

Automobile Sales
(841)

Average Vehicle Trip Ends vs: 1000 Sq. Feet Gross Floor Area
On a: Weekday,
P.M. Peak Hour of Generator

Number of Studies: 25
Average 1000 Sq. Feet GFA: 31
Directional Distribution: 47% entering, 53% exiting

Trip Generation per 1000 Sq. Feet Gross Floor Area

Average Rate	Range of Rates	Standard Deviation
2.80	0.89 - 5.41	1.91

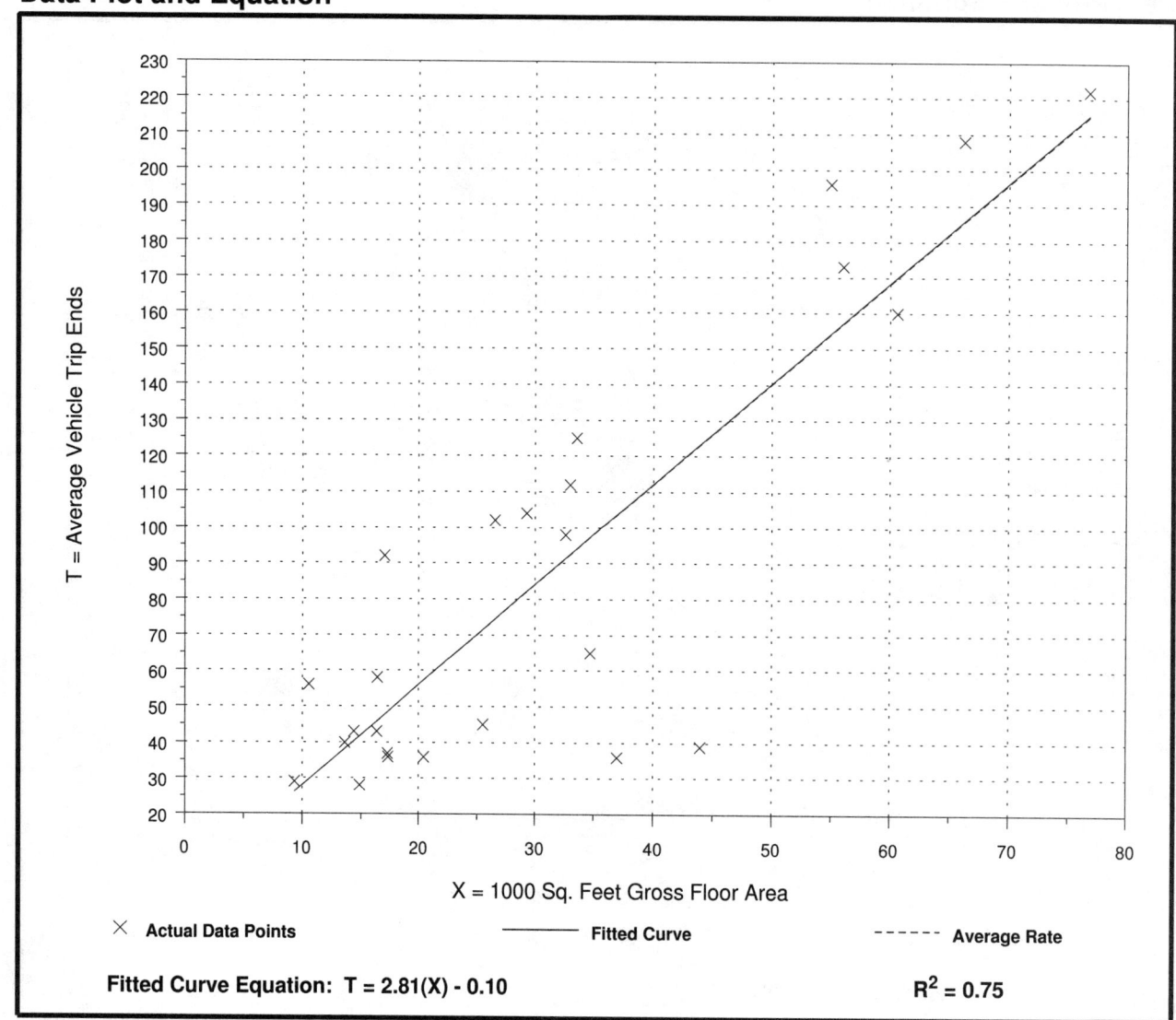

Data Plot and Equation

Fitted Curve Equation: $T = 2.81(X) - 0.10$ $\qquad R^2 = 0.75$

Automobile Sales
(841)

Average Vehicle Trip Ends vs: 1000 Sq. Feet Gross Floor Area
On a: Saturday

Number of Studies: 4
Average 1000 Sq. Feet GFA: 30
Directional Distribution: 50% entering, 50% exiting

Trip Generation per 1000 Sq. Feet Gross Floor Area

Average Rate	Range of Rates	Standard Deviation
29.74	15.47 - 52.24	16.58

Data Plot and Equation

Caution - Use Carefully - Small Sample Size

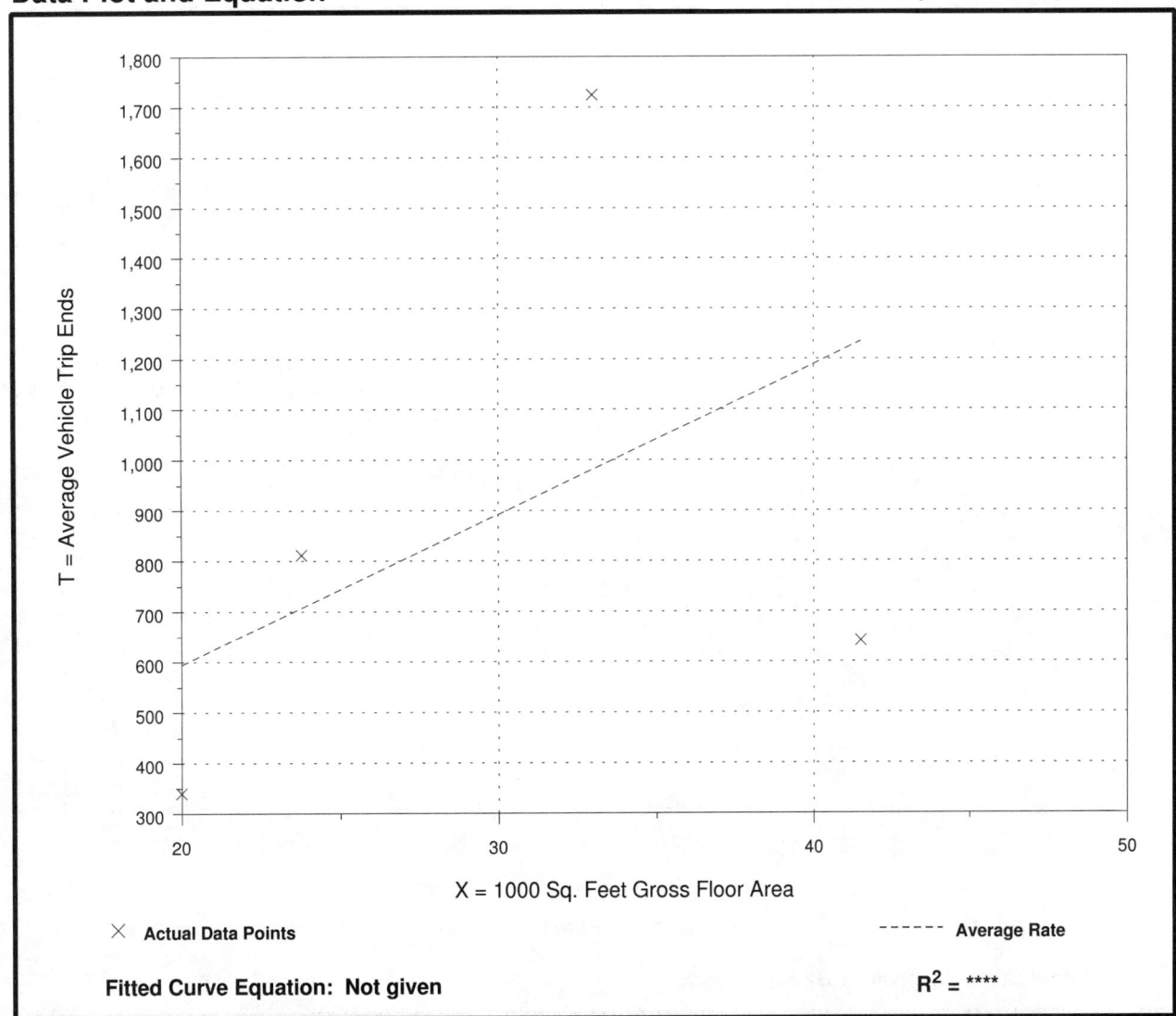

[Scatter plot: X = 1000 Sq. Feet Gross Floor Area (20–50), T = Average Vehicle Trip Ends (300–1,800). Actual Data Points marked with X; dashed line = Average Rate.]

Fitted Curve Equation: Not given $R^2 = ****$

Automobile Sales
(841)

Average Vehicle Trip Ends vs: 1000 Sq. Feet Gross Floor Area
On a: Saturday,
Peak Hour of Generator

Number of Studies: 4
Average 1000 Sq. Feet GFA: 21
Directional Distribution: 50% entering, 50% exiting

Trip Generation per 1000 Sq. Feet Gross Floor Area

Average Rate	Range of Rates	Standard Deviation
4.02	1.41 - 5.64	2.58

Data Plot and Equation

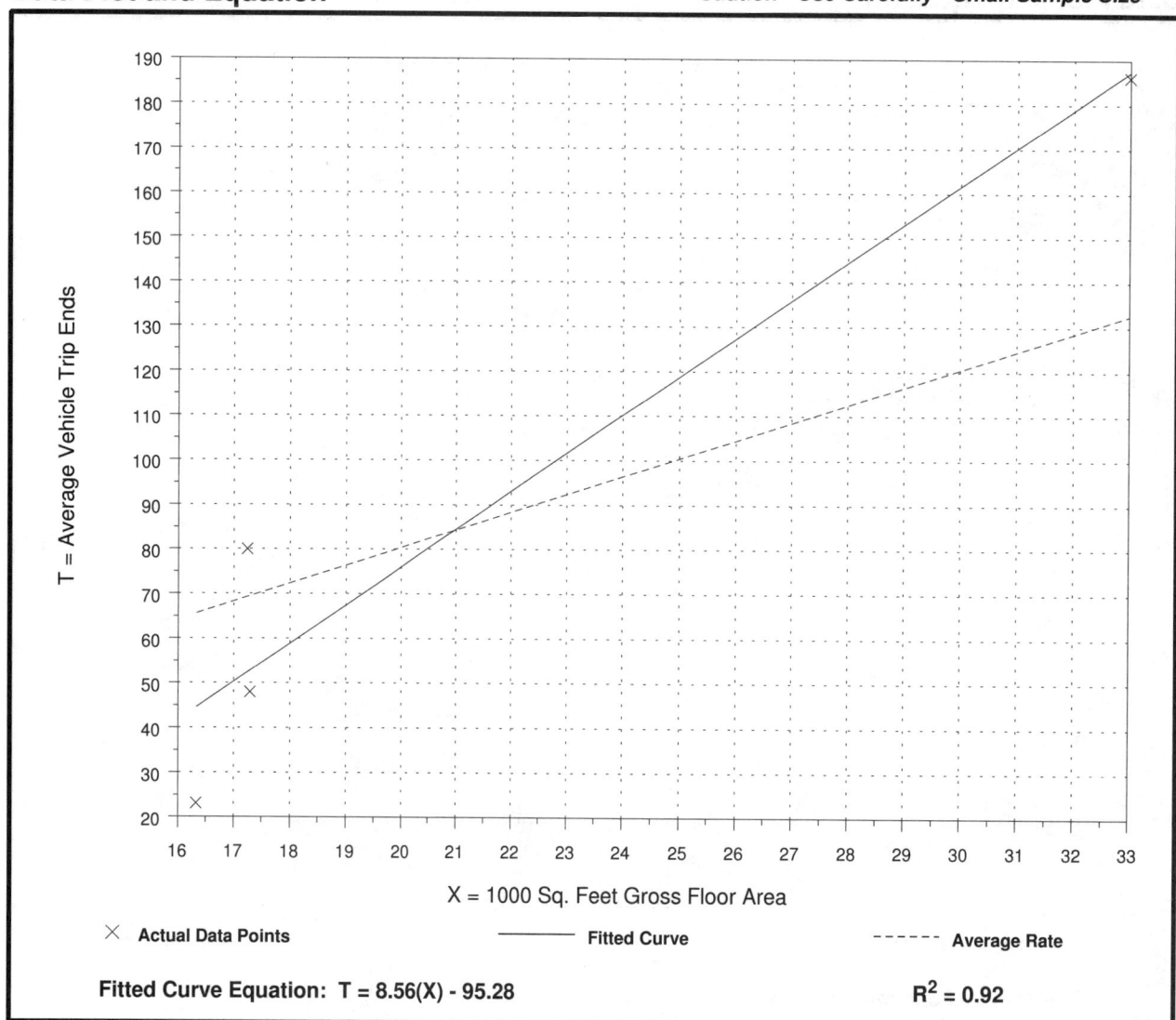

Caution - Use Carefully - Small Sample Size

Fitted Curve Equation: T = 8.56(X) - 95.28 $R^2 = 0.92$

Automobile Sales
(841)

Average Vehicle Trip Ends vs: 1000 Sq. Feet Gross Floor Area
On a: Sunday

Number of Studies: 4
Average 1000 Sq. Feet GFA: 30
Directional Distribution: 50% entering, 50% exiting

Trip Generation per 1000 Sq. Feet Gross Floor Area

Average Rate	Range of Rates	Standard Deviation
13.62	7.82 - 21.73	7.16

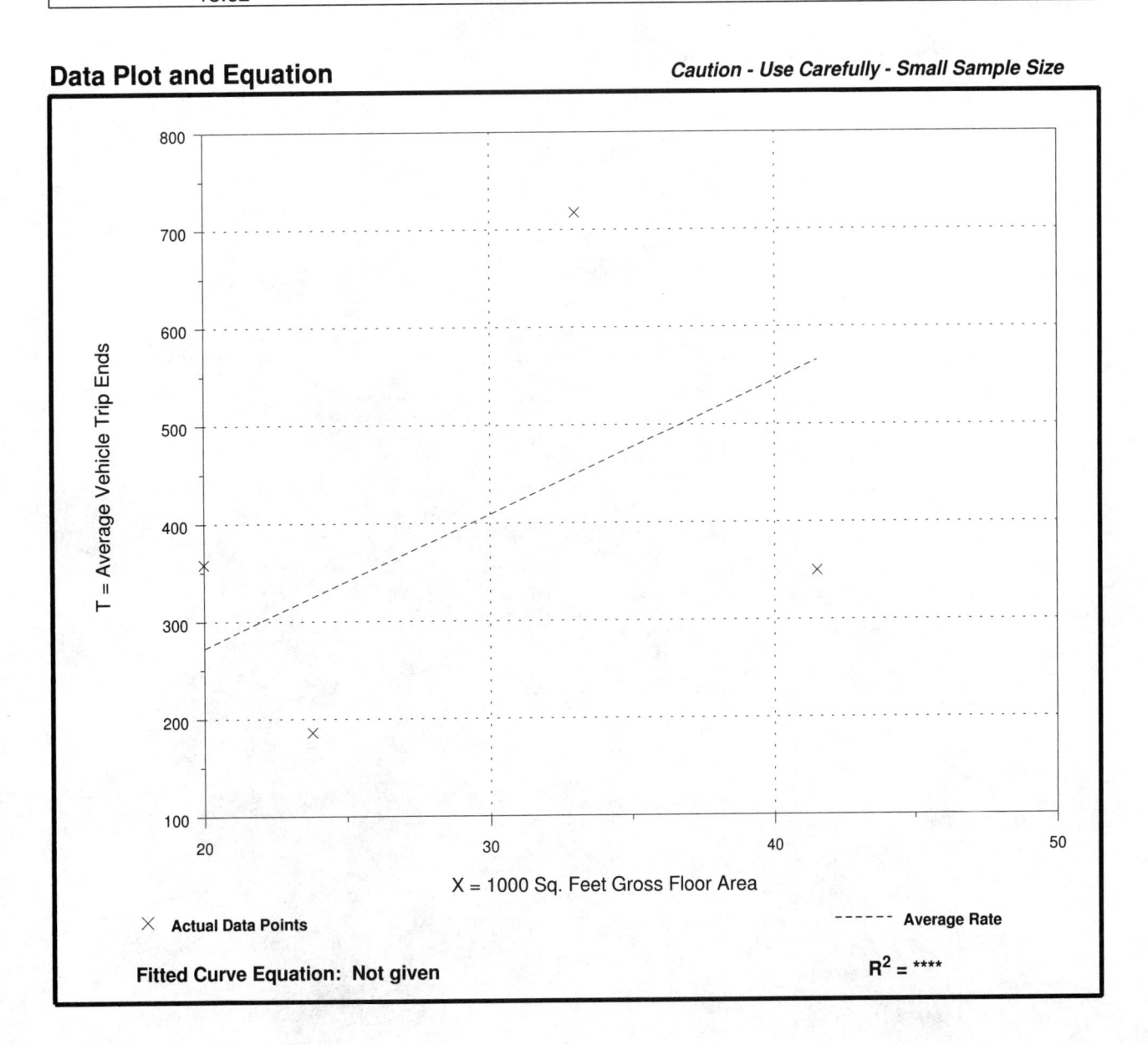

Data Plot and Equation

Caution - Use Carefully - Small Sample Size

Fitted Curve Equation: Not given $R^2 = ****$

Land Use: 842
Recreational Vehicle Sales

Description

A recreational vehicle (RV) sales dealership is a free-standing facility that specializes in the sales of new RVs. Recreational vehicle services, parts and accessories sales and substantial used RV sales may also be available. Some RV dealerships may also include boat sales and servicing. Automobile sales (Land Use 841) is a related use.

Additional Data

The sites were surveyed in 2007 in Florida.

Source Number

721

Recreational Vehicle Sales
(842)

Average Vehicle Trip Ends vs: 1000 Sq. Feet Gross Floor Area
On a: Weekday,
Peak Hour of Adjacent Street Traffic,
One Hour Between 4 and 6 p.m.

Number of Studies: 2
Average 1000 Sq. Feet GFA: 7
Directional Distribution: 36% entering, 64% exiting

Trip Generation per 1000 Sq. Feet Gross Floor Area

Average Rate	Range of Rates	Standard Deviation
2.54	2.20 - 3.67	*

Data Plot and Equation

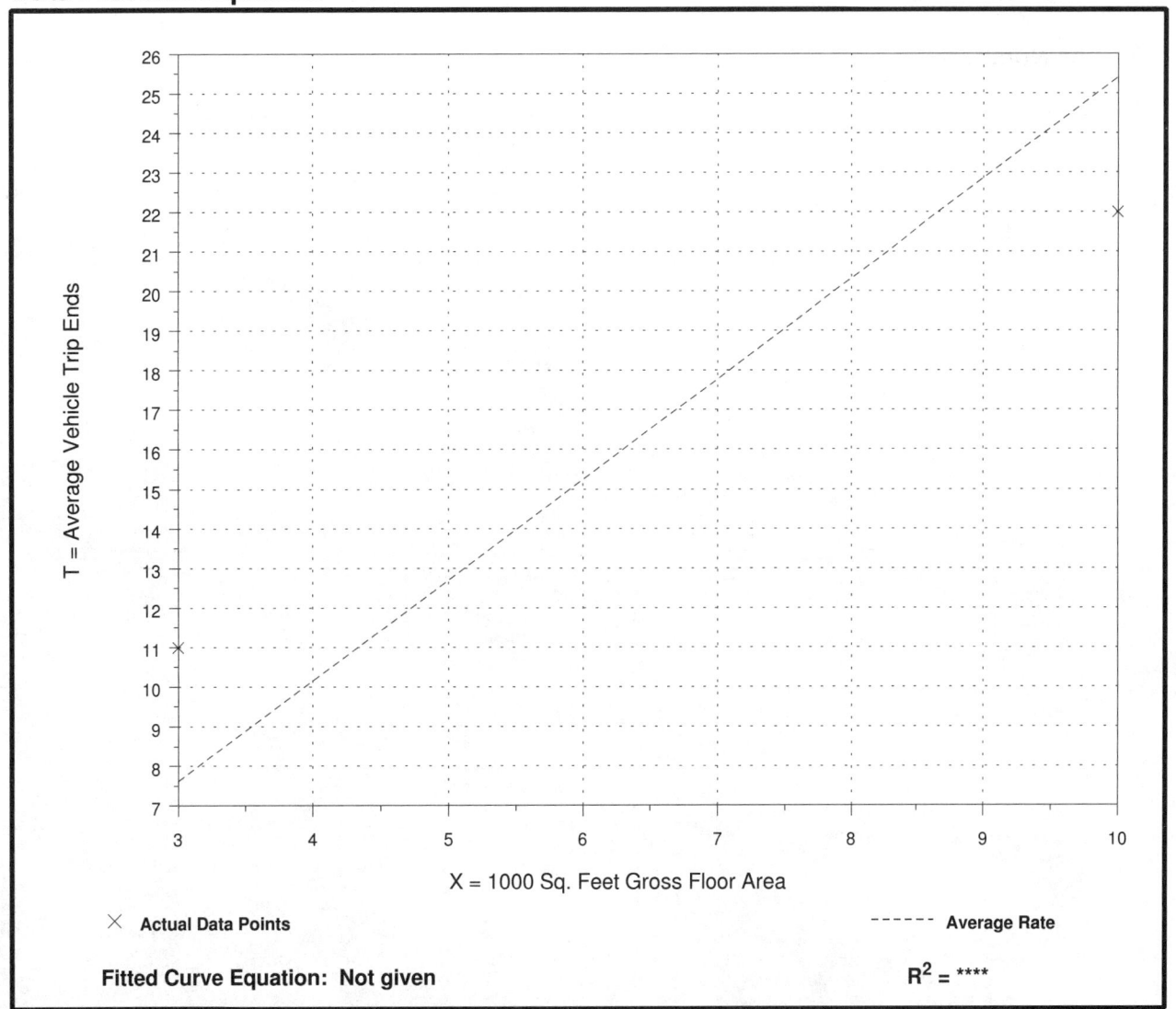

Caution - Use Carefully - Small Sample Size

Fitted Curve Equation: Not given $R^2 = ****$

Land Use: 843
Automobile Parts Sales

Description

Automobile parts sales facilities specialize in the sale of automobile parts for maintenance and repair. Items sold at these facilities include spark plugs, oil, batteries and a wide range of automobile parts. These facilities are not equipped for on-site vehicle repair. Tire store (Land Use 848), tire superstore (Land Use 849) and automobile parts and service center (Land Use 943) are related uses.

Additional Data

The sites were surveyed in the 1990s and 2000s in Florida and New Hampshire.

Source Numbers

436, 439, 618

Land Use: 843
Automobile Parts Sales
Independent Variables with One Observation

The following trip generation data are for independent variables with only one observation. This information is shown in this table only; there are no related plots for these data.

Users are cautioned to use data with care because of the small sample size.

Independent Variable	Trip Generation Rate	Size of Independent Variable	Number of Studies	Directional Distribution
1,000 Square Feet Gross Floor Area				
Saturday Peak Hour of Generator	11.10	7.3	1	51% entering, 49% exiting

Automobile Parts Sales
(843)

Average Vehicle Trip Ends vs: 1000 Sq. Feet Gross Floor Area
On a: Weekday

Number of Studies: 5
Average 1000 Sq. Feet GFA: 8
Directional Distribution: 50% entering, 50% exiting

Trip Generation per 1000 Sq. Feet Gross Floor Area

Average Rate	Range of Rates	Standard Deviation
61.91	42.17 - 70.67	12.97

Data Plot and Equation

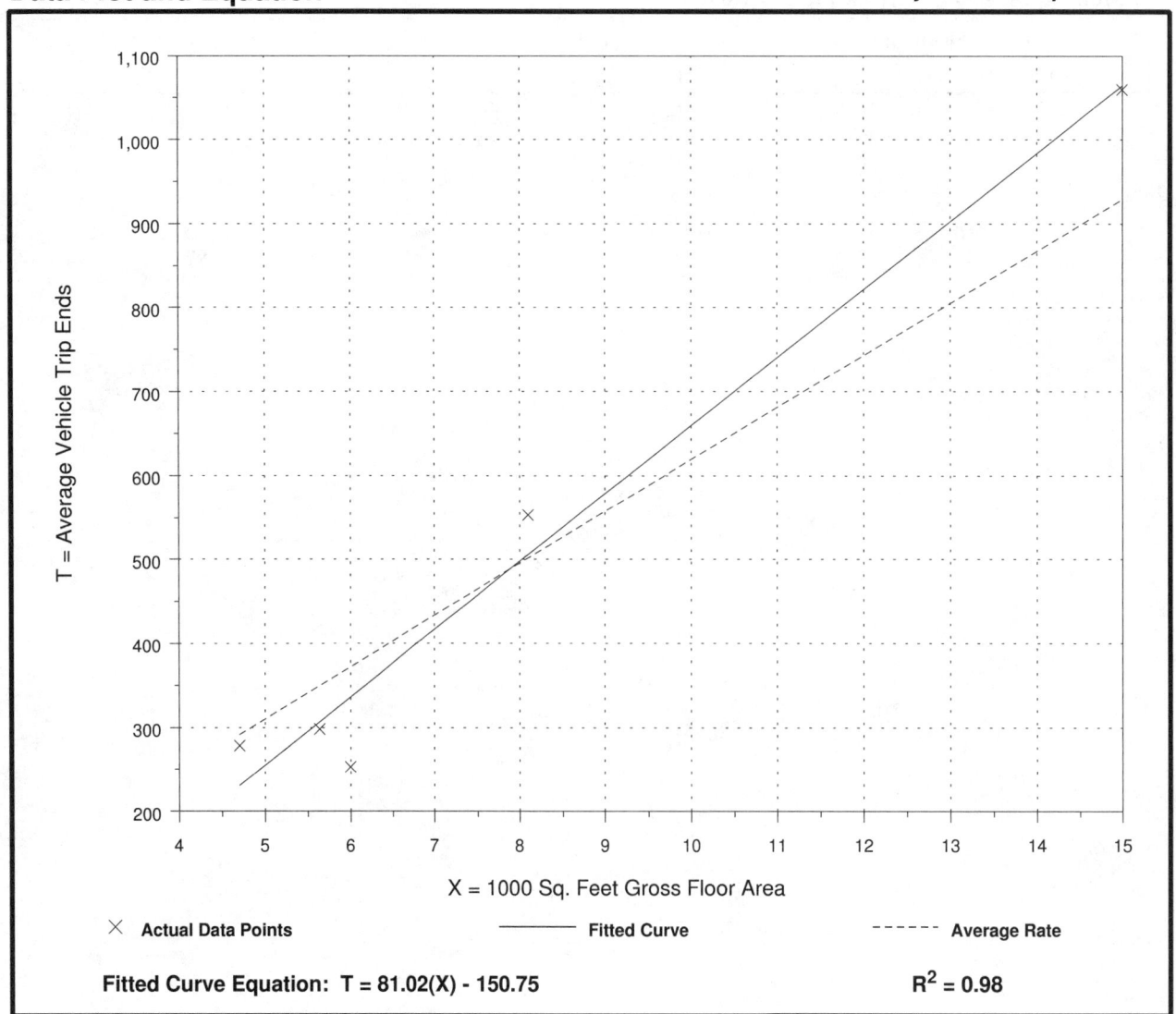

Caution - Use Carefully - Small Sample Size

Fitted Curve Equation: T = 81.02(X) - 150.75 $R^2 = 0.98$

Automobile Parts Sales
(843)

Average Vehicle Trip Ends vs: 1000 Sq. Feet Gross Floor Area
On a: Weekday,
        Peak Hour of Adjacent Street Traffic,
        One Hour Between 7 and 9 a.m.

Number of Studies: 5
Average 1000 Sq. Feet GFA: 8
Directional Distribution: Not available

Trip Generation per 1000 Sq. Feet Gross Floor Area

Average Rate	Range of Rates	Standard Deviation
2.21	1.17 - 2.55	1.49

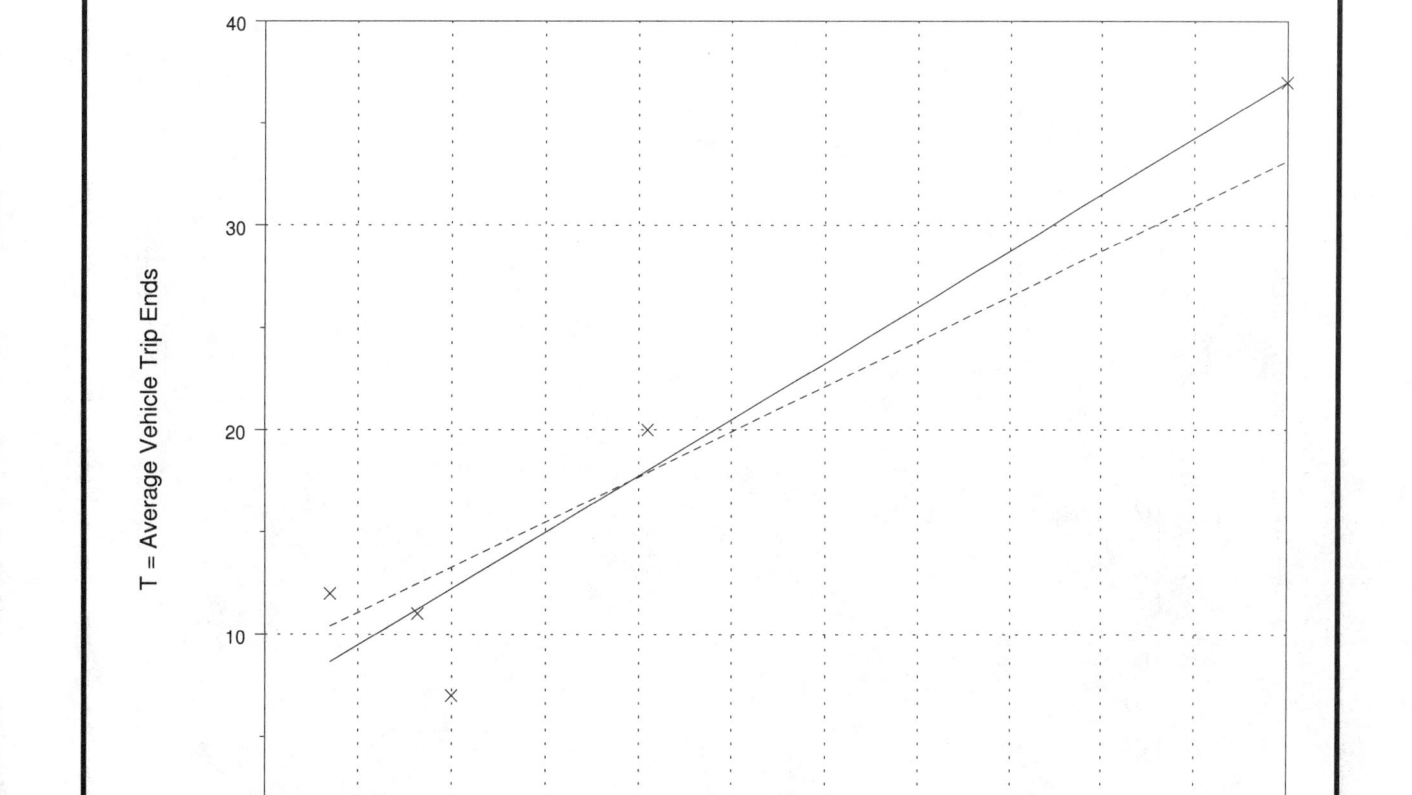

Data Plot and Equation

Caution - Use Carefully - Small Sample Size

Fitted Curve Equation: $T = 2.76(X) - 4.34$ $R^2 = 0.93$

Automobile Parts Sales
(843)

Average Vehicle Trip Ends vs: 1000 Sq. Feet Gross Floor Area
On a: Weekday,
Peak Hour of Adjacent Street Traffic,
One Hour Between 4 and 6 p.m.

Number of Studies: 5
Average 1000 Sq. Feet GFA: 8
Directional Distribution: 49% entering, 51% exiting

Trip Generation per 1000 Sq. Feet Gross Floor Area

Average Rate	Range of Rates	Standard Deviation
5.98	3.83 - 6.87	2.57

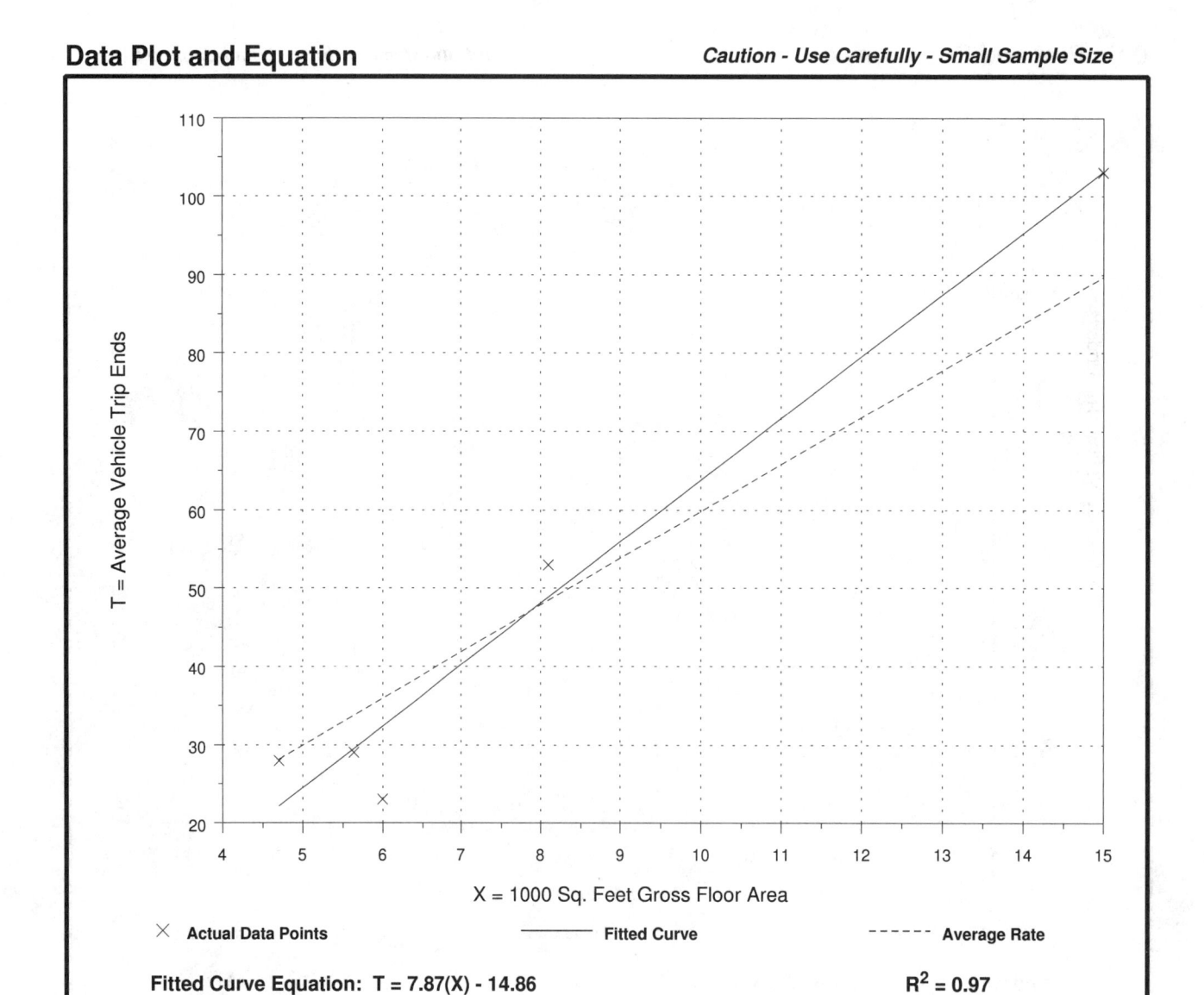

Data Plot and Equation

Caution - Use Carefully - Small Sample Size

Fitted Curve Equation: $T = 7.87(X) - 14.86$ $R^2 = 0.97$

Automobile Parts Sales
(843)

Average Vehicle Trip Ends vs: 1000 Sq. Feet Gross Floor Area
On a: Weekday,
A.M. Peak Hour of Generator

Number of Studies: 5
Average 1000 Sq. Feet GFA: 8
Directional Distribution: Not available

Trip Generation per 1000 Sq. Feet Gross Floor Area

Average Rate	Range of Rates	Standard Deviation
4.41	2.47 - 5.80	2.39

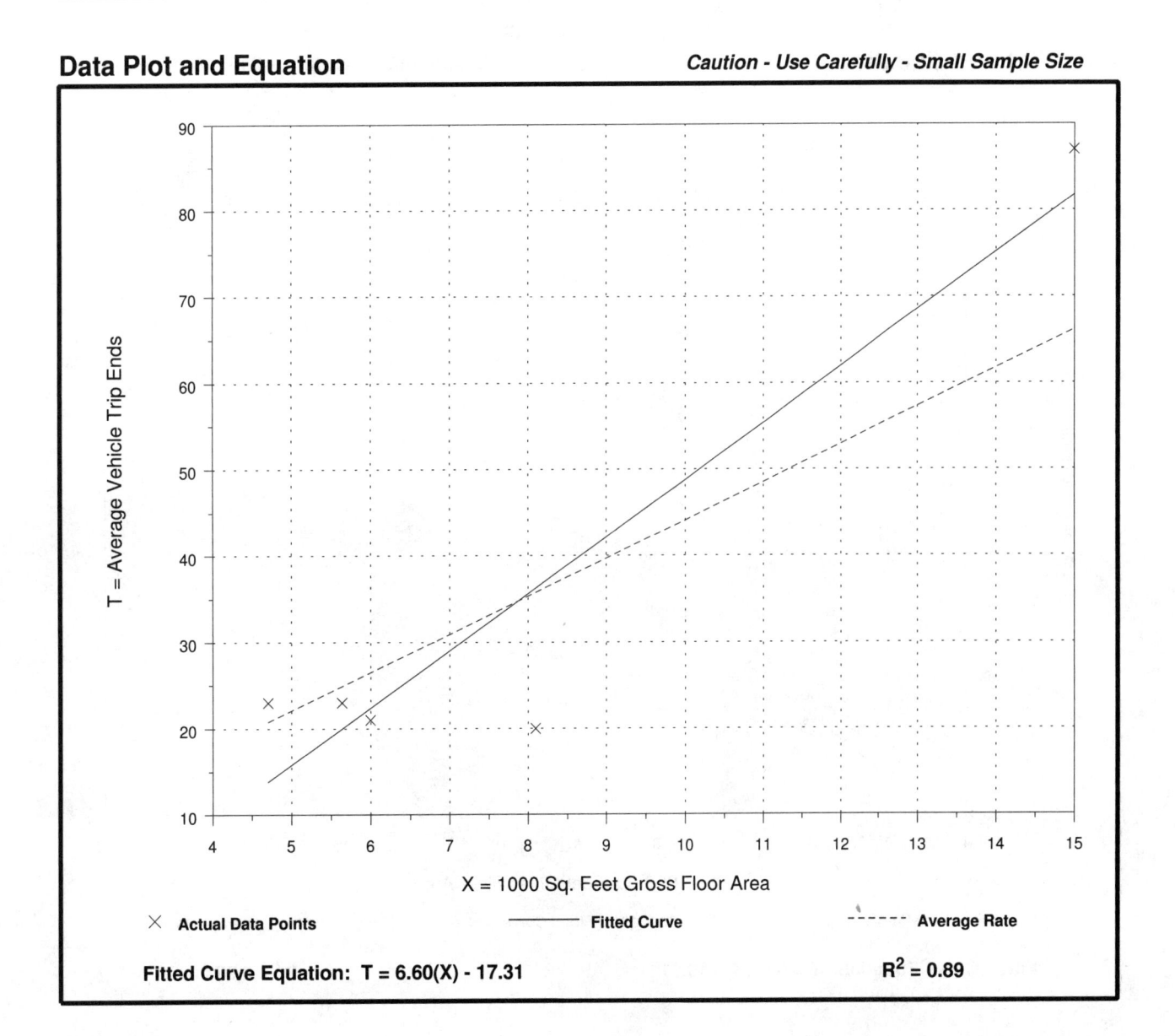

Data Plot and Equation

Caution - Use Carefully - Small Sample Size

Fitted Curve Equation: T = 6.60(X) - 17.31 $R^2 = 0.89$

Automobile Parts Sales
(843)

Average Vehicle Trip Ends vs: 1000 Sq. Feet Gross Floor Area
On a: Weekday,
P.M. Peak Hour of Generator

Number of Studies: 5
Average 1000 Sq. Feet GFA: 8
Directional Distribution: 51% entering, 49% exiting

Trip Generation per 1000 Sq. Feet Gross Floor Area

Average Rate	Range of Rates	Standard Deviation
6.44	4.33 - 7.60	2.72

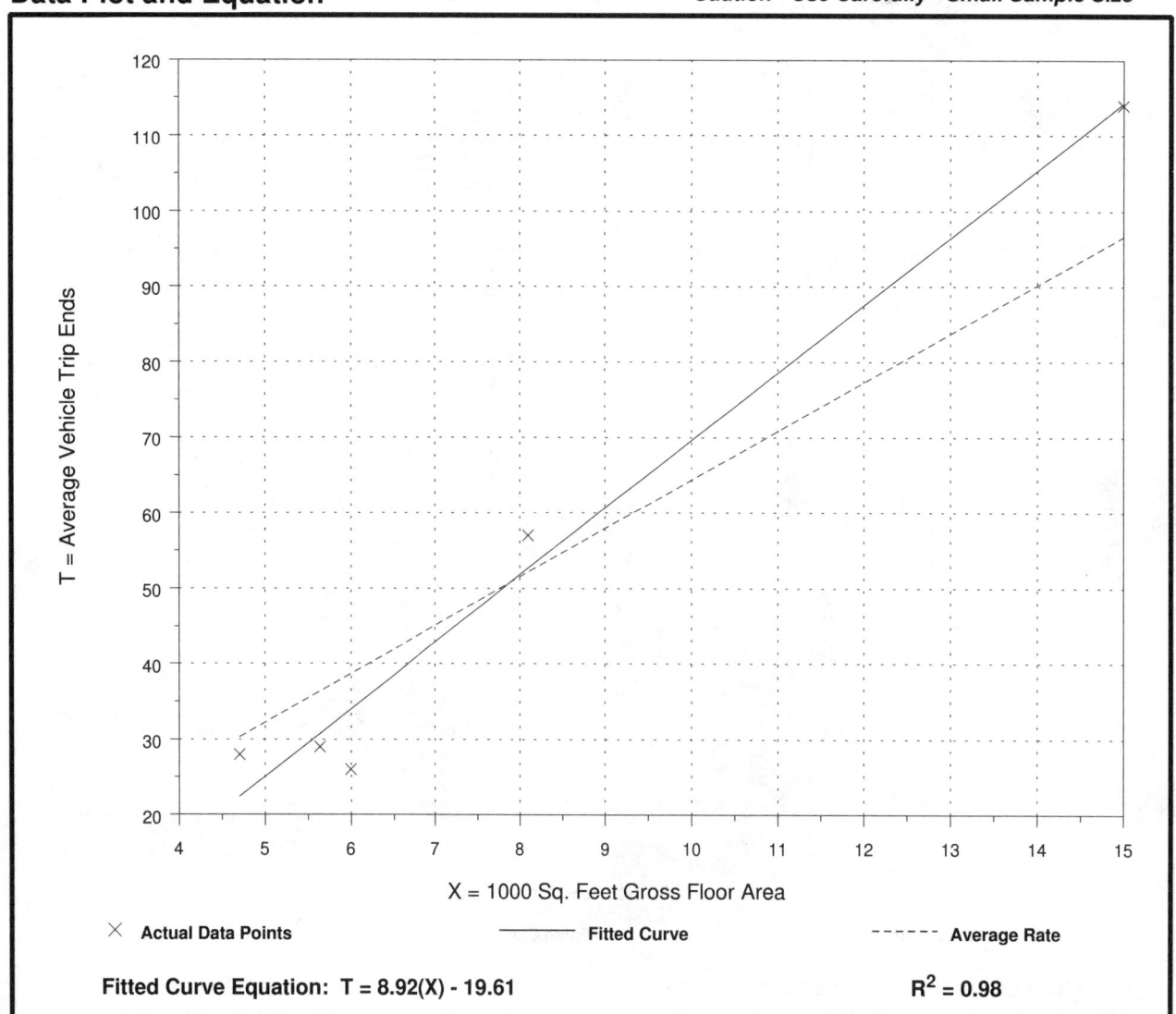

Data Plot and Equation

Caution - Use Carefully - Small Sample Size

Fitted Curve Equation: T = 8.92(X) - 19.61 $R^2 = 0.98$

Land Use: 848
Tire Store

Description

A tire store's primary business is the sale and marketing of tires for automotive vehicles. Services offered by these stores usually include tire installation and repair, as well as other automotive maintenance or repair services and customer assistance. These stores generally do not contain large storage or warehouse areas. Automobile parts sales (Land Use 843), tire superstore (Land Use 849) and automobile parts and service center (Land Use 943) are related uses.

Additional Data

The sites were surveyed between the 1990s and the 2000s in Florida, New York, Oregon, Pennsylvania, New Jersey and South Dakota.

Source Numbers

328, 359, 438, 555, 571, 583, 599

Tire Store
(848)

Average Vehicle Trip Ends vs: Service Bays
On a: Weekday,
Peak Hour of Adjacent Street Traffic,
One Hour Between 7 and 9 a.m.

Number of Studies: 9
Average Num. of Service Bays: 8
Directional Distribution: 64% entering, 36% exiting

Trip Generation per Service Bay

Average Rate	Range of Rates	Standard Deviation
2.10	0.80 - 4.80	1.71

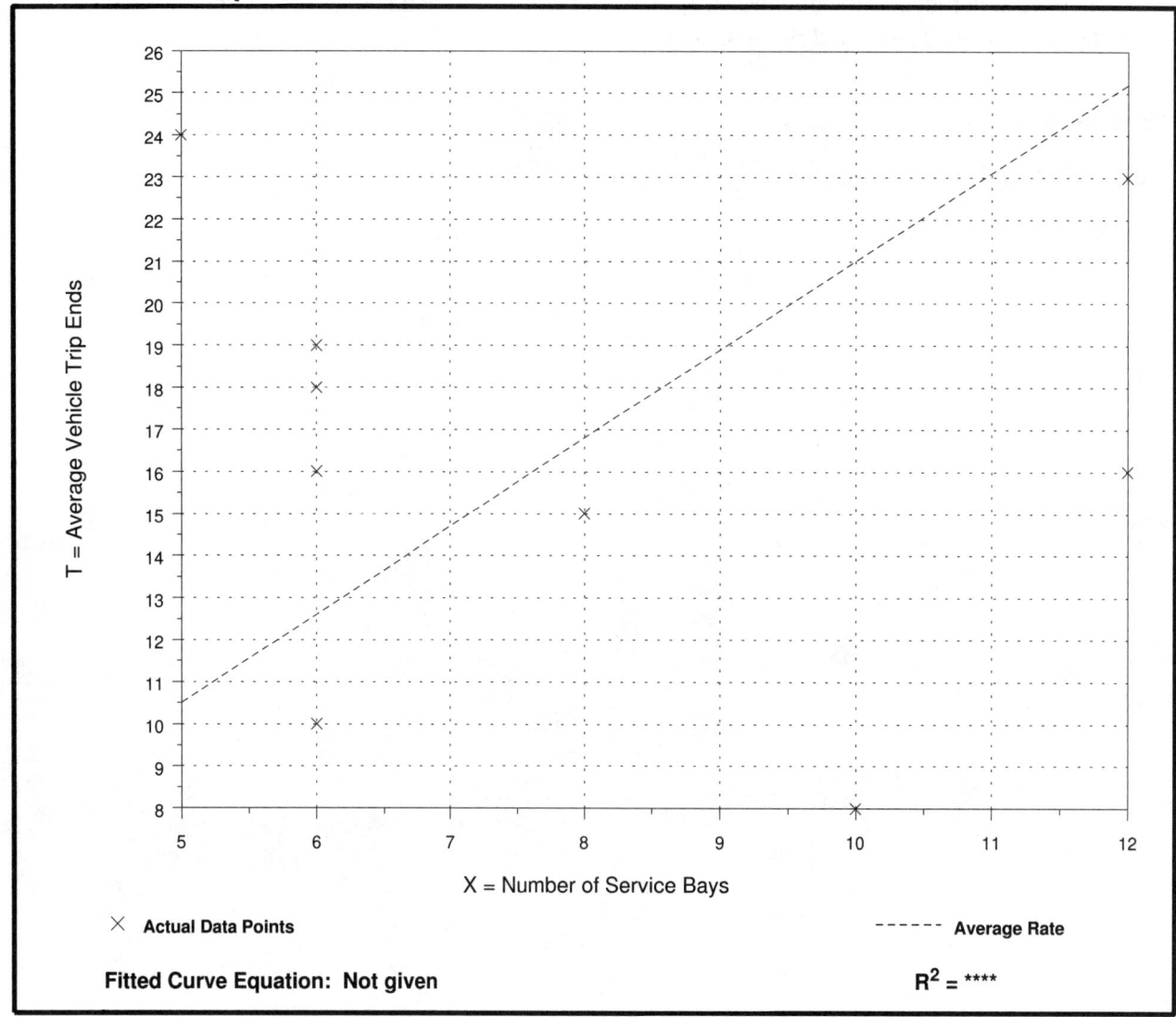

Data Plot and Equation

Fitted Curve Equation: Not given $R^2 = ****$

Tire Store
(848)

Average Vehicle Trip Ends vs: Service Bays
On a: Weekday,
Peak Hour of Adjacent Street Traffic,
One Hour Between 4 and 6 p.m.

Number of Studies: 9
Average Num. of Service Bays: 8
Directional Distribution: 42% entering, 58% exiting

Trip Generation per Service Bay

Average Rate	Range of Rates	Standard Deviation
3.54	2.00 - 6.67	2.27

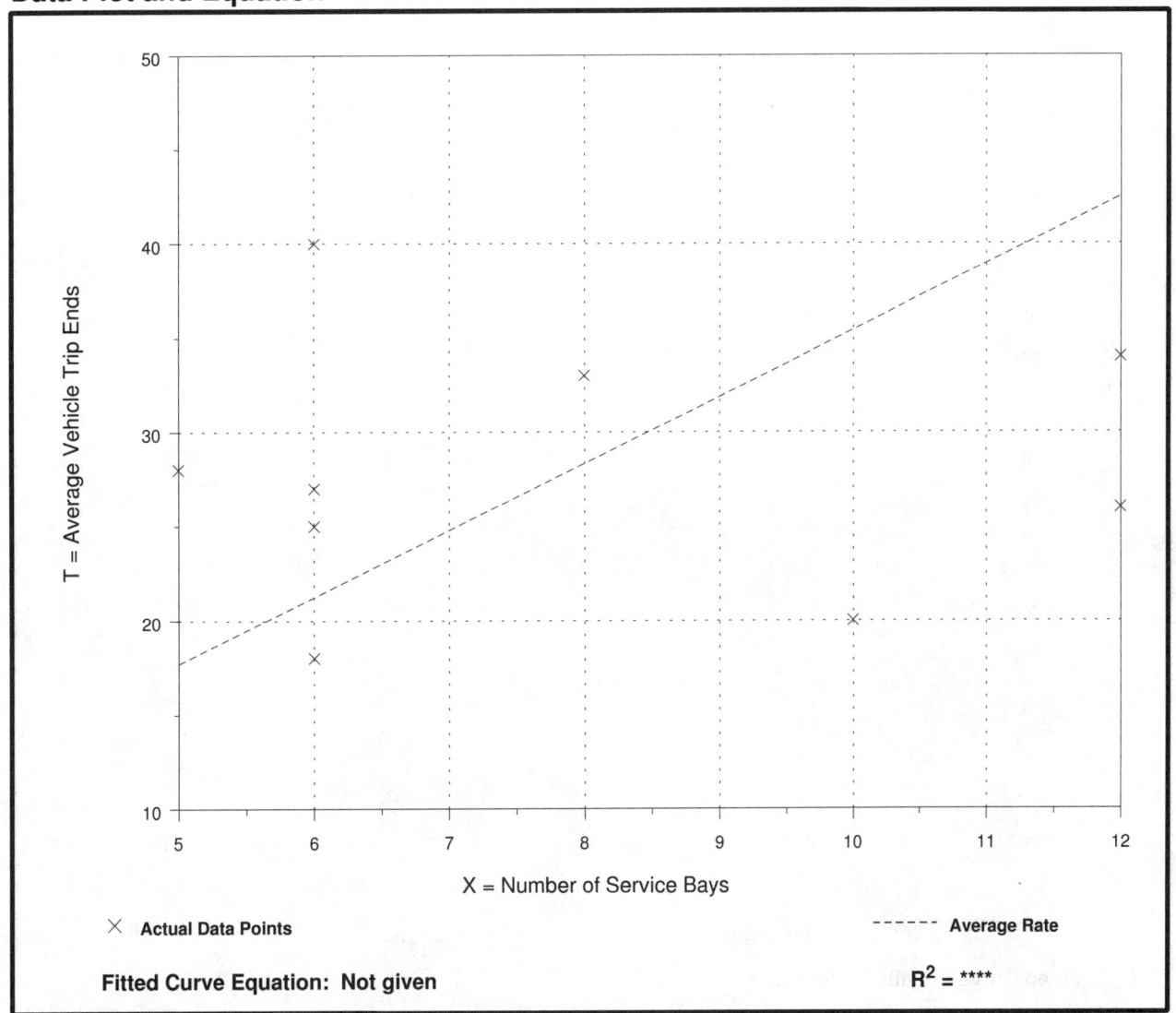

Data Plot and Equation

Fitted Curve Equation: Not given R^2 = ****

Tire Store
(848)

Average Vehicle Trip Ends vs: Service Bays
On a: Weekday,
A.M. Peak Hour of Generator

Number of Studies: 3
Average Num. of Service Bays: 6
Directional Distribution: 53% entering, 47% exiting

Trip Generation per Service Bay

Average Rate	Range of Rates	Standard Deviation
5.12	3.50 - 6.00	2.45

Data Plot and Equation

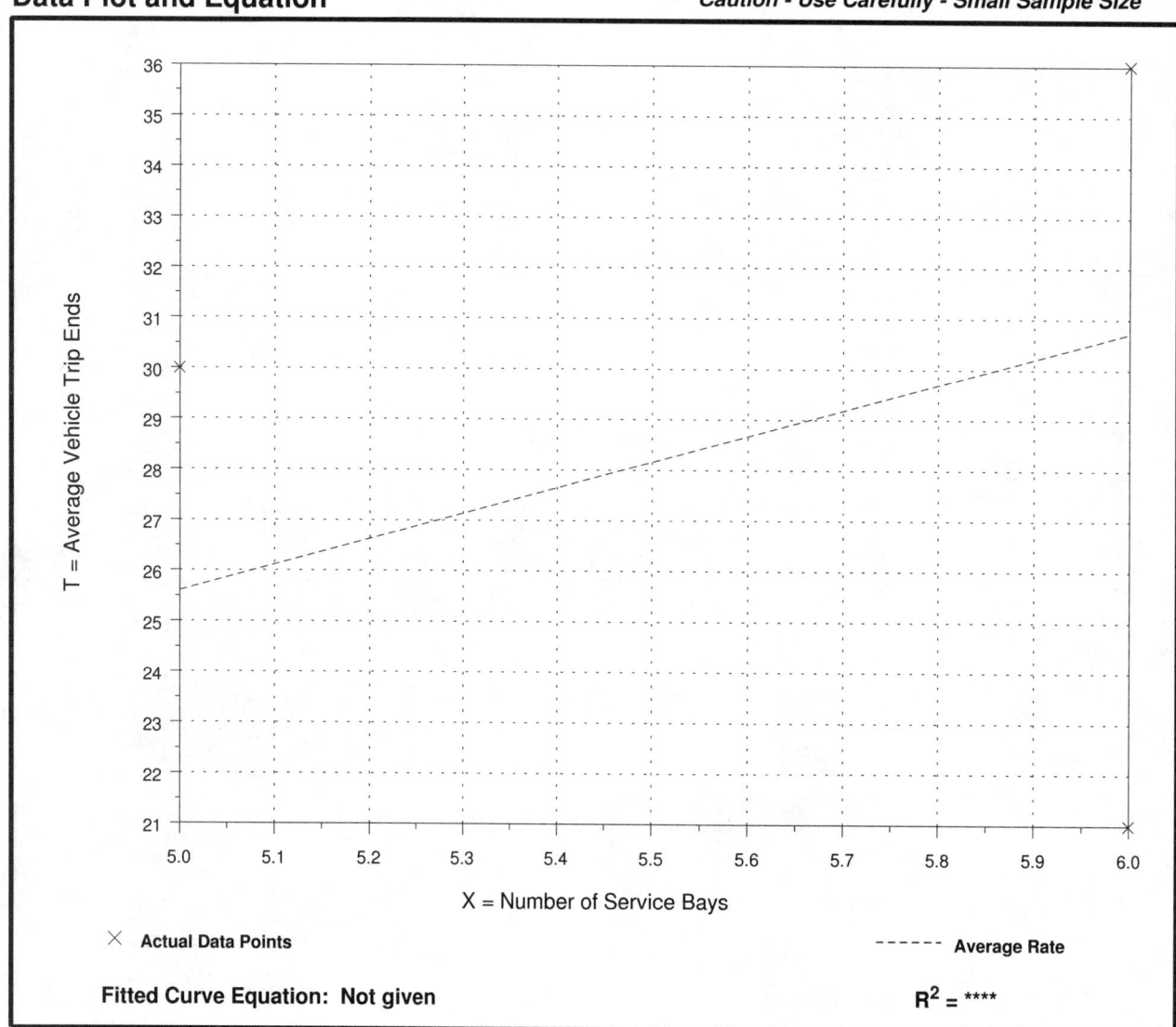

Caution - Use Carefully - Small Sample Size

×   Actual Data Points   - - - - - Average Rate

Fitted Curve Equation: Not given R^2 = ****

Tire Store
(848)

Average Vehicle Trip Ends vs: Service Bays
On a: Weekday,
P.M. Peak Hour of Generator

Number of Studies: 3
Average Num. of Service Bays: 6
Directional Distribution: 46% entering, 54% exiting

Trip Generation per Service Bay

Average Rate	Range of Rates	Standard Deviation
5.65	3.33 - 8.00	3.00

Data Plot and Equation

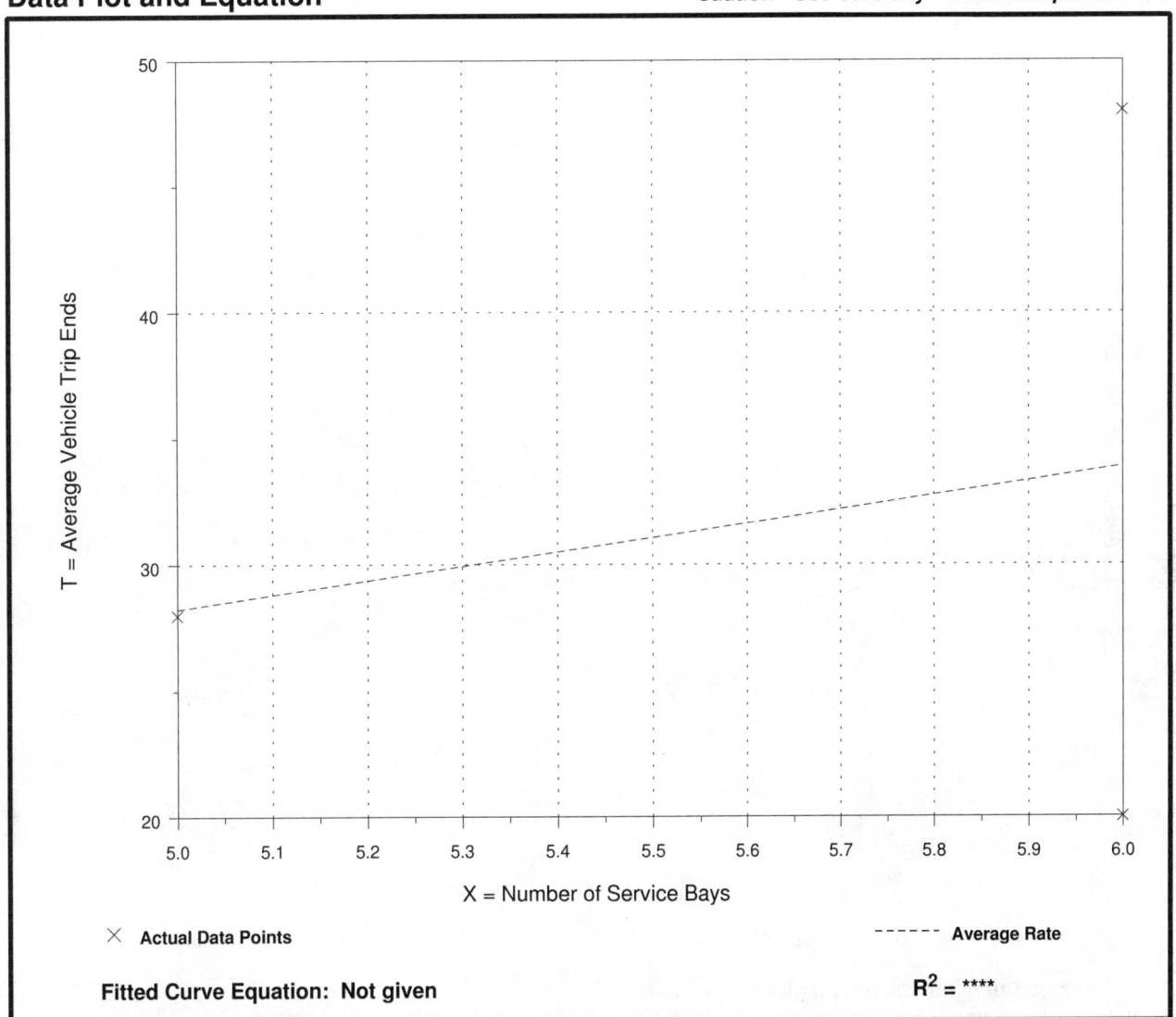

Caution - Use Carefully - Small Sample Size

X   Actual Data Points   ----- Average Rate

Fitted Curve Equation: Not given $R^2 = ****$

Tire Store
(848)

Average Vehicle Trip Ends vs: Service Bays
On a: Saturday,
Peak Hour of Generator

Number of Studies: 3
Average Num. of Service Bays: 10
Directional Distribution: 43% entering, 57% exiting

Trip Generation per Service Bay

Average Rate	Range of Rates	Standard Deviation
2.47	1.20 - 3.38	1.78

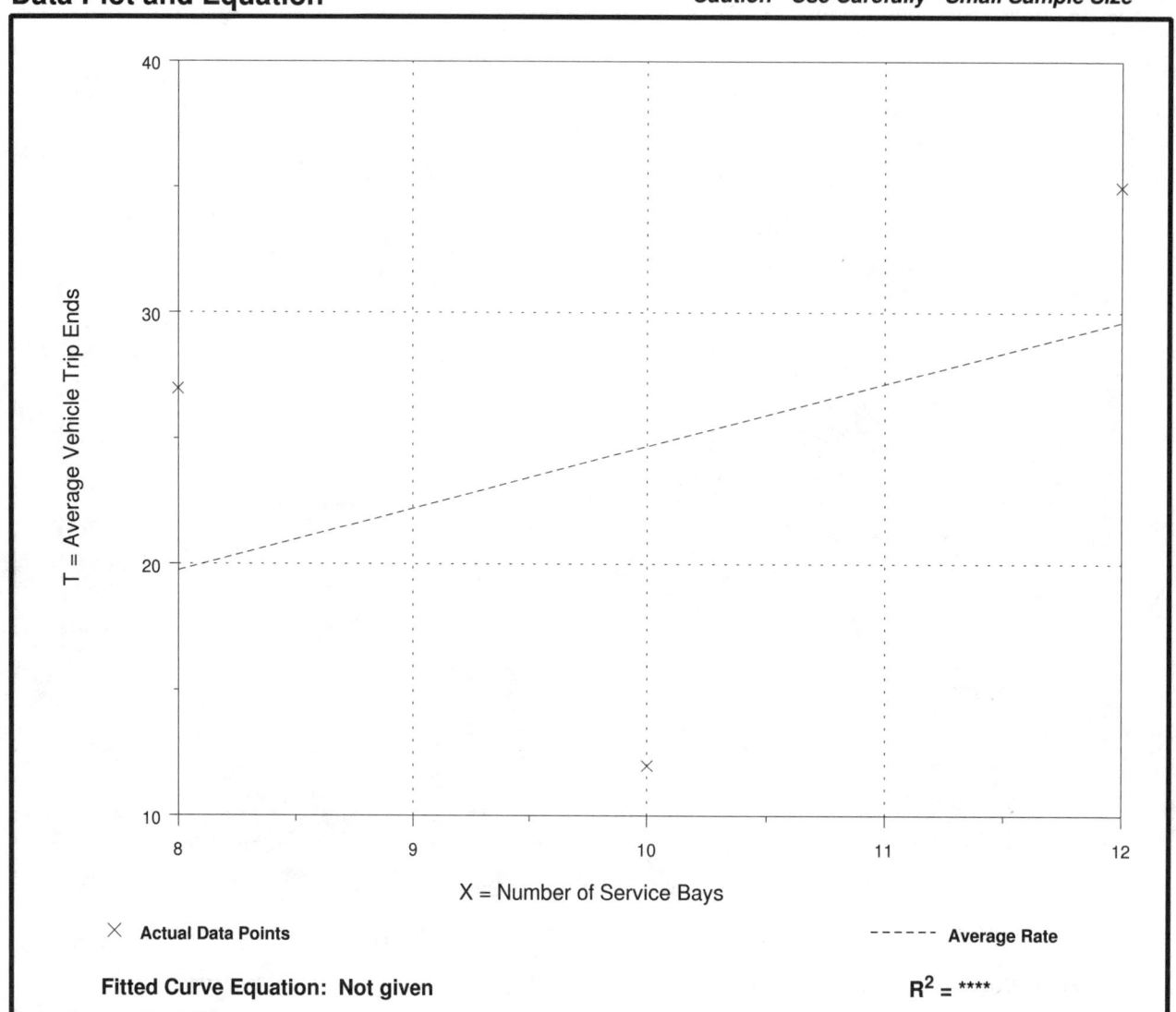

Data Plot and Equation

Caution - Use Carefully - Small Sample Size

Fitted Curve Equation: Not given $R^2 =$ ****

Tire Store
(848)

Average Vehicle Trip Ends vs: 1000 Sq. Feet Gross Floor Area
On a: Weekday

Number of Studies: 5
Average 1000 Sq. Feet GFA: 5
Directional Distribution: 50% entering, 50% exiting

Trip Generation per 1000 Sq. Feet Gross Floor Area

Average Rate	Range of Rates	Standard Deviation
24.87	19.40 - 36.02	7.41

Data Plot and Equation

Caution - Use Carefully - Small Sample Size

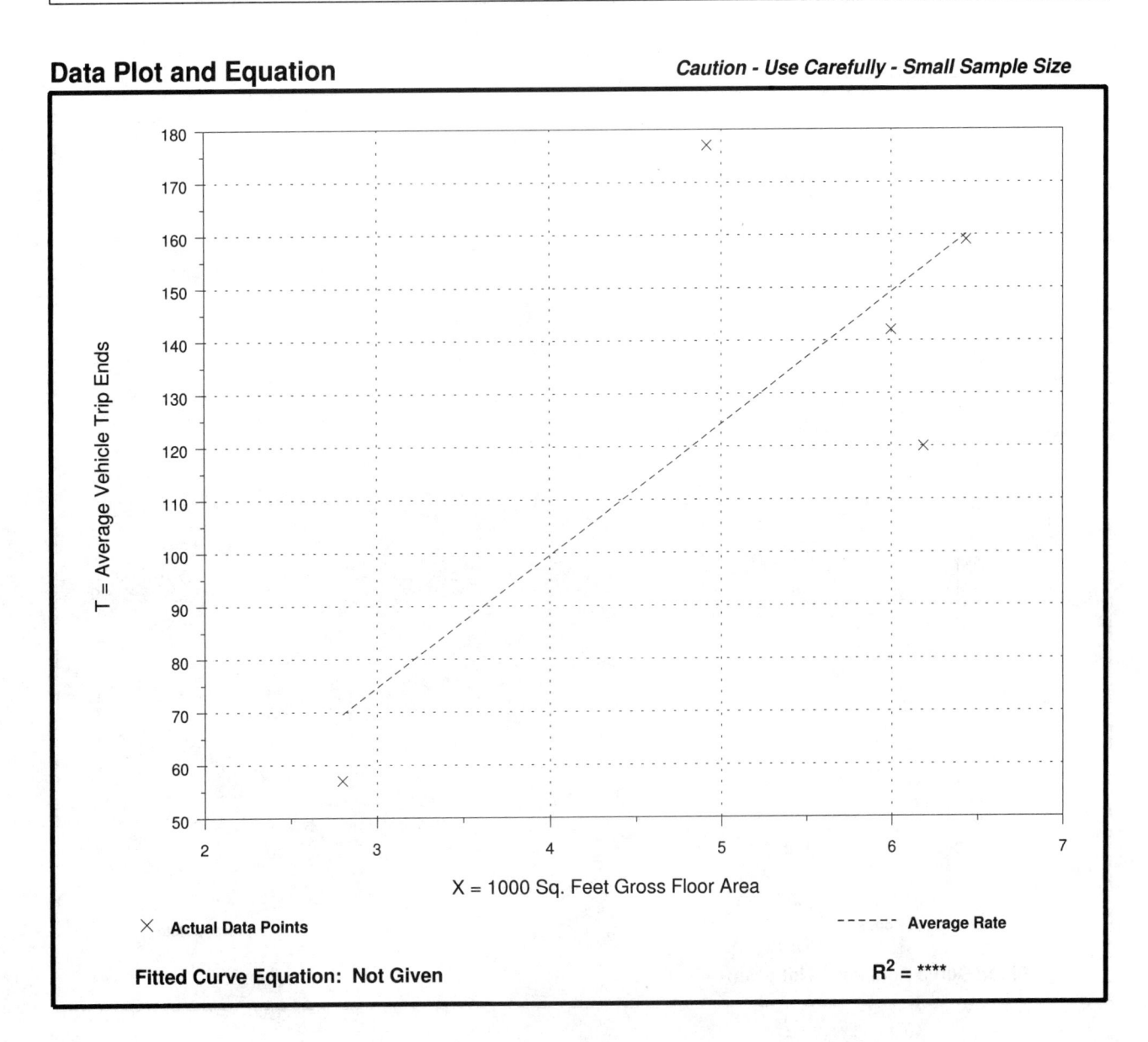

Fitted Curve Equation: Not Given $R^2 = ****$

Tire Store
(848)

Average Vehicle Trip Ends vs: 1000 Sq. Feet Gross Floor Area
On a: Weekday,
Peak Hour of Adjacent Street Traffic,
One Hour Between 7 and 9 a.m.

Number of Studies: 13
Average 1000 Sq. Feet GFA: 5
Directional Distribution: 63% entering, 37% exiting

Trip Generation per 1000 Sq. Feet Gross Floor Area

Average Rate	Range of Rates	Standard Deviation
2.89	0.97 - 6.98	2.10

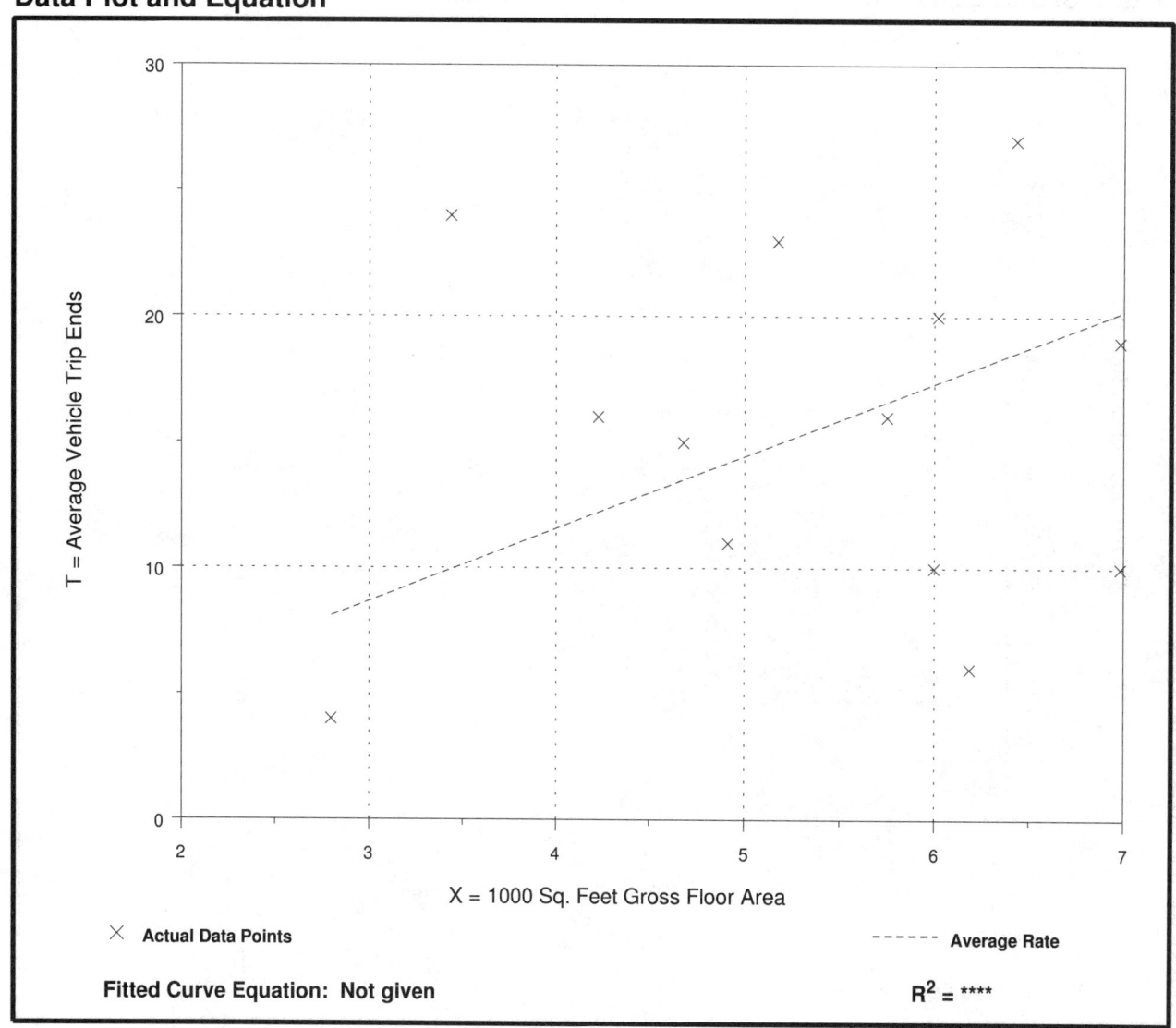

Data Plot and Equation

Fitted Curve Equation: Not given $R^2 = ****$

Tire Store
(848)

Average Vehicle Trip Ends vs: 1000 Sq. Feet Gross Floor Area
On a: Weekday,
Peak Hour of Adjacent Street Traffic,
One Hour Between 4 and 6 p.m.

Number of Studies: 16
Average 1000 Sq. Feet GFA: 6
Directional Distribution: 43% entering, 57% exiting

Trip Generation per 1000 Sq. Feet Gross Floor Area

Average Rate	Range of Rates	Standard Deviation
4.15	1.46 - 8.14	2.55

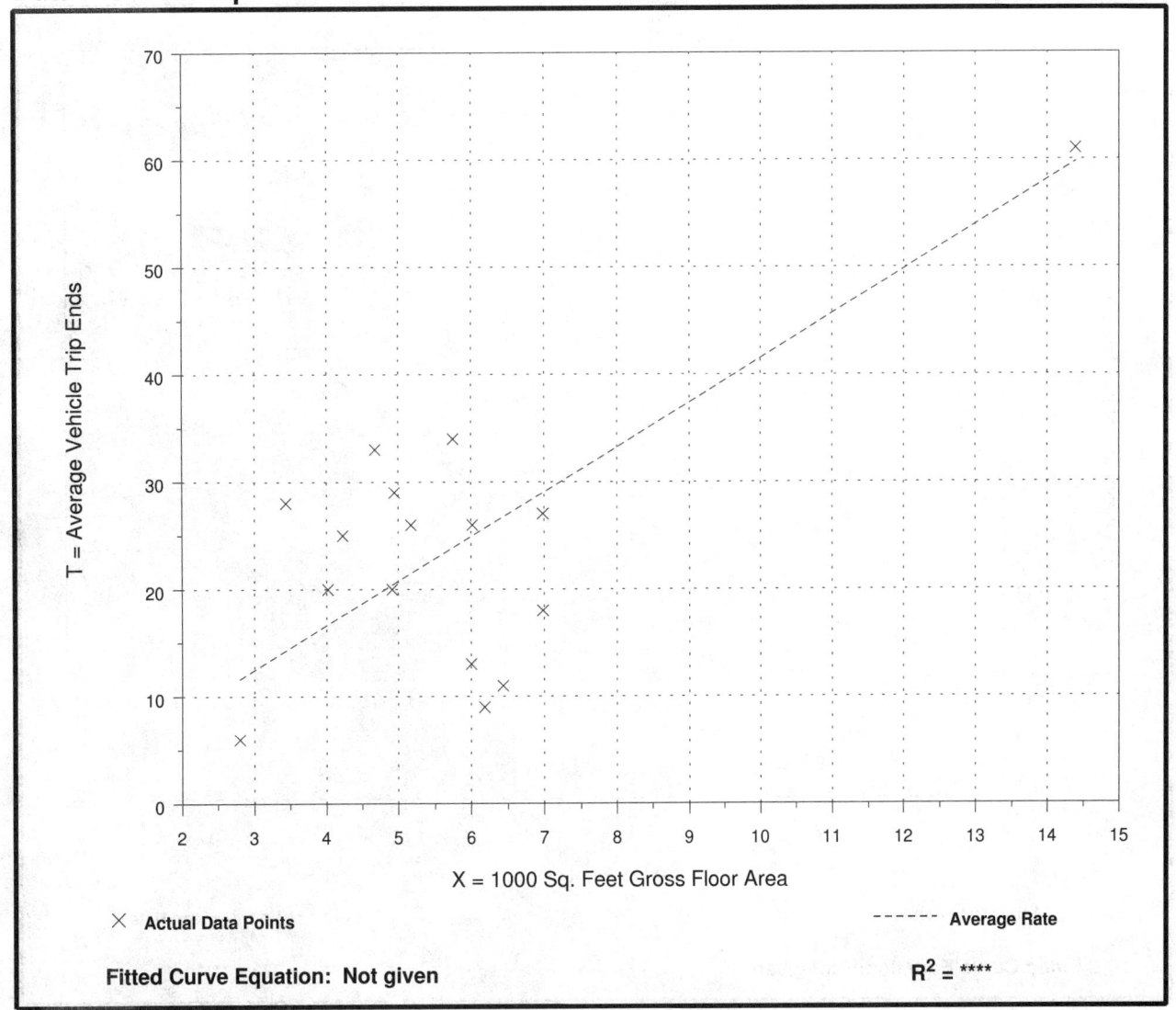

Data Plot and Equation

Fitted Curve Equation: Not given $R^2 = ****$

Tire Store
(848)

Average Vehicle Trip Ends vs: 1000 Sq. Feet Gross Floor Area
On a: Weekday,
A.M. Peak Hour of Generator

Number of Studies: 7
Average 1000 Sq. Feet GFA: 5
Directional Distribution: 50% entering, 50% exiting

Trip Generation per 1000 Sq. Feet Gross Floor Area

Average Rate	Range of Rates	Standard Deviation
3.45	1.62 - 8.72	2.55

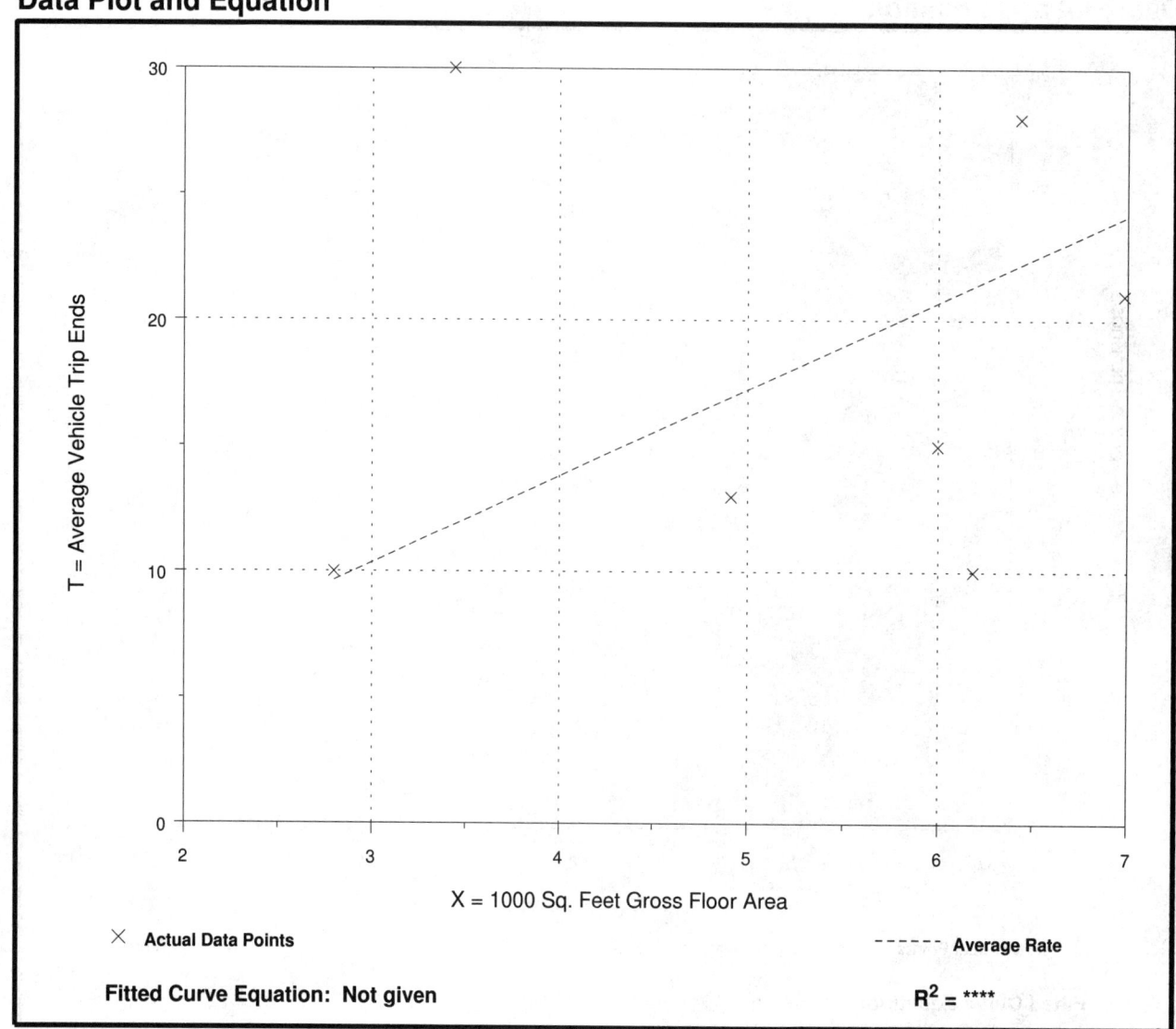

Data Plot and Equation

Fitted Curve Equation: Not given R^2 = ****

Tire Store
(848)

Average Vehicle Trip Ends vs: 1000 Sq. Feet Gross Floor Area
On a: Weekday,
P.M. Peak Hour of Generator

Number of Studies: 7
Average 1000 Sq. Feet GFA: 5
Directional Distribution: 42% entering, 58% exiting

Trip Generation per 1000 Sq. Feet Gross Floor Area

Average Rate	Range of Rates	Standard Deviation
3.26	1.62 - 8.14	2.41

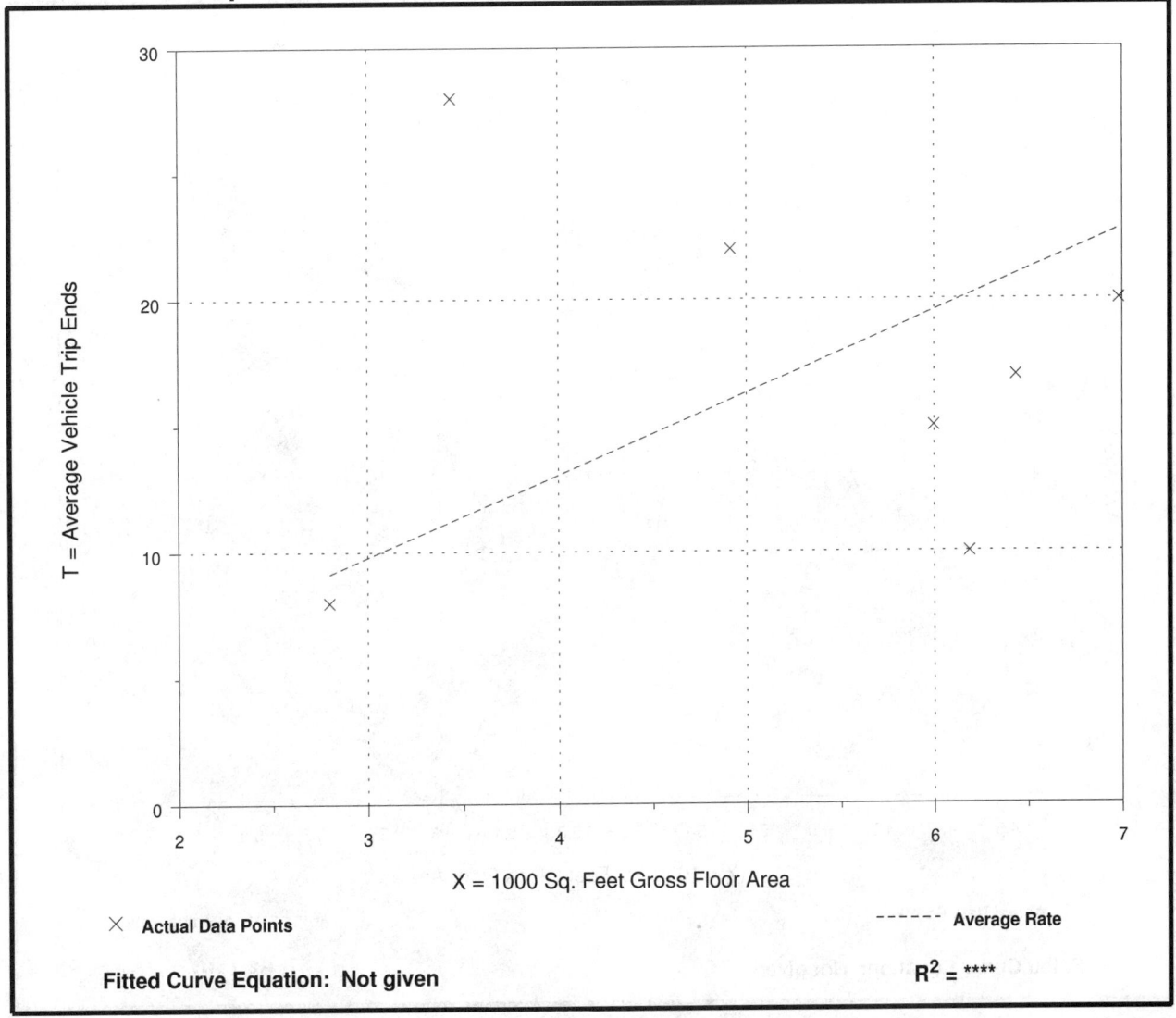

Data Plot and Equation

Fitted Curve Equation: Not given $R^2 = ****$

Tire Store
(848)

Average Vehicle Trip Ends vs: 1000 Sq. Feet Gross Floor Area
On a: Saturday,
Peak Hour of Generator

Number of Studies: 3
Average 1000 Sq. Feet GFA: 5
Directional Distribution: 47% entering, 53% exiting

Trip Generation per 1000 Sq. Feet Gross Floor Area

Average Rate	Range of Rates	Standard Deviation
5.05	3.49 - 6.09	2.43

Data Plot and Equation

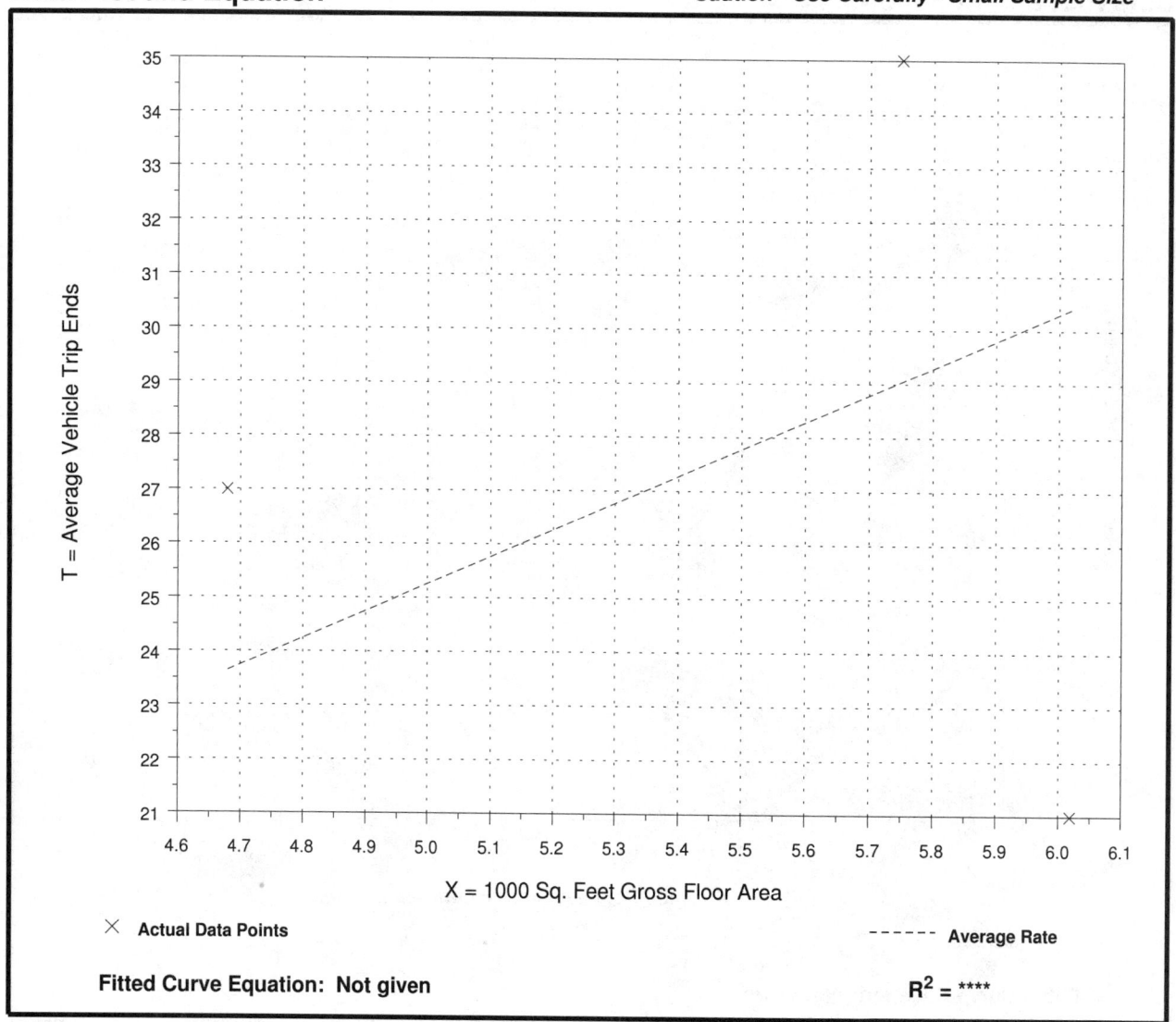

Caution - Use Carefully - Small Sample Size

Fitted Curve Equation: Not given $R^2 = ****$

Tire Store
(848)

Average Vehicle Trip Ends vs: Employees
On a: Weekday,
Peak Hour of Adjacent Street Traffic,
One Hour Between 7 and 9 a.m.

Number of Studies: 7
Avg. Number of Employees: 5
Directional Distribution: 65% entering, 35% exiting

Trip Generation per Employee

Average Rate	Range of Rates	Standard Deviation
3.24	1.88 - 6.00	2.10

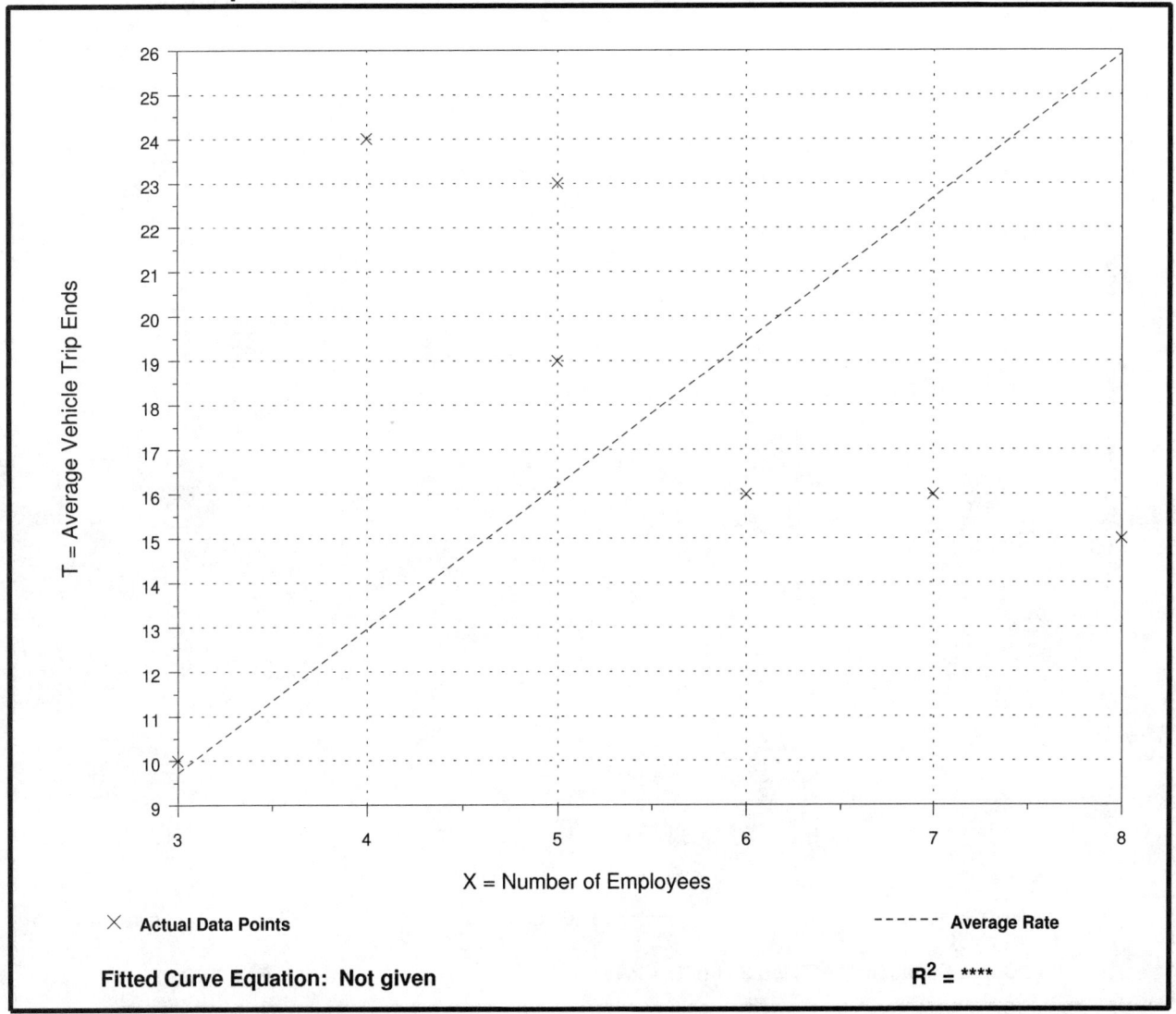

Data Plot and Equation

Fitted Curve Equation: Not given R^2 = ****

Tire Store
(848)

Average Vehicle Trip Ends vs: Employees
On a: Weekday,
Peak Hour of Adjacent Street Traffic,
One Hour Between 4 and 6 p.m.

Number of Studies: 7
Avg. Number of Employees: 5
Directional Distribution: 42% entering, 58% exiting

Trip Generation per Employee

Average Rate	Range of Rates	Standard Deviation
5.03	4.13 - 7.00	2.23

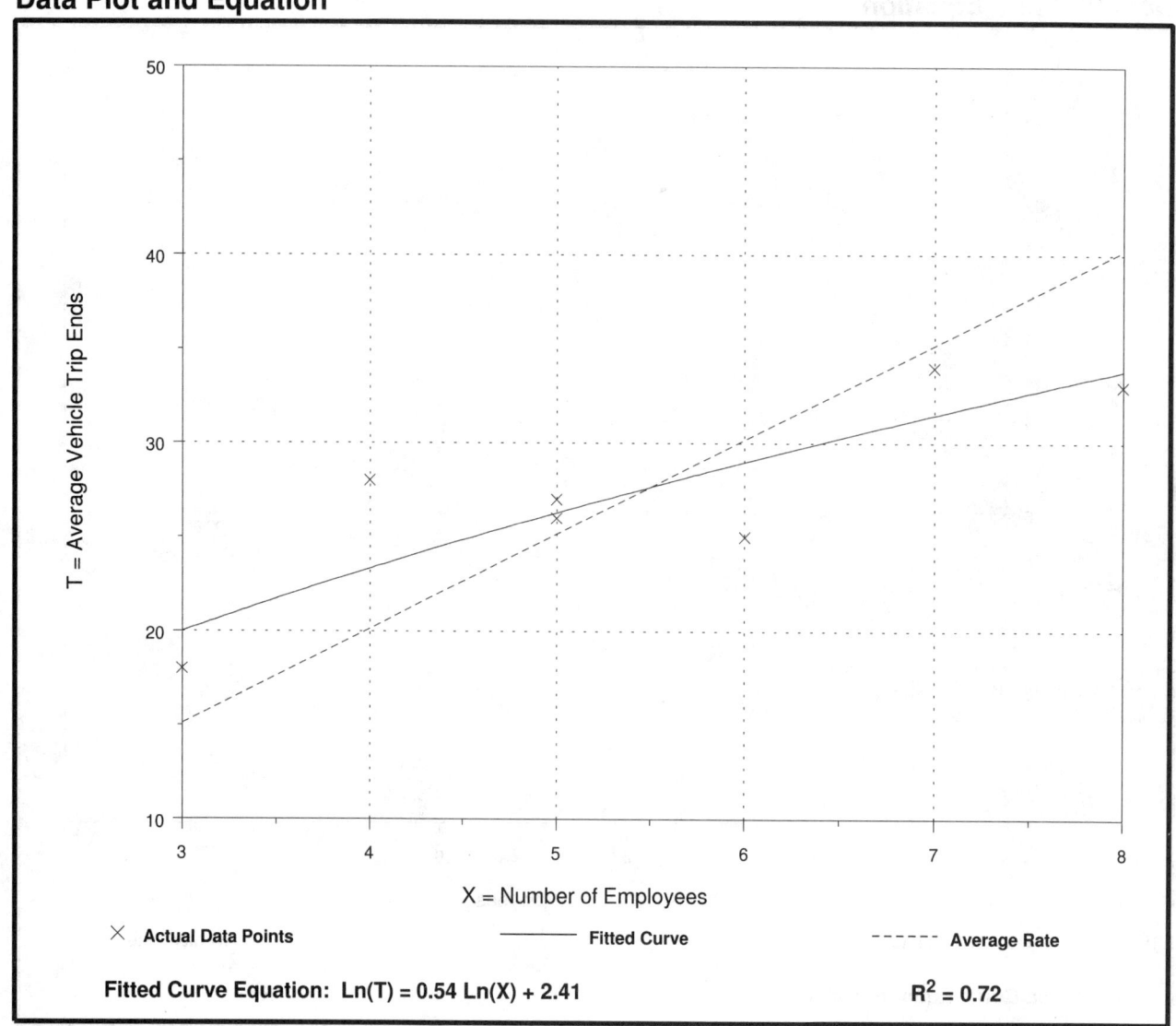

Data Plot and Equation

Fitted Curve Equation: $Ln(T) = 0.54\ Ln(X) + 2.41$ $R^2 = 0.72$

Tire Store
(848)

Average Vehicle Trip Ends vs: Employees
On a: Weekday,
A.M. Peak Hour of Generator

Number of Studies: 2
Avg. Number of Employees: 4
Directional Distribution: 47% entering, 53% exiting

Trip Generation per Employee

Average Rate	Range of Rates	Standard Deviation
7.29	7.00 - 7.50	*

Data Plot and Equation

Caution - Use Carefully - Small Sample Size

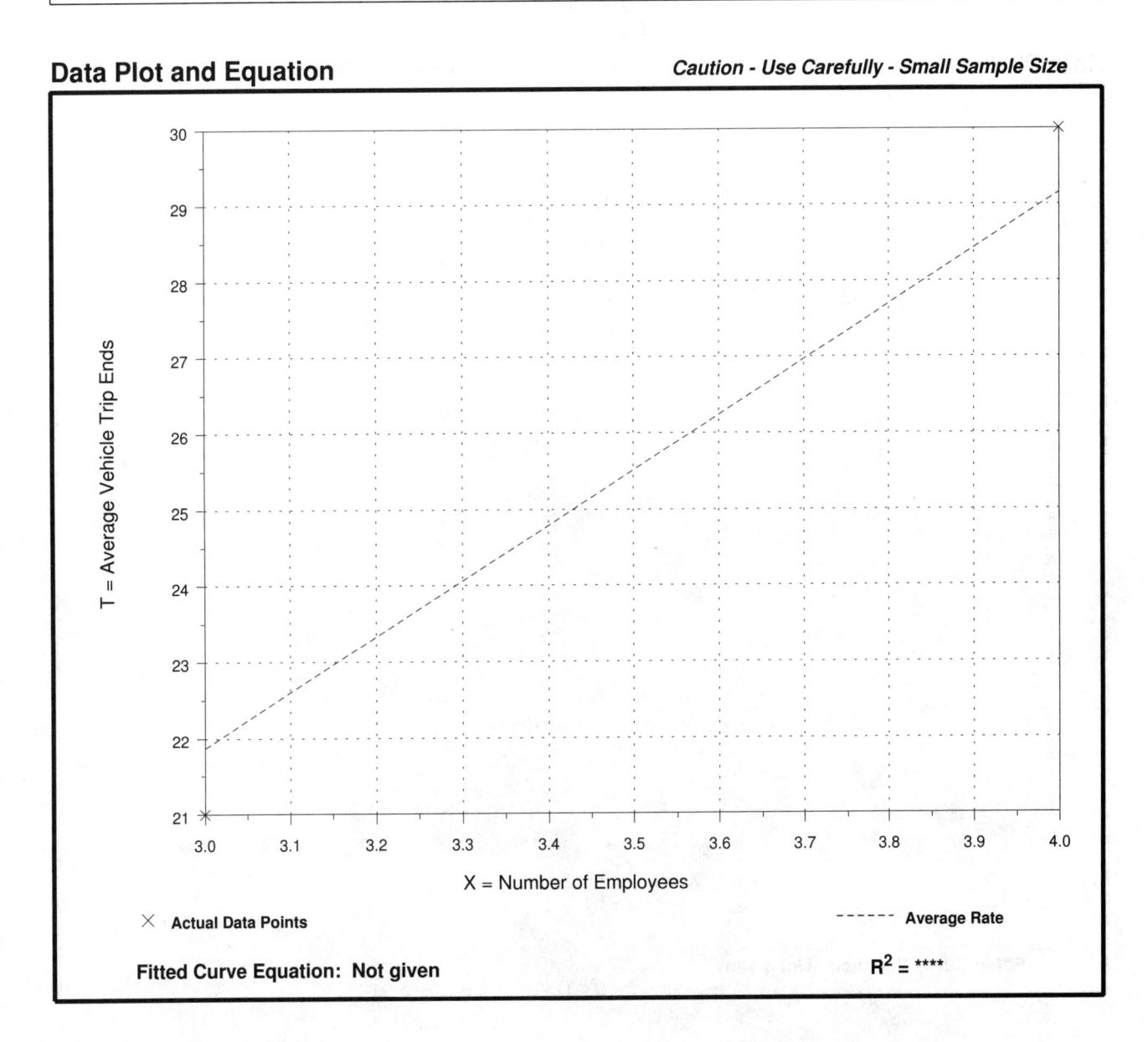

Fitted Curve Equation: Not given

$R^2 = $ ****

Tire Store
(848)

Average Vehicle Trip Ends vs: Employees
On a: Weekday,
P.M. Peak Hour of Generator

Number of Studies: 2
Avg. Number of Employees: 4
Directional Distribution: 38% entering, 62% exiting

Trip Generation per Employee

Average Rate	Range of Rates	Standard Deviation
6.86	6.67 - 7.00	*

Data Plot and Equation

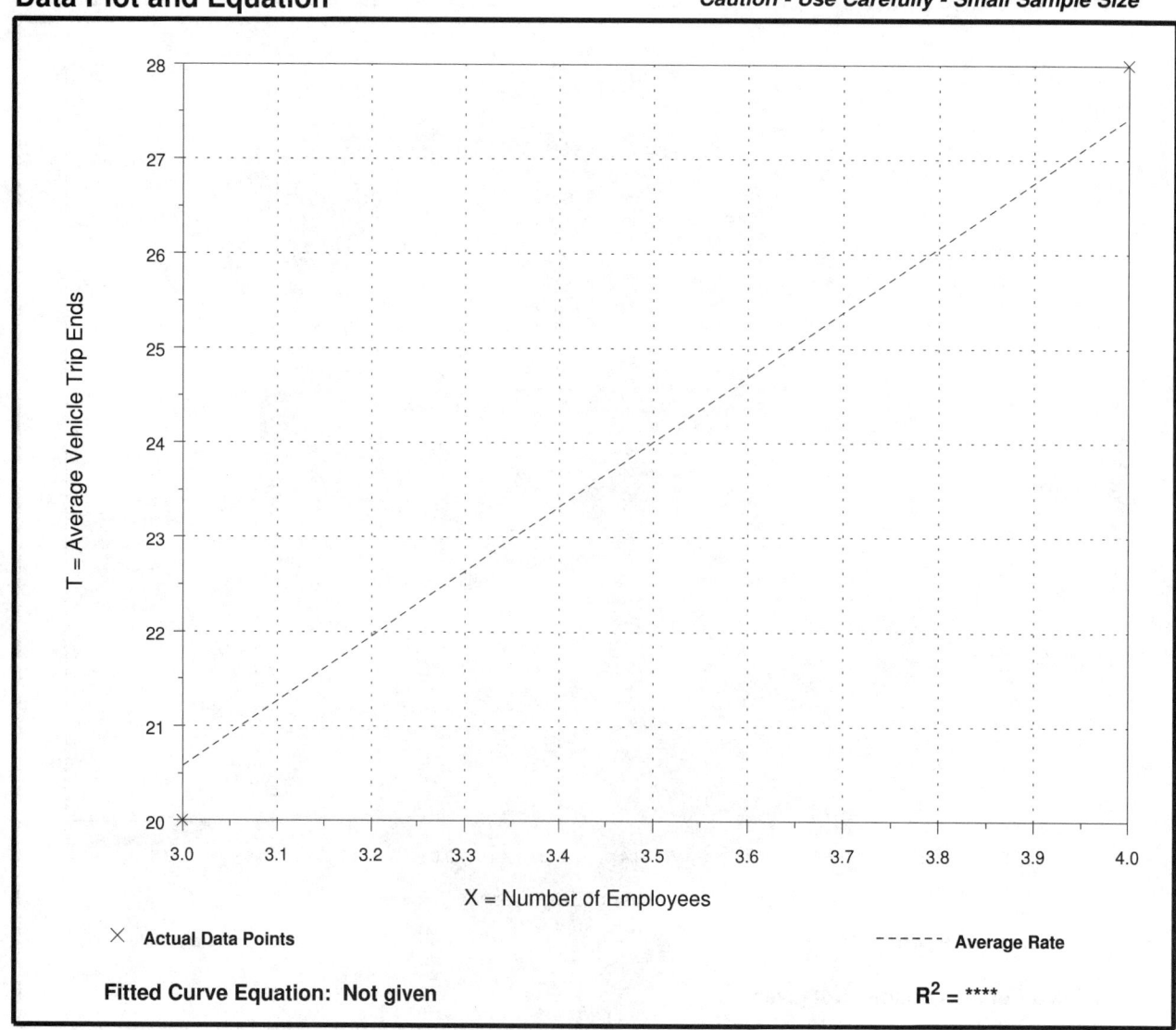

Caution - Use Carefully - Small Sample Size

Fitted Curve Equation: Not given $R^2 = ****$

Tire Store
(848)

Average Vehicle Trip Ends vs: Employees
On a: Saturday,
Peak Hour of Generator

Number of Studies: 2
Avg. Number of Employees: 8
Directional Distribution: 44% entering, 56% exiting

Trip Generation per Employee

Average Rate	Range of Rates	Standard Deviation
4.13	3.38 - 5.00	*

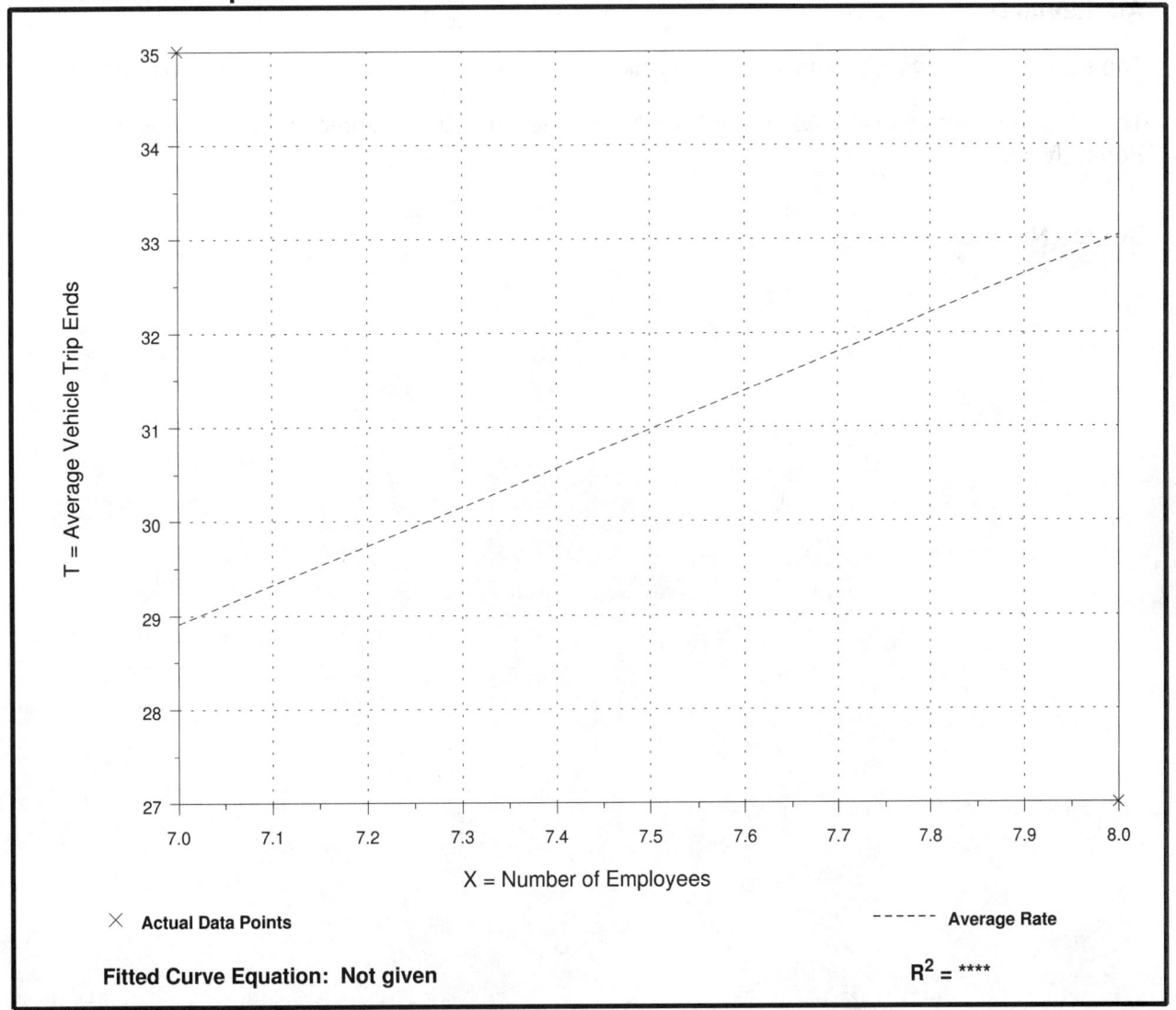

Data Plot and Equation

Caution - Use Carefully - Small Sample Size

Fitted Curve Equation: Not given $R^2 = ****$

Land Use: 849
Tire Superstore

Description

Tire superstores are warehouse-like facilities with the primary function of selling and installing tires for automobiles and small trucks. Other services provided may include automotive maintenance functions, such as wheel alignment or shock and brake service, and customer services. A tire display, customer waiting lounge, restroom facilities, staff office space and significant storage area are also provided. General mechanical repairs and bodywork are usually not conducted at these facilities. Automobile parts sales (Land Use 843), tire store (Land Use 848) and automobile parts and service center (Land Use 943) are related uses.

Additional Data

The superstores surveyed for this land use typically provided storage for approximately 5,000 tires.

The sites were surveyed in 1992 and 1993 in Massachusetts, New Hampshire, New Jersey and Pennsylvania.

Source Number

416

Tire Superstore
(849)

Average Vehicle Trip Ends vs: Service Bays
On a: Weekday

Number of Studies: 12
Average Num. of Service Bays: 9
Directional Distribution: 50% entering, 50% exiting

Trip Generation per Service Bay

Average Rate	Range of Rates	Standard Deviation
30.55	20.62 - 40.00	8.03

Data Plot and Equation

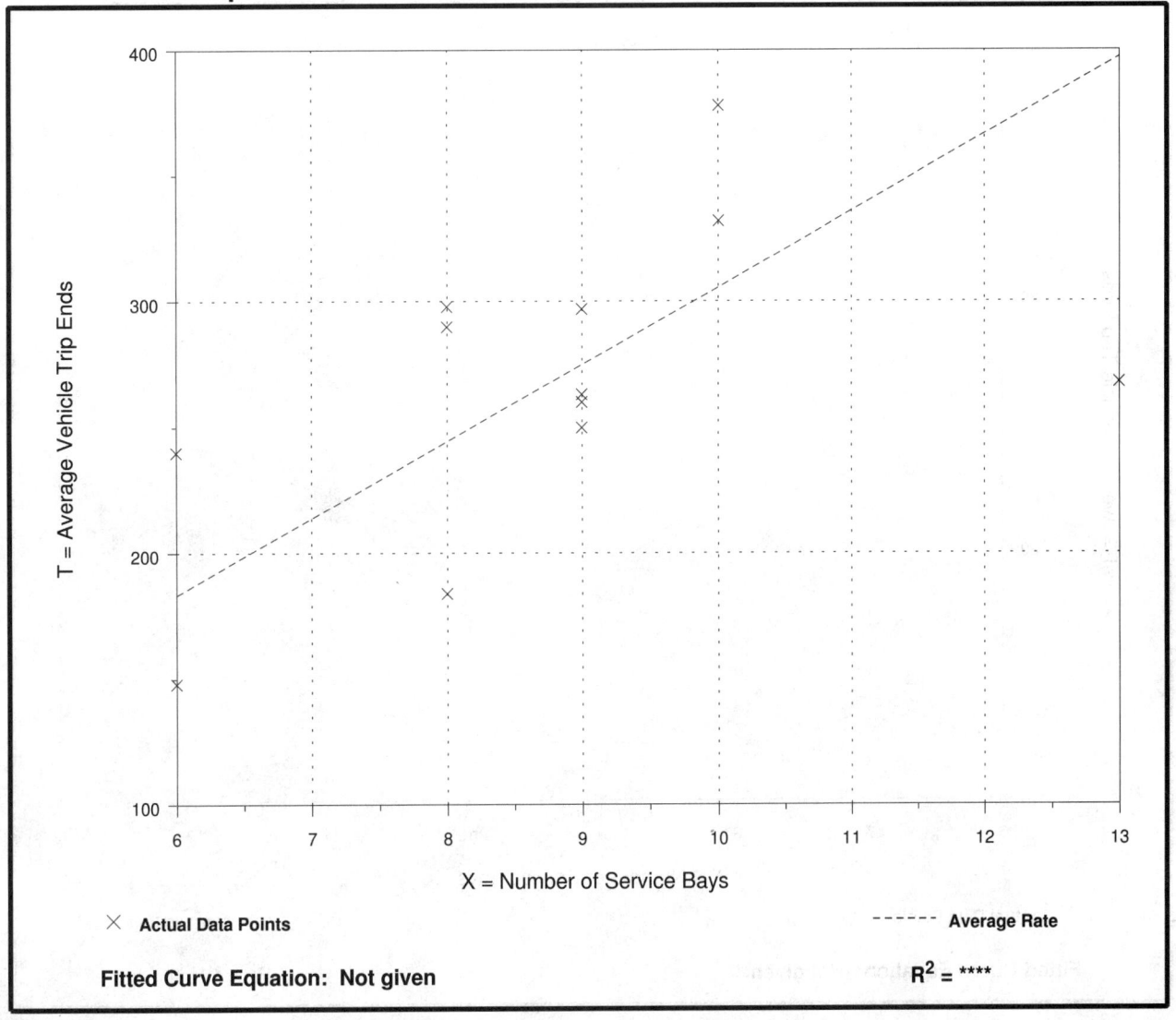

Fitted Curve Equation: Not given $R^2 = ****$

Tire Superstore
(849)

Average Vehicle Trip Ends vs: Service Bays
On a: Weekday,
Peak Hour of Adjacent Street Traffic,
One Hour Between 7 and 9 a.m.

Number of Studies: 11
Average Num. of Service Bays: 9
Directional Distribution: 65% entering, 35% exiting

Trip Generation per Service Bay

Average Rate	Range of Rates	Standard Deviation
2.01	1.11 - 4.00	1.58

Data Plot and Equation

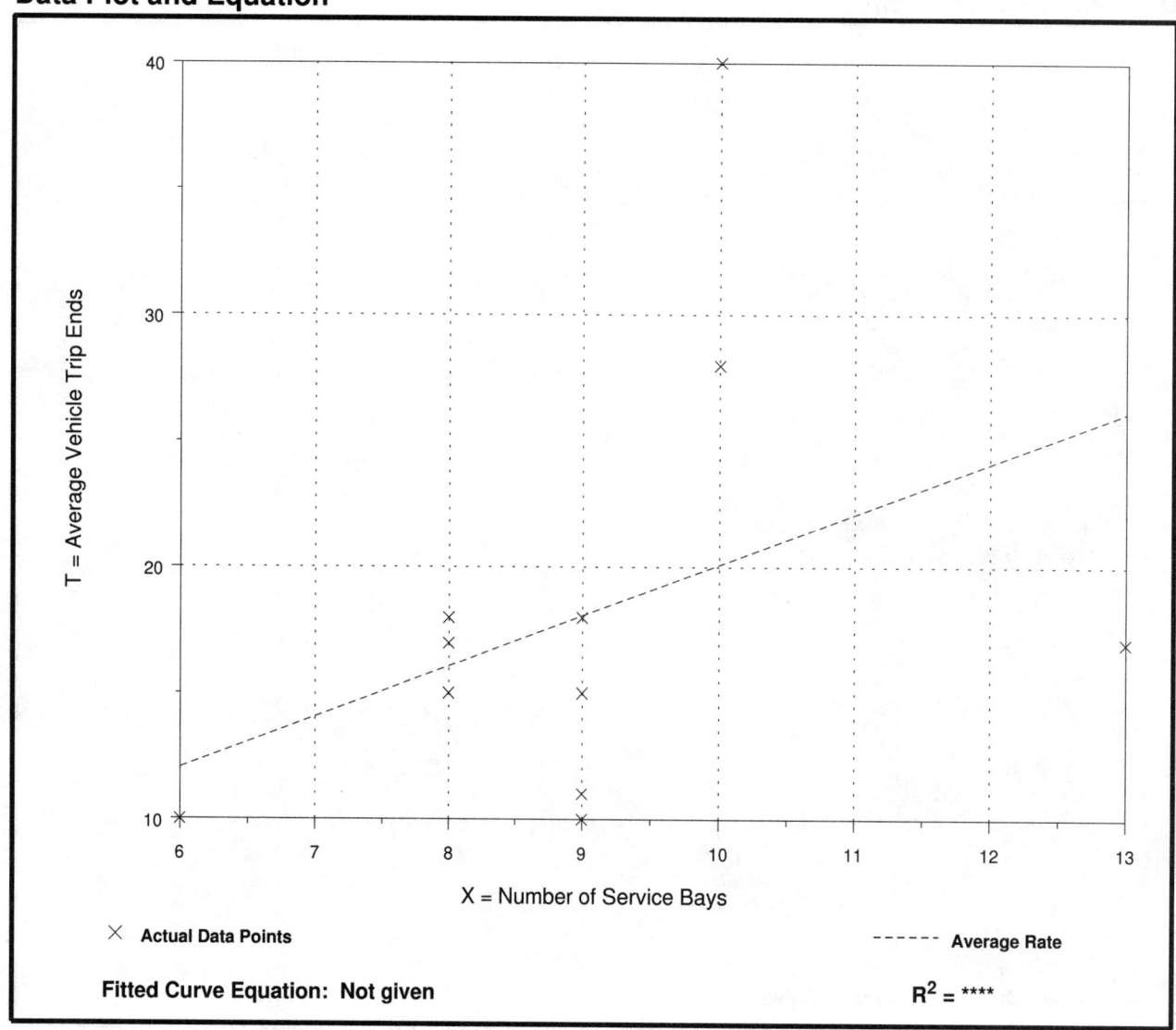

Fitted Curve Equation: Not given $R^2 = ****$

Tire Superstore
(849)

Average Vehicle Trip Ends vs: Service Bays
On a: Weekday,
Peak Hour of Adjacent Street Traffic,
One Hour Between 4 and 6 p.m.

Number of Studies: 12
Average Num. of Service Bays: 9
Directional Distribution: 47% entering, 53% exiting

Trip Generation per Service Bay

Average Rate	Range of Rates	Standard Deviation
3.17	2.25 - 4.67	1.85

Data Plot and Equation

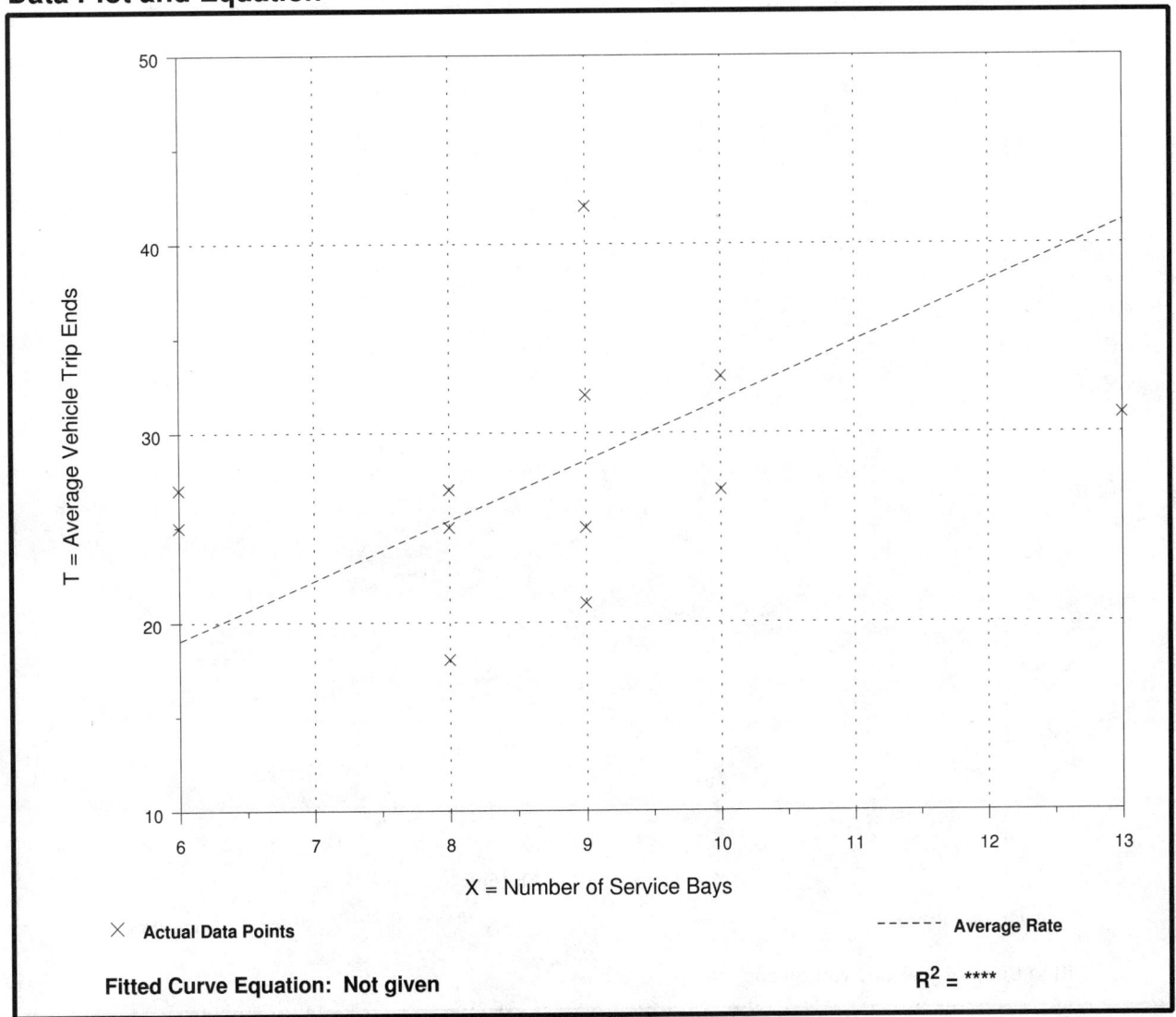

[Plot: T = Average Vehicle Trip Ends vs. X = Number of Service Bays, with Actual Data Points and Average Rate line]

Fitted Curve Equation: Not given $R^2 = ****$

Tire Superstore
(849)

Average Vehicle Trip Ends vs: Service Bays
On a: Weekday,
A.M. Peak Hour of Generator

Number of Studies: 12
Average Num. of Service Bays: 9
Directional Distribution: 56% entering, 44% exiting

Trip Generation per Service Bay

Average Rate	Range of Rates	Standard Deviation
3.63	2.23 - 5.00	2.03

Data Plot and Equation

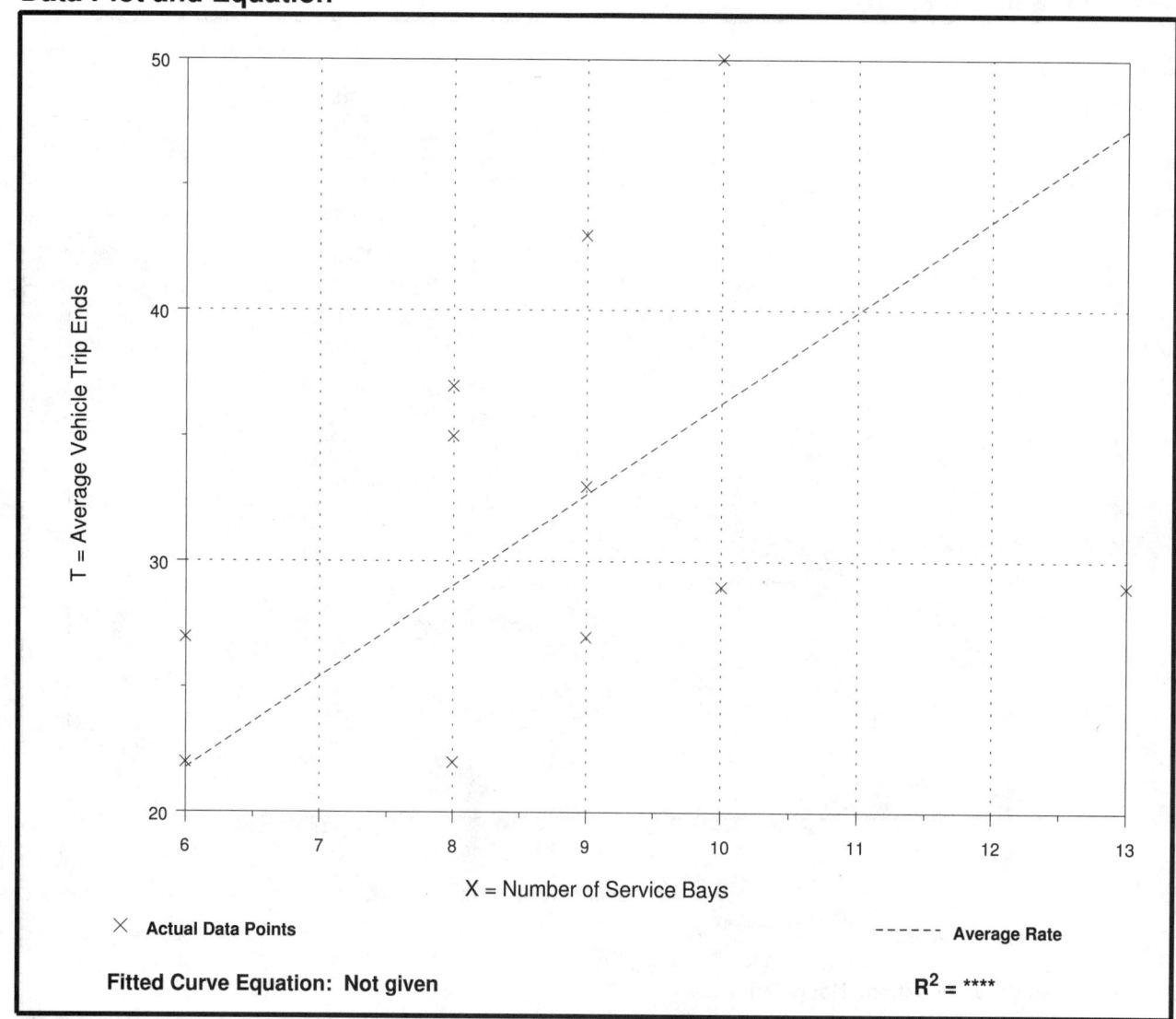

Fitted Curve Equation: Not given $R^2 = ****$

Tire Superstore
(849)

Average Vehicle Trip Ends vs: Service Bays
On a: Weekday,
P.M. Peak Hour of Generator

Number of Studies: 12
Average Num. of Service Bays: 9
Directional Distribution: 49% entering, 51% exiting

Trip Generation per Service Bay

Average Rate	Range of Rates	Standard Deviation
3.87	2.38 - 6.17	2.12

Data Plot and Equation

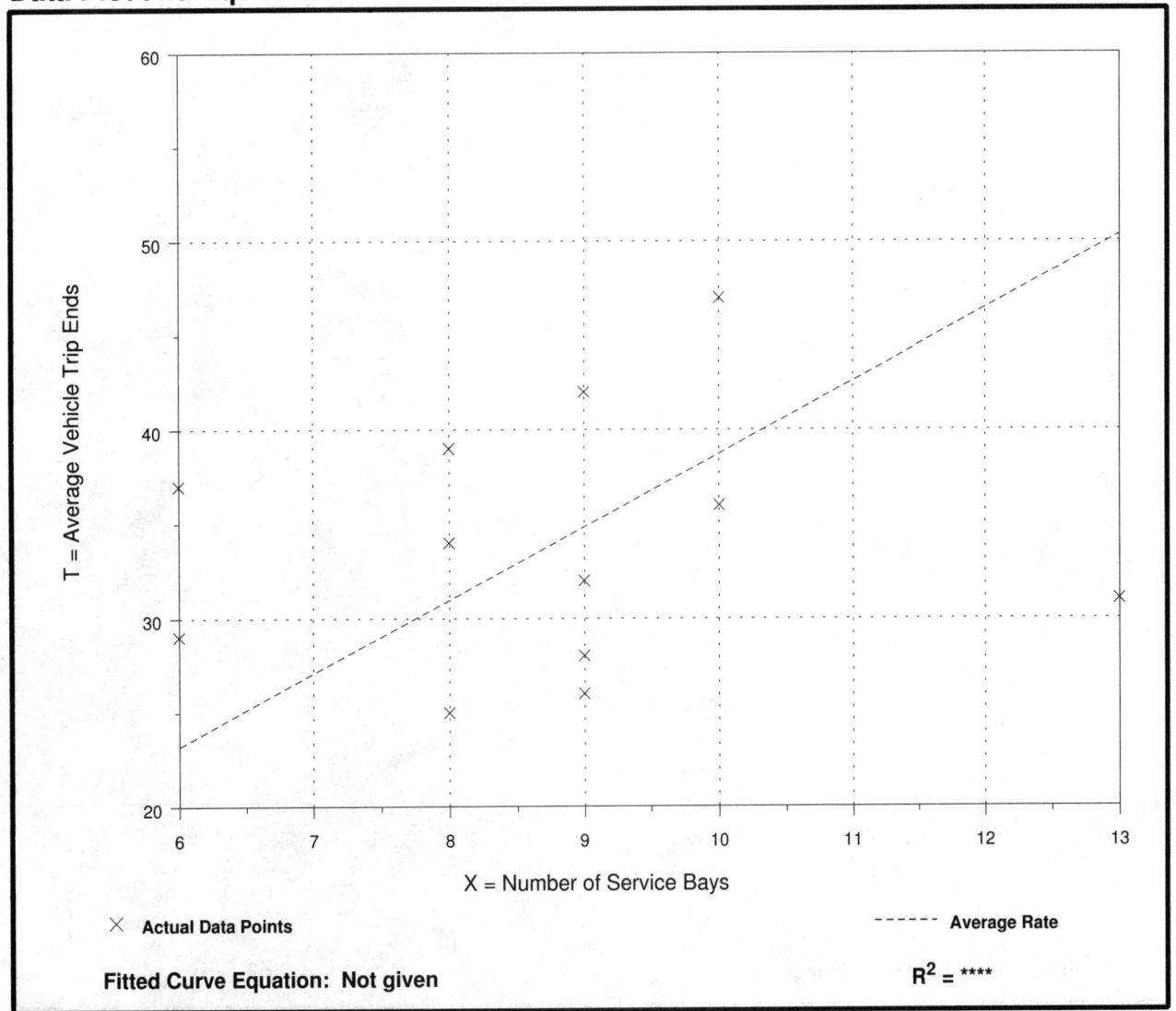

X   Actual Data Points   ----- Average Rate

Fitted Curve Equation: Not given $R^2 = $ ****

Tire Superstore
(849)

Average Vehicle Trip Ends vs: Service Bays
On a: Saturday

Number of Studies: 12
Average Num. of Service Bays: 9
Directional Distribution: 50% entering, 50% exiting

Trip Generation per Service Bay

Average Rate	Range of Rates	Standard Deviation
28.55	14.50 - 44.11	10.61

Data Plot and Equation

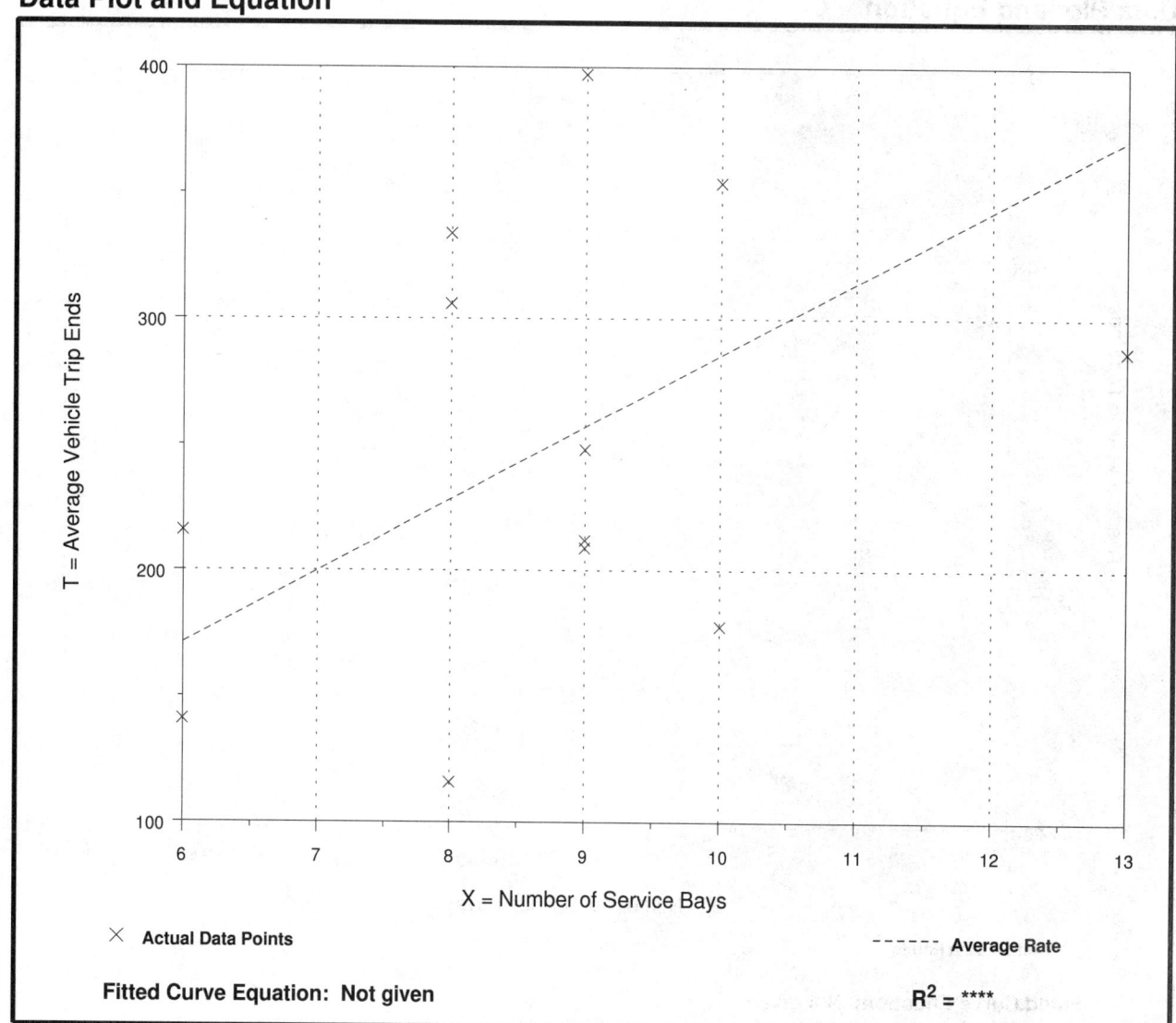

Fitted Curve Equation: Not given $R^2 = ****$

Tire Superstore
(849)

Average Vehicle Trip Ends vs: Service Bays
On a: Saturday,
Peak Hour of Generator

Number of Studies: 11
Average Num. of Service Bays: 9
Directional Distribution: 47% entering, 53% exiting

Trip Generation per Service Bay

Average Rate	Range of Rates	Standard Deviation
4.76	3.69 - 6.50	2.26

Data Plot and Equation

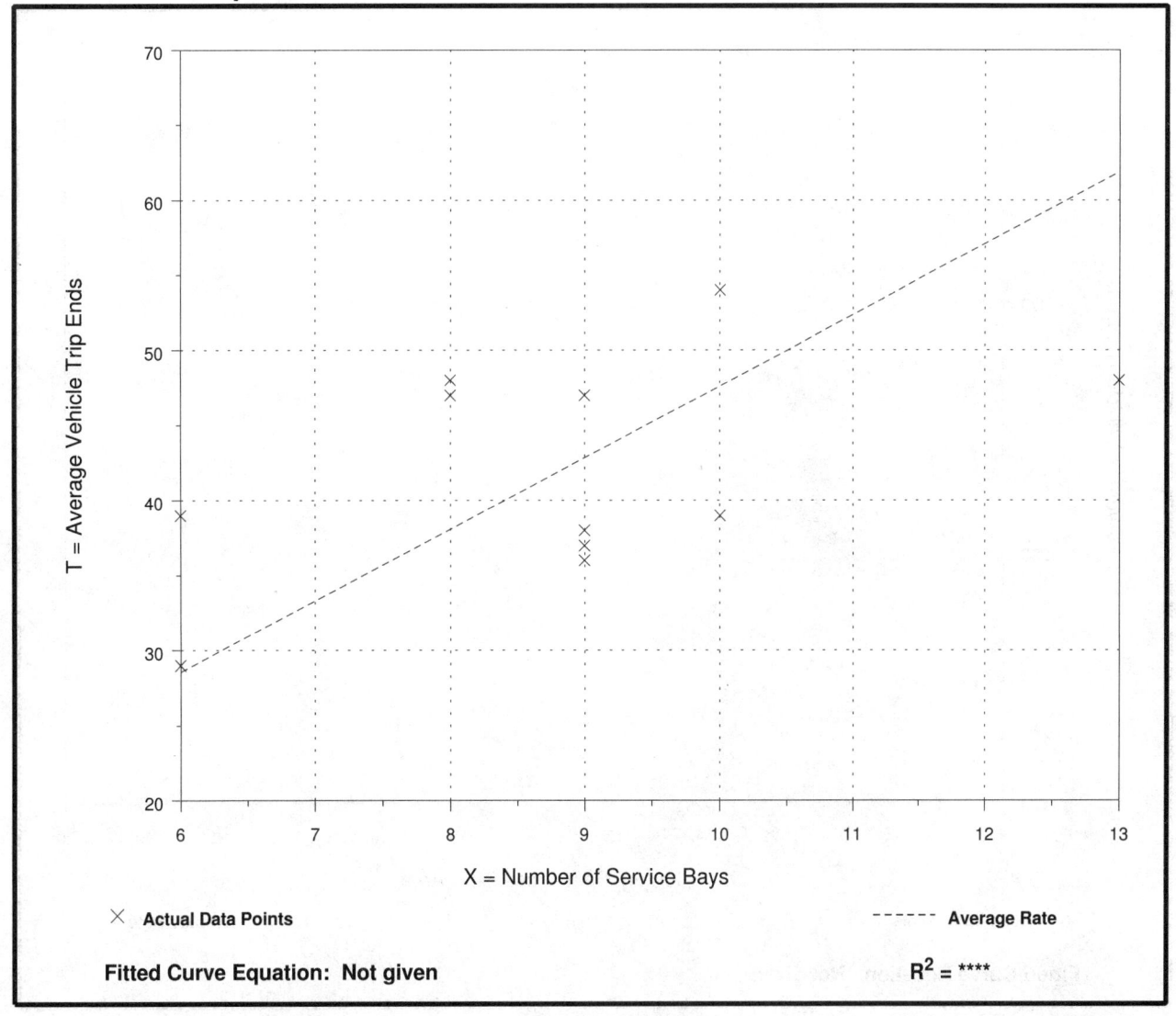

Fitted Curve Equation: Not given $R^2 = ****$

Tire Superstore
(849)

Average Vehicle Trip Ends vs: 1000 Sq. Feet Gross Floor Area
On a: Weekday

Number of Studies: 12
Average 1000 Sq. Feet GFA: 13
Directional Distribution: 50% entering, 50% exiting

Trip Generation per 1000 Sq. Feet Gross Floor Area

Average Rate	Range of Rates	Standard Deviation
20.36	14.06 - 27.35	6.10

Data Plot and Equation

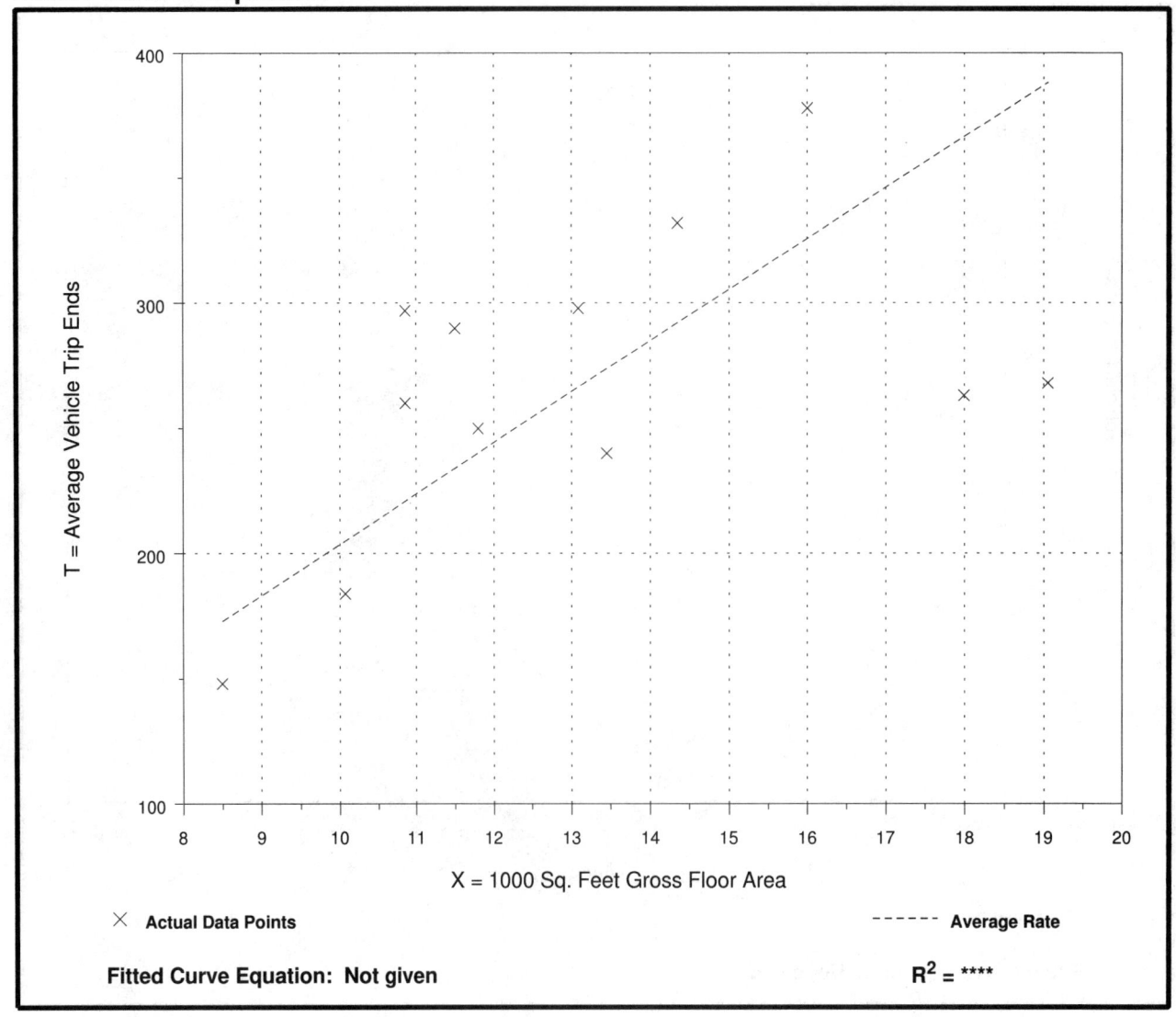

Fitted Curve Equation: Not given R^2 = ****

Tire Superstore
(849)

Average Vehicle Trip Ends vs: 1000 Sq. Feet Gross Floor Area
On a: Weekday,
Peak Hour of Adjacent Street Traffic,
One Hour Between 7 and 9 a.m.

Number of Studies: 11
Average 1000 Sq. Feet GFA: 14
Directional Distribution: 65% entering, 35% exiting

Trip Generation per 1000 Sq. Feet Gross Floor Area

Average Rate	Range of Rates	Standard Deviation
1.34	0.74 - 2.50	1.24

Data Plot and Equation

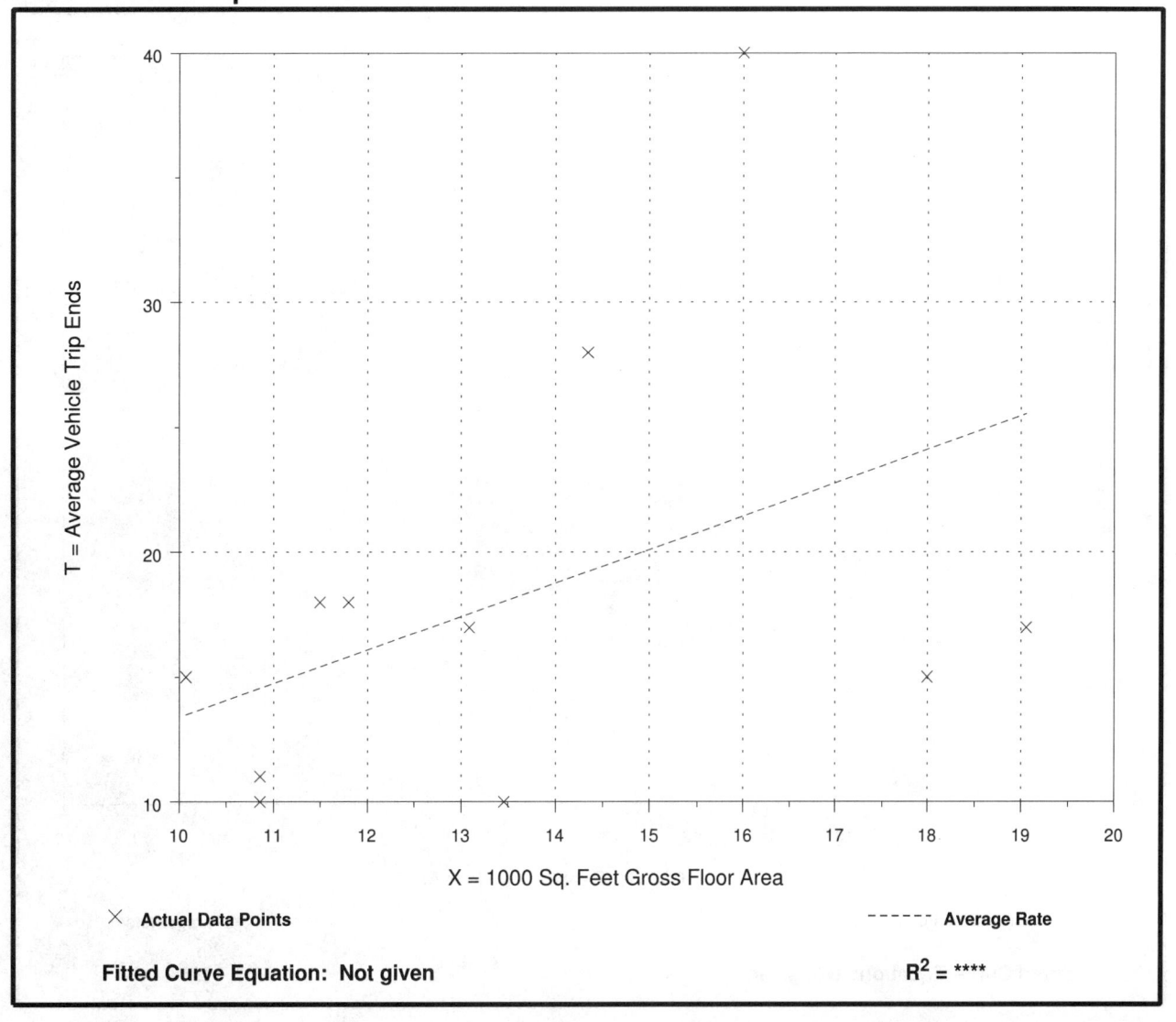

Fitted Curve Equation: Not given $R^2 = ****$

Tire Superstore
(849)

Average Vehicle Trip Ends vs: 1000 Sq. Feet Gross Floor Area
On a: Weekday,
Peak Hour of Adjacent Street Traffic,
One Hour Between 4 and 6 p.m.

Number of Studies: 12
Average 1000 Sq. Feet GFA: 13
Directional Distribution: 47% entering, 53% exiting

Trip Generation per 1000 Sq. Feet Gross Floor Area

Average Rate	Range of Rates	Standard Deviation
2.11	1.57 - 2.95	1.46

Data Plot and Equation

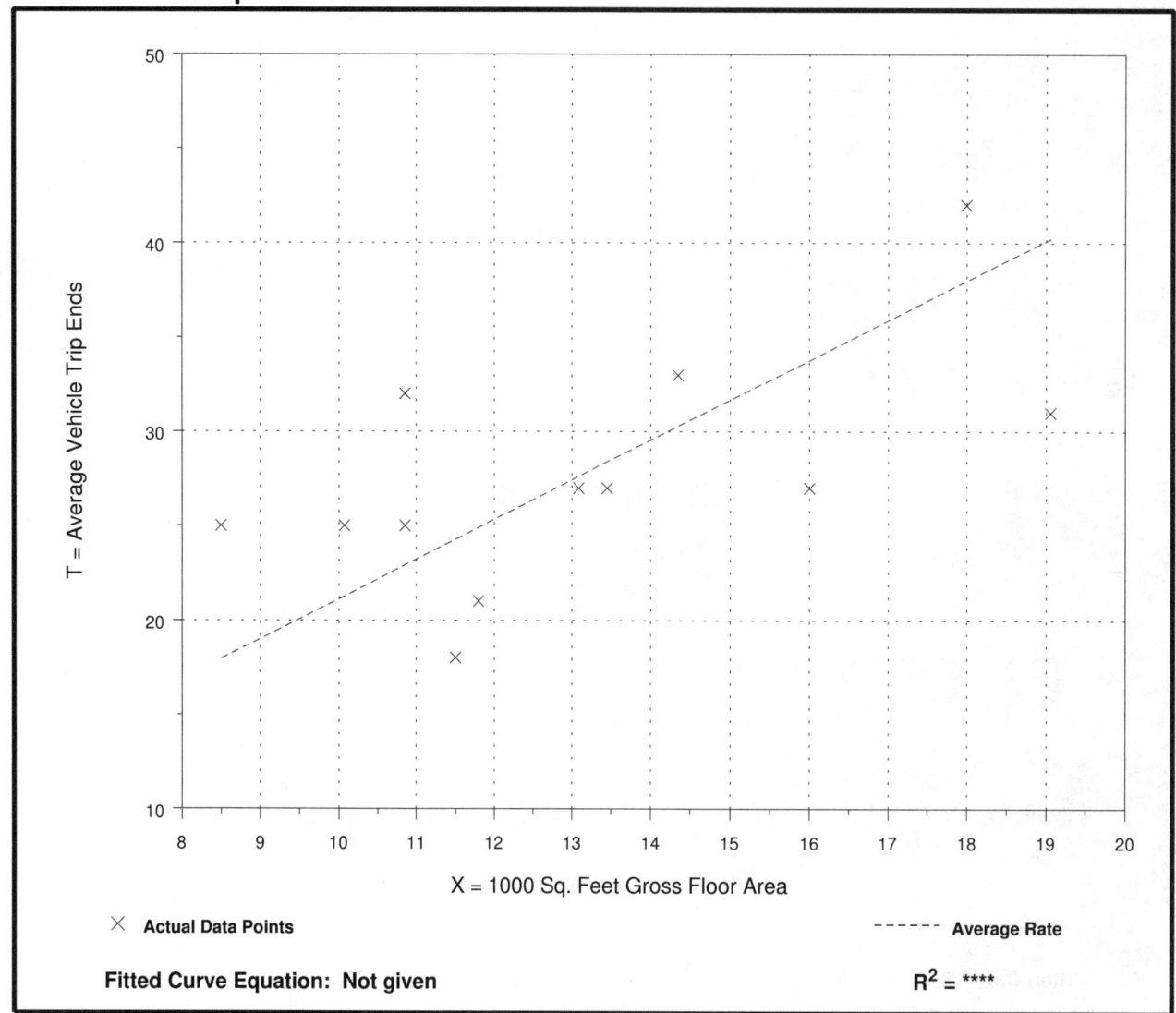

Fitted Curve Equation: Not given $R^2 = ****$

Tire Superstore
(849)

Average Vehicle Trip Ends vs: 1000 Sq. Feet Gross Floor Area
On a: Weekday,
A.M. Peak Hour of Generator

Number of Studies: 12
Average 1000 Sq. Feet GFA: 13
Directional Distribution: 56% entering, 44% exiting

Trip Generation per 1000 Sq. Feet Gross Floor Area

Average Rate	Range of Rates	Standard Deviation
2.42	1.50 - 3.96	1.66

Data Plot and Equation

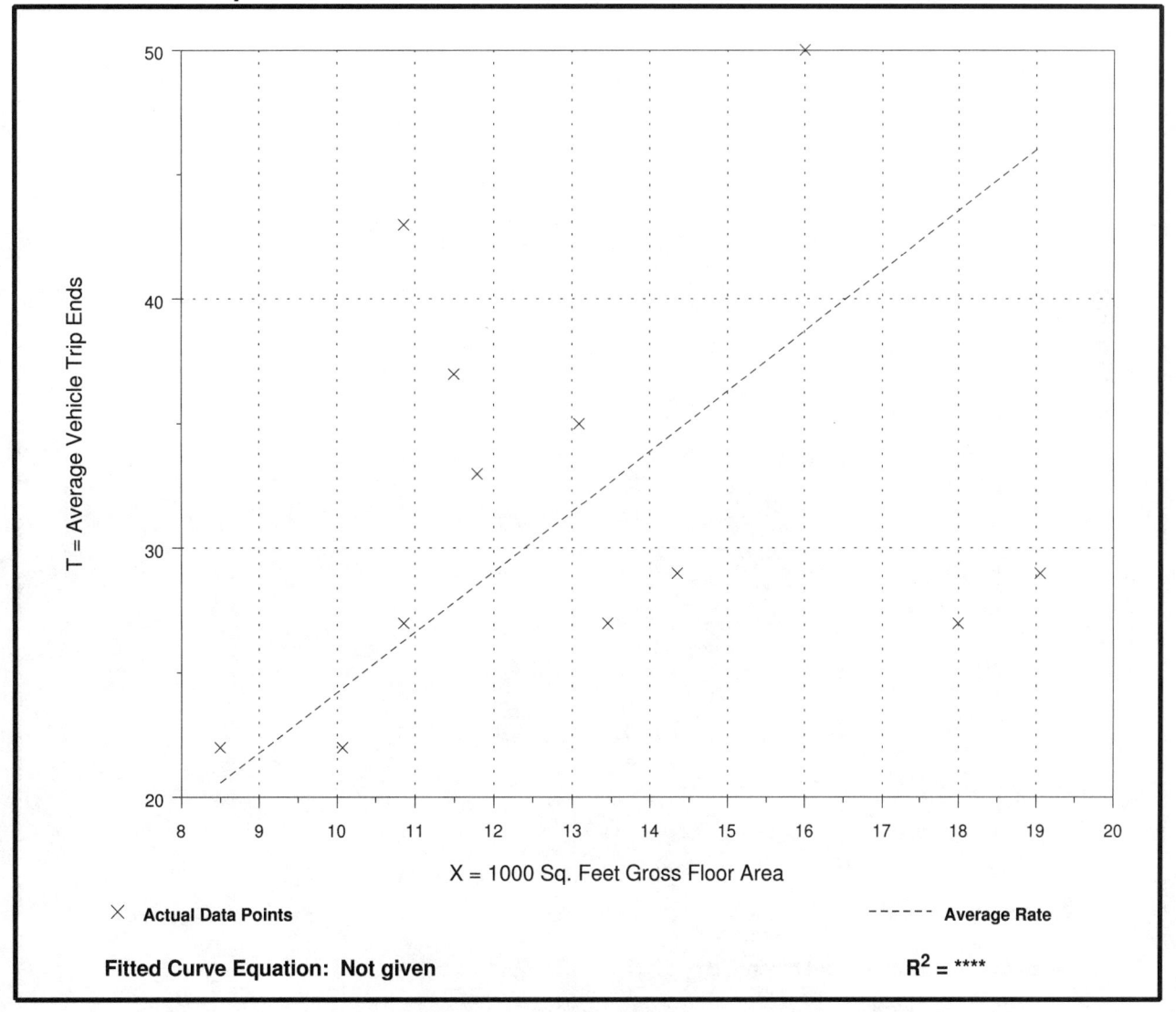

X = 1000 Sq. Feet Gross Floor Area
T = Average Vehicle Trip Ends

× Actual Data Points
----- Average Rate

Fitted Curve Equation: Not given $R^2 = ****$

Tire Superstore
(849)

Average Vehicle Trip Ends vs: 1000 Sq. Feet Gross Floor Area
On a: Weekday,
P.M. Peak Hour of Generator

Number of Studies: 12
Average 1000 Sq. Feet GFA: 13
Directional Distribution: 49% entering, 51% exiting

Trip Generation per 1000 Sq. Feet Gross Floor Area

Average Rate	Range of Rates	Standard Deviation
2.58	1.63 - 3.41	1.61

Data Plot and Equation

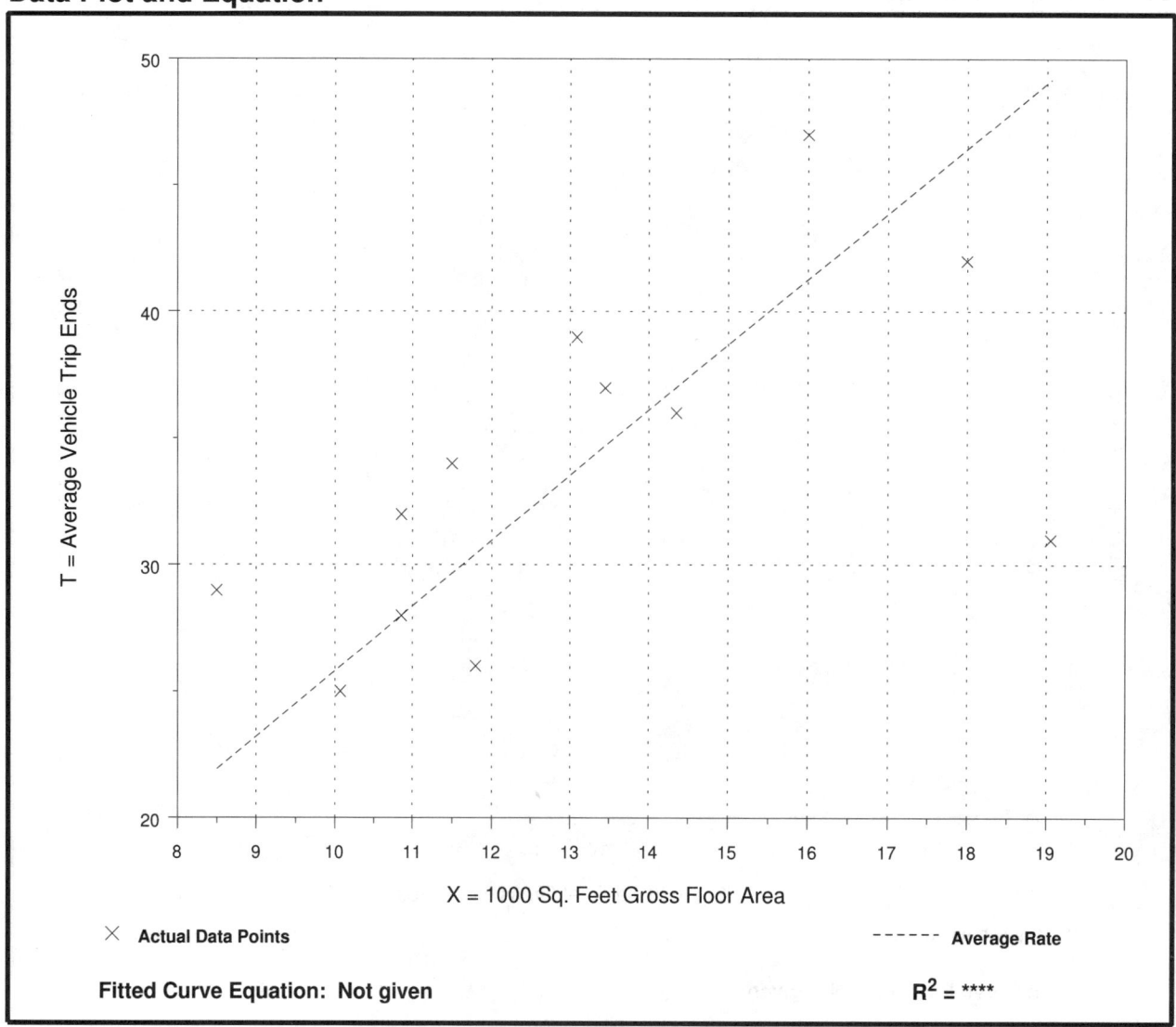

Fitted Curve Equation: Not given $R^2 = ****$

Tire Superstore
(849)

Average Vehicle Trip Ends vs: 1000 Sq. Feet Gross Floor Area
On a: Saturday

Number of Studies: 12
Average 1000 Sq. Feet GFA: 13
Directional Distribution: 50% entering, 50% exiting

Trip Generation per 1000 Sq. Feet Gross Floor Area

Average Rate	Range of Rates	Standard Deviation
19.03	11.13 - 36.56	8.31

Data Plot and Equation

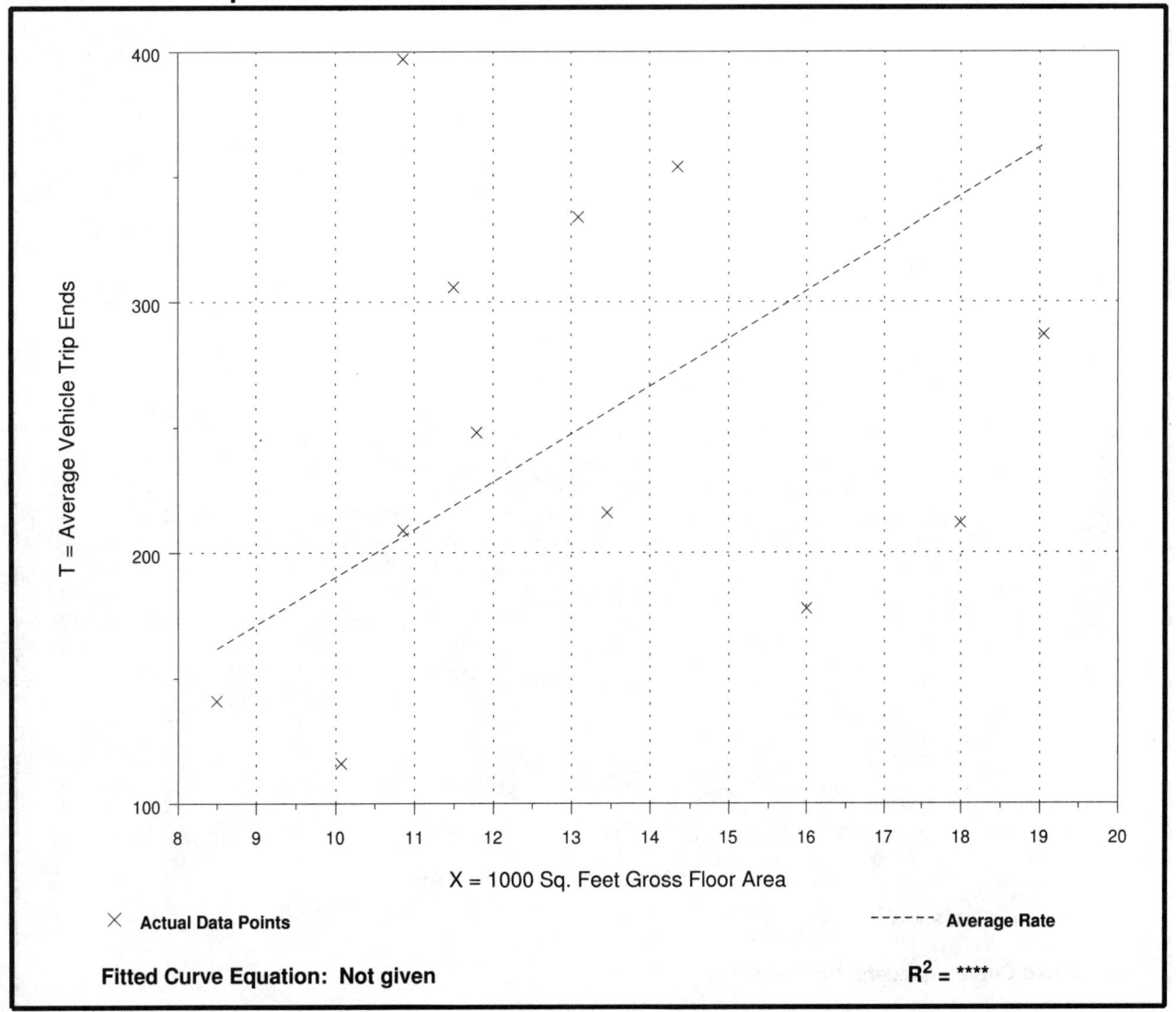

Fitted Curve Equation: Not given R^2 = ****

Tire Superstore
(849)

Average Vehicle Trip Ends vs: 1000 Sq. Feet Gross Floor Area
On a: Saturday,
Peak Hour of Generator

Number of Studies: 11
Average 1000 Sq. Feet GFA: 13
Directional Distribution: 47% entering, 53% exiting

Trip Generation per 1000 Sq. Feet Gross Floor Area

Average Rate	Range of Rates	Standard Deviation
3.13	2.06 - 4.59	1.90

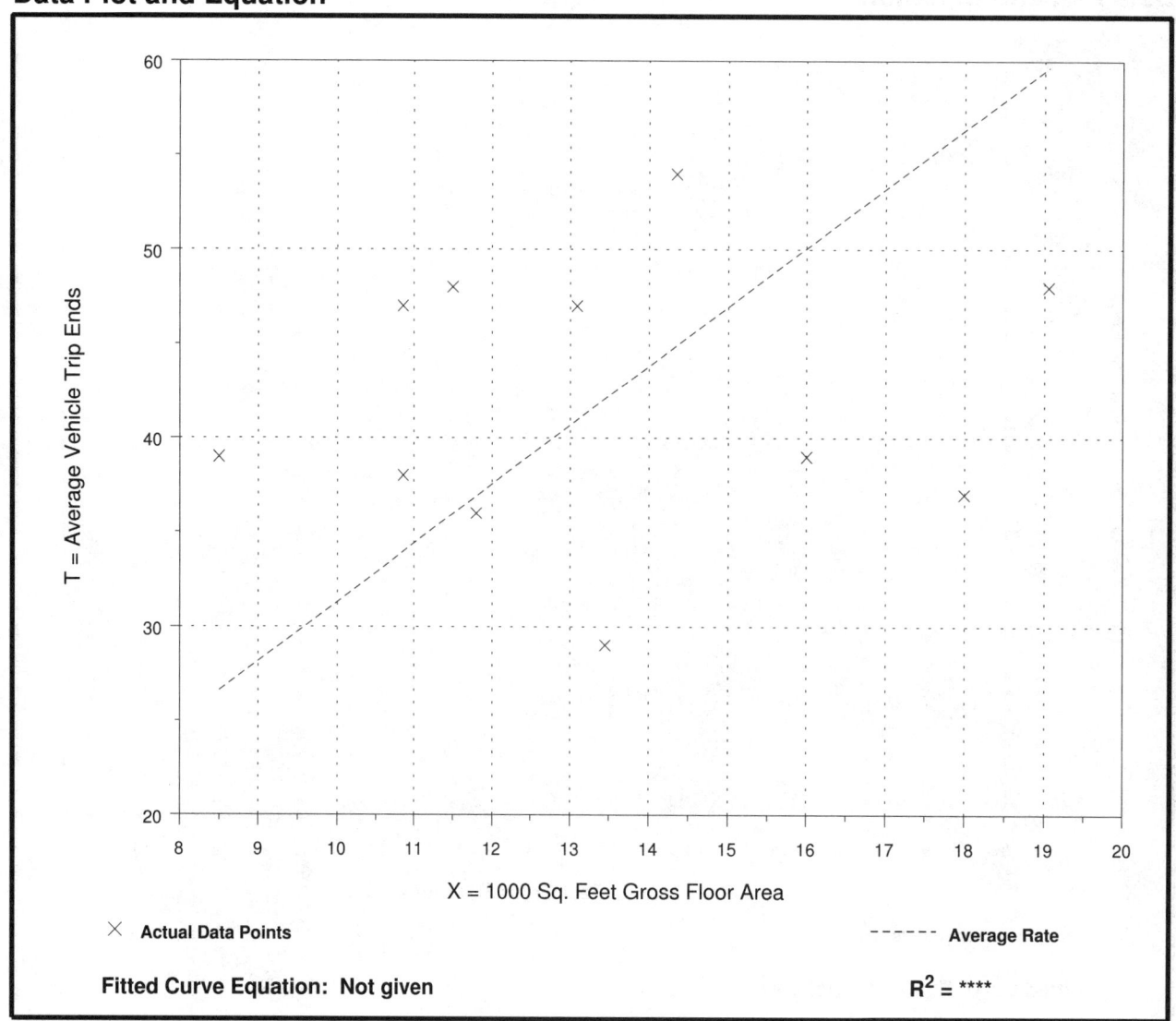

Data Plot and Equation

Fitted Curve Equation: Not given $R^2 = ****$

Land Use: 850
Supermarket

Description

Supermarkets are free-standing retail stores selling a complete assortment of food, food preparation and wrapping materials, and household cleaning items. Supermarkets may also contain the following products and services: ATMs, automobile supplies, bakeries, books and magazines, dry cleaning, floral arrangements, greeting cards, limited-service banks, photo centers, pharmacies and video rental areas. Some facilities may be open 24 hours a day. Discount supermarket (Land Use 854) is a related use.

Additional Data

Caution should be used when applying daily trip generation rates for supermarkets, as the database contains a mixture of facilities with varying hours of operation. Future data submissions should specify hours of operation of a site.

Specialized Land Use Data

One study provided data on a supermarket in Oregon that also carried clothing, footwear, bedding, furniture, jewelry, beauty products, electronics, toys, lumber and garden supplies. The secondary products offered at this supermarket varied from the other stores in this land use; therefore, the information collected for this facility is presented in the following table and was excluded from the data plots. The weekday morning and afternoon peak hours of the generator at this site were between 8:45 a.m. and 9:45 a.m. and between 4:45 p.m. and 5:45 p.m., respectively. The Saturday and Sunday peak hours of the generator were between 3:00 p.m. and 4:00 p.m. and between 12:45 p.m. and 1:45 p.m., respectively.

Independent Variable	Trip Generation Rate	Size of Independent Variable	Number of Studies	Directional Distribution
1,000 Square Feet Gross Floor Area				
Weekday A.M. Peak Hour of Generator	4.21	78	1	Not available
Weekday P.M. Peak Hour of Generator	10.13	78	1	Not available
Saturday Peak Hour of Generator	10.91	78	1	Not available
Sunday Peak Hour of Generator	9.83	78	1	Not available

Source: 746

The sites were surveyed between the 1960s and the 2000s throughout the United States.

Source Numbers

2, 4, 5, 72, 98, 203, 213, 251, 273, 305, 359, 365, 438, 442, 447, 448, 514, 520, 552, 577, 610, 716, 746

Land Use: 850
Supermarket
Independent Variables with One Observation

The following trip generation data are for independent variables with only one observation. This information is shown in this table only; there are no related plots for these data.

Users are cautioned to use data with care because of the small sample size.

Independent Variable	Trip Generation Rate	Size of Independent Variable	Number of Studies	Directional Distribution
Employees				
Weekday	87.82	44	1	50% entering, 50% exiting

Supermarket
(850)

Average Vehicle Trip Ends vs: 1000 Sq. Feet Gross Floor Area
On a: Weekday

Number of Studies: 4
Average 1000 Sq. Feet GFA: 39
Directional Distribution: 50% entering, 50% exiting

Trip Generation per 1000 Sq. Feet Gross Floor Area

Average Rate	Range of Rates	Standard Deviation
102.24	68.65 - 168.88	31.73

Data Plot and Equation

Caution - Use Carefully - Small Sample Size

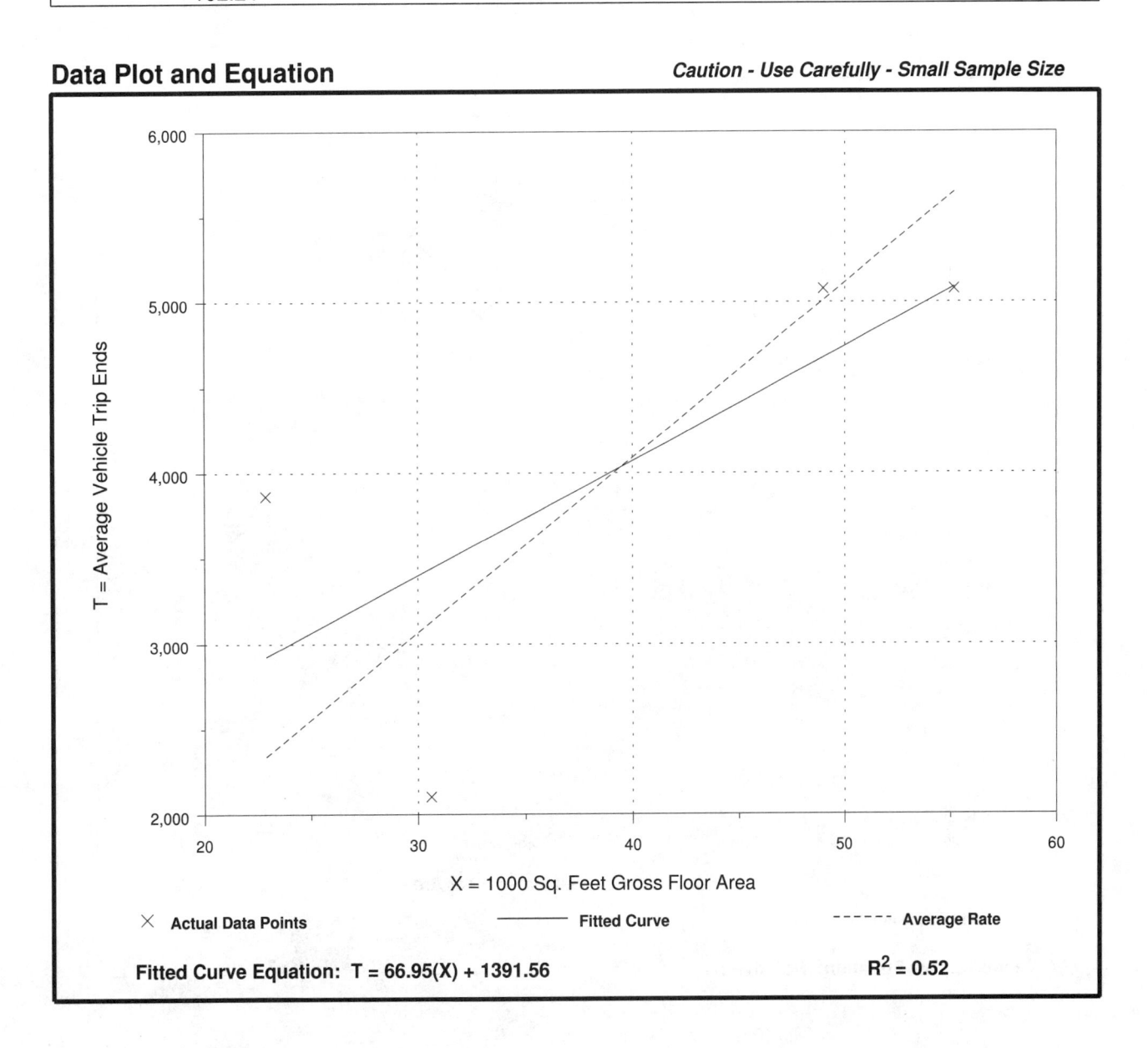

Fitted Curve Equation: T = 66.95(X) + 1391.56 $R^2 = 0.52$

Supermarket
(850)

Average Vehicle Trip Ends vs: 1000 Sq. Feet Gross Floor Area
On a: Weekday,
Peak Hour of Adjacent Street Traffic,
One Hour Between 7 and 9 a.m.

Number of Studies: 13
Average 1000 Sq. Feet GFA: 37
Directional Distribution: 62% entering, 38% exiting

Trip Generation per 1000 Sq. Feet Gross Floor Area

Average Rate	Range of Rates	Standard Deviation
3.40	1.00 - 7.78	2.64

Data Plot and Equation

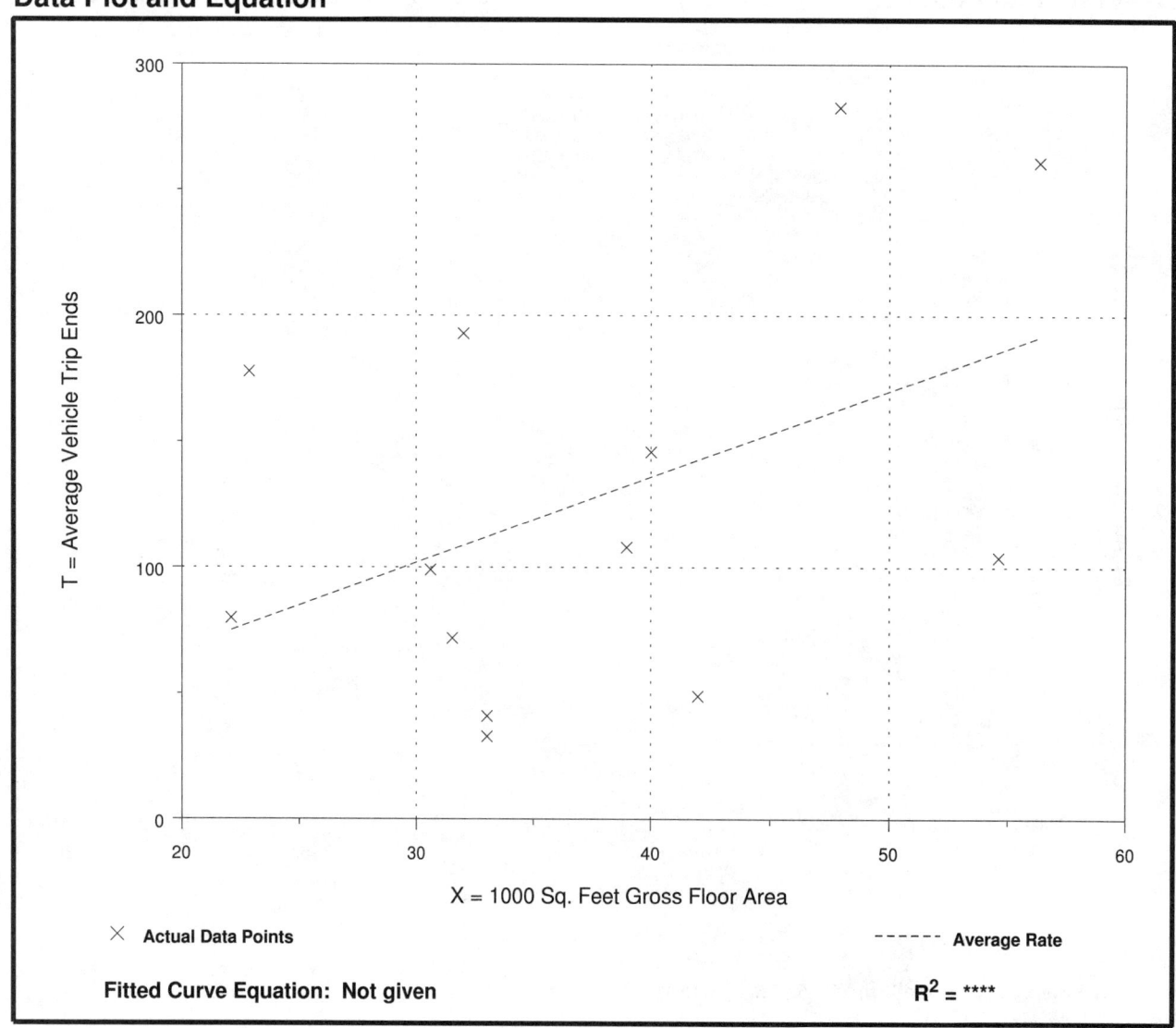

Fitted Curve Equation: Not given $R^2 = ****$

Supermarket
(850)

Average Vehicle Trip Ends vs: 1000 Sq. Feet Gross Floor Area
On a: Weekday,
Peak Hour of Adjacent Street Traffic,
One Hour Between 4 and 6 p.m.

Number of Studies: 62
Average 1000 Sq. Feet GFA: 56
Directional Distribution: 51% entering, 49% exiting

Trip Generation per 1000 Sq. Feet Gross Floor Area

Average Rate	Range of Rates	Standard Deviation
9.48	3.53 - 20.29	4.81

Data Plot and Equation

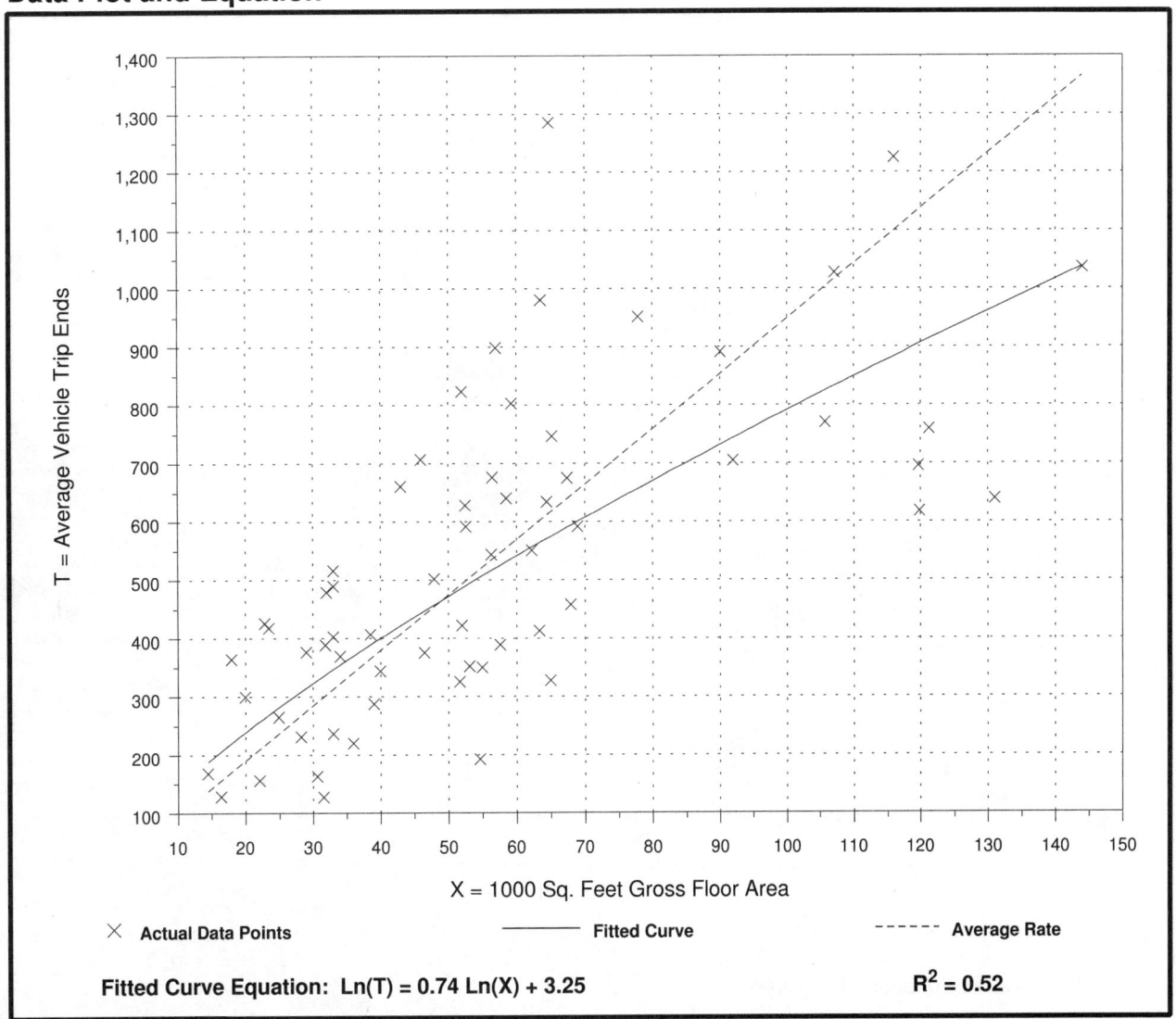

Fitted Curve Equation: $Ln(T) = 0.74\ Ln(X) + 3.25$ $R^2 = 0.52$

Trip Generation, 9th Edition • Institute of Transportation Engineers

Supermarket
(850)

Average Vehicle Trip Ends vs: 1000 Sq. Feet Gross Floor Area
On a: Weekday,
A.M. Peak Hour of Generator

Number of Studies: 12
Average 1000 Sq. Feet GFA: 39
Directional Distribution: 52% entering, 48% exiting

Trip Generation per 1000 Sq. Feet Gross Floor Area

Average Rate	Range of Rates	Standard Deviation
7.07	2.28 - 12.67	4.18

Data Plot and Equation

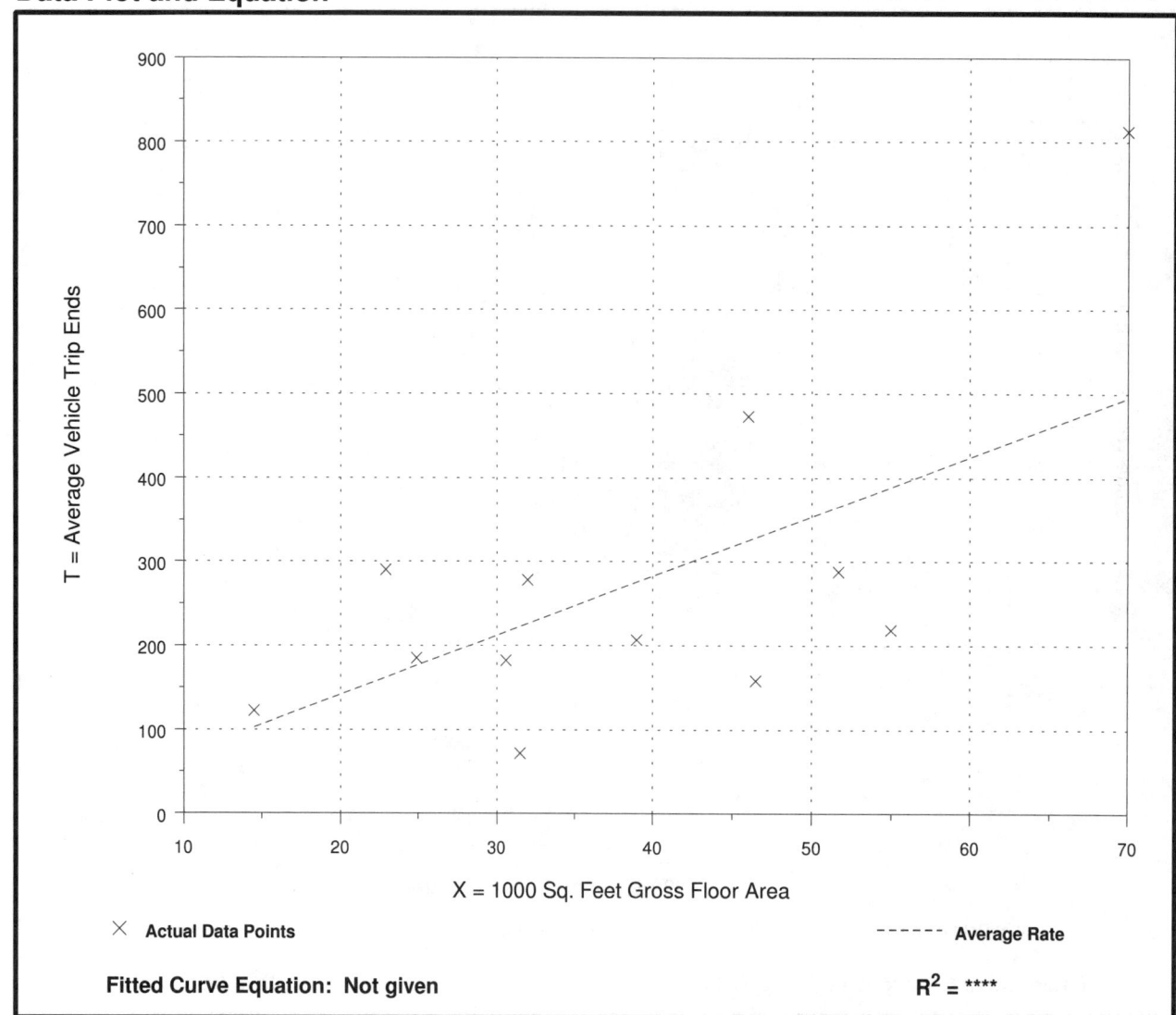

Fitted Curve Equation: Not given $R^2 = ****$

Supermarket
(850)

Average Vehicle Trip Ends vs: 1000 Sq. Feet Gross Floor Area
On a: Weekday,
P.M. Peak Hour of Generator

Number of Studies: 13
Average 1000 Sq. Feet GFA: 45
Directional Distribution: 52% entering, 48% exiting

Trip Generation per 1000 Sq. Feet Gross Floor Area

Average Rate	Range of Rates	Standard Deviation
8.37	4.55 - 18.62	4.80

Data Plot and Equation

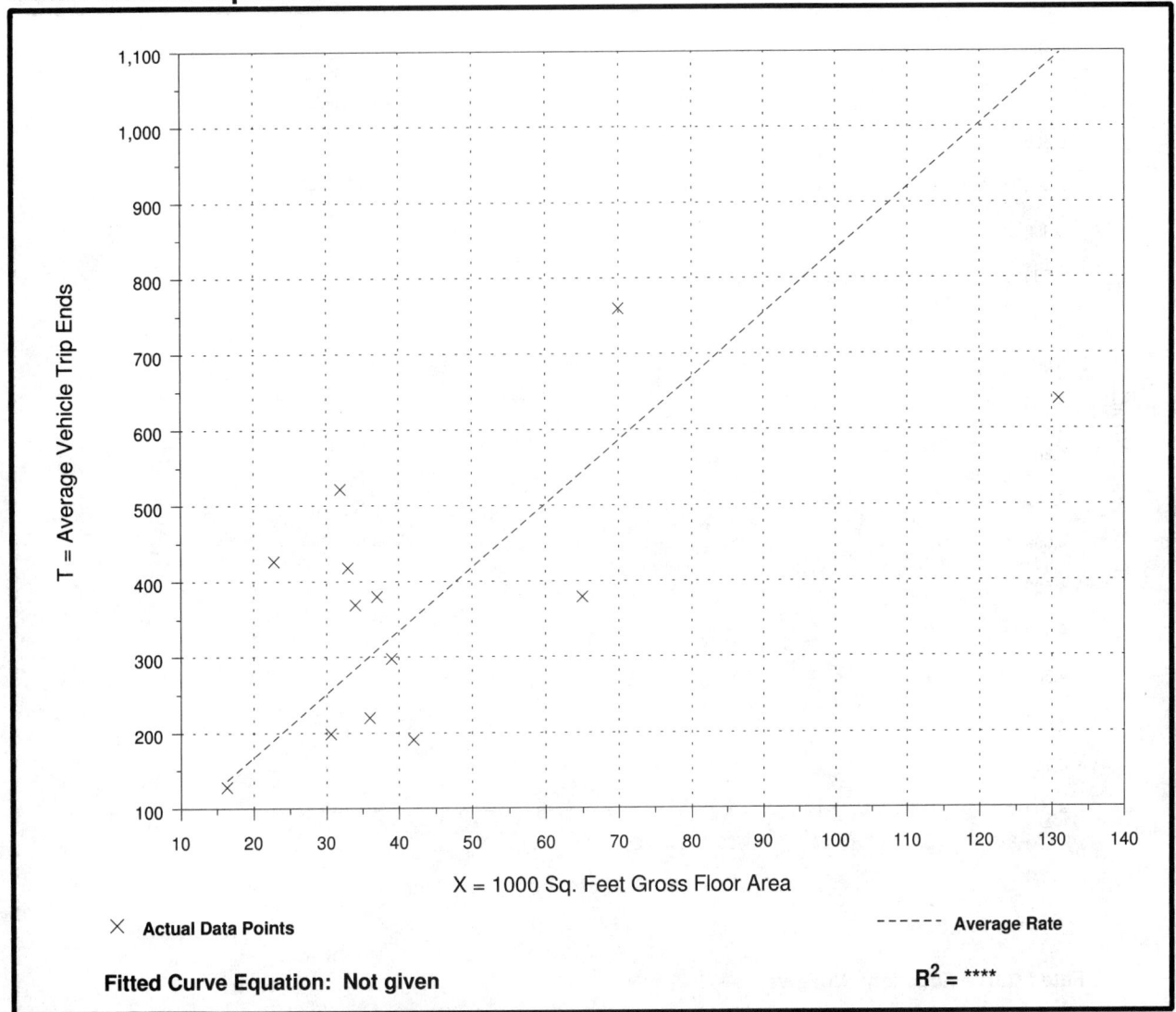

Fitted Curve Equation: Not given $R^2 = ****$

Supermarket
(850)

Average Vehicle Trip Ends vs: 1000 Sq. Feet Gross Floor Area
On a: Saturday

Number of Studies: 2
Average 1000 Sq. Feet GFA: 27
Directional Distribution: 50% entering, 50% exiting

Trip Generation per 1000 Sq. Feet Gross Floor Area

Average Rate	Range of Rates	Standard Deviation
177.59	168.41 - 190.43	*

Data Plot and Equation

Caution - Use Carefully - Small Sample Size

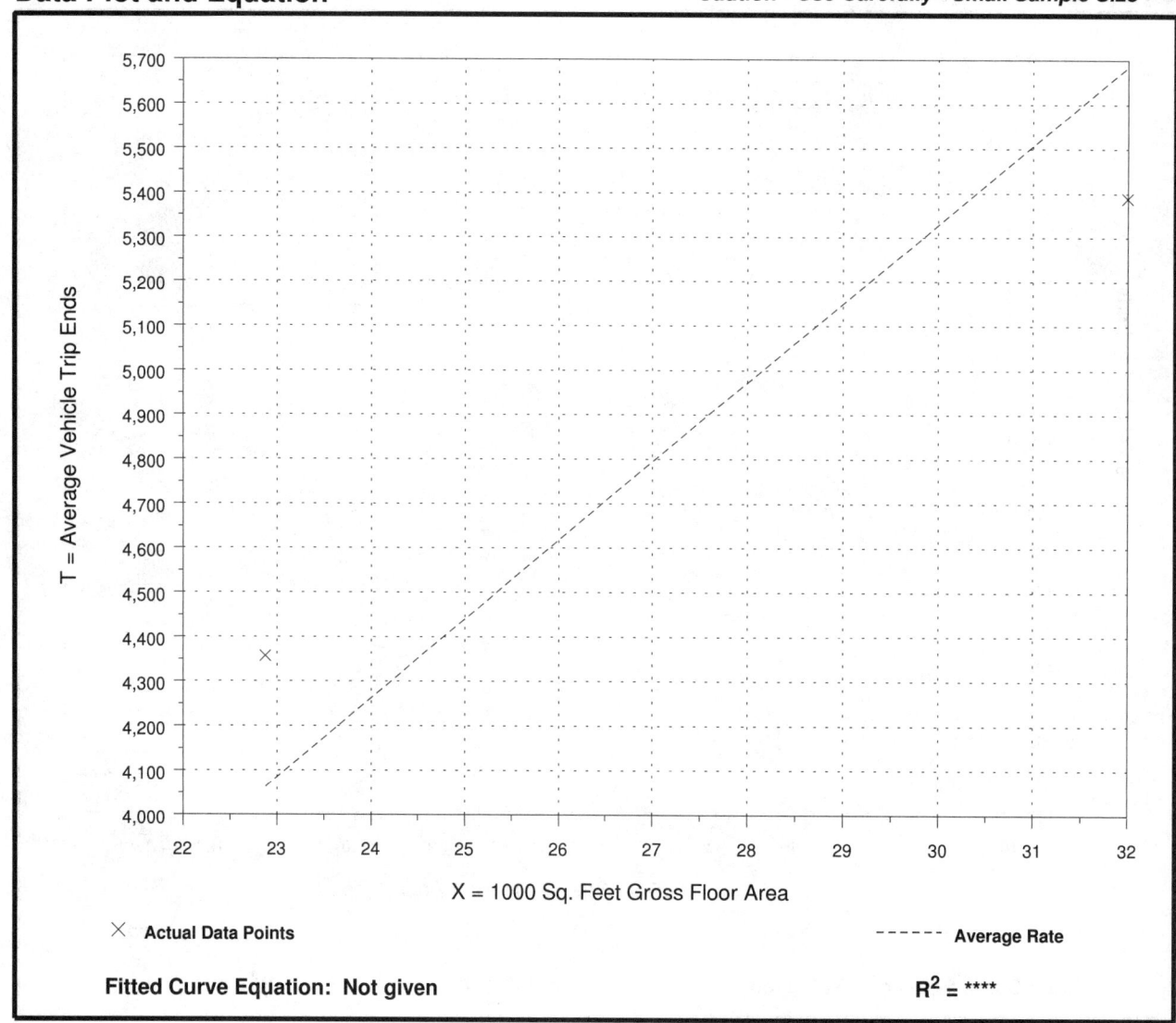

Fitted Curve Equation: Not given $R^2 = ****$

Supermarket
(850)

Average Vehicle Trip Ends vs: 1000 Sq. Feet Gross Floor Area
On a: Saturday,
Peak Hour of Generator

Number of Studies: 34
Average 1000 Sq. Feet GFA: 67
Directional Distribution: 51% entering, 49% exiting

Trip Generation per 1000 Sq. Feet Gross Floor Area

Average Rate	Range of Rates	Standard Deviation
10.65	5.78 - 22.60	4.88

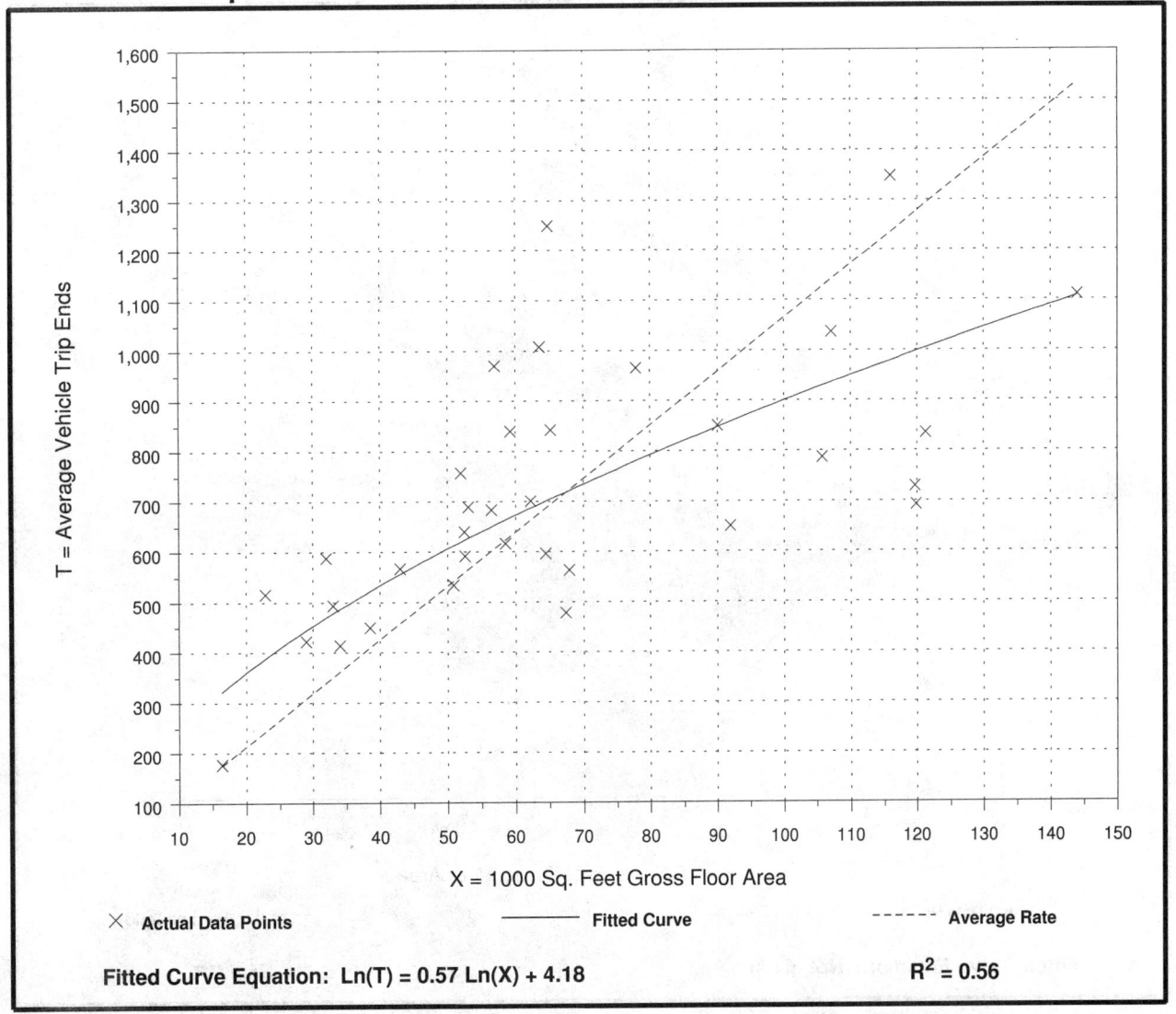

Data Plot and Equation

Fitted Curve Equation: $Ln(T) = 0.57 \, Ln(X) + 4.18$ $R^2 = 0.56$

Supermarket
(850)

Average Vehicle Trip Ends vs: 1000 Sq. Feet Gross Floor Area
On a: Sunday

Number of Studies: 2
Average 1000 Sq. Feet GFA: 27
Directional Distribution: 50% entering, 50% exiting

Trip Generation per 1000 Sq. Feet Gross Floor Area

Average Rate	Range of Rates	Standard Deviation
166.44	150.52 - 177.81	*

Data Plot and Equation

Caution - Use Carefully - Small Sample Size

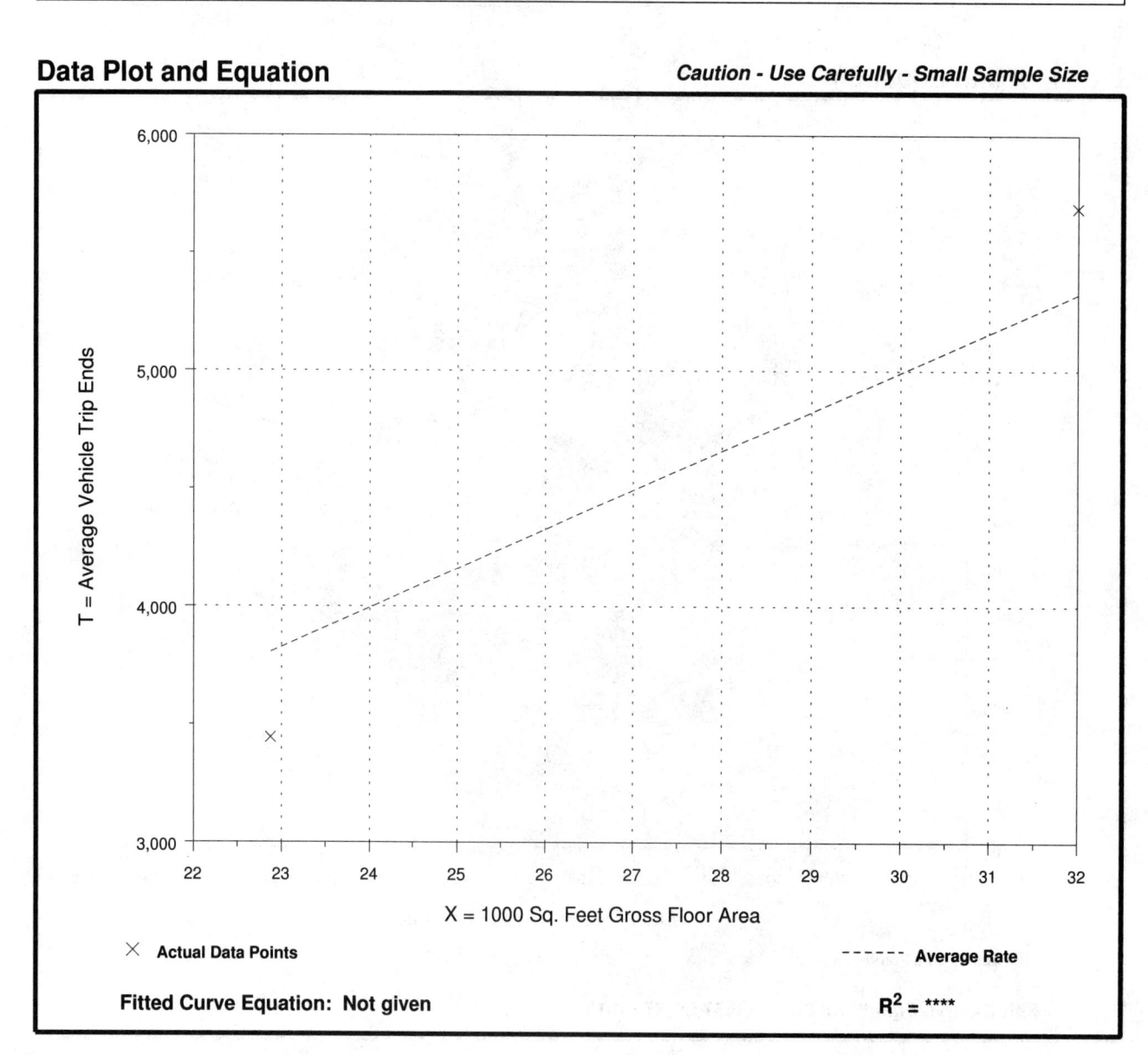

X Actual Data Points

- - - - - Average Rate

Fitted Curve Equation: Not given $R^2 = ****$

Supermarket
(850)

Average Vehicle Trip Ends vs: 1000 Sq. Feet Gross Floor Area
On a: Sunday,
Peak Hour of Generator

Number of Studies: 2
Average 1000 Sq. Feet GFA: 27
Directional Distribution: Not available

Trip Generation per 1000 Sq. Feet Gross Floor Area

Average Rate	Range of Rates	Standard Deviation
18.93	17.79 - 19.75	*

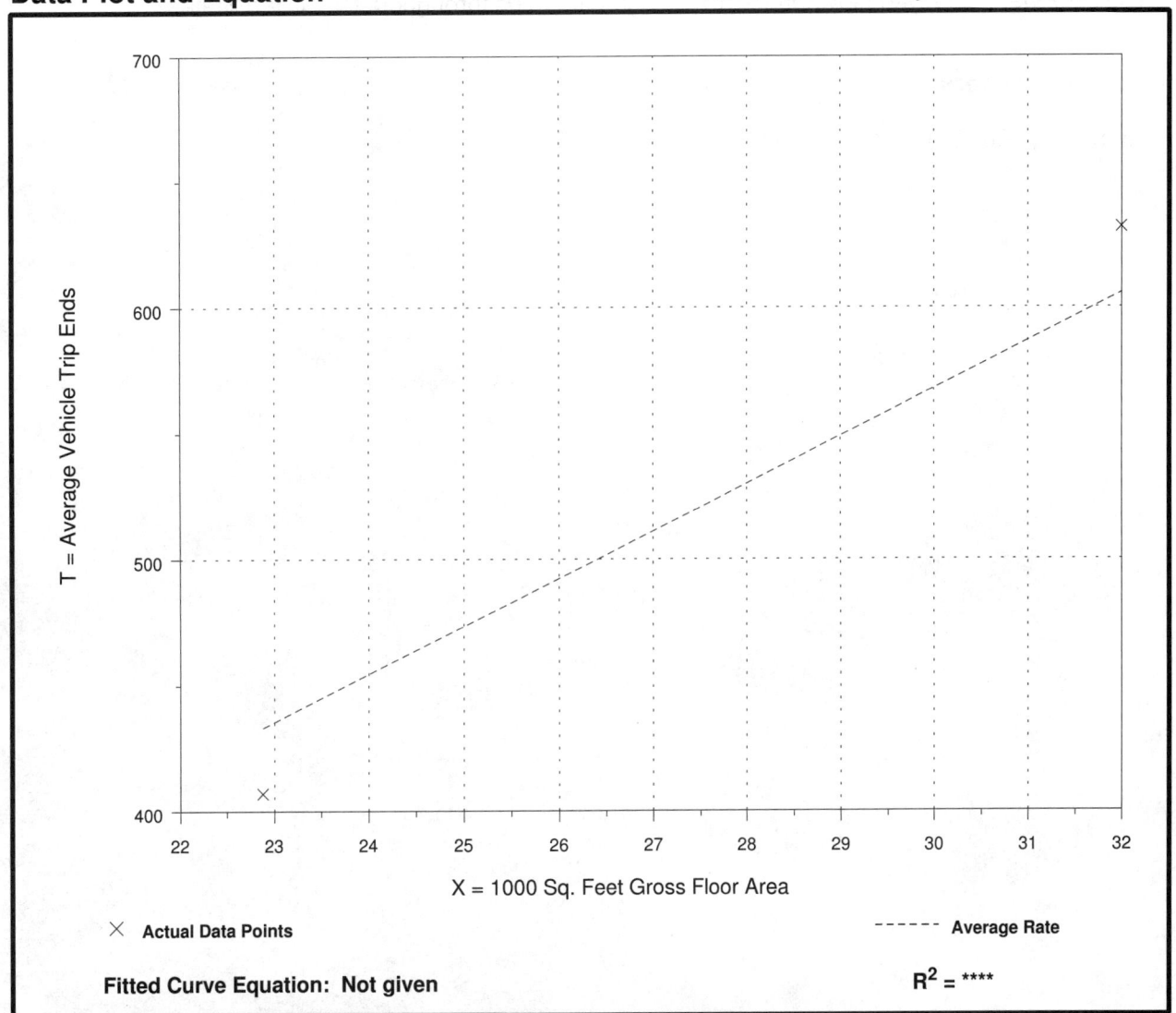

Data Plot and Equation

Caution - Use Carefully - Small Sample Size

Fitted Curve Equation: Not given $R^2 = ****$

Land Use: 851
Convenience Market (Open 24 Hours)

Description

The convenience markets in this classification are open 24 hours per day. These markets sell convenience foods, newspapers, magazines and often beer and wine; they do not have gasoline pumps. Convenience market (open 15-16 hours) (Land Use 852), convenience market with gasoline pumps (Land Use 853), gasoline/service station with convenience market (Land Use 945) and gasoline/service station with convenience market and car wash (Land Use 946) are related uses.

Additional Data

The sites were surveyed between the 1970s and the 2000s throughout the United States.

Source Numbers

78, 87, 90, 168, 213, 237, 542, 550

Convenience Market (Open 24 Hours)
(851)

Average Vehicle Trip Ends vs: 1000 Sq. Feet Gross Floor Area
On a: Weekday

Number of Studies: 8
Average 1000 Sq. Feet GFA: 2
Directional Distribution: 50% entering, 50% exiting

Trip Generation per 1000 Sq. Feet Gross Floor Area

Average Rate	Range of Rates	Standard Deviation
737.99	330.00 - 1438.00	336.24

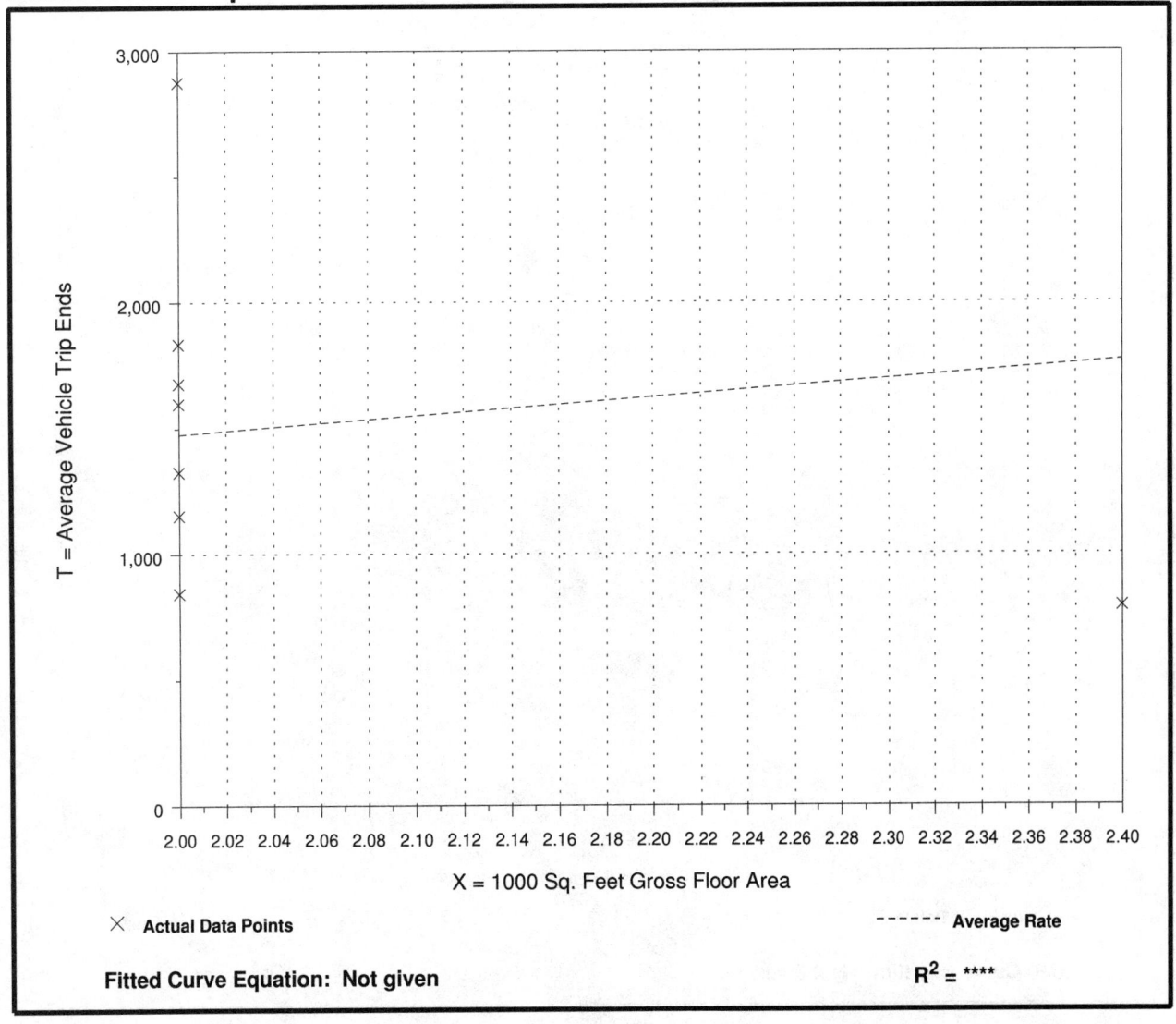

Data Plot and Equation

Fitted Curve Equation: Not given $R^2 = ****$

Trip Generation, 9th Edition • Institute of Transportation Engineers

Convenience Market (Open 24 Hours)
(851)

Average Vehicle Trip Ends vs: 1000 Sq. Feet Gross Floor Area
On a: Weekday,
Peak Hour of Adjacent Street Traffic,
One Hour Between 7 and 9 a.m.

Number of Studies: 33
Average 1000 Sq. Feet GFA: 3
Directional Distribution: 50% entering, 50% exiting

Trip Generation per 1000 Sq. Feet Gross Floor Area

Average Rate	Range of Rates	Standard Deviation
67.03	16.67 - 138.48	33.78

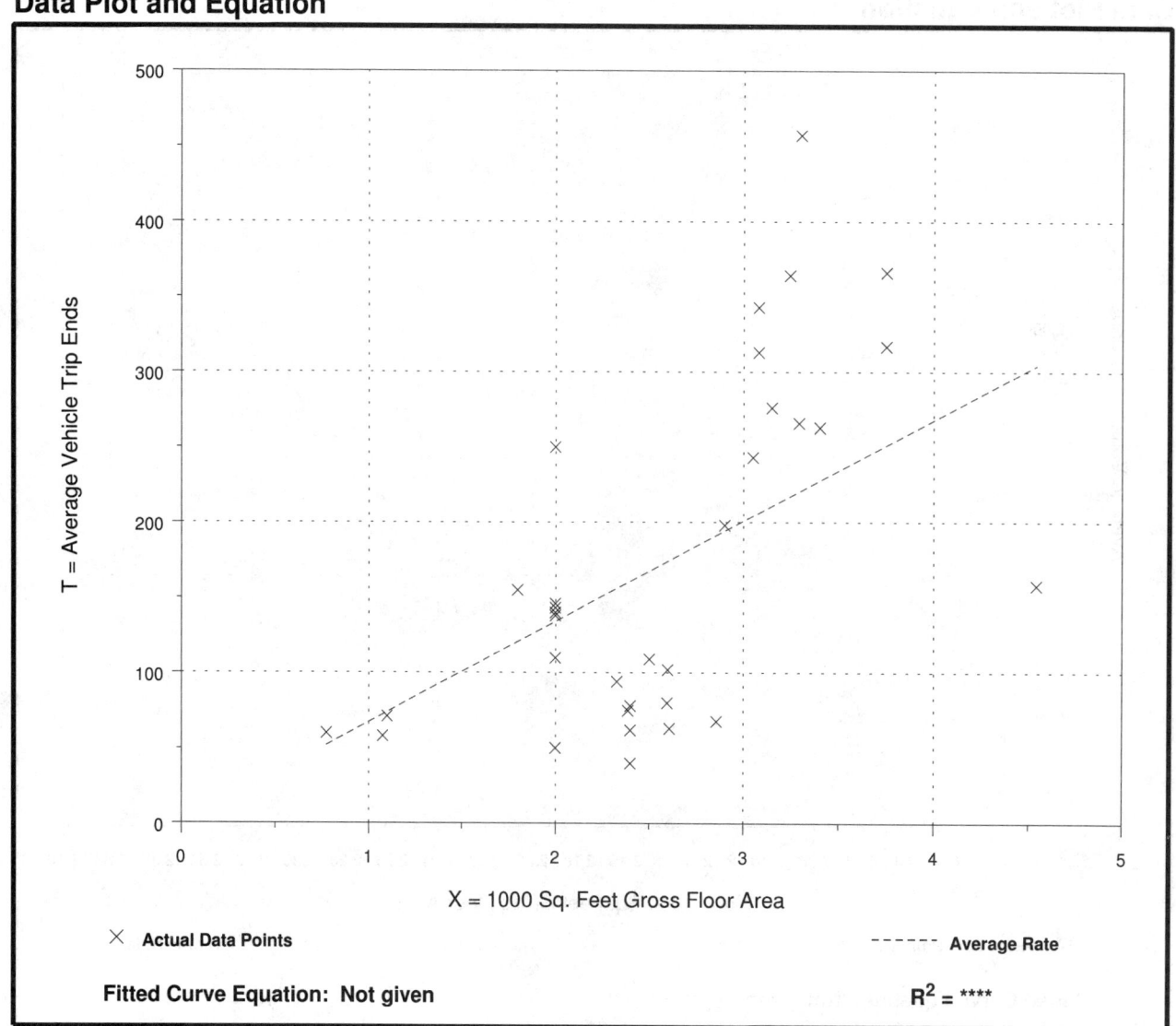

Data Plot and Equation

Fitted Curve Equation: Not given $R^2 = ****$

Convenience Market (Open 24 Hours)
(851)

Average Vehicle Trip Ends vs: 1000 Sq. Feet Gross Floor Area
On a: Weekday,
Peak Hour of Adjacent Street Traffic,
One Hour Between 4 and 6 p.m.

Number of Studies: 33
Average 1000 Sq. Feet GFA: 3
Directional Distribution: 51% entering, 49% exiting

Trip Generation per 1000 Sq. Feet Gross Floor Area

Average Rate	Range of Rates	Standard Deviation
52.41	16.67 - 97.88	21.41

Data Plot and Equation

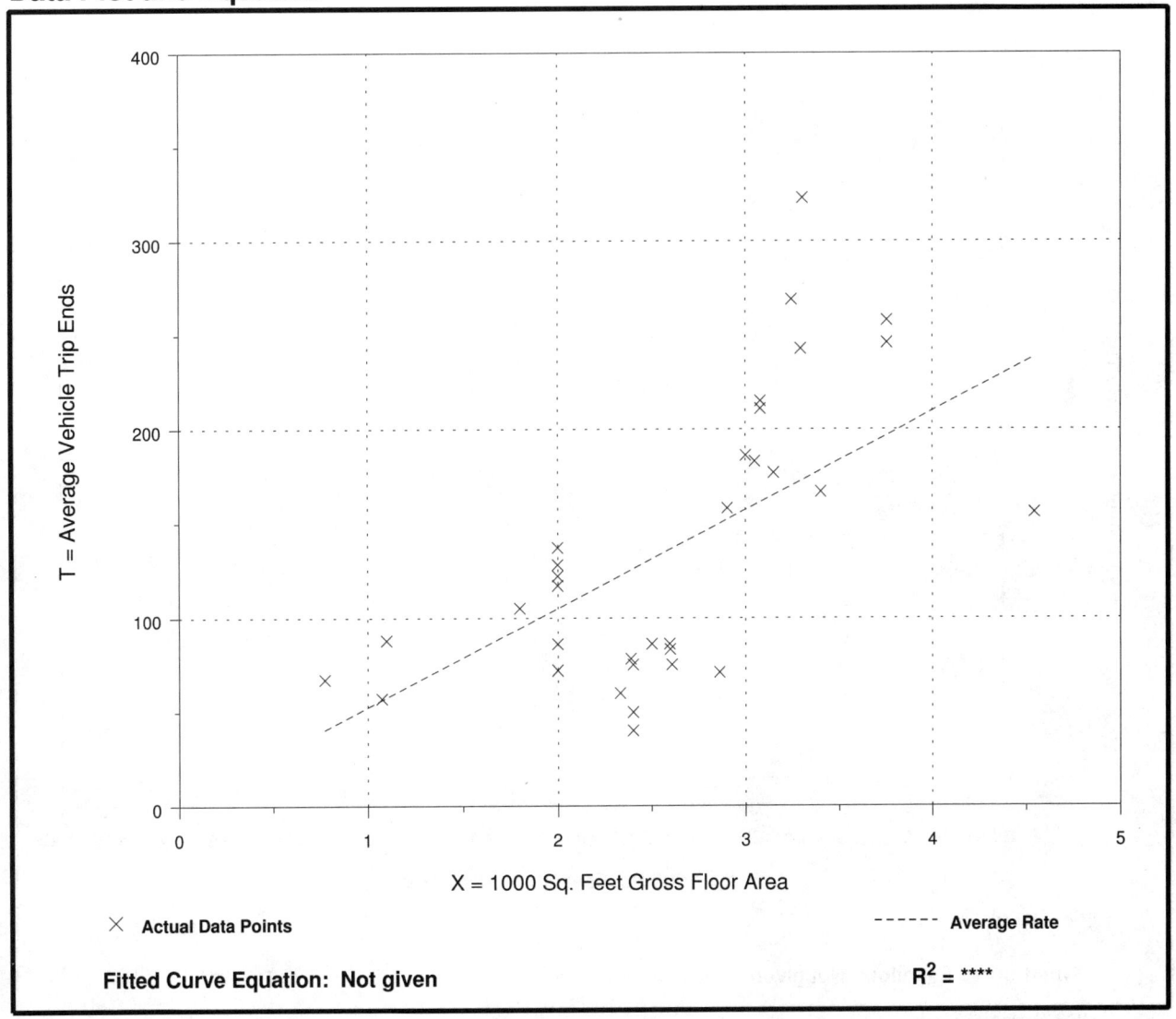

X   Actual Data Points   ----- Average Rate

Fitted Curve Equation: Not given $R^2 = ****$

Trip Generation, 9th Edition • Institute of Transportation Engineers

Convenience Market (Open 24 Hours)
(851)

Average Vehicle Trip Ends vs: 1000 Sq. Feet Gross Floor Area
On a: Weekday,
A.M. Peak Hour of Generator

Number of Studies: 11
Average 1000 Sq. Feet GFA: 2
Directional Distribution: 49% entering, 51% exiting

Trip Generation per 1000 Sq. Feet Gross Floor Area

Average Rate	Range of Rates	Standard Deviation
73.10	25.00 - 139.50	30.67

Data Plot and Equation

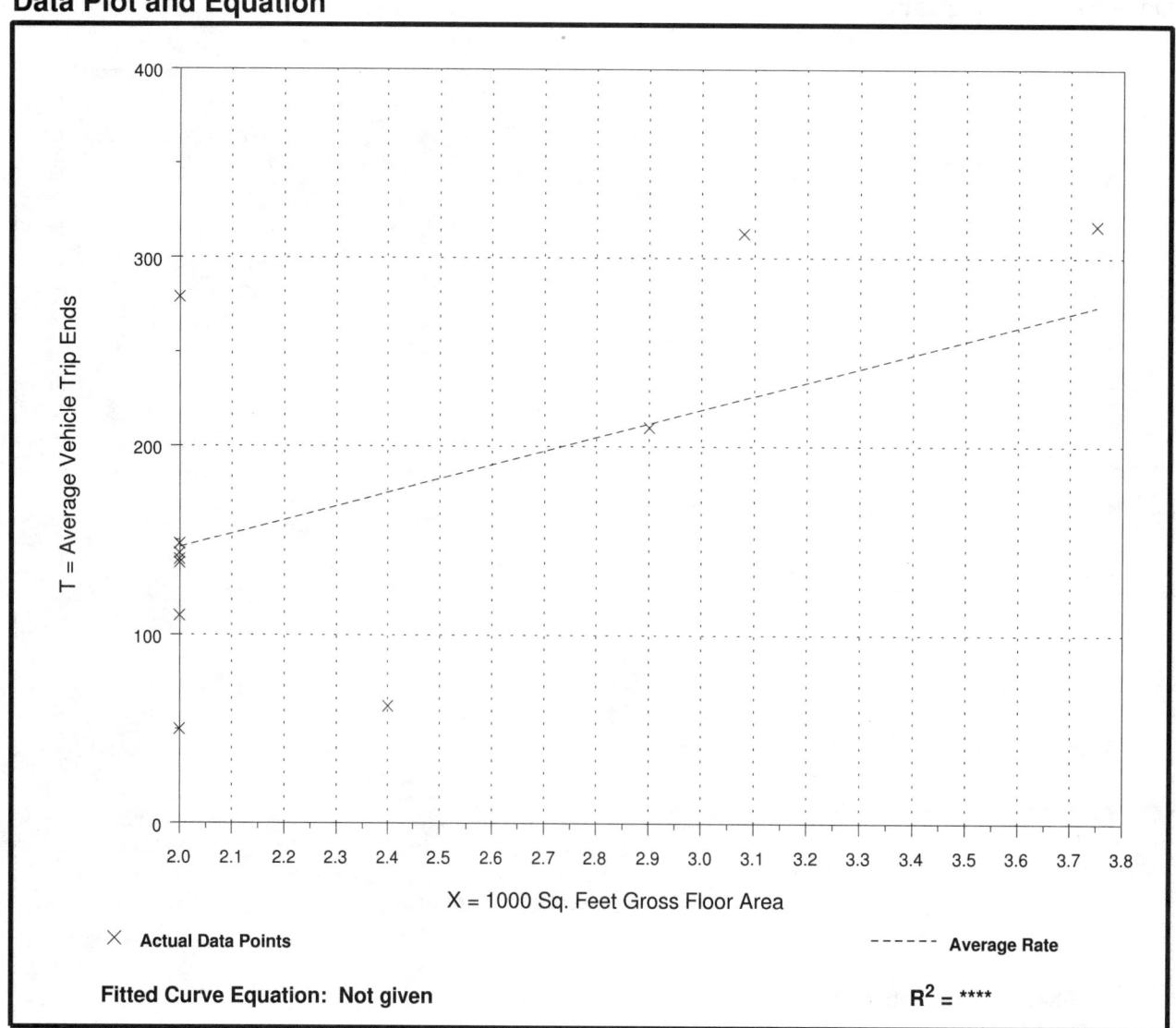

Fitted Curve Equation: Not given $R^2 = ****$

Convenience Market (Open 24 Hours)
(851)

Average Vehicle Trip Ends vs: 1000 Sq. Feet Gross Floor Area
On a: Weekday,
P.M. Peak Hour of Generator

Number of Studies: 9
Average 1000 Sq. Feet GFA: 2
Directional Distribution: 52% entering, 48% exiting

Trip Generation per 1000 Sq. Feet Gross Floor Area

Average Rate	Range of Rates	Standard Deviation
53.42	20.83 - 79.00	19.25

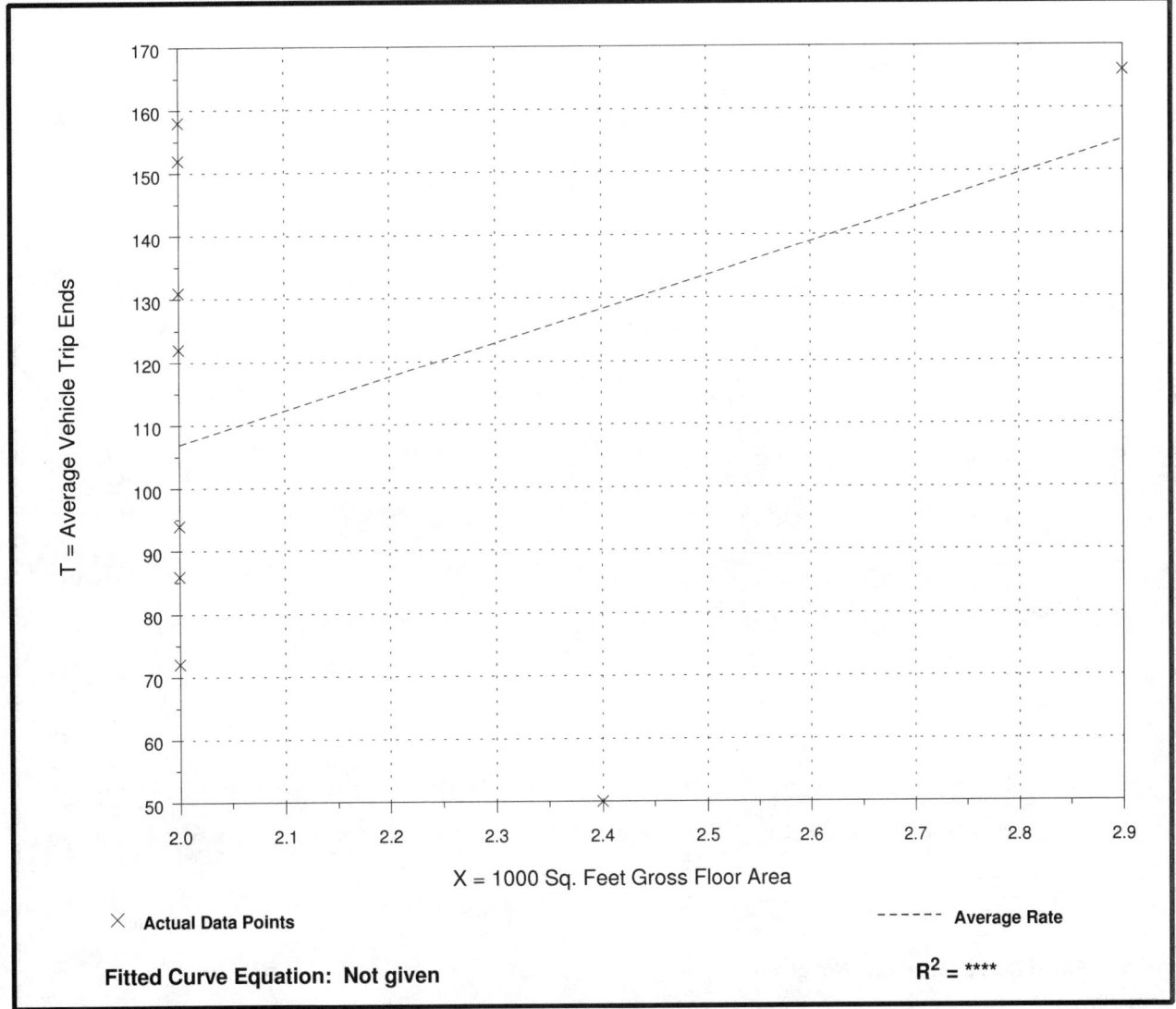

Data Plot and Equation

Fitted Curve Equation: Not given $R^2 = ****$

Convenience Market (Open 24 Hours)
(851)

Average Vehicle Trip Ends vs: 1000 Sq. Feet Gross Floor Area
On a: Saturday

Number of Studies: 4
Average 1000 Sq. Feet GFA: 2
Directional Distribution: 50% entering, 50% exiting

Trip Generation per 1000 Sq. Feet Gross Floor Area

Average Rate	Range of Rates	Standard Deviation
863.10	310.42 - 1627.00	511.99

Data Plot and Equation

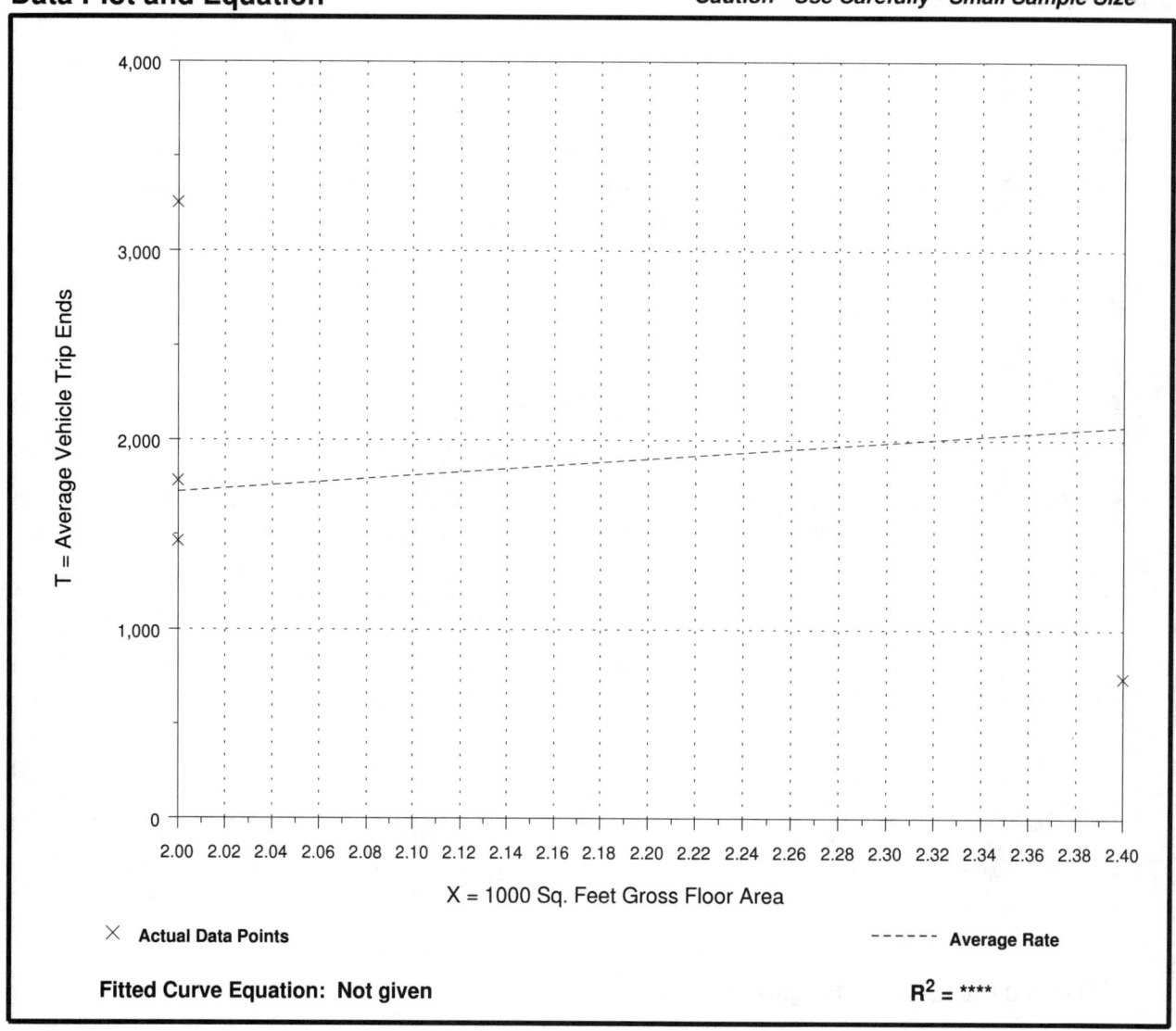

Caution - Use Carefully - Small Sample Size

Fitted Curve Equation: Not given $R^2 = ****$

Convenience Market (Open 24 Hours)
(851)

Average Vehicle Trip Ends vs: 1000 Sq. Feet Gross Floor Area
On a: Saturday,
Peak Hour of Generator

Number of Studies: 16
Average 1000 Sq. Feet GFA: 3
Directional Distribution: 50% entering, 50% exiting

Trip Generation per 1000 Sq. Feet Gross Floor Area

Average Rate	Range of Rates	Standard Deviation
77.11	26.67 - 122.73	27.79

Data Plot and Equation

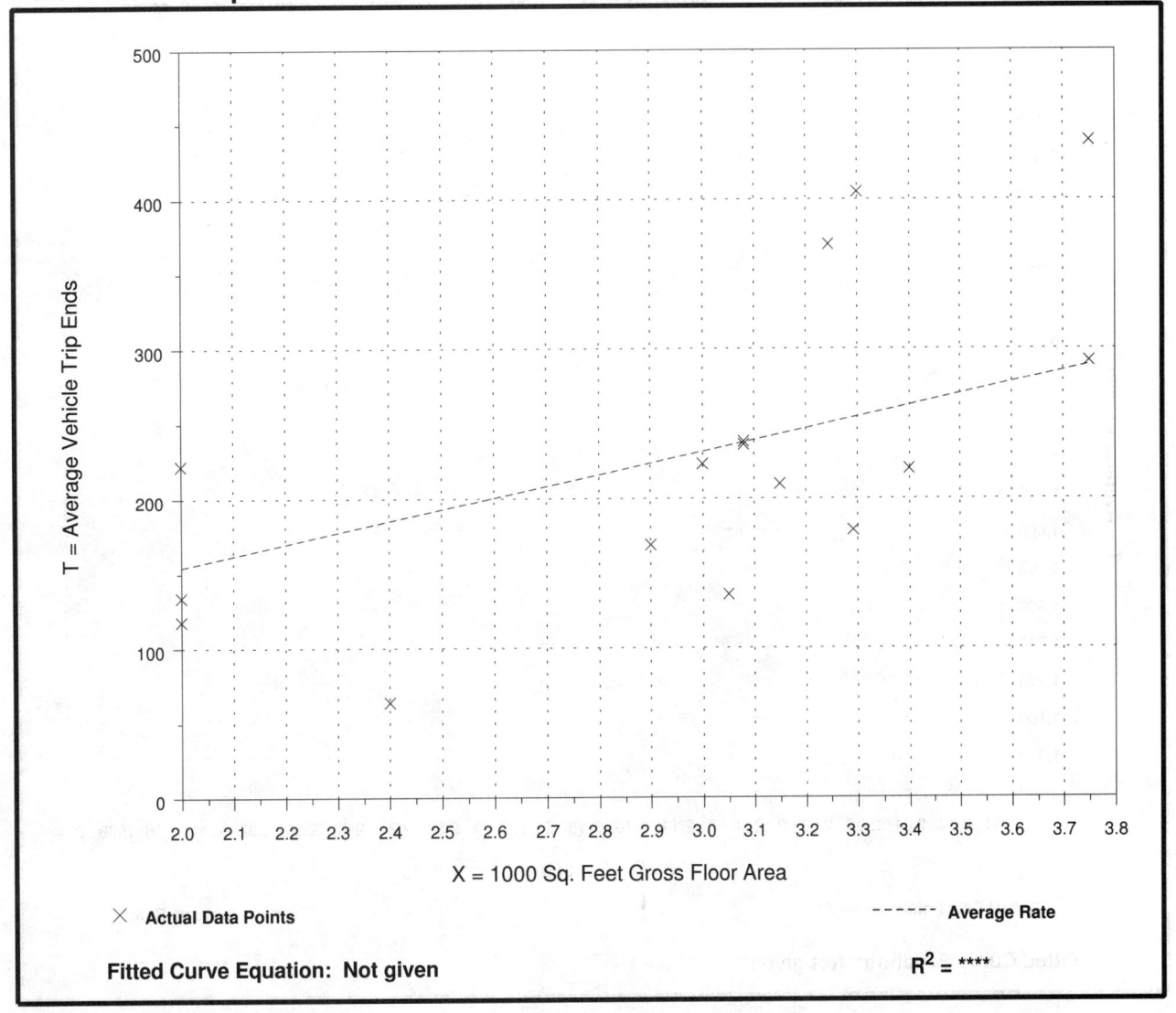

Fitted Curve Equation: Not given $R^2 = ****$

Convenience Market (Open 24 Hours)
(851)

Average Vehicle Trip Ends vs: 1000 Sq. Feet Gross Floor Area
On a: Sunday

Number of Studies: 4
Average 1000 Sq. Feet GFA: 2
Directional Distribution: 50% entering, 50% exiting

Trip Generation per 1000 Sq. Feet Gross Floor Area

Average Rate	Range of Rates	Standard Deviation
758.45	401.67 - 1416.00	415.86

Data Plot and Equation

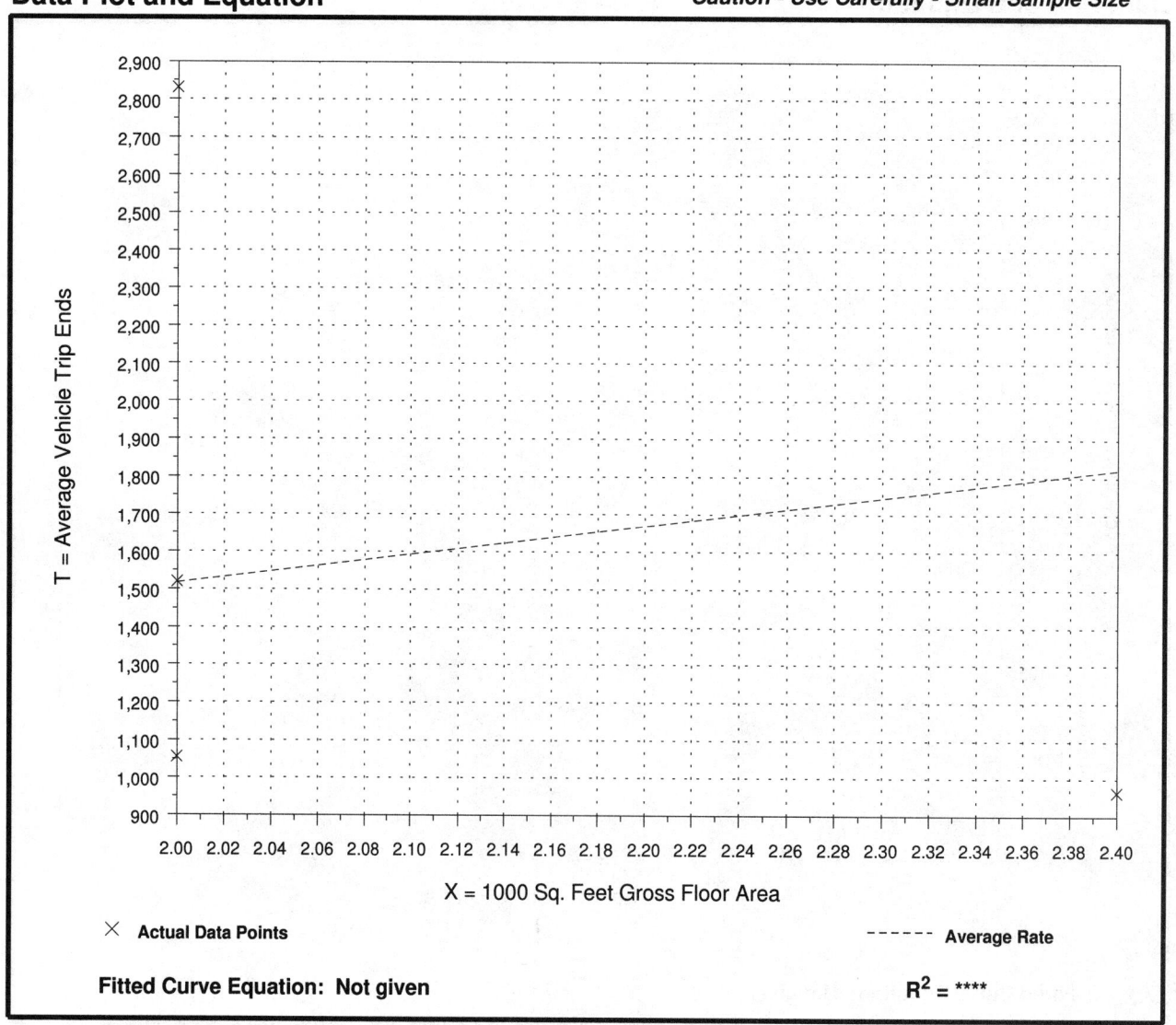

Caution - Use Carefully - Small Sample Size

Fitted Curve Equation: Not given $R^2 = ****$

Convenience Market (Open 24 Hours)
(851)

Average Vehicle Trip Ends vs: 1000 Sq. Feet Gross Floor Area
On a: Sunday,
Peak Hour of Generator

Number of Studies: 4
Average 1000 Sq. Feet GFA: 2
Directional Distribution: 47% entering, 53% exiting

Trip Generation per 1000 Sq. Feet Gross Floor Area

Average Rate	Range of Rates	Standard Deviation
65.00	41.67 - 109.00	28.49

Data Plot and Equation

Caution - Use Carefully - Small Sample Size

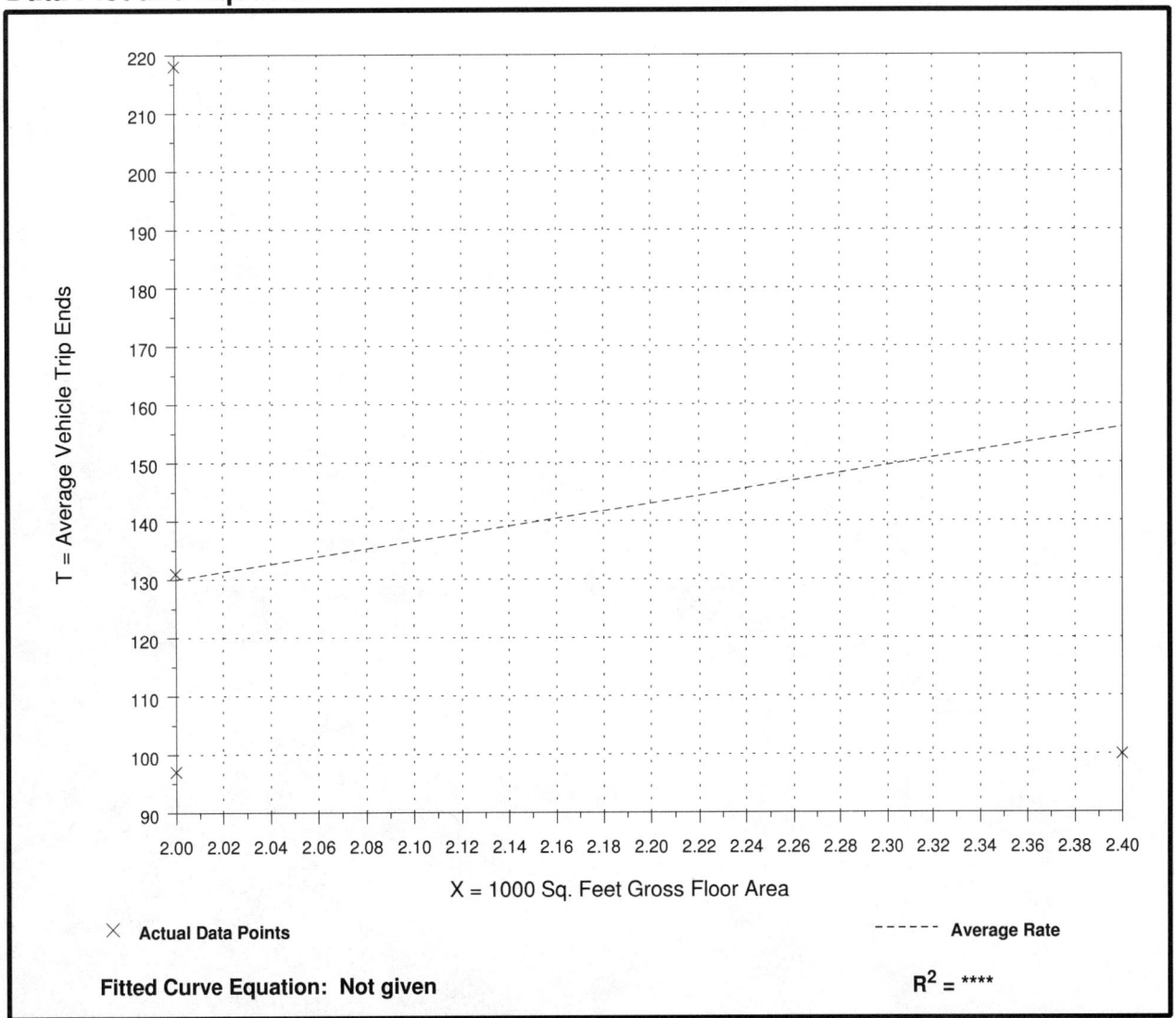

× Actual Data Points   ------ Average Rate

Fitted Curve Equation: Not given $R^2 = ****$

Land Use: 852
Convenience Market (Open 15-16 Hours)

Description

The convenience markets in this classification are open 15-16 hours per day. These markets sell convenience foods, newspapers, magazines and often beer and wine; they do not have gasoline pumps. Convenience market (open 24 hours) (Land Use 851), convenience market with gasoline pumps (Land Use 853), gasoline/service station with convenience market (Land Use 945) and gasoline/service station with convenience market and car wash (Land Use 946) are related uses.

Additional Data

The sites were surveyed in the 1980s in Pennsylvania and Oregon.

Source Numbers

253, 282

Convenience Market (Open 15-16 Hours)
(852)

Average Vehicle Trip Ends vs: 1000 Sq. Feet Gross Floor Area
On a: Weekday,
Peak Hour of Adjacent Street Traffic,
One Hour Between 7 and 9 a.m.

Number of Studies: 5
Average 1000 Sq. Feet GFA: 3
Directional Distribution: 50% entering, 50% exiting

Trip Generation per 1000 Sq. Feet Gross Floor Area

Average Rate	Range of Rates	Standard Deviation
31.02	12.92 - 72.00	24.36

Data Plot and Equation

Caution - Use Carefully - Small Sample Size

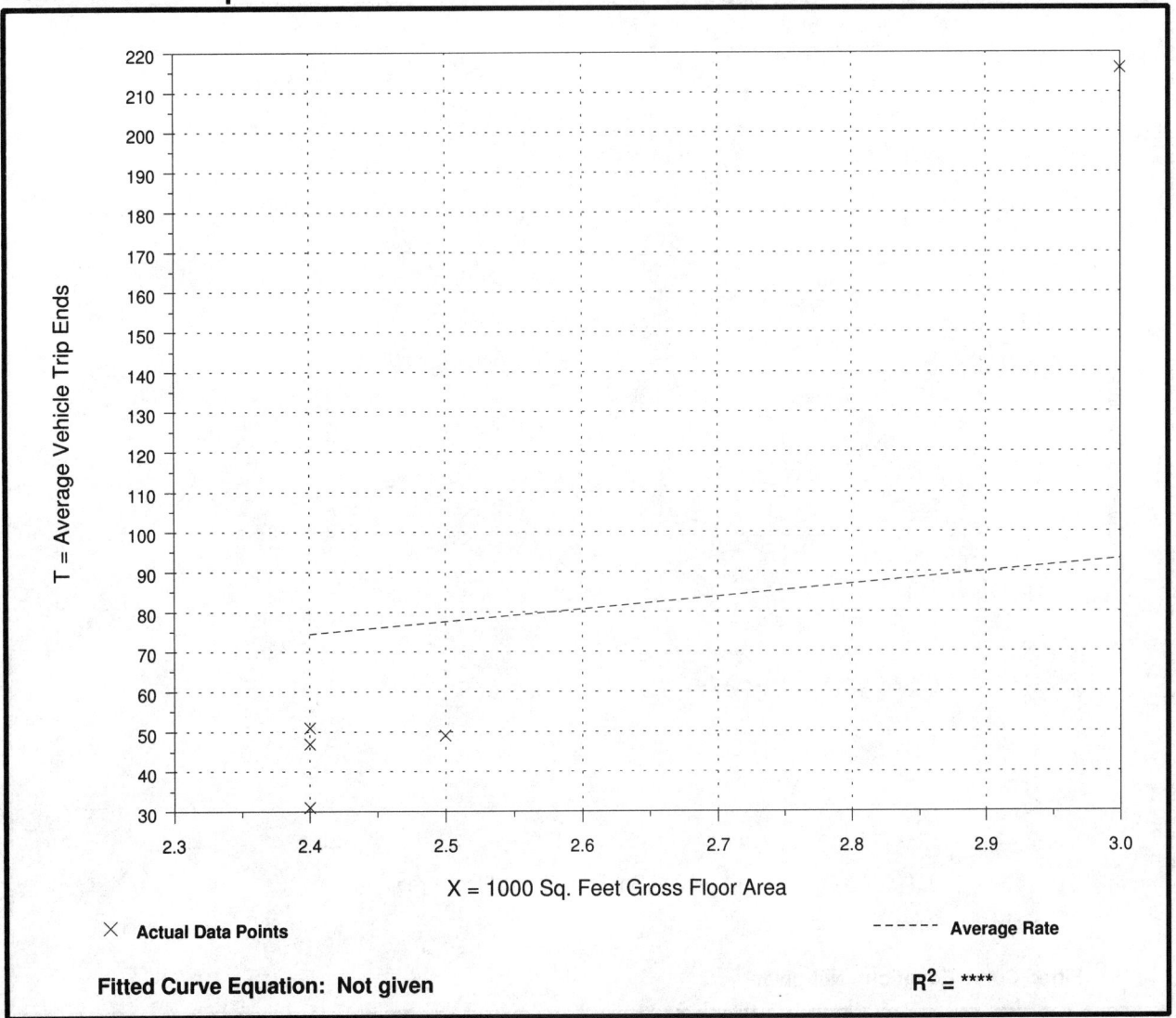

Fitted Curve Equation: Not given $R^2 = ****$

Convenience Market (Open 15-16 Hours)
(852)

Average Vehicle Trip Ends vs: 1000 Sq. Feet Gross Floor Area
On a: Weekday,
Peak Hour of Adjacent Street Traffic,
One Hour Between 4 and 6 p.m.

Number of Studies: 5
Average 1000 Sq. Feet GFA: 3
Directional Distribution: 49% entering, 51% exiting

Trip Generation per 1000 Sq. Feet Gross Floor Area

Average Rate	Range of Rates	Standard Deviation
34.57	15.83 - 56.67	17.61

Data Plot and Equation

Caution - Use Carefully - Small Sample Size

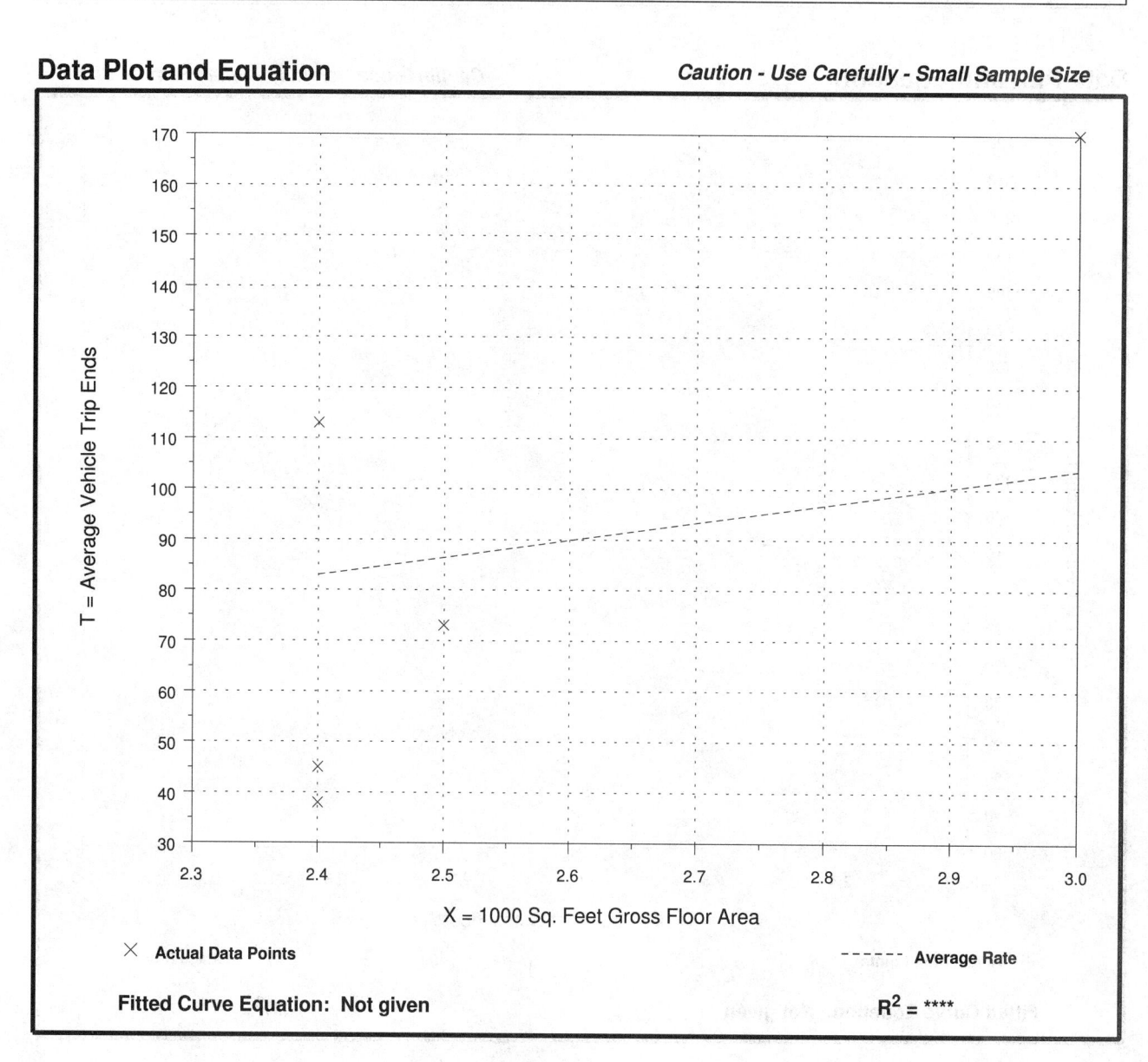

X Actual Data Points

------ Average Rate

Fitted Curve Equation: Not given $R^2 = ****$

Convenience Market (Open 15-16 Hours)
(852)

Average Vehicle Trip Ends vs: 1000 Sq. Feet Gross Floor Area
On a: Weekday,
A.M. Peak Hour of Generator

Number of Studies: 5
Average 1000 Sq. Feet GFA: 3
Directional Distribution: 51% entering, 49% exiting

Trip Generation per 1000 Sq. Feet Gross Floor Area

Average Rate	Range of Rates	Standard Deviation
32.60	13.75 - 72.00	23.64

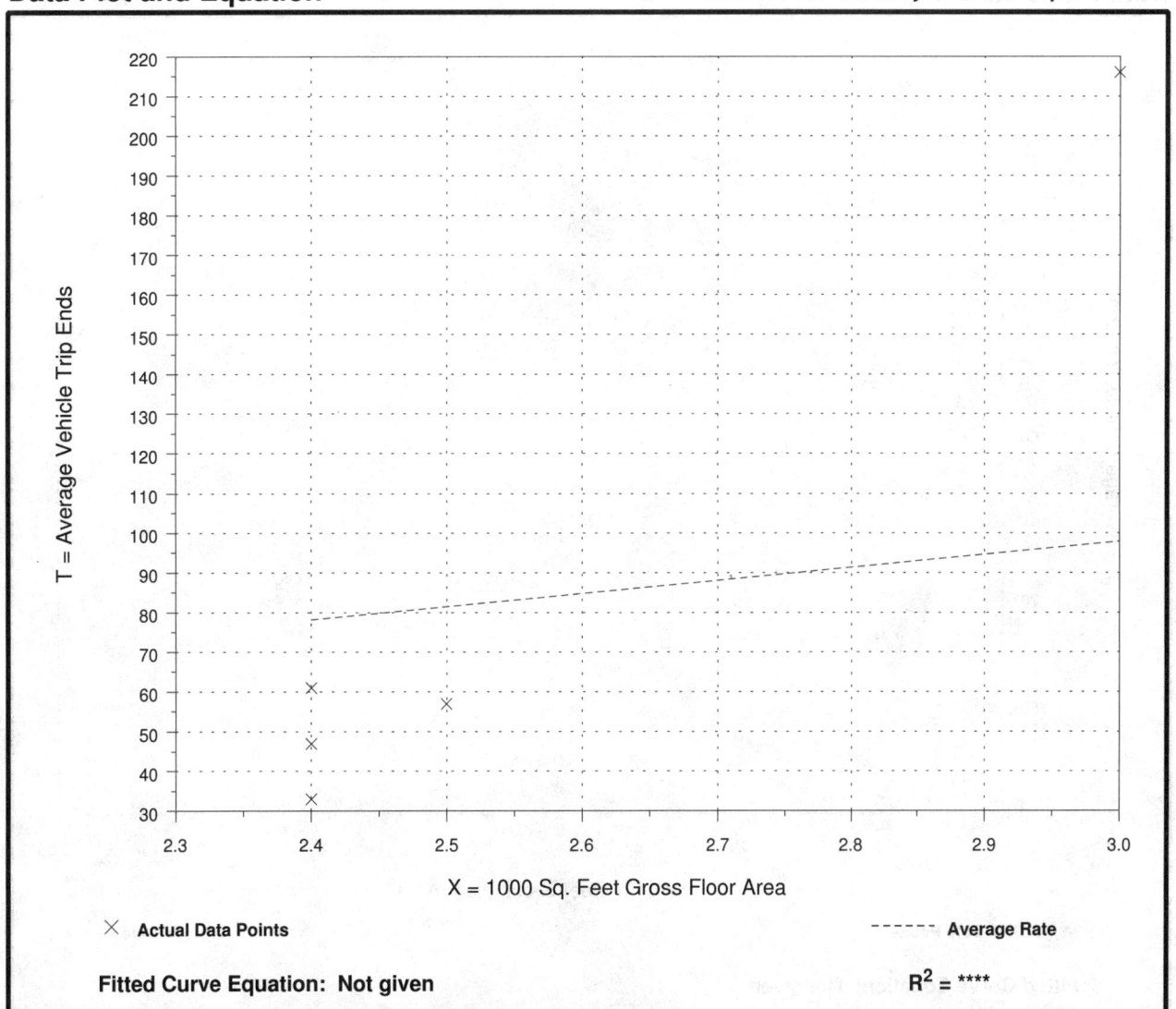

Data Plot and Equation

Caution - Use Carefully - Small Sample Size

Fitted Curve Equation: Not given $R^2 = ****$

Convenience Market (Open 15-16 Hours)
(852)

Average Vehicle Trip Ends vs: 1000 Sq. Feet Gross Floor Area
On a: Weekday,
P.M. Peak Hour of Generator

Number of Studies: 5
Average 1000 Sq. Feet GFA: 3
Directional Distribution: 49% entering, 51% exiting

Trip Generation per 1000 Sq. Feet Gross Floor Area

Average Rate	Range of Rates	Standard Deviation
36.22	15.83 - 56.67	16.69

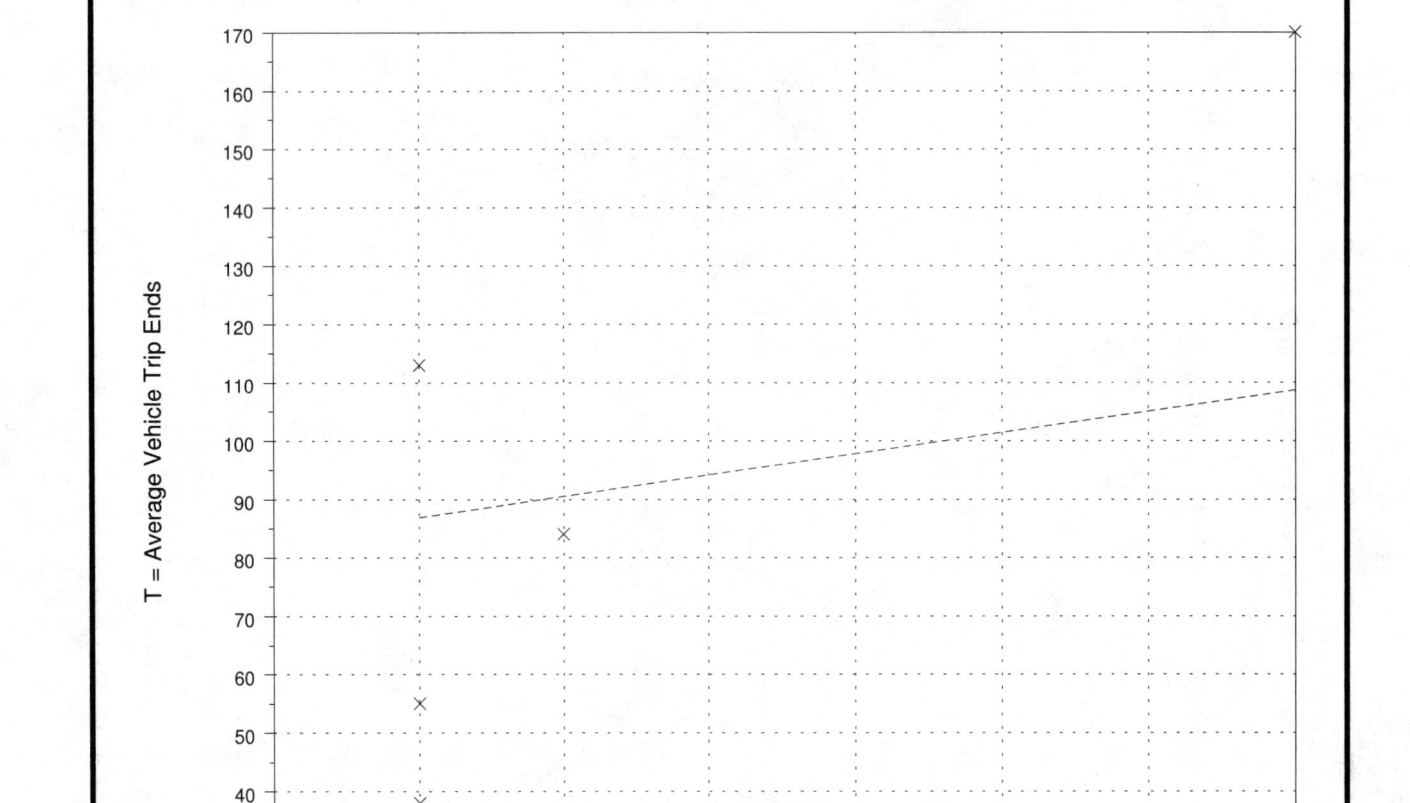

Data Plot and Equation

Caution - Use Carefully - Small Sample Size

Fitted Curve Equation: Not given $R^2 = ****$

Land Use: 853
Convenience Market with Gasoline Pumps

Description

The convenience markets surveyed sell gasoline, convenience foods, newspapers, magazines and often beer and wine. This land use includes convenience markets with gasoline pumps where the primary business is the selling of convenience items, not the fueling of motor vehicles. Convenience market (open 24 hours) (Land Use 851), convenience market (open 15-16 hours) (Land Use 852), gasoline/service station (Land Use 944), gasoline/service station with convenience market (Land Use 945) and gasoline/service station with convenience market and car wash (Land Use 946) are related uses.

Additional Data

The independent variable, vehicle fueling positions, is defined as the maximum number of vehicles that can be fueled simultaneously.

The weekday peak hours of the generator typically coincided with the peak hours of the adjacent street traffic.

The sites were surveyed between the 1980s and the 2000s throughout the United States.

Source Numbers

237, 245, 274, 276, 278, 300, 340, 350, 351, 352, 359, 366, 617, 618, 728

Convenience Market with Gasoline Pumps
(853)

Average Vehicle Trip Ends vs: Vehicle Fueling Positions
On a: Weekday

Number of Studies: 10
Average Vehicle Fueling Positions: 4
Directional Distribution: 50% entering, 50% exiting

Trip Generation per Vehicle Fueling Position

Average Rate	Range of Rates	Standard Deviation
542.60	370.25 - 701.00	113.52

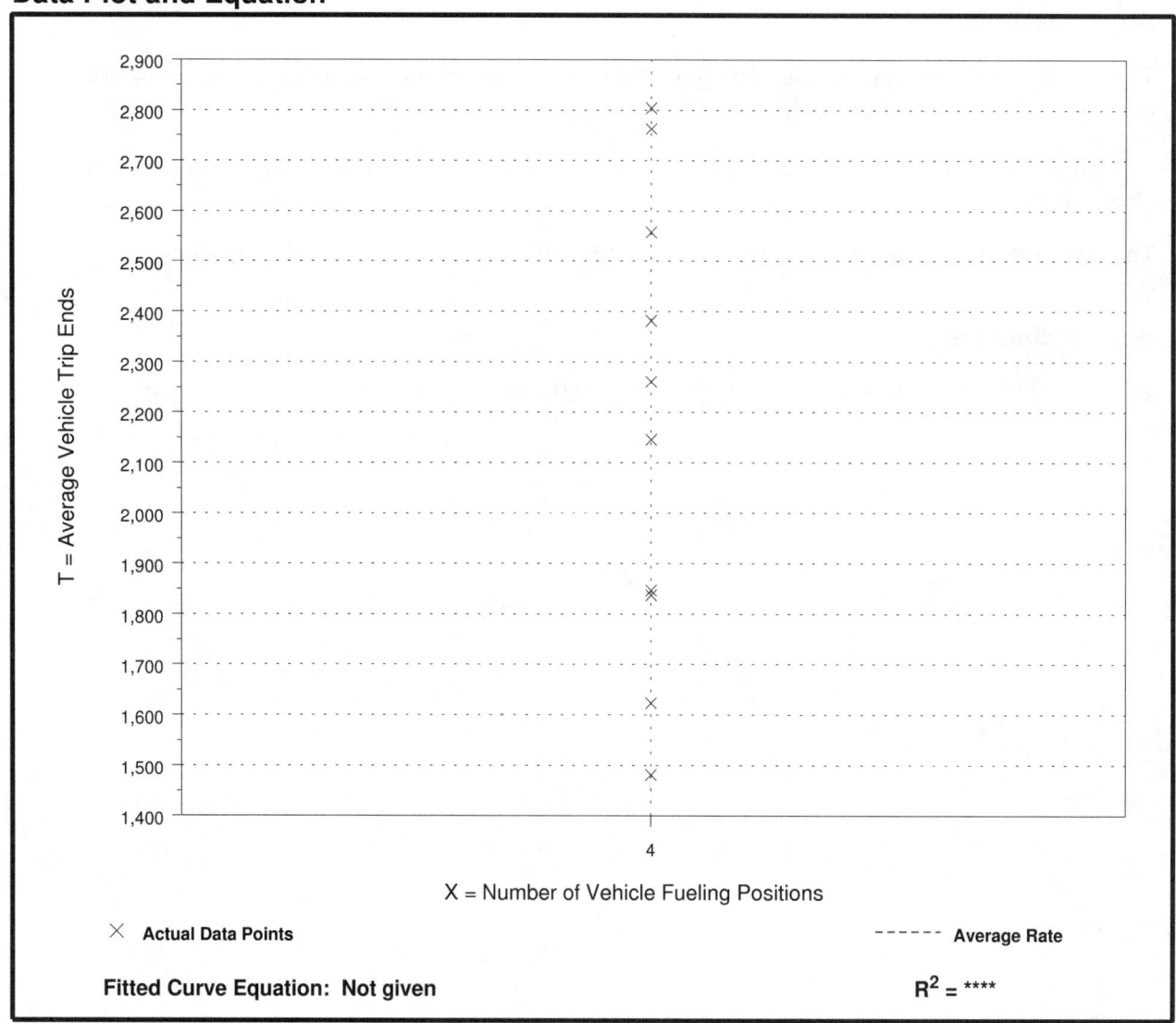

Data Plot and Equation

× Actual Data Points   ------ Average Rate

Fitted Curve Equation: Not given $R^2 =$ ****

Convenience Market with Gasoline Pumps
(853)

Average Vehicle Trip Ends vs: Vehicle Fueling Positions
On a: Weekday,
Peak Hour of Adjacent Street Traffic,
One Hour Between 7 and 9 a.m.

Number of Studies: 28
Average Vehicle Fueling Positions: 7
Directional Distribution: 50% entering, 50% exiting

Trip Generation per Vehicle Fueling Position

Average Rate	Range of Rates	Standard Deviation
16.57	5.40 - 47.00	11.34

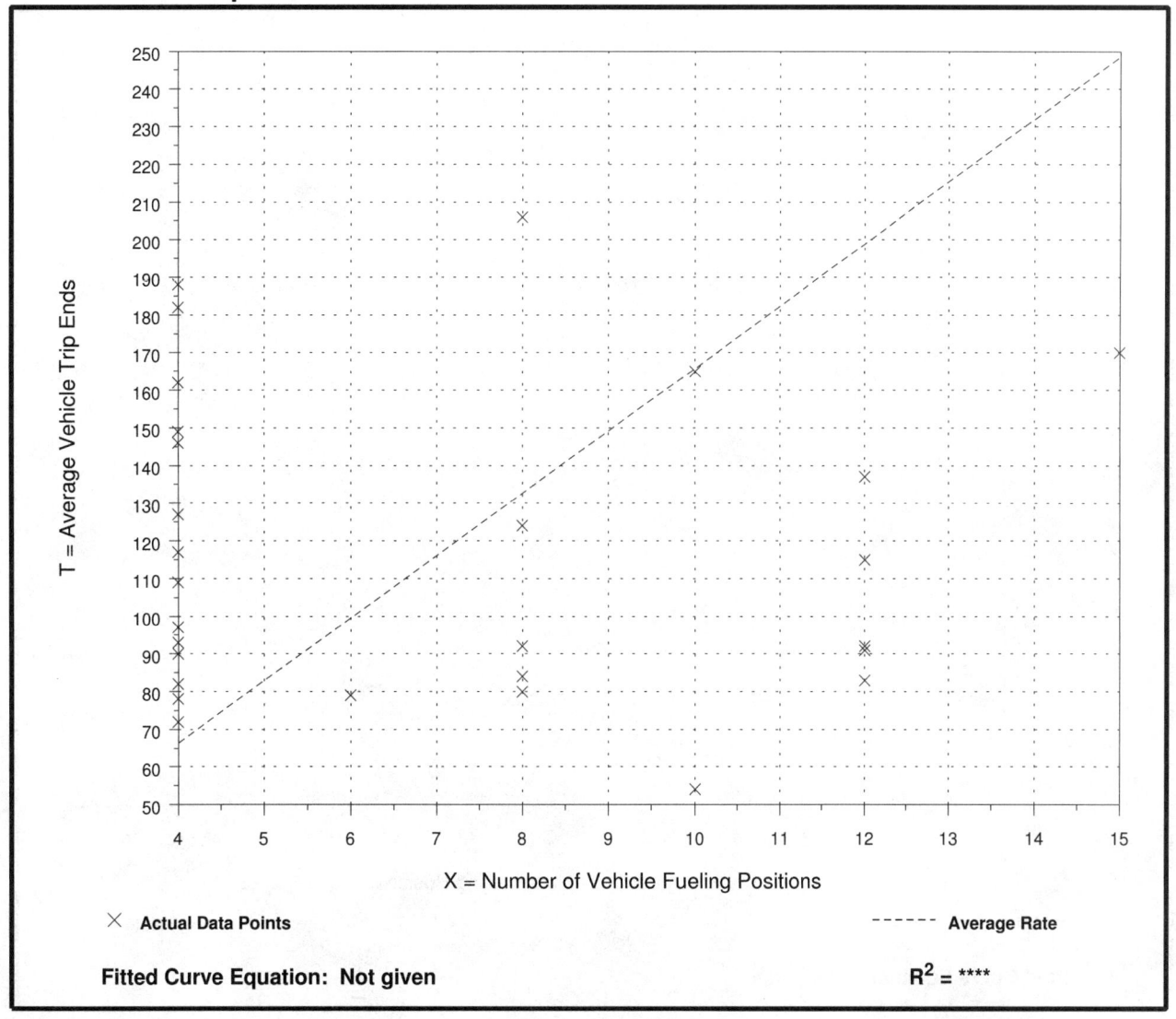

Data Plot and Equation

X = Number of Vehicle Fueling Positions
T = Average Vehicle Trip Ends

× Actual Data Points
------ Average Rate

Fitted Curve Equation: Not given $R^2 = ****$

Trip Generation, 9th Edition • Institute of Transportation Engineers

Convenience Market with Gasoline Pumps
(853)

Average Vehicle Trip Ends vs: Vehicle Fueling Positions
On a: Weekday,
Peak Hour of Adjacent Street Traffic,
One Hour Between 4 and 6 p.m.

Number of Studies: 54
Average Vehicle Fueling Positions: 8
Directional Distribution: 50% entering, 50% exiting

Trip Generation per Vehicle Fueling Position

Average Rate	Range of Rates	Standard Deviation
19.07	5.53 - 75.50	11.93

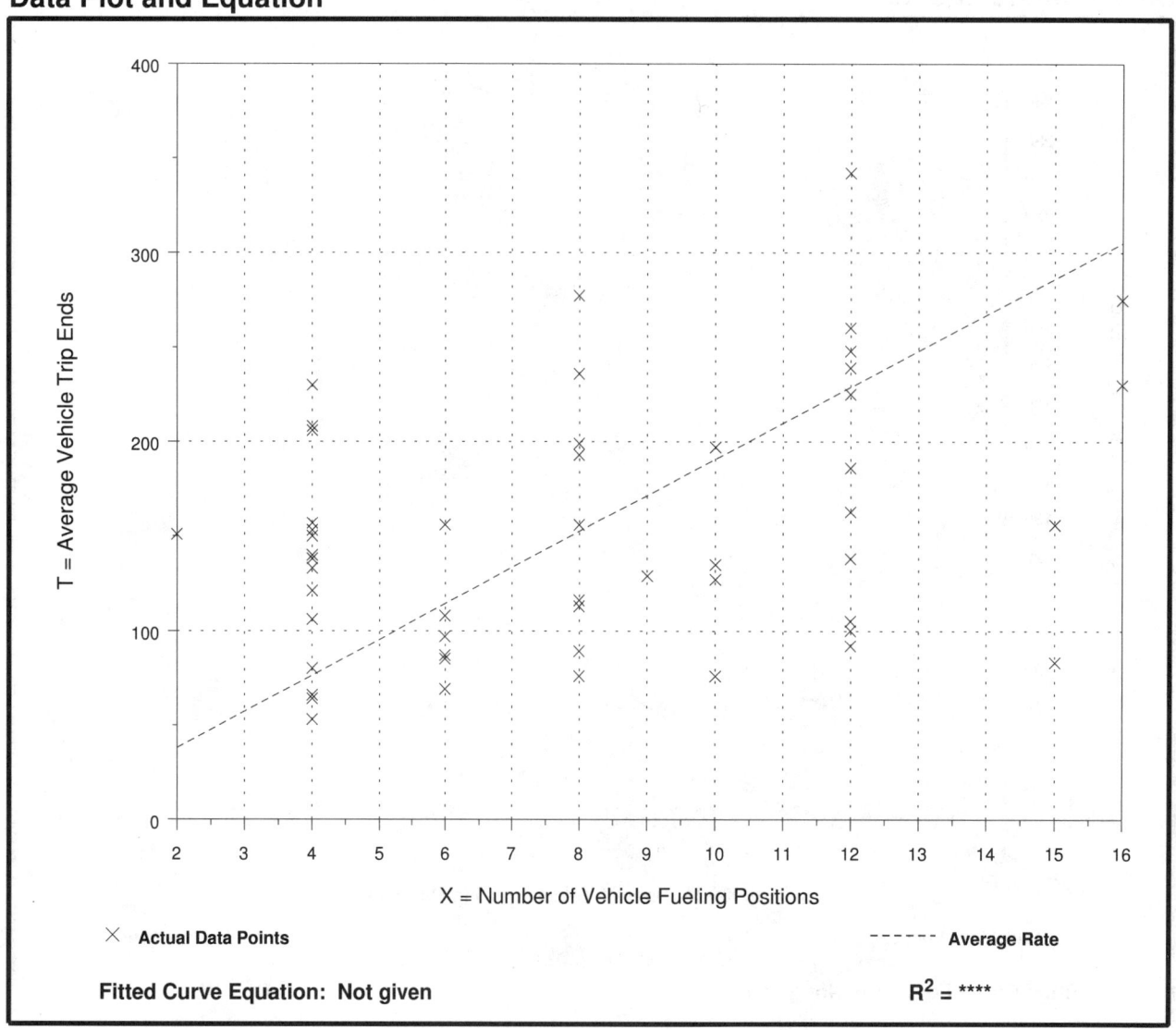

Data Plot and Equation

X = Number of Vehicle Fueling Positions

× Actual Data Points   ----- Average Rate

Fitted Curve Equation: Not given $R^2 = ****$

Convenience Market with Gasoline Pumps
(853)

Average Vehicle Trip Ends vs: Vehicle Fueling Positions
On a: Weekday,
A.M. Peak Hour of Generator

Number of Studies: 26
Average Vehicle Fueling Positions: 7
Directional Distribution: 50% entering, 50% exiting

Trip Generation per Vehicle Fueling Position

Average Rate	Range of Rates	Standard Deviation
17.03	6.92 - 47.00	11.39

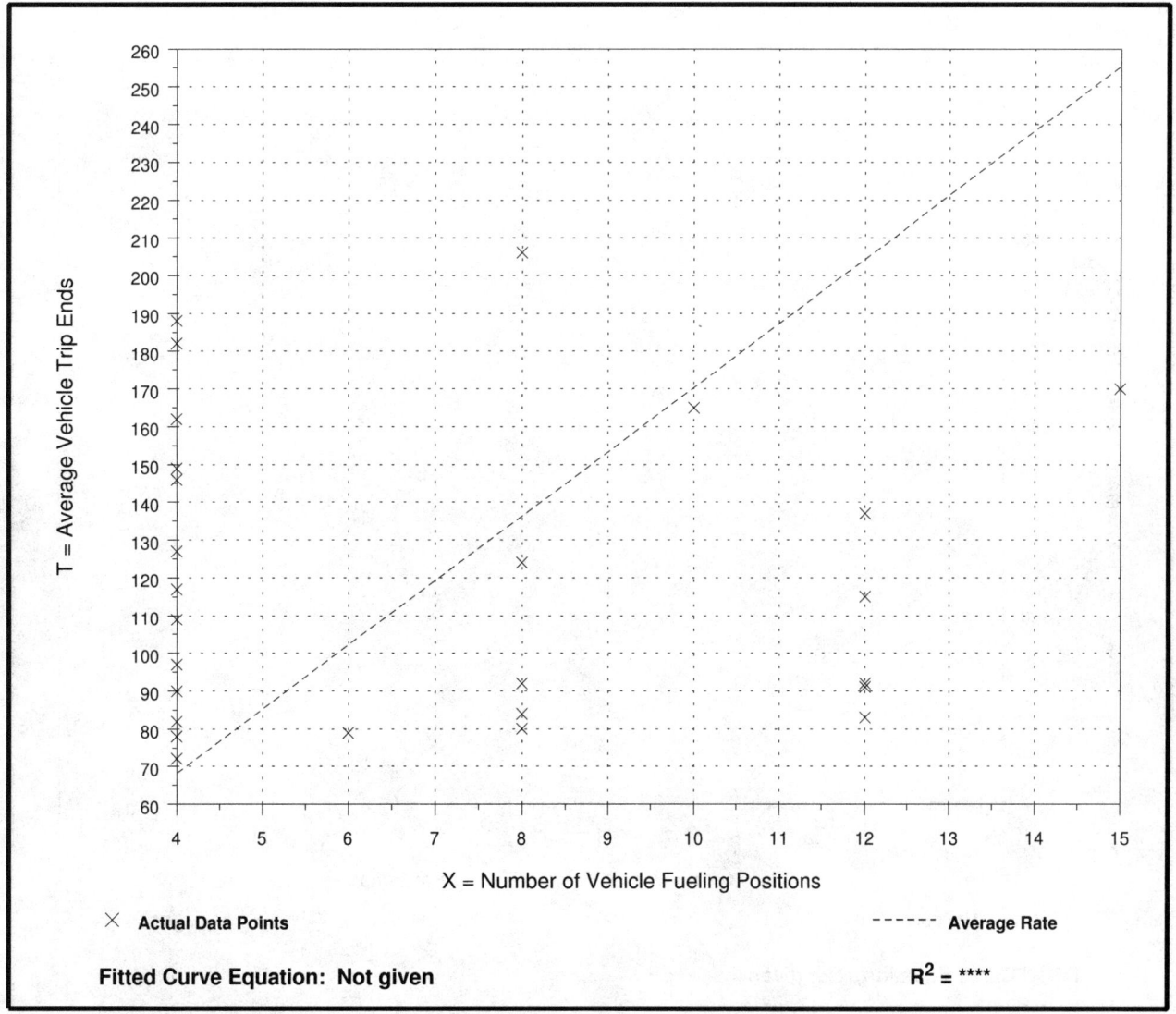

Data Plot and Equation

Fitted Curve Equation: Not given $R^2 = ****$

Convenience Market with Gasoline Pumps
(853)

Average Vehicle Trip Ends vs: Vehicle Fueling Positions
On a: Weekday,
P.M. Peak Hour of Generator

Number of Studies: 46
Average Vehicle Fueling Positions: 8
Directional Distribution: 50% entering, 50% exiting

Trip Generation per Vehicle Fueling Position

Average Rate	Range of Rates	Standard Deviation
19.98	7.60 - 75.50	12.23

Data Plot and Equation

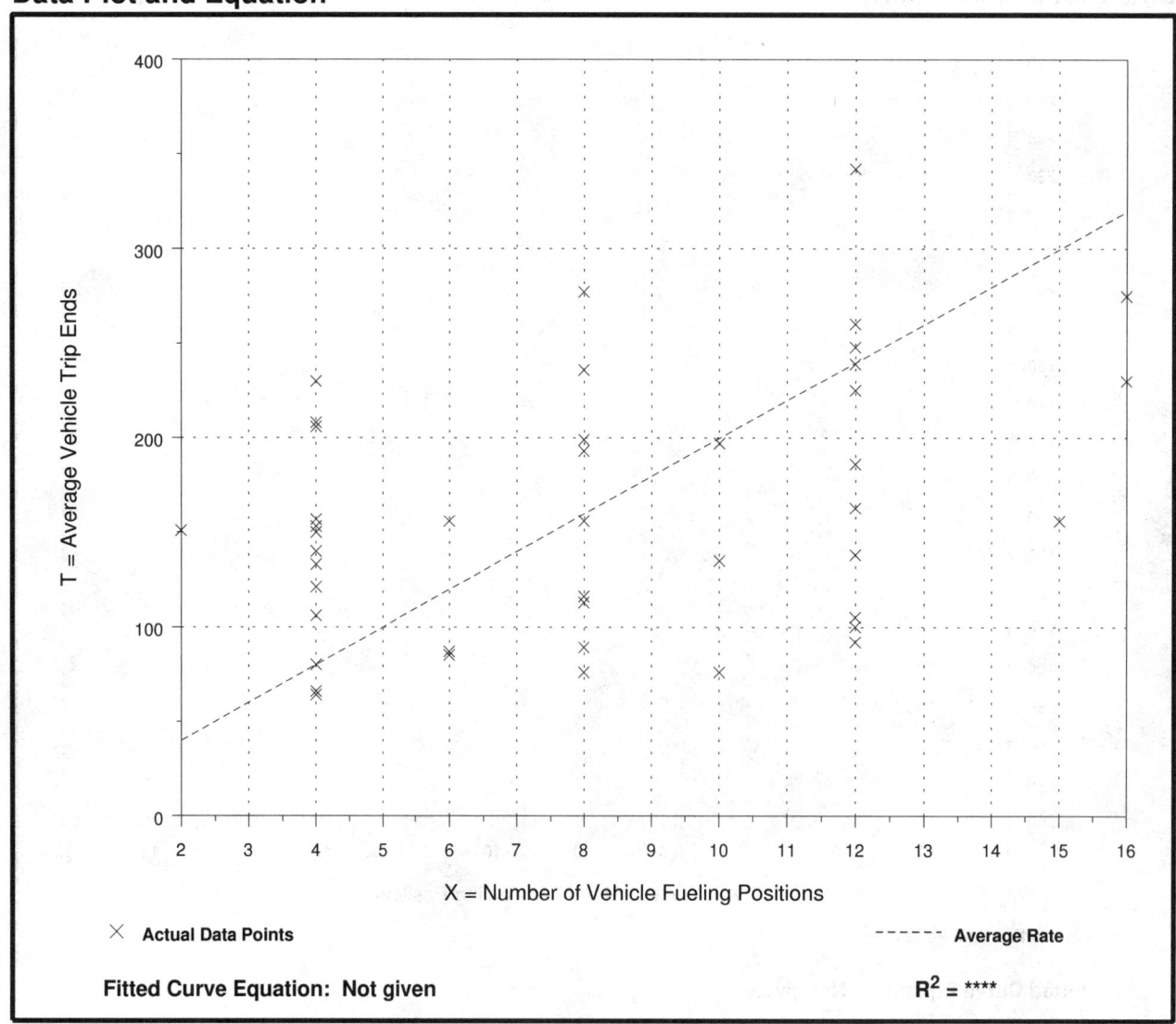

Fitted Curve Equation: Not given $R^2 = ****$

Convenience Market with Gasoline Pumps
(853)

Average Vehicle Trip Ends vs: Vehicle Fueling Positions
On a: Saturday

Number of Studies: 3
Average Vehicle Fueling Positions: 11
Directional Distribution: 50% entering, 50% exiting

Trip Generation per Vehicle Fueling Position

Average Rate	Range of Rates	Standard Deviation
204.47	166.88 - 545.00	88.93

Data Plot and Equation

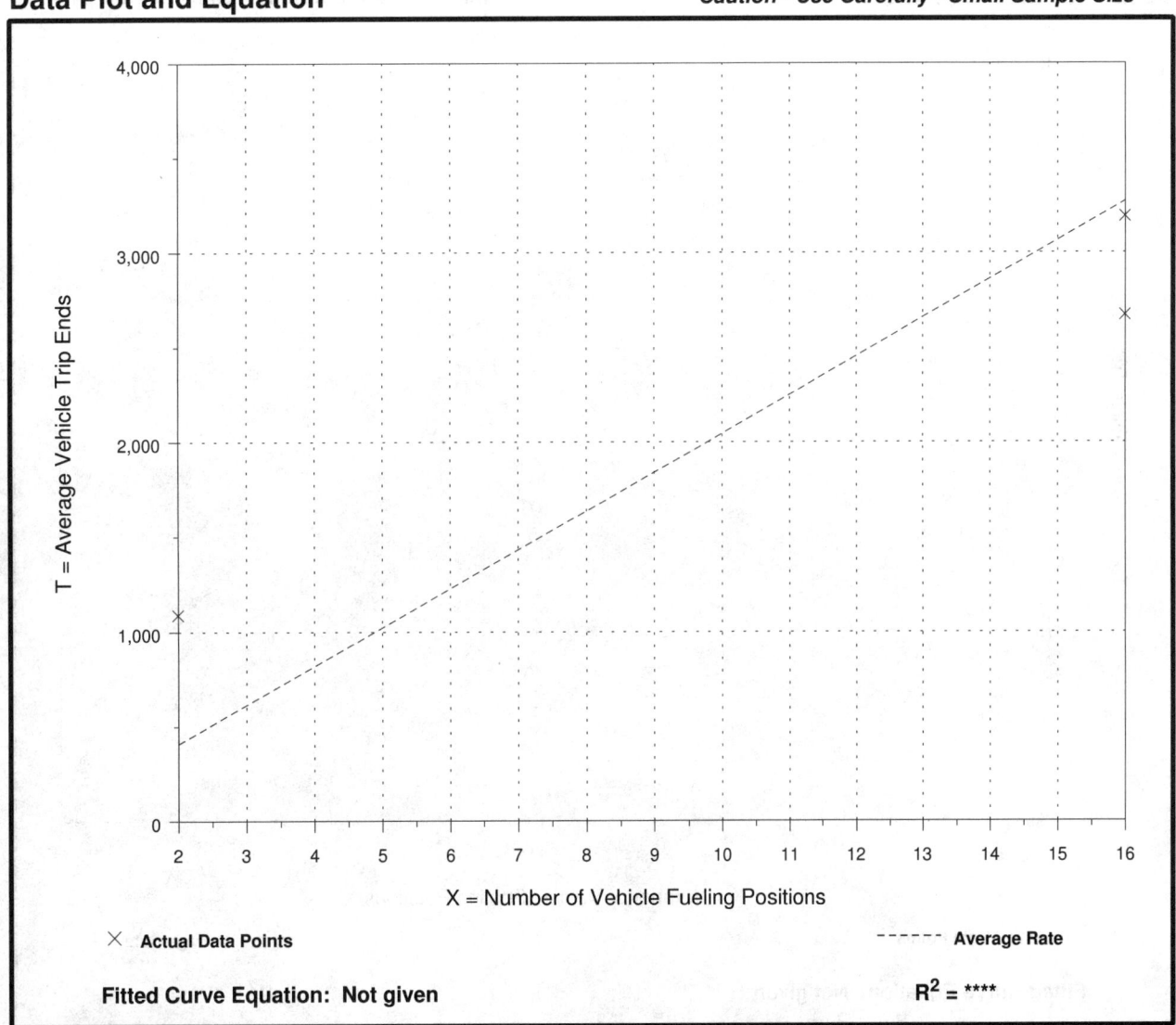

Caution - Use Carefully - Small Sample Size

Fitted Curve Equation: Not given $R^2 = ****$

Convenience Market with Gasoline Pumps
(853)

Average Vehicle Trip Ends vs: Vehicle Fueling Positions
On a: Saturday,
Peak Hour of Generator

Number of Studies: 2
Average Vehicle Fueling Positions: 11
Directional Distribution: 51% entering, 49% exiting

Trip Generation per Vehicle Fueling Position

Average Rate	Range of Rates	Standard Deviation
10.00	9.80 - 10.17	*

Data Plot and Equation

Caution - Use Carefully - Small Sample Size

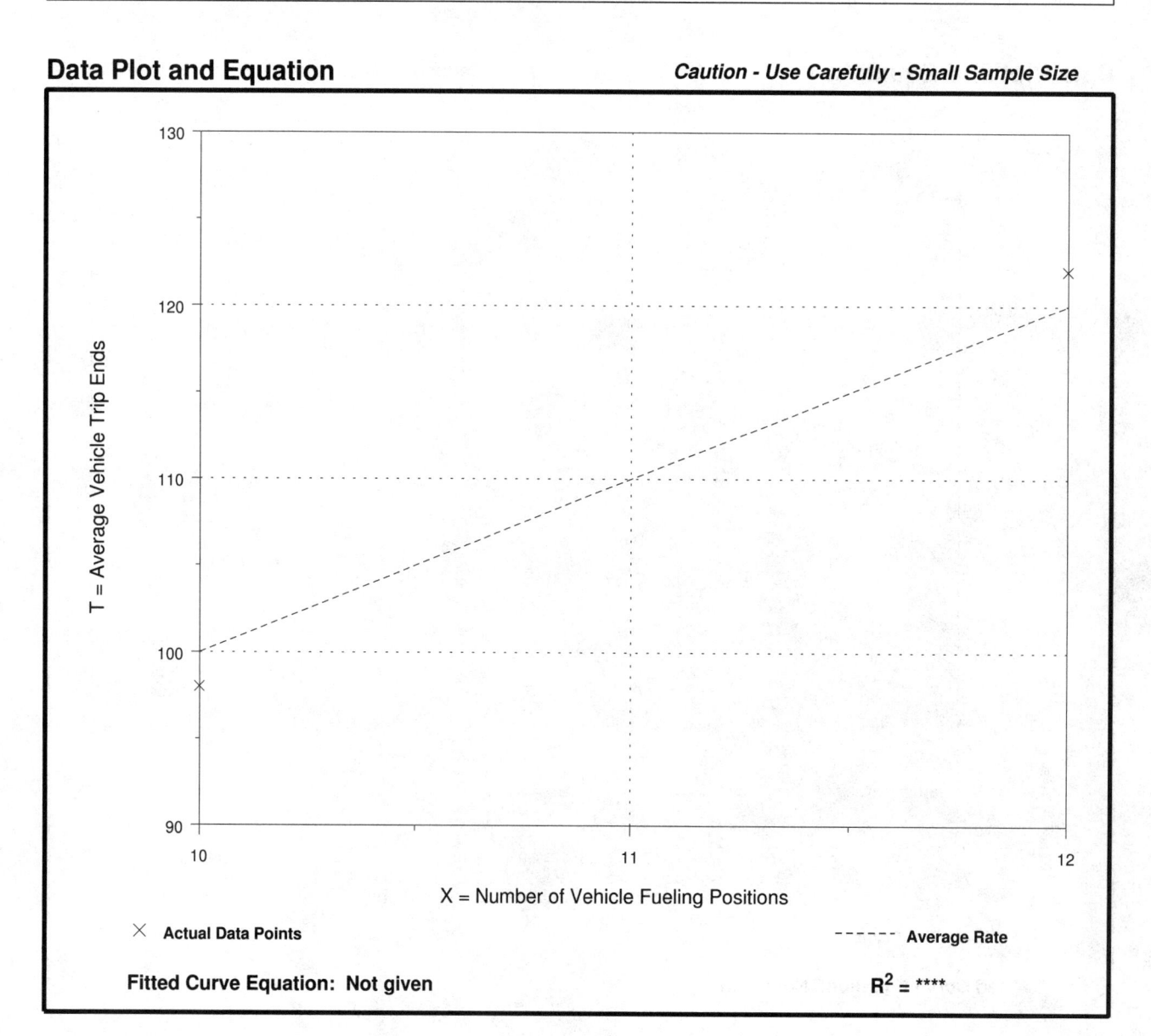

Fitted Curve Equation: Not given $R^2 = ****$

Convenience Market with Gasoline Pumps
(853)

Average Vehicle Trip Ends vs: Vehicle Fueling Positions
On a: Sunday

Number of Studies: 3
Average Vehicle Fueling Positions: 11
Directional Distribution: 50% entering, 50% exiting

Trip Generation per Vehicle Fueling Position

Average Rate	Range of Rates	Standard Deviation
166.88	146.75 - 439.00	70.19

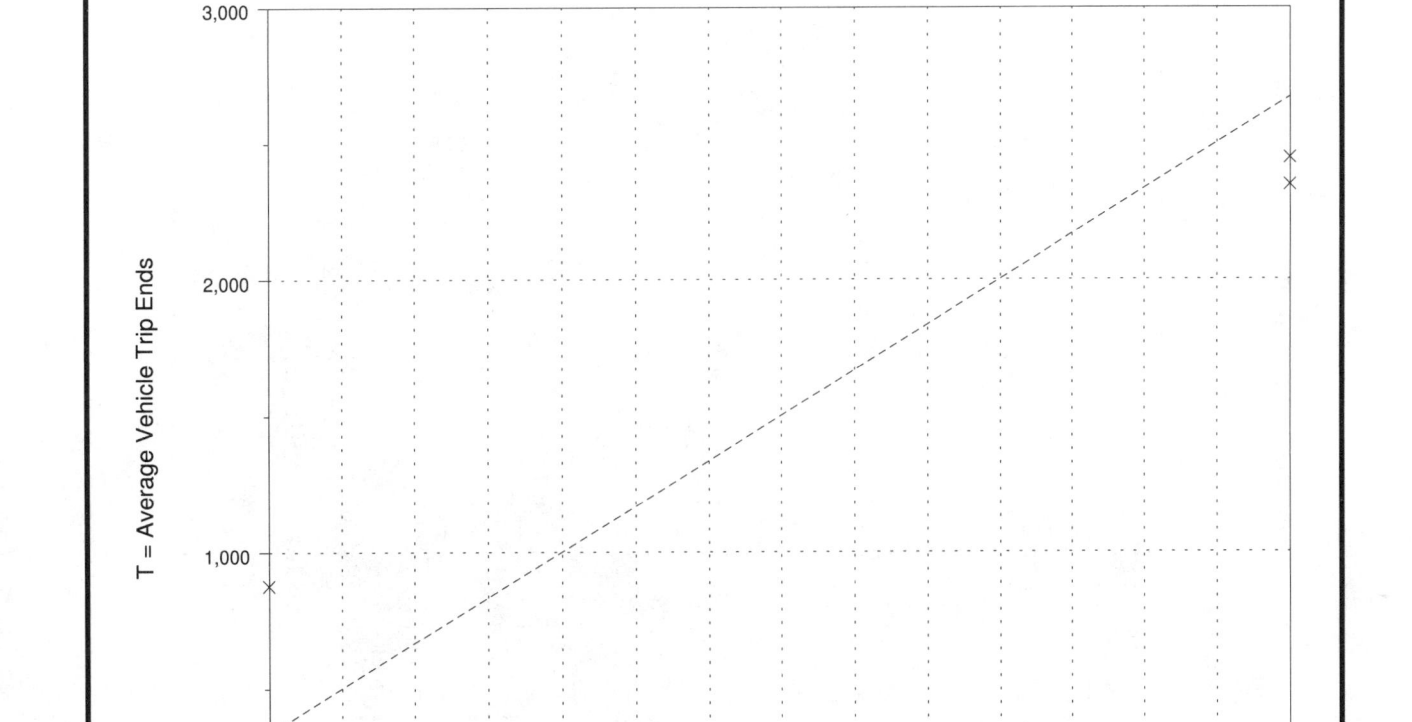

Data Plot and Equation

Caution - Use Carefully - Small Sample Size

Fitted Curve Equation: Not given $R^2 = ****$

Convenience Market with Gasoline Pumps
(853)

Average Vehicle Trip Ends vs: 1000 Sq. Feet Gross Floor Area
On a: Weekday

Number of Studies: 10
Average 1000 Sq. Feet GFA: 3
Directional Distribution: 50% entering, 50% exiting

Trip Generation per 1000 Sq. Feet Gross Floor Area

Average Rate	Range of Rates	Standard Deviation
845.60	578.52 - 1084.72	163.67

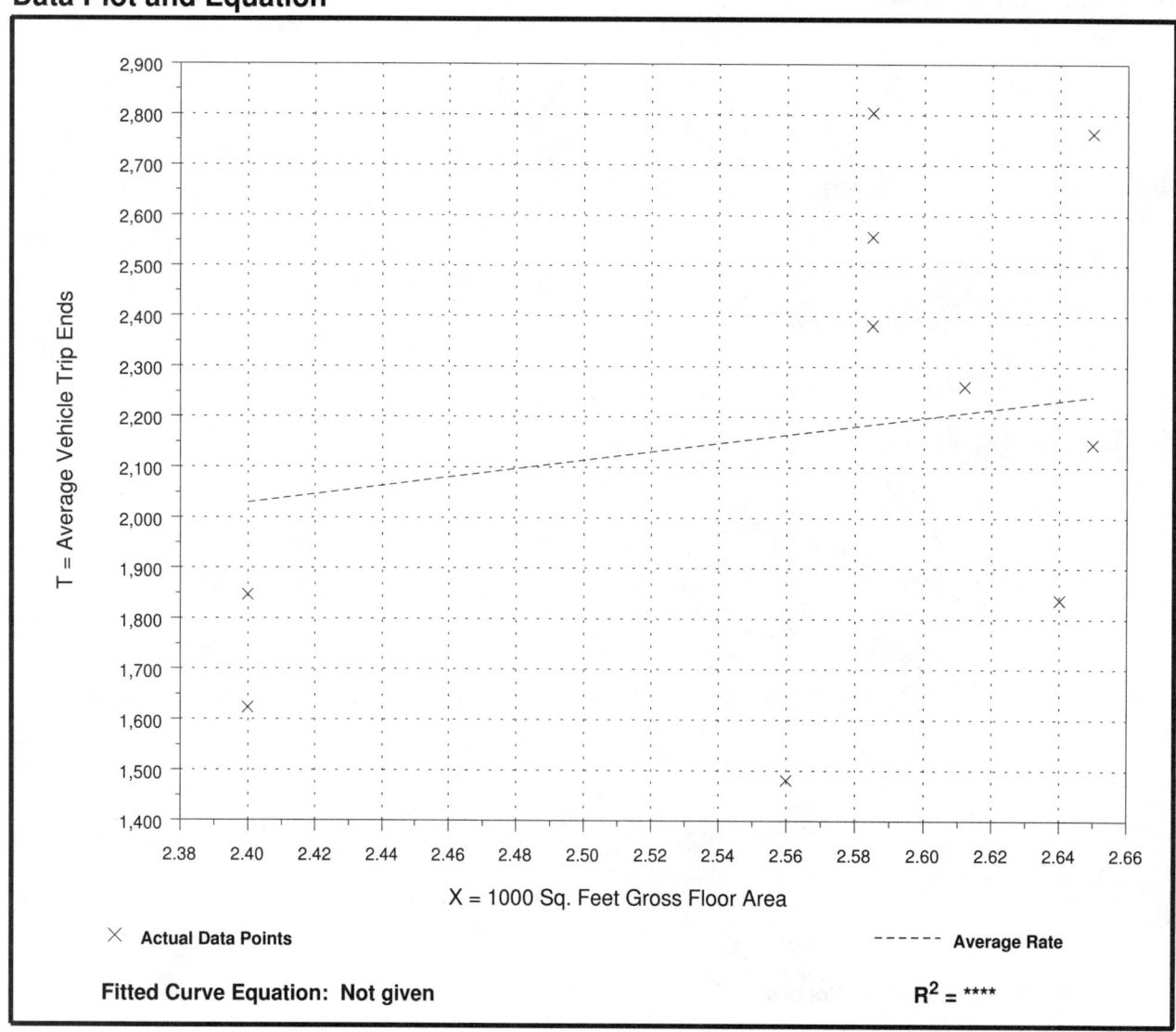

Data Plot and Equation

Fitted Curve Equation: Not given $R^2 = ****$

Convenience Market with Gasoline Pumps
(853)

Average Vehicle Trip Ends vs: 1000 Sq. Feet Gross Floor Area
On a: Weekday,
Peak Hour of Adjacent Street Traffic,
One Hour Between 7 and 9 a.m.

Number of Studies: 53
Average 1000 Sq. Feet GFA: 3
Directional Distribution: 50% entering, 50% exiting

Trip Generation per 1000 Sq. Feet Gross Floor Area

Average Rate	Range of Rates	Standard Deviation
40.92	11.67 - 119.29	20.75

Data Plot and Equation

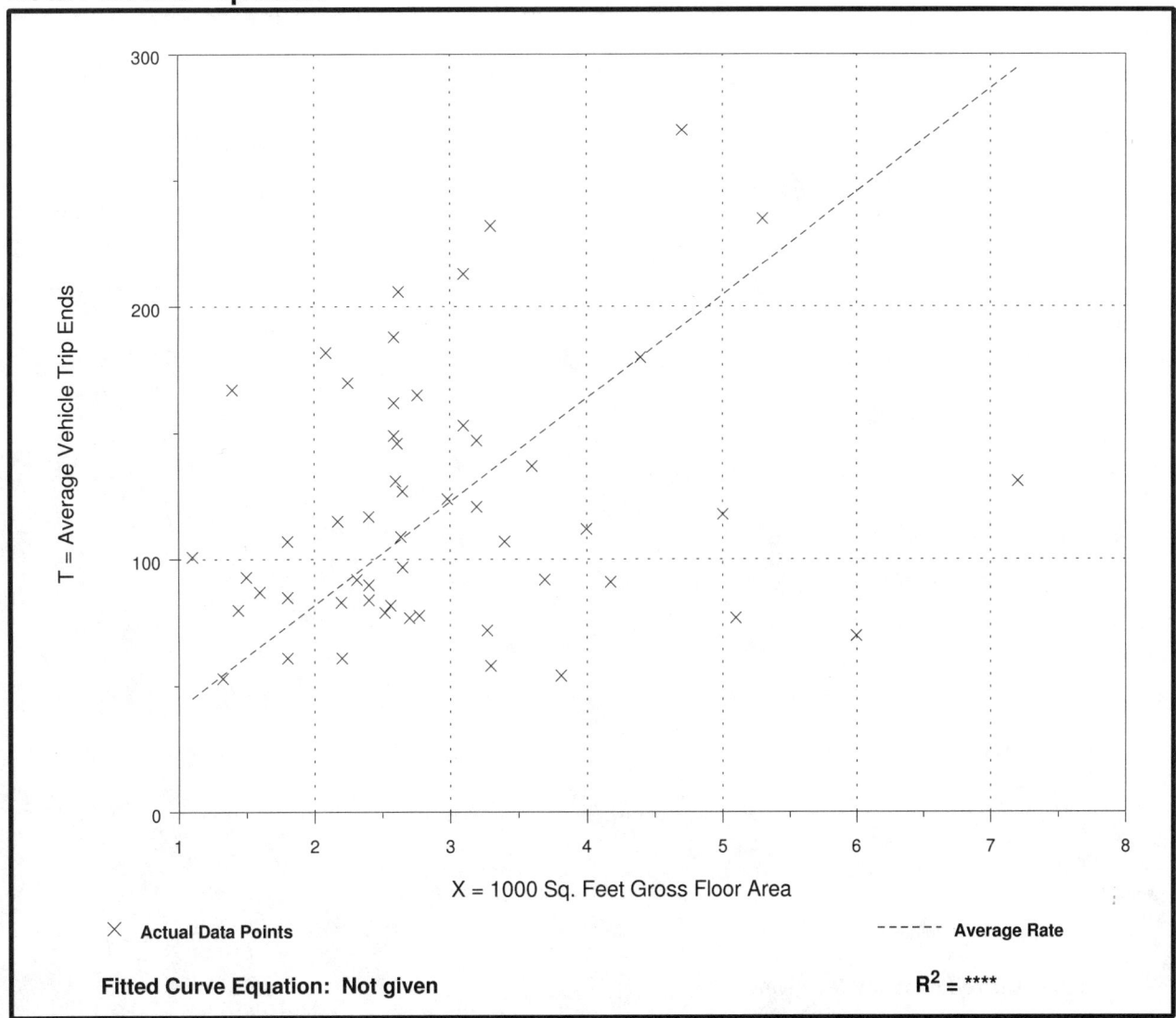

Fitted Curve Equation: Not given $R^2 = ****$

Convenience Market with Gasoline Pumps
(853)

Average Vehicle Trip Ends vs: 1000 Sq. Feet Gross Floor Area
On a: Weekday,
Peak Hour of Adjacent Street Traffic,
One Hour Between 4 and 6 p.m.

Number of Studies: 78
Average 1000 Sq. Feet GFA: 3
Directional Distribution: 50% entering, 50% exiting

Trip Generation per 1000 Sq. Feet Gross Floor Area

Average Rate	Range of Rates	Standard Deviation
50.92	13.53 - 292.89	32.15

Data Plot and Equation

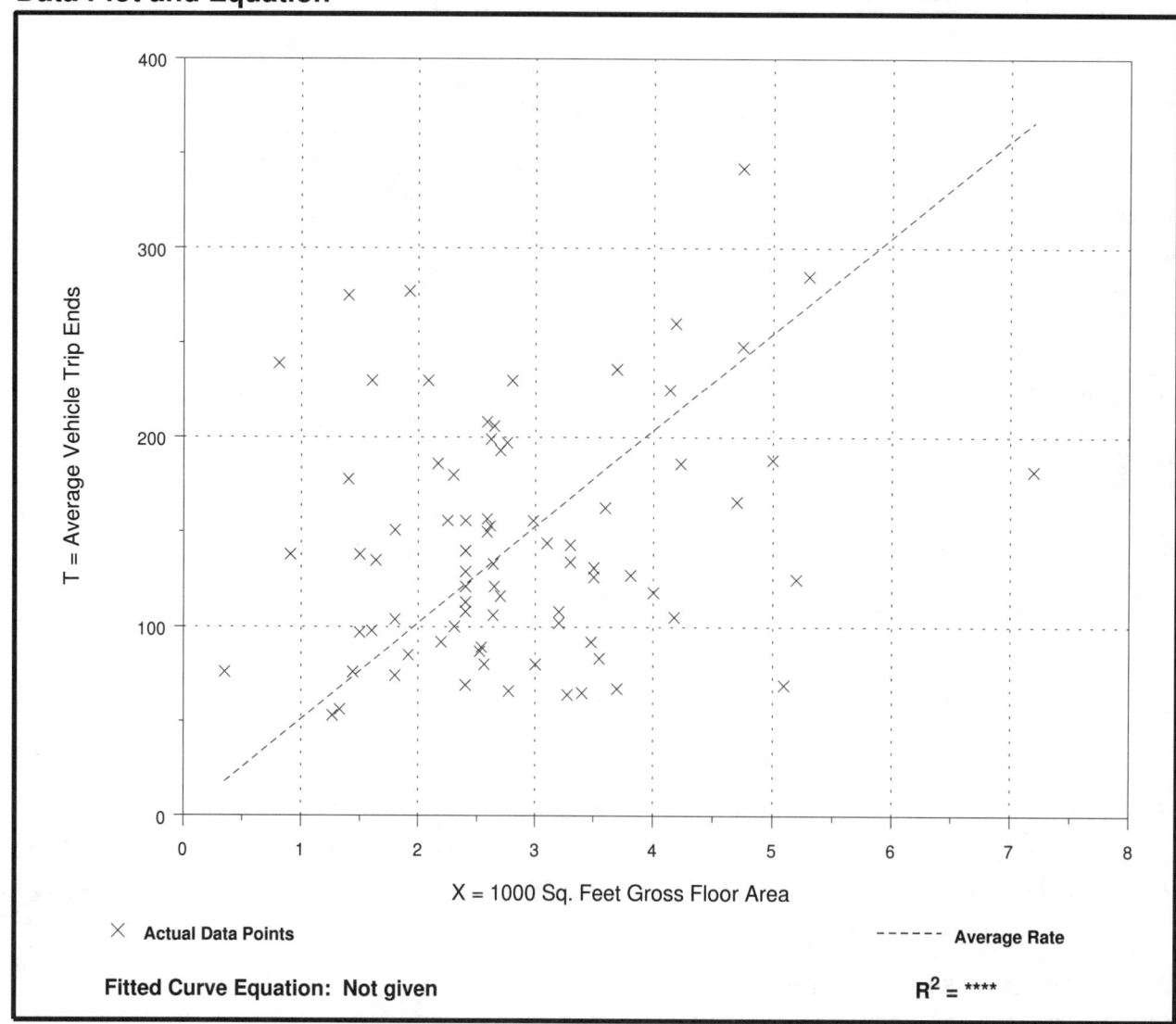

Fitted Curve Equation: Not given $R^2 = ****$

Convenience Market with Gasoline Pumps
(853)

Average Vehicle Trip Ends vs: 1000 Sq. Feet Gross Floor Area
On a: Weekday,
A.M. Peak Hour of Generator

Number of Studies: 34
Average 1000 Sq. Feet GFA: 3
Directional Distribution: 50% entering, 50% exiting

Trip Generation per 1000 Sq. Feet Gross Floor Area

Average Rate	Range of Rates	Standard Deviation
42.86	20.45 - 99.23	18.85

Data Plot and Equation

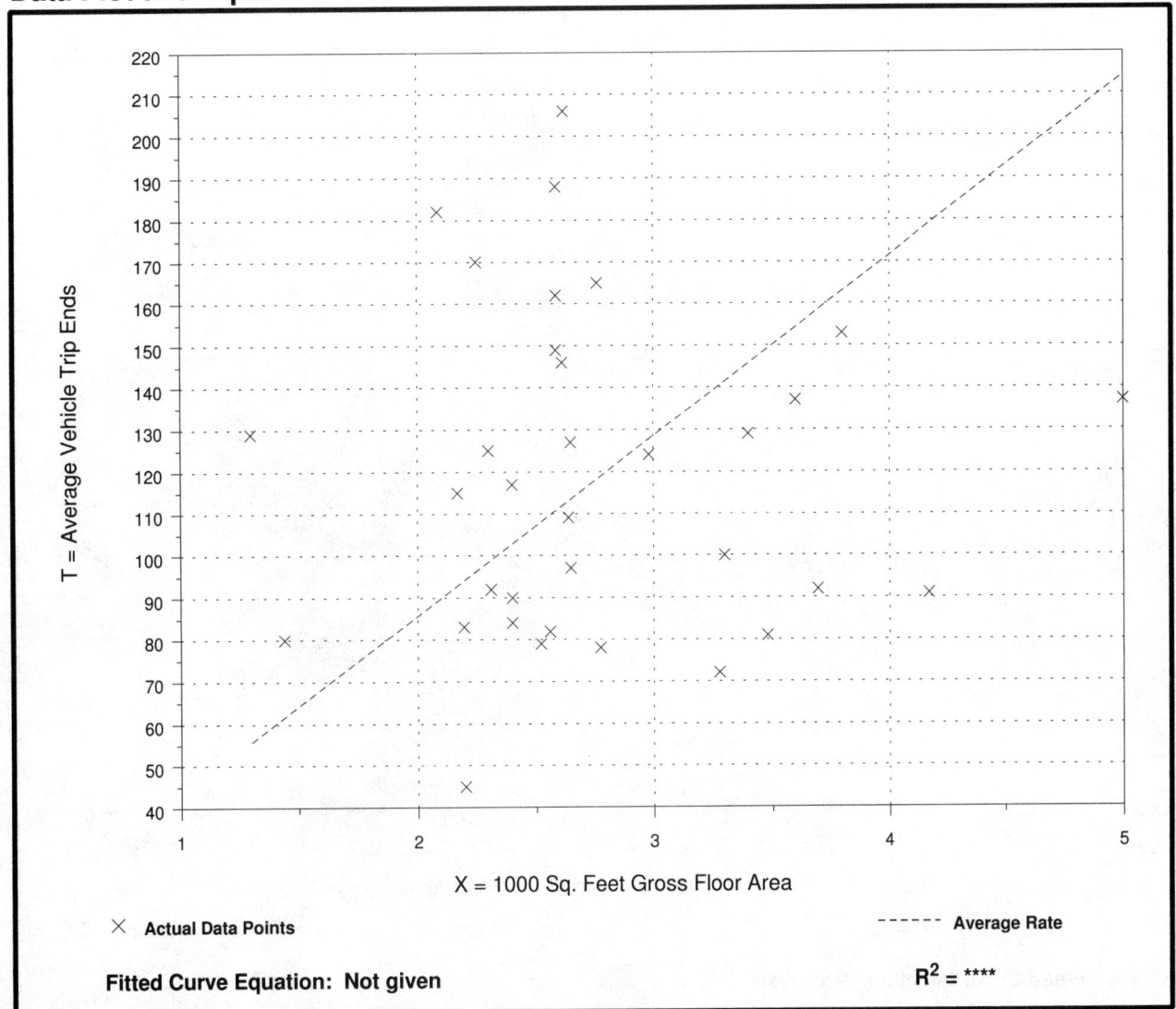

Fitted Curve Equation: Not given $R^2 = ****$

Convenience Market with Gasoline Pumps
(853)

Average Vehicle Trip Ends vs: 1000 Sq. Feet Gross Floor Area
On a: Weekday,
P.M. Peak Hour of Generator

Number of Studies: 46
Average 1000 Sq. Feet GFA: 3
Directional Distribution: 50% entering, 50% exiting

Trip Generation per 1000 Sq. Feet Gross Floor Area

Average Rate	Range of Rates	Standard Deviation
62.57	19.54 - 292.89	36.32

Data Plot and Equation

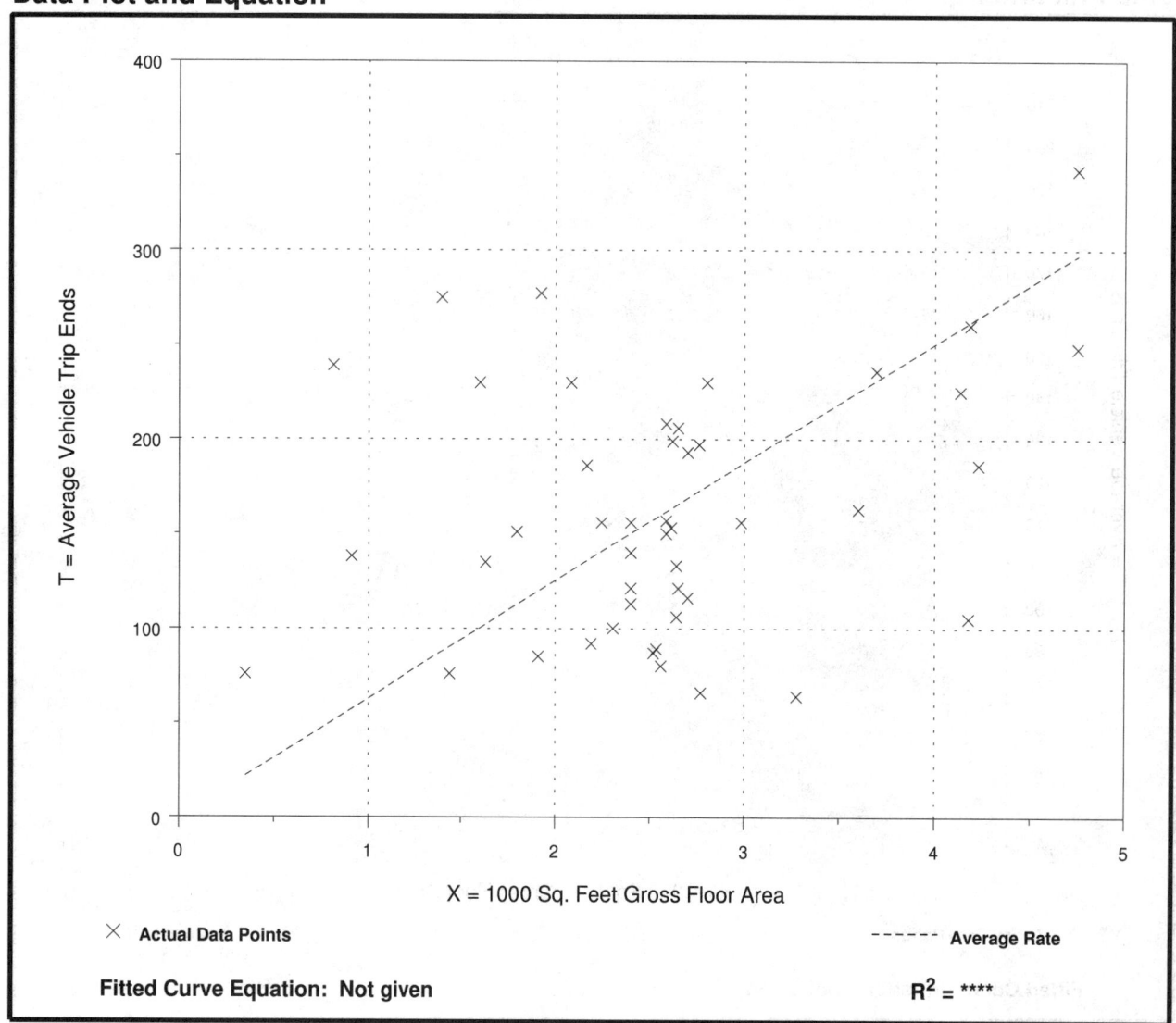

Fitted Curve Equation: Not given $R^2 = ****$

Convenience Market with Gasoline Pumps
(853)

Average Vehicle Trip Ends vs: 1000 Sq. Feet Gross Floor Area
On a: Saturday

Number of Studies: 3
Average 1000 Sq. Feet GFA: 2
Directional Distribution: 50% entering, 50% exiting

Trip Generation per 1000 Sq. Feet Gross Floor Area

Average Rate	Range of Rates	Standard Deviation
1448.33	605.56 - 1995.00	735.17

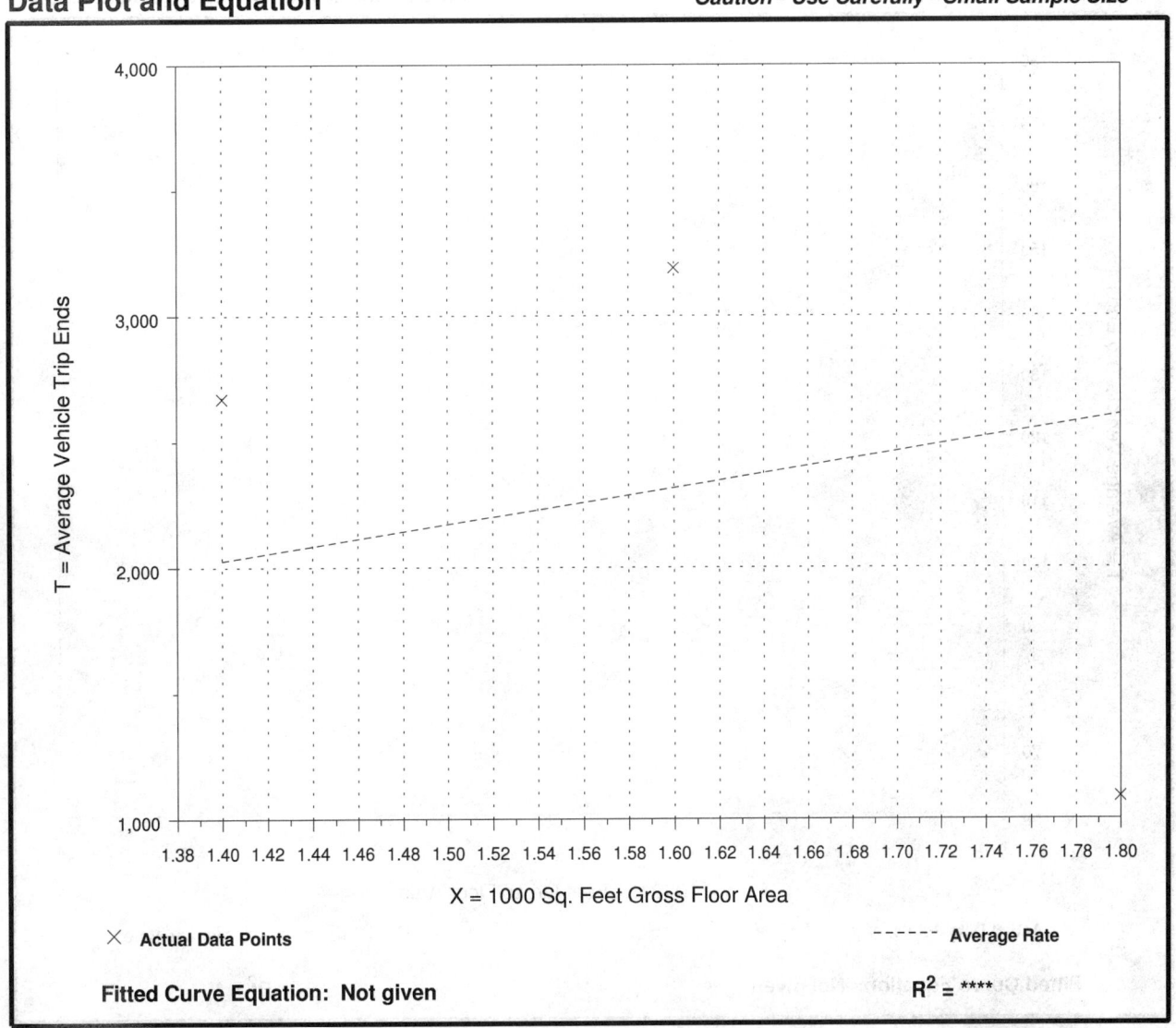

Data Plot and Equation

Caution - Use Carefully - Small Sample Size

Fitted Curve Equation: Not given $R^2 =$ ****

Convenience Market with Gasoline Pumps
(853)

Average Vehicle Trip Ends vs: 1000 Sq. Feet Gross Floor Area
On a: Saturday,
Peak Hour of Generator

Number of Studies: 3
Average 1000 Sq. Feet GFA: 2
Directional Distribution: 51% entering, 49% exiting

Trip Generation per 1000 Sq. Feet Gross Floor Area

Average Rate	Range of Rates	Standard Deviation
45.94	25.70 - 88.17	26.05

Data Plot and Equation

Caution - Use Carefully - Small Sample Size

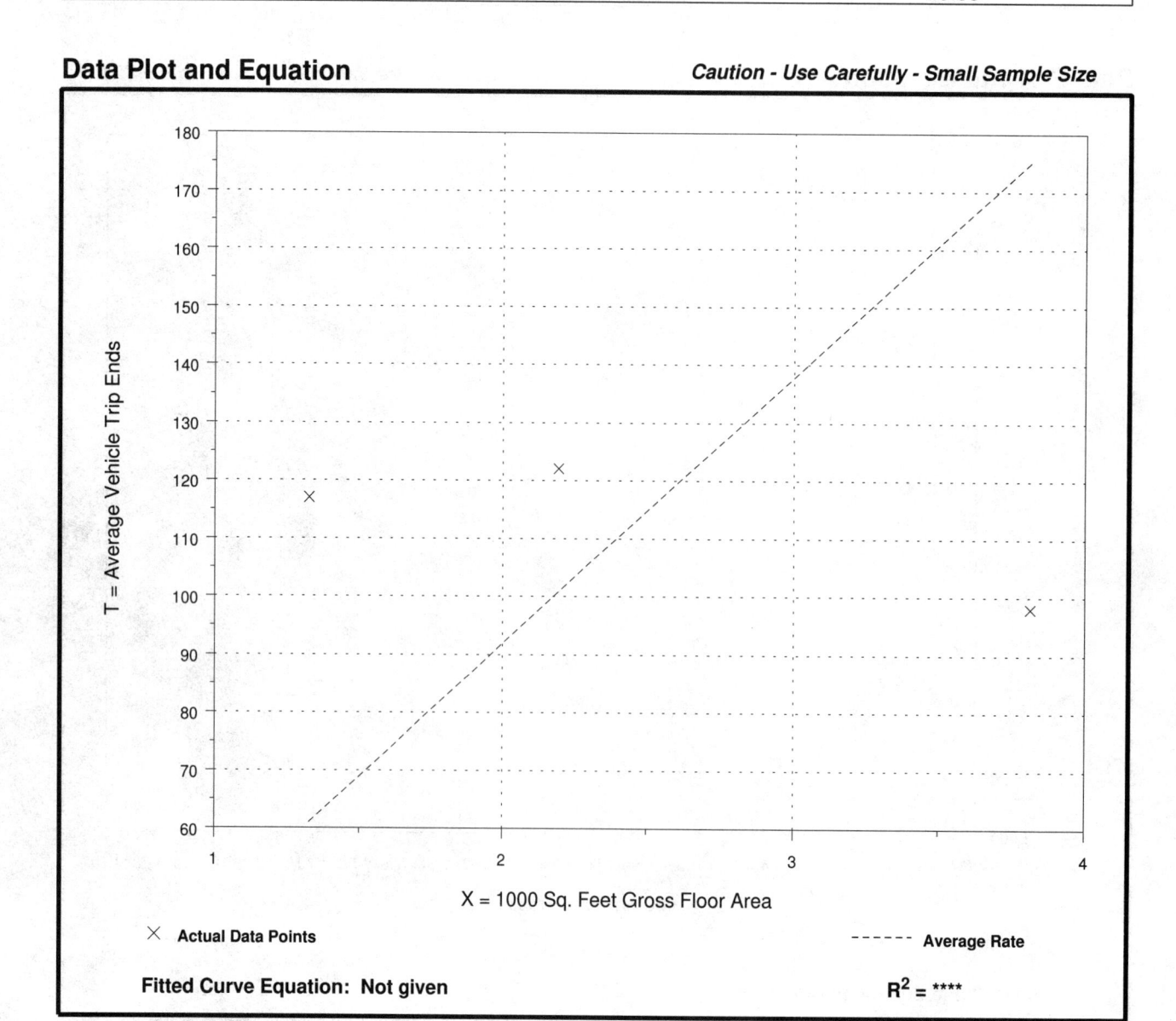

X Actual Data Points
----- Average Rate

Fitted Curve Equation: Not given $R^2 = ****$

Convenience Market with Gasoline Pumps
(853)

Average Vehicle Trip Ends vs: 1000 Sq. Feet Gross Floor Area
On a: Sunday

Number of Studies: 3
Average 1000 Sq. Feet GFA: 2
Directional Distribution: 50% entering, 50% exiting

Trip Generation per 1000 Sq. Feet Gross Floor Area

Average Rate	Range of Rates	Standard Deviation
1182.08	487.78 - 1677.14	608.38

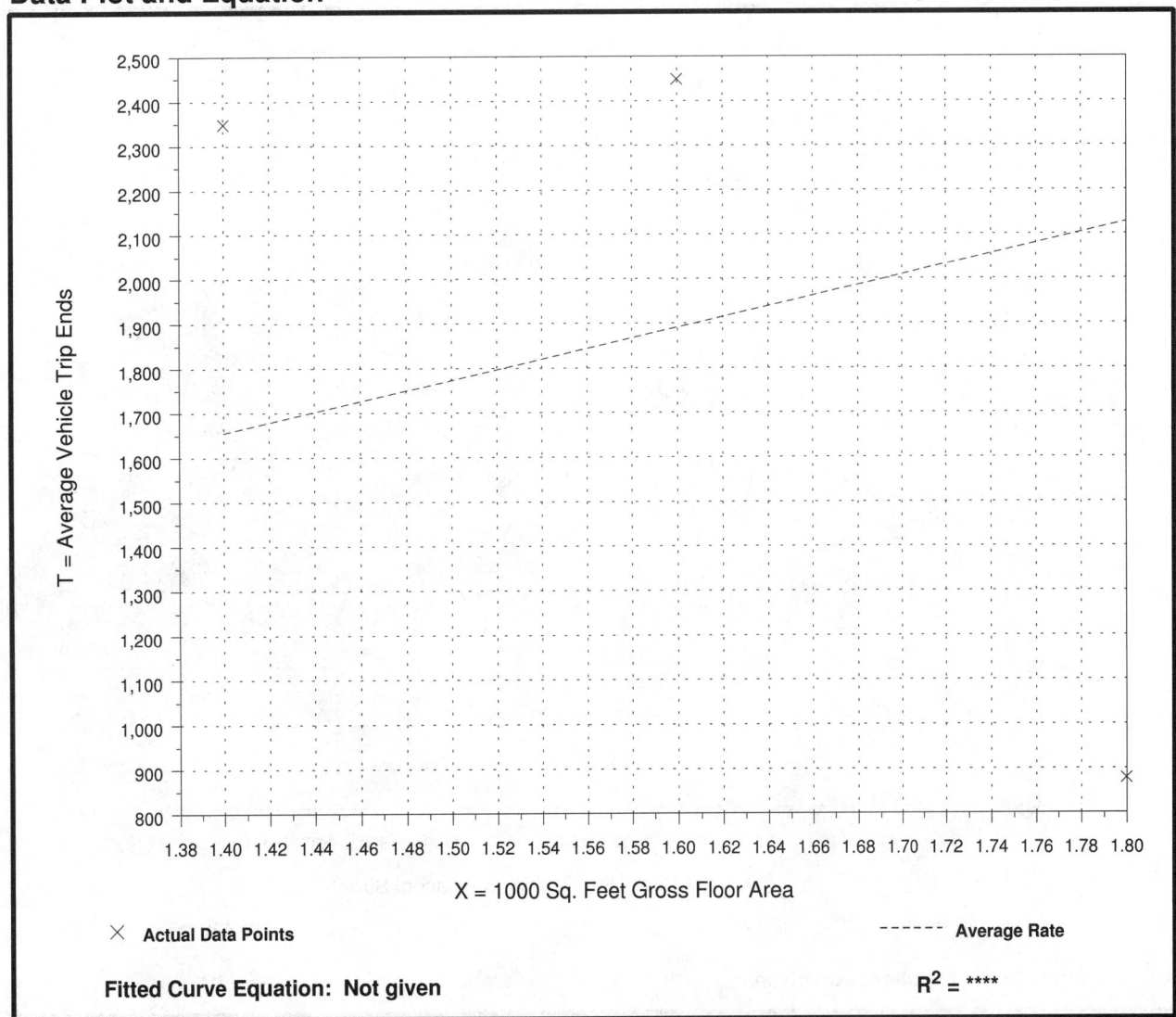

Data Plot and Equation

Caution - Use Carefully - Small Sample Size

X Actual Data Points   ----- Average Rate

Fitted Curve Equation: Not given $R^2 = ****$

Convenience Market with Gasoline Pumps
(853)

Average Vehicle Trip Ends vs: A.M. Peak Hour Traffic on Adjacent Street
On a: Weekday,
Peak Hour of Adjacent Street Traffic,
One Hour Between 7 and 9 a.m.

Number of Studies: 11
Avg. A.M. Peak Hr. Traf. on Adj. Street: 1,146
Directional Distribution: 51% entering, 49% exiting

Trip Generation per AM Peak Hour Traffic on Adjacent Street

Average Rate	Range of Rates	Standard Deviation
0.10	0.04 - 0.22	0.31

Data Plot and Equation

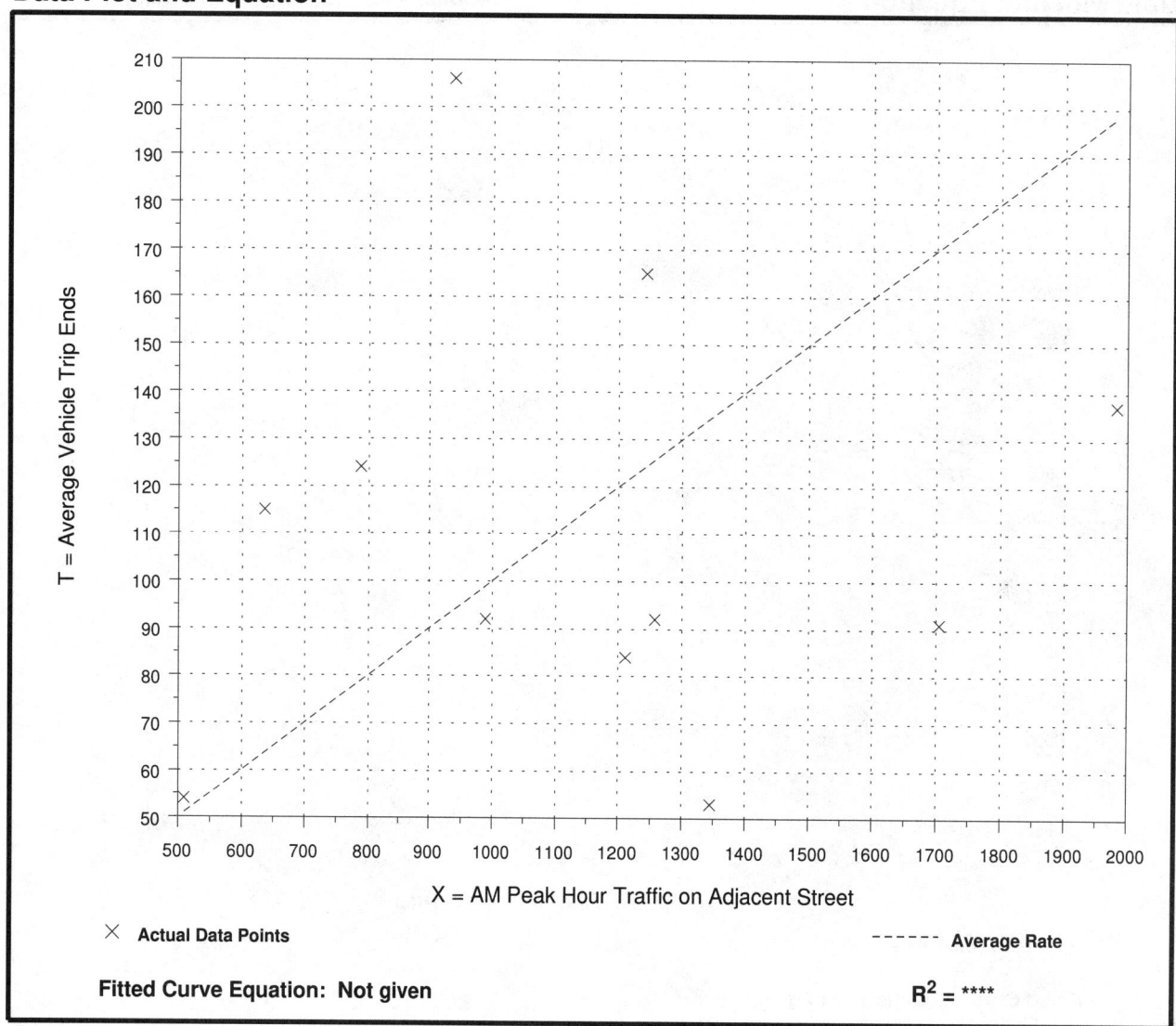

Fitted Curve Equation: Not given $R^2 =$ ****

Convenience Market with Gasoline Pumps
(853)

Average Vehicle Trip Ends vs: P.M. Peak Hour Traffic on Adjacent Street
On a: Weekday,
Peak Hour of Adjacent Street Traffic,
One Hour Between 4 and 6 p.m.

Number of Studies: 11
Avg. P.M. Peak Hr. Traf. on Adj. Street: 1,968
Directional Distribution: 50% entering, 50% exiting

Trip Generation per P.M. Peak Hour Traffic on Adjacent Street

Average Rate	Range of Rates	Standard Deviation
0.08	0.02 - 0.23	0.28

Data Plot and Equation

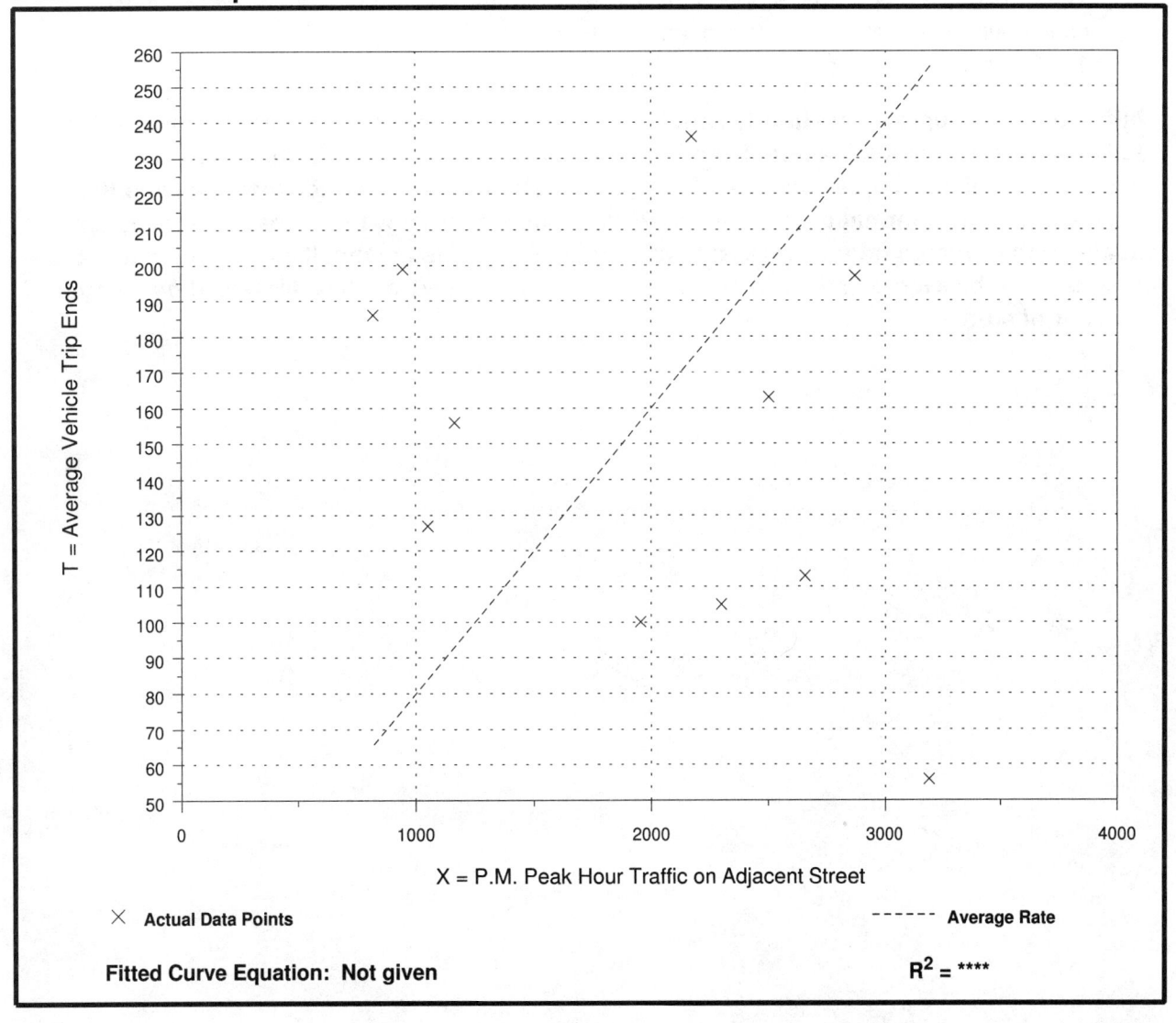

Fitted Curve Equation: Not given $R^2 = ****$

Land Use: 854
Discount Supermarket

Description

Discount supermarkets are free-standing retail stores selling a complete assortment of food (often in bulk), food preparation and wrapping materials, and household cleaning and servicing items at discounted prices. Some facilities may be open 24 hours a day. Supermarket (Land Use 850) is a related use.

Additional Data

Peak hours of the generator—
The weekday A.M. peak hour varied between 10:00 a.m. and 12:00 p.m. The weekday P.M. peak hour varied between 3:00 p.m. and 6:00 p.m. The weekend peak hour varied between 11:30 a.m. and 5:00 p.m.

Information on approximate hourly variation in discount supermarket traffic is shown in the following table. It should be noted, however, that the information contained in this table is based on a limited sample size. Therefore, caution should be exercised when applying the data. Also, some information provided in the table may conflict with the results obtained by applying the average rate or regression equations. When this occurs, it is suggested that the results from the average rate or regression equations be used, as they are based on a larger number of studies.

| Hourly Variation in Discount Supermarket Traffic ||||||
| Time | Average Weekday[a] || Average Saturday[b] || Average Sunday[c] ||
	Percent of 24-Hour Entering Traffic	Percent of 24-Hour Exiting Traffic	Percent of 24-Hour Entering Traffic	Percent of 24-Hour Exiting Traffic	Percent of 24-Hour Entering Traffic	Percent of 24-Hour Exiting Traffic
6 a.m.–7 a.m.	0.6	0.3	0.4	0.2	0.4	0.1
7 a.m.–8 a.m.	1.0	0.6	0.8	0.2	0.5	0.3
8 a.m.–9 a.m.	2.3	1.0	3.0	1.0	1.2	0.5
9 a.m.–10 a.m.	6.0	3.2	6.6	3.9	4.3	2.1
10 a.m.–11 a.m.	6.7	6.1	8.5	6.8	7.9	6.0
11 a.m.–12 p.m.	6.7	6.8	10.0	9.3	9.7	9.2
12 p.m.–1 p.m.	7.8	7.7	8.6	9.5	12.4	11.8
1 p.m.–2 p.m.	8.4	7.7	10.8	9.2	12.9	12.6
2 p.m.–3 p.m.	8.2	8.5	9.8	10.5	14.7	11.9
3 p.m.–4 p.m.	8.8	9.0	9.2	9.7	12.0	14.2
4 p.m.–5 p.m.	9.4	9.4	8.9	10.3	12.0	13.1
5 p.m.–6 p.m.	8.4	9.5	7.5	9.0	7.3	10.7
6 p.m.–7 p.m.	7.7	7.6	5.9	7.3	1.8	4.9
7 p.m.–8 p.m.	7.0	7.3	4.1	5.3	0.6	0.6
8 p.m.–9 p.m.	5.3	6.6	2.8	3.5	0.5	0.4
9 p.m.–10 p.m.	3.3	5.1	1.2	2.4	0.1	0.5
10 p.m.–6 a.m.	2.4	3.6	1.9	1.9	1.7	1.1

Sites ranged in size from 63,000 to 127,000 square feet gross floor area
[a] Source numbers – 221 and 236; based on three studies
[b] Source number – 221; based on one study
[c] Source number – 221; based on one study

The sites were surveyed between the late 1980s and the 2000s throughout the United States.

Caution should be used when applying daily trip generation rates for discount supermarkets, as the database contains a mixture of facilities with varying hours of operation. Future data submissions should specify the hours of operation of the study site.

Source Numbers

221, 236, 440, 537, 566, 734, 738

Discount Supermarket
(854)

Average Vehicle Trip Ends vs: 1000 Sq. Feet Gross Floor Area
On a: Weekday

Number of Studies: 11
Average 1000 Sq. Feet GFA: 80
Directional Distribution: 50% entering, 50% exiting

Trip Generation per 1000 Sq. Feet Gross Floor Area

Average Rate	Range of Rates	Standard Deviation
90.86	68.66 - 127.13	19.14

Data Plot and Equation

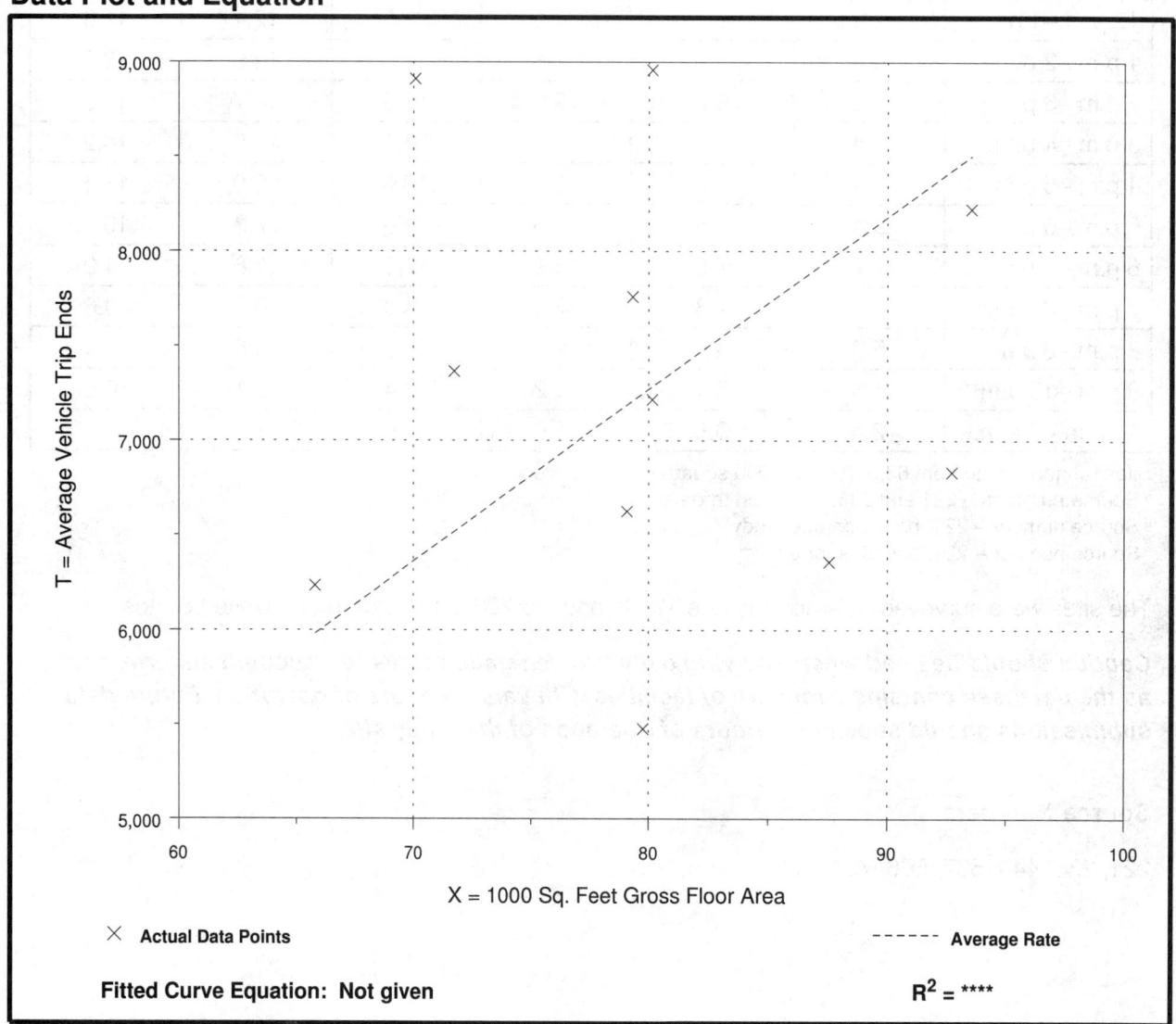

[Plot: T = Average Vehicle Trip Ends vs. X = 1000 Sq. Feet Gross Floor Area]

Fitted Curve Equation: Not given $R^2 = ****$

Discount Supermarket
(854)

Average Vehicle Trip Ends vs: 1000 Sq. Feet Gross Floor Area
On a: Weekday,
Peak Hour of Adjacent Street Traffic,
One Hour Between 7 and 9 a.m.

Number of Studies: 17
Average 1000 Sq. Feet GFA: 84
Directional Distribution: 58% entering, 42% exiting

Trip Generation per 1000 Sq. Feet Gross Floor Area

Average Rate	Range of Rates	Standard Deviation
2.53	1.03 - 4.80	1.84

Data Plot and Equation

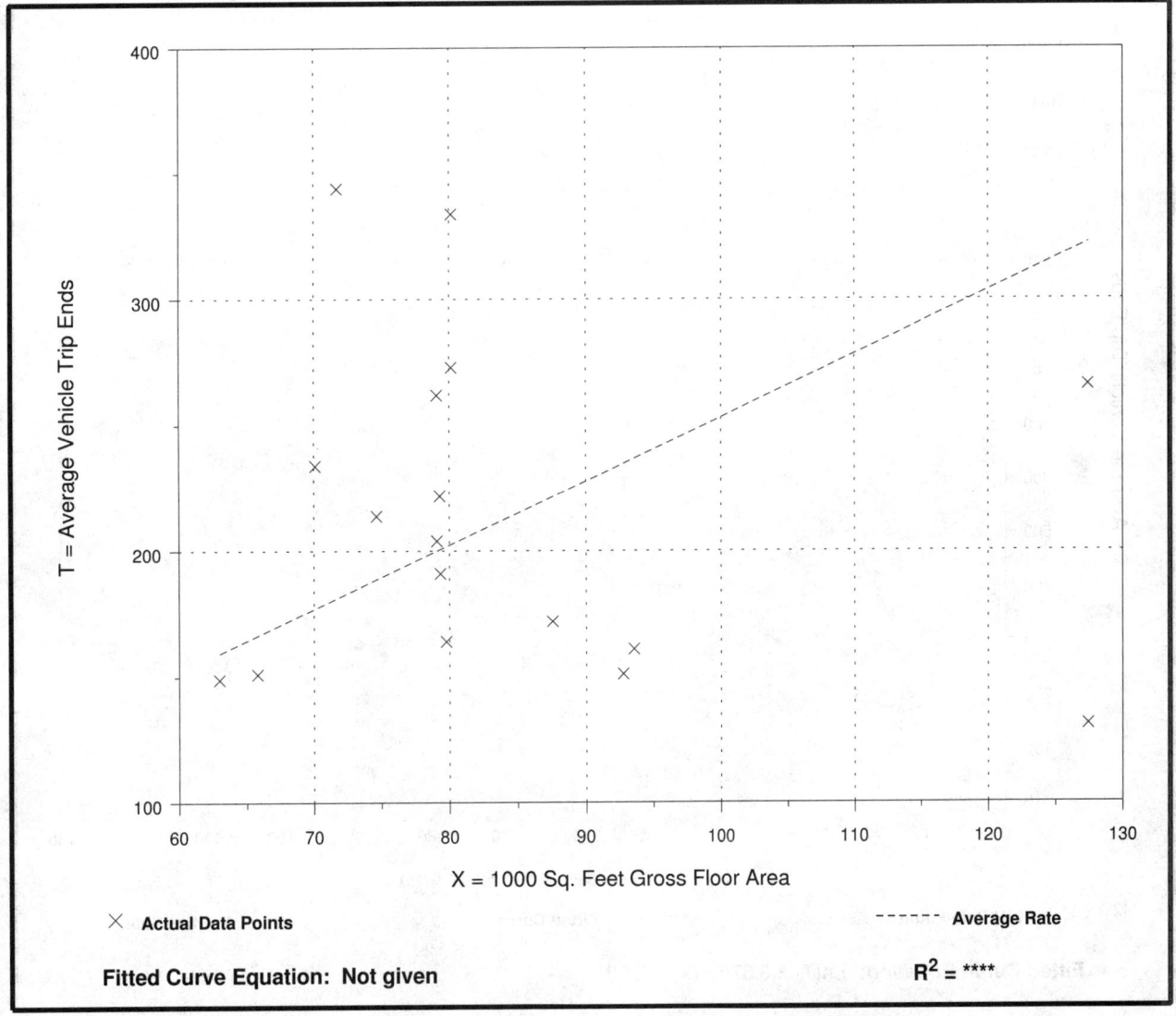

Fitted Curve Equation: Not given $R^2 = ****$

Discount Supermarket
(854)

Average Vehicle Trip Ends vs: 1000 Sq. Feet Gross Floor Area
On a: Weekday,
Peak Hour of Adjacent Street Traffic,
One Hour Between 4 and 6 p.m.

Number of Studies: 19
Average 1000 Sq. Feet GFA: 77
Directional Distribution: 50% entering, 50% exiting

Trip Generation per 1000 Sq. Feet Gross Floor Area

Average Rate	Range of Rates	Standard Deviation
8.34	5.67 - 12.18	3.33

Data Plot and Equation

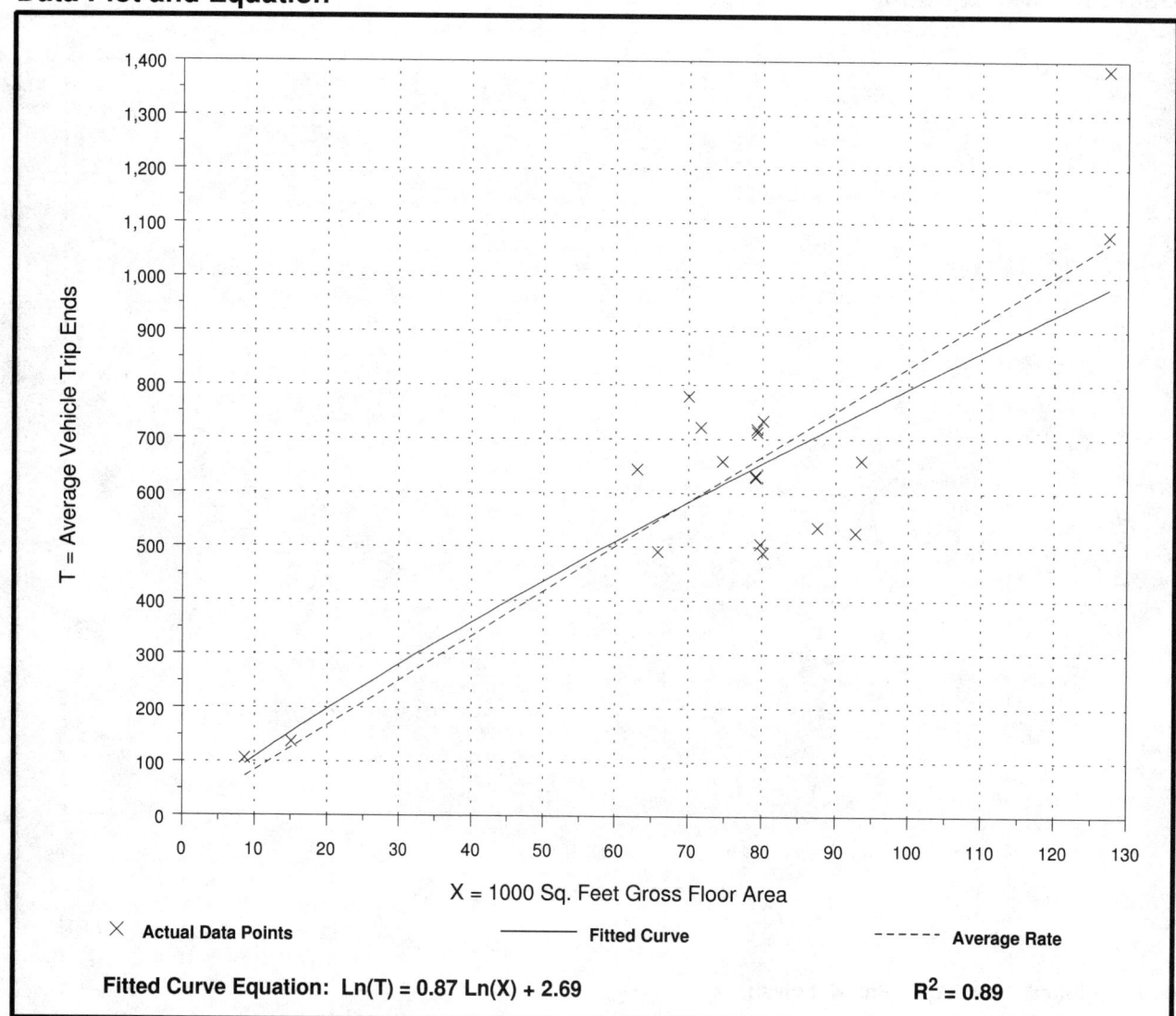

Fitted Curve Equation: $Ln(T) = 0.87\ Ln(X) + 2.69$ $R^2 = 0.89$

Discount Supermarket
(854)

Average Vehicle Trip Ends vs: 1000 Sq. Feet Gross Floor Area
On a: Weekday,
A.M. Peak Hour of Generator

Number of Studies: 7
Average 1000 Sq. Feet GFA: 94
Directional Distribution: 51% entering, 49% exiting

Trip Generation per 1000 Sq. Feet Gross Floor Area

Average Rate	Range of Rates	Standard Deviation
6.33	4.78 - 7.92	2.74

Data Plot and Equation

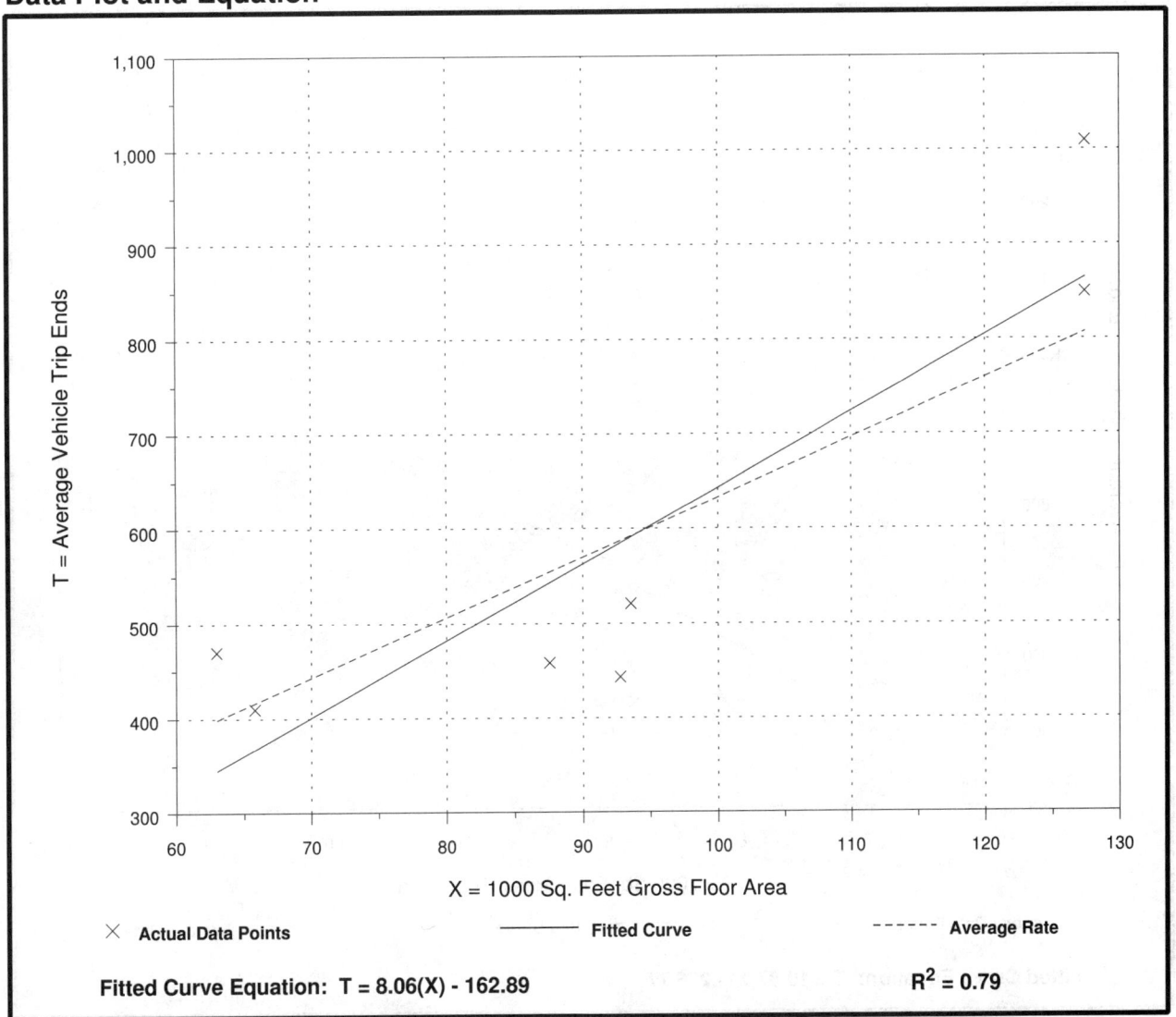

Fitted Curve Equation: $T = 8.06(X) - 162.89$ $R^2 = 0.79$

Discount Supermarket
(854)

Average Vehicle Trip Ends vs: 1000 Sq. Feet Gross Floor Area
On a: Weekday,
P.M. Peak Hour of Generator

Number of Studies: 7
Average 1000 Sq. Feet GFA: 94
Directional Distribution: 49% entering, 51% exiting

Trip Generation per 1000 Sq. Feet Gross Floor Area

Average Rate	Range of Rates	Standard Deviation
8.13	5.67 - 10.85	3.41

Data Plot and Equation

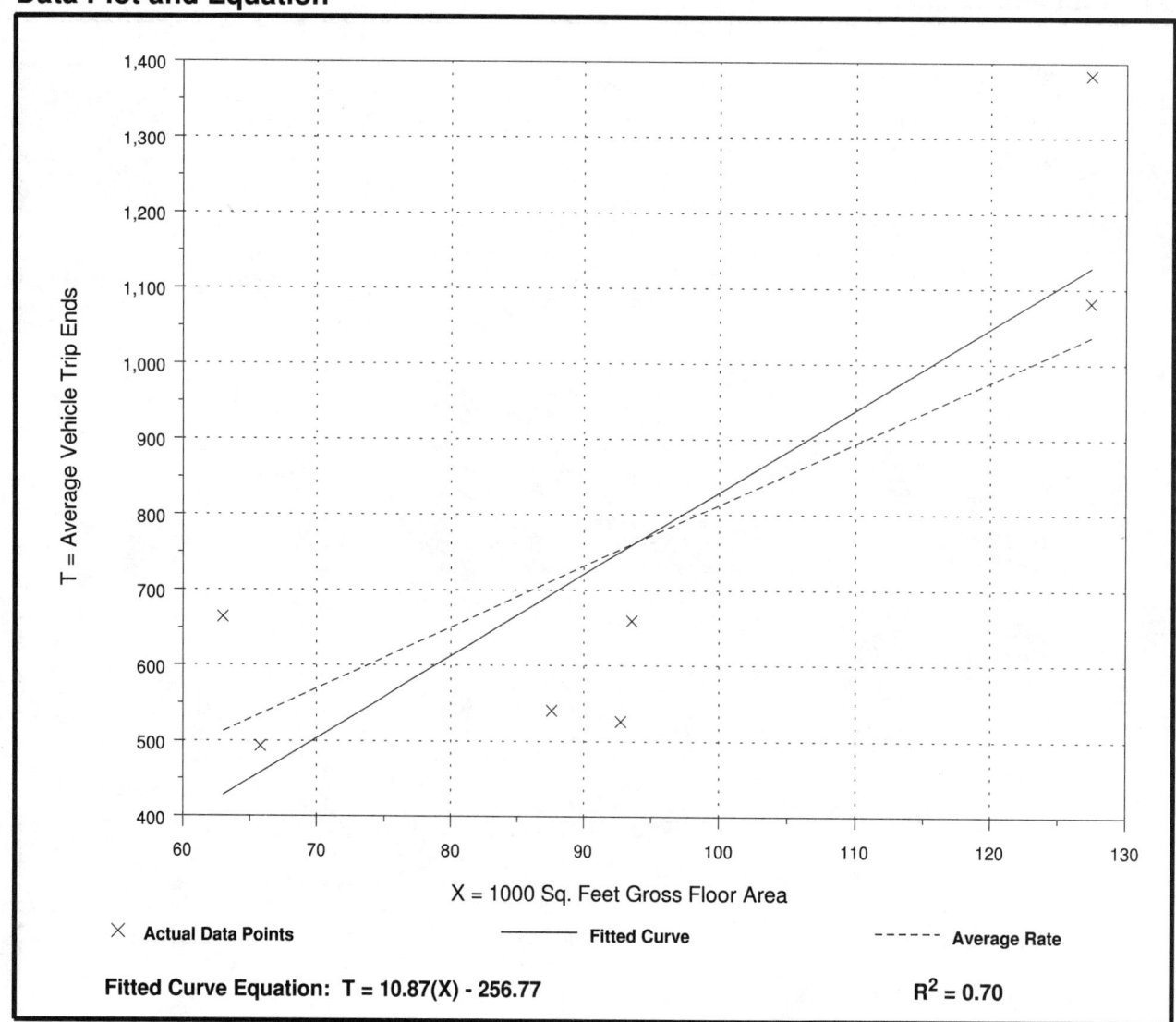

Fitted Curve Equation: $T = 10.87(X) - 256.77$ $R^2 = 0.70$

Discount Supermarket
(854)

Average Vehicle Trip Ends vs: 1000 Sq. Feet Gross Floor Area
On a: Saturday

Number of Studies: 15
Average 1000 Sq. Feet GFA: 83
Directional Distribution: 50% entering, 50% exiting

Trip Generation per 1000 Sq. Feet Gross Floor Area

Average Rate	Range of Rates	Standard Deviation
111.85	86.08 - 152.26	21.63

Data Plot and Equation

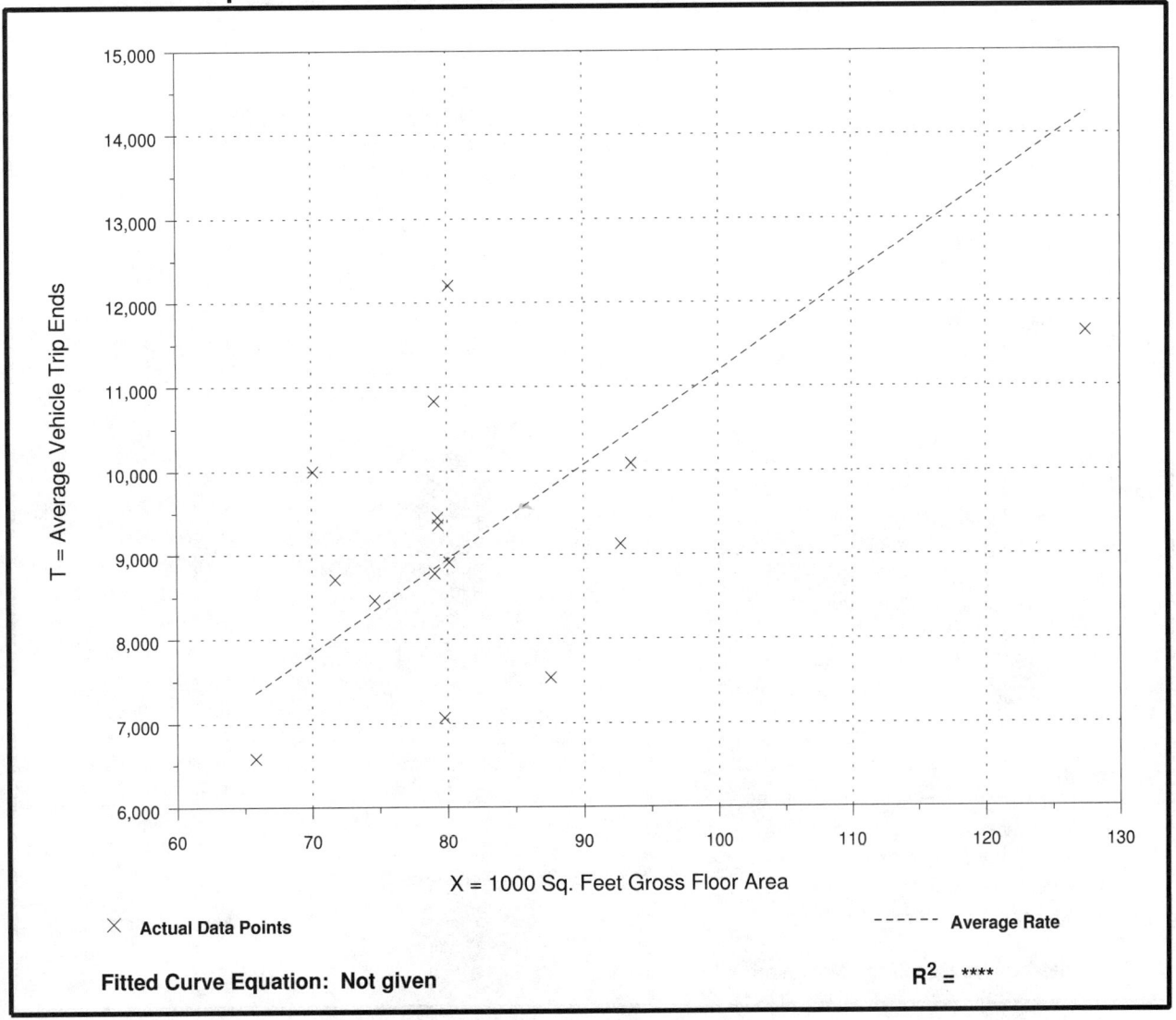

Fitted Curve Equation: Not given R^2 = ****

Discount Supermarket
(854)

Average Vehicle Trip Ends vs: 1000 Sq. Feet Gross Floor Area
On a: Saturday,
Peak Hour of Generator

Number of Studies: 16
Average 1000 Sq. Feet GFA: 81
Directional Distribution: 50% entering, 50% exiting

Trip Generation per 1000 Sq. Feet Gross Floor Area

Average Rate	Range of Rates	Standard Deviation
9.65	6.87 - 12.63	3.57

Data Plot and Equation

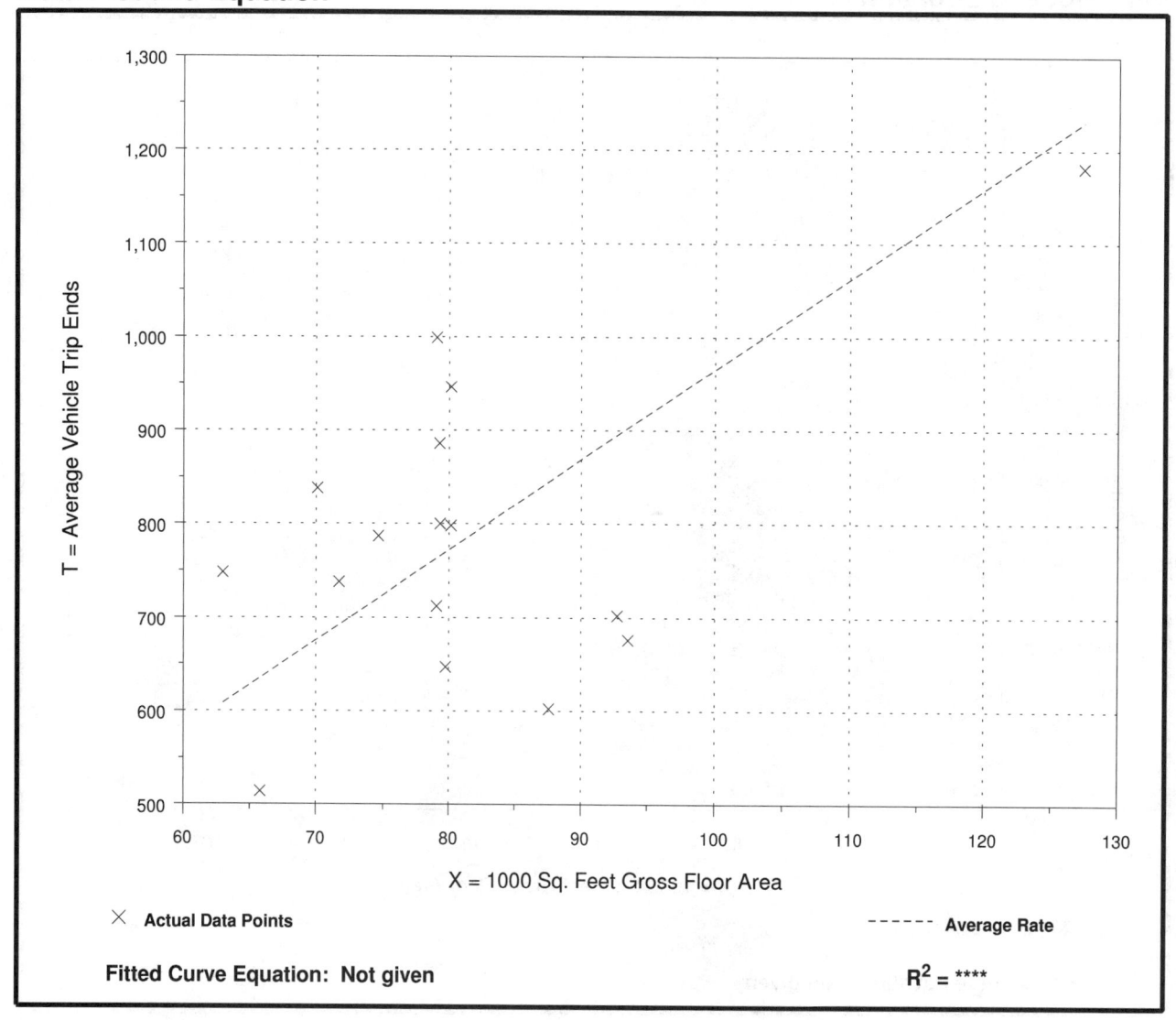

Fitted Curve Equation: Not given R^2 = ****

Discount Supermarket
(854)

Average Vehicle Trip Ends vs: 1000 Sq. Feet Gross Floor Area
On a: Sunday

Number of Studies: 13
Average 1000 Sq. Feet GFA: 84
Directional Distribution: 50% entering, 50% exiting

Trip Generation per 1000 Sq. Feet Gross Floor Area

Average Rate	Range of Rates	Standard Deviation
99.75	65.39 - 130.20	19.78

Data Plot and Equation

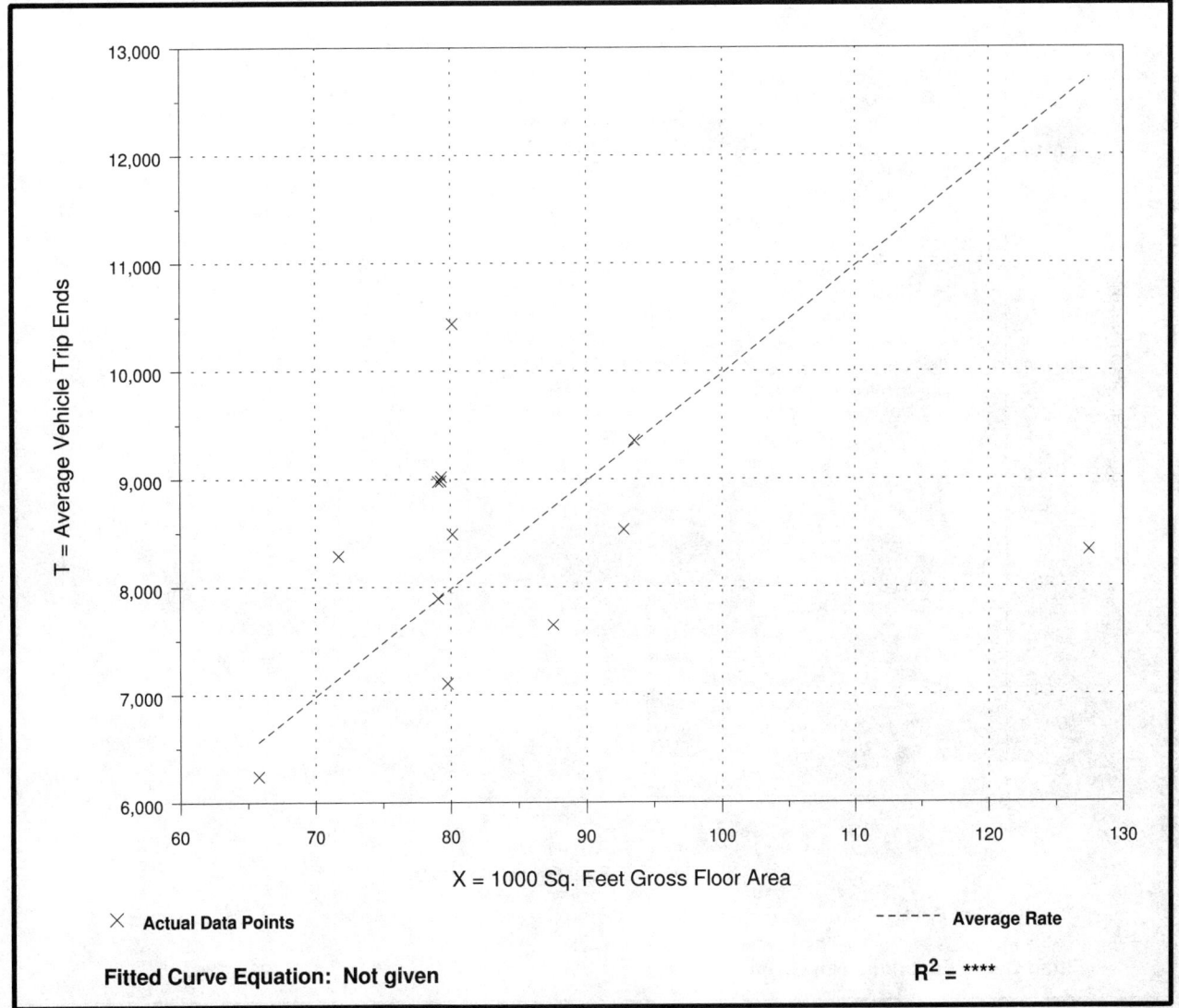

Fitted Curve Equation: Not given $R^2 = ****$

Discount Supermarket
(854)

Average Vehicle Trip Ends vs: 1000 Sq. Feet Gross Floor Area
On a: Sunday,
Peak Hour of Generator

Number of Studies: 5
Average 1000 Sq. Feet GFA: 93
Directional Distribution: 53% entering, 47% exiting

Trip Generation per 1000 Sq. Feet Gross Floor Area

Average Rate	Range of Rates	Standard Deviation
7.85	7.39 - 8.70	2.84

Data Plot and Equation

Caution - Use Carefully - Small Sample Size

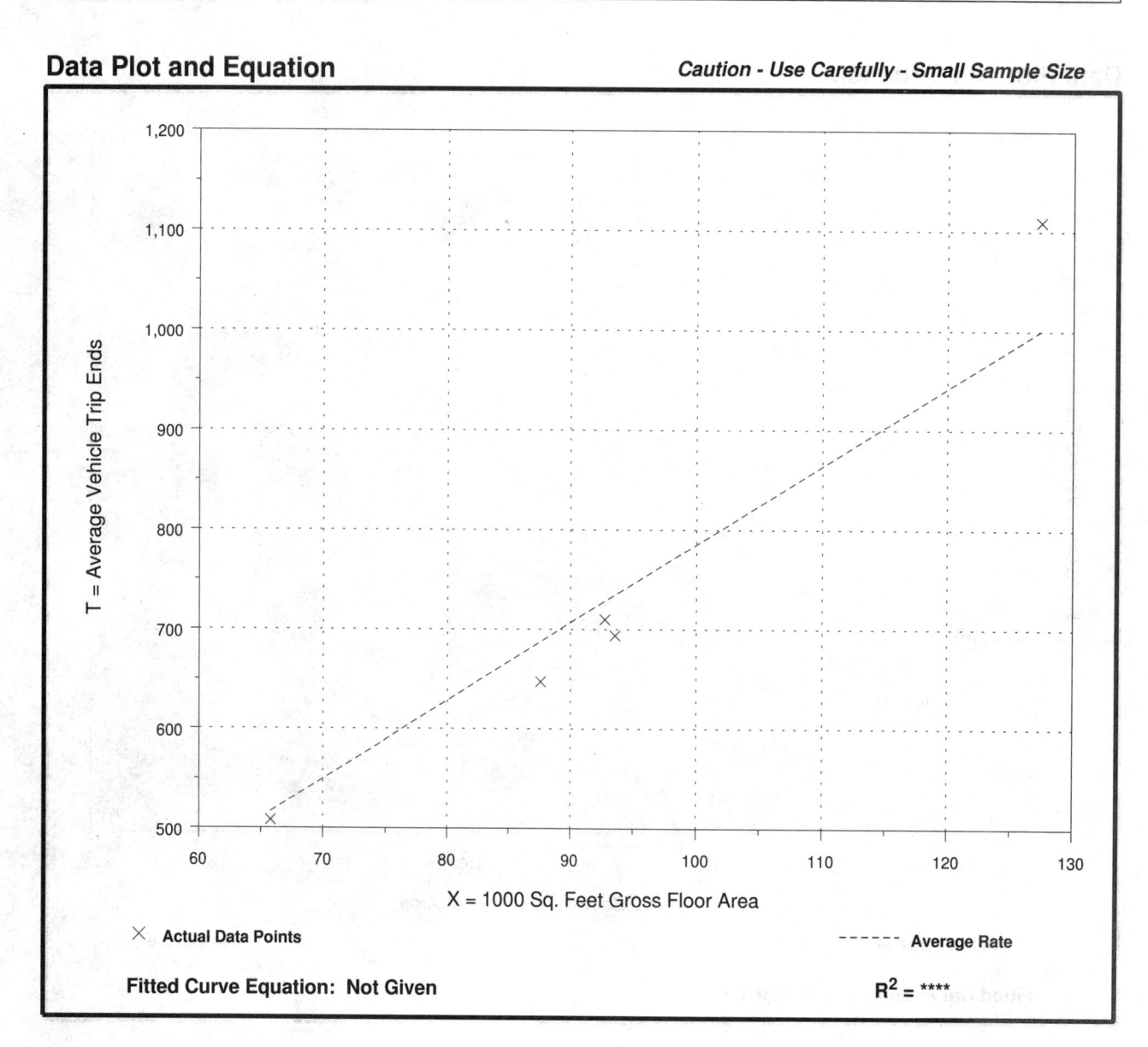

Fitted Curve Equation: Not Given $R^2 = ****$

Discount Supermarket
(854)

Average Vehicle Trip Ends vs: Employees
On a: Weekday

Number of Studies: 4
Avg. Number of Employees: 171
Directional Distribution: 50% entering, 50% exiting

Trip Generation per Employee

Average Rate	Range of Rates	Standard Deviation
40.36	33.33 - 48.09	8.84

Data Plot and Equation

Caution - Use Carefully - Small Sample Size

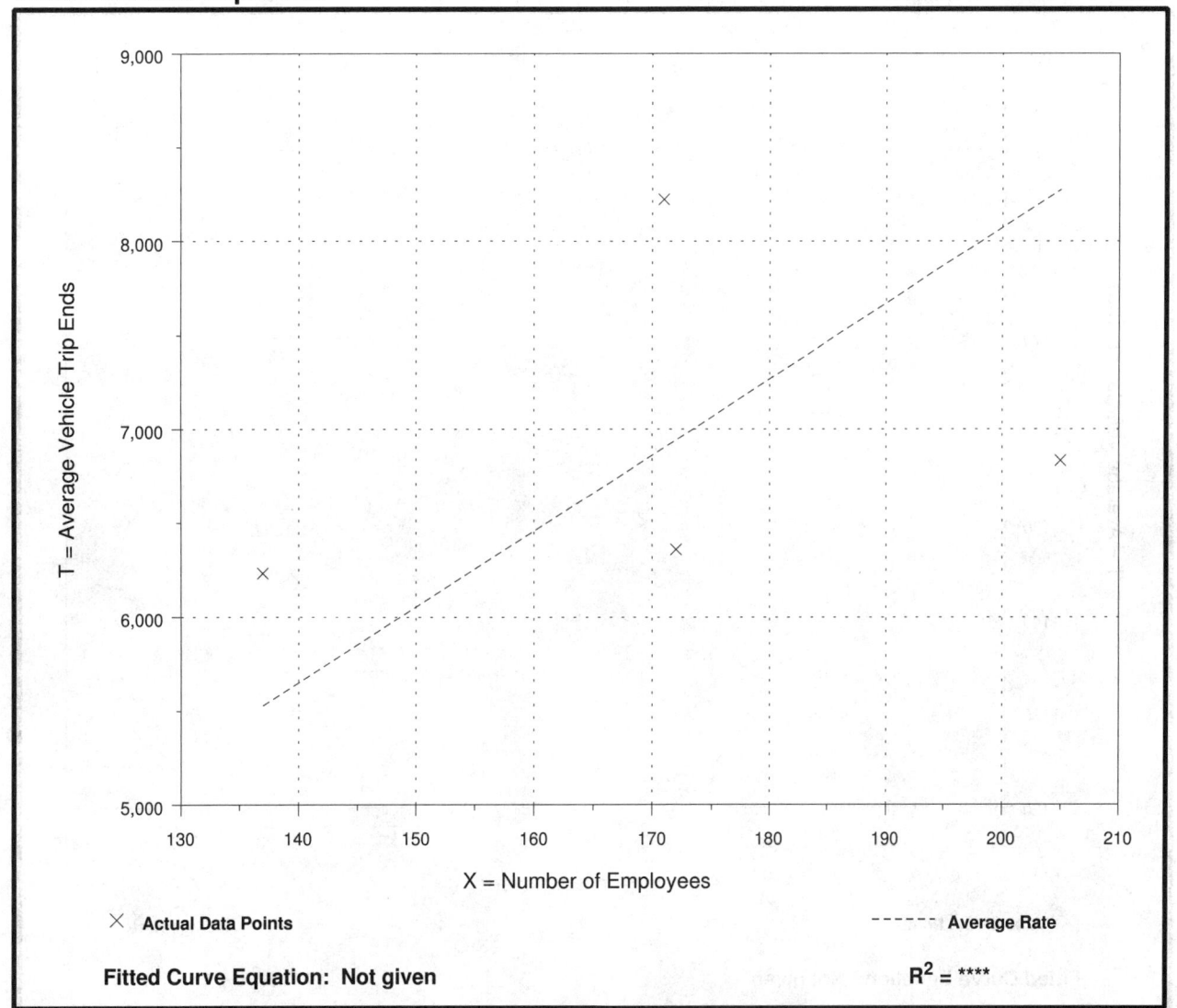

Fitted Curve Equation: Not given $R^2 = ****$

Discount Supermarket
(854)

Average Vehicle Trip Ends vs: Employees
On a: Weekday,
Peak Hour of Adjacent Street Traffic,
One Hour Between 7 and 9 a.m.

Number of Studies: 4
Avg. Number of Employees: 171
Directional Distribution: 54% entering, 46% exiting

Trip Generation per Employee

Average Rate	Range of Rates	Standard Deviation
0.93	0.74 - 1.10	0.97

Data Plot and Equation

Caution - Use Carefully - Small Sample Size

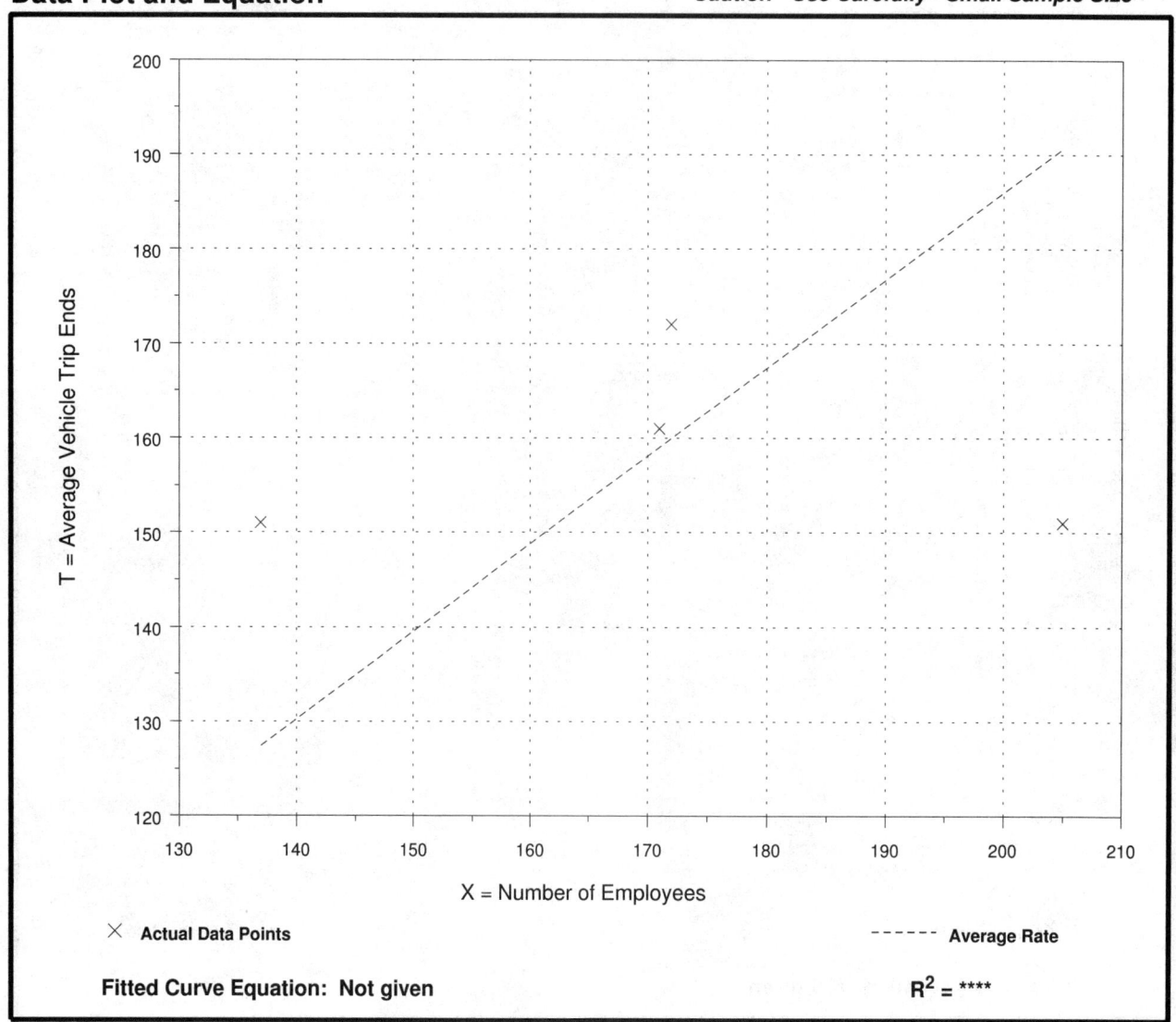

X Actual Data Points
----- Average Rate

Fitted Curve Equation: Not given $R^2 = ****$

Discount Supermarket
(854)

Average Vehicle Trip Ends vs: Employees
On a: Weekday,
Peak Hour of Adjacent Street Traffic,
One Hour Between 4 and 6 p.m.

Number of Studies: 4
Avg. Number of Employees: 171
Directional Distribution: 50% entering, 50% exiting

Trip Generation per Employee

Average Rate	Range of Rates	Standard Deviation
3.23	2.57 - 3.86	1.86

Data Plot and Equation

Caution - Use Carefully - Small Sample Size

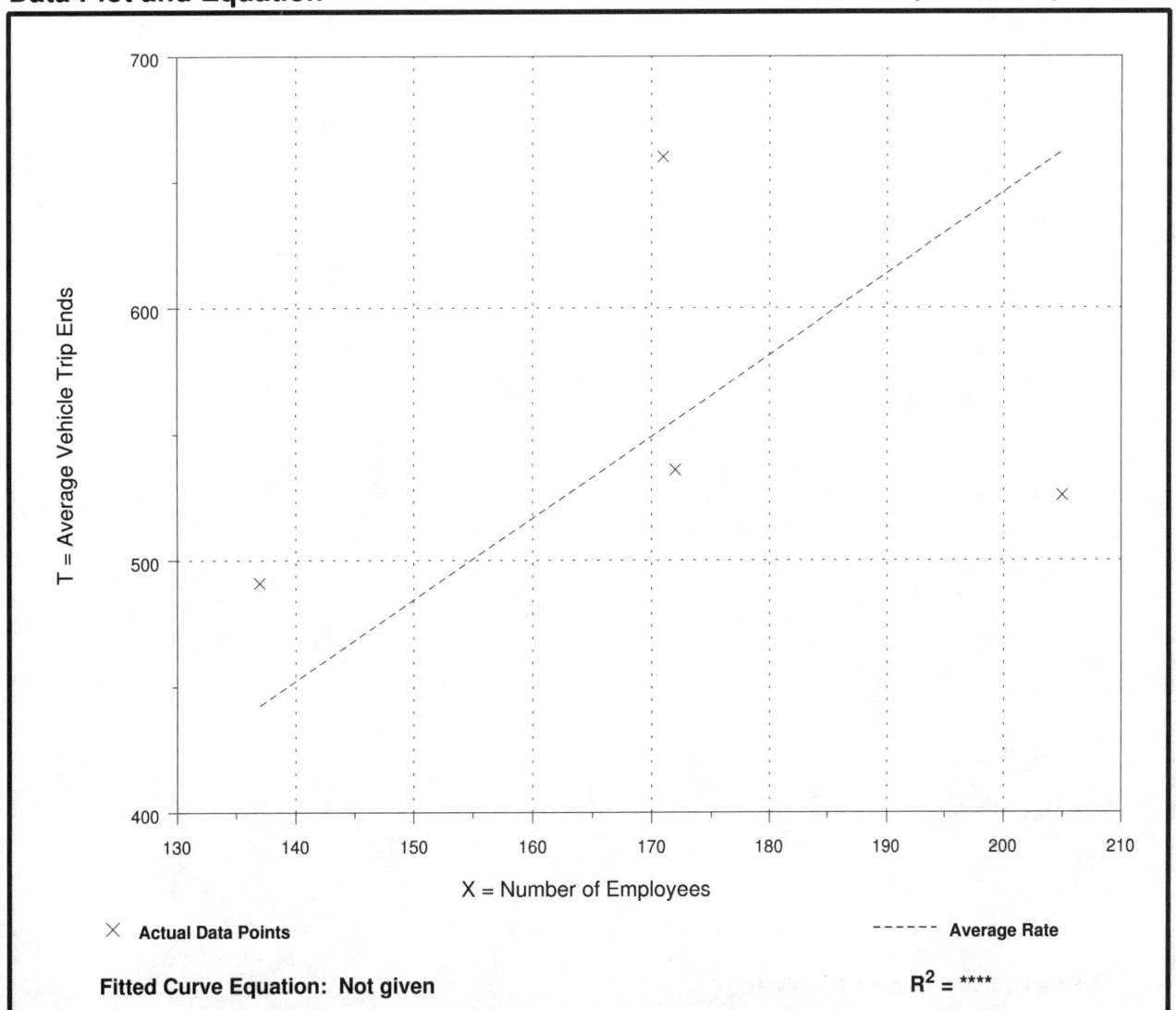

Fitted Curve Equation: Not given $R^2 = ****$

Discount Supermarket
(854)

Average Vehicle Trip Ends vs: Employees
On a: Weekday,
A.M. Peak Hour of Generator

Number of Studies: 4
Avg. Number of Employees: 171
Directional Distribution: 51% entering, 49% exiting

Trip Generation per Employee

Average Rate	Range of Rates	Standard Deviation
2.67	2.16 - 3.05	1.67

Data Plot and Equation

Caution - Use Carefully - Small Sample Size

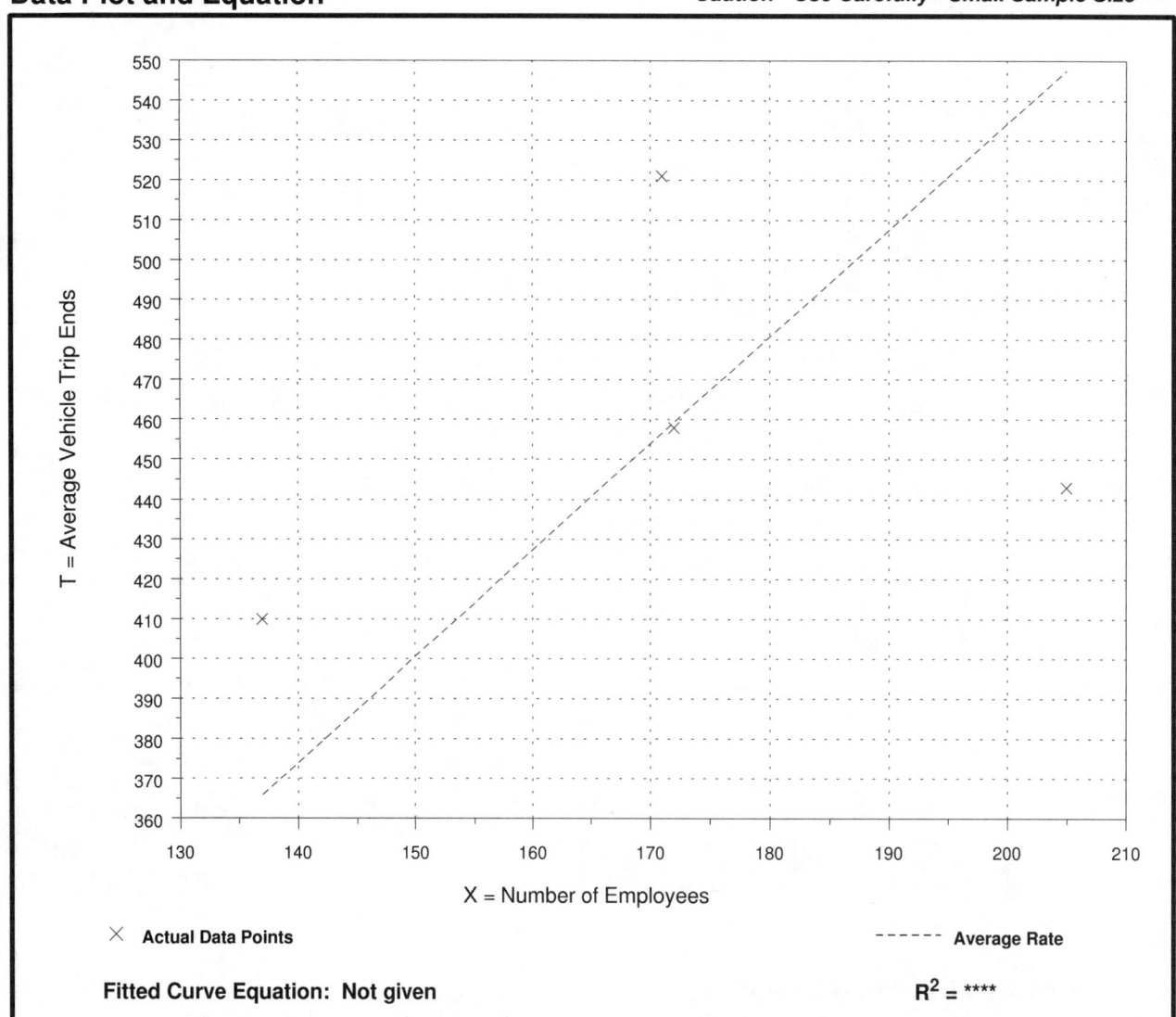

Fitted Curve Equation: Not given $R^2 = ****$

Discount Supermarket
(854)

Average Vehicle Trip Ends vs: Employees
On a: Weekday,
P.M. Peak Hour of Generator

Number of Studies: 4
Avg. Number of Employees: 171
Directional Distribution: 49% entering, 51% exiting

Trip Generation per Employee

Average Rate	Range of Rates	Standard Deviation
3.24	2.57 - 3.86	1.87

Data Plot and Equation

Caution - Use Carefully - Small Sample Size

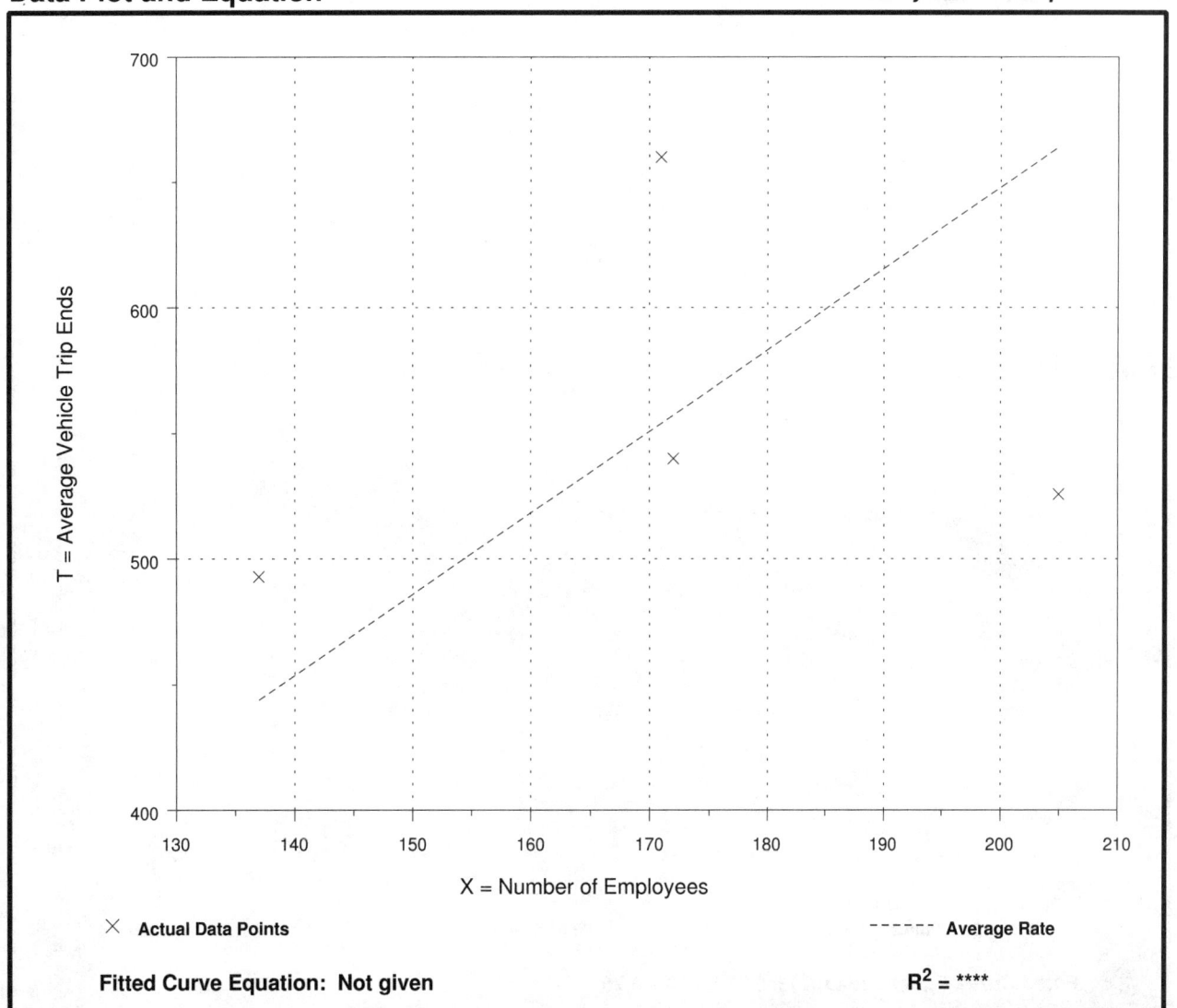

Fitted Curve Equation: Not given $R^2 = ****$

Trip Generation, 9th Edition • Institute of Transportation Engineers

Discount Supermarket
(854)

Average Vehicle Trip Ends vs: Employees
On a: Saturday

Number of Studies: 4
Avg. Number of Employees: 171
Directional Distribution: 50% entering, 50% exiting

Trip Generation per Employee

Average Rate	Range of Rates	Standard Deviation
48.66	43.83 - 58.98	9.28

Data Plot and Equation

Caution - Use Carefully - Small Sample Size

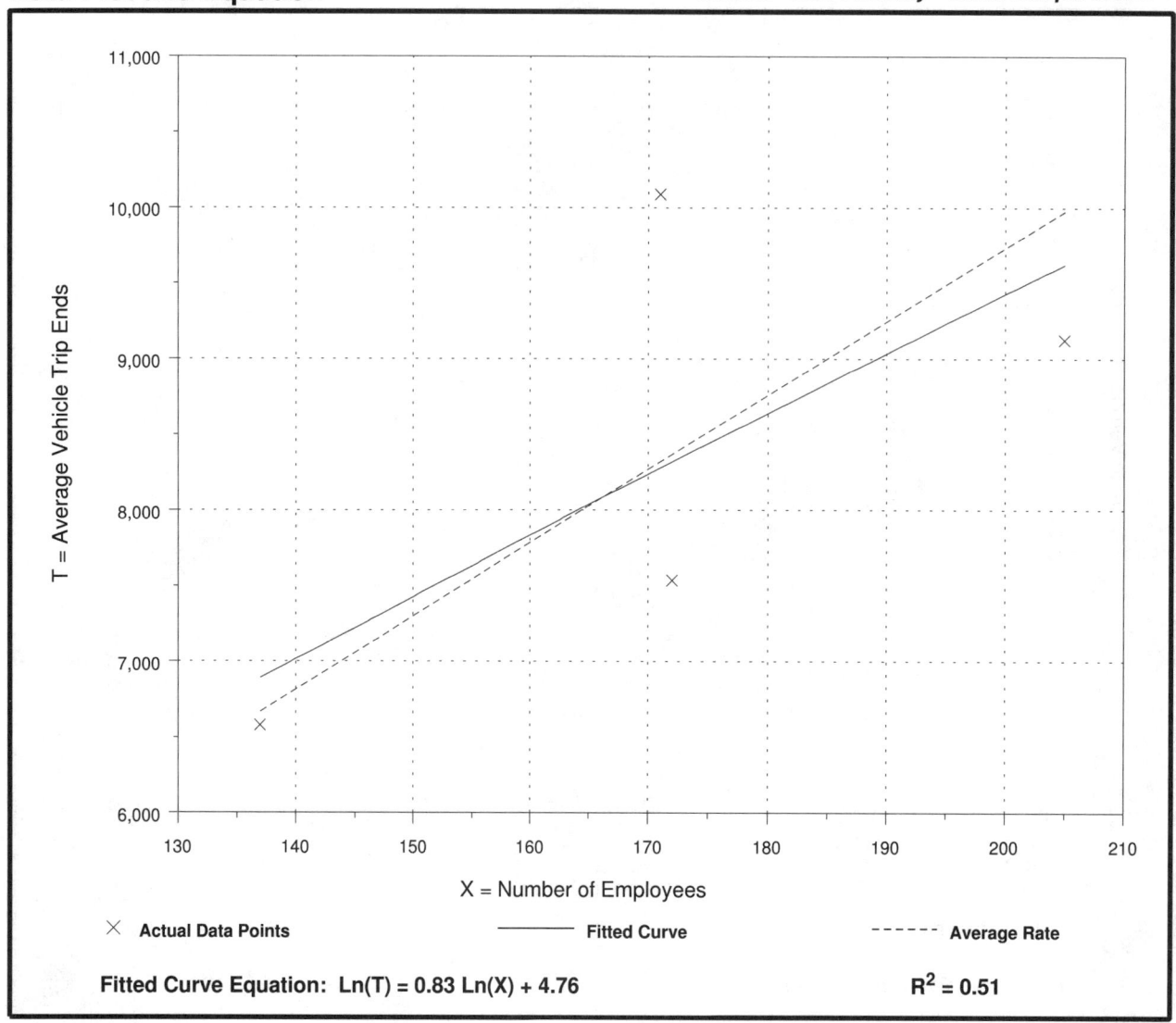

Fitted Curve Equation: $Ln(T) = 0.83\ Ln(X) + 4.76$ $R^2 = 0.51$

Discount Supermarket
(854)

Average Vehicle Trip Ends vs: Employees
On a: Saturday,
Peak Hour of Generator

Number of Studies: 4
Avg. Number of Employees: 171
Directional Distribution: 51% entering, 49% exiting

Trip Generation per Employee

Average Rate	Range of Rates	Standard Deviation
3.64	3.42 - 3.95	1.92

Data Plot and Equation

Caution - Use Carefully - Small Sample Size

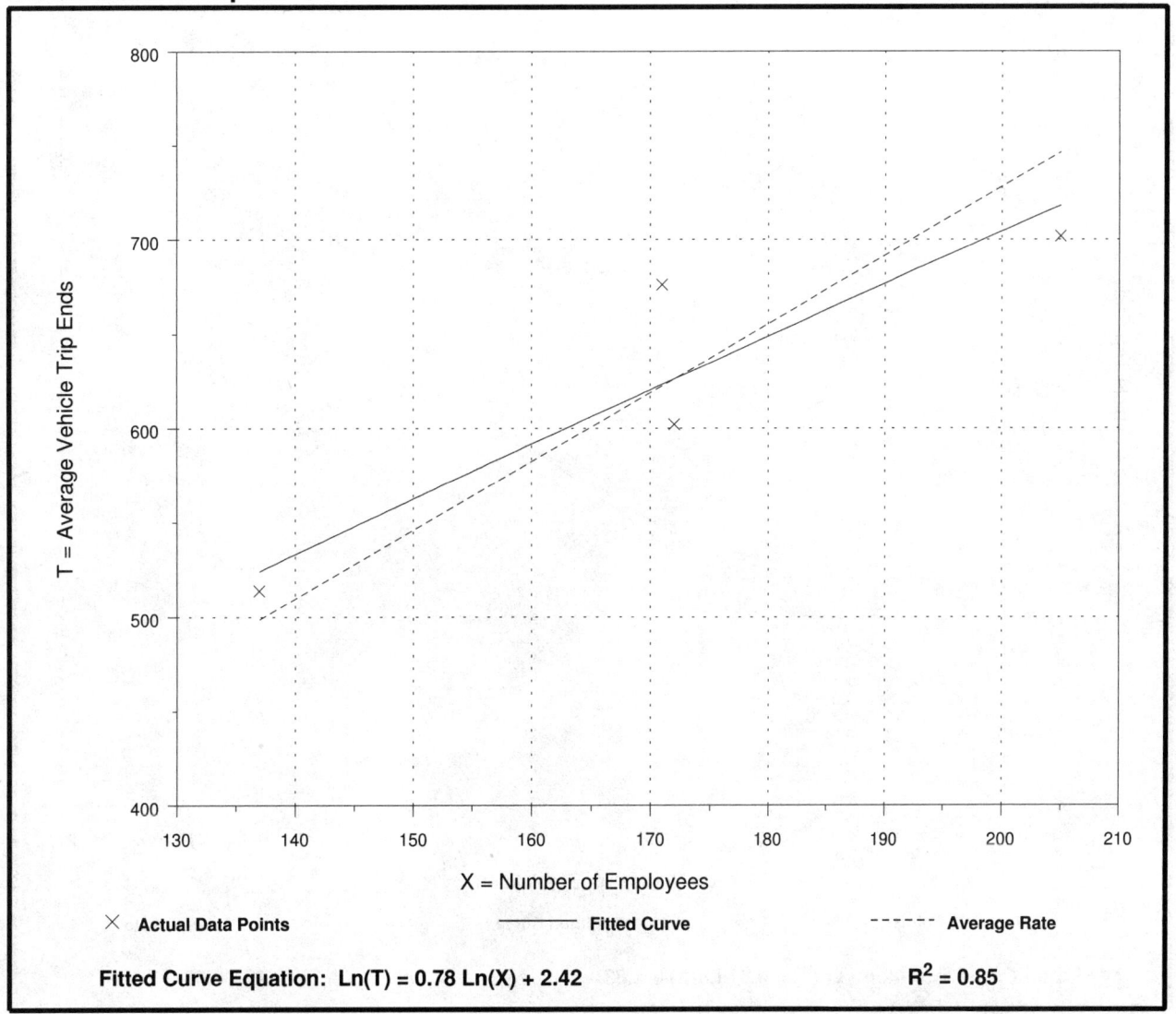

Fitted Curve Equation: $Ln(T) = 0.78 \, Ln(X) + 2.42$ $R^2 = 0.85$

Discount Supermarket
(854)

Average Vehicle Trip Ends vs: Employees
On a: Sunday

Number of Studies: 4
Avg. Number of Employees: 171
Directional Distribution: 50% entering, 50% exiting

Trip Generation per Employee

Average Rate	Range of Rates	Standard Deviation
46.39	41.62 - 54.72	8.45

Data Plot and Equation

Caution - Use Carefully - Small Sample Size

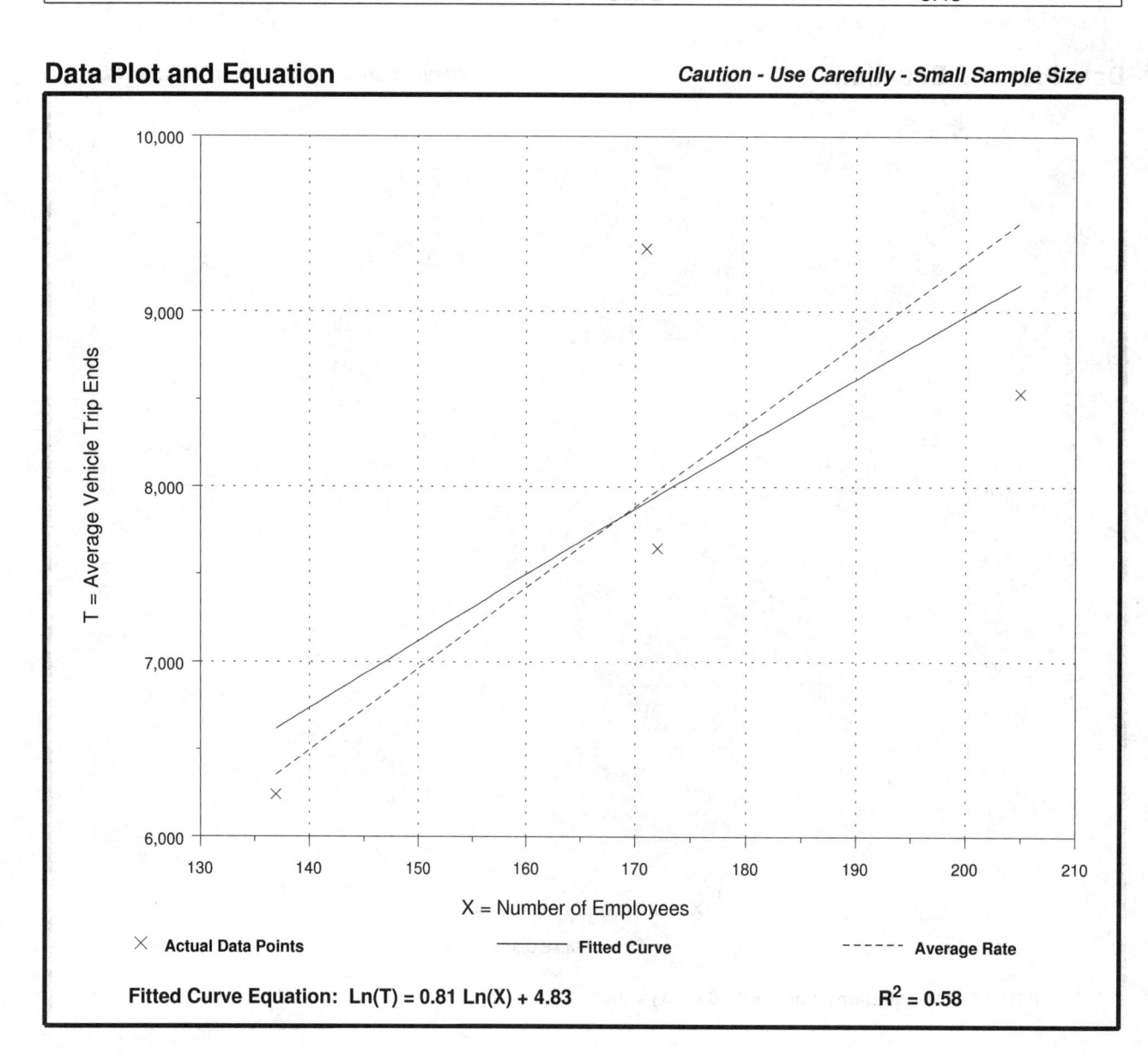

× Actual Data Points   —— Fitted Curve   ----- Average Rate

Fitted Curve Equation: $Ln(T) = 0.81 \, Ln(X) + 4.83$ $R^2 = 0.58$

Discount Supermarket
(854)

Average Vehicle Trip Ends vs: Employees
On a: Sunday,
Peak Hour of Generator

Number of Studies: 4
Avg. Number of Employees: 171
Directional Distribution: 52% entering, 48% exiting

Trip Generation per Employee

Average Rate	Range of Rates	Standard Deviation
3.74	3.46 - 4.06	1.94

Data Plot and Equation

Caution - Use Carefully - Small Sample Size

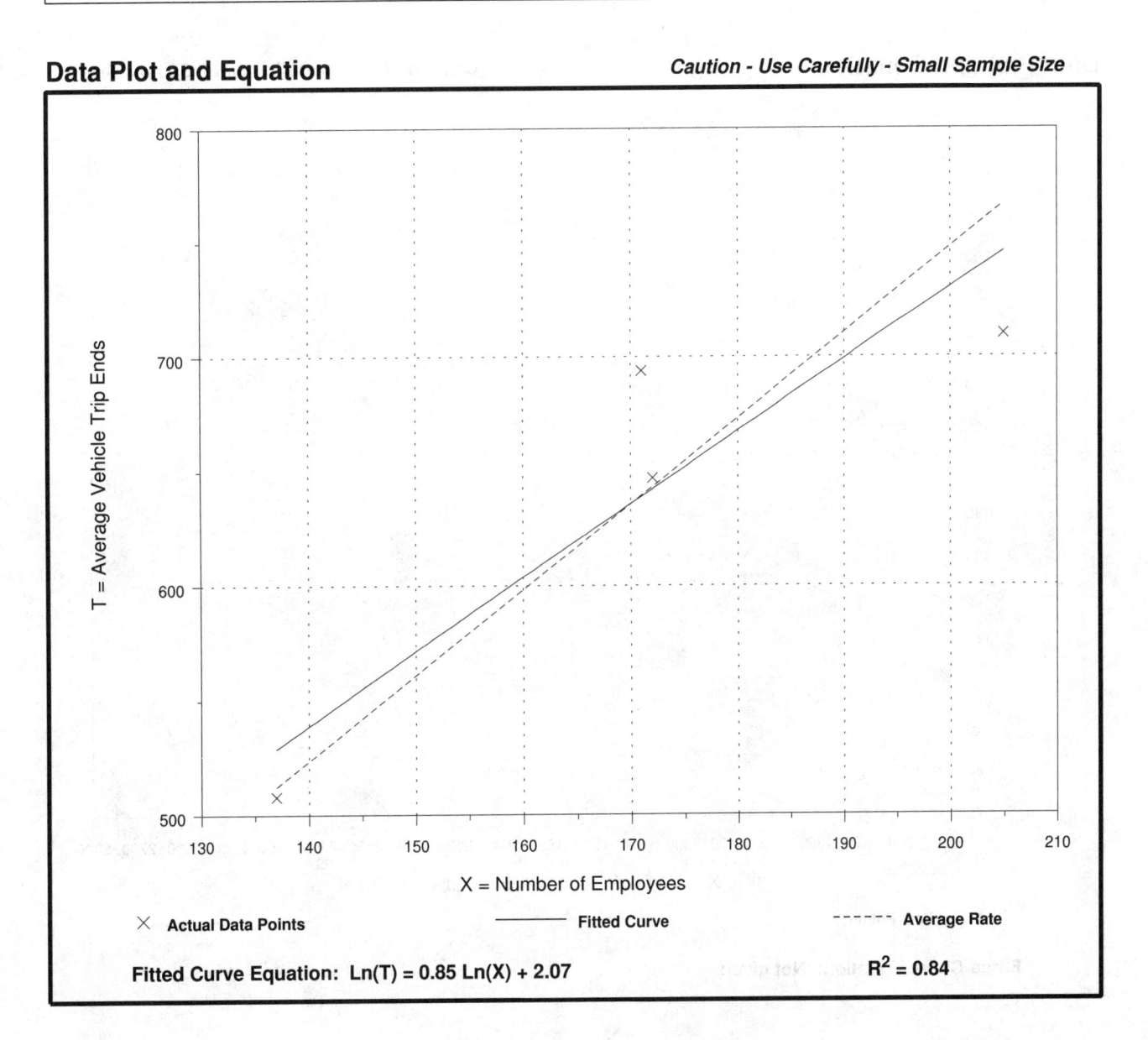

Fitted Curve Equation: $Ln(T) = 0.85 \, Ln(X) + 2.07$ $R^2 = 0.84$

Discount Supermarket
(854)

Average Vehicle Trip Ends vs: A.M. Peak Hour Traffic on Adjacent Street
On a: Weekday,
Peak Hour of Adjacent Street Traffic,
One Hour Between 7 and 9 a.m.

Number of Studies: 4
Avg. A.M. Peak Hr. Traf. on Adj. Street: 2,004
Directional Distribution: 54% entering, 46% exiting

Trip Generation per AM Peak Hour Traffic on Adjacent Street

Average Rate	Range of Rates	Standard Deviation
0.08	0.06 - 0.15	0.28

Data Plot and Equation

Caution - Use Carefully - Small Sample Size

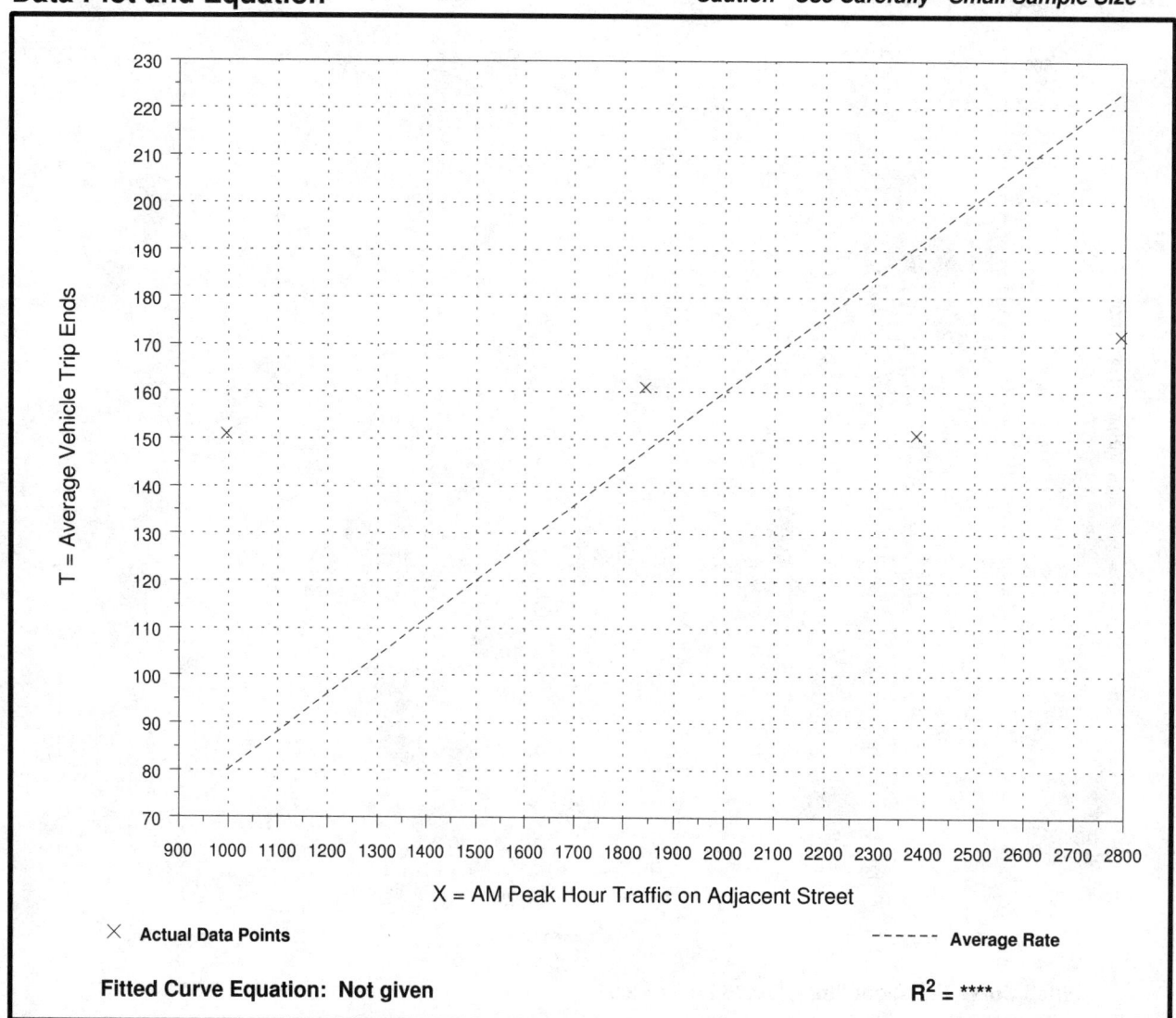

[Scatter plot with X = AM Peak Hour Traffic on Adjacent Street (900-2800) on x-axis and T = Average Vehicle Trip Ends (70-230) on y-axis. Four actual data points shown with dashed Average Rate line.]

X = Actual Data Points   ----- Average Rate

Fitted Curve Equation: Not given $R^2 = ****$

1708   *Trip Generation*, 9th Edition • Institute of Transportation Engineers

Discount Supermarket
(854)

Average Vehicle Trip Ends vs: P.M. Peak Hour Traffic on Adjacent Street
On a: Weekday,
Peak Hour of Adjacent Street Traffic,
One Hour Between 4 and 6 p.m.

Number of Studies: 4
Avg. P.M. Peak Hr. Traf. on Adj. Street: 2,579
Directional Distribution: 50% entering, 50% exiting

Trip Generation per P.M. Peak Hour Traffic on Adjacent Street

Average Rate	Range of Rates	Standard Deviation
0.21	0.17 - 0.27	0.46

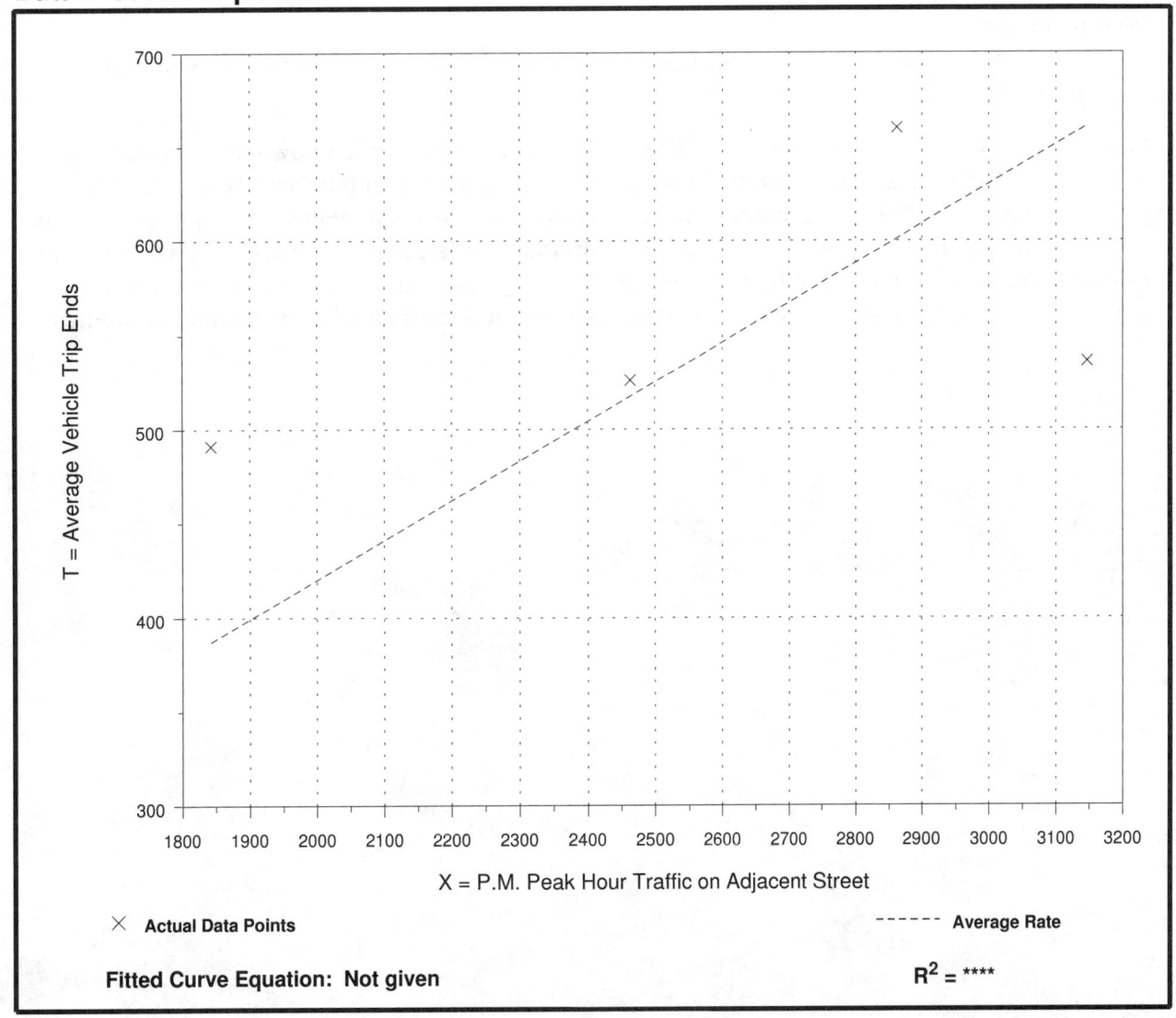

Data Plot and Equation

Caution - Use Carefully - Small Sample Size

[Scatter plot: X = P.M. Peak Hour Traffic on Adjacent Street (1800–3200); T = Average Vehicle Trip Ends (300–700). Four data points with dashed Average Rate line.]

X   Actual Data Points   - - - - Average Rate

Fitted Curve Equation: Not given R^2 = ****

Land Use: 857
Discount Club

Description

A discount club is a discount store or warehouse where shoppers pay a membership fee in order to take advantage of discounted prices on a wide variety of items such as food, clothing, tires and appliances; many items are sold in large quantities or bulk. Some sites may include on-site fueling pumps.

Additional Data

Based on a limited sample, the average vehicle occupancy was 1.45 persons per automobile.

Peak hours of the generator—
    The weekday peak hour varied between 11:00 a.m. and 2:00 p.m. The weekend peak hour varied between 12:00 p.m. and 3:00 p.m.

Information on approximate hourly variation in discount club traffic is shown in the following table. It should be noted, however, that the information contained in this table is based on a limited sample size. Therefore, caution should be exercised when applying the data. Also, some information provided in the table may conflict with the results obtained by applying the average rate or regression equations. When this occurs, it is suggested that the results from the average rate or regression equations be used, as they are based on a larger number of studies.

Hourly Variation in Discount Club Traffic						
Time	Average Weekday[a]		Average Saturday[b]		Average Sunday[c]	
	Percent of 24-Hour Entering Traffic	Percent of 24-Hour Exiting Traffic	Percent of 24-Hour Entering Traffic	Percent of 24-Hour Exiting Traffic	Percent of 24-Hour Entering Traffic	Percent of 24-Hour Exiting Traffic
6 a.m.–7 a.m.	0.9	0.3	0.3	0.1	0.2	0.1
7 a.m.–8 a.m.	1.1	0.6	0.5	0.3	0.1	0.3
8 a.m.–9 a.m.	1.1	0.7	1.2	0.5	0.0	0.0
9 a.m.–10 a.m.	2.6	1.8	3.9	1.6	0.4	0.3
10 a.m.–11 a.m.	5.2	3.6	7.0	4.2	1.3	2.3
11 a.m.–12 p.m.	8.5	5.7	11.0	7.9	5.2	4.9
12 p.m.–1 p.m.	11.4	9.5	12.6	10.8	18.0	12.2
1 p.m.–2 p.m.	10.4	11.0	12.9	11.8	17.2	16.4
2 p.m.–3 p.m.	9.0	10.2	13.2	12.5	20.5	16.5
3 p.m.–4 p.m.	8.2	9.4	10.9	12.6	16.6	15.2
4 p.m.–5 p.m.	7.9	8.6	9.4	11.3	12.7	16.2
5 p.m.–6 p.m.	8.3	7.9	6.7	9.8	5.0	10.0
6 p.m.–7 p.m.	10.1	8.2	5.0	7.1	1.7	4.1
7 p.m.–8 p.m.	9.5	9.7	3.4	4.8	0.3	1.0
8 p.m.–9 p.m.	3.7	9.9	1.0	3.5	0.3	0.2
9 p.m.–10 p.m.	0.7	2.0	0.3	0.9	0.1	0.1
10 p.m.– 6 a.m.	1.4	0.9	0.7	0.3	0.4	0.2

Sites ranged in size from 101,000 to 131,000 square feet gross floor area
[a] Source numbers – 333 and 345; based on five studies
[b] Source numbers – 333 and 345; based on five studies
[c] Source number – 333; based on one study

The sites were surveyed between the 1980s and the 2000s throughout the United States.

To assist in the future analysis of this land use, it is important to collect and include information on the presence of vehicle fueling stations in trip generation data submissions.

Source Numbers

212, 333, 344, 345, 346, 424, 438, 445, 580, 584, 715, 719

Discount Club
(857)

Average Vehicle Trip Ends vs: 1000 Sq. Feet Gross Floor Area
On a: Weekday

Number of Studies: 19
Average 1000 Sq. Feet GFA: 112
Directional Distribution: 50% entering, 50% exiting

Trip Generation per 1000 Sq. Feet Gross Floor Area

Average Rate	Range of Rates	Standard Deviation
41.80	25.44 - 78.02	14.35

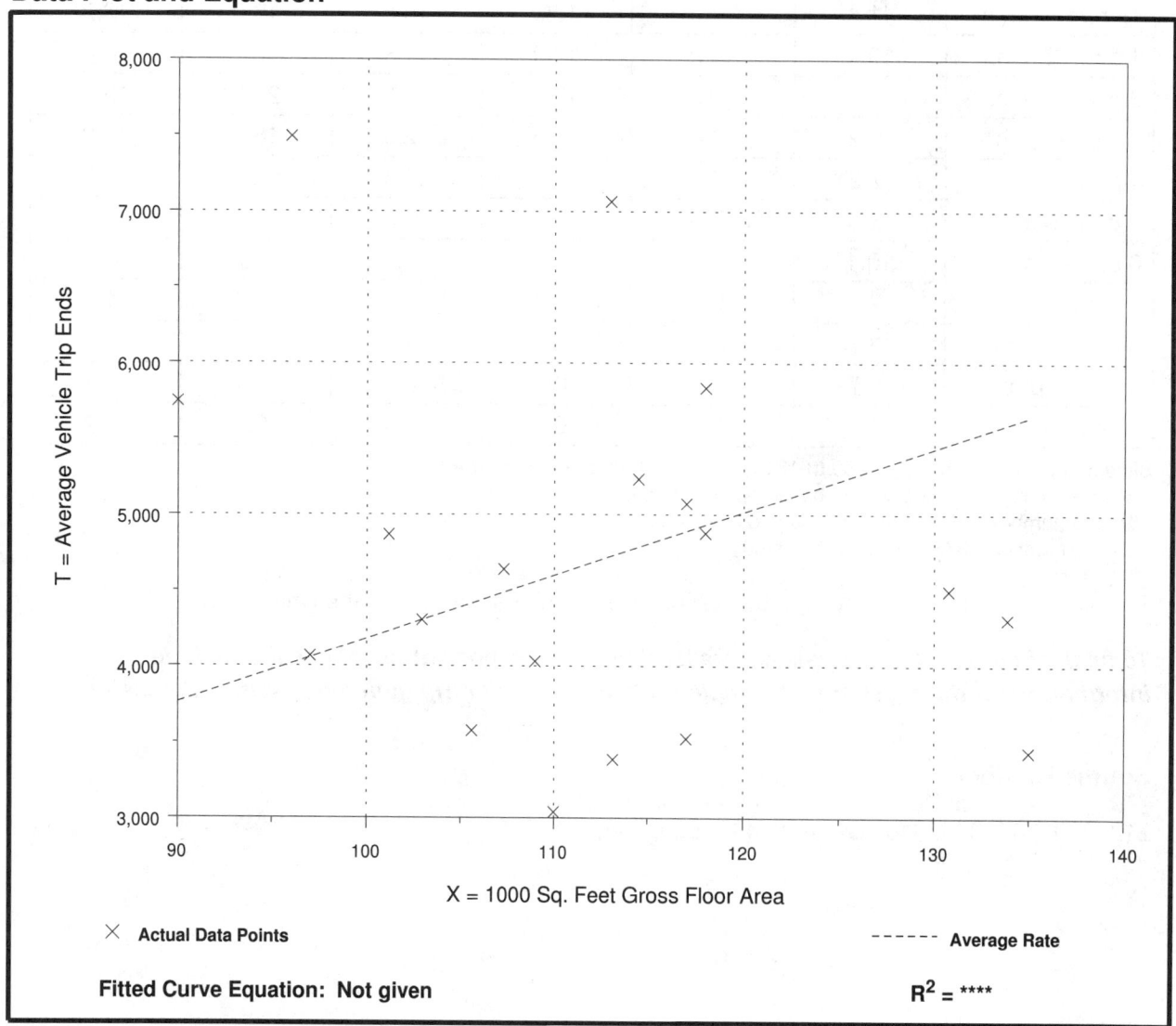

Data Plot and Equation

Fitted Curve Equation: Not given $R^2 =$ ****

Discount Club
(857)

Average Vehicle Trip Ends vs: 1000 Sq. Feet Gross Floor Area
On a: Weekday,
Peak Hour of Adjacent Street Traffic,
One Hour Between 7 and 9 a.m.

Number of Studies: 4
Average 1000 Sq. Feet GFA: 121
Directional Distribution: 70% entering, 30% exiting

Trip Generation per 1000 Sq. Feet Gross Floor Area

Average Rate	Range of Rates	Standard Deviation
0.49	0.30 - 1.07	0.76

Data Plot and Equation

Caution - Use Carefully - Small Sample Size

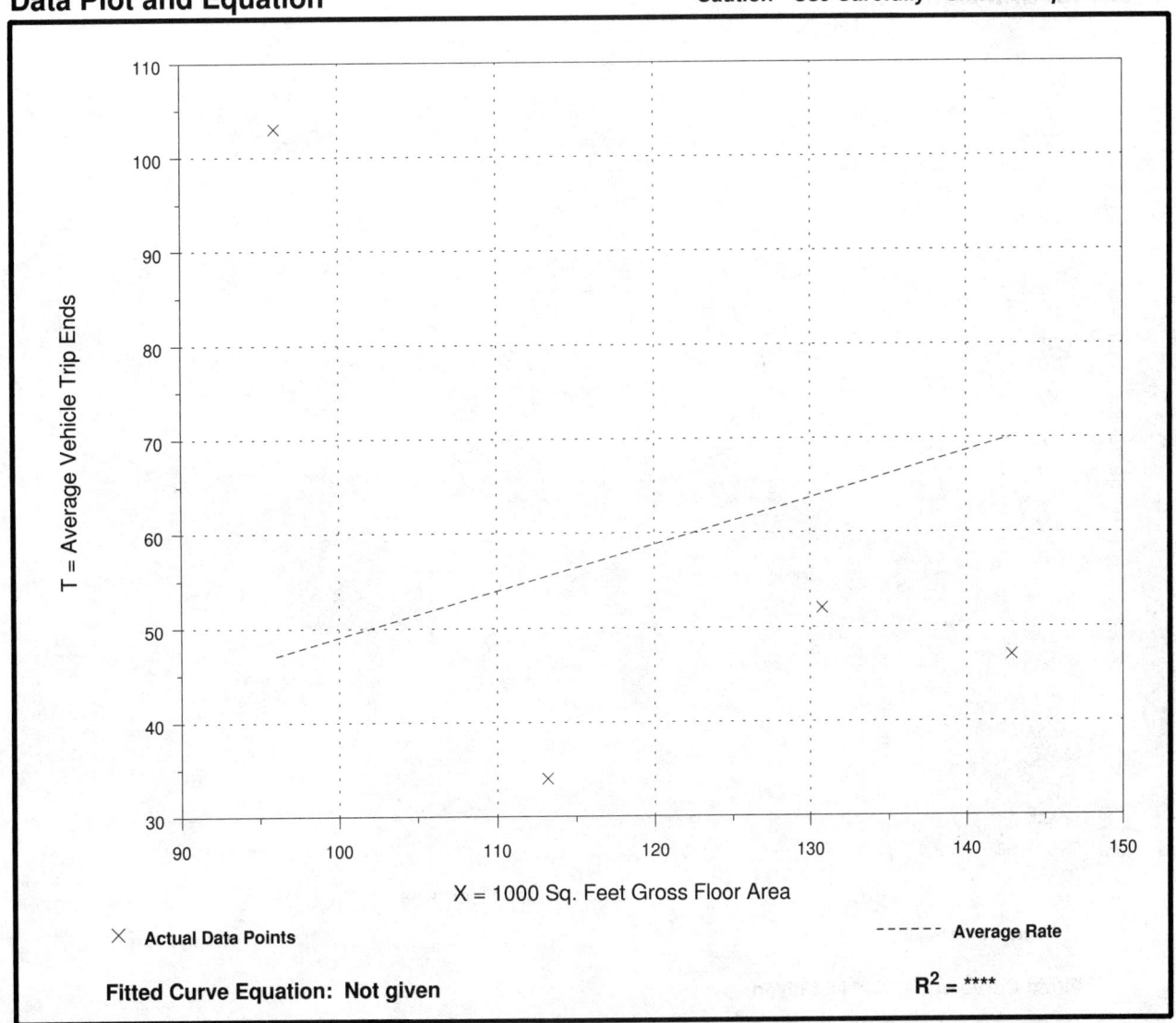

Fitted Curve Equation: Not given $R^2 = ****$

Discount Club
(857)

Average Vehicle Trip Ends vs: 1000 Sq. Feet Gross Floor Area
On a: Weekday,
Peak Hour of Adjacent Street Traffic,
One Hour Between 4 and 6 p.m.

Number of Studies: 31
Average 1000 Sq. Feet GFA: 120
Directional Distribution: 50% entering, 50% exiting

Trip Generation per 1000 Sq. Feet Gross Floor Area

Average Rate	Range of Rates	Standard Deviation
4.18	1.85 - 8.13	2.65

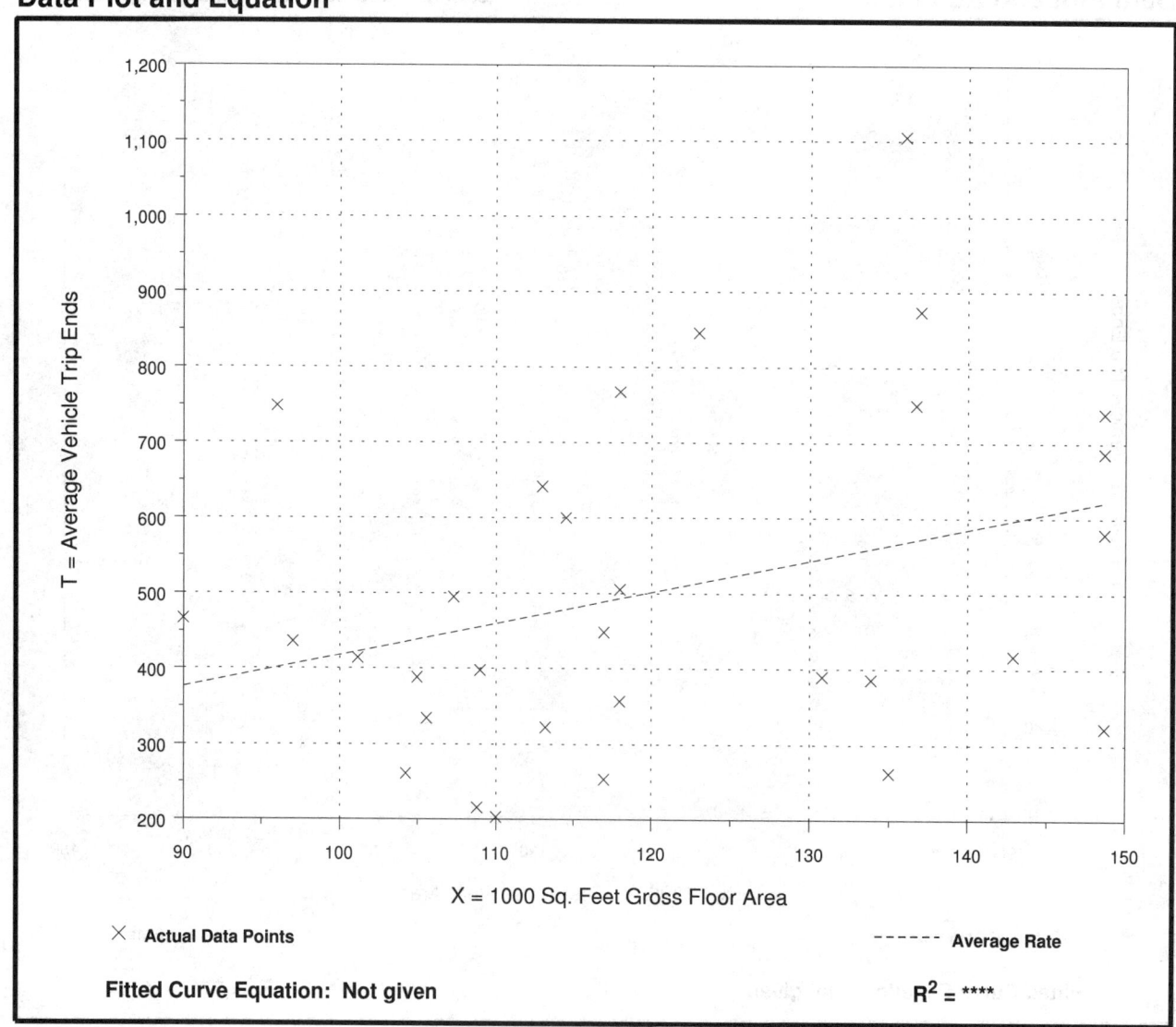

Data Plot and Equation

Fitted Curve Equation: Not given $R^2 =$ ****

Discount Club
(857)

Average Vehicle Trip Ends vs: 1000 Sq. Feet Gross Floor Area
On a: Weekday,
A.M. Peak Hour of Generator

Number of Studies: 11
Average 1000 Sq. Feet GFA: 116
Directional Distribution: 52% entering, 48% exiting

Trip Generation per 1000 Sq. Feet Gross Floor Area

Average Rate	Range of Rates	Standard Deviation
3.37	0.79 - 6.34	2.35

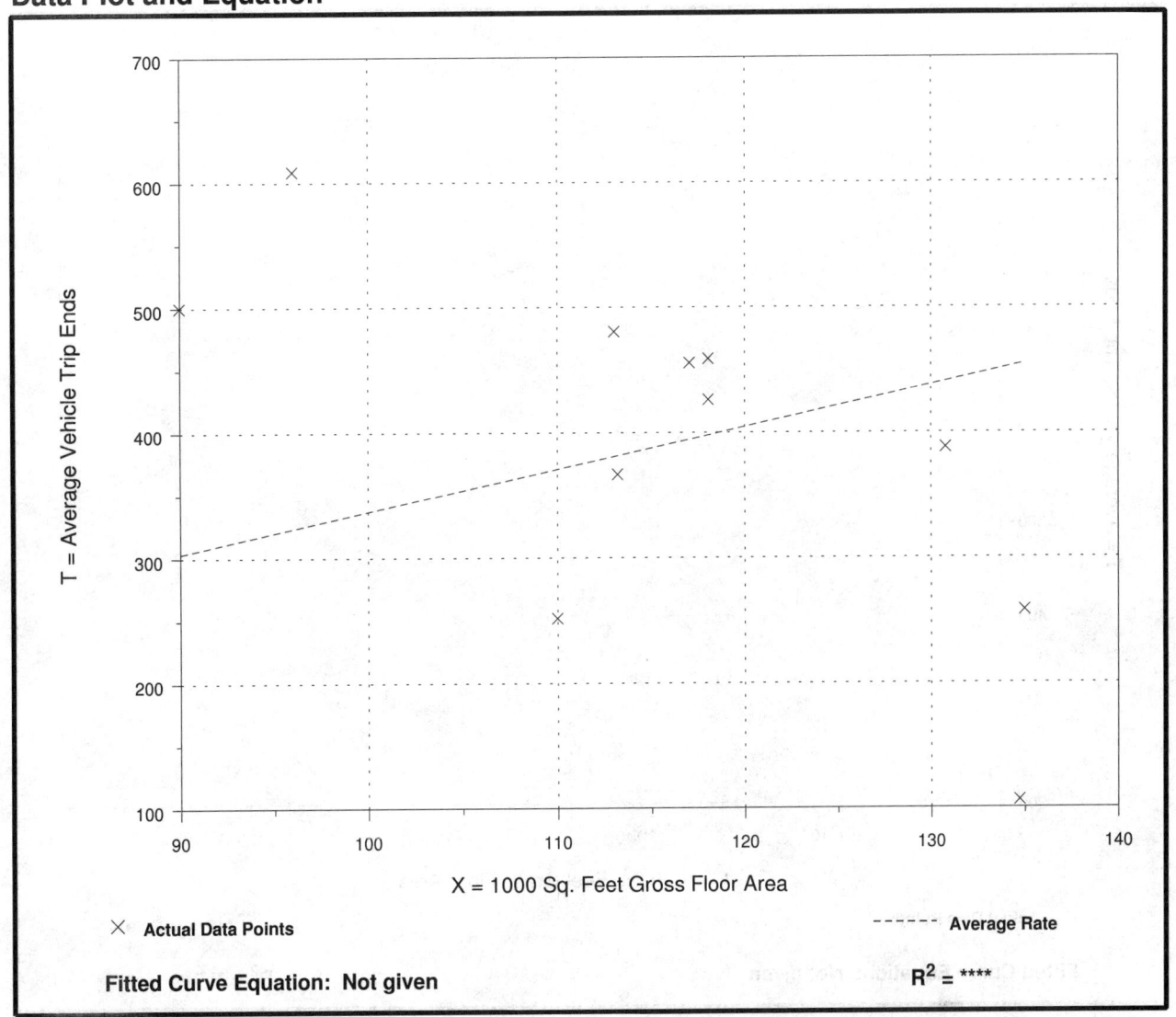

Data Plot and Equation

Fitted Curve Equation: Not given $R^2 = ****$

Discount Club
(857)

Average Vehicle Trip Ends vs: 1000 Sq. Feet Gross Floor Area
On a: Weekday,
P.M. Peak Hour of Generator

Number of Studies: 21
Average 1000 Sq. Feet GFA: 113
Directional Distribution: 50% entering, 50% exiting

Trip Generation per 1000 Sq. Feet Gross Floor Area

Average Rate	Range of Rates	Standard Deviation
4.63	2.42 - 9.67	2.80

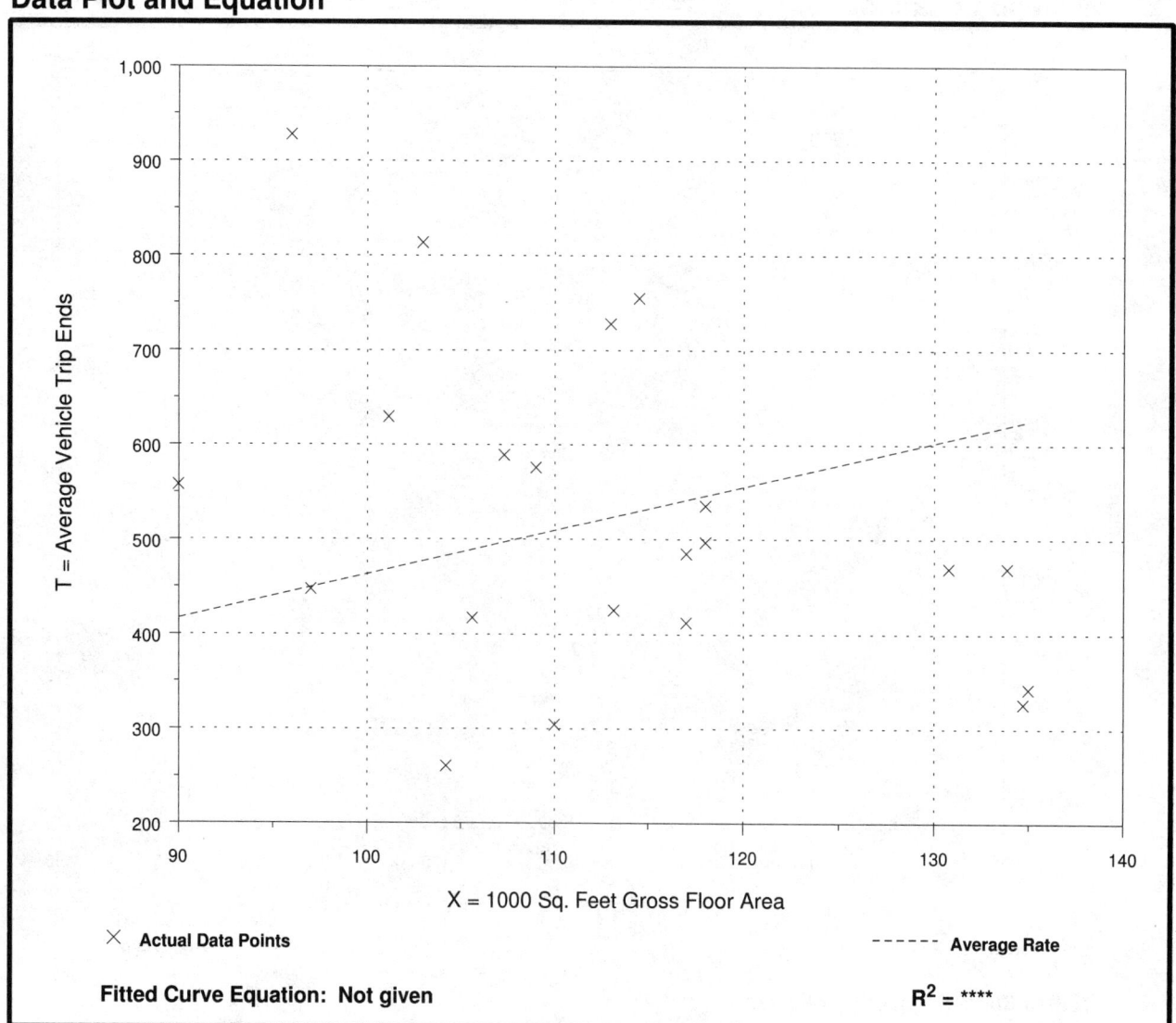

Data Plot and Equation

T = Average Vehicle Trip Ends vs. X = 1000 Sq. Feet Gross Floor Area

X Actual Data Points
----- Average Rate

Fitted Curve Equation: Not given $R^2 = ****$

Discount Club
(857)

Average Vehicle Trip Ends vs: 1000 Sq. Feet Gross Floor Area
On a: Saturday

Number of Studies: 16
Average 1000 Sq. Feet GFA: 113
Directional Distribution: 50% entering, 50% exiting

Trip Generation per 1000 Sq. Feet Gross Floor Area

Average Rate	Range of Rates	Standard Deviation
53.75	31.96 - 82.43	16.47

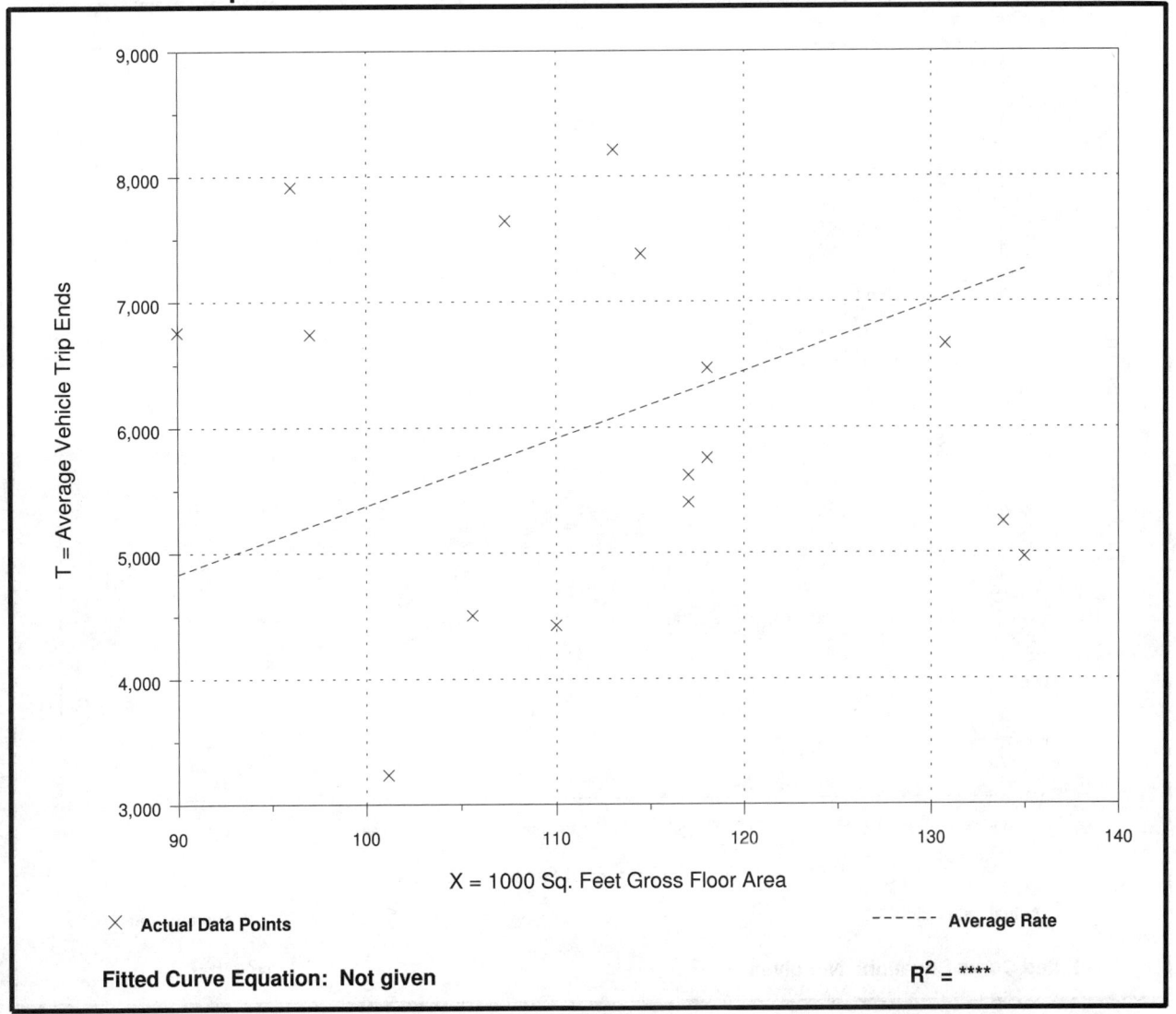

Data Plot and Equation

Fitted Curve Equation: Not given $R^2 = ****$

Discount Club
(857)

Average Vehicle Trip Ends vs: 1000 Sq. Feet Gross Floor Area
On a: Saturday,
Peak Hour of Generator

Number of Studies: 23
Average 1000 Sq. Feet GFA: 123
Directional Distribution: 49% entering, 51% exiting

Trip Generation per 1000 Sq. Feet Gross Floor Area

Average Rate	Range of Rates	Standard Deviation
6.37	3.79 - 12.52	3.46

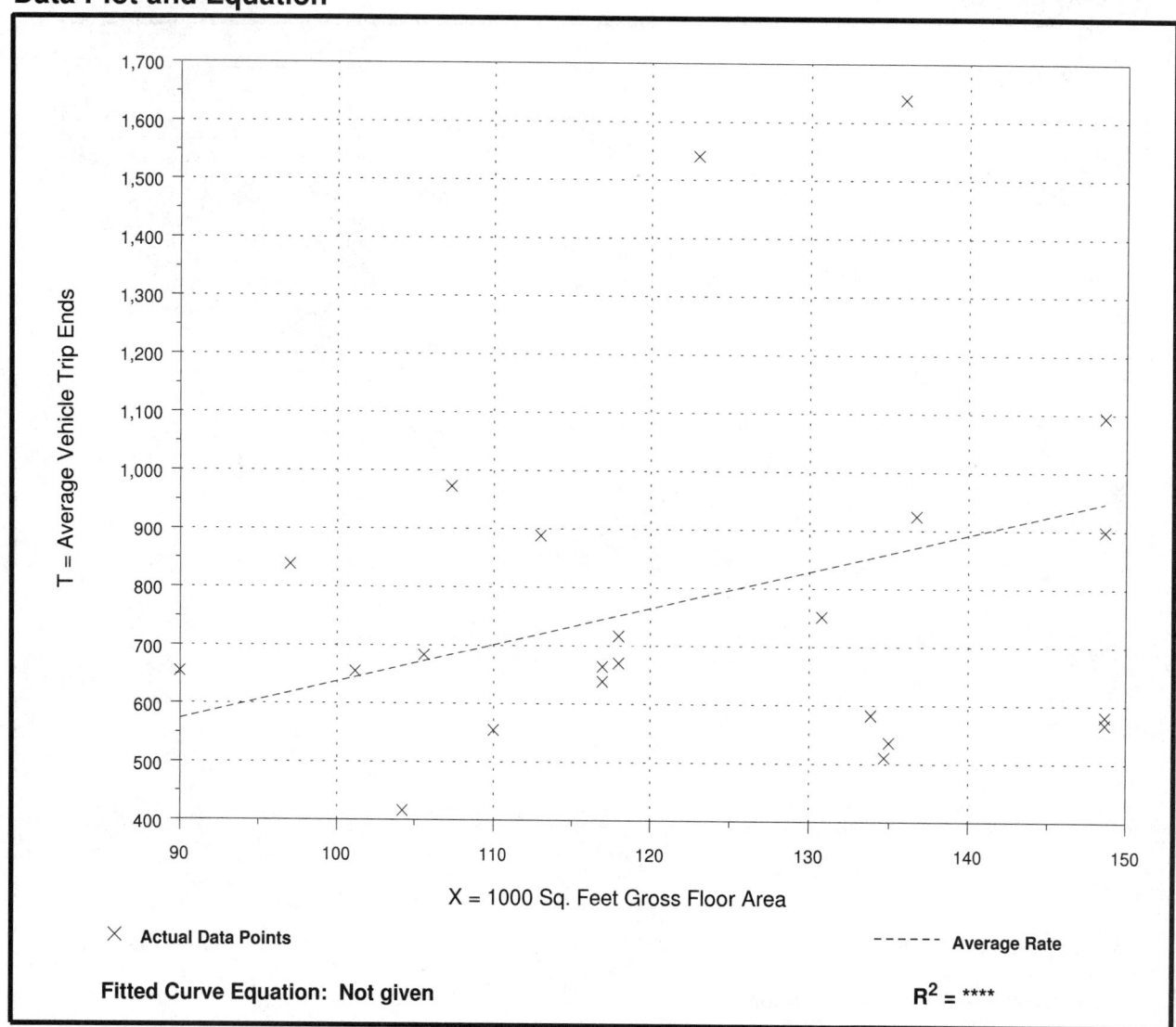

Data Plot and Equation

Fitted Curve Equation: Not given $R^2 = ****$

Discount Club
(857)

Average Vehicle Trip Ends vs: 1000 Sq. Feet Gross Floor Area
On a: Sunday

Number of Studies: 8
Average 1000 Sq. Feet GFA: 112
Directional Distribution: 50% entering, 50% exiting

Trip Generation per 1000 Sq. Feet Gross Floor Area

Average Rate	Range of Rates	Standard Deviation
33.67	17.17 - 61.79	14.48

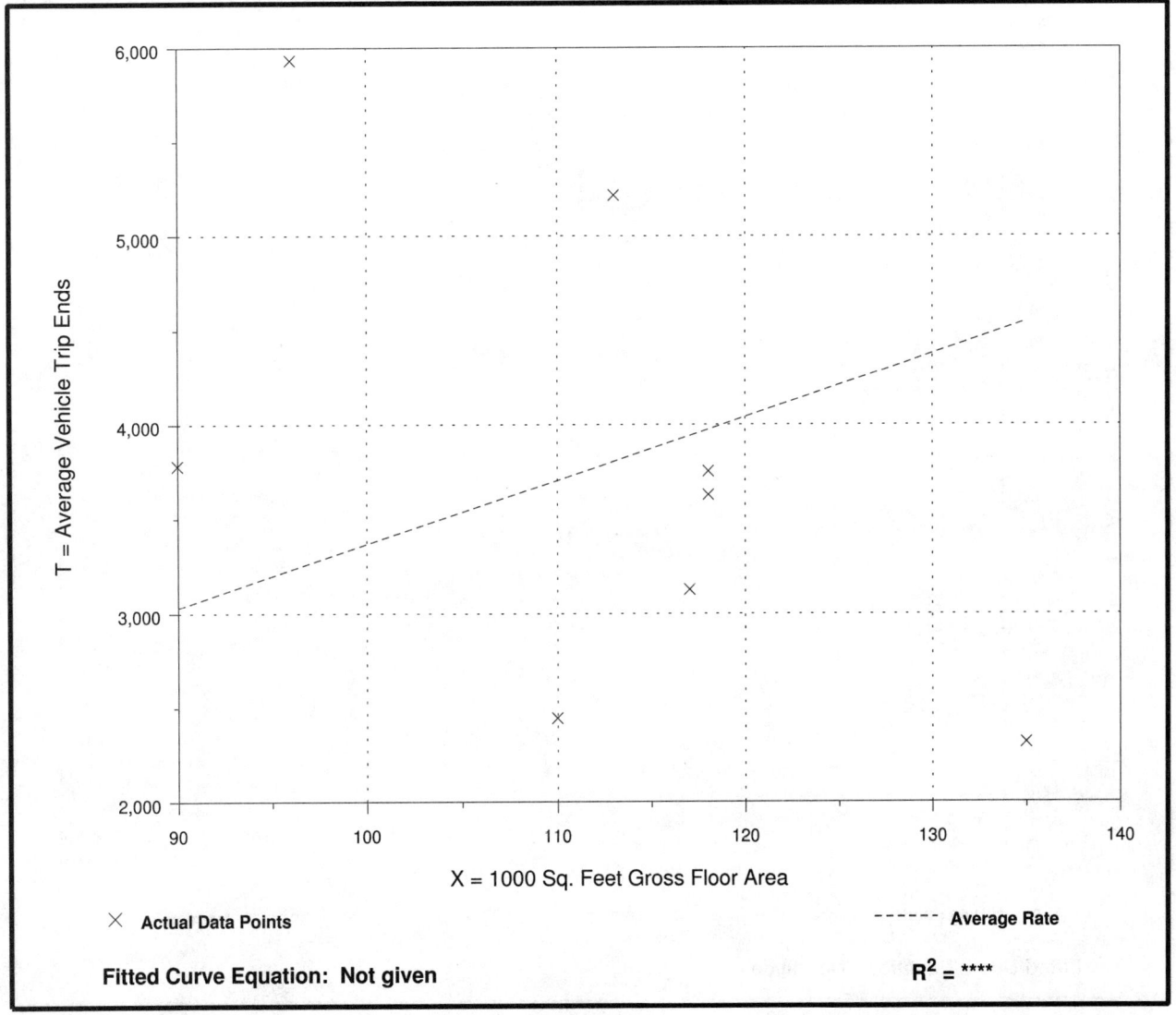

Data Plot and Equation

Fitted Curve Equation: Not given $R^2 = ****$

Discount Club
(857)

Average Vehicle Trip Ends vs: 1000 Sq. Feet Gross Floor Area
On a: Sunday,
Peak Hour of Generator

Number of Studies: 7
Average 1000 Sq. Feet GFA: 114
Directional Distribution: 50% entering, 50% exiting

Trip Generation per 1000 Sq. Feet Gross Floor Area

Average Rate	Range of Rates	Standard Deviation
5.62	3.24 - 7.58	2.75

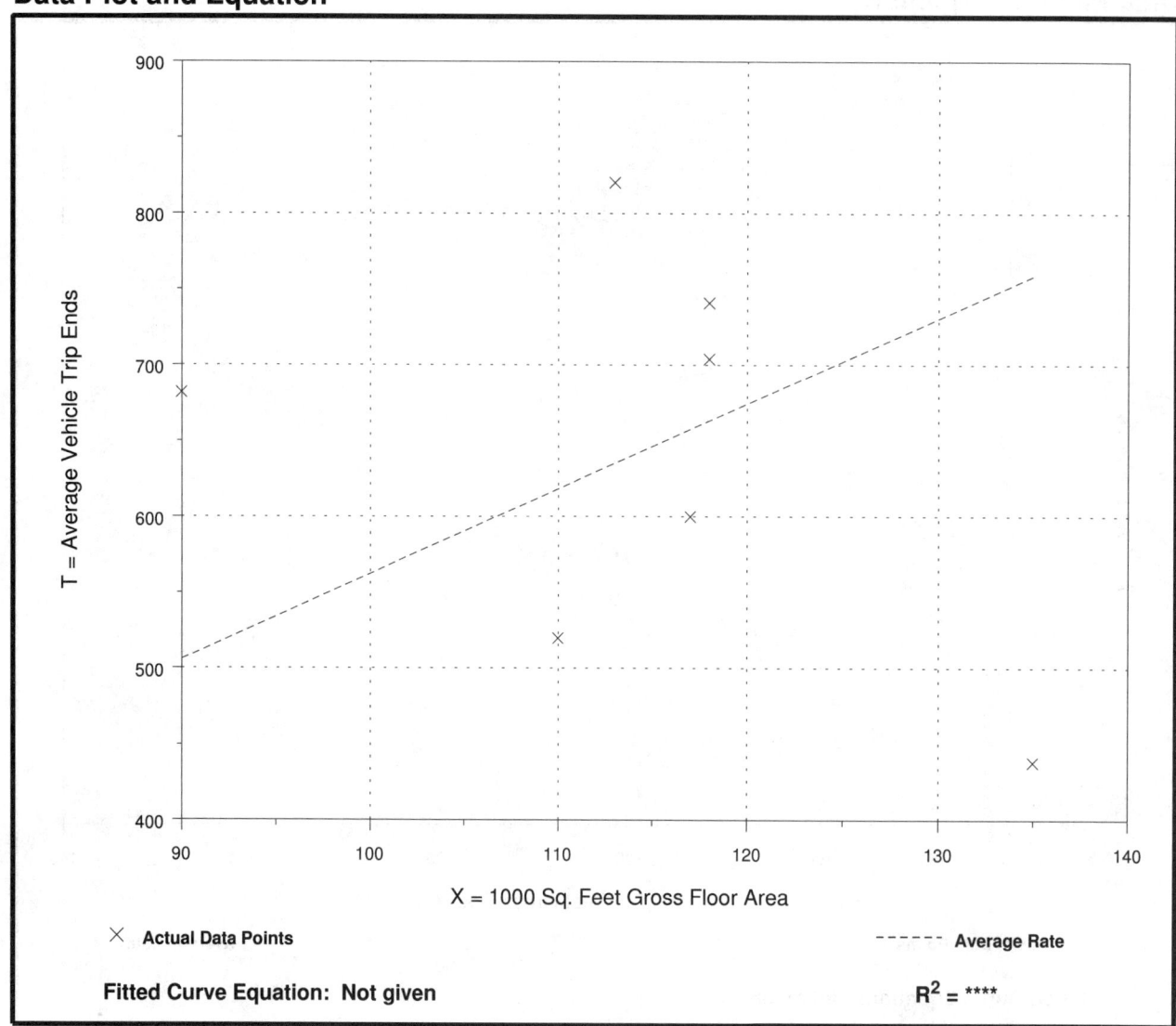

Data Plot and Equation

Fitted Curve Equation: Not given $R^2 = ****$

Discount Club
(857)

Average Vehicle Trip Ends vs: Employees
On a: Weekday

Number of Studies: 10
Avg. Number of Employees: 152
Directional Distribution: 50% entering, 50% exiting

Trip Generation per Employee

Average Rate	Range of Rates	Standard Deviation
32.21	24.76 - 49.79	8.44

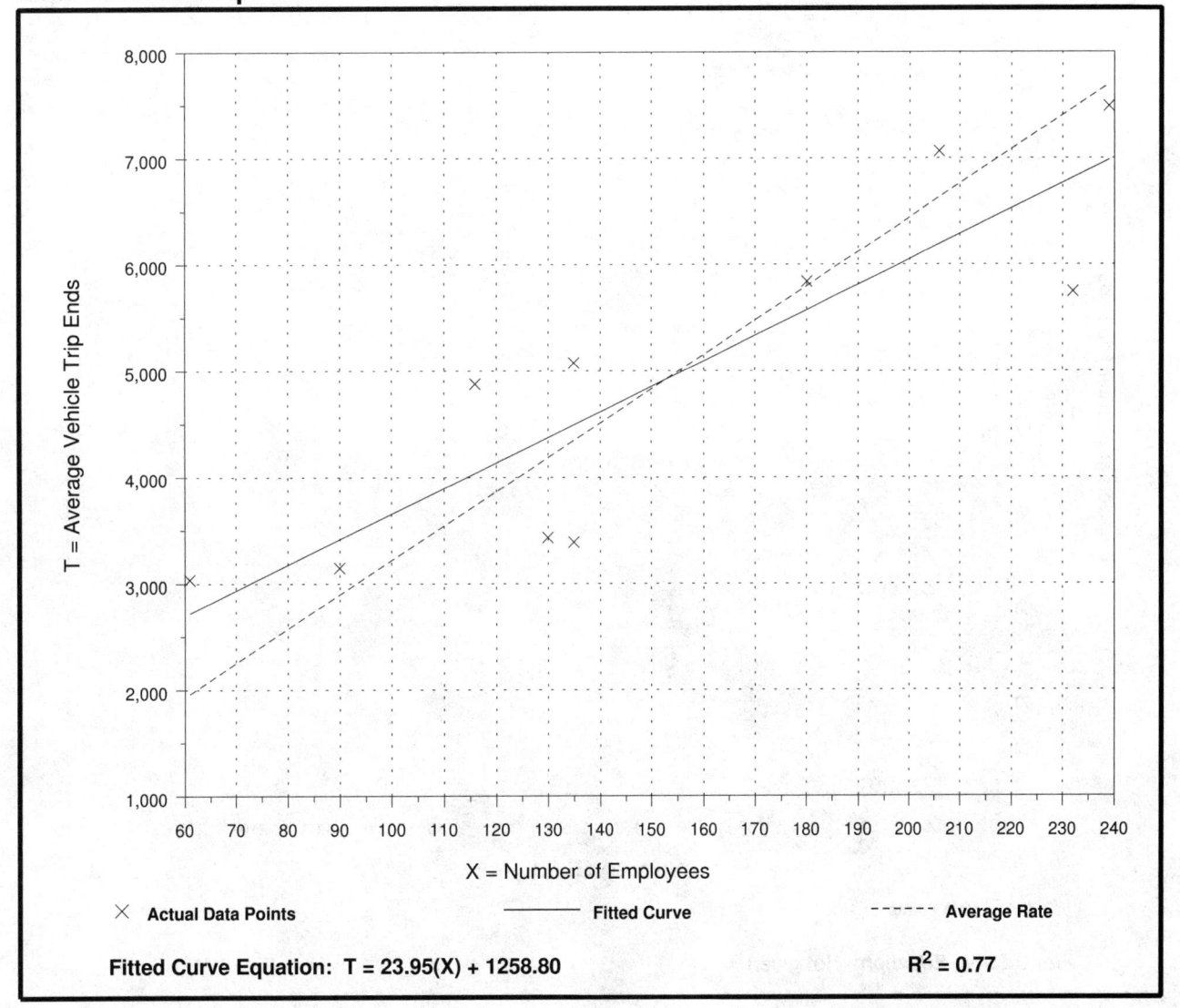

Data Plot and Equation

Fitted Curve Equation: T = 23.95(X) + 1258.80 $R^2 = 0.77$

Discount Club
(857)

Average Vehicle Trip Ends vs: Employees
On a: Weekday,
Peak Hour of Adjacent Street Traffic,
One Hour Between 7 and 9 a.m.

Number of Studies: 3
Avg. Number of Employees: 155
Directional Distribution: 77% entering, 23% exiting

Trip Generation per Employee

Average Rate	Range of Rates	Standard Deviation
0.36	0.25 - 0.43	0.61

Data Plot and Equation

Caution - Use Carefully - Small Sample Size

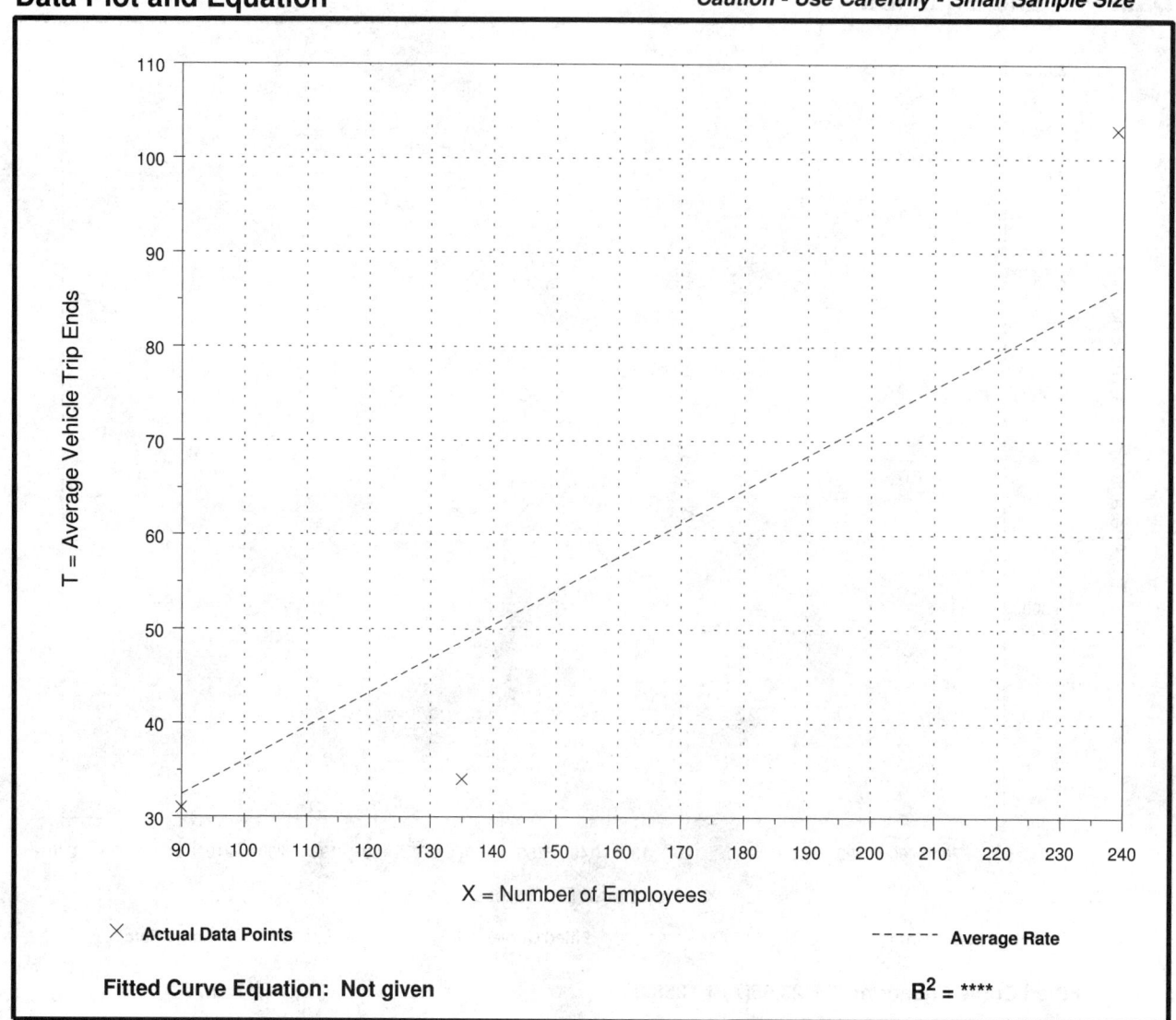

Fitted Curve Equation: Not given $R^2 = ****$

Discount Club
(857)

Average Vehicle Trip Ends vs: Employees
On a: Weekday,
Peak Hour of Adjacent Street Traffic,
One Hour Between 4 and 6 p.m.

Number of Studies: 10
Avg. Number of Employees: 152
Directional Distribution: 48% entering, 52% exiting

Trip Generation per Employee

Average Rate	Range of Rates	Standard Deviation
2.79	2.01 - 3.34	1.74

Data Plot and Equation

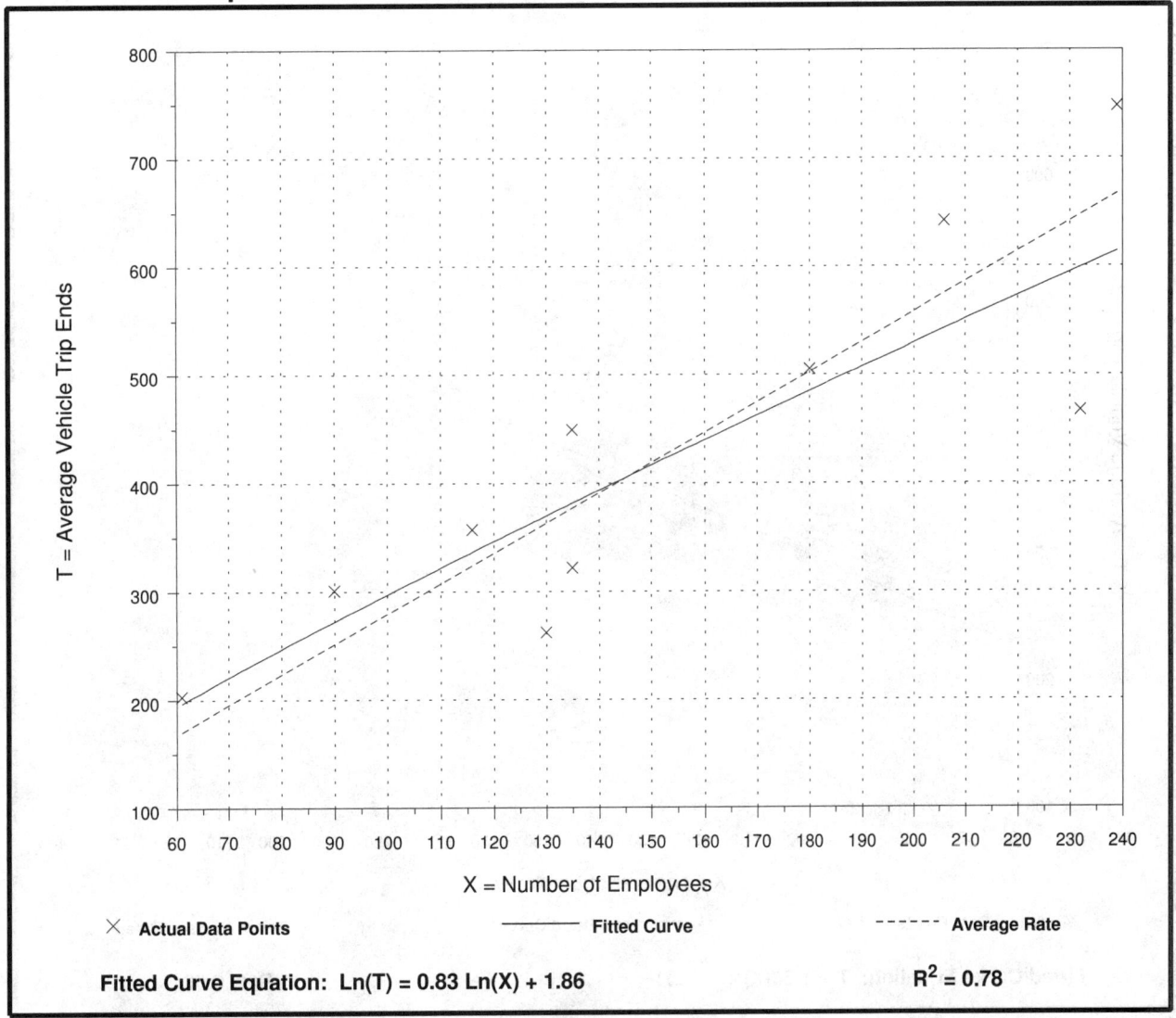

Fitted Curve Equation: $Ln(T) = 0.83 \, Ln(X) + 1.86$ $R^2 = 0.78$

Discount Club
(857)

Average Vehicle Trip Ends vs: Employees
On a: Weekday,
A.M. Peak Hour of Generator

Number of Studies: 10
Avg. Number of Employees: 152
Directional Distribution: 51% entering, 49% exiting

Trip Generation per Employee

Average Rate	Range of Rates	Standard Deviation
2.61	1.97 - 4.13	1.73

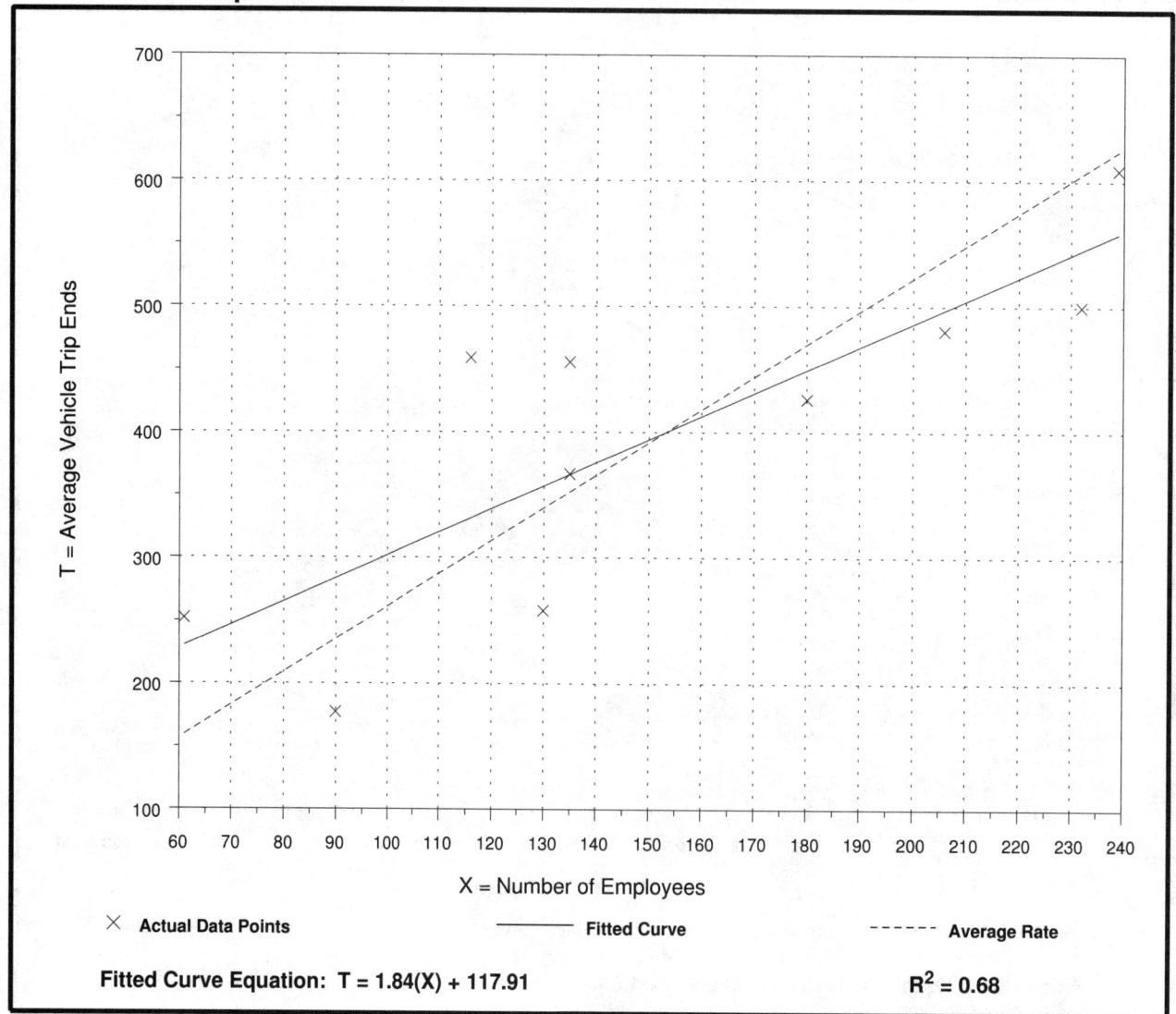

Data Plot and Equation

Fitted Curve Equation: T = 1.84(X) + 117.91 $R^2 = 0.68$

Discount Club
(857)

Average Vehicle Trip Ends vs: Employees
On a: Weekday,
P.M. Peak Hour of Generator

Number of Studies: 10
Avg. Number of Employees: 152
Directional Distribution: 47% entering, 53% exiting

Trip Generation per Employee

Average Rate	Range of Rates	Standard Deviation
3.36	2.41 - 4.98	1.94

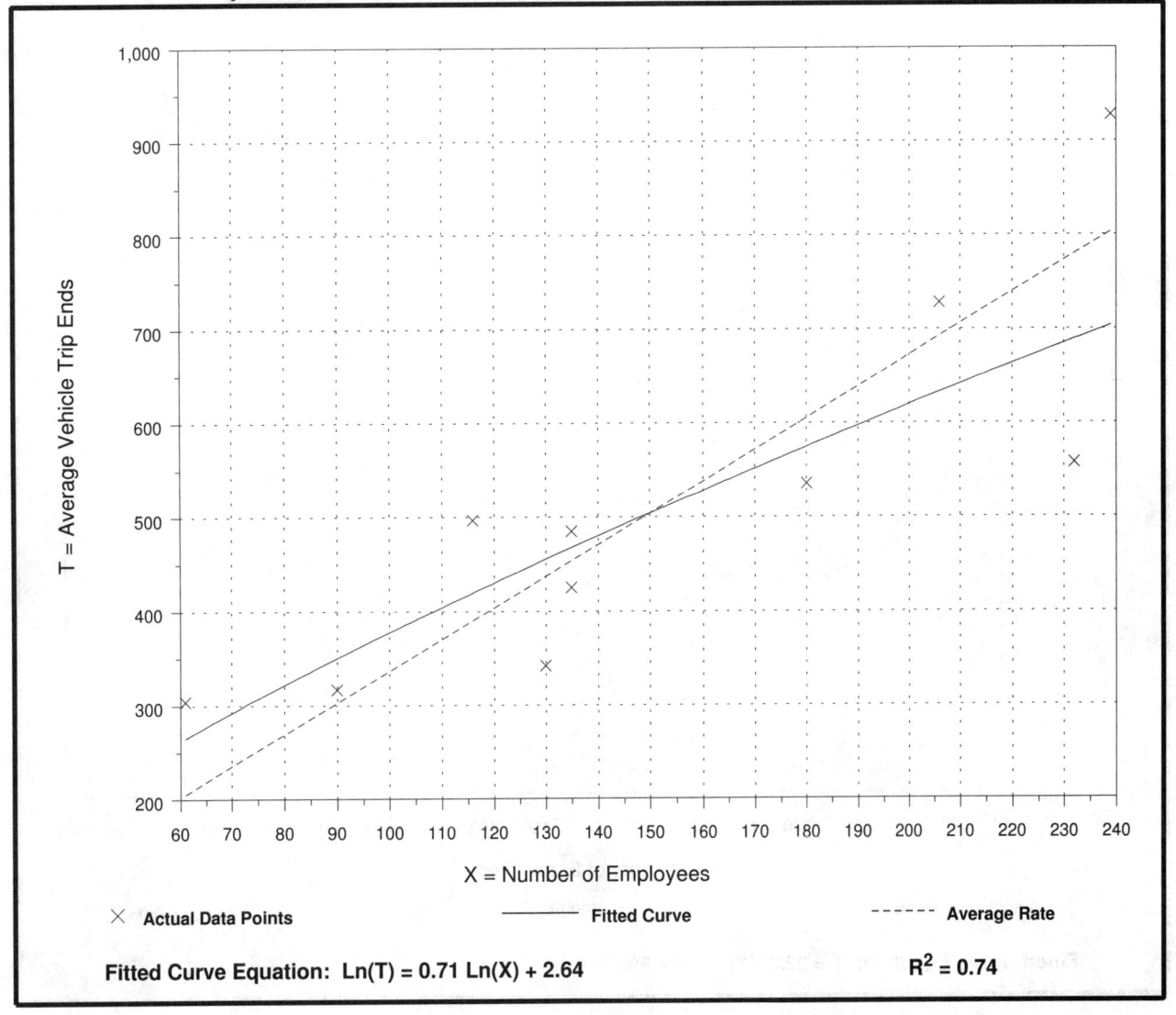

Data Plot and Equation

Fitted Curve Equation: $Ln(T) = 0.71\ Ln(X) + 2.64$ $R^2 = 0.74$

Discount Club
(857)

Average Vehicle Trip Ends vs: Employees
On a: Saturday

Number of Studies: 9
Avg. Number of Employees: 154
Directional Distribution: 50% entering, 50% exiting

Trip Generation per Employee

Average Rate	Range of Rates	Standard Deviation
38.44	29.14 - 72.56	11.03

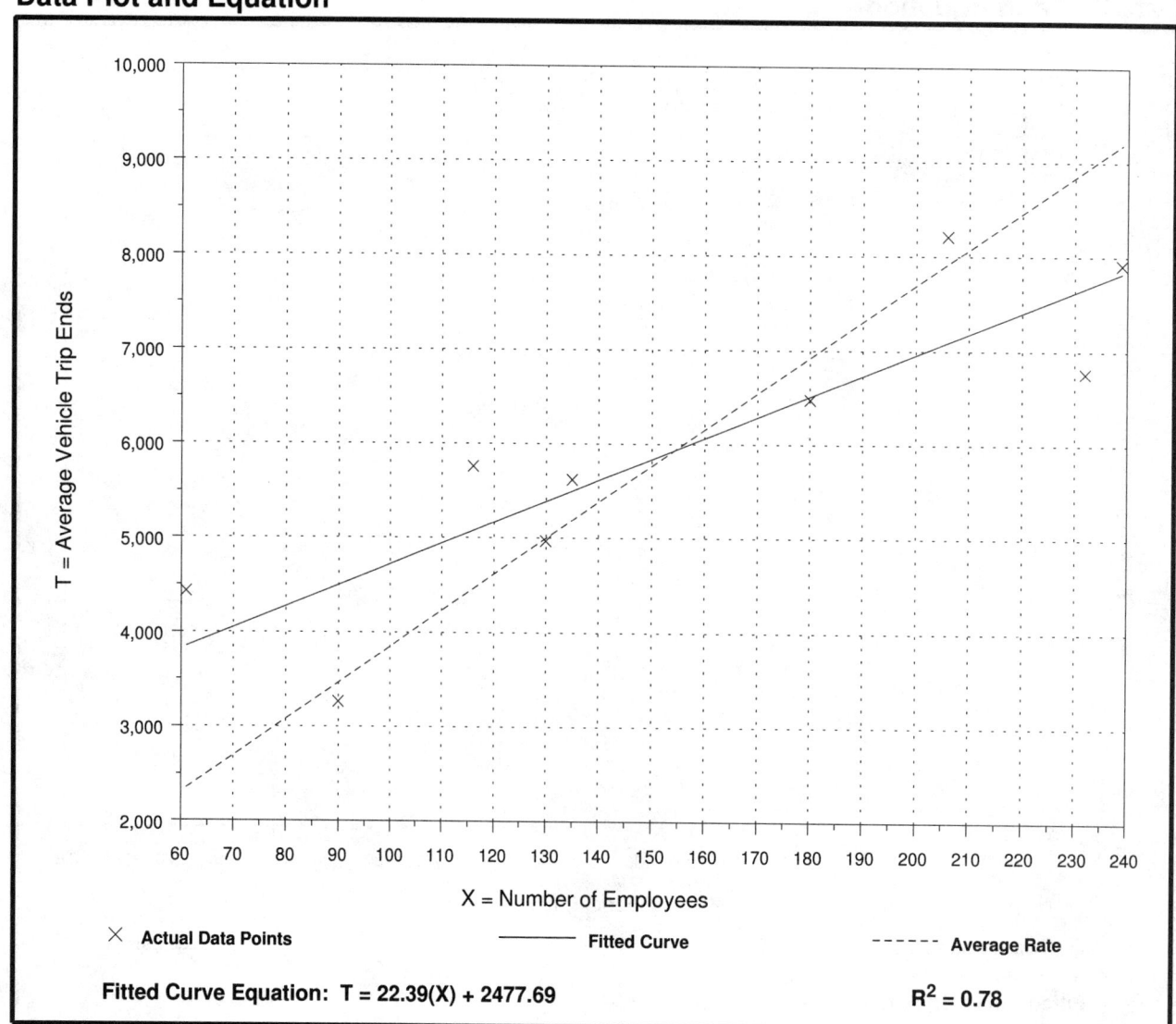

Data Plot and Equation

Fitted Curve Equation: T = 22.39(X) + 2477.69 $R^2 = 0.78$

Discount Club
(857)

Average Vehicle Trip Ends vs: Employees
On a: Saturday,
Peak Hour of Generator

Number of Studies: 8
Avg. Number of Employees: 144
Directional Distribution: 49% entering, 51% exiting

Trip Generation per Employee

Average Rate	Range of Rates	Standard Deviation
4.48	2.82 - 9.10	2.57

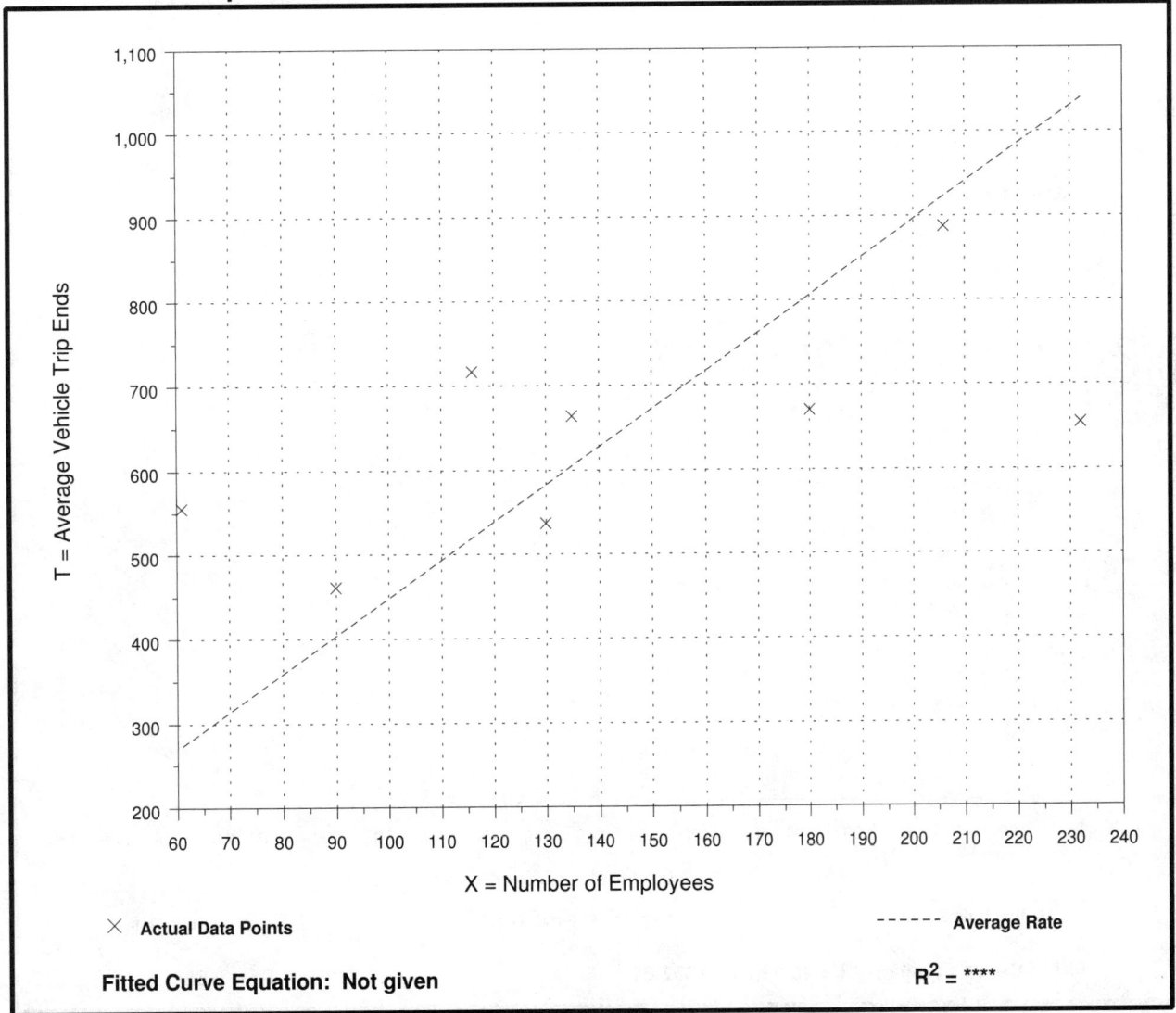

Data Plot and Equation

Fitted Curve Equation: Not given $R^2 = ****$

Discount Club
(857)

Average Vehicle Trip Ends vs: Employees
On a: Sunday

Number of Studies: 9
Avg. Number of Employees: 154
Directional Distribution: 50% entering, 50% exiting

Trip Generation per Employee

Average Rate	Range of Rates	Standard Deviation
23.35	16.30 - 40.08	7.48

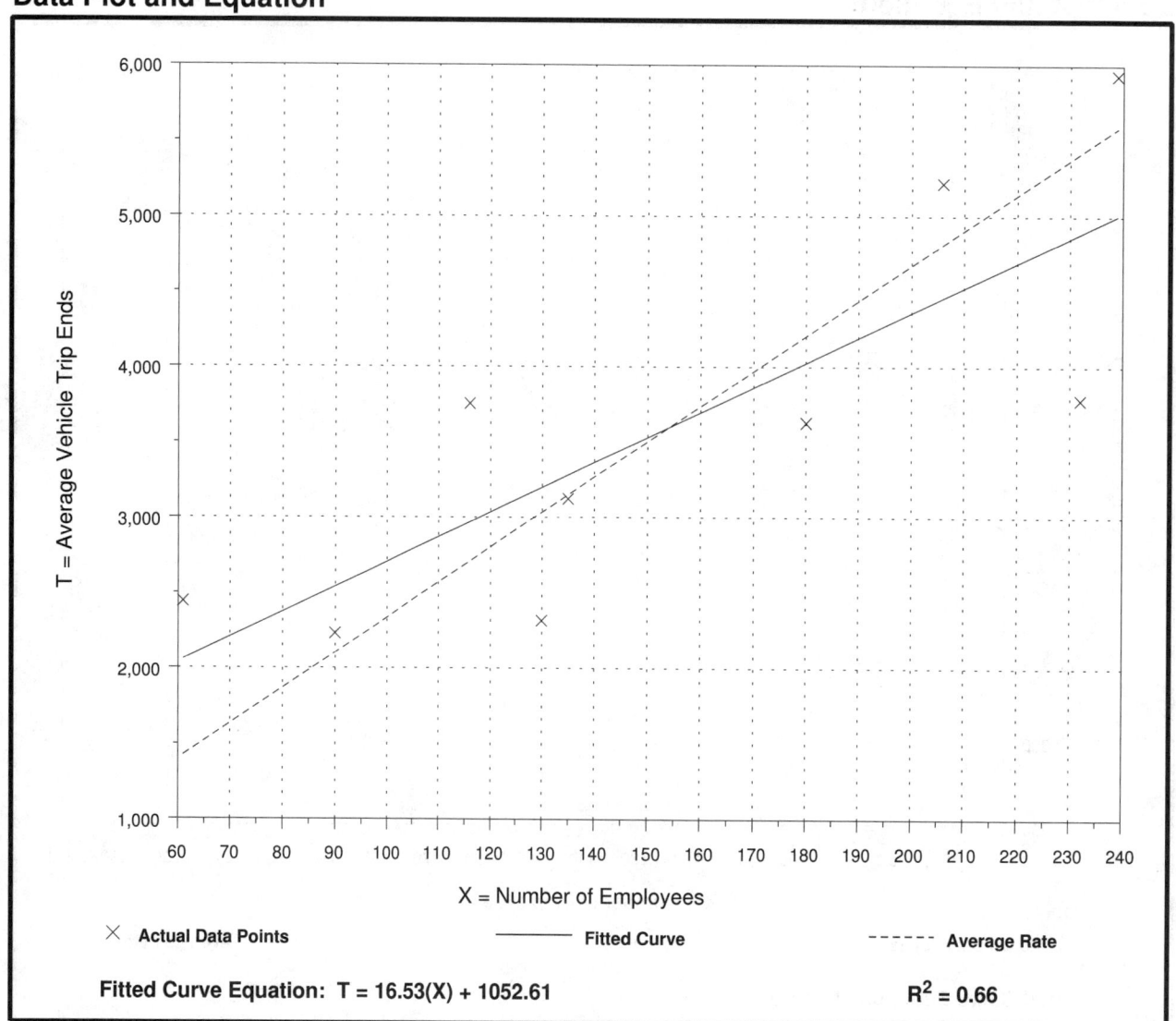

Data Plot and Equation

Fitted Curve Equation: T = 16.53(X) + 1052.61 $R^2 = 0.66$

Discount Club
(857)

Average Vehicle Trip Ends vs: Employees
On a: Sunday,
Peak Hour of Generator

Number of Studies: 8
Avg. Number of Employees: 144
Directional Distribution: 51% entering, 49% exiting

Trip Generation per Employee

Average Rate	Range of Rates	Standard Deviation
4.28	2.94 - 8.52	2.48

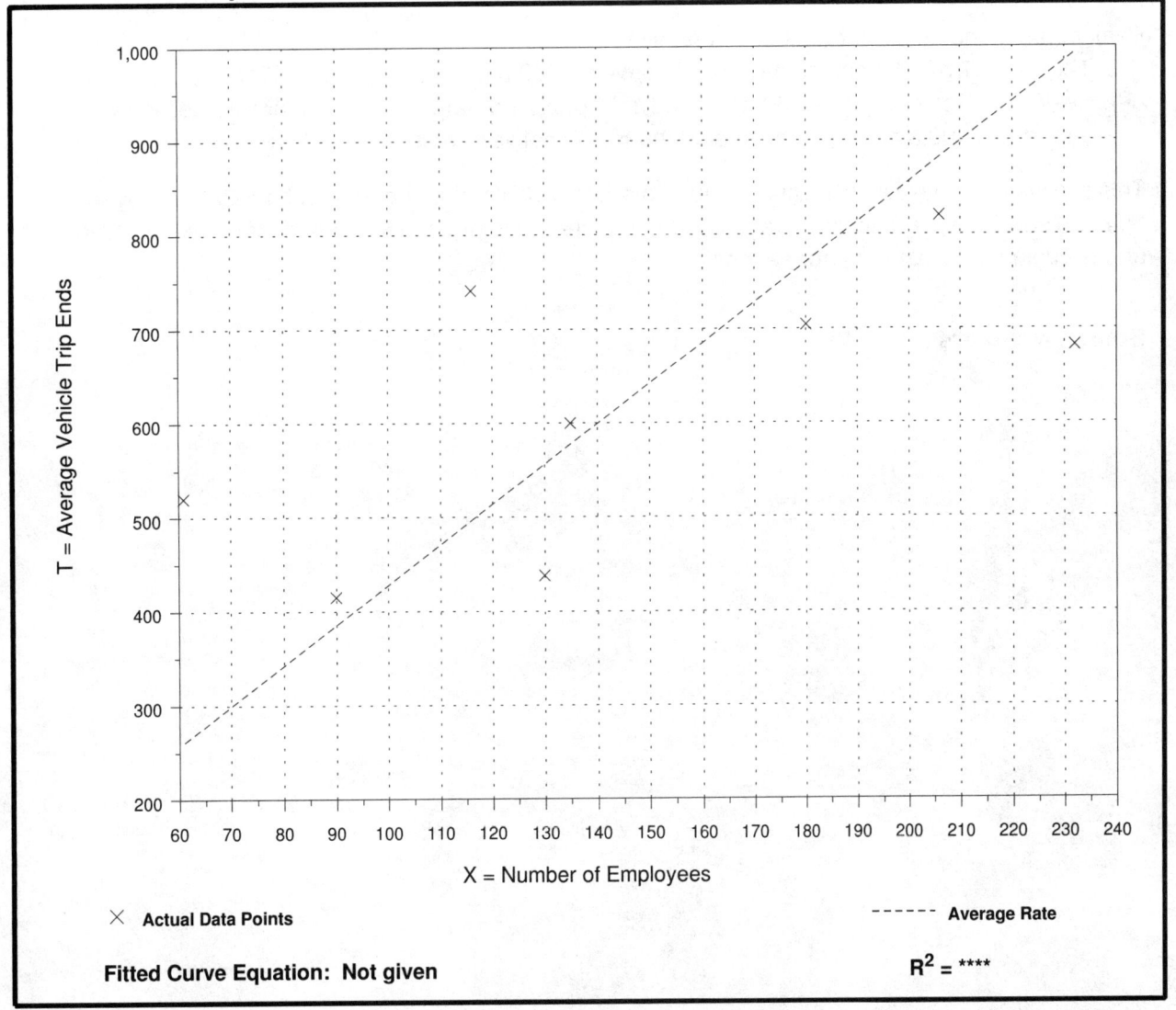

Data Plot and Equation

Fitted Curve Equation: Not given $R^2 = ****$

Land Use: 860
Wholesale Market

Description

Wholesale markets generally include large storage and distribution areas for receiving goods and shipping these goods to places such as grocery stores and restaurants. Generally, these markets are characterized by little drive-in business; truck deliveries and pick-ups take place at all hours of the day.

Additional Data

Truck trips accounted for 30 percent of the total traffic at the site.

Average vehicle occupancy for the site was 1.21 persons per automobile.

Peak hours of the generator (based on one site)—
    The weekday A.M. peak hour occured between 9:00 a.m. and 10:00 a.m. The weekday P.M. peak hour was between 12:00 noon and 1:00 p.m. The Saturday peak hour was between 10:00 a.m. and 11:00 a.m. The Sunday peak hour was between 5:00 p.m. and 6:00 p.m.

This site was surveyed in 1973 at a produce market in California. The market had 250 employees, 304,920 square feet gross floor area, 392,040 square feet of truck parking and maneuvering area, and occupied 16 acres (nine for parking).

Source Numbers

88, 599

Land Use: 860
Wholesale Market
Independent Variables with One Observation

The following trip generation data are for independent variables with only one observation. This information is shown in this table only; there are no related plots for these data.

Users are cautioned to use data with care because of the small sample size.

Independent Variable	Trip Generation Rate	Size of Independent Variable	Number of Studies	Directional Distribution
Employees				
Weekday	8.21	250	1	50% entering, 50% exiting
Weekday A.M. Peak Hour of Adjacent Street Traffic	0.61	250	1	Not available
Weekday P.M. Peak Hour of Adjacent Street Traffic	0.26	250	1	Not available
Weekday A.M. Peak Hour of Generator	0.70	250	1	Not available
Weekday P.M. Peak Hour of Generator	0.64	250	1	Not available
Saturday	1.94	250	1	50% entering, 50% exiting
Saturday Peak Hour of Generator	0.22	250	1	Not available
Sunday	2.80	250	1	50% entering, 50% exiting
Sunday Peak Hour of Generator	0.28	250	1	Not available
1,000 Square Feet Gross Floor Area				
Weekday	6.73	305	1	50% entering, 50% exiting
Weekday A.M. Peak Hour of Generator	0.58	305	1	Not available
Weekday P.M. Peak Hour of Generator	0.52	305	1	Not available
Saturday	1.59	305	1	50% entering, 50% exiting
Sunday	2.30	305	1	50% entering, 50% exiting
Sunday Peak Hour of Generator	0.23	305	1	Not available

Land Use: 860
Wholesale Market

Independent Variables with One Observation

Independent Variable	Trip Generation Rate	Size of Independent Variable	Number of Studies	Directional Distribution
Acres				
Weekday	128.25	16	1	50% entering, 50% exiting
Weekday A.M. Peak Hour of Adjacent Street Traffic	9.56	16	1	Not available
Weekday P.M. Peak Hour of Adjacent Street Traffic	4.06	16	1	Not available
Weekday A.M. Peak Hour of Generator	11.00	16	1	Not available
Weekday P.M. Peak Hour of Generator	9.94	16	1	Not available
Saturday	30.38	16	1	50% entering, 50% exiting
Saturday Peak Hour of Generator	3.44	16	1	Not available
Sunday	43.81	16	1	50% entering, 50% exiting
Sunday Peak Hour of Generator	4.31	16	1	Not available

Wholesale Market
(860)

Average Vehicle Trip Ends vs: 1000 Sq. Feet Gross Floor Area
On a: Weekday,
Peak Hour of Adjacent Street Traffic,
One Hour Between 7 and 9 a.m.

Number of Studies: 2
Average 1000 Sq. Feet GFA: 210
Directional Distribution: 67% entering, 33% exiting

Trip Generation per 1000 Sq. Feet Gross Floor Area

Average Rate	Range of Rates	Standard Deviation
0.51	0.50 - 0.55	*

Data Plot and Equation

Caution - Use Carefully - Small Sample Size

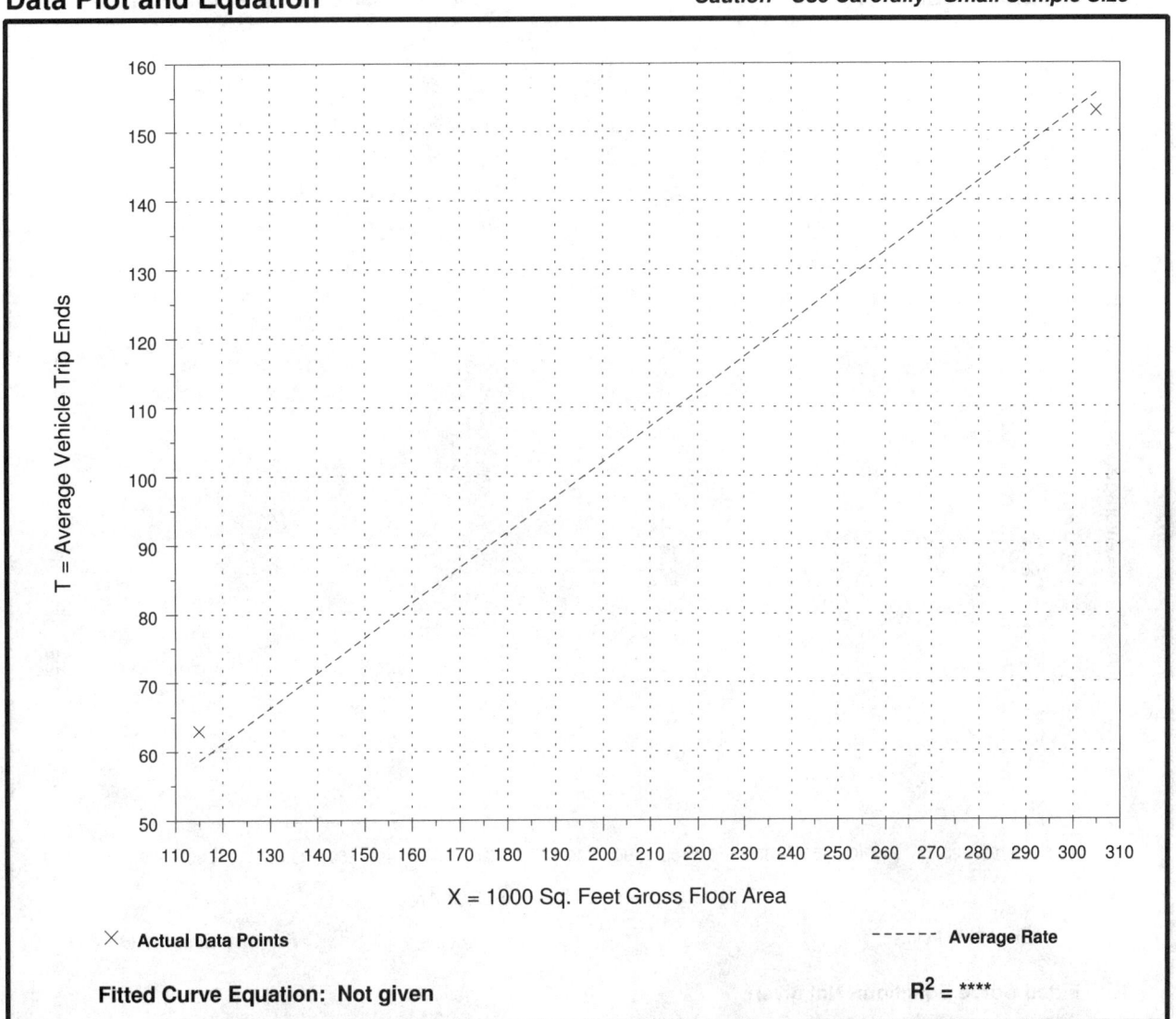

Fitted Curve Equation: Not given $R^2 =$ ****

Wholesale Market
(860)

Average Vehicle Trip Ends vs: 1000 Sq. Feet Gross Floor Area
On a: Weekday,
Peak Hour of Adjacent Street Traffic,
One Hour Between 4 and 6 p.m.

Number of Studies: 3
Average 1000 Sq. Feet GFA: 178
Directional Distribution: 53% entering, 47% exiting

Trip Generation per 1000 Sq. Feet Gross Floor Area

Average Rate	Range of Rates	Standard Deviation
0.88	0.21 - 1.78	1.21

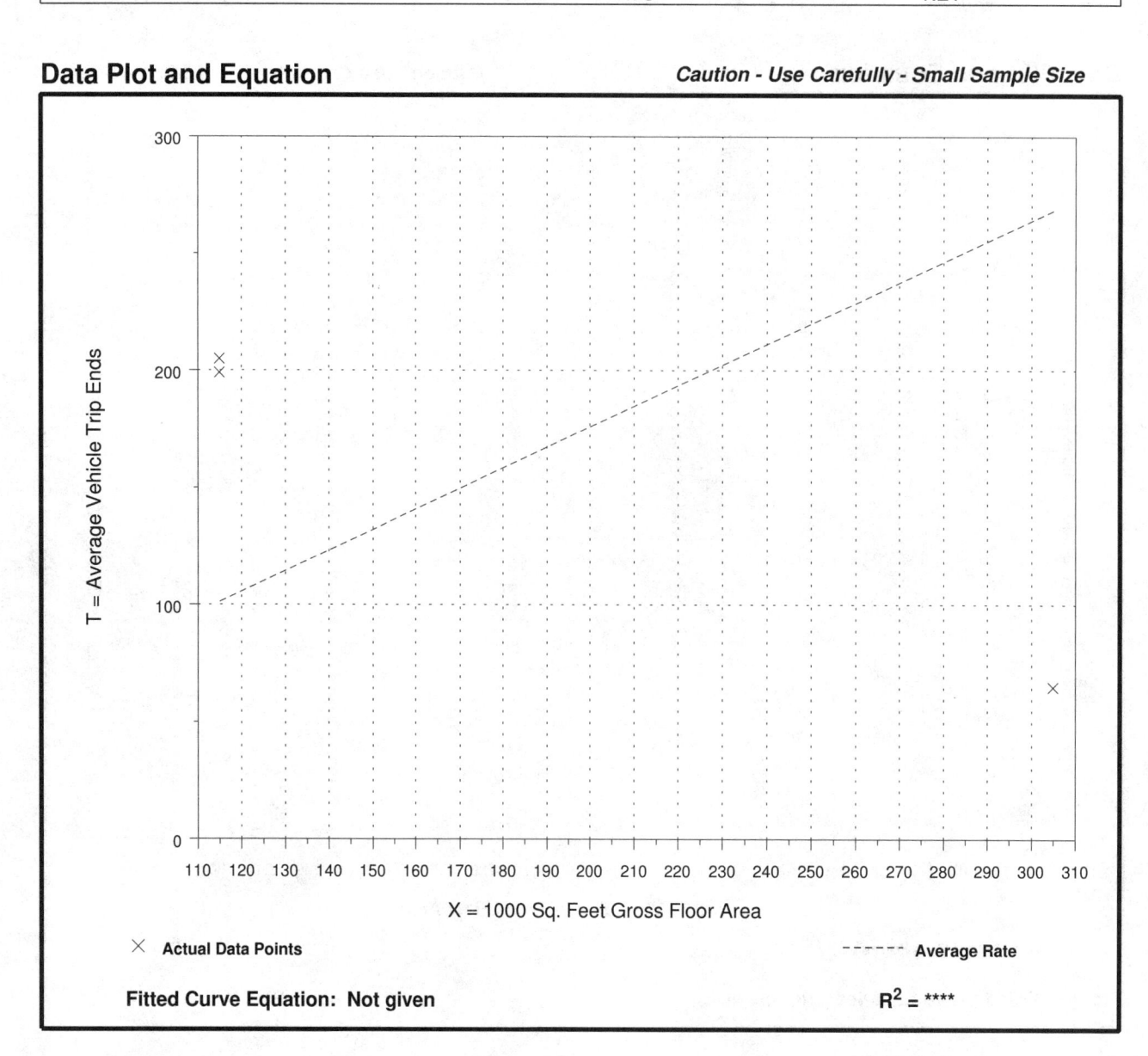

Data Plot and Equation

Caution - Use Carefully - Small Sample Size

Fitted Curve Equation: Not given $R^2 = ****$

Land Use: 861
Sporting Goods Superstore

Description

Sporting goods superstores are free-standing facilities. These stores generally offer a variety of customer services and centralized cashiering and specialize in the sale of athletic and outdoor-oriented merchandise. These stores typically maintain long store hours 7 days a week. Examples of items sold in these stores include outdoor/athletic clothing, sports equipment, shoes and hunting/boating/fishing gear. Some may also carry automotive supplies. Sporting goods superstores are sometimes also found as separate parcels within a retail complex, with or without their own dedicated parking.

Additional Data

Peak hours of the generator—
    The Saturday peak hour varied between 12:00 p.m. and 2:00 p.m.

The sites were surveyed in the 2000s in Massachusetts, Nevada and New Hampshire.

Source Numbers

618, 745, 747

Land Use: 861
Sporting Goods Superstore
Independent Variables with One Observation

The following trip generation data are for independent variables with only one observation. This information is shown in this table only; there are no related plots for these data.

Users are cautioned to use data with care because of the small sample size.

Independent Variable	Trip Generation Rate	Size of Independent Variable	Number of Studies	Directional Distribution
1,000 Square Feet Gross Floor Area				
Weekday A.M. Peak Hour of Adjacent Street Traffic	0.25	295	1	80% entering, 20% exiting

Sporting Goods Superstore
(861)

Average Vehicle Trip Ends vs: 1000 Sq. Feet Gross Floor Area
On a: Weekday,
Peak Hour of Adjacent Street Traffic,
One Hour Between 4 and 6 p.m.

Number of Studies: 5
Average 1000 Sq. Feet GFA: 93
Directional Distribution: 48% entering, 52% exiting

Trip Generation per 1000 Sq. Feet Gross Floor Area

Average Rate	Range of Rates	Standard Deviation
1.84	1.11 - 4.69	1.82

Data Plot and Equation

Caution - Use Carefully - Small Sample Size

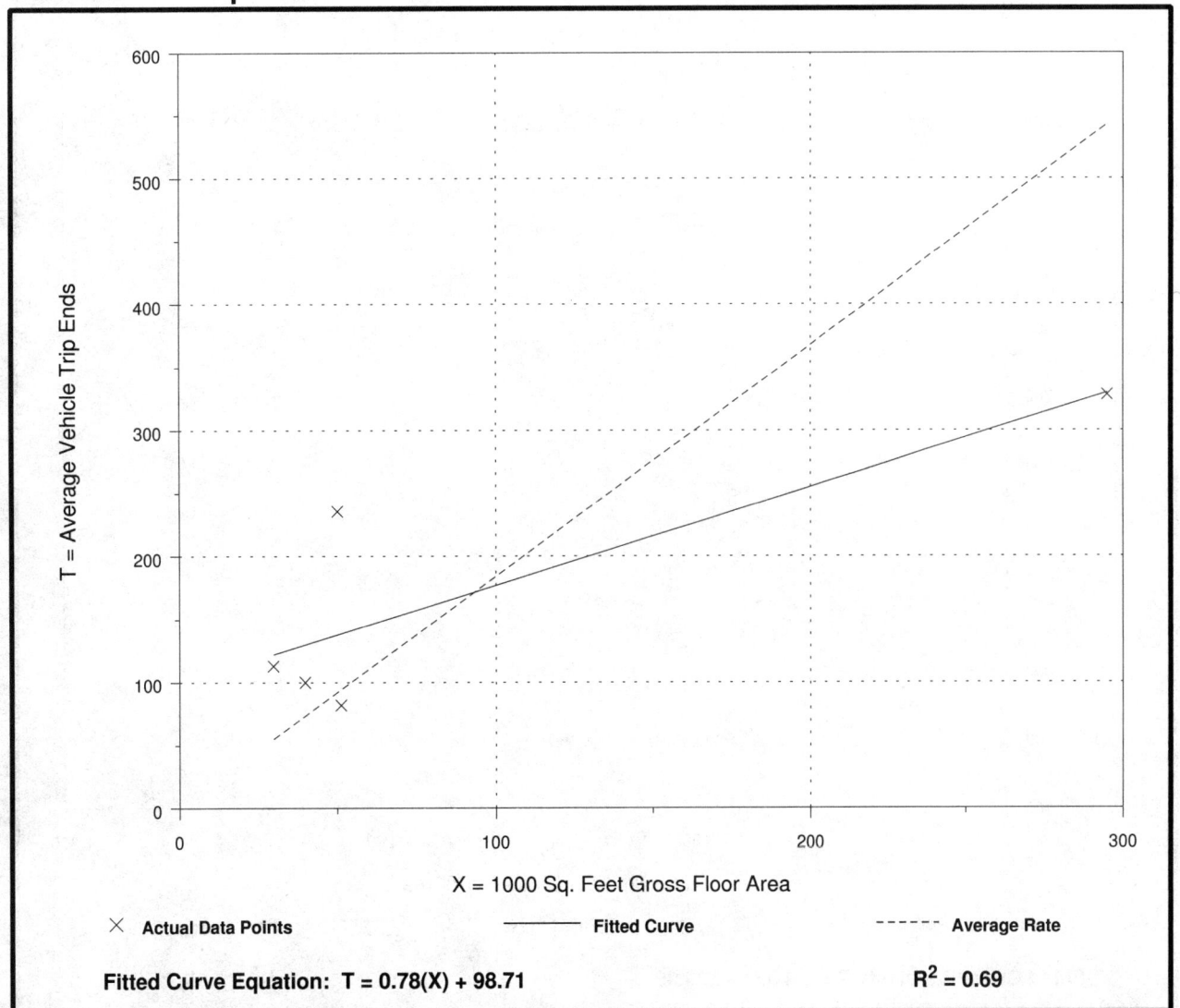

Fitted Curve Equation: $T = 0.78(X) + 98.71$ $R^2 = 0.69$

Sporting Goods Superstore
(861)

Average Vehicle Trip Ends vs: 1000 Sq. Feet Gross Floor Area
On a: Saturday,
Peak Hour of Generator

Number of Studies: 5
Average 1000 Sq. Feet GFA: 93
Directional Distribution: 51% entering, 49% exiting

Trip Generation per 1000 Sq. Feet Gross Floor Area

Average Rate	Range of Rates	Standard Deviation
3.84	2.19 - 9.83	3.30

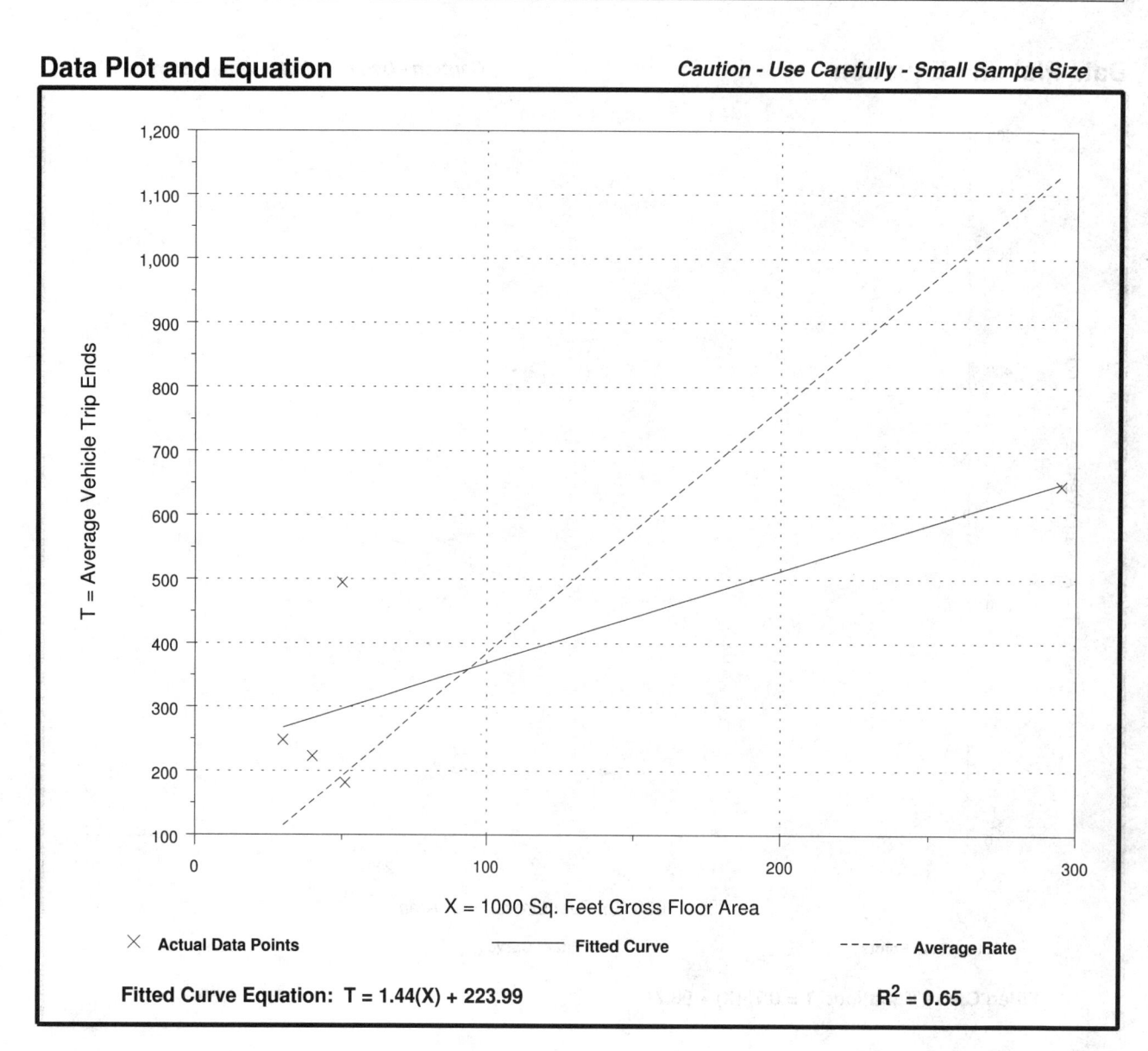

Data Plot and Equation

Caution - Use Carefully - Small Sample Size

Fitted Curve Equation: T = 1.44(X) + 223.99 $R^2 = 0.65$

Land Use: 862
Home Improvement Superstore

Description

Home improvement superstores are free-standing facilities that specialize in the sale of home improvement merchandise. These stores generally offer a variety of customer services and centralized cashiering. Home improvement superstores typically maintain long store hours 7 days a week. Examples of items sold in these stores include lumber, tools, paint, lighting, wallpaper and paneling, kitchen and bathroom fixtures, lawn equipment, and plant and garden accessories. The stores included in this land use are often the only ones on the site, but they can also be found in mutual operation with a related or unrelated garden center. Home improvement superstores are sometimes found as separate parcels within a retail complex, with or without their own dedicated parking. The buildings contained in this land use usually range in size from 50,000 to 200,000 square feet gross floor area. This land use does not include interior design stores. Building materials and lumber store (Land Use 812) and hardware/paint store (Land Use 816) are related uses.

Additional Data

Peak hours of the generator—
   The weekday A.M. peak hour varied between 10:00 a.m. and 12:00 p.m. The weekday P.M. peak hour varied between 12:00 p.m. and 5:00 p.m. The weekend peak hour varied between 12:00 p.m. and 3:00 p.m.

Outside storage areas are not included in the overall gross floor area measurements. However, if storage areas are located within the principal outside faces of the exterior walls, they are included in the overall gross floor area of the building.

Garden centers contained within the principal outside faces of the exterior building walls were included in the gross square floor areas reported. Outdoor or fenced-in areas outside the principal faces of the exterior building walls were excluded. Please refer to Volume 1, User's Guide, for a more detailed definition of gross floor area.

The sites were surveyed between the 1980s and the 2000s throughout the United States.

To assist in the future analysis of this land use, it is important to collect and include information on the presence and size of garden centers, outdoor fenced-in space and service stations in trip generation data submissions.

Source Numbers

126, 376, 434, 437, 507, 616, 617, 728, 731

Home Improvement Superstore
(862)

Average Vehicle Trip Ends vs: 1000 Sq. Feet Gross Floor Area
On a: Weekday

Number of Studies: 19
Average 1000 Sq. Feet GFA: 135
Directional Distribution: 50% entering, 50% exiting

Trip Generation per 1000 Sq. Feet Gross Floor Area

Average Rate	Range of Rates	Standard Deviation
30.74	18.35 - 55.34	10.01

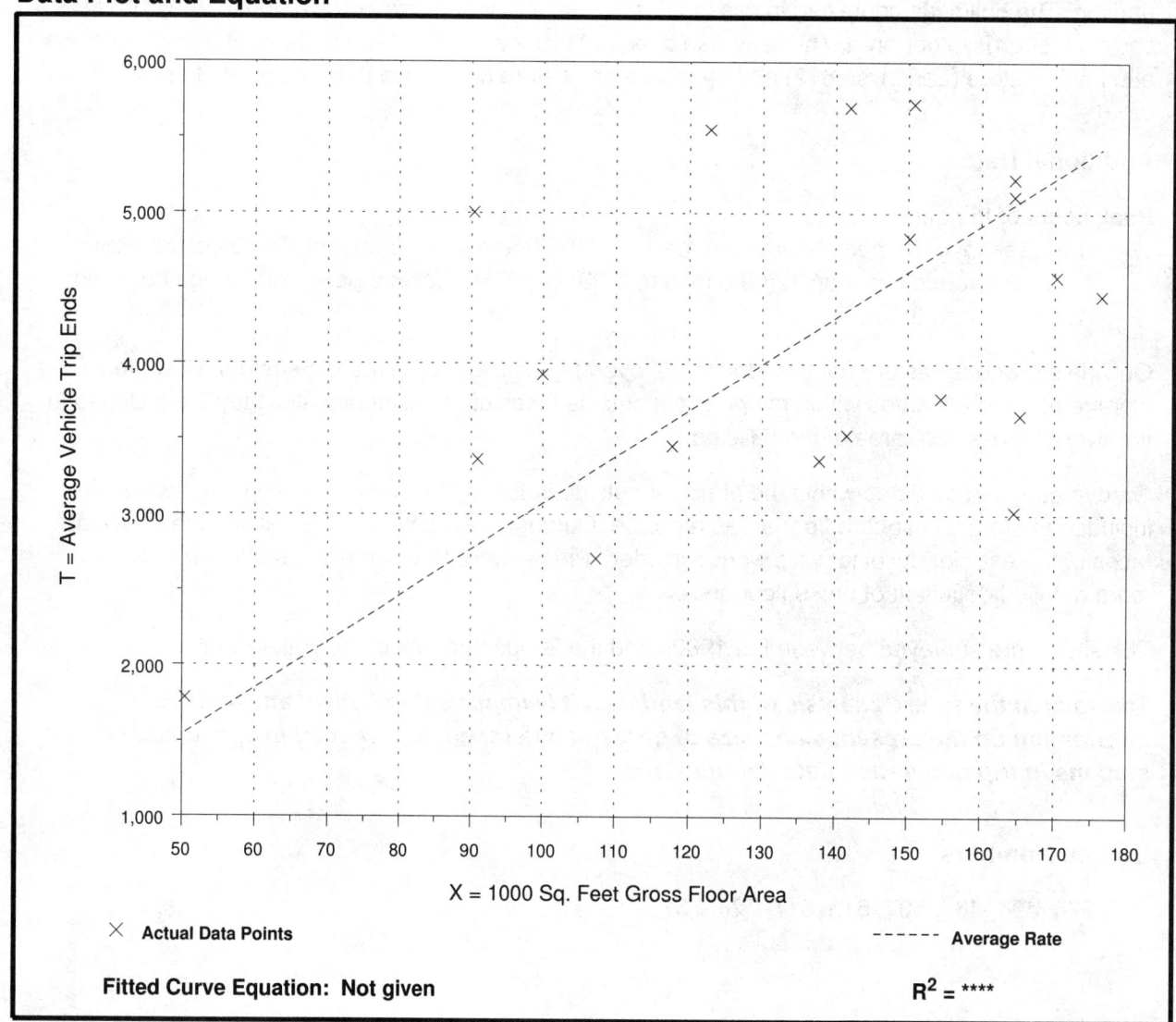

Data Plot and Equation

Fitted Curve Equation: Not given $R^2 = ****$

Home Improvement Superstore
(862)

Average Vehicle Trip Ends vs: 1000 Sq. Feet Gross Floor Area
On a: Weekday,
Peak Hour of Adjacent Street Traffic,
One Hour Between 7 and 9 a.m.

Number of Studies: 46
Average 1000 Sq. Feet GFA: 135
Directional Distribution: 57% entering, 43% exiting

Trip Generation per 1000 Sq. Feet Gross Floor Area

Average Rate	Range of Rates	Standard Deviation
1.49	0.32 - 4.16	1.42

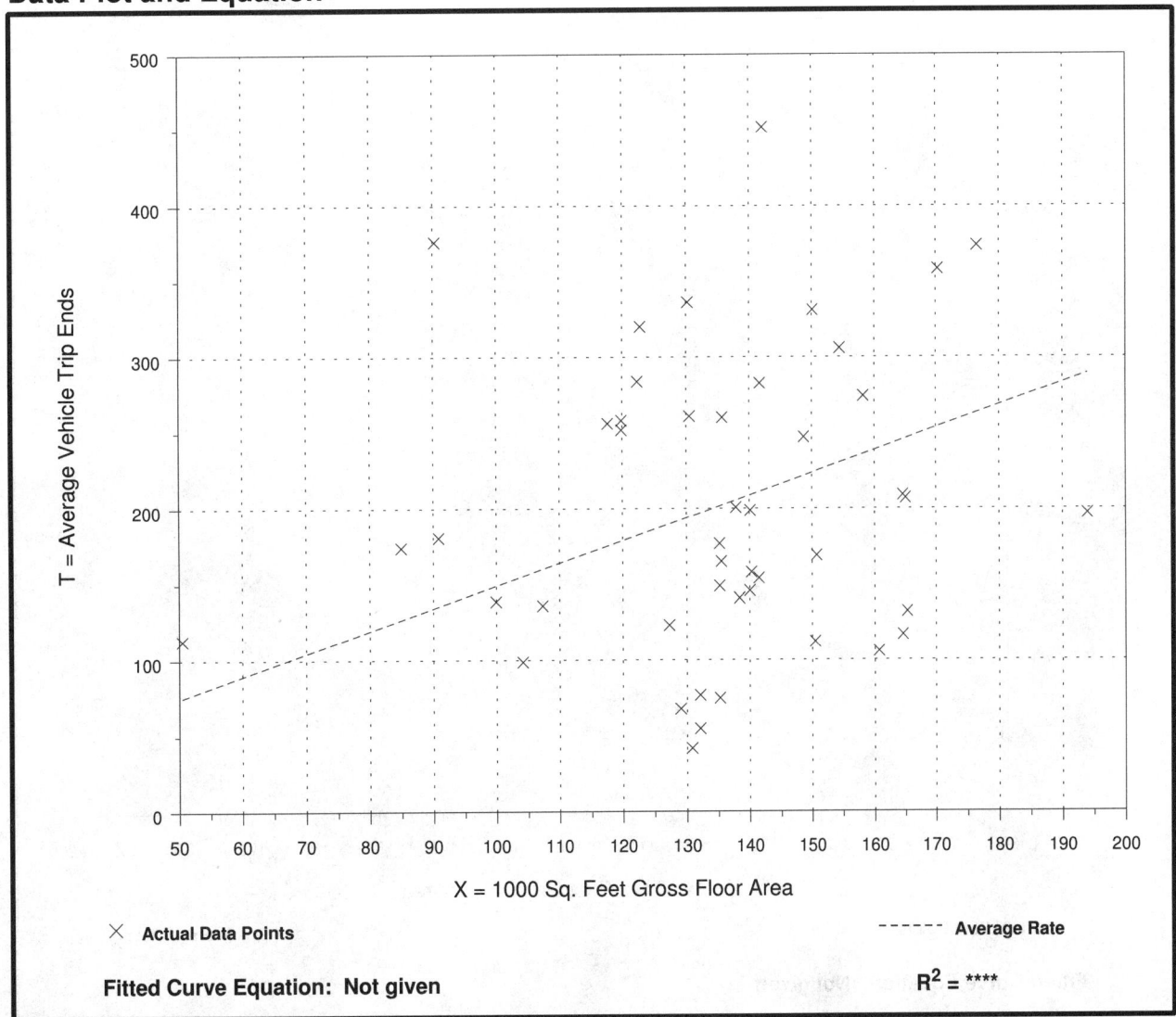

Data Plot and Equation

Fitted Curve Equation: Not given $R^2 =$ ****

Home Improvement Superstore
(862)

Average Vehicle Trip Ends vs: 1000 Sq. Feet Gross Floor Area
On a: Weekday,
Peak Hour of Adjacent Street Traffic,
One Hour Between 4 and 6 p.m.

Number of Studies: 51
Average 1000 Sq. Feet GFA: 135
Directional Distribution: 49% entering, 51% exiting

Trip Generation per 1000 Sq. Feet Gross Floor Area

Average Rate	Range of Rates	Standard Deviation
2.33	1.20 - 4.34	1.69

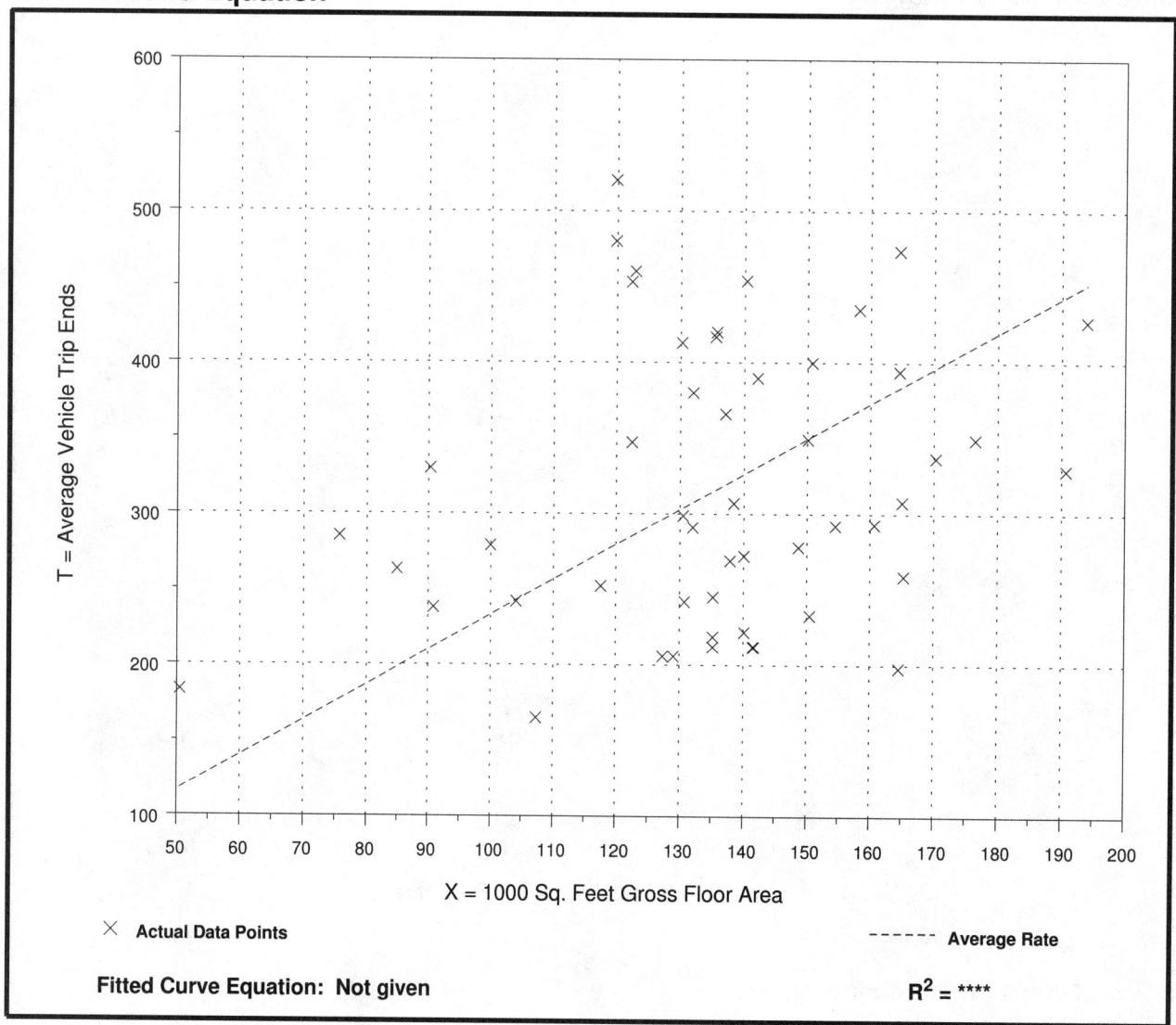

Data Plot and Equation

Fitted Curve Equation: Not given $R^2 = ****$

Home Improvement Superstore
(862)

Average Vehicle Trip Ends vs: 1000 Sq. Feet Gross Floor Area
On a: Weekday,
A.M. Peak Hour of Generator

Number of Studies: 23
Average 1000 Sq. Feet GFA: 139
Directional Distribution: 52% entering, 48% exiting

Trip Generation per 1000 Sq. Feet Gross Floor Area

Average Rate	Range of Rates	Standard Deviation
2.57	1.46 - 5.31	1.84

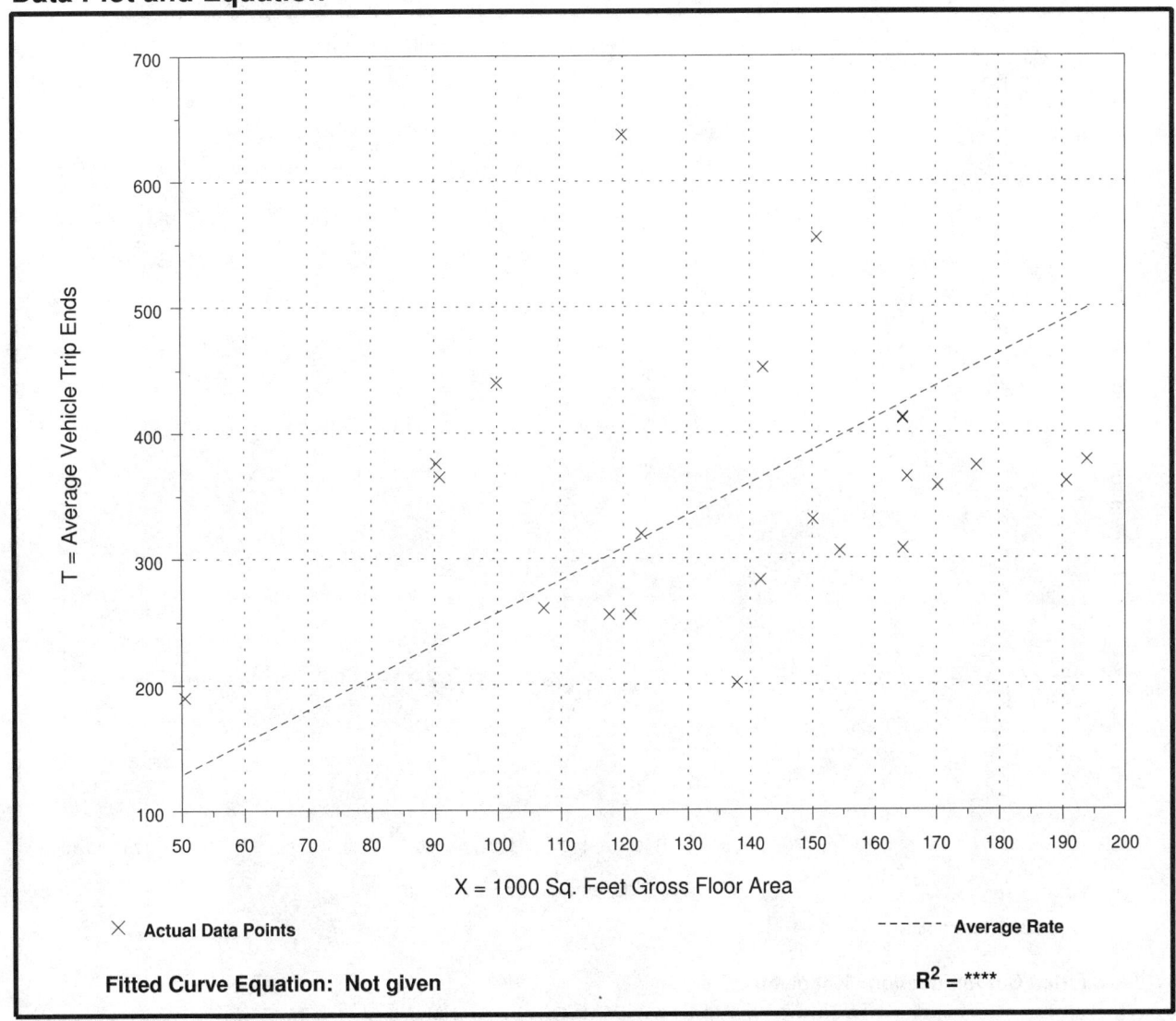

Data Plot and Equation

X   Actual Data Points   ----- Average Rate

Fitted Curve Equation: Not given $R^2 = ****$

Home Improvement Superstore
(862)

Average Vehicle Trip Ends vs: 1000 Sq. Feet Gross Floor Area
On a: Weekday,
P.M. Peak Hour of Generator

Number of Studies: 20
Average 1000 Sq. Feet GFA: 134
Directional Distribution: 52% entering, 48% exiting

Trip Generation per 1000 Sq. Feet Gross Floor Area

Average Rate	Range of Rates	Standard Deviation
3.17	1.96 - 5.89	2.06

Data Plot and Equation

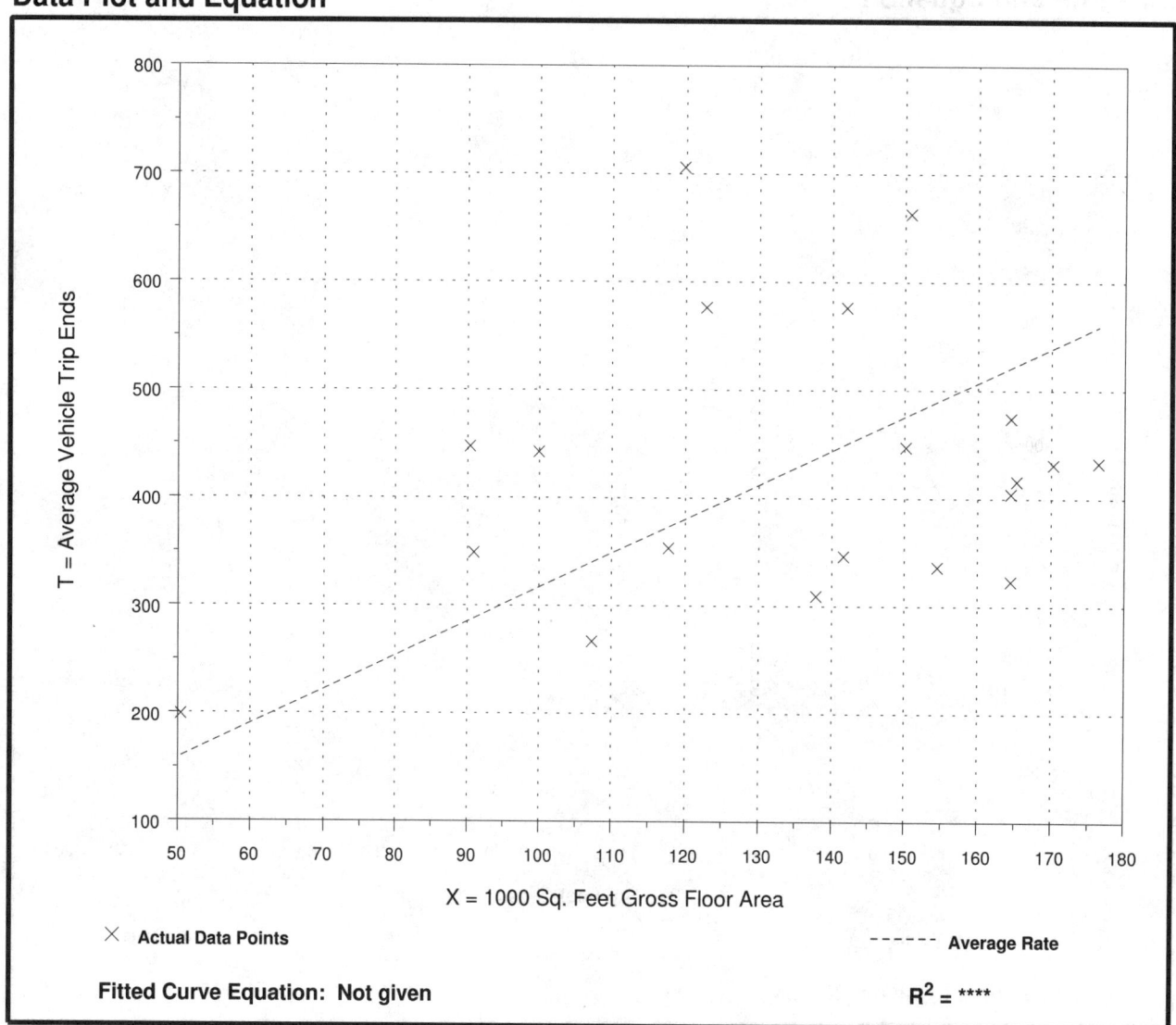

× Actual Data Points   ----- Average Rate

Fitted Curve Equation: Not given $R^2 = ****$

Home Improvement Superstore
(862)

Average Vehicle Trip Ends vs: 1000 Sq. Feet Gross Floor Area
On a: Saturday

Number of Studies: 3
Average 1000 Sq. Feet GFA: 99
Directional Distribution: 50% entering, 50% exiting

Trip Generation per 1000 Sq. Feet Gross Floor Area

Average Rate	Range of Rates	Standard Deviation
56.72	34.77 - 73.12	16.32

Data Plot and Equation

Caution - Use Carefully - Small Sample Size

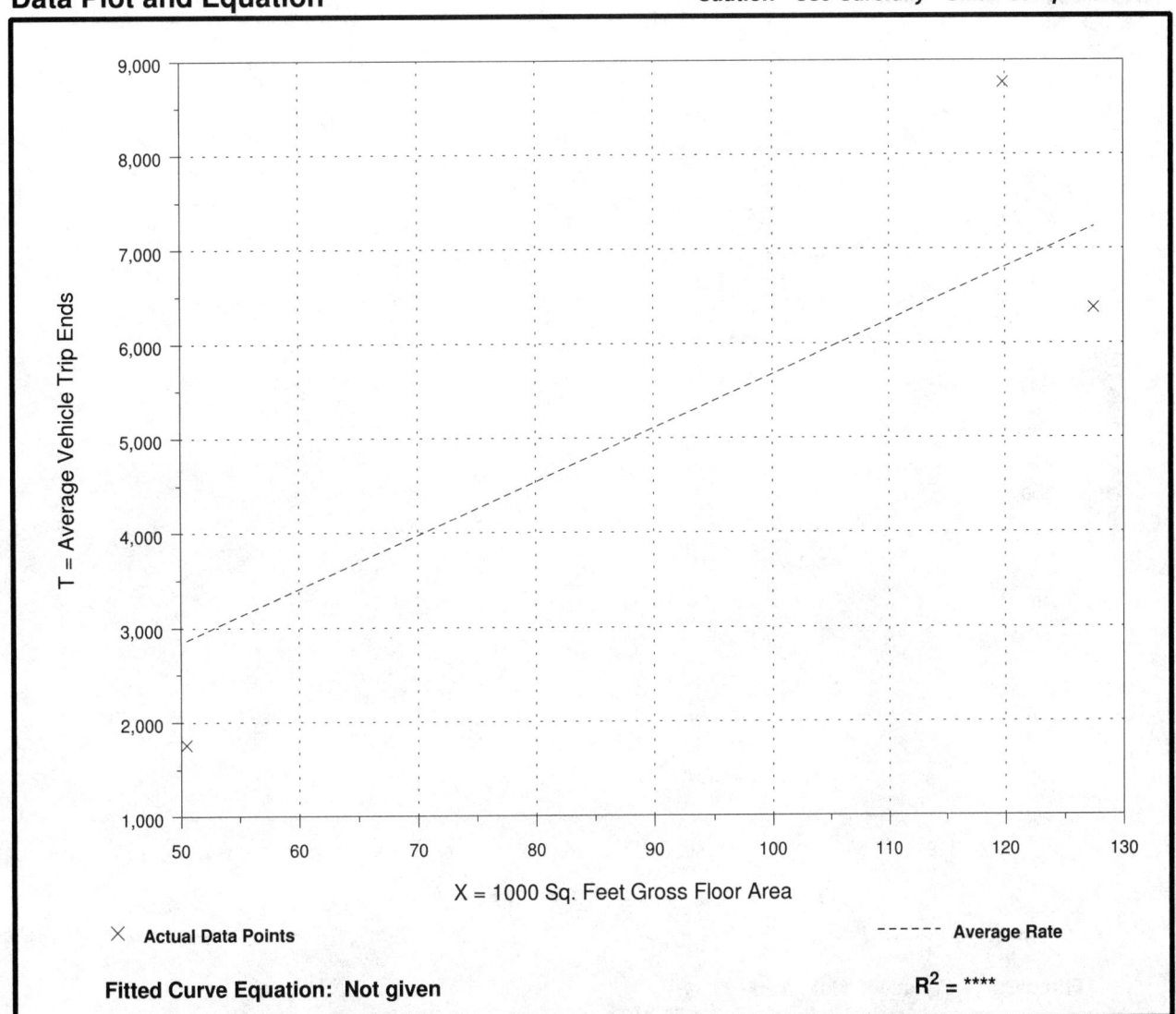

Fitted Curve Equation: Not given

$R^2 = ****$

Home Improvement Superstore
(862)

Average Vehicle Trip Ends vs: 1000 Sq. Feet Gross Floor Area
On a: Saturday,
Peak Hour of Generator

Number of Studies: 31
Average 1000 Sq. Feet GFA: 129
Directional Distribution: 51% entering, 49% exiting

Trip Generation per 1000 Sq. Feet Gross Floor Area

Average Rate	Range of Rates	Standard Deviation
4.51	2.63 - 7.28	2.44

Data Plot and Equation

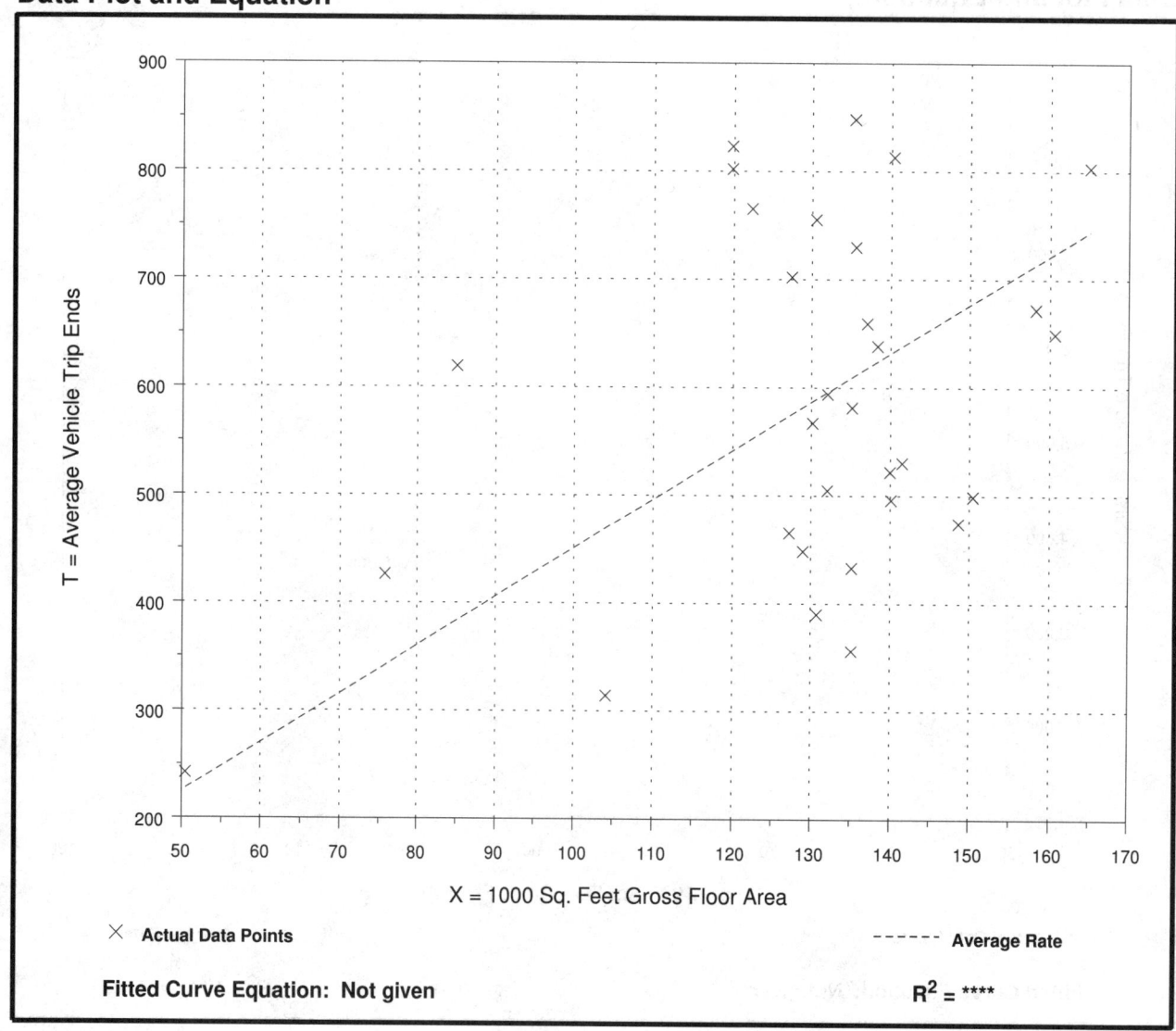

Fitted Curve Equation: Not given $R^2 = ****$

Home Improvement Superstore
(862)

Average Vehicle Trip Ends vs: 1000 Sq. Feet Gross Floor Area
On a: Sunday

Number of Studies: 2
Average 1000 Sq. Feet GFA: 85
Directional Distribution: 50% entering, 50% exiting

Trip Generation per 1000 Sq. Feet Gross Floor Area

Average Rate	Range of Rates	Standard Deviation
55.80	20.93 - 70.49	*

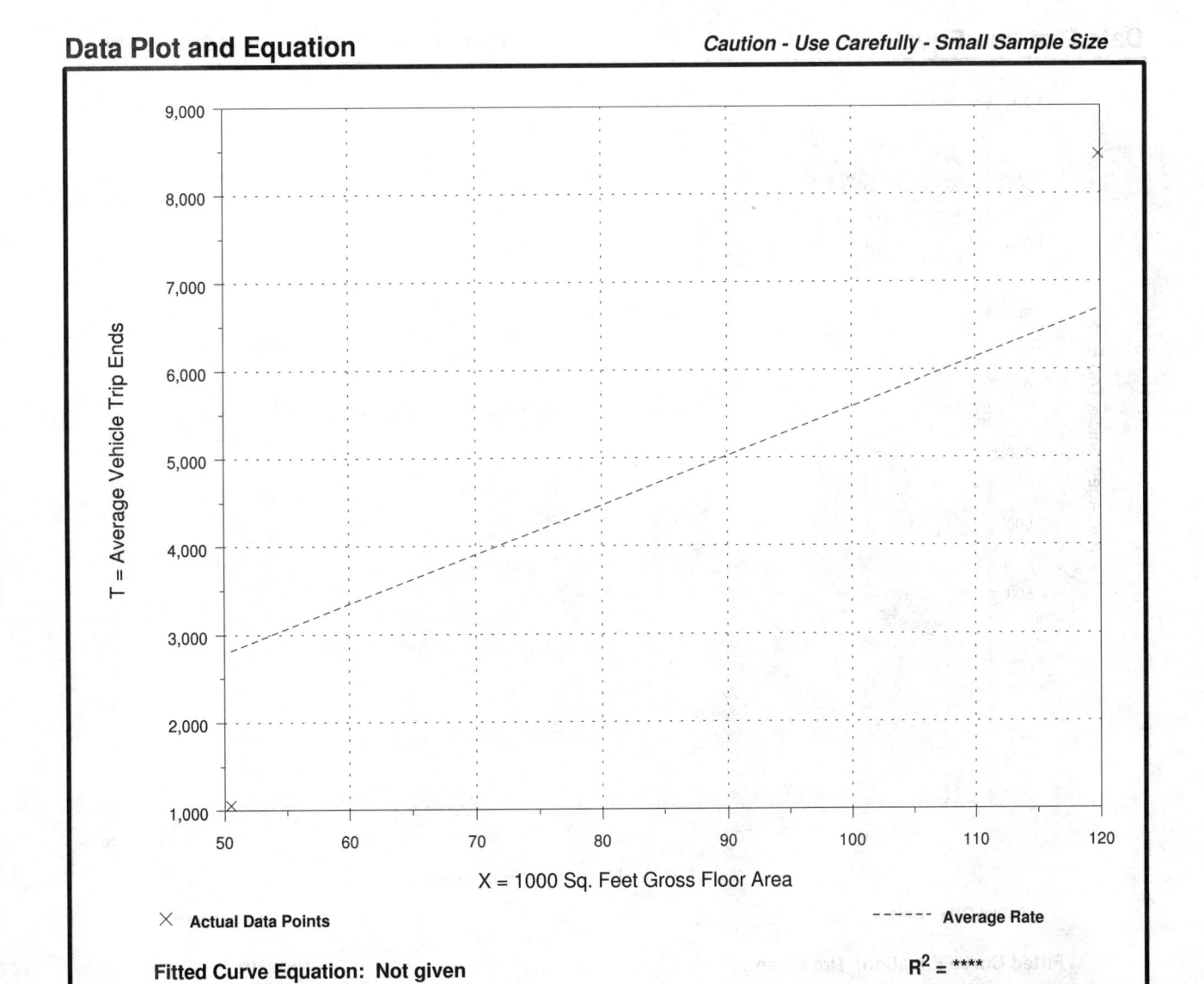

Data Plot and Equation

Caution - Use Carefully - Small Sample Size

Fitted Curve Equation: Not given $R^2 = ****$

Home Improvement Superstore
(862)

Average Vehicle Trip Ends vs: 1000 Sq. Feet Gross Floor Area
On a: Sunday,
Peak Hour of Generator

Number of Studies: 2
Average 1000 Sq. Feet GFA: 85
Directional Distribution: 54% entering, 46% exiting

Trip Generation per 1000 Sq. Feet Gross Floor Area

Average Rate	Range of Rates	Standard Deviation
8.03	4.20 - 9.64	*

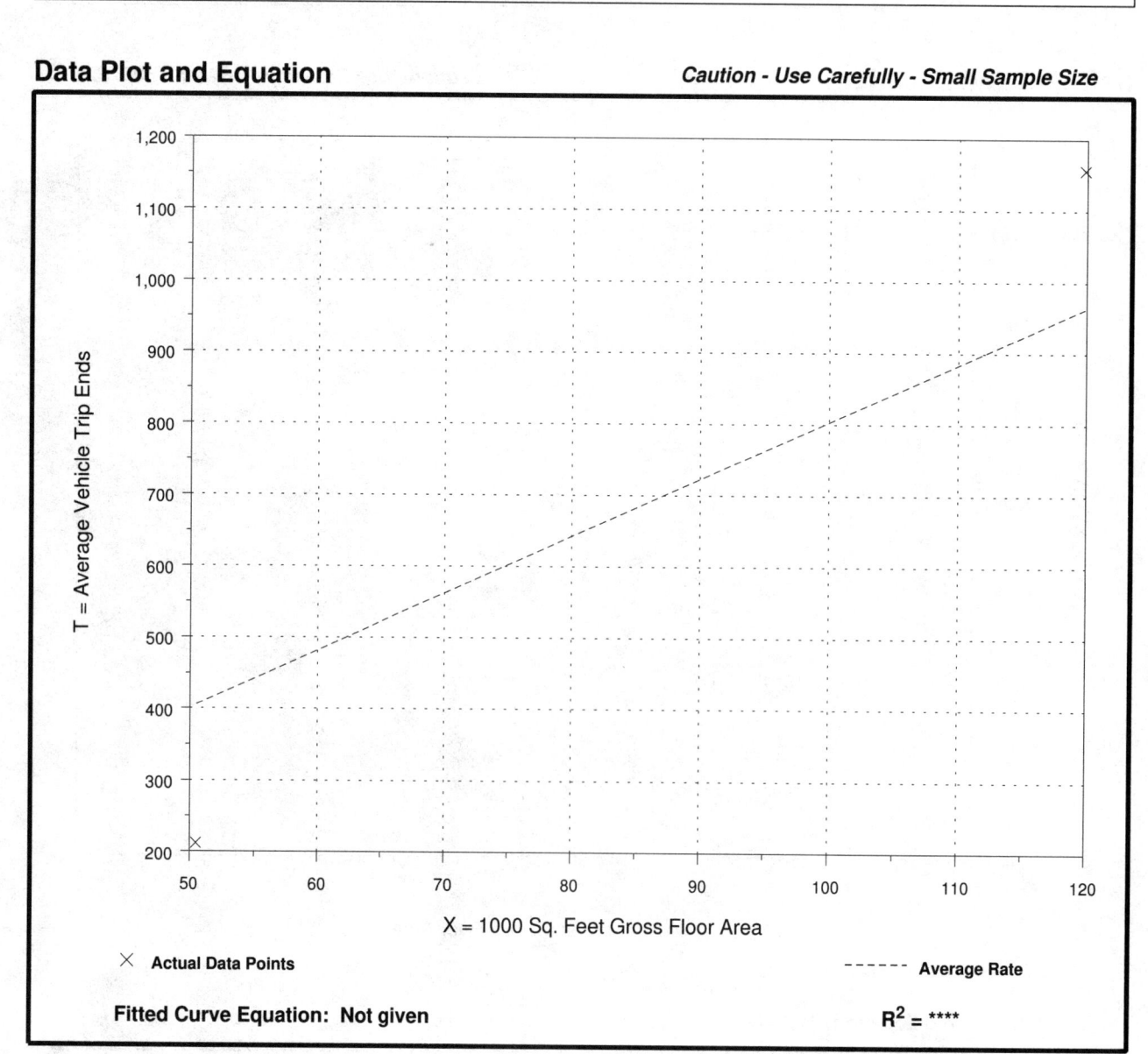

Data Plot and Equation

Caution - Use Carefully - Small Sample Size

× Actual Data Points
----- Average Rate

Fitted Curve Equation: Not given $R^2 = ****$

Land Use: 863
Electronics Superstore

Description

Electronics superstores are free-standing facilities that specialize in the sale of electronic merchandise. These facilities generally offer a variety of customer services and centralized cashiering. Electronics superstores typically maintain long store hours 7 days a week. Examples of items sold in these stores include televisions, audio and video players and recorders, software, telephones, computers and general electronic accessories. Major home appliances may also be sold at these facilities. Electronics superstores are sometimes found as separate parcels within a retail complex, with or without their own dedicated parking.

Additional Data

The sites were surveyed in 1994 and 1995 in urban areas of Florida.

Source Numbers

436, 437

Land Use: 863
Electronics Superstore
Independent Variables with One Observation

The following trip generation data are for independent variables with only one observation. This information is shown in this table only; there are no related plots for these data.

Users are cautioned to use data with care because of the small sample size.

Independent Variable	Trip Generation Rate	Size of Independent Variable	Number of Studies	Directional Distribution
1,000 Square Feet Gross Floor Area				
Weekday A.M. Peak Hour of Adjacent Street Traffic	0.28	46	1	Not available

Electronics Superstore
(863)

Average Vehicle Trip Ends vs: 1000 Sq. Feet Gross Floor Area
On a: Weekday

Number of Studies: 3
Average 1000 Sq. Feet GFA: 37
Directional Distribution: 50% entering, 50% exiting

Trip Generation per 1000 Sq. Feet Gross Floor Area

Average Rate	Range of Rates	Standard Deviation
45.04	33.74 - 59.17	13.69

Data Plot and Equation

Caution - Use Carefully - Small Sample Size

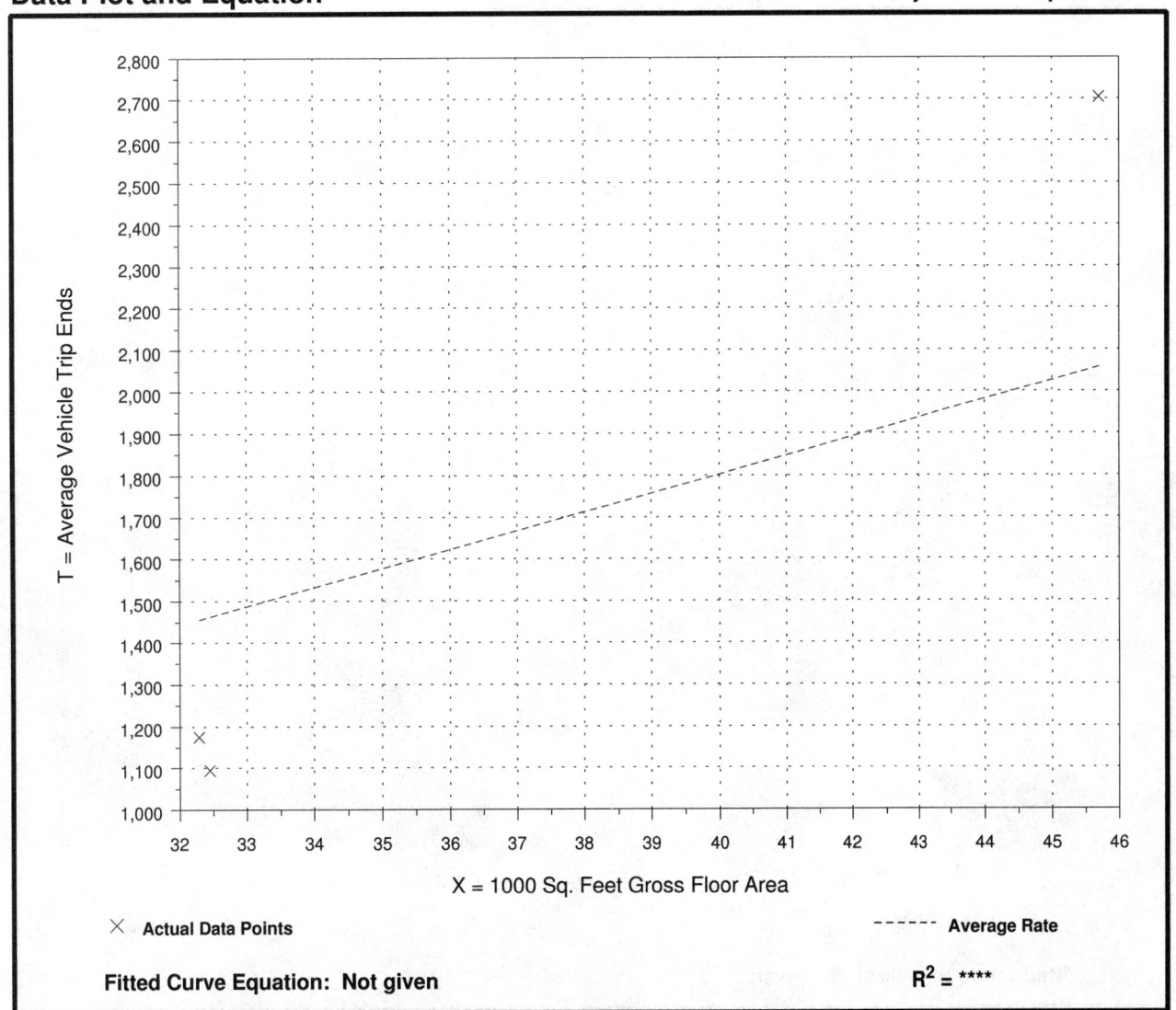

Fitted Curve Equation: Not given $R^2 = ****$

Trip Generation, 9th Edition • Institute of Transportation Engineers

Electronics Superstore
(863)

Average Vehicle Trip Ends vs: 1000 Sq. Feet Gross Floor Area
On a: Weekday,
Peak Hour of Adjacent Street Traffic,
One Hour Between 4 and 6 p.m.

Number of Studies: 3
Average 1000 Sq. Feet GFA: 37
Directional Distribution: 49% entering, 51% exiting

Trip Generation per 1000 Sq. Feet Gross Floor Area

Average Rate	Range of Rates	Standard Deviation
4.50	3.45 - 5.78	2.37

Data Plot and Equation

Caution - Use Carefully - Small Sample Size

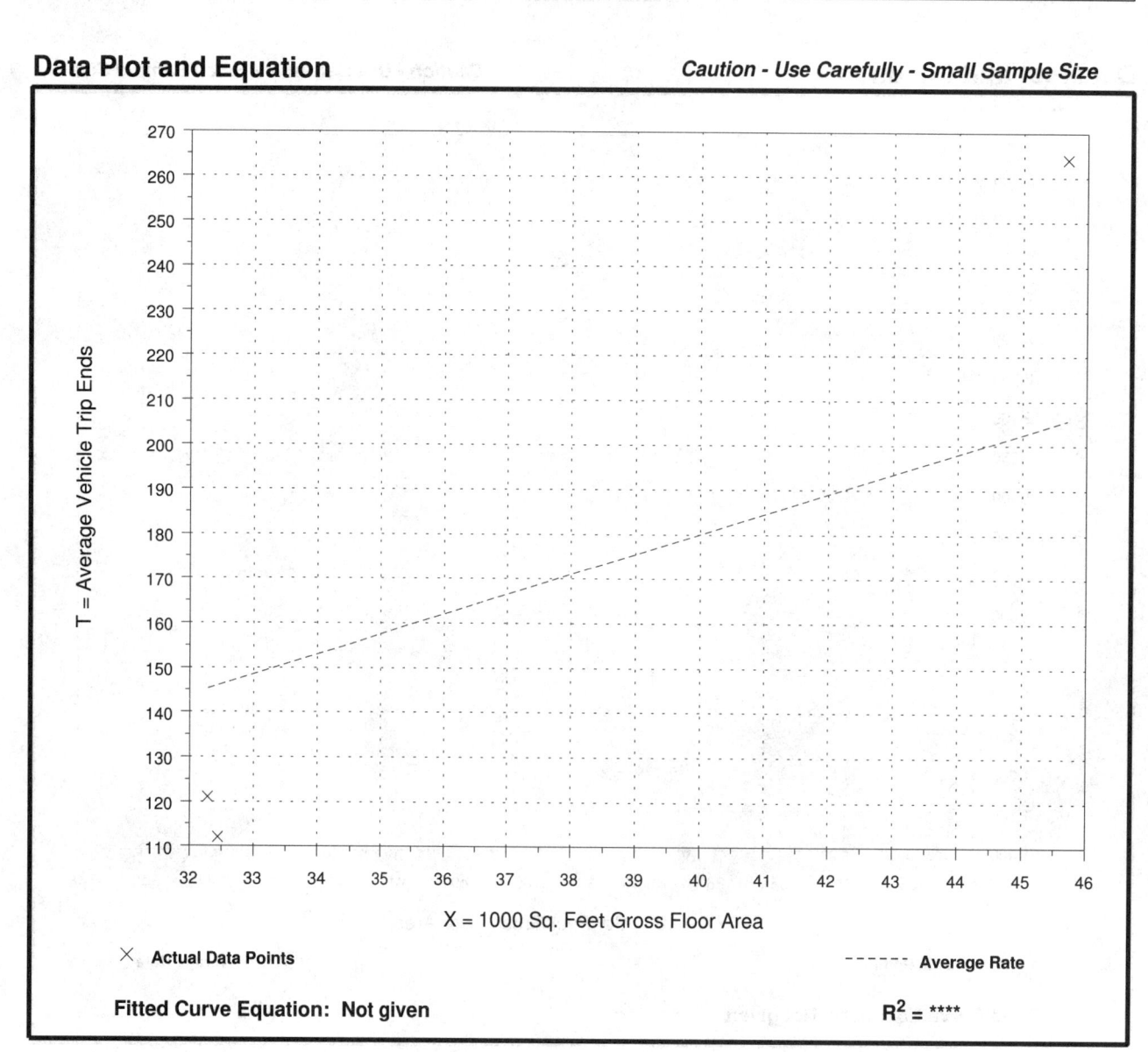

Fitted Curve Equation: Not given $R^2 = ****$

Electronics Superstore
(863)

Average Vehicle Trip Ends vs: 1000 Sq. Feet Gross Floor Area
On a: Weekday,
A.M. Peak Hour of Generator

Number of Studies: 3
Average 1000 Sq. Feet GFA: 37
Directional Distribution: 53% entering, 47% exiting

Trip Generation per 1000 Sq. Feet Gross Floor Area

Average Rate	Range of Rates	Standard Deviation
3.46	2.91 - 4.18	1.94

Data Plot and Equation

Caution - Use Carefully - Small Sample Size

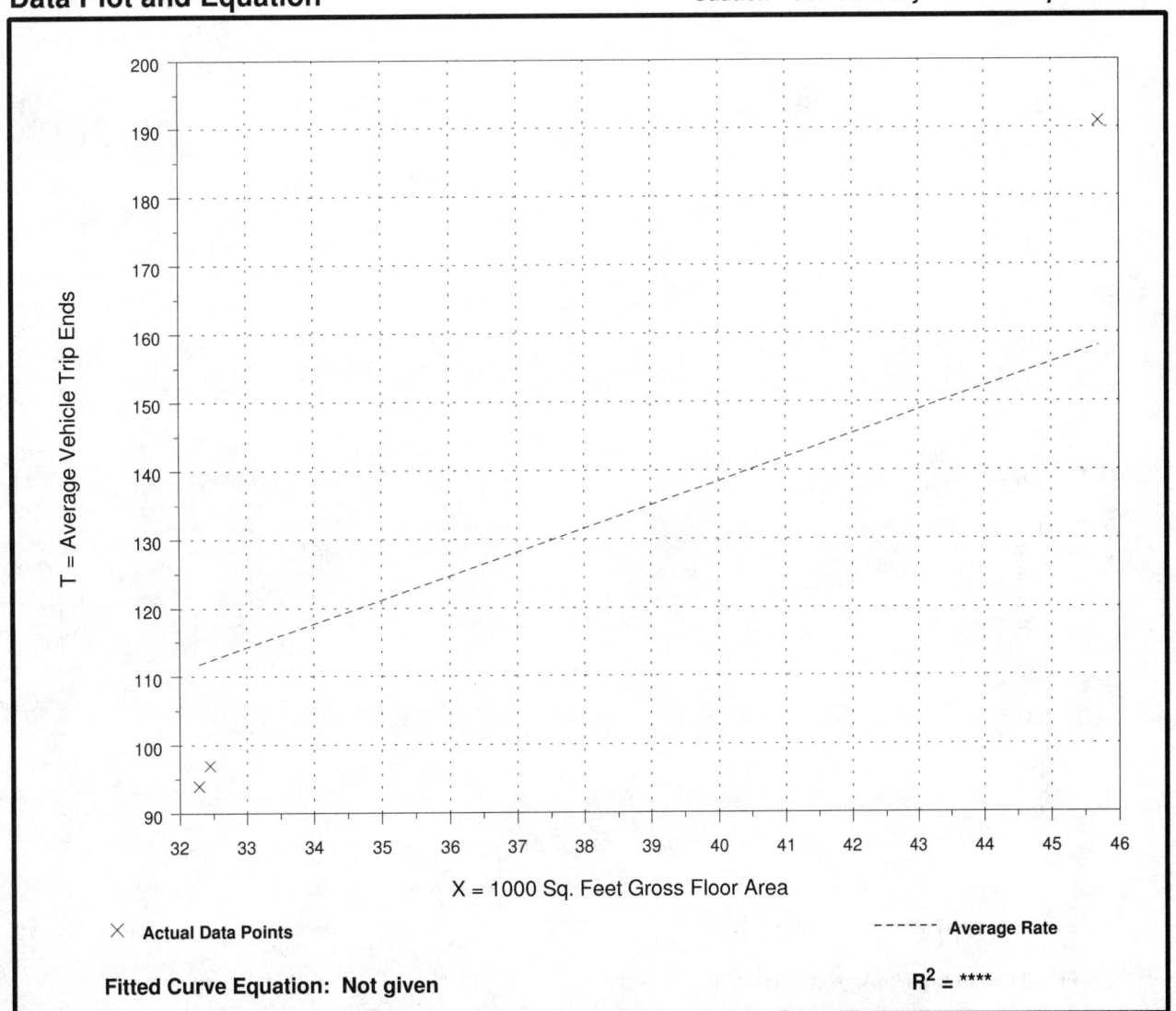

Fitted Curve Equation: Not given $R^2 = ****$

Electronics Superstore
(863)

Average Vehicle Trip Ends vs: 1000 Sq. Feet Gross Floor Area
On a: Weekday,
P.M. Peak Hour of Generator

Number of Studies: 3
Average 1000 Sq. Feet GFA: 37
Directional Distribution: 49% entering, 51% exiting

Trip Generation per 1000 Sq. Feet Gross Floor Area

Average Rate	Range of Rates	Standard Deviation
4.50	3.45 - 5.78	2.37

Data Plot and Equation

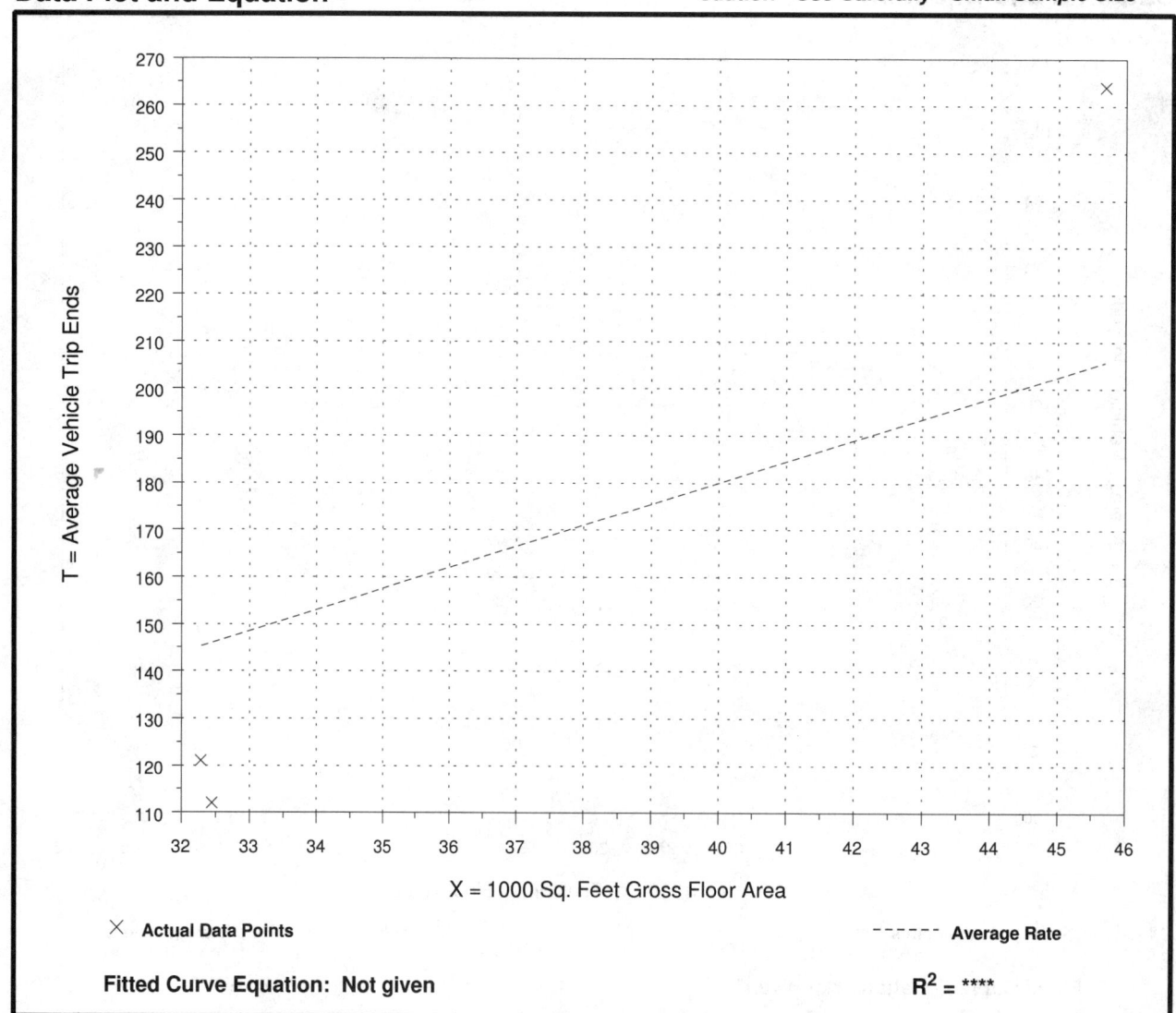

Caution - Use Carefully - Small Sample Size

Fitted Curve Equation: Not given $R^2 = ****$

Land Use: 864
Toy/Children's Superstore

Description

Toy/children's superstores are free-standing facilities that specialize in the sale of child-oriented merchandise. These facilities generally offer a variety of customer services and centralized cashiering. Toy/children's superstores typically maintain long store hours 7 days a week. Examples of items sold in these stores include board and video games, toys, bicycles/tricycles, wagons, outdoor play equipment and school supplies. Some may also carry children's clothing. Toy/children's superstores are sometimes found as separate parcels within a retail complex, with or without their own dedicated parking.

Additional Data

Peak hours of the generator—
    The weekday P.M. peak hour varied between 4:00 p.m. and 6:00 p.m. The Saturday peak hour varied between 12:00 p.m. and 2:00 p.m.

The sites were surveyed in 1993 in Maryland.

Source Number

376

Toy/Children's Superstore
(864)

Average Vehicle Trip Ends vs: 1000 Sq. Feet Gross Floor Area
On a: Weekday,
Peak Hour of Adjacent Street Traffic,
One Hour Between 4 and 6 p.m.

Number of Studies: 2
Average 1000 Sq. Feet GFA: 46
Directional Distribution: 50% entering, 50% exiting

Trip Generation per 1000 Sq. Feet Gross Floor Area

Average Rate	Range of Rates	Standard Deviation
4.99	4.99 - 5.00	*

Data Plot and Equation

Caution - Use Carefully - Small Sample Size

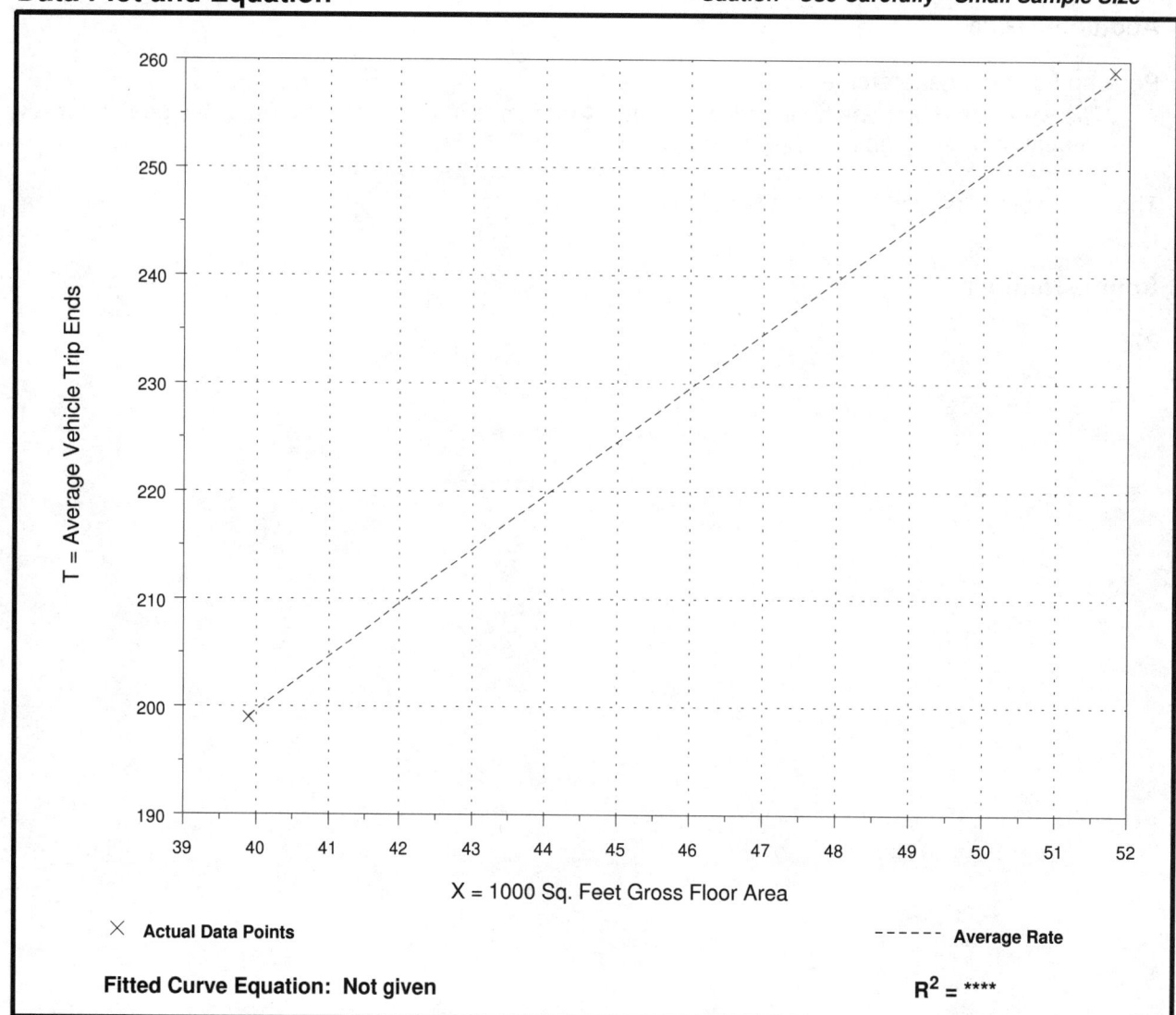

Fitted Curve Equation: Not given $R^2 = ****$

Toy/Children's Superstore
(864)

Average Vehicle Trip Ends vs: 1000 Sq. Feet Gross Floor Area
On a: Saturday,
Peak Hour of Generator

Number of Studies: 2
Average 1000 Sq. Feet GFA: 46
Directional Distribution: 46% entering, 54% exiting

Trip Generation per 1000 Sq. Feet Gross Floor Area

Average Rate	Range of Rates	Standard Deviation
5.53	4.66 - 6.20	*

Data Plot and Equation

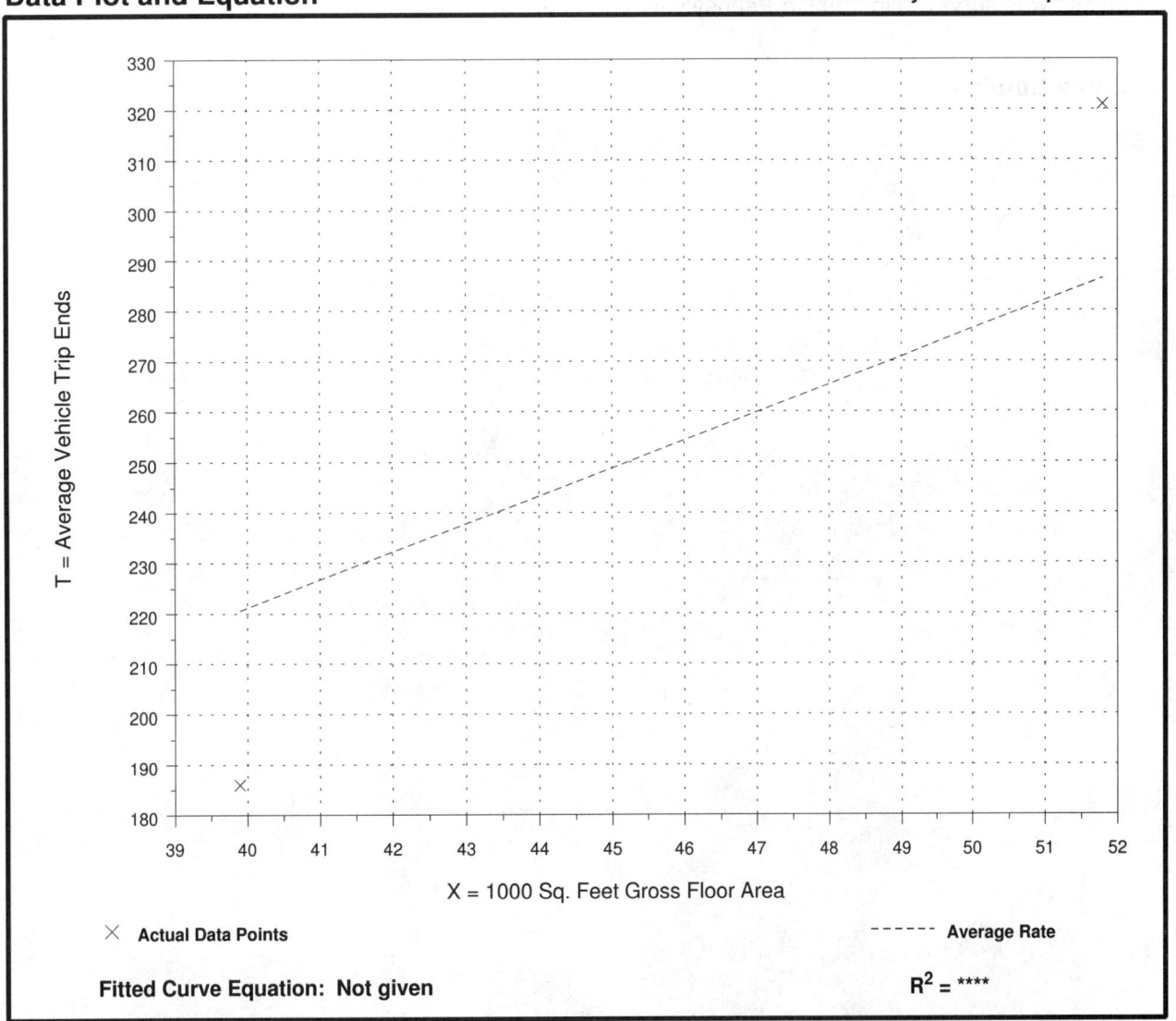

Caution - Use Carefully - Small Sample Size

[Plot: T = Average Vehicle Trip Ends vs. X = 1000 Sq. Feet Gross Floor Area. Actual data points at approximately (40, 186) and (52, 322). Dashed line represents Average Rate.]

× Actual Data Points ----- Average Rate

Fitted Curve Equation: Not given $R^2 = ****$

Land Use: 865
Baby Superstore

Description

Baby superstores are free-standing facilities that specialize in the sale of juvenile merchandise. These facilities generally offer a variety of customer services and centralized cashiering. Baby superstores typically maintain long store hours 7 days a week. Examples of items sold in these stores include furniture, strollers, infant safety products, clothing and toys. Baby superstores are sometimes found as separate parcels within a retail complex, with or without their own dedicated parking.

Additional Data

The site was surveyed in 1999 in Pennsylvania.

Source Number

571

Land Use: 865
Baby Superstore
Independent Variables with One Observation

The following trip generation data are for independent variables with only one observation. This information is shown in this table only; there are no related plots for these data.

Users are cautioned to use data with care because of the small sample size.

1,000 Square Feet Gross Floor Area

Independent Variable	Trip Generation Rate	Size of Independent Variable	Number of Studies	Directional Distribution
Weekday P.M. Peak Hour of Adjacent Street Traffic	1.82	45	1	50% entering, 50% exiting
Saturday Peak Hour of Generator	3.73	45	1	53% entering, 47% exiting

Land Use: 866
Pet Supply Superstore

Description

Pet supply superstores are free-standing facilities that specialize in the sale of pets and pet supplies, food and accessories. These facilities generally offer a variety of customer services, have centralized cashiering and maintain long store hours 7 days a week. Pet supply superstores are sometimes found as separate parcels within a retail complex, with or without their own dedicated off-street parking.

Additional Data

The Saturday peak hour of the generator varied between 12:00 p.m. and 2:00 p.m.

The sites were surveyed in 1999 and the 2000s in New York and New Jersey.

Source Numbers

555, 610

Land Use: 866
Pet Supply Superstore
Independent Variables with One Observation

The following trip generation data are for independent variables with only one observation. This information is shown in this table only; there are no related plots for these data.

Users are cautioned to use data with care because of the small sample size.

Independent Variable	Trip Generation Rate	Size of Independent Variable	Number of Studies	Directional Distribution
1,000 Square Feet Gross Floor Area				
Weekday P.M. Peak Hour of Generator	2.19	27	1	50% entering, 50% exiting

Pet Supply Superstore
(866)

Average Vehicle Trip Ends vs: 1000 Sq. Feet Gross Floor Area
On a: Weekday,
Peak Hour of Adjacent Street Traffic,
One Hour Between 4 and 6 p.m.

Number of Studies: 2
Average 1000 Sq. Feet GFA: 24
Directional Distribution: 50% entering, 50% exiting

Trip Generation per 1000 Sq. Feet Gross Floor Area

Average Rate	Range of Rates	Standard Deviation
3.38	2.19 - 4.96	*

Data Plot and Equation

Caution - Use Carefully - Small Sample Size

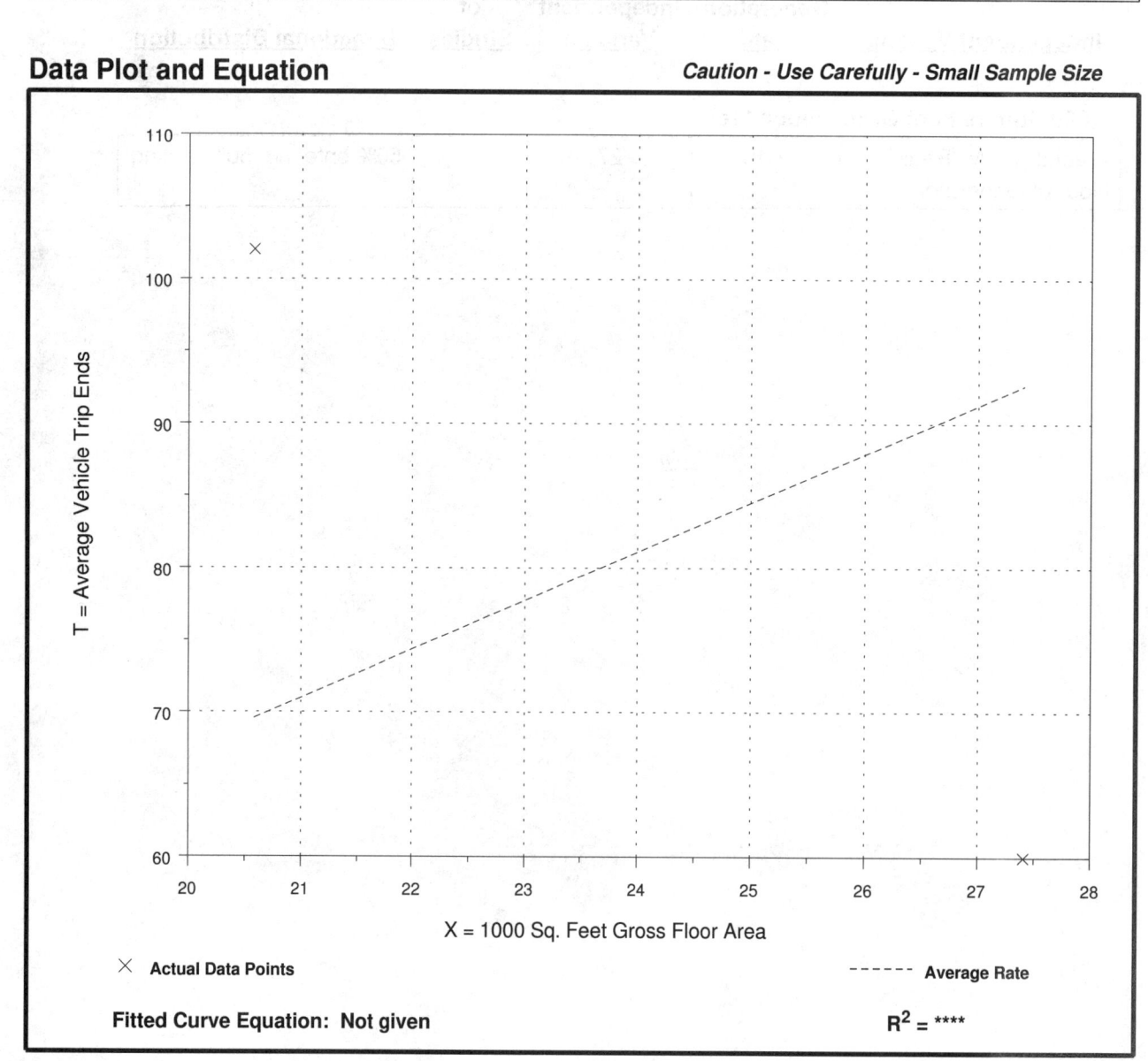

X Actual Data Points
----- Average Rate

Fitted Curve Equation: Not given $R^2 = ****$

Pet Supply Superstore
(866)

Average Vehicle Trip Ends vs: 1000 Sq. Feet Gross Floor Area
On a: Saturday,
Peak Hour of Generator

Number of Studies: 2
Average 1000 Sq. Feet GFA: 24
Directional Distribution: 50% entering, 50% exiting

Trip Generation per 1000 Sq. Feet Gross Floor Area

Average Rate	Range of Rates	Standard Deviation
6.98	3.90 - 11.08	*

Data Plot and Equation

Caution - Use Carefully - Small Sample Size

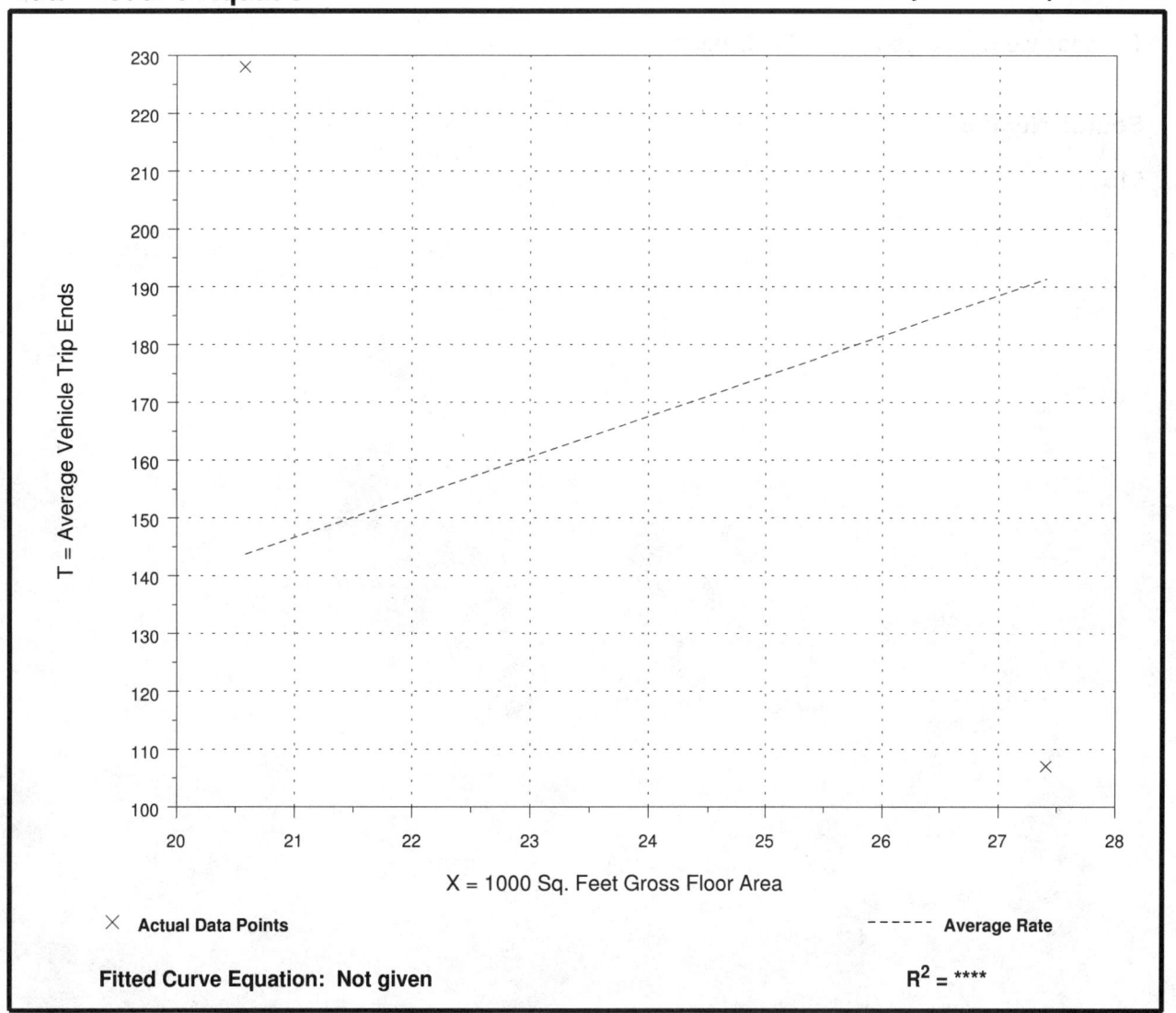

Fitted Curve Equation: Not given $R^2 = ****$

Land Use: 867
Office Supply Superstore

Description

Office supply superstores are free-standing facilities that specialize in the sale of office equipment and supplies including computers, paper, furniture and desk accessories. These facilities may offer a variety of business services including administrative, communications, custom printing and Internet services. The stores generally have centralized cashiering and maintain long store hours 7 days a week. Office supply superstores are sometimes found as separate parcels within a retail complex, with or without their own dedicated off-street parking.

Additional Data

The sites were surveyed in 1999 in Oregon.

Source Number

515

Office Supply Superstore
(867)

Average Vehicle Trip Ends vs: 1000 Sq. Feet Gross Floor Area
On a: Weekday,
Peak Hour of Adjacent Street Traffic,
One Hour Between 4 and 6 p.m.

Number of Studies: 2
Average 1000 Sq. Feet GFA: 26
Directional Distribution: 53% entering, 47% exiting

Trip Generation per 1000 Sq. Feet Gross Floor Area

Average Rate	Range of Rates	Standard Deviation
3.40	2.27 - 4.55	*

Data Plot and Equation

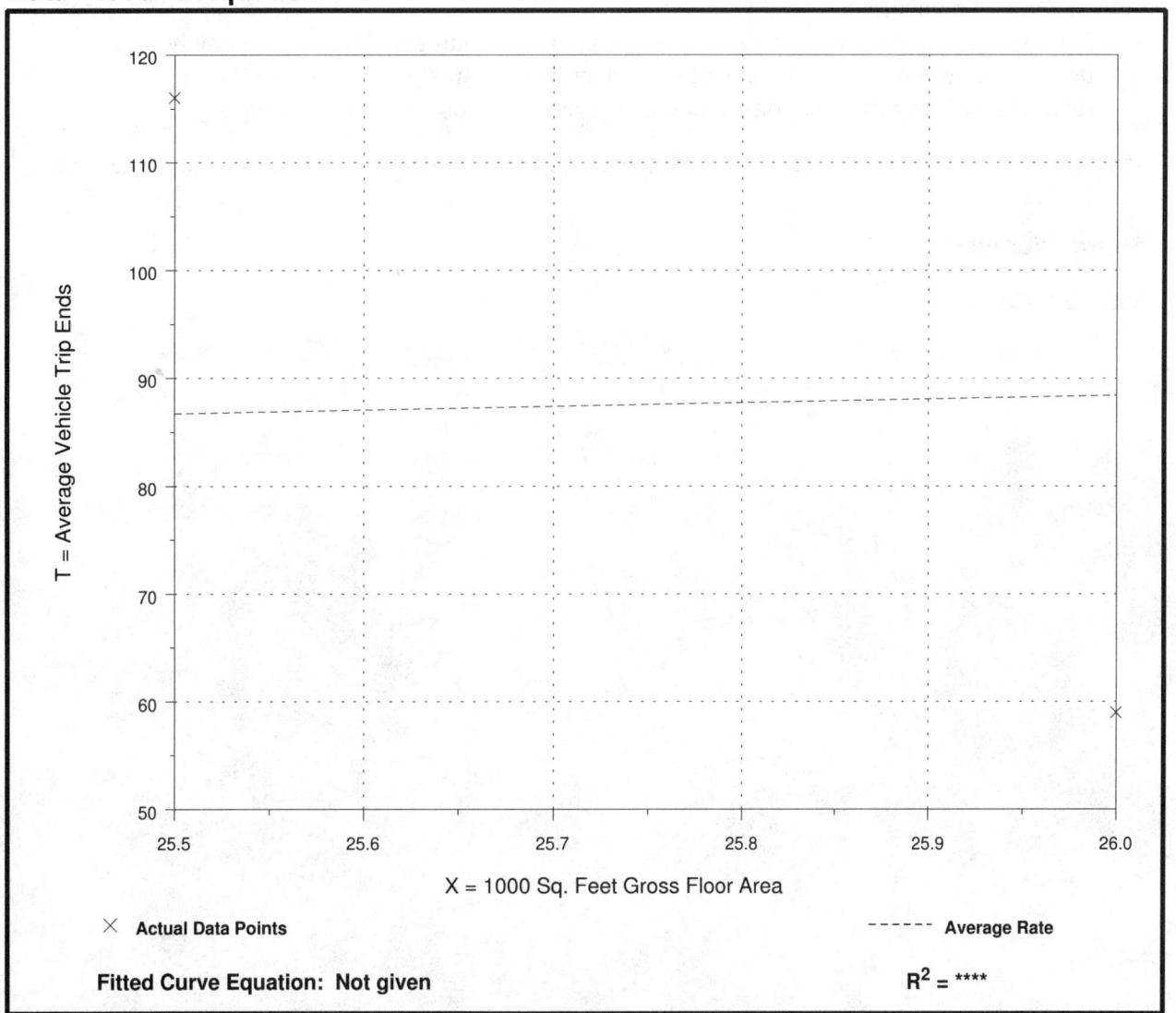

Caution - Use Carefully - Small Sample Size

Fitted Curve Equation: Not given $R^2 = ****$

Land Use: 868
Book Superstore

Description

Book superstores are free-standing facilities that specialize in the sale of books. Some stores may also include audio/video sales. Some book superstores have small cafés as ancillary facilities. The stores generally have centralized cashiering and maintain long store hours 7 days a week. Book superstores are sometimes found as separate parcels within a retail complex, with or without their own dedicated off-street parking.

Additional Data

Peak hours of the generator—
    For the one site from which data were available, the weekday A.M. and P.M. peak hours occurred between 11:00 a.m. and 12:00 p.m. and between 7:00 p.m. and 8:00 p.m., respectively. The Saturday peak hour varied between 11:00 a.m. and 1:00 p.m.

The sites were surveyed in the late 1990s and the 2000s in New Hampshire, New Jersey and New York.

Source Numbers

555, 562, 734

Land Use: 868
Book Superstore

Independent Variables with One Observation

The following trip generation data are for independent variables with only one observation. This information is shown in this table only; there are no related plots for these data.

Users are cautioned to use data with care because of the small sample size.

Independent Variable	Trip Generation Rate	Size of Independent Variable	Number of Studies	Directional Distribution
1,000 Square Feet Gross Floor Area				
Weekday	143.53	13	1	50% entering, 50% exiting
Weekday A.M. Peak Hour of Adjacent Street Traffic	1.27	13	1	Not available
Weekday A.M. Peak Hour of Generator	10.66	13	1	Not available
Weekday P.M. Peak Hour of Generator	13.99	13	1	Not available

Book Superstore
(868)

Average Vehicle Trip Ends vs: 1000 Sq. Feet Gross Floor Area
On a: Weekday,
Peak Hour of Adjacent Street Traffic,
One Hour Between 4 and 6 p.m.

Number of Studies: 2
Average 1000 Sq. Feet GFA: 11
Directional Distribution: 52% entering, 48% exiting

Trip Generation per 1000 Sq. Feet Gross Floor Area

Average Rate	Range of Rates	Standard Deviation
15.82	12.88 - 19.53	*

Data Plot and Equation

Caution - Use Carefully - Small Sample Size

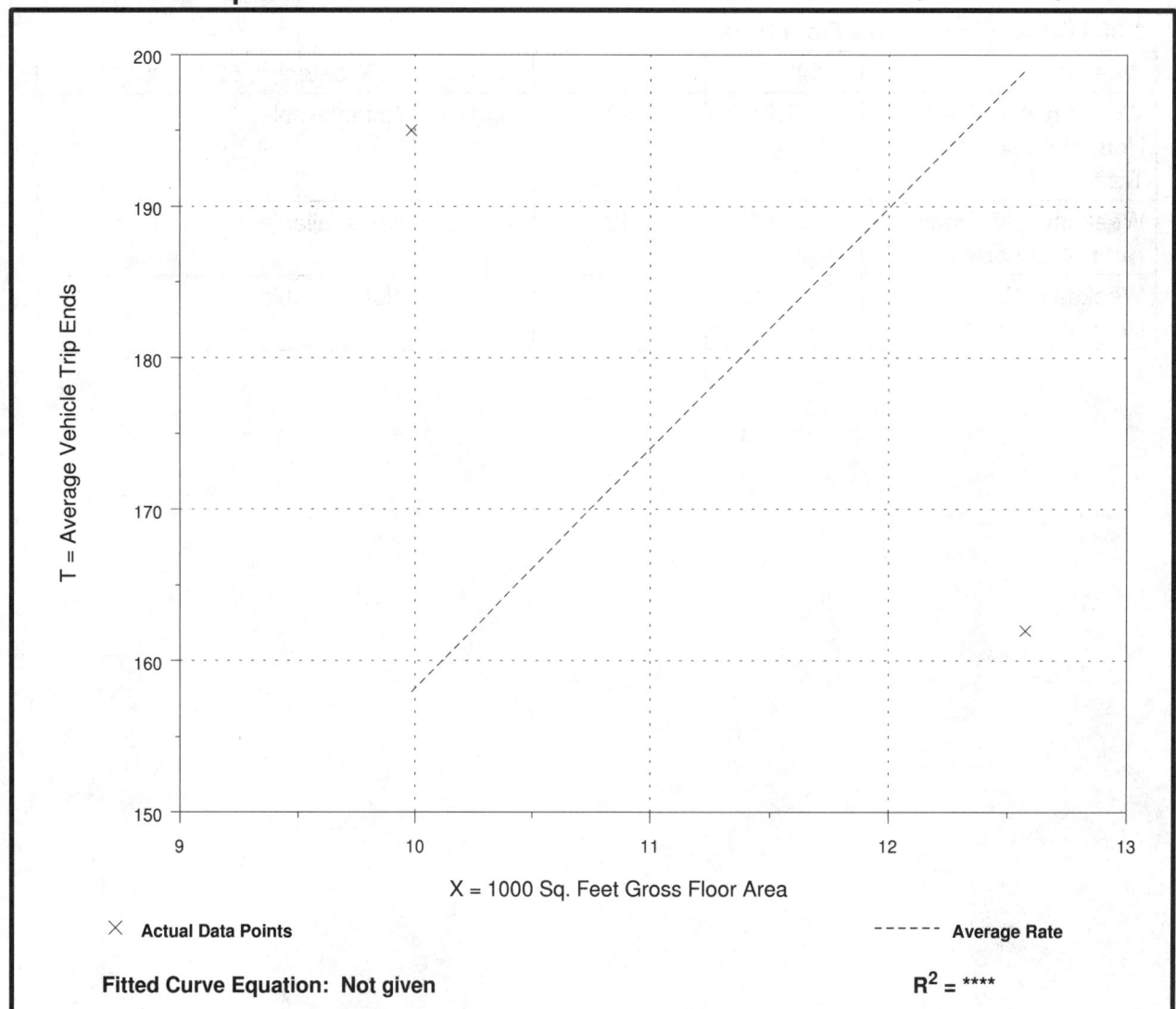

Fitted Curve Equation: Not given $R^2 = ****$

Book Superstore
(868)

Average Vehicle Trip Ends vs: 1000 Sq. Feet Gross Floor Area
On a: Saturday,
Peak Hour of Generator

Number of Studies: 2
Average 1000 Sq. Feet GFA: 15
Directional Distribution: 53% entering, 47% exiting

Trip Generation per 1000 Sq. Feet Gross Floor Area

Average Rate	Range of Rates	Standard Deviation
21.30	19.05 - 26.04	*

Data Plot and Equation

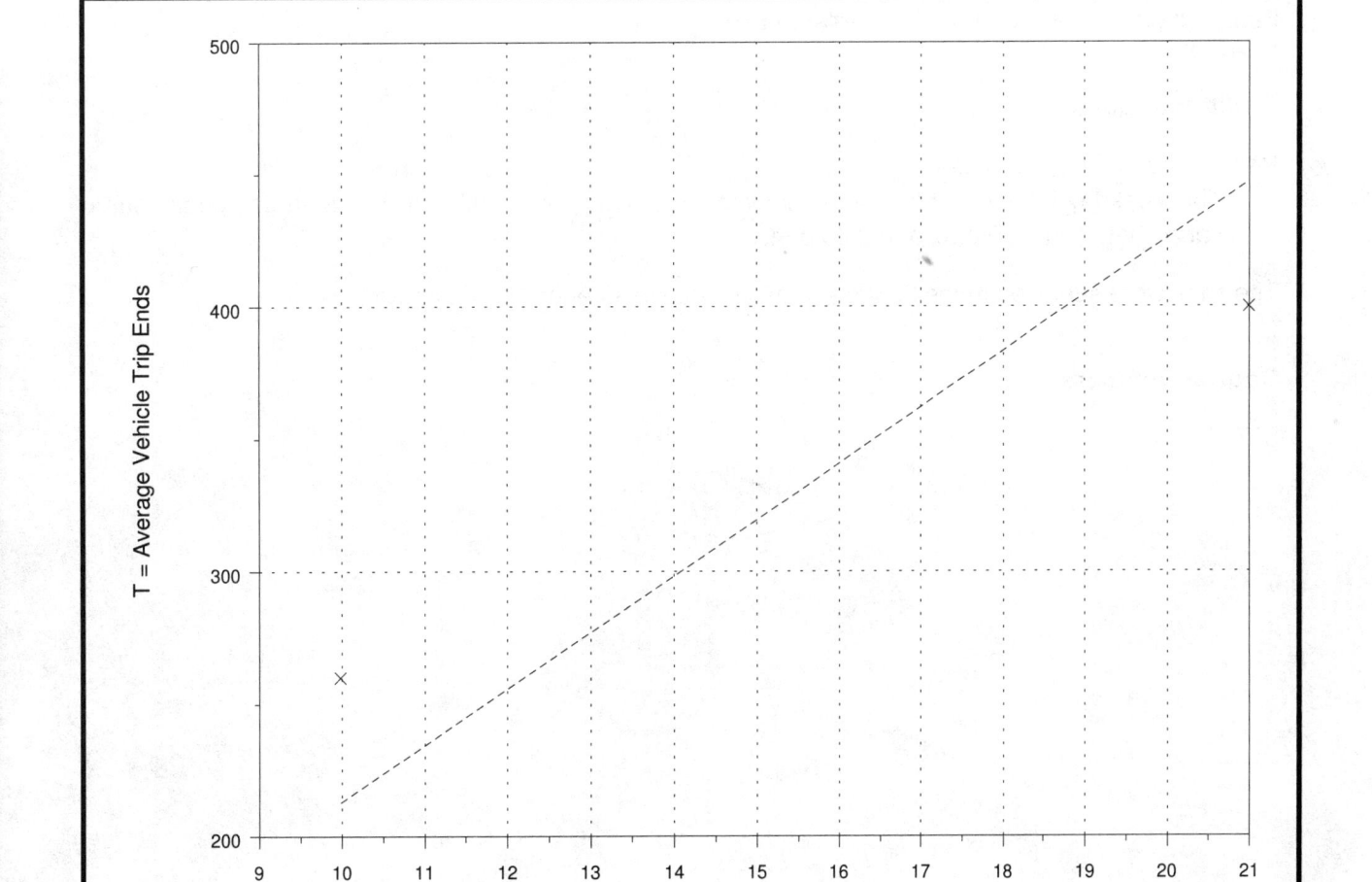

Caution - Use Carefully - Small Sample Size

Fitted Curve Equation: Not given $R^2 = ****$

Land Use: 869
Discount Home Furnishing Superstore

Description

Discount home furnishing superstores are free-standing facilities that sell an extensive variety of home furnishings and accessories. These facilities typically have showrooms that display products, many of which require assembly. The superstores are typically larger than 100,000 gross square feet in size. These superstores maintain self-serve, on-site inventories of their products within the facilities; customers pick up most of their items from these inventory locations prior to completing their purchases. Some may include convenience services, such as small restaurants and children's play areas. The stores generally have centralized cashiering and maintain long store hours 7 days a week. Discount home furnishing superstores are sometimes found as separate parcels within a retail complex, with or without their own dedicated off-street parking. Furniture store (Land Use 890) is a related use.

Additional Data

Peak hours of the generator—
    The weekday P.M. peak hour varied between 12:00 p.m. and 2:00 p.m. The Saturday peak hour varied between 1:00 p.m. and 3:00 p.m.

The sites were surveyed in the 2000s in Massachusetts, Oregon and Pennsylvania.

Source Numbers

577, 650, 733

Discount Home Furnishing Superstore
(869)

Average Vehicle Trip Ends vs: 1000 Sq. Feet Gross Floor Area
On a: Weekday

Number of Studies: 8
Average 1000 Sq. Feet GFA: 316
Directional Distribution: 50% entering, 50% exiting

Trip Generation per 1000 Sq. Feet Gross Floor Area

Average Rate	Range of Rates	Standard Deviation
20.00	12.01 - 47.81	9.70

Data Plot and Equation

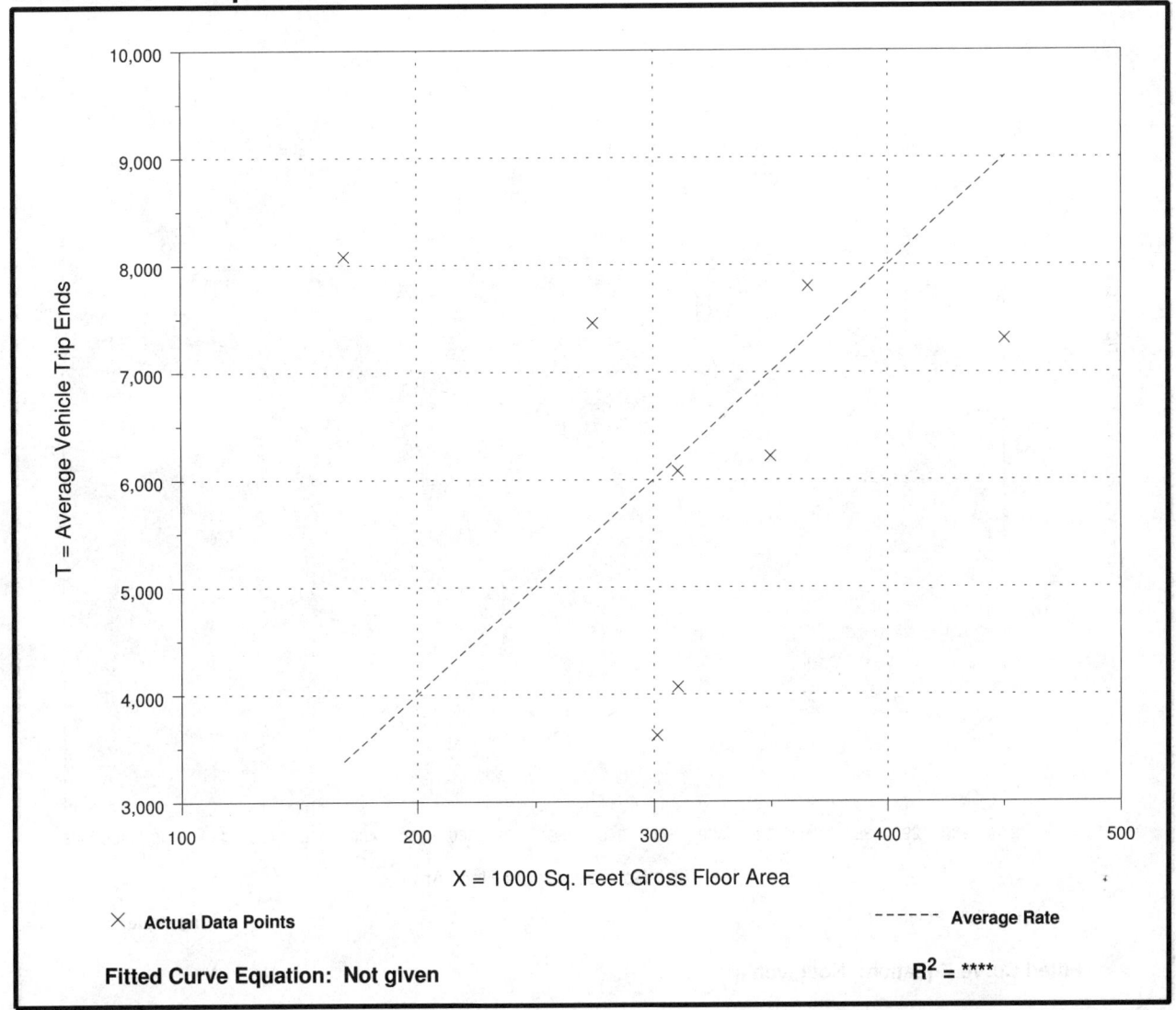

X — Actual Data Points
----- Average Rate

Fitted Curve Equation: Not given $R^2 = ****$

Discount Home Furnishing Superstore
(869)

Average Vehicle Trip Ends vs: 1000 Sq. Feet Gross Floor Area
On a: Weekday,
Peak Hour of Adjacent Street Traffic,
One Hour Between 7 and 9 a.m.

Number of Studies: 7
Average 1000 Sq. Feet GFA: 338
Directional Distribution: 64% entering, 36% exiting

Trip Generation per 1000 Sq. Feet Gross Floor Area

Average Rate	Range of Rates	Standard Deviation
0.57	0.16 - 1.00	0.81

Data Plot and Equation

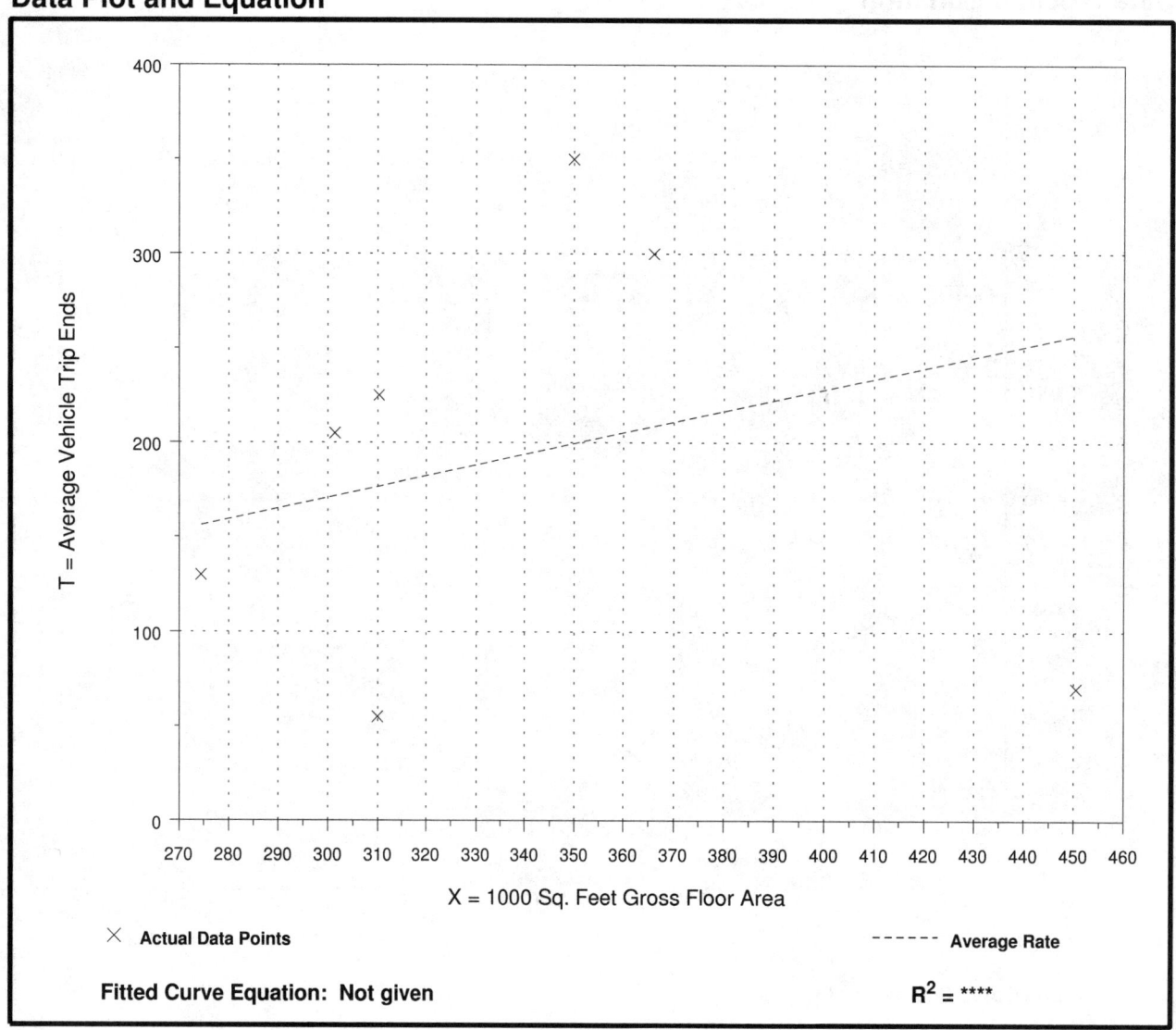

Fitted Curve Equation: Not given $R^2 = ****$

Discount Home Furnishing Superstore
(869)

Average Vehicle Trip Ends vs: 1000 Sq. Feet Gross Floor Area
On a: Weekday,
Peak Hour of Adjacent Street Traffic,
One Hour Between 4 and 6 p.m.

Number of Studies: 8
Average 1000 Sq. Feet GFA: 316
Directional Distribution: 53% entering, 47% exiting

Trip Generation per 1000 Sq. Feet Gross Floor Area

Average Rate	Range of Rates	Standard Deviation
1.57	0.94 - 4.01	1.48

Data Plot and Equation

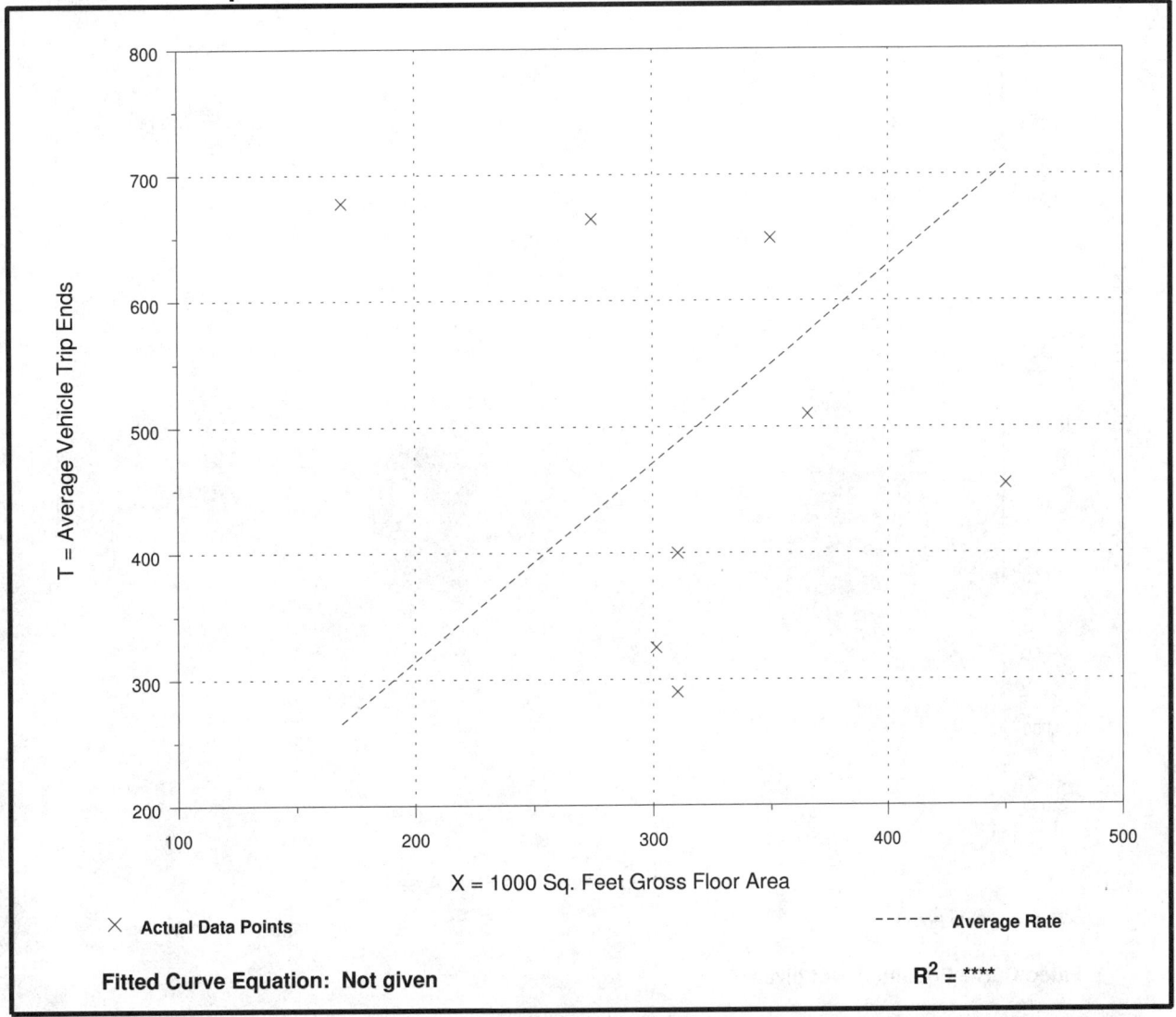

Fitted Curve Equation: Not given $R^2 = ****$

Discount Home Furnishing Superstore
(869)

Average Vehicle Trip Ends vs: 1000 Sq. Feet Gross Floor Area
On a: Saturday

Number of Studies: 8
Average 1000 Sq. Feet GFA: 316
Directional Distribution: 50% entering, 50% exiting

Trip Generation per 1000 Sq. Feet Gross Floor Area

Average Rate	Range of Rates	Standard Deviation
33.29	17.39 - 70.01	16.09

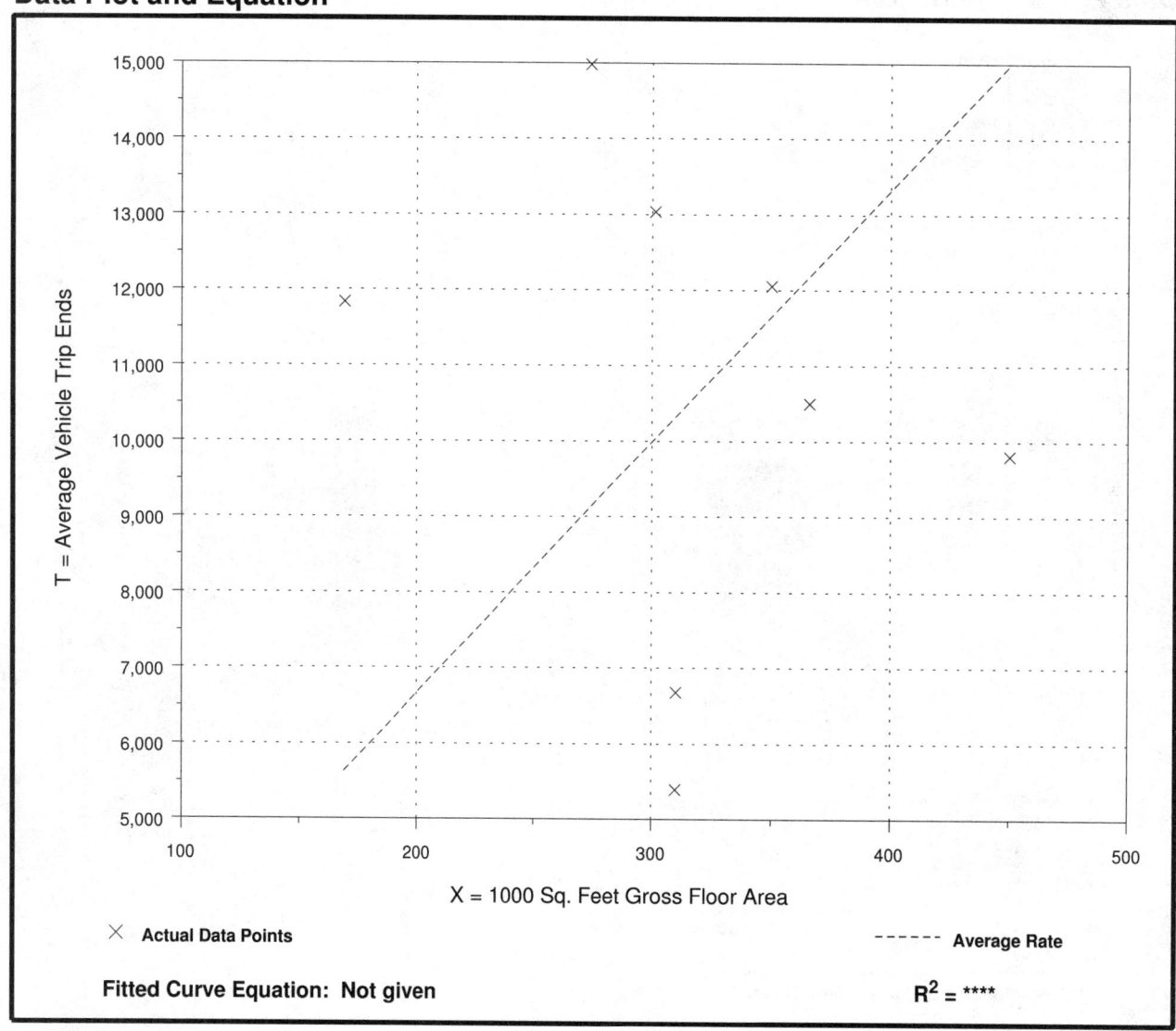

Data Plot and Equation

Fitted Curve Equation: Not given $R^2 = ****$

Discount Home Furnishing Superstore
(869)

Average Vehicle Trip Ends vs: 1000 Sq. Feet Gross Floor Area
On a: Saturday,
Peak Hour of Generator

Number of Studies: 9
Average 1000 Sq. Feet GFA: 312
Directional Distribution: 54% entering, 46% exiting

Trip Generation per 1000 Sq. Feet Gross Floor Area

Average Rate	Range of Rates	Standard Deviation
3.29	1.44 - 6.19	2.38

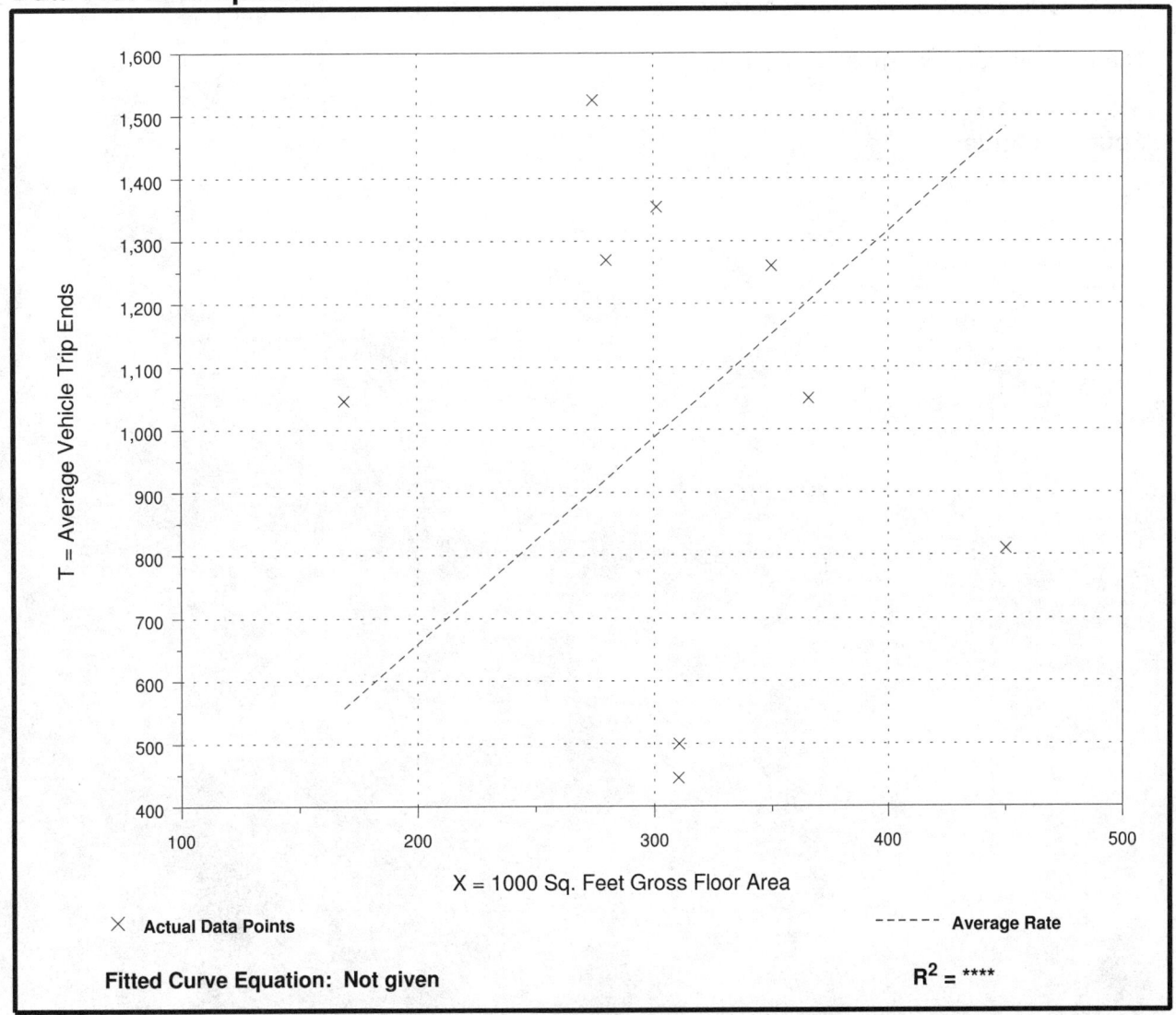

Data Plot and Equation

Fitted Curve Equation: Not given $R^2 = ****$

Land Use: 872
Bed and Linen Superstore

Description

Bed and linen superstores are free-standing facilities that specialize in the sale of bedding and bath merchandise. These facilities generally offer a variety of customer services and centralized cashiering. Bed and linen superstores typically maintain long hours 7 days a week. Examples of items sold in these stores include bedding, home textiles, home accessories, home furnishings, kitchenware and luggage. Department store (Land Use 875) is a related use.

Additional Data

The Saturday peak hour of the generator occurred between 2:00 p.m. and 3:00 p.m.

The site was surveyed in 2005 in Oregon.

Source Number

612

Land Use: 872
Bed and Linen Superstore
Independent Variables with One Observation

The following trip generation data are for independent variables with only one observation. This information is shown in this table only; there are no related plots for these data.

Users are cautioned to use data with care because of the small sample size.

Independent Variable	Trip Generation Rate	Size of Independent Variable	Number of Studies	Directional Distribution
1,000 Square Feet Gross Floor Area				
Weekday P.M. Peak Hour of Adjacent Street Traffic	2.22	32	1	41% entering, 59% exiting
Saturday Peak Hour of Generator	6.97	32	1	51% entering, 49% exiting

Land Use: 875
Department Store

Description

Department stores are free-standing facilities that specialize in the sale of a wide range of products including apparel, footwear, home products, bedding and linens, luggage, jewelry and accessories. These stores typically maintain long hours of operations 7 days a week. Free-standing discount store (Land Use 815), bed and linen superstore (Land Use 872) and apparel store (Land Use 876) are related uses.

Additional Data

Peak hours of the generator—
The weekday A.M. peak hour occurred between 11:00 a.m. and 12:00 p.m. The weekday P.M. peak hour varied between 12:30 p.m. and 5:00 p.m. The Saturday peak hour varied between 1:00 p.m. and 5:00 p.m.

Information on approximate hourly variation in department store traffic is shown in the following table. It should be noted, however, that the information contained in this table is based on a limited sample size. Therefore, caution should be exercised when applying the data. Also, some information provided in the table may conflict with the results obtained by applying the average rate or regression equations. When this occurs, it is suggested that the results from the average rate or regression equations be used, as they are based on a larger number of studies.

Hourly Variation in Department Store Traffic				
Time	Average Weekday[a]		Average Saturday[b]	
	Percent of 24-Hour Entering Traffic	Percent of 24-Hour Exiting Traffic	Percent of 24-Hour Entering Traffic	Percent of 24-Hour Exiting Traffic
6 a.m.–7 a.m.	0.9	0.8	0.1	0.5
7 a.m.–8 a.m.	2.1	1.2	0.8	0.4
8 a.m.–9 a.m.	2.9	1.9	2.4	1.5
9 a.m.–10 a.m.	4.6	3.3	4.9	4.2
10 a.m.–11 a.m.	6.7	5.1	7.5	5.5
11 a.m.–12 p.m.	9.0	7.3	8.9	8.5
12 p.m.–1 p.m.	10.1	10.0	9.1	9.1
1 p.m.–2 p.m.	9.4	10.1	11.2	9.5
2 p.m.–3 p.m.	7.8	8.6	11.2	10.1
3 p.m.–4 p.m.	7.6	8.5	10.5	11.3
4 p.m.–5 p.m.	7.8	6.7	9.0	10.1
5 p.m.–6 p.m.	8.2	7.3	7.8	8.2
6 p.m.–7 p.m.	7.1	7.0	5.9	6.3
7 p.m.–8 p.m.	6.5	7.3	4.8	5.1
8 p.m.–9 p.m.	4.5	6.3	3.1	4.3
9 p.m.–10 p.m.	2.1	4.3	1.3	2.8
10 p.m.–6 a.m.	2.7	4.3	1.5	2.6

Sites ranged in size from 95,000 to 127,000 square feet gross floor area
[a] Source number – 604; based on five studies
[b] Source number – 604; based on three studies

The sites were surveyed in the 2000s in California and Pennsylvania.

Source Numbers

604, 719

Department Store
(875)

Average Vehicle Trip Ends vs: 1000 Sq. Feet Gross Floor Area
On a: Weekday

Number of Studies: 5
Average 1000 Sq. Feet GFA: 104
Directional Distribution: 50% entering, 50% exiting

Trip Generation per 1000 Sq. Feet Gross Floor Area

Average Rate	Range of Rates	Standard Deviation
22.88	16.64 - 30.94	7.01

Data Plot and Equation

Caution - Use Carefully - Small Sample Size

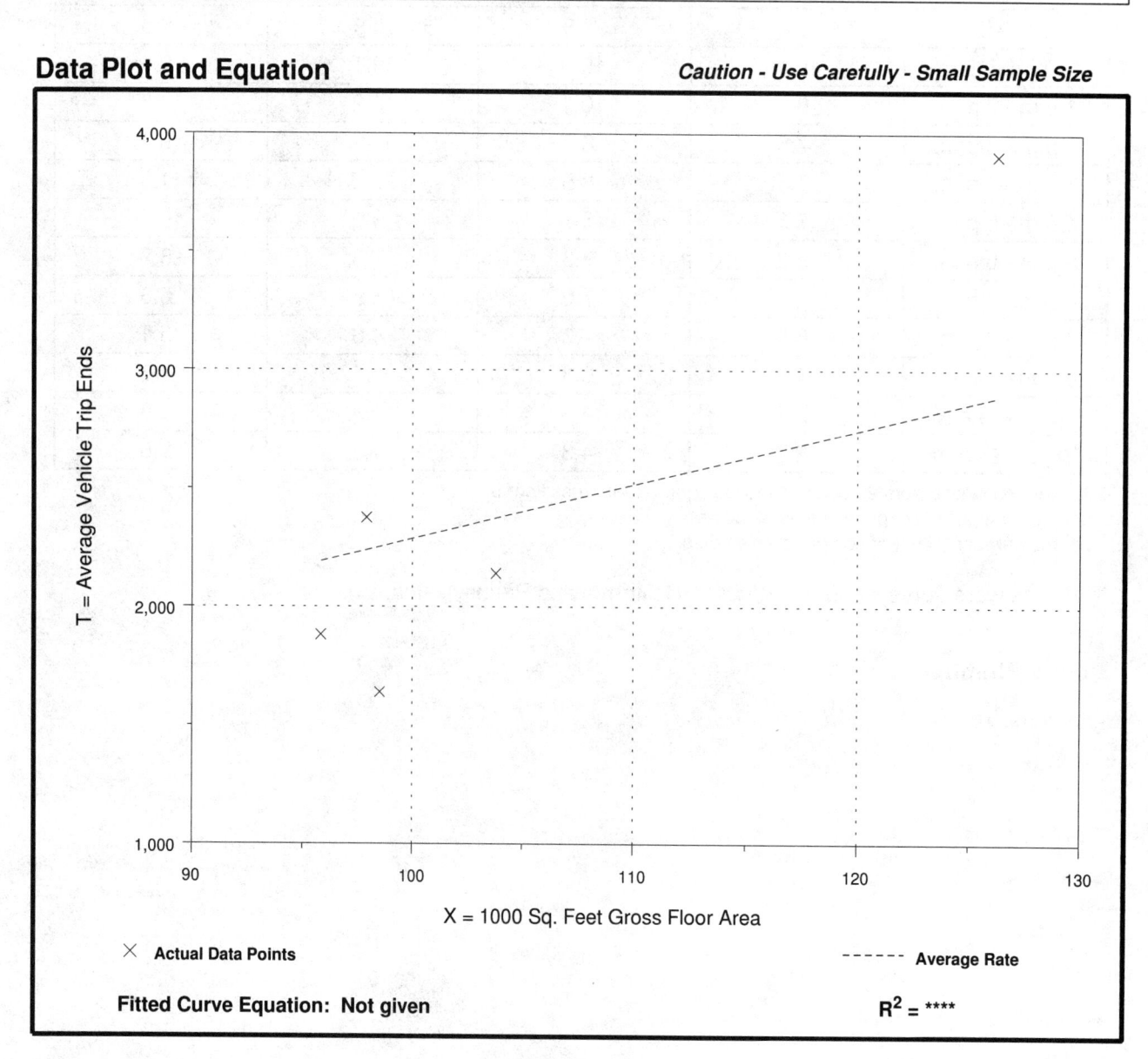

X   Actual Data Points
- - - - - Average Rate

Fitted Curve Equation: Not given $R^2 = ****$

Department Store
(875)

Average Vehicle Trip Ends vs: 1000 Sq. Feet Gross Floor Area
On a: Weekday,
Peak Hour of Adjacent Street Traffic,
One Hour Between 7 and 9 a.m.

Number of Studies: 7
Average 1000 Sq. Feet GFA: 102
Directional Distribution: 64% entering, 36% exiting

Trip Generation per 1000 Sq. Feet Gross Floor Area

Average Rate	Range of Rates	Standard Deviation
0.58	0.20 - 1.30	0.84

Data Plot and Equation

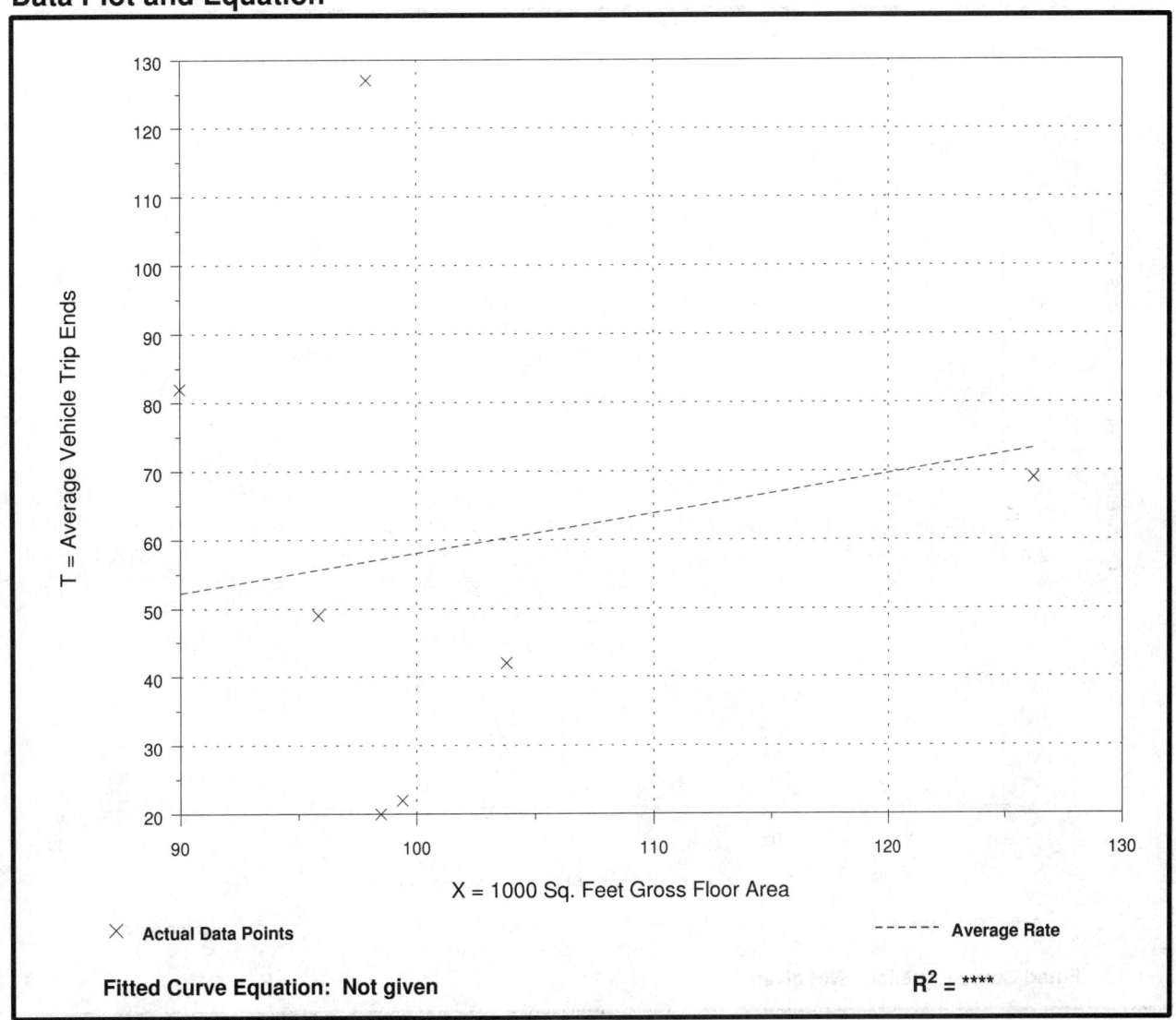

Fitted Curve Equation: Not given $R^2 =$ ****

Department Store
(875)

Average Vehicle Trip Ends vs: 1000 Sq. Feet Gross Floor Area
On a: Weekday,
Peak Hour of Adjacent Street Traffic,
One Hour Between 4 and 6 p.m.

Number of Studies: 10
Average 1000 Sq. Feet GFA: 100
Directional Distribution: 51% entering, 49% exiting

Trip Generation per 1000 Sq. Feet Gross Floor Area

Average Rate	Range of Rates	Standard Deviation
1.87	1.31 - 2.80	1.46

Data Plot and Equation

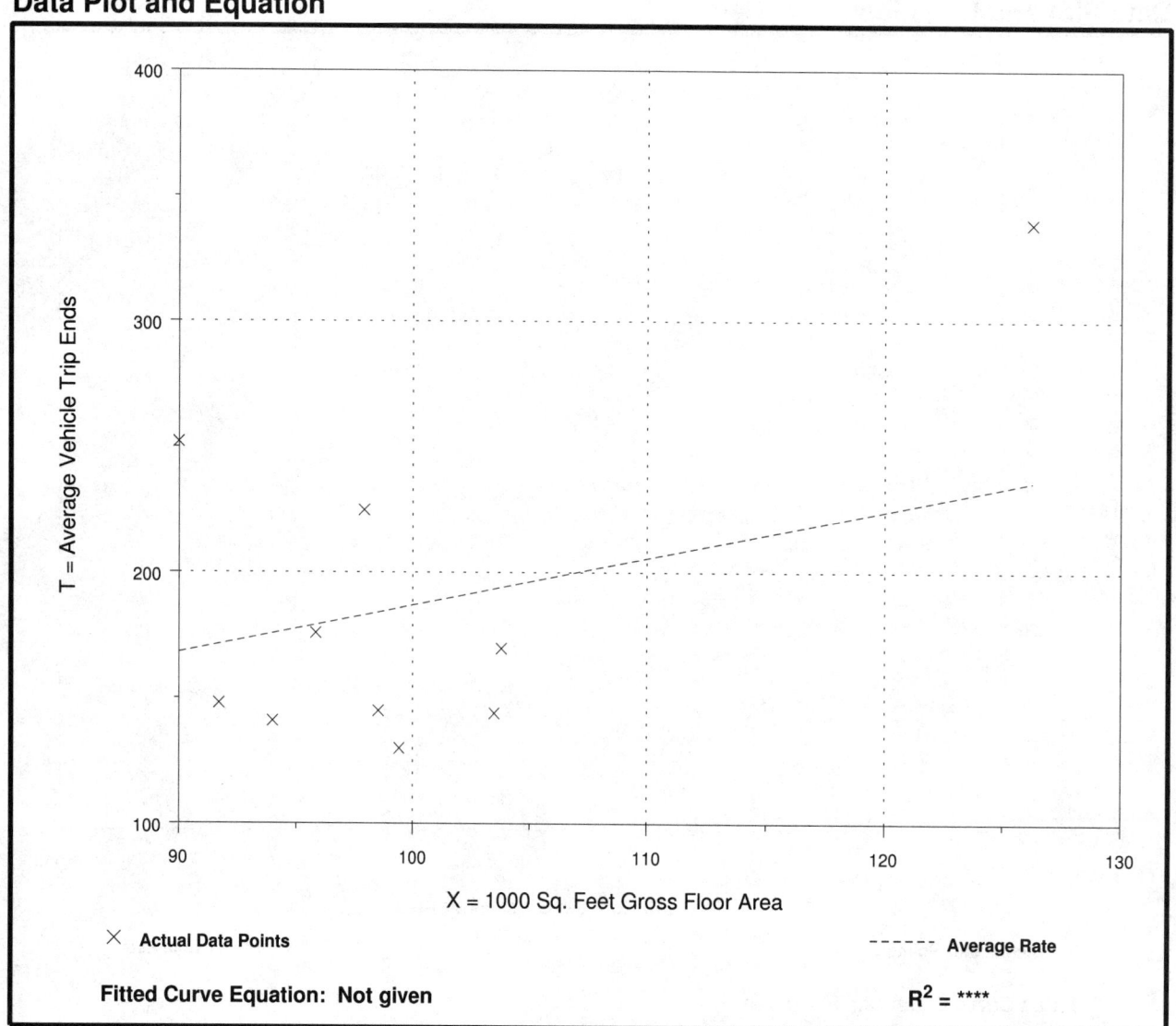

Fitted Curve Equation: Not given $R^2 = ****$

Department Store
(875)

Average Vehicle Trip Ends vs: 1000 Sq. Feet Gross Floor Area
On a: Weekday,
A.M. Peak Hour of Generator

Number of Studies: 5
Average 1000 Sq. Feet GFA: 104
Directional Distribution: 51% entering, 49% exiting

Trip Generation per 1000 Sq. Feet Gross Floor Area

Average Rate	Range of Rates	Standard Deviation
2.14	1.24 - 3.40	1.65

Data Plot and Equation

Caution - Use Carefully - Small Sample Size

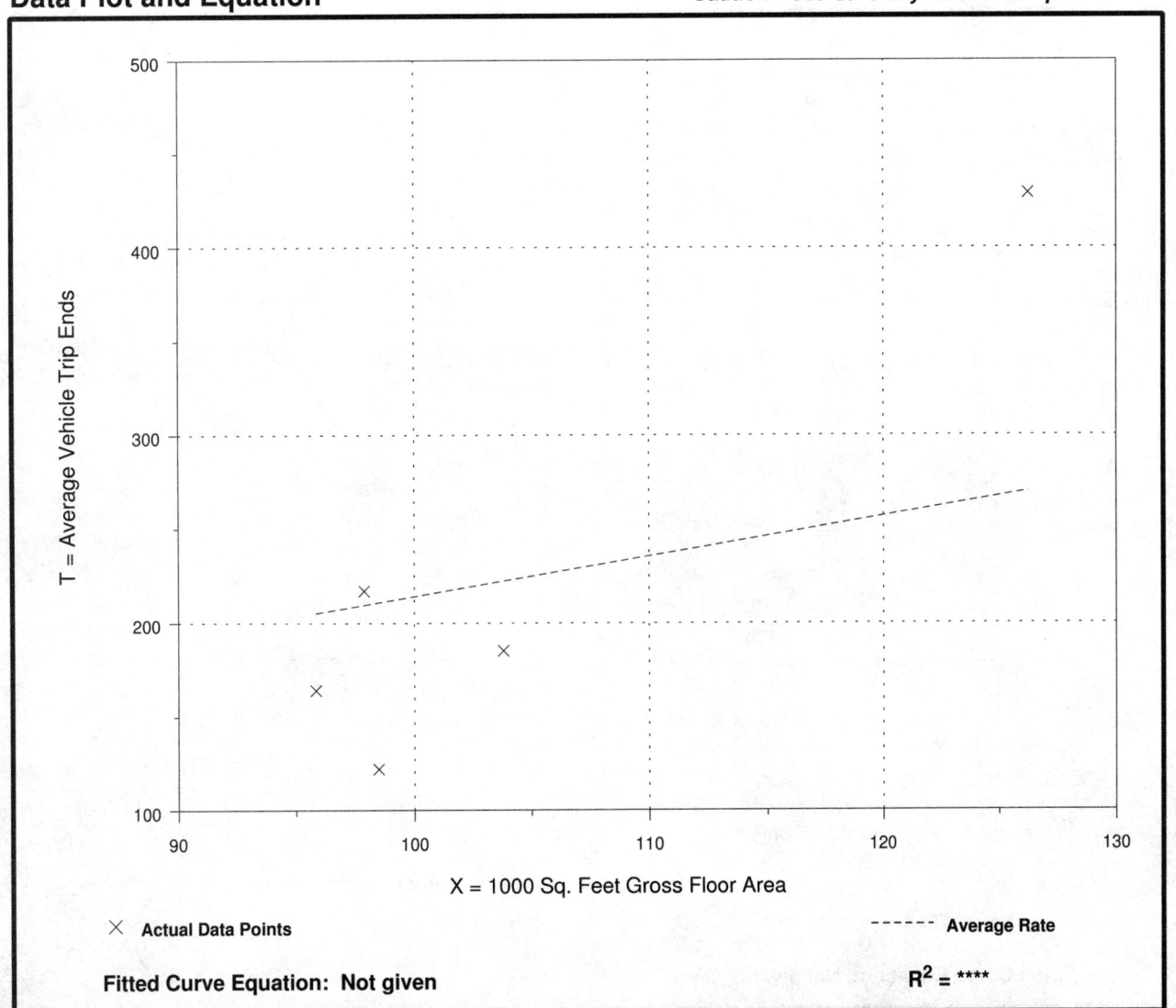

Fitted Curve Equation: Not given $R^2 = ****$

Department Store
(875)

Average Vehicle Trip Ends vs: 1000 Sq. Feet Gross Floor Area
On a: Weekday,
P.M. Peak Hour of Generator

Number of Studies: 5
Average 1000 Sq. Feet GFA: 104
Directional Distribution: 45% entering, 55% exiting

Trip Generation per 1000 Sq. Feet Gross Floor Area

Average Rate	Range of Rates	Standard Deviation
2.81	1.68 - 4.70	2.00

Data Plot and Equation

Caution - Use Carefully - Small Sample Size

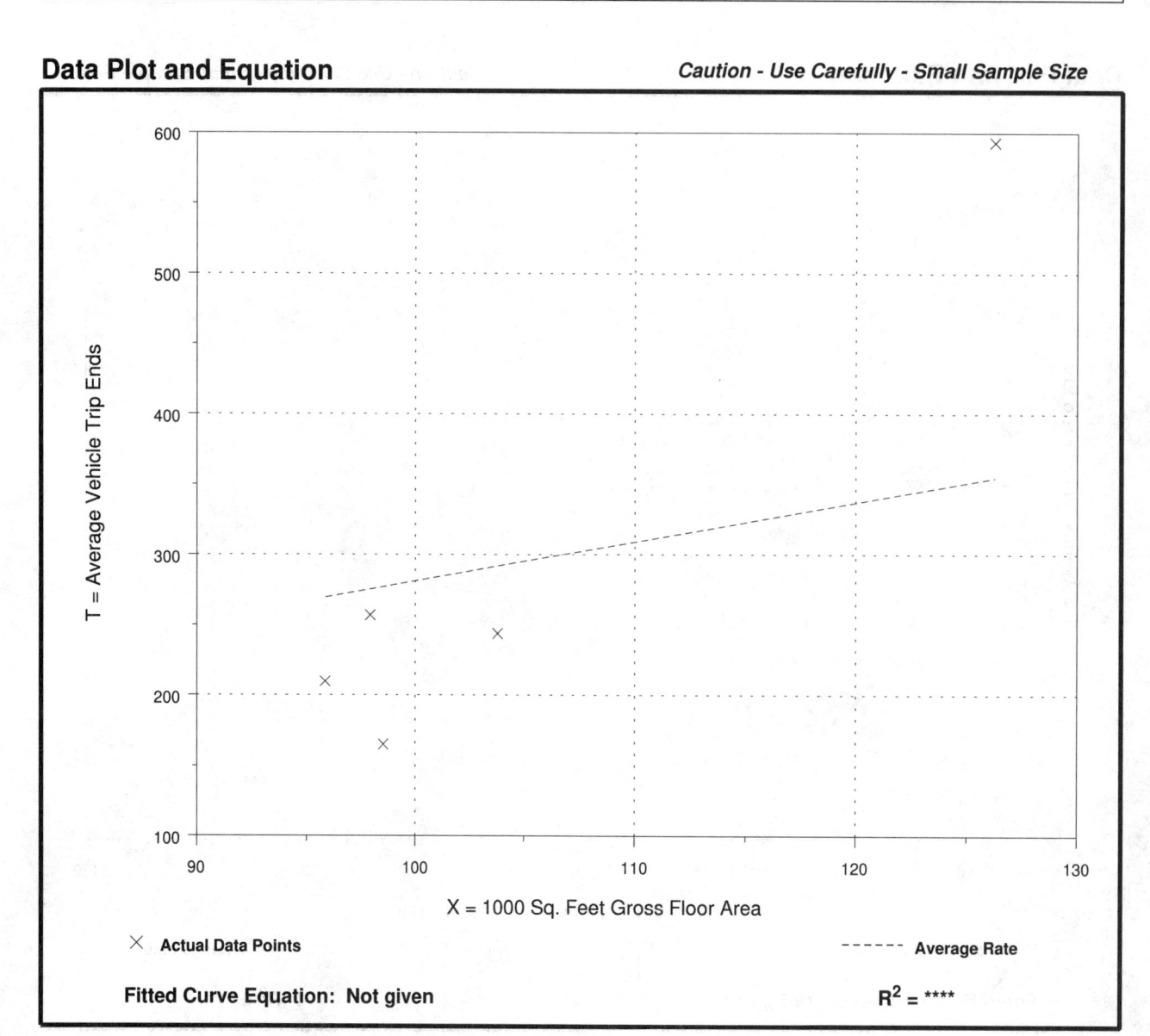

Fitted Curve Equation: Not given $R^2 = ****$

Department Store
(875)

Average Vehicle Trip Ends vs: 1000 Sq. Feet Gross Floor Area
On a: Saturday

Number of Studies: 3
Average 1000 Sq. Feet GFA: 97
Directional Distribution: 50% entering, 50% exiting

Trip Generation per 1000 Sq. Feet Gross Floor Area

Average Rate	Range of Rates	Standard Deviation
25.40	21.73 - 31.83	6.80

Data Plot and Equation

Caution - Use Carefully - Small Sample Size

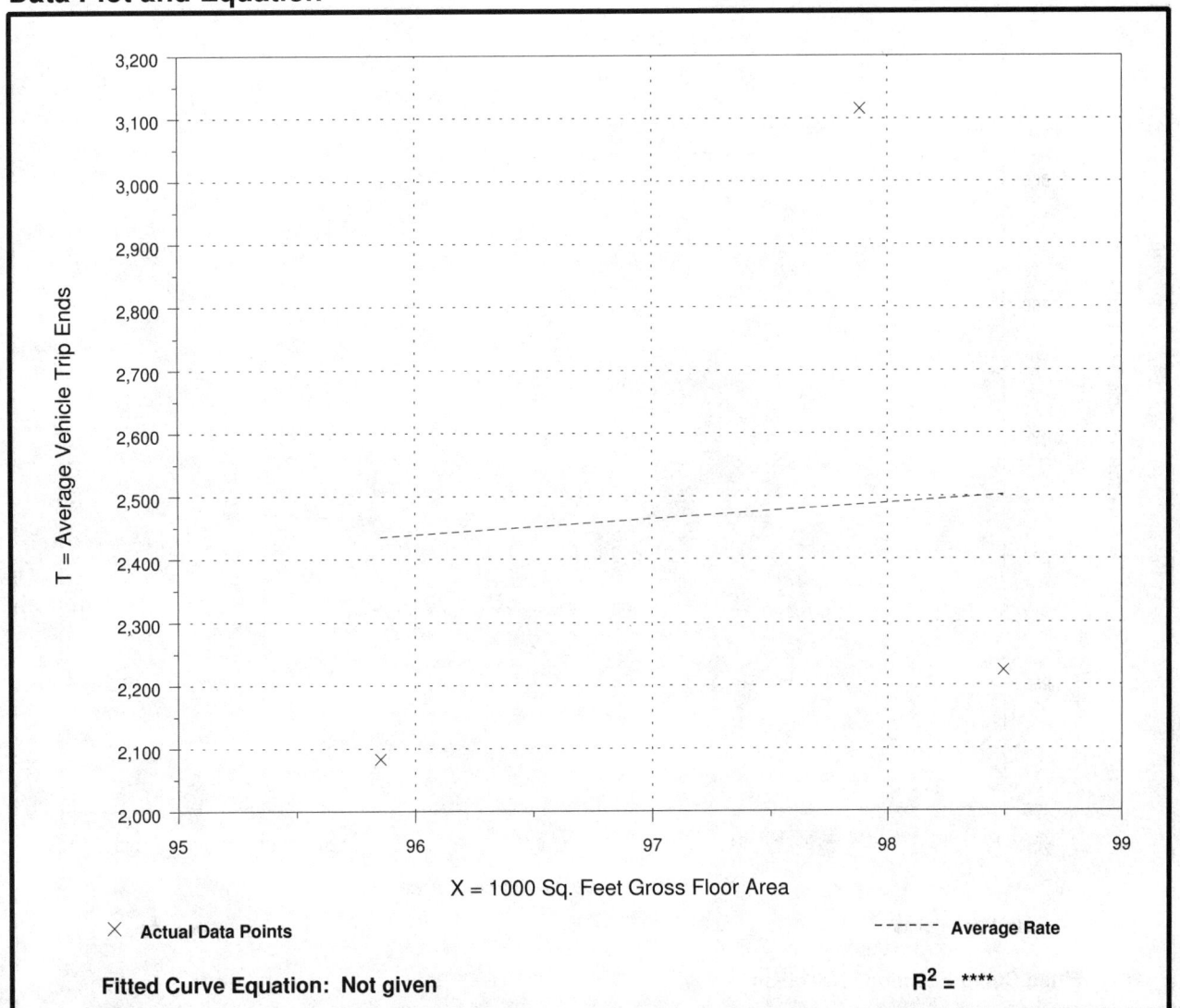

Fitted Curve Equation: Not given

$R^2 = ****$

Department Store
(875)

Average Vehicle Trip Ends vs: 1000 Sq. Feet Gross Floor Area
On a: Saturday,
Peak Hour of Generator

Number of Studies: 8
Average 1000 Sq. Feet GFA: 96
Directional Distribution: 53% entering, 47% exiting

Trip Generation per 1000 Sq. Feet Gross Floor Area

Average Rate	Range of Rates	Standard Deviation
3.32	1.68 - 6.76	2.29

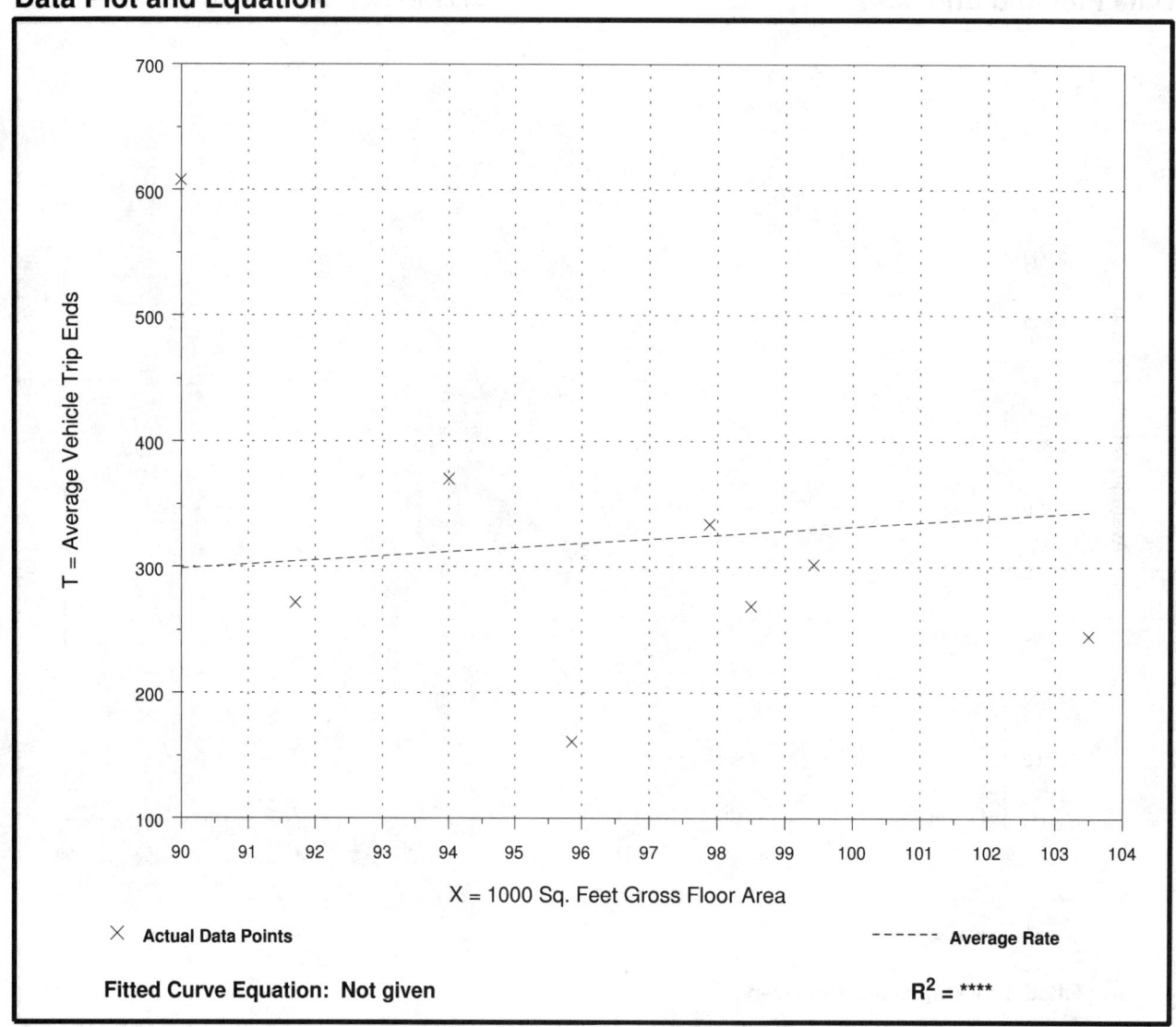

Data Plot and Equation

Fitted Curve Equation: Not given $R^2 = ****$

Land Use: 876
Apparel Store

Description

An apparel store is an individual store specializing in the sale of clothing. Department store (Land Use 875) is a related use.

Additional Data

The sites were surveyed in 1984 and 1994 in Florida and Vermont.

Source Numbers

210, 439

Land Use: 876
Apparel Store
Independent Variables with One Observation

The following trip generation data are for independent variables with only one observation. This information is shown in this table only; there are no related plots for these data.

Users are cautioned to use data with care because of the small sample size.

Independent Variable	Trip Generation Rate	Size of Independent Variable	Number of Studies	Directional Distribution
1,000 Square Feet Gross Floor Area				
Weekday	66.40	5	1	50% entering, 50% exiting
Weekday A.M. Peak Hour of Adjacent Street Traffic	1.00	5	1	80% entering, 20% exiting
Weekday A.M. Peak Hour of Generator	4.80	5	1	54% entering, 46% exiting

Apparel Store
(876)

Average Vehicle Trip Ends vs: 1000 Sq. Feet Gross Floor Area
On a: Weekday,
Peak Hour of Adjacent Street Traffic,
One Hour Between 4 and 6 p.m.

Number of Studies: 7
Average 1000 Sq. Feet GFA: 5
Directional Distribution: 50% entering, 50% exiting

Trip Generation per 1000 Sq. Feet Gross Floor Area

Average Rate	Range of Rates	Standard Deviation
3.83	1.50 - 6.80	2.81

Data Plot and Equation

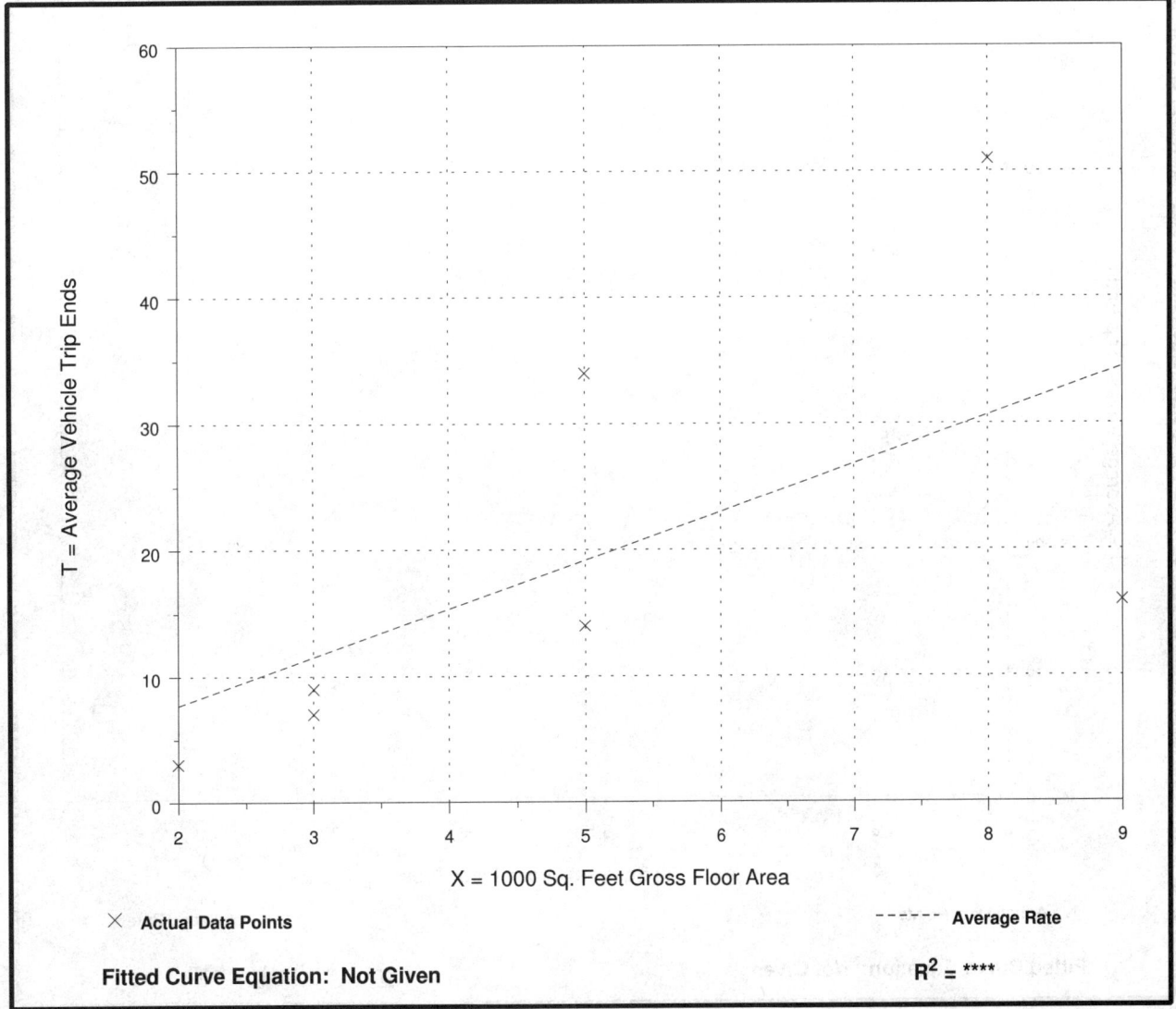

Fitted Curve Equation: Not Given $R^2 = ****$

Apparel Store
(876)

Average Vehicle Trip Ends vs: 1000 Sq. Feet Gross Floor Area
On a: Weekday,
P.M. Peak Hour of Generator

Number of Studies: 7
Average 1000 Sq. Feet GFA: 5
Directional Distribution: 50% entering, 50% exiting

Trip Generation per 1000 Sq. Feet Gross Floor Area

Average Rate	Range of Rates	Standard Deviation
4.20	1.78 - 6.80	2.73

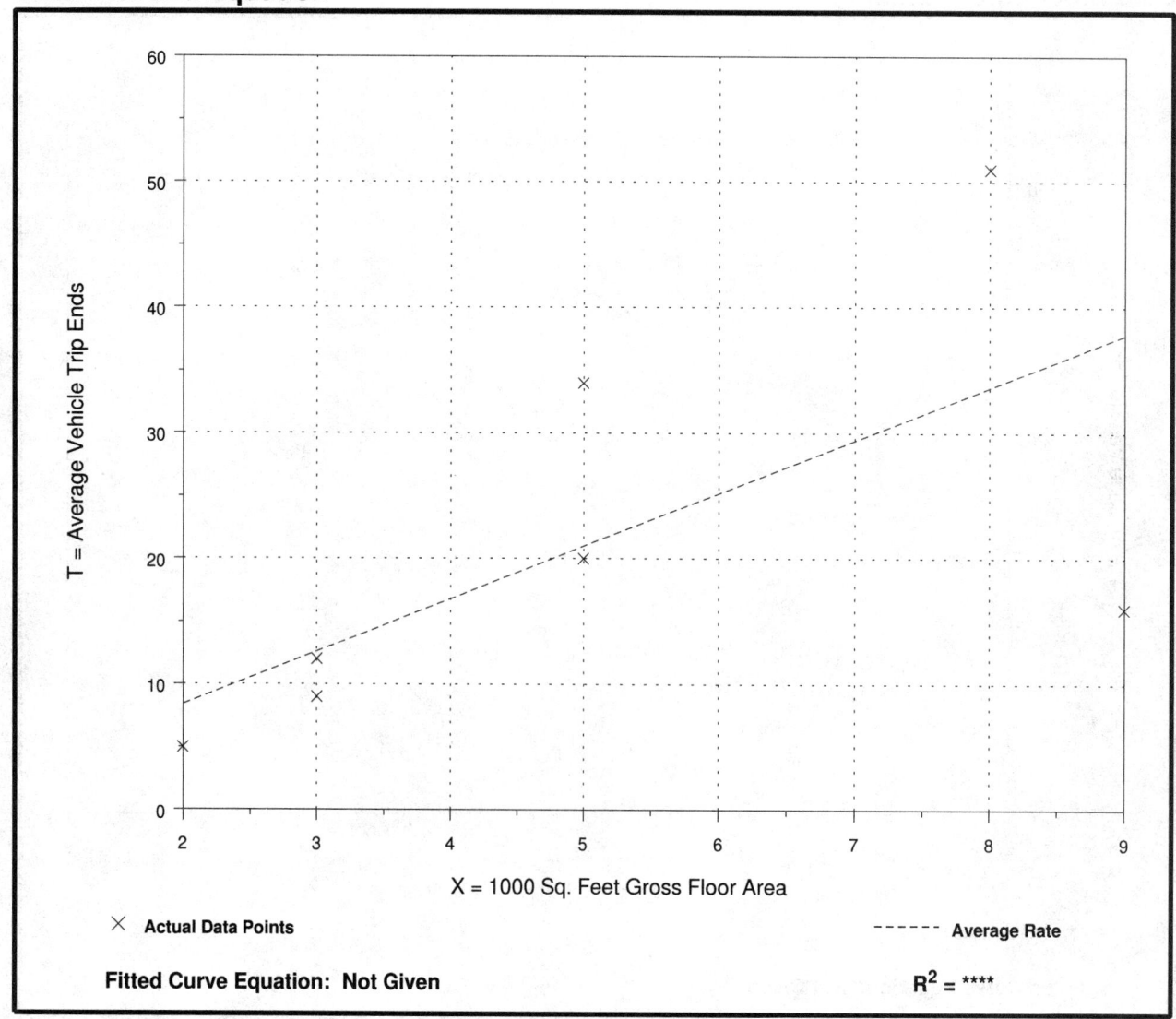

Data Plot and Equation

Fitted Curve Equation: Not Given $R^2 = ****$

Land Use: 879
Arts and Crafts Store

Description

Arts and crafts stores are free-standing facilities that sell art, framing, wall décor and seasonal merchandise. These stores may provide in-store arts and crafts classes. Arts and crafts stores are sometimes found as separate parcels within a retail complex, with or without their own dedicated off-street parking.

Additional Data

The sites were surveyed in the 2000s in Pennsylvania.

Source Number

571

Land Use: 879
Arts and Crafts Store
Independent Variables with One Observation

The following trip generation data are for independent variables with only one observation. This information is shown in this table only; there are no related plots for these data.

Users are cautioned to use data with care because of the small sample size.

Independent Variable	Trip Generation Rate	Size of Independent Variable	Number of Studies	Directional Distribution

1,000 Square Feet Gross Floor Area

Weekday	56.55	20	1	50% entering, 50% exiting
Weekday A.M. Peak Hour of Generator	4.65	20	1	49% entering, 51% exiting
Weekday P.M. Peak Hour of Generator	6.85	20	1	51% entering, 49% exiting

Arts and Crafts Store
(879)

Average Vehicle Trip Ends vs: 1000 Sq. Feet Gross Floor Area
On a: Weekday,
Peak Hour of Adjacent Street Traffic,
One Hour Between 4 and 6 p.m.

Number of Studies: 2
Average 1000 Sq. Feet GFA: 20
Directional Distribution: 46% entering, 54% exiting

Trip Generation per 1000 Sq. Feet Gross Floor Area

Average Rate	Range of Rates	Standard Deviation
6.21	4.90 - 7.55	*

Data Plot and Equation

Caution - Use Carefully - Small Sample Size

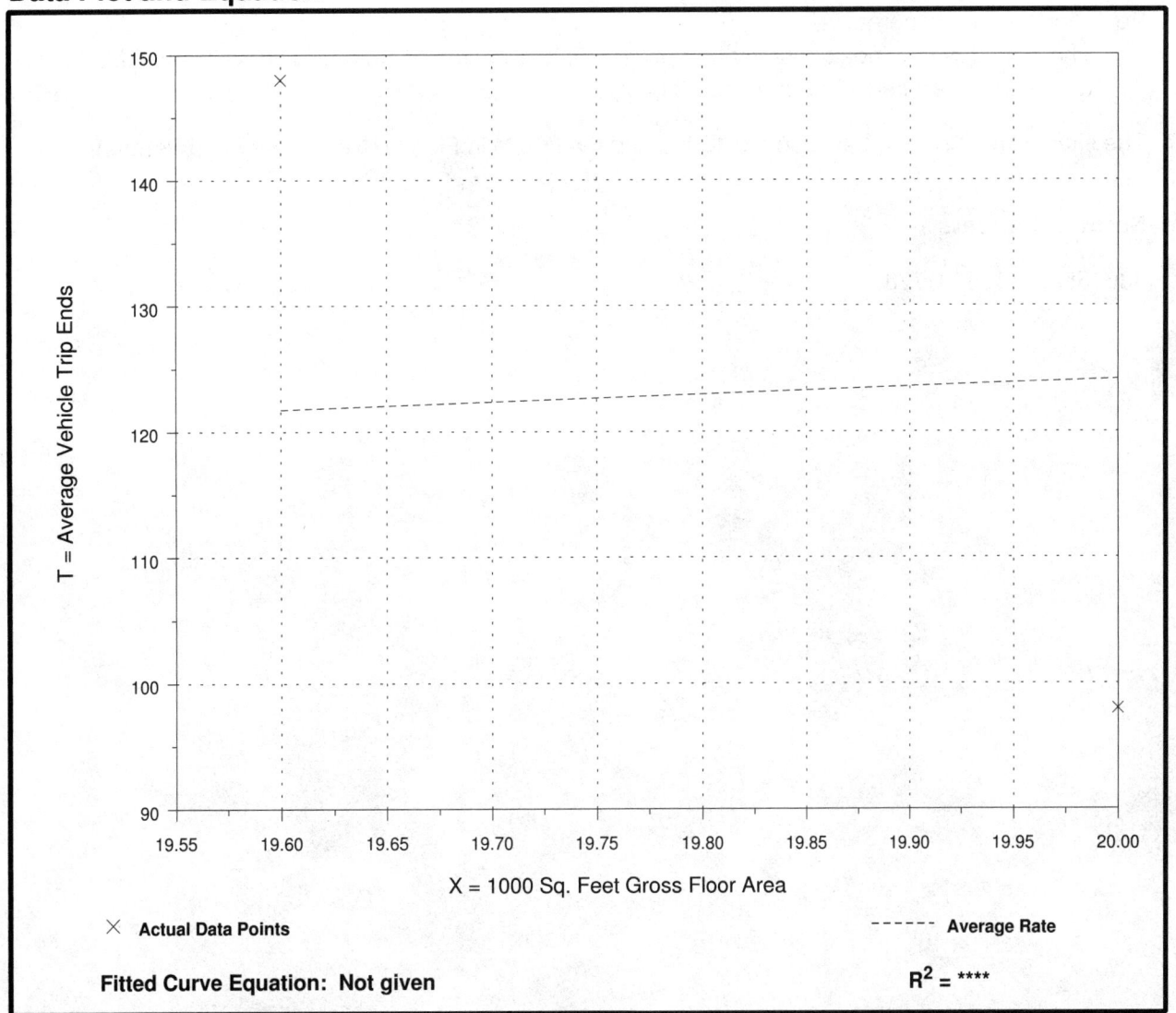

Fitted Curve Equation: Not given $R^2 = ****$

Land Use: 880
Pharmacy/Drugstore without Drive-Through Window

Description

Pharmacies/drugstores are retail facilities that primarily sell prescription and non-prescription drugs. These facilities may also sell cosmetics, toiletries, medications, stationery, personal care products, limited food products and general merchandise. The drug stores in this category do not contain drive-through windows. Pharmacy/drugstore with drive-through window (Land Use 881) is a related use.

Additional Data

Peak hours of the generator—
　　The weekday A.M. peak hour varied between 9:00 a.m. and 11:00 a.m. The weekday P.M. peak hour varied between 12:00 p.m. and 3:00 p.m.

The sites were surveyed between the 1990s and the 2000s in Florida, New Jersey and Vermont.

Source Numbers

436, 550, 551, 573, 728

Pharmacy/Drugstore without Drive-Through Window
(880)

Average Vehicle Trip Ends vs: 1000 Sq. Feet Gross Floor Area
On a: Weekday

Number of Studies: 6
Average 1000 Sq. Feet GFA: 11
Directional Distribution: 50% entering, 50% exiting

Trip Generation per 1000 Sq. Feet Gross Floor Area

Average Rate	Range of Rates	Standard Deviation
90.06	81.00 - 106.50	12.24

Data Plot and Equation

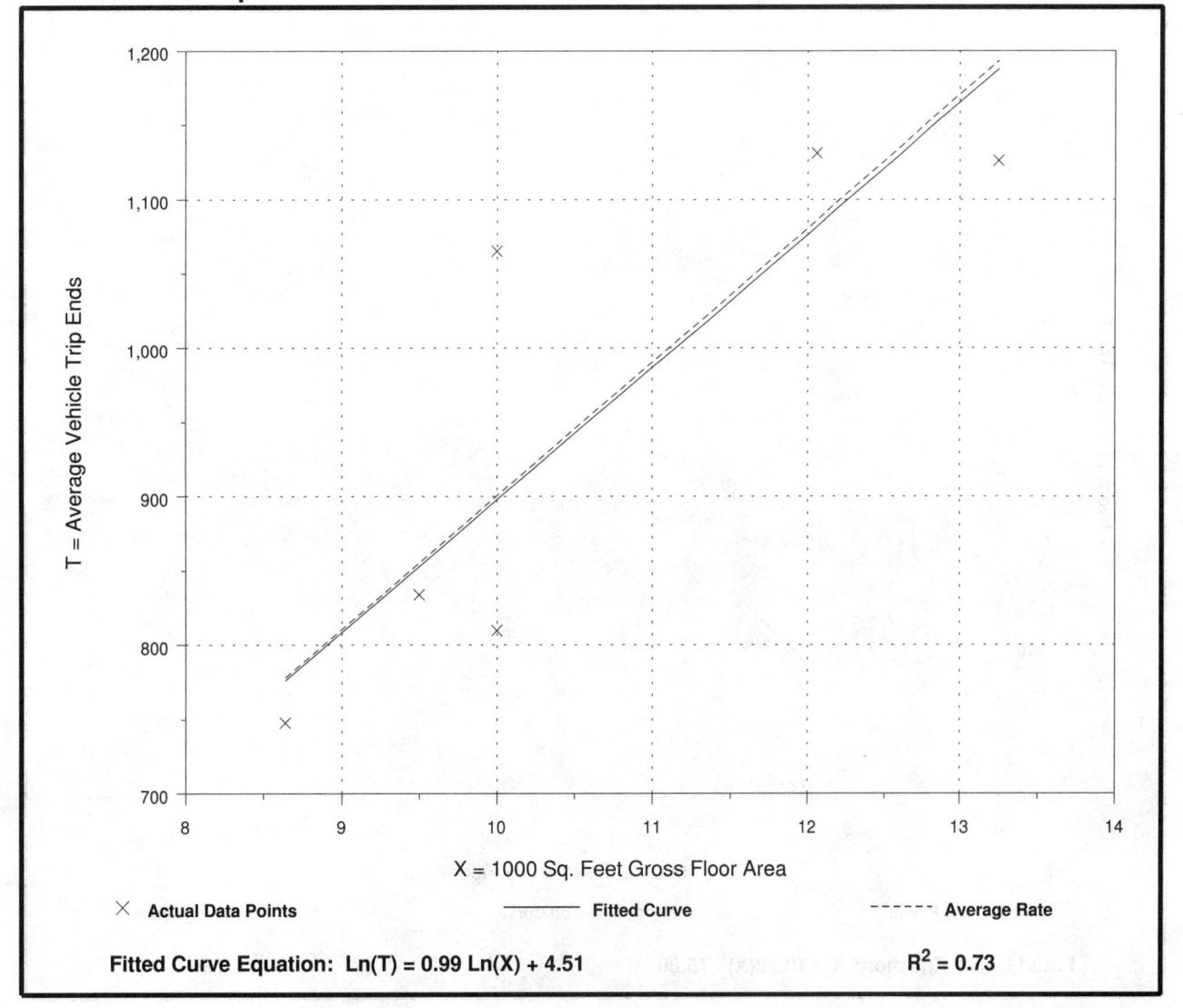

Fitted Curve Equation: $Ln(T) = 0.99\, Ln(X) + 4.51$ $R^2 = 0.73$

Pharmacy/Drugstore without Drive-Through Window
(880)

Average Vehicle Trip Ends vs: 1000 Sq. Feet Gross Floor Area
On a: Weekday,
Peak Hour of Adjacent Street Traffic,
One Hour Between 7 and 9 a.m.

Number of Studies: 7
Average 1000 Sq. Feet GFA: 10
Directional Distribution: 65% entering, 35% exiting

Trip Generation per 1000 Sq. Feet Gross Floor Area

Average Rate	Range of Rates	Standard Deviation
2.94	1.17 - 4.30	2.02

Data Plot and Equation

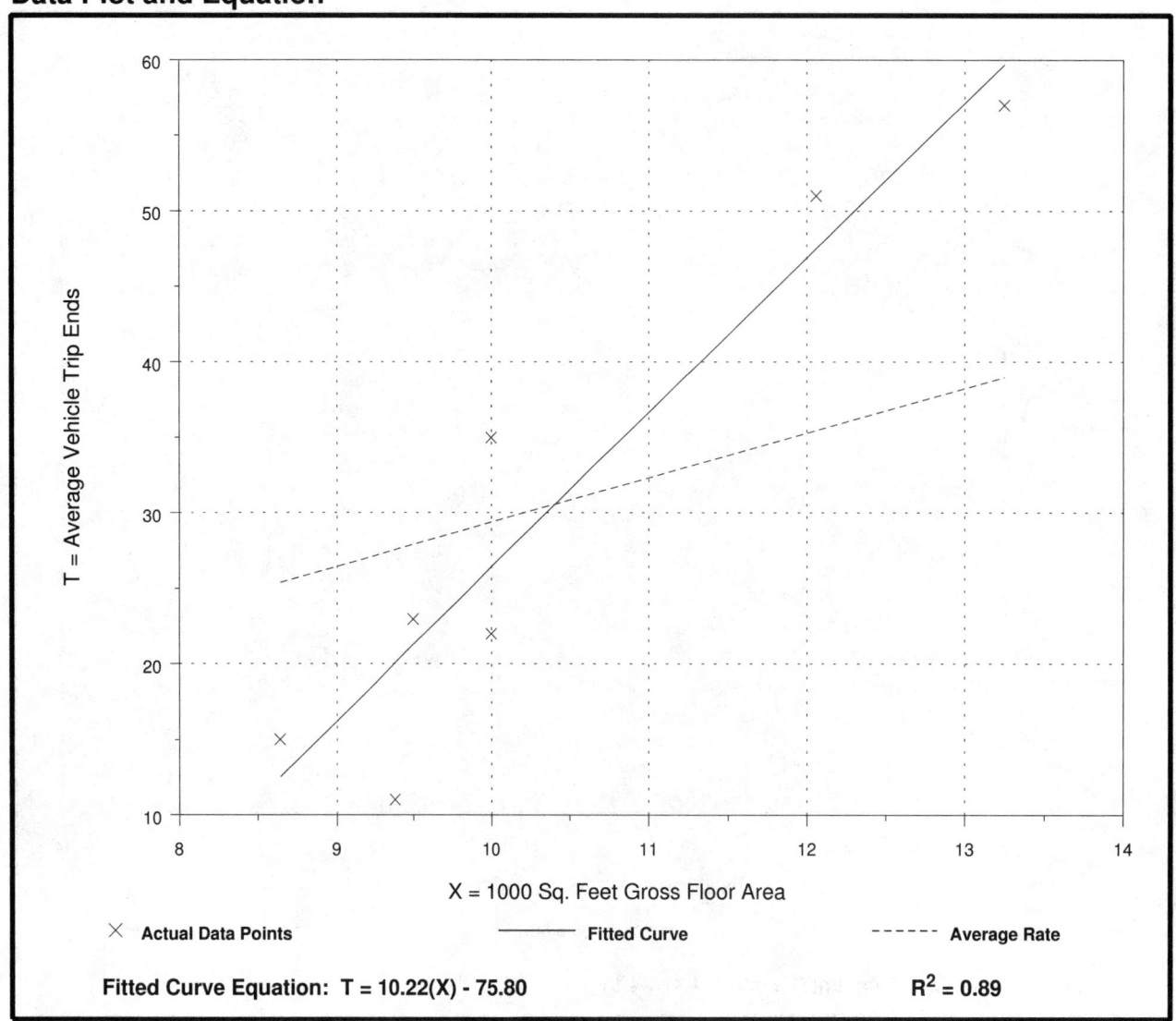

Fitted Curve Equation: T = 10.22(X) - 75.80 $R^2 = 0.89$

Pharmacy/Drugstore without Drive-Through Window
(880)

Average Vehicle Trip Ends vs: 1000 Sq. Feet Gross Floor Area
On a: Weekday,
Peak Hour of Adjacent Street Traffic,
One Hour Between 4 and 6 p.m.

Number of Studies: 12
Average 1000 Sq. Feet GFA: 10
Directional Distribution: 49% entering, 51% exiting

Trip Generation per 1000 Sq. Feet Gross Floor Area

Average Rate	Range of Rates	Standard Deviation
8.40	5.10 - 11.70	3.51

Data Plot and Equation

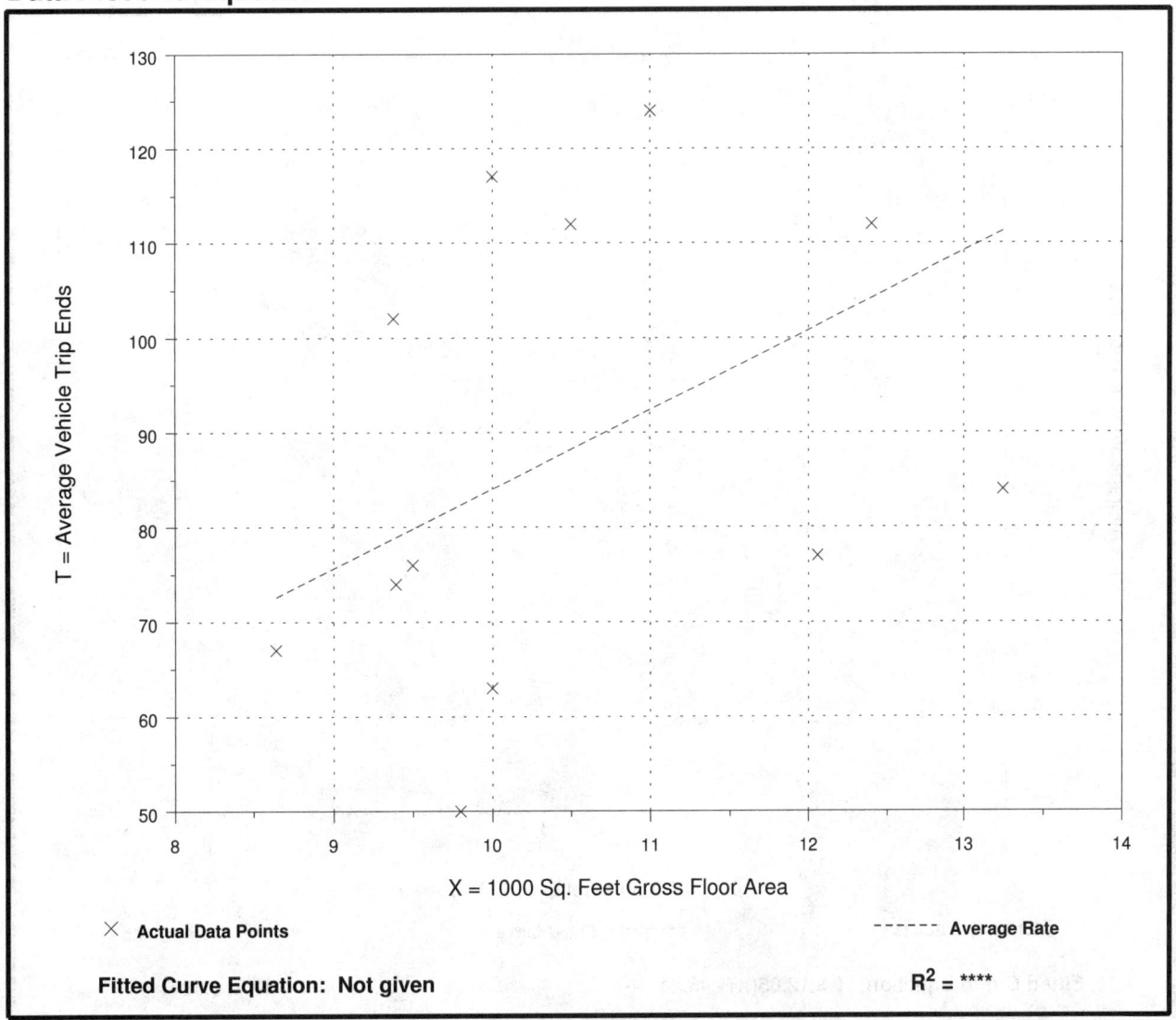

Fitted Curve Equation: Not given $R^2 = ****$

Pharmacy/Drugstore without Drive-Through Window
(880)

Average Vehicle Trip Ends vs: 1000 Sq. Feet Gross Floor Area
On a: Weekday,
A.M. Peak Hour of Generator

Number of Studies: 8
Average 1000 Sq. Feet GFA: 11
Directional Distribution: 50% entering, 50% exiting

Trip Generation per 1000 Sq. Feet Gross Floor Area

Average Rate	Range of Rates	Standard Deviation
7.71	6.02 - 8.83	2.81

Data Plot and Equation

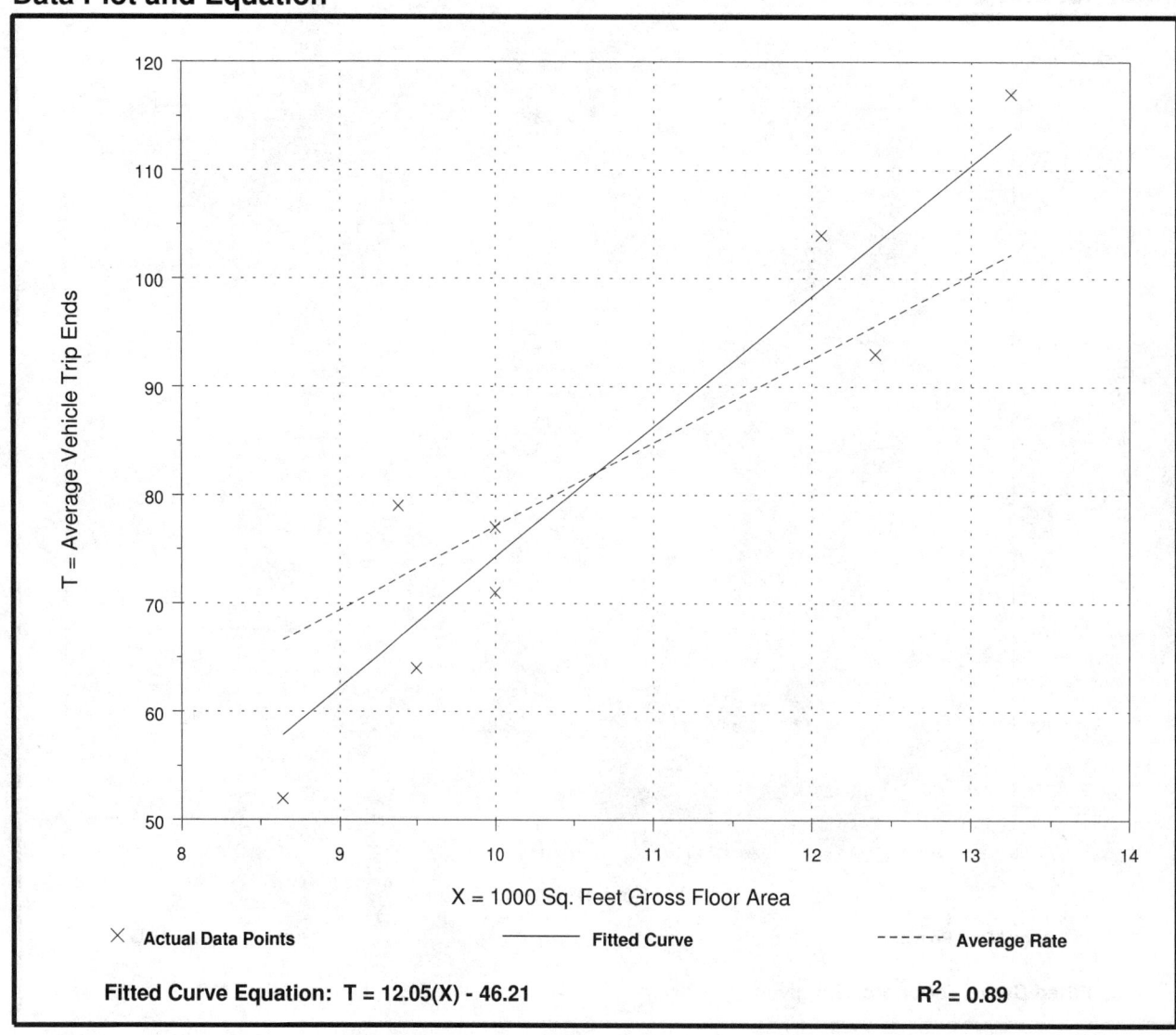

Fitted Curve Equation: T = 12.05(X) - 46.21 $R^2 = 0.89$

Pharmacy/Drugstore without Drive-Through Window
(880)

Average Vehicle Trip Ends vs: 1000 Sq. Feet Gross Floor Area
On a: Weekday,
P.M. Peak Hour of Generator

Number of Studies: 7
Average 1000 Sq. Feet GFA: 11
Directional Distribution: 50% entering, 50% exiting

Trip Generation per 1000 Sq. Feet Gross Floor Area

Average Rate	Range of Rates	Standard Deviation
11.07	7.47 - 24.00	6.63

Data Plot and Equation

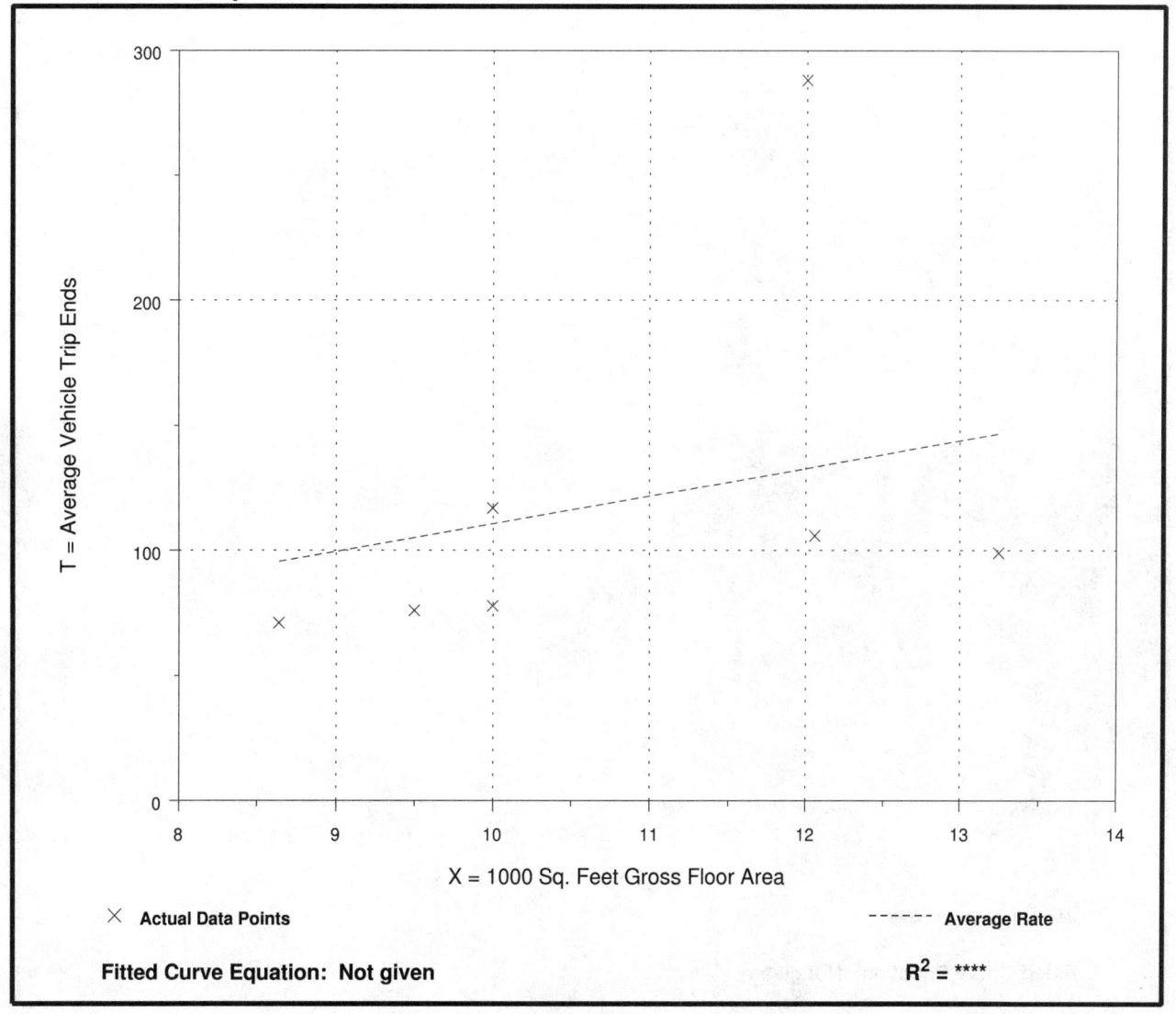

Fitted Curve Equation: Not given $R^2 = ****$

Pharmacy/Drugstore without Drive-Through Window
(880)

Average Vehicle Trip Ends vs: 1000 Sq. Feet Gross Floor Area
On a: Saturday,
Peak Hour of Generator

Number of Studies: 3
Average 1000 Sq. Feet GFA: 10
Directional Distribution: 49% entering, 51% exiting

Trip Generation per 1000 Sq. Feet Gross Floor Area

Average Rate	Range of Rates	Standard Deviation
10.68	9.26 - 13.09	3.65

Data Plot and Equation

Caution - Use Carefully - Small Sample Size

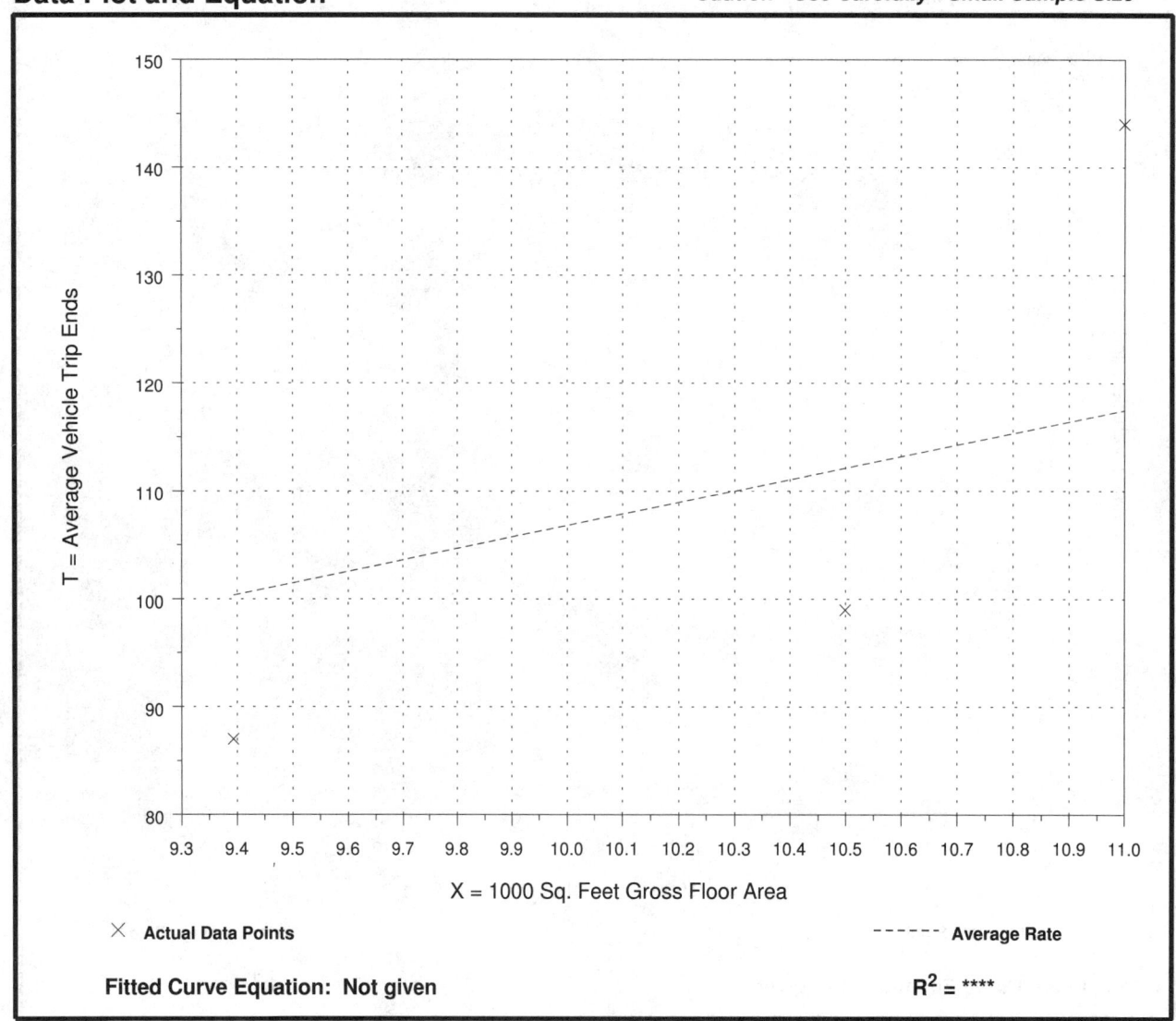

Fitted Curve Equation: Not given $R^2 = ****$

Land Use: 881
Pharmacy/Drugstore with Drive-Through Window

Description

Pharmacies/drugstores are retail facilities that primarily sell prescription and non-prescription drugs. These facilities may also sell cosmetics, toiletries, medications, stationery, personal care products, limited food products and general merchandise. The drug stores in this category contain drive-through windows. Pharmacy/drugstore without a drive-through window (Land Use 880) is a related use.

Additional Data

Several studies indicated that they had two drive-through windows.

Peak hours of the generator—
    The weekday A.M. peak hour varied between 8:00 a.m. and 12:00 p.m. The weekday P.M. peak hour varied between 12:00 p.m. and 6:00 p.m. The weekend peak hour varied between 12:00 p.m. and 7:00 p.m.

The sites were surveyed between the 1990s and the 2000s in California, Colorado, Florida, Minnesota, New Hampshire, New Jersey, New York, Pennsylvania, Vermont and Wisconsin.

To assist in the future analysis of this land use, it is important that the number of drive-through lanes at the study site be reported.

Source Numbers

369, 418, 436, 547, 550, 552, 563, 568, 573, 599, 621, 716, 727, 728, 734

Pharmacy/Drugstore with Drive-Through Window
(881)

Average Vehicle Trip Ends vs: 1000 Sq. Feet Gross Floor Area
On a: Weekday

Number of Studies: 10
Average 1000 Sq. Feet GFA: 13
Directional Distribution: 50% entering, 50% exiting

Trip Generation per 1000 Sq. Feet Gross Floor Area

Average Rate	Range of Rates	Standard Deviation
96.91	67.09 - 133.45	21.59

Data Plot and Equation

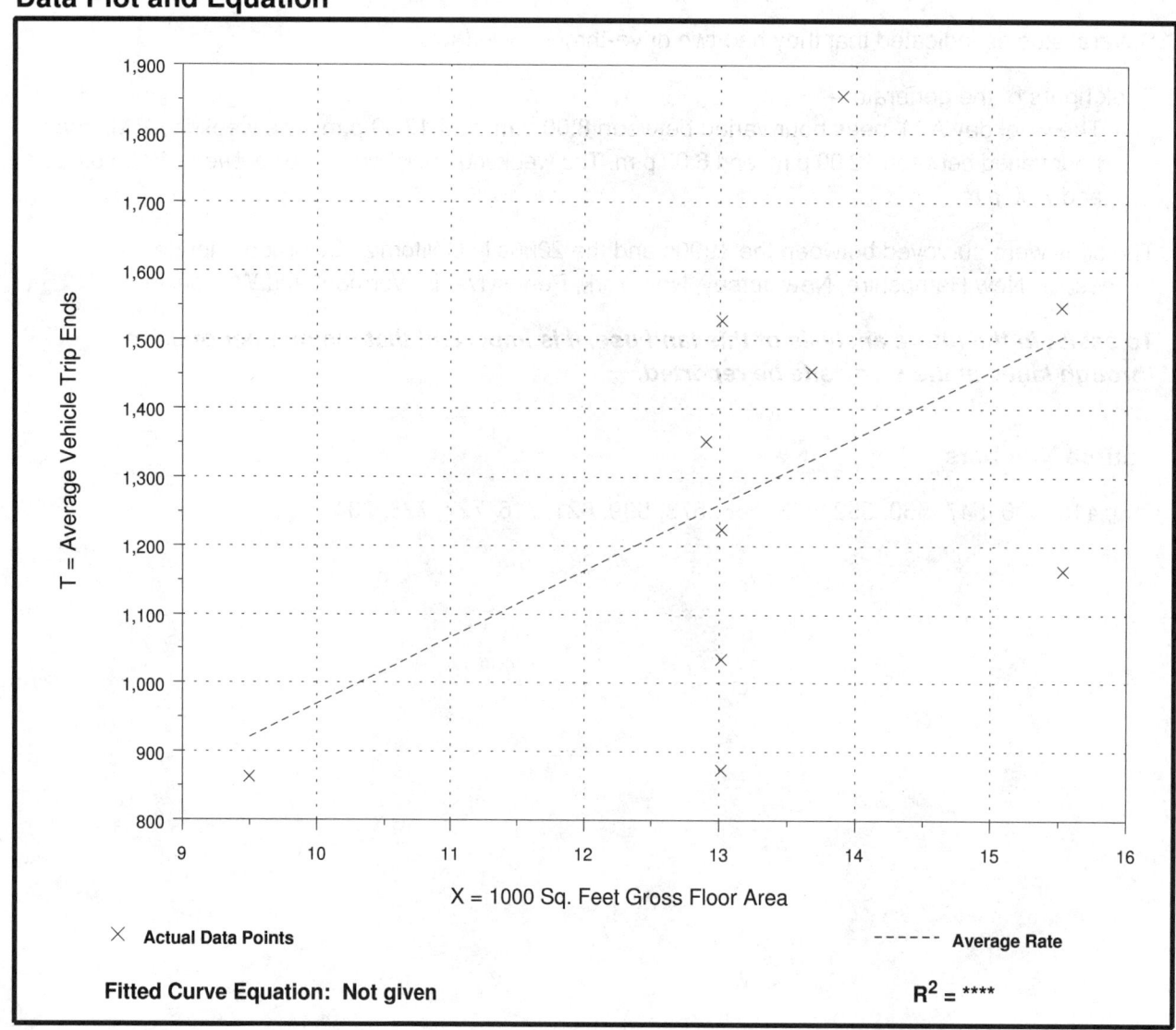

Fitted Curve Equation: Not given $R^2 = ****$

Pharmacy/Drugstore with Drive-Through Window
(881)

Average Vehicle Trip Ends vs: 1000 Sq. Feet Gross Floor Area
On a: Weekday,
Peak Hour of Adjacent Street Traffic,
One Hour Between 7 and 9 a.m.

Number of Studies: 13
Average 1000 Sq. Feet GFA: 13
Directional Distribution: 52% entering, 48% exiting

Trip Generation per 1000 Sq. Feet Gross Floor Area

Average Rate	Range of Rates	Standard Deviation
3.45	1.93 - 5.92	2.10

Data Plot and Equation

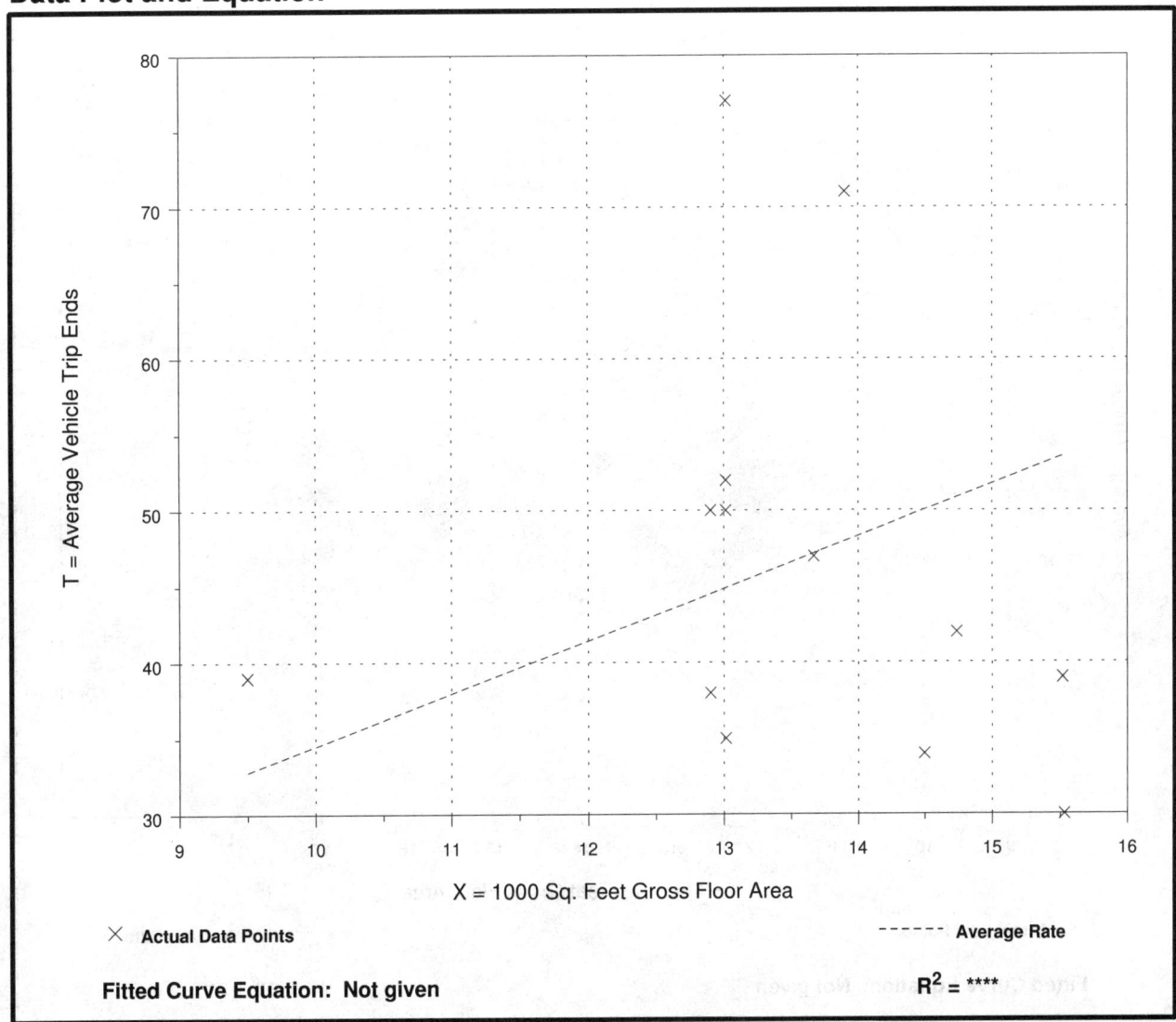

Fitted Curve Equation: Not given $R^2 = ****$

Pharmacy/Drugstore with Drive-Through Window
(881)

Average Vehicle Trip Ends vs: 1000 Sq. Feet Gross Floor Area
On a: Weekday,
Peak Hour of Adjacent Street Traffic,
One Hour Between 4 and 6 p.m.

Number of Studies: 31
Average 1000 Sq. Feet GFA: 14
Directional Distribution: 50% entering, 50% exiting

Trip Generation per 1000 Sq. Feet Gross Floor Area

Average Rate	Range of Rates	Standard Deviation
9.91	4.85 - 20.43	5.04

Data Plot and Equation

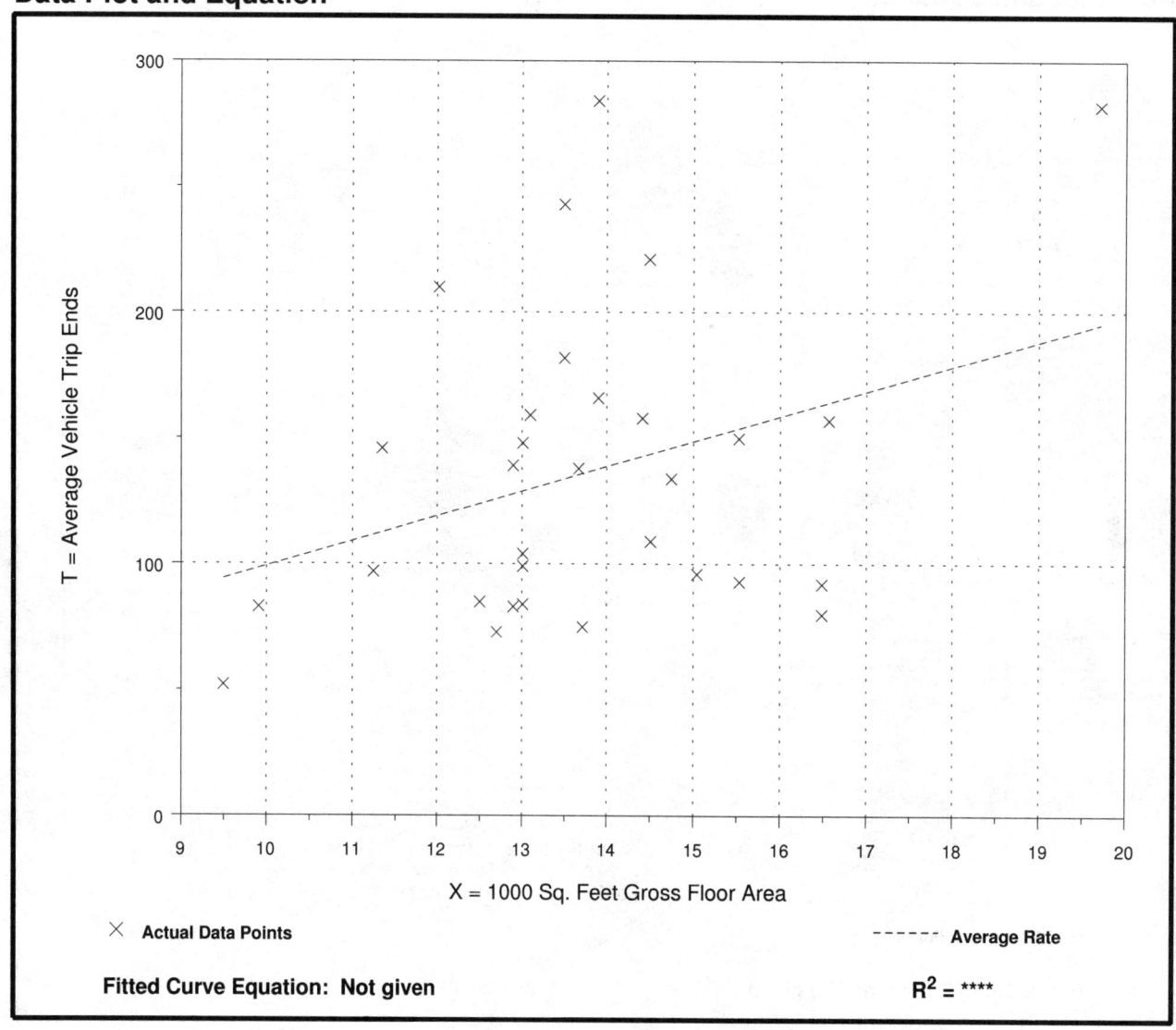

Fitted Curve Equation: Not given $R^2 = ****$

Pharmacy/Drugstore with Drive-Through Window
(881)

Average Vehicle Trip Ends vs: 1000 Sq. Feet Gross Floor Area
On a: Weekday,
A.M. Peak Hour of Generator

Number of Studies: 17
Average 1000 Sq. Feet GFA: 13
Directional Distribution: 50% entering, 50% exiting

Trip Generation per 1000 Sq. Feet Gross Floor Area

Average Rate	Range of Rates	Standard Deviation
8.36	5.07 - 14.69	3.64

Data Plot and Equation

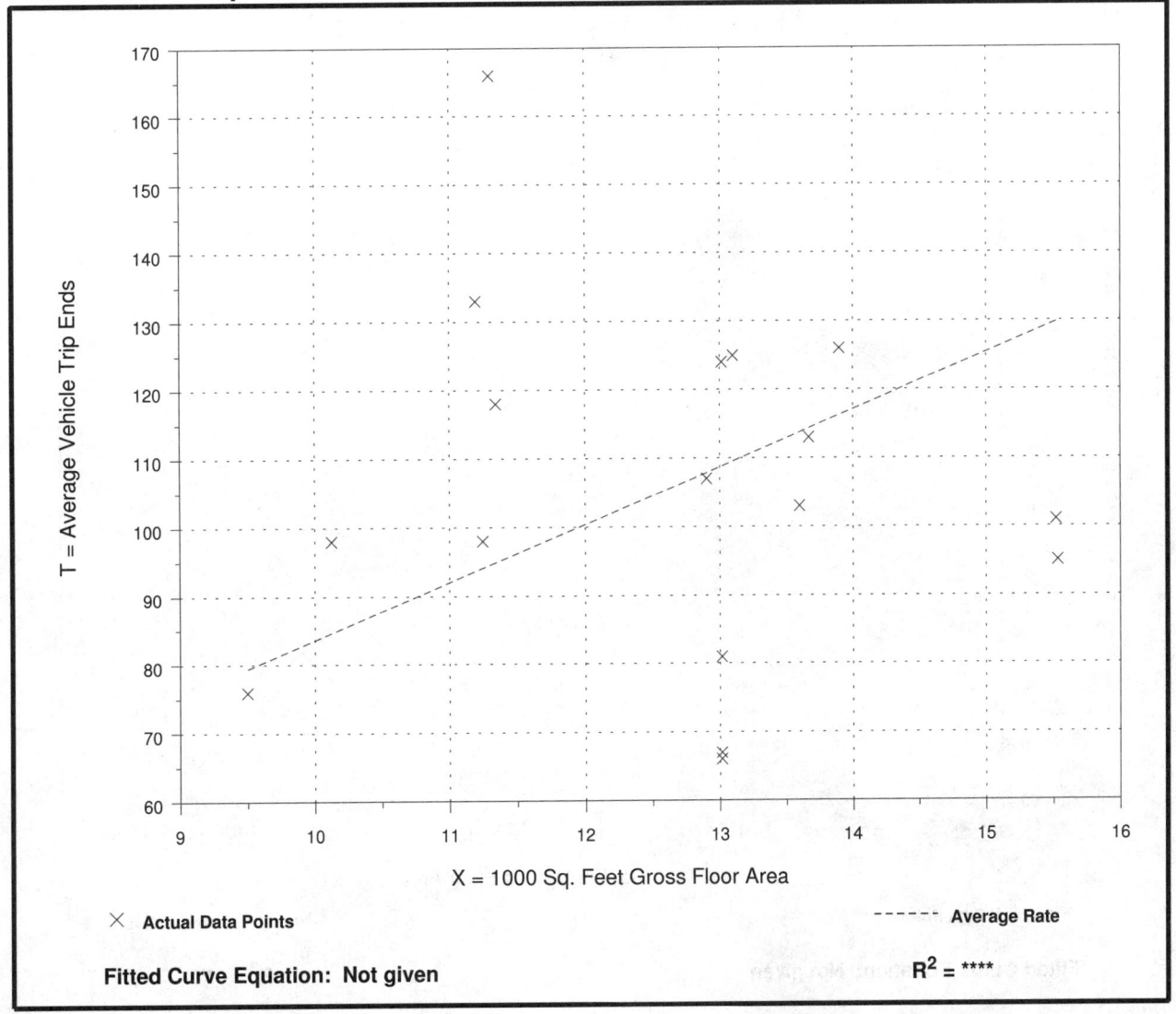

Fitted Curve Equation: Not given $R^2 = ****$

Pharmacy/Drugstore with Drive-Through Window
(881)

Average Vehicle Trip Ends vs: 1000 Sq. Feet Gross Floor Area
On a: Weekday,
P.M. Peak Hour of Generator

Number of Studies: 9
Average 1000 Sq. Feet GFA: 13
Directional Distribution: 50% entering, 50% exiting

Trip Generation per 1000 Sq. Feet Gross Floor Area

Average Rate	Range of Rates	Standard Deviation
9.72	6.50 - 13.48	4.04

Data Plot and Equation

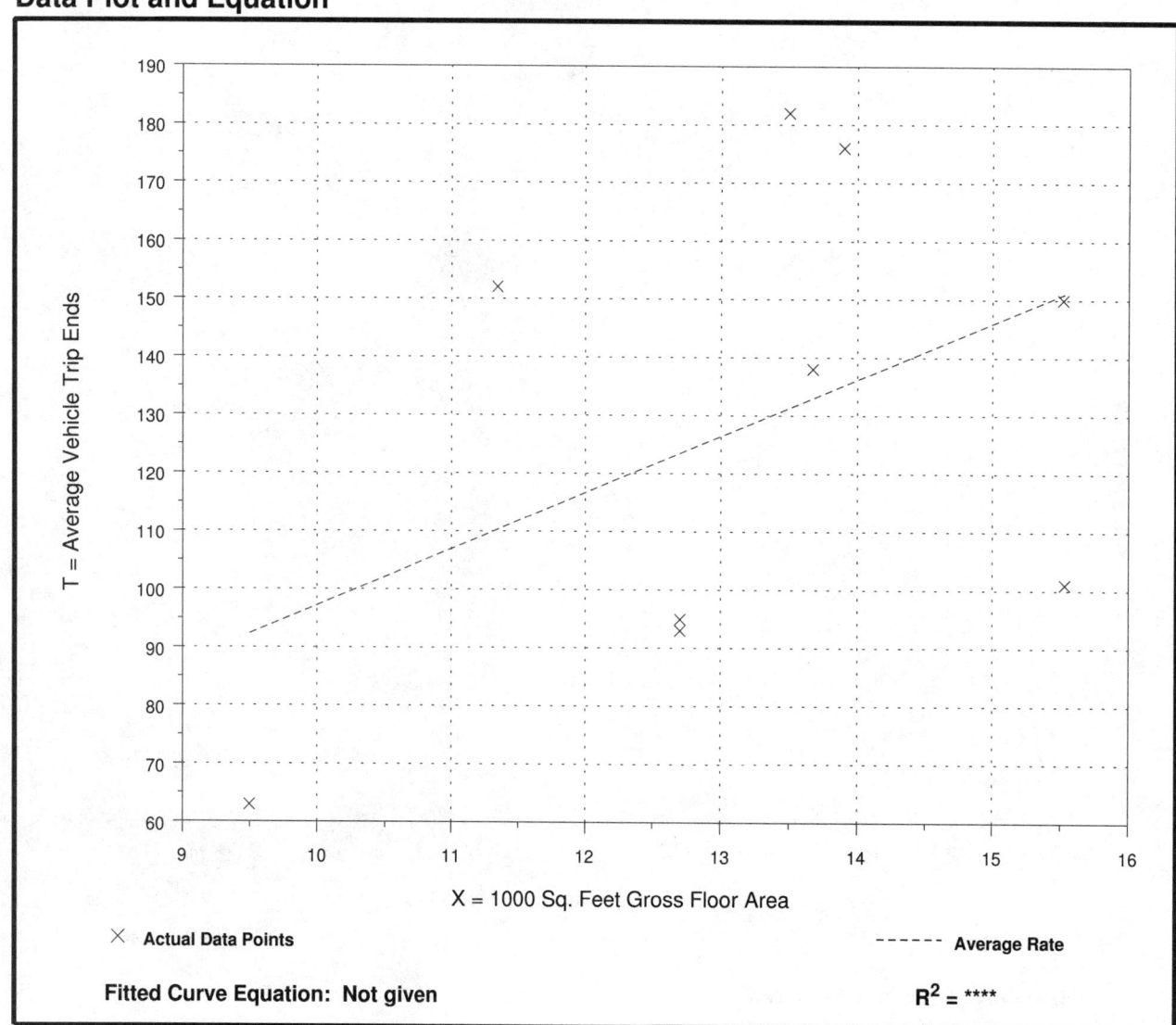

Fitted Curve Equation: Not given $R^2 = ****$

Pharmacy/Drugstore with Drive-Through Window
(881)

Average Vehicle Trip Ends vs: 1000 Sq. Feet Gross Floor Area
On a: Saturday,
Peak Hour of Generator

Number of Studies: 14
Average 1000 Sq. Feet GFA: 14
Directional Distribution: 49% entering, 51% exiting

Trip Generation per 1000 Sq. Feet Gross Floor Area

Average Rate	Range of Rates	Standard Deviation
8.20	4.31 - 11.40	3.57

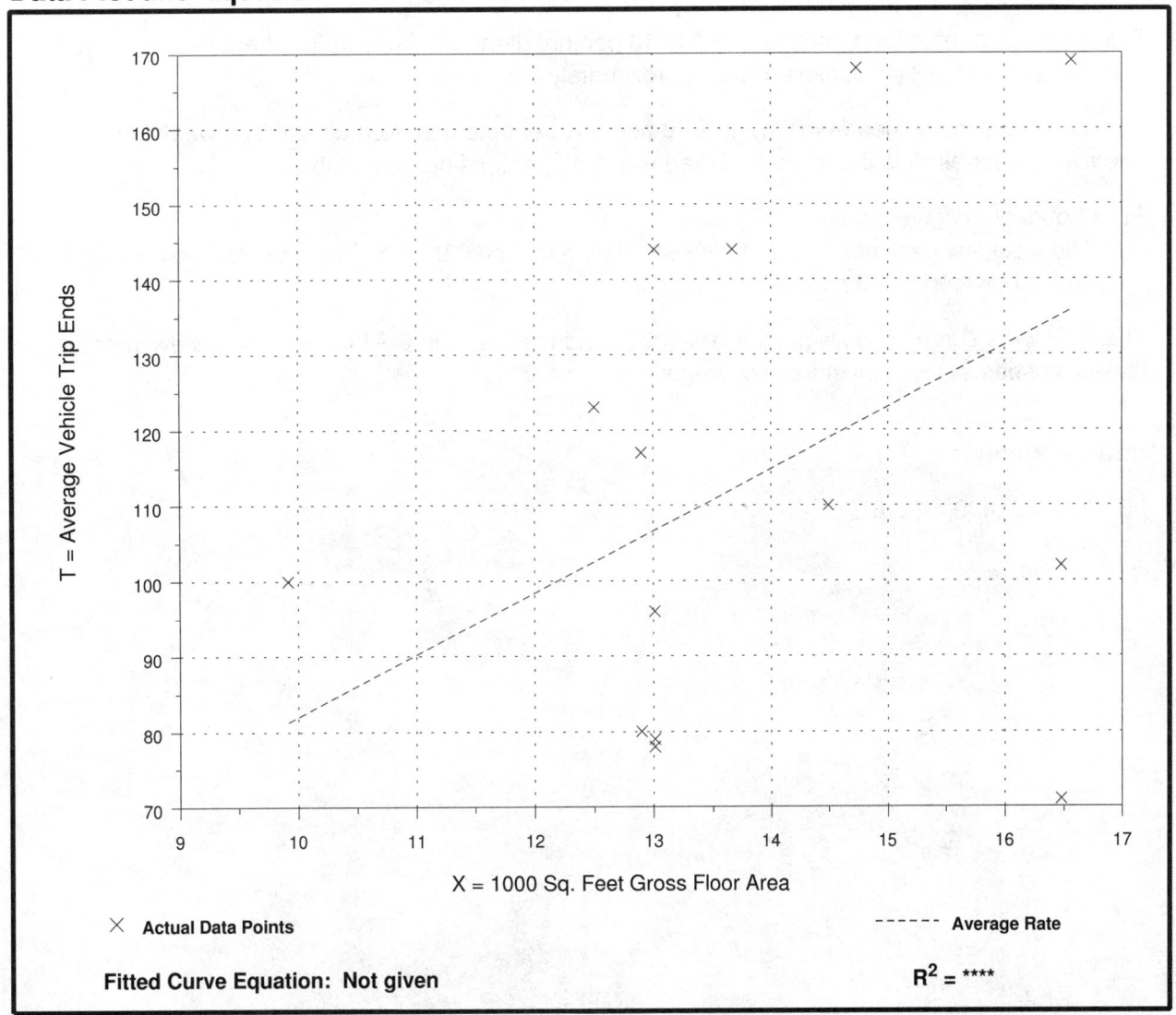

Data Plot and Equation

Fitted Curve Equation: Not given $R^2 = ****$

Trip Generation, 9th Edition • Institute of Transportation Engineers

Land Use: 890
Furniture Store

Description

A furniture store is a full-service retail facility that specializes in the sale of furniture and often carpeting. Furniture stores are generally large and may include storage areas. The sites surveyed included both traditional retail furniture stores and warehouse stores with showrooms. Although some home accessories may be sold, furniture stores primarily focus on the sale of pre-assembled furniture. A majority of items sold at these facilities must be ordered for delivery. Discount home furnishing superstore (Land Use 869) is a related use.

Additional Data

Truck trips accounted for approximately 1 to 13 percent of the weekday traffic at the sites surveyed. The average for the sites surveyed was approximately 5 percent.

Vehicle occupancy ranged from 1.12 to 2.00 persons per automobile on an average weekday. The average for all sites that were surveyed was 1.42 persons per automobile.

Peak hours of the generator—
    The weekday peak hour varied between 10:00 a.m. and 6:00 p.m. The weekend peak hour varied between 12:00 p.m. and 5:00 p.m.

The sites were surveyed between the late 1970s and the 2000s in California, Florida, New York, Massachusetts, New Hampshire and Oregon.

Source Numbers

95, 126, 280, 439, 532, 617

Furniture Store
(890)

Average Vehicle Trip Ends vs: 1000 Sq. Feet Gross Floor Area
On a: Weekday

Number of Studies: 13
Average 1000 Sq. Feet GFA: 69
Directional Distribution: 50% entering, 50% exiting

Trip Generation per 1000 Sq. Feet Gross Floor Area

Average Rate	Range of Rates	Standard Deviation
5.06	0.70 - 15.35	4.38

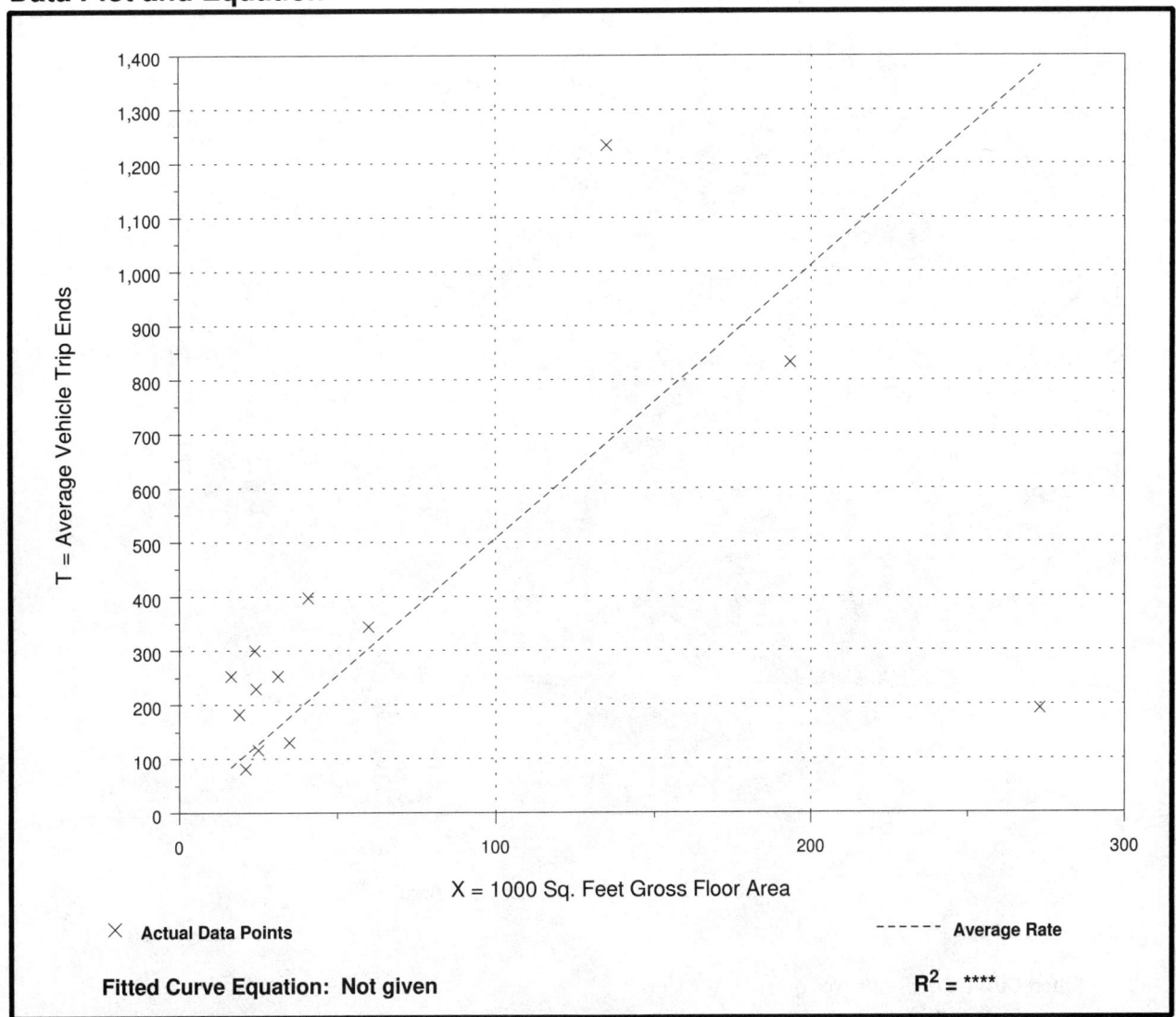

Data Plot and Equation

X Actual Data Points ----- Average Rate

Fitted Curve Equation: Not given $R^2 = ****$

Furniture Store
(890)

Average Vehicle Trip Ends vs: 1000 Sq. Feet Gross Floor Area
On a: Weekday,
Peak Hour of Adjacent Street Traffic,
One Hour Between 7 and 9 a.m.

Number of Studies: 16
Average 1000 Sq. Feet GFA: 64
Directional Distribution: 69% entering, 31% exiting

Trip Generation per 1000 Sq. Feet Gross Floor Area

Average Rate	Range of Rates	Standard Deviation
0.17	0.03 - 0.45	0.44

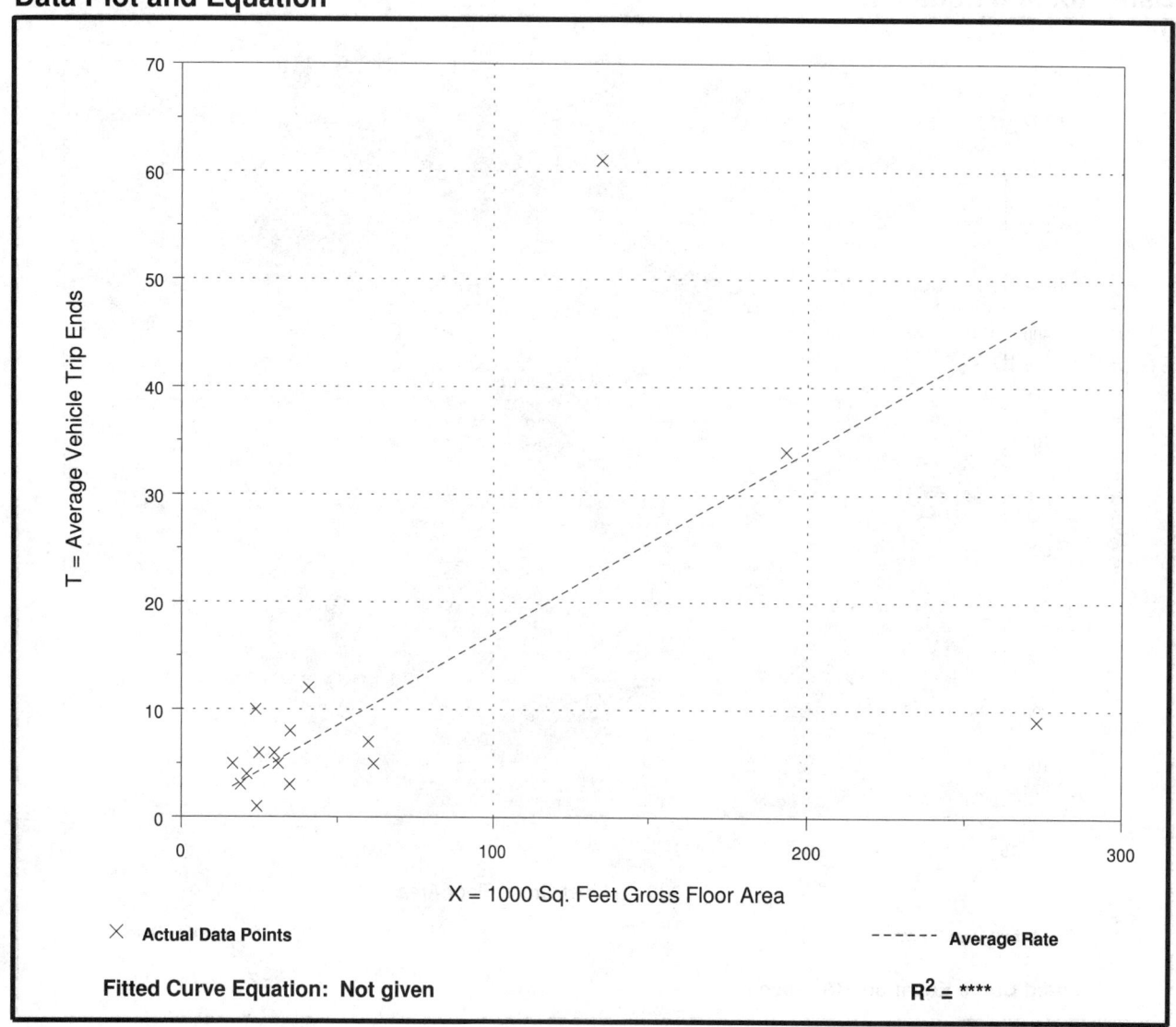

Data Plot and Equation

Fitted Curve Equation: Not given $R^2 = ****$

Furniture Store
(890)

Average Vehicle Trip Ends vs: 1000 Sq. Feet Gross Floor Area
On a: Weekday,
Peak Hour of Adjacent Street Traffic,
One Hour Between 4 and 6 p.m.

Number of Studies: 19
Average 1000 Sq. Feet GFA: 69
Directional Distribution: 48% entering, 52% exiting

Trip Generation per 1000 Sq. Feet Gross Floor Area

Average Rate	Range of Rates	Standard Deviation
0.45	0.06 - 1.70	0.74

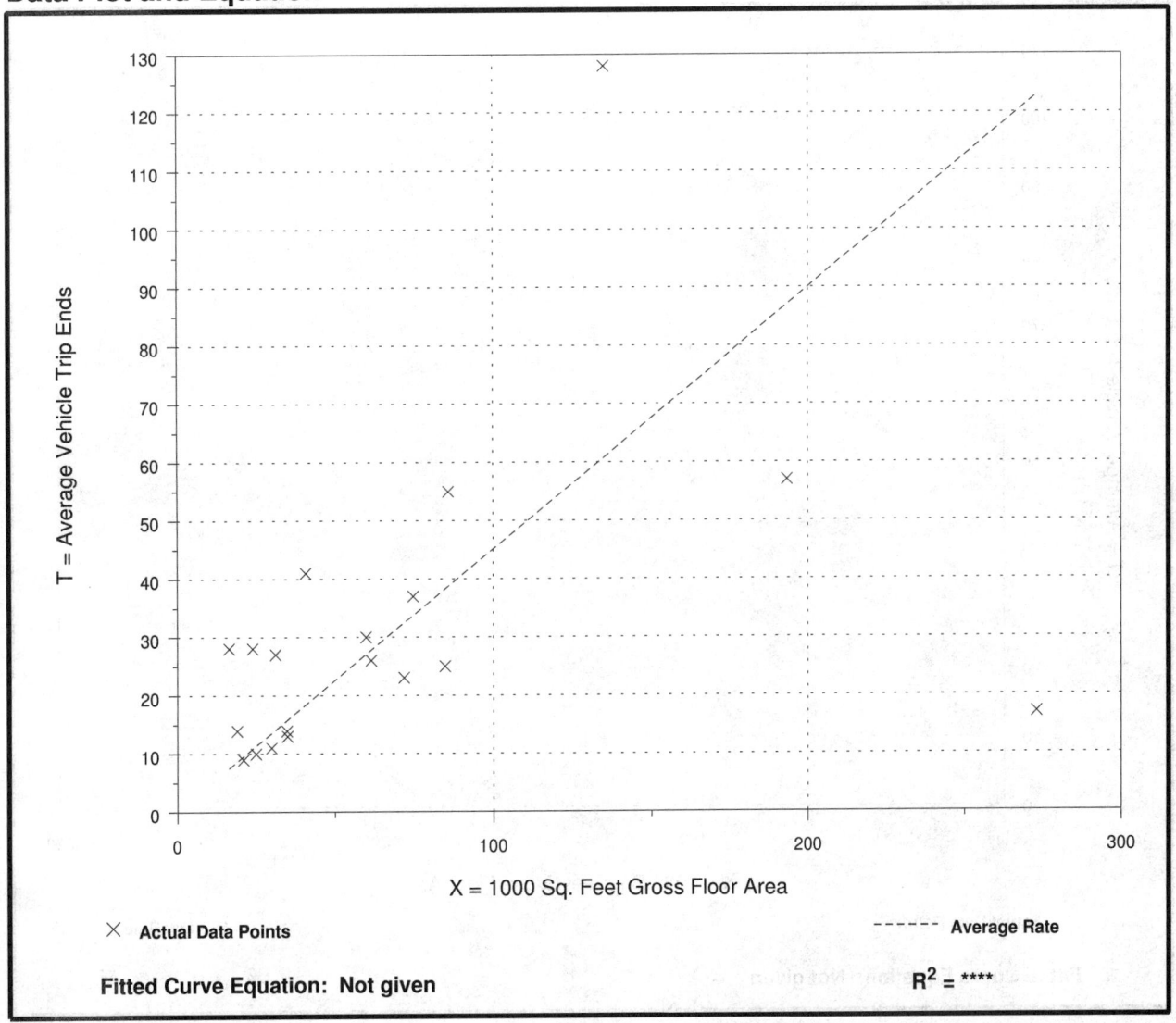

Data Plot and Equation

Fitted Curve Equation: Not given $R^2 = ****$

Furniture Store
(890)

Average Vehicle Trip Ends vs: 1000 Sq. Feet Gross Floor Area
On a: Weekday,
A.M. Peak Hour of Generator

Number of Studies: 16
Average 1000 Sq. Feet GFA: 64
Directional Distribution: 63% entering, 37% exiting

Trip Generation per 1000 Sq. Feet Gross Floor Area

Average Rate	Range of Rates	Standard Deviation
0.40	0.09 - 1.17	0.68

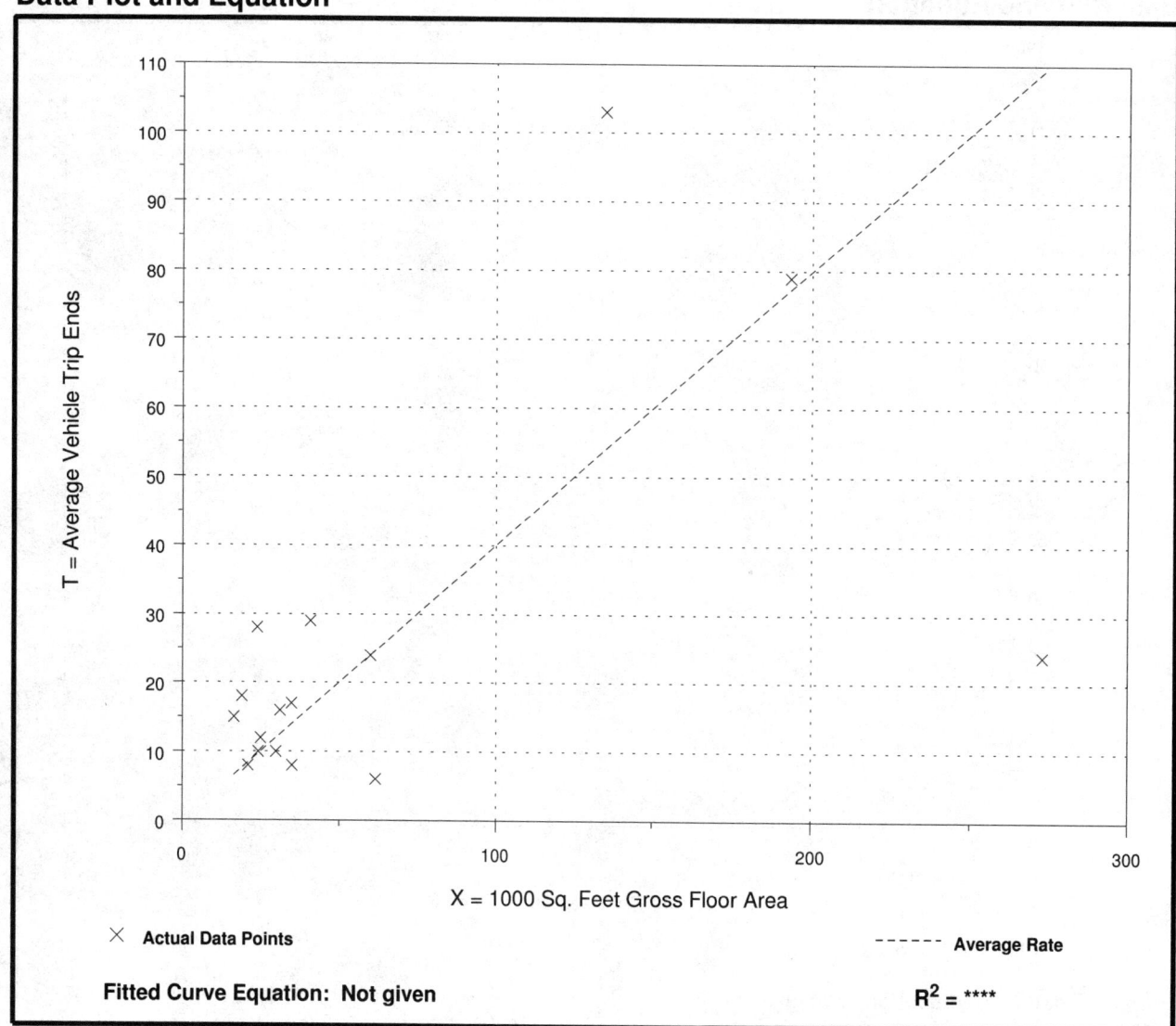

Data Plot and Equation

X   Actual Data Points   - - - - - Average Rate

Fitted Curve Equation: Not given R^2 = ****

Furniture Store
(890)

Average Vehicle Trip Ends vs: 1000 Sq. Feet Gross Floor Area
On a: Weekday,
P.M. Peak Hour of Generator

Number of Studies: 16
Average 1000 Sq. Feet GFA: 67
Directional Distribution: 50% entering, 50% exiting

Trip Generation per 1000 Sq. Feet Gross Floor Area

Average Rate	Range of Rates	Standard Deviation
0.53	0.09 - 1.70	0.81

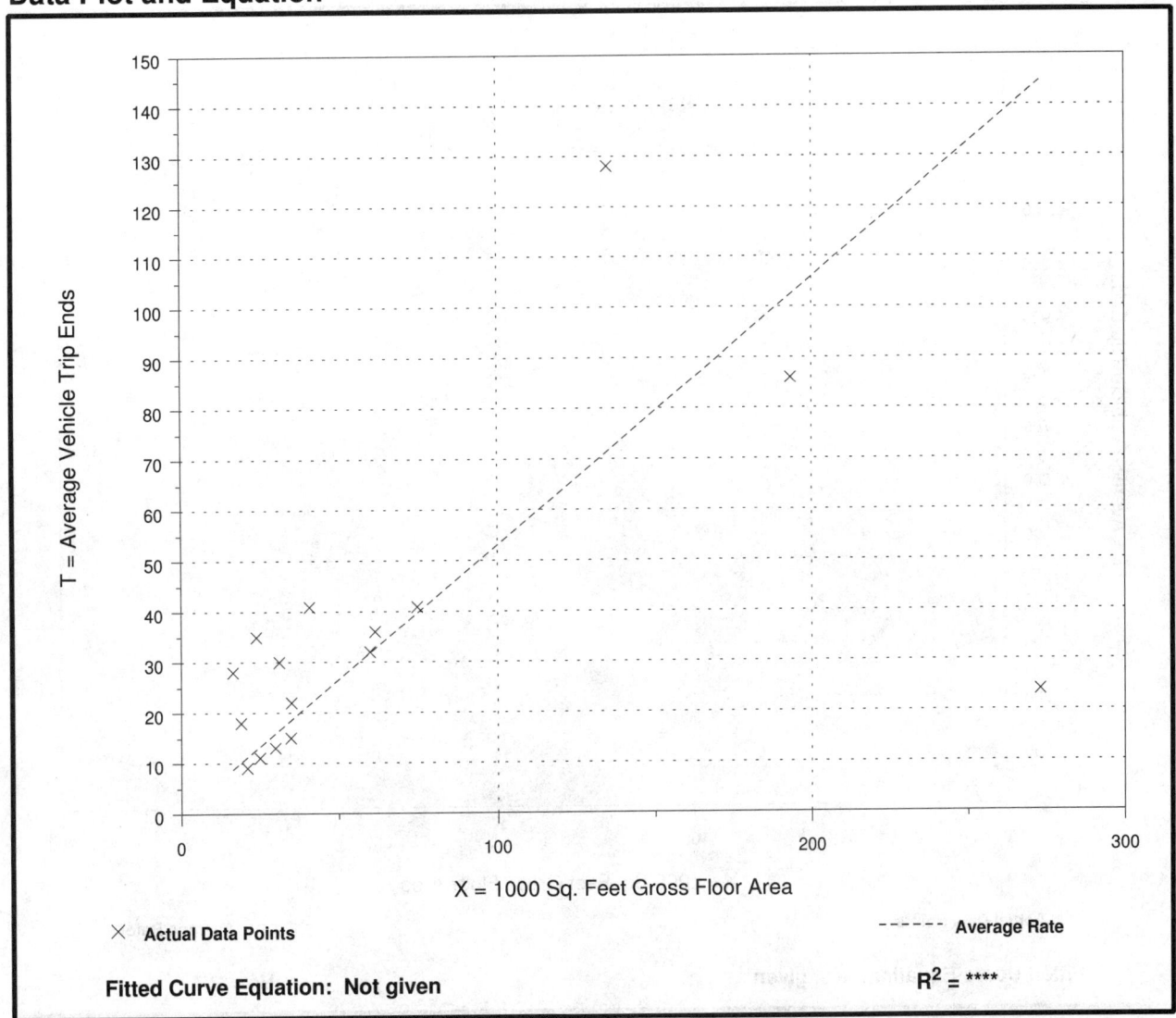

Data Plot and Equation

Fitted Curve Equation: Not given $R^2 = ****$

Furniture Store
(890)

Average Vehicle Trip Ends vs: 1000 Sq. Feet Gross Floor Area
On a: Saturday

Number of Studies: 9
Average 1000 Sq. Feet GFA: 87
Directional Distribution: 50% entering, 50% exiting

Trip Generation per 1000 Sq. Feet Gross Floor Area

Average Rate	Range of Rates	Standard Deviation
4.94	0.78 - 13.96	4.26

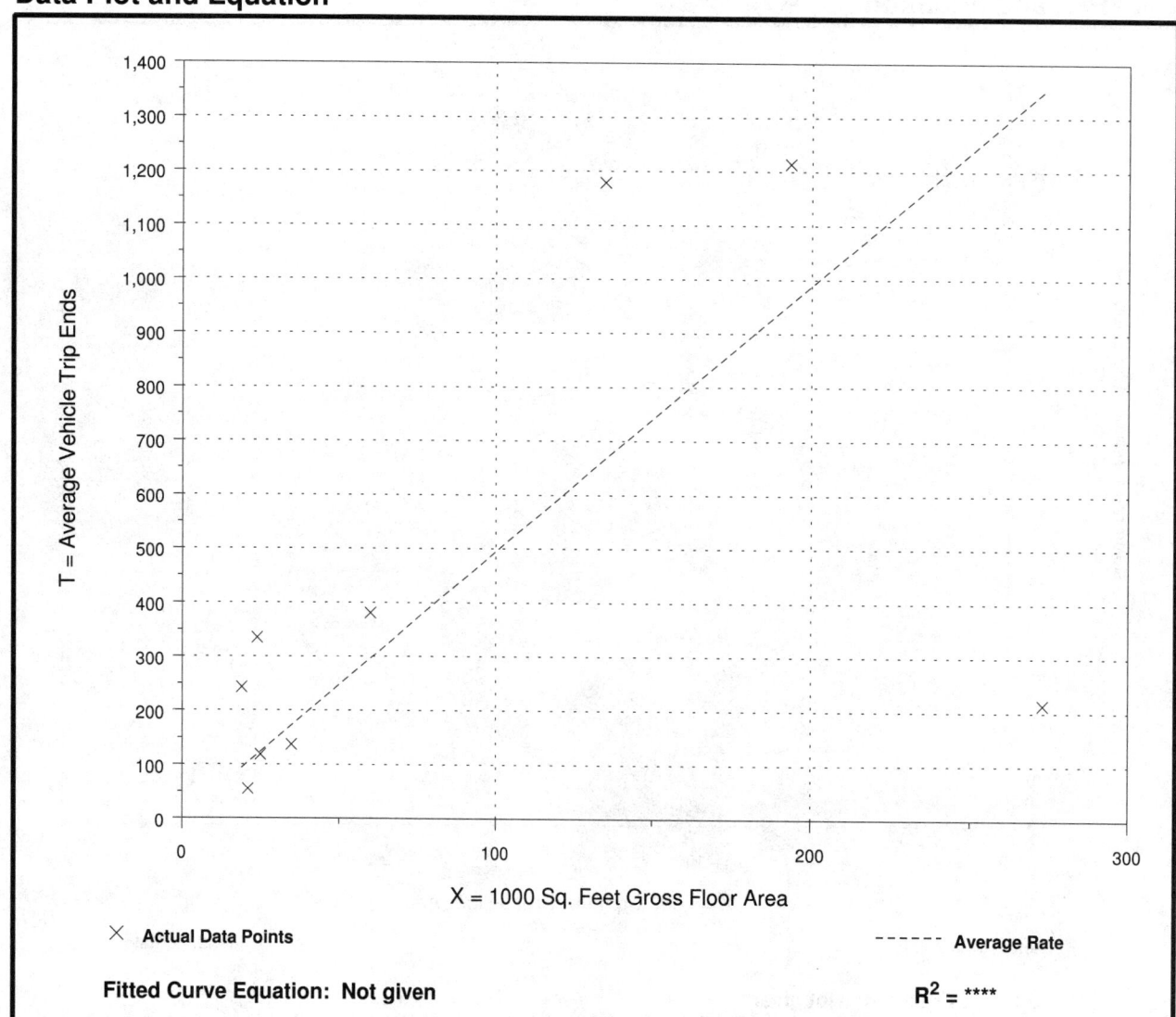

Data Plot and Equation

Fitted Curve Equation: Not given $R^2 =$ ****

Furniture Store
(890)

Average Vehicle Trip Ends vs: 1000 Sq. Feet Gross Floor Area
On a: Saturday,
Peak Hour of Generator

Number of Studies: 16
Average 1000 Sq. Feet GFA: 77
Directional Distribution: 54% entering, 46% exiting

Trip Generation per 1000 Sq. Feet Gross Floor Area

Average Rate	Range of Rates	Standard Deviation
0.95	0.15 - 2.79	1.13

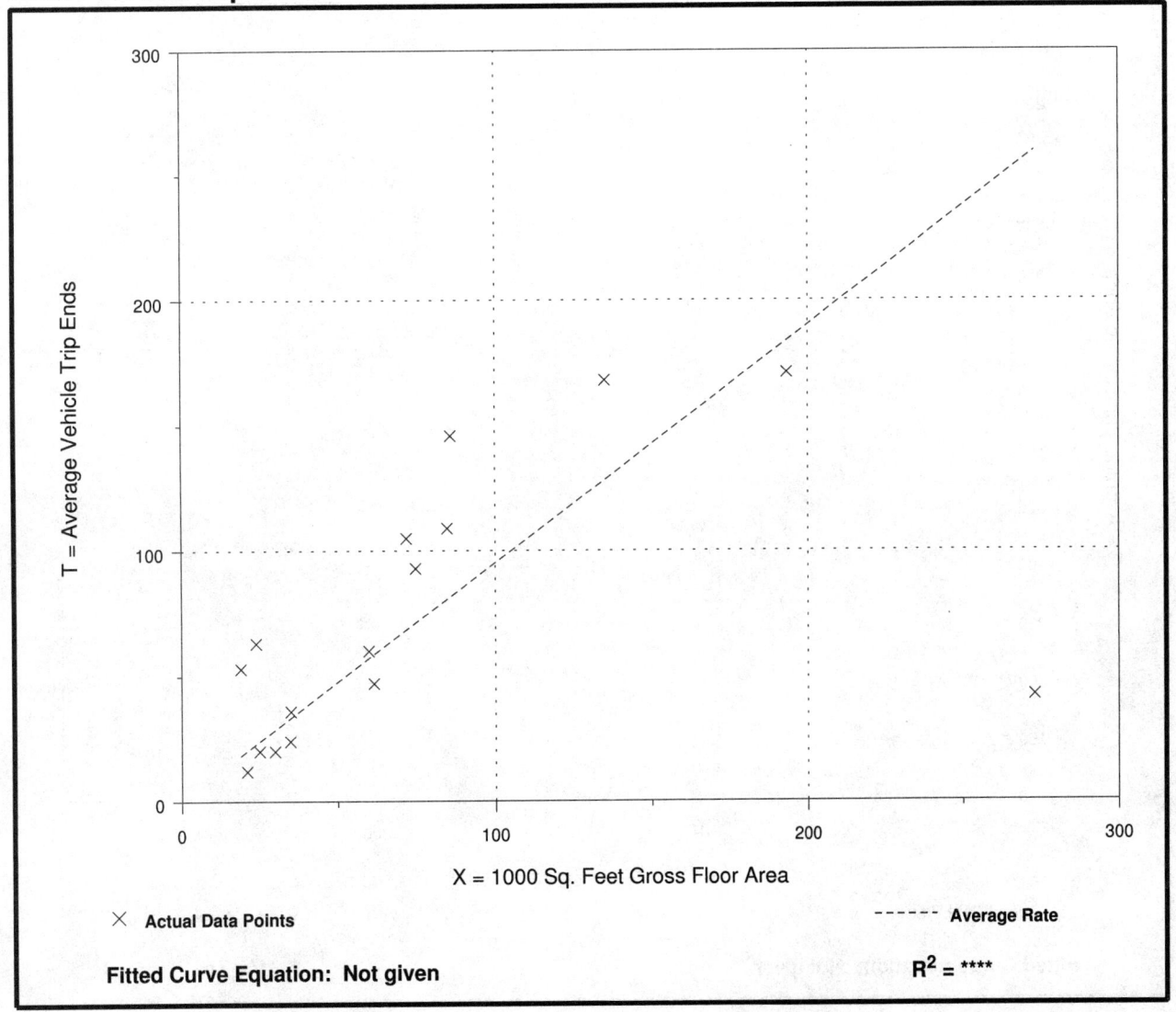

Data Plot and Equation

Fitted Curve Equation: Not given $R^2 = ****$

Furniture Store
(890)

Average Vehicle Trip Ends vs: 1000 Sq. Feet Gross Floor Area
On a: Sunday

Number of Studies: 9
Average 1000 Sq. Feet GFA: 87
Directional Distribution: 50% entering, 50% exiting

Trip Generation per 1000 Sq. Feet Gross Floor Area

Average Rate	Range of Rates	Standard Deviation
4.64	0.14 - 14.17	4.28

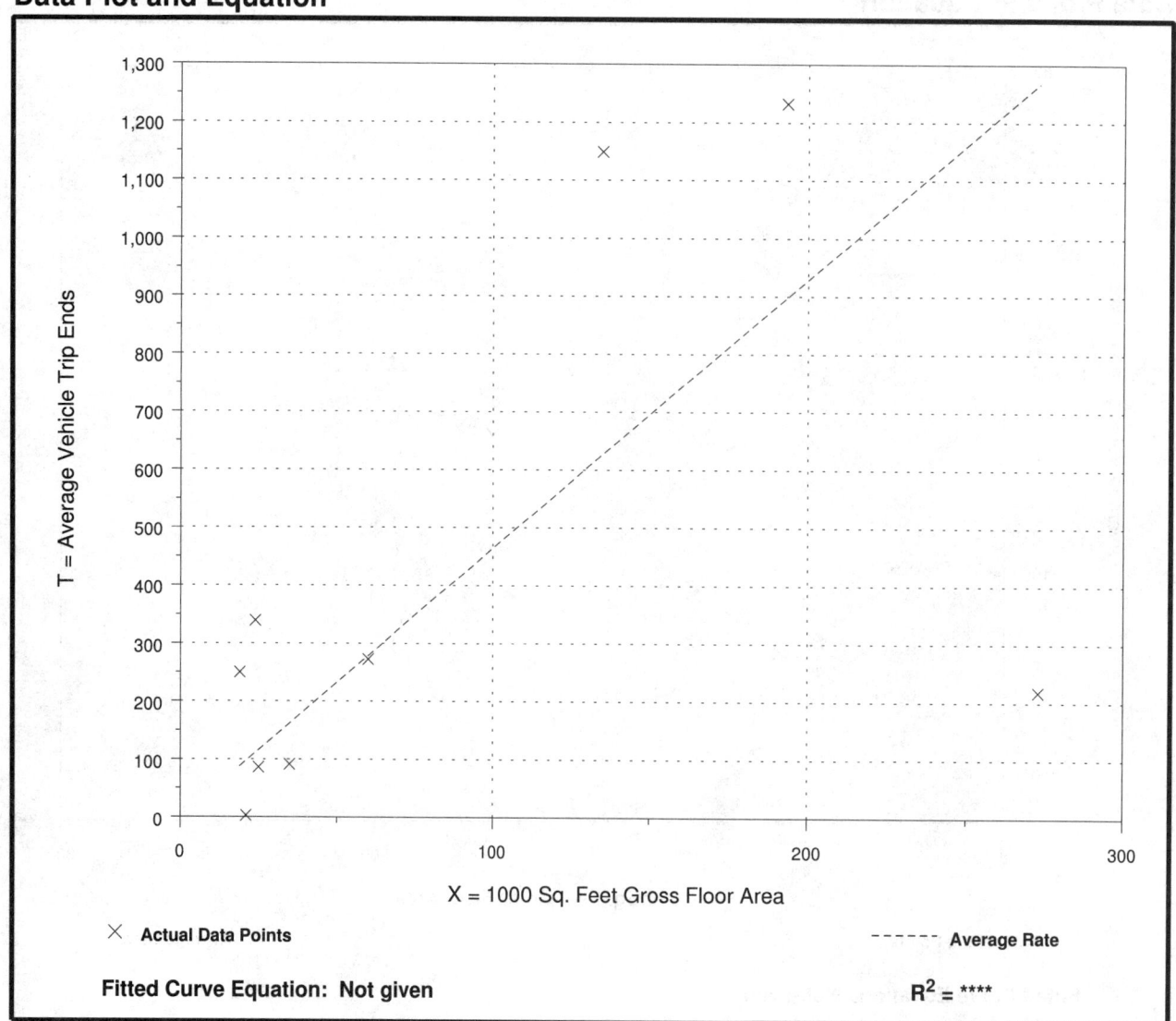

Data Plot and Equation

× Actual Data Points   ----- Average Rate

Fitted Curve Equation: Not given $R^2 = ****$

Furniture Store
(890)

Average Vehicle Trip Ends vs: 1000 Sq. Feet Gross Floor Area
On a: Sunday,
Peak Hour of Generator

Number of Studies: 9
Average 1000 Sq. Feet GFA: 87
Directional Distribution: Not available

Trip Generation per 1000 Sq. Feet Gross Floor Area

Average Rate	Range of Rates	Standard Deviation
0.92	0.10 - 3.42	1.18

Data Plot and Equation

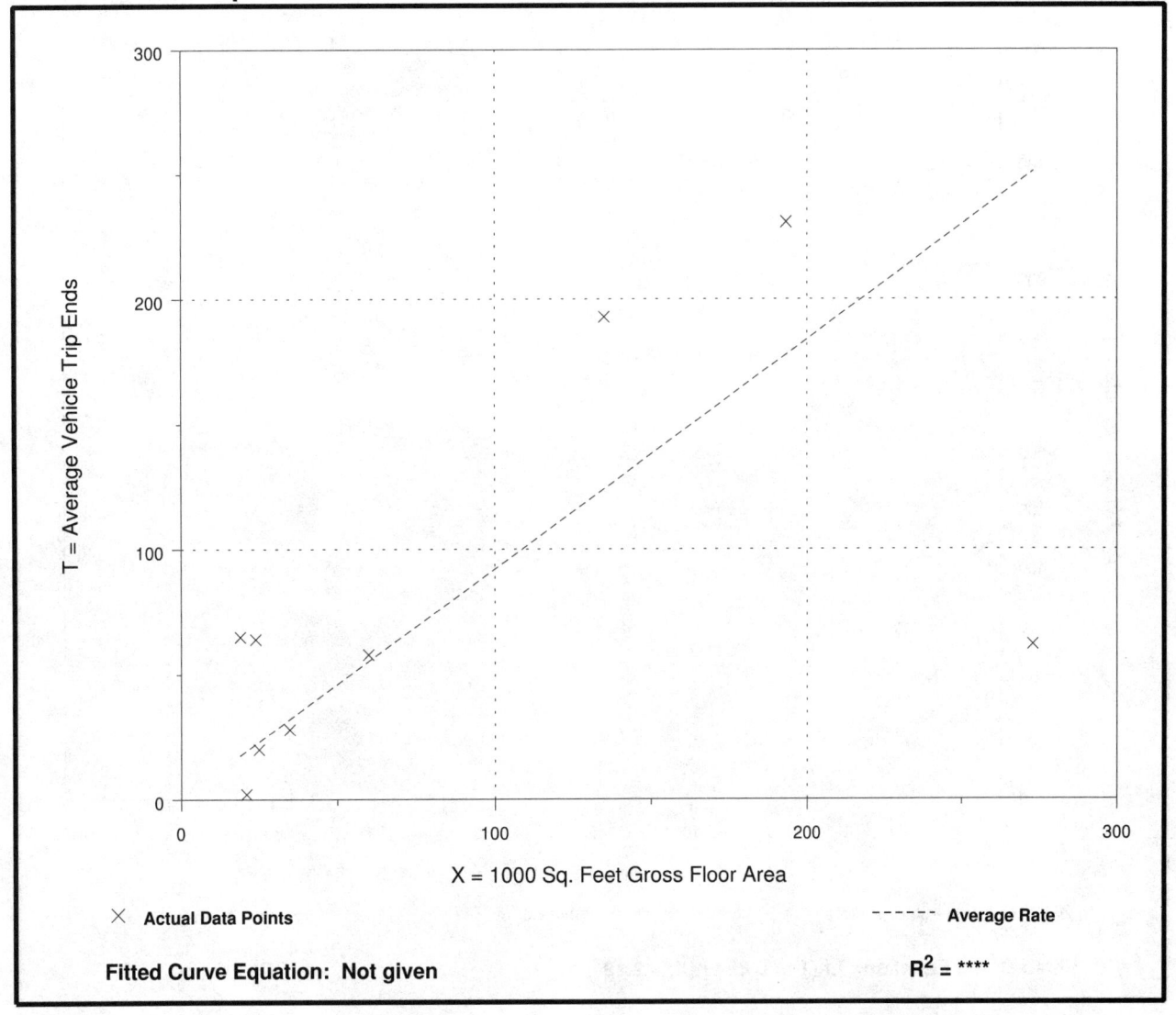

Fitted Curve Equation: Not given $R^2 = ****$

Furniture Store
(890)

Average Vehicle Trip Ends vs: Employees
On a: Weekday

Number of Studies: 8
Avg. Number of Employees: 33
Directional Distribution: 50% entering, 50% exiting

Trip Generation per Employee

Average Rate	Range of Rates	Standard Deviation
12.19	5.80 - 30.00	7.22

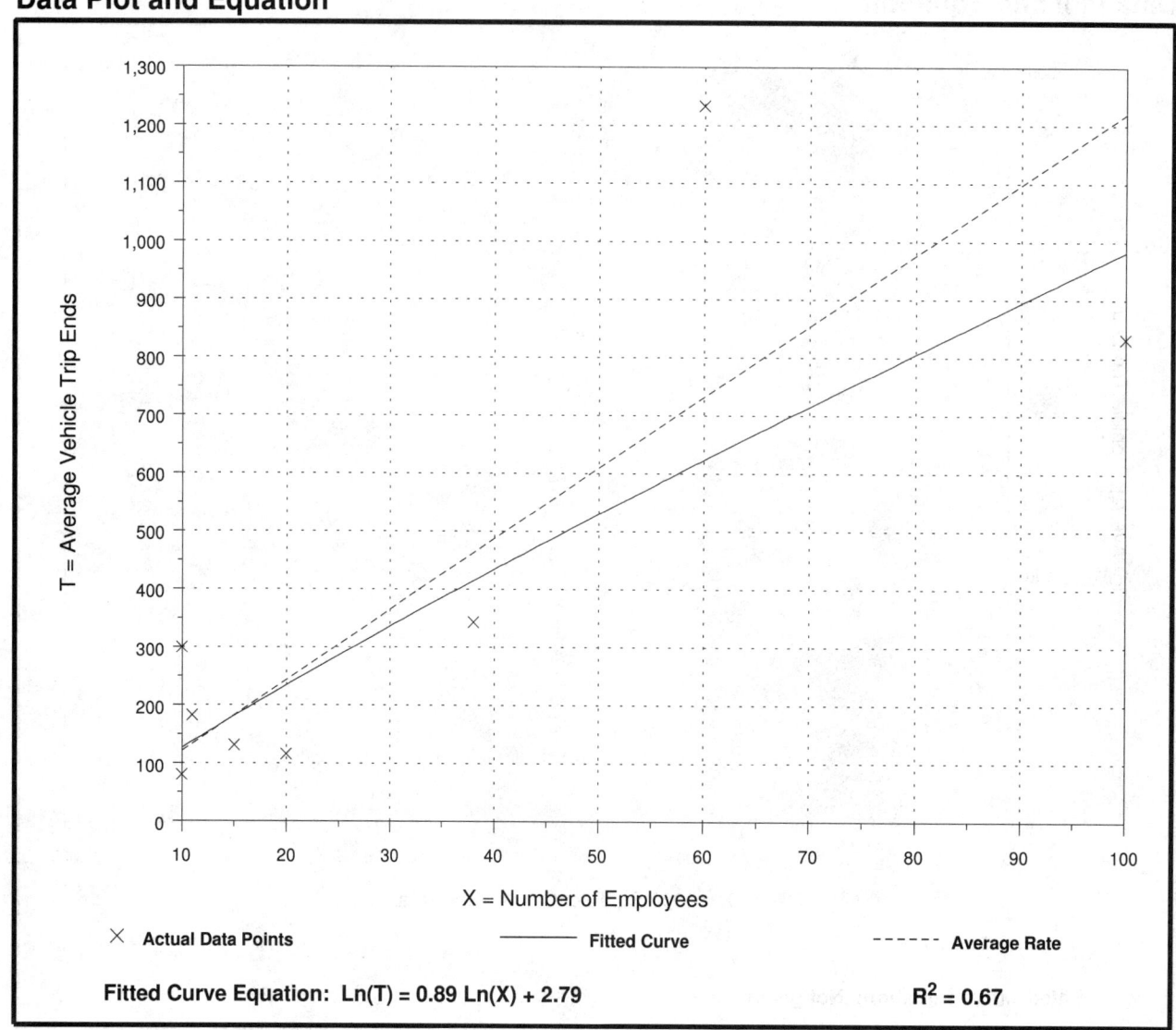

Data Plot and Equation

Fitted Curve Equation: $Ln(T) = 0.89\ Ln(X) + 2.79$ $R^2 = 0.67$

Furniture Store
(890)

Average Vehicle Trip Ends vs: Employees
On a: Weekday,
Peak Hour of Adjacent Street Traffic,
One Hour Between 7 and 9 a.m.

Number of Studies: 8
Avg. Number of Employees: 33
Directional Distribution: Not available

Trip Generation per Employee

Average Rate	Range of Rates	Standard Deviation
0.48	0.18 - 1.02	0.76

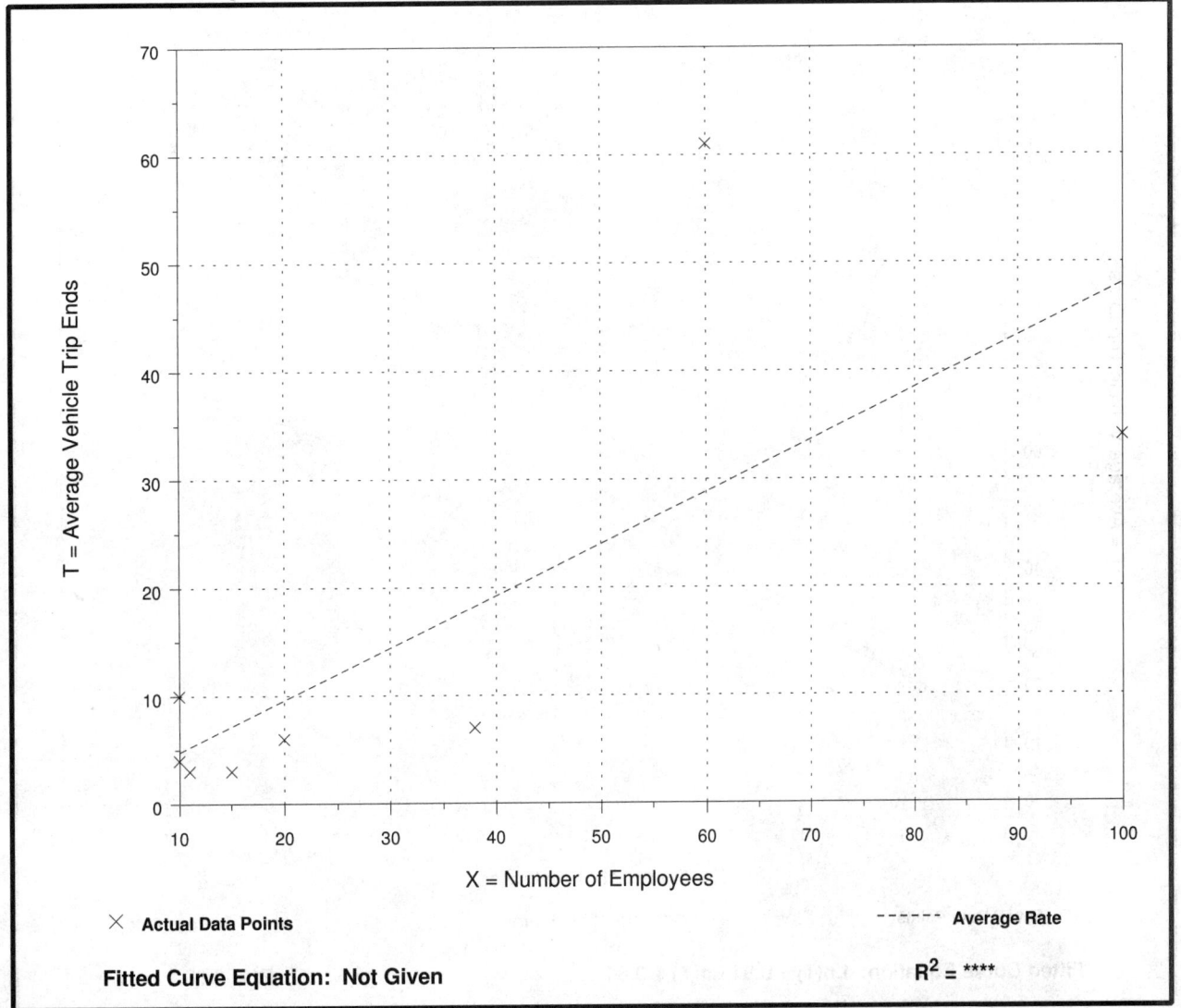

Data Plot and Equation

Fitted Curve Equation: Not Given $R^2 = ****$

Furniture Store
(890)

Average Vehicle Trip Ends vs: Employees
On a: Weekday,
Peak Hour of Adjacent Street Traffic,
One Hour Between 4 and 6 p.m.

Number of Studies: 8
Avg. Number of Employees: 33
Directional Distribution: Not available

Trip Generation per Employee

Average Rate	Range of Rates	Standard Deviation
1.10	0.50 - 2.80	1.25

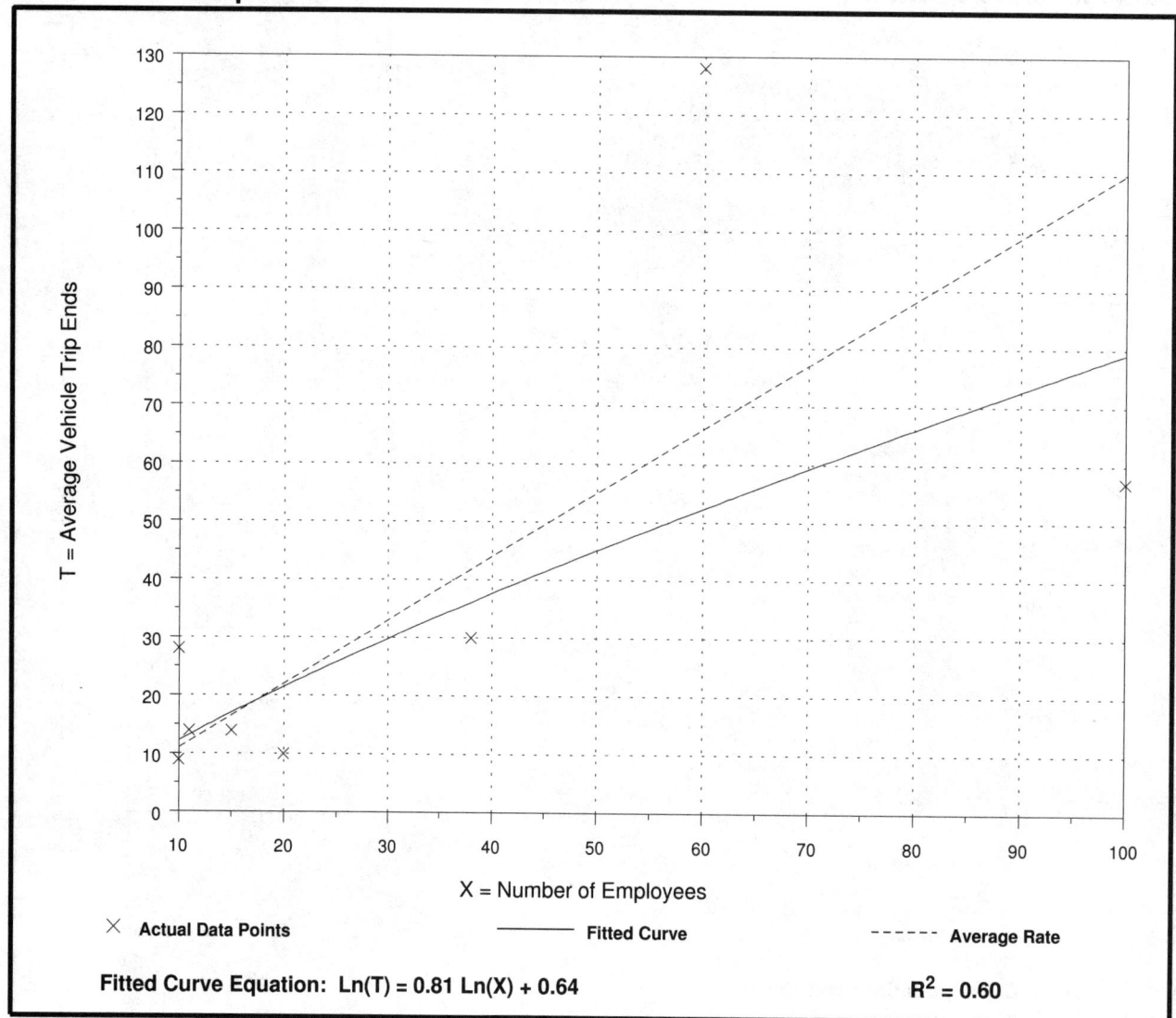

Data Plot and Equation

Fitted Curve Equation: $Ln(T) = 0.81\ Ln(X) + 0.64$ $R^2 = 0.60$

Furniture Store
(890)

Average Vehicle Trip Ends vs: Employees
On a: Weekday,
A.M. Peak Hour of Generator

Number of Studies: 8
Avg. Number of Employees: 33
Directional Distribution: Not available

Trip Generation per Employee

Average Rate	Range of Rates	Standard Deviation
1.09	0.60 - 2.80	1.17

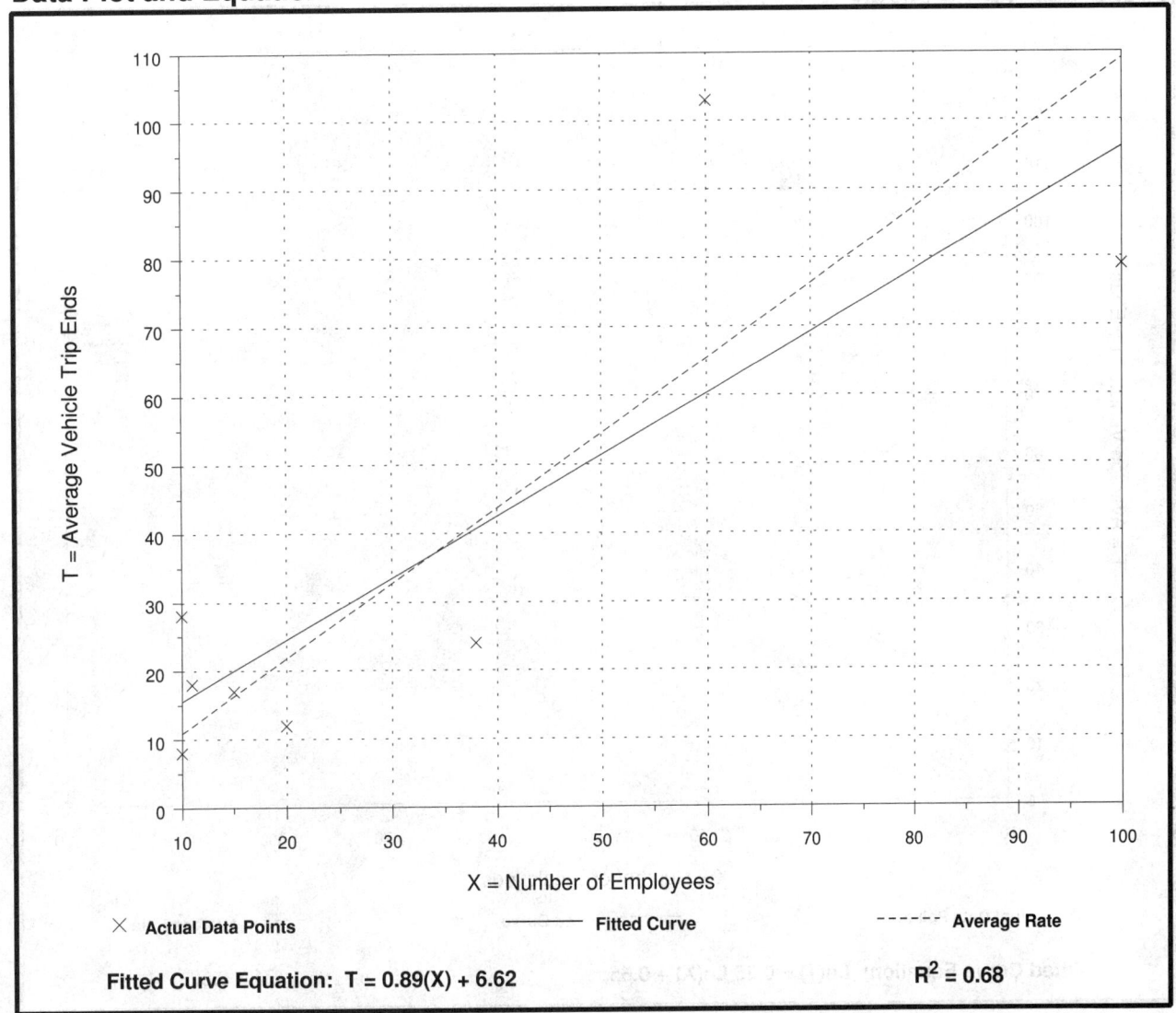

Data Plot and Equation

Fitted Curve Equation: T = 0.89(X) + 6.62 $R^2 = 0.68$

Furniture Store
(890)

Average Vehicle Trip Ends vs: Employees
On a: Weekday,
P.M. Peak Hour of Generator

Number of Studies: 8
Avg. Number of Employees: 33
Directional Distribution: Not available

Trip Generation per Employee

Average Rate	Range of Rates	Standard Deviation
1.27	0.55 - 3.50	1.32

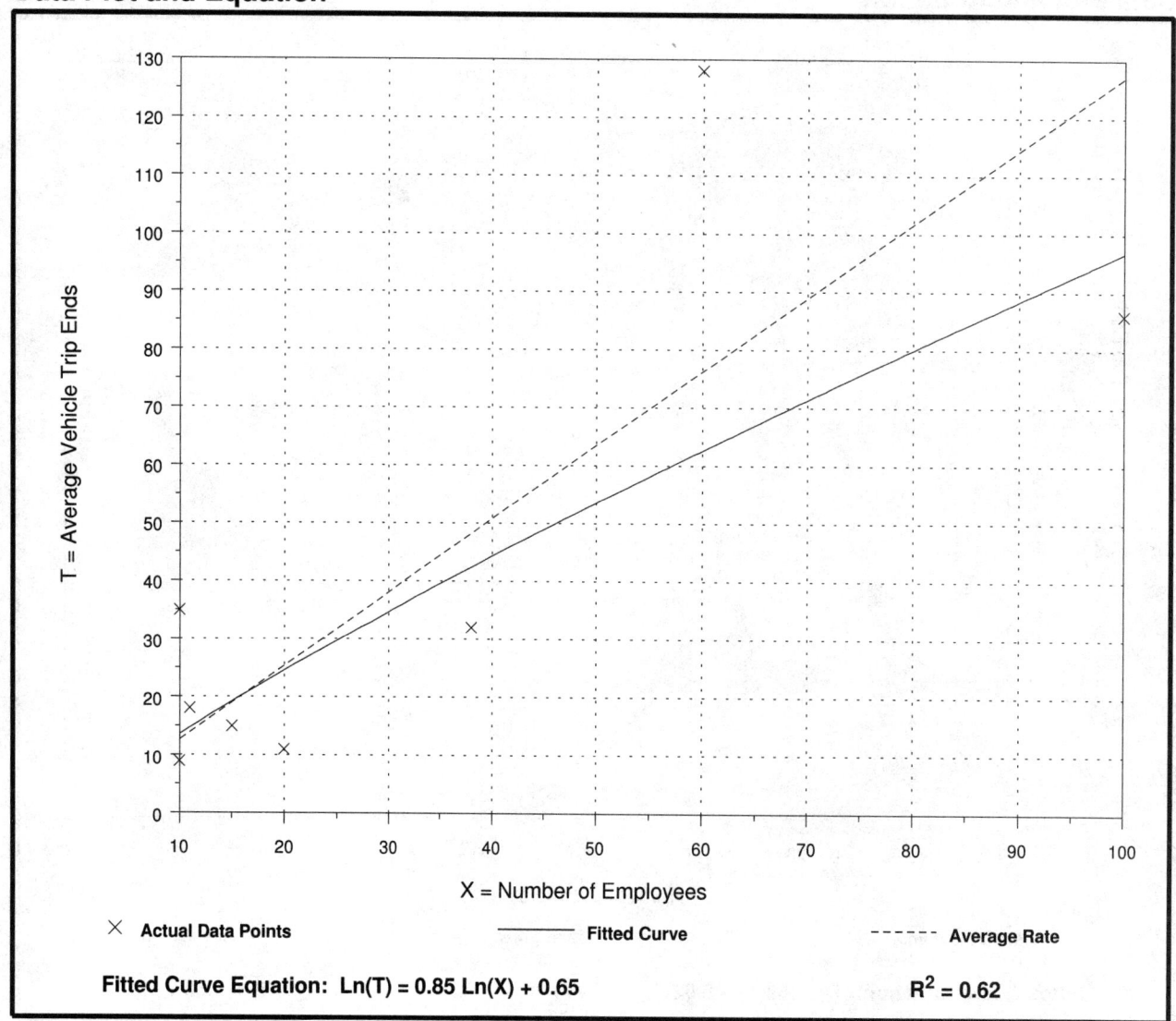

Data Plot and Equation

Fitted Curve Equation: $Ln(T) = 0.85\ Ln(X) + 0.65$ $R^2 = 0.62$

Furniture Store
(890)

Average Vehicle Trip Ends vs: Employees
On a: Saturday

Number of Studies: 8
Avg. Number of Employees: 33
Directional Distribution: 50% entering, 50% exiting

Trip Generation per Employee

Average Rate	Range of Rates	Standard Deviation
13.87	5.50 - 33.50	7.11

Data Plot and Equation

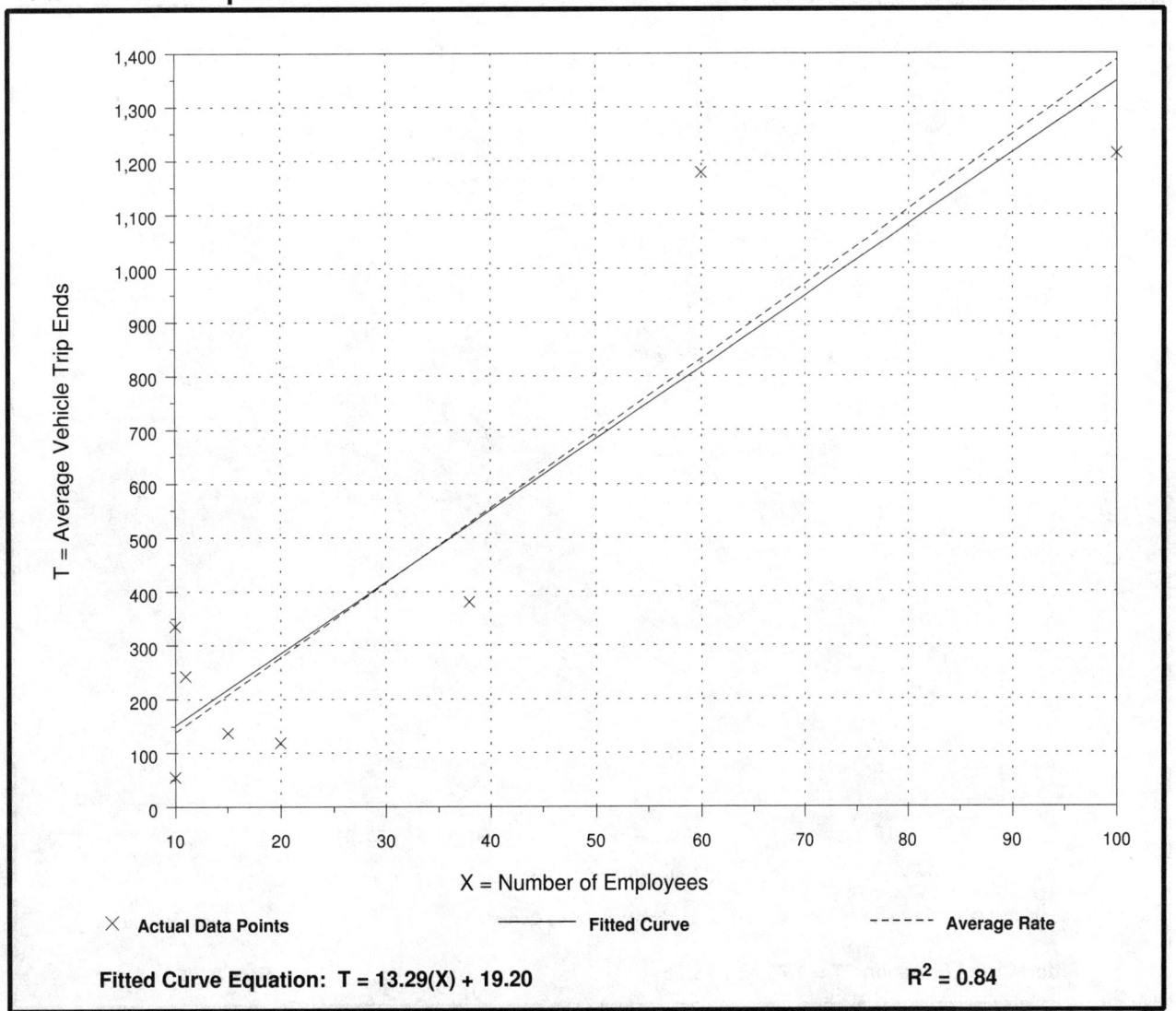

Fitted Curve Equation: $T = 13.29(X) + 19.20$ $R^2 = 0.84$

Furniture Store
(890)

Average Vehicle Trip Ends vs: Employees
On a: Saturday,
Peak Hour of Generator

Number of Studies: 8
Avg. Number of Employees: 33
Directional Distribution: Not available

Trip Generation per Employee

Average Rate	Range of Rates	Standard Deviation
2.16	1.00 - 6.30	1.85

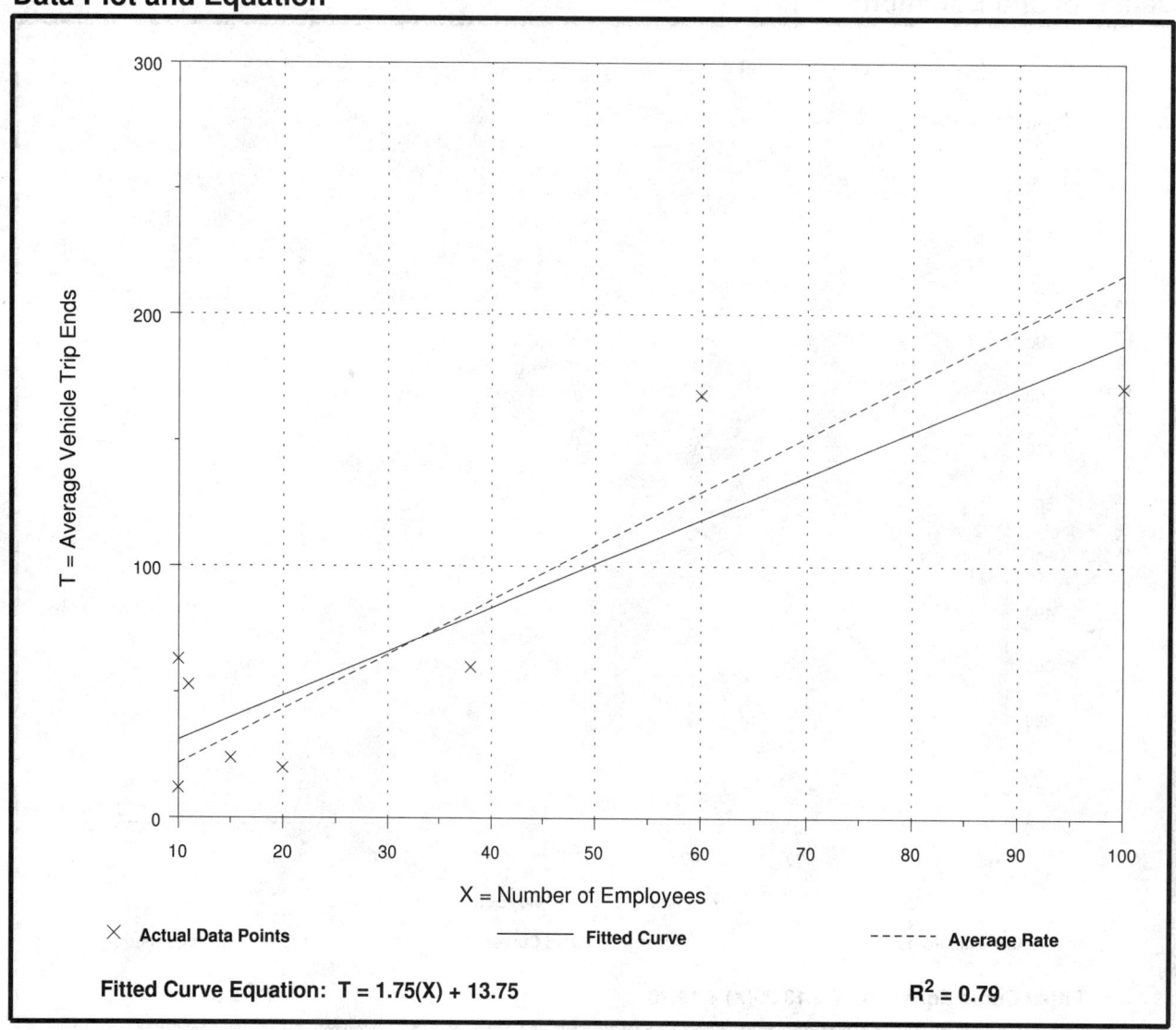

Data Plot and Equation

Fitted Curve Equation: $T = 1.75(X) + 13.75$ $R^2 = 0.79$

Furniture Store
(890)

Average Vehicle Trip Ends vs: Employees
On a: Sunday

Number of Studies: 8
Avg. Number of Employees: 33
Directional Distribution: 50% entering, 50% exiting

Trip Generation per Employee

Average Rate	Range of Rates	Standard Deviation
12.97	0.30 - 34.00	7.84

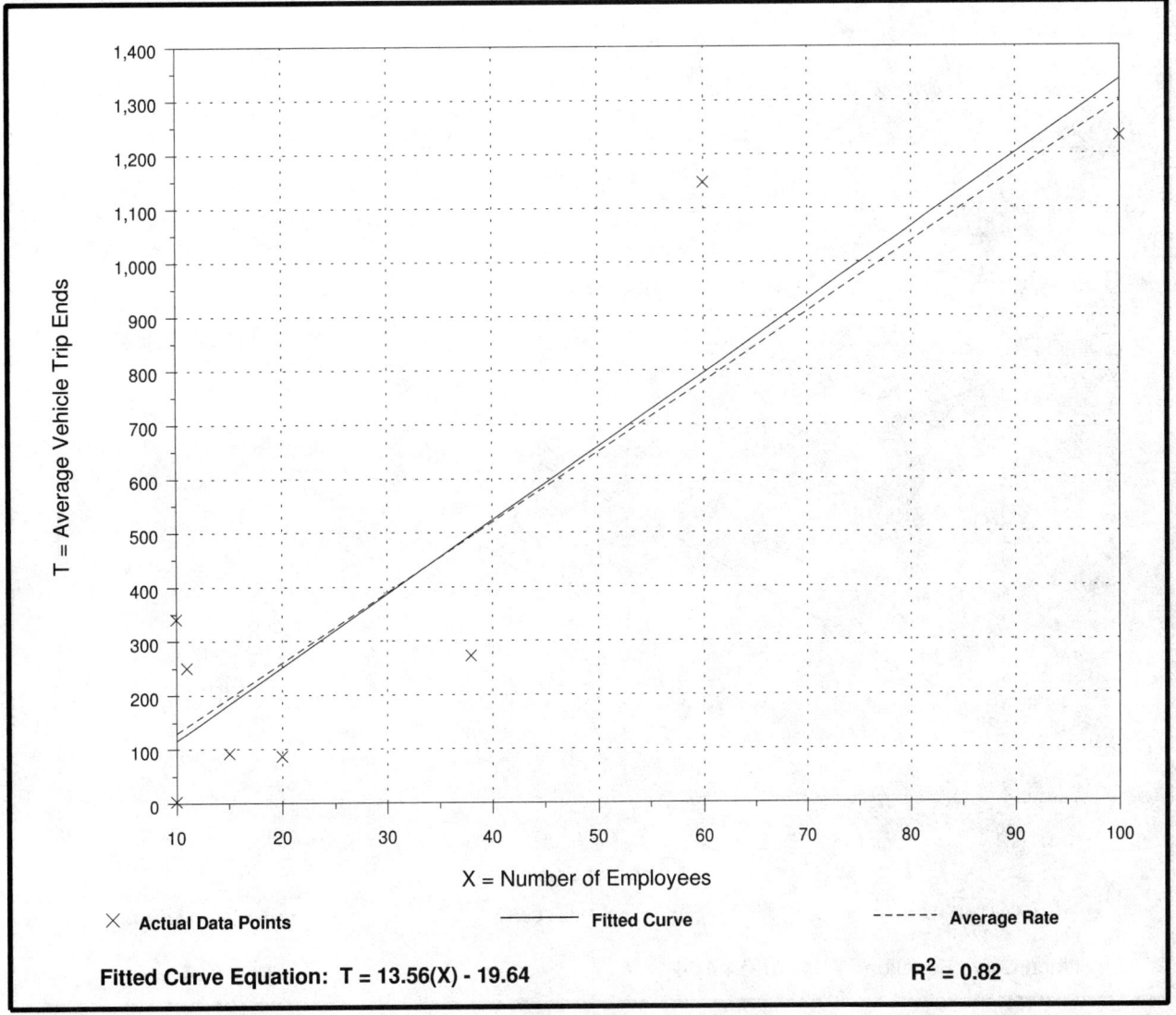

Data Plot and Equation

Fitted Curve Equation: $T = 13.56(X) - 19.64$ $R^2 = 0.82$

Furniture Store
(890)

Average Vehicle Trip Ends vs: Employees
On a: Sunday,
Peak Hour of Generator

Number of Studies: 8
Avg. Number of Employees: 33
Directional Distribution: Not available

Trip Generation per Employee

Average Rate	Range of Rates	Standard Deviation
2.50	0.20 - 6.40	2.04

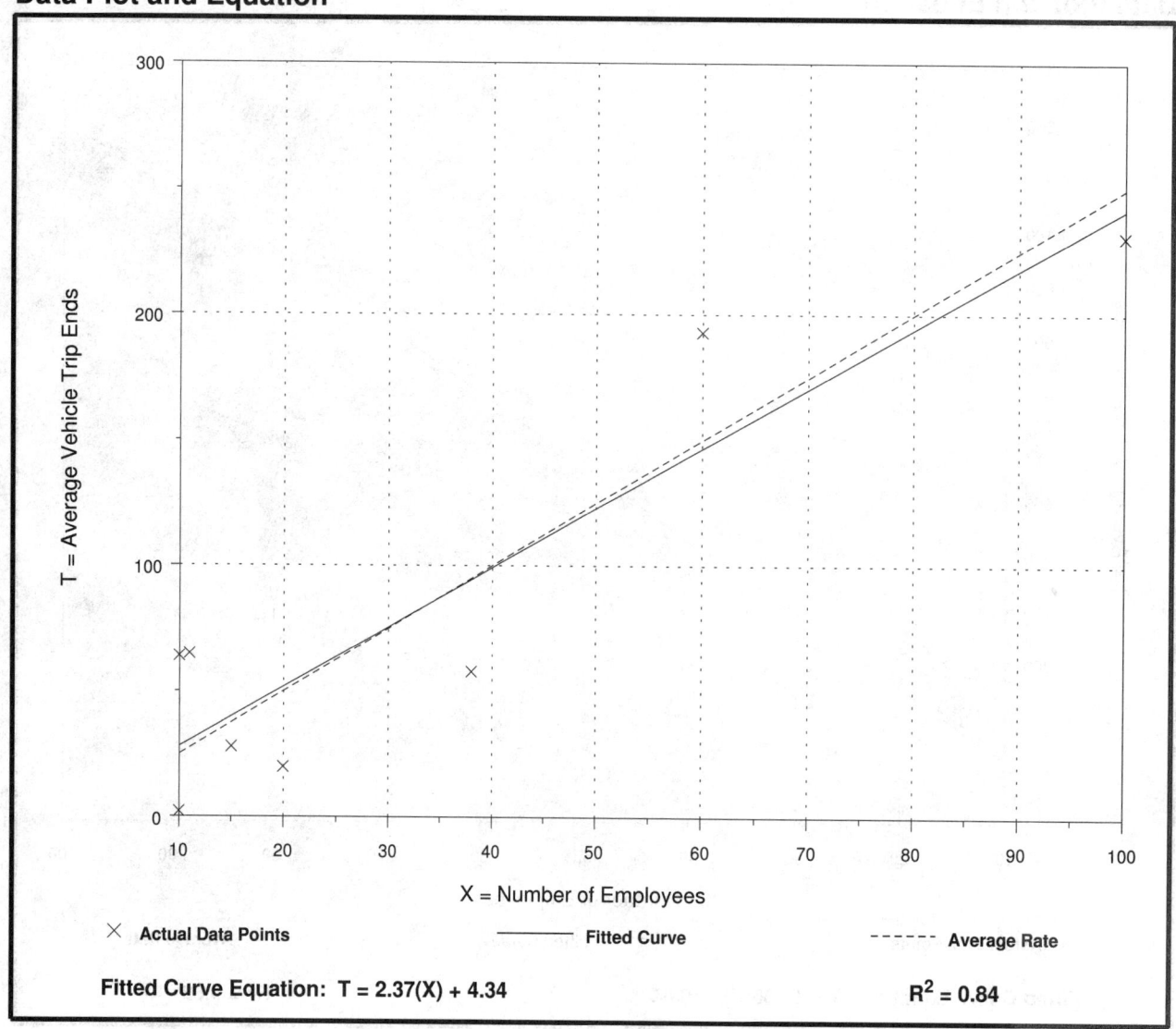

Data Plot and Equation

Fitted Curve Equation: T = 2.37(X) + 4.34 $R^2 = 0.84$

Land Use: 896
DVD/Video Rental Store

Description

DVD/video rental stores are businesses specializing in the rental of home movies and video games. Movies and video games may also be available for purchase. DVD/video rental stores typically maintain long store hours and are usually open 7 days a week.

Additional Data

Friday trip generation was typically higher than other weekdays. Caution should be exercised when using these data, as they contain studies taken from Friday as well as other weekdays. One study indicated that a site generated approximately two times as many trips during the Friday P.M. peak hour of adjacent street traffic as during the Thursday P.M. peak hour of adjacent street traffic.

The sites were surveyed in the 1990s in Pennsylvania, Utah and Wisconsin.

Source Numbers

372, 387, 399, 432

Land Use: 896
DVD/Video Rental Store
Independent Variables with One Observation

The following trip generation data are for independent variables with only one observation. This information is shown in this table only; there are no related plots for these data.

Users are cautioned to use data with care because of the small sample size.

Independent Variable	Trip Generation Rate	Size of Independent Variable	Number of Studies	Directional Distribution
1,000 Square Feet Gross Floor Area				
Weekday P.M. Peak Hour of Generator	31.54	7	1	50% entering, 50% exiting
Saturday Peak Hour of Generator	26.92	7	1	46% entering, 54% exiting

DVD/Video Rental Store
(896)

Average Vehicle Trip Ends vs: 1000 Sq. Feet Gross Floor Area
On a: Weekday,
Peak Hour of Adjacent Street Traffic,
One Hour Between 4 and 6 p.m.

Number of Studies: 6
Average 1000 Sq. Feet GFA: 5
Directional Distribution: 46% entering, 54% exiting

Trip Generation per 1000 Sq. Feet Gross Floor Area

Average Rate	Range of Rates	Standard Deviation
13.60	6.94 - 20.15	6.13

Data Plot and Equation

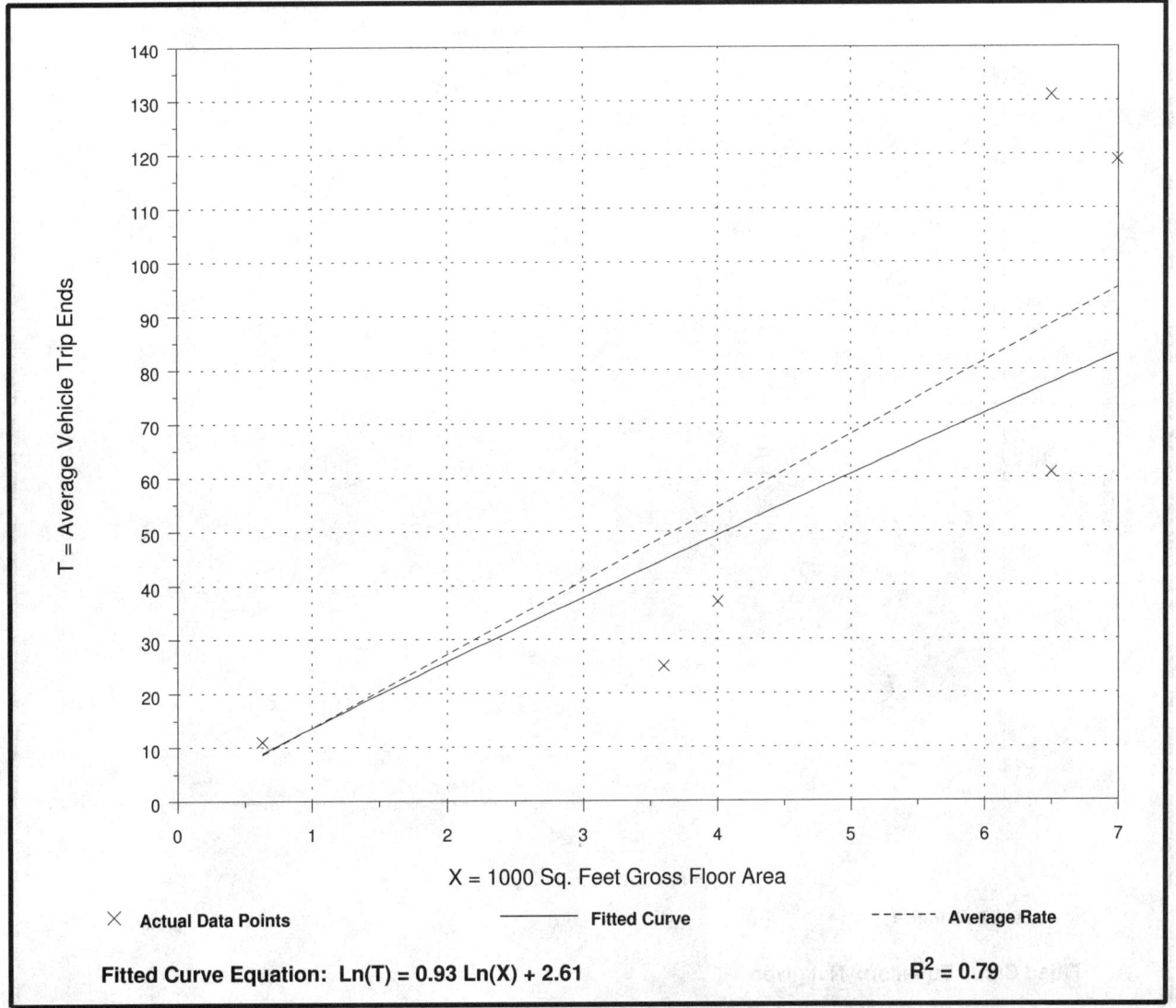

Fitted Curve Equation: $Ln(T) = 0.93\ Ln(X) + 2.61$ $R^2 = 0.79$

DVD/Video Rental Store
(896)

Average Vehicle Trip Ends vs: Employees
On a: Weekday,
Peak Hour of Adjacent Street Traffic,
One Hour Between 4 and 6 p.m.

Number of Studies: 2
Avg. Number of Employees: 4
Directional Distribution: Not available

Trip Generation per Employee

Average Rate	Range of Rates	Standard Deviation
6.00	5.50 - 6.17	*

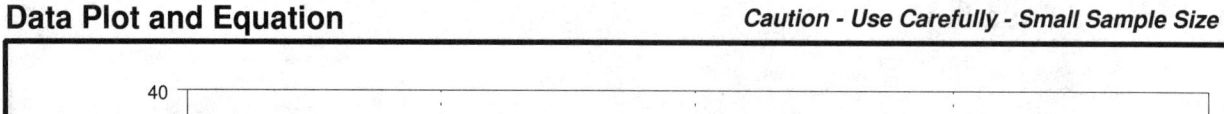

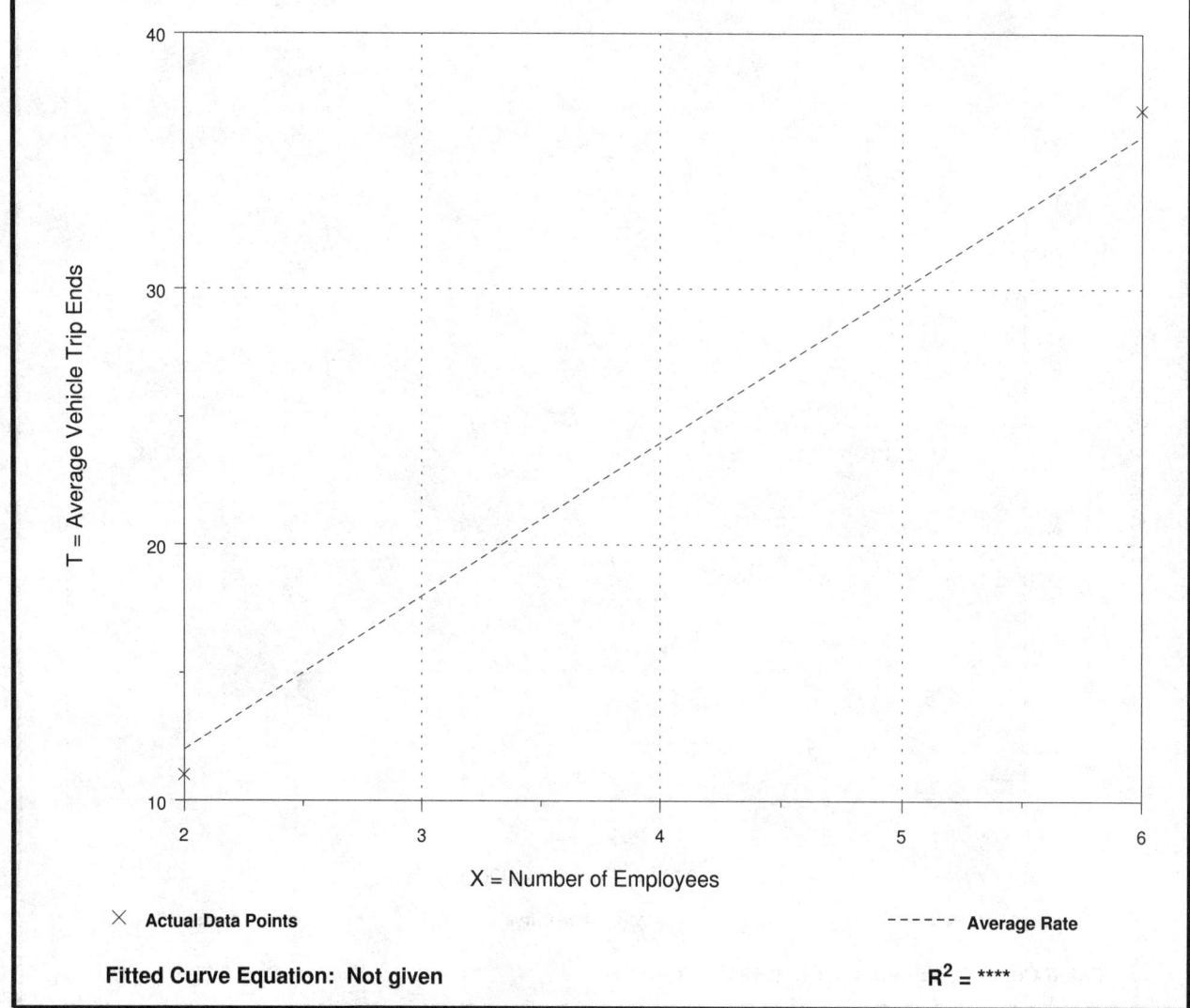

Data Plot and Equation

Caution - Use Carefully - Small Sample Size

Fitted Curve Equation: Not given $R^2 = ****$

Land Use: 897
Medical Equipment Store

Description

A medical equipment store is a free-standing building that specializes in the sales of medical equipment, such as hospital beds, physical therapy equipment and rehabilitation equipment. Equipment rental options may also be available at some stores.

Additional Data

Peak hours of the generator—
> The A.M. and P.M. weekday peak hours of generator coincided with the peak hours of the adjacent street traffic and occurred between 7:45 a.m. and 8:45 a.m. and between 4:45 p.m. and 5:45 p.m., respectively.

The site was surveyed in 2007 in Florida.

Source Number

721

Land Use: 897
Medical Equipment Store
Independent Variables with One Observation

The following trip generation data are for independent variables with only one observation. This information is shown in this table only; there are no related plots for these data.

Users are cautioned to use data with care because of the small sample size.

Independent Variable	Trip Generation Rate	Size of Independent Variable	Number of Studies	Directional Distribution
1,000 Square Feet Gross Floor Area				
Weekday	6.00	15	1	50% entering, 50% exiting
Weekday A.M. Peak Hour of Adjacent Street Traffic	1.17	15	1	88% entering, 12% exiting
Weekday P.M. Peak Hour of Adjacent Street Traffic	1.24	15	1	11% entering, 89% exiting
Weekday A.M. Peak Hour of Generator	1.17	15	1	88% entering, 12% exiting
Weekday P.M. Peak Hour of Generator	1.24	15	1	11% entering, 89% exiting

Land Use: 911
Walk-in Bank

Description

Walk-in banks are generally free-standing buildings with their own parking lots. These banks do not have drive-in lanes but usually contain non-drive-through automatic teller machines (ATMs). Drive-in bank (Land Use 912) is a related use.

Additional Data

To reflect changes in travel patterns resulting from recent technological advances in the banking industry, data from years prior to the year 2000 have been removed from this land use. The elimination of these data resulted in lower trip generation rates for most time periods presented.

Peak hours of the generator—
The weekday P.M. peak hour varied between 4:00 p.m. and 5:30 p.m.

The sites were surveyed in the 2000s in California.

To assist in the future analysis of this land use, it is important that Friday data be collected and reported separately from weekday data. It is also important to specify the date and month of the data collection period.

Source Number

594

Walk-in Bank
(911)

Average Vehicle Trip Ends vs: 1000 Sq. Feet Gross Floor Area
On a: Weekday,
Peak Hour of Adjacent Street Traffic,
One Hour Between 4 and 6 p.m.

Number of Studies: 3
Average 1000 Sq. Feet GFA: 5
Directional Distribution: 44% entering, 56% exiting

Trip Generation per 1000 Sq. Feet Gross Floor Area

Average Rate	Range of Rates	Standard Deviation
12.13	2.00 - 24.15	10.96

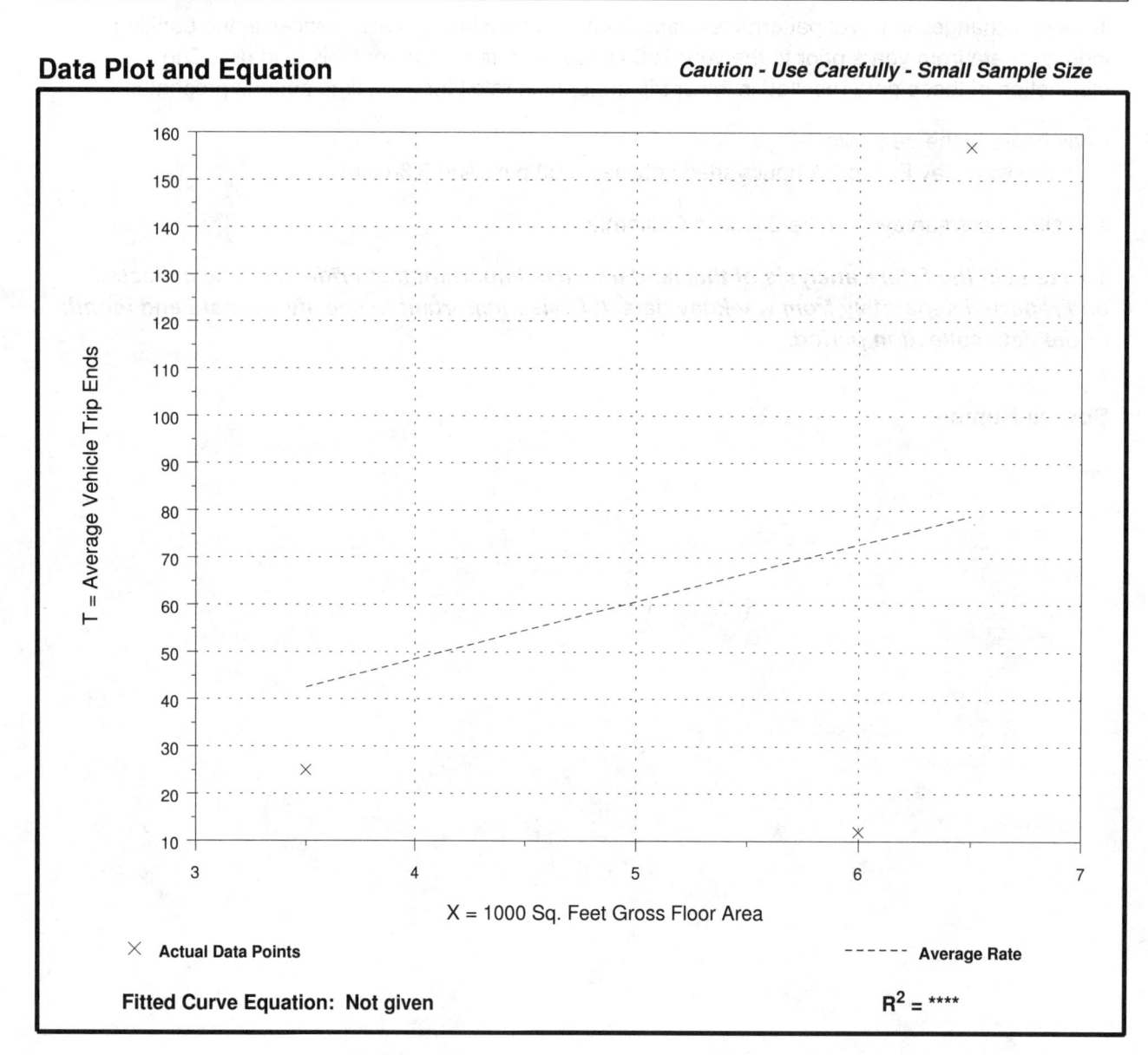

Data Plot and Equation

Caution - Use Carefully - Small Sample Size

Fitted Curve Equation: Not given $R^2 = ****$

Land Use: 912
Drive-in Bank

Description

Drive-in banks provide banking facilities for motorists who conduct financial transactions from their vehicles; many also serve patrons who walk into the building. The drive-in lanes may or may not provide automatic teller machines (ATMs). Walk-in bank (Land Use 911) is a related use.

Additional Data

To reflect changes in travel patterns resulting from recent technological advances in the banking industry, data from years prior to the year 2000 have been removed from this land use. The elimination of these data resulted in substantially lower trip generation rates for most time periods presented.

The independent variable, drive-in lanes, refers to all lanes at a banking facility used for financial transactions, including ATM-only lanes.

Peak hours of the generator—
    The weekday A.M. peak hour varied between 8:00 a.m. and 12:00 p.m. The weekday P.M. peak hour varied between 12:00 p.m. and 6:00 p.m. The weekend peak hour varied between 9:00 a.m. and 1:30 p.m.

The sites were surveyed in the 2000s throughout the United States.

To assist in the future analysis of this land use, it is important that Friday data be collected and reported separately from weekday data. It is also important to specify the date and month of the data collection period and the number of drive-through lanes that are open at the time of the study.

Specialized Land Use Data

One study provided data on a drive-in bank with an office on the second floor. The size and scale of this site differs considerably from those contained in this land use. Therefore, the information collected for this facility is presented in the following table and was excluded from the data plots.

Independent Variable	Trip Generation Rate	Size of Independent Variable	Number of Studies	Directional Distribution
Weekday A.M. Peak Hour of Adjacent Street Traffic	3.55	15,200	1	69% entering, 31% exiting
Weekday P.M. Peak Hour of Adjacent Street Traffic	5.57	15,200	1	44% entering, 56% exiting
Weekday A.M. Peak Hour of Generator	3.55	15,200	1	69% entering, 31% exiting
Weekday P.M. Peak Hour of Generator	5.57	15,200	1	44% entering, 56% exiting

1,000 Square Feet Gross Floor Area

Source: 656

Source Numbers

553, 555, 573, 577, 600, 624, 626, 629, 630, 637, 656, 657, 710, 724, 728

Drive-in Bank
(912)

Average Vehicle Trip Ends vs: Employees
On a: Weekday

Number of Studies: 2
Avg. Number of Employees: 9
Directional Distribution: 50% entering, 50% exiting

Trip Generation per Employee

Average Rate	Range of Rates	Standard Deviation
30.94	28.89 - 33.00	*

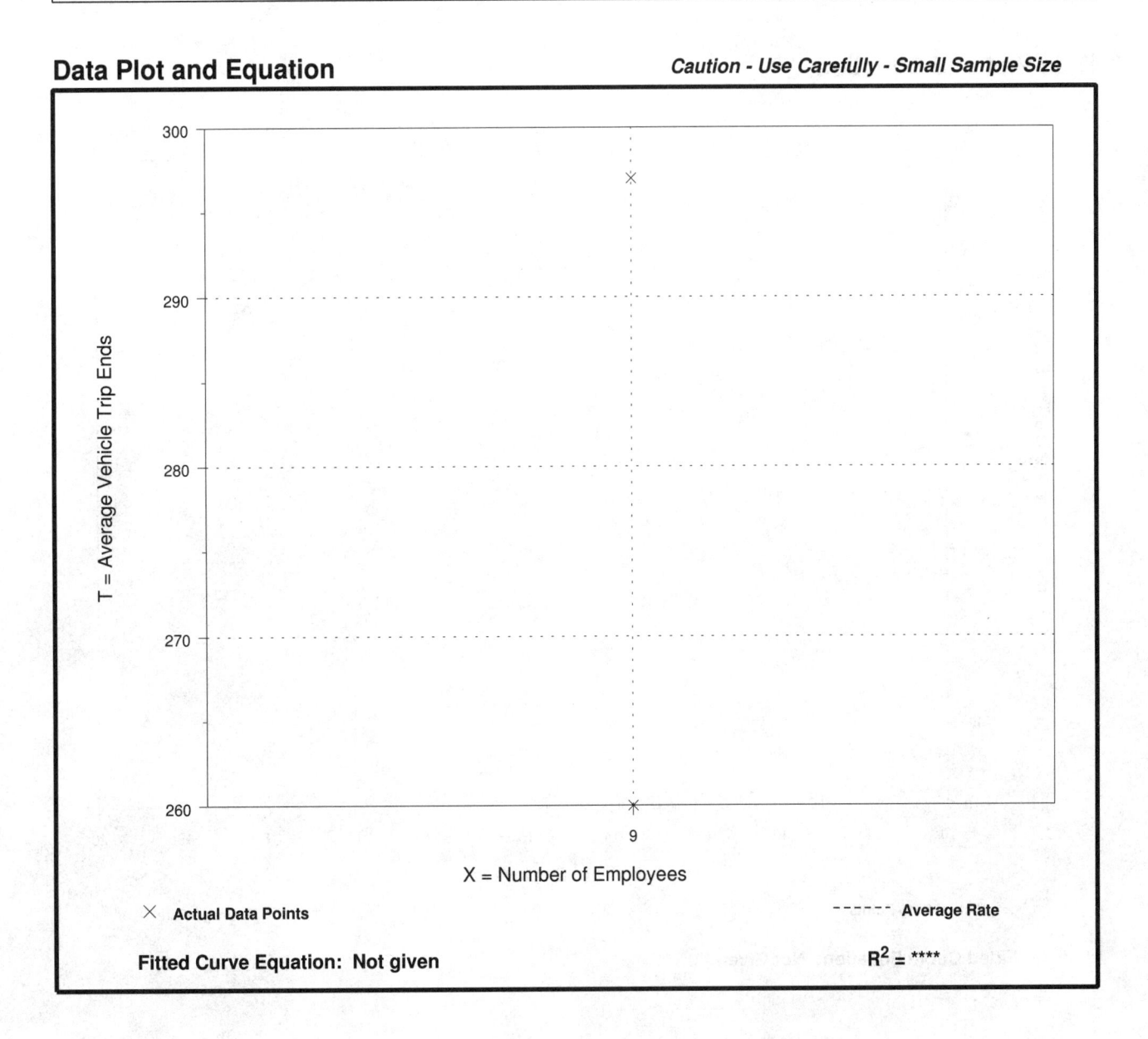

Data Plot and Equation

Caution - Use Carefully - Small Sample Size

Fitted Curve Equation: Not given

$R^2 = ****$

Drive-in Bank
(912)

Average Vehicle Trip Ends vs: Employees
On a: Weekday,
Peak Hour of Adjacent Street Traffic,
One Hour Between 7 and 9 a.m.

Number of Studies: 7
Avg. Number of Employees: 15
Directional Distribution: 62% entering, 38% exiting

Trip Generation per Employee

Average Rate	Range of Rates	Standard Deviation
2.63	1.30 - 4.72	1.98

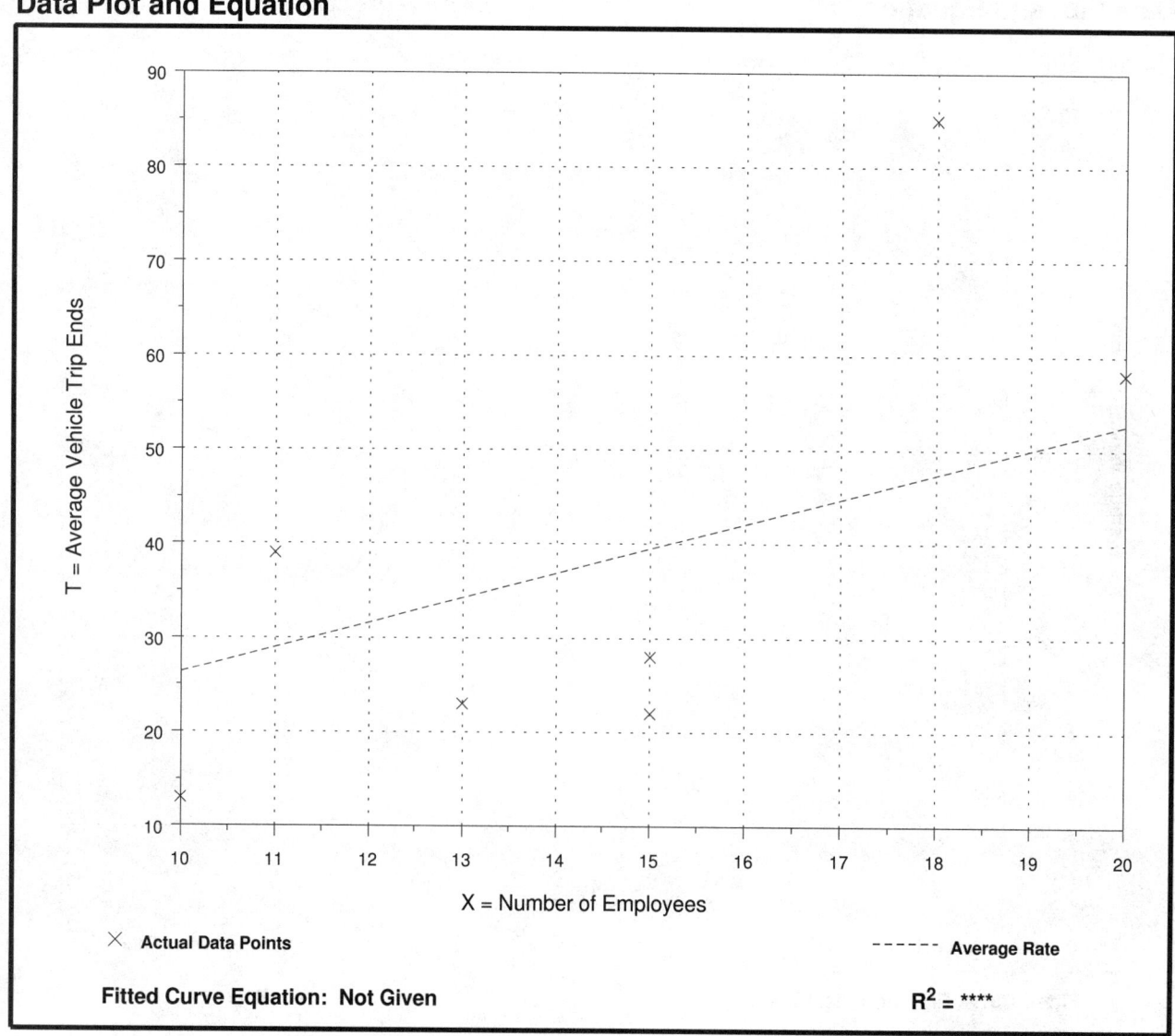

Data Plot and Equation

Fitted Curve Equation: Not Given $R^2 = ****$

Drive-in Bank
(912)

Average Vehicle Trip Ends vs: Employees
On a: Weekday,
Peak Hour of Adjacent Street Traffic,
One Hour Between 4 and 6 p.m.

Number of Studies: 11
Avg. Number of Employees: 13
Directional Distribution: 48% entering, 52% exiting

Trip Generation per Employee

Average Rate	Range of Rates	Standard Deviation
5.42	1.70 - 8.90	2.93

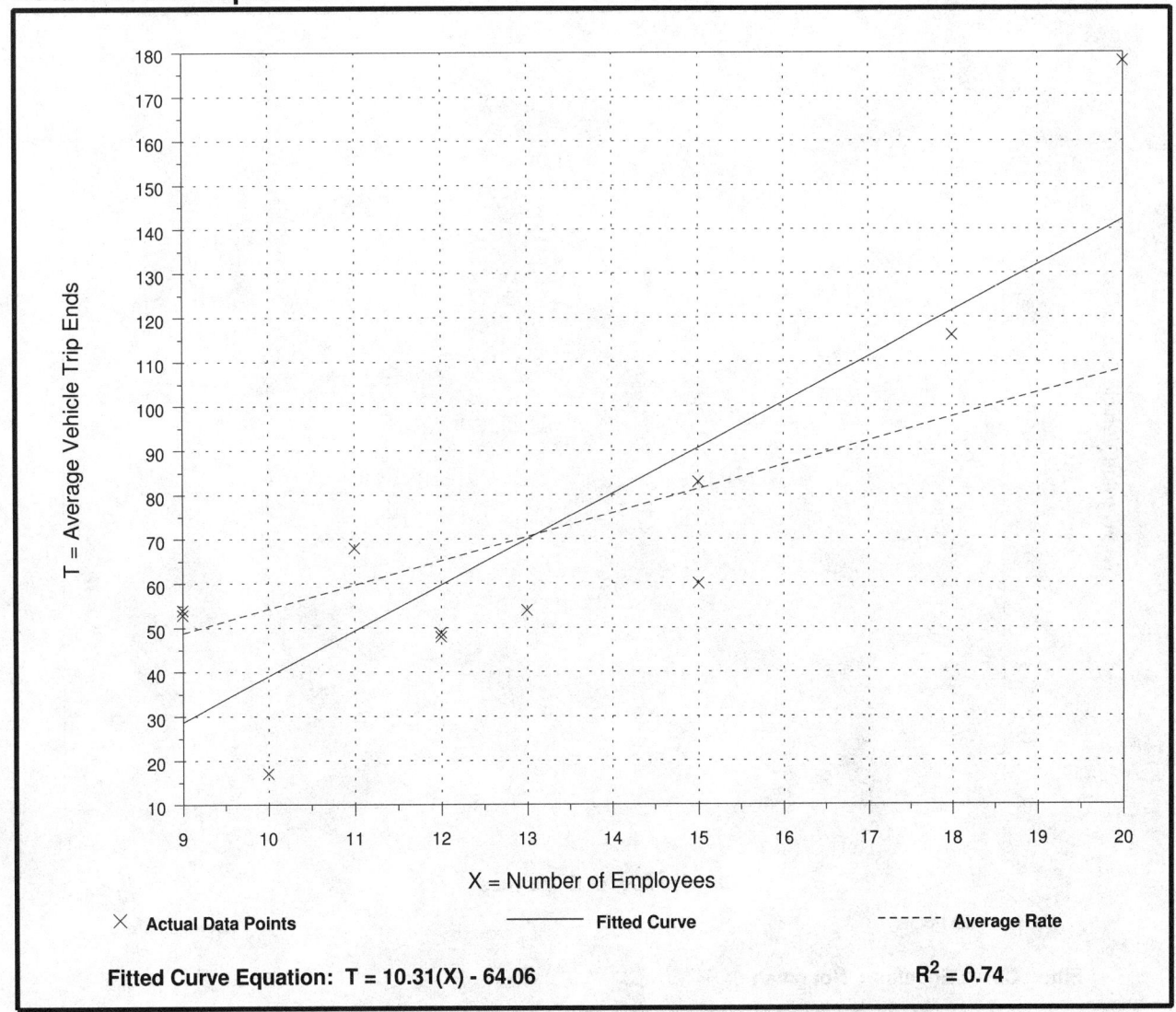

Data Plot and Equation

X Actual Data Points   —— Fitted Curve   ----- Average Rate

Fitted Curve Equation: $T = 10.31(X) - 64.06$ $R^2 = 0.74$

Trip Generation, 9th Edition • Institute of Transportation Engineers   1839

Drive-in Bank
(912)

Average Vehicle Trip Ends vs: Employees
On a: Weekday,
A.M. Peak Hour of Generator

Number of Studies: 2
Avg. Number of Employees: 11
Directional Distribution: 54% entering, 46% exiting

Trip Generation per Employee

Average Rate	Range of Rates	Standard Deviation
2.48	1.30 - 3.55	*

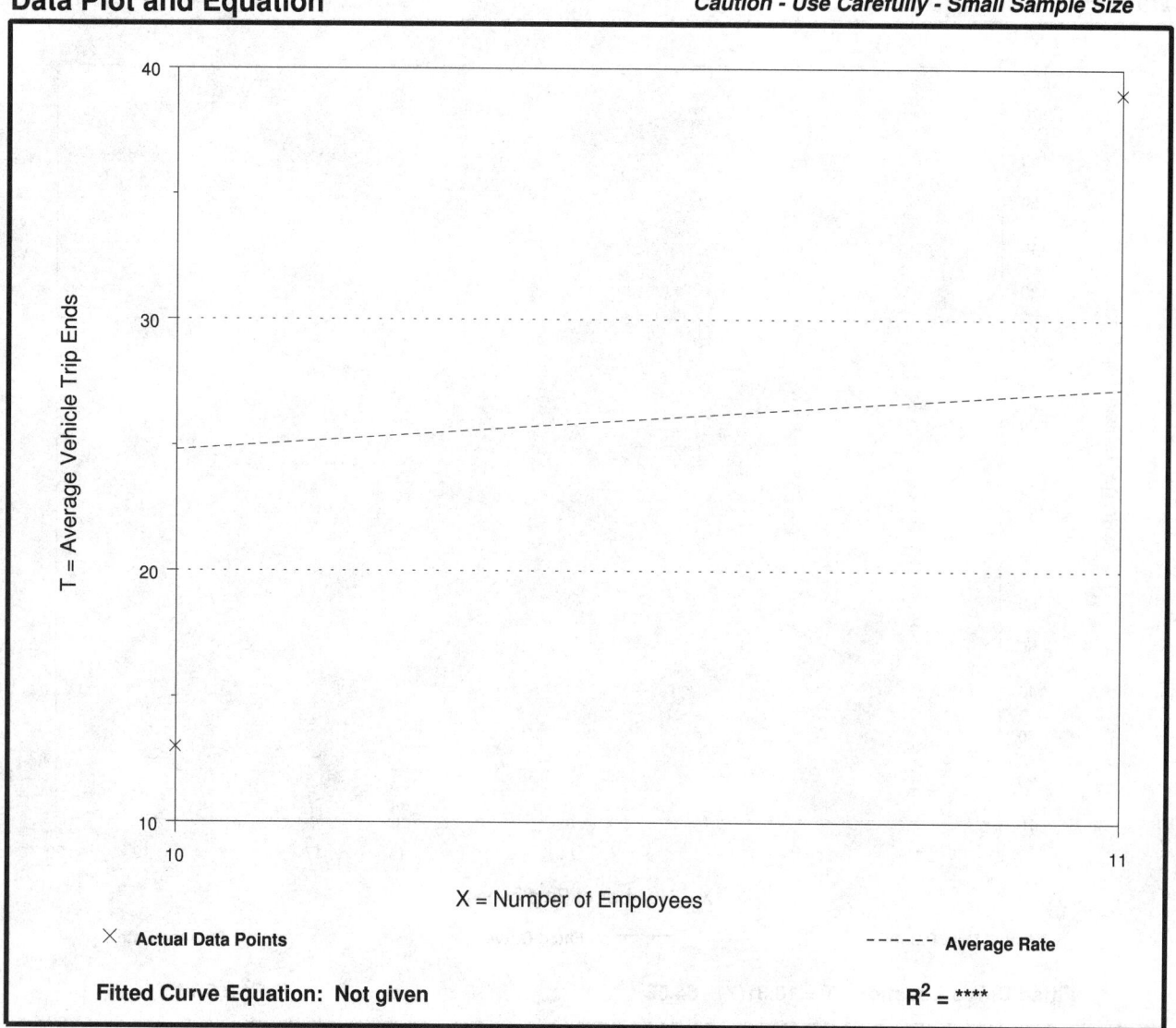

Data Plot and Equation

Caution - Use Carefully - Small Sample Size

Fitted Curve Equation: Not given $R^2 = ****$

Drive-in Bank
(912)

Average Vehicle Trip Ends vs: Employees
On a: Weekday,
P.M. Peak Hour of Generator

Number of Studies: 2
Avg. Number of Employees: 11
Directional Distribution: 41% entering, 59% exiting

Trip Generation per Employee

Average Rate	Range of Rates	Standard Deviation
4.71	3.10 - 6.18	*

Data Plot and Equation

Caution - Use Carefully - Small Sample Size

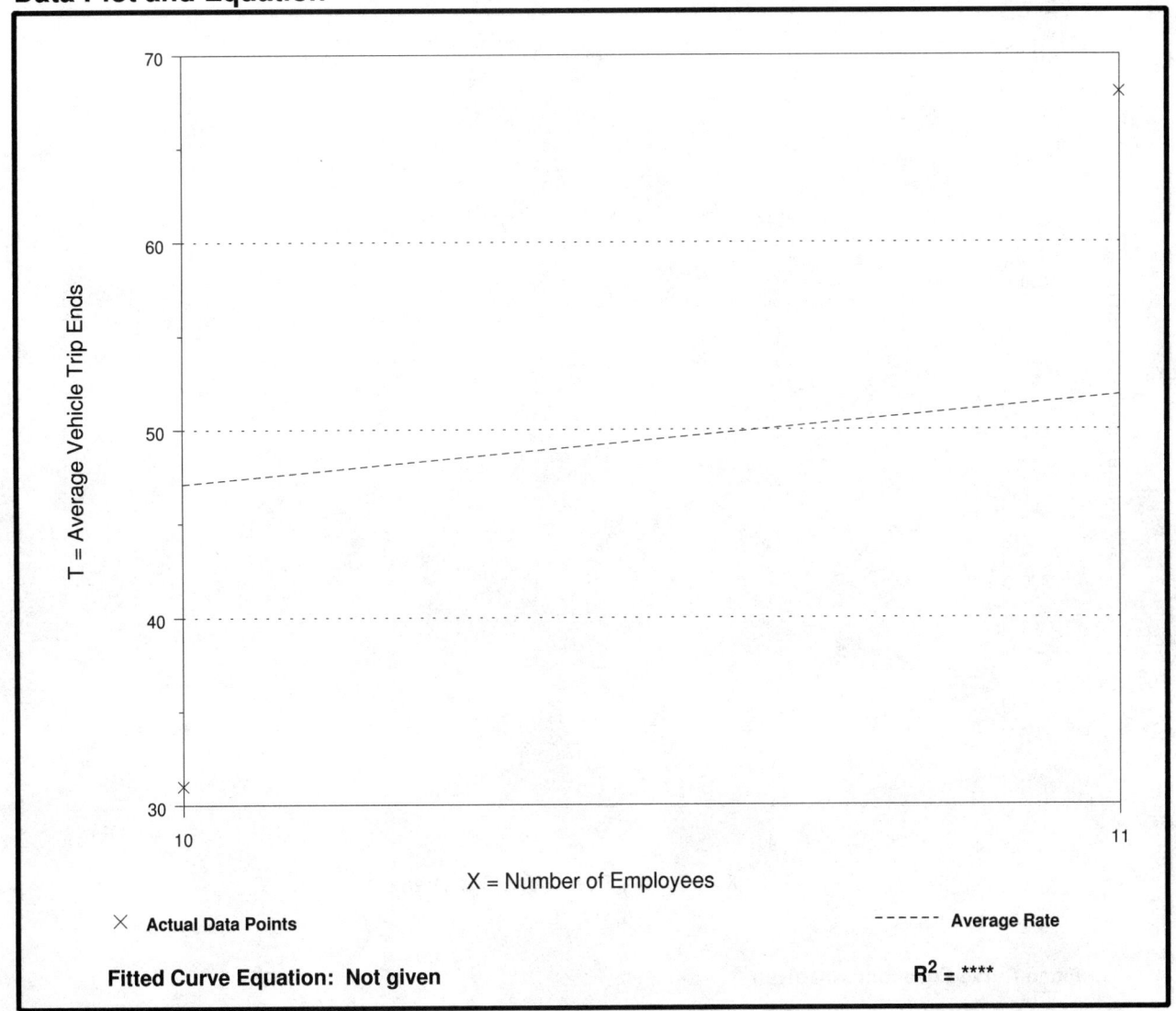

Fitted Curve Equation: Not given $R^2 = ****$

Drive-in Bank
(912)

Average Vehicle Trip Ends vs: 1000 Sq. Feet Gross Floor Area
On a: Weekday

Number of Studies: 7
Average 1000 Sq. Feet GFA: 3
Directional Distribution: 50% entering, 50% exiting

Trip Generation per 1000 Sq. Feet Gross Floor Area

Average Rate	Range of Rates	Standard Deviation
148.15	68.23 - 407.21	97.36

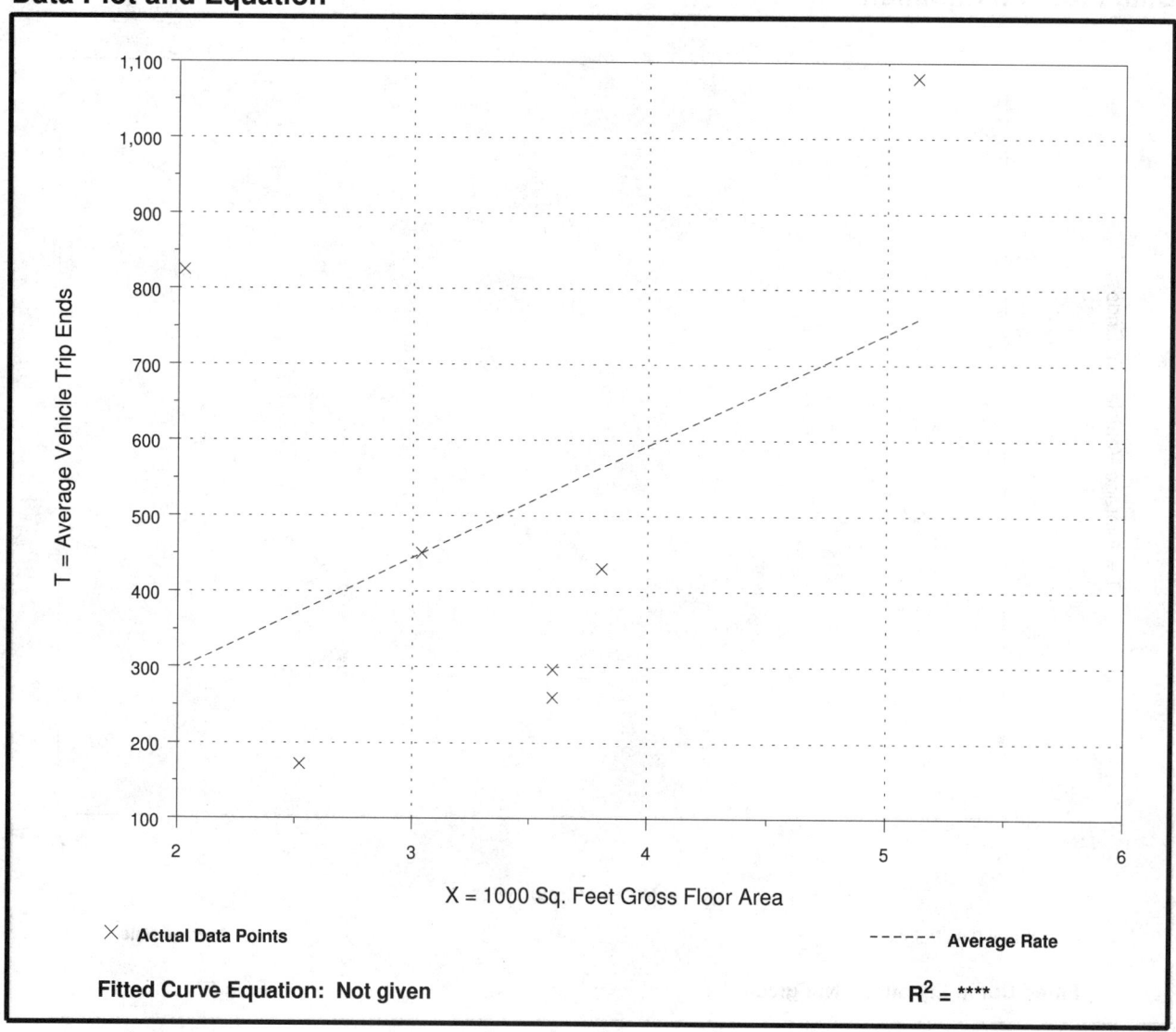

Data Plot and Equation

Fitted Curve Equation: Not given $R^2 = ****$

Drive-in Bank
(912)

Average Vehicle Trip Ends vs: 1000 Sq. Feet Gross Floor Area
On a: Weekday,
Peak Hour of Adjacent Street Traffic,
One Hour Between 7 and 9 a.m.

Number of Studies: 31
Average 1000 Sq. Feet GFA: 4
Directional Distribution: 57% entering, 43% exiting

Trip Generation per 1000 Sq. Feet Gross Floor Area

Average Rate	Range of Rates	Standard Deviation
12.08	3.71 - 29.40	6.88

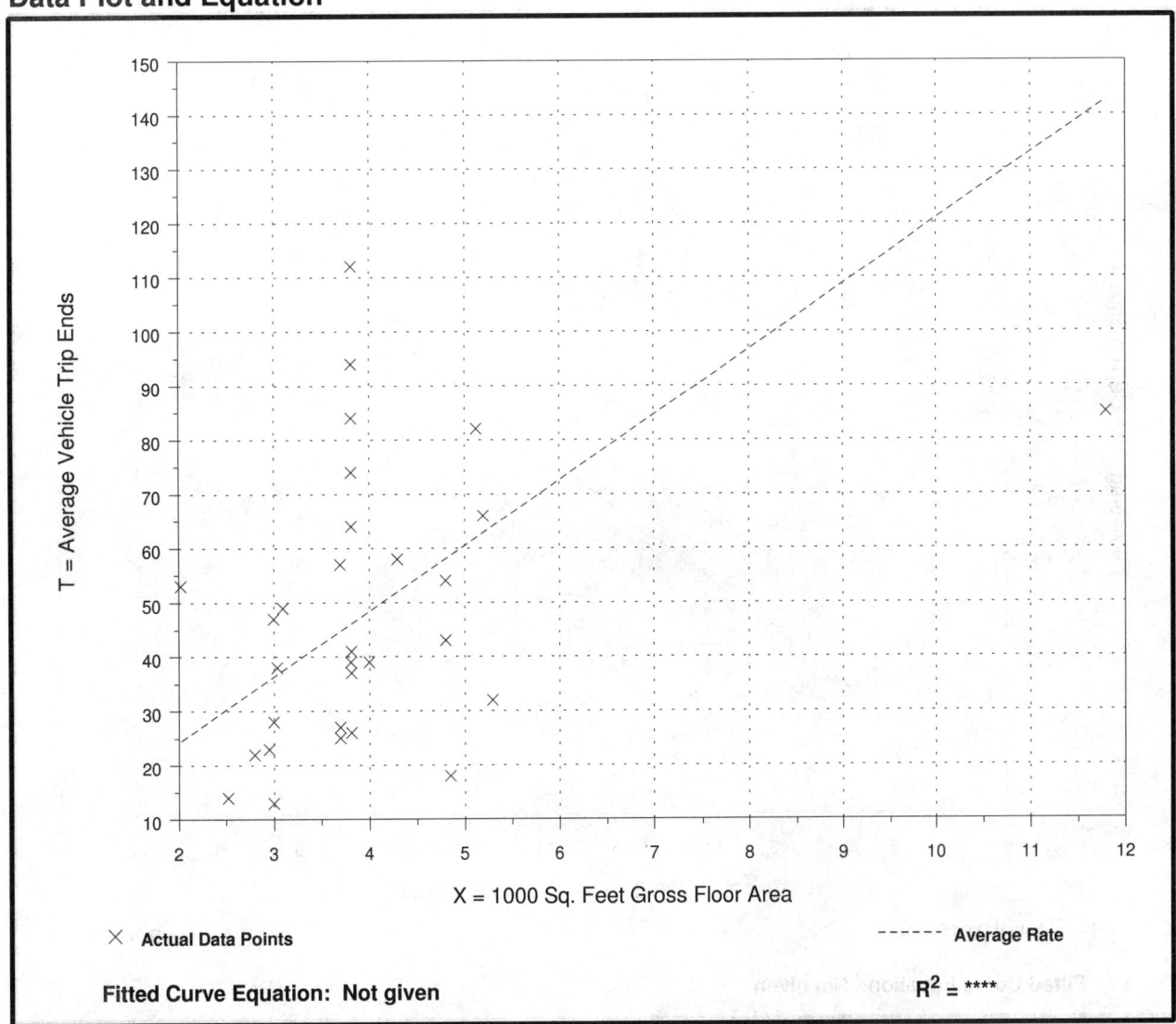

Data Plot and Equation

Fitted Curve Equation: Not given $R^2 = ****$

Drive-in Bank
(912)

Average Vehicle Trip Ends vs: 1000 Sq. Feet Gross Floor Area
On a: Weekday,
Peak Hour of Adjacent Street Traffic,
One Hour Between 4 and 6 p.m.

Number of Studies: 102
Average 1000 Sq. Feet GFA: 4
Directional Distribution: 50% entering, 50% exiting

Trip Generation per 1000 Sq. Feet Gross Floor Area

Average Rate	Range of Rates	Standard Deviation
24.30	3.09 - 109.68	16.24

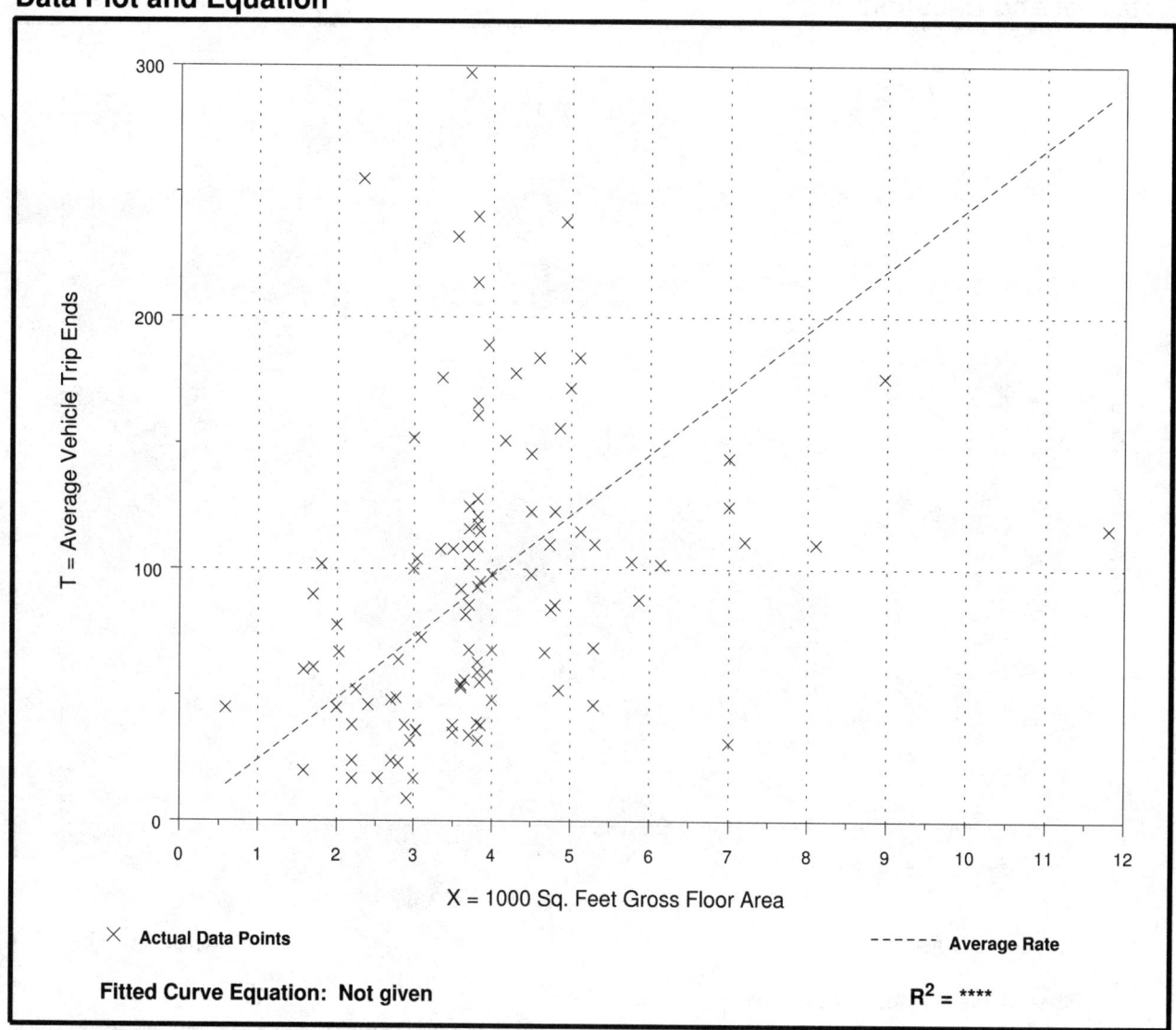

Data Plot and Equation

Fitted Curve Equation: Not given $R^2 = ****$

Drive-in Bank
(912)

Average Vehicle Trip Ends vs: 1000 Sq. Feet Gross Floor Area
On a: Weekday,
A.M. Peak Hour of Generator

Number of Studies: 39
Average 1000 Sq. Feet GFA: 4
Directional Distribution: 53% entering, 47% exiting

Trip Generation per 1000 Sq. Feet Gross Floor Area

Average Rate	Range of Rates	Standard Deviation
17.57	4.33 - 46.89	10.68

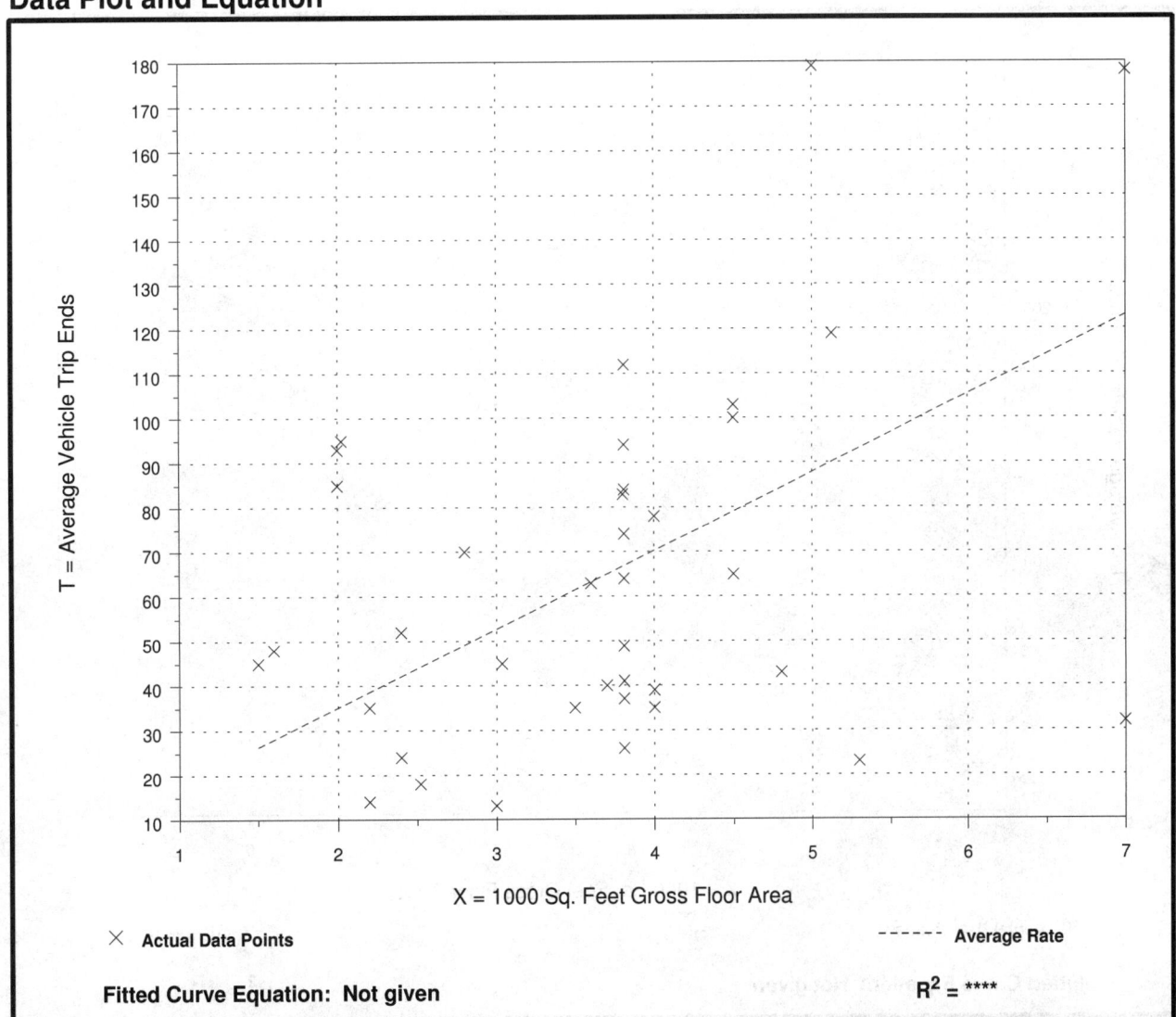

Data Plot and Equation

Fitted Curve Equation: Not given $R^2 = ****$

Drive-in Bank
(912)

Average Vehicle Trip Ends vs: 1000 Sq. Feet Gross Floor Area
On a: Weekday,
P.M. Peak Hour of Generator

Number of Studies: 47
Average 1000 Sq. Feet GFA: 4
Directional Distribution: 51% entering, 49% exiting

Trip Generation per 1000 Sq. Feet Gross Floor Area

Average Rate	Range of Rates	Standard Deviation
26.69	7.14 - 68.50	14.58

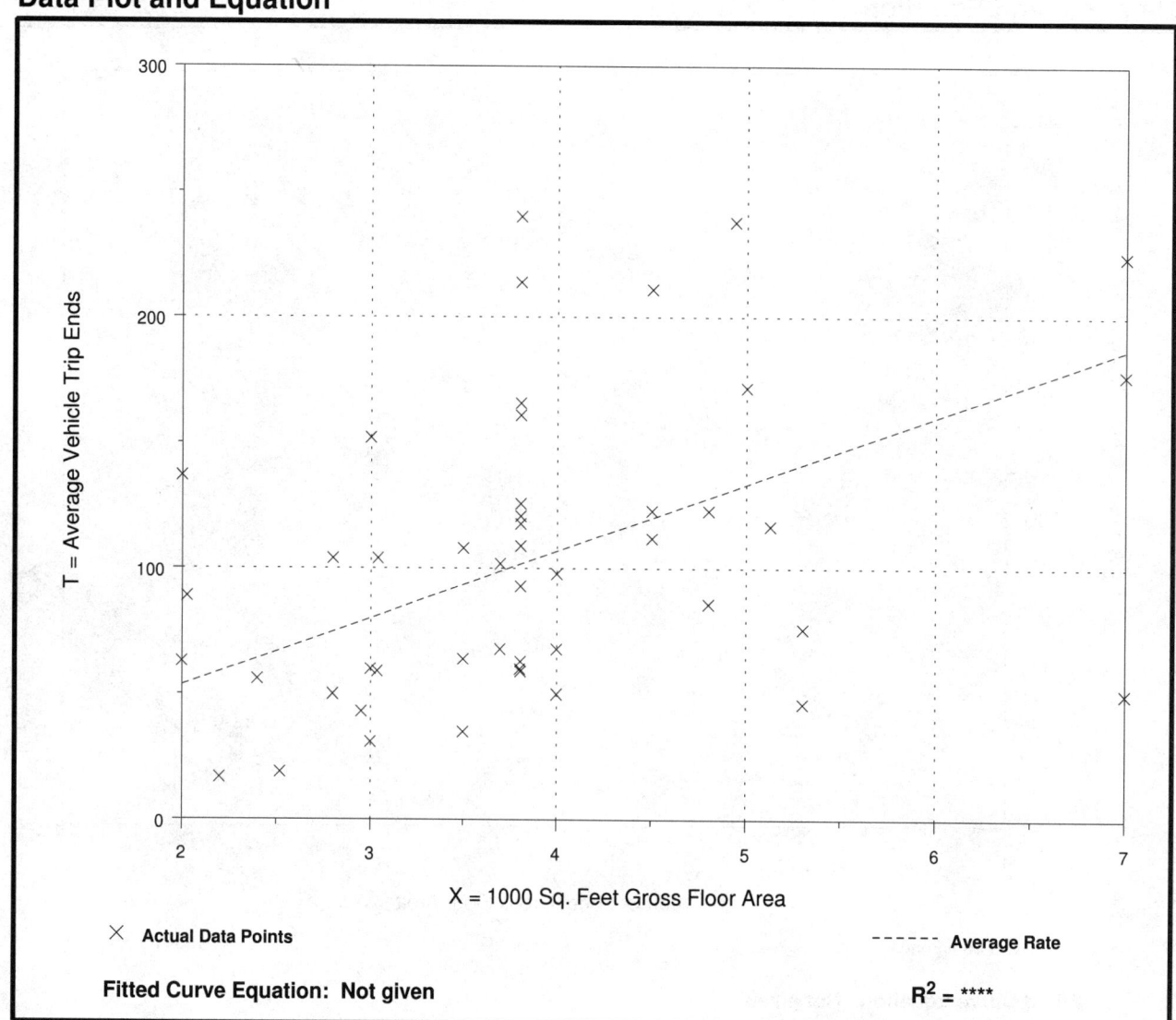

Data Plot and Equation

[Scatter plot: X = 1000 Sq. Feet Gross Floor Area (2 to 7); T = Average Vehicle Trip Ends (0 to 300); X marks = Actual Data Points; dashed line = Average Rate]

Fitted Curve Equation: Not given $R^2 = ****$

Drive-in Bank
(912)

Average Vehicle Trip Ends vs: 1000 Sq. Feet Gross Floor Area
On a: Saturday

Number of Studies: 5
Average 1000 Sq. Feet GFA: 3
Directional Distribution: 50% entering, 50% exiting

Trip Generation per 1000 Sq. Feet Gross Floor Area

Average Rate	Range of Rates	Standard Deviation
86.32	42.44 - 171.27	36.65

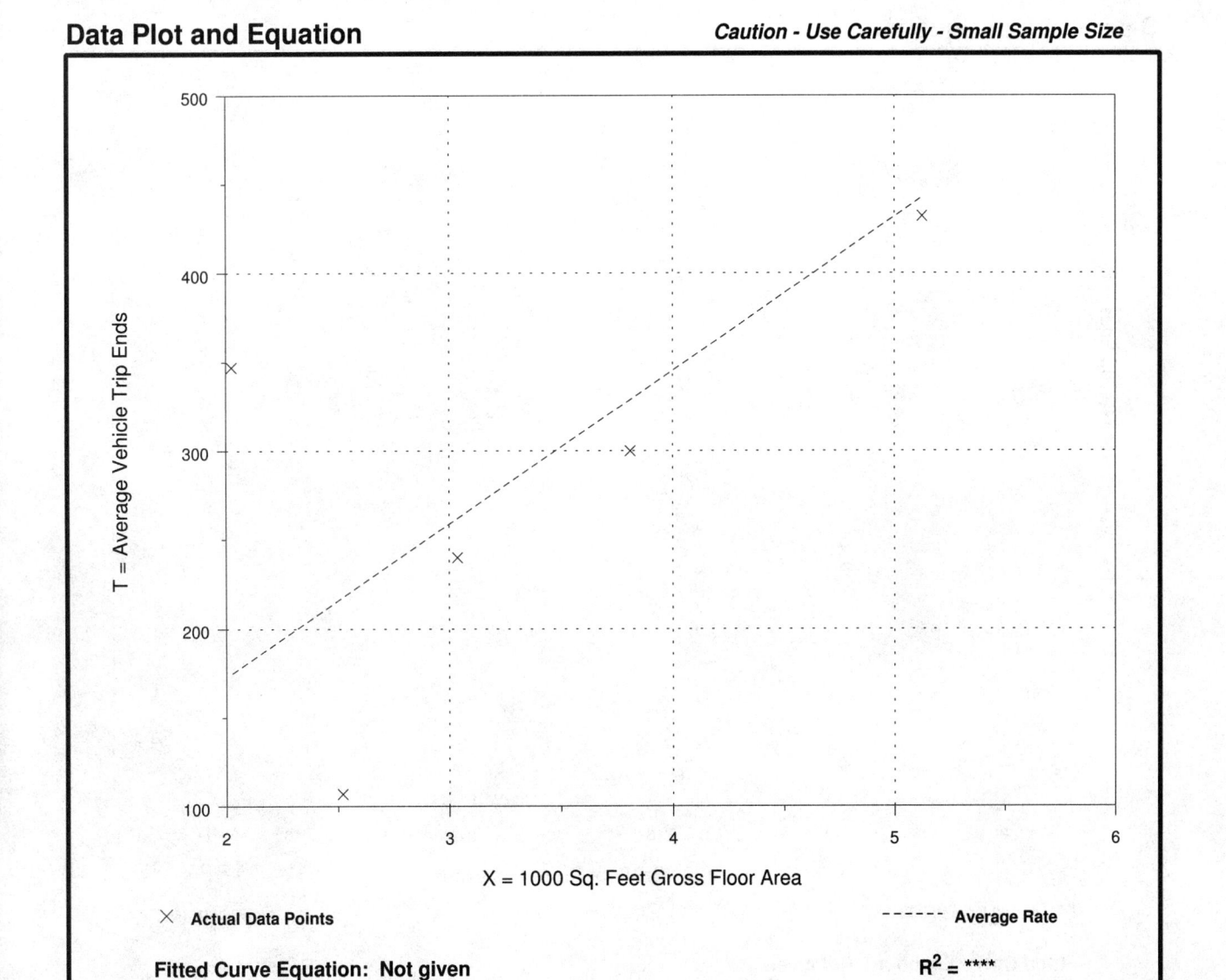

Data Plot and Equation

Caution - Use Carefully - Small Sample Size

[Plot: T = Average Vehicle Trip Ends vs. X = 1000 Sq. Feet Gross Floor Area]

× Actual Data Points   ----- Average Rate

Fitted Curve Equation: Not given $R^2 = ****$

Drive-in Bank
(912)

Average Vehicle Trip Ends vs: 1000 Sq. Feet Gross Floor Area
On a: Saturday,
Peak Hour of Generator

Number of Studies: 41
Average 1000 Sq. Feet GFA: 4
Directional Distribution: 51% entering, 49% exiting

Trip Generation per 1000 Sq. Feet Gross Floor Area

Average Rate	Range of Rates	Standard Deviation
26.31	7.17 - 107.00	15.79

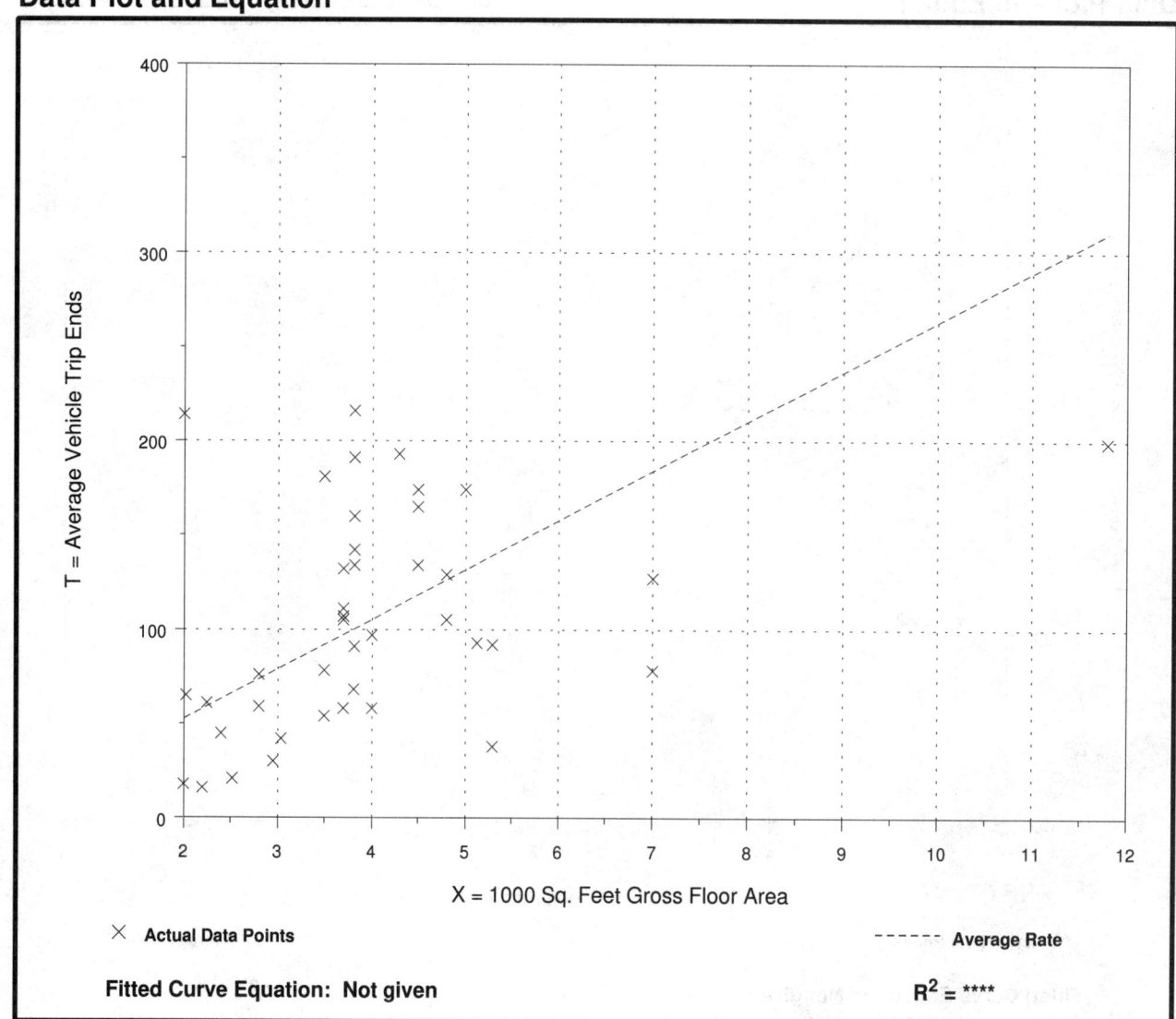

Data Plot and Equation

Fitted Curve Equation: Not given $R^2 = ****$

Drive-in Bank
(912)

Average Vehicle Trip Ends vs: 1000 Sq. Feet Gross Floor Area
On a: Sunday

Number of Studies: 5
Average 1000 Sq. Feet GFA: 3
Directional Distribution: 50% entering, 50% exiting

Trip Generation per 1000 Sq. Feet Gross Floor Area

Average Rate	Range of Rates	Standard Deviation
31.90	23.40 - 69.10	15.45

Data Plot and Equation

Caution - Use Carefully - Small Sample Size

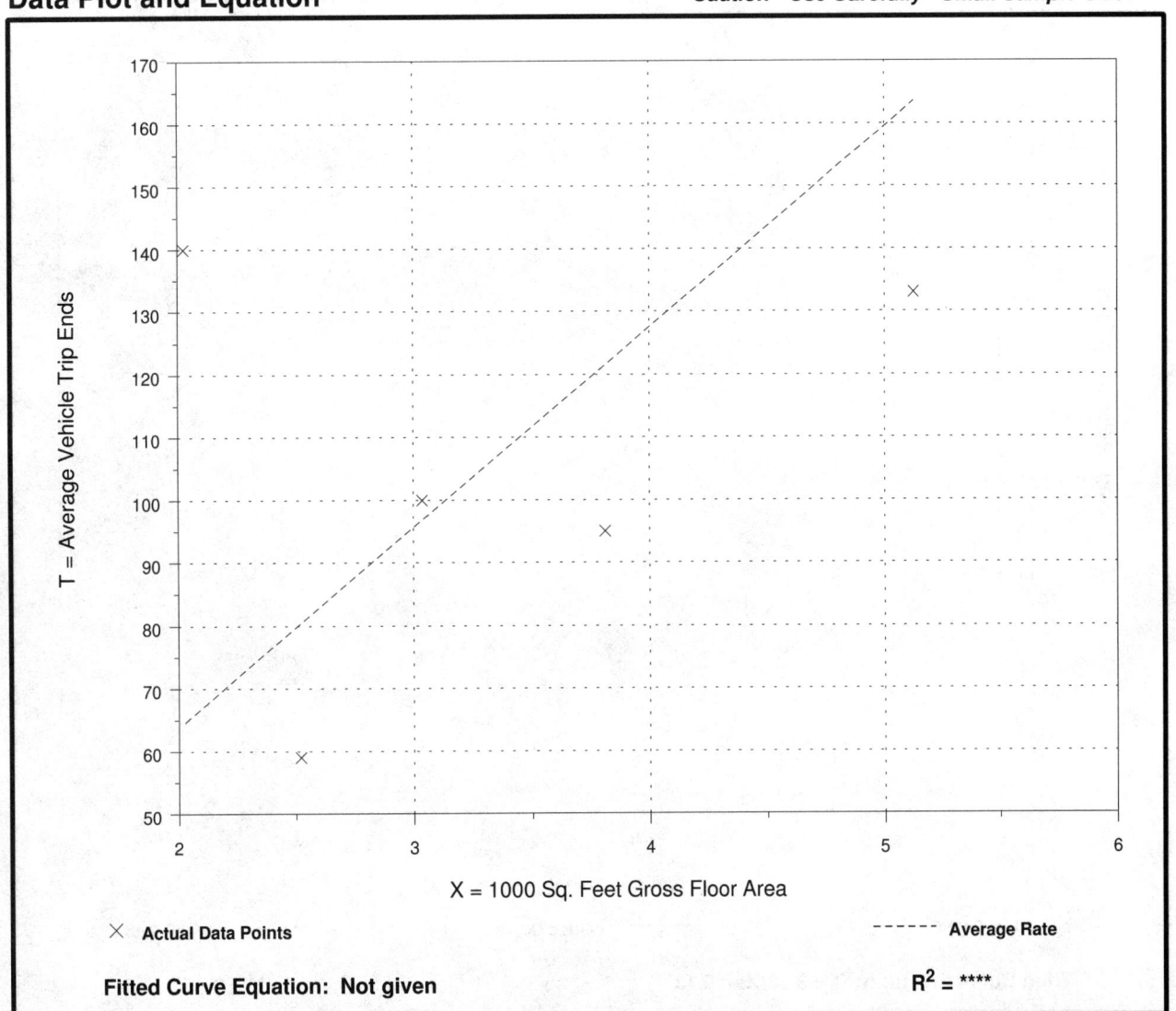

X   Actual Data Points   ------ Average Rate

Fitted Curve Equation: Not given R^2 = ****

Drive-in Bank
(912)

Average Vehicle Trip Ends vs: 1000 Sq. Feet Gross Floor Area
On a: Sunday,
Peak Hour of Generator

Number of Studies: 5
Average 1000 Sq. Feet GFA: 3
Directional Distribution: Not available

Trip Generation per 1000 Sq. Feet Gross Floor Area

Average Rate	Range of Rates	Standard Deviation
4.78	3.68 - 7.40	2.17

Data Plot and Equation

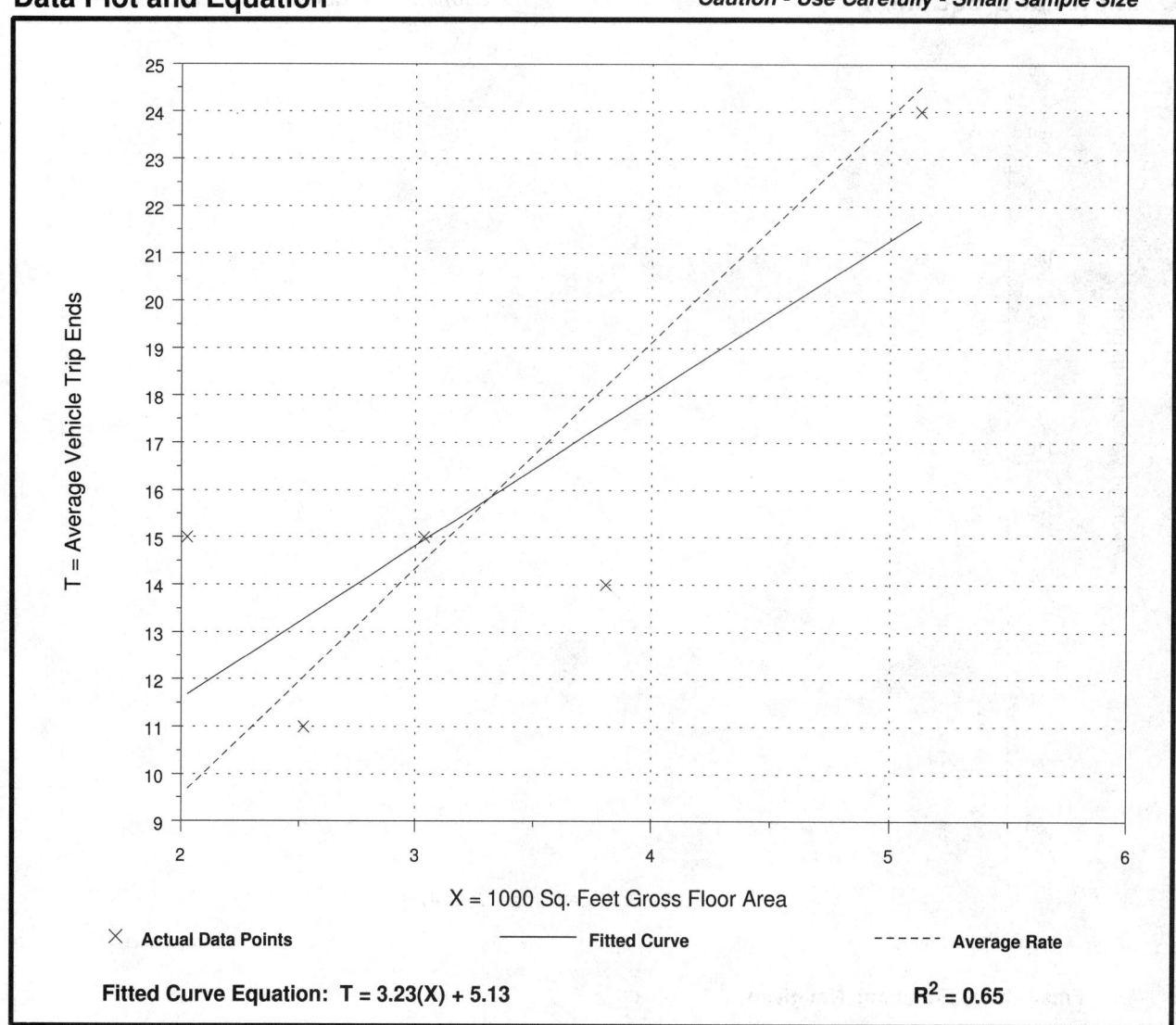

Caution - Use Carefully - Small Sample Size

Fitted Curve Equation: $T = 3.23(X) + 5.13$ $R^2 = 0.65$

Drive-in Bank
(912)

Average Vehicle Trip Ends vs: Drive-in Lanes
On a: Weekday

Number of Studies: 2
Avg. Number of Drive-in Lanes: 2
Directional Distribution: 50% entering, 50% exiting

Trip Generation per Drive-in Lane

Average Rate	Range of Rates	Standard Deviation
139.25	130.00 - 148.50	*

Data Plot and Equation

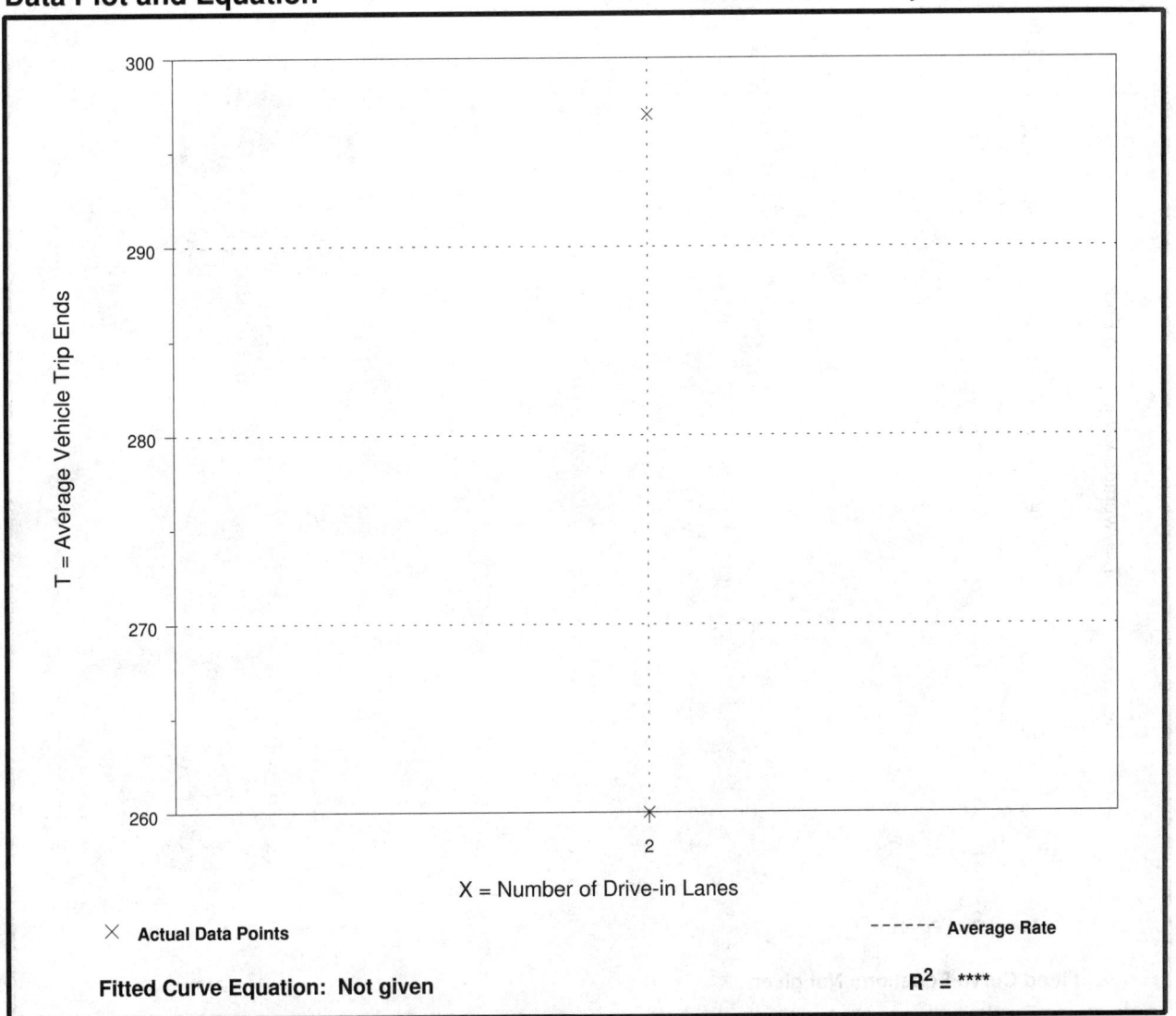

Caution - Use Carefully - Small Sample Size

Fitted Curve Equation: Not given $R^2 = ****$

Drive-in Bank
(912)

Average Vehicle Trip Ends vs: Drive-in Lanes
On a: Weekday,
Peak Hour of Adjacent Street Traffic,
One Hour Between 7 and 9 a.m.

Number of Studies: 18
Avg. Number of Drive-in Lanes: 4
Directional Distribution: 60% entering, 40% exiting

Trip Generation per Drive-in Lane

Average Rate	Range of Rates	Standard Deviation
9.29	4.33 - 45.00	6.03

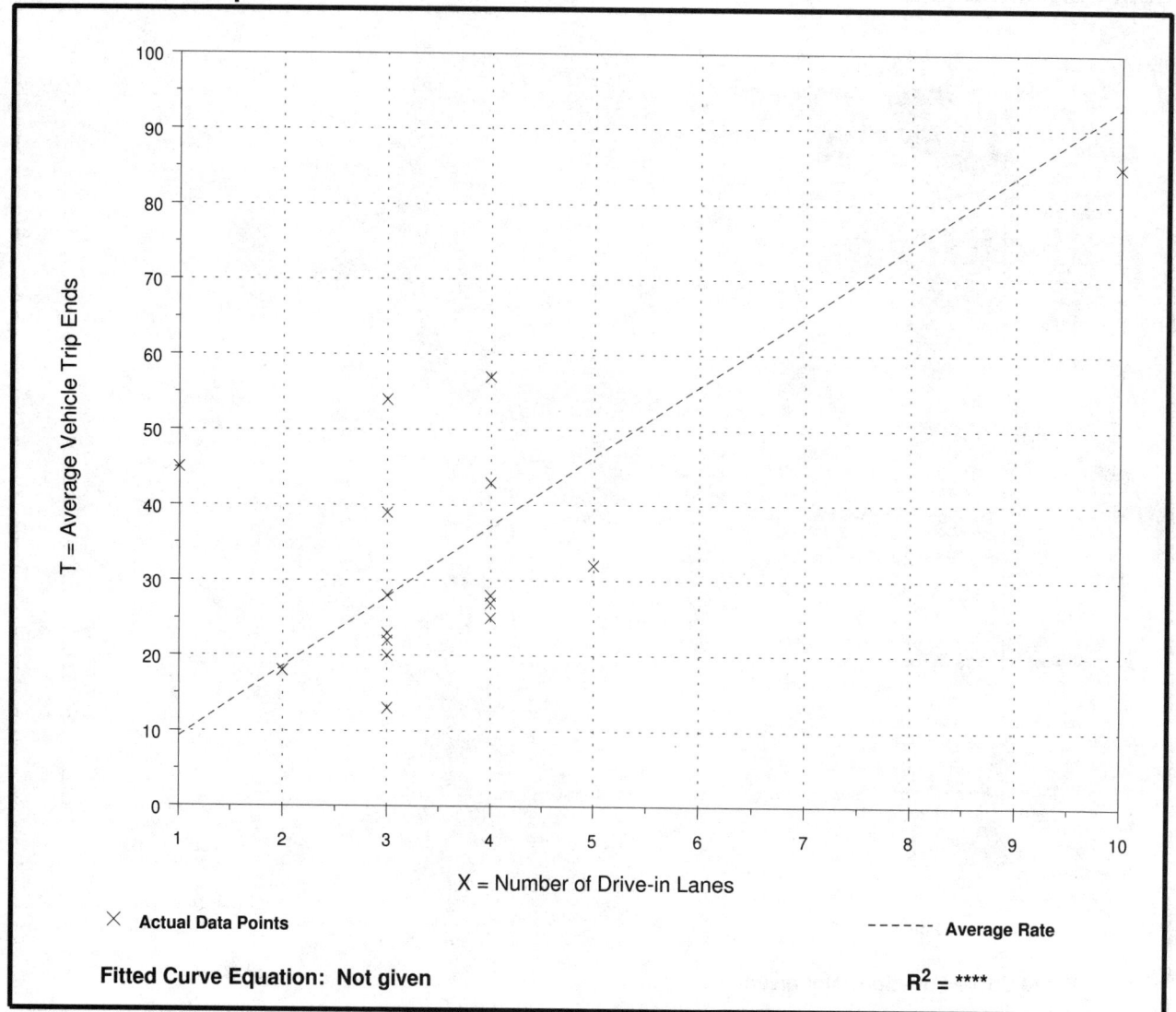

Data Plot and Equation

T = Average Vehicle Trip Ends
X = Number of Drive-in Lanes

× Actual Data Points
------ Average Rate

Fitted Curve Equation: Not given $R^2 = ****$

Drive-in Bank
(912)

Average Vehicle Trip Ends vs: Drive-in Lanes
On a: Weekday,
Peak Hour of Adjacent Street Traffic,
One Hour Between 4 and 6 p.m.

Number of Studies: 85
Avg. Number of Drive-in Lanes: 3
Directional Distribution: 49% entering, 51% exiting

Trip Generation per Drive-in Lane

Average Rate	Range of Rates	Standard Deviation
33.24	3.00 - 176.00	24.48

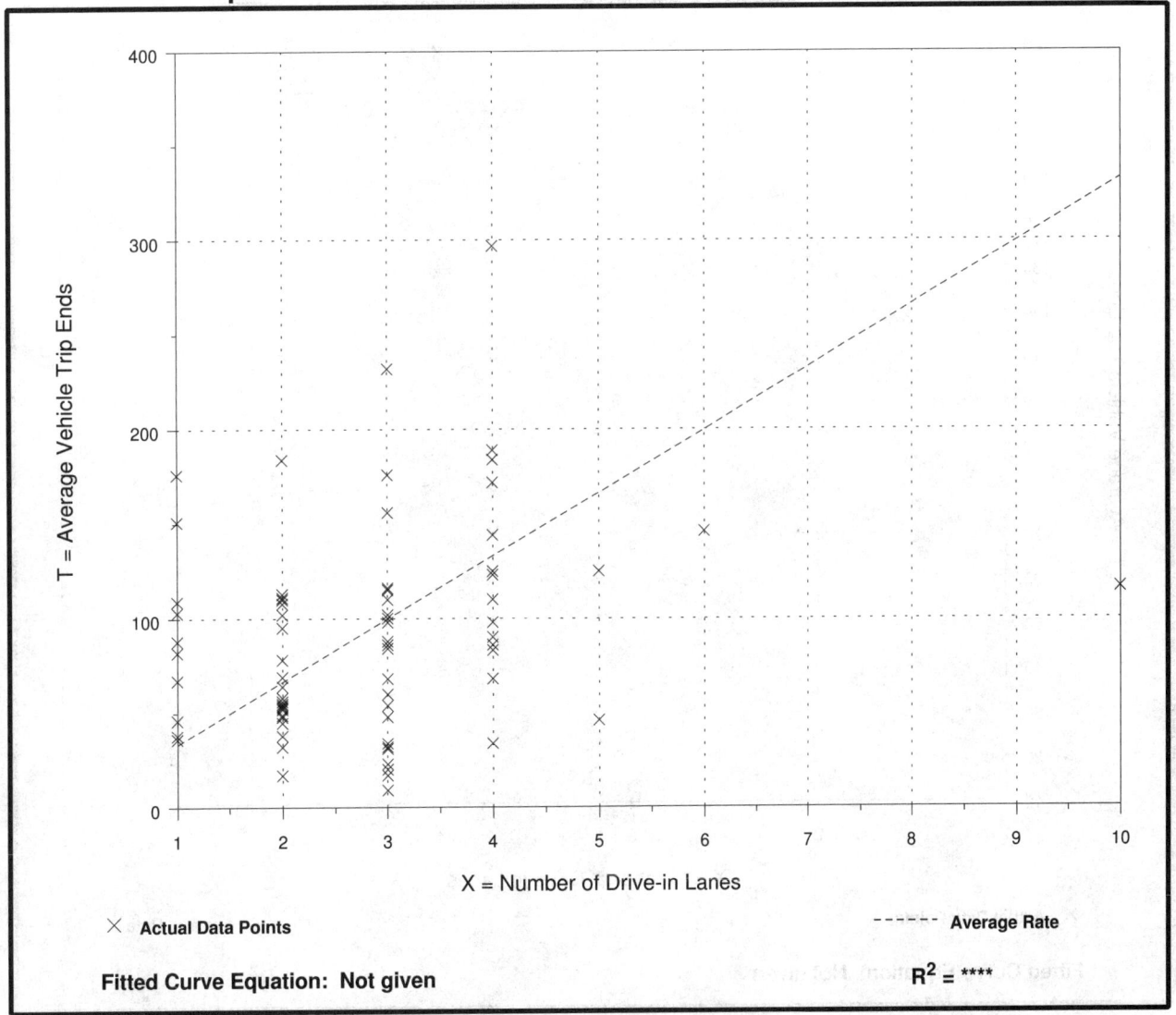

Data Plot and Equation

Fitted Curve Equation: Not given $R^2 =$ ****

Drive-in Bank
(912)

Average Vehicle Trip Ends vs: Drive-in Lanes
On a: Weekday,
A.M. Peak Hour of Generator

Number of Studies: 19
Avg. Number of Drive-in Lanes: 3
Directional Distribution: 56% entering, 44% exiting

Trip Generation per Drive-in Lane

Average Rate	Range of Rates	Standard Deviation
21.64	4.33 - 52.00	13.89

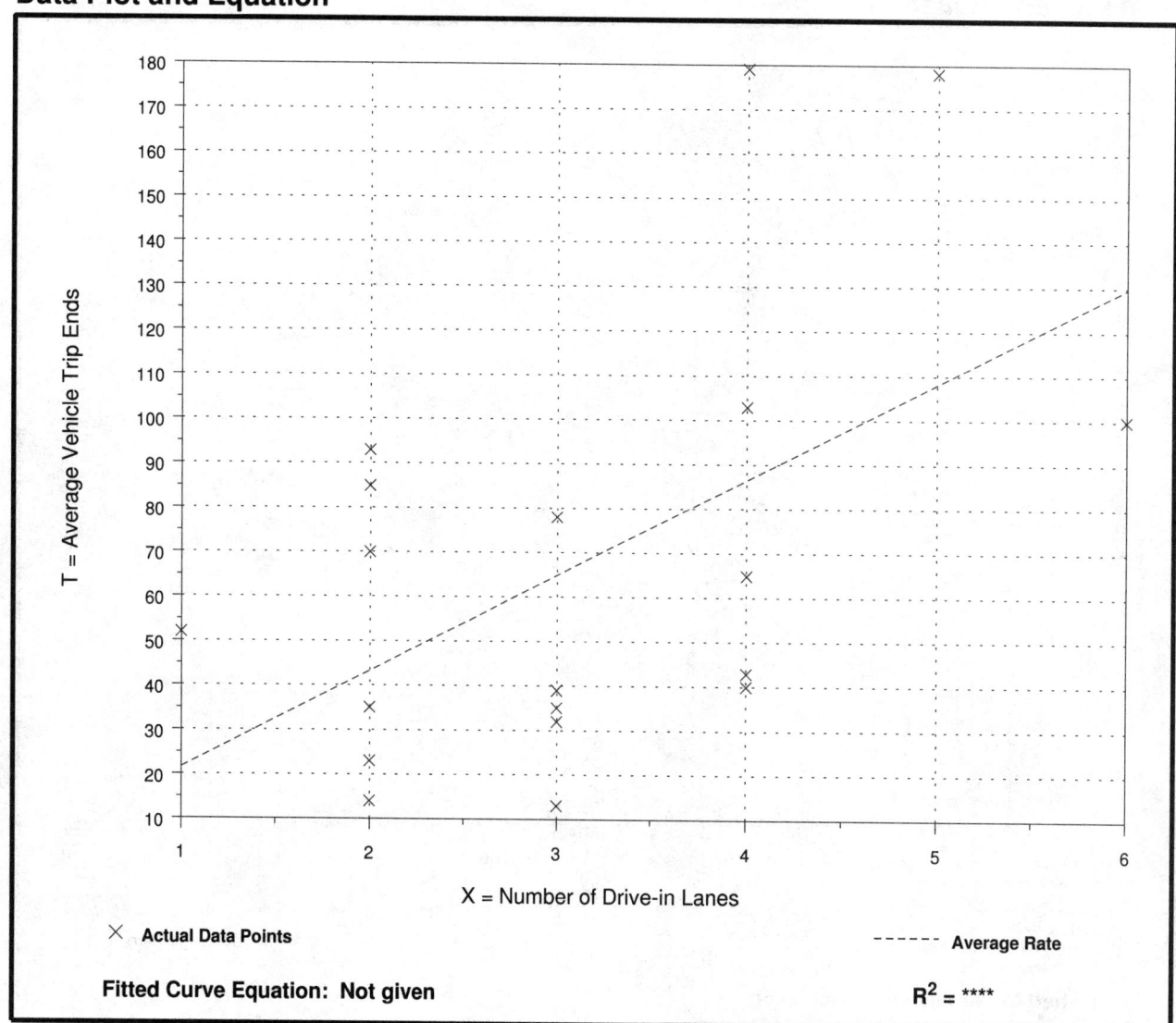

Data Plot and Equation

× Actual Data Points   ------ Average Rate

Fitted Curve Equation: Not given $R^2 = ****$

Drive-in Bank
(912)

Average Vehicle Trip Ends vs: Drive-in Lanes
On a: Weekday,
P.M. Peak Hour of Generator

Number of Studies: 26
Avg. Number of Drive-in Lanes: 3
Directional Distribution: 49% entering, 51% exiting

Trip Generation per Drive-in Lane

Average Rate	Range of Rates	Standard Deviation
29.05	8.50 - 68.50	14.73

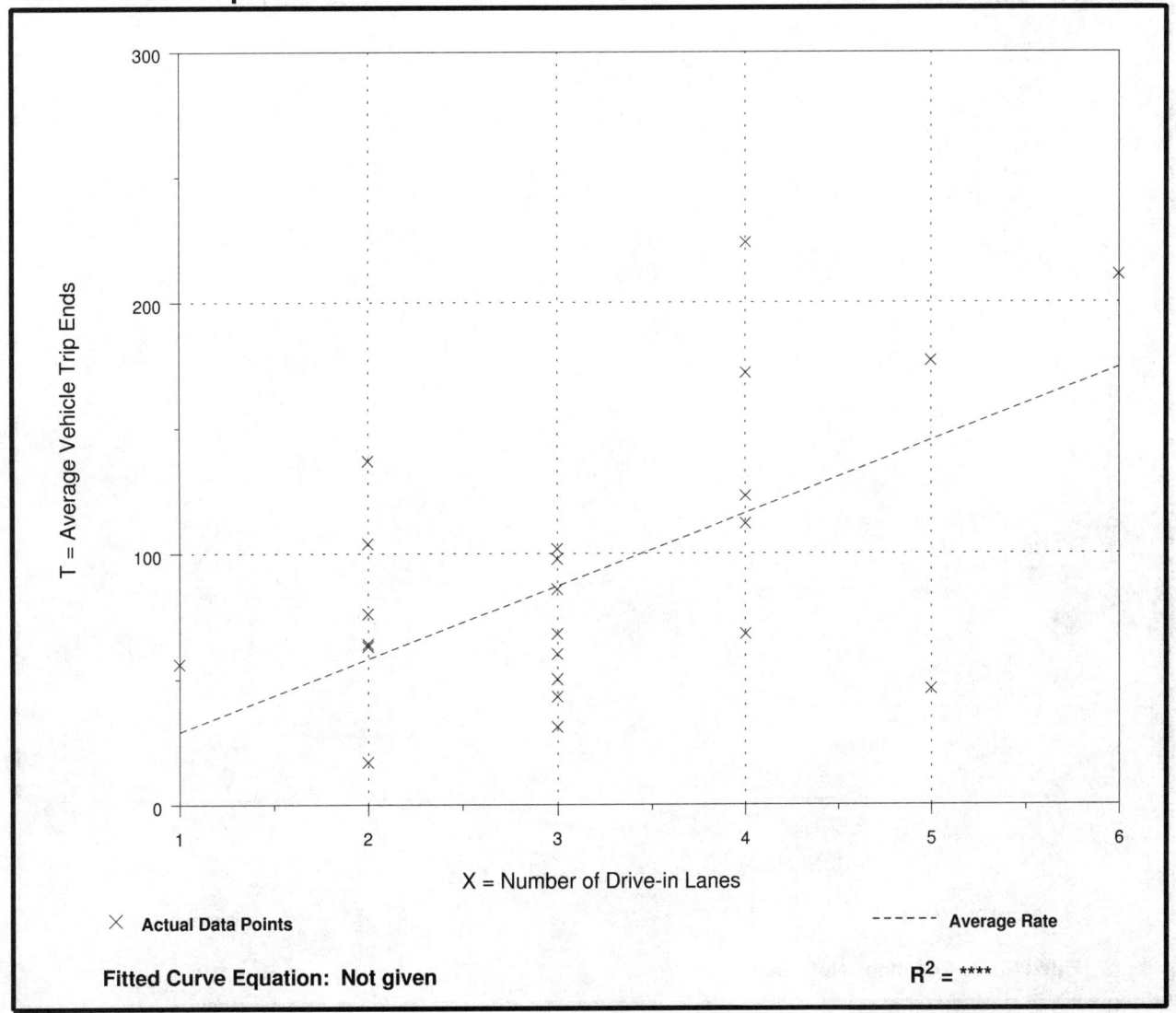

Data Plot and Equation

Fitted Curve Equation: Not given $R^2 = ****$

Drive-in Bank
(912)

Average Vehicle Trip Ends vs: Drive-in Lanes
On a: Saturday,
Peak Hour of Generator

Number of Studies: 26
Avg. Number of Drive-in Lanes: 4
Directional Distribution: 49% entering, 51% exiting

Trip Generation per Drive-in Lane

Average Rate	Range of Rates	Standard Deviation
28.78	7.60 - 107.00	16.77

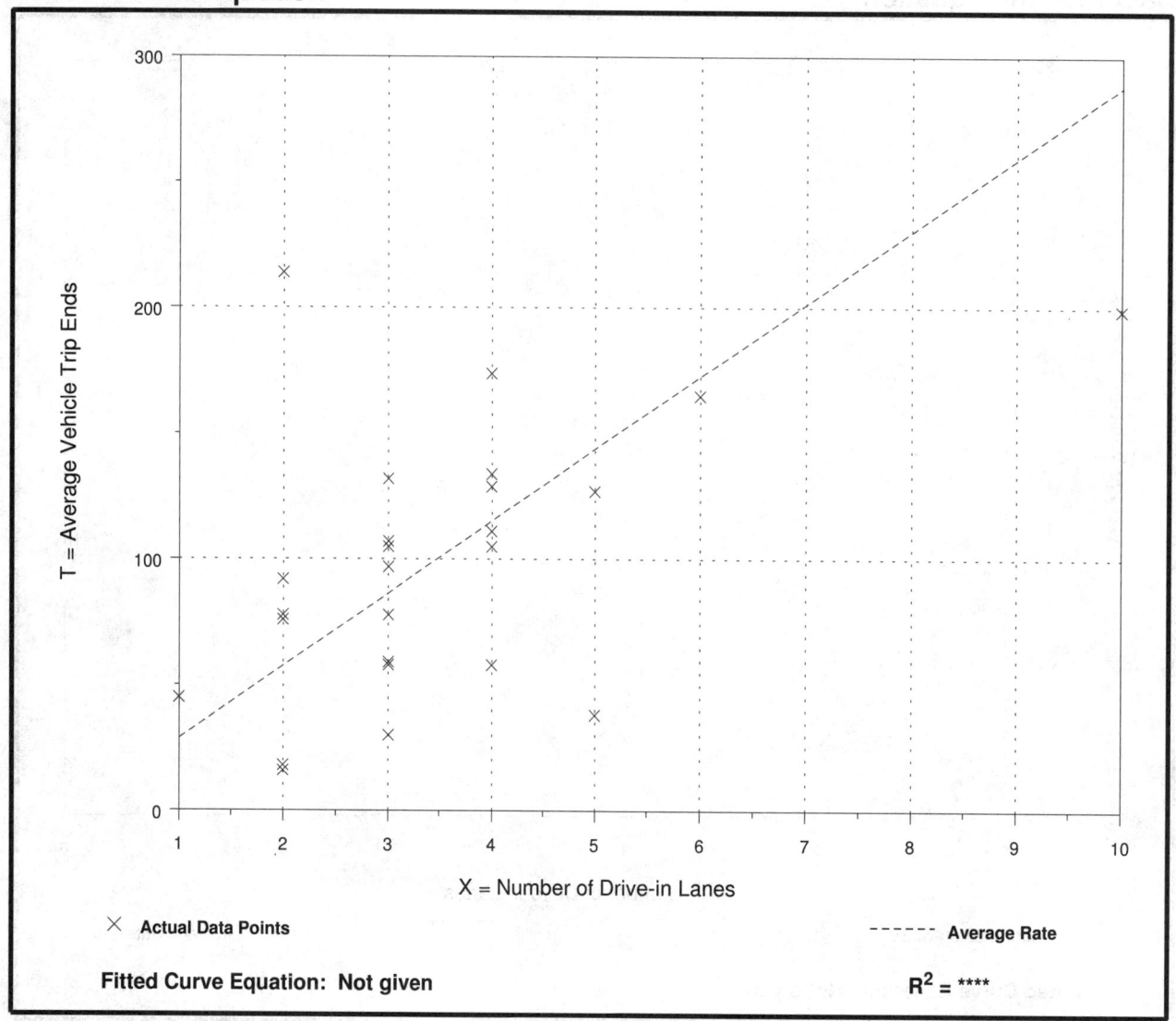

Data Plot and Equation

Fitted Curve Equation: Not given $R^2 = ****$

Land Use: 918
Hair Salon

Description

Hair salons are facilities that specialize in cosmetic and beauty services including hair cutting and styling, skin and nail care, and massage therapy. Hair salons may also contain spa facilities.

Additional Data

The surveyed site had 15 parking spaces.

The site was surveyed in 2007 in New York.

Source Number

586

Land Use: 918
Hair Salon
Independent Variables with One Observation

The following trip generation data are for independent variables with only one observation. This information is shown in this table only; there are no related plots for these data.

Users are cautioned to use data with care because of the small sample size.

Independent Variable	Trip Generation Rate	Size of Independent Variable	Number of Studies	Directional Distribution
1,000 Square Feet Gross Floor Area				
Weekday A.M. Peak Hour of Adjacent Street Traffic	1.21	4	1	100% entering, 0% exiting
Weekday P.M. Peak Hour of Adjacent Street Traffic	1.45	4	1	17% entering, 83% exiting
Weekday A.M. Peak Hour of Generator	1.21	4	1	100% entering, 0% exiting
Weekday P.M. Peak Hour of Generator	1.93	4	1	38% entering, 62% exiting
Saturday Peak Hour of Generator	5.08	4	1	36% entering, 64% exiting

Land Use: 920
Copy, Print and Express Ship Store

Description

Copy, print and express ship stores are facilities that offer a variety of copying, printing, binding and shipping services. Retail sales of a limited range of office-related items including packing and shipping supplies are also commonly available. Technology services, such as computer rental and wireless Internet may also be provided. Copy, print and express ship stores typically maintain long store hours 7 days a week. Some stores may be open 24 hours a day.

Additional Data

Peak hours of the generator—
    The weekday A.M. peak hour occurred between 10:30 a.m. and 11:30 p.m. The weekday P.M. peak hour occurred between 3:30 p.m. and 4:30 p.m.

The average vehicle occupancy rate for the A.M. and P.M. peak hour of the adjacent street at the site where information was submitted was 1.12 and 1.21, respectively. The average vehicle occupancy rate for the peak hour of the generator was 1.16 at the site surveyed.

The site was surveyed in 2007 in Texas.

Source Number

608

Land Use: 920
Copy, Print and Express Ship Store
Independent Variables with One Observation

The following trip generation data are for independent variables with only one observation. This information is shown in this table only; there are no related plots for these data.

Users are cautioned to use data with care because of the small sample size.

Independent Variable	Trip Generation Rate	Size of Independent Variable	Number of Studies	Directional Distribution
1,000 Square Feet Gross Floor Area				
Weekday A.M. Peak Hour of Adjacent Street Traffic	2.78	4	1	75% entering, 25% exiting
Weekday P.M. Peak Hour of Adjacent Street Traffic	7.41	4	1	44% entering, 56% exiting
Weekday A.M. Peak Hour of Generator	8.10	4	1	51% entering, 49% exiting
Weekday P.M. Peak Hour of Generator	12.27	4	1	43% entering, 57% exiting
Employees				
Weekday A.M. Peak Hour of Adjacent Street Traffic	1.50	8	1	75% entering, 25% exiting
Weekday P.M. Peak Hour of Adjacent Street Traffic	4.00	8	1	44% entering, 56% exiting
Weekday A.M. Peak Hour of Generator	4.38	8	1	51% entering, 49% exiting
Weekday P.M. Peak Hour of Generator	6.63	8	1	43% entering, 57% exiting

Land Use: 925
Drinking Place

Description

A drinking place contains a bar, where alcoholic beverages and food are sold, and possibly some type of entertainment, such as music, television screens, video games, or pool tables. Establishments that specialize in serving food but also have bars are not included in this land use.

Additional Data

The sites were surveyed in 1987, 1995 and 1997 in Colorado, Oregon and South Dakota.

Source Numbers

291, 358, 583

Drinking Place
(925)

Average Vehicle Trip Ends vs: 1000 Sq. Feet Gross Floor Area
On a: Weekday,
Peak Hour of Adjacent Street Traffic,
One Hour Between 4 and 6 p.m.

Number of Studies: 12
Average 1000 Sq. Feet GFA: 4
Directional Distribution: 66% entering, 34% exiting

Trip Generation per 1000 Sq. Feet Gross Floor Area

Average Rate	Range of Rates	Standard Deviation
11.34	3.73 - 29.98	8.04

Data Plot and Equation

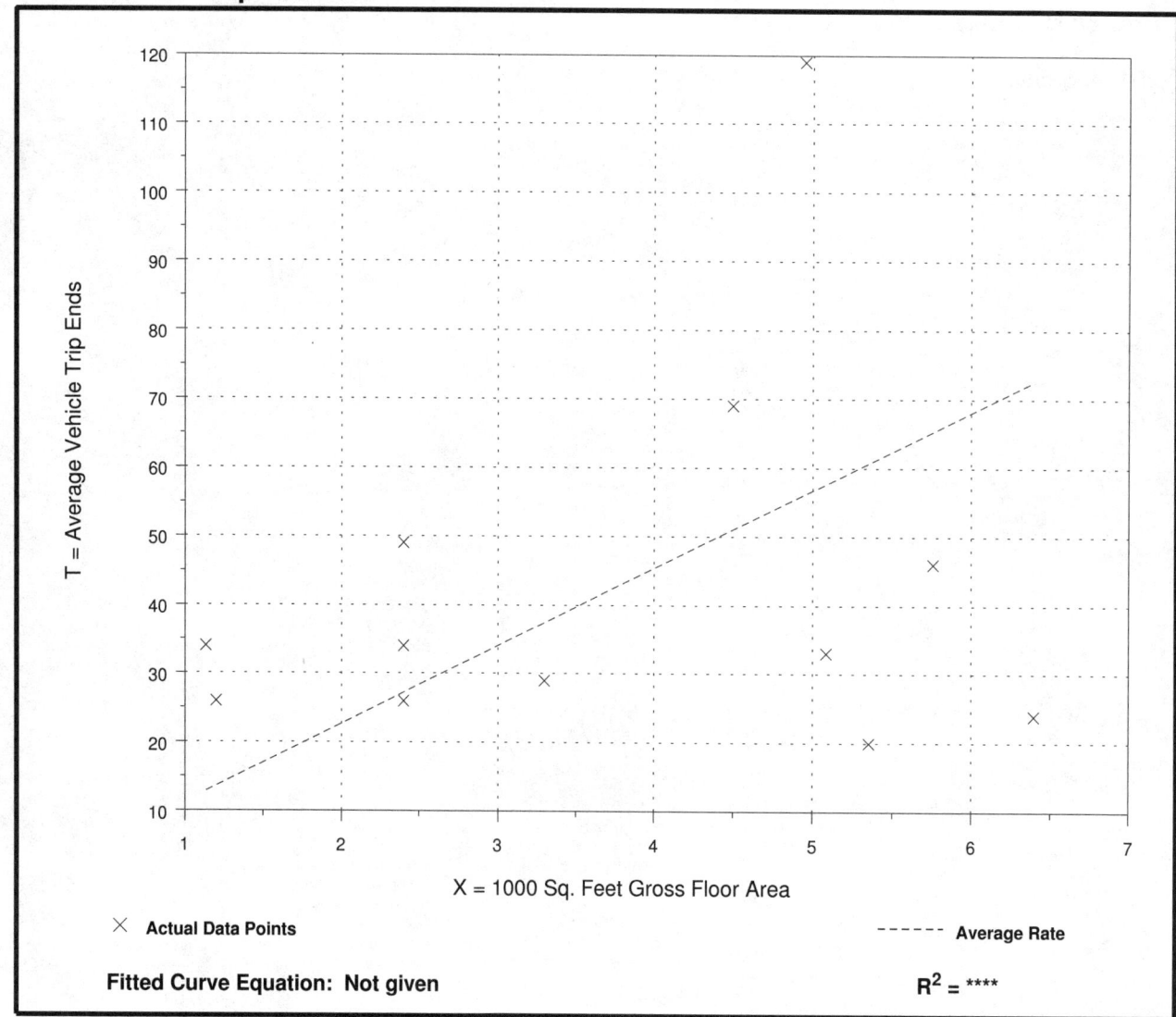

Fitted Curve Equation: Not given $R^2 = ****$

Drinking Place
(925)

Average Vehicle Trip Ends vs: 1000 Sq. Feet Gross Floor Area
On a: Weekday,
P.M. Peak Hour of Generator

Number of Studies: 8
Average 1000 Sq. Feet GFA: 3
Directional Distribution: 68% entering, 32% exiting

Trip Generation per 1000 Sq. Feet Gross Floor Area

Average Rate	Range of Rates	Standard Deviation
15.49	3.73 - 29.98	8.63

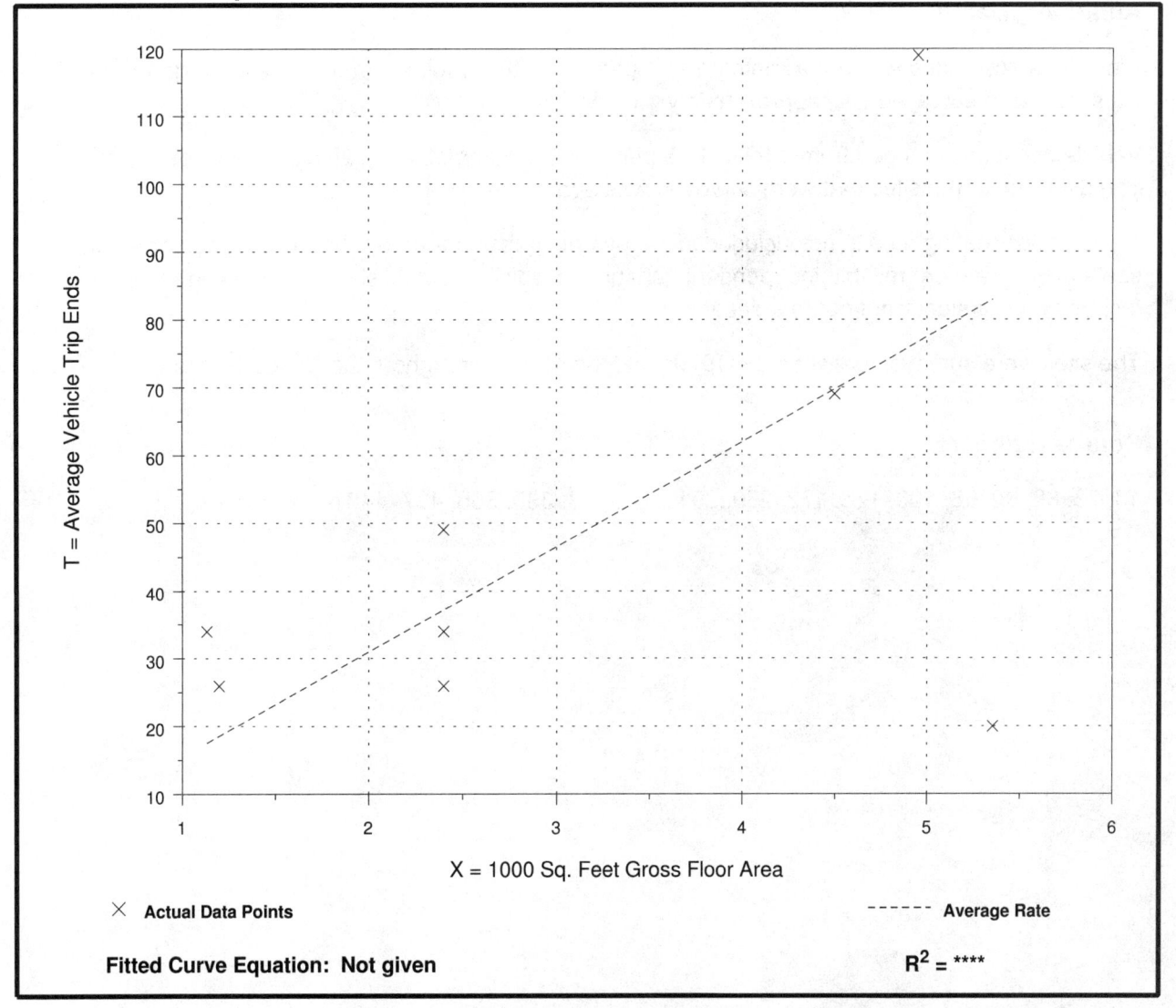

Data Plot and Equation

Fitted Curve Equation: Not given $R^2 = ****$

Land Use: 931
Quality Restaurant

Description

This land use consists of high quality, full-service eating establishments with typical duration of stay of at least one hour. Quality restaurants generally do not serve breakfast; some do not serve lunch; all serve dinner. This type of restaurant often requests and sometimes requires reservations and is generally not part of a chain. Patrons commonly wait to be seated, are served by a waiter/waitress, order from menus and pay for meals after they eat. While some of the study sites have lounge or bar facilities (serving alcoholic beverages), they are ancillary to the restaurant. High-turnover (sit-down) restaurant (Land Use 932) is a related use.

Additional Data

Truck trips accounted for approximately 1 to 4 percent of the weekday traffic. The average for the sites that were surveyed was approximately 1.6 percent.

Vehicle occupancy ranged from 1.59 to 1.98 persons per automobile on an average weekday. The average for the sites that were surveyed was approximately 1.78.

The outdoor seating area is not included in the overall gross floor area. Therefore, the number of seats may be a more reliable independent variable on which to establish trip generation rates for facilities having significant outdoor seating.

The sites were surveyed between the 1970s and the 1990s throughout the United States.

Source Numbers

13, 73, 88, 90, 98, 100, 126, 172, 260, 291, 301, 338, 339, 368, 437, 440

Quality Restaurant
(931)

Average Vehicle Trip Ends vs: 1000 Sq. Feet Gross Floor Area
On a: Weekday

Number of Studies: 15
Average 1000 Sq. Feet GFA: 9
Directional Distribution: 50% entering, 50% exiting

Trip Generation per 1000 Sq. Feet Gross Floor Area

Average Rate	Range of Rates	Standard Deviation
89.95	33.41 - 139.80	36.81

Data Plot and Equation

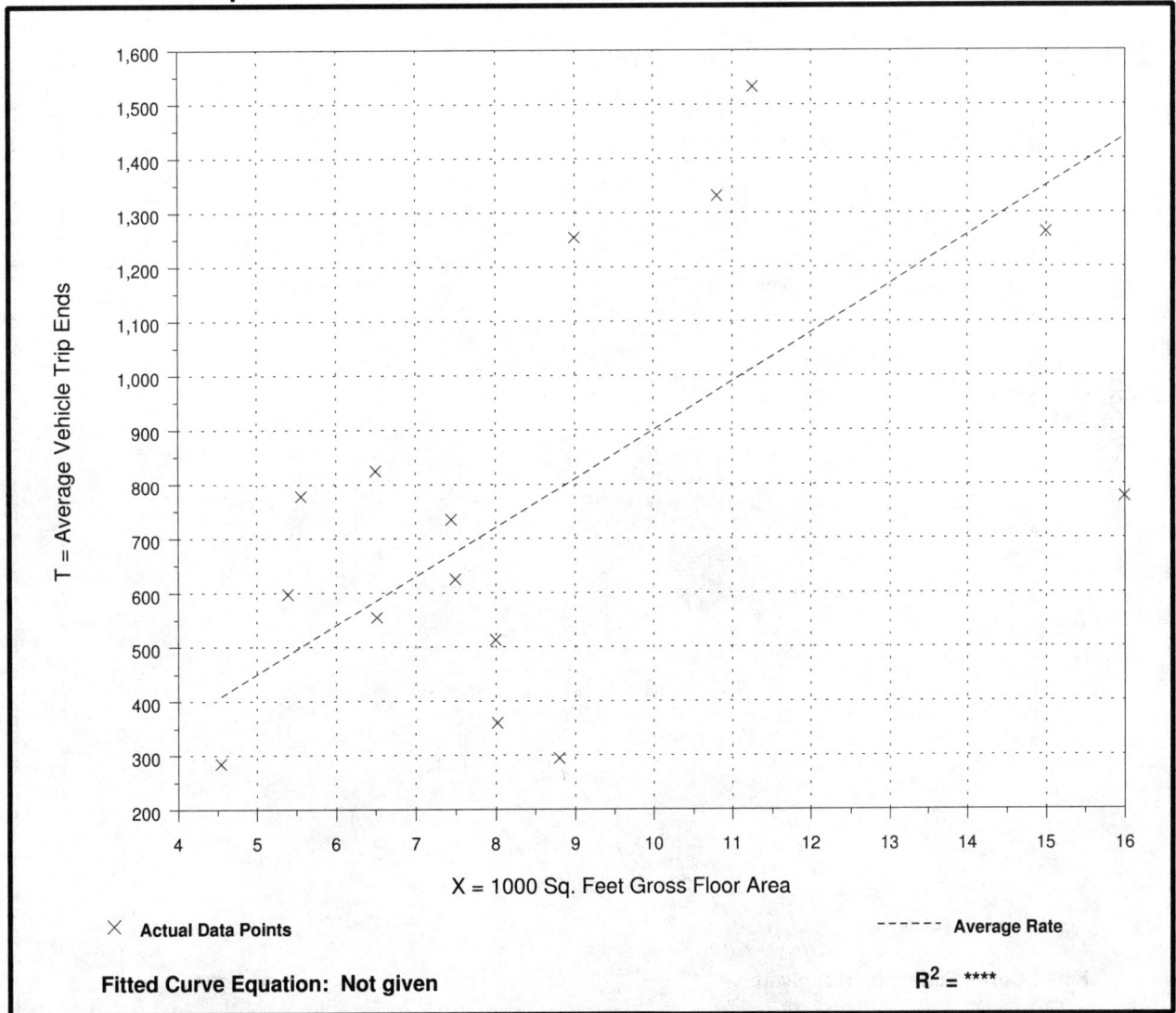

Fitted Curve Equation: Not given $R^2 = ****$

Quality Restaurant
(931)

Average Vehicle Trip Ends vs: 1000 Sq. Feet Gross Floor Area
On a: Weekday,
Peak Hour of Adjacent Street Traffic,
One Hour Between 7 and 9 a.m.

Number of Studies: 11
Average 1000 Sq. Feet GFA: 9
Directional Distribution: Not available

Trip Generation per 1000 Sq. Feet Gross Floor Area

Average Rate	Range of Rates	Standard Deviation
0.81	0.25 - 1.60	0.93

Data Plot and Equation

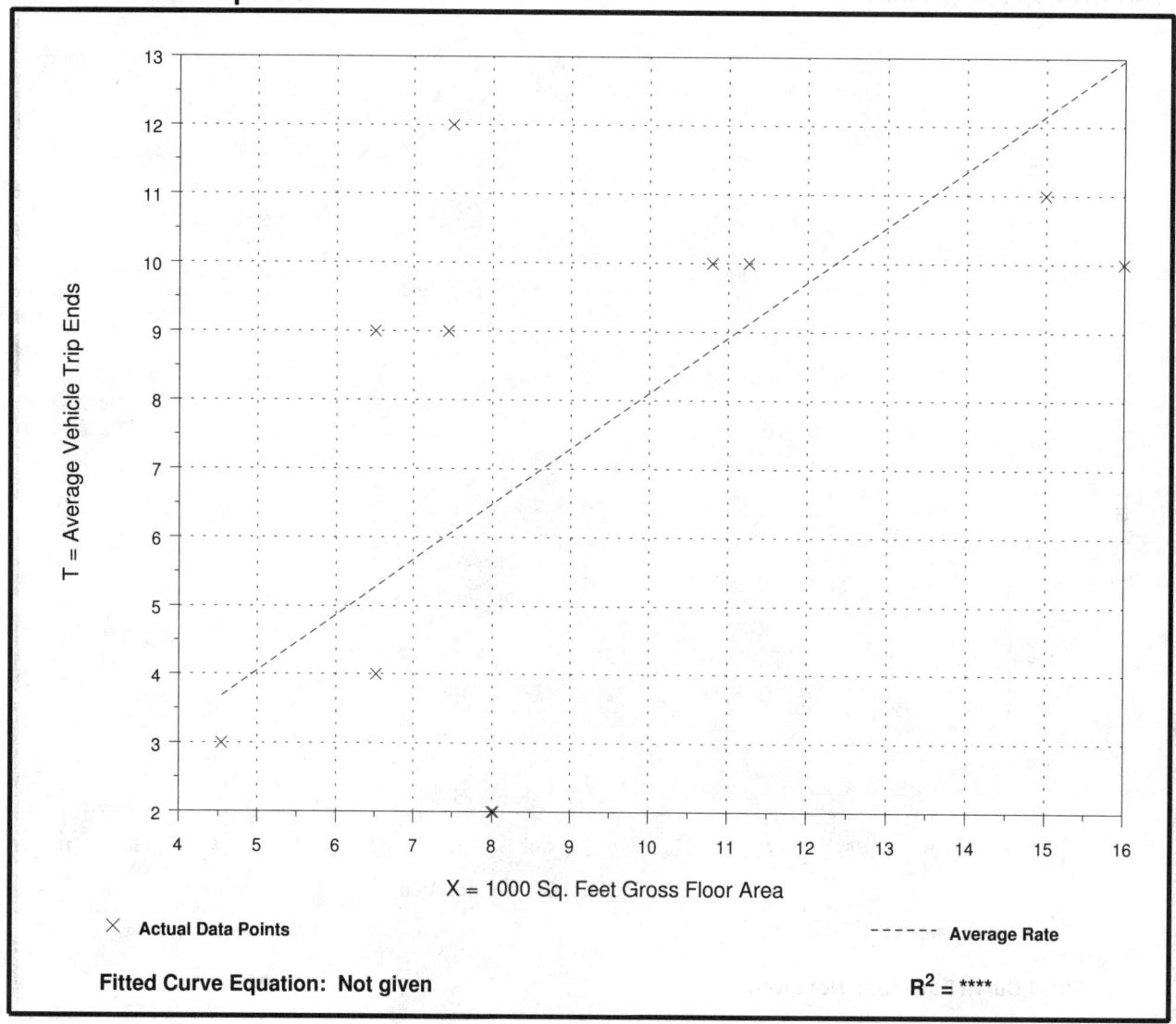

Fitted Curve Equation: Not given $R^2 =$ ****

Quality Restaurant
(931)

Average Vehicle Trip Ends vs: 1000 Sq. Feet Gross Floor Area
On a: Weekday,
Peak Hour of Adjacent Street Traffic,
One Hour Between 4 and 6 p.m.

Number of Studies: 24
Average 1000 Sq. Feet GFA: 9
Directional Distribution: 67% entering, 33% exiting

Trip Generation per 1000 Sq. Feet Gross Floor Area

Average Rate	Range of Rates	Standard Deviation
7.49	2.42 - 18.64	4.89

Data Plot and Equation

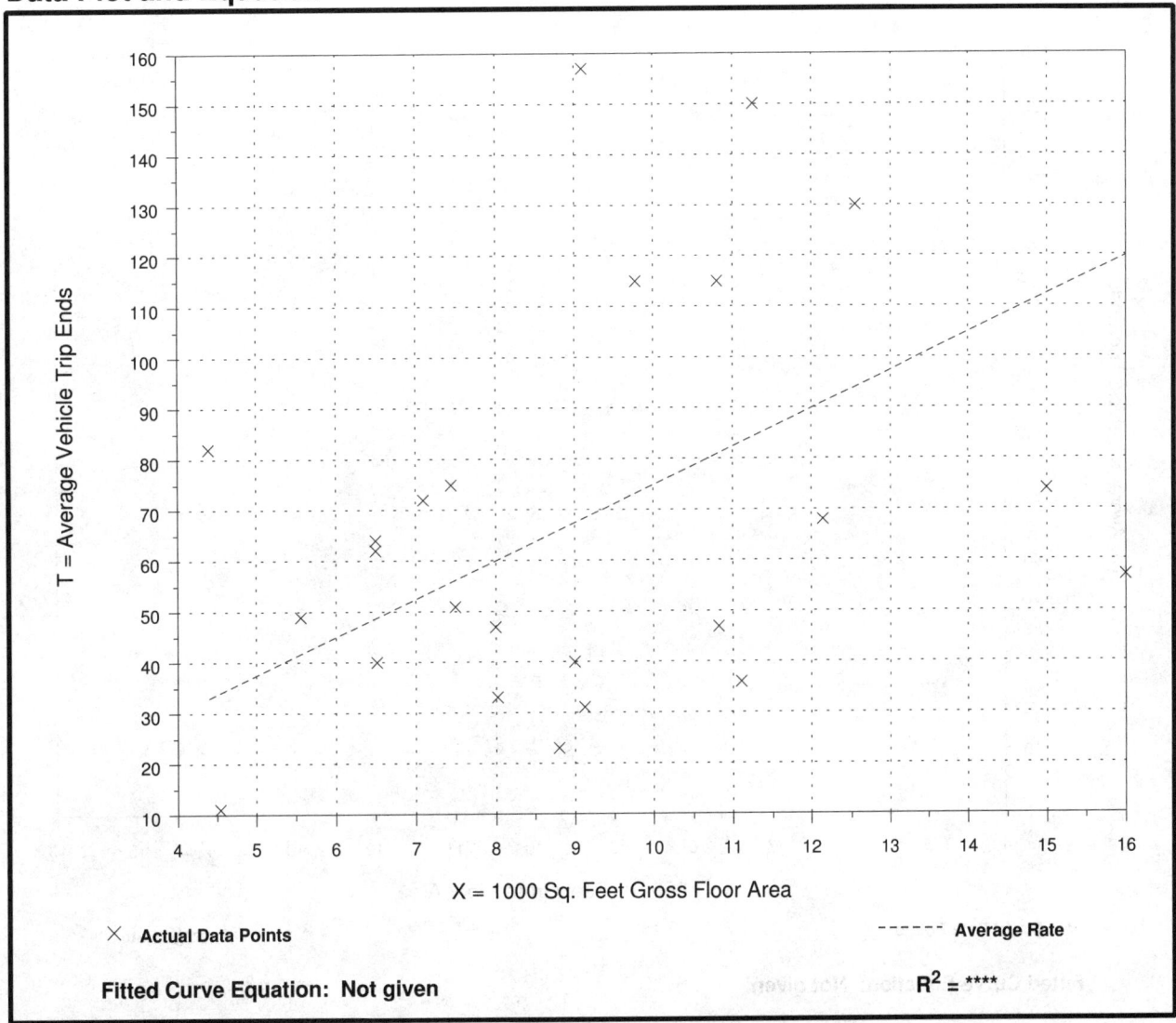

X = Actual Data Points ------ Average Rate

Fitted Curve Equation: Not given $R^2 = ****$

Quality Restaurant
(931)

Average Vehicle Trip Ends vs: 1000 Sq. Feet Gross Floor Area
On a: Weekday,
A.M. Peak Hour of Generator

Number of Studies: 14
Average 1000 Sq. Feet GFA: 9
Directional Distribution: 82% entering, 18% exiting

Trip Generation per 1000 Sq. Feet Gross Floor Area

Average Rate	Range of Rates	Standard Deviation
5.57	0.87 - 10.37	3.79

Data Plot and Equation

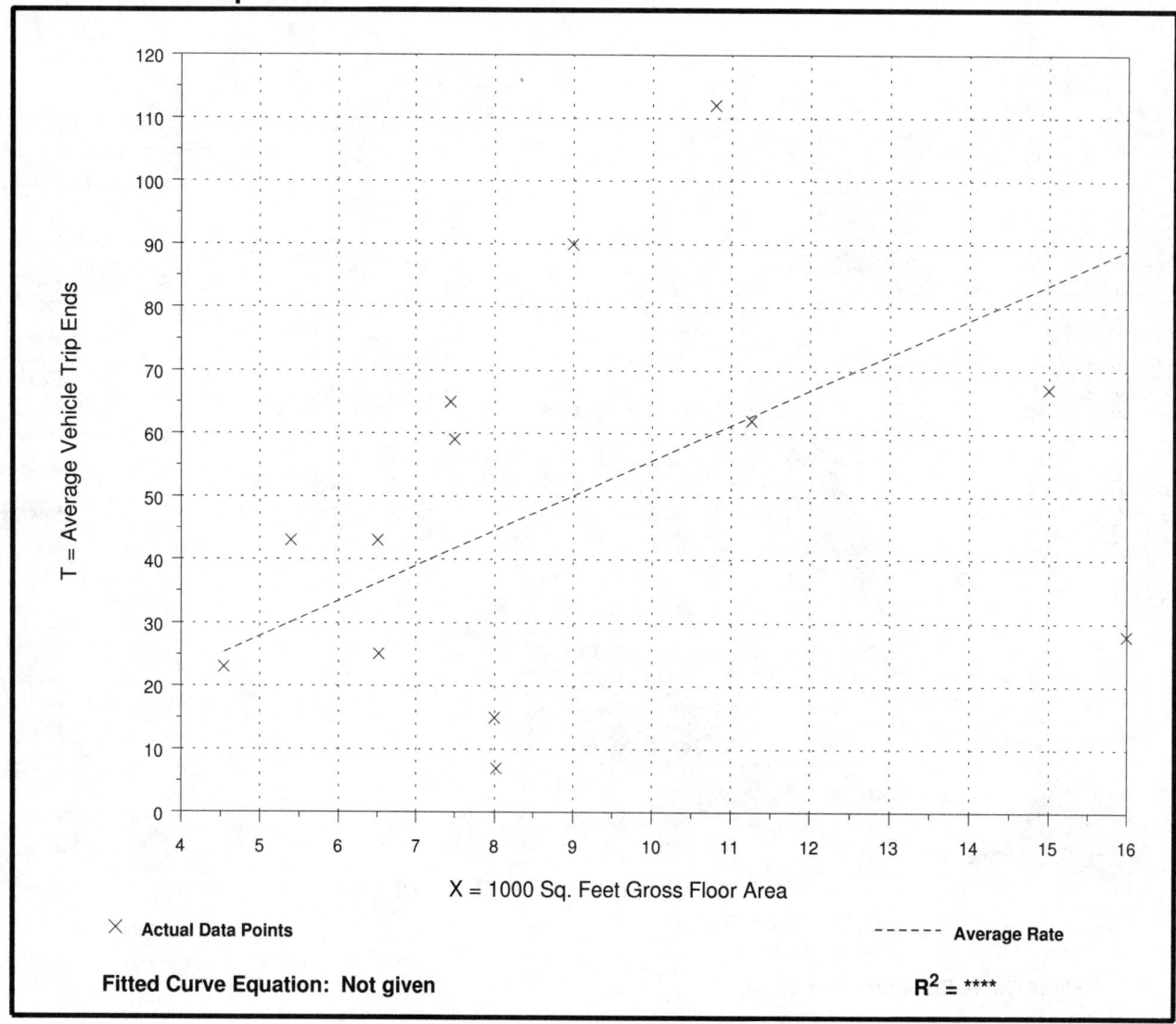

Fitted Curve Equation: Not given $R^2 = ****$

Quality Restaurant
(931)

Average Vehicle Trip Ends vs: 1000 Sq. Feet Gross Floor Area
On a: Weekday,
P.M. Peak Hour of Generator

Number of Studies: 16
Average 1000 Sq. Feet GFA: 9
Directional Distribution: 62% entering, 38% exiting

Trip Generation per 1000 Sq. Feet Gross Floor Area

Average Rate	Range of Rates	Standard Deviation
9.02	3.24 - 15.89	4.55

Data Plot and Equation

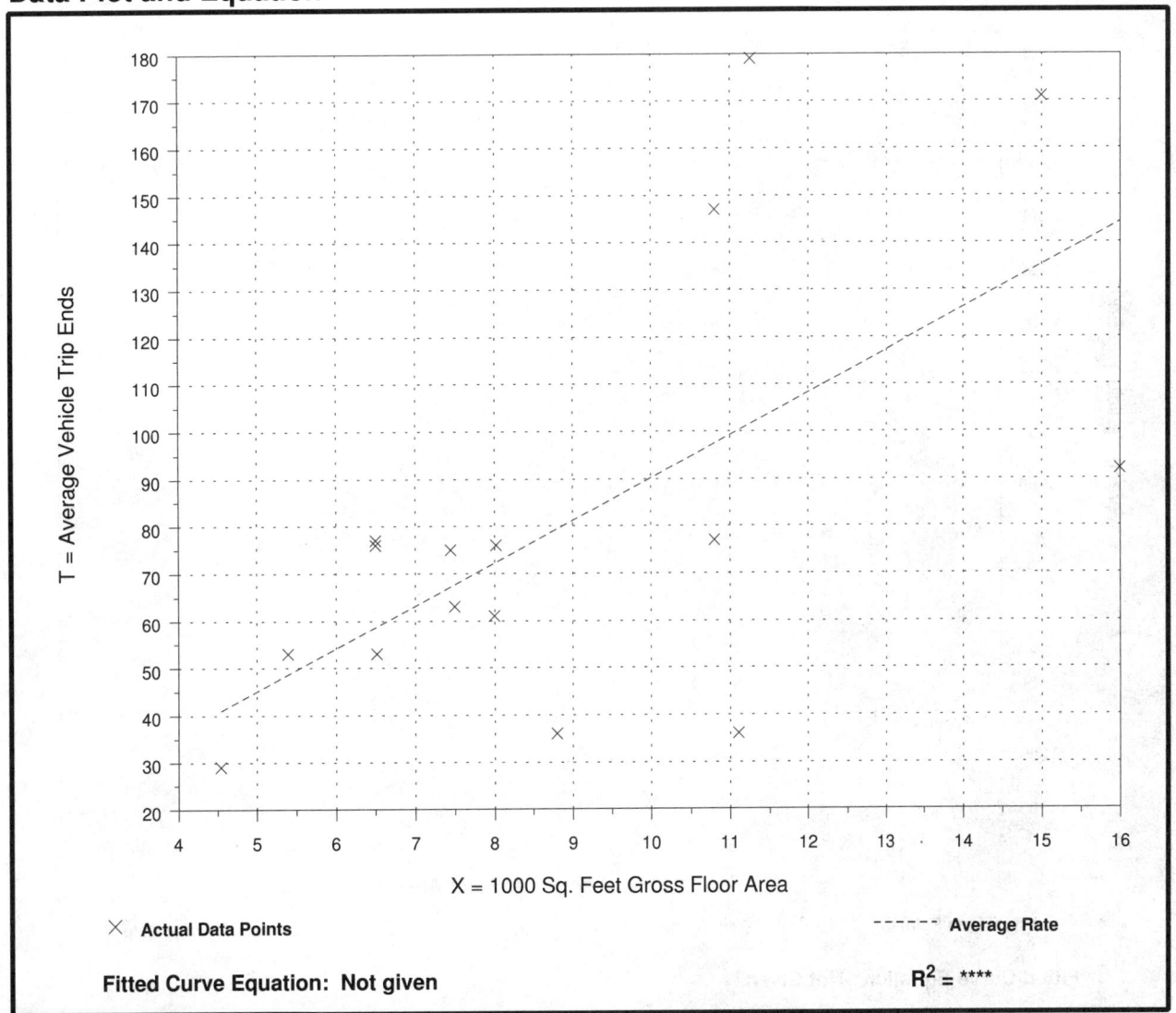

X Actual Data Points

------ Average Rate

Fitted Curve Equation: Not given $R^2 =$ ****

Quality Restaurant
(931)

Average Vehicle Trip Ends vs: 1000 Sq. Feet Gross Floor Area
On a: Saturday

Number of Studies: 11
Average 1000 Sq. Feet GFA: 9
Directional Distribution: 50% entering, 50% exiting

Trip Generation per 1000 Sq. Feet Gross Floor Area

Average Rate	Range of Rates	Standard Deviation
94.36	53.63 - 156.67	34.42

Data Plot and Equation

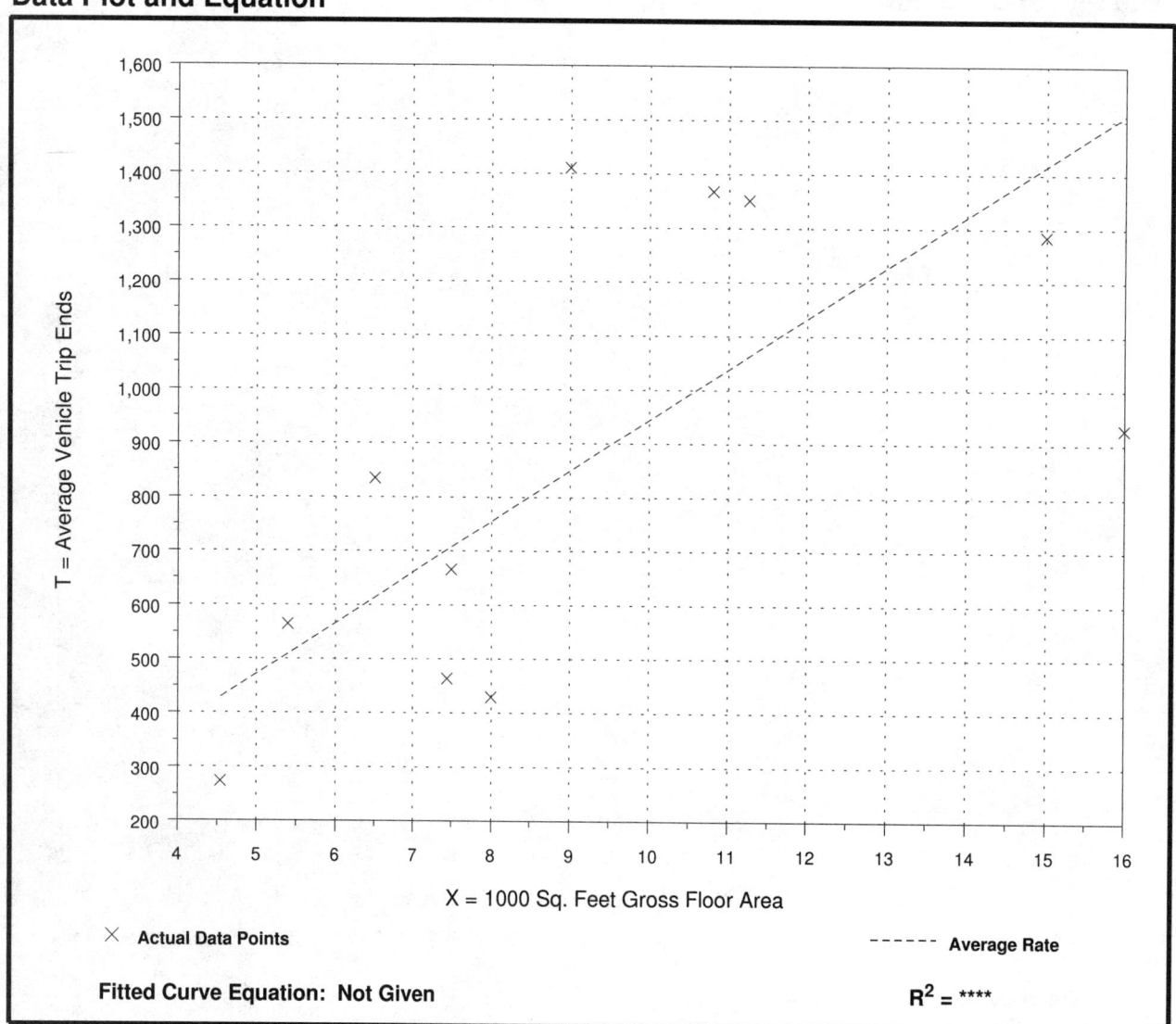

Fitted Curve Equation: Not Given $R^2 = ****$

Quality Restaurant
(931)

Average Vehicle Trip Ends vs: 1000 Sq. Feet Gross Floor Area
On a: Saturday,
Peak Hour of Generator

Number of Studies: 11
Average 1000 Sq. Feet GFA: 9
Directional Distribution: 59% entering, 41% exiting

Trip Generation per 1000 Sq. Feet Gross Floor Area

Average Rate	Range of Rates	Standard Deviation
10.82	5.75 - 15.28	4.38

Data Plot and Equation

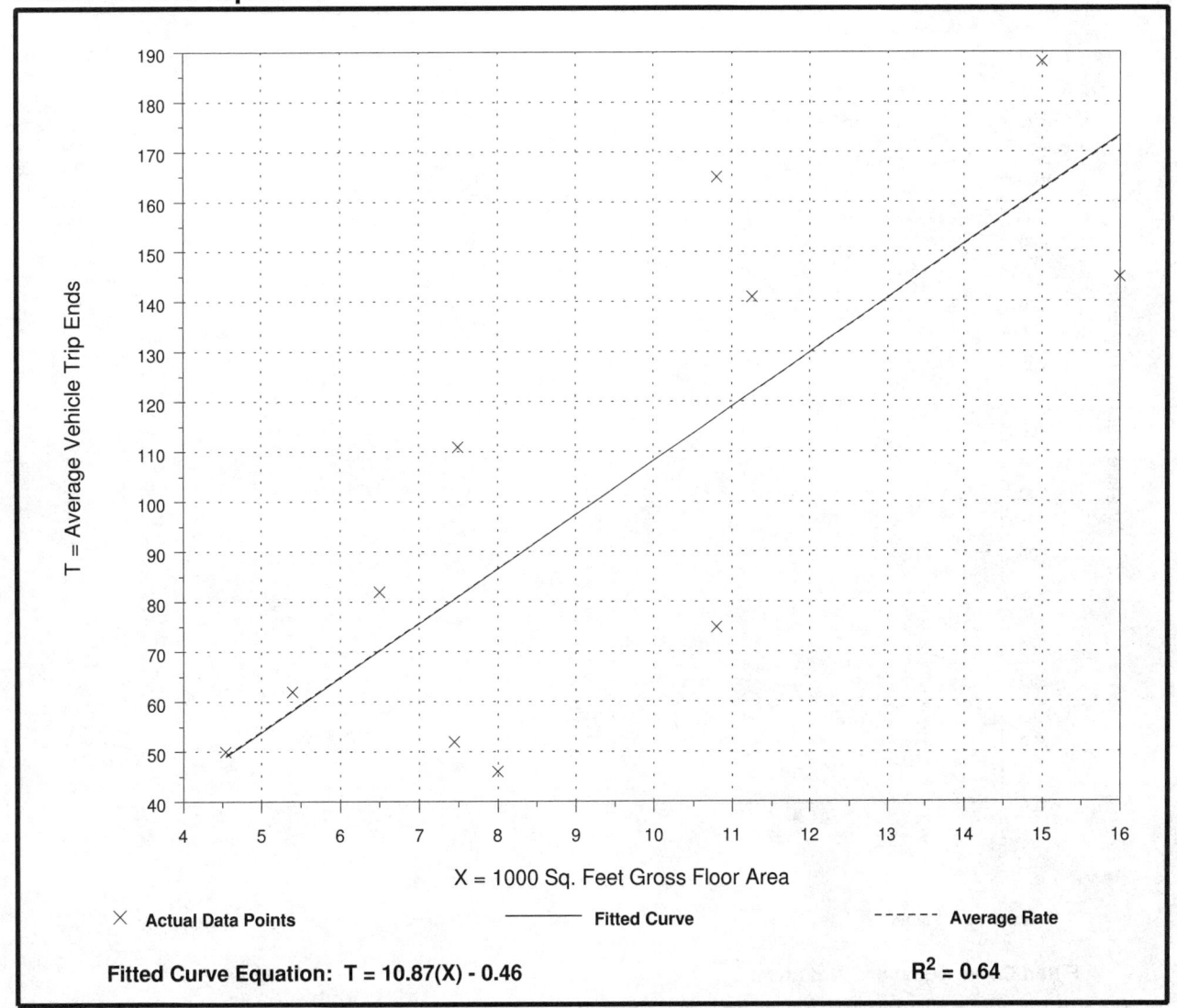

Fitted Curve Equation: T = 10.87(X) - 0.46 $R^2 = 0.64$

Quality Restaurant
(931)

Average Vehicle Trip Ends vs: 1000 Sq. Feet Gross Floor Area
On a: Sunday

Number of Studies: 11
Average 1000 Sq. Feet GFA: 9
Directional Distribution: 50% entering, 50% exiting

Trip Generation per 1000 Sq. Feet Gross Floor Area

Average Rate	Range of Rates	Standard Deviation
72.16	34.09 - 137.78	32.35

Data Plot and Equation

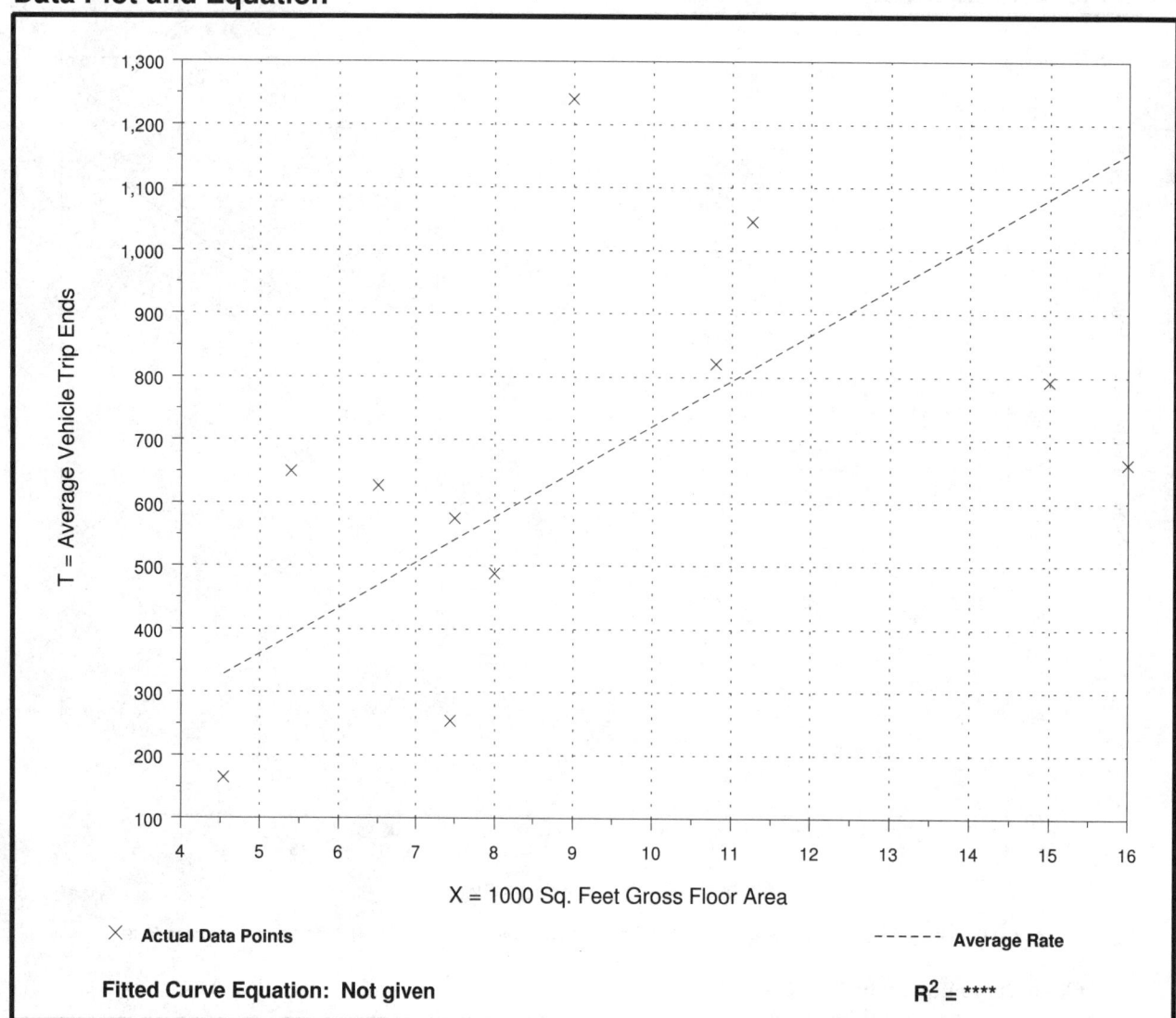

Fitted Curve Equation: Not given $R^2 = ****$

Quality Restaurant
(931)

Average Vehicle Trip Ends vs: 1000 Sq. Feet Gross Floor Area
On a: Sunday,
Peak Hour of Generator

Number of Studies: 10
Average 1000 Sq. Feet GFA: 9
Directional Distribution: 63% entering, 37% exiting

Trip Generation per 1000 Sq. Feet Gross Floor Area

Average Rate	Range of Rates	Standard Deviation
8.38	4.56 - 12.07	3.88

Data Plot and Equation

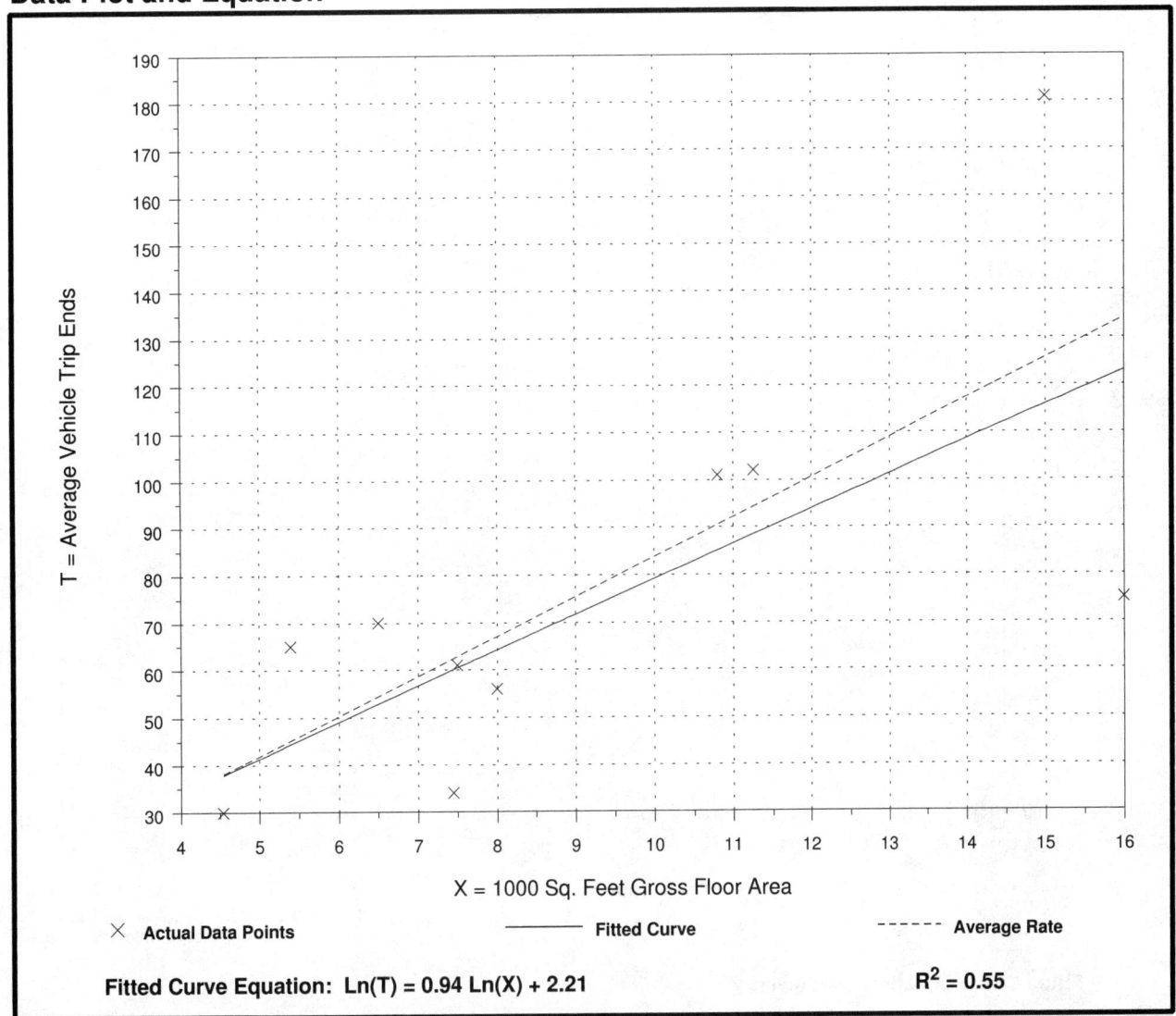

Fitted Curve Equation: $Ln(T) = 0.94 Ln(X) + 2.21$ $R^2 = 0.55$

Quality Restaurant
(931)

Average Vehicle Trip Ends vs: Seats
On a: Weekday

Number of Studies: 11
Average Number of Seats: 309
Directional Distribution: 50% entering, 50% exiting

Trip Generation per Seat

Average Rate	Range of Rates	Standard Deviation
2.86	1.77 - 5.50	1.96

Data Plot and Equation

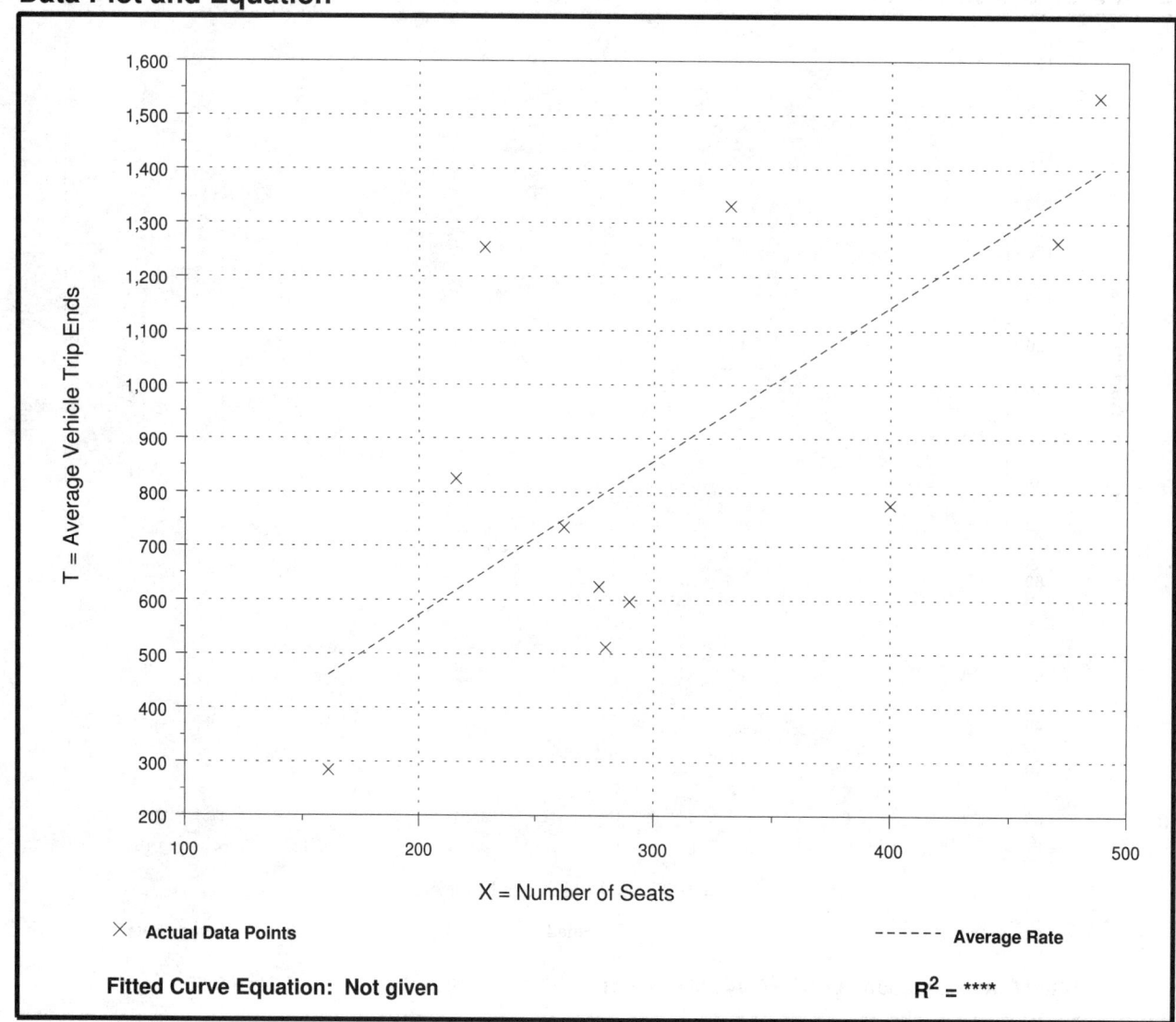

X Actual Data Points

-------- Average Rate

Fitted Curve Equation: Not given

$R^2 = ****$

Quality Restaurant
(931)

Average Vehicle Trip Ends vs: Seats
On a: Weekday,
Peak Hour of Adjacent Street Traffic,
One Hour Between 7 and 9 a.m.

Number of Studies: 9
Average Number of Seats: 321
Directional Distribution: Not available

Trip Generation per Seat

Average Rate	Range of Rates	Standard Deviation
0.03	0.01 - 0.04	0.16

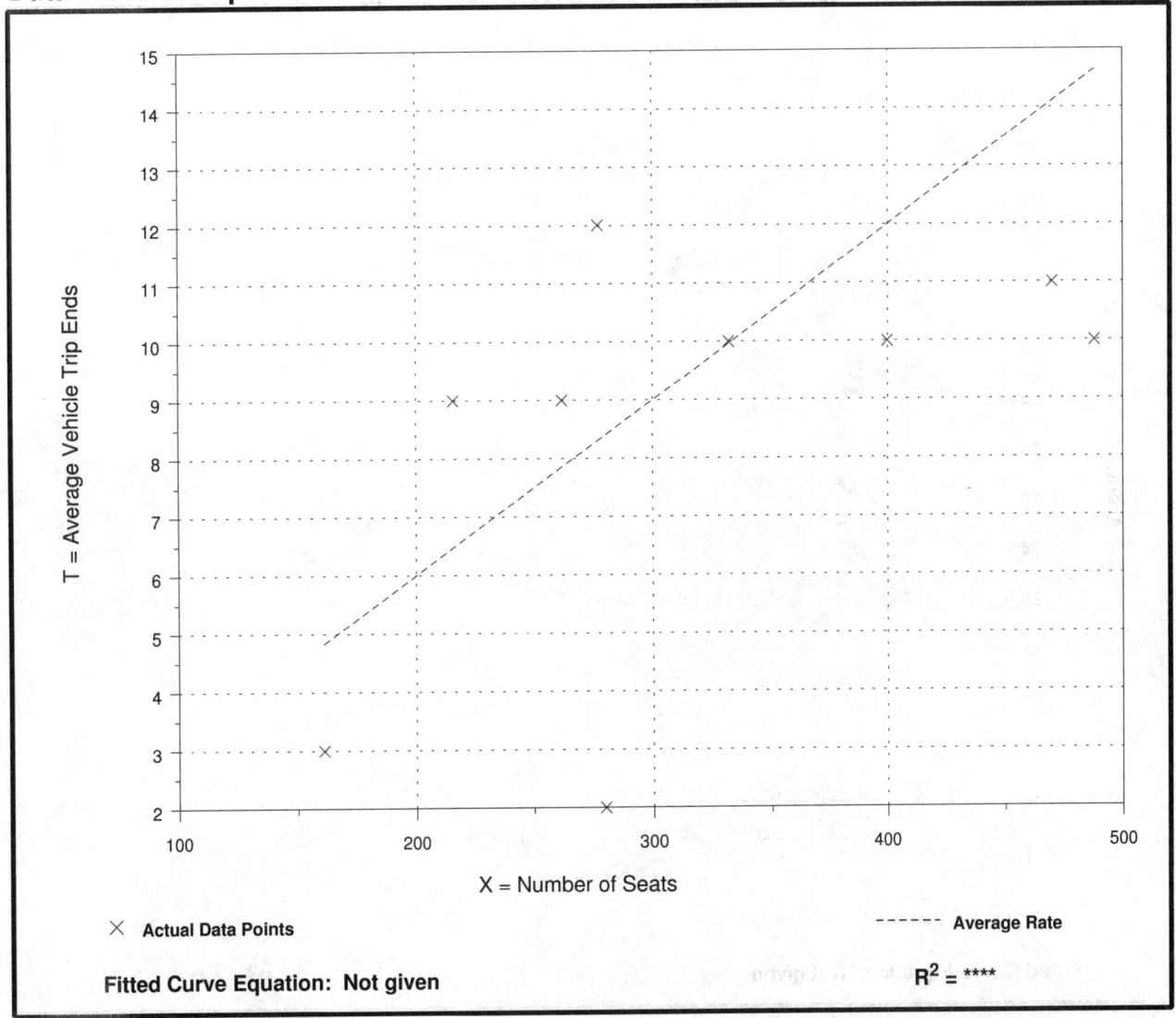

Data Plot and Equation

X Actual Data Points   ------ Average Rate

Fitted Curve Equation: Not given $R^2 = ****$

Trip Generation, 9th Edition • Institute of Transportation Engineers

Quality Restaurant
(931)

Average Vehicle Trip Ends vs: Seats
On a: Weekday,
Peak Hour of Adjacent Street Traffic,
One Hour Between 4 and 6 p.m.

Number of Studies: 15
Average Number of Seats: 326
Directional Distribution: 67% entering, 33% exiting

Trip Generation per Seat

Average Rate	Range of Rates	Standard Deviation
0.26	0.07 - 0.50	0.52

Data Plot and Equation

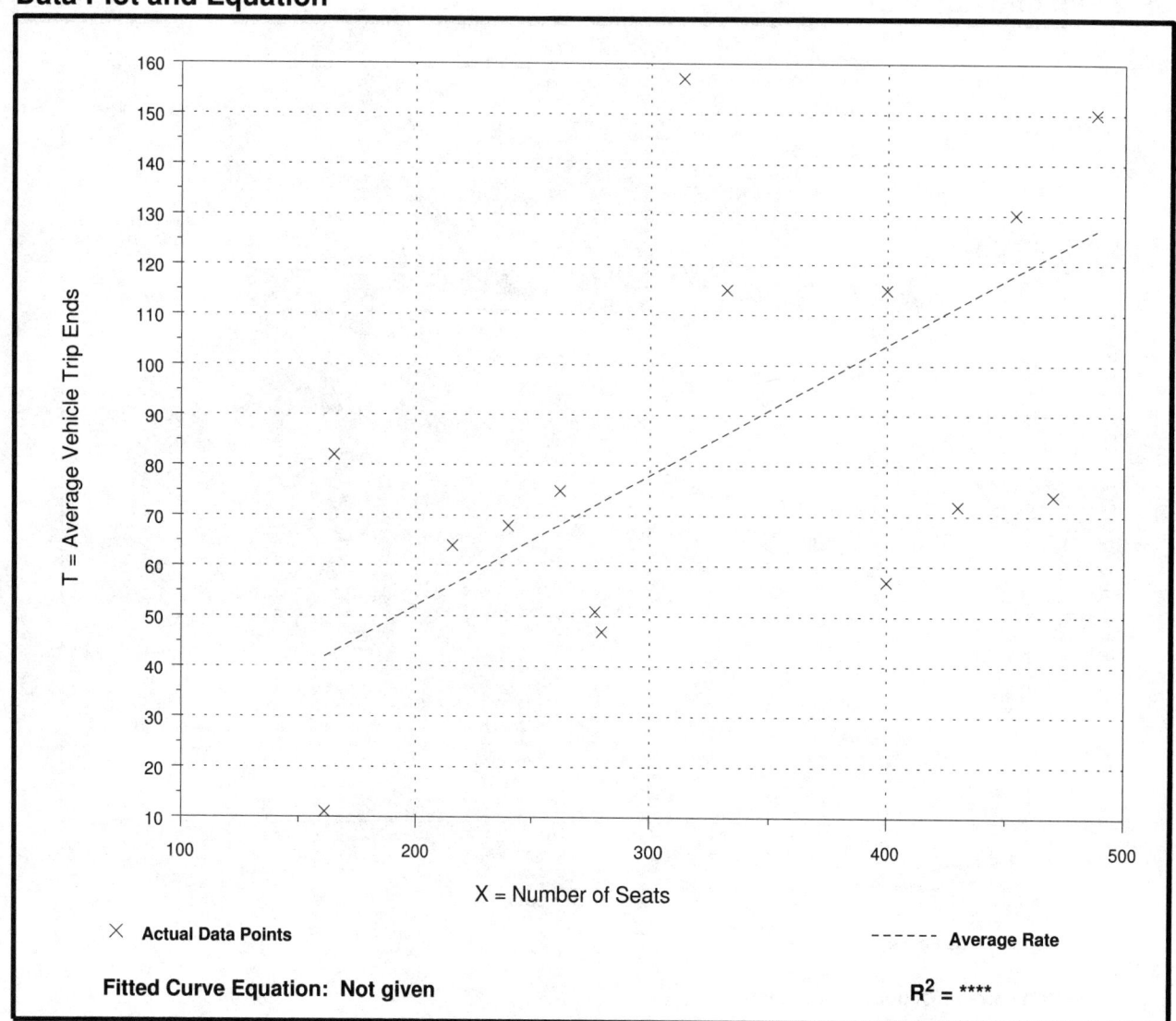

Fitted Curve Equation: Not given $R^2 = ****$

Quality Restaurant
(931)

Average Vehicle Trip Ends vs: Seats
On a: Weekday,
A.M. Peak Hour of Generator

Number of Studies: 10
Average Number of Seats: 318
Directional Distribution: 70% entering, 30% exiting

Trip Generation per Seat

Average Rate	Range of Rates	Standard Deviation
0.16	0.05 - 0.34	0.41

Data Plot and Equation

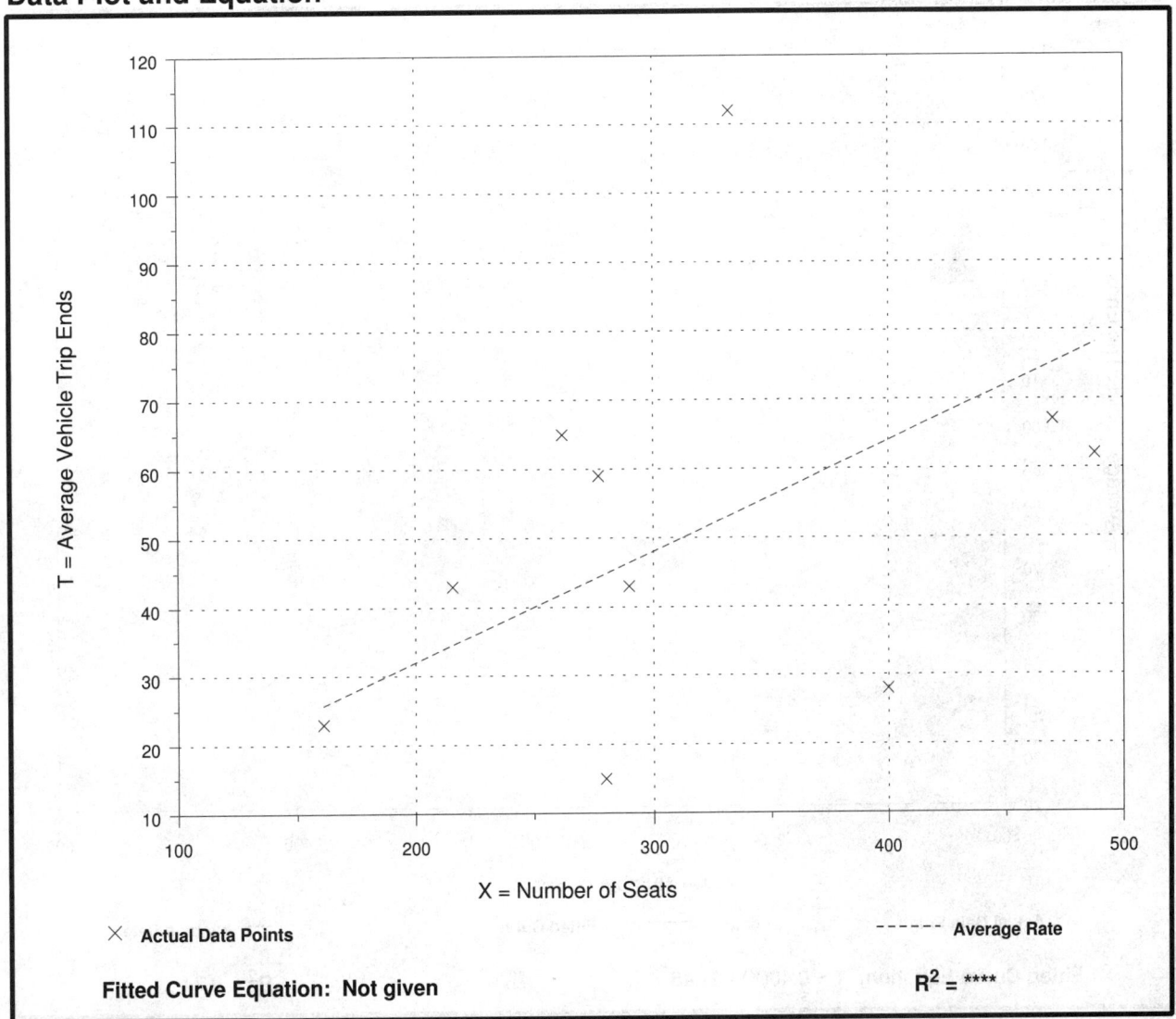

× Actual Data Points   ----- Average Rate

Fitted Curve Equation: Not given $R^2 = ****$

Trip Generation, 9th Edition • Institute of Transportation Engineers

Quality Restaurant
(931)

Average Vehicle Trip Ends vs: Seats
On a: Weekday,
P.M. Peak Hour of Generator

Number of Studies: 11
Average Number of Seats: 308
Directional Distribution: 59% entering, 41% exiting

Trip Generation per Seat

Average Rate	Range of Rates	Standard Deviation
0.30	0.18 - 0.44	0.56

Data Plot and Equation

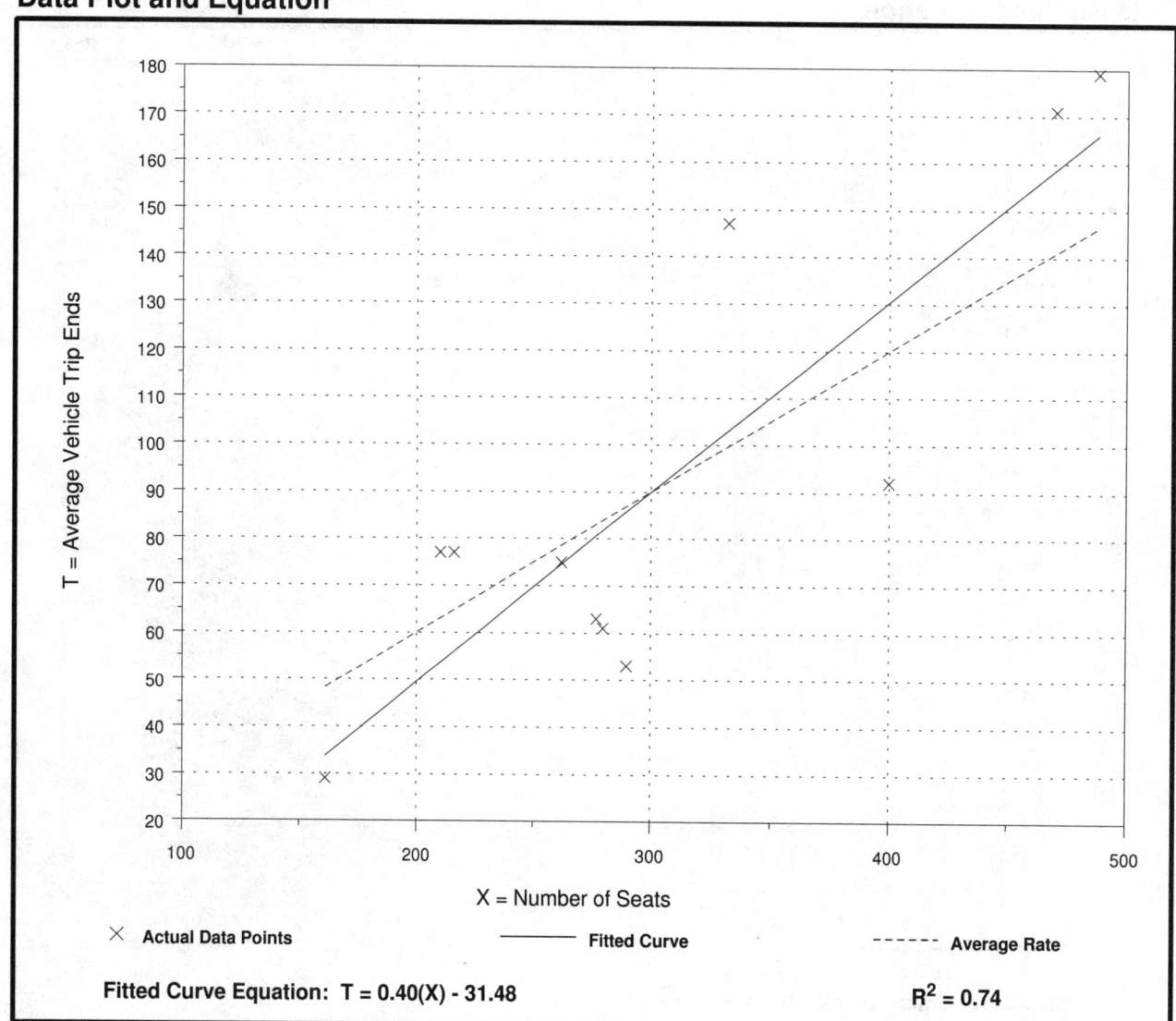

Fitted Curve Equation: $T = 0.40(X) - 31.48$ $R^2 = 0.74$

Quality Restaurant
(931)

Average Vehicle Trip Ends vs: Seats
On a: Saturday

Number of Studies: 11
Average Number of Seats: 309
Directional Distribution: 50% entering, 50% exiting

Trip Generation per Seat

Average Rate	Range of Rates	Standard Deviation
2.81	1.53 - 6.18	2.05

Data Plot and Equation

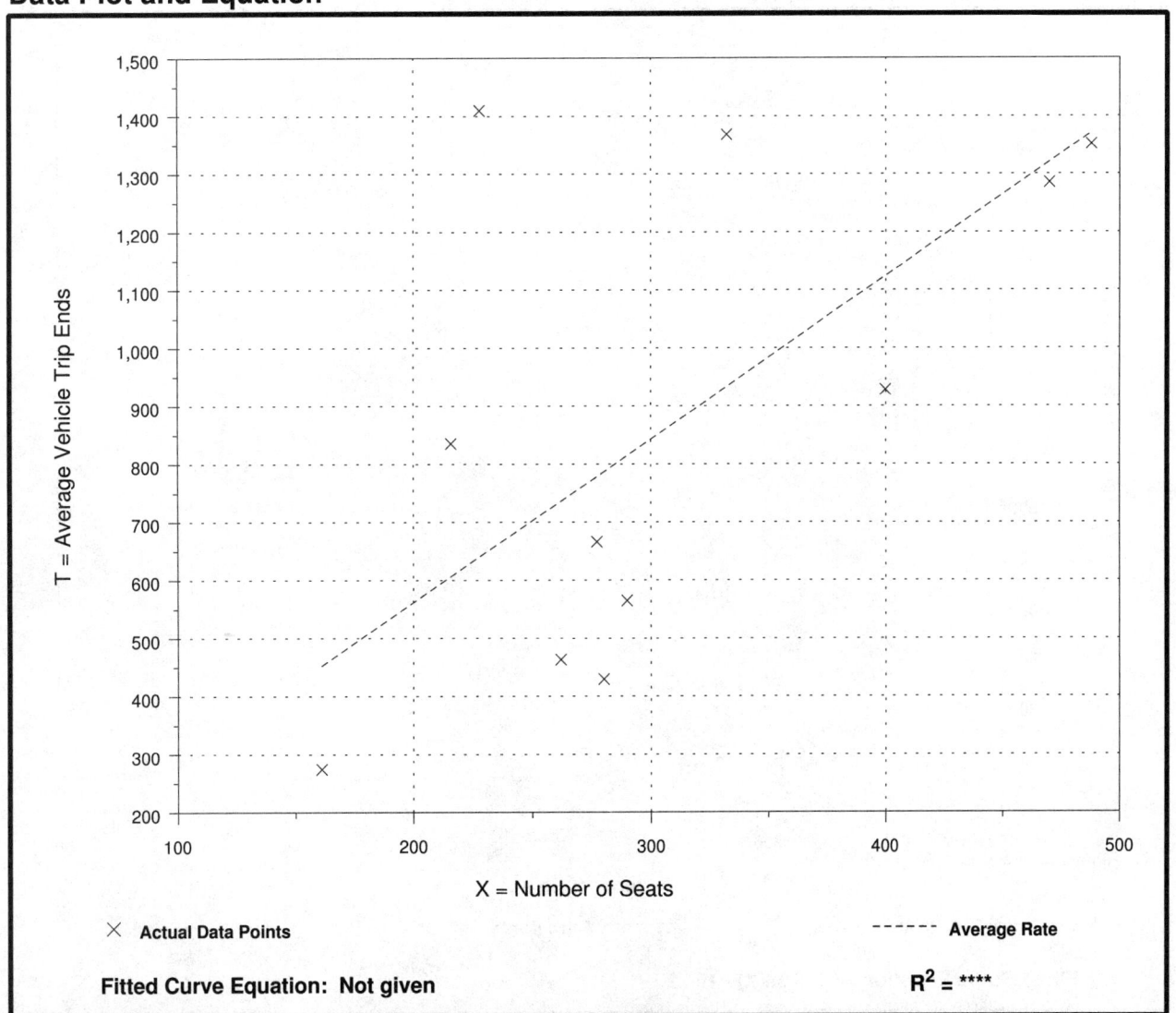

Fitted Curve Equation: Not given $R^2 = ****$

Quality Restaurant
(931)

Average Vehicle Trip Ends vs: Seats
On a: Saturday,
Peak Hour of Generator

Number of Studies: 11
Average Number of Seats: 308
Directional Distribution: 59% entering, 41% exiting

Trip Generation per Seat

Average Rate	Range of Rates	Standard Deviation
0.33	0.16 - 0.50	0.58

Data Plot and Equation

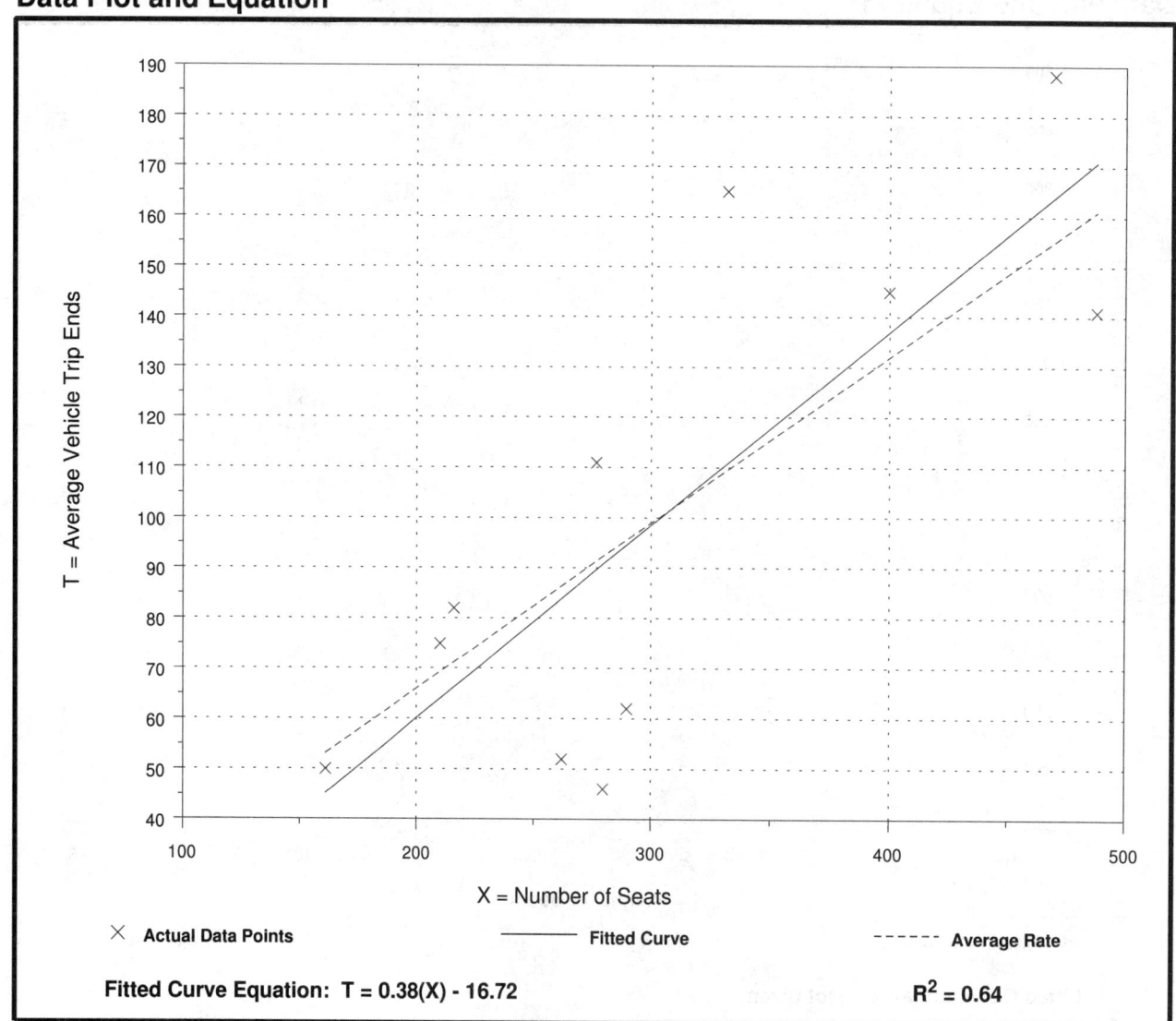

Fitted Curve Equation: T = 0.38(X) - 16.72 $R^2 = 0.64$

Quality Restaurant
(931)

Average Vehicle Trip Ends vs: Seats
On a: Sunday

Number of Studies: 11
Average Number of Seats: 309
Directional Distribution: 50% entering, 50% exiting

Trip Generation per Seat

Average Rate	Range of Rates	Standard Deviation
2.15	0.97 - 5.44	1.78

Data Plot and Equation

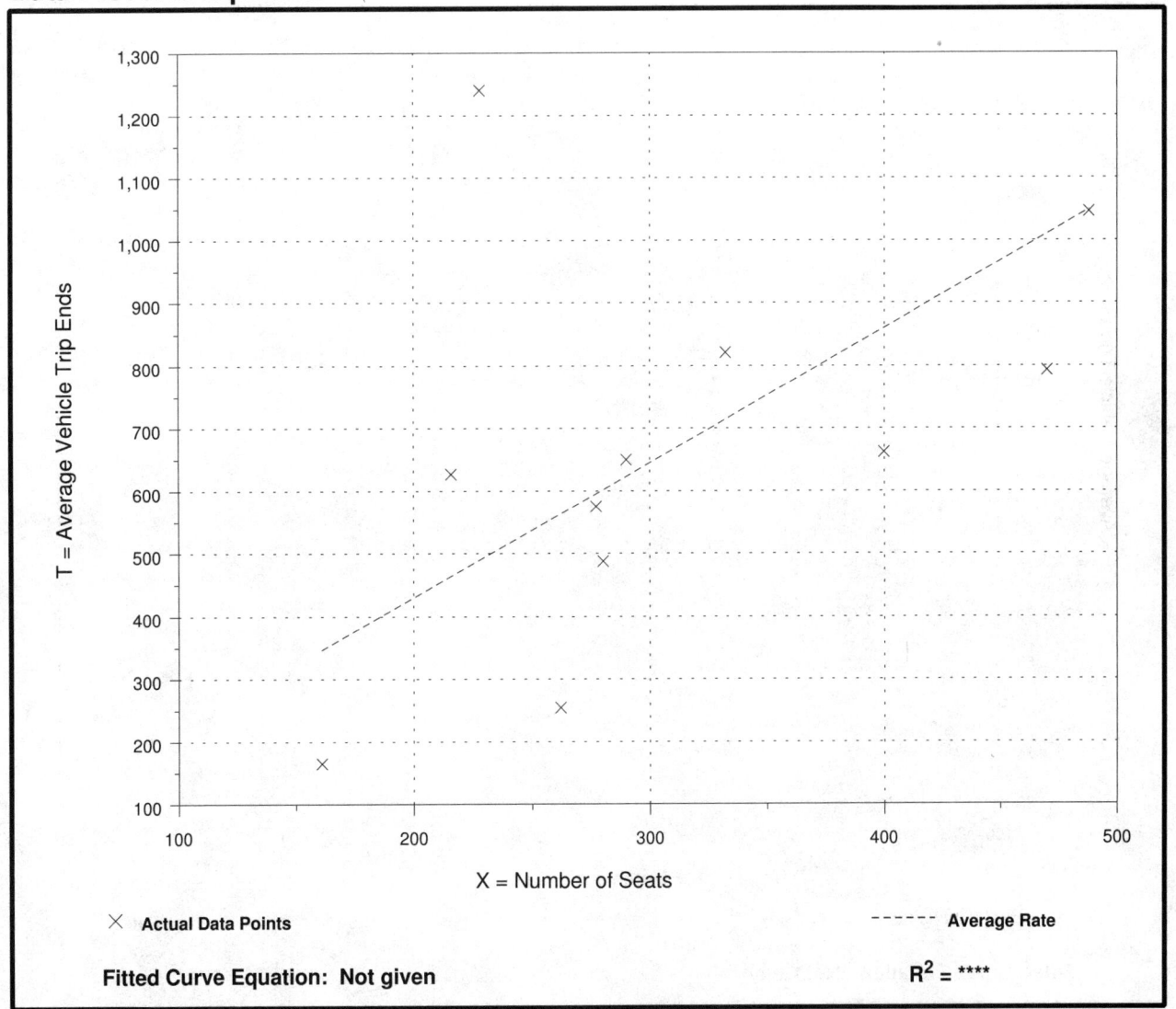

X Actual Data Points   ----- Average Rate

Fitted Curve Equation: Not given $R^2 = ****$

Quality Restaurant
(931)

Average Vehicle Trip Ends vs: Seats
On a: Sunday,
Peak Hour of Generator

Number of Studies: 10
Average Number of Seats: 318
Directional Distribution: 63% entering, 37% exiting

Trip Generation per Seat

Average Rate	Range of Rates	Standard Deviation
0.24	0.13 - 0.39	0.50

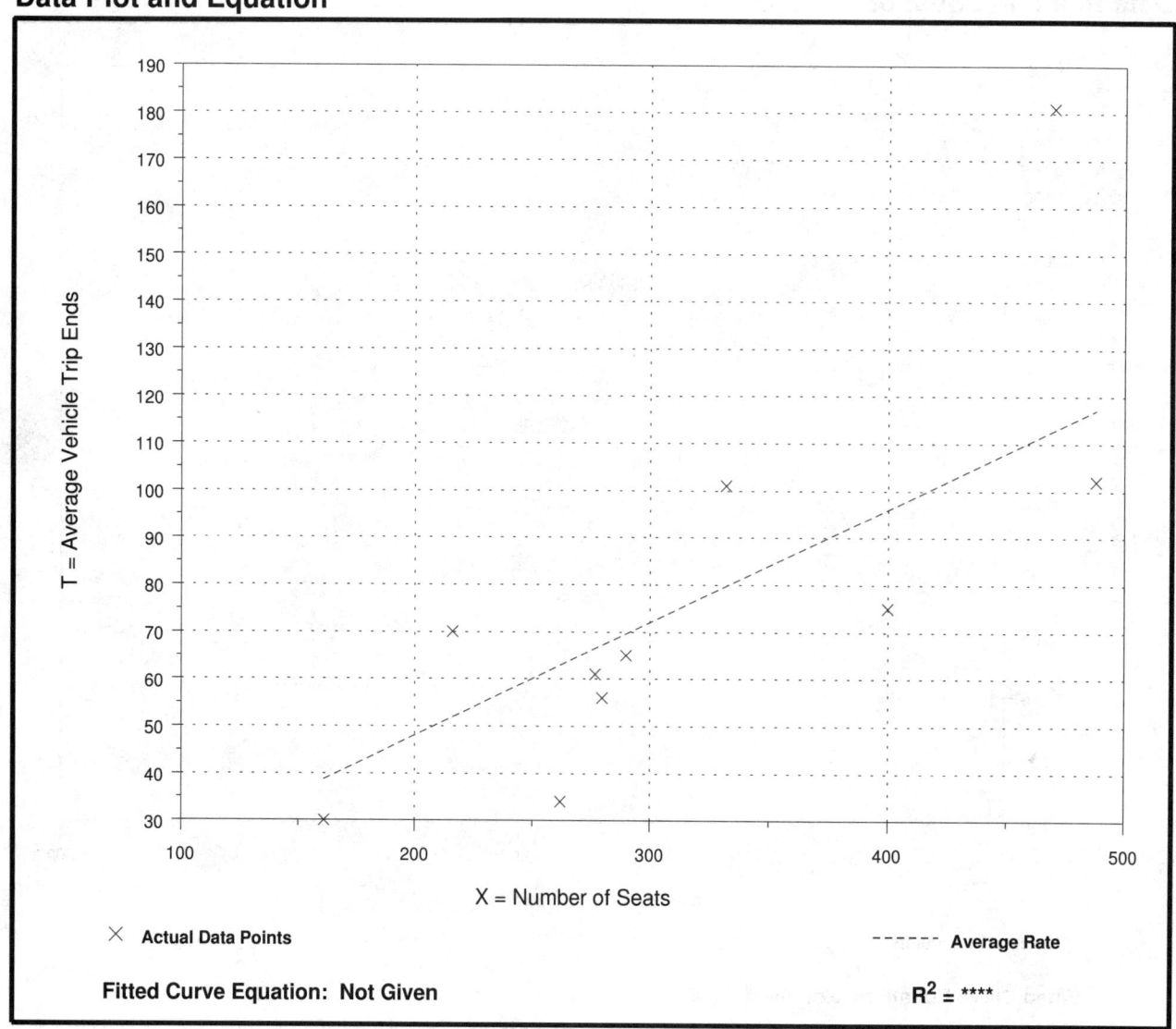

Data Plot and Equation

Fitted Curve Equation: Not Given $R^2 = ****$

Land Use: 932
High-Turnover (Sit-Down) Restaurant

Description

This land use consists of sit-down, full-service eating establishments with typical duration of stay of approximately one hour. This type of restaurant is usually moderately priced and frequently belongs to a restaurant chain. Generally, these restaurants serve lunch and dinner; they may also be open for breakfast and are sometimes open 24 hours per day. These restaurants typically do not take reservations. Patrons commonly wait to be seated, are served by a waiter/waitress, order from menus and pay for their meal after they eat. Some facilities contained within this land use may also contain a bar area for serving food and alcoholic drinks. Quality restaurant (Land Use 931), fast-food restaurant without drive-through window (Land Use 933), fast-food restaurant with drive-through window (Land Use 934) and fast-food restaurant with drive-through window and no indoor seating (Land Use 935) are related uses.

Additional Data

Users should exercise caution when applying statistics during the A.M. peak periods, as the sites contained in the database for this land use may or may not be open for breakfast. In cases where it was confirmed that the sites were not open for breakfast, data for the A.M. peak hour of the adjacent street traffic were removed from the database.

Information on approximate hourly variation in high-turnover (sit-down) restaurant traffic is shown in the following table. It should be noted, however, that the information contained in this table is based on a limited sample size. Therefore, caution should be exercised when applying the data. Also, some information provided in the table may conflict with the results obtained by applying the average rate or regression equations. When this occurs, it is suggested that the results from the average rate or regression equations be used, as they are based on a larger number of studies.

Time	Hourly Variation in High-Turnover (Sit-Down) Restaurant Traffic					
	Average Weekday[a]		Average Saturday[b]		Average Sunday[c]	
	Percent of 24-Hour Entering Traffic	Percent of 24-Hour Exiting Traffic	Percent of 24-Hour Entering Traffic	Percent of 24-Hour Exiting Traffic	Percent of 24-Hour Entering Traffic	Percent of 24-Hour Exiting Traffic
6 a.m.–7 a.m.	1.5	0.8	0.9	0.6	0.1	0.4
7 a.m.–8 a.m.	3.0	1.7	2.2	1.0	0.9	1.3
8 a.m.–9 a.m.	3.6	2.3	4.1	2.8	1.7	0.1
9 a.m.–10 a.m.	4.1	2.7	4.1	3.5	1.4	1.2
10 a.m.–11 a.m.	3.3	3.2	4.6	3.7	2.3	4.2
11 a.m.–12 p.m.	7.4	3.8	4.6	4.0	5.5	2.6
12 p.m.–1 p.m.	8.6	6.6	5.1	3.6	8.8	3.9
1 p.m.–2 p.m.	4.8	8.6	4.4	4.3	6.6	8.2
2 p.m.–3 p.m.	3.2	5.5	3.8	4.3	5.9	5.1
3 p.m.–4 p.m.	3.0	4.0	3.6	3.5	8.7	7.2
4 p.m.–5 p.m.	5.6	4.5	4.5	4.0	10.0	8.4
5 p.m.–6 p.m.	9.7	4.6	7.1	4.3	12.4	10.5
6 p.m.–7 p.m.	10.7	7.9	9.9	6.7	11.3	10.0
7 p.m.–8 p.m.	9.5	9.0	8.5	7.3	8.7	9.3
8 p.m.–9 p.m.	7.7	9.0	8.1	8.5	5.9	8.0
9 p.m.–10 p.m.	4.9	8.6	6.5	7.3	4.2	7.5
10 p.m.–6 a.m.	9.4	17.2	18.0	30.6	5.6	12.1

Sites ranged in size from 4,500 to 21,000 square feet gross floor area
[a] Source numbers – 13, 88, 126, 507 and The Traffic Group, Inc.; based on seven studies
[b] Source numbers – 13, 88, 126 and The Traffic Group, Inc.; based on five studies
[c] Source numbers – 13, 88 and 126; based on three studies

Vehicle occupancy ranged from 1.39 to 1.69 persons per automobile on an average weekday. The average for the sites surveyed was approximately 1.52.

Five sites submitted for inclusion in this land use indicated the presence of an on-site pick-up window. From the limited data sample, it does not appear that the presence of a pick-up window had a significant impact on trip generation.

The outdoor seating area is not included in the overall gross floor area. Therefore, the number of seats may be a more reliable independent variable on which to establish trip generation rates for facilities having significant outdoor seating.

The sites were surveyed between the 1960s and the 2000s throughout the United States.

Source Numbers

2, 4, 5, 72, 90, 100, 126, 269, 275, 280, 300, 301, 305, 338, 340, 341, 358, 384, 424, 432, 437, 438, 444, 507, 555, 577, 589, 617, 618, 728

High-Turnover (Sit-Down) Restaurant
(932)

Average Vehicle Trip Ends vs: 1000 Sq. Feet Gross Floor Area
On a: Weekday

Number of Studies: 14
Average 1000 Sq. Feet GFA: 7
Directional Distribution: 50% entering, 50% exiting

Trip Generation per 1000 Sq. Feet Gross Floor Area

Average Rate	Range of Rates	Standard Deviation
127.15	73.51 - 246.00	41.77

Data Plot and Equation

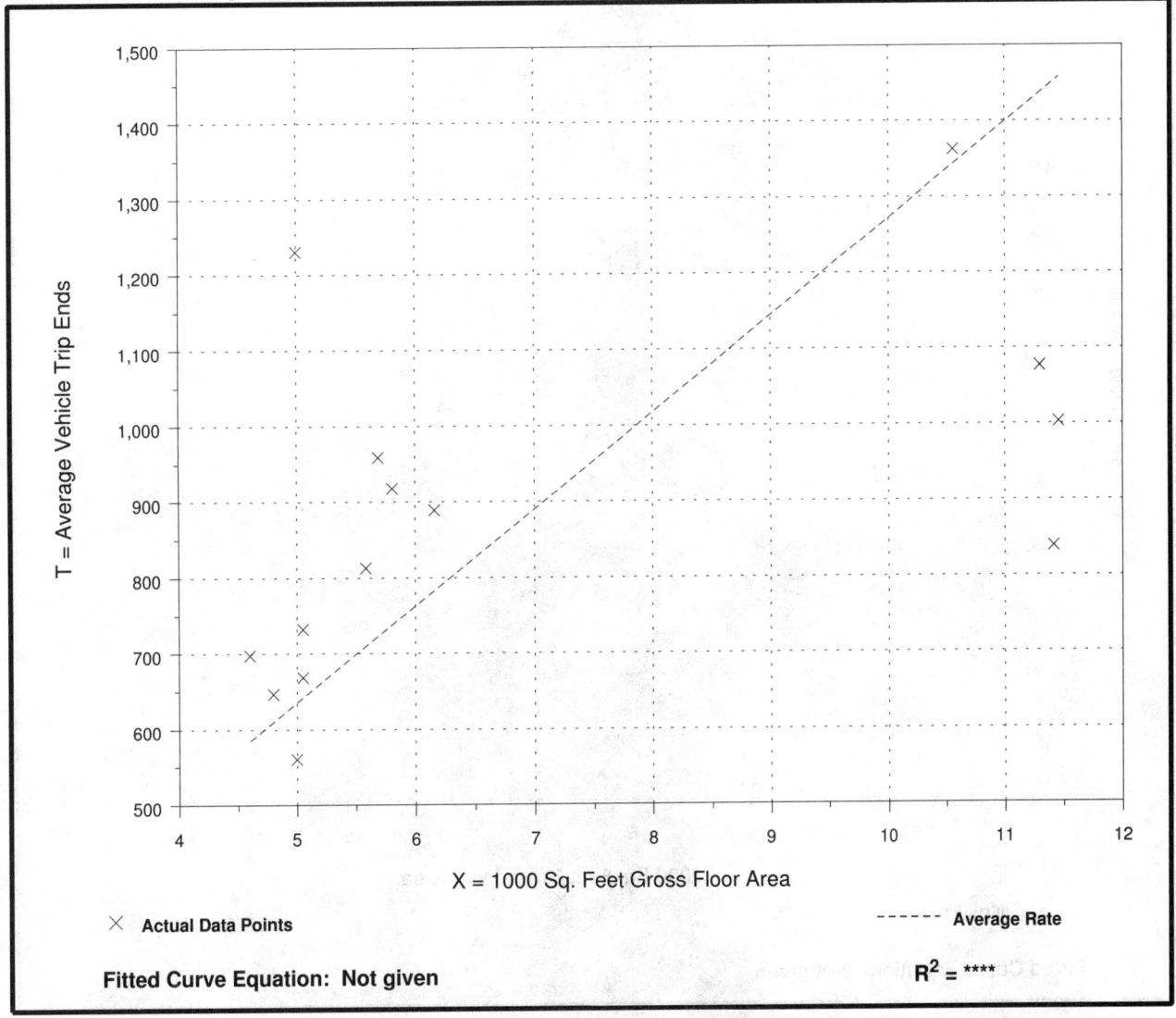

Fitted Curve Equation: Not given $R^2 = ****$

High-Turnover (Sit-Down) Restaurant
(932)

Average Vehicle Trip Ends vs: 1000 Sq. Feet Gross Floor Area
On a: Weekday,
Peak Hour of Adjacent Street Traffic,
One Hour Between 7 and 9 a.m.

Number of Studies: 24
Average 1000 Sq. Feet GFA: 6
Directional Distribution: 55% entering, 45% exiting

Trip Generation per 1000 Sq. Feet Gross Floor Area

Average Rate	Range of Rates	Standard Deviation
10.81	2.32 - 25.60	6.59

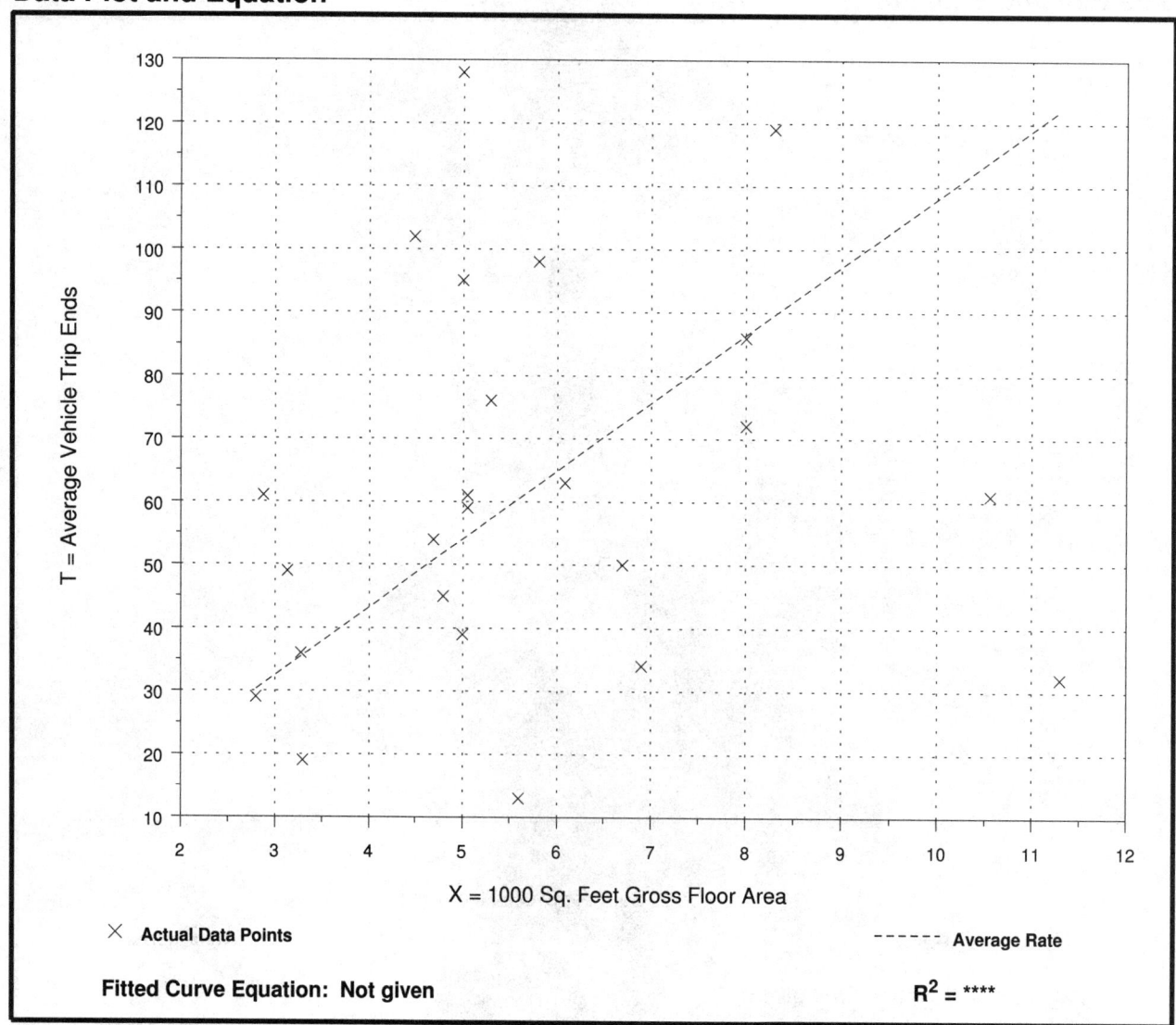

Data Plot and Equation

Fitted Curve Equation: Not given $R^2 = ****$

High-Turnover (Sit-Down) Restaurant
(932)

Average Vehicle Trip Ends vs: 1000 Sq. Feet Gross Floor Area
On a: Weekday,
Peak Hour of Adjacent Street Traffic,
One Hour Between 4 and 6 p.m.

Number of Studies: 60
Average 1000 Sq. Feet GFA: 6
Directional Distribution: 60% entering, 40% exiting

Trip Generation per 1000 Sq. Feet Gross Floor Area

Average Rate	Range of Rates	Standard Deviation
9.85	0.92 - 62.00	8.54

Data Plot and Equation

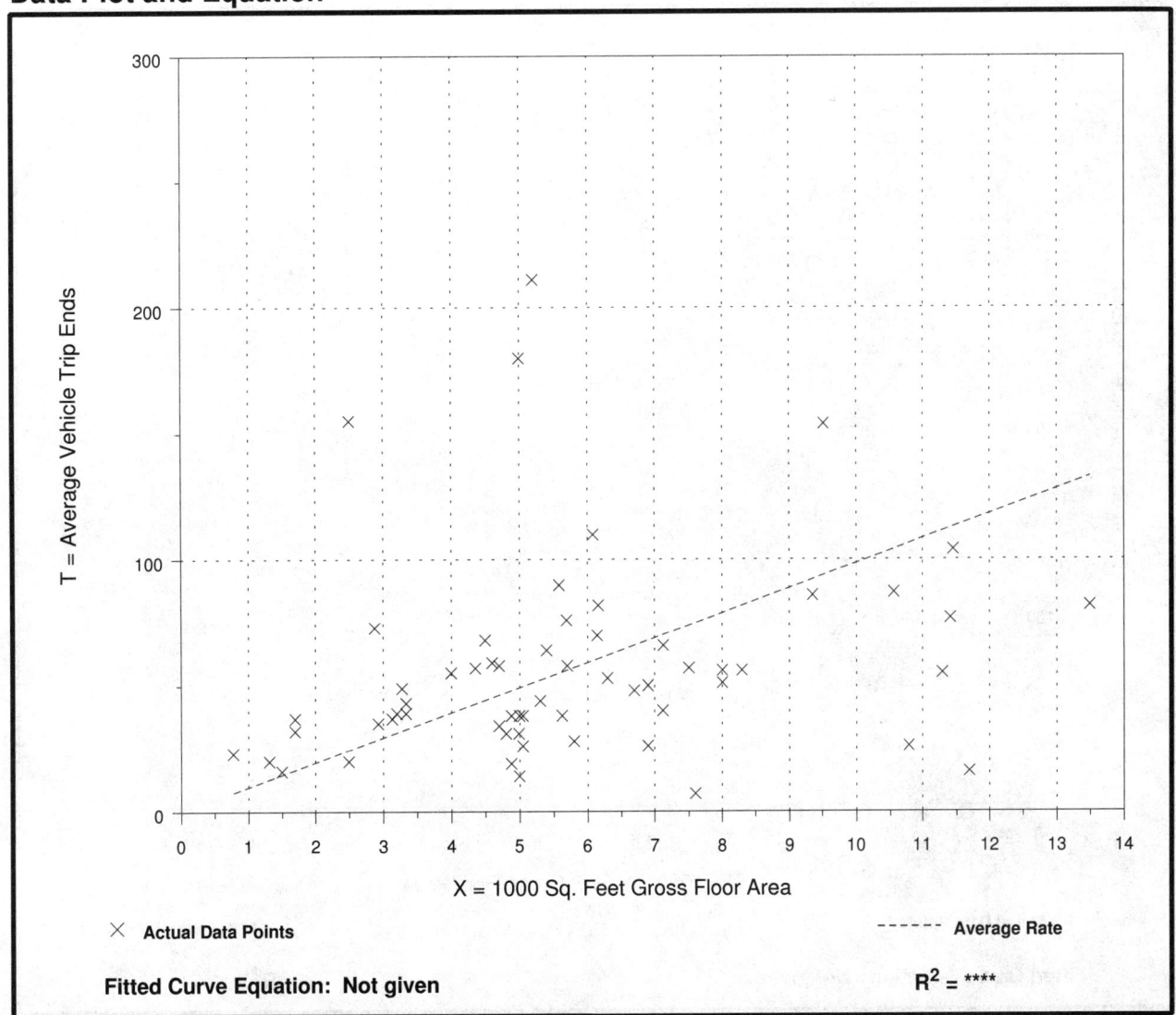

Fitted Curve Equation: Not given $R^2 = ****$

High-Turnover (Sit-Down) Restaurant
(932)

Average Vehicle Trip Ends vs: 1000 Sq. Feet Gross Floor Area
On a: Weekday,
A.M. Peak Hour of Generator

Number of Studies: 25
Average 1000 Sq. Feet GFA: 7
Directional Distribution: 53% entering, 47% exiting

Trip Generation per 1000 Sq. Feet Gross Floor Area

Average Rate	Range of Rates	Standard Deviation
13.33	3.00 - 54.09	9.44

Data Plot and Equation

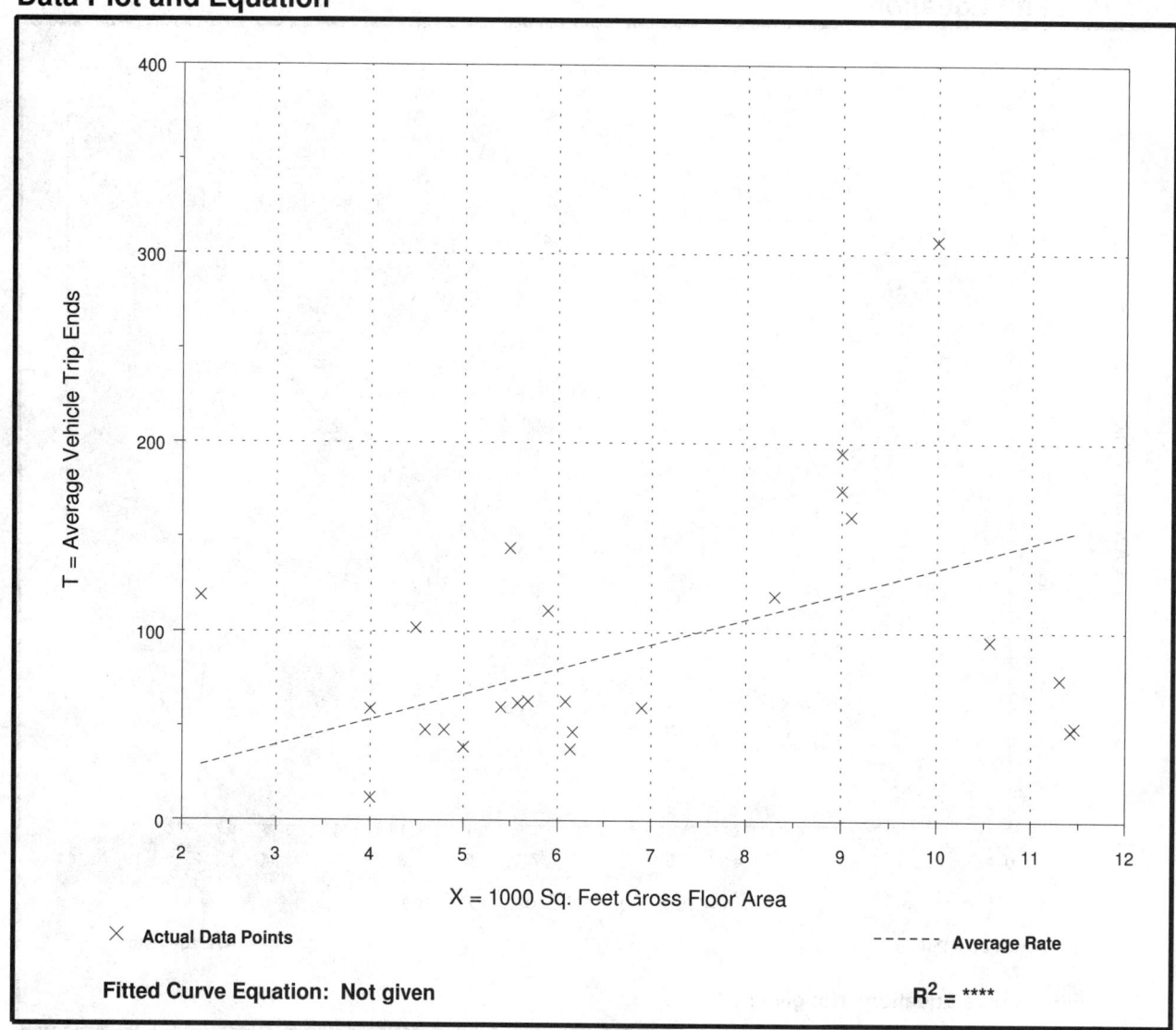

Fitted Curve Equation: Not given $R^2 = ****$

High-Turnover (Sit-Down) Restaurant
(932)

Average Vehicle Trip Ends vs: 1000 Sq. Feet Gross Floor Area
On a: Weekday,
P.M. Peak Hour of Generator

Number of Studies: 31
Average 1000 Sq. Feet GFA: 5
Directional Distribution: 54% entering, 46% exiting

Trip Generation per 1000 Sq. Feet Gross Floor Area

Average Rate	Range of Rates	Standard Deviation
18.49	5.60 - 69.20	13.32

Data Plot and Equation

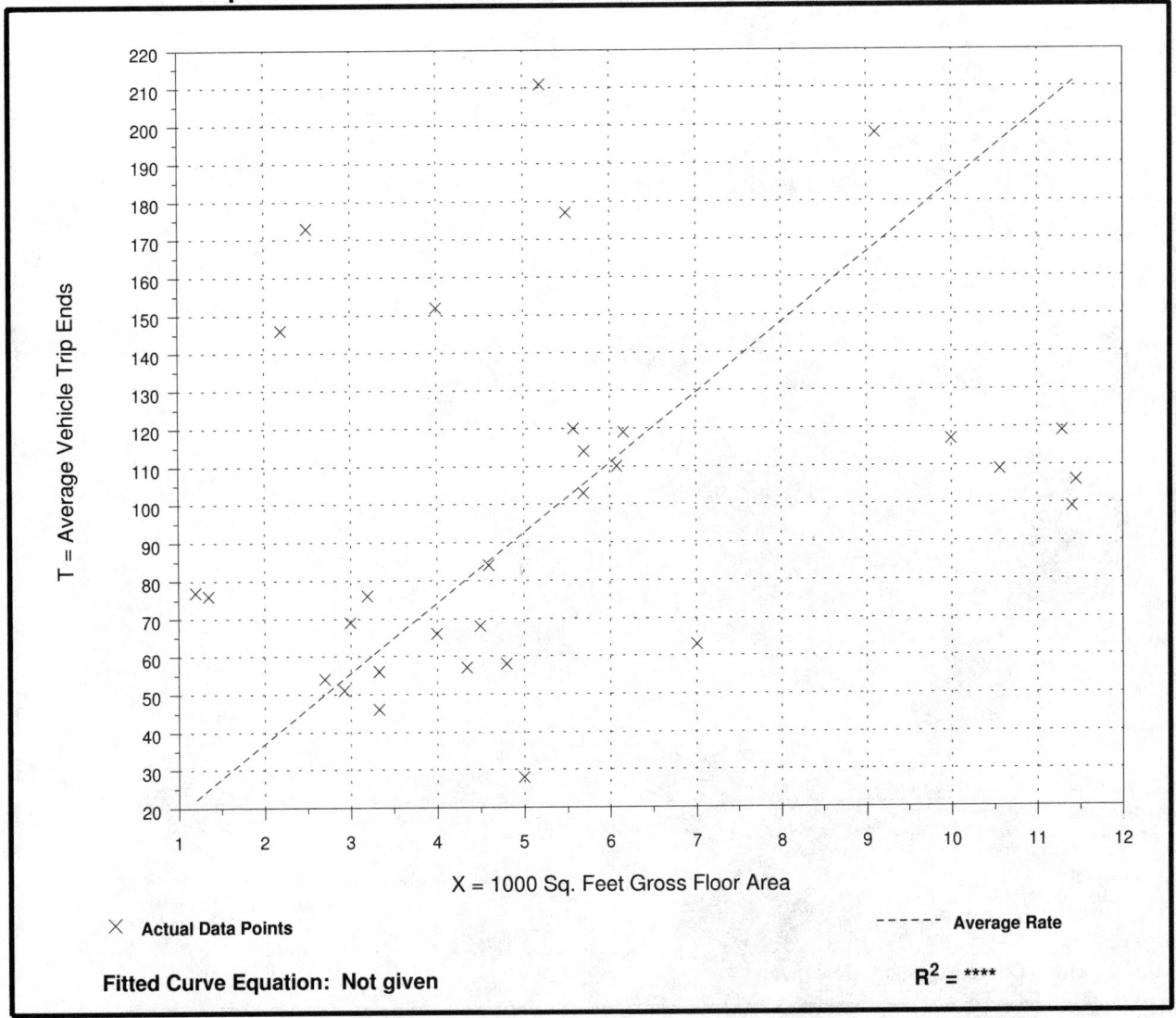

Fitted Curve Equation: Not given $R^2 = ****$

High-Turnover (Sit-Down) Restaurant
(932)

Average Vehicle Trip Ends vs: 1000 Sq. Feet Gross Floor Area
On a: Saturday

Number of Studies: 2
Average 1000 Sq. Feet GFA: 5
Directional Distribution: 50% entering, 50% exiting

Trip Generation per 1000 Sq. Feet Gross Floor Area

Average Rate	Range of Rates	Standard Deviation
158.37	144.60 - 172.71	*

Data Plot and Equation

Caution - Use Carefully - Small Sample Size

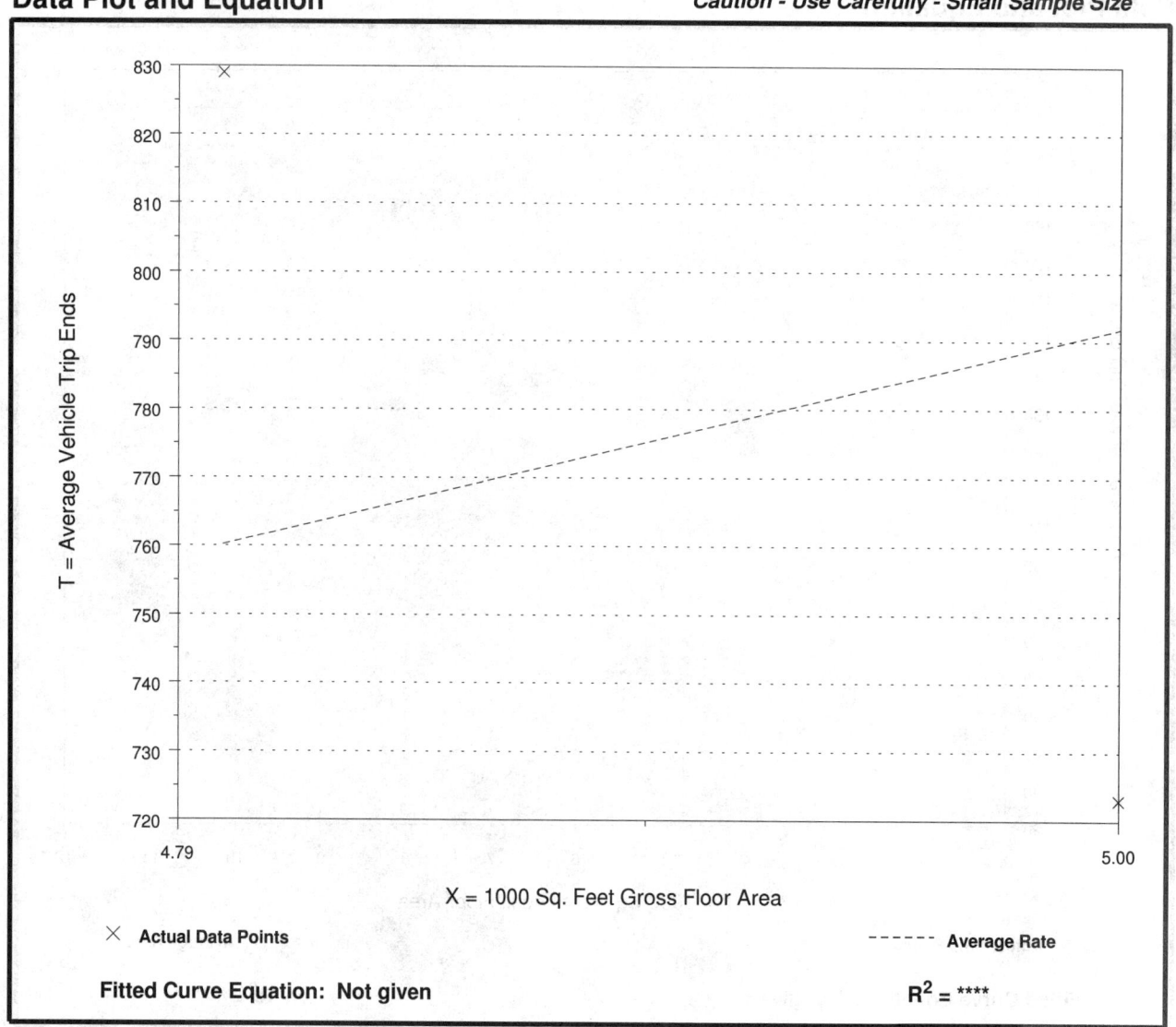

Fitted Curve Equation: Not given

$R^2 = ****$

High-Turnover (Sit-Down) Restaurant
(932)

Average Vehicle Trip Ends vs: 1000 Sq. Feet Gross Floor Area
On a: Saturday,
Peak Hour of Generator

Number of Studies: 8
Average 1000 Sq. Feet GFA: 4
Directional Distribution: 53% entering, 47% exiting

Trip Generation per 1000 Sq. Feet Gross Floor Area

Average Rate	Range of Rates	Standard Deviation
14.07	4.44 - 50.40	12.19

Data Plot and Equation

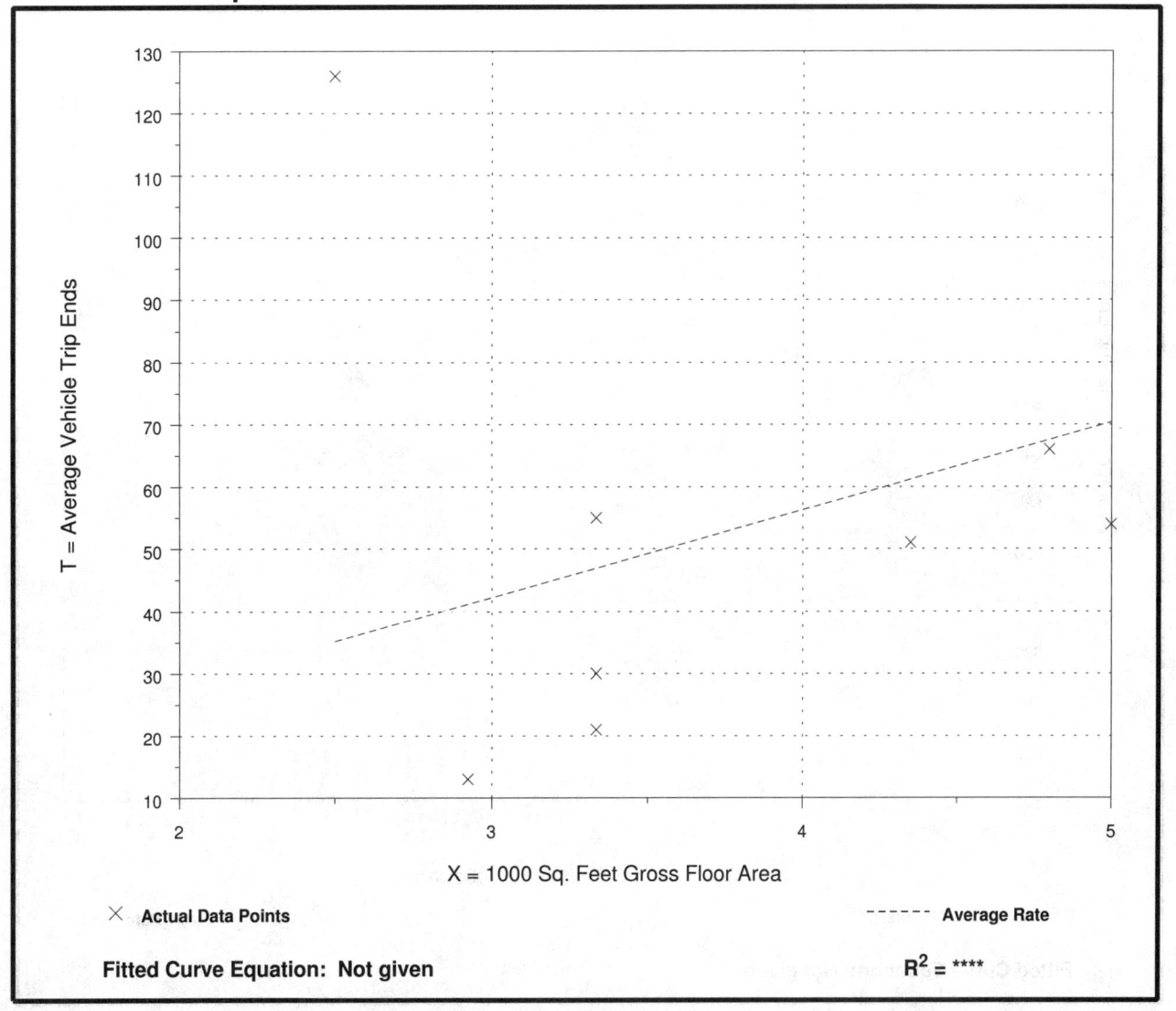

Fitted Curve Equation: Not given $R^2 = ****$

High-Turnover (Sit-Down) Restaurant
(932)

Average Vehicle Trip Ends vs: 1000 Sq. Feet Gross Floor Area
On a: Sunday

Number of Studies: 2
Average 1000 Sq. Feet GFA: 5
Directional Distribution: 50% entering, 50% exiting

Trip Generation per 1000 Sq. Feet Gross Floor Area

Average Rate	Range of Rates	Standard Deviation
131.84	119.38 - 143.80	*

Data Plot and Equation

Caution - Use Carefully - Small Sample Size

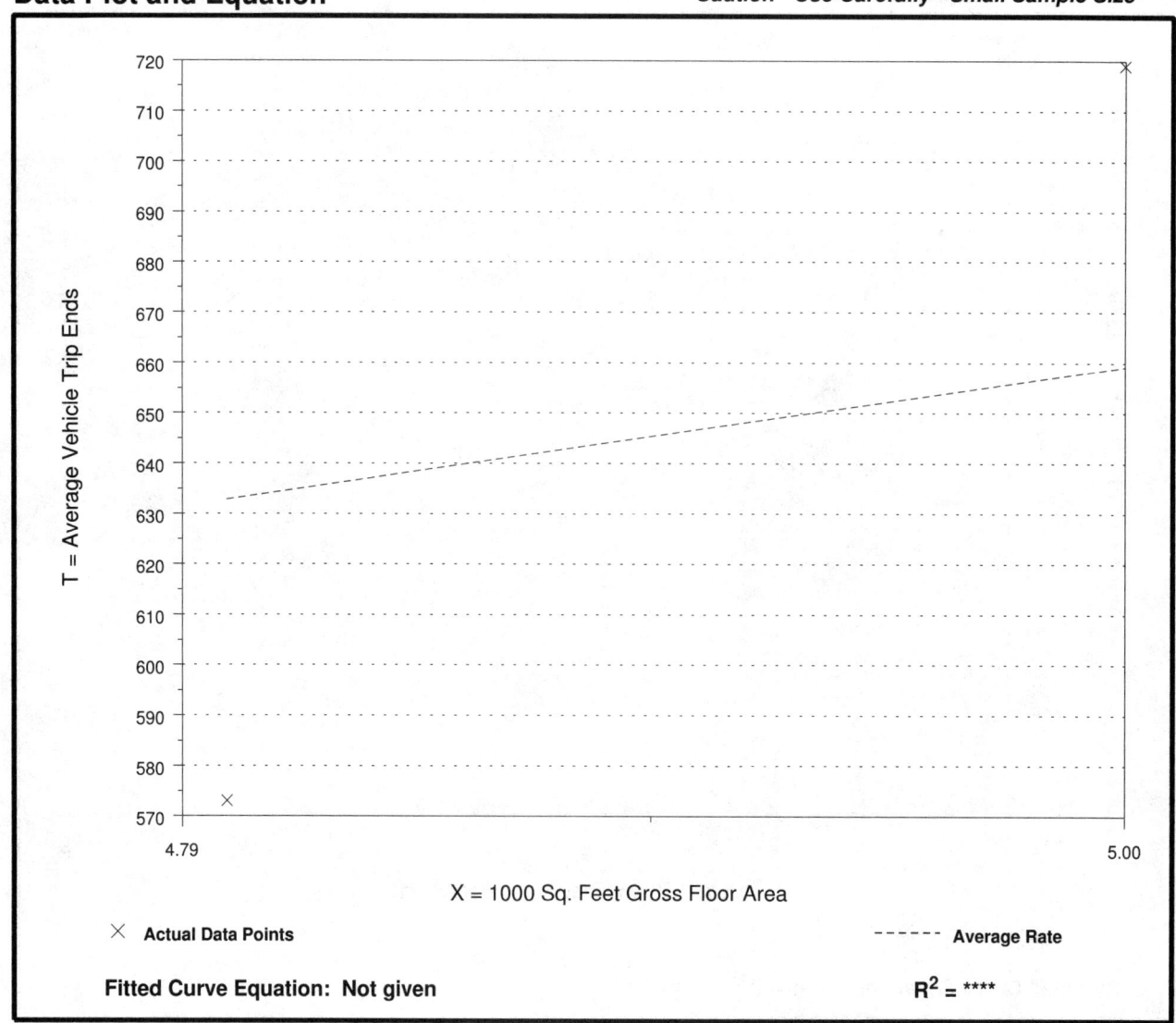

Fitted Curve Equation: Not given

R^2 = ****

High-Turnover (Sit-Down) Restaurant
(932)

Average Vehicle Trip Ends vs: 1000 Sq. Feet Gross Floor Area
On a: Sunday,
Peak Hour of Generator

Number of Studies: 3
Average 1000 Sq. Feet GFA: 4
Directional Distribution: 55% entering, 45% exiting

Trip Generation per 1000 Sq. Feet Gross Floor Area

Average Rate	Range of Rates	Standard Deviation
18.46	9.79 - 43.20	13.74

Data Plot and Equation

Caution - Use Carefully - Small Sample Size

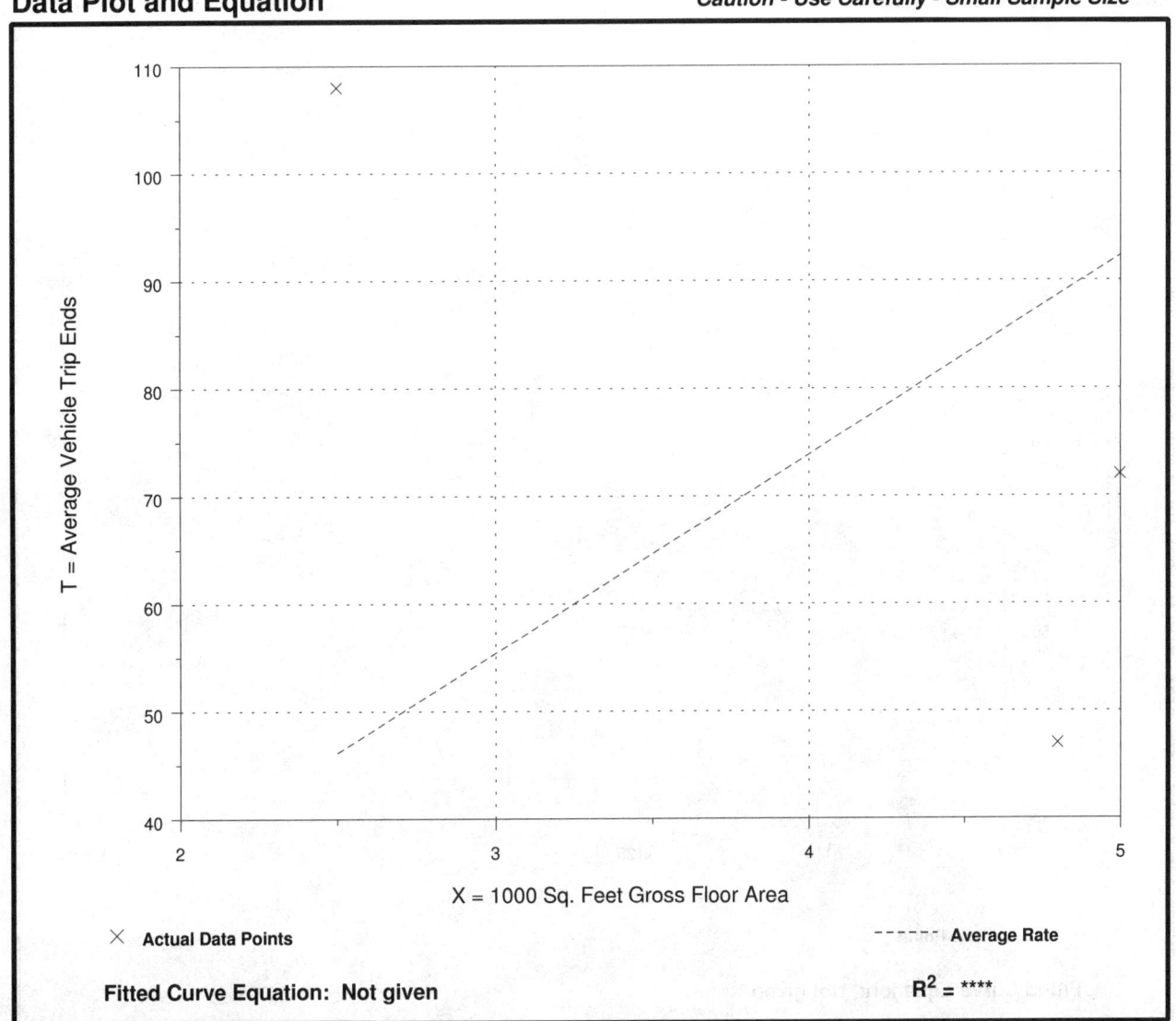

Fitted Curve Equation: Not given

R^2 = ****

High-Turnover (Sit-Down) Restaurant
(932)

Average Vehicle Trip Ends vs: Seats
On a: Weekday

Number of Studies: 2
Average Number of Seats: 125
Directional Distribution: 50% entering, 50% exiting

Trip Generation per Seat

Average Rate	Range of Rates	Standard Deviation
4.83	4.37 - 5.49	*

Data Plot and Equation

Caution - Use Carefully - Small Sample Size

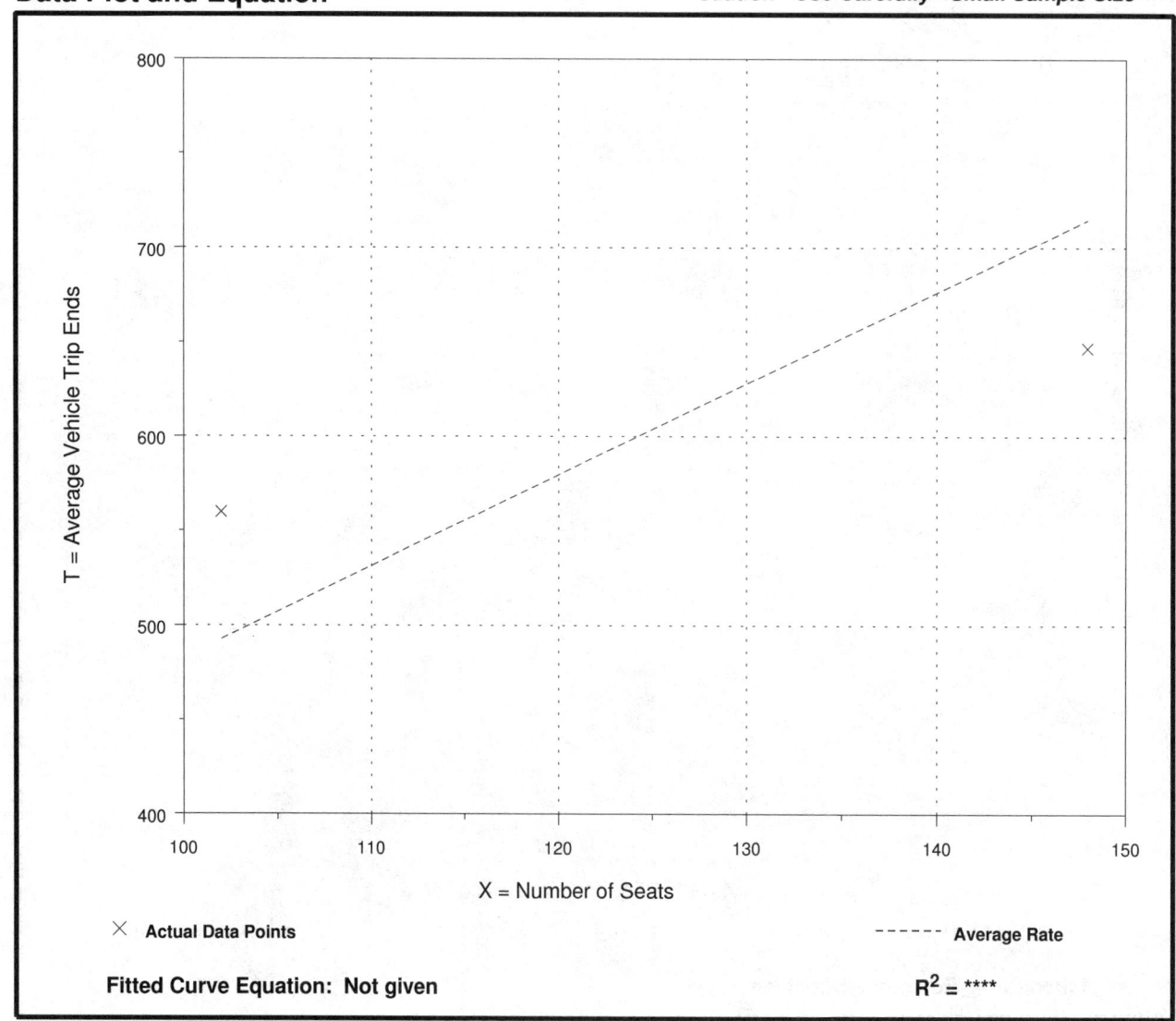

Fitted Curve Equation: Not given

R^2 = ****

High-Turnover (Sit-Down) Restaurant
(932)

Average Vehicle Trip Ends vs: Seats
On a: Weekday,
Peak Hour of Adjacent Street Traffic,
One Hour Between 7 and 9 a.m.

Number of Studies: 10
Average Number of Seats: 150
Directional Distribution: 52% entering, 48% exiting

Trip Generation per Seat

Average Rate	Range of Rates	Standard Deviation
0.47	0.30 - 0.76	0.70

Data Plot and Equation

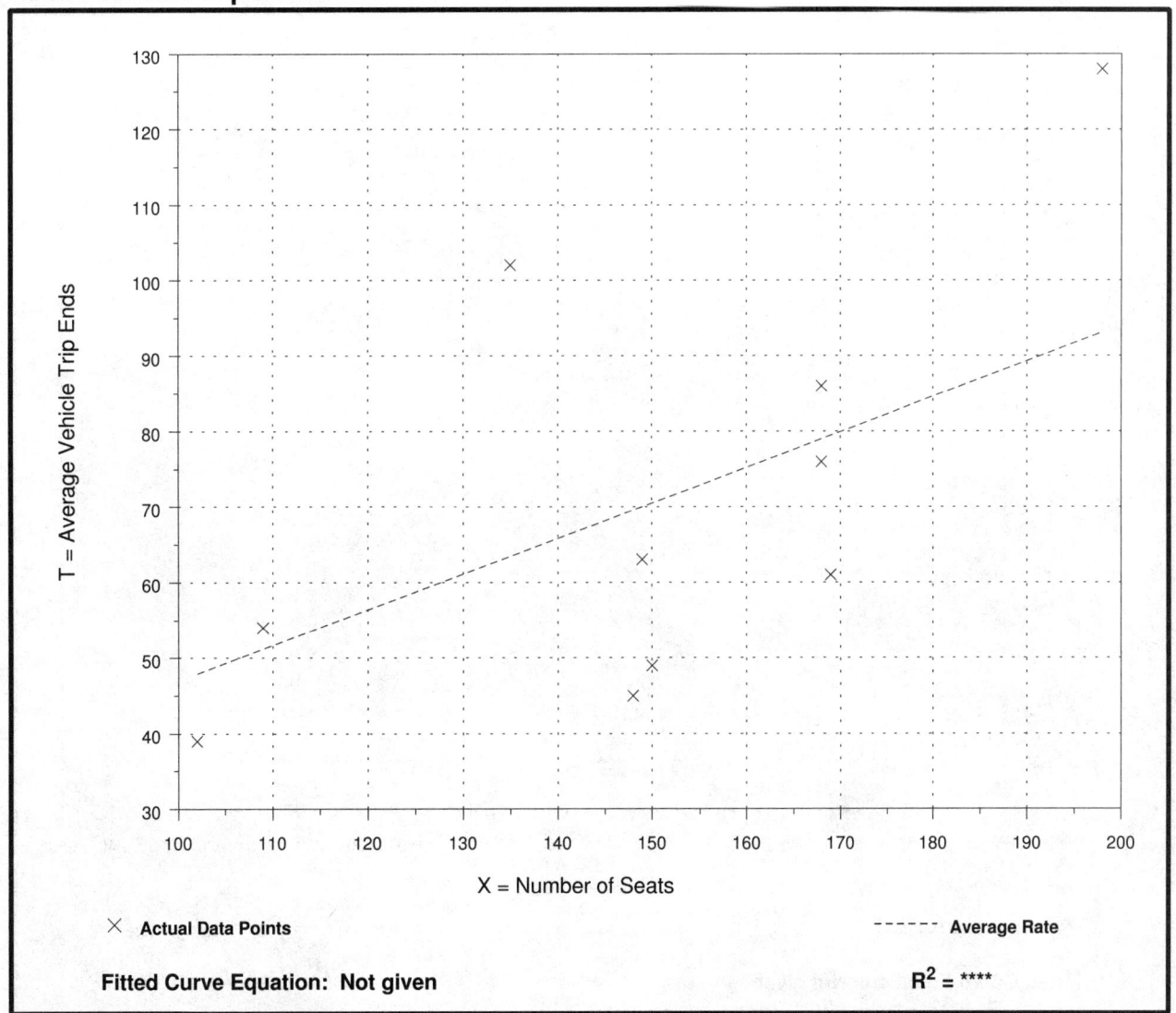

X   Actual Data Points   ----- Average Rate

Fitted Curve Equation: Not given R^2 = ****

High-Turnover (Sit-Down) Restaurant
(932)

Average Vehicle Trip Ends vs: Seats
On a: Weekday,
Peak Hour of Adjacent Street Traffic,
One Hour Between 4 and 6 p.m.

Number of Studies: 18
Average Number of Seats: 139
Directional Distribution: 57% entering, 43% exiting

Trip Generation per Seat

Average Rate	Range of Rates	Standard Deviation
0.41	0.14 - 1.73	0.73

Data Plot and Equation

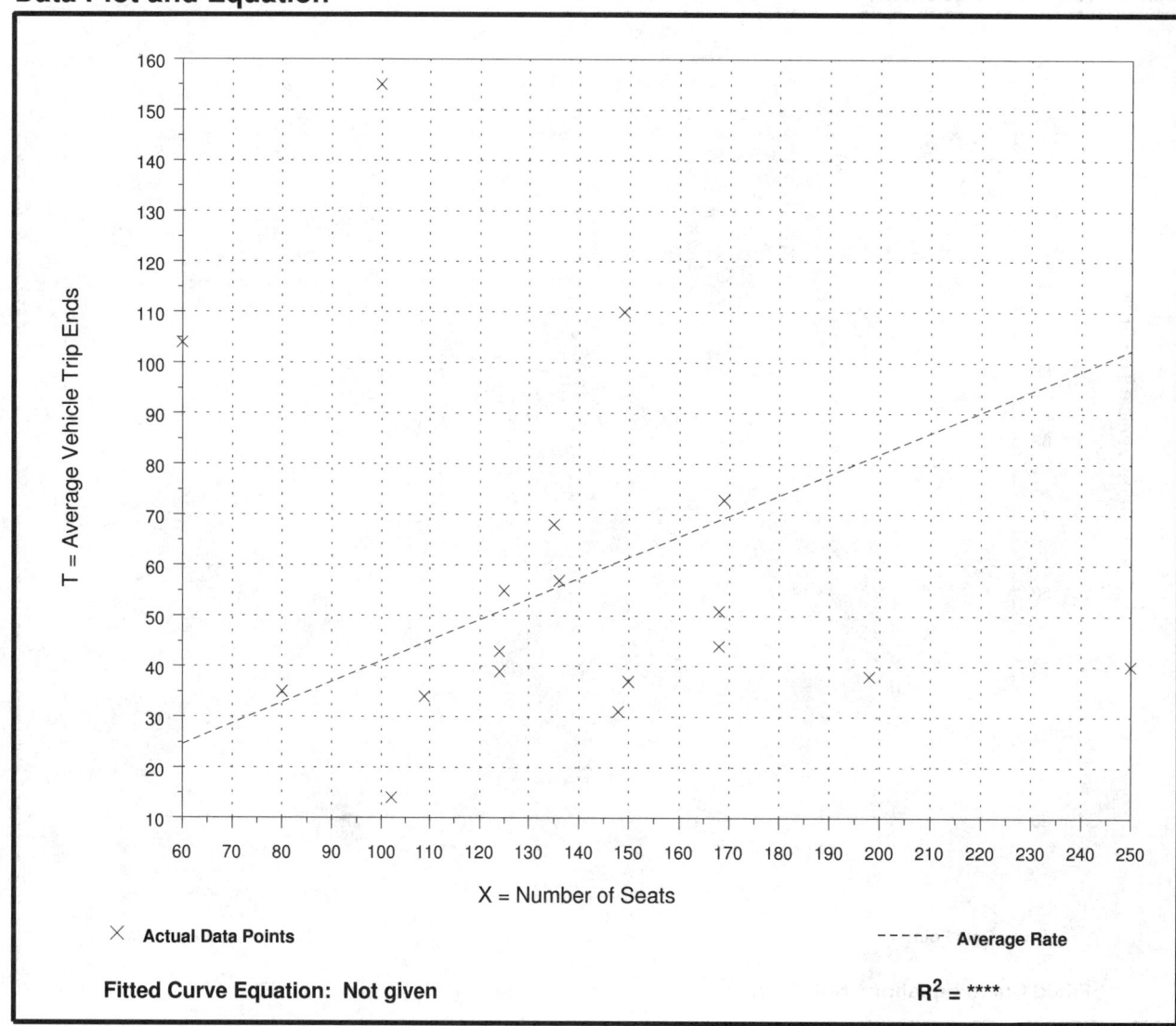

Fitted Curve Equation: Not given $R^2 = ****$

High-Turnover (Sit-Down) Restaurant
(932)

Average Vehicle Trip Ends vs: Seats
On a: Weekday,
A.M. Peak Hour of Generator

Number of Studies: 8
Average Number of Seats: 143
Directional Distribution: 58% entering, 42% exiting

Trip Generation per Seat

Average Rate	Range of Rates	Standard Deviation
0.60	0.10 - 1.70	0.86

Data Plot and Equation

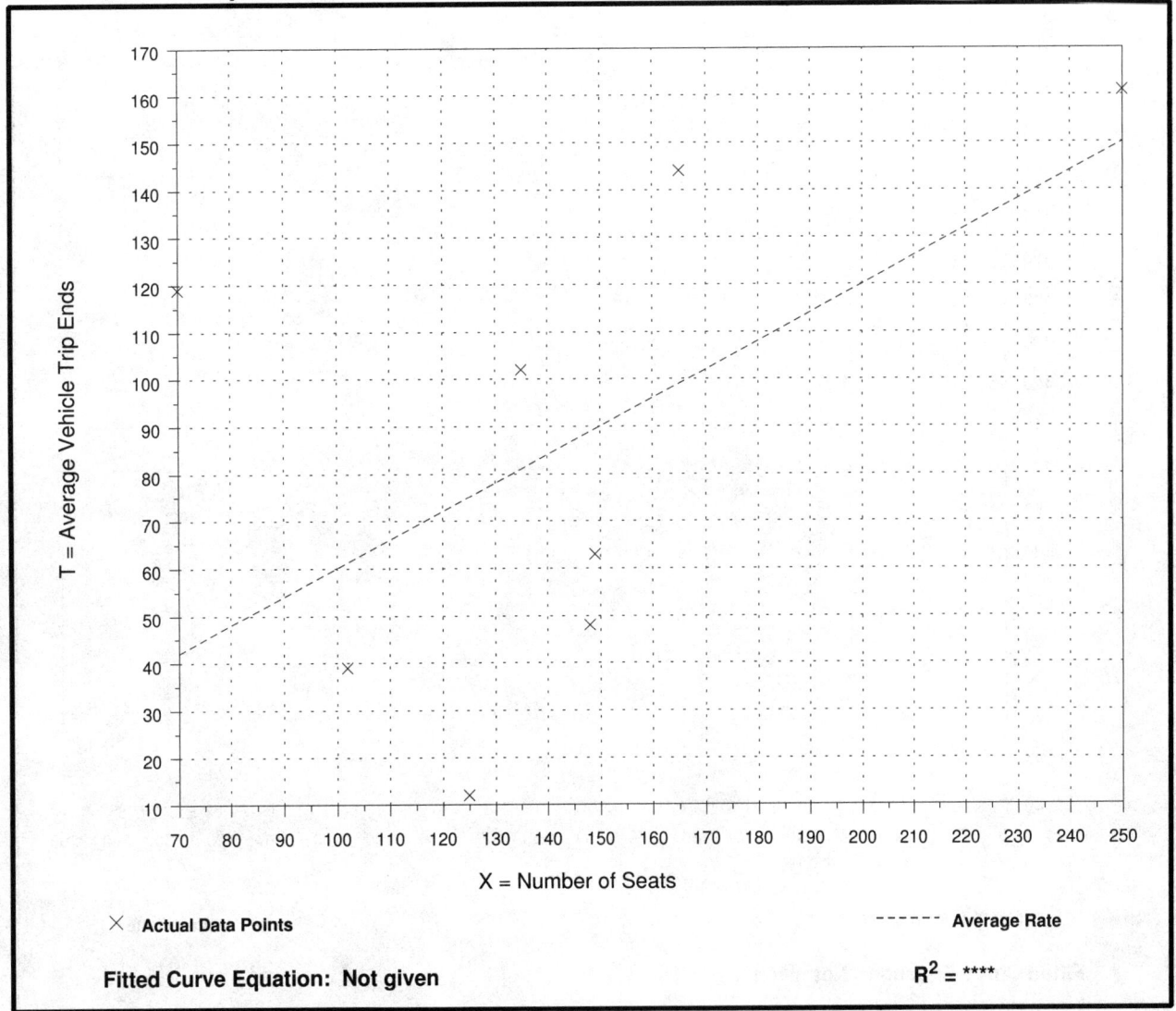

Fitted Curve Equation: Not given $R^2 = ****$

High-Turnover (Sit-Down) Restaurant
(932)

Average Vehicle Trip Ends vs: Seats
On a: Weekday,
P.M. Peak Hour of Generator

Number of Studies: 13
Average Number of Seats: 131
Directional Distribution: 52% entering, 48% exiting

Trip Generation per Seat

Average Rate	Range of Rates	Standard Deviation
0.72	0.27 - 2.09	0.96

Data Plot and Equation

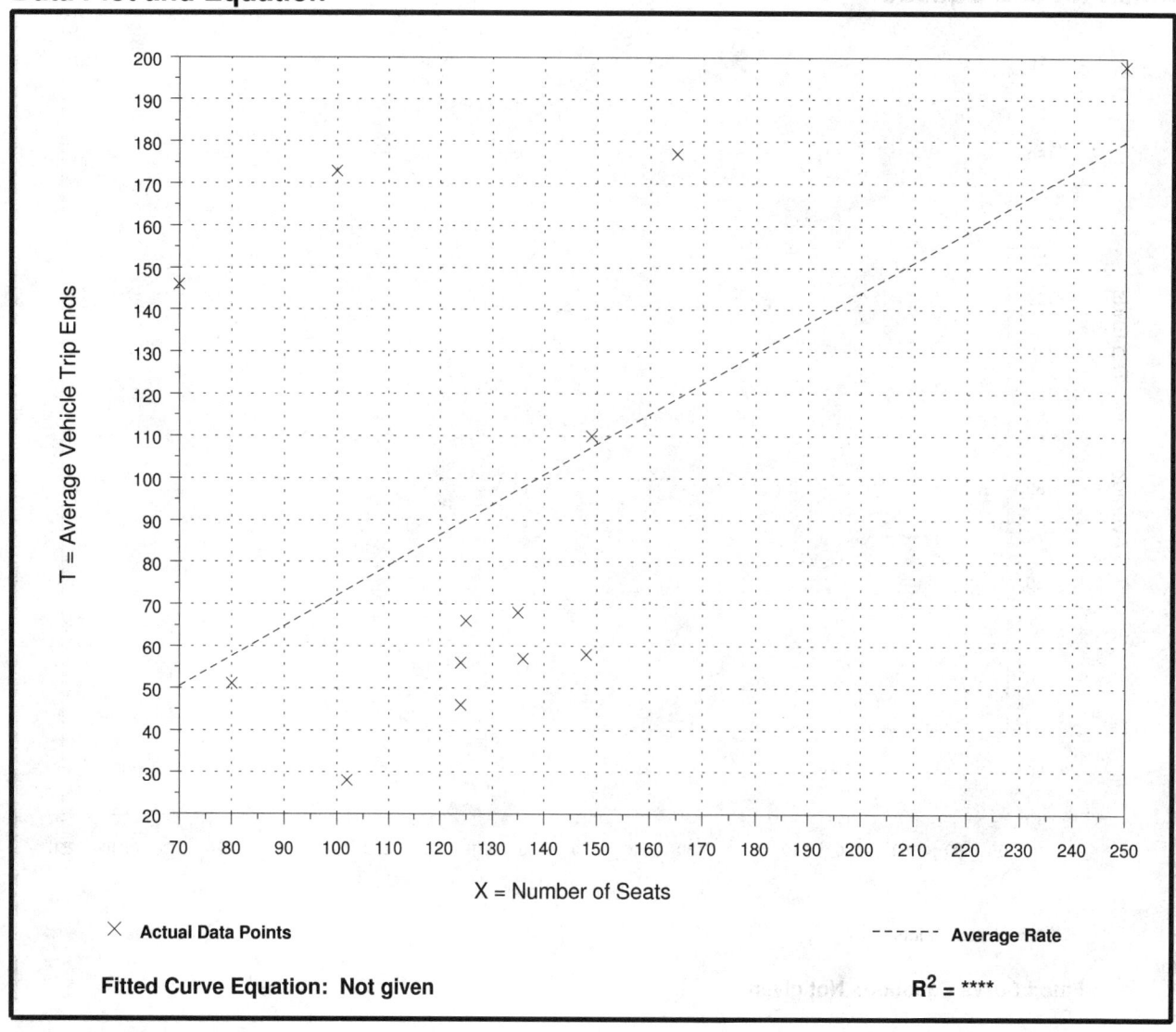

X = Actual Data Points ----- Average Rate

Fitted Curve Equation: Not given $R^2 = ****$

High-Turnover (Sit-Down) Restaurant
(932)

Average Vehicle Trip Ends vs: Seats
On a: Saturday

Number of Studies: 2
Average Number of Seats: 125
Directional Distribution: 50% entering, 50% exiting

Trip Generation per Seat

Average Rate	Range of Rates	Standard Deviation
6.21	5.60 - 7.09	*

Data Plot and Equation

Caution - Use Carefully - Small Sample Size

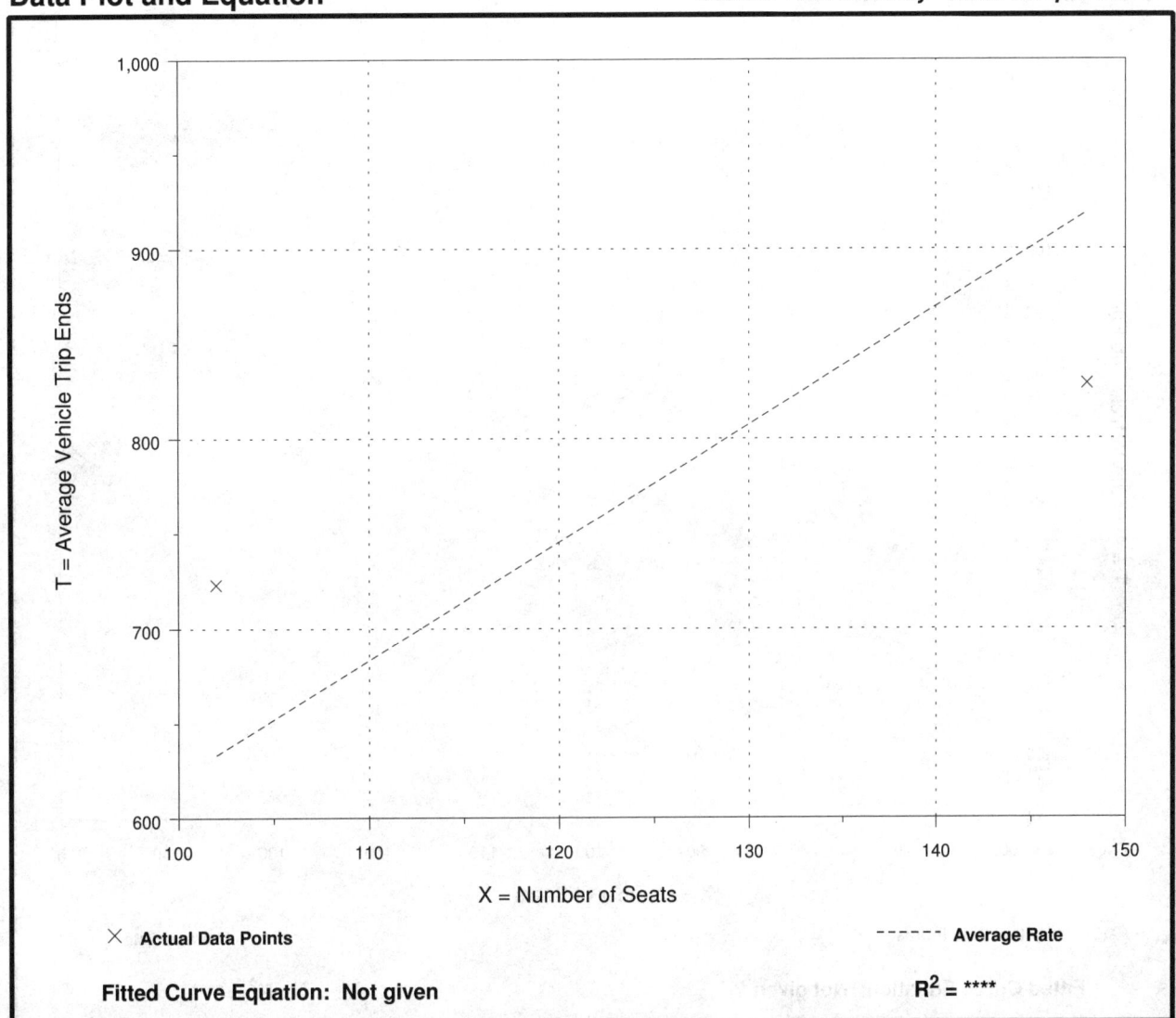

[Plot: T = Average Vehicle Trip Ends vs X = Number of Seats, with two actual data points (≈102, 720) and (≈148, 830), dashed line showing Average Rate]

X   Actual Data Points   ----- Average Rate

Fitted Curve Equation: Not given R^2 = ****

Trip Generation, 9th Edition • Institute of Transportation Engineers

High-Turnover (Sit-Down) Restaurant
(932)

Average Vehicle Trip Ends vs: Seats
On a: Saturday,
Peak Hour of Generator

Number of Studies: 9
Average Number of Seats: 111
Directional Distribution: 53% entering, 47% exiting

Trip Generation per Seat

Average Rate	Range of Rates	Standard Deviation
0.53	0.16 - 1.88	0.86

Data Plot and Equation

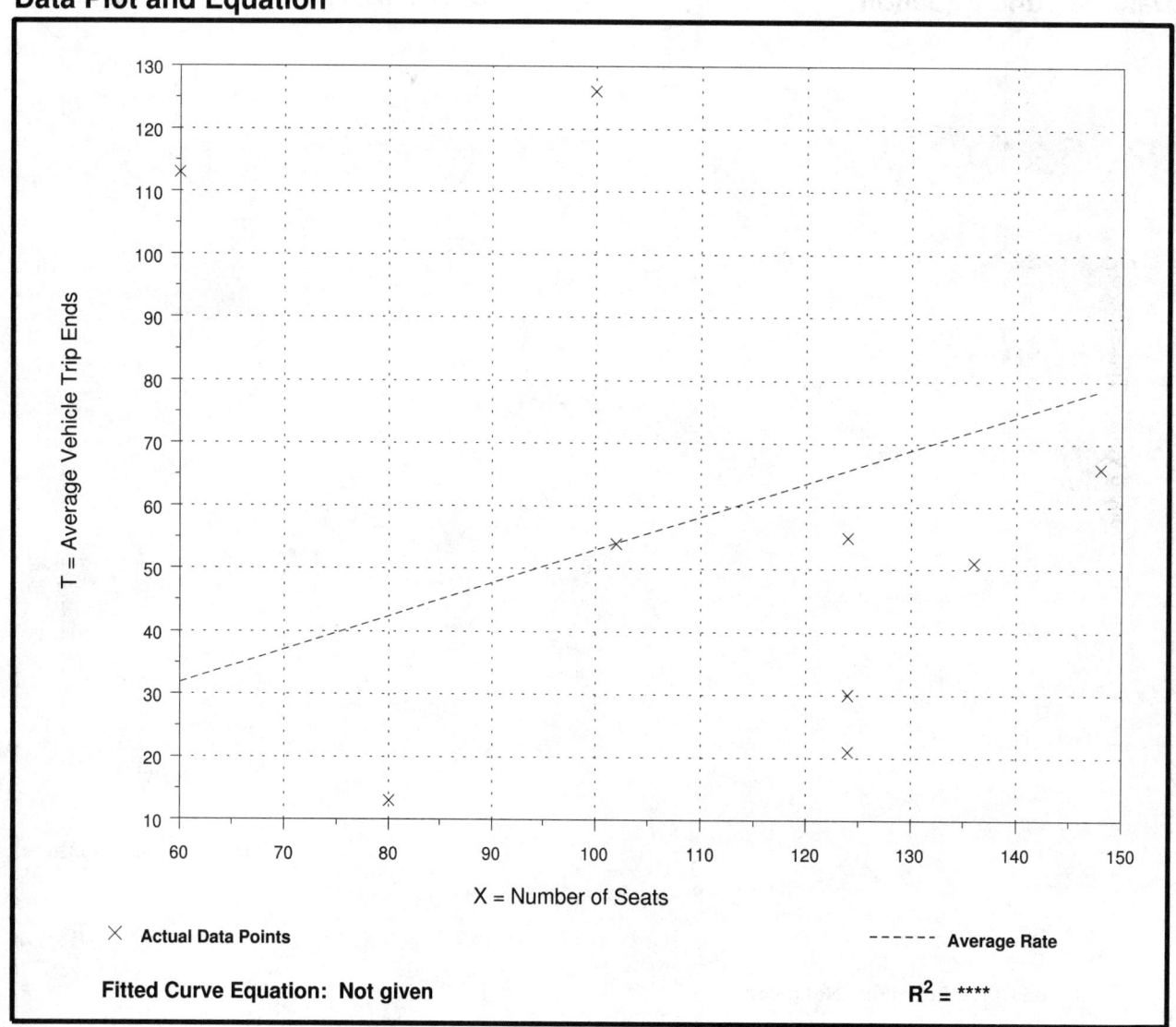

Fitted Curve Equation: Not given

R^2 = ****

High-Turnover (Sit-Down) Restaurant
(932)

Average Vehicle Trip Ends vs: Seats
On a: Sunday

Number of Studies: 2
Average Number of Seats: 125
Directional Distribution: 50% entering, 50% exiting

Trip Generation per Seat

Average Rate	Range of Rates	Standard Deviation
5.17	3.87 - 7.05	*

Data Plot and Equation

Caution - Use Carefully - Small Sample Size

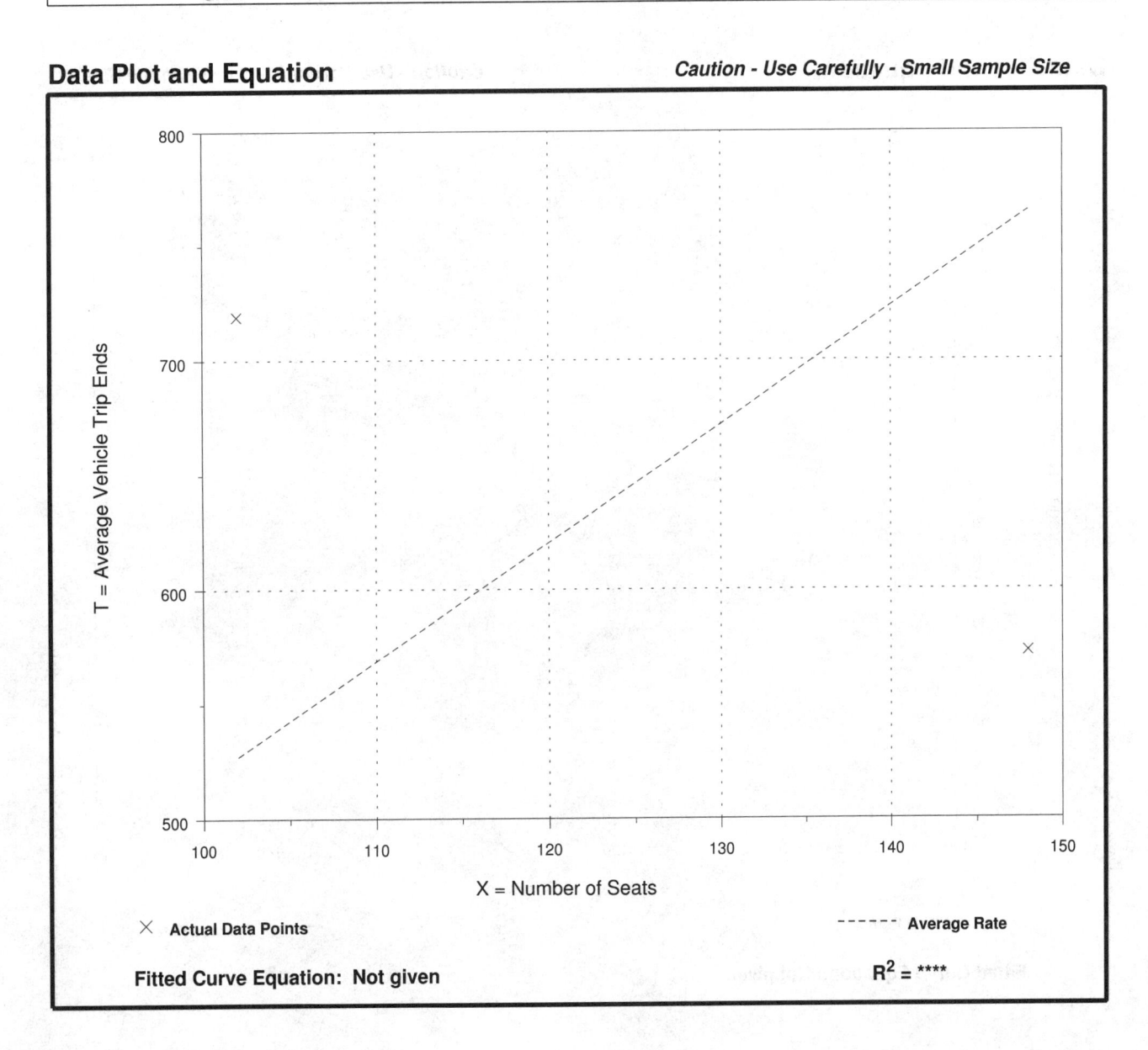

Fitted Curve Equation: Not given

R^2 = ****

High-Turnover (Sit-Down) Restaurant
(932)

Average Vehicle Trip Ends vs: Seats
On a: Sunday,
Peak Hour of Generator

Number of Studies: 3
Average Number of Seats: 117
Directional Distribution: 55% entering, 45% exiting

Trip Generation per Seat

Average Rate	Range of Rates	Standard Deviation
0.65	0.32 - 1.08	0.86

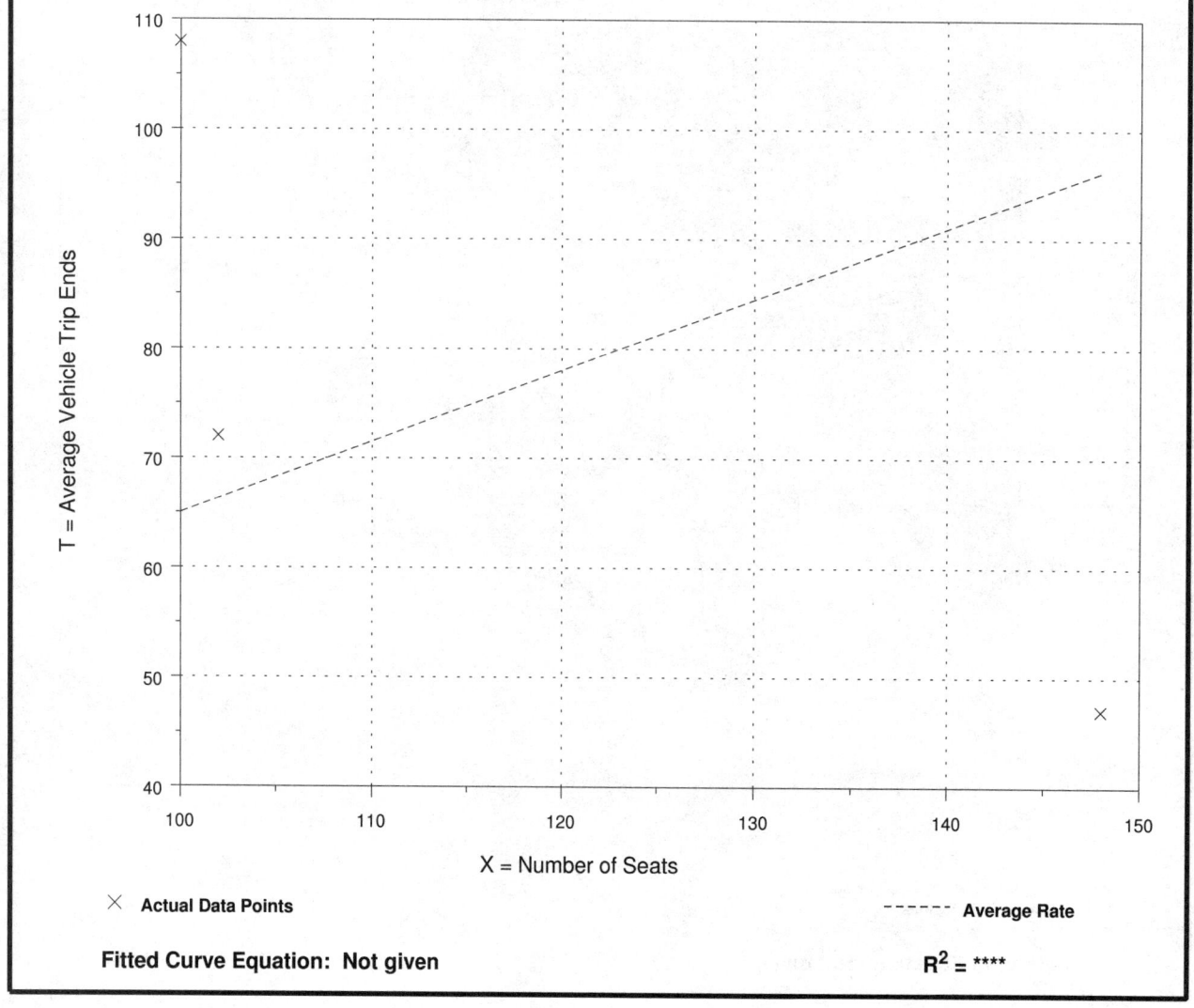

Data Plot and Equation

Caution - Use Carefully - Small Sample Size

Fitted Curve Equation: Not given $R^2 = ****$

Land Use: 933
Fast-Food Restaurant without Drive-Through Window

Description

This land use includes fast-food restaurants without drive-through windows. This type of restaurant is characterized by a large carry-out clientele, long hours of service (some are open for breakfast, all are open for lunch and dinner, some are open late at night or 24 hours per day) and high turnover rates for eat-in customers. These limited-service eating establishments do not provide table service. Patrons generally order at a cash register and pay before they eat. High-turnover (sit-down) restaurant (Land Use 932), fast-food restaurant with drive-through window (Land Use 934) and fast-food restaurant with drive-through window and no indoor seating (Land Use 935) are related uses.

Additional Data

Although the facilities in this land use may or may not be open for breakfast, it was confirmed that the data provided during the A.M. peak periods were strictly for facilities that were open during this time frame.

The outdoor seating area is not included in the overall gross floor area. Therefore, the number of seats may be a more reliable independent variable on which to establish trip generation rates for facilities having significant outdoor seating.

The sites were surveyed between the 1980s and the 2000s throughout the United States.

It has been speculated that hamburger restaurants may generate trips at a higher rate than other types of fast-food restaurants. The database was tested in an attempt to verify this assumption; the data neither verified nor disproved it. Future research is needed in this area.

Specialized Land Use Data

Current industry trends have resulted in the emergence of several new fast-food restaurants without drive-through windows that specialize in the sale of very specific food items. The trip generation characteristics of these facilities differ from the facilities typically contained in this land use; their sizes, trip generation rates and peak hour of service vary considerably. Another notable difference in these land uses is that they are typically not stand-alone facilities; these restaurants are generally located in small shopping centers. Therefore, the information collected for these facilities is presented in the following table and was excluded from the data plots.

Table: Yogurt Shop

Independent Variable	Trip Generation Rate	Size of Independent Variable	Number of Studies	Directional Distribution
1,000 Square Feet Gross Floor Area				
Weekday P.M. Peak Hour of Street Adjacent Traffic	15.12	0.86	1	38% entering, 62% exiting
Weekday P.M. Peak Hour of Generator	18.60	0.86	1	50% entering, 50% exiting

Source: 414

Source Numbers

163, 247, 278, 319, 342, 414

Land Use: 933
Fast-Food Restaurant without Drive-Through Window
Independent Variables with One Observation

The following trip generation data are for independent variables with only one observation. This information is shown in this table only; there are no related plots for these data.

Users are cautioned to use data with care because of the small sample size.

Independent Variable	Trip Generation Rate	Size of Independent Variable	Number of Studies	Directional Distribution
1,000 Square Feet Gross Floor Area				
Weekday	716	1	1	50% entering, 50% exiting
Saturday	696	1	1	50% entering, 50% exiting
Saturday Peak Hour of Generator	54.55	5	1	49% entering, 51% exiting
Sunday	500	1	1	50% entering, 50% exiting
Seats				
Weekday	42.12	17	1	50% entering, 50% exiting
Weekday P.M. Peak Hour of Adjacent Street Traffic	2.13	30	1	64% entering, 36% exiting
Weekday P.M. Peak Hour of Generator	6.59	17	1	52% entering, 48% exiting
Saturday	40.94	17	1	50% entering, 50% exiting
Sunday	29.41	17	1	50% entering, 50% exiting

Fast-Food Restaurant without Drive-Through Window
(933)

Average Vehicle Trip Ends vs: 1000 Sq. Feet Gross Floor Area
On a: Weekday,
Peak Hour of Adjacent Street Traffic,
One Hour Between 7 and 9 a.m.

Number of Studies: 2
Average 1000 Sq. Feet GFA: 3
Directional Distribution: 60% entering, 40% exiting

Trip Generation per 1000 Sq. Feet Gross Floor Area

Average Rate	Range of Rates	Standard Deviation
43.87	42.00 - 45.45	*

Data Plot and Equation

Caution - Use Carefully - Small Sample Size

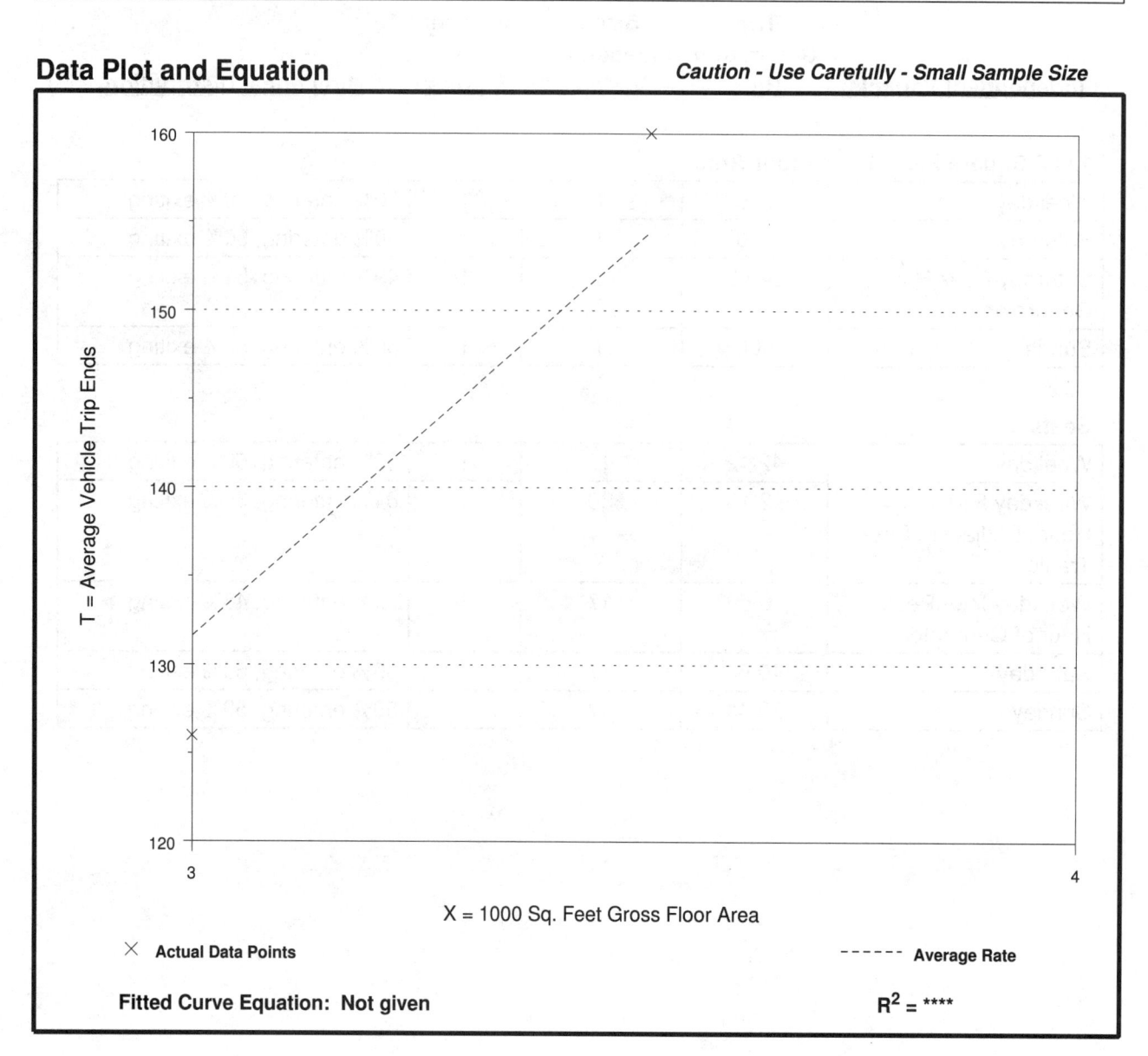

Fitted Curve Equation: Not given $R^2 = ****$

Fast-Food Restaurant without Drive-Through Window
(933)

Average Vehicle Trip Ends vs: 1000 Sq. Feet Gross Floor Area
On a: Weekday,
Peak Hour of Adjacent Street Traffic,
One Hour Between 4 and 6 p.m.

Number of Studies: 4
Average 1000 Sq. Feet GFA: 4
Directional Distribution: 51% entering, 49% exiting

Trip Generation per 1000 Sq. Feet Gross Floor Area

Average Rate	Range of Rates	Standard Deviation
26.15	14.28 - 36.36	10.51

Data Plot and Equation

Caution - Use Carefully - Small Sample Size

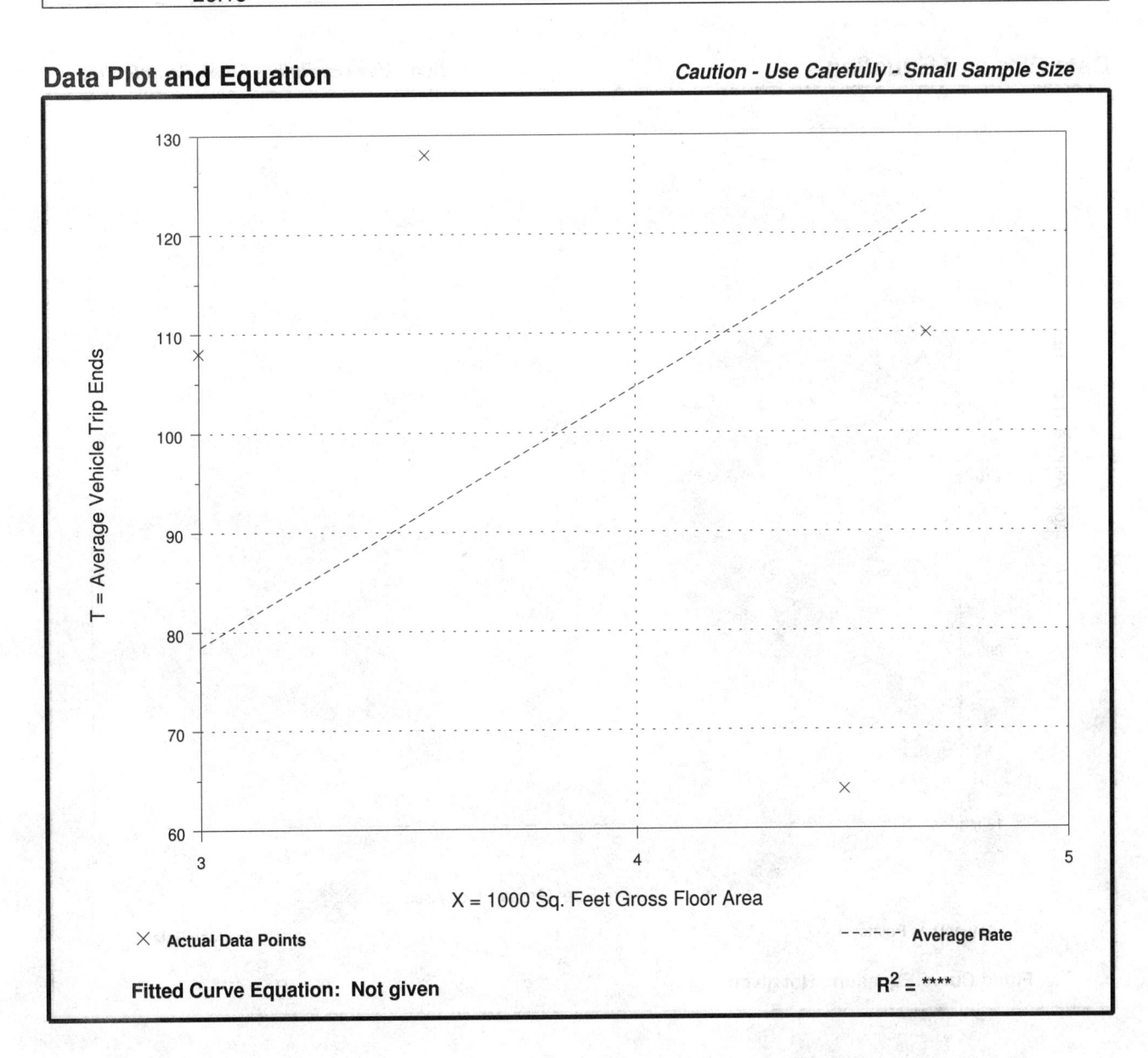

X Actual Data Points
------ Average Rate

Fitted Curve Equation: Not given $R^2 = ****$

Fast-Food Restaurant without Drive-Through Window
(933)

Average Vehicle Trip Ends vs: 1000 Sq. Feet Gross Floor Area
On a: Weekday,
A.M. Peak Hour of Generator

Number of Studies: 2
Average 1000 Sq. Feet GFA: 3
Directional Distribution: 52% entering, 48% exiting

Trip Generation per 1000 Sq. Feet Gross Floor Area

Average Rate	Range of Rates	Standard Deviation
63.50	59.38 - 68.33	*

Data Plot and Equation

Caution - Use Carefully - Small Sample Size

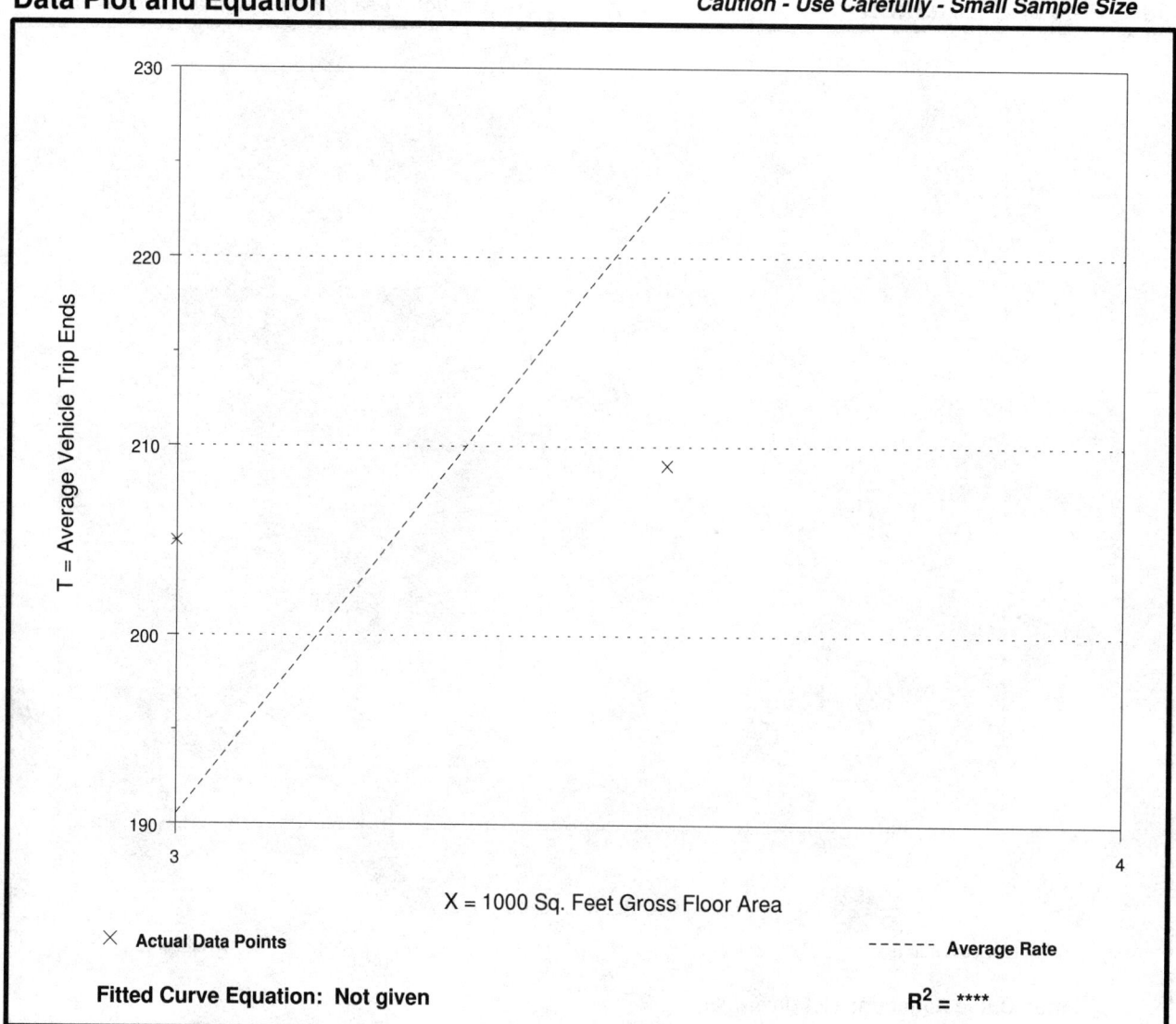

Fitted Curve Equation: Not given $R^2 = ****$

1908   *Trip Generation*, 9th Edition • Institute of Transportation Engineers

Fast-Food Restaurant without Drive-Through Window
(933)

Average Vehicle Trip Ends vs: 1000 Sq. Feet Gross Floor Area
On a: Weekday,
P.M. Peak Hour of Generator

Number of Studies: 5
Average 1000 Sq. Feet GFA: 3
Directional Distribution: 51% entering, 49% exiting

Trip Generation per 1000 Sq. Feet Gross Floor Area

Average Rate	Range of Rates	Standard Deviation
52.40	29.05 - 112.00	19.86

Data Plot and Equation

Caution - Use Carefully - Small Sample Size

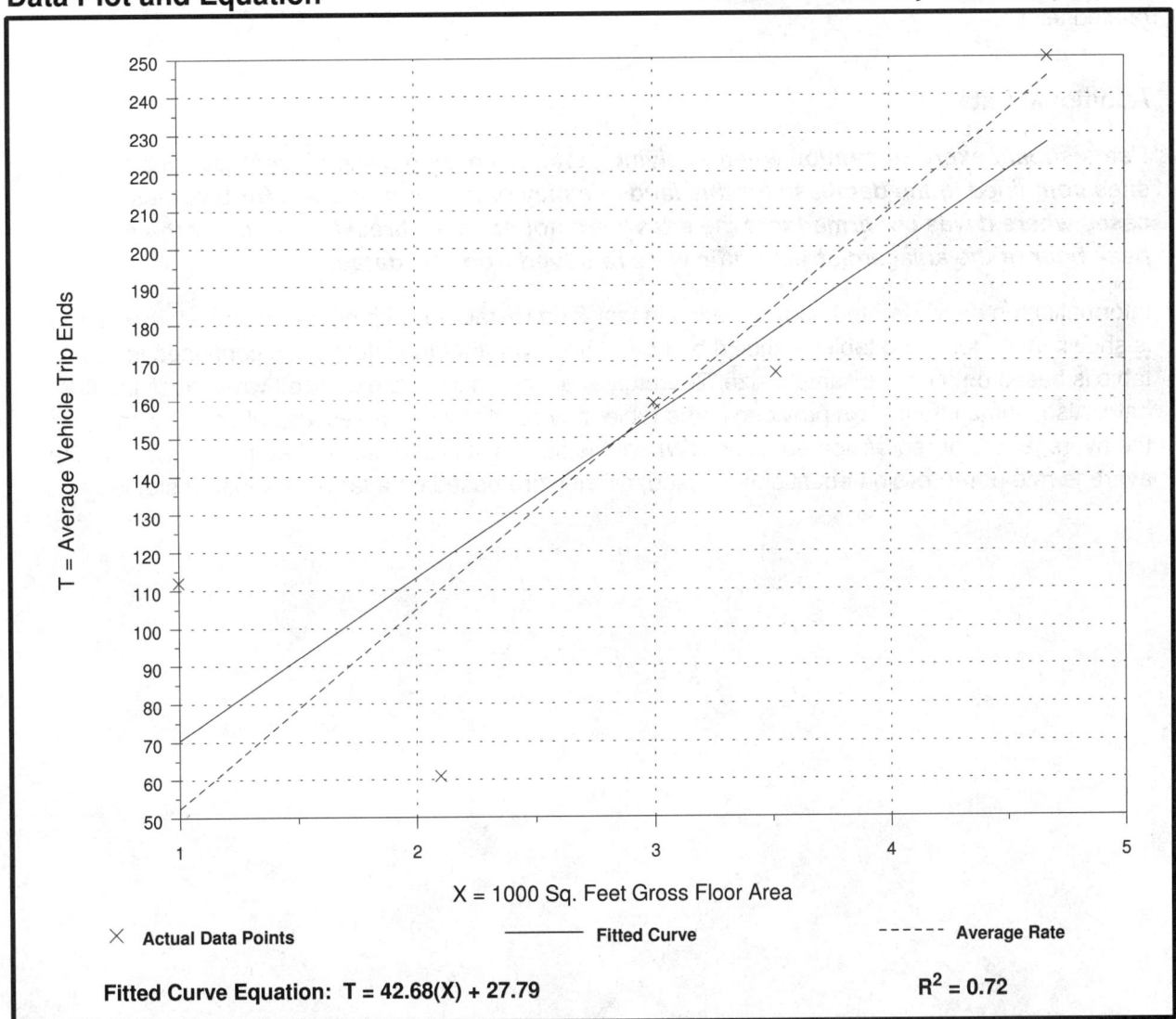

Fitted Curve Equation: $T = 42.68(X) + 27.79$ $R^2 = 0.72$

Land Use: 934
Fast-Food Restaurant with Drive-Through Window

Description

This category includes fast-food restaurants with drive-through windows. This type of restaurant is characterized by a large drive-through clientele, long hours of service (some are open for breakfast, all are open for lunch and dinner, some are open late at night or 24 hours per day) and high turnover rates for eat-in customers. These limited-service eating establishments do not provide table service. Non-drive-through patrons generally order at a cash register and pay before they eat. High-turnover (sit-down) restaurant (Land Use 932), fast-food restaurant without drive-through window (Land Use 933) and fast-food restaurant with drive-through window and no indoor seating (Land Use 935) are related uses.

Additional Data

Users should exercise caution when applying statistics during the A.M. peak periods, as the sites contained in the database for this land use may or may not be open for breakfast. In cases where it was confirmed that the sites were not open for breakfast, data for the A.M. peak hour of the adjacent street traffic were removed from the database.

Information on approximate hourly variation in fast-food restaurant with drive-through window traffic is shown in the following table. It should be noted, however, that the information contained in this table is based on a limited sample size. Therefore, caution should be exercised when applying the data. Also, some information provided in the table may conflict with the results obtained by applying the average rate or regression equations. When this occurs, it is suggested that the results from the average rate or regression equations be used, as they are based on a larger number of studies.

Hourly Variation in Fast-Food Restaurant with Drive-Through Window Traffic		
	Average Weekday	
Time	Percent of 24-Hour Entering Traffic	Percent of 24-Hour Exiting Traffic
6 a.m.–7 a.m.	1.1	1.1
7 a.m.–8 a.m.	3.1	3.8
8 a.m.–9 a.m.	3.3	4.3
9 a.m.–10 a.m.	3.5	3.7
10 a.m.–11 a.m.	4.5	4.2
11 a.m.–12 p.m.	8.1	8.0
12 p.m.–1 p.m.	14.9	14.4
1 p.m.–2 p.m.	12.1	12.4
2 p.m.–3 p.m.	7.6	8.0
3 p.m.–4 p.m.	6.1	5.9
4 p.m.–5 p.m.	5.2	5.3
5 p.m.–6 p.m.	5.9	5.8
6 p.m.–7 p.m.	7.0	6.4
7 p.m.–8 p.m.	6.9	6.1
8 p.m.–9 p.m.	5.4	5.0
9 p.m.–10 p.m.	3.8	3.9
10 p.m.–6 a.m.	1.5	1.7

Sites ranged in size from 2,900 to 4,300 square feet gross floor area
Source: The Traffic Group, Inc.; based on four studies

The outdoor seating area is not included in the overall gross floor area. Therefore, the number of seats may be a more reliable independent variable on which to establish trip generation rates for facilities having significant outdoor seating.

One site indicated that a two-story play area and a video arcade were included in the gross floor area.

The sites were surveyed between the 1980s and the 2000s throughout the United States.

It has been speculated that hamburger restaurants may generate trips at a higher rate than other types of fast-food restaurants. The database was tested in an attempt to verify this assumption; the data neither verified nor disproved it. Future research is needed in this area.

Source Numbers

163, 164, 168, 180, 181, 241, 245, 278, 294, 300, 301, 319, 338, 340, 342, 343, 358, 389, 438, 502, 552, 555, 577, 583, 584, 617, 640, 641, 704, 715, 728

Fast-Food Restaurant with Drive-Through Window
(934)

Average Vehicle Trip Ends vs: 1000 Sq. Feet Gross Floor Area
On a: Weekday

Number of Studies: 21
Average 1000 Sq. Feet GFA: 3
Directional Distribution: 50% entering, 50% exiting

Trip Generation per 1000 Sq. Feet Gross Floor Area

Average Rate	Range of Rates	Standard Deviation
496.12	195.98 - 1132.92	242.52

Data Plot and Equation

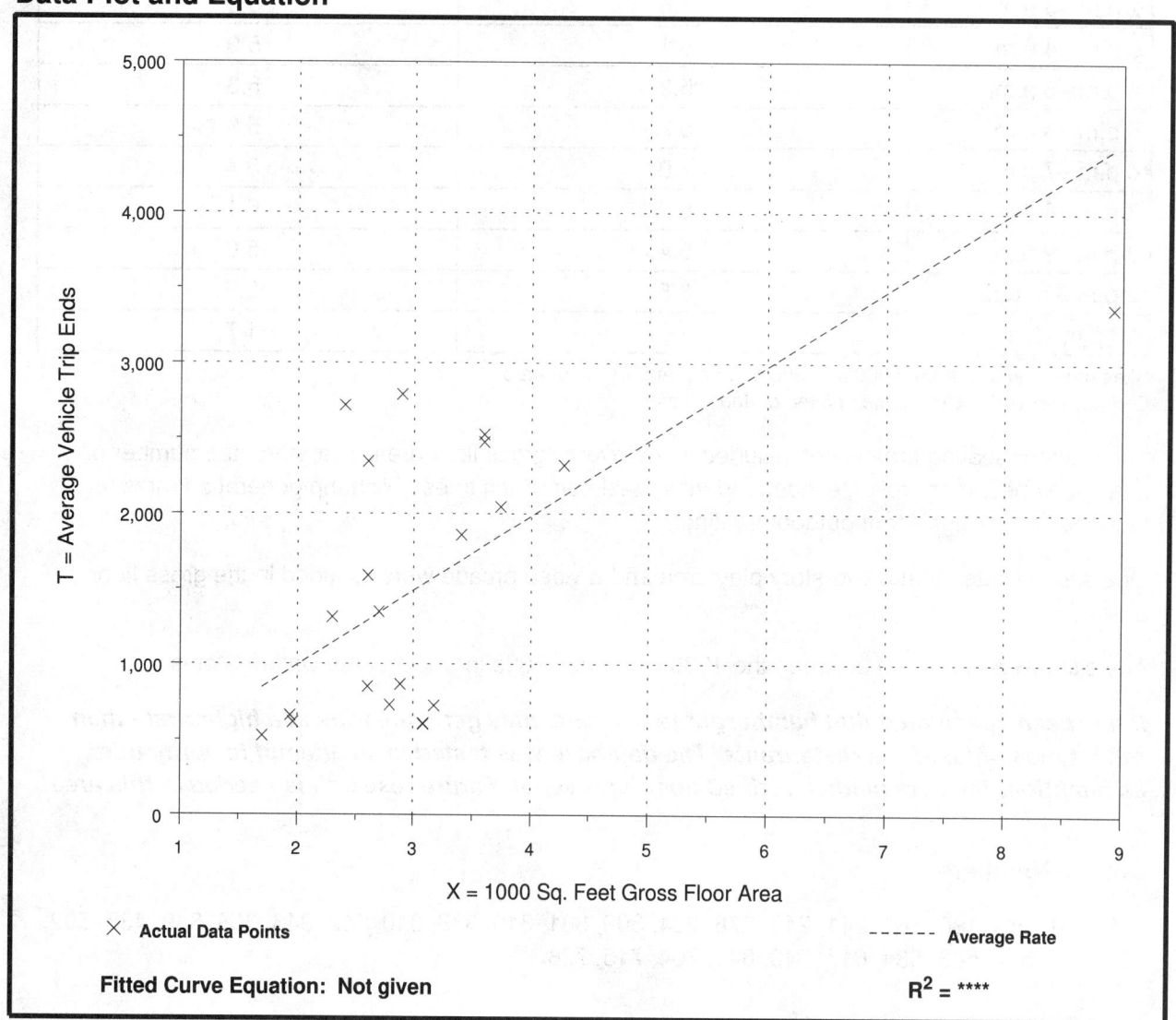

Fitted Curve Equation: Not given $R^2 = ****$

Fast-Food Restaurant with Drive-Through Window
(934)

Average Vehicle Trip Ends vs: 1000 Sq. Feet Gross Floor Area
On a: Weekday,
Peak Hour of Adjacent Street Traffic,
One Hour Between 7 and 9 a.m.

Number of Studies: 75
Average 1000 Sq. Feet GFA: 4
Directional Distribution: 51% entering, 49% exiting

Trip Generation per 1000 Sq. Feet Gross Floor Area

Average Rate	Range of Rates	Standard Deviation
45.42	1.02 - 163.33	28.63

Data Plot and Equation

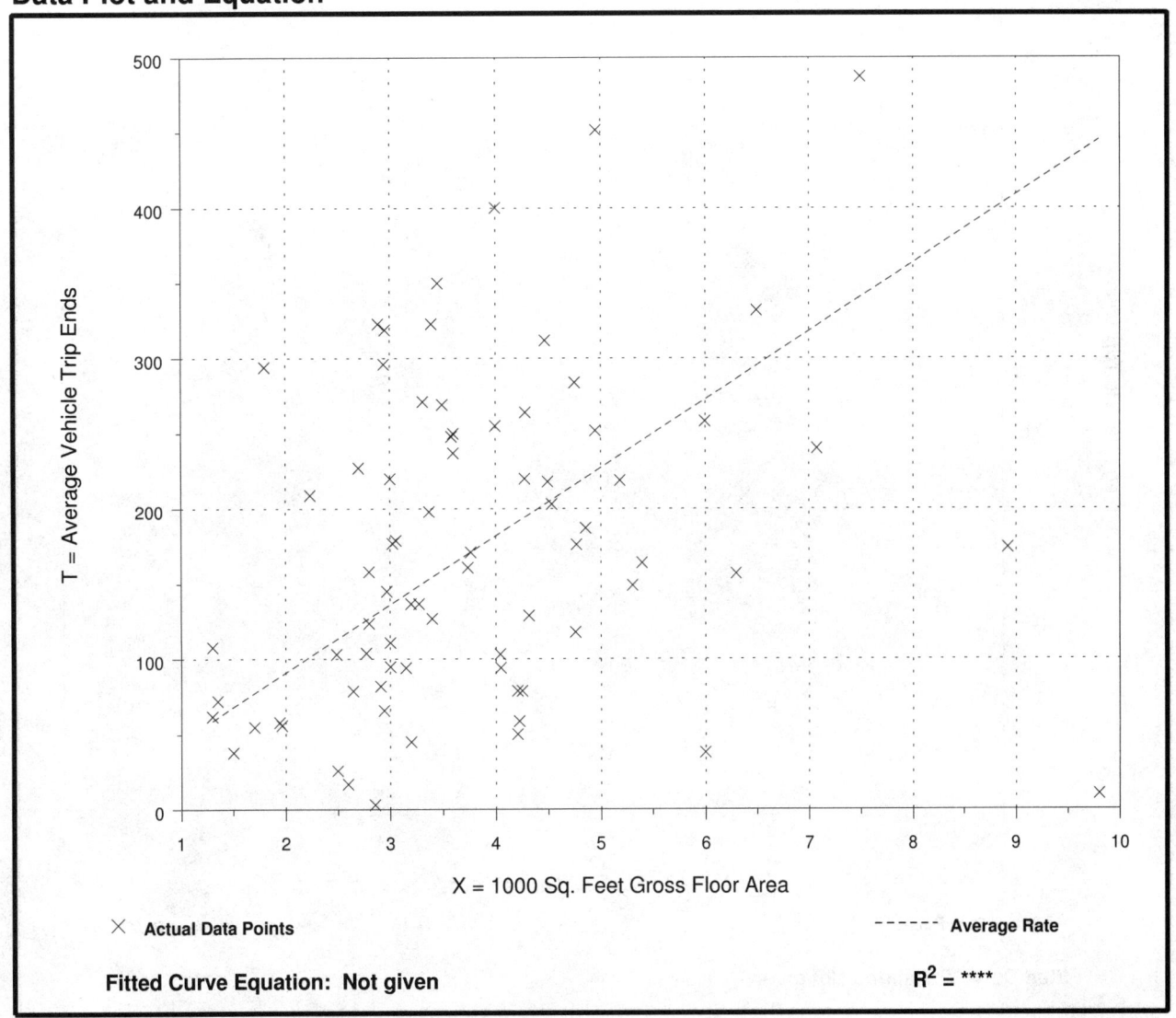

Fitted Curve Equation: Not given $R^2 = ****$

Fast-Food Restaurant with Drive-Through Window
(934)

Average Vehicle Trip Ends vs: 1000 Sq. Feet Gross Floor Area
On a: Weekday,
Peak Hour of Adjacent Street Traffic,
One Hour Between 4 and 6 p.m.

Number of Studies: 132
Average 1000 Sq. Feet GFA: 3
Directional Distribution: 52% entering, 48% exiting

Trip Generation per 1000 Sq. Feet Gross Floor Area

Average Rate	Range of Rates	Standard Deviation
32.65	7.96 - 117.15	19.73

Data Plot and Equation

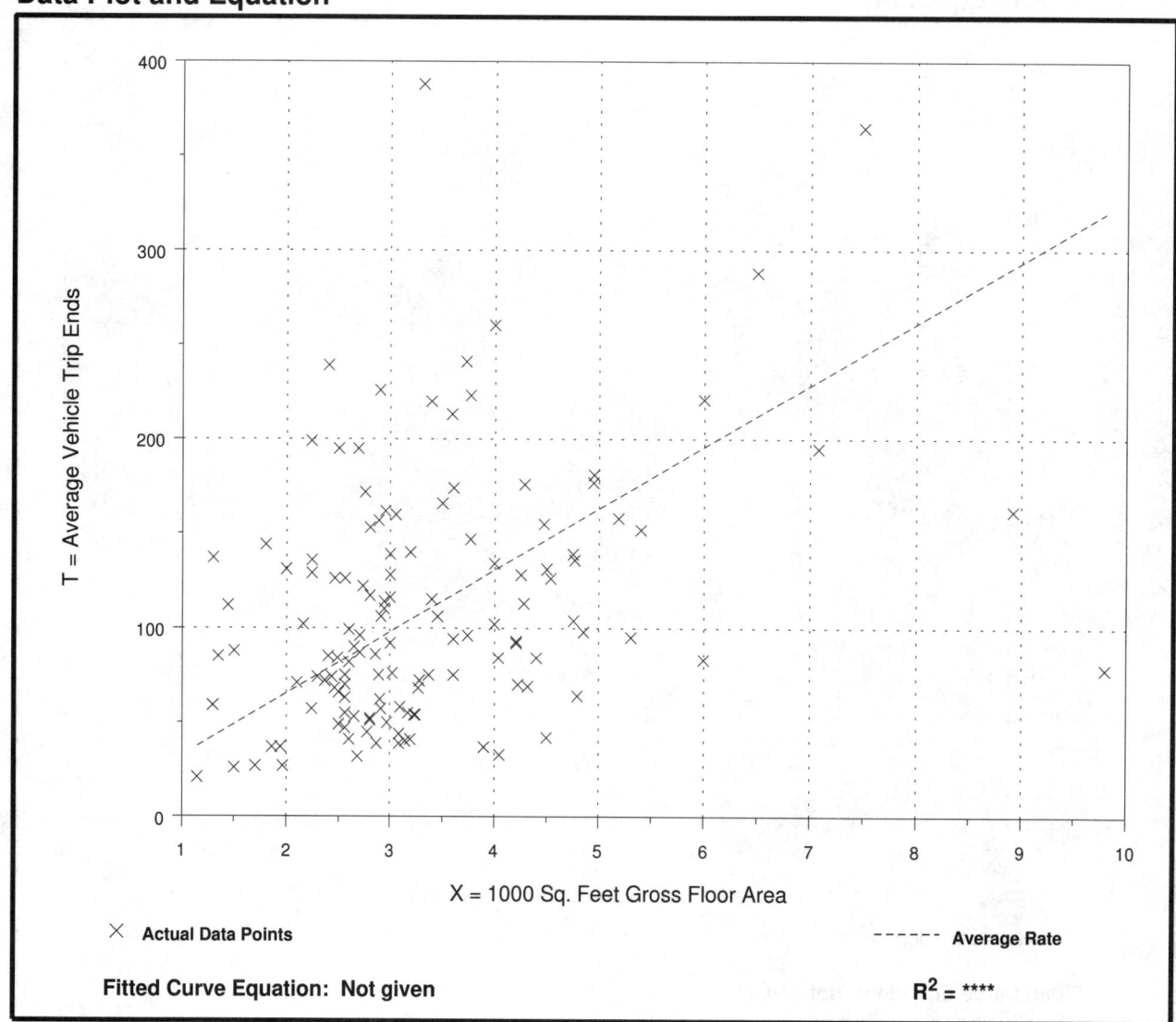

Fitted Curve Equation: Not given $R^2 = ****$

Fast-Food Restaurant with Drive-Through Window
(934)

Average Vehicle Trip Ends vs: 1000 Sq. Feet Gross Floor Area
On a: Weekday,
A.M. Peak Hour of Generator

Number of Studies: 65
Average 1000 Sq. Feet GFA: 4
Directional Distribution: 51% entering, 49% exiting

Trip Generation per 1000 Sq. Feet Gross Floor Area

Average Rate	Range of Rates	Standard Deviation
53.61	12.96 - 163.33	26.27

Data Plot and Equation

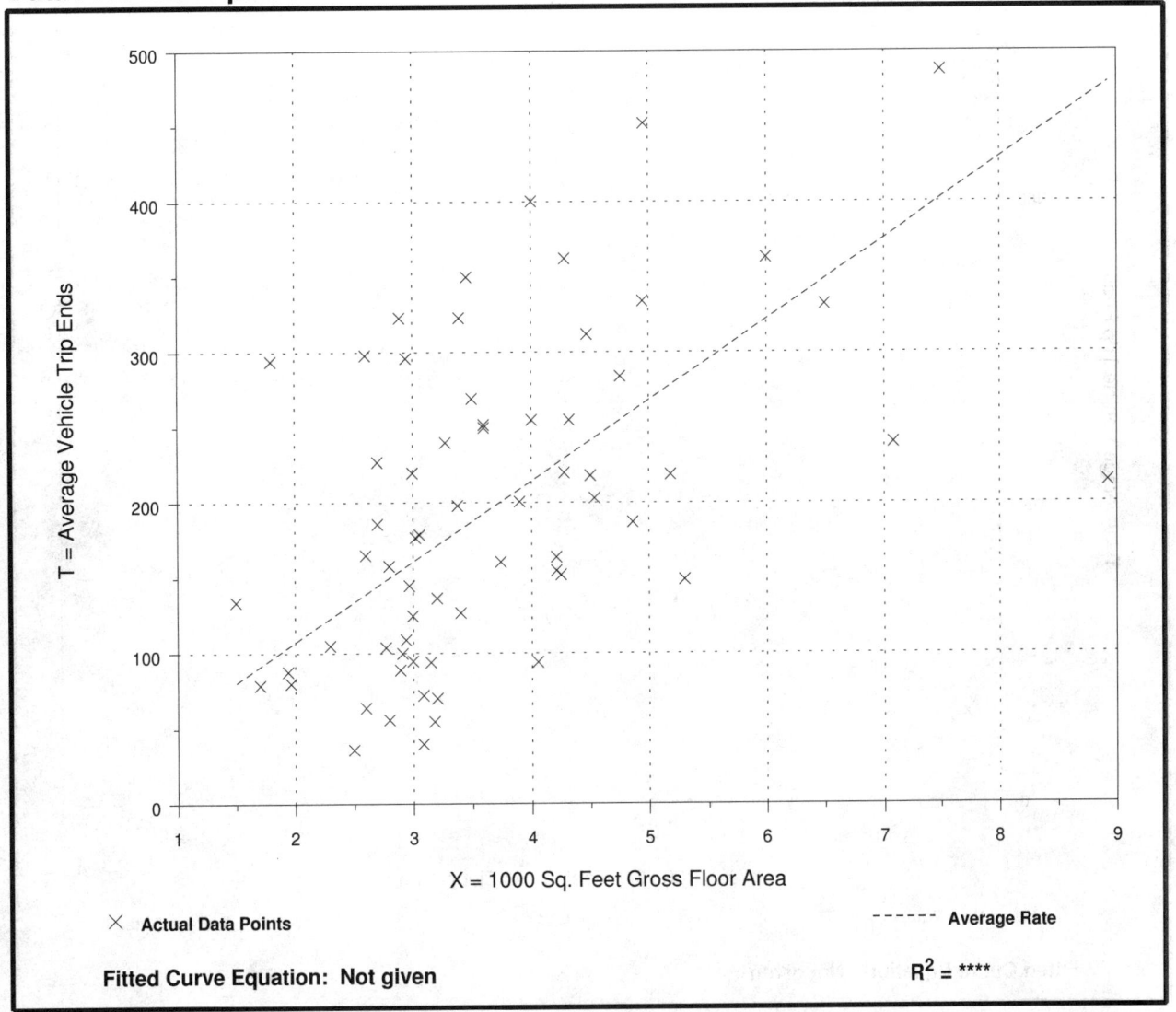

Fitted Curve Equation: Not given $R^2 = ****$

Trip Generation, 9th Edition • Institute of Transportation Engineers

Fast-Food Restaurant with Drive-Through Window
(934)

Average Vehicle Trip Ends vs: 1000 Sq. Feet Gross Floor Area
On a: Weekday,
P.M. Peak Hour of Generator

Number of Studies: 81
Average 1000 Sq. Feet GFA: 4
Directional Distribution: 52% entering, 48% exiting

Trip Generation per 1000 Sq. Feet Gross Floor Area

Average Rate	Range of Rates	Standard Deviation
47.30	13.33 - 158.46	25.52

Data Plot and Equation

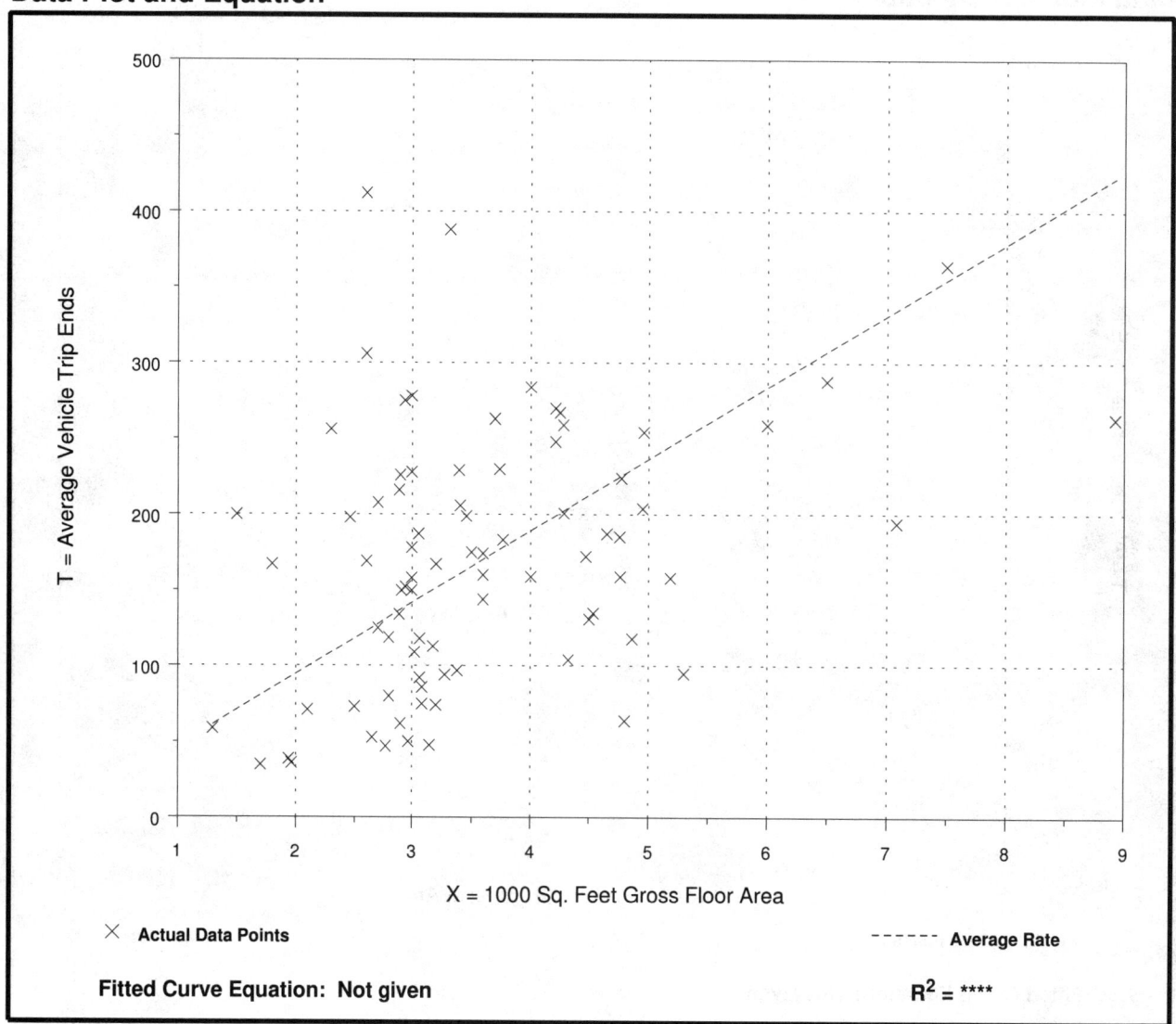

Fitted Curve Equation: Not given $R^2 = ****$

Fast-Food Restaurant with Drive-Through Window
(934)

Average Vehicle Trip Ends vs: 1000 Sq. Feet Gross Floor Area
On a: Saturday

Number of Studies: 11
Average 1000 Sq. Feet GFA: 3
Directional Distribution: 50% entering, 50% exiting

Trip Generation per 1000 Sq. Feet Gross Floor Area

Average Rate	Range of Rates	Standard Deviation
722.03	338.92 - 1405.00	295.62

Data Plot and Equation

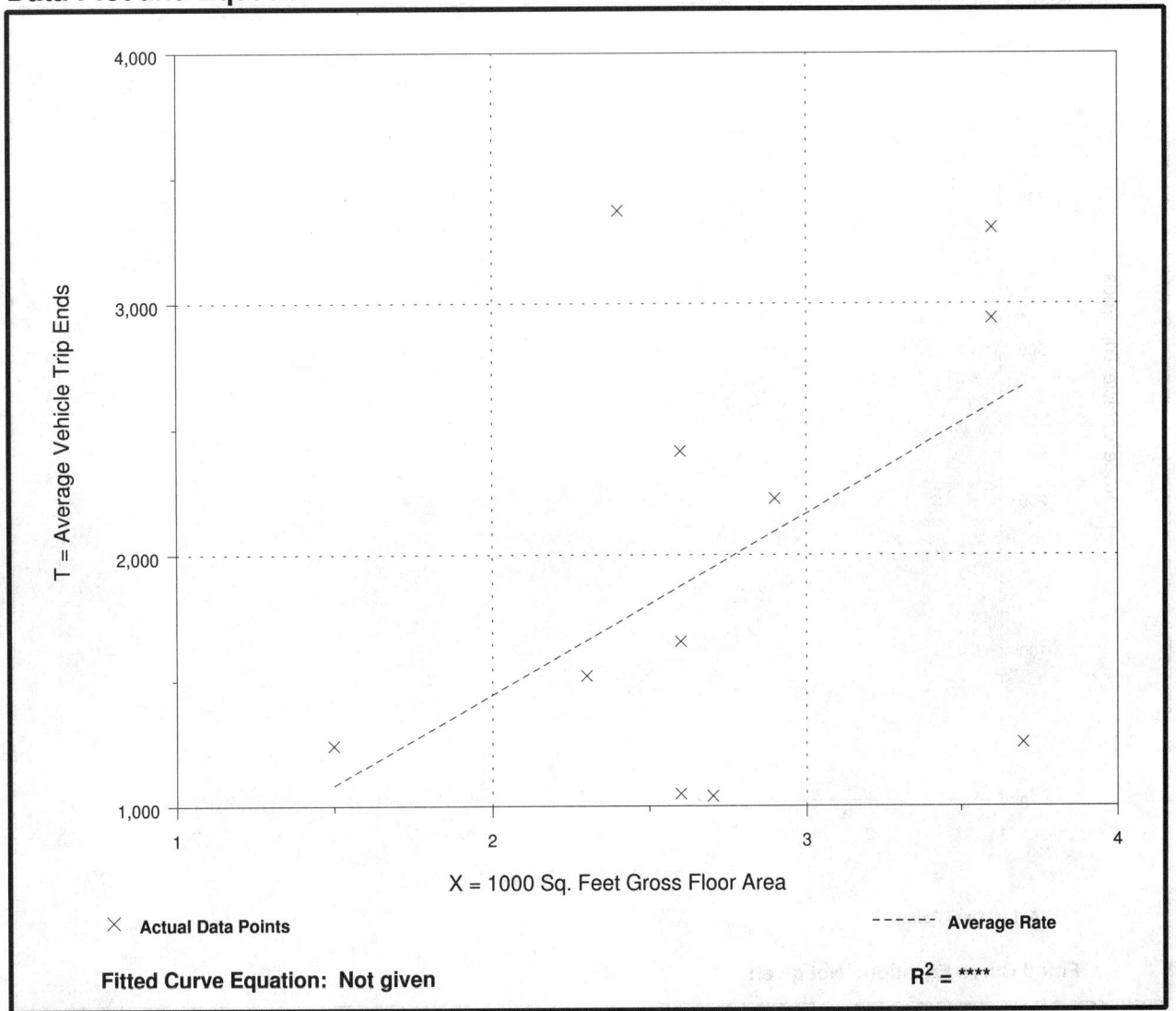

X   Actual Data Points   - - - - - Average Rate

Fitted Curve Equation: Not given $R^2 =$ ****

Fast-Food Restaurant with Drive-Through Window
(934)

Average Vehicle Trip Ends vs: 1000 Sq. Feet Gross Floor Area
On a: Saturday,
Peak Hour of Generator

Number of Studies: 41
Average 1000 Sq. Feet GFA: 4
Directional Distribution: 51% entering, 49% exiting

Trip Generation per 1000 Sq. Feet Gross Floor Area

Average Rate	Range of Rates	Standard Deviation
59.00	19.21 - 122.49	22.89

Data Plot and Equation

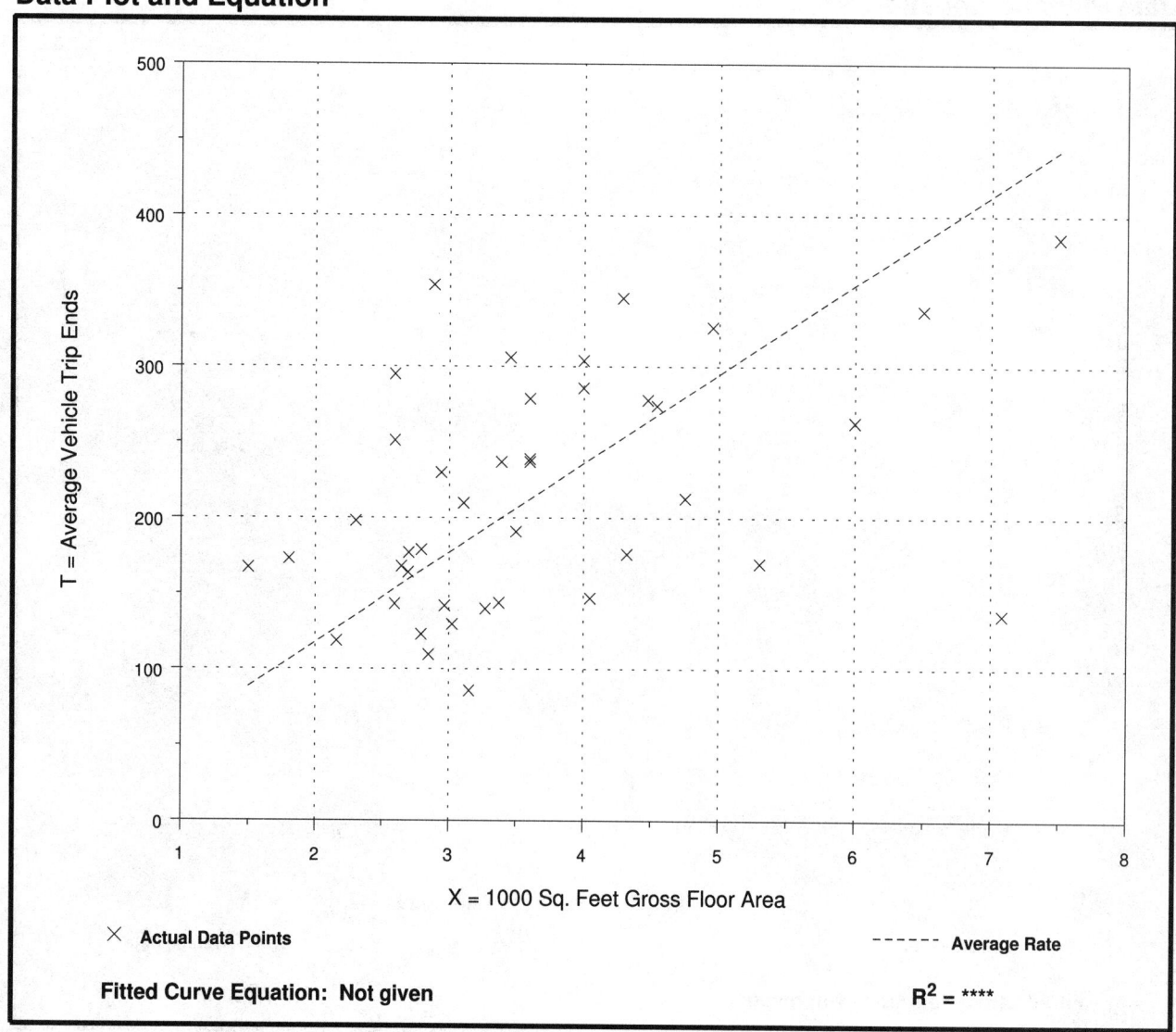

X Actual Data Points   ----- Average Rate

Fitted Curve Equation: Not given $R^2 = ****$

Fast-Food Restaurant with Drive-Through Window
(934)

Average Vehicle Trip Ends vs: 1000 Sq. Feet Gross Floor Area
On a: Sunday

Number of Studies: 11
Average 1000 Sq. Feet GFA: 3
Directional Distribution: 50% entering, 50% exiting

Trip Generation per 1000 Sq. Feet Gross Floor Area

Average Rate	Range of Rates	Standard Deviation
542.72	225.41 - 950.00	206.86

Data Plot and Equation

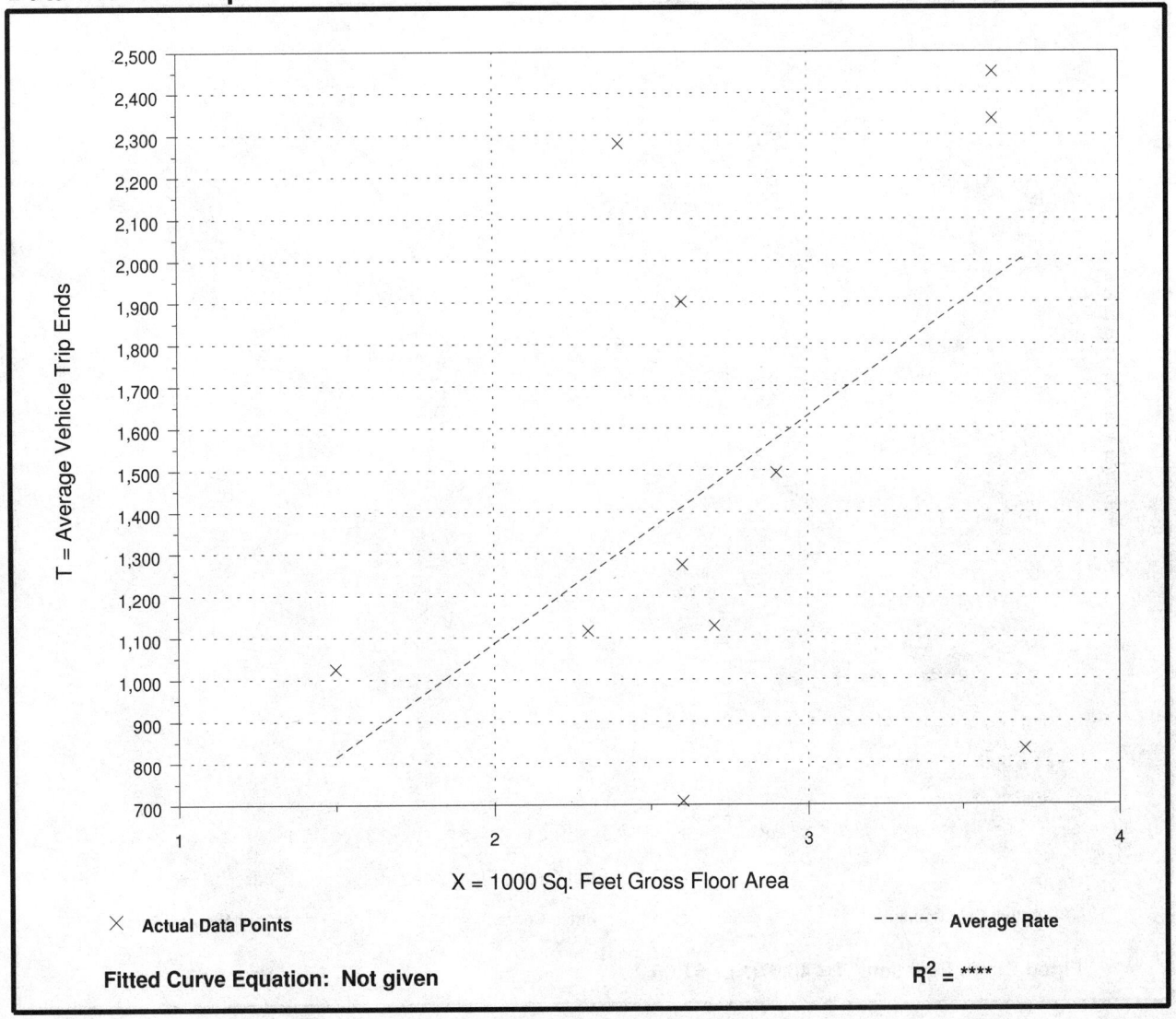

T = Average Vehicle Trip Ends
X = 1000 Sq. Feet Gross Floor Area

× Actual Data Points
------ Average Rate

Fitted Curve Equation: Not given $R^2 = ****$

Trip Generation, 9th Edition • Institute of Transportation Engineers

Fast-Food Restaurant with Drive-Through Window
(934)

Average Vehicle Trip Ends vs: 1000 Sq. Feet Gross Floor Area
On a: Sunday,
Peak Hour of Generator

Number of Studies: 5
Average 1000 Sq. Feet GFA: 2
Directional Distribution: 48% entering, 52% exiting

Trip Generation per 1000 Sq. Feet Gross Floor Area

Average Rate	Range of Rates	Standard Deviation
72.74	60.00 - 93.33	11.95

Data Plot and Equation

Caution - Use Carefully - Small Sample Size

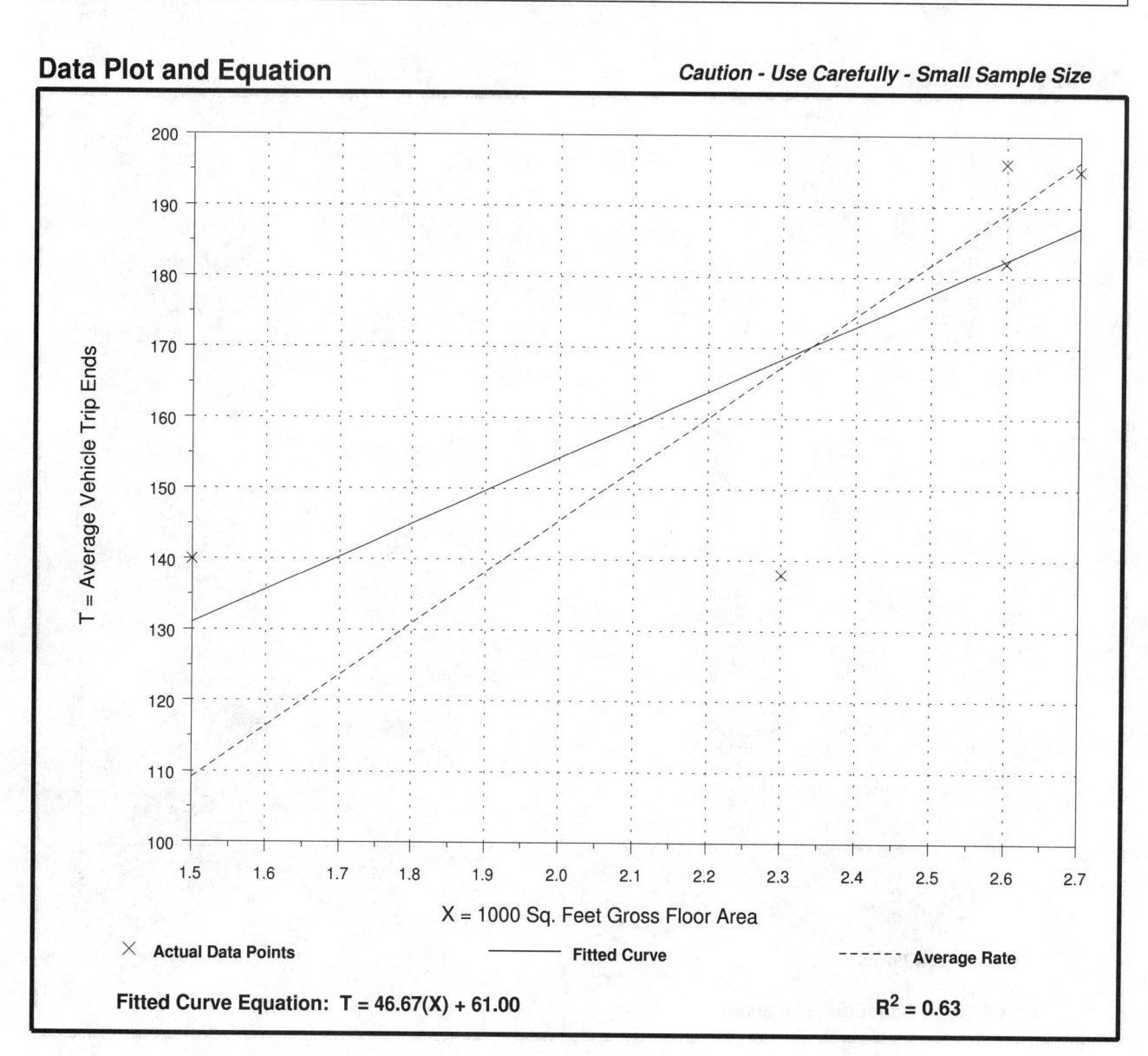

Fitted Curve Equation: $T = 46.67(X) + 61.00$ $\qquad R^2 = 0.63$

Trip Generation, 9th Edition • Institute of Transportation Engineers

Fast-Food Restaurant with Drive-Through Window
(934)

Average Vehicle Trip Ends vs: Seats
On a: Weekday

Number of Studies: 10
Average Number of Seats: 109
Directional Distribution: 50% entering, 50% exiting

Trip Generation per Seat

Average Rate	Range of Rates	Standard Deviation
19.52	8.88 - 35.78	9.97

Data Plot and Equation

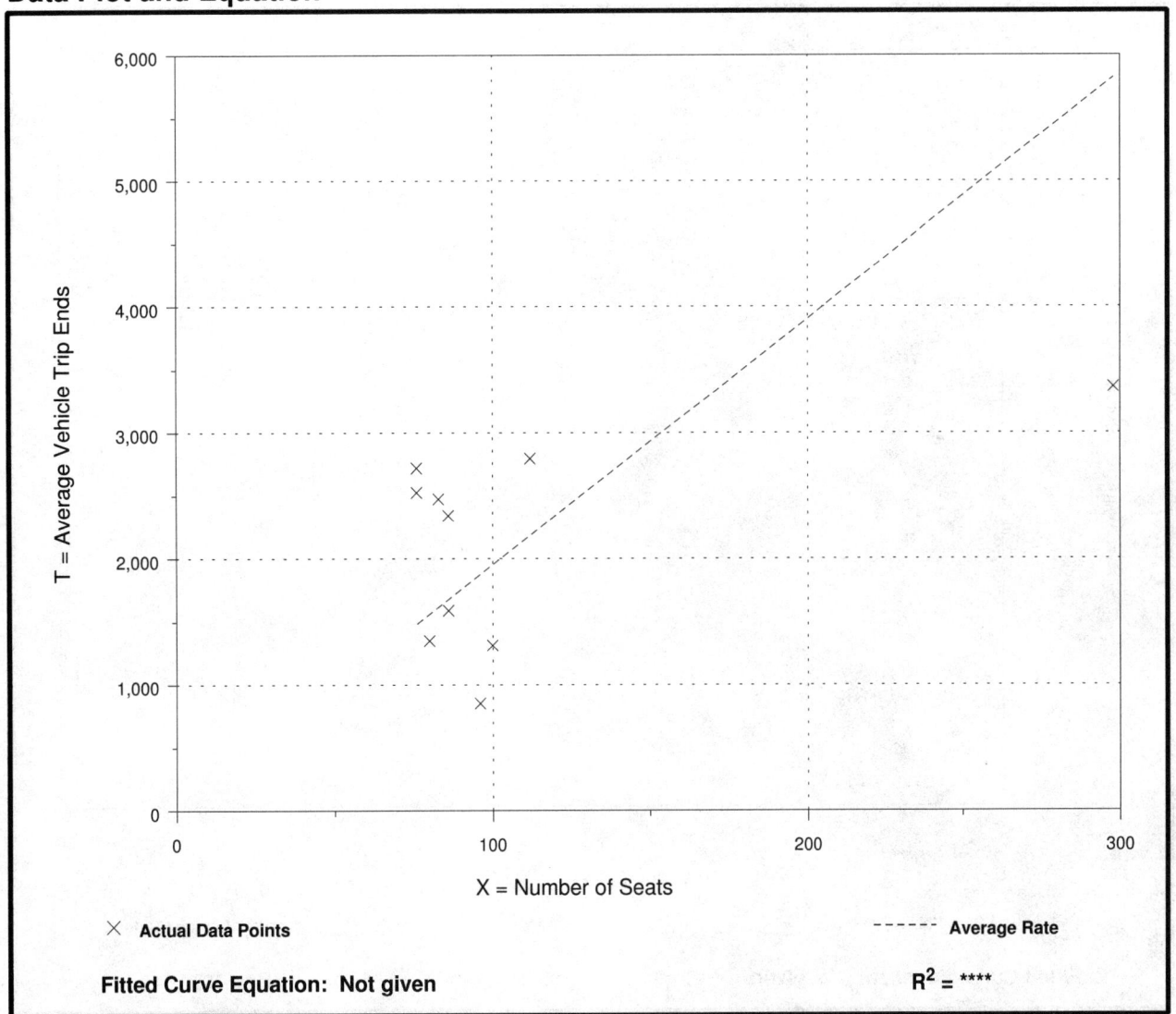

Fitted Curve Equation: Not given $R^2 = ****$

Fast-Food Restaurant with Drive-Through Window
(934)

Average Vehicle Trip Ends vs: Seats
On a: Weekday,
Peak Hour of Adjacent Street Traffic,
One Hour Between 7 and 9 a.m.

Number of Studies: 12
Average Number of Seats: 124
Directional Distribution: 53% entering, 47% exiting

Trip Generation per Seat

Average Rate	Range of Rates	Standard Deviation
1.27	0.05 - 3.12	1.40

Data Plot and Equation

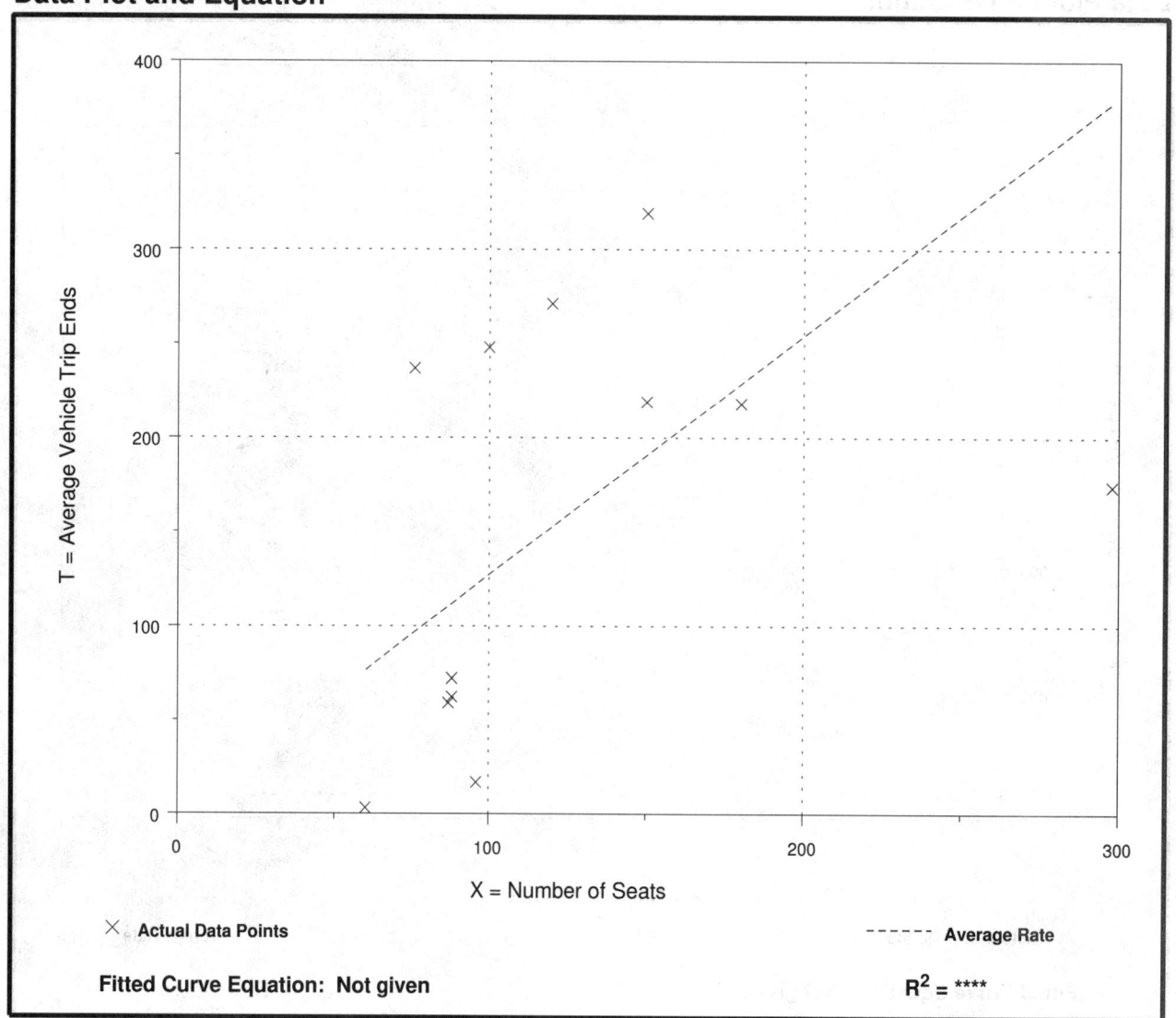

Fitted Curve Equation: Not given $R^2 = ****$

Fast-Food Restaurant with Drive-Through Window
(934)

Average Vehicle Trip Ends vs: Seats
On a: Weekday,
Peak Hour of Adjacent Street Traffic,
One Hour Between 4 and 6 p.m.

Number of Studies: 28
Average Number of Seats: 105
Directional Distribution: 53% entering, 47% exiting

Trip Generation per Seat

Average Rate	Range of Rates	Standard Deviation
0.95	0.26 - 3.23	1.17

Data Plot and Equation

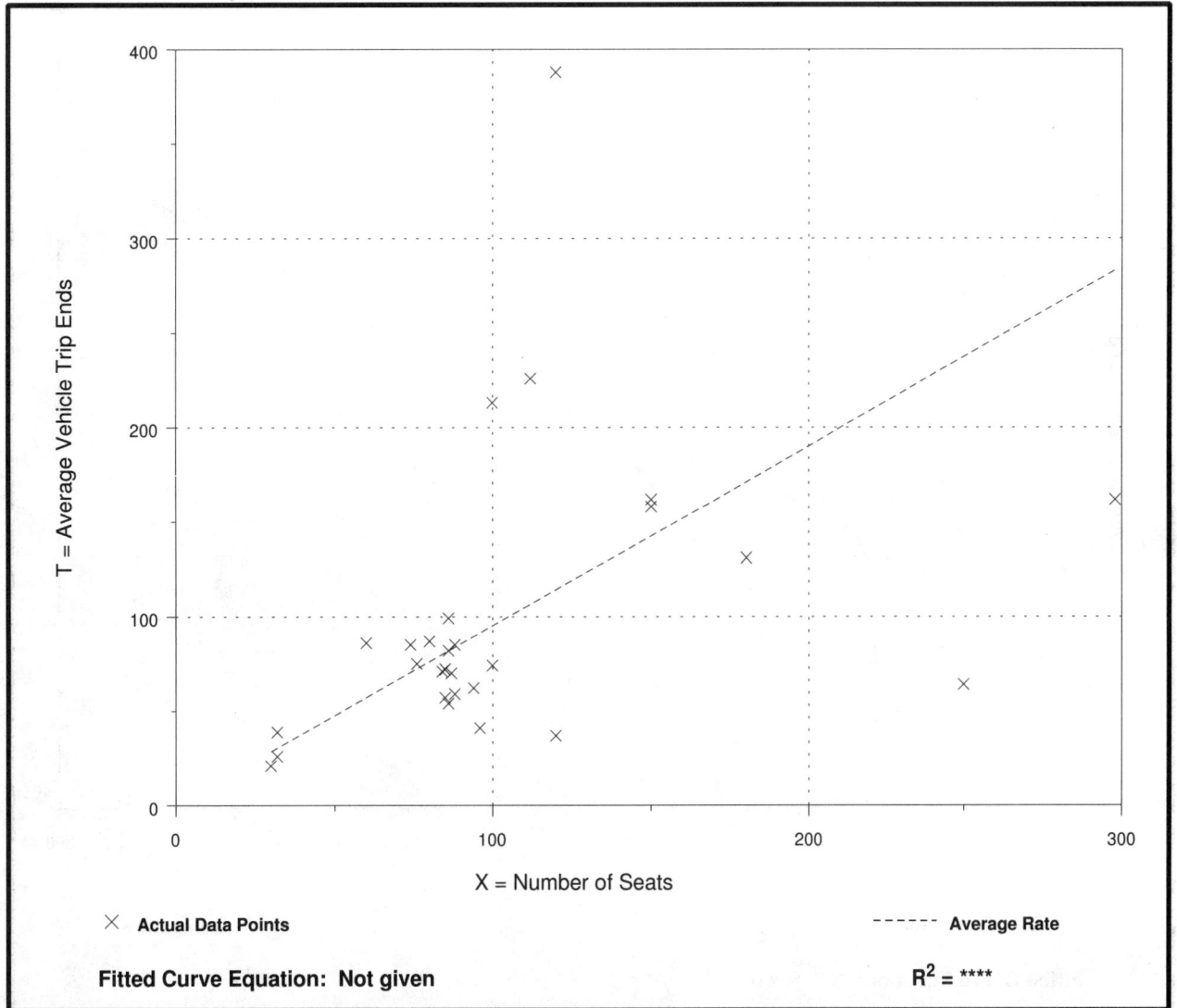

× Actual Data Points   ------ Average Rate

Fitted Curve Equation: Not given $R^2 =$ ****

Fast-Food Restaurant with Drive-Through Window
(934)

Average Vehicle Trip Ends vs: Seats
On a: Weekday,
A.M. Peak Hour of Generator

Number of Studies: 9
Average Number of Seats: 128
Directional Distribution: 53% entering, 47% exiting

Trip Generation per Seat

Average Rate	Range of Rates	Standard Deviation
1.49	0.67 - 3.47	1.51

Data Plot and Equation

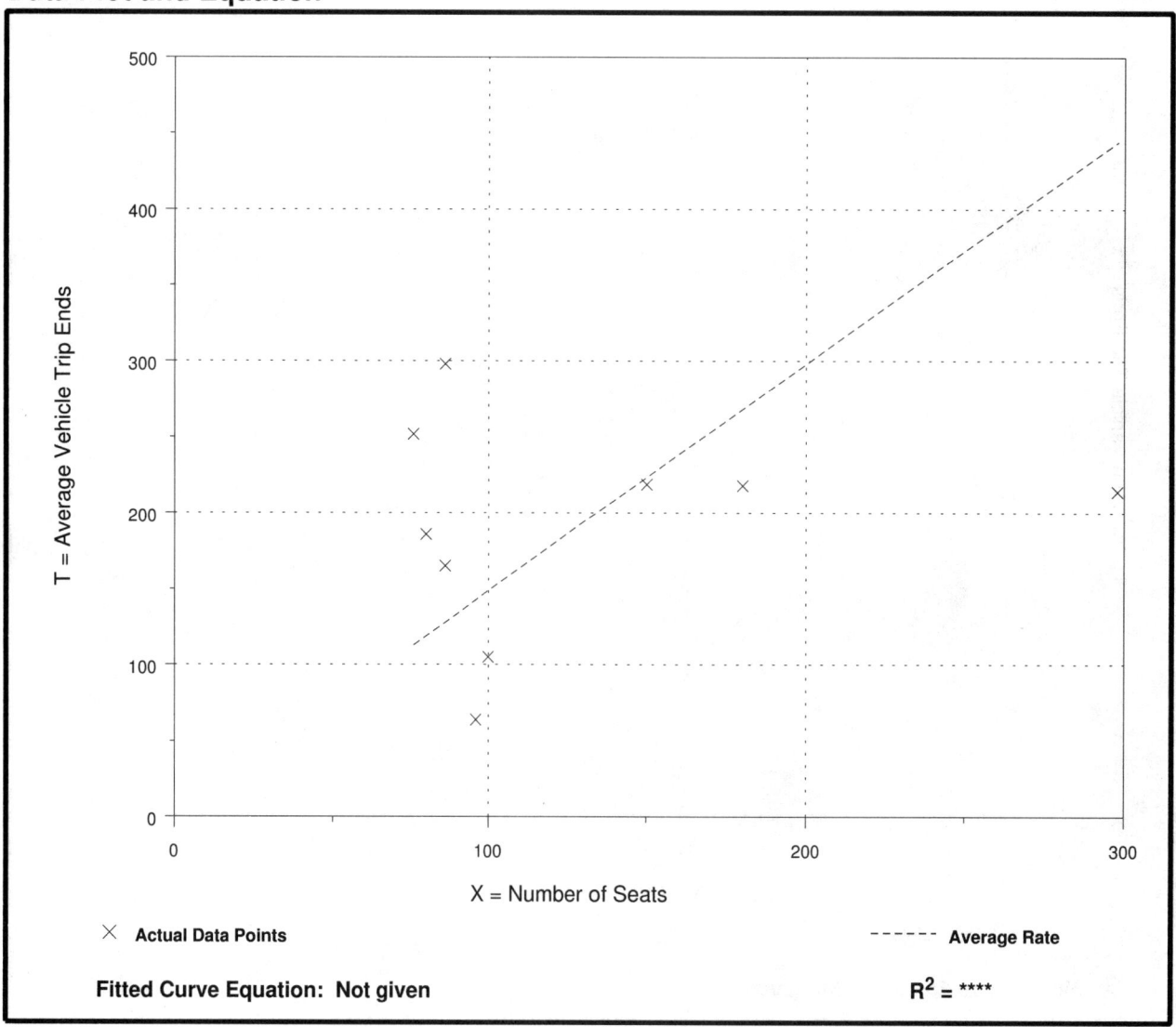

Fitted Curve Equation: Not given $R^2 = ****$

Fast-Food Restaurant with Drive-Through Window
(934)

Average Vehicle Trip Ends vs: Seats
On a: Weekday,
P.M. Peak Hour of Generator

Number of Studies: 18
Average Number of Seats: 119
Directional Distribution: 50% entering, 50% exiting

Trip Generation per Seat

Average Rate	Range of Rates	Standard Deviation
1.62	0.26 - 4.79	1.73

Data Plot and Equation

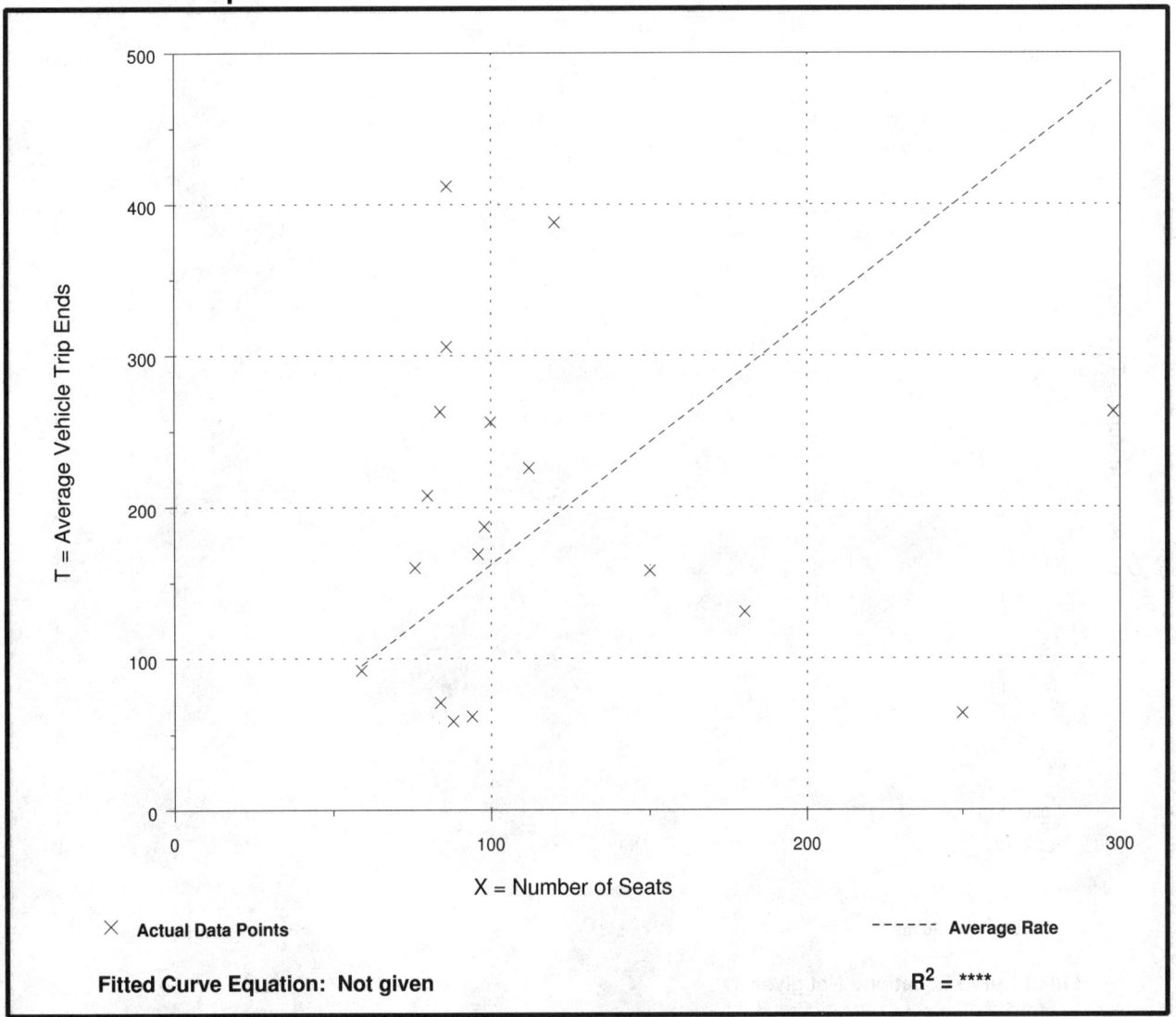

T = Average Vehicle Trip Ends
X = Number of Seats

× Actual Data Points
----- Average Rate

Fitted Curve Equation: Not given R^2 = ****

Trip Generation, 9th Edition • Institute of Transportation Engineers   1925

Fast-Food Restaurant with Drive-Through Window
(934)

Average Vehicle Trip Ends vs: Seats
On a: Saturday

Number of Studies: 10
Average Number of Seats: 88
Directional Distribution: 50% entering, 50% exiting

Trip Generation per Seat

Average Rate	Range of Rates	Standard Deviation
23.64	10.94 - 44.37	12.52

Data Plot and Equation

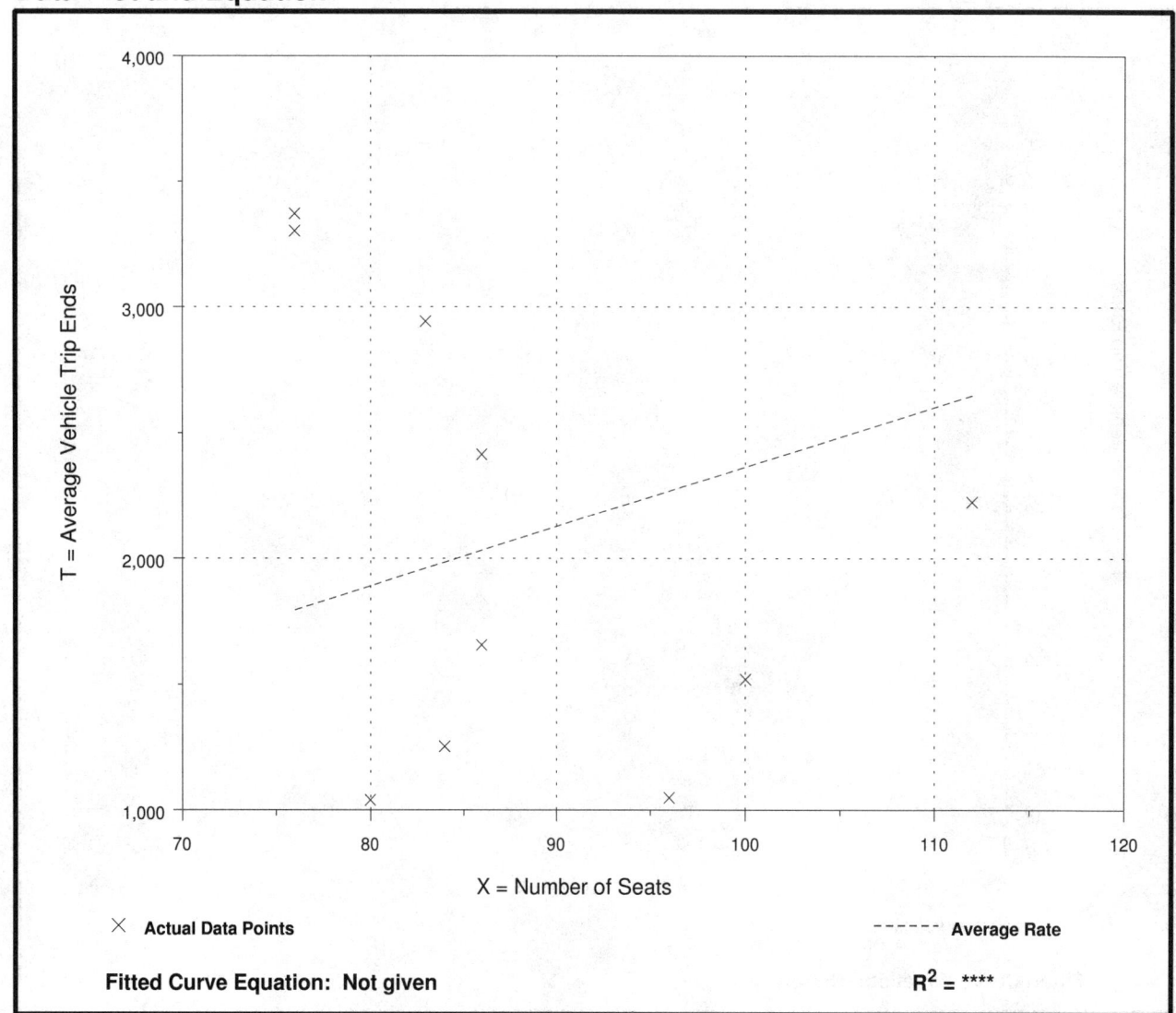

[Scatter plot: X = Number of Seats (70–120), T = Average Vehicle Trip Ends (1,000–4,000). Dashed line shows Average Rate.]

Fitted Curve Equation: Not given $R^2 =$ ****

Fast-Food Restaurant with Drive-Through Window
(934)

Average Vehicle Trip Ends vs: Seats
On a: Saturday,
Peak Hour of Generator

Number of Studies: 7
Average Number of Seats: 83
Directional Distribution: 51% entering, 49% exiting

Trip Generation per Seat

Average Rate	Range of Rates	Standard Deviation
2.39	1.49 - 3.43	1.69

Data Plot and Equation

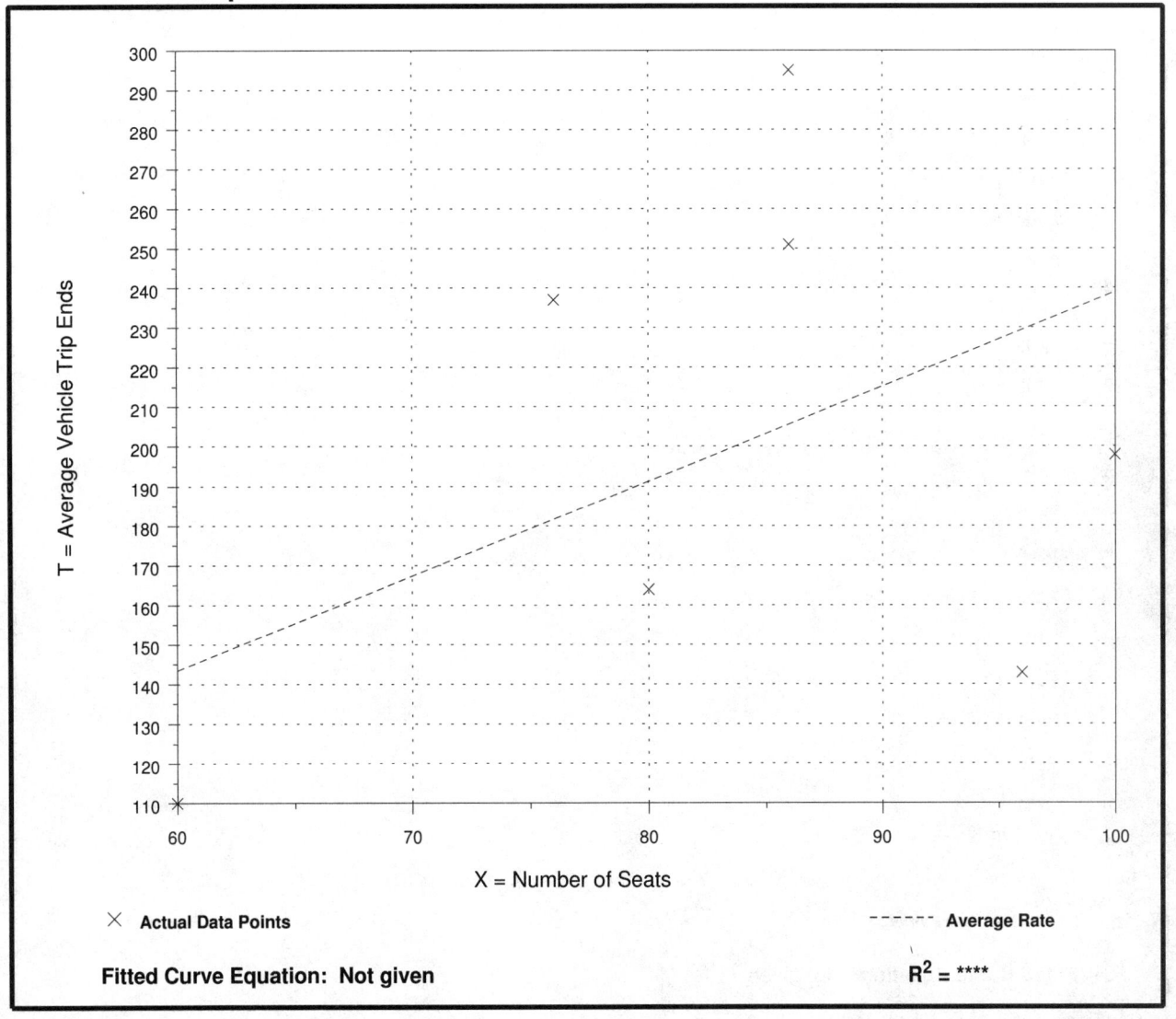

Fitted Curve Equation: Not given $R^2 = ****$

Fast-Food Restaurant with Drive-Through Window
(934)

Average Vehicle Trip Ends vs: Seats
On a: Sunday

Number of Studies: 10
Average Number of Seats: 88
Directional Distribution: 50% entering, 50% exiting

Trip Generation per Seat

Average Rate	Range of Rates	Standard Deviation
17.67	7.39 - 32.25	9.38

Data Plot and Equation

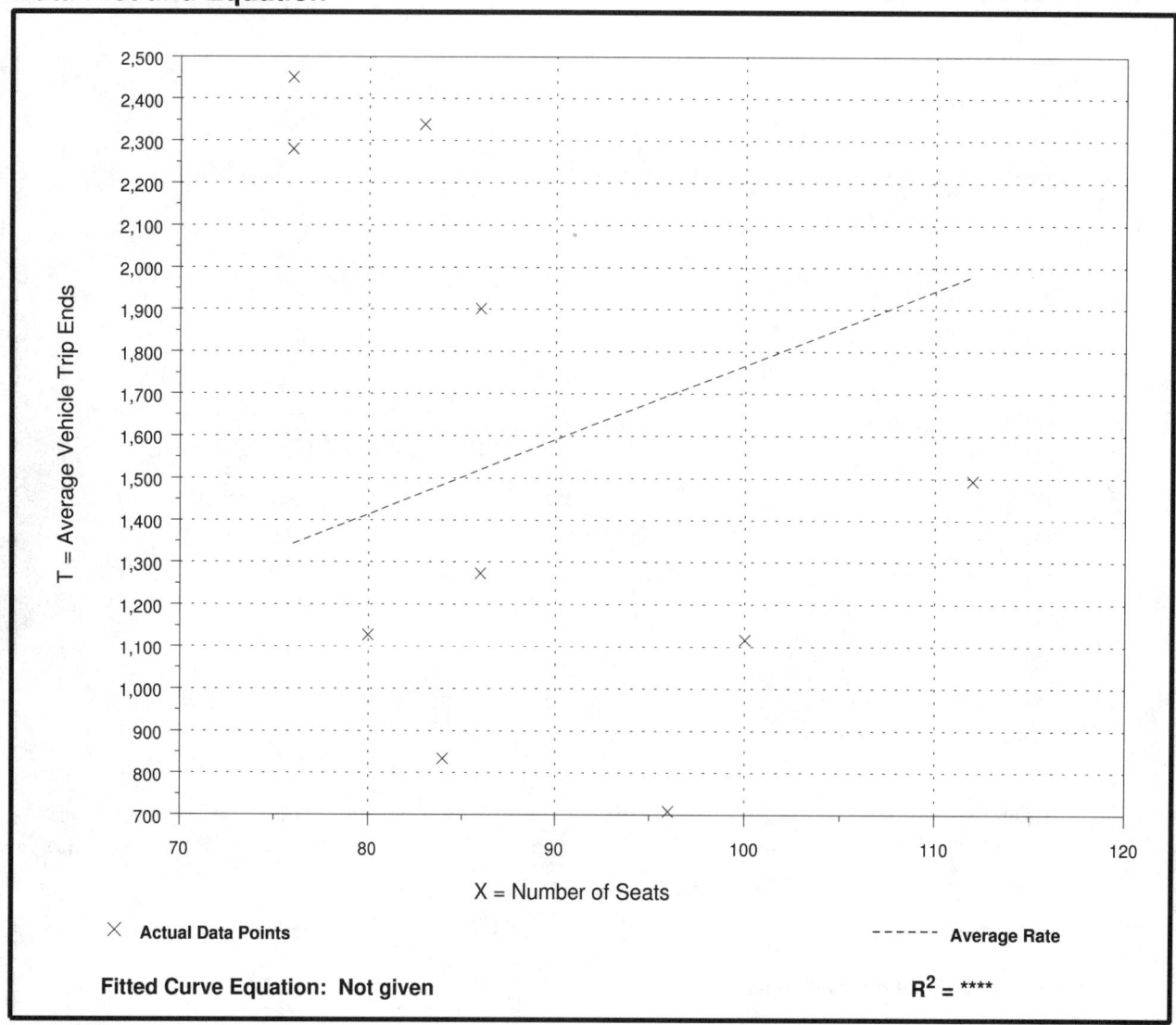

Fitted Curve Equation: Not given $R^2 = ****$

Fast-Food Restaurant with Drive-Through Window
(934)

Average Vehicle Trip Ends vs: Seats
On a: Sunday,
Peak Hour of Generator

Number of Studies: 4
Average Number of Seats: 88
Directional Distribution: 49% entering, 51% exiting

Trip Generation per Seat

Average Rate	Range of Rates	Standard Deviation
2.02	1.38 - 2.44	1.48

Data Plot and Equation

Caution - Use Carefully - Small Sample Size

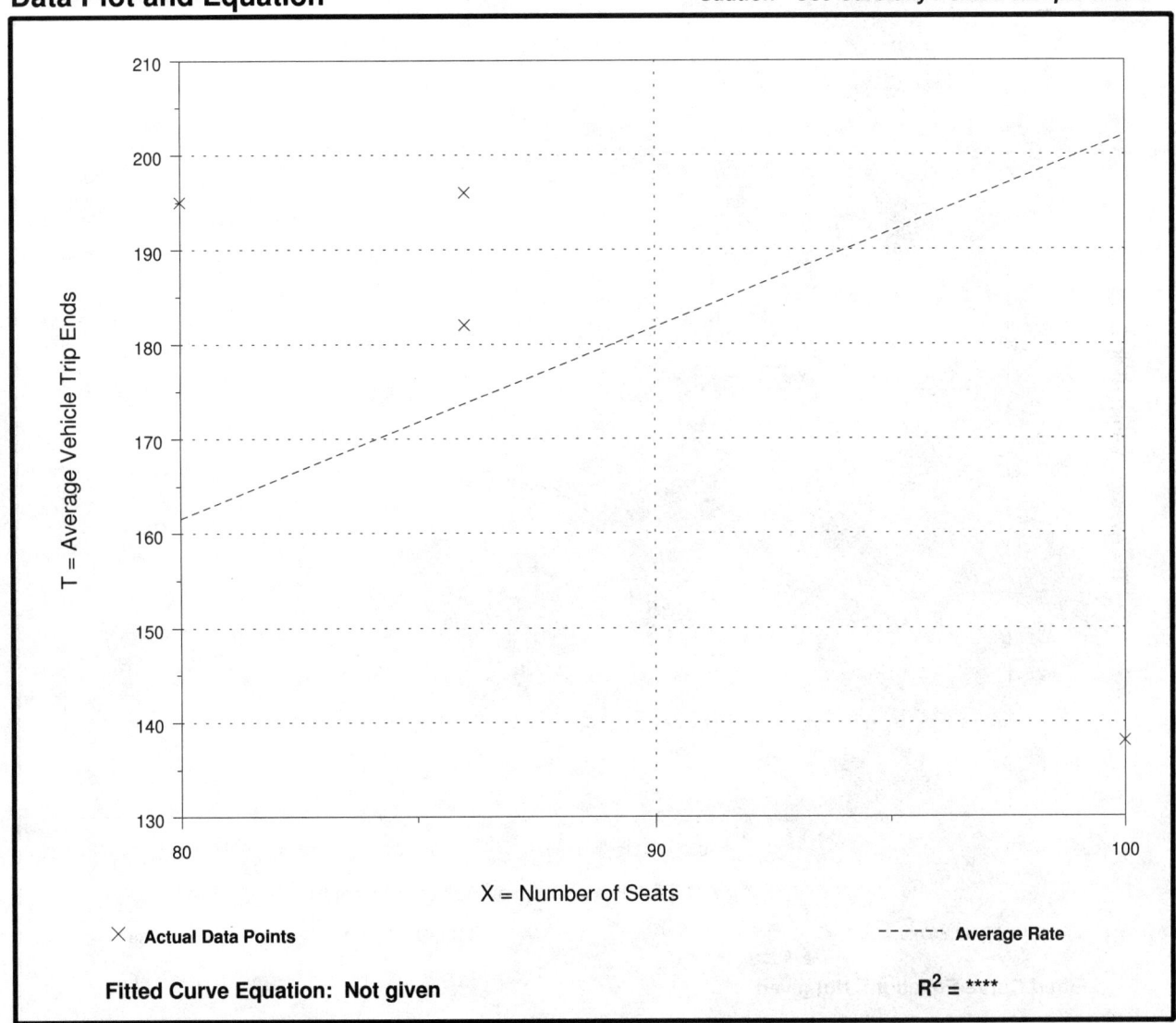

× Actual Data Points   ------ Average Rate

Fitted Curve Equation: Not given $R^2 = ****$

Fast-Food Restaurant with Drive-Through Window
(934)

Average Vehicle Trip Ends vs: A.M. Peak Hour Traffic on Adjacent Street
On a: Weekday,
Peak Hour of Adjacent Street Traffic,
One Hour Between 7 and 9 a.m.

Number of Studies: 5
Avg. A.M. Peak Hr. Traf. on Adj. Street: 1,351
Directional Distribution: 54% entering, 46% exiting

Trip Generation per AM Peak Hour Traffic on Adjacent Street

Average Rate	Range of Rates	Standard Deviation
0.12	0.05 - 0.57	0.36

Data Plot and Equation

Caution - Use Carefully - Small Sample Size

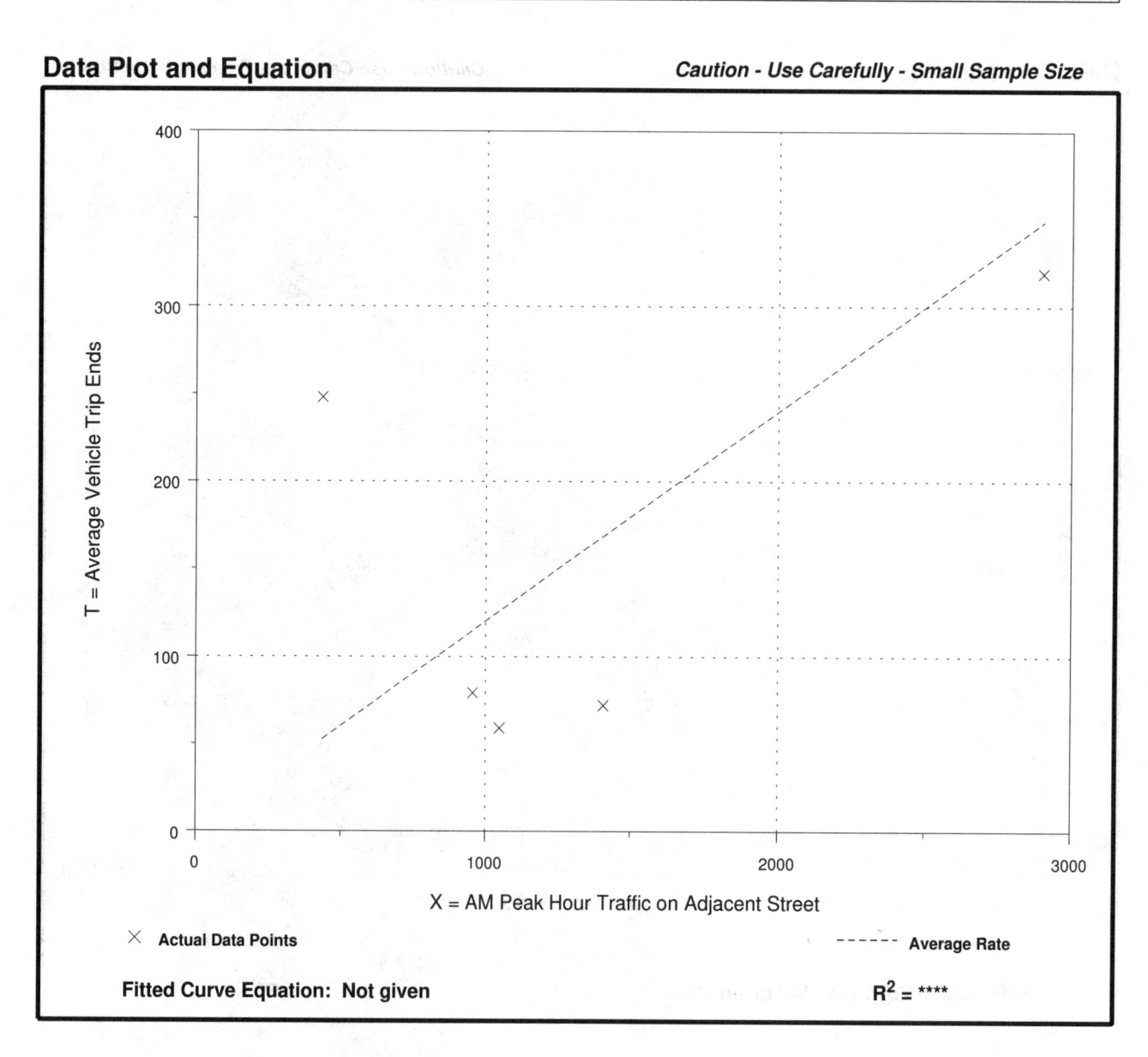

[Scatter plot: X = AM Peak Hour Traffic on Adjacent Street (0 to 3000); T = Average Vehicle Trip Ends (0 to 400). Dashed line shows Average Rate. × marks Actual Data Points.]

Fitted Curve Equation: Not given R^2 = ****

1930   *Trip Generation*, 9th Edition • Institute of Transportation Engineers

Fast-Food Restaurant with Drive-Through Window
(934)

Average Vehicle Trip Ends vs: P.M. Peak Hour Traffic on Adjacent Street
On a: Weekday,
Peak Hour of Adjacent Street Traffic,
One Hour Between 4 and 6 p.m.

Number of Studies: 14
Avg. P.M. Peak Hr. Traf. on Adj. Street: 2,199
Directional Distribution: 53% entering, 47% exiting

Trip Generation per P.M. Peak Hour Traffic on Adjacent Street

Average Rate	Range of Rates	Standard Deviation
0.04	0.02 - 0.32	0.21

Data Plot and Equation

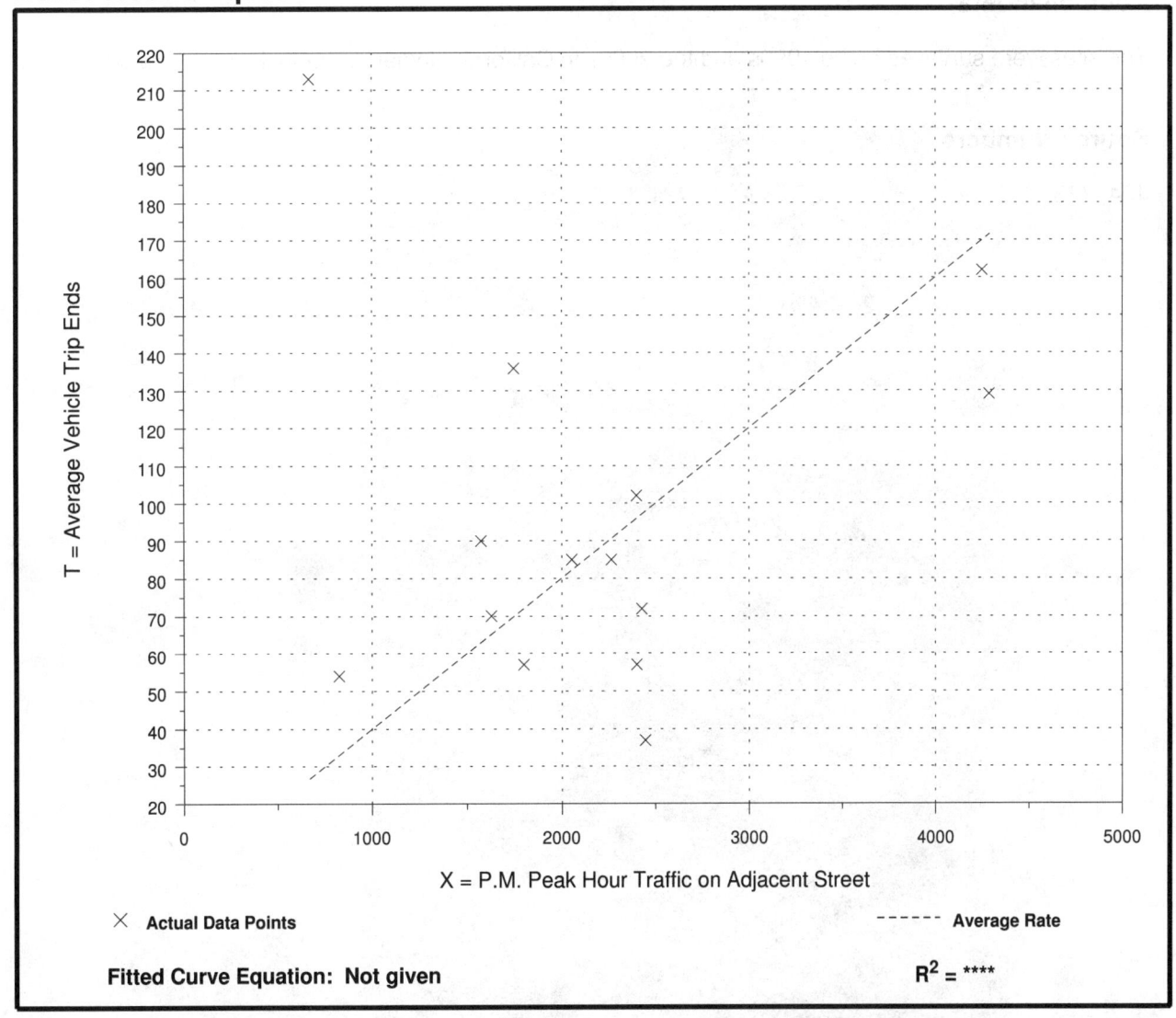

Fitted Curve Equation: Not given $R^2 = ****$

Land Use: 935
Fast-Food Restaurant with Drive-Through Window and No Indoor Seating

Description

This category includes fast-food restaurants with drive-through service only. These facilities typically have very small building areas and may provide a limited amount of outside seating. These limited-service eating establishments usually do not provide table service. High-turnover (sit-down) restaurant (Land Use 932), fast-food restaurant without drive-through window (Land Use 933) and fast-food restaurant with drive-through window (Land Use 934) are related uses.

Additional Data

The sites were surveyed in the 1990s and the 2000s in California, Indiana, Kentucky and New Jersey.

Source Numbers

404, 713, 720

Land Use: 935
Fast-Food Restaurant with Drive-Through Window and No Indoor Seating
Independent Variables with One Observation

The following trip generation data are for independent variables with only one observation. This information is shown in this table only; there are no related plots for these data.

Users are cautioned to use data with care because of the small sample size.

Independent Variable	Trip Generation Rate	Size of Independent Variable	Number of Studies	Directional Distribution
1,000 Square Feet Gross Floor Area				
Weekday A.M. Peak Hour of Adjacent Street Traffic	24.43	1.8	1	47% entering, 53% exiting

Fast-Food Restaurant with Drive-Through Window and No Indoor Seating
(935)

Average Vehicle Trip Ends vs: 1000 Sq. Feet Gross Floor Area
On a: Weekday,
Peak Hour of Adjacent Street Traffic,
One Hour Between 4 and 6 p.m.

Number of Studies: 6
Average 1000 Sq. Feet GFA: 1
Directional Distribution: 51% entering, 49% exiting

Trip Generation per 1000 Sq. Feet Gross Floor Area

Average Rate	Range of Rates	Standard Deviation
44.99	27.78 - 191.56	38.88

Data Plot and Equation

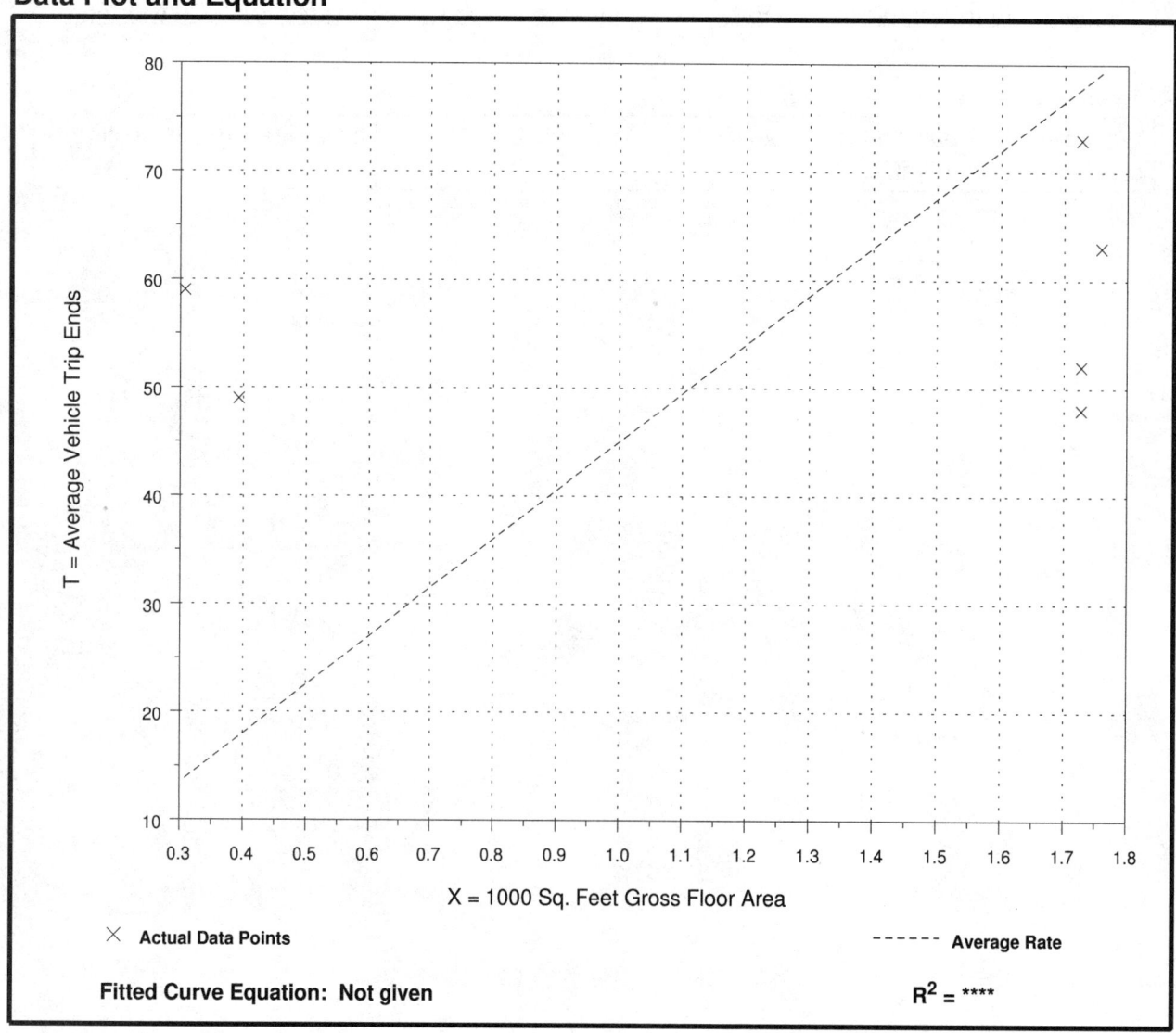

× Actual Data Points   ----- Average Rate

Fitted Curve Equation: Not given $R^2 = ****$

Land Use: 936
Coffee/Donut Shop without Drive-Through Window

Description

This land use includes single-tenant coffee and donut restaurants without drive-through windows. Freshly brewed coffee and a variety of coffee-related accessories are the primary retail products sold at these sites. They may also sell other refreshment items, such as donuts, bagels, muffins, cakes, sandwiches, wraps, salads and other hot and cold beverages. Some sites may also sell newspapers, music CDs and books. The coffee and donut shops contained in this land use typically hold long store hours (over 15 hours) with an early morning opening. Also, limited indoor seating is generally provided for patrons; however, table service is not provided. Coffee/donut shop with drive-through window (Land Use 937), coffee/donut shop with drive-through window and no indoor seating (Land Use 938), bread/donut/bagel shop without drive-through window (Land Use 939) and bread/donut/bagel shop with drive-through window (Land Use 940) are related uses.

Additional Data

Many of the facilities in this land use were located within a shopping center or as an outparcel to a shopping center.

It should be noted that those stores specializing in the sale of coffee (Land Uses 936-938) generated higher trip generation rates than those specializing in other products (Land Uses 939-940).

The sites were surveyed between the 1990s and the 2000s in California, Massachusetts, Minnesota, New Hampshire, New Jersey, New York, Pennsylvania and Vermont.

Specialized Land Use Data

Current industry trends have resulted in the emergence of several coffee/donut shops combined with other types of restaurants. The trip generation characteristics of these facilities differ from the facilities typically contained in this land use, as their sizes, trip generation rates and peak hour of service vary considerably. Therefore, the information collected for these facilities is presented in the following tables and was excluded from the data plots.

Table 1: Donut and Ice Cream Shop

Independent Variable	Trip Generation Rate	Size of Independent Variable	Number of Studies	Directional Distribution
1,000 Square Feet Gross Floor Area				
Weekday P.M. Peak Hour of Adjacent Street Traffic	20.00	2.4	1	50% entering, 50% exiting
Weekday P.M. Peak Hour of Generator	21.67	2.4	1	52% entering, 48% exiting

Source: 563

Table 2: Donut and Sandwich Shop

1,000 Square Feet Gross Floor Area

Weekday A.M. Peak Hour of Adjacent Street Traffic	59.75	4.0	1	49% entering, 51% exiting
Weekday P.M. Peak Hour of Adjacent Street Traffic	13.00	4.0	1	52% entering, 48% exiting
Weekday P.M. Peak Hour of Generator	27.75	4.0	1	51% entering, 49% exiting

Source: 563

Source Numbers

555, 563, 571, 594, 617, 618, 621, 728

Coffee/Donut Shop without Drive-Through Window
(936)

Average Vehicle Trip Ends vs: 1000 Sq. Feet Gross Floor Area
On a: Weekday,
Peak Hour of Adjacent Street Traffic,
One Hour Between 7 and 9 a.m.

Number of Studies: 15
Average 1000 Sq. Feet GFA: 2
Directional Distribution: 51% entering, 49% exiting

Trip Generation per 1000 Sq. Feet Gross Floor Area

Average Rate	Range of Rates	Standard Deviation
108.38	54.29 - 254.50	47.90

Data Plot and Equation

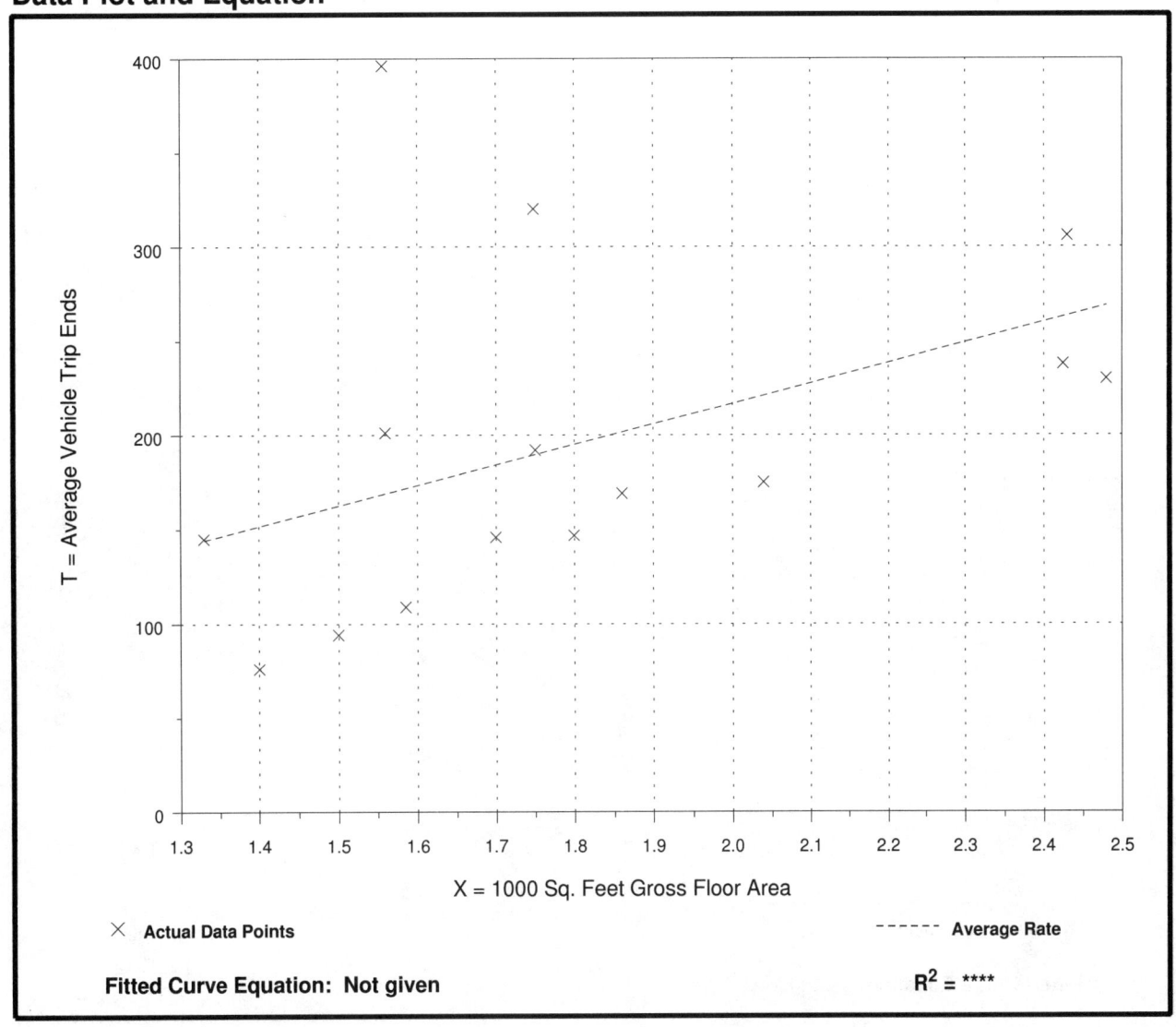

× Actual Data Points   ------ Average Rate

Fitted Curve Equation: Not given $R^2 =$ ****

Trip Generation, 9th Edition • Institute of Transportation Engineers

Coffee/Donut Shop without Drive-Through Window
(936)

Average Vehicle Trip Ends vs: 1000 Sq. Feet Gross Floor Area
On a: Weekday,
Peak Hour of Adjacent Street Traffic,
One Hour Between 4 and 6 p.m.

Number of Studies: 8
Average 1000 Sq. Feet GFA: 2
Directional Distribution: 50% entering, 50% exiting

Trip Generation per 1000 Sq. Feet Gross Floor Area

Average Rate	Range of Rates	Standard Deviation
40.75	25.00 - 74.55	14.42

Data Plot and Equation

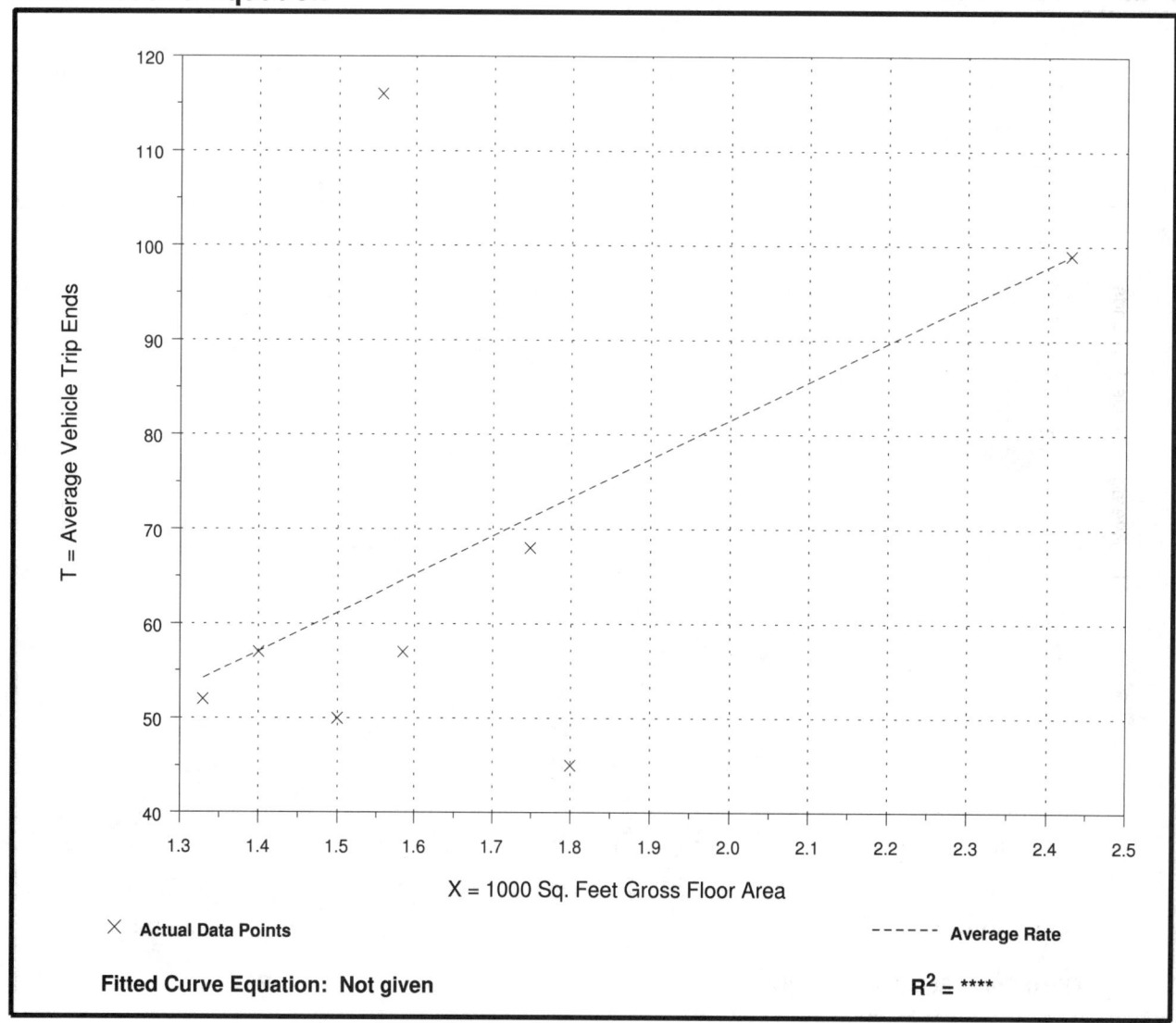

Fitted Curve Equation: Not given $R^2 = ****$

Coffee/Donut Shop without Drive-Through Window
(936)

Average Vehicle Trip Ends vs: 1000 Sq. Feet Gross Floor Area
On a: Weekday,
A.M. Peak Hour of Generator

Number of Studies: 5
Average 1000 Sq. Feet GFA: 2
Directional Distribution: 51% entering, 49% exiting

Trip Generation per 1000 Sq. Feet Gross Floor Area

Average Rate	Range of Rates	Standard Deviation
64.21	49.82 - 112.78	21.09

Data Plot and Equation

Caution - Use Carefully - Small Sample Size

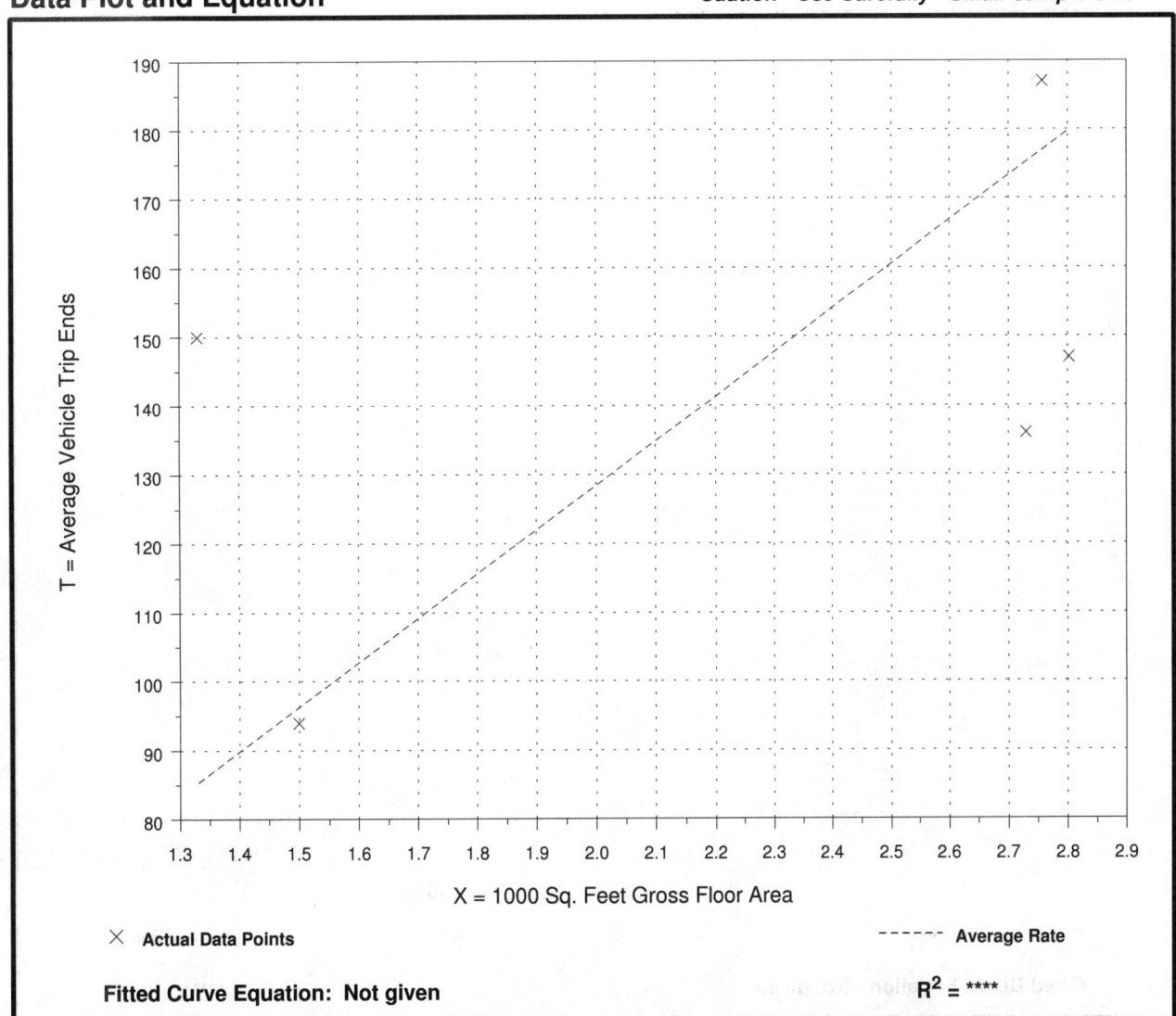

Fitted Curve Equation: Not given $R^2 = ****$

Coffee/Donut Shop without Drive-Through Window
(936)

Average Vehicle Trip Ends vs: 1000 Sq. Feet Gross Floor Area
On a: Weekday,
P.M. Peak Hour of Generator

Number of Studies: 5
Average 1000 Sq. Feet GFA: 2
Directional Distribution: 49% entering, 51% exiting

Trip Generation per 1000 Sq. Feet Gross Floor Area

Average Rate	Range of Rates	Standard Deviation
25.81	18.19 - 39.10	8.08

Data Plot and Equation

Caution - Use Carefully - Small Sample Size

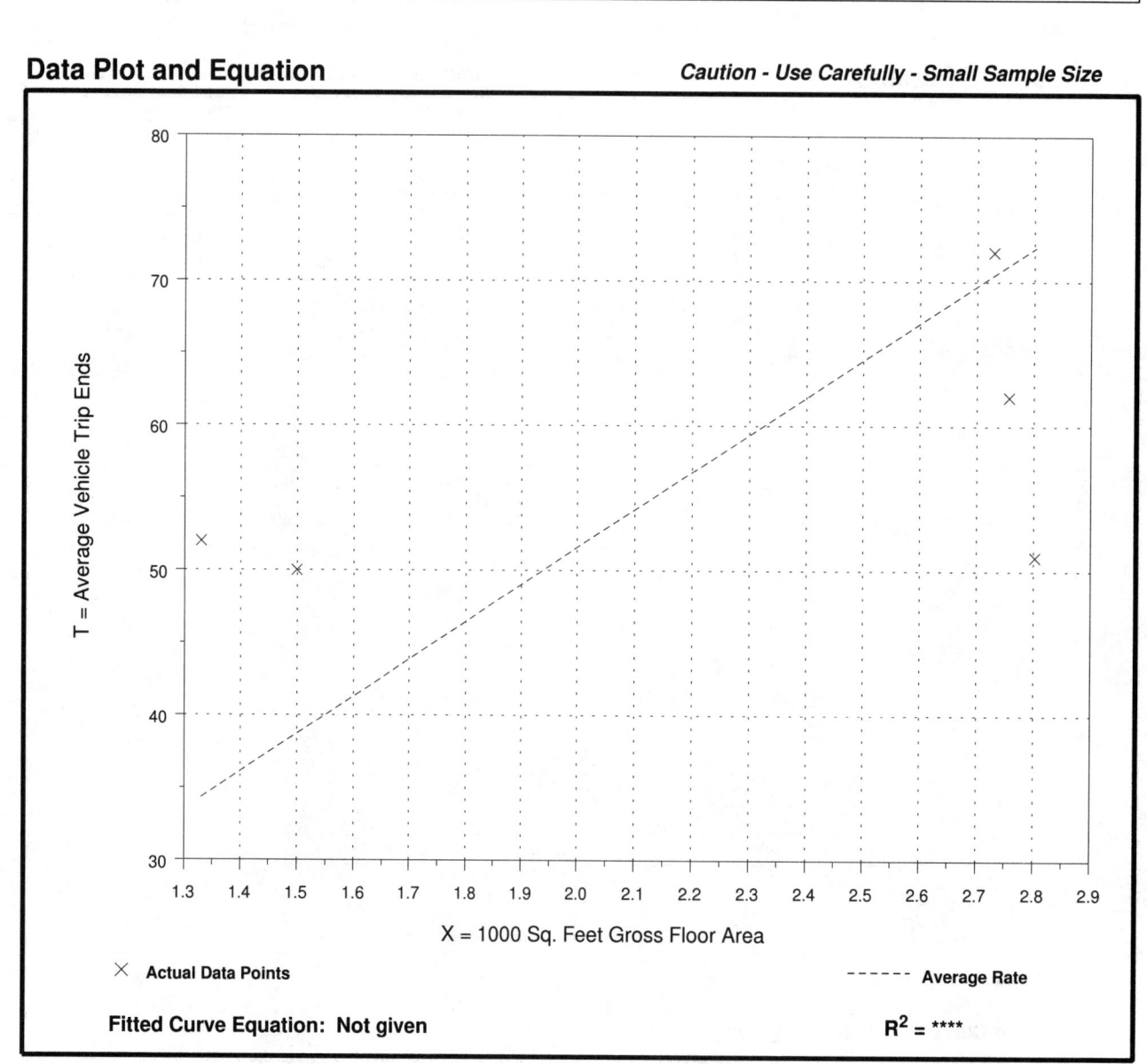

Fitted Curve Equation: Not given $R^2 = ****$

Coffee/Donut Shop without Drive-Through Window
(936)

Average Vehicle Trip Ends vs: 1000 Sq. Feet Gross Floor Area
On a: Saturday,
Peak Hour of Generator

Number of Studies: 4
Average 1000 Sq. Feet GFA: 2
Directional Distribution: 48% entering, 52% exiting

Trip Generation per 1000 Sq. Feet Gross Floor Area

Average Rate	Range of Rates	Standard Deviation
65.96	36.67 - 116.97	31.10

Data Plot and Equation

Caution - Use Carefully - Small Sample Size

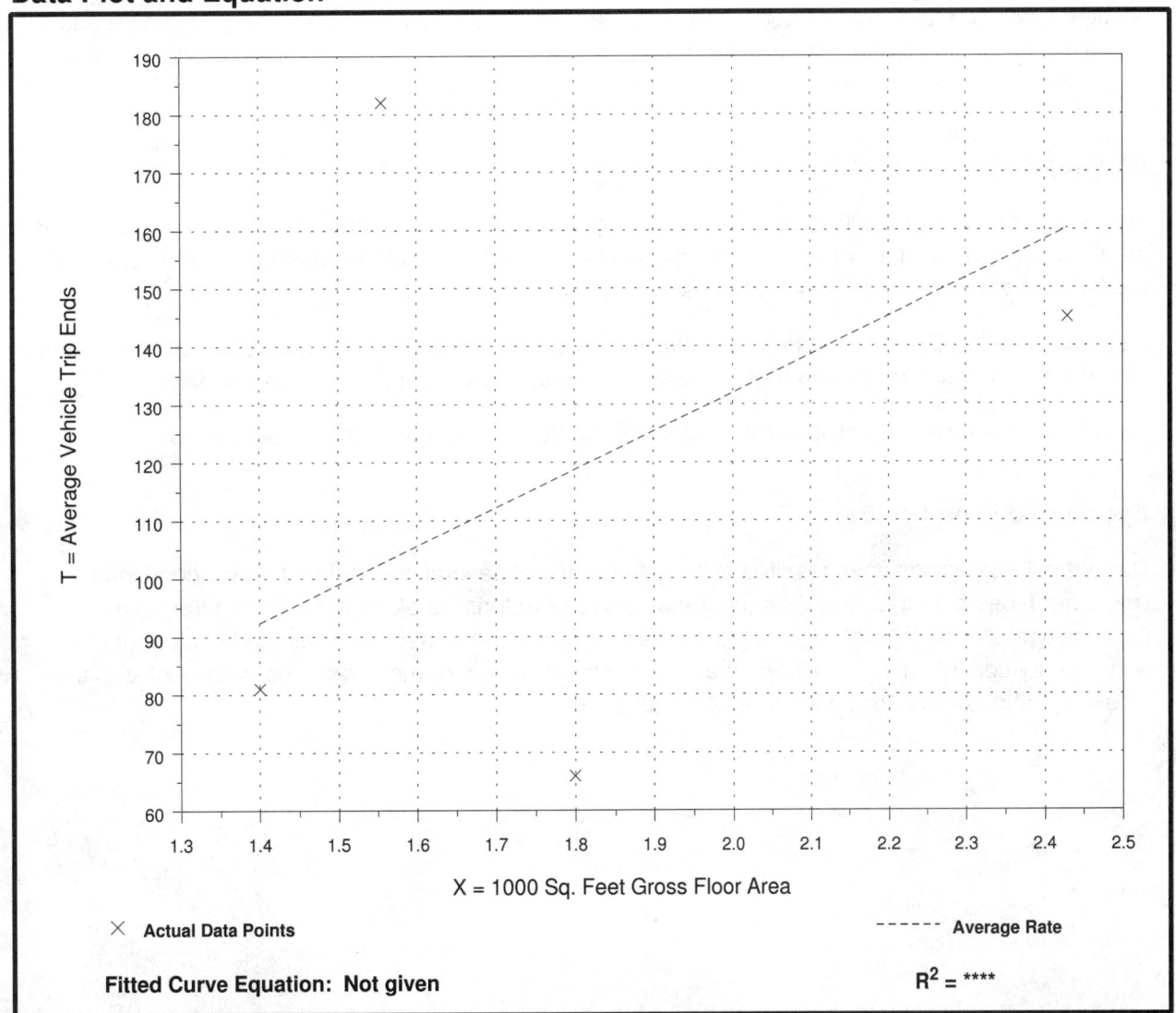

× Actual Data Points ------ Average Rate

Fitted Curve Equation: Not given R^2 = ****

Trip Generation, 9th Edition • Institute of Transportation Engineers

Land Use: 937
Coffee/Donut Shop with Drive-Through Window

Description

This land use includes single-tenant coffee and donut restaurants with drive-through windows. Freshly brewed coffee and a variety of coffee-related accessories are the primary retail products sold at these sites. They may also sell other refreshment items, such as donuts, bagels, muffins, cakes, sandwiches, wraps, salads and other hot and cold beverages. Some sites may also sell newspapers, music CDs and books. The coffee and donut shops contained in this land use typically hold long store hours (over 15 hours) with an early morning opening. Also, limited indoor seating is generally provided for patrons; however, table service is not provided. Coffee/donut shop without drive-through window (Land Use 936), coffee/donut shop with drive-through window and no indoor seating (Land Use 938), bread/donut/bagel shop without drive-through window (Land Use 939) and bread/donut/bagel shop with drive-through window (Land Use 940) are related uses.

Additional Data

Most of the facilities in this land use were in free-standing buildings in retail shopping areas. Some of the facilities were located within a shopping center or as an outparcel to a shopping center. Some of the facilities shared parking areas with one or more other businesses.

It should be noted that those stores specializing in the sale of coffee (Land Uses 936-938) generated higher trip generation rates than those specializing in other products (Land Uses 939-940).

The sites were surveyed between the 1990s and the 2000s throughout the United States.

Specialized Land Use Data

Current industry trends have resulted in the emergence of several coffee/donut shops combined with other types of restaurants. The trip generation characteristics of these facilities differ from the facilities typically contained in this land use, as their sizes, trip generation rates and peak hour of service vary considerably. Therefore, the information collected for these facilities is presented in the following table and was excluded from the data plots.

Table: Donut and Ice Cream Shop with Drive-Through Window

Independent Variable	Trip Generation Rate	Size of Independent Variable	Number of Studies	Directional Distribution
1,000 Square Feet Gross Floor Area				
Weekday A.M. Peak Hour of Adjacent Street Traffic	128.77	3.3	1	52% entering, 48% exiting
Weekday P.M. Peak Hour of Adjacent Street Traffic	71.54	3.3	1	52% entering, 48% exiting

Source : 617

Source Numbers

593, 594, 599, 615, 617, 618, 621, 622, 635, 639, 712, 714, 725, 726, 728

Coffee/Donut Shop with Drive-Through Window
(937)

Average Vehicle Trip Ends vs: 1000 Sq. Feet Gross Floor Area
On a: Weekday

Number of Studies: 2
Average 1000 Sq. Feet GFA: 2
Directional Distribution: 50% entering, 50% exiting

Trip Generation per 1000 Sq. Feet Gross Floor Area

Average Rate	Range of Rates	Standard Deviation
818.58	734.34 - 869.00	*

Data Plot and Equation

Caution - Use Carefully - Small Sample Size

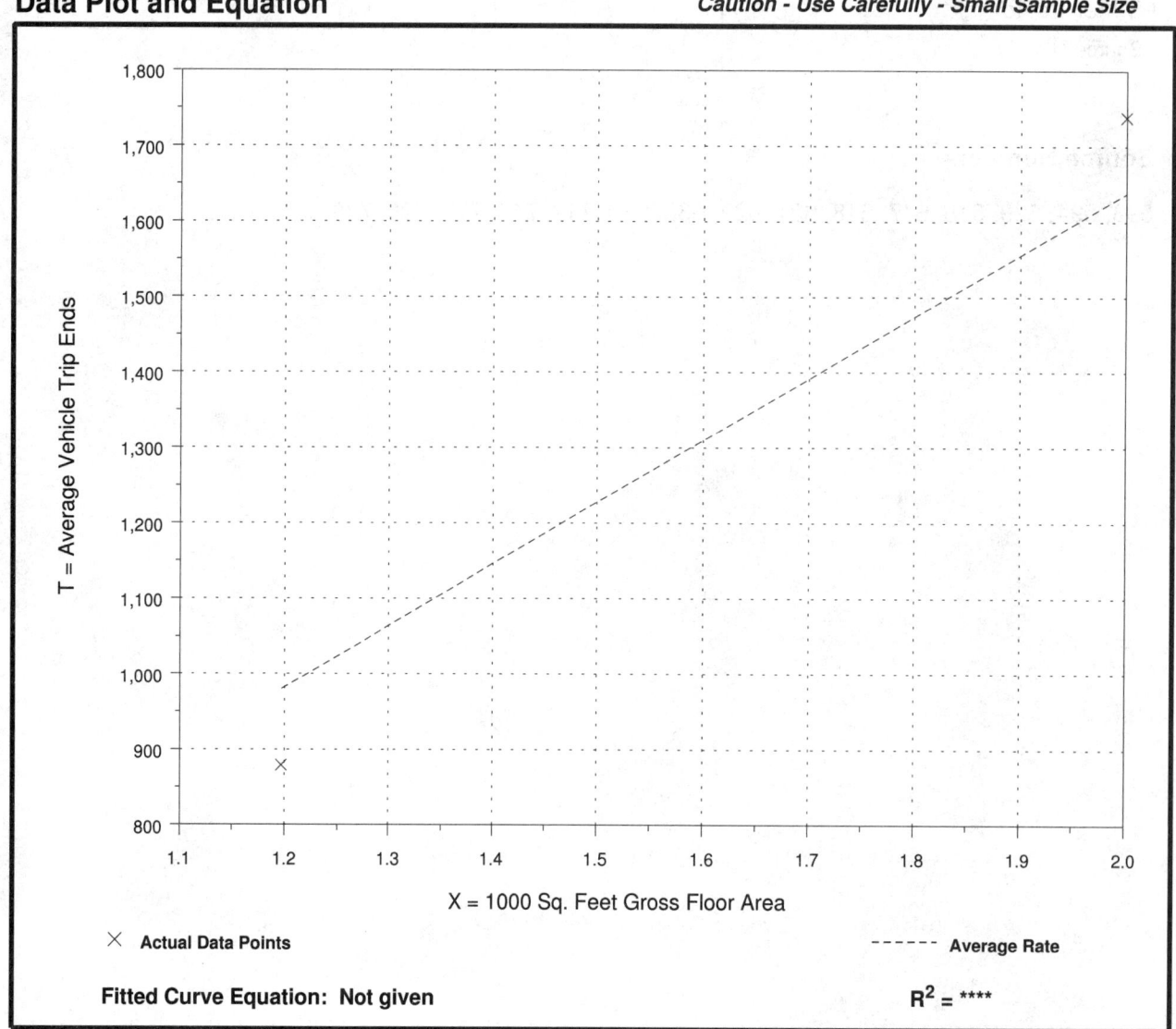

Fitted Curve Equation: Not given $R^2 = ****$

Coffee/Donut Shop with Drive-Through Window
(937)

Average Vehicle Trip Ends vs: 1000 Sq. Feet Gross Floor Area
On a: Weekday,
Peak Hour of Adjacent Street Traffic,
One Hour Between 7 and 9 a.m.

Number of Studies: 43
Average 1000 Sq. Feet GFA: 2
Directional Distribution: 51% entering, 49% exiting

Trip Generation per 1000 Sq. Feet Gross Floor Area

Average Rate	Range of Rates	Standard Deviation
100.58	18.23 - 349.41	49.38

Data Plot and Equation

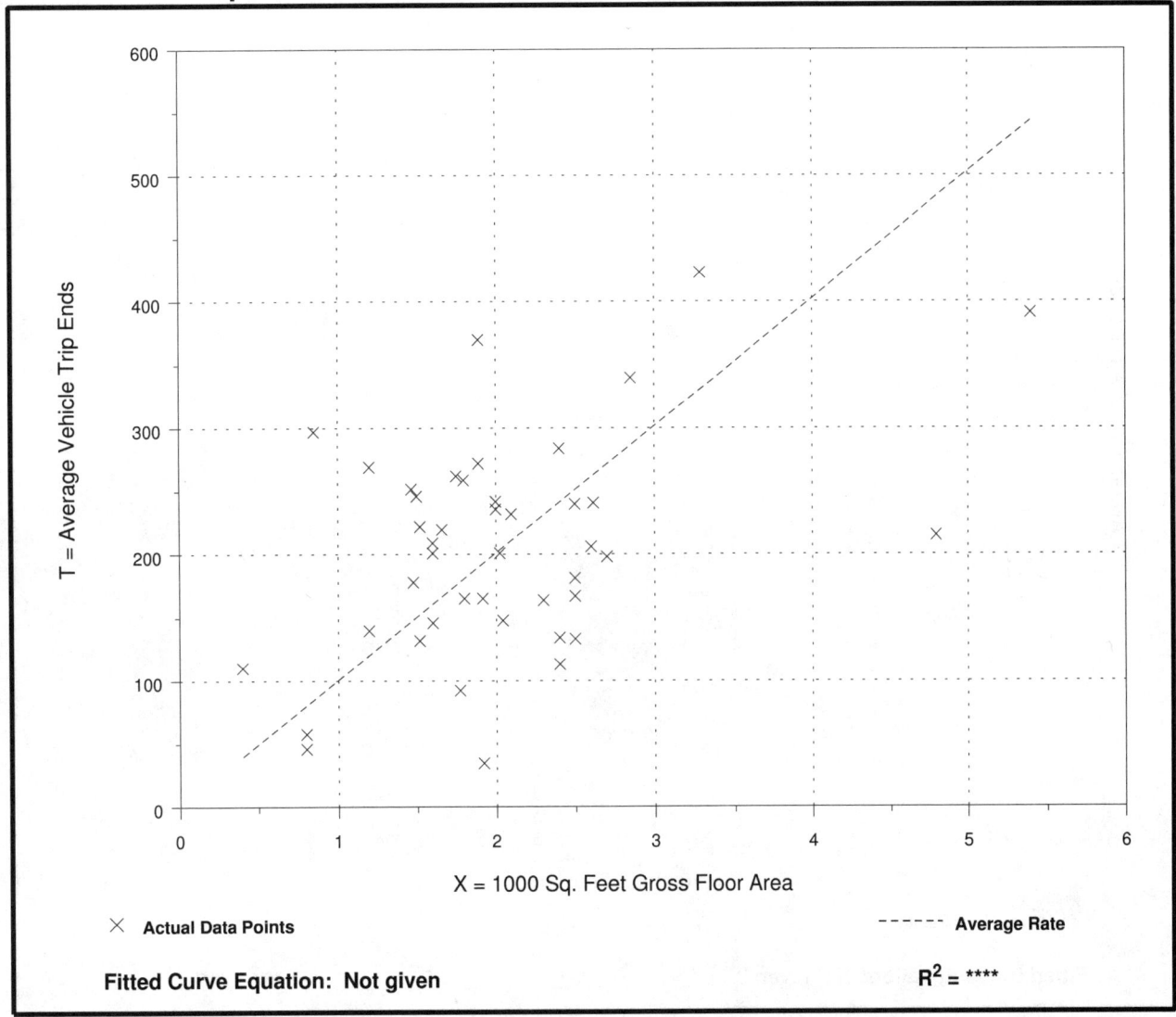

Fitted Curve Equation: Not given $R^2 = ****$

Coffee/Donut Shop with Drive-Through Window
(937)

Average Vehicle Trip Ends vs: 1000 Sq. Feet Gross Floor Area
On a: Weekday,
Peak Hour of Adjacent Street Traffic,
One Hour Between 4 and 6 p.m.

Number of Studies: 24
Average 1000 Sq. Feet GFA: 2
Directional Distribution: 50% entering, 50% exiting

Trip Generation per 1000 Sq. Feet Gross Floor Area

Average Rate	Range of Rates	Standard Deviation
42.80	2.08 - 90.00	18.06

Data Plot and Equation

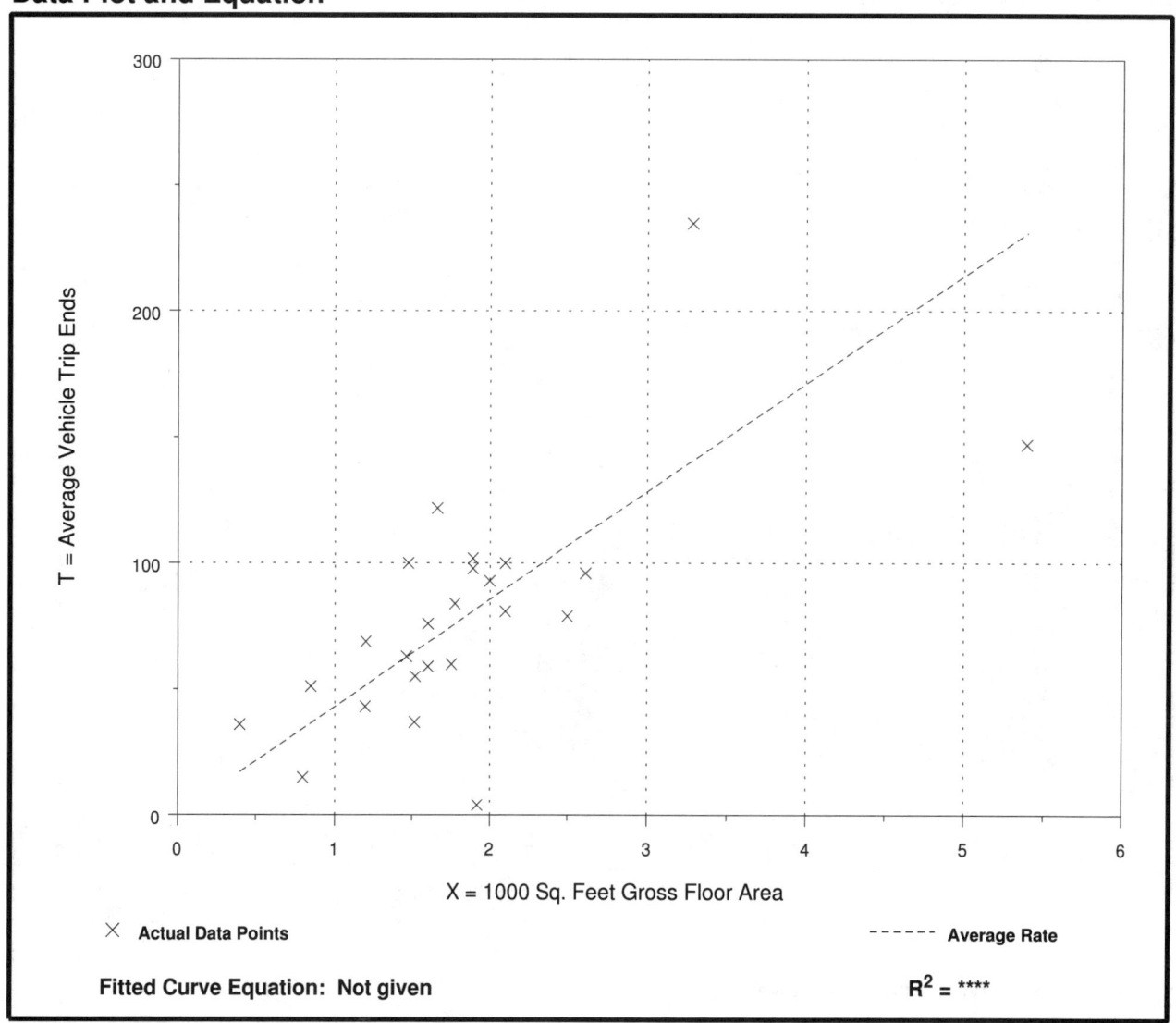

Fitted Curve Equation: Not given $R^2 = ****$

Coffee/Donut Shop with Drive-Through Window
(937)

Average Vehicle Trip Ends vs: 1000 Sq. Feet Gross Floor Area
On a: Weekday,
A.M. Peak Hour of Generator

Number of Studies: 20
Average 1000 Sq. Feet GFA: 2
Directional Distribution: 49% entering, 51% exiting

Trip Generation per 1000 Sq. Feet Gross Floor Area

Average Rate	Range of Rates	Standard Deviation
101.40	18.23 - 275.00	45.90

Data Plot and Equation

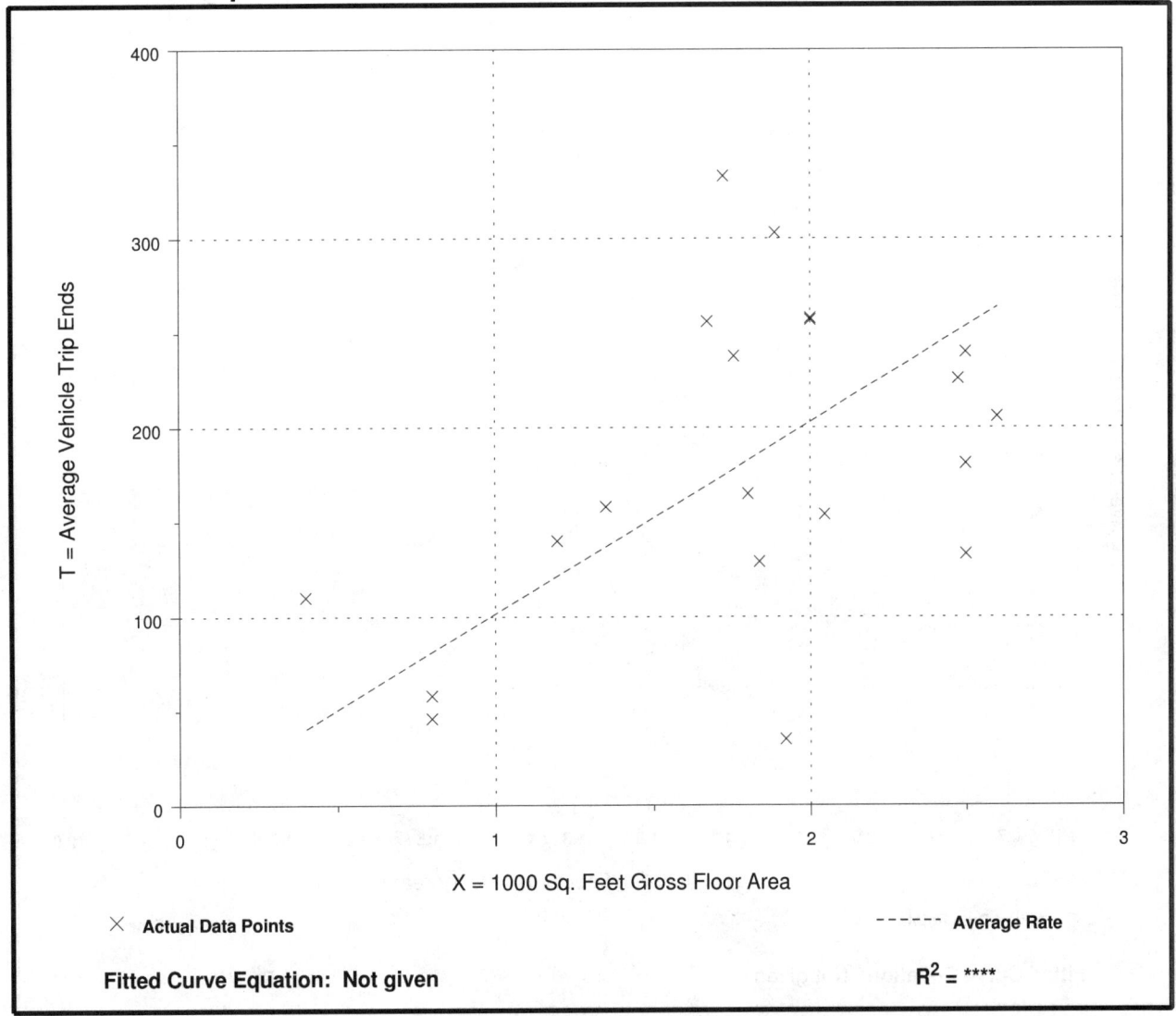

Fitted Curve Equation: Not given $R^2 = ****$

Coffee/Donut Shop with Drive-Through Window
(937)

Average Vehicle Trip Ends vs: 1000 Sq. Feet Gross Floor Area
On a: Weekday,
P.M. Peak Hour of Generator

Number of Studies: 8
Average 1000 Sq. Feet GFA: 2
Directional Distribution: 51% entering, 49% exiting

Trip Generation per 1000 Sq. Feet Gross Floor Area

Average Rate	Range of Rates	Standard Deviation
36.16	2.08 - 60.50	19.50

Data Plot and Equation

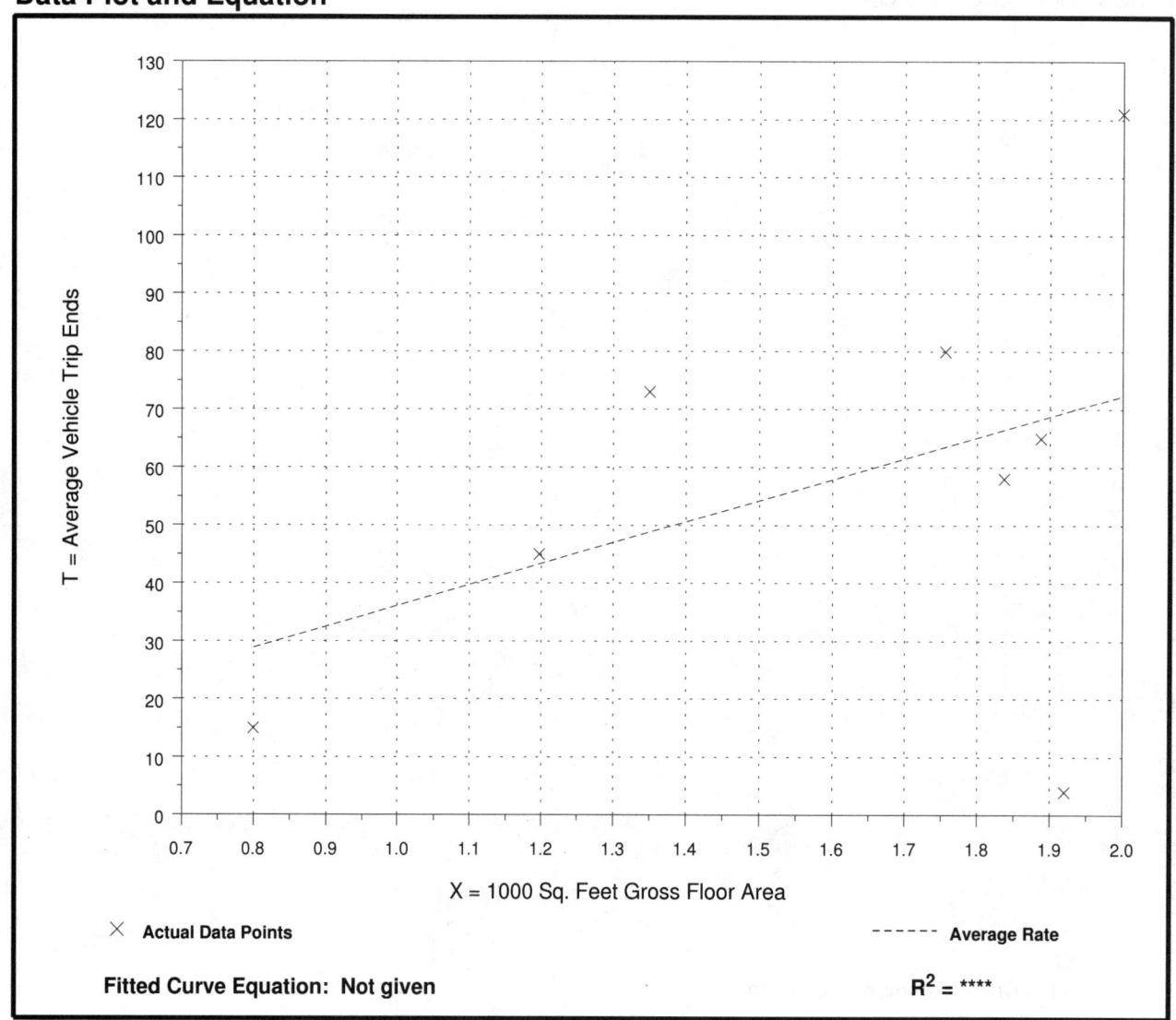

Fitted Curve Equation: Not given $R^2 = ****$

Coffee/Donut Shop with Drive-Through Window
(937)

Average Vehicle Trip Ends vs: 1000 Sq. Feet Gross Floor Area
On a: Saturday,
Peak Hour of Generator

Number of Studies: 9
Average 1000 Sq. Feet GFA: 2
Directional Distribution: 50% entering, 50% exiting

Trip Generation per 1000 Sq. Feet Gross Floor Area

Average Rate	Range of Rates	Standard Deviation
84.52	48.33 - 137.62	33.38

Data Plot and Equation

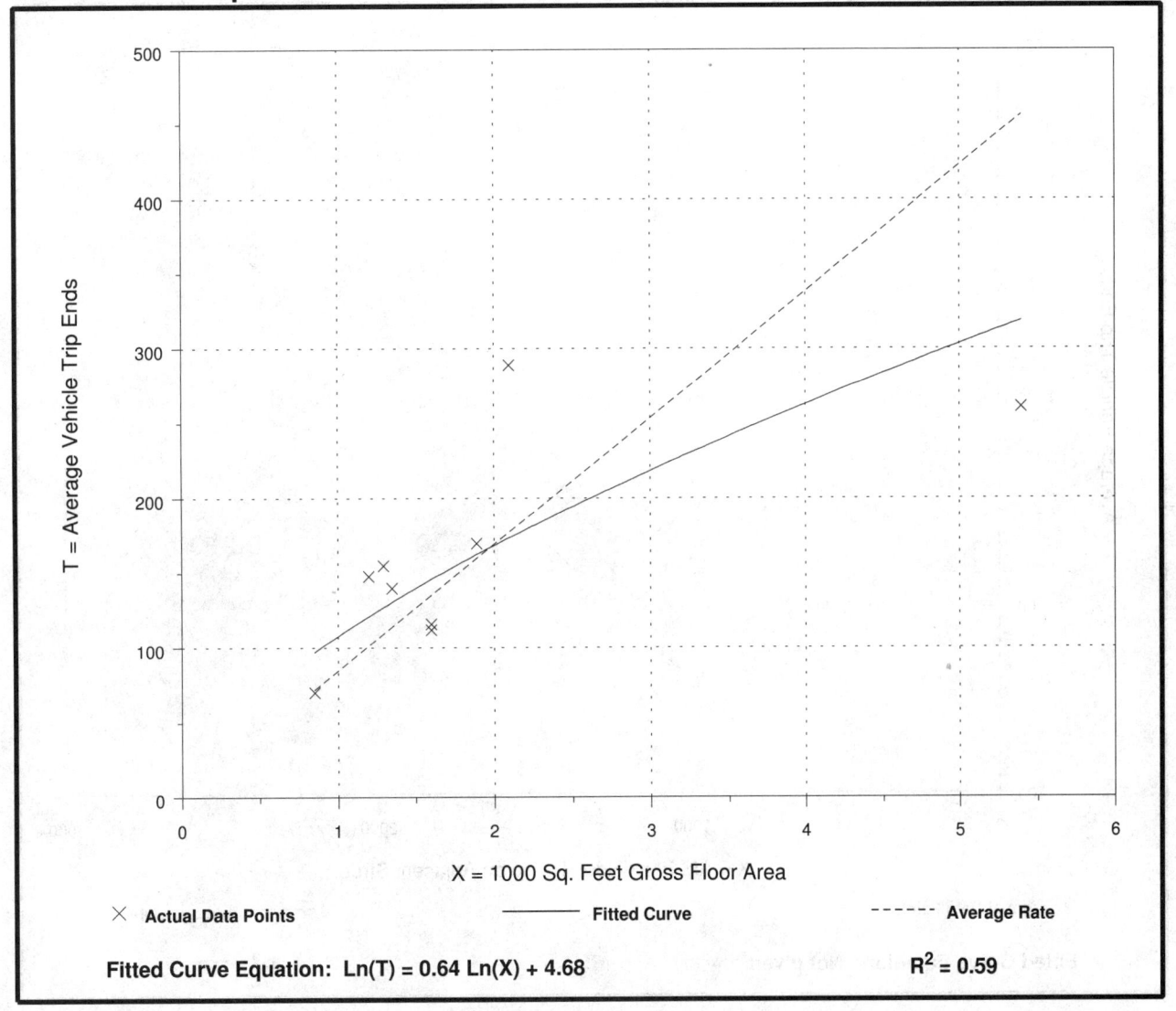

Fitted Curve Equation: $Ln(T) = 0.64 Ln(X) + 4.68$

$R^2 = 0.59$

Coffee/Donut Shop with Drive-Through Window
(937)

Average Vehicle Trip Ends vs: A.M. Peak Hour Traffic on Adjacent Street
On a: Weekday,
Peak Hour of Adjacent Street Traffic,
One Hour Between 7 and 9 a.m.

Number of Studies: 9
Avg. A.M. Peak Hr. Traf. on Adj. Street: 1,778
Directional Distribution: 51% entering, 49% exiting

Trip Generation per AM Peak Hour Traffic on Adjacent Street

Average Rate	Range of Rates	Standard Deviation
0.16	0.05 - 0.29	0.41

Data Plot and Equation

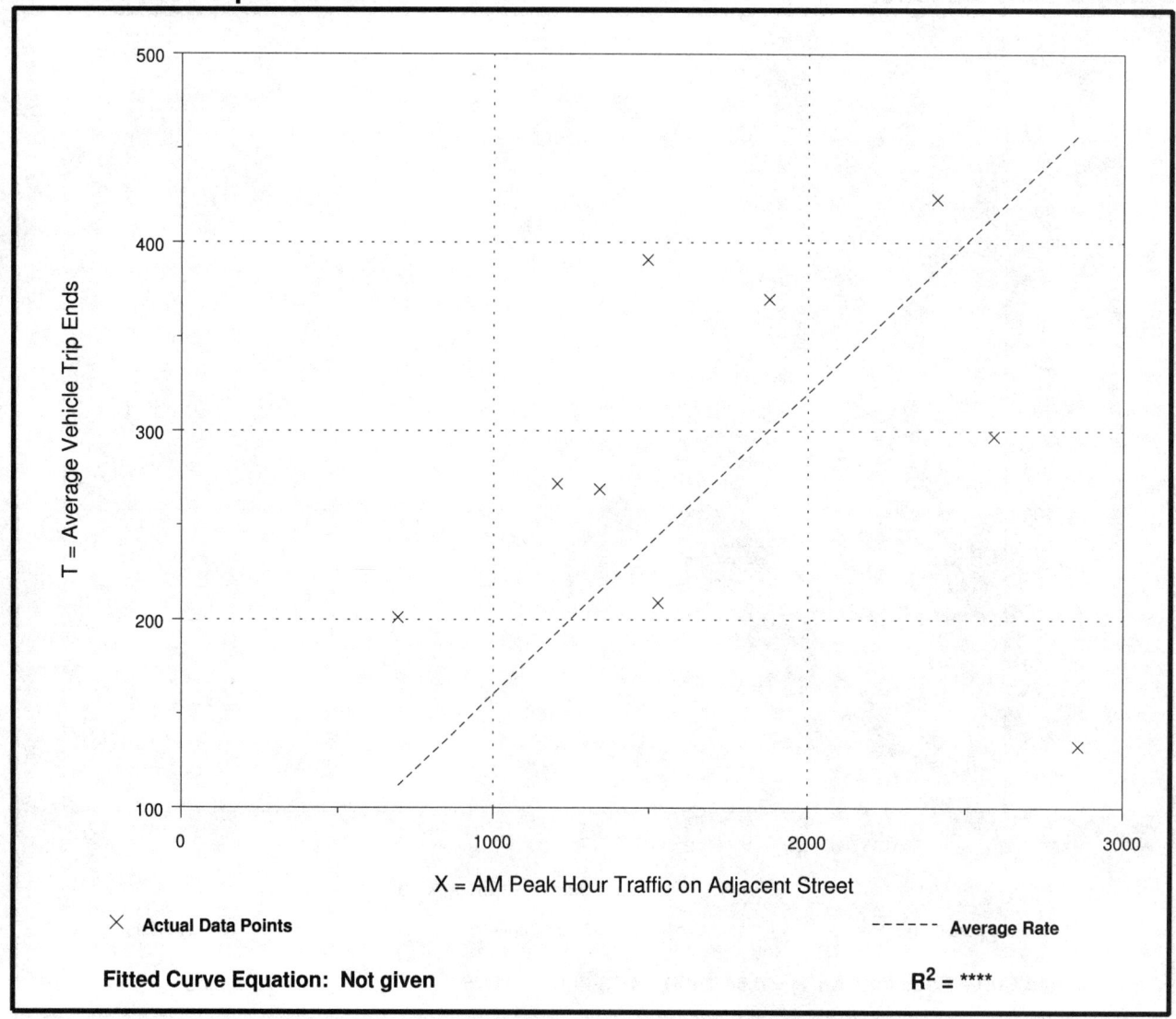

Fitted Curve Equation: Not given $R^2 = ****$

Coffee/Donut Shop with Drive-Through Window
(937)

Average Vehicle Trip Ends vs: P.M. Peak Hour Traffic on Adjacent Street
On a: Weekday,
Peak Hour of Adjacent Street Traffic,
One Hour Between 4 and 6 p.m.

Number of Studies: 9
Avg. P.M. Peak Hr. Traf. on Adj. Street: 1,903
Directional Distribution: 51% entering, 49% exiting

Trip Generation per P.M. Peak Hour Traffic on Adjacent Street

Average Rate	Range of Rates	Standard Deviation
0.05	0.02 - 0.09	0.24

Data Plot and Equation

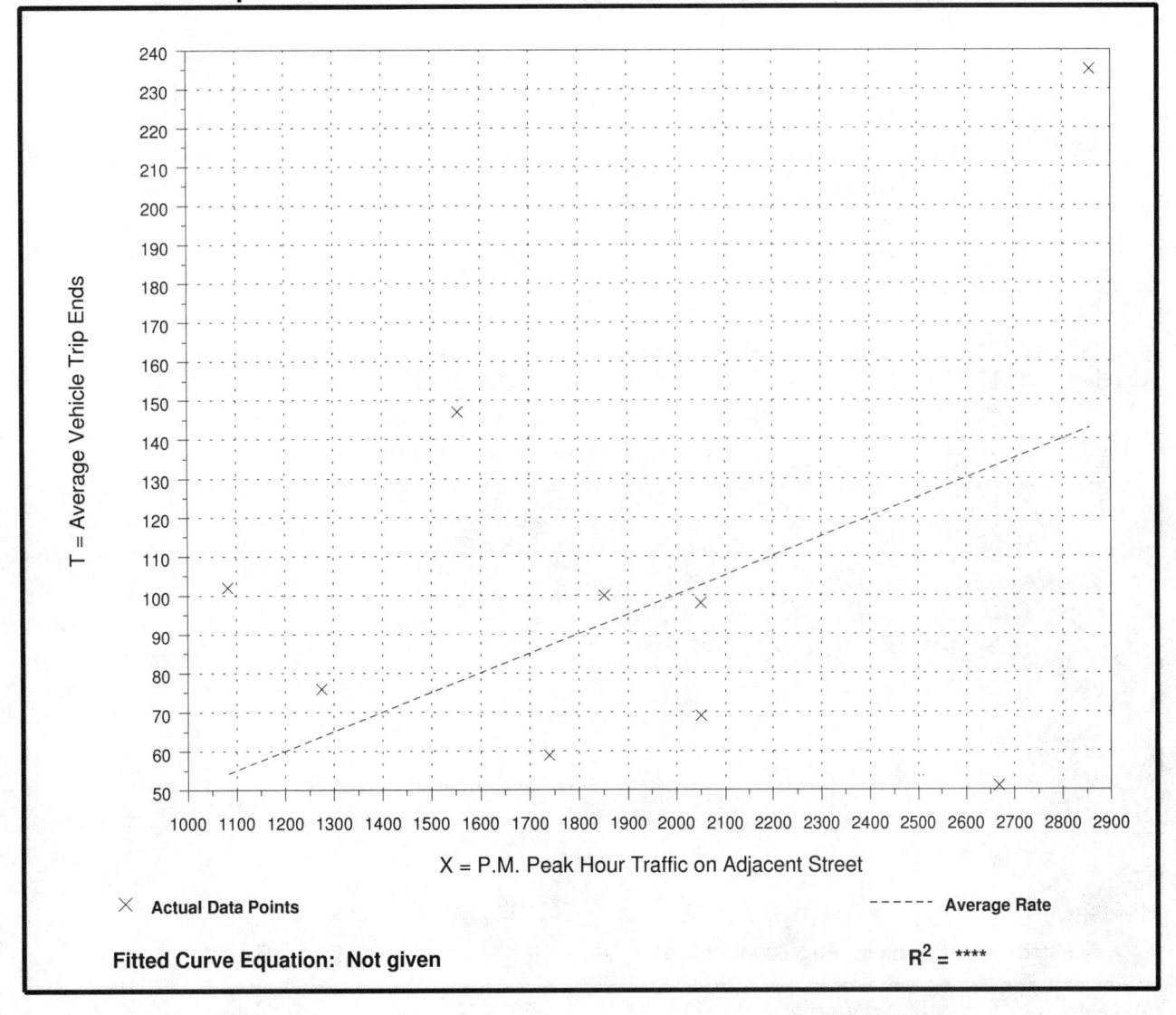

Fitted Curve Equation: Not given $R^2 = ****$

Coffee/Donut Shop with Drive-Through Window
(937)

Average Vehicle Trip Ends vs: Seats
On a: Weekday,
Peak Hour of Adjacent Street Traffic,
One Hour Between 7 and 9 a.m.

Number of Studies: 4
Average Number of Seats: 13
Directional Distribution: 51% entering, 49% exiting

Trip Generation per Seat

Average Rate	Range of Rates	Standard Deviation
4.70	2.69 - 7.25	2.57

Data Plot and Equation

Caution - Use Carefully - Small Sample Size

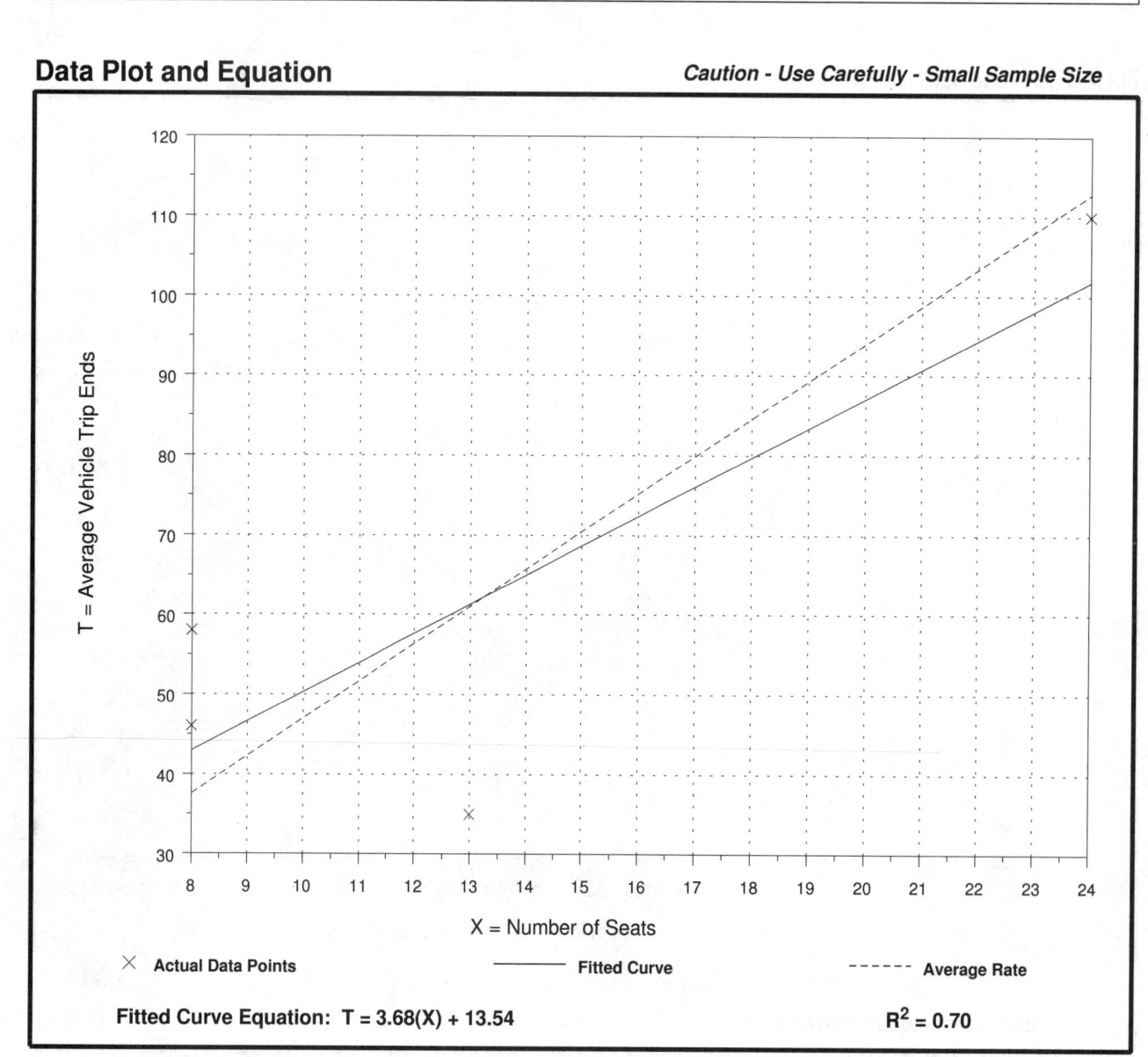

Fitted Curve Equation: T = 3.68(X) + 13.54 $R^2 = 0.70$

Coffee/Donut Shop with Drive-Through Window
(937)

Average Vehicle Trip Ends vs: Seats
On a: Weekday,
Peak Hour of Adjacent Street Traffic,
One Hour Between 4 and 6 p.m.

Number of Studies: 3
Average Number of Seats: 15
Directional Distribution: 45% entering, 55% exiting

Trip Generation per Seat

Average Rate	Range of Rates	Standard Deviation
1.22	0.31 - 1.88	1.24

Data Plot and Equation

Caution - Use Carefully - Small Sample Size

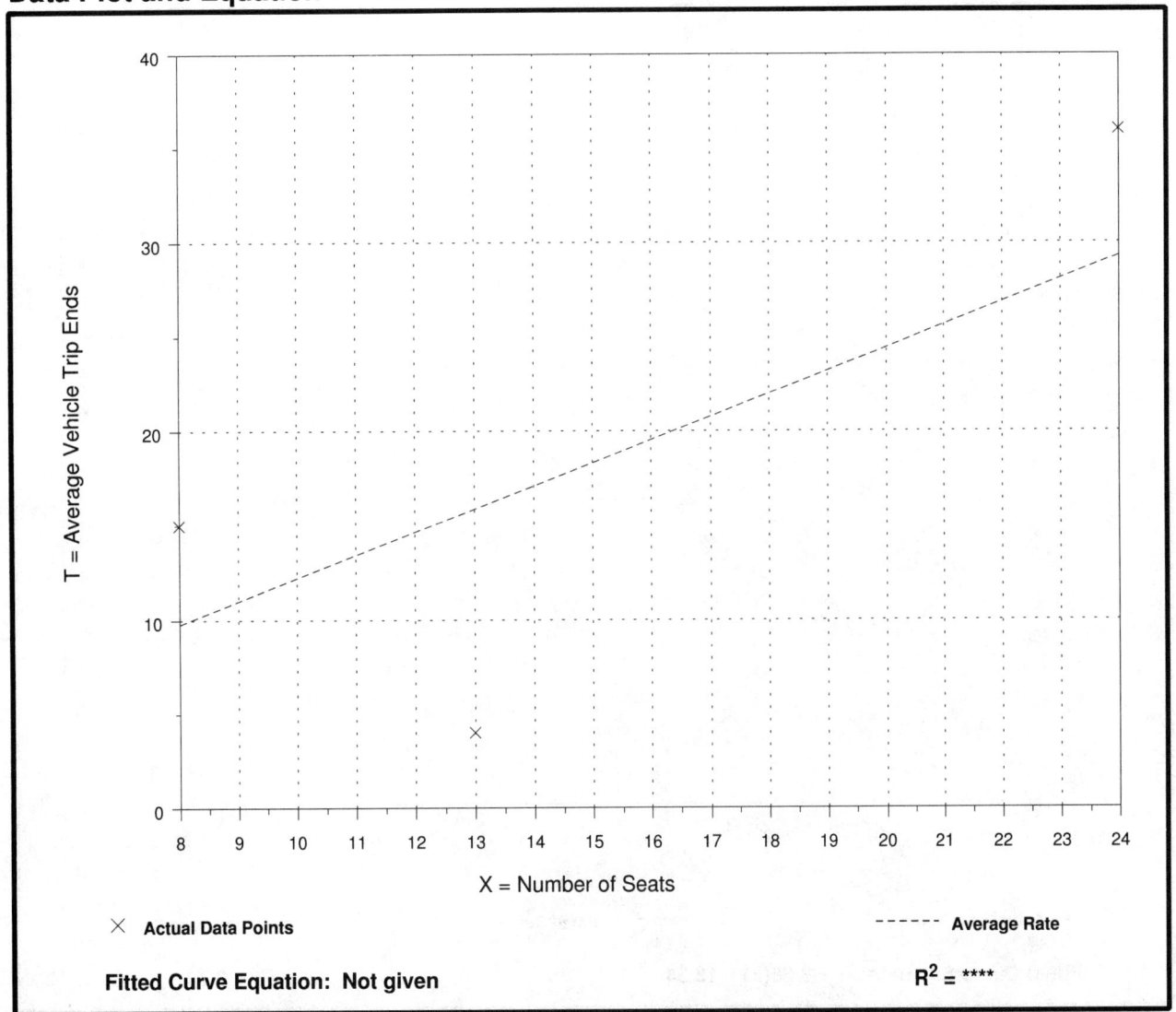

Fitted Curve Equation: Not given $R^2 = ****$

Coffee/Donut Shop with Drive-Through Window
(937)

Average Vehicle Trip Ends vs: Seats
On a: Weekday,
A.M. Peak Hour of Generator

Number of Studies: 4
Average Number of Seats: 13
Directional Distribution: 51% entering, 49% exiting

Trip Generation per Seat

Average Rate	Range of Rates	Standard Deviation
4.70	2.69 - 7.25	2.57

Data Plot and Equation

Caution - Use Carefully - Small Sample Size

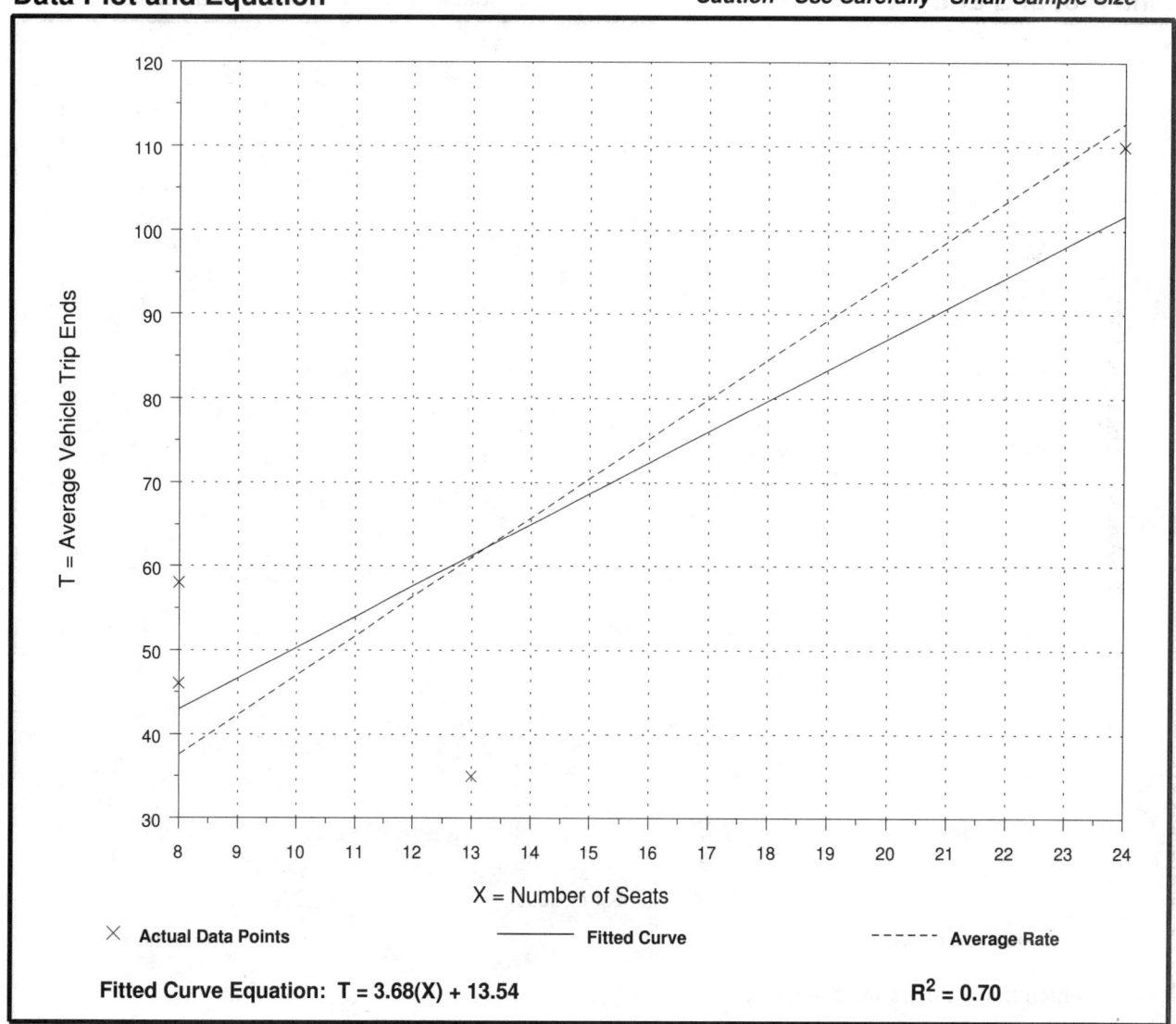

Fitted Curve Equation: T = 3.68(X) + 13.54 $R^2 = 0.70$

Coffee/Donut Shop with Drive-Through Window
(937)

Average Vehicle Trip Ends vs: Seats
On a: Weekday,
P.M. Peak Hour of Generator

Number of Studies: 2
Average Number of Seats: 11
Directional Distribution: 53% entering, 47% exiting

Trip Generation per Seat

Average Rate	Range of Rates	Standard Deviation
0.90	0.31 - 1.88	*

Data Plot and Equation

Caution - Use Carefully - Small Sample Size

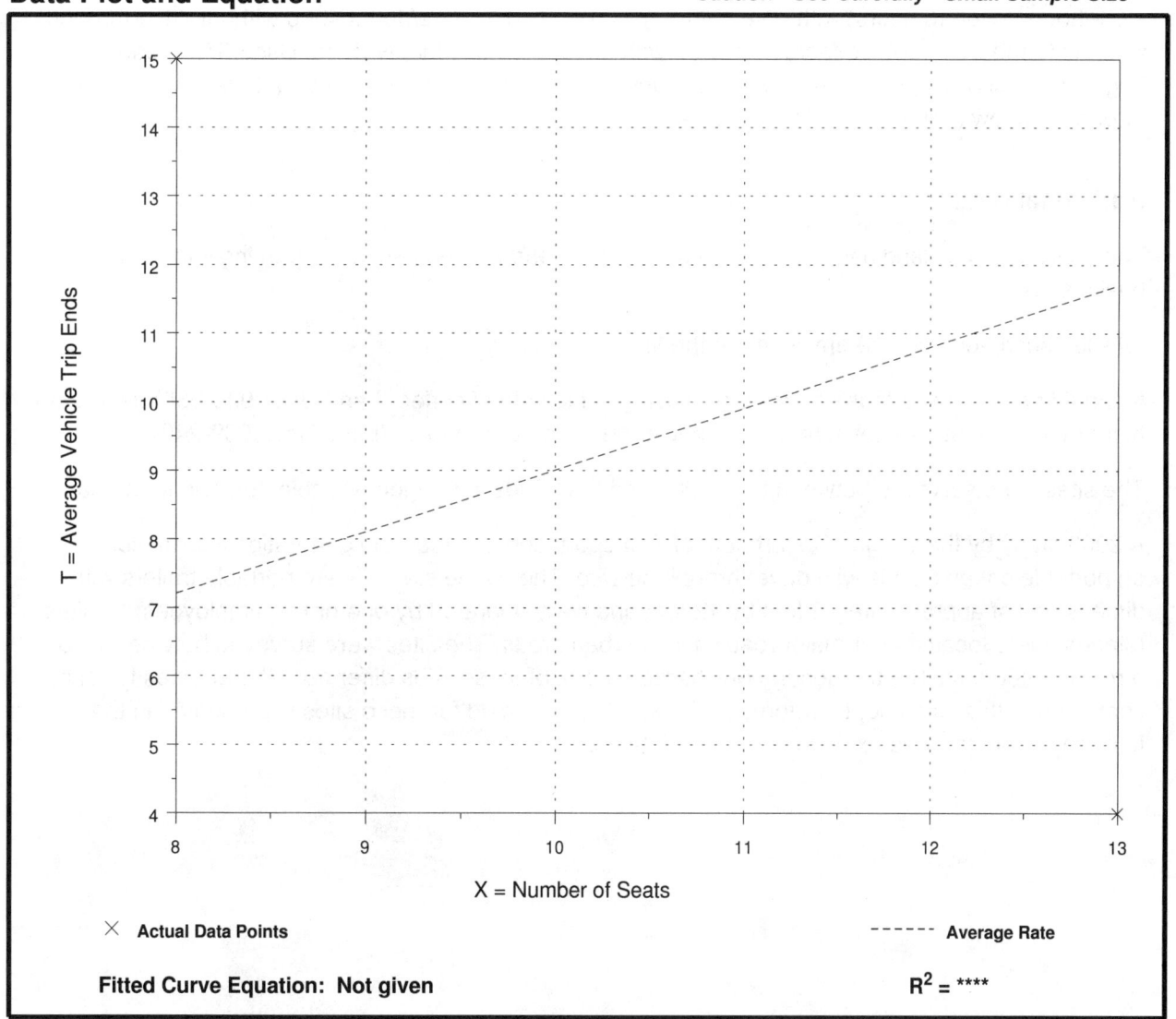

Fitted Curve Equation: Not given $R^2 = ****$

Land Use: 938
Coffee/Donut Shop with Drive-Through Window and No Indoor Seating

Description

This land use includes single-tenant coffee and donut restaurants with drive-through windows. Freshly brewed coffee and a variety of coffee-related accessories are the primary retail products sold at these sites. They may also sell other refreshment items, such as donuts, bagels, muffins, cakes, sandwiches, wraps, salads and other hot and cold beverages. Some sites may also sell newspapers, music CDs and books. The coffee and donut shops contained in this land use typically hold long store hours (over 15 hours) with an early morning opening. Coffee/donut shop without drive-through window (Land Use 936), coffee/donut shop with drive-through window (Land Use 937), bread/donut/bagel shop without drive-through window (Land Use 939) and bread/donut/bagel shop with drive-through window (Land Use 940) are related uses.

Additional Data

The facilities in this land use are typically stand-alone and are located in the parking lots of other businesses.

All sites surveyed had one employee at the time of the survey.

It should be noted that those stores specializing in the sale of coffee (Land Uses 936-938) generated higher trip generation rates than those specializing in other products (Land Uses 939-940).

The sites were surveyed between the 1990s and the 2000s in Oregon, Washington and Montana.

A 2003 study by the Oregon Department of Transportation provided trip generation information on portable coffee stands with drive-through service. The coffee stands were portable trailers with dimensions of approximately 8 feet by 12 feet and were operated by one or two employees. All sites (stands) were located near major roadways in urban areas. The sites were surveyed between 7:00 a.m. and 9:00 a.m. The trip generation characteristics of these sites differ from the facilities typically contained in this land use; therefore, the information collected for these sites is presented in the following table and was excluded from the data plots.

Table: Portable Coffee Stand with Drive-Through Window

Independent Variable	Trip Generation Rate	Size of Independent Variable	Number of Studies	Directional Distribution

1,000 Square Feet Gross Floor Area

Independent Variable	Trip Generation Rate	Size of Independent Variable	Number of Studies	Directional Distribution
Weekday A.M. Peak Hour of Adjacent Street Traffic	396.67	0.1	9	50% entering, 50% exiting
Weekday P.M. Peak Hour of Adjacent Street Traffic	71.54	3.3	1	52% entering, 48% exiting

Source : 755

1,000 Average Daily Traffic (ADT) on Adjacent Street

Independent Variable	Trip Generation Rate	Size of Independent Variable	Number of Studies	Directional Distribution
Weekday A.M. Peak Hour of Adjacent Street Traffic	1.29	31	9	50% entering, 50% exiting

Source : 755

A.M. Peak Hour Traffic on Adjacent Street

Independent Variable	Trip Generation Rate	Size of Independent Variable	Number of Studies	Directional Distribution
Weekday A.M. Peak Hour of Adjacent Street Traffic	0.02	1,865	9	50% entering, 50% exiting

Source : 755

Source Numbers

514, 644, 755

Coffee/Donut Shop with Drive-Through Window and No Indoor Seating
(938)

Average Vehicle Trip Ends vs: 1000 Sq. Feet Gross Floor Area
On a: Weekday

Number of Studies: 3
Average 1000 Sq. Feet GFA: 0.1
Directional Distribution: 50% entering, 50% exiting

Trip Generation per 1000 Sq. Feet Gross Floor Area

Average Rate	Range of Rates	Standard Deviation
1800.00	1400.00 - 2340.00	*

Data Plot and Equation

Caution - Use Carefully - Small Sample Size

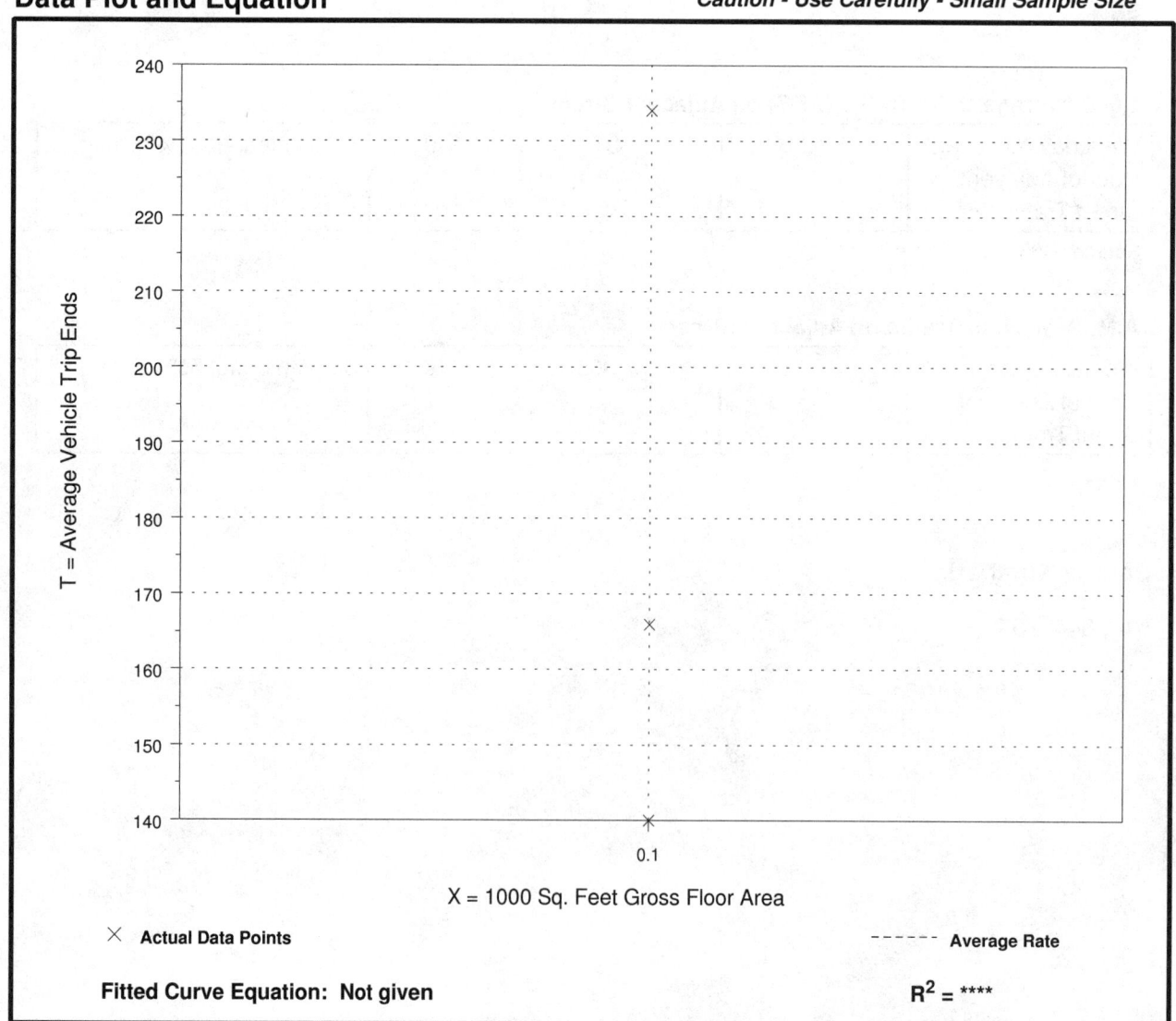

X = 1000 Sq. Feet Gross Floor Area
T = Average Vehicle Trip Ends

× Actual Data Points
------ Average Rate

Fitted Curve Equation: Not given $R^2 = ****$

Coffee/Donut Shop with Drive-Through Window and No Indoor Seating
(938)

Average Vehicle Trip Ends vs: 1000 Sq. Feet Gross Floor Area
On a: Weekday,
Peak Hour of Adjacent Street Traffic,
One Hour Between 7 and 9 a.m.

Number of Studies: 9
Average 1000 Sq. Feet GFA: 0.1
Directional Distribution: 50% entering, 50% exiting

Trip Generation per 1000 Sq. Feet Gross Floor Area

Average Rate	Range of Rates	Standard Deviation
303.33	100.00 - 600.00	*

Data Plot and Equation

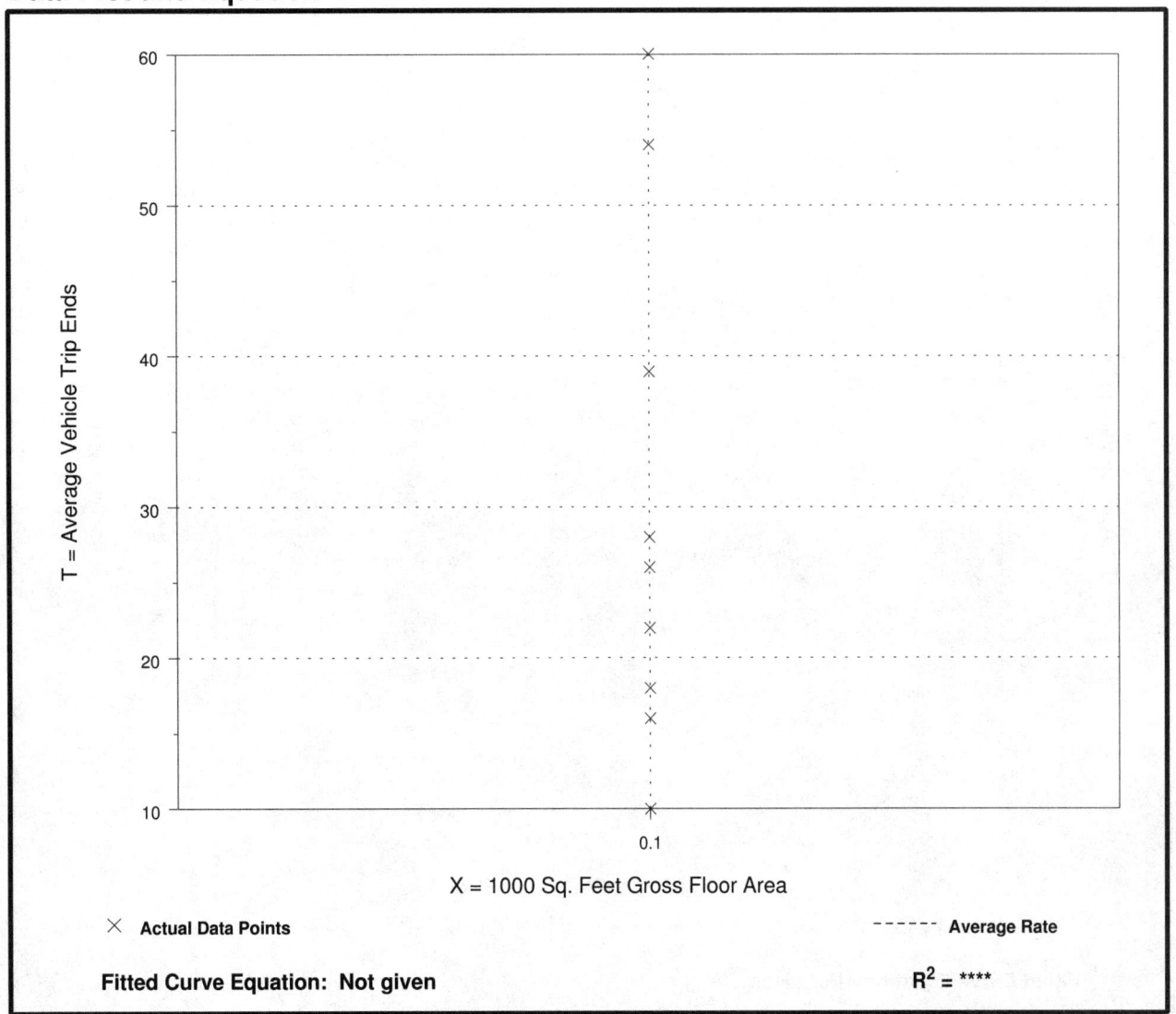

X = Actual Data Points

----- Average Rate

Fitted Curve Equation: Not given $R^2 =$ ****

Coffee/Donut Shop with Drive-Through Window and No Indoor Seating
(938)

Average Vehicle Trip Ends vs: 1000 Sq. Feet Gross Floor Area
On a: Weekday,
Peak Hour of Adjacent Street Traffic,
One Hour Between 4 and 6 p.m.

Number of Studies: 4
Average 1000 Sq. Feet GFA: 0.1
Directional Distribution: 50% entering, 50% exiting

Trip Generation per 1000 Sq. Feet Gross Floor Area

Average Rate	Range of Rates	Standard Deviation
75.00	50.00 - 100.00	12.91

Data Plot and Equation

Caution - Use Carefully - Small Sample Size

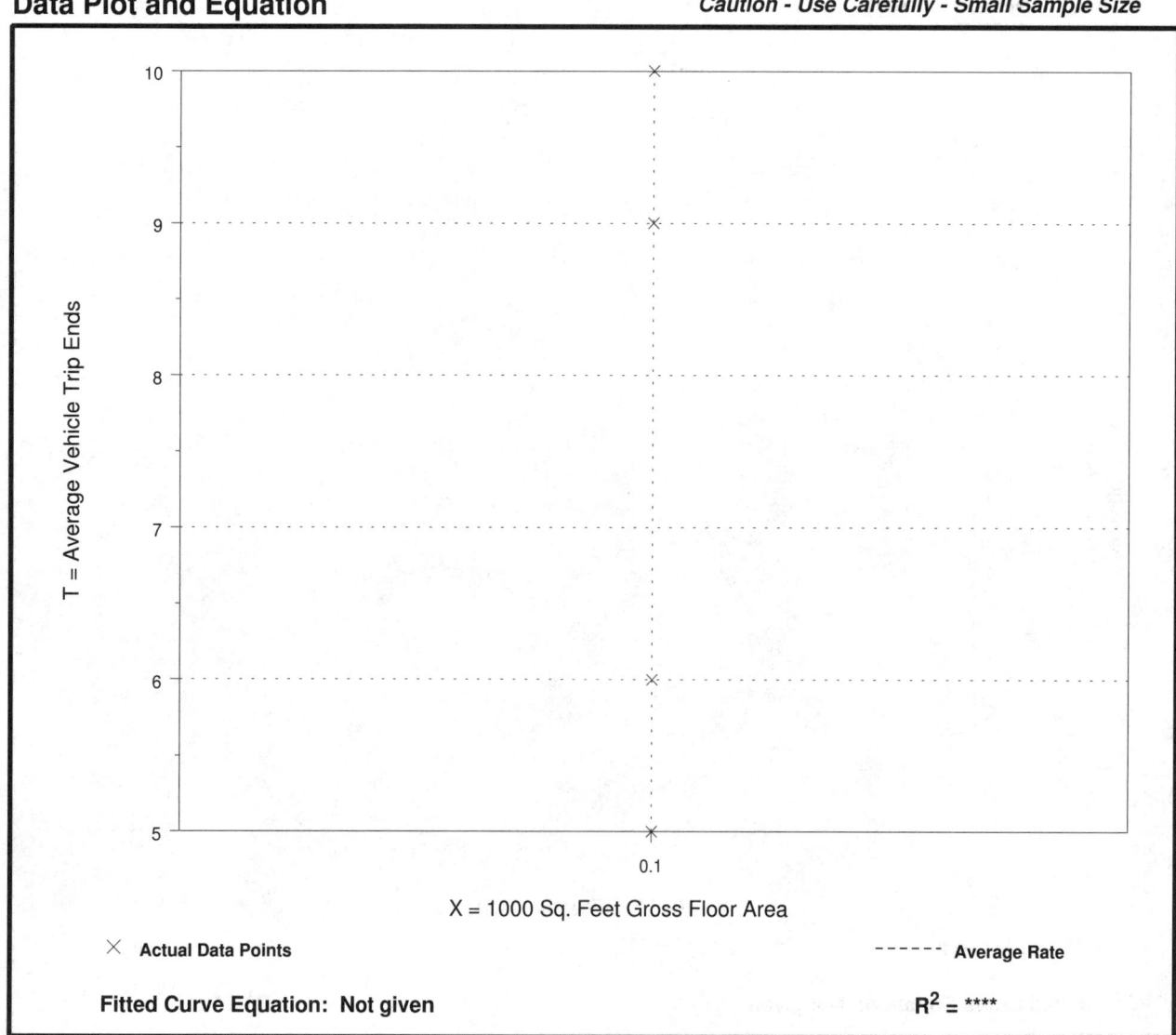

Fitted Curve Equation: Not given $R^2 = ****$

Coffee/Donut Shop with Drive-Through Window and No Indoor Seating
(938)

Average Vehicle Trip Ends vs: 1000 Sq. Feet Gross Floor Area
On a: Weekday,
A.M. Peak Hour of Generator

Number of Studies: 9
Average 1000 Sq. Feet GFA: 0.1
Directional Distribution: 50% entering, 50% exiting

Trip Generation per 1000 Sq. Feet Gross Floor Area

Average Rate	Range of Rates	Standard Deviation
310.00	100.00 - 600.00	*

Data Plot and Equation

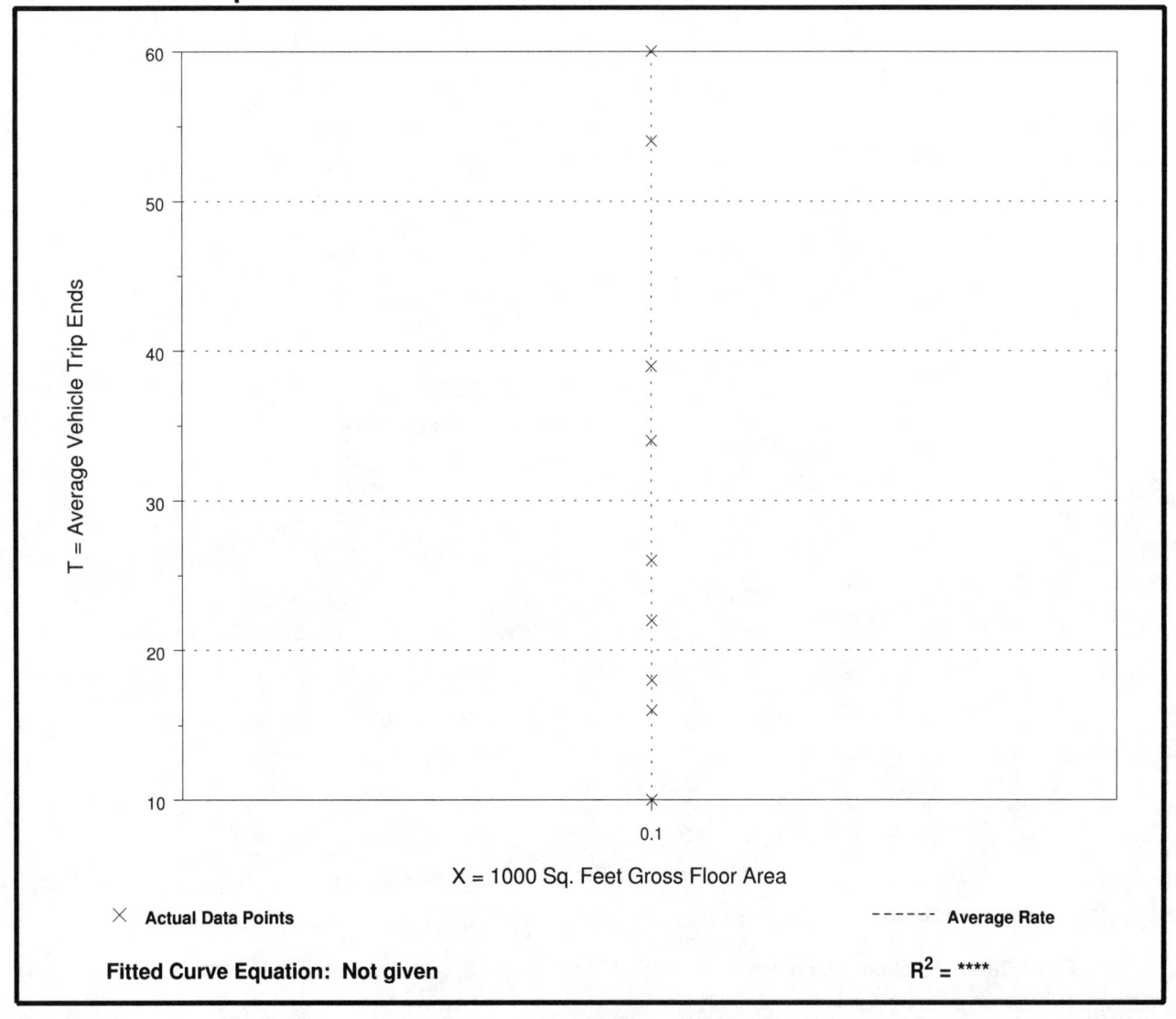

Fitted Curve Equation: Not given $R^2 = ****$

Coffee/Donut Shop with Drive-Through Window and No Indoor Seating
(938)

Average Vehicle Trip Ends vs: 1000 Sq. Feet Gross Floor Area
On a: Weekday,
P.M. Peak Hour of Generator

Number of Studies: 5
Average 1000 Sq. Feet GFA: 0.1
Directional Distribution: 50% entering, 50% exiting

Trip Generation per 1000 Sq. Feet Gross Floor Area

Average Rate	Range of Rates	Standard Deviation
96.00	50.00 - 150.00	*

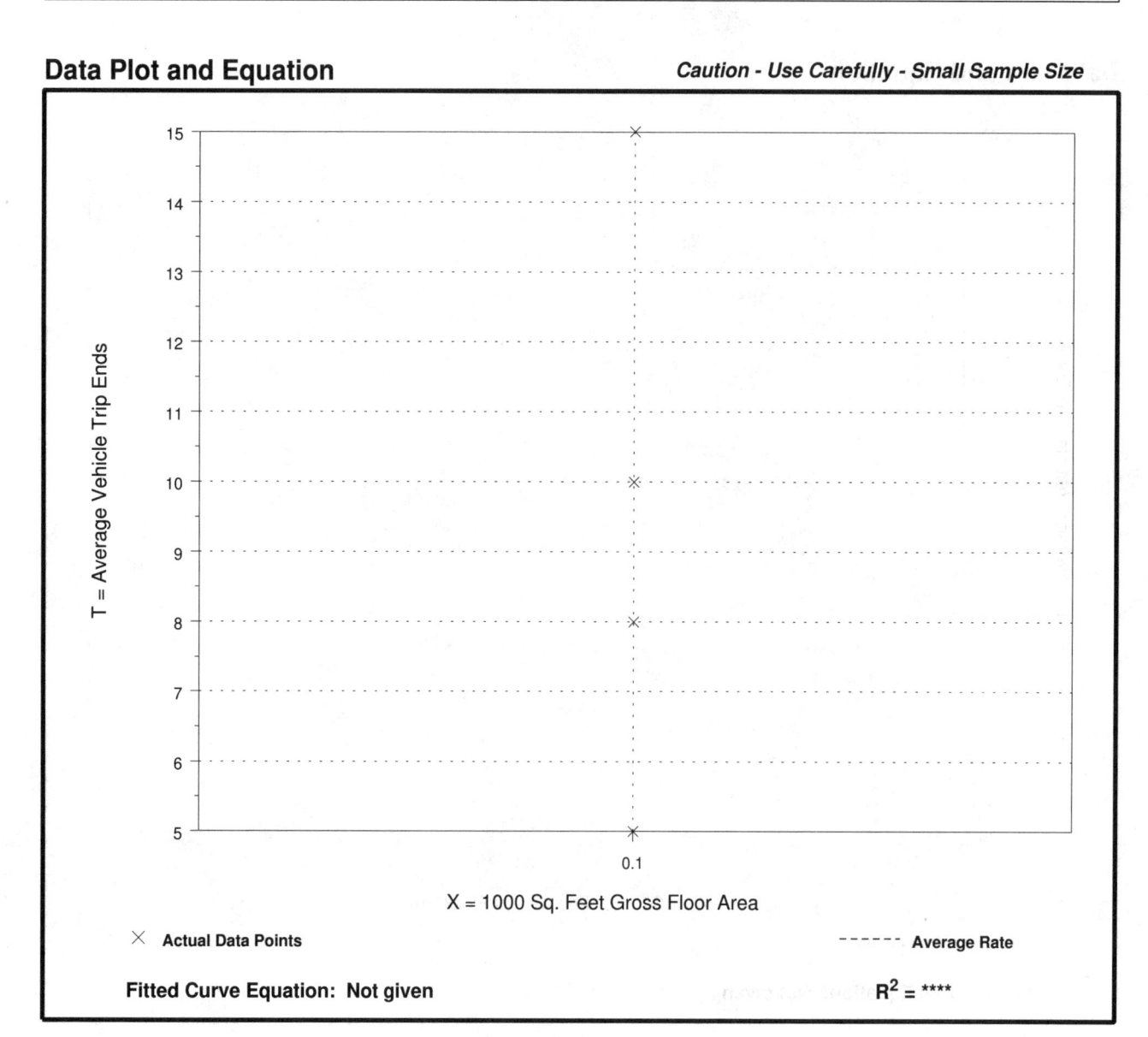

Data Plot and Equation

Caution - Use Carefully - Small Sample Size

Fitted Curve Equation: Not given $R^2 = ****$

Land Use: 939
Bread/Donut/Bagel Shop without Drive-Through Window

Description

This land use includes single-tenant bread, donut and bagel shops without drive-through windows. The sites surveyed specialize in producing and selling a variety of breads, donuts and bagels as the primary products sold. Some sites offer a breakfast menu. They may also sell other refreshment items, such as coffee, tea, soda, or other hot and cold beverages. Limited indoor seating is generally available at the sites surveyed. Coffee/donut shop without drive-through window (Land Use 936), coffee/donut shop with drive-through window (Land Use 937), coffee/donut shop with drive-through window and no indoor seating (Land Use 938) and bread/donut/bagel shop with drive-through window (Land Use 940) are related uses.

Additional Data

It should be noted that those stores specializing in the sale of coffee (Land Uses 936-938) generated higher trip generation rates than those specializing in other products (Land Uses 939-940).

The sites were surveyed in the 2000s in New Jersey and New York.

Source Numbers

551, 555

Land Use: 939
Bread/Donut/Bagel Shop without Drive-Through Window
Independent Variables with One Observation

The following trip generation data are for independent variables with only one observation. This information is shown in this table only; there are no related plots for these data.

Users are cautioned to use data with care because of the small sample size.

Independent Variable	Trip Generation Rate	Size of Independent Variable	Number of Studies	Directional Distribution
1,000 Square Feet Gross Floor Area				
Weekday A.M. Peak Hour of Adjacent Street Traffic	70.22	1.1	1	47% entering, 53% exiting
Weekday P.M. Peak Hour of Adjacent Street Traffic	28.00	2.0	1	50% entering, 50% exiting

Bread/Donut/Bagel Shop without Drive-Through Window
(939)

Average Vehicle Trip Ends vs: 1000 Sq. Feet Gross Floor Area
On a: Saturday,
Peak Hour of Generator

Number of Studies: 2
Average 1000 Sq. Feet GFA: 2
Directional Distribution: 52% entering, 48% exiting

Trip Generation per 1000 Sq. Feet Gross Floor Area

Average Rate	Range of Rates	Standard Deviation
48.87	33.78 - 56.62	*

Data Plot and Equation

Caution - Use Carefully - Small Sample Size

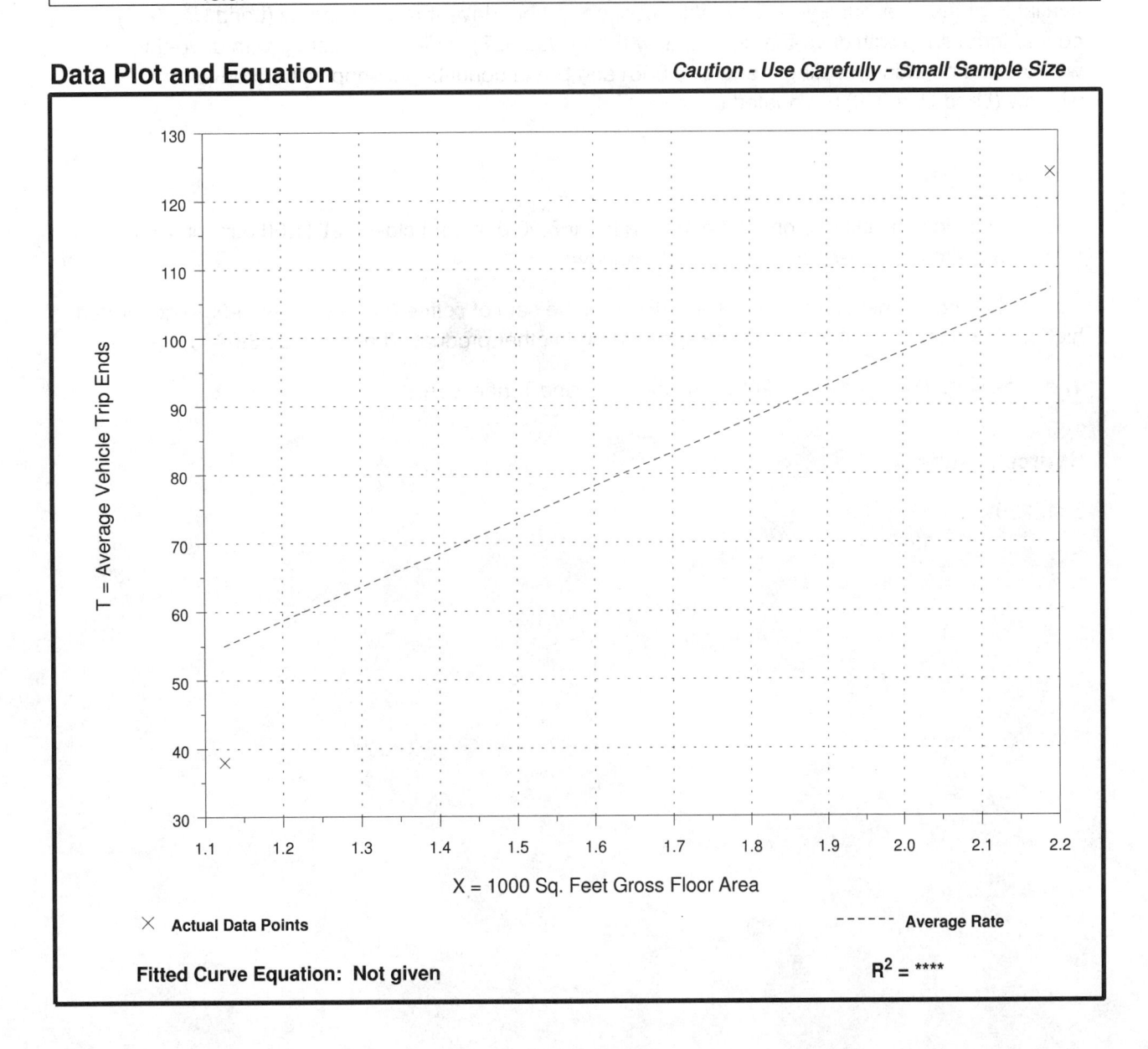

Fitted Curve Equation: Not given $R^2 = ****$

Land Use: 940
Bread/Donut/Bagel Shop with Drive-Through Window

Description

This land use includes single-tenant bread, donut and bagel shops with drive-through windows. The sites surveyed specialize in producing and selling a variety of breads, donuts and bagels as the primary products sold. Some sites offer a breakfast menu. They may also sell other refreshment items, such as coffee, tea, soda, or other hot and cold beverages. Limited indoor seating is generally available at the sites surveyed. Coffee/donut shop without drive-through window (Land Use 936), coffee/donut shop with drive-through window (Land Use 937), coffee/donut shop with drive-through window and no indoor seating (Land Use 938) and bread/donut/bagel shop without drive-through window (Land Use 939) are related uses.

Additional Data

The facilities in this land use opened at 5:00 a.m. or 6:00 a.m. and closed at 11:00 p.m. or 12:00 a.m. Some sites had 24-hour drive-through windows.

It should be noted that those stores specializing in the sale of coffee (Land Uses 936-938) generated higher trip generation rates than those specializing in other products (Land Uses 939-940).

The sites were surveyed in the 2000s in California and Tennessee.

Source Numbers

594, 725

Bread/Donut/Bagel Shop with Drive-Through Window
(940)

Average Vehicle Trip Ends vs: 1000 Sq. Feet Gross Floor Area
On a: Weekday,
Peak Hour of Adjacent Street Traffic,
One Hour Between 7 and 9 a.m.

Number of Studies: 5
Average 1000 Sq. Feet GFA: 4
Directional Distribution: 50% entering, 50% exiting

Trip Generation per 1000 Sq. Feet Gross Floor Area

Average Rate	Range of Rates	Standard Deviation
38.60	34.25 - 52.00	7.89

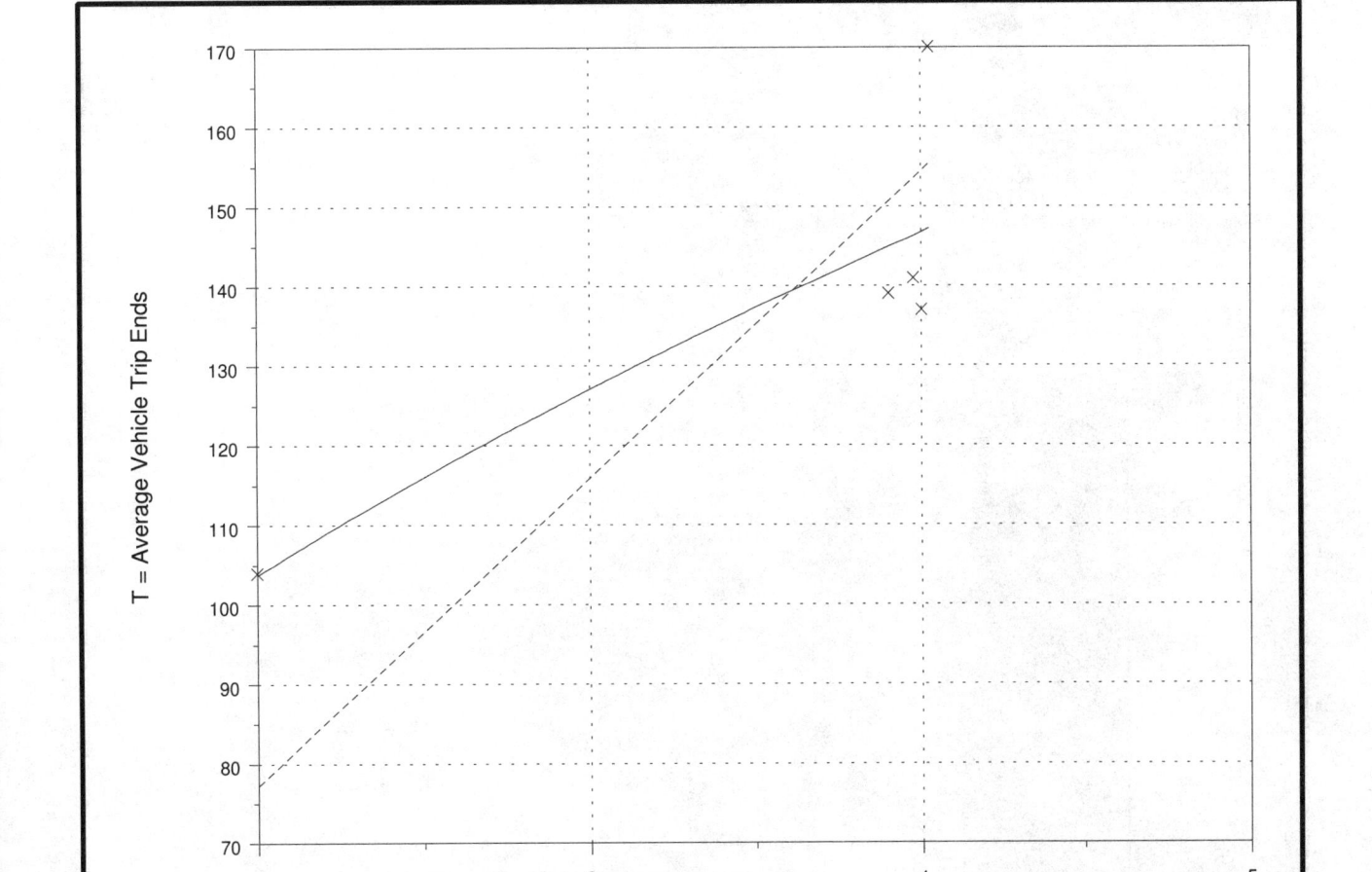

Data Plot and Equation

Caution - Use Carefully - Small Sample Size

Fitted Curve Equation: $Ln(T) = 0.50 \, Ln(X) + 4.29$ $R^2 = 0.77$

Bread/Donut/Bagel Shop with Drive-Through Window
(940)

Average Vehicle Trip Ends vs: 1000 Sq. Feet Gross Floor Area
On a: Weekday,
Peak Hour of Adjacent Street Traffic,
One Hour Between 4 and 6 p.m.

Number of Studies: 5
Average 1000 Sq. Feet GFA: 4
Directional Distribution: 49% entering, 51% exiting

Trip Generation per 1000 Sq. Feet Gross Floor Area

Average Rate	Range of Rates	Standard Deviation
18.99	14.50 - 25.90	5.50

Data Plot and Equation

Caution - Use Carefully - Small Sample Size

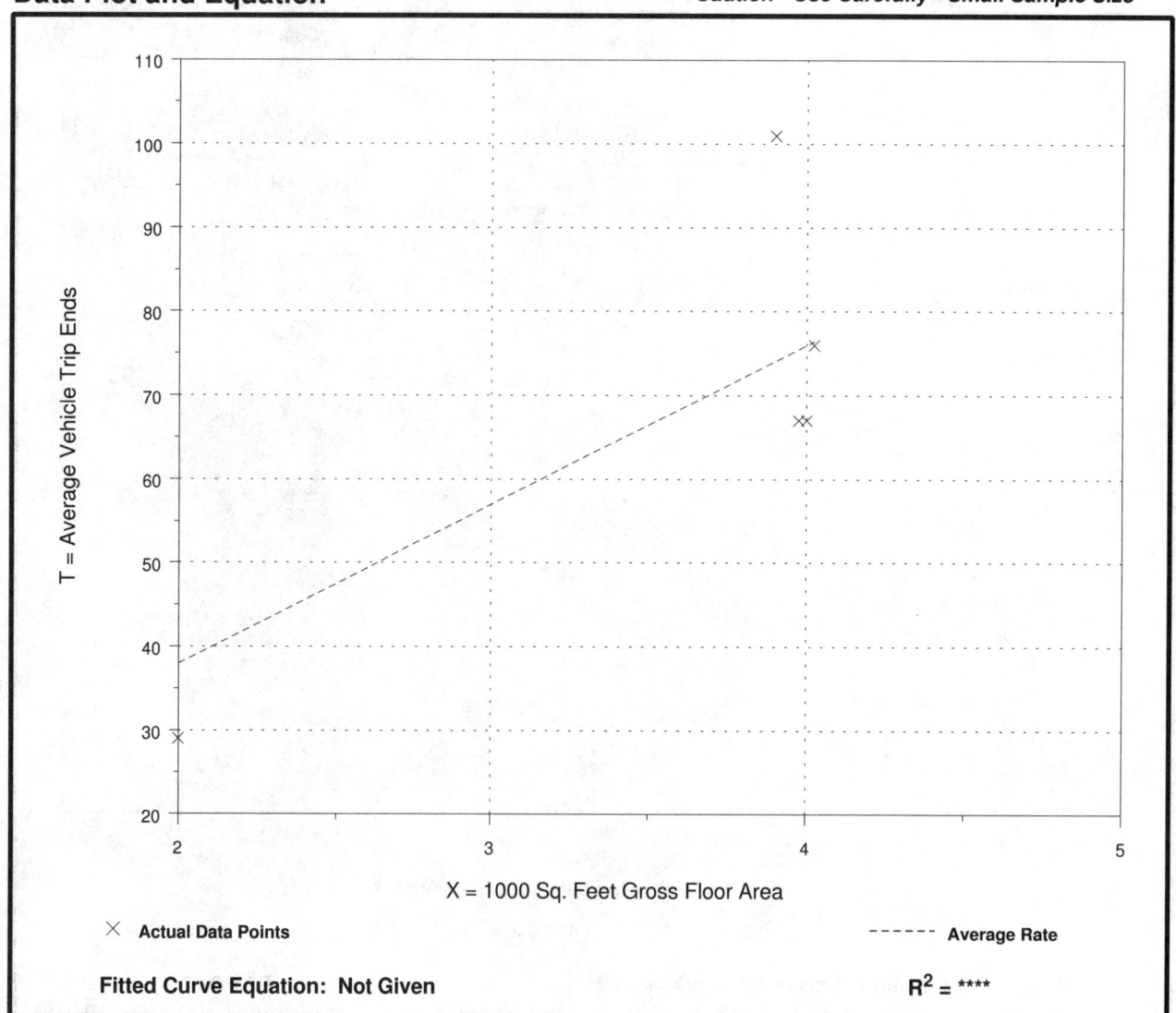

[Scatter plot: X = 1000 Sq. Feet Gross Floor Area (2 to 5); T = Average Vehicle Trip Ends (20 to 110). Data points at approximately (2, 29), (4, 67), (4, 67), (4, 76), (3.9, 101). Dashed line shows Average Rate.]

× Actual Data Points   ----- Average Rate

Fitted Curve Equation: Not Given $R^2 = ****$

Land Use: 941
Quick Lubrication Vehicle Shop

Description

A quick lubrication vehicle shop is a business where the primary activity is to perform oil change services for vehicles. Other ancillary services provided may include preventative maintenance, such as fluid and filter changes. Automobile repair service is generally not provided. Automobile care center (Land Use 942) and automobile parts and service center (Land Use 943) are related uses.

Additional Data

For the purpose of this land use, the independent variable, servicing positions, is defined as the maximum number of vehicles that can be serviced simultaneously.

The sites were surveyed in the 1990s in California and Washington.

Source Numbers

362, 441

Land Use: 941
Quick Lubrication Vehicle Shop
Independent Variables with One Observation

The following trip generation data are for independent variables with only one observation. This information is shown in this table only; there are no related plots for these data.

Users are cautioned to use data with care because of the small sample size.

Servicing Positions

Independent Variable	Trip Generation Rate	Size of Independent Variable	Number of Studies	Directional Distribution
Weekday	40.0	2	1	50% entering, 50% exiting
Weekday A.M. Peak Hour of Adjacent Street Traffic	3.0	2	1	67% entering, 33% exiting
Weekday A.M. Peak Hour of Generator	4.0	2	1	50% entering, 50% exiting
Saturday	42.0	2	1	50% entering, 50% exiting
Saturday Peak Hour of Generator	7.0	2	1	50% entering, 50% exiting
Sunday	28.0	2	1	50% entering, 50% exiting
Sunday Peak Hour of Generator	4.5	2	1	56% entering, 44% exiting

Quick Lubrication Vehicle Shop
(941)

Average Vehicle Trip Ends vs: Servicing Positions
On a: Weekday,
Peak Hour of Adjacent Street Traffic,
One Hour Between 4 and 6 p.m.

Number of Studies: 8
Avg. Num. of Servicing Positions: 3
Directional Distribution: 55% entering, 45% exiting

Trip Generation per Servicing Position

Average Rate	Range of Rates	Standard Deviation
5.19	3.00 - 10.00	2.96

Data Plot and Equation

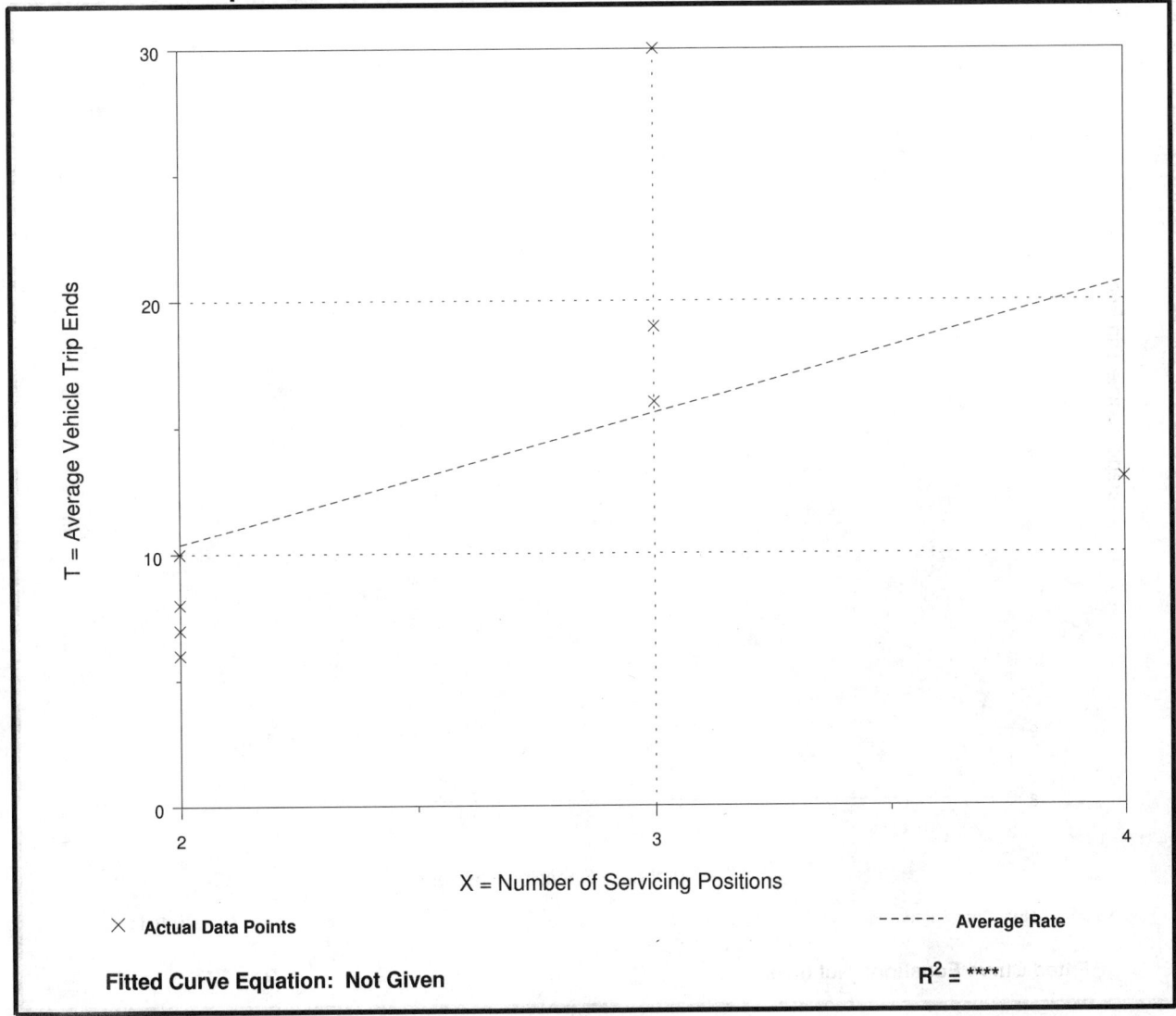

Fitted Curve Equation: Not Given $R^2 = ****$

Quick Lubrication Vehicle Shop
(941)

Average Vehicle Trip Ends vs: Servicing Positions
On a: Weekday,
P.M. Peak Hour of Generator

Number of Studies: 6
Avg. Num. of Servicing Positions: 3
Directional Distribution: 55% entering, 45% exiting

Trip Generation per Servicing Position

Average Rate	Range of Rates	Standard Deviation
4.60	3.25 - 6.00	1.97

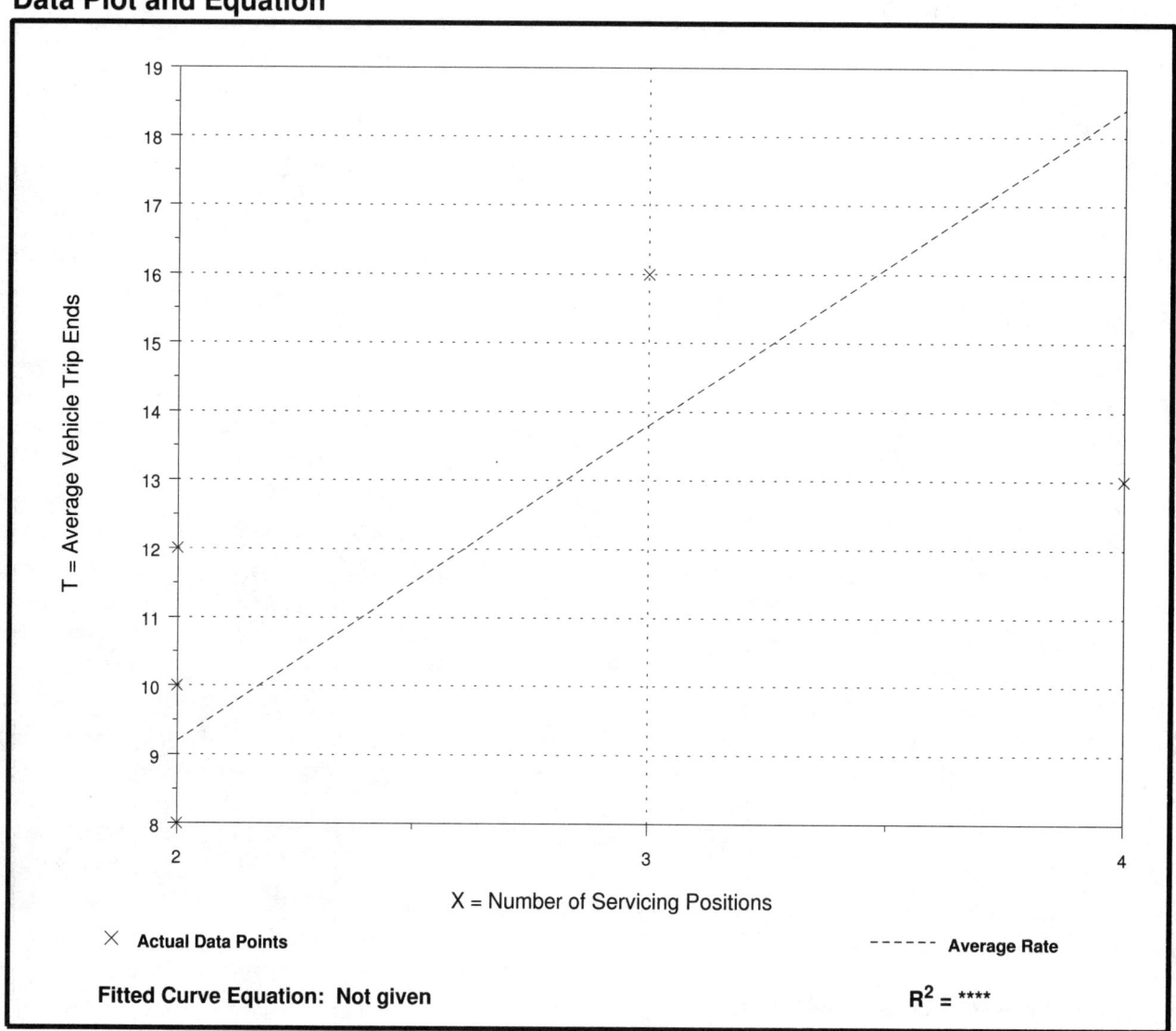

Data Plot and Equation

Fitted Curve Equation: Not given $R^2 = ****$

Land Use: 942
Automobile Care Center

Description

An automobile care center houses numerous businesses that provide automobile-related services, such as repair and servicing, stereo installation and seat cover upholstering. Quick lubrication vehicle shop (Land Use 941) and automobile parts and service center (Land Use 943) are related uses.

Additional Data

The P.M. peak hour of the generator typically coincided with the peak hour of the adjacent street traffic.

The sites were surveyed in 1988, 1989 and 1994 in California, Maryland and Florida.

Source Numbers

267, 273, 439, 715

Land Use: 942
Automobile Care Center
Independent Variables with One Observation

The following trip generation data are for independent variables with only one observation. This information is shown in this table only; there are no related plots for these data.

Users are cautioned to use data with care because of the small sample size.

Independent Variable	Trip Generation Rate	Size of Independent Variable	Number of Studies	Directional Distribution
Employees				
Weekday A.M. Peak Hour of Adjacent Street Traffic	1.00	44	1	68% entering, 32% exiting
Weekday P.M. Peak Hour of Adjacent Street Traffic	1.43	44	1	Not available
Weekday A.M. Peak Hour of Generator	1.00	44	1	68% entering, 32% exiting
Weekday P.M. Peak Hour of Generator	1.43	44	1	Not available
Saturday	8.23	44	1	50% entering, 50% exiting
Sunday	1.34	44	1	50% entering, 50% exiting
Service Stalls				
Weekday A.M. Peak Hour of Adjacent Street Traffic	1.52	29	1	68% entering, 32% exiting
Weekday P.M. Peak Hour of Adjacent Street Traffic	2.17	29	1	Not available
Weekday A.M. Peak Hour of Generator	1.52	29	1	68% entering, 32% exiting
Weekday P.M. Peak Hour of Generator	2.17	29	1	Not available
Weekday A.M. Peak Hour of Adjacent Street Traffic	1.52	29	1	68% entering, 32% exiting
Weekday P.M. Peak Hour of Adjacent Street Traffic	2.17	29	1	Not available
Weekday A.M. Peak Hour of Generator	1.52	29	1	68% entering, 32% exiting
Weekday P.M. Peak Hour of Generator	2.17	29	1	Not available
Saturday	12.48	29	1	50% entering, 50% exiting
Sunday	2.03	29	1	50% entering, 50% exiting

Automobile Care Center
(942)

Average Vehicle Trip Ends vs: 1000 Sq. Feet Occ. Gr. Leasable Area
On a: Weekday,
Peak Hour of Adjacent Street Traffic,
One Hour Between 7 and 9 a.m.

Number of Studies: 6
Average 1000 Sq. Feet OGLA: 17
Directional Distribution: 66% entering, 34% exiting

Trip Generation per 1000 Sq. Feet Occ. Gr. Leasable Area

Average Rate	Range of Rates	Standard Deviation
2.25	1.20 - 5.29	1.99

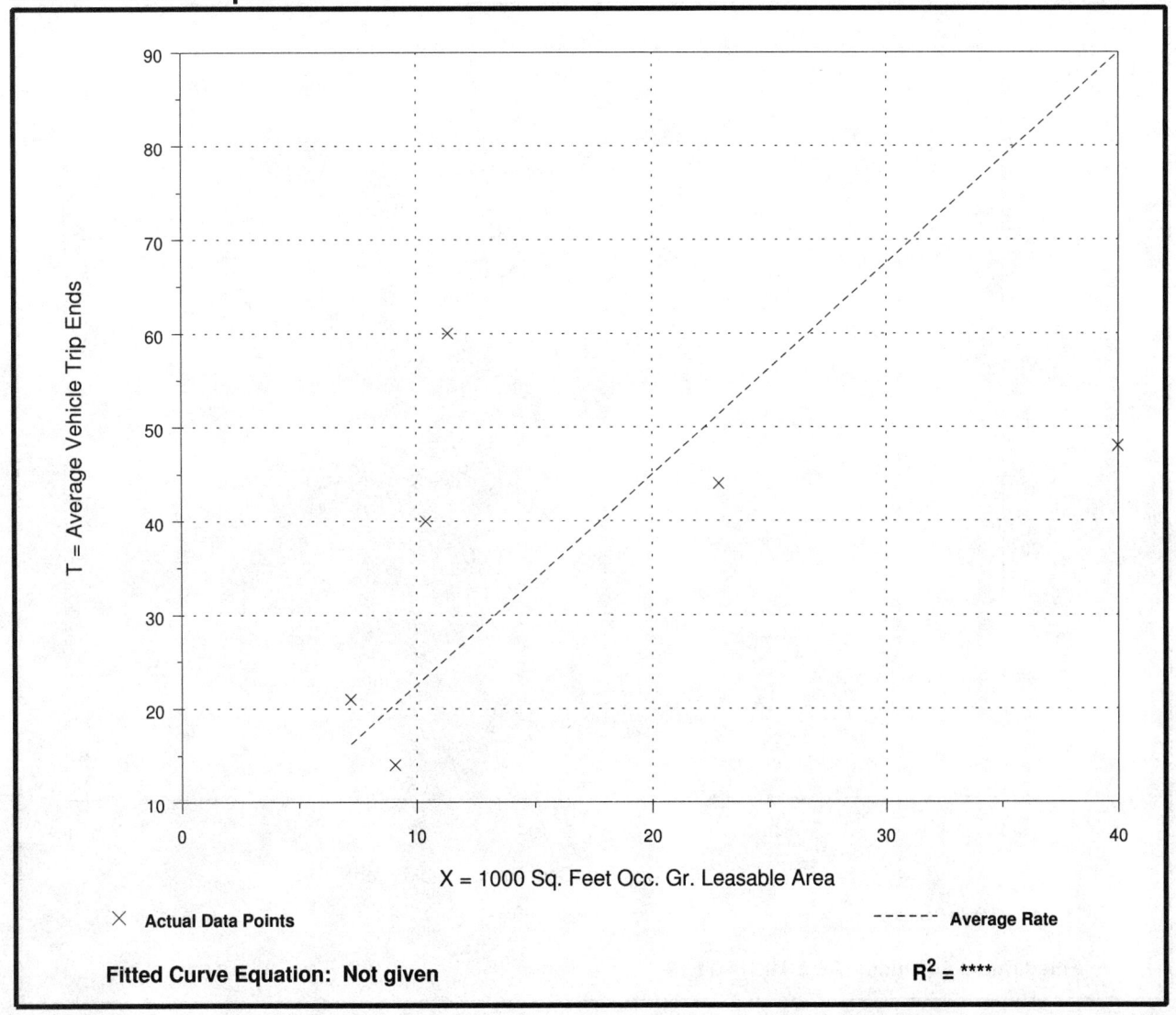

Data Plot and Equation

Fitted Curve Equation: Not given $R^2 = ****$

Automobile Care Center
(942)

Average Vehicle Trip Ends vs: 1000 Sq. Feet Occ. Gr. Leasable Area
On a: Weekday,
Peak Hour of Adjacent Street Traffic,
One Hour Between 4 and 6 p.m.

Number of Studies: 6
Average 1000 Sq. Feet OGLA: 17
Directional Distribution: 48% entering, 52% exiting

Trip Generation per 1000 Sq. Feet Occ. Gr. Leasable Area

Average Rate	Range of Rates	Standard Deviation
3.11	1.87 - 5.64	1.98

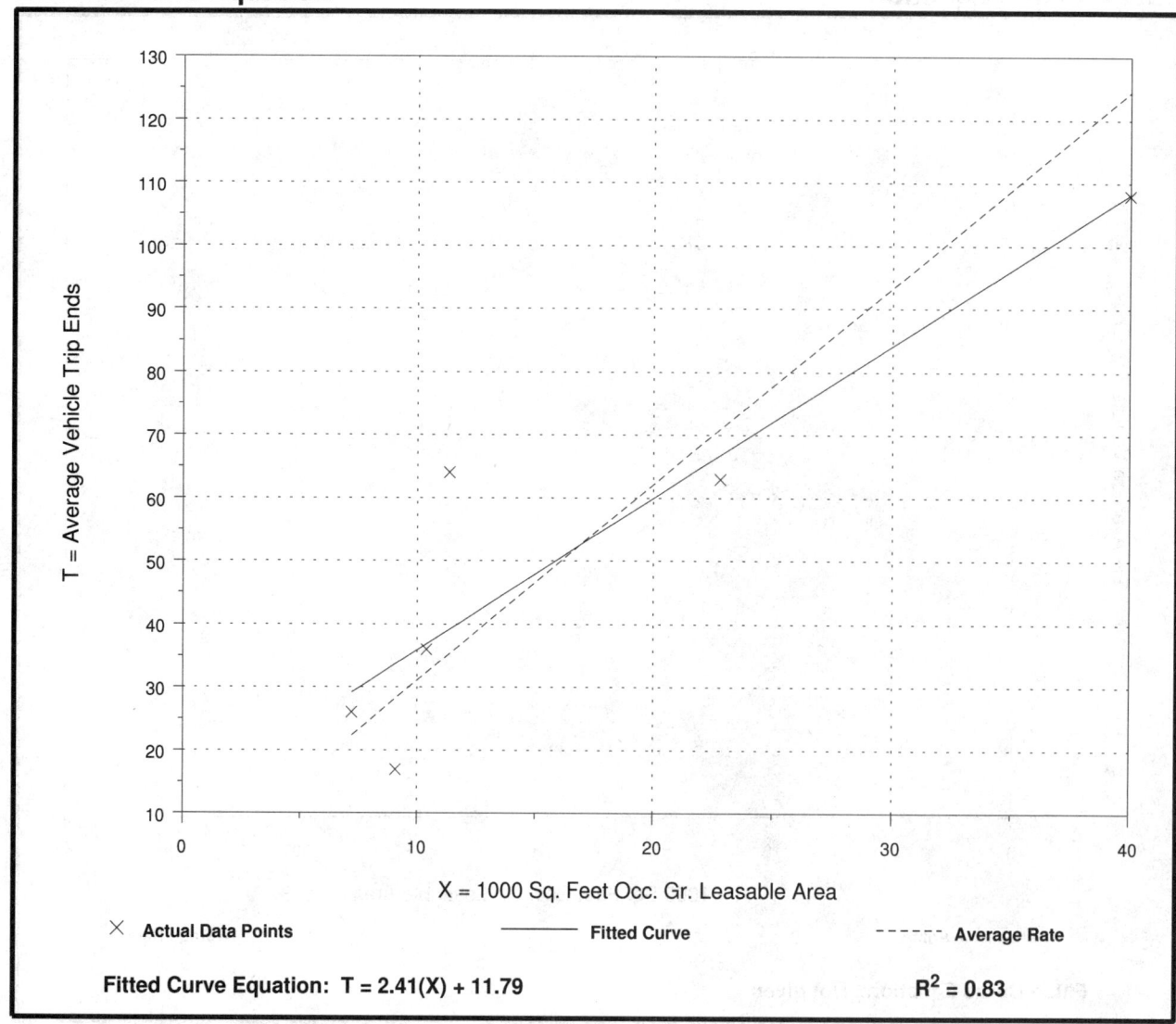

Data Plot and Equation

Fitted Curve Equation: $T = 2.41(X) + 11.79$ $R^2 = 0.83$

1976   *Trip Generation*, 9th Edition • Institute of Transportation Engineers

Automobile Care Center
(942)

Average Vehicle Trip Ends vs: 1000 Sq. Feet Occ. Gr. Leasable Area
On a: Weekday,
A.M. Peak Hour of Generator

Number of Studies: 6
Average 1000 Sq. Feet OGLA: 17
Directional Distribution: 56% entering, 44% exiting

Trip Generation per 1000 Sq. Feet Occ. Gr. Leasable Area

Average Rate	Range of Rates	Standard Deviation
2.83	1.93 - 5.73	2.05

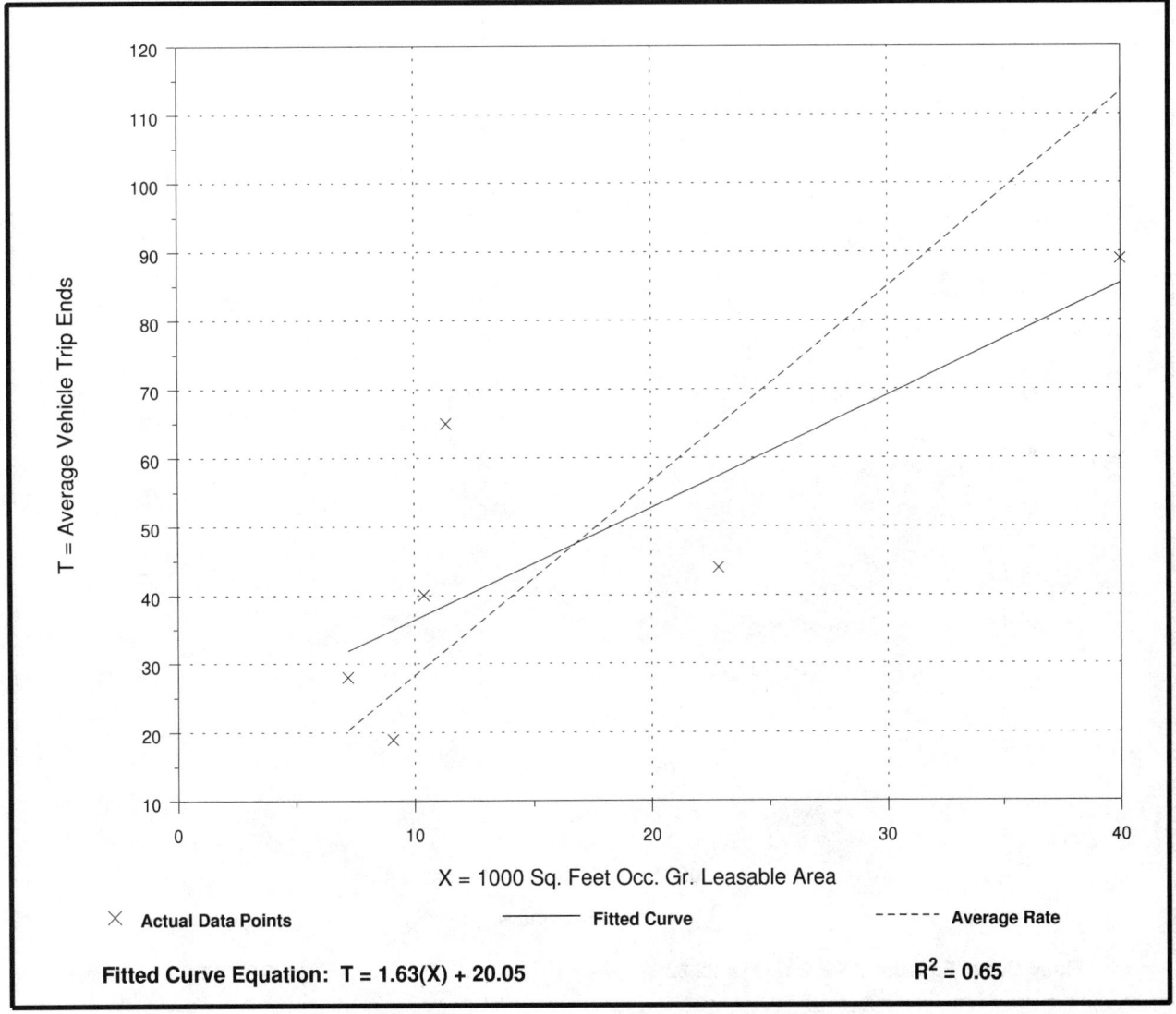

Data Plot and Equation

Fitted Curve Equation: $T = 1.63(X) + 20.05$ $R^2 = 0.65$

Automobile Care Center
(942)

Average Vehicle Trip Ends vs: 1000 Sq. Feet Occ. Gr. Leasable Area
On a: Weekday,
P.M. Peak Hour of Generator

Number of Studies: 6
Average 1000 Sq. Feet OGLA: 17
Directional Distribution: 49% entering, 51% exiting

Trip Generation per 1000 Sq. Feet Occ. Gr. Leasable Area

Average Rate	Range of Rates	Standard Deviation
3.51	2.75 - 7.14	2.29

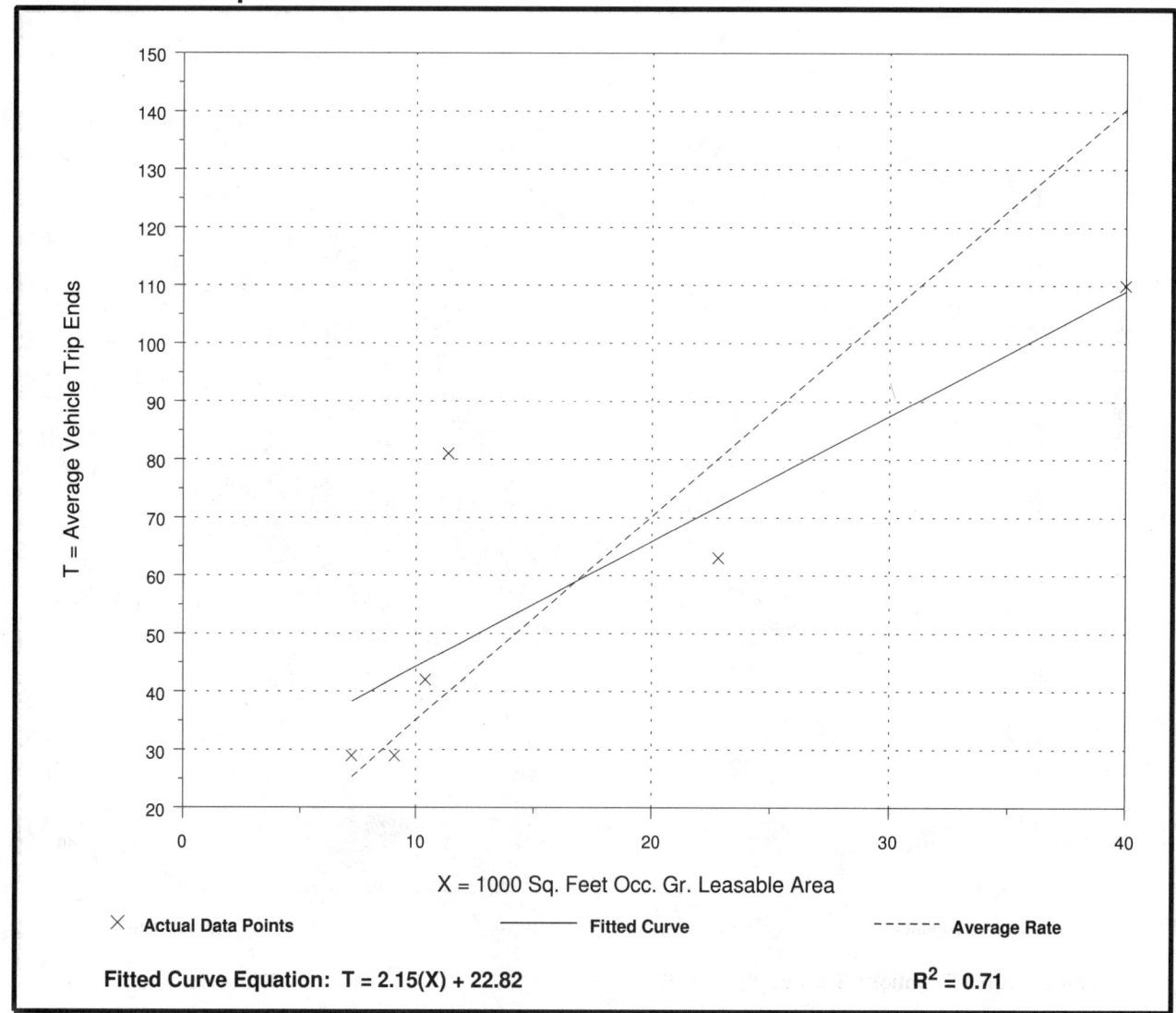

Data Plot and Equation

Fitted Curve Equation: $T = 2.15(X) + 22.82$ $R^2 = 0.71$

Automobile Care Center
(942)

Average Vehicle Trip Ends vs: 1000 Sq. Feet Occ. Gr. Leasable Area
On a: Saturday

Number of Studies: 2
Average 1000 Sq. Feet OGLA: 31
Directional Distribution: 50% entering, 50% exiting

Trip Generation per 1000 Sq. Feet Occ. Gr. Leasable Area

Average Rate	Range of Rates	Standard Deviation
23.72	15.86 - 28.20	*

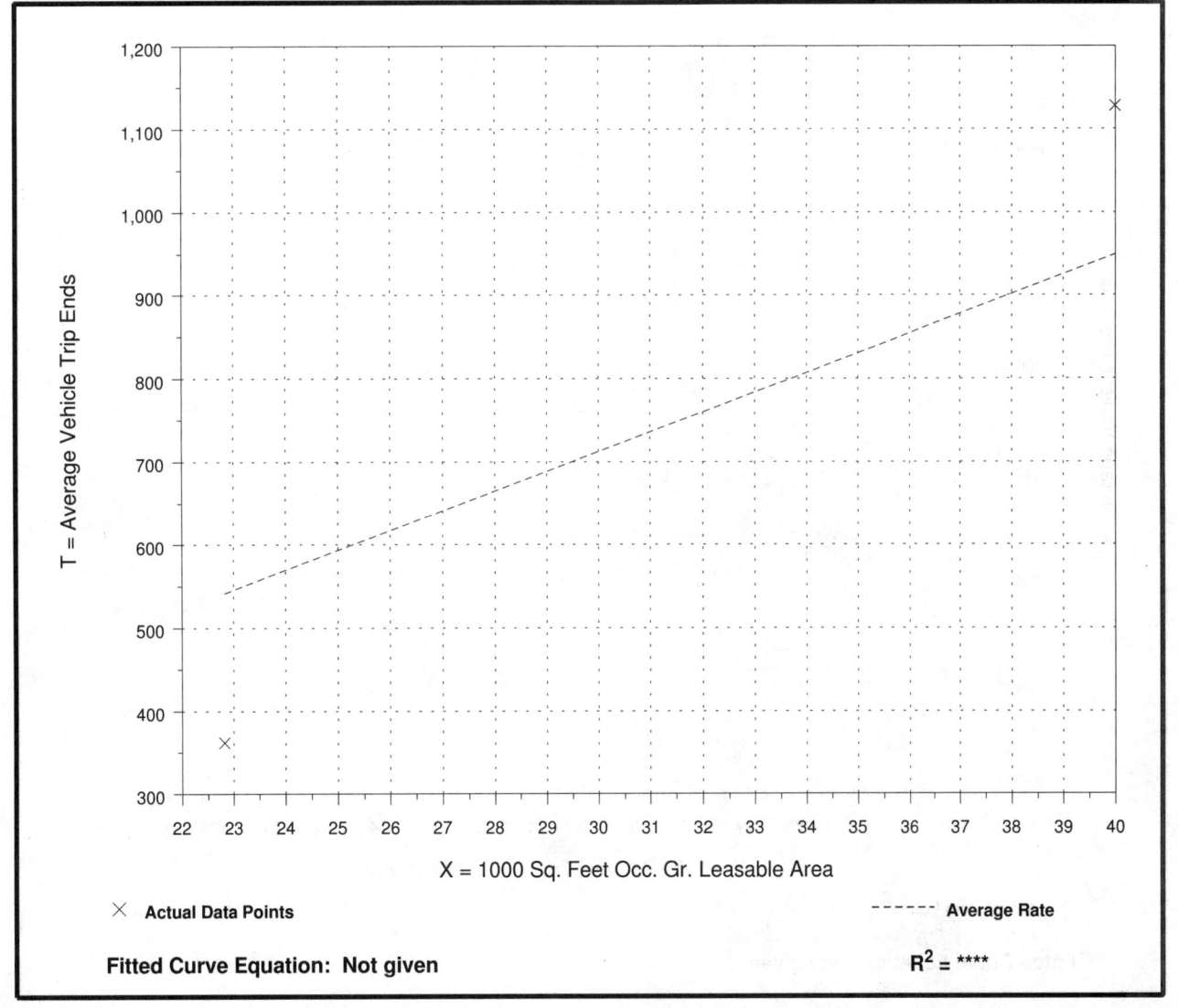

Data Plot and Equation

Caution - Use Carefully - Small Sample Size

Fitted Curve Equation: Not given $R^2 = ****$

Automobile Care Center
(942)

Average Vehicle Trip Ends vs: 1000 Sq. Feet Occ. Gr. Leasable Area
On a: Sunday

Number of Studies: 2
Average 1000 Sq. Feet OGLA: 31
Directional Distribution: 50% entering, 50% exiting

Trip Generation per 1000 Sq. Feet Occ. Gr. Leasable Area

Average Rate	Range of Rates	Standard Deviation
11.88	2.59 - 17.18	*

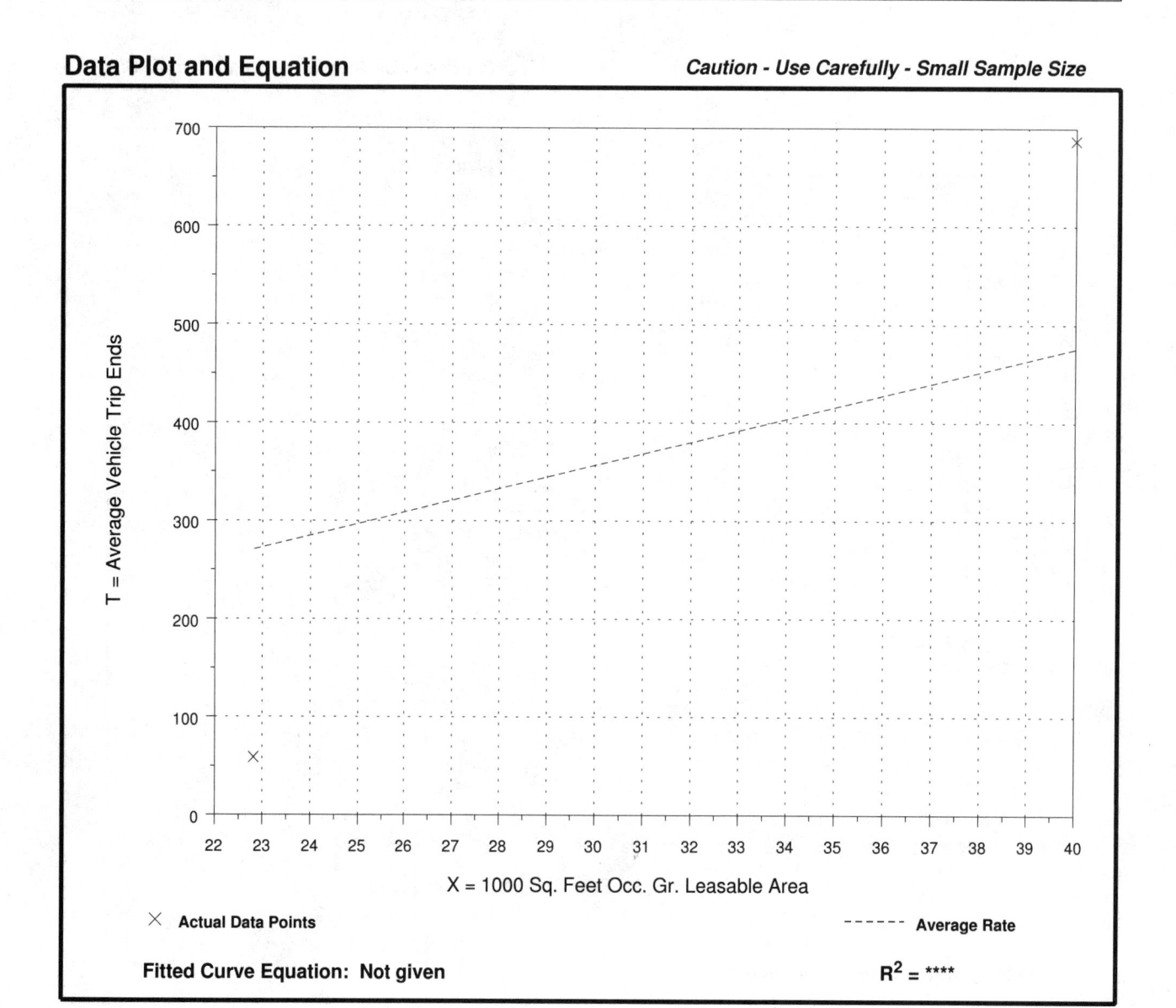

Data Plot and Equation

Caution - Use Carefully - Small Sample Size

Fitted Curve Equation: Not given $R^2 = ****$

Land Use: 943
Automobile Parts and Service Center

Description

Automobile parts and service centers sell automobile parts for do-it-yourself maintenance and repair including tires, batteries, oil and sparks plugs. The stores may also sell automobile parts to retailers and repair facilities. Automobile parts and service centers also provide a full array of on-site services for various automobiles. These facilities provide centralized cashiering and maintain long hours 7 days per week. Automobile parts and service centers are sometimes found as separate parcels within a retail complex. Automobile parts sales (Land Use 843), tire store (Land Use 848), tire superstore (Land Use 849), quick lubrication vehicle shop (Land Use 941) and automobile care center (Land Use 942) are related uses.

Additional Data

The site was surveyed in 2000 in New York.

Source Number

555

Land Use: 943
Automobile Parts and Service Center
Independent Variables with One Observation

The following trip generation data are for independent variables with only one observation. This information is shown in this table only; there are no related plots for these data.

Users are cautioned to use data with care because of the small sample size.

Independent Variable	Trip Generation Rate	Size of Independent Variable	Number of Studies	Directional Distribution
1,000 Square Feet Gross Floor Area				
Weekday P.M. Peak Hour of Adjacent Street Traffic	4.46	19	1	42% entering, 58% exiting
Saturday Peak Hour of Generator	6.61	19	1	46% entering, 54% exiting

Land Use: 944
Gasoline/Service Station

Description

This land use includes gasoline/service stations where the primary business is the fueling of motor vehicles. These service stations may also have ancillary facilities for servicing and repairing motor vehicles. Service stations are generally located at intersections or interchanges. Service stations with convenience stores and car washes are not included in this land use. Convenience market with gasoline pumps (Land Use 853), gasoline/service station with convenience market (Land Use 945), gasoline/service station with convenience market and car wash (Land Use 946) and truck stop (Land Use 950) are related uses.

Additional Data

For the purpose of this land use, the independent variable, vehicle fueling positions, is defined as the maximum number of vehicles that can be fueled simultaneously.

Gasoline/service stations in this land use include "pay-at-the-pump" and traditional fueling stations.

The weekday peak hours of the generator typically coincided with the peak hours of the adjacent street traffic.

The sites were surveyed between the 1970s and the 2000s throughout the United States.

Specialized Land Use Data

A 2006 study provided data on four private fuel facilities in Florida. These facilities provide self-fuel service, but are not open for use by the general public. To use the services offered at the facility, a pre-established membership account is required. Information collected for these sites is presented in the following table and was excluded from the data plots.

Independent Variable	Average Trip Generation Rate	Average Size of Independent Variable	Number of Studies	Directional Distribution
Vehicle Fueling Positions				
Weekday A.M. Peak Hour of Adjacent Street Traffic	1.32	9	4	49% entering, 51% exiting
Weekday P.M. Peak Hour of Adjacent Street Traffic	0.82	9	4	46% entering, 54% exiting

Source: 721

Source Numbers

347, 349, 355, 440, 444, 445, 540, 551, 552, 583, 599, 721

Gasoline/Service Station
(944)

Average Vehicle Trip Ends vs: Vehicle Fueling Positions
On a: Weekday

Number of Studies: 6
Average Vehicle Fueling Positions: 8
Directional Distribution: 50% entering, 50% exiting

Trip Generation per Vehicle Fueling Position

Average Rate	Range of Rates	Standard Deviation
168.56	73.00 - 306.00	71.19

Data Plot and Equation

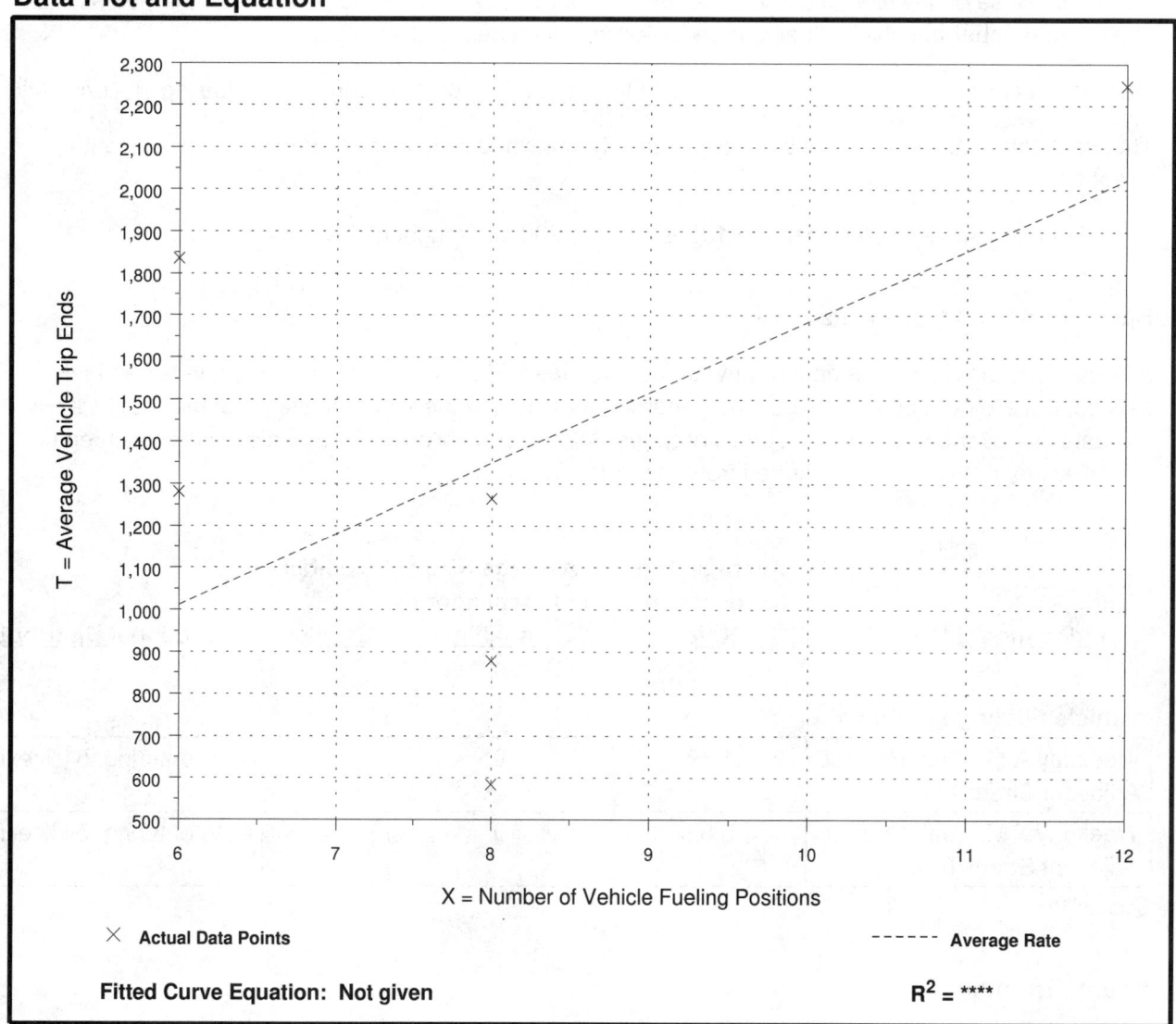

X = Number of Vehicle Fueling Positions
T = Average Vehicle Trip Ends

× Actual Data Points
- - - - - Average Rate

Fitted Curve Equation: Not given $R^2 =$ ****

Gasoline/Service Station
(944)

Average Vehicle Trip Ends vs: Vehicle Fueling Positions
On a: Weekday,
Peak Hour of Adjacent Street Traffic,
One Hour Between 7 and 9 a.m.

Number of Studies: 15
Average Vehicle Fueling Positions: 7
Directional Distribution: 51% entering, 49% exiting

Trip Generation per Vehicle Fueling Position

Average Rate	Range of Rates	Standard Deviation
12.16	7.33 - 17.50	4.29

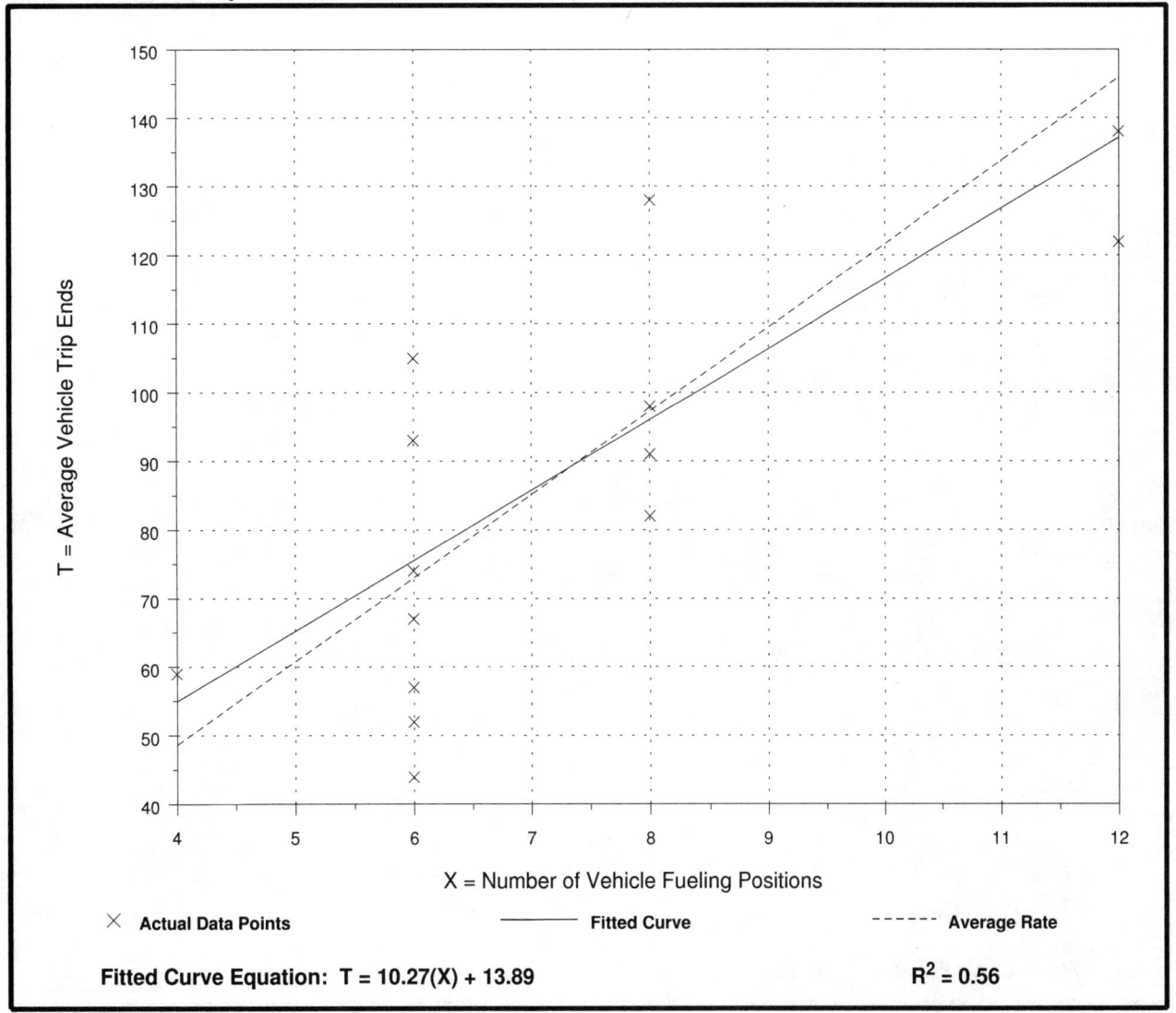

Data Plot and Equation

Fitted Curve Equation: $T = 10.27(X) + 13.89$ $R^2 = 0.56$

Gasoline/Service Station
(944)

Average Vehicle Trip Ends vs: Vehicle Fueling Positions
On a: Weekday,
Peak Hour of Adjacent Street Traffic,
One Hour Between 4 and 6 p.m.

Number of Studies: 28
Average Vehicle Fueling Positions: 8
Directional Distribution: 50% entering, 50% exiting

Trip Generation per Vehicle Fueling Position

Average Rate	Range of Rates	Standard Deviation
13.87	5.00 - 27.33	6.65

Data Plot and Equation

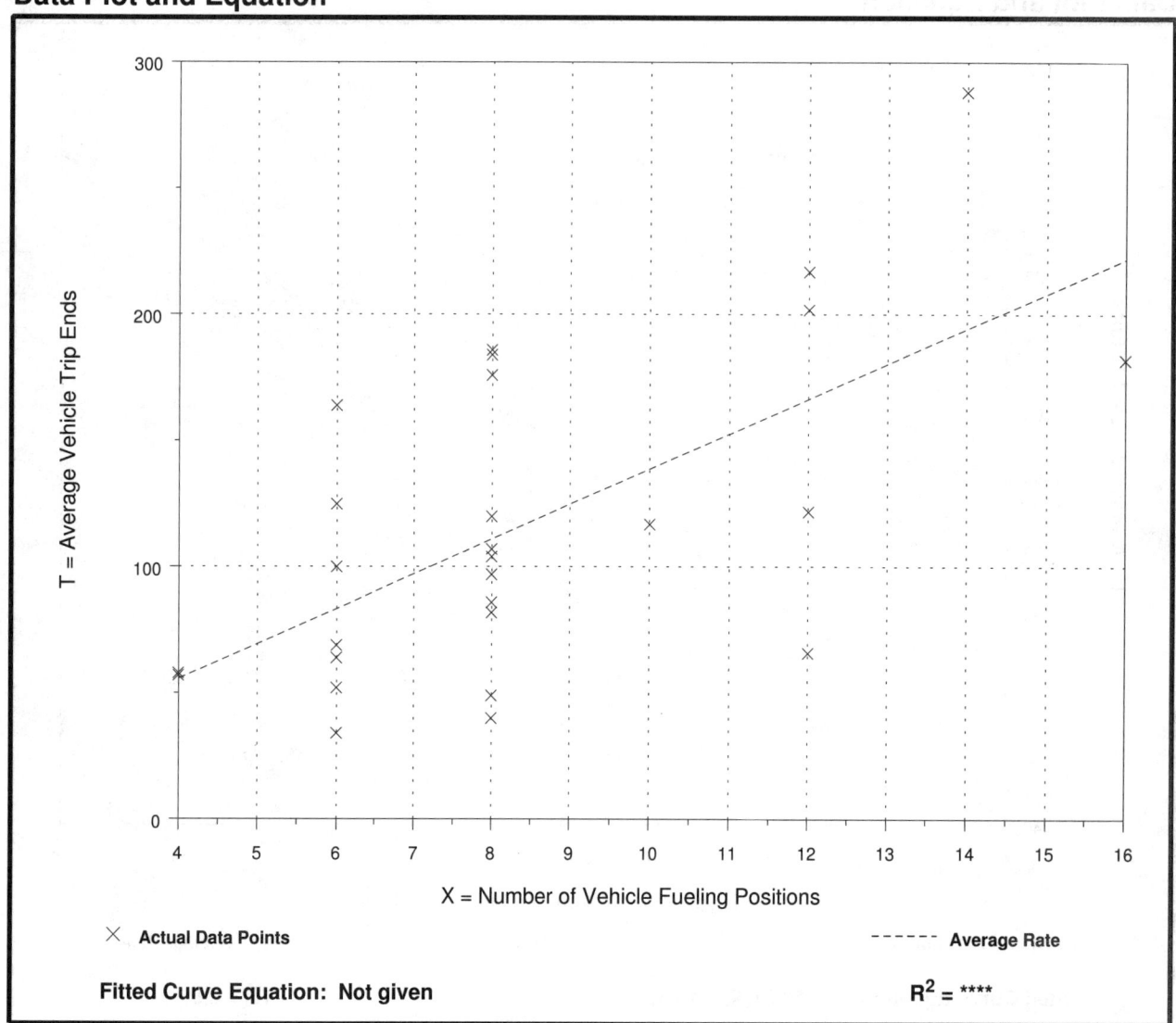

Fitted Curve Equation: Not given $R^2 = ****$

Gasoline/Service Station
(944)

Average Vehicle Trip Ends vs: Vehicle Fueling Positions
On a: Weekday,
A.M. Peak Hour of Generator

Number of Studies: 13
Average Vehicle Fueling Positions: 8
Directional Distribution: 50% entering, 50% exiting

Trip Generation per Vehicle Fueling Position

Average Rate	Range of Rates	Standard Deviation
12.58	7.33 - 17.50	4.55

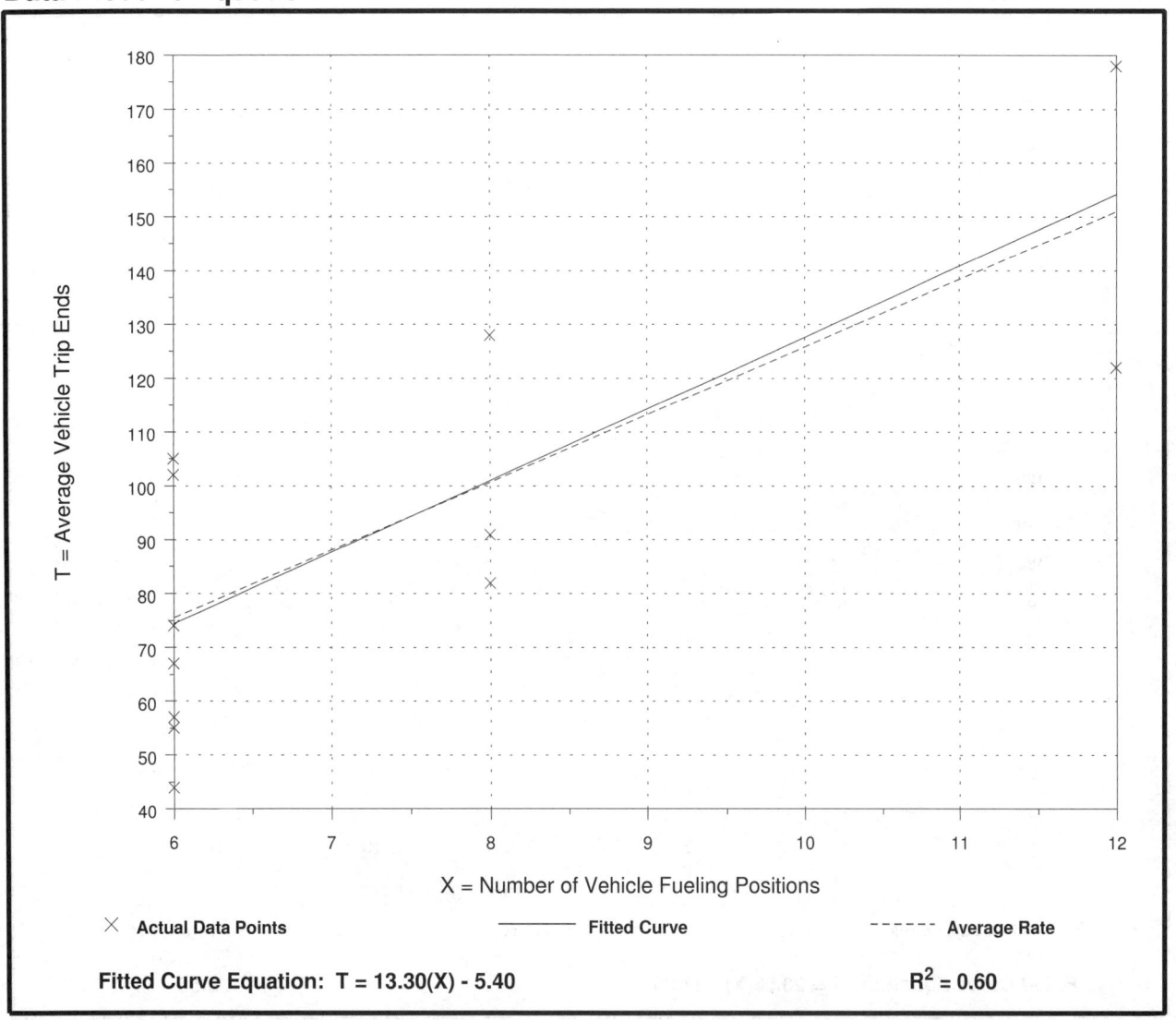

Data Plot and Equation

Fitted Curve Equation: T = 13.30(X) - 5.40 $R^2 = 0.60$

Gasoline/Service Station
(944)

Average Vehicle Trip Ends vs: Vehicle Fueling Positions
On a: Weekday,
P.M. Peak Hour of Generator

Number of Studies: 14
Average Vehicle Fueling Positions: 8
Directional Distribution: 50% entering, 50% exiting

Trip Generation per Vehicle Fueling Position

Average Rate	Range of Rates	Standard Deviation
15.65	6.83 - 29.33	6.62

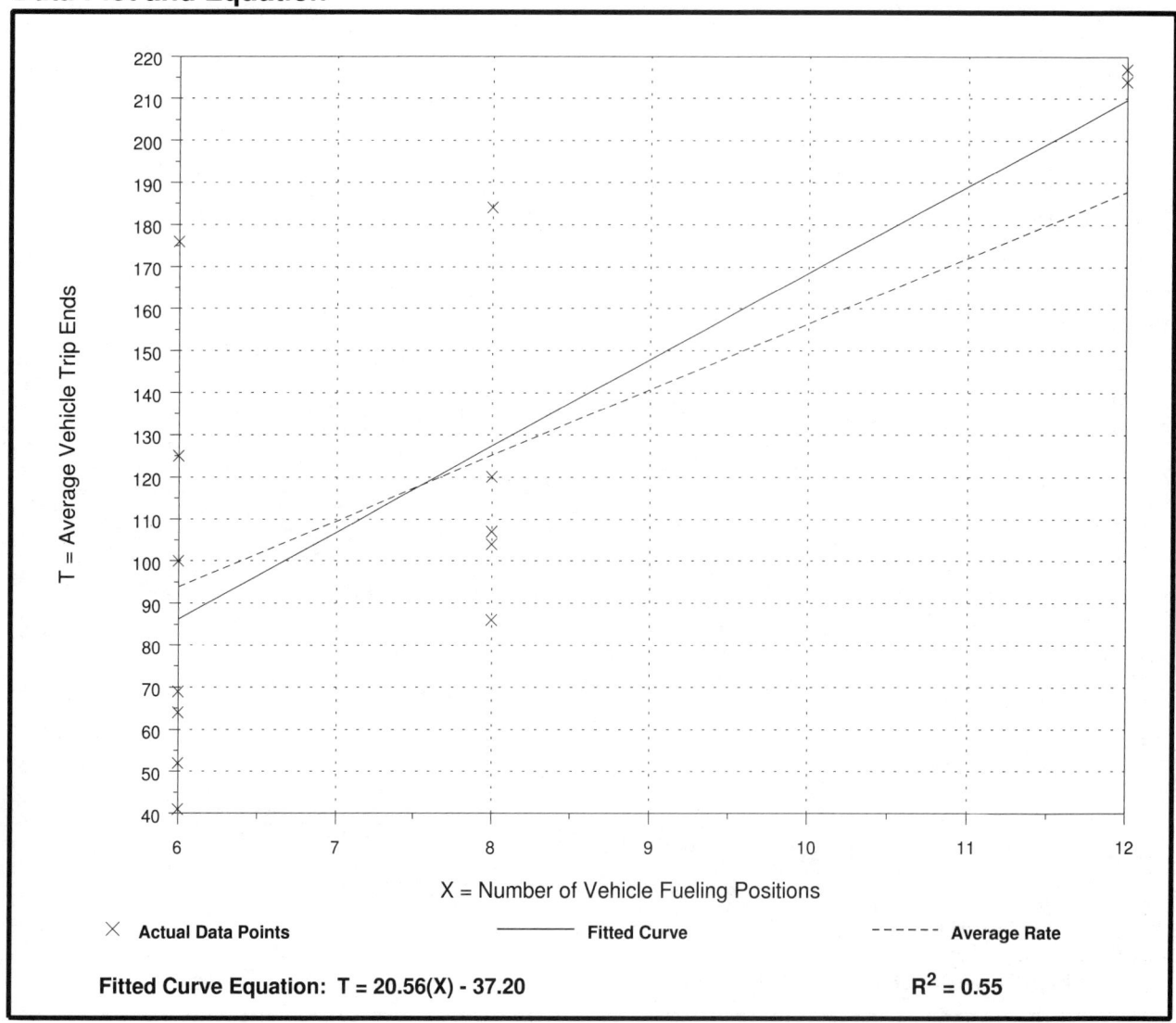

Data Plot and Equation

Fitted Curve Equation: $T = 20.56(X) - 37.20$ $\qquad R^2 = 0.55$

Gasoline/Service Station
(944)

Average Vehicle Trip Ends vs: A.M. Peak Hour Traffic on Adjacent Street
On a: Weekday,
Peak Hour of Adjacent Street Traffic,
One Hour Between 7 and 9 a.m.

Number of Studies: 6
Avg. A.M. Peak Hr. Traf. on Adj. Street: 1,941
Directional Distribution: 50% entering, 50% exiting

Trip Generation per AM Peak Hour Traffic on Adjacent Street

Average Rate	Range of Rates	Standard Deviation
0.04	0.03 - 0.07	0.21

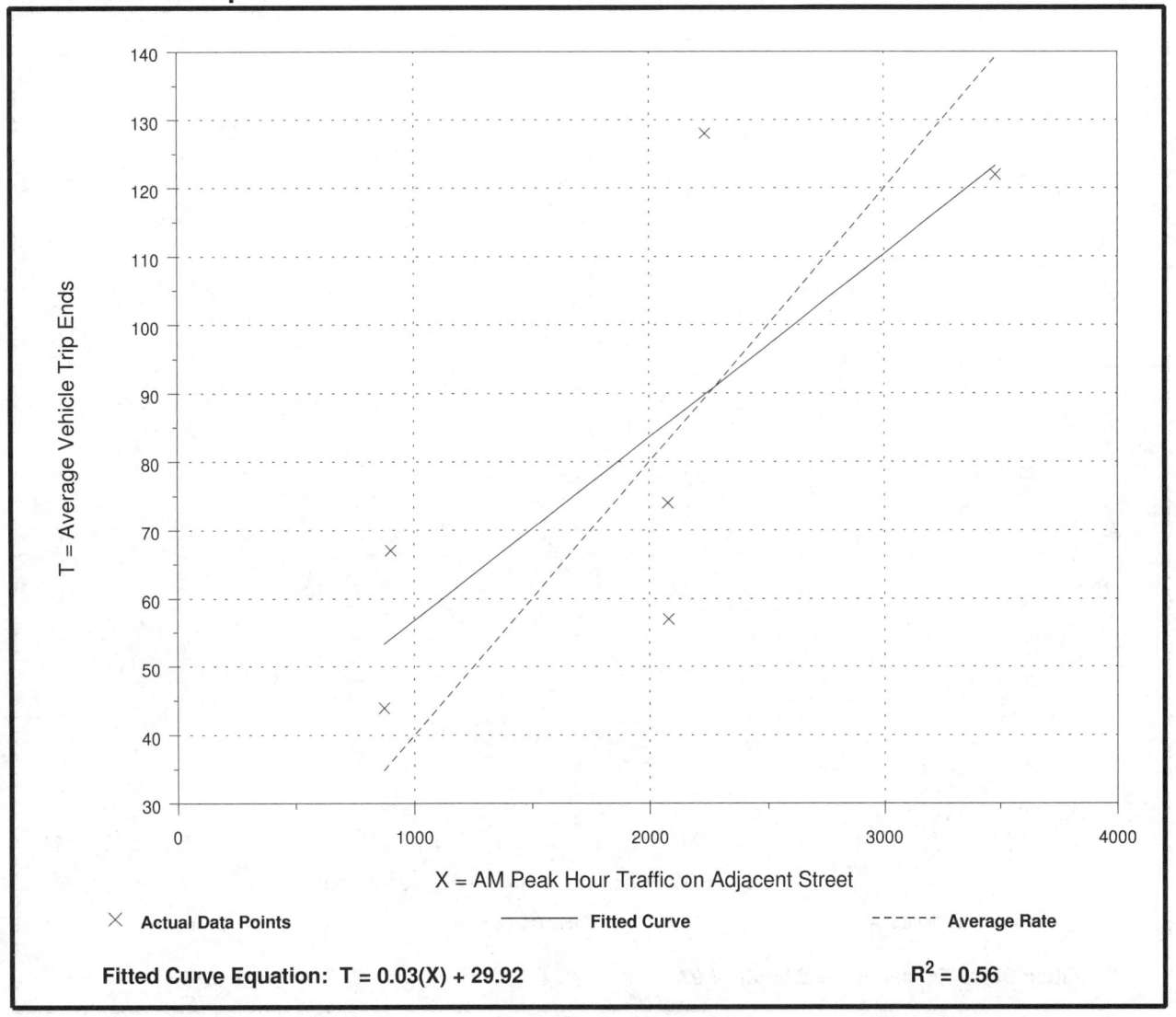

Data Plot and Equation

Fitted Curve Equation: T = 0.03(X) + 29.92 $R^2 = 0.56$

Gasoline/Service Station
(944)

Average Vehicle Trip Ends vs: P.M. Peak Hour Traffic on Adjacent Street
On a: Weekday,
Peak Hour of Adjacent Street Traffic,
One Hour Between 4 and 6 p.m.

Number of Studies: 6
Avg. P.M. Peak Hr. Traf. on Adj. Street: 2,302
Directional Distribution: 50% entering, 50% exiting

Trip Generation per P.M. Peak Hour Traffic on Adjacent Street

Average Rate	Range of Rates	Standard Deviation
0.05	0.03 - 0.06	0.21

Data Plot and Equation

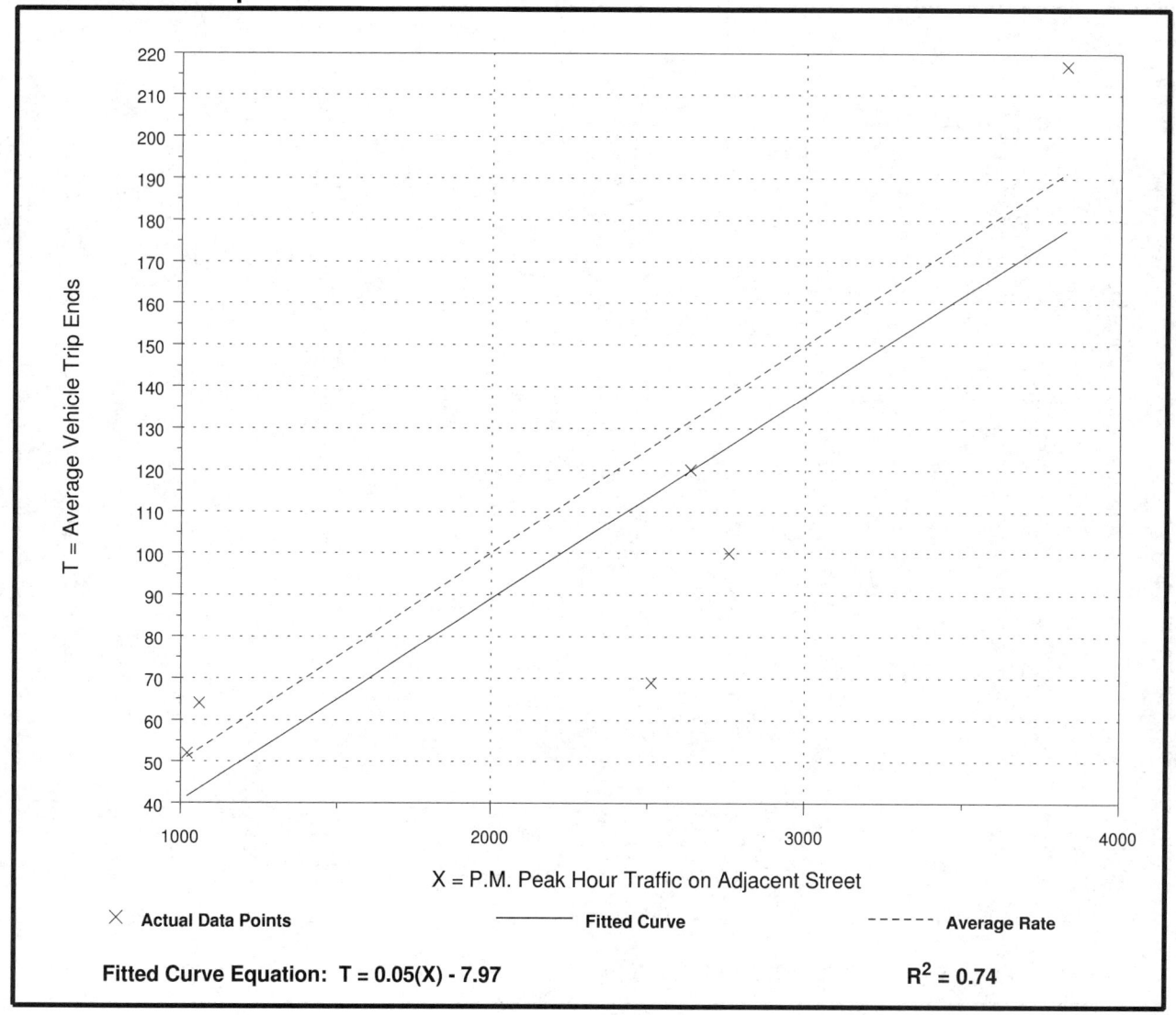

Fitted Curve Equation: $T = 0.05(X) - 7.97$ $R^2 = 0.74$

Land Use: 945
Gasoline/Service Station with Convenience Market

Description

This land use includes gasoline/service stations with convenience markets where the primary business is the fueling of motor vehicles. These service stations may also have ancillary facilities for servicing and repairing motor vehicles. Some commonly sold convenience items are newspapers, coffee or other beverages and snack items that are usually consumed in the car. These service stations are generally located at intersections or interchanges. This land use does not include stations with car washes. Convenience market (open 24 hours) (Land Use 851), convenience market (open 15-16 hours) (Land Use 852), convenience market with gasoline pumps (Land Use 853), gasoline/service station (Land Use 944), gasoline/service station with convenience market and car wash (Land Use 946) and truck stop (Land Use 950) are related uses.

Additional Data

For the purpose of this land use, the independent variable, vehicle fueling positions, is defined as the maximum number of vehicles that can be fueled simultaneously.

Gasoline/service stations in this land use include "pay-at-the-pump" and traditional fueling stations.

The weekday peak hours of the generator typically coincided with the peak hours of the adjacent street traffic.

The sites were surveyed between the late 1980s and the 2000s throughout the United States, with many conducted in New England.

Source Numbers

221, 255, 288, 347, 350, 351, 355, 440, 631, 718

Gasoline/Service Station with Convenience Market
(945)

Average Vehicle Trip Ends vs: Vehicle Fueling Positions
On a: Weekday

Number of Studies: 11
Average Vehicle Fueling Positions: 12
Directional Distribution: 50% entering, 50% exiting

Trip Generation per Vehicle Fueling Position

Average Rate	Range of Rates	Standard Deviation
162.78	90.67 - 299.50	68.16

Data Plot and Equation

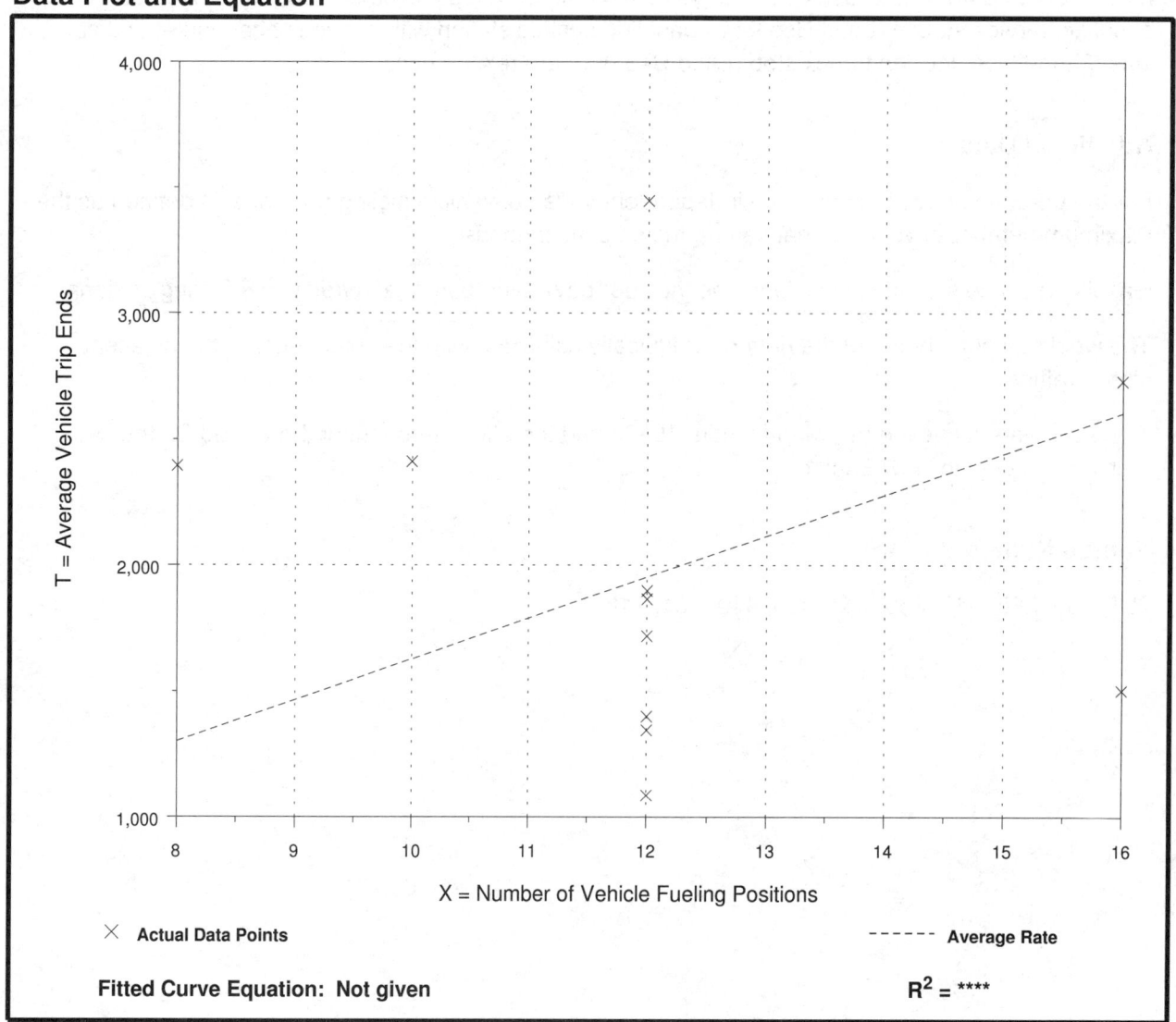

Fitted Curve Equation: Not given $R^2 = ****$

1992 *Trip Generation*, 9th Edition • Institute of Transportation Engineers

Gasoline/Service Station with Convenience Market
(945)

Average Vehicle Trip Ends vs: Vehicle Fueling Positions
On a: Weekday,
Peak Hour of Adjacent Street Traffic,
One Hour Between 7 and 9 a.m.

Number of Studies: 36
Average Vehicle Fueling Positions: 11
Directional Distribution: 50% entering, 50% exiting

Trip Generation per Vehicle Fueling Position

Average Rate	Range of Rates	Standard Deviation
10.16	3.50 - 33.40	6.01

Data Plot and Equation

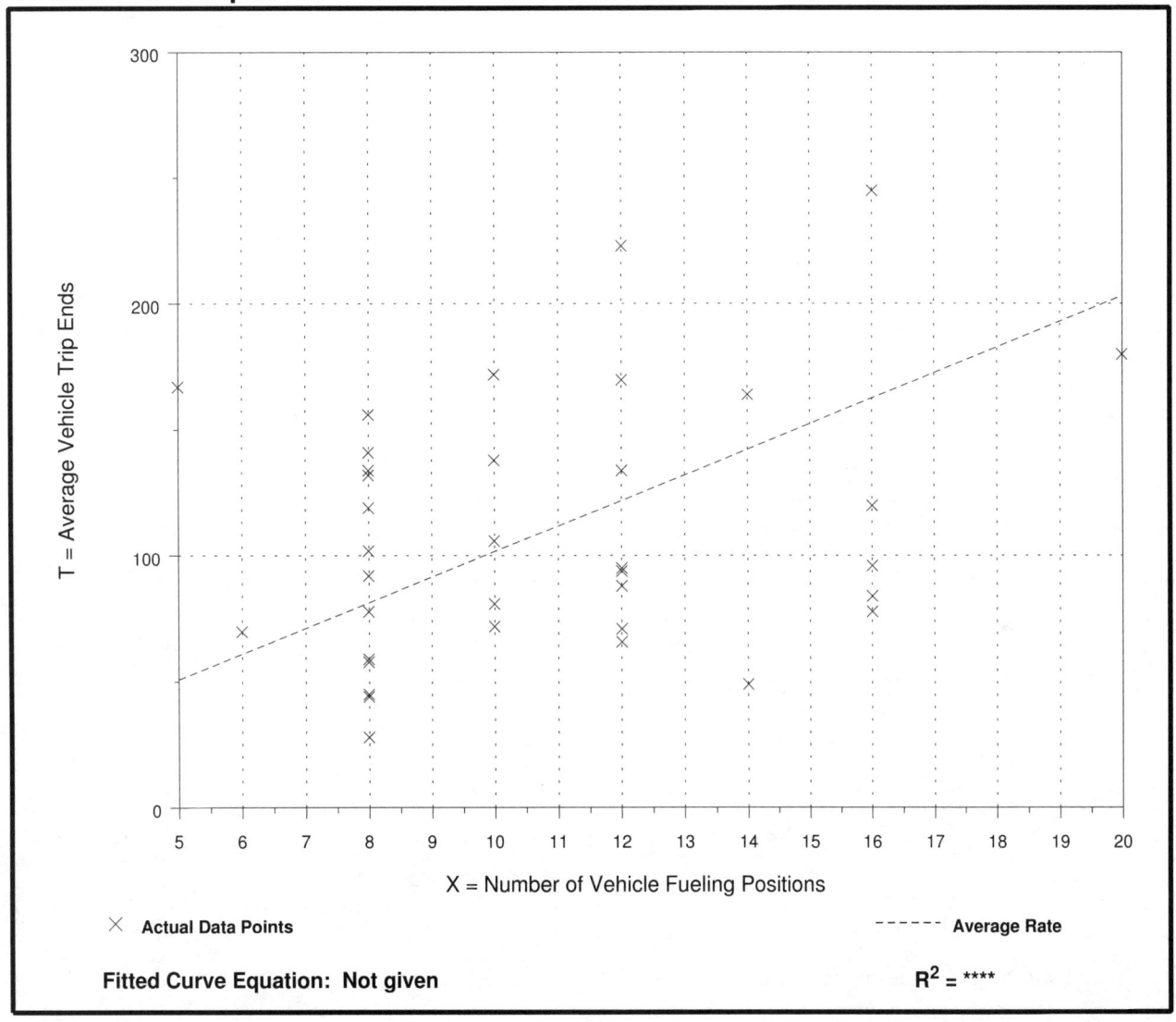

Fitted Curve Equation: Not given $R^2 = ****$

Gasoline/Service Station with Convenience Market
(945)

Average Vehicle Trip Ends vs: Vehicle Fueling Positions
On a: Weekday,
Peak Hour of Adjacent Street Traffic,
One Hour Between 4 and 6 p.m.

Number of Studies: 39
Average Vehicle Fueling Positions: 10
Directional Distribution: 50% entering, 50% exiting

Trip Generation per Vehicle Fueling Position

Average Rate	Range of Rates	Standard Deviation
13.51	4.25 - 57.80	7.91

Data Plot and Equation

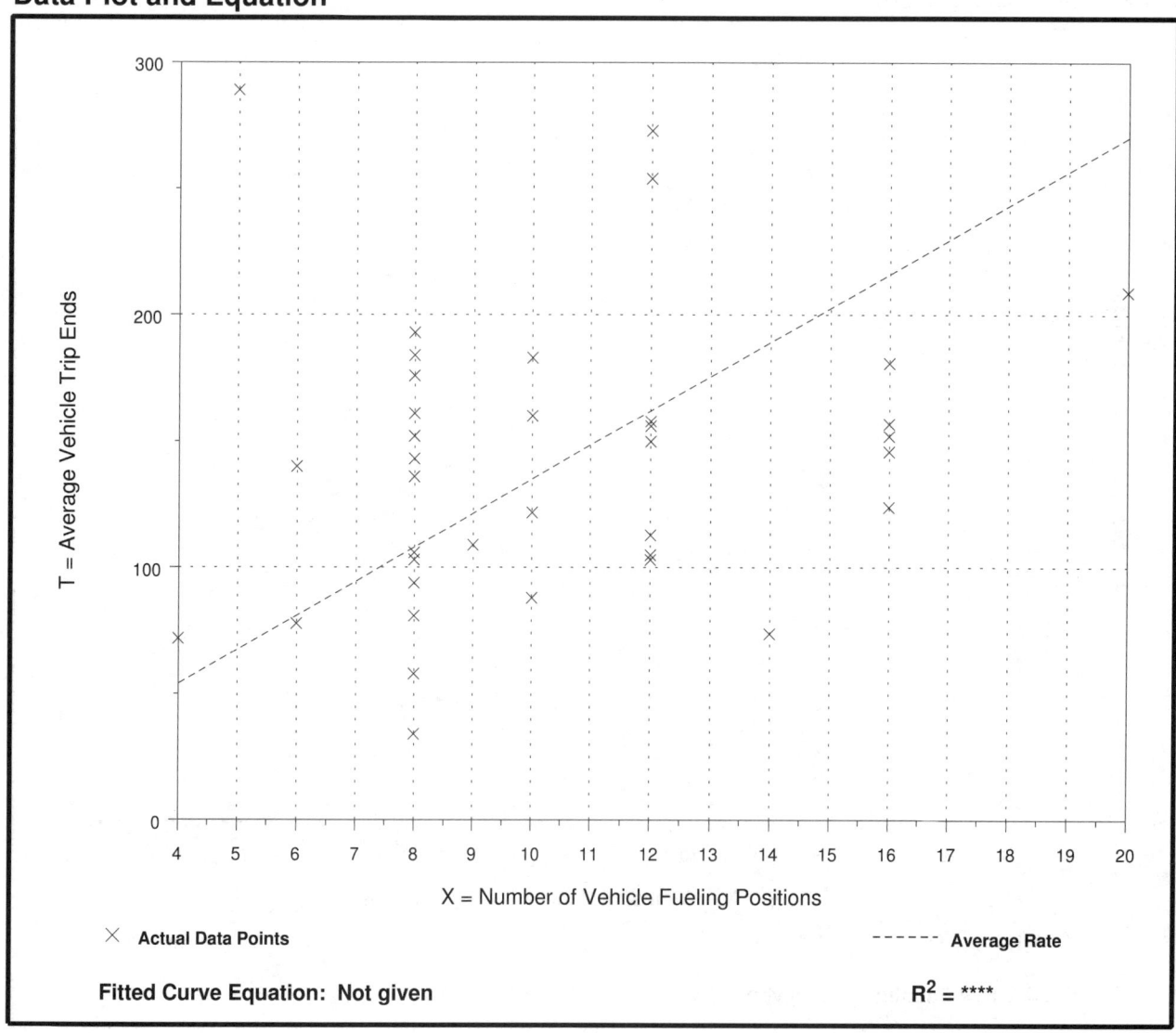

Fitted Curve Equation: Not given $R^2 = ****$

1994   *Trip Generation*, 9th Edition • Institute of Transportation Engineers

Gasoline/Service Station with Convenience Market
(945)

Average Vehicle Trip Ends vs: Vehicle Fueling Positions
On a: Weekday,
A.M. Peak Hour of Generator

Number of Studies: 35
Average Vehicle Fueling Positions: 11
Directional Distribution: 50% entering, 50% exiting

Trip Generation per Vehicle Fueling Position

Average Rate	Range of Rates	Standard Deviation
10.56	3.50 - 33.40	6.15

Data Plot and Equation

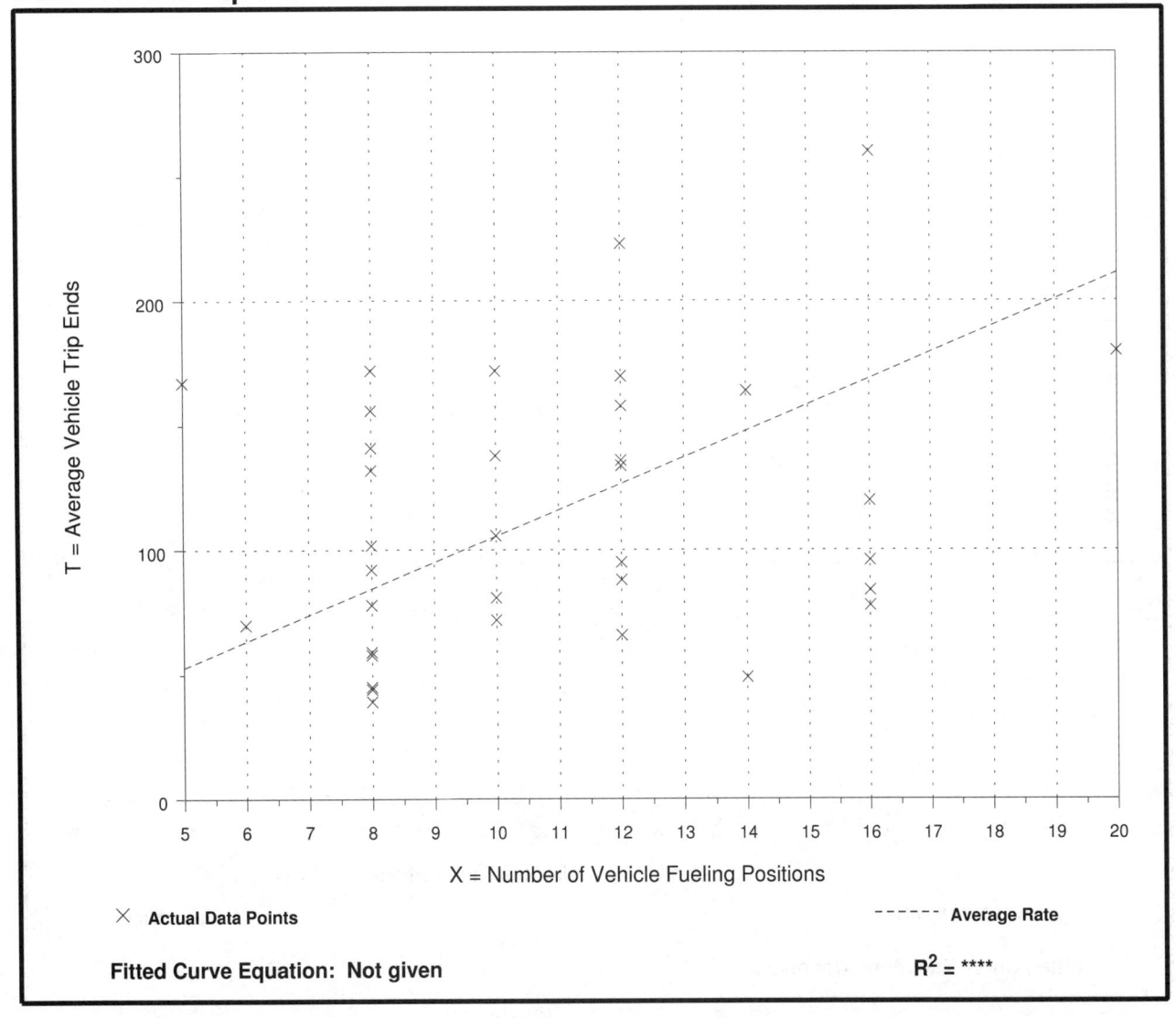

Fitted Curve Equation: Not given $R^2 = ****$

Gasoline/Service Station with Convenience Market
(945)

Average Vehicle Trip Ends vs: Vehicle Fueling Positions
On a: Weekday,
P.M. Peak Hour of Generator

Number of Studies: 37
Average Vehicle Fueling Positions: 10
Directional Distribution: 50% entering, 50% exiting

Trip Generation per Vehicle Fueling Position

Average Rate	Range of Rates	Standard Deviation
13.57	4.25 - 57.80	7.94

Data Plot and Equation

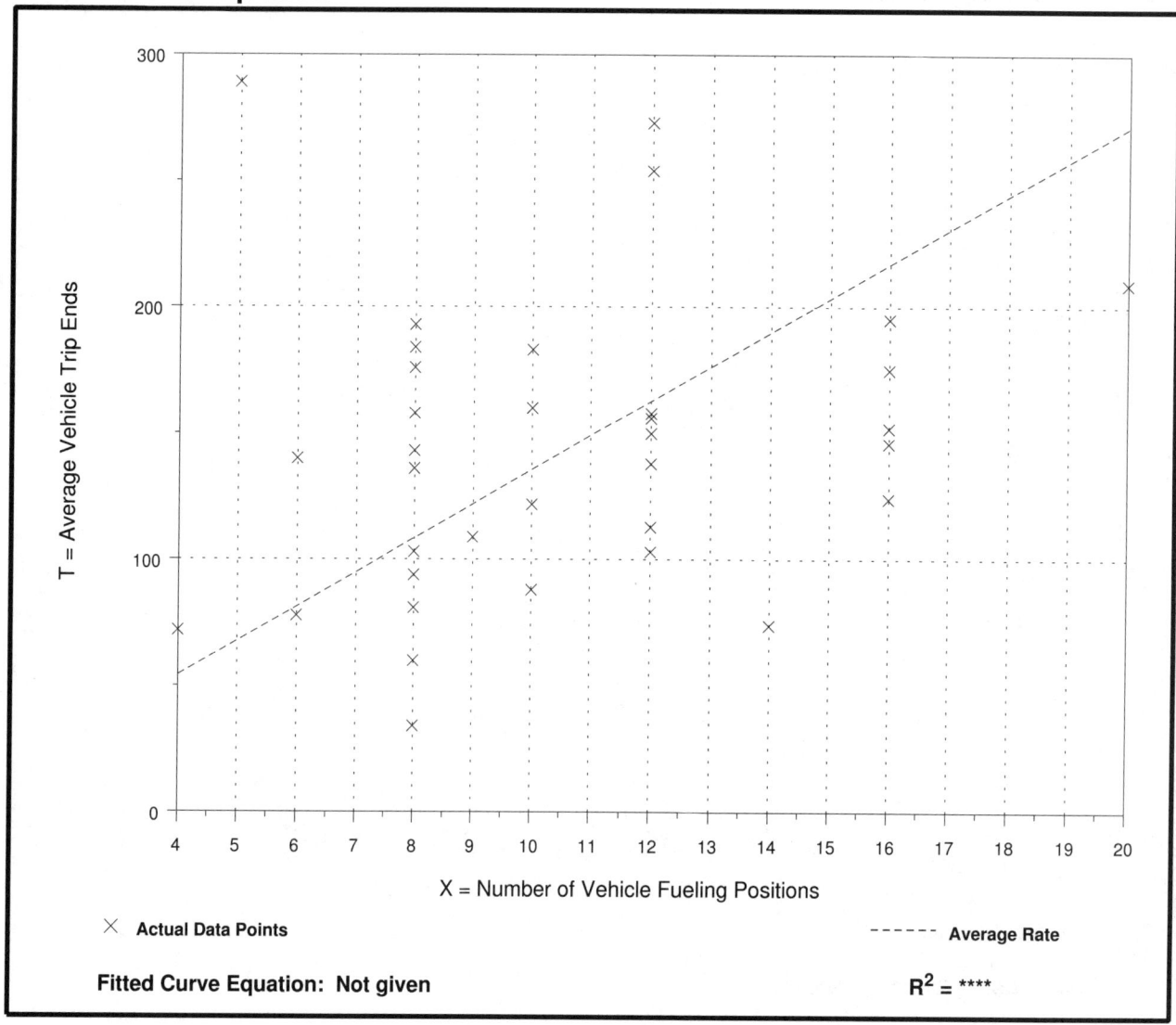

X = Number of Vehicle Fueling Positions
T = Average Vehicle Trip Ends

× Actual Data Points ----- Average Rate

Fitted Curve Equation: Not given $R^2 =$ ****

Gasoline/Service Station with Convenience Market
(945)

Average Vehicle Trip Ends vs: 1000 Sq. Feet Gross Floor Area
On a: Weekday,
Peak Hour of Adjacent Street Traffic,
One Hour Between 7 and 9 a.m.

Number of Studies: 30
Average 1000 Sq. Feet GFA: 1
Directional Distribution: 51% entering, 49% exiting

Trip Generation per 1000 Sq. Feet Gross Floor Area

Average Rate	Range of Rates	Standard Deviation
82.13	25.00 - 276.60	59.13

Data Plot and Equation

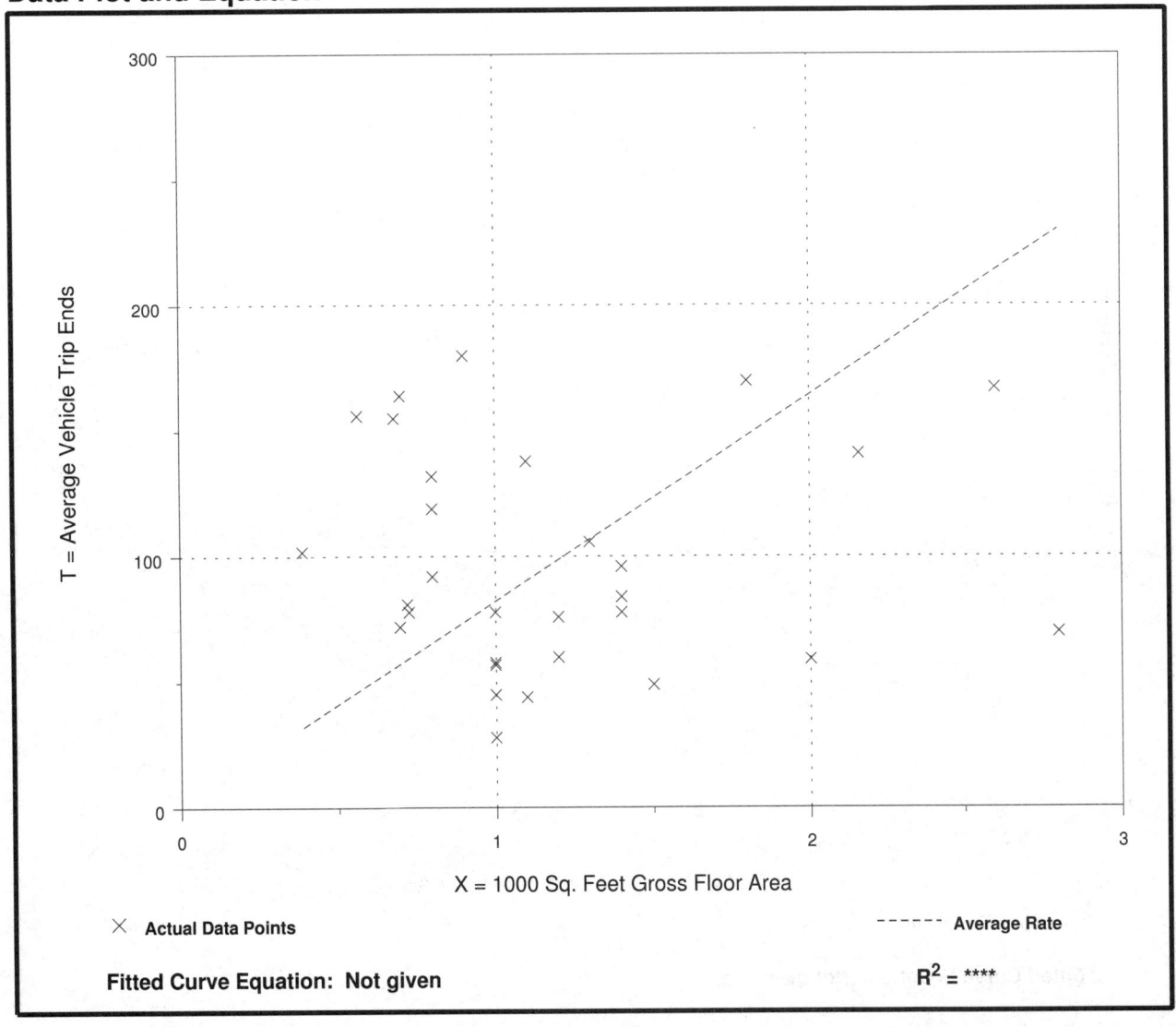

Fitted Curve Equation: Not given $R^2 = ****$

Gasoline/Service Station with Convenience Market
(945)

Average Vehicle Trip Ends vs: 1000 Sq. Feet Gross Floor Area
On a: Weekday,
Peak Hour of Adjacent Street Traffic,
One Hour Between 4 and 6 p.m.

Number of Studies: 34
Average 1000 Sq. Feet GFA: 1
Directional Distribution: 50% entering, 50% exiting

Trip Generation per 1000 Sq. Feet Gross Floor Area

Average Rate	Range of Rates	Standard Deviation
97.47	27.86 - 451.28	65.13

Data Plot and Equation

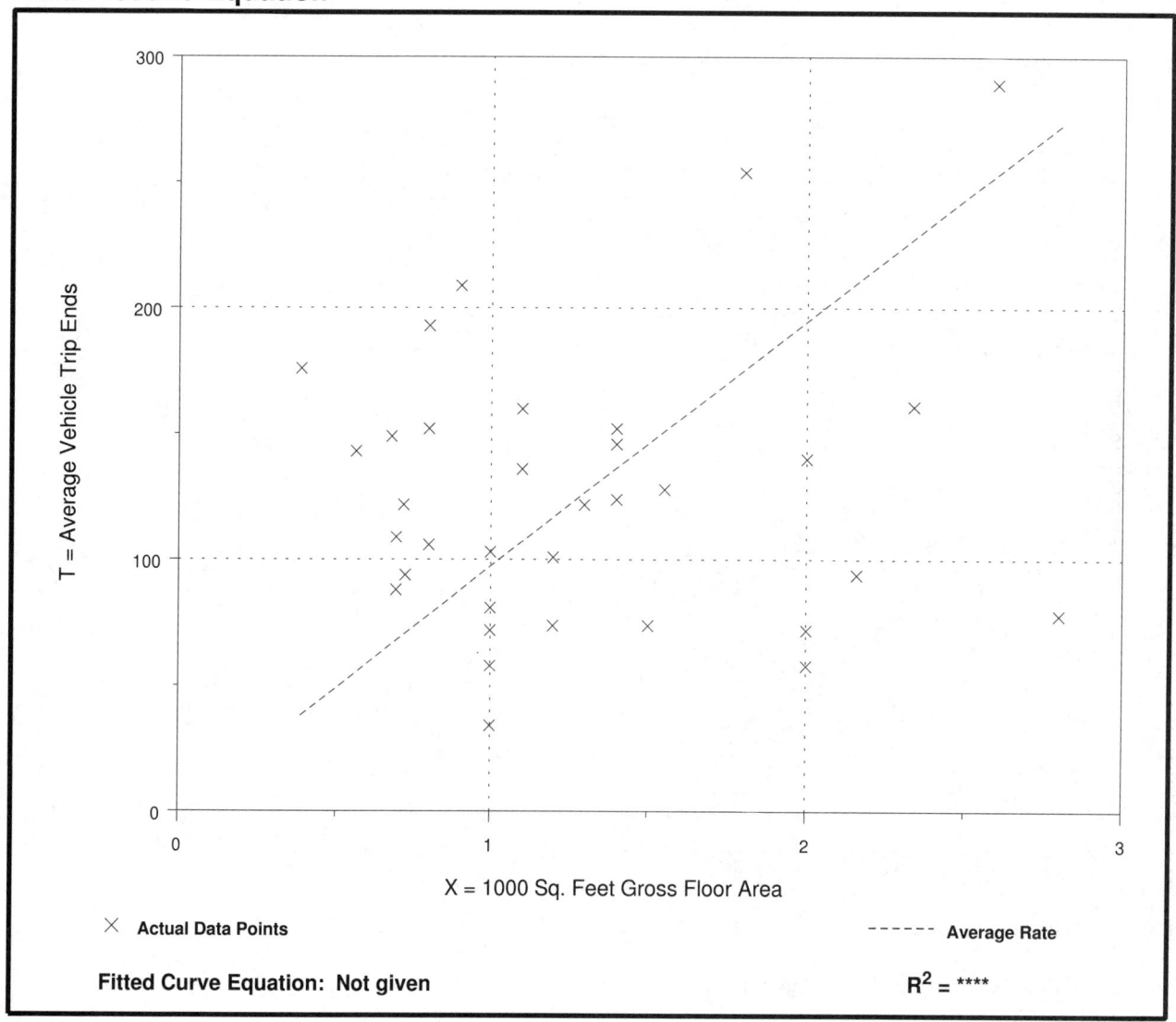

Fitted Curve Equation: Not given $R^2 = ****$

Gasoline/Service Station with Convenience Market
(945)

Average Vehicle Trip Ends vs: 1000 Sq. Feet Gross Floor Area
On a: Weekday,
A.M. Peak Hour of Generator

Number of Studies: 28
Average 1000 Sq. Feet GFA: 1
Directional Distribution: 51% entering, 49% exiting

Trip Generation per 1000 Sq. Feet Gross Floor Area

Average Rate	Range of Rates	Standard Deviation
78.06	25.00 - 276.60	55.34

Data Plot and Equation

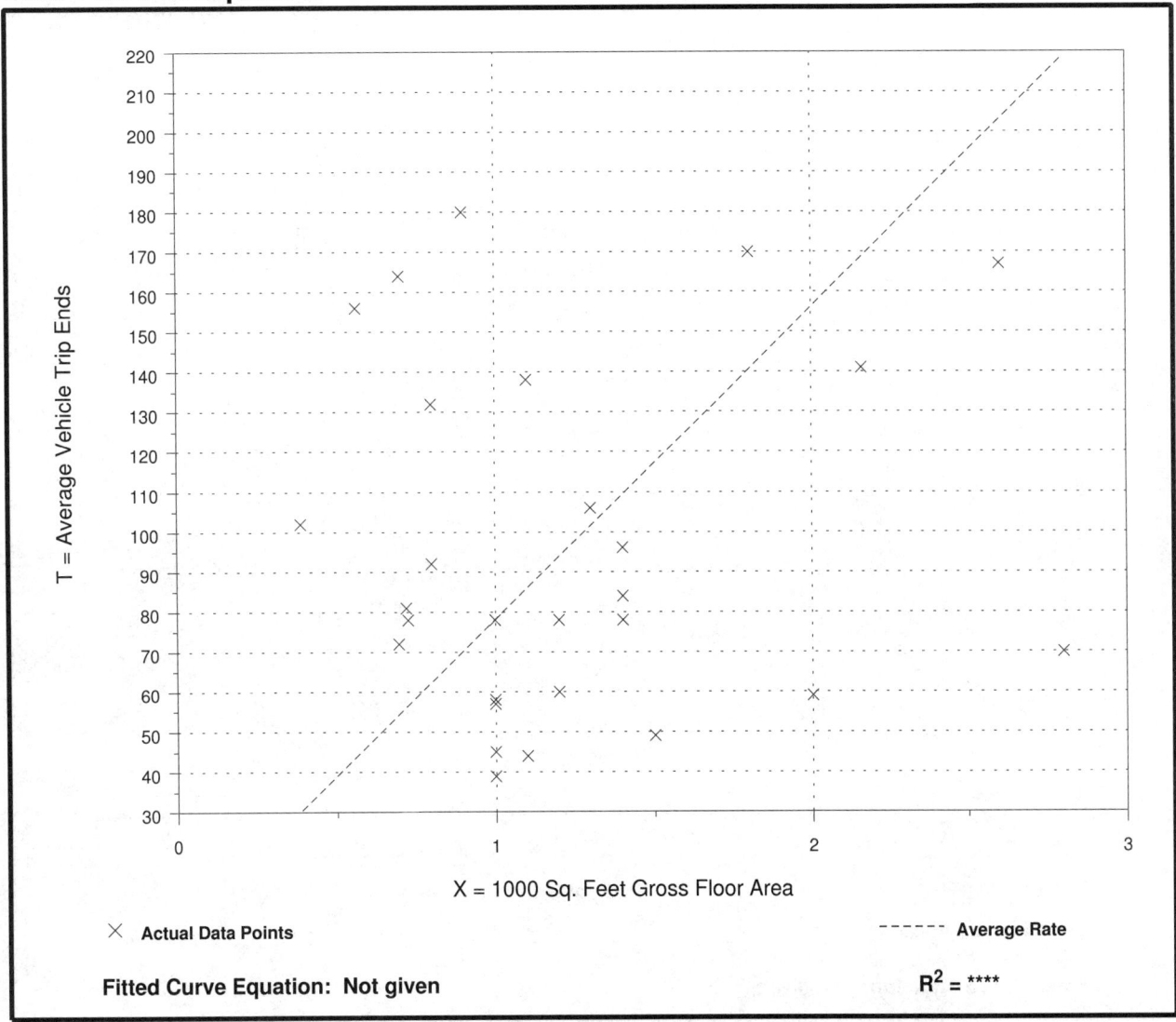

X = 1000 Sq. Feet Gross Floor Area
T = Average Vehicle Trip Ends

× Actual Data Points
----- Average Rate

Fitted Curve Equation: Not given $R^2 = ****$

Trip Generation, 9th Edition ● Institute of Transportation Engineers   1999

Gasoline/Service Station with Convenience Market
(945)

Average Vehicle Trip Ends vs: 1000 Sq. Feet Gross Floor Area
On a: Weekday,
P.M. Peak Hour of Generator

Number of Studies: 30
Average 1000 Sq. Feet GFA: 1
Directional Distribution: 50% entering, 50% exiting

Trip Generation per 1000 Sq. Feet Gross Floor Area

Average Rate	Range of Rates	Standard Deviation
97.14	27.86 - 451.28	67.23

Data Plot and Equation

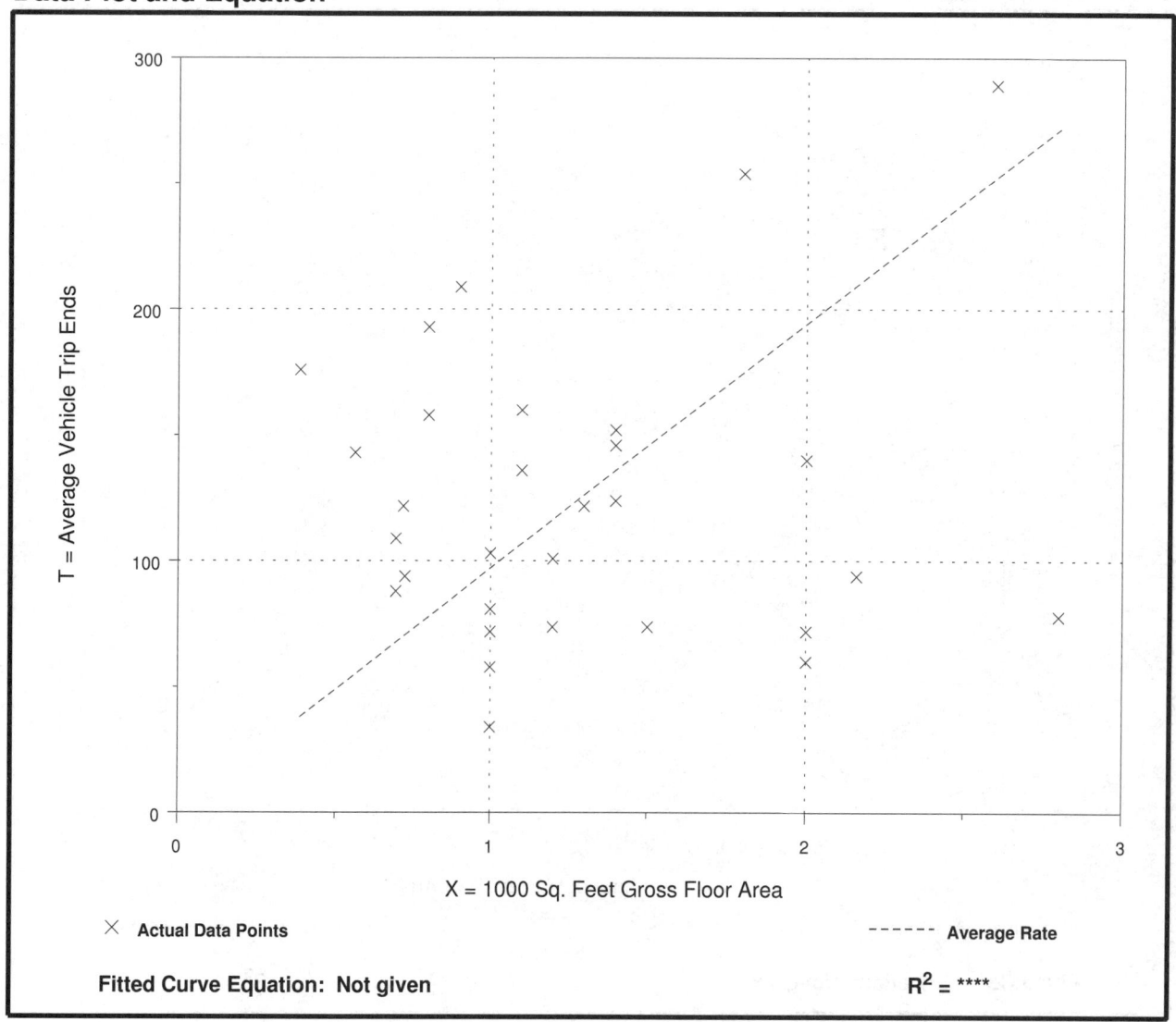

Fitted Curve Equation: Not given $R^2 = ****$

Gasoline/Service Station with Convenience Market
(945)

Average Vehicle Trip Ends vs: A.M. Peak Hour Traffic on Adjacent Street
On a: Weekday,
Peak Hour of Adjacent Street Traffic,
One Hour Between 7 and 9 a.m.

Number of Studies: 8
Avg. A.M. Peak Hr. Traf. on Adj. Street: 2,649
Directional Distribution: 52% entering, 48% exiting

Trip Generation per AM Peak Hour Traffic on Adjacent Street

Average Rate	Range of Rates	Standard Deviation
0.04	0.01 - 0.12	0.20

Data Plot and Equation

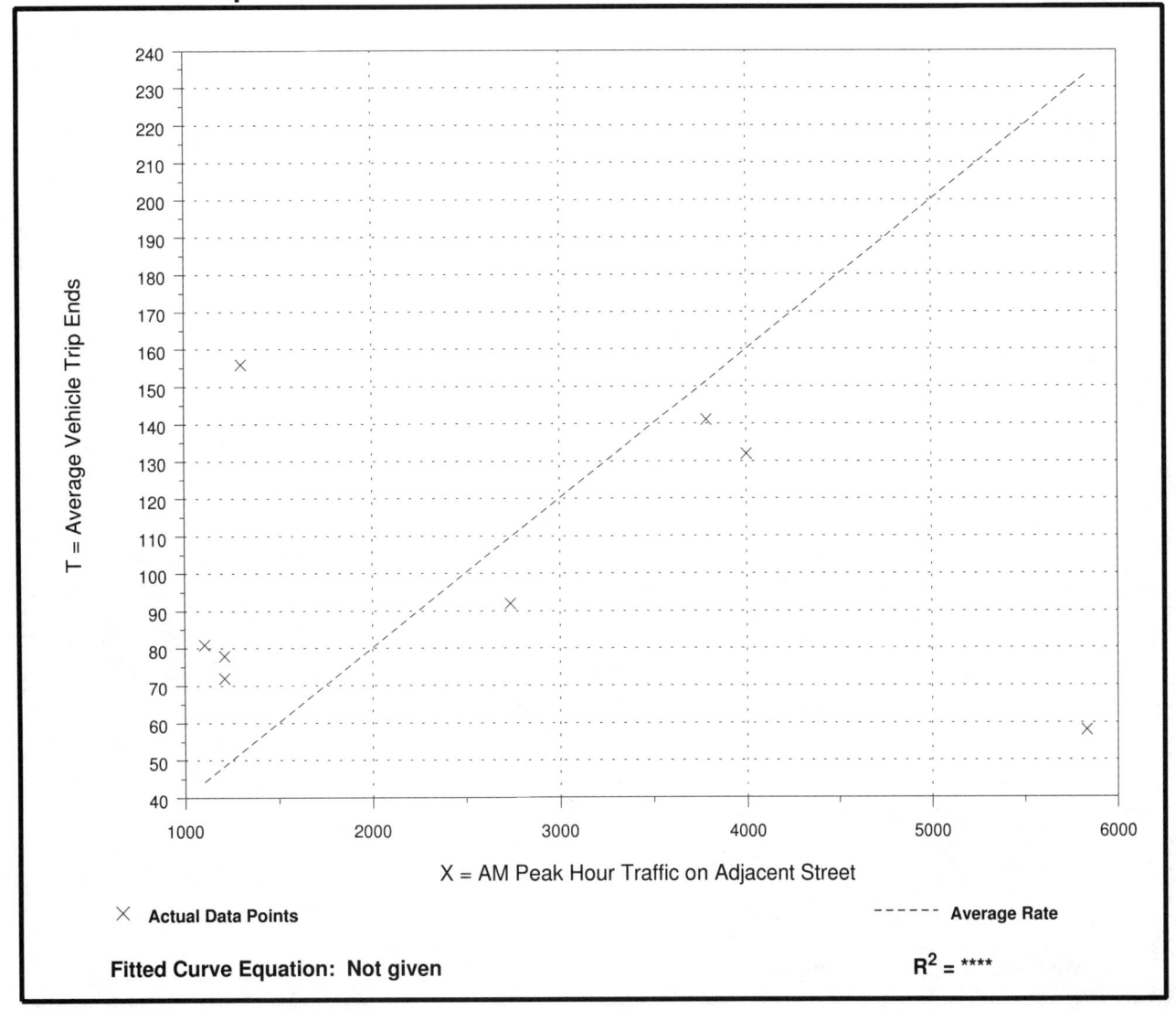

Fitted Curve Equation: Not given $R^2 = ****$

Trip Generation, 9th Edition • Institute of Transportation Engineers

Gasoline/Service Station with Convenience Market
(945)

Average Vehicle Trip Ends vs: P.M. Peak Hour Traffic on Adjacent Street
On a: Weekday,
Peak Hour of Adjacent Street Traffic,
One Hour Between 4 and 6 p.m.

Number of Studies: 9
Avg. P.M. Peak Hr. Traf. on Adj. Street: 3,257
Directional Distribution: 51% entering, 49% exiting

Trip Generation per P.M. Peak Hour Traffic on Adjacent Street

Average Rate	Range of Rates	Standard Deviation
0.04	0.01 - 0.10	0.20

Data Plot and Equation

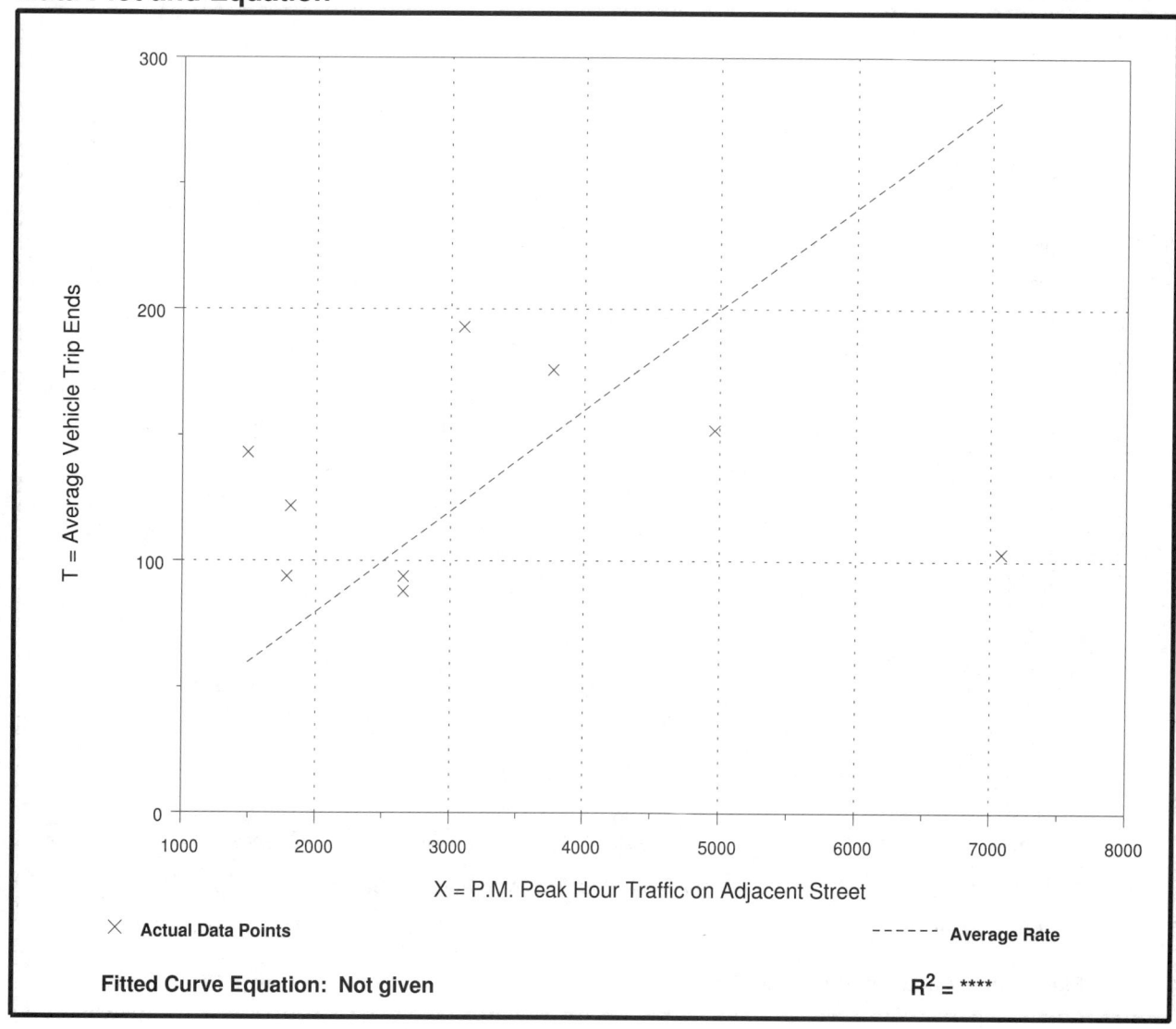

Fitted Curve Equation: Not given $R^2 = ****$

Land Use: 946
Gasoline/Service Station with Convenience Market and Car Wash

Description

This land use includes gasoline/service stations with convenience markets and car washes where the primary business is the fueling of motor vehicles. They may also have ancillary facilities for servicing and repairing motor vehicles. These service stations are generally located at intersections or interchanges. Convenience market (open 24 hours) (Land Use 851), convenience market (open 15-16 hours) (Land Use 852), convenience market with gasoline pumps (Land Use 853), gasoline/service station (Land Use 944) and gasoline/service station with convenience market (Land Use 945) are related uses.

Additional Data

For the purpose of this land use, the independent variable, vehicle fueling positions, is defined as the maximum number of vehicles that can be fueled simultaneously.

Gasoline/service stations in this land use include "pay-at-the-pump" and traditional fueling stations.

The sites were surveyed between the late 1980s and the 2000s throughout the United States.

Source Numbers

334, 340, 347, 348, 355, 385, 440, 445, 540, 580, 586, 617

Gasoline/Service Station with Convenience Market and Car Wash
(946)

Average Vehicle Trip Ends vs: Vehicle Fueling Positions
On a: Weekday

Number of Studies: 10
Average Vehicle Fueling Positions: 12
Directional Distribution: 50% entering, 50% exiting

Trip Generation per Vehicle Fueling Position

Average Rate	Range of Rates	Standard Deviation
152.84	97.20 - 250.88	45.80

Data Plot and Equation

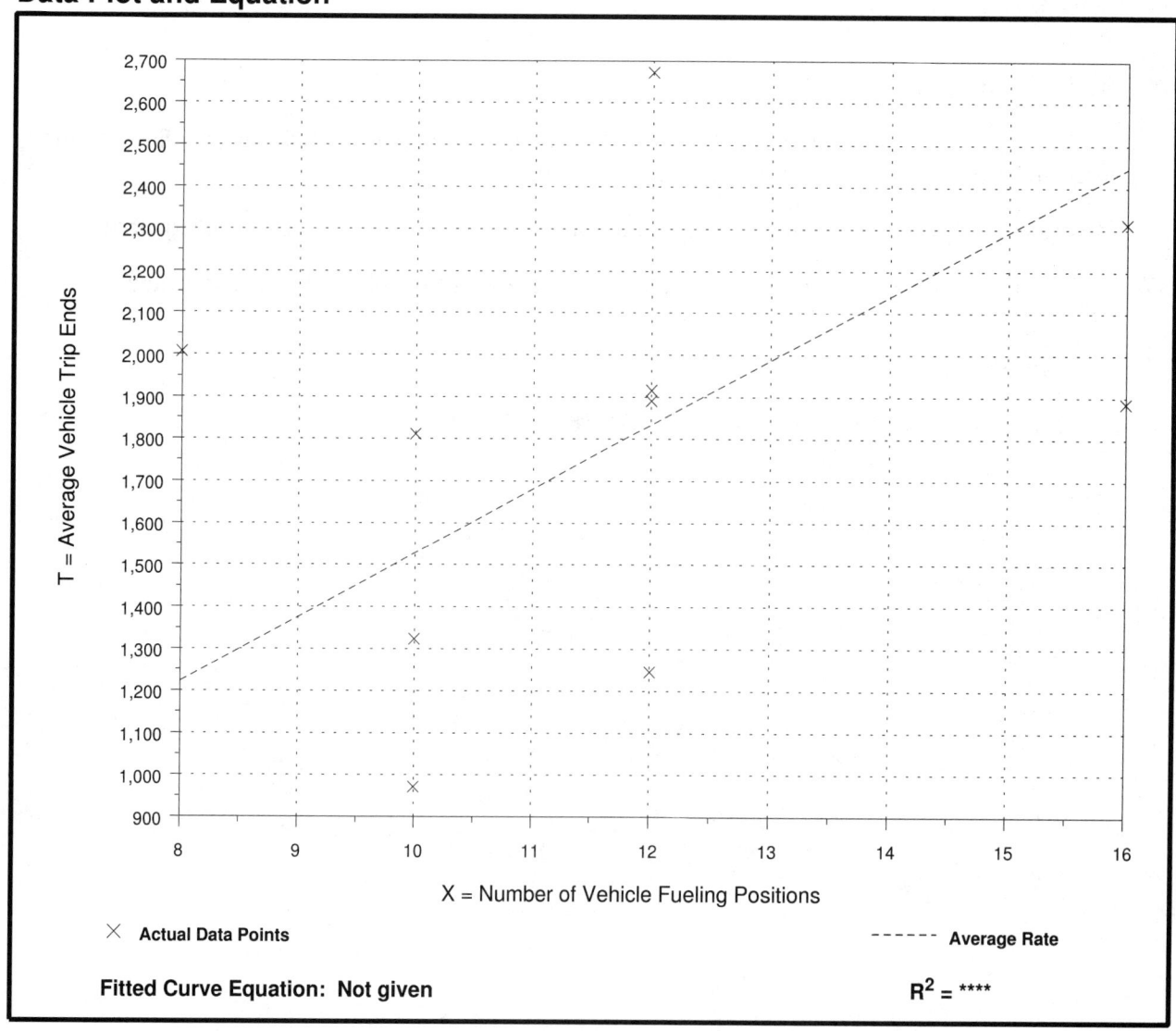

Fitted Curve Equation: Not given $R^2 = ****$

Gasoline/Service Station with Convenience Market and Car Wash
(946)

Average Vehicle Trip Ends vs: Vehicle Fueling Positions
On a: Weekday,
Peak Hour of Adjacent Street Traffic,
One Hour Between 7 and 9 a.m.

Number of Studies: 22
Average Vehicle Fueling Positions: 11
Directional Distribution: 51% entering, 49% exiting

Trip Generation per Vehicle Fueling Position

Average Rate	Range of Rates	Standard Deviation
11.84	4.33 - 29.00	6.95

Data Plot and Equation

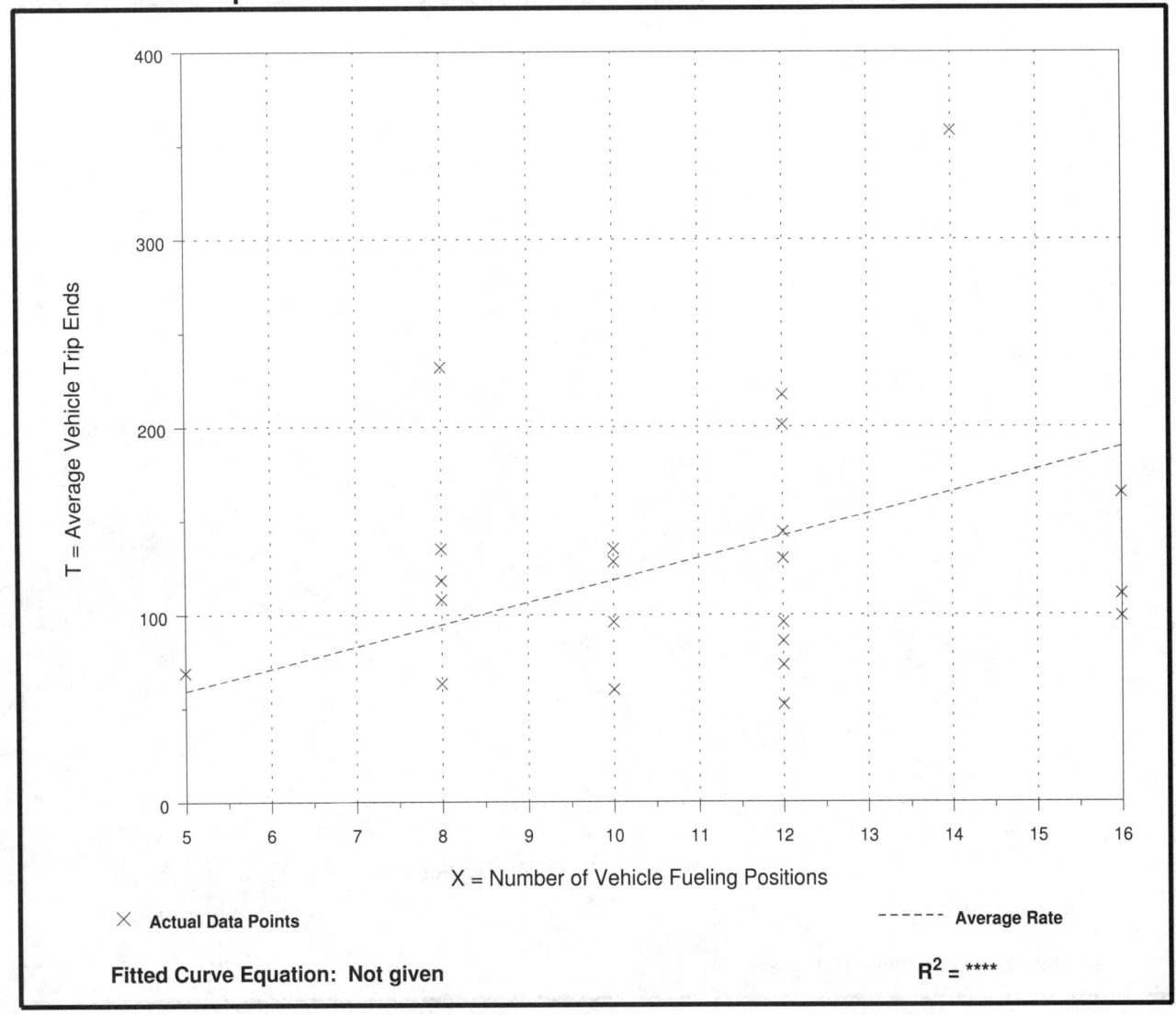

Fitted Curve Equation: Not given $R^2 = ****$

Gasoline/Service Station with Convenience Market and Car Wash
(946)

Average Vehicle Trip Ends vs: Vehicle Fueling Positions
On a: Weekday,
Peak Hour of Adjacent Street Traffic,
One Hour Between 4 and 6 p.m.

Number of Studies: 30
Average Vehicle Fueling Positions: 11
Directional Distribution: 51% entering, 49% exiting

Trip Generation per Vehicle Fueling Position

Average Rate	Range of Rates	Standard Deviation
13.86	7.00 - 29.63	6.31

Data Plot and Equation

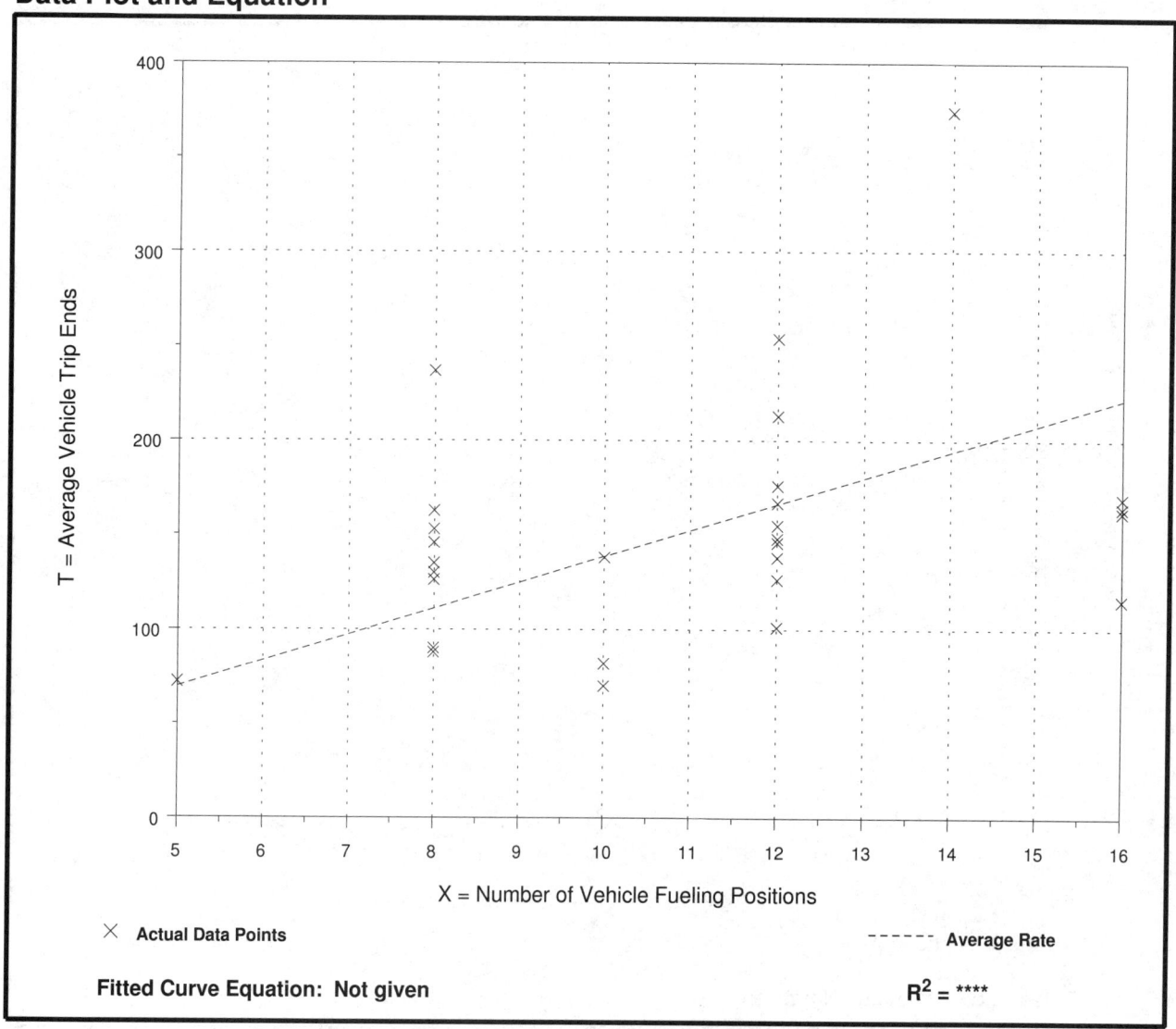

Fitted Curve Equation: Not given $R^2 = ****$

Gasoline/Service Station with Convenience Market and Car Wash
(946)

Average Vehicle Trip Ends vs: Vehicle Fueling Positions
On a: Weekday,
A.M. Peak Hour of Generator

Number of Studies: 20
Average Vehicle Fueling Positions: 11
Directional Distribution: 51% entering, 49% exiting

Trip Generation per Vehicle Fueling Position

Average Rate	Range of Rates	Standard Deviation
13.32	6.08 - 29.00	6.60

Data Plot and Equation

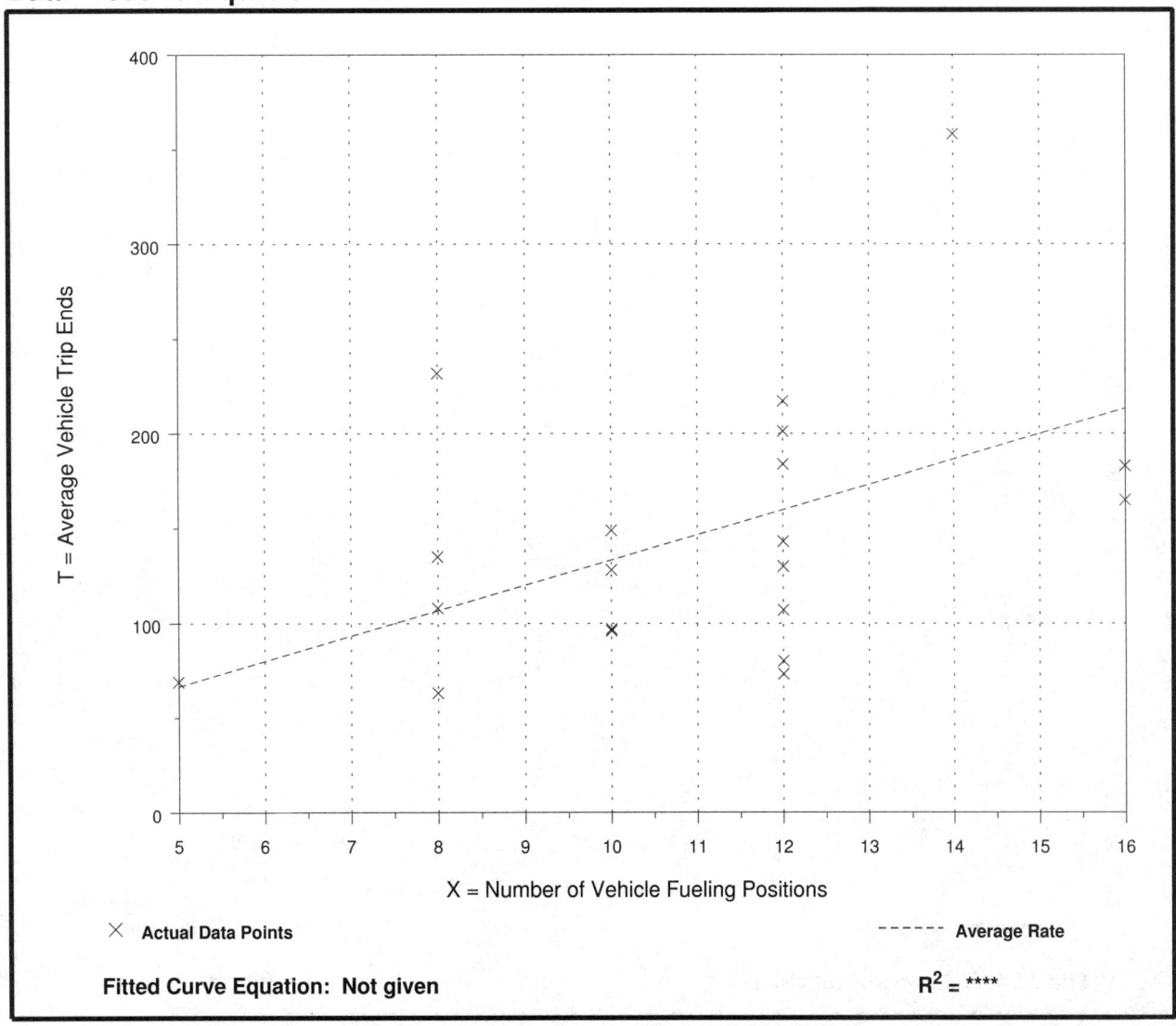

[Scatter plot: X = Number of Vehicle Fueling Positions (5 to 16); T = Average Vehicle Trip Ends (0 to 400). Dashed line shows Average Rate.]

Fitted Curve Equation: Not given $R^2 = ****$

Gasoline/Service Station with Convenience Market and Car Wash
(946)

Average Vehicle Trip Ends vs: Vehicle Fueling Positions
On a: Weekday,
P.M. Peak Hour of Generator

Number of Studies: 22
Average Vehicle Fueling Positions: 11
Directional Distribution: 50% entering, 50% exiting

Trip Generation per Vehicle Fueling Position

Average Rate	Range of Rates	Standard Deviation
14.52	7.00 - 26.71	5.87

Data Plot and Equation

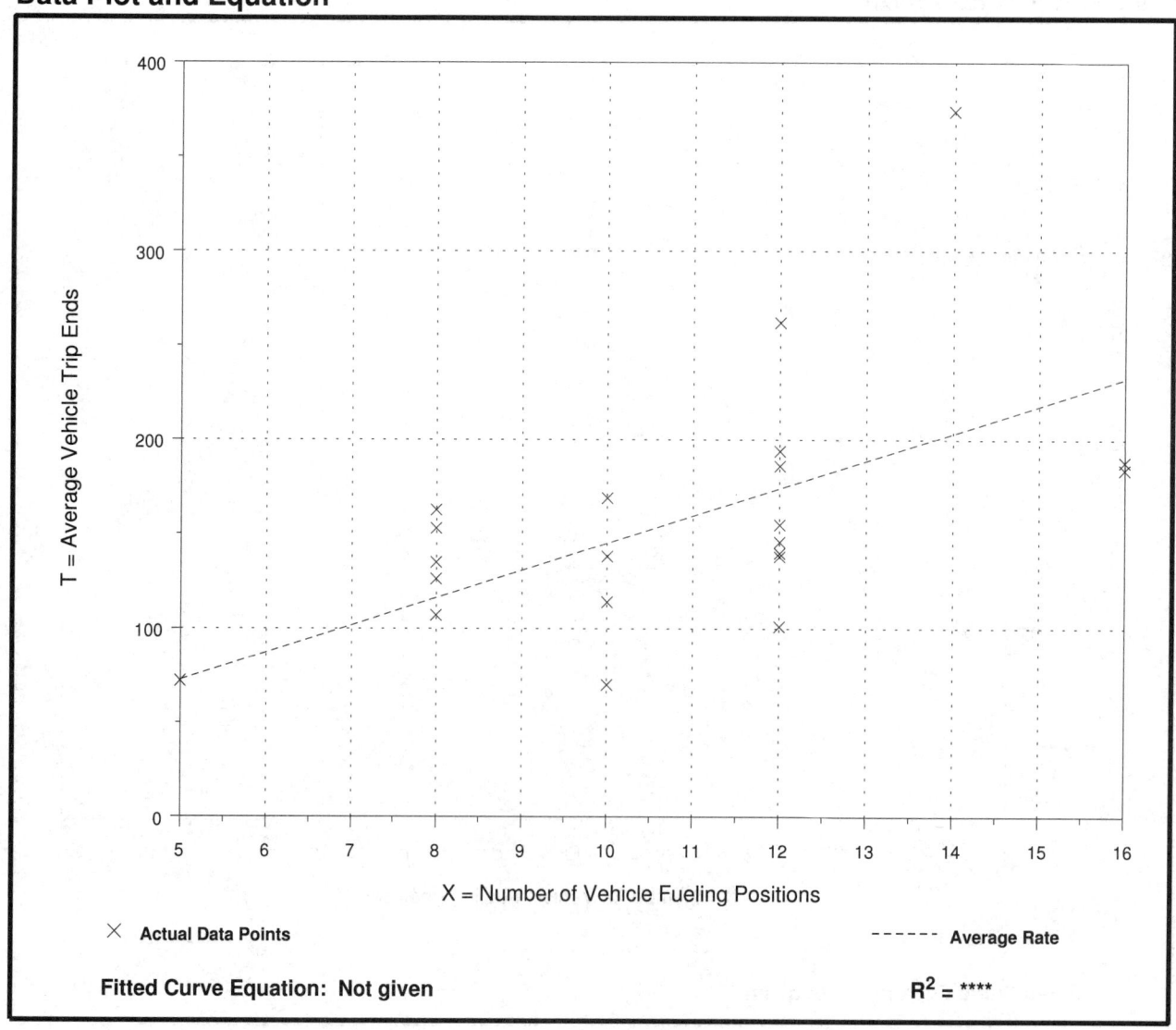

Fitted Curve Equation: Not given $R^2 = ****$

Gasoline/Service Station with Convenience Market and Car Wash
(946)

Average Vehicle Trip Ends vs: Vehicle Fueling Positions
On a: Saturday,
Peak Hour of Generator

Number of Studies: 5
Average Vehicle Fueling Positions: 11
Directional Distribution: 50% entering, 50% exiting

Trip Generation per Vehicle Fueling Position

Average Rate	Range of Rates	Standard Deviation
19.46	10.42 - 32.71	9.38

Data Plot and Equation

Caution - Use Carefully - Small Sample Size

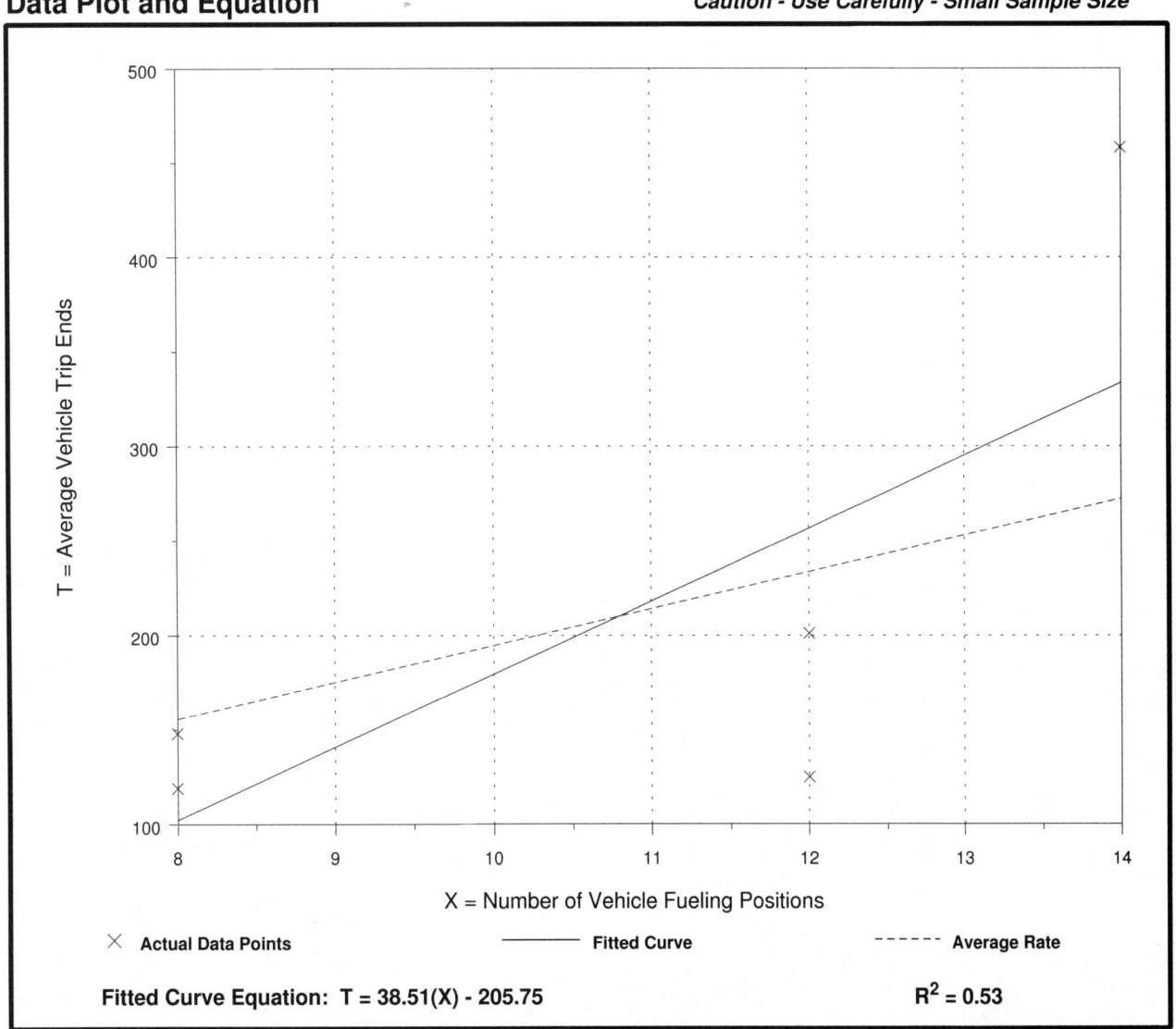

Fitted Curve Equation: T = 38.51(X) - 205.75 $R^2 = 0.53$

Trip Generation, 9th Edition • Institute of Transportation Engineers

Land Use: 947
Self-Service Car Wash

Description

Self-service car washes allow manual cleaning of vehicles by providing stalls to park and wash vehicles. Automated car wash (Land Use 948) is a related use.

Additional Data

The sites were surveyed between the late 1960s and the 2000s in Colorado, New Jersey and South Dakota.

Source Numbers

171, 178, 358, 359, 550

Land Use: 947
Self-Service Car Wash
Independent Variables with One Observation

The following trip generation data are for independent variables with only one observation. This information is shown in this table only; there are no related plots for these data.

Users are cautioned to use data with care because of the small sample size.

Independent Variable	Trip Generation Rate	Size of Independent Variable	Number of Studies	Directional Distribution
Wash Stalls				
Weekday	108.00	5	1	50% entering, 50% exiting
Weekday A.M. Peak Hour of Generator	8.00	5	1	50% entering, 50% exiting
Weekday P.M. Peak Hour of Generator	8.00	5	1	50% entering, 50% exiting
Saturday	132.80	5	1	50% entering, 50% exiting

Self-Service Car Wash
(947)

Average Vehicle Trip Ends vs: Wash Stalls
On a: Weekday,
Peak Hour of Adjacent Street Traffic,
One Hour Between 4 and 6 p.m.

Number of Studies: 6
Average Num. of Wash Stalls: 7
Directional Distribution: 51% entering, 49% exiting

Trip Generation per Wash Stall

Average Rate	Range of Rates	Standard Deviation
5.54	4.00 - 8.00	2.67

Data Plot and Equation

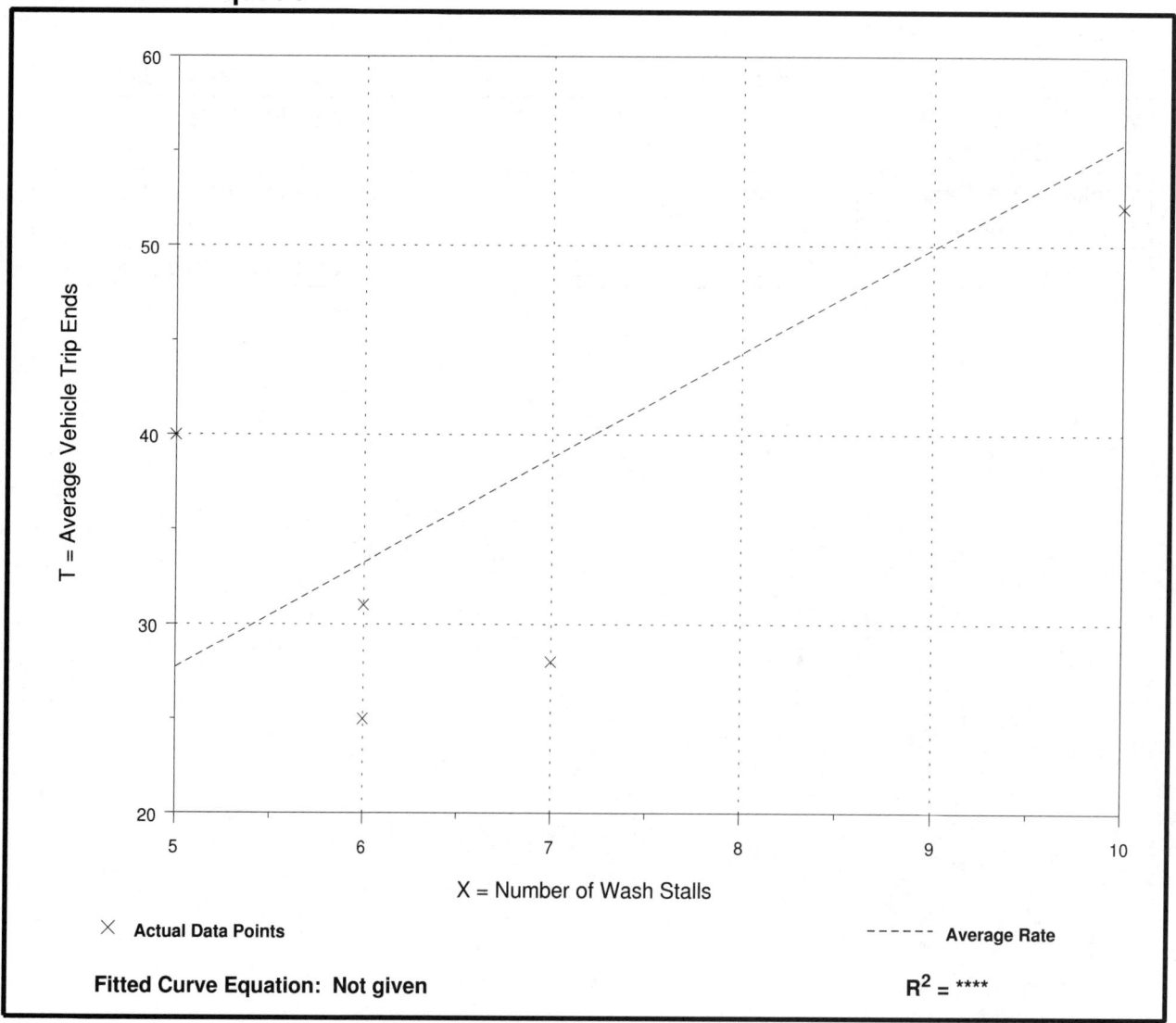

Fitted Curve Equation: Not given $R^2 = ****$

Self-Service Car Wash
(947)

Average Vehicle Trip Ends vs: Wash Stalls
On a: Saturday,
Peak Hour of Generator

Number of Studies: 2
Average Num. of Wash Stalls: 5
Directional Distribution: 50% entering, 50% exiting

Trip Generation per Wash Stall

Average Rate	Range of Rates	Standard Deviation
20.60	11.20 - 30.00	*

Data Plot and Equation

Caution - Use Carefully - Small Sample Size

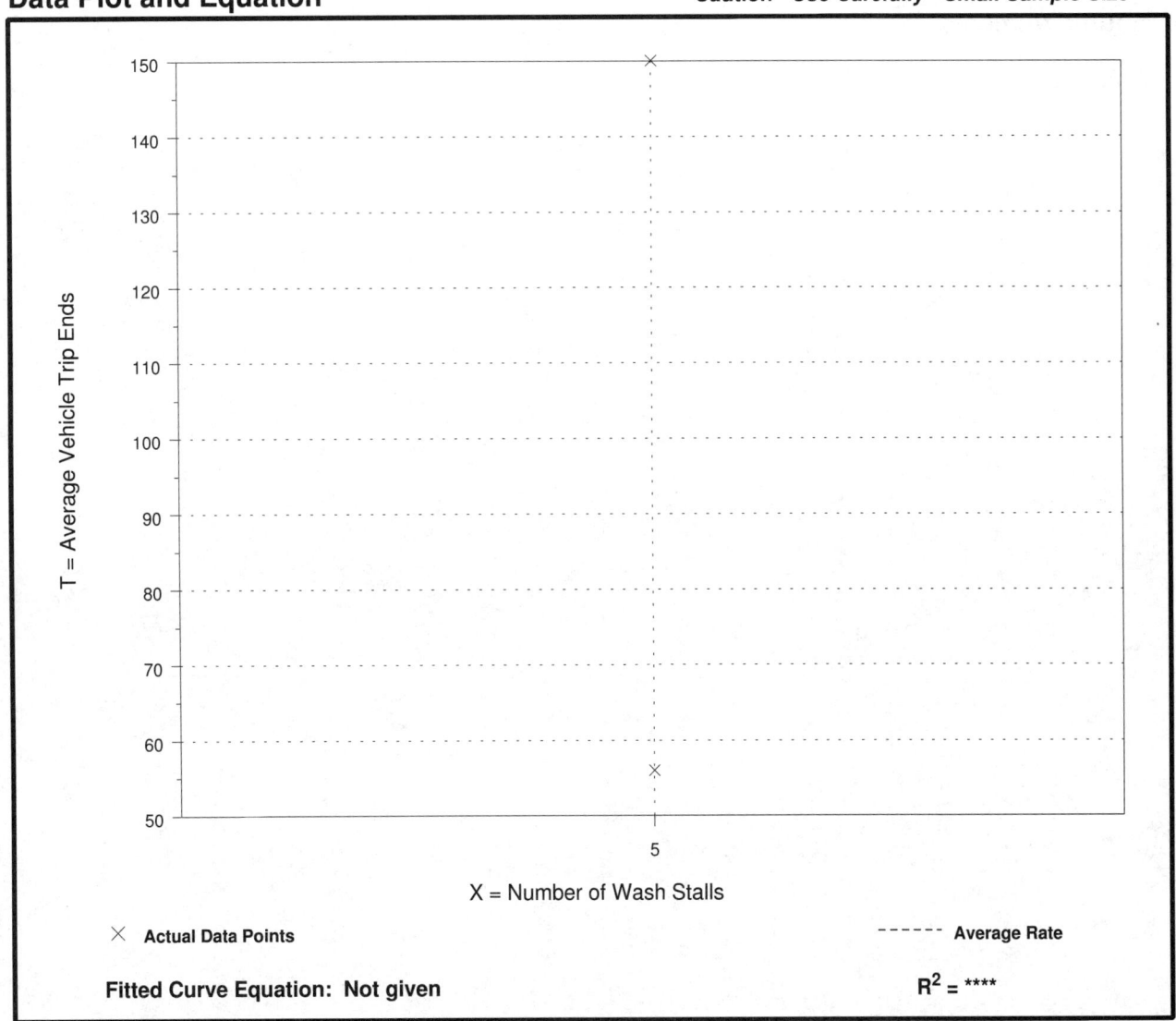

Fitted Curve Equation: Not given $R^2 = ****$

Land Use: 948
Automated Car Wash

Description

Automated car washes are facilities that allow for the mechanical cleaning of the exterior of vehicles. Manual cleaning and car detailing services may also be available at these facilities. Self-service car wash (Land Use 947) is a related use.

Additional Data

The sites were surveyed in the 2000s in New Jersey, New York and Washington.

Source Numbers

552, 555, 585, 599

Land Use: 948
Automated Car Wash
Independent Variables with One Observation

The following trip generation data are for independent variables with only one observation. This information is shown in this table only; there are no related plots for these data.

Users are cautioned to use data with care because of the small sample size.

Independent Variable	Trip Generation Rate	Size of Independent Variable	Number of Studies	Directional Distribution

1,000 Square Feet Gross Floor Area

Weekday P.M. Peak Hour of Adjacent Street Traffic	14.12	2	1	50% entering, 50% exiting
Saturday Peak Hour of Generator	14.12	2	1	50% entering, 50% exiting

Wash Stalls

Saturday Peak Hour of Generator	41	1	1	46% entering, 54% exiting

Land Use: 950
Truck Stop

Description

Truck stops are facilities located on or near major roadways and provide refuelling, food and other services to motorists and truck drivers. These facilities typically contain convenience stores, showers, restaurants and on-site truck parking spaces. Gasoline/service station (Land Use 944) and gasoline/service station with convenience market (Land Use 945) are related uses.

Additional Data

The sites were surveyed in 2006 in Florida.

To assist in the future analysis of this land use, it is important that the number of gasoline and diesel pumps at the study site be reported.

Source Number

721

Truck Stop
(950)

Average Vehicle Trip Ends vs: 1000 Sq. Feet Gross Floor Area
On a: Weekday,
Peak Hour of Adjacent Street Traffic,
One Hour Between 4 and 6 p.m.

Number of Studies: 3
Average 1000 Sq. Feet GFA: 20
Directional Distribution: 52% entering, 48% exiting

Trip Generation per 1000 Sq. Feet Gross Floor Area

Average Rate	Range of Rates	Standard Deviation
13.63	11.60 - 15.75	3.98

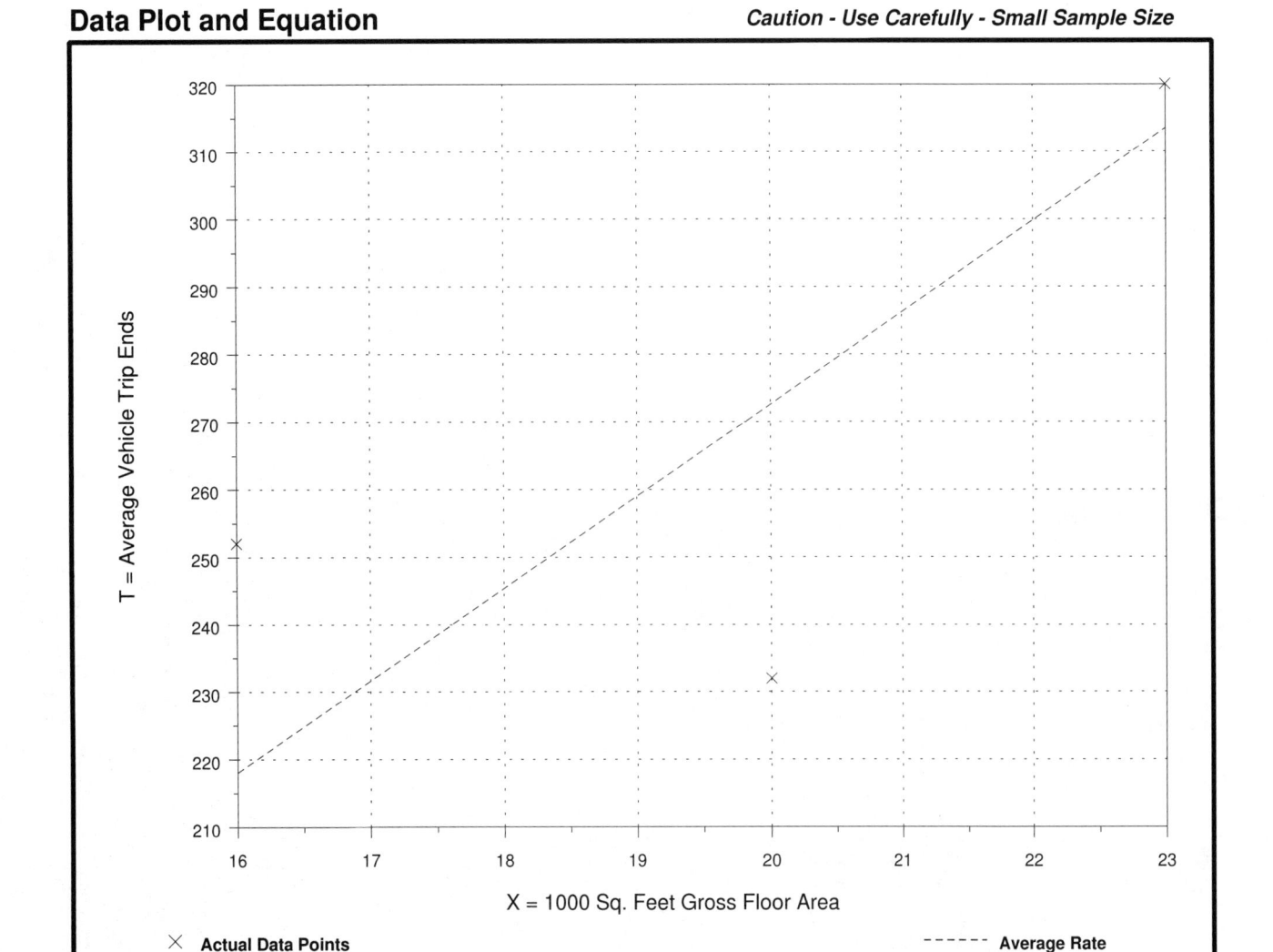

Data Plot and Equation

Caution - Use Carefully - Small Sample Size

Fitted Curve Equation: Not given

$R^2 = ****$